2016 7th Power Electronics and Drive Systems Technologies Conference (PEDSTC 2016)

Tehran, Iran
16 – 18 February 2016

IEEE Catalog Number: CFP1611J-POD
ISBN: 978-1-5090-0376-1

Copyright © 2016 by the Institute of Electrical and Electronics Engineers, Inc
All Rights Reserved

Copyright and Reprint Permissions: Abstracting is permitted with credit to the source. Libraries are permitted to photocopy beyond the limit of U.S. copyright law for private use of patrons those articles in this volume that carry a code at the bottom of the first page, provided the per-copy fee indicated in the code is paid through Copyright Clearance Center, 222 Rosewood Drive, Danvers, MA 01923.

For other copying, reprint or republication permission, write to IEEE Copyrights Manager, IEEE Service Center, 445 Hoes Lane, Piscataway, NJ 08854. All rights reserved.

***This publication is a representation of what appears in the IEEE Digital Libraries. Some format issues inherent in the e-media version may also appear in this print version.**

IEEE Catalog Number: CFP1611J-POD
ISBN (Print-On-Demand): 978-1-5090-0376-1
ISBN (Online): 978-1-5090-0375-4

Additional Copies of This Publication Are Available From:

Curran Associates, Inc
57 Morehouse Lane
Red Hook, NY 12571 USA
Phone: (845) 758-0400
Fax: (845) 758-2633
E-mail: curran@proceedings.com
Web: www.proceedings.com

Table of Content

001 "Evaluation of Various Dynamic Modeling Methods of a Circumferential-Flux Hysteresis Motors"
Abolfazl Halvaei Niasar, Arsham Ghanbari   1-6

002 "Design and Analysis of a Compound SRM and PM Relay for Automatic Drilling tools"
O. Safdarzadeh, E. Afjei, M. Nezamabadi, A. Mosallanejad   7-12

003 "Investigation of the Effect of Eccentricity in Flux Switching Permanent Magnet Machines"
Mostafa Ahmadi Darmani, Seyyed Mehdi Mirimani   13-18

004 "Force Calculation with Complete Modeling of End-Effect in Single-Sided Linear Induction Motors"
Amir Zare-Bazghaleh, Esmaeil Fallah-Chulabi, Seyed Hamid Shahalami   19-23

005 "Comparing the Effect of Using New Hysteresis Material on the Performance of a Hystersis Motor"
Mohammad Mohammadi Firozjaee, Abolfazl Vahedi, Mostafa Sanikhani   24-28

006 "Eccentricity Fault Detection of A Salient-Pole Synchronous Machine Using Modified Winding Function Approaches and Finite Element Method"
Azadeh Doulatshah, Peyman Naderi   29-34

007 "Sensitivity Analysis to Improve the Design Process of an Induction Motor Using MEC"
Reza Haghighian, Mohamad Reza Mazinanian, Abolfazl Vahedi   35-40

008 "A Precise Method to Detect the Accurate Rotor Position in BLDC Motors at Standstill Condition"
O. Safdarzadeh, M. Nezamabadi, E. Afjei   41-47

009 "Application of Multiband Hysteresis Modulation in Field Oriented Control Based IPMSM Drive Fed by Asymetrical Multilevel Cascaded H-Bridge Inverter"
Abdolhossein Ejlali, Davood Arab Khaburi, Javad Soleimani   48-52

010 "Grid Synchronization of Brushless Doubly Fed Induction Generator Using Direct Torque Control"
Ramtin Sadeghi, Seyed M. Madani, M.R. Agha Kashkooli   53-57

011 "Improved Direct Torque Control of DFIG with Reduced Torque and Flux Ripples at Constant Switching Frequency"
M.R. Agha Kashkooli, Seyed M. Madani, Ramtin Sadeghi   58-63

012 "Using Extended Kalman Filter and Adaptive Filter for Sensorless Predictive Torque Control of PM-Assissted Synchronous Reluctance Motor"
Ali Sarajian, Davood Arab Khaburi, Marco Rivera   64-69

013 "Implementation of Drirect Torque Control Method on Brushless Doubly Fed Induction Machines in Unbalenced Situations"
Ali Ghaffarpour, Farhad Barati, Hashem Oraee   70-75

014 "Diagnosis and Compensation of Amplitude Imbalance, Imperfect Quadrant and Offset in Resolver Signals"
N. Noori, D.A. Khaburi   76-81

015 "Fixed-Point Computing with Variable Fraction Length for Motion Profile Generation"
S.M. Dehghan, M. Mansourian   82-86

016 "A New Hardware Device to Simulate the Movement of Electric Train Wheel on Rail"
Sajad Sadr, Davood Arab Khaburi, M. Rivera   87-92

017 "Design and Simulation of Hybrid Electrical Energy Storage (HEES) for Esfahan Urban Railway to Store Regenerative Braking Energy"
Hasan Moghbeli, Hossein Hajisadeghian, Mahdi Asadi   93-98

018 "Analysis of Electromechanical Model of Traction System with Single Inverter Dual Induction Motor"
Pooya Ghani, Mohammad Arasteh, Hamid Reza Tayebi   99-104

019 "Electromagnetic Force Analysis in Linear Induction Motors, Considering End Effect"
Abbas Shiri   105-110

020 "Electro Pump Modeling Using Laboratory System Data"
Pooya Samanipour, Javad Poshtan   111-115

021 "Family of Single-Switch-Soft Switching PWM Converters with Single Magnetic Core"
Moretza Esteki, Mehdi Mohammadi, Ehsan Adib, Hosein Farzanehfard   116-122

022 "A New Switched Boost Inverter Using Transformer Suitable for the Microgrid-Connected PV with High Boost Ability"
Masoud Ghodsi, Seyed Masoud Barakati   123-128

023 "A Family of Single Phase Converters with Reduced Number of Components and Leakage Current Elimination in Photovoltaic Systems"
Masoumeh Adham Haghighpour, Nima Salehiyan Zandi, Ali Mostaan, Alfred Baghramian   129-133

024 "A High Step-Down DC-DC Converter with Low Switch Voltage Stress and Extremely Low Output Current Ripple"
Morteza Esteki, Nasrin Einabadi, Ehsan Adib, Hosein Farzanehfard   134-139

025 "AC Voltage Regulator Based on AC/AC Buck Converter"
M.R. Hajimoradi, H. Mokhtari   140-146

026 "An Interleaved High Step-Up DC-DC Converter with Low Input Current Ripple"
Mahmoud Fekri, Hosein Farzanehfard, Ehsan Adib   147-152

027 "New Asymmetrical Commutation Cell for Multilevel Inverters with Reduced Number of Components"
Mahdi Vizheh, Mohammad Rezanejad, Emad Samadaei   153-158

028 "A Novel Single Switch High Gain DC-DC Converter Employing Coupled Inductor and Diode Capacitor"
Masoume Amirbande, Keyvan Yari, Mojtaba Forouzesh, Alfred Baghramian   159-164

029 "A New Topology for Cascaded Multilevel Inverters with Reduced Number of Power Electronic Switches"
Ebrahim Babaei, Maryam Sarbanzadeh, Mohammad Ali Hosseinzadeh, Concettina Buccella   165-170

030 "Ultra Step-Up DC-DC Converter Based On Three Windings Coupled Inductor"
S.M. Salehi, S.M. Dehghan, S. Hasanzadeh   171-176

031 "A New Switched-Inductor Quasi-Z-Source Inverter"
Masoud Ghodsi, S. Masoud Barakati   177-184

032 "A New Basic Unit for Cascaded Multilevel Inverters with Reduced Number of Power Electronic Devices"
Ebrahim Babaei, Mohammad Ali Hosseinzadeh, Maryam Sarbanzadeh, Carlo Cecati   185-190

033 "Coupled Inductor Based Boost DC/DC Converter"
Simin Sakhavati, Ebrahim Babaei   191-196

034 "A New Current-Fed High Step-Up Quasi-Resonant DC-DC Converter with Voltage Quadrupler"
Sina Salehi Dobakhshari, Seyed Hamid Fathi, Amin Banaiemoqadam,   197-203
Javad Shokrollahi Moghani

035 "Design and Implementing of a Novel Resonant Switched-Capacitor Converter for Improving Balancing Speed of Lithium-Ion Battery Cells"
Shahin Goodarzi, Reza Beiranvand, Reza Rezaii, Mohammad Amin Abolhasani,   204-210
Mustafa Mohamadian

036 "DC-DC Converter for Energy Loss Compensation and Maximum Frequency Limitation in Capacitive Deionization Systems"
Hamed Mehrabian-Nejad, Babak Farhangi, Shahrokh Farhangi, Sadegh Vaez-Zadeh   211-216

037 "Implementation of The First Commercial Medium Power Active Front End Transformerless Uninterruptible Power Supply Made In Iran"
Mustafa Mohamadian, Adib Abrishamifar, Mehdi Shahrdad, Masoud Arefian, Mahdi Fazeli   217-221

038 "Solar Charger System With LED Driver Using Capacitor-Less Multi-Port Converter"
S.M. Dehghan, M. Alinaghizadeh Ardestani   222-228

039 "A High Performance Bi-directional AC-DC Converter for Charge Equalization"
Nima Tashakor, Ebrahim Farjah, Seyed Reza Khayam Hosseini, Teymoor Ghanbari   229-234

040 "Control of Super-Capacitor SOC in a Railway Transit Network"
Elham Rahimi , Ali Dastfan, Saeed Ahmadi   235-240

041 "Zero-Voltage-Transition with Dual Resonant Tank for Bridgeless Boost PFC Rectifier with Low Current Stress"
Farzad Yazdani, Farzad Tahami   241-247

042 "Z-Source DG-Active Filter"
S.A. Saremi Hasari, S. Soori, A. Salemnia, S. Khosrogorji   248-252

043 "An Adaptive Recursive Discrete Fourier Transform Technique for the Reference Current Generation of Single-Phase Shunt Active Power Filters"
Mohammad-Sadegh Karbasforooshan, Mohammad Monfared   253-260

044 "A Z-Source Railway Static Power Conditioner for Power Quality Improvement"
Hossein Mahdinia Roudsari, Alireza Jalilian, Sadegh Jamali   261-267

045 "Parallel Operation of Series-Parallel Uninterruptible Power Supplies"
M. Shahbazi, M. Mohamadian, A. Yazdian Varjani   268-272

046 "Sliding Mode Control of DC-Link Capacitors Voltages of a NPC 4-Wire Shunt Active Power Filter with Selective Harmonic Extraction Method"
M. Asadi, H. Ebrahimirad   273-278

047 "Sizing of Power Electronics EMC Filters Using Design by Optimization Methodology"
J.L. Schanen, A. Baraston, M. Delhommais, P. Zanchetta   279-284

048 "Improving Battery Performance in Hybrid Energy Storage System of PMSG Wind Turbine by Variable Filter Cut off Frequency"
Mohammad Eydi, Javad Farhang, Behzad Asaei, Babak Farhangi   285-290

049 "Modified Local Voltage Controller Design of Inverter-based DGs in a Microgrid"
Hadi Hosseini Kordkheili, Mahdi Banejad   291-296

050 "Developing a New Fault Location Topology for DC Microgrid Systems"
Reza Kheirollahi, Ehsan Dehghanpour   297-301

051 "Three-Phase PFC rectifier with High Efficiency and Low Cost for Small PM Synchronous Wind Generators"
Ghasem Rezazadeh, Farzad Tahami, Hamed Valipour   302-307

052 "A High Step-Up Switched-Capacitor Converter with Zero Current Switching Technique for Using in Solar System Applications"
Seyed Mohammad Mousavi, Reza Rezaii, Reza Beiranvand, Ali Yazdian Varjani   308-313

053 "Space Vector PWM Algorithm for Three-Phase 16-Level Class B-2 Converter"
O. Salari, M.J. Mojibian, M. Tavakoli Bina   314-319

054 "An Interleaved High-Power Two-Switch Flyback Inverter with a Fast and Robust Maximum Power Point Tracker"
Saleh Mohammadi, H. Abootorabi Zarchi   320-325

055 "An Improved Method for Power Management and Voltage Control of PV Unit in DC Microgrid"
S. Soori, S.A. Saremi Hasari, A. Salemnia, S. Khosrogorji   326-331

056 "A New DPC Method For Single VSC Based DFIG Under Unbalanced Grid Voltage Condition"
Ali Izanlo, S. Asghar Gholamian, Mohammad Verij Kazemi   332-337

057 "Energy Management of Dual-Source Propelled Electric Vehicle using Fuzzy Controller Optimized via Genetic Algorithm"
S. Khoobi Arani, A. Halvaei Niasar, A. Haji Zadeh   338-343

058 "Three Phase Photovoltaic Grid-Tied Inverter Based on Feed-Forward Decoupling Control Using Fuzzy-PI Controller"
Faramarz Karbakhsh, G.B. Gharehpetian, Jafar Milimonfared, Armin Teymoori 344-348

059 "Analysis of the Boost Converter Under the DCM Condition to Reduce the MIC Volume to Mitigate Partial Shading Effects in PV Arrays"
Reza Rezaii, Mohammad Amin Abolhasani, Ali Yazdian Varjani, Reza Beiranvand 349-355

060 "A Noninvasive On-line Failure Prediction Technique for Aluminum Electrolytic Capacitors in Photovoltaic Grid-connected Inverters"
Amir Sepehr, Mehdi Saradarzadeh, Shahrokh Farhangi 356-361

061 "Single Stage DC-AC Boost Converter"
Ali Nahavandi, Mehdi Roostaee, Mohammad Reza Azizi 362-366

062 "A Comparison Between Buck and Boost Topologies as Module Integrated Converters To Mitigate Partial Shading Effects on PV Arrays"
Mohammad Amin Abolhasani, Reza Rezaii, Reza Beiranvand, Ali Yazdian Varjani 367-372

063 "Output Power Smoothing of PMSG-Based Wind Energy Conversion System Equipped with Matrix Converter"
Koosha Mehdizadegan, Karim Abbaszadeh 373-377

064 "Control of Storage System in Series Collection Grid in Offshore Wind Farm for Limiting DC/DC Converter Overvoltaging"
Diana Florez Rodriguez, Ehsan Enferad, Christophe Saudemont 378-383

065 "Analysis and Control of Single-Phase Converters for Integration of Small-Scaled Renewable Energy Sources into the Power Grid"
Majid Mehrasa, Mohamad Rezanejhad, Edris Pouresmaeil, João P.S. Catalão, Sasan Zabihi 384-389

066 "Comparison of Single Loop Based Control Strategies for a Grid Connected Inverter in a Photovoltaic System"
Mohammad Hossein Mahlooji, Hamid Reza Mohammadi, Mohsen Rahimi 390-395

067 "THD Minimization in Variable Input Cascaded H-Bridge Multi-level Inverters via State Table"
Reza Emamalipour, Behzad Asaei, Babak Farhangi 396-402

068 "New Switching Strategy for Single-Phase Multilevel Quasi-Z-Source Inverter"
Saman Radman, Mostafa Shahnazari, Hamidreza Toodeji 403-408

069 "Reliability Evaluation of Two-Stage Interleaved Boost Converter Interfacing PV Panels Based on Mode of Use"
Farid Hamzeh Aghdam, Mehrdad Tarafdar Hagh, Mehdi Abapour 409-414

070 "A Comparative Study of Different Multilevel Converter Topologies for High Power Photovoltaic Applications"
A. Delavari, I. Kamwa, A. Zabihinejad 415-420

071 "New Topology to Reduce Leakage Current in Three-Phase Transformerless Grid-Connected Photovoltaic Inverters"
Ramin Rahimi, Babak Farhangi, Shahrokh Farhangi 421-426

072 "Efficiency Optimization and Power Management in a Stand-Alone Photovoltaic (PV) Water Pumping System"
Behzad Mirshekarpour, S. Alireza Davari　　　　　　　　　　427-433

073 "Increasing the Battery Life of the PMSG Wind Turbine by Improving Power Division of the Hybrid Energy Storage System"
Mohammad Eydi, Javad Farhang, Behzad Asaei, Reza Emamalipour　　　　434-439

074 "Fast and Simple Open-Circuit Fault Detection Method for Interleaved DC-DC Converters"
Mahmoud Shahbazi, Mohammad Reza Zolghadri, Saeed Ouni　　　　440-445

075 "A New Hybrid Method of MPPT for Photovoltaic Systems Based on FLC and Three Point-Weight Methods"
M. Bahrami, M. Zandi, R. Gavagsaz, B. Nahid-Mobarakeh, S. Pierfederici　　　446-450

076 "Stabilizing a Photovoltaic Plant Power Output by Employing an Auxiliary Power Source"
Naier Mahdinejad, Luiz Machado, Ricardo Nicolau Nassar Koury, Ramon Molina Valle　　451-456

077 "The Effect of Energy Efficiency Increase on a PV Plant with a Non-conventional Energy Storage System Income"
Ali Moallemi, Luiz Machado, Ricardo Nicolau Nassar Koury, Fabricio Jose Pacheco Pujatti　　457-463

078 "Signal Flow Graph Modeling and Disturbance Observer based Output Voltage Regulation of an Interleaved Boost Converter"
Majid Abbasi, Ahmad Afifi, Mohammad Reza Alizadeh Pahlavani　　　　464-469

079 "Design Procedure and Experimental Validation of a 30 kVA DSP-Based PWM Rectifier"
Mohammad Pichan, Adib Abrishamifar, Amir Mirzabayati, Mehdi Fazeli　　　470-475

080 "Virtual Flux Based Direct Power Control of a Three-Phase Rectifier Connected to an LCL Filter with Sensorless Active Damping"
Mehran Maghamizadeh, S. Hamid Fathi　　　　　　　　　476-481

081 "Dynamic Formation of Time-Dependent Duty Cycles for a Three-Phase Boost-Type DC-AC Converter Based on Averaging Model"
B. Eskandari, J. Javidi Hagh, J. Shojaee, M. Tavakoli Bina　　　　482-485

082 "A Modified SVM Switching Pattern for Z-Source Inverter"
Mostafa Abarzadeh, Hossein Fathi kivi, Hossein Madadi Kojabadi　　　　486-491

083 "Reduced Size Single-Phase PHEV Charger with Output Second-Order Voltage Harmonic Elimination Capability"
Hamid Rezaie, Hassan Rastegar, Mohammad Pichan　　　　　　492-497

084 "A Novel Application of H_∞ Robust Controller for a Single Phase Inverter in Uninterruptible Power Supply"
Ahmad Irani, Mahdi Sojoodi, Mustafa Mohamadian　　　　　　498-503

085 "Design and Implementation of an FPGA-based Real-Time Simulator for H-Bridge Converter"
Morteza Rezaei Larijani, Mohammad-Reza Zolghadri, Mahmoud Shahbazi　　　504-510

086 "A New SVM-based Voltage Balancing method for Five-Level NPC Inverter"
Pouria Qashqai, Abdolreza Sheikholeslami, Hani Vahedi, Kamal Al-Haddad   511-516

087 "Model Predictive Control of Classic Bidirectional DC-DC Converter for Battery Applications"
A. Pirooz, R. Noroozian   517-522

088 "A Novel Method for Real-time Selective Harmonic Elimination in Five-Level Converters"
Adib Abrishamifar, Mohammad Arasteh, Farzad Golshan   523-528

089 "Elimination of Low Order Harmonics in Nine-level Cascaded H-bridge Converter"
Adib Abrishamifar, Mohammad Arasteh, Farzad Golshan   529-534

090 "Robust Control of the DC-DC Ćuk Converter in Discontinuous Conduction Mode"
Vadood Hajbani, Mahdi Salimi, Jafar Soltani   535-540

091 "Indirect Voltage Regulation of Double Input Y-Source DC-DC Converter Based on Sliding Mode Control"
Soheil Ahmadzadeh, Gholamreza Arab Markadeh   541-546

092 "Flatness-Based Control Method: A Review of its Applications to Power Systems"
M. Soheil-Hamedani, M. Zandi, R. Gavagsaz-Ghoachani, B. Nahid-Mobarakeh   547-552

093 "Space Vector PWM Method for Two-Phase Three-Leg Inverters"
Maedeh Mirazimi, Ebrahim Babaei, Mohammad Bagher Banna Sharifian   553-558

094 "Improved Equations of Switching Loss and Conduction Loss in SPWM Multilevel Inverters"
Abolfazl Babaie, Bagher Karami, Adib Abrishamifar   559-564

095 "High Efficiency Wireless Power Transfer System Design for Circular Magnetic Structures"
A. Ramezani1, Sh. Farhangi, H. Iman-Eini ,B. Farhangi   565-570

096 "Minimum Weight Wireless Power Transfer Coil Design"
Adel Moradi, Farzad Tahami, Amirreza Poorfakhraei   571-576

097 "A New Pulsed Power Generator Topology for Corona Discharge"
Mohammad Kebriaei, Abolfazl HalvaeiNiasar, Abbas Ketabi   577-581

098 "A Review of Predictive Control Techniques for Matrix Converters - Part I"
M. Rivera, P. Wheeler, A. Olloqui, D.A. Khaburi   582-588

099 "A Review of Predictive Control Techniques for Matrix Converters - Part II"
M. Rivera, P. Wheeler, A. Olloqui, D.A. Khaburi   589-595

100 "Optimized Current Control of Vienna Rectifier Using Finite Control Set Model Predictive Control"
Ali R. Izadinia, Hamid R. Karshenas   596-601

101 "Predictive Control of a Five-Level NPC Inverter Using a Three-Phase Coupled Inductor"
Seyed Saeed Fazel, Hamid Reza Piryaei   602-607

102 "A Predictive Control Strategy for a Single-Phase AC-AC Converter"
M. Rivera, S. Rojas, P. Wheeler, J. Rodriguez   608-613

103 "Predictive Torque Control of a Permanent Magnet Synchronous Motor Fed by a Matrix Converter without Weighting Factor"
Mohsen Siami, Hamed Kiani Savadkoohi, Alireza Abbaszadeh, D. A. Khaburi, Jose Rodriguez,   614-619
Marco Rivera

104 "One Step Model Predictive Control of Five Level ANPC Permanent Magnet Motor Drive"
Mohammad Niliyan, Mustafa Mohamadian, Ali Yazdian Varjani   620-625

105 "Improved Direct Torque Control of Induction Motor with the Model Predictive Solution"
Mohammad Reza Nikzad, Seyed omid Ahmadi, Behzad Asaei   626-630

106 "Minimum Slope Model Predictive Control with Double Margins"
Reza Fotouhi, Hui Fang, Ralph Kennel   631-635

107 "The Challenges of Predictive Control to Reach Acceptance in the Power Electronics Industry"
Margarita Norambuena, Cristian Garcia, Jose Rodriguez   636-640

108 "Distributed Secondary Control in DC Microgrids with Low-Bandwidth Communication Link"
Saeed Peyghami-Akhuleh, Hossein Mokhtari, Poh Chiang Loh, Frede Blaabjerg   641-645

109 "Voltage Unbalance and Harmonic Compensation in Microgrids by Cooperation of Distributed Generators and Active Power Filters"
Mohammad M. Hashempour, Mehdi Savaghebi, Juan C. Vasquez   646-651

110 "Method for Load Sharing and Power Management in a Hybrid PV/Battery Source Islanded Microgrid"
Yaser Karimi, Hashem Oraee, Josep M. Guerrero, Juan C. Vasquez, Mehdi Savaghebi   652-657

7th Power Electronics, Drive Systems & Technologies Conference (PEDSTC 2016)
16-18 Feb. 2016, Iran University of Science and Technology, Tehran, Iran

Evaluation of Various Dynamic Modeling Methods of a Circumferential-Flux Hysteresis Motors

Abolfazl Halvaei Niasar
Department of Electrical & Computer Engineering
University of Kashan, Kashan, Iran
halvaei@kashanu.ac.ir

Arsham Ghanbari
Department of Electrical Engineering
University of Allame Feiz Kashani, Kashan, Iran
arshamghanbari@gmail.com

Abstract—To analyze hysteresis motor performance from startup to synchronism, we need to a valid dynamic model. This paper studies and evaluates various dynamic modeling methods for hysteresis motor. Modeling based on constant impedance for the rotor, variable impedance based on load angle for the rotor, and variable impedance as well as dependent voltage source for the rotor is assessed. These models are implemented in Simulink for a given high-speed, circumferential-flux hysteresis motor and their dynamic responses, attributes and defects are explored. Unlike the hysteresis motor with constant impedance modeling for the rotor, the hysteresis motor with variable impedance modeling based on hysteresis delay angle and equivalent voltage source for the rotor will reach the synchronism at full load condition. Also we discover by this model the variation of lag angle would stay under 90 degree (ideal degree of hysteresis lag angle) and at the synchronism would be equal to nominal value of the typical motor. The simulation results obtained by variable impedance modeling based on load angle for the rotor will show us the motor reaches synchronism but the value of load angle is beyond 90° and almost equal to 180°. Finally, some suggestions are given to improve of dynamic model.

Index Terms—Hysteresis Motor; Dynamic Modeling; B-H Loop; circumferential-flux.

I. INTRODUCTION

The hysteresis motor can be regarded as a self-starting synchronous machine, which has the same structure as a squirrel cage asynchronous motor, unlike to induction motors (IMs), semi-hard materials used in the rotor of hysteresis motors have wider B-H loops than soft magnetic materials used in IMs and so, it leads to more complicated model. It is often used for small power application that needs a very smooth torque at high speed such as gas centrifuge and gyroscope [1]. Relatively low starting current, simple and strong structure, uniform torque-speed characteristic and also low power factor, low efficiency and hunting phenomenon are some of hysteresis motor advantages and disadvantages [2]. There are two main structures for hysteresis motors: cylindrical and disk types. Cylindrical hysteresis motors are

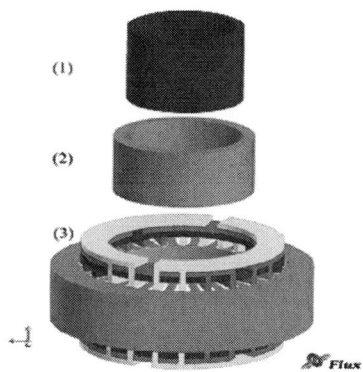

Fig. 1. Structure of circumferential-flux type of hysteresis motor
(1) Rotor support, (2) Rotor (3) Stator

classified in two types of circumferential-flux and radial-flux. In circumferential-flux hysteresis motor, the rotor ring is mounted on the non-magnetic material as support. The magnetic field lines in the rotor are mostly circumferential to the ring. This type of hysteresis motor is mostly used in industrial applications [3]. Fig. 1 shows structure of a circumferential-flux hysteresis motor.

Mathematical modeling of electrical motors is essential to verify the design procedure and study and understand the transient behavior of machine [4]. Dynamic modeling of a hysteresis motor is relatively similar to an induction machine, but the modeling of hysteresis ring in the rotor of hysteresis motor makes it quite different from the asynchronous motor. The basic concept used in dynamic model of hysteresis motor comes from steady-state model. There has been a little contribution in the literature for dynamic modeling of hysteresis motors. Dynamic modeling with constant parameters using various approximation of major B-H loop of magnetic material used in rotor ring is very common. Hysteresis motor using this model at no load condition will reach the synchronism [5]. But there is a significant deficiency at this kind of modeling. As it will be discussed in the next part, by using this model the motor under full load condition could not reach the synchronism. There are a few papers

978-1-5090-0376-1/16 $31.00 © 2016 IEEE

which consider the variation of B-H loops especially for hysteresis motors.

In this paper, various dynamic models of hysteresis motors are studied. Modeling of the stator in proposed dynamic model are the same as stator model in induction motors. The difference between various models comes from how they model the hysteresis (B-H) loop of the rotor. In this manner, B-H loop of the rotor's magnetic material is often assumed as elliptical shape, and so, the rotor is modeled with inductive impedance. The proposed rotor models can be classified into three groups: (1) Modeling based on constant impedance (2) Modeling based on variable impedance, (3) Modeling based on variable impedance in series with a dependent voltage source on the rotor circuit. In the rest of the paper, mentioned modeling methods are used for a cylindrical hysteresis motors and the dynamic responses are simulated in Simulink and the results are analyzed.

II. DYNAMIC MODEL BASED ON CONSTANT IMPEDANCE FOR THE ROTOR

Dynamic model with constant impedance of the rotor use the major B-H loop of hysteresis material of the rotor [6-8]. In this manner, B-H loop of the rotor's magnetic material is assumed as elliptical shape, and so, the rotor is modeled with inductive impedance. Reported models have just considered major B-H loop of rotor that is related to nominal stator voltage [6]. The main problem of this modeling method is the presence of speed slip under load variations, whereas, this motor is a synchronous motor type. Fig. 2 shows the dynamic model of hysteresis motor with constant impedance for rotor in rotating dq reference frame. The value of rotor's resistance and reactance are derived from elliptical B-H of hysteresis ring. Fig. 3 shows the rotor speed ω_r for a typical high-speed motor with given parameters at the appendix under no-load and full load conditions. In no-load case, motor reaches synchronism and slip in steady state is equal to zero as we expected from results obtained by former works [6-8]. But under load or full-load conditions, there is always a slip in steady state performance of the motor. For this case at full load, ω_r is 0.9276 p.u which means the hysteresis motor would

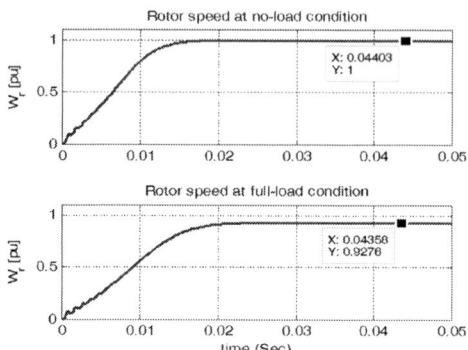

Fig. 3. Rotor speed ω_r (p.u) of hysteresis motor for model with constant impedance in rotor at no-load and full-load conditions

act just like an IM with 0.0724 slip. It's the most important deficiency of this kind of modeling for hysteresis motor.

III. DYNAMIC MODEL BASED ON VARIABLE IMPEDANCE FOR THE ROTOR DEPENDENT TO LOAD ANGLE

Dynamic model with variable impedance has been proposed by Nitao in [9] that they have developed a dynamic model for hysteresis motor in stationary dq coordinates based on the model by Miyairi and Kataoka [10]. In proposed work the resistance and reactance related to B-H curve are considered with respect to load angle (δ) as:

$$R_{hr} = \omega_b \frac{m K_\omega^2 N_\omega^2 V_r \mu}{\pi^2 r_r^2} \sin \delta \qquad (1)$$

$$L_{hr} = \frac{m K_\omega^2 N_\omega^2 V_r \mu}{\pi^2 r_r^2} \cos \delta \qquad (2)$$

Where the load angle δ obeys the equation:

$$\frac{d\delta}{dt} = \left(\frac{2}{P}\right)(s) = \left(\frac{2}{P}\right)(\omega_b - \omega_r) \qquad (3)$$

Subject to the condition:

$$|\delta| \le \delta_{max} \qquad (4)$$

The author has considered this constraint similar to constrain for value of lag angle β that should be less than β_{max} of major loop for variation of B-H loop. Fig. 4 shows the

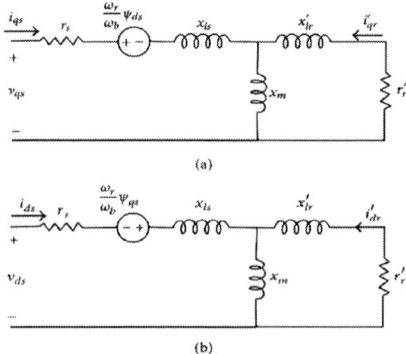

(a)

(b)

Fig. 2. Dynamic model of hysteresis motor with constant impedance for the rotor in rotating dq reference frame (a) q axis circuit (b) d axis circuit [7]

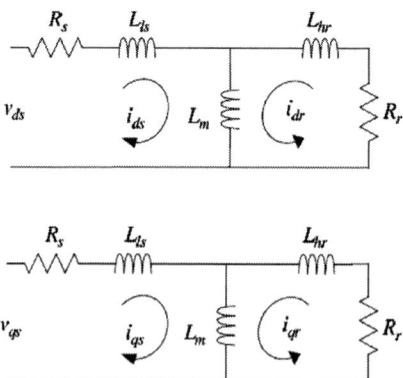

Fig. 4. Dynamic model of hysteresis motor with variable impedance for the rotor in stationary dq reference frame [9]

978-1-5090-0376-1/16 $31.00 © 2016 IEEE

electrical equivalent circuit for dynamic modeling of hysteresis motor in dq stationary reference frame presented in [9]. The rotor resistance (R_r) in this circuit is $R_{hr}||R_e$, which R_e represents the eddy current resistance.

Proposed modeling technique has been successful to reach the motor to synchronism under load variations and is able to modify the rotor's equivalent impedance with lag angle. For better analyze of this kind of modeling we've done two different tests, the first one with δ_{max} constraint and the other one without it. We have simulated some other hysteresis motors with $\delta_{max} < 90°$; but the motors under load torque cannot reach synchronism. For this purpose, δ_{max} should be increased to 180°. Fig. 5 shows the simulation results with constrained load angle δ_{max} to 0.7 rad (or 40°). This constraint is corresponding to lag angle β_{max} due to major B-H loop. The motor speed cannot pass 0.88 pu. For the higher value of δ_{max} the speed can reach to higher value, but not to synchronism.

Fig. 6 shows the simulation results with unconstrained load angle δ_{max}, in which it can varies between 0° to 180° ($\delta = \pi$ rad). It seems the speed reaches to synchronism and load angle converges to 180°. But, as we know the β_{max} cannot reach to 90° in hysteresis motor because 90° is corresponding to ideal rectangular B-H loop [2,3]. Moreover, load angle δ more than 90° in generalized theory of electrical machines for steady-state condition is meaningless. Furthermore, there are some strange simulation results for presented model by [8] such as developing constant instantaneous torque during acceleration and oscillation in steady state in torque that cannot satisfy the theoretical behavior of hysteresis motors.

IV. DYNAMIC MODEL BASED ON VARIABLE IMPEDANCE VOLTAGE SOURCE FOR THE ROTOR

This modeling method has been presented in [11,12] for a disk-type hysteresis motor. In proposed model, the hysteresis material of rotor can be replaced by two phase balanced windings with the same number of turns of the stator in dq reference frame as shown in Fig. 7. The eddy current effect is modeled by resistance R_e, and hysteresis loss or power are represented by R_h. Parameter α is the angle between d axis and phase 'a' axis of the stator, which is called hysteresis delay angle. Also, two induced voltages V_{dr} and V_{qr} are considered in the rotor circuit for temporary magnet property due to hysteresis loop. The rotor resistance and reactance related to

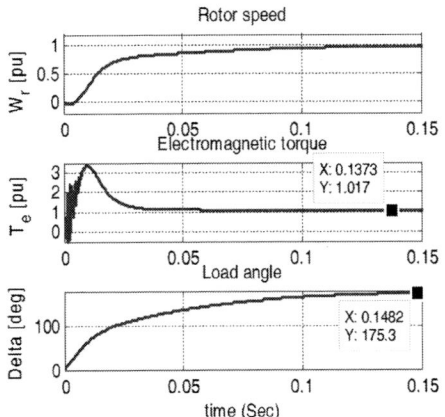

Fig. 6. The variations of speed and load (lag) angle of hysteresis motor with modeling based on variable impedance for rotor [9] at full-load condition without δ_{max} constraint

hysteresis loop are modeled as functions of delay angle α and can be obtained from structural parameters of disc type hysteresis motor as [12]:

$$R_h = \frac{4mf\left(K_w N_{ph}\right)^2 t_r (R_o - R_i) K_{sf} B_q}{1000 R_{av} H_p} \sin\alpha \qquad (5)$$

$$R_e = \frac{12\rho l_h}{A_h 10^4} \qquad (6)$$

$$r_r = \frac{(R_h R_e)}{(R_e + sR_h)} \qquad (7)$$

$$X_{lr} = X_h = \frac{R_h}{\tan\alpha} \qquad (8)$$

The hysteresis delay angle (α) in above equations is related to hysteresis loops of rotor and these loops depend on the stator voltage amplitude. Larger voltage amplitude leads to wider B-H loop and greater delay angle and vice-versa. For calculation the real value of rotor parameters and magnetizing part, the operational delay angle α or hysteresis loop should be identified. For this purpose, an initial B-H curve (or B_q) is chosen and the stator voltage is calculated from the steady-state equivalent circuit shown in Fig. 8 and then is compared with real stator voltage. The value of B_q is modified by a

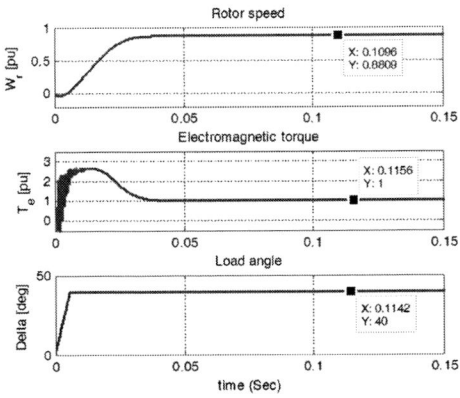

Fig. 5. The variations of speed and load (lag) angle of hysteresis motor with modeling based on variable impedance for rotor [9] at full-load condition with $\delta_{max} = 40°$ constraint

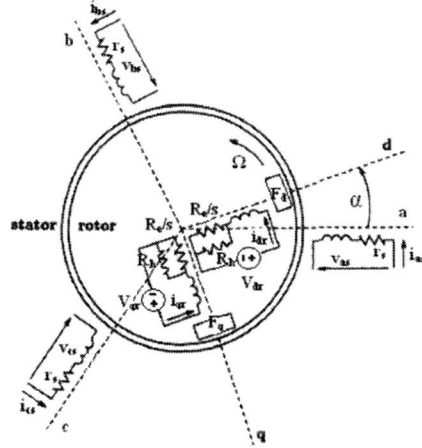

Fig. 7. The Dynamic model schematic of hysteresis motor in [12]

978-1-5090-0376-1/16 $31.00 © 2016 IEEE

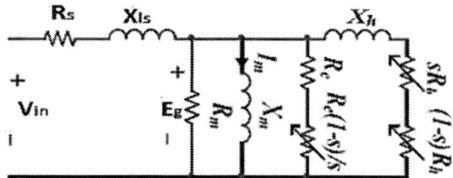

Fig. 8. Steady State electrical equivalent circuit of hysteresis motors [5,11]

proper step size and procedure is repeated until the voltage difference falls into desired tolerance. Proposed simulation results in [12] indicate that the motor can reach to synchronism under any sudden change in the load. The hysteresis motor voltage equations in synchronously rotating dq reference frame can be written as [12]:

$$V_{dq} = R \cdot I_{dq} + \frac{d\Lambda_{dq}}{dt} + E_{dq} \qquad (9)$$

It can be expanded as:

$$
\begin{bmatrix} V_{ds} \\ V_{qs} \\ V_{dr} \\ V_{qr} \end{bmatrix} =
\begin{bmatrix} R_s & 0 & 0 & 0 \\ 0 & R_s & 0 & 0 \\ 0 & 0 & R_r & 0 \\ 0 & 0 & 0 & R_r \end{bmatrix}
\begin{bmatrix} I_{ds} \\ I_{qs} \\ I_{dr} \\ I_{qr} \end{bmatrix}
$$
$$
+ \frac{1}{\omega_b}
\begin{bmatrix} X_{ls} & 0 & X_m & 0 \\ 0 & X_{ls} & 0 & X_m \\ X_m & 0 & X_{lr} & 0 \\ 0 & X_m & 0 & X_{lr} \end{bmatrix}
\times \frac{d}{dt}
\begin{bmatrix} I_{ds} \\ I_{qs} \\ I_{dr} \\ I_{qr} \end{bmatrix} \qquad (10)
$$
$$
+ \frac{\omega_r}{\omega_b}
\begin{bmatrix} 0 & -X_{ls} & 0 & -X_m \\ X_{ls} & 0 & X_m & 0 \\ 0 & 0 & 0 & 0 \\ 0 & 0 & 0 & 0 \end{bmatrix}
\begin{bmatrix} I_{ds} \\ I_{qs} \\ I_{dr} \\ I_{qr} \end{bmatrix}
$$

Developed electromagnetic torque can be expressed in term of flux and current as:

$$T_{em} = \frac{3}{2}\frac{P}{2}(\lambda_{ds} i_{qs} - \lambda_{qs} i_{ds}) \qquad (11)$$

And the mechanical speed in per unit is obtained from:

$$T_{em} - T_{mech} - T_{damp} = 2H \frac{d\omega_r}{dt} \qquad (12)$$

A. Determination of Operational B-H loop and Related Lag angle β

The parameters of the equivalent circuit of hysteresis motor are function of hysteresis lag angle β that should be identified.

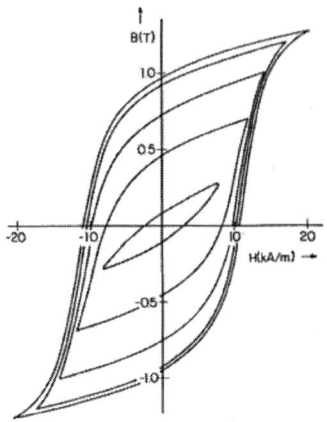

Fig. 9. B-H characteristics for a typical magnetic material used in hysteresis motors

In [9], it has been considered equal to load angle δ that we proved it is correct. The lag angle β is directly related to B-H curves of the rotor materials. B-H curves for a given material have the nonlinear shapes as shown in Fig. 9. The major B-H loop is corresponding to the rated stator voltage of the motor and with applying the lower voltage minor B-H loops are created. In [12], the operational B-H curve and so related lag angle β are identified using steady-state model of hysteresis motor and through a recursive algorithm as shown in Fig. 10.

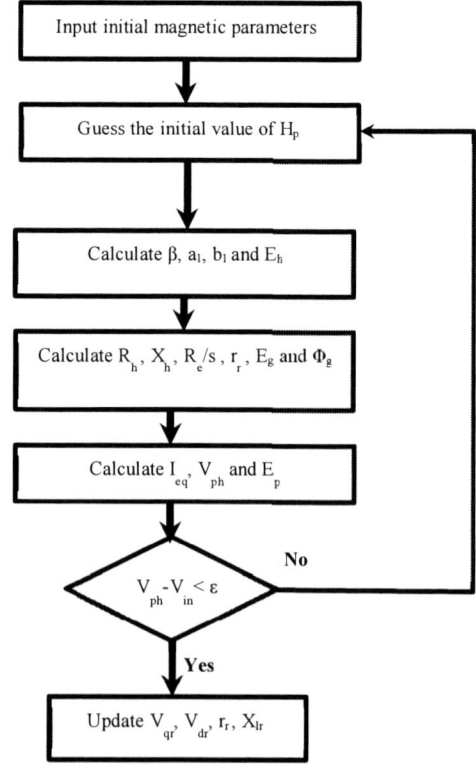

Fig.10. Flowchart for calculation of the rotor's equivalent circuit parameters [12]

For this purpose, B-H curves of rotor's material are supposed as elliptical curves. It means only fundamental harmonic of magnetic field intensity of rotor is considered as:

$$B(t) = a_1 \cos(\omega t) + b_1 \sin(\omega t) = B_q \cos(\omega t - \beta) \qquad (13)$$

Firstly an initial value is considered for B_q that means an initial B-H curve is chosen. So, lag angle β is calculated and by that the rotor's circuit parameters are computed. Using these parameters the input voltage (V_{in}) of the stator is calculated from steady-state circuit shown in Fig. 8. Afterwards, V_{in} is compared with real applied voltage to the stator (V_{ph}). If the difference falls to desired tolerance ε, the calculated β is the operational lag angle. Otherwise the initial guess for B_q or H_P changes with a proper step size.

For a given circumferential-flux, high speed hysteresis motor with parameters given in Table I, some simulations are carried out based on the proposed modeling method in [12]. Fig. 11 shows the motor behavior under full load and applying the nominal voltage of stator. As shown, the motor's speed reach to synchronism under full load and developed torque is equal to load torque (1 pu). The current and power factor of motor are 0.29 A and 0.31 at steady state that are near to their value in experiment. Fig. 12 shows the variation of lag angle β and rotor's resistance R_h and reactance X_h that converges to 41.6°, 173.7 Ω and 195.3 Ω respectively. The initial value for flux intensity H_p in Fig. 10 has been chosen 3000 A/m. These rotor's parameters remain constant, if the stator voltage doesn't change.

None of previous models can reach the motor to synchronous speed under load. Proposed dynamic model satisfies many aspects of hysteresis motor behavior. Hunting phenomena and variation of power factor by changing of stator voltage can be shown by this model.

TABLE I
RATED SPECIFICATIONS AND PARAMETERS OF USED CIRCUMFERENTIAL-FLUX HYSTERESIS MOTOR

Symbol	Quantity	Value	Dimension
V_{rated}	line rated voltage	380	V
T_{rated}	rated torque	0.01	N.m
f_{rated}	rated frequency	1000	Hz
ω	supply angular frequency	60×1000	rpm
m	number of phase	3	
P	number of poles	2	
J	shaft inertia moment	3×10^{-4}	kg.m^2
R_s	stator resistance	60	Ω/ph
R_c	stator core loss equivalent resistance	10580	Ω/ph
R_h	rated rotor hysteresis resistance	173	Ω/ph
R_e	eddy current resistance of rotor	223	Ω/ph
X_{ls}	stator leakage reactance	78	Ω/ph
X_m	rated magnetizing reactance	165	Ω/ph
X_h	rated rotor hysteresis reactance	195	Ω/ph
l_h	axial length of hysteresis ring	25	mm
g	air gap length	1	mm
t_r	thickness of hysteresis ring	1	mm
β	hysteresis lag angle	42	deg

V. MAGNETIC BEHAVIOR OF ROTOR HYSTERESIS MATERIAL DUE TO LOAD TORQUE CHANGING

Among previous models, the modeling based on variable impedance and voltage source satisfies more aspects of hysteresis motor theory. However, this model and all earlier models cannot predict the operational B-H loop or hysteresis lag angle, while load torque changes. In other words, to calculate of hysteresis resistance R_h, they have used a constant hysteresis lag angle [6-8] or varied with stator voltage [11,12]. This fact can be explained as follows. When the rotor turns at synchronous speed, the load torque is equal to output (or developed electromagnetic) torque and a specific B-H loop is experienced. The area of this loop is proportional to output torque T_{em} that for a circumferential-flux hysteresis motor in steady state can be obtained from [6]:

$$T_{em} = \frac{mP}{2}\pi(B_m H_c)r_h t_r l_h \sin\beta \qquad (14)$$

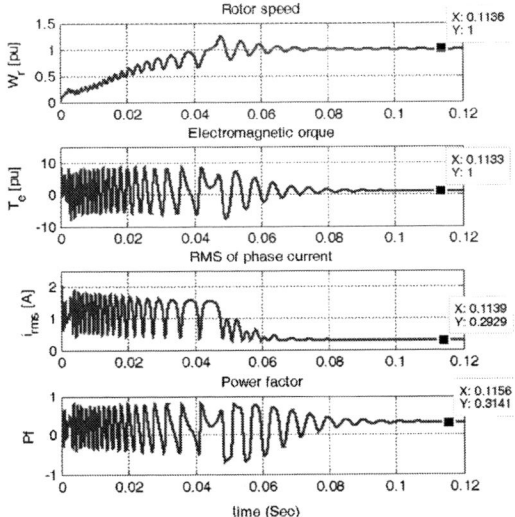

Fig. 11. Speed, torque, RMS current, and power factor of hysteresis motor under full-load and nominal supply voltage conditions based on variable impedance and voltage source for rotor [12]

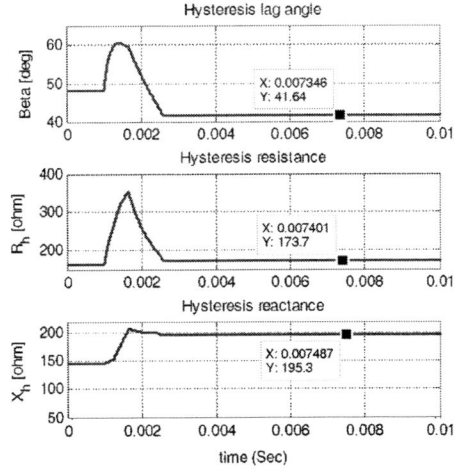

Fig. 12. Lag angle β, rotor's resistance (R_h) and reactance (X_h) of hysteresis motor under full-load and nominal supply voltage conditions based on variable impedance and voltage source for rotor [12]

It means that, with change the load torque and then output torque, the lag angle β varies. Fig. 13 shows two B-H loops of rotor's material corresponding to two value of load torque [13-15]. The stator voltage for both cases remains constant. Both real and elliptical representations for B-H loops are shown. The variation of load torque with developed model described in section IV is simulated. As shown in Fig. 14, load torque changes from 1 pu to 0.5 pu and then goes back to 1 pu. The motor remains in synchronous speed and developed torque in steady state follows the load torque. But the load angle β and so the rotor parameters remain constant. So, this model cannot predict the value of lag angle under load changing.

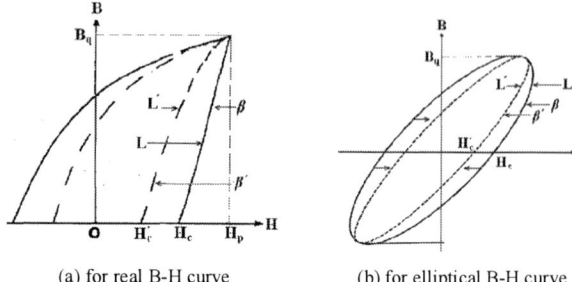

(a) for real B-H curve   (b) for elliptical B-H curve

Fig. 13. The change of hysteresis loops of hysteresis motor for two different load torque, (L' loop corresponds to lower load torque)

VI. CONCLUSIONS

An investigation based on various dynamic models for this type motor were studied and assessed. Comparison of the results obtained from three case studies in this paper validates, the modeling based on variable impedance as well as dependent source voltage can incorporate the impact of B-H loops of hysteresis motor, and satisfied many aspects of the motor such as synchronism at full load condition and maintaining the variation of lag angle under 90°. However, it cannot predict and model the influence of load torque variations, whereas the operational B-H loop not only changes with changing stator voltage, but also is affected with variation of load torque. This fact was described at the end part of paper in details. Modeling method based on constant impedance for rotor leads to a linear, time invariant for hysteresis motor. It is a simple way, but it cannot satisfy the essential features of hysteresis motor such as synchronism at full load condition. With modeling based on variable impedance using load angle, a linear, time varying model is obtained, but it cannot show the impact of B-H loops and the variation of stator voltage. It cannot model the influence of hysteresis material of the rotor. Consequently, a perfect dynamic model of hysteresis motor must consider the variations of voltage and load to identify the correct lag angle.

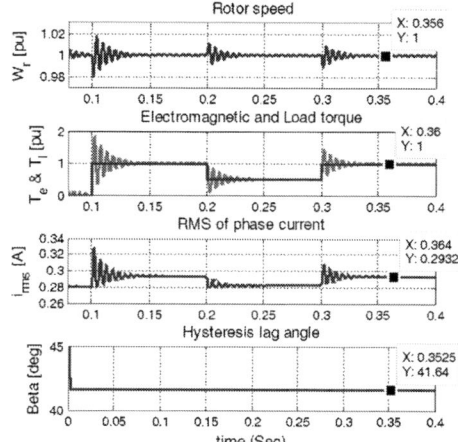

Fig. 14. Speed, torque, RMS current, and power factor of hysteresis motor under full-load and nominal supply voltage conditions

REFERENCES

[1] B.R. Teare, "Theory of Hysteresis Motor Torque", *AIEE Transactions*, vol. 59, pp. 907-912, 1940.

[2] M.A. Copeland and G.R. Slemon, "An Analysis of the Hysteresis Motor: part-I-Analysis of the Idealized Machine", *IEEE Trans. on Power Apparatus and System*, vol. 82, no. 65, pp. 34-42, 1963.

[3] M.A. Copeland and G.R. Slemon, "An Analysis of the Hysteresis Motor: part-II-The Circumferential-flux Machine", *IEEE Trans. on Power Apparatus and Systems*, vol. 83, pp. 619-625, June 1964.

[4] C.M. Ong, *Dynamic Simulations of Electric Machinery*, Prentice-Hall Inc., 1998.

[5] M.A. Rahman, "Analytical Models for Poly phase Hysteresis Motor", *IEEE Trans. on Power Apparatus and Systems*, vol. PAS-92, no. 1, pp. 237-242, 1973.

[6] M.A. Rahman and A.M. Osheiba, "Dynamic Performance Prediction of Poly Phase Hysteresis Motors", *IEEE Trans. on Industry Applications*, vol. 26, no.6, pp. 1026-1033, 1990.

[7] O. M. Badeeb, "Investigation of the Dynamic Performance of Hysteresis Motors using MATLAB/SIMULINK", *Journal of Electrical Engineering*, vol. 56, no. 3-4, pp.106-109, 2005.

[8] A. Halvaei Niasar, M. Zare and H. Moghbelli, "Dynamic Modeling and Simulation of a Super-High-Speed Circumferential-Flux Hysteresis Motor," *Journal of Engineering*, vol. 2013, Article ID 898634, pp. 1-7, January 2013.

[9] J. Nitao, E. T. Scharlemann, and B. A. Kirkendall, "Equivalent Circuit Modeling of Hysteresis Motors", U.S. Department of Energy by Lawrence Livermore National Laboratory Report no. LLNL-TR-416493, July 2009.

[10] S. Miyairi and T. Kataoka, "A Basic Equivalent Circuit of the Hysteresis Motor", *IEE Journal, Japan*, vol. 85, pp. 41-50, 1965.

[11] A. Darabi and H. Lesani, "Modeling and Optimum Design of Disk-Type Hysteresis Motors", *IEEE International Conference on Electrical Machines and Systems (ICEMS)*, pp. 998-1002, October 2007.

[12] A. Darabi, T. Ghanbari, M. Rafiei, H. Lesani, and M. Sanati-Moghadam, "Dynamic Performance Analysis of Hysteresis Motors by a Linear Time-Varying Model", *Iranian Journal of Electrical & Electronic Engineering*, vol. 4, no. 4, pp. 202-215, 2008.

[13] T. Ishikawa and T. Kataoka, "V curve of hysteresis motor", *IEE Proceedings-B, Electric Power Applications, Electric Power Applications*, vol. 138, no. 3, pp. 137-141, May 1991.

[14] S. F. Rabbi, M. A. Rahman, "Equivalent Circuit Modeling of an Interior Permanent Magnet Hysteresis Motor", *The 2014 Canadian Conference on Electrical and Computer Engineering*, 2014, pp. 1-5.

[15] S. F. Rabbi, M. Halloran, T. LeDrew, A. Matchem and M. A. Rahman, "Modeling and V/F control of a Hysteresis Interior Permanent Magnet Motor", IEEE International Advance Computing Conference (IACC), 2014, pp. 1-8.

7th Power Electronics, Drive Systems & Technologies Conference (PEDSTC 2016)
16-18 Feb. 2016, Iran University of Science and Technology, Tehran, Iran

Design and Analysis of a Compound SRM and PM Relay for Automatic Drilling tools

O. Safdarzadeh
Department of Electrical Engineering
Shahid Beheshti Univ. G. C
Tehran, Iran
omidSafdarzadeh@chmail.ir

E. Afjei
Department of Electrical Engineering
Shahid Beheshti Univ. G. C
Tehran, Iran
e-afjei@sbu.ac.ir

M. Nezamabadi
Department of Electrical Engineering
Shahid Beheshti Univ. G. C
Tehran, Iran
m_nezamabadi@sbu.ac.ir

A. Mosallanejad
Department of Electrical Engineering
Shahid Beheshti Univ. G. C
Tehran, Iran
mosallanejad@pwut.ac.ir

Abstract— Drilling is performed manually by drill tables or automated multi-motor systems. In this paper a specific machine is designed to be used as an automatic drilling tool. The machine moving part can move rotationally and linearly. The automated multi-motor drilling system can be substituted by this machine. The rotary motion is based on a 6/4 switched reluctance motor (SRM) while the linear force pushing the rotor toward the objective surface is also produced by specific operation of the SRM. The linear force pulling the rotor back to its initial position is provided by an innovative design of permanent magnet (PM) relay. Due to the lack of active coil the whole system efficiency is improved. Theoretical equations governing the machine are derived and the machine 3-D model is defined for finite element analysis. The results are extracted to validate the derived equations and indicate the feasibility of implementation.

Keywords—SRM; PM relay; compound machine; drilling tool; automated systems;

I. INTRODUCTION

Utilizing automated systems instead of manual systems have been extended in various industries. One of the tools used in numerous applications is drilling machine [1]. It is employed in various applications from automobile industry to PCB manufacturing systems [2].Some kinds of these machines are used by hand and some others are automatic [3]. The manual models which are mainly comprised of a simple motor, a gear box and a mechanical table require a human user who provides the pushing force as shown in Fig. 1_a [4], [5]. In automatic models all the drilling process is performed by machines without human direct control as shown in Fig. 1_b [4]. Automatic models require at least two separate motors and a complex controller [6]. One part of this system is a rotary motor rotating the drill bit independently. The other part, which is responsible for linear motion, is introduced in two different mechanisms. First mechanism consists of a linear motor which directly makes the linear force. A rotary motor

with a mechanical system for converting the rotational torque to linear force is used in the second mechanism [7].

Manual drilling machines are seen less in high-tech industry because of requiring a human user. That increases not only the cost production but also the drilling error. Instead of manual machines, precise automatic drilling systems are used widely even though cost of initial investment, maintenance issues and big volume occupation are the obstacles in front of the industries [8].

In this paper an innovative design is presented as a drilling machine consisting two main parts. The first part is a SRM which its rotor moves linearly along the rotational axis in addition to conventional rotation. In another words, exciting the stator windings moves the rotor linearly and rotationally. The SRM operation is discussed completely in section II.

a   b
Fig. 1.   Different types of drilling systems

Second part of the design consists of a cylindrical laminated steel rotor moving along the rotational axis and a stator equipped with permanent magnets. Rotor of the relay shares the same shaft with the rotor of the SRM (Fig. 2) therefore the both rotors together are called "rotor-set" in this

978-1-5090-0376-1/16 $31.00 © 2016 IEEE

paper as a solid integrated object. The machine has two operational modes. In ON-mode, turning on the SRM moves the rotor-set downward and performs the drilling action. After that in OFF-mode, turning off the windings, pulls the rotor-set back to the initial position by PM relay. The design of PM relay is explained in section II.

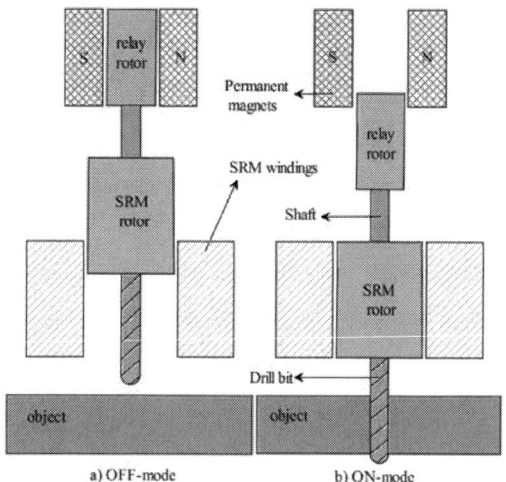

Fig. 2.   The machine operation modes

II. THE MACHINE DESIGN

A. Switched reluctance motor

As mentioned earlier, SRM is the first part of the machine. SRM rotor moves along the rotational axis in addition to its conventional rotation. In the OFF-mode in which none of the stator windings are excited, only a part of the rotor is inside the stator. Exciting the stator windings not only rotates the rotor but also pulls it downward due to the reduction of reluctance in both direction between the rotor and the stator as shown in Fig. 3.

For designing the motor, 6/4 pattern is chosen arbitrarily. Other patterns affect the machine specifications but not the operation principles. A 12 volt power supply and 100 watt as the output power is assumed. 0.77 tesla and 11000 amperes are selected for specific magnetic and electrical loadings, respectively. According to [9] the coefficient of the output power equation can be derived from (1).

$$C_0 = \pi^2 B_{av}\, ac = \pi^2 \times 0.77 \times 11000 = 84000 \qquad (1)$$

Where, B_{av} is specific magnetic loading, ac is specific electrical loading and C_0 is coefficient of output power equation.

Output power equation is derived from (2).

$$Q = C_0 D^2 L\ n \qquad (2)$$

Where, Q is output power, D is diameter of air gap, L is length of air gap and n is nominal rotational speed in rotation per second (r/s⁻¹) [9]. Substituting $n = 28$ r/s⁻¹ and (1) in (2) results in (3).

$$D^2 L = \frac{100 \times 10^{-3}}{84 \times 28} = 4.25 \times 10^{-5} \qquad (3)$$

According to the constraints of the target application L and D could be determined. Nevertheless assuming $L/D = 1.5$ results in $D = 35mm$ and $L = 53mm$. Mechanical width of the rotor and the stator teeth are chosen 30 and 30 degree, respectively. Minimum and maximum width limits are derived from (4) and (5), respectively.

$$w_r\ , w_s \ge \frac{2\pi}{q N_r} = \frac{360°}{3 \times 4} \qquad (4)$$

$$w_r\ , w_s \le \frac{2\pi}{q N_r} = \frac{360°}{4} \qquad (5)$$

Fig. 3 shows the structure of the SRM based on the performed calculations. As shown in Fig. 3 force (F) and torque (□) refer to linear and rotational forces, respectively.

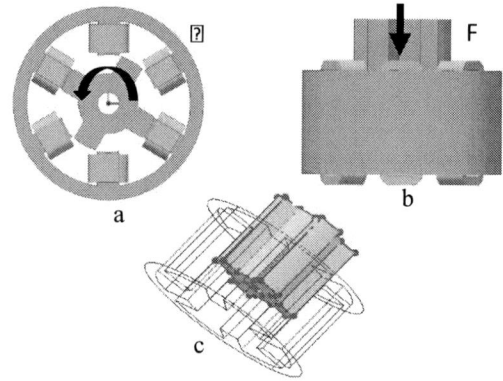

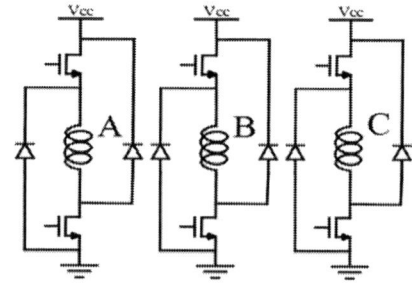

Fig. 3.   The SRM structure

Fig. 4 shows a conventional driver for SRMs which is utilized for the motor control [10]. There is a half bridge circuit for each phase consists of two switches and two diodes. The structure is able to return the exciting energy from energized winding to power supply when the phase is turned off.

Fig. 4.   The SRM driver

Exciting pattern selected here is shown in **Fig. 5**. The excitation pattern has periods of 90 mechanical degrees divided by three equal sections. In each section one phase is excited and the other phases are off.

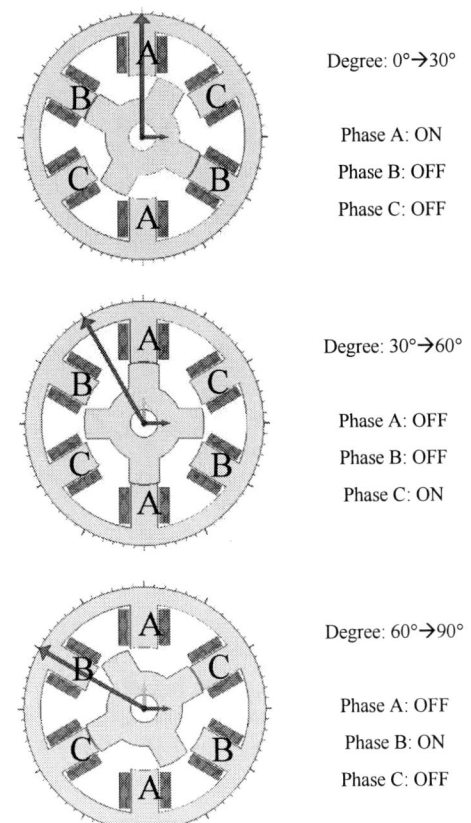

Degree: 0°→30°

Phase A: ON

Phase B: OFF

Phase C: OFF

Degree: 30°→60°

Phase A: OFF

Phase B: OFF

Phase C: ON

Degree: 60°→90°

Phase A: OFF

Phase B: ON

Phase C: OFF

Fig. 5.   The excitation pattern

As mentioned before, the rotor rotates while it translates along the axial direction. The torque and force which move the rotor is extracted from reluctance variation. One of the rotor and stator teeth is shown in Fig. 6.

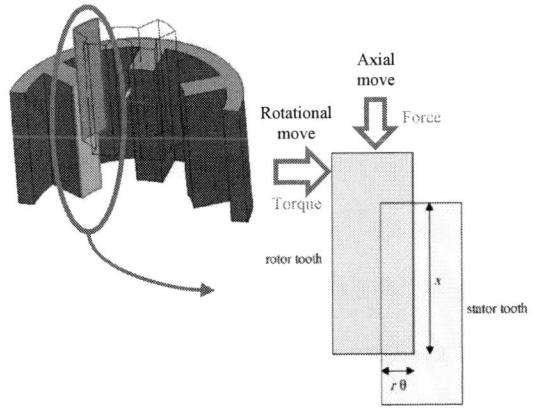

Fig. 6.   The produced torque and force in the SRM

According to the parameters in Fig. 6, the air gap reluctance can be obtained from (6). Where R, l_g, x, r and θ

are the reluctance, radial air gap distance, axial distance variation, air gap radius and angle variation respectively.

$$R = \frac{l}{\mu_0 A} = \frac{l_g}{\mu_0 \, x \, r \, \theta} \tag{6}$$

Therefore the inductance is derived from (7) where N is the turns of the winding.

$$L = \frac{N^2}{R} = \frac{N^2 \mu_0 \, x \, r \, \theta}{l_g} \tag{7}$$

Due to conservation of energy the torque and the force are derived from (8) and (9), respectively.

$$\tau_\theta = \frac{1}{2} \frac{dL}{d\theta} i^2 = \frac{N^2 \mu_0 \, x \, r}{2 l_g} i^2 \tag{8}$$

$$F_x = \frac{1}{2} \frac{dL}{dx} i^2 = \frac{N^2 \mu_0 \, r \, \theta}{2 l_g} i^2 \tag{9}$$

B. Permanent magnet relay

Second part of the machine is a PM relay operating without any windings. As shown in **Fig. 7**, a laminated cylinder works as the translating rotor for the relay. The stator is a steel tube equipped by two permanent magnets which one of them is magnetized radially inward and the other is magnetized radially outward as shown in **Fig. 7**.

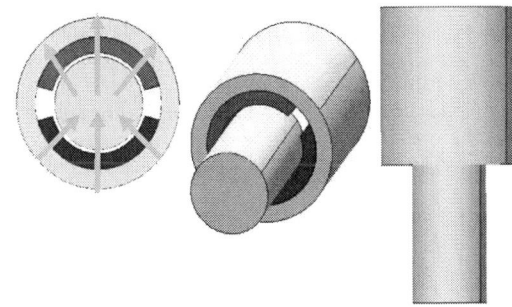

Fig. 7.   The PM relay structure

As mentioned earlier, in ON-mode, SRM moves the rotor-set downward. Due to the decreasing overlapping area between permanent magnets and the rotor of the relay, some energy is stored in the permanent magnets. In OFF-mode the energy is used to produce a linear force pulling up the rotor-set to the first position.

To study the relay operation, the relay is divided in to non-overlapping (A) and overlapping (B) sections as shown in **Fig. 8.**

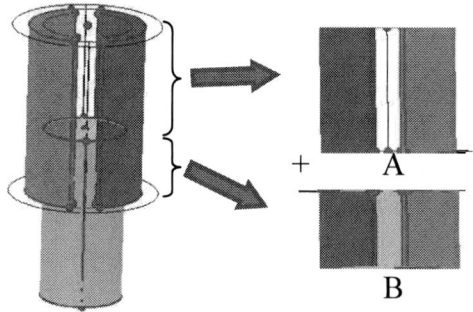

Fig. 8.   The sections of the relay

To find the total energy stored in PMs an equal static magnetic load line for magnets shown in **Fig. 9**. Line A and B represent static magnetic load lines for the magnets in section A and B, respectively[11].

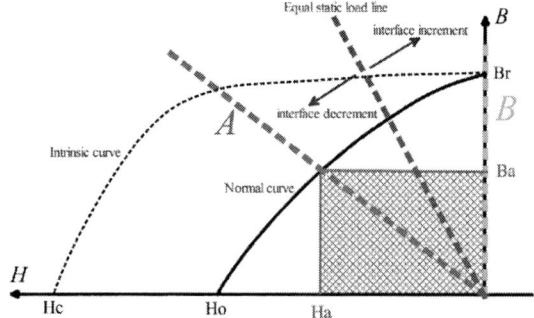

Fig. 9.   Stored energy in the permanent magnets

As shown in Fig. 9, varying the overlapping area between the rotor and the magnets of the stator rotates the static magnetic load line around the origin. Intersection of the static magnetic load line and the normal demagnetization curve indicates the operating point of the relay magnets. According to [12], stored energy in the permanent magnet is derived from (10).

$$w = VBH \qquad (10)$$

Where, V, B and H are volume, flux density and intensity of the permanent magnets, respectively. Reduction of the overlapping area between the rotor and the magnets is equivalent to increasing the volume of section A and decreasing the volume of section B. According to (11), this translation raises the overall stored energy in PMs.

$$w = w_A + w_B = V_A B_A H_A + V_B B_B H_B$$
$$H_B \cong 0 \Rightarrow w = V_A B_A H_A \qquad (11)$$

In OFF-mode, relay magnets produce upward linear force to reduce their stored energy. This force is presented in (14) derived from (12) and (13).

$$\Delta w = F.\Delta x \qquad (12)$$

$$\Delta (V_A B_A H_A) = B_A H_A \Delta V_A = B_A H_A S \Delta x \qquad (13)$$

$$\Rightarrow F = B_A H_A S \qquad (14)$$

Where, S and F are the cross-section of the magnets area and the linear force.

Finally, according to the target application, dimensions and materials are extracted from (14). The machine parameters in this paper are summarized in TABLE I.

TABLE I.   THE MAIN MACHINE CHARACTERISTICS

SRM poles pattern	6/4
Air gap's radius	17.5 mm
Air gap's lenght	35 mm
Each winding turns	50
Magnet material	Ceramic ferrit 4
Magnet length	30 mm
Megnat cross-section area	85 mm^2
Relay's rotor radius	8.5 mm

Type and dimensions of the PMs depend on the weight of the rotor set (in vertical operation as shown in Fig. 2) which in this paper is 180 grams.

III. FINITE ELEMENT ANALYSIS OF THE MACHINE

According to TABLE I. the machine model is imported in a finite element method (FEM)-aided software. For simulating the machine, FEM is applied to calculate the torque and the force. The rotor-set is rotated in three linear positions as shown in **Fig. 10**. Since at any moment just one phase is excited, the results which are shown for phase-A, represent the whole machine. The rotational angle ($\Delta\theta$) varies from 0 degree (un-alignment for phase A) to 30 degrees (full-alignment for phase A).

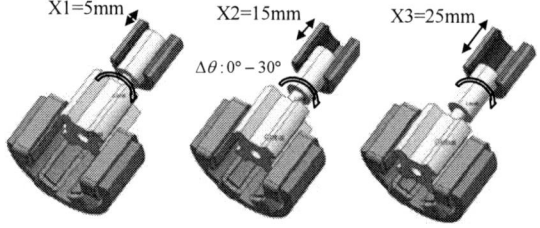

Fig. 10.   The rotor-set positions

Fig. 11 shows the torque produced by SRM for the positions described in Fig. 10.

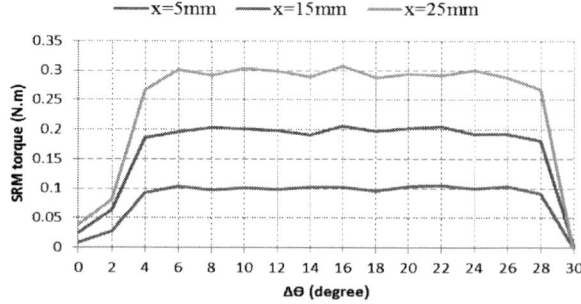

Fig. 11.   The produced SRM torque

Similarly linear force produced by SRM is shown in **Fig. 12** for the positions described in Fig. 10.

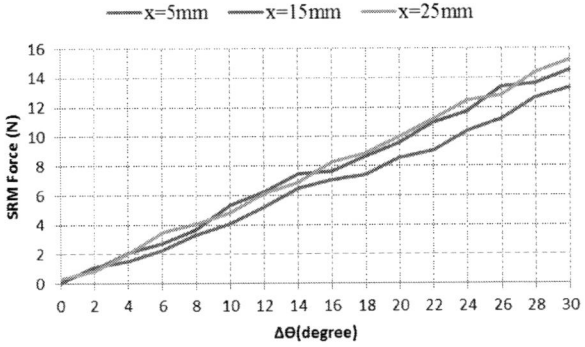

Fig. 12.   The produced SRM force

Fig. 13 shows the flux density distribution for two different linear alignments at the beginning of rotary alignment.

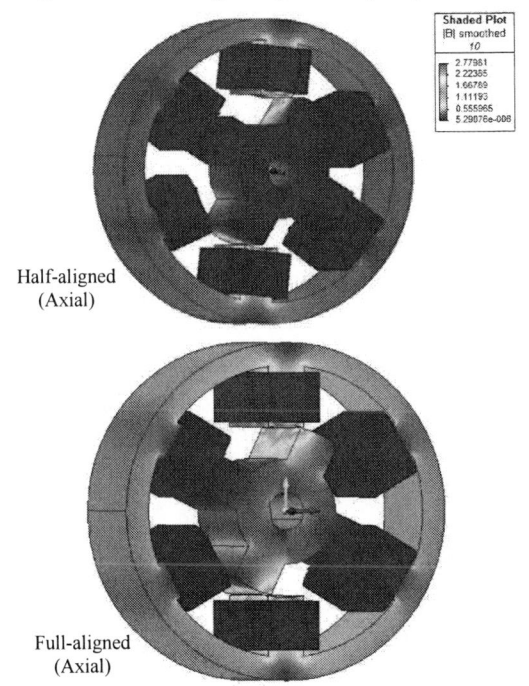

Fig. 13.   FEM analysis for Flux distribution in the machine

Fig. 14 shows linear force produced by the PM relay for the linear displacement span.

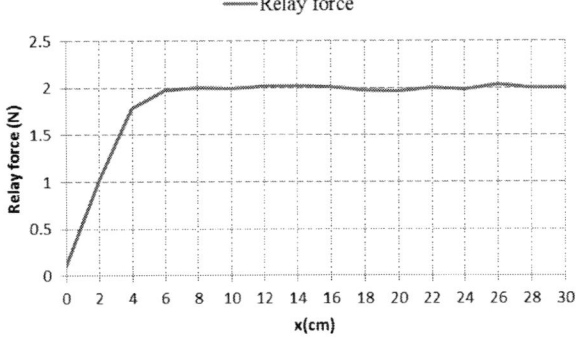

Fig. 14.   The Relay force

Complete structure of the machine is demonstrated in **Fig. 15.**

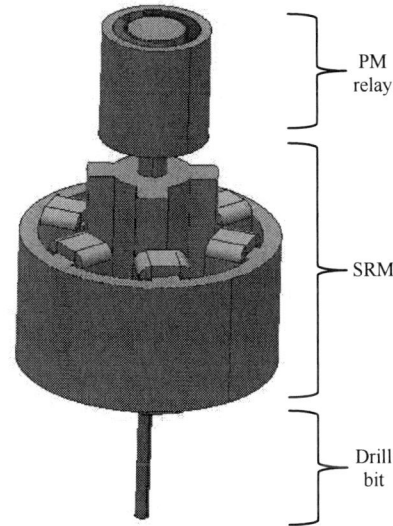

Fig. 15.   The whole construction of the machine

IV. CONCLUSION

In this paper a new machine for drilling purposes is presented. This machine is comprised of a switched reluctance motor and a permanent magnet relay. The relay operates without windings so the machine efficiency increases especially in OFF-mode. Based on parameters presented in TABLE I. 0.3 (N.m.) and 12 (N) are obtained from SRM structure in ON-mode approximately. The machine dimensions could be changed to reach the desired torque and force according to the target application. Also the windings excitation algorithm and driver topology could be specified to fulfill the application. Simplicity of the machine in comparison with the existent drilling systems [4] can gain financial benefits for manufacturers.

978-1-5090-0376-1/16 $31.00 © 2016 IEEE

References

[1] C. J. Singer, E. J. Holmyard, A. R. Hall, and E. P. H. S. T. A. R. Hall, *A History of Technology, V1: From Early Times to Fall of Ancient Empires*: Literary Licensing, LLC, 2011.

[2] M. Daumas, *A History of Technology & Invention: The origins of technological civilization*: Crown Publishers, 2006.

[3] J. E. Landmeyer, *Introduction to Phytoremediation of Contaminated Groundwater: Historical Foundation, Hydrologic Control, and Contaminant Remediation*: Springer Netherlands, 2011.

[4] STROJIMPORT. (2015). *DRILLING MACHINES*. Available: http://www.strojimport.com/drilling-machines/

[5] R. S. Woodbury, *Studies in the History of Machine Tools*: M.I.T. Press, 1972.

[6] A. J. Lynch and C. A. Rowland, *The History of Grinding*: Society for Mining, Metallurgy, and Exploration, 2005.

[7] J. J. Mohr, S. Sengupta, and S. F. Slater, *Marketing of High-technology Products and Innovations*: Prentice Hall, 2009.

[8] B. Guo and G. Liu, *Applied Drilling Circulation Systems: Hydraulics, Calculations and Models*: Elsevier Science, 2011.

[9] E. S. Hamdi, *Design of small electrical machines*: Wiley, 1994.

[10] R. Krishnan, *Switched Reluctance Motor Drives: Modeling, Simulation, Analysis, Design, and Applications*: CRC Press, 2001.

[11] H. C. Lovatt and P. Watterson, "Energy stored in permanent magnets," *Magnetics, IEEE Transactions on*, vol. 35, pp. 505-507, 1999.

[12] T. D. Nguyen, K.-J. Tseng, S. Zhang, and H. T. Nguyen, "A novel axial flux permanent-magnet machine for flywheel energy storage system: design and analysis," *Industrial Electronics, IEEE Transactions on*, vol. 58, pp. 3784-3794, 2011.

7th Power Electronics, Drive Systems & Technologies Conference (PEDSTC 2016)
16-18 Feb. 2016, Iran University of Science and Technology, Tehran, Iran

Investigation of the Effect of Eccentricity in Flux Switching Permanent Magnet Machines

Mostafa Ahmadi Darmani
Faculty of Electrical and Computer Engineering
Islamic Azad University Science and Research Branch
Tehran, Iran
mostafa.ahmadi.d@gmail.com

Seyyed Mehdi Mirimani
Faculty of Electrical and Computer Engineering
Babol University of Technology
Babol, Iran
mirimani@nit.ac.ir

Abstract—This paper studies a 10/12 Flux Switching Permanent Magnet (FSPM) machine under static eccentricity fault to assess the performance of the machine under different operating conditions. Two-dimensional finite element method (FEM) is used to model the complicated structure of the machine in order to obtain accurate results. Furthermore, finite element analysis (FEA) is used to obtain air-gap flux density, machine's torque and the force between stator and rotor under different degrees of static eccentricity. Furthermore, Fourier analysis is performed to study magnetic force and torque profiles. To the best awareness of the authors, the effects of the eccentricity of FSPM machines have not been studied previously.

Keywords— *Eccentricity;Flux Switching Permanent Magnet Machines(FSPM); Finite Element Analysis (FEA)*

I. INTRODUCTION

Permanent-magnet (PM) brushless machines have been extensively used for several applications because of their high efficiency and high torque/power density [1]. Stator-PM machines include Doubly Salient PM (DSPM) machine, Flux Reversal PM (FRPM) machine, and Flux Switching Permanent Magnet (FSPM) machines. In this kind of machines, PMs can be cooled simply and the PMs will not be exposed to the centrifugal forces of a rotating rotor [2]-[5]. FSPM machines have complicated structure in comparison with other stator-PM machines. This type of machines is used as one of high frequency inductor generators [6], [7]. Furthermore, FSPM machines are used in aerospace applications [8], high-speed applications [9], wind turbine application [10] and hybrid electric vehicles [11]. Over the last few years, several publications have concentrated their interest to design, analysis, and improve the FSPM structure in order to attain the desired requirements [12]-[16]. Moreover, some literatures have been done to improve the performance of FSPM machines especially reducing cogging torque [17]-[19] and back-EMF harmonics [20].

Generally, there are numerous types of fault which may occur in electrical machines which are categorized into two main groups according to their causes, external faults and internal faults. Furthermore, these faults are consisted in electrical faults and mechanical faults. Among these deficiencies, bearing faults and eccentricity faults are the most common failures in electrical machines. Static or dynamic

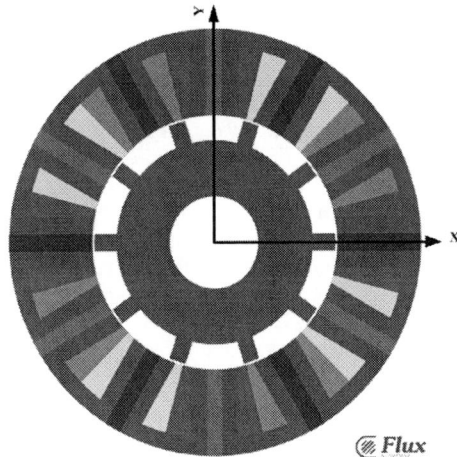

Fig. 1. Schematic of proposed FSPM machine

eccentricity faults are among the mechanical faults that are due to manufacturing imprecision of mechanical parts such as unbalanced mass, bearing deficiency, misalignment and excessive tolerances which cause a vibration and ultrasound noise [21]. There is no literature discussing the effects of the eccentricity on an FSPM and this paper is the first step in such an analysis for FSPM.

In this paper, a three phase 12/10 flux switching permanent magnet motor under static eccentricity fault is simulated at different percentage of static eccentricity using two-dimensional (2D) finite element analysis (FEA). Flux density distribution of the machine is calculated and the output torque is acquired. In addition, the force between rotor and stator of the machine and torque profile with 40% eccentricity is derived and compared with a faultless motor.

II. MOTOR SPECIFICATIONS

The proposed FSPM machine is a three-phase 12/10 stator/rotor pole motor, as shown in Fig. 1. Major parameters of the machine are given in Table I.

III. FIINITE ELEMENT SIMULATION

In order to study eccentricity in FSPM machines, it is essential to use an accurate modeling of all machine details such as geometry and non-linear properties of the magnetic materials. Analytical methods are complicated when precise evaluation of saturation effects is needed.

978-1-5090-0376-1/16 $31.00 © 2016 IEEE

TABLE I.   MAJOR PARAMETERS OF FSPM MACHINE

Quantity	Value
Stator pole numbers (N_s)	12
Rotor pole numbers (N_r)	10
Outer diameter of stator (D_{so})	90mm
Active axial length (L_{st})	25mm
Air-gap length (g)	0.5mm
Rotor pole width (L_{pr})	4mm
Outer diameter of rotor (D_{or})	55mm
PM thickness (L_{PM})	3.6mm
Stator tooth width (L_{st})	3.6mm
Stator back iron thickness (y)	3.6mm
Number of turns per phase (N_{ph})	72
Rated current (I_a)	14A
Speed (N_s)	400rpm

This issue can be overcome by using FEA which can apply precise numerical approach. The finite element model of the FSPM machine is created in CEDRAT Flux package in order to study and analysis static eccentricity. Fig. 1 shows the schematic of proposed FSPM machine. The machine has concentrated winding with twelve coils which have star connection. The coils of each phase are connected in series and supplied by three-phase sinusoidal current waveforms as it is shown in Fig. 2. The circuit is coupled to magnetic domain in this study. Two different coordinate systems are defined for each of rotor and stator. Stator coordinate system is fixed and rotor coordinate system varies along positive direction of x-axis in order to generate eccentricity and the rotor rotates around globalist own coordinate system.

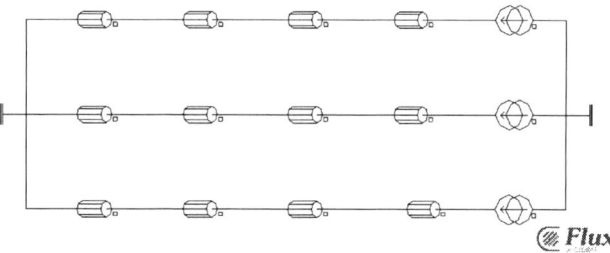

Fig. 2.   Circuit diagram of proposed FSPM machine

IV.   STATIC ECCENTRICITY IN FSPM MACHINES

Bearing faults are the most regular sources of failure in electric machines that causes ball fatigue which is the main result of machine vibration. These vibrations in the air-gap are considered as non-uniform air-gap which is called eccentricity. Therefore, it can conclude that bearing faults change the air-gap balance like eccentricity faults.

When there is a non-uniform air-gap between rotor and stator, the axis of stator, the axis of rotor and rotation axis of rotor are misaligned. Eccentricity can be categorized into three types, static eccentricity, dynamic eccentricity and mixed eccentricity. In the static eccentricity, the rotor rotates around its own symmetrical axis, but the axis does not coincide with the stator center as shown in Fig. 3.

In [22], [23] the definition of Static Eccentricity Factor (SEF) is proposed which can be defined as follows:

$$SEF = \frac{r}{g} \times 100 \qquad (1)$$

Where "r" is the offset between the rotor and the stator axes and "g" is the radial air-gap length in faultless condition.

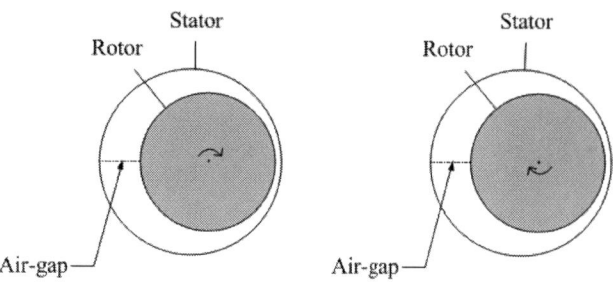

Fig. 3.   Schematic representation of static eccentricity

V.   EFFECT OF ECCENTRICITY ON THE FLUX DENSITY

A. Flux Density Distribution

Flux density distribution of the faultless motor and the motor with 40% static eccentricity are illustrated in Figs. 4 and Fig. 5 respectively. As mentioned in part III, the static eccentricity has happened along x-axis. It is observed that by decreasing the air-gap length in right side of the machine, machine's reluctance in x-direction and consequently amplitude of magnetic flux density growths to 2.911 T in hotspots. However, on the opposite side, the air-gap length of the machine goes up and consequently reluctance and amplitude of magnetic flux density declines to 2.21 Tesla. So, it can be concluded that the static eccentricity causes asymmetric magnetic flux density distribution and harmonic components in the air-gap field.

B. Air-gap Flux Density

The air-gap flux density is one of the crucial features in electrical machines because every change in this characteristic will affect other characteristics of motor. On the other hand, the air-gap flux density shows the machine condition. Thus, by monitoring air-gap flux density it is possible to foresee and diagnosis. This characteristic in static eccentricity which is non-uniform air-gap length should be analyzed. Air-gap flux density in the radial direction is computed using 2D-FEM. Fig. 6 shows the normal component of air-gap flux density computed by 2D-FEM simulation for both faultless motor and motor with SEF=40%.

By comparing the results it can be concluded that static eccentricity has high impact on air-gap flux density of the motor. At the right side of the machine where the air-gap has the smallest length, the air-gap flux density has the peak values and on the other side where the air-gap has the maximum length, the air-gap flux density has the minimum value. This happens because the flux path reluctance is depended on air-gap length and thus varying small value of air-gap has a great impact on the value of air-gap flux density. In Fig. 7 maximum value of air-gap flux density for various values of SEF is shown. It can be seen that the static eccentricity lead to increase the maximum air-gap flux density.

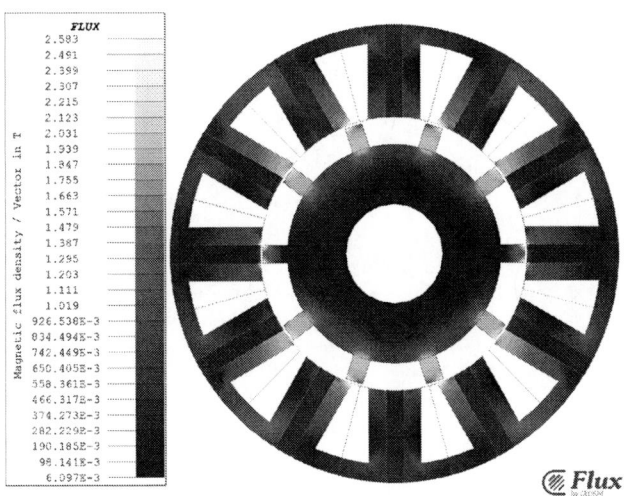

Fig. 4. Flux disturbution of FSPM motor with SEF=0%

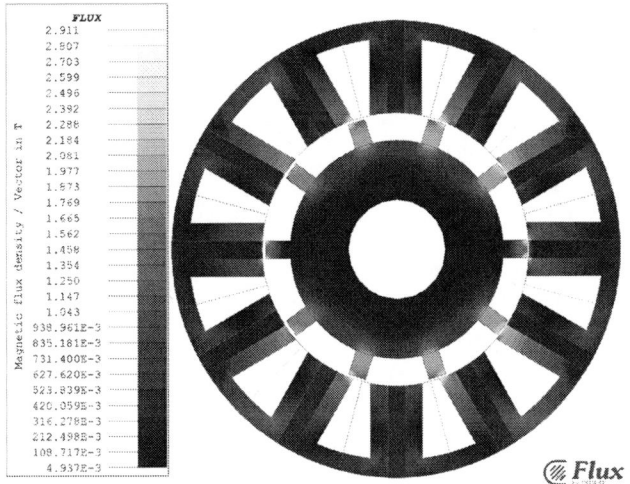

Fig. 5. Flux disturbution of FSPM motor with SEF=40%

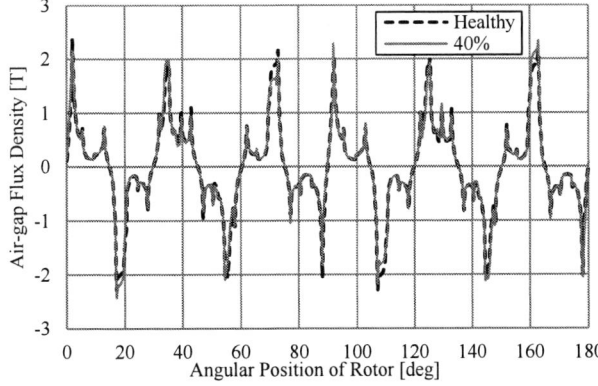

Fig. 6. Air-gap flux density of the faultless and motor with 40% eccentricity

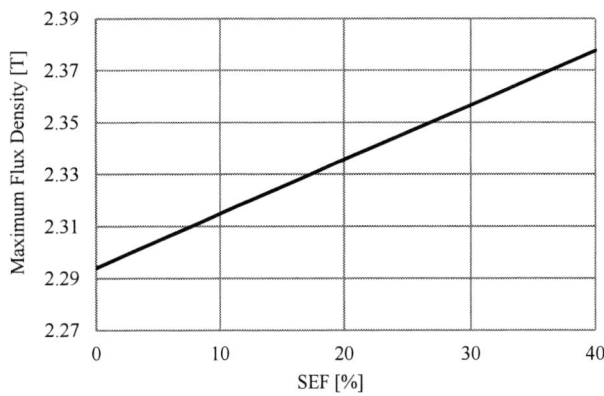

Fig. 7. Maximum air-gap flux density versus different values of SEF

VI. EFFECT OF ECCENTRICITY ON MOTOR FORCE CHARACTERISTIC

Electrical machines eccentricity is an important consequence of bearing deterioration. Eccentricity disrupts the magnetic field and the balance between the magnetic forces of the machine and consequently resulting in unbalanced magnetic force. This unbalanced force places more loads on the bearing which can cause an increase in vibration at certain frequencies specified by the motor configuration [24].

The machine that is studied in this paper is a radial flux FSPM motor. Generally, when rotor and stator place in one centerline and the air-gap between the rotor and stator is uniformed, the resultant of acting forces between rotor and stator of radial flux machines is near zero. But, when there is an eccentricity fault, the resultant of acting forces between rotor and stator is not near to zero any more. FEM simulation shows that the magnetic force between the rotor and the stator is periodic and the period of this force can be expressed as follows:

$$\tau = \frac{2\pi}{N_r} rad = \frac{360}{N_r} deg \qquad (2)$$

Fig. 8 indicates the effect of the static eccentricity on the magnetic force of the faultless motor and the motor with 40% eccentricity. This magnetic force is the result of interaction between the air-gap flux density and stator and rotor teeth. The exact nature of the produced force waves is a function of motor dimensions and stator and rotor pole numbers combinations. Furthermore, it can be seen that the motor force is increased by increasing SEF.

Fig.9 (a) and Fig.9 (b) indicates the results of Fourier analysis of magnetic force performed for SEF=0% and SEF=40%, respectively. It is observed that the magnetic force between stator and rotor produces extra 1st, 2nd, 3rd, 4th harmonics orders. For the machine with SEF=40% the amplitude of magnetic force content has a significant value in comparison with the machine with SEF=0% which leads to motor condition deterioration.

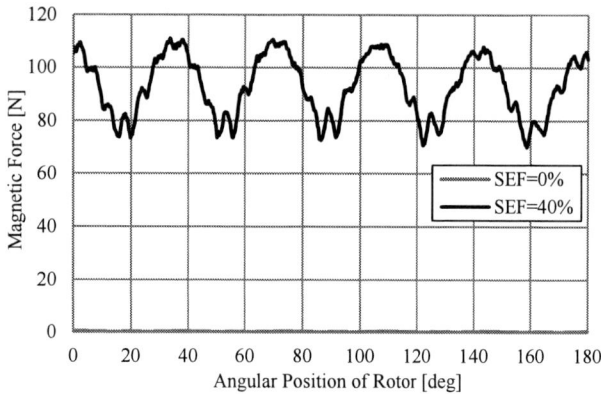

Fig. 8. A period of radial forces between rotor and stator at 40% eccentricity

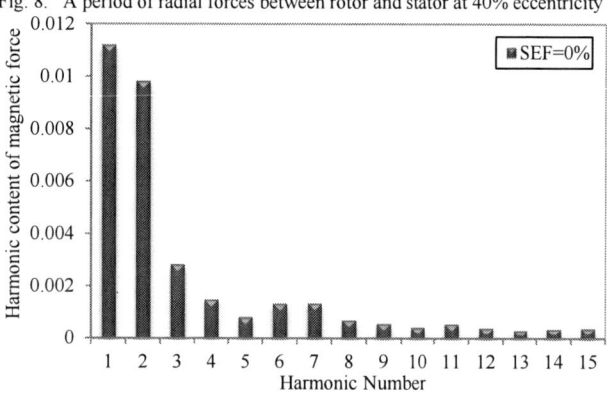

(a)

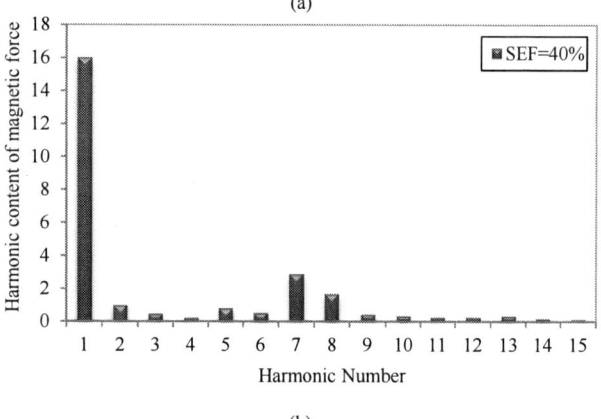

(b)

Fig. 9. Harmonic of magnetic force between stator and rotor, a): SEF=0%, b): SEF=40%

Moreover, as it is indicated in Fig. 10, the mean value of magnetic force increases by increasing the SEF.

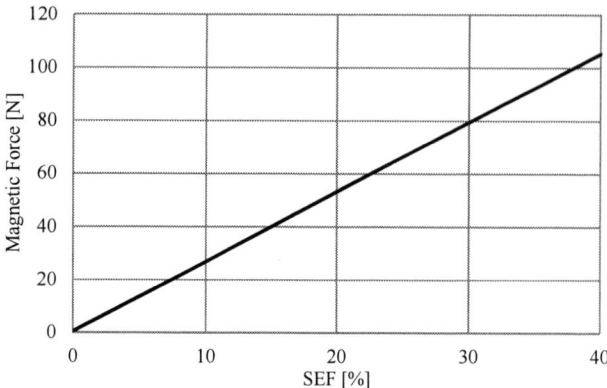

Fig. 10. Magnetic force versus different values of SEF

VII. EFFECT OF ECCENTRICITY ON MOTOR TORQUE CHARACTERISTIC

As it was shown in section VI, static eccentricity increases the amount of forces between stator and rotor. This non-uniformed air-gap may cause an undesirable unbalanced torque and bearings defect [22]. Usually, the Maxwell stress tensor method is used to calculate the instantaneous torque of machines. In symmetric condition, vectorial sum of the radial components of the acting force is equal to zero and only tangential torque causes to produce torque. However, in asymmetric condition, the rotor is gravitated to stator, so the radial components of the acting force have not been in a balanced condition any more. Thus, it should be taken into account in calculation of torque [25].

The presented method of calculating the machine's torque is valid for doubly salient machines too. But, according to complex structure of FSPM machines, it is really difficult to calculate the machine's torque based on analytical methods. Hence FEA is used to obtain machine's torque. Fig. 11 shows the waveform of torque which is produced in faultless and faulty machine with SEF=40%. FEA is used in order to study this torque under different values of SEF. The machine in faultless condition has a little lower torque ripple in comparison with the machine with SEF=40%. Also by increasing the SEF the mean value of the machine torque is reduced which is indicated in Fig. 12.

Fourier analysis of torque profile is performed for both SEF=0% and SEF=40% and the results are shown in Fig. 13. As it can be seen there is an increment in the torque 7th harmonic order by increasing the SEF.

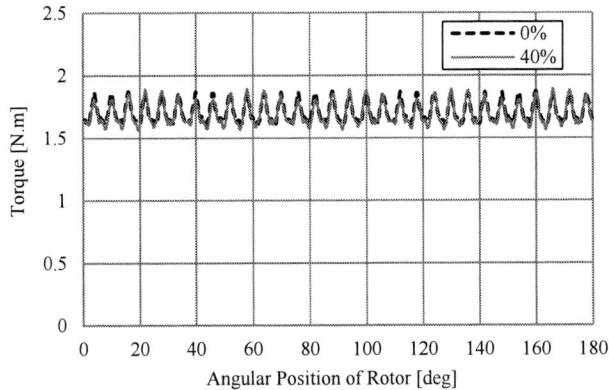

Fig. 11. Torque of the faultless and motor with SEF=40%

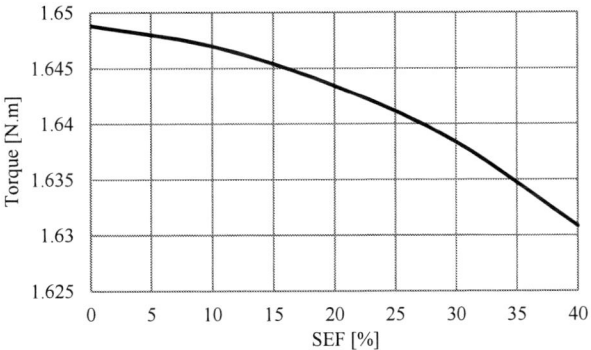

Fig. 12. Variation of fundamental torque with different values of SEF

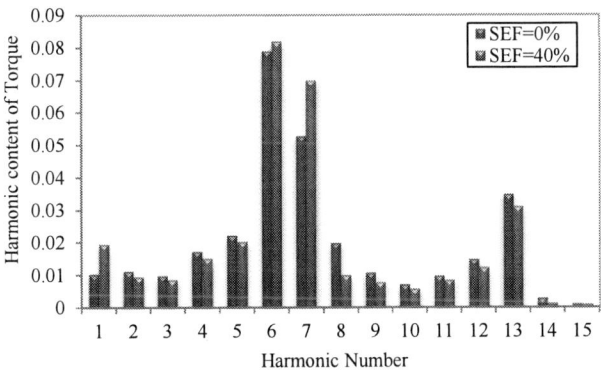

Fig. 13. Harmonic of torque in the faultless motor and motor with SEF=40%

VIII. CONCLUSION

The effect of static eccentricity on an FSPM motor has been studied through 2D-FEA. Use of this precise modeling makes it possible to attain required characteristics.

The results of FEM model show that static eccentricity deforms air-gap magnetic field and consequently leads to produce an unbalanced magnetic force between rotor and stator. Moreover, it is discovered that the produced magnetic force has a remarkable value. Furthermore, it is observed that the eccentricity does not vary the output torque profile significantly. Also, the machine torque has elevated 7th

harmonic order which can be used for fault detection based on machine vibration analysis. Also magnetic force between rotor and stator is significantly elevated which leads to machine condition deterioration. These facts have not been studied before and the investigation can be useful for future studies.

REFERENCES

[1] Yangsheng Chen; Haodong Yang; Ze Ying Han, "Investigation of electromagnetic vibration of permanent magnet brushless machines," in *Electrical Machines and Systems, 2008. ICEMS 2008. International Conference on* , vol., no., pp.621-626, 17-20 Oct. 2008

[2] Ming Cheng; Wei Hua; Jianzhong Zhang; Wenxiang Zhao, "Overview of Stator-Permanent Magnet Brushless Machines," in *Industrial Electronics, IEEE Transactions on* , vol.58, no.11, pp.5087-5101, Nov. 2011

[3] Chunhua Liu; Chau, K.T.; Jiang, J.Z.; Niu, S., "Comparison of Stator-Permanent-Magnet Brushless Machines," in *Magnetics, IEEE Transactions on*, vol.44, no.11, pp.4405-4408, Nov. 2008

[4] Yuefeng Liao; Liang, F.; Lipo, T.A., "A novel permanent magnet motor with doubly salient structure," in *Industry Applications Society Annual Meeting, 1992., Conference Record of the 1992 IEEE* , vol., no., pp.308-314 vol.1, 4-9 Oct. 1992

[5] Deodhar, R.P.; Andersson, S.; Boldea, I.; Miller, T.J.E., "The flux-reversal machine: a new brushless doubly-salient permanent-magnet machine," in *Industry Applications, IEEE Transactions on* , vol.33, no.4, pp.925-934, Jul/Aug 1997

[6] Rauch, S.E.; Johnson, L.J., "Design Principles of Flux-Switch Alternators," in *Power Apparatus and Systems, Part III. Transactions of the American Institute of Electrical Engineers* , vol.74, no.3, pp., Jan. 1955

[7] Zhu, Z.Q., "Switched flux permanent magnet machines — Innovation continues," in *Electrical Machines and Systems (ICEMS)*, 2011 International Conference on , vol., no., pp.1-10, 20-23 Aug. 2011

[8] Sanabria-Walter, C.; Polinder, H.; Ferreira, J.A., "High-Torque-Density High-Efficiency Flux-Switching PM Machine for Aerospace Applications," in *Emerging and Selected Topics in Power Electronics, IEEE Journal of* , vol.1, no.4, pp.327-336, Dec. 2013

[9] Thomas, A.S.; Zhu, Z.Q.; Jewell, G.W., "Comparison of flux switching and surface mounted permanent magnet generators for high-speed applications," in *Electrical Systems in Transportation, IET* , vol.1, no.3, pp.111-116, September 2011

[10] Akuru, U.B.; Kamper, M.J., "Comparative advantage of flux switching PM machines for medium-speed wind drives," in *Domestic Use of Energy (DUE), 2015 International Conference on the* , vol., no., pp.149-154, March 31 2015-April 1 2015

[11] Wei Hua; Gan Zhang; Ming Cheng, "Investigation and Design of a High-Power Flux-Switching Permanent Magnet Machine for Hybrid Electric Vehicles," in *Magnetics, IEEE Transactions on* , vol.51, no.3, pp.1-5, March 2015

[12] Zhu, Z.Q.; Chen, J.T., "Advanced Flux-Switching Permanent Magnet Brushless Machines," in *Magnetics, IEEE Transactions on* , vol.46, no.6, pp.1447-1453, June 2010

[13] Zhu, Z.Q.; Pang, Y.; Howe, D.; Iwasaki, S.; Deodhar, R.; Pride, A., "Analysis of electromagnetic performance of flux-switching permanent-magnet Machines by nonlinear adaptive lumped parameter magnetic circuit model," in *Magnetics, IEEE Transactions on* , vol.41, no.11, pp.4277-4287, Nov. 2005

[14] Sulaiman, E.; Kosaka, T.; Matsui, N., "Design optimization of 12Slot-10Pole hybrid excitation flux switching synchronous machine with 0.4kg permanent magnet for hybrid electric vehicles," in *Power Electronics and ECCE Asia (ICPE & ECCE), 2011 IEEE 8th International Conference on* , vol., no., pp.1913-1920, May 30 2011-June 3 2011

[15] Jurca, F.N.; Martis, C., "Optimal design of a flux-switching permanent magnet machine for small power automotive applications," in *Power Electronics, Electrical Drives, Automation and Motion (SPEEDAM), 2014 International Symposium on* , vol., no., pp.439-443, 18-20 June 2014

[16] Chen, J.T.; Zhu, Z.Q., "Winding Configurations and Optimal Stator and Rotor Pole Combination of Flux-Switching PM Brushless AC Machines," in *Energy Conversion, IEEE Transactions on* , vol.25, no.2, pp.293-302, June 2010

[17] Zhu, Z.Q.; Thomas, A.S.; Chen, J.T.; Jewell, G.W., "Cogging Torque in Flux-Switching Permanent Magnet Machines," in *Magnetics, IEEE Transactions on* , vol.45, no.10, pp.4708-4711, Oct. 2009. doi: 10.1109/TMAG.2009.2022050

[18] Daohan Wang; Xiuhe Wang; Sang-Yong Jung, "Reduction on Cogging Torque in Flux-Switching Permanent Magnet Machine by Teeth Notching Schemes," in *Magnetics, IEEE Transactions on* , vol.48, no.11, pp.4228-4231, Nov. 2012

[19] Zhu, Z.Q.; Chen, J.T.; Pang, Y.; Howe, D.; Iwasaki, S.; Deodhar, R., "Analysis of a Novel Multi-Tooth Flux-Switching PM Brushless AC Machine for High Torque Direct-Drive Applications," in *Magnetics, IEEE Transactions on* , vol.44, no.11, pp.4313-4316, Nov. 2008

[20] Bobba, D.; Li, Y.; Sarlioglu, B., "Harmonic Analysis of Low Stator Slot and Rotor Pole Combination FSPM Machine Topology for High Speed," in *Magnetics, IEEE Transactions on* , vol.PP, no.99, pp.1-1

[21] Mirimani, S.M.; Vahedi, A.; Marignetti, F., "Effect of static eccentricity in back-EMF of Axial Flux Permanent Magnet Machines," in *Power Electronics, Electrical Drives, Automation and Motion (SPEEDAM), 2012 International Symposium on* , vol., no., pp.464-468, 20-22 June 2012

[22] Mirimani, S.M.; Vahedi, A.; Marignetti, F., "Effect of Inclined Static Eccentricity Fault in Single Stator-Single Rotor Axial Flux Permanent Magnet Machines," in *Magnetics, IEEE Transactions on*, vol.48, no.1, pp.143-149, Jan. 2012

[23] Mirimani, S.M.; Vahedi, A., "Effects of static eccentricity in axial flux permanent magnet machines," in *Power Electronic & Drive Systems & Technologies Conference (PEDSTC), 2010 1st* , vol., no., pp.311-315, 17-18 Feb. 2010

[24] Kim, U.; Lieu, D.K., "Effects of magnetically induced vibration force in brushless permanent-magnet motors," in *Magnetics, IEEE Transactions on* , vol.41, no.6, pp.2164-2172, June 2005

[25] Katsuyoshi, E; Kazurou, H; Yoshiyuki, I.; Yoshiyuki, T., "A calculation of torque in motors considering rotor eccentricity", Electrical Engineering in Japan, Volume 132, Issue 4, pages 53–61, September 20

Force Calculation with Complete Modeling of End-Effect in Single-Sided Linear Induction Motors

Amir Zare-Bazghaleh
Department of Electrical Engineering
Langroud Branch
Islamic Azad University
Langroud, Iran.
amir.zarebazghaleh27@gmail.com

Esmaeil Fallah-Chulabi
Department of Electrical Engineering
University of Guilan
Rasht, Iran.
fallah_e@guilan.ac.ir

Seyed Hamid Shahalami
Department of Electrical Engineering
University of Guilan,
Rasht, Iran.
shahalami@guilan.ac.ir

Abstract— **Longitudinal end-effect (LEE) is an important phenomenon which affects the Linear Induction Motor, adversely. This phenomenon leads to some problems such as considerable changes in equivalent circuit (EC) parameters by varying speed, non-uniform field distribution along the air-gap and so, weaker output performance comparing to rotary induction motors (RIMs). This paper proposes to include LEE of Single-Sided Linear Induction Motor (SLIM) in traction and normal forces calculation completely, using the Poynting theorem. By help of the method it will be possible to include the effect of LEE even at entry and exit area of the motor. The analysis will be carried out by help of one-dimensional field analysis (1-D) method. Simulation results based on the proposed method will be evaluated by experimental measurements from Canadian Institute of Guided Ground Transportation (CIGGT).**

Keywords— *Linear induction motor, Poynting theorem, Longitudinal end effect.*

I. INTRODUCTION

Nowadays linear motors are widely used to develop linear motion directly, where, the elimination of mechanical converters, results in better out-put performance of motor [1]. Single-sided linear induction motors (SLIMs) are widely used to produce linear motion in various applications, especially in medium and high speed transportation [2].

Special structure of SLIMs causes some baneful phenomena. Longitudinal End Effect (LEE) is the most important phenomenon which makes the SLIM analysis different from its rotary counterpart.

Several technics can be used to analysis the SLIMs such as one and two-dimensional field analysis, numerical technics such as finite elements (FE) and EC methods [3-10].

In reference [11], using the average eddy current of secondary and energy conversion balance theorem the conventional EC of RIM is modified. But, slip change causes considerable change in EC parameters of SLIMs and the parameters cannot be obtained by no-load and locked rotor tests, as described in [11]. Therefore, reference [2], according to the result in [11], has used the FE to estimate the EC parameters due to slip change.

Reference [12] suggested a series EC of SLIM in which the LEE is brought in to account by help of three different impedances, each one models the effect of fundamental, forward and backward waves in the air-gap.

In another research [13], four coefficients K_r, K_x, C_r, and C_x are used to modify secondary resistance, and magnetizing reactance in a conventional EC model of RIM.

One of the important aspects of SLIMs design is to reduce the effects of specific phenomena on the motor output performance. So, it's necessary to analysis and measure the phenomena intensity.

In reference [14], LIM is analyzed by 3-dimensional FE to study the effective secondary reaction plate parameters on the transverse edge effect.

In [4] and [15] entry force to fundamental force ratio is used to measure the LEE intensity.

Reference [16], has analyzed effects of design parameters on motor out-put performances. In the mentioned paper, the ratio between the air-gap active power considering LEE and air-gap active power neglecting LEE is used to measure LEE intensity.

All of the mentioned papers tried to include LEE in the analysis process accurately but, all of them ignored some aspects of LEE. In this paper, it is proposed to include LEE in SLIMs analysis, completely. In order to gain this goal, we use the 1-D method and Poynting Vector to calculate the power flow to the all area of the motor, even to the entry and exit area. LEE is included into the field analysis using two boundary conditions which are based on the continuity of flux and power densities at the boundaries between the air-gap, entry and exit areas. Saturation Effect (SE) and TEE is considered by help of proper coefficients [17], [18]. Experimental measurements from Canadian Institute of guided ground transformation (CIGGT) validates the simulation method.

II. THE PROPOSED SLIM MODEL

The model the SLIM can be seen in fig.1. In order to simplify the analysis some assumptions are necessary to be made:

1) The primary core is ideal.
2) All layers are infinitely long in the ±X Direction.
3) The reference frame is attached to the primary.
4) Changes due to change in time and position are sinusoidal.
5) The primary winding and secondary eddy currents are rreplaced by infinitesimally thin current sheets.
6) Currents flow only on Z-direction.
7) Z-directed variations can be ignored.

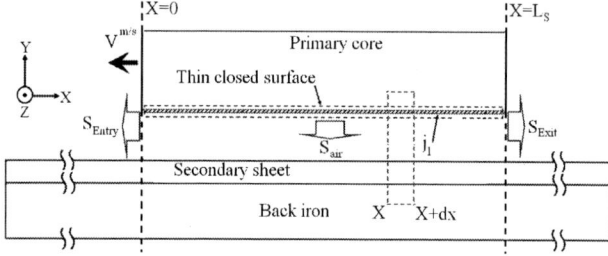

Figure 1- One-dimensional field analysis model

8) skin-effect, Edge-effect, air-gap leakage, back-iron saturation, and the effect of slotting of the primary are included into the analysis by suitable correction coefficients.

Applying Ampere's law and Ohm's law to the model shown in fig.1 result in differential equation of air-gap flux density as following:

$$g_{ei}\frac{\partial^2 B_y}{\partial x^2} - \sigma_e \mu_0 v \frac{\partial B_y}{\partial x} - \sigma_e \mu_0 \frac{\partial B_y}{\partial t} = \mu_0 \frac{\partial J_1}{\partial x} \quad (1)$$

where:

$$J_1 = J_m e^{j(\omega t - \pi x/\tau)} \quad (2)$$

So, the resulting air-gap flux density in $0 \leq x \leq L_s$ becomes:

$$B_{ya} = \left(B_0 e^{-j\frac{\pi x}{\tau}} + B_1 e^{\xi_1 x} + B_2 e^{\xi_2 x} \right) e^{j\omega t} \quad (3)$$

$$B_0 = jJ_m \mu_0 / [g_{ei}\beta(1 + jsG_e)] \quad (4)$$

$$J_m = \sqrt{2} m N_{ph} K_d I / L_s \quad (5)$$

$$\xi_1 = -\frac{1}{\alpha_1} - j\frac{\pi}{\tau_e} \quad (6)$$

$$\xi_2 = \frac{1}{\alpha_2} + j\frac{\pi}{\tau_e} \quad (7)$$

where [9]:

$$\alpha_1 = \frac{2g_{ei}}{g_{ei}X - \mu_0 \sigma_e v}, \quad \alpha_2 = \frac{2g_{ei}}{g_{ei}X + \mu_0 \sigma_e v} \quad (8)$$

$$\tau_e = \frac{2\pi}{Y} \quad (9)$$

$$\sqrt{\left(\frac{\mu_0 \sigma_e v}{g_{ei}}\right)^2 + 4j\frac{\omega \mu_0 \sigma_e}{g_{ei}}} = X + jY \quad (10)$$

SE and TEE are included in the model using two correction coefficients as following [17]:

$$\sigma_{1e} = \sigma_{fe}/k_{fe} \quad (11)$$

$$\sigma_{2e} = \sigma_{Al} \times k_{Al} \quad (12)$$

In order to adjust the equations (1)-(10) for SLIMs, the equivalent resistivity of secondary plates can be defined as following:

$$\sigma_e = \sigma_{2e}d + \sigma_{1e}d_{Fe} \quad (13)$$

Furthermore, the equivalent air-gap should be modified as following:

$$g_{ei} = K_l K_c K_s (g_m + d) \quad (14)$$

Where:

$$K_s = 1 + \frac{\mu_0}{\mu_i \delta (g_m + d) K_c \beta^2} \quad (15)$$

$$\delta = 1/\sqrt{\pi s f \sigma_i \mu_i} \quad (16)$$

By applying the eq.1 to the entry and exit areas ($x \leq 0$, $x \geq L_s$), and considering the fact that there is no primary MMF here, the resulting flux density become [17]:

$$B_{y\,Entry} = B_1' e^{-\left(\frac{x}{\alpha_1} + j\frac{\pi x}{\tau_e}\right)} e^{j\omega t}, \qquad for\ x \leq 0 \quad (17)$$

$$B_{y\,Exit} = B_2' e^{\left(\frac{x}{\alpha_2} + j\frac{\pi x}{\tau_e}\right)} e^{j\omega t}, \qquad for\ x \geq L_s \quad (18)$$

Using the magnetic flux density, and through $\nabla \times E = \partial B/\partial t$, it is possible to calculate the electric field intensities as following:

$$E_{za} = \left(-\frac{\tau\omega}{\pi} B_0 e^{-j\frac{\pi x}{\tau}} + j\frac{\omega}{\xi_1} B_1 e^{\xi_1 x} + j\frac{\omega}{\xi_2} B_2 e^{\xi_2 x} \right) e^{j\omega t} \quad (19)$$

$$E_{z\,Entry} = j\frac{\omega}{\xi_1} B_1' e^{\xi_1 x} e^{j\omega t} \quad (20)$$

$$E_{z\,Exit} = j\frac{\omega}{\xi_2} B_2' e^{\xi_2 x} e^{j\omega t} \quad (21)$$

The magnetic flux density coefficients, B_1, B_1', B_2, and B_2' should be determined by help of four boundary conditions. Two boundary surfaces, $x = 0$, and $x = L_S$, are shown in fig.1.

Two of the boundary conditions will be given by continuity of magnetic flux density at $x = 0$, and $x = L_S$:

$$B_{ya}\big|_{x=0} = B_{y\,Entry}\big|_{x=0} \quad (22)$$

$$B_{ya}\big|_{x=L_S} = B_{y\,Exit}\big|_{x=L_S} \quad (23)$$

And the continuity of power density (Poynting Vector) at these boundaries will give the third and fourth boundary conditions [19]:

$$\frac{1}{2\mu_0} E_{za}\hat{a}_z \times B^*_{ya}\hat{a}_y\big|_{x=0} = \frac{1}{2\mu_0} E_{z\,Entry}\hat{a}_z \times B^*_{y\,Entry}\hat{a}_y\big|_{x=0} \quad (24)$$

$$\frac{1}{2\mu_0} E_{za}\hat{a}_z \times B^*_{ya}\hat{a}_y\big|_{x=L_S} = \frac{1}{2\mu_0} E_{z\,Exit}\hat{a}_z \times B^*_{y\,Exit}\hat{a}_y\big|_{x=L_S} \quad (25)$$

The equations (22)-(25) forms a system of nonlinear equations, which gives the field coefficients at every operating slip.

III. POWER FLOW IN SLIM AND PROPULSION FORCE

Considering the model of fig.1 the power flow can be divided into three parts:

1) Air-gap power (The power which flows from primary to secondary through the air-gap), S_{air}.

2) The power which flows from primary to the entry area, S_{Entry}.

3) The power which flows from primary to the exit area, S_{Exit}.

As it is shown in fig. 1 the out flow of power from the imaginary thin closed surface, **A**, can be written as following [19]:

$$\oiint_A \mathbf{P}.d\mathbf{A} = -\frac{1}{2}\oiiint_V \mathbf{j}_1^* .\mathbf{E}\, dV$$

$$-\frac{\partial}{\partial t}\left(\frac{1}{4}\oiiint_V \varepsilon_0 |\mathbf{E}|^2 dV\right) \tag{26}$$

$$-\frac{\partial}{\partial t}\left(\frac{1}{4}\oiiint_V \mu_0 |\mathbf{H}|^2 dV\right)$$

The left hand side of eq. 26 represents the integral of Poynting vector on $\mathbf{A}$. The integral result is nonzero just on the surface located in the air-gap. So, the result of the integral gives the power which flows to the secondary through the air-gap.

The first term in right hand side of the eq. 26 represents the power generated by a source inside V. V is the Volume enclosed by the surface $\mathbf{A}$. The second and third terms in the right hand side represent the rate of increase of electric and magnetic stored energies, respectively. Because of the negligible volume of V, these terms equals zero. So, the air-gap power is [19]:

$$S_{air} = -0.5 W_{se}\int_0^{L_S} E_{za}J_1^*\, dx = \overbrace{\frac{B_0 J_m L_S W_{se}\omega\tau}{2\pi}}^{S_0}$$

$$+ j\overbrace{\int_0^{L_S} J_m \frac{\omega}{\xi_1} B_1 e^{\left(\xi_1 + j\frac{\pi}{\tau}\right)x}\, dx}^{S_f} \tag{27}$$

$$+ j\overbrace{\int_0^{L_S} J_m \frac{\omega}{\xi_2} B_2 e^{\left(\xi_2 + j\frac{\pi}{\tau}\right)x}\, dx}^{S_b}$$

where, S_0, S_f, and S_b are the air-gap power respectively due to fundamental, forward and backward waves.

However, finding the surfaces in entry and exit areas, in which the flux density is zero, is greatly complicated [12], but the power entered to the areas can be calculated, simply. The only power source in this field analysis case is the equivalent current sheet. The power flows to the entry and exit areas through the boundary surfaces located at $x = 0$ and $x = L_S$. So, the flowed power to the entry and exit areas can be calculated by the minus surface integral of Poynting vector on the boundary surfaces located $x = 0$ and $x = L_S$, respectively. So [19]:

$$S_{Entry} = -\oiint_{X=0} \mathbf{P}.d\mathbf{A} = -\frac{1}{2\mu_0}\iint_{X=0}\left(B^*_{y\,Entry}.E_{z\,Entry}\right)dA \tag{28}$$

$$S_{Exit} = -\frac{1}{2\mu_0}\iint_{X=L_S}\left(B^*_{y\,Exit}.E_{z\,Exit}\right)dA \tag{29}$$

and the propulsion force become:

$$F_p = \frac{real(S_0)}{V_s} + \frac{real(S_f + S_b + S_{entry} + S_{exit})}{V_s} \tag{30}$$

in which the first term in right hand side is the fundamental wave force and the second term is LEE force.

In order to calculate the normal force, co-energy equation is written as following:

$$W_f' = \overbrace{\frac{3V_{ph}}{\omega}\overbrace{I_1}^{i}}^{\lambda} - \overbrace{\frac{Im\{S_t\}}{2\omega}}^{W_f} \tag{31}$$

In which, V_{ph} is the per-phase voltage of SLIM:

$$V_{ph} = \frac{S_t}{3I_1} \tag{32}$$

By a differential change, Δy, in the air-gap length, the change in co-energy value becomes:

$$\Delta W_f' = \frac{3\left(V_{ph}(g+\Delta y) - V_{ph}(g)\right)}{\omega}$$

$$-\frac{Im\{S_t(g+\Delta y) - S_t(g)\}}{2\omega} \tag{33}$$

$$-\frac{Re\{S_t(g+\Delta y) - S_t(g)\}}{2\omega}$$

It is necessary to be noted that the change of air-gap length result, in changes in the value of traction force and ohmic loss in addition to variation of per-phase voltage, $\left(V_{ph}(g+\Delta y) - V_{ph}(g)\right)$, and stored magnetic energy, $Im\{S_t(g+\Delta y) - S_t(g)\}$. So, the third part of (27) is the energy due to change in air-gap length that caused to variation of traction force and ohmic loss, $Re\{S_t(g+\Delta y) - S_t(g)\}$.

As mentioned before, there are five travelling waves in the introduced model described in Fig. 1. So, each wave results in its own energy, co-energy and force. Then, the values could be superimposed on top of each other to calculate the resultant value. Thus, the resultant normal force could be implied as following:

$$\overbrace{\frac{\Delta W_f'}{\Delta y}}^{F_n} = \overbrace{\frac{\Delta W_{f0}'}{\Delta y}}^{F_{n0}} + \overbrace{\frac{\Delta W_{ff}'}{\Delta y}}^{F_{nf}} + \overbrace{\frac{\Delta W_{fb}'}{\Delta y}}^{F_{nb}} + \overbrace{\frac{\Delta W_{f\,Entry}'}{\Delta y}}^{F_{n\,en}} + \overbrace{\frac{\Delta W_{f\,Exit}'}{\Delta y}}^{F_{n\,ex}} \tag{34}$$

In which, $\Delta W_{f0}'$, $\Delta W_{ff}'$, $\Delta W_{fb}'$, $\Delta W_{f\,Entry}'$ and $\Delta W_{f\,Exit}'$ are the changes in co-energy value respectively due to fundamental, forward, backward, entry and exit waves. F_n represents resultant normal force and F_{n0}, F_{nf}, F_{nb}, $F_{n\,en}$ and $F_{n\,ex}$ are normal forces due to fundamental, forward, backward, entry and exit waves, respectively.

IV. LONGITUDINAL END EFFECT FACTOR

As mentioned before, LEE is the most important consequence of finite length of primary and relative motion between primary and secondary [20], [21].
Therefore, inclusion of relative speed using (1), finite geometry of primary by help of (23)-(25) and the power flow to the entry and exit areas using (28) and (29) make it possible to include LEE in the proposed model, completely. LEE causes some problem such as phases unbalance, braking force, and reduction in motor efficiency.

In this paper, the LEE power to actual input power ratio is introduced as LEE factor:

978-1-5090-0376-1/16 $31.00 © 2016 IEEE

TABLE I
DESIGN DATA OF THE CASE STUDY SLIM

Design Variable	Value
Number of poles	6
Number of phases	3
Number of slots per pole per phase	3
Air-gap length(g_m) (mm)	15
Conducting plate thickness(d) (mm)	2.5
Secondary back-iron thickness(h_{Fe}) (mm)	25
Primary width (m)	0.101
Conducting plate width (m)	0.179
Back-iron width (m)	0.111
Pole pitch (m)	0.25 m
Coil pitch (m)	0.1944
Nominal primary current (A)	200
Nominal frequency (Hz)	40 Hz
Conducting plate material	Aluminum 1100 alloy
Back-iron material	516 grade 17 mild steel

$$K_{lee} = \frac{S_f + S_b + S_{entry} + S_{exit}}{S_{air} + S_{entry} + S_{exit}} \quad (35)$$

V. SIMULATION RESULTS

To show the validity of the suggested method, the simulation results are compared to the experimental results from the SLIM tested by CIGGT. The motor design data is presented in table I [17], [18]. The out-put performance of the motor is calculated for input frequencies between 5-40 Hz at constant excitation current of 200 A.

Fig. 2 and 3 show the change of motor propulsion and normal forces due to change in motor speed. It can be seen that simulation results are close enough to the experimental results. It's necessary to be noted that the force-speed curves obtained by the suggested method are closer to the experimental results than the curves obtained in [17], [18]. Furthermore, LEE force-speed curves indicate that the direction and value of the forces due to LEE depend on slip frequency, in-put frequency and consequently synchronous speed.

By help of the method change of the secondary back-iron permeability with speed could be plotted as it can be seen in fig. 4.

The real part, imaginary part and magnitude of the Y-component of magnetic flux density in the air-gap, entry and exit areas for the input frequency of 40 Hz are shown fig. 5 (a), (b) and (c), respectively. Flux distributions are plot for three different slip, $s=1$, $s=0.25$ and $s=0.05$. It can be seen that flux density distribution is not so non-uniform along the air-gap, at $s=1$. But, with slip reduction the flux density in the air-gap is reduced at the entry area and is reflected at exit area. This phenomenon causes LEE. Fig. 6 shows the variation of the presented LEE factor, K_{lee}, versus motor speed. It's seen that as the motor speed increases from zero the factor goes up almost in all input frequencies, except for 5 Hz, in which the motor speed is too low and there is no considerable LEE.

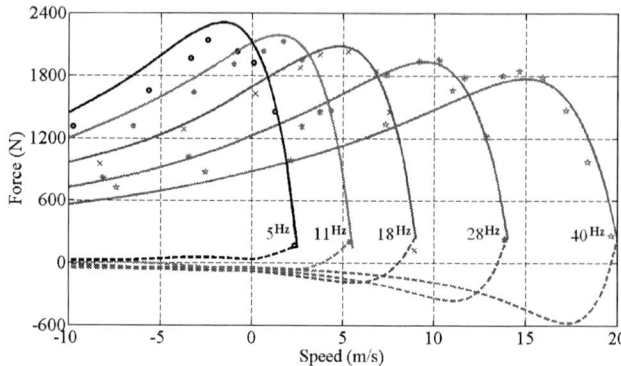

Figure 2- Forces versus motor speed. (Continued lines indicate calculated propulsion forces, dashed lines indicate forces due to LEE, shapes indicate experimental results).

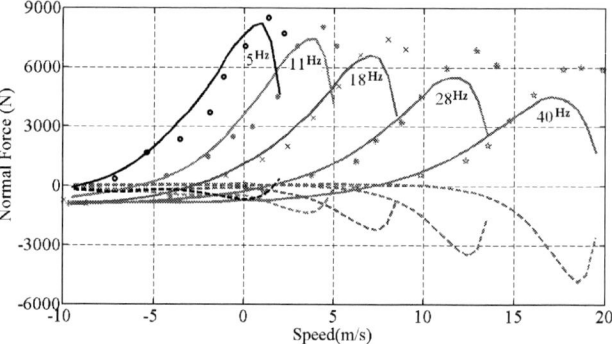

Figure 3- Normal forces versus motor speed. (Continued lines indicate calculated normal forces, dashed lines indicate normal forces due to LEE, shapes indicate experimental results)

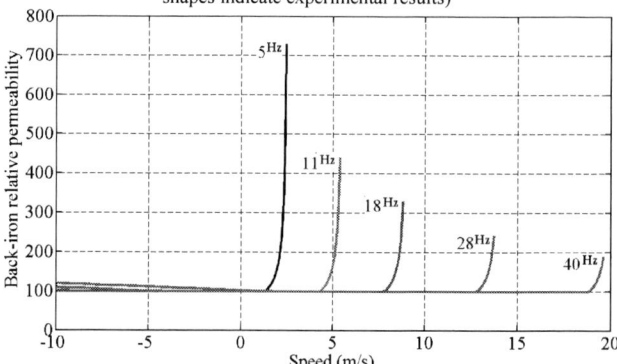

Figure 4- Permeability of secondary back-iron versus motor speed.

VI. CONCLUSION

The paper suggested a new method to include LEE in the SLIMs analysis. 1-D field analysis method is used to have a comprehensive description of electromagnetic waves in the air-gap, entry and exit areas. Then by help of the Poynting theorem the power flow to the all areas of the motor was gained. Using the obtained power flow it was possible to calculate the propulsion force versus motor speed. The ratio between the LEE power and actual input power was introduced as LEE factor. In the presented model LEE was taken into account by help of four boundary conditions.

Neglecting static LEE [11], backward traveling waves [4],[16],[22] and the power transmitted to the entry and exit areas [12],[13] may bring some errors. So, the main advantage of the suggested method is that it is possible to include LEE in the calculation procedure, completely.

Comparison between the calculated propulsion force-speed curves and the measured forces shows the acceptable accuracy of the suggested model.

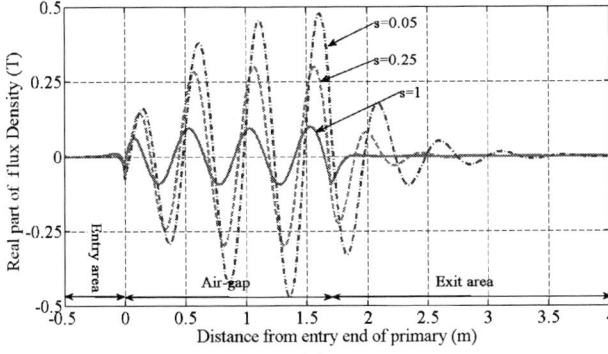

(a)

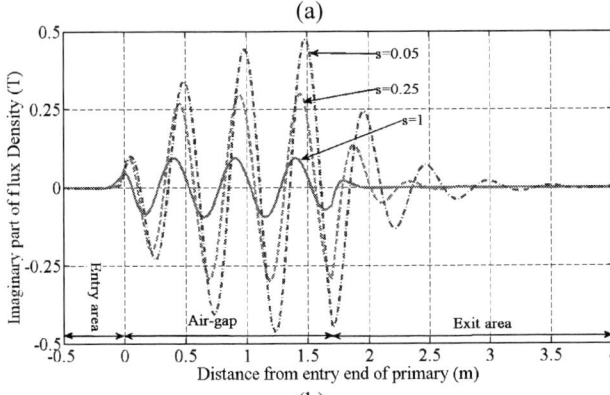

(b)

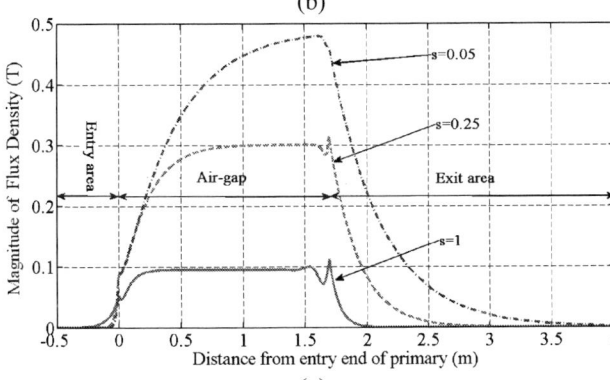

(c)

Figure 5- Flux density distribution

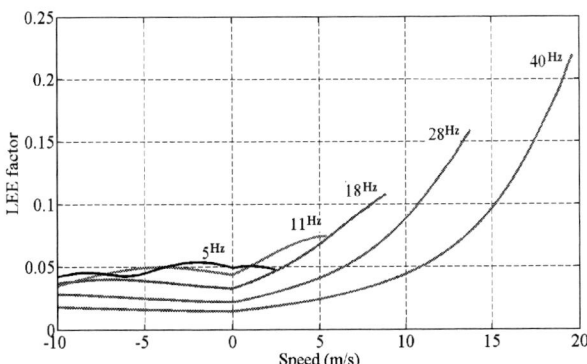

Figure 6- LEE factors versus motor speed.

VII.   REFERENCES

[1] J. F. Gieras, *Linear Induction Drives.* Oxford, U.K.: New Clarendon Press, 1994.

[2] M. Mirsalim, A. Doroudi,; J.S. Moghani, "Obtaining the operating characteristics of linear induction motors: a new approach" *IEEE Trans. Magn.* Vol.38, pp:1365–1370. Issue:2, March 2002.

[3] W. Xu, G. Sun, Y, Li,; "Research on Performance Characteristics of Linear Induction Motor" *IEEE Industrial electronics and applications conf.* 2007 pp.86-88.

[4] R.C. Creppe, J.A.C. Ulson, J.F. Rodrigues,; "Influence of Design Parameters on Linear Induction Motor End Effect" *IEEE Trans. Ener.Conv .vol.23 no.2* pp. 358-362, June. 2008.

[5] Y. Nozaki, T. Koseki, E. Masada; "Analysis of Linear Induction Motors for HSST and Linear Metro using Finite Difference Method" *Proc. of the 5th International Symposium on Linear Drives for Industry Applications,* pp: 168-171, LDIA2005.

[6] Byung-Jun Lee, Dae-Hyun Koo, and Yun-Hyun Cho, "Investigation of linear induction motor according to secondary conductor structure" *IEEE Trans. Magn.,* vol. 45, no. 6, pp. 2839–2842, June. 2009.

[7] Yuxing Zhang; Weiming Ma; Junyong Lu; Zhaolong Sun; Jin Xu; Weibo Li, "A new approach to research the transverse edge effect in linear induction motor considering the edge fringing flux" *IEEE Trans. Magn.,* vol. 47, no. 11, pp. 4660–4668, November. 2011.

[8] Byeong-Hwa Lee; Kyu-Seob Kim; Jung-Pyo Hong; Jung-Ho Lee, "Optimum shape design of single-sided linear induction motors using response surface methodology and finite-element method" *IEEE Trans. Magn.,* vol. 47, no. 10, pp. 3657–3660, October. 2011.

[9] S. Yamamura, H. Ito, and Y. Ishulawa, "Theories of the linear induction motor and compensated linear induction motor," *IEEE Trans. Power App. Syst.,* vol. PAS-91, no. 4, pp. 1700–1710, Dec. 1971.

[10] D. Li, W. Li, J. Fang, X. Zhang, J. Cao, "Performance Evaluation of A Low-Speed Single-Side HTS Linear Induction Motor Used for Subway System" *IEEE Trans. Magn.* In press, 2013.

[11] J. Duncan, "Linear induction motor—Equivalent circuit model," *Proc. Inst. Elec. Eng.,* pt. B, vol. 130, no. 1, 1983.

[12] Wei Xu; Jian Guo Zhu; Yongchang Zhang; Yaohua Li; Yi Wang; Youguang Guo, "An improved equivalent circuit model of a single-sided linear induction motor," *IEEE Trans. Veh. Technol.,* vol. 59, no. 5, pp. 2277–2289, June. 2010.

[13] Wei Xu; Jian Guo Zhu; Yongchang Zhang; Zixin Li; Yaohua Li; Yi Wang; Youguang Guo; Yongjian Li; "Equivalent circuits for single-sided linear induction motors" *IEEE Trans. Indus.* vol.46, no.6, pp. 2410-2423, Dec. 2010.

[14] Sung Gu Lee; Hyung-Woo Lee; Sang-Hwan Ham; Chang-Sung Jin; Hyun-June Park; Ju Lee; "Influence of the construction of secondary reaction plate on the transverse edge effect in linear induction motor" *IEEE Trans. Magn.,* vol. 45, no. 6, pp. 2815–2818, June. 2009.

[15] Tong Yang; Libing Zhou; Langru Li, "Influence of design parameters on end effect in long primary double-sided linear induction motor," *IEEE Trans. plasma science,* vol. 39, no. 1, pp. 192–197, Jan. 2011.

[16] A.Z. Bazghaleh; M.R. Naghashan; M.R. Meshkatoddini, "Optimum design of single-sided linear induction motors for improved motor performance" *IEEE Trans. Magn.,* vol. 46, no. 11, pp. 3939–3947, Nov. 2010.

[17] R. M. Pai, I. Boldea, and S. A. Nasar, "A complete equivalent circuit of a linear induction motor with sheet secondary," *IEEE Trans. Magn.,* vol. 24, pp. 639–654, Jan. 1988.

[18] J. Faiz, H. Jafari; "Accurate Modeling of Single-Sided Linear Induction Motor Considers End Effect and Equivalent Thickness" *IEEE Trans. Magn.* vol.36, no.5, pp:3785-3790, Sep 2000.

[19] A. Zare-Bazghaleh, M. R. Naghashan, and A. Khodadoost "Derivation of equivalent circuit parameters for single-sided linear induction motors" *IEEE Trans. Plasma Sci.* vol.43, no.10, pp. 3637-3644, Oct. 2015.

[20] A. H. Selçuk, H. Kürüm, "Investigation of End Effects in Linear Induction Motors by Using the Finite-Element Method" *IEEE Trans. Magn.,* vol. 44, no. 7, pp. 1791–1795, July. 2008.

[21] E. Amiri, E. Mendrela, "A Novel Equivalent Circuit Model of Linear Induction Motors" *IEEE Trans. Magn.* In press, 2013.

[22] J. F. Gieras, G. E. Dawson, and A. R. Eastham, "A new longitudinal end effect factor for linear induction motors," *IEEE Trans. Energy Conversion,* vol. EC-2, pp. 152–159, 1987.

7th Power Electronics, Drive Systems & Technologies Conference (PEDSTC 2016)
16-18 Feb. 2016, Iran University of Science and Technology, Tehran, Iran

Comparing the Effect of Using New Hysteresis Material on the Performance of a Hystersis Motor

Mohammad Mohammadi
Firozjaee
Electrical Engineering Department
of Iran University of Science &
Technology
Center of Excellence for Power
Systems Automation and Operation
Tehran, Iran
m_firozjaee@elec.iust.ac.ir

Abolfazl Vahedi
Electrical Engineering Department
of Iran University of Science &
Technology
Center of Excellence for Power
Systems Automation and Operation
Tehran, Iran
avahedi@iust.ac.ir

Mostafa Sanikhani
Electrical Engineering Department
of Iran University of Science &
Technology
Center of Excellence for Power
Systems Automation and Operation
Tehran, Iran
m_sanikhani@elec.iust.ac.ir

Abstract— The hysteresis motor is the self-starting synchronism motor. From the time of its start until it reaches synchronous speed, the motor produces a synchronous torque and an ideally flat speed/torque characteristic. The rotor of a hysteresis motor is a cylindrical tube of high hysteresis loss permanent magnet material without winding or slots. It is known that the hysteresis ring material magnetic characteristics plays an important role in flux distribution that can influence the output torque, terminal current and the efficiency of this motor. Regarding this issue, in this study effect of using new material for hysteresis ring on the performance characteristics of a radial flux hysteresis motor is investigated. Finally, the simulation of hysteresis motor is done in order to extract the output values of motor is done using 2D- Finite Element Model.

Keywords— Hysteresis Motor, Semi-hard Magnetic Material, Finite Element Analysis

I. INTRODUCTION

The hysteresis motor is a synchronous machine that produces torque by magnetic hysteresis of the rotor material, and it has been used in applications requiring a precisely constant speed [1, 2]. The hysteresis motors are widely used in small motor applications because of its distinct advantageous such as soft torque, simple construction with conventional poly phase stator windings structure (Fig. 1) and self-starting torque during the run up and synchronization period. It has no rotor slots and then it has low noise in operation. These advantageous make hysteresis motor suitable for some applications such as gyros, centrifuges, pumps, timing and recording equipment, in which constant torque, constant speed, and quite operation are required [3, 4].

Hysteresis characteristics of the magnetic material in hysteresis motors plays an important role in determining the hysteresis characteristics of motor, thus the proper use of these material is an important factor in determining the performance of the motor. It is known the magnetic characteristics of the motor could be easily affected by the air gap length and structure dimensions variations [5, 6]. Then fixed structure used for the simulation of the motor with different material for accurate comparison. Some analytical methods have been

proposed to analyze the motor, but these methods seem to have complexities and limits to get good results [7]. These problems are overcome by applying numerical methods. Finite element method allows a precise analysis of magnetic devices taking into account geometric details and magnetic nonlinearity [8].

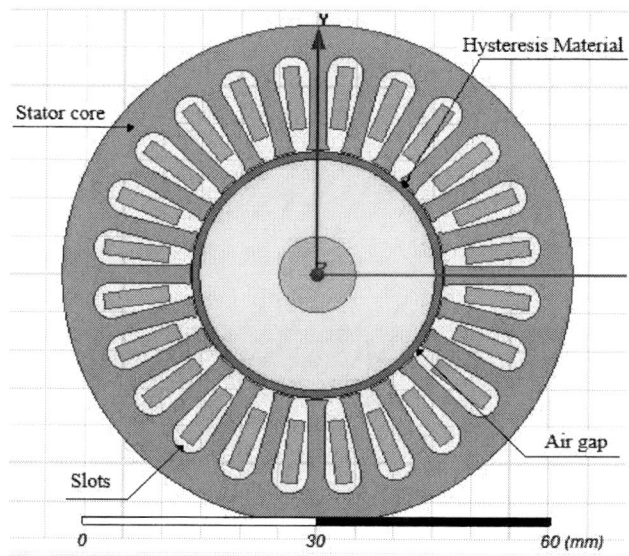

Fig. 1: cross section of a hysteresis motor

In this paper, new semi hard magnetic material has been used and the performance characteristics of hysteresis motor through a 2D finite element analysis (FEA) are provided. Such simulation is based upon Maxwell's field equations considering the case of radial flux type machine at synchronous speed [8]. For better understanding of the simulation the equivalent circuit diagram has been introduced and necessitate for the simulation is considered. Numerical model has been implanted by Ansys Maxwell v. 16.0. At the end, comparison taking place between other conventional materials that used for hysteresis ring of the hysteresis motor with the newest one.

II. EQUIVALENT CIRCUIT DIAGRAM

As for all machine analysis the hysteresis motor can be represented by the equivalent magnetic circuit (Fig. 2). F_α Represents the spatial distribution given by:

$$F_\alpha = (\frac{m}{2})(\frac{N_s I_m}{P})\sin(wt - P\alpha) \qquad (1)$$

Where N_s is the number of turns per phase, m the number of motor phases and I_m the magnitude of the current in each phase. The m.m.f. F_α pushes the magnetic flux $_\alpha\Phi$ across the air gap g with the magnetic reluctance R_g. Leakage flux $_\alpha\Phi\text{-}\Phi_{h\alpha}$ has a path of reluctance R_p and the useful flux $\Phi_{h\alpha}$ across the rotor sleeve is responsible for the material magnetization. By simplicity the hysteresis loop of the rotor material it is represented by an ideally square loop and this approximation can be compensated by addition of a reluctance R_0 to R_g. Expressions for R_0, R_g, R_p and equivalent magnetic permeability μ_p are displayed in the diagram.

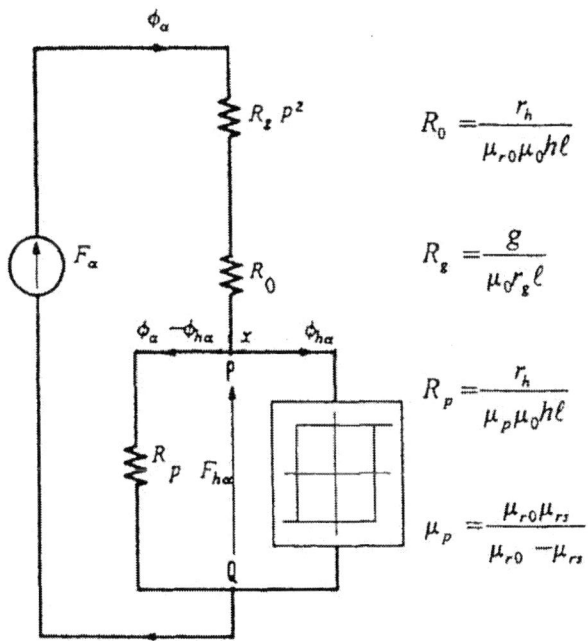

Fig. 2: equivalent circuit diagram

The resulting motor torque general equation is:

$$T = (\frac{\pi}{2})R_{eq}^{-1}(\frac{r_h H_c}{\sqrt{2}})^2 EC \sin\theta \qquad (2)$$

Where:

$$R_{eq} = R_p(\frac{R_0 + R_g P^2}{R_p + R_0 + R_g P^2}) \qquad (3)$$

$$C = (\frac{mN_s I_m}{4r_h H_c})(\frac{R_p}{R_p + R_0 + R_g P^2}) \qquad (4)$$

Represents the ratio of the maximum m.m.f. $N_s I_m$ of the stator to the m.m.f. $r_h H_c$ necessary to produce the coactivity H_c of the rotor material. In (2) the parameter E and torque angle θ depend on the normalized stator m.m.f., C, given by (4) and are plotted in figure 5 [9]. As it can be seen from (2) that the motor constant torque and speed depends on machine configuration as well as on magnetic properties of the rotor material, then we do our simulation by constant machine structure and configuration to gain trustworthy results.

III. INTRODUCTION TO NEW SEMI HARD MAGNETIC MATERIALS

Much of the focus of industry is on high coercivity permanent magnets and very low coercivity soft magnetic steels. A less glamorous, but not less important set of materials provide modest coercivity for applications such as brakes and hysteresis coupled devices. Many semi-hard materials are also malleable and therefore capable of being formed and of being machined with standard metal-working tools. These malleable alloys can also be extruded into wire, rods and stamped into other forms [10].

There are so many different type of semi hard magnets in the market. In this paper we focus on three newest introduced semi hard magnets. These material are Cobalt steel, Magnetoflex 35U and Hysteresis Alloy 2J9. These materials have a proper characteristic to be used in hysteresis machines.

TABLE I.   PROPERTIES OF SEMI-HARD MAGNETIC ALLOYS

Name	Composition	H_c (KAm^{-1})	B_r	$BH_m(\frac{mW.s}{cm^3})$
36% Cobalt steel	Co,Cr,W,C	19.9	0.9	7.56
Magnetoflex 35U	Co,Fe,V	27	0.83	12
Hysteresis Alloy 2J9	C,Si,P,S,Mn,Co,V,Ni	8.5	1.25	8.55

IV. HYSTERESIS LOOP APPROXIMATION

In this study, a complex permeability is used to predict the hysteresis loop in the inclined ellipse shape in order of simulation in Maxwell software. The complex permeability is a useful tool for dealing with magnetic effect. In this study, a complex permeability is used to predict the hysteresis loop in the inclined ellipse to find and use these hysteresis materials characteristics. It is well known when a sinusoidal field intensity wave (H(wt)), is applied to the hysteresis material, a corresponding flux density wave (B(wt)) will be produced. These quantities react top each other with some lag time. This lag presented in angle γ, then H and B would be:

$$H(wt) = H_m . \sin(wt)$$
$$B(wt) = B_m . \sin(wt - \gamma) \qquad (5)$$

By math operation the sin wave can be expressed through the time phasors:

$$H = H_m e^{iwt}$$
$$B = B_m . e^{j(wt-\gamma)} \qquad (6)$$

Regarding to the fact that in this simulation the whole hysteresis loop can be introduced based on complex permeability, a hysteresis loop in the shape of an inclined ellipse is adopted [11].

$$\left| \mu_r \right| = \frac{\mu}{\mu_0} = \frac{B_{max}}{\mu_0 H_{max}} \qquad (7)$$

$$\gamma = \arcsin(\frac{H_c}{H_{max}}) \qquad (8)$$

$$\mu_r^{'} = \mu_r . \cos(\gamma) \qquad (9)$$

$$\mu_r^{"} = \mu_r . \sin(-\gamma) \qquad (10)$$

Where μ_r is the relative Permeability, $\mu_r^{'}$ and $\mu_r^{"}$ are the real and imaginary part of complex permeability.

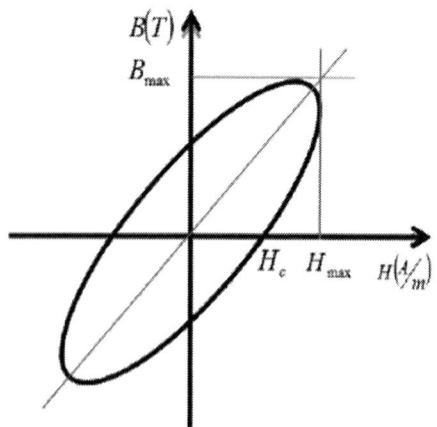

Fig. 3: inclined hysteresis loop approximation

Based on mentioned method, quantities for simulation of each material have been obtained. These quantities have been used in Ansys maxwell software in order of simulation. This method of approximation is easy and also has very close results [12, 8].

TABLE II.   Relative Permeability and Magnetic Loss Tangent OF SEMI-HARD MAGNETIC ALLOYS

Name	μ_r	γ
Cobalt steel	35	0.5
Magnetoflex 35U	24.44	1.05
Hysteresis Alloy 2J9	103.61	1.08

V.   SIMULATION RESULTS

This paper presents the results of using the finite element method in the design of Radial flux hysteresis motor by using new semi-hard magnetic materials. Based on the above respects, finite element simulation has been done for each of the materials on hysteresis motor and the torque in each of the situation has been shown to have a good insight for comparison. The design features of a motor are the same in all of the simulation and as has been shown in Table III. Fig. 3 shows the mesh diagram of motor that has been built by the software to predict the magnetic flux density in desired regions.

TABLE III.   MOTOR FEATURES

Quantity	Value
Line RMS voltage(v)	200
Rated phase current(A)	3.6
Frequency(Hz)	400
Speed(rpm)	12000
Phase connection	Y
Pole pairs	2
Air gap length (mm)	0.22
Ring thickness (mm)	1
Outer diameter of stator (mm)	33
Inner diameter of stator (mm)	16.5
Number of slots	24
Number of turns per coil	200
Fill factor	0.5
Phase resistance at 400Hz(Ω)	3.25

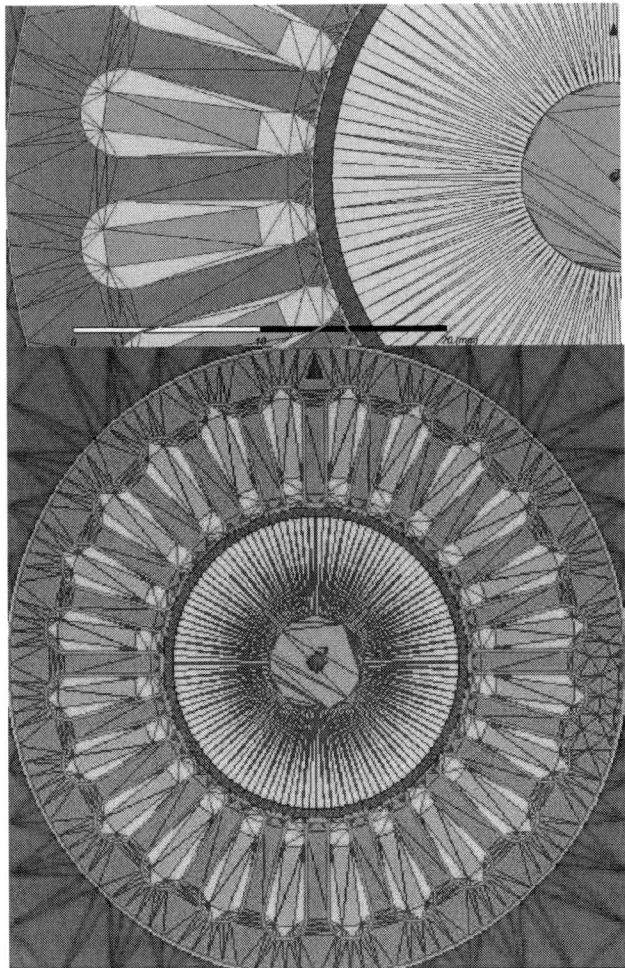

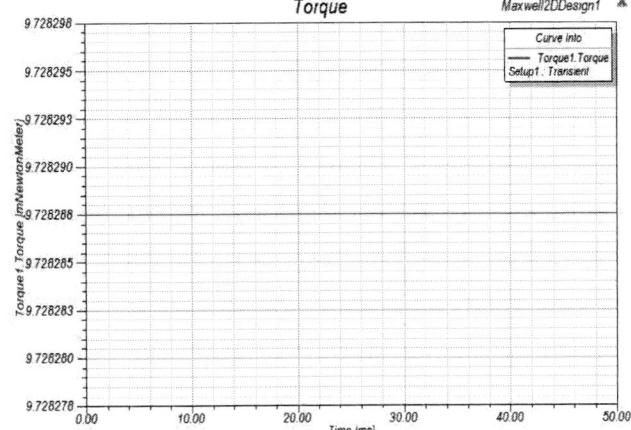

Fig. 6: torque scheme of the motor in transient state using Cobalt steel

Magnetoflex 35U

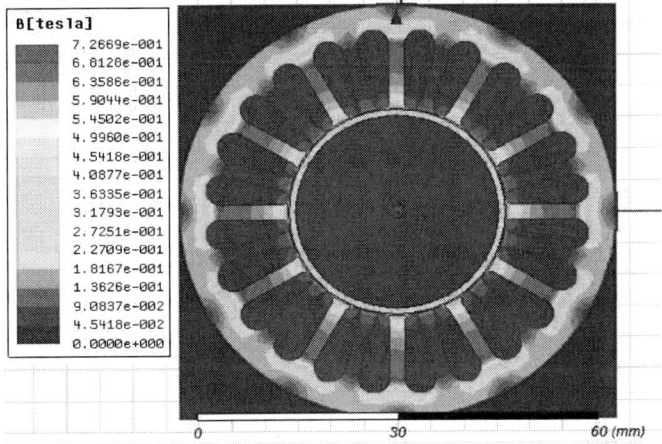

Fig. 4: Mesh diagram of simulated motor

Fig. 7: distribution of the circumferential flux using Magnetoflex 35U

36% Cobalt steel

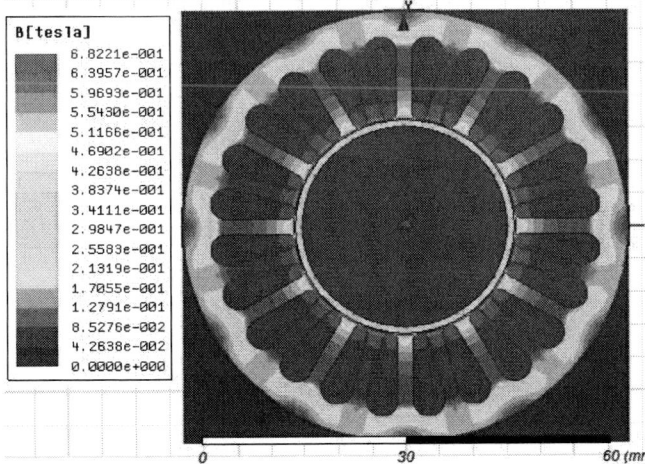

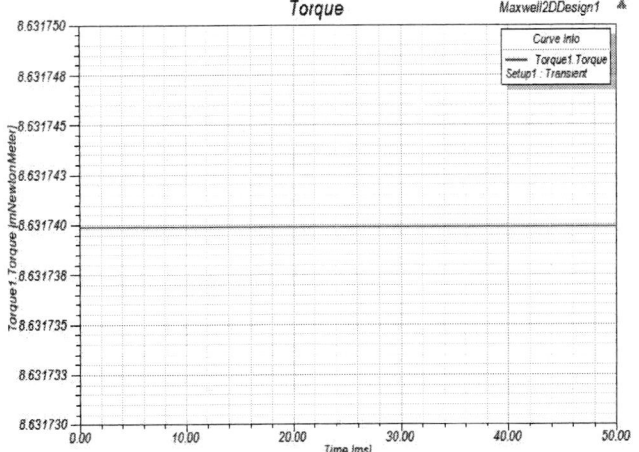

Fig. 5: distribution of the circumferential flux using Cobalt steel

Fig. 8: torque scheme of the motor in transient state using Magnetoflex 35U

Hysteresis Alloy 2J9

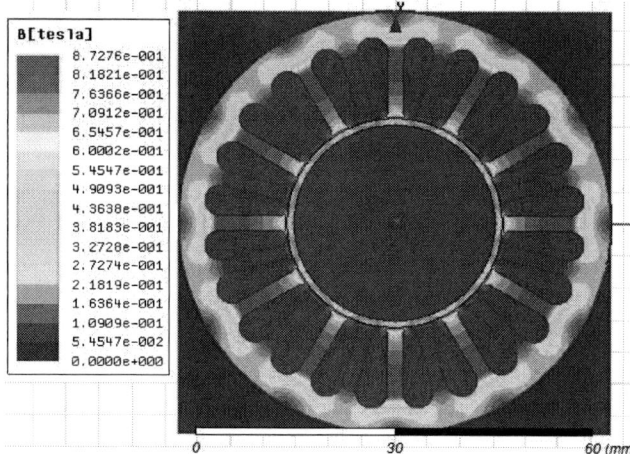

Fig. 9: distribution of the circumferential flux using Hysteresis Alloy 2J9

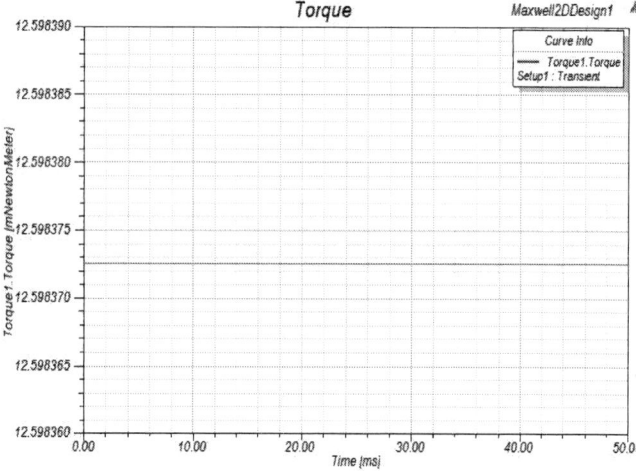

Fig. 10: torque scheme of the motor in transient state using Hysteresis Alloy 2J9

TABLE IV.   OUTPUT TORQUE OF THE HEYSTERESIS MACHINE BY EACH OF THE MATERIALS

Name	Torque(mNewtone Meter)
Cobalt steel	9.728288
Magnetoflex 35U	8.631740
Hysteresis Alloy 2J9	12.598373

VI. CONCLUSIONS

In this paper, finite element analysis has been done to investigate the effect of using new hysteresis material to reach higher output power (torque). In order of comparing, each of the materials have been used in hysteresis ring of the same hysteresis motor. Results at transient state of the hysteresis motor can be seen in table IV.

As it's evident from table IV by using Hysteresis Alloy 2J9 maximum performance can be reached. Hysteresis characteristics of this material bring out the most suitable hysteresis loop and as a result the highest air gap magnetic flux density between these material that result in more torque, more output power and also higher efficiency for the motor. It should be noted that determining the best material for the hysteresis ring of hysteresis motor depend on other factor too, such as mechanical and chemical and thermal characteristics, that could be the subject of future works. But this work focus on choosing the best material in order to reach the maximum torque and efficiency in these motors.

REFERENCES

[1] Kataoka, T., T. Ishikawa, and T. Takahashi, *Analysis of a hysteresis motor with overexcitation.* Magnetics, IEEE Transactions on, 1982. **18**(6): p. 1731-1733.

[2] Qin, R. and R. Zhong, *The starting process analysis of hysteresis motors.* Small and special electrical machines, China, 1986(4): p. 25-31.

[3] Rahman, M.A. and R. Qin, *Starting and synchronization of permanent magnet hysteresis motors.* Industry Applications, IEEE Transactions on, 1996. **32**(5): p. 1183-1189.

[4] Rahman, M.A., M. Copeland, and G.R. Slemon, *An Analysis of the Hysteresis Motor Part III: Parasitic Losses.* Power Apparatus and Systems, IEEE Transactions on, 1969(6): p. 954-961.

[5] Sedagati, A. and A. Vahedi. *Effects of parameters design on the characteristics of hysteresis motor.* in *Electrical Machines and Systems, 2003. ICEMS 2003. Sixth International Conference on.* 2003.

[6] Rajagopal, K., *Design of a compact hysteresis motor used in a gyroscope.* Magnetics, IEEE Transactions on, 2003. **39**(5): p. 3013-3015.

[7] Kim, H.-K., S.-K. Hong, and H.-K. Jung, *Analysis of hysteresis motor using finite element method and magnetization-dependent model.* Magnetics, IEEE Transactions on, 2000. **36**(4): p. 685-688.

[8] Mirimani, S., et al., *Electromagnetic Analysis of Hysteresis Synchronous Motor Based on Complex Permeability Concept.* Iranian Journal of Electrical and Electronic Engineering, 2013. 9(2): p. 88-93.

[9] Hadjipanayis, J.E.M.a.G., *Some aspect of the design and operation of the polyphase hysteresis motor.* Magnetic Hysteresis in Novel Magnetic Materials, 1997. 338: p. 875.

[10] Bozorth, R.M., *Ferromagnetism.* Ferromagnetism, by Richard M. Bozorth, pp. 992. ISBN 0-7803-1032-2. Wiley-VCH, August 1993., 1993. **1**.

[11] Modarres, M., A. Vahedi, and M. Ghazanchaei. *Effect of Air gap variation on characteristics of an Axial flux hysteresis motor.* in *Power Electronic & Drive Systems & Technologies Conference (PEDSTC), 2010 1st.* 2010. IEEE.

[12] Modarres, M., A. Vahedi, and M. Ghazanchaei, *Study on axial flux hysteresis motors considering airgap variation.* Journal of Electromagnetic Analysis and Applications, 2010. 2010.

7th Power Electronics, Drive Systems & Technologies Conference (PEDSTC 2016)
16-18 Feb. 2016, Iran University of Science and Technology, Tehran, Iran

Eccentricity Fault Detection of A Salient-Pole Synchronous Machine Using Modified Winding Function Approaches and Finite Element Method

Azadeh Doulatshah
Department of Electrical Engineering
Borujerd Branch, Islamic Azad University
Borujerd, Iran
azadehdoulatshah@yahoo.com

Peyman Naderi
Faculty of Electrical & Computer
Shahid Rajaee University
Tehran, Iran
p.naderi@srttu.edu

Abstract— in this paper, a 6 salient-pole 3-phase synchronous machine with specified dimensions and characteristics is modeled and simulated using modified winding function approaches and finite element method. In this model, all non-ideal effects are taken into account. That they include the machine geometry, slots effects and saturation, non-linear behavior due to ferromagnetic materials, type of winding connection and winding distribution and static and dynamic eccentricity between the stator and rotor. That is able to consider most of the important features of a salient pole machine. In order to analyze the behavior of machine in normal conditions and faulty, Fast Fourier Transform (FFT) has been used for eccentricity fault identification. Torque and stator current spectra of the SynM under different condition are computed and a novel approach is presented to detect static and dynamic eccentricity in salient-pole SynM based on vibration torque and currents signature.

Keywords—Eccentricity; Finite element method; Modified winding function approaches; Vibration torque; Currents signature

I. INTRODUCTION

Synchronous machines are the most important and valuable devices in power systems. Consequence of many electrical and mechanical faults occurring during the operation of electrical machines is the eccentricity between the rotor and stator. Eccentricity fault is categorized into three general groups: static eccentricity (SE), dynamic eccentricity (DE) and mixed eccentricity (ME) [1]. Fault diagnosis can produce significant cost saving to allow the scheduling of preventive maintenance, thereby preventing extensive downtime periods caused by extensive failure. There are three important factors for precise fault diagnosis in synchronous machines: modeling approach, processing method and feature extraction [1]. The essential stage of any reliable fault diagnosis method is precise modeling of the machine under different fault conditions and the simulation of the healthy machine is the first stage of this modeling procedure. Modeling faulty electrical machines requires long computation using techniques that are normally more complicated than of healthy machines. To date, Finite Element Method (FEM) and Winding Function Approaches (WFA) have been used for modeling, simulation and analysis of electrical machines with different types of fault. During the

last decade, some research investigated these phenomena, with various works published accordingly. In old research, the eccentricity fault was considered by Joksimovic et al. [2]; however, for a three phase induction machine, the slot opening effect was missed. Faiz et al; [3-6] in various studies, eccentricity faults for induction machines and PMSM were studied using FEM. Also, in [7] was presented comparison between finite-element analysis and winding function theory for inductances and torque calculation of a Synchronous Reluctance Machine. An FEM-based eccentricity fault diagnosis for induction machines was proposed in [8] by Nandy et al. in which fault diagnosis has been performed using current signatures; in that study, the authors focused only on hybrid eccentricity, and static and dynamic eccentricity was not addressed individually. In [9], a review was performed on induction machine fault diagnosis. It is known that the WFT and the FEM are two powerful methods for numerical analysis of electrical machines.

It is obvious that most of the above-mentioned research in the fault-diagnostic area have been presented for induction machines. Therefore, other types of machines, such as SynMs, have less been investigated.

In this research, a 6 salient-pole 72-slot SynM under healthy and eccentricity fault conditions is modeled and simulated using MWFA and FEM. This model, all non-ideal effects, including the machine geometry, slots effects and saturation, non-linear behavior due to ferromagnetic materials, type of winding connection and winding distribution and static and dynamic eccentricity are taken into account. Mathematical-based models and numerical methods using MWFA and Two-dimensional (2D) Finite-Element method have been utilized to compute torque and stator currents in the faulty SynM, and then torque and stator currents spectra have been employed to introduce an index for SE and DE fault diagnosis. Then a novel approach is presented to detect SE and DE in salient-pole SynM based on vibration torque and currents signature.

II. ECCENTRICITY FAULT

An eccentricity fault usually occurs due to ball bearing erosion and machine obsolescence, and consequently, will be

978-1-5090-0376-1/16 $31.00 © 2016 IEEE

produced an un-uniformed air gap. It has three conditions: static, dynamic, and hybrid eccentricity. The third condition is a combination of static and dynamic conditions, as shown in Fig1. The air-gap function for both dynamic and static eccentricity is written in Eqs. (1-3).

$$f_e(\theta) = g_0[1 + \delta.Cos(\theta - \varepsilon)] \tag{1}$$

$$\theta = \begin{cases} \theta_s & for \quad Static \quad Eccentricity \\ \theta_r & for \quad Dynamic \quad Eccentricity \end{cases} \tag{2}$$

$$\theta_r = \theta_s - \theta_m - tg^{-1}\left(\frac{\delta.g_0}{r}\right).Cos(\theta_s) \tag{3}$$

Where $f_e(\theta)$ is whole air-gap function, g_0 gap length between teeth and rotor saliency θ_s , θ_r mechanical angle of stator and rotor, θ_m mechanical position of rotor. Also δ, ε are eccentricity coefficient and denotes center of eccentricity, respectively.

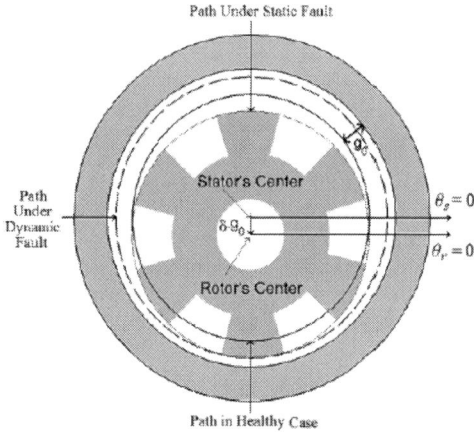

Fig. 1. Eccentricity fault schematic.

A. Static Eccentricity

Static eccentricity is the case where the rotor rotation does not change the air gap length minima position as shown in Figs.2a and 2b. [10]

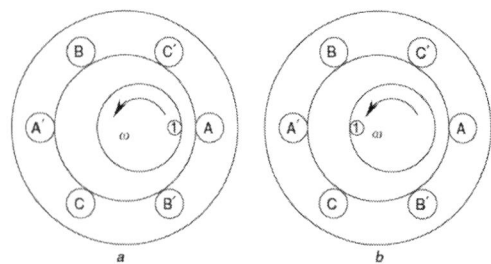

Fig. 2. Rotor bar position in machine with static eccentricity, (a) Initial position , (b) position aftar half rotor revolution.

B. Dynamic Eccentricity

Dynamic eccentricity is another faulty operating regime of the electrical machine. In this case, the position of the air gap length minima moves along with the rotor as shown in Figs. 3a and 3b.[10]

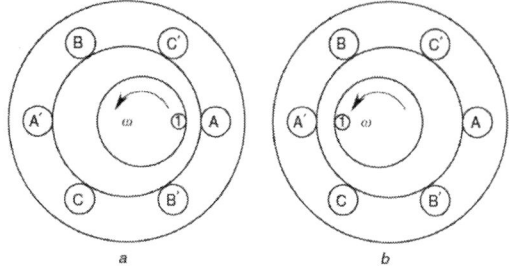

Fig. 3. Rotor bar position in machine with dynamic eccentricity, (a) Initial position , (b) position aftar half rotor revolution.

III. CASE STUDY FOR HEALTHY AND FAULTY STUDYING

It was considering a 3-phase 6-pole salient-pole Synchronous machine with 72 slots in stator and has 2 layer windings in order to decrease effects due to space harmonics: as shown in Fig4 and its listed parameters are subsequently presented in Table I. As shown in Fig4, the reference of the stator place ($\theta_s = 0$) and the reference of the rotor place ($\theta_r = 0$) are related to the center of winding (A) and the maximum air-gap length, respectively.

TABLE I.   MACHINE PROPERTIES FOR SIMULATION

Symbol	Parameter	Value
l	Machine length	250mm
r	Stator radius	75mm
P	number of poles	6
β	Pole saliency	30°
τ	Angle of slot opening	2.2°
g_0	Constant part of air gap	0.5mm
N_s	Turn per pole	160 turns
R_a	Stator winding resistance	0.2 Ω
R_f	Field winding resistance	0.1 Ω
l_{ls}	Stator winding's leakage inductance	0.01H
l_{lf}	Field winding's leakage inductance	0.02H

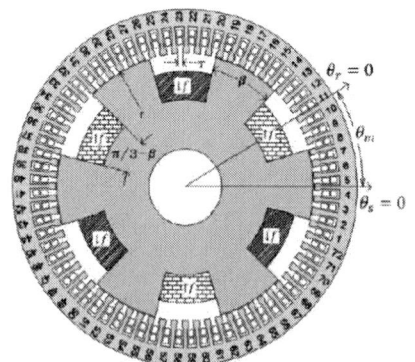

Fig. 4. Schematic of a salient-pole 3-phase SynM with 2 layer windings.

IV. MODIFIED WINDING FUNCTION APPROACHES

The non-sinusoidal winding function and slot opening number are the two most important reasons for space harmonic production in electrical machines. The sinusoidal winding function is an ideal feature of a machine that cannot be assumed for an excellent model. In fact, limitation of number of slots is one of the real factors that cause winding and magnetic-motive-force (MMF) functions to be non-sinusoidal [11]. It is noted that each 6-pole machine with 72 slots and 160 turns per pole is equivalent to a machine with 20 turns per slot. Now, the turn function for the stator and field winding can be considered as shown in Fig 5. The turn function as depicted in Fig 5 is based on the WFT [2], which will be used for modeling as follows.

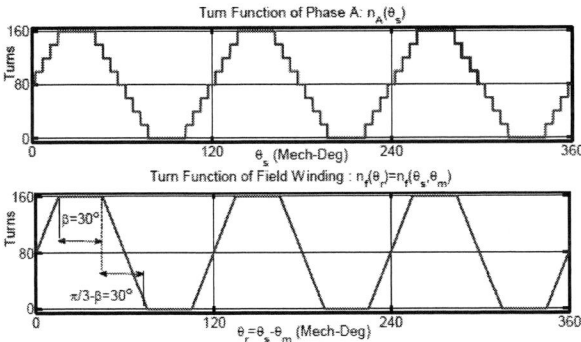

Fig. 5. Turn function of 6-pole 72-slots SynM shown in Fig 4.

Slot opening geometry is regarded as another non-ideal feature of a real machine. To increase the accuracy of modeling, the proper function should be assigned to the rotor poles and slot opening [11]. A relative figure is shown in Figs 6, and relevant mathematical modeling are done according to which were first proposed in [7]. Figs 7, 8 show air gap function including rotor saliency and stator slots effect.

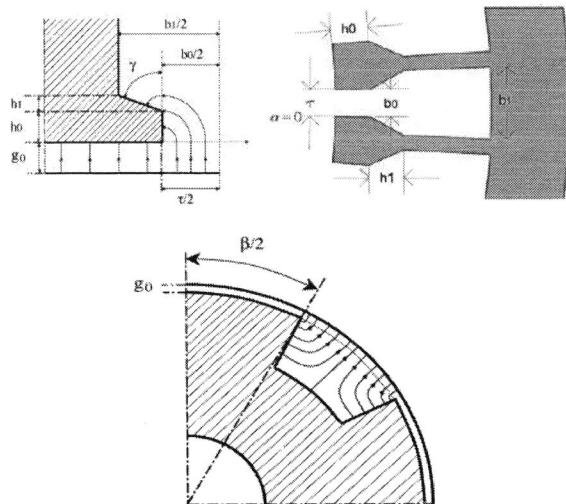

Fig. 6. Geometric structure of the stator slot and rotor poles and Flux lines distribution due to the stator slots and rotor saliency.

And the next Figs show Inverse of the air gap function in healthy condition, static eccentricity and dynamic eccentricity in four given positions of rotor.

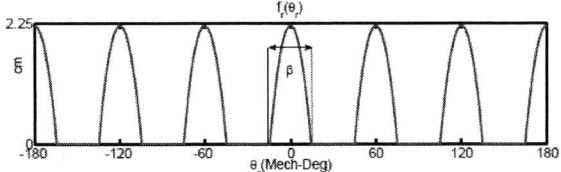

Fig. 7. Air gap function due to rotor poles.

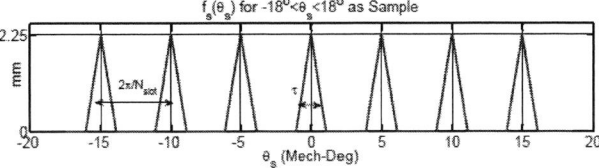

Fig. 8. Air gap function due to stator slot.

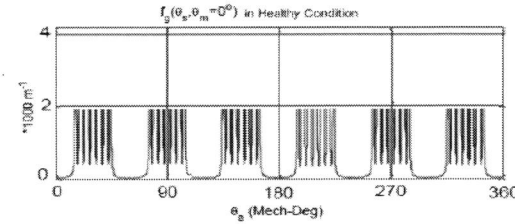

Fig. 9. Inverse of air gap function in healthy condition.

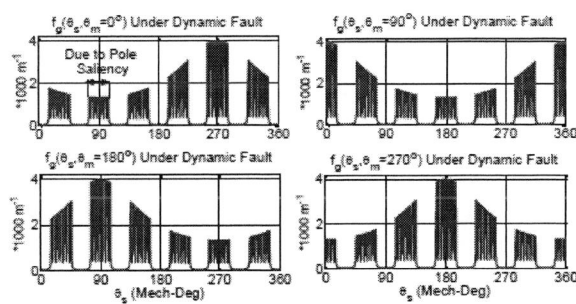

Fig. 10. Inverse of air gap function in dynamic eccentricity in 4 given position of rotor.

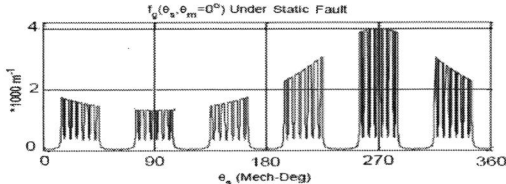

Fig. 11. Inverse of air gap function in static eccentricity.

The MWFA is used for calculating the machine inductances as written in Eq.(4-6), while n_i, n_j turn function of phases (as shown in Fig 5 for Example phase A), and f_g^{-1} inverse of air gap function consists of three parts for rotor poles, slot opening and conventional air gap.

$$L_{ij}(\theta_m) = \mu_0 rl \left[\begin{array}{c} \int_0^{2\pi} n_i(\theta_s,\theta_m) n_j(\theta_s,\theta_m) . f_g^{-1}(\theta_s,\theta_m) d\theta_s \\ -2\pi \langle M_i(\theta_s,\theta_m) \rangle \langle M_j(\theta_s,\theta_m) \rangle \langle f_g^{-1}(\theta_s,\theta_m) \rangle \end{array} \right] \quad (4)$$

$$\langle f_g^{-1}(\theta_s,\theta_m) \rangle = \frac{1}{2\pi} \int_0^{2\pi} f_g^{-1}(\theta_s,\theta_m) d\theta_s \quad (5)$$

$$\langle M_k(\theta_s,\theta_m) \rangle = \frac{1}{2\pi \langle f_g^{-1}(\theta_s,\theta_m) \rangle} \int_0^{2\pi} n_K(\theta_s,\theta_m) f_g^{-1}(\theta_s,\theta_m) d\theta_s \quad (6)$$

In this paper, a 50% eccentricity fault (δ=0.5) is considered with ε=90°, and in order to simulate performance of a SynM in healthy and faulty condition, all self and mutual inductances are computed at 721 rotor angular positions. Then dynamic equations of SynM are solved. To analyze torque and stator currents of healthy and faulty machine, a 60-HZ, 220Vrms, 3-phase voltage source is considered. Many signal processing tools have been so far utilized in order to detect faults in electrical machines. In this paper, Fast Fourier Transform (FFT) has been used for eccentricity fault identification in SynM. Therefor torque and stator current spectra of the SynM under different condition are computed. Torque and current curve with its power spectral density (PSD) at 1200rpm are obtained. Stator current in healthy condition and torque in healthy and faulty condition is shown in Fig12,13, respectively.

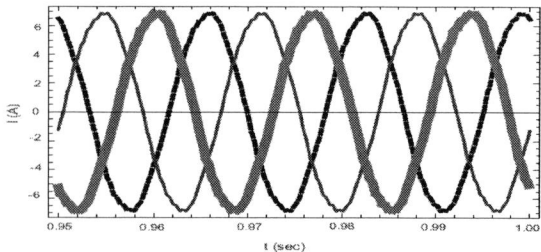

Fig. 12. Stator current in healthy condition (MWFA).

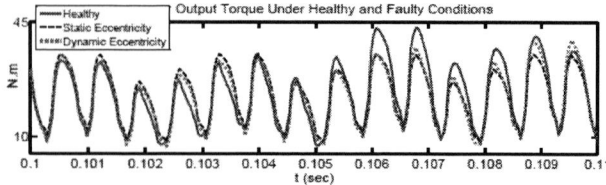

Fig. 13. Torque in healthy and faulty condition (MWFA).

V. FINITE ELEMENT METHOD

FEM is a technique based on magnetic field analysis that takes into account the magnetic circuit geometry, spatial distribution of stator windings and rotor bars, existing slots around air gap, nonlinear behavior of ferromagnetic materials, and different types of rotor and stator faults [3]. By calculating the magnetic field distribution, other quantities such as magnetic flux density, inductances of different windings, induced voltage waveform, and electromagnetic torque can be obtained. Although, FEM gives accurate result, however, this method is time consuming.

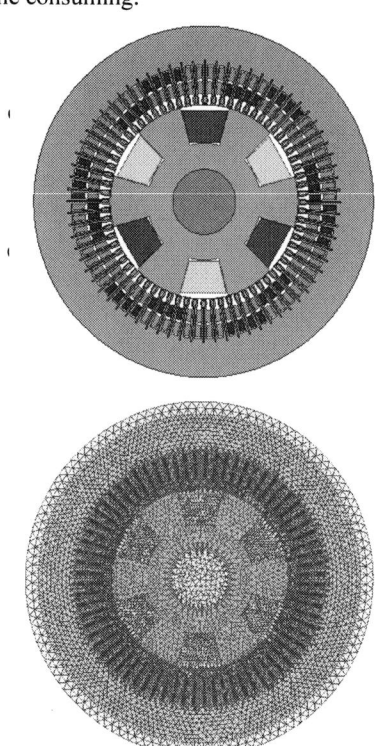

Fig. 14. Geometric configuratins of the modelled SynM and using 2D FEM.

In this paper, now the same analysis is verified by the FEM. The 2D scheme of the simulated SynM has been shown in Fig.14. In this modeling, non-linear characteristic of the SynMs, stator and rotor cores, spatial harmonics due to the stator slots and non-uniformity due to eccentricity fault are taken in to account. The stator and rotor laminated M19-24G sheets and there are 72 slots in the stator filled with copper. The details of the simulated machine are according to Table I. In order to modeling of saturation is used real B-H curve in the simulation. Also to reduce the losses due to eddy currents and core laminations, in 2D modeling, conductivity of rotor and stator's sheets is zero. A 220-V, 60-Hz 3-phase voltage source has been considered as a power supply for both methods and time-stepping FEM is used in this model. It should be noted that the machine was made to function as a generator.

A. Modelling of eccentricity fault

In this modeling is modeled non-uniformity of the air gap due to static and dynamic eccentricity in different degrees. In the SE case, the rotor is displaced from the center of the stator and rotor is rotated around its own center. Although in the DE fault the rotor is moved from the center of the stator, rotor is rotated around the center of the stator [4]. Fig15 shows the lines flux and electromagnetic field density in healthy and faulty condition.

Fig. 15. Lines flux density and magnetic field density plots (right to left) healthy, 50% SE, 50% DE.

With respect to Fig 15, the change in the magnitude of flux density with eccentricity is significant and clearly distinguishable. And shows the modeling of machine under SE and DE is correct. It is clear that there is a marginal increase in the magnitude of flux density and magnetic field density under 50% SE and DE condition as compared to the healthy condition.

In this part, has been extracted the output of simulation contains stator currents and torque curves in healthy and faulty condition after passing transient mode.

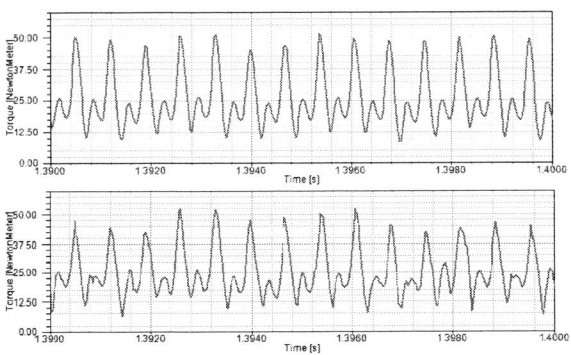

Fig. 16. Torque (steady state) in healthy and faulty condition (FEM)

To study the effect of the eccentricity faults, the simulation is performed in three parts as healthy, static, and dynamic eccentricity cases. In each part, a torque and current frequency scan is performed for a machine, as shown in Figs 18-23.then harmonics components of stator currents and torque are show in next Figs. In figures, DDE and DSE denote to "due to dynamic eccentricity" and "due to static eccentricity," respectively.

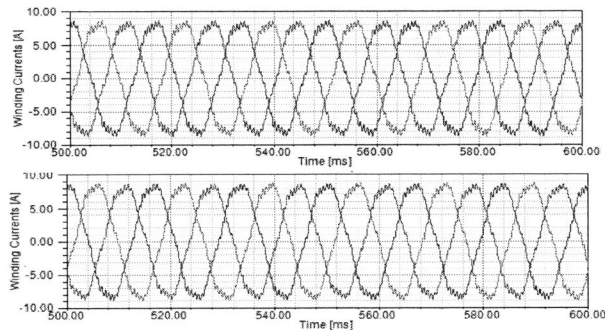

Fig. 17. Stator Current (steady state) in healthy and faulty condition (FEM)

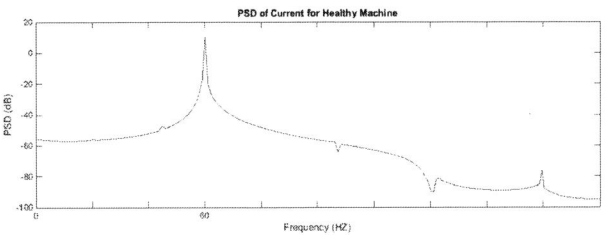

Fig. 18. PSD of stator currents in healthy condition.

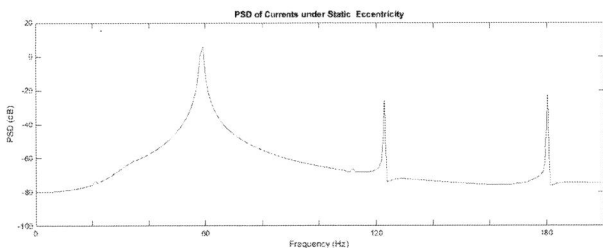

Fig. 19. PSD of stator currents under 50% SE

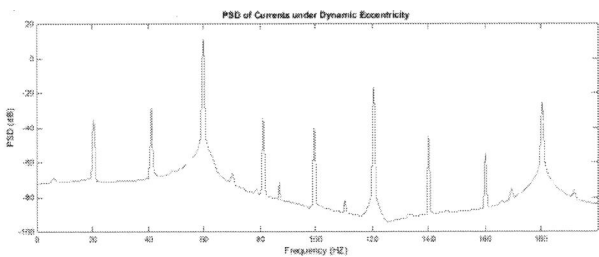

Fig. 20. PSD of stator currents under 50% DE

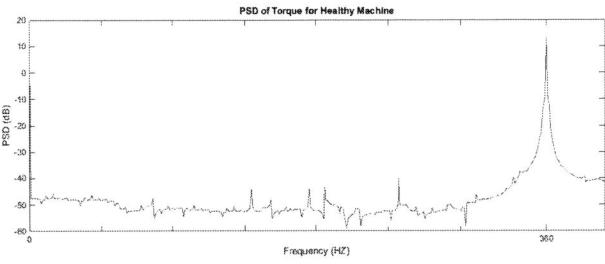

Fig. 21. PSD of Torque in healthy condition

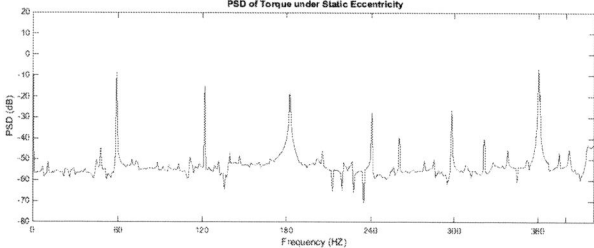

Fig. 22. PSD of Torque under 50% SE

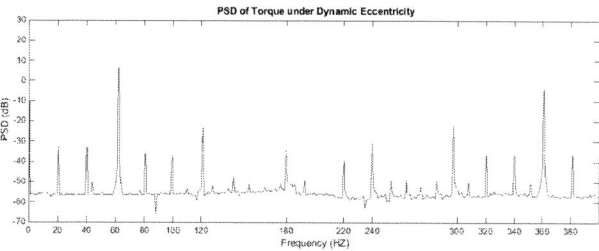

Fig. 23. PSD of Torque under 50% DE

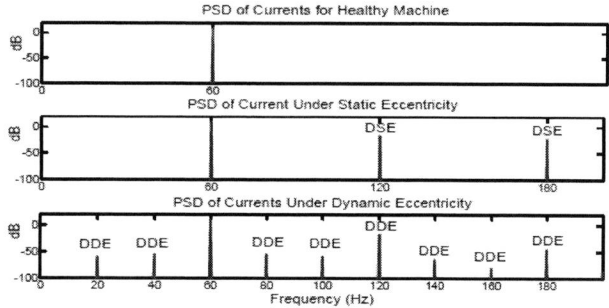

Fig. 24. Harmonic components of current in healthy and 50% faulty condition

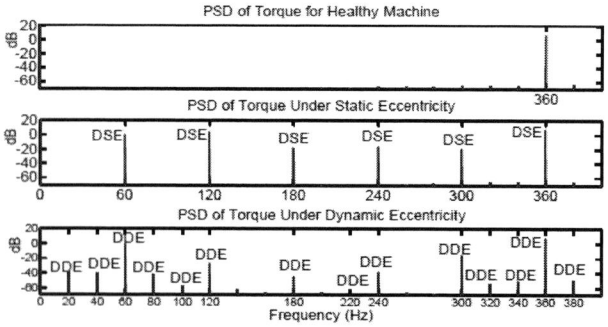

Fig. 25. Harmonic components of torque in healthy and 50% faulty condition

According to Figs 24, 25, the eccentricity fault could be observed from the torque and current signature for both static and dynamic cases.Comparing the healthy and faulty condition, frequency components of current and torque production are same. Frequency components of the SE with Eq.7 and frequency components of the DE with Eq.8 are shown.

$$f_{i\&\tau-Static} = k.f_s \qquad (7)$$

$$f_{i\&\tau-Dynamic} = k.\frac{1}{3}f_s \qquad (8)$$

VI. CONCLUSION

In this article, a 6 salient-pole 72-slot SynM under healthy and eccentricity fault conditions was investigated using MWFA and FEM. This model, are taken in to account all non-ideal effects. That they include the machine geometry, slots effects and saturation, non-linear behavior due to ferromagnetic materials, type of winding connection and winding distribution, static and dynamic eccentricity. It can consider most of the important features of a salient-pole machine. To study the effect of the eccentricity faults, torque and stator currents are computed, and then a novel approach is presented to detect static and dynamic eccentricity in salient-pole SynM based on vibration torque and currents signature.

REFERENCES

[1] B. M. Ebrahimi, M. Etemadzadeh, J. Faiz,"Dynamic eccentricity fault diagnosis in round rotor synchronous motors", Journal of Energy Concertion and Management, Vol. 52, 2011.

[2] G. M. Jocsimovic, D. J. Penman, and N. Arthur, "Dynamic Simulation of Dynamic Eccentricity in Induction Machine- Winding Function Theory", IEEE Transaction on Energy Conversion, Vol. 15, No.2, June2000.

[3] J. Faiz, B. M. Ebrahimi, B. Akin, and H.A. Toliyat, "Finite Element Transient Analysis of Induction Motors Under Mixed Eccentricity Fult",IEEE Transaction on Magnetics, Vol. 44, No.1, January 2008.

[4] B.M. Ebrahimi, J. Faiz, B.N. Araabi, "Pattern identification for eccentricity fault diagnosis in permanent magnet synchronous motors using stator current monitoring", IET Electric Appl, Vol.4, Issue.6,2010.

[5] B.M. Ebrahimi, J. Faiz, " Magnetic field and vibration monitoring in permanent magnet synchronous motors under eccentricity fault", IET Electric Power Appl, Vol.6, Issue.1, 2012.

[6] B. M. Ebrahimi, J. Faiz, "Diagnosis and performance analysis o threephase permanent magnet synchronous motor with static, dynamic and mixed eccentricity", IET electric Power Appl., Vol. 4, No.1, 2010.

[7] T. Lubin, T. Hamiti, H. Razik, and A. Rezzoug, "Comparison Between Finite-Element Analysis and Winding Function Theory for Inductances and Torque Calculation of a Synchronous Reluctance Machine. IEEE Transaction on Magnetics, Vol. 43, No. 8, August 2007.

[8] Nandy, S., Ahmed, S., Toliat, H. A., and Bharadwaj, R. M., "Selection criteria of inductionmachines for speed-sensorless drive applications," IEEE Trans. Ind., Vol. 39, No. 3, pp. 704–712, May/June 2003.

[9] Bellini, A., Filippeti, F., Tassoni, C., and Capolino, G. A., "Advances in diagnostic technique for induction machines," IEEE Trans. Ind., Vol. 55, No. 12, pp. 4109–4126, December 2008.

[10] G. M. Jocsimovic, "Dynamic Simulation of cage Induction Machine with air gap eccentricity", IEE Proc-Electr Power Appl, Vol. 152, No.4, July2005.

[11] P.Naderi, "Eccentricity Fault Diagnosis and Torque Ripple Analysis of a Four-pole Synchronous Reluctance Machine in Healthy and Faulty Conditions", Electric Power Components and Systems, 43(11):1236–1245, 2015.

7th Power Electronics, Drive Systems & Technologies Conference (PEDSTC 2016)
16-18 Feb. 2016, Iran University of Science and Technology, Tehran, Iran

Sensitivity Analysis to Improve the Design Process of an Induction Motor Using MEC

Reza Haghighian
School of Electrical Engineering
Iran University of Science & Technology
Tehran, Iran
haghighianreza@gmail.com

Mohamad Reza Mazinanian
School of Electrical Engineering
Iran University of Science & Technology
Tehran, Iran
mrm.mazinanian@gmail.com

Abolfazl Vahedi
School of Electrical Engineering
Iran University of Science & Technology
Tehran, Iran
avahedi@iust.ac.ir

Abstract— **Design of any system in particular electrical motors requires knowledge of all aspects of design, So Designing is a complex and unique process. Usually, after the initial design of electrical motors, modelling is integral part of designing. Due to acceptable accuracy and high speed, the Magnetic Equivalent Circuit (MEC) model used for modeling. Due to the high accuracy, finite element model (FEM) is appropriated to validation. In this paper, First modeling of 18.3 kW, 380 V, 3000 rpm induction motor based on magnetic equivalent circuit is done. Then by finite element model validation is performed. Next, Sensitivity analysis of important motor characteristics including torque, power factor and so on to the geometrical and electrical design variables is presented. For example simulations indicate that to have better torque and efficiency in motor should be changed number of turns, length and rotor tooth width.**

Keywords— Magnetic equivalent circuit; MEC; Sensitivity analysis; Finite Element Method; FEM

I. INTRODUCTION

From past to the present induction motors a large portion of the world's electricity consumption is allocated. Induction machines due to many advantages in many applications first choice. Statistics show that more than 90 percent of the motors from this type are used in industry [1]. The advantages of this motors compared to the other types that are more robust and easier maintenance. For this reason, in certain applications, such as wells, grinding and etc. are applied. Always modeling in many various fields, especially electrical machines are used. So in 1970 [2], finite element model based on a small element has been applied to electrical machine modeling. Then despite acceptable accuracy this model, Due to the complex and time-consuming calculations in 1980, returning to analytical model based on empirical formula and lumped parameters for the simplicity of calculation, despite low accuracy was considered [3]. Then in late 1980s magnetic equivalent circuit (MEC) for electric machine was presented [4]. This model is based on a reluctance network, Evaluates the performance of the electric machine. Magnetic equivalent circuit in terms of accuracy and

speed of response is compromise between the finite element model and analytical model [3].

A list of criteria and the corresponding qualitative ratings are presented in Table I ('+' denotes an advantage, 'o' a neutral rating, and '-'a disadvantage). An analytical model has advantages in computational effort, and Parameterization. FEM despite reasonable accuracy, the calculations are Time-consuming and complex. MEC in terms of accuracy and speed of response is compromise between the finite element model and analytical model [3].

TABLE I.   COMPARISON OF THREE METHODS OF MODELING [3]

index	Analytical	FEA	MEC
accuracy	--	+	o
Computational complexity	++	-	o
Parameterization	+	o	+
Simplicity of implementation	o	o	o

This article is divided in three parts, the first part was MEC modeling in brief and second part MEC modeling validation with finite element model is done. The last part, a sensitivity analysis of the induction motor output such as: torque, efficiency, stator current and power factor, is carried out to the design parameters. Using sensitivity analysis can be change design variables to improve motor performance.

II. MEC MODELING

This section MEC modeling briefly based on electrical and geometric parameters motor is implemented. Permeance structure generation, the relationship between of the MEC model with electrical and geometrical parameters will be determined [5]. The theory of the MEC-modeling in detail at the [6] source is expressed. Permeance values of different parts of the motor can be calculated using the following equation:

978-1-5090-0376-1/16 $31.00 © 2016 IEEE

$$p = \frac{\mu s}{l} \qquad (1)$$

So that p, μ, S and l are permeance, permeability, cross sectional area, and length of an element, respectively. In a rotating equipment, air-gap permeance depends on the angle of the rotating element (rotor).

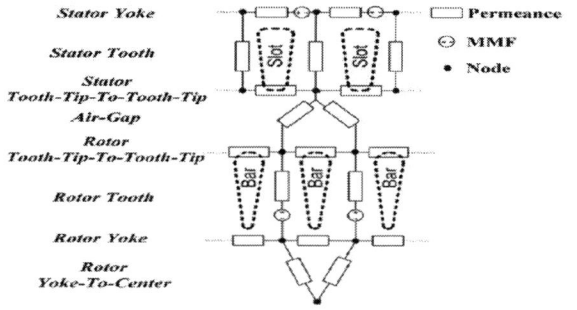

Fig. 1. *Permeance network model for a portion of a squirrel cage induction machine [7].*

Rotor and stator permeance network model for a induction motor is depicted in Figs. 1. In these figure the node potentials and the location of permeances clearly shown. The stator MMF source is placed in the yoke. Rotor sources is caused by the current passing through the bars.

The stator nodal equations can be written as:

$$A_{sb-sb}M_{sb} + A_{sb-st}M_{st} + A_{sb-si}i_{qd0} = 0 \qquad (2)$$

$$A_{st-sb}M_{sb} + A_{st-st}M_{st} + A_{st-rt}M_{rt} = 0 \qquad (3)$$

The rotor nodal equations are given as:

$$A_{rt-st}M_{st} + A_{rt-rt}M_{rt} + A_{rt-st}M_{rb} + A_{rt-ri}i_r = 0 \qquad (4)$$

$$A_{rb-rt}M_{rt} + A_{rb-rb}M_{rb} + A_{rb-ri}i_r = 0 \qquad (5)$$

Flux linkage equations of stator and rotor may be expressed as:

$$A_{s\lambda-sb}M_{sb} + A_{s\lambda-st}M_{st} = \lambda_{qd0} \qquad (6)$$

$$A_{r\lambda-rt}M_{rt} + A_{r\lambda-rb}M_{rb} + A_{r\lambda-ri}i_r = \lambda_r^T \qquad (7)$$

Equations (2) to (7) can be mixed, and a non-linear system equation as follows formed.

$$AX = b \qquad (8)$$

So that A, X, and b are partitioned matrices of the form

$$A = \begin{bmatrix} A_{sb-sb} & A_{sb-st} & 0 & 0 & A_{sb-si} & 0 \\ A_{st-sb} & A_{st-st} & A_{st-rt} & 0 & 0 & 0 \\ 0 & A_{rt-st} & A_{rt-rt} & A_{rt-rb} & 0 & A_{rt-ri} \\ 0 & 0 & A_{rb-rt} & A_{rb-rb} & 0 & A_{rb-ri} \\ A_{s\lambda-sb} & A_{s\lambda-st} & 0 & 0 & 0 & 0 \\ 0 & 0 & A_{r\lambda-rt} & A_{r\lambda-rb} & 0 & A_{r\lambda-ri} \end{bmatrix} \qquad (9)$$

$$X = \begin{bmatrix} M_{sb}^T & M_{st}^T & M_{rt}^T & M_{rb}^T & i_{qd0}^T & i_r^T \end{bmatrix}^T \qquad (10)$$

$$b = \begin{bmatrix} 0 & 0 & 0 & 0 & \lambda_{qd0}^T & \lambda_r^T \end{bmatrix} \qquad (11)$$

In the system, zeros is the expression of matrix, and subscripts "sb", "st", "rt", "rb", "si", "ri", "$s\lambda$" and "$r\lambda$" denote association with the potential of stator base, potential of stator tooth, potential of rotor tooth, potential of rotor base, stator current, rotor current, stator flux linkage, and rotor flux linkage, respectively. Non-linear equation with any system such as Gauss Seidel and Newton Raphson method is solved. For this work, an iterative method was used. The Flowchart of the iterative method for MEC Solution is shown in Fig. 2 and the algorithm proceeds as follows.

1. Calculate constant resistances, permeances and leakage inductances and set initial X matrix to zero.
2. According to X matrix, determine unknown parameters such as stator and rotor currents and magnetic potentials.
3. After calculation of stator and rotor flux, b matrix formed.
4. Depending on the angle rotor, air-gap permeances is determined.
5. According to node potentials, the field intensity of the various branches is calculated and non-linear permittivity is determined.
6. By identification A matrix and b matrix, calculate X matrix. If convergence was done. The algorithm has the answer and go to Step 7; otherwise the algorithm goes to Step 5. In (12), X_k and X_{k-1} represents current and previous of X matrix.

$$\frac{|X_k - X_{k-1}|}{|X_{k-1}|} \langle 0.01 \qquad (12)$$

7. Determine X matrix and after that calculate torque, rotor speed and position. So stator to rotor permeances corrected.

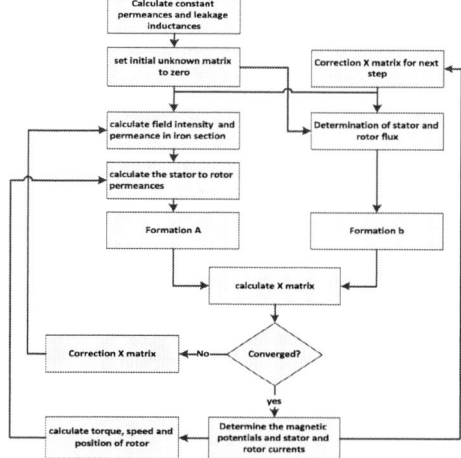

Fig. 2. Flowchart of the iterative method for MEC Solution

In this paper sensitivity analysis some important design variables than output parameters for an induction motor with three-phase, two poles, 3000 [rpm], 18.3 [kW] is done.

Geometric and electrical characteristics of the selective induction motor are respectively shown in tables II and III. Shape of the rotor and the stator tooth are also shown in Figs. 3 and 4.

TABLE II.   GENERAL CHARACTERISTICS OF THE SELECTIVE INDUCTION MACHINE

Parameter	Value
nominal Power	18.3 Kw
line-line voltage	380 V
operating frequency	50 Hz
stator winding connection	delta
rated power factor	0.8
application	Driving ESP
number of poles	2

TABLE III.   SPECIFICATIONS OF THE SELECTIVE INDUCTION MOTOR

Parameter	Value
shaft diameter	90 mm
rotor diameter	113.8 mm
air-gap length	0.6 mm
stator inner diameter	115 mm
stator outer diameter	220 mm
axial length	1430 mm
number of rotor slots	14
number of stator slots	18
number of slot per pole per phase	3
stator slot type	semi-closed
rotor slot type	semi-closed
lamination material	M270-35A
shaft material	Monel-500
stator winding material	copper
rotor bar material	copper
number of conductor per slot	8
number of turns per coil	4
number of coils per phase	6
coils connection type	series

Stator slot design (mm)

Ssd	*19.7*
Stw	*10.3*
Wsst	*9.7*
Wssb	*17.2*
Wstt	*5*
Stft	*0.5*

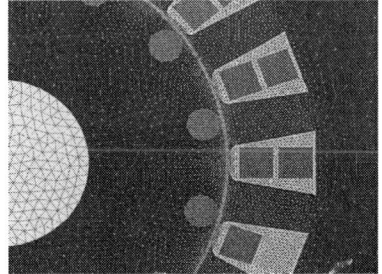

Fig. 3. *Stator geometry*

Rotor slot design (mm)

rsd	*8.4*
rtw	*14.9*
Wrtt	*2.1*
rtft	*0.5*

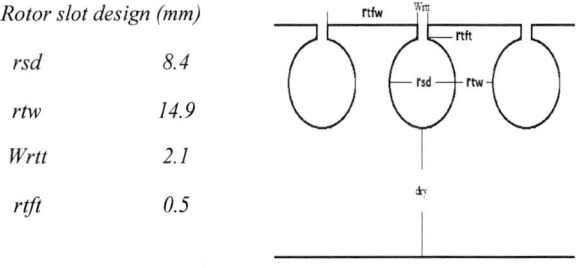

Fig. 4. *Rotor geometry*

III. MEC VALIDATION MODEL WITH FEM

In order to validate and determine the accuracy of the MEC modeling, the finite element method is selected. In this section, the motor using FEM modeled and some results are compared with MEC.

As is clear from Fig.5 approximately highest flux density in the stator teeth is 0.9.

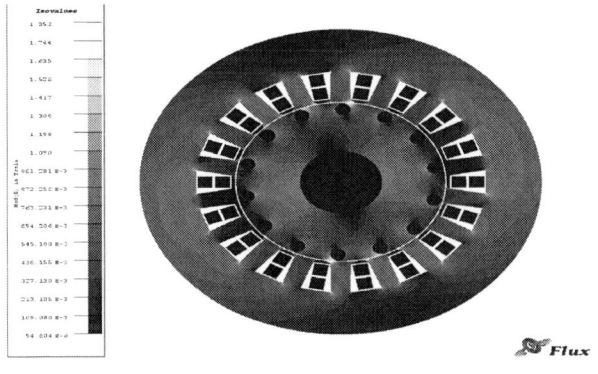

Fig. 5. flux density

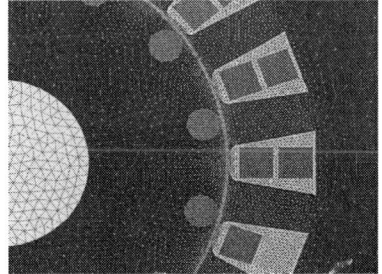

Fig. 6. The rotor and stator slot and teeth

The speed and current waveforms of both MEC and FEM models in Fig.7 to Fig.10 is clear.

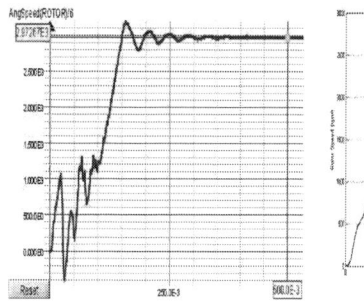

Fig. 7. Speed in FEM model

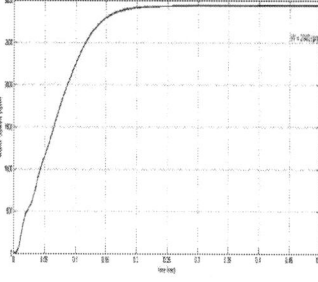

Fig. 8. Speed in MEC model

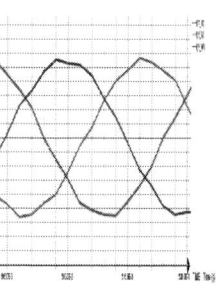

Fig. 9. Phase current waveform in FEM model

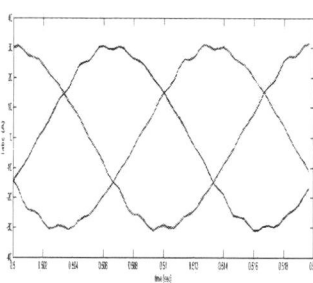

Fig. 10. Phase current waveform in MEC model

power factor, increases the drawn current and decline the torque.

Fig. 16 and Fig. 17, represent Rtfw and Rtw variations to output variables, respectively. Increasing Rtfw and Rtw causing to reduce the motor performance. Results show that the torque, efficiency, power factor are decreasing, it causes the rotor tooth face width and rotor tooth width increase. And also stator current increases.

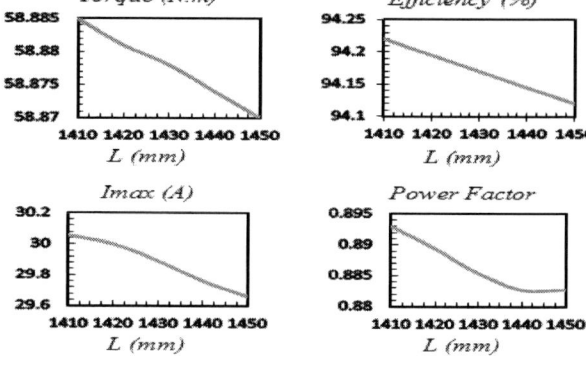

Fig. 11. Variations output variables to axial length (L) Variatiog

IV. SENSITIVITY ANALYSIS

A. Sensitivity

After MEC modeling, variations of design parameters to output variables is evaluated (sensitivity analysis). The selective induction motor is a three-phase, two poles, 3000 [rpm], 18.3 [kW].

This section is dedicated to defining the effective parameters, i. e. axial length (L), air gap length (g), number of turns of the stator winding (Nph), stator tooth face width (Stfw), stator tooth width (Stw), rotor tooth face width (Rtfw), rotor tooth width (Rtw), Output variables include the torque, efficiency, power factor and stator current.

Fig. 11 shows the sensitivity of the outputs to the axial length of the motor. With increasing length, all output variables are reduced.

Fig. 12 represents analyzes of the air-gap length. The torque and stator current are increasing as well as the air-gap length increase and although the efficiency and power factor decreases.

In Fig. 13, the sensitivity analysis of the number of stator turns on output variables are visible. Increasing Nph thereby reducing torque, efficiency and stator current and also the higher power factor.

Fig. 14 represent variations of stator tooth face width's. As shown, increase in Stfw, reduced the torque and efficiency leads.

In Fig. 15, represent Stw Variations to output variables, as it is clear, enhancement Stw, despite better the efficiency and

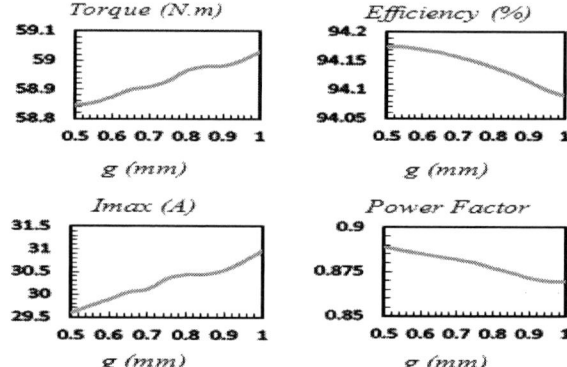

Fig. 12. Air-gap Variations the output parameters

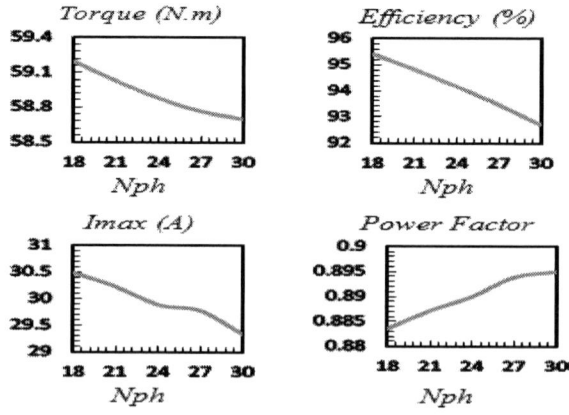

Fig. 13. Sensitivity analysis due to the number of stator winding turns (Nph)

978-1-5090-0376-1/16 $31.00 © 2016 IEEE

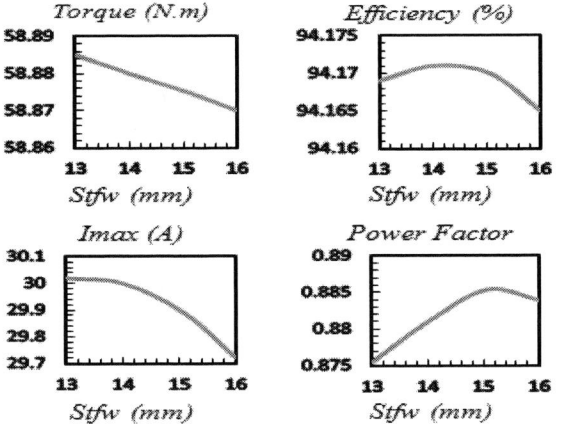

Fig. 14. Sensitivity analysis due to the stator tooth face width (Stfw)

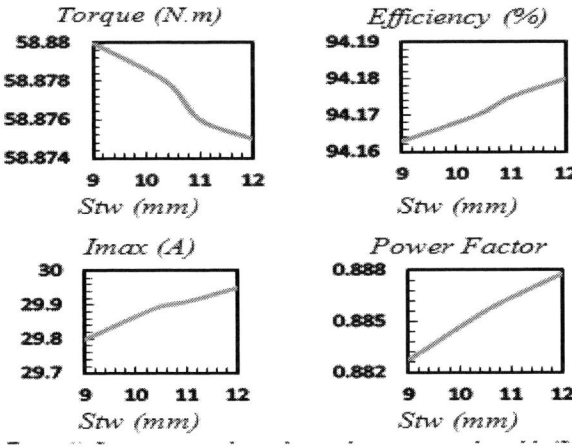

Fig. 15. Sensitivity analysis due to the stator tooth width (Stw)

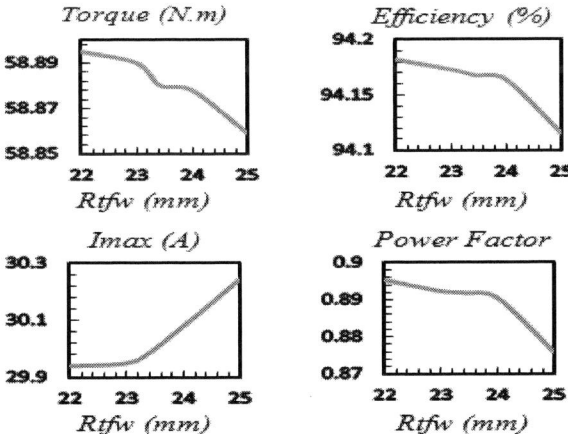

Fig. 16. Variations of the rotor tooth face width (Rtfw)

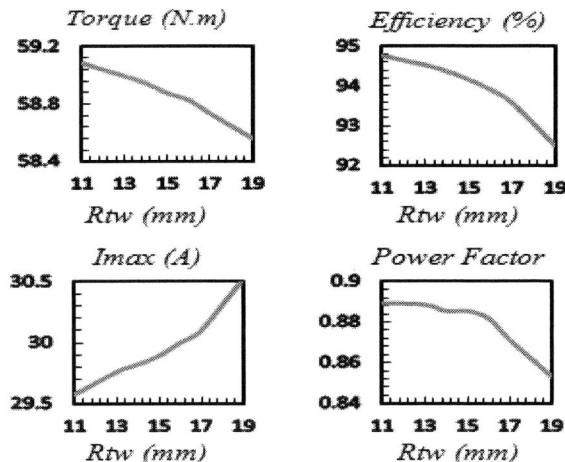

Fig. 17. Variations of sthe rotor tooth width (Rtw)

B. All Sensitivities

In order to better compare of sensitivity analysis by using (13) normalized output parameters to all of design parameters are calculated.

$$S_I^o = \frac{\Delta o / o_{av}}{\Delta I / I_{av}} \tag{13}$$

Fig. 18 compares the sensitivities of all design variables for each output parameters.

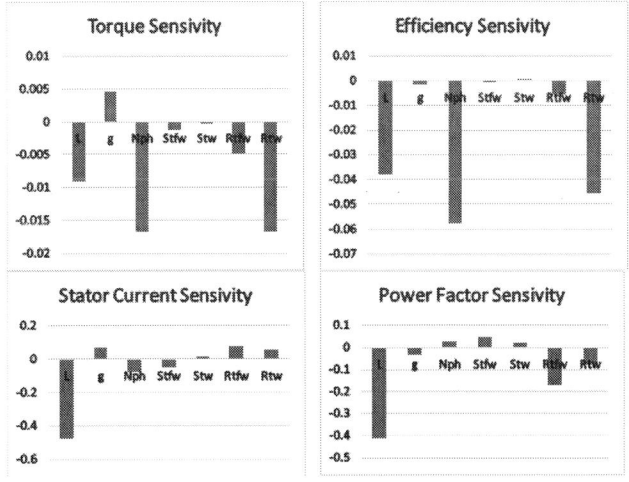

Fig. 18. The sensitivities of all outputs to all design parameters

V. CONCLUSION

In this paper, variations of the torque, efficiency, stator current and power factor due to the geometrical and electrical variables is presented. Sensitivity analysis not only in the design process is useful, but is a prerequisite in motor optimization. Results show that to have better torque and efficiency in motor should be changed number of turns, length and rotor teeth width. In order to have the best current performance (copper losses) length of the motor must be changed. And finally decreasing the length and rotor and stator flange parameters suited for improving the power factor.

References

[1] F. J. T. E. Ferreira and A. T. de Almeida, "Method for in-fieldevaluation of the stator winding connection of three-phase induction motors to maximize efficiency and power factor," IEEE Trans. Energy Conv., vol. 21, pp. 370-379, Jun. 2006.

[2] Amrhein, Marco, and Philip T. Krein. "Magnetic Equivalent Circuit Simulations of Electrical Machines for Design Purposes." Electric Ship Technologies Symposium, 2007. ESTS'*07. IEEE International.*

[3] V. Ostovic, Dynamics of Saturated Electric Machines. New York: Springer-Verlag, 1989.

[4] Amrhein, Marco, and Philip T. Krein. "Magnetic equivalent circuit modeling of induction machines design-oriented approach with extension to 3-D." *Electric Machines & Drives Conference, 2007. IEMDC'07. IEEE International.* Vol. 2.

[5] S. D. Sudhoff, B. T. Kuhn, K. A. Corzine, and B. T. Branecky, "Magnetic equivalent circuit modeling of induction motors," IEEE Trans. Energy Convers., vol. 22, no. 2, pp. 259–270, Jun. 2007.

[6] Marco Amrhein, Philip T. Krein, "Induction Machine Modeling Approach Based on 3-D Magnetic Equivalent Circuit Framework," IEEE Trans. Energy Convers., vol. 25, no. 2, pp. 259–270, Jun. 2010.

[7] Babak Asghari, Venkata Dinavahi, "Experimental Validation of a Geometrical Nonlinear Permeance Network Based Real-Time Induction Machine Model," IEEE Trans. Industrial Electronics, vol. 59, no. 11, NOVEMBER 2012.

7th Power Electronics, Drive Systems & Technologies Conference (PEDSTC 2016)
16-18 Feb. 2016, Iran University of Science and Technology, Tehran, Iran

A Precise Method to Detect the Accurate Rotor Position in BLDC Motors at Standstill Condition

O. Safdarzadeh
Department of Electrical
Engineering
Shahid Behesti Univ. G. C
Tehran, Iran
Safdarzadeh.ieee@chmail.ir

M. Nezamabadi
Department of Electrical
Engineering
Shahid Behesti Univ. G. C
Tehran, Iran
m_nezamabadi@sbu.ac.ir

E. Afjei
Department of Electrical
Engineering
Shahid Behesti Univ. G. C
Tehran, Iran
e-afjei@sbu.ac.ir

Abstract— **Sensorless startup methods in brushless DC (BLDC) motor drives have not been perfectly improved. Detecting the rotor initial position is a problematic step in these methods. In this paper a new method based on a probabilistic algorithm is proposed to find the position of the rotor at standstill condition. A set of diagnostic pulses are injected to stator windings for an unknown rotor position due to the inductance of each phase, different current responses are received. These responses are analyzed by an innovative algorithm, which is based on probability theory, to extract the rotor position in accuracy of 15 electrical degrees which is four times more accurate in comparison with earlier methods [9]. Due to the convergence feature of the used algorithm, the final result is completely faultless. This recognition method is very fast due the simplicity of calculations. A prototype driver is implemented with low-cost microcontroller and experimental results are demonstrated to validate the proposed method.**

Keywords—Brushless DC Motor; Standstill condition; Sensorless control; Active probing method;

I. INTRODUCTION

Utilization of the BLDC motors instead of their older counterparts, brushed DC motors, in recent decades has motivated the scientists to obtain optimized performance for these devices. Several advantages have been introduced for BLDC motors such as high speed [1], robust performance [2], high efficiency, etc. These advantages are mainly achieved by the absence of brushes in the motor structure [3].

In brushed DC motors, detecting the position of rotor to excite the proper windings is performed by commutator but in BLDC motors other mechanisms have to be employed. According to the utilized mechanism, two categories for BLDC drives have been introduced. The first category is based on position sensors which are installed on the motor to detect the rotor position [4]. But using sensors requires additional wiring, precise installation, frequent adjustment [5] and limited operating temperature [6].

In order to eliminate the problems of position sensors, sensorless drives have been introduced. In sensorless drives the motor electrical variables are measured to detect the rotor position. Back electro motive force (back-emf) produced in

stator windings, can be measured from the non-excited windings whenever the rotor rotates. In [7], [8] back-emf has been used to detect the rotor position. Since the back-emf depends directly on rotor rotational speed, the main defect of these methods is inability to detect the rotor position in standstill condition. Since most conventional drives operation are based on the back-emf thus one of the challenges in sensorless control is detecting the rotor position at standstill condition. To solve this problem a group of methods called active probing have been proposed [9]. these methods are based on the injection of diagnostic signal to a non-excited winding, receiving the electrical response from that winding or the other windings, and then analyzing the responses to find the rotor position. This methods are ideal in standstill condition since all the windings are non-excited. In [10] a set of voltage pulses are injected to the motor winding then voltage of neutral point and current of DC bus are measured to extract the initial rotor position but accuracy of this method is limited in 30 electrical degrees.

In this paper diagnostic signals are applied to stator windings, responses are received and a new algorithm is presented to extract the rotor position from the responses.

The effect of saturation on permanent magnet and iron core characteristic are studied in section II. In section III the procedure to obtain the response signals is explained and responses are analyzed. The proposed probabilistic algorithm for extracting the rotor position are discussed in section IV in detail. The experimental results are shown in section V which verify the effectiveness of the proposed method and finally the conclusion is presented in section VI.

II. MAGNET AND IRON CORE CHARACTERISTICS

Permanent magnets (PMs) store magnetic energy. They can produce magnetic flux without any excitation system. Storage of magnetic energy in PM can be evacuated if a high contrary external magneto motive force (mmf) is applied to it. PMs magnetic characteristic is illustrated in fig1. In the normal operating interval of PMs in machines the slope of ($\Delta B/\Delta H$), which shows PM permeability, is about $\mu_0 = 4\pi \times 10^{-7}$.

978-1-5090-0376-1/16 $31.00 © 2016 IEEE

Since high external mmf can demagnetize the PM, in standard utilization of PM, rang of external mmf is limited in a way that PM can keep its storage magnetic energy. Therefore the approximate range of tolerable external mmf is shown in Fig. 1 as the operating interval.

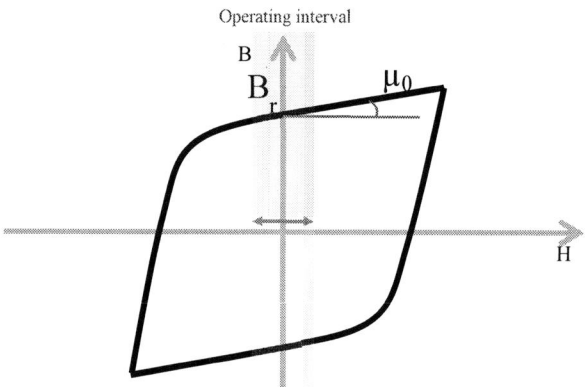

Fig. 1. PM characteristic and approximated operating interval

The stator of motor is comprised of laminated steel. Produced flux in the iron core is highly sensitive to the externally applied mmf.

According to ampere principle, $Ni = \int H.dl$, applying current to stator windings produces magnetic field intensity (H), this mmf changes iron core operating point to a new point as shown in Fig. 2. Regardless of hysteresis effect, without external mmf there is no magnetic flux density (B) in the iron core. Applying an external mmf to the iron core increases linearly the B in linear region but after passing the soft saturation region increasing the H does not raise the B in the iron core like before as depicted in Fig. 2.

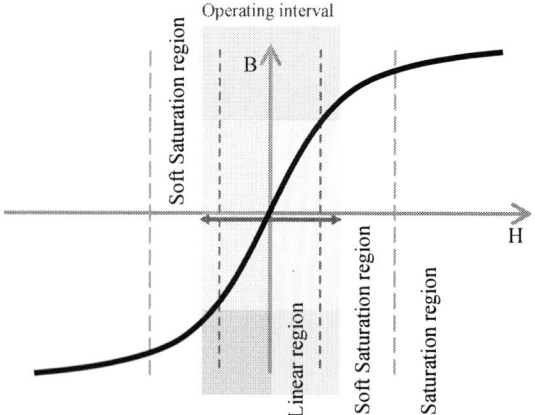

Fig. 2. Iron core characteristic

A simplified version of BLDC motor is illustrated in Fig. 3. The motor is comprised of a non-salient two-pole rotor and a three-pole stator with a winding on each pole. It is assumed that exciting each winding produce the magnetic flux in the direction shown in Fig. 3. This direction is called winding

vector and the direction of the magnetic field in the PMs of the rotor is called rotor vector in this paper.

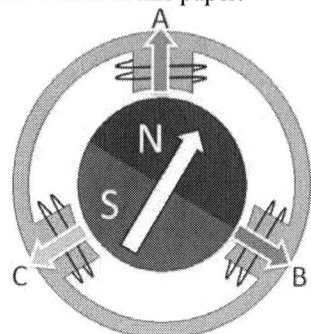

Fig. 3. The simplified BLDC motor with the rotor vector and the stator winding vectors

By rotating the rotor, the flux linkage density in winding A produce by rotor vector can be expressed by $B_r|A|\cos(\theta)$ in which B_r is residual flux density of the rotor magnets, $|A|$ is normalized vector of winding A and θ is the angle between them. So in different rotor positions, different flux densities are expected. Consequently the operating point of the stator pole A changes as shown in Fig. 2. For using the maximum capability of magnetic materials, BLDC motors are designed to operate close to soft saturation regions (Fig. 2). In another words, PMs are designed in a way that they can put the operating point of the stator iron core close to the soft saturation region.

III. RESPONSE SIGNALS ANALYSIS

To obtain response signals, voltage pulses called diagnostic pulses (Fig. 4) are applied to selected winding terminals. The amplitudes of the resulting currents called response signals are received from the selected winding.

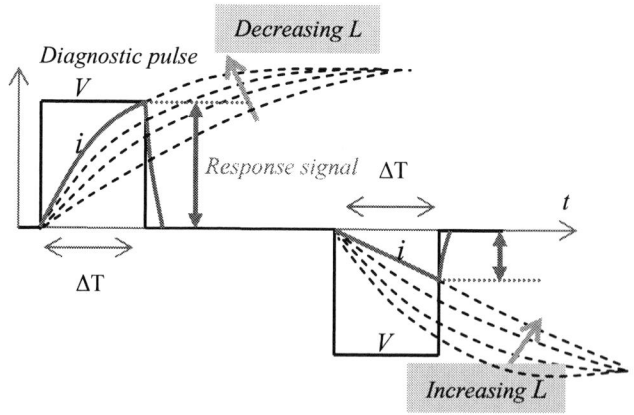

Fig. 4. The diagnostic pulses and the response signals

By injecting a diagnostic pulse to the winding A, the current increases according to (1).

978-1-5090-0376-1/16 $31.00 © 2016 IEEE

$$i(t) = \frac{V_{dc}}{R}\left(1 - e^{-\frac{R}{L}t}\right) \qquad (1)$$

Since the duration of the diagnostic pulse in comparison with the time constant of the winding is negligible the winding resistivity effect can be neglected so the current can be expressed as (2).

$$i = \frac{1}{L}\int_0^t v \qquad (2)$$

As discussed earlier, depending on the rotor vector direction, there is an initial flux in the stator poles. If the direction of produced flux by the diagnostic pulse is in the same direction of the initial flux, the net flux density becomes bigger and the operating point goes toward the saturation region as shown in Fig. 5. According to (3) the corresponding inductance of winding A becomes smaller due to the saturation effect. According to (2) this results in larger amplitude of the response signals from the applied diagnostic pulses as shown in Fig. 4.

$$L = \frac{\Delta\varphi}{\Delta I} = \frac{v \times \Delta t}{\Delta I} = \frac{(v_{dc} - R \times i) \times \Delta t}{\Delta I} \approx \frac{v_{dc} \times \Delta t}{\Delta I} \qquad (3)$$

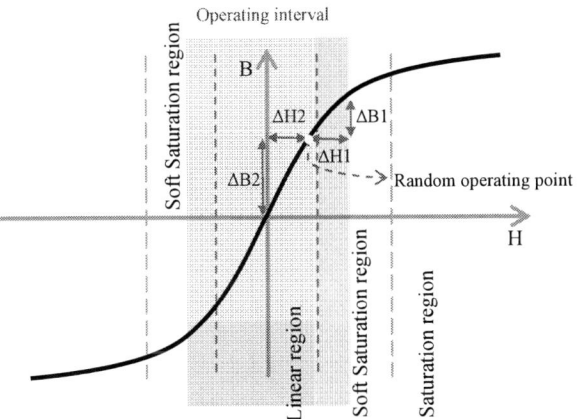

Fig. 5. Effect of diagnostic pulses on iron core characteristic

But if the direction of produced flux by the diagnostic pulse is in the opposite direction of the initial flux in the stator pole the net flux density becomes smaller and the operating point goes backward from the saturation region as shown in Fig. 5. This corresponds to bigger inductance for the winding A according to (3) and results in smaller amplitude of response signals as shown in Fig. 4.

In the simplified BLDC model (Fig. 3) which has star windings connection, diagnostic pulses with same duration are injected to every two phase windings while the third phase winding is left unconnected therefore six states are created as shown in Fig. 6.

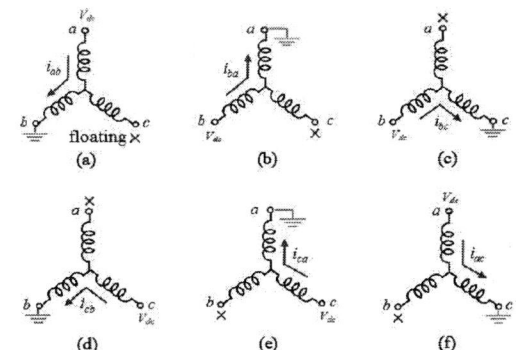

Fig. 6. The windings excitation modes

The net vector of the winding vectors for each excitation mode is shown in Fig. 7 in dashed vector. As discussed ealier, alignment of the rotor vector and the stator net vector can be indicated by the amplitude of the response signals. It should be notified that the same winding vectors in Fig. 3 and Fig. 7 are depicted by same color.

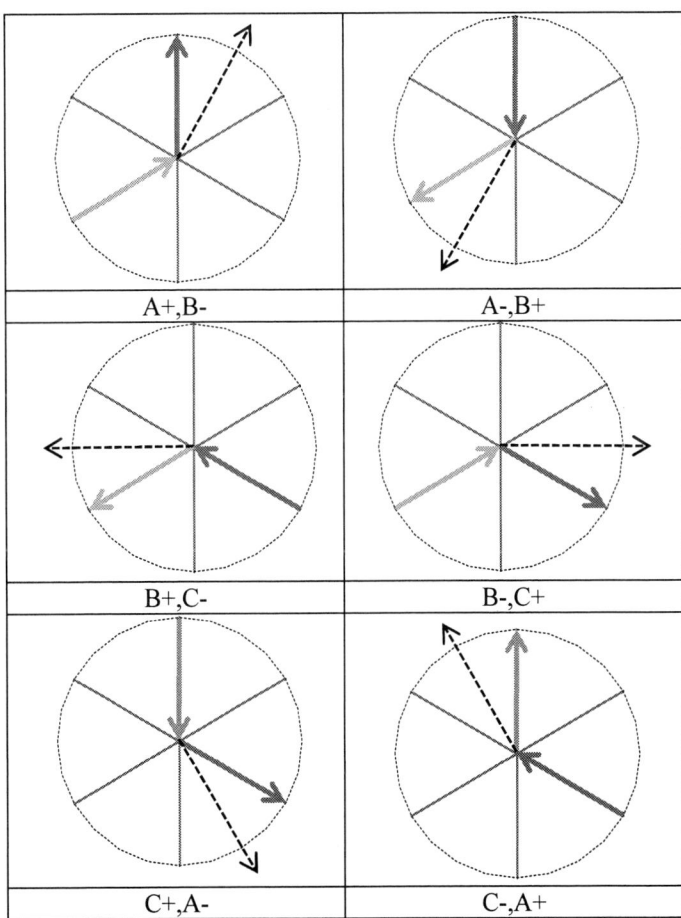

Fig. 7. The winding Net vector in each excitation mode

IV. PROPOSED PROBABILISTIC ALGORITHM

As described earlier, depending on the rotor vector position response signals, named in Table. 1, have different amplitudes. For an example, S1 (A+, B-) stands for the amplitude of the response signal produce by connecting phase A to +V_{dc} and phase B to ground.

TABLE 1. THE RESPONSE SIGNALS

S1 (A+,B-)	S2 (A-,B+)
S3 (B+,C-)	S4 (B-,C+)
S5 (C+,A-)	S6 (C-,A+)

In Fig. 8 the signal responses (TABLE 1.) are plotted in one electrical turn of the rotor vector then in TABLE 2. , They are sorted according to their amplitudes. It is clear that at the points of intersection the signals order changes. Based on the signals order distinct states are defined. Sorting signals for one electrical turn of the rotor vector results in 24 sectors as shown in TABLE 2. Each sector has 15 electrical degrees duration. According to the signals orders shown in TABLE 2. the proposed algorithm finds the section which the rotor vector points to it.

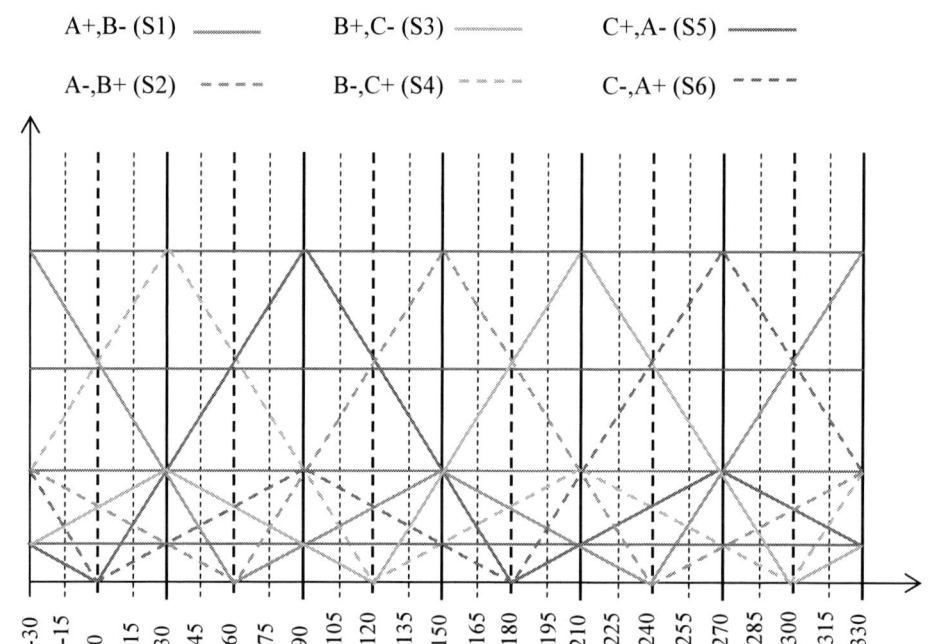

Fig. 8. The signal responses for the duration of one electrical turn

TABLE 2. PATTERNS OF THE SORTED SIGNALS FOR THE IDEAL MODEL

degree	-30_-15	-15_0	0_15	15_30	30_45	45_60	60_75	75_90	90_105	105_120	120_135	135_150	150_165	165_180	180_195	195_210	210_225	225_240	240_255	255_270	270_285	285_300	300_315	315_330
Signals order	S1	S1	S4	S4	S4	S5	S5	S5	S5	S5	S2	S2	S2	S2	S3	S3	S3	S3	S6	S6	S6	S6	S1	S1
	S4	S4	S1	S5	S5	S4	S4	S2	S2	S2	S5	S5	S3	S3	S2	S2	S6	S6	S3	S3	S1	S1	S6	S6
	S2	S2	S3	S3	S3	S6	S6	S6	S6	S6	S1	S1	S1	S1	S4	S4	S4	S4	S5	S5	S5	S5	S2	S2
	S6	S3	S2	S1	S6	S3	S3	S4	S4	S1	S6	S3	S5	S4	S1	S6	S2	S5	S4	S1	S3	S2	S5	S4
	S3	S6	S5	S6	S1	S2	S2	S1	S1	S4	S3	S6	S4	S5	S6	S1	S5	S2	S1	S4	S2	S3	S4	S5
	S5	S5	S6	S2	S2	S1	S1	S3	S3	S3	S4	S4	S6	S6	S5	S5	S1	S1	S2	S2	S4	S4	S3	S3

Due to the signals intersections in the border of the sectors, extracting the rotor position is the main challenge when the rotor vector points to the border of two sectors. In addition unforeseen feature of iron core such as hysteresis, external disturbances such as temperature variation and measurement error may affect the extracted results so a robust algorithm is required to confidently find the true rotor position.

The first part of the proposed algorithm is shown in fig. 9. After the injection of diagnostic pulses and sorting the response signals two conditions may occur; if the pattern of sorted signals exists in the predefined patterns of the ideal model (Table. 2), the rotor position is extracted otherwise the procedure continues in the second part of the algorithm as shown in Fig. 10. Difference between every two signals has to be calculated as shown in Table. 3. Therefore every pair of signals has a calculated difference which is called DIFF here. As an example it is assumed that the response signals, for a random rotor position, have order as below:

$$S1 \Rightarrow S4 \Rightarrow S3 \Rightarrow S2 \Rightarrow S5 \Rightarrow S6$$

This pattern doesn't exist in the predefined patterns (TABLE 2.) so DIFFs have to be calculated and sorted from smallest to largest. It is supposed that the order of DIFFs after sorting is as below: (D6, D15, D2, D12, D11, D9, D8, D14, D13, D5, D4, D7, D10, D1 and D3).
By this way, paired signals are sorted according to their DIFFs as follows: {(S2, S3), (S5, S6), (S1, S3), (S3, S6) and …}

The signals in the first pair (in this example: S2, S3) will be exchanged with each other. Therefore new pattern of the sorted signals will be created:

$$S1 \Rightarrow S4 \Rightarrow S2 \Rightarrow S3 \Rightarrow S5 \Rightarrow S6$$

This new pattern will be compared with patterns of TABLE 2, if this new pattern exist in TABLE 2. , the rotor position is extracted, else the second paired signals (in this example: S5, S6) have to be involved in the process of permutation. The second pair exchange will be applied for both state of first pair condition (S2, S3 or S3, S2) and result will be compared with predefined patterns and this process will continue until finding a pattern matched with the predefined patterns to extract the rotor position.

In this example by applying second exchange, the new patterns are as follow:

$$S1 \Rightarrow S4 \Rightarrow S3 \Rightarrow S2 \Rightarrow S6 \Rightarrow S5$$

$$S1 \Rightarrow S4 \Rightarrow S2 \Rightarrow S3 \Rightarrow S6 \Rightarrow S5$$

The second new pattern exists in the sector #2 of predefined patterns therefore the rotor vector position is found.
This algorithm is based on the fact that the probability for sorting signals incorrectly increases whenever the response signals become closer to each other.
It should be highlighted that if the rotor vector points to the sector border there are always some response signals which are far from signals intersection area that forces the process to track the answer by iterations, in the other words this feature avoid the process to converge to the incorrect result, so the process is always converged to true rotor position. Fig. 10 shows the details.

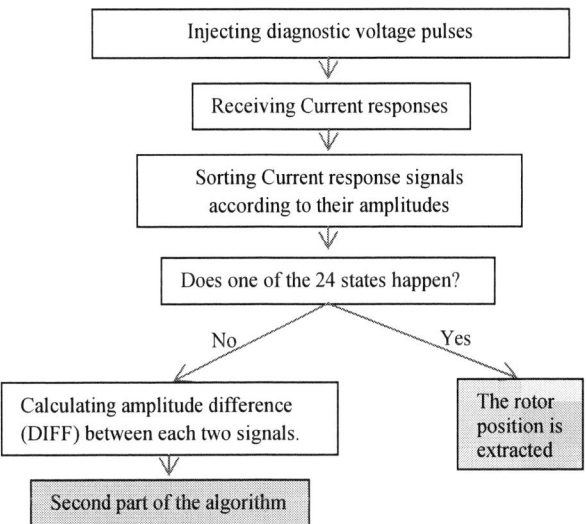

Fig. 9.   First part of the proposed algorithm

TABLE 3.   CALCULATED DIFFERENCES

D1=|S1-S2|	D6=|S2-S3|	D10=|S3-S4|	D13=|S4-S5|	D15=|S5-S6|
D2=|S1-S3|	D7=|S2-S4|	D11=|S3-S5|	D14=|S4-S6|	
D3=|S1-S4|	D8=|S2-S5|	D12=|S3-S6|		
D4=|S1-S5|	D9=|S2-S6|			
D5=|S1-S6|				

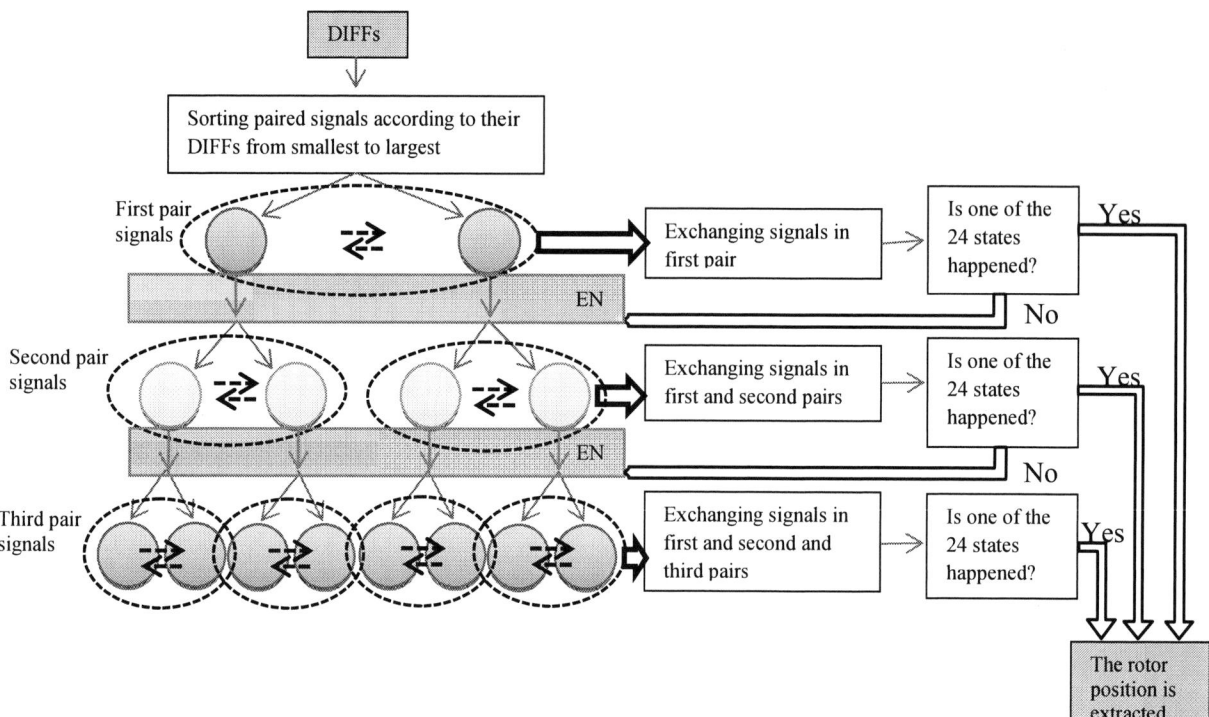

Fig. 10. Second part of the proposed algorithm

V. EXPERIMENTAL RESULTS

The proposed method is implemented by a low cost 8-bit microcontroller and a simple three phase converter as shown in fig.11.

Fig. 11. The implemented driver

The circuit schematic is shown in fig. 12. The current responses are measured simply by resistors which are connected to the source of the lower MOSFETs. To minimize measuring error, it is suggested that every response signal is measured six times and average value of two middle signals are used as the response signal.

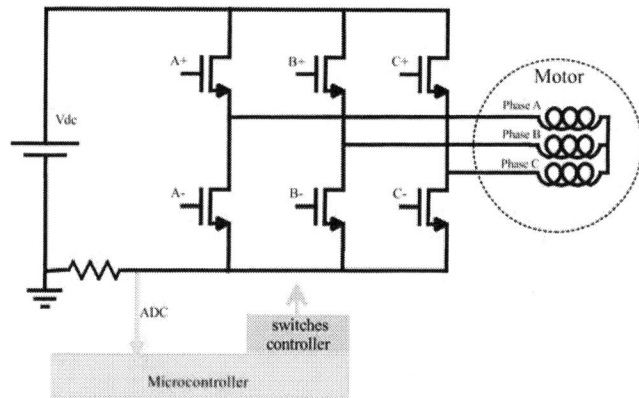

Fig. 12. The schematic circuit

For a random position of rotor, Current responses are measured as shown in fig. 13. The rotor position was extracted in sector1 by analyzing the responses in proposed algorithm.

The duration of the diagnostic pulses are chosen short enough (about 20us) in the way, the amplitude of current responses does not exceed the half of rated current. It is why the rotor stands at its initial position during the process of detection. In the other words, due to the short duration of diagnostic pulses the given energy is too small to rotate the rotor.

978-1-5090-0376-1/16 $31.00 © 2016 IEEE   46

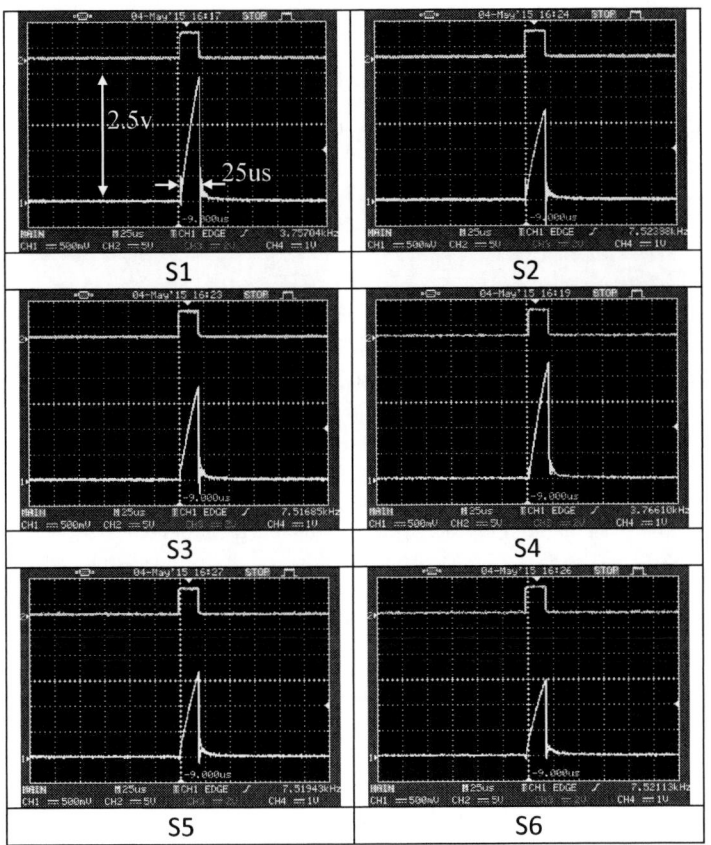

Fig. 13. The response signals for a random rotor position

VI. CONCLUSION

This paper proposes a new algorithm to find the rotor position in standstill condition with high accuracy. The algorithm is based on analyzing current responses obtained from diagnostic pulses.

Since all response signals are compared together the implemented driver could be applied to other BLDC motors. All the diagnostic pulse injection modes are used so the whole diagnostic potential is utilized and it results in 15 degree accuracy in rotor position detecting while conventional methods regardless to the challenges in sectors border could obtain 60 degree accuracy in rotor position detecting.

REFERENCES

[1] Wenjuan Zhu, Xiangyu Yang, Zhiyong Lan, "Structure Optimization Design of High-Speed BLDC Motor Using Taguchi Method, Electrical and Control Engineering (ICECE), 2010 International Conference on, pp. 4247 – 4249, 2010.

[2] Kamal, M.M., Mathew, L., Chatterji, S., "Speed control of brushless DC motor using intelligent controllers," Students Conference on Engineering and Systems (SCES), pp. 1-5, 2014

[3] Priya B., Krishnakumar V., "Performance improvement of brushless DC motor," International Conference on Energy Efficient Technologies for Sustainability (ICEETS), pp. 524 – 529, 2013

[4] Jinglin Liu, Lianghui Dong, Gang Wang, "Research on brushless DC motor based on new type electromagnetic position sensor," 15th International Conference on Electrical Machines and Systems (ICEMS), pp. 1 – 5, 2012

[5] Tsotoulidis, S., Safacas, A., "Side-effects of Hall sensors misplacement on BLDC motor drive operation," International Conference on Electrical Machines (ICEM), pp. 1825 – 1830, 2014

[6] Cassat, A., Espanet, C., Wavre, N., "BLDC motor stator and rotor iron losses and thermal behavior based on lumped schemes and 3-D FEM analysis," IEEE Transactions on Industry Applications, vol. 39, pp. 1314 – 1322, 2003

[7] Ma Xiaohan, Wang Xiaolin, Deng Zhiquan, Zhou PengFei, Zhao Yao, "Position sensorless starting method of BLDC motor based on SVPWM and stator magneto motive force control," 39th Annual Conference of the IEEE Industrial Electronics Society, IECON, pp. 3054 – 3059, 2013

[8] Fei Yang, Chenguang Jiang, Taylor, A., Hua Bai, Kotrba, A., Yetkin, A., Gundogan, A., "Design of a High-Efficiency Minimum-Torque-Ripple 12-V/1-kW Three-Phase BLDC Motor Drive System for Diesel Engine Emission Reductions," IEEE Transactions on Vehicular Technology, vol. 63, pp. 3107 – 3115, 2014

[9] Chiao-Chien Lin, Ying-Yu Tzou, "Robust Startup Control of Sensorless PMSM Drives with Self-Commissioning," International Power Electronics Conference (IPEC-Hiroshima - ECCE-ASIA), pp. 3072 – 3078, 2014

[10] Champa, Somsiri, Wipasuramonton, Nakmahachalasint, "Initial Rotor Position Estimation for Sensorless Brushless DC Drives," Industry Applications, IEEE Transactions on , vol.45, no.4, pp.1318,1324, July-aug. 2009

7th Power Electronics, Drive Systems & Technologies Conference (PEDSTC 2016)
16-18 Feb. 2016, Iran University of Science and Technology, Tehran, Iran

Application of Multiband Hysteresis Modulation in Field Oriented Control Based IPMSM Drive Fed by Asymetrical Multilevel Cascaded H-Bridge Inverter

Abdolhossein Ejlali, Member, IEEE
Department of electrical Engineering, Mehran Branch
Islamic Azad university
Mehran, Iran
abdolhossein.ejlali@gmail.com

Davood Arab Khaburi, and Javad Soleimani, Member, IEEE
Centre of Excellence for Power Systems Automation &
Operation, Dept. Electrical Engineering
Iran University of Science and Technology, IUST
Tehran, Iran
khaburi@iust.ac.ir, javad.soleimani@gmail.com

Abstract— **Recently, Inner permanent magnet synchronous Machines are widely used in several applications such as hybrid electric vehicles (HEVs), starters/alternators, power steering, and air conditioning motors. This paper presents multiband hysteresis modulation to Field Oriented Controlled (FOC) IPMSM fed by a 9-level asymmetrical Cascaded Half Bridge (CHB) inverter. For this reason, hysteresis controllers have 9-band current error. The CHB is more attractive choice due to its modular structure and higher number of output voltage level for the same number of semiconductors in comparison with symmetrical CHB. The results show several advantages of this method such as high accuracy, fast dynamic response, insensitivity to parameter variations and simple implementation.**

Keywords—IPMSM; Multiband Hysteresis Controller; CHB Inverter; FOC.

I. INTRODUCTION

Due to the unique merits of IPMSMs such as; high torque per volume ratio, high efficiency, good power factor and rapid dynamic performance, these motors have reached high attentions, especially in railway industries. Furthermore, due to the existence of magnetic saliency, IPMSMs offer a very good startup torque, that can be a good trait for the control system in some applications like traction and crane applications [1].

DTC method, proposed by M. Depenbrock and I. Takahashi in 1985, and has been applied to PMSM since 1990's. This method is based on two hysteresis units, which will control torque and flux independently, and leads to high dynamic performances. DTC method has several advantages such as good torque response in transient operating conditions, simple structure, robustness to the variations of motor parameters [2]. However, with inherent characteristics of DTC, this method leads to torque ripples that limit the usage of DTC method for some applications especially at low speeds.

Field Oriented Control (FOC) method proposed by F.Blaschke is based on stator currents control. This method

can be implemented using hysteresis current controllers by comparison of current error with hysteresis band of controller, which has several benefits such as high accuracy, fast dynamic response, insensitivity to parameters variation and simple implementation [3].

Evolving power converter technology, leads to extension of researches on application of new converters such as matrix converters and multilevel inverters in electrical drives [4-5].

However, schemes which produce the waveforms with higher number of voltage levels, have several merits such as better harmonic spectrum, low Electro-Magnetic Interference (EMI), nearly sinusoidal output, low dv/dt and high efficiency [6-7].

Different topologies of multilevel inverter are diode clamped, flying capacitor and Cascaded Half Bridge (CHB) which the first one, needs complex PWM control and the second one, has large size because of necessity of more capacitor. However, CHB is more attractive choice due to its modular structure and higher number of output voltage level for the same number of semiconductors [8-9].

Among the switching methods of multi level inverters, Sinusoidal Pulse Width Modulation (SPWM) and Space Vector Modulation (SVM) are more popular. SVM method uses three voltage vectors to produce reference voltage. This method needs a metrology to determine the three nearest space vectors while carrier based method is simpler and the number of computations are less than SVM. Also, the used PWM method for Multilevel inverters which called multi-carrier based PWM technique needs multiple carrier [10].

Several literatures have reported the multiband hysteresis modulation method used to control the multilevel inverters. For example, In [11] a method has been used for calculation of hysteresis band limits. In [12] a generalized multiband hysteresis modulation has been proposed for multilevel inverter controlled system.

978-1-5090-0376-1/16 $31.00 © 2016 IEEE

In this paper, multiband hysteresis modulation has been used in field oriented controlled IPMSM fed by a 9-level asymmetrical CHB inverter. In this modulation technique, hysteresis controllers have 9-band current error. In the remind of this paper, the IPMSM model and basic principle of DTC will be presented in section II. In section III, CHB topologies will be described and voltage generation, based on multiband method will be explained. Section IV of this paper presents simulation results which shows effectiveness of control system by employing 9-level CHB.

II. IPMSM MODEL AND FIELD ORIENTED CONTROL

A. IPMSM Model

Voltage equations of IPMSM in the synchronous rotating reference frame are [13]:

$$\begin{cases} u_d = Ri_d + p\lambda_d - \omega\lambda_q \\ u_q = Ri_q + p\lambda_q + \omega\lambda_d \end{cases} \quad (1)$$

where

$$\begin{cases} \lambda_d = L_d i_d + \lambda_{pm} \\ \lambda_q = L_q i_q \end{cases} \quad (2)$$

Where u_d, u_q, i_d and i_q are the d-q axis stator voltages and currents, respectively; R is stator resistance; p is differential coefficient; λ_d, λ_q are the d-q axis stator magnetic flux, respectively; ω is electrical motor speed; L_d, L_q are the d-q axis stator inductances, respectively; λ_{pm} is coupling flux linkage of rotor on stator. The electromagnetic torque is given by:

$$T_e = \frac{3}{2} P[\lambda_{pm} + (L_d - L_q)i_d]i_q \quad (3)$$

where P is the number of pole pairs.

B. Field Oriented Control Method

FOC method proposed by Blaschke is based on stator currents control which is implemented in synchronous reference frame for simplification of control strategy [14]. The dq components of stator currents can be compared to dq reference currents, the error value will used in PI controllers to produce reference voltages for switching strategy. Another scheme is transforming the dq reference currents to abc form and comparing to real stator currents. In this scheme, errors are given to hysteresis controllers to produce switching signals. Fig.1 shows the FOC method for multilevel inverter fed IPMSM.

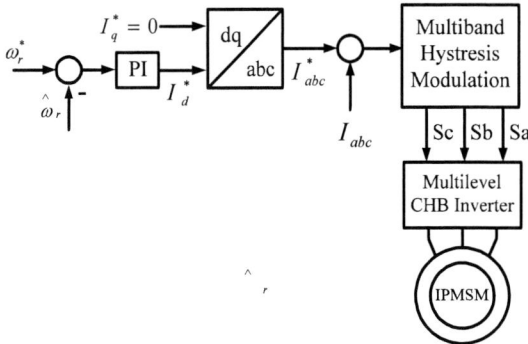

Fig. 1. Field Oriented Control for multilevel inverter fed IPMSM

III. MULTILEVEL CASCADED H-BRIDGE INVERTER

Fig. 2 shows the Cascaded H-bridge inverter structure which consists of 3 arms and each arm has more than one series connected to H-cell. This structure has several advantages such as high output voltage level production in comparison with other multilevel inverters and its modular topology which composed of identical cells.

Fig. 3 shows one arm of the inverter fed from DC voltage sources V_{dc1} and V_{dc2}. If all sources have the same value of the voltage, the inverter cells of each arm will generate a similar voltage steps. Such inverters are called symmetrical multilevel inverters. In asymmetrical inverters, H-cells are fed by unequal voltages and their output voltages have different voltage steps.

DC voltage sources selection in an asymmetric inverter is carried out with a factor equal to 2 or 3 [15]. If the inverter has N number of H-bridges and the amplitude of DC voltage has been chosen by the factor 2, the number of output voltage levels will be:

$$n=2^{N+1}-1 \quad (1)$$

If the factor equals 3, the number of output voltage levels is:

$$n=3^N \quad (2)$$

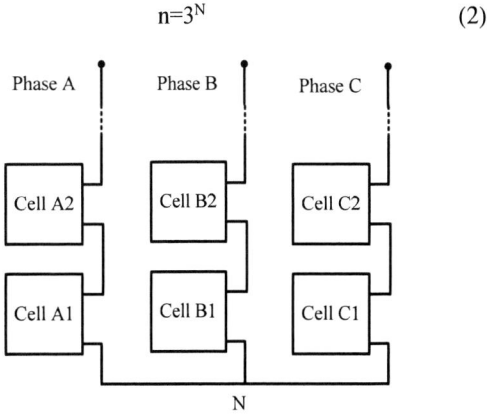

Fig. 2. Three phase cascaded H-bridge multilevel inverter

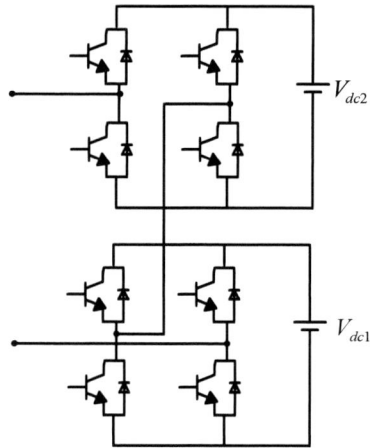

Fig. 3. One arm of cascaded H-bridge multilevel inverter

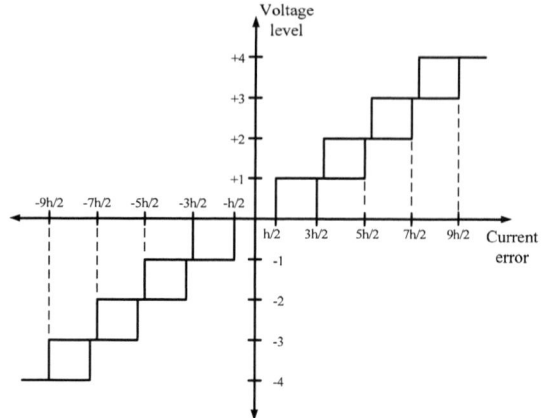

Fig. 4. Multiband Hysteresis controller for 9-level CHB inverter

In this paper, an asymmetrical cascaded H-bridge inverter with 2 cells in each arm has been used. The voltage of second cell in each arm is 3 times greater than v_{dc1}. Therefore, 9 level voltage will appear in output. Table I shows the state of each stages in one inverter arm and generated output voltage levels. The output voltage of each stage can be positive, negative or zero.

IV. MULTIBAND HYSTERESIS OF CASCADED H-BRIDGE MULTILEVEL INVERTER

There are many different approaches to switch power electronic devices used in multilevel cascaded H-bridge inverters. Several literatures have reported Multiband hysteresis modulation method used for multilevel inverter control.

This method is based on extension of hysteresis current control used for conventional inverter. The method needs 9-level hysteresis controllers to generate multilevel output voltage. Fig. 4 shows the characteristic of a multiband hysteresis controller. As it can be seen, two parameters have to be determined for this controller: hysteresis band h and dead-zone δ which has been defined to avoid overlapping.

TABLE I. RELATION BETWEEN STAGE STATE AND OUTPUT VOLTAGE LEVEL

Stage 1	Stage 2	Output Voltage
-	-	-4Vdc
0	-	-3Vdc
+	-	-2Vdc
-	0	-Vdc
0	0	0
+	0	+Vdc
-	+	+2Vdc
0	+	+3Vdc
+	+	+4Vdc

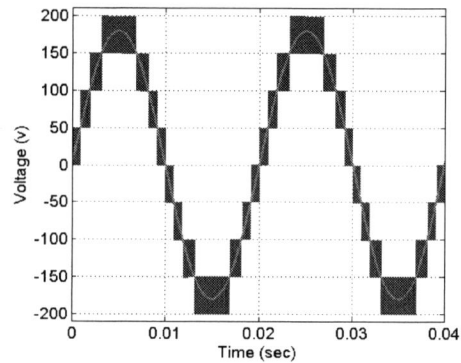

Fig. 5. Output voltage of 9-level CHB

The multilevel CHB shown in Fig. 2 and Fig. 3 is simulated using multiband hysteresis modulation. Fig. 5 shows the output voltage of 9-level CHB, which is used as the voltage source for field oriented controlled IPMSM drive system in this paper.

V. SIMULATION RESULTS

Direct torque control of IPMSM fed by a 9-level CHB, is simulated in MATLAB/SIMULINK. The motor parameters has been listed in Table II. The DC link voltages in stage 1 and stage 2 of CHB are 50 and 150 volts respectively. In the simulation, hysteresis band h is 0.05A. Fig.6 shows speed control response of IPMSM fed in FOC method.

TABLE II. MOTOR PARAMETER

Parameter1	Numerical Value
PM Flux	0.1688
Stator resistance	0.0918
Armature inductance (H)	0.000975
Pole pairs	4
Inertia	0.003945

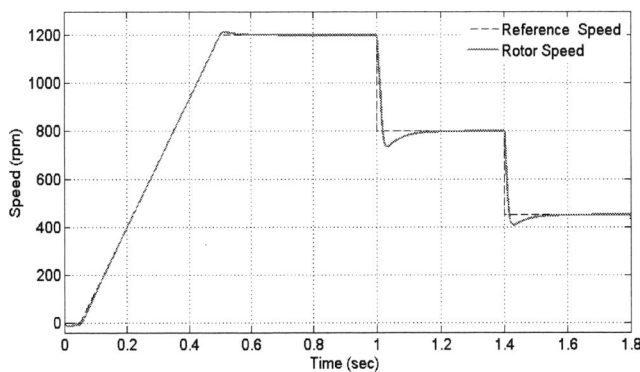

Fig. 6.   Speed control response of IPMSM in FOC method

The reference speed at t=0.05sec increases from zero to 1200 rpm at t=0.5sec, then, at t=1sec and t=1.4 sec will decrease to 800rpm and 450rpm respectively. Fig. 7 depicts the torque response of control system which has feasible ripple. The results show that the used method has fast dynamic response and error speed can be neglected.

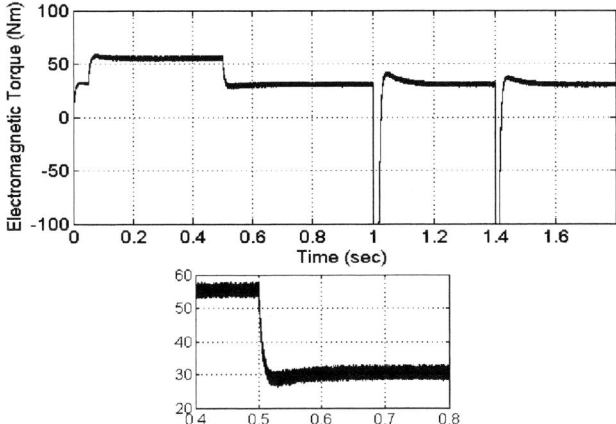

Fig. 7.   Torque response of FOC method applied to IPMSM

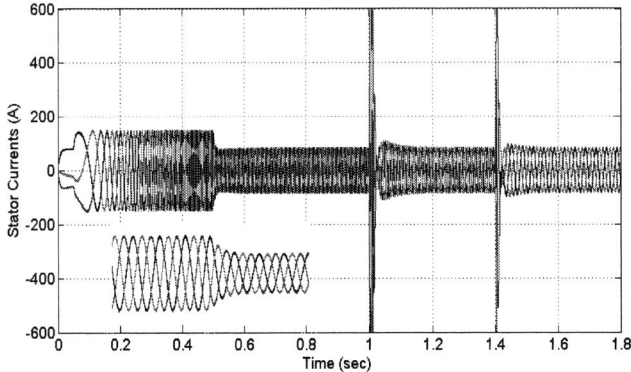

Fig. 8.   Three phase current of IPMSM in FOC method applied to IPMSM fed by 9-level CHB inverter

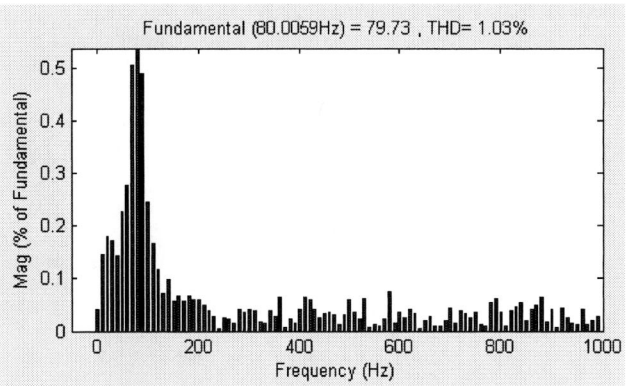

Fig. 9.   Harmonic spectrum of stator current

As it can be seen in Fig. 8, all phases currents have sinusoidal shape and the total harmonic distortion (THD) of stator currents shown in Fig. 9 is 1.03 which is the result of increment of voltage levels given to motor by Asymmetrical multilevel CHB.

VI.   CONCLUSION

In this paper multiband hysteresis modulation has been applied to field oriented controlled IPMSM fed by 9-level asymmetrical CHB inverter. The CHB is more attractive choice for inverter stage of drive systems due to its modular structure. Higher number of output voltage level for the same number of semiconductors hysteresis controllers have several benefits such as high accuracy, fast dynamic response, insensitivity to parameter variations and simple implementation. Speed characteristic shows good dynamic response and torque ripple has low value. The stator current shape ha nearly sinusoidal shape.

REFERENCES

[1]   B. Majidi, J. Milimonfared, K. Malekian, "Performance improvement of direct torque controlled Oct.2008. interior permanent magnet synchronous motor drive by considering magnetic saturation," Power Electronics and Motion Control Conference, 2008. EPE-PEMC 2008. 13th , vol., no., pp.763-768, 1-3 Sept. 2008.

[2]   Islam, D.; Reza, C.M.F.S.; Mekhilef, S., "DTC-IM drive with 5-level hybrid cascaded h-bridge inverter," in Industrial Electronics and Applications (ICIEA), 2014 IEEE 9th Conference on , vol., no., pp.332-337, 9-11 June 2014.

[3]   Gautam, S.; Gupta, R., "Switching Frequency Derivation for the Cascaded Multilevel Inverter Operating in Current Control Mode Using Multiband Hysteresis Modulation," in IEEE Transactions on Power Electronics, vol. 29, no. 3, pp. 1480-1489, March 2014.

[4]   D. Xiao and M.F. Rahman, "Direct Torque Control of an Interior Permanent Magnet Synchronous Machine fed by a Direct AC-AC Converter", Power Electronics and Motion Control Conference, 2006. IPEMC '06. CES/IEEE 5th International Volume 3, pp:1 - 6, Aug. 2006.

[5]   Abdul Kadir, M.N.; Mekhilef, S.; Hew Wooi Ping, "Direct Torque Control Permanent Magnet Synchronous Motor drive with asymmetrical multilevel inverter supply," in Internatonal Conference on Power Electronics, 2007. ICPE '07. 7th , pp.1196-1201, 22-26 Oct. 2007.

[6]   Gholinezhad, J.; Noroozian, R., "Application of cascaded H-bridge multilevel inverter in DTC-SVM based induction motor drive," in Power Electronics and Drive Systems Technology (PEDSTC), 2012 3rd , pp.127-132, 15-16 Feb. 2012.

[7] Mortezaei, A.; Azli, N.A.; Idris, N.R.N.; Mahmoodi, S.; Nordin, N.M., "Direct torque control of induction machines utilizing 3-level cascaded H-Bridge Multilevel Inverter and fuzzy logic," in IEEE Applied Power Electronics Colloquium (IAPEC), pp.116-121, 18-19 April 2011.

[8] Hosseini, S.H.; Ahmadi, M.; Zadeh, S.G., "Reducing the output harmonics of cascaded H-bridge multilevel inverter for Electric Vehicle applications," in 8th International Conference on Electrical Engineering/Electronics, Computer, Telecommunications and Information Technology (ECTI-CON), pp.752-755, 17-19 May 2011.

[9] Brovanov, S.V.; Egorov, S.D.; Dubkov, I.S., "Study on the efficiency at h-bridge cascaded VSI for different PWM methods," in Micro/Nanotechnologies and Electron Devices (EDM), 2014 15th International Conference of Young Specialists on , vol., no., pp.440-443, June 30 2014-July 4 2014.

[10] R. Gupta, A. Ghosh, and A. Joshi, "Multi-band hysteresis modulation and switching characterization for sliding mode controlled cascaded multilevel inverter", IEEE Trans on Ind Electron.,vol.57, no.7 ,Jul.2010.

[11] H. Mao, X. Yang, Z. Chen, and Z. Wang, "A hysteresis current controller for single-phase three-level voltage source inverters," IEEE Trans on Power Electron, vol.27, no.7 ,pp.3330–3339, Jul.2012.

[12] Gupta, R.; Ghosh, A.; Joshi, A., "Multiband Hysteresis Modulation and Switching Characterization for Sliding-Mode-Controlled Cascaded Multilevel Inverter," i, IEEE Transactions on Industrial Electronics, vol.57, no.7, pp.2344-2353, July 2010.

[13] L. Tang, L. Zhong, M. F. Rahman and Y. Hu, "A Novel Direct Torque Control for Interior Permanent Magnet Synchronous Machine Drive System with Low Ripple in Flux and Torque and Fixed Switching Frequency," IEEE Trans. Power Electron., vol. 19, pp. 346-354, March 2004.

[14] Lakshmi, G.S.; Kumar, P.V.; Sudeepika, P., "Field Oriented Control of IPMSM using diode-clamped multilevel inverter," in 2015 International Conference on Power and Advanced Control Engineering (ICPACE), vol., no., pp.355-360, 12-14 Aug. 2015.

[15] Khoucha, F.; Lagoun, M.S.; Kheloui, A.; El Hachemi Benbouzid, M., "A Comparison of Symmetrical and Asymmetrical Three-Phase H-Bridge Multilevel Inverter for DTC Induction Motor Drives," in, IEEE Transactions on Energy Conversion, vol.26, no.1, pp.64-72, March 2011.

7th Power Electronics, Drive Systems & Technologies Conference (PEDSTC 2016)
16-18 Feb. 2016, Iran University of Science and Technology, Tehran, Iran

Grid Synchronization of Brushless Doubly Fed Induction Generator Using Direct Torque Control

Ramtin Sadeghi, Seyed M. Madani, M. R. Agha Kashkooli

Department of Electrical Engineering

University of Isfahan

Isfahan, Iran

r.sadeghi@eng.ui.ac.ir, m.madani@eng.ui.ac.ir, mr.kashkooli.1987@ieee.org

Abstract— **A direct torque control (DTC) method is developed for grid synchronization of Brushless Doubly Fed Induction Generator (BDFIG). The presented method and the dynamic model of BDFIG are conferred in a reference frame aligned to the stator of control machine. The proposed controller has high performance, i.e., grid connection with minimum current overshoot and fast dynamic response, during the BDFIG loading. The robustness and efficiency of the proposed controller are verified by both simulation and experimental tests.**

Keywords—Brushless Doubly Fed Induction Generator (BDFIG); Direct Torque Control (DTC); Grid Synchronization.

I. INTRODUCTION

The Doubly-Fed Induction Generator (DFIG) constitutes over 50% of installed Wind Energy Conversion Systems (WECS). However, conventional DFIG needs brushes and slip-rings in order to connect a power converter to the rotor windings. Elimination of the brushes and slip-rings increases WECS robustness reduces its maintenance costs, because there is no need to replace or inspect these elements. For this reason, the Brushless Doubly-Fed Induction Generator (BDFIG) is a suitable replacement for widely used conventional DFIG [1].

Since 1902, different control algorithms for BDFIG are presented [2]. In [3], the method's purpose is to maintain stability and reactive power control independent of speed. Nonlinear relationship between the values of current, torque and reactive power with slip and flux is the weakness of this study. [4] mentions the limits of the BDFIG control methods, due to their complex structures. Thus, direct control methods are used to control this machine. Also, they analyze the behavior of this machine under unbalance voltage, and improve the machine power quality. In their study, the rotor current is utilized to control the machine, which is not accessible in real BDFIG. This is the disadvantage of the method.

In [5], Direct torque control of DFIG is presented. In this method, torque and rotor flux references are changed into frequency and voltage references, for synchronizing DFIG to grid. Other control strategy for grid-connection of DFIG by directly controlling both rotor flux and virtual torque and of the generator, is presented in [6].

The Cascaded Doubly Fed Induction Generator (CDFIG) is a model for BDFIG. CDFIG consists of two DFIGs that are connected together both mechanically and electrically. One of the DFIGs is considered as Power Machine (PM) which its stator is connected to the main grid. The other DFIG is entitled Control Machine (CM) and its stator supplied from a back-to-back converter.

In this paper, dynamic model of CDFIG is expressed in the CM stator reference frame. Then DTC method for this machine is presented. This study develops DTC strategy for grid synchronization of the BDFIG based on current stator of CM. Experimental results are confirmed the feasibility and accuracy of the proposed method.

II. DYNAMIC MODEL OF DOUBLY FED INDUCTION GENERATOR

The dynamic model of the CDFIG is consisted of known models of two DFIM which are connected together; shown in figure 1.

CDFIG can be expressed in four different reference frame that are shown in Fig2. The stationary reference frame viewed from the stator of PM is designated by D_{sp}/Q_{sp}. The D_{rp} $(D_{rc})/Q_{rp}$ (Q_{rc}) axes represent the rotor reference frame. It should be noted that the rotors of power and control machines rotating at the same electrical angular speed of $\frac{P_p}{2}\omega_m$. Finally, the frame aligned to the stator of CM is shown by D_{sc}/Q_{sc} axes and rotates at the electrical angular speed of $\frac{P_c+P_p}{2}\omega_m$; where ω_m is mechanical angular speed of machine. Also P_c and P_p are pole number of CM and PM, respectively. This figure

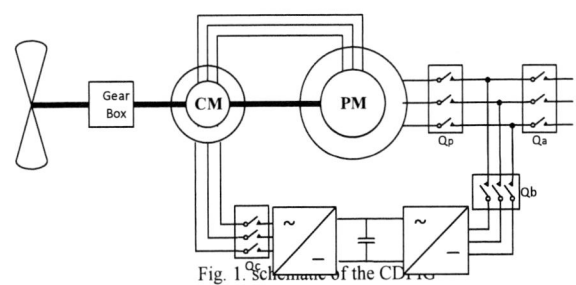

Fig. 1. Schematic of the CDFIG

978-1-5090-0376-1/16 $31.00 © 2016 IEEE

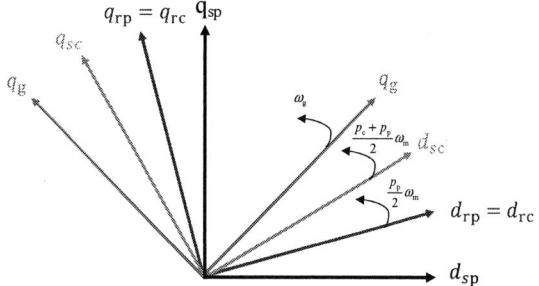

Fig. 2. Four different reference frame for CDFIG

shows that the CM rotates with respect to the PM. This rotation is not a physical concept, but the Stator of the CM rotates electrically from the stator of the PM point of view. [3, 7].

Dynamic equations of the machine are written in the CM stator reference frame:

$$\begin{bmatrix} \vec{v}_{sp} \\ \vec{v}_{rp} \end{bmatrix} = \begin{bmatrix} -R_{sp}\,\vec{i}_{sp} + \dfrac{(d\vec{\varphi}_{sp})}{dt} + j\omega_p\,\vec{\varphi}_{sp} \\ R_{rp}\,\vec{i}_{rp} + \dfrac{(d\vec{\varphi}_{rp})}{dt} + j(\omega_p - \omega_{rp})\,\vec{\varphi}_{rp} \end{bmatrix} \quad (1)$$

$$\begin{bmatrix} \vec{\varphi}_{sp} \\ \vec{\varphi}_{rp} \end{bmatrix} = \begin{bmatrix} -L_{sp} & L_{mp} \\ -L_{mp} & L_{rp} \end{bmatrix} \begin{bmatrix} \vec{i}_{sp} \\ \vec{i}_{rp} \end{bmatrix} \quad (2)$$

$$\begin{bmatrix} \vec{v}_{sc} \\ \vec{v}_{rc} \end{bmatrix} = \begin{bmatrix} R_{sc}\,\vec{i}_{sc} + \dfrac{(d\vec{\varphi}_{sc})}{dt} + j\omega_c\,\vec{\varphi}_{sc} \\ -R_{rc}\,\vec{i}_{rc} + \dfrac{(d\vec{\varphi}_{rc})}{dt} + j(\omega_p - \omega_{rc})\,\vec{\varphi}_{rc} \end{bmatrix} \quad (3)$$

$$\omega_c = \omega_p - \frac{p_p + p_c}{2}\omega_m \quad (4)$$

$$\omega_{rp} = \omega_{rc} = \frac{p_p}{2}\omega_m \quad (5)$$

$$\begin{bmatrix} \vec{\varphi}_{sc} \\ \vec{\varphi}_{rc} \end{bmatrix} = \begin{bmatrix} L_{sc} & -L_{mc} \\ L_{mc} & -L_{rc} \end{bmatrix} \begin{bmatrix} \vec{i}_{sc} \\ \vec{i}_{rc} \end{bmatrix} \quad (6)$$

Where ω_p is electrical angular speed of PM, also v, φ and i denote the voltage, flux and current space vectors respectively. The sp, rp, rc and sc subscripts stand for the stator of the PM, the rotor of the PM, the rotor of the CM and the stator of the CM quantities, respectively.

The dynamic equivalent circuit for the machine is shown in figure (3).

III. GRID SYNCHRONISATION PROCESS

Soft and fast grid synchronization of variable speed wind turbines is offered as one of the main advantages over fixed-speed wind turbine technology. A squirrel cage induction generator is usually utilized in fixed-speed wind turbines. In the same way as in an induction machine, grid connection is achieved by a soft starter. First the generator is driven by the wind turbine to near synchronous speed, then the soft starter connects the generator to the grid by applying a reduced voltage that is gradually increased. The main problem of this method is a greatly distorted current during connection and also that synchronization is only probable at synchronous speed.

Main advantage of variable-speed wind turbines is grid synchronization that is possible at any operational speed. The most common case is grid synchronization from a stop or idling wind turbine state.

First the circuit breaker (Q_a, Q_p, Q_b and Q_c that are shown figure (1)) are open, keeping the electrical CDFIG disconnected from the grid. As wind speed flowing on the wind turbine overcomes the cut in wind speed, the connection process begins. In these conditions, the wind turbine develops a mechanical torque and the rotational speed increases until 220 rpm (cut-in rotational speed). At this specific moment the grid synchronization starts.

The conditions for synchronization are similar to conventional synchronous machine: voltage magnitude, frequency and voltage phase at the generator terminals has to be the similar to the grid.

These conditions should be checked and reached as fast as possible, because electrical braking torque isn`t applied when the wind turbine continues accelerating. Nevertheless, synchronization takes less than one second, and because of the high inertia of the wind turbine, it can be considered that speed is constant during this process.

Synchronization process are as follows:

First, the main circuit breaker, Q_a, is closed and immediately after that, the converter circuit breaker, Q_b is

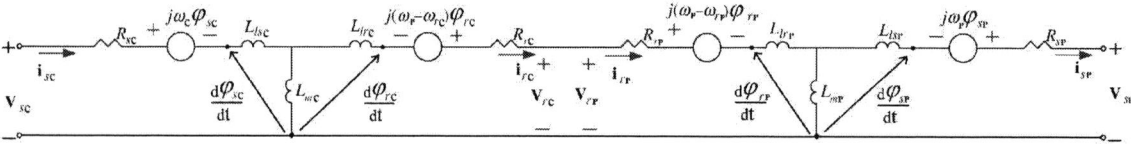

Fig. 3. Dynamic equivalent for CDFIG

978-1-5090-0376-1/16 $31.00 © 2016 IEEE

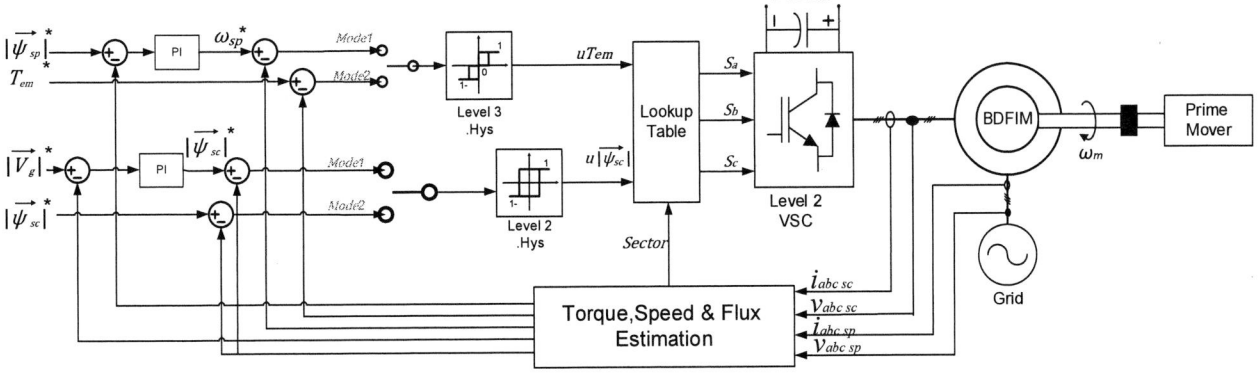

Fig.4. Proposed DTC diagram.

done. At this time, the DC link capacitor charges and the converter is directly connected to the grid. The DC link voltage controller rises the DC link voltage to its reference level (580 V).

Following, the CM stator circuit breaker, Q_c, is closed and the DTC of CM converter operates. This control method has modified for synchronization and its principles are described in next section.

In the end, when synchronization conditions are established, the PM stator circuit breaker is closed, and the CDFIG is connected to the grid with zero stator currents. Then, standard DTC is applied and generator can be controlled.

IV. DTC Synchronization Control Method

In figure 4, a block diagram of the control method is shown. The torque control method for CDFIG was proposed in [8]. For CDFIG grid synchronization, DTC algorithm is modified and switching table of this method is similar to the DTC switching table, but changing external references.

First, CM stator flux reference is obtained from an outer PI controller. This controller is used to control PM stator voltage similar to synchronous machine that open circuit voltage is controlled by rotor excitation. In this way, the first condition for grid synchronization, PM voltage equal to grid voltage, can be obtained.

Second, torque reference is changed into a frequency reference. The idea is that in open circuit torque is equal to zero, i.e., torque angle is always zero as CM stator flux and PM stator flux are on the same direction. Though, this control loop is responsible for rotary CM stator flux, so if a PM stator frequency reference is given, CM stator flux will rotate to its slip frequency. In this way, the second condition for synchronization, PM frequency equal to grid frequency, can be achieved. Finally, an outer phase controller is placed that sets up the frequency reference, so frequency is changed until phase difference between grid and the machine is zero. The last condition for synchronization, PM phase equal to grid phase, can be obtained. Vector theory is utilized to obtain mentioned condition as follows:

$$\vec{v}_{sp} = v_{sp\_a} + v_{sp\_b}e^{j2\pi/3} + v_{sp\_c}e^{-j2\pi/3}$$

(7)

Then, magnitude, phase and frequency are calculated as follow:

$$V_{sp} = \sqrt{V_{sp\_d}^2 + V_{sp\_q}^2}$$

(8)

$$\varphi_{sp} = \arctan\left(\frac{V_{sp\_q}}{V_{sp\_d}}\right)$$

(9)

$$\omega_s = \frac{d\varphi_{sp}}{dt}$$

(10)

V. Experimental Results

The performance of the grid synchronization is validated through experimental results i.e., soft and fast generator synchronization and loading. Table I demonstrates The parameters of both power and control machine parameters. In figure 5, the experimental setup is shown. The controller is implemented on TMS320F2812. The actual synchronization process takes less than 200 ms and speed is constant at cut-in speed, 220 rpm, during the whole process due to the high inertia of the machine.

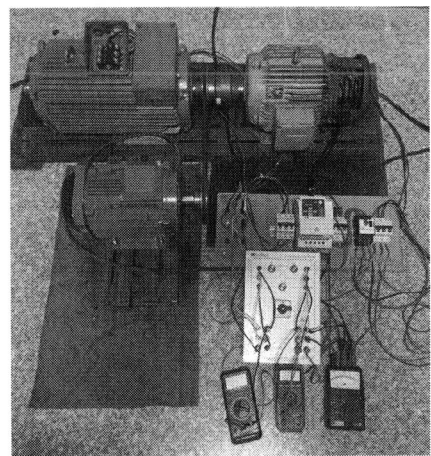

Fig 5. Experimental setup

At t=0 ms, main and converter circuit breaker are closed, then the DC link voltage controller will rise DC voltage level from precharge value to reference value, Fig. 6. Voltage is stabilized at 200 ms. approximately. In order to eliminate the phase difference between PM stator and grid voltages, the voltage reference enforced to be equal to grid (described in Section II). A phase difference is generated between two voltage vectors when they are not collinear. This fact has been verified by experiment as shown in figure 6.

In the proposed torque control loop, a voltage step of 1 p.u is applied. It is understandable how the voltage difference has an impact on the phase difference between the stator and grid voltages. When the voltage reference is equal to grid voltage, the difference voltage is obligated to be null. Thus, these two waveforms overlap. Figure 7 show Grid (U_g) and PM stator (U_{sp}) line-to-line voltages at grid-connection moment.

The grid connection i.e. control response is smooth and fast, as predictable which is illustrated in Fig. 9. Definitely, at the connection instant there is no current inrushs.

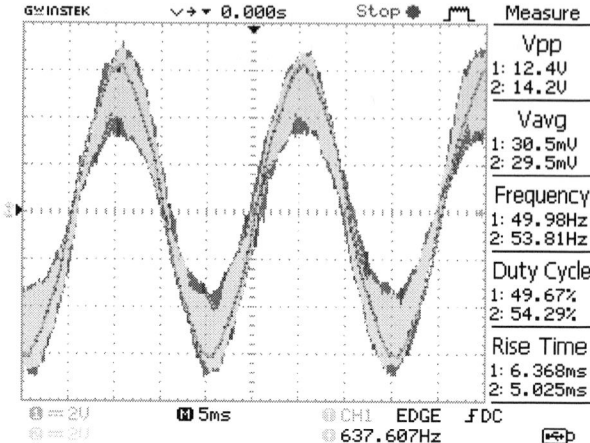

Fig7. Grid (U_g) and PM stator (U_{sp}) line-to-line voltages at grid-connection moment

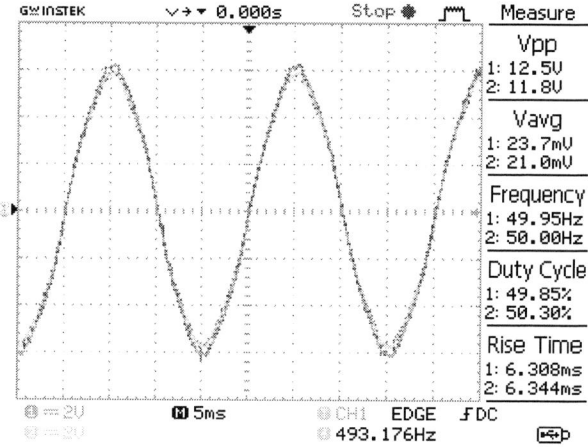

Fig. 8. grid and PM voltage at the grid-connection instant.

TABLE I.   Experimental Parameters

PM Rated Power	**Sn**	3 kW
PM Pole Pairs	**p**	3
PM Nominal Frequency	**f_b**	50 Hz
PM Rotor Resistance	**r_r**	0.125 Ω
PM Stator Line to Line Voltage	**Vs**	380 Vrms
PM Rotor Voltage	**Vr**	120 Vrms
CM Rated Power	**Sn**	4 kW
CM Pole Pairs	**p**	2
CM Nominal Frequency	**f_b**	50 Hz
CM Rotor Resistance	**r_r**	0.1 Ω
CM Stator Line to Line Voltage	**Vs**	380 Vrms
CM Rotor Voltage	**Vr**	150 Vrms

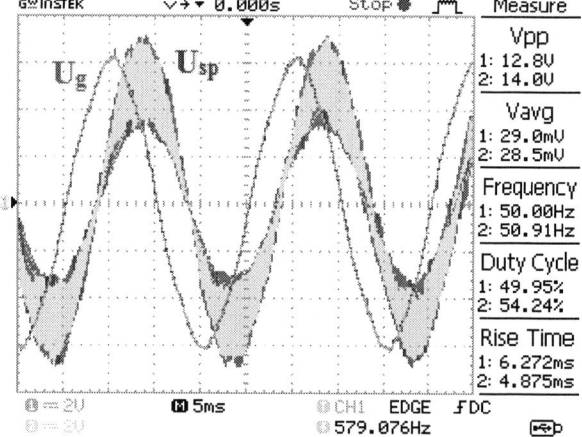

Fig 6. Line-to-line voltages of grid (U_g) and PM stator (U_{sp})

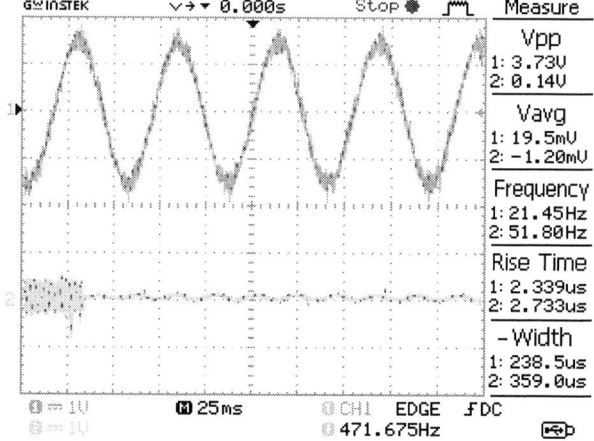

Fig. 9. CM and PM current at the grid-connection instant.

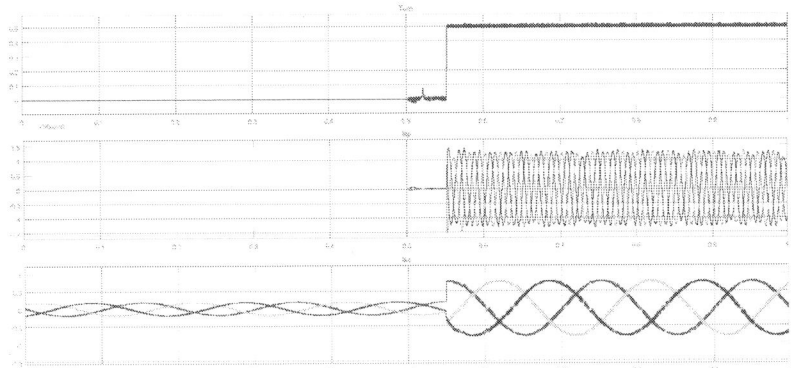

Fig.10. Simulation results of transition from no-load mode to grid-connection mode.

A. Transition from The No-Load Mode to The Grid-Connected Mode

The following definitions are presented for simplification:

Mode 1) Through this operation mode, the PM stator terminals are not connected to the grid and the synchronization process is in the course of execution. This mode is mentioned earlier as the no-load mode.

Mode 2) In this operation mode the stator windings of power machine are connected to the grid. (grid connected mode.)

Then, the transition between these two modes of operation in the proposed method is explained as follow:

1) The voltage estimator added for the synchronizing in mode 1 is swapped to the CDFIG electromagnetic torque reference necessary for the DTC algorithm in mode 2.

2) The substitution from the CM stator flux reference in mode 1 to the other reference based on the required stator reactive power command; as mentioned in [8].

In figure 4, the switching procedure is shown. Changing the proposed CM flux reference in mode 1 to the one in mode 2 is identical to applying a step in the CM stator flux reference. Also, the estimation of CM stator flux is absolutely similar in these two modes.

Simulations through MATLAB/Simulink environment validate the effectiveness of the proposed control method.

In figure 10, at t = 0.5 (s) the stator of PM is connected to the grid. After a quarter of cycle (5 ms), a torque reference step equal to 0.5 p.u. is applied to the torque control loop. During this time, there is no load on the CDFIG which results that stator current of PM and the CDFIG torque are both remain zero.

At t = 0.505 (s), the control algorithm changes from the grid connection to the electromagnetic torque one. It must take into account that in his moment, torque reference is set to zero. This is the reason of smooth switching without any adverse effect on the efficiency of the proposed method. Until t = 0.505 (s), the magnetizing current of CM stator is imposed in a way that guarantees the suitable voltage induces in PM stator which has the same amplitude, frequency and phase of grid voltage. Figure 10 shows that proposed method achieves soft grid connection and transition between different operation modes without any current inrush and overshoots.

VI. CONCLUSION

A new control method for BDFIG synchronization is proposed. The proposed control method utilizes a modified DTC scheme. Voltage Phase, frequency and voltage magnitude controllers are employed to assure synchronization conditions, using the presented DTC algorithm. The presented method results in a soft and fast synchronization.

Once synchronization is achieved, BDFIG load is controlled by the standard DTC method. The high performance of the synchronization method is experimentally validated.

REFERENCES

[1] R. Cárdenas, R. Peña, P. Wheeler, J. Clare, A. Muñoz and A. Sureda, "Control of a wind generation system based on a Brushless Doubly-Fed Induction Generator fed by a matrix converter," Electric Power Systems Research, Volume 103, pp. 49-60, October 2013.

[2] P. C. Roberts, "A Study of Brushless Doubly-Fed (Induction) Machines," PhD dissertation, University of Cambridge, 2005.

[3] J. Hu, J. Zhu and D. G. Dorrell, "A New Control Method of Cascaded Brushless Doubly Fed Induction Generators Using Direct Power Control," IEEE TRANSACTIONS ON ENERGY CONVERSION, vol.29, no.3, pp.771-779, Sept. 2014.

[4] M. N. Hashemnia, F. Tahami, P. Tavner and S. Tohidi, "Steady-state analysis and performance of a brushless doubly fed machine accounting for core loss," IET Electr. Power Appl., vol. 7, Iss. 3, pp. 170–178, 2013

[5] S. A. Gomez and J. L. R. Amenedo, "Grid Synchronisation of Doubly Fed Induction Generators using Direct Torque Control," in *IECON 02* (IEEE 2002 28th Annual Conference of the Industrial Electronics Society), vol.4, pp.3338-3343 vol.4, 5-8 Nov. 2002

[6] J. Hu, J. Zhu, Y. Zhang, G. Platt, Q. Ma and D. G. Dorrell, "Direct Virtual Torque Control for Doubly Fed Induction Generator Grid Connection," IEEE TRANSACTIONS ON POWER ELECTRONICS, VOL. 28, NO. 7, JULY 2013.

[7] S. M. Madani, R. Sadeghi, B. Kharkan and M. Abadi, "Improving The LVRT Capability of The DFIG-Based Wind Turbines During Fault," Power Electronics," Drives Systems & Technologies Conference (PEDSTC), pp. 503 – 508, 3-4 Feb. 2015.

[8] Y. Zhang and J. Zhu, "Direct torque control of cascaded brushless doubly fed induction generator for wind energy applications," Electric Machines & Drives -Conference (IEMDC), 2011 IEEE International, pp. 741-746, 15-18 May 2011.

7th Power Electronics, Drive Systems & Technologies Conference (PEDSTC 2016)
16-18 Feb. 2016, Iran University of Science and Technology, Tehran, Iran

Improved Direct Torque Control of DFIG with Reduced Torque and Flux Ripples at Constant Switching Frequency

M. R. Agha Kashkooli, Seyed M. Madani, Ramtin Sadeghi

Department of Electrical Engineering

University of Isfahan

Isfahan, Iran

mr.kashkooli.1987@ieee.org, m.madani@eng.ui.ac.ir, r.sadeghi@eng.ui.ac.ir

Abstract— **This paper presents a modified hysteretic controller to improve Direct Torque Control (DTC) of the Doubly-Fed Induction Generator (DFIG) to achieve constant switching frequency and reduced torque and flux ripples. The proposed controller generates appropriate voltage vector with variable magnitude and angle, in order to reduce those ripples. The Space Vector Modulation is employed to apply the reference voltage vector to rotor windings. Other advantages of DTC methods such as fast dynamic response and simple coordinate transforms are preserved. The experimental results verify the advantages of this method compared to conventional DTC.**

Keywords—Doubly Fed Induction Generator (DFIG); Direct Torque Control (DTC); Space Vector Modulation (SVM); Hysteretic Controller

I. Introduction

Doubly fed induction generators (DFIGs) are commonly used in modern wind power generation systems due to their variable speed operation, low-converter cost, and reduced power losses, compared to other solutions such as fixed speed induction generators or synchronous generators with full sized converters. Usually, control of grid-connected DFIGs are based on either stator voltage [1] or stator/rotor flux oriented [2] - [4] called vector control (VC). These schemes decompose the rotor current into two decoupled components in phase with torque and flux, in the synchronous reference frame. Control of torque and flux is then achieved by regulating the rotor current components, using proportional-integral (PI) controllers. A drawback of these control schemes is that their performance are highly rely on the tuning of the PI and machine parameters such as stator and rotor inductances and resistances. Thus, performance may degrade when actual machine parameters deviate from values used in the control system.

Considering discrete nature of voltage source inverters, direct torque control (DTC), as an alternative to the vector control for induction machine and DFIG, is proposed in [5] and [6]. The DTC strategy provides direct torque regulation of the machine, which reduces the complexity of the VC strategy and minimizes the use of machine parameters. Initially, the basic DTC method directly controls the torque and flux by selecting voltage vectors from a predefined lookup table (LUT) based on the stator or rotor flux and torque information. One main problem is that the converter switching frequency varies with torque/flux hysteresis controller's bandwidth, which significantly complicates current filter designs and results in obvious torque pulsations.

To overcome these drawbacks, Predictive Direct Torque Control is proposed in [7]. To achieve constant switching frequency, two consecutive active vector applied to rotor followed by a zero vector in a fixed switching period. The order of these three vectors are defined by an optimal switching lookup table to provide constant switching frequency. The duration of each active vector is calculated based on cost function minimizations of torque and flux ripple. As a result, ripples are minimized but the duration calculations becomes complex. A comparative study of VC, DTC and DPC is presented in [8] for different conditions. Other efforts have been made in [9] and [10] to improve the performance of predictive direct control of DFIG by injecting three active vectors.

Variable structure or sliding-mode control (SMC) is an effective scheme for nonlinear systems with uncertainties [11]. The design principles of SMC and its applications in electrical drive systems were primarily presented in [12]. It features strong robustness, disturbance rejection, and fast dynamic response, but the controlled variables may be impacted by chattering effect. A SMC-based DTC drive for induction machine is proposed in [13] and [14] combined with PI regulator. It is named linear and variable structure control, which employs a switching and linear component, and has hybrid behavior. These schemes, also require synchronous transformation frame, which needs the angle of stator flux vector. A similar controller based on sliding mode applied to a high power DFIG in [15].

This paper, first introduces the state space model of DFIG and the SVM technique. Then, an improved DTC method is proposed. Furthermore, the proposed controller is simulated and validated by an experimental setup during balanced grid condition. Compared to conventional DTC, the proposed controller offers constant switching frequency and less torque and flux ripples.

978-1-5090-0376-1/16 $31.00 © 2016 IEEE

II. Dynamic Model of Doubly Fed Induction Generator

The per unit state-space representation of DFIG presented in the rotor reference frame is: [16]

$$\frac{d}{dt}\begin{bmatrix} \vec{i}_s \\ \vec{i}_r \end{bmatrix} = \frac{\omega_b}{\sigma L_s L_r}\begin{bmatrix} r_s L_r + j\omega_r L_s L_r & j\omega_r L_M L_r \\ -r_s L_M - j\omega_r L_M L_r & -j\omega_r L_M^2 \end{bmatrix}\begin{bmatrix} \vec{i}_s \\ \vec{i}_r \end{bmatrix} +$$
$$\frac{\omega_b}{\sigma L_s L_r}\begin{bmatrix} L_r & -L_M \\ -L_M & L_s \end{bmatrix}\begin{bmatrix} \vec{v}_s \\ \vec{v}_r \end{bmatrix} \quad (1)$$

$$\frac{d\omega_r}{dt} = \frac{\omega_b}{2H}(T_{em} - T_L) \quad (2)$$

Where:
v_s : stator voltage vector
v_r : rotor voltage vector
i_s : stator current vector
i_r : rotor current vector
L_s : stator magnetizing inductance
L_r : rotor magnetizing inductance
L_m : mutual inductance
ω_r : rotor angular speed
ω_b : base angular speed
H : inertia constant
σ : dispersion coefficient

The electromagnetic torque is governed by

$$T_{em} = \frac{X_m}{\sigma X_s X_r}\,\text{Im}\{\vec{\psi}_r^* \cdot \vec{\psi}_s\} = \frac{X_m}{\sigma X_s X_r}\,|\vec{\psi}_s||\vec{\psi}_r|\sin\delta \quad (3)$$

Where ψ_s and ψ_r are stator and rotor flux vectors, δ is angle between them.

III. Space Vector Modulation

Space vector pulse width modulation is a popular PWM technique for three-phase voltage-source inverters (VSI), applicable to control of AC induction and permanent-magnet synchronous machines.

A. Basic Principles

The structure of a typical three-phase VSI is shown in Fig.1 (a). The relationship between the switching state vector $[S_a, S_b, S_c]^t$ and the phase (line-to-neutral) output voltage vector $[V_a, V_b, V_c]^t$ is given by (4).

$$\begin{bmatrix} V_a \\ V_b \\ V_c \end{bmatrix} = \frac{V_{DC}}{3}\begin{bmatrix} 2 & -1 & -1 \\ -1 & 2 & -1 \\ -1 & -1 & 2 \end{bmatrix}\begin{bmatrix} S_a \\ S_b \\ S_c \end{bmatrix} \quad (4)$$

Where V_{DC} is the DC link voltage.

There are eight possible combinations of "on" and "off" states for the three upper power switches. The results of which are six non-zero and two zero vectors as shown in Fig.1 (b). The non-zero vectors are form axes of a hexagonal. The angle between any adjacent two non-zero vectors is 60 degrees. The zero vectors apply zero voltage to a three-phase load. These eight vectors are called the Basic Space Vectors. The objective of SVM technique is to generate an arbitrary voltage vector by combining the basic space vectors. To achieve this, for every PWM period the output voltage vector is approximately defined by (5).

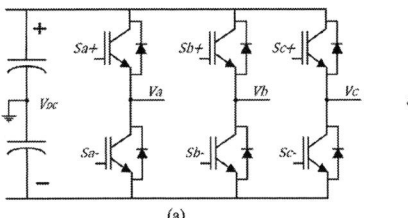 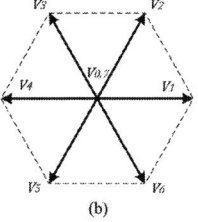

Fig. 1. Three phase VSI: (a)Power Circuit, (b) Basic Space Vectors

$$\vec{v}_{out}(n\,T_{PWM}) = \frac{1}{T_{PWM}}(T_1\vec{v}_k + T_2\vec{v}_{k\pm1} + T_0\vec{v}_{0,7}) \quad (5)$$

In the above equation, v_k and $v_{k\pm1}$ are two adjacent non-zero basic space vectors and T_1, T_2 are their duration time, respectively. The sum of T_1 and T_2 should be less than the PWM period. Assume the angle between v_{out} and v_k is α, the duration times are obtained as follows

$$T_1 = \sqrt{2}\,T_{PWM}|\vec{v}_{out}|\cos(\alpha + \pi/6) \quad (6)$$

$$T_2 = \sqrt{2}\,T_{PWM}|\vec{v}_{out}|\sin(\alpha) \quad (7)$$

$$T_{0,7} = T_{PWM} - (T_1 + T_2) \quad (8)$$

B. Switching Patterns

There are two possible arrangements for the order of v_k, $v_{k\pm1}$ and $v_{0,7}$ in each PWM period. Two symmetric switching patterns that can be implemented on TI's C2000 DSP are called Software-Determined and Hardware-Implemented Switching which are discussed in this section. [17]

In the software determined switching pattern the switching order is fixed among the three PWM channels for each sector, by the means of a program. Other feature of this technique are:

- Each PWM channel switches twice per every PWM period except when the duty cycle is 0% or 100%.
- Every PWM period starts and ends with v_0.

The period of v_0 is the same as v_7 in each PWM period.

On the other hand, the built-in SV PWM hardware module uses the basic space vectors in an order that results minimum switching. In the following some remarks about this switching technique is mentioned.

- There is always an upper switch that stays constant for the entire PWM period.
- The switch that stay unchanged for the entire sector, has no dead band effect. As a result, the dead band will affect the three phase outputs unevenly, which leads to harmonics in the inverter output.

Fig. 2 shows the switching state of upper switches, for both techniques when the output voltage vector is located in the first sector. The hardware implemented pattern reduces the switching loss and brings lower stress on semiconductor devices. On the other hand, the software determined pattern has better performance when harmonics are unfavorable.

IV. Proposed Direct Torque Controller

Generally the rotor windings of a DFIG are connected via a back to back converter to the grid for wind energy applications. In this paper, Rotor Side Converter (RSC) is studied and the

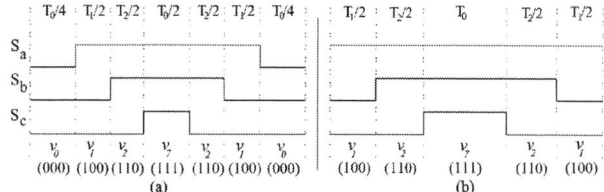

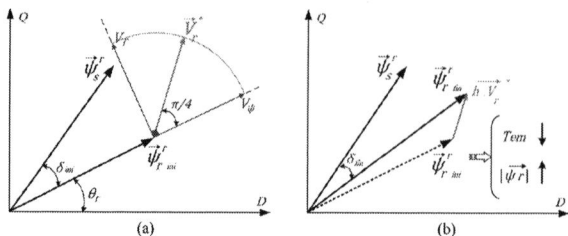

Fig. 2. Switching patterns: (a) Software determined, (b) Hardware Implemented

Fig. 4. Reference Rotor Voltage Vector Selection

DC link voltage is assumed to be constant. The overall control block diagram is illustrated in Fig3.

Equation (3) shows that the torque depends on machine parameters, and rotor and stator fluxes magnitude and the angle between them. Neglecting the stator resistance, stator flux is directly determined by integration of grid voltage, and is not affected by rotor variable. On the other hand, the rotor flux can be fully controlled via RSC. By keeping the magnitude of rotor flux within a desired range, the machine's electromagnetic torque becomes directly related to δ.

To present the method, V_ϕ and V_T must be introduced. V_ϕ is a voltage vector aligned with rotor flux. Applying this vector can control the magnitude of the rotor flux ψ_r, without significant impact on δ. Meanwhile V_T is perpendicular to the rotor flux and affect δ directly. Therefore the resultant of these two vectors is chosen as the rotor reference voltage vector in order to control torque and flux simultaneously. Fig. 4 (a) illustrates the strategy of selecting the proper reference voltage vector. The effect of a given vector applied during one sampling period (h) can be further analyzed in Fig. 4 (b).

TABLE I. Reference Voltage Vector Angle Selection

		uT_{em}	
		-1	1
$u\psi_r$	-1	$\theta_r + 3\pi/4$	$\theta_r + 5\pi/4$
	1	$\theta_r + \pi/4$	$\theta_r + 7\pi/4$

Hysteretic controllers with zero band detect variation of torque and flux immediately. Depending on the outputs of them, the angle of the reference vector is determined by Table I. These four possible vectors is better to be symmetric so the angle between rotor flux vector and applied voltage vector is multiple of $\pi/4$. Thereupon the proposed controller can work in both motor mode and generator mode. Also the speed of rotor does not affect the performance of controller.

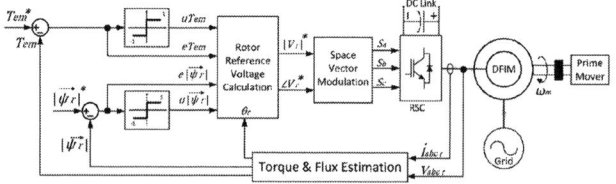

Fig. 3. Proposed DTC block diagram

Since the space vector modulation is capable of produce any arbitrary voltage vector, magnitude of the reference vector can be modified too. So the controller uses two different values for magnitudes of the reference vector. In the steady state operation of the machine when torque and flux errors are slightly small, applying a voltage vector less than one per unit is capable of reducing ripples

Hence, when the errors of torque and flux are less than the maximum allowable ripples, SVM is commanded to generate a vector with magnitude of 0.5 per unit. In other conditions such as changes in command torque and flux values or occurring a voltage dip in the grid, reference voltage magnitude is set to 1.0 per unit to retain the good dynamic response. Consequently, the proposed controller provides the appropriate magnitude and angle of the reference rotor voltage then SVM block generate the gate pulses for two level voltage source converter.

V. TORQUE AND FLUX ESTIMATION

In the proposed DTC, all the calculations are executed in rotor reference frame with direct component aligned with phase 'a' of rotor windings. Assuming balanced condition and neglecting zero sequence currents, the simplified Clarke transformation applied to the three phase rotor currents is:

$$\begin{bmatrix} i_{Dr} \\ i_{Qr} \end{bmatrix} = \frac{2}{3} \begin{bmatrix} 1 & -1/2 & -1/2 \\ 0 & \sqrt{3}/2 & \sqrt{3}/2 \end{bmatrix} \begin{bmatrix} i_{ar} \\ i_{br} \\ i_{cr} \end{bmatrix} \quad (9)$$

Since the rotor winding of a typical DFIG has star connection and its neutral point is unavailable, the line to line voltages should be used in transformation matrix. [2]

$$\begin{bmatrix} v_{Dr} \\ v_{Qr} \end{bmatrix} = \frac{3}{2} \begin{bmatrix} 2/\sqrt{3} & 1/\sqrt{3} \\ 0 & 1 \end{bmatrix} \begin{bmatrix} v_{abr} \\ v_{bcr} \end{bmatrix} \quad (10)$$

Direct and quadrature component of rotor flux is estimated by a simple integrator and the only machine parameter which is used is rotor resistance [2]:

$$\varphi_{Dr} = \int (v_{Dr} - r_r i_{Dr}) \, dt \quad (11)$$

$$\varphi_{Qr} = \int (v_{Qr} - r_r i_{Qr}) \, dt \quad (12)$$

The sampling frequency of rotor voltage should be at least twice of the switching frequency, according to the Nyquist-Shannon theorem. In this case, the sampling frequency is barely enough to accurately estimate the rotor voltage. So, a

978-1-5090-0376-1/16 $31.00 © 2016 IEEE

first order low pass filter is used to smooth the PWM waveform. As a result, the sampling is performed without aliasing and rotor fluxes are estimated precisely.

The per unit torque is calculated with only rotor variables as follows

$$T_{em} = \varphi_{Qr} \cdot i_{Dr} - \varphi_{Dr} \cdot i_{qr} \qquad (13)$$

Consequently, the estimation process needs no speed sensor and no measurements in the stator side. Moreover, the dq-transformation is simple like conventional DTC.

VI. EXPERIMENTAL RESULTS

The performance of the proposed DTC is evaluated on a 3kW DFIG and experimental results are compared with conventional DTC method. The machine parameters, sampling time, switching frequency and desired maximum ripples are listed in Table II. The controller is implemented on TMS320F2812 and the software determined SVM technique is chosen to minimize harmonics in stator current. Fig. 5 shows the experimental setup. The different components including F2812 DSP, voltage source inverter, low-pass filter, DC Link, voltage sensors, current sensors, prime mover and DFIG are designated from (1) to (8) respectively, in Fig. 5.

TABLE II.   Experimental Parameters

Rated Power	S_n	3 kW
Pole Pairs	p	3
Nominal Frequency	f_b	50 Hz
Rotor Resistance	r_r	0.125 Ω
Stator Line to Line Voltage	V_s	380 Vrms
Rotor Voltage	V_r	120 Vrms
DC Link Voltage	V_{dc}	160 V
Sampling Time per channel	T_s	4.8 μSec
SVM Switching Frequency	f_{sw}	4 kHz
LPF Cutoff Frequency	f_c	1 kHz
Maximum Torque Ripple	H_T	5 %

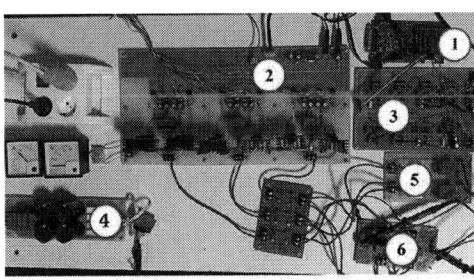

Fig. 5. Experimental Setup

Maximum Torque Ripple	H_φ	2%

A. Performance During Balanced Grid Condition

The first experiment is done during balanced grid condition, when the machine is operating as a generator at constant sub-synchronous speed of $\omega_{m\,pu} = 0.75$. In the middle of the test, torque reference is reversed to its positive nominal value and the machine operates as a motor. During this experiment, the rotor flux reference kept constant at its nominal value. This study can demonstrate the steady state and dynamic performance of the proposed controller versus the conventional DTC. The results are shown in Fig. 6.

The dynamic response of the proposed controller is similar to the conventional DTC which is obvious in Fig. 6 (a) and (d). Meanwhile, the comparison of Fig. 6 (b) and (e) along with Fig. 6 (f) indicates that in the steady state, the proposed comptroller has smaller torque and flux ripples.

Finally, the magnitude of rotor voltage is shown in Fig 6(c). Based on proposed control rule, this magnitude is taken equal to 0.5 pu, during the steady state in both motor and generator modes. When a step change occurs in torque command and the error of torque is greater than the maximum allowable ripple, the rotor voltage magnitude is raised to 1 pu, in order to keep the DTC dynamic response.

The Total Harmonic Distortion (THD) of injected current to the grid by a DFIG in important from power quality point of

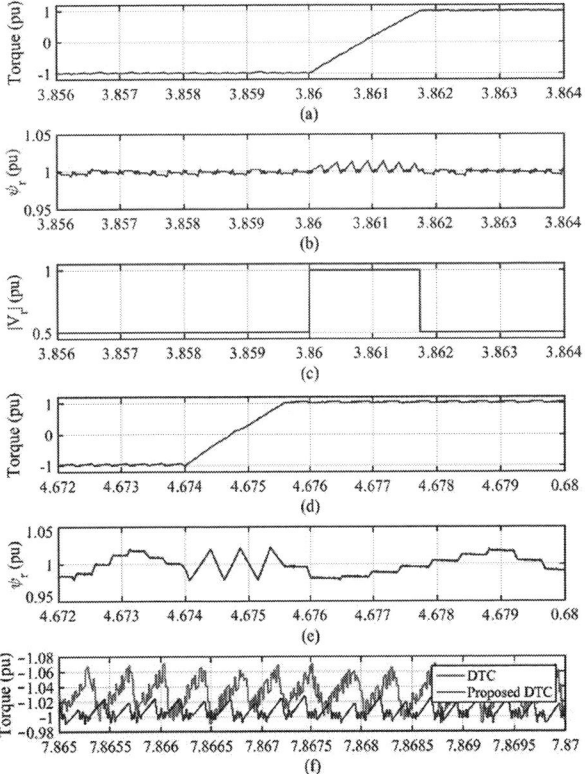

Fig. 6. Experimental dynamic and steady state response of proposed and conventional DTC

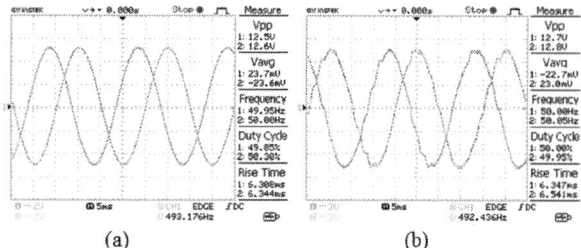

(a)   (b)

Fig. 7. Stator Currents, (a) Proposed DTC, (b) Conventional DTC

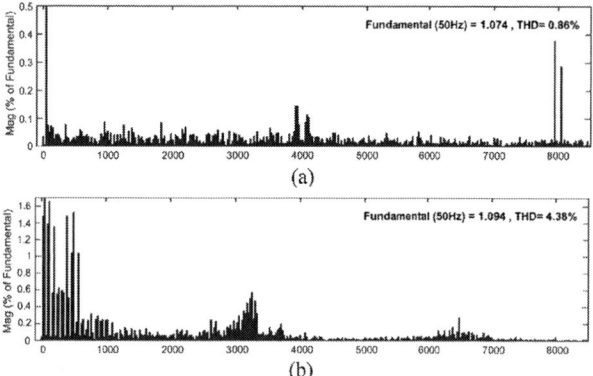

Fig. 8. FFT analysis of stator current, (a) Proposed DTC, (b) Conventional DTC

view. So the stator currents are measured in proposed controller and conventional DTC, then a Fast Fourier Transform (FFT) analysis is performed. Fig 7 shows the output of stator current sensors resulted by both controller. Their respective FFT analysis of per unit stator currents of these two controllers is illustrated in Fig. 8. The proposed controller has lower Total Harmonic Distortion (THD) and constant switching frequency behavior. Furthermore, the SVM switching frequency of 4 kHz and its second harmonic (8 kHz) have the greatest magnitudes. These magnitudes depend on the SVM switching frequency, unlike the conventional DTC where the frequency of the dominant harmonics are determined by the hysteresis bandwidth. Therefore in conventional DTC, reducing the torque and flux ripples is only possible by

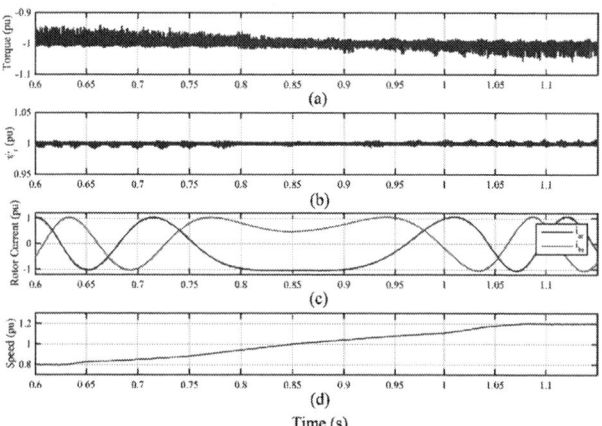

Time (s)

Fig. 9. Variable speed operation: (a) Electromagnetic torque, (b) Rotor currents, (c) Mechanical speed

decreasing the hysteresis bandwidth however causes more THD. As an advantage, in the proposed method torque and flux ripples reduced without affecting the stator current THD.

B. Variable Speed Operation

The main application of DFIG and its associated controller is in variable speed wind turbine. So the controller must capable of handling both sub-synchronous and hyper-synchronous speed operation. In this experiment, the mechanical speed increased from 800 rpm to 1200 rpm while the torque reference is kept constant at its nominal value. The DFIG torque tracks its reference when the rotor speed is near synchronous speed, which is illustrated in Fig 9. In this situation, the rotor current became DC as predicted by steady state analysis of induction machine where

$$f_r = s \cdot f_s \tag{14}$$

VII. CONCLUSION

In this paper, a simple and effective DTC for DFIG is proposed, to achieve constant switching frequency and reduce torque and flux ripples. The rotor voltage vector is selected according to hysteretic controller outputs, which has variable angle and magnitude. Then, the SVM technique is used to apply this voltage vector to the rotor windings via a two-level VSI.

The method does not use synchronous reference frame transformation, speed sensor, and PI regulators. However, it results good steady-state and dynamic performance. In addition to this, the proposed controller is capable of operating at variable speed, from sub-synchronous to hyper-synchronous speeds, which makes it suitable for wind power applications. The implementation of the control algorithm does not require powerful digital processors, due to its computational simplicity. The performance of the proposed controller is compared to conventional DTC. Simulation and experimental results for a 3kW DFIG confirms the effectiveness of the proposed DTC.

REFERENCES

[1] Muller, S.; Deicke, M.; De Doncker, R.W., "Doubly fed induction generator systems for wind turbines," in *Industry Applications Magazine, IEEE* , vol.8, no.3, pp.26-33, May/Jun 2002

[2] Pena, R.; Clare, J.C.; Asher, G.M., "Doubly fed induction generator using back-to-back PWM converters and its application to variable-speed wind-energy generation," in *Electric Power Applications, IEE Proceedings -* , vol.143, no.3, pp.231-241, May 1996

[3] Madani, S.M.; Sadeghi, R.; Kharkan, B.; Abadi, M., "Improving the LVRT capability of the DFIG-based wind turbines during fault," in *Power Electronics, Drives Systems & Technologies Conference (PEDSTC), 2015 6th* , vol., no., pp.503-508, 3-4 Feb. 2015

[4] Amiri, N.; Madani, S.M.; Lipo, T.A.; Zarchi, H.A., "An Improved Direct Decoupled Power Control of Doubly Fed Induction Machine Without Rotor Position Sensor and With Robustness to Parameter Variation," in *Energy Conversion, IEEE Transactions on* , vol.27, no.4, pp.873-884, Dec. 2012

[5] Takahashi, I.; Ohmori, Y., "High-performance direct torque control of an induction motor," in *Industry Applications, IEEE Transactions on* , vol.25, no.2, pp.257-264, Mar/Apr 1989

978-1-5090-0376-1/16 $31.00 © 2016 IEEE

[6] Buja, G.S.; Kazmierkowski, M.P., "Direct torque control of PWM inverter-fed AC motors - a survey," in *Industrial Electronics, IEEE Transactions on* , vol.51, no.4, pp.744-757, Aug. 2004

[7] Abad, G.; Rodriguez, M.A.; Poza, J., "Two-Level VSC Based Predictive Direct Torque Control of the Doubly Fed Induction Machine With Reduced Torque and Flux Ripples at Low Constant Switching Frequency," in *Power Electronics, IEEE Transactions on* , vol.23, no.3, pp.1050-1061, May 2008

[8] Tremblay, E.; Atayde, S.; Chandra, A., "Comparative Study of Control Strategies for the Doubly Fed Induction Generator in Wind Energy Conversion Systems: A DSP-Based Implementation Approach," in*Sustainable Energy, IEEE Transactions on* , vol.2, no.3, pp.288-299, July 2011

[9] Xiangjie Liu; Xiaobing Kong, "Nonlinear Model Predictive Control for DFIG-Based Wind Power Generation," in *Automation Science and Engineering, IEEE Transactions on* , vol.11, no.4, pp.1046-1055, Oct. 2014

[10] Yongchang Zhang; Jiefeng Hu; Jianguo Zhu, "Three-Vectors-Based Predictive Direct Power Control of the Doubly Fed Induction Generator for Wind Energy Applications," in *Power Electronics, IEEE Transactions on* , vol.29, no.7, pp.3485-3500, July 2014

[11] K.D. Young, Ü. Özgüner, *Variable structure systems, sliding mode and nonlinear control.* London, Springer, 1999

[12] Utkin, V.I., "Sliding mode control design principles and applications to electric drives," in *Industrial Electronics, IEEE Transactions on* , vol.40, no.1, pp.23-36, Feb 1993

[13] Lascu, C.; Boldea, I.; Blaabjerg, F., "Direct torque control of sensorless induction motor drives: a sliding-mode approach," in *Industry Applications, IEEE Transactions on* , vol.40, no.2, pp.582-590, March-April 2004

[14] Lascu, C.; Boldea, I.; Blaabjerg, F., "A Class of Speed-Sensorless Sliding-Mode Observers for High-Performance Induction Motor Drives," in *Industrial Electronics, IEEE Transactions on* , vol.56, no.9, pp.3394-3403, Sept. 2009

[15] Yung-Tsai Weng; Yuan-Yih Hsu, "Sliding mode regulator for maximum power tracking and copper loss minimization of a doubly fed induction generator," in *Renewable Power Generation, IET* , vol.9, no.4, pp.297-305, 5 2015

[16] G. Abad, J. L´opez, M. A. Rodr´ıguez, L. Marroyo, and G. Iwanski, *Doubly Fed Induction Machine: Modeling and Control for Wind Energy Generation.* New York: Wiley, Sep. 2011

[17] Yu, Z., Application Report SPRA524: Space-Vector PWM with TMS320C24x Using H/W & S/W Determined Switching Patterns, Texas Instruments, 1999

7th Power Electronics, Drive Systems & Technologies Conference (PEDSTC 2016)
16-18 Feb. 2016, Iran University of Science and Technology, Tehran, Iran

Using Extended Kalman Filter and Adaptive Filter for Sensorles Predictive Torque Control of PM-Assissted Synchronous Reluctance Motor

Ali Sarajian[1], Davood Arab Khaburi[1], Marco Rivera[2]

[1]Department of Electrical Engineering, Iran University of Science and Technology, Tehran, Iran

[2]Department of Industrial Technologies, Universidad de Talca, Talca, Chile

asarajian@elec.iust.ac.ir, khaburi@iust.ac.ir, marcoriv@utalca.cl

Abstract—**The Permanent Magnet-Assissted Synchronous Reluctance Motor (PMA-SynRM) drive has become one of the most interesting replacements for the high efficiency variable speed drive. Herein, sensorless predictive torque control of a PMA-SynRM with non-sinusoidal back electromotive force (Back-EMF) is introduced. In order to control PMA-SynRM, finite control set-model predictive control (FCS-MPC) is implemented by means of a two-level inverter. Furthermore, an improved form of FCS-MPC, i.e., direct mean torque control (DMTC), is utilized as a second method to control PMA-SynRM. For improving the sensorless the combination of Extended Kalman Filter (EKF), Adaptive Filter (AF) and quadrature Phase-Locked Loop (PLL) are used for better estimation of the non-sinusoidal back EMF, elimination of the high order harmonics, and the accurate estimation of position and speed rotor, respectively. The simulation result in effectively minimizing torque ripples compared to conventional FCS-MPC. The outcomes of the observer simulation are successfully guaranteed the accurate estimation of speed and rotor position.**

Keywords—*Permanent magnet-assissted synchronous reluctance motor, extended kalman filter; adaptive filter; predictive torque control; sensorless control*

I. INTRODUCTION

Permanent-Magnet Assisted Synchronous Reluctance Machines (PMA-SynRMs) are known for application in high-efficiency adjustable speed drives (ASDs). These ASDs utilize electromagnetic torque consisting reluctance and the permanent magnet components to achieve this purpose. The amount of magnet is small in comparison with the conventional Interior PM Motors. Compared with the conventional Synchronous reluctance machines, the PMA-SynRMs increase efficiency and power density. Altogether, the good characteristics of SynRMs and PM motors are gathered in PMA-SynRMs [1] .

Conventional direct torque control (DTC) is a modern control method that illustrates a good dynamic behavior. This method selects the active voltage vector (AVV) and the zero voltage vector (ZVV) by using a switching table [2]. However, this method has some disadvantages. The most important ones is considerable torque ripple. To tackle this problem, a direct mean torque control (DMTC) and then an improved algorithm of DMTC were proposed for SynRM

[3]. Moreover, Predictive DTC [4] is proposed methods for reducing the torque ripple. Finite control set predictive torque control (FCS-PTC) is an effective method. The FCS-PTC method calculates all possible voltage vectors within one sampling interval and selects the best one by using an optimization cost function [5]. The PTC method along with DMTC method have been used in SynRM [6] and PMSM [7]. To use predictive control, we need accurate model of machine. The PTC method unlike the DTC method is not inherently sensorless [8]. The main source of torque pulsation in the machine is existence of the harmonic components of the air gap flux [9]. In order to estimate the air gap flux, sliding mode observer [10] and extended kalman filter [11] have been used for IMs. Inaccurate estimate of rotor position is due to the harmonics in the air gap flux. So, in order to eliminate the harmonic components, an adaptive filter should be used. Adaptive filter detects the fifth and seventh harmonic components based on recursive least square (RLS) algorithm [12]. In addition, to reduce the influence of errors in estimating position of the rotor, the phase lock loop (PLL) should be used [13].

Sensorless vector-control for PMA–SynRM using sliding mode observer has been presented in [14]. Also, Sensorless direct torque and flux control with space vector modulation (DTFC–SVM) for PMA–SynRM has been successfully implemented [15]. This paper presents sensorless FCS-PTC method with improved DMTC for achieving low torque ripples of PMA-SynRM based on extended kalman filter, adaptive filter and PLL. Simulation results using Matlab/Simulink show the validity of the proposed control method and observer scheme as well as the excellent dynamic response of the electromagnetic torque.

II. MODEL AND OPERATION OF PMA-SYNRM

A. PMA-SynRM model

The voltage equation of a PMA-SynRM in the synchronous reference frame can be expressed as follows:

$$u_{qs} = R_s i_{qs} + \frac{d\psi_{qs}}{dt} + \omega_e \psi_{ds}$$
$$u_{ds} = R_s i_{ds} + \frac{d\psi_{ds}}{dt} - \omega_e \psi_{qs}$$

(1)

978-1-5090-0376-1/16 $31.00 © 2016 IEEE   64

Where R_S is stator resistance and ω_e is electrical rotor angular speed, u_{ds} and u_{qs} are the d and q-axes voltages, and ids and iqs are the d and q-axes currents. Ψ_{qs} and Ψ_{ds} are stator flux components in d and q-axes. Rotor nonsinusoidal flux distribution produces two new components: Ψ_{qm} and Ψ_{dm} that are the rotor flux linkages in d and q axes, respectively. Stator flux equations of the PMA-SynRM in the d-q reference frame are expressed as follows:

$$\psi_{qs} = L_q i_{qs} - \psi_m - \psi_m \times \sum_{n=1}^{\infty} \left(K_{6n-1} + K_{6n+1} \right) \cos\left(6n\theta \right) = L_q i_{qs} - \psi_{qm} \quad (2)$$

$$\psi_{ds} = L_d i_{ds} - \psi_m \times \sum_{n=1}^{\infty} \left(K_{6n-1} + K_{6n+1} \right) \sin\left(6n\theta \right) = L_d i_{ds} - \psi_{dm} \quad (3)$$

If the back EMF harmonics higher than sixth order are negligible, the back EMF equations i.e., e_{ds} and e_{qs} that are the d and q-axes back EMFs, are given as fallow:

$$e_{qs} = \omega_e \psi_{qm} = E_1 + E_6 \cos\left(6\theta \right) \quad (4)$$

$$e_{ds} = \omega_e \psi_{dm} = E_6 \sin\left(6\theta \right) \quad (5)$$

where E_1 and E_6 are the amplitude of the fundamental component and sixth order harmonic of the back EMFs, in the synchronous reference frame due to the nonsinusoidal rotor flux, respectively. According to (2)-(5), (1) can be rearranged as

$$u_{qs} = R_s i_{qs} + L_q \frac{di_{qs}}{dt} - \frac{d\psi_{qm}}{dt} + \omega_e L_d i_{ds} + e_{ds}$$
$$u_{ds} = R_s i_{ds} + L_d \frac{di_{ds}}{dt} + \frac{d\psi_{dm}}{dt} - \omega_e L_q i_{qs} + e_{qs} \quad (6)$$

where L_d and L_q are d and q-axes inductances, respectively. Since the rotor flux can not vary sharply, the derivative of rotor flux components are zero $(d\psi_{qm}/dt = 0, d\psi_{dm}/dt = 0)$.

The electromagnetic torque is

$$T_e = \frac{3}{2} \frac{p}{2} \left(\psi_{ds} i_{qs} - \psi_{qs} i_{ds} \right)$$
$$= \frac{3}{2} \frac{p}{2} \left(\psi_{dm} i_{qs} - \psi_{qm} i_{ds} + \left(L_d - L_q \right) i_{ds} i_{qs} \right) \quad (7)$$

B. Improved DMTC

Direct mean torque control (DMTC) was introduced to control the mean value of the torque at reference value [6]. Figure. 1 shows a typical cycle of operation of the improved DMTC. In this method, the interval switching time are divided into two segments including AVV and ZVV. Formerly, when AVV applies the torque increment occurs at the beginning cycle. Following, the reaching to virtual hysteresis width ΔT, ZVV applies and the torque decreases. In the steady state, the value of T_n at the beginning of a cycle should be equal to its value T_{n+1} at the end. The hatch area in Fig. 1, shows the torque response during the cycle of n. The value of the torque at the end of the cycle T_{n+1} can be described directly by the following formula:

$$T_{n+1} = T_n + \frac{dT_{AVV}}{dt} t_{AVV} + \frac{dT_{ZVV}}{dt} \left(T_s - t_{AVV} \right) = T^* - \frac{1}{2} \Delta T \quad (8)$$

The virtual hysteresis width ΔT can be expressed

$$\Delta T = \frac{dT_{AVV}}{dt} t_{AVV} = -\frac{dT_{ZVV}}{dt} \left(T_s - t_{AVV} \right) \quad (9)$$

(10) can be derived from inserting (8) to (9); thus:

$$t_{AVV} = \frac{T^* - T - \frac{1}{2}\Delta T - \frac{dT_{ZVV}}{dt} T_s}{\frac{dT_{AVV}}{dt} - \frac{dT_{ZVV}}{dt}} \quad , \quad t_{ZVV} = T_s - t_{AVV} \quad (10)$$

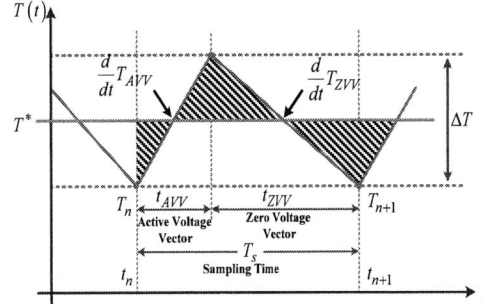

Fig. 1. Typical operation cycle of DMTC in steady state.

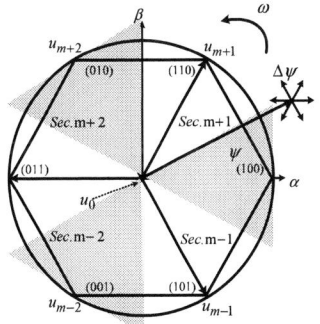

Fig. 2. Voltage space phasor, sectors and stator flux variation.

As can be seen in Figure. 2, in two-level inverter, the six possible AVVs have different effects on the flux as well as on the torque. With respect to improved DMTC, it is assumed that the stator flux vector (Ψ) is situated in the mth sector (m =1,...,6) of the α–β plane (Figure. 2) for positive rotation ($\omega > 0$). The two most favorable voltage vectors are preselected, and thus calculating t_{AVV} for both of them. For increasing the value of ψ, the AVVs u_{m+1} can be selected, while selecting u_{m+2} results in decreasing its magnitude. It should be noted that the value of T increases irrelevant from the number of sectors.

C. calculation of the slopes of the torque

By differentiating (7), dT/dt can be determined by

$$\frac{dT}{dt} = \frac{3}{2} \frac{P}{2} \left(\psi_{qm} \frac{di_{ds}}{dt} + \psi_{dm} \frac{di_{qs}}{dt} \right.$$
$$\left. + \left(L_{ds} - L_{qs} \right) \left(i_{ds} \frac{di_{qs}}{dt} + i_q \frac{di_{ds}}{dt} \right) \right) \quad (11)$$

From (11) and (6) and by supposing that a ZVV is exerted to the motor, the torque decreases, and its time derivative is obtained as follows:

$$\frac{dT_{ZVV}}{dt} = \frac{3}{2} \frac{P}{2} \left[\psi_{qm} \left(\frac{-R_s i_{ds} + \omega_e L_q i_{qe} - \omega_e \psi_{qm}}{L_d} \right) \right.$$
$$-\psi_{dm} \left(\frac{-R_s i_{qs} - \omega_e L_d i_{ds} - \omega_e \psi_{dm}}{L_q} \right) + \left(L_d - L_q \right) \times \quad (12)$$
$$\left. \left(i_{ds} \frac{-R_s i_{qs} - \omega_e L_d i_{ds} - \omega_e \psi_{dm}}{L_q} + i_{qs} \frac{-R_s i_{ds} + \omega_e L_q i_{qs} - \omega_e \psi_{qm}}{L_d} \right) \right]$$

978-1-5090-0376-1/16 $31.00 © 2016 IEEE

In addition, for an AVV, the torque increases with the time derivative

$$
\frac{dT_{AVV}}{dt} = \frac{3}{2}\frac{P}{2}\left[\psi_{qm}\left(\frac{u_{ds}}{L_d}\right) - \psi_{dm}\left(\frac{u_{qs}}{L_q}\right) + \\ \left(L_d - L_q\right)\times\left(i_{ds}\frac{u_{qs}}{L_q} + i_{qs}\frac{u_{ds}}{L_d}\right)\right] + \frac{dT_{ZVV}}{dt} \tag{13}
$$

In order to decide which voltage space phasor has to be applied to the machine, a second criterion is necessary, and the flux has to be taken into account.

D. Torque and flux prediction

In order to decide the best voltage vector, the FCS-PTC approach can be applied. In this approach a cost function that is the evaluation criterion to decide which AVV is the best to be applied. the cost function is defined as fallows

$$
J_m = \frac{1}{2}\left(\left|T_{e,k+1} - T^*\right|^2 + Q\left\|\psi_{s,k+1}\right|^2 - \psi_s^*\right|^2\right) \tag{14}
$$

where $T_{e,k+1}$ and $\psi_{s,k+1}$ are the predicted torque and stator flux T^* and ψ^* are the torque and flux references. Q is a weighting factor that determines the importance of flux control compared to torque control. Since the d–q components of the stator current space phasor and stator flux space phasor can be predicted at the end of the switching interval as (15) and (16) in discrete form from the voltages (1) and (6)

$$
i_{ds,n+t_{AVV}} = i_{ds,n} + t_{AVV}/L_d\left(u_d - R_s i_{ds} + \omega_e L_q i_{qs} - e_{qs}\right) \\
i_{qs,n+t_{AVV}} = i_{qs,n} + t_{AVV}/L_q\left(u_q - R_s i_{qs} - \omega_e L_d i_{ds} + e_{ds}\right) \tag{15}
$$

$$
\psi_{ds,n+t_{AVV}} = \psi_{ds,n} + t_{AVV}\left(u_d - R_s i_{ds} + \omega_e \psi_{qs}\right) \\
\psi_{qs,n+t_{AVV}} = \psi_{qs,n} + t_{AVV}\left(u_q - R_s i_{qs} - \omega_e \psi_{ds}\right) \tag{16}
$$

then the components of the torque space phasor at the end of the same cycle can be predicted by

$$
T_{n+t_{AVV}} = \frac{3}{2}\frac{p}{2}\left[\psi_{qm}i_{ds,n+t_{AVV}} - \psi_{dm}i_{qs,n+t_{AVV}} \\ + \left(L_d - L_q\right)i_{ds,n+t_{AVV}}i_{qs,n+t_{AVV}}\right] \tag{17}
$$

With respect to the prediction of the torque and stator flux for every AVV and following by substituting them into the cost function, the AVV that minimizes the cost function is selected as the best voltage vector.

E. Modeling of EKF Estimator for back EMF Estimation

Kalman filtering is an optimal stochastic approach to state estimation and filtering in linear systems [11]. For nonlinear systems, the state space equation can be written in the following form:

$$
x(t) = f\left(x(t)\right) + Bu(t) + \sigma(t) \\
\delta y(t) = Hx(t)\delta x(t) + \mu(t) \tag{18}
$$

where $x(t)$, $u(t)$, and $y(t)$ are the states, inputs, and outputs of the system, respectively. The system noise $\sigma(t)$ and measurement noise $\mu(t)$ are supposed to be zero mean and white with Gaussian distributions of covariances $Q(t)$ and $R(t)$, respectively. Once a nominal solution to the nonlinear (18) has been obtained, the linearized perturbation equations of the system are as follows:

$$
\delta x(t) = F\left(x(t)\right)\delta x(t) + B\delta u(t) + \sigma(t) \tag{19}
$$

$$
\delta y(t) = H\delta x(t) + \mu(t) \tag{20}
$$

where the Jacobian and output matrices are defined respectively as follows:

$$
F\left(x(t)\right) = \frac{\partial f}{\partial x}\bigg|_{x=x(t)} \tag{21}
$$

$$
H\left(x(t)\right) = \frac{\partial h}{\partial x}\bigg|_{x=x(t)} \tag{22}
$$

Using (6), the PMA-SynRM voltage equations and the flux linkage equations can be expressed in the rotor reference frames as the following equations

$$
\begin{cases}
\dfrac{di_d}{dt} = \dfrac{u_d}{L_d} - \dfrac{R_s}{L_d}i_d + \dfrac{L_q}{L_d}\omega_e i_q - \dfrac{\omega_e}{L_d}\psi_{qm} \\[2mm]
\dfrac{di_q}{dt} = \dfrac{u_q}{L_q} - \dfrac{R_s}{L_q}i_q - \dfrac{L_d}{L_q}\omega_e i_d - \dfrac{\omega_e}{L_q}\psi_{dm} \\[2mm]
\dfrac{d\psi_{dm}}{dt} = 0 \\[2mm]
\dfrac{d\psi_{qm}}{dt} = 0
\end{cases} \tag{23}
$$

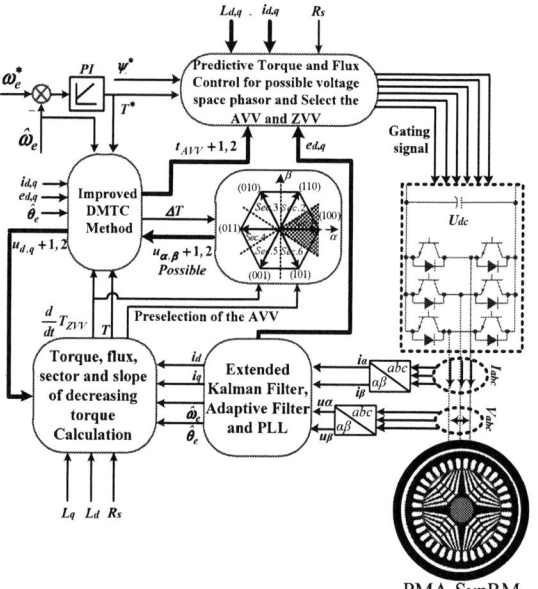

Fig. 3. block diagram of the proposed PTC for the PMA-SynRM.

F. Adaptive Filter with PLL for Compensation of Position Harmonic Error

Following the estimation of the back EMF, speed and position of the rotor based on the arc tangent function is calculated from the below equations [16]:

$$
\theta_e(n) = -\tan^{-1}\left(\hat{e}_\beta(n)/\hat{e}_\alpha(n)\right) \tag{24}
$$

$$
\omega_e(n) = -\frac{d}{dt}\tan^{-1}\left(e_\beta(n)/e_\alpha(n)\right) \tag{25}
$$

where $\hat{e}_\alpha$ and $\hat{e}_\beta$ are the estimation back EMF in α and β-axes, respectively. It should be noted (24) and (25) suffer from the lack of the accuracy due to appearance of harmonic in air gap flux. In order to reduce the effect of estimation error, PLL was used as a replacement of arc tangent function. Because the back EMF changes in different speeds, the input estimation back EMF for the PLL is normalized [17]. The

978-1-5090-0376-1/16 $31.00 © 2016 IEEE

normalized error position of signal is calculated from the below equation as follows:

$$\varepsilon = \frac{1}{\sqrt{\hat{e}_\alpha^2 + \hat{e}_\beta^2}} \left[\hat{e}_\alpha \cos\left(\hat{\theta}_e\right) - \hat{e}_\beta \sin\left(\hat{\theta}_e\right) \right] \quad (26)$$

Therefore, the estimation of rotor position acquired from the PLL can be achieved as follows:

$$\theta_e(n) = -\tan^{-1}\left(\hat{e}_\beta(n)/\hat{e}_\alpha(n)\right) \quad (27)$$

Considering the dominant fifth and seventh harmonic components of back EMF generated by air gap flux harmonics, the phase difference between actual rotor position and estimated rotor position of the PMA-SynRM can be expressed as [17]:

$$\varepsilon \approx \theta_e - \hat{\theta}_e + e_6 \sin\left(6\omega_e t + \theta_{e6}\right) \quad (28)$$

where e_6 is the amplitude of the equivalent sixth back EMF harmonic and θ_{e6} is the initial phase.

It is noticeable that there is an additive error in the phase difference due to the sixth back EMF harmonic. However, due to the widely bandwidth of the PLL, utilizing the type-II system will partially eliminate the sixth harmonic. In addition, the difficulty of design of the type-II system parameters to eliminate the sixth harmonic without influencing on the fundamental components is due to variation of the speed. Therefore, an adaptive filter (AF) with PLL is used for the effectiveness compensation of harmonics in back EMF [17].

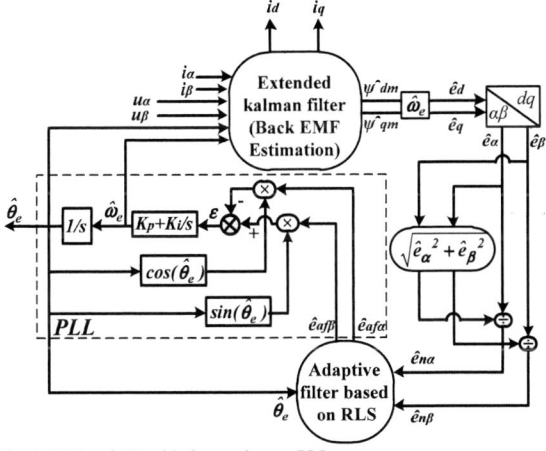

Fig. 4. EKF and AF with the quadrature PLL.

Figure. 5 depicts the theory of the harmonic eliminating on the base of the adaptive noise-cancelling technique with two orthogonal references for the position estimator. $p(n)$ denotes the primary input which is comprised of a given signal and an additional harmonic signal. The two orthogonal signals, i.e., the first row and the second row of matrix $r(n)$, denote the harmonic references that the high-order harmonic components are converged to them. $w_1(n)$ and $w_2(n)$ are the adjustable filter coefficients corresponding to the harmonic references. The sum of these two weighted references form the output of filter $h(n)$ that indicates the harmonic estimation. The fundamental component of desired signal is acquired from the error signal $e(n)$ resulted from deducting $h(n)$ from $p(n)$. Correspondingly, making the filter output $h(n)$ to converge to the real high-order harmonic components is the objective of adaptive filter. Adjusting the filter

coefficients is done by the RLS algorithm because of its fast convergence speed [16].

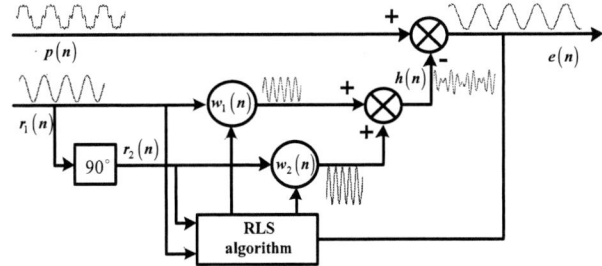

Fig. 5. Concept of the harmonic detection based on the adaptive noise cancelling.

The normalized estimations of back EMF, i.e., $\hat{e}_{n\alpha}$ and $\hat{e}_{n\beta}$, that are comprised of the fifth and seventh harmonic components, are multiplied by $\sin(5\theta_e)$, $\cos(5\theta_e)$ and $\sin(7\theta_e)$, $\cos(7\theta_e)$, respectively (θ_e is the estimation of rotor position from the PLL output) are used for producing the AF harmonic references. $w_{ji}(n)$ and $k_{ji}(n)$ are coefficients and gains of the filter. The coefficients of filter can be adjusted by the RLS algorithm in a way that the fifth and seventh harmonic estimations converge to their actual values. The desirable fundamental components of the normalized back EMF is obtained as fallow:

$$\hat{e}_{af\alpha}(n) = \hat{e}_{n\alpha} - \sum_{i=1}^{2}\sum_{j=1}^{2} w_{ji}(n) r_{ji}(n) \quad (29)$$

$$\hat{e}_{af\beta}(n) = \hat{e}_{n\beta} - \sum_{i=3}^{4}\sum_{j=1}^{2} w_{ji}(n) r_{ji}(n) \quad (30)$$

where $r_{ji}(n)$ imposes the harmonic reference signals and is given by the below matrix:

$$r(n) = \begin{bmatrix} \sin\left(5\hat{\theta}_e\right) & \sin\left(7\hat{\theta}_e\right) & \sin\left(5\hat{\theta}_e\right) & \sin\left(7\hat{\theta}_e\right) \\ \cos\left(5\hat{\theta}_e\right) & \cos\left(7\hat{\theta}_e\right) & \cos\left(5\hat{\theta}_e\right) & \cos\left(7\hat{\theta}_e\right) \end{bmatrix} \quad (31)$$

The meanings of $w_{ji}(n)$ is the estimation of the corresponding harmonic component magnitude, and it can be adjusted online according to the specification of the harmonic reference by the equations as follows:

$$w_{ji}(n) = w_{ji}(n-1) + k_{ji}(n)\hat{e}_{af\alpha}(n) \ j \ \& \ i = 1,2 \quad (32)$$

$$w_{ji}(n) = w_{ji}(n-1) + k_{ji}(n)\hat{e}_{af\beta}(n) \ j = 1,2 \ \& \ i = 3,4 \quad (33)$$

The gain vector $k_{ji}(n)$ can be calculated as follows:

$$k_{ji}(n) = \frac{\phi_{ji}(n)}{\lambda + r_{ji}(n)\phi_{ji}(n)} \qquad j = 1,2 \ \& \ i = 1,2,3,4 \quad (34)$$

where $\lambda \in (0,1)$ is the forgetting factor. With $\lambda = 1$, the conventional method of least squares can be achieved. The presence of λ is to guarantee that the data in the far apart past have less effect on the convergence. $\Phi_{ji}(n)$ is the intermediate variables that is obtained as follows:

$$\phi_{ji}(n) = P_{ji}(n-1)r_{ji}(n) \qquad j = 1,2 \ \& \ i = 1,2,3,4 \quad (35)$$

The inverse of the autocorrelation matrix $P_{ji}(n)$ can be achieved as

$$P_{ji}(n) = \lambda^{-1}\left(P_{ji}(n-1) - k_{ji}(n)\phi_{ji}(n)\right)$$
$$j = 1,2 \ \& \ i = 1,2,3,4 \tag{36}$$

The initial values of $w_{ji}(n)$ and $P_{ji}(n)$ are selected as

$$P_{ji}(0) = \sigma \ , \ w_{ji}(0) = 0 \qquad j = 1,2 \ \& \ i = 1,2,3,4 \tag{37}$$

It should be noticed that a smaller value of λ improves the tracking capability, but the stability of the RLS algorithm can be influenced. In this application, λ is set at 0.99995 and σ is chosen 0.001.

III. SIMULATION RESULTS

In order to validate the proposed method to control of PMA-SynRM based kalman filter and adaptive filter, some simulations are done using Matlab\Simulink software. Fig. 3 and Fig. 4 shows the predictive torque control scheme of the sensorless PMA-SynRM drive based on a EKF and AF with PLL. The parameters of the PMA-SynRM used in this paper are shown in TABLE 1.

TABLE I.   SIMULATION PARAMETERS

Parameter	Value
p	4
R_S	1.25[Ω]
L_d	49.801[mH]
L_q	17.901[mH]
ψ_r	0.48[wb]
J_m	0.0012[kg.m^2]
B	0.0001[N.m.s]
ω_n	6000[rpm]
T_n	3.7[N.m]

Fig. 6 shows the actual and estimated d-axes and q-axes flux linkage rotor. The motor rotates at 6000 rpm with rated load. The dotted lines shown in all figures express estimated results. At steady state, the mean value of the flux linkage estimation error at the d-axes and q-axes are 0.0308% and 0.0463%, respectively. From the presented result of kalman filter, it can be seen that the kalman filter is good performance and accurate.

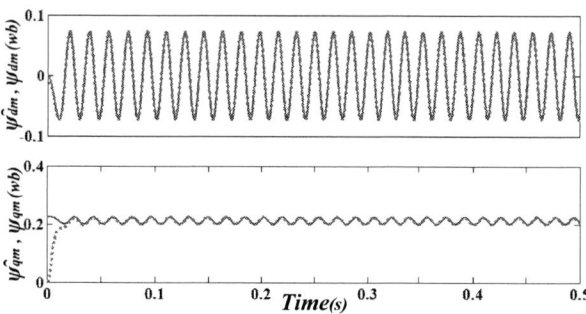

Fig. 6. Actual and stimated d-axes and q-axes flux linkage rotor.

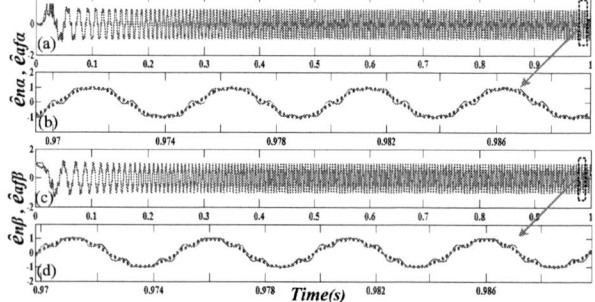

Fig. 7. The α-axes and β-axes normalized back EMF estimation and output of Adaptive Filter.

Fig. 7-8 show the comparison of the α-axes and β-axes back EMF estimate with AF at 6000 rpm with rated load. The normalized back EMF estimate $\hat{e}_{n\alpha}$ and $\hat{e}_{n\beta}$ directly obtained from the EKF, the fifth and seventh harmonic compensation $\hat{e}_{5\alpha}$, $\hat{e}_{7\alpha}$, and the back EMF estimate with the compensation $\hat{e}_{af\alpha}$ and $\hat{e}_{af\beta}$ through the AF are given. By comparison, it can be seen that the back EMF waveform gets more sinusoidal with the proposed AF using the RLS algorithm, and the fifth and seventh harmonic distortions are effectively removed.

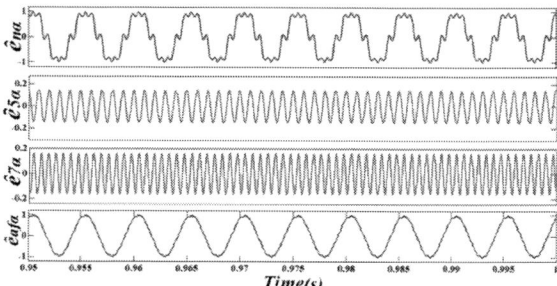

Fig. 8. Obtained back EMF harmonic waveform by using the AF.

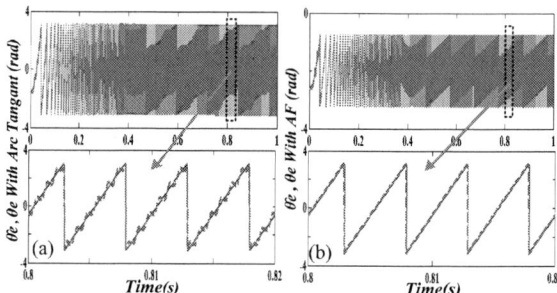

Fig. 9. Comparison of the actual and estimated position at 6000 rpm. a) with arc tangent function. b) with the AF using the RLS algorithm.

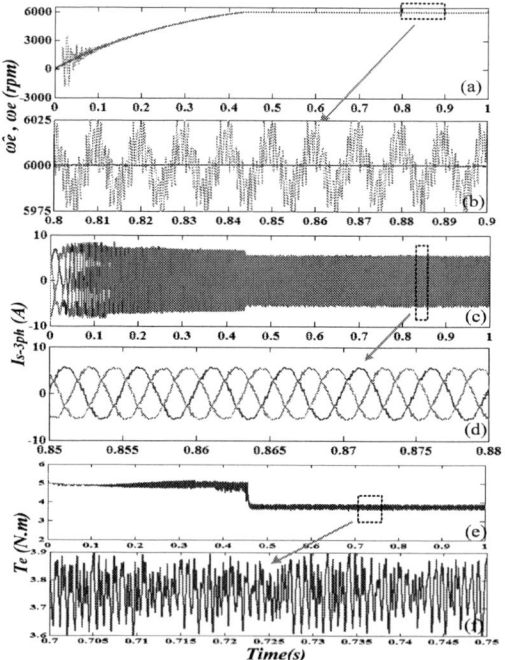

Fig. 10. Nominal torque step response of the improved predictive control simulation result.

Fig. 9 shows the comparison of the estimated rotor position with arc tangent function and with the AF at 6000 rpm with rated load. It can be seen from Fig. 11(a), the estimated rotor position contains the sixth harmonic component obviously. Note that the estimated position gets relatively smooth and the sixth harmonic component is eliminated. At steady state, the mean value of the position estimation error is less than 1%.

Figs. 10 show the speed, stator current and torque responses of the proposed PTC method, respectively, when a torque step is imposed. The torque step is set to its nominal value (3.7 N.m) while the load torque is rated load. The speed is kept close to the nominal value. The total harmonic distortion (THD) for steady state stator current is 4.3% and the mean error for speed estimation is 0.59%. As expected, torque ripple will be reduced predominantly if the proposed method is utilized. The torque ripple for proposed PTC is 8%.

IV. CONCLUSION

High performance applications require torque smoothness. Presence of harmonic components in PMA-SynRM back-EMF is a source of torque pulsations. In this paper, a sensorless predictive torque control of PMA-SynRM with nonsinusoidial back EMF (EMF) based on Extended Kalman Filter and Adaptive Filter has been proposed. The main goal of this paper is to accurately estimate the position and speed rotor and to reduce the torque ripple when FCS-MPC is implemented by means of the two-level inverter. The effectiveness of the proposed method has been verified accurately at a 2.3 kW PMA-SynRM sensorless drive. The simulation results demonstrate that the observer scheme including EKF, AF and PLL make an effective estimation of back EMF, the speed and position of the rotor. In both of nominal and low speed, the proposed method proved the promising results in the point of view of precise torque and speed control as well as the excellent dynamic response of the electromagnetic torque.

References

[1] P. Niazi, H. a. Toliyat, D. H. Cheong, and J. C. Kim, "A low-cost and efficient permanent-magnet-assisted synchronous reluctance motor drive," IEEE Trans. Ind. Appl., vol. 43, no. 2, pp. 542–550, 2007.

[2] F. Wang, Z. Zhang, S. Davari, R. Fotouhi, D. Arab Khaburi, J. Rodriguez, and R. Kennel, "An Encoderless Predictive Torque Control for an Induction Machine with a Revised Prediction Model and EFOSMO," IEEE Trans. Ind. Electron., vol. Early Acce, no. 12, pp. 6635–6644, 2014.

[3] E. Flach, "Improved algorithm for direct mean torque control of an induction motor," in Proc. PCIM, Nuremberg, Germany, 1998, pp. 261–267.

[4] M. Siami, S. A. Gholamian, M. Yousefi, "A Comparative Study Between Direct Torque Control and Predictive Torque Control for Axial Flux Permanent Magnet Synchronous Machines," Journal of Electrical Engineering, vol. 64, no. 6, pp. 346-353, Dec, 2013.

[5] M. Siami, D. A. Khaburi, A. Abbaszadeh and J. Rodriguez, "Robustness Improvement of Predictive Current Control Using Prediction Error Correction for Permanent Magnet Synchronous Machines,"

IEEE Trans. Ind. Electron., Accepted for publication, DOI: 10.1109/TIE.2016.2521734.

[6] R. Morales-caporal, M. Pacas, "A Predictive Torque Control Of Synchronous Reluctance Machine Taking Into Account the Magnetic Cross Saturation," IEEE Trans. Ind. Electron., vol. 54, no. 2, pp. 1161–1167, 2007.

[7] M. Siami, D. A. Khaburi, M. Yousefi and J. Rodriguez "Improved predictive torque control of a permanent magnet synchronous motor fed by a matrix converter," in Proc. 6th IEEE Power Electron., Drive Syst. Technol. Conf., 2015, pp. 369 – 374.

[8] S. A. Davari, D. A. Khaburi, and R. Kennel, "An improved FCS-MPC algorithm for an induction motor with an imposed optimized weighting factor," IEEE Trans. Power Electron., vol. 27, no. 3, pp. 1540–1551, 2012.

[9] X. Xiao and C. M. Chen, "Reduction of Torque Ripple Due to Demagnetization in PMSM Using Current Compensation," IEEE Trans. Appl. Supercond., vol. 20, no. 3, pp. 1068–1071, 2010.

[10] M. Tursini, "Adaptive sliding-mode observer for speed-sensorless control of induction motors," IEEE Trans. Ind. Appl., vol. 36, no. 5, pp. 1380–1387, 2000.

[11] S. Bolognani, L. Tubiana, and M. Zigliotto, "Extended Kalman filter tuning in sensorless PMSM drives," IEEE Trans. Ind. Appl., vol. 39, no. 6, pp. 1741–1747, 2003.

[12] L. Qian, D. A. Cartes, and H. Li, "An Improved Adaptive Detection Method for Power Quality Improvement," Industry Applications, IEEE Transactions on, vol. 44, no. 2. pp. 525–533, 2008.

[13] L. Tong, X. Zou, S. Feng, Y. Chen, Y. Kang, Q. Huang, and Y. Huang, "An SRF-PLL-based sensorless vector control using the predictive deadbeat algorithm for the direct-driven permanent magnet synchronous generator," IEEE Trans. Power Electron., vol. 29, no. 6, pp. 2837–2849, 2014.

[14] A. K. Chakali, H. a. Toliyat, and H. Abu-Rub, "Observer-based sensorless speed control of PM-assisted SynRM for direct drive applications," IEEE Int. Symp. Ind. Electron., no. 3, pp. 3095–3100, 2010.

[15] I. Boldea, C. I. Pitic, C. Lascu, G. D. Andreescu, L. Tutelea, F. Blaabjerg, and P. Sandholdt, "DTFC-SVM motion-sensorless control of a PM-assisted reluctance synchronous machine as starter-alternator for hybrid electric vehicles," IEEE Trans. Power Electron., vol. 21, no. 3, pp. 711–719, 2006.

[16] P. Guglielmi, M. Pastorelli, G. Pellegrino, and A. Vagati, "Position-Sensorless Control of Synchronous Reluctance Motor," Ind. Appl. IEEE Trans., vol. 40, no. 2, pp. 615–622, 2004.

[17] G. Wang, T. Li, G. Zhang, X. Gui, D. Xu, "Position Estimation Error Reduction Using Recursive-Least-Square Adaptive Filter for Model-Based Sensorless Interior Permanent-Magnet Synchronous Motor Drives," IEEE Trans. Ind. Electron., vol. 61, no. 9, pp. 5115–5125, 2014.

7th Power Electronics, Drive Systems & Technologies Conference (PEDSTC 2016)
16-18 Feb. 2016, Iran University of Science and Technology, Tehran, Iran

Implementation of Drirect Torque Control Method on Brushless Doubly Fed Induction Machines in Unbalenced Situations

Ali Ghaffarpour
Electrical Engineering Department
Sharif University of Technology
Tehran, Iran
Ghaffarpour_ali@ee.sharif.edu

Farhad Barati
Materials and Energy Research
Center
Karaj, Iran

Hashem Oraee
Electrical Engineering Department
Sharif University of Technology
Tehran, Iran

Abstract— This paper presents a new method for implementing the direct torque control (DTC) on the brushless doubly fed induction machines (BDFM) in presence of turbulence in voltage grid. It is supposed irrupts in three phase voltage grid suddenly. Effects of these events on closed-loop operation are inevitable and in some cases could lead to system instability. It presumed that grid voltages disturbance effect on back-to-back converter's outputs is negligible. Control of a complex machine is problematic. Hence, firstly has explained performance of the control method on BDFM, Proposed a new method for reducing turbulence effects of system operation afterward. Simulation results on BDFM model with prototype parameters in MATLAB/Simulink shows the validity of the presented method.

Keywords—Brushless Doubly Fed Induction Machine. Direct torque Control, Harmonic distortions, Total Harmonics Disturbance

Nomenclature

$PW/CW/RW$	Power/Control/Rotor Windings
P_1/P_2	The pole pair of PW/CW
T_e	Electromagnetic Torque
T_{ep}/T_{ec}	Electromagnetic Torque related to PW/CW
T_m	Mechanical Torque
ω_{a1}/ω_{a2}	Angular frequency related to PW/CW
ω_p/ω_c	Synchronies angular frequency of PW/CW
$\psi_p/\psi_c/\psi_r$	Flux vector of PW/CW/rotor
$V_p/V_c/V_r$	Voltages of PW/CW/rotor
$i_p/i_c/i_r$	Currents of PW/CW/rotor
$L_p/L_c/L_r$	Self- inductance of PW/CW/rotor
Rp / Rc / Rr	Resistance of PW/CW/rotor
L_{ph}/L_{ch}	Coupling inductances between PW/CW and rotor
V_{rp}/V_{rc}	Voltage of rotor induced by PW/CW
i_{rp}/i_{rc}	Current of rotor induced by PW/CW
k_p/k_c	Constants coefficients in the torque equation related to PW/CW
ψ_{rp}/ψ_{rc}	Rotor fluxes induced by PW/CW
ω_r	Angular speed of rotor
J	rotor specify inertia
$\dot{Z}$	Derivative of Z to time (dZ/dt)
$*$	Conjugate Operator
$Im\{z\}$	Imaginary part of z
$|z|$	Amplitude of z

I. INTRODUCTION

As illustrated in Fig 1, The Brushless DFIG, also known as the Brushless Doubly-Fed Machine (BDFM), is one of a class of doubly-fed generators. Doubly-fed generators are machines with two accessible three phase windings to/from which electrical power can be fed/extracted. Generally, one winding is connected to the mains or grid and therefore has a fixed frequency (Power Winding), and the other is supplied from a fractionally rated power electronics converter (Control Winding). Nested loop rotor structure that shown in Fig 2 gives important property to BDFM.

Thanks to the specific design of rotor, the CW can modify and control the rotor current which is being induced by PW. This is achieved by an electromagnetic cross-coupling effect between the two stator windings through the rotor [1].

The reduced system cost due to fractional size of the converter is an encouragement to electric drive manufacturers to utilize the BDFM. The brushless operation and thereby reduced manufacturing and operational costs as well as improved reliability have led to interests in the BDFM as a replacement for the doubly fed (slip-ring) induction generator

978-1-5090-0376-1/16 $31.00 © 2016 IEEE

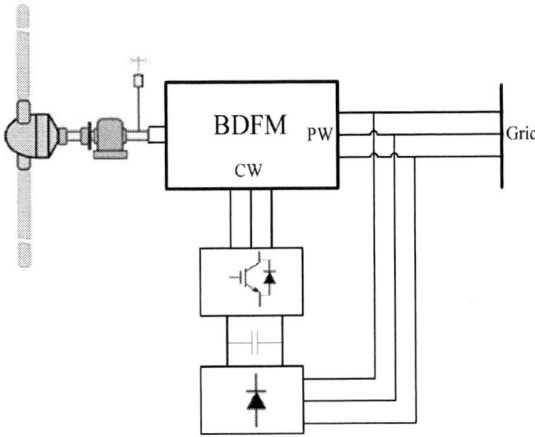

Fig. 1. BDFM as a wind field generator

Fig. 2. Nested loop rotor of BDFM

which is currently widely used in the wind power industry [2]. Compared to the doubly fed induction machine (DFIM), the BDFIM has the advantage of higher reliability due to the elimination of brush gears. This advantage will become even more significant as more wind turbines are built offshore [3]. The brushless doubly-fed machine can also be used as a motor in variable speed drive applications, making it the ideal choice in pumping, as well as many other applications. Therefore, the BDFIM shows great potential in some industrial applications, for example wind power generation and variable speed drives [4]. Furthermore, it shows commercial potential due to the relatively lower demand of the converter capacity when compared to conventional induction machines driven by fully rated converters. Currently, the research of the BDFIM focuses on BDFIM control and design optimization [5]

To date, there have been several attempts to propose control strategies on BDFM based on conventional methods that tested on induction machine previously. With respect to Vector Control, rotor-flux-oriented VC was achieved based on a dual-synchronous reference frame [3]. However, this control scheme suffers from complicated control structure. In other studies, the power-winding-flux-oriented VC has been implemented based on the BDFIM vector model in a common synchronous reference frame of the power winding, achieving new levels in the control of the BDFIM [4]

The DTC strategy for induction machines was first proposed by Depenbrocket and Takahashi et al., after the appearance of VC [6], [7]. Unlike VC, DTC does not require rotating coordinate transformations to achieve decoupled control of the flux linkage and torque. Additionally, DTC is not overly sensitive to machine parameters. Despite its simplicity, DTC permits good torque control under steady-state and transient operating conditions. Therefore, since its appearance, DTC has received much attention and interest [3].

Implementing DTC on BDFM suggested firstly in [8]. Other methods based on DTC proposed for BDFM control in [4],[9]. However there is no published research result for control of BDFM in unbalanced grid situation until now. Disruptive harmonics in power grid can led to interrupts in back-to-back

converter that is outside of this paper discussion and ignored.

II. BDFM MODEL

A static two-axis model can be obtained by generic reference frame d-q model that proposed in [9]. The generic reference frame d-q model is composed of two subsystems. Each subsystem contained a set of stator windings (PW or CW) and the corresponding rotor windings. The set of equations of the PW or CW subsystem are written in two different arbitrary reference frames with the Angular frequency ω_{a1} and ω_{a2}, which related to each pole-pair distribution. This leads to a couple of equations describing the dynamics of two independent rotor vectors which correspond to two different dynamics reference frames. If $\omega_{a1} = \omega_{a2} = 0$, the static model will obtain.

The flux equations, electromagnetic torque, and voltage equations could express by (1), (2) and (3) respectively.

$$
\begin{aligned}
\vec{\psi}_p &= L_p \vec{i}_p + L_{hp} \vec{i}_{rp} + L_{hp} e^{j\omega_N t} \vec{i}_{rc}^* \\
\vec{\psi}_c &= L_c \vec{i}_c + L_{hc} \vec{i}_{rc} + L_{hc} e^{j\omega_N t} \vec{i}_{rp}^* \\
\vec{\psi}_{rp} &= L_{hp} \vec{i}_p + L_r \vec{i}_{rp} \\
\vec{\psi}_{rc} &= L_{hc} \vec{i}_c + L_r \vec{i}_{rc} \\
\omega_N &= (p_1 + p_2)\omega_r
\end{aligned} \tag{1}
$$

$$
\begin{aligned}
T_e &= T_{ep} + T_{ec} \\
T_{ep} &= \frac{3}{2} P_p \operatorname{Im}\left\{ \vec{\psi}_p^* \vec{i}_p \right\} \\
T_{ec} &= \frac{3}{2} P_c \operatorname{Im}\left\{ \vec{\psi}_c^* \vec{i}_c \right\}
\end{aligned} \tag{2}
$$

$$\vec{V}_p = R_p \vec{i}_p + \frac{d\vec{\psi}_p}{dt}$$

$$\vec{V}_c = R_c \vec{i}_c + \frac{d\vec{\psi}_c}{dt} \qquad (3)$$

$$\vec{V}_{rp} = R_r \vec{i}_{rp} + \frac{d\vec{\psi}_{rp}}{dt} - jp_1\omega_r\vec{\psi}_{rp} = 0$$

$$\vec{V}_{rc} = R_r \vec{i}_{rc} + \frac{d\vec{\psi}_{rc}}{dt} - jp_2\omega_r\vec{\psi}_{rc} = 0$$

Mechanical equation of BDFM like all rotary machines, can expressed based on Newton's second law.

$$T_e - T_m = J\,\dot{\omega} \qquad (4)$$

BDFM model based on mentioned equations created in MATLAB/Simulink. Table 1 contains the parameters of model that are accordance with measured values of a prototype BDFM.

III. IMPELEMENTING THE DTC ON BDFM

To control the BDFM Torque in accordance with DTC algorithm the electromagnetic torque equation of machine must represents in (5) format. This equations congruous the theory that segregating rotor flux to two components. Fig.3 shows these components in BDFM air-gap. Applying each of 6 voltage vector may led to variation of the CW fluxes absolute or pharos. Thus a part of BDFM torque can controlled in a specific range [5].

$$T_{ep} = k_p |\psi_p||\psi_{rp}|\sin\delta_p \qquad (5)$$

$$T_{ec} = k_c |\psi_c||\psi_{rc}|\sin\delta_c$$

The closed loop system works successfully in normal situation. Response of this system to the step in reference torque is shown in Fig.5. As illustrated the BDFM torque trace the reference torque whereas CW fluxes amplitude is approximately constant. If mentioned system connected to an unbalanced system, can't work prosperity again. Fig.6. shows the operation of closed loop system in the unbalanced situation. Between seconds 0.2 and 0.3 simulated an unstable voltage grid for the system and electromagnetic torque has an uncontrolled manner. Total harmonic distortion of grid is shown in Fig 10.

TABLE I. PARAMETERS OF PROTOTYPE BDFM

Parameter	Value	Parameter	Value
PW pole-pairs	3	L_{ch}(mH)	4
CW pole-pairs	1	L_r(mH)	0.048
RW pole-pairs	4	$R_p(\Omega)$	3.2
Natural speed	750 rpm	$R_c(\Omega)$	5.32
PW rated Power	3 kw	$R_r(m\Omega)$	0.173
CW rated Power	1.5 kw	L_{sp}(mH)	292
L_{ph}(mH)	2.16	L_{sc}(mH)	642

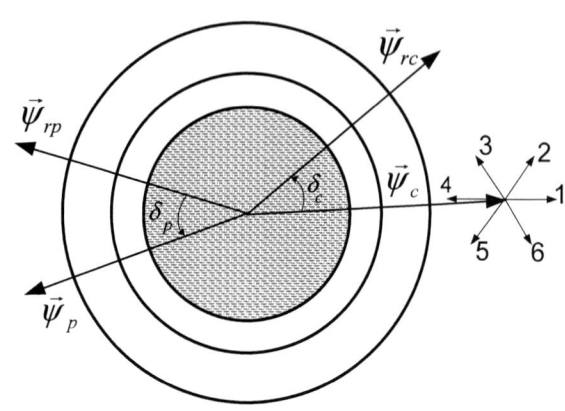

Fig. 3. The fluxes of BDFM components

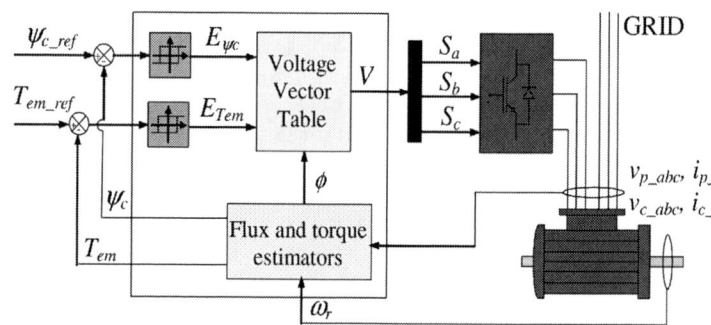

Fig. 4. Structure of evaluating DTC strategy on BDFM

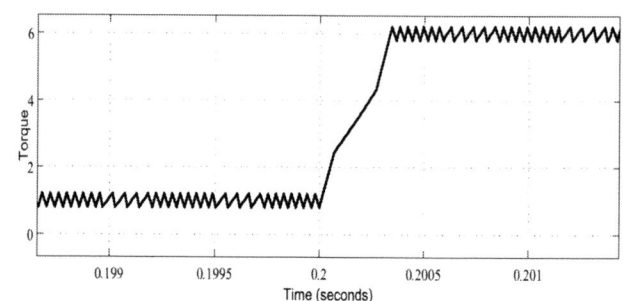

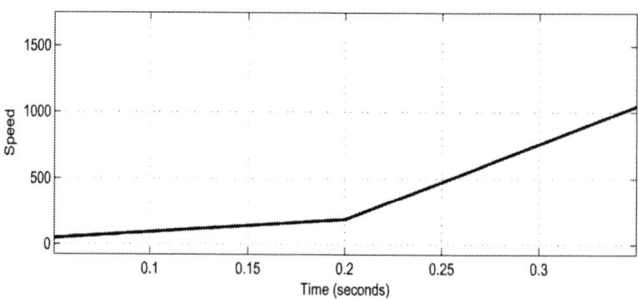

Fig. 5. Trque(N.m) and speed(rpm) step response of closed loop system with conventional DTC in normal situation

978-1-5090-0376-1/16 $31.00 © 2016 IEEE

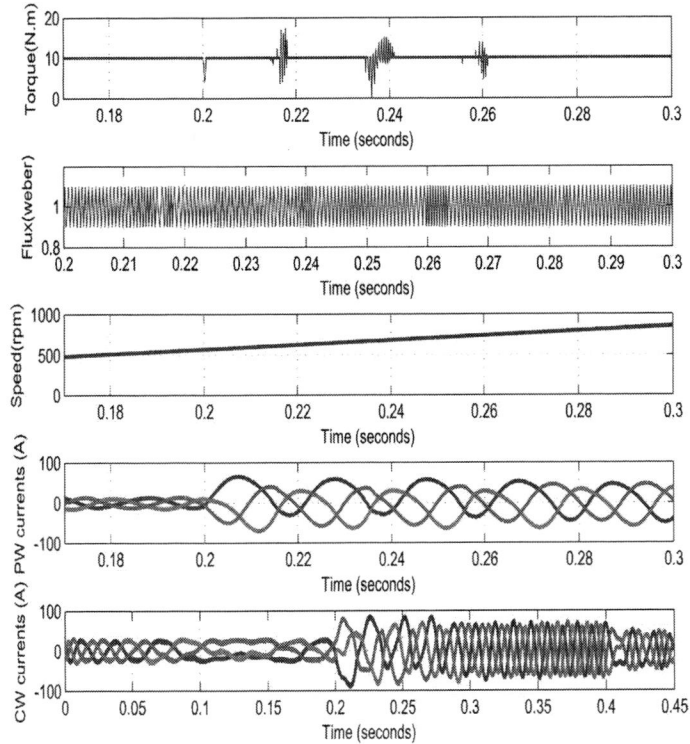

Fig. 6. Torque, CW flux, speed, PW currents and CW currents of closed loop system with conventional DTC in unbalanced situation

IV. PROPOSED DTC STRUCTURE

One of important problems of conventional DTC strategy is related to the dead time of switches. However, we haven't any control on hysteresis comparators outputs. Thus some signals that sent to switches are too closed and practically the switches can't operate ideally.

In proposed method, hysteresis bands of comparators are so tiny and they are sending switching signals to driver almost constantly. By sampling of these signals with specific frequency, dead time problem would solve spontaneously.

It seems obvious that the accuracy of systems output depends on the sampling frequency. Whatever this frequency increases the out-put electromagnetic torque and CW flux amplitude will trace the reference torque and reference flux respectively.

The total harmonic distortion, or THD, of a signal is a measurement of the harmonic distortion present and is defined as the ratio of the sum of the powers of all harmonic components to the power of the fundamental frequency which is 50 Hz here. When the main performance criterion is the "purity" of the original sine wave (in other words, the contribution of the original frequency with respect to its harmonics), the measurement is most commonly defined as the ratio of the RMS amplitude of a set of higher harmonic frequencies to the RMS amplitude of the first harmonic, or fundamental, frequency so in proposed method THD calculate according to (6).

$$THD = \frac{\sqrt{V_2^2 + V_3^2 + V_4^2 + V_4^2 + \cdots}}{V_1} \tag{6}$$

As illustrated in Fig 7, THD level assigns the sampling frequency. If the harmonic distortion of grid voltage increases, the controller leads will growth the frequency to reach better output signals to switches.

The frequency controller acts as a quantic function, in other words, the sampling frequency doesn't change permanently and have only 3 predesigned values, in our case 10 KHz, 20 KHz, 30 KHz.

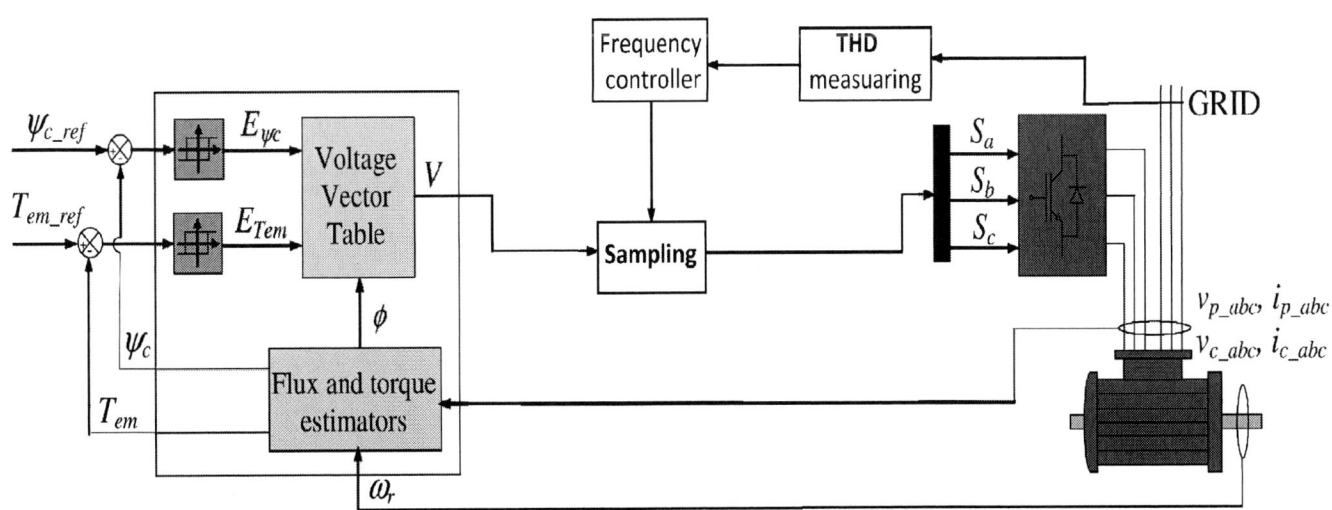

Fig. 7. Block diagram of proposed method direct torque control on BDFM

V. SIMULATION RESULTS

All simulation done on MATLAB/Simulink model with the parameters are mentioned in Table 1. First of all, simulation results of the proposed method on normal situations shown in Fig 8. Ripples in electromagnetic torque are decreased significantly by increase the sampling frequency. Fig 9 shows the voltage vectors that applied to BDFM in 10 KHz and 25 KHz.

The next step of simulation is to test the proposed method verses conventional method in same situation that the grid have some destructive harmonics. For the three phases of grid THD level has been shown in Fig 10. Simulation results of suggested method in this situation illustrated in Fig 11.

VI. CONCLUSION

This paper presented a structure of applying a standard electrical machine drive method (DTC) on a fairly complex machine (BDFM). The standard method has some problems in working on BDFM. However, presented a new structure for solve some of this problems. In this method a sampler in specific frequency gives the output signals of hysteresis comparators to switches. The sampling frequency controlled based on THD level of grid and by delimitate of controller output the dead time problem would solve spontaneously.

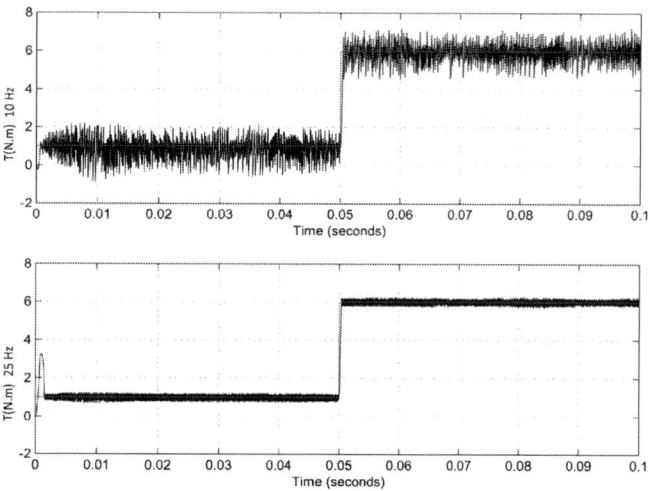

Fig. 8. Electromagnetic torque of BDFM in step response of closed loop system with conventional DTC in normal situation

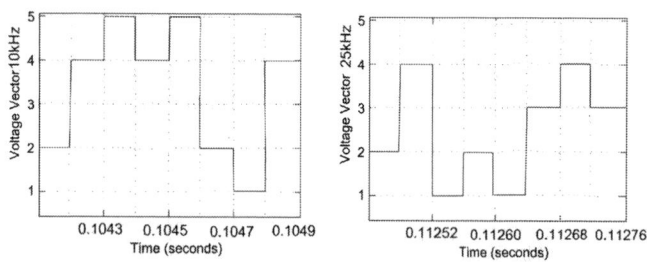

Fig. 9. Voltage vector numbers that applied to the machine the simulations with different frequencies

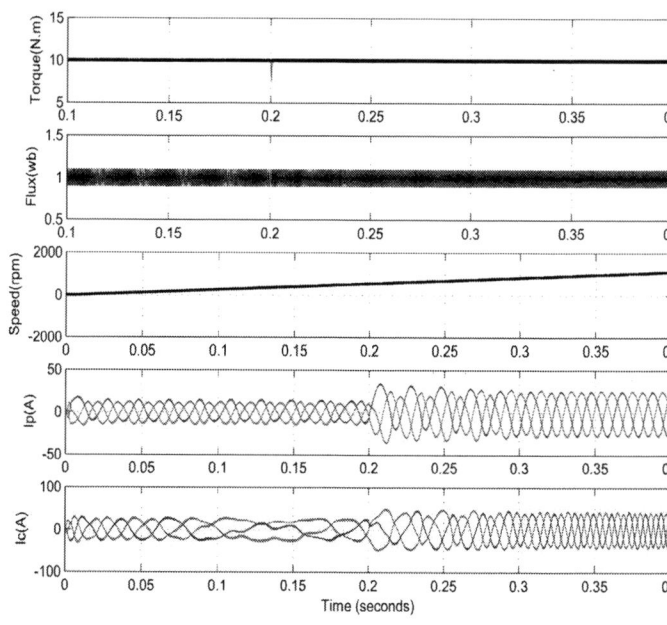

Fig. 10. THD level of phases in simulation

Fig. 11. Torque, CW flux, speed, PW currents and CW currents of closed loop system with proposed DTC in unbalanced situation

REFERENCES

[1] Roberts PC. A study of brushless doubly-fed (induction) machines. PhD. thesis, University of Cambridge, September 2004

[2] Barati F, Mohan R Mc, Shao S, Abdi E, and Oraee H. Generalized Vector Control for Brushless Doubly Fed Machines With Nested-Loop Rotor. IEEE transactions on industrial electronics, Vol. 60, No. 6, June 2013.

[3] R. Zhao, A Zhang; Y. Ma; X Wang, " The Dynamic Control of Reactive Power for the Brushless Doubly Fed Induction Machine With Indirect Stator-Quantities Control Scheme" IEEE Trans Power Electronics, vol. 30., 2015

[4] A. Zhang, X. Wang, W. Jia and Yun Ma "Indirect Stator-Quantities Control for the Brushless Doubly Fed Induction Machine" IEEE Trans Power Electronics, vol. 29. 3, March 2014

[5] A. Ghaffarpour, F. Barati and H. Oraee " Direct Torque control of Brushless Doubly Fed Induction Machine " The international confrance in new research of electrical engineering, Tehran, Iran, 2015

[6] Takahashi I, Noguchi T. A new quick-response and high efficiency control strategy of an induction motor. IEEE Trans Ind.Appl 1986; IA-22(5):820–7.

[7] Depenbrock, M.; Ruhr-Univ. Bochum, West Germany" Direct self-control (DSC) of inverter-fed induction" IEEE Transaction on Power Electronics, vol. 3, 1989

[8] W. Brassfield, Spee R, Habetler TG. Direct torque control for brushless doubly fed machines. IEEE Trans Ind Appl 1996;32(5):1098–104

[9] J. Poza, E. Oyarbide, and D. Roye, "New vector control algorithm for brushless doubly-fed machines," in Proc. IEEE 28th Annul. Conf. Ind.Electron. Soc., 2002, pp. 1138–1143.

[10] A.E. Fitzgerald, Jr. Charles Kingsley, and Stephen D. Umans. Electric Machinery. Electrical Engineering. McGraw-Hill, New York, 4 editions, 1983.

7th Power Electronics, Drive Systems & Technologies Conference (PEDSTC 2016)
16-18 Feb. 2016, Iran University of Science and Technology, Tehran, Iran

Diagnosis and Compensation of Amplitude Imbalance, Imperfect Quadrant and Offset in Resolver Signals

N. Noori
Dispatching Deputy
Iran Grid Management Company
Tehran, Iran
noori@igmc.ir

D. A. Khaburi
Electrical Engineering Department
Iran University of Science and Technology
Tehran, Iran
khaburi@iust.ac.ir

Abstract— **In this paper, a new method for detecting some errors in resolver signals is presented. Sin and cosine signals in resolver to digital converter's (RDC's) input might have errors, such as amplitude imbalance, imperfect quadrant or DC offset. Like other sensors some defects in mechanical structure and electrical or magnetic circuits cause to these errors in resolver output signals. The amplitude and phase of resolver output signals and consequently calculated angle are affected by these errors at any moment. Resolver output signals are more affected in their peak points than the other points. These peak points are easier and more accurate than the other points for analyzing, too. It is obvious that any error has unique effects on the signals. Therefore, by analyzing the effect on signals' peak points the error type and the value can be detected. To avoid wrong detection, error signal is defined, which detects the per unit signals deviation from the unit circle. Also, for error detection, there is no need to interrupt converter's operation. The proposed method is simulated in Matlab/Simulink environment, and the obtained results validate its effectiveness. Also, it is simulated on TMS320F2812 DSP using CCS 3.3, and the results are presented. Furthermore, by implementing the present RDC on motor controller DSP, it will be a low-cost approach.**

Keywords— *Error Detection, Error Compensation, Amplitude Imbalance, Imperfect Quadrant, DC Offset, Resolver.*

I. INTRODUCTION

Resolvers, as angular position sensors, are commonly used in electrical motor's control loop; they are also used in some industries, such as radars, robots, process automation, electrical brake system and military applications. They measure the absolute rotor angle and are able to work in industrial, harsh and noisy environments.

Usually a resolver to digital converter calculates the absolute rotor angle. Many hardware and software techniques have been proposed until now to perform this conversion. Some commercially available closed loop converters use the phase locked loop method [1]- [2]. Some other methods have been proposed to increase the accuracy of calculated angle. The software-based method proposed in [3] provides a method for calculating the rotor angle using angle tracking observer

(ATO). This method is implemented on the DSP which is used for motor control. So, this method is accurate an low-cost.

In practice, since there are some restrictions, the resolvers are not constructed ideally. For example windings are not distributed completely sinusoidal, or the magnetic structure of the rotor and stator are not perfectly uniform. Besides, during the installation process of resolver, acceleration fault may be occurred, or during the resolver operation, motor leakage flux can affect the resolver, or turn to turn short circuit may be occurred. These non-ideal characteristics cause to imperfect signals. Common faults in resolver output signals are amplitude imbalance, imperfect quadrant, harmonics, and DC offset.

In 1989, "D.C. Hanselman" analyzed the amplitude imbalance, imperfect quadrant, inductance harmonics, DC component imperfect cancelation and quadrant component imperfect rejection errors, and calculated their effects on RDC's output [4]. In a later research, he studied imperfect quadrant and inductance harmonics, and proposed some methods to reduce these errors [5]. These techniques require adding two new winding in resolver structure to reduce inductance harmonic effects. In 1995, "L. A. Knox" studied capacitance loading problem in long connector cables between transmitter and receiver resolvers [6]. In another study, an RDC was proposed and implemented, which was programmed to calculate the rotor angle and detect four faults [7].

In 2004, "A. Bunte" and "S. Binke" analyzed resolvers and sinusoidal encoders output's errors in complex numbers. As a result, they proposed a method to diagnose and remove these faults [8]. In another research, a resolver is used as a sensor in an electromagnetic brake system, and calibration method was used to reduce the resolver effects [9]. In 2009, "S. H. Hwang" and et al. analyzed the amplitude imbalance and imperfect quadrant errors and their effects on motor control process [10]. They proposed a method to compensate these errors using PMSM's direct component current in synchronous frame. Effects of PMSM's flux leakage on resolver's rotor and stator fluxes have been analyzed in [11]. They proposed a magnetic shield to remove motor leakage flux effects. Although, some improved methods have been presented by "K. Masaki" and et

978-1-5090-0376-1/16 $31.00 © 2016 IEEE

al. in 2000, [12], "D. W. Brown" in 2008, [13], "K. C. Kim" in 2010, [14], and "S. H. Hwang" in 2011.

In [15], the author has studied the effect of uneven harmonics, short circuit conditions in input and output windings, and rotor eccentricity. The VR resolver and conventional RDC are cosimulated, and resolver behavior is analyzed by 2D-FEM. In [16], the rotor position extracted from unbalanced resolver output signals are studied based on the quarter which they expected to be, and each fault is detected separately using the resolver position and resolver output signals.

Some mentioned methods require interrupting the process and then an operator detects the faults. Furthermore, some changes should be applied in RDC to compensate the errors. Some methods diagnose resolver errors by analyzing motor signals, so motor errors can affect the result, and some of them are so complicated and need high processing ability. The aim of this work is to eliminate these deficiencies. The proposed method in this paper is able to diagnose and compensate one or more errors; it has high accuracy; and RDC will diagnose and compensate faults intelligently, without interrupting resolver application.

II. RESOLVER

At first, an ideal resolver operation and the output signals production should be analyzed. Then, the effects of errors on these signals will be introduced. Resolver's structure is formed with two quadrant output windings (see Fig.1). The input and output signals are given by equations (1) and (2).

$$V_{ref} = V_m \cos \omega t \tag{1}$$

$$\begin{cases} V_{SIN} = V_1 \sin \theta \cos \omega t \\ V_{COS} = V_1 \cos \theta \cos \omega t \end{cases} \tag{1}$$

Where V_{ref} is the input excitation signal; ω (rad/sec) and V_m(V) are the angular velocity and amplitude of the input signal; V_{SIN} (V) and V_{COS} (V) are induced voltages in COS and SIN output windings; V_1 (V) is the amplitude of the induced output voltages and θ (rad) is the rotor angle, respectively.

When an AC current, directly or indirectly, is applied to these inductances, variable sinusoidal voltages are induced in output windings. These signals are encoded by a RDC to represent the position angle in digital form. In PLL method, a tracking angle (φ (rad)) traces the real angle. The difference between them will be very small, if the tracking angle is so close to the real angle. So, the shaft's angular position can be calculated continuously and without delay.

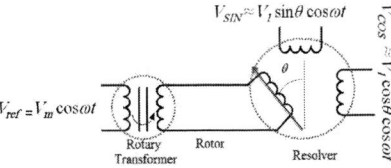

Fig. 1. Rotary transformer and brushless resolver Schematic [3].

A. Rotary Transformer Modeling

The equivalent circuit presented for rotary transformer is like single phase transformer's equivalent circuit. Considering the small rotor current value, the circuit, given below, can present the rotary transformer model. The values of rotary transformer parameters, presented in [17], are given in Table I. These values are obtained from several tests on a brushless resolver. Corresponding to Fig.2, the equations of rotary transformer model are:

$$V_r = L_{mt}(di/dt) \tag{3}$$

$$i = ((R_{st} + R'_{rt})V_{ref})/((R_{st} + R'_{rt})^2 + (L_{lst} + L'_{lrt} + L_{mt})^2 \omega^2)$$
$$- ((L_{lst} + L'_{lrt} + L_{mt})/((R_{st} + R'_{rt})^2 + (L_{lst} + L'_{lrt} + L_{mt})^2 \omega^2)).((dV_{ref})/dt). \tag{4}$$

Where V_r (V) and i (A) are the induced voltage in the rotor winding of rotary transformer and the input current. L_{lst} (H), R_{st}(Ω), L_{mt} (H), L_{lrt}' (H) and R_{rt}' (Ω) are stator leakage inductance and resistance, the mutual inductance, and the rotor leakage inductance and resistance in the stator side.

B. Brushless Resolver Modeling

Equations (5) to (8) are used to model a brushless resolver. Resolver's parameters are given in Table II [17]. By using the last section's model, V_r will be obtained; rotary transformer's output voltage is applied to the resolver. The rotor current can be calculated and the resolver's output signals will be obtained using following equations:

$$\begin{cases} V_r = R_r i_r + L_r(di/dt) \\ V_r = V'_r \cos(\omega t + \psi) \end{cases} \tag{5}$$

$$i = \frac{V'_r}{R_r^2 + (L_{rr}\omega)^2}[R_r \cos(\omega t + \psi) + (L_r\omega)\sin(\omega t + \psi)] \tag{6}$$

$$\begin{cases} L_{SIN} = L_0 + L_1 \sin \theta \\ L_{COS} = L_0 + L_1 \cos \theta \end{cases} \tag{7}$$

$$\begin{cases} V_{SIN} = (L_0 + L_1 \sin \theta)(di/dt) = V_0 \cos \omega t + V_1 \sin \theta \cos \omega t \\ V_{COS} = V_0 \cos \omega t + V_1 \cos \theta \cos \omega t \end{cases} \tag{8}$$

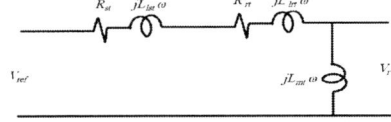

Fig. 2. Rotary transformer equivalent circuit [17].

TABLE I. ROTARY TRANSFORMER PARAMETERS

Parameter	Symbol	Unit	Value
Primary Series Resistance	R_{st}	Ohm	16.3
Primary Linkage Inductance	L_{lst}	mH	0.39
Mutual Inductance	L_{mt}	mH	4.9
Secondary Series Resistance	R_{rt}'	Ohm	12.98
Secondary Linkage Inductance	L_{lrt}'	mH	0.39

TABLE II.   RESOLVER PARAMETERS (TRANSFERRED TO STATOR SIDE)

Parameter	Symbol	Unit	Value
Stator Series Resistance	R_s	Ohm	40
Stator Linkage Inductance	L_{ls}	mH	0.2
Mutual Inductance	L_m	mH	2.089
Rotor Series Resistance	$R_r{}'$	Ohm	19
Rotor Linkage Inductance	$L_{lr}{}'$	mH	0.2

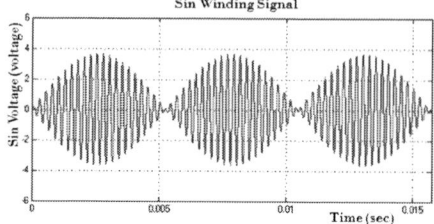

Fig. 3.   SIN output voltage.

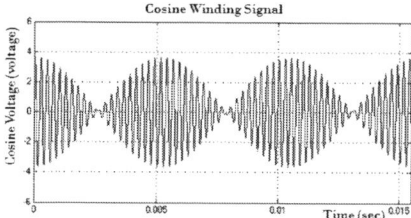

Fig. 4.   COS output voltage.

Where $V_r{}'$ (V) and ψ (rad) are the amplitude and phase of the rotor voltage; R_r (Ω) and L_{rr} (H) are the resistance and inductance of the rotor winding; L_{SIN} (H) and L_{COS} (H) are the mutual inductance between the rotor and SIN output windings and between the rotor and COS windings; L_0 (H) and L_1 (H) are the DC component and main component of the mutual inductance; and i (A) is the rotor current, respectively.

Equation (2) is an accurate approximation of resolver signals. An equivalent circuit for the brushless resolver is like transformer with two output windings; this circuit is used to model the resolver and its parameters are presented in Table II. Where L_{ls} (Ω), R_S (Ω), L_m (Ω), $L_{lr}{}'$ (H) and R_r (Ω) are the stator windings' leakage inductance and resistance, the mutual inductance amplitude between the rotor and stators windings, the leakage inductance and resistance of the rotor winding, respectively. Fig.3 and Fig.4 represent the output SIN and COS winding signals. The mentioned model and parameters are used to model the resolver and the rotary transformer.

III. RESOLVER FAULTS

In a real system, the resolver output signals have non-ideal characteristics, such as amplitude imbalance, imperfect quadrant, inductance harmonics, exciting signal's distortion and noise signals; position information is distorted significantly because of these non-idealities. These non-idealities might occur during operation, installation or construction of the resolver.

Among these errors, amplitude imbalance, offset and imperfect quadrant are more noticeable than others; so, these three faults are studied in this paper.

A. Offset

It was assumed that first terms in (8), the dc components, are canceled before processing output signals by RDC. If this component is not canceled perfectly, a position error will be generated. To study this error, RDC's input signals are written as below (they are represented after demodulation).

$$\begin{cases} V_{\sin} = \gamma + \sin\theta \\ V_{\cos} = \gamma + \cos\theta \end{cases} \tag{9}$$

Where V_{sin} (p.u.) and V_{cos} (p.u.) are the SIN and COS output voltages per unit after demodulation; the base voltage is V_1 and γ (p.u.) represents the remained DC component, which is not canceled, $\gamma=0$ and $\gamma=1$ are corresponding to perfect cancellation and no cancellation, respectively.

Where V_{sin} (p.u.) and V_{cos} (p.u.) are the SIN and COS output voltages per unit after demodulation; the base voltage is V_1 and γ (p.u.) represents the remained DC component, which is not canceled, $\gamma=0$ and $\gamma=1$ are corresponding to perfect cancellation and no cancellation, respectively.

B. Amplitude Imbalance

An amplitude imbalance is caused by an unbalance excitation of two output windings, non-equal output inductances, different turn ratios or nonlinear analogue devices such as Op-Amps and low-pass filters. Considering the amplitude imbalance, resolver's output signals are:

$$\begin{cases} V_{\sin} = \sin\theta \\ V_{\cos} = (1+\alpha)\cos\theta. \end{cases} \tag{10}$$

Where α (p.u.) presents the imbalance rate, and COS output contains the amplitude imbalance.

C. Imperfect Quadrant

When two output inductances are not exactly in $\pi/2$ radians phase difference, imperfect quadrant fault exists. Mathematical expression of this fault is presented in the following equations:

$$\begin{cases} V_{\sin} = \sin\theta \\ V_{\cos} = \cos(\theta + \beta). \end{cases} \tag{11}$$

Where β (rad) is the imperfect quadrant's angle value.

IV. FAULT DIAGNOSIS

Considering the importance of fault detecting and its compensating, it is necessary to have an accurate fault diagnosis. The diagnosing method proposed in this paper is based on the peaks value of the signals. To avoid incorrect fault diagnosis, the Error Signal is defined and used.

In the ideal condition, normalized output signals must trace unit circle. Actually the error signal reveals the deviation of normalized output signals from the unit circle. When this deviation becomes more than a defined threshold value, an error existence will be announced.

A. Fault Existence Detecting

V_{sin} and V_{cos} signals are checked continuously, as mentioned earlier. Equation (12) introduces the error signal. If the error signal exceeds a defined value, an error has been occurred.

$$error\ signal = |1 - (V_{sin}^2 + V_{cos}^2)| \tag{12}$$

B. Proposed Method

The signal's peaks are more affected than the other parts of signal by the mentioned faults. The peak of each signal occurs in the other signal's zero crossing point. Therefore, it is necessary to detect the signals' zero crossing point, and to sample the complementary signal's value at this time for fault diagnosis. This method diagnoses the errors' size and type regardless of rotor speed before fault occurrence, and just uses the SIN and COS signals.

C. Offset Diagnosis

A dc offset on the desired signal can be calculated by analyzing the signals' peaks. The difference value between the absolute value of the positive and negative peaks of each signal indicates twice its offset. In this case, there is no need to compare two SIN and COS signals, so, if both signals include offset error, both of them can be calculated.

$$\begin{cases} |\ positive\ peak\ | - |\ negative\ peak\ | = 2\ (offset\ amplitude) \\ \gamma = (|positive\ peak| - |negative\ peak\ |)/2 \qquad ,|\gamma| < 1 \end{cases} \tag{13}$$

D. Amplitude Imbalance Diagnosing

When an amplitude imbalance occurs, the peak value of the signal will differ from its ideal value. By comparing two signals' peaks, it will be detected that which one is reduced, and the size of amplitude imbalance can be calculated. (Considering the carried out researches, amplitude imbalance commonly means amplitude reduction.)

$$\alpha = -|(|(V_{sin})_{peak}| - |(V_{cos})_{peak}|)| \tag{14}$$

E. Imperfect Quadrant Diagnosing

To detect the imperfect quadrant fault, the sum and subtraction of two input signals are used. These two signals are quadrant even if imperfect quadrant fault occurs in the original signals, and their peaks will be as functions of "$\beta/2$". Following equations show how to extract the β value.

$$\begin{cases} \hat{V}_s = V_{sin} + V_{cos} \\ \hat{V}_c = V_{sin} - V_{cos} \end{cases} \tag{15}$$

$$\begin{cases} \hat{V}_{s\ peak} = 2\cos((\beta/2) + (\pi/4)) \\ \qquad = \sqrt{2} \times (\cos(\beta/2) - \sin(\beta/2)) \\ \hat{V}_{c\ peak} = -2\cos((\beta/2) - (\pi/4)) \\ \qquad = -\sqrt{2} \times (\cos(\beta/2) + \sin(\beta/2)) \end{cases} \tag{16}$$

$$\beta = (\hat{V}_{s\ peak} + \hat{V}_{c\ peak})/\sqrt{2} \tag{17}$$

Where $\hat{V}_s$, $\hat{V}_c$ (p.u.) are the sum and subtract of two original signal, respectively; $\hat{V}_{s\ peak}$ is the peak value of $\hat{v}_s$, and $\hat{V}_{c\ peak}$ is the peak value of $\hat{v}_c$. Usually the value of β is small, so the trigonometric approximations can be used to simplify the above equations and to calculate the value of β accurately.

In the proposed method, to detect the faults correctly, priorities are defined for faults. The first priority is for DC offset, then amplitude imbalance and finally the imperfect quadrant. Faulty signals are compensated by priority as well, and the compensated signals are used to detect the next fault. The last produced signals will be processed to calculate the rotor angular position.

V. COMPENSATION

A. Offset Compensation

To compensate the offset error, we can simply subtract the offset value (γ) from the under fault signal. For example, you can see the following equation.

$$\sin\theta = V_{sin} - \gamma = (\sin\theta + \gamma) - \gamma = \sin\theta \tag{18}$$

B. Amplitude Imbalance Compensation

If there is a detected amplitude imbalance fault, it can be compensated as below:

$$\sin\theta = V_{sin} \times \frac{1}{(1+\alpha)} = \frac{(1+\alpha)\sin\theta}{(1+\alpha)} = \sin\theta \tag{19}$$

C. Imperfect Quadrant Compensation

Compensation of imperfect quadrant is not as simple as amplitude imbalance and offset compensation. For this purpose, it is proposed to use the trigonometric equations and Taylor series approximation. β is small and using Taylor series approximation, we have:

$$\cos\theta = [V_{cos} + (\sin\theta)\beta]/(1 - \beta^2/2) \tag{20}$$

VI. SIMULATION RESULTS

In this section, simulation results in Matlab/Simulink environment for three mentioned faults are presented separately. Fig. 5, Fig. 6 and Fig. 7 show the ideal and actual signals, detected fault value, and the peaks of processed signals for 0.1 (p.u.) sin offset, 0.1 (p.u.) cosine amplitude imbalance and 0.1 (rad) imperfect quadrant occurred at 0.2 (sec), respectively. Fig. 8 shows these signal when all three errors occurred at 0.2 (sec).

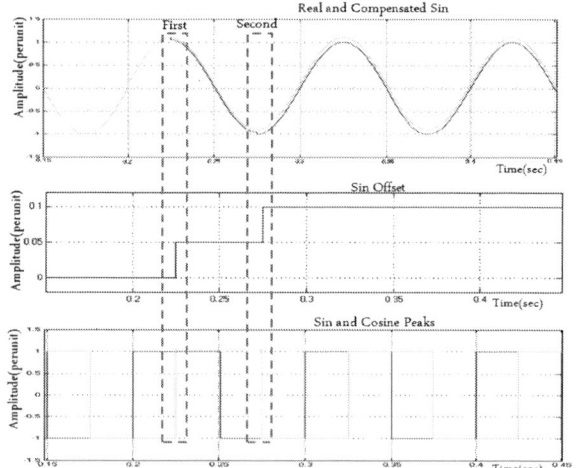

Fig. 5. The ideal and actual SIN signal before and after the error occurrence (the above figure), Detected sin offset error (the middle figure), and the sin and cosine peak's before and after fault occurrence. (Error is implied at 0.2 (sec) and the rotor speed is 20π (rad/sec)).

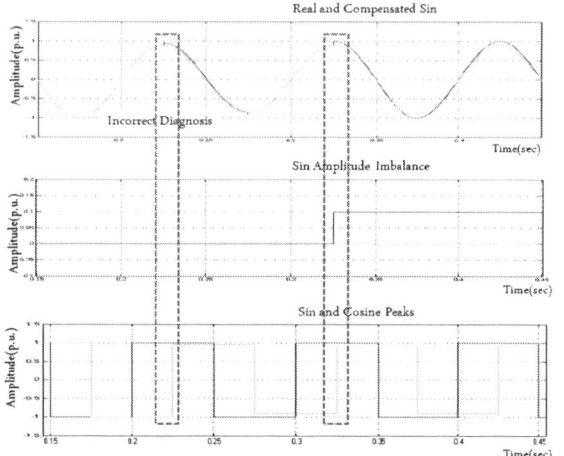

Fig. 6. The ideal and actual COS signals before and after implying 0.1 COS amplitude imbalance (the above figure), the value of COS amplitude imbalance detected by proposed algorithm (the middle figure), and the SIN and COS peaks' values. (Error is implied at 0.2 (sec) and the rotor speed is 20π (rad/sec))

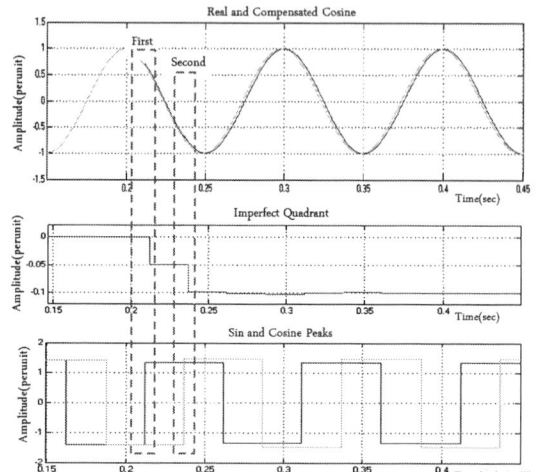

Fig. 7. The ideal and actual COS signals before and after applying 0.1 (rad) imperfect quadrant (the above figure), The value of detected imperfect quadrant error (middle figure), The peaks of $\cos(\theta-\pi/4)$ and $\sin(\theta-\pi/4)$ (the imperfect error occurred at 0.2 (sec) and the rotor speed is 20π (rad/sec)).

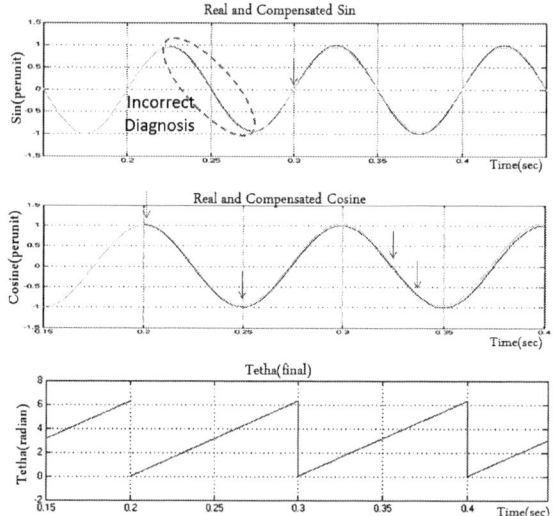

Fig. 8. The compensated and real SIN and COS signals before and after implying 3 faults: 0.05 SIN amplitude imbalance, 0.05 COS offset, 0.05 (rad) imperfect quadrant, the above figure and the middle figure, respectively, calculated rotor angle, the last figure.

This method is simulated in CCS 3.3 environment. The DSP which is used for simulation is TMS320F2812 series of Texas Instrument production. These simulation results are shown in Fig. 9, Fig. 10 and Fig. 11, for 0.1 (p.u.) sin offset, 0.1 (p.u.) sin amplitude imbalances and 0.1 (rad) imperfect quadrant, respectively. These faults were applied to the original signals one cycle after RDC had started to work.

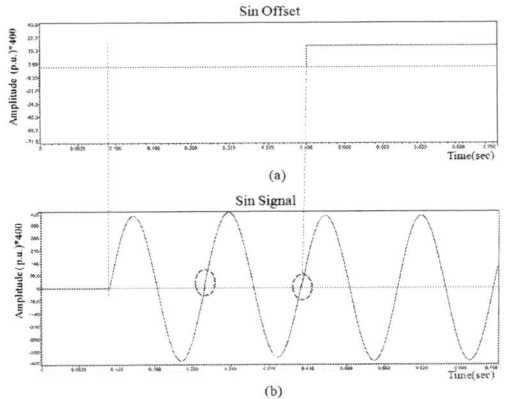

Fig. 9. Detected sin offset error (the above figure), the actual SIN signal before and after the error occurrence. (CCS 3.3 Simulation)

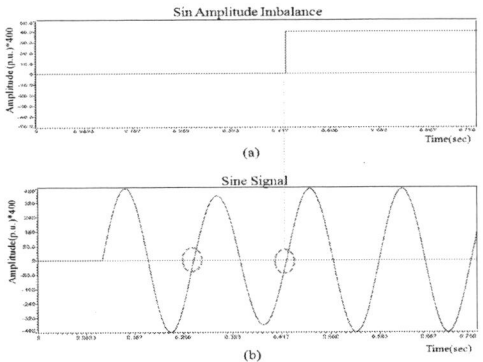

Fig. 10. a) Detected sin amplitude imbalance error (the above figure), b) The actual SIN signal before and after the error occurrence (CCS3.3 simulation)

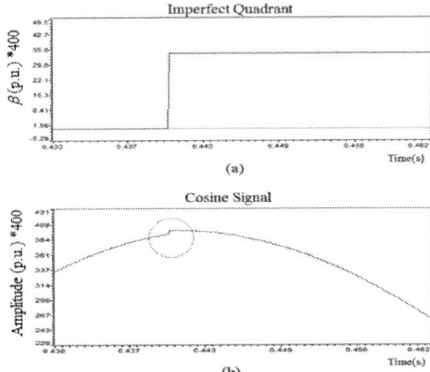

Fig. 11. a) The value of detected imperfect quadrant error (the above figure), b) The actual COS signals before and after compensating 0.1(rad) imperfect quadrant. (CCS 3.3 simulation)

VII. CONCLUSION

The proposed method is able to diagnose and compensate the resolver's common faults: DC offset, amplitude imbalance, and imperfect quadrant. The offset error can be detected for both signals. Furthermore, two or three and even four simultaneous error also can be detected and compensated accurately, as shown in Simulink results. It is noticeable that

compensation and diagnosing are implemented with defined priority. This RDC is low-cost because it can be implemented on motor controller DSP and is able to do these tasks intelligently, which is the outstanding property of this work.

REFERENCES

[1] C. Bruguier, G. Champenois and J. P. Rognon, "ID-Model Control of a Synchronnous Motor Without Position and Speed Sensor," in *IEEE Conference*, 1995.

[2] H. Madadi Kojabadi, L. Chang and R. Doriaswami, "Recent Progress in Sensoreless Vector-Controlled Introduction Motor Drives," in *IEEE, Proceeding of the Large Engineering System Conference Power Engineering*, 2002.

[3] D. Arab Khaboori, "Software-Based Resolver-to-Digital Converter for DSP-Based Drives Using an Improved Angle-Tracking Observer," *IEEE Trans. Instrumentation and Measurement*, vol. 61, no. 4, pp. 922-929, April, 2012.

[4] D. Hancelman, "Resolver Signal Requirements for High Accuracy Resolver-to-Digital Conversion," in *IEEE Conf. Industrial Electronics Society*, 1989.

[5] D. C. Hanselman, "Technique for Improving Resolver-to-Digital Conversion Accuracy," *IEEE Trans. on Industrial Electronics*, vol. 38, pp. 501-504, 1991.

[6] L. A. Knox, "Resolver Function Error Versus "R-C" Loading," *IRE Trans. Component Parts*, vol. 4, no. 1, pp. 44-60, 1955.

[7] A. Murray, B. Hare and A. Hirao, "Resolver Position Sensing System wiyh Integrated Fault Detection for Automotive Applications," in *Proceeding of IEEE, Sennsors*, 2002.

[8] A. Bünte and S. Beineke, "High-Performance Speed Measurement by Suppression of Systematic Resolver and Encoder Errors," *IEEE Trans. on Industrial Electronics*, vol. 51, no. 1, pp. 49-53, Feb. 2004.

[9] R. Hoseinnezhad, A. Bab-Hadiashar and P. Harding, "Calibration of Resolver Sensors in Electromechanical Braking Systems: A Modified Recursive Weighted Least-Squares Approach," *IEEE Trans. Industrial Electronics*, vol. 54, no. 2, pp. 1052-1060, 2007.

[10] S. H. Hwang, Y. H. Kwon, J. M. Kim and J. S. Oh, "Compensation of Position Error due to Amplitude Imbalance in Resolver Signals for PMSM Drives," *IEEE Journal of Power Electronics*, vol. 9, no. 5, pp. 1827-1831, Sep, 2009.

[11] K.-C. Kim, C. S. Jin and J. Lee, "Magnetic Shield Design Between Interior Permanent Magnet Synchronous Motor and Sensor for Hybrid Electric Vehicle," *IEEE Transaction on Magnetic*, vol. 45, no. 6, pp. 2835-2838, June, 2009.

[12] K. Massaki, K. Kitazawa, H. Mimura, M. Nirei, K. Tsuchimichi, H. Wakiwaka and H. Yamada, "Magnetic Field Analysis of a Resolver with a skewed and eccentric rotor," *Elsevier Trans. on Sensor and Actuators*, 2000.

[13] D. W. Brown, D. L. Edwards, G. Georgoulas, B. B. Zhang and G. J. Vachtsevanost, "Real-Time Fault Detection and Accommodation for COTS Resolver Position Sensors," in *IEEE International Conf. on Prognostics and Health Management*, 2008.

[14] K. C. Kim, S. J. Hwang, K. Y. Sung and Y. S. Kim, "A study on the Fault Diagnosis Analysis of Variable Reluctance Resover for Electric Vehicle," in *IEEE Conf. Sensors*, Korea, 2010.

[15] C. K. Ki, "Analysis on the Charateristics of Variable Reluctance Resolver Considering Uneven Magnetic Fields," *IEEE Transaction on Magnetics*, vol. 49, no. 7, pp. 3858-3861, 2013.

[16] J. J. Moon, H. J. Heo, W. S. Im and J. M. Kim, "Classification and Compensation of Amplitude Imbalance and Imperfect Quadrature in Resolver Signals," in *Power Electronics and Applications*, Lappeenranta, 2014.

[17] D. Arab-Khabur, F. Tootoonchian and Z. Nasiri-Gheidar, "Parameter Identification of a Brushless Resolver Using Charge Response of Stator Current," *IJEEE*, pp. 42-52, 2007.

7th Power Electronics, Drive Systems & Technologies Conference (PEDSTC 2016)
16-18 Feb. 2016, Iran University of Science and Technology, Tehran, Iran

Fixed-Point Computing with Variable Fraction Length for Motion Profile Generation

S.M. Dehghan
Faculty of Electrical and Computer Engineering,
Qom University of Technology, Qom, Iran

M. Mansourian
Faratavan Automation Company, Tehran, Iran

Abstract—This paper proposes a new fixed-point computing method for motion profile generation. When parameters have wide ranges, floating-point computing usually is used for accurate computation; however, this method increases execution time. In the proposed method, by considering input data, optimum fraction lengths are determined for different parameters. In this way, the fixed-point computing offers an accurate computation in a minimum execution time. Experimental results show the performance of the proposed method.

Keywords—Motion control; motion profile; fixed-point; floating-point.

I. INTRODUCTION

Motion control has a wide range of applications in industry, such as the machines for semiconductor manufacturing, automatic assembly of printed circuit boards, the digital mechatronic systems (e.g. hard disk drive and optical data storage systems) and CNC machining. A motion control system includes many functions such as position and velocity profile generation, interpolation, inverse kinematics, and proportional–integral–derivative (PID) controller; however generation of motion profile is a critical problem in the motion control area especially when s-curve acceleration/deceleration method is applied in motion profile to achieve fast and precise motion with the reduction of the residual vibration [1-9].

For precise control, floating-point computing is required which lead to long computing time especially when a fixed-point processor is used. On the other hand when a wide range acceleration and deceleration time is required, the fixed-point computing offers different accuracies for different acceleration and deceleration times that it can cause the error in the motion profile generation.

To meet the desired accuracy and speed, high-speed processors, multi-processor systems, floating-point-based processors, and the field-programmable gate array (FPGA) can be used in a motion control system. However these solutions can increase cost, complexity, size and power consumption [10-13].

Aim of this paper is to achieve a simple and low cost but fast and accurate motion control system. As a remedy, this paper proposes a fixed-point computing method with variable radix point. In the proposed method considering selected acceleration, deceleration and s-curve times, optimum placement of point for all parameters such as acceleration (positive and negative) and jerk is calculated, and then motion profile is generated.

The paper discusses the principles of motion profile first. Then floating-point and fixed-point methods are described in section III. The proposed method is presented in section IV. Finally the performance of the proposed method is experimentally validated in section V.

II. MOTION PROFILE

Fig. 1 shows the curves for a third order polynomial s-curve model. If maximum velocity (V_{max}), acceleration time (T_a), deceleration time (T_d), acceleration jerk time (T_{Ja}), and deceleration jerk time (T_{Jd}) are selected by the user, maximum acceleration (A_{max}), maximum deceleration (D_{max}), maximum acceleration jerk (J_{a-max}), and value maximum deceleration jerk (J_{d-max}) can be calculated by:

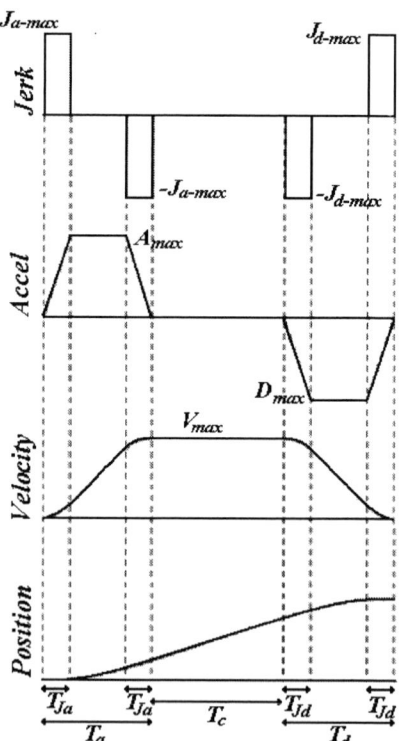

Fig. 1. S-Curve motion profiles.

$$A_{max} = V_{max} / (T_a - 2 T_{Ja}) \qquad (1)$$

$$D_{max} = V_{max} / (T_d - 2 T_{Jd}) \qquad (2)$$

$$J_{a\text{-}max} = A_{max} / T_{Ja} \qquad (3)$$

$$J_{d\text{-}max} = D_{max} / T_{Jd} \qquad (4)$$

In discrete domain, instantaneous acceleration (A), velocity (V) and position (P) are determined by:

$$A[n+1] = A[n] + J[n+1] \times T_S \qquad (5)$$

$$V[n+1] = V[n] + A[n] \times T_S + J[n+1] \times T_S^2 / 2 \qquad (6)$$

$$P[n+1] = P[n] + V[n] \times T_S + A[n] \times T_S^2 / 2 + J[n+1] \times T_S^3 / 6 \quad (7)$$

where J is instantaneous jerk, T_S is computing cycle time and n is discrete time index.

III. COMPUTING METHOD

There is a variety of ways to store and represent numbers on microprocessors. However floating-point and fixed-point formats are the standard format for decimal numbers [14]. Similar to scientific notation, in floating-point format a number multiplied by a base number raised to some power. 32-bit single-precision floating-point format is a floating-point representation defined by IEEE 754 [14]. This format consists of three parts: the sign bit, the exponent, and the fraction. The division of the three parts is shown in Fig. 2 and the number is determined by:

$$\text{number} = (-1)^{\text{sign}} \times (1 + \text{fraction} \times 2^{-23}) \times 2^{\text{exponent} - 127} \quad (8)$$

In the fixed-point format, a specific radix point is chosen (Fig. 3). The bits to the left of the radix point are called the integer bits. The bits to the right of the radix point are called the fractional bits. The number is determined by:

$$\text{number} = \text{integer} + \text{fractional} \times 2^{-m} \qquad (9)$$

where the m is length of fraction part. Advantage of the fixed-point format is simple arithmetic principles and consequently faster computation especially on processors without a floating-point unit (FPU). Disadvantages of the fixed-point format are limited range of values and different precise ratio in whole range. For example for an unsigned number with range value of 1 to 16383, allowable maximum fraction length is 18 bits in a 32-bit format. In this case precise is 2^{-18}, however precise ratio changes from 2^{-18} to 2^{-32} in whole range.

In the motion profile generation, when wide ranges for the acceleration and the jerk are required, using the conventional fixed-point representation leads to error in velocity and position values especially when small values are selected for acceleration and jerk times.

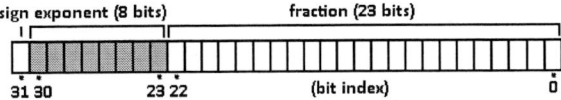

Fig. 2. Floating-point representation.

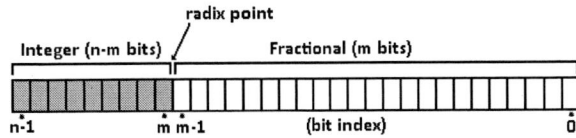

Fig. 3. Fixed-point representation.

IV. PROPOSED METHOD

Instead of the conventional fixed-point method, this paper proposes a variable radix-point fixed-point method. In the proposed method for critical parameters, an optimum fraction length is calculated. In the case of the motion profile, at first, fraction length for the time and the velocity and the position are determined (P_{Time}, $P_{Velocity}$ and $P_{Position}$). Then using (1)-(4), optimum fraction length for the maximum acceleration, the maximum deceleration, the maximum acceleration jerk and the maximum deceleration jerk are calculated:

A. Maximum acceleration (Fig. 4)

Maximum velocity is shifted to the left by sum of word length (B_{Word}) and fraction length of time (P_{Time}) and then is divided over $T_a - 2T_{Ja}$. Then, the obtained data is stored in a memory with two word length and number of unused bits in the left-side (B_{Unused}) is calculated. Finally the fraction length of the acceleration is calculated by:

$$P_{Acc} = P_{Unused} + P_{Velocity} + P_{Acc\text{-}max} - B_{Word} \qquad (10)$$

where $P_{Acc\text{-}max}$ is allowable maximum fraction length for the acceleration.

B. Maximum deceleration (Fig. 5)

Similar to the acceleration, optimum fraction length can be calculated for the deceleration. Where shifted maximum speed is divided over $T_d - 2T_{Jd}$ and the fraction length of the deceleration is calculated by:

$$P_{Dec} = P_{Unused} + P_{Velocity} + P_{Dec\text{-}max} - B_{Word} \qquad (11)$$

where $P_{Dec\text{-}max}$ is allowable maximum fraction length for the deceleration.

C. Maximum acceleration jerk (Fig. 6)

Maximum acceleration is shifted to left by sum of word length (B_{Word}) and fraction length of time (P_{Time}) and then is divided over T_{Ja}. Then obtained data is stored in a memory with two word length and number of unused bits in the left-side (B_{Unused}) is calculated. Finally the fraction length of the acceleration jerk is calculated by:

$$P_{Ja} = P_{Unused} + P_{Acc} + P_{Ja\text{-}max} - B_{Word} \qquad (12)$$

where $P_{Ja\text{-}max}$ is allowable maximum fraction length for the acceleration jerk.

978-1-5090-0376-1/16 $31.00 © 2016 IEEE

D. Maximum deceleration jerk (Fig. 7)

Similar to acceleration jerk, optimum fraction length can be calculated for deceleration jerk. Where shifted maximum deceleration is divided over T_{Jd} and the fraction length of the deceleration jerk is calculated by:

$$P_{Jd} = P_{Unused} + P_{Dec} + P_{Jd-max} - B_{Word} \qquad (13)$$

where P_{Jd-max} is allowable maximum the fraction length for the deceleration jerk.

The equations (5)-(7) should be rewritten for the proposed method:

$$A[n+1] = A[n] + (J[n+1] \times T_S) >> (P_{Time} + P_{Jerk} - P_{Acc/Dec}) \quad (14)$$

$$V[n+1] = V[n] + (A[n] \times T_S) >> (P_{Time} + P_{Acc/Dec} - P_{Velocity}) + \\ (J[n+1] \times T_S^2 / 2)) >> (2P_{Time} + P_{Jerk} - P_{Velocity}) \qquad (15)$$

$$P[n+1] = P[n] + (V[n] \times T_S) >> (P_{Time} + P_{Velocity} - P_{Position}) + \\ (A[n] \times T_S^2 / 2) >> (2P_{Time} + P_{Acc/Dec} - P_{Position}) + \\ (J[n+1] \times T_S^3 / 6) >> (3P_{Time} + P_{Jerk} - P_{Position}) \qquad (16)$$

where $P_{Acc/Dec}$ and P_{Jerk} is determined considering the sign of the acceleration and the jerk using the algorithm shown in Fig. 8.

$$Temp1 = \frac{V_{max} << (B_{Word} + P_{Time})}{T_a - 2T_{Ja}}$$
$$Temp2 = 1 << (2B_{Word} - 1)$$

$for(B_{Unused} = 0; B_{Unused} < (B_{Word} - P_{Speed}); B_{Unused} + +)$
$\quad if((Temp1 \ \& \ Temp2) == 0) \ Temp2 = Temp2 >> 1$
$\quad else \ break$
end

$$A_{max} = Temp1 >> (2B_{Word} - B_{Unused} - P_{Acc-max})$$
$$P_{Acc} = B_{Unused} + P_{Speed} + P_{Acc-max} - B_{Word}$$

Fig. 4. Algorithm for calculation of optimum fraction length of acceleration.

$$Temp1 = \frac{V_{max} << (B_{Word} + P_{Time})}{T_d - 2T_{Jd}}$$
$$Temp2 = 1 << (2B_{Word} - 1)$$

$for(B_{Unused} = 0; B_{Unused} < (B_{Word} - P_{Speed}); B_{Unused} + +)$
$\quad if((Temp1 \ \& \ Temp2) == 0) \ Temp2 = Temp2 >> 1$
$\quad else \ break$
end

$$D_{max} = Temp1 >> (2B_{Word} - B_{Unused} - P_{Dec-max})$$
$$P_{Dec} = B_{Unused} + P_{Speed} + P_{Dec-max} - B_{Word}$$

Fig. 5. Algorithm for calculation of optimum fraction length of deceleration.

$$Temp1 = \frac{A_{max} << (B_{Word} + P_{Time})}{T_{Ja}}$$
$$Temp2 = 1 << (2B_{Word} - 1)$$

$for(B_{Unused} = 0; B_{Unused} < (B_{Word} - P_{Acc}); B_{Unused} + +)$
$\quad if((Temp1 \ \& \ Temp2) == 0) \ Temp2 = Temp2 >> 1$
$\quad else \ break$
end

$$J_{a-max} = Temp1 >> (2B_{Word} - B_{Unused} - P_{Ja-max})$$
$$P_{Ja} = B_{Unused} + P_{Acc} + P_{Ja-max} - B_{Word}$$

Fig. 6. Algorithm for calculation of optimum fraction length of acceleration jerk.

$$Temp1 = \frac{D_{max} << (B_{Word} + P_{Time})}{T_{Ja}}$$
$$Temp2 = 1 << (2B_{Word} - 1)$$

$for(B_{Unused} = 0; B_{Unused} < (B_{Word} - P_{Acc}); B_{Unused} + +)$
$\quad if((Temp1 \ \& \ Temp2) == 0) \ Temp2 = Temp2 >> 1$
$\quad else \ break$
end

$$J_{d-max} = Temp1 >> (2B_{Word} - B_{Unused} - P_{Jd-max})$$
$$P_{Jd} = B_{Unused} + P_{Dec} + P_{Jd-max} - B_{Word}$$

Fig. 7. Algorithm for calculation of optimum fraction length of deceleration jerk.

$if (A[n] > 0 \ || \ (A[n] == 0 \ \&\& \ J[n+1] >= 0))$
$\{$
$\qquad P_{Acc/Dec} = P_{Acc}$
$\qquad P_{Jerk} = P_{Ja}$
$\}$
$else$
$\{$
$\qquad P_{Acc/Dec} = P_{Dec}$
$\qquad P_{Jerk} = P_{Jd}$
$\}$

Fig. 8. Algorithm for calculation of $P_{Acc/Dec}$ and P_{Jerk}.

V. EXPERIMENTAL RESULTS

To verify the validity of the proposed method, the digital control board of Faratavan Universal Servo Drive (FUSD) was used for the implementation of the floating point, the conventional fixed-point and the proposed fixed-point methods. The digital control board includes two fixed-point digital signal processors (TI TMS320F2810) for motor control and position control and an Altera Cyclone II FPGA for peripheral and communication handling (Fig. 9). The motion control algorithm was implemented in the position control processor. Generated profiles were transferred to personal computer using Ethernet link of the control board and monitored and stored by Faratavan Fast Monitoring software.

The table I shows the parameters for the fixed-point methods. Two scenarios were considered for experimental tests which can be seen in Table II. First scenario leads to large value for acceleration and jerk, however these values in the second scenario are small. Fraction bits in the fixed-point methods during these scenarios are shown in Table III. Fraction lengths for the conventional method were selected in a manner to support the values in the both tests.

Fig. 10 and Fig. 11 show the curves calculated by the DSP for the Test 1 and the Test 2, respectively. The curves for all three methods are almost the same. However there is a little difference between the values, especially in the position values (Table IV). While the final position in the Test 1 for all three methods is equal to the expected value, the final position in the Test 2 has error. Produced error in the conventional fixed point method is considerable (-2182 unit), while errors in floating point and proposed fixed point method are 78 and -152 units, respectively. Table IV also shows maximum execution time for different methods. It can be seen that execution time of the floating point method is four times more than of the fixed point methods (39.6 vs. 7.6 μs)

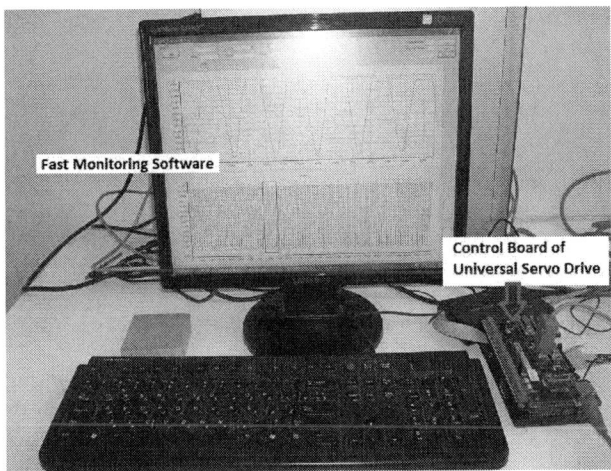

Fig. 9. Experimental set-up for motion profiles generation.

TABLE I. EXPERIMENTAL PARAMETERS FOR THE FIXED-POINT METHODS.

Parameters	Length (bits)
B_{Word}	32
P_{Speed}	15
P_{Time}	10
$P_{Acc-max}$	31
$P_{Dec-max}$	31
P_{Ja-max}	31
P_{Jd-max}	31

TABLE II. EXPERIMENTAL PARAMETERS FOR TEST1 AND TEST2.

Parameters	Value	
	Test1	Test2
V_{max}	50000	500
T_a	10	900
T_d	10	100
T_{Ja}	2	180
T_{Jd}	2	20
T_c	10	500

TABLE III. FRACTION BITS IN THE FIXED-POINT METHODS.

Parameters	Length (bits)			
	Proposed Fixed Point		Conventional Fixed Point	
	Test1	Test2	Test1	Test2
P_{Acc}	18	28	18	18
P_{Dec}	18	25	18	18
P_{Ja}	19	31	19	19
P_{Jd}	19	29	19	19

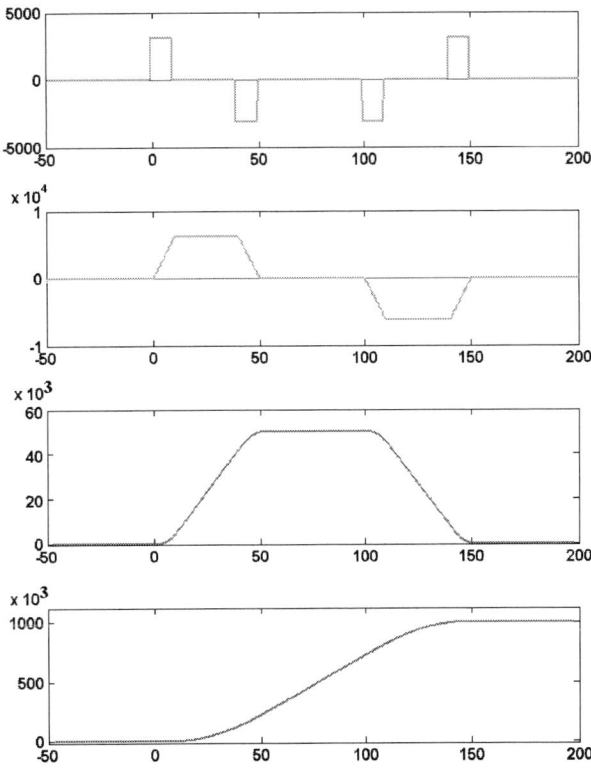

Fig. 10. Experimental s-curve motion profiles for Test1. (Jerk (Top), Acceleration, Velocity, Position (Bottom)).

I. CONCLUSION

A fixed-point computing method with variable fraction bits for motion profile generation was proposed in this paper. When a wide range is considered for the parameters such as the acceleration and deceleration times in a motion control application, using the fixed-point computation decreases accuracy and using the floating-point computation increases computation time. In the proposed method, considering input data for acceleration and deceleration times and maximum velocity, optimum fraction lengths are determined for positive and negative accelerations and acceleration and deceleration jerks. Experimental results showed while execution time of the proposed fixed point method is as low as the conventional fixed-pint method, its computation error is almost similar to that of the floating point method.

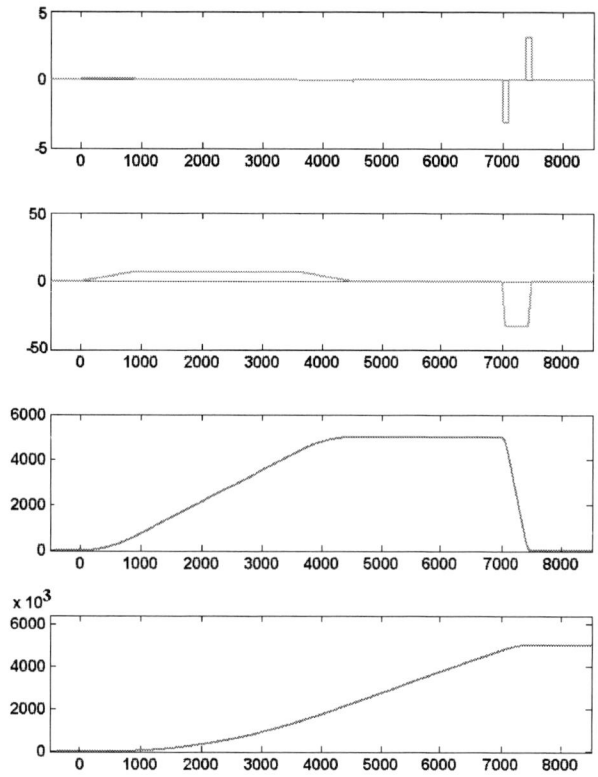

Fig. 11. Experimental s-curve motion profiles for Test2. (Jerk (Top), Acceleration, Velocity, Position (Bottom)).

TABLE IV. FINAL POSITION AND MAXIMUM EXECUTION TIME IN DIFFERENT METHODS.

Method	Final Position		Maximum Execution Time [μs]
	Test1	Test2	
Expected	1000000	5000000	-
Floating Point	1000000	5000087	39.6
Conventional Fixed Point	999999	4997818	7.84
Proposed Fixed Point	999999	4999848	7.84

REFERENCES

[1] Kim Doang Nguyen, I-Ming Chen, Teck-Chew Ng, "Planning Algorithms for S-curve Trajectories," *IEEE/ASME International Conference on Advanced Intelligent Mechatronics, Zurich, Switzerland,* 2007.

[2] Chang-Wan Ha, Keun-Ho Rew, Kyung-Soo Kim, and Soohyun Kim, "Tuning the S-curve motion profile in short distance case," *American Control Conference (ACC), Washington, DC, USA,* 2013.

[3] Sang-Yun Lee, Chang Soon Kang, Chang-Ho Hyun, and Mignon Park, "S-curve profile switching method using fuzzy system for position control of DC motor under uncertain load," *International Conference on Control, Automation and Systems (ICCAS), Kintex, Korea,* 2012.

[4] A. Shahzadeh, A. Khosravi, S. Nahavandi, "Path Planning for CNC Machines Considering Centripetal Acceleration and Jerk," *IEEE International Conference on Systems, Man, and Cybernetics (SMC), Manchester, UK,* 2013.

[5] Keun-Ho Rew and Kyung-Soo Kim, "Using Asymmetric S-Curve Profilefor Fast and Vibrationless Motion," *International Conference on Control, Automation and Systems, Seoul, Korea,* 2007.

[6] Kuijing Zheng , Li Cheng, "Adaptive S-curve Acceleration/Deceleration Control Method," *World Congress on Intelligent Control and Automation, Chongqing, China,* 2008.

[7] Chang-Wan Ha, Keun-Ho Rew and Kyung-Soo Kim "A Complete Solution to Asymmetric S-curve Motion Profile: Theory & Experiments," *International Conference on Control, Automation and Systems, Seoul, Korea,* 2008.

[8] Keun-Ho Rew, and Kyung-Soo Kim, "A Closed-Form Solution to Asymmetric Motion Profile Allowing Acceleration Manipulation," *IEEE Transactions on Industrial Electronics,* vol. 57, no. 7, July 2010.

[9] Fengshan Zou, Daokui Qu, Fang Xu, "Asymmetric S-curve Trajectory Planning for Robot Point-to-point Motion," *International Conference on Robotics and Biomimetics, Guilin, China,* 2009.

[10] U. Komin, K. Tanta-ngai, "DSP-Based Motion Controller development for Milling Machine," *SICE Annual Conference, Tokyo, Japan,* 2011.

[11] Jung Uk Cho, Quy Ngoc Le, and Jae Wook Jeon, "An FPGA-Based Multiple-Axis Motion Control Chip," *IEEE Transactions on Industrial Electronics,* vol. 56, no. 3, 2009.

[12] Ying-Shieh Kung, Ping-Hang Huang, Fong-Chin Su, and Tain-Song Chen, "Realization of an FPGA-based Motion control system for electric standing wheelchairs," *IEEE Symposium on Industrial Electronics and Applications (ISIEA), Langkawi, Malaysia,* 2011.

[13] Jun Tang, and Xiaojuan Zhao, "Research on Software Design of DSP Motion Controller," *International Conference on MEMS, NANO, and Smart Systems (ICMENS), Dubai, UAE,* 2009 .

[14] "Comparing Floating-Point and Fixed-Point Implementations on ADI Blackfin Processors with LabVIEW," *National Instrument Company,* 2012, http://www.ni.com/white-papers/

7th Power Electronics, Drive Systems & Technologies Conference (PEDSTC 2016)
16-18 Feb. 2016, Iran University of Science and Technology, Tehran, Iran

A New Hardware Device to Simulate the Movement of Electric Train Wheel on Rail

Sajad Sadr
Iran University of Science and
Technology
Tehran, Iran
sajadsadr@iust.ac.ir

Davood Arab Khaburi
Iran University of Science and
Technology
Tehran, Iran
khaburi@iust.ac.ir

M. Rivera
Universidad de Talca
Curico, CHILE
marcoesteban@gmail.com

Abstract—In all of the control speed systems one of the main factor to analysis the performance of the system, is the speed control in the acceleration mode. The enhancing speed control in acceleration mode is like magnifying the brake performance. The goal of both speed control system in acceleration mode and braking system, is reducing the time duration of speed changes. Braking control systems is common between, cars, trucks, motorcycles, trains, etc. For this reason several project to improve the performance of the braking systems have been done. The most famous of this advances is Anti-lock Braking System (ABS). The base of ABS, is using the maximum of the friction between tire and road surface. In order to reducing energy consumption, train wheels and rails are built from iron. Since wheel and rails are built from iron, the friction between them is weak. Hence when the propulsion system tries to push forward the train, the driven will began to rotation and the speed of the driven wheel is more than the train actual speed. The name of this phenomena is wheel slip. As described in this paper wheel slip has a significant impact on the adhesion between wheel and rail. In the other hand train acceleration is function of the adhesion between wheel and rail. Therefore in order to improve the train acceleration, the maximum of the adhesion should be utilized. For these reasons in this paper a hardware simulator to implement and test the adhesion control methods is proposed.

Keywords—adhesion coefficient; electric drive; electric train; hardware simulator ; wheel slip

NOMENCLATURE

θ	Slop angle [rad]
λ	Wheel slip
λ_m	Wheel slip of maximum adhesion coefficient
μ	Adhesion coefficient
μ_m	Maximum adhesion coefficient
τ_a	Adhesion torque
τ_m	Driving torque [N.m]
ω_{wh}	Angular speed of train driven wheel [rad/s]
F_a	Adhesion force [N]
F_m	Driving force [N]
F_r	Train rolling resistance [N]
g	Earth gravity [m/s²]
J	Moment of inertia [kg.m²]
M	Mass of train [kg]
N	Normal force [N]

r	Wheel radius [m]
t	Time [sec]
v_s	Slip speed [m/s]
v_t	Train actual speed [m/s]
v_{wh}	Speed of train driven wheel [m/s]

I. INTRODUCTION

The main advantage of rail transportation is the low friction between wheels and rails [1, 2]. By a constant energy because of this low friction, the train can move more than a car. Because the friction between wheels and rails is much lower than the friction between car tires and the road asphalt surface. On the other hand this low friction in the railway leads to wheel slip and wheel skid phenomenon. Wheel slip occurs in the train acceleration while wheel skid takes place in the train braking. Urban trains compare to inter-city trains have more number of start-up and brake. So importance of slip/skid control in urban trains is more prominent than inter-city trains. Also high speed trains must reach to their reference speed in the minimum time. As shown in this article wheel slip control has a significant impact on the train acceleration. Therefore high speed trains need slip control system too. Wheel slip control in two field can be interesting for researchers: A. wheel slip control when the train is in the braking mode. B. wheel slip control when the train is in the acceleration mode.

A. Wheel Slip Control When the Train is in the Braking Mode

Wheel slip control in the braking mode is common between trains and cars. Because in the both of these two types of vehicle, brake acceleration is so important and wheel slip control has a significant impact on the brake acceleration. Hence extensive research on wheel slip control systems in the braking mode of the vehicles have been done [3-11]. Some of the main research in this field refer to Anti-lock Braking System (ABS) in all types of vehicles [3-5], nonlinear control and robust control of the wheel slip in the braking mode [6], separate controls for front and rear wheels during braking [7], adaptive control in the braking mode [8] and fuzzy control in the Anti-lock Braking Systems [9]. For wheel slip software modeling in the braking mode there are two models. First of them is a static model based on Beam model which its application is mostly in modeling cars braking systems. The second one is dynamic model based on Bristle model which its application is mostly modeling trains braking systems [10-11].

978-1-5090-0376-1/16 $31.00 © 2016 IEEE

For studying on the wheel slip control in the baking mode a laboratory test bench has been presented in [12]. This valuable hardware device is only designed for investigation on braking methods of the trains and cannot be used for studying on the acceleration mode of the train.

B. Wheel Slip Control When the Train is in the Acceleration Mode

One of the factor that has significant impact on the adhesion coefficient between wheels and rails is the wheel slip. The train acceleration is a function of the adhesion coefficient. So changes in the wheel slip cause changes in the train acceleration. Hence some study on the wheel slip control have been done. In many of these study, to reduce the acceleration time, their suggestion is preventing wheel slip [13-17]. The base of these system in acceleration mode is like Anti-lock Braking Systems in the braking mode. By prevention the wheel slip, can be guaranteed that the wheelspin could not occur. But as will be shown the maximum of the acceleration achieves at an indefinite value of the wheel slip. In addition of the wheel slip, the adhesion coefficient is affected by other factors like physical conditions of wheels and rails [1, 2 and 18]. Therefore other effective methods to improving the train acceleration can be like enhancing the adhesion coefficient by spraying sands on the rail [19], or spraying special liquid on the rail [20]. In the method of spraying sands on the rail, size and substance of the sands have the main impact on the adhesion coefficient improvement [19].

C. Paper Contribution and Organization.

As will be shown, the wheel slip has a significant impact on the train acceleration. As described in part "A" of introduction, a valuable laboratory test bench to study on the braking mode of the train has been presented in [12]. In the contrary of the braking mode there is no hardware device to study on the wheel slip control in the acceleration mode. Therefore the contribution of this paper is:

• Introducing a hardware device which simulates the movement of electric train wheel on the rail. This hardware is designed to study on the adhesion between wheel and rail.

The structure of the paper continuation is as follows: in Section II, the wheel slip and adhesion phenomenon has been described. Section III, is dedicated to equations on train movement on rail. In Section IV, the simulator of the train movement on the rail has been presented, and its structure has been described. By some experimental test on the proposed simulator, principles of the device performance is described in Section V. And at the end, Section VI, belongs to the paper conclusion.

II. THE WHEEL SLIP AND ADHESION CONCEPTS

Before describing the structure and performance of the proposed hardware system, it's necessary to define the concepts of wheel slip and adhesion. For this propose, suppose a simple model of train movement, like Fig. 1. Train speed (v_t) is the train actual speed. Concept of slip speed (v_s) is defined as difference between train speed (v_t) and linear speed of train driven wheel (v_{wh}):

$$v_s = v_{wh} - v_t \qquad (1)$$

Linear speed of the train driven wheel is obtained by multiplying the rotation speed of the driven wheel (ω_{wh}) in driven wheel radius (r):

$$v_{wh} = r.\omega_{wh} \qquad (2)$$

Wheel slip (λ) is defined as result of dividing the slip speed to the train actual speed:

$$\lambda = \frac{v_s}{v_t} = \frac{v_{wh} - v_t}{v_t} \qquad (3)$$

Adhesion coefficient (μ) can be defined as the surface friction coefficient between train wheel and rail, in the contact place. In fact, the reason of train movement on the rail is the adhesion between wheels and rail. In other words, when the driven wheel is rotated by the motor torque, the driven wheel tries to push back the rail. At this moment the adhesion causes the reaction force from the rail which leads to train movement.

Fig. 2, shows a curve of adhesion versus wheel slip curve. This curve is for a typical condition and as will be described in Figs. 3 and 4, some factors have impact on it. Regard to Figs. 2, 3 and 4 when the wheel slip is near to zero the adhesion coefficient is near to zero too. At first by increasing the wheel slip, the adhesion coefficient will be enhanced, up to its maximum. After a wheel slip value, wheel slip increasing leads to adhesion coefficient decreasing. The name of this point is wheel slip of maximum adhesion coefficient. As will be described this point does not have a certain or constant value. When the wheel slip is between zero (0) and wheel slip of maximum adhesion coefficient (λ_m), wheel slip and adhesion coefficient have a direct relationship together. This region is

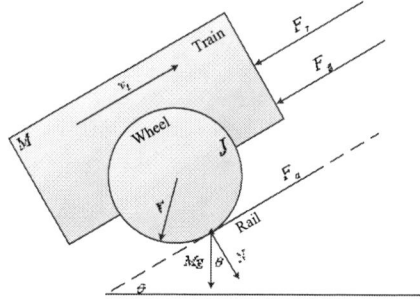

Fig. 1. Simplified model of train, for train movement study

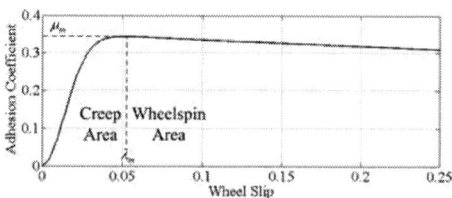

Fig. 2. Adhesion coefficient versus wheel slip curve

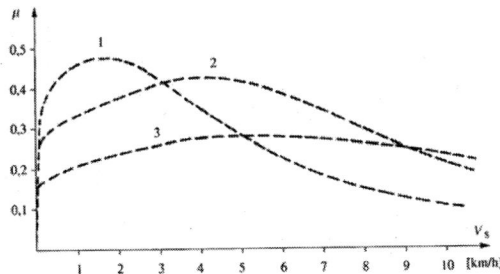

Fig. 3. Adhesion coefficient versus wheel slip curves in difference situations; 1. Rail is clean and dry, 2. There are moisture and dew on rail, 3. There are leaves and oil on rail. [2]

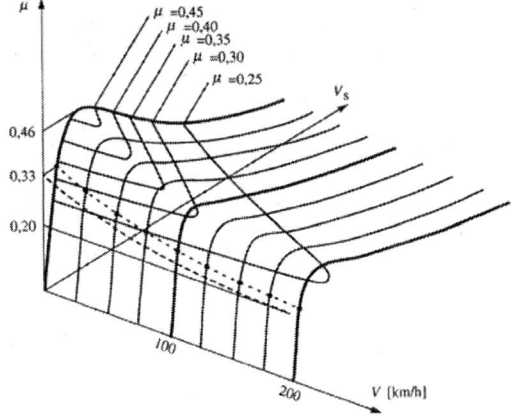

Fig. 4. Impact of the train speed on adhesion coefficient versus wheel slip curve. [2]

named *Creep Area*. Creep area has a stable situation. Because increasing wheel slip leads to increasing adhesion coefficient which enhances the train acceleration. If the wheel slip is between more than wheel slip of maximum adhesion coefficient (λ_m), wheel slip and adhesion coefficient will have an inverse relationship together. This region is named *Wheelspin Area*. Wheelspin area has an unstable situation. Because in this area increasing wheel slip leads to decreasing adhesion coefficient, which diminish the train acceleration. Several factors have impact on the adhesion versus wheel slip curve. The most important of this factors are descried in follow.

The first factor which has impact on friction and adhesion, is surface conditions of contact places. When the wheel and rail are dry and clean, adhesion coefficient between them has its maximum value. In the other hand existence of leaves, dust, moisture, oil, etc. on the rail lead to reducing in adhesion coefficient. The impact of different wheel and rail surface conditions on the adhesion versus wheel slip curve is illustrated in Fig. 3.

In addition of wheel and rail surface condition, the train speed has significant impact on the adhesion versus wheel slip curve. By rising the train speed, both of adhesion coefficient and wheel speed of maximum adhesion coefficient, will be reduced. Therefore can be said that, adhesion coefficient and

wheel speed of maximum adhesion coefficient have an inverse relationship with the train speed. The impact of train speed on the adhesion versus wheel slip curve is demonstrated in Fig. 4.

III. EQUATIONS OF TRAIN MOVEMENT ON THE RAIL

When the traction motor tries to move the train on the rail, difference forces will be appeared. This forces are illustrated in Fig. 1. The adhesion force (F_a) tries to move the train, while rolling resistance (F_r) and gravity resistance (F_g) are against this force. When the traction motor rotate the driven wheel, in order to accelerate the train, wheel slip wheel be occurred. The equations of train movements is as below [1, 2, 21 and 22]:

$$M \frac{d}{dt} v_t = F_a - F_r - F_g \tag{4}$$

$$F_a = \mu N \tag{5}$$

$$F_g = MgSin(\theta) \tag{6}$$

$$F_r = (k_0 + k_1 v_t + k_2 v_t^2)Mg \tag{7}$$

$$J \frac{d}{dt} \omega_{wh} = \tau_m - \tau_a \tag{8}$$

$$\tau_a = F_a r \tag{9}$$

$$\tau_m = F_m r \tag{10}$$

The factor of k_0, k_1 and k_2 are functions of train mass, number of wagons and train shape.

As shown in train motion equations like (5), adhesion coefficient has a main role in train movement, but there is no straight equation to calculate the adhesion coefficient. Furthermore as described and illustrated in section II, adhesion coefficient does not have a certain and fixed relationship to wheel slip. Hence to determine the adhesion coefficient, an observer should be used. But the performance of the observer in simulation is difference with implementation. By testing the new methods on a test bench the accuracy of them can be improved. The importance of wheel slip control and adhesion control is described in [23]. For this propose, testing the new methods of train slip control on a hardware device, a simulator has been designed. This simulator, can simulate the train movement on rail, which is described in next section.

IV. DESCRIBING THE SIMULATOR STRUCTURE AND ITS PERFORMANCE

The motion always measured by a coordinate system. The speed and movement is defined in comparison of coordinate system place and speed. For example if the place of coordinate system is fixed in the train, in all of the time, the speed of the train is zero. When the train moves, in that coordinate system, the speed of train is zero and the rail speed has a non-zero value. This rail speed, that coordinate system, is exactly the train speed in a stationary coordinate system. This note is used in the proposed simulator.

A. Structure and Components of Proposed Hardware Simulator

In this device there are two steel wheels. One wheel plays the role of the train driven wheel, while other wheel plays the role of the train rail. These two steel wheels are installed in an iron frame. In order to simulate the condition of wheel and rail these two wheels are tangent to each other, at their contact place. The structure and components of this simulator are shown in Fig. 5. The wheel which plays the role of the train driven wheel, in one side coupled with a shaft encoder and in other side coupled with a 3-phase induction motor. This shaft encoder senses the speed of the driven wheel. The induction motor is as train traction motor. On the other hand, the wheel which plays the role of the train rail, in one side coupled with a shaft encoder and in other side coupled with a 3-phase synchronous generator. This shaft encoder senses the speed of the rail. As described, this speed is exactly the train actual speed. The synchronous generator is connected to a parallel series of electrical resistance, which are built from light bulbs. In order to control the load of the synchronous generator, there are simple switches between light bulbs and the synchronous generator. The synchronous generator is coupled with that wheel which plays the role of the rail, therefore its speed is the train speed. The train load consist of some factors like train and wagon weight, shape, etc. that produce the resistance torque. In the other hand, the electrical load of this generator, produce the resistances torque against generator speed. Therefore this load, which consists of light bulbs, is playing role of the train load.

B. Implementation the Train Speed Control Methods

Besides these motors and shaft encoders, in order to implement the speed control methods, a microprocessor is needed. This simulator uses a TMS320F2812 Digital Signal Processor (DSP) in the microprocessor part. The microprocessor part of the proposed simulator and its accessory is depicted in Fig. 6. With use of the microprocessor, all kinds of the power electronics and drive strategies like [24-25] can be implemented on this simulator. The main application of the proposed simulator is implementation and test of the future

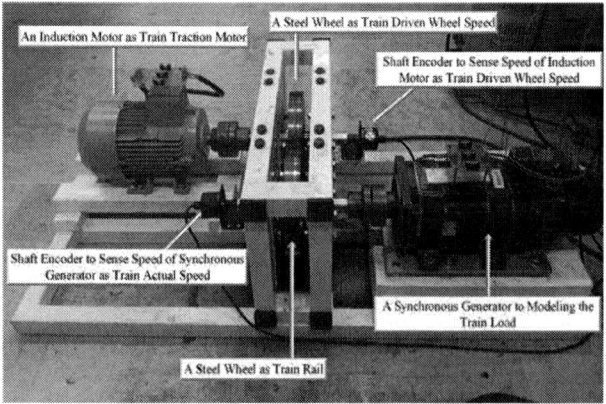

Fig. 6. Main part of the proposed simulator and its component

adhesion control methods. Independent of the adhesion control strategy the inner control loop of the system is speed control of induction motors. This induction motor speed control methods, can be Field Oriented Control (FOC) [1, 28 and 29], Direct Torque Control (DTC) [28, 29], Predictive Control [30-33], Scalar Control [28, 29], etc. As described in the next section, in order to depiction the application of the proposed simulator, several train speed control methods has been implemented on this simulator.

V. IMPLEMENTING SEVERAL TRAIN SPEED CONTROL METHODS ON THE PROPOSED SIMULATOR

As has been described in introduction section, the value of adhesion coefficient has significant effect on the train acceleration. Maximizing the train acceleration is an aim of the train speed control systems. Therefore the goal of the adhesion control methods is use of maximum adhesion coefficient in train acceleration. Train speed control systems may be equipped with adhesion control subsystem, or not. Both of this situation are discussed in this paper. In all of the experimental test, all of the conditions are like together. As discussed in the previous section, the generator speed is as train actual speed, and induction motor speed is as train driven wheel speed. Hence the generator speed is as train speed, reference speed is defined to generator speed. In all of this paper test, reference speed of the generator is 1500 rpm.

A. Implementation the Train Speed Control Methods which are not Equipped with Adhesion Control

When there is no adhesion control in the train speed control, the operation point cannot be at the maximum point of the adhesion. This operation point could be in creep area or wheelspin area, Fig. 2.

1) Operation in creep area: In the adhesion versus wheel slip curve, corresponding torque of the maximum adhesion point, is maximum adhesion torque. Assume that the driving torque of the induction motor is limited to a value less than the maximum adhesion torque. Therefore in acceleration time the operation point cannot reach to maximum adhesion

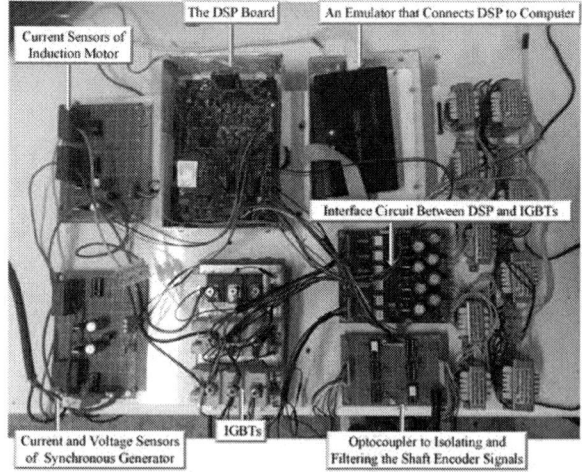

Fig. 5. The microprocessor part of the proposed simulator and its accessory

point. In this situation operation point is in the creep area. The advantage of this strategy is wheel slip limitation. When the wheel slip is limited up to low value, the amortization of wheels and rail will be low. In the contrary low adhesion coefficient lead to low acceleration. Disadvantage of limited driving torque to control speed of the traction motor, is low acceleration. Meaning that train needs a lot of time to reach its reference. This situation experimental results are illustrated in Figs. 7 and 8. By operation in creep area the acceleration time is about 87 sec. As illustrated in Fig. 8, at the beginning of the acceleration time, the generator speed is zero. The speed control system tries to increase the speed of induction motor up to the speed reference. Therefore at first, the wheel slip is high. By rising the speed of the induction motor, because of limited driving torque, the wheel slip will be reduced. As depicted in Fig.8, after around t = 10 sec, up to the end of the acceleration time, t = 87 sec, the wheel slip is near to zero.

2) *Operation in wheelspin area:* Assume that the driving torque of the induction motor is not limited to maximum adhesion torque and the only limitation of it, is thermal limitations of the induction motor. Also assume that maximum torque of the induction motor is greater than the maximum adhesion torque. Therefore in the acceleration time, the operation point reaches to maximum adhesion point and it will be transferred to wheelspin area. Regard to Figs. 2, 3 and 4, in all conditions, absolute value of slope of adhesion versus wheel slip curve in wheelspin area is much lower than absolute value of slope of adhesion versus wheel slip curve in creep area. This points means that, in comparison with operation in creep area, by operation in wheelspin area, speed control system can use a greater value of adhesion coefficient which leads to a greater acceleration. Therefore it can be said that, the advantage of operation in wheelspin area is enhancement of train acceleration. But this high value of acceleration obtained beside high value of wheel slip. High value of wheel slip causes significant amortization of wheels and rails. In the other word disadvantage of operation in wheelspin area is significant amortization of wheels and rails, which leads to damage on wheels and rails. This situation experimental results are illustrated in Figs. 7 and 8. By operation in wheelspin area, the acceleration time is about 15 sec. As illustrated in Fig. 8, at the beginning of the acceleration time, the generator speed is zero. The speed control system tries to rise the speed of induction motor up to the generator reference speed. Therefore at first, the wheel slip is high. By rising the speed of the induction motor, because of that the driving torque is unlimited, the wheel slip will be high too. As depicted in Fig.8, in all of the acceleration time, between t = 0 sec and t = 15 sec, the wheel slip is too high.

B. Implementation the Train Speed Control Method which is Equipped with Adhesion Control

As shown in previous part advantage of operation in creep area is low amortization of wheels and rails and disadvantage of it, is low train acceleration. In contrary, advantage of operation in wheelspin area is high value train acceleration and disadvantage of it, is significant amortization of wheels and rail. Therefore it can be said that the speed control system needs a tradeoff between enhancement of train acceleration and reduction of wheels and rails amortization. This operation point is exactly on the maximum adhesion point. Because by this

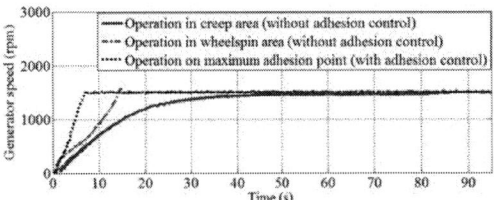

Fig. 7. Generator speeds (as train actual speed) with and without adhesion control, in acceleration mode.

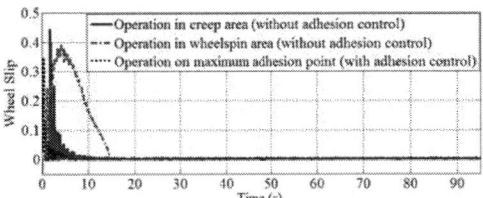

Fig. 8. Wheel slip with and without adhesion control, in acceleration mode.

value of wheel slip, the adhesion coefficient is maximized. If the speed control of train equipped with adhesion control, the operation point will be on the maximum of the adhesion versus wheel slip curve, Fig. 2. This situation experimental results are illustrated in Figs. 7 and 8. By operation in wheelspin area the acceleration time is about 7 sec. As illustrated in Fig. 8, at the beginning of the acceleration time, the generator speed is zero. The speed control system tries to rise the speed of induction motor up to the generator reference speed. Therefore at first, the wheel slip is high and by rising the speed of the induction motor, because of that, the driving torque is under control of the adhesion control system, the wheel slip will be controlled around the wheel slip of maximum adhesion coefficient. This value of wheel slip in this experimental test is between 0.02 and 0.04. As depicted in Figs. 7 and 8, by operation on maximum adhesion point the acceleration time is minimized and the wheel slip is optimized.

VI. CONCLUSION

As discussed in this paper, improving the acceleration of electric trains is one of the main goals of train designer. The train acceleration has straight relationship to the adhesion coefficient and wheel slip has a significant impact on the adhesion coefficient. Therefore in order to improve the train acceleration, the wheel slip should be controlled. The desired wheel slip in acceleration time, is the wheel slip of the maximum adhesion coefficient. But the problem is that the adhesion versus wheel slip does not have a fixed curve. Also the adhesion coefficient is a non-measurable quantity. Hence the new adhesion methods should be tested on a similar condition of real train. The proposed hardware device by simulating the wheel and rail condition, provide an appropriate simulator to testing the new adhesion control methods. The contracture of the proposed simulator described in this paper. In order to demonstrate the application and performance of the proposed device, several experimental test is done. The experimental tests consist of speed control systems, which are equipped and not equipped with adhesion control subsystem is discussed in this article. The experimental test illustrate the appropriate performance of the proposed simulator.

REFERENCES

[1] A. Steimel, Electric Traction - Motion Power and Energy Supply: Basics and Practical Experience, Oldenbourg, 2010.

[2] J.-M. Allenbach, P. Chapas, M. Comte, and R. Kaller, Traction électrique, polytechniques et universitaires romandes, 2008.

[3] L. Weng-Ching, L. Chun-Liang, H. Ping-Min, and W. Meng-Tzong, "Realization of Anti-Lock Braking Strategy for Electric Scooters", IEEE Transactions on Industrial Electronics, vol. 61, pp. 2826-2833, 2014.

[4] M. Tanelli, L. Piroddi, and S. M. Savaresi, "Real-time identification of tire-road friction conditions", Control Theory & Applications, IET, vol. 3, pp. 891-906, 2009.

[5] J. S. Lin and W. E. Ting, "Nonlinear control design of anti-lock braking systems with assistance of active suspension", Control Theory & Applications (IET), vol. 1, pp. 343-348, 2007.

[6] H.-o. Yamazaki, Y. Karino, T. Kamada, M. Nagai, and T. Kimura. (2007). "Effect of Wheel-Slip Prevention Based on Sliding Mode Control Theory for Railway Vehicles", Quarterly Report of Railway Technical Research Institute, vol. 48, pp. 22-29, 2007.

[7] N. Mutoh, Y. Hayano, H. Yahagi, and K. Takita, "Electric Braking Control Methods for Electric Vehicles With Independently Driven Front and Rear Wheels" IEEE Transactions on Industrial Electronics. vol. 54, pp. 1168-1176, 2007.

[8] J. J. Choi, S. H. Park, and J. S. Kim, "Dynamic Adhesion Model and Adaptive Sliding Mode Brake Control System for the Railway Rolling Stocks", Part F: J. Rail and Rapid Transit, vol. 221, pp. 313-320, 2007.

[9] P. Khatun, C. M. Bingham, N. Schofield, and P. H. Mellor, "Application of Fuzzy Control Algorithms for Electric Vehicle Antilock Braking/Traction Control Systems", IEEE Transactions on Vehicular Technology, vol. 52, pp. 1356-1364, 2003.

[10] S. H. Park, J. S. Kim, J. J. Choi, and H.-o. Yamazaki, "Modeling and Control of Adhesion Force in Railway Rolling Stocks", IEEE Control Systems Magazin vol. 28, pp. 44-58, 2008.

[11] H.-o. Yamazakiy, M. Nagai, and T. Kamada, "A Study of Adhesion Force Model for Wheel Slip Prevention Control". Jsme International Journal, vol. 47, pp. 496-501, 2004.

[12] P. Khatun, C. M. Bingham, N. Schofield, and P. H. Mellor, "An experimental laboratory bench setup to study electric vehicle antilock braking/traction systems and their control," Presented at IEEE 56th Vehicular Technology Conference, Vol. 3, pp. 1490-1494, 2002.

[13] T. Watanabe, "Anti-slip Readhesion Control with Presumed Adhesion Force. - Method of Presuming Adhesion Force and and Running Test Results of High-speed Shinkansen Train", Quarterly Report of Railway Technical Research Institute, vol. 41, pp. 32-36, 2000.

[14] K. Ohishi, S. Kadowaki, Y. Smizu, T. Sano, S. Yasukawa, and T. Koseki, "Anti-slip Readhesion Control of Electric Commuter Train Based on Disturbance Observer Considering Bogie Dynamics", presented at the 32nd Annual Conference on IEEE Industrial Electronics, 2006.

[15] Y. Shimizu, K. Ohishi, T. Sano, S. Yasukawa, and T. Koseki, "Anti-slip/skid Re-adhesion Control Based on Disturbance Observer Considering Bogie Vibration," Power Conversion Conference - PCC '07, Nagoya, 2-5April, 2007.

[16] M. Yamashita and T. Watanbe, "A Readhesion Control Method without Speed Sensor for Electric Railway Vehicles", Electric Machines and Drives Conference .EMDC'03, 1-4 June, 2003.

[17] M. Yamashita and T. Watanbe, "A Readhesion Control Method without Speed Sensor for Electric Railway Vehicles", Quarterly Report of Railway Technical Research Institute, vol. 45, pp. 85-89,2005.

[18] W. Zhang, J. Chen, X. Wu, and X. Jin, "Wheel/Rail Adhesion and Analysis by Using Full Scale Roller Rig", Wear, vol. 253, pp. 82-88, 2002.

[19] O. Arias-Cuevas, Z. Li, and R. Lewis, "A Laboratory Investigation on the Influence of the Particle Size and Slip During Sanding on the Adhesion and Wear in the Wheel–Rail Contact", Wear, vol. 271, pp. 14-24, 2011.

[20] M. Tomeoka, N. Kabe, M. Tanimotob, E. Miyauchib, and M. Nakatac, "Friction Control Between Wheel and Rail by Means of on-Board Lubrication", Wear, vol. 253, pp. 124-129, 2002.

[21] A. Nayal, S. P. Gupta, and S. P. Singh, "Performance Analysis of DC Motor Drive in Electric Traction with Wheel Slip Control", Journal of the Institution of Engineers, vol. 87, pp. 55-60, 2006.

[22] K. Wei, J. Zhao, and T. Q. Xiaojie YouZheng, "Development of a Slip and Slide Simulator for Electric Locomotive Based on Inverter-Controlled Induction Motor", 4th IEEE Conference on Industrial Electronics and Applications., Xi'an, 25-27 May, 2009.

[23] S. Sadr, D.A.Khaburi, M.Namazi, A.Shiri, D. Esmaeil Moghadam, "Modeling of wheel and rail slip and demonstration of the benefit of maximum adhesion control in train propulsion system", IEEE 23rd International Symposium on Industrial Electronics (ISIE), pp.847-852, 1-4 June, 2014.

[24] M. Aleenejad, H. Mahmoudi, P. Moamaei, R. Ahmadi, "A New Fault-Tolerant Strategy Based on a Modified Selective Harmonic Technique for Three Phase Multilevel Converters", IEEE Transactions on Power Electronics, unpublished.

[25] M. Aleenejad, P. Moamaei, H. Mahmoudi, R. Ahmadi, "Unbalanced Selective Harmonic Elimination for fault-tolerant operation of three phase multilevel Cascaded H-bridge inverters", Applied Power Electronics and Exposition Conference (APEC), IEEE, pp.1589-1594, 15-19 Mar, 2015.

[26] M. Aleenejad, H. Iman-Eini, S. Farhangi, "Modified space vector modulation for fault-tolerant operation of multilevel cascaded H-bridge inverters", Power Electronics, IET , vol.6, no.4, pp.742-751, April, 2013.

[27] H. Mahmoudi, M.J. Lesani, D. Arab khabouri, "Online fuzzy tuning of weighting factor in model predictive control of PMSM", 13th Iranian Conference on Fuzzy Systems (IFSC), pp.1-5, 27-29 Aug, 2013.

[28] R. Krishnan, "Electric motor drives: modeling, analysis, and control", Prentice Hall, 2001.

[29] B. K. Bose, "Power Electronics and Motor Drives Advances and Trends', Elsevier, 2006.

[30] S. Alireza Davari, D. A. Khaburi, W. Fengxiang, and R. M. Kennel, "Using Full Order and Reduced Order Observers for Robust Sensorless Predictive Torque Control of Induction Motors," IEEE Transactions on Power Electronics, vol. 27, pp. 3424-3433, 2012.

[31] S. Alireza Davari, D. A. Khaburi, and R. Kennel, "An Improved FCS-MPC Algorithm for an Induction Motor With an Imposed Optimized Weighting Factor" IEEE Transactions on Power Electronics, vol. 27, pp. 1540-1551, 2012.

[32] W. Fengxiang, Z. Zhenbin, S. Alireza Davari, R. Fotouhi, D. Arab Khaburi, J. Rodriguez, et al., "An Encoderless Predictive Torque Control for an Induction Machine With a Revised Prediction Model and EFOSMO" IEEE Transactions on Industrial Electronics, vol. 61, pp. 6635-6644, 2014.

[33] S. A. Davari, D. A. Khaburi, F. Wang, and R. Kennel, "Robust sensorless predictive control of induction motors with sliding mode voltage model observer", Turkish Journal of Electrical Engineering & Computer Sciences, vol. 21, pp. 1539-1552, 2013.

7th Power Electronics, Drive Systems & Technologies Conference (PEDSTC 2016)
16-18 Feb. 2016, Iran University of Science and Technology, Tehran, Iran

Design and Simulation of Hybrid Electrical Energy Storage (HEES) for Esfahan Urban Railway to Store Regenerative Braking Energy

Hasan Moghbeli
Dept. of Electrical Engineering
Arak University of Technology
Arak, Iran
Hamoghbeli@yahoo.ir

Hossein Hajisadeghian
Dept. of Electrical Engineering
Shahid Rajaee University
Tehran, Iran
Hossein.hajisadegian@gmail.com

Mahdi Asadi
Dept. of Electrical Engineering
Arak University of Technology
Arak, Iran
Asadi@iust.ac.ir

Abstract— **This paper deals with design and simulation of a hybrid electrical energy storage (HEES) for Esfahan urban railway under regenerative braking condition. The HEES presented in this paper, is comprised of battery and supercapacitor. The capacity of the supercapacitor and battery is calculated based on regenerative braking energy from each train considering other trains acceleration, normal traveling, and braking. This paper presents an algorithm and its control system for DC/DC converters in order to store regenerative braking energy in supercapacitor-battery, and to feed auxiliary system. Simulation results carried out by MATLAB/Simulink confirm the effective performance of the presented algorithm and control system.**

Keywords— *supercapacitor, battery, hybrid electrical energy storage, regenerative braking*

I. INTRODUCTION

Among all types of energy sources, fossil fuels are the most desirable where this kind of energy is going to be finished. Some issues like global warming and environmental pollution are the effects of using fossil fuels [1]. It is important to find ways to reduce energy consumption and reuse wasted energy. Regeneration of energy reduces the total energy cost and increases efficiency. Several studies have shown that application of regenerative braking in urban rail systems could potentially reduce their net energy consumption from 10 to 45 percent, depending on the characteristics of each system [2]. In Iran Based on calculation done for line 3 of Tehran metro the 24% of energy can be returned back to electrical system [1].
Regenerative braking in trains converts kinetic energy into electrical energy while braking process [3]. A typical induction machine generally holds motoring, generating and braking modes. During motoring mode, the direction of motor rotating speed agrees with the direction of torque. On contrary, when the direction of rotating speed opposes the direction of torque, the electric machine enters the generating mode. For a railway vehicle, during the regenerative braking (generating mode), the torque reduces the motor speed and generates electric power. Regenerative braking energy (RBE) can be converted by power electronic devices into electric energy. There are three main following methods for using regenerative

braking energy. Otherwise, the regenerative energy can be converted into heat using a large resistance bank referred to as "dynamic braking" [4]:

A. Returning Energy Back to DC Network

If regenerative braking power was enough qualified, it can be sent to DC network. Other nearby trains must be in acceleration mode to use this instantaneous power. Otherwise line voltage increases and protection relays trip. Fig. 1 is a snapshot of system while Train 2 is in braking mode. Energy flows to network and it is used by Trains 1 and 3 during acceleration mode [2].

B. Returning Energy Back to AC Power Grid

In this method a DC/AC convertor is required to change regenerative braking energy to a high quality fixed voltage and frequency energy. Producing this low THD power needs new technologies like multi-level converters. Moreover usually there is a long distance between traction power system (TPS) and trains. So the transmission costs increases.
Fig.2 shows a model of AC traction motor and its electrical power utility consists of power grid, diode rectifier, main DC-AC inverter, braking resistor and Silicon-Carbide (SiC) support converter (SC). Two series diodes are also applied in order to prevent circulating currents [5].
Two basic functions of the SC converter are:

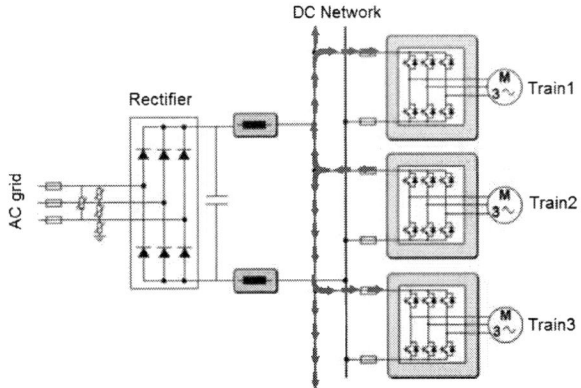

Fig.1. Regenerative braking energy sent back to DC network

978-1-5090-0376-1/16 $31.00 © 2016 IEEE

- Active filtering and harmonics compensation.
- Energy regeneration in regenerative breaking mode.

C. Storing in Hybrid electrical energy storage (HEES)

The best way to deal with regenerative braking energy is to store energy in HEES like chemical batteries, supercapacitor[1] (ECDL capacitor) and flywheel [2].A comparison between battery, supercapacitor and flywheel, used in metro application is done in [6]. The main idea for hybridization of these energy storage systems are factors like power density and energy density. Supercapacitor-battery, supercapacitor-flywheel, flywheel-battery and supercapacitor-flywheel-Battery are four different combinations of HEES which can be used. In this paper because of some reasons like long lifetime, lower size and weight, ease to control, ease to calculate state of charge of elements, supercapacitor-battery HEES is implemented for Esfahan urban railway. First, calculation of energy and capacity is done for supercapacitor and battery based on real data. Then a control system is defined for system combination. Based on results of simulation, control system is examined. To drive and control induction motor DTC method introduced in [7] is used in simulation.

II. SUPERCAPACITOR AND BATTERY MODELING

A. Supercapacitor

Supercapacitor or electrochemical double layer capacitor (EDLC) stores electrical energy in its original type on their plates. Supercapacitor can be modeled similar to conventional capacitors. In low frequencies the first order equivalent circuit of a supercapacitor is shown in Fig.3. This model consists of an ideal capacitor (C), internal resistor (R_s), series inductance (L) and a parallel resistor (R_p). L is usually very small and in DC condition is omitted [8]. The efficient storing energy in Supercapacitor is calculated as

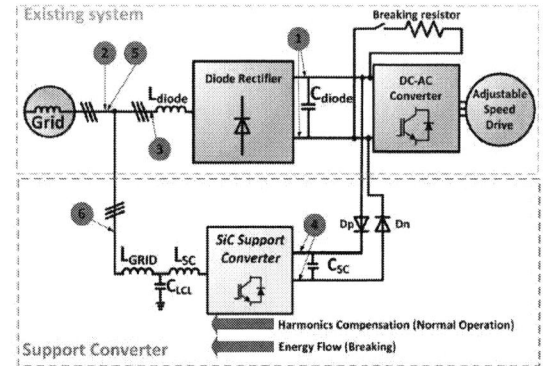

Fig.2 Regenerative braking energy sent back to AC grid

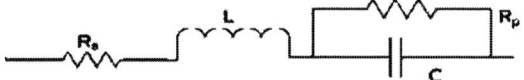

Fig.3. First order model of supercapacitor

---

[1] Also known as Ultracapacitor

$$W_{eff} = \frac{1}{2} C(V^2_{max} - V^2_{min}) \tag{1}$$

Where C is capacitance, V_{max} is nominal voltage and V_{min} is minimum voltage of suppercapacitor. State of charge (SOC) is energy stored in supercapacitor over its nominal energy. It can also be calculated as below

$$SOC_C(t) = u.V^2_C.100 + SOC_C(t_0) \tag{2}$$

Which u is a gain and depends on supercapacitor structure. SOC (t_0) is initial value of state of charge.

B. Battery

Chemical battery stores electrical energy in the form of chemical potential energy. Batteries have some advantages such as high energy density, low self-discharging rate, low main cost, long lifetime and some disadvantages like low power density and high dependency on temperature. Today Lead-Acid, Ni-Cd and Li-ion batteries are commonly used in railway and traction applications. Fig.4 (a) shows conventional equivalent circuit of a battery which consists of an ideal internal voltage power supply (V_0) and a series resistor (R_0). Fig.4 (b) is proposed model of a battery. In this model V_0 is function of SOC and RC parallel circuit shows dynamic model of battery [9]. Energy stored in a battery is expressed as below

$$E = VQ_B \tag{3}$$

Where E is in kWh, Q_B is capacity of battery in Ah and V is nominal voltage of battery. SOC relationship with current of battery is expressed as below

$$SOC_B(t) = \frac{1}{Q_B}\int_{t0}^{t} i(T)dT + SOC_B(t_0) \tag{4}$$

Which $SOC_B(t_0)$ is initial value of SOC.

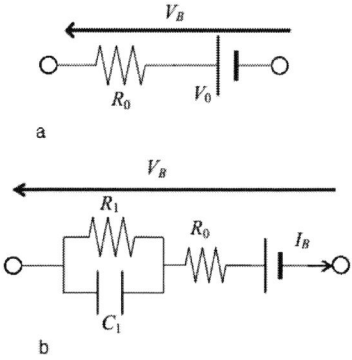

Fig.4 (a) Conventional model of battery. (b) Proposed model of a battery.

III. DESIGN OF SUPERCAPACITOR-BATTERY HEES

A. Technical and economical study of Esfahan urban railway

In line 1 of Esfahan urban railway organization there are 20 stations and 28 trains which move in forward and return path. These trains in conditions of braking and downhill generate power. Table 1 shows calculation for consumed and regenerative energy of each train during toward 20 stations [11]. Total travel time is 4120 s.

Fig.5 shows regenerative braking power for 20 stations. Maximum power regeneration happens at station S-17.

Fig.5 is only for one train. Because of time difference between two trains, to obtain other train diagrams it is necessary to shift time axis by 150 s. By adding up all of these 28 shifted diagrams total regenerative power diagram can be obtain as illustrated in Fig.6. The total consuming and regenerative braking power for individual train is calculated and summarized in Table 2.

In Iran, metro power systems are usually connected to 20kV distribution feeders. The price of power in this voltage is 1.89 cents (641 Rial) per kWh in mid load [11]. Due to this, total costs and benefits are shown in Table 2. A large amount of money saving can be obtained by optimal use of regenerative braking.

B. Design

Combination of HEES can be designed as stationary or onboard. In stationary form, HEES is placed on stations. HEES receives energy of all trains which brake in station. In [1] stationary designation is done. Unlike in onboard configuration, HEES is placed on train set. The objective of this paper is to design supercapacitor ESS as stationary and battery as onboard. This method has following advantages

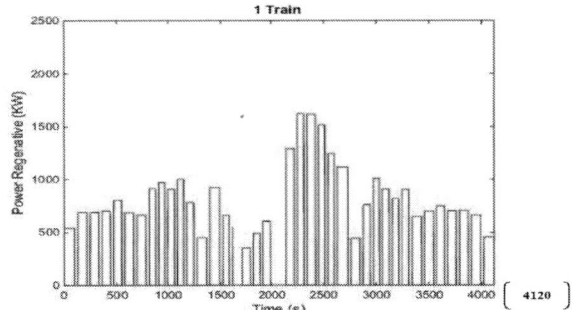

Fig. 5. Regenerative braking power in line 1

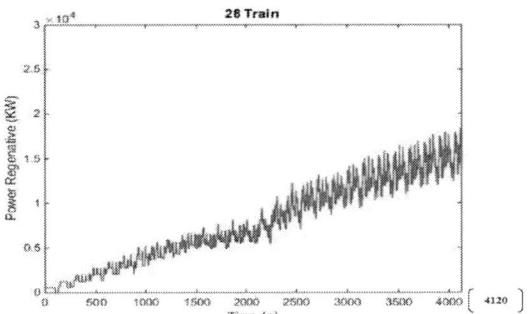

Fig. 6. Total regenerative braking power of trains

- Extra energy in supercapacitore which is not able to store in battery, can be used in traction, power supply (UPS) and internal usage of station like illustration.
- It reduces space and weight.
- It increases System reliability by using redundant supercapacitor set.

TABLE 1. Energy and supercapacitor storage calculation

Station	RBE of forward path (kWh)	RBE of return path (kWh)	Consumed energy of forward path(kWh)	Consumed energy of return path(kWh)	Number of serried cells of super capacitor	Number of parallel strings of super capacitor	Total net weight of super capacitor (kg)
S-1	0	12.5	0	24.6	480	13	3182.4
S-2	14.9	16	20.1	27.5	480	16	3916.8
S-3	16.5	16.1	23	26.1	480	18	4406.4
S-4	15.9	16	22.8	26.7	480	17	4161.6
S-5	16	15.8	22.5	23.6	480	16	3916.8
S-6	17	16.7	24.9	16.1	480	17	4161.6
S-7	16.4	16.2	28.4	22.1	480	17	4161.6
S-8	16.4	16.6	25.6	24.9	480	17	4161.6
S-9	16.3	13.7	24.3	21.5	480	17	4161.1
S-10	16.2	15.9	18.2	23.2	480	17	4161.6
S-11	15.9	15.7	22.2	22.1	480	17	4161.6
S-12	15.9	14.4	21.1	21.8	480	17	4161.6
S-13	14.6	10.6	20.1	18.4	480	15	3672
S-14	11.3	31	17.5	34.7	480	31	7588.8
S-15	25.2	19.7	38.9	16.1	480	26	6364.8
S-16	11.2	25.7	25.3	15.1	480	26	6364.8
S-17	10	35.5	29.6	15.7	480	36	8812.8
S-18	8.1	28.4	46.9	14.9	480	29	7099.2
S-19	9.2	27.6	35	16.6	480	28	6854.4
S-20	13	0	39.8	0	480	11	2692.8

TABLE.2. Total energy and annual prices

Title	1train	28 train
Consumed energy(kWh)	923.5	25858
Regenerative braking energy(kWh)	644.1	18034.8
Consumed power(kW)	806.9	22593.2
Regenerative braking power(kW)	562.8	15758.4
Annual total consumed energy price ($)	171,428	4,800,000
Annual total regenerative braking energy benefits ($)	43,214	1,210,000

Maximum voltage of the supercapacitor should be lower than the network voltage, in order to simplify the design of DC/DC converter [1]. The voltage of DC power grid is 1500 V but supercapacitor voltage is 1200 V. Also nominal voltage of battery is 200 V. This voltage level is made by a buck converter.

To save regenerative braking energy in stations, series and parallel configuration of supercapacitor is used. In this case BCAP3400 ultracapacitors of MAXWELL company are used. Nominal voltage of this cell is 2.85 V and its capacity is 3400 F. With considering a margin voltage the operating voltage of each cell in practice is 2.5 V.

To reach 1200 V, 480 supercapacitor cells must become in series.To compensate extra capacity parallel strings must be used. Table 1 shows calculations for all 20 stations. Between forward and return path maximum regenerative braking energy is chosen. Net weight of each cell is 0.52 kg which due to number of cells total weight is obtained and shown in Table 1.

Batteries are placed onboard because of connecting to auxiliary system. Today NiCd batteries are commonly used in metro applications. These batteries must be under charge for approximately an hour. Due to 70 minutes of train travel time battery pack can be charged during a complete trip. If battery stores 10% of suppercapacitor energy during a braking cycle, by adding up 10% energy of each station (40 stations), total energy for storing in battery will be 64.4 kWh. If nominal voltage of battery pack assumed to be 200 V the total capacity of battery pack due to (3) would be 322 Ah. KPL1000 is a NiCd battery cell from QUALMEGA company. Output voltage and capacity of this cell is 1.2 v and 1000Ah. By putting 167 cells in series and 53 strings in parallel, battery pack can be constructed.

IV. SIMULATION

A. System Toplogy

There are two topologies for supercapacitor-Battery HEES; active cascade system and parallel system. Fig.7 shows active cascade system which is usually used because of these reasons:

- A lower voltage battery can be used because of using buck converter.
- Because the voltage level decreases in two steps, a softer current is received by battery.

- Battery can be always full charged.

Parallel system is shown in Fig.8. In this system because energy must to send back to grid, bidirectional DC/DC convertors are used.

Power circuit is illustrated in Fig.9. Two buck converters are added to step down voltage for each energy storage system. Braking resistor is implemented in condition of overcharge of supercapacitor and failure [13]. In Esfahan urban railway two motors of each vehicle are connected to one inverter by reverse phase sequence.

Fig.10 shows Simulink-MATLAB model of supercapacitor-Battery HEES. In this configuration regenerative braking energy is modeled by an induction motor fed by an inverter which is controlled by DTC. Tree buck DC/DC converters are controlled by feedback PI controller to regulate output voltage. Supercapacitor and battery characteristics are chosen based on part III calculations.

B. Control Method

The algorithm used to control supercapacitor-battery HEES is shown in Fig.11. This system works based on torque sign.

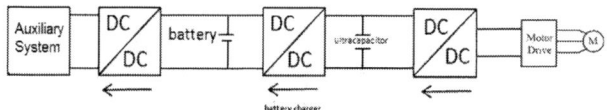

Fig.7 Cascade connection of supercapacitor-battery HEES

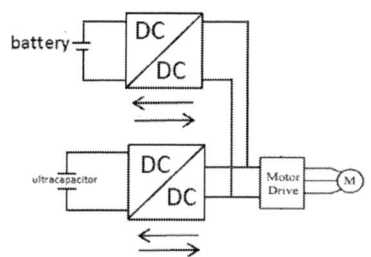

Fig.8 Parallel connection of supercapacitor-battery HEES

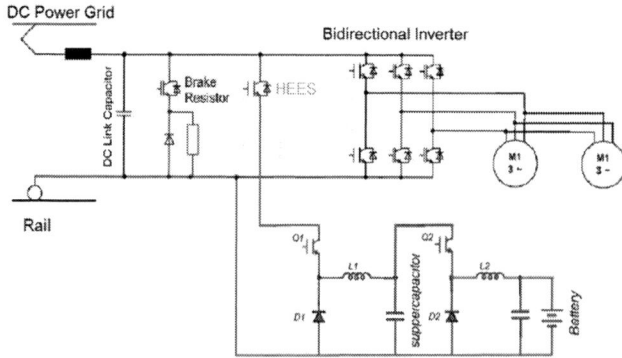

Fig.9 Power circuit of traction and HEES system

978-1-5090-0376-1/16 $31.00 © 2016 IEEE

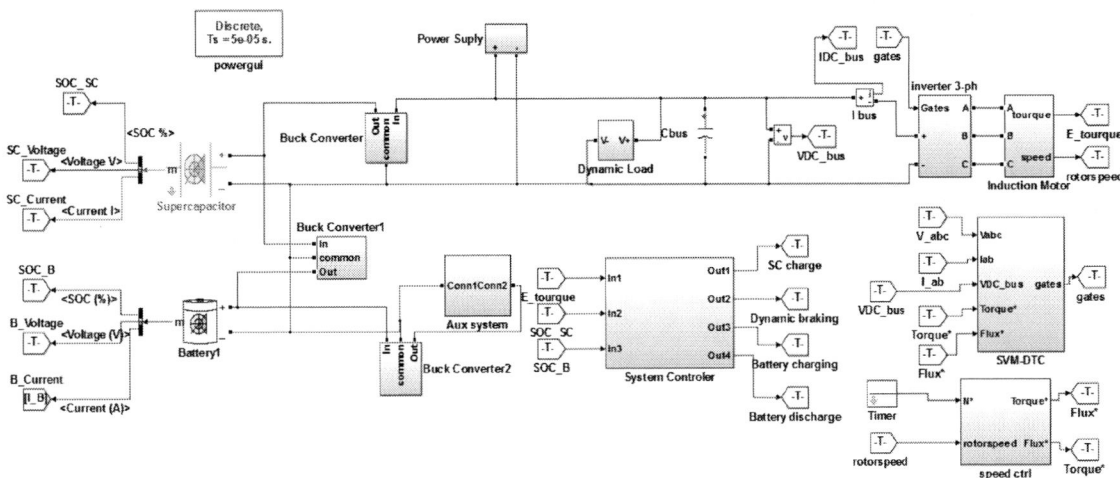

Fig.10. Block diagram of traction and HEES in MATLAB

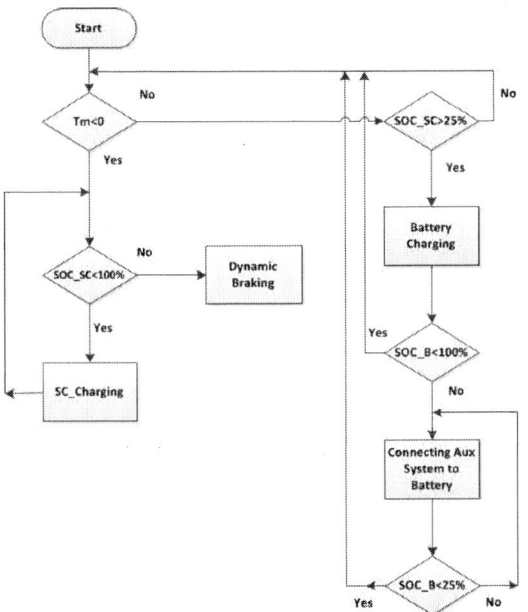

Fig11. Control algorithm of HEES system

It expresses regenerative braking mode when it is negative and motoring mode when it is positive. Battery doesn't have permission to charge in condition of regenerative braking while supercapacitor is being charged. Also SOC variation of the supercapacitors will be kept between 25% and 100%. Hence the voltage variation will be between 50% and 100% of its maximum voltage. It is because to lower stress on power switches due to voltage difference [1]. This is also true for battery which is connected to load. In condition of supercapacitor overcharge and failure, energy is sent to dynamic braking.

C. Simulation Resaults

Braking starts at t = 2.7 s. It happens by putting -1000N torque on induction motor. Fig 12 shows rotor speed, electromagnetic torque and stator current of induction motor.

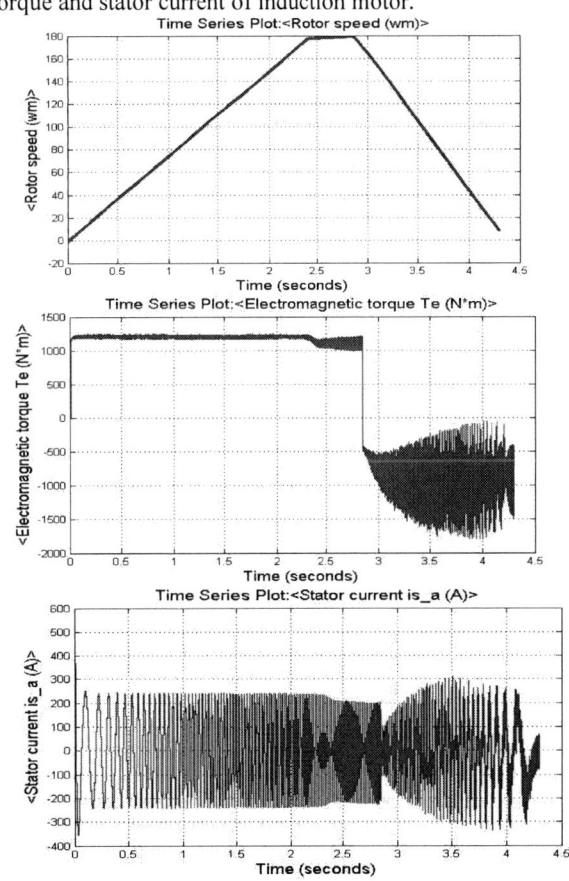

Fig12. Speed, torque and stator current of induction motor

978-1-5090-0376-1/16 $31.00 © 2016 IEEE   97

In braking mode reference speed for DTC drive is zero. It can be seen after 1.2 s Induction motor stops. Initial SOC value for battery is 0% and for supercapacitor is 96%. Variation in battery and supercapacitor parameters are shown in Fig.13 and Fig.14 for two cycles.

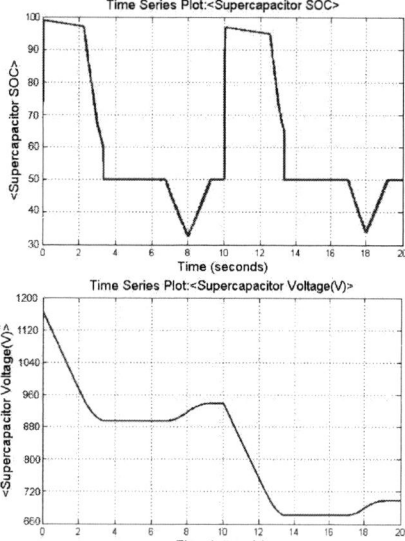

Fig13. Supercapacitor SOC and Voltage variation

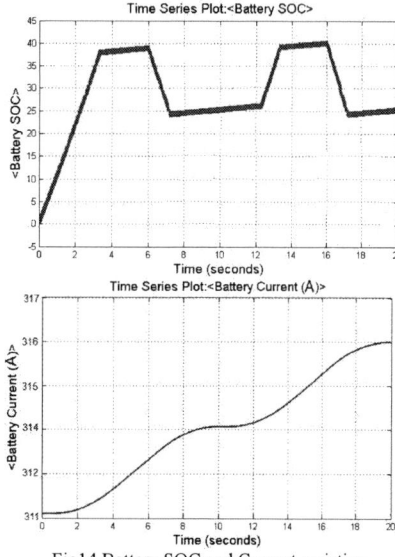

Fig14.Battery SOC and Current variation

As it is obvious the SOC of battery and supercapacitor is kept between 25 to100 percent.

V. CONCLUSION

In this paper three different strategies for storing regenerative braking energy (RBE) were discussed. The optimum and practical solution is to use hybrid electrical energy storage (HEES). Due to important factors like high power and high energy density four different configurations were offered. The best choice for line 1 of Esfahan urban railway is supercapacitor-battery HEES. Two stationary and onboard methods were discussed and designed. To design, BCAP3400 supercapacitor cells and NiCd KPL1000 battery cells were used. Finally a supercapacitor-battery HEES was simulated. Investment on using regenerative braking energy brings about 1,210,000 $ profit for a year which in comparison to 4,800,000 $ annual consumed cost saves about 25% of money.

ACKNOWLEDGMENT

The authors would like to acknowledge Ghatar Shahri Esfahan Organization for providing data and financial supports in this research. Also the authors would like to acknowledge Mr. Ali Salimi for his contribution in obtaining regenerative braking energy.

REFERENCES

[1] R. Teymourfar, B. Asaei, H. Iman-Eini, and R. Nejati Fard, "Stationary super-capacitor energy storage system to save regenerative braking energy in metro line,", energy conversion and management, vol 56(2012) ,pp. 206-214, january 2012.

[2] A. Gonzalez-Gil, R. Palacin, and P. Batty, "Sustainable urban rail systems: Strategies and technologies for optimal management of regenerative braking energy", Energy Conversion and Management,vol 75, pp. 374–388, June 2013.

[3] A. Adinolfi, R. Lamedica, C. Modesto, A. Prudenzi , and S. Vimercati, " Experimental assessment of energy saving due to trains regenerative braking in an electrified subway line,", IEEE Transactions on Power Delivery, Vol. 13, pp. 1536-1542, No. 4, October 1998.

[4] S. Lu, P. Weston, S. Hillmansen, H.B. Gooi, and C. Roberts, "Increasing the Regenerative Braking Energy for Railway Vehicles,", IEEE transaction on intelligent transportation system, Vol 15, pp. 2506 – 2515, Nov. 2014.

[5] S. Piasecki, J. Rąbkowski, G. Wrona, T. Płatek, "SiC-Based Support Converter for Passive Front-End AC Drive Applications,", 39th Annual Conference of the IEEE, pp. 6010 – 6015, Nov. 2013.

[6] S. Koohi-Kamali, V.V. Tyagi, N.A. Rahim, N.L. Panwar, and H. Mokhlis, "Emergence of energy storage technologies as the solution for reliable operation of smart power systems: A review,", Renewable and Sustainable Energy Reviews, vol 25, pp.135–165, May 2013.

[7] H. Moghbeli, M. Zarei, S.S. Mirhoseini, "Transient and steady states analysis of traction motor drive with regenerative braking and using modified direct torque control (SVM-DTC), ", The 6th International Power Electronics Drive Systems and Technologies Conference (PEDSTC 2015),pp. 615 – 620, Feb.2015 Tehran,Iran.

[8] H. Douglas, P. Pillay, "Sizing ultracapacitors for hybrid electric vehicles, ", Industrial Electronics Society, 31st Annual Conference of IEEE, 6-10 Nov. 2005.

[9] K. Moon-Young, K. Jong-Woo, J. Chol-Ho, C. Shin-Young, M. Goon-Woo , "Automatic charge equalization circuit based on regulated voltage source for series connected lithium-ion batteries, ", 8th International Conference on Power Electronics and ECCE Asia (ICPE & ECCE), 2011 IEEE, p.p 2248 – 2255 , May 30 2011-June 3 2011.

[10] AW3-Esfahan train Operation line Volt - Esfahan train traction. (Documents of Ghatar Shahri Esfahan Organization).

[11] "http://www.moe.gov.ir." .

[12] S. Koohi-Kamali, V.V. Tyagi, N.A. Rahim, N.L. Panwar, and H.Mokhlis, "Emergence of energy storage technologies as the solution for reliable operation of smart power systems: A review,", Renewable and Sustainable Energy Reviews, vol 25, pp.135–165, May 2013.

[13] M. Steiner, M. Klohr, and S. Pagiela," Energy Storage System with UltraCaps on Board of Railway Vehicles, ", European Conference on Power Electronics and Applications, pp. 1-10, Sept.2007

7th Power Electronics, Drive Systems & Technologies Conference (PEDSTC 2016)
16-18 Feb. 2016, Iran University of Science and Technology, Tehran, Iran

Analysis of Electromechanical Model of Traction System with Single Inverter Dual Induction Motor

Pooya Ghani, Mohammad Arasteh, Hamid Reza Tayebi

Industrial Power Supply Research Group
Iranian Research Institute for Electrical Engineering (IRIEE), ACECR,
Tehran, Iran
Pooya.ghani@gmail.com

Abstract—The purpose of this paper is to analyze an electromechanical model of a bogie, which is a main unit of traction system. Electrical model consists of parallel-connected dual induction motors, fed by a single inverter. Conventional and weighted vector controls are used as motor speed control method. Mechanical model is considered train body, wheel-rail contact, gearbox, drag force etc. This model is established using Simscape library of MATLAB. The comparison of the control methods are obtained through simulations.

Keywords— *weighted vector control; traction drive; electromechanical model; dual induction motor*

I. INTRODUCTION

Many software can used for simulating complex power systems. However, in many cases it is preferred to simulate electrical system along with mechanical and sophisticated control systems. Using Simscape library of MATLAB, all of these parts can simulate together. Traction system consists of complex mechanical and electrical components. Study of these components together is very important to investigate motor control methods. This paper analyses the dynamic model of the combined electrical, mechanical and control systems of a bogie of traction system to investigate control strategies.

The total model of a bogie as main unit of traction system is presented. In railway traction systems, the traction forces are transmitted to bogie through the contact of the wheels with the rail and the transmitted forces depend on the adhesion coefficient (μ). The principal factor for an unbalanced load in a railway traction system is the friction force change, which depends on the variation of the adhesion coefficient. Adhesion coefficient also depends on the slip velocity, rail condition, train load, train speed etc. Slip velocity (v_{slip}) is defined in (1) where r is wheel radius, ω is angular velocity of wheel and V_{train} is transitional speed of train body. Fig.1 shows adhesion coefficient of rail-wheel according to slip velocity [1]. The adhesive force (F) is defined as (2) where M is weight of train. Adhesion coefficient is highly non-linear parameter, as fig.1 can be divided into two parts: stable and unstable area. In the stable area, the traction effort is equal to the adhesive effort, thus propulsive force is well transmitted to the body. In an unstable area, since the adhesion coefficient is low friction force is low. Therefore, slip velocity increases and system is

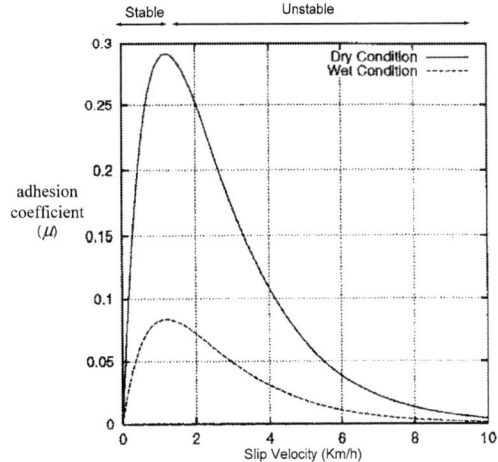

Fig. 1. adhesion coefficient.

unstable. As shown in Fig.1, the maximum value of adhesion coefficient (μ) is quite reduced when rail is wet. When friction force changes suddenly, controller must maintain stability and performance [2].

$$v_{\text{slip}} = \text{r} \times \omega - V_{train} \qquad (1)$$

$$F = \mu Mg \qquad (2)$$

Most applications require multiple motors according to the desired power. These multiple motors are generally coupled mechanically, so the drive control of these systems, is challengeable. For urban railway and subway traction system, the single inverter multi motor drive system has been used widely. The advantages for this structure are low cost, light weight and compactness [3],[4]. Due to safety problems and performance of induction motors, adhesion loss of wheel is considered as a challenge for these traction systems [5],[6].

Recently, researchers of single inverter multi motor drive field focus on vector control methods. In conventional methods, the control is applied to a virtual bigger motor rather than two parallel motors. Therefore, to control flux and torque through the d and q axes respectively, averages of the variables are used in controller. Sensorless conventional

978-1-5090-0376-1/16 $31.00 © 2016 IEEE

vector control of single inverter, dual induction motor are presented in [7],[8]. A vector control method based on average differential control (ADC) for single inverter, dual induction motor is presented in [9]. It can control the total torque of two induction motors. In additional to decrease the behavior deviation between the motors, the differential parameters are used.

Some researchers presented the drive controls based on a weighted vector control strategy. The control is applied to a virtual motor chosen by the weighting coefficients. The author of [10] proposed a weighted voltage vector control strategy as an anti-slip strategy, which actively inhibits the slip of two induction motors. To improve the control performance of unbalanced load, [11] proposed a control strategy based on the weighted vector control. The motors' speed and torque are used to calculate the weight value, and then is distributed to the weight control of dual induction motors. This method is implemented on irrelevant motors.

In this paper, weighted and conventional vector control methods are implemented as control strategy of electromechanical model of traction system and performance of methods is investigated.

The system modeling is introduced first. Then, the control structures are presented. Thereafter, the simulation results of the system are presented and response of control methods to changing condition is discussed.

II. ELECTROMECHANICALL MODELING

Modeling includes two parts: electrical and mechanical modeling. Speed sensors and ideal torque sources create a relation between two parts. In the following, the details of modeling are described. The model is similar to practical system and all of the model components chosen from Simscape library of MATLAB. Simscape library consists of SimPowerSystems, SimDriveline, Mechanical and Physical Signals libraries. SimPowerSystems is used for electrical modeling and other libraries are used for mechanical

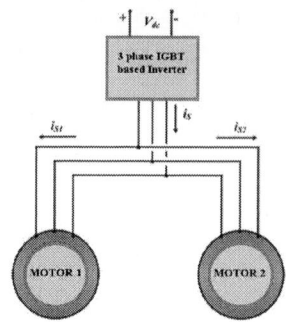

Fig. 2.   Electrical model of single inverter dual induction motor.

modeling.

A. Electrical Modeling

Electrical modeling part of a bogie model considers two parallel induction motors fed by single inverter. Electrical model is shown in fig.2. Three-leg IGBT-base inverter with ideal dc source is used as drive. Space vector modulation (SVM) is used for inverter gate commands. Load torque or shaft speed can be used as mechanical input of the motor. In this paper, speed of shaft is mechanical input of the motor, which comes from mechanical part of modeling. In addition, produce torque of motor is mechanical output, which goes to mechanical part of modeling.

B. Mechanical Modeling

In mechanical part of modeling of a bogie, two axles that coupled to motors using gearboxes and shafts, bogie body, wheel-rail contact, drag force, wind speed and road incline angle are considered. Mechanical model is shown in fig.3. In mechanical modeling, two types of movement are considered: rotational movement and transitional movement. As shown in fig.3 produced torque of motor using ideal torque source provided rotational power of mechanical model. Angular velocity sensor provided mechanical input of electrical modeling part. The power passes through shaft model, gearbox model and wheel-axle model. Wheel-axle model

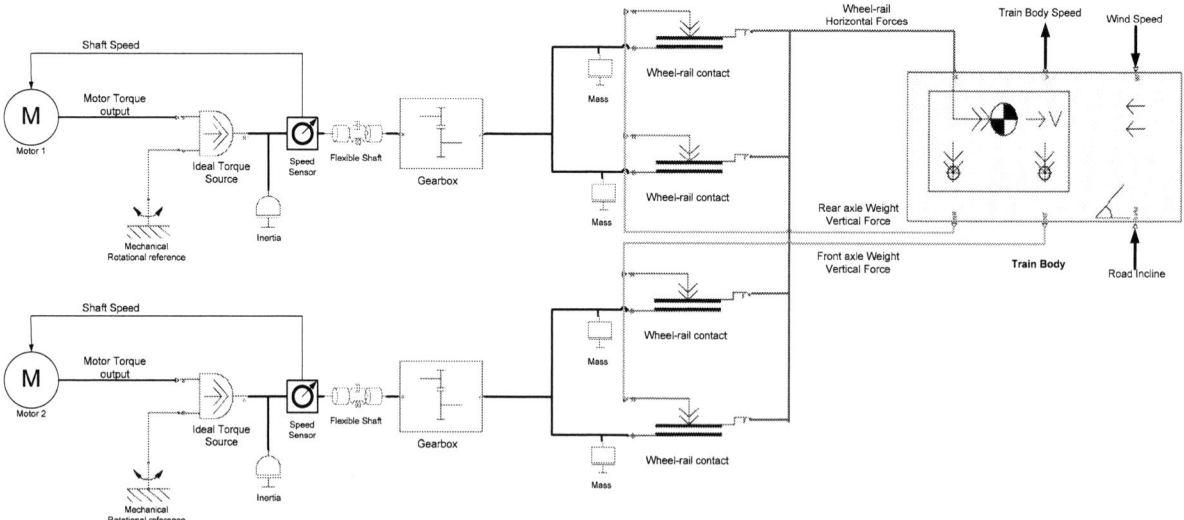

Fig. 3.   Mechanical model of the bogie.

changes rotational quantities to transitional quantities. Output of wheel-axle model is applied to mass of wheel and wheel-rail contact model. Weight vertical force is applied to Wheel-rail contact model and value of adhesion coefficient is chosen from lookup table proportional of slip speed as shown in fig.1. Last object of mechanical modeling is train body model, which is chosen from SimDriveline library. Four wheel-rail contact models transmit force to body of train. Train body model considers mass of train, drag coefficient, wind velocity, center of gravity and road incline angle. It calculates weight vertical force and body velocity.

III. CONTROL METHODS

Due to nature of train, working cycle, fast acceleration and declaration is necessary. Vector control can achieve a quick and accurate motor torque control which assurance fast acceleration and declaration. In this section, two vector control methods, which are based on direct field oriented control, are compared.

A. Weighted Vector Control Method

In this method [11], rotor flux is estimated using flux observer then the flux linkage of dual induction motors is reoriented, after that, the control expressions of weighted rotor flux linkage and weighted torque of dual motors are derived.

According to fig.4 the weighted value of rotor flux linkage, stator current and angular velocity are calculated respectively in new synchronous reference frame by (3), (4) and (5). The d axle of new frame is aligned with $\overline{\Psi}_r^e$. Superscript "e" represents synchronous reference frame.

$\overline{\Psi}_r^e$: Weighted rotor flux.

$\overline{\Psi}_{dr}^e$: d component of $\overline{\Psi}_r^e$.

Ψ_{r1}^e, Ψ_{r2}^e: Rotor flux of motor 1, 2.

$\Delta\Psi_r^e$: Difference of rotor flux.

$\Delta\Psi_{qr}^e, \Delta\Psi_{dr}^e$: q, d component of $\Delta\Psi_r^e$.

$\overline{i}_s^e$: Weighted stator current.

i_{s1}^e, i_{s2}^e: Stator current of motor 1, 2.

$\overline{i}_{qs}^{e*}, \overline{i}_{ds}^{e*}$: q, d component of $\overline{i}_s^e$.

i_{qs1}^e, i_{qs2}^e: q component of i_{s1}^e, i_{s2}^e.

Δi_s^e: Difference of stator current.

$\Delta i_{qs}^e, \Delta i_{ds}^e$: q, d component of Δi_s^e.

$\overline{\omega}_r$: Weighted angular velocity.

ω_{r1}, ω_{r2}: rotor angular velocity of motor 1, 2.

ω_e: Synchronous speed.

$\Delta\omega_r$: Difference of angular velocity.

L_r, R_r: Rotor inductance, Rotor resistance.

L_m: Excitation inductance.

$S_r: R_r/L_r$

T_s: Sum of produce torques.

T_1, T_2: Produce torque of motor 1, 2.

n_p: Pole pairs.

k_m: Weight value.

k_w, k_u: Intermediate variables.

θ_a: Weighted rotor flux angle.

$$\begin{cases} \overline{\Psi}_r^e = k_m\Psi_{r1}^e + (1-k_m)\Psi_{r2}^e \\ \Delta\Psi_r^e = \Psi_{r2}^e - \Psi_{r1}^e \end{cases} \qquad (3)$$

$$\begin{cases} \overline{i}_s^e = k_m i_{s1}^e + (1-k_m)i_{s2}^e \\ \Delta i_s^e = i_{s2}^e - i_{s1}^e \end{cases} \qquad (4)$$

$$\begin{cases} \overline{\omega}_r = k_m\omega_{r1} + (1-k_m)\omega_{r2} \\ \Delta\omega_r = \omega_{r2} - \omega_{r1} \end{cases} \qquad (5)$$

In synchronous reference frame, the rotor flux oriented state equation of each induction motor can be described by:

$$\begin{cases} \dfrac{d\Psi_{r1}^e}{dt} + \left(S_r + j(\omega_e - \omega_{r1})\right)\Psi_{r1}^e = L_m S_r i_{s1}^e \qquad (6) \\ \dfrac{d\Psi_{r2}^e}{dt} + \\ \left(S_r + j(\omega_e - \omega_{r2})\right)\Psi_{r2}^e = L_m S_r i_{s2}^e \end{cases}$$

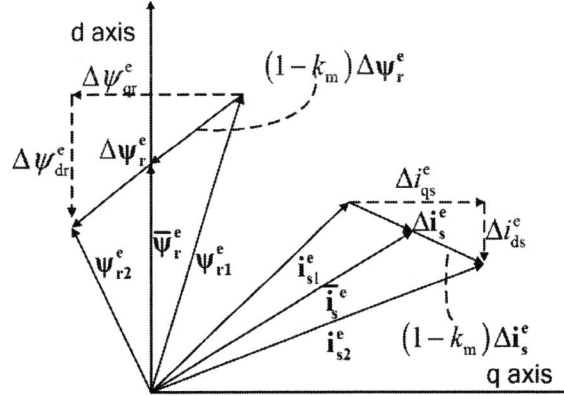

Fig. 4.   Weighted vector diagram of dual induction motor.

The summation torque of dual induction motors is shown in (7). By defining (8) and substation (3),(4) and (5) in (6) after simplification, the expression of the q and d components of stator reference current are achieved as (9) [11]. The q component is used to control the summation torque and the d component of stator reference current is used to control the excitation current of dual induction motors.

$$T_s = T_1 + T_2 = \frac{3}{2} n_p \frac{L_m}{L_r} (\Psi^e_{r1} \times i^e_{s1} + \Psi^e_{r2} \times i^e_{s2}) \qquad (7)$$

$$\begin{cases} k_w = 2(k_m)^2 - 2k_m + 1 \\ k_u = 2k_m - 1 \end{cases} \qquad (8)$$

$$\begin{cases} \bar{i}^{e*}_{ds} = \frac{\bar{\Psi}^e_{dr}}{L_m} + \frac{(k_u \Delta\omega_r - (k_w)^2)\Delta\omega_r \Delta\Psi^e_{qr}}{2 S_r L_m} \\ \\ \bar{i}^{e*}_{qs} = \dfrac{\frac{T_s}{\frac{3}{2} n_p \frac{L_m}{L_r}} - k_w(\bar{\Psi}^e_{dr}\Delta i^e_{qs} - \bar{i}^e_{ds}\Delta\Psi^e_{qr}) - k_u(\Delta\Psi^e_{dr}\Delta i^e_{qs} - \Delta i^e_{ds}\Delta\Psi^e_{qr})}{2\bar{\Psi}^e_{dr} + k_w \Delta\Psi^e_{dr}} \end{cases} \qquad (9)$$

The weight value k_m is calculated by load torques. Since motors' torques proportional to q component of stator current, (10) is used for calculation k_m [12]. Block diagram of this method is shown in fig.5.

$$k_m = \frac{T_1}{T_1 + T_2} \cong \frac{i^e_{qs1}}{i^e_{qs1} + i^e_{qs2}} \qquad (10)$$

B. Conventional Vector Control Method

In conventional vector control method, despite deference between loads of each motor, average value of variables of motors are used in controller. Therefore, it is sufficed that, average of fluxes, speeds and currents be applied to direct field oriented control. It means that in weighted control method k_m is considered 0.5.

IV. SIMULATION RESULT AND DISCOUSTION

In this section, the bogie total system is simulated using Simscape library of MATLAB. System parameters are shown in Table I. Some parameters are obtained from Tehran metro line 1 [13].

TABLE I. SYSTEM PARAMETERS

No.	Parameter	Value
1	Motor rated power	180 kw
2	Moment of interia	3 kg.m²
3	Flexible shaft stiffness	2×10⁵ N.m/rad
4	Gearbox ratio	5 (speed reducer)
5	Wheel raduis	0.5 m
6	Body wieght	20 ton
7	Wheel weight	10 kg

In this simulation, Drag factor coefficient is 0.4, bogie velocity set point is 5m/s and adhesion coefficient is taken from fig.1. Train motion starts from stop condition and zero incline of road. Acceleration time takes up to 3.8s. At t=4s incline of road increases to 8 degree and at t=6s adhesion coefficient curve for front axle wheels, according to fig.1, changes from dry condition to wet condition. This situation can take place when train goes out of the tunnel in rainy day. Suddenly front axle wheels enter to wet rail area and rear axle wheels are still in dry area.

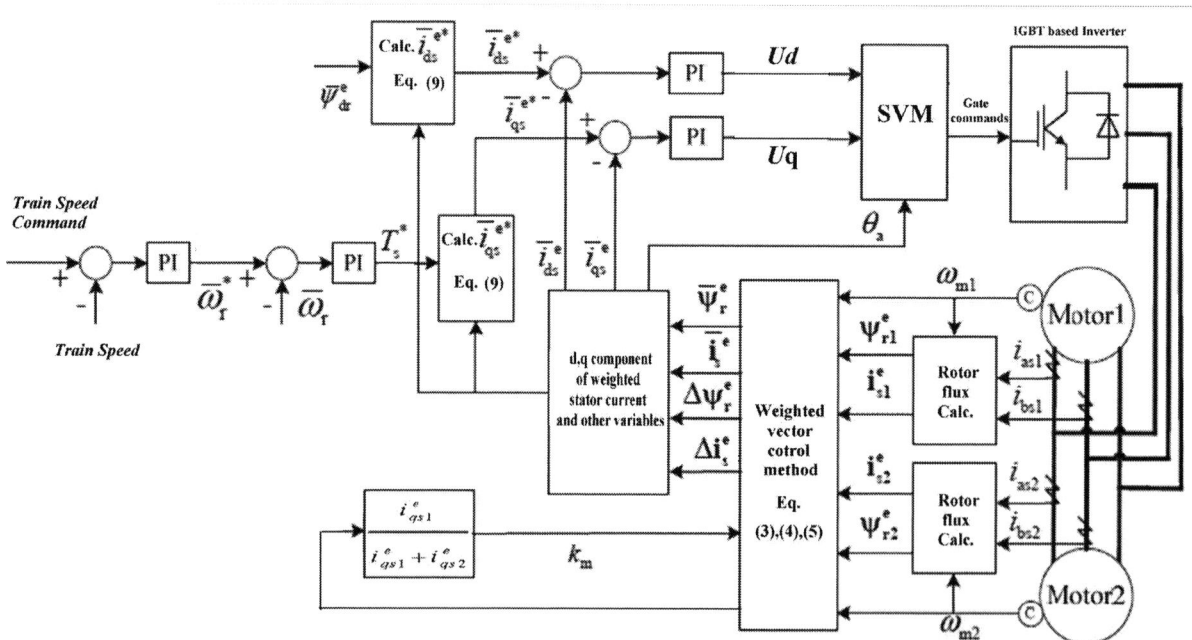

Fig. 5. Weighted vector control block diagram.

978-1-5090-0376-1/16 $31.00 © 2016 IEEE

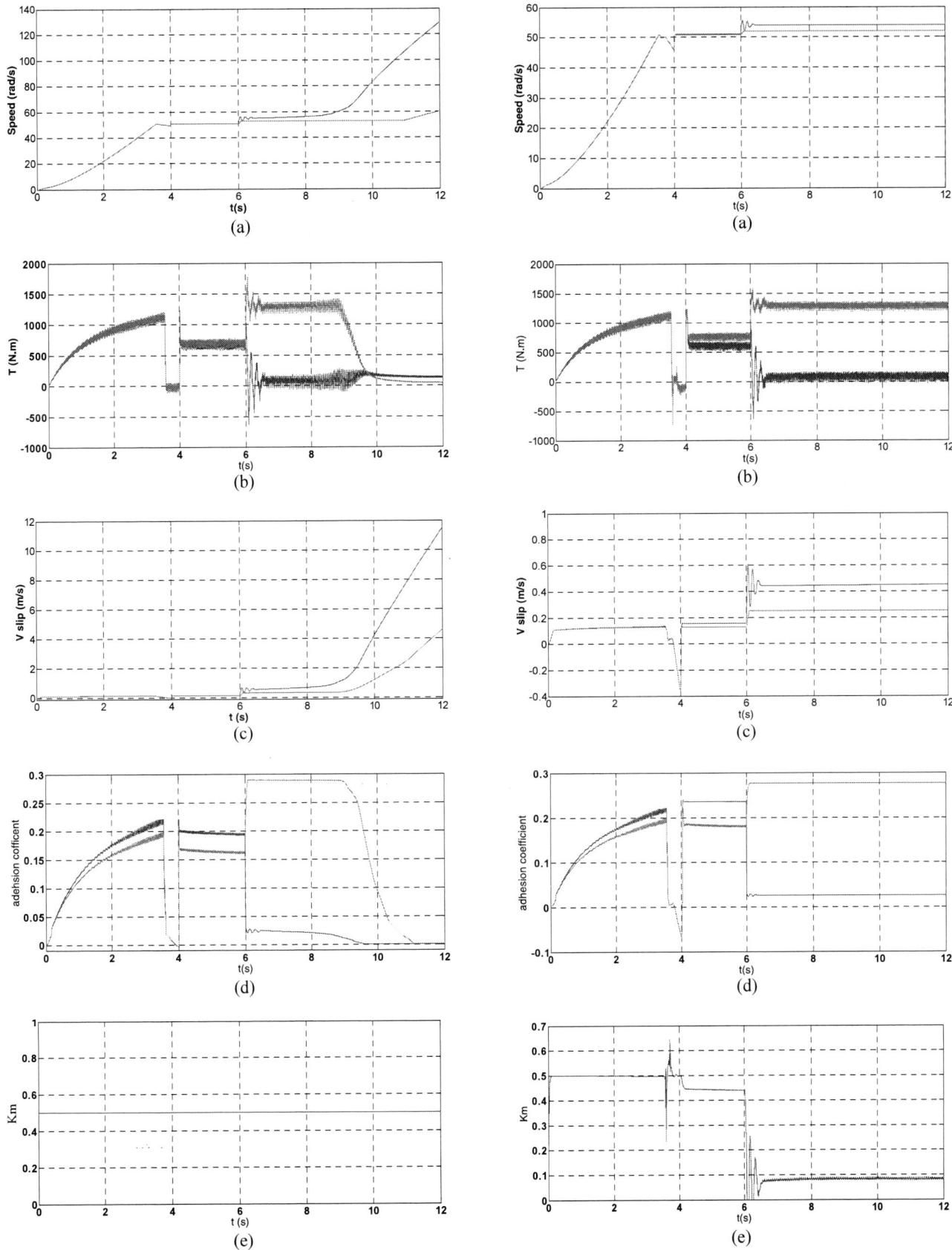

Fig. 6   Conventional vector control: (a) motor speed, (b) motor torque, (c) slip velocity, (d) rail-wheel adhesion coefficient, (e) k_m parameter

Fig.7   Weighted vector control: (a) motor speed, (b) motor torque, (c) slip velocity, (d) rail-wheel adhesion coefficient, (e) km parameter

Adhesion coefficient of front axle wheels decreases. In This situation, response of two control methods is investigated. Fig.6 and Fig.7 in order show system parameters of conventional vector control method and weighted vector control method. System parameters consist of (a) motor speed (rad/s), (b) motor torque (N.m), (c) slip velocity (m/s), (d) rail-wheel adhesion coefficient, (e) k_m parameter. Motor 1 and motor 2 in order are front and rear axle actuator. Velocity of bogie body is shown in fig.8 and fig.9 in order for conventional and weighted vector control method. As shown in fig.6 after t=6s speed of two motors is increased. Since transmitted friction force to bogie body is decreased, motor torque causes rotor and coupled parts accelerate. In addition, bogie body velocity decreases therefore motor speed reference increases. Motor speed increase and bogie body velocity decrease causes slip velocity increases. Due to fig.1, this process continues until slip velocity enters to unstable region. After t=9s system is in unstable region. Since in conventional vector control method mean of parameters is used in controller (k_m=0.5), influence of motor1, disturbs control of motor 2. Finally system is unstable, all wheels slip and bogie body speed decreases (fig.8).

As shown in fig.7 after t=6s, according load of motors k_m is changed below 0.1, therefore, vector control is adapted with new condition. Torque of motor 1 changes adequate with friction force and torque of motor 2 increases to remain bogie body velocity near set point. System finds new work point, which provides appropriate adhesion coefficient for both axle wheels.

Using weighted vector control velocity of bogie body is kept constant, stability is provided and dynamic performance is enhanced. Notice that, re-adhesion and anti slip control are not used in controller. Weighted vector control has inherent anti slip control.

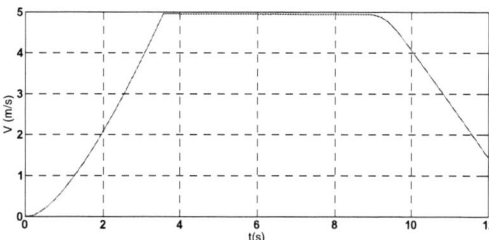

Fig.8   Train velocity with conventional control method

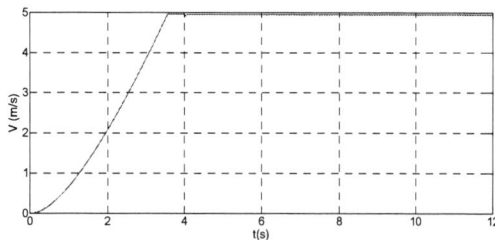

Fig.9   Train velocity with weighted control method

V.   CONCLUSION

In this paper, total modeling of the train bogie is presented. Mechanical, electrical and system control parts are combined. This model can be used for investigate motors control methods in various train conditions.

Conventional and weighted vector control methods for single inverter dual induction motor are simulated and compared. Simulation results show that in the case of adhesion loss of one axle wheels, weighted vector control has better stability and performance than conventional vector control. The comparison shows that weighted vector control method has inherent anti slip feature and it can be enhanced for single inverter dual motor, traction application.

REFERENCES

[1]   Ishikawa, Yoshiki, and Atsuo Kawamura. "Maximum adhesive force control in super high speed train." In Power Conversion Conference-Nagaoka 1997., Proceedings of the, vol. 2., 1997, pp. 951-954.

[2]   Achour, Tahar, Mustapha Debbou, and Maria Pietrzak-David. "Control strategy of a dual induction motor: Anti-slip control application." *Control Engineering Practice* 36, 2015, pp. 58-71.

[3]   P. Gratzfeld and H. C. Skudelny, "Dynamic performance of two parallel-connected induction machines for traction drives," in IEEEIAS Conf. Rec., 1984 Annu. Mtg., pp. 287-295.

[4]   P. M.Kelecy and R. D. Lorenz, "Control methodology for single inverter, parallel connected dual induction motor drives for electric vehicles", in Proc. IEEE PESC'94, 1994, pp. 987–991.

[5]   N. K. Mutoh, T. and Takita, K., "Driving characteristics of an electric vehicle system with independently driven front and rear wheels," IEEE Trans. Ind. Electron., vol. 53, no.3, Jun. 2006, pp. 803-813.

[6]   Jalali, Kiumars, Kai Bode, Steve Lambert, and John McPhee. "Design of an Advanced Traction Controller for an Electric Vehicle Equipped with Four Direct Driven In-Wheel Motors." SAE international journal of passenger cars-electronic and electrical systems, 2008, pp. 211-219.

[7]   K. Matsuse, Y. Kouno, H. Kawai, S. Yokomizo, "A speed sensorless vector control method of parallelconnected dual induction motor fed by a single inverter," IEEE Trans. Ind. App., vol. 38, pp. 1566–1571, Nov./Dec.2002

[8]   J. Nishimura, K. Oka, K. Matsuse, "A Method of Speed Sensorless Vector Control of Parallel-Connected Dual Induction Motors by a Single Inverter with a Rotor Flux Control", Proceeding of International Conference on Electrical Machines and Systems Seoul, Korea , Oct. 2007.

[9]   P. M.Kelecy and R. D. Lorenz, "Control methodology for single inverter, parallel connected dual induction motor drives for electric vehicles", in Proc. IEEE PESC'94, 1994, pp. 987–991.

[10] A. Bouscayrol, M. Pietrzak-David, P. Delarue, R. Peña-Eguiluz, P. Vidal, X. Kestelyn, "Weighted Control of Traction Drives with Parallel-Connected AC Machines", IEEE trans. on ind. elec., VOL. 53, NO. 6, DEC. 2006.

[11] Xu, Fei, Liming Shi, and Yaohua Li. "The weighted vector control of speed-irrelevant dual induction motors fed by the single inverter." Power Electronics, IEEE Transactions on 28, no. 12, 2013, pp. 5665-5672.

[12] Qi, Long, Chenchen Wang, and Xiaojie You. "Study of Speed-Sensorless Weighted Vector Control of Parallel Connected Induction Motors Drive." InElectronics and Application Conference and Exposition (PEAC), 2014 International, 2014, pp. 553-559.

[13] https://www.metro.tehran.ir

7th Power Electronics, Drive Systems & Technologies Conference (PEDSTC 2016)
16-18 Feb. 2016, Iran University of Science and Technology, Tehran, Iran

Electromagnetic Force Analysis in Linear Induction Motors, Considering End Effect

Abbas Shiri

Department of Electrical Engineering
Shahid Rajaee Teacher Training University
Tehran, Iran
Abbas.shiri@srttu.edu

Abstract—**Along with different phenomena, the end effect phenomenon deteriorates the performance of linear induction motors (LIMs). The end effect produces a braking force and reduces the thrust of the motor. In this paper, using Ampère and Maxwell basic equations, the air-gap flux density of the motor is calculated. Then, the electromagnetic force of LIM is calculated by using 1-D and 2-D field analysis. In order to calculate the braking force produced by the end effect, a finite length motor is considered. The forces due to entrance and exit end effect and the force without end effect are separately calculated. To validate the proposed methods, the obtained results are compared with those reported in literature.**

Keywords- linear induction motor; electromagnetic forc; entrance end effect; exit end effect;1-D and 2-D analysis

I. INTRODUCTION

Linear induction motors (LIMs) have been widely used in many applications, especially, in high speed transportation systems [1]-[4]. To analyze these motors, different models have been developed in literature. The equivalent circuit model is suitable for design purposes [5]-[10]; However, for calculation and prediction of the forces and the performance of the motor, electromagnetic field theories have been employed [11]-[17]. In these researches, by using field theories, the air-gap flux density is calculated. The electromagnetic forces produced by the LIMs have been calculated in [18]-[20]. In these references, the long motors are investigated. So, the end effect phenomenon has not been considered. In [21], the forces of laterally displaced secondary LIMs have been investigated. The end effect phenomenon produces braking force and degrades the performance of the motor [22]-[24]. In this paper, 1-D and 2-D analysis are carried out by using the Ampère and Maxwell equations. Then, net electromagnetic forces as well as the entry and exit end effect force are derived, separately. Finally, the results of the calculation are compared with the results of the methods appeared in literature.

II. 1-DIMENTIONAL MODELLING

A side view of a LIM is shown in Fig. 1. As seen in this figure, the primary of the motor consists of an iron layer as well as excitation layer (current sheet model is used) and the secondary sheet which is made of a simple aluminum sheet. The model is chosen in such a way that it can be applicable for both single-sided and double-sided linear induction motors. In Fig. 1, j_1 is the primary current density and j_2 is the

secondary sheet current density which both of them are in z-direction. The machine width in z-direction is considered 1 meter. Ampère circuit law is employed to calculate the magnetic field in the air-gap. Considering the closed rectangle in Fig. 1 and employing Ampère law, the following equation is derived [25]:

$$g \frac{\partial H}{\partial x} = j_1 + j_2 \qquad (1)$$

Where H is the air-gap field intensity and g is the magnetic air-gap length. Regarding the relation between field intensity and the flux density, (1) can be written as:

$$\frac{\partial B}{\partial x} = \frac{\mu_0}{g} (j_1 + j_2) \qquad (2)$$

In the above equation, B is the air-gap flux density and μ_0 is the magnetic permeability of the vacuum. Due to time-variable property of the flux density and also the relative motion between the primary and the secondary of the motor, a voltage induces in the secondary which is equal to [24]:

$$\frac{\partial e_2}{\partial x} = \frac{\partial B}{\partial t} + v \frac{\partial B}{\partial x} \qquad (3)$$

Where v is the relative speed between the primary and the secondary of the motor. In the 1-D analysis, the flux density has only y-component; this means that the secondary leakage is negligible; so, the induced voltage e_2 drops only in the secondary resistance. Thus, the following equation holds:

$$e_2 = j_2 \rho_s \qquad (4)$$

Where ρ_s is the surface resistivity of the secondary sheet which is given by:

$$\rho_s = \rho / d \qquad (5)$$

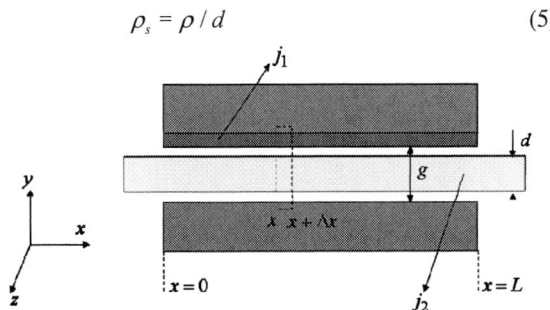

Figure 1. Side view of a linear induction motor

978-1-5090-0376-1/16 $31.00 © 2016 IEEE   105

In the above equation, ρ is the secondary volume resistivity and d is its width. By using equations (1)-(5), the following differential equation is derived for the air-gap magnetic flux density:

$$\frac{g}{\mu_0}\frac{\partial^2 B}{\partial x^2} - \frac{v}{\rho_s}\frac{\partial B}{\partial x} - \frac{1}{\rho_s}\frac{\partial B}{\partial t} = \frac{\partial j_1}{\partial x} \quad (6)$$

The motor is supplied by the following input current:

$$j_1(x,t) = J_1 e^{j\frac{\pi}{\tau}(v_s t - x)} \quad (7)$$

Where τ is the pole pitch and v_s is linear synchronous speed which is equal to:

$$v_s = 2f\tau \quad (8)$$

The air-gap magnetic flux density is derived by solving (6):

$$B(x,t) = K_p e^{j\frac{\pi}{\tau}(v_s t - x)} + K_1 e^{-\frac{x}{\alpha_1}} e^{j(\omega t - \frac{\pi}{\tau_e}x)} + K_2 e^{+\frac{x}{\alpha_2}} e^{j(\omega t + \frac{\pi}{\tau_e}x)} \quad (9)$$

Employing appropriate initial conditions, coefficients K_p, K_1 and K_2 are derived as follows [24]:

$$K_p = \frac{J_1}{\sqrt{(\frac{\pi g}{\mu_0 \tau})^2 + [\frac{1}{\rho_s}(v_s - v)]^2}} e^{j\delta_s} \quad (10)$$

$$K_1 = B_s[-\frac{B}{A}\frac{q_2}{q_3}\frac{e^{q_3 L} - 1}{e^{q_2 L} - 1} - \frac{q_2}{q_1}\frac{e^{q_1 L} - 1}{e^{q_2 L} - 1}] + J_1\frac{\rho_s}{v.A}\frac{q_2}{q_3}\frac{e^{q_3 L} - 1}{e^{q_2 L} - 1} \quad (11)$$

$$K_1 = \frac{B}{A}B_s - \frac{\rho_s}{v.A}J_1 \quad (12)$$

Different coefficients in the above equations are as follows [24]:

$$\delta_s = tan^{-1}\frac{\pi\rho_s g}{\mu_0 \tau(v_s - v)} \quad (13)$$

$$q_1 = -j\frac{\pi}{\tau} \quad (14)$$

$$q_2 = -\frac{1}{\alpha_1} - j\frac{\pi}{\tau_e} \quad (15)$$

$$q_3 = +\frac{1}{\alpha_2} + j\frac{\pi}{\tau_e} \quad (16)$$

$$A = [1 - \frac{\rho_s g}{v\mu_0}q_3] - \frac{q_2}{q_3}[1 - \frac{\rho_s g}{v\mu_0}q_2]\frac{e^{q_3 L} - 1}{e^{q_2 L} - 1} \quad (17)$$

$$B = [\frac{\rho_s g}{v\mu_0}q_1 - 1] + \frac{q_2}{q_1}[1 - \frac{\rho_s g}{v\mu_0}q_2]\frac{e^{q_1 L} - 1}{e^{q_2 L} - 1} \quad (18)$$

$$\alpha_1 = \frac{2\rho_s g}{\rho_s g A_1 - \mu_0 v} \quad (19)$$

$$\alpha_2 = \frac{2\rho_s g}{\rho_s g A_1 + \mu_0 v} \quad (20)$$

$$\tau_e = \frac{2\pi}{B_1} \quad (21)$$

$$A_1 + jB_1 = \sqrt{(\frac{v\mu_0}{\rho_s g})^2 + 4j\frac{\omega\mu_0}{\rho_s g}} \quad (22)$$

THRUST CALCULAATION

The mean thrust produced by the motor is derived as [17]:

$$F_T = \frac{1}{2}W_s\int_0^L Re[j_1 * (x,t)B(x,t)]dx \quad (23)$$

Where $j*(x,t)$ is the complex conjugate of $j(x,t)$. Using (7) and (9) in (23) yields:

$$F_T = \frac{1}{2}W_s J_1[L|K_p|\cos\delta_s + A_1|K_1| + A_2|K_2|] \quad (24)$$

In the above equation, different coefficients are given as follows:

$$A_1 = \frac{1}{1 + (\frac{1}{\alpha_1 q_4})^2}[\frac{1}{q_4}(e^{-\frac{L}{\alpha_1}}\sin(q_4 L + \delta_{K_1}) - \sin(\delta_{K_1})) - \frac{1}{\alpha_1 q_4^2}(e^{-\frac{L}{\alpha_1}}\cos(q_4 L + \delta_{K_1}) - \cos(\delta_{K_1}))] \quad (25)$$

$$A_2 = \frac{1}{1 + (\frac{1}{\alpha_2 q_5})^2}[\frac{1}{q_5}(e^{\frac{L}{\alpha_2}}\sin(q_5 L + \delta_{K_2}) - \sin(\delta_{K_2})) + \frac{1}{\alpha_2 q_5^2}(e^{+\frac{L}{\alpha_2}}\cos(q_5 L + \delta_{K_2}) - \cos(\delta_{K_2}))] \quad (26)$$

$$K_1 = |K_1|e^{j\delta_{K_1}} \quad (27)$$

$$K_2 = |K_2|e^{j\delta_{K_2}} \quad (28)$$

$$q_4 = \pi/\tau - \pi/\tau_e \quad (29)$$

$$q_5 = \pi/\tau + \pi/\tau_e \quad (30)$$

III. 2-DIMENTIONAL MODELLING

In 1-D analysis, it is assumed that the magnetic field has only y-component; while in reality, it has components in three directions; i.e. x, y and z-directions. However, the component in z direction is negligible. So, in this section 2-D analysis is carried out to calculate the air-gap magnetic flux of the machine. To do this, the Maxwell equations are employed. For a substance in an isotropic environment, moving with the speed of V, the governing differential equation for the vector magnetic potential is derived as [24]:

$$\nabla^2 A = \mu\sigma\left[\frac{\partial A}{\partial t} - V\times(\nabla\times A)\right] \quad (31)$$

The above equation is a general equation for the magnetic vector potential. Fig. 2 illustrates the linear induction motor for 2-D analysis. In this figure, the motor is divided into three regions. Region 1 is the primary iron core, region 2 is the secondary sheet and region 3 is the air-gap. In this study, the current sheet concept is employed for analysis. It means that the primary windings as well as the slots and the teeth are replaced with an infinitely thin sheet carrying the same current (j_1) as the windings.

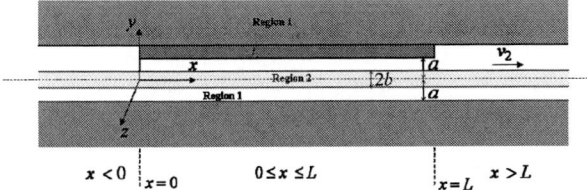

Figure 2. Side view of a linear induction motor for 2-D analysis

Regarding the direction of the current j_1 (z-direction), B and H have components in only x and y directions. Also, the magnetic vector potential A has component only in z-direction. Using (31) and (32), the magnetic flux density is derived as (33).

$$B = \nabla \times A \qquad (32)$$

$$B = a_x \frac{\partial A_z}{\partial y} - a_y \frac{\partial A_z}{\partial x} \qquad (33)$$

Employing (31), the field equations for three regions are derived as follows:

$$\frac{\partial^2 A_1}{\partial x^2} + \frac{\partial^2 A_1}{\partial y^2} = 0 \qquad (34)$$

$$\left[\frac{\partial^2 A_2}{\partial x^2} + \frac{\partial^2 A_2}{\partial y^2}\right] a_z = \mu_2 \sigma_2 \left[\frac{\partial A_2}{\partial t} a_z + v_2 \frac{\partial A_2}{\partial x} a_z\right] \qquad (35)$$

$$\frac{\partial^2 A_3}{\partial x^2} + \frac{\partial^2 A_3}{\partial y^2} = 0 \qquad (36)$$

In the above equations, the indices denote the different regions and v_2 is the speed of the secondary sheet.

As mentioned before, the magnetic vector potential have component in only z-direction and it is a function of x, y. Also, it is sinusoidal function of t, so, we can write:

$$A(x,y,t) = A(x,y)e^{j\omega t} \qquad (37)$$

Using (37) in (35) the following equation is derived:

$$\frac{\partial^2 A_2}{\partial x^2} + \frac{\partial^2 A_2}{\partial y^2} = \mu_2 \sigma_2 \left[j\omega A_2 + v_2 \frac{\partial A_2}{\partial x}\right] \qquad (37)$$

To solve the above equation, Fourier transform is employed. The Fourier transform of $A_2(x,y)$ is defined as:

$$\tilde{A}_2(x,y) = \int_{-\infty}^{+\infty} A_2(x,y) e^{-j\zeta x} dx \qquad (38)$$

Using (38) in (37), the following equation is derived:

$$\frac{d^2 \tilde{A}_2}{dy^2} = \gamma^2 \tilde{A}_2 \qquad (39)$$

Where

$$\gamma^2 = \zeta^2 + j\omega\mu_2\sigma_2 + j\zeta\mu_2\sigma_2 v_2 \qquad (39)$$

Similarly, applying Fourier transform, (34) and (36) become:

$$\frac{d^2 \tilde{A}_1}{dy^2} = \zeta^2 \tilde{A}_1 \qquad (40)$$

$$\frac{d^2 \tilde{A}_3}{dy^2} = \zeta^2 \tilde{A}_3 \qquad (41)$$

Due to the structure of the machine, the following functions are suggested for the different equations (40), (39) and (41):

$$\tilde{A}_1 = K_1 e^{-\zeta y} \qquad (42)$$

$$\tilde{A}_2 = K_2 \cosh(\gamma y) \qquad (43)$$

$$\tilde{A}_3 = K_3 e^{\zeta y} + K_4 e^{-\zeta y} \qquad (44)$$

Coefficients $K_1 - K_4$ which have been derived from the boundary conditions, are as follows:

$$K_1 = \frac{2\zeta \tilde{J}_1}{\mu_3 |A|} \qquad (45)$$

$$K_2 = \frac{K_1}{2}\{[\cosh(\gamma b) + \frac{\mu_3}{\mu_1}\frac{\gamma}{\zeta}\sinh(\gamma b)]e^{-\zeta b}e^{2\zeta a} + [\cosh(\gamma b) - \frac{\mu_3}{\mu_1}\frac{\gamma}{\zeta}\sinh(\gamma b)]e^{\zeta b}\} \qquad (46)$$

$$K_3 = \frac{K_1}{2}[\cosh(\gamma b) + \frac{\mu_3}{\mu_1}\frac{\gamma}{\zeta}\sinh(\gamma b)]e^{-\zeta b} \qquad (47)$$

$$K_4 = \frac{K_1}{2}[\cosh(\gamma b) - \frac{\mu_3}{\mu_1}\frac{\gamma}{\zeta}\sinh(\gamma b)]e^{\zeta b} \qquad (48)$$

In (45), $|A|$ is equal to:

$$|A| = [-\frac{\gamma}{\mu_2}\sinh(\gamma b)][(\frac{1}{\mu_1} + \frac{1}{\mu_3})\zeta e^{\zeta a}e^{-\zeta b} + (-\frac{1}{\mu_1} + \frac{1}{\mu_3})\zeta e^{-\zeta a}e^{\zeta b}]$$
$$+ [\frac{\zeta}{\mu_2}\cosh(\gamma b)][-(\frac{1}{\mu_1} + \frac{1}{\mu_3})\zeta e^{\zeta a}e^{-\zeta b} + (-\frac{1}{\mu_1} + \frac{1}{\mu_3})\zeta e^{-\zeta a}e^{\zeta b}] \qquad (49)$$

Now, inversely transforming the Fourier transforms, one can calculate the vector potentials in different regions:

$$A_1(x,y) = \frac{1}{2\pi}\int_{-\infty}^{+\infty}(K_3 e^{2\zeta a} + K_4)e^{-\zeta y}e^{j\zeta x}d\zeta \quad , \quad |y| \ge a \qquad (50)$$

$$A_2(x,y) = \frac{j2J_1}{2\pi\mu_3}\int_{-\infty}^{+\infty}(e^{-j(k+\zeta)L}-1)\frac{\cosh(\gamma b)}{(k+\zeta)H_1(\zeta)}e^{j\zeta x}d\zeta \quad , \quad -b \le y \le b \qquad (51)$$

$$A_3(x,y) = \frac{j2J_1}{2\pi\mu_3}\int_{-\infty}^{+\infty}(e^{-j(k+\zeta)L}-1)\frac{G(\zeta,y)}{(k+\zeta)H_1(\zeta)}e^{j\zeta x}d\zeta \quad , \quad b \le |y| \le a \qquad (52)$$

Where

$$H_1(\zeta) = |A| / \zeta \qquad (53)$$

$$G(\zeta,y) = \cosh(\gamma b)\cosh(y-b)\zeta + \frac{\mu_3}{\mu_2}\frac{\gamma}{\zeta}\sinh(\gamma b)\sinh(y-b)\zeta \qquad (54)$$

Equations (50), (51) and (52) can be solved by residue theorem. For the thrust calculation, we need the air-gap flux density. So, the air-gap magnetic vector potential in region 3, A_3 is calculated. Thus, (52) is calculated as:

$$A_3(x,y) = \frac{j2J_1}{2\pi\mu_3}[(2\pi j)\sum_{l=1}^{n}\frac{P(\zeta_l)}{H'(\zeta_l)} + \pi j(\frac{G(-k,y)}{H_1(-k)}e^{-jkx})] \qquad (55)$$

where

$$P(\zeta) = (e^{-j(k+\zeta)L}-1)\frac{G(\zeta,y)}{(k+\zeta)} \qquad (56)$$

If the primary iron magnetic permeability is assumed to be infinity ($\mu_1 \to \infty$), then, the air-gap magnetic vector potential for different positions along x-axis are derived as [24]:

$$A_3(x,y) = \mu_3 J_1 \frac{(e^{-j(\zeta_0'+k)L}-1)}{(\zeta_0'+k)} \frac{G(\zeta_0',y)}{H'(\zeta_0')} e^{j\zeta_0'x} \quad , \quad x<0 \tag{57}$$

$$A_3(x,y) = \mu_3 J_1 [e^{-jkx}\frac{G(-k,y)}{H(-k)} + \frac{e^{j\zeta_0 x}}{(\zeta_0+k)}\frac{G(\zeta_0,y)}{H'(\zeta_0)} + \frac{e^{-j(\zeta_0'+k)L}}{(\zeta_0'+k)}\frac{G(\zeta_0',y)}{H'(\zeta_0')}e^{j\zeta_0'x}] \ , \ 0<x<L \tag{58}$$

$$A_3(x,y) = -\mu_3 J_1 \frac{(e^{-j(\zeta_0'+k)L}-1)}{(\zeta_0+k)} \frac{G(\zeta_0,y)}{H'(\zeta_0)} e^{j\zeta_0 x} \quad , \quad x>L \tag{59}$$

In the above equations, only the two dominant poles (ζ_0 and ζ_0') of the integrand in (55) is considered. By using (33) and (57)-(59), the y and x-components of the air-gap flux density are derived as:

$$B_{3y}(x,y) = -j\zeta_0'\mu_3 J_1 \frac{(e^{-j(\zeta_0'+k)L}-1)}{(\zeta_0'+k)} \frac{G(\zeta_0',y)}{H'(\zeta_0')} e^{j\zeta_0'x} \ , \ x<0 \tag{60}$$

$$B_{3y}(x,y) = j\mu_3 J_1 [ke^{-jkx}\frac{G(-k,y)}{H(-k)} - \frac{\zeta_0 e^{j\zeta_0 x}}{(\zeta_0+k)}\frac{G(\zeta_0,y)}{H'(\zeta_0)} - \zeta_0'\frac{e^{-j(\zeta_0'+k)L}}{(\zeta_0'+k)}\frac{G(\zeta_0',y)}{H'(\zeta_0')}e^{j\zeta_0'x}], \ 0<x<L \tag{61}$$

$$B_{3y}(x,y) = j\zeta_0\mu_3 J_1 \frac{(e^{-j(\zeta_0'+k)L}-1)}{(\zeta_0+k)} \frac{G(\zeta_0,y)}{H'(\zeta_0)} e^{j\zeta_0 x} \ , \ x>L \tag{62}$$

$$B_{3x}(x,y) = \mu_3 J_1 \frac{(e^{-j(\zeta_0'+k)L}-1)}{(\zeta_0'+k)H'(\zeta_0')} \frac{\partial G(\zeta_0',y)}{\partial y} e^{j\zeta_0'x} \ , \ x<0 \tag{63}$$

$$B_{3x}(x,y) = \mu_3 J_1 [\frac{e^{-jkx}}{H(-k)}\frac{\partial G(-k,y)}{\partial y} + \frac{e^{j\zeta_0 x}}{(\zeta_0+k)H'(\zeta_0)}\frac{\partial G(\zeta_0,y)}{\partial y} + \frac{e^{-j(\zeta_0'+k)L}}{(\zeta_0'+k)H'(\zeta_0')}\frac{\partial G(\zeta_0',y)}{\partial y}e^{j\zeta_0'x}], 0<x<L \tag{64}$$

$$B_{3x}(x,y) = -\mu_3 J_1 \frac{(e^{-j(\zeta_0'+k)L}-1)}{(\zeta_0+k)H'(\zeta_0)} \frac{\partial G(\zeta_0,y)}{\partial y} e^{j\zeta_0 x} \ , \ x>L \tag{65}$$

THRUST CALCULATION

The thrust of the motor is calculated by [17]:

$$F_T = \frac{1}{2}W_s \int_0^L \mathrm{Re}[j_1^*(x,t)b_y(x,y,t)]dx \tag{66}$$

Where $j*(x,t)$ is the complex conjugate of $j(x,t)$ which is given by (7). Also, $b_y(x,y,t)$ equals to:

$$b_y(x,y,t) = B_y(x,y)e^{j\omega t} \tag{67}$$

Where $B_y(x,y)$ is given by (61). Because, the motor current density is zero for $x<0$ and $x>L$ and the thrust in this zones are zero. Using (7) and (61) in (66), the thrust produced by the motor is derived as:

$$F_T = \frac{1}{2}W_s J_1^2 \{\mathrm{Re}[jk\mu_3 L\frac{G(-k,y)}{H(-k)}] + \mathrm{Re}[\zeta_0\mu_3 \frac{G(\zeta_0,y)}{(\zeta_0+k)^2 H'(\zeta_0)}(1-e^{j(\zeta_0+k)L})]$$
$$+ \mathrm{Re}[\zeta_0'\mu_3 \frac{e^{-j(\zeta_0'+k)L}}{(\zeta_0'+k)^2}\frac{G(\zeta_0',y)}{H'(\zeta_0')}(1-e^{j(\zeta_0'+k)L})]\} \tag{68}$$

It is seen that the thrust in (68) and also in (24) are composed of three components: net thrust without end effect, thrust due to entry end effect and thrust due to exit end effect.

IV. CALCULATION RESULTS

In this section, the flux density and the produced thrust of the motor are calculated by using 1-D and 2-D analysis and compared with the results of references [18]-[20] for validation. The specifications of the studied motor are illustrated in Table I.

TABLE I.   THE STUDIED MOTORS SPESIFICATIONS

Specification	Value
Primary current sheet density, A/m	50000
Synchronous speed, m/s	10
Secondary sheet thickness, mm	5
Air-gap length, mm	15
Slip	0.1
Input frequency, Hz	50
Motor length, m	0.5
Motor width, m	0.12

Fig. 3 shows the air-gap flux density with and without end effect which is obtained by 1-D and 2-D analysis. It is seen that the flux densities obtained by two methods are approximately the same in case without end effect. To better comparison, a part of the curve is magnified in the figure. Also, the results of the calculation by the two methods are close enough to each other in case of considering the end effect. The differences in the flux density at the entrance and exit ends are because of the different initial and boundary conditions used for solving the field equations. Fig. 4 illustrates the different components of the thrust produced by the motor versus the motor slip (only the results of 2-D analysis are shown). It is seen that the force due to exit end effect is negligible while the force due to entry end effect has large value and reduces the total thrust of the machine. In order to investigate the effect of the secondary sheet thickness on the thrust of the motor, the latter is calculated for different secondary sheet thicknesses which are shown in Fig. 5. It is seen that decreasing the secondary sheet thickness increases the maximum thrust. Also, the latter occurs in the large values of the motor slip.

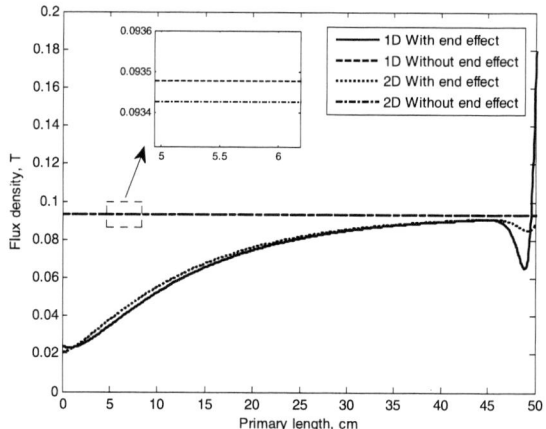

Figure 3.   Comparison of the air-gap flux density calculated by 1-D and 2-D models

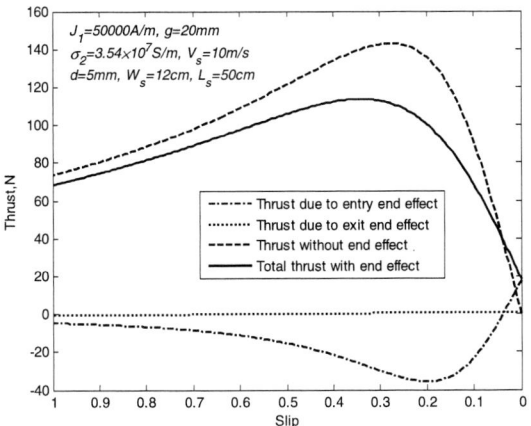

Figure 4. The thrust versus the motor slip with and without end effect (2-D analysis)

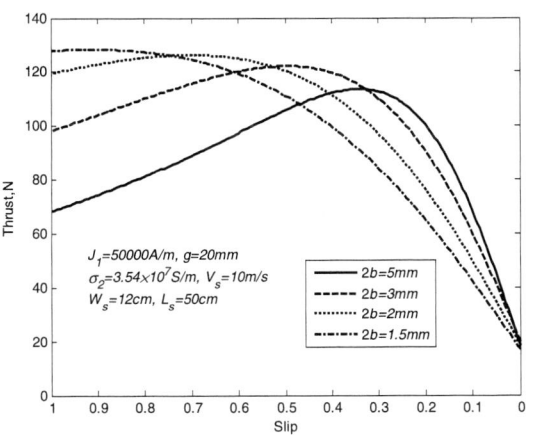

Figure 5. The effect of the the secondary sheet thikness on thrust with end effect (2-D analysis)

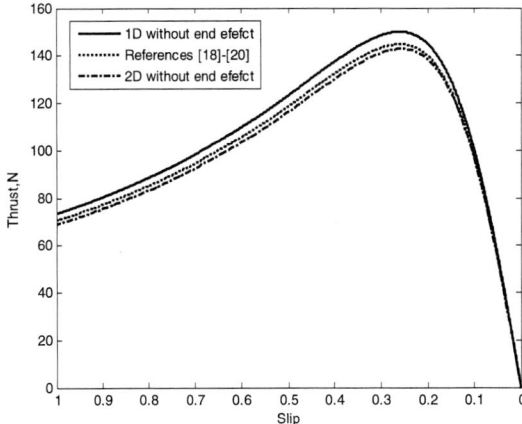

Figure 6. Comparison of the thrust obtained by 1-D and 2-D method(without end effect) with references [18]-[20]

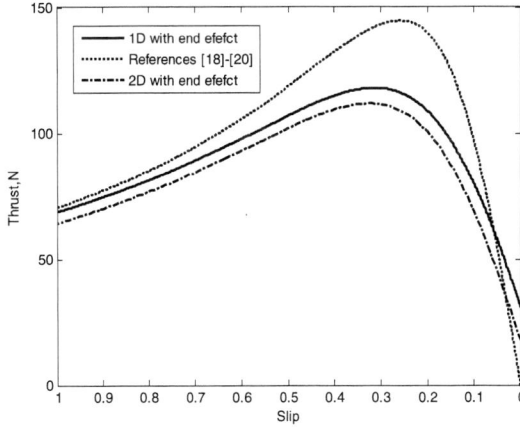

Figure 7. Comparison of the thrust obtained by 1-D and 2-D method(with end effect) with references [18]-[20]

V. VALIDATION

To validate the proposed methods, the results of the calculations are compared with those reported in [18]-[20]. Fig. 6 shows the 1-D and 2-D calculation results of the thrust without considering end effect and the results of [18]-[20]. In [18]-[20], the length of the motor is supposed infinitely long. It means that the end effect have been ignored in calculations. As it is seen in Fig. 6, the results of [18]-[20] are close enough to the results of the proposed 2-D analysis confirming the latter method. Although it is seen a difference between the results of 1-D model and the results of the 2-D and also the results of [18]-[20], the former are acceptable for many practical applications. It should be mentioned that the calculation procedure of the 1-D analysis is very simple in comparison with the 2-D method. In Fig. 7, the results of the thrust calculations considering end effect are presented. As it is seen in this figure, the thrust obtained by 2-D analysis is slightly less than that of obtained by 1-D method. Besides having less air-gap flux density in 2-D analysis, increasing the flux density in 1-D analysis at the exit end of the motor is the reason of this phenomenon.

VI. CONCLUSION

Electromagnetic forces produced by linear induction motor are calculated by using Maxwell field equations. As it is seen from the results, the end effect phenomenon produces braking force at two ends of the motor. The results of the calculations show that the exit end effect force is negligible and can be ignored in many practical applications, while, the entry end effect force is considerable and degrade the performance of the motor. The latter produces a braking force that acts as an external load. The comparison between the results of the calculations with the results of the methods reported in literature confirms the accuracy of the proposed methods.

VII. REFERENCES

[1] B. N. Turman, B. M. Marder, G. J. Rohwein, D. P. Aeschliman J. B. Keliey, M. Owan, R. M. Zimmerman, "The Pulsed Linear Induction Motor Concept for High-Speed Trains", Sandia National Laboratories, the United States Department of Energy, 1995.

[2] Kanji Wako, Kazuo Sawada, "Railway Technology Today", Japan Railway & Transport Review, October 2000.

[3] E. R. Laithwaite "Induction Machines for Special Purposes", Chemical Publishing Company, Inc., New York, 1966.

[4] R. Hellinger and P. Mnich, "Linear motor-powered transportation: history, present status, and future outlook", Proceedings of IEEE, Vol. 97, No. 11.

[5] S. Yamamura, H. Ito and Y. Ishikawa, "Theories of the linear induction motor and compensated linear induction motor", IEEE Transactions on Power Apparatus and Systems, Vol. PAS-91, No. 4, pp.: 1700-1710, 1972.

[6] M. Iwamoto, S. Sakabe, K. Itani, K. Kitagawa and G. Utsumi, "Experimental and theoretical study of high-speed single-sided linear induction motors", Proceedings of IEE, Vol. 128, Part B, No. 6, pp.: 306-312 November 1981.

[7] C. H. Lee and C. Y. Chin, "A theoretical analysis of linear induction motors", IEEE Transactions on Power Apparatus and Systems, Vol. PAS-98, No. 2, pp.: 679-688, March/April 1979.

[8] S. A. Nasar and L. del Cid, "Certain approaches to the analysis of single-sided linear induction motors", Proceedings of IEE, Vol. 120, No. 4, pp.: 477-483 April 1973.

[9] I. Boldea and S. A. Nasar, "Simulation of high-speed linear-induction-motor end effects in low-speed tests", Proceedings of IEE, Vol. 121, No. 9, pp.: 961-964 September 1974.

[10] B. T. Ooi, "A generalized machine theory of the linear induction motor", IEEE Transactions on Power Apparatus and Systems, Vol. PAS-92, No. 4, pp.: 1252-1259, 1973.

[11] M. Poloujadoff, B. Morel and A. Bolopion, "Simultaneous consideration of finite length and finite width of linear induction motors", IEEE Transactions on Power Apparatus and Systems, Vol. PAS-99, No. 3, pp.: 1172-1180, May/June 1973.

[12] J. Duncan, "Linear induction motor-equivalent-circuit model", IEE Proceedings B, Electric Power Applications, Vol. 130, No. 1, pp.: 51-57, January 1983.

[13] R. M. Pai, I. Boldea, and S. A. Nasar, "A complete equivalent circuit of a linear induction motor with sheet secondary", IEEE Trans. Magn., Vol. 24, No. 1, January 1988.

[14] W. Xu, J. Zhu, Y. Guo, Y. Wang, Y. Zhang and L. Tan, "Equivalent circuits for single-sided linear induction motors", IEEE Energy Congress and Exposition, pp.: 1288-1295, 2009.

[15] W. Xu, Y. Guo, Y. Zhang, Y. Li, Y. Wang, and Y. Guo, "An improved equivalent circuit model of a single-sided linear induction motor", IEEE Transactions on Vehicular Technology, Vol. 59, No. 5, pp.: 2277-2289, June 2010.

[16] W. Xu, J. G. Zhu, Y. Zhang, Z. Li, Y. Li, Y. Wang, Y. Guo and Y. Li, "Equivalent circuits for single-sided linear induction motors", IEEE Transactions on Industry Applications, Vol. 46, No. 6, pp.: 2410-2423, November/December 2010.

[17] W. Xu, G. Sun, G. Wen, Z. Wu and P. K. Chu, "Equivalent circuit derivation and performance analysis of a single-sided linear induction motor based on the winding function theory", IEEE Transactions on Vehicular Technology, Vol. 61, No. 4, pp.: 1515-1525, May 2012.

[18] B. T. Ooi and D C. White, "Traction and normal forces in the linear induction motor", IEEE Transactions on Power Apparatus and Systems, Vol. PAS-89, No. 4, April 1970.

[19] Z. Yang, J. Zhao and T. Q. Zheng, "A novel traction and normal forces study for the linear induction motor", IEEE International Conference on Electrical Machines and Systems, pp.: 3474-3477, 2008.

[20] L. Shi, H. Zhang and K. Wang, "Analysis of forces in combined levitation-and-propulsion SLIM system", IEEE International Conference on Electrical Machines and Systems, pp.: 2960-2963, 2008.

[21] H. Bolton, "Forces in induction motors with laterally asymmetric sheet secondary", Proceedings of IEE, Vol. 177, No. 12, December 1970.

[22] A. Shiri and A. Shoulaie, "Design optimization and analysis of single-sided linear induction motor, considering all Phenomena", IEEE Transactions on Energy Conversion, Vol. 27, No. 2, June 2012.

[23] A. Shiri and A. Shoulaie, "End effect braking force reduction in high-speed single-sided linear induction machine", Energy Conversion and Management, Elsevier, Volume 61, 2012.

[24] A. Shiri, "Design and optimization of linear induction motor, considering end effect", PhD Desertation, Iran University of Science and Technology, Feb. 2013.

[25] D. K. Cheng, Field and wave electromagnetics. Eddison Wesley, Massachusetts 1983.

7th Power Electronics, Drive Systems & Technologies Conference (PEDSTC 2016)
16-18 Feb. 2016, Iran University of Science and Technology, Tehran, Iran

Electro Pump Modeling Using Laboratory System Data

Pooya Samanipour
Faculty of Electrical Engineering
Iran University of Science and Technology
Tehran, Iran
samanipour.pooya@gmail.com

Javad Poshtan
Faculty of Electrical Engineering
Iran University of Science and Technology
Tehran, Iran
jposhtan@iust.ac.ir

Abstract —**pumps are widely used in various industries and they are mostly driven by induction motors. For purposes such as system control, monitoring and model-based fault detection, a dynamic model of the motor-pump behavior is necessary. We used a nonlinear model to describe the behavior of an electro pump. The main problem to model the pump are the unknown parameters of hydraulic part of the pump. In this paper, two approaches are proposed to identify the hydraulic part of the pump. First, laboratory system data are used to train a neural network which identifies a black-box model for the hydraulic part of the pump. In the second approach, the parameters are estimated via least squares using laboratory system data. The main advantage of the proposed methods is that the impeller vane angel need not be measured to model the hydraulic part of the pump. Finally, the model is simulated with both proposed methods, and results are evaluated on a laboratory system. These results show the model describes the electro pump dynamic with good precision.**

Keywords— electro pump; modeling; parameter estimation; least square error; neural networks;

I. INTRODUCTION

For different purposes such as fault detection and system control, a model which describes the dynamic behavior of an electro pump is necessary. The electro pump plays an important role in different industries therefore the behavior of electro pumps have been described employing different models. [1], [2] proposed a linear model for the electro pump which just performs well around the operating point. [2] used least square error to estimate pump parameters for fault detection purposes. However, [3] proposed a nonlinear model in which the velocity triangle is used to extract pump parameters. This is difficult to perform and may introduce substantial error into angel measuring. In [4] the model is presented for a couple of synchronous motor and centrifugal pump in steady state, and equivalent electrical circuits are used for modeling.

In this paper a nonlinear model is derived to describe the dynamic behavior of a motor-pump. Equations are presented in the state space form so that the model is suitable for different applications such as model-based fault detection and monitoring. One of the main challenges in motor-pump modeling are the unknown parameters of the hydraulic part of the pump. Impeller type determines these parameters. In this paper two approaches are suggested to identify the hydraulic part of the pump, and the main advantage of both approaches is that the impeller vane angel need not be

measured. First, an MLP neural network is used, and the neural network is trained using the laboratory system data. The network identifies the hydraulic part of the pump as a black-box. In the second approach, the pump parameters are obtained via least squares estimation using laboratory system data. Finally the presented model is simulated with both approaches, and simulation results are evaluated on a laboratory system. Measured volume flow and pump pressure are compared with simulated signals, and results show that the proposed model is able to successfully describe the laboratory system with good precision. Also the neural network showed better precision than least squares in modeling the hydraulic part of the pump.

II. MODELING

In this section a model is presented for a motor-pump set. The electro pump is composed of four subsystems. The first subsystem is the induction motor, the second one is the mechanical part of the pump, the third subsystem is the hydraulic part of the pump, and the last one is the hydraulic application of the pump. The induction motor produces mechanical energy, and the hydraulic part of the pump converts the mechanical energy to hydraulic energy. Subsystems equations are extracted in section A and B, and the final model is described in section C. also, in section D an observer is designed to estimate states and rotational speed of the induction motor. A block diagram of the motor-pump set is depicted in Fig.1 [5].

A. Induction motor model

One of the major approaches in modeling induction motors is first to obtain the equations in *abc* form, and then a transformation is used to map them to the *dq0* coordinate. Induction motor state equations can be describes as [6]:

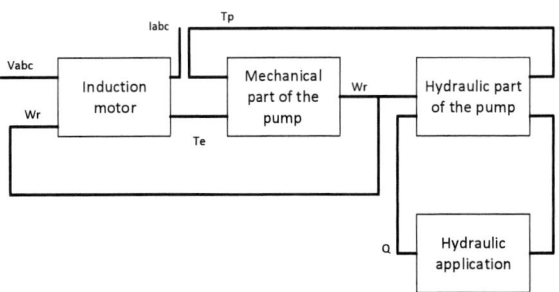

Fig. 1. Motor pump structure

978-1-5090-0376-1/16 $31.00 © 2016 IEEE

$$p\psi_{qs} = \omega_b\left[V_{qs} - \frac{\omega}{\omega_b}\psi_{ds} + \frac{r_s}{X_{ls}}\left(\psi_{mq} - \psi_{qs}\right)\right] \quad (1)$$

$$p\psi_{ds} = \omega_b\left[V_{ds} + \frac{\omega}{\omega_b}\psi_{qs} + \frac{r_s}{X_{ls}}\left(\psi_{md} - \psi_{ds}\right)\right]$$

$$p\psi_{0s} = \omega_b\left[V_{0s} - \frac{r_s}{X_{ls}}\left(\psi_{0s}\right)\right]$$

$$p\psi'_{qr} = \omega_b\left[-\frac{\omega - \omega_r}{\omega_b}\psi'_{dr} + \frac{r'_r}{X'_{lr}}\left(\psi_{mq} - \psi'_{qr}\right)\right]$$

$$p\psi'_{qr} = \omega_b\left[-\frac{\omega - \omega_r}{\omega_b}\psi'_{dr} + \frac{r'_r}{X'_{lr}}\left(\psi_{mq} - \psi'_{qr}\right)\right]$$

$$p\psi'_{0r} = \omega_b\left[\frac{r'_r}{X'_{lr}}\psi'_{0r}\right]$$

where

$$\psi_{mq} = X_{aq}\left(\frac{\psi_{qs}}{X_{ls}} + \frac{\psi'_{qr}}{X'_{lr}}\right) \quad (2)$$

$$\psi_{md} = X_{ad}\left(\frac{\psi_{qs}}{X_{ls}} + \frac{\psi'_{qr}}{X'_{lr}}\right)$$

$$X_{ad} = X_{aq} = \left(\frac{1}{X_M} + \frac{1}{X_{ls}} + \frac{1}{X'_{lr}}\right)^{-1}$$

In (1) and (2) s and r refer to stator and rotor circuits respectively. In these equations p is the derivative operator, $\psi_{q,d,0}$ is $dq0$ flux linkage, $r_{s,r}$ is stator and rotor resistance, X_m is magnetizing reactance, and X_l is leakage reactance. Also, $V_{q,d}$ is power supply voltage in dq0 coordinate, and $\omega, \omega_r, \omega_b$ are angular velocity, rotating circuit angular velocity and base electrical rotating velocity [6].

B. Pump model

As it mentioned earlier, the pump converts mechanical energy into hydraulic energy. The mechanical part of the pump transforms the mechanical energy which is produced by the induction motor to the hydraulic part of the pump. The mechanical part of the pump can be described using newton's law as follows [3]:

$$J\frac{d\omega_r}{dt} = T_e - B\omega_r - T_p \quad (3)$$

In (3) T_e is the electromagnetic torque which is produced by the motor, J is the moment of inertia of the electro pump, and T_p is pump torque. It is assumed that the friction has a linear relation with angular speed and B is the friction coefficient.

The hydraulic part of the pump includes the input and the output of the pump, impeller and diffuser. Impeller is the rotational component that induces rotational speed to fluid. Speed converts to static pressure in diffuser. Thus, the pressure difference among impeller eye and volute exit is obtained and this pressure difference is used in pump equations [3].

In this section, both the pump torque and pressure are introduced as a function of flow and angular velocity. A block diagram of the hydraulic part of the pump is illustrated in Fig.2. Torque and pressure can be described as the following:

$$T_p = f_t(Q, \omega_r) \quad (4)$$

$$H = f_H(Q, \omega_r)$$

In these equations Q is flow, and H is pressure. In major references, f_t and f_H in (4) are described by the following equations [3]:

$$H = \rho g(-a_{h2}Q^2 + a_{h1}\omega_r Q + a_{h0}\omega_r^2) \quad (5)$$

$$T_p = -a_{t2}Q^2 + a_{t1}\omega_r Q + a_{t0}\omega_r^2 \quad (6)$$

Equation (5) describes the pump pressure, and equation (6) describes the produced torque by the pump. In these equations $a_{h0}, a_{h1}, a_{h2}, a_{t0}, a_{t1}$ and a_{t2} are the unknown parameters. In these equations some terms are neglected.

As it can be seen in Fig.1, flow is the input of the hydraulic part of the pump. To determine the fluid flow, it is necessary to the model the hydraulic application of the pump. If the application of the pump is to transports fluid from lower to higher level in a pipeline, the related equation can be described as [2]:

$$h = a_F\frac{dQ}{dt} + h_{rr}Q^2 + h_{static} \quad (7)$$

In (7) h is the delivery head, h_{rr} is the resistant coefficient of the pipeline, and $a_F = l/gA$ where l is the pipeline length and A is the pipe cross sectional area. Also h_{static} is the height of the higher level. If the pump transports the fluid in a closed pipe circuit, then $\frac{dQ}{dt} = 0$ and $h_{static} = 0$ in steady-state. The hydraulic application of the pump can be simply modeled due to laboratory system as [2]:

$$h = h_{rr}Q^2 \quad (8)$$

Now the pump model is complete and it can be used in the electro pump state-space equations.

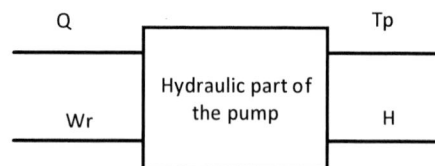

Fig. 2. Hydraulic part of the pump

C. Final model

In this section equations presented above are rewritten in a state-space form as the following:

$$\dot{x} = f(x) + Gu + m(x, Q) \tag{9}$$

$$y = h(x, Q) \tag{10}$$

In (9) state vector x includes φ and ω_r which represent stator and rotor flux in the dq0 frame, and rotational speed respectively. u is the input vector, and includes power supply voltages in dq0 frame. Stator currents and pump pressure are measured, and both are considered as system outputs. $m(x, Q)$ is employed to describe pump torque in (3). Thus, state vector, input vector and measurement vector can be describes as:

$$x = [\psi_{qs} \ \psi_{ds} \ \psi_{qr} \ \psi_{dr} \ \omega_r] \tag{11}$$

$$u = \begin{bmatrix} v_{sd} & v_{sq} \end{bmatrix} \tag{12}$$

$$y = \begin{bmatrix} i_{sd} & i_{sq} & H_p \end{bmatrix}^T \tag{13}$$

As a result, $f(x), G, m$ and h can be described as the following [3].

$$f(x) = \begin{pmatrix} \omega_b \left[-\frac{\omega}{w_b} \psi_{ds} + \frac{r_s}{X_{ls}} \left(\psi_{mq} - \psi_{qs} \right) \right] \\ \omega_b \left[\frac{\omega}{w_b} \psi_{qs} + \frac{r_s}{X_{ls}} (\psi_{md} - \psi_{ds}) \right] \\ \omega_b \left[-\frac{\omega - \omega_r}{\omega_b} \psi'_{dr} + \frac{r'_r}{X'_{lr}} (\psi_{mq} - \psi'_{qr}) \right] \\ \omega_b \left[\frac{\omega - \omega_r}{\omega_b} \psi'_{qr} + \frac{r'_r}{X'_{lr}} (\psi_{md} - \psi'_{dr}) \right] \\ T_e/J - B\omega_r/J \end{pmatrix} \tag{14}$$

$$G = \begin{pmatrix} \omega_b & 0 \\ 0 & \omega_b \\ 0 & 0 \\ 0 & 0 \\ 0 & 0 \end{pmatrix}, h(x, w) = \begin{pmatrix} i_{sd} \\ i_{sq} \\ f_H(Q, \omega_r) \end{pmatrix} \tag{15}$$

$$m(x, Q) = \begin{pmatrix} 0 \\ 0 \\ 0 \\ 0 \\ \frac{-1}{J} f_T(Q, \omega_r) \end{pmatrix} \tag{16}$$

The first four equations in (14) are the induction motor equations presented in (1), and the last equation is related to the mechanical part of the pump as described in (3).

D. Induction motor speed estimation

As it can be seen, in (14), (15) and (16) f_T, f_H and T_e are used. As it mentioned earlier, f_T and f_H are functions of angular speed. T_e is a function of fluxes and stator currents. Angular speed and fluxes are state variables and stator currents are measured variables. As a result, a state observer is required to determine the state variables values.

In general, induction motor state equations are as follows:

$$\frac{d}{dt}\begin{bmatrix} \psi_s \\ \psi_r \end{bmatrix} = \begin{bmatrix} A_{11} & A_{12} \\ A_{21} & A_{22} \end{bmatrix} \begin{bmatrix} \psi_s \\ \psi_r \end{bmatrix} + \begin{bmatrix} B_1 \\ 0 \end{bmatrix} V_s = Ax + BV_s \tag{17}$$

$$i_s = Cx \tag{18}$$

Consequently, observer equations for (17) and (18) can be described as:

$$\frac{d}{dt}\hat{x} = A\hat{x} + BV_s + L(\hat{i}_s - i_s) \tag{19}$$

$$\hat{i}_s = C\hat{x} \tag{20}$$

where A and B are the main system matrices, L is the observer gain, and $\hat{x}$ is the estimated state.

According to Lyapunov equation and observer stability condition, the estimated speed is given by [7]:

$$\hat{\omega}_r = K_P (e_{ids}\hat{\psi}_{qr} - e_{iqs}\hat{\psi}_{dr}) + K_I \int (e_{ids}\hat{\psi}_{qr} - e_{iqs}\hat{\psi}_{dr}) dt \tag{21}$$

In (21), e_{ids} and e_{iqs} are current estimation errors and k_p and k_i coefficients are obtained experimentally on a trial and error basis. Eventually, electromagnetic torque is obtained from estimated states in the following form[7].

$$T_e = \frac{3}{2}\frac{P}{2}\frac{1}{\omega_b}(\hat{\psi}'_{qr}\hat{i}'_{dr} - \hat{\psi}'_{dr}\hat{i}'_{qr}) = \frac{3}{2}\frac{P}{2}\frac{1}{\omega_b}(\hat{\psi}_{ds}\hat{i}_{qs} - \hat{\psi}_{qs}\hat{i}_{ds}) \tag{22}$$

III. PUMP PARAMETERS

In the final model (9) to (16), there are some unknown parameters to describe torque and pressure. These parameters are determined according to the pump impeller type. Velocity triangle is used to estimate these parameters in [8]. Velocity triangle uses impeller vane angels which are hard to measure correctly. In this paper, a heuristic method is proposed to identify the hydraulic part of the pump. Two approaches are considered, in the first approach, neural networks are used to model the hydraulic part of the pump as a black-box. The neural network is trained using laboratory system data. In the second approach, least square error is used to estimate parameters using pressure, flow, speed and torque.

As mentioned earlier, torque and pressure are functions of the angular speed and flow. As a result, in the first approach, flow and speed are considered as the inputs, while pressure and torque are the outputs of the neural network. Initial structure of the identifier is selected to be a feedforward neural network with five hidden layers. The hidden layer transfer function is assumed to be a sigmoid function, and output layer transfer function is a linear activation function. In this neural network, Levenberg-Marquardt algorithm is used to optimize and update network weights. Flow and pressure are measured, and speed is estimated according to (21). Also torque is obtained by (3). Fig.3 shows the structure of the neural network.

Therefore the neural network outputs are obtained as follows:

$$H = \sum_{j=1}^{5} w_{1j} \frac{1}{1+\exp\left(-(\sum_{i=1}^{2} w_{ij} u_i - b_j)\right)} - b_{11} \quad (23)$$

$$T_p = \sum_{j=1}^{5} w_{2j} \frac{1}{1+\exp(-(\sum_{i=1}^{2} w_{ij} u_i - b_j))} - b_{21} \quad (24)$$

In (23) and (24) b_i and w_{ij} are the neural network biases and weights respectively. To train the neural network, data are collected from laboratory system in different operating points to provide a rich data-base.

In the second approach, least square error is used to estimate pump parameters. Flow and pressure are measured, speed is estimated with rotor and stator currents according to (21), and pump torque can be calculated as follows:

$$T_p \omega_r = HQ \quad (25)$$

In (25) some terms such as friction are neglected. Finally, parameters can be estimated using linear regression as below:

$$H = \phi^T \Theta_H \quad (26)$$

$$T_p = \phi^T \Theta_T \quad (27)$$

In (26) and (27) data vector and parameters vector are defined as:

$$\phi^T = [\omega^2 \ \omega Q \ Q^2] \quad (28)$$

$$\Theta_H^T = [a_{h0} \ a_{h1} \ a_{h2}] \quad (29)$$

$$\Theta_T^T = [a_{t0} \ a_{t1} \ a_{t2}] \quad (30)$$

Flow, pressure and torque change in different operating points along the pump characteristic curves. In (28), the data vector contains laboratory system data in different operating points, then parameter vector are estimated according to (26) and (27) using least squares. In [2], least square error is used to estimate the pump parameters for fault detection purposes.

IV. RESULT AND DISCUSSION

The extracted model is evaluated on a laboratory system consisting of a 3KW induction motor and a 3-stage centrifugal pump. Data is transferred to a computer via an ADVANTECH PCI-1711 data card. Maximum frequency of data collection is 10000 samples per second. Voltage, current, flow and pressure of the pump are measured by related sensors.

Rotational speed is estimated by (21) where $K_p = .5$ and $K_i = 50$ are chosen by trial and error. To obtain the induction motor parameters, different tests are carried out on the induction motor such as no-load test, DC test and blocked-rotor test.

The pump parameters are obtained to complete the model by (26) and (27). These parameters are shown in TABLE I.

The parameters are extracted in a way which is compatible with the pump pressure-flow and power-flow curves.

To evaluate the extracted model precision, simulation results are compared to laboratory system data. In Fig.4 the laboratory pump pressure signal is compared with the simulated one. As it can be seen, the simulated signals converge to measured signal after 2 seconds, with small errors shown in Fig5. Also the neural network has better precision compare to least squares. A rich data-base can help to reduce errors. As a result, the extracted model has reliable result and model behavior is close enough to the laboratory system.

Also both measured flow and simulated flow signals can be compared in Fig.6. Error is small and simulation results converge to measured data after 2 seconds. Simulation error is shown in Fig.7.

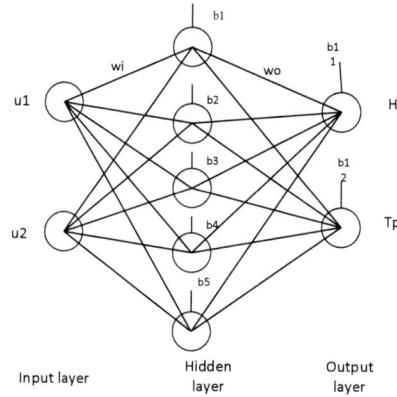

Fig. 3. Neural network structure

TABLE I. CENTRIFUGAL PUMP PARAMETERS

	a_0	a_1	a_2
H	0.0299	65.3831	-4.57e+07
T_p	-5.36e-05	328.8306	-7.02e+06

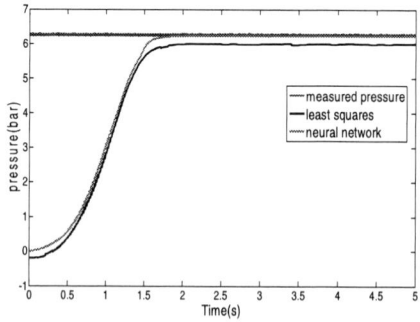

Fig. 4. Measured pressure and simulated signals

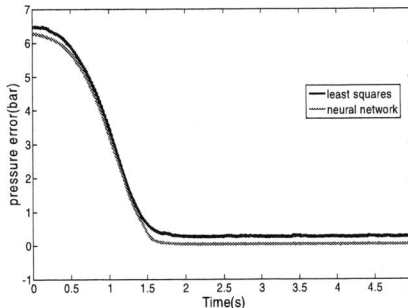

Fig. 5. Pressure error

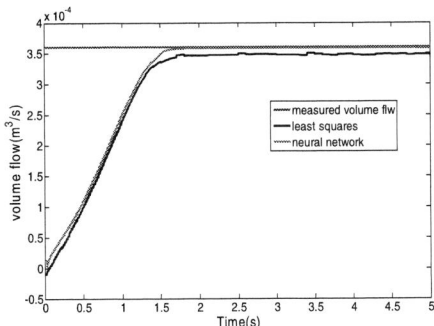

Fig. 6. Measured volume flow and simulated signals

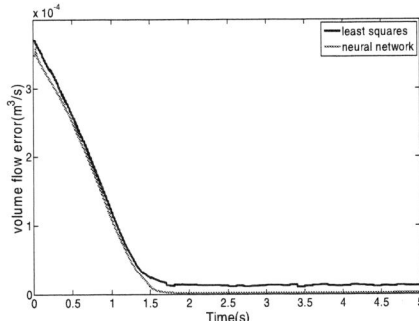

Fig. 7. Volume flow error

V. CONCLUSION

In this paper a nonlinear model is presented for an electro pump. Dynamic equations are derived in state space form which are suitable for some applications such as control and fault diagnosis. The main challenge in this modeling are parameters of the hydraulic part of the pump. To obtain these parameters the measurement of impeller vane angle is normally required. In this paper two methods are suggested to model the hydraulic part of the pump using laboratory system data to avoid measuring impeller vane angel. A neural network is used to identify the hydraulic part of the pump as a black-box, and Laboratory data is used to train the neural network. Also laboratory data is used to estimate the unknown parameters via least square error.

Finally suggested methods were applied to a laboratory motor-pump. Simulation results are compared with the laboratory system data which showed that the both methods are capable to model the hydraulic part of the pump. Also the neural network showed better precision than least square in modeling the hydraulic part of the pump.

REFRENCES

[1] G. Janevska, "Mathematical Modeling of Pump System," in *Proceedings in EIIC-The 2nd Electronic International Interdisciplinary Conference*, 2013.

[2] R. Isermann, *Fault-diagnosis applications: model-based condition monitoring: actuators, drives, machinery, plants, sensors, and fault-tolerant systems*: Springer Science & Business Media, 2011.

[3] C. Kallesøe, *Fault detection and isolation in centrifugal pumps*: Videnbasen for Aalborg UniversitetVBN, Aalborg UniversitetAalborg University, Det Teknisk-Naturvidenskabelige FakultetThe Faculty of Engineering and Science, Institut for Elektroniske SystemerDepartment of Electronic Systems, 2005.

[4] P. Gogolyuk, V. Lysiak, and I. Grinberg, "Mathematical modeling of a synchronous motor and centrifugal pump combination in steady state," in *Power Systems Conference and Exposition, 2004. IEEE PES*, 2004, pp. 1444-1448.

[5] C. S. KallesØe, V. Cocquempot, and R. Izadi-Zamanabadi, "Model based fault detection in a centrifugal pump application," *Control Systems Technology, IEEE Transactions on*, vol. 14, pp. 204-215, 2006.

[6] P. C. Krause, O. Wasynczuk, S. D. Sudhoff, and S. Pekarek, *Analysis of electric machinery and drive systems* vol. 75: John Wiley & Sons, 2013.

[7] H. Kubota, K. Matsuse, and T. Nakano, "DSP-based speed adaptive flux observer of induction motor," *IEEE Transactions on Industry Applications*, vol. 29, pp. 344-348, 1993.

[8] A. T. Sayers, *Hydraulic and compressible flow turbomachines*: McGraw-Hill, 1990.

7th Power Electronics, Drive Systems & Technologies Conference (PEDSTC 2016)
16-18 Feb. 2016, Iran University of Science and Technology, Tehran, Iran

Family of Single-Switch-Soft Switching PWM Converters with Single Magnetic Core

Moretza Esteki
Department of Electrical
and Computer Engineering
Isfahan University of Technology
Isfahan, Iran
m.esteki@ec.iut.ac.ir

Mehdi Mohammadi
Department of Electrical
and Computer Engineering
University of British Columbia
Vancouver, BC, Canada
mehdi.mohammadi.m@ieee.org

Ehsan Adib, Hosein Farzanehfard
Department of Electrical
and Computer Engineering
Isfahan University of Technology
Isfahan, Iran
ehsan.adib@cc.iut.ac.ir,hosein@cc.iut.ac.ir

Abstract—This paper introduces a family of single switch soft switching PWM converters where only one magnetic element is employed. The proposed converters within the family have the ability to provide soft switching conditions for all semiconductor elements. As a result, the problems related to the reverse recovery of diodes are excluded. The lossless passive snubber circuit in this family can be applied to a wide variety of switching power converters. Among the converters within the family, the theoretical analysis is provided for the buck converter, since the basic performance of the lossless snubber circuit is the same for the converters. To validate the theoretical analysis, a 120W prototype buck converter is implemented experimentally. The experimental results show that the proposed converter improves the converter's efficiency by over 2% compared to a conventional buck converter.

Keywords— soft switching; PWM converters; zero current switching (ZCS); zero voltage switching (ZVS)

I. INTRODUCTION

These days, switching power converters are being employed in many applications such as driving light emitting diodes (LEDs) that can be used as lighting systems [1], power factor corrector systems [2], fuel cell interface systems [3], photovoltaic systems [4], and battery chargers [5]. There is an overwhelming interest to design and implement these systems in a smaller volume and weight, while the overall efficiency should be maintained or improved [6]. Generally, increasing the switching frequency is the most effective way to reduce the weight and size of power converters. However, its unwanted side effect is increasing the switching losses. Generally, for PWM converters to decrease the switching losses, lossless snubber circuits are used to provide soft switching conditions. Providing soft switching conditions in high switching frequency converters has several merits, including reduction of switching losses and electromagnetic interference (EMI) which consequently can increase the converter power density [7-8]. In general, lossless snubber circuits are categorized as active and passive snubbers. In active snubbers, an auxiliary switch is used to control the operation of the snubber circuit. Thus, active snubber employs an additional switch which requires proper control and gate

drive circuits [9] where the appropriate time to drive the converter switches should be detected suitably [10]. Moreover, due to the presence of additional switch, the switching losses may increase [11]. In some topologies both the converter and auxiliary switches require floating gate signals which complicate the driving circuit [12] -[13]. Also, only in some of these methods, the auxiliary switch can be driven with the bootstrap technique which reduces the complexity of the gate drive circuit [14] -[15]. If the bootstrap technique cannot be applied to a floating switch, a pulse transformer or Opto-coupler should be used which adversely affects the power conversion density. In converters with a high switching frequency, pulse transformers are not effective due to their leakage inductance and also application of Opto-couplers can contribute to the converter cost. In general, the only parasitic component which passive methods cannot recover its energy is the output capacitor of the converter switch. Since this capacitor is very small, its losses compared to other switching losses in the converter are not important at low and moderate switching frequencies [16]. In [17], soft switching conditions are obtained with an active auxiliary cell, but the auxiliary switch is not turned on under zero voltage switching (ZVS) condition and so its output parasitic capacitor discharges into the auxiliary switch after turn on. Despite the disadvantages of active methods, some of these methods are able to provide ZVS and zero current switching (ZCS) conditions at turn on and off instants, respectively. Turning the converter switch on under ZVS would recover capacitive turn on losses [18].

In passive approach, no extra switch is used and to obtain soft switching conditions and only passive elements, including inductors, capacitors and diodes are used. Since, these components have lower failure rates in comparison to active elements, passive methods are more reliable [19]. In [20], a lossless passive snubber circuit is introduced for the double ended flyback converter. The passive snubber provides ZVS and ZCS conditions for the switches at turn off and turn on moments, respectively. Although, the soft switching condition is achieved for all semiconductor elements, once the converter switches are on, the currents of the two snubber inductors freewheel through the converter switches and snubber diodes which contributes to circulating energy losses. Also, in [21]

978-1-5090-0376-1/16 $31.00 © 2016 IEEE

a simple snubber cell is applied to a half bridge interleaved flyback converter. Although, the soft switching condition is provided and the converter output current is continuous in the view point of the output capacitor, two magnetic cores are engaged which increases the weight and size of the converter. In [22] a turn off lossless passive snubber circuit is suggested for boost converter. In the power path two diodes exist which leads to high conduction losses. Also, some of the snubber circuit components are omitted in [23]. The number of the snubber circuit components in [23] is lower than its counterpart in [22], but when the converter switch is off, the output current passes through three diodes. This causes the conduction losses to be more than the converter in [22]. In [24] a family of soft switching PWM converters is introduced which uses the concept of coupled inductors. Because, one of the coupled inductors of the snubber circuit is in series with the converter switch, the switch is turned off under semi ZVS condition. Moreover, the coupling coefficient of the coupled inductors is not unity in practice and hence on each side the leakage inductors exist. Thus, at turn off instant, a ringing occurs due to the resonance between the output capacitor of the converter switch and the leakage inductance of the coupled inductors. The introduced snubber circuit in [25] can be applied for a boost converter. In this converter, when the converter switch is on, the current through the snubber inductors which are coupled together, freewheels through the converter switch until it is turned off. Thus, a portion of the snubber circuit energy is wasted in the switch. The inductor of the snubber circuit suggested in [26], uses the magnetic core of the main converter transformer. So, the snubber cell does not require an additional core for the snubber inductor. Also, a clamp capacitor at the secondary side of transformer clamps the voltage stress of the output diode.

In this paper, a family of non-isolated soft switching PWM converters is introduced. The snubber circuit uses the main converter's magnetic core. Thus, the suggested method to provide soft switching conditions is cost effective and also no additional inductors are required. The proposed snubber circuit is able to provide ZVS and ZCS conditions for the converter switch at turn off and turn on instants respectively. All semiconductor elements are soft switched in the proposed converter. The operation of a soft switched buck converter using the proposed snubber is discussed in details. The paper is arranged in 6 sections. In section II, the operation of the proposed soft switching buck converter is explained in details. In section III, the design procedure of the proposed snubber circuit is explained. Section IV presents the experimental results obtained from a prototype of a buck converter with the suggested passive snubber. In section V, other soft switching converters using the proposed snubber are introduced.

II. OPERATING PRINCIPLES

In Fig. 1, the proposed soft (switching) single switch (SSS) buck converter is shown. This converter consists of S_1, D_1, L_1 and C_o in addition to D_1, D_2, D_3, C_S and the two coupled inductors. It is noticeable that the inductors of the snubber cell are coupled with the converter output filter inductor. In order to simplify the description of the proposed converter, the following assumptions are made:

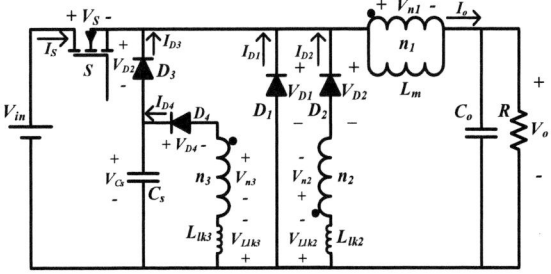

Fig. 1. Proposed PWM SSS Single magnetic core buck converter.

1) All components are ideal.

2) The output capacitor is large enough so that the output voltage can be considered constant.

In the proposed converter the coupling coefficient of three coupled inductors is not unitary and therefore, series leakage inductors L_{lk1}, L_{lk2} and L_{lk3} are considered. These leakage inductors provide ZCS condition for the converter switch. The proposed converter in each switching period has six operating modes. Fig. 2 shows the equivalent circuits of each operating mode and in Fig. 3, the key theoretical waveforms of the proposed buck converter are shown. Before the first mode, it is assumed that the converter switch S is off, D_2 is on and other semiconductor elements are off. Also, the voltage of C_S is zero.

Mode 1 [t_0-t_1]: At t_0, S is turned on under ZCS condition due to the leakage inductors L_{lk1}, L_{lk2} and L_{lk3}. Since the currents through L_{lk1} and L_{lk2} are equal before Mode 1, D_2 remains on in this mode. So, I_{Llk2} starts to decrease from I_o to zero and I_S increases from zero to I_o linearly. Also D_4 turns on under ZCS which causes a resonance to occur between C_S and L_{lk3}. Important equations of this mode are as follows:

$$I_S(t) = (\frac{n_3}{n_1})^2.(\frac{L_m.}{L_m + L_{lk1}})^2.\frac{(V_{in} - V_o)}{Z_1}.\sin \omega_1(t - t_0)$$
$$+ \frac{(L_m + L_{lk1}).V_{in} + (V_{in} - V_o).\frac{n_2}{n_1}.L_m}{L_{lk_2}(L_{lk1} + L_m.(1 + (\frac{n_2}{n_1})^2))}(t - t_0) \qquad (1)$$

$$I_{D_2}(t) = I_0 - \frac{(L_m + L_{lk1}).V_{in} + (V_{in} - V_o).\frac{n_2}{n_1}.L_m}{L_{lk_2}(L_{lk1} + L_m.(1 + (\frac{n_2}{n_1})^2))}(t - t_0) \qquad (2)$$

$$I_{D_4}(t) = \frac{n_3}{n_1}.\frac{(V_{in} - V_o)}{Z_1}.(\frac{L_m}{L_m + L_{lk_1}}).\sin \omega_1(t - t_0) \qquad (3)$$

$$V_{C_s}(t) = \frac{n_3}{n_1}.(V_{in} - V_o).(\frac{L_m}{L_m + L_{lk_1}})(1 - \cos \omega_1(t - t_0)) \qquad (4)$$

Where

$$I_0 = I_o(t = t_0) \qquad (5)$$

978-1-5090-0376-1/16 $31.00 © 2016 IEEE   117

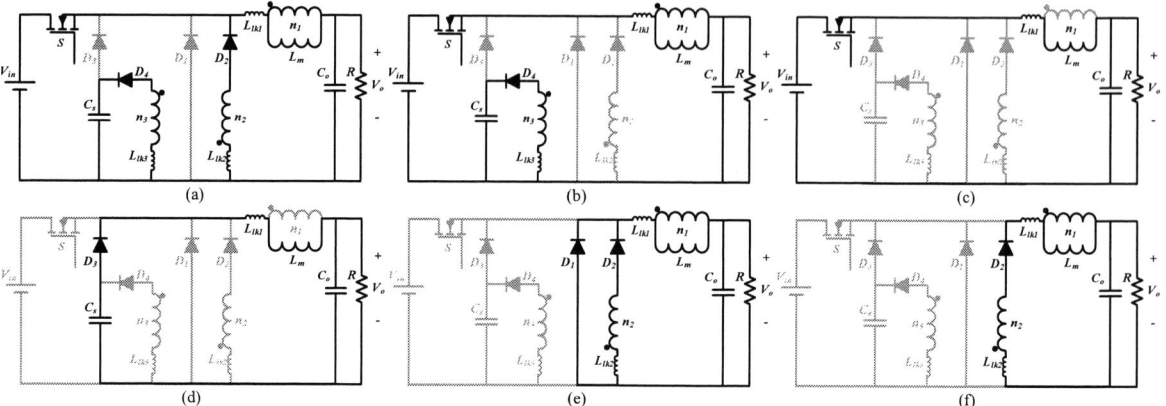

Fig. 2. The equivalent circuit of each operating mode of the proposed converter. (a) mode 1, (b) mode 2, (c) mode 3, (d) mode 4, (e) mode 5, and (f) mode 6.

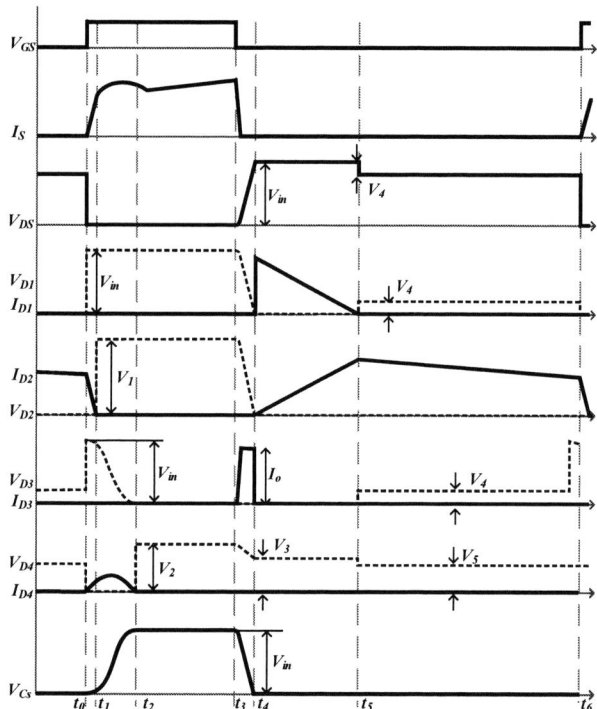

Fig. 3. Key waveforms of the proposed PWM SSS buck converter.

$$\omega_1 = \frac{1}{\sqrt{L_{eq}.C_s}} \qquad (6)$$

$$Z_1 = \sqrt{\frac{L_{eq}}{C_s}} \qquad (7)$$

$$L_{eq} = L_{lk3} + (\frac{n_3}{n_1})^2 . \frac{L_m.L_{lk1}}{L_m + L_{lk1}} \qquad (8)$$

When the current of D_2 reaches zero, D_2 turns off under ZCS and this mode ends.

Mode 2 [t_1-t_2]: At t_2, D_2 turns off under zero current switching condition and its voltage will increase to the

following value

$$V_1 = V_{in} + (V_{in} - V_o).\frac{n_2}{n_1}.\frac{L_m}{L_m + L_{lk1}} \qquad (9)$$

In this mode, L_m is charged and C_s continues its resonance with L_{lk3}. The equations of V_{Cs} and I_{D4} are the same as the pervious mode. The equation of switch current is as follows:

$$\begin{aligned} I_S(t) = (\frac{n_3}{n_1})^2 . (\frac{L_m.}{L_m + L_{lk1}})^2 . \frac{(V_{in} - V_o)}{Z_1}.\sin \omega_1 (t - t_1) \\ + \frac{V_{in} - V_o}{L_m + L_{lk1}}.(t - t_1) + I_1 \end{aligned} \qquad (10)$$

Where

$$I_1 = I_s(t = t_1) \qquad (11)$$

After half of the resonance period, the voltage of C_S has reached V_{in} and D_4 turns off.

Mode 3 [t_2-t_3]: In this mode, the converter operates similar to the conventional hard switching buck converter. All diodes are off and L_m is being charged. The voltage across D_4 during this mode is V_2.

$$V_2 = V_{in} - (V_{in} - V_o).\frac{n_3}{n_1}.\frac{L_m}{L_m + L_{lk1}} \qquad (12)$$

Mode 4 [t_3-t_4]: At the beginning of this mode the converter switch S is turned off under ZVS condition due to C_s, and D_3 turns on. Since the converter operates in continuous conduction mode (CCM), I_o is assumed almost constant in this mode and thus, C_s is being discharged by I_o. When C_s is discharged completely, D_1 turns on and this mode ends. Important equation of this mode is as follows:

$$V_{C_s}(t) = V_{in} - \frac{I_o}{C_s}(t - t_3) \qquad (13)$$

Mode 5 [t_4-t_5]: In this mode D_1 and D_2 conduct. During this interval, I_o is considered to be constant, also I_{Llk2} increases from zero to I_o and I_{D1} decreases from I_o to zero. During this mode,

978-1-5090-0376-1/16 $31.00 © 2016 IEEE   118

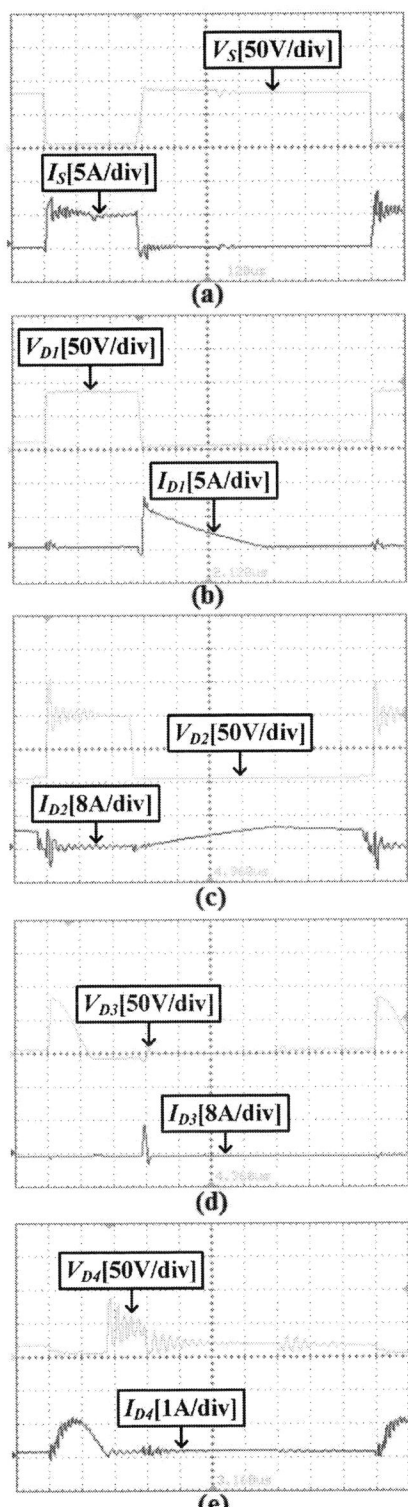

Fig. 4. Experimental results of the proposed converter; Voltage and current of (a) switch, (b) diode D_1, (c) diode D_2, (d) Diode D_3, and (e) diode D_4 (time scale 1µs/div).

Fig. 5. A picture of the power stage of the implemented prototype converter.

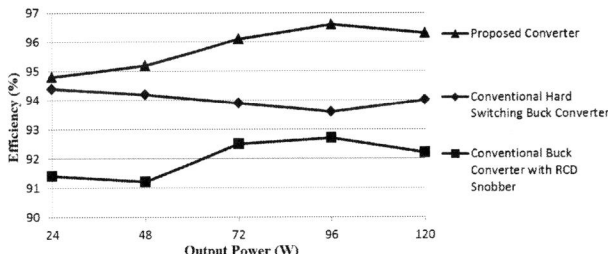

Fig. 6. Efficiency comparison of the conventional hard switching buck converter, conventional buck converter with RCD snubber and the proposed converter.

the voltage across D_4 is V_3.

$$V_3 = V_o . \frac{n_3}{n_1} . \frac{L_m}{L_m + L_{lk1}} \qquad (14)$$

When I_{D1} decreases to zero, this mode ends.

Mode 6 [t_5-t_6]: when I_{D1} reaches zero, D_1 turns off and this mode begins. During this mode, L_m transfers its energy to the output. At the beginning of this mode, the voltages across diodes D_1 and D_4 are V_4 and V_5, respectively.

$$V_4 = \frac{(\frac{n_2}{n_1}+1).L_m.V_o.\frac{n_3}{n_1}}{L_{lk1}+L_{lk2}+L_m.(\frac{n_2}{n_1}+1)^2} \qquad (15)$$

$$V_5 = \frac{L_{lk2}.V_o + \frac{n_2}{n_1}.(\frac{n_2}{n_1}+1).L_m.V_o}{L_{lk1}+L_{lk2}+L_m.(\frac{n_2}{n_1}+1)^2} \qquad (16)$$

III. DESIGN CONSIDERATION

The voltage conversion ratio of the proposed converter is

$$G = \frac{D}{1+(1-D).\dfrac{n_1}{n_1+n_2}} \qquad (17)$$

In order to design the proposed converter, it is assumed that the turn ratio of n_1/n_2 is large and the converter's conversion ratio is considered equal to the conventional buck converter. The output filter capacitor and the magnetizing inductor Lm which acts as the output filter inductor can be designed similar to the conventional converter. To obtain ZVS condition for the converter switch, C_S should limit the slope of the converter switch voltage at turn off instant [27]. So, this capacitor can be designed using the following equation [27]:

$$C_s = \frac{I_s . t_f}{2 \Delta V_{C_s}} \quad (18)$$

where, I_s is the switch current before turning the converter switch off, t_f is the switch current fall time, ΔV_{Cs} is the voltage across the switch when the switch current reaches zero. Using equation (4), and by considering the fact that V_{Cs} reaches V_{in} at the end of Mode 2, n_3/n_1 can be obtained.

$$\frac{n_3}{n_1} = \frac{V_{in}}{2.(V_{in} - V_o).(\frac{L_m}{L_m + L_{lk1}})} \quad (19)$$

The leakage inductor L_{lk2} provides ZCS at turn-on for the converter switch. For this purpose, L_{lk2} can be obtained using the equation bellow [27]:

$$L_{lk2} = \frac{V_{L_{lk2}} . t_r}{2 . \Delta I_s} \quad (20)$$

where, t_r is the switch current rise time, V_{Llk2} is voltage across L_{lk2} and ΔI_s is the variation of the switch current in the meantime that the voltage of the switch decrease to zero and the switch turns on .Also the leakage inductor L_{lk3} can be obtained from (7) and (18) as follows:

$$L_{lk3} = \frac{1}{\omega_1^2} - (\frac{n_3}{n_1})^2 . L_m . (1 - \frac{L_m}{L_m + L_{lk1}}) \quad (21)$$

Where, ω_1 is determined according to the minimum switch on-time.

$$\omega_1 = \frac{2\pi f_{sw}}{D} \quad (22)$$

Where $\underline{D}$ is the minimum duty cycle of the switch and f_{sw} is the switching frequency.

IV. EXPERIMENTAL RESULTS

In order to validate the theoretical analysis, a 120W prototype of the proposed buck converter with output voltage of 24V is implemented. The input voltage of the converter varies from 40V to 70V and the switching frequency is 100 KHz. The other parameters of the implemented prototype buck converter are shown in Table I. The magnetic core EI-3329 is used with n_1/n_2 =6 and n_1/n_3=1.2. The experimental waveforms of the implemented prototype are shown in Fig. 4. As can be observed from Fig 4.a, the converter switch is turned on under ZCS and turned off under ZVS condition. In Fig 4.b shows that D_1 turns on and off under ZVS and ZCS, respectively. Also, as can be observed from Fig 4.c and Fig. 4.e, D_2 and D_4 turn on and off under ZCS. It is shown in Fig 4.d that D_3 turns on and off under ZVS condition. Therefore, the soft switching condition is achieved for all semiconductor components. Fig 5 shows a picture of the power stage of the implemented prototype buck converter and Fig 6 shows the efficiency diagrams of the proposed buck converter, the

TABLE I
PARAMETERS OF THE PROTOTYPE CONVERTER

components	specification
L_m	$340\,\mu H$
n_1/n_2	6
n_1/n_3	1.2
L_{lk1}	$25\,\mu H$
L_{lk2}	$3\,\mu H$
L_{lk3}	$20\,\mu H$
C_s	$10\,\mu F$
Converter Switch	IRF540
D_1 and D_2	BYV32-200
D_3 and D_4	MUR115
Output capacitor	$47\,\mu F$

TABLE II
COMPARISON OF SSS CONVERTERS

Symbol	Converter Proposed in [11]	Converter proposed in [24]	Proposed Converter
Number of snubber diodes	4	1	3
Number of snubber capacitors	2	1	1
Number of snubber cores	2	1	0
Peak switch voltage (V)	102	108	70
Peak switch current (A)	7.7	6.1	7.3
Simulation efficiency (%)	95.4	95.3	96.5

conventional hard switching buck converter and the soft switching buck converter which uses RCD snubber circuit. Fig. 6 shows that the soft switching method has increased the converter efficiency as compared to the hard switching counterpart. Also, in Table II a quantitative comparison is performed between the proposed converter and other soft switching converters. Simulation results from PSPICE software are used to compare the proposed converter with other soft switching converters in Table II.

V. TOPOLOGY DERIVATION

The proposed soft switching method is extendable to other topologies. Fig. 7 shows SSS boost, buck-boost, Cuk, SEPIC and Zeta converters. In all of these converters, no additional magnetic core is used. In the converters in Fig. 7, L_{Lk2} is responsible for providing ZCS condition at turn on instant for the converter switch.

VI. CONCLUSION

In this paper, a family of single switch soft switching DC-DC converters which use only one magnetic core is introduced. The proposed method not only provides soft switching condition for the main converter switch, but also for all semiconductor elements as well. The presented experimental results justify the theoretical analysis. The experimental results show that the proposed technique increases the conversion efficiency as compared to the conventional converters.

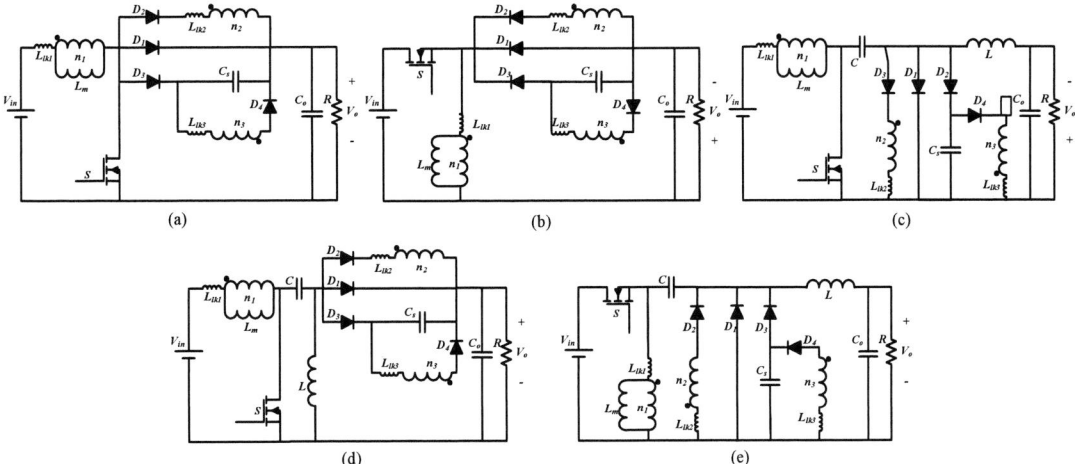

Fig. 7. Topology variations of the proposed converter (a) Boost, (b) Buck-boost (c) Cuk, (d) SEPIC, and (e) Zeta.

REFERENCES

[1] J.-K. Kim, J.-B. Lee, G.-W. Moon, "Isolated Switch-Mode Current Regulator With Integrated Two Boost LED Drivers," IEEE Trans. Ind. Electron., vol. 61, no. 9, pp. 4649-4653, Sep. 2014.

[2] H. S. Athab, D. D.-C. Lu, A. Yazdani, B. Wu, "An Efficient Single-Switch Quasi-Active PFC Converter with Continuous Input Current and Low DC-Bus Voltage Stress," *IEEE Trans. Ind. Electron.*, vol. 61, no. 4, pp. 1735-1749, Apr. 2014.

[3] N. Molavi, M. Esteki, E. Adib, H. Farzanehfard, "High step-up/down DC-DC bidirectional converter with low switch voltage stress, " *Power Electronics, Drives Systems & Technologies Conference (PEDSTC)*, 2015 6th, pp.162-167, 3-4 Feb. 2015.

[4] I.-O. Lee, G.-W. Moon, "Half-Bridge Integrated ZVS Full-Bridge Converter with Reduced Conduction Loss for Electric Vehicle Battery Chargers," *IEEE Trans. Ind. Electron.*, vol. 61, no. 8, pp. 3978-3988, Aug. 2014.

[5] M. Esteki, E. Adib, and H. Farzanehfard, "Soft switching interleaved PWM buck converter with one auxiliary switch," *Electrical Engineering (ICEE), 2014 22nd Iranian Conf.*, pp.232,237, 20-22 May 2014.

[6] N. Altintas, A. F. Bakan, I. Aksoy, "A Novel ZVT-ZCT-PWM Boost Converter," *IEEE Trans. Power Electron.*, vol. 29, no. 1, pp. 256-265, Jan. 2014.

[7] Y.-W. Kim, J.-H. Kim, K.-Y. Choi, B.-S. Suh, R.-Y. Kim, "A Novel Soft-Switched Auxiliary Resonant Circuit of a PFC ZVT-PWM Boost Converter for an Integrated Multichip Power Module Fabrication," *IEEE Trans. Ind. Applications,* vol. 49, no. 6, pp. 2802-2809, Nov. 2013.

[8] M. Mohammadi, E. Adib, M.R. Yazdani, "Family of Soft-Switching Single-Switch PWM Converters with Lossless Passive Snubber," *IEEE Trans. Ind. Electron.*, vol., no, pp., Nov. 2014.

[9] C. A. Gallo, F. L. Tofoli, J. A. C. Pinto, "A Passive Lossless Snubber Applied to the AC-DC Interleaved Boost Converter," *IEEE Trans. Power Electron.*, vol. 25, no. 3, pp. 775-785, Mar. 2010.

[10] J. L. Russi, V. F. Montagner, M. L. da Silva Martins, H. L. Hey, "A Simple Approach to Detect ZVT and Determine Its Time of Occurrence for PWM Converters," *IEEE Trans. Ind. Electron.*, vol. 60, no. 7, pp. 2576-2585, Jul. 2013.

[11] R. T. H. Li, H. S.-H. Chung, "A Passive Lossless Snubber Cell with Minimum Stress and Wide Soft-Switching Range," *IEEE Trans. Power Electron.*, vol. 25, no. 7, pp. 1725-1738, Jul. 2010.

[12] E. Adib, H. Farzanehfard, "Family of Soft-Switching PWM Converters with Current Sharing in Switches," *IEEE Trans. Power Electron.*, vol. 24, no. 4, pp. 979-985, Apr. 2009.

[13] M. Esteki, E. Adib, H. Farzanehfard, and S. A. Arshadi, "Auxiliary circuit for zero voltage transition interleaved PWM buck converter," *IET Power Electron.*, Aug. 2015, in press.

[14] E. Adib, H. Farzanehfard, "Zero-Voltage-Transition PWM Converters with Synchronous Rectifier," *IEEE Trans. Power Electron.*, vol. 25, no. 1, pp. 105-110, Jan. 2010.

[15] K.-B. Park, G.-W. Moon, M.-J. Youn, "Two-Switch Active-Clamp Forward Converter With One Clamp Diode and Delayed Turnoff Gate Signal," *IEEE Trans. Ind. Electron.*, vol. 58, no. 10, pp. 4768-4772, Oct. 2011.

[16] K. M. Smith, K. M. Smedley, "Properties and Synthesis of Passive Lossless Soft-Switching PWM Converters," *IEEE Trans. Power Electron.*, vol. 14, no. 5, pp. 890-899, Sep. 1999.

[17] L. Chen, H. Hu, Q. Zhang, A. Amirahmadi, I. Batarseh, "A Boundary-Mode Forward-Flyback Converter With an Efficient Active LC Snubber Circuit," *IEEE Trans. Power Electron.*, vol. 29, no. 6, pp. 2944-2958, Jun. 2014.

[18] B. Akın, "An Improved ZVT-ZCT PWM DC-DC Boost Converter with Increased Efficiency," *IEEE Trans. Power Electron.*, vol. 29, no. 4, pp. 1919-1926, Apr. 2014.

[19] K. M. Smith, K. M. Smedley, "Engineering Design of Lossless Passive Soft Switching Methods for PWM Converters—Part I: With Minimum Voltage Stress Circuit Cells," *IEEE Trans. Power Electron.*, vol. 16, no. 3, pp. 336-344, May 2001.

[20] M. Mohammadi, E. Adib, H. Farzanehfard, "Lossless Passive Snubber for Double Ended Flyback Converter with Passive Clamp Circuit," *IET Power Electron.*, vol. 7, Iss. 2, pp. 245-250, 2014.

[21] M. Mohammadi, E. Adib, "Lossless Passive Snubber for Half Bridge Interleaved Flyback Converter," *IET Power Electron.*, vol. 7, Iss. 6, pp. 1475-1481, 2014.

[22] K. Fujiwara, H. Nomura, "A Novel Lossless Passive Snubber for Soft-Switching Boost-Type Converters," *IEEE Trans. Power Electron.*, vol. 14, no. 6, pp. 1065-1069, Nov. 1999.

[23] M. Mohammadi, E. Adib, "Reducing Turn off Losses with A Passive Lossless Snubber for Boost Converter," *Power Electronics, Drive Systems and Technologies Conference (PEDSTC)*, 2014 5th, pp. 385-389, 5-6 Feb. 2014.

[24] M. R. Amini, H. Farzanehfard, "Novel Family of PWM Soft-Single-Switched DC-DC Converters with Coupled Inductors," *IEEE Trans. Ind. Electron.*, vol. 56, no. 6, pp. 2108-2114, Jun. 2009.

[25] T. Zhan, Y. Zhang, J. Nie, Y. Zhang, Z. Zhao, "A Novel Soft-Switching Boost Converter With Magnetically Coupled Resonant Snubber," *IEEE Trans. Power Electron.*, vol. 29, no. 11, pp. 5680-5687, Nov. 2014.

[26] C. Vartak, A. Abramovitz, K. M. Smedley, "Analysis and Design of Energy Regenerative Snubber for Transformer Isolated Converters," *IEEE Trans. Power Electron.*, vol. 29, no. 11, pp. 6030-6040, Nov. 2014.

[27] John G. Kassakian, Martin F. Schlecht, and George C. Verghese, 'Principles of Power Electronics' (Paperback), 1991.

978-1-5090-0376-1/16 $31.00 © 2016 IEEE

7th Power Electronics, Drive Systems & Technologies Conference (PEDSTC 2016)
16-18 Feb. 2016, Iran University of Science and Technology, Tehran, Iran

A New Switched Boost Inverter Using Transformer Suitable for the Microgrid-Connected PV with High Boost Ability

Masoud Ghodsi
University sistan and baluchestan
Zahedan, Iran
ghodsi.masoud.2012@gmail.com

Seyed Masoud Barakati
University sistan and baluchestan
Zahedan, Iran
smbaraka@gmail.com

Abstract— Switched boost inverter (SBI) topology exhibits advantage similar to Z-source inverter (ZSI) with fewer passive components; however it requires more active component. In this study a new class of single-stage high inversion gain inverters based on transformers is proposed. Switched trans boost inverter (STBI) is built by replacing inductor in switched boost inverter (SBI) with a transformer and two diodes. Three topologies are presented for STBIs namely ripple input current STBI (rSTBI), discontinuous input current STBI (dSTBI) and DC-linked type STBI. The proposed inverters produce high voltage gain when the turn ratio is larger than 1. Thus, the size, weight can reduce. These kind of inverters could use in applications photovoltaic and other renewable systems, where a high voltage gain is usually requested and weight and size important. In this paper presents the operating principle and compare them with SBI, ZSI, SL-ZSI, Trans quasi and TZ source inverter. The simulation verified that the proposed inverters have high voltage gain.

Keywords—switched trans boost inverter (STBI) ; shoot-through state; non-shoot-through state; transformer; high inversion gain.

I. Introduction

Voltage source inverter (VSI) is two-stage power conversion. Since we need that the peak AC output voltage has higher than the DC input source, an additional DC/DC boost converter used. Thus, the size, weight, cost and complexity increased and is desirable to eliminate if possible. Two switches in one phase cannot turn on simultaneously. Because it will cause a short circuit of the voltage DC source and damage the devices. To solve the problem a dead time exits between the turn on and turn off of the two power switches in one phase. But quality of the output voltage waveform is reduced. In 2002, a inverter proposed that provides features that cannot be yield with conventional VSI and it can be overcome the above limitations mentioned. This inverter called Z-source inverter (ZSI) as shown in fig. 1. It has two capacitors and two inductors connected in X-sharp and buck-boost capability with single stage power conversion. Both upper and lower switches in one phase turn on simultaneously to boost the output voltage and not need to dead time therefore the system reliability increased. In addition to active switch states and zero switch states, it has

extra shoot-through states. The zero states can be replaced by the shoot-through states partially or completely. Depending on the desired voltage boosting, ZSI can boost the output voltage using duty cycle D for shoot-through [1]. To date, many research worked on ZSI. Some paper focused on modeling and control [2]-[3], applications [4]-[6] pulse width modulation control [7].

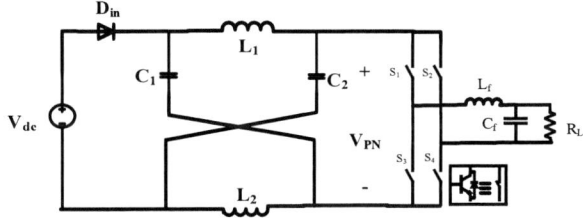

Fig. 1. the conventional Z-source inverter [1].

The boost factor in switched boost inverter is

$$B_{ZSI} = \frac{V_{pn}}{V_{in}} = \frac{1}{1 - 2\left(\frac{T_0}{T}\right)} = \frac{1}{1 - 2D} \quad (1)$$

B is boost factor, T_0 is the shoot-through interval during the switching period T and D is the duty cycle of shoot-through state. When D=[0-0.5] the boost factor varies between 1 and ∞, due to parasitic effects, the infinite value of B is not achievable. In conventional ZSI, current drawn from the DC source is discontinuous.

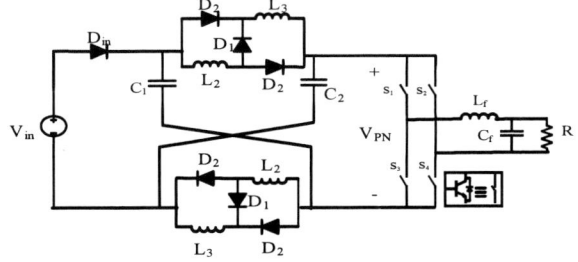

Fig. 2. Switched inductor Z-source inverter [8].

978-1-5090-0376-1/16 $31.00 © 2016 IEEE

Because of the modulation index M is limited by M≤1-D, shoot-through duty ratio decrease with the increase of M, so using a low M result in increasing the total harmonic distortion (THD). Thus, may not be suitable in full cell and solar cell. We attend to focused on the ZSIs which in order to improve the output power quality and increase the boost factor using a low D.

Fig. 2 shows the switched inductor Z-source inverter (SL-ZSI), where two inductors are replaced by a SL cell to obtain a high voltage gain [8]. The boost factor of the SL-ZSI is

$$B_{SL-ZSI} = \frac{V_{pn}}{V_{in}} = \frac{1+D}{1-3D} \qquad (2)$$

But in this structure the number of passive components increased. Fig. 3 shows the trans-quasi Z-source inverter. In order to obtain high voltage gain, two inductors are replaced by a transformer with a turn ratio of N:1 [9].

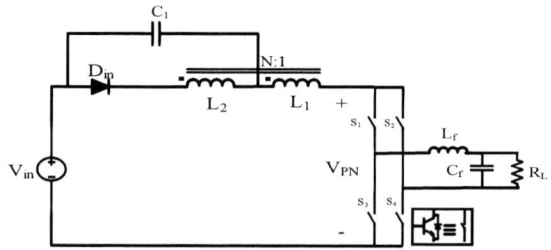

Fig. 3.   Trans-quasi Z-source inverter [9].

The boost factor in trans-quasi Z-source inverter is

$$B_{trans-quasi-ZSI} = \frac{V_{pn}}{V_{in}} = \frac{1}{1-(N+1)D} \qquad (3)$$

When N>1, a high boost inversion ability can be obtained. Fig. 4 shows the TZ-source inverter. In order to obtain very high voltage gain applies two transformers to the conventional ZSI [10].

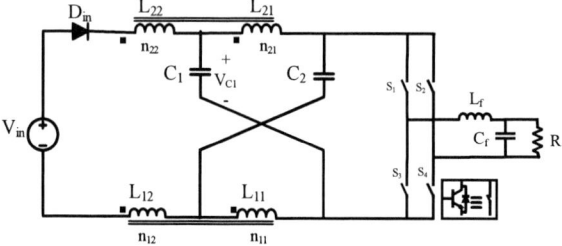

Fig. 4.   TZ-source inverter [10].

The boost factor in TZ-source inverter is

$$B_{TZ} = \frac{V_{pn}}{V_{in}} = \frac{1}{1-(2+N_1+N_2)} \qquad (4)$$

N_1, N_2 are turn ratio the transformers T_1, T_2 respectively. Due to use two transformers it increase the cost and weight and not suitable for applications such as micro inverter where weight, size and cost are important and this structure should not be

compared with other structures. Therefore for such as applications switched boost inverters (SBIs) have been proposed [11].

As shown in Fig. 5 SBI has a lower number of passive component and one more active switch than ZSI. Similar to the ZSI, SBI has shoot-through state. The SBI consist of two diodes, one inductor, one capacitor and one active switch.

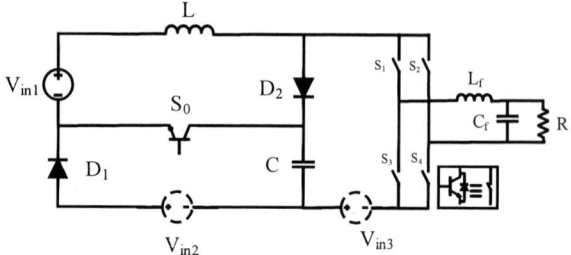

Fig. 5.   Switched boost inverter

The boost factor in switched boost inverter is

$$B_{SBI} = \frac{V_{pn}}{V_{in}} = \frac{1-D}{1-2D} \qquad (5)$$

But SBI has drawbacks boost factor is low, voltage across capacitor is high and current from the source is discontinuous. To solve the aforementioned drawbacks DC input can be placed in various locations as shown in Fig. 5 [11].

In this paper, by replacing inductor in SBI with a transformer and two diodes, new single stage high boost voltage inverters presented. By inserting DC source into switched boost cell various topologies are built. The proposed inverters called switched trans boost inverter (STBIs). The proposed inverters produce a very high boost voltage gain when turn ratio of transformer is higher than 1. Operating principles, analysis and comparison with ZSI, SL-ZSI, trans-guasi ZSI with N=1,2 and SBI are presented. The proposed STBI topologies suitable for applications such as micro inverter where a low input voltage must be inverted to a high AC output voltage.

II. CIRCUIT ANALYSIS OF THE PROPOSED STBIS

Fig. 6 shows the proposed switched trans boost inverters (STBIs), which consist of one isolated two-winding transformer, four diodes (D_1, D_2, D_A, D_S), one capacitor (C) and one active switch (S_0). By inserting DC source (V_{in}) into three locations shown that in Fig. 6, these topologies are built. Inductor of the conventional SBI is replaced by a isolated two-winding transformer and two diodes. The operating principal of the proposed inverters are similar to those of the conventional ZSI and SBI. These kind of inverters have the extra shoot-through states beside two zero states and two active states for single phase. The operation principal can be simplified into shoot-through and non-shoot-through state. We assumption all component used in Fig. 6 are ideal without parasitic, conduction and losses. The DC input voltage can be placed in three locations as shown in Fig. 6. In shoot-through state, D_1 and D_A are off and transformer secondary windings are disconnect, whereas D_S and S_0 are on. In this state

978-1-5090-0376-1/16 $31.00 © 2016 IEEE   124

capacitor C is discharged, while transformer primary windings stores energy from both capacitor C and DC input source. During this state, both the upper and lower switches of any phase legs are turned simultaneously. Figs. 7a, c and e show the schematics of STBI during this state. In non-shoot-through state, D_S and S_0 are off whereas D_1 and D_2 are on, so transformer secondary and primary windings in series, In this state capacitor C is charged, while windings transfer energy from the input DC source to the load. Figs. 7b, d and f show the schematics of STBI during this state. In following the operation principle for each structure will be described.

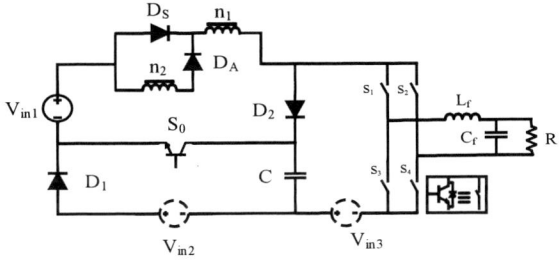

Fig. 6. Proposed switched trans boost inverters (STBIs)

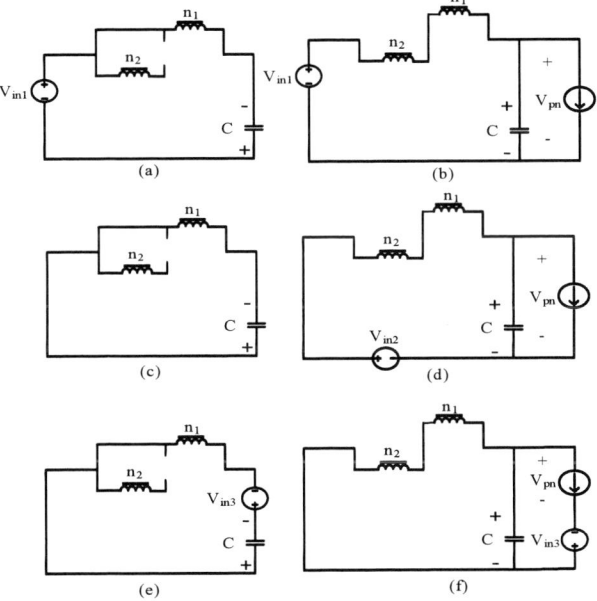

Fig. 7. Equivalent circuits when DC input source location V_{in1}, V_{in2} and V_{in3} : (a), (c), (e) shoot-through states when DC input source located in V_{in1}, V_{in2} and V_{in3} respectively, (b), (d), (f) non-shoot-through states when DC input source located in V_{in1}, V_{in2} and V_{in3} respectively.

A. Operating principles when input DC source located in V_{in1}

When the DC source located in V_{in1}, the source current is the current flows to the transformer winding Therefore a ripple appears on the input current of this topology denoted as a ripple input current STBI (rSTBI). This inverter does not share

the negative point between the DC input source and the inverter. Fig. 7(a) shows the equivalent circuit of the rSTBI in shoot-through state. In shoot-through state, we acquire

$$\begin{cases} V_{n1-shoot} = V_{in1} + V_C \\ V_{n2-shoot} = N(V_{in1} + V_C) \end{cases} \quad (6)$$

The turn ratio of the transformer is defined as $N=n_2/n_1$. Fig. 7(b) shows the equivalent circuit of the rSTBI in non-shoot-through state. In non-shoot-through state, we acquire

$$\begin{cases} V_{n1-active} = \dfrac{V_{in1} - V_C}{1+N} \\ V_{n2-active} = \dfrac{N(V_{in1} - V_C)}{1+N} \end{cases} \quad (7)$$

Applying the volt-second balance principle to n_1, n_2 from (6) and (7) yield

$$V_C = V_{pn} = BV_{in1} = \frac{1+ND}{1-D(N+2)}V_{in1} \quad (8)$$

The boost factor of the proposed inverter B is defined by

$$B = \frac{1+ND}{1-D(N+2)} \quad (9)$$

The boost factor of the rSTBI with N=1 is similar to SL-ZSI.

B. Operating principles when input DC source located in V_{in2}

When the DC source located in V_{in2}, input current is discontinuous since connect in series to diode; and this topology is called discontinuous input current STBI (dSTBI). Fig. 7(c) shows the equivalent circuit of the dSTBI in shoot-through state. In shoot-through state, we acquire

$$\begin{cases} V_{n1-shoot} = V_C \\ V_{n2-shoot} = N(V_C) \end{cases} \quad (10)$$

Fig. 7(d) shows the equivalent circuit of the dSTBI in non-shoot-through state. In non-shoot-through state, the obtained equations are similar to (7).
Applying the volt-second balance principle to n_1, n_2 from (7) and (10) yield

$$V_C = V_{pn} = \frac{1-D}{1-D(N+2)}V_{in2} = BV_{in2} \quad (11)$$

The boost factor of the proposed inverter B is defined by

$$B = \frac{1-D}{1-D(N+2)} \quad (12)$$

C. Operating principles when input DC source located in V_{in3}

When the DC source located in V_{in3}, voltage stress on capacitor reduced and DC source current is the current that flows to the bridge since the DC source connect in series to DC bus, so this topology is called DC-linked type STBI. Fig. 7(e) shows the equivalent circuit of the DC-linked type STBI in shoot-through state. In shoot-through state, the obtained equations are similar to (6).
Fig. 7(f) shows the equivalent circuit of the DC-linked type STBI in non-shoot-through state. In non-shoot-through state, we acquire,

$$\begin{cases} -V_{in3} + V_{n1-active} + V_{n2-active} + V_{pn} = 0 \\ V_C = V_{pn} - V_{in3} \end{cases} \quad (13)$$

Applying the volt-second balance principle to n_1, n_2 from (6) and (13) yield

$$\begin{cases} V_C = \dfrac{(N+3)}{1+(N+2)D} \\ V_{pn} = \dfrac{1-D}{1-D(N+2)} V_{in3} = B V_{in3} \end{cases} \quad (14)$$

The boost factor of the proposed inverter B is defined by

$$B = \frac{1-D}{1-D(N+2)} \quad (15)$$

The boost factor of the dSTBI is similar to DC-linked type STBI. Fig. 8 shows the boost factor versus duty cycle for SBI, ZSI conventional and trans-quasi ZSI, rSTBI, dSTBI, DC-linked type STBI with turn ratios N=1 and 2.

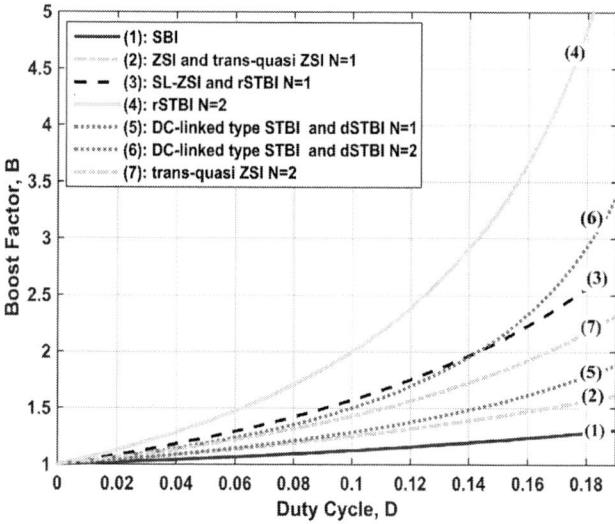

Fig. 8. Comparison of Boost factor between proposed STBIs, SBI, ZSI conventional and trans-quasi ZSI with turn ratios N=1 and 2

III. PWM CONTROL FOR THE SBIs

A PWM control for SBIs must be modified to control the shoot-through state. Three PWM control methods simple boost, maximum boost and maximum constant boost control are presented in [12], [13]. In this paper from simple boost control used. In simple boost control, the shoot-through time per switching period is kept constant. To generate a control signal for the S_0, a straight line (V_p), whose amplitude is equal to the peak value of the $V_{control}$ used and compared to triangle waveform with double frequency and half of the amplitude of that of V_{tri}. V_{tri}, is a high frequency triangle waveform with amplitude -1 to 1. To generate control signals for switches S_1

to S_4, V_{tri}, compared with two control waveform, $V_{control}$ and $-V_{control}$ then inserted into the control signal of S_0 through OR logic gate [14]. Fig. 9 shows this method control for single phase SBIs.

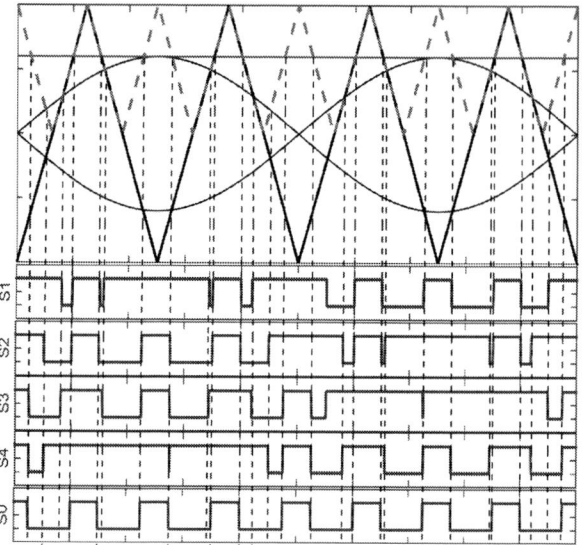

Fig. 9. Simple boost control for single phase SBIs

The voltage gain G of the Z-source inverter can be expressed as

$$G = \frac{\hat{V}_o}{V_{in}/2} = MB \quad (16)$$

Where $\hat{V}_o$ is the output peak phase voltage V_{in} is the input DC source, M is the modulation index and B is the boost factor. In simple boost control method the maximum duty cycle of the shoot-through state is 1-M. Thus, the voltage gain of rSTBI, DC-linked type STBI and dSTBI rewritten as

$$\begin{cases} G_{rSTBI} = \dfrac{N(1-M)+1}{1-(1-M)(N+2)} \\ G_{DC-linked\ STBI} = G_{dSTBI} = \dfrac{M}{1-(1-M)(N+2)} \end{cases} \quad (17)$$

Fig. 10 shows the voltage gain versus modulation index of the SBI, ZSI, SL-ZSI and trans-quasi ZSI, rSTBI, dSTBI and DC-liked type STBI with turn ratios 1 and 2 using the simple boost control. As shown in Fig. 10 using the same modulation the proposed rSTBI compared with dSTBI and DC-linked type provides higher voltage gain. In other hand, voltage gain of the proposed rSTBI with N=2 the highest in compersion with other topologis and uses a higher modulation index to improve the output quality. At all modulation indexes, DC-linked type STBI with dSTBI is the same and SL-ZSI, rSTBI with N=1 is the same.

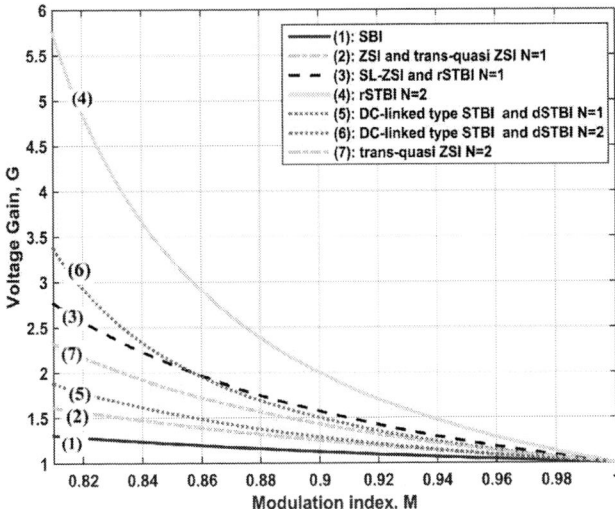

Fig. 10. Voltage gain versus modulation index of the different topologies

IV. SIMULATION

Proposed topology rSTBI selected for the simulation. Simulation studies are performed in PSIM. The parameters were C=470 μF, L_f=3 mH, C_f=20 μF and R=10 Ω. The switching frequency was 10 kHz and input DC source was 30 V. Simple boost control was used. The magnetic inductance was 980 μH. To verify the properties of the rSTBI turn ratios of the transformer N=1 and 2 used within various conditions. The simulations worked done in three cases. In case 1, turn ratio of transformer of 2 with high modulation index and low shoot-through duty cycle was used. In case 2, turn ratio of transformer of 2 with low modulation index and high shoot-through duty cycle was used. In case 3, transformer with N=2 is replaced by transformer with N=1 and rest of parameters are similar to case 2.

A. Case1: M=0.85, D=0.15, Vin=30, N=2

Fig. 11 shows the simulation results for rSTBI in case 1. In this case capacitor voltage is boosted to 97.5 V, the DC-linked voltage is boosted to 97.5V, the peak DC-linked voltage is the same as the capacitor voltage, peak output voltage is 81 V, voltage across active switch (V_{so}) is 110 V and the peak shoot-through current across the main power circuit (I_{sh}) is 24 A. A ripple appears on the input current (I_{in}) of this topology denoted as a ripple input current TSBI (rSTBI).

B. Case2: M=0.8, D=0.2, Vin=30, N=2

Fig. 12 shows the simulation results for rSTBI in case 2. In this case capacitor voltage is boosted to 234 V, the DC-linked voltage is boosted to 234 V, peak output voltage is 170, voltage across active switch (V_{so}) is 236 V and peak shoot-through current across the main power circuit (I_{sh}) is 90 A

C. Case3: M=0.8, D=0.2, Vin=30, N=1

Fig. 13 shows the simulation results for rSTBI in case 3. In this case capacitor voltage is boosted to 90 V, the DC-linked

voltage is boosted to 90 V, peak output voltage is 72 V, voltage across active switch (V_{so}) is 101 V and peak shoot-through current across the main power circuit (I_{sh}) is 15 A. The transformer in simulation was modeled by an ideal transformer without leakage inductance.

Leakage inductance caused decreasing boost factor in practical. The simulation results indicate the proposed inverter achieve high boost inversion using a very low D or high M. case 2 compared to case 3, shows by increasing the turn ratio of the transformer peak AC output voltage is increased.

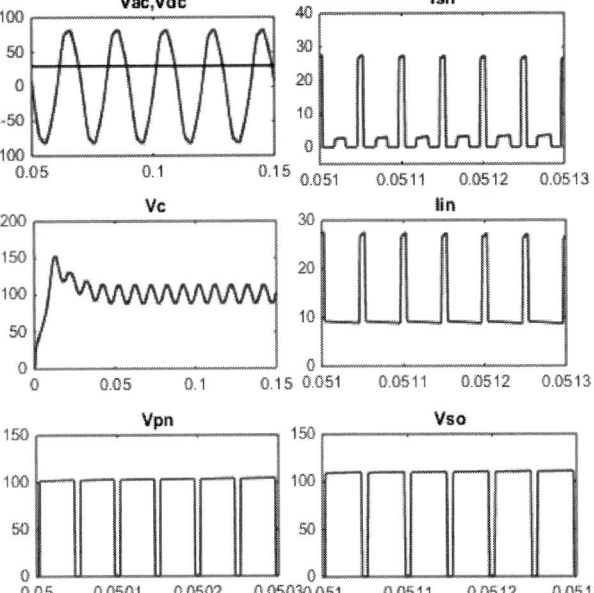

Fig. 11. Simulation results of rSTBI in case1 when M=0.85, D=0.15, Vin=30, N=2

V. CONCLUSION

A new class of switched boost inverters have been proposed. The proposed inverters used a transformer by replacing inductor in switched boost inverter. By inserting DC input voltage in the various locations of STBI, rSTBI, dSTBI and DC-linked type are built. The proposed inverters have the following main advantage

1) By increasing the turn ratio of transformer a high boost factor can be obtained, 2) Under the same voltage gain, turn ratio of transformer in STBIs are smaller than trans-quasi ZSI, 3) In comparison SL-ZSI, lower number of passive component and one more active switch. This may be lead to reduction in the size, weight and cost, 4) Continuous input current and larger boost factor can be obtained in the rSTBI, 5) DC-linked type STBI has a lower voltage stress on capacitor, 6) To obtain the same voltage gain rSTBI uses a modulation index higher than other topologies. Thus, the quality of the output waveform is improved.

The proposed inverters could use in applications photovoltaic and other renewable systems, where a high voltage gain is usually requested and cost, weight, size are important. It should be noted Transformer in STBI should be designed with low leakage inductance to result in a practice which is closer to the result of simulation.

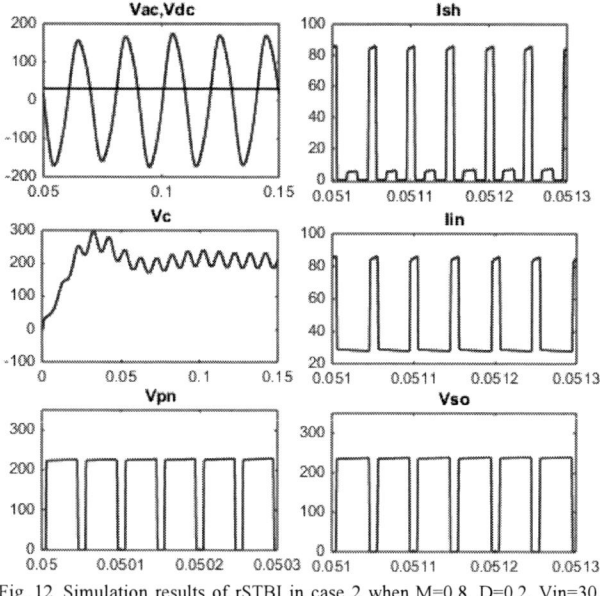

Fig. 12. Simulation results of rSTBI in case 2 when M=0.8, D=0.2, Vin=30, N=2

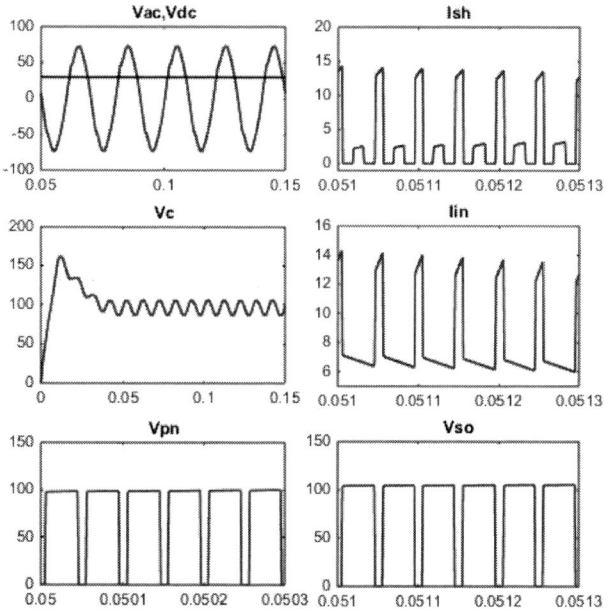

Fig. 13. Simulation results of rSTBI in case 2 when M=0.8, D=0.2, Vin=30, N=2

REFERENCES

[1] F. Z. Peng, "Z-Source Inverter," IEEE Trans. Ind. Applicat., vol. 39, no. 2, pp. 504-510, Mar./Apr. 2003.

[2] J. B. Liu, J. G. Hu, and L. Y. Xu, "Dynamic modeling and analysis of Z-source converter-derivation of ac small signal model and design-oriented analysis," IEEE Trans. Power Electron., vol. 22, no. 5, pp. 1786–1796, Sep. 2007.

[3] Y. Tang, S. J. Xie, C. H. Zhang, and Z. G. Xu, "Improved Z-source inverter with reduced capacitor voltage stress and soft-start capability," IEEE Trans. Power Electron., vol. 24, no. 2, pp. 409–415, Feb. 2009.

[4] F. Z. Peng, M. Shen, and K. Holland, "Application of Z-source inverter for traction drive of fuel cell-battery hybrid electric vehicles," IEEE Trans. Power Electron., vol. 22, no. 3, pp. 1054–1061, May 2007.

[5] M. Hanif, M. Basu, and K. Gaughan, "Understanding the operation of a Z-source inverter for photovoltaic application with a design example," IET Power Electron., vol. 4, no. 3, pp. 278–287, Mar. 2011.

[6] Y. Huang, M. Shen, F. Z. Peng, and J. Wang, "Z-source inverter for residential photovoltaic systems," IEEE Trans. Power Electron., vol. 21, no. 6, pp. 1776–1782, Nov. 2006.

[7] P.C. Loh, D. M. Vilathgamuwa, Y. S. Lai, G. T. Chua and Y. Li, "Pulse-Width Modulation of Z-Source Inverters," IEEE Trans. Power Electron., vol. 20, no. 6, pp. 1346-1355, Nov. 2005.

[8] M. Zhu, K. Yu and F. L. Luo, "Switched Inductor Z-Source Inverter," IEEE Trans. Power Electron., vol. 25, no. 8, pp. 2150-2158, Aug. 2010

[9] W. Qian, F. Z. Peng and H. Cha, "Trans-Z-Source Inverters," IEEE Trans. Power Electron., vol. 26, no. 12, pp. 3453-3463, Dec. 2011.

[10] M. K. Nguyen, Y. C. Lim and Y. G. Kim, "TZ-Source Inverters," IEEE Trans. Ind. Electron., vol. 60, no. 12, pp. 5686-5695, Dec. 2013.

[11] M. K. Nguyen, T. V. Le, S. J. Park, Y. C. Lim and J. Y. Yoo "Class of high boost inverters based on switched-inductor structure", IET Power Electron., vol. 8, no. 5, pp. 750 – 759, Apr 2015.

[12] F. Z. Peng, M. Shen and Z. Qian, "Maximum Boost Control of the Z-Source Inverter," IEEE Trans. Power Electron., vol. 20, no. 4, pp. 833-838, July 2005.

[13] M. Shen, J. Wang, A. Joseph, F. Z. Peng, L. M. Tolbert and D. J. Adams, "Maximum Constant Boost Control of The Z-Source Inverter," in Proc. 39th ISA Annual Meeting, pp. 142-147, Oct. 2004.

[14] A. Ravindranath, S. K. Mishra, and A. Joshi, "Analysis and PWM Control of Switched Boost Inverter", IEEE Trans. Ind. Electron., vol. 60, no. 12, pp. 5593-5602, Dec. 2013.

7th Power Electronics, Drive Systems & Technologies Conference (PEDSTC 2016)
16-18 Feb. 2016, Iran University of Science and Technology, Tehran, Iran

A Family of Single Phase Converters with Reduced Number of Components and Leakage Current Elimination in Photovoltaic Systems

MasoumehAdhamHaghighpour[1], NimasalehiyanZandi[2], Ali Mostaan[2] and Alfred Baghramian[3]

(1):Islamic Azad University, Lahijan Branch, Lahijan, Iran. Email:adham_e85@yahoo.com

(2):Iranian Central Oil Field Company (ICOFC), Tehran, Iran. Email: nszandi@yahoo.com , ali_8457@yahoo.com,

(3): University of Guilan, Department of the electrical engineering, Rasht, Iran. Email: baghramian2000@yahoo.com

Abstract–**A family of single phase converters with reduced number of components is introduced in this paper. There are two switches, one coupled inductors and one capacitor in their structure, therefore the cost is reduced considerably. The proposed converters can be used as four quadrant DC/DC converter when fixed duty cycle (D) is applied to switches. On the other hand, they can be used as single phase inverter when the duty cycle is changed sinusoidal with time. In addition, the input voltage and the output load share the common ground. Therefore the leakage current is eliminated naturally in proposed converters when the photovoltaic source is applied as the input source. The performance of the proposed converter is validated with theoretical and simulation results using MATLAB/SIMULINK.**

Keywords—DC/ACinverter, coupled inductors, Z source network, leakage current, photovoltaic systems

I. INTRODUCTION

In recent years because of the global warning and air pollution problems many investigationshave been done on green energy sources such as wind turbine, fuel cell and photovoltaic systems and etc [1]. Many renewable energy sources such as full cells and photovoltaic modules only produced DC voltage; therefore a DC/AC inverter should be utilized in order to convert the DC input voltage into AC voltage. Grid connected inverters are divided in two major groups: 1) isolated inverters and 2) non-isolated inverters [2]. Isolated inverters use the line frequency or high frequency transformers for electrical isolation between input source and output. From weight and size viewpoint, the high frequency transformers have advantages in comparison to line frequency transformer. Also, isolated inverters have higher gain and safely advantages. However they are more expensive and their efficiency is lower. In other hand, when the grid voltage is low or power level is below 20kw the isolation is not a requirement. In addition non-isolated inverters have low cost and higher efficiency in compare with their isolated counterpart. Particularly the non-isolated single phase inverters become popular in recent years. Unfortunately in photovoltaic system the PV modules are not ideal and there is leakage capacitor between the PV module and ground. This leakage capacitor leads to leakage current if the PV module and grid do not share the common ground.

This leakage current should be lower than 200mA for safety consideration. Traditionally, full bridge and half bridge inverters can be used in photovoltaic system in order to invert the DC input voltage to AC voltage and inject the power AC to grid. However in full bridge inverters with unipolar modulation, the common mode voltage is not constant and fluctuates with high frequency. Therefore high leakage current flows between grid and PV terminal that may be lead to safety problems. Using bi polar modulation the common mode voltage can be keep to constant value. However the output AC current has higher ripple and large output filter inductors are required that can be increased complexity and cost. On the other hand in half bridge inverter the input voltage should be double value of the input voltage to obtain the same output voltage in compare with full bridge inverter. Therefore, half bridge inverter requires more PV module in compare with full bridge inverter. Also with same output voltage the voltage stress on switches is two times in half bridge inverter that requires power switches with high voltage stress that raises the cost [3]. In recent years many efforts has been done to decrease or eliminate the leakage current in non-isolated single phase grid connected inverters and many topologies based on full bridge inverters using unipolar modulation have been introduced in literature[4-11]. The leakage current can be reduced considerable using H5 topology [4]. This inverter has very simple structure and realize by adding one power switch to full bridge inverter. In this structure, the grid is isolated from the PV source during the zero states, however because of switches non-ideal characteristics the common mode voltage is not constant and leakage current is not suppressed completely. Also, three switches are conducting during the active states that increase the power loss in compare with full bridge inverter. HERIC topology has introduced in [5] and its operation is quite similar to H5, but two switches are conducting in every switching state that decreases the power losses. Similar topologies have been introduced in [6-7]. Another effective approach to suppress the leakage current is clamping to common mode voltage to half value of the input voltage using two clamp capacitors.OH5 [8], FB-DCBP [9],HBZVR [10] and HBZVR-D [11] topologies are utilized this method to suppress the leakage current. Although the common

978-1-5090-0376-1/16 $31.00 © 2016 IEEE

voltage is fixed to half value of the input voltage and the leakage current is reduced significantly below of its maximum allowable value, at least six power switches are required in this method that increase the cost and complexity. Also, in order to clamp the common mode voltage, two electrolyte capacitors are required that decrease the system reliability. In [12], a novel single phase inverters bases on Z source inverter was presented. In this topology, the input voltage and load, share common ground, therefore the leakage current is suppress automatically without using the clamping method. However, there are two inductors and two capacitors in this inverter.

In this paper, two single phase inverters are introduced. Similar to semi-Z source inverter [12], the leakage current is suppressed automatically and clamping capacitors are not required. But, there are two coupled inductors that are wounded in one core and only one capacitor that can be lead to lower occupied space and lower cost. These converters have been presented by one of this paper authors in [13] as a four quadrant DC/DC converter. This study expands these converters application as a single phase dc/ac inverter that the input voltage and grid share the common ground, therefore the leakage current in PV system is eliminated naturally.

This paper is organized as follow: in section II, the proposed converters are introduced and their voltage gain in steady state is obtained. It is shown that proposed converters can work as a four quadrant DC/DC converter with fixed duty cycle (D) and as a single phase inverter when the duty cycle is changed sinusoidal with time. Simulation results are presented in section III and finally, the conclusion is followed in section IV.

II. PROPOSED CONVERTERS STRUCTURE

A. Analysis in steady state

Recently, Trans Z source [13] and Γ-Z source inverter [14] have been introduced to decrease the number of components and increase the voltage gain in traditional Z-source inverter. As shown in Fig. 1. The proposed converters that are derived from Z-source inverter and $\Gamma-Z$ source inverter are shown in Fig.2a and Fig.2b, respectively. In order to these converters work as dc/ac inverter, two bi-directional switches (e.g. MOSFET and its anti-parallel diode) should be utilized.

With fixed duty cycle (D) [13], these converters can work as a four quadrant dc/dc converter and the output voltage can be positive or negative that depends on duty cycle. In this section, the first topology is analyzed in steady state. There are twomodes in one switching period that are shown in Fig. 3a and Fig.3b respectively. The duty cycle is defined D for S1 and (1-D) for S2.

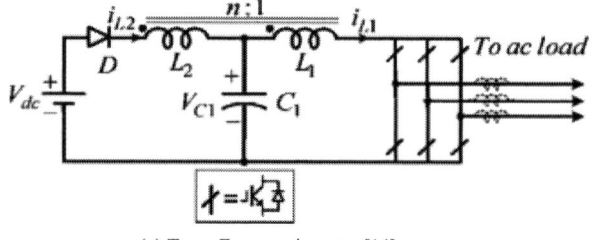

(a) Trans Z source inverter [14]

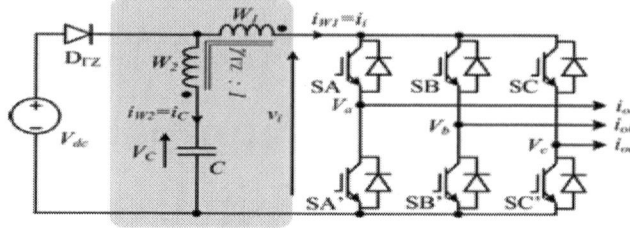

(b) Γ-Z source inverter [15]
Fig.1.Trans Z source and Γ-Z source inverters

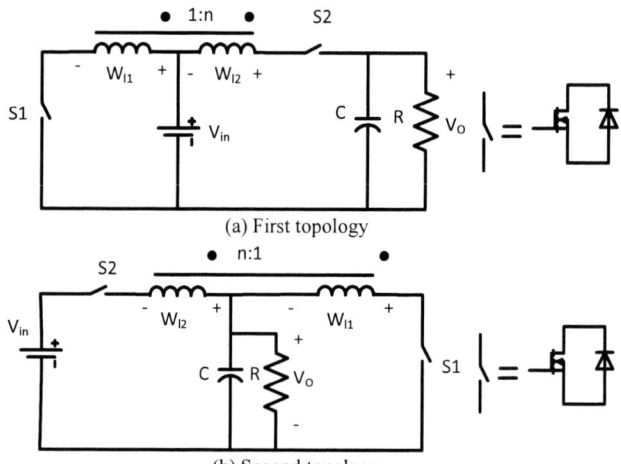

(a) First topology

(b) Second topology
Fig.2. proposed single phase converters/inverters

A): *steady state analysis of the first topology*

In order to analyze the converter in steady state, the coupled inductors are modeled with an ideal transformer and a magnetic inductance (L_m) that is in parallel with transformer primary side. The effect of the leakage inductance has been neglected in this study.

When S1 is turn on, the voltage across the magnetic inductance is equal to input voltage.

$$V_{Lm} = V_{W1} = V_{in} \qquad (1)$$

In mode 2, when S2 is turn on the voltage across L_m is

$$V_{Lm} = V_{W1} = \frac{V_O - V_{in}}{n} \qquad (2)$$

Using the volt-second balance across the magnetic inductance during a switching cycle we have

$$DV_{in} + \frac{(1-D)(V_O - V_{in})}{n} = 0 \qquad (3)$$

Therefore

$$\frac{V_O}{V_{in}} = \frac{1 - D(1+n)}{1-D} \qquad (4)$$

Using (4) the output voltage can be positive or negative that depends on transformer turn ratio (n) and duty cycle (D) .The voltage gain of this converter versus duty cycle (D) for different transformer turn ratio is shown in Fig.4 [13]. The output voltage always is lower than the input voltage when the output voltage is positive. However, the output voltage can be higher or lower than input voltage when the output voltage is negative. Also, the negative output voltage gain can be raised by increase the transformer turn ratio. If n=1, using (4)the voltage gain can be written as

$$\frac{V_O}{V_{in}} = \frac{1 - 2D}{1-D} \qquad (5)$$

The voltage gain of the converter 1 with fixed duty cycle under n=1 is equal to other four quadrant DC/DC converters that are proposed in [16-17], however the number of components are lower in proposed converter. Under n=1, the output voltage is positive while D>0.5 negative while D<0.5. The operation of the first converter with D<0.5 and D>0.5 is shown in Fig. 5. The operation as a four quadrant DC/DC converter is one advantage in compare with conventional DC/DC converters such as buck, boost, etc.

It can be shown that the voltage gain for second converter is [13].

$$\frac{V_O}{V_{in}} = \frac{1-n(1-D)}{(1-n)(1-D)} \quad (6)$$

The voltage gain versus transformer turn ratio is shown in Fig.6.Similar the previous topology, this converter work only in buck mode when positive output voltage is required. However, it can operate in buck-boost mode when the output voltage is negativeand the gain can be increased by lowering the transformer turn ratio. Under n=2, the voltage gain can be obtain using (5). It should be mentioned in order to obtain high voltage gain; it is not a requirement high transformer turn ratio that leads to lower leakage inductance. However, leakage inductance can lead to voltage spike across the components. The leakage inductance effect and design the snubber circuit to suppress the voltage spike can be a subject for future work.

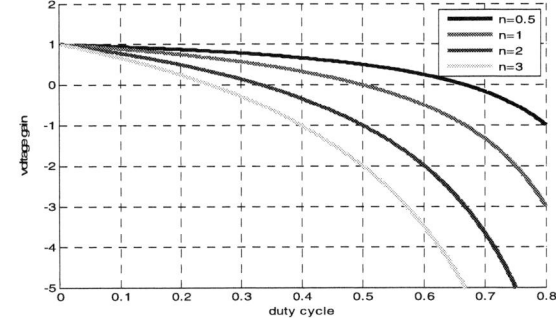

Fig.4.Voltage gain of the first topology versus the transformer turn ratio [13]

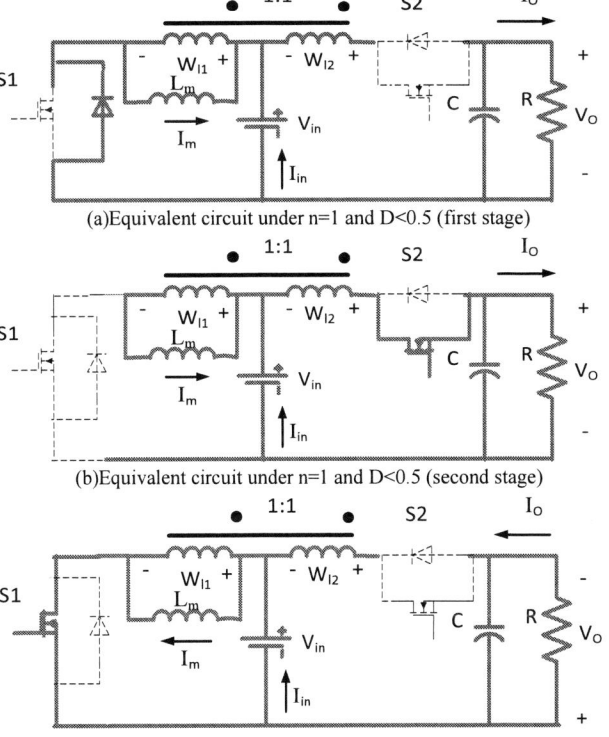

(a)Equivalent circuit under n=1 and D<0.5 (first stage)

(b)Equivalent circuit under n=1 and D<0.5 (second stage)

(c) Equivalent circuit under n=1 and D>0.5 (first stage)

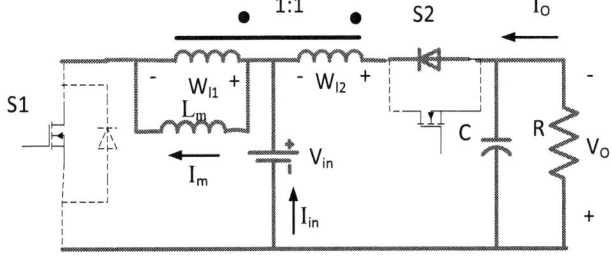

(d) Equivalent circuit under n=1 and D>0.5 (second stage)

Fig.5. Equivalent circuits of first topology under n=1

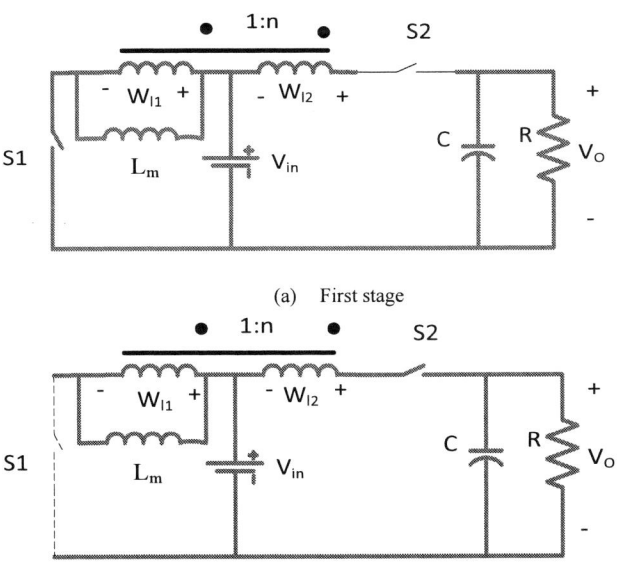

(a)   First stage

(b)   Second stage

Fig. 3.Equivalent circuits of first topology in one switching cycle

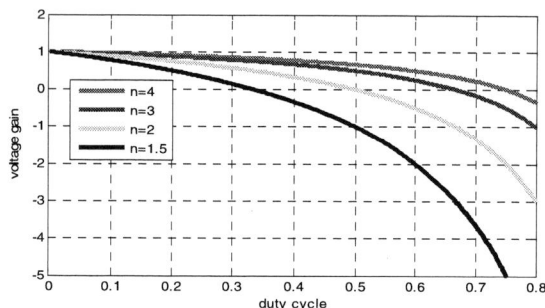

Fig.6. Voltage gain of second topology versus duty cycle under different transformer turn ratio [13]

B): DC/AC operation of the proposed converters

In the previous section, it is shown that the output voltage in the proposed converters can be positive or negative depending on duty cycle or transformer turn ratio, therefore these converters can be operated as a DC/AC inverter if the duty cycle changessinusoidal with time. It should be notice that these inverters can be operate only in buck mode, because according Fig. 4 or Fig. 6, the output voltage is lower than the output voltage if positive output voltage is required.

In order to the first converter operate as a DC/AC inverter the output voltage should be

$$V_o = MV_{in}sin\omega t \quad (7)$$

Where,ω is the angular frequency of the output voltage and M is the voltage gain that varies between zero and one, depending on required output voltage. Substituting (7) into (4), the time variant duty cycle can be obtained as

$$D = \frac{1 - Msin(\omega t)}{1 + n - Msin(\omega t)} \quad (8)$$

$$1 - D = \frac{n}{1 + n - Msin(\omega t)} \quad (9)$$

If n=1, we have

$$D = \frac{1 - Msin(\omega t)}{2 - Msin(\omega t)} \quad (10)$$

$$1 - D = \frac{1}{2 - Msin(\omega t)} \quad (11)$$

In above equations, D is duty cycle for S_1 and (1-D) is duty cycle for S_2. It is clear that for inverting operation the duty cycle is not fixed and is changed sinusoidal with time.

Using similar method, it can be shown the duty cycle for S_1 and S_2 in second converter can be calculated from (12) and (13), respectively in order to obtain sinusoidal output voltage.

$$D = \frac{Msin(\omega t) - nMsin(\omega t) - 1 + n}{Msin(\omega t) - Mnsin(\omega t) + n} \quad (12)$$

$$1 - D = \frac{n - 1}{Msin(\omega t) - Mnsin(\omega t) + n} \quad (13)$$

If n=2, the duty cycle for S_1 and S_2 can be obtained using (9) and (10), respectively.

Using Fig. 3, the current ripple of the magnetic inductor and voltage ripple of the output capacitor in first topology under n=1 and second topology under n=2 can be obtained using (13) and (14), respectively

$$\Delta i_L = \frac{DV_{in}}{L_m f_s} = \frac{1 - Msin(\omega t)}{2 - Msin(\omega t)} \times \frac{V_{in}}{L_m f_s} \quad (14)$$

$$\Delta V_C = \frac{(1-D)V_O}{RCf_s} = \frac{MV_{in}sin(\omega t)}{RCf_s(2 - Msin(\omega t))} \quad (15)$$

The above equations can be used to select the magnetic inductance and the output capacitorvalues.

Also, the voltage stress on switches is

$$V_{S1} = V_{S2} = 2V_{in} - V_O = V_{in}\big(2 - Msin(\omega t)\big) (16)$$

From (16), it is clear that under worst condition (M=1), the maximum voltage stress on switches is three time of the input voltage that should be considered in circuit design procedure.

III. SIMULATION RESULTS

In order to verify the theoretical results, the first topology with n=1 is simulated using MALTAB/SIMULINK under fixed duty cycle (the converter operates as DC/DC converter) and time variant duty cycle according to (8), when the converter works as a DC/AC inverter. The circuits parameters that was used in simulation are

1- The input voltage is 400V.
2- The magnetic inductance is $400\mu H$
3- The output capacitor is $22\mu F$
4- The switching frequency is 10KHz

Fig.7-9 show the output voltage under D=0.25, 0.5 and 0.75, respectively

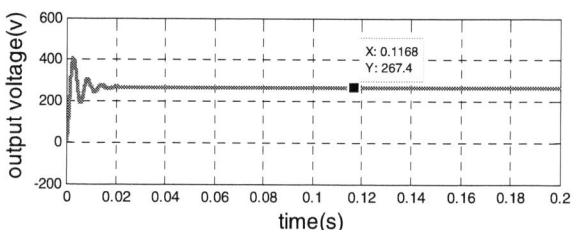

Fig.7. output voltage under D=0.25

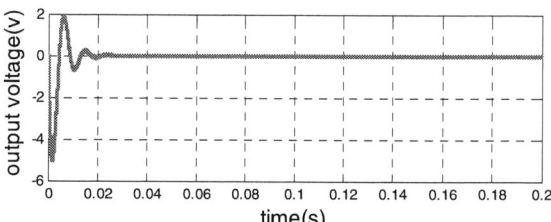

Fig.8. Output voltage under D=0.5

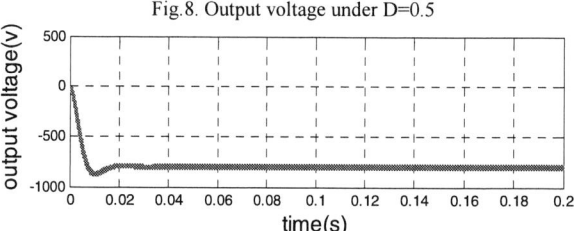

Fig.9. Output voltage under D=0.75

From Fig. 7-9, it is obvious that the output voltage can be positive (D = 0.25), zero (D = 0.5) or negative (D = 0.75) that depends on duty cycle. Therefore this converter can be used as a four quadrant DC/DC converter using bi-directional switches. Also, the output voltages in steady state are in consistent with theoretical results that can be calculated using (4).

Fig. 10, shows the output voltage of the proposed converter under M=0.5 and$\omega = 314 \, rad/s$ when an inductive load (R=10Ω and L=1mH) is connected to the converter output. In

978-1-5090-0376-1/16 $31.00 © 2016 IEEE

this case the peak value of the AC output voltage is 200 V that is in agreement with theoretical analysis.

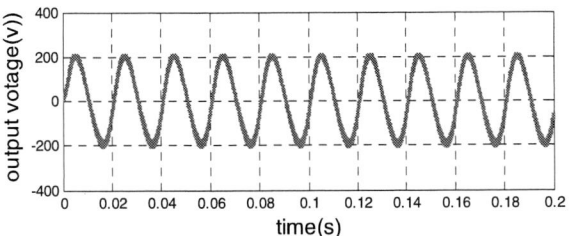

Fig.10. AC output voltage under time variant duty cycle using (8) with inductive load

From Fig.10 it is clear that the proposed converter can generate AC output voltage with appropriate time variant duty cycle that can be obtained using (8).

Fig. 11 shows the grid voltage, grid current and leakage current when the proposed converter is connected to single phase AC gird with V_{rms}=110 V.

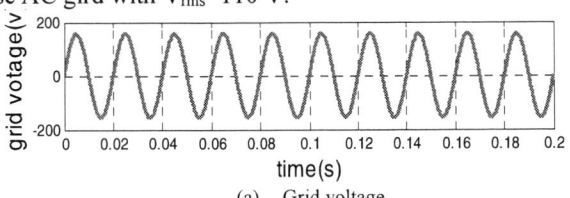

(a) Grid voltage

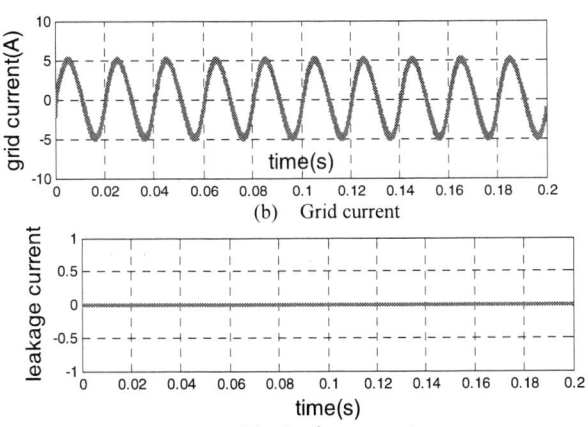

(b) Grid current

(c) Leakage current

Fig.11. Simulation results with V_{grid}=110V_{rms}

From Fig.11 it is clear that the proposed converter can inject the sinusoidal current to grid while the leakage current is zero because the input voltage and grid voltage share the common ground.

IV. CONCLUSION

Families of single phase converters are introduced in this paper. The proposed converters have very simple structure. There are two couple inductors and one capacitor in their structure that is lower in compare with similar topologies. The output voltage can be positive, zero or negative that depends on duty cycle and transformer turn ratio. Therefore, the proposed converters can be used as a four quadrant DC/DC converter using bi-directional switches in some applications such as DC motor control. Also, the output voltage can be sinusoidal using appropriate time variant duty cycle. In the proposed converters the input voltage and the

grid voltage share the common ground. Therefore the leakage current is eliminated without any modification on these converters. Therefore, they can be used in photovoltaic cells that the leakage current is the major problem. The performance of the proposed converters is verified using the simulation results. The effects of the leakage inductance and snubber circuit design can be a subject for future study.

REFERENCES

[1]- F, Blaabjerg and Y, Yang. "Power Electronics –The Key Technology for Renewable Energy Systems" Ninth International Conference on Ecological Vehicles and Renewable Energies (EVER), 2014, pp 1-10

[2]- S. B. Kjaer, J. K. Pedersen, and F. Blaabjerg, "A review of single-phase grid-connected inverters for photovoltaic modules," IEEE Trans. Ind. Appl., vol. 41, no. 5, pp. 1292–1306,

[3]-O. Lopez, F. D. Freijedo, A. G. Yepes, P. Fernandez-Comesaa, J. Malvar,R. Teodorescu, and J. Doval-Gandoy, "Eliminating ground current in a transformer less photovoltaic application," IEEE Trans. Energy conversion., vol. 25, no. 1, pp. 140–147, Mar. 2010.

[4]- M. Victor, F. Greizer, S. Bremicker, and U. H¨ubler, "Method of convertinga direct current voltage from a source of direct current voltage, more specifically from a photovoltaic source of direct current voltage, into a alternating current voltage," U.S. Patent 7 411 802, Aug. 12,2008.

[5]- S. Heribert, S. Christoph, and K. Jurgen, " Inverter for transforming a DC voltage into an AC current or an AC voltage," Europe Patent 1 369 985(A2), May 13, 2003.

[6]- . Yang, W. Li, Y. Gu, W. Cui, and X. He, "Improved transformer less inverter with common-mode leakage current elimination for a photovoltaicgrid-connected power system," IEEE Trans. Power Electron., vol. 27, no. 2, pp. 752–762, Feb. 2012.

[7]- L. Zhang, K. Sun, Y. Xing, and M. Xing, "H6 transformer less full bridgePV grid-tied inverters," IEEE Trans. Power Electron., vol. 29, no. 3,pp. 1229–1238, Mar. 2014.

[8]- H. Xiao, S. Xie, Y. Chen, and R. Huang, "An optimized transformer less photovoltaic grid-connected inverter," IEEE Trans. Ind. Electron., vol. 58, no. 5, pp. 1887–1895, May 2011.

[9]- R. Gonzalez, J. Lopez, P. Sanchis, and L. Marroyo, "Transformer less inverter for single-phase photovoltaic systems," IEEE Trans. Power Electron.,vol. 22, no. 2, pp. 693–697, Mar. 2007.

[10]- T. Kerekes, R. Teodorescu, P. Rodriguez, G. Vazquez, and E. Aldabas, "A new high-efficiency single-phase transformer less PV inverter topology,"IEEE Trans. Ind. Electron., vol. 58, no. 1, pp. 184–191, Jan. 2011.

[11]- Freddy, T., Rahim, N.A., Hew, W.P., Che, H.S.: 'Comparison and analysis of single-phase transformer less grid-connected PV inverters', IEEE Trans. Power Electron., 2014, 29, pp. 5358–5369

[12]- D. Cao, S. Jiang, X. Yu and F. Z. Peng, " Low cost semi Z source inverter for single phase photovoltaic system" IEEE. Tran Power Electron, 2011, Vol.26, No. 12, pp 3514- 3523

[13]- A. Mostaan and M. Soltani, " A family of four quadrant DC/DC converters with reduced number of components" . in proc, IEEE. International Telecommunication Energy Conference (INTELEC 2015), PP 1-6

[14]- W. Qian, F. Z. Peng and H. Cha, " Trans Z-source inverters," IEEE Trans Power Electron, vol 26, no.11, pp 3453-3463, Nov 2011.

[15]- P. C. Loh, D. Li and F. Blaabjerg, " Γ-Z source inverters, " IEEE Trans Power Electron, vol 28, no.11, pp 4880-4884, Nov 2013

[16]-Y. Berkovich, B. Axelrod, S. Tapuchi, and A. Ioinovici, A family Of Four-Quadrant, PWM DC-DC Converters," in proc. IEEE PowerElectronics Specialists Conference, 2007.(PESC 2007). IEEE, 2007, pp. 1878-1883.

[17]- D. Cao and F. Z. Peng. "A family of Z source and semi Z source DC/DC converters."AppliedPower Electronics Conference And Exposition (APEC). IEEE, 2009, PP 1097-1101

7th Power Electronics, Drive Systems & Technologies Conference (PEDSTC 2016)
16-18 Feb. 2016, Iran University of Science and Technology, Tehran, Iran

A High Step-Down DC-DC Converter with Low Switch Voltage Stress and Extremely Low Output Current Ripple

Morteza Esteki, Nasrin Einabadi, Ehsan Adib, Hosein Farzanehfard

Department of Electrical and Computer Engineering
Isfahan University of Technology
Isfahan, Iran
m.esteki@ec.iut.ac.ir, n.einabadi@ec.iut.ac.ir, ehsan.adib@cc.iut.ac.ir, hosein@cc.iut.ac.ir

Abstract—In this paper, an interleaved high step-down DC–DC converter with low switch voltage stress is proposed. The proposed converter provides an extended duty cycle for switches and also, the voltage stress across two switches and two didoes is one fourth of the input voltage and for the other semiconductor elements is smaller than half of the input voltage. This voltage stress is much lower than that of conventional interleaved buck which makes it possible to use switches and diodes with lower voltage rating. As a results both switching and conduction losses, can be reduced and consequently the overall efficiency can be improved. Another advantage of the proposed converter is its extremely low output current ripple which requires an additional small inductor. All these benefit are obtained without applying additional stress on active components or using transformers. The simulation results based on a 480W, 400 to 48 V dc/dc prototype verify the effectiveness of the theoretical analysis.

Keywords— DC-DC converter; low switch voltage stress; step-down conversion ratio.

I. INTRODUCTION

Nowadays, the tendency to use high performance high step-down DC-DC converters, in applications such as VRMs for microprocessors and battery chargers, has increased [1]-[5]. For non-isolated low output current ripple applications, interleaved buck converter (IBC) is an excellent choice due to its simple structure and control [5]–[11]. Although, the conventional IBC shown in Fig. 1 has some advantages, like current sharing capability between modules, current ripple cancellation and fast transient response, but it suffers from few disadvantages. First of all, voltage stress of switches and diodes of IBC is equal to the input voltage which is an important disadvantage in high input voltage applications. So, high voltage semiconductor devices should be used, where they suffer from high on resistance, high forward voltage drop, high output capacitor and high cost. High voltage across switches and diodes, before turn-on and after turn-off causes high switching losses and high losses related to reverse recovery of diodes [12]. Another problem of IBC in high step down applications is small operating Duty cycle (D). Extremely small value for D causes very short regulation interval, particularly at high switching frequencies [13]. Moreover, small D results in high current stress of circuit elements.

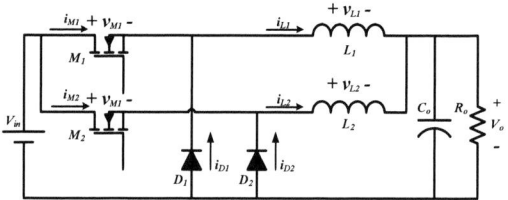

Fig. 1.   The Conventional IBC.

In addition to the mentioned increment of losses, input current ripple and input filter size increases, too. The main drawback of the conventional converters is the required high inductor value to guarantee low output current ripple which results in high inductor losses [14]. In order to reduce the current ripple, parallel converters with interleaved control are an attractive technique [15]. But still in conventional IBC there is a tradeoff between the inductor filter size and the output current ripple, switching frequency and the losses. In other words, in conventional IBC in order to decrease current ripple, inductor size or switching frequency should increase but by increasing inductor size, converter size and inductor losses increase. On the other hand, by increasing switching frequency, switching losses would increase.

To solve these problems, various solutions have been proposed. In [16], a coupled inductor buck converter is introduced, which uses the turns ratio of coupled inductors to extend the duty cycle of switches in high step down conversion ratio applications. Also, by adjusting coupled inductors turns ratio, the current stress of switches are reduced, but, the structure is complex. Another IBC with low switching losses and improved step-down conversion ratio is presentenced in [17]. The converter switches and diodes voltage stress is half of the input voltage, just after turn-on and before turn-off instants. This also means that the switching losses have decreased, but one of switches voltage rating should be more than the input voltage. Also, in [17], despite using interleaved structure in the proposed converter, the input current ripple has not improved. In [18], an interleaved high step-down converter is proposed. One fourth of conventional IBC conversion ratio is the DC-DC gain of the converter which is an important advantage. However, the converter has four switches, which three of them

978-1-5090-0376-1/16 $31.00 © 2016 IEEE   134

must be rated more than half of the input voltage. By using coupled inductors in interleaved buck converter and some simplification an interesting topology is introduced in [19], which has proper duty cycles for converter switches in high step-down applications. However the topology is not suitable for high input voltage high step-down applications because of its high voltage stress for switches which is much more than the input voltage. In [20]-[22] three-level buck converters are introduced to reduce the voltage stress of semiconductors components. However, to perform interleaving, many components are required and complexity of structure and control increases. A three level buck converter is introduced in [23]. The voltage stress of its switches and diodes are half of its input voltage but the conversion ratio is still similar to the conventional buck converter.

This paper introduces an IBC topology, which is suitable for high input voltage, high step-down, non-isolated applications with low output current and continuous input current. At steady state, the voltage stress across the switches is much lower than the input voltage and thus, switching and capacitive turn on losses are reduced. The voltage stress of the freewheeling diodes is also much lower than that of the conventional IBC so fast schottky diode can be used to reduce reverse-recovery problems. The conversion ratio of the proposed converter is lower than the conventional IBC, thus the switches duty cycles can be extended which reduces peak current of switches. The output current ripple of the converter is considerably low. The design considerations are presented in section III. In section IV, the simulation results are presented to validate the theoretical analysis.

II. POROPSED CONVERTER AND ITS OPERATIONAL PRINCIPLES

As shown in Fig. 2, the proposed converter is composed of four switches, four diodes, two filter inductors and 3 capacitors. It has eight distinct operating modes in one switching period and functions at $D<0.5$. In order to simplify the analysis of the proposed converter, following assumptions are made

- All components are assumed to be ideal.
- Capacitors C_{b1} and C_{b2} are large enough, so their voltage variations can be ignored.

Equivalent circuit of the proposed converter for each operating mode is shown in Fig 3. Also, the converter typical waveforms are shown in Fig 4.

Mode 1 [t_0-t_1]: In this mode M_2, M_3 and M_4 are off. Inductor L_1 has two charging loops during this interval which are C_1-M_1-C_{b1}-D_1-L_1-L_o-C_o and V_{in}-M_1-C_{b1}-D_1-L_1-C_2. Also, C_2 is charging through V_{in}-M_1-C_{b1}-D_1-L_1-C_2 path while C_1 is discharging through C_1-M_1-C_{b1}-D_1-L_1-L_o-C_o path. Diodes D_3 and D_4 are conducting L_2 current to output. Important equations of this mode are as follows:

$$v_{L1}(t) = V_{in} - V_{Cb1} - V_{C2} = V_{C1} - V_{Cb1} - V_o - v_{Lo}(t) \quad (1)$$

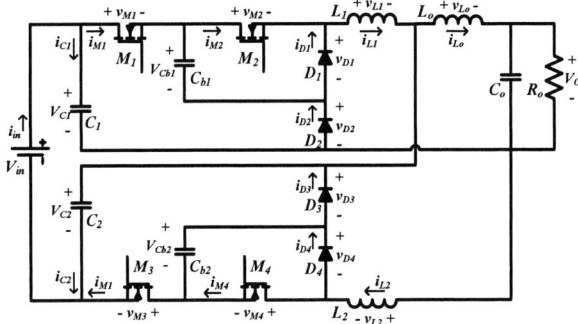

Fig. 2. The proposed high step-down DC-DC converter.

$$v_{L2}(t) = -V_o - v_{Lo}(t) \quad (2)$$

$$v_{Lo}(t) = V_{C1} + V_{C2} - V_{in} - V_o \quad (3)$$

Mode 2 [t_1-t_2]: In this mode all switches are off and all diodes are conducting. Also, C_1 and C_2 are being charged through V_{in}-C_1-D_2-D_1-L_1-C_2 path. The voltage of L_o is constant and same as mode 1. Inductors currents i_{L1} and i_{L2} are decreasing linearly. The important equations expressed as follows:

$$v_{L1}(t) = v_{L2}(t) = -V_o - v_{Lo}(t) \quad (4)$$

$$v_{Lo}(t) = V_{C1} + V_{C2} - V_{in} - V_o \quad (5)$$

Mode 3 [t_2-t_3]: Mode 3 begins when M_4 turns on at t_2. Capacitor C_{b2} starts to discharge through C_{b2}-D_3-L_o-C_o-L_2-M_4 path. The current of L_1 (i_{L1}) passes through D_1 and D_2. During this mode V_{Lo} is constant, and i_{L2} increases linearly. The converter equations in this mode are as follows:

$$v_{L1}(t) = -V_o - v_{Lo}(t) \quad (6)$$

$$v_{L2}(t) = V_{Cb2} - V_o - v_{Lo}(t) = V_{in} + V_{Cb2} - V_{C1} - V_{C2} \quad (7)$$

$$v_{Lo}(t) = V_{C1} + V_{C2} - V_{in} - V_o \quad (8)$$

Mode 4 [t_3-t_4]: When M_4 is turned off this mode begins. Operation of converter in this mode is the same as mode 2, in which all switches are off and all diodes are conducting. Note that voltage of L_o in this mode is like the pervious modes.

Mode 5 [t_4-t_5]: At t_4, M_2 turns on and this mode begins. Inductor L_1 is charging through C_{b1}-M_2-L_1-L_o-C_o-D_2 and V_{in}-C_1-D_2- C_{b1}-M_2-L_1-C_2-V_{in}. The current of L_2 (i_{L2}) passes through D_1 and D_2. Important equations of this interval can be expressed as follows:

$$v_{L1}(t) = V_{Cb1} - V_o - v_{Lo}(t) = V_{in} + V_{Cb1} - V_{C1} - V_{C2} \quad (9)$$

$$v_{L2}(t) = -V_o - v_{Lo}(t) \quad (10)$$

978-1-5090-0376-1/16 $31.00 © 2016 IEEE

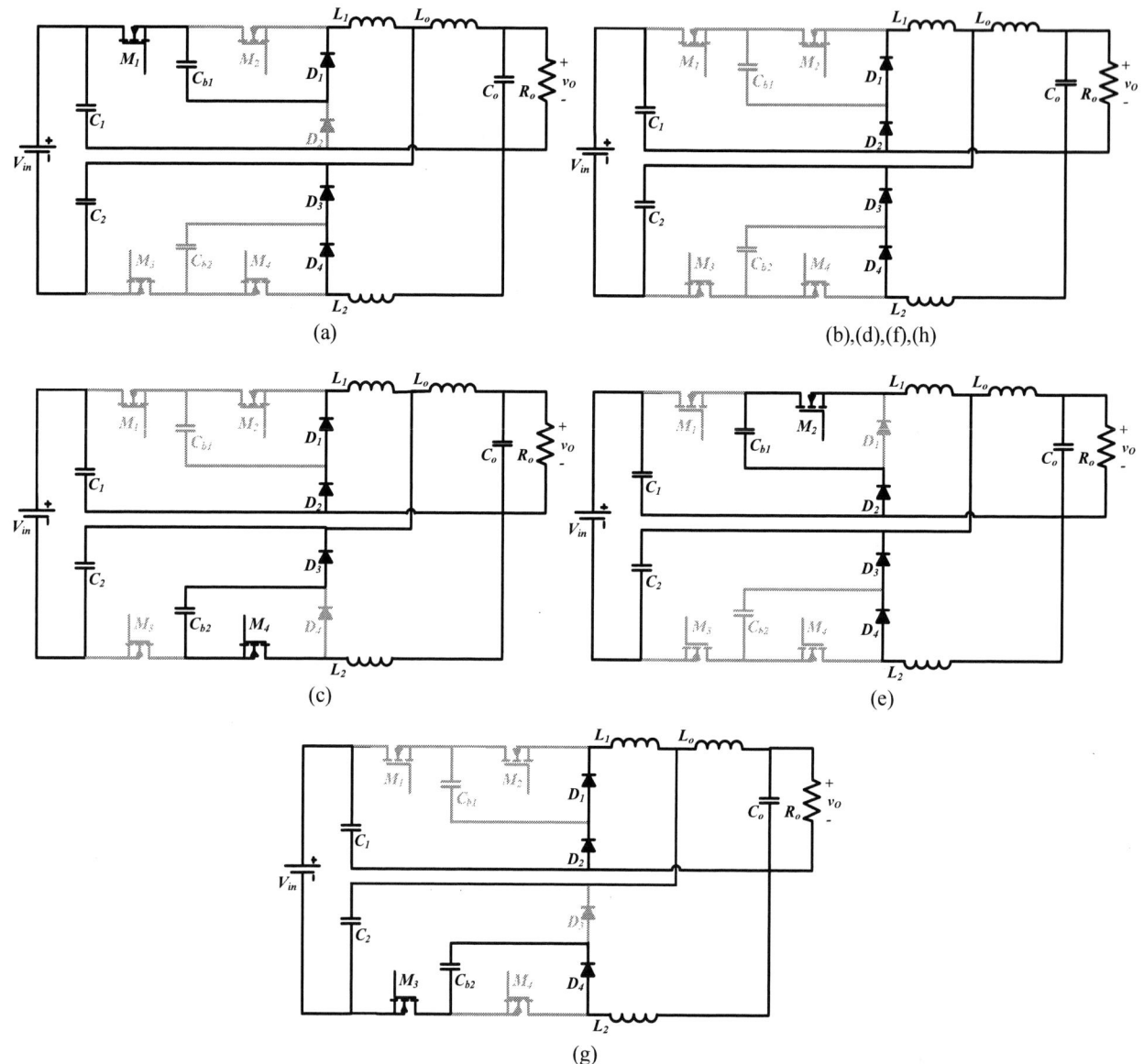

Fig. 3. Equivalent circuit of each operating mode: (a) Mode 1, (b) Mode 2, (c) Mode 3, (d) Mode 4, (e) Mode 5, (f) Mode 6, (g) Mode 7, and (h) Mode 8.

$$v_{L2}(t) = -V_o - v_{Lo}(t) \tag{10}$$

$$v_{Lo}(t) = V_{C1} + V_{C2} - V_{in} - V_o \tag{11}$$

Mode 6 [t_5-t_6]: This mode begins when M_2 is turned off at t_5 which is similar to mode 2 and 4.

Mode 7 [t_6-t_7]: At t_6 M_3 is turned on and V_{Cb2} is placed across M_4. Current of L_1, i_{L1} passes through D_1 and D_2, and i_{L2} increase linearly through two loops, one V_{in}-C_1-L_2-D_4-C_{b2}-M_3

and the other C_2-L_o-C_o-L_2-D_4-C_{b2}-M_3. The equations of this mode are as follows:

$$v_{L1}(t) = -V_o - v_{Lo}(t) \tag{12}$$

$$v_{L2}(t) = V_{in} - V_{Cb2} - V_{C1} = V_{C2} - V_{Cb2} - V_o - v_{Lo}(t) \tag{13}$$

$$v_{Lo}(t) = V_{C1} + V_{C2} - V_{in} - V_o \tag{14}$$

Mode 8 [t_7-t_8]: The proposed converter operation in this interval is like the second mode of operation and v_{Lo} is like (5).

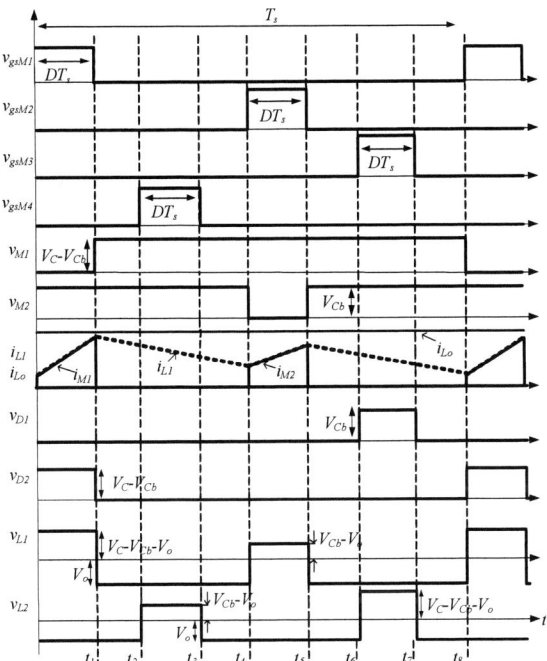

Fig. 4. Typical key waveforms of the proposed converter.

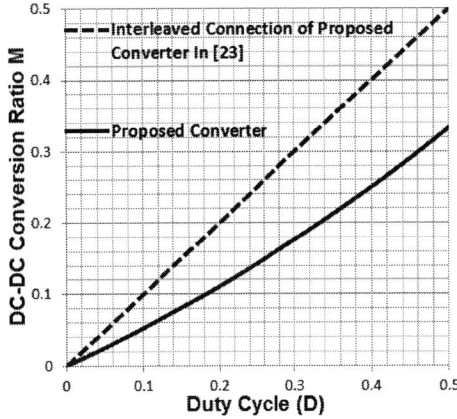

Fig. 5. Comprison of DC- DC conversion ratio between the proposed converter and the inteleaved connection of the converter in [23].

III. DESINGN CONSIDERATION

The following equation describes the Volt-Second-Balance (VSB) of an inductor.

$$\int_T v_{L_o} \cdot dt = 0 \tag{15}$$

According to the pervious section equations, V_{Lo} in all operating modes is similar and thus, to confirm (15), V_{Lo} must be equal to zero.

$$v_{L_o}(t) = 0 \tag{16}$$

$$v_{L_o}(t) = L_o \frac{di_{L_o}(t)}{dt} \tag{17}$$

Therefore, the output current ripple is zero in this converter.

A. DC-DC Conversion ratio

By using the principle VSB, the conversion ratio of the proposed converter can be derived. The VSB equation for L_1 and L_2 can be expressed as follows:

$$\int_{T_{sw}} v_{L1}(t) \cdot dt = [(V_{in} - V_{Cb1} - V_{C2}) \cdot D \tag{18}$$
$$+ (-V_o) \cdot (1 - 2.D) + (V_{Cb1} - V_o) \cdot D] = 0$$

$$\int_{T_{sw}} v_{L2}(t) \cdot dt = [(V_{in} - V_{Cb2} - V_{C1}) \cdot D \tag{19}$$
$$+ (-V_o) \cdot (1 - 2.D) + (V_{Cb2} - V_o) \cdot D] = 0$$

By simplifying (18) and (19), (20) and (21) are obtained respectively.

$$(V_{in} - V_{C2})D = (1 - D)V_o \tag{20}$$

$$(V_{in} - V_{C1})D = (1 - D)V_o \tag{21}$$

And form (20) and (21)

$$V_C = V_{C1} = V_{C2} \tag{22}$$

Thus, by substituting (20), in (3)

$$V_C = \frac{V_{in} + V_o}{2} \tag{23}$$

And also by performing similar procedure

$$V_{Cb} = V_{Cb1} = V_{Cb2} \tag{24}$$

And

$$V_{Cb} = \frac{V_{in}}{4} \tag{25}$$

Form (20) or (21) and (22), DC-DC conversion ratio of the proposed converter is

$$M = \frac{V_O}{V_{in}} = \frac{D}{2 - D} \tag{26}$$

Fig. 5 shows a comparison between the proposed converter and the interleaved connection of the converter in [23]. As shown in Fig. 5, the step down ability of the proposed converter is much better.

B. Voltage Stress of semiconductor components

The voltage stress of M_1 and M_3 can be obtained form

$$V_{M1} = V_{M3} = V_C - V_{Cb} = \frac{2 + D}{4.(2 - D)} . V_{in} \tag{27}$$

Also, the voltage stress of M_2 and M_4 is

$$V_{M2} = V_{M4} = V_{Cb} = \frac{V_{in}}{4} \tag{28}$$

978-1-5090-0376-1/16 $31.00 © 2016 IEEE

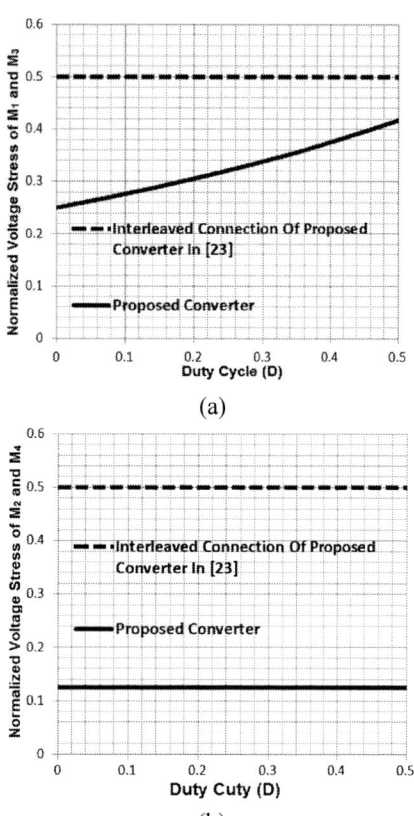

(a)

(b)

Fig. 6. A comprison of switches voltage stress in the proposed converter and in the interleaved connection of the converter in [23]. (a) voltage stress of M_1 and M_3. (b) voltage stress of M_2 and M_4.

Fig. 6 shows a comparison of switches voltage stress between the interleaved connection of converter introduced in [23] and the proposed converter. Also, the voltage stress of D_1 and D_4 is similar to (28) and the voltage stress of D_2 and D_3 is like (27). As it is obvious from Fig. 6, voltage stress of switches and diodes are much smaller than the interleaved connection of converter in [23].

C. Inductor design and output current ripple

Fig. 4 shows the voltage and current waveforms of the inductors. From this figure, the current ripple of each inductor can be expressed as following

$$\Delta i_L = \frac{V_O (1 - 2D).T_S}{L} \quad (29)$$

According to (29), and considering (16), variation of L_o current, regardless of its amount, is zero. So the output current ripple is extremely low.

$$v_L = L.\frac{di(t)}{dt} \quad (30)$$

D. Capacitors Design

In order to design capacitors, the following equation can be obtained for input current by using *KCL* law in the proposed converter

$$i_{in}(t) = i_{L1} + i_{L2} - i_{Lo} \quad (31)$$

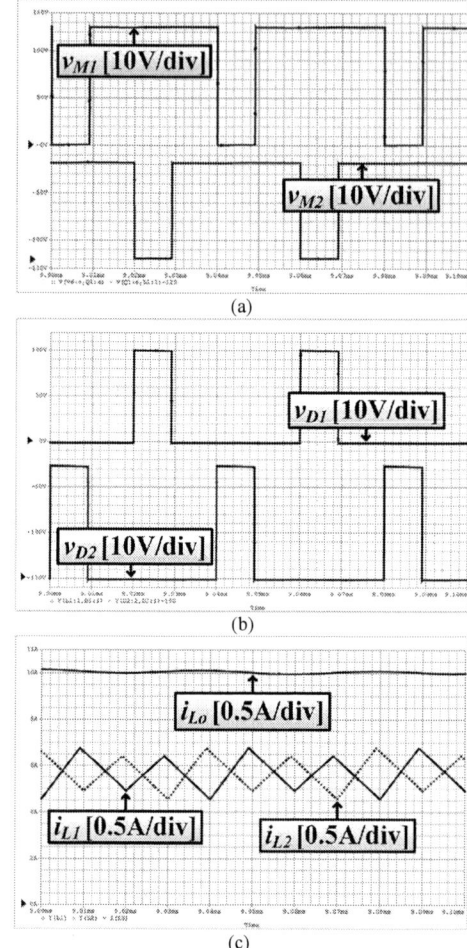

(a)

(b)

(c)

Fig. 7. Simulation results of the poroposed converter. (time scale 1μs/ div) (a) Voltage of switches (b), Voltage of Diodes, (c) current of L_1 , L_2 and L_o.

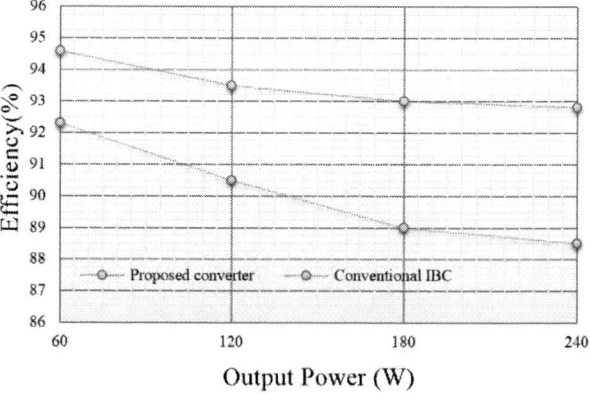

Fig. 8. Efficinecy comparison of the proposed converter and the conventional IBC by simulation.

So, the equation below is obtained by assuming $P_{in} = P_{out}$.

$$C_1 = C_2 = C = \frac{1}{\Delta V_C} \int_{T_S} i_C.dt = \frac{1}{\Delta V_C}.I_O.\frac{D.(1-2.D)}{2-D}.T_S \quad (32)$$

978-1-5090-0376-1/16 $31.00 © 2016 IEEE   138

Also, C_b can be designed by

TABLE I. PARAMETERS OF THE PROPOSED CONVERTER

Component	Specification
Input Voltage (V)	400
Output Voltage (V)	48
Output Power (W)	480
Switches M_1, M_2, M_3, M_4	IRF640
Diodes D_1, D_2, D_3, D_4	MBR20-150
Inductors L_1 and L_2 (µH)	300
Inductor L_o (µH)	10
Capacitors C_1 and C_2 (µF)	220
Capacitors C_{b1} and C_{b2} (µF)	100
Capasitor C_o (µF)	10

$$C_{b1} = C_{b2} = C_b = \frac{1}{\Delta V_{C_b}} \int_{T_S} i_{C_b} . dt = \frac{1}{\Delta V_{C_b}} . I_o . \frac{D}{2-D} . T_S \quad (33)$$

I. SIMULATION RESULTS

A prototype converter is simulated with PSPICE to verify the theoretical analysis. The prototype converter has 400 V input voltage and 48V, 10A output. Switching frequency is 25 kHz. L_1 and L_2 are designed for CCM condition. Table I shows the parameters of proposed converter. Simulation results are presented in Fig. 7. Fig. 7.a shows the voltage stress of switches M_1, M_2, M_3 and M_4. According to this figure, voltage stress of M_2 and M_4 is a quarter of input voltage and voltage stress of M_1 and M_3 is approximately $V_{in}/4$. Also, Fig. 7.b shows the voltage stress of diodes. Voltage stress of D_1 and D_4 is similar to voltage stress of M_2, M_4 and voltage stress of D_1 and D_3 is just like M_1, and M_3. In Fig. 8.c, currents of inductors L_1, L_2 and L_o are shown. According to Fig. 7.c, L_o current ripple is extremely low.

II. CONCLUSIONS

In this paper a high step-down DC-DC converter is proposed. The proposed converter has many advantages like improved step-down conversion ratio, extremely low output current ripple and low switching losses. Also, the voltage stress of semiconductor components in the proposed converter is much smaller than the conventional interleaved buck converter. All these benefits are obtained without applying any additional voltage or current stress on the components. The simulation results based on a 400V to 48V-10A prototype validates the theoretical analysis.

REFERENCES

[1] H. R. E. Larico and I. Barbi, "Three-Phase push-pull DC-DC converter: analysis, design, and experimentation," *IEEE Trans. Ind. Electron.*, vol. 59, pp. 4629-4636, 2012.

[2] K. Sun, L. Zhang, Y. Xing, and J. M. Guerrero, "A Distributed Control Strategy Based on DC Bus Signaling for Modular Photovoltaic Generation Systems With Battery Energy Storage," *IEEE Trans. Ind. Electron.*, vol. 26 no. 10 pp. 3032-3045, Oct. 2010.

[3] Pit-Leong Wong; Peng Xu; B. Yang and F.C., Lee, "Performance improvements of interleaving VRMs with coupling inductors," *IEEE Trans. Power Electron.*, vol.16, no.4, pp.499, 507, Jul 2001.

[4] M. Pahlevaninezhad, J. Drobnik, P. K. Jain, and A. Bakhshai, "A load adaptive control approach for a zero-voltage switching DC/DC converter used for electric vehicles," *IEEE Trans. Ind. Electron.*, vol. 59 no. 2 pp. 920-933, Feb. 2012.

[5] M. Esteki, E. Adib, and H. Farzanehfard, "Soft switching interleaved PWM buck converter with one auxiliary switch," *Electrical Engineering (ICEE), 2014 22nd Iranian Conf.*, pp.232,237, 20-22 May 2014.

[6] M. Esteki, B. Poorali, E. Adib and H. Farzanehfard, "Interleaved buck converter with continuous input current, extremely low output current ripple, low switching losses, and improved step-down conversion ratio," *IEEE Trans. Ind. Electron.*, vol.62, no.8, pp.4769-4776, Aug. 2015.

[7] M. Ilic, and D. Maksimovic, "Interleaved zero-current-transition buck converter," *IEEE Trans. industry App.*, vol.43, no.6, pp.1619,1627, Nov.-dec. 2007.

[8] M. Esteki, E. Adib, H. Farzanehfard, and S. A. Arshadi, "Auxiliary circuit for zero voltage transition interleaved PWM buck converter," *IET Power Electron.*, Aug. 2015, in press.

[9] J.-B. Baek; W.-I. Choi and B.-H. Cho, "Digital adaptive frequency modulation for bidirectional DC–DC converter," *IEEE Trans. Ind. Electron.*, vol.60, no.11, pp.5167, 5176, Nov. 2013.

[10] A. orrell, M. Castilla, J. Miret, J. Matas, and L. Garcia de Vicuna, "Control design for multiphase synchronous buck converters based on exact constant resistive output impedance," *IEEE Trans. Ind. Electron.*, vol.60, no.11, pp.4920,4929, Nov. 2013.

[11] N. Molavi, M. Esteki, E. Adib and H. Farzanehfard, "High step-up/down DC-DC bidirectional converter with low switch voltage stress," *The 6th Power Electronics, Drives Systems & Technologies Conf. (PEDSTC)*, vol., no., pp.162-167, 3-4 Feb. 2015.

[12] Xiong Du; Luowei Zhou; Heng-Ming Tai, "Double-frequency buck converter," *IEEE Trans. Ind. Electron.*, vol.56, no.5, pp.1690,1698, May 2009.

[13] Yao, K.; Mao Ye; Ming Xu; Lee, F.C., "Tapped-inductor buck converter for high-step-down DC-DC conversion," *IEEE Trans. Power Electron.*, vol.20, no.4, pp.775, 780, July 2005.

[14] Garcia, J.; Calleja, A.J.; López rominas, E.; Gacio Vaquero, D.; Campa, L., "Interleaved buck converter for Fast PWM dimming of high-brightness LEDs," *IEEE Trans. Power Electron.*, vol.26, no.9, pp.2627,2636, Sept. 2011.

[15] Forest, F.; Labouré, E.; Meynard, T.A.; Huselstein, J.-J., "Multicell interleaved flyback using intercell transformers," *IEEE Trans. Power Electron.*, vol.22, no.5, pp.1662,1671, Sept. 2007.

[16] K. Yao, Q. Yang, X. Ming and F. C. Lee, "A novel winding-coupled buck converter for high-frequency, high-step-down DC-DC conversion," *IEEE Trans. Power Electron.*, vol.20, no.5, pp.1017, 1024, Sept. 2005.

[17] Il-Oun Lee; Shin-Young Cho; Gun-Woo Moon, "Interleaved buck converter having low switching losses and improved step-down conversion ratio," *IEEE Trans. Power Electron.*, vol.27, no.8, pp.3664,3675, Aug. 2012.

[18] Ching-Tsai Pan; Chen-Feng Chuang; Chia-Chi Chu, "A novel transformerless interleaved high step-down conversion ratio DC–DC converter with low switch voltage stress, " *IEEE Trans. Ind. Electron.*, vol.61, no.10, pp.5290,5299, Oct. 2014.

[19] Peng Xu; Jia Wei; Lee, F.C., "Multiphase coupled-buck converter-a novel high efficient 12 V voltage regulator module," *IEEE Trans. Power Electron.*, vol.18, no.1, pp.74,82, Jan 2003.

[20] Xinbo Ruan; Bin Li; Qianhong Chen; Siew-Chong Tan; Tse, C.K., "Fundamental considerations of three-level DC–DC converters: topologies, analyses, and control," *IEEE Trans. Circuits and Systems I: Regular Papers*, vol.55, no.11, pp.3733,3743, Dec. 2008.

[21] Y. Zhang, J.-T. Sun and Y.-F. Wang, "Hybrid boost three-level DC–DC converter with high voltage gain for photovoltaic generation systems," *IEEE Trans. Power Electron.*, vol.28, no.8, pp.3659, 3664, Aug. 2013.

[22] S.-S. Lee, "Step-down converter with efficient ZVS operation with load variation," *IEEE Trans. Ind. Electron.*, vol.61, no.1, pp.591, 597, Jan. 2014.

[23] X. Ruan, J. Wei and Y. Xue, "Three-level converters with the input and output sharing the ground," *Power Electronics Specialist Conference, 2003. PESC '03. 2003 IEEE 34th Annual*, vol.4, no., pp.1919, 1923 vol.4, 1June 2003.

7th Power Electronics, Drive Systems & Technologies Conference (PEDSTC 2016)
16-18 Feb. 2016, Iran University of Science and Technology, Tehran, Iran

AC Voltage Regulator Based on AC/AC Buck Converter

M. R. Hajimoradi
Electrical Engineering Department
Sharif University of Technology
Tehran, Iran
hajimoradi@ee.sharif.edu

H. Mokhtari
Electrical Engineering Department
Sharif University of Technology
Tehran, Iran
mokhtari@sharif.edu

Abstract— **This paper introduces a high efficiency AC voltage regulator based on an AC/AC buck converter cascaded by a transformer in series with the input voltage. The AC/AC converter uses an overlap time in the gate signals to solve the commutation problem. Non-use of any snubber circuits and current sensors leads to lower cost, smaller size and simpler hardware. The converter generates only the compensation term which results in smaller switches and, thus, lower cost. Simulation and experimental results verify the performance of the proposed topology.**

Keywords— *Ac Voltage Regulator, AC Chopper, AC/AC Buck Converter, Power Electronics, Pulse Width Modulation*

I. INTRODUCTION

Expanding the use of electronic equipment in various aspects of life from one hand and increasing the sensitivity of the electronic appliances to the power system quality from the other hand, make the power quality improvement an important issue for the utilities and the end users. Voltage sags and swells are known as the most common and destructive power quality problems in the power system [1]. AC voltage regulators, based on solid state power switches are the main devices used for increasing the reliability of the sensitive loads operation against power quality problems. One of the proposed solutions to compensate voltage deviations is the use of a dynamic voltage restorer (DVR) with indirect conversion of AC to AC voltage and a DC link capacitor [2]. But the dc-electrolytic capacitors are very sensitive to temperature and are expected to have higher failure rates at higher temperatures [3]. An indirect AC to AC conversion uses more power switches; hence, has more loss and less efficiency as compared to direct AC/AC conversion [1].

An AC/AC converter which was first introduced as an "Intelligent Transformer" in 1996 [4], offers several advantages such as high power density, multi-functionality, intelligence and environmental protection [5]. Direct conversion of AC to AC and removing the

dc-link and its bulky electrolytic capacitor can lead to higher efficiency, lower cost, smaller size, longer life and increased reliability [6]. As a result, a transition from double conversion converters (like back-to-back) to direct AC/AC converter (like matrix converters) is in progress.

An AC/AC high frequency converter has the drawback of commutation problem. A circuit diagram of an AC/AC buck converter is shown in Fig. 1. During the commutation, if a dead-time is inserted between the gating signals, an open circuit in the inductor current path causes voltage spikes. On the other hand, applying overlapping PWM signals leads to a short circuit among the source terminals.

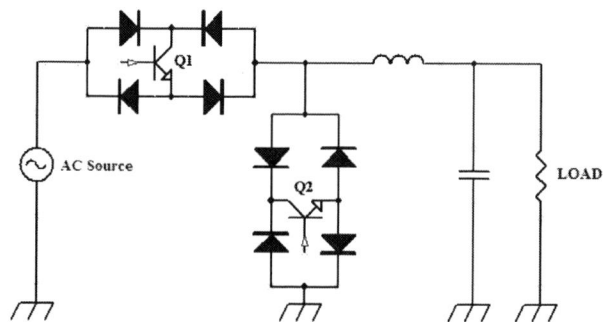

Fig. 1. A prototype AC/AC buck converter

Several topologies have been introduced as an AC/AC buck converter or an AC chopper. These topologies either use snubber circuits [7-10] or resonance circuits for soft-switching (ZVS or ZCS) [11, 12]. Soft switching AC/AC converters have some restrictions on the load power factor [13]. In some practices, selective switching patterns based on current and voltage polarities are proposed to overcome the commutation problem [14, 15].

This paper uses a new configuration for a direct

978-1-5090-0376-1/16 $31.00 © 2016 IEEE   140

AC/AC buck converter to generate a duty proportional AC voltage. The generated voltage is added in phase or 180° out of phase with the input voltage by the use of an isolating transformer to compensate voltage sags or swells. The converter takes the required compensation power directly from the grid and does not need any storage element in its structure. Another benefit of the series connected AC voltage regulator is that, only the required power for compensating the deviations is transferred through the converter, therefor the power circuit has reduced ratings and stresses as compared to the ones that handle all of the load power.

The remainder of the paper is organized as follows. In the following section the proposed topology is introduced. The elements of the power circuit and the way it works are explained, and the rules governing the system are expressed. In section III, the feedback loop and control strategy with simulation results of the voltage regulator are presented. All of the theoretical findings are confirmed and validated by the experimental tests performed using a 1 KVA prototype AC/AC voltage regulator presented in section IV. Finally, this paper is summarized in section V.

II. THE PROPOSED TOPOLOGY

As mentioned earlier, the commutation problem is the most determinative factor in the expansion of use and power ratings of direct AC/AC converters. In this paper, a new class of bidirectional semiconductor power switch connections is used which eliminates the commutation problem. This configuration which is shown in Fig. 2 is a development to the classic AC/AC buck configuration of Fig. 1.

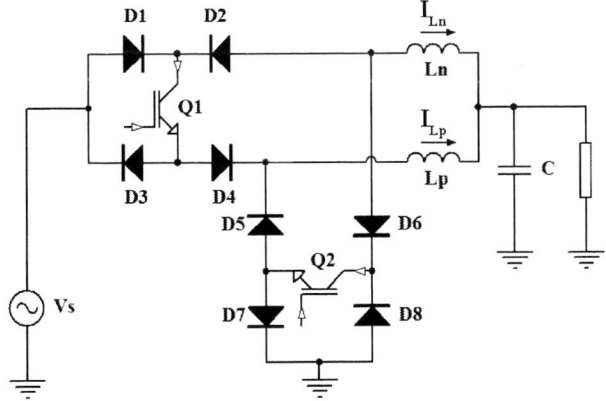

Fig. 2. The new class of bidirectional switches connection used in an AC/AC buck converter

In the new configuration, the inductor of the filter is divided into two parts, and the combination of each 4-

quadrant switches and their connections are reconfigured. A simple complementary PWM is applied to Q_1 and Q_2. To keep the current path of the inductors always closed, an overlap is inserted in the gating signals of the power switches during the switch transition. In other words, at the end of the duty cycle, prior to turning off Q_1, Q_2 is turned on, and after a while (equal to the overlap time) Q_1 turns off. This pattern is repeated reversely at the end of the period when Q_1 is going to be triggered once again.

Based on the state of the PWM gating signals, three operation modes exist as described below.

A. Q_1 is on and Q_2 is off

In this mode, the sign of $I_{Ln}+I_{Lp}$ determines which elements are conducting. In Fig. 3 and Fig. 4, the two scenarios are shown and the active components are highlighted.

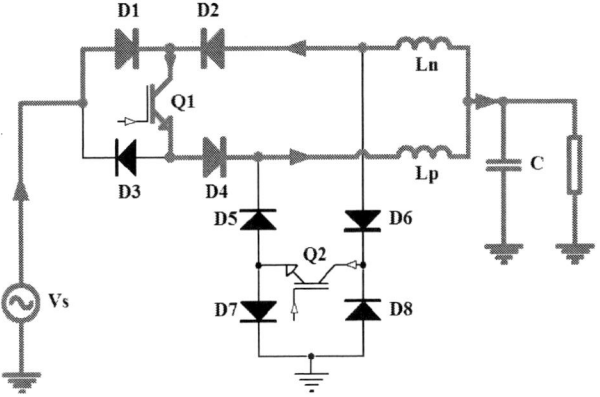

Fig. 3. Current path when Q_1 is on and $I_{Ln}+I_{Lp}>0$

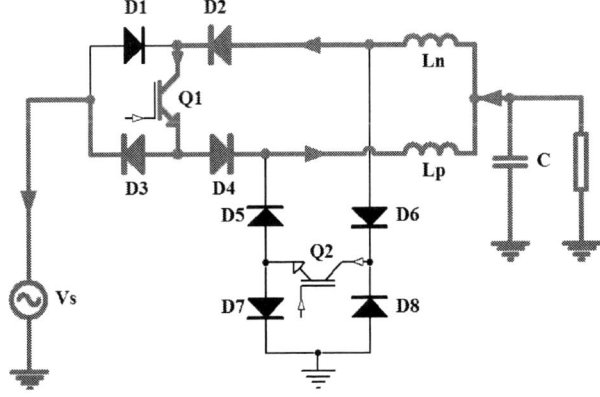

Fig. 4. Current path when Q_1 is On and $I_{Ln}+I_{Lp}<0$

B. Both Q_1 and Q_2 are on

During the overlap period, both Q1 and Q2 are on. Depending on the polarity of the source voltage, two paths exist for the currents of L_n and L_p. When $V_S>0$ the

978-1-5090-0376-1/16 $31.00 © 2016 IEEE   141

current path is shown in Fig. 5. Fig. 6 depicts the current flow when $V_S<0$.

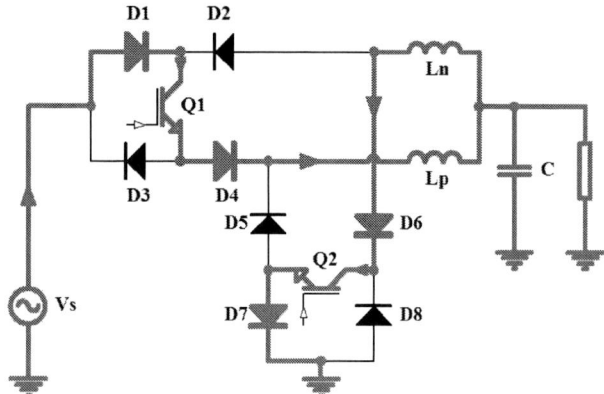

Fig. 5. Current path during the overlap period, when $V_S>0$

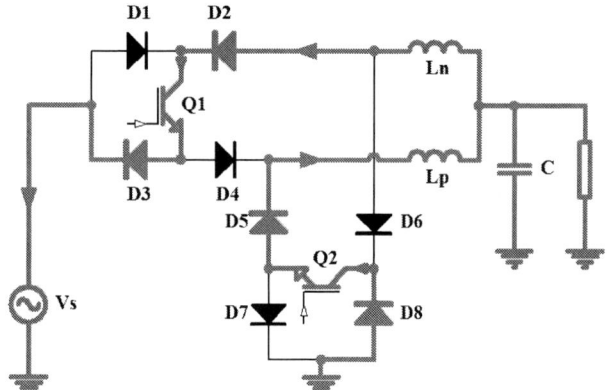

Fig. 6. Current path during the overlap period, when $V_S<0$

C. Q1 is off and Q2 is on

In this mode, the same as the first mode, the sign of the sum of I_{Ln} and I_{Lp} is the determining factor for the current path. The current flows for positive and negative values are shown in Fig. 7 and Fig. 8 respectively by the highlighted components.

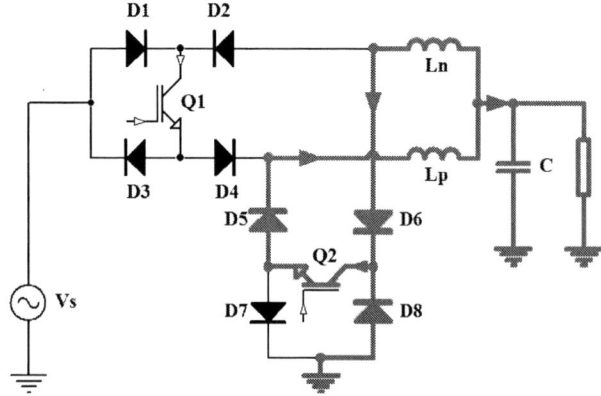

Fig. 7. Current path when Q_2 is on and $I_{Ln}+I_{Lp}>0$

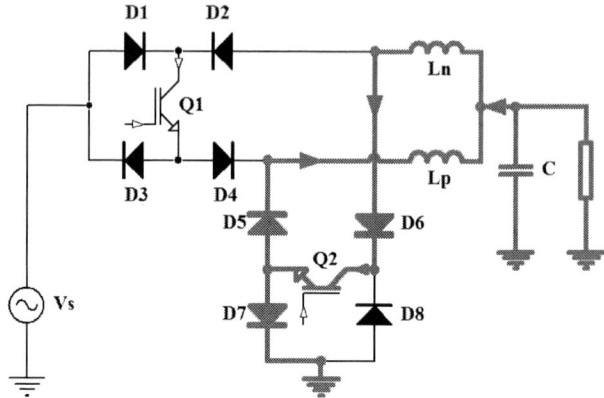

Fig. 8. Current path when Q_2 is on and $I_{Ln}+I_{Lp}<0$

Another important concern about the new AC/AC buck converter is that the maximum switching frequency is limited by the peak value of the source voltage (V_m), the forward voltage drop across the power switches and the diodes (V_Q and V_D) and the duration of the overlap time (T_{OVLP}). This relationship can be expressed by

$$f_S \leq \frac{(V_Q+2V_D)}{\left(\frac{4}{\pi}V_m-2(V_Q+2V_D)\right)T_{OVLP}} \tag{1}$$

If the switching frequency is set more than the threshold frequency determined by (1), the amplitude of the inductors currents increases by time, resulting in system instability. This is due to the fact that by increasing the switching frequency, the stored energy in the inductors during the overlap time becomes more than the inductors energy loss due to the forward voltage drops across the power switches and the diodes.

In the AC/AC buck converter of Fig. 2, the familiar conversion ratio of the DC/DC buck converters still exists and can be expressed as

$$V_C = DV_S \tag{2}$$

where V_C is the capacitor voltage (load voltage), D is the duty ratio of the applied PWM signals and V_S is the source voltage.

Considering the conversion ratio of (2), the buck converter of Fig. 2 is only capable of compensating the voltage swells. In order to add the capability of compensating voltage sags and swells and benefit from the advantages of series voltage compensation, an isolating transformer with two primary windings is cascaded by the AC/AC buck converter as demonstrated in Fig. 9. The configuration is similar to an auto-transformer.

978-1-5090-0376-1/16 $31.00 © 2016 IEEE

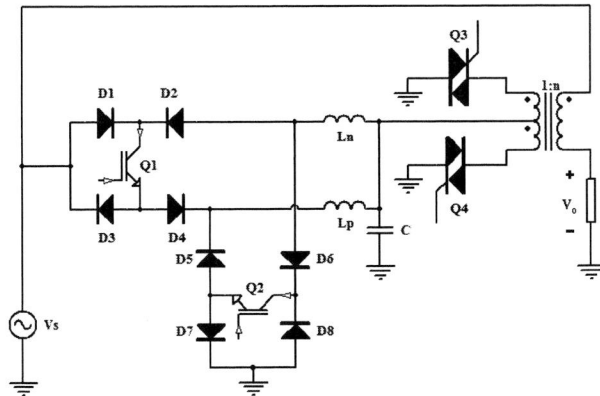

Fig. 9. The proposed topology for AC/AC voltage regulator

The output voltage of the buck converter is connected to the middle terminal of the primary side windings. During the normal operation of the converter, one of the silicon controlled rectifiers (SCR) is always on. If Q_3 is triggered, the output voltage of the isolating transformer is added to the source voltage to compensate voltage sags, and if Q_4 is gated, voltage swells are regulated. In order to prevent any short circuit, after turning on the required switch (Q3 or Q4) and stopping the gate signals of the other switch (Q4 or Q3) and prior to applying PWM pulses to the buck converter, a short delay (at least one half-cycle) should be elapsed to ensure that the SCR is turned off naturally. This schema is illustrated in Fig. 10.

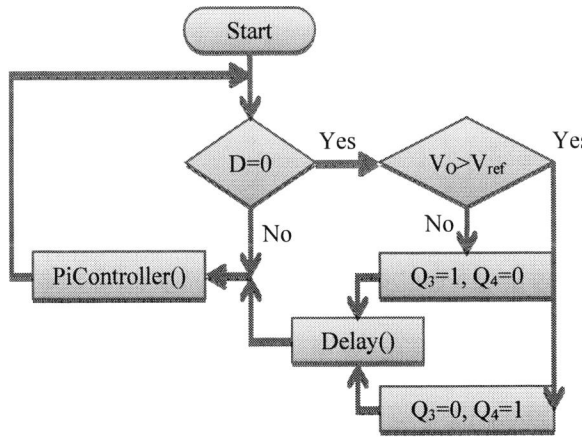

Fig. 10. Flowchart of transition from sag to swell compensation or vice versa

In addition, in the case of an overcurrent or short circuit detection, the duty ratio of the AC/AC buck converter is set to zero and both Q_3 and Q_4 should be triggered to protect the power circuit of the converter from any damage. In this situation, a series connected protective fuse could operate automatically and saves the system.

For the AC voltage regulator shown in Fig. 9, the input-output voltage relation can be expressed by

$$V_O = V_S(1 \pm nD) \tag{3}$$

where n is the turns ratio of the transformer and $\pm$ represents the voltage sag or voltage swell regulation selected by Q_3 or Q_4.

III. CONTROL SYSTEM AND SIMULATION RESULTS

In this part, the feedback loop of the new AC/AC voltage regulator is explained, and the simulation results of the proposed circuit are presented. Simulations are carried out for a 1 KW system using PSCAD/EMTDC and the simulation parameters are given in Table I.

TABLE I. SYSTEM PARAMETERS FOR SIMULATION STADIES

Parameter	Value
L_n, L_p	500 µH
C	10 µF, 400V
$V_{CE,SAT}$ of Q_1, Q_2	1 V
Forward voltage drop of D_1-D_8	1 V
T_{OVLP}	0.5 µSec
Transformer turns ratio (n)	50/170

Considering the parameters listed in Table I and the inequality of (1), the maximum allowed switching frequency for the peak of 368V in the AC supply is 12.983 KHz. Therefore, a switching frequency of 12 KHz will guarantee operation in the safe area.

One of the most effective factors in the quality of voltage regulation, from the fast response performance point of view, is the delay time of the error detecting unit.

Error detection based on an RMS calculation in the most effective form takes about one-fourth of a cycle to detect the voltage deviations. Another way is to use the simple trigonometric principle of

$$V_n^2 \sin^2(\omega t) + V_n^2 \cos^2(\omega t) = V_n^2 \tag{4}$$

If the first term is substituted with the instantaneous

output voltage, the dc feedback value is generated, and the deviation from the set point can produce the error signal of the PI controller. This scheme is described in [16]. During the voltage sag compensation (Q_3=1 and Q_4=0), any increase in the duty ratio of the PWM pulses increases the output voltage. In contrast, when the voltage swells are regulated (Q_3=0 and Q_4=1), increasing the duty factor causes the output voltage to be decreased. This behavior affects the error signal calculation and makes two following different modes.

- In the voltage sag compensation mode:

$$e = V_{ref} - V_{fb} \tag{5}$$

- In the voltage swell compensation mode:

$$e = V_{fb} - V_{ref} \tag{6}$$

where e is the error signal and V_{ref} and V_{fb} are defined as:

$$V_{ref} = V_n^2 \tag{7}$$

$$V_{fb} = V_O^2 + V_n^2 \cos^2(\omega t) \tag{8}$$

where V_O and V_n are the instantaneous output voltage and the nominal output voltage amplitude, respectively.

The first interesting issue to be verified by simulation is the switching frequency threshold. The results of two situations are demonstrated in Fig. 11. If the switching frequency is set lower than the threshold frequency given by (1), the inductors currents are convergent as shown in Fig. 11-a. Increasing the switching frequency beyond the limit, will cause instabilities and divergence of the currents as presented in Fig. 11-b. The divergence is finite and is limited by the parasitic resistance of the inductors, switches, diodes, paths, etc., in the real power circuit.

The normal operation of the system in regulating the mains voltage deviations and the output voltage transients in the case of sudden source voltage change are illustrated in Fig. 12 and Fig. 13.

In Fig. 12, the performance of the system in case of a 30V voltage jump (from 180V to 210V) is shown. The voltage deviation is detected and compensated within about one half-cycle.

In the second try, a 70V voltage drop is applied to the source. The PI controller detects the voltage deviation and reduces the duty ratio of the PWM to zero. In this case, the voltage swell is changed to voltage sag. Therefore, it is required to turn off Q4 and trigger Q3.

This is done upon achieving the duty factor to zero and initiates a delay of 10 ms to make sure Q4 is turned off naturally. During the delay time, PWM pulses are stopped, and the output voltage is equal to the mains voltage. At the end of this state, the PI controller starts its normal operation and regulates the load voltage to the desired level. This scenario is depicted in Fig. 13. In this case, the voltage regulation takes place within about two cycles that can be an acceptable time for compensating of 70V voltage drop.

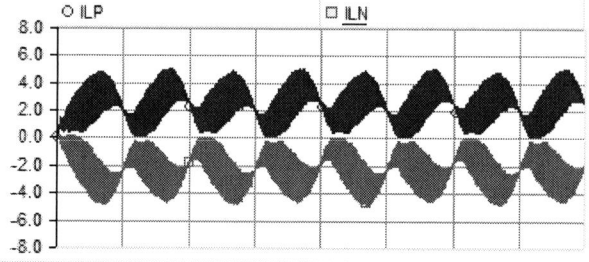

(a)

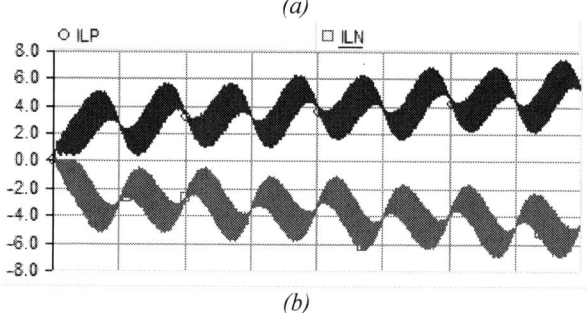

(b)

Fig. 11. The inductors current waveforms. (a) The currents are convergent. (b) The currents are divergent.

In the next section, the experimental results support the analytical and simulation findings of the proposed regulator.

IV. EXPERIMENTAL RESULTS

In this part, the experimental results of the proposed regulator based on the new configuration of an AC/AC buck converter are presented. A prototype with 1 KW nominal power is setup as a proof of concept.

The parameters of the system are the same as those listed in Table I. In practice, STGW40NC60V and MUR3060 are chosen as the IGBT and diodes of the power circuit and the part number of the SCR is BTA16. The switching frequency is set to 12 KHz and the overlap time is set to 0.5 µs.

In the experimental tests, the input power is delivered by a programmable/arbitrary waveform AC power source (CHROMA 61501), and the waveforms are recorded by a PC based digital oscilloscope.

978-1-5090-0376-1/16 $31.00 © 2016 IEEE

In Fig. 14 and Fig. 15, a 25% voltage drop and rise occur in the source. The error detection unit generates the error signal, and the PI controller regulates the output voltage. The disturbance is compensated after about one cycle.

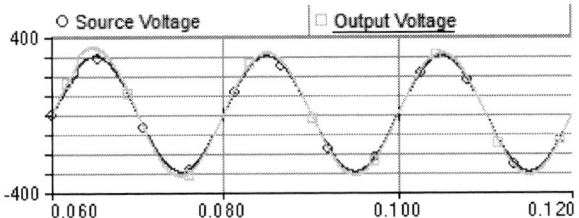

Fig. 12. Output voltage transient in a sudden voltage jump.

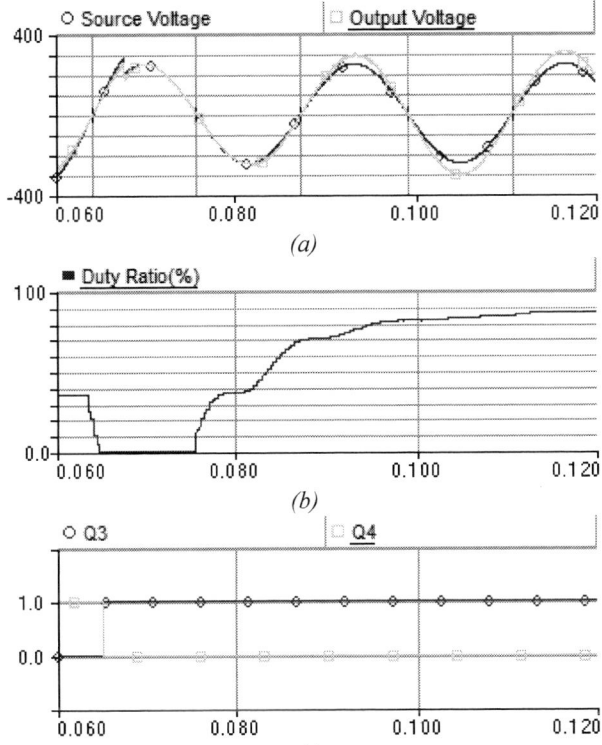

Fig. 13. Suddenly change of voltage swell to voltage sag. (a) Input/output voltages. (b) The graph of changes in duty ratio. (c) Trigger pulses of Q3 and Q4

The efficiency of the proposed voltage regulator is also measured using a CHAUVIN ARNOUX C.A 8334 power quality analyzer. For this test, the source voltage is set to 190V and the load voltage is regulated at 220V. The load type is pure resistive and has three 300W steps. The load steps are connected one by one to the output terminal. The results of the measurements are summarized in Table 2.

According to the data of Table 2, the regulator has a

good efficiency and the power loss is very low. In practice, the heat radiators of the power switches and diodes are quite cool without any forced cooling system such as a fan.

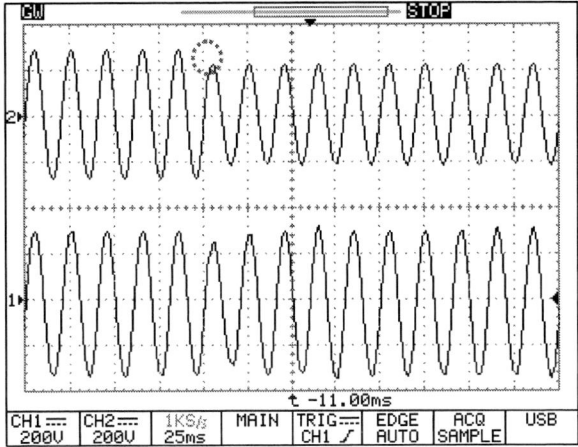

Fig. 14. Input (CH2) and output (CH1) waveforms for a 240V to 185V sudden voltage drop.

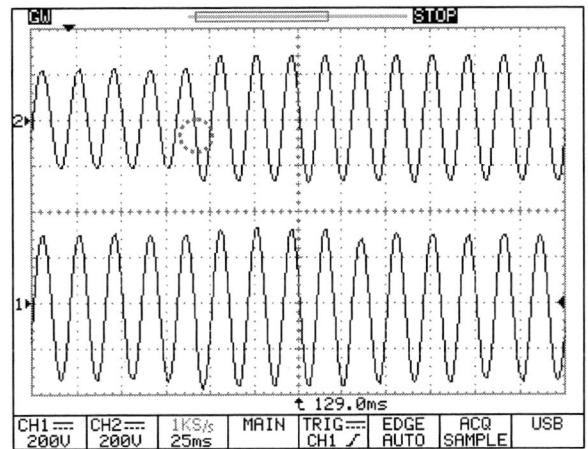

Fig. 15. Input (CH2) and output (CH1) waveforms for a 185V to 240V sudden voltage rise.

TABLE 2. EFFICIENCY MEASUREMENTS DATA

Load Power	Efficiency
305W	93.8%
607W	96.0%
911W	96.8%

V. CONCLUSION

This paper has introduced a high efficiency AC voltage regulator based on an AC/AC buck converter cascaded by a transformer to produce an isolated voltage in series with the source voltage. The AC/AC buck converter introduced a new configuration in the power circuit and uses an overlap time in the gate signals to solve the commutation problem. No need of any snubber circuits and current sensors lead to lower cost, smaller size and simpler hardware. The converter generates only the compensation term which results in smaller switches, and thus, lower cost. The design and concept of the converter has also been studied in this paper, and the performance of the regulator has been evaluated through simulation and experimental results.

REFERENCES

[1] Park, C. Y., Kwon, J. M., & Kwon, B. H. 'Automatic voltage regulator based on series voltage compensation with ac chopper', IET J. Power Electronics, 2012, 5, (6), pp. 719 – 725.

[2] Jimichi, T., Fujita, H., Akagi, H. 'A dynamic voltage restorer equipped with a high-frequency isolated DC–DC converter', Industry Applications, IEEE Transactions on, 47, (1), pp. 169-175.

[3] Amirabadi, M., Baek, J., Toliyat, H. 'Bidirectional Soft Switching Series AC-Link Inverter', Industry Applications, IEEE Transactions on, 51, (3), pp. 2312 – 2320.

[4] Harada, K., Anan, F., Yamasaki, K., Jinno, M., Kawata, Y., Nakashima, T.: 'Intelligent Transformer' Proc . of the IEEE Power Electronics Specialist Conference, Baveno, Jun 1996, 2, pp. 1337-1341.

[5] Ma Huasheng, Zhao Liang, Zhang Bo, Zheng Jianchao.:'AC/AC Buck Converter Instantaneous Value Control Modeling, System Analyzing and Designing', Proc. of the IEEE Electrical Machines and Systems Conference, Nanjing, Sept 2005, 2, pp. 1391 – 1395.

[6] R. Moghe, R.P. Kandula, A. Iyer, D. Divan, 'Losses in Medium-Voltage Megawatt-Rated Direct AC/AC Power Electronics Converters', IEEE Transactions on Power Electronics, 2015, 30, (7), pp. 3553-3562.

[7] Rosas-Caro, J.C., Mancilla-David, F., Ramirez-Arredondo, J.M., Bakir, A.M.: 'Two-switch three-phase ac-link dynamic voltage restorer', IET J. Power Electronics, 2012, 5, (9) , pp. 1754-1763.

[8] Jothibasu, S., Mishra, M.K., 'An Improved Direct AC–AC Converter for Voltage Sag Mitigation', IEEE Transactions on Industrial Electronics, 2015, 62, (1), pp. 21- 29.

[9] Subramanian, S., Mishra, Mahesh Kumar, 'Interphase AC–AC Topology for Voltage Sag Supporter', IEEE Transactions on Power Electronics, 2010, 25, (2), pp. 514-518.

[10] Kwon, J.-M., Kim, K.-t., Kwon, B.: 'Instant voltage compensator based on a three-leg converter', IET J. Power Electronics, 2013, 6, (8), pp. 1618 – 1625.

[11] Chien-Ming Wang, Chang-Hua Lin, Ching-Hung Su, Shih-Yuan Chang, 'A Novel Single-Phase Soft-Switching AC Chopper without Auxiliary Switches', IEEE Transactions on Power Electronics, 2011, 26, (7), pp. 2041-2048.

[12] Minh-Khai Nguyen, Young-cheol Lim, Yong-Jae Kim: 'A Modified Single-Phase Quasi-Z-Source AC–AC Converter',IEEE Transactions on Power Electronics, 2012, 27, (1), pp. 201-207.

[13] Amirabadi, M., Toliyat, H.A., Alexander, W.C.: 'Single-stage soft-switching ac-link AC-AC and DC-AC buck-boost converters with unrestricted load power factor', 29th Conf. on Power Electronics (APEC), Fort Worth, TX , March 2014, pp. 1268 – 1275.

[14] Khan, M. M., Rana, A., Fei Dong: 'Improved ac/ac choppers-based voltage regulator designs', IET J. Power Electronics, 2014, 7, (8), pp. 1989 – 2000.

[15] Soeiro, T.B., Petry, C.A., dos S.Fagundes, J.C., Barbi, I.: 'Direct AC–AC Converters Using Commercial Power Modules Applied to Voltage Restorers', IEEE Transactions on Industrial Electronics, 2011, 58, (1), pp. 278-288.

[16] Hajimoradi, M. R., Karimi, E., Mokhtari, H., Yazdian, A., 'Performance improvement of a double stage switch mode AC voltage regulator', 3rd conf. on Power Electronics and Drive Systems Technology (PEDSTC), February 2012, pp. 181-186.

7th Power Electronics, Drive Systems & Technologies Conference (PEDSTC 2016)
16-18 Feb. 2016, Iran University of Science and Technology, Tehran, Iran

An Interleaved High Step-Up DC-DC Converter With Low Input Current Ripple

Mahmoud Fekri, Hosein Farzanehfard, Ehsan Adib
Department of Electrical and Computer Engineering
Isfahan University of Technology
Isfahan, Iran
m.fekri@ec.iut.ac.ir, hosein@cc.iut.ac.ir, e.adib@cc.iut.ac.ir.

Abstract— A new interleaved high step-up DC-DC converter is presented in this paper. Coupled inductors with series secondary windings are used in order to minimize the input current ripple and extend the voltage gain. Passive clamp capacitors are used to absorb the leakage energy and recycle it to the output. Thus, the voltage stress across switches is low and efficiency is improved. Due to charge- balance of clamp capacitor, the converter has inherent current sharing capability between two interleaved legs without any extra control circuit. All the mentioned advantages are joined in a simple structure with low component count. The converter structure and its operational modes are explained. A 40V input, 320V output, 130W prototype is implemented and examined in order to confirm the performance of the converter.

Keywords—High step-up, interleaved, coupled inductor, passive clamp.

I. INTRODUCTION

Global warming, greenhouse effect, environmental pollutions and problems associated with fossil fuels have resulted in major research toward new energy sources, among which solar energy and fuel cells are of promising candidates [1-3]. But unfortunately the output voltage produced by Photovoltaic (PV) modules and fuel cells is not high enough to directly supply an inverter, hence, a DC-DC converter with high conversion ratio is indispensable in such systems. It is desirable for such a converter to efficiently provide high voltage gain, while having simple structure, and low cost. Reducing voltage stress across power switches and diodes yields in low cost and low loss semiconductor devices, which in turn improves overall efficiency. Having low input current ripple is a valuable merit for a high step-up converter, since, large input current ripple leads to large volume and high cost of input capacitor due to high current rating of input stage. Moreover, low current ripple is desirable in photovoltaic applications for better Maximum Power Point Tracking (MPPT).

Boost converter is the simplest structure used for increasing the voltage level. But obtaining large voltage gain from boost converter results in large operating duty cycle, instability, high voltage stress across semiconductor devices and severe reverse recovery problems [4]. Isolated structures could be used to attain high voltage gain by adjusting transformer turn ratio [5], [6], but problems such as large

volume, weight and losses due to leakage inductance are the disadvantages of isolated converters [7]. Hence, several non-isolated structures are proposed. Switched capacitor technique could be used to increase the output voltage [8],[9], but higher voltage ratios require several switched capacitor cells and leads to complexity. Furthermore, switches would suffer large transient currents [10]. Some converters are developed utilizing coupled inductors [11-13], in which turn ratio of the coupled inductors could be employed to step up the gain. However, active or passive snubbers are required to suppress the voltage spikes across the switches caused by leakage inductance [14-16].

For higher power density and lower input current ripple, interleaved structures are suggested. In [17], an interleaved converter is presented in which cuk-type converter is integrated to achieve higher gain. Input current ripple is low as well as the voltage stress across the power switches, but voltage gain is not high enough. An interleaved voltage quadrupler is proposed in [18], in which automatic current sharing capability is achieved, and voltage across switches is a quarter of the output voltage, however, the voltage gain is limited. In [19] a Zero Voltage Transition (ZVT) interleaved converter is proposed. A built-in transformer is used in order to extend the voltage gain and reduce the voltage stress of the switches. Soft switching condition is provided by active clamp scheme so that both turn-on and turn-off of the main and the auxiliary switches are performed under ZVS condition. But gate signals and control mechanism are more complex with two additional switches, and voltage gain is not very large. In [20] the voltage gain of the converter in [19] is doubled by adding two capacitors in series with secondary windings of the built-in transformer, and the voltage stress of the switches is halved, but, the complexity problem due to additional switches still exists. The active clamp switches in [20] are replaced by diodes in [21] to form a passive clamp network along with capacitors in order to make the structure simpler. However, extra control circuitry is needed for maintaining current sharing between input inductors. In [22] an interleaved converter using Winding Cross Coupled Inductors (WCCIs) is presented, in which the voltage gain is improved and switches endure low voltage stress due to WCCIs. Furthermore, because of leakage inductance of the WCCIs, switches turn on under Zero Current Switching (ZCS) condition, and reverse-recovery problem of the output diodes is alleviated. However, WCCIs

978-1-5090-0376-1/16 $31.00 © 2016 IEEE   147

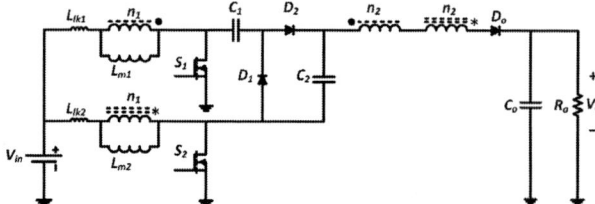

Fig. 1. The schematic of the proposed converter

have complex structure. Similar winding complexity problem exists in converters proposed in [23] and [24] in which three-winding coupled inductors with voltage multiplier cells are employed in order to achieve higher voltage gain. The leakage energy is recycled to the output by passive clamp technique, and leakage inductance helps soft turn-on of the switches. The voltage stress across switches is low in both converters, but, the input current is pulsating in [24].

An interleaved high step-up converter is presented in this paper in which coupled inductors technique is used in order to minimize the input current ripple as well as achieving high voltage gain. Leakage inductance energy is absorbed by clamp capacitors and recycled to the output, so, in addition to suppressing voltage spikes across the switches, the clamp capacitors would further improve the voltage gain. The proposed converter possesses automatic current sharing mechanism, so, there is no need for extra current-sharing control circuitry. All the aforementioned advantages are achieved using a very simple structure with low component count. The converter structure and its operating principles are explained, and voltage gain as well as voltage stress across semiconductor components are calculated. Finally the experimented results are illustrated in order to verify the circuit performance.

II. PROPOSED CONVERTER AND OPERATION PRINCIPLE

The schematic of the proposed converter is shown in Fig. 1. In order to minimize the input current ripple, two coupled inductors are inserted in each legs of the interleaved converter, and the secondary windings are positioned in series with each other and also in series with the output. So, in addition to developing the output voltage, when one of the switches is turned off, the same current drop in input would be compensated by the current induced by the secondary side, therefore, the input current ripple would be dramatically reduced. The energy stored in the leakage inductance of the coupled inductors is absorbed by the clamp capacitors, and is transferred to the output which contributes to the voltage gain, efficiency improvement and suppression of voltage spikes across switches. Duty cycle should be greater than 0.5 for proper operation of the clamp circuit which is usual in high step up converters.

As can be observed from Fig. 1, the coupled inductors are modeled by an ideal transformer with a magnetizing inductor and an equivalent leakage inductance. L_{m1} and L_{m2} are magnetizing inductors, L_{lk1} and L_{lk2} are the leakage inductors, S_1 and S_2 are MOSFET switches, C_1 and C_2 are the clamp capacitors while D_1 and D_2 are the clamp diodes, and

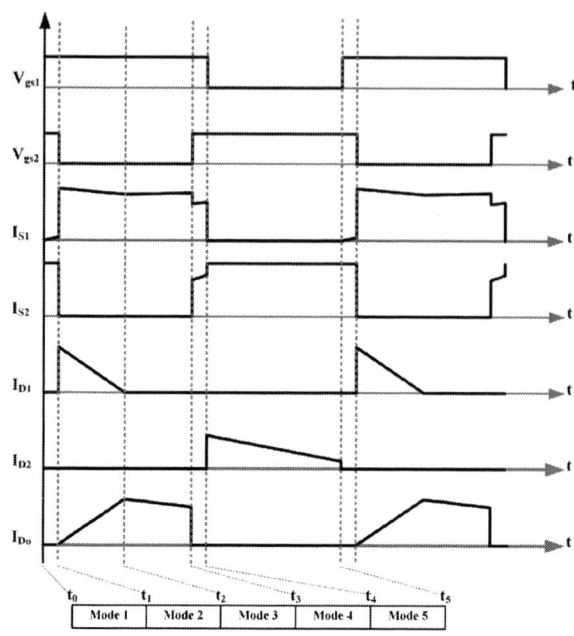

Fig. 2. Key waveforms of the proposed converter

C_o and D_o are the output capacitor and the output diode, respectively. N is defined as n_2/n_1

The proposed converter operates in Continuous Conduction Mode (CCM). There is 180° phase shift between switches S_1 and S_2, and duty cycle is slightly greater than 0.5. A complete operating cycle of the converter consists of five modes, which are explained as follows. The key waveforms are shown in Fig. 2. , and the equivalent circuit of each mode is illustrated in Fig. 3.

Mode 1 [t_0, t_1]: Both switches, S_1 and S_2, are on. All diodes, D_1, D_2 and D_o are off, and there is no energy transfer from input to output. Magnetizing inductances, L_{m1} and L_{m2}, are charging from the input source, and currents through the leakage inductors, L_{lk1} and L_{lk2} are increasing. This mode ends when S_2 turns off.

Mode 2 [t_1, t_2]: When S_2 turns off, D_1 and D_o start to conduct and the energy stored in L_{m2} through secondary winding and C_2 is transferred to the output. So the L_{m2} current reduces. Also, the energy stored in the leakage inductor L_{lk2} is transferred to C_1 through D_1, and the voltage across S_2 is clamped at V_{C1} level. The leakage inductance current reduces to the secondary side current. When I_{Llk2} equals to the secondary side current, I_{sec}, the current flowing through D_1 reaches zero, and D_1 turns off under ZCS condition. This mode ends when D_1 turns off.

Mode 3 [t_2, t_3]: D_1 and D_2 are off, and D_o is still conducting, so still energy is transferred from L_{m2} and C_2 to the output. Hence, I_{Lm2} is still decreasing with the falling rate equal to the previous stage, so, the secondary side current, I_{sec}, is declining with the same ratio. This mode ends when S_2 is switched on.

978-1-5090-0376-1/16 $31.00 © 2016 IEEE   148

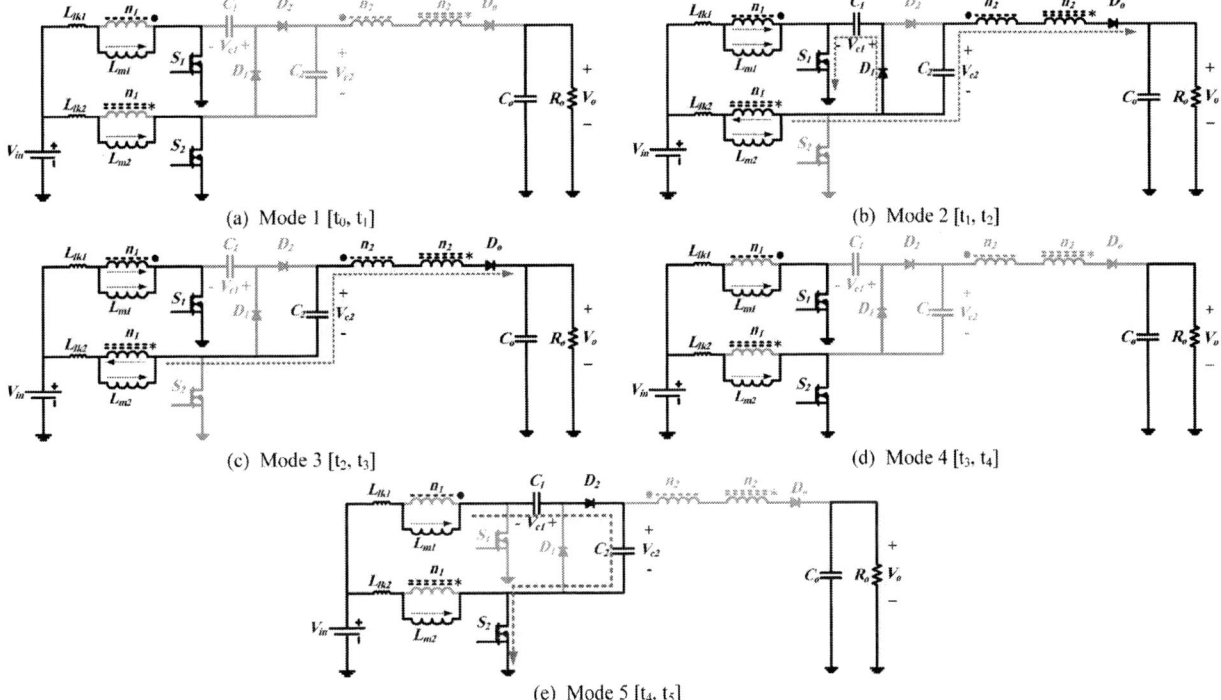

(a) Mode 1 [t₀, t₁]   (b) Mode 2 [t₁, t₂]

(c) Mode 3 [t₂, t₃]   (d) Mode 4 [t₃, t₄]

(e) Mode 5 [t₄, t₅]

Fig. 3. Operational modes of the proposed converter

Mode 4 [t₃, t₄]: As S_2 turns on, the leakage inductance current increases to the magnetizing inductance current and then D_o turns off and there is no energy transfer from the primary side to the output. Since there is no current in the secondary side, I_{sec} is equal to zero, hence, I_{Lm} and I_{Llk} are identical. Both magnetizing inductors, L_{m1} and L_{m2}, are storing energy from the input terminal. This mode ends when S_1 is switched on.

Mode 5 [t₄, t₅]: When S_1 turns off, D_2 starts to conduct, and the energy stored in leakage inductor, L_{lk1}, along with the energy stored in the clamp capacitor C_1 are transferred to C_2, so the current through L_{lk1}, which is equal to I_{Lm1}, is decreasing. D_o is off and the primary side is disconnected from the secondary one, and C_o is still providing the output energy. This mode ends when S_1 turns on and D_2 turns off.

III. STEADY STATE ANALYSIS

The analysis is based on CCM operation of the converter, and in order to simplify the calculations, some assumptions are made:

1. All components are considered to be ideal.

2. All capacitors are large enough so that the voltage across them is considered constant during a period.

A. Voltage gain

In this section the converter voltage gain is obtained. It can be shown by analytical analysis that the effect of the leakage inductance on the voltage gain is negligible, which also is confirmed by the simulation results.

The turn ratio of the coupled inductors is defined as (1)

$$N = \frac{n_2}{n_1} \qquad (1)$$

Clamp capacitor C_1 functions the same as the output capacitor in a conventional boost converter, so its voltage could be calculated as in (2)

$$V_{C1} = \frac{1}{1-D}V_{in} \qquad (2)$$

Where duty cycle is denoted by D. C_2 would be charged two times higher than V_{C1} and its voltage is derived from

$$V_{C2} = \frac{2}{1-D}V_{in} \qquad (3)$$

From voltage-second balance of L_{m1} and L_{m2} and substituting V_{C1} and V_{C2} from (2) and (3) the voltage gain could be derived from

$$\frac{V_o}{V_{in}} = \frac{3+N}{1-D} \qquad (4)$$

978-1-5090-0376-1/16 $31.00 © 2016 IEEE   149

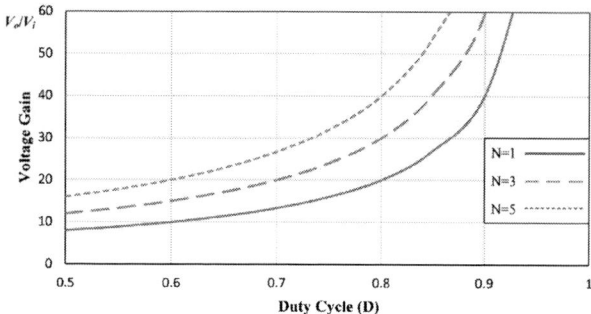

Fig. 4. Voltage gain versus duty cycle for different amounts of N.

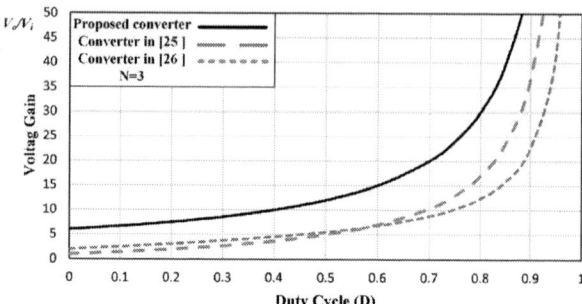

Fig. 5. The voltage gain comparison between the proposed converter and converters presented in [25] and [26]

From (4), it can be concluded that the proposed converter possesses a good step-up gain, which could be further extended by adjusting the coupled inductors turns ratio, N, as can be seen in Fig. 4. A comparison between the voltage gain and turns ratio is plotted in Fig. 4. So, higher voltage gain could be obtained with a moderate duty cycle.

B. Voltage Stress of the semiconductor devices

The voltage-second balance of the magnetizing inductors, L_{m1} and L_{m2}, forces the voltage across switches to be the same as the output voltage of a conventional boost converter with same operating duty cycle

$$V_{S1} = V_{S2} = \frac{1}{1-D}V_{in} = \frac{1}{3+N}V_o \qquad (5)$$

The voltage across diodes D_1 and D_2 can be derived as below

$$V_{D1} = V_{D2} = \frac{2}{1-D}V_{in} = \frac{2}{3+N}V_o \qquad (6)$$

The voltage stress across the output diode D_o is calculated from (7)

$$V_{Do} = V_o + \frac{N-2}{1-D}V_{in} = \frac{1+2N}{3+N}V_o \qquad (7)$$

Assuming $N=1$ the aforementioned equations would be as following:

The voltage gain is derived from (8)

$$\frac{V_o}{V_{in}} = \frac{4}{1-D} \qquad (8)$$

From (8) can be observed that even with unity turn ratio, the proposed converter provides voltage gain four times larger than the conventional boost converter. Thus, the voltage stress across switches would be as (9)

$$V_{S1} = V_{S2} = \frac{V_o}{4} \qquad (9)$$

The voltage stress of the switches, as stated in (9) is a quarter of the output voltage, so switches with lower $R_{DS(ON)}$ could be employed to improve efficiency.

The voltage stress across diodes D_1, D_2 and D_o is obtained from (10) and (11)

$$V_{D1} = V_{D2} = \frac{1}{2}V_o \qquad (10)$$

$$V_{Do} = \frac{3}{4}V_o \qquad (11)$$

In Fig. 5, the voltage gain of the proposed converter is compared with converters proposed in [25] and [26]. The turns ratio of the coupled inductors, N, is equal to three in this comparison. As can be observed from Fig. 5, the proposed converter reaches higher gain ratios via simpler structure when comparing with the structures presented in [25] and [26].

IV. EXPERIMENTAL RESULTS

In order to validate the theoretical analysis of the proposed converter, a 40V input 320V output 130W laboratory prototype of the proposed converter is implemented. The switching frequency is 50 kHz and for the better ripple cancellation, the operating duty cycle of the interleaved switches is selected about 0.5. Circuit elements used are listed in Table I.

TABLE I
SELECTED PARTS FOR THE PROTOTYPE

Symbol	Quantity	
L_{m1}, L_{m2}	Magnetizing inductances	300 µH
L_{lk1}, L_{lk2}	Leakage inductances	11 µH
N (n_2/n_1)	Turns ratio	1
S_1, S_2	Power MOSFETs	IRFP260
C_1, C_2	Clamp capacitors	22 µF/200V
C_o	Output capacitor	47 µF/400V
D_1, D_2	Clamp diodes	MUR460
D_o	Output diode	STTH 806DTI

978-1-5090-0376-1/16 $31.00 © 2016 IEEE

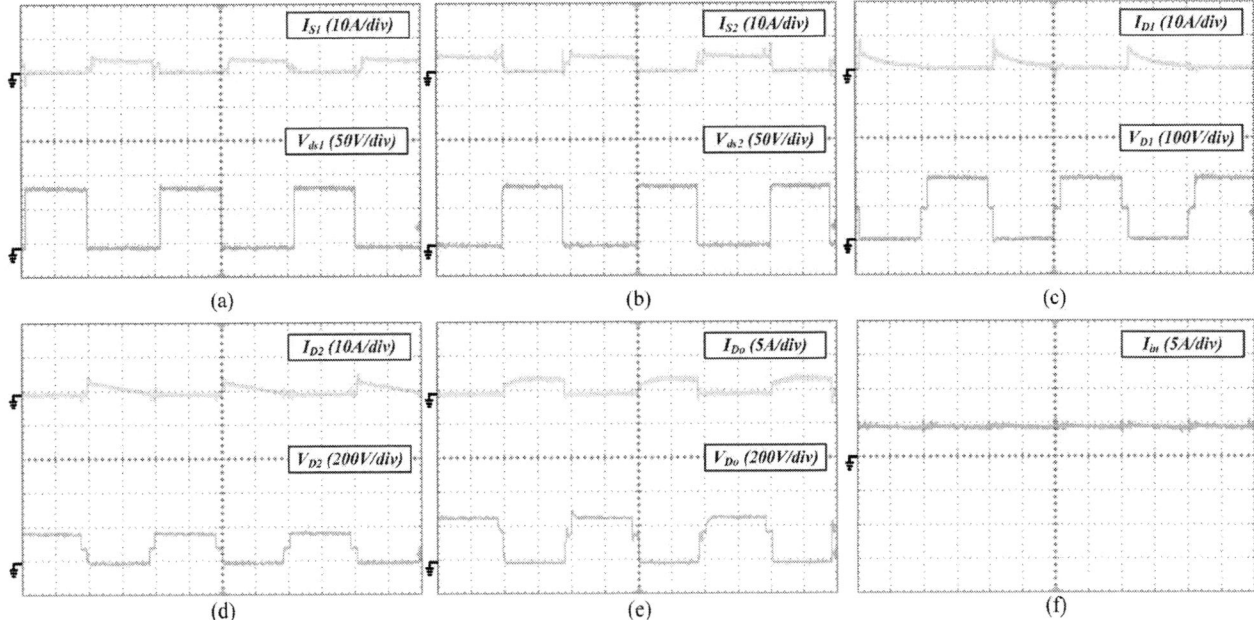

Fig. 6. Experimental results: (a) V_{ds1}, I_{S1}. (b) V_{ds2}, I_{S2}. (c) I_{D1}, V_{D1}. (d) I_{D2}, V_{D2}. (e) I_{Do}, V_{Do}. (f) I_{in}

There is 180° phase shift between switches S_1 and S_2. Fig. 6(a) and (b), show the voltage across drain-source of the switches S_1 and S_2, respectively, as well as the currents flowing through them. The voltage across both switches is clamped to almost 80V which is a quarter of the high 320V output voltage. So, switches with lower voltage ratings and on- resistance can be used which leads to lower conduction losses and better efficiency. The voltage stress across clamp diodes D_1 and D_2, and their currents are shown in Fig. 6(c) and (d), respectively. The voltage across these clamp diodes is half of the output. Since D_1 turns off naturally, the reverse recovery problem associated with it is alleviated. The voltage across the output diode D_o and its current is shown in Fig. 6(e). The leakage inductance of the secondary windings of the coupled inductors helps alleviation of the reverse recovery problem of the output diode. The measured input current of the converter is shown in Fig. 6(f). The measured $\Delta I/I$ is about %5 which has resulted from 0.2A ripple of the input current, and confirms low input current ripple characteristic of the converter which is desirable in photovoltaic applications.

The efficiency of the proposed converter for different load conditions, while considering all circuit parasitics, is shown in Fig. 7. Maximum efficiency achieved is %96.3 at 100W load condition.

V. CONCLUSION

A new interleaved high step-up converter is presented in this paper. Coupled inductors are used to increase the voltage gain, and secondary windings of the coupled inductors are positioned in series to minimize the input current ripple. The leakage inductance energy is absorbed and recycled to the output by clamp capacitors, so voltage spikes across switches are suppressed and the voltage gain is further improved. The proposed converter has high voltage gain which is needed in renewable energy applications with a simple structure and high efficiency. The voltage stress across switches is a quarter of the output voltage, and the input current ripple is very low. The experimental results presented confirm the performance of the proposed converter for high step-up applications.

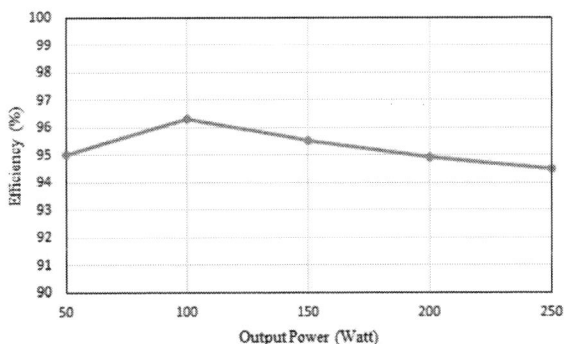

Fig. 7. Efficiency of the proposed converter for different loads

REFERENCES

[1] Y. Bo, L. Wuhua, Z. Yi, and H. Xiangning, "Design and analysis of a grid-connected photovoltaic power system," *IEEE Trans. Power Electron.*, vol. 25, pp. 992-1000, 2010.

[2] J. Ke, R. Xinbo, Y. Mengxiong, and X. Min, "A hybrid fuel cell power system," *IEEE Trans. Ind. Electron.*, vol. 56, pp. 1212-1222, 2009.

[3] D. Velasco de la Fuente, Rodri, x, C. L. T. guez, Garcera, x, *et al.*, "Photovoltaic power system with battery backup with grid-connection and islanded operation capabilities," *IEEE Trans. Ind. Electron.*, vol. 60, pp. 1571-1581, 2013.

[4] L. Wuhua and H. Xiangning, "Review of nonisolated high-step-up dc/dc converters in photovoltaic grid-connected applications," *IEEE Trans. Ind. Electron.*, vol. 58, pp. 1239-1250, 2011.

[5] D. Vinnikov and I. Roasto, "Quasi-z-source-based isolated dc/dc converters for distributed power generation," *IEEE Trans. Ind. Electron.*, vol. 58, pp. 192-201, 2011.

[6] M. Nymand and M. A. E. Andersen, "High-efficiency isolated boost dc-dc converter for high-power low-voltage fuel-cell applications," *IEEE Trans. Ind. Electron.*, vol. 57, pp. 505-514, 2010.

[7] W. Gang, R. Xinbo, and Y. Zhihong, "Nonisolated high step-up dc-dc converters adopting switched-capacitor cell," *IEEE Trans. Ind. Electron.*, vol. 62, pp. 383-393, 2015.

[8] C. L. Wei and M. H. Shih, "Design of a switched-capacitor dc-dc converter with a wide input voltage range," *IEEE Trans. Circuits Syst. I, Reg. Papers*, vol. 60, pp. 1648-1656, 2013.

[9] Q. Wei, C. Dong, J. G. Cintron-Rivera, M. Gebben, D. Wey, and P. Fang Zheng, "A switched-capacitor dc-dc converter with high voltage gain and reduced component rating and count," *IEEE Trans. Ind. Appl*, vol. 48, pp. 1397-1406, 2012.

[10] C. Hyuntae, M. Ciobotaru, J. Minsoo, and V. G. Agelidis, "Performance of medium-voltage dc-bus pv system architecture utilizing high-gain dc-dc converter," *IEEE Trans. Sustainable Energy*, vol. 6, pp. 464-473, 2015.

[11] Y. Lung-Sheng, L. Tsorng-Juu, L. Hau-Cheng, and C. Jiann-Fuh, "Novel high step-up dc-dc converter with coupled-inductor and voltage-doubler circuits,". *IEEE Trans. Ind. Electron*, vol. 58, pp. 4196-4206, 2011.

[12] H. Xuefeng and G. Chunying, "A high voltage gain dc-dc converter integrating coupled-inductor and diode-capacitor techniques," *IEEE Trans. Power Electron.*, vol. 29, pp. 789-800, 2014.

[13] Z. Yi, L. Wuhua, and H. Xiangning, "Single-phase improved active clamp coupled-inductor-based converter with extended voltage doubler cell," *IEEE Trans. Power Electron* vol. 27, pp. 2869-2878, 2012.

[14] K. I. Hwu and Y. T. Yau, "High step-up converter based on coupling inductor and bootstrap capacitors with active clamping,", *IEEE Trans. Power Electron.* vol. 29, pp. 2655-2660, 2014.

[15] M. Tao, Y. Shuai, B. Hongqi, and W. Guo, "A family of multilevel passive clamp circuits with coupled inductor suitable for single-phase isolated full-bridge boost PFC converter," *IEEE Trans. Power Electron.*, vol. 29, pp. 4348-4356, 2014.

[16] W. Tsai-Fu, L. Yu-Sheng, J. C. Hung, and C. Yaow-Ming, "Boost converter with coupled inductors and buck-boost type of active clamp," *IEEE Trans. Ind. Electron.*, vol. 55, pp. 154-162, 2008.

[17] P. Ching-Tsai and L. Ching-Ming, "A high-efficiency high step-up converter with low switch voltage stress for fuel-cell system applications," *IEEE Trans. Ind. Electron.*, vol. 57, pp. 1998-2006, 2010.

[18] P. Ching-Tsai, C. Chen-Feng, and C. Chia-Chi, "A novel transformer-less adaptable voltage quadrupler dc converter with low switch voltage stress," *IEEE Trans. Power Electron.*, vol. 29, pp. 4787-4796, 2014.

[19] W. Li, W. Li, and X. He, "Zero-voltage transition interleaved high step-up converter with built-in transformer," *IET Power Electron.* vol. 4, pp. 523-531, 2011.

[20] L. Weichen, X. Xin, L. Chushan, L. Wuhua, and H. Xiangning, "Interleaved high step-up ZVT converter with built-in transformer voltage doubler cell for distributed pv generation system," *IEEE Trans. Power Electron.*, vol. 28, pp. 300-313, 2013.

[21] L. Wuhua, L. Weichen, X. Xin, H. Yihua, and H. Xiangning, "High step-up interleaved converter with built-in transformer voltage multiplier cells for sustainable energy applications," *IEEE Trans. Power Electron.* vol. 29, pp. 2829-2836, 2014.

[22] L. Wuhua, Z. Yi, W. Jiande, and H. Xiangning, "Interleaved high step-up converter with winding-cross-coupled inductors and voltage multiplier cells," *IEEE Trans. Power Electron.*, vol. 27, pp. 133-143, 2012.

[23] T. Nouri, S. H. Hosseini, E. Babaei, and J. Ebrahimi, "Interleaved high step-up dc-dc converter based on three-winding high-frequency coupled inductor and voltage multiplier cell," *IET Power Electron.*, vol. 8, pp. 175-189, 2015.

[24] T. Kuo-Ching, C. Jyun-Ze, L. Jang-Ting, H. Chi-Chih, and Y. Tzu-Hsiang, "High step-up interleaved forward-flyback boost converter with three-winding coupled inductors," *IEEE Trans. Power Electron.*, vol. 30, pp. 4696-4703, 2015.

[25] S. Dwari and L. Parsa, "An efficient high-step-up interleaved dc-dc converter with a common active clamp," *IEEE Trans. Power Electron.*, vol. 26, pp. 66-78, 2011.

[26] L. Ching-Ming, P. Ching-Tsai, and C. Ming-Chieh, "High-efficiency modular high step-up interleaved boost converter for dc-microgrid applications," *IEEE Trans. Ind. Appl.*, vol. 48, pp. 161-171, 2012.

New Asymmetrical Commutation Cell for Multilevel Inverters with Reduced Number of Components

Mahdi Vizheh, Mohammad Rezanejad
Department of Electrical Engineering
Mazandaran University of Science and Technology
Babol, Iran
mahdi.vizheh@gmail.com
mohamad.rezanezhad@gmail.com

Emad Samadaei
Department of Computer and Electrical Engineering
Babol University of Technology
Babol, Iran
e.samadaei@stu.nit.ac.ir

Abstract— **This paper proposes a new topology of multilevel inverters with asymmetrical commutation cell that consists of three sources and six switches to synthesize asymmetrical output voltage level. This category of topologies is typically used in voltage AC generation. The structure provides multilevel waveform utilizing reduced number of components as compared to the conventional topologies. The performance and operation of the proposed inverter is introduced in two modes: 1) six level asymmetrical, 2) eleven level symmetrical output voltage. A comparison between the proposed and conventional structures is established. The proposed inverter is simulated in MATLAB/Simulink software. To validate proper operation of the proposed circuit, experimental setups are implemented in the laboratory in two modes.**

Keywords—Multilevel inverters; Voltage source inverter; Reduced components; Power electronic

I. INTRODUCTION

Multilevel inverters (MLI) have been attracted more attentions in recent years because of their noticeable features such as high power quality, low harmonic components, reduced electromagnetic interference and lower switching losses. These inverters are employed to create more output voltage levels, so that the output voltage quality is enhanced and also harmonic components are reduced [1-3]. The basic concept of power conversion in multilevel inverters is based on a series connection of switching components with several lower DC voltage sources to synthesize a staircase voltage waveform [4]. Low voltages rate operation is known as an important feature of the multilevel inverter. Therefore, switching losses and voltage stress on power semiconductor devices will be decreased [5].

The best known multilevel structures are Neutral-Point Clamped Inverter (NPC), Flying Capacitor (FC) and Cascaded H-bridge Inverter (CHB), which are called as a basic topology. NPC was the first multilevel inverter that designed in 1970s. The main disadvantage of NPC topology is that the required number of clamping diodes is quite high and for higher number of voltage levels this topology will be impractical due to this fact [6, 7]. Voltage balancing is another problem of this topology, which can be solved by adding an additional balancing circuit or implementing more complex control

methods, but it makes this topology unusual for high voltage applications. A similar topology to the NPC topology is FC but with the difference, that FC topology uses clamping capacitor instead of diodes to hold the voltages to the desired values [8]. FC topology also has the same drawback that is voltage balancing, which makes it inappropriate for high-voltage applications [6, 9]. CHB differs in several ways from NPC and FC in how to achieve the voltage waveform [10]. It uses cascaded full-bridge inverters with separate DC sources, in a modular setup, to create the voltage waveform. Compared to the NPC and FC the CHB is suitable for high-voltage applications and requires fewer components, every voltage level requires the same amount of components, and there is no voltage balancing problem because of using DC voltage sources, however; sometimes providing DC voltage sources are become more difficult than voltage balancing [11].

In recent years, a variety of multilevel inverters have been introduced in order to reduce the number of overall power semiconductor devices, cost or present a simpler structure [12]. In fact, designing a new structure with less power semiconductor devices is one of the main challenges in MLI. A number of recent structures were presented with this fact [13, 14]. In recent topologies a modularized setup of sub-modules is used, which have found a lot of attention in industrial applications. Being able to produce high output voltages with rather less harmonic in output could be mentioned as their considerable features. [15, 16] were introduced by this purpose. In [17-19] the positive and negative levels both can be make together by themselves; however, it cannot be done in [20, 21], so in these kinds of structures using an H-bridge to build up an AC output voltage. In this condition, some of switches must be able to tolerate total output voltage. This issue limits application of this kind of inverter in high voltages.

The aim of this paper is to propose a new topology of multilevel inverters with asymmetrical commutation cell that consists of three sources and six switches to synthesize asymmetrical output voltage level. This sort of topologies is typically used in High Voltage DC (HVDC) transmission and high voltage AC generation [22]. In this strategy, as shown in Fig. 1, upper and lower modules are complemented. The complemented asymmetrical modules (e.g. 1 & 1'; 2 & 2'; ...., n & n') are connected to each other to create a symmetrical output waveform. For example, upper modules create 3 positive and 2 negative levels while lower (complemented) modules generate 2 positive and 3 negative levels, so 11 level symmetrical output voltages can be achieved in the output.

In the following, the structure and operation of the proposed topology is described and compared with basic multilevel inverters. In addition, the proposed structure is simulated in MATLAB/Simulink software and two prototypes are implemented in the laboratory. The accuracy and validity have been demonstrated through experimental and simulation results on 6 level and 11 level inverter.

II. PROPOSED TOPOLOGY

The proposed topology is shown in Fig. 2. It consists of three sources and six switches to synthesize an asymmetrical 6-level output voltage waveform. In other word, different paths are provided by switches for sources, so that various values of them are available in the output, which create the output voltage waveform. Extendibility is one of the best features of this topology. Fig. 3. shows the proposed extended structure. The number of output levels can be increased by duplicating the sub-modules that is specified in the figure.

The value of all DC sources is considered equal. All the available relations for extension are indicated in Table I. Where N_{level}, $N_{switch,diode}$, N_{source}, $V_{o,max}$ and m are the number of output voltage levels, number of switches and diodes, number of the sources, the maximum output Voltage and number of duplicated sources, respectively. According to Fig. 2, duplication of sub-module A leads to add the source V_a and switches S_a and $S_{a'}$ in which to the left side while duplication of sub-module B leads to add the sources V_b and $V_{b'}$ and switches S_b, $S_{b'}$ and $S_{b''}$ in which to the right side. For each duplication of sub-module A and B, the output level will be increased to 2 and 4 levels, respectively.

The performance of the proposed structure is completely discussed in the following. All possible switching states and levels are shown in Table II. From Table II, switches $S_{a'}$ and S_b are turned on to create $3V_{DC}$ voltage in the output. Therefore, through the direction that provided by these switches, positive sum of the three DC sources is given in the output. The switches $S_{a'}$ and $S_{b'}$ are triggered to achieve $2V_{DC}$ voltage level, so that the positive sum of the sources V_a and $V_{b'}$ appear in the output. The switches $S_{a'}$ and $S_{b''}$ are switched on, so that the positive value of V_a appears in the output and creates V_{DC} voltage level. The zero level can be defined by switching S_a and S_b. To achieve $-V_{DC}$ voltage level, the switches S_a and $S_{b'}$ are turned on to have the negative value of V_b in the output. The switches S_a and $S_{b''}$ are switched on, so that the negative sum of the sources V_b and $V_{b'}$ appears in the output and create $-2V_{DC}$ voltage level.

III. COMPARISON OF THE PROPOSED TOPOLOGY

In this paper, the aim is to reduce the number of components, especially switches, which is known as a main challenge in MLI. Table III shows the comparison of the proposed topology with conventional topologies. In this case, two modules are connected complementary as shown in Fig. 1, to synthesize a symmetrical output waveform, and then have compared with conventional topologies.

According to table III, the proposed structure uses only N_{Level} +1 switches when conventional topologies need $2(N_{Level}-1)$ to provide the same number of output voltage levels. In other word, the proposed structure approximately uses half switches less than conventional. For example, for 11-level output voltage, the proposed structure uses 12 switches when conventional topologies use 20 switches. However, the structure utilizes 1 more sources, the number of switches are reduced significantly. Generally, the proposed structure requires lower components than other topologies, especially for higher number of voltage.

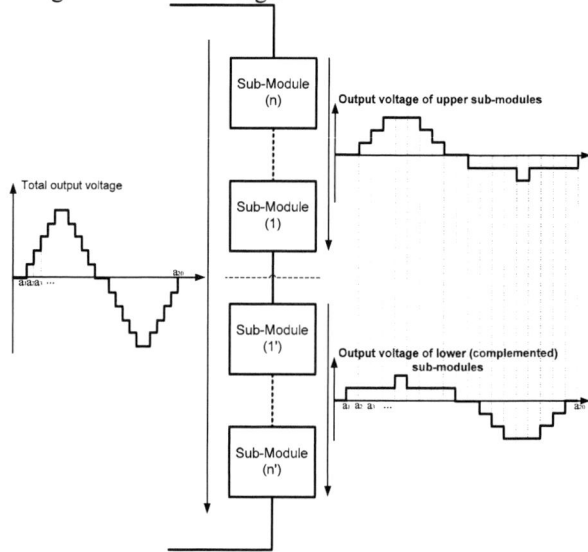

Fig. 1. Basic structure of modular multilevel converter for high voltage generation

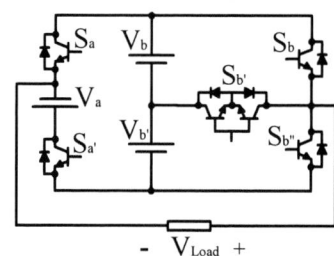

Fig. 2. The proposed topology of multilevel inverter

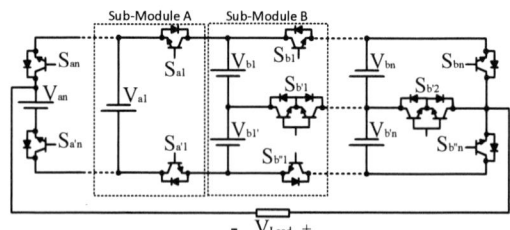

Fig. 3. The proposed extended topology of multilevel inverter

TABLE I.   DIFFERENT PARAMETERS CONSIDERED FOR PROPOSED EXTENDED TOPOLOGY

Components		Proposed Methods			
		duplication of sub-module A		duplication of sub-module B	
N_{level}	Positive level	2m+6	m+3	4m+6	2m+3
	Negative level		m+2		2m+2
$N_{switche,diode}$		2m+6		4m+6	
N_{source}		m+3		2m+3	
$V_{o,max}$		$(m+3)V_{dc}$		$(2m+3) V_{dc}$	

TABLE II.   SWITCHES STATES FOR A PROPOSED 6-LEVEL TOPOLOGY

State	Switch states					V_{Load}
	S_a	$S_{a'}$	S_b	$S_{b'}$	$S_{b''}$	
1	0	1	1	0	0	$3V_{dc}$
2	0	1	0	1	0	$2V_{dc}$
3	0	1	0	0	1	V_{dc}
4	1	0	1	0	0	0
5	1	0	0	1	0	$-V_{dc}$
6	1	0	0	0	1	$-2V_{dc}$

TABLE III.   COMPARISON OF POWER COMPONENT REQUIREMENTS AMONG CONVENTIONAL TOPOLOGTIES

Components	Structures			
	NPC	FC	CHB	Proposed Topology
switches	$2N_{Level}-2$	$2N_{Level}-2$	$2N_{Level}-2$	$N_{Level}+1$
diodes	$(4N_{Level}-7)^{a}$	$2N_{Level}-2$	$2N_{Level}-2$	$N_{Level}+1$
capacitor	0	$N_{Level}-2$	0	0
DC links	$(N_{Level}-1)/2$	$(N_{Level}-1)/2$	$(N_{Level}-1)/2$	$(N_{Level}+1)/2$
total num. of components	$(13N_{Level}-19)/2$	$(13N_{Level}-13)/2$	$(9N_{Level}-9)/2$	$(5N_{Level}+5)/2$

a. Total required clamping and main diodes

IV. SIMULATION AND EXPERIMENTAL RESULTS

Simulation and experimental results are presented in this section in order to validate the operation of the proposed structure. In this paper, two different modes of the proposed topology have been studied. The simulation and experimental results of the proposed 6-level and 11-level topology are described. The proposed structure is simulated in MATLAB/Simulink software. In addition, to validate the accuracy of the proposed structure two prototypes are implemented in the laboratory. The prototype built using IGBT SGP15N120 and diode MUR860. In order to control the multilevel inverter, the switching table of the inverter is implemented in the AVR microcontroller, which generates the gate signals of the switches. The simulation and experimental study has been carried out under a series R-L connection load (R= 100 Ω and L= 950 mH) to create 50 Hz sinusoidal voltage waveform.

Several techniques of switching modulation have been presented for multilevel inverters [23-25]. in this paper, Selective Harmonic Elimination (SHE) has been used [26, 27]. A voltage waveform made up by a (2N + 1)-level inverter is shown in Fig. 4., where the transition angles change according

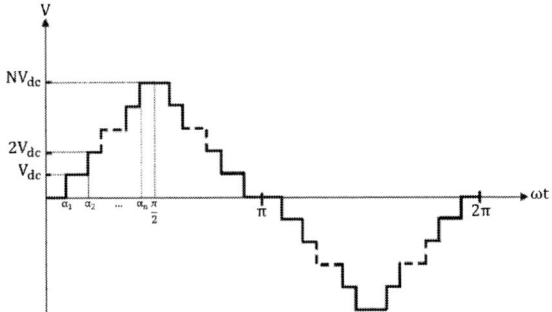

Fig. 4.   Generalised (2N+1)-level output voltage waveform

to the voltage level variations. The Fourier series expansion of the voltage waveform is expressed by:

$$V(\omega t)=\sum_{n=1}^{\infty} U_n Sin(k\theta) \qquad (1)$$

Where U_n is the amplitude of the harmonic of order k and is obtained as follows

$$U_n = \frac{4V_{dc}}{k\pi}\sum_{k=1}^{N} Cos(k\theta_1) \qquad k=1,3,5,... \qquad (2)$$

NV_{dc} represents the number of DC sources of the multilevel inverter. The switching angles α_1, α_2, ..., and α_n must satisfy the following restriction

$$0 \leq \alpha_1 \leq \alpha_2 \leq ... \leq \alpha_n < \frac{\pi}{2} \qquad (3)$$

The purpose of SHE-PWM technique for a Given desired fundamental voltage amplitude, U_1, is to determine the switching angles α_1, α_2, ..., and α_n, in order to adjust the fundamental component to its desired amplitude and the (N–1) harmonics are omitted. The standard formulation of SHE-PWM problem is expressed by the following equations

$$\sum_{i=1}^{N} Cos(\alpha_i) - NM = 0$$
$$\sum_{i=1}^{N} Cos(3\alpha_i) = 0$$
$$\sum_{i=1}^{N} Cos(5\alpha_i) = 0 \qquad (4)$$
$$.$$
$$.$$
$$\sum_{i=1}^{N} Cos(k\alpha_n) = 0$$

M is the modulation index and is defined as

$$M = \frac{U_1}{(4NV_{dc})/\pi} \qquad 0 < M \leq 1 \qquad (5)$$

Solving the N transcendental equations in (4) attains the switching angels. According to the above mentioned, the standard SHE-PWM technique is implemented for the proposed symmetrical 11-level inverter, which generate 11-level stepped voltage waveform. Based on the equation sets of (4), the switching angles are calculated to adapt the amplitude of the ac-side voltage and to eliminate the harmonics of order 3, 5, 7, 9, 11 for 11-level inverter. The computed switching angels for the proposed 11-level inverter are collected in Table IV.

First, the proposed 6-level topology is described. The photo of prototype is shown in Fig. 5. In this case, the magnitude of DC sources is equal to 50v. Therefore, the maximum output voltage is 150v. The output voltage and current of simulation and experimental results are shown in Fig. 6 and Fig. 7, respectively. The results are satisfactory and also quite similar to each other, which confirm the performance of the proposed structure. It should be mentioned that the voltage of the resistor is used to measure output current.

In regard to Fig. 1, two modules are connected complementary to create a symmetrical output waveform as shown in Fig. 8. For this state, the magnitude of all DC sources is equal to 50v. Therefore, the maximum output voltage is 250v. The switching patterns are indicated in Table V. The output voltage and current of simulation and experimental results of proposed 11-level structure are shown in Fig. 9 and Fig. 10, respectively. The results are satisfactory and also quite similar to each other, which validates the performance of the proposed structure. The THD values of output voltage and current based on simulation and experimental are calculated 7.51%, 0.80%, 7.62% and 1.10%, respectively.

TABLE IV. SWITCHES STATES FOR A PROPOSED 11-LEVEL TOPOLOGY

α_i	α_1	α_2	α_3	α_4	α_5
Angles	8.594	13.178	30.218	45.782	67.452

TABLE V. SWITCHES STATES FOR A PROPOSED 6-LEVEL TOPOLOGY

State	S1	S2	S3	S4	S5	S6	S7	S8	S9	S10	V_{Load}
1	0	1	0	1	0	1	0	0	0	1	250v
2	0	1	0	1	0	0	0	1	0	1	200v
	0	1	1	0	0	1	0	0	0	1	
3	0	1	0	1	0	0	1	0	0	1	150v
	0	1	1	0	0	0	1	0	0	1	
	0	1	0	0	1	1	0	0	0	1	
4	0	1	1	0	0	0	1	0	0	1	100v
	0	1	0	0	1	0	0	1	0	1	
	0	1	0	1	0	1	0	0	1	0	
	1	0	0	1	0	1	0	0	0	1	
5	0	1	0	0	0	1	0	1	0	1	50v
	0	1	1	0	0	1	0	0	1	0	
	0	1	0	1	0	0	0	1	1	0	
	1	0	1	0	0	1	0	0	0	1	
	1	0	0	1	0	0	0	1	0	1	
6	1	0	0	1	0	0	1	0	0	1	0
	1	0	1	0	0	0	0	1	0	1	
	1	0	0	0	1	1	0	0	0	1	
	0	1	0	0	1	1	0	0	1	0	
	0	1	1	0	0	0	0	1	1	0	
	0	1	0	1	0	0	1	0	1	0	
7	1	0	1	0	0	0	1	0	0	1	-50v
	1	0	0	0	1	0	0	1	0	1	
	1	0	0	1	0	1	0	0	1	0	
	0	1	0	0	1	0	0	1	1	0	
	0	1	1	0	0	0	1	0	1	0	
8	1	0	0	0	1	0	1	0	0	1	-100v
	1	0	1	0	0	1	0	0	1	0	
	1	0	0	1	0	0	0	1	1	0	
	0	1	0	0	1	0	1	0	1	0	
9	1	0	0	0	1	1	0	0	1	0	-150v
	1	0	1	0	0	0	0	1	1	0	
	1	0	0	1	0	0	1	0	1	0	
10	1	0	0	0	1	0	0	1	1	0	-200v
	1	0	1	0	0	0	1	0	1	0	
11	1	0	0	0	1	0	1	0	1	0	-250v

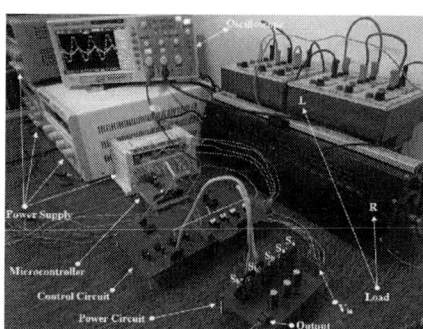

Fig. 5. Photo of prototype

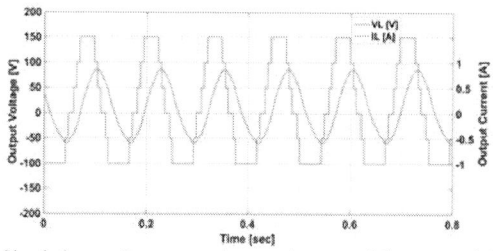

Fig. 6. Simulation results, output voltage and current of the proposed 6-level inverter

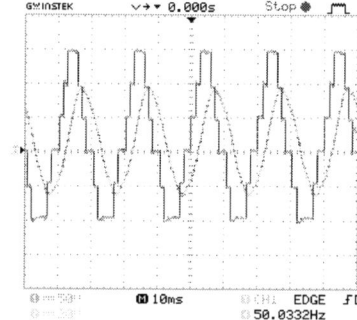

Fig. 7. Experimental results, output voltage and current of the proposed 6-level inverter

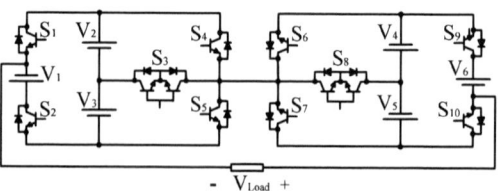

Fig. 8. The proposed 11-level topology of multilevel inverter

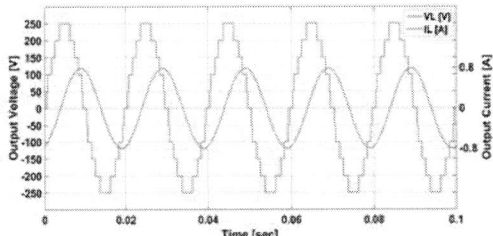

Fig. 9. Simulation results, output voltage and current of the proposed 11-level inverter

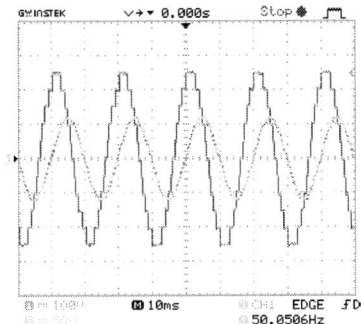

Fig. 10. Experimental results, output voltage and current of the proposed 11-level inverter

V. CONCOLUSION

In this paper, a new asymmetric structure has proposed that can synthesize more output voltage levels with reduced number of components. The proposed main module is consisting of three sources and six switches to generate an asymmetrical 6 level output voltage. Also it can be extended easily, so that the output voltage levels can be increased significantly. Therefore, the proposed structure is beneficial with high quality output. The proposed 6-level and 11-level structures both are simulated in MATLAB/Simulink software. Moreover, two experimental setups have implemented in the laboratory. The results show low harmonics with standard that confirm the theoretical analysis and perfectly similar to simulation results that demonstrate the superiority and validity of the structure.

REFERENCES

[1] Franquelo, L.G.; Rodriguez, J.; Leon, J.I.; Kouro, S.; Portillo, R.; Prats, M.A.M., "The age of multilevel converters arrives," in Industrial Electronics Magazine, IEEE , vol.2, no.2, pp.28-39, June 2008

[2] Jang-Hwan Kim; Seung-Ki Sul; Enjeti, P.N., "A Carrier-Based PWM Method With Optimal Switching Sequence for a Multilevel Four-Leg Voltage-Source Inverter," in Industry Applications, IEEE Transactions on , vol.44, no.4, pp.1239-1248, July-aug. 2008

[3] Rodriguez, J.; Bernet, S.; Bin Wu; Pontt, J.O.; Kouro, S., "Multilevel Voltage-Source-Converter Topologies for Industrial Medium-Voltage Drives," in Industrial Electronics, IEEE Transactions on , vol.54, no.6, pp.2930-2945, Dec. 2007

[4] Babaei, E., "A Cascade Multilevel Converter Topology With Reduced Number of Switches," in Power Electronics, IEEE Transactions on , vol.23, no.6, pp.2657-2664, Nov. 2008

[5] Rodriguez, J.; Franquelo, L.G.; Kouro, S.; Leon, J.I.; Portillo, R.C.; Prats, M.A.M.; Perez, M.A., "Multilevel Converters: An Enabling Technology for High-Power Applications," in Proceedings of the IEEE , vol.97, no.11, pp.1786-1817, Nov. 2009

[6] Rodriguez, J.; Jih-Sheng Lai; Fang Zheng Peng, "Multilevel inverters: a survey of topologies, controls, and applications," in Industrial Electronics, IEEE Transactions on , vol.49, no.4, pp.724-738, Aug 2002

[7] Rodriguez, J.; Bernet, S.; Steimer, P.K.; Lizama, I.E., "A Survey on Neutral-Point-Clamped Inverters," in Industrial Electronics, IEEE Transactions on , vol.57, no.7, pp.2219-2230, July 2010

[8] Meynard, T.A.; Foch, H., "Multi-level conversion: high voltage choppers and voltage-source inverters," in Power Electronics Specialists Conference, 1992. PESC '92 Record., 23rd Annual IEEE , vol., no., pp.397-403 vol.1, 29 Jun-3 Jul 1992

[9] Jih-Sheng Lai; Fang Zheng Peng, "Multilevel converters-a new breed of power converters," in Industry Applications, IEEE Transactions on , vol.32, no.3, pp.509-517, May/Jun 1996

[10] Sivakumar, K.; Das, A.; Ramchand, R.; Patel, C.; Gopakumar, K., "A Hybrid Multilevel Inverter Topology for an Open-End Winding Induction-Motor Drive Using Two-Level Inverters in Series With a Capacitor-Fed H-Bridge Cell," in Industrial Electronics, IEEE Transactions on , vol.57, no.11, pp.3707-3714, Nov. 2010

[11] Malinowski, M.; Gopakumar, K.; Rodriguez, J.; Pérez, M.A., "A Survey on Cascaded Multilevel Inverters," in Industrial Electronics, IEEE Transactions on , vol.57, no.7, pp.2197-2206, July 2010

[12] Gupta, K.K.; Ranjan, A.; Bhatnagar, P.; Kumar Sahu, L.; Jain, S., "Multilevel Inverter Topologies With Reduced Device Count: A Review," in Power Electronics, IEEE Transactions on , vol.31, no.1, pp.135-151, Jan. 2016

[13] Shalchi Alishah, R.; Nazarpour, D.; Hosseini, S.H.; Sabahi, M., "Novel Topologies for Symmetric, Asymmetric, and Cascade Switched-Diode Multilevel Converter With Minimum Number of Power Electronic Components," in Industrial Electronics, IEEE Transactions on , vol.61, no.10, pp.5300-5310, Oct. 2014

[14] Ebrahimi, J.; Babaei, E.; Gharehpetian, G.B., "A New Topology of Cascaded Multilevel Converters With Reduced Number of Components for High-Voltage Applications," in Power Electronics, IEEE Transactions on , vol.26, no.11, pp.3109-3118, Nov. 2011

[15] Kangarlu, M.F.; Babaei, E.; Sabahi, M., "Cascaded cross-switched multilevel inverter in symmetric and asymmetric conditions," in Power Electronics, IET , vol.6, no.6, pp.1041-1050, July 2013

[16] Babaei, E.; Kangarlu, M.F.; Sabahi, M., "Extended multilevel converters: an attempt to reduce the number of independent DC voltage sources in cascaded multilevel converters," in Power Electronics, IET , vol.7, no.1, pp.157-166, January 2014

[17] Gupta, K.K.; Jain, S., "Multilevel inverter topology based on series connected switched sources," in Power Electronics, IET , vol.6, no.1, pp.164-174, Jan. 2013

[18] Babaei, E.; Laali, S.; Alilu, S., "Cascaded Multilevel Inverter With Series Connection of Novel H-Bridge Basic Units," in Industrial Electronics, IEEE Transactions on , vol.61, no.12, pp.6664-6671, Dec. 2014

[19] Babaei, E.; Laali, S., "Optimum Structures of Proposed New Cascaded Multilevel Inverter With Reduced Number of Components," in Industrial Electronics, IEEE Transactions on , vol.62, no.11, pp.6887-6895, Nov. 2015

[20] Najafi, E.; Yatim, A.H.M., "Design and Implementation of a New Multilevel Inverter Topology," in Industrial Electronics, IEEE Transactions on , vol.59, no.11, pp.4148-4154, Nov. 2012

[21] Shalchi Alishah, R.; Nazarpour, D.; Hosseini, S.H.; Sabahi, M., "New hybrid structure for multilevel inverter with fewer number of components for high-voltage levels," in Power Electronics, IET , vol.7, no.1, pp.96-104, January 2014

[22] Nami, A.; Jiaqi Liang; Dijkhuizen, F.; Demetriades, G.D., "Modular Multilevel Converters for HVDC Applications: Review on Converter Cells and Functionalities," in Power Electronics, IEEE Transactions on, vol.30, no.1, pp.18-36, Jan. 2015

[23] Poh Chiang Loh; Holmes, D.G.; Lipo, T.A., "Implementation and control of distributed PWM cascaded multilevel inverters with minimal harmonic distortion and common-mode voltage," in Power Electronics, IEEE Transactions on , vol.20, no.1, pp.90-99, Jan. 2005

[24] Wenxi Yao; Haibing Hu; Zhengyu Lu, "Comparisons of Space-Vector Modulation and Carrier-Based Modulation of Multilevel Inverter," in Power Electronics, IEEE Transactions on , vol.23, no.1, pp.45-51, Jan. 2008

[25] Chiasson, J.N.; Tolbert, L.M.; McKenzie, K.J.; Zhong Du, "Control of a multilevel converter using resultant theory," in Control Systems Technology, IEEE Transactions on , vol.11, no.3, pp.345-354, May 2003

[26] Zhong Du; Tolbert, L.M.; Chiasson, J.N., "Active harmonic elimination for multilevel converters," in Power Electronics, IEEE Transactions on , vol.21, no.2, pp.459-469, March 2006

[27]Blasko, V., "A Novel Method for Selective Harmonic Elimination in Power Electronic Equipment," in Power Electronics, IEEE Transactions on , vol.22, no.1, pp.223-228, Jan. 2007

7th Power Electronics, Drive Systems & Technologies Conference (PEDSTC 2016)
16-18 Feb. 2016, Iran University of Science and Technology, Tehran, Iran

A Novel Single Switch High Gain DC-DC Converter Employing Coupled Inductor and Diode Capacitor

Masoume Amirbande, Keyvan Yari, Mojtaba Forouzesh, Alfred Baghramian

Department of Electrical Engineering
University of Guilan
Rasht, Iran
Masoume.amirbande@yahoo.com, yari.keyvan1369@gmail.com, m.forouzesh.ir@ieee.org, alfred@guilan.ac.ir

Abstract— **This paper presents a novel single switch high step-up DC-DC converter with high efficiency feature. Due to using coupled inductors and switched capacitor techniques the proposed converter achieves high voltage gain without large duty cycle. In spite of the fact that voltage gain is high, voltage stress across all elements has been reduced. Moreover the input current of the proposed converter is continues and the stored energy in leakage inductance is recycled to the output. These characteristics with low voltage rating components make the proposed converter very suitable for renewable applications. All the operational modes and steady state analysis in CCM are discussed in detail. Finally, some simulations have been done to verify the performance of the proposed converter in Pspice/Orcad.**

Keywords— high voltage gain, reduced voltage stress, coupled inductors, DC-DC.

I. INTRODUCTION

The intensive utilization of fossil fuels causes serious pollution on the environment. In order to eliminate effects of pollution, renewable energy sources, such as photovoltaic (PV) systems, fuel cells, wind energy, etc., have been studied and developed. The renewable energy resources are clean. They are effective for various applications like electronic controllers, hybrid electric vehicles. However, the main impediment of this sources is low output voltage. Therefore, the need of converters which have high voltage gain ratio is explicit [1].

Many converters have been proposed recently in order to increase voltage gain ratio. The converter in [2] uses capacitors and switches in order to attain high voltage gain. However, this topology have some defects such as using several switches which complicate control and implementation of the proposed converter and voltage across circuit elements is high. The switched inductor and voltage multiplier functions are used to achieve high voltage gain in [3]. In addition high voltage gain is achieved with appropriate duty cycle. The aforementioned converters have higher voltage gain compared with the simple boost converter, however they aren't appropriate for high step-up applications. The major drawback of using transformer and coupled inductors in DC-DC converters is the stored energy in their leakage inductance. This stored energy increases voltage spikes across power switches so that clamp circuits are introduced to solve the problem of storing energy in transformer and coupled inductors leakage inductance. The

interleaved converter with high voltage gain and active clamp circuit is proposed in [4]. The proposed converters in [5-6] use coupled inductors to reach high voltage gain without extreme duty cycle in addition the stored energy in the leakage inductance is recycled to the output. With all the aforementioned advantages the major flaw of these converter is discontinues input current characteristics.

Coupled inductors and switched capacitors techniques are used in order to increase voltage gain of DC-DC converters. The proposed converters in [7-11] use this method and the voltage stresses across elements are reduced. The converter which is introduced in [12] is a combination of flyback and boost converter to attain higher voltage gain. Transformer and voltage Multiplier Cell has been implemented in [13] to increase voltage gain. Besides voltage across circuit elements has been reduced and because of zero current switching, power losses of semiconductor devices are decreased and the efficiency is increased.

In the proposed converter, the major purpose is to obtain high voltage gain meanwhile keeping the voltage stress of components as low as possible. By decreasing in voltage stress on switch, capacitors and diodes, size and cost of elements are reduced. Using coupled inductors in this converter increases voltage gain. In addition, efficiency of the proposed converter increases by employing quasi-resonant operation. The resonance circuit between the leakage inductance of coupled inductors and resonance capacitor helps to alleviate the semiconductor losses. Fig. 1 shows proposed converter. The rest of this paper is formed as follows. In section II, describe operation principle of the proposed converter. Section III, illustrates voltage gain and voltage stresses across elements derivation. In order to validate the performance of the proposed converter some

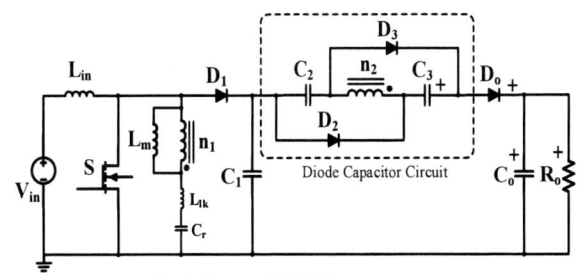

Fig.1. Proposed DC-DC converter

978-1-5090-0376-1/16 $31.00 © 2016 IEEE

simulations have been done and their results are presented in section IV. Finally, section V concludes this paper.

II. DESCRIPTION OF CIRCUIT OPERATON

According to the Fig. 1, the proposed converter consist of one switch, two coupled inductors, five capacitors, four diodes and an input filter inductor. In order to simplify the analysis of the proposed converter following assumptions have been used:

1) The switch S (in conducting mode) and other circuit elements are ideal except leakage inductance L_{lk} of coupled inductors.

2) The capacitors are large enough in order to consider voltage across them in one switching cycle is constant.

3) The secondary to primary turns ratio of the coupled inductors represent by $N = \frac{n_2}{n_1}$

The characteristic waveforms of this converter in one switching cycle are illustrated in Fig. 2.

A. Mode 1 $[t_0 - t_1]$:

At $t = t_0$ the switch S is turned ON and its turn-ON current is increased from zero. The magnetizing inductance L_m receives energy from resonance capacitor C_r and input source V_{in} gives energy to the input inductor L_{in}. The diode D_1 is reverse biased and the diodes D_2, D_3 are forward biased. The magnetizing inductance L_m transfers its energy through the coupled inductors to charge the capacitors C_2, C_3. In addition the output diode D_o turns ON naturally and the capacitors C_1 ، C_2 و C_3 discharge their energy to the output capacitor C_o and load. The coupling coefficient of the coupled inductors is considered $k = \frac{L_m}{L_m + L_{lk}}$ and following equations can be written in this time interval

$$V_{Lm} = \frac{L_m}{L_m + L_{lk}} V_{Cr} = k V_{Cr} \qquad (1)$$
$$V_{lk} = (k - 1) V_{in} \qquad (2)$$
$$V_{L2} = -V_{C2,C3} \qquad (3)$$

B. Mode 2 $[t_1 - t_2]$:

Before t_1 the switch S is in the turn-ON state. At $t = t_1$ the diodes D_2 and D_3 are turned OFF naturally. Moreover the leakage inductance L_{lk} and the resonance capacitor C_r begin to resonant. Other conditions are the same as first time interval. At the end of this time interval the output diode D_o turns OFF naturally which minimizes the output diode reverse recovery losses. Following equation can be expressed in this time interval

$$V_{C1} + V_{C2} + V_{C3} = V_{out} \qquad (4)$$

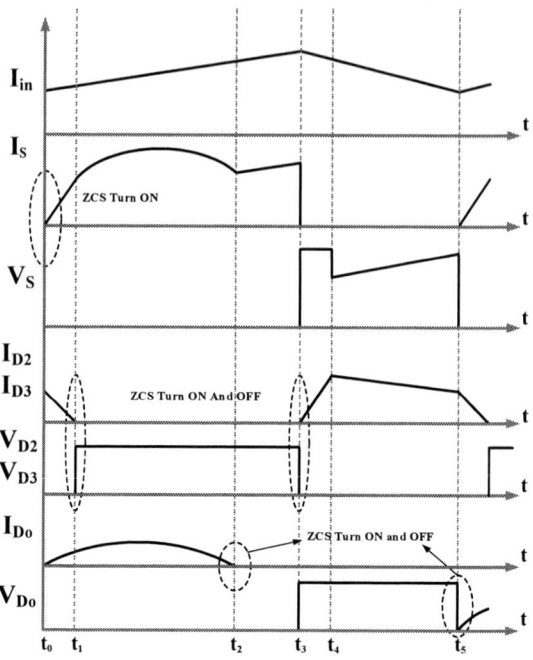

Fig. 2. The characteristic waveforms of the proposed converter

C. Mode 3 $[t_2 - t_3]$:

At this time interval the switch S is still ON. At $t = t_2$ the leakage inductance L_{lk} and the resonance capacitor C_r resonance ends. The magnetizing inductance L_m stores energy and its current increases linearly. In this mode the output capacitor C_o supplies the load.

D. Mode 4 $[t_3 - t_4]$:

At $t = t_3$ the switch S is turned OFF and the diodes D_1, D_2 and D_3 are turned ON. Meanwhile the stored energy in the input inductor L_{in} charges the resonance capacitor C_r and the capacitor C_1 through the diode D_1. The magnetizing inductance L_m transfers its energy through the coupled inductors to charge the capacitors C_2 and C_3. In this mode the output capacitor C_o supplies the load. Following equations can be found in this time interval

$$V_{Lin} = V_{in} - V_{C1} \qquad (5)$$
$$V_{L2} = -V_{C2,3} \qquad (6)$$

E. Mode 5 $[t_4 - t_5]$:

At $t = t_4$ the diode D_1 turns OFF naturally. The diodes D_2 and D_3 are still ON and the stored energy in the magnetizing inductance L_m charges the capacitors C_2 and C_3. This time interval ends when the switch S is turned on again.

Figure 3 shows all time intervals of the proposed converter in one switching cycle.

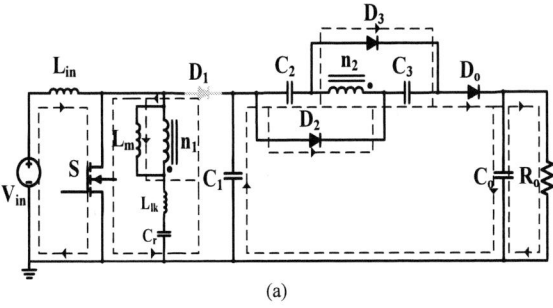

(a)

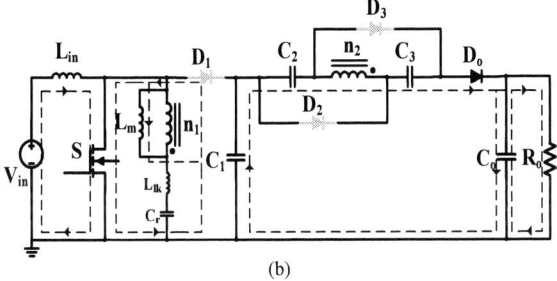

(b)

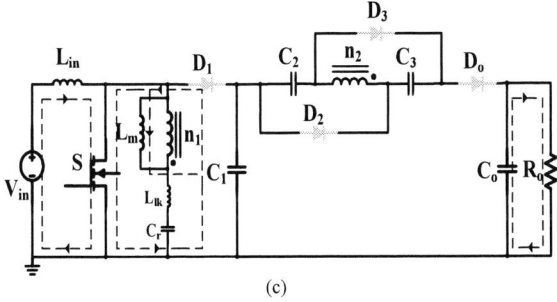

(c)

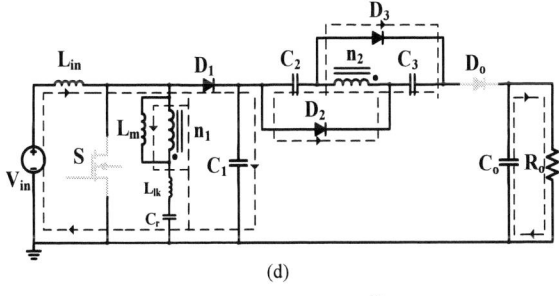

(d)

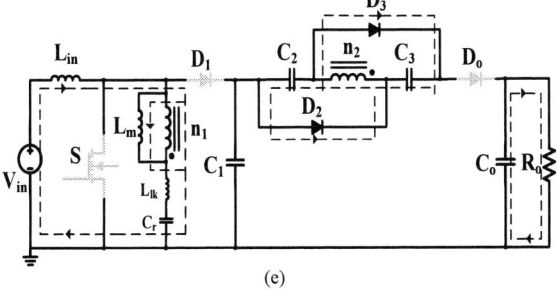

(e)

Fig. 3. Operating modes of the proposed DC-DC converter (a) mode 1, (b) mode 2, (c) mode 3, (d) mode 4 and (e) mode 5

III. DRIVATION OF VOLTAGE GAIN AND VOLTAGE STRESS ACROSS ELEMENTS

By applying KVL in the input loop of the proposed converter the following equation can be derived

$$V_{Cr(avg)} = V_{in} \qquad (7)$$

In order to simplify steady state analysis voltage ripple across the resonance capacitior C_r is neglected and the coupling coefficinet is considred ideal (k=1). From (1), (2) and (5) the voltage across V_{C1} is

$$V_{C1} = \frac{kDV_{in} - DV_{in} + V_{in}}{1 - D} = \frac{V_{in}}{1 - D} \qquad (8)$$

In (8) D is the duty cycle of the switch S. From (6) and applying volt-second law across the secondary winding of the coupled inductors voltage stress across C_2 and C_3 can be written as

$$V_{C2} = V_{C3} = \frac{kND}{1 - D} V_{in} = \frac{ND}{1 - D} V_{in} \qquad (9)$$

And the voltage gain of the proposed converter can be expressed as

$$M = \frac{V_o}{V_{in}} = \frac{1 + Nk(1 + D) + D(k - 1)}{1 - D} \qquad (10)$$
$$= \frac{1 + N(1 + D)}{1 - D}$$

Due to equality of voltage across diode D_1 and switch S with voltage across C_1 during switch OFF time, the voltage across D_1 and S are

$$V_{SW} = V_{D1} = V_{C1} = \frac{V_{in}}{1 - D} \qquad (11)$$

Voltage across diodes D_2, D_3 and D_o is equal and can be written as

$$V_{Do} = V_{D2} = V_{D3} = \frac{N}{1 - D} V_{in} \qquad (12)$$

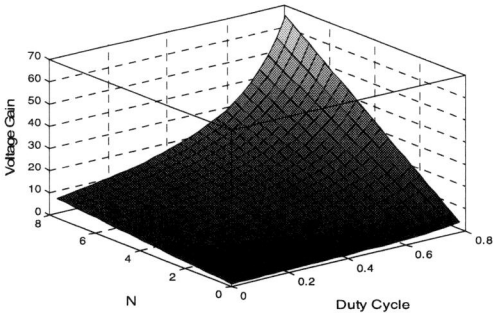

Fig. 4. A 3D plot for the voltage gain of the proposed converter versus both duty cycle and turns ratio

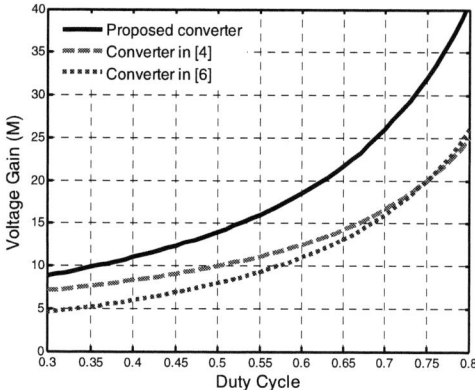

Fig. 5. A comparison between voltage gain of the proposed converter and some other mentioned high step up converters.

It can been seen by increasing the duty cycle of the switch, the voltage stress of switch and diodes increases. Whereas, voltage stress across the diodes increases by increasing turns ratio of the coupled inductor too. Hence, a compromise should be conducted in order to achieve demanded gain while maintaining a balance between the voltage stress of switch S and diodes. Fig. 4 illustrates a 3D plot for the voltage gain of the proposed DC-DC converter in which the effect of both duty cycle and turns ratio is illustrated at the same time. Besides, a comparison between the voltage gain of the proposed converter and some mentioned high gain converter is depicted in Fig. 5. Apparently the proposed converter achieves a higher voltage gain than the other mentioned converter in the entire duty cycle range.

IV. SIMULATION RESULTS

In order to verify the proposed converter performance the simulations have been done in PSpice/Orcad. The parameters, which have been used for the simulations, are

TABLE I. SIMULATION PARAMETERS

Parameter	Value
Output power	200 W
Input voltage (V_{in})	25 V
Output voltage (V_{out})	400 V
Switching Frequency (f_s)	50 KHz
Capacitors C_1, C_2, C_3	100 uF
Turns ratio ($N = n_2/n_1$)	4
Magnetizing inductor (L_m)	300 uH
Leakage Inductance (L_{lk})	1.5 uH
Resonance Capacitor (C_r)	5 uF
Output Capacitor (C_o)	220uF
Input Inductor (L_{in})	200 uH

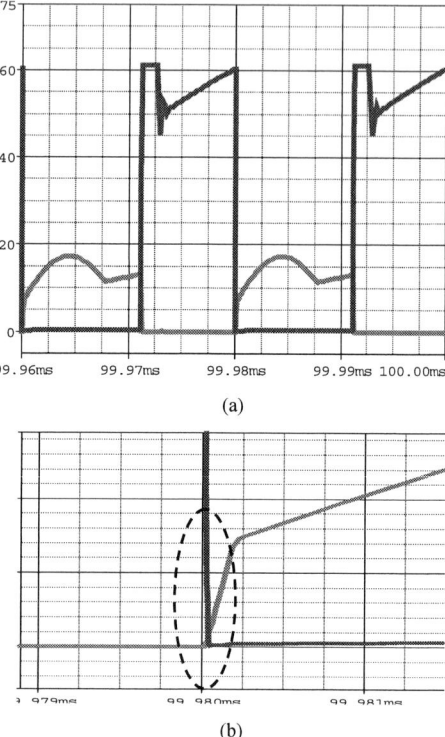

Fig. 6. (a) Switch voltage (red) and current (green) waveforms and (b) ZCS condition at switch turn ON

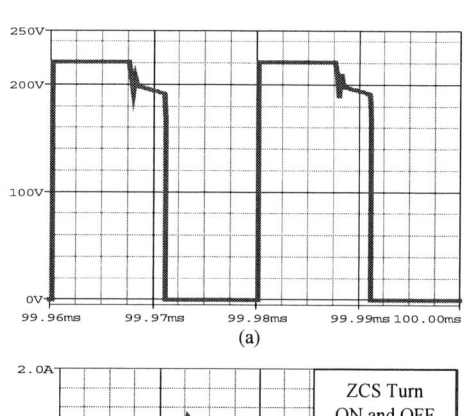

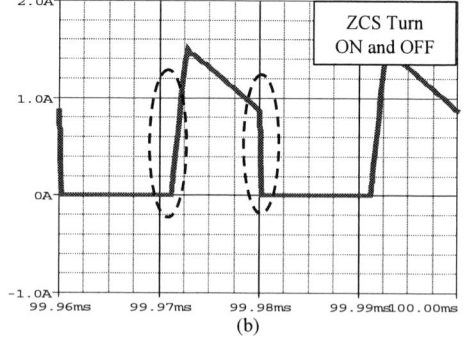

Fig. 7. (a) Voltage stress across D_2 and D_3 and (b) current through D_2 and D_3

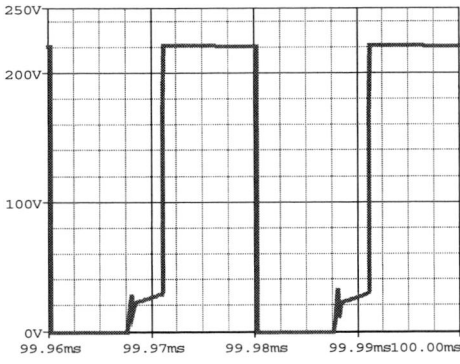

(a)

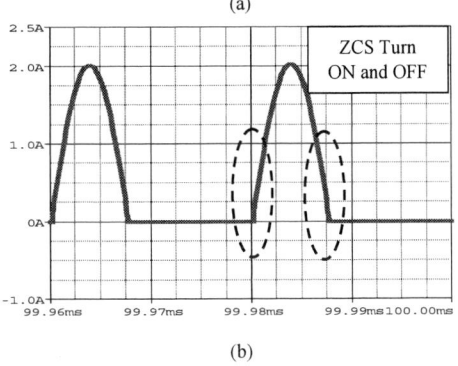

ZCS Turn
ON and OFF

(b)

Fig. 8. (a) Voltage stress across D_o and (b) current through D_o

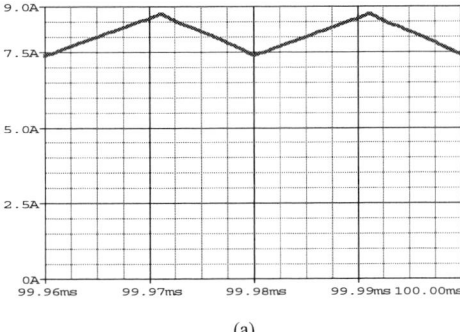

(a)

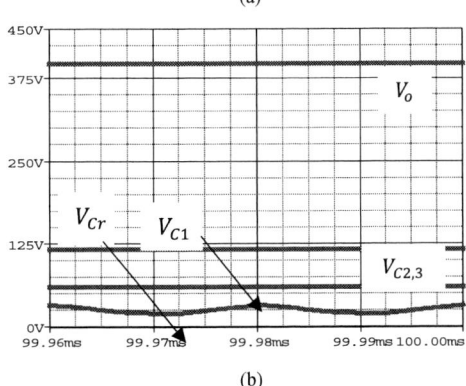

V_o

V_{Cr} V_{C1}

$V_{C2,3}$

(b)

Fig. 9. (a) Input current and (b) output voltage and voltage across C_1, C_2, C_3 and C_r.

shown in Table I.

Current and voltage waveforms of the switch S are shown in Fig. 6.(a). From Fig 6.(a) it can been seen voltage stress across the switch is less than 100v so that a switch with low $R_{ds(on)}$ can be used to implement the proposed converter.

Fig. 6.(b) represents zero current switching (ZCS) of the proposed converter, so that all the turn ON losses are alleviated. Fig. 7.(a) illustrates voltage stress of diodes D_2, D_3 and Fig. 7.(b) shows diodes D_2 and D_3 current stresses. Fig. 8.(a) depicts voltage stress of output diode D_o and Fig. 8.(b) represents the output diode current stress. It can be seen that voltage stresses of all diodes are less than output voltage. Since output diode D_o and diodes D_2, D_3 are turned ON and OFF under ZCS condition, causing diminished reverse recovery problem of the diodes.

Fig. 9.(a) depicts the proposed converter input current. Output voltage and voltage stresses across the capacitors C_1, C_2, C_3 and C_r have been shown in Fig. 9.(b). It is obvious that soft switching feature for switch S and diodes reduces losses and increases the proposed converter efficiency. Voltage stresses of all the components have been reduced so the proposed converter can be implemented with low cost and volume. Clearly all the simulation results are in accordance with the calculated values from the steady state analysis.

V. CONCLUSION

In this paper, a novel DC-DC converter with high voltage gain and reduced voltage stress across the switch and other circuit elements has been proposed. In addition, Soft switching feature of the proposed converter increases its efficiency and alleviates all the diodes reverse recovery losses. Operation principles and steady sates analysis have been demonstrated. According to the steady state analysis and simulation results voltage stresses across all elements have been reduced so the proposed converter can be implemented with low cost and volume. All the aforementioned features make the proposed converter suitable for high power conversion.

REFRENCES

[1] Blaabjerg, F.; Iov, F.; Terekes, T.; Teodorescu, R.; Ma, K., "Power electronics - key technology for renewable energy systems," in Power Electronics, Drive Systems and Technologies Conference (PEDSTC), 2011 2nd , vol., no., pp.445-466, 16-17 Feb. 2011.

[2] Oded Abutbul, Amir Gherlitz, Yefim Berkovich, Member and Adrian Ioinovici, "Step-Up Switching-Mode Converter With High Voltage Gain Using a Switched-Capacitor Circuit," IEEE Trans. On Circuits and Syetems., vol. 50, no. 8, pp. 1098 - 1102, August 20.

[3] Mahajan Sagar Bhaskar Ranjana, Nandyala SreeramulaReddy, Repalle Kusala Pavan Kumar, "A Novel Non-Isolated Switched Inductor Floating Output DC-DC Multilevel Boost Converter For Fuelcell Applications," Electrical, Electronics and Computer Science (SCEECS), 2014 IEEE Students' Conference on, Bhopal, pp. 1-5, 1-2 March. 2014.

[4] High Step-Up Active-Clamp Converter With Input-Current Doubler and Output-Voltage Doubler for Fuel Cell Power Systems," IEEE Trans. On Power Electronics, vol. 24, no. 1, January 2009.

[5] Rong-Jong Wai and Kun-Huai Jheng, "High-Efficiency Single-Input Multiple-Output DC–DC Converter," IEEE Trans. On Power Electronics, vol. 28, no. 2, February 2013.

[6] Lung-Sheng Yang, Tsorng-Juu Liang, Hau-Cheng Lee and Jiann-Fuh Chen, "Novel High Step-Up DC–DC Converter With Coupled-Inductor and Voltage-Doubler Circuits," IEEE Trans. On Industrial Electronics, vol. 58, no. 9, September 2011.

[7] K. Yari, M. Forouzesh, and A. Baghramian, "A novel high voltage gain DC-DC converter with reduced components voltage stress," in Proc. PEDSTC, pp. 173-177, February. 2015.

[8] Yi-Ping Hsieh, Jiann-Fuh Chen, Tsorng-Juu Liang and Lung-Sheng Yang, "Novel High Step-Up DC–DC Converter for Distributed Generation System," IEEE Trans. On Industrial Electronics, vol. 60, no. 4, April 2013.

[9] V. T. Liu and L. J. Zhang, "Design of high efficiency Boost-Forward-Flyback converters with high voltage gain," in Proc. ICCA, pp. 1061-1066, June 2014.

[10] Yi-Ping Hsieh, Tsorng-Juu Liang, and Lung-Sheng Yang, "A Novel High Step-Up DC–DC Converter for a Microgrid System," IEEE Trans. On Power Electronics, vol. 26, no. 4, April 2011.

[11] Yi-Ping Hsieh, Tsorng-Juu (Peter) Liang and Lung-Sheng Yang, "Novel High Step-Up DC–DC Converter With Coupled-Inductor and Switched-Capacitor Techniques for a Sustainable Energy System," IEEE Trans. On Power Electronics, vol. 26, no. 12, December 2011.

[12] Shih-Ming Chen, Tsorng-Juu Liang, Lung-Sheng Yang and Jiann-Fuh Chen, "A Cascaded High Step-Up DC–DC Converter With Single Switch for Microsource Applications," IEEE Trans. On Power Electronics, vol. 26, no. 4, April 2011.

[13] Yan Deng, Qiang Rong, Wuhua Li, Yi Zhao, Jianjiang Shi and Xiangning He, "Single-Switch High Step-Up Converters With Built-In Transformer Voltage Multiplier Cell." IEEE Trans. On Power Electronics vol. 27, no. 8, August 2012.

7th Power Electronics, Drive Systems & Technologies Conference (PEDSTC 2016)
16-18 Feb. 2016, Iran University of Science and Technology, Tehran, Iran

A New Topology for Cascaded Multilevel Inverters with Reduced Number of Power Electronic Switches

Ebrahim Babaei[1], *Member, IEEE*, Maryam Sarbanzadeh[2], Mohammad Ali Hosseinzadeh[3], Concettina Buccella[4], *Senior Member, IEEE*

[1&2] Faculty of Electrical and Computer Engineering, University of Tabriz, Tabriz, Iran

[3] Department of Electronic Engineering, University of Applied Science and Technology, Jajarm Branch, Jajarm, Iran

[4] Department of Information Engineering, Computer Science and Mathematics, University of L'Aquila, L'Aquila, Italy

E-mails: e-babaei@tabrizu.ac.ir; may14_1368@yahoo.com; m.a_hosseinzadeh@yahoo.com; concettina.buccella@univaq.it

Abstract—**In this paper, a new basic unit for cascaded multilevel inverter is proposed. The proposed basic unit consists of *m* cells. Each cell can generate three levels. The proposed multilevel inverter is based on series connection of several basic units. The proposed multilevel inverter is investigated at both symmetric and asymmetric topologies. In order to generate all voltage levels at the output, four algorithms to determine the magnitudes of dc voltage sources are proposed. Reduction of number of power switches, driver circuits and IGBTs are some advantages of the proposed cascaded multilevel inverter. Finally, to verify the correctness operation of the proposed inverter, the simulation results of a 27-level inverter based on proposed topology by using PSCAD/EMTDC software are used.**

Keywords—*Basic unit; symmetric and asymmetric multilevel inverters; cascaded multilevel inverter*

I. INTRODUCTION

Multilevel inverters include an array of power semiconductors and dc voltage sources to generate voltages with stepped waveforms [1]. Recently, there is a good attention to multilevel inverters as new kind of converters. Compared to two-level inverter configurations, the multilevel inverters have significant advantages which mainly include: higher power quality, more electromagnetic capability, lower switching losses, voltage capability, higher efficiency, lower dv/dt, and lower THD [2-3]. The multilevel inverters are basically divided into three categories: cascaded H-Bridge multilevel inverters (CHBMLs) [4], diode clamped multilevel inverters [5], and flying capacitor multilevel inverters [6]. The diode-clamped multilevel inverter, also called neutral point clamped (NPC), can be considered the first generation of multilevel inverter [7]. The CHBMLs use series connection of H-bridge cells with isolated dc voltage sources. From the view point of magnitudes of the dc voltage sources the CHBMLs can be divided into two groups of symmetric and asymmetric topologies. In the symmetric topology, the magnitudes of all of dc voltage sources are equal. This characteristic gives good modularity, but the number of the switching devices is rapidly increased by increasing the number of output voltage levels. In order to increase the number of output voltage levels, the magnitudes of dc voltage sources are selected to be different, these topologies are called

asymmetric [8-9]. Reducing the stress on power switches and fault-tolerant operations, extendibility, modularization, simplicity of control and high reliability are some of main advantages of these inverters in comparison with two other basic topologies of multilevel inverters. However, high number of required dc voltage sources and semiconductor power switches by increasing the number of generated output levels are disadvantageous of this inverter [10-12].

In this paper, in order to increase the number of output levels by using lower number of power electronic switches a new basic unit is proposed. Then, using series connection of these basic units, a new cascaded multilevel inverter is proposed. In order to generate all positive and negative levels at the output, four algorithms to determine the magnitude of dc voltage sources are proposed. The proposed inverter is compared with several conventional cascaded multilevel inverters to investigate its advantages and disadvantages. Finally, the accuracy performance of the proposed inverter is reconfirmed by using simulation results on proposed 27-level inverter.

II. PROPOSED BASIC UNIT

The proposed basic unit for cascaded multilevel inverter is shown in Fig. 1. As shown in this figure, the proposed basic unit consists of m number of three-level inverters that each of they called one cell. Each cell consist of two dc voltage sources and three power electronic switches of S_1, S_1' and S_1'' (S_1 and S_1' the unidirectional and S_1'' is the bidirectional power switches). The magnitude of dc voltage source is equal V_j and cells are connected in series by using one power electronic switch. Each cell generates three levels of 0, V_j and $-V_j$. The switches S_j, S_j' and S_j'' are turned on to generate positive voltage level of V_j, negative voltage level of $-V_j$ and zero level, respectively. It is clear that the switches S_j, S_j' and S_j'' cannot be turned on simultaneously because a short circuit across the dc voltage V_j would be produced. The proposed basic unit can be generate a stepped waveform with positive level, negative level and zero voltage level with no needing to the H-bridge inverter at the output. As shown in this Fig. 1, the

978-1-5090-0376-1/16 $31.00 © 2016 IEEE

unidirectional and bidirectional power switches are used in the basic unit. Each unidirectional power switch consists of an IGBT with an anti-parallel diode as a common emitter configuration that is able to conduct current in both direction and block voltage in one polarity. While, the bidirectional power switches include of two IGBTs with two anti-parallel diodes that conducts current in both direction and blocks voltage with both positive and negative polarities. As a result, each of unidirectional and bidirectional switches need on driver circuits. In the basic unit, the power switches S_1, S_1', T_1, S_2, S_2', T_2,..., S_m, S_m', T_{m-1} are unidirectional switches while the power switches of S_1'', S_2'', ... , S_m'' are bidirectional power switches. In addition, in order to generate same at steps output, the magnitudes of dc voltage sources have to be considered equally. In this topology, if the power switches of S_1'', S_2'', ... , S_m'' are turned on simultaneously, the output voltage will be zero. When the power switches of S_1, S_2,..., S_m turned on simultaneously the inverter will generate the maximum positive voltage ($\sum_{i=1}^{m} V_i$) and when the power switches S_1', S_2',..., S_m' turned on simultaneously, the inverter will generate the maximum negative output voltage level ($-\sum_{i=1}^{m} V_i$). Table I shows the switching patterns and generated output levels for different combinations of on and off states of switches in the proposed basic unit. As it is obvious from this Table, the proposed basic unit is able to generate positive and negative voltage levels at the output. In this Table 1, and 0 indicate the on and off states of switches, respectively. As it is obvious from this Table, the basic unit is able to generate $2m^2 + 1$ levels at the output. Moreover, in order to generate all voltage levels at the output except the zero level, only two power switches in different operating modes are turned on.

In the proposed basic unit, the number of generated output voltage levels ($N_{level,unit}$), power switches ($N_{switch,unit}$), IGBTs ($N_{IGBT,unit}$), driver circuits ($N_{driver,unit}$), dc voltage sources ($N_{source,unit}$) and the maximum output voltage ($V_{o,max,unit}$) are calculated as follows, respectively:

$$N_{level,unit} = 2m^2 + 1 \tag{1}$$

$$N_{switch,unit} = 4m - 1 \tag{2}$$

$$N_{IGBT,unit} = 5m - 1 \tag{3}$$

$$N_{driver,unit} = 4m - 1 \tag{4}$$

$$N_{source,unit} = 2m \tag{5}$$

$$V_{o,max,unit} = \sum_{i=1}^{m} V_i \tag{6}$$

In above equations, m is number of used cells in the proposed basic unit.

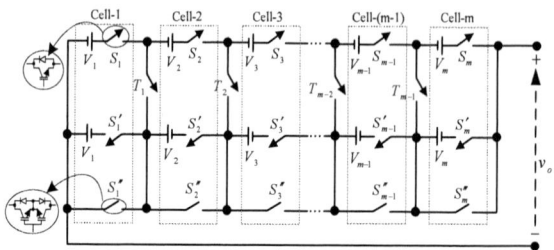

Fig. 1. The proposed basic unit.

TABLE I. THE GENERATED OUTPUT LEVELS IN DIFFERENT SWITCHING PATTERNS

v_o	S_1	S_1'	S_1''	S_2	S_2'	S_2''	...	S_m	S_m'	S_m''	T_1	T_2	...	T_{m-2}	T_{m-1}
0	0	0	1	0	0	1	...	0	0	1	0	0	...	0	0
	0	0	1	0	0	1		0	0	1	0	0		0	0
V_1	1	0	0	0	0	1	...	0	0	1	1	0	...	0	0
$-V_1$	0	0	0	0	0	1	...	0	0	1	0	0	...	0	0
V_2	0	0	1	1	0	0	...	0	0	1	1	1	...	0	0
$-V_2$	0	0	1	0	1	0	...	0	0	1	0	0	...	0	0
$V_1 - V_2$	1	0	0	0	1	0	...	0	0	1	1	0	...	0	0
$-(V_1 - V_2)$	0	1	0	1	0	0	...	0	0	1	1	1	...	0	0
$V_1 + V_2$	1	0	0	1	0	0	...	0	0	1	0	1	...	0	0
$-(V_1 + V_2)$	0	1	0	0	1	0	...	0	0	1	0	0	...	0	0
⋮	⋮	⋮	⋮	⋮	⋮	⋮	...	⋮	⋮	⋮	⋮	⋮	...	⋮	⋮
$\sum_{i=1}^{m} V_i$	1	0	0	1	0	0	...	1	0	0	0	0	...	0	0
$-\sum_{i=1}^{m} V_i$	0	1	0	0	1	0	...	0	1	0	0	0	...	0	0

III. PROPOSED CASCADE MULTILEVEL INVERTER

To generate high number of levels, it is possible to use k basic units in series as shown in Fig. 2. For this topology, the number of power switches (N_{switch}), IGBTs (N_{IGBT}), driver circuits (N_{driver}) and dc voltage sources (N_{source}) are calculated as follows, respectively:

$$N_{switch} = \sum_{i=1}^{k} (4m_i - 1) \tag{7}$$

$$N_{IGBT} = \sum_{i=1}^{k} (5m_i - 1) \tag{8}$$

$$N_{driver} = \sum_{i=1}^{k} (4m_i - 1) \tag{9}$$

$$N_{source} = \sum_{i=1}^{k} 2m_i \tag{10}$$

In order to generate the maximum number of output levels for a fixed number of components, it is necessary to consider the equal number of cells in each basic unit. In other word:

$$m_1 = m_2 = \cdots = m_k = m \tag{11}$$

In this condition, the proposed topology consists of km cells and the $k(4m-1)$ power switches. The different output voltage levels can be determined by combinations of switching states of units. If proper magnitudes for dc voltage sources are selected, then the output voltage of the inverter can be obtained between $-\sum_{i=1}^{m}\sum_{j=1}^{k}V_{i,j}$ and $\sum_{i=1}^{m}\sum_{j=1}^{k}V_{i,j}$.

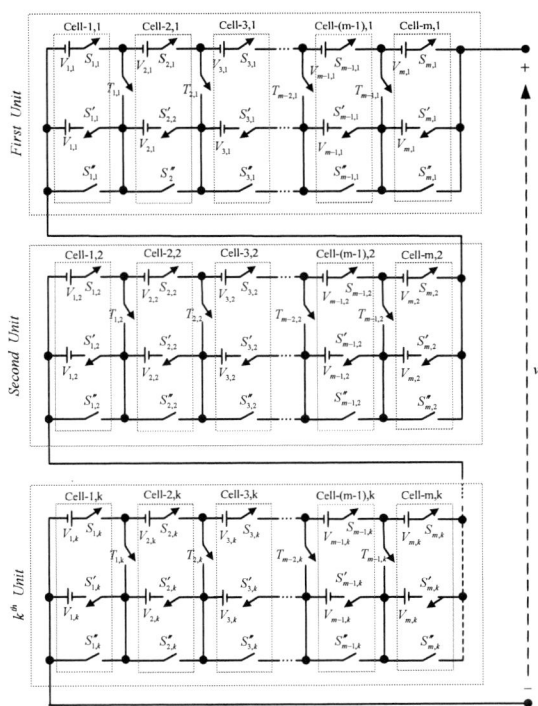

Fig. 2. The proposed cascaded multilevel inverter.

By applying (11), the equations of (7) to (10) are rewritten as follows:

$$N_{switch} = k(4m-1) \tag{12}$$

$$N_{IGBT} = k(5m-1) \tag{13}$$

$$N_{driver} = k(4m-1) \tag{14}$$

$$N_{source} = k(2m) \tag{15}$$

$$N_{variety} = 2km \tag{16}$$

$$V_{o,\max} = \sum_{i=1}^{m}\sum_{j=1}^{k}V_{i,j} \tag{17}$$

In the proposed inverter, the number of generated levels is completely depended on the magnitudes of used dc voltage sources. In next sub-sections, four different algorithms to determine the magnitudes of dc voltage sources are proposed.

A. First Proposed Algorithm
In the first proposed algorithm, the magnitudes of all dc voltage sources are considered equally. In other word:

$$V_{1,1} = V_{2,1} = \cdots = V_{m,1} = V_{dc} \tag{18}$$

$$V_{1,2} = V_{2,2} = \cdots = V_{m,2} = V_{dc} \tag{19}$$

$$\vdots$$

$$V_{1,k} = V_{2,k} = \cdots = V_{m,,k} = V_{dc} \tag{20}$$

The proposed inverter based on this algorithm is known as symmetric cascaded multilevel inverter. In this condition, the number of generated output levels and maximum magnitude of output voltage are obtained as follows:

$$N_{level} = 2mk+1 \tag{21}$$

$$V_{o,\max} = kmV_{dc} \tag{22}$$

B. Second Proposed Algorithm
In the second proposed algorithm, the magnitudes of dc voltage sources are considered as follows:

$$V_{1,1} = V_{2,1} = \cdots = V_{m,1} = V_1 \tag{23}$$

$$V_{1,2} = V_{2,2} = \cdots = V_{m,2} = V_2 \tag{24}$$

$$\vdots$$

$$V_{1,k} = V_{2,k} = \cdots = V_{m,k} = V_k \tag{25}$$

Now, the magnitudes of dc voltage sources V_1 , V_2 ,..., V_k are determined as follows:

$$V_1 = V_{dc} \tag{26}$$

$$V_2 = V_1 + mV_1 = (m+1)V_{dc} \tag{27}$$

$$\vdots$$

$$V_k = V_{k-1} + mV_{k-1} = (m+1)^{k-1}V_{dc} \tag{28}$$

For the second algorithm, the maximum magnitude of output voltage and the number of output levels are obtained as follows:

$$V_{o,\max} = m\sum_{j=1}^{k}V_j = [(m+1)^k - 1]\,V_{dc} \tag{29}$$

$$N_{level} = 2(m+1)^K - 1 \tag{30}$$

C. Third Proposed Algorithm
The third method to determine the magnitudes of dc voltage sources is as binary form as follows:

First unit:

$$V_1 = V_{dc} \qquad (31)$$

$$V_{i,1} = 2^{i-1}V_{1,1} \qquad i = 2,3,\ldots,m \qquad (32)$$

Second unit:

$$V_{1,2} = V_{1,1} + \sum_{i=1}^{m} V_{i,1} \qquad (33)$$

$$V_{i,2} = 2^{i-1}V_{1,2} \qquad i = 2,3,\ldots,m \qquad (34)$$

$$\vdots$$

k^{th} unit:

$$V_{1,k} = V_{1,1} + \sum_{i=1}^{m}\sum_{j=1}^{k-1} V_{i,j} \qquad (35)$$

$$V_{i,k} = 2^{i-1}V_{1,k} \qquad i = 2,3,\ldots,m \qquad (36)$$

The proposed inverter based on this algorithm is known as asymmetric cascaded inverter. According to Fig. 2 and above mentioned equations, the number of output levels and maximum amplitude of the output voltage are obtained as follows:

$$N_{level} = 2^{mk+1} - 1 \qquad (37)$$

$$V_{o,\max} = \sum_{i=1}^{m}\sum_{j=1}^{k} V_{i,j} = (2^{mk}-1)V_{dc} \qquad (38)$$

D. Fourth Proposed Algorithm
The fourth method to determine the magnitude of dc voltage sources is as trinary form as follows:

First unit:

$$V_1 = V_{dc} \qquad (39)$$

$$V_{i,1} = 3^{i-1}V_{1,1} \qquad i = 2,3,\ldots,m \qquad (40)$$

Second unit:

$$V_{1,2} = V_{1,1} + \sum_{i=1}^{m} V_{i,1} \qquad (41)$$

$$V_{i,2} = 3^{i-1}V_{1,2} \qquad i = 2,3,\ldots,m \qquad (42)$$

$$\vdots$$

k^{th} unit:

$$V_{1,k} = V_{1,1} + \sum_{i=1}^{m}\sum_{j=1}^{k-1} V_{i,j} \qquad (43)$$

$$V_{i,k} = 3^{i-1}V_{1,k} \qquad i = 2,3,\ldots,m \qquad (44)$$

The proposed inverter based on this algorithm is known as asymmetric cascaded inverter. According to Fig. 2 and above mentioned equations, the maximum amplitude of output voltage and number of generated output levels are calculated as follows:

$$V_{o,\max} = \sum_{i=1}^{m}\sum_{j=1}^{k} V_{i,j} = \left(\frac{3^{mk}-1}{2}\right)V_{dc} \qquad (45)$$

$$N_{level} = 3^{mk} \qquad (46)$$

As it is obvious, the proposed cascaded inverter based on the fourth proposed algorithm is able to generate higher number of output levels in comparison with the other proposed algorithms. It is also possible to propose other algorithms to determine the magnitudes of dc voltage sources. This feature shows the high flexibility of the proposed inverter.

IV. COMPARISON OF THE PROPOSED INVERTER WITH CONVENTIONAL MULTILEVEL INVERTERS

In this section, the proposed cascaded multilevel inverter is compared with several conventional cascaded multilevel inverters. This comparison is done from different points of view as the number of IGBTs, power switches and dc voltage sources. In this comparison, the proposed cascaded multilevel inverter based on forth algorithm is indicated by P_1. The CHBML with two asymmetric algorithms presented in [13] are indicated by R_1 and R_2, respectively. (R_1 for $V_j = 2^{j-1}V_{dc}$ $(j = 1,2,\cdots,n)$ and R_2 for $V_j = 3^{j-1}V_{dc}$ $(j = 1,2,\cdots,n)$). The presented asymmetric multilevel inverter in [14] is indicated by R_3 ($V_{1,1} = V_{dc}$, $V_{2,1} = 2V_{dc}$, $V_{2,j} = 2V_{1,j}$ $(j = 1,2,\cdots,n-1)$). In addition, three other algorithms have been introduced for these inverters in [15-17]. These algorithms are also used in asymmetric state. For the presented cascaded multilevel inverter in [15], the asymmetric algorithm is considered by R_4 ($V_{1,1} = V_{2,1} = V_{dc}$, $V_{1,j} = V_{2,j} = 3^{j-1}$ $(j = 1,2,\cdots,n)$). In this comparison, the presented cascaded multilevel inverters in [16] and [17] are considered by R_5 and R_6, respectively. (R_5 for $V_{1,1} = V_{2,1} = V_{dc}$, $V_{1,j} = V_{2,j} = 3^{j-1}V_{dc}$ $(j = 2,3,\cdots,n)$ and R_6 for $V_{1,j} = 4^{j-1}V_{dc}$, $V_{2,j} = 3(4^{j-1}V_{dc})$ $(j = 1,2,\cdots,n)$). Fig. 3 indicates all of the above-mentioned topologies. Fig. 4 (a) compares the number of IGBTs of the proposed topology with the above-mentioned cascaded multilevel inverters. As it is obvious from this figure, the proposed cascaded multilevel inverter needs lower number of IGBTs than other presented multilevel inverters in the references except the presented multilevel inverter by R_2. It is pointed out that the bidirectional and unidirectional power switches are used in the proposed cascaded multilevel inverter. As mentioned before, the number of IGBTs is as same as the number of power diodes. As a result, the proposed inverter also needs minimum number of power diodes than other conventional cascaded multilevel inverters.

978-1-5090-0376-1/16 $31.00 © 2016 IEEE

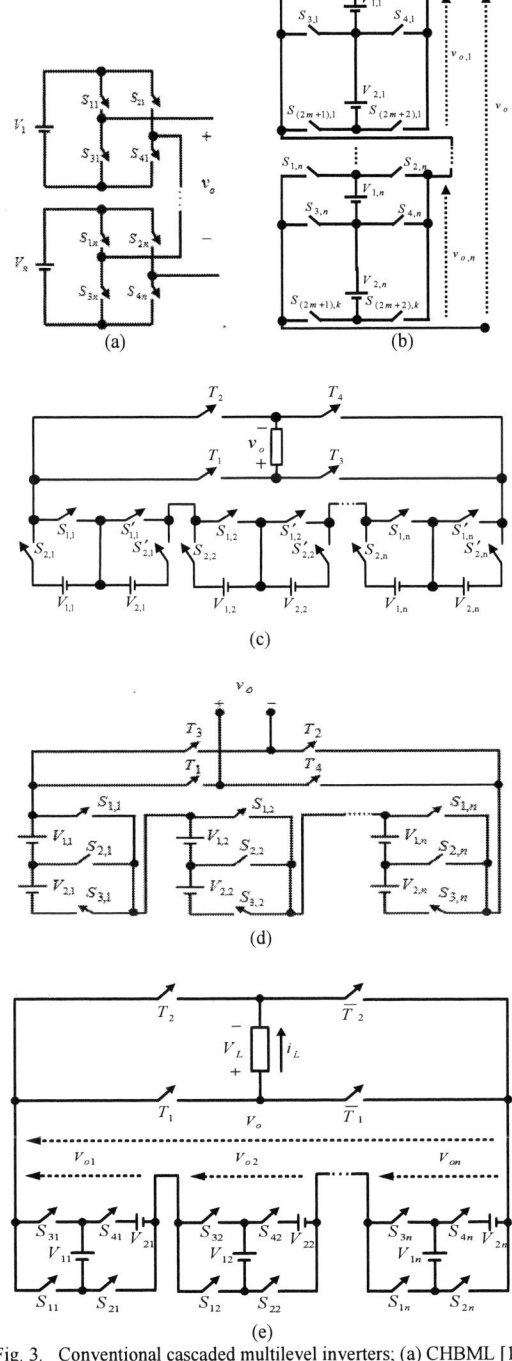

Fig. 3. Conventional cascaded multilevel inverters; (a) CHBML [13] ($R_1 - R_2$); (b) presented topology in [14] (R_3); (c) presented topology in [15] (R_4); (d) presented topology in [16] (R_5); (e) presented topology in [17] (R_6).

The number of power switches in the proposed cascaded multilevel inverter is compared with above-mentioned inverters. This comparison is shown in Fig. 4(b). As it is

obvious from this figure, the proposed inverter needs lower number of power switches than other presented multilevel inverters. As it is mentioned previously, all unidirectional and bidirectional power switches require a driver circuit. Therefore, the proposed inverter also needs lower number of driver circuit than other multilevel inverters. Fig. 4(c) compares the number of dc voltage sources in the proposed topology with the above mentioned cascaded multilevel inverters. It is obvious that the proposed topology requires higher number of dc voltage sources than other mentioned topologies except the presented multilevel inverter by R_4 .

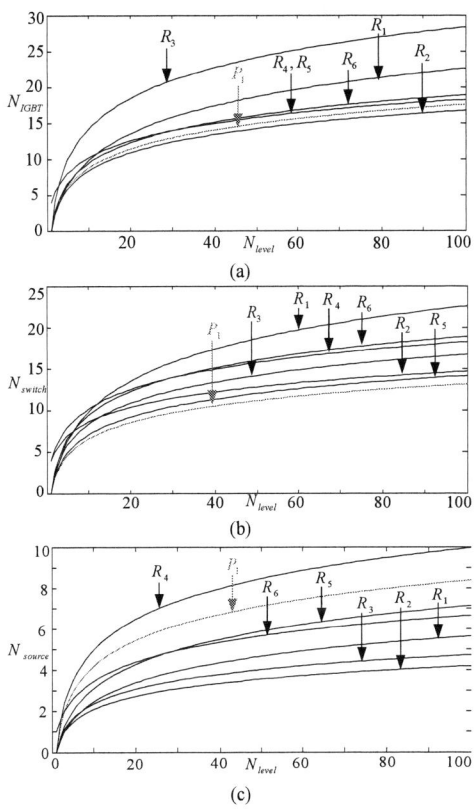

Fig. 4. Comparison results; (a) variation of N_{IGBT} versus N_{level} ; (b) variation N_{switch} of versus N_{level} ; (c) variation of N_{source} versus N_{level} .

V. SIMULATION RESULTS

In this section, the ability of the proposed multilevel inverter to generate all voltage levels is reconfirmed by using simulation results on a 27-level inverter. The simulation results are obtained by using PSCA/EMTDC software program. The fundamental frequency control method is used to control the proposed multilevel inverter. In the simulation, the load is assumed as $R-L$ load with $R = 107\Omega$ and $L = 55\,mH$ and the output voltage frequency is assumed $50Hz$. According to Fig. 5, the proposed multilevel inverter consists of two basic units. Two cells in first unit and one cell in second unit are used. To determine the magnitudes of dc voltage sources, the fourth algorithm is used. Based on this algorithm, the magnitudes of dc voltage sources in the first unit are $V_{1,1} = V_{dc}$ and $V_{1,1} = V_{dc}$ and in the second unit is $V_{1,2} = 9\,V_{dc}$. The value of V_{dc} is assumed $10V$. Considering these values the peak value of output voltage will be $130V$. Figs. 6(a), 6(b) shows the output voltage of first and second basic units, respectively. The sum of output voltages of first and second basic units creates the output voltage of the inverter which is shown in Fig. 6(c). As this figure shows, the inverter can generate all of the expected voltage levels. Fig. 6(d) shows the output current waveform. Considering the waveforms of current and voltage, there is a phase displacement between them. This is resulted from the inductive characteristic of the load.

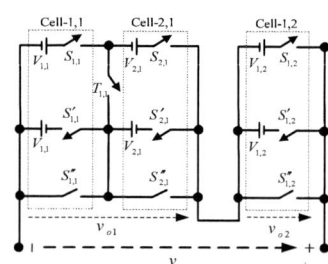

Fig. 5.   27-level inverter based on proposed topology.

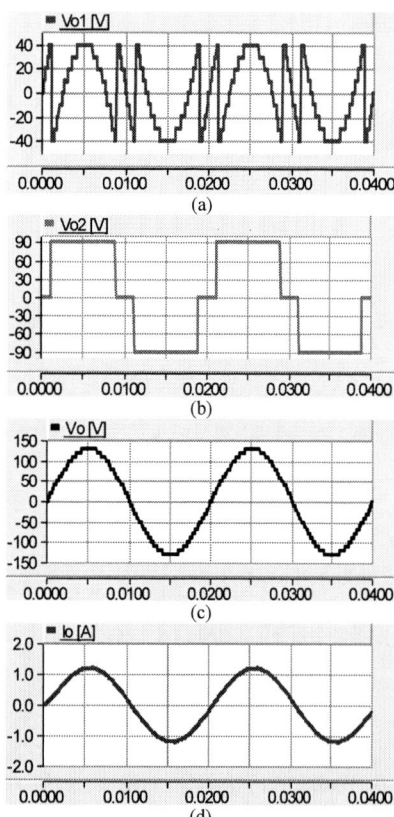

Fig. 6.   Simulation results; (a) output voltage of first basic unit; (b) output voltage of second basic unit; (c) output voltage; (d) output current.

VI. CONCLUSION

In this paper, a new basic unit for cascaded multilevel inverters is proposed. Then, by series connection of the basic units, the proposed multilevel inverter is proposed. Then, four different algorithms are proposed to determine the magnitudes of dc voltage sources. This inverter uses lower number of power switches, diodes and driver circuits for a specific number of voltage levels in comparison with many of conventional multilevel inverters that have been presented in literature. Finally, all of the obtained theoretical issues are reconfirmed by using simulation results on a proposed 27-level inverter based on fourth proposed algorithm.

REFERENCES

[1] J. Rodríguez, J.S. Lai, and F.Z. Peng, "Multilevel inverters: A survey of topologies, controls, and applications," *IEEE Trans. Ind. Electron.*, vol. 49, no. 49, pp. 724-738, Aug. 2002.

[2] J. Rodriguez, B. Wu, S. Bernet, J. Pontt, and S. Kouro, "Multilevel voltage source converter topologies for industrial medium voltage drives," *IEEE Trans. Ind. Electron.*, vol. 54, pp. 2930-2945, Dec. 2007.

[3] M.N.A. Kadir, S. Mekhilef, and H.W. Ping, "Voltage vector control of a hybrid three-stage eighteen-level inverter by vector decomposition," *IET Trans. Power Electron.*, vol. 3, no. 4, pp. 601–611, Jul. 2010.

[4] G.P. Adam, O. Anaya-Lara, G.M. Burt, D. Telford, B.W. Williams, and J.R. McDonald, "Modular multilevel inverter: pulse width modulation and capacitor balancing technique," *IET Power Electron.*, vol. 3, no. 5, pp. 702–715, 2010.

[5] I. Nabaei, Takahashi, and H. Akagi, "A neutral-point clamped PWM inverter," *IEEE Trans. Ind. Appt.*, vol. 1A-17, no. 5, pp. 518-523, Sep. 1981.

[6] T.A. Meynard and H. Foch, "Multilevel conversion: high voltage choppers and voltage source inverters," in *Proc. PESC,* 1992, Spain, pp.397-403.

[7] J. Li, S. Bhattacharya, and A.Q. Huang, "A new nine-level active NPC (ANPC) converter for grid connection of large wind turbines for distributed generation, " *IEEE Trans. Power Electron.*, vol. 26, no. 3, pp. 961–972, 2011

[8] J. Dixon, J. Pereda, C. Castillo, and S. Bosch, "Asymmetrical multilevel inverter for traction drives using only one dc supply," *IEEE Trans. Ind. Electron.*, vol. 59, no. 8, pp. 3736–3743, Oct. 2010.

[9] A.L. Batschauer, S.A. Mussa, and M.L. Heldwein, "Three-phase hybrid multilevel inverter based on half-bridge modules," *IEEE Trans. Ind. Electron.*, vol. 59, no. 2, pp. 668-678, Feb. 2012.

[10] J. Pereda and J. Dixon, "Cascaded multilevel converters: optimal asymmetries and floating capacitor control," *IEEE Trans. Ind. Electron.*, vol. 60, no. 11, pp. 4784-4793, Nov. 2013.

[11] M.A. Hosseinzadeh, E. Babaei, and M. Sabahi, "Back-to-back stacked multicell converter," in *Proc. PEDSTC*, 2012, Iran, pp. 410-415.

[12] E. Babaei, M. Farhadi, and M.A. Hosseinzadeh, "Asymmetrical multilevel converter topology with reduced number of components," *IET Power Electron.*, vol. 6, no. 6, pp. 1188-1196, Jan. 2013.

[13] M. Manjrekar and T.A. Lipo, "A hybrid multilevel inverter topology for drive application," in *Proc. APEC*, 1998, pp. 523-529.

[14] E. Babaei, S.H. Hosseini, G.B. Gharehpetian, M.T. Haque, and M. Sabahi, "Reduction of dc voltage sources and switches in asymmetrical multilevel converters using a novel topology," *Electr. Power Syst. Res.*, vol. 77, no. 8, pp. 1073–1085, Jun. 2007.

[15] M. Farhadi Kangarlu and E. Babaei, "A generalized cascaded multilevel inverter using series connection of sub-multilevel inverters," *IEEE Trans. Power Electron.*, vol. 28, no. 2, pp. 625-636, Feb. 2013.

[16] S. Laali, E. Babaei, and M.B. Bannae Sharifian, "Reduction the number of power electronic devices of a cascaded multilevel inverter based on new general topology," *Journal of Operation and Automation in Power Engineering (JOAPE)*, vol. 2, no. 2, pp. 81-90, Summer/Fall 2014.

[17] M. Sarbanzadeh, E. Babaei, and S. Laali, "A new basic unit for cascaded multilevel inverters with reduced power switches," in *Proc. ITC-CSCC*, 2014, Phuket, Thailand, pp. 46-49.

7th Power Electronics, Drive Systems & Technologies Conference (PEDSTC 2016)
16-18 Feb. 2016, Iran University of Science and Technology, Tehran, Iran

Ultra Step-Up DC-DC Converter Based On Three Windings Coupled Inductor

S.M. Salehi, S.M. Dehghan, S. Hasanzadeh
Faculty of Electrical and Computer engineering
Qom University of Technology, Qom, Iran
dehghan@qut.ac.ir

Abstract—In this paper, an ultra-step-up DC-DC converter using a three-winding coupled inductor and a voltage multiplier cell are proposed to achieve high step-up voltage. The important advantages of the proposed converter are as: Only one active switch is required without extreme duty cycle that high voltage is attained, The voltage stress on the active switch and diodes are reduced by increasing the turns ratio of the secondary windings of the coupled inductor, and the energy of the leakage inductance of the coupled inductor is recycled by switched capacitor and directly transfers to the load. The operation principle and steady state analysis of continuous conduction mode are used to determine voltage gain and switch voltage stress. Also the extension of the proposed converter to attain more step-up voltage gain are presented. The simulation results are used to verify the performance and validity.

Keywords—DC-DC boost converter; ultra step-up; coupled inductor; three windings

I. INTRODUCTION

Recently, because of global warming, environmental contamination and the shortage of resources such as oil, coal and natural gas which do not seem to suffice according to the growing of the global energy demand. So more and more researches focus on exploring renewable energy sources, such as the photovoltaic (PV) sources, the wind energy, the fuel cells (FC) and etc. These energy systems are sources low voltage, so DC-DC converters with steep voltage ratio are usually used to boost voltage to generate AC utility voltage [1]-[3]. The conventional boost converters cannot provide a high voltage gain, even using an extreme duty cycle. Also the voltage stress of the switches in these converters is equal to the output voltage. Thus high rating switches should be selected. In recent years many topologies have been proposed to achieve high steep voltage ratio. Recent researches on converters for high voltage gain systems have included the switched-inductor and switched-capacitor type [4]-[7], and the voltage-lift type [8],[9]. Some converters, that are the combination of conventional boost converter and any other type of converters [10]-[14], are developed to achieve a high step-up voltage ratio. Also many converters use the coupled-inductor technique for a considerably high voltage gain and reduce the serious reverse-recovery problem [15],[16]. Meanwhile, the influence of the leakage energy of the coupled inductor will cause the high voltage spike on active switches when the switches are turned off. A small resistor and

capacitor of the snubber can be used to dissipate this leakage energy and suppress the voltage spike. In this way, the power conversion will be decreased [16]. As an alternative solution, using an active clamp technique for recycling the leakage energy can achieve high efficiency [17]. This technique increases the complexity of control and the part count of circuit. Thus the power conversion efficiency and voltage gain are restrained by the parasitic effect of the components. The charging current of switched-capacitor reduces the inrush current of the switching-capacitor using leakage inductance of the coupled inductor, and the leakage inductor energy is recycled and directly output to the load which increases the power conversion efficiency [18].

The voltage multiplier cell (VMC) techniques can extend the voltage conversion ratio. A high step-up converter with a three-winding high-frequency coupled inductor and VMC with recycling the leakage inductor energy have been proposed in [19]-[22]. These converters have high step-up voltage gain and because of leakage inductor energy recycling, suitable rating switches can be selected. Thus the efficiency is increased.

In this paper, a new dc-dc converter with an ultra step-up voltage gain conversion and low semiconductor voltage stress is proposed using a switched-coupled-inductor and a voltage multiplier. Also using a hybrid switched-capacitor technique for providing a higher step-up voltage gain is presented. The circuit configuration of the proposed converter is shown in Fig. 1. The reminder of this paper is organized as follows. Section II presents operating principles of the proposed converter. Section III described the steady state analysis, calculating voltage gain and voltage stresses on switches. The extension of the proposed converter to attain more step-up voltage gain is presented in section IV.

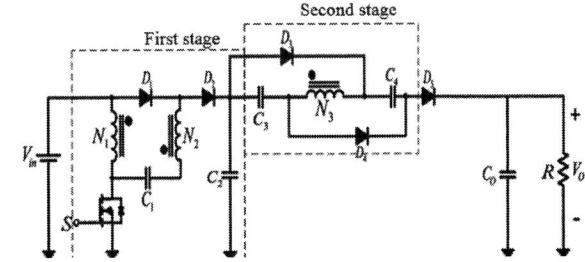

Fig. 1. Circuit configuration of the porposed converter

978-1-5090-0376-1/16 $31.00 © 2016 IEEE

II. OPERATING PRINCIPLE OF THE PORPOSED CONVERTER

The operation principles for the continuous conduction mode (CCM) are explained in this section. It is assumed that all devices are ideal, the converter is at steady state and inductor and capacitors are very large enough so that inductor current and capacitor voltages are constant during one switching cycle and the magnetizing inductance considered at primary side of coupled inductor. The converter consist of a DC input voltage V_{in}, a active switch S, a coupled inductor, five capacitors and five diodes. The coupled inductor can be modeled by an ideal transformer with primary winding, two secondary windings and a magnetizing inductance L_m. ' • ' notates for the windings of coupled inductor. The primary of coupled inductor has N_1 turns, its secondary windings have N_2 and N_3 turns. Turns ratio N_{21} and N_{31} of the coupled inductor are equal to $N_{21} = N_2/N_1$ and $N_{31} = N_3/N_1$, respectively. The converter is divided to two stages. The first stage of the converter is shown in Fig.1 included of two diodes D_1 and D_2, switched capacitor C_1, clamp capacitor C_2 and coupled inductor. The second stage of the converter uses a VMC; the VMC is made of two blocking capacitors (C_3, C_4) and two regenerative diodes (D_3, D_4). D_5 is output diode and C_O is the output capacitor.

The coupled inductor plays the role of energy storage and a transfer device. The magnetizing inductor L_m of the coupled inductor is equivalent to the input inductor of a conventional boost converter. Switching capacitor C_1 obtains energy from input source V_{in} and secondary winding N_2 and then releases it to clamp capacitor C_2 and VMC through output rectifier diode D_2. The typical waveforms of several keys components in the CCM operation are shown in Fig. 2. Circuit topology in each interval and the current paths are illustrated Fig. 3.

The four steady operating states are described as follows:

Mode 1 $[t_0 - t_1]$: At $t = t_0$ the active switch S is turned on. In this time interval, the diodes D_1, D_2 and D_5 are in turn-off state and the diodes D_3 and D_4 are in turn-on state. During this mode the energy is being stored in magnetizing inductor L_m from the DC input voltage and energy stored in coupled inductor by secondary winding N_3 is released into the capacitors C_3 and C_4. The current path shown in Fig. 3(a). This operating mode ends when I_{Lm} reaches to source current.

Mode 2 $[t_1 - t_2]$: At $t = t_1$, diode D_1 is being conducted. The diodes D_2 and D_5 are in off state. The diodes D_3 and D_4 are conducted and capacitors C_3 and C_4 are continually charging in the same manner. The energy stored in L_m from the DC input voltage is continually charging in the same manner, too. During this mode, switched capacitor C_1 receives energy from the input source and secondary winding N_2. The charging current from input source V_{in} flows to switched capacitor C_1 through diode D_1 in series with the secondary winding N_2 of coupled inductor. This operating mode ends when active switch S is turned off at $t = t_2$. Fig. 3(b) shows the current flow path of this mode operation.

Mode 3 $[t_2 - t_3]$: Active switch S and diodes D_1, D_2 and D_4 are turned off, but diodes D_2 and D_5 are conducted. During this mode, energy is being released through the series-connected path including clamp capacitor C_2, blocking capacitors C_3 and C_4, secondary winding N_3 and diode D_5 to charge capacitor C_O and load R. The energy of secondary winding N_3 is coupled from magnetizing inductor L_m at the primary side of the coupled inductor. This operating mode ends when I_{Lm} reaches to $(1/N_{31})I_3$. The current-flow path of this interval is illustrated in Fig. 3(c).

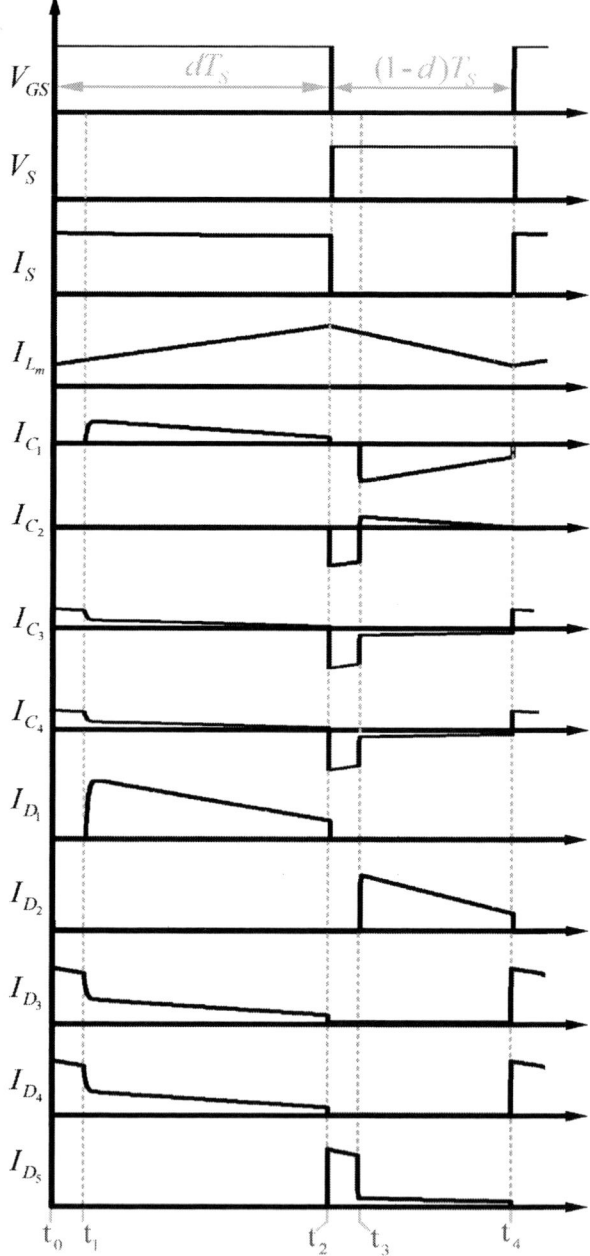

Fig. 2. Characteristic waveform of the porposed converter

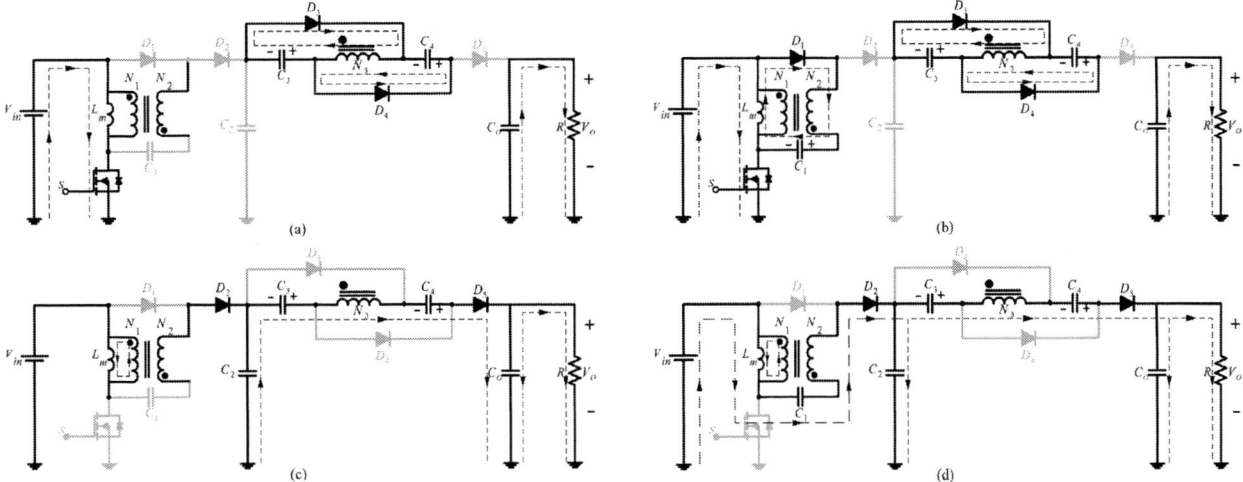

Fig. 3. Operating modes of the porposed converter. (a) Mo1de 1. (b) Mode 2. (c) Mode 3. (d) Mode 4.

Mode 4 $[t_3 - t_4]$: At $t = t_3$ the current through switched capacitor C_1 reverses its direction. The coupled inductor, DC input power source and capacitors C_1, C_3 and C_4 are in series to transfer the energy to the output capacitor C_O and load. Also the energy of clamp capacitor C_2 released. This operating mode ends when active switch is turned on at $t = t_4$ and beginning of the next switching period. The circuit topology and path of current are shown in Fig. 3(d).

III. STEADY STATE ANALYSIS OF THE PORPOSED CONVERTER

The steady state only takes the CCM operating into consideration, and for the simplicity of the analysis, leakage inductances of coupled inductor at the primary and secondary sides are neglected.

A. Step-Up Gain

When the active switch S is turned on, the voltage across the magnetizing inductor L_m can be denoted via (1) as

$$V_{N1} = V_{in} \qquad (1)$$

Moreover, the voltage across the secondary windings of coupled inductor V_{N2} and V_{N3} are N_{21} and N_{31} times of V_{Lm}, respectively. It can be represented via (2) and (3) as

$$V_{N2} = N_{21}V_{N1} \qquad (2)$$

$$V_{N3} = N_{31}V_{N3} \qquad (3)$$

Because the series voltages of V_{Lm} and V_{N2} charge switched capacitor C_1, the voltage across C_1 can be described via (4) as

$$V_{C1} = N_{21}V_{in} + V_{in} \qquad (5)$$

Also, the secondary winding N_3 is parallel with blocking capacitors C_3 and C_4, i.e.

$$V_{C3} = V_{C4} = N_{31}V_{in} \qquad (6)$$

Using volt-second balance for primary winding N_1 of the coupled inductor yields

$$< V_{N1} >_{T_S} = dT_S N_{N1}^{dT_S} + d'T_S V_{N1}^{d'T_S} = 0 \qquad (7)$$

$$V_{N1} = \frac{d}{1-d} V_{in} \qquad (8)$$

where $\langle V_{N1} \rangle_{T_S}$ denotes the average value of the primary winding voltage and d' denotes the $(1-d)$ in one cycle period.

When the active switch S is turned off, using KVL in Fig. 2(c) and Fig. 2(d) one can derive the voltage across clamp capacitor C_2 and output capacitor C_O as

$$V_{C2} = V_{in} + V_{N1} + N_{C1} + N_{21}V_{N1} \qquad (9)$$

$$V_O = V_{C2} + V_{C3} + N_{31}V_{N1} + V_{C4} \qquad (10)$$

Substituting V_{C1} and V_{N1} from (5) and (8) into (9) yields the voltage clamp capacitor C_2 as

$$V_{C2} = V_{in} \frac{2 + N_{21} - d}{1 - d} \qquad (11)$$

From (2)-(11), the voltage gain of the proposed converter in CCM operation is obtained as follows

$$M_{CCM} = \frac{V_O}{V_{in}} = \frac{2 + N_{21} + 2N_{31} - dN_{31} - d}{1 - d} \qquad (12)$$

The voltage transfer gain curve of the proposed converter by substituting $N_{21} = N_{31} = 2 - 5$ into (12) is depict in Fig. 4. Moreover Fig. 5 gives a comparison between voltage gains of the proposed converter and three other converters presented in [18],[21] and [22].

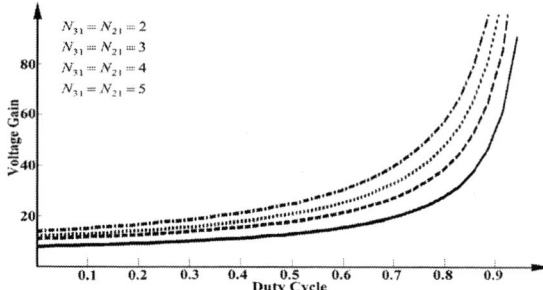

Fig. 4. Voltage gain of the porposed converter

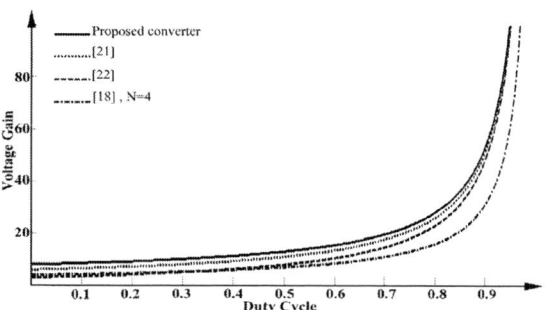

Fig. 5. Voltage gain of the porposed converter and other converters [18], [21] and [22]

B. Voltage Stress Analysis

The voltage stress of the active switch is given by

$$V_S = \frac{1}{2+N_{21}+2N_{31}-dN_{31}-d} V_O \qquad (13)$$

It can be concluded that seen that voltage stress of the active switch is determined by the turns ratio of the built-in transformer, duty cycle and the output voltage. As the turns ratio increases, the switch voltage stress decreases, which makes the low-voltage rated active switch with low resistance available to reduce the conduction losses.

The voltage stress of diodes D_1 and D_2 are as follows

$$V_{D1} = V_{D2} = \frac{N_{21}+1}{2+N_{21}+2N_{31}-dN_{31}-d} V_O \qquad (14)$$

When the active switch S is turn-off, the diodes D_3 and D_4 are reverse biased. Therefore, the voltage stresses of diodes D_3 and D_4 are as follows

$$V_{D3} = V_{D4} = V_{D5} = \frac{N_{31}}{2+N_{21}+2N_{31}-dN_{31}-d} V_O \qquad (15)$$

Equations (13)-(15) can be illustrated to determine the maximum voltage stress on each device. The voltage stress on each switch is shown in Fig. 6.

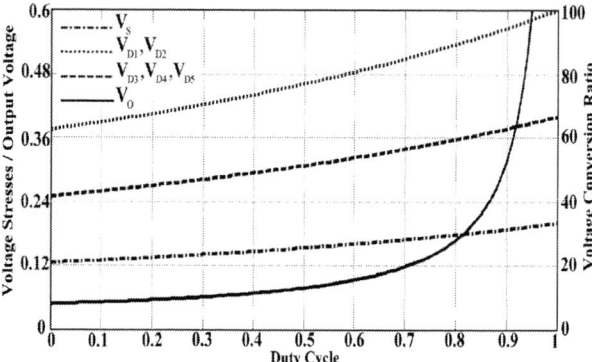

Fig. 6. Voltage stresses on avtive switch and diodes

IV. EXTENSION OF THE PROPOSED CONVERTER

In this section an extension of the proposed converter using a multiwindings coupled inductor, with each secondary winding inserted into a switched capacitor voltage multiplier circuit is added to the proposed converter, as shown in fig. 7. The output voltage of extension of this generalized extension topology is presented in (16)

$$V_O = \frac{(2+N_{21}-d)+\sum_{i=1}^{K} N_{3-i}(2-d)}{1-d} V_{in} \qquad (16)$$

Also, the voltage gain is obtained as (17), where K is the number of secondary windings N_3 of the coupled inductor, and each secondary winding N_3 of the coupled inductor has a similar turn ratio.

$$V_O = \frac{(2+N_{21}-d)+KN_{31}(2-d)}{1-d} V_{in} \qquad (17)$$

Fig. 7. Extension of the proposed converter using multiwinding coupled inductor and switched capacitor

V. SIMULATION RESULTS

In this section, the performance and validity of the proposed converter is verified by simulations. The specification of the proposed converter is shown in Table I. considering (11) and (12), the output voltage V_O and clamp capacitor voltage V_{C2} should be 85V and 155V, respectively. Simulation results are shown in Fig. 8 verify these values.

Fig. 9 shows the maximum reverse voltages across the diodes as $V_{D1} = V_{D2} = 75$ and $V_{D3} = V_{D4} = V_{D5} = 50$ which are equal to their expected values from (14) and (15). In regard to (13) the voltage across active switch should be 25V that it can be seen in Fig. 10. Current of magnetizing inductance of the coupled inductor is shown in Fig. 11. According to Fig. 11, operating mode of the converter should be CCM mode.

TABLE I. SPECIFICATION OF THE PROPOSED CONVERTER

Input voltage, V_{in}	10 V
Output voltage, V_O	155 V
Rated output power, P_O	240 W
Switching frequency, f_s	50 KHz
Load, R	100 Ω
Duty cycle, d	0.6
Coupled inductor turns ratio	1:2:2
Magnetizing inductor, L_m	23μH
C_1, C_3, C_4	48μF
C_2	30μF
C_O	20μF

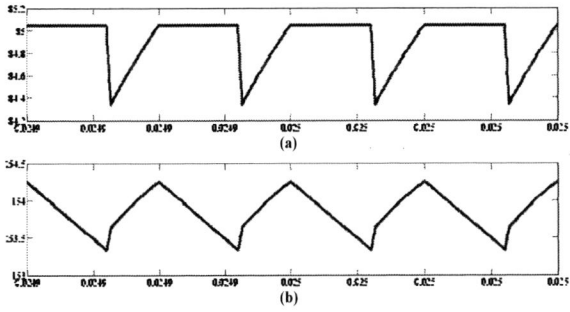

Fig. 8. (a) capacitor clamp voltage, (b) output voltage

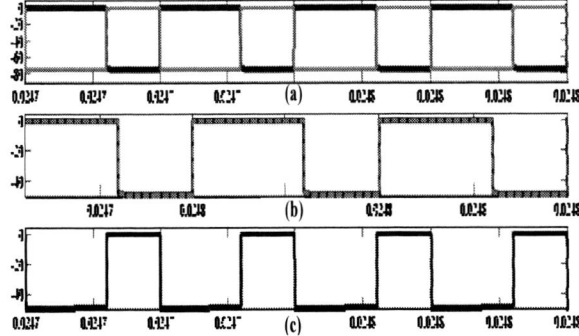

Fig. 9. The voltage across diodes, (a) V_{D1} and V_{D2}, (b) V_{D3} and V_{D4}, (c) V_{D5}

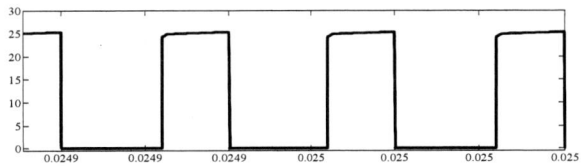

Fig. 10. The voltage across active switch, V_s

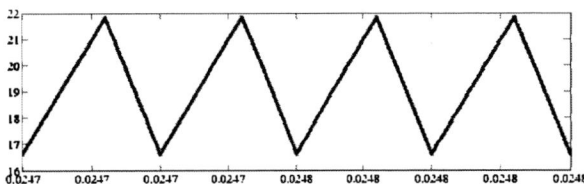

Fig. 11. The current of magnetizing inductor, I_{Lm}

VI. CONCLUSION

This paper has developed an ultra step-up converter based on three windings coupled inductor and VMC. Principles and steady state analysis of the proposed converter in CCM are presented in order to calculate voltage gain and switches stress. The voltage gain of the proposed converter is compared with other similar converters. The voltage gain is increased without using an extreme duty cycle. The voltage across active switch and diodes are decreased by turns ratio of coupled inductor increase, thus lower rating switches can be selected. Recycling the energy of the leakage inductance of the coupled inductor and passing it directly to the load, causes voltage gain increase. Extended proposed converter and its analysis are also mentioned. Computer simulation results validate the theoretical analysis of the proposed converter.

REFERENCES

[1] Kou-Ching Tseng, Jang-Ting Lin and Chi-Chih Hung "High step-up converter with three-winding coupled inductor for fuel cell energy source application," *IEEE Trans. Power Electron.*, vol. 30, pp. 574-581, Feb. 2015.

[2] Antonio Alisson Alencar Freitas, Fernando Lessa Tofoli, Edilson Mineiro Sa Junior, Sergio Daher and Fernando Luiz Marcelo Antunes "High-voltage gain dc-dc boost converter with coupled inductors for photovoltaic systems," Power Electron., IET, vol.8, no. 10,pp.1885-1892, Apr. 2015.

[3] Amir Parastar and Jul-Ki Seok "High-gain resonant switched capacitor cell-based dc/dc converter for offshore wind energy systems," IEEE Trans. Power Electron., in press.

[4] Yu Tang, Dongjin Fu, Ting Wang, and Zhiwei Xu, "Hybrid switched-inductor converters for high step-up conversion," IEEE Trans. Ind. Electron., in press.

[5] Boris Axelrod, Yefim Berkovich, and Adrian Ioinovici, "Switched-capacitor/switched-inductor structures for getting transformerlees hybrid dc-dc pwm converters" IEEE Trans. Circuit and Systems, vol.55, no.2, pp. 687-696, Mar. 2008.

[6] Yu Tang, Ting Wang, and Dongjin Fu "Multicell switched-inductor/switched-capacitor combined Active-network converters," IEEE Trans. Power Electtron, vol. 30, no. 4, Apr. 2015.

978-1-5090-0376-1/16 $31.00 © 2016 IEEE

[7] Ka Wai Eric Cheng and Yuan-mao Ye " Duality approach to the study of switched-inductor power converters and its higher-order variations," Power Electron., IET, vol. 8, no. 4, pp. 489-496, Apr. 2015.

[8] F. L. Luo, "Six self –lift DC-DC converters, voltage lift technique," IEEE Trans. Ind. Electron., vol. 48, no. 6, pp. 1268-1272, Dec. 2001.

[9] F. L. Luo, and H.Ye, "Positive output super-lift converters," IEEE Trans. Ind. Electron., vol. 18, no. 1, pp. 105-113, Jan. 2003.

[10] Mustafa A. Al-Saffar, Esam H.Ismail, Ahmad J. Sabzali, and Abbas A.Fardoun, " An improved topology of sepic converter with reduced output voltage ripple." IEEE Trans. Power Electtron, vol. 23, no. 5, pp. 2377-2386, Sep. 2008.

[11] Yefim Berkovich, Boris Axelrod, Rotem Madar, Avraham Twina "Improved luo converter modifications with increasing voltage ratio," Power Electron., IET, vol.8, no. 2, pp.202-212, Apr. 2015.

[12] Xiaotian Zhang, Timothy C. Green, "The modular multilevel converter for high step-up ratio dc-dc conversion," IEEE Trans. Ind. Electron., in press.

[13] Gang Wu, Xinbo Ruan, Zhihong Ye, "Non-isolated high step-up dc-dc converters adopting switched-capacitor cell," IEEE Trans. Ind. Electron., in press.

[14] Abbas A. Fardoun, and Esam H. Ismail, "Ultra step-up dc-dc converter with reduced switch stress" IEEE Trans. Ind. Appl., vol. 46, no. 5, pp. 105-113, Oct. 2010.

[15] X. Hu, and C. Gong, "A high voltage gain DC-DC converter integrating coupled-inductor and diode capacitor techniques," IEEE Trans. Power Electtron, vol. 29, no. 2, pp. 789-800, Feb. 2014.

[16] W. Li, C. Xu, H. Yu, Y.Gu, and X. He, "Analysis, design and implementation of isolated bidirectional converter with winding-cross-coupled inductors for high step-up and high step-down conversion system," Power Electron., IET, vol.7, no. 1, pp.67-77, Jan. 2014.

[17] J. Zhang, X. Huang, X. Wu, and Z. Qian, "A high efficiency flyback converter with new active clamp technique," IEEE Trans. Power Electtron, vol. 25, no. 7, pp. 1775-1785, Jul. 2010.

[18] Shih-Ming Chen, Man-Long Lao, Yi-hsun Hsieh, "A novel switched-coupled-inductor DC-DC step-up converter and its derivatives" IEEE Trans. Ind. Appl., vol. 51, no. 1, pp. 309-313, Feb. 2015.

[19] Tohid Nouri, Seyed Hossein Hosseini, Ebrahim Babei, Jaber Ebrahimi, "Interleaved high step-up DC-DC converter based on three-winding high-frequency coupled inductor and coltage multiplier cell" Power Electron., IET, vol.8, no. 2, pp.175-189, Apr. 2015

[20] M.Khalilzadeh, M.Mahdipour, K. Abbaszadeh, "High Step-up DC-DC Converter Based on Three-Winding Coupled Inductor" The 6th International Power Electronics Drive Systems and Technologies Conference (PEDSTC), pp. 195-200, February 2015

[21] Rong-Jong Wai, Chung-You Lin, Rou-Yong Duan, and Yung-Ruei Chang "High-Efficiency DC-DC ConverterWith High Voltage Gain and Reduced Switch Stress", IEEE Trans. Ind Electtron, vol. 54, no. 1, pp. 354-364, Feb. 2007

[22] Shih-Kuen Changchien, Tsorng-Juu Liang, Jiann-Fuh Chen, and Lung-Sheng Yang "Novel High Step-Up DC–DC Converter for Fuel Cell Energy Conversion System" IEEE Trans. Ind Electtron, vol. 57, no. 6, pp. 2007-2017, Jun. 2010.

7th Power Electronics, Drive Systems & Technologies Conference (PEDSTC 2016)
16-18 Feb. 2016, Iran University of Science and Technology, Tehran, Iran

A New Switched-Inductor Quasi-Z-Source Inverter

Masoud Ghodsi

Department of Electrical Engineering
University of Sistan and Baluchestan
Zahedan, Iran
ghodsi.masoud@pgs.usb.ac.ir

S. Masoud Barakati

Department of Electrical Engineering
University of Sistan and Baluchestan
Zahedan, Iran
smbaraka@ece.usb.ac.ir

Abstract— Z-source inverter (ZSI) is an attractive power converter both in industry and academic area. In this paper, a new switched-inductor network, called extended switched-inductor quasi-Z-source inverter (ESL-qZSI), is proposed for Z-source inverters. From topology point of view, the presented network has an additional inductor besides three diodes, in comparison with the switched-inductor quasi Z-source inverter (SL-qZSI). The suggested inverter has a dc source ground point, continuous input current, and no startup inrush current. The capacitor voltage stress of the proposed voltage inverter with the same input and output voltages is less than SL-qZSI and SL-ZSI which provides lower the dc-link voltage in comparison with SL-qZSI. For the same input source and modulation index, the proposed inverter has a higher voltage gain with respect to the SL-qZSI.

Keywords— Z-source inverter (ZSI), switched-inductor quasi-Z-source inverter (SL-qZSI), shoot-through mode, maximum boost control.

I. INTRODUCTION

Bidirectional energy conversion between three-phase ac and dc sources is achieved by voltage source inverter (VSI) or current source inverter (CSI), conventionally. Those cannot perform as a buck-boost inverter. In VSI the output voltage is limited to below input voltage and in CSI the output voltage is greater than the input voltage. In order to deal with this drawback an additional boost or buck dc-dc converter is needed. However, using two-stage power conversion will result in high cost and low efficiency [1], [2]. The Z-Source inverter (ZSI) can provide buck-boost ability as compared to VSI and CSI. In VSI, shoot-through problem destroy the switching devices, so dead time method employs to prevent from short-circuit the upper and lower devices of each phase leg; but it will cause waveform distortion. In ZSI the upper and lower devices in a leg can be turned on simultaneously, which eliminates the dead time and improves the reliability and will lead to reducing of the output waveform distortion [3]. Various ZSI topologies have been presented in many papers recently. Some papers are focused on type of control and modulation that are suitable for ZSI [4]-[10], application of ZSI, such as motor drive, uninterrupted power supply, renewable energy [11]-[17], and other papers have focused on improving different topologies [18]-[30]. The traditional Z-source inverter topology has the following drawbacks: a. the current drawn from the source is discontinuous, b. cannot suppress the inrush

current and resonance between the capacitors and inductors at startup, c. high-voltage capacitors, which may increase the volume and cost, d. does not share the ground point of dc source with the inverter. In order to solve the aforementioned problems, quasi-Z-source inverters (qZSIs) were proposed [4]. Recent papers used a very high modulation index in order to improve the main output power quality and overcome the boost limitations and increase the boost factor. These topologies are suitable for applications such as fuel cell and solar cell. These topologies introduce an impendence network including the inductors, capacitors, diodes, and transformer to produce a high output ac voltage for the main power circuit from a very low input dc voltage.

The ZSIs transformer-based, such as T-source [27], TZ-source [28], and Trans-Z-source [29] inverters extend the boost ability without adding extra components, only by increasing the transformer turn ratio. In diode/capacitors-assisted quasi-ZSIs [23] and an inductor-capacitor-capacitor-transformer ZSI (LCCT-ZSI) [12] which each one of them add diode, inductor, and capacitor in order to overcome the boost limitations. Applying switched-capacitors (SC), switched-inductors (SL), hybrid switched-capacitor/switched inductors structures, of the high boost gain dc-dc converters are introduced to combine with the traditional ZSI and qZSI for providing a strong step-up inversion [32]-[35].

In this paper, by adding three diodes and one inductor to the switched-inductor quasi ZSI topology, a new type of inverter, called the extended switched inductor quasi-Z-source inverter (ESL-qZSI), is introduced. The proposed inverter possesses higher boost voltage inversion ability than ZSI, SL-qZSI and reduces the voltage and current stress compared to SL-ZSI for the same input voltage and modulation index. The operating principle and simulation results of the proposed inverter are compared with SL-ZSI and SL-qZSI .

The rest of the paper is organized as follows. In section II, the original ZSI, qZSI and switched inductor Z-source inverter topologies are explained. In Section III, the topology, operation principle, PWM control method of the proposed network will be discussed in detail. Section IV is dedicated to present the simulation results. Finally, the conclusions of this paper is discussed and presented in section V.

978-1-5090-0376-1/16 $31.00 © 2016 IEEE

II. REVIEW OF Z-SOURCE INVERTER TOPOLOGIES

The original topology of ZSI shown in Fig. 1. Due to use two inductors and two capacitors in an "X" configuration, both power switches of a phase-leg can turn on simultaneously without damaging the inverter. This additional shoot-through mode is added to the switching modes in order to boost voltage.

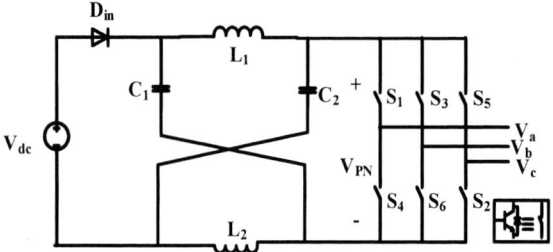

Fig. 1.   The original Z-source inverter [3].

The original ZSI does not share the ground point of dc source with the main circuit inverter and the current drawn from the source is discontinuous so capacitor bank at the front end is used to protect the energy source and avoid current discontinuity. This problem limits the application of ZSI. To solve these problems, quasi ZSI (qZSI) inverter was offered [5], as depicted in Fig. 2. In qZSI the voltage stress on the component is much lower in comparison with that of the conventional ZSI.

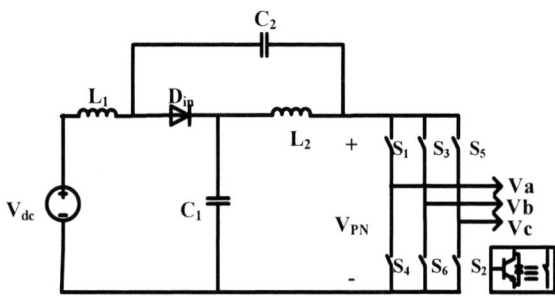

Fig. 2.   Quasi Z-source inverter [4].

The conversion relation between dc-link voltage across the inverter bridge, V_{PN} and the input source voltage V_{dc} can be stated as follow:

$$B = \frac{V_{PN}}{V_{dc}} = \frac{1}{1-2(T_0/T)} = \frac{1}{1-2D}, \qquad (1)$$

where T_0 is the interval of the shoot-through mode during the switching period T and $D=T_0/T$ is the duty cycle of the shoot-through for each cycle and B is the boost factor. Equation (1), shows that is limited to the range from value zero to maximum value 0.5. Value of high D will result in high THD value and low modulation index value.

Enhancement techniques such as switched inductor (SL), switched capacitor (SC) are used in dc-dc converter [35]. By usage of the SL concept in the conventional Z-source inverter will experience higher output power quality, high voltage conversion ratios and lower D.

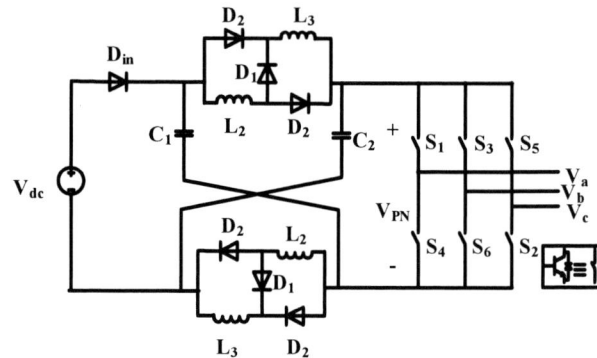

Fig. 3.   Switched inductor Z-source inverter [31].

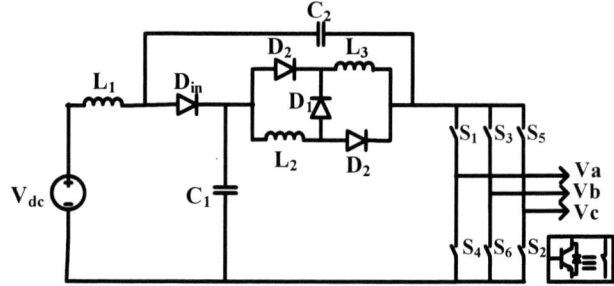

Fig. 4.   Switched inductor quasi Z-source inverter [32].

SL-ZSI topology is displayed in Fig. 3. The boost factor of this inverter is increased to

$$B = \frac{1+T_0/T}{1-3(T_0/T)} = \frac{1+D}{1-3D}. \qquad (2)$$

Nevertheless, SL-ZSI has a several defects, it does not share dc ground point between the source and inverter, the current drawn from the source is also discontinuous, and it cannot suppress the inrush current introduced at the startup that can lead to destruction of the devices. In order to deal with the aforesaid problems, SL-qZSI is proposed. Fig. 4 shows SL-qZSI. Boost factor of this inverter is increased to

$$B = \frac{1+D}{1-2D-D^2}. \qquad (3)$$

III. PROPOSED TOPOLOGY

The proposed inverter, has higher boost voltage inversion ability than ZSI, SL-qZSI, causes reduction of the voltage stress on the capacitors and lower current through the inductors in comparison with SL-ZSI and SL-qZSI with the same voltage gain. The idea multicell switched-inductor used in Z-source inverter conventional [32] but in this paper it used in quasi Z-source topology.

978-1-5090-0376-1/16 $31.00 © 2016 IEEE   178

A. Characteristics of Circuit

Fig. 5 shows the proposed extended switched inductor quasi Z-source inverter (ESL-qZSI). It consists of four inductors, two capacitors, and seven diodes. It is created by adding three diodes and one inductor to the SL-qZSI.

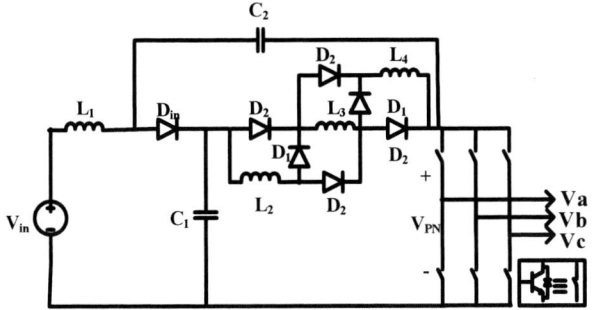

Fig. 5. Proposed switched inductor quasi Z-source inverter.

The main characteristics of the proposed topology are as follows:

- Unlike the SL-ZSI topology, no inrush current flows to the main circuit at startup,

- The input dc current is continuous,

- Lower peak shoot-through current compared to the SL-ZSI topology,

- Lower capacitor voltage stress compared to SL-ZSI and SL-qZSI,

- It does share a dc ground point between the input dc source and main circuit inverter which will cause safety and will have lower EMI,

- Using more passive components than SL-qZSI is main drawback of the proposed topology.

B. Circuit Operation

Switching modes of the main circuit and the operating principle of the proposed inverter are similar to conventional ZSI. Like the ZSI, the proposed ESL-qZSI has an extra shoot-through zero states and the six active states and two conventional zero states. The shoot-through zero mode can be realized by short-circuiting both the upper and lower switching devices of each of the phase legs, any two phase legs, or all three phase legs. For simplicity the proposed circuit has two operating modes; shoot-through mode and non-shoot-through mode.

Fig. 6(a) shows the equivalent circuit in the non-shoot-through mode, in this mode the proposed inverter has six active states and two zero states of the inverter main circuit. Inductors L_2, L_3, L_4 connected in series because diodes D_1 and D_{in} are turned on while diodes D_2 are turned off.

During this term, both inductors and dc source energies are transferring to the inverter bridge and the capacitors are charging, so one can obtain

$$V_{L1} = V_{C1} - V_{dc}, \qquad (4)$$

$$V_{L2non} = -V_{L3non} - V_{L4non} + V_{C2}, \qquad (5)$$

$$V_{L3non} = -V_{L2non} - V_{L4non} + V_{C2}, \qquad (6)$$

$$V_{L4non} = -V_{L2non} - V_{L3non} + V_{C2}, \qquad (7)$$

$$V_{PN} = V_{C1} + V_{C2}. \qquad (8)$$

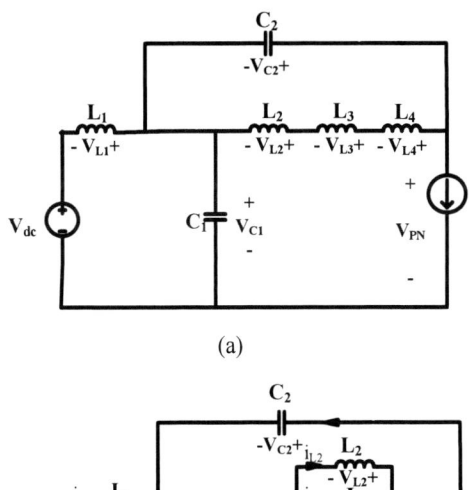

(a)

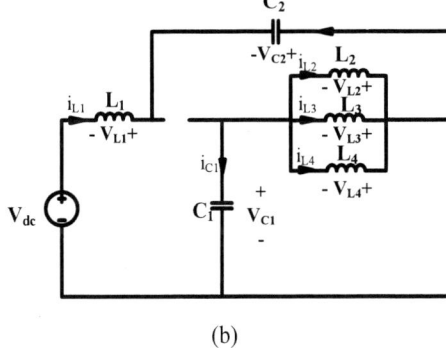

(b)

Fig. 6. Operating modes: (a) non-shoot-through and (b) shoot-through mode.

Fig. 6 (b) shows the equivalent circuit in the shoot-through operation mode. In this mode, the inverter side is shorted by both the upper and lower switching devices of any phase leg. The inductors L_2, L_3, L_4 are connected in parallel because diodes D_1 and D_{in} are turned off while diodes D_2 are turned on. During this term, capacitors C_1 and C_2 are discharged, whereas inductors L_1, L_2, L_3 and L_4 are charged. The relations for this term can be obtained as follow

$$V_{L1} = -V_{dc} - V_{C2}, \qquad (9)$$

$$V_{L2} = V_{L3} = V_{L4} = -V_{C1}, \qquad (10)$$

$$I_{C1} = -I_{L1} - I_{L2} - I_{L3}, \qquad (11)$$

$$I_{C2} = -I_{L1}. \qquad (12)$$

978-1-5090-0376-1/16 $31.00 © 2016 IEEE

By applying the volt-second balance principle to the inductors L_2, L_3, L_4 from (4)-(7) to (9), (10) one can deduce

$$D(-V_{C1}) + (1-D)(V_{L2non}) = 0, \qquad (13)$$

hence,

$$V_{L2non} = V_{L3non} = V_{L4non} = \frac{DV_{C1}}{1-D}, \qquad (14)$$

substituting (14) into (5), it can be expressed as follow

$$V_{C2} = V_{L3non} + V_{L2non} + V_{L1non} = \frac{3D}{1-D}V_{C1}. \qquad (15)$$

By applying the volt-second balance principle to L_1 from (9), (4) and (15), relations can be stated as follow

$$D(-V_{dc} - V_{C2}) + (1-D)(-V_{C1} - V_{dc}) = 0, \qquad (16)$$

$$V_{C1} = \frac{1-D}{1-2D-2D^2}V_{dc}; \quad V_{C2} = \frac{3D}{1-2D-2D^2}V_{dc}. \qquad (17)$$

The peak dc-link voltage crosses the inverter main circuit is expressed in (8) and can be rewritten as

$$V_{PN} = \frac{1+2D}{1-2D-2D^2}V_{dc}, \qquad (18)$$

the boost factor of the proposed ESL-qZSI is defined by

$$B = \frac{1+2D}{1-2D-2D^2}. \qquad (19)$$

Fig. 7 shows the curves of the boost factor B against the duty cycle for topologies SL-ZSI, SL-qZSI and ESL-qZSI. The boost ability of the ESL-qZSI is higher than SL-qZSI and lower than of the SL-ZSI for the same D. The proposed inverter in comparison with SL-qZSI at the same B uses a smaller shoot-through duty cycle.

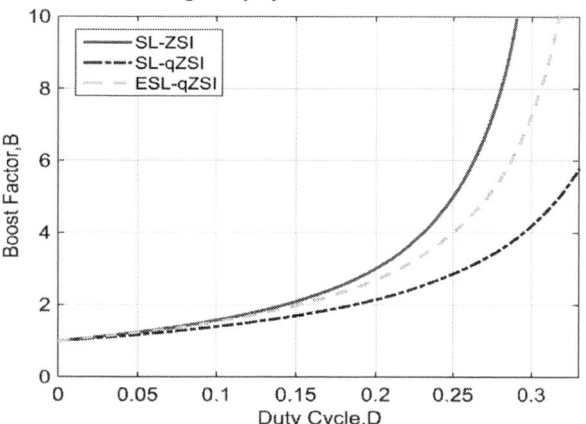

Fig. 7. Comparison of the boost ability of ESL-qZSI; SL-ZSI [31]; SL-qZSI [32] topologies.

C. PWM Switching Strategy for the Proposed Inverter

To control the proposed ESL-qZSI, all basic PWM switching strategies can be employed. These strategies are presented in detail in[5]-[7]. In this paper, the maximum boost control strategy is applied to control the proposed inverter.

The maximum boost control maintains six active states unchanged and turns all zero states into shoot-through states. The duty cycle of the shoot-through mode, varies each cycle and it results in ripple in the current through the inductor and voltage across the capacitor.

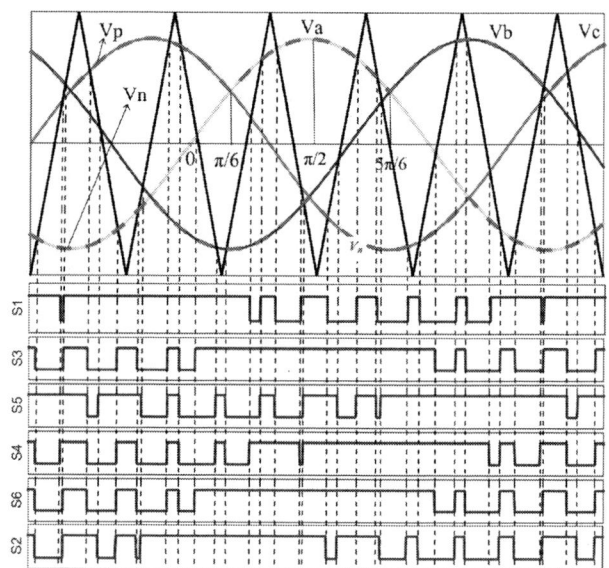

Fig. 8. PWM Signal from Maximum Boost Control

As illustrated in Fig. 8, envelops V_P, V_n are employed to realize the shoot through duty ratio (D_0). V_P is equal upper envelop of the three-phase sinusoidal reference voltages while V_n is the negative of the V_P. When the triangular carrier waveforms are greater than the upper envelope V_P, or lower than the bottom envelop V_n, the circuit turns into shoot-through mode, otherwise it operates just as traditional carrier based PWM. The pulse generation of the three phase leg switches are shown in Fig. 8. By turning all zero state into shoot-through states, the Z-source inverter achieves the maximum boost, minimizes the voltage stress, and make the duty ratio as large as possible, however, the shoot-through duty cycle repeats every $\pi/3$ and the voltage boost is related to the shoot-through duty ratio, so ripple in shoot-through duty ratio will result in ripple in the current through the inductor and voltage across the capacitor. As discussed in [6], the current ripple through the inductor becomes

$$\Delta I_L = \frac{V_{pk2pk}}{6\omega L} = \frac{(2\sqrt{3}-3)M\pi V_{dc}}{24(3\sqrt{3}M - \pi)\omega L}, \qquad (20)$$

V_{pk2pk} is voltage ripple across the inductor and $L_1 = L_2 = L$. Therefore the inductor has to be large for low output frequency in order to limit the current ripple. In this method, maximum D which is as mentioned in [5] is limited to

$$D = \frac{T_0}{T} = \frac{2\pi - 3\sqrt{3}M}{2\pi}, \qquad (21)$$

M, is the modulation index. Substituting (21) into (19) where the boost factor is

$$B = \frac{6\pi^2 - 6\sqrt{3}\pi M}{-6\pi^2 + 18\sqrt{3}\pi M - 27M^2}, \qquad (22)$$

the peak value of the phase voltage form the inverter output is

$$\hat{V}_{ph} = MB\frac{V_{dc}}{2}, \qquad (23)$$

the voltage gain is defined by

$$G = MB = \frac{6\pi^2 M - 6\sqrt{3}M^2\pi}{-6\pi^2 + 18\sqrt{3}\pi M - 27M^2}. \qquad (24)$$

Fig. 9 indicates the voltage conversion ratios versus the modulation index under maximum boost control. Using the same modulation index, the proposed ESL-qZSI provides higher voltage boost inversion than SL-qZSI. Therefore for the same voltage conversion ratio, the proposed inverter used a higher modulation index to improve the inverter output quality.

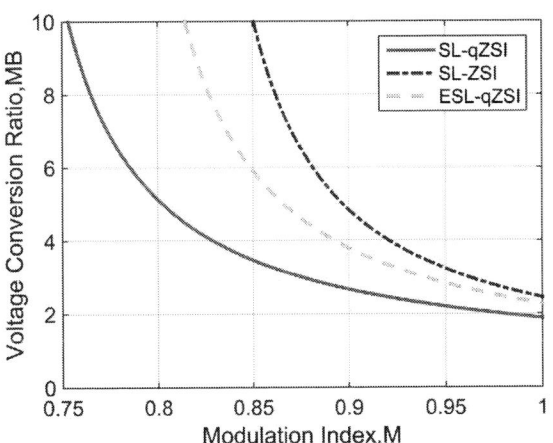

Fig. 9. Voltage conversion ratios versus the modulation index for ESL-qZSI; SL-ZSI [31]; SL-qZSI [32] topologies under the maximum boost control.

D. Comparison of Current and Voltage Stress

In the power inverters, the current and voltage stresses varied under different control and load conditions. Thus, for simulation the maximum boost control method is used. Fig. 10 indicates the simplified equivalent circuit of the impedance-type power inverters SL-ZSI, SL-qZSI, and ESL-qZSI. An inductive load impedance (Z_l) is connected in parallel with switch S. i_l and v_l are the instantaneous load current and voltage, I_l and V_l are the average current and voltage, respectively, during a switching cycle in the steady state.

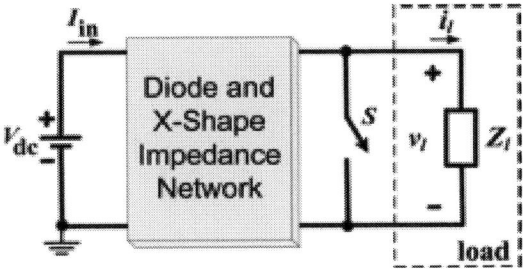

Fig. 10. Simplified equivalent circuit of the power inverter: ESL-qZSI, SL-qZSI, SL-ZSI.

By applying the steady state analysis method proposed in [10] for the equivalent circuit (Fig. 10), the voltage and current stresses of the main components are obtained and tabulated in Table I.

Table I compares the governing equations of SL-ZSI, SL-qZSI, and ESL-qZSI for the same D and V_{dc} using maximum boost control method. For the same D and V_{dc} the voltage stresses and current stresses of the proposed topology are higher than those of SL-qZSI, because ESL-qZSI boost factor is greater than SL-qZSI and has voltage gain more than SL-qZSI. I_{sh} is the peak shoot-through current across the main power circuit.

IV. SIMULATION

The simulation results of the ESL-qZSI with SL-ZSI and SL-qZSI are compared. The simulation parameters $L_1 = L_2 = L_3 = L_4 = 1mH$, $C_1 = C_2 = 1000$ μF, $L_f = 1$ mH, $C_f = 22$ μF, diode drop voltage Vf = 1.5 V and R = 30 Ω/phase are chosen. The switching frequency was 5 kHz; the input voltage was 20 Vdc. The maximum boost control method is used. According to Fig. 9, in gain G=3, M = 0.962 for the SL-ZSI, M = 0.876 for the SL-qZSI, and M = 0.939 for ESL-qZSI. Table II indicates theoretical results for three inverters in the same input dc voltage and phase output voltage. Figs. 11, 12 and 13 show the simulation results for the SL-ZSI, SL-qZSI and ESL-qZSI when M = 0.962, M = 0.876 and M = 0.939, respectively. From Fig. 11 we can see capacitor voltage (V_{C1}, V_{C2}) is boosted to 34 V; the dc-link voltage (V_{pn}) is boosted to 50 V and peak shoot-through current (I_{sh}) is 11.1 A, in the steady state. From Fig. 12 we can see capacitors voltage C_1 and C_2 are boosted to 38 V and 25 V respectively; the dc-link voltage (V_{pn}) is boosted to 63 V and peak shoot-through current (I_{sh}) is 11 A, in the steady state. From Fig. 13 we can see capacitors voltage C_1 and C_2 are boosted to 32 V and 20 V respectively; the dc-link voltage (V_{pn}) is boosted to 54 V and peak shoot-through current (I_{sh}) is 12 A, in the steady state. In SL-ZSI, the huge inrush current flows through Din, C_1, C_2 at startup but in SL-qZSI and ESL-qZSI no current flows to the main circuit at startup.

TABLE I. CURRENT AND VOLTAGE STRESSES IN THE SAME D AND V_{DC} CASE

	ESL-qZSI	SL-qZSI [32]	SL-ZSI [31]
$\hat{v}_{ph}$	$\dfrac{3\pi^2 M - 3\sqrt{3}M^2\pi}{-6\pi^2 + 18\sqrt{3}\pi M - 27M^2}V_{dc}$	$\dfrac{4\pi^2 M - 3\sqrt{3}\pi M^2}{-8\pi^2 + 24\sqrt{3}\pi M - 27M^2}V_{dc}$	$\dfrac{4\pi M - 3\sqrt{3}M^2}{18\sqrt{3}M - 8\pi}V_{dc}$
V_C	$V_{C1} = \dfrac{1-D}{1-2D-2D^2}V_{dc}$ $V_{C2} = \dfrac{3D}{1-2D-2D^2}V_{dc}$	$V_{C1} = \dfrac{1-D}{1-2D-D^2}V_{dc}$ $V_{C2} = \dfrac{2D}{1-2D-D^2}V_{dc}$	$\dfrac{1-D}{1-3D}V_{dc}$
V_{PN}	$\dfrac{1+2D}{1-2D-2D^2}V_{dc}$	$\dfrac{1+D}{1-2D-D^2}V_{dc}$	$\dfrac{1+D}{1-3D}V_{dc}$
I_L	$I_{L1} = \dfrac{(1+2D)(1-D)^2}{1-2D-2D^2}\dfrac{V_{PN}}{R}$ $I_{L2,3,4} = \dfrac{(1-D)^2}{1-2D-2D^2}\dfrac{V_{PN}}{R}$	$I_{L1} = \dfrac{(1+D)(1-D)^2}{1-2D-D^2}\dfrac{V_{PN}}{R}$ $I_{L2,3} = \dfrac{(1-D)^2}{1-2D-D^2}\dfrac{V_{PN}}{R}$	$\dfrac{(1-D)^2}{1-3D}\dfrac{V_{PN}}{R}$
I_{Din}	$I_{L2} + I_{L1} - I_l$	$I_{L2} + I_{L1} - I_l$	$2I_L - I_l$
I_{sh}	$I_{L1} + 3I_{L2}$	$I_{L1} + 2I_{L2}$	$4I_L$
I_l	$(1-D)V_{pn}/R_L$	$(1-D)V_{pn}/R_L$	$(1-D)V_{pn}/R_L$

TABLE II. THEORETICAL RESULTS FOR THE SAME GAIN G=3

	ESL-qZSI	SL-qZSI [32]	SL-ZSI [31]
M	0.939	0.876	0.962
D	0.223	0.275	0.204
B	3.18	3.40	3.10
V_{PN}	63.6	68	62
V_{C1}	34.18	38.73	41.03
V_{C2}	29.43	29.38	41.03
V_{ph}	29.86	29.78	29.82

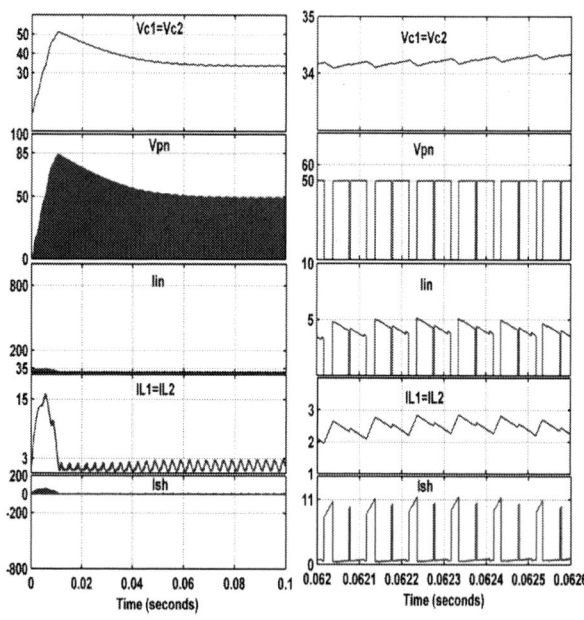

Fig. 11. SL-ZSI with M=0.962. From top to bottom : capacitors voltage: V_{c1}, V_{c2}; dc-link voltage: V_{pn}; input current: I_{in}, inductor current: I_{L1}, I_{L2} and shoot-through current: I_{sh}.

As shown in Figs. 11, 12 and 13 due to diode in series with the dc voltage source, input current in SL-ZSI is discontinuous while in SL-qZSI and ESL-qZSI it is continuous. For producing the same gain voltage G = 3, proposed ESL-qZSI has lower capacitor voltage stresses compared with the SL-ZSI and SL-qZSI. As the proposed inverter uses a modulation index higher than SL-qZSI, the dc-link voltage (V_{pn}) in proposed inverter is lower than SL-qZSI.

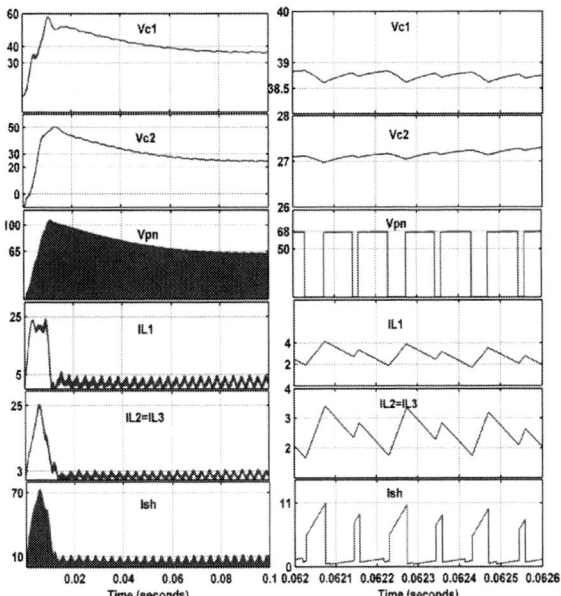

Fig. 12. SL-qZSI with M=0.876. From top to bottom : capacitors voltage: V_{c1},V_{c2}; dc-link voltage: Vpn; inductors current: I_{L1}, I_{L2}, I_{L3}, I_{L4} and shoot-through current: I_{sh}.

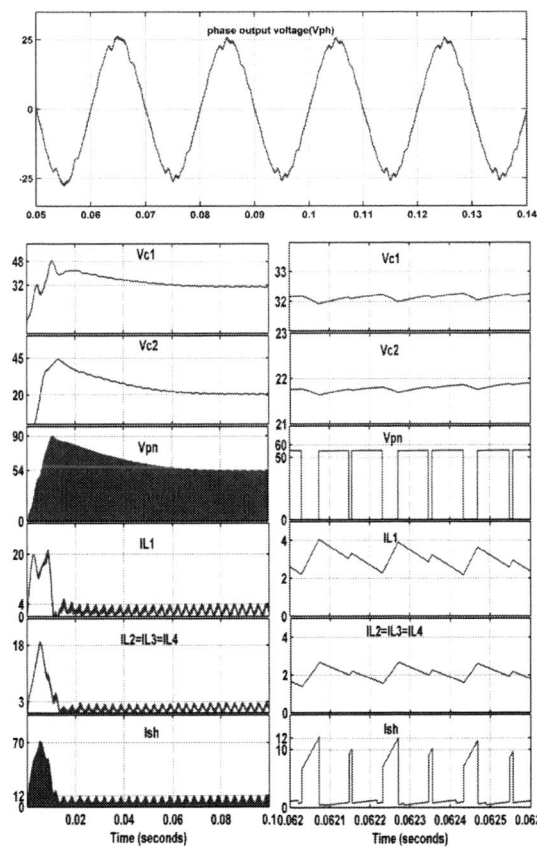

Fig. 13. ESL-qZSI with M=0.939. From top to bottom: From top to bottom : phase output voltage: V_{ph}; capacitors voltage: V_{c1},V_{c2}; dc-link voltage: V_{pn}; inductors current: I_{L1}, I_{L2}, I_{L3}, I_{L4} and shoot-through current: Ish.

V. CONCLUSIONS

This paper presented a new impedance network for switched inductor Z-source inverter called the ESL-qZSI. The proposed inverter has one inductor and three diodes more than switched inductor quasi- Z-source inverter (SL-qZSI). The recommended inverter has following main characteristics:

• Continuous input current,

• Sharing the dc source ground point,

• Suppressing the startup inrush current,

• High boost voltage inversion ability.

Moreover, for the same voltage gain the proposed inverter has dc-link voltage (V_{pn}) lower than SL-qZSI and reduced voltage stress on capacitors compared with SL-ZSI and SL-qZSI. inrush current lower than SL-ZSI. For same modulation index and same input dc source the proposed ESL-qZSI provides higher voltage boost inversion (MB) than SL-qZSI.The offered ESL-qZSI is applicable for PV and fuel cell applications, where a high ac output voltage are need. The presented inverter can be applied to various power conversion applications.

REFERENCES

[1] T. Kerekes, R. Teodorescu, M. Liserre, C. Klumpner, and M. Sumner, "Evaluation of three-phase transformerless photovoltaic inverter topologies," IEEE Trans. Power Electron., vol. 24, no. 9, pp. 2202–2211, Sep. 2009.

[2] F. Gao, P. C. Loh, R. Teodorescu, and F. Blaabjerg, "Diode-assisted buck- boost voltage-source inverters," IEEE Trans. Power Electron., vol. 24, no. 9, pp. 2057–2064, Sep. 2009.

[3] F. Z. Peng, "Z-Source Inverter," IEEE Trans. Ind. Applicat., vol. 39, no. 2, pp. 504-510, Mar./Apr. 2003.

[4] J. Anderson and F. Z. Peng, "Four Quasi-Z-Source Inverters," in Proc. PESC 2008, pp. 2743-2749, June 2008.

[5] F. Z. Peng, M. Shen and Z. Qian, "Maximum Boost Control of the Z-Source Inverter," IEEE Trans. Power Electron., vol. 20, no. 4, pp. 833-838, Jul. 2005.

[6] M. Shen, J. Wang, A. Joseph, F. Z. Peng, L. M. Tolbert and D. J Adams, "Constant Boost Control of the Z-Source Inverter to Minimize Current Ripple and Voltage Stress," IEEE Trans. Ind. Applicat., vol. 42, no. 3, pp. 770-778, May/June 2003.

[7] M. Shen, J. Wang, A. Joseph, F. Z. Peng, L. M. Tolbert and D. J Adams, "Maximum Constant Boost Control of the Z-Source Inverter," in Proc. IEEE IAS Annual Meeting 2004, Oct. 2004.

[8] E. C. dos Santos Jr., E. P. X. P. Filho, A. C. Oliveria and E. R. C. da Silva, "Hybrid Pulse Width Modulation for Z-Source Inverters," in Proc. IEEE ECCE 2010, pp. 2888-2892, Sep. 2010.

[9] U. S. Ali and V. Kamaraj, "A Novel Space Vector PWM for Z-Source Inverter," in Proc.1st Int. Conf. on Electr. Energy Systems, pp. 82-85, Jan. 2011.

[10] J. B. Liu, J. G. Hu, and L. Y. Xu, "Dynamic modeling and analysis of Z- source converter-derivation of AC small signal model and design-oriented analysis," IEEE Trans. Power Electron., vol. 22, no. 5, pp. 1786–1796, Sep. 2007.

[11] P. C. Loh, D. M. Vilathgamuwa, Y. S. Lai, G. T. Chua and Y. Li, "Pulse-Width Modulation of Z-Source Inverters," IEEE Trans. Power Electron., vol. 20, no. 6, pp. 1346-1355, Nov. 2005.

[12] M. Adamowicz, R. Strzelecki, F. Z. Peng, J. Guzinski and H. A. Rub, "High Step-Up Continuous Input Current LCCT-Z-Source Inverters for Fuel Cells," in Proc. ECCE 2011, pp. 2276-2282, Sept. 2011.

978-1-5090-0376-1/16 $31.00 © 2016 IEEE

[13] Y. Zhou, L. Liu and H. Li, "A High-Performance Photovoltaic Module-Integrated Converter (MIC) Based on Cascaded Quasi-Z-Source Inverters (qZSI) Using eGaN FETs," IEEE Trans. Power Electron., vol. 28, no. 6, pp. 2727-2738, Jun. 2013.

[14] D. Cao, S. Jiang, X. Yu and F. Z. Peng, "Low-Cost Semi-Z-Source Inverter for Single-Phase Photovoltaic Systems," IEEE Trans. Power Electron., vol. 26, no. 12, pp. 3514-3523, Dec. 2011.

[15] J. Liu, S. Jiang, D. Cao and F. Z. Peng, "Sliding-Mode Control of Quasi-Z-Source Inverter with Battery for Renewable Energy System," in Proc. IEEE ECCE 2011, pp. 3665-3671, Sep. 2011.

[16] H. A. Rub, A. Iqbal, Sk. M. Ahmed, F. Z. Peng, Y. Li and B. Ge, "Quasi-Z-Source Inverter-Based Photovoltaic Generation System With Maximum Power Tracking Control Using ANFIS," IEEE Trans. Sustainable Energy, vol. 4, no. 1, pp. 11-20, Jan. 2013.

[17] Y. Liu, B. Ge, H. A. Rub and F. Z. Peng, "An Effective Control Methods for Quasi-Z-Source Cascade Multilevel Inverter Based Grid-tie Single Phase Photovoltaic Power System," IEEE Trans. Ind. Informatics, (accepted for future publication, Aug. 2013).

[18] D. Li, P. C. Loh, M. Zhu, F. Gao and F. Blaabjerg, "Enhanced-Boost Z-Source Inverters with Alternate-Cascaded Switched-and Tapped-Inductor Cells," IEEE Trans. Ind. Electron., vol. 60, no. 9, pp. 3567-3578, Sep. 2013.

[19] D. Cao, S. Jiang, X. Yu and F. Z. Peng, "Low Cost Single-Phase Semi-Z-Source Inverter," in Proc. APEC 2011, pp. 429-436, March 2011.

[20] P. C. loh, F. Gao and F. Blaabjerg, "Embedded EZ-Source Inverter," IEEE Trans. Ind. App., vol. 46, no. 1, pp. 256-267, Jan./Feb. 2010.

[21] F. Zhang, F. Z. Peng and Z. Qian, "Z-H Converter," in Proc. PESC 2008, pp. 1004-1007, June 2008.

[22] P. C. Loh, N. Duan, C. Liang, F. Gao and F. Blaabjerg, "Z-Source B4 Inverter," in Proc. PESC 2007, pp. 1363-1369, June 2007.

[23] C. J. Gajanayake, F. L. Luo, H. B. Gooi, P. L. So and L. K. Siow, "Extended-Boost Z-Source Inverters," IEEE Trans. Power Electron., vol. 25, no. 10, pp. 2642-2652, Oct. 2010.

[24] L. Huang, M. Zhang, L. Hang, W. Yao and Z. Lu, "A Family of Three-Switch Three-State Single-Phase Z-Source Inverters," IEEE Trans. Power Electron., vol. 28, no. 5, pp. 2317-2329, May 2013.

[25] P. C. Loh, D. Li and F. Blaabjerg, "T-Z-Source Inverters," IEEE Trans. Power Electron. (letters), vol. 28, no. 11, pp. 4880-4884, Nov. 2013.

[26] R. Strzelecki, M. Adamowicz, N. Strzelecka and W. Bury, "New type T-source inverter," in Proc. CPE 2009, pp. 191-195, May 2009.

[27] W. Qian, F. Z. Peng and H. Cha, "Trans-Z-Source Inverters," IEEE Trans. Power Electron., vol. 26, no. 12, pp. 3453-3463, Dec. 2011.

[28] M. K. Nguyen, Y. C. Lim and Y. G. Kim, "TZ-Source Inverters," IEEE Trans. Ind. Electron., vol. 60, no. 12, pp. 5686-5695, Dec. 2013.

[29] M. Adamowicz, R. Strzelecki, F. Z. Peng, J. Guzinski and H.A. Rub, "New Type LCCT-Z-Source Inverters," in Proc. EPE 2011, pp. 1-10, Sept. 2011.

[30] M. Zhu, K. Yu and F. L. Luo, "Switched Inductor Z-Source Inverter," IEEE Trans. Power Electron., vol. 25, no. 8, pp. 2150-2158, Aug. 2010.

[31] M. K. Nguyen, Y. C. Lim and G. B. Cho, "Switched-Inductor Quasi-Z-Source Inverter," IEEE Trans. Power Electron., vol. 26, no. 11, pp. 3183-3191, Nov 2011.

[32] D. Li, P. C. Loh, M. Zhu, F. Gao and F. Blaabjerg, "Generalised Multicell Switched-inductor and Switched-Capacitor Z-Source Inverters," IEEE Trans. Power Electron., vol. 28, no. 2, pp. 837-848, Feb. 2013.

[33] M. K. Nguyen, Y. C. Lim and J. H. Choi, "Two Switched-Inductor Quasi-Z-Source Inverters," IET Power Electron., vol. 5, Iss. 7, pp. 1017-1025, 2012.

[34] H. Itozakura and H. Koizumi, "Embedded Z-Source Inverter with Switched Inductor," in Proc. IECON 2011, pp. 1342-1347, Nov. 2011.

[35] B. Axelrod, Y. Berkovich, and A. Ioinovici, "Switched-capacitor/switched-inductor structures for getting transformerless hybrid dc-dc PWM converters," IEEE Trans. Circ. Syst. I: Fundamental Theory Appl., vol. 55, no. 2, pp. 687–696, Mar. 2008.

7th Power Electronics, Drive Systems & Technologies Conference (PEDSTC 2016)
16-18 Feb. 2016, Iran University of Science and Technology, Tehran, Iran

A New Basic Unit for Cascaded Multilevel Inverters with Reduced Number of Power Electronic Devices

Ebrahim Babaei[1], *Member, IEEE*, Mohammad Ali Hosseinzadeh[2], Maryam Sarbanzadeh[3], Carlo Cecati[4], *Senior Member, IEEE*

[1&3]Faculty of Electrical and Computer Engineering, University of Tabriz, Tabriz, Iran
[2]Department of Electronic Engineering, University of Applied Science and Technology, Jajarm Branch, Jajarm, Iran
[4]Department of Information Engineering, Computer Science and Mathematics, University of L'Aquila, L'Aquila, Italy
E-mails: e-babaei@tabrizu.ac.ir; m.a_hosseinzadeh@yahoo.com; may14_1368@yahoo.com; carlo.cecati@univaq.it

Abstract— In this paper, a new topology for asymmetrical cascaded multilevel inverter is proposed. The proposed topology consists of series connection of several basic units. Reduction of number of power switches, driver circuits, IGBTs and dc voltage sources are some advantages of the proposed topology in comparison with the conventional cascaded multilevel inverters. In order to generate all output voltage levels, a new algorithm to determine the magnitudes of dc voltage sources is proposed. Finally, to verify the performance of the proposed inverter, the simulation results by using PSCAD/EMTDC software on a 33-level single-phase inverter are used.

Keywords—Asymmetric multilevel inverters; cascaded multilevel inverter; Basic unit

I. INTRODUCTION

Multilevel inverters due to many advantages such as high quality output waveform, reducing lower order harmonics, lower switching losses and better electromagnetic interference are used in different applications [1-4]. The conventional topologies are divided into three main groups: the neutral point clamped (NPC) multilevel inverter, flying capacitor (FC) multilevel inverter and cascaded H-bridge (CHB) multilevel inverter [5-9]. Among the multilevel inverter topologies, the cascaded multilevel inverters have attracted more attention mainly because of simple structure and easily of extending to higher voltage levels. Cascaded multilevel inverters synthesize a desired voltage output based on a series connection of power cells. In recent years, researchers have presented many various topologies of multilevel inverters for different purposes [10-15]. Most of the presented topologies try to reduce the number of components. One of these topologies is the modular multilevel inverter [16]. This topology is simpler than the cascaded four-switch H-bridge-based inverter and has several advantages [17]. However, the topology does not consider reduction in the number of components. The multilevel inverter presented in [18] is based on symmetric topology and uses series/parallel connection of the dc voltage sources. This topology uses lower number of switches in comparison with the symmetric CHB. The topologies presented in [13] and [14] consider reduction in the components. These topologies are basically based on asymmetric topologies; hence, the used dc

voltage sources have different values. However, the number of switching devices still remains high in these topologies.

In this paper, first a new topology for basic unit for multilevel inverter is proposed which utilizes lower number of power switches and dc voltage sources. Then, series connection of the basic units is proposed as a generalized multilevel inverter. It is aimed to increase the number of output voltage levels and reduce the number of power switches, driver circuits and total cost of the inverter. In order to generate all positive and negative levels at the output, a new algorithm to determine the magnitudes of dc voltage sources are proposed. The proposed inverter is compared with several conventional cascaded multilevel inverters to investigate its advantages. Finally, the accuracy performance of the proposed inverter is reconfirmed by using simulation results on 33-level proposed inverter.

II. PROPOSED BASIC UNIT

In this paper, a new basic unit for asymmetrical cascaded multilevel inverter is proposed. The proposed basic unit for cascaded multilevel inverter is shown in Fig. 1. As shown in this figure, the proposed basic unit consists of m dc voltage sources that their magnitudes can be different values. Each basic unit consists of $m + 5$ power switches. The power switches used in the basic unit inverter are unidirectional and bidirectional types. The power switches S_1, S_3, S_4, S_6, S_7, ..., and S_{m+5} are unidirectional and two switches S_2 and S_5 are bidirectional. There are different circuit configurations for bidirectional power switches. In this paper, the common emitter structure with one gate driver for a switch is used. Each unidirectional power switch consists of an IGBT with an anti-parallel diode that is able to conduct current in both direction and block voltage in one polarity. While, the bidirectional power switch includes of two IGBTs with Two anti-parallel diodes that conducts current in both direction and blocks voltage in two positive and negative polarities. In the condition, each bidirectional power switch needs one driver circuit. In the proposed basic unit, if the power switches of S_1, S_4 or S_2, S_5 or S_3, S_6 turned on simultaneously, the output voltage will be zero and if the power switches of S_2,

978-1-5090-0376-1/16 $31.00 © 2016 IEEE   185

S_6 with each of the switches $S_7,...,$ S_{m+5} turned on simultaneously, the inverter generates positive voltage levels. In this condition, if S_1, S_{m+5}, S_6 turned on simultaneously, the maximum positive output voltage level (V_1+V_m) will be generated. To generate the negative output voltage level, the power switches S_3 and S_4 with one of switches $S_7,...,$ S_{m+5} must be turned on simultaneously. In this condition, if the power switches S_3, S_{m+5}, S_4 are turned on simultaneously, the maximum negative output voltage level $-(V_1+V_m)$ will be generated. Table I shows the switching patterns and generated output levels for the proposed basic unit. As it is obvious from this Table, the proposed basic unit is able to generate positive and negative voltage levels at the output. In this Table, 1 and 0 indicate the turned on and off states of switches, respectively. As obvious from this Table, the basic unit is able to generate $4m-1$ levels at the output. Moreover, in order to generate all voltage levels at the output except the zero level, only three power switches in different operating modes are turned on. In the proposed basic unit, the number of power switches $(N_{switch,unit})$, IGBTs $(N_{IGBT,unit})$, driver circuits $(N_{driver,unit})$, dc voltage sources $(N_{source,unit})$, maximum output voltage $(V_{o,max,unit})$ and generated output voltage levels $(N_{level,unit})$ are calculated as follows, respectively:

$$N_{switch,unit} = \begin{cases} 2(m+1), & \text{for } m=1,2 \\ m+5, & \text{for } m>2 \end{cases} \quad (1)$$

$$N_{IGBT,unit} = \begin{cases} 4m, & \text{for } m=1,2 \\ m+7, & \text{for } m>2 \end{cases} \quad (2)$$

$$N_{driver,unit} = \begin{cases} 2(m+1), & \text{for } m=1,2 \\ m+5, & \text{for } m>2 \end{cases} \quad (3)$$

$$N_{source,unit} = m \quad (4)$$

$$V_{o,max,unit} = V_1 + V_m = (2m-1)V_{dc} \quad (5)$$

$$N_{level,unit} = 4m-1 \quad (6)$$

In above equations, m is the number of used dc voltage sources in the basic unit.

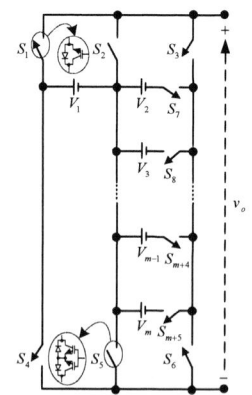

Fig. 1.   The proposed basic unit.

TABLE I.   DIFFERENT OUTPUT LEVELS

v_o	S_1	S_2	S_3	S_4	S_5	S_6	S_7	S_8	$\cdots$	S_{m+3}	S_{m+4}	S_{m+5}
0	0	0	0	0	0	0	1	1	$\cdots$	0	0	0
	0	0	0	0	1	1	0	0	$\cdots$	0	0	0
	0	0	1	1	0	0	0	0		0	0	0
V_1	0	0	0	1	0	0	0	1	$\cdots$	0	0	0
$-V_1$	0	0	0	0	1	0	1	0	$\cdots$	0	0	0
V_2	0	0	1	0	0	0	1	0	$\cdots$	0	0	0
$-V_2$	0	0	0	1	0	1	0	0	$\cdots$	0	0	0
V_1+V_2	0	1	1	0	0	0	0	1	$\cdots$	0	0	0
$-(V_1+V_2)$	0	0	1	1	0	0	1	0	$\cdots$	0	0	0
V_1+V_3	1	0	0	0	0	1	0	1	$\cdots$	0	0	0
$-(V_1+V_3)$	0	0	1	1	0	0	0	1	$\cdots$	0	0	0
$\vdots$	$\vdots$	$\vdots$	$\vdots$	$\vdots$	$\vdots$	$\vdots$	$\vdots$	$\vdots$	$\vdots$	$\vdots$	$\vdots$	$\vdots$
V_1+V_m	1	0	0	0	0	1	0	0	$\cdots$	0	0	1
$-(V_1+V_m)$	0	0	1	1	0	0	0	0	$\cdots$	0	0	1

As an example, considering (6) and Table I for generation of 11-level, the proposed basic unit needs three dc voltage source ($m=3$) with the values of V_1, $2V_1$ and $4V_1$ as shown in Fig. 2(a). The typical 11-level output voltage is shown in Fig. 2(b).

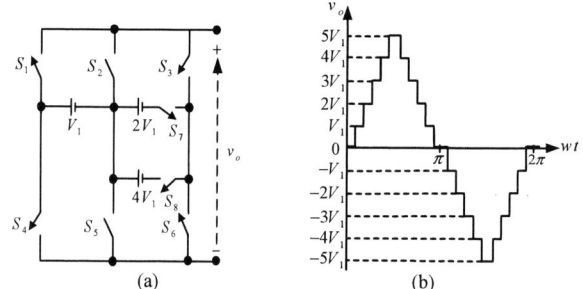

(a)   (b)

Fig. 2.   (a) 11-level basic unit based on the proposed basic unit; (b) typical output voltage of the 11-level inverter.

III. PROPOSED CASCADED MULTILEVEL INVERTER

It is possible to use series connection of n numbers of the proposed basic unit to create a new cascaded multilevel inverter Fig. 3 shows the proposed cascaded multilevel inverter. For this topology, the number of power switches (N_{switch}), IGBTs (N_{IGBT}), driver circuits (N_{driver}) and dc voltage sources (N_{source}) are calculated as follows, respectively:

$$N_{switch} = \begin{cases} 2(m_1+m_2+\cdots+m_n)+2n, & \text{for } m=1,2 \\ m_1+m_2+\cdots+m_n+5n, & \text{for } m>2 \end{cases} \quad (7)$$

$$N_{IGBT} = \begin{cases} 4(m_1+m_2+\cdots+m_n), & \text{for } m=1,2 \\ m_1+m_2+\cdots+m_n+7n, & \text{for } m>2 \end{cases} \quad (8)$$

$$N_{driver} = \begin{cases} 2(m_1 + m_2 + \cdots + m_n) + 2n, & \text{for } m = 1,2 \\ m_1 + m_2 + \cdots + m_n + 5n, & \text{for } m > 2 \end{cases} \quad (9)$$

$$N_{source} = \sum_{i=1}^{n} m_i = m_1 + m_2 + \cdots + m_n \quad (10)$$

In order to generate the maximum number of output levels for a fixed number of components, it is necessary to consider the equal number of dc voltage sources in each basic unit. In other word:

$$m_1 = m_2 = \cdots = m_n = m \quad (11)$$

The different output voltage levels can be determined by combination of switching states of basic units. If proper values for the dc voltage sources are selected, then the output voltage of the inverter can be obtained between $\sum_{j=1}^{n}(V_{1,j} + V_{m,j})$ and

$-\sum_{j=1}^{n}(V_{1,j} + V_{m,j})$.

Considering (11), the equations (7) to (10) can be rewritten as follows:

$$N_{switch} = \begin{cases} 2n(m+1), & \text{for } m = 1,2 \\ n(m+5), & \text{for } m > 2 \end{cases} \quad (12)$$

$$N_{IGBT} = \begin{cases} 4nm, & \text{for } m = 1,2 \\ n(m+7), & \text{for } m > 2 \end{cases} \quad (13)$$

$$N_{driver} = \begin{cases} 2n(m+1), & \text{for } m = 1,2 \\ n(m+5), & \text{for } m > 2 \end{cases} \quad (14)$$

$$N_{sourse} = nm \quad (15)$$

$$N_{variety} = nm \quad (16)$$

$$V_{o,\max} = \sum_{j=1}^{n}(V_{1,j} + V_{m,j}) \quad (17)$$

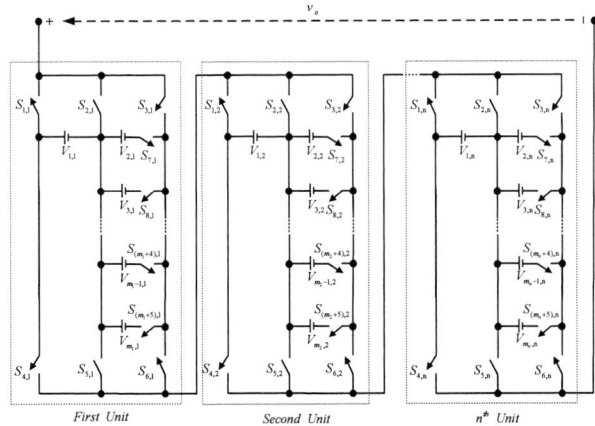

First Unit *Second Unit* n^{th} *Unit*

Fig. 3. The proposed cascaded multilevel inverter.

In the proposed inverter, the number of generated output levels is completely depended on the magnitudes of dc voltage sources. At the following, a new algorithm to determine the magnitudes of dc voltage sources is proposed.

For the first basic unit, the values of dc voltage sources are considered as follows:

$$V_1 = V_{dc} \quad (18)$$

$$V_{i,1} = 2(i-1)V_{1,1} \qquad i = 2,3,\ldots,m_1 \quad (19)$$

For the second basic unit, the values of dc voltage sources are determined as follows:

$$V_{1,2} = V_{dc} + 2(V_{o,\max 1}) = (4m_1 - 1)V_{dc} \quad (20)$$

$$V_{i,2} = 2(i-1)V_{1,2} \qquad i = 2,3,\ldots,m_2 \quad (21)$$

The values of the dc voltage sources in the third basic unit are assumed as following:

$$V_{1,3} = V_{dc} + 2(V_{o,\max 1} + V_{o,\max 2}) = (4m_1 - 1)(4m_2 - 1)V_{dc} \quad (22)$$

$$V_{i,3} = 2(i-1)V_{1,3} \qquad i = 2,3,\ldots,m_3 \quad (23)$$

The $k-th$ basic unit has dc voltage sources with the following values:

$$V_{1,j} = V_{dc} + 2(V_{o,\max 1} + V_{o,\max 2} + \cdots + V_{o,\max(j-1)})$$
$$= \prod_{j=1}^{n-1}(4m_j - 1)V_{dc} \quad (24)$$

$$V_{i,j} = 2(i-1)V_{1,j} \qquad i = 2,3,\ldots,m_j \quad (25)$$

According to this algorithm, the number of generated output levels (N_{level}) and maximum amplitude of output voltage source are ($V_{o,\max}$) obtained as follows:

$$V_{o,\max} = \sum_{j=1}^{n}(2m_j - 1)V_{1,j} \quad (26)$$

$$N_{level} = (4m_1 - 1) \times (4m_2 - 1) \times \cdots \times (4m_n - 1)$$
$$= \prod_{j=1}^{n}(4m_j - 1) \quad (27)$$

It is also possible to propose other algorithms to determine the magnitudes of dc voltage sources.

IV. COMPARISON RESULTS

Recently, some new topologies with reduced number of switches have been presented for multilevel inverters. In this section, the proposed cascaded multilevel inverter is compared with several conventional cascaded multilevel inverters. This comparison is done from different points of view such as the number of IGBTs, power switches and dc voltage sources. In this comparison, the proposed cascaded multilevel inverter based on new proposed algorithm is indicated by P_1. The CHB multilevel inverter with binary algorithm is indicated by R_1. (R_1 for $V_1 = 2^{j-1}V_{dc}$ ($j = 1, 2, \cdots, n$)) [5]. The presented cascaded multilevel inverter in [19] is considered by R_2. (R_2 for $V_j = (m+1)^{j-1}V_{dc}$ $j = 1, 2, \cdots, n$). In addition, for the presented inverter in [13], [20] and [21] three different algorithms have been considered that are indicated by R_3, R_4 and R_5, respectively, as follows:

R_3 for $V_j = (m^{j-1})V_{dc}$ ($j = 1, 2, \cdots, n$)

R_4 for $V_{1,j} = 0.5V_{2,j} = V_{3,j} = 2^{j-1}V_{dc}$ ($j = 1, 2, \cdots, n$)

R_5 for $V_{1,1} = V_{1,2} = V_{dc}$, $V_{1,3} = 2V_{1,1} = 2V_{dc}$,

$V_{1,i} = 2^{i-4} \times 5 V_{dc}$ ($i = 4, \cdots, m$), $\ldots$,

$V_{n,1} = V_{n,2} = V_{dc} \times (5 \times 2^{m-2} - 1)^{n-1}$,

$V_{n,3} = 2V_{dc} \times (5 \times 2^{m-2} - 1)^{n-1}$, $\ldots$,

$V_{n,i} = V_{dc}(2^{i-4} \times 5)(5 \times 2^{m-2} - 1)^{n-1}$ ($i = 4, \ldots, m$))

Fig.4 shows all of the above-mentioned topologies. Fig. 5 (a) compares the number of IGBTs of the proposed topology with the above-mentioned cascaded multilevel inverters. As it is obvious from this figure, the proposed cascaded inverter needs lower number of IGBTs than other presented inverters in the references except the presented inverter by R_5. It is pointed out that the bidirectional and unidirectional power switches are used in the proposed cascaded inverter. As mentioned before, the number of IGBTs is as same as the number of power diodes. As a result, the proposed inverter also needs minimum number of power diodes than other conventional cascaded inverters. The number of power switches in the proposed cascaded inverter is compared with above-mentioned inverters. This comparison is shown in Fig. 5(b). As it is obvious from this figure, the proposed multilevel inverter needs lower number of power switches in comparison with other presented multilevel inverters in literature. As it is mentioned previously, all unidirectional and bidirectional power switches require a driver circuit. Therefore, the proposed inverter also needs lower number of driver circuit than other inverters. Fig. 4(c) compares the number of dc voltage sources in the proposed topology with the above-mentioned cascaded multilevel inverters. It is obvious that the proposed topology requires lower number of dc voltage sources than other mentioned topologies. It is clear from above comparisons that the proposed cascaded inverter is able to generate higher number of output levels with lower number of power switches, IGBTs, diodes, driver circuits and dc voltage sources than above mentioned topologies.

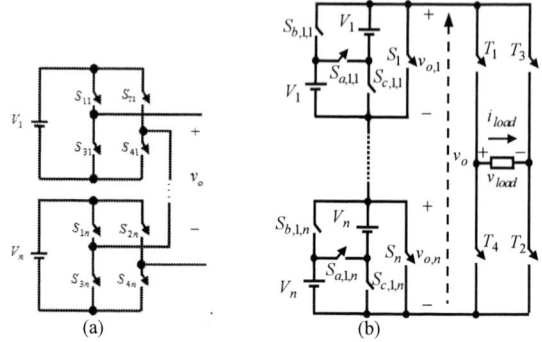

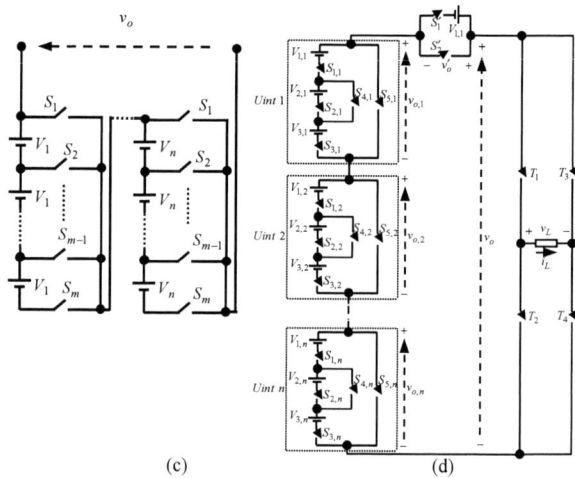

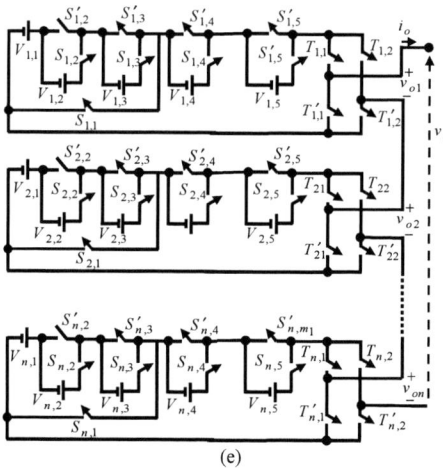

Fig. 4. Conventional cascaded multilevel inverters; (a) CHB multilevel inverter [5] (R_1); (b) presented topology in [19] (R_2); (c) presented topology in [13] (R_3); (d) presented topology in [20] (R_4); (e) presented topology in [21] (R_5).

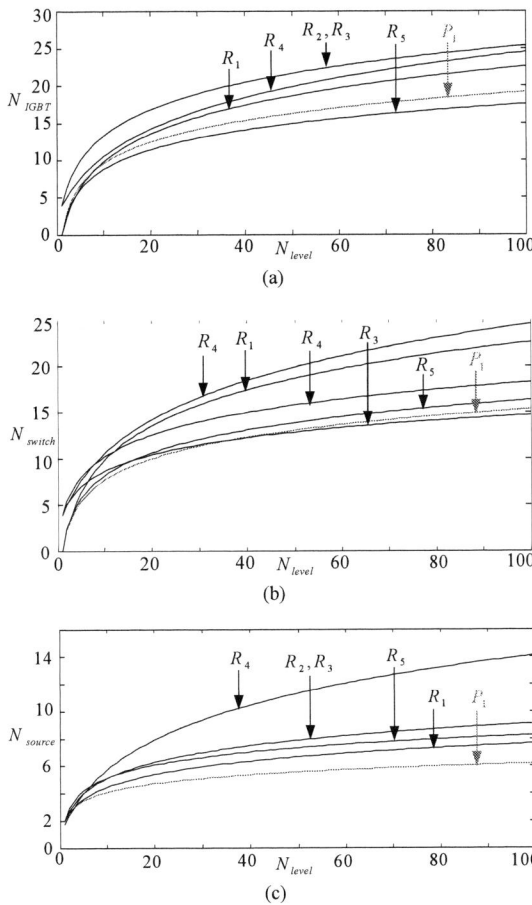

Fig. 5. (a) variation of N_{IGBT} versus N_{level} ; (b) variation N_{switch} of versus N_{level} ; (c) variation N_{source} of versus N_{level} .

V. SIMULATION RESULTS

In this section, the ability of the proposed inverter to generate all voltage levels is reconfirmed by using simulation results on a 33-level single phase inverter. The simulation results are obtained by using PSCA/EMTDC software program. There are various control methods for controlling the multilevel inverters. In this paper, the fundamental frequency control method is used. According to Fig. 6 two basic units are cascaded to generate 33 levels. The proposed cascaded multilevel inverter has four unequal dc voltage sources. Based on the proposed algorithm, the values of dc voltage sources are $V_{1,1} = V_{dc}$, $V_{2,1} = 2V_{dc}$, $V_{3,1} = 4V_{dc}$, and $V_{1,2} = 11V_{dc}$. The value of dc voltage source V_{dc} is assumed of $10V$. Considering these values, the peak value of output voltage will be $160V$. In the simulation, the frequency of the output voltage is assumed $50Hz$. and the load is considered R–L with the value of $R = 107\Omega$ and $L = 55mH$.

Figs. 7(a), 7(b) shows the output voltage of the first basic unit and the second basic unit, respectively. The sum of the output voltages of the first and the second basic units is the output voltage of the inverter which is shown in Fig. 7 (c). As this figure shows, the inverter can generate all of the expected voltage levels. Fig. 7(d) shows the output current waveform. Regarding the waveforms of output current and voltage, there is a phase displacement between them. This is resulted from the inductive characteristic of the load.

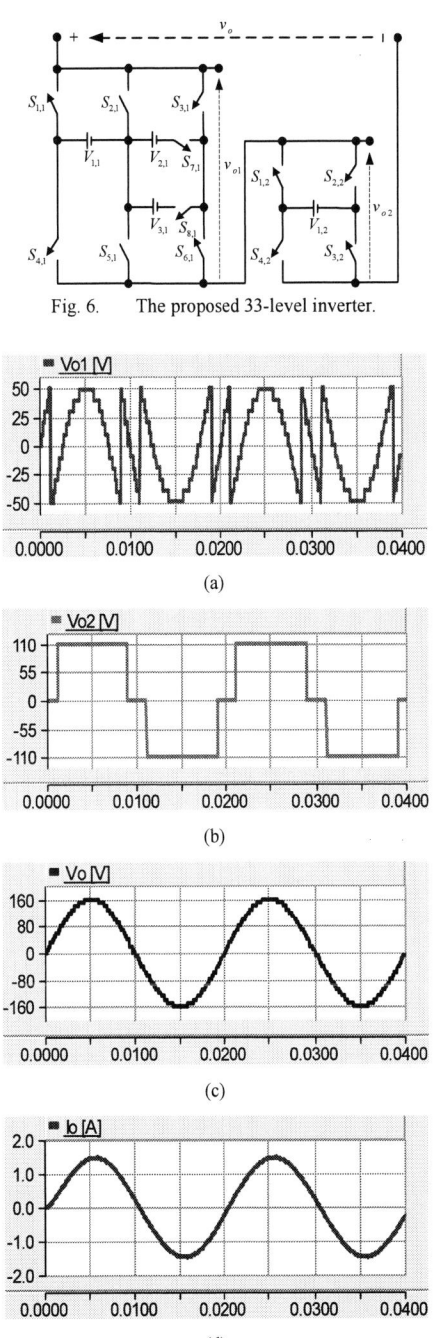

Fig. 6. The proposed 33-level inverter.

Fig. 7. Simulation results ; (a) output voltage of the first basic unit; (b) output voltage of the second basic unit; (c) output voltage; (d) output current.

978-1-5090-0376-1/16 $31.00 © 2016 IEEE

VI. CONCLUSION

In this paper, a new asymmetric basic unit was proposed for cascaded multilevel inverters. The proposed multilevel inverter based on the basic unit is connected in series. Moreover, a new algorithm is proposed to determine the magnitude of dc voltage sources. The proposed cascaded multilevel inverter uses lower number of power electronic switches, IGBTs, power diode, driver circuit and dc voltage sources for a same number of voltage levels in comparison with the conventional multilevel inverters. The simulation results obtained from a 33-level inverter are presented to verify the correctness operation of the proposed topology.

REFERENCES

[1]   J. Rodriguez, B. Wu, S. Bernet, J. Pontt, and S. Kouro, "Multilevel voltage source converter topologies for industrial medium voltage drives," *IEEE Trans. Ind. Electron.*, vol. 54, pp. 2930-2945, Dec. 2007.

[2]   M.N.A. Kadir, S. Mekhilef, and H. W. Ping, "Voltage vector control of a hybrid three-stage eighteen-level inverter by vector decomposition, " *IET Trans. Power Electron.*, vol. 3, no. 4, pp. 601–611, Jul. 2010.

[3]   M. Farhadi Kangarlu and E. Babaei, "A generalized cascaded multilevel inverter using series connection of sub-multilevel inverters," *IEEE Trans. Power Electron.*, vol. 28, no. 2, pp. 625-636, Feb. 2013.

[4]   M.A. Eltawil and Z. Zhao, "Grid-connected photovoltaic power systems: Technical and potential problems—A review," *Renewable and Sustainable Energy Reviews*, vol. 14, no. 1, pp. 112-129, Jan 2010.

[5]   M. Manjrekar and T.A. Lipo, "A hybrid multilevel inverter topology for drive application," in *Proc. APEC*, 1998, pp. 523-529

[6]   M.A. Hosseinzadeh, E. Babaei, and M. Sabahi, "Back-to-back stacked multicell converter," in *Proc. PEDSTC*, 2012, Tehran, Iran, PP. 410-415.

[7]   E. Babaei, M. Farhadi Kangarlu, and M.A. Hosseinzadeh, "Asymmetrical multilevel converter topology with reduced number of components," *IET Power Electron.*, vol. 6, no. 6, pp. 1188–1196, Jan. 2013.

[8]   A.L. Batschauer, S.A. Mussa, and M.L. Heldwein, "Three-phase hybrid multilevel inverter based on half-bridge modules," *IEEE Trans. Ind. Electron.*, vol. 59, no. 2, pp. 668-678, Feb. 2012.

[9]   F. Carnielutti, H. Pinheiro, and C. Rech, "Generalized carrier based modulation strategy for cascaded multilevel converters operating under fault conditions," *IEEE Trans. Ind. Electron.*, vol. 59, no. 2, pp. 679-689, Feb. 2012.

[10] E. Babaei, M. Farhadi Kangarlu, and M. Sabahi, "Extended multilevel converters: An attempt to reduce the number of independent dc voltage sources in cascaded multilevel converters," *IET Power Electron.*, vol. 7, no. 1, pp. 157-166, Jan. 2014.

[11] M. Malinowski, K. Gopakumar, J. Rodriguez, and M. Perez, "Survey on cascaded multilevel inverters," *IEEE Trans. Ind. Electron.*, vol. 3 no. 57, pp. 2197–2206, 2010.

[12] M. Farhadi Kangarlu and E. Babaei, "A generalized cascaded multilevel inverter using series connection of sub-multilevel inverters," *IEEE Trans. Power Electron.*, vol. 28, no. 2, pp. 625-636, Feb. 2013.

[13] E. Babaei, "A cascade multilevel converter topology with reduced number of switches," *IEEE Trans. Power Electron.*, vol. 23, no. 6, pp.2657–2664, 2008.

[14] E. Babaei, S.H. Hosseini, G.B. Gharehpetian, M. Tarafdar Haque, and M. Sabahi, "Reduction of dc voltage sources and switches in asymmetrical multilevel converters using a novel topology," *Electric Power Syst. Res.*, vol. 77, no.8, pp. 1073–1085, 2007.

[15] W.K. Choi and F.S. Kang, "H-bridge based multilevel inverter using PWM switching functions," *Proc. INTELEC*, 2009, pp. 1–5.

[16] A. Lesnicar and R. Marquardt, "An innovative modular multilevel converter topology suitable for a wide power range," in *proc. IEEE Power Tech. Conf.*, 2003.

[17] G.P. Adam, O. Anaya-Lara, G.M. Burt, D. Telford, B.W. Williams, and J. R. McDonald, "Modular multilevel inverter: pulse width modulation and capacitor balancing technique," *IET Power Electron.*, vol. 3, no. 5, pp. 702–715, 2010.

[18] Y. Hinago and H. Koizumi, "A single phase multilevel inverter using switched series/parallel dc voltage sources," *IEEE Trans. Ind. Electron.*, vol. 58, no. 8, pp. 2643–2650, Aug. 2010.

[19] E. Babaei , S. Sheermohammadzadeh Gowgani, and M. Sabahi, "Modified multilevel inverters using series and parallel connection of dc voltage sources, " *Arabian Journal for Science and Engineering*, Vol. 39, no. 4, pp 3077-3094, April 2014.

[20] E. Babaei, S. Laali, and Z. Bayat, "A single-phase cascaded multilevel inverter based on a new basic unit with reduced number of power switches," *IEEE Trans. Ind. Electron.*, vol. 62, no. 2, pp. 922-929, Feb. 2015

[21] E. Babaei, A. Dehqan, and M. Sabahi. "A new topology for multilevel inverter considering its optimal structures," *Electric Power Syst. Res.*, vol. 103, no. 8, pp. 145–156, 2013.

7th Power Electronics, Drive Systems & Technologies Conference (PEDSTC 2016)
16-18 Feb. 2016, Iran University of Science and Technology, Tehran, Iran

Coupled Inductor Based Boost DC/DC Converter

Simin Sakhavati
Department of Electrical Engineering,
Shabestar Branch, Islamic Azad University,
Shabestar, Iran
E-mail: simin.sakhavati@iaushab.ac.ir

Ebrahim Babaei, *Member, IEEE*
Faculty of Electrical and Computer Engineering,
University of Tabriz,
Tabriz, Iran
E-mail: e-babaei@tabrizu.ac.ir

Abstract—The boost dc-dc converters are used to increase the magnitude of dc voltage. The renewable energy sources such as photovoltaic and fuel cell generate low magnitudes of voltage. To solve this problem, it is necessary to use a boost dc/dc converter. In this paper, a new boost dc-dc converter by using coupled inductor is proposed. In the proposed converter, by choosing the suitable value of the coupled inductor coefficient, it is possible to obtain minimum ouptut ripple. In the proposed converter as same as the conventional boost dc-dc converters by changing the value of duty cycle, the output voltage could be controlled. Finally, the accuracy performance of the proposed converter is reconfirmed thruogh the simulation results in PSCAD/EMTDC software.

Keywords—*Boost dc-dc converter; coupled inductor; renewable energy sources*

I. INTRODUDTION

The dc-dc converters directly convert the dc voltage from one level to a specific level. Up to now, different types of dc-dc converters such as buck, boost, buck-boost, Cuk and Sepic converters have been presented in literature. These converters have most important effects on industries, electrical vehicles, electrical machines, renewable systems and etc. [1-6]. In addition, according to worse effects of greenhouse gases, environmental pollution, leakage of fossil fuels and high necessity to energy, the attention to renewable energy sources are increased. Moreover, the renewable energy sources such as photovoltaic and fuel cell generate dc voltage levels with low magnitude that leads to use boost dc-dc converter to increase the voltage levels [7].
In the boost converter, to reduce the input current ripple the value of inductor can be increased [8]. But increasing the inductor value leads to a slow dynamic response of the converter and also the weight of the converter is increased [9]. By proper designing the input filter, the input current ripple can be reduced. However, high values of filters increase the weight and size of the converter [10-11]. The coupled inductor based converters can reduce the input current ripple [12-15]. Moreover, by using tapped inductor, the input current ripple can be eliminated [16-17]; however, the tapped inductor increases the failure rate of inductor.
In this paper, a new boost dc-dc converter based on coupled inductors is proposed. In this converter, a capacitor connected in series to the secondary winding of the coupled inductor is used to store energy and transmit it to the load. Moreover, the coupling coefficient can be used to control the magnitude of the input current ripple. As a result, it is possible to eliminate the input current ripple by choosing a suitable value of coupling coefficient. This converter also consists of a diode at the output of the converter to avoid of reverse current and a capacitor that is connected in parallel to the load. This capacitor is used to reduce the output voltage ripple and store energy as a storage element.
In this paper, the topology of the proposed converter is presented. Then, the performance of the proposed converter in different operating modes is analyzed. In addition, the equations of the current ripple of the primary winding and the voltage gain of converter are calculated. At the end, the simulation results of the proposed boost converter by using PSCAD/EMTDC software are used to reconfirm the correctness of obtained theoretical analysis.

II. PROPOSED BOOST DC-DC CONVERTER

Fig. 1 shows the proposed boost dc-dc converter. This converter consists of a coupled inductor with low leakage inductance in comparison with the conventional converters. In this coupled inductor, the secondary winding is connected in series to a capacitor. The coupled inductor is used to develop the converter operating range, and increase the efficiency and reliability. In order to generate different output voltage levels, it is possible to change turns ratio, inductance value, coupling coefficient and time interval of turning on of the switch. The secondary winding of the coupled inductor is series with a capacitor. In order to decrease the output voltage ripple, a large is connected in parallel with the load. In the proposed boost converter, the power switch of S and diode D are used as a circuit switching elements and rectification. This converter consists of two operating modes with equivalent circuits shown in Fig. 2. The coupled inductor model can be expressed by the following mathematical equations:

$$v_{LP} = L_P \frac{di_{LP}}{dt} + M \frac{di_{LS}}{dt} \qquad (1)$$

$$v_{LS} = M \frac{di_{LP}}{dt} + L_S \frac{di_{LS}}{dt} \qquad (2)$$

where L_p and M are the primary winding inductance and mutual inductance of the coupled inductor, respectively. In

978-1-5090-0376-1/16 $31.00 © 2016 IEEE

addition, $\frac{di_{LP}}{dt}$ and $\frac{di_{LS}}{dt}$ are the current differential of the primary and secondary windings' currents of the coupled inductor, respectively. The mutual inductance is equal to:

$$M = K\sqrt{L_P L_S} \qquad (3)$$

where K is the coupled inductor coefficient and L_S is the self-inductance of the secondary winding.

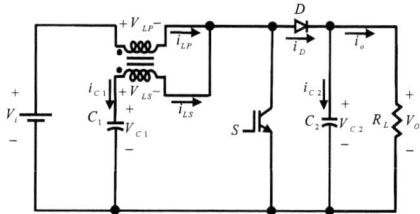

Fig. 1. Proposed boost dc-dc converter.

A. The First Operating Mode ($0 \leq t < t_1$)

In this operating mode, the power switch S is turned on and the diode D is turned off. According to Fig. 2(a), by turning on the switch, the voltage value of the primary winding is equal to the value of voltage source. Therefore, the primary winding current of the coupled inductor is increased from its minimum value (I_{LP1}) to its maximum value (I_{LP2}) in steady state. As a result, the induced voltage in the secondary winding of the coupled inductor will be positive. In this condition, the secondary winding current is increased. Moreover, in this operating mode, C_1 is charged by the induced voltage in the secondary winding because of series-connected of C_1 to the secondary winding. In the first operating mode, the stored energy in C_2 is transmitted to the load. As a result, the current of C_2 is decreased from its maximum value (I_{C2}) to its minimum value (I_{C1}). As shown in Fig. 2(a), the primary (v_{LP}) and secondary (v_{LS}) windings' voltages are calculated as follows:

$$v_{LP} = V_i \qquad (4)$$
$$v_{LS} = v_{C1} = V_i \qquad (5)$$

Considering (1), (3) and (4), the ripple current of the primary winding (Δi_{LP}) in the first operating mode is equal to:

$$\Delta i_{LP} = I_{LP2} - I_{LP1} = \frac{V_i - K\sqrt{L_P L_S}\dfrac{di_{LS}}{dt}}{L_P} t_1 \qquad (6)$$

According to (6), in order to decrease the primary winding current ripple, it is possible to increase K, and or L_P and L_S. Moreover, it is also possible to reduce t_1 that leads to increase the switching losses.

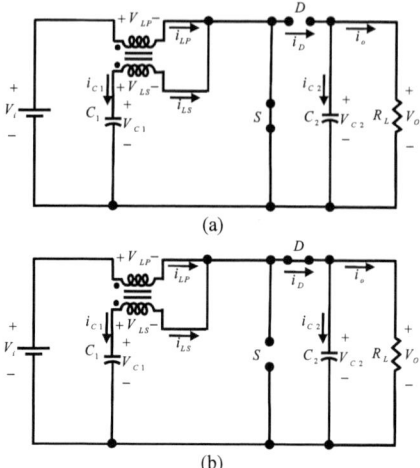

(a)

(b)

Fig. 2. Equivalent circuits of the proposed converter; (a) first operating mode; (b) second operating mode.

B. Second Operating Mode ($t_1 \leq t < t_2$)

In this operating mode, the power switch of S is turned off and the diode D is turned on. According to Fig. 2(b) and by applying KVL to the external loop, a negative value for voltage v_{LP} is obtained that leads to reduce the current of the primary winding. In addition, this negative voltage value is induced in the secondary winding and causes to decrease the current of the secondary winding that is equal to current of the capacitor C_1. In other word, the stored energies in the capacitor C_1 and coupled inductor are transmitted to the load that leads to increase output voltage. Moreover, the voltage value of C_2 is positive, so, the capacitor C_2 is charged and its energy is transmitted to the load. As shown in Fig. 2(b), the primary (v_{LP}) and secondary (v_{LS}) windings' voltages are calculated as follows:

$$v_{LP} = V_i - V_o \qquad (7)$$
$$v_{LS} = v_{C1} - V_o = V_i - V_o \qquad (8)$$

At the time of t_2 that the switch is turned on, the value of coupled inductor current is reached I_{LP1} (minimum value of current). Therefore, the current ripple in the second operating mode is obtained as follows:

$$\Delta i_{LP} = I_{LP2} - I_{LP1} = \frac{-V_i + V_o + M\dfrac{di_{LS}}{dt}}{L_P}(t_2 - t_1) \qquad (9)$$

According to (3) and (9), it is clear that in the second operating mode to decrease the current ripple of the primary winding, it is enough to increase the values of L_P, L_S and or coupled inductor coefficient. It is also possible to decrease the value of t_2 but this method leads to increase the switching losses.

978-1-5090-0376-1/16 $31.00 © 2016 IEEE

Fig. 3 shows the voltage and current waveforms of the used devices in the proposed converter.

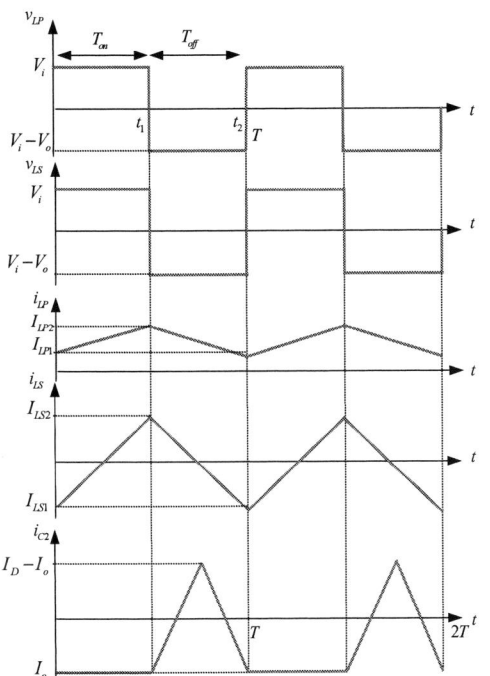

Fig. 3. Voltage and current waveforms in the for the proposed converter.

C. Transformer Model of the Coupled Inductor

The coupled inductor can be modeled by leakage inductance (L_K), magnetized inductance (L_M) and an ideal transformer by turns ratio of (N) as shown in Fig. 4(a). The turns ratio of the transformer is defined as follows:

$$N = \frac{N_S}{N_P} \tag{10}$$

where, N_P and N_S denote the primary and secondary turns, respectively.

The effective conversion ratio (N_e) is defined as follows:

$$N_e = \sqrt{\frac{L_S}{L_P}} \tag{11}$$

The following equation is always valid between N_e and N:

$$N_e = NK \tag{12}$$

Considering (3), (11) and (12), we have:

$$M = K\sqrt{L_P L_S} = K^2 N L_P \tag{13}$$

By replacing (13) into (1) and (2), the following equations can be obtained:

$$v_{LP} = L_P(1-K^2)\frac{di_{LP}}{dt} + \frac{1}{N}v_{LS} \tag{14}$$

$$v_{LS} = K^2 N v_{LP} + (KN)^2(1-K^2)L_P\frac{di_{LS}}{dt} \tag{15}$$

According to (15), it is possible to consider the equivalent circuit of the coupled inductor as shown in Fig. 4(a). In this figure, $(1-K^2)L_P$ and K^2L_P are the leakage and magnetized inductances, respectively. This circuit is a transformer model for the coupled inductor that its leakage inductance is transferred to only in one side. Here, the coupled inductor is modeled by a magnetized inductance, ideal transformer with conversion ratio of N and a leakage inductance.

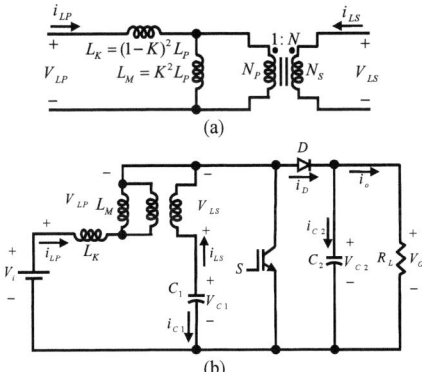

(b)
Fig. 4. (a) Transformer model of coupled inductor; (b) equivalent circuit of the proposed converter by transformer model of coupled inductor.

D. Gain Voltage Calculation

The value of duty cycle (D) is calculated as follows:

$$D = \frac{T_{on}}{T} \tag{16}$$

where T_{on}, and T are the time interval of turning on of the switch and switching, respectively.

By replacing the transformer model of the coupled inductor (Fig. 4(a)) in Fig. 1, the equivalent circuit shown in Fig. 4(b) is obtained. Based on the voltage balance law, the average value of inductance voltage is equal to zero as follows:

$$\int_0^T V_L = 0 \tag{17}$$

Considering Fig. 4(a) and by applying (17) for one of L_M, L_k and L_P, the following equation is obtained:

$$DV_i + (1-D)(V_i - V_o) = 0 \tag{18}$$

By applying the voltage balance law for the primary and secondary windings, the voltage gain of the proposed converter is obtained as follows:

$$\frac{V_o}{V_i} = \frac{1}{(1-D)} \quad (19)$$

Equation (19) shows that the proposed converter is a boost dc-dc converter and has the same voltage gain as the conventional boost converter.

By applying the voltage balance law on secondary winding, the following equation is obtained:

$$v_{C1} = (1-D)V_o \quad (20)$$

Considering (19) and (20), the following equation is resulted:

$$v_{C1} = V_i \quad (21)$$

III. INPUT CURRENT RIPPLE ELIMINATION

The proposed converter consists of simple topology that is more like to the conventional dc-dc converter. The control method of the proposed topology is same as the conventional topology. As a result, in the proposed converter, the control complexity for the switch is not considered. In comparison with the conventional boost converter, the proposed topology can eliminate the ripple of input current. At the following, the aim is to obtain the condition to eliminate the input current ripple.

Considering (1), (2), (4) and (5), the following equation can be written for the first operating mode:

$$L_P \frac{di_{LP}}{dt} + M \frac{di_{LS}}{dt} = M \frac{di_{LP}}{dt} + L_S \frac{di_{LS}}{dt} = V_i \quad (22)$$

By simplification the above equation, it is resulted:

$$\frac{di_{LP}}{dt} = \frac{L_S - M}{L_S L_P - M^2} V_i \quad (23)$$

Considering (1), (2), (7) and (8), the following equation can be written for secondary operating mode:

$$L_P \frac{di_{LP}}{dt} + M \frac{di_{LS}}{dt} = M \frac{di_{LP}}{dt} + L_S \frac{di_{LS}}{dt} = V_i - V_o \quad (24)$$

The above equation can be rewritten as follows:

$$\frac{di_{LP}}{dt} = \frac{L_S - M}{L_S L_P - M^2} (V_i - V_o) \quad (25)$$

Considering (3), (23) and (25), it is resulted that the condition of zero value for input current is given by

$$L_S = M = K\sqrt{L_S L_P} \quad (26)$$

From (26), the following condition is calculated:

$$K = \sqrt{\frac{L_S}{L_P}} \quad (27)$$

Considering (11), (12) and (27), it is clear that the above condition for transformer model of coupled inductor is obtained if $N = 1$ and in this condition we have $N_e = K$.

IV. SIMULATION RESULTS

In order to verify the accuracy performance of the proposed converter, the simulation results in PSCAD/EMTDC software are used. The input voltage is assumed $200V$ and the capacitance of the capacitors C_1 and C_2 are considered $100\mu F$ and $220\mu F$, respectively. The switching frequency and the duty cycle are also assumed $25kHz$ and $D = 66.67\%$, respectively. To prove the validity of the given theories, several values of inductances for primary and secondary windings, and also coupled inductor coefficient are considered. For the first case study, the following values are considered:

$$R = 90\Omega \,;\; L_P = L_S = 150\,\mu H \,;\; K = 0.8 \,;\; N_P = 200$$

Considering (13) and (16), the following values are obtained for coupled inductor:

$$L_M = K^2 L_P = 96\,\mu H$$

$$L_K = (1-K^2)L_P = 54\,\mu H$$

$$N_e = \sqrt{\frac{L_s}{L_P}} = \sqrt{\frac{150}{150}} = 1$$

$$N = \frac{N_e}{K} = \frac{1}{0.8} = 1.25$$

$$N_S = N \times N_P = 1.25 \times 200 = 250$$

Considering (19), the voltage gain of the converter is calculated as follows:

$$\frac{V_o}{V_i} = \frac{1}{(1-D)} = \frac{1}{(1-0.6667)} = 3$$

Fig. 5 shows the voltage and current waveforms of the converter. As it is obvious from Figs. 5(a) and 5(b), by turning on the power switch of S, the voltage across the primary and secondary windings is equal to the input voltage that is $200V$. When the switch is turned off, according to (7) and (8) the voltage across the primary and secondary windings is equal

$-400V$. As shown in Fig. 5(c), during the turn on time interval of switch S the current of the primary winding is linearly increased and by turning off the switch S, due to the constant negative value of primary winding voltage, the primary winding current is linearly decreased. The same conditions are valid for secondary winding current as shown in Fig. 5 (d). The current waveform of the capacitor C_2 is shown in Fig. 5(e). As it is obvious from Fig. 5(e), in the first operating mode, the current value of the capacitor C_2 is $-6A$. In the second operating mode, the current value of the capacitor C_2 is $i_D - I_o$. Fig. 5(f) shows the output voltage of the converter. It is obvious from this figure that the output voltage is equal to $600V$ and this value proves the correctness of (19).

For the first case study considering the values of $L_P = L_S = 150\,\mu H$ and $K = 0.8$, it is clear that these values are not verified in (27). So, as shown in Fig. 5(c), there is considerable value for primary winding current ripple (in this case approximately 20A). As mentioned before, the main advantage of the proposed converter is to eliminate the ripple of input current if the condition given in (27) is valid. To prove the correctness of (27), the second case study is investigated. For the second case study, the following values are considered:

$$R = 90\Omega \,;\ L_P = 150\,\mu H \,;\ K = 0.95 \,;\ N_P = 200$$

Considering (27), the value of L_S is calculated as follows:

$$L_S = K^2 L_P = 0.95^2 (150) = 135.375\,\mu H$$

According to (10), (11) and (12), the following values are calculated:

$$N_e = \sqrt{\frac{L_S}{L_P}} = \sqrt{\frac{135.375}{150}} = 0.95$$

$$N = \frac{N_e}{K} = \frac{0.95}{0.95} = 1$$

$$N_S = N_P N = 200 \times 1 = 200$$

The simulation results for the second case study are shown in Fig. 6. Comparing Figs. 5 and 6, it is obvious that there is main difference between the primary and secondary windings' currents in both case studies. As shown in Fig. 6(c), the input current ripple is approximately 0.4A. The same value for the first case study is 20A. These results show that by taking the condition given in (27), it is possible to reach zero ripple in input current.

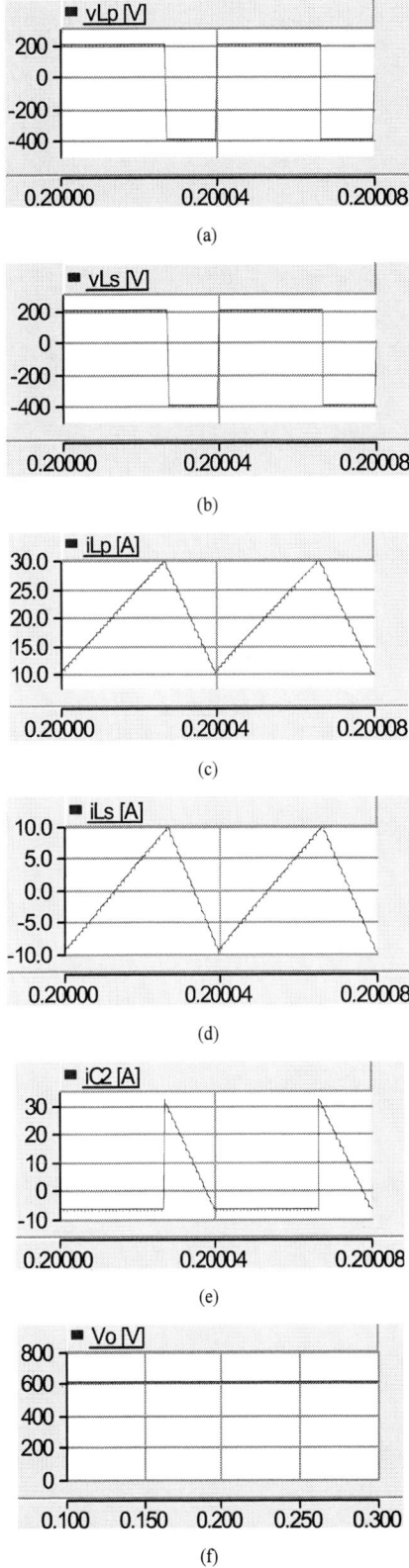

Fig. 5. The simulation results for the first case study $(K = 0.8)$.

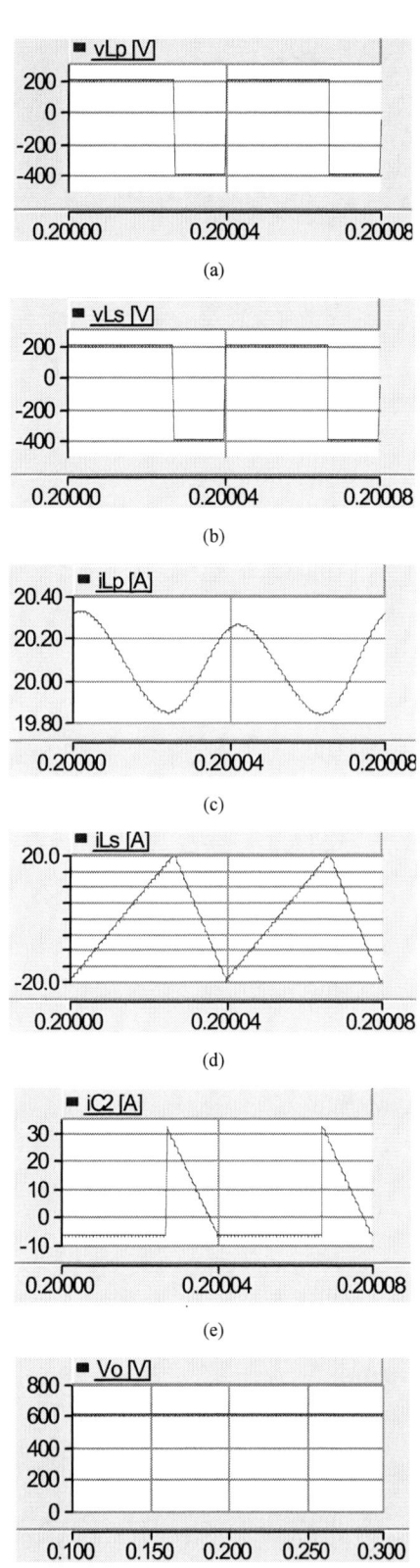

(a)

(b)

(c)

(d)

(e)

(f)

Fig. 6. The simulation results for the second case study $(K = 0.95)$.

V. CONCLUSION

In this paper, a new boost dc-dc converter is proposed. This converter is same as the conventional boost converter however the coupled inductor is used instead of the inductor in the convnetional converter. In the proposed converter, the secondary winding of the coupled inductor is connected in series to a capacitor. High numbers of changable parameters to reduce inductor current ripple is one of the main advantages of the proposed converter that leads to select optimm value to control current ripple and avoid of increasing losses. In other word, it is possible to choose a suitable value for coupling coefficient to get zero ripple in input current of converter. The correctness and accuracy performance of the proposed converter is verified through the simulation results in PSCAD/EMTDC software.

REFERENCES

[1] E. Babaei, M.E. Seyed Mahmoodieh, and M. Sabahi, "Investigating buck dc-dc converter operation in different operational modes and obtaining the minimum output voltage ripple considering filter size," *Journal of Power Electronics*, vol. 11, no. 6, pp. 793-800, Nov. 2011.

[2] I.-D. Kim, J.-Y. Kim, E.-C. Nho, and H.-G. Kim, "Analysis and design of a soft-switched PWM Sepic dc-dc converter," *Journal of Power Electronics*, vol. 10, no. 5, pp.461-467, Sep. 2010.

[3] E. Babaei and M.E. Seyed Mahmoodieh "Systematical method of designing the elements of the Cuk converter," *International Journal of Electrical Power and Energy Systems*, vol. 55, pp. 351-361, Feb. 2014.

[4] H. Cha, R. Ding, Q. Tang, and Z. Peng, "Design and development of high power dc-dc converter for metro vehicle system," *IEEE Trans. Ind. Appl.*, vol. 44, no. 6, pp. 1795-1804, Nov./Dec. 2008.

[5] E. Babaei, M.E. Seyed Mahmoodieh, and H. Mashinchi Mahery, "Operational modes and output voltage ripple analysis and design considerations of buck-boost dc-dc converters," *IEEE Trans. Ind. Electron.*, vol. 59, no. 1, pp. 381-391, Jan. 2012.

[6] E. Babaei and M.E. Seyed Mahmoodieh, "Calculation of output voltage ripple and design considerations of Sepic converter," *IEEE Trans. Ind. Electron.*, vol. 61, no. 3, pp. 1213-1222, March 2014.

[7] Y. Zhou and W. Huang, "Single-stage boost inverter with coupled Inductor," *IEEE Trans. Power Electron.*, vol. 27, no. 4, April 2012.

[8] R. Erickson and D. Maksimovic, Fundamentals of Power Electronics, 2nd ed. Boston, MA, USA: Kluwer, 2001, pp. 42–45.

[9] S. Maniktala, Switching Power Supplies A to Z. Amsterdam, The Netherlands: Elsevier, pp. 51–54.

[10] J. M. Simonelli and D.A. Torrey, "Input filter design considerations for boost- derived high power-factor converters," in *Proc. APEC*, 1992, vol. 2, pp. 186–192.

[11] R.S. Balog and P.T. Krein, "Coupled-Inductor Filter: A Basic Filter Building Block," *IEEE Trans. Power Electron*, vol. 28, no. 1, pp. 537-546, Jan. 2013.

[12] T. Nouri, S.H. Hosseini, E. Babaei and J. Ebrahimi, "Interleaved high step-up dc-dc converter based on three-winding high-frequency coupled inductor and voltage multiplier cell," *IET Power Electron.*, vol. 8, no. 2, pp. 175-189, Feb. 2015.

[13] Y. Gu and D. Zhang "Interleaved boost converter with ripple cancellation network," *IEEE Trans. Power Electron.*, vol. 28, no. 8, pp. 3860 -3868, 2013.

[14] J.J. Lee and B.H. Kwon, "Active-clamped ripple-free dc/dc converter using an input-output coupled inductor," *IEEE Trans. Ind. Electron.*, vol. 55, no. 4, pp. 1842–1854, Apr. 2008.

[15] J.J. Lee and B.H. Kwon, "DC-DC converter using a multiple-coupled inductor for low output voltages," *IEEE Trans. Ind. Electron.*, vol. 54, no. 1, pp. 467–478, Feb. 2007.

[16] J. Rosas-Caro "A transformer-less high-gain boost converter with input current ripple cancelation at a selectable duty cycle," *IEEE Trans. Ind. Electron.*, vol. 60, no. 10, pp. 4492 -4499, 2013

[17] Y. Gu, D. Zhang, and Z. Zhao, "Input current ripple cancellation technique for Boost converter using tapped inductor," *IEEE Trans. Ind. Electron.*, vol. 61, no. 10, pp. 5323-5333, Oct. 2014.

978-1-5090-0376-1/16 $31.00 © 2016 IEEE

A New Current-Fed High Step-Up Quasi-Resonant DC-DC Converter with Voltage Quadrupler

Sina Salehi Dobakhshari*

Seyed Hamid Fathi*

Amin Banaiemoqadam*

Javad Shokrollahi Moghani*

*Dept. of Electrical Engineering, Amirkabir University of Technology
Tehran, Iran

Abstract— **A new high step-up DC-DC converter with voltage quadrupler (HCVQ) for Fuel Cell (FC) application is presented in this paper. This converter employs two Quasi-Resonant (QR) and Pulse Width Modulation (PWM) schemes, and the converter's performance in these modes is compared. In both schemes, switches work under Zero Voltage Switching (ZVS) condition, and output diodes work under Zero Current Switching (ZCS). As a result of incorporating the resonant operation for main switch in QR scheme, its turn-off voltage and current is lower than PWM scheme, so total switching loss is reduced, and switches have approximately the same conduction losses. The proposed converter in QR scheme operates in Discontinuous Conduction Mode (DCM). Also it has a higher voltage gain and efficiency in comparison with the one using PWM scheme. Moreover, this converter is compared with conventional half-bridge current-fed converter with voltage doubler (HCVD). Finally, two 150 W HCVQ with QR and PWM schemes and one HCVQ are simulated to validate the theoretical and simulation results.**

Keywords— *Current-fed, fuel cells, isolated boost converter, quasi-resonant dc–dc converter, zero-voltage switching (ZVS). Introduction (*Heading 1*)*

I. INTRODUCTION

In recent years, renewable energy resources such as Photovoltaic (PV) and fuel cell have gained significant importance in energy markets. Since these power sources generate dysfunctional voltage and current, a dc-dc converter is needed to regulate the voltage to a suitable level [1-9].

PWM converters have the problems of voltage and current spikes across the switches due to capacitance and inductance leakage of transformer windings and reverse recovery of diodes [12]. In order to eliminate the spikes, active clamp circuit can be utilized [13], [8-9]. Furthermore, these converters have a lower switching frequency than resonant ones which causes lower efficiency and bulky magnetic components.

To bring fuel cells into application, an isolated voltage-fed or current-fed boost converter is needed. In [1-3] step-up voltage-fed dc-dc converters are used for fuel cells application, but due to large input current ripple of these converters, the catalyst's life time in fuel cells will decline. In voltage-fed converters an additional LC filter is required which is placed

across the fuel cell to reduce input current ripple resulting in lower efficiency. As discussed in [4], current-fed converters are better suited for fuel cell applications than voltage-fed ones. These converters represent lower input current ripple and transformer's turn-ratio, higher voltage gain, and lower switching conduction loss.

There are three kinds of current-fed converters: push-pull, full-bridge, and half-bridge. In the proposed full-bridge converter, an active clamp circuit is employed [12] which absorbs voltage spikes across the switches. However, the current of each switch is twice as the input current. In contrast with full-bridge converters, the push-pull converter in [5] has higher efficiency, but voltage stress of switches are higher. Although push-pull converters have simple structures, implementation of center-tapping in high current side of transformer is difficult. On the other hand, in full-bridge converters, switching frequency of additional switch is twice as switching frequency of the main switch. The proposed converter in [6] is a current-fed full-bridge converter with hard switching the switching frequency of which is 10kHz, causing magnetic components of bigger size. To achieve ZVS, in [13] many extra components are added which have caused lower efficiency.

A HCVD is proposed in [14] as a bidirectional converter for Plug-in Hybrid Electric Vehicle (PHEV) applications. Also, in [15] the HCVD operates as a bidirectional converter in medium power. Both converters suffer from high current stress in switches. A proposed half-bridge resonant converter in [7] is utilized for photovoltaic (PV) applications, but this converter is controlled by frequency control method which causes high Electro Magnetic Interference (EMI).

In this paper a half-bridge current-fed converter with voltage quadrupler (HCVQ) is proposed. A comparison between voltage doubler and quadrupler has been conducted in [16]. By utilizing voltage quadrupler at output, transformer's turn-ratio reduces. Although voltage quadrupler has one more diodes and capacitors compared to voltage doubler, it has lower transformer turn-ratio which totally reduces the converter losses. Furthermore, the proposed converter losses as well as voltage and current stress of semiconductors in PWM and QR schemes are compared. Since switching loss is a major portion of converters losses, by operating in QR scheme, switches current stresses will reduce using resonant operation. Also, an active clamp circuit is used to absorb voltage spikes across the switches which is caused by leakage inductance of high frequency transformer. Additionally, proposed converter is controlled with Pulse Width Modulation (PWM) method. Consequently, the proposed converter works in constant frequency, which reduces EMI [17].

II. OPERATION PRINCIPLE AND CONVERTER ANALYSIS

A HCVD converter and the proposed HCVQ converter are shown in Fig. 1. Input boost circuit is comprised of inductance L_B and switch S_1. Switch S_2 and capacitor C_2 make active clamp circuit. In secondary side of transformer of HCVQ, a voltage quadrupler is made by C_3, C_4, D_1, D_2, D_3 and D_4. Load resistance is modeled as R_L and "n" is transformer turn-ratio defined as $n= N_2/ N_1$. Operation principle of the proposed converter is discussed in PWM and QR schemes. Leakage inductance of transformer is modeled as L_k which makes resonance circuit with capacitor C_1 in Quasi-Resonant mode.

A. PWM scheme

The key voltage and current waveforms for some specific components of the proposed converter in PWM and QR scheme are plotted in Fig. 2(a) and (b), respectively. It is assumed that circuit input current and voltages of all capacitors are constant. The operation of the proposed converter is described through six intervals of one switching cycle the operating circuits of which are shown in Fig.3.

Interval 1 [t_0- t_1]: At t_0, S_2 turns off, and its current charges and discharges the parasitic capacitor of S_2 and S_1, respectively. Then, the current flows through the body diode of switch S_1. In secondary side of transformer, one half of secondary current rings through D_1 and C_3 charging C_3. the other half feeds the load through C_4 and D_4 discharging C_4. This interval finishes when body diode current reaches zero, but before that, the gate pulse should be applied to switch S_1 in order to achieve ZVS.

Interval 2 [t_1- t_2]: During this interval, the current flows through the switch S_1 which is turned ON in previous interval. The other components work the same as previous interval.

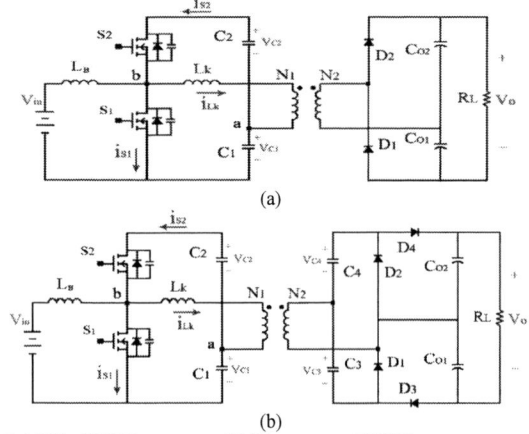

(a)

(b)
Fig. 1. (a) The HCVD converter. (b) The proposed HCVQ converter.

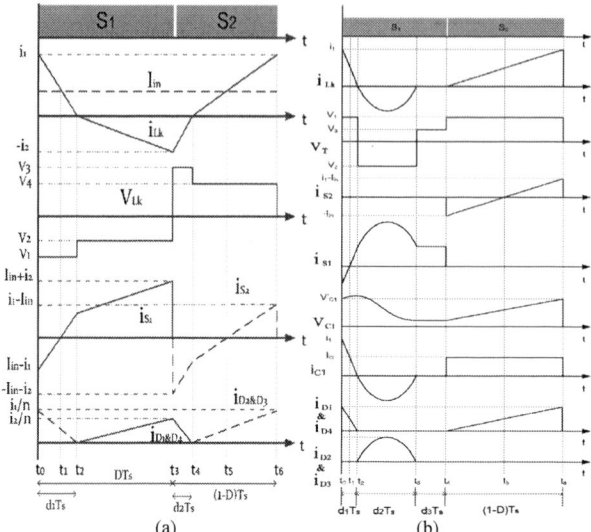

(a)   (b)
Fig. 2. The key voltage and current waveforms of the proposed converter in (a) PWM scheme, (b) QR scheme

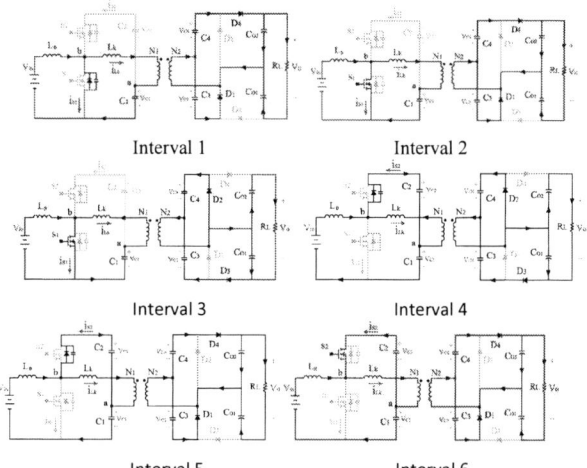

Interval 1   Interval 2

Interval 3   Interval 4

Interval 5   Interval 6

Fig. 3. Operation intervals of the proposed converter in PWM scheme.

Interval 3 [t_2- t_3]: At t_2 ,direction of leakage inductance current changes, so output current of diodes D_1 and D_4 reaches to zero, and diodes D_2 and D_3 start to conduct. Consequently, capacitors C_4 and C_3 start charging and discharging, respectively. Moreover, voltage of capacitor C_4 is applied to primary side by means of transformer, causing leakage inductance current to rise. By removing gate pulse of switch S_1, this interval finishes.

Interval 4 [t_3- t_4]: At t_3, current of switch S_1 flows through the body diode of switch S_2 and charges capacitor C_2. Simultaneously, C_4 starts to feed the output through D_4,in the meanwhile, C_3 discharges through D_1. In this interval gate pulse can be applied to switch S_2 so as to achieve ZVS. Voltage of leakage inductance will be:

$$V_3=V_{C2}+V_{C4}/n \qquad (1)$$

978-1-5090-0376-1/16 $31.00 © 2016 IEEE   198

Since this voltage is negative, current of leakage inductance decreases to zero.

Interval 5 [t_4- t_5]: At t_4, current of leakage inductance changes direction and starts rising. At the output, diodes D_2 and D_3 go OFF, and diodes D_1 and D_3 start conducting. So capacitors C_3 and C_4 charges and discharges, respectively. In this interval, voltage of leakage inductance will be:

$$V_4 = V_{C2} - V_{C3}/n \qquad (2)$$

Interval 6 [t_5- t_6]: During this interval, current of S_2 changes direction and flows through the switch. The other components operate the same as the previous interval.

B. Quasi-Resonant (QR) scheme

The key voltage and current waveforms for some specific components of the proposed converter in this mode are plotted in Fig.2(b). The operation of the proposed converter is the same as PWM scheme, but in interval 4 it goes in DCM.

III. THEORETICAL ANALYSIS

A. Voltage Gain in PWM Scheme

According to equivalent circuit of the proposed converter in Fig. 3(a), (b), (e) and (f), applying KVL to capacitors C_{O2}, C_3 and C_4 yields:

$$V_{C3} + V_{C4} = V_{Co2} \qquad (3)$$

And in Fig. 3(c):

$$V_{C3} + V_{C4} = V_{Co1} \qquad (4)$$

$$V_{Co1} + V_{Co2} = V_O \qquad (5)$$

from (3), (4), and (5):

$$V_{C3} + V_{C4} = V_{Co1} = V_{Co2} = \frac{V_O}{2}. \qquad (6)$$

As shown in Fig. 3, during the time $(1-D+d_1-d_2)T_S$ voltage of transformer is V_{C3}, and during the time $(D-d_1+d_2)T_S$, it's V_{C4}. By ignoring the value of d_1-d_2, volt-second balance in transformer yields:

$$DV_{C4} = (1 - D)V_{C3}. \qquad (7)$$

In intervals 5 and 6, the voltage across leakage inductance is $V_{C3} + V_{C2}$, So i_1 can be obtained as follow:

$$\Delta i_{Lk} = \frac{(1-D)T_s}{L_k}\left(V_{C2} - \frac{V_{C3}}{n}\right) = i_1, \qquad (8)$$

by applying KVL:

$$V_{in} = \langle V_{LB} \rangle + \langle V_{Lk} \rangle + \langle V_T \rangle + V_{C1}, \qquad (9)$$

Where V_T is voltage of transformer ($\langle \alpha \rangle$ means average of α). Since in steady state average voltage of inductors and transformer are zero, so:

$$V_{in} = V_{C1}. \qquad (10)$$

Also, it can be said that:

$$V_{in} = \langle V_{S1} \rangle = V_{C1} \qquad (11)$$

and:

$$\langle V_{S2} \rangle = V_{C2} \qquad (12)$$

$$V_{S1} = \frac{\langle V_{S1} \rangle}{1-D} = \frac{V_{in}}{1-D} \qquad (13)$$

$$V_{S2} = \frac{\langle V_{S2} \rangle}{D}. \qquad (14)$$

Where V_{S2} and V_{S1} are voltage of switches S_2 and S_1 in off state, respectively. In PWM mode, switches voltages in off state are equal ($V_{S1} = V_{S2}$), so from (12), (13) and (14):

$$V_{C2} = \frac{DV_{in}}{1-D}. \qquad (15)$$

Since, average current of diodes D_4 or D_3 is equal to output current, so according to Fig. 2(b), output current will be (d_1 is neglected):

$$I_O = \langle i_{D4} \rangle = \frac{1}{2} \cdot \frac{i_1}{2n}(1 - D)T_S. \qquad (16)$$

Substituting (8) in (16) yields:

$$I_O = \frac{1}{4n}\frac{(1-D)^2 T_s}{L_k}\left(\frac{D}{1-D}V_{in} - \frac{DV_O}{2n}\right) = \frac{V_o}{R_L} \qquad (17)$$

So voltage gain can be obtained as:

$$M = \frac{D}{\frac{4nL_k f_s}{(1-D)R_L} + \frac{D(1-D)}{2n}}. \qquad (18)$$

3.1 Voltage Gain of Quasi-Resonant Scheme

Voltage of transformer during the time $(1-D+d_1)T_S$ is V_{C4}, and during the time $d_2 T_S$ it is V_{C3}. Also, during the time $d_3 T_S$, voltage of capacitor C_1 is reflected to secondary side of the transformer. From t_4 to t_6, constant input current charges capacitor C_1:

$$\Delta V_{C1} = \frac{I_{in}.\Delta t}{C_1} = \frac{I_{in}.(1-D)T_S}{C_1}. \qquad (19)$$

Also, from t_0 to t_3, voltage of capacitor C_1 decreases to V'_{C1} by resonance. By ignoring d_1 compared with D, average voltage of this capacitor can be calculated as follow:

$$\langle V_{C1} \rangle = V'_{C1} + \frac{1}{2}\Delta V_{C1}(1 - D + d_2) = V_{in}. \qquad (20)$$

Substituting (19) in (20) yields:

$$V'_{C1} = V_{in} - \frac{I_{in}}{2C_1 f_s}(1 - D)(1 - D + d_2). \qquad (21)$$

By applying volt-second balance to the transformer, the following relation is achieved:

$$nV'_{C1}d_3 - (1 - D)V_{C3} + d_2 V_{C4} = 0. \qquad (22)$$

From (6) and (22):

$$V_{C3} = \frac{\frac{d_2 V_O}{2} + nd_3 V'_{C1}}{1-D+d_2}, \qquad (23)$$

$$V_{C4} = \frac{\frac{V_O}{2}(1-D) - nd_3 V'_{C1}}{1-D+d_2}. \qquad (24)$$

During interval 3, voltage of capacitor C_4 is applied to C_1 and L_k by means of transformer. Relevant voltage of capacitor C_1 resonances around V_{C4}/n, so:

$$\frac{V_{C4}}{n} = \frac{\Delta V_{C1}}{2} + V'_{C1}. \qquad (25)$$

By substituting (19) and (24) in (25):

$$\frac{I_{in}(1-D)}{2C_1 f_s} = \frac{V_O(1-D)}{2n(1-D+d_2)} - \frac{V'_{C1}}{1-D+d_2}. \qquad (26)$$

From (21) and (26):

$$\frac{I_{in}(1-D)}{2C_1 f_s} = \frac{V_O(1-D)}{2n(1-D+d_2)} - \frac{V_{in}}{1-D+d_2} + \frac{I_{in}(1-D)}{2C_1 f_s}, \qquad (27)$$

$$M = \frac{V_o}{V_{in}} = \frac{2n}{1-D}. \qquad (28)$$

Voltage gain against duty cycle in both PWM and QR schemes is plotted in Fig. 4. According to (18) and (28) and Fig. 4, voltage gain sensitivity to leakage inductance in QRs is less than that in PWMs.

978-1-5090-0376-1/16 $31.00 © 2016 IEEE

B. ZVS Conditions in PWM Scheme

According to the first interval in Fig. 2 ZVS, current of switch S_1 is:

$$I_{ZVS,S1} = I_{in} - i_1. \tag{29}$$

At time t_0, this current charges and discharges parasitic capacitors of switches S_1 and S_2, respectively. As a result, to ensure ZVS occurrence, the following condition should be satisfied:

$$\frac{1}{2}L_B I_{in}^2 - \frac{1}{2}L_k i_1^2 > \frac{1}{2}C_{oss,total}\left(\frac{V_{in}}{1-D}\right)^2. \tag{30}$$

The first term of the above condition is much bigger than the second one, so the condition can be simplified as follow:

$$\frac{1}{2}L_B I_{in}^2 > \frac{1}{2}C_{oss,total}\left(\frac{V_{in}}{1-D}\right)^2, \tag{31}$$

where $C_{oss,total}$ is equivalent capacitor of switches parasitic capacitor.

During the fourth interval, ZVS current of switch S_2 is:

$$I_{ZVS,S2} = -I_{in} - i_2. \tag{32}$$

This current is always negative, so ZVS occurs over the whole load range.

C. Voltage and Current Stresses of Switches

In PWM scheme, according to (13) and (14), voltage stress of switches in turn on and off states are equal to $V_{in}/(1-D)$. As shown in Fig. 2, turn-off current of switches can be obtained as:

$$i_{S1,turn-OFF} = i_2 + I_{in} = I_{in} + \frac{\frac{V_O(1-D)}{2n}-V_{in}}{L_k f_s}D. \tag{33}$$

Since average current of D_1 and D_2 are equal, ignoring $d_1 T_S$ and $d_2 T_S$ leads to:

$$\frac{1}{2}\frac{i_2}{n}DT_S = \frac{1}{2}\frac{i_1}{n}(1-D)T_S \tag{34}$$

According to (34):

$$i_2 = \frac{2(1-D)}{D}I_{in} \tag{35}$$

$$i_{S1,turn-OFF} = I_{in} + \frac{2(1-D)}{D}I_{in} \cong 2I_{in}, \tag{36}$$

$$i_{S2,turn-OFF} = i_1 - I_{in} = \frac{\frac{V_O(1-D)}{2n}-V_{in}}{L_k f_s}D - I_{in} \cong I_{in} \tag{37}$$

In QR scheme, by applying KVL to switches as well as capacitors C_1 and C_2, and as Fig. 2(b) shows, turn-on and OFF voltages of switches can be obtained as:

$$V_{S2,turn-OFF} = \frac{V_{in}}{1-D} + \frac{I_{in}}{2C_1 f_s}(1-D)(1+D-d_2) \tag{38}$$

$$V_{S1,turn-OFF} = \frac{V_{in}}{1-D} - \frac{I_{in}}{2C_1 f_s}(1-D)(1-D+d_2) \tag{39}$$

Also, based on Fig. 2(b), turn-off currents of switches are equal to I_{in}. So, in QR scheme, turn-off voltage and current of S_1 is less than PWM scheme. Consequently, turn-off loss in QR scheme is much lower than that in PWM. Switch S_2 in both schemes has the same turn-off turn, but turn-off voltage in QR scheme is a little more.

D. Comparison between QR and PWM schemes.

Switching losses can be calculated from the following equations [18]:

$$P_{turn\,OFF} = \frac{1}{2}V_{turn\,ON}i_{turn\,off}\left(\frac{t_{tr}}{T_S}\right), \tag{40}$$

$$P_{turn\,ON} = \frac{1}{2}V_{turn\,ON}i_{turn\,ON}\left(\frac{t_{tr}}{T_S}\right), \tag{41}$$

$$P_{conduction} = R_{on}I_{rms}^2, \tag{42}$$

$$P_{switching} = P_{turn\,OFF} + P_{turn\,ON} + P_{conduction}. \tag{43}$$

Due to ZVS, turn-on loss is neglected, yielding to:

$$P_{switching} = \frac{1}{2}V_{turn\,ON}i_{turn\,off}\left(\frac{t_{tr}}{T_S}\right) + R_{on}I_{rms}^2. \tag{44}$$

A comparison between QR, resonant, and PWM Schemes is shown in Fig. 5 in different resonance frequencies. Assumptions are as followed:

Switching frequency of switch S_1: $f_{s1} = \frac{1}{DT_s}$ $f_{s1} > f_{r1}$

Switching frequency of switch S_2: $f_{s2} = \frac{1}{(1-D)T_s}$

Resonant frequency of switch S_1: $f_{r1} = \frac{1}{2\pi\sqrt{L_k C_1}}$

Resonant frequency of switch S_2: $f_{r2} = \frac{1}{2\pi\sqrt{L_k C_2}}$

As shown in Fig. 5 (a), turn-off current of S_1 in PWM Scheme ($f_{s1} \gg f_{r1}$) is as much as QR Scheme ($f_{s1} < f_{r1}$), so switching loss of S_1 in PWM scheme will be considerably higher than that in QR scheme. However, current waveform of switch S_2 would be as shown in Fig. 5 (b), which is similar to PWM Scheme, and also, would have the same turn-off current. According to Fig. 5 (c), if resonant time period of S_2 becomes lower than switching period ($f_{s2} < f_{r2}$), its turn-off current will decrease. But if load becomes less than nominal value, controller circuit will decrease the duty cycle (D; consequently, current of S_2 will have enough time to resonate completely, making resonant current negative, and S_2 turns off by ZCS as is shown in Fig. 5 (d). Therefore, S_2 turns on by ZVS and turns off by ZCS. It can be seen that in this condition total switching loss decreases dramatically, but negative turn-off current in S_2 causes reverse recovery loss in body diode of S_2. Moreover, ZVS condition for the other switch will be lost.

As a result, resonant frequency for switch S_2 (f_{s2}) has to be much lower than $1/(1-D_{min})T_S$, where D_{min} is duty cycle at minimum load. To decrease switching loss of S_2, it is better to operate just like PWM scheme as shown in Fig. 5 (b).

IV. SIMULATION RESULTS

The proposed converter is simulated in both PWM and QR schemes in Pspice OrCAD, and component values are represented in Table I. Simulation results are shown in Fig. 6. Current waveform of leakage inductances are shown in Fig. 6 (a). Here the differences between QR and PWM schemes are clearly shown. In Fig. 6 (b) current of switch S1 in PWM and QR scheme are shown. Turn-off current of S_1 in QR scheme is 6.5A and in PWM scheme is 12.5A. As mentioned before, turn-off current of S_1 in QR scheme is dramatically reduced. Current of switch S_2 is shown in Fig. 6 (c), as discussed in section 3.5 and 3.6, these currents are almost the same, and their turn-off current is 6.5A. Voltage waveform of S_1 and S_2 in QR scheme are depicted in Fig. 6 (d) and (e), respectively.

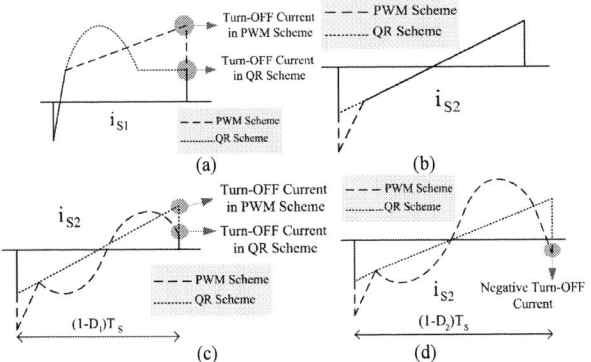

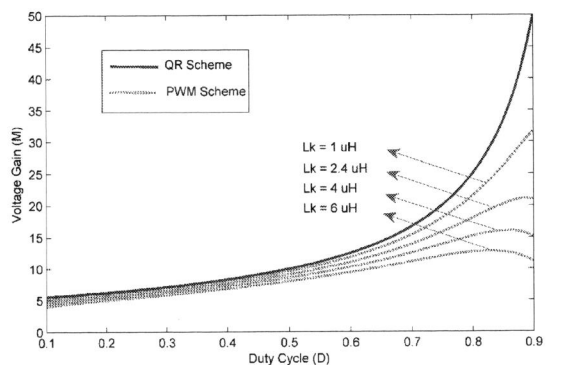

Fig. 5. A comparison of switch's turn-off current in some different states of resonance. (a) PWM ($f_{S1} \gg f_{r1}$) and QR ($f_{S1} < f_{r1}$) scheme for S_1. (b) PWM ($f_{S1} \gg f_{r1}$) and QR ($f_{S1} < f_{r1}$) scheme for S_2. (c) PWM ($f_{S1} \gg f_{r1}$) and Resonant ($f_{S1} = f_{r1}$) scheme for S_2 (d) PWM ($f_{S1} \gg f_{r1}$) and Resonant ($f_{S1} = f_{r1}$) scheme for S_2.

Fig. 6. Voltage gain of the proposed converter in PWM and QR schemes according to duty cycle

Table I Proposed converter's performance and components value

Components		Values
Input Voltage		24 V
Output Voltage		380 V
Output power		150 W
Switching Frequency		100 KHz
Capacitors C_{O1}-C_{O2}		470 uF 250 V
Capacitor C_1	PWM	47uF
	QR	1 uF
Capacitors C_2-C_4		47 uF
Input Inductance		270uH
Leakage Inductance		1.2uH

Turn-off voltage of S_1 and S_2 is 65v and 90v, respectively. While in PWM scheme turn-off voltage of the switches are 80v, and their waveform are as a square wave (that is why they are not shown in here). The voltage change of switches in QR scheme is caused by capacitor C_1 alternative voltage. Although this swing in QR scheme increases turn-off voltage in S_2, decreases in S_1. Current of diodes D1 and D4 and also

D2 and D3 are shown in Fig. 6 (d) and (e), respectively. Needless to say, in these figures, di/dt of diodes in QR scheme is lower than PWM scheme. Maximum current of diodes D1 and D4 in both schemes is 2.2A, but for diodes D2 and D3 in QR scheme it's 2A, and in PWM scheme it's 0.6A. Efficiency of the proposed converter in both schemes and HCVD is plotted in Fig. 7.

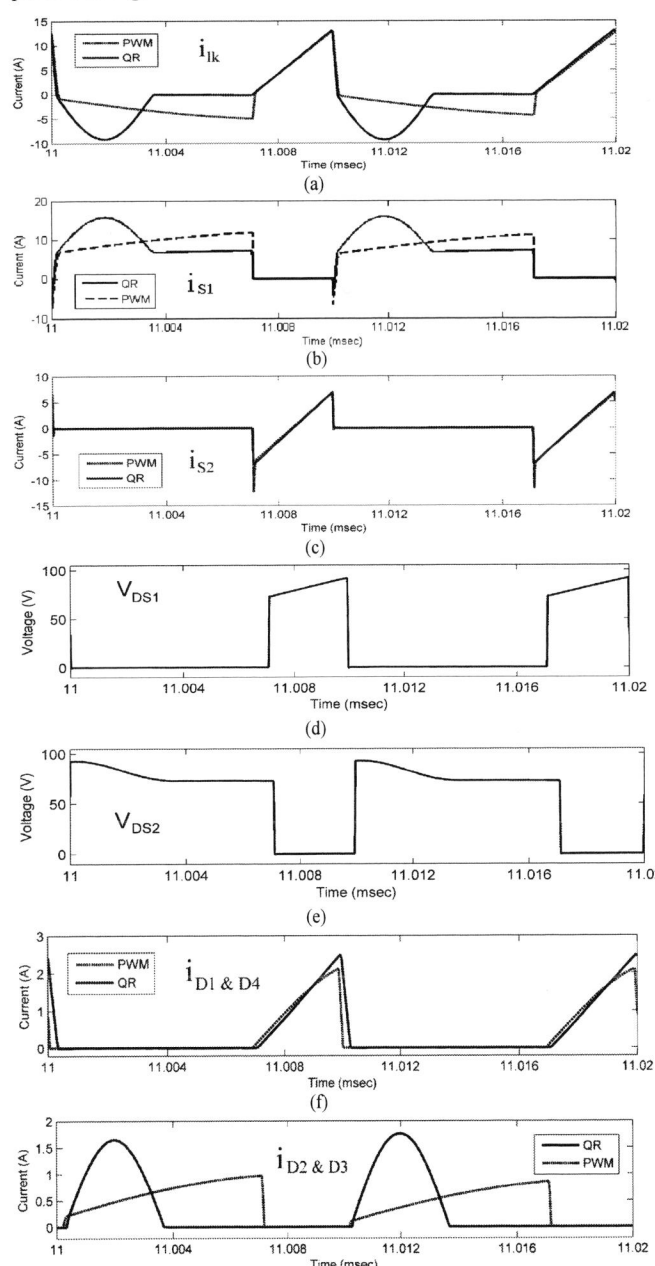

Fig. 6. Simulation results in both PWM and QR schemes. (a) Leakage inductance's current. (b) Current of switch S_1. (c) Current of switch S_2. (d) Voltage of switch S_1. (e) voltage of switch S_2. (f) Current of diodes D_1 and D_4. (g) Current of diodes D_2 and D_3.

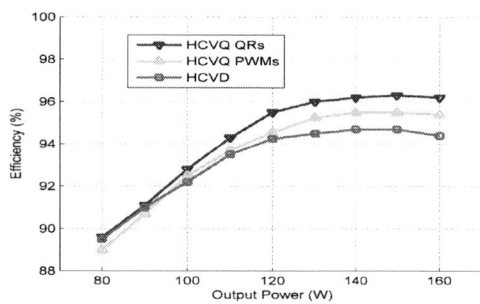

Fig. 7 Efficiency Comparison.

I. Conclusion

A high step-up current-fed DC-DC converter with voltage quadrupler for fuel cell application is presented in PWM and QR schemes. The two schemes in which the converter operates are compared in this paper. Also, this proposed converter is compared with conventional high step-up half-bridge current-fed converter with voltage doubler (HCVD). Although the proposed converter has two diodes and two capacitors more than HCVD, it has halved transformer turns ratio which results in lower loss, more efficiency, and a smaller converter. Moreover, the proposed converter in QR scheme has the following benefits over the PWM schem:

- Turn-off current and voltage of main switch over whole load range is minimized, which decreases switching loss.
- Lower di/dt in output diodes.
- More voltage gain by employing resonance features.

Having employed these features, the proposed converter in QR scheme achieved an efficiency of 96.3 % at full load. In addition, in PWM scheme and HCVD, 95.4 % and 94.5 % efficiency is achieved, respectively.

References

[1] J. L. Duarte, M. Hendrix, and M. G. Simoes, "Three-port bidirectional converter for hybrid fuel cell systems," IEEE Trans. Power Electron., vol. 22, no. 2, pp. 480–487, Mar. 2007.

[2] H. Tao, A. Kotsopoulos, J. L. Duarte, and M. A. M Hendrix, "Transformer coupled multiport ZVS bidirectional dc–dc converter with wide input range," IEEE Trans. Power Electron., vol. 23, no. 2, pp. 771–781, Mar. 2008.

[3] R. Sharma and H. Gao, "Low cost high efficiency dc–dc converter for fuel cell powered auxiliary power unit of a heavy vehicle,"IEEE Trans. Power Electron., vol. 21, no. 3, pp. 587–591, May 2006.

[4] J. Wang, F. Z. Peng, J. Anderson, A. Joseph, and R. Buffenbarger, "Low cost fuel cell converter system for residential power generation," IEEE Trans. Power Electron., vol. 19, no. 5, pp. 1315–1322, Sep. 2004.

[5] J.-M. Kwon and B.-H. Kwon, "High step-up active-clamp converter with input-current doubler and output-voltage doubler for fuel cell power systems," IEEE Trans. Power Electron., vol. 24, no. 1, pp. 108–115, Jan. 2009.

[6] X. Kong and A. M. Khambadkone, "Analysis and implementation of a high efficiency, interleaved current-fed full bridge converter for fuel cell system,"IEEE Trans. Power Electron., vol. 22, no. 2, pp. 543–550, Mar. 2007.

[7] B. York, W Yu and J. Sh Lai, "An Integrated Boost Resonant Converter for Photovoltaic Applications," IEEE Trans. on Power Electron. , VOL. 28, NO. 3, pp.1199-1207, March 2013.

[8] R. L Andersen, and I. Barbi, "A ZVS-PWM Three-Phase Current-Fed Push–Pull DC–DC Converter" IEEE Trans. Ind. Electron., Vol. 60, No. 3, pp. 838-847, March 2013.

[9] P. Xuewei, P. R. Prasanna, and A. K.Rathore, "Magnetizing-Inductance-Assisted Extended Range Soft-Switching Three-Phase AC-Link Current-Fed DC/DC Converter for Low DC Voltage Applications" IEEE Trans. Power Electron., vol. 28, no. 7, pp. 3317–3328, July 2013.

[10] S. S. Lee, S. W. Rhee, and G. W. Moon, "Coupled inductor incorporated boost half-bridge converter with wide ZVS operation range,"IEEE Trans. Ind. Electron., vol. 56, no. 7, pp. 2505–2512, Jul. 2009.

[11] Sh.-J Chen, K.-Shan, S.-P. Yang, and M.-F. Cho, "Analysis and implementation of an interleaved series input parallel output active clamp forward converter" IET Power Electronics, vol. 6, no. 4, pp. 774-781, April 2013.

[12] U.R. Prasanna, A.K. Rathore, "Extended Range ZVS Active-Clamped Current-Fed Full-Bridge Isolated DC/DC Converter for Fuel Cell Applications: Analysis, Design, and Experimental Results" IEEE Transactions on Industrial Electron., Vol. 60, Issue: 7, pp.2661-2672, July 2013.

[13] A. J. Mason, D. J. Tschirhart, and P. K. Jain, "New ZVS phase shift modulated full-bridge converter topologies with adaptive energy storage for SOFC application," IEEE Trans. Power Electron., vol. 23, no. 1, pp. 332–342, Jan. 2008.

[14] H. Sangtaek and D. Divan, "Bi-directional dc/dc converters for plug-in hybrid electric vehicle (PHEV) applications," inProc. IEEE Appl. Power Electron. Conf. Expo (APEC), 2008, pp. 1–5.

[15] H, Li, F, Z, Peng and J. S. Lawler, "A Natural ZVS Medium-Power Bidirectional DC–DC Converter With Minimum Number of Devices," IEEE Transactions on Industry Applications, vol. 39, no. 2, pp. 525-535, March/April 2003.

[16] Y. Zhao, X. Xiang, W. Li, X. He and Ch. Xia, "Advanced Symmetrical Voltage Quadrupler Rectifiers for High Step-Up and High Output-Voltage Converters" IEEE Trans. Power Electron., vol. 28, no. 4, pp. 1622–1631, April. 2013.

[17] W. J. Lee, S. W. Choi,C. E. Kim, and G. W. Moon, "A New PWM-Controlled Quasi-Resonant Converter for a High Efficiency PDP Sustaining Power Module," IEEE Trans. on Power Electron. , Vol 23, no.4, pp.1782-1790, July 2008.

[18] R. Erickson and D. Maksimovic, Fundamentals of Power Electronics, 2004 :Kluwer

This page intentionally left blank.

7th Power Electronics, Drive Systems & Technologies Conference (PEDSTC 2016)
16-18 Feb. 2016, Iran University of Science and Technology, Tehran, Iran

Design and Implementing of a Novel Resonant Switched-Capacitor Converter for Improving Balancing Speed of Lithium-Ion Battery Cells

Shahin Goodarzi[1], Reza Beiranvand[2], Reza Rezaii[3], Mohammad Amin Abolhasani[4], Mustafa Mohamadian[5]

Department of Electrical and Computer Engineering

Tarbiat Modares University

Tehran, Iran

s.goodarzi@modares.ac.ir[1], beiranvand@modares.ac.ir[2], reza.rezaii@modares.ac.ir[3],
m.abolhassani@modares.ac.ir[4], mohamadian@modares.ac.ir[5]

Abstract— Active topologies based on the Switched-Capacitor converters (SCC) are used for battery cell balancing because they eliminate magnetic elements effectively. In addition, these converters have easier implementation, integration capability, and reduced size. Despite of all these benefits, SCCs suffer from some disadvantages such as increased number of active switches, edge current, low balancing speed, and high switching losses. In this paper, a chain resonant SCC is introduced which can realize soft switching condition at zero current, limiting edge current, and increasing the balancing speed to overcome the problems of SCCs. To confirm the operation of the proposed converter, a chain resonant SCC with the capacity of 2150 mAh, the nominal voltage of 3.6 V, and the switching frequency of 50 kHz is simulated in MATLAB/SIMULINK, and implemented in the laboratory. The simulation and implementation results of the converter approve the correctness of the converter performance.

Keywords— Lithium-Ion battery cell, switched capacitor converter (SCC), resonant power conversion.

I. INTRODUCTION

Lithium-Ion battery is one of the most used rechargeable batteries in industry due to its valuable advantages such as high energy density and low self-discharge rate [1]-[14]. In order to prepare the required load voltage, several batteries are connected in series, so they are charged and discharged in series connection at a same time [3]. In this case, repeating the charge and discharge processes causes some mismatches in the batteries due to unavoidable differences in batteries' chemical and electrical characteristics such as: production variety and unbalanced aging and temperature distribution. Lithium-Ion battery cells should be protected from overvoltage and under voltage rates. So, there should be a balancing circuit to avoid the unpredicted events. The balancing circuit removes these mismatches, significantly, enhances the total capacity, improves the efficiency, and increases the battery lifetime [4], [5]. Different balancing circuits have been introduced in the literature, which are categorized in two groups: 1) passive, and 2) active.

The passive balancing methods have disadvantages such as energy loss and thermal problems [2]. To overcome these problems active balancing circuits have been introduced which active elements such as inductors and capacitors are employed in their constructions.

Usually, two types of active converters are used for balancing battery cells: 1) Switched capacitor based converters, 2) Inductor based converters. Different circuits based on inductor have been introduced in [5]-[8]. But, these circuits have large sizes, high cost, and excess loss. Therefore, switched-capacitor topologies have been introduced for balancing battery cells owing to the fact that they do not use heavy electromagnetic elements like inductors and transformers, and are much compact and easily implemented [1],[12], [13]. Studying these converters, it is found that SCCs have two problems in the process of balancing battery cells: 1) Balancing battery cells needs more time in comparison with the inductor based converters, especially when the number of cells is increased [1]. 2) Switching loss is high due to hard switching [2]. 3) The current has excessive spike.

To increase speed of the SCCs, these converters are studied in the literature. The double-tiered switched capacitor converter has been introduced to increase the balancing speed [1], [11]. This method reduces the balancing time, strikingly, by employing additional bridging. However, due to the different gradient of ambient temperature, the risk of unbalancing between outer cells will increase normally [1]. In addition, this method is useless in reaching much higher balancing speed. To improve the performance by increasing the speed especially between outer cells, two circuits with chain structure have been studied in [1]. Decrease of the farthermost cells to half is one of the advantages of this circuit [1]. Therefore, the cells balancing speed are significantly increased by this chain structure.

As mentioned above, other main problem of the SCCs is due to limitation on accessible size of storage elements and dependence of their sizes on switching frequency causes to increase switching frequency, and it will increase the switching loss because switches due to hard switching. To overcome this problem, resonant SCC is used, in practice. In [1]-[14], Different converters based on the resonant SCC have been presented. The performance analysis of complex SCCs is introduced in [1]-[3]. Specifically, a resonant SCC has been used for voltage balancing of series-connected capacitor in [2].

Although SCC with Chain Structure that presented in [1] has high cells balancing speed, there are some problems such as

978-1-5090-0376-1/16 $31.00 © 2016 IEEE

high number of switches and hard switching that increases the switching loss. For these reasons, Zero-Current Switching (ZCS) SCC with Chain Structure is presented in this paper and applied to a series battery-cell string or super capacitor string to achieve zero-voltage gap. Here, there is a very small inductor in series with the capacitor in resonant circuit to limit the current spike, and achieve zero-current switching. Therefore, zero voltage gap among cells, limited spike current, overcoming the problem of switching loss, and improving the cell balancing speed are advantages of this converter. Moreover, at the same time, it avoids the existence of awkward large inductor. So, it is more suitable for series battery string or super capacitor string.

In Sec. II slow balancing speed of conventional resonant SCCs is described. In Sec. III, the design of the proposed system is presented, and the soft switching in the chain structure is investigated. Finally, conclusion is made in Sec. VI.

II. ANALYSIS AND EXPLANATION OF SLOW BALANCING SPEED OF THE CONVENTIONAL RESONANT SCCS

Conventional resonant SCC made for balancing battery cells are presented in Fig. 1(a). There are four switches in this converter which are controlled simultaneously with a constant switching frequency. Also, they are connected to up and down with a duty cycle of 45% by considering a dead time interval.

Based on how switches are connected, the operation of the circuit is divided into two operating modes, as illustrated in Fig. 1(b) and (c). When the voltages of two adjacent cells are different, the charge transfer between these cells is conducted by the common capacitor (C_{ab}) to balancing the cells. The charge transfer in this converter is done cell by cell. So, the

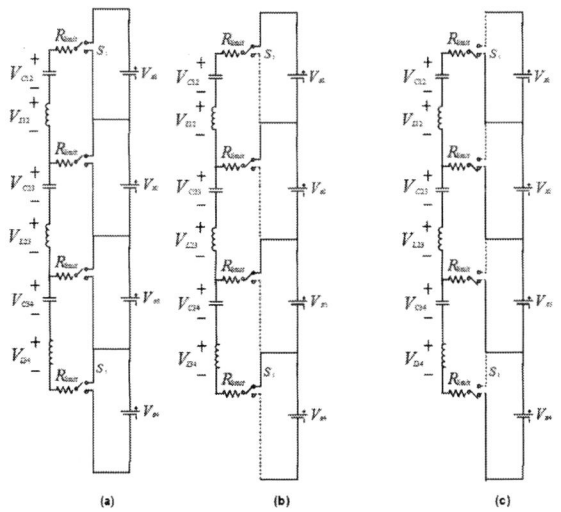

Fig. 1. Conventional resonant SCC [2] (a) circuit description, (b) state A, and (c) state B.

balancing speed is slow especially when the number of cells is increased.

Equivalent circuits of two switching modes are plotted in Fig. 1 (b) and (c) by considering the battery's inner impedance (r_b) equivalent series resistor of the common capacitor, and

resonant inductor (r_c). Since common capacitor between two cells would become connected to upper and lower cells with a unique duty cycle, voltage of the common capacitor (V_{Cab}) would be calculated as follows:

In the first switching subinterval the capacitor voltage is

$$V_{C_{ab}} = V_{B_a} - V_{L_{ab}} \tag{1}$$

For the second subinterval the capacitor voltage expression will be:

$$V_{C_{ab}} = V_{B_b} - V_{L_{ab}}$$
$$\tag{2}$$

By averaging the capacitor voltage:

$$<V_{C_{ab}}>_{T_s} = \frac{1}{T}\left(\int_0^{\frac{T}{2}} V_{C_{ab}} dt + \int_{\frac{T}{2}}^{T} V_{C_{ab}} dt\right)$$
$$= \left(\int_0^{\frac{T}{2}} (V_{Ba} - V_{Lab}) dt + \int_{\frac{T}{2}}^{T} (V_{Bb} - V_{Lab}) dt\right) \tag{3}$$

Since inductor current value is equal to its value at the end of period:

$$i_l(t) - i_l(0) = \frac{1}{T}\int_0^T (V_L) dt = 0 \rightarrow <V_L>_{T_s} = 0 \tag{4}$$

Combination of (1), (2), and (3) yields:

$$<V_{C_{ab}}>_{T_s} = \frac{(V_{B_a} + V_{B_b})}{2} \tag{5}$$

Each cell is connected to the common capacitor just in one subinterval. So, the average current entering the batteries is calculated as follows:

$$[i_{Ba}]_{avg} = \frac{\frac{(V_{Ba} + V_{Bb})}{2} - \left(\frac{1}{T}\int_0^T V_{Lab} dt\right) - V_{Ba}}{R_{eq}} = \frac{V_{Bb} - V_{Ba}}{2 \times R_{eq}} \tag{6}$$

In the second switching subinterval, batteries average current is equal to:

$$[i_{Bb}]_{avg} = \frac{\frac{(V_{Ba} + V_{Bb})}{2} - \left(\frac{1}{T}\int_0^T V_{Lab} dt\right) - V_{Bb}}{R_{eq}} = \frac{V_{Ba} - V_{Bb}}{2 \times R_{eq}} \tag{7}$$

Where, D is duty cycle and R_{eq} is summation of the switch conduction resistor (R_{sw}) and inrush current limiting resistor (R_{limit}). According to (6) and (7) current of the cell is related to the equal resistor and the existing voltage difference between two cells. For ease of calculation it is assumed that the voltage of the most upper cell is different from the other cells as follows:

$$V_{B1} > V_{B2} = V_{B3} = V_{B4} \tag{8}$$

Considering Fig. 1, we can write:

$$\begin{bmatrix} V_{B1}-V_{C12}-\left(\dfrac{1}{T}\displaystyle\int_0^T (V_{L12})\,dt\right) \\[2mm] V_{B2}-V_{C23}-\left(\dfrac{1}{T}\displaystyle\int_0^T (V_{L23})\,dt\right) \\[2mm] V_{B3}-V_{C34}-\left(\dfrac{1}{T}\displaystyle\int_0^T (V_{L34})\,dt\right) \\[2mm] 0 \end{bmatrix} = \begin{bmatrix} (V_{B1}-V_{B2})/2 \\ 0 \\ 0 \\ 0 \end{bmatrix}$$

$$= \begin{bmatrix} 2R_{eq}+r_p & -R_{eq} & 0 & 0 \\ -R_{eq} & 2R_{eq}+r_p & -R_{eq} & 0 \\ 0 & -R_{eq} & 2R_{eq}+r_p & 0 \\ 0 & 0 & 0 & 0 \end{bmatrix} \cdot \begin{bmatrix} i_{1A} \\ i_{2A} \\ i_{3A} \\ i_{4A} \end{bmatrix} \qquad (9)$$

$$\begin{bmatrix} 0 \\[2mm] V_{B2}-V_{C12}-\left(\dfrac{1}{T}\displaystyle\int_0^T (V_{L12})\,dt\right) \\[2mm] V_{B3}-V_{C23}-\left(\dfrac{1}{T}\displaystyle\int_0^T (V_{L23})\,dt\right) \\[2mm] V_{B4}-V_{C34}-\left(\dfrac{1}{T}\displaystyle\int_0^T (V_{L34})\,dt\right) \end{bmatrix} = \begin{bmatrix} 0 \\ (V_{B2}-V_{B1})/2 \\ 0 \\ 0 \end{bmatrix}$$

$$= \begin{bmatrix} 0 & 0 & 0 & 0 \\ 0 & 2R_{eq}+r_p & -R_{eq} & 0 \\ 0 & -R_{eq} & 2R_{eq}+r_p & -R_{eq} \\ 0 & 0 & -R_{eq} & 2R_{eq}+r_p \end{bmatrix} \cdot \begin{bmatrix} i_{1B} \\ i_{2B} \\ i_{3B} \\ i_{4B} \end{bmatrix} \qquad (10)$$

Here, $r_p = r_b + r_c$. Generally, the batteries balancing current is expressed as follows, which N is the number of battery cells [1]:

$$\begin{bmatrix} I_{B1} \\ I_{B2} \\ . \\ . \\ I_{BN-1} \\ I_{BN} \end{bmatrix} = \begin{bmatrix} i_{1A}+i_{1B} \\ i_{2A}+i_{2B} \\ . \\ . \\ i_{(N-1)A}+i_{(N-1)B} \\ i_{NA}+i_{NB} \end{bmatrix} = \frac{1}{|i_{NB}|} \begin{bmatrix} X(N-1) \\ X(N-2)-X(N-1) \\ . \\ . \\ X(1)-X(2) \\ 0-X(1) \end{bmatrix} \qquad (11)$$

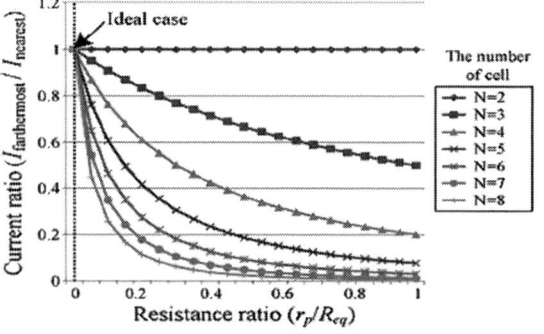

Fig. 2. Balancing current ratio between the nearest cell and the farthermost cell according to the number of cells and resistance ratio [1].

$$X(N)=\alpha^{N-1}-(N-2)\alpha^{N-3}+\left(\sum_{K=1}^{N-4}K\right)\alpha^{N-5}-\left(\sum_{M=1}^{N-6}\sum_{K=1}^{M}K\right)\alpha^{N-7}+\ldots \qquad (12)$$

$$\alpha=2+\frac{r_p}{R_{sw}},\ and \sum_{k=1}^{m}k=0\,(m<1) \qquad (13)$$

In Fig. 2, the ratio of the balancing current of farthest cell to closest cell is depicted in terms of the number of cells and resistance ratio (r_p/R_{eq}) [1]. In Fig. 2, by increasing the number of cells and the resistance ratio, current of the farthest cell is reduced, so in spite of its good efficiency and limited edge current, the conventional resonant switched capacitor balancing circuit has a slow balancing speed.

III. THE PROPOSED RESONANT SWITCHED CAPACITOR CHAIN CONVERTER

As explained in Sec. II, despite of higher balancing speed of the switched capacitor chain converter, it has some problems such as high edge current, high switching loss, and low efficiency due to its hard switching and large number of switches. Also, the resonant SCC has lower balancing speed despite of its high efficiency, low switching loss, and low edge current. Thus, resonant switched capacitor chain structure is proposed which has low switching loss due to ZCS, high efficiency, edge current prevention, and high balancing speed in comparison with the resonant SCC. In this topology, switches are turned off under the ZCS condition. Efficiency of this converter is inversely proportional to the rate of power, because, the higher resonant current produces higher resistance loss. Thus, in normal conditions this circuit will be used in average power rate to achieve highest efficiency. It should be mentioned that this converter can be used in high power rating with changing resonant tank parameters if much higher balancing speed is needed.

The balancing circuit of the proposed converter consists of two sections: 1) Switches network and 2) Resonant tank network which are shown in Fig. (3).

Resonant tank consisting resonant capacitor and inductor is designed for saving energy, and switches network for connection to resonant tank.

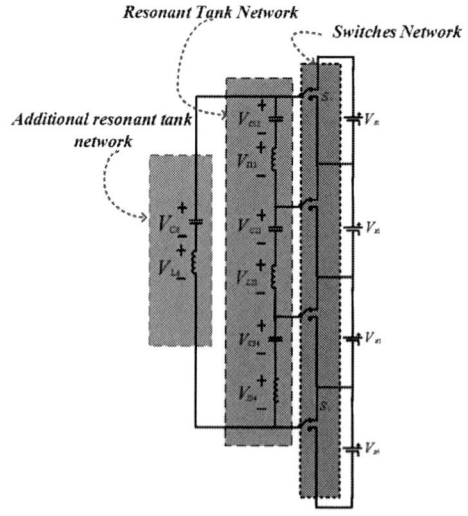

Fig. 3. Proposed converter with an additional resonant tank network.

978-1-5090-0376-1/16 $31.00 © 2016 IEEE

A. Circuit operation modes

To simplify design and analyses of the circuit and its operation modes, it is assumed that there are two battery cells in the string, and the upper cell has higher voltage than the lower cell, as shown in Fig. 4. Also, waveforms of the proposed circuit are illustrated in Fig. 5. It should be mentioned that analyzing and designing for n battery cells is based on the aforementioned analysis, and the initial conditions for inductor and capacitor is equal to zero.

1) Saving energy mode [t_0-t_1]:

Switches S_1 and S_3 are turned on in this operating mode and the resonant tank is connected to the upper battery cell. The current will flow through the resonant capacitor which causes the capacitor voltage to increase until it reaches to $2v_{B1}$. The equivalent circuit of this operating mode is given in Fig. 4 (a). When the resonant inductor's current, (i_{lr}), reaches zero, S_1 and S_2 will be turned off under the ZCS condition. The resonant inductor current and resonant capacitor voltage typical waveforms have been plotted in Fig. 5. The converter is described by the following equations:

$$i_{lr}(t) = v_{B1}\sqrt{\frac{c_r}{l_r}} \sin\left(\omega_r(t-t_0)\right) \tag{14}$$

$$v_{cr}(t) = v_{B1}\left(1-\cos\left(\omega_r(t-t_0)\right)\right) \tag{15}$$

$$\Delta t = t_1 - t_0 = \pi\sqrt{l_r c_r} \quad , \quad \omega_r = \frac{1}{\sqrt{l_r c_r}} \tag{16}$$

2) Releasing energy mode [t_1-t_2]:

Switches S_2 and S_4 will be turned on in this mode and the resonant tank will be connected to the lower battery cell. Then, the current will flow through the lower cell from the resonant capacitor c_r. The amplitude of the resonant current will decrease in comparison with the saving energy mode, owing to the fact that the quality factor of the circuit is not infinite, the voltage of the resonant capacitor will decrease until it reaches zero. The equivalent circuit of this mode is drawn in Fig. 4 (b). When the resonant inductor current become zero, S_2 and S_4 will be turned off, so same as the previous mode they are turned off by soft switching at zero current. The waveforms of this mode are presented in Fig. (5), and its equations are:

$$i_{lr}(t) = (2v_{B1} - v_{B2})\sqrt{\frac{c_r}{l_r}} \sin\left(\omega_r(t-t_1)\right) \tag{17}$$

$$\Delta t = t_2 - t_1 = \pi\sqrt{l_r c_r} \quad , \quad \omega_r = \frac{1}{\sqrt{l_r c_r}} \tag{18}$$

B. Designing of elements of the resonant tank

For designing with high quality factor, primary waveform will be similar to a sine wave, and by choosing the circuit's switching frequency equal to the resonant frequency, soft switching in zero current can be achieved. Therefore, according to the equations (19) and (20), the value of elements of resonant tank i.e., resonant inductor and capacitor can be calculated.

$$Q = \frac{2\pi f_r l_r}{R_c} \tag{19}$$

$$f_s \leq f_r = \frac{1}{2\pi\sqrt{l_r c_r}} \tag{20}$$

C. Balancing Speed Comparisons

1) Average of the Steps for Charge Transfer

A comparison between resonant SCC with chain structure and conventional resonant SCC is presented. In the proposed structure, when the power rating is equal to the conventional converter, average steps of charge transfer is a suitable factor for balancing speed of battery cells. Average steps of charge transfer are calculated by the needed steps for transferring energy from one cell to the other. For simplicity of analysis it is assumed that each battery is composed of N series cells, and N is an odd number [1], [2].

Average steps of charge transfer are calculated by the following formula [1], [4]:

$$Step_{average} = \frac{\sum Step_{ij-min}}{\# of\ cases} \tag{21}$$

In which $Step_{ij}$ is the minimum steps needed for charge transfer from cell number i to j, and # of cases is the number of whole steps.

2) Comparison between Averages of Steps

Minimum steps for charge transfer from cell number i to the cell number j in conventional resonant SCC is displayed in

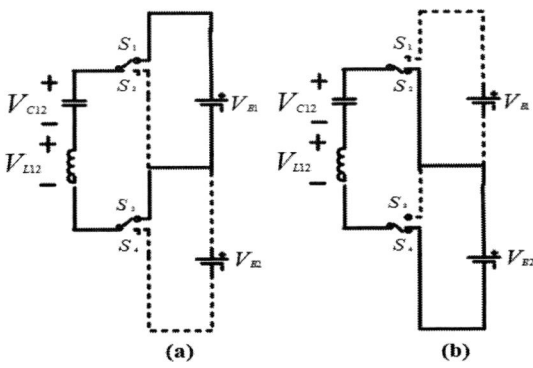

Fig. 4. Circuit operation of. (a) store mode, (b) release mode.

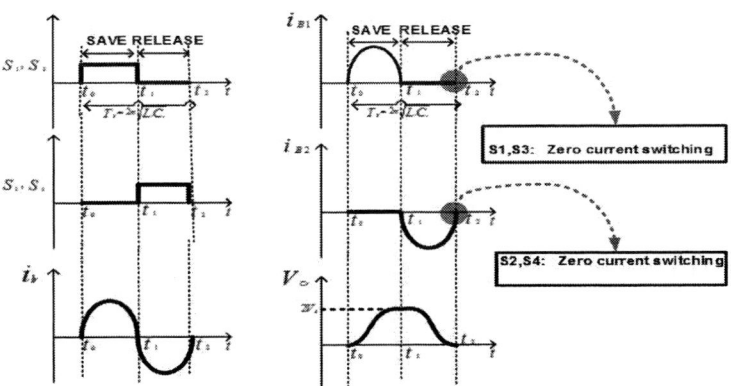

Fig. 5. Operational waveforms of the proposed circuit.

978-1-5090-0376-1/16 $31.00 © 2016 IEEE

Table I. According to this table the whole number of steps and the total number of minimum steps is calculated as follows [1]:

$$\sum Step_{ij-\min} = 2\left(\sum_{K=1}^{N-1}\sum_{i=1}^{K} i\right) = \sum_{K=1}^{N-1} K(K+1) = \frac{N(N-1)(N+1)}{3} \quad (22)$$

$$X = \frac{N-1}{2} \ , \# of \ cases = N(N-1)$$

Thus, based on these equations, the average of steps for the structure of conventional resonant SCC is:

$$Step_{average} = \frac{(N+1)}{3} \quad (23)$$

In Table II the minimum steps for charge transfer from cell number i to the cell number j in the proposed converter is illustrated. According to this table the whole number of steps and the total number of minimum steps is calculated as follows[1]:

$$\sum Step_{ij-\min} = 2\left(N\sum_{K=1}^{X} K\right) = N*X(X+1) = N\left(\frac{N-1}{2}\right)\left(\frac{N+1}{2}\right) \quad (24)$$

$$X = \frac{N-1}{2}, \# of \ cases = N(N-1)$$

Thus, based on these equations, the average of steps for the structure of the proposed converter is:

$$Step_{average} = \frac{(N+1)}{4} \quad (25)$$

In Fig. 6 the average steps for charge transfer in each structure is depicted, and it can be seen that the proposed converter have lower average steps than the conventional converter. Thus, the balancing speed in the proposed converter is higher than the conventional converter.

IV. SIMULATION RESULTS

The simulation of the proposed resonant SCC for three battery cells is done due to comparison between the conventional and proposed converter's balancing behavior. For the proposed converter the switching frequency of 50 kHz is considered, the upper cell has an unbalanced voltage of 3.56 volts, and the other cells' voltages is 3.28 volts. So, the voltage difference between the unbalanced cell and other cells is equal to 280 mV. In the dynamic model, batteries are considered as a capacitor in series with a resistor, and the capacitor is polarized for modeling the transient response in the charge and discharge duration.

To reduce the time of simulation and according to the fact that time constant of the resistor and the paralleled capacitor is very short, they are neglected. Thus, battery cells are modeled as capacitors with large capacitance (50 mF). The characteristics of the circuit elements are shown in Table III. For validating the performance of the proposed balancing circuit, several situations are studied in simulations which are presented hereafter. Also, the simulation results for both of conventional and proposed system is delivered.

A. Balancing Speed Comparison

The simulation results for two converters are depicted in Fig. 7. In a specified time like 10 ms, the maximum and minimum voltage difference is measured, and for the conventional and proposed system are respectively 49 mV and 19 mV. This value is very less in the proposed converter rather than the conventional one. Thus, the balancing speed in the

TABLE I. Minimum steps for charge transfer of conventional resonant switched capacitor converter [1].

To cell / From cell	B_1	B_2	...	B_X	...	B_{N-2}	B_{N-1}	B_N
B_1		1	...	X-1	...	N-3	N-2	N-1
B_2	1		...	X-2	...	N-4	N-3	N-2
...	...	...	...	...		...	...	...
B_{N-1}	N-2	N-3	...	X	...	1		1
B_N	N-1	N-2	...	X+1	...	2	1	

TABLE II. Minimum steps for charge transfer of the proposed converter[1].

To cell / From cell	B_1	B_2	...	B_X	...	B_{N-2}	B_{N-1}	B_N
B_1		1	...	X-1	...	3	2	1
B_2	1		...	X-2	...	4	3	2
...	...	...		...		...	...	...
B_{N-1}	2	3	...	X	...	1		1
B_N	1	2		X	...	2	1	

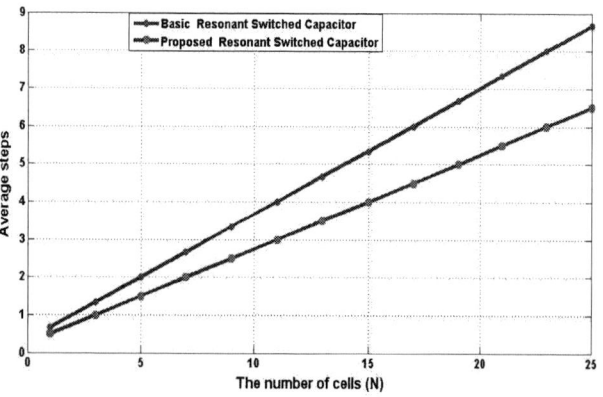

Fig. 6. Average steps for charge transfer in each structure.

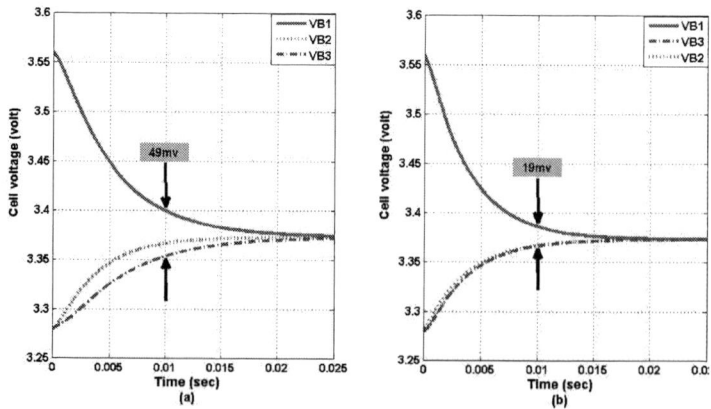

Fig. 7. MATLAB simulation results. (a) Conventional, (b) Proposed.

proposed converter is very higher than the conventional converter.

The comparison between voltage differences as a function of time of balancing is depicted in Fig. 8. In the conventional converter, for reaching the 50 *mV* voltage difference the running time is 10 *ms*. But in the proposed converter this time is equal to 7 *ms*.

B. Achieving Soft Switching in Zero Current

The resonant tank current and gate to source voltage of the third switch is illustrated in Fig. 9. As mentioned in the previous section, by adding an inductor in series with the resonant capacitor, zero current switching can be achieved.

V. EXPERIMENTAL RESULTS

To verify the simulation results, several experimental tests have been done. Power switches IRFP460 MOSFETs are used in the circuit. Value of the resonant inductors is equal to 10 *μH*, and their cores are *RM6*. The polyester capacitors are used for resonant capacitor due to their acceptable response in high frequencies, and their capacitances are 1 *μF*. As shown in Fig. 10, for firing and controlling 6 power switches in the power circuit, there should be 6 gate driver signals and 6 isolated power supplies because all source pin of these switches have not common points. Thus, HCPL3120 is employed to isolate the power section from signals and driving the switches. Besides, for supplying these ICs, the MAU215 isolated power

regulator is used. The second circuit used in this balancing converter is control board which is composed of an ARM-CORTEX-STM32F407 digital controller. The whole system is depicted in Fig. 10.

A. Balancing speed comparison

According to Fig. 11, in a specified time like 70 minutes, maximum and minimum measured voltage difference in conventional converter is 15 *mV*, and in the proposed converter is 0.4 *mV*, which is very smaller than the conventional one. Voltage difference as a function of time spent on balancing is shown in Fig. 12. When voltage difference reaches to 150 *mV*, 21 minutes is needed in the conventional method, while in the proposed method this time is reduced to 10 minutes.

B. Realizing the ZCS condition

Resonant tank current and gate-source voltage of the third switch is drawn in Fig. 13, which illustrates realizing the ZCS, clearly.

VI. CONCLUSIONS

To achieve high balancing speed and realizing ZCS condition resonant SCC with chain structure has been proposed. Also, its different operation modes have been described, in detail. The converter has been analyzed, mathematically, too. Soft switching is achieved by adding an inductor in series with the resonant tank's capacitor. In addition, it has been proved that the proposed converter has higher balancing speed than the conventional converters.

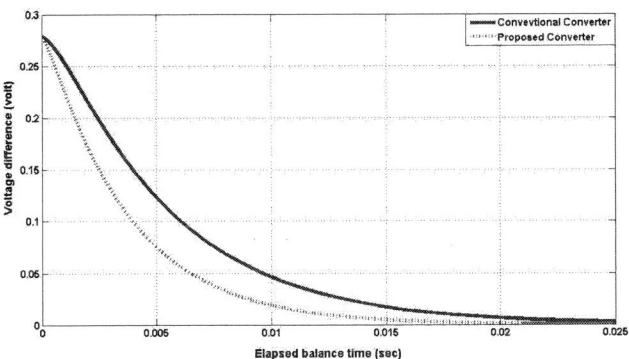

Fig. 8.   Comparison between voltage differences.

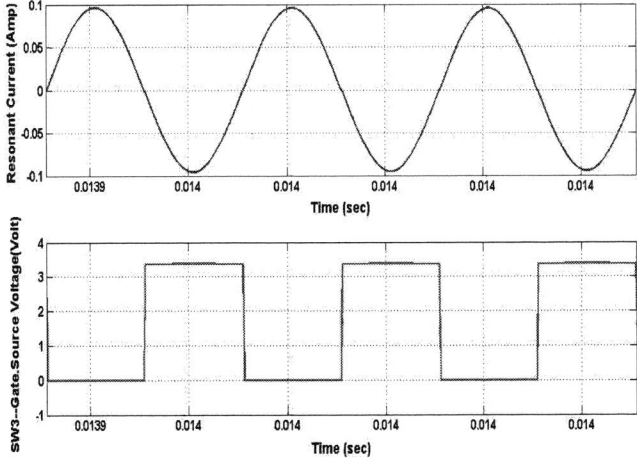

Fig. 9.   MATLAB simulation waveforms.

TABLE III. VALUES OF THE CIRCUIT COMPONENTS

Parameter	Value
Frequency	50 kHz
Equivalent resistance=r_b+r_c+r_{sw}	12 mΩ
Battery Capacitance	50 mF
Resonant Capacitor	1 μF
Resonant Inductor	10μH

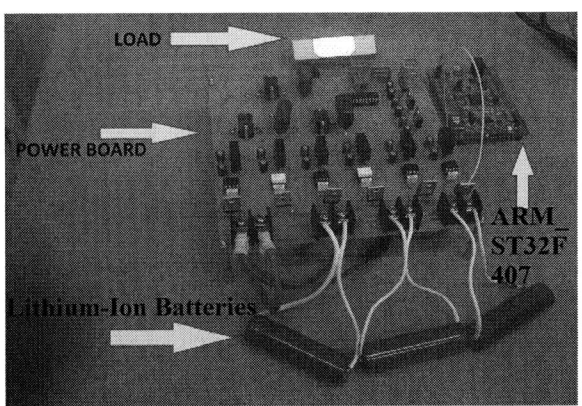

Fig. 10.   Prototype converter for balancing 3 lithium-ion battery cells experimental setup.

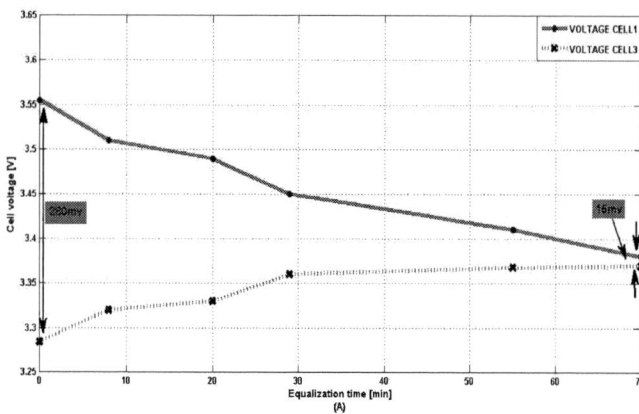

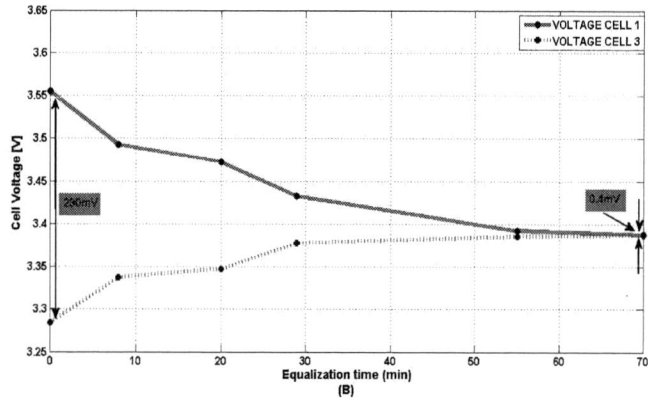

Fig. 11. Experimental results (a) conventional converter and (b) the proposed converter

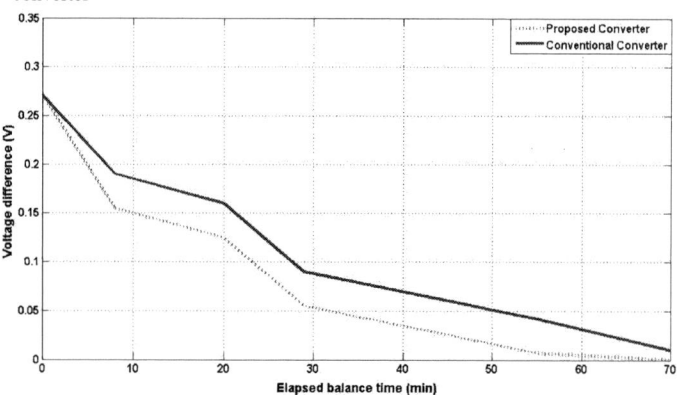

Fig. 12. Experimental results of comparison between voltage differences.

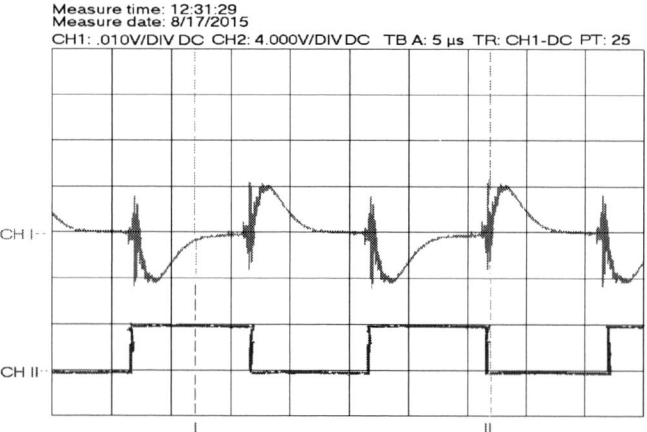

Fig. 13. Experimental waveforms of the prototype converter.

REFERENCES

[1]. Kim, Moon-Young, et al. "A chain structure of switched capacitor for improved cell balancing speed of lithium-ion batteries." *Industrial Electronics, IEEE Transactions on* 61.8 (2014): 3989-3999.

[2]. Lee, Kyung-min, Yoo-chae Chung, Chang-Hyeon Sung, and Bongkoo Kang. "Active Cell Balancing of Li-Ion Batteries using LC Series Resonant Circuit." *2015 IEEE.* IEEE,2015.

[3]. Wilkins, Steven, et al. "Optimised battery capacity utilisation within Battery Management Systems." *Ecological Vehicles and Renewable Energies (EVER), 2015 Tenth International Conference on.* IEEE, 2015.

[4]. Kim, Moon-Young, et al. "A new cell-to-cell balancing circuit with a center-cell concentration structure for series-connected batteries." *ECCE Asia Downunder (ECCE Asia), 2013 IEEE.* IEEE, 2013.

[5]. Phung, Thanh Hai, Alvaro Collet, and Jean-Christophe Crebier. "An optimized topology for next-to-next balancing of series-connected lithium-ion cells." *Power Electronics, IEEE Transactions on* 29.9 (2014): 4603-4613.

[6]. Kim, Moon-Young, Jun-Ho Kim, and Gun-Woo Moon. "Center-cell concentration structure of a cell-to-cell balancing circuit with a reduced number of switches." *Power Electronics, IEEE Transactions on* 29.10 (2014): 5285-5297.

[7]. Yuanmao, Ye, K. W. E. Cheng, and Y. P. B. Yeung. "Zero-current switching switched-capacitor zero-voltage-gap automatic equalization system for series battery string." Power Electronics, IEEE Transactions on 27.7 (2012): 3234-3242.

[8]. Daowd, Mohamed, et al. "Passive and active battery balancing comparison based on MATLAB simulation." Vehicle Power and Propulsion Conference (VPPC), 2011 IEEE. IEEE, 2011.

[9]. Emadi, Ali, Young Joo Lee, and Kaushik Rajashekara. "Power electronics and motor drives in electric, hybrid electric, and plug-in hybrid electric vehicles." Industrial Electronics, IEEE Transactions on 55.6 (2008): 2237-2245.

[10]. Active Cell Balancing Methods for Li-Ion Battery Management ICs using the ATA6870.

[11]. Baughman, Andrew C., and Mehdi Ferdowsi. "Double-tiered switched-capacitor battery charge equalization technique." Industrial Electronics, IEEE Transactions on 55.6 (2008): 2277-2285.

[12]. Park, Hong-Sun, et al. "Two-stage cell balancing scheme for hybrid electric vehicle Lithium-ion battery strings." Power Electronics Specialists Conference, 2007. PESC 2007. IEEE. IEEE, 2007.

[13]. Lee, Yuang-Shung, and Guo-Tian Cheng. "Quasi-resonant zero-current-switching bidirectional converter for battery equalization applications." Power Electronics, IEEE Transactions on 21.5 (2006): 1213-1224.

[14]. Kuhn, B.T., Pitel, G.E. and Krein, P.T., 2005, September. Electrical properties and equalization of lithium-ion cells in automotive applications. In *Vehicle Power and Propulsion, 2005 IEEE Conference* (pp. 5-pp). IEEE.

7th Power Electronics, Drive Systems & Technologies Conference (PEDSTC 2016)
16-18 Feb. 2016, Iran University of Science and Technology, Tehran, Iran

DC-DC Converter for Energy Loss Compensation and Maximum Frequency Limitation in Capacitive Deionization Systems

Hamed Mehrabian-Nejad[1], Babak Farhangi[2], Shahrokh Farhangi[3], Sadegh Vaez-Zadeh[4]

Department of Electrical and Computer Engineering, University of Tehran, Tehran, Iran.

Emails: [1] mehrabian@ut.ac.ir, [2] b.farhangi@ece.ut.ac.ir, [3] farhangi@ut.ac.ir, [4] vaezs@ut.ac.ir

Abstract— Currently, shortage of the water resources is a global issue. Water desalination is a solution that can be used to solve the water shortage problem. Capacitive deionization (CDI) is an effective water desalination method. CDI offers relatively low energy consumptions and low initial costs. This paper improves energy exchange between CDI units used in efficient CDI systems. Traditionally, hysteresis control is employed for transferring energy through power converters; this control method induces switching losses. Converter losses cause CDI systems not to operate in nominal voltage. In this paper, a buck converter and a CDI selector are proposed to compensate the losses. The maximum switching frequency is limited to prevent high switching losses. Simulation results validate the proposed scheme.

Keywords— *Water desalination systems, Capacitive deionization (CDI), buck-boost converter, buck converter, hysteresis control, high switching losses.*

I. INTRODUCTION

Nowadays, due to population growth, water misuse, and management problems, the usable water resources are significantly limited. Water scarcity defines as a lack of sufficient water resources. According to United Nation (UN) research, one seventh of world's population do not access to clean drinking water. Desalination of brackish water and seawater is a technology to address water shortage. Desalination processes require energy for separation of salt from brackish water or seawater [1,2]. There are several methods to desalinate water such as Reverse Osmosis (RO), distillation, electrodialysis and Capacitive Deionization (CDI) [3]. In comparison with other methods, CDI consumes less energy. For instance, RO technology uses 4 kWh/m³ for water desalination while CDI reduces this amount to 1 kWh/m³. Considering environmental pollution, distillation systems consume fossil fuel as a source of energy that it can produce high density of CO^2 emissions. In contrast, CDI can be supplied by electricity that is produced through photovoltaic cells [4,5].

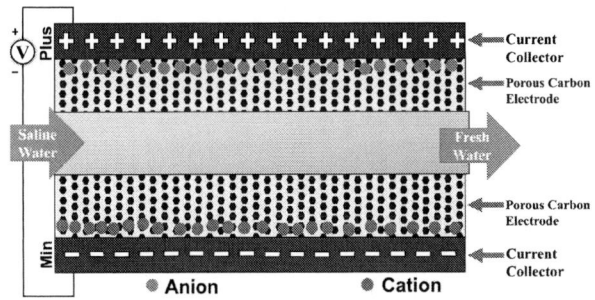

Fig. 1. Purification mode by applying a DC voltage in CDI system

CDI systems can use membrane for desalination which is called membrane capacitive deionization (MCDI). By adding membrane to CDI, system consumes less energy. However the price of maintenance and water pressure can increase. Therefore, CDI system can be considered without any membrane for desalination and trying to save energy with power converter [6].

CDI method is based on two operating mode which is called purification and purging modes. Fig.1 shows one cell of CDI. As it can be seen in this figure, the DC voltage is applied to the CDI electrode for removing the ions when water flows between two plates. This mode continues up to the ions saturation of the plates and is called purification mode. The purging mode starts by charging the other electrode or transferring the stored energy from one CDI cell to another. In parallel with this action, output valve changes from desalinated water to waste water and the feeding water purges the saturated electrodes [7]. The energy stored in CDI cells during water desalination process can be recovered through power converters. Desalination system efficiency is improved by this energy recovery technique. Unidirectional and bidirectional power converters are suitable for this application [8-21]. In [22, 23], a buck-boost converter is proposed that can transfer energy between two CDI cells.

978-1-5090-0376-1/16 $31.00 © 2016 IEEE

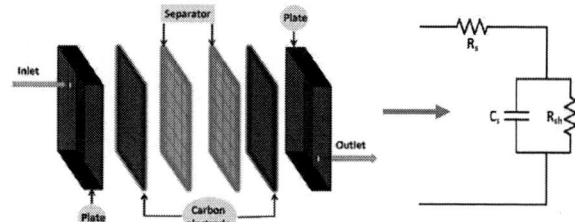

Fig. 2. Electrical model of CDI application

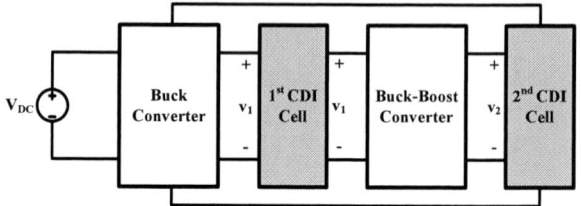

Fig. 3. General converter of the CDI system

A mathematical algorithm and hysteresis control have been developed for buck-boost converter in [24,25]. In these papers, with mode changing, power converter losses are not compensated. Moreover, by applying hysteresis control, frequency of the converter will increase that leads to high switching losses.

In this paper, a buck-boost converter is applied to transfer energy between two CDI cells; additionally, a buck converter is proposed to compensate losses that is caused by buck-boost converter and internal resistance of CDI system. Also, a CDI selector is added to change CDI cell during different mode of system operation. Finally, the amount of inductor increases to control and limit the maximum switching frequency. The next section will discussed about CDI electric models and its formula. Power converters for energy transferring is defined in section III. Section IV and V describe controller design and simulation results, respectively. Finally conclusions are given in section VI.

II. CDI MODELS

As mentioned in the introduction, CDI is able to store energy. This is due to the existence of two electrodes and a brine water between them that plays a role of the medium. This structure presents a supercapacitor. In other words, CDI is modeled as a supercapacitor as shown in Fig.2. C_{Cell} is capacitance of the cell that can be defined as:

$$C_{Cell} = \frac{\varepsilon A}{d} \tag{1}$$

Where, ε is the permeability of input water (brine water), A is surface area of the plates and d is the distance between the plates. With increasing water salinity of, cell capacitance will increase due to variation in ε. R_s is the series resistance of CDI that the smaller series resistance is preferred. This resistance is formed by resistance of wires and electrodes.

R_{sh} is shunt resistor of CDI that relates to the leakage current of electrodes and desire to be as high as possible.

III. POWER CONVERTER

The function of power converter for this plant is to transfer energy from DC voltage source to CDI cell and also between two CDI cells. Fig.3 presents the proposed system that includes the buck and the buck-boost converters. To obtain equation of frequency variations, each converter is

analyzed in this section.

A. Buck converter

Generally, this converter is used to step down the input voltage of the system. For CDI application, buck converter can be used as an energy compensator and initial charge source. As it can be seen in Fig.4, when buck converter's switch is conducting, the buck inductor current can be calculated as:

$$i_1 = i_{1\min} + \frac{1}{L_1}\int v_1(t)d(t) \tag{2}$$

Due to the nature of supercapacitor, voltage $v_1(t)$ across first supercapacitor (1st CDI cell model) is constant during each cycle. Hence, when the inductor current reaches its maximum value, the elapsed time t_{on1} can be estimated as:

$$i_{1\max} = i_{1\min} + \frac{1}{L_1}(V_{DC} - v_1)t_{on1} \Rightarrow t_{on1} = \frac{\Delta i_1 L_1}{(V_{DC} - v_1)} \tag{3}$$

Where, i_1, L_1 are value of inductor and inductor's current of the buck converter, respectively. When the diode of buck converter is conducting, the feeding source is closed by switch and the time's equation can be extracted as:

$$t_{off1} = \frac{\Delta i_1 L_1}{v_1} \tag{4}$$

By considering equation (3) and (4), the frequency variations of the buck converter can be expressed as:

$$f_1(t) = \frac{1}{t_{on1} + t_{off1}} = \frac{V_{DC}}{\Delta i_1 L_1 (V_{DC} - v_1)v_1} \tag{5}$$

v_1 varies with time. Thus, in control system, if Δi_1 is constant, frequency varies with time too.

B. Buck-Boost converter

Bidirectional buck-boost converter can transfer energy between two CDI cells. In first subinterval of buck-boost, inductor current can be calculated as equation (3). Voltage of the supercapacitor is constant during switching period. Therefore, the elapsed time t_{on2} of buck-boost converter can be expressed as:

$$i_{2\max} = i_{2\min} + \frac{1}{L_2}(v_2)t_{on2} \Rightarrow t_{on2} = \frac{\Delta i_2 L_2}{v_2} \tag{6}$$

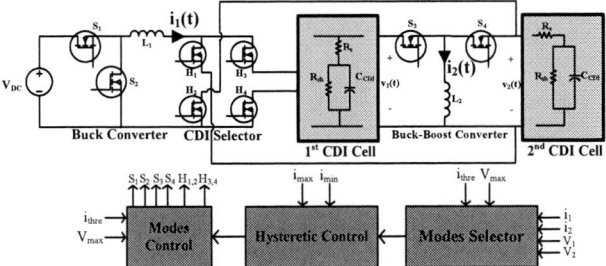

Fig. 4. Proposed model of CDI for compensating losses

TABLE I: Switching pattern of CDI system

Mode Numbers	Transferring Energy from	S_1	S_2	S_3	S_4	H_1,H_2	H_3,H_4
1	DC to Cell$_1$	HC*	HC	0	0	0	1
2	DC to Cell$_2$	HC	HC	0	0	1	0
3	Cell$_1$ to Cell$_2$	0	0	HC	HC	0	0
4	Cell$_2$ to Cell$_1$	0	0	HC	HC	0	0

HC* = Hysteretic Control

Where i_2 , L_2 are the value of inductor and inductor current of the buck-boost converter, respectively. For second subinterval, the turn-off time can be extracted as:

$$t_{off2} = \frac{\Delta i_2 L_2}{V_2} \qquad (7)$$

Equations (6) and (7) lead to obtain frequency variations of the buck-boost converter. Hence,

$$f_2(t) = \frac{1}{t_{on2} + t_{off2}} = \frac{V_1 + V_2}{\Delta i_2 L_2 V_1 V_2} \qquad (8)$$

According to equation (8), frequency of buck-boost converter can be limited by current difference Δi_2 .

IV. CONTROLER DESIGN

Electrical model of the CDI system is illustrated in Fig.4. It includes buck converter, buck-boost converter, CDI selector, DC voltage source, and two CDI cells. Table 1 shows how switches operate during different operating modes. Hysteresis controller is used that operates at a variable frequency. According to mode 1 switching pattern, S_3 and S_2 operate in base of the hysteresis control. In first subinterval, DC source commences to charging the buck inductor and S_1 is on, while S_2 is off. After its current reaches to $i_{1\max}$ S_1 is off, S_2 is on and the current falls down. By reaching current of inductor to $i_{1\min}$, switching strategy will change again to previous mode. The difference between these two current limits is:

$$\Delta i_1 = i_{1\max} - i_{1\min} \qquad (9)$$

The cycle of buck converter continues until CDI voltage reaches to $V_{\max}$. This procedure also will carry out in buck-boost converter to transfer energy from Cell$_1$ to Cell$_2$. The transition time between the operating modes is determined by monitoring the inductor current i_{thre} . The control block diagram is shown in Fig. 5. As system starts to operate,

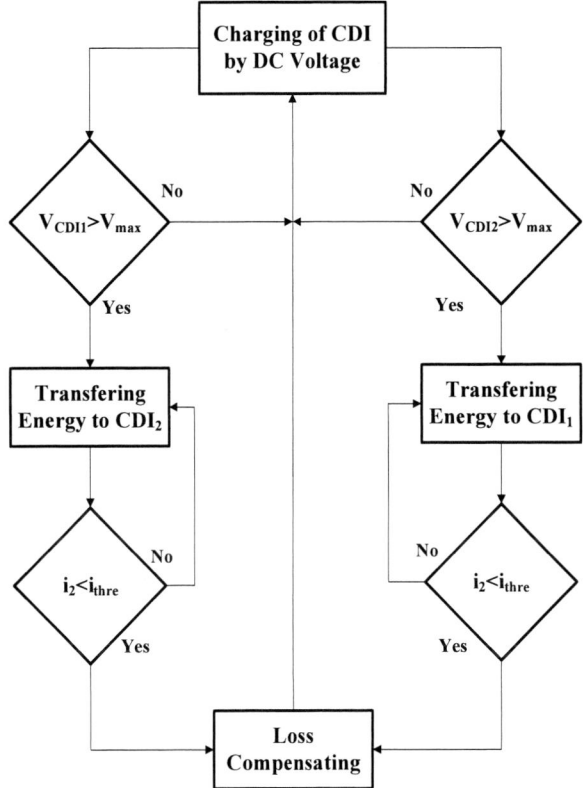

Fig. 5. Control algorithm by considering loss compensation

energy transfers from DC source to the first CDI through CDI selector. Then, When voltage of the first cell reaches to $V_{\max}$, mode of the system will change and the stored energy in that cell will transfer to the second CDI cell. This operation continues until current of buck-boost inductor reaches to i_{thre} . At this time, the stored energy in the first CDI cell is fully transferred to the second cell; the operating mode should be changed again. Afterwards, through CDI selector and buck converter, the energy loss of buck-boost converter is compensated. This will continue up to point that the CDI voltage reaches to $V_{\max}$. This cycle repeats until adequate water is provided.

V. SIMULATION RESULTS

In this section, the proposed control structure is simulated in MATLAB SIMULINK environment. As discussed, the CDI physical cell is replaced by its electrical model. Parameters of the system are listed in Table II. CDI parameters are for a 100 mm×100 mm graphite electrodes covered with 100μm of nanoporous carbon [19]. System operation in modes 2 and 3 are shown in simulation case studies. Energy is transferred between two CDI cells in mode 3 that is shown in Fig. 6a and in mode 2, the loss is compensated by buck converter that is shown in Fig. 6b.

978-1-5090-0376-1/16 $31.00 © 2016 IEEE   213

TABLE II: Main Parameter of converters and CDI model [19]

Converter Parameter	Value	Description
$L_{1,2}(mH)$	6.5	Inductance of both
$Rs(\Omega)$	0.05	Series Resistance of CDI
$R_{sh}(\Omega)$	50	Parallel Resistance of CDI
$C_{CDI}(F)$	2	Capacitance of CDI
$V_{max,CDI}(V)$	1.2	Maximum Voltage of CDI
$i_{thre}(A)$	0.8	Current threshold of mode changing
$[i_{max1},i_{min1}](A)$	[1.15,0.85]	Current difference of buck converter
$[i_{max2},i_{min2}](A)$	[1.05,0.95]	Current difference of buck-boost converter
$V_{DC}(V)$	10	A Voltage of DC Source
$T_s(S)$	1e-05	Sample Time of Simulation

Fig.7 shows voltage and current waveforms of the CDI cells. Simulation commences from mode 3 (transferring energy between cells) and continues up to (t=10.5 s) that is in mode 2. Inductors currents in both converters are plotted in Fig. 8. As it can be seen in Fig. 8a, the difference of current Δi_1 in buck converter increase up to 0.3 A while current difference of buck-boost converter Δi_2 is 0.1 A. This is due to limit the maximum frequency variations by using equation (5). The frequency variations of this system and the one which is presented in [18] are illustrated in Fig.9. As it can be seen, the frequency variation in Fig. 9a is much higher than Fig.9b. This is due to the increasing of current difference Δi_2 and inductor value L that is calculated by equation (8). Therefore, instead of increasing frequency that increases the converter losses, the converters design parameters Δi_2, L are modified. This modification is used for buck-boost converter. Finally, maximum switching frequency of the converter can be evaluated with current difference as seen in Fig.10. According to this figure, the maximum frequency is reduced by increasing the current difference. This figure is plotted according to ($L_2 = 6.5\,mH$). Therefore, a small increases in current difference lead to remarkably reduce switching frequency.

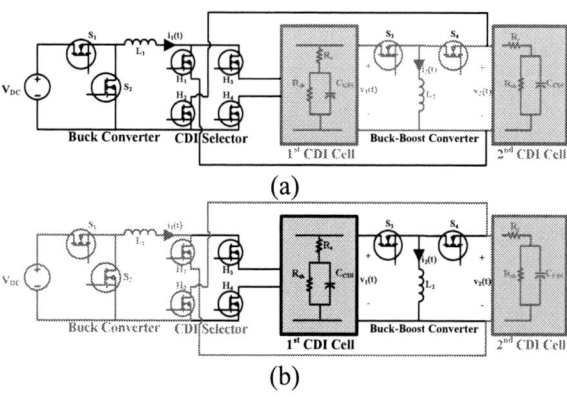

(a)

(b)

Fig. 6. Operation of CDI during a) transferring energy in mode 3 and b) compensating energy in mode 2

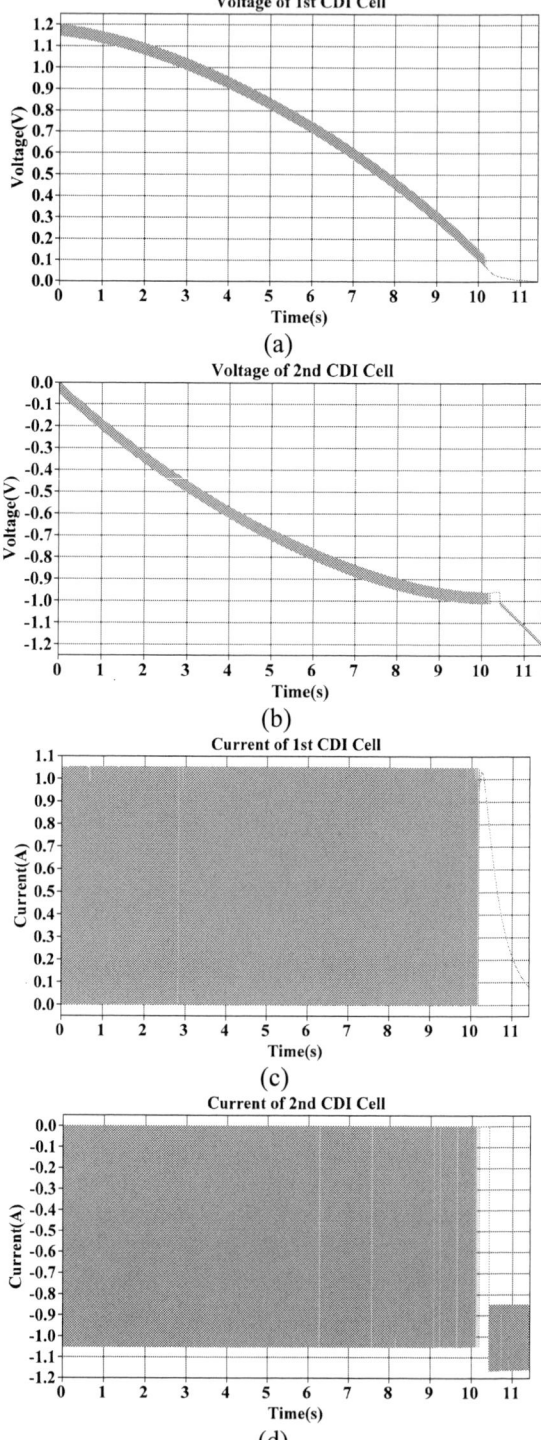

(a)

(b)

(c)

(d)

Fig. 7. Waveform of CDI system during operations of mode 3 and mode 2 a) voltage of first CDI cell b) voltage of second CDI cell c) current of first CDI cell d) current of second CDI cell

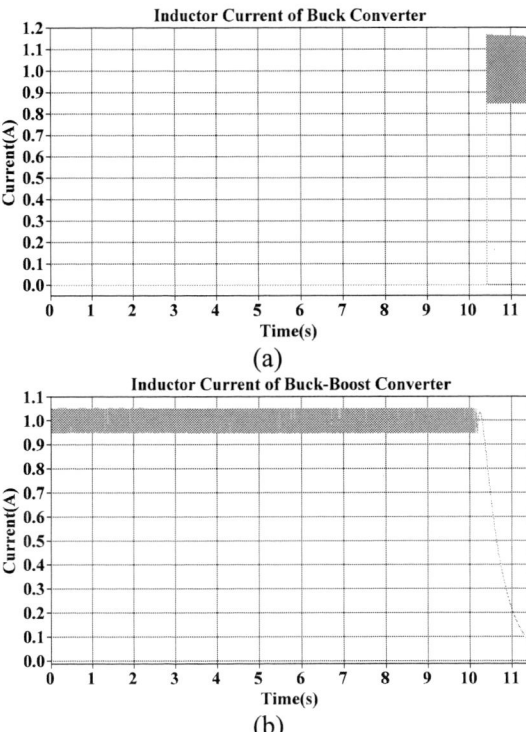

(a)

(b)

Fig. 8. Current of inductor a) buck converter b) buck-boost converter

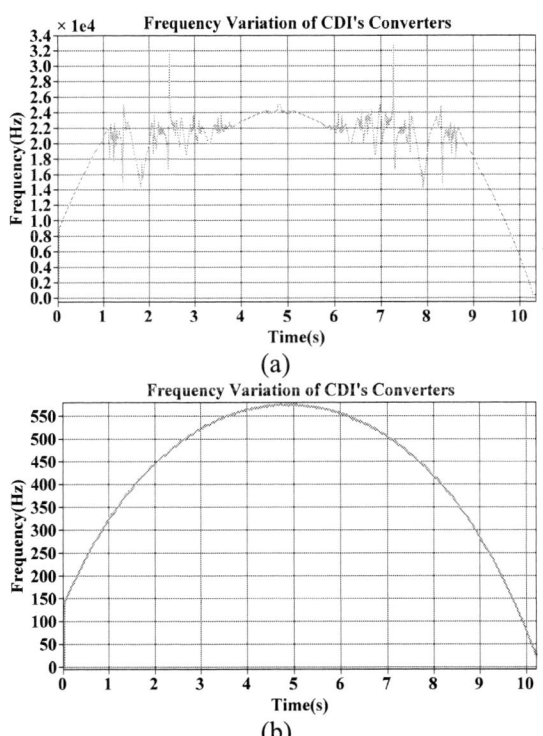

(a)

(b)

Fig. 9. Variation of frequency that is presented a) in [18] b) in this paper

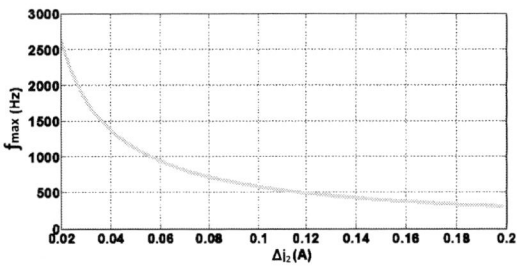

Fig. 10. Maximum switching frequency of buck-boost converter versus difference current Δi_2

VI. CONCLUSION

The capacitive deionization (CDI) is an attractive water desalination technique due to its capability to store and recover energy. A buck converter align with CDI selector was proposed in this paper to compensate losses during CDI process. The proposed CDI system was mathematically analyzed to derive the maximum switching. A system design approach was proposed to design the switching frequency bandwidth within the desired range. The modified design results in reducing the high frequency power converter losses. The proposed system was verified through simulations.

REFERENCES

[1] H. Mehrabian-Nejad, B. Farhangi and Sh. Farhangi, "Application of PV and Solar Energy in Water Desalination System," in Proc. 2nd Int. Conf. and Exhibition on Solar Energy, 2015, ICESE-1078.

[2] S. A. Kalogirou, "Chapter eight - Solar Desalination Systems," Solar Energy Engineering, S. A. Kalogirou, ed., pp. 421-468, Boston: Academic Press, 2009.

[3] Y. Oren, "Capacitive deionization (CDI) for desalination and water treatment-past, present and future (a review)," Desalination, vol. 228, no. 1–3, pp. 10–29, Aug. 1. 2008.

[4] A.M.Pernia, M.J.Prieto; F.Nuno, F.J.Alvarez-Gonzalez, J.A.Martinez, "Improving energy recovery in CDI systems," in Industry Applications Society Annual Meeting, 2013 IEEE , vol., no., pp.1-8, 6-11 Oct. 2013.

[5] A.M.Pernía, F.J.Alvarez-González, J.Díaz, P.J.Villegas, F.Nuño, "Optimum Peak Current Hysteresis Control for Energy Recovering Converter in CDI Desalination".Energies 2014, 7, 3823-3839.

[6] R. Zhao, P. M. Biesheuvel, and A. van der Wal, "Energy consumption and constant current operation in membrane capacitive deionization," Energy & Environmental Science, vol. 5, no. 11, pp. 9520-9527, 2012.

[7] M. Alkuran and M. Orabi, "Utilization of a buck–boost converter and the method of segmented capacitors in a CDI water purification system," in Proc. 12th Int. Middle-East Power Syst. Conf., 2008, pp. 470–474.

[8] H. Mehrabian-Nejad, S. Mohammadi and B. Farhangi, "Novel control method for reducing EMI in shunt active filters with level shifted random modulation," in Power Electronics, Drives Systems & Technologies Conference (PEDSTC), 2015 6th , vol., no., pp.585-590, 3-4 Feb. 2015.

[9] V.Samavatian, A.Radan, "A High Efficiency Input/Output Magnetically Coupled Interleaved Buck–Boost Converter With Low Internal Oscillation for Fuel-Cell Applications: CCM Steady-State Analysis," in Industrial Electronics, IEEE Transactions on , vol.62, no.9, pp.5560-5568, Sept. 2015.

[10] F.Caricchi, F.Crescimbini, F.Giulii Capponi, L.Solero, "Study of bi-directional buck-boost converter topologies for application in electrical vehicle motor drives," in Applied Power Electronics Conference and Exposition, 1998. APEC '98. Conference Proceedings 1998., Thirteenth Annual , vol.1, no., pp.287-293 vol.1, 15-19 Feb 1998.

[11] R. Rahimi, B. Farhangi, and Sh. Farhangi, " New Topology to Reduce Leakage Current in Three-Phase Transformerless Grid-Connected Photovoltaic Inverters ," in Proc. Power Electronics Drive Systems and Technologies Conference (PEDSTC), 7th 2016, pp. 1-6, 16-18 Feb. 2016.

[12] B. Farhangi and H. A. Toliyat, "Modeling Isolation Transformer Capacitive Components in a Dual Active Bridge Power Conditioner," in *Energy Conversion Congress and Exposition (ECCE), 2013 IEEE*, 2013, pp. 5476-5480.

[13] B. Farhangi and H. Toliyat, "A Novel Vehicular Integrated Power System Realized with Multi-port Series Ac Link Converter," in *Applied Power Electronics Conference and Exposition, 2015. APEC 2015. 30th Annual IEEE*, 2015, pp. 1353-1359.

[14] B. Farhangi and H. A. Toliyat, "Piecewise Linear Model for Snubberless Dual Active Bridge Commutation," *Industry Applications, IEEE Transactions on,* vol. 51, pp. 4072-4078, 2015.

[15] B. Farhangi and H. A. Toliyat, "Piecewise linear modeling of snubberless dual active bridge commutation," in *Energy Conversion Congress and Exposition (ECCE), 2014 IEEE*, 2014, pp. 2065-2071.

[16] B. Farhangi, "A novel modified deadbeat controller for vehicle to grid application," in *Power Electronics, Drives Systems & Technologies Conference (PEDSTC), 2015 6th*, 2015, pp. 47-52.

[17] B. Farhangi, H. A. Toliyat, and A. Balaster, "High impedance grounding for onboard plug-in hybrid electric vehicle chargers," in *Power Engineering, Energy and Electrical Drives (POWERENG), 2013 Fourth International Conference on*, 2013, pp. 609-613.

[18] B. Farhangi, and H. A. Toliyat, "Modeling and Analyzing Multi-Port Isolation Transformer Capacitive Components for Onboard Vehicular Power Conditioners," Industrial Electronics, IEEE Transactions on, Accepted for Publication.

[19] E. Afshari, R. Rahimi, B. Farhangi, Sh. Farhangi, "Analysis and Modification of the Single Phase Transformerless FB-DCB Inverter Modulation for Injecting Reactive Power," in Proc. Energy Conversion (CENCON), 2015 IEEE Conference on , pp.1-6, 19-20 Oct. 2015.

[20] R. Rahimi, E. Afshari, B. Farhangi, Sh. Farhangi, "Optimal placement of additional switch in the photovoltaic single-phase grid-connected transformerless full bridge inverter for reducing common mode leakage current", in Proc. Energy Conversion (CENCON), 2015 IEEE Conference on, pp. 1-5, 19-20 Oct, 2015.

[21] A. Sepehr, M. Saradarzadeh, and Sh. Farhangi, " A Noninvasive On-line Failure Prediction Technique for Aluminum Electrolytic Capacitors in Photovoltaic Grid-connected Inverters," in Proc. Power Electronics Drive Systems and Technologies Conference (PEDSTC), 7th 2016, pp. 1-6, 16-18 Feb. 2016.

[22] P. Długołęcki, and A. van der Wal, "Energy Recovery in Membrane Capacitive Deionization," Environmental Science & Technology, vol. 47, no. 9, pp. 4904-4910, 2013/05/07, 2013.

[23] M. Alkuran, M. Orabi, and N. Scheinberg, "Highly efficient Capacitive De-Ionization (CDI) water purification system using a buck-boost converter." pp. 1926-1930.

[24] A. M. Pern´ıa, J. G. Norniella, J. A. Mart´ın-Ramos, J. D´ıaz, and J. A. Mart´ınez, "Up–down converter for energy recovery in a CDI desalination system," IEEE Trans. Power Electron., vol. 27, no. 7, pp. 3257–3265, Jul. 2012.

[25] A.M.Pernia, F.J.Alvarez-Gonzalez, M.A.J.Prieto, P.J.Villegas and F.Nuno, "New Control Strategy of an Up–Down Converter for Energy Recovery in a CDI Desalination System," IEEE Trans. Power Electron , vol.29, no.7, pp.3573-3581, July 2014.

978-1-5090-0376-1/16 $31.00 © 2016 IEEE

7th Power Electronics, Drive Systems & Technologies Conference (PEDSTC 2016)
16-18 Feb. 2016, Iran University of Science and Technology, Tehran, Iran

Implementation of The First Commercial Medium Power Active Front End Transformerless Uninterruptible Power Supply Made In Iran

Mustafa Mohamadian[1], Adib Abrishamifar[2], Mehdi Shahrdad[3], Masoud Arefian[3], Mahdi Fazeli[3]

1:Dept. of Elec. & Comp.Eng.,Tarbiat Modares University,Tehran, Iran, mohamadian@modares.ac.ir
2: School of Elec. Eng.,.,Iran University of Science & Technology,Tehran, Iran
3: Electr. Converters & Power Syst. Dept., ACECR

Abstract— Today there are numerous companies in Iran manufacturing Uninterruptible Power Supplies (UPS). However, in medium and high power range they manufacture online double conversion transformer based UPSs, with thyristor controlled rectifier. In this paper the power topology, control algorithm and experimental results of the first commercial active front end 30 KVA transformerless UPS in Iran is presented.

Keywords— Uninterruptible Power Supply; transformerless; Active Front End;Standard UPS;

I. INTRODUCTION

Interruption of many manufacturing processes can cause serious loss or damage. Today UPSs are a crucial part of modern industrial systems to prevent interruptions due to power loss or low power quality.

In many critical applications with sensitive loads, it is imperative to use transformer based UPSs. Transformer based UPS is proven technology that improves reliability and provides fault isolation.

But today, transformerless UPS [1] market is growing rapidly. This is due to the benefits such as smaller footprint, higher efficiency and lower cost.

Transformerless UPS is now becoming a mature technology. However, there is no research paper or document describing the power topology and control block diagram of one particular commercial Transformerless UPS.

In this paper the power topology, control algorithm and performance of first commercial 30 KVA Transformerless UPS prototyped in Electr. Converters & Power Syst. Dept., ACECR is elaborated.

Section II of this paper compares transformerless UPSs with conventional ones. In section III, the control hardware is discussed. In section IV the PWM rectifier topology and hardware are described. Section V discusses the bidirectional DC/DC converter followed by DC bus voltage balancing scheme. The inverter control algorithm is explained in section VII. Final section demonstrates the performance of the UPS.

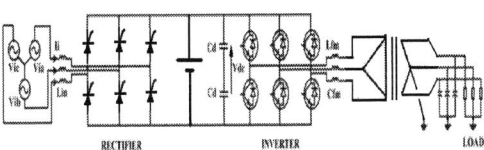

Fig. 1: Transformer based UPS

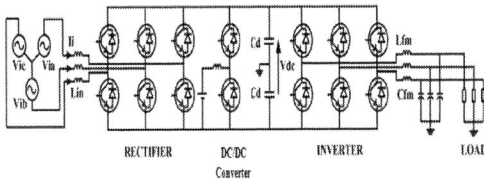

Fig. 2: Transformerless UPS

II. TRANSFORMER BASED VERSUS TRANSFORMERLESS UPS

A. Transformer Based UPS

The power topology of a typical transformer based UPS is shown in Fig. 1. The front end is a thyristor controlled rectifier. In normal operation, the firing angle is adjusted to keep the condition of the battery, connected to the DC bus, ready for power loss operation. The input current Total Harmonic Distortion (THD) is normally not within standard limit, but this can be remedied by 12 pulse rectifiers and passive filters with costumer demand. Compared to a transformerless UPS, due to the isolation transformer transfer ratio, the DC bus voltage or the number of batteries can be kept low without a DC/DC converter.

The inverter does keep the load voltage at standard level with low THD. In order to keep the transformer loss at

978-1-5090-0376-1/16 $31.00 © 2016 IEEE

minimum, the switching frequency is also kept low typically

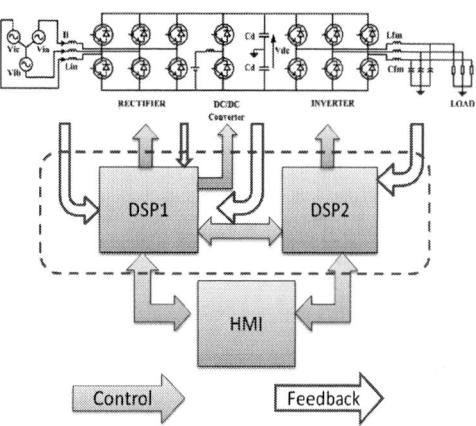

Fig. 3: Control hardware diagram

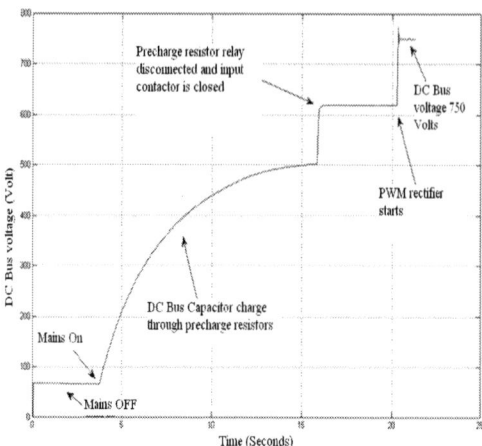

Fig. 4: Experimental results :PWM rectifier start up

3-5 kHz. Achieving an acceptable dynamic response and low THD is a challenging task in this condition. Space Vector modulation is the conventional switching algorithm to exploit maximum output voltage from the DC bus.

B. Transformerless UPS

The power topology of the transformerless UPS [2] used in this application is shown in Fig. 2. The rectifier is a Pulse Width Modulation (PWM) rectifier, keeping the input current THD under 5%. The battery is connected to the DC bus through a DC/DC converter. This is required if the number of series connected batteries is low because the DC bus voltage has to be high enough, typically 750 V, for the inverter to produce required output AC voltage.

In contrast to transformer based UPS that the load neutral is connected to the transformer secondary neutral point, in this configuration the load neutral is connected to the DC bus midpoint. Hence Space Vector modulation is not possible for the inverter and conventional sinusoidal PWM is used in this case.

It is seen that in transformerless configuration, the UPS foot print and cost can be reduced considerably due to lack of isolation transformer. Also the full load efficiency is improved and switching frequency can be increased, typically 8-16 kHz, which improves dynamic response and helps reduction of output voltage THD. But with eliminating the transformer, the galvanic isolation and fault current limitation which is a requirement for critical applications is lost.

III. CONTROL HARDWARE

Fig. 3 shows the control, hardware diagram. Two Digital Signal Processors (DSP) are used for UPS control.

DSP1 is responsible for rectifier and DC/DC converter control. Hence three groups of feedbacks, input voltages and currents, battery voltage and current and DC bus upper and lower capacitor voltages are fed to DSP1 board. As shown in

this figure, DSP1 generates the gate control signals for PWM rectifier and DC/DC converter. It also communicates with DSP2, Human Machine Interface (HMI) and monitors mains voltage quality including voltage magnitude and frequency.

DSP2 only controls the inverter and it also is responsible for switching the load between the UPS output and mains bypass. It also has a communication link with DSP1 and HMI. Load voltages and currents are fed back to DSP2 for output voltage control. DSP2 is also responsible for keeping the load voltage synchronized with mains voltage. Hence switching between the load and mains happens with minimum load voltage distortion.

IV. PWM RECTIFIER CONTROL

As mentioned before, DSP1 is responsible for PWM rectifier control. Its main responsibility is to keep the DC bus voltage at constant 750 V level. Fig. 4 shows the PWM rectifier startup procedure. At time 4s, the mains voltage is on and this is realized by DSP1. Hence the precharge resistor relay is activated. At time 16 s, when the DC bus voltage is at

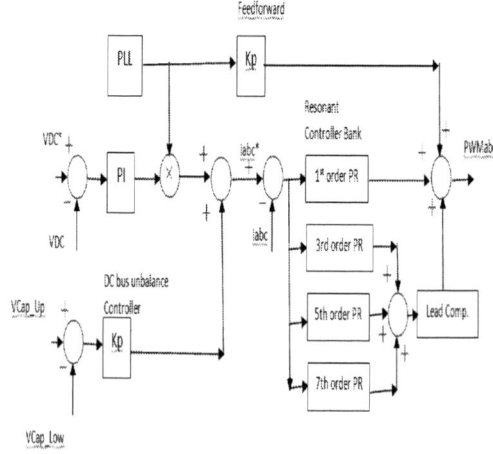

Fig. 5: PWM rectifier Control block diagram

978-1-5090-0376-1/16 $31.00 © 2016 IEEE

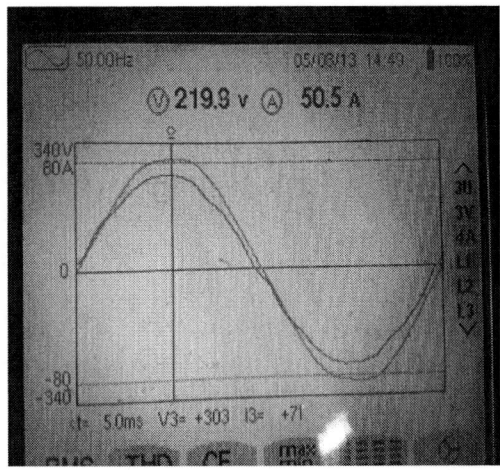

Fig. 6: PWM rectifier main input voltage and current experimental results. Input current THD=2.4%

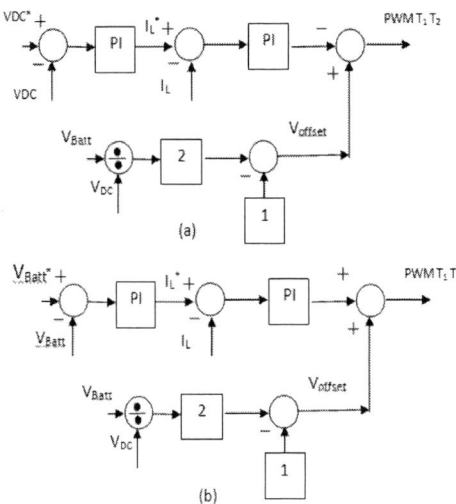

(a)

(b)

Fig. 8: DC/DC converter control block diagram (a): Boost (battery discharge). (b): Buck (battery charge)

500V, the precharge resistor relay is disconnected and mains input contactor is closed. At this point the DC bus capacitors are charged through rectifier diodes up to 620V. This procedure prevents the inrush DC bus capacitor current. At time 21s, the PWM rectifier switching starts and the DC bus voltage is boosted to 750V.

The PWM rectifier control block diagram is shown in Fig. 5. The target is to keep the DC bus voltage at reference target while keeping the input current in phase with mains voltage. A Proportional Integrator (PI) controller multiplied by Phase Locked Loop (PLL) output and added with a DC bus unbalance controller, generates the ref current for phase currents (Iabc*). A Proportional Resonant (PR) [3] filter bank with 1st, 3rd,5th and 7th harmonic resonant frequency is required to reduce the input current THD. The lead compensator improves the system stability. Also a feed forward term is added to improve dynamic response.

The experimental results for the PWM rectifier input current and mains voltage at 30kW linear load is shown in Fig. 6. It is shown that input current is in phase with mains voltage. The input current THD is 2.4% which is within standard limit.

V. BIDIRECTIONAL DC/DC CONVERTER OPERATION

The bidirectional DC/DC converter is shown in Fig. 7. In normal operation, the converter charges the battery from DC bus V_{dc} with buck operation of IGBT T_1 and diode D_2. During power failure, the battery V_{Batt} charges the DC bus V_{dc} with

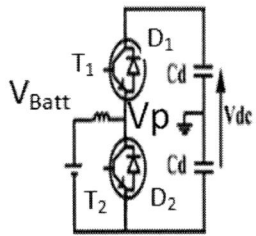

Fig. 7: Bidirectional DC/DC Converter

boost operation of IGBT T_2 and diode D_1. IGBTs T_1 and T_2 are complementary. When T_1 is on T_2 is off and vice versa. This ensures continuous conduction of inductance current.

The converter control block diagram which is a conventional PI control with internal inductance current control loop is shown in Fig. 8. V_{offset} is the limit modulation index value where the converter changes from buck operation to boost operation or vice versa.

Fig. 9 depicts the DC/DC converter experimental results. In the beginning, the UPS is at normal operation. The DC bus voltage is at 750 V and the battery is charging with a small floating voltage current. At time 4 S, the mains voltage is disconnected. Immediately the DC/DC switches from buck to boost operation and 45 A is drawn from the battery and the battery voltage drops from 650 V to 525 V. At time 19.5 S the

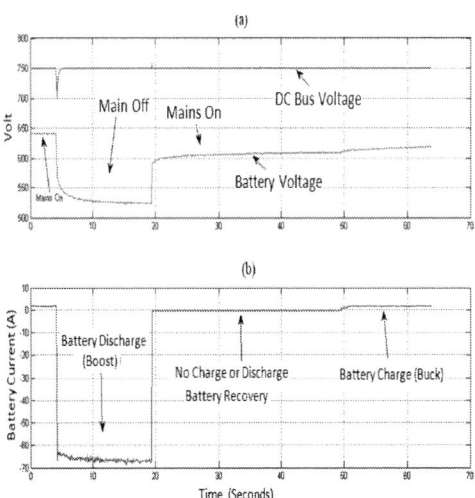

Fig. 9: Experimental results (a): DC bus voltage and battery voltage. (b): Battery current

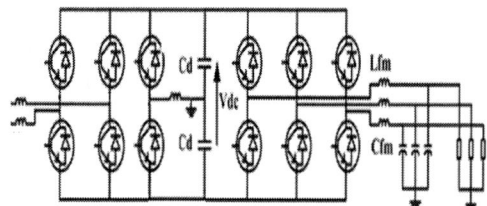

Fig. 10: UPS power topology during power failure. (One PWM rectifier leg is connected to midpoint of DC bus). (Battery and DC/DC converter is not shown in this figure)

mains voltage is connected back and PWM rectifier supplies the DC bus. During the battery recovery time there is no charge or discharge. At time 50 S, the converter starts the buck operation again, charging the battery.

VI. DC BUS VOLTAGE BALANCE IN POWER FAILURE

When the mains power is available, the DC bus mid-point voltage balance is provided by PWM rectifier controller as shown in Fig. 5. In case of power failure, the PWM rectifier is not active. In this situation, the input contactor is disconnected and an auxiliary contactor connects PWM rectifier leg C and its input inductance to the DC bus midpoint. This leg is now responsible for DC bus mid-point voltage balance. This configuration is shown in Fig. 10. The DC bus mid-point voltage balance control block diagram is dual loop with PI controller as shown in Fig. 11.

VII. INVERTER OPERATION

The inverter control block diagram implemented in abc reference frame is shown in Fig. 10 [4]. The reference voltage using a proportional and resonant filter bank filter [3] with capacitor current feed forward plus load current generates the inductance reference current. Inductance reference current internal loop compared to capacitor current internal loop convenes parallel operation of UPSs. A proportional controller plus the reference voltage feed forward is used as PWM commands for the three phases.

VIII. MISCELLANOUS

In a commercial UPS there are many more details that has to be considered such as protection, Human Machine Interface (HMI), Static Transfer Switch (STS), inductance design, parts

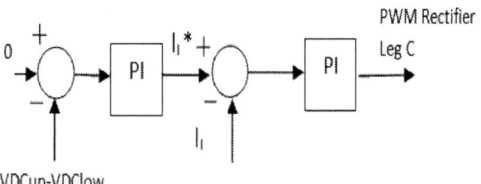

Fig. 11: DC bus voltage balance controller during power failure

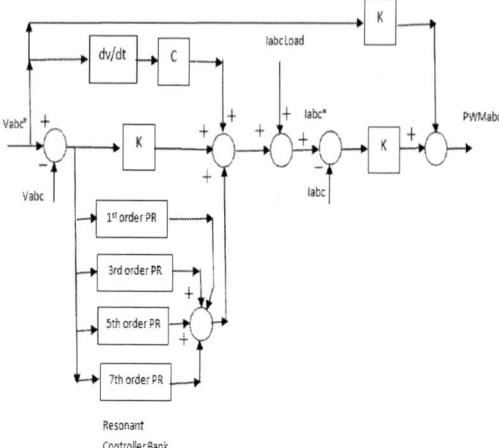

Fig. 12: Inverter Control block diagram

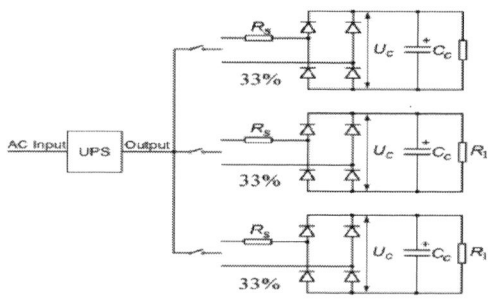

Fig. 13: Nonlinear load test: Three single phase rectifiers 27 kVA

placement, cooling and so on.

IX. UPS PERFORMANCE

UPS response to nonlinear load, dynamic behavior and transfer characteristics are demonstrated in this section.

A. Nominal nonlinear load

Three single phase rectifiers with a total apparent power of 27 kVA and 17 kW is used for nonlinear test as depicted in Fig. 13. Phase A load voltage and current is shown in Fig. 14. The measured voltage THD is 2.8% and voltage magnitude regulation is 0.45% .

B. Dynamic Response

In Fig. 15, in the beginning, the UPS is at no load condition when at time 44 ms, a step linear load of 30 kW is applied to the UPS. The load voltage suddenly drops 28%, but in first positive cycle it reaches a voltage drop of 4% and in less than 100 ms the load voltage is 1 PU. This response meets the IEC 62040-3 Classification I UPS standard.

Also removal of the 30 kW linear load is shown in Fig. 16. Load is disconnected at time 49 ms and current reaches zero at time 51 ms. First the load voltage is increased to 1.06 PU but after 18 ms the load voltage is normal. Finally in Fig. 17 it is shown that load voltage remains within standard level with mains power failure at time 45 ms.

X. CONCLUSIONS

Power topology, control block diagrams and performance of a commercial active front end transformerless UPS was presented in this paper. Experimental results verified the performance of the UPS. The input current THD was measured 2.4% with almost unity power factor. The load voltage THD with nonlinear load wad 2.8% with steady state voltage regulation of 0.45%. Also the dynamic tests indicated that the designed UPS qualified as a classification I Uninterruptible Power Supply according to IEC 62040-3 standard.

Acknowledgment

The authors would like to thank the financial support from ACECR. Also the authors are grateful to Mr. Erfanian, Mr. Afghani, Mr. Rajabi and Mr. Nazari for their support.

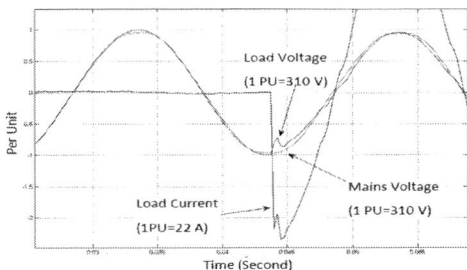

Fig. 15: Experimental results: 30kW linear step load, Phase A, load voltage and current and mains voltage

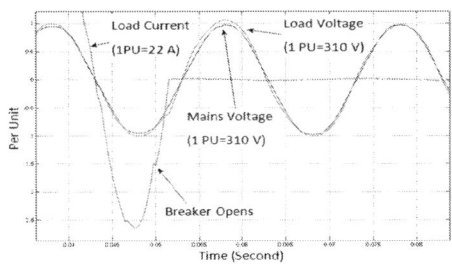

Fig. 16: Experimental results: 30kW linear load removal, Phase A, load voltage and current and mains voltage

References

[1] E.-H. Kim J.-M. Kwon B.-H. Kwon"Transformerless three-phase on-line UPS with high performance", IET Power Electronics, V.2, Issue 2, PP.103-112, April 2009

[2] I.Günel, B.Üstüntepe and A. M. Hava, "Modern Transformerless Uninterruptable Power Supply (UPS) Systems", International Conference on Electrical and Electronics Engineering, Bursa, 5-8 Nov. 2009, pp. 316-320.

[3] D. N. Zmood and D. G. Holmes,"Stationary Frame Current Regulation of PWM Inverters With Zero Steady-State Error", IEEE Trans. on Power Electronics, Vol. 18, No. 3, MAY 2003

[4] M.J. Ryan, W.E. Brumsickle, R.D. Lorenz, "Control topology options for single-phase UPS inverters", IEEE Trans. on Industry Applications, V.33, Issue 2, PP.493-501, Mar/Apr 1997

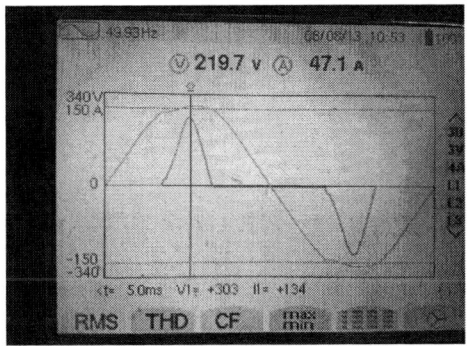

Fig. 14: Experimental results: Three single phase rectifier, Phase A, load voltage and current

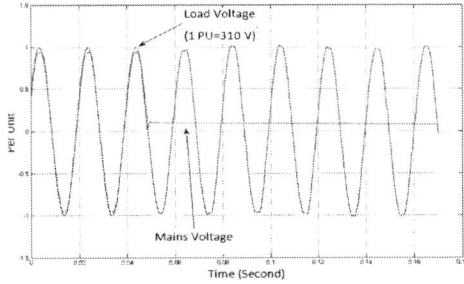

Fig. 17: Experimental results: Mains power failure while supplying a 30kW linear load l, Phase A, load voltage and mains voltage

7th Power Electronics, Drive Systems & Technologies Conference (PEDSTC 2016)
16-18 Feb. 2016, Iran University of Science and Technology, Tehran, Iran

Solar Charger System With LED Driver Using Capacitor-Less Multi-Port Converter

S.M. Dehghan
Faulty of Electrical and Computer Engineering,
Qom University of Technology, Qom, Iran
dehghan@qut.ac.ir

M. Alinaghizadeh Ardestani
Faulty of Electrical and Computer Engineering,
Qom University of Technology, Qom, Iran
ardestani@qut.ac.ir

Abstract—**In this paper, a three-port converter, without capacitor in the common bus, is used for a solar charger application. Two ports of the converter are connected to a solar panel and a battery bank. Third port is used as a LED driver with constant current. This paper analyzes the used converter and proposes special switching, control and design approaches for the mentioned application. The performance of the proposed configuration and approaches are surveyed using simulation results.**

Keywords—Solar Charger; Photovoltaic; Maximum Power Point Tracking (MPPT); LED Driver; Multi-Port DC/DC Converter.

I. INTRODUCTION

In the recent years, because of environmental and economic issues, renewable energy sources have gained attention. Among these energy sources, solar energy due to its special features, has been interested for many of governments, companies and research institutes [1-4].

Solar panels have been widely used for public lighting applications. Also combination of solar cells and LED lamps has made an efficient system. Such system, also requires an energy storage, and therefore uses several power converters to control power flow [4-6] (Fig. 1).

In many researches, in order to decreasing size and cost, increasing efficiency and dynamic response, and implementing centralized control , using a multi-port converter has been proposed instead of several conventional two-port converters [7-13].

A bidirectional multi-port converter without capacitor in bus common has been presented in [14] (Fig. 2). This converter is formed by the connection of several bidirectional non-isolated converters. In this converter, capacitor is removed in connection point (common bus). This, despite of decreasing cost, can increase complexity of control of converter.

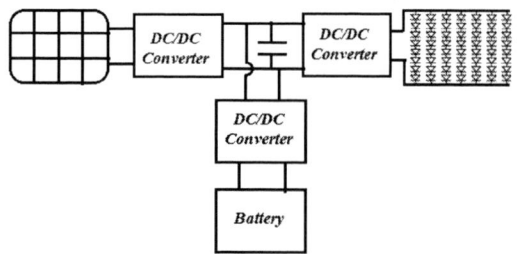

Fig. 1. Conventional solar charger system.

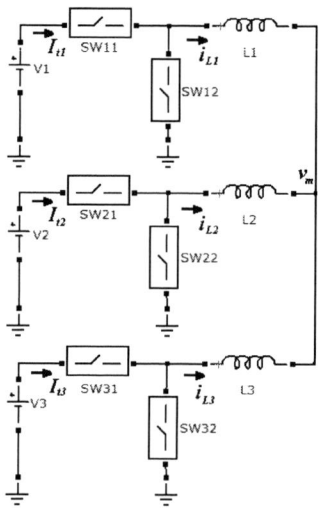

Fig. 2. Multi-port converter without capacitor in common bus.

In this paper, the presented converter in [14], are used for a solar charger system together with LED driver. In addition, along with complete analysis of converter, novel switching and control strategies are proposed. Also, considering the mentioned application, a design approach is proposed for choosing control parameters.

II. ANALYSIS OF CONVERTER

The converter shown in Fig. 2, has six solid-state switches. Given that state of switches of each port are opposite of each other, eight switching combination is possible. TABLE 1 and Fig. 3 show state of switches and equivalent circuit of converter during these eight combinations, respectively.

The voltages of the common bus and the inductors in various states can be expressed as following:

$$v_{m\_D1} = N_1 V_1 \tag{1}$$

$$v_{L1\_D1} = V_1 - v_{m\_D1} \tag{2}$$

$$v_{L2\_D1} = -v_{m\_D1} \tag{3}$$

$$v_{L3\_D1} = -v_{m\_D1} \tag{4}$$

$$v_{m\_D2} = N_1 V_1 + N_2 V_2 \tag{5}$$

$$v_{L1\_D2} = V_1 - v_{m\_D2} \tag{6}$$

$$v_{L2\_D2} = V_2 - v_{m\_D2} \tag{7}$$

978-1-5090-0376-1/16 $31.00 © 2016 IEEE

TABLE I. STATE OF SOLID STATE SWITCHES IN VARIOUS SWITCHING COMBINATIONS

	SW11	SW21	SW31	SW12	SW22	SW32
ST1	ON	OFF	OFF	OFF	ON	ON
ST2	ON	ON	OFF	OFF	OFF	ON
ST3	OFF	ON	OFF	ON	OFF	ON
ST4	OFF	ON	ON	ON	OFF	OFF
ST5	OFF	OFF	ON	ON	ON	OFF
ST6	ON	OFF	ON	OFF	ON	OFF
ST7	ON	ON	ON	OFF	OFF	OFF
ST8	OFF	OFF	OFF	ON	ON	ON

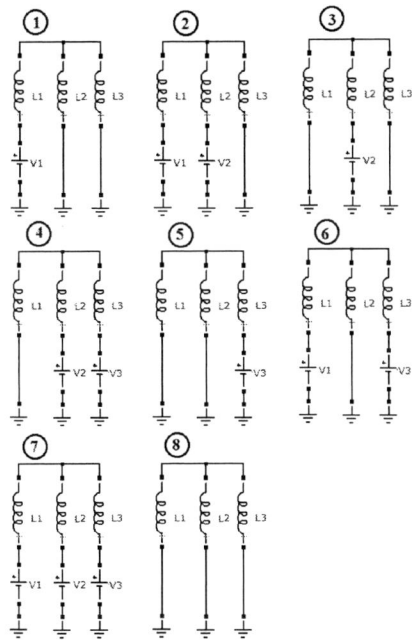

Fig. 3. Equivalent circuit of converter in various switching combinations.

$$v_{L3\_D2} = -v_{m\_D2} \tag{8}$$

$$v_{m\_D3} = N_2 V_2 \tag{9}$$

$$v_{L1\_D3} = -v_{m\_D3} \tag{10}$$

$$v_{L2\_D3} = V_2 - v_{m\_D3} \tag{11}$$

$$v_{L3\_D3} = -v_{m\_D3} \tag{12}$$

$$v_{m\_D4} = N_2 V_2 + N_3 V_3 \tag{13}$$

$$v_{L1\_D4} = -v_{m\_D4} \tag{14}$$

$$v_{L2\_D4} = V_2 - v_{m\_D4} \tag{15}$$

$$v_{L3\_D4} = V_3 - v_{m\_D4} \tag{16}$$

$$v_{m\_D5} = N_3 V_3 \tag{17}$$

$$v_{L1\_D5} = -v_{m\_D5} \tag{18}$$

$$v_{L2\_D5} = -v_{m\_D5} \tag{19}$$

$$v_{L3\_D5} = V_3 - v_{m\_D5} \tag{20}$$

$$v_{m\_D6} = N_1 V_1 + N_3 V_3 \tag{21}$$

$$v_{L1\_D6} = V_1 - v_{m\_D6} \tag{22}$$

$$v_{L2\_D6} = -v_{m\_D6} \tag{23}$$

$$v_{L3\_D6} = V_3 - v_{m\_D6} \tag{24}$$

$$v_{m\_D7} = N_1 V_1 + N_2 V_2 + N_3 V_3 \tag{25}$$

$$v_{L1\_D7} = V_1 - v_{m\_D7} \tag{26}$$

$$v_{L2\_D7} = V_2 - v_{m\_D7} \tag{27}$$

$$v_{L3\_D7} = V_3 - v_{m\_D7} \tag{28}$$

$$v_{m\_D8} = 0 \tag{29}$$

$$v_{L1\_D8} = 0 \tag{30}$$

$$v_{L2\_D8} = 0 \tag{31}$$

$$v_{L3\_D8} = 0 \tag{32}$$

where v_{m_Dj} and v_{Li_Dj} are instantaneous voltage of common bus and i-th inductor during j-th switching time interval. In the above equations, N_1, N_2, and N_3 are defined as:

$$N_1 = \frac{L_2 L_3}{L_1 L_2 + L_2 L_3 + L_3 L_1} \tag{33}$$

$$N_2 = \frac{L_1 L_3}{L_1 L_2 + L_2 L_3 + L_3 L_1} \tag{34}$$

$$N_3 = \frac{L_1 L_2}{L_1 L_2 + L_2 L_3 + L_3 L_1} \tag{35}$$

If one of the ports is disconnected (both related switches are turned off), the value of related inductor in above equations should be considered as infinite. For example, when second port is disconnected, it can be written that:

$$N_1 = \frac{L_3}{L_1 + L_3} \tag{36}$$

$$N_2 = 0 \tag{37}$$

$$N_3 = \frac{L_1}{L_1 + L_3} \tag{38}$$

Considering (1-32), the average voltage of the common bus can be expressed as:

$$V_m = \sum_{j=1}^{8} D_j v_{m\_Dj} = D_{eq1} N_1 V_1 + D_{eq1} N_2 V_2 + D_{eq3} N_3 V_3 \tag{39}$$

where

$$D_{eq1} = D_1 + D_2 + D_6 + D_7 \tag{40}$$

$$D_{eq2} = D_2 + D_3 + D_4 + D_7 \tag{41}$$

$$D_{eq3} = D_4 + D_5 + D_6 + D_7 \tag{42}$$

The average voltage of inductors should be zero, thus:

$$\sum_{j=1}^{8} D_j v_{L1\_Dj} = D_{eq1} V_1 - \sum_{j=1}^{8} D_j v_{m\_Dj} = D_{eq1} V_1 - V_m = 0 \tag{43}$$

$$\sum_{j=1}^{8} D_j v_{L2\_Dj} = D_{eq2} V_2 - \sum_{j=1}^{8} D_j v_{m\_Dj} = D_{eq2} V_2 - V_m = 0 \tag{44}$$

$$\sum_{j=1}^{8} D_j v_{L3\_Dj} = D_{eq3} V_3 - \sum_{j=1}^{8} D_j v_{m\_Dj} = D_{eq3} V_3 - V_m = 0 \tag{45}$$

In regard to (43), (44), and (45), it can be written that:

$$V_m = D_{eq1} V_1 = D_{eq1} V_2 = D_{eq3} V_3 \tag{46}$$

In the other hand, the average current of ports are equal to:

$$I_{T1} = D_1 i_{L1\_D1} + D_2 i_{L1\_D2} + D_6 i_{L1\_D6} + D_7 i_{L1\_D7} \tag{47}$$

$$I_{T2} = D_2 i_{L2\_D2} + D_3 i_{L2\_D3} + D_4 i_{L2\_D4} + D_7 i_{L1\_D7} \tag{48}$$

$$I_{T3} = D_4 i_{L3\_D4} + D_5 i_{L3\_D5} + D_6 i_{L3\_D6} + D_7 i_{L3\_D7} \tag{49}$$

where i_{Li_Dj} are instantaneous current of i-th inductor during j-th switching time interval.

Considering various switching combinations, it is obvious that there are seven control parameters (duty cycles of switching combinations). However for full control of a three-port converter, three control parameters are sufficient. Therefore four switching combinations can be ignored for simplicity. These combinations can be selected considering different goals. Also more switching combinations may be considered to enhance dynamic performance and optimum utilization.

III. PROPOSED SWITCHING

In this paper, the switching combinations 1, 3, 5, and 8 (from TABLE I) are proposed for making a new switching algorithm. In this case, it can be written that:

$$V_m = D_1 V_1 = D_3 V_2 = D_5 V_3 \tag{50}$$

Given that in each of combinations 1, 3, and 5, only one of the ports is connected (only one of switches SW11, SW21, and SW31 are ON), switching method shown in Fig. 4 is proposed. In the proposed method, signals D1, D1+D3, and D1+D3+D5 are compared with a carrier signal. Output of first comparator makes switching signal G11. The switching signal G21 is generated by the logical AND of output of second comparator and the logical NOT output of first comparator. In a similar manner, the logical AND of output of third comparator and the logical NOT of output of second comparator makes switching signal G31. The switching signals G12, G22, and G32 are generated by the logical NOT of signals G11, G21, and G31, respectively. Fig. 5 shows the implementation of the proposed switching method.

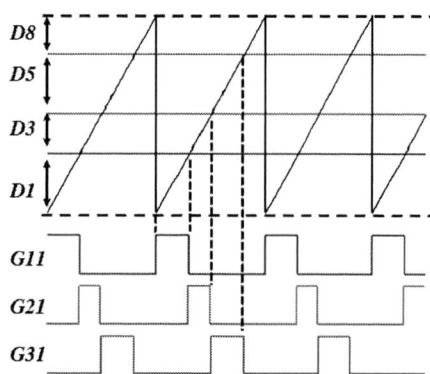

Fig. 4. The proposed switching method based on carrier wave.

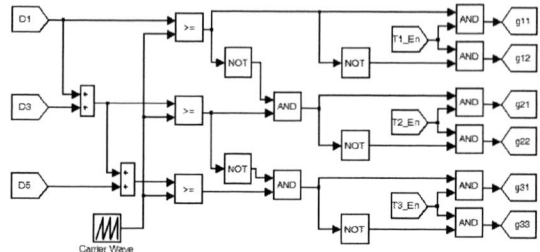

Fig. 5. The implementation of the proposed switching method.

IV. CURRENT OF INDUCTORS

Analysis of input and inductor currents can be helpful to verify proper operation of the converter. If the proposed switching method shown in Fig. 4 is used, from (47), (48), and (49), currents of the ports can be expressed as:

$$I_{T1} = D_1 i_{L1\_D1} \tag{51}$$

$$I_{T2} = D_3 i_{L2\_D3} \tag{52}$$

$$I_{T3} = D_5 i_{L3\_D5} \tag{53}$$

The average current of first inductor during time interval of switching combinations 3, 5, and 8, can be written as:

$$i_{L1\_D3} = i_{L1\_D1} + \frac{D_1 v_{L1\_D1} + D_3 v_{L1\_D3}}{2L_1 f_s} \tag{54}$$

$$i_{L1\_D5} = i_{L1\_D3} + \frac{D_3 v_{L1\_D3} + D_5 v_{L1\_D5}}{2L_1 f_s} \tag{55}$$

$$i_{L1\_D8} = i_{L1\_D5} + \frac{D_5 v_{L1\_D5} + D_8 v_{L1\_D8}}{2L_1 f_s} \tag{56}$$

where f_s is switching frequency. With substituting (2), (10), (18), and (30) in the above equations, it can be written that:

$$i_{L1\_D3} = i_{L1\_D1} + \frac{D_1 V_1 - D_1 N_1 V_1 - D_3 N_2 V_2}{2L_1 f_s} \tag{57}$$

$$i_{L1\_D5} = i_{L1\_D3} + \frac{-D_3 N_2 V_2 - D_5 N_3 V_3}{2L_1 f_s} \tag{58}$$

$$i_{L1\_D8} = i_{L1\_D5} + \frac{-D_5 N_3 V_3}{2L_1 f_s} \tag{59}$$

Considering (50), the above equations can be rewritten as:

$$i_{L1\_D3} = i_{L1\_D1} + \frac{N_3 V_m}{2L_1 f_s} \tag{60}$$

$$i_{L1\_D5} = i_{L1\_D1} - \frac{N_2 V_m}{2L_1 f_s} \tag{61}$$

$$i_{L1\_D8} = i_{L1\_D1} - \frac{(N_2 + N_3) V_m}{2L_1 f_s} \tag{62}$$

Therefore total average current of first inductor can be calculated as:

$$I_{L1} = D_1 i_{L1_{D1}} + D_3 i_{L1_{D3}} + D_5 i_{L1_{D5}} + D_8 i_{L1_{D8}} \tag{63}$$

$$I_{L1} = i_{L1\_D1} + \left(D_3 N_3 - D_5 N_2 - D_8 (N_2 - N_3) \right) \frac{V_m}{2L_1 f_s} \tag{64}$$

In a similar approach, for second inductor, it can be written that:

978-1-5090-0376-1/16 $31.00 © 2016 IEEE

$$i_{L2\_D1} = i_{L2\_D3} - \frac{N_3 V_m}{2L_1 f_s} \tag{65}$$

$$i_{L2\_D5} = i_{L2\_D3} + \frac{N_1 V_m}{2L_1 f_s} \tag{66}$$

$$i_{L1\_D8} = i_{L1\_D1} + \frac{(N_1 - N_3) V_m}{2L_1 f_s} \tag{67}$$

$$I_{L2} = i_{L2\_D3} + (-D_1 N_3 + D_5 N_1 + D_8 (N_1 - N_3)) \frac{V_m}{2L_1 f_s} \tag{68}$$

Also for third inductor, it can be written that:

$$i_{L3\_D1} = i_{L3\_D5} + \frac{N_2 V_m}{2L_1 f_s} \tag{69}$$

$$i_{L3\_D3} = i_{L3\_D5} - \frac{N_1 V_m}{2L_1 f_s} \tag{70}$$

$$i_{L1\_D8} = i_{L3\_D5} + \frac{(N_1 + N_2) V_m}{2L_1 f_s} \tag{71}$$

$$I_{L3} = i_{L3\_D5} + (D_1 N_2 - D_3 N_1 + D_8 (N_1 + N_2)) \frac{V_m}{2L_1 f_s} \tag{72}$$

V. CONTROL STRATEGY

In solar charger application together with LED driver, there are different operation modes. Therefore controller structure should change in regard to operation mode. Usually when the solar cell can generate energy, lighting is not required. However here it is assumed that lighting may be always required (for example for indoor lighting):

1- Solar panel does not generate energy, and state of charge (SOC) of battery is low. In this case, the load is disconnected to avoid the damage of the battery. It is also possible that SOC is acceptable, but load connection is not required.

2- Solar panel does not generate energy, but state of charge (SOC) of battery is acceptable. In this case, the load is supplied with constant current (LED lamps are usually supplied with constant current).

3- Solar panel generates energy and the load is connected. However load power is more than solar panel power, therefore the battery provides some portion of the load power. In this case, maximum power point tracking (MPPT) algorithm should be applied.

4- Solar panel generates energy. The load is disconnected (because of no need to it or low SOC of battery). The SOC of the battery is less than 70% and the bulk charging mode is applicable. In this case, the MPPT algorithm should be applied.

5- Solar panel generates energy and the load is connected. The SOC of the battery is less than 70% and the bulk charging mode is applicable. In this case , the MPPT algorithm should be applied.

6- This mode is similar to Mode 4. However if MMPT algorithm is applied, the charging current of the battery is more the allowable maximum current. Therefore the battery is charged with constant current.

7- This mode is similar to Mode 5. However if MMPT algorithm is applied, the charging current of the battery is more the allowable maximum current. Therefore the battery is charged with constant current.

8- Solar panel generates energy, but the load is disconnected . The SOC of the battery is more than 70% and the absorption charging mode is applicable. In this case , the voltage of battery should be kept constant.

9- Solar panel generates energy, the load is connected . The SOC of the battery is more than 70% and the absorption charging mode is applicable. In this case , the voltage of battery should be kept constant.

It is obvious that if the load and solar panel don't work simultaneously, the operation modes 3, 5, 7 , and 9 don't exist.

In the rest of the paper, it is assumed that the battery, solar panel, and LED module are connected to ports 1, 2, and 3, respectively. Given that the battery voltage is approximately constant, D1 factor can be always selected constant (Fig. 6). For this purpose, a desired value can be chosen for common bus voltage (This value should be less than is minimum possible voltage of all ports). D1 factor is determined based on rated voltage of the battery and chosen common bus voltage. Note that in the operation mode 1, the battery is disconnected, thus the port 1 is disable and D1 is zero (Fig. 6).

D3 factor may be determined by solar panel voltage controller (MPPT) (in the operation modes 3, 4, and 5), the battery current controller (in constant current charging mode) (in the operation modes 6 and 7), or the battery voltage controller (in constant voltage charging mode) (in the operation modes 8 and 9) (Fig. 7). In the operation modes 1 and 2, the port 2 is disable and D3 is zero.

LED drier should provide a constant current. Therefore D5 factor is determined using a current controller (in the operation modes 2, 3, 5, 7, and 9) (Fig. 8). Note that in the operation modes 1, 4, 6, and 8, the port 3 is disable and D5 is zero.

VI. DESIGN APPROACH

It is assumed that V_{load}, $V_{b-\min}$, $V_{b-\max}$, $V_{pv-\min}$, and $V_{pv-\max}$ are load voltage, minimum battery voltage, maximum battery voltage, minimum solar panel voltage, and maximum solar panel voltage, respectively. if $D1$ is considered constant, in regard to (50), it can be written that:

$$\frac{V_{b-\min}}{V_{pv-\max}} D_1 \le D_3 \le \frac{V_{b-\max}}{V_{pv-\min}} D_1 \tag{73}$$

$$\frac{V_{b-\min}}{V_{load}} D_1 \le D_5 \le \frac{V_{b-\max}}{V_{load}} D_1 \tag{74}$$

In order to obtain maximum values for $D_{3-\min}$ and $D_{5-\min}$ (for using slower switches), maximum possible value should be selected for $D1$. For this purpose, following condition is required:

$$D_1 + D_{3-\max} + D_{5-\max} = 1 \tag{75}$$

Considering (73) and (74), (75) can be rewritten as:

$$D_1 + \frac{V_{b-\max}}{V_{pv-\min}} D_1 + \frac{V_{b-\max}}{V_{load}} D_1 = 1 \tag{76}$$

Thus:

$$D_1 = \frac{1}{1 + \frac{V_{b-\max}}{V_{pv-\min}} + \frac{V_{b-\max}}{V_{load}}} \tag{77}$$

If the load and solar panel don't work simultaneously, $D1$ can be calculated using following equation:

$$D_1 + \max(D_{3-\max} + D_{5-\max}) = 1 \tag{78}$$

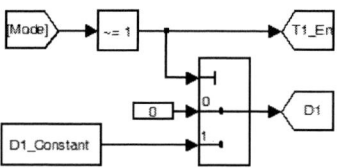

Fig. 6. Block diagram of the controller for the port connected to the battery.

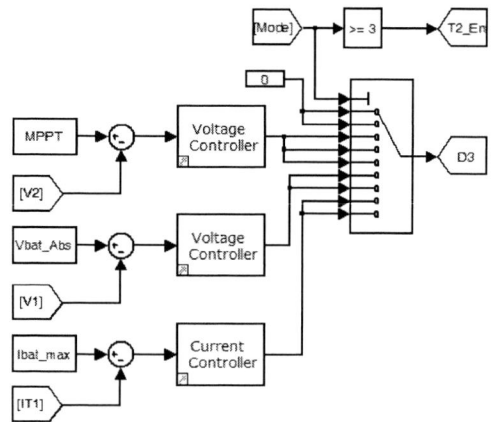

Fig. 7. Block diagram of the controller for the port connected to the solar panel.

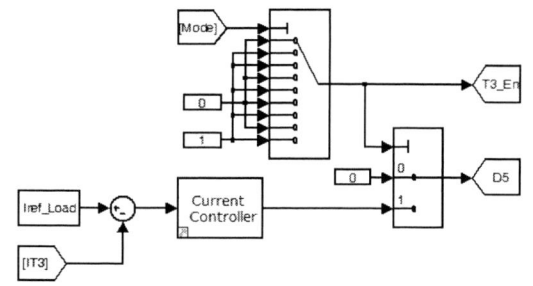

Fig. 8. Block diagram of the controller for the port connected to the LED module.

VII. Simulation

TABLE II shows simulation parameters. In regard to (77) and TABLE II, D1 is selected as 0.35. To verify the performance of the proposed configuration, operation modes 2, 3, and 7 are simulated in this section. The validity and performance of converters in these modes, would show that the converter can work in the all operation modes, properly.

At first, second operation mode is simulated. In regard to TABLE II, it can be written that: $N_1 = N_3 = 0.5$ and $N_2 = 0$. In this case, it is assumed that the battery voltage is equal to 12V. Fig. 9 shows simulation results. D5 is determined by current controller and its value is 0.105 at steady state. As it can be seen from Fig. 9, the voltage and current of the load are 40V and 1A respectively. The common bus voltage at different switching combinations, regarding (1), (17), and (29), is 6, 20, and 0, and its average is 4.2 V. In regard to (53), i_{L3_D5} is equal to 9.53A. Also considering (69) and (71), i_{L3_D1} and i_{L3_D8} are 9.53A and 7.43A, respectively. Regarding (72), total average current of third inductor is equal to 8.39A. In this simulation, magnitude of current of first and third inductors are same.

TABLE II. SIMULATION PARAMETERS

	Parameters	Value	Unit
Battery	Rated Voltage	12	V
	Maximum voltage	15	V
	Minimum voltage	10	V
	Voltage Absorption Mode	14	V
	Maximum Charging Current	5	A
Solar Panel	Open circuit voltage	21	V
	Short Circuit Current	7	A
	Maximum voltage	17	V
	Minimum voltage	10.12	V
LED Driver	Voltage	40	V
	Current	1	A
Converter	L1	10	uH
	L2	10	uH
	L3	10	uH
	Switching Frequency	50	kHz

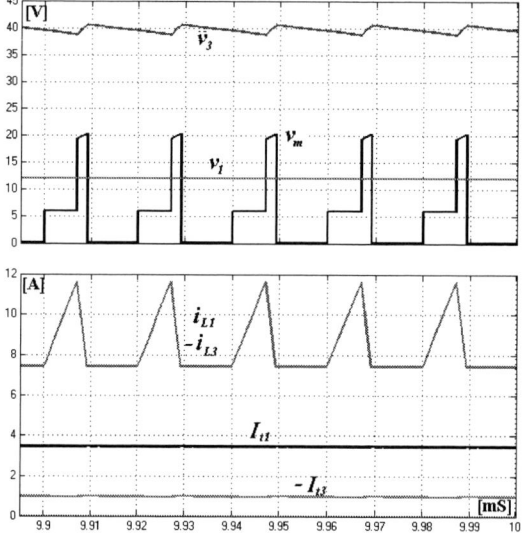

Fig. 9. Simulation results of second operation mode: Voltage of first and third ports and common bus (Top figures). current of first and third inductors , average current of first and third ports (Bottom figures).

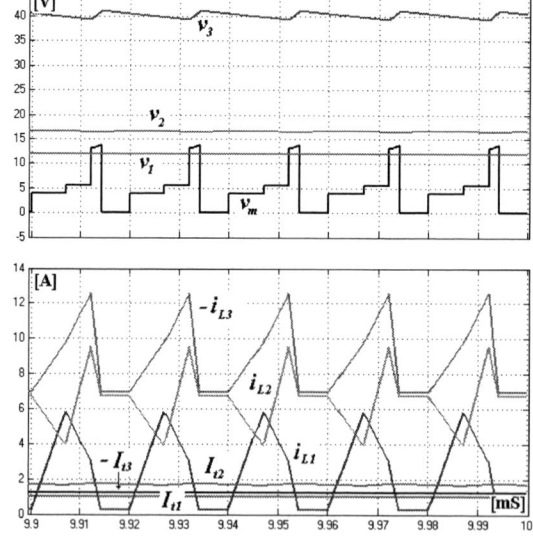

Fig. 10. Simulation results of third operation mode: Voltage of first, second and third ports and common bus (Top figures). current of first, second and third inductors , average current of first, second and third ports(Bottom figures).

978-1-5090-0376-1/16 $31.00 © 2016 IEEE

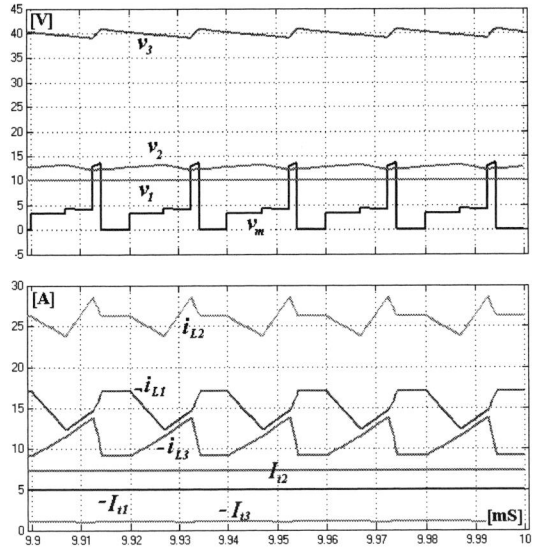

Fig. 11. Simulation results of seventh operation mode: Voltage of first, second and third ports and common bus (Top figures). current of first, second and third inductors , average current of first, second and third ports(Bottom figures).

Then third operation mode was simulated. Regarding TABLE II, it can be written that $N_1 = N_2 = N_3 = 1/3$. It was assumed that the battery voltage is 12V and reference voltage of solar panel for MPPT is 16.5V. Fig. 10 shows simulation results. D3 and D5 are determined by voltage and current controllers, respectively and their value s are 0.254 and 0.105 at steady state. As it can be seen from Fig. 10, the voltage and current of the load are 40V and 1A, respectively. Common bus voltage at different switching combinations, regarding (1), (9), (17), and (29), is 4, 5.5, 13.33, and 0, and its average is 4.2 V. In regard to (53), i_{L3_D5} is equal to 9.53A. Also considering (69), (70) and (71), i_{L3_D1}, i_{L3_D3} and i_{L3_D8} are 8.13A, 10.93A and 6.73A, respectively. Regarding (72), total average current of third inductor is equal to 8.58A. In this simulation, total average current of second port (I_{t2}) is equal to 1.75A that means generation 28.875W power by solar panel. Shortage of power (11.125W) is provided by the battery. total average current of the battery is 0.927A that proves this point.

As last test, seventh operation mode was simulated. It this simulation, it is assumed that the battery voltage is equal to 10V. Fig. 11 shows simulation results. D3 and D5 are determined by two current controllers, and their values are 0.2785 and 0.0875 at steady state. As it can be seen from Fig. 11, the voltage and current of the load are 40V and 1A, respectively. The solar panel voltage is 12,7V. Common bus voltage at different switching combinations, regarding (1), (9), (17), and (29), is 3.33, 4.18, 13.33, and 0, and its average is 3.5 V. In regard to (53), i_{L3_D5} is equal to 11.42A. Also considering (69), (70) and (71), i_{L3_D1}, i_{L3_D3} and i_{L3_D8} are 10.26A, 12.26A and 9.1A, respectively. Regarding (72), total average current of third inductor is equal to 10.68A. In this simulation, total average current of second port (I_{t2}) is equal to 7.16A that means generation 90W power by solar panel. 50W of power generated by solar panel is consumed to charge battery. Total average current of the battery is 5A that proves this point.

VIII. CONCLUSION

In this paper, a three-port converter was used for solar charger application with lighting load. The used converter is based on the connection of three conventional two-port converters with non-constant common bus. Therefore the common bus is capacitor-less. In this paper, various possible switching modes of the converter were analyzed. Based on the analysis, the control, switching and parameter design strategies were proposed for the motioned application. In the proposed method, four of eight switching combinations were selected. Using three of these four modes, voltage of all ports can be controlled as a function of common bus voltage, independently (Forth mode, zero switching mode, does not effect on voltage of ports).Also based on the proposed method, a pulse width modulation approach with three inputs, corresponding with duty cycles of the mentioned switching modes, was presented. Simulation results showed the performance and validity of the proposed configuration.

REFERENCES

[1] JP Benner, "Photovoltaics gaining greater visibility", *IEEE Spectrum*, Vol.36, Issue 9, Sep.1999 ,pp. 34-42 .

[2] DE Carlson, "Recent Advances in Photovoltaics," *Proceedings of the Intersociety Engineering Conference on Energy Conversion*, 1995.

[3] Ibrahim, O.; Yahaya, N.Z.; Saad, N., "Single phase inverter with wide-input voltage range for solar photovoltaic application," *IEEE International Conference on Environment and Electrical Engineering (EEEIC), Italy, Rome*, 2015..

[4] J.M. Ho, & C.C. Lou, "The Design and Implementation of Stand-Alone Solar Power LED Lighting Systems," *Recent Researches in Circuits, Systems, Electronics, Control & Signal Processing, Athens, Greece*, pp. 66-69, 2011.

[5] A. Gulati, and S. NVNS, "MPPT Solar Charger with Integrated LED Drive," *PowerPSoC Application Note*, www.cypress.com.

[6] F. da Rocha, Sa. Antonia, Silva, Martins Carlos E.A, Geraldo E.; Marques, Edson R., "A step-up switched-capacitor converter for LEDs applied to photovoltaic systems," *IEEE International Symposium on Industrial Electronics (ISIE), Armação dos Búzios, Brazil*, 2015, Pages: 1159 - 1165

[7] A. Wechsler, B. C. Mecrow, D. J. Atkinson, et al. "Condition monitoring of DC-Link capacitors in aerospace drives," *IEEE Transactions on Industry Applications*, vol. 48, no. 6, pp. 1866-1874, 2012.

[8] Li W, Xiao J, Zhao Y, He X, "PWM plus phase angle shift (PPAS) control scheme for combined multiport DC-DC converters," *IEEE Transactions on Power Electronics*, vol. 27, no. 3, pp. 1479-1489, 2012.

[9] L. Wang , Z. Wang and H. Li., "Asymmetrical duty cycle control and decoupled power flow design of a three-port bidirectional DC-DC converter for fuel cell vehicle application,", *IEEE Transactions on Power Electronics*, vol. 27, no. 2, pp. 891 -904, 2012.

[10] N. Katayama, and S. Kogoshi, "Frequency characteristic of a fuel cell-EDLC hybrid power source system with a multi-port bidirectional dc-dc converter," *IEEE International Conference on Power Electronics and ECCE Asia (ICPE ECCE), Jeju, Korea* , 2011, pp. 2561–2564.

[11] B. Zhao, Q. Yu and W. Sun, "Extended-Phase-Shift Control of Isolated Bi-directional DC-DC Converter for Power Distribution in Microgrid," *IEEE Transactions on Power Electronics*, vol. 27, no. 11, pp. 4667–4680, 2012.

[12] H. Wu, R. Chen, J. Zhang, Y. Xing, H. Hu and H. Ge, "A family of Three-Port Half-Bridge Converters for a Stand-Alone Renewable Power System," *IEEE Transactions on Power Electronics*, vol. 26, no. 9, pp. 2697-2706, 2011.

[13] H. Wu, K. Sun, S. Ding and Y Xing. "Topology Derivation of Nonisolated Three-Port DC–DC Converters From DIC and DOC," *IEEE Transactions on Power Electronics*, vol. 28, no. 7, pp. 3297–3307, 2013.

[14] Junjun Zhang, Hongfei Wu, Jun Huang, Yan Xing, Xudong Ma, "A Novel Multi-Port Bidirectional Converter for Interfacing Distributed DC Micro-Grid," *IEEE International Symposium on Industrial Electronics (ISIE), Istanbul, Turkey*, 2014.

978-1-5090-0376-1/16 $31.00 © 2016 IEEE

7th Power Electronics, Drive Systems & Technologies Conference (PEDSTC 2016)
16-18 Feb. 2016, Iran University of Science and Technology, Tehran, Iran

A High Performance Bi-directional AC-DC Converter for Charge Equalization

Nima Tashakor, Ebrahim Farjah and Seyed Reza
Khayam Hosseini
School of Electrical and Computer Engineering
Shiraz University
Shiraz, Iran
Nima.mpe@gmail.com

Teymoor Ghanbari
School of Advanced Technologies
Shiraz University
Shiraz, Iran
Ghanbarih@shirazu.ac.ir

Abstract— **To achieve high voltage and power in propulsion systems, cascade battery stacks are essential. Because of small discrepancies in charge or discharge characteristics, even between identical batteries, voltage imbalance would happen in cascaded battery stacks which get worse over the time with battery cell degradation. In this paper, a multilevel bi-directional battery charger with two operation modes for cascaded battery stack is developed. Grid to Vehicle (G2V) mode is an Electric Vehicle (EV) battery charger with reactive power control, controllable input current and charge equalization over cascade battery cells. Vehicle to Grid (V2G) mode is considered as an auxiliary energy storage device with active and reactive power control. The proposed charge equalization control uses each voltage cell as a state of charge indicator to prevent over charge or discharge of the cascaded cells which is very harmful for battery cell lifetime.**

Keywords— *cascade battery stack; Electric Vehicle; AC-DC multi-level converter; Grid to Vehicle (G2V); Vehicle to Grid (V2G), charge equalization.*

I. INTRODUCTION

Although there is a growing market for electric and plug-in hybrid vehicles, Electric Vehicles (EV) and Plug-in Hybrid Electric Vehicles (PHEV) technologies need more researches to become completely developed. Three important obstacles for EVs to achieve complete public acceptance are EV batteries' high cost and low lifetime, Complication of chargers, and lack of wide spread charging infrastructures [1]. Another drawback is the effect of EV battery chargers on distribution network because of high harmonic injection due to battery charging procedure [2].

Due to increasing demands for EVs, there is a vital requirement for a high power, high efficiency charger with features like: reactive power control, possibility of active power injection (V2G), low harmonic [3] and active charge equalization of cascaded battery stacks. Charge equalization between battery cells plays an essential role in extending EV battery lifetime and preventing battery cells from over charge or discharge.

There are two main charging methods: inductive and conductive. Despite recent progress in inductive charging, conductive charging is still the most popular method for EV battery charging. Conductive chargers can be divided in two groups: unidirectional and bi-directional chargers. Furthermore, they can be categorized to on-board and off-board battery chargers with single phase, three phase or high voltage dc interfaces as input source [1], [3].

There are three standard level of charge for EV battery chargers: level 1 is the slowest level that utilizes 120 V/15 A outlet and usually takes up to 4-8 hours for complete charge; level 2 is the primary method for home and public units that uses 208 to 240 V and up to 80 A; and level 3 or dc fast charging is for charging stations [1].

In [4], an on-board single-phase EV charger with reactive power control is proposed. The high voltage battery cells are charged with constant current but there is no charge equalization control.

In [5], a hybrid multi-level converter topology for EV applications is proposed. The proposed converter achieves multi-level voltage via bypassing and/or connecting battery cells. Although this converter could be used as battery charger, battery cells lifetime is decreased due to the high ripple of charging current. Furthermore, due to increased switching, switching loss is increased.

In [6], a diode-clamped series resonant converter with resonant valley compensator is proposed. Due to the Zero Current Switching (ZCS) control, switching loss is decreased. Increased elements and complex control of the converter are the main issues of this topology. [7] Also uses a resonant converter with Zero voltage switching (ZVS) capability and simpler structure, but the problem of complex control still remains.

A DC/DC converter for charge equalization is proposed in [8]. Charge equalization process is obtained using buck-boost operation. Although number of the elements is higher than usual, control strategy is simple.

In [9], an active charge equalizer for cascaded lithium battery cells using a Super-Capacitor (SC) as intermediate device is proposed. The system initially stores charge drawn from a cell in the SC and then the charge is moved to another

978-1-5090-0376-1/16 $31.00 © 2016 IEEE

cell with the lower state of charge. One problem is the time consuming process of charge equalization. To increase equalization speed, a multi-layer charge equalization control could is proposed in [10].

In this paper, a two stage multi-level converter topology and control is developed. The preferred topology for connecting three battery cells in series for realizing charge equalization procedure, and proposed control algorithm for charge equalization with constant current mode is explained. Although, number of the cascaded battery cells could be much higher than three cells. Simulation results using MATLAB software verify the performance of the developed topology and control algorithm. Battery cells are charged by a DC current with current ripple around 2% of nominal reference current.

II. TWO STAGE BI-DIRECTIONAL CONVERTER

Fig.1 represents two stage bidirectional converter topology and preferred the connection for cascaded battery cells. It is composed of two power stages. A full-bridge AC-DC bidirectional converter is the first stage and a DC-DC buck/boost converter is the second stage of power conversion.

Since most on-board chargers are in mid power level, they need to be compatible with home installations. A single phase AC power grid is considered for input power. However, the topology and control algorithm can be easily modified for three phase input sources.

A. G2V Mode

During this mode, AC-DC bidirectional converter operates as an active rectifier with sinusoidal current absorption and controlled power factor. The buck/boost converter operates as a buck converter in constant current or constant voltage charging mode.

1) Full-Bridge AC-DC Bidirectional Converter Control

Basic requirement of the full bridge in this mode of operation is sinusoidal current absorption in order to reduce filter size and harmonic distortion in Point of Common Coupling (PCC). Since according to IEC 61000-3-2 standard, it is mandatory for the controller to be synchronized with power grid voltage. Hence, First a Phased-Locked Loop

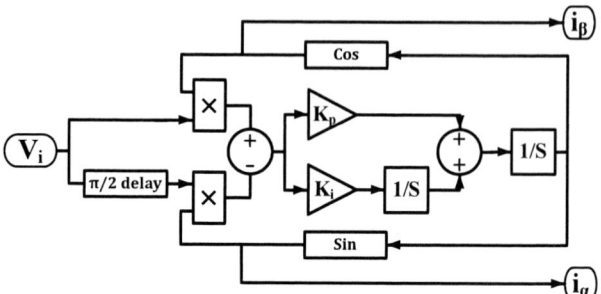

Fig. 2. PLL Control Block Diagram

algorithm is implemented using [11], [12]. In Fig. 2, block diagram of a single phase PLL is illustrated. The output signals of PLL have unity amplitude. When PLL is synchronized, signals pllα and pllβ are the direct and quadrature components of the grid fundamental voltage, as is requested by the standard. These signals are used to synthesize the source reference current.

Reference current of the full-bridge converter is composed of two components. First one is the active power needed for charging of traction batteries which is accomplished by controlling DC link voltage. Second component is the reactive power needed to control charger power factor, which can be injected back or absorbed. Since the converter is designed to work with a maximum apparent power, the calculated active or reactive power should be limited to a maximum to protect the device. Fig. 3 shows the control block diagram of the reference current is*.

Power delivered to the traction batteries should be constant, therefore, a dc link capacitor is used as an intermediate energy storage device. In order to have constant power delivered to the traction, the full-bridge converter is responsible to stabilize average DC link voltage [4].

After synthesizing i_s^*, (1) is used to calculate the reference voltage for unipolar PWM controller with switching frequency of 20 kHz to control the DC link voltage of C1. The control block diagram of the full-bridge is shown in Fig. 4.

$$V_F = V_S - L\ (di_s^*/dt + di_{err}/dt) \qquad (1)$$

Where i_{err} is calculated using (2).

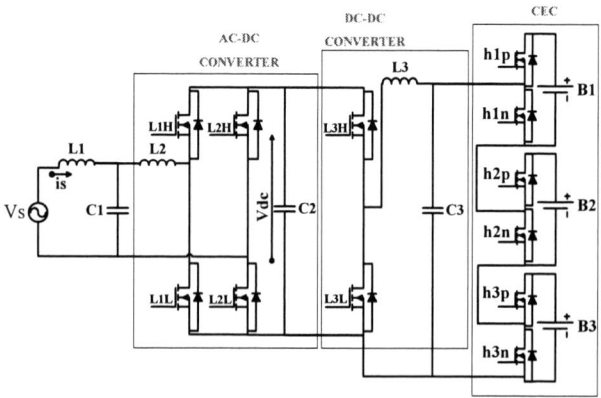

Fig. 1. Battery Charger Circuit Topology

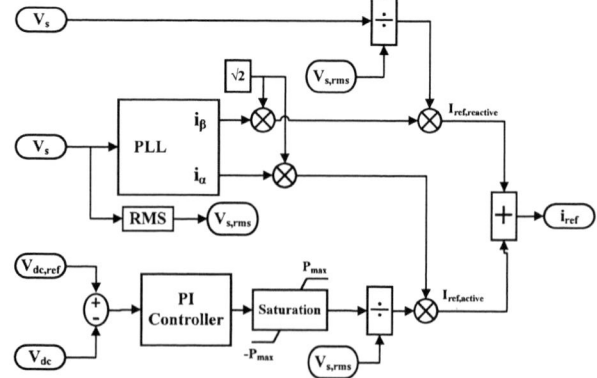

Fig. 3. Block diagram relevant to control of $i_{s,ref}$

978-1-5090-0376-1/16 $31.00 © 2016 IEEE   230

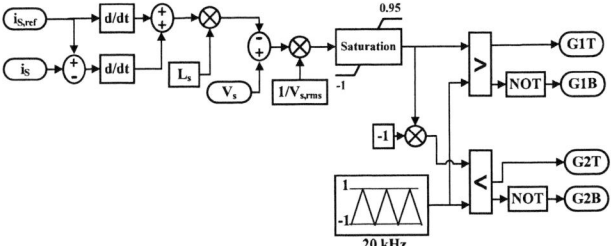

Fig. 4.   AC-DC Full-Bridge Control Block Diagram

$$i_{err} = i_s{}^* - i_s \qquad (2)$$

For practical implementation of the obtained reference voltage, a predictive current control algorithm could be used [4].

2) DC-DC Converter-Buck Operation

Since in charging mode of operation, DC link voltage is always higher than traction batteries voltage, DC-DC converter works in Buck mode.

For increasing battery cells lifetime, two mode of charging is used. When battery voltage is lower than the rated, a constant current mode is used and whenever the battery voltage is equal to or higher than rated, a constant voltage charging mode is utilized. When battery voltage is rated value or higher, battery won't draw too much current and thus, constant voltage mode is not harmful for the battery. On the other hand, when battery voltage is lower than the nominal voltage, if allowed battery will draw too much current which would harm the battery or lower its lifetime. As shown in Fig. 5, the control algorithm for DC-DC buck converter uses voltage or current error as input to PI controller which adjusts duty cycle of PWM modulator with a carrier frequency equal to 40 kHz.

B. V2G Mode

In this mode of operation, the battery stack returns its stored energy back to the grid. The main requirement in this mode is a synchronized sinusoidal current on the AC side with

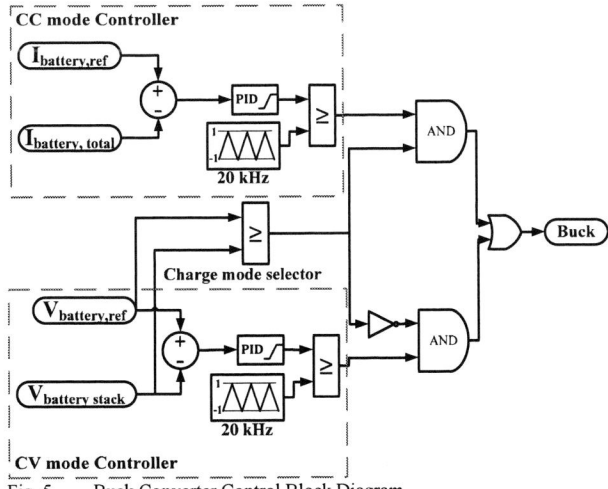

Fig. 5.   Buck Converter Control Block Diagram

controllable active and reactive power.

1) AC-DC Full-Bridge Control

In this mode, AC-DC full-bridge converter is used as inverter with sinusoidal output current and reactive control.

Procedure of obtaining reference current in this mode is quite similar to the G2V mode. The only difference is that here both active and reactive components of the reference current are external input parameters determined by the power grid controller. Switching pattern for IGBT switches is obtained by predictive current control algorithm through unipolar PWM modulation with a 20 kHz triangular carrier.

2) DC-DC Converter Controller

DC link voltage must be greater than the voltage amplitude of the power grid to make power transfer from the DC link capacitor possible. Therefore, in this mode, DC-DC converter works as a boost converter. Since during short periods of time, the traction batteries voltage does not fluctuate, delivering reference power to the grid with constant discharge current is possible. During the discharge period, as the traction batteries voltage decreases, reference discharge current of the traction batteries increases in an inverse proportion to obtain constant power injection. The control algorithm for obtaining switching pattern is shown in Fig. 6.

III. CASCADE BATTERY STACKS CONNECTION AND CONTROL ALGORITHM

Power ratings needed for propulsion drive system in EVs makes cascade battery stacks essential. But even though identical cells are used in a battery stack, because of the small differences in chemical characteristics or charge and discharge patterns, there would be some voltage unbalance. This can cause over charge and discharge in case of cascaded battery stacks and circulating current in case of parallel battery cells.

As shown in Fig.1, each battery cell is connected in parallel with two bidirectional switches. With the upper switch on and the lower one off, the battery cell is connected to the circuit and when the lower switch is on and the upper one off, the battery cell is bypassed from the circuit, and hence the battery is neither charged nor discharged.

The control algorithm for bypassing or connecting battery cells to the circuit can be determined by state of charge (SOC) or output voltage of the cells. In simulations, both SOC control and output voltage control are possible, but since determining SOC of the battery cell is hard in practical implementations, battery cell voltage is used as input for control algorithm in simulations.

978-1-5090-0376-1/16 $31.00 © 2016 IEEE   231

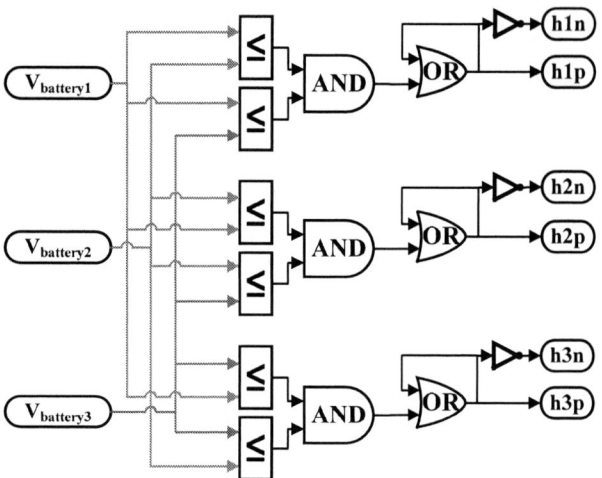

Fig. 7. Proposed control algorithm for charge equalization of the battery cells

Fig. 7 shows the proposed control algorithm for charge equalization of the battery cells. First, the battery cell with the lowest state of charge is identified as the primary cell and VTBref (reference voltage for buck converter) is chosen as the cell nominal voltage. The primary cell is charged until its voltage is equal to the second cell with the lowest voltage. Then the second battery cell is connected to the circuit and both cells are charged together, and so on.

Because of small differences in charge and discharge characteristics, even after primary charge equalization in the charging process, still some state of charge differences occurs during the charging process. Since some batteries have lower capacity than the others, with equal current, they are charged faster. So the secondary charge equalization is to bypass those cells that have reached their rated value, to prevent them from over charge.

The discharge process in V2G operation is quite the same. At first all of the fully charged battery cells are in series. V2G mode starts and gradually, battery cell voltages start to decrease with respect to their capacity, until the first battery cell has reached its discharge limit. At this point to avoid over discharge, battery which is discharged to its limit is bypassed. Other battery cells are still connected, until the second battery cell has reached its limit and so on.

It should be noted that every time that a battery is switched on or bypassed, VTBref is changed, but the nominal charging and discharging current remain constant.

IV. DESIGN CONSIDERATIONS

Table I shows some design specifications of the battery charger. Filter elements are designed using those specifications.

In AC-DC full-bridge converter, AC current ripple depends on the DC link voltage, switching frequency (f_s) and coupling inductance (L_1). According to (3) for a 2% ripple, coupling inductance will be 5mH:

$$L1 = (v_{DC}/4i_{ripple}f_s) \qquad (3)$$

TABLE I.
BATTERY CHARGER SPECIFICATIONS

Parameters	Value	Unit
Input AC Voltage (RMS)	230	V
AC Input Frequency	50	Hz
Maximum Input AC Current (RMS)	20	A
Maximum Input AC Current Ripple	0.1	A
Output DC Voltage Range	400	V
Output DC Voltage Ripple	8	V
Maximum Output DC Current	10	A
Maximum Output DC Current Ripple	.1	A
Maximum Output Power	5	kW

To obtain a constant power in battery charging or discharging, a suitable DC link capacitor is needed to be used as an intermediate storage device. Sizing of the element is very important for voltage ripple. According to (4), DC link capacitor can be calculated considering a 2% dc ripple in a maximum of 400 V DC link voltage which would be 3.6 mF:

$$C = (2V_sI_s/\omega V_{ripple}V_{DC}) \qquad (4)$$

Considering a constant charging current for battery cells, for 2% DC current ripple in battery side, a 500 mH inductance is needed.

V. SIMULATION RESULTS

MATLAB 7 software is chosen for simulation studies. Series conduction resistance of IGBT switches are considered equal to 1mΩ and snubber resistance equal to 100 kΩ. Measuring devices are considered ideal. Battery cell characteristics are listed in Table II.

As shown in Table II, different Battery types with slightly different nominal voltages and capacities are chosen, to verify the proposed charge equalization control.

Simulation results for G2V mode of operation are shown in Fig.8 to Fig.12. As shown in Fig. 8, AC side input current is sinusoidal and synchronous with input voltage. To eliminate high frequency components of the input current, a LC filter can be used.

As shown in Fig. 9 and 10, first the cell with the lowest voltage (cell 2) is charged by constant current and all the other

TABLE II
BATTERY CELLS SPECIFICATIONS

Battery Cell	Battery Type	Nominal Voltage	Rated capacity	Initial SOC	Internal Resistance
Cell 1	Lead-Acid	100 V	100 Ah	70 %	0.01 Ω
Cell 2	Li-Ion	99 V	99 Ah	65 %	0.01 Ω
Cell 3	Ni-Cad	102 V	102 Ah	75 %	0.0099 Ω

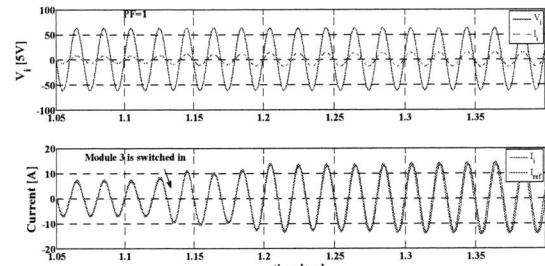

Fig. 8.   Simulation results for AC side parameters

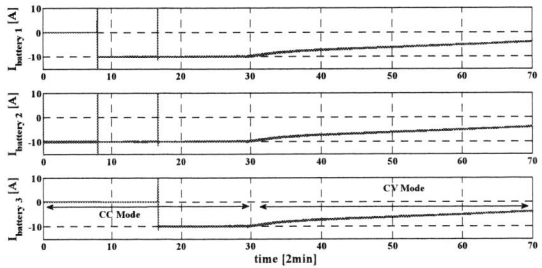

Fig. 9.   Simulation results for Battery cell currents.

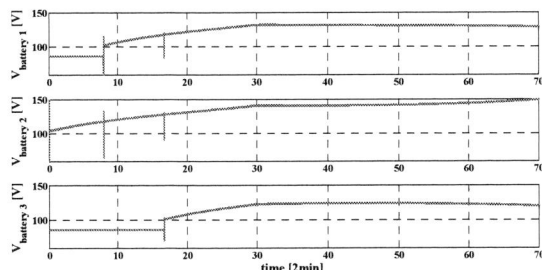

Fig. 10.   Simulation results for Battery cell Voltages

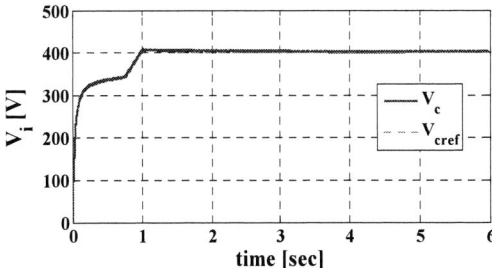

Fig. 11.   Simulation results for DC link voltages

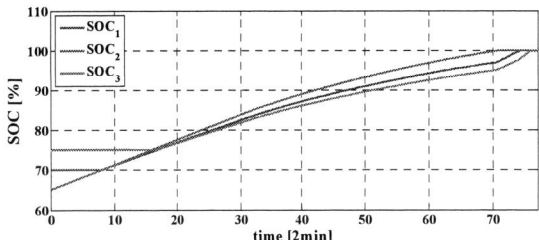

Fig. 12.   Simulation Results for State of Charge

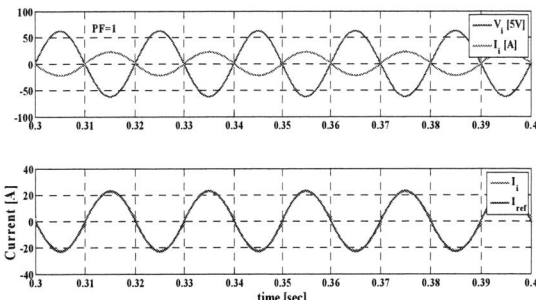

Fig. 13.   AC side simulation results during V2G

batteries are bypassed from the charger.

The moment that SOC of cell 2 reaches cell 1 (the second lowest SOC), h1p switch is switched on and h1n is off. Cell 1 is connected in series with cell 2 and both of them are charged with constant current until their terminal voltage reaches the next cell. As shown in Fig. 8 maximum charge current ripple in the steady state is 1 A and with increasing number of the connected cells, current ripple would decrease.

Each time a cell is switched connected or bypassed from the circuit, reference battery voltage changes accordingly, but the DC link reference voltage is constant through the whole process. Fig. 11 shows that DC link voltage ripple is less than 2 %. Fig. 12 shows battery cells' state of charge. For V2G mode, simulation results are shown in Fig. 13 to Fig.15. In this mode, battery cells should discharge with constant current and the ac side current must be sinusoidal and in phase with voltage. Fig.13 shows AC side output voltage and current. AC side output currents are sinusoidal and in phase with AC side voltage.

Battery side discharge current is shown in Fig. 14. Discharge current ripple is less than 8%. In order to deliver

constant power to the grid, discharge reference current increases as the batteries terminal voltage is decreased.

In this mode of operation DC link ripple is increased because of the increased discharge current of the battery cells. If the injected power to the grid is decreased, so does the DC link voltage ripple (see Fig. 15). During all operation modes, THD is than 3%. Furthermore, as mentioned before, here reactive power can be controlled. In Fig. 16, AC side voltage and current with PF=0.9 lead is shown. As it is obvious, current is completely sinusoidal.

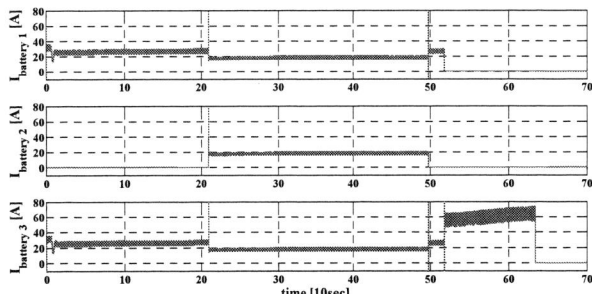

Fig. 14.   Battery Current simulation results during V2G operation

978-1-5090-0376-1/16 $31.00 © 2016 IEEE   233

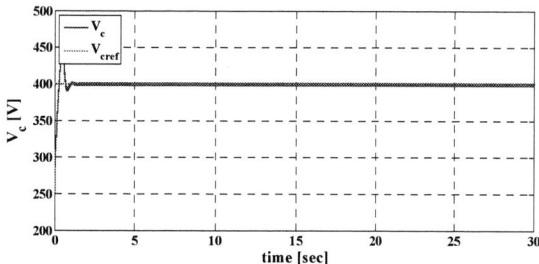

Fig. 15. DC link voltage.

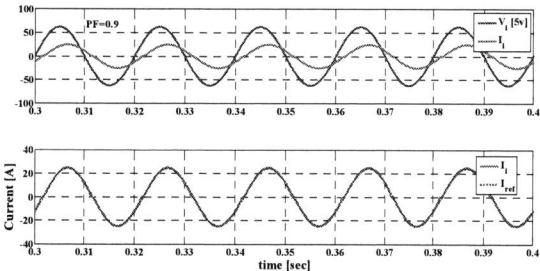

Fig. 16. AC side voltage and current during G2V with PF=0.9

VI. CONCLUSION

In this paper, a high efficiency AC-DC front-end bi-directional converter with charge equalization is developed. The converter is suitable for EV battery charger and DG battery bank AC-DC interface applications. Minimum switching loss for charge equalization condition, high efficiency, and reactive power control are the main advantages of the converter. Therefore, power quality control and power factor correction can be realized by the developed converter. Simulation results verify the performance of the developed converter.

REFERENCES

[1] M. Yilmaz, P. T. Krein, "Review of battery charger topologies, charging power levels, and infrastructure for plug-in electric and hybrid vehicles," IEEE Transactions on Power Electronics, vol 28, no. 5, may 2013.

[2] C. Chan, K. T. Chau, "An Overview of Power electronics on EVs," IEEE Transactions on Industrial Electronics, vol. 44, no. 1, pp.3-13, February 1997.

[3] Y. Du, X. Zhou, S. Bai, S. Lukic, A. Huang, "Review of non-isolated bi-directional DC-DC converters for plug-in hybrid electric vehicle charge station application at Municipal parking decks, " IEEE Applied Power Electronics Conference (APEC), USA, pp. 1145–1151, February 2010.

[4] G. Pinto, V. Monteiro, H. Goncalves, J. L. Afonso, "On-board reconfigurable battery charger for electric vehicles with traction-to-auxiliary mode," IEEE Transactions, Vehicular Technlogy, vol.63, pp 1104–1116, March 2014.

[5] Z. Zheng, K. Wang, L. Xu, Y. Li, "A hybrid cascaded multi-level converter for battery energy management applied in electric vehicles," IEEE Transactions. Power Electronics, vol. 29, pp. 3537–3546, October 2013.

[6] J. Y. Lee, Y. D. Yoon, J. W. Kang, "A Single-Phase Battery Charger Design for LEV Based on DC-SRC with Resonant Valley-Fill Circuit," IEEE Trans. Industrial Electronics (TIE), 2014, unpublished.

[7] M. Pahlevaninezhad, J. Drobnik, P. K. Jain , and A. Bakhshi, "A Load Adaptive Control Approach for a Zero-Voltage-Switching DC/DC Converter Used for Electric Vehicles," IEEE Trans. Industrial Electronics., vol. 59, no. 2, pp. 920–933, Feb. 2012.

[8] S. Yarlagadda, T. T. Hartley, I. Husain"A Battery Management System Using an Active Charge Equalization Technique Based on a DC/DC Converter Topology," IEEE Trans. Industrial Electronics, vol. 49, No. 6, pp. 2720–2729, Nov./Dec. 2013.

[9] F. Baronti, G. Fantechi, R. Roncella. R. Saletti, "High-Efficiency Digitally Controlled Charge Equalizer for series-Connected Cells Based on Switching Converter and Super-Capacitor,"IEEE Trans. Industrial Informatics, vol. 9, No. 2, pp. 1139–1147, May. 2013.

[10] Dong, Y. Li, Y. Han, "Parallel Architecture for Battery Charge Equalization," IEEE Trans. Power Electronics, 2014, unpublished.

[11] H. Carneiro, L. F. Monteiro, J. L. Afonso, "Comparisons between Synchronizing Circuits to Control Algorithms for Single-Phase Active Converters," IEEE Industrial Electronics (IECON '09), pp. 3229–3234, November 2009.

[12] L. G. B. Rolim, D. R. da Costa, M. Aredes, "Analysis and Software Implementation of a RobustSynchronizing PLL Circuit Based on the pq Theory," IEEE Transactions. Industrial Electronics, vol. 53, no.6, pp. 1919–1926, November 2006.

7th Power Electronics, Drive Systems & Technologies Conference (PEDSTC 2016)
16-18 Feb. 2016, Iran University of Science and Technology, Tehran, Iran

Control of Super-Capacitor SOC in a Railway Transit Network

Elham Rahimi , Ali Dastfan and Saeed Ahmadi
Department of Electrical and Robotic Engineering
University of Shahrood
Shahrood, Iran
elhamrahimi_1367@yahoo.com, dastfan@ieee.org, saeedahmadi35@yahoo.com

Abstract— **This paper proposes a control strategy for Energy Storage System (ESS) in railway transit network. The system includes a super-capacitor (SC) and a DC-DC convertor. The super-capacitor is used for energy storing and regulating the DC line voltage which is achieved by a bidirectional DC-DC convertor. First the structure of the metro network is introduced, then the model of ESS and its control strategy are explained. An important feature of this simple and effective control strategy is the use of appropriate reference current for charging and discharging of SC. Simulation in MATLAB/Simulink showed that DC bus voltage fluctuation is limited, which is acceptable, and the super-capacitor voltage changes are kept within the allowable range and total input energy of substations is reduced considerably.**

Keywords— Supercapacitor; Energy storage system; Railway transit network; regenerative braking; energy saving

I. INTRODUCTION

Nowadays, the urban rail transit network has attracted the attention of researchers due to some advantages, including safety, energy saving, faster transportation, being more environment-friendly and high capacity[1].

One of the major means for energy saving is using the regenerative braking energy. In this way the kinetic energy is converted to electrical energy, and then it is used in next acceleration period [2]. Three possible conditions may appear:
1- One train is accelerating while the other one is braking [3].
2- Braking energy returns to the network.
3- Braking energy is saved in Energy Storage System (ESS).

Since rectifiers in most metro network are unidirectional, so regenerative braking energy cannot be returned to the supply network [4, 5]. Additionally in [6, 7] it is recommended that using reversible substations is not completely appropriate due to the high cost of new equipment.

In subway network, the super-capacitors (SC) are used as storage systems which is a suitable choice for energy storing, because of its high power density and durability against frequent charging and discharging due to the repeated start–stop of the vehicles over a day.

In [4, 8] regenerative braking current value is calculated in each substation, then suitable ESS configuration is proposed for each station, but in [8] the proposed model is more accurate. In [9] benefits of using different ESS for both stationary and on-board system have been discussed but equipment sizing, positioning and control algorithm are not considered.

As the state of charge (SOC) variation of the super-capacitor should be between 25% and 100% [4], in this paper a simple control method for managing the SOC of the super-capacitor has been proposed and used. Our focus is on stationary storage systems.

In Section II subway network structure will be introduced. The ESS and its control system will be discussed in section III. In section IV, the simulation results of control system, on DC bus voltage and the amount of energy savings in the metro network will be presented. Finally, the conclusion of the paper is given in section V.

II. SPECIFICATIONS OF THE NETWORK

The metro network model consists of stationary ESS, vehicle, unidirectional substations and connecting lines. These components are described in the following subsections and shown in Fig.1.

A. Traction substation and connecting lines model

Ideal DC voltage sources with an internal resistance are used to describe the substation. The connecting lines are modeled as electric resistors. Due to the fact that trains are moving between two stations, the resistance between the train and the previous or next station is time-dependent. These values are shown in Fig. 1 and are calculated as [4]:

$$R' = \frac{kx(t)}{1000} \tag{1}$$

$$R'' = \frac{k(D - x(t))}{1000} \tag{2}$$

Where R' is the line resistance between the train and the previous station (Ω/km), R'' is the line resistance between the train and the next station (Ω/km), k represents the resistive coefficient, $x(t)$ is the distance between the train and the last station, D is the distance between the first and next stations (km).

B. Vehicle model

A controlled current source is used to simulate the train that consumes power during the acceleration time and delivers it during the regeneration time. Traction and braking power drawn and given to the DC bus are defined by (3) and (5) respectively.

$$P_{t\,raction} = \frac{(m_t \times a + F_r) \times v}{\eta_g \times \eta_m \times \eta_i} + P_a \qquad (3)$$

$$I_{t\,rain} = \frac{P_{traction}}{V_{bus}}$$
$$(4)$$

$$P_{braking} = (m_t \times a + F_r) \times v\,(\eta_g \times \eta_m \times \eta_i) - P_a \qquad (5)$$

$$I_{t\,rain} = \frac{P_{braking}}{V_{bus}} \qquad (6)$$

The resistive forces are calculated by (7):

$$F_r = (0.0064 \times m_t + 1.056) + (0.00014 \times m_t \times v + 0.00046 \times v^2) \qquad (7)$$

Where m_t is the total mass of the train with passengers, a is acceleration, v is the velocity, η_g, η_m, η_i represent gear box, motor and inverter efficiencies respectively. P_a is the power that is required for lighting and air-conditioning (auxiliary power). Equations (4) and (6) demonstrate the vehicle behavior.

C. Energy storage system model

The ESS model includes the super-capacitors and DC-DC converter.
The advantages of ESS are [10]:
- energy saving.
- improve power quality.
- dc bus voltage regulation.

III. ESS DESIGNING

A. Supercapacitor Design

The SC model consists of various components:
- A capacitor (C)
- An equivalent parallel resistor (EPR); representing the self-discharging loss.
- An equivalent series resistor (ESR); modeling the ohmic loss [11].

The maximum energy that an ESS can store corresponds to the sum of the potential and kinetic energy. Equation 8 is used to calculate the ESS capacitance regardless of potential energy:

$$C = \frac{m_t \times v^2}{V_{sc}^2} \qquad (8)$$

Table 1 show the parameters of Maxwell BCAP3000P270 super-capacitor which is used in this paper.

B. Controller Design

Since there is no direct connection between the forward and backward paths, two half-bridge bi-directional DC-DC converters have been utilized. The structure of DC-DC converters are shown in Fig. 1. Also r_1, r_2, r_3 and r_4 are rail electric resistors. In charging mode, they operate similarly to a buck converter and in discharging mode they operate like a boost converter.

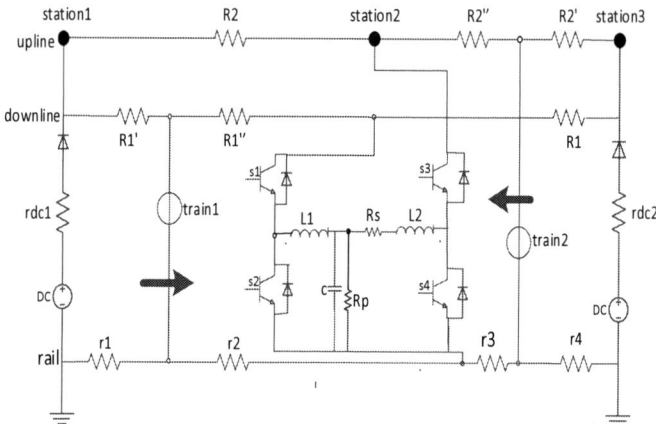

Fig. 1. Modeling of the metro network with 3 stations

When the upper switch (S1) is turned on, inductor voltage (v_l) is equal to the difference between Super-capacitor voltage (v_{sc}) and bus voltage (v_{bus}) and when bottom switch (S2) is on, v_l is v_{sc}. As the duty ratio of the bottom switch is d, the average inductor voltage is obtained as (9) [12].

$$V_l = V_{sc} - V_{bus} \times (1-d) \qquad (9)$$

A compensator is used to remove non-linear part and linearize the system. For this purpose, the following control law has been suggested [14].

$$d = (V_{bus} - V_{sc} + V_l) \div V_{bus} \qquad (10)$$

It is worth mentioning that duty ratio is limited between 0.05 and 0.95.

The closed loop transfer function can be written as in (11).

$$T(s) = \frac{I_{sc}}{I_{sc-ref}} = \frac{K_p\,(s + K_i/K_p)}{L(S^2 + \frac{K_p}{L}S + \frac{K_i}{L})} \qquad (11)$$

By considering given damping ratio ζ and crossover frequency f_c (crossover frequency must be much lower than f_s), k_p and k_i are calculated according to equations (13) and (14) [14].

TABLE I   Maxwell BCAP3000P270 Characteristics

Parameters	Unit	Value
Capacitance	(F)	3000
Rated voltage	(v)	2.7
Surge voltage	(v)	2.8
ESR,DC	(mΩ)	0.29
Leakage current	(mA)	5.2
Vsc_min	(v)	300
Vsc_max	(v)	600
Maximum Continuous Current	(A$_{rms}$)	210

$$\omega_C = \frac{K_p}{L} \tag{12}$$

$$k_p = 2\zeta\omega_n L \tag{13}$$

$$k_i = \omega_n{}^2 L \tag{14}$$

- *DC/DC converter controller*

The control strategy of the converter is based on a current feedback loop to regulate the super-capacitor current in Fig(2). The reference current (I_{sc-ref}) consists of two components: I_{sc-l} and I_{sc-re} ($I_{sc-ref} = I_{sc-l} - I_{sc-re}$). I_{sc-l} is determined based on the power balance calculation and assuming converters without losses, as shown in (15) [14].

$$I_{sc-l} = \frac{(V_{bus} \times I_L)}{V_{sc}} \tag{15}$$

Where I_L is load current.
I_{sc-re} manages the SOC of the super-capacitor as shown in (16) [14].

$$I_{sc\_re} = k_1 \left(V_{scmax} - V_{sc} \right)^{k2} \tag{16}$$

k_1 and k_2 are two positive parameters that have an effect on SOC of the super-capacitor and so should be appropriately selected so that super-capacitor voltage does not exceed the permitted range and the v_{sc} is 300 V at full load. By considering k_2 value 10, k_1 value can be calculated.

IV. SIMULATION RESULTS

In this part, several simulations were carried out by MATLAB/Simulink software to ensure the effectiveness of the control strategy. The network model includes three stations and two trains. One train moves in forward and the other in backward path. Moreover, it is supposed that the ESS is set in station 2 (Fig. 1). Tables 2 and 3 display the parameters of simulated DC-DC converter and network respectively.

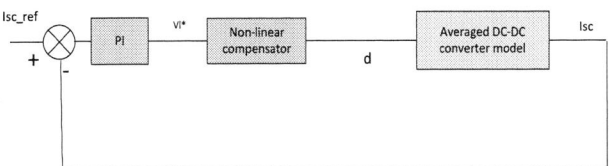

Fig.2. Super-capacitor current control system

TABLE II   Parameters of simulated dc-dc converter

Parameters	Unit	Value
L	μH	10
f$_s$ (switching frequency)	(kHz)	10
K$_p$	-	0.01
K$_i$	-	3.2283
ζ		0.88
f$_C$ (crossover frequency)	(kHz)	1

TABLE III   Parameters of network

Parameters		Unit	Value
Distance between stations		(m)	1000
Rail electric resistance		(mΩ/km)	33
Contact wire resistance		(mΩ/km)	6
Substation internal resistance		(mΩ)	18
η_g	gear box efficiency	(%)	95
η_m	motor efficiency	(%)	90
η_i	inverter efficiency	(%)	97
P_a	auxiliary power	(kW)	70
	maximum train power	(Kw)	500
v	maximum speed	(m/s)	19.44
a	maximum acceleration	(m/s^2)	1
R	braking resistor	(Ω)	1.42

The speed of two trains at different times are represented in Fig. (3). As shown in Figs (3), when train 1 brakes in the forward path, train 2 starts its movement in the backward path. The model included powering, cruising and braking regimes, which are shown in Fig (4). Train 1 starts moving from the first station to the second in the forward path and after 10 seconds, stops at station 2, it then starts its movement again. Simulation has been done from second 55 to 130.

978-1-5090-0376-1/16 $31.00 © 2016 IEEE   237

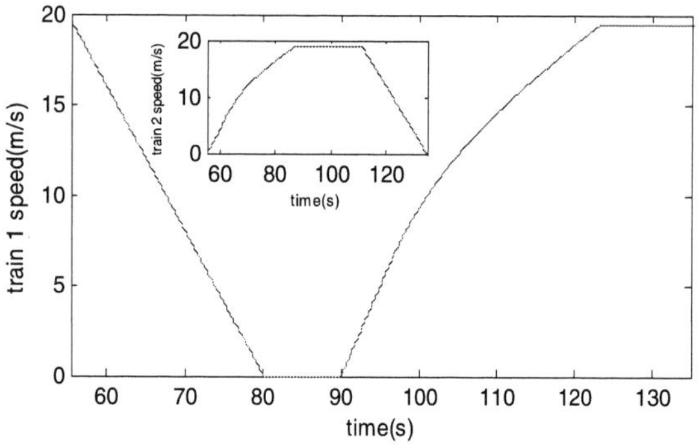

Fig.3. Train one speed

The super-capacitor voltage is shown in Fig (5). The regenerative braking energy of train 1 is stored in ESS, which is used during the acceleration of train 2. With ESS considered, in the first 18 seconds less energy will be drawn out of substations than that are shown in Figs (6) and (7).

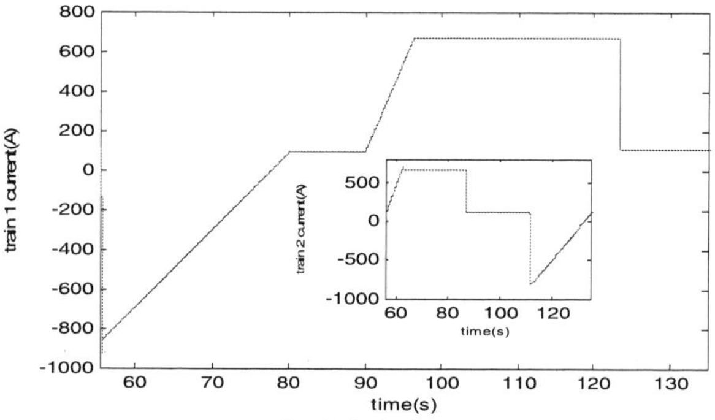

Fig.4. Train 1 current

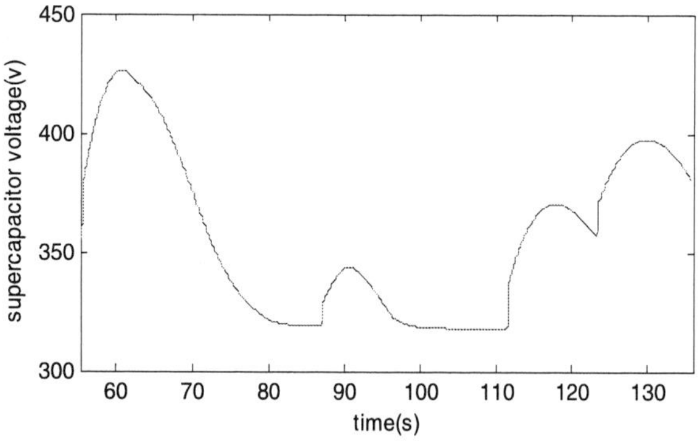

Fig.5. Super-capacitor voltage

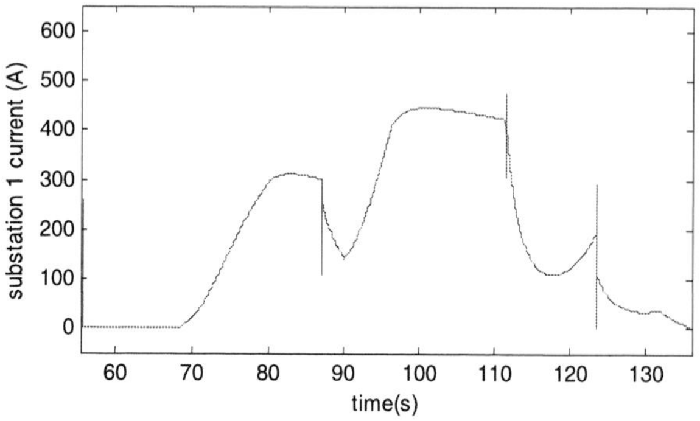

Fig.6. Substation 1 current with ESS

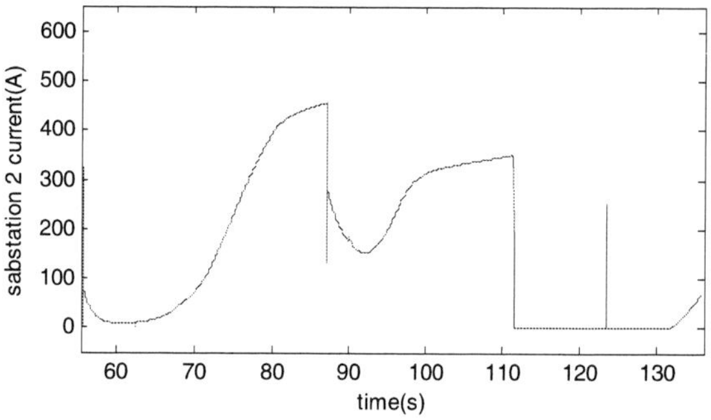

Fig.7. Substation 2 current with ESS

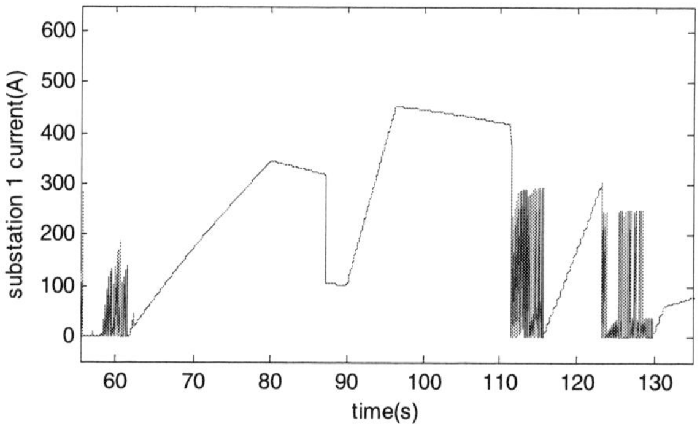

Fig.8. Substation 1 current without ESS

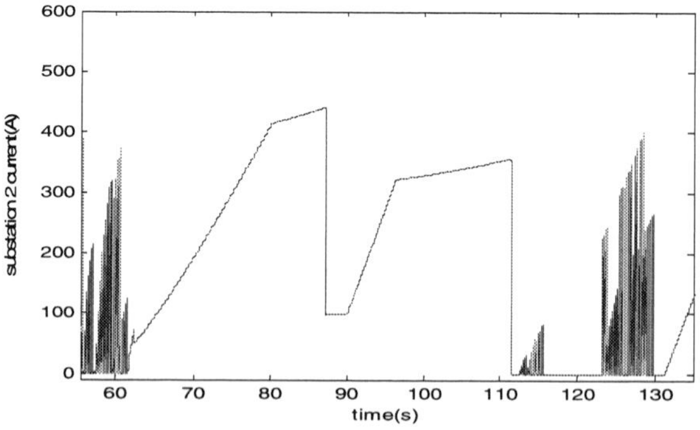

Fig.9. Substation 2 current without ESS

Comparing the states with and without ESS, the energy drawn out of substations would be equal to 4.7 and 5.42 kWh respectively. The energy dissipated in the braking resistors would be 0.63kwh without ESS and it drops to an almost negligible amount in case of using an ESS.

Bus 1 and 2 voltages with and without considering ESS are shown in Figs (10) to (13) respectively. By comparing Figs (10) and (12) we can observe that from second 55.6 to 70 in Fig (10), bus voltages are 750V but at the same time in Fig (12), bus voltages are reduced. Generally dc bus voltage variation is limited to 10% that is acceptable. Similarly to Figs (11) and (13).

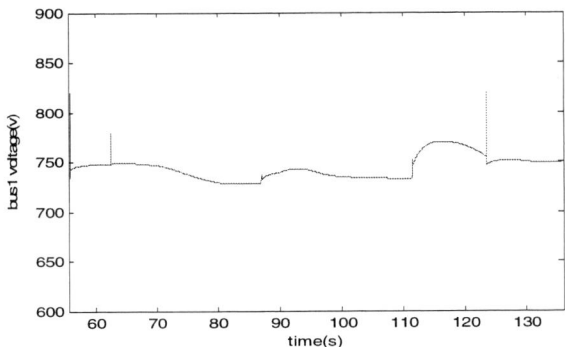

Fig.10. Bus 1 voltage with ESS

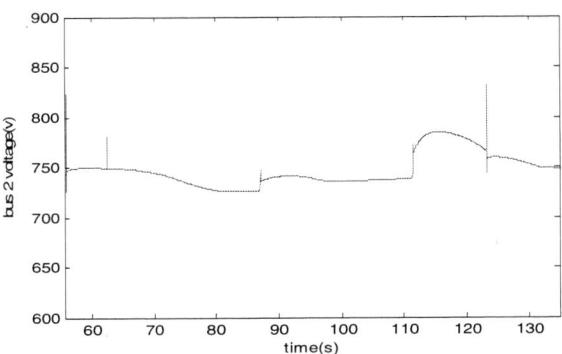

Fig.11. Bus 2 voltage with ESS

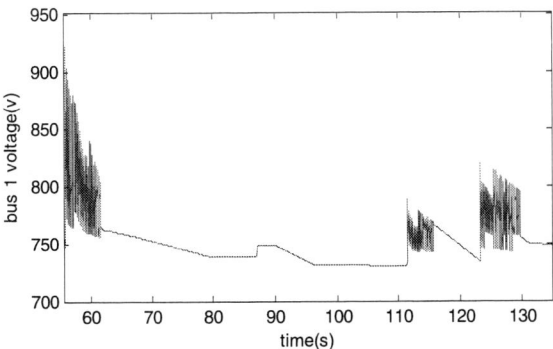

Fig.12. Bus 1 voltage without ESS

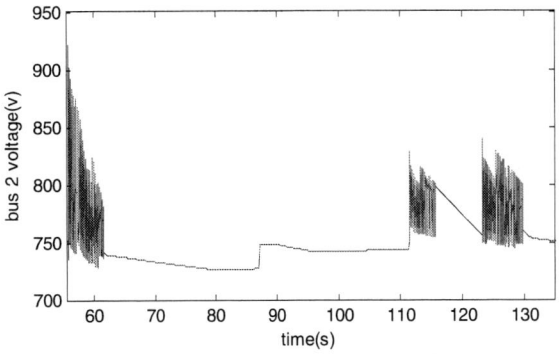

Fig.13. Bus 2 voltage without ESS

V. CONCLUSION

In this paper an efficient control method is used to control the super-capacitor SOC. By comparing results both with and without considering ESS, we found out that by considering ESS, The amount of energy which is drawn from substations can be reduced from 5.42 KWh to 4.7 KWh. Moreover, energy dissipated in the braking resistor is reduced from 0.63 KWh to a negligible amount, showing that the control strategy has a significant effect in suppressing the voltage fluctuation. The DC bus voltage variation is also limited to 10% which is acceptable.

REFERENCES

[1]   Z. Gao, J. Fang, Y. Zhang, L. Jiang, D. Sun, and W. Guo, "Control of urban rail transit equipped with ground-based supercapacitor for energy saving and reduction of power peak demand," *International Journal of Electrical Power & Energy Systems*, vol. 67 pp. 439-447, 2015.

[2]   X.-l. Chen, D.-q. Liang and W.-d. Zhang, "Braking energy recovery for electric traction based on super-capacitor and Bidirectional DC-DC converter," in *Power Electronics and Motion Control Conference (IPEMC), 2012 7th International*, 2012, pp. 879-883.

[3]   F. Ciccarelli, D. Iannuzzi, and P. Tricoli, "Control of metro-trains equipped with onboard supercapacitors for energy saving and reduction of power peak demand," *Transportation Research Part C: Emerging Technologies*, vol. 24, pp. 36-49, 2012.

[4]   R. Teymourfar, B. Asaei, and H. Iman-Eini, "Stationary super-capacitor energy storage system to save regenerative braking energy in a metro line," *Energy Conversion and Management*, vol. 56, pp. 206-214, 2012.

[5]   M. Steiner, M. Klohr, and S. Pagiela, "Energy storage system with ultracaps on board of railway vehicles," in *Power Electronics and Applications, 2007 European Conference on*, 2007, pp. 1-10.

[6]   D. Cornic, "Efficient recovery of braking energy through a reversible dc substation," in *Electrical Systems for Aircraft, Railway and Ship Propulsion (ESARS), 2010*, 2010, pp. 1-9.

[7]   H .Chuang, "Optimisation of inverter placement for mass rapid transit systems by immune algorithm," *IEE Proceedings-Electric Power Applications*, vol. 152, pp. 61-71, 2005.

[8]   G. Zongyu, F. Jianjun, Y. Zhang, and S. Di, "Control strategy for wayside supercapacitor energy storage system in railway transit network," *Journal of Modern Power Systems and Clean Energy*, vol. 2, pp. 181-190, 2014.

[9]   F. Foiadelli, M. Roscia, and D. Zaninelli, "Optimization of storage devices for regenerative braking energy in subway systems," in *Power Engineering Society General Meeting, 2006. IEEE*, 2006, p. 6 pp.

[10] X. Shen, S. Chen, G. Li, Y. Zhang, X. Jiang, and T. T. Lie, "Configure methodology of onboard supercapacitor array for recycling regenerative braking energy of URT vehicles," *Industry Applications, IEEE Transactions on*, vol. 49, pp. 1678-1686, 2013.

[11] A. Masmoudi, A. Abdelkafi, and L. Krichen, "Electric power generation based on variable speed wind turbine under load disturbance," *Energy*, vol. 36, pp. 5016-5026 .2111,

978-1-5090-0376-1/16 $31.00 © 2016 IEEE

[12]   K. Inoue, K. Ogata, and T. Kato, "A Control Method of a Regenerative Power Storage System for Electric Machinery," in *Power Electronics Specialists Conference, 2006. PESC'06. 37th IEEE*, 2006, pp. 1-5.

[13]   D. Iannuzzi, F. Ciccarelli, and D. Lauria, "Stationary ultracapacitors storage device for improving energy saving and voltage profile of light transportation networks," *Transportation Research Part C: Emerging Technologies*, vol. 21, pp. 321-337, 2012.

[14]   R. Todd, D. Wu, J. dos Santos Girio ,M. Poucand, and A. Forsyth, "Supercapacitor-based energy management for future aircraft systems," in *Applied Power Electronics Conference and Exposition (APEC), 2010 Twenty-Fifth Annual IEEE*, 2010, pp. 1306-1312.

7th Power Electronics, Drive Systems & Technologies Conference (PEDSTC 2016)
16-18 Feb. 2016, Iran University of Science and Technology, Tehran, Iran

Zero-Voltage-Transition with Dual Resonant Tank for Bridgeless Boost PFC Rectifier with Low Current Stress

Farzad Yazdani
Dept. Electrical Engineering
Sharif University of technology
Tehran, Iran
Yazdani_farzad@ee.sharif.edu

Farzad Tahami
Dept. Electrical Engineering
Sharif University of technology
Tehran, Iran
Tahami@sharif.edu

Abstract— In this paper, a novel zero-voltage-switching topology for power factor correction bridgeless boost rectifier is proposed. By employing an improved switch cell, ZVS for all the main switches is achieved without additional current or voltage stress. The zero-current-switching is also provided for the auxiliary switch. In all modes of operation of this converter, the input current flows through only two semiconductors. Therefore, conduction loss of this converter is as low as the conventional parent converter. With these features, the efficiency of the proposed converter is very high. A 500 W prototype is designed and simulated to verify the system performance.

Index Terms— Power factor correction, zero-voltage-switching, Bridgeless Boost rectifier

I. INTRODUCTION

Nowadays, as the number of the switching power supplies used in industry and home increases, the reduction of the harmonic and power factor improvement are receiving much increasing attention to satisfy the harmonic standard such as IEC 61000-3-2 to improve the system efficiency [1]. For this purpose, the PFC rectifiers are promoted and widely used [2-5]. Among various approaches, the boost converter in continuous conduction mode (CCM) with the average current control is the most popular topology in power factor correction converters [6-9].

Bridgeless boost PFC rectifiers are employed because of their higher efficiency .The conventional bridgeless PFC rectifier always has two semiconductors in power flow path so has less conduction losses than the conventional PFC converter with three semiconductors in power-flow path [2]. This PFC converter is shown in Fig.1. Reducing the volume and weight of a converter is another issue that should be considered. Thus, the converter switching frequency should be increased. Increasing the switching frequency results in higher switching losses, which makes the employing of soft-switching techniques essential. Consequently, AC-DC power supplies with higher power density and improved efficiency can be obtained. Zero-voltage-transition (ZVT) and Zero-current-transition (ZCT) are soft-switching techniques, whereas the desirable features of the parent PWM converters remain [10-17]. Various soft-switching techniques are applied to the bridgeless PFC rectifier [18-26].

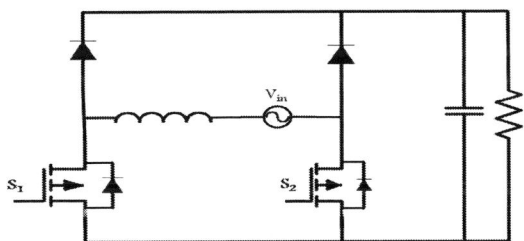

Fig. 1. The bridgeless boost PFC rectifier

In most of the ZVS topologies, voltage and current stresses of the main switches are increased [21-25]. The proposed converter has a PFC bridgeless boost rectifier and an auxiliary switch cell that provides ZVS for all the semiconductors. Also zero-current-switching is obtained for the auxiliary switch. The theoretical analysis of the proposed converter is presented in the next section. The design consideration are explained and the simulation results of the proposed converter are shown.

II. THE PROPOSED ZVS BRIDGELESS BOOST PFC

The power stage of the proposed ZVS bridgeless Boost PFC rectifier is shown in Fig.2 [8]. The circuit is divided in two sections. The first section is the bridgeless Boost PFC rectifier composed of S_1, S_2, D_1, D_2, L_{in} and C_{out}. The second section is the soft-switching cell shown in the dotted box that provides the ZVS conditions for the main switches. When the main switch is turned off, the snubber capacitor (C_s) is charged to V_{out}. So before turning on the main switch, this energy must been taken from C_s. By turning on S_a in the soft-switching cell, the main switch can be turned on in zero-voltage and the switching loss is reduced.

The proposed converter has eight operation modes during one switching cycle. To simplify the analysis, the input inductor (L_{in}) is assumed large enough to be considered as a dc current source (I_{in}) and the output capacitor (C_{out}) is assumed to be large enough to be considered as a constant voltage (V_{out}). There are two symmetrical cycles in circuit operation, positive and negative input cycles. The equivalent circuit in positive input cycle is illustrated in Fig.3. Before The first mode (Fig. 3.a), it is assumed that:

- The diode D_1 and body diode of S_2 is conducting

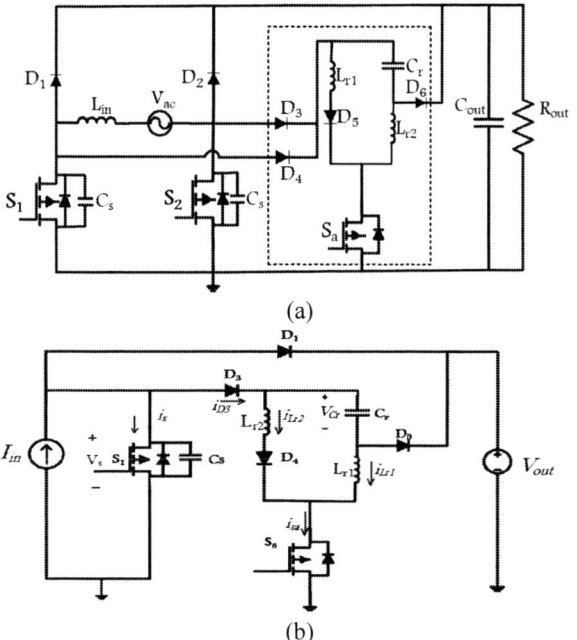

(a)

(b)

Fig. 2. (a) Proposed rectifier topology (b) Equivalent circuit of the proposed rectifier topology during steady state.

- All other semiconductor elements are off
- The voltage of C_r and the currents of L_{r1} and L_{r2} are zero
- C_s is charged to input voltage V_{in}

1) Mode 1: [t_0-t_1] (Fig. 3.b)

The first mode begins by turning S_a on. Due to existence of inductors L_{r1} and L_{r2} in the auxiliary switch path, ZCS condition is achieved in turning on. By conducting of D_1, D_3 and D_4, the voltage on L_{r2} jumped to V_{out}. Therefore, the current of L_{r2} increases linearly. The resonance between C_r and L_{r1} starts. Equations in this mode are:

$$i_{Lr2}=\frac{V_{out}}{L_{r2}}t \tag{1}$$

$$i_{Lr1}=\frac{V_{out}}{Z_r}\sin(\omega_r t) \tag{2}$$

$$V_{Cr}=V_{out}(1-\cos(\omega_r t)) \tag{3}$$

$$\omega_r=\frac{1}{\sqrt{L_{r1}C_r}} \tag{4}$$

$$Z_r=\sqrt{\frac{L_{r1}}{C_r}} \tag{5}$$

This mode ends when sum of currents of L_{r1} and L_{r2} equals to I_{in}. The current of D_1 reaches zero and it is turned off in zero-current.

2) Mode 2: [t_1-t_2] (Fig. 3.c)

After turning off D_1, C_s starts to resonate with the resonant tank. In this mode, it is assumed that L_{r2} is large enough that i_{Lr2} approximately constant during this mode. Thus, equations in this mode are:

$$i_{Lr2}=i_{Lr2}(t_1) \tag{6}$$

$$V_s=V_{Cr}+L_{r1}\frac{di_{Lr1}}{dt} \tag{7}$$

$$i_{Lr2}=C_r\frac{dV_{Cr}}{dt} \tag{8}$$

$$I'_{in}=I_{in}-i_{Lr2}=C_s\frac{dV_{Cs}}{dt}+i_{Lr1} \tag{9}$$

By solving the above differential equations the following results are obtained:

$$i_{Lr1}=\frac{V_{out}-V_{Cr}(t_1)}{Z_r}\sin(\omega'_r(t-t_1))+i_{Lr1}(t_1) \tag{10}$$

$$V_{Cr}=\left(\frac{C_r}{C_r+C_s}\right)V_{Cr}(t_1)+(V_{out}-V_{Cr}(t_1))\left(\frac{C_r}{C_r+C_s}\right)\times\cos(\omega'_r(t-t_1)) \tag{11}$$

$$\omega'_r=\frac{1}{\sqrt{L_{r1}C_r\|C_s}} \tag{12}$$

$$Z_r=\sqrt{\frac{L_{r1}}{C_r\|C_s}} \tag{13}$$

The voltage of C_s must reach to zero in this mode to achieve soft-switching for S_1. From equation (11) the condition for achieving ZVS is found as:

$$V_{Cr}(t_1)\leq\frac{V_{out}}{2} \tag{14}$$

3) Mode 3: [t_2-t_3] (Fig. 3.d)

When the voltage of C_s reaches to zero, the body diode of S_1 is turned on. This provides the ZVS condition for S_1. The equations in this mode are:

$$i_{Lr2}=i_{Lr2}(t_1) \tag{15}$$

$$i_{Lr1}=i_{Lr1}(t_2)\cos(\omega'_r(t-t_2))-\frac{V_{Cr}(t_2)}{Z'_r}\sin(\omega'_r(t-t_2)) \tag{16}$$

$$V_{Cr}=V_{Cr}(t_2)\cos(\omega'_r(t-t_2))-i_{Lr1}(t_2)Z'_r\sin(\omega'_r(t-t_2)) \tag{17}$$

4) Mode 4: [t_3-t_4] (Fig. 3.e)

This mode is started by S_1 conducting, the voltage across L_{r2} is zero. Therefore, its current is constant and the resonance between C_r and L_{r1} continues. After one half resonance cycle this current is negative in direction so the auxiliary switch current (I_{sa}) decreases. The equations in this mode are similar to the previous mode.

5) Mode 5: [t_4-t_5] (Fig. 3.f)

I_{sa} reaches zero and it can be turned off in zero-current. Also, D_3 is turned off in zero-current. There is no path for resonance current to flow in the main switch so the extra current stress does not exist. A resonance between C_r and the L_{r1} and L_{r2} starts through D_4 .This mode continues until the resonance current reaches zero and D_4 is turned off. Capacitor

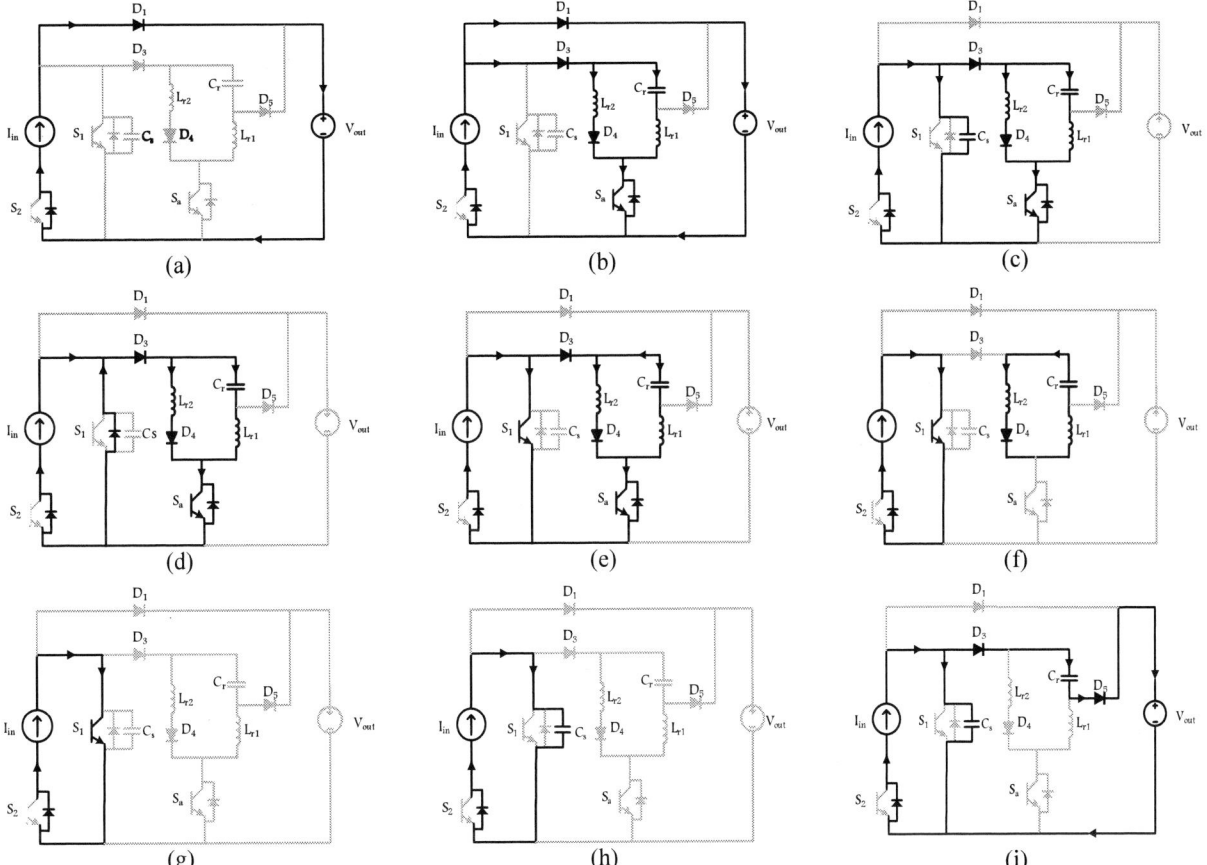

Fig. 3. Equivalent circuit of each mode

C_r is charge to negative value. The equations of the resonance currents and voltage are:

$$i_{Lr1}(t)=-i_{Lr2}(t)=i_{Lr2}(t_1)\cos\left(\omega''_r(t-t_5)\right)-\frac{V_{Cr}(t5)}{Z''_r}\sin\left(\omega''_r(t-t_5)\right) \quad (18)$$

$$V_{Cr}=V_{Cr}(t_5)\cos\left(\omega''_r(t-t_5)\right)-i_{Lr2}(t_1)Z''_r\sin\left(\omega''_r(t-t_5)\right) \quad (19)$$

$$Z''_r=\sqrt{\frac{L_{r1}+L_{r2}}{C_r}} \quad (20)$$

$$\omega''_r=\frac{1}{\sqrt{(L_{r1}+L_{r2})C_r}} \quad (21)$$

6) Mode 6: [t_5-t_6] (Fig. 3.g)

This mode is similar to the energizing mode of basic boost PFC rectifier. It continues until S_1 is turned off.

7) Mode 7: [t_6-t_7] (Fig. 3.h)

At the beginning, S_1 is turned off. The source current (I_{in}) begins to charge C_s. At the end of this mode, when the voltage of C_s is reached to difference between V_{out} and voltage of C_r. Finally, D_3 is turned on.

8) Mode 8: [t_7-t_8] (Fig. 3.i)

By conducting of D_3 and D_5, capacitor C_r is discharged linearly and C_s is charged to V_{out}. After charge of C_s, D_1 is turned on in zero-voltage and a next cycle starts.

III. CONTROL OF THE PROPOSED ZVS PFC RECTIFIER

Fig.2 shows the power stage of the proposed PFC rectifier consists of two main switches that have the same gate signal. The control circuit regulates the input current in phase with input voltage for unity power factor operation and also, regulates the output voltage. For this purpose, the control stage has two control loop. The Inner loop is used for the input current regulation and the outer loop is utilized for the output

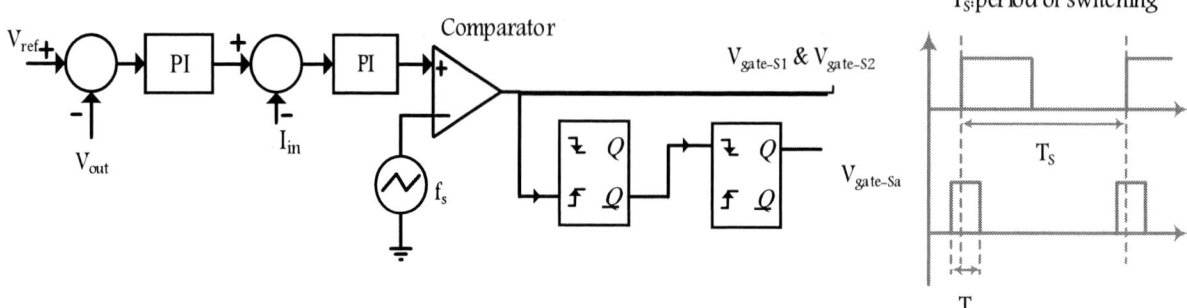

Fig. 4. Control circuit block diagram and timing chart

voltage regulation. From the main switch signal and mono-stable IC, the gate signal of the auxiliary switch is generated. The control circuit block diagram and the timing chart is shown in Fig. 4.

IV. DESIGN PROCEDURE AND EXAMPLE

The design procedure and an example of the proposed converter are described as follows. The design procedure has two steps as follows:

Step1: Designing of the resonance elements:

For making sure that the ZVS condition is achieved for the main switch in mode 2, the peak of auxiliary current is considered to be two times of peak input current:

$$\frac{V_{out}}{Z_{r1}}=2\times I_{inmax} \qquad (22)$$

With this assumption the equation (14) is satisfied. The resonance frequency is chosen 10 times higher than the switching frequency. Therefore, based on this two equations, the capacitor C_r and inductor L_{r1} can be obtained.

$$f_{res}=\frac{1}{2\pi\sqrt{L_{r1}C_r}}=10\times f_{sw} \qquad (23)$$

Therefore, based on these two equations, the capacitor C_r and inductor L_{r1} can be obtained. The inductor L_{r2} has been used to provide ZCS condition for the auxiliary switch S_a and in analysis we assume that its inductance is greater than the inductance of L_{r1}. By this assumption, the value of inductance of L_{r2} is chosen by:

$$L_{r2}=4L_{r1} \qquad (24)$$

Step2: Designing of bridgeless Boost PFC elements

The input inductor L_{in} and output capacitor C_{out} are selected to minimize the input-current ripple and output-voltage ripple. The design method of input inductor and output capacitor can be obtained like the conventional boost PFC rectifier [27].

TABLE I
SPECIFICATION OF THE PROPOSED CONVERTER

Parameters	Value/Model
Output voltage, V_{out}	300Vdc
Output power, P_{out}	500W
Input voltage, V_{in}	100-150V_{rms}
Efficiency, η	0.9
Switching frequency, f_s	50KHz

TABLE II
PROPOSED CONVERTER PARAMETERS

Parameters	Value/Model
Input inductor, L_{in}	3mH
Resonant inductor, L_{r1}	8μH
Resonant inductor, L_{r2}	32μH
Resonant capacitor, C_r	13nf
Snubber capacitor, C_s	5nf
Output capacitor, C_{out}	1mf
Output resistor, R_{out}	180Ω -1kΩ
Switches, S_1, S_2, S_a	IRFP460
All diodes	BYT08-600

V. SIMULATION RESULTS

The specification of the prototype converter has been tabulated in Table I. With these values, the peak input current value ($I_{in,max}$) is:

$$I_{in,max}=\frac{\sqrt{2}P_{out}}{\eta V_{in,min}}=7.86(A)$$

Therefore, with equations (22), (23) and (24) we can obtain the resonance parameters shown in table II. To verify the system performance, the PFC rectifier is simulated in PSIM at 50 KHz. The waveforms of the voltage and current of the ZVS-PWM PFC rectifier at the rated power are shown in Fig .5.

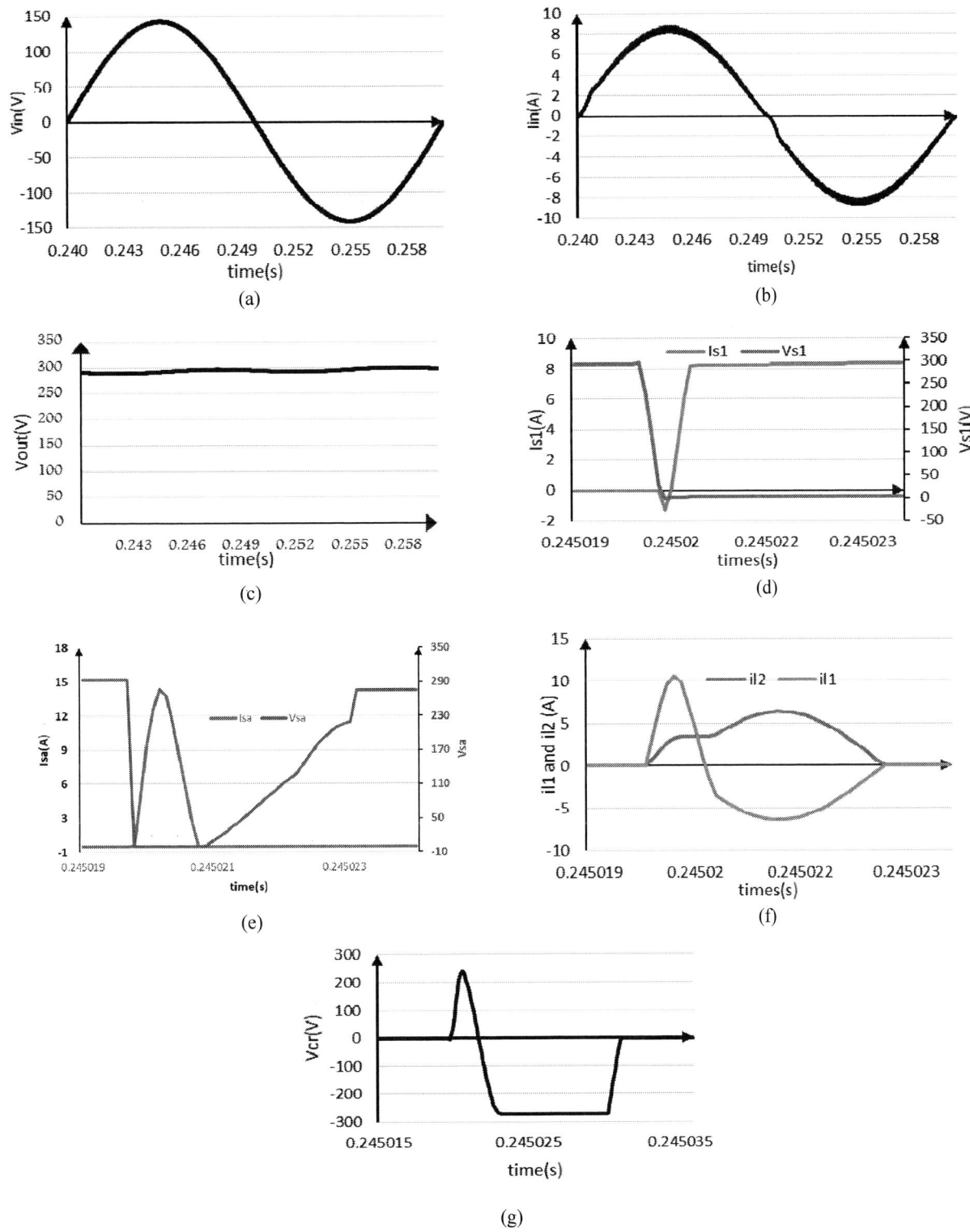

Fig. 5. Simulation results of the proposed converter at rated power and V_{in}=100 Vrms - (a) input voltage (b) input current (c) output voltage (d) voltage and current of the main switch (e) voltage and current of the auxiliary switch (f) currents of L_{r1} and L_{r2} (g) voltage of C_r

It can be seen that the main switches (S_1,S_2) are turned on witch ZVS and turned off at a reduced voltage. Also, ZCS is obtained for the auxiliary switch. With the help of the soft-switching technique, the efficiency of the PFC rectifier is increased by 2.5% and 2% at the minimum and maximum ranges of the input voltage, as shown in Fig. 6.

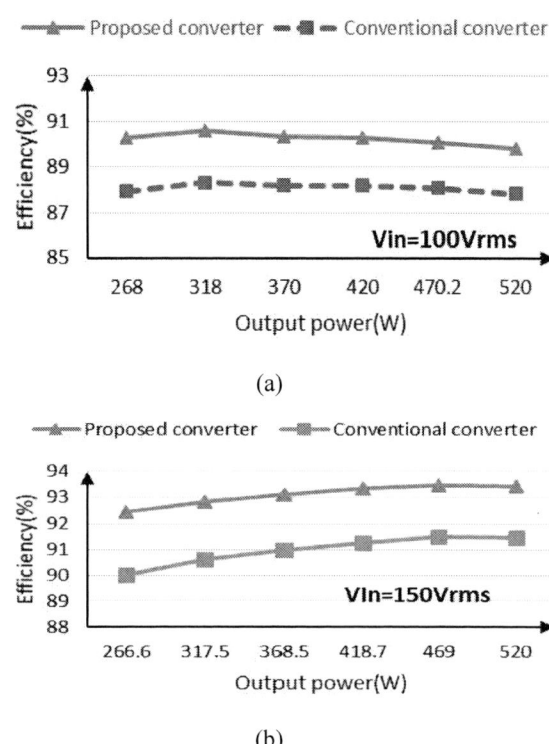

(a)

(b)

Fig. 6. Efiiciency comparison of the proposed converter and the conventional brodgless boost PFC rectifier (a) at Vin=100 Vrms (b) at Vin=150 Vrms

VI. CONCLUSION

In this paper, a novel ZVS bridgeless boost PFC rectifier was presented. The operation of the proposed converter has been theoretically analyzed, and verified by simulation. With the elimination of the switching losses, the efficiency of the PFC rectifier is improved by at least 2% over the hard-switched bridgeless boost PFC rectifier, without increasing the voltage stress of the main switches.

VII. REFERENCES

[1] IEC 61000-3-2, "International Electro technical ommission" Geneva,Switzerland, 1998.

[2] Jang, Yungtaek, and Milan M. Jovanovic. "Bridgeless buck PFC rectifier." In Applied Power Electronics Conference and Exposition (APEC), 2010 Twenty-Fifth Annual IEEE, pp. 23-29. IEEE, 2010.

[3] Mahdavi, M. Farzanehfard, H. "Bridgeless SEPIC PFC Rectifier With Reduced Components and Conduction

Losses," IEEE IEEE Transaction on Industrial Electronics, vol. 58, no. 9, pp. 4153-4160, 2011.

[4] Sahid, M. R., A. H. M. Yatim, and Taufik. "A new AC-DC converter using bridgeless SEPIC." In IECON 2010-36th Annual Conference on IEEE Industrial Electronics Society, pp. 286-290. IEEE, 2010.

[5] A. J. Sabzali, E. H. Ismail,et al. "New Bridgeless DCM Sepic and Cuk PFC Rectifiers With Low Conduction and Switching Losses," IEEE Transaction on Industrial Electronics, vol. 47, no. 2, pp. 873-881, 2011.

[6] F. Musavi, W. Eberle ,et al. "A High-Performance Single-Phase Bridgeless Interleaved PFC Converter for Plug-in Hybrid," IEEE Transaction on Industrial Application, vol. 47, no. 4, pp. 183-1143, 2011.

[7] Muhammad, K., and D. Lu. "ZCS Bridgeless Boost PFC Rectifier Using Only Two Active Switches." IEEE IEEE Transaction on Industrial Electronics, vol. 62, no. 5, pp. 2795-2806, 2015.

[8] Jovanovic M, M. Yungtaek Jang, "Performance Evaluation of Bridgeless PFC Boost Rectifier" IEEE Transaction on Power Electronics, vol. 23, no. 3, pp. 1381-1390, 2008.

[9] Jovanovic, M., Jang, Y, "State-of-the-art, single-phase, active PFC techniques for high-power applications" IEEE IEEE Transaction on Industrial Electronics, vol. 52, no. 3, pp. 701-708, 2005.

[10] Wang, Chien-Ming. "A new family of zero-current-switching (ZCS) PWM converters." Industrial Electronics, IEEE Transactions on 52, no. 4 (2005): 1117-1125.

[11] Taheri, M.; Milimonfared, J.; Bayat, H.; Riazmontazer, H.; Noroozi, A., "Analysis and design of a new soft switching interleaved converter using an integrated transformer," Power Electronics, Drive Syst. and Technologies Conference (PEDSTC), Feb. 2011., pp.98,103, 16-17.

[12] Bellar, Maria D., Tzong-Shiann Wu, Aristide Tchamdjou, Javad Mahdavi, and M. Ehsani. "A review of soft-switched DC-AC converters." Industry Applications, IEEE Transactions on 34, no. 4 (1998): 847-860.

[13] Mousavi, Ahmad, Pritam Das, and Gerry Moschopoulos. "A comparative study of a new ZCS DC–DC full-bridge boost converter with a ZVS active-clamp converter." Power Electronics, IEEE Transactions on 27, no. 3 (2012): 1347-1358.

[14] Adib, Ehsan, and Hosein Farzanehfard. "Analysis and design of a zero-current switching forward converter with simple auxiliary circuit." Power Electronics, IEEE Transactions on 27, no. 1 (2012): 144-150.

[15] Taheri, M., J. Milimonfared, H. Bayat, H. Riazmontazer, and A. Noroozi. "Analysis and design of a new choke less interleaved ZVS forward-flyback converter." In Power Electronics, Drive Systems and Technologies Conference (PEDSTC), 2011 2nd, pp. 81-86. IEEE, 2011.

[16] Adib, Ehsan, and Hosein Farzanehfard. "Family of zero-voltage transition pulse width modulation converters with low auxiliary switch voltage stress."Power Electronics, IET 4, no. 4 (2011): 447-453.

[17] Riazmontazer, H., J. S. Moghani, M. Taheri, and H. Bayat. "An isolated bi-directional dc-dc converter with zero current switching in LV Side and zero voltage switching in HV side." International Review of Electrical Engineering6, no. 2 (2011).

[18] Mahdavi, M. Farzanehfard, H. "Zero-Current-Transition Bridgeless PFC Without Extra Voltage and Current Stress," IEEE Transaction on Industrial Electronics, vol. 56, no. 7,

pp. 2540-2547, July 2009.

[19] Amini, M.R. Mahdavi, M. Emrani, A. Farzanehfard, H. "Soft switching bridgeless PFC with reduced conduction losses and no stresses," IET Power Electronics, vol. 5, no. 3, pp. 334-340, 2012.

[20] Wang, C. M. "A novel ZCS-PWM PF preregulator with reduced conduction losses," IEEE Transaction on industrial electronics, vol. 52, no. 3, pp. 689-700, July 2005.

[21] Wang, C. M. "ZCS-PWM Boost Rectifier with High Power Factor And Low Conduction Losses," IEEE Transaction on Aerospace And Electronic Systems, vol. 40, no. 2, pp. 650-660, April 2004.

[22] Mahdavi, M. Farzanehfard, H. "A new soft switching bridgeless PFC without any extra switch" in Int. Rev. Electr., Eng., 2008.

[23] de Souza, A.F., Barbi, I. "A new ZVS-PWM unity PF rectifier with reduced conduction losses," IEEE Tranaction on Power Electronics, vol. 10, no. 6, pp. 746-752, 1995.

[24] Tsai, Hsien-Yi, Tsun-Hsiao Hsia, Dan Chen. "A novel soft-switching bridgeless power factor correction circuit," in Power Electronics And Applications, Alborg, Denmark, 2007.

[25] Haghi, R. Zolghadri, M. Beiranvand, R. "Novel Zero-Voltage-Switching Bridgeless PFC Converter," Journal of Power Electronics, vol. 13, no. 1, pp. 40-50, Jan 2013.

[26] Hsien-Yi Tsai, Tsun-Hsiao Hsia, and Dan Chen, "A Family of Zero-Voltage-Transition Bridgeless PFC Circuits With a ZCS Auxiliary Switch," IEEE Transaction on Industrial Electronics, vol. 58, no. 5, pp. 1848-1855, 2011

[27] Unitrode product and applications handbook 1995-1996 ,pp. 10-303-10-322

978-1-5090-0376-1/16 $31.00 © 2016 IEEE

7th Power Electronics, Drive Systems & Technologies Conference (PEDSTC 2016)
16-18 Feb. 2016, Iran University of Science and Technology, Tehran, Iran

Z-Source DG-Active Filter

S. A. Saremi Hasari[1], S. Soori[2], A. Salemnia[3], S. Khosrogorji[4]

Department of Electrical and Computer Engineering

Shahid Beheshti University

Tehran, Iran

[1]A_Saremi@sbu.ac.ir, [2]sepehr.soori70@gmail.com, [3]Salemnia@pwut.ac.ir, [4]Soheyl_Gorji1991@ace.sbu.ac.ir

Abstract- **Nowadays with respect to growth of sensitive loads in power grid, power quality improvement is undeniable. One of power quality improvement techniques is using distribution generation interfaces as an active filter (DG-active filter). Because of some benefits of DG-active filters, such as active and reactive power injection, harmonic current injection and power factor improvement, they are attracting attentions. In this paper, a z-source inverter (ZSI) is used to connect a DG to grid and compensating reactive power and harmonic components of local load current. This structure is named Z-source DG-active filter in this paper. ZSI operation modes are introduced and the method of harmonic reference extraction and harmonic injection are explained. The performance of the presented method is verified using simulations carried out in MATLAB/SIMULINK software.**

Keywords: power quality improvement, DG-active filter, ZSI converter.

I. INTRODUCTION:

Nowadays with respect to new issues in power grid such as increasing of sensitive loads in power grid or increasing of people's knowledge about power quality's issues, improvement of grid's power quality is unavoidable. In general, power quality improvement has bilateral benefits for electricity consumers and producers.

In power grid, decrement of power quality generally occurs because of some factors such as lightning, presence of nonlinear loads, which is the most common factor in power quality decrement, and so on. Nonlinear loads drown a current from grid, which not only has fundamental harmonic but also has other harmonics such as 5th harmonic or 7th harmonic. Harmonic current causes harmonic voltage on main feeder which causes other loads, which are connected to main feeder, experience low power quality.

In order to increase power quality in grids by omitting unwanted harmonics from load's current, DG active filter has been investigated. In DG active filter, renewable source not only delivers active power to the grid but also tries to compensate load's reactive power and injects local unwanted load's harmonics to grid so that drown current from grid be sinusoidal and in phase with grid's voltage. By DG active filter, grid's power factor will be become nearly one and it can be said that grid will deliver only active power to loads that helps to free line capacity.

In general, for connecting renewable source to grid two types of converters are used: a rectifier which extracts maximum power from renewable source and an inverter which makes AC voltage in terminals of DG. Each converter has six IGBT switch. Rectifier type can be replaced by diode bridge using Z-source inverter and because of boost ability in Z-source converter, DC link of DG with low voltage could be connected to power grid [1].

From other aspect, voltage and current THD in Z-source output are less than conventional PWM converters [2].

Z-source converter not only can compensate reactive power and harmonics components of loads, also, it can be used to compensate voltage sag in grid [3]. With respect to [4-6] because of ability of Z-source converter in power quality improvement and low amount of output THD, it can be used for motor drive purposes. Conventional drive's defects such as high input current THD and sensitivity to input voltage sag are improved in ZSI converters.

Z-source converter is usually controlled by regulating voltage of capacitor in X-shape network. In [7], a wind turbine is connected to grid using a Z-source converter and its active power is delivered to grid. In that Z-source, a bidirectional converter is used to regulate capacitor voltage in X-shape network. The main defect of this control method is complexity of that and using unnecessary elements for controlling capacitor voltage.

In [8], it is tried to produce a voltage reference to control the active power that is delivered to grid, using PLL and dq0-transformation. This control method can't compensate unwanted load's current harmonic.

In other references like [9-11] other controlling methods for grid connected Z-source are introduced which harmonic and reactive power compensation are not considered in this references.

In DC Microgrids, DC link voltage can be affected by sudden load increment or decrement which causes low power quality in inverter output. ZSI inherent characteristics make it able to overcome DC link voltage fluctuation. If DC Microgrid be in fault situation like sudden load changing, shoot trough value can be changed so that X-shape output voltage be kept constant. This causes higher power quality in DC Microgrid.

In this paper, it is tried to improve power quality of grid, using a Z-source inverter that connects a PV cell to grid and transmits its power to grid. This structure is named Z-source DG-active filter in this paper. An effective control method is proposed for identifying reactive power and harmonic components of load's current and compensating them. This method isn't sensitive to fluctuation of DC link voltage of renewable sources.

This paper is organized as follows: proposed ZSI will be described in section II, proposed harmonic compensation method will be introduced in section III, simulation results will be analyzed in section IV and finally, conclusion will be presented in section V.

978-1-5090-0376-1/16 $31.00 © 2016 IEEE

II. Proposed ZSI

Proposed ZSI has been shown in Fig. 1. As said in [1] to have normal operation in ZSI, its controlling system should divided into decoupled part:

- Control system for regulating capacitor voltage in impedance network.

- Control system for producing sinusoidal voltage on inverter terminals.

As said in [1], there are two operation modes in z-source inverters:

- Shoot trough: in this operation mode switches in inverter legs are shorted. In this mode capacitor voltage is boosted.

- Non-shoot trough: in this mode inverter operated in normal mode to produce sine wave in output terminal.

In non-shoot trough mode according to Fig. 1, using KVL's law it can be written:

$$V_L = V_{DC} - V_C \qquad (1)$$

In shoot trough mode according to Fig. 1, using KVL's law it can be written:

$$V_{in\_inverter} = V_C - V_L \qquad (2)$$

Equation (3) could be extracted from (1) and (2) as follows:

$$V_{in\_inverter} = 2V_C - V_{DC} \qquad (3)$$

Which $V_{in\_inverter}$ is maximum amount of input voltage of inverter, V_C is the X-shape capacitance voltage and V_{DC} is renewable output voltage.

Using (3) it can be said that if capacitor voltage in impedance network be regulated at a fixed value, input voltage of inverter will be regulated and then output voltage of inverter will be regulated, so minimum distortion will be appear in output voltage.

Relation between shoot trough duty cycle ($D_{shoot\ trough}$) and boosting factor (B) can be stated as follows [1]:

$$B = \frac{V_C}{V_{DC}} = \frac{1}{1 - 2D_{shoot\ trough}} \qquad (4)$$

$$D_{shoot\ trough} = \frac{T_{shoot\ trough}}{T_{switching}} = \frac{1-B}{1-2B} \qquad (5)$$

In Eq. (5), $T_{shoot\ trough}$ is shoot trough duration and $T_{switching}$ is switching duration. By using Equations (4, 5), it could be said that, if shoot trough time increases, boosting factor and capacitor voltage will increase, too.

The method for applying shoot trough times has been described in Fig. 2 according to [12]. In this method shoot trough level ($V_{SC} = 1 - D_{shoot\ trough}$) must be greater than modulation ratio (m).

Proposed control system for regulating capacitor voltage has been shown in Fig. 3. In this control system, using capacitor's reference voltage and DG terminal voltage, shoot trough duty cycle ($D_{shoot\ trough}$) is calculated at first. Then, for regulating capacitor voltage, it is changed with respect to capacitor voltage error. An advantage of proposed control method is that it can regulate inverter input voltage when DC input voltage is changed.

III. Power Injection and Compensation of Reactive and Harmonic Components of Load Current

The designed Z-source DG-active filter could do three actions as follows:

1- Injecting DG generated active power to grid.

2- Compensating reactive power of load and improving power factor of grid current.

3- Compensating harmonic components of load current and decreasing THD of grid current.

Control system must include following parts for accurate compensation:

- Producing an in phase voltage with grid voltage in point of common coupling (PCC), for preventing circulating current between DG and grid. For this propose a PLL and voltage measurement unit could be used.

- Harmonic components of load current extraction from load current and injecting them to PCC. Fast Fourier transform (FFT) has been used for doing this action, in this paper.

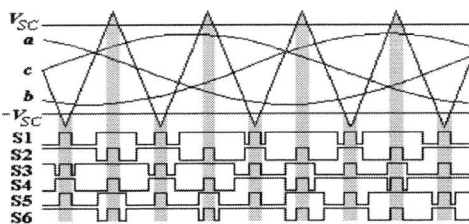

Fig. 2. The method for extracting shoot trough time [12]

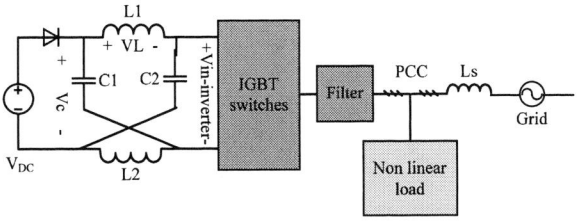

Fig. 1. Grid connected ZSI

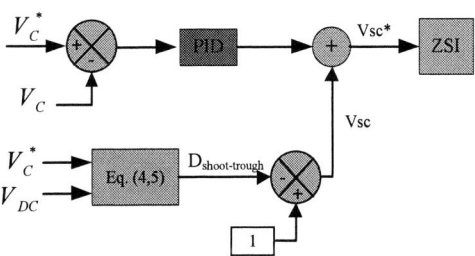

Fig. 3. Capacitor voltage control system

For injecting desired harmonic component of load current to PCC:

- First, output filter's impedance for desired harmonic component of load current must be calculated.

- Then, voltage drop across filter impedance for desired harmonic component could be calculated.

- In final step, inverter output reference voltage is extracted by adding calculated voltage drop to PCC desired harmonic voltage.

This technique has been shown briefly in (6). In this equation, I_{hn} is phasor of desired harmonic current, $V_{inv,hn}$ is phasor of inverter output reference voltage in desired harmonic, $V_{PCC,hn}$ is phasor of desired harmonic component of V_{PCC} and $Z_{filter,hn}$ is output filter impedance for desired harmonic.

$$I_{hn} = \frac{V_{inv,hn} - V_{PCC,hn}}{Z_{filter,hn}} \qquad (6)$$

Because of small amount of grid impedance, voltage of PCC is approximately equal to grid voltage and it is supposed in this paper that voltage grid is purely sinusoidal and free of harmonic components.

In this paper, harmonic compensation system will compensate 5st, 7st, 11st and 13st harmonic components of load current. Harmonic compensation method block diagram for 5st harmonic has been shown in Fig. 4. In this figure the signs "| |" and "∠" are used for showing amplitude and phase of a phasor, respectively. As can be seen, phase and amplitude of desired harmonic component of load current is extracted by FFT. Then the suitable voltage drop across filter impedance is calculated. This voltage drop could generate the desired harmonic current.

Efficiency of purposed method is related to amounts of filter components. For solving this problem a closed loop control has been added to harmonic compensation system in Fig. 4. By means of dq0 transform and PI controller, steady state error is forced to nearly zero.

As it had been said earlier, other important duty of designed DG-active filter is compensating the reactive power of load and improving the power factor of grid current. This action is illustrated in Fig. 5. In this compensation method at first, magnitude of reactive current will be calculated by identifying phase and amplitude of main component of load current. Then the suitable voltage drop across filter impedance is calculated. This voltage drop could generate the desired reactive current.

But, the main duty of designed DG-active filter is injecting the DG active power to grid. This has been shown in Fig. 6. As can be seen in this figure, the voltage drop across filter impedance is calculated appropriately.

Finally, as shown in Fig. 7, equivalent voltage drop across filter impedance could be calculated by summing calculated voltage drops. The reference voltage for inverter is extracted by summing equivalent voltage drop and voltage of PCC. Because of small amount of grid impedance, voltage of PCC is approximately equal to grid voltage. Grid voltage is free of harmonic components.

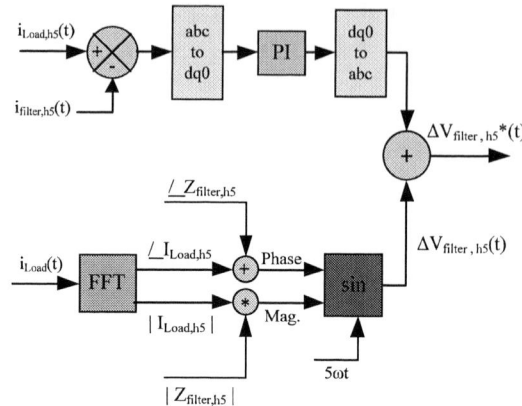

Fig. 4. Load harmonic component compensation for 5th harmonic

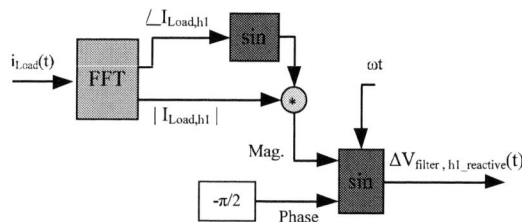

Fig. 5. Load reactive power compensation (PF improvement)

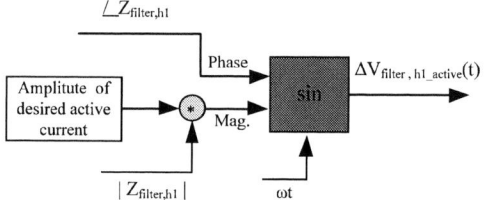

Fig. 6. Active power injecting

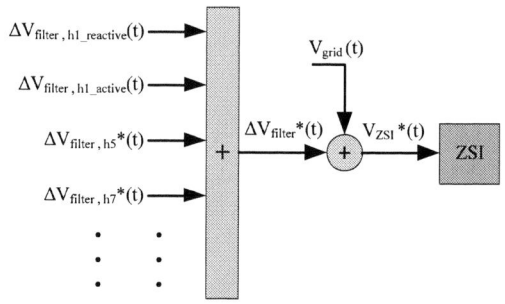

Fig. 7. Extracting inverter output reference voltage

IV. SIMULATION RESULTS

A grid like Fig. 1 has been assumed for simulation. In this combination, a diode rectifier is supposed for non-linear load. DG will be connected to utility grid at t=0.05s. Parameters of supposed grid have been adjusted as Table I.

For showing the ability of desired DG active filter three scenarios will be studied in this section.

First scenario: As it said earlier, one of important advantages of ZSI is that it could control the x shape capacitor voltage so that, maximum amount of input voltage of inverter could be constant. For using this advantage scenario is designed so that, DG output voltage decreases 10% at t=0.2 and then at t=0.4 increases to its previous amount. It could be seen in Fig. 8 that capacitor voltage is kept nearly reference voltage (400v) without considering the fluctuation of V_{DC}.

Second scenario: In this scenario generated active power of DG is zero and harmonic compensation and PF improvement will be studied. At first, DG is not connected to grid and capacitor voltage control system begins to regulate its voltage. Then, at t=0.05s converter is connected to grid and started to compensate the reactive power.

Grid current and converter current are shown in Fig. 9. This figure shows that by connecting the converter to grid, harmonic and reactive components of load current are compensated by ZSI and grid's current shape becomes sinusoidal. Grid's voltage and current are shown in Fig. 10. As can be seen, after compensation, current and voltage of grid are in phase and grid current THD is decreased, noticeably.

Grid 5th harmonic current in line 1 has been shown in Fig. 11, before and after compensation. As can be seen this component is omitted after compensation. FFT of grid current is shown in Fig. 12. As can be seen, THD of grid current has been changed from 18.07%, before compensation, to 3.12%, after compensation. Decrement in THD is noticeable.

Third scenario: In this scenario ZSI not only operates as active filter but also injects generated power of DG to grid, too. In this scenario DG injects two different active current which is shown in Fig. 13 and Fig. 14. As can be seen, magnitude of grid current is decreased by increasing the magnitude of injected current.

Table I. Parameters of supposed grid

Parameter	Value
V_{grid}	310V / 50Hz
L_s	0.001 mH
L_1	0.2 mH
C_1	400 uF
L_{Filter}	10 mH
R_{Filter}	0.2 Ω
P_{load}	4 kW
Q_{load}	2 kVAR
$Vc1$ reference	400 v DC
V_{DC}	200v

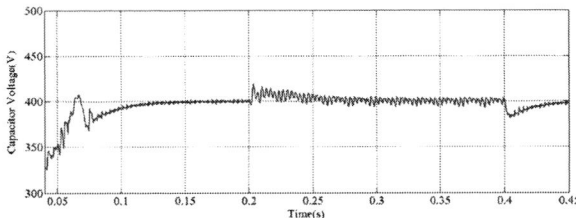

Fig. 8. Capacitor voltage

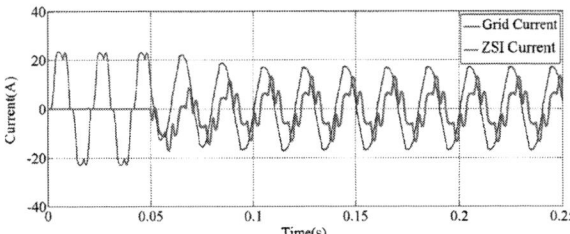

Fig. 9. Grid current and converter current before and after compensation

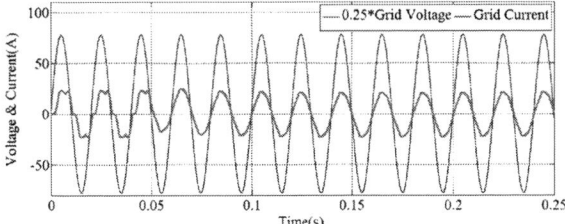

Fig. 10. Grid's current and voltage

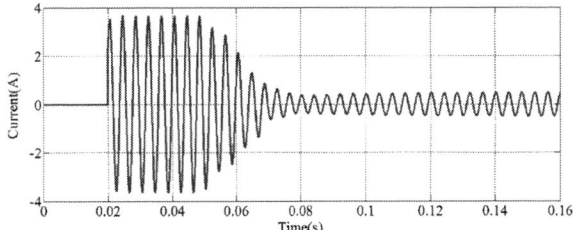

Fig. 11. Fifth harmonic of grid current before and after compensation

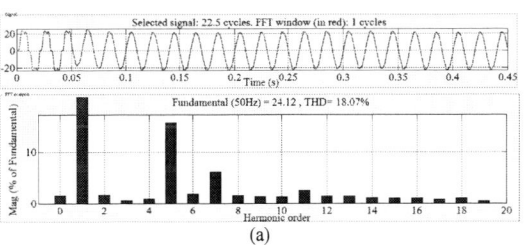

(a)

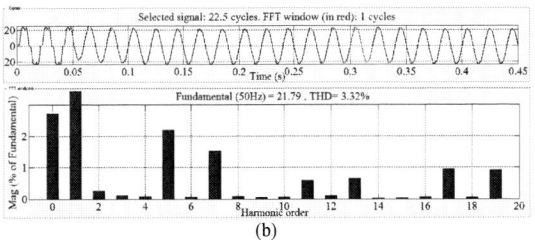

(b)

Fig. 12. Grid's current THD (a) befor compesation (b) after compensatio

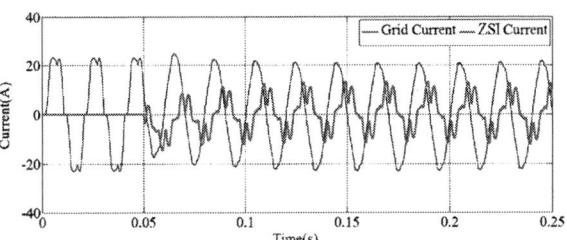

Fig. 13. Grid and ZSI current (DG active current magnitude is 5A)

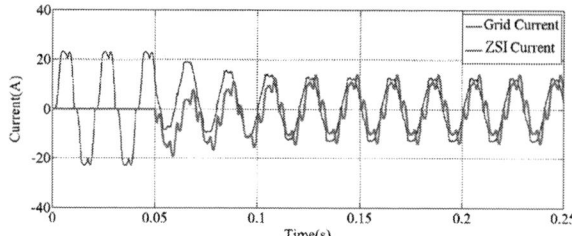

Fig. 14. Grid and ZSI current (DG active current magnitude is 10A)

V. CONCLUSION

The operation of a DG-active filter that is constructed using ZSI has been investigated in this paper. Using ZSI is useful for increasing voltage level of DG without any need of boost converter that is usual in conventional interfaces. The suggested DG-active filter has two important duties: injecting the generated power of DG to grid and improving power quality without considering the fluctuation of DC voltage of DG. ZSI operation and its control system were described and a new algorithm for extracting and injecting reactive and harmonic components of load current was explained.

For future work, the operation of suggested interface could be investigated in islanded mode.

REFERENCES

[1] P. Fang Zheng, "Z-source inverter," *Industry Applications, IEEE Transactions on,* vol. 39, pp. 504-510, 2003.

[2] L. Poh Chiang, D. M. Vilathgamuwa, L. Yue Sen, C. Geok Tin, and L. Yunwei, "Pulse-width modulation of Z-source inverters," *Power Electronics, IEEE Transactions on,* vol. 20, pp. 1346-1355, 2005.

[3] D. M. Vilathgamuwa, C. J. Gajanayake, P. C. Loh, and Y. W. Li, "Voltage Sag Compensation With Z-Source Inverter Based Dynamic Voltage Restorer," in *Industry Applications*

Conference, 2006. 41st IAS Annual Meeting. Conference Record of the 2006 IEEE, 2006, pp. 2242-2248.

[4] P. Fang Zheng, A. Joseph, W. Jin, S. Miaosen, C. Lihua, P. Zhiguo, *et al.*, "Z-source inverter for motor drives," *Power Electronics, IEEE Transactions on,* vol. 20, pp. 857-863, 2005.

[5] F. Z. Peng, Y. Xiaoming, F. Xupeng, and Q. Zhaoming, "Z-source inverter for adjustable speed drives," *Power Electronics Letters, IEEE,* vol. 1, pp. 33-35, 2003.

[6] D. Xinping, Q. Zhaoming, Y. Shuitao, C. Bin, and P. Fangzheng, "A New Adjustable-Speed Drives (ASD) System Based on High-Performance Z-Source Inverter," in *Industry Applications Conference, 2007. 42nd IAS Annual Meeting. Conference Record of the 2007 IEEE*, 2007, pp. 2327-2332.

[7] Z. Alnasir and M. Kazerani, "Standalone SCIG-based wind energy conversion system using Z-source inverter with energy storage integration," in *Electrical and Computer Engineering (CCECE), 2014 IEEE 27th Canadian Conference on*, 2014, pp. 1-6.

[8] M. G. Simoes, B. K. Bose, and R. J. Spiegel, "Design and performance evaluation of a fuzzy-logic-based variable-speed wind generation system," *Industry Applications, IEEE Transactions on,* vol. 33, pp. 956-965, 1997.

[9] R. Bharanikumar, R. Senthilkumar, and A. Nirmal Kumar, "Impedance Source Inverter for Wind Turbine Driven Permanent Magnet Generator," in *Power System Technology and IEEE Power India Conference, 2008. POWERCON 2008. Joint International Conference on*, 2008, pp. 1-7.

[10] K. Liu, J. Zou, X. Fu, X. Jiang, and F. Xu, "A new Flywheel Energy Storage System (FESS) Using Z-Source Inverter," in *Electromagnetic Field Computation (CEFC), 2010 14th Biennial IEEE Conference on*, 2010, pp. 1-1.

[11] W. Xiaoyu, D. M. Vilathgamuwa, K. J. Tseng, and C. J. Gajanayake, "Controller design for variable-speed permanent magnet wind turbine generators interfaced with Z-source inverter," in *Power Electronics and Drive Systems, 2009. PEDS 2009. International Conference on*, 2009, pp. 752-757.

[12] Q.-N. Trinh and H.-H. Lee, "Z-source inverter based grid connected for PMSG wind power system," in *Strategic Technology (IFOST), 2010 International Forum on*, 2010, pp. 145-150.

7th Power Electronics, Drive Systems & Technologies Conference (PEDSTC 2016)
16-18 Feb. 2016, Iran University of Science and Technology, Tehran, Iran

An Adaptive Recursive Discrete Fourier Transform Technique for the Reference Current Generation of Single-Phase Shunt Active Power Filters

Mohammad-Sadegh Karbasforooshan, *Student Member, IEEE,* and Mohammad Monfared, *Senior Member, IEEE*

Department of Electrical Engineering, Faculty of Engineering
Ferdowsi University of Mashhad
Mashhad, Iran
m.karbasforooshan.1991@ieee.org, m.monfared@um.ac.ir

Abstract—**This paper proposes a novel adaptive recursive discrete Fourier transform (ARDFT) technique for the reference current generation of single-phase shunt active power filters (APFs). The suggested method is robust to input frequency changes and exactly extracts the reference current of the APF. Modeling of the converter system and design procedure of the control parameters are presented in this paper. To confirm the theoretical results, simulation results are provided. These results show effectiveness and excellent performance of the suggested technique.**

Keywords—Active power filter (APF); discrete Fourier transform (DFT); recursive Fourier transform; phase locked loop (PLL).

I. INTRODUCTION

Nowadays, the utilization of nonlinear loads, such as CFLs, LEDs, computers, electronic drives and so on is more increased in the grid. These loads cause power quality problems, additional losses, stability problems and create resonance, fault in the protection systems, electromagnetic interference (EMI), damage to the power system equipment and etc. To solve these problems, many type of compensators are introduced. Passive filters are the first compensators that introduced for the current harmonics elimination and the power factor improvement. Although passive filters have advantages of simplicity and low cost, but due to their drastic limitations, always have been used with cautions. Active power filters (APFs) are proposed to be used in distribution systems, which have the capability of eliminating the whole undesired harmonics and improving the power factor by compensating the reactive power, simultaneously [1]-[4].

Reference current generation is the most important part of the control system of APFs. Up to now, many different methods for the reference current generation of APFs are proposed in literature. These methods can be categorized into time-domain and frequency-domain techniques [5]-[7]. Fourier transform method is the most known reference current generation technique in the frequency-domain. This method provides high accuracy in harmonic detection and is used in single-phase and three-phase systems. Despite of many advantages of Fourier transform technique, this method suffers

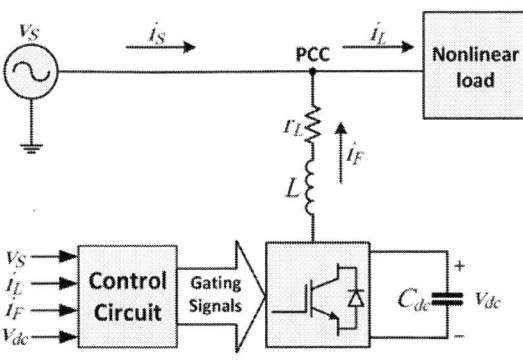

Fig. 1. Single-phase shunt active power filter.

from high computational burden and high sensitivity to frequency changes of the measured signal [8]-[17].

This paper proposes a new frequency adaptive recursive discrete Fourier transform (ARDFT) structure to overcome the problems of sensitivity to input frequency changes. The paper is organized as follows: in section II, the modeling of a single-phase shunt APF and its controller parameters design are described. In section III, the proposed technique for the APF is explained. Section IV is devoted to simulation of the system with the proposed reference current generation technique. Finally, conclusion is coming in section V.

II. MODELING OF THE SINGLE-PHASE SHUNT APF AND CONTROLLERS DESIGN

A. System Modeling

Fig. 1 shows the power and control system of a single-phase shunt APF. According to this, the APF is intended to inject a compensating current, so that the grid current will be an in-phase sinusoidal waveform with grid voltage in point of common coupling (PCC). The power circuit of APF is composed of a single-phase full-bridge inverter, a DC-link capacitor and an inductor. The grid voltage includes harmonic contents and the nonlinear load is a single-phase diode rectifier that in DC-side sees a resistor parallel with a capacitor.

978-1-5090-0376-1/16 $31.00 © 2016 IEEE

TABLE I.   SYSTEM PARAMETERS

Parameter	Symbol	Value
DC-link voltage	V_{dc}	380 V
Grid voltage	V_S	220 V_{rms}
Inverter rating	S	2 kVA
Filter inductance	L	1 mH
ESR of the inductance	r_L	0.25 Ω
DC-link capacitance	C_{dc}	2200 uF
Grid frequency	f	50 Hz
Switching/sampling frequency	f_{sw}/f_{samp}	12.8 kHz

Table I listed system parameters. The following equation shows the inductor voltage:

$$v_F = r_L i_F + L \frac{di_F}{dt} + v_S \qquad (1)$$

where L and r_L are inductance and resistance of the inductor. Equation (1) can be rewritten as

$$\frac{di_F}{dt} = \frac{1}{L}\left(v_F - r_L i_F - v_S\right) \qquad (2)$$

By applying Laplace operator to (2), the transfer function of the filter current, $i_F(s)$ based on $v_F(s)$ and $v_S(s)$ is obtained as

$$I_F(s) = \frac{1}{Ls + r_L}\left(V_F(s) - V_S(s)\right) \qquad (3)$$

Also, according to Fig. 1, the filter capacitor current based on average model method is [1]

$$C_{dc}\frac{d\hat{v}_{dc}}{dt} + \frac{\hat{v}_{dc}}{r_{dc}} = d\hat{i}_F \qquad (4)$$

where r_{dc} represent the inverter losses. Also, the transfer function of $i_F(s)$ based on $d(s)$ assuming $v_S=0$ can be readily obtained as

$$\left.\frac{\hat{i}_F(s)}{\hat{d}(s)}\right|_{\hat{v}_S(s)\equiv 0} = \frac{2V_{dc}}{Ls} \qquad (5)$$

The transfer function of $i_S(s)$ to $v_{dc}(s)$ is

$$\frac{\hat{v}_{dc}(s)}{\hat{i}_S(s)} = \frac{V_S}{2C_{dc}V_{dc}} \cdot \frac{1}{s + \left(\dfrac{2}{r_{dc}C_{dc}}\right)} \qquad (6)$$

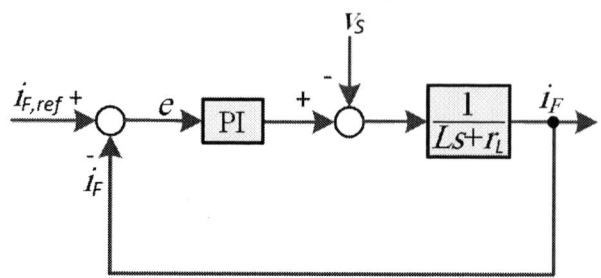

Fig. 2. Inner filter current control loop.

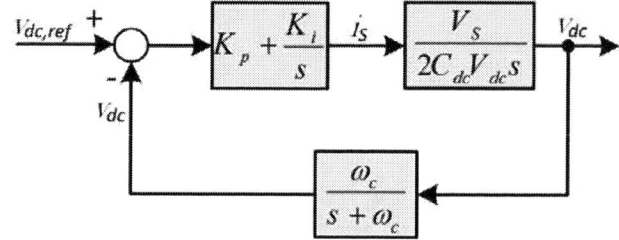

Fig. 3. Outer DC-link voltage control loop.

Therefore, by assuming that the power losses is equal to zero (r_{dc} tends to infinity), (6) simplifies to

$$\frac{\hat{v}_{dc}(s)}{\hat{i}_S(s)} = \frac{V_S}{2C_{dc}V_{dc}s} \qquad (7)$$

B. Controller Parameter Design

The filter current controller is a simple proportional regulator. The inner filter current loop is shown in Fig. 2. Therefore, the closed loop transfer function of the inner loop is obtained as

$$G_I(s) = \frac{i_F(s)}{i_{F,ref}(s)} = \frac{P}{Ls + \left(r_L + P\right)} \qquad (8)$$

By considering -3dB attenuation of (8) at the current bandwidth frequency, one has

$$\frac{P^2}{\left(r_L + P\right)^2 + \left(L\omega_{bi}\right)^2} = \frac{1}{2} \qquad (9)$$

So, the current loop proportional gain, P is obtained as

$$P = r_L + \sqrt{2r_L^2 + L^2\omega_{bi}^2} \qquad (10)$$

In this case, the current control bandwidth is considered 1000 Hz which is in the range of ten times the grid frequency (500 Hz) and one-tenth the switching frequency (1280 Hz).

To control the DC-link voltage of APF, we consider a proportional-integral (PI) controller. Also, a low-pass filter for the disturbance and noise rejection of the DC-link voltage is used. Fig. 3 shows the outer voltage control loop of APF. The closed-loop transfer function of the outer loop is

$$G_{Vdc}(s) = \frac{v_{dc}(s)}{v_{dc,ref}(s)}$$
$$= \frac{V_S\left(K_p s + K_i\right)}{2C_{dc}V_{dc}s^3 + 2C_{dc}V_{dc}\omega_c s^2 + K_p V_S \omega_c s + K_i V_S \omega_c} \quad (11)$$

Tuning the PI controller is a compromise between accessible control bandwidth and the loop stability [18]. The effect of the integral part of PI controller around the crossover and bandwidth frequency can be neglected. So, equation (11) without K_i simplified to

$$\left.\frac{v_{dc}(s)}{v_{dc,ref}(s)}\right|_{K_i \equiv 0} = \frac{V_S K_p}{2C_{dc}V_{dc}s^2 + 2C_{dc}V_{dc}\omega_c s + K_p V_S \omega_c} \quad (12)$$

By considering -3dB attenuation of (12) at the voltage bandwidth frequency, the proportional gain of the DC-link voltage is calculated as

$$K_p = \frac{2C_{dc}V_{dc}\omega_{bv}\left(\omega_{bv} + \omega_c\right)}{V_S \omega_c} \quad (13)$$

The voltage bandwidth frequency is always selected bellow one-tenth the grid frequency. Therefore, the voltage bandwidth is selected as $\omega_{bv} = 15$ rad/s. On the other hand, characteristic equation of the system from (11) is

$$2C_{dc}V_{dc}s^3 + 2C_{dc}V_{dc}\omega_c s^2 + K_p V_S \omega_c s + K_i V_S \omega_c = 0 \quad (14)$$

By applying Routh-Hurwitz stability criterion to (14), the following condition is obtained.

$$K_i < K_p \omega_c \quad (15)$$

So, according to (15), the integral part of DC-link voltage controller is selected in the range $[0, K_p\omega_c]$.

III. ADAPTIVE RECURSIVE DISCRETE FOURIER TRANSFORM TECHNIQUE

Fourier transform is one of the amplitude and phase detection methods of harmonics. The Fourier series integral in time-domain is [8]:

$$<x>_h (t) = \frac{1}{T}\int_{t-T}^{t} x(\tau)e^{jh\omega\tau}d\tau \quad (16)$$

The above equation determines harmonic coefficients of the Fourier series of the signal. The discrete form of (16) can be written as

$$<x>_h [n] = \frac{1}{N}\sum_{k=n-N+1}^{n} x[k]e^{j\frac{2\pi hk}{N}} \quad (17)$$

Equation (17) is the direct or non-recursive calculation way of discrete-time Fourier transform. This equation implies that for the calculation of amplitude and phase, a whole AC cycle information is required. Equation (17) can be rewritten as

$$<x>_h [n]$$
$$= \frac{1}{N}\sum_{k=n-N+1}^{n} x[k]\left(\cos\left(\frac{2\pi hk}{N}\right) + j\sin\left(\frac{2\pi hk}{N}\right)\right) \quad (18)$$

Because of high calculation requirement of direct method, the recursive technique is already proposed. The following subsections, focus on conventional recursive method and proposed adaptive recursive method.

A. Conventional Recursive Discrete Fourier Transform

To extract the fundamental component of a signal, the recursive discrete Fourier transform (RDFT) is proposed as follows

$$\begin{cases} C_j = C_{j-1} + \frac{1}{N}\cos\left(\frac{2\pi j}{N}\right)\left(x_j - x_{j-N}\right) \\ S_j = S_{j-1} + \frac{1}{N}\sin\left(\frac{2\pi j}{N}\right)\left(x_j - x_{j-N}\right) \end{cases} \quad (19)$$

where N is the number of samples per one cycle, and S_j and C_j are the real and imaginary part of the Fourier transform. Fig. 4 shows the corresponding block diagram of equation (19). The main advantage of recursive method over the direct method is the high memory and calculation space saving. But the main limitation of the conventional recursive method is the high sensitivity to frequency changes of the input signal. Such that, the truncation, accumulation and rounding errors lead to instability of the output signal in presence of a small frequency deviation of the input signal. To clarify the situation, assume that the input signal of the Fourier transform is a pure sinusoidal waveform with 10V amplitude and 50Hz frequency. Suddenly, the frequency changed to 49.5Hz (shown in Fig. 5). The output amplitude and phase of the Fourier transform is shown in Fig. 6. As can be seen, the amplitude and phase of the output signal is unstable following the frequency change.

Therefore, the recursive method, despite the fewer memory occupation, has the major problem of instability in case of frequency changes.

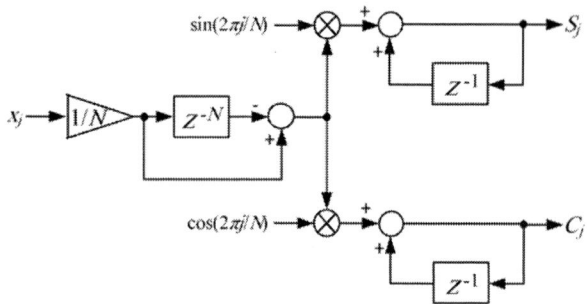

Fig. 4. Conventional recursive discrete Fourier transform.

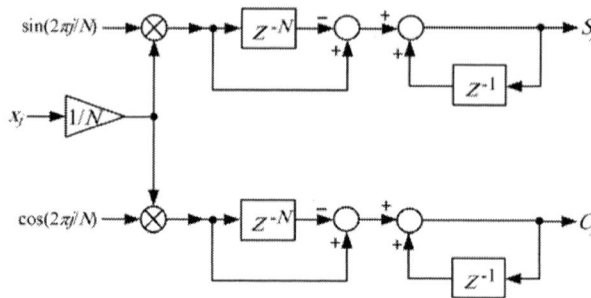

Fig. 7. Improved recursive discrete Fourier transform.

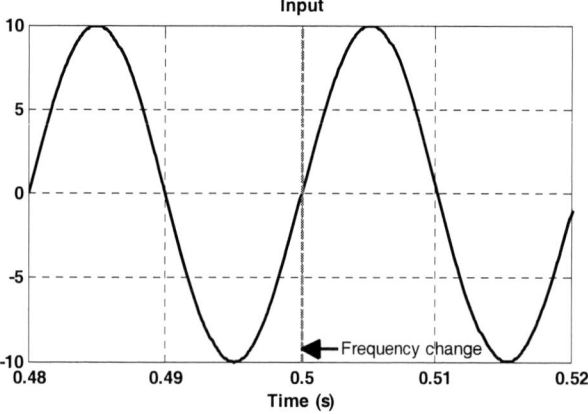

Fig. 5. A pure sinusoidal waveform with 50Hz frequency, frequency changes to 49.5Hz at t=0.5 s.

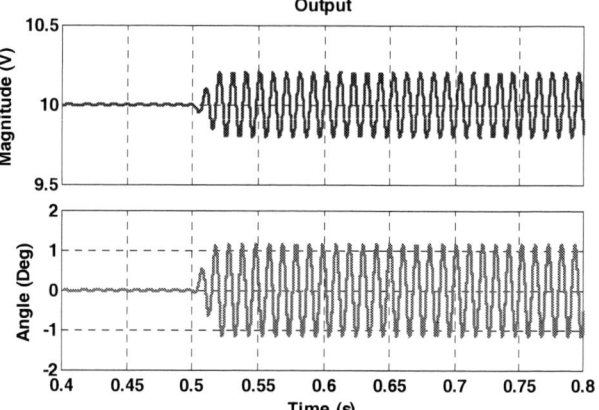

Fig. 8. Output amplitude and phase of the improved recursive discrete Fourier transform for frequency change from 50Hz to 49.5Hz at t=0.5 s.

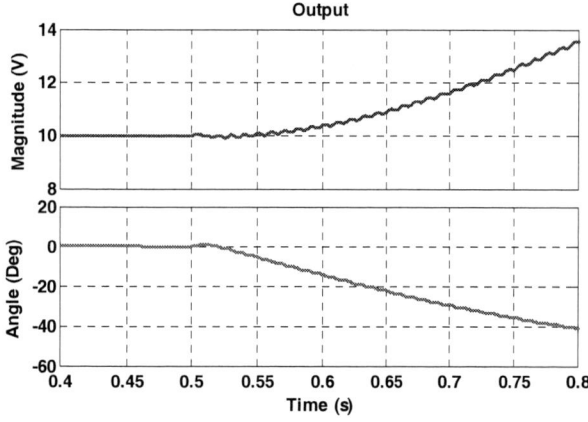

Fig. 6. Output amplitude and phase of the conventional recursive discrete Fourier transform for frequency change from 50Hz to 49.5Hz at t=0.5 s.

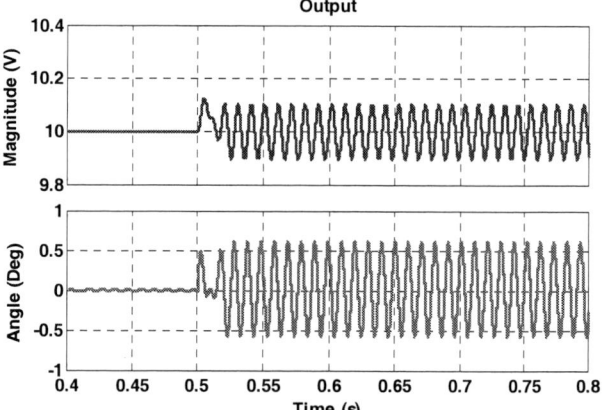

Fig. 9. Output amplitude and phase of the improved recursive discrete Fourier transform with updating N for frequency change from 50Hz to 49.5Hz at t=0.5 s.

B. Improved Recursive Discrete Fourier Transform

The following equations show the improved recursive discrete Fourier transform method

$$\begin{cases} C_j = C_{j-1} + \left(P_j - P_{j-N} \right) \\ S_j = S_{j-1} + \left(M_j - M_{j-N} \right) \end{cases} \qquad (20)$$

where P_j and M_j are equal to $(1/N)x_j \cos(2\pi j/N)$ and $(1/N)x_j \sin(2\pi j/N)$, respectively. Fig. 7 shows the corresponding block diagram of this technique. Except adding two P_j and M_j calculations, the improved technique has no calculations more than the conventional method. In order to present the performance of this technique, the same sinusoidal waveform of Fig. 6 is applied to the improved technique and the output amplitude and phase are reported in Fig. 8. As shown in this figure, the output signal of the improved technique   remains

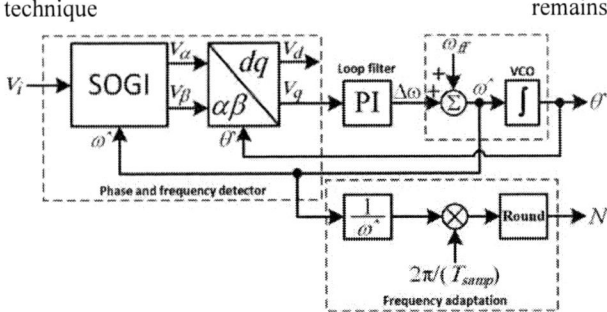

Fig. 10. Phase locked loop.

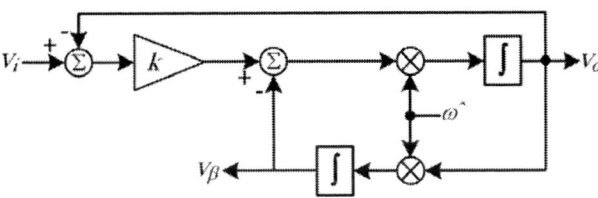

Fig. 11. Second order generalized integrator.

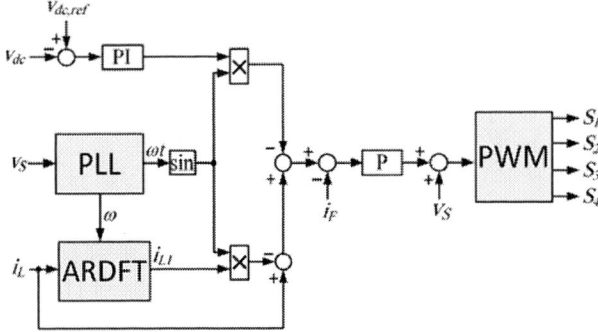

Fig. 12. Overall proposed control block diagram of the single-phase shunt active power filter.

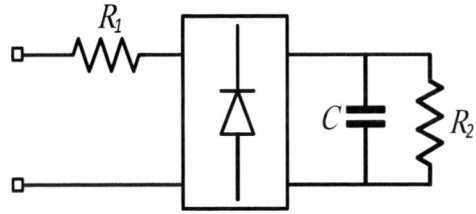

Fig. 13. Structure of the nonlinear load.

stable after the frequency change and only oscillates in a narrow band. These oscillations originates from the variations N as a result of the frequency change of the input signal. In the proposed technique, N is estimated continually by a phase locked loop (PLL) and its real value is used in the Fourier transform. The output result of applying sinusoidal waveform of Fig. 6 to the input of the ARDFT is shown in Fig. 9. In fact, for 10kHz sampling frequency and 50Hz fundamental frequency, the value of N was equal to 200 which by changing the frequency, the actual value of N is 202.02. By comparing Fig. 8 and Fig. 9, it can be seen that oscillations around the steady-state value have decreased. The remained oscillations are caused by the rounding error of N for the new frequency.

C. Phase Locked Loop and Frequency Adaptation

In this paper, for the calculation of the grid frequency and update of N, a second order generalized integrator (SOGI) based PLL is used that has many advantages, such as simple implementation, low computational burden, high accuracy, high speed and robust performance to disturbances. In fact, with having a PLL, two goals of grid phase and frequency extraction are fulfilled. Fig. 10 shows the PLL block diagram to extract the grid phase and frequency [5]. In this figure, v_i, θ, ω and ω_{ff} are the grid voltage, extracted phase, extracted frequency and nominal grid frequency, respectively. To calculate α and β component of the grid voltage, the SOGI block is used, which its structure is shown in Fig. 11. The SOGI provides the necessary signals for the PLL, and at the same time, creates a pure sinusoidal voltage for the reference current generation by attenuating the grid voltage harmonics and noises. The PLL provides required frequency for appropriate performance of the ARDFT technique and has an important role in increasing the system robustness against grid frequency changes.

IV. SIMULATION RESULTS

The overall control block diagram of the APF is shown in Fig. 12. According to Fig. 12, firstly the grid voltage phase and frequency are extracted by the PLL. Then, the ARDFT block calculates the fundamental component of the load current and by multiplying it by the voltage phase, the grid reference current is constructed. On the other hand, the difference between the DC-link voltage and its reference passes through a PI controller and by multiplying the controller output signal by the voltage phase, the filter power losses is calculated. The difference of filter current with its reference passes through a simple proportional gain. Also, in the proposed method, a feedforward path including grid voltage is added to the control signal to improve system dynamics. Finally, the resulting control signal enters PWM block and gating signals of inverter switches are produced. Table I shows the system parameters. The nominal value of N is $N=12800/50=256$.

In this section, different test on system in MATLAB/Simulink are done. It should be noted that the grid voltage has 5% third harmonic and 3% fifth harmonic and the nonlinear load consists of a single-phase diode rectifier as shown in Fig. 13 with $R_1=1\ \Omega$, $R_2=10\ \Omega$ and $C=1000\ uF$. The grid inductance is 20uH.

First, the steady-state response of the system is shown in Fig. 14. In this figure, the grid voltage, grid current, load current, filter current and DC-link voltage are shown. While the grid voltage THD is equal to 5.72% and load current THD is equal to 46.64%, the grid current THD is obtained equal to 2.36%. The filter current follows its reference properly and the maximum current amplitude is 25A. Also, the DC-link voltage by 2.6% ripple oscillates around the 380V reference.

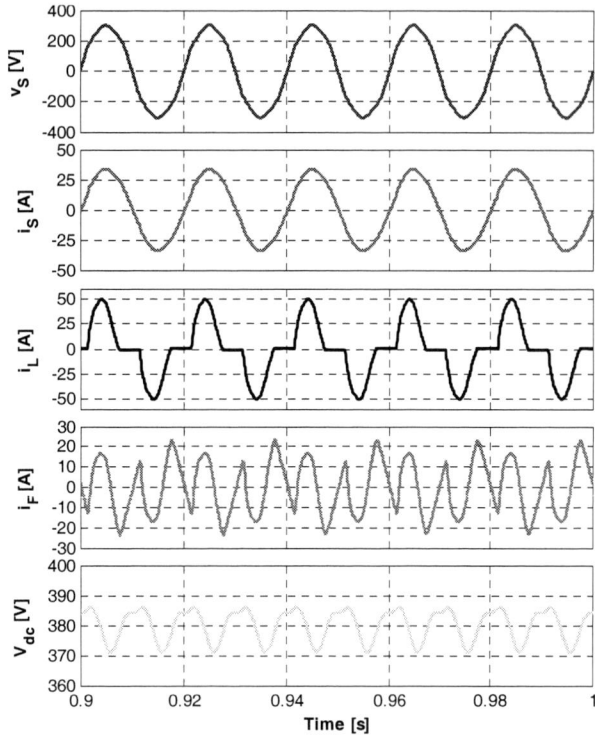

Fig. 14. Steady-state waveforms: grid voltage, grid current, load current, filter current and DC-link voltage.

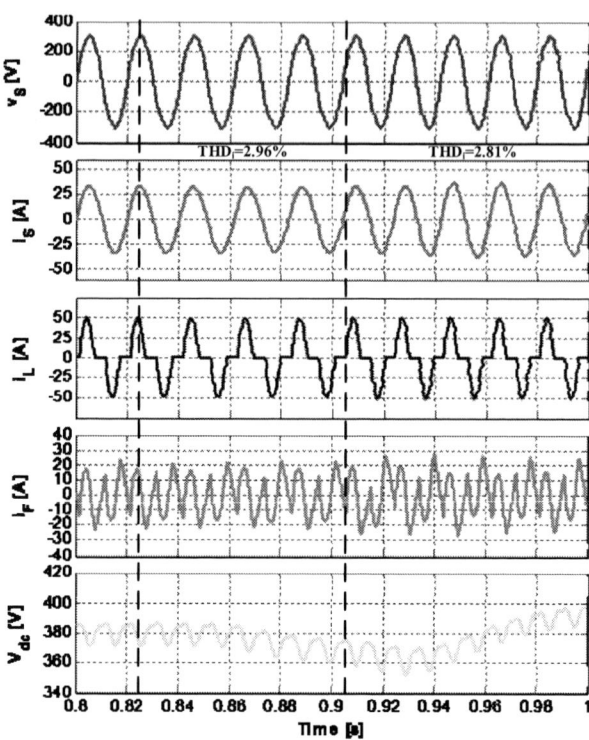

Fig. 16. Transient waveforms in response to grid voltage frequency changes: grid voltage, grid current, load current, filter current and DC-link voltage.

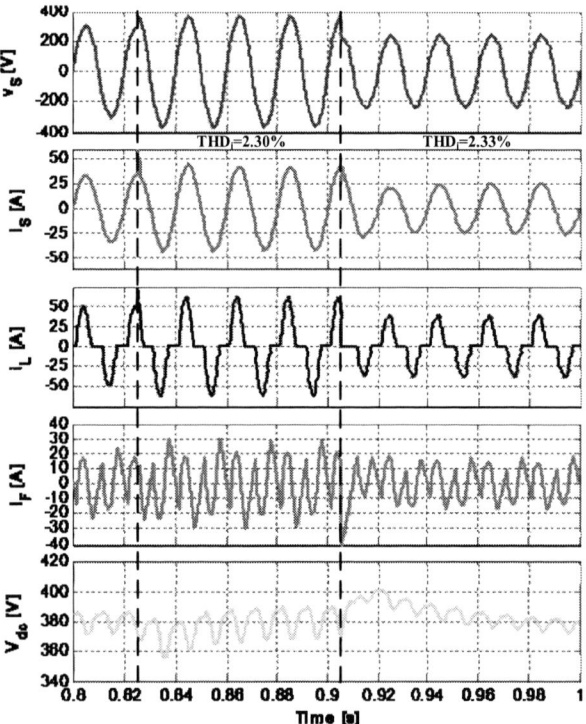

Fig. 15. Transient waveforms in response to grid voltage amplitude changes: grid voltage, grid current, load current, filter current and DC-link voltage.

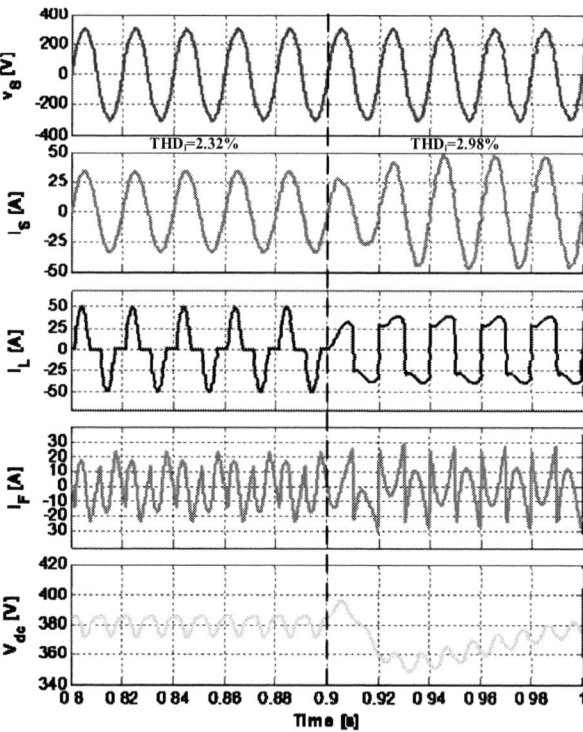

Fig. 17. Transient waveforms in response to sudden load change: grid voltage, grid current, load current, filter current and DC-link voltage.

978-1-5090-0376-1/16 $31.00 © 2016 IEEE

In the next test, the transient behavior of the system against grid voltage amplitude and grid frequency changes is analyzed. Figs. 15 and 16 show the output result of these tests. In Fig. 15, the grid voltage amplitude jumped 20% at t=0.825 s and the fell 50% at t=0.905 s. In Fig. 16, the grid voltage frequency changes from 50Hz to 47.5Hz at t=0.825 s and from 47.5Hz to 52.5Hz at t=0.905 s. As can be seen in Fig. 15, a little change in the grid current and DC-link voltage occurred at the moment of sudden change of grid voltage amplitude, but the current control system acts quickly and the current attain to its steady-state in less than one cycle. According to Fig. 16, the grid frequency change has no effect on the control system and reference current generation performance, and this shows excellent and robust performance of the proposed technique. It should be noted that in two above tests, the grid current THD remains in [2% - 3%] range.

In the final test, a nonlinear load consists of a single-phase rectifier along with a 5Ω resistor series with a 30mH inductor in DC-side enters to the system at t=0.9 s and previous load is disconnected. The output result of this test is shown in Fig. 17. Despite the load change from capacitive to inductive, the grid current THD remains 3% and accurate performance of the reference current generation is well done.

V. CONCLUSION

This paper proposed an adaptive recursive discrete Fourier transform technique for the reference current generation of single-phase shunt APFs. The proposed technique has advantages of simplicity, high accuracy in current extraction, less calculation burden than non-recursive Fourier transform and robust performance to the frequency changes. To confirm the theoretical achievements, simulation results are presented. The results show excellent performance of the control system and the reference current generation in steady-state and transient conditions.

REFERENCES

[1] S. Rahmani, K. Al-Haddad, F. Fnaiech, and P. Agarwal, "Modified PWM with indirect current control technique applied to a single-phase shunt active power filter topology," *Can. J. Electr. Comput. Eng.*, vol. 31, no. 3, pp. 135–144, 2006.

[2] F. Pottker de Souza and I. Barbi, "Single-phase active power filters for distributed power factor correction," *2000 IEEE 31st Annu. Power Electron. Spec. Conf. Conf. Proc. (Cat. No.00CH37018)*, vol. 1, no. c, pp. 500–505, 2000.

[3] S. Rahmani, K. Al-Haddad, and H. Y. Kanaan, "A comparative study of shunt hybrid and shunt active power filters for single-phase applications: Simulation and experimental validation," *Math. Comput. Simul.*, vol. 71, no. 4–6, pp. 345–359, Jun. 2006.

[4] J. Miret, M. Castilla, J. Matas, J. M. Guerrero, and J. C. Vasquez, "Selective harmonic-compensation control for single-phase active power filter with high harmonic rejection," *IEEE Trans. Ind. Electron.*, vol. 56, no. 8, pp. 3117–3127, Aug. 2009.

[5] S. Golestan, M. Monfared, and J. M. Guerrero, "Second order generalized integrator based reference current generation method for single-phase shunt active power filters under adverse grid conditions," in *4th Annual International Power Electronics, Drive Systems and Technologies Conference (PEDSTC)*, pp. 510-517, 2013.

[6] M. I. M. Montero, E. R. Cadaval, and F. B. Gonzalez, "Comparison of control strategies for shunt active power filters in three-phase four-wire systems," *IEEE Trans. Power Electron.*, vol. 22, no. 1, pp. 229–236, Jan. 2007.

[7] B. N. Singh, V. Khadkikar, and A. Chandra, "Generalised single-phase p-q theory for active power filtering: simulation and DSP-based experimental investigation," *IET Power Electron.*, vol. 2, no. 1, pp. 67–78, Jan. 2009.

[8] M. Dogruel and H. H. Çelik, "Harmonic control arrays method with a real time application to periodic position control," *IEEE Trans. Control Syst. Technol.*, vol. 19, no. 3, pp. 521–530, May 2011.

[9] K. Borisov and H. Ginn, "A computationally efficient RDFT-based reference signal generator for active compensators," *IEEE Trans. Power Deliv.*, vol. 24, no. 4, pp. 2396–2404, Oct. 2009.

[10] M. S. Reza, M. Ciobotaru, and V. G. Agelidis, "A recursive DFT based technique for accurate estimation of grid voltage frequency," in *IECON 2013 - 39th Annual Conference of the IEEE Industrial Electronics Society*, pp. 6420–6425, 2013.

[11] H. L. Ginn, "CPC based converter control for systems with non-ideal supply voltage," in *2010 International School on Nonsinusoidal Currents and Compensation*, pp. 117–122, 2010.

[12] L. Asiminoael, F. Blaabjerg, and S. Hansen, "Detection is key - harmonic detection methods for active power filter applications," *IEEE Ind. Appl. Mag.*, vol. 13, no. 4, pp. 22–33, Jul. 2007.

[13] H. L. Ginn and G. Chen, "Digital control method for grid-connected converters supplied with nonideal voltage," *IEEE Trans. Ind. Informatics*, vol. 10, no. 1, pp. 127–136, Feb. 2014.

[14] S. A. Gonzalez, R. Garcia-Retegui, and M. Benedetti, "Harmonic computation technique suitable for active power filters," *IEEE Trans. Ind. Electron.*, vol. 54, no. 5, pp. 2791–2796, Oct. 2007.

[15] S. Lai, S. Lei, C. Chang, C. Lin, and C. Luo, "Low computational complexity, low power, and low area design for the implementation of recursive DFT and IDFT algorithms," *IEEE Trans. Circuits Syst. II Express Briefs*, vol. 56, no. 12, pp. 921–925, Dec. 2009.

[16] H. A. Darwish and M. Fikri, "Practical considerations for recursive DFT implementation in numerical relays," *IEEE Trans. Power Deliv.*, vol. 22, no. 1, pp. 42–49, Jan. 2007.

[17] J. M. Maza-Ortega, J. A. Rosendo-Macias, A. Gomez-Exposito, S. Ceballos-Mannozzi, and M. Barragan-Villarejo, "Reference current computation for active power filters by running DFT techniques," *IEEE Trans. Power Deliv.*, vol. 25, no. 3, pp. 1986–1995, Jul. 2010.

[18] M. Monfared, S. Golestan, and J. M. Guerrero, "Analysis, design and experimental verification of a synchronous reference frame voltage control for single-phase inverters," *IEEE Trans. Ind. Electron.*, vol. 61, no. 1, pp. 258-269, Jan. 2014.

978-1-5090-0376-1/16 $31.00 © 2016 IEEE   260

A Z-Source Railway Static Power Conditioner for Power Quality Improvement

Hossein Mahdinia Roudsari, Alireza Jalilian, Sadegh Jamali
School of Electrical Engineering
Centre of Excellence for Power System Automation and Operation
Iran University of Science and Technology
Narmak, Tehran, 1684613114, IRI

Abstract— Electrified railway system is a huge single phase load which imposes some serious power quality problems such as negative sequence current (NSC), harmonics, low power factor, pantographs arcs, resonance, electromagnetic interference, etc. to power grid. Several compensation strategies and configurations such as RPC, APQC, HBRPC, and etc. have been introduced up to now. The main problem corresponding to these compensators is their high nominal ratings. In this paper a new topology of railway static power compensator is presented. In proposed configuration, a Z-source converter is employed to benefit the pros of such converter in electrified railway system, especially it's higher electrical efficiency and lower switching losses. Theoretical calculations and simulation results verify the effectiveness of the proposed configuration.

Keywords—Power quality; Z-source Converter; Harmonic; Negative Sequence Current; Impedance matching Transformer

I. INTRODUCTION

Electrified railway system is a huge single phase load in electric power system which impose serious power quality problems such as current and voltage unbalance, harmonics, low power factor, pantograph arcing, resonance, electromagnetic interference (EMI), and etc. These Power quality problems have been investigated in many researches in the literature. A comprehensive review on power quality issues of electrified railway system has been made in [1]. In the paper compensation strategies are classified and compared. In [2] power-quality impact assessment for high-speed railway with high-speed trains has been studied using train timetable of trains. Many researchers used power electronic converters to overcome negative sequence current (NSC), harmonics, and low power factor (PF) in the same time using an appropriate control system. Several compensation strategies and configurations such as RPC, APQC, HBRPC, and etc. have been introduced since 1993 up to now. The evolution of these converters has been briefly reviewed here.

For the first time in 1993 Mochinaga *et al.* presented the Railway static Power Conditioner (RPC) which is being the base compensator for all the researches made on this field [3]. A RPC consist of two single phase back-to-back converter which can shift active and reactive power between traction substation feeders. Morimoto *et al.* in 2002 and Uzaka *et al.* in similar researches utilized RPC with a balancing Scott transformer in order for voltage fluctuation in the high-voltage

side of the traction transformer [4-6]. Luo *et al.*in 2011 proposed a RPC with three-phase V/v transformers with a compensation strategy to eliminate the negative-sequence and harmonic currents in the high-speed train traction systems [7]. They published another article focused on the control system of converter [8]. In the paper a dual-loop control strategy for maintaining the DC link voltage in a constant value is presented.

Active power quality conditioner (APQC) which is introduced for the first time in 2004 by Sun, consists of a three-leg converter and a Scott balancing transformer [9]. The main advantage of APQC in comparison with RPC is the reduced number of switches, higher reliability, and ease of control. In [10] a control method for APQC presented in [9] has been proposed which uses only capacitor voltage control, and it is claimed that there is no need to detection methods for harmonic, reactive and negative sequence currents. A constant DC voltage based control strategy proposed in [11] for an active power quality compensator (APQC) used in electrified railways. The proposed strategy consists of only an I-PD based constant DC capacitor voltage control with an added moving average type low-pass filter (LPF). In [12] a simpler configuration similar to APQC is proposed by Maghsood *et al.* which a control strategy based on FBD method is used in the paper. The control strategy is independent to the type and connection of the traction transformer.

For the first time a half-bridge inverter based on APQC presented in [13]. The APQC comprised of three legs of power electronic switches which two of them are the main legs and produce the compensating current, and the third one switches control the voltage of two capacitors placed in the middle of half-bridge. Ma *et al.* proposed a half-bridge-converter-based railway static power conditioner (HBRPC) which consists of two half-bridge converters connected by two capacitors in series [14]. In the paper a balanced voltage control developed to control the DC voltages of two capacitors. Jafari *et al.* utilized the half-bridge configuration with a control strategy based on modified instantaneous reactive power theory in [15].

One of the most important issues incorporated with above converters is their high power rating. Many researches focused on reduction of the power rating of these converters. In [16] the rating of RPC has been calculated. The results showed that in worst case scenario the rating of the RPC in 115% of the apparent power consumed in one of the traction transformer feeders. In [17], based on the decoupling of NSC and reactive current, two priority control strategies to achieve flexible compensation targets is presented to realize the optimal utilization of the power capacity of PRC. Dai *et al.* in order to moderate the power rating of the RPC in a Co-Phase electrified railway system used a RC network in series with RPC [18]. Iranian researchers (Ghasemmi *et al.*) utilized TSC in parallel with one of the traction feeders to reduce the rating of the RPC [16]. It is obvious that using current balancing transformers such as Scott, Le-Blanc, Woodbridge, impedance matching (IM) transformers, etc. reduce the power rating of the converters installed in parallel with traction substations.

All of the presented configurations of power quality compensators use voltage source inverters (VSI) as a current controlled current source. VSIs cannot work in boost condition because VSI cannot produce an AC voltage more than DC link voltage. On the other hand in all of the converters mentioned before, by simultaneously turning on the switches placed on a same leg, DC link short circuited and may damage the converter. All of these disadvantages have been solved in Z-Source converter originally presented by Peng in 2003 [20]. Impedance networks cover the entire of electric power conversion consists of DC-DC, DC-AC, AC-DC, and AC-AC converters in a wide range of applications. These converters widely used to overcome the limitations and problems of the traditional voltage source, current source as well as various classical buck-boost, unidirectional, and bidirectional converter topologies [21]. Various converter topologies have been reported in the literature such as variable drive systems [22], grid-connected photovoltaic systems[23-25], electrical vehicles [26, 27], etc. These extensive uses of Z-Source converter make it a good candidate for using it as a compensator in traction substations.

In this paper a new topology of railway static power compensator is presented. In proposed configuration, a Z-source converter is employed to benefit the pros of such converter in electrified railway system, especially it's higher electrical efficiency and lower switching losses. In order to reduce the rating of proposed Z-Source Railway static Power Conditioner (ZSRPC) an IM balancing transformer has been considered in traction substation (TSS). The simulation results verified the performance of the proposed configuration. In following section the configuration of proposed compensator is illustrated, and in section III the operating principle of the ZSRPC is explained. The forth section focused on the control system of the ZSRPC. In order to validate the operating principle of the proposed ZSRPC simulation results from MATLAB/Simulink Presented in V, and finally in section VI conclusion is drawn.

II. CONFIGURATION OF PROPOSED COMPENSATOR

In this section the configuration of proposed ZSRPC is presented and details of the scheme are discussed. An IM transformer is used as traction transformer. IM is a commonly used traction transformer especially in China railway system. The transformer produces orthogonal voltages in its output. Unlike other balancing transformers such as Le-Blanc, Woodbridge, and Scott, the primary windings of IM transformer are star-connected and the neutral point is grounded. The grounded star connection in the primary side reduces the required insulation level in the primary side. The output orthogonal voltages of IM transformer is obtained as:

$$\begin{bmatrix} V_\alpha \\ V_\beta \end{bmatrix} = \frac{1}{2K} \begin{bmatrix} \sqrt{3}-1 & 0 & -2 \\ 1-\sqrt{3} & 2 & 0 \end{bmatrix} \begin{bmatrix} V_{AB} \\ V_{BC} \\ V_{CA} \end{bmatrix} = \frac{\sqrt{3}V}{\sqrt{2}K} \begin{bmatrix} e^{-\frac{\pi}{12}} \\ e^{\frac{7\pi}{12}} \end{bmatrix} \quad (1)$$

where V is the phase to ground voltage of the primary side of traction transformer, and K is turn ratio of primary to secondary windings of IM transformer (n_1/n_2). The vector diagram of the output voltages of IM transformer and line voltages of primary side of the transformer are depicted in Fig. 1. In Fig. 1 for ease of understanding and similarity between vector diagram and circuit diagram of IM, the origin of vector diagram (0°) is shifted 150°.

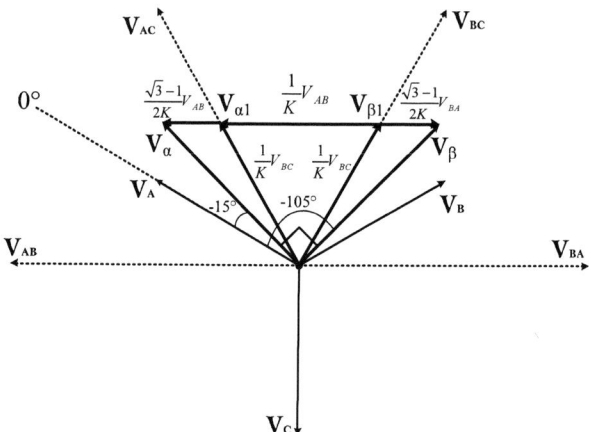

Fig. 1. Vector diagram of primary and secondary voltages of IM transformer

As mentioned before and as can be seen in above figure output voltages (V_α, V_β) are orthogonal. V_α and V_β phase angle are -15° and -105° respectively. Primary side currents (I_A, I_B, I_C) can be obtained as follows:

$$\begin{bmatrix} I_A \\ I_B \\ I_C \end{bmatrix} = \frac{1}{2\sqrt{3}K} \begin{bmatrix} (\sqrt{3}+1) & -(\sqrt{3}-1) \\ -(\sqrt{3}-1) & (\sqrt{3}+1) \\ -2 & -2 \end{bmatrix} \cdot \begin{bmatrix} I_\alpha \\ I_\beta \end{bmatrix} \quad (2)$$

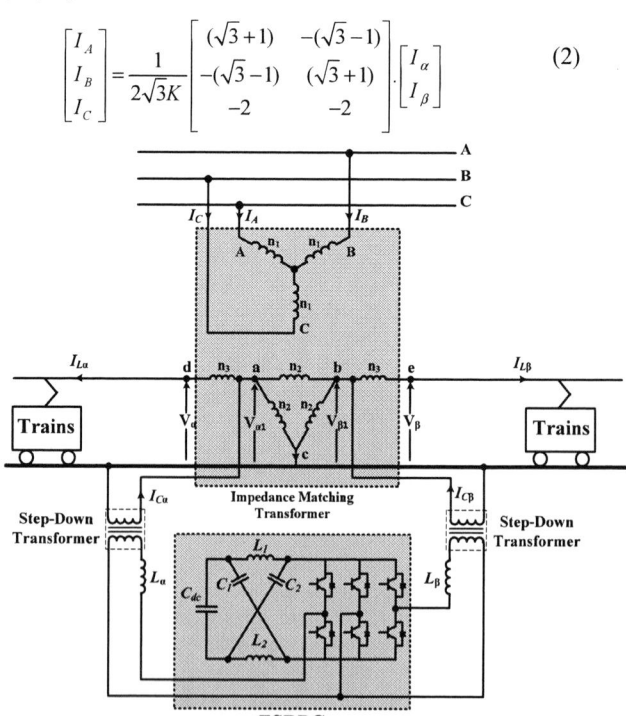

Fig. 2. Proposed Configuration (ZSRPC)

978-1-5090-0376-1/16 $31.00 © 2016 IEEE

where I_α and I_β are the output currents of IM transformer. If the magnitude of I_α equals to the magnitude of I_β primary side currents will be the same and there will be no NSC in grid side.

If IM transformer designed in a way that the "a" and "b" terminals are accessible in order to reduce the nominal voltage of the compensator, It can be connected to these nodes. As can be seen in Fig. 2 ZSRPC is connected to "a" and "b" terminals instead of to "d" and "e" terminals. This leads to decreasing the nominal voltage of the compensator, and also changes the control system of the converter. In the following sections detailed operating principle and control system of the proposed ZSRPC is discussed.

III. OPERATING PRINCIPLE OF THE ZSRPC

ZSRPC which is shown in Fig. 2 contains a Z-source converter (ZSC). The ZSC comprised of a three-phase converter and an Impedance network. The impedance network contains inductances L_1 and L_2 and capacitors C_1 and C_2. The ZSC acts like a current controlled current source, which injects compensating current into the "a" and "b" terminals of IM. The left and right leg of the converter injects α and β phases compensating current and the middle leg absorb the return current from "c" terminal of the IM transformer. Fig. 3 shows the equivalent circuit diagram of the three-phase Z-Source converter.

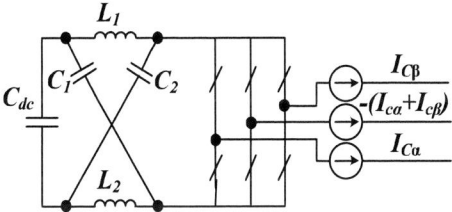

Fig. 3. Equivalent circuit diagram of the three-phase Z-Source converter

Assuming the same value for inductances ($L_1=L_2$) and the same value for capacitors ($C_1=C_2$) Z-source network is a symmetrical lattice (Eq. 3) [20].

$$v_{L1}=v_{L2}=v_L \quad , \quad V_{C1}=V_{C2}=V_C \qquad (3)$$

In a switching cycle with period T, during T_0 shoot-through period (equivalent circuit diagram of the ZSC in shoot-through mode and in non-shoot-through mode depicted in Fig. 4), the output voltage will be zero and the capacitor and inductor voltages in shoot-through mode are calculated as follows:

$$\begin{cases} v_{L1}=V_{C1} \\ v_{L2}=V_{C2} \end{cases} \Rightarrow Symmetrical \Rightarrow \begin{cases} v_{L1}=v_{L2}=v_L \\ V_{C1}=V_{C2}=V_C \end{cases} \Rightarrow V_C=v_L \qquad (4)$$

$$\Rightarrow V_d=2V_C=2v_L \quad , \quad V_i=0 \qquad (5)$$

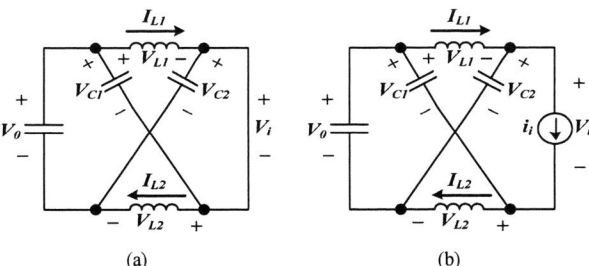

(a)   (b)

Fig. 4. Equivalent circuit diagram of ZSC in: a) Shoot-through mode b) Non Shoot-through mode

If the ZSC works on non-shoot-through mode output voltage can be obtained from equation below

$$v_L=V_0-V_C \quad , \quad V_d=V_0 \quad , \quad V_i=V_C-v_L=2V_C-V_0 \qquad (6)$$

V_0 is the average DC voltage of DC link and $T=T_0+T_1$. Average voltages of the inductor in a switching period (T) must be zero:

$$V_L=\overline{v_L}=\frac{T_0 V_C+T_1.(V_0-V_C)}{T}=0 \qquad (7)$$

$$\frac{V_C}{V_0}=\frac{T_1}{T_1-T_0} \qquad (8)$$

Consequently the average output voltage of the impedance network (input voltage of the three phase bridge) is:

$$V_i=\overline{v_i}=\frac{T_0.0+T_1.(2V_C-V_0)}{T}=\frac{T_1}{T_1-T_0}V_0=V_C \qquad (9)$$

It can be realized that in steady state the average output voltage equals to average voltage of capacitors in impedance network. DC link peak voltage can be calculated as:

$$\hat{v}_i=V_C-v_L=2V_C-V_0=\frac{T}{T_1-T_0}V_0=\frac{T}{(T-T_0)-T_0}V_0=BV_0 \qquad (10)$$

$$B=\frac{T}{T_1-T_0}=\frac{1}{1-2\frac{T_0}{T}}\geq 1 \qquad (11)$$

B is the boost factor established from shoot-through mode. On the other hand assuming modulation index M maximum output voltage in the ac side of the bridge can be written as:

$$\overline{V_{ac}}=M.B.\frac{V_i}{2} \qquad (12)$$

$$B_B=M.B=(0\sim\infty) \qquad (13)$$

B_B coefficient (Buck-Boost coef.) can be determined by M and B. B_B theoretically can change between zero and infinite. Considering above relations the average voltage of the capacitors in impedance network is calculated as follows:

$$V_{C1}=V_{C2}=V_C=\frac{1-\frac{T_0}{T}}{1-2\frac{T_0}{T}}V_0 \qquad (14)$$

In the next section the control system of the proposed configuration briefly explained.

IV. CONTROL SYSTEM

In this section a comprehensive overview of the control system corresponding to ZSRPC is presented. In the first place the procedure of reference current generation considering new connection is presented. After illustrating the block diagram of the reference current generation the current and voltage control parts of the control system of ZSRPC are projected.

A. Refernce Current Generation

In order to avoid losing generality, both feeders are considered resistive-inductive, and the harmonic components created by traction drives are considered as current controlled harmonic source. The fundamental frequency component of α and β feeder currents lag their corresponding voltages by $\theta_{L\alpha}$ and $\theta_{L\beta}$ degrees respectively (Eq. 15 and Eq. 16).

$$i_{L\alpha\_(1)}(\omega t)=I_{L\alpha 1}Sin(\omega t-\frac{\pi}{12}-\theta_{L\alpha}) \qquad (15)$$

$$i_{L\beta\_(1)}(\omega t) = I_{L\beta 1}Sin(\omega t - \frac{7\pi}{12} - \theta_{L\beta})$$ (16)

Considering the harmonic component of traction loads of both sides, $I_{L\alpha}$ and $I_{L\beta}$ are as:

$$i_{L\alpha}(\omega t) = I_{p\alpha}sin(\omega t - \frac{\pi}{12}) - I_{q\alpha}\cos(\omega t - \frac{\pi}{12}) + ...$$

$$...+ \sum_{n=2}^{\infty} I_{h\alpha}(n\omega).sin(n\omega t + \theta_{Lh\alpha})$$ (17)

$$i_{L\beta}(\omega t) = I_{p\beta}sin(\omega t - \frac{7\pi}{12}) - I_{q\beta}\cos(\omega t - \frac{7\pi}{12}) + ...$$

$$...+ \sum_{n=2}^{\infty} I_{h\beta}(n\omega).sin(n\omega t + \theta_{Lh\beta})$$ (18)

By installing the ZSRPC in "a" and "b" terminals, the control system which was used in earlier papers does not work here. It is necessary to regenerate the reference currents. The equivalent circuit diagram of ZSRPC is depicted in Fig. 5.

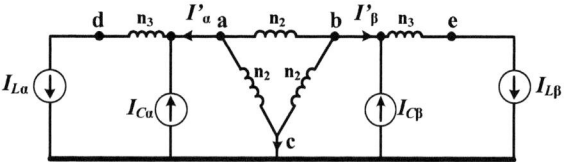

Fig. 5. Equivalent circuit diagram of ZSRPC

In order to calculate the grid side currents it is necessary to write down the mmf equations corresponding to traction loads ($I_{L\alpha}$ and $I_{L\beta}$), compensating currents ($I_{C\alpha}$ and $I_{C\beta}$). Considering the connection of the IM transformer the mmf relationships are as follows:

$$n_1 I_A = n_2(I_{L\alpha} - I_{C\alpha}) + n_3 I_{L\alpha} - n_3 I_{L\beta}$$
$$n_1 I_B = n_2(I_{L\beta} - I_{C\beta}) - n_3 I_{L\alpha} + n_3 I_{L\beta}$$
$$n_1 I_C = n_2(I_{C\alpha} + I_{C\beta} - I_{L\alpha} - I_{L\beta})$$ (19)

If $N_2 = n_2/n_1 = (\sqrt{3}-1)/(2\sqrt{3}K)$ and $N_3 = n_3/n_1 = 1/(\sqrt{3}K)$, then:

$$\begin{bmatrix} I_A \\ I_B \\ I_C \end{bmatrix} = \begin{bmatrix} (N_2 + N_3) & -N_3 & -N_2 & 0 \\ -N_3 & (N_2 + N_3) & 0 & -N_2 \\ -N_2 & -N_2 & N_2 & N_2 \end{bmatrix} \cdot \begin{bmatrix} I_{L\alpha} \\ I_{L\beta} \\ I_{C\alpha} \\ I_{C\beta} \end{bmatrix}$$ (20)

Using inverse Fortescue transformation, zero, positive and negative sequence components of grid side currents can be obtained as (a=1<120°):

$$\begin{bmatrix} I_0 \\ I_+ \\ I_- \end{bmatrix} = \frac{1}{3}\begin{bmatrix} 1 & 1 & 1 \\ 1 & a & a^2 \\ 1 & a^2 & a \end{bmatrix}\begin{bmatrix} I_A \\ I_B \\ I_C \end{bmatrix} = M.\begin{bmatrix} I_{L\alpha} \\ I_{L\beta} \\ I_{C\alpha} \\ I_{C\beta} \end{bmatrix}$$ (21)

$$M = \begin{bmatrix} 0 & 0 & 0 & 0 \\ N_2(1-a^2)+N_3(1-a) & N_2(a-a^2)+N_3(a-1) & N_2(a^2-1) & N_2(a^2-a) \\ N_2(1-a)+N_3(1-a^2) & N_2(a^2-a)+N_3(a^2-1) & N_2(a-1) & N_2(a-a^2) \end{bmatrix}$$ (22)

Using (21) and (22) the NSC of the grid side three-phase currents is as (23). In order to remove NSC I_ must be zero.

$$I_- = \left[N_2(1-a) + N_3(1-a^2)\right]I_{L\alpha} + ...$$
$$\left[N_2(a^2-a) + N_3(a^2-1)\right]I_{L\beta} + ...$$
$$\left[N_2(a-1)\right]I_{C\alpha} + \left[N_2(a-a^2)\right]I_{C\beta} = 0$$ (23)

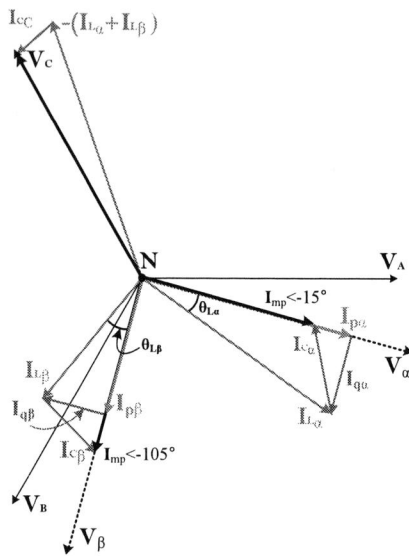

Fig. 6. Vector diagram of the traction load currents and compensating currents

Equation (23) involves four terms which are traction loads ($I_{L\alpha}$ and $I_{L\beta}$) and compensating currents ($I_{C\alpha}$ and $I_{C\beta}$). Using the concept presented in [8] each compensating currents comprise of an active and a reactive part. The vector diagram of train's load and compensating currents is shown in Fig. 6.

B. Current Control

In this part a modified hysteresis current control consist of four hysteresis band (two upper bands and two lower bands) to encompass the shoot-through mode of the ZSC is presented (Fig. 7). The first upper hysteresis band and first lower hysteresis band works such as a common bipolar hysteresis band, but the second upper band and second lower band have been considered in order to apply the shoot-through mode in switching of the IGBTs. If the actual current exceeds from first upper band and reaches the second upper band, one of shoot-through mode takes place. This concept can occur for the lower bands either.

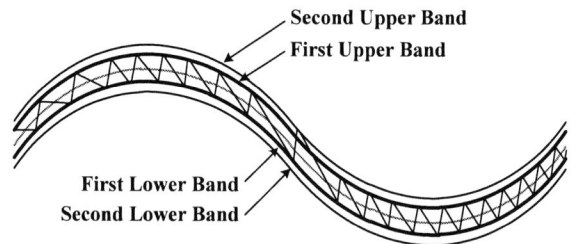

Fig. 7. Four band hystersis current control

TABLE I. shows all of switching states of the ZSC which is used as compensator. Four band hysteresis current control can determines which state apply to six switches.

C. Voltage Control

The output voltage of the impedance network must be kept constant in order to ensure the correct performance of the compensator. In this paper a PI controller used to control the input voltage of three-phase Bridge is shown in Fig. 2.

TABLE I. SWITCHING STATES FOR ZSRPC

Switching States	Switches						Output Voltage
	S1	*S2*	*S3*	*S4*	*S5*	*S6*	
Non Shoot-Through	1	0	0	0	0	1	Determined Voltage
	1	1	0	0	0	0	
	0	1	1	0	0	0	
	0	0	1	1	0	0	
	0	0	0	1	1	0	
	0	0	0	0	1	1	
Zero State Non Shoot-Through	1	0	1	0	1	0	Zero
	0	1	0	1	0	1	
Shoot-Through State	1	×	×	1	×	×	Zero
	×	×	1	×	×	1	Zero
	×	1	×	×	1	×	Zero
	1	1	×	1	1	×	Zero
	1	×	1	1	×	1	Zero
	×	1	1	×	1	1	Zero
	1	1	1	1	1	1	Zero

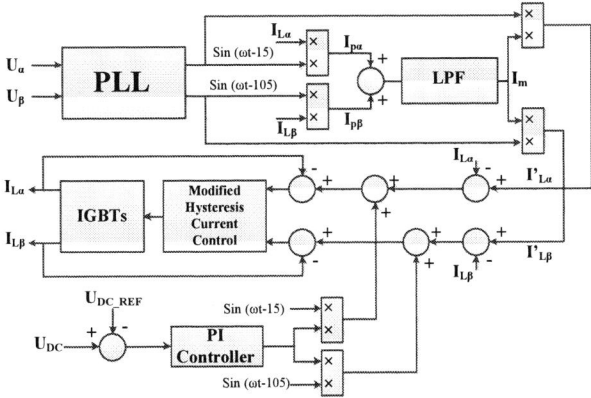

Fig. 8. Overall control system of the ZSRPC

The overall block diagram of control system of ZSRPC illustrated in Fig. 8. The reference DC voltage is a constant value that when actual DC voltage (V_i in Fig. 4) differ from U_{DC_REF} PI controller inject an active current into the reference current in order to stabilize the output voltage of the impedance network.

V. SIMULATION RESULTS

In order to validate the performance of the proposed ZSRPC simulation studies are performed in the MATLAB/Simulink. Electrified traction substation shown in Fig. 2 has been implemented in Simulink environment. A 230 kV/27.5 kV IM traction transformer supplies the overhead catenary system. The traction transformer is grounded in neutral point of primary star connection. For considering the reactive power consumed and harmonic generated by traction loads a Thyristor controlled rectifier with two RL loads is considered in the load model (Fig. 9).

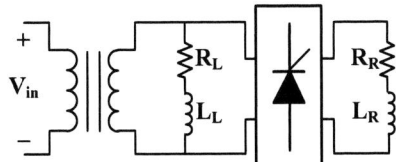

Fig. 9. Traction load model

ZSRPC connects to IM transformer using two step-down transformers. The nominal voltages of the HV and LV windings of the transformers are 22.5 kV and 1.5 kV respectively. Two 4.7 mH inductors placed in series with converter. These inductances play an important role in performance of the ZSRPC. Two 1mH inductor and two 1.6 µF capacitors placed in impedance network, and a 0.2 mF capacitor used as C_{DC}.

In order to evaluate the performance of the proposed ZSRPC two scenarios presented. The power quality indices related to these two scenarios are written in TABLE II.

TABLE II. SIMULATION SCENARIOS

Scenarios	Power Quality Indices		
	Unbalance ($\zeta_{\alpha\beta}=|I_\beta|/|I_\alpha|$)	*Power Factor (PF)*	*Total Harmonic Distortion (THD)*
Scenario 1	0.5	0.85	13%
Scenario 2	0	0.85	13%

A. Case 1

The simulation time is considered 0.5 (sec) and the ZSRPC inserted in parallel with traction transformer. The simulation results illustrated in Fig. 10-12. Fig. 10 shows the grid side three phase current which have been unbalanced until the ZSRPC inserted into the system. After t=0.25 (s) and after a transient response the three phase currents have been balanced and the NSC has been minimized. NSC values reduces from 25% to below the standard margin (5%) and the final amount of current unbalance reaches to 2.31%.

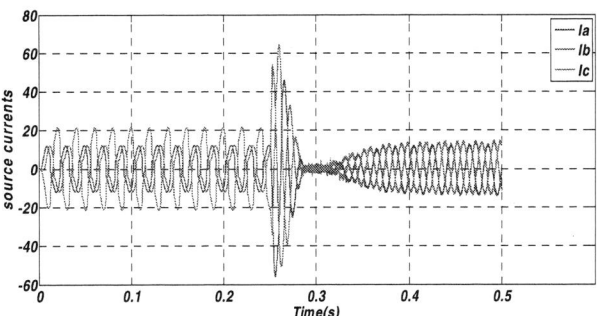

Fig. 10. Grid side three phase currents (Case 1)

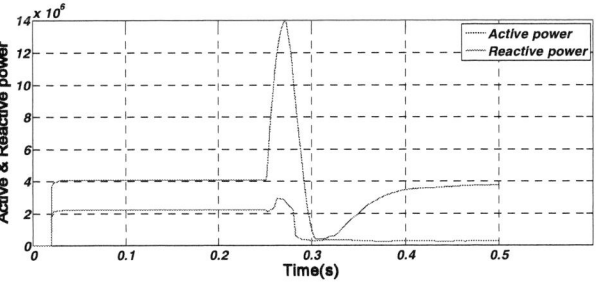

Fig. 11. Active and reactive power calculated in primary side of traction substation (Case 1)

Fig. 11 shows the active and reactive power calculated in primary side of traction substation. The ZSRPC after entering to the system, reduces the reactive power consumed by traction loads and improved the power factor from 0.85 to 0.997. ZSRPC improve the NSC, power factor and also reduce the harmonic components of grid side three-phase current. Harmonic components of grid side currents before and after compensation illustrated in Fig. 12. THD before and after compensation equals 18.83% and 4.39% respectively and these results show the suitable performance of the ZSRPC.

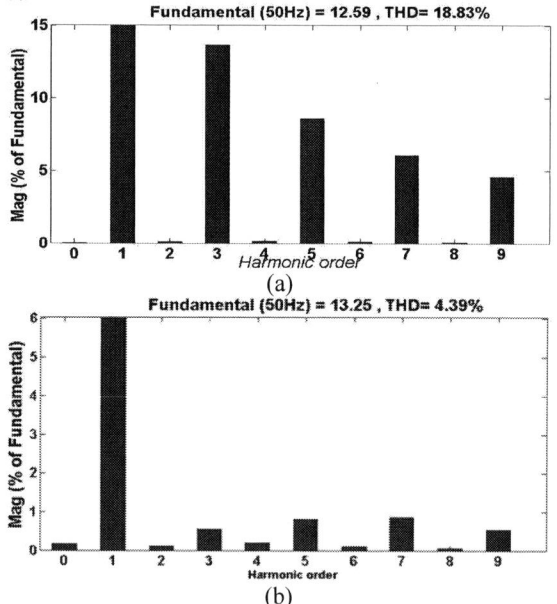

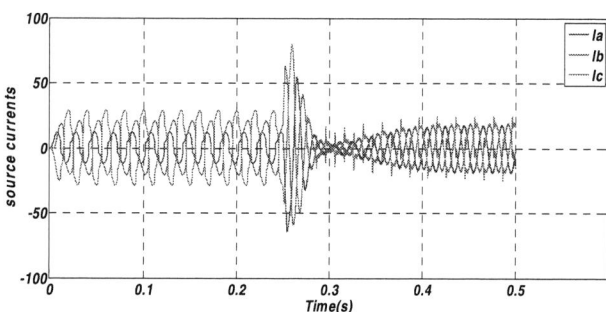

Fig. 12. Harmonic components of grid side currents (Case 1). a) Before compensation b) after compensation

B. Case 2

Same as case 1 simulation time is considered 0.5 (sec) and the ZSRPC inserted in parallel with traction transformer in t=0.25 (s). The grid side three-phase current illustrated in Fig. 13. NSC precentage before and after compensation is 50% and 2.94% respectively.

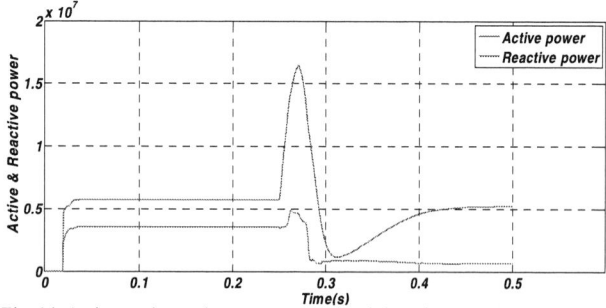

Fig. 13. Grid side three phase currents (Case 2)

Fig. 14. Active and reactive power calculated in primary side of traction substation (Case 2)

Fig. 14 shows the active and reactive power consumed in the traction power supply system. The results shows that the power factor increses from 0.85 to 0.994.

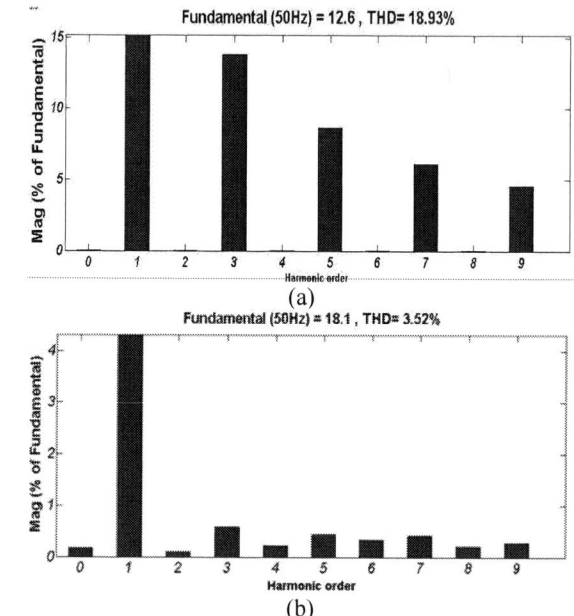

Fig. 15. Harmonic components of grid side currents (Case 2). a) Before compensation b) after compensation

Fig. 15 shows that THD before and after compensation equals 18.93% and 3.52% respectively and these results show the suitable performance of the ZSRPC.

VI. CONCLUSION

In this paper a novel configuration for electrified railway static power conditioner based on Z-Source converter is presented. In order to reduce the rating of proposed Z-Source Railway static Power Conditioner (ZSRPC) an IM balancing transformer has been considered in traction substation (TSS). In order to reduce the rating of the ZSRPC converter connected to the middle terminal of the IM transformer. The Current control and voltage control proportional to new configuration have been modified. The simulation results in MATLAB/Simulink environment verified the performance of the proposed configuration.

REFERENCES

[1] S. M. M. Gazafrudi, A. Tabakhpour Langerudy, E. F. Fuchs, and K. Al-Haddad, "Power Quality Issues in Railway Electrification: A Comprehensive Perspective," *IEEE Transactions on Industrial Electronics,* vol. 62, pp. 3081-3090, 2015.

[2] H. Hu, Z. He, X. Li, K. Wang, and S. Gao, "Power-Quality Impact Assessment for High-Speed Railway Associated With High-Speed Trains Using Train Timetable; Part I: Methodology and Modeling," *IEEE Transactions on Power Delivery,* vol. PP, pp. 1-1, 2015.

[3] Y. Mochinaga, Y. Hisamizu, M. Takeda, T. Miyashita, and K. Hasuike, "Static power conditioner using GTO converters for AC electric railway," in *Power Conversion Conference,* 1993, pp. 641-646.

[4] H. Morimoto, M. Ando, Y. Mochinaga, T. Kato, J. Yoshizawa, T. Gomi, T. Miyashita, S. Funahashi, M. Nishitoba, and S. Oozeki, "Development of railway static power conditioner used at substation for Shinkansen," in *Proceedings of the Power Conversion Conference,* 2002, pp. 1108-1111 vol.3.

[5] T. Uzuka, S. Ikedo, and K. Ueda, "A static voltage fluctuation compensator for AC electric railway," in *Power Electronics Specialists Conference (PESC)*, 2004, pp. 1869-1873 Vol.3.

[6] T. Uzuka, S. Ikedo, K. Ueda, Y. Mochinaga, S. Funahashi, and K. Ide, "Voltage fluctuation compensator for Shinkansen," *IEEJ Transactions on Power and Energy*, vol. 125, pp. 885-892, 2005.

[7] L. An, W. Chuanping, J. Shen, S. Zhikang, and M. Fujun, "Railway Static Power Conditioners for High-speed Train Traction Power Supply Systems Using Three-phase V/V Transformers," *IEEE Transactions on Power Electronics*, vol. 26, pp. 2844-2856, 2011.

[8] L. An, M. Fujun, W. Chuanping, D. Shi Qi, Q. C. Zhong, and S. Zhikang, "A Dual-Loop Control Strategy of Railway Static Power Regulator Under V/V Electric Traction System," *IEEE Transactions on Power Electronics*, vol. 26, pp. 2079-2091, 2011.

[9] S. Zhuo, J. Xinjian, Z. Dongqi, and Z. Guixin, "A novel active power quality compensator topology for electrified railway," *IEEE Transactions on Power Electronics* vol. 19, pp. 1036-1042, 2004.

[10] T. Tanaka, N. Ishikura, and E. Hiraki, "A novel simple control method of an active power quality compensator used in electrified railways with constant DC voltage control," in *Industrial Electronics Conference (IECON)*, 2008, pp. 502-507.

[11] N. Ishikura, E. Hiraki, and T. Tanaka, "A constant DC voltage control based strategy for an active power quality compensator used in electrified railways with improved response," in *35th Annual Conference of IEEE on Industrial Electronics*, 2009, pp. 3199-3204.

[12] I. Maghsoud, A. Ghassemi, S. Farshad, and S. S. Fazel, "Current balancing, reactive power and harmonic compensation using a traction power conditioner on electrified railway system," in *21st Iranian Conference on Electrical Engineering (ICEE)*, 2013, pp. 1-6.

[13] W. Tint Soe, Y. Baba, E. Hiraki, T. Tanaka, and M. Okamoto, "A half-bridge inverter based Active Power Quality Compensator using a constant DC capacitor voltage control for electrified railways," in *7th International Power Electronics and Motion Control Conference*, 2012, pp. 314-320.

[14] M. Fujun, L. An, X. Xianyong, X. Huagen, W. Chuanping, and W. Wen, "A Simplified Power Conditioner Based on Half-Bridge Converter for High-Speed Railway System," *IEEE Transactions on Industrial Electronics*, vol. 60, pp. 728-738, 2013.

[15] H. J. Kaleybar, S. Farshad, M. Asadi, and A. Jalilian, "Multifunctional control strategy of Half-Bridge based Railway Power Quality Conditioner for Traction System," in *13th International Conference on Environment and Electrical Engineering (EEEIC)*, 2013, pp. 207-212.

[16] A. Ghassemi, S. S. Fazel, I. Maghsoud, and S. Farshad, "Comprehensive study on the power rating of a railway power conditioner using thyristor switched capacitor," *IET Electrical Systems in Transportation*, vol. 10, pp. 1–10, 2014.

[17] W. Yingdong, J. Qirong, and Z. Xiujuan, "An optimal control strategy for power capacity based on railway power static conditioner," in *Asia Pacific Conference on Circuits and Systems*, 2008, pp. 236-239.

[18] N. Y. Dai, K. W. Lao, M. C. Wong, and C. K. Wong, "Hybrid power quality conditioner for co-phase power supply system in electrified railway," *Power Electronics, IET*, vol. 5, pp. 1084-1094, 2012.

[19] S. Zhuo, J. Xinjian, Z. Dongqi, and Z. Guixin, "A novel active power quality compensator topology for electrified railway," *IEEE Transactions on Power Electronics*, vol. 19, pp. 1036-1042, 2004.

[20] P. Fang Zheng, "Z-source inverter," *IEEE Transactions on Industry Applications*, vol. 39, pp. 504-510, 2003.

[21] Y. P. Siwakoti, P. Fang Zheng, F. Blaabjerg, L. Poh Chiang, and G. E. Town, "Impedance-Source Networks for Electric Power Conversion Part I: A Topological Review," *IEEE Transactions on Power Electronics*, vol. 30, pp. 699-716, 2015.

[22] F. Z. Peng, Y. Xiaoming, F. Xupeng, and Q. Zhaoming, "Z-source inverter for adjustable speed drives," *IEEE Power Electronics Letters*, vol. 1, pp. 33-35, 2003.

[23] A. Das, D. Lahiri, and A. K. Dhakar, "Residential solar power systems using Z - source inverter," in *IEEE Region 10 Conference TENCON*, 2008, pp. 1-6.

[24] H. Yi, S. Miaosen, F. Z. Peng, and W. Jin, "Z-Source Inverter for Residential Photovoltaic Systems," *IEEE Transactions on Power Electronics*, vol. 21, pp. 1776-1782, 2006.

[25] S. Kouro, J. I. Leon, D. Vinnikov, and L. G. Franquelo, "Grid-Connected Photovoltaic Systems: An Overview of Recent Research and Emerging PV Converter Technology," *IEEE Industrial Electronics Magazine*, vol. 9, pp. 47-61, 2015.

[26] P. Fang Zheng, S. Miaosen, and K. Holland, "Application of Z-Source Inverter for Traction Drive of Fuel Cell—Battery Hybrid Electric Vehicles," *IEEE Transactions on Power Electronics*, vol. 22, pp. 1054-1061, 2007.

[27] S. Miaosen, A. Joseph, H. Yi, F. Z. Peng, and Q. Zhaoming, "Design and Development of a 50kW Z-Source Inverter for Fuel Cell Vehicles," in *CES/IEEE 5th International Power Electronics and Motion Control Conference*, 2006, pp. 1-5.

978-1-5090-0376-1/16 $31.00 © 2016 IEEE

7th Power Electronics, Drive Systems & Technologies Conference (PEDSTC 2016)
16-18 Feb. 2016, Iran University of Science and Technology, Tehran, Iran

Parallel Operation of Series-Parallel Uninterruptible Power Supplies

M.Shahbazi, M. Mohamadian, A. Yazdian Varjani

Faculty of Electrical and Computer Engineering

Tarbiat Modares University

Tehran, Iran

m.shahbazi@modares.ac.ir,mohamadian@modares.ac.ir,yazdian@modares.ac.ir

Abstract-With growing demand for electric power with high power quality and high availability for sensitive loads, a kind of Uninterruptible Power Supplies (UPS) named series-parallel have been introduced. Based on its structure, this topology could show both good power-quality indices –like low input current THD- and high power availability to the load. In fact, this topology utilizes a direct path called Delta path to reach high efficiencies. Parallel operation of conventional UPS is discussed in the literature, but there are no documents available regarding parallel operation of series-parallel UPS yet. This paper, proposes a method for parallel operation of this kind of UPS.

Keywords—UPS; Parallel operation; Master-Slave; Load Sharing

I. INTRODUCTION

With the increased concerns on quality and cost of energy, the power industry is experiencing changes toward utilizing more reliable energy sources [1]. Uninterruptible power supplies (UPSs), which could deliver power with minimized loss of load power seen as a good candidate to achieve acceptable levels of reliability. In other hand, increase in power level demand put some limitations on this solution, because the nature of this power supplies which have the basic volt-ampere limitation of semiconductor switches[2].

To overcome this issue, parallel operation of UPSs, is proposed as a solution to achieve higher power levels. A review of literature [1],[2],[8]-[15] indicates that in conventional topologies of uninterruptible power supplies, there is no basic difference in parallel operation of inverters and these power supplies.

In case of series-parallel UPSs, which involves presence of current-controlled voltage source inverters (VSIs), and voltage-controlled VSIs at same UPS unit, conventional solutions could be inconvenient. This work proposes a solution for this problem.

In section II a brief description of conventional topologies of UPSs have been presented, and describe the operational control strategy of series-parallel UPS. Then an appropriate control strategy for parallel operation has been selected.

In section IV the selected control strategy is modified to meet limitations of this application. Simulation results are shown in section V to validate parallel operation of the series-parallel UPS. Finally, conclusions are presented in section VI.

II. TOPOLOGIES OF UNINTERRUPTIBLE POWER SUPPLIES

There are two different topologies for operation of uninterruptible power supplies, named "double conversion" and "line-interactive" [3]-[5]. Each one has its advantages and disadvantages which make them applicable for specific load-grid characteristics, like difference in efficiencies and power quality conditioning capabilities in presence of grid voltage.

In both topologies a combination of two inverters connected "back to back" in order with common DC-link. This DC-link stores required energy for operation under circumstance which the grid voltage is not available or not suitable to be delivered to the load.

In line-interactive topology, main path deliver power directly to the load and the inverter-rectifier pack remains disconnected from load until main supply fails. That is why this scheme is sometimes called "off-line".

In contrast, the double conversion or on-line topology uses a solid-state switched rectifier, to charge the energy storage (which could be batteries or flywheel or even superconductor media) from grid all the time. Load is supplied by means of an inverter with appropriate AC waveform.

The nature of this strategy which implies continuous operation of both rectifier/inverter units make it more reliable and could use the rectifier unit as a power quality issue compensator, but implies lower efficiency due to switching and conduction loss of rectifier/inverter combination.

A different topology which combines characteristics of previous topologies to reach higher levels of efficiency and power quality conditioning capabilities at the same time is presented in [4], [5].

This difference is in the presence of a direct path between grid and load which delivers the major part of power directly from grid to the load without any kind of conversion through semiconductors.

This could be done by utilizing two independent inverters in the structure of this UPS named as "series" and "shunt" inverter, respectively. This combination gives some degree of freedom in control strategies.

978-1-5090-0376-1/16 $31.00 © 2016 IEEE   268

III. STRATEGIES OF CONTROL

Authors suggest two strategies to control this kind of UPS [4]-[7]. The difference is in predefined duties for each inverter under normal operational condition. In the lack of grid voltage or under unacceptable power quality condition, both strategies use only the shunt inverter to supply the load.

A series-parallel is shown in Fig. 1. It is seen that this topology employs an inverter directly connected to the load named "Shunt" and another, one which is connected in series with the direct "Delta" path, named "series".

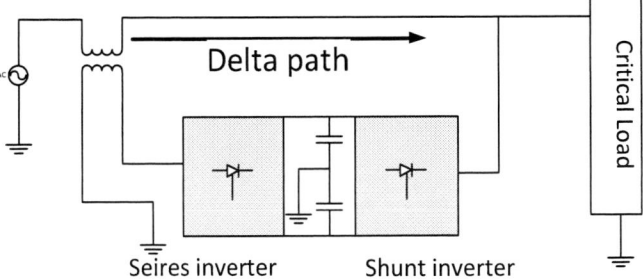

Figure 1. Basic structure of series-parallel UPS and the "delta-path".

In the normal operational condition, because each inverter acts independently, one could deliver the required reactive power to the load without loading the UPS feeder, and the other inverter could compensate the power quality issues of the grid. These control strategies are described below.

A. strategy I

In this strategy shown in Fig. 2, the series inverter could be modeled as an ideal voltage source, which compensates input voltage non-idealities. Shunt inverter also could be modeled as an ideal current source to compensate load current non-ideal behavior. [4],[7].

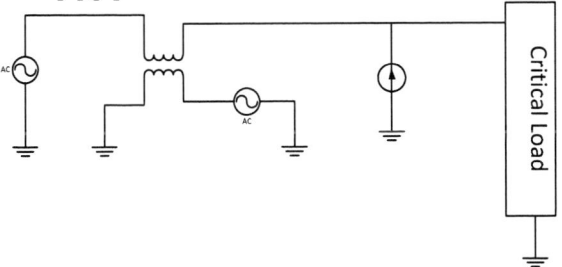

Figure 2. " first strategy"

This strategy whose schematic diagram shown in Fig. 2, implies high speed response for both inverters to inject desired voltage to the connection point and secondary winding of the series transformer. It also needs to use high quality thin lamination of transformer core, capable of transferring the high frequency power waveforms to the other side [6]. This strategy

comes from the known operation principle of Unified Power Quality Conditioner (UPQC).

B. strategy II

In this approach [5],[6] the UPS dictates the desired current-voltage wave forms instead of trying to compensate non-idealities made by system elements.

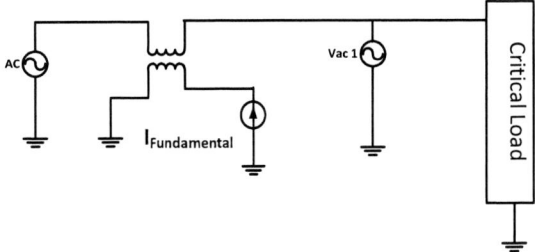

Figure 3. " Second Strategy"

As Fig. 3 illustrates, in this strategy we force the voltage wave form of the load to be sinusoidal by a shunt voltage-source inverter, combined with a passive filter to ensure acceptable power quality of delivered power to the load. In this diagram, V_{ac1} is used to show that only the first order wave-form is present in shunt inverter.

It also utilizes a series-connected inverter via transformer to force the UPS input current in-phase with input voltage and keep it pure sinusoidal too. Like shunt inverter the $I_{fundamental}$ refers to the first order harmonic of the load current.

This control strategy has the advantage of better transient response because there is no need to change the control strategy of shunt inverter in case of emergency and possibility to use less passive parts in filter design. In this paper the second strategy is used.

IV. PARALLEL CONNECTION OF DELTA UPS UNITS

Authors propose different parallel operation methods in literature [6]-[15]. In this paper we chose the Master-Slave strategy. The data exchange bus has the role of transferring the data from the master unit to the slave unit. It is seen that the two units have two independent controllers which only uses the other unit's data to generate the required control signals, and no central controller exists.

In this Fig. 4 three control boxes are present:

A. Gating Pattern Controller

This block utilizes a simple PWM method. It uses two major input signals to generate the gating pattern for both series and shunt inverters in each unit. The V_{Sref} is used generate shunt inverter PWM signals. This controller use V_{Pref} to define the series inverter PWM signals.

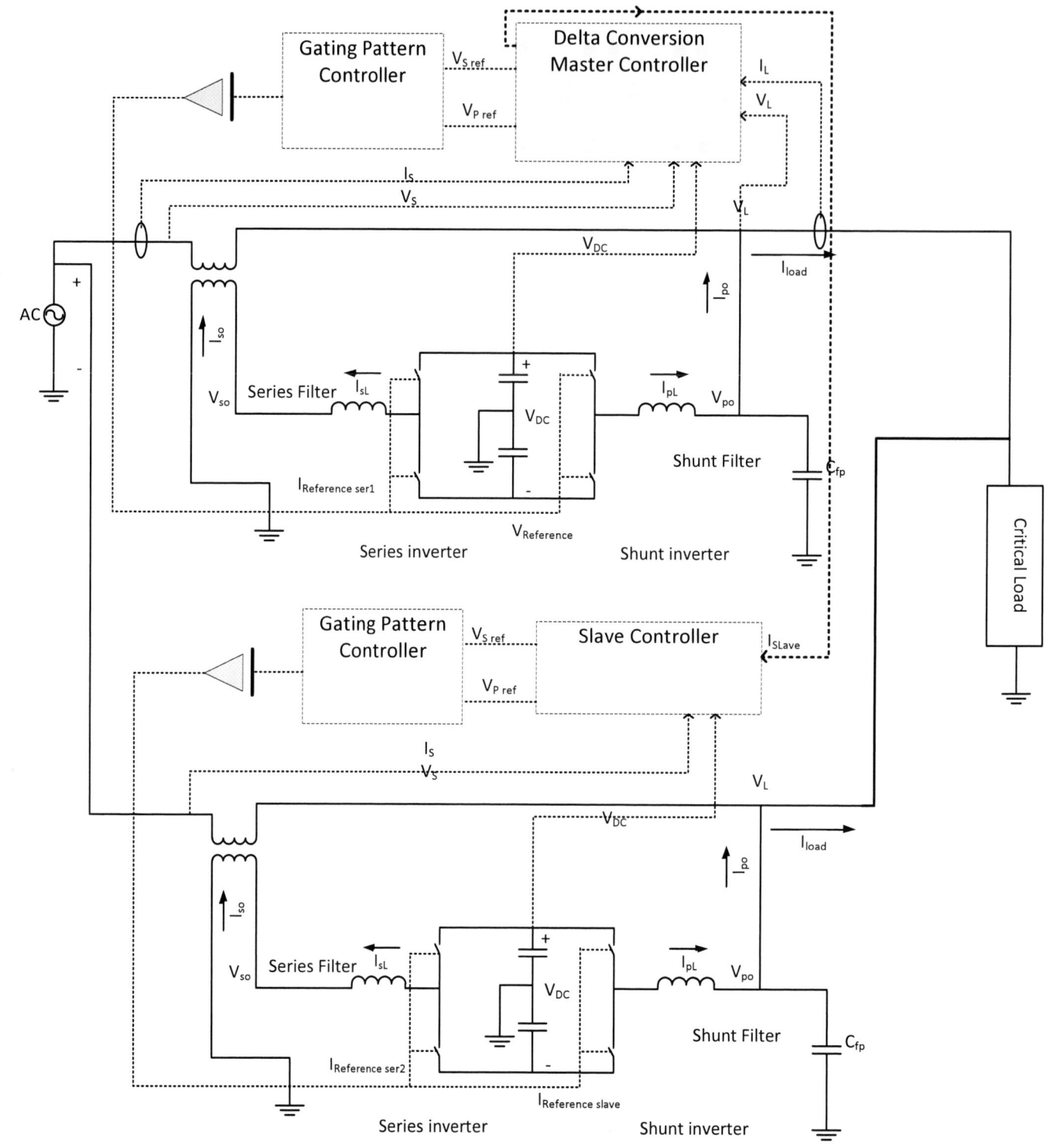

Figure 4. Schematic one-phase diagram of parallel UPSs

B. Delta Conversion Master Controller

As shown in "Fig.5", this block uses V_S from input side of the series inverter transformer to determine the frequency and phase of desired current. It also employs a PI controller to adjust the amplitude of series inverter current by using feedback from input current (I_S), DC-Link voltage (V_{DC}) and Load Voltage (V_L). Load current (I_L) is also used in this block diagram to generate the current reference for the series inverter.

This block generates the voltage reference for the series inverter (V_{Sref}) with desired phase. It also determines the required active power for master unit series controller (V_{Pref}) and current share of slave units (I_{Slave}).

978-1-5090-0376-1/16 $31.00 © 2016 IEEE   270

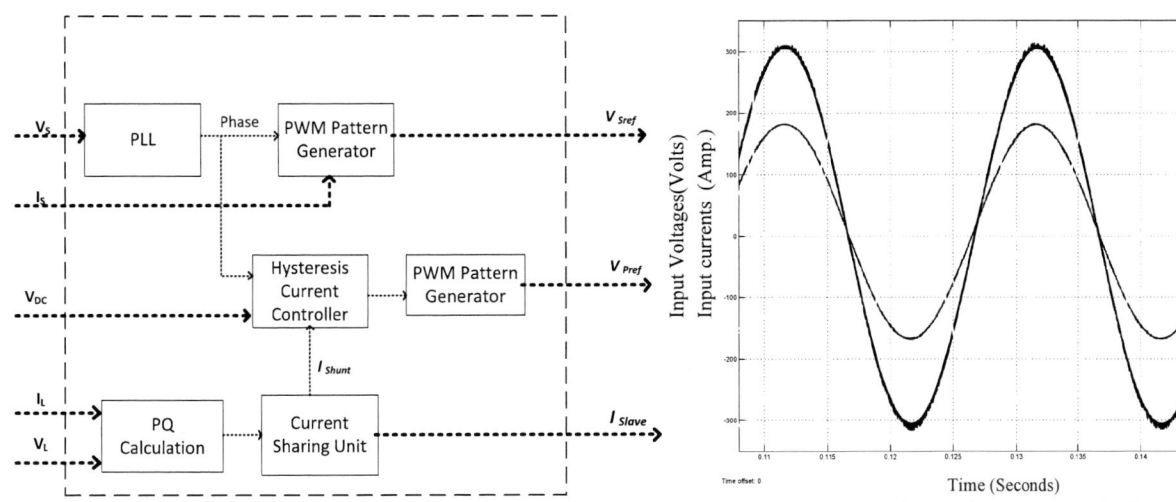

Figure 5. Master UPS Controller

C. Slave Controller

Slave controller has some features of master controller like DC-Link voltage (V_{DC}) controller. Schematic diagram of this controller have been shown in "Fig.6". It utilizes a feedback loop from master unit and tracking the phase and frequency of the master series inverter by another feedback loop from input voltage (V_S). But it uses an external current reference signal from master unit (I_{Slave}) to determine active power reference (V_{Pref}) like conventional master-slave controllers [6]-[8].

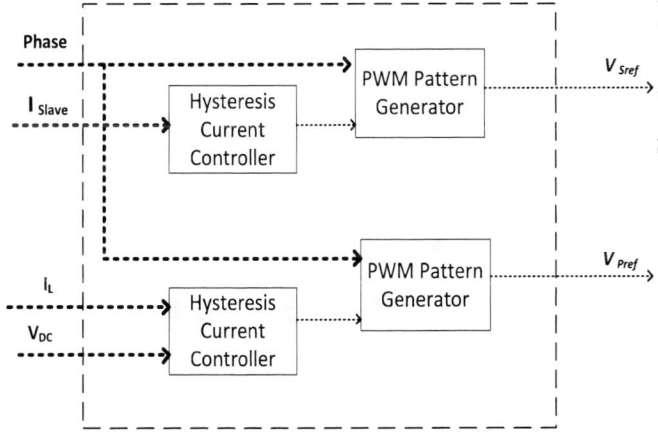

Figure 6. Master UPS Controller

V. SIMULATION RESULTS

Here the two units feed a non-linear load. To illustrate the load sharing capability of control strategy, we assume that one unit has 50 percent more capacity than the other unit.

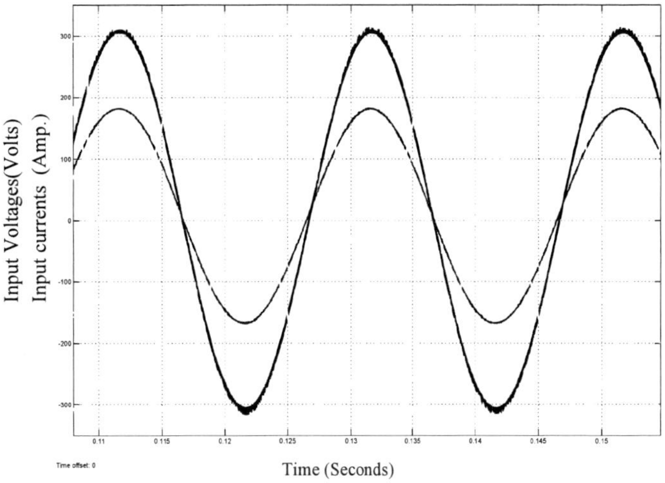

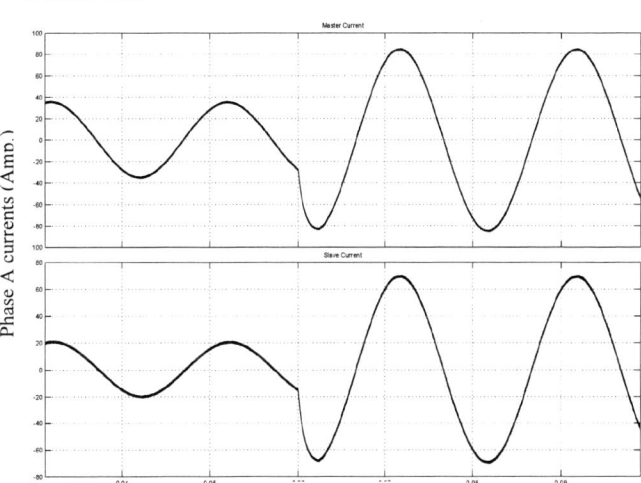

Figure 7. input current and voltage

To ensure proper operation of both UPSs, at first we show the input voltage and currents drawn from the grid. As shown in "Fig. 7", the current and voltages are in phase and current wave form is pure sinusoidal. This proves the correct operation of each unit.

Figure 8. Phase-A input currents of both UPSs during load change

To verify the appropriate operation of parallel controller phase "A" currents are shown together. As illustrated in "Fig. 8", the currents of both units are in-phase with each other and share the common load proportional to their ratings. It is seen that one UPS with higher capacity, handles 50 percent more current than other as expected. Implemented Master-Slave strategy also could handle a sudden 10 percent change in overall load too.

As shown in "Fig. 8" the 1:1.5 ratio of UPS capacities is also seen in steady state after sudden load change.. It is also seen the current waveforms are both sinusoidal in "Fig. 9".

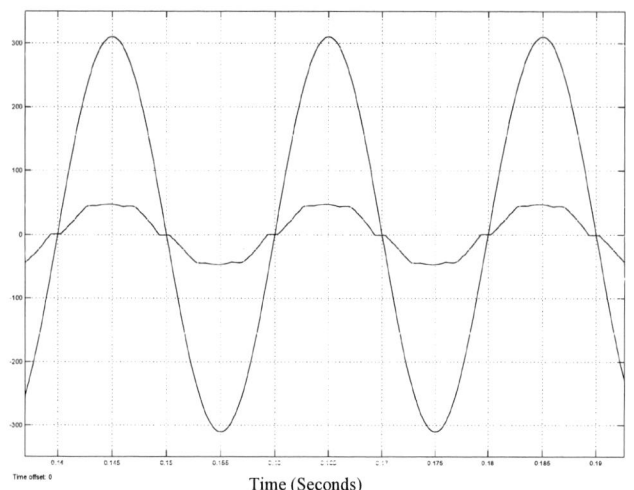

Time (Seconds)

Figure 9. Phase-A output current-voltage of UPSs

VI. CONCLUSION

Parallel operation of conventional UPSs, have been studied in many aspects by researchers, but nothing said about parallel operation of series-parallel inverter till now. In this paper two UPS units have been connected in parallel to share a common load. Some modifications have been made on well-known "Master-Slave" strategy to make it appropriate for this case. Proper operation has been shown by simulation and dynamic response of the whole system has been investigated too.

It was shown that input current drawn from ac grid, was sinusoidal and in-phase with input voltage. Each unit has sinusoidal output voltage in spite of non-linear nature of the load.

Load current divides between units, proportional to each unit nominal power. To investigate dynamic performance of the system, sudden change in load power was studied. In this situation each unit could keep its portion of the newly changed overall load.

REFERENCES

[1] J. Rocabert, A. Luna, f. Blaabjerg, P. Rodriguez, "Control of Power Converters in A Microgrids", *IEEE transactions on power electronics"*,Nn.11, Issue27, pp. 4734-4749, Nov 2012.

[2] J. M. Guerrero, L. Hang , J. Uceda" ,Control of Distributed Uninterruptible Power Supply Systems ",*IEEE TRANSACTIONS ON INDUSTRIAL ELECTRONICS:* ,No. 8 Isuue 55 ,pp. 2845-2858, AUGUST 2008.

[3] A. Nasiri, "Digital Control of Three-Phase Series-Parallel Uninterruptible Power Supply Systems", *IEEE transactions on power electronics*, vol. 22, issue 4, pp. 1116 – 1127, July 2007.

[4] S. B. Bekiarov and A. Emadi, "Uninterruptible power supplies: Classification, operation, dynamics, and control," in Proc. *IEEE APEC*, 2002, pp. 597–604.

[5] A. Nasiri, Zhong Nie, S.B. Bekiarov, A.Emadi, "An On-Line UPS System With Power Factor Correction and Electric Isolation Using BIFRED Converter", *IEEE transactions on power electronics*, vol. 55, issue 2, pp. 722 - 730, February 2008.

[6] M. Niroomand, H.R. Karshenas," Review and comparison of control methods for uninterruptible power supplies", *Drive Systems and Technologies Conference (PEDSTC), 2010 1st*, pp. 18 -23, Tehran, Iran , February 2010.

[7] Nasiri, A., Bekiarov, S.B., Emadi, A., "Reduced parts three-phase series-parallel UPS system with active filter capabilities*", Industry Applications Conference, 2003. 38th IAS Annual Meeting*, Conference Record of the Year: 2003, Volume: 2, Pages: 963 - 969 vol.2, 2003.

[8] S.M. Dehghan, A.A. Ahmad, R. Lourakzadegan, M. Fazeli, M. Mohamadian, A. Abrishamifar ,"A high performance controller for parallel operation of three-phase UPSs powering unbalanced and nonlinear loads Power Electronics", *Drive Systems and Technologies Conference (PEDSTC), 2011 2nd*, pp. 433 -438, Tehran, Iran , February 2011.

[9] W.-C. Lee, T.-K. Lee, S.-H. Lee, K.-H. Kim, D.-S. Hyun, and I.-Y. Suh, "A master and slave control strategy for parallel operation of three phase UPS systems with different ratings," in Proc. *IEEE APEC*, 2004, pp. 456–462.

[10] Y. J. Cheng and E. K. K. Sng, "A novel communication strategy for decentralized control of paralleled multi-inverter systems," *IEEE Trans.Power Electron.*, vol. 21, no. 1, pp. 148–156, Jan. 2006.

[11] X. Sun, Y.-S. Lee, and D. Xu, "Modeling, analysis, and implementation of parallel multi-inverter system with instantaneous average-current-sharing scheme," *IEEE Trans. Power Electron.*, vol. 18, no. 3, pp. 844–856,May 2003.

[12] Z. He, Y. Xing, and Y. Hu, "Low cost compound current sharing control for inverters in parallel operation," in Proc. *IEEE PESC*, 2004, pp. 222–227.

[13] S. J. Chiang, C. H. Lin, and C. Y. Yen, "Current limitation control technique for parallel operation of UPS inverters," in Proc. *IEEE PESC*, 2004, pp. 1922–1926.

[14] M. N. Marwali, J.-W. Jung, and A. Keyhani, "Control of distributed generation systems—Part II: Load sharing control," *IEEE Trans. Power Electron.*, vol. 19, no. 6, pp. 1551–1561, Nov. 2004.

[15] H. Deng, R. Oruganti, and D. Srinivasan, "A simple control method for high-performance UPS inverters through output-impedance reduction," *IEEE Trans. Ind. Electron.*, vol. 55, no. 2, pp. 888–898, Feb. 2008.

7th Power Electronics, Drive Systems & Technologies Conference (PEDSTC 2016)
16-18 Feb. 2016, Iran University of Science and Technology, Tehran, Iran

Sliding Mode Control of DC-Link Capacitors Voltages of a NPC 4-Wire Shunt Active Power Filter with Selective Harmonic Extraction Method

M. Asadi
Dept. of electrical engineering
Arak University of Technology (AUT)
Arak, Iran
m.asadi@arakut.ac.ir

H. Ebrahimirad
Niroo Research Institute (NRI)
Tehran, Iran
hebrahimirad@nri.ac.ir

M. S. Mousavi, A. Jalilian
Dept. of electrical engineering
Centre of Excellence for Power System Automation and Operation
Iran University of Science and Technology (IUST)
Tehran, Iran
mahdi_mousavi@elec.iust.ac.ir, jalilian@iust.ac.ir

Abstract—this paper presents a sliding mode voltage controller in order to regulate the dc-link voltages of a three level neutral-point-clamped (NPC) inverter of a shunt active power filter. A Lyapunov function is used to determine sliding mode control laws that guarantees the stability of the control system. Also a selective harmonic extraction method is employed in the control system to compensate desired harmonic components of the load currents. For validation of the proposed control system an experimental set-up has been prepared. The control system hardware is based on DSP-TMS320F2812. The experimental results confirms the performance of the proposed control scheme.

Keywords— dc-link controller; Active Power Filter; Three level invertr; sliding mode control

I. INTRODUCTION

Shunt active power filters (SAPF) are considered as one of the effective solutions for power quality problems in the power distribution networks [1], [2]. Two level and multi-level voltage source inverters are widely used in the SAPF topology. In the multi-level inverter topologies, the power losses and switching frequency is lower than the two level inverters [3-7]. For these reasons, this paper employs a three-level NPC inverter for the SAPF structure.

Several current control methods, such as PI and resonant controller, Lyapunov function control, adaptive robust control, state space variables control and etc. have been proposed for SAPFs [8-12]. The effectiveness of the current control methods depends on the performance of the dc-link voltage controller. The most of the papers use a conventional PI regulator to control the dc-link voltage [8-12]. In order to improve the dc-link voltage controller performance, some methods have been proposed in the literatures, such as

regulating capacitor's energy in dc-link [13], adaptive dc-link control [14] and sliding mode based PI controller [15].

In this paper a new dc-link voltage controller based on sliding mode control method, is proposed for three level NPC inverter of a SAPF. The proposed controller regulates and balances the dc-link voltages without making overshoot on the SAPF start-up. In this method, stability of the dc-link control loop is guaranteed. Furthermore a selective harmonic extraction method is presented. Each selected harmonic is separately detected and compensated in its own reference frame. In high power applications this feature is very useful, because in this conditions the control system selectively compensates only the most harmful harmonic components. Effectiveness of the proposed dc-link voltage controller and the selective harmonic extraction method is verified with a 400V laboratory prototype of SAPF.

II. STRUCTURE OF A SHUNT APF

The configuration of a shunt APF is illustrated in fig.1. In This structure, a three level neutral-point-clamped voltage source inverter is connected in parallel to the three phase network at the point of common coupling (PCC) through the L_f inductors. Also the dc-link mid- point is connected to the grid's neutral wire. The nonlinear load is a three phase diode rectifier with RL in dc-side. These types of loads are widely used at the front-ends of industrial ac drives and inject harmonic currents into the networks, which have odd orders: $6n \circ 1(n = 1 \circ 2 \circ 3 \ \triangleright \triangleright \triangleright)$ of the fundamental frequency. The

SAPF must generate the harmonic currents to compensate the harmonic components that are produced by the nonlinear load.

978-1-5090-0376-1/16 $31.00 © 2016 IEEE

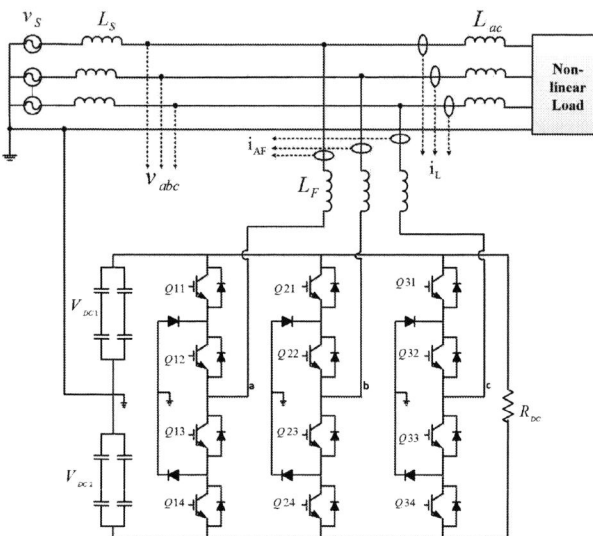

Fig. 1. Three level NPC inverter based SAPF

III. CONTROL SYSTEM

In this section, two parts of the control system are explored. The first part of control system is employed to control output currents of the inverter to compensate the harmonic components of the source currents. The second part of control system is for regulation of the dc-link voltage.

A. Selective harmonic extraction control method

In order to extract the harmonic components of the load currents, the load measured currents are transformed into the $\alpha\beta$ reference frame as follows:

$$
\begin{bmatrix} i_{L\alpha} \\ i_{L\beta} \\ i_{Lo} \end{bmatrix} = \sqrt{\frac{2}{3}} \begin{bmatrix} 1 & -1/2 & -1/2 \\ 0 & \sqrt{3}/2 & -\sqrt{3}/2 \\ 1/\sqrt{2} & 1/\sqrt{2} & 1/\sqrt{2} \end{bmatrix} \begin{bmatrix} i_{La} \\ i_{Lb} \\ i_{Lc} \end{bmatrix} \tag{1}
$$

Where, i_{Labc} are the load currents as shown in fig.1. in this control system, harmonic component (h) is transformed to 'dq' coordinating system with $\theta_h = h\theta_s$ and by following equation:

$$
\begin{bmatrix} i_{Ld,h} \\ i_{Lq,h} \end{bmatrix} = \begin{bmatrix} \cos(h\theta_s) & \sin(h\theta_s) \\ -\sin(h\theta_s) & \cos(h\theta_s) \end{bmatrix} \begin{bmatrix} i_{L\alpha} \\ i_{L\beta} \end{bmatrix} \tag{2}
$$

Where, h is the harmonic order and θ_s is angular frequency of PCC voltage. θ_s is obtained from phase locked loop (PLL) as shown in fig.2.

Equation (2) is employed for the harmonic orders which is desired to compensate. Regarding to the fig.2 in order to extract the harmonic components of the currents the output of the transforms are passed through low pass filters (LPF). Then, the outputs of the LPFs are transformed to ABC coordinating system as $i_{h,ABC}$ as shown in fig.2.

The output of the dc-link voltage controller which will be explained in the next section, is subtracted from the extracted harmonics i_{Lh_abc}. The achieved result is compared with the SAPF actual currents and finally the inverter switching pulses are obtained by employing the hysteresis current control.

B. Sliding mode based dc-link voltage controller

Fig.3 shows the dc-link equivalent circuit. According to this figure, the state space equations are expressed as follows.

$$
C\frac{dV_1}{dt} + \frac{V_1 + V_2}{R} = I_1 \Rightarrow C\dot{x}_1 + \frac{x_1 + x_2}{R} = I_1
$$

$$
C\frac{dV_2}{dt} + \frac{V_1 + V_2}{R} = -I_2 \Rightarrow C\dot{x}_2 + \frac{x_1 + x_2}{R} = -I_2 \tag{3}
$$

$$
\Rightarrow \begin{bmatrix} \dot{x}_1 \\ \dot{x}_2 \end{bmatrix} = -\frac{1}{RC} \begin{bmatrix} 1 & 1 \\ 1 & 1 \end{bmatrix} \begin{bmatrix} x_1 \\ x_2 \end{bmatrix} + \frac{1}{C} \begin{bmatrix} 1 & 0 \\ 0 & -1 \end{bmatrix} \begin{bmatrix} I_1 \\ I_2 \end{bmatrix}
$$

Where, the capacitor voltages V_1 and V_2 is choose as the state variables x_1 and x_2. The errors between the state variables and their reference values, is defined as follows:

$$
e_1 = x_{ref1} - x_1 = V_1^* - V_1
$$

$$
e_2 = x_{ref2} - x_2 = V_2^* - V_2 \tag{4}
$$

Where, V_1^* and V_2^* are the dc-link reference voltages. State space equations in (3) with the reference values is written as:

$$
\begin{bmatrix} \dot{x}_{ref1} \\ \dot{x}_{ref2} \end{bmatrix} = -\frac{1}{RC} \begin{bmatrix} 1 & 1 \\ 1 & 1 \end{bmatrix} \begin{bmatrix} x_{ref1} \\ x_{ref2} \end{bmatrix} + \frac{1}{C} \begin{bmatrix} 1 & 0 \\ 0 & -1 \end{bmatrix} \begin{bmatrix} I_{ref1} \\ I_{ref2} \end{bmatrix} \tag{5}
$$

By subtracting (5) from (3), the following relations are obtained

$$
\begin{bmatrix} \dot{e}_1 \\ \dot{e}_2 \end{bmatrix} = -\frac{1}{RC} \begin{bmatrix} 1 & 1 \\ 1 & 1 \end{bmatrix} \begin{bmatrix} e_1 \\ e_2 \end{bmatrix} + \frac{1}{C} \begin{bmatrix} -1 & 0 \\ 0 & 1 \end{bmatrix} \begin{bmatrix} \Delta i_1 \\ \Delta i_2 \end{bmatrix} \tag{6}
$$

Where,

$$
\Delta i_1 = I_1 - I_{ref1} \ , \ \Delta i_2 = I_2 - I_{ref2} \tag{7}
$$

978-1-5090-0376-1/16 $31.00 © 2016 IEEE

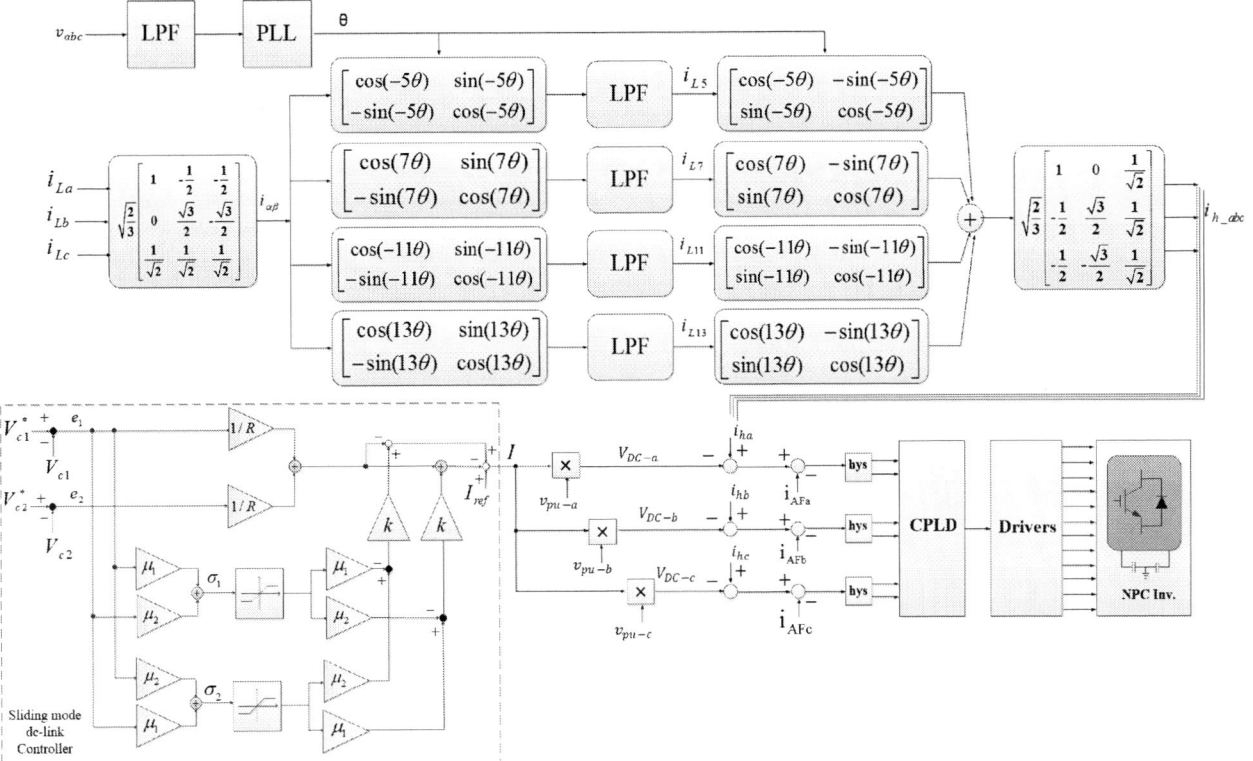

Fig. 2. Block diagram of the proposed control system

In order to obtain sliding mode control laws for (6), a sliding surface vector is considered as below:

$$\sigma = \begin{bmatrix} \sigma_1 \\ \sigma_2 \end{bmatrix} = \begin{bmatrix} \mu_1 & \mu_2 \\ \mu_2 & \mu_1 \end{bmatrix} \begin{bmatrix} e_1 \\ e_2 \end{bmatrix} , \quad \mu_1 > \mu_2 > 0 \qquad (8)$$

The μ_1 and μ_2 are the gains of the sliding mode controller. The Constraint $\mu_1 > \mu_2 > 0$ is necessary to achieve a positive matrix. The Lyapunov function for the sliding surface vector σ is defined as follows:

$$V(X) = \frac{1}{2}\sigma\sigma^T \qquad (9)$$

The system is globally stable if and only if the derivative of the Lyapunov function is negative :

$$\dot{V} < 0 \Rightarrow \sigma\dot{\sigma} < 0 \qquad (10)$$

The time derivative $\dot{\sigma}$ can be expressed as

$$\dot{\sigma} = \begin{bmatrix} \mu_1 & \mu_2 \\ \mu_2 & \mu_1 \end{bmatrix} \begin{bmatrix} \dot{e}_1 \\ \dot{e}_2 \end{bmatrix} \qquad (11)$$

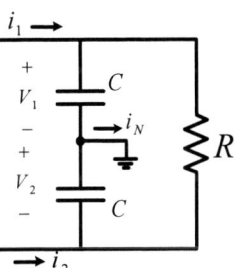

Fig. 3. DC-link equivalent circuit

Therefore, the Lyapunov function derivative becomes negative if the following switching law is satisfied:

$$\begin{bmatrix} \mu_1 & \mu_2 \\ \mu_2 & \mu_1 \end{bmatrix} \begin{bmatrix} \dot{e}_1 \\ \dot{e}_2 \end{bmatrix} = -\begin{bmatrix} sgn(\sigma_1) \\ sgn(\sigma_2) \end{bmatrix} \qquad (12)$$

Where, $sgn(\sigma)$ is a mathematical function that returns the values 1 for positive inputs and -1 for negative inputs. Equation (12) can be reformed as:

$$\mu(AX + BU) = -\begin{bmatrix} sgn(\sigma_1) \\ sgn(\sigma_2) \end{bmatrix} \qquad (13)$$

Therefore, the matrix U became:

$$U = -B^{-1}\mu^{-1}\begin{bmatrix} sgn(\sigma_1) \\ sgn(\sigma_2) \end{bmatrix} - B^{-1}AX \qquad (14)$$

Where:

$$\mu = \begin{bmatrix} \mu_1 & \mu_2 \\ \mu_2 & \mu_1 \end{bmatrix}, A = -\frac{1}{RC}\begin{bmatrix} 1 & 1 \\ 1 & 1 \end{bmatrix}, X = \begin{bmatrix} e_1 \\ e_2 \end{bmatrix},$$

$$\qquad\qquad\qquad (15)$$

$$B = +\frac{1}{C}\begin{bmatrix} -1 & 0 \\ 0 & 1 \end{bmatrix}, U = \begin{bmatrix} \Delta i_1 \\ \Delta i_2 \end{bmatrix}$$

Finally, the proposed controller relation is obtained as follows:

$$\begin{bmatrix} \Delta i_1 \\ \Delta i_2 \end{bmatrix} = k\begin{bmatrix} -\mu_1 & \mu_2 \\ -\mu_2 & \mu_1 \end{bmatrix}\begin{bmatrix} sgn(\sigma_1) \\ sgn(\sigma_2) \end{bmatrix} + \frac{1}{R}\begin{bmatrix} -1 & -1 \\ 1 & 1 \end{bmatrix}\begin{bmatrix} e_1 \\ e_2 \end{bmatrix} \qquad (16)$$

Where, k is calculated as below:

$$k = -\frac{C}{\mu_1^2 - \mu_2^2} \qquad (17)$$

Equation (16) has been implemented in the control system as shown in fig.3. Also this figure, shows the block diagram of the proposed selective harmonic extraction control method.

IV. EXPERIMENTAL RESULTS

The proposed selective harmonic compensation control method and the sliding mode dc-link voltage controller is implemented on a laboratory prototype of the SAPF that is shown in fig.4. It is implemented by a DSP-TMS320F2812.

The SAPF parameters is given in Table I.

TABLE I.   SYSTEM PARAMETERS

Grid frequency and voltage RMS(line to line)	220 V(RMS) , 50Hz
Grid impedance	$R_s = 0.1\Omega$, $L_s = 0.1mH$
dc-link voltage	300V for each capacitor V_{dc}=600 volts
dc-link capacitor and resistance	C=15000 μF, R=600 Ω
SAPF inductance	$L_F = 10$ mH
Load impedance	$R_L = 32\ \Omega$, $L_L = 176mH$

Fig.5 shows the grid currents without compensating in phases a, b and c respectively. As seen in this figure, the

source current before compensation is a distorted waveform. Fig.6 shows the grid currents after compensating using the SAPF with applying the proposed control system. As can be seen in this figure, after compensating the current waveform became almost sinusoidal.

The dc-link voltages at the start-up of the SAPF is shown in fig.7. This figure demonstrates that not only the voltages of the dc-link capacitors are fixed in their references in the steady state, at the starting transients the voltages not have any overshoot. Furthermore it can be noticed from fig.7 that the capacitors voltages are balanced regarding to the proposed control strategy.

Fig.8 shows the harmonic spectrum of the source currents before compensation from 0 to 2000 Hz. It can be seen from this figure, the source currents are distorted and containing all the harmonic orders (6k±1). Harmonic spectrum of the compensated source currents are shown in fig.9 from 0 to 2000 Hz. The selected harmonic orders in the control method, i.e. the 5, 7, 11 and 13th harmonics, in the fig.9 have lower magnitude than the figure 8.

Fig. 4.   Laberatory prototype of SAPF

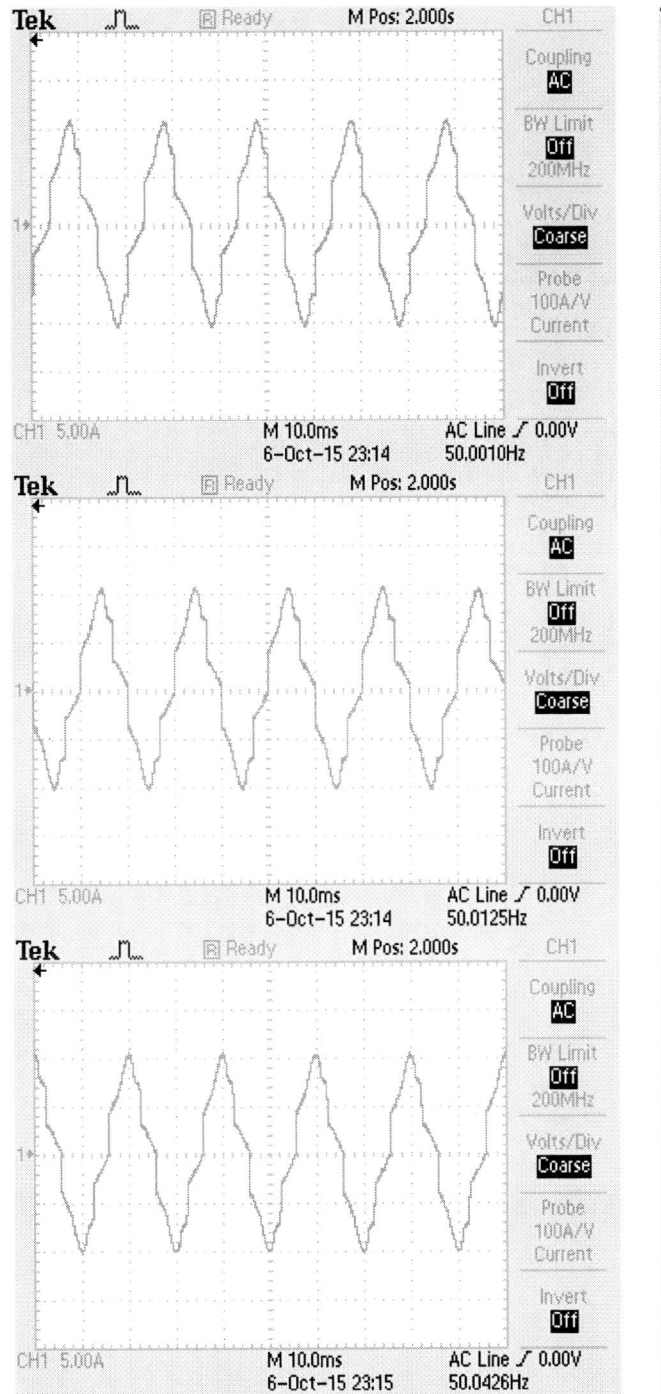

Fig. 5. Source currents waveform of phases "a", "b" and "c" before compensation

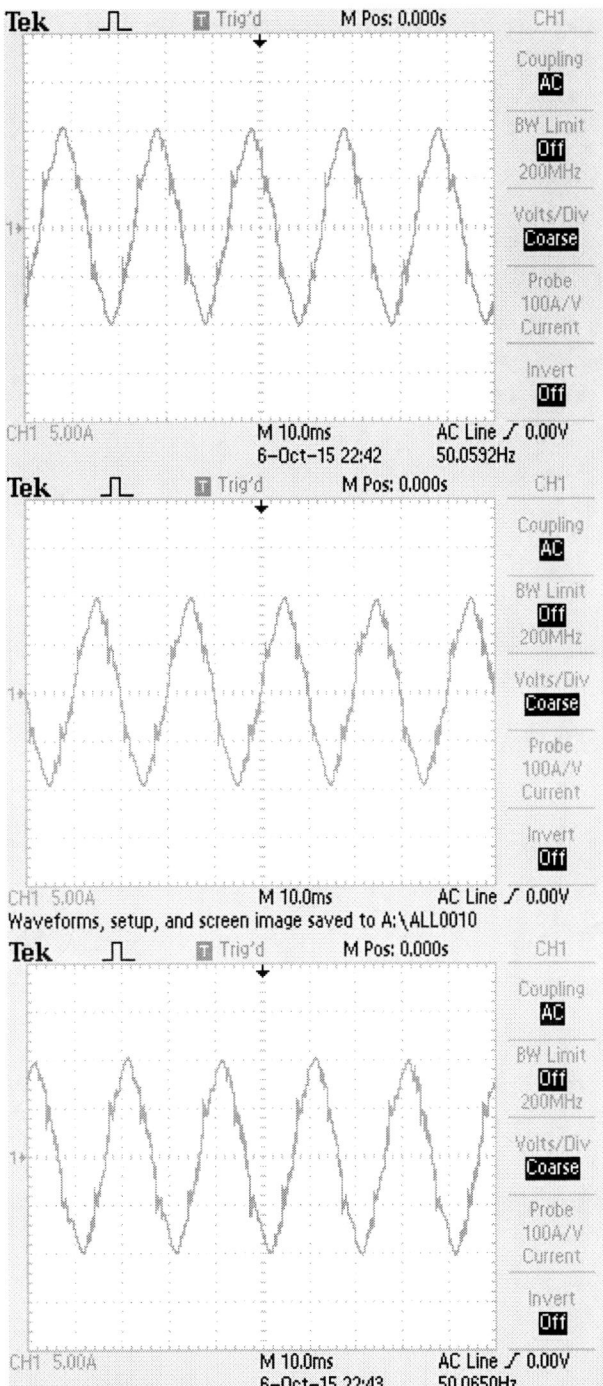

Fig. 6. Source currents waveform of phases "a", "b" and "c" before compensation

V. CONCLUSION

In this paper a sliding mode control method has been proposed to regulate the dc-link capacitor voltages of the NPC 4-wire SAPF. Furthermore a selective harmonic extraction method was employed in order to compensate desired harmonic components of the load currents. The laboratory prototype of the SAPF system has been developed to verify the effectiveness of the proposed system control. The experimental results confirm that the dc-link voltages is fixed in its reference value without making overshoot at the system start-up. Also according to the results, selected harmonic components of the load currents are cancelled in the source currents.

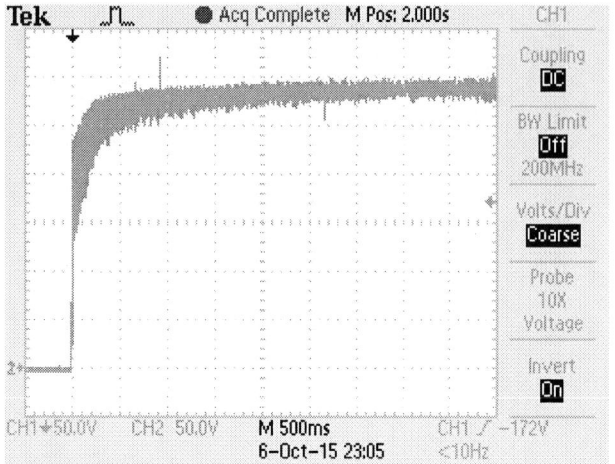

Fig. 7. DC-link capacitor voltages of the SAPF at the start-up

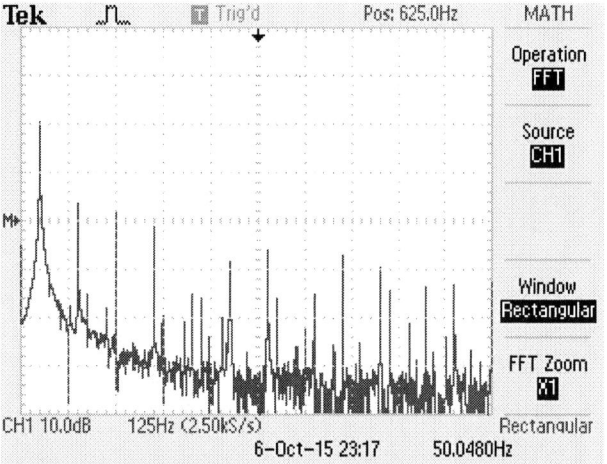

Fig. 8. harmonic spectrum of the source currents before compensation

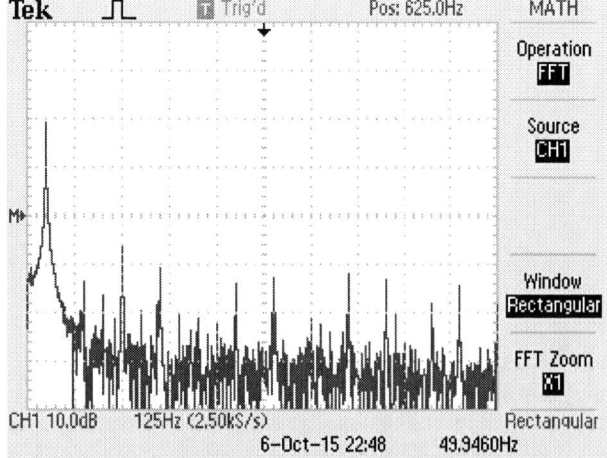

Fig. 9. harmonic spectrum of the source currents before compensation

REFERENCES

[1] H. Akagi, "Active Harmonic Filters. Proceedings of the IEEE vol 93, no. 12, December 2005.

[2] B. Singh, K. Al-Haddad, "A Review of Active Filters for Power Quality Improvement" IEEE Transaction on Industrial Electronics, vol. 46, no. 5, October. 1999.

[3] Singh, B., Verma, V., Chandra, A., & Al-Haddad, K. "Hybrid filters for power quality improvement". In *Generation, Transmission and Distribution, IEE Proceedings*. Vol. 152, No. 3. May 2005.

[4] Rodriguez, Jose, Jih-Sheng Lai, and Fang Zheng Peng. "Multilevel inverters: a survey of topologies, controls, and applications." *IEEE Transactions on Industrial Electronics,* vol.49 no.4, 2002.

[5] Sangeetha, B., and K. Geetha. "Performance of multilevel shunt active filter for smart grid applications." *International Journal of Electrical Power & Energy Systems* 63, 2014.

[6] M. Asadi, A. Jalilian ," Three-Level NPC Inverter Control System of Hybrid Active Power Filter by Modulation Ratios of Switching Functions ", 17th Electric Power Distribution Conference, EPDC 2012.

[7] O. Vodyakho, and Chris C. Mi. "Three-level inverter-based shunt active power filter in three-phase three-wire and four-wire systems." *IEEE Transactions on Power Electronics,* vol.24 no.5, 2009.

[8] C. Lascu, L. Asiminoaei, I. Boldea, and F. Blaabjerg, "High performance current controller for selective harmonic compensation in active power filters," IEEE Transaction Power Electronic., vol. 22, no. 5, Sep. 2007.

[9] S. Rahmani, A. Hamadi and K. Al-Haddad, "A Lyapunov-Function-Based Control for a Three-Phase Shunt Hybrid Active Filter" IEEE Transactions on Industrial Electronics, Vol. 59, No. 3, March 2012.

[10] M. Asadi, A. Jalilian , "Lyapunov control method of shunt active power filter." *Power Electronics, 5th Drive Systems and Technologies Conference (PEDSTC), 2014.*

[11] R. de Araujo Ribeiro, C. de Azevedo and R. de Sousa "A Robust Adaptive Control Strategy of Active Power Filters for Power-Factor Correction, Harmonic Compensation, and Balancing of Nonlinear Loads" IEEE Transactions on Power Electronics, Vol. 27, No. 2, February 2012.

[12] M. Asadi, A. Jalilian, "An Improved Current Control Method of Shunt Active Power Filter Based on State-Space Variables Under Asymmetrical and Non-Sinusiodal Conditions ", The 5 th International Power Engineering and Optimization Conference (PEOC02011), Shah Alam, Selangor, Malaysia : 6-7 June 2011.

[13] Mishra, M. K., & Karthikeyan, K. "A fast-acting DC-link voltage controller for three-phase DSTATCOM to compensate AC and DC loads" IEEE Transactions on Power Delivery, Vol. 24, No. 4, October 2009.

[14] Lam, Chi-Seng, et al. "Design and performance of an adaptive low-DC-voltage-controlled LC-hybrid active power filter with a neutral inductor in three-phase four-wire power systems." IEEE Transactions on Industrial Electronics, Vol. 61, No.6, June 2014.

[15] R. L. A. Ribeiro, T. O. A. Rocha, R. M. Sousa, E. C. dos Santos Jr, and A. M. N. Lima, "A Robust DC-Link Voltage Control Strategy to Enhance the Performance of Shunt Active Power Filters without Harmonic Detection Schemes " IEEE Transactions on Industrial Electronics, June 2014.

7th Power Electronics, Drive Systems & Technologies Conference (PEDSTC 2016)
16-18 Feb. 2016, Iran University of Science and Technology, Tehran, Iran

Sizing of Power Electronics EMC Filters Using Design by Optimization Methodology

JL.Schanen, A.Baraston, M.Delhommais
G2ELab
University Grenoble Alps
CS 90624, 38031 Grenoble CEDEX 1
jean-luc.schanen@g2elab.grenoble-inp.fr

P.Zanchetta
University of Nottingham, UK
D.Boroyevitch
CPES Virginia Tech, USA

Abstract—**This paper proposes a synthesis of EMC filter design method for power electronics converters. It starts with the description of the legacy approach, using the usual Common Mode / Differential Mode decomposition, and underlines the need of symmetry and the associated limits. Then an illustration of a design by optimization process is provided in the case of a simple switching cell. Finally, a full system composed of a PFC rectifier is provided, using EMC filters on both AC and DC sides. This example requires a design by optimization, since the two filters exhibit strong interactions.**

Keywords—Power Electronics, EMC, Design by Optimization

I. INTRODUCTION

Power Electronics can be seen as the major "enabling technology", which initiated the "More Electrical" trend of the XXI century. However, if the switching mode power conversion allows high efficiency and easy control of the electrical power, it also generates high frequency disturbances, especially due to harmonics of absorbed currents and voltage, and also to stray currents induced in the ground. The mitigation of these stray effects is mandatory in modern systems where all becomes "all electronic", with higher susceptibility to external noise. The "ElectroMagnetic Compatibility" vocabulary has been launched in the early 90's [1], and is today a mandatory step in converter design. EMC filters are a significant part of the weight and cost of a power converter, very often between 25% and 40% of the total converter [2]. The specificity of EMC filters of Power Electronics converter is that the design has to account for the high voltages and currents operated in the converter, therefore, design constraints are not only the attenuation of noise, but also limited losses and saturation for filter components. With the development of power electronics in embedded networks, as aircraft or automotive applications, the weight reduction becomes clearly a major challenge that has to be addressed. For instance, the first studies of More Electrical Aircraft have led to the conclusion that it was not so interesting in comparison with conventional hydraulic systems, due to the added weight of power electronics converters [3]. In this context, the EMC filters are obviously under concern, and have to be designed according to a minimum weight. Two parallel strategies can be defined: either technological integration, to reduce the filter weight by using new materials and technologies [4-5] or design by optimization, by trying to use at best given technologies and materials. This second

approach will be proposed in this paper. Part II will start with the conventional way to design filters, and remind the basics notions used in EMC filter design. Part III will introduce the design by optimization concept for EMC filters, applied to a simple switching cell. Finally Part IV will present a full converter including two filters, which have to be designed simultaneously due to a strong interaction. The major advantage of design by optimization will be underlined in this case.

II. CONVENTIONAL FILTER DESIGN

A. Common Mode Differential Mode Separation

EMC investigations and filter design widely uses the well-known Common Mode (CM) Differential Mode (DM) separation. The mathematical definition of this specific EMC basis is reminded hereafter, according to the notations of Fig.1.

$$\begin{bmatrix} V_{DM} \\ I_{DM} \\ V_{CM} \\ I_{CM} \end{bmatrix} = [P] \cdot \begin{bmatrix} V_1 \\ V_2 \\ I_1 \\ I_2 \end{bmatrix} \text{ with } [P] = \begin{bmatrix} 1 & -1 & 0 & 0 \\ 0 & 0 & \frac{1}{2} & \frac{-1}{2} \\ \frac{1}{2} & \frac{1}{2} & 0 & 0 \\ 0 & 0 & 1 & 1 \end{bmatrix} \quad (1)$$

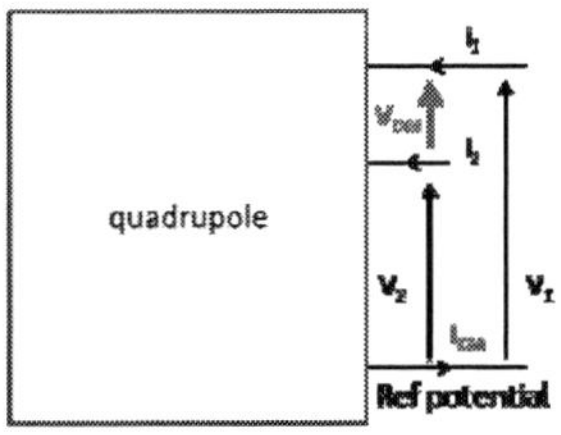

Fig. 1.   Conventional quadrupole notations and link with CM/DM basis

The main difference with conventional electrical dipole representation used by electrical engineers is to take into account the reference potential and the associated current I_{CM}. V_{DM} is simply the natural voltage used between wires, and if $I_{CM}=0$, I_{DM} becomes I_1. As illustrated in [6], the main interest of this CM/DM representation is that it allows solving the quadrupole equations using two uncoupled systems, corresponding to a dipole for CM and another dipole for DM,

978-1-5090-0376-1/16 $31.00 © 2016 IEEE

provided that the electrical circuit is symmetrical. This can be simply illustrated in Fig.2 and equation (2) & (3). The link between V_1, V_2 and I_1, I_2 is clearly coupled since it is a quadrupole, and the expression of the same relation in the DM/CM basis underlines that for a symmetrical circuit (i.e. $Z_1 = Z_2$), the two equations (3) in DM and CM become decoupled.

$$\begin{bmatrix} I_1 \\ I_2 \end{bmatrix} = \begin{bmatrix} \frac{1}{Z_{12}} + \frac{1}{Z_1} & -\frac{1}{Z_{12}} \\ -\frac{1}{Z_{12}} & \frac{1}{Z_{12}} + \frac{1}{Z_2} \end{bmatrix} \cdot \begin{bmatrix} V_1 \\ V_2 \end{bmatrix} \qquad (2)$$

$$\begin{bmatrix} I_{DM} \\ I_{CM} \end{bmatrix} = \begin{bmatrix} \frac{1}{4Z_1} + \frac{1}{4Z_2} + \frac{1}{2Z_{12}} & \frac{1}{2Z_1} - \frac{1}{2Z_2} \\ \frac{1}{2Z_1} - \frac{1}{2Z_2} & \frac{1}{Z_1} + \frac{1}{Z_2} \end{bmatrix} \cdot \begin{bmatrix} V_{DM} \\ V_{CM} \end{bmatrix} \quad (3)$$

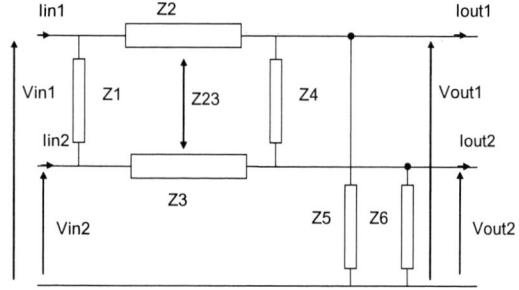

Fig. 2. Conventional quadrupole notations

In a more general case developed in [6], it is clearly shown that for a symmetrical link illustrated in Fig.3 (i.e. Z5 = Z6 and Z2 = Z3), the relations between input and output becomes decoupled if expressed in the CM/DM basis.

Fig. 3. Electrical link representation (from [6]). The relation between input and output becomes decoupled if symmetrical and expressed in CM/DM basis

Therefore, the CM/DM representation allows simplifying the study of the ElectroMagnetic Interferences (EMI), by solving only dipoles instead of quadrupoles, but it is crucial to keep **symmetry** of the electrical circuit.

B. Filter topologies

EMC filters for power electronics usually use simple second order filters. The topology of these filters depends on both converter and line impedances [7].

For DM filter, the topology supposes a low impedance on the converter side and a high one on the line side. Therefore, the topology is the one of Fig.4a. For CM filter, the converter impedance is supposed to by high, and the line one can be quite low, leading to the topology of Fig.4b.

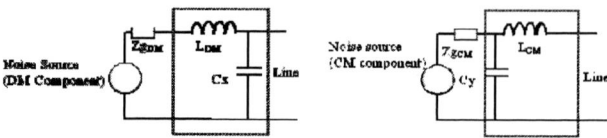

Fig. 4. a) typical DM filter   b) typical CM filter

The filter attenuation is defined by Eq. (4)

$$AdB = 20.Log\left(\frac{V_{line\_without\_Filter}}{V_{line\_with\_Filter}} \right) \qquad (4)$$

With some assumption on converter impedances in DM and CM ($Z_{gDM} = 0$, $Z_{gCM} \gg Z_{line}$), both attenuations for DM and CM filters are given by Eq. (5) and Eq.(6) (Z_N being the Line impedance, represented by a Line Impedance Stabilization Network)

$$A_{DM} = 20.Log\left(1 + \frac{L_{DM}}{2.Z_N} \cdot s + L_{DM} \cdot Cx \cdot s^2 \right) \qquad (5)$$

$$A_{CM} = 20.Log\left(1 + \frac{Z_N}{2} \cdot Cy \cdot s + L_{CM} \cdot Cy \cdot s^2 \right) \qquad (6)$$

C. Design methodology

The legacy approach consists in measuring the EMI on the LISN, in the CM/DM basis, without any filter. These disturbances are compared to the required standard, in order to determine the desired attenuation on the whole frequency range. Then the cut-off frequency of the filters, both in DM and CM, are adjusted to provide the needed attenuation. This design process is illustrated in Fig. 5 in the case of CM. In this figure, a 2kW Buck converter has been considered (parameters given in Table I) and simulated using Pspice with a quite precise ElectroMagnetic environment (stray inductances, stray capacitances, Pspice models of components).

Once the cut off frequency has been found, the determination of the filters elements L and C depends on technological considerations. Very often, most of the filter weight and volume come from the magnetics [8], and therefore capacitances are increased. For Cx (DM), the limitation may be the bandwidth, due to esl (equivalent series inductor) which may be too large for big capacitances. For Cy (CM), safety issue arises due to the connection of these capacitors to metallic parts which are accessible to the users. In our example, Cx has been limited to 6.8µF, and Cy (between power line and chassis) to 28nF, according to EN

501178, Annex A.5.2.8.2 table A1 (DC Voltage = 400V). These quite high values of capacitance lead to small values of magnetics (what was the objective). The discussion in part D will underline some limitations of this choice.

Last, the implementation on the CM and DM filters on the converters power lines has to guarantee the symmetry rule, therefore, the final scheme is the one depicted in Fig.6. The values of the line elements account for the splitting of each filter among the two lines, as well as for the coupling of magnetics between the two lines. This magnetic coupling is interesting to reduce the size of CM filter, and also allows avoiding the influence of DM inductance on CM one and vice versa. At this point, all filter components are supposed to be ideal, including the coupling. Obviously this will not be the case in real life, but it will not affect the global design of the filter, since these stray effects only impact the high frequency behavior. Only the unit magnetic coupling is too optimistic, but it can easily be taken into account when realizing the components, by modifying the desired inductances, accounting for the leakage inductances.

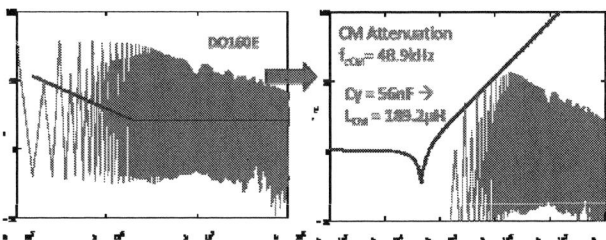

Fig. 5. Illustration of the desired attenuation and the choice of cut-off frequency in the case of CM for the Buck converter under study.

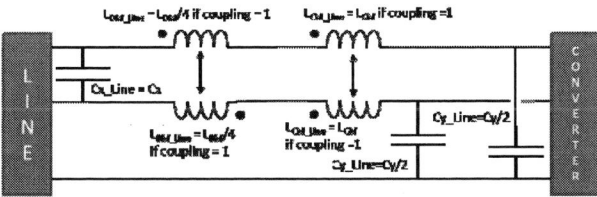

Fig. 6. Implementation of the DM and CM filter on the power lines, accounting for the necessary symmetry and the coupling of magnetics between lines.

D. Limitations of the Conventional Method

The legacy design method is based on the symmetrical realization of the filter. If, for instance the DM filter is not equally split among the two power lines, despite the good inductor and capacitor values for achieving the filter, the circulation of CM current generates different voltage drops on each part of the DM inductor, what creates additional DM noise. Moreover, the conventional design method supposes that the optimum corresponds to a maximization of the capacitors. In the considered example, this is not true. Indeed, the global weight of capacitors is evaluated to 50g when the filter mass is 54g ! There is thus another possible optimum to be found with increased inductances and lower capacitance.

Also, all attenuations have been obtained under the assumption that the converter impedances are either zero or very large in DM or CM. When trying to obtain better performances in terms of weight for EMC filter, all these assumptions will perhaps no more be valid, and design by optimization must be more generic. This will be detailed in the next section.

TABLE I. CONVERTER AND FILTER PARAMETERS

Power Converter			
Power	2kW		
DC Input Voltage	400V		
DC Output Current	10A		
Switching Frequency	100kHz		
Duty Cycle	0.5		
Input Capacitor	Cin= 400µF, esl = 10nH, esr=0.2Ω		
EMC Filter			
F_{cDM} = 37.7kHz F_{cCM} = 48.9kHz			
CM/DM basis		Power Line Values	
L_{DM}	2.6µH	L_{DM_Line}	650nH – Coupling -1
Cx	6.8µF	C_{x_Line}	6.8µF
L_{CM}	189.2µH	L_{CM_Line}	189.2µH Coupling 1
Cy	56nF	C_{y_Line}	28nF

III. DESIGN BY OPTIMIZATION IN A SIMPLE CASE

A. EMC Model of a Switching Cell

The time domain simulation can be used to investigate the EMI from a power converter, as shown in the previous part. However, simulation time, memory space, convergence issues are often associated with this kind of simulations, and this is definitely not adapted to optimization. Therefore, the model of the switching cell in the frequency domain is preferred [9]. The idea of this modeling method is to replace the switching devices by equivalent sources, reproducing the voltage and current discontinuities. For the considered Buck converter, the equivalent circuit is given in Fig.7: the current source corresponds to the switch current and the voltage source to its voltage. These sources can be expressed in the frequency domain, either by a FFT of the switch signals (simulated or measured), or by approximating these waveforms with simple slopes and performing the Laplace transform of these sources. Then, the converter is associated with the filter to be designed and the LISN, and the disturbances can immediately be evaluated using the formal resolution of the equivalent circuit of Fig.7 in the frequency domain.

B. Optimization Principle

The design methodology starts from technological description of the components : EMC filter inductors and capacitors are characterized by geometrical data and material properties, or interpolated from manufacturer datasheets. Then, electrical quantities, such as capacitance, inductance, but also maximum flux, maximum current density or rms current, … are obtained thanks to analytical formulas. Voltage and current ripple are also evaluated, as well as EMC behavior using the equivalent circuit of Fig.7-bottom. Other additional constraints such as the cooling system of the semiconductors can be added, but will not be developed in this paper. The full optimization process is illustrated in Fig. 8. All models are implemented in a specific optimization framework developed in our lab [10].

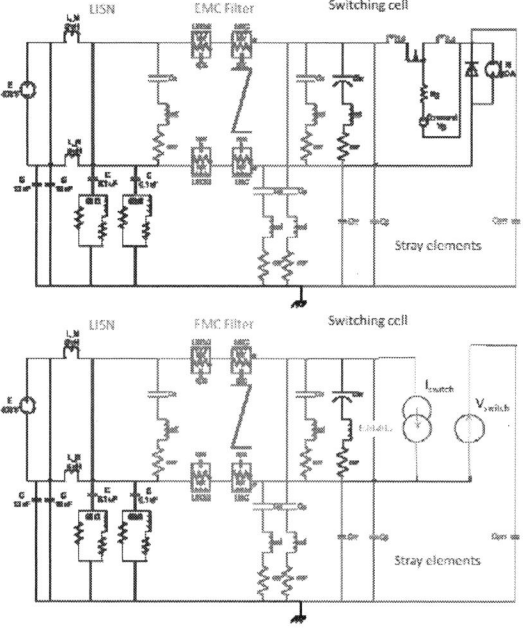

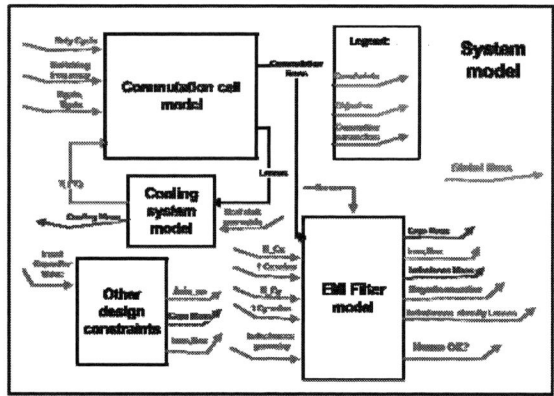

Fig. 7. Top: buck converter model - Bottom: Equivalent circuit for EMC studies.

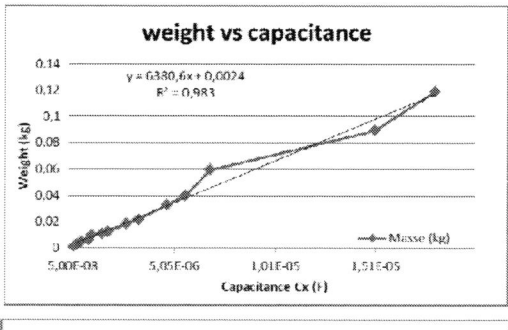

Fig. 8. Design methodology.

In reference to Fig.8, the cooling system has not been considered. The capacitor model is simply an interpolation of manufacturer datasheets, providing the weight of the capacitors as a function of the capacitance, for both technologies Cx and Cy (Fig. 9). Several capacitors can be associated in parallel if needed. As explained previously, the inductors are described using technological variables: a toroidal core is used, with homothetic variation (changing its

heights –one of the design parameters– changes all other size), the turn number is also a design variable, as well as the wire diameter The magnetic material is fixed. Permeability has been chosen at 60 for DM inductor, and 5000 for CM one. To avoid any saturation, the maximum flux in the core is a constraint during the optimization process. For the Common Mode inductor, the peak flux results both from peak CM current and non-perfect coupling between the two wirings (leakage flux). The peak Common Mode current is evaluated using the equivalent frequency of the MOSFET voltage slope [4], and the leakage effect is taken into account using [11].

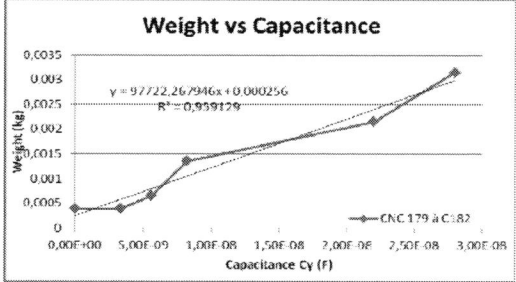

Fig. 9. Top: Interpolation for Weight vs Capacitance. Top: Cx – Bottom Cy

C. Results

The optimization of the proposed buck converter has been performed to respect all design criteria (maximum current density, saturation …), including the EMC standard DO160E. The main optimization results are given in Table II. As stated in the discussion of the previous section, the optimal design does not anymore result in increasing the capacitors, since it has been shown that the magnetics were very small in the previous design. It is worth noting that the cut off frequencies for DM and CM has not changed a lot in comparison with the legacy design. Indeed, the considered converter is symmetrical; therefore, the mode separation assumption (what is the basis for legacy approach) is valid. Also the non-perfect coupling of the CM inductor replaces the DM filter (the leakage inductor of the designed CM filter is evaluated to 38µH, which is sufficient for filtering the DM noise.

D. Discussion

The Design by Optimization performed in this simple application example shows that the optimum design does not correspond to the legacy approach. Of course, in this simple case, other approach may have provided more or less the same results, but it should be noticed that we kept the system simple in order to present the methodology. For instance the converter has been symmetrical, what avoids any mode coupling. Furthermore, the presented methodology only focused on EMC filter design, keeping all other parameters constant. The full methodology also applies by changing the switching frequency, the driver parameters, the input capacitor, … but it is out of the scope of this paper. Next section will illustrate how the method can also be applied to a more complex topology: a three phases PFC rectifier with both input and output filters.

TABLE II.   EMC FILTER LEGACY DESIGN AND OPTIMIZATION RESULTS

Legacy design		Optimization Results	
L_{DM_Line}	650nH – Coupling -1	L_{DM_Line}	38µH (CM leakage)
C_{x_Line}	6.8µF	C_{x_Line}	467nF
L_{CM_Line}	189.2µH Coupling 1	L_{CM_Line}	479µH Coupling 0.92 (computed)
C_{y_Line}	28nF	C_{y_Line}	14.6nF
Weight 55g		Weight: 13g	

IV. APPLICATION TO PFC RECTIFIER

A. Studied Converter

In comparison with the previous example, the power range is completely different. We keep the aircraft application, but we try to design a converter between the HV AC grid and a HV DC grid. According to the system optimization presented in [12], the AC network voltage is fixed to 350V-400Hz and the DC voltage to 990V. The power is 120kW. The switching frequency is fixed to 40kHz. The circuit is given in Fig. 10. It is worth noting that we decided to insert two different LISN, one at the input and one at the output. Indeed, even if it is not described in the DO160, we wanted to fix the EMC environment of this converter, and there is no specific load to plug on the DC side. Therefore, two LISN are the best way to provide a known environment for the EMC study of the converter. The boost inductors are fixed (51.2µH) according to current ripple, the output capacitor (13.6µF) according to voltage ripple. Some stray capacitances are added in order to be representative of CM noise generated by the converter. The floating points of the PFC legs (100pF) and the DC bus stray capacitance (300pF) are thus considered. The filter topology on the AC side and on the DC side are shown on Fig. 10.

B. Manual Filter design

Designing both input and output filters is not an easy task. Indeed, the C_{y_AC} capacitors on the AC side may originate an additional CM current, especially if the boost inductors are non-perfect (i.e. exhibit a stray parallel capacitance). Therefore, if increasing these C_{y_AC} reduces the disturbances measured on the AC LISN, it may increase the noise on the DC one. In the same idea, increasing the C_{y_DC} capacitors reduces the high frequency impedance on the DC side and thus may increase the CM current on the AC side [12-13]. Therefore, both DC and AC filters have coupled effects on both noise on DC and AC LISN, and therefore should be designed simultaneously. To perform the design, we fixed all capacitors to the maximum allowed values and used the trial and error method with a time simulation software, varying the CM and DM inductors (DC and AC) until we fulfilled the DO160F on both AC and DC sides. Table 3 shows the obtained results.

TABLE III.   EMC FILTER MANUAL DESIGN FOR PFC

AC side		DC side	
L_{DM_AC}	350µH	L_{DM_DC}	614µH
C_{x_AC}	1µF	C_{x_DC}	500nF
L_{CM_AC}	200mH	L_{CM_DC}	180mH
C_{y_AC}	8nF	C_{y_DC}	8nF
Total Weight: 98.88kg			

C. EMC equivalent circuit and optimization models

In order to use the Design by Optimization method, we have to provide models for all physical phenomena to be taken into account. We used exactly the same models as in the previous section (just adapting the boundaries for the core size, and the manufacturer datasheets for capacitor interpolation).

Regarding EMC model, we used the same approach as presented in section III, representing all current/voltage variations by equivalent sources. As a result, three voltage sources are replacing the three bottom switches of the three PFC legs, whereas one single current source allows reconstructing the DC current exciting the DC output capacitor and the DC circuit (including the DC filter, stray elements and DC LISN). The equivalent circuit is depicted in Fig. 12.

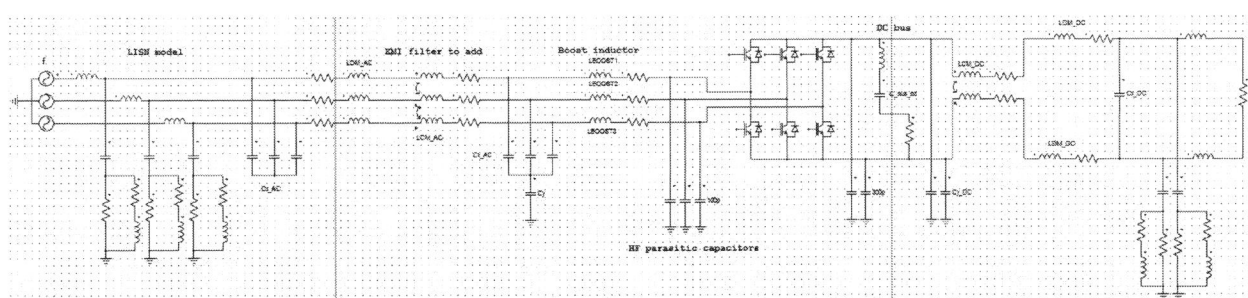

Fig. 10. Circuit description of the PFC with the two LISN and the two filters

D. Optimization Results

Results are provided in Table IV. To facilitate comparison, we kept the capacitors identical and just optimized the inductors. Obviously, a full optimization is feasible. The optimization result shows an improved global weight. The DM inductors are so small that they are part of the Common Mode inductors (using the leakage). As shown in Fig.11, the EMC constraints are respected, on both DC and AC side, and the EMC model used (CADES in Fig. 11) exhibit a good agreement with the time simulation software. With this result, the Design by Optimization method is validated on a complex case, and shows a quick convergence to more or less the same optimal point as the one obtained after a long and boring trial and error method. Furthermore, additional degrees of freedom can be considered, as filter capacitors, switching frequency… what would lead to more global results.

TABLE IV.   EMC FILTER MANUAL DESIGN FOR PFC

EMC Filter			
AC side		DC side	
L_{DM_AC}	486µH (leakage of CM)	L_{DM_DC}	540µH (leakage of CM)
C_{x_AC}	1µF	C_{x_DC}	500nF
L_{CM_AC}	236mH	L_{CM_DC}	170mH
C_{y_AC}	8nF	C_{y_DC}	8nF
Total Weight: 90.36kg			

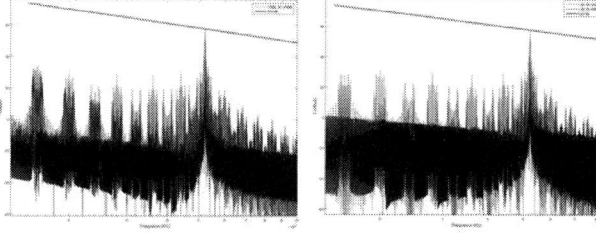

Fig. 11. AC and DC noise in comparison with the DO160F standard after optimal design of the filter

V. CONCLUSION

Design by Optimization method has been presented in the case of EMC filter design, and compared to legacy approach. This new way of design is promising, since it allows avoiding all usual assumptions, such as "capacitors are negligible compared to magnetics", or the usual CM/DM separation, which may be wrong for non-symmetrical topologies. It does also not consider any assumption for converter impedance. EMC models in the frequency domain are useful for this task, using other analytical models for taking into account other physical constraints associated with components design. All these models have been implemented in a specific framework, dedicated to optimization. This avoids many programming efforts and expertise in optimization, in order to conduct such a design method.

References

[1] D. N. Heirman, "A History of the Evolution of EMC Regulatory Bodies and Standards", 16th International Zurich Symposium on Electromagnetic Compatibility, February 2005.

[2] Boroyevich, Dushan; Zhang, Xuning; Bishinoi, Hemant; Burgos, Rolando; Mattavelli, Paolo; Wang, Fred, "Conducted EMI and Systems Integration", CIPS 2014 8th International Conference, Nürenberg, 2014

[3] EU project MOET: http://www.transport-research.info/project/more-open-electrical-technologies

[4] F.Mesmin, "Materiaux magnetiques et solutions innovantes de filtrage CEM pour applications aeronautiques", PHD dissertation, University Grenoble Alps, 28 sept 2012 (in French)

[5] Ali, M.; Labouré, E.; Costa, F.; Revol, B., "Design of a Hybrid Integrated EMC Filter for a DC–DC Power Converter", IEEE Transactions on Power Electronics, 2012, Volume: 27, Issue: 11

[6] De Oliveira Thomas, Mandray Sylvain, Guichon Jean-Michel, Jean-Luc Schanen, Adrian Perregaux "Reduction of conducted EMC using busbar stray elements" IEEE APEC'09, Feb. 2009, Washington DC, USA

[7] M. L. Nave, 3Power Line Filter Design for Switch-Mode Power Supplies", Van Nostrand Reinold, 1991, pp. 210

[8] Yoann Y. Maillet, "High-Density Discrete Passive EMI Filter Design for Dc-Fed Motor Drives", MsC dissertation Virginia Polytechnic Institute and State University, August 2008

[9] B. Revol, J. Roudet, J.L. Schanen, P.Loizelet, "EMI study of a three phase inverter-Fed Motor Drives", IEEE trans on IAS, Vol 47 n° 1 January/Feb 2011, pp 223 - 231

[10] http://www.vesta-system.fr/en/products/vestacades/

[11] M. J. Nave, "On modeling the common mode inductor," in Proc. IEEE Int. Symp. Electromagn. Compat., 1991, pp. 452–457

[12] B.Wen, X.Zhang, F.Effah, A.Baraston, P.Zanchetta, D.Boroyevich, JL.Schanen, R.Burgos, P.Wheeler, A.Tardy, "Integrated Design by Optimization of Electrical Power Systems for More Electric Aircraftl", MEA 2015, February 3-5 2015, Toulouse, France

[13] Jettanasen, C.; Costa, F.; Vollaire, C.; Revol, B.; Morel, F., "Measurements and Simulation of Common Mode Conducted Noise Emissions in Adjustable-Speed AC Drive Systems", EMC Zurich 2009

[14] Tallam, R.M.; Skibinski, G.L.; Shudarek, T.A.; Lukaszewski, R.A., "Integrated differential-mode and common-mode filter to mitigate the effects of long motor leads on AC drives", ECCE 2010 Pages: 838 - 845

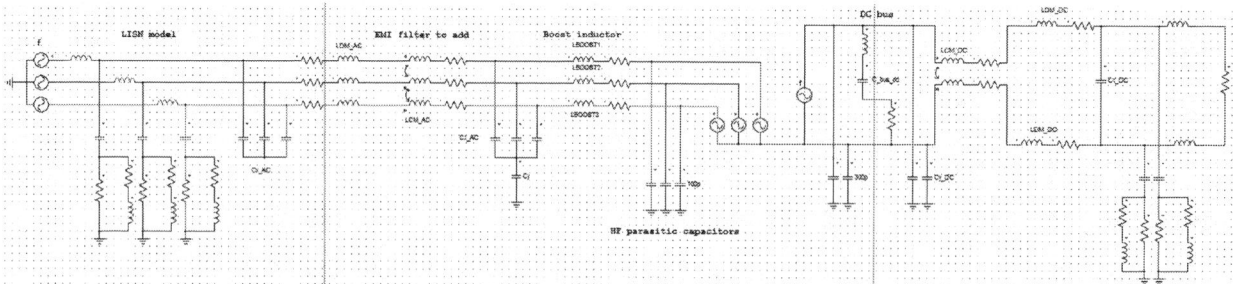

Fig. 12. Equivalent circuit of the PFC for EMC modelling

7th Power Electronics, Drive Systems & Technologies Conference (PEDSTC 2016)
16-18 Feb. 2016, Iran University of Science and Technology, Tehran, Iran

Improving Battery Performance in Hybrid Energy Storage System of PMSG Wind Turbine by Variable Filter Cut off Frequency

Mohammad Eydi, Javad Farhang, Behzad Asaei, Babak Farhangi

School of Electrical and Computer Engineering, Faculty of Engineering, University of Tehran, Tehran, Iran

Abstract—Nowadays, the control of hybrid battery and capacitor energy storage system is one of the most important problems in output power smoothing of wind turbines with Permanent Magnet generators. In this paper, a novel control method is proposed, which not only takes account the system limitations in terms of operation region, but also depending on the capacitor's capabilities to compensate the fluctuation in each region, distribute the power between the battery and capacitor. In the proposed method, if the capacitor's charge is in a condition that can compensate the low frequency fluctuation of output power, the cut-off frequency, is reduced to optimize the utilization of the capacitor and reduce the utilization of the battery. By this method, batteries current has a smoother waveform and lower dissipation, result in battery's improved performance and higher lifetime. The proposed method is simulated using MATLAB/SIMULINK environment. Simulation results prove the advantages of this method.

Keywords— Hybrid energy storage system; Permanent magnet synchronous generator; Power fluctuation; Power smoothing; Wind turbine.

I. INTRODUCTION

The rising temperature of the planet and environmental problems in one hand, and increasing price of fuels in other hand, caused a shift of a lot of attentions from fossil fuels to renewable energy sources. Renewable energy systems are pollution and cost free. Wind and Solar energy are two of the most important types of renewable energy. Typically, wind turbines have much higher power ratings comparing to solar systems, so that today we have operating wind turbines with power ratings of several megawatts. These turbines usually fall into two categories based on their generators: The turbines with Doubly-Fed Induction Generator (DFIG) which are simple, robust and have low maintenance and repair costs [1,2], and the turbines with Permanent Magnet Synchronous Generator (PMSM). Lower acoustic noise, full control of the system to operate under failure and extract maximum power [3], elimination of gear box [4], feasibility of active and reactive power control [5], elimination of DC excitation [6], simpler drive circuit, higher reliability, and higher efficiency [7] are among the advantages of PMSM wind turbines. The aforementioned features along with the grid requirements cause a growing trend to utilizing PMSM turbines alongside active rectifier [8].

The renewable energy sources produce oscillating power due to their random nature (wind velocity and solar irradiation). If the power injected by renewables to the grid is negligible, doesn't harm the stability of the grid, but if is significant, can cause oscillation in grid frequency and even instability of the grid.

There are some grid requirements to ensure the stability of the grid voltage and frequency, which deny power sources to increase or decrease their output power drastically. For instance, the maximum allowed slope changes for output power of wind turbines in Germany and Japan are 10 and 2 percent per minute, respectively [9,10]. If the input power increases with a higher slope than maximum allowed, the control system must limit the increase slope of output power to the maximum allowed and transfer the excessive power to a storage unit. Also, if the input power decreased with a higher slope than the maximum allowed, the storage unit must provide the extra power to limit the decreasing slope of the output power. The energy storage system should have fast dynamic response and high capacity to deal with rapid or long period power variations.

There are many researches on energy storage systems. In [11], flywheels are utilized with induction machine to improve the power quality. This is also achieved by utilizing static synchronous compensators and batteries as storage unit in [12]. In [13], only batteries are considered as energy storage system, and an additional capacitor bank is introduced in [14].

A capacitor storage unit with high power rating needs a high number of capacitors. Also, in a battery storage unit,

978-1-5090-0376-1/16 $31.00 © 2016 IEEE   285

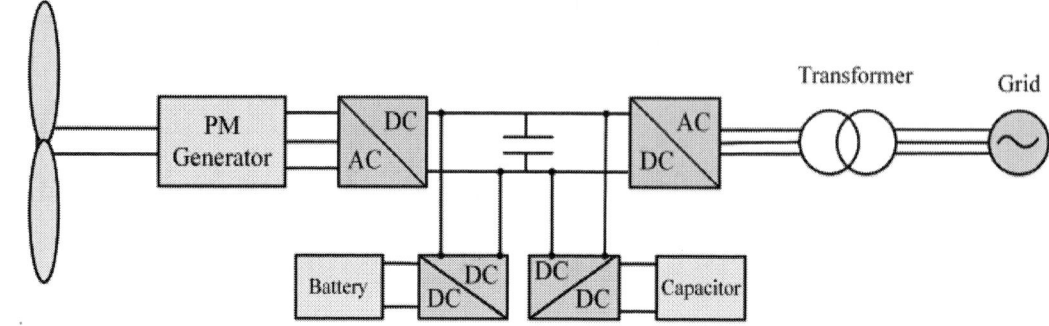

Fig.1. Overall structure of the system

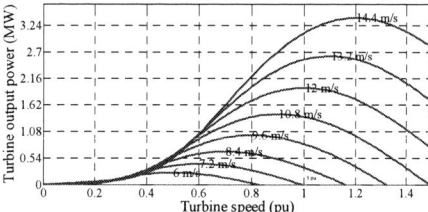

Fig.2. The turbine's output power against turbine speed for different wind velocities

fast dynamic is nearly impossible. So, to achieve improved response, the combination of batteries and capacitor is suitable to exploit the fast dynamic of capacitors and high energy density of batteries [15, 16].

II. MODEL OF SYSTEM

A. Turbine model

According to fig. 1, the wind energy is transferred through the turbine to the PMSG. The amount of this energy depends on wind velocity and turbine speed. The turbine's output power diagram vs. turbine speed for different wind velocities are illustrated in fig. 2. As can be seen from the figure, to extract the maximum output power, the turbine speed can be change using an active rectifier, so that the output power can be placed on its peek for any wind velocity. This is usually done when the peek output power is lower than the nominal power of the turbine. If the peak power is higher than the nominal power, the power produced by the turbine is controlled by controlling the blade angles. The relation between turbine's torque and power is described as [17]:

$$P_{\mathrm{w}} = \frac{1}{2} C_{\mathrm{p}}(\lambda, \beta) \rho \pi R^2 V_{\mathrm{w}}{}^3 \tag{1}$$

$$T_{\mathrm{w}} = \frac{1}{2} C_{\mathrm{p}}(\lambda, \beta) \rho \pi R^3 \frac{V_{\mathrm{w}}{}^2}{\lambda} \tag{2}$$

where V_{w} is the wind velocity, ρ is the air density, R is the blades' radius, λ is the ratio of the blades' tip velocity to the wind velocity, and c_{p} is the turbine's power factor which can be obtained from [18]:

$$C_{\mathrm{p}}(\lambda, \beta) = 0.73 \left(\frac{151}{\lambda_{\mathrm{i}}} - 0.85\theta - 0.002\theta^{2.14} - 13.2 \right) e^{\frac{-18.4}{\lambda_{\mathrm{i}}}} \tag{3}$$

And:

$$\lambda_{\mathrm{i}} = \frac{1}{\frac{1}{\lambda - 0.02\theta} - \frac{0.003}{\theta^3 + 1}} \tag{4}$$

B. AC/DC/AC converter

To extract the maximum power, the turbine must be set to a specific speed. Since the generator and turbine are coupled, turbine speed is a factor of generator speed. Thus, the turbine speed can be controlled by controlling the generator speed. So, the converter must adjust the generator speed to the value provided by the Maximum Power Point Tracker (MPPT). There are different methods to control the generator speed. One of the most important methods is using active rectifier and control in rotating frame. The equation of the system in rotating frame is described as:

$$V_{\mathrm{d}} = -R_s I_{\mathrm{d}} - L_{\mathrm{d}} \frac{dI_{\mathrm{d}}}{dt} + \omega_r L_q I_q \tag{5}$$

$$V_{\mathrm{q}} = -R_s I_{\mathrm{q}} - L_{\mathrm{q}} \frac{dI_{\mathrm{q}}}{dt} - \omega_r L_d I_d + \omega_r \Psi_f \tag{6}$$

where V_{d} and V_{q} are stator's d and q axis voltages, R_s is the stator resistor, L_{d} and L_{q} are stator's d and q axis inductances, I_{d} and I_{q} are stator's d and q axis current, and ω_r is the angular velocity of the rotor.

The output power and electromagnetic torque of the machine can be calculated from:

$$P = \frac{3}{2} \left(V_{\mathrm{d}} I_{\mathrm{d}} + V_{\mathrm{q}} I_{\mathrm{q}} \right) \tag{7}$$

$$T = -\frac{3}{2} P_n (\Psi_f I_q + (L_d - L_q) I_d I_q) \tag{8}$$

Where P_n is the number of pair pole's. Since the values of L_{d} and L_{q} are slightly different, the torque equation can be simplified as:

$$T = -\frac{3}{2} P_n \Psi_f I_{\mathrm{q}} \tag{9}$$

Eq. (9) shows that the output torque has a linear relation to the q-axis current[19].

978-1-5090-0376-1/16 $31.00 © 2016 IEEE

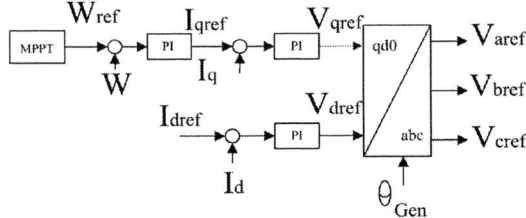

Fig.3. Control system of PMSG

Fig.3 illustrates the control system of PMSG. As can be seen from the figure, the generator speed is compared to speed reference provided by MPPT, and the resulting error is used to generate the q-axis current reference by means of a PI controller. The d-axis current reference is usually fixed to zero to reduce the power dissipation.

The second part of this converter, transfer power from DC bus to the grid. This converter is controlled so that the grid voltage is regulated and the frequency variation in weak grids are avoided. The converter controls the voltage by absorbing and injecting the reactive power from and to the grid and controls the frequency by changing the amount of active power injected to the grid. All of these controls are executed in the synchronous rotating frame. The voltage and injected current equations are:

$$V_{dinv} = RI_d + L\frac{dI_d}{dt} - L\omega_s I_q + V_d \tag{10}$$

$$V_{qinv} = RI_q + L\frac{dI_q}{dt} + L\omega_s I_d + V_q \tag{11}$$

Assuming the grid voltage is aligned with the d axis, the power transferred to the grid can be calculated from:

$$P = \frac{3}{2}V_d I_d \tag{10}$$

$$Q = \frac{3}{2}V_d I_q \tag{11}$$

So we can control the active and reactive power by controlling the d and q axis current that inject to the grid[20]. The control system of active and reactive powers is illustrated in Fig. 4.

C. DC/DC converters

When the input power is higher than the output power, the DC link voltage increased, and when the input power is lower than the output power, the DC link voltage is decreased. One

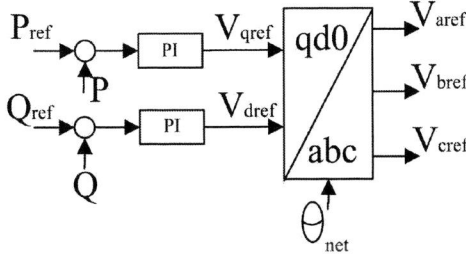

Fig.4. Control system of inverter's output power

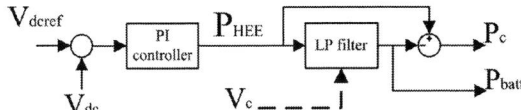

Fig.5. Power distribution algorithm

solution for this problem is using a hybrid energy storage system. This system contains, battery and capacitor energy sources, that each of them connected to the DC link via a DC/DC converter.

The DC link voltage is compared to the reference value. The error signal is controlled using a PI controller. The output of PI controller is the power that must be provided by energy storage system. This signal is filtered by means of a Low-Pass Filter (LPF) to form the reference power required from batteries . The remaining power must be provided by capacitors. So, the fast dynamic power is provided by capacitors and the static power is provided by batteries .Fig.5 illustrate this algorithm [15].

1) Battery: The selection of appropriate battery type depends on its characteristics. The selected battery must have a long lifecycle, high efficiency, low weight, low cost, and high power density [21]. Nowadays, Li-Ion batteries attract a lot of attentions due to their higher energy and power density, higher efficiency, lower self-discharge, and higher lifetime.

Batteries have operating temperature range. The battery's temperature increases with increased dissipation and can cause exceeding temperature range and even damage to the battery. Electrical equivalent model of the battery shows how the temperature increases. Fig.6 illustrates the electrical model a battery which comprises of a voltage source, a series resistor, and a resistor and capacitor mesh. This resistor and capacitor mesh describes the dynamic behavior of the battery [22]. The R_self is also used to describe the battery's discharge in open circuit [23].

The increasing temperature of the battery have a linear relation to the series resistor and RMS value of battery's current. If the RMS current is reduced, the power dissipation reduces and the battery's lifetime increases.

Another factor in battery's lifetime is the current waveform. Battery's lifetime increases with less amounts of peak and variations in current. For instance, constant and pulsed currents with same amount of DC value have different effects on the battery's lifetime. So, if we can manage to

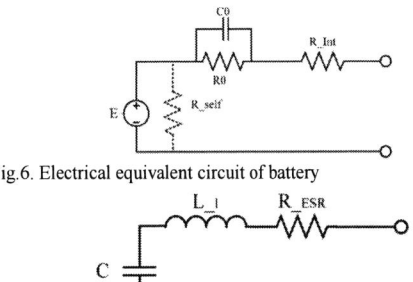

Fig.6. Electrical equivalent circuit of battery

Fig.7. Equivalent circuit of and ultracapacitor

978-1-5090-0376-1/16 $31.00 © 2016 IEEE   287

reduce the peek current, the lifetime is increased.

2)Ultracapacitors: Ultracapacitors have much higher power densities than batteries and absorb or inject large amount of power from or to the grid. Ultracapacitors usually have a capacity of thousands times higher than typical capacitors [24]. Like batteries, capacitors also have operating temperature range. For instance, ultracapacitors' operating range is usually from -40°C to 70°C. Fig.7 illustrate the equivalent circuit of an ultracapacitor.

Fig.7 shows that it is the series resistor that cause power dissipation in an ultracapacitor. This resistor ranges is from hundreds of uΩ to hundreds of mΩ. The value of series resistor changes with temperature, but it can be assumed constant over a long temperature range.

III. PROPOSED METHOD

As mentioned earlier, distribution of the power between battery and ultracapacitor is done by means of a low-pass filter. A relatively low cut-off frequency around 0.5 Hz is usually chosen to filter the power waveform. This filtered power is then provided by battery and the remaining power with higher frequency oscillation is provided by ultracapacitor. Note that in this method, condition of the system irrelevant to the power distribution process and this leads to inefficient exploitation of the energy storage unit. Therefore, in this paper a new method is proposed in which the cut-off frequency is changed based on the condition of ultracapacitor bank. If the ultracapacitor is in a condition that can compensate lower frequency fluctuation, cut-off frequency of the filter is reduced, otherwise, cut-off frequency is back to its constant value. This leads to better waveform of battery current and hence lower power dissipation. Depending on the ultracapacitors condition, we can define 3 operating regions.

- **High Capacity for Discharging (HCD)**: The capacitor has stored a lot of energy in this region and in case that it injects power to the dc bus, if needed, has the ability to supply the lower frequency fluctuations. However, if it absorbs power from the dc bus, its voltage will rise and may not be able to compensate the high frequency fluctuations, which should be supplied by the capacitor in normal mode

- **High Capacity for Charging (HCC)**: In this region, the ultracapacitor have a low amount of stored energy and is mostly discharged. This means the system can absorb low frequency power fluctuation, but if to inject this power fluctuation, the possible future fluctuation can't be dealt with. Therefore, in this region also the cut-off frequency must remain constant.

- **Medium Capacity for Charge and Discharge (MCCD)**: In this region, not only ultracapacitor can absorb low frequency fluctuations, but also can inject low frequency power fluctuation to the grid. Whatever we reach the center of this region, this capability is intensified. So, we have the lowest cut-off frequency in the center of this

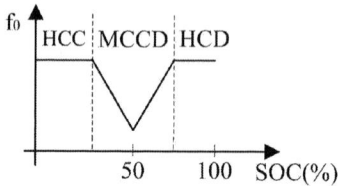

Fig.8. Cut-off frequency vs. SOC of ultracapacitor

region. From center to HCD and HCC, the cut-off frequency is increased toward its constant value. The cut-off frequency is changing continuously in this region.

The diagram of cut-off frequency for different ultracapacitor's State of Charges (SOCs) is illustrated in Fig. 8.

IV. SIMULATION RESULTS

For further analysis, the proposed method is simulated in MATLAB/Simulink environment. The general parameters are

Table I   Simulation parameters

Generator parameters	
Rated output power	2MW
Resistance	50uΩ
d axis Inductance	3.75mH
q axis Inductance	5.5mH
Number of Pair Pole	11
Field Flux	136.25 V.s/rad
Equivalent Inertia	10000 Kg.m^2
Other Parameters	
DC link voltage	8kV
Battery Capacity	50kVAh
HESS Capacitance	0.2F

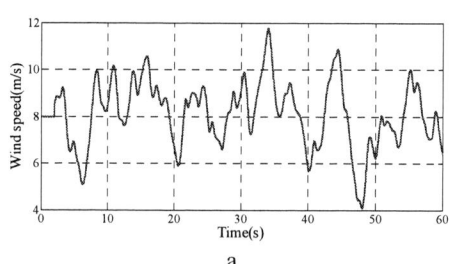

a

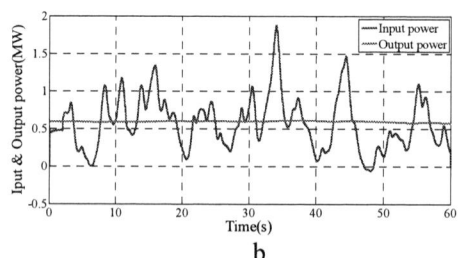

b

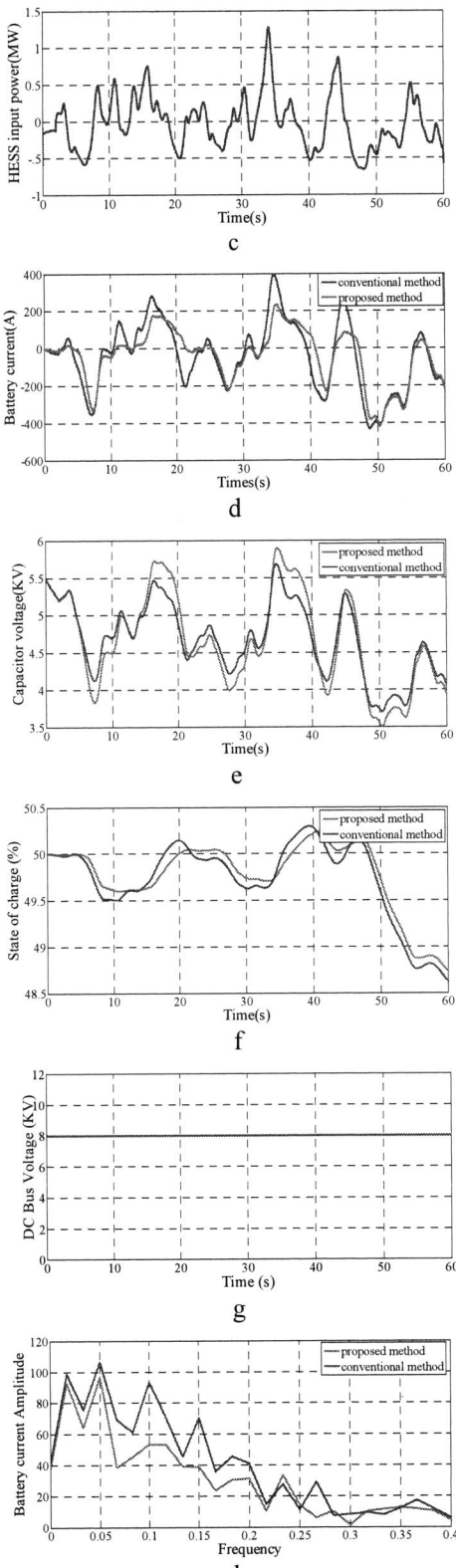

Fig.9. Simulation results. (a) wind speed,(b)Input & output power, (c)HESS input power,(d) battery current,(e) capacitor voltage,(f) battery SOC,(g)DC bus voltage,(h) battery current amplitude

depicted in Table (1)[17],[25].

In simulation, the Fig.9(a) is used as wind speed. Fig.9(b) illustrates the DC link input power (or obtained power from turbine) and the output power. From the fig.9(b) can be realized that the input power increases with the wind velocity increment. The Germany's standard is used to calculate the injected power to the grid [9]. According to this standard, the maximum allowed slope of changing power for wind systems is 10% per minute of the nominal power. Therefore, the output power (power injected to the grid) can be as Fig.9(b). According to fig.9(b), the injected power to the grid is nearly constant due to the limited changes slope. Fig.9(c) illustrates the injected power to the energy storage system. Fig.9(d) illustrates the injected current to the battery pack for both conventional and proposed methods. When the ultracapacitor voltage is in MCCD region (around 5000 volts), the peak current of the battery is lower in the proposed method than the conventional. But, for some periods of time such as the 28th second or between 50 and 60 s, the battery's currents are the same for both method, because the cut-off frequencies are constant and equal. The ultracapacitor bank voltage is illustrated in Fig.9(e). As can be seen, the ultracapacitor bank voltage changes in the proposed method is much higher than the conventional method. For instance, in t=17 s. From fig.9(e), this variation in voltage is due to passing peak current in the period from t=13.5 s to t=17s. Fig.9(f) illustrates the SOC of the ultracapacitor bank. The variations of SOC in the proposed method is lower than the conventional method. This means fewer charging and discharging of the battery. Fig.9(g) illustrates the DC link voltage in the proposed method. According to the figure, the proposed method can appropriately control the DC link voltage. Fig.9(h) illustrates the battery's current waveform versus frequency. As can be seen from the figure, the current in proposed method is lower than the conventional. The RMS value of current in the conventional method is 171.64 A, while the RMS value of current in the proposed method is 145.5 A, which means 28% reduction in power dissipation.

V. CONCLUSION

In this paper, a new control method is introduced for regulation of the output power of a PMSG wind turbine using a hybrid energy storage system. In this method, the cut-off of frequency of the LPF (used for distribution of power between battery and ultracapacitor) is change according to SOC of the ultracapacitor bank. Simulation results in MATLAB/Simulink environment is used for comparison between proposed and conventional method. Using the proposed method, battery's peak current is reduced. Also, the dynamic response of the system is become faster due to more utilization of the ultracapacitor. Simulation results show that for applied wind profile, the RMS value of battery current is reduced by 15%, thus, a Considerable reduction of 28% in power dissipation.

REFERENCES

[1]. X.Yuan,'' A Set of Multilevel Modular Medium-Voltage High Power Converters for 10-MW Wind Turbines'' IEEE TRANSACTIONS ON SUSTAINABLE ENERGY

[2]. G.Tian, Sh.Wang, G. Liu,''Design and Realization of STATCOM for Power Quality improvement of Wind Turbine with Squirrelcage Induction Generator'' IEEE 7th International Power Electronics and Motion Control Conference June 2012 pp1985-1989

[3]. G. Michalke, A. D. Hansen, and T. Hartkopf, "Control strategy of a variable speed wind turbine with multipole permanent magnet synchronous generator," presented at the 2007 Eur. Wind Energy Conf. Exhib., Milan,Italy, May 7–10, 2007.

[4]. Y. Chen, P. Pillary, and A. Khan, "PM wind generator topologies," *IEEE Trans. Ind. Appl.*, vol. 41, no. 6, pp. 1619–1626, Nov./Dec. 2005

[5]. G. Michalke, A. D. Hansen, and T. Hartkopf, "Control strategy of a variablespeed wind turbine with multipole permanent magnet synchronousgenerator," presented at the 2007 Eur.Wind Energy Conf. Exhib., Milan,Italy, May 7–10, 2007.

[6]. H. Polinder, S.W. H. de Haan,M. R. Dubois, and J. Slootweg, "Basic operation principles and electrical conversion systems of wind turbines," presented at the Nordic Workshop Power Ind. Electron., Trondheim, Norway, Jun. 14–16, 2004.

[7]. A. Grauers, "Efficiency of three wind energy generator systems," *IEEE Trans. Energy Convers.*, vol. 11, no. 3, pp. 650–657, Sep. 1996.

[8]. S. Jˉockel, "High energy production plus built in reliability—The new Vensys 70/77 gearless wind turbines in the 1.5 MW class," presented at the 2006 Eur.Wind Energy Conf., Athens, Greece, Feb. 27–Mar. 2, 2006.

[9]. Grid Code Regulations for High and Extra High Voltage, E.ON. NetzGmbh, Bayreuth, Germany, Report ENENARHS2006, Apr. 1, 2006.

[10]. Eltra specifications for connecting wind farms to the transmission network 2000, Eltra doc. no. 74174. [Online]. Available http://www.eltra.dk

[11]. R. Cardenas, R. Pena, G. Asher, and J. Clare, "Power smoothing in wind generation systems using a sensorless vector controlled induction machine driving a flywheel," *IEEE Transaction. Energy Convers.*, vol. 19, no. 1, pp. 206–216, Mar. 2004.

[12]. S. M. Muyeen, M. H. Ali, R. Takahashi, T. Murata, and J. Tamura, "Wind generator output power smoothing and terminal voltage regulation by using STATCOM/ESS," in *Proc. IEEE PowerTech. 2007 Conf.*, Lausanne,Switzerland, Jul., pp. 1232–1237, Paper 258.

[13]. M. EL Mokadem, C. Nichita, P. Reghem, and B. Dakyo, "Wind diesel system for DC bus coupling with battery and flywheel storage," presented at the XVII Int. Conf. Electr. Mach. (ICEM 2006), Chania, Greece, Paper ID 699.

[14]. S. M. Muyeen RionTakahashi, Toshiaki Murataand Junji Tamura "Integration of an Energy Capacitor System With a Variable-Speed Wind Generator" IEEE Transaction on energy conversion vol24,no.3,september2009,pp740-749.

[15]. W.Li,G. Joós, and Jean Bélanger,''Real-Time Simulation of a Wind Turbine Generator Coupled With a Battery Supercapacitor Energy Storage System''IEEE TRANSACTIONS ON INDUSTRIAL ELECTRONICS, VOL. 57, NO. 4, APRIL 2010.pp.1137-1145

[16]. Wei Li, G. Joos, "A power electronic interface for a battery supercapacitor hybrid energy storage system for wind applications,"
IEEE Power Electronics Specialists Conference, 2008.PESC2008, pp.1762-1768.

[17]. A. Uehara, A. Pratap, T.Goya,T.Senjyu, A.Yona,N.Urasaki,and T. Funabashi,''A Coordinated Control Method to Smooth Wind Power Fluctuations of a PMSG-Based WECS '' IEEE TRANSACTIONS ON ENERGY CONVERSION, VOL. 26, NO. 2, JUNE 2011,pp.550-558

[18]. J. G. Slootweg, S. W. H. de Haan, H. Polinder, and W. L. Kling,'' General Model for Representing Variable SpeedWind Turbines in Power System Dynamics Simulations '' IEEE TRANSACTIONS ON POWER SYSTEMS, VOL. 18, NO. 1, FEBRUARY 2003.pp .144-151

[19]. M.E. Haque, M. Negnevitsky, K.M. Muttaqi, , "A novel control strategy for a variable speed wind turbine with a permanent magnet synchronous generator," iEEE Trans. On Industry Applications, vol. 46, no. I, January/February 2010.

[20]. S.Li, T.A. Haskew, RP.Swatloski, and W.Gathings, ''Optimal and Direct-Current Vector Control of Direct-Driven PMSG Wind Turbines '',IEEE TRANSACTIONS ON POWER ELECTRONICS, VOL. 27, NO. 5, MAY 2012.pp.2325-2337

[21]. M.ehsani,Y.geo, sebastien.e.gay and A.emadi,''Modern Electric, Hybrid Electric and Full Cell Vehicles''

[22]. X. Hu, S. Li, and H. Peng, "A comparative study of equivalent circuit models for Li-ion batteries," *Journal of Power Sources*, vol. 198, pp. 359-367, Jan 15 2012.

[23]. E. Manla, A. Nasiri, C. H. Rentel, and M. Hughes, "Modeling of zinc/bromide energy storage for vehicular applications," *IEEE Trans. Ind.Electron.*, vol. 57, no. 2, pp. 624–632, Feb. 2010.

[24]. Xie Hailian, Lennart Angquist, Hans-Peter Nee, " Design study of a converter interface interconnecting energy stotage with DC link Statcom" iEEE Trans. On Power Delivery vol. 26, no.4,pp.2676-2686, Oct. 2011.

[25]. A.Esmaili, B.Novakovic, A.Nasiri, and O. Abdel-Baqi,'' A Hybrid System of Li-Ion Capacitors and Flow Battery for Dynamic Wind Energy Support '',IEEE TRANSACTIONS ON INDUSTRY APPLICATIONS, VOL. 49, NO. 4, JULY/AUGUST 2013.pp.1649-1657

Modified Local Voltage Controller Design of Inverter-Based DGs in a Microgrid

Hadi Hosseini Kordkheili

School of Electrical Engineering, University of Shahrood,
Shahrood, Iran
Hadi.h.k@ieee.org

Mahdi Banejad

School of Electrical Engineering, University of Shahrood,
Shahrood, Iran
m.banejad@shahroodut.ac.ir

Abstract—This paper concentrates on a multi-loop hierarchical controllers for a microgrid system in order to design a modified local voltage controller with a proper power sharing performance. The proposed controller which contains inner current and voltage loops, is based on dynamic model of microgrid system. This controller regulates the switching functions of interfaced-inverters and the reference values of DG currents. Also, a new droop controller is introduced to enhance the performance of the designed voltage control loop based on the measured fundamental current components of microgrid. In addition to improving the dynamic operation of the proposed controller (by the considering nonlinear loads), the extraction and analysis of harmonic components of PCC voltages are accomplished. The performance of the proposed control technique has been verified in simulation using MATLAB/SIMULINK environment. The designed controller provides a good transient performance. This suitable performance is maintained specially during non-linear load changes.

Keywords— Microgrid; Primary controller; Inner control loops; Droop Control; Hierarchical Control.

I. INTRODUCTION

Electric distribution networks are in a transition era in which their structure and issues are changing dramatically. The main characteristic of all modern distribution networks is the existence of renewable energy resources (RES) which creates new concepts, network architectures and challenges, along with their wonderful and undeniable advantages.

One of the most important concepts, suggested based on RES, is Microgrid (MG). MGs are small-scale, medium or low voltage distribution networks, which are used to provide power for electrical loads of small communities. A MG is an active distribution network, as it is a combination of renewable distributed generation (DG) and different loads in the distribution voltage levels [1].

MGs contain DGs which require interface power electronics converters (dc/ac or ac/dc/ac) which their control is the main challenge of MGs operation. The converter should provide voltage and current during any changes and disturbances e.g. load changes. At the same time, it should also be able to control voltage magnitude and frequency as well as proper power sharing between DG units.

Different approaches have been investigated in the recent years to properly control the voltage of MGs. The concept of hierarchical control has been adopted as an important and effective approach. In this approach, four control levels are generally considered, each has its own control responsibility as well as a supervision task on lower levels by providing set points for them. The hierarchical levels have been introduced as inner control loops, primary control, secondary control and tertiary control [2]. Inner control loops and primary control are local controllers of each DG unit. The main objective of inner control loops is the regulation of output voltages and currents. Besides, the primary control level is responsible for proper power sharing. Droop control approach has been usually applied to fulfill the latter. In [3, 4], the design of power sharing control loop has been investigated using active and reactive power droop methods. A nested controller has been used with an inner voltage loop, a virtual impedance loop and a droop based power sharing outer loop. In [5, 6], the concept of generalized droop control (GDC) in inverter-based DGs has been developed to decouple the active and reactive power control process.

In [6, 7], a multi-function control strategy has been presented for stable operation of DG units based on a direct Lyapunov control (DLC) theory. A Multi-objective control technique for integration of distributed generation with and without using a phase locked loop (PLL) has also been reported in [8, 9]. In [11], a passivity based control strategy has been developed to provide compensation of active and reactive powers and harmonic current components of the loads as well as injection of the maximum available power from DG units to the grid.

In this paper, the voltage of inverter-based DGs in a MG system is controlled by multi-loop controllers. Based on a dynamic model of MG system, the inner current and voltage loops are included in the proposed controller. In order to enhance the performance of the designed voltage control loop, a new droop controller is presented using low pass filters (LPF) and DG unit currents. Also, the extraction and analysis of harmonic components of PCC voltages are performed to improve the dynamic operation of the proposed controller in operating with nonlinear loads and their intermittency. The paper is organized as follows: Firstly, the dynamic model of the proposed MG system is achieved. Then, the inner currents and voltage control loops are designed. The new droop control loop is presented in the next section. The harmonic component extraction and analysis of PCC voltages are performed as the last part of design process of the controller. Finally, simulation, analysis of the results and conclusion are given.

II. MG STRUCTURE AND MODEL

The MG is usually a weak distribution network. During the utility grid intentional or non-intentional outages, the disconnection from the main network is occurred and an islanded low voltage MG begins to operate. In this circumstance, the MG includes a number of loads supplied by specific DGs such as photovoltaic, wind, and so on. Normally in this case, there is no major generating unit or a dominant power supply in this network to compensate voltages and frequency deviations. Each DG unit has a limited capability to supply its local loads as well as the MG. Therefore, a control system should be used to compensate the deviations which is going to be discussed in the following subsections. It is also assumed that the capacities of DG units are larger than load demands.

A. Proposed MG Structure

Fig. 1 shows the schematic configuration of the proposed MG structure. Two locally load connected DGs are in the MG with some common load. The MG is disconnected from utility grid at PCC.

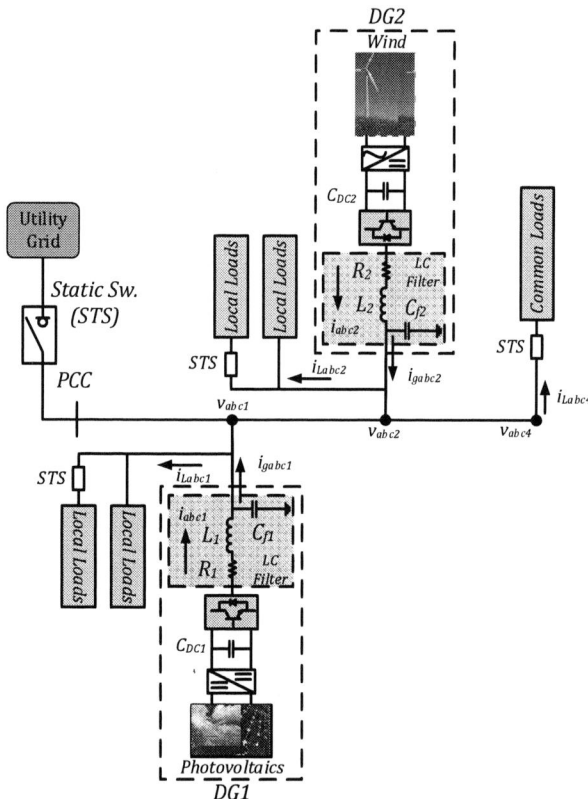

Fig. 1.   Schematic configuration of the proposed MG structure

In this figure, the interfacing system between RES and MG includes a dc-link, inverter, an equivalent LC filter, local loads and the proposed primary control system. Voltage and current signs as well as notations of parameters are indicated in the diagram. The LC filter represents the equivalent resistance, inductance and capacitance of the ac filter, coupling transformer, and connection cables.

B. Dynamic Model of the System

In order to design an appropriate control plan, the dynamic model of the proposed MG system have to be obtained. By considering Fig.1 and the single-phase equivalent circuit of each interface filter in Fig.2, the dynamic model of the proposed MG system in *a-b-c* frame can be expressed as follows:

$$L_i \frac{di_{ki}}{dt} + R_i i_{ki} + u_{eq_{ki}} v_{DC_i} + v_{ki} = 0$$

$$C_{fi} \frac{dv_{ki}}{dt} = i_{fki} \qquad (1)$$

$$C_{DCi} \frac{dv_{DCi}}{dt} + \left(u_{eq_{ai}} i_{ai} + u_{eq_{bi}} i_{bi} + u_{eq_{ci}} i_{ci} \right) - i_{DCi} = 0$$

where the indices "*i*" and "*k*" are employed for the DG unit numbers and *a-b-c* frame representation, respectively.

The ability of sine reference tracking is greatly depended on the closed loop bandwidth which may affect the speed of controller as well as undesired significant errors. One of widely used and accepted solutions is to design a controller in *d-q* rotating frame in which all variables are DC quantities and the design of controllers is much easier. Using *a-b-c/d-q* transformation, the following dynamic model can be written in *d-q* rotating frame:

$$L_i \frac{di_{di}}{dt} + R_i i_{di} - \omega L_i i_{qi} + v_{di} + u_{eq_{di}} v_{DC_i} = 0 \qquad (2)$$

$$L_i \frac{di_{qi}}{dt} + R_i i_{qi} + \omega L_i i_{di} + v_{qi} + u_{eq_{qi}} v_{DC_i} = 0 \qquad (3)$$

$$C_{fi} \frac{dv_{di}}{dt} - \omega C_{fi} v_{qi} - i_{fdi} = 0 \qquad (4)$$

$$C_{fi} \frac{dv_{qi}}{dt} + \omega C_{fi} v_{di} - i_{fqi} = 0 \qquad (5)$$

$$C_{DC_i} \frac{dv_{DC_i}}{dt} - u_{eq_{di}} i_{di} - u_{eq_{qi}} i_{qi} - i_{DC_i} = 0 . \qquad (6)$$

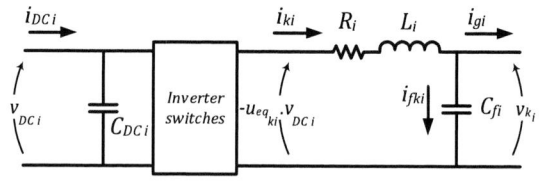

Fig. 2.   Single phase equivalent circuit of each DG's interface filter

where i_{abci} and i_{fabci} are output currents of AC filter which are transformed to i_{dqi} and i_{fdqi}, respectively, and u_{eqdqi} are the equivalent switching state functions in *d-q* rotating frame, v_{DCi} is the voltage across the dc-link capacitor (C_{DCi}) and ω is the angular frequency in *Rad/s*.

III. CONTROL TECHNIQUE

It is clear that the MG should operate appropriately to provide the required output voltage magnitude and frequency as well as proper active and reactive power sharing. These can be achieved by designing suitable control loops proposed in the following sections. Our main control objectives are

978-1-5090-0376-1/16 $31.00 © 2016 IEEE

accurate active and reactive power sharing, firm reference tracking, and accurate magnitude and frequency regulation of DG unit output voltages.

A. Proposed Inner Control Loops

In the inverter, each output current should track a reference current which is achieved by a proper compensator. The state-space model associated with current in the system can be derived from (2) and (3) as follows:

$$\begin{bmatrix} \dot{i}_{di} \\ \dot{i}_{qi} \end{bmatrix} = \frac{1}{L_i} \begin{bmatrix} -R_i & \omega L_i \\ -\omega L_i & -R_i \end{bmatrix} \begin{bmatrix} i_{di} \\ i_{qi} \end{bmatrix} - \frac{v_{DCi}}{L_i} \begin{bmatrix} u_{eq_{di}} \\ u_{eq_{qi}} \end{bmatrix} - \frac{1}{L_i} \begin{bmatrix} v_{di} \\ v_{qi} \end{bmatrix}. \tag{7}$$

These two equations show two current control loops which are not completely decoupled from each other. Thus, the coupled terms should be eliminated by some decoupling methods. Also, when the system is starting from a zero state condition, a significant undershoot may occur due to non-zero amount of v_{abc}. To overcome these problems, some additional compensation is adopted. The equivalent switching functions can be written from (2) and (3) as:

$$u_{eq_{di}} = -v_{DCi}^{-1} \left(L_i \frac{di_{di}}{dt} + R_i i_{di} - \omega L_i i_{qi} + v_{di} \right) \tag{8}$$

$$u_{eq_{qi}} = -v_{DCi}^{-1} \left(L_i \frac{di_{qi}}{dt} + R_i i_{qi} + \omega L_i i_{di} + v_{qi} \right). \tag{9}$$

To compensate the coupling terms as well as the grid possible voltage startup transient, (8) and (9) can be modified using (10) and (11). This also eliminates the undesired terms:

$$u_{eq_{di}} = -v_{DCi}^{-1} \left(m_{di} - \omega L_i i_{qi} + v_{di} \right) \tag{10}$$

$$u_{eq_{qi}} = -v_{DCi}^{-1} \left(m_{qi} + \omega L_i i_{di} + v_{qi} \right). \tag{11}$$

The terms m_d and m_q are control inputs that operate in two independent control loops to perfectly control the output currents of DG units in d-q frame. These can be obtained by two various PI controllers (G_{dqi}) with k_{Pdqi} and k_{Idqi} coefficients as,

$$m_{di} = k_{Pdi}.e_{di} + k_{Idi}.\int (e_{di})dt \tag{12}$$

$$m_{qi} = k_{Pqi}.e_{qi} + k_{Iqi}.\int (e_{qi})dt \tag{13}$$

where ($e_{dqi} = i_{dqi\_ref} - i_{dqi}$) are the errors between the measured and reference currents of DG units. Fig. 3 shows the above mentioned inner current control loop for each DG unit.

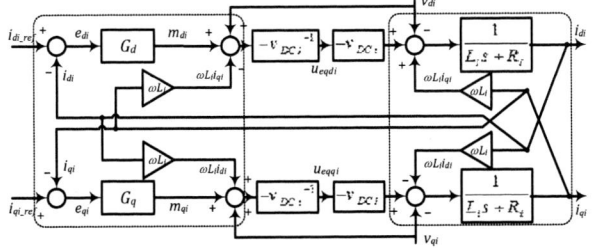

Fig. 3.   Current control loops model for each DG

B. Proposed Voltage Control Loops

The above mentioned current control loops are so important in active and reactive power injection of each DG unit by changing the current reference commands. In order to generate proper current references, equations (4) and (5) are used which lead to the following equations:

$$\begin{bmatrix} \dot{v}_{di} \\ \dot{v}_{qi} \end{bmatrix} = \frac{1}{C_{fi}} \begin{bmatrix} 0 & \omega C_{fi} \\ -\omega C_{fi} & 0 \end{bmatrix} \begin{bmatrix} v_{di} \\ v_{qi} \end{bmatrix} + \frac{1}{C_{fi}} \begin{bmatrix} i_{fdi} \\ i_{fqi} \end{bmatrix}. \tag{14}$$

Again, (14) shows two coupled voltage control loops. They can be decoupled by some feed-forward compensations, similar to the current loops. The currents i_{dqi} can be calculated as follows:

$$i_{di} = i_{fdi} + i_{gdi} = C_{fi} \frac{dv_{di}}{dt} - \omega C_{fi} v_{qi} + i_{gdi} \tag{15}$$

$$i_{qi} = i_{fqi} + i_{gqi} = C_{fi} \frac{dv_{qi}}{dt} + \omega C_{fi} v_{di} + i_{gqi}. \tag{16}$$

The d-q decoupled loops with PI regulators can be achieved by introducing n_d and n_q control inputs for voltage control:

$$i_{di\_ref} = n_{di} - \omega C_{fi} v_{qi} + i_{gdi} \tag{17}$$

$$i_{qi\_ref} = n_{qi} + \omega C_{fi} v_{di} + i_{gqi} \tag{18}$$

As mentioned in the current loop, the terms n_{dqi} is obtained from a PI controller (G_{vdqi}) with k_{Pvdqi} and k_{Ivdqi} coefficients:

$$n_{di} = k_{Pvdi} e_{vdi} + k_{Ivdi}.\int (e_{vdi})dt \tag{19}$$

$$n_{qi} = k_{Pvqi} e_{vqi} + k_{Ivqi}.\int (e_{vqi})dt \tag{20}$$

where ($e_{vdqi} = v_{dqi\_ref} - v_{dqi}$) is the error between the measured and reference voltages. The voltage loop is illustrated in Fig. 4 along with the block of inner current loop.

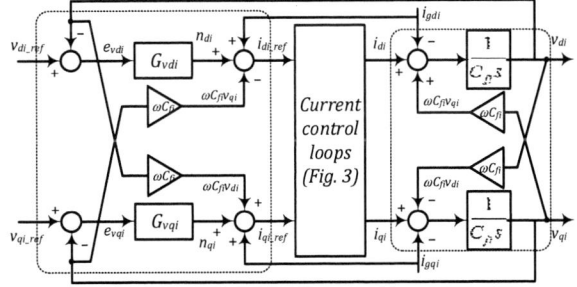

Fig. 4.   Voltage control loops model for each DG

The above mentioned control loops are generally called inner control loops which is the heart of hierarchical control structure of a MG. The voltage reference commands must be generated properly to complete the local controller of each DG unit. This can be accomplished through a droop-based primary controller.

C. Proposed Primary Controller

In order to design the proposed primary controller, a new droop curve is employed in this section. The conventional (P-f) and (Q-v) droop equations are expressed as follows:

$$f_{i1} = f_{0i} - \alpha_i (P_{i1} - P_{0i1}) \tag{21}$$

$$V_{i1} = V_{0i} - \beta_i (Q_{i1} - Q_{0i1}) \tag{22}$$

where f_{i1}, V_{i1}, f_{0i} and V_{0i} are the fundamental harmonic component of generated frequency and voltage commands for inner control loops and their desired reference quantities, respectively. Droop coefficients (α_i and β_i), output powers (P_{i1}, Q_{i1}) and desired reference powers (P_{0i1} and Q_{0i1}) are also used for each DG unit in main frequency. The equations of (21) and (22) can be rewritten as,

$$f_{i1} = f'_{0i} - \alpha_i P_{i1}, \qquad V_{i1} = V'_{0i} - \beta_i Q_{i1} \tag{23}$$

where

$$f'_{0i} = f_{0i} + \alpha_i P_{0i1}, \qquad V'_{0i} = V_{0i} + \beta_i Q_{0i1}. \tag{24}$$

On the other hand, the instantaneous output active and reactive powers of DG units are calculated as follows:

$$P_i = 1.5 \times (v_{di}.i_{di} + v_{qi}.i_{qi}) \tag{25}$$

$$Q_i = 1.5 \times (v_{qi}.i_{di} - v_{di}.i_{qi}) \tag{26}$$

With an acceptable approximation and by considering v_{dqi1} and i_{dqi1} as currents and voltages of DG units in fundamental frequency, (25) and (26) can be rewritten as:

$$P_{i1} = 1.5 v_{di1}.i_{di1}, \qquad Q_{i1} = -1.5 v_{di1}.i_{qi1}. \tag{27}$$

By substituting (27) into (23), the proposed droop equations based on currents of DG units are obtained as:

$$f_{i1} = f'_{0i} - \alpha'_i i_{di1} \tag{28}$$

$$V_{i1} = V'_{0i} + \beta'_i i_{qi1} \tag{29}$$

where $\alpha'_i = 1.5 v_{di1}.\alpha_i$ and $\beta'_i = 1.5 v_{di1}.\beta_i$. Fig. 5 shows the current-based frequency and voltage droop curves, respectively. Therefore, the voltage magnitude and frequency can be controlled by fundamental current components of each DG unit in d-q frame.

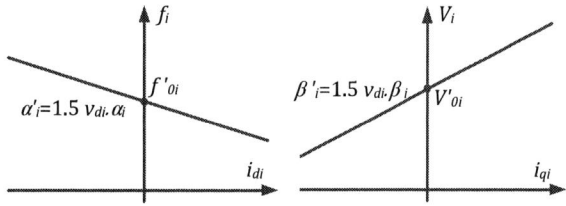

Fig. 5. Current-based frequency and voltage droop curves of each DG

D. Uncertainty of output voltage harmonics in DG units

By considering nonlinear loads for both local and common loads in the proposed MG system, each DG unit should generate the currents including:

1) Total q-component of the respective load currents;

2) All or part of maximum capacity of d-component of DG currents at fundamental frequency;

3) Total harmonic d-components of the respective loads.

All mentioned points are considered in the voltage control loop as given in (17) and (18). Since the proposed droop controller is derived in fundamental frequency, the different harmonic components of PCC voltages should be considered in voltage control loop. Thus, the d and q components of PCC voltages can be expressed as,

$$v_{dqi} = V_{dqi1} + \sum_{n=2}^{\infty} v_{dqih}. \tag{30}$$

The terms V_{dqi1} and $\sum v_{dqih}$ are fundamental components and total harmonic components of PCC voltages in dq rotating frame that should be compensated by each DG unit. The total harmonic components of the voltages can be extracted by the use of a low pass filter (LPF) as,

$$\sum_{n=2}^{\infty} v_{dqih} = (1 - LPF) v_{dqi}. \tag{31}$$

In order to decrease the negative effects of nonlinear loads, which cause PCC voltages to be distorted significantly in the proposed MG system, Eq. (31) is added to voltage control loop as depicted in Fig. 6.

IV. SIMULATION AND RESULTS

In order to verify the proposed control technique, the MG system shown in Fig. 1 is simulated in MATLAB/SIMULINK environment. The overall structure of the proposed droop controller is illustrated in Fig. 6 and the MG parameters are given in Table I.

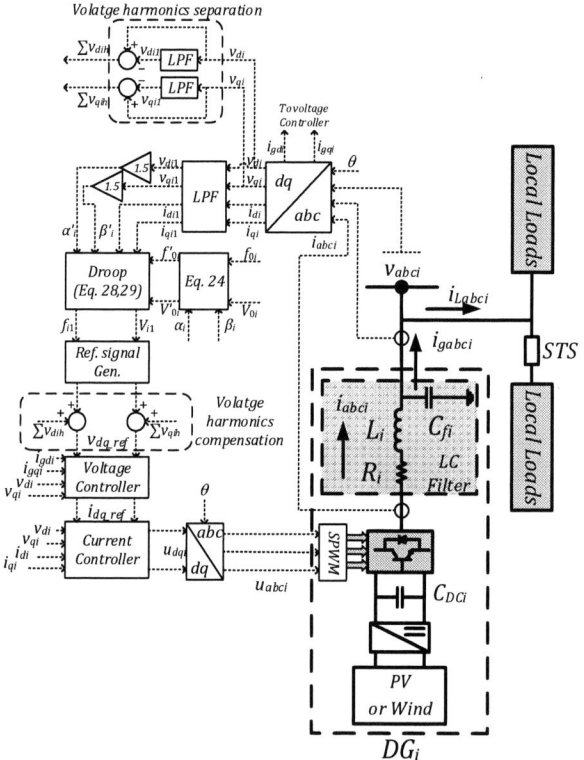

Fig. 6. Local controllers of each DG

The scenario of the proposed MG simulation is schemed as the following. Firstly, both DG units are working in steady state and supply their respective nonlinear loads. Then, at t=0.15 sec, a variable local nonlinear load is connected to PCC of each DG unit. The rated values of the local nonlinear loads are presented in Table I. Finally at t=0.3 sec, a common nonlinear load is linked to the PCC of the MG system and DG units are responsible to supply active and reactive power of the loads. It is assumed that the DG units can completely supply the required active and reactive power of their respective loads in entire simulation process. The flowchart of MG control process is illustrated in Fig. 7.

978-1-5090-0376-1/16 $31.00 © 2016 IEEE

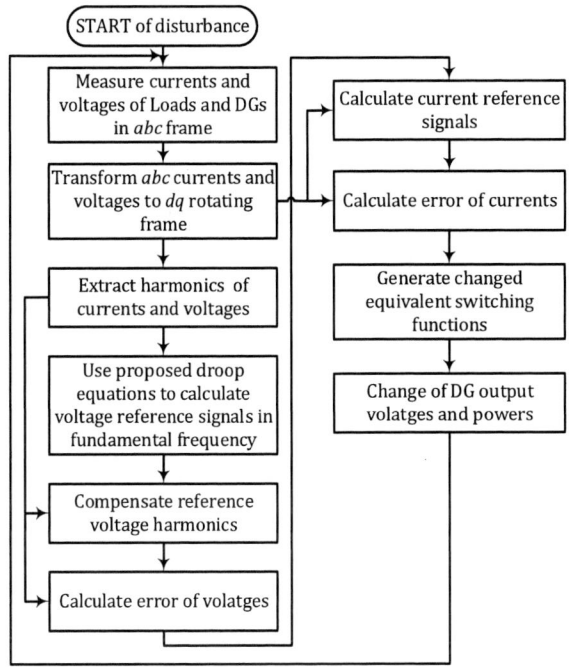

Fig. 7. Flowchart of MG control process

TABLE I. MICROGRID PARAMETERS

Parameters	Values	Parameters	Values
Microgrid ratings		**Base local loads** (a rectifier supplying RL loads)	
Voltage	220/380 V	R_{lbDGi}	40 Ω
DC-link voltage	900	L_{lbDGi}	10 mH
Fundamental frequency	50 Hz	**intermittent local loads** (a rectifier supplying RL loads)	
Switching frequency	10 kHz	R_{lDGi}	30 Ω
DGi		L_{lDGi}	10 mH
R_i	0.1 Ω	**intermittent common load** (a rectifier supplying RL load)	
L_i	45 mH	R_{lc}	30 Ω
C_{fi}	120 μF	L_{lc}	10 mH
C_{DCi}	2200 μF		
Rated power	15kW+j10kVAR		

A. Active and reactive power sharing evaluation

The active and reactive powers of DG units and nonlinear loads are shown in Fig. 8. As can be seen in this figure, the DG units are able to generate active power for their local load including both fundamental and harmonic components in steady state with acceptable transient time. According to this figure, the DG units consume the reactive power generated by ac filter to keep the PCC voltages at its desired value.

After local load changes at t=0.15, the DG units are regulated by the proposed droop controller to supply the required active power as depicted in Fig. 8. It leads to an appropriate operation for DG units in tracking their reference currents.

The reactive power of DG units is changed to reach the desired sinusoidal voltages at PCC. By connecting common load to PCC at t=0.3 sec, the DG units appropriately supply the specified active power of the load (that is $P_{DG1}=P_{lc}/3$,

$P_{DG2}=2P_{lc}/3$). Also, the reactive power of common load is supplied so that PCC voltages remain at desired range.

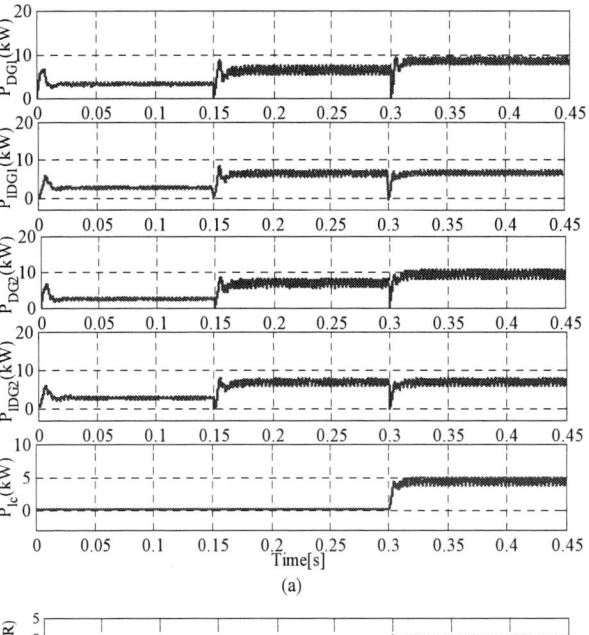

(a)

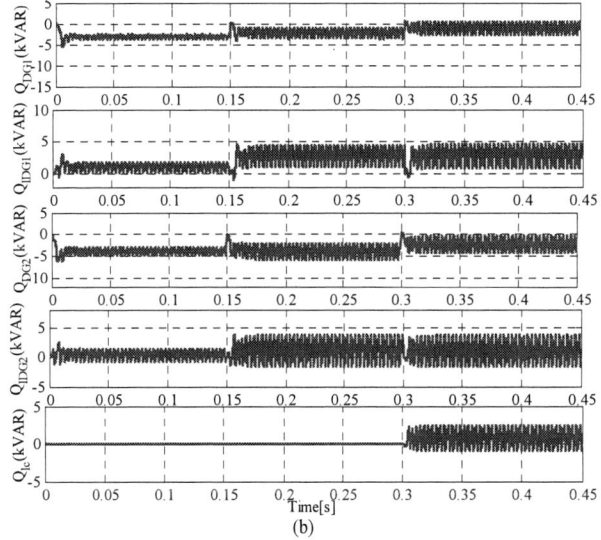

(b)

Fig. 8. Active and reactive powers variation of DGs (P_{DG}, Q_{DG}), local loads (P_{IDG}, Q_{IDG}) and common loads (P_{lc}, Q_{lc}); (a) active powers (b) reactive powers

B. Frequency and voltage magnitude assessment of PCC and DG units output voltages

In this section, the output frequency and voltage magnitude of PCC voltages within the proposed MG system are completely analyzed. As it can be seen from Fig. 9, the PCC voltages acceptably maintain their magnitude and frequency around the reference value with slight fluctuations during the whole process of simulation.

The magnitude of PCC voltages obtains the desired amount with a small transient time period as illustrated in Fig 10. Also, three phase PCC voltages is demonstrated in this figure. The figure shows that the proposed droop control technique has a good ability to adjust the DG units for generating balanced and sinusoidal three phase voltages at PCC.7

978-1-5090-0376-1/16 $31.00 © 2016 IEEE   295

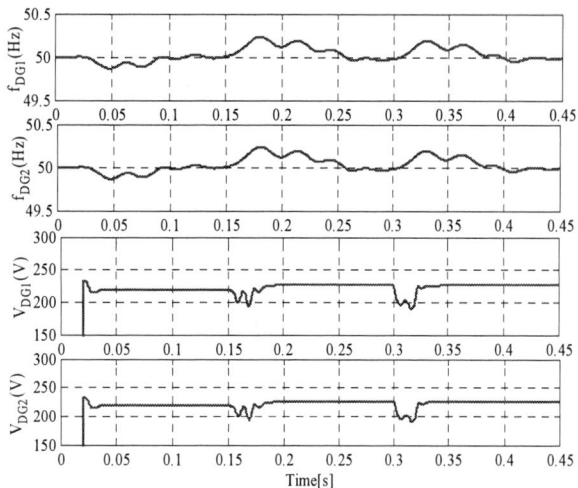

Fig. 9. Frequency and voltage variation of DGs

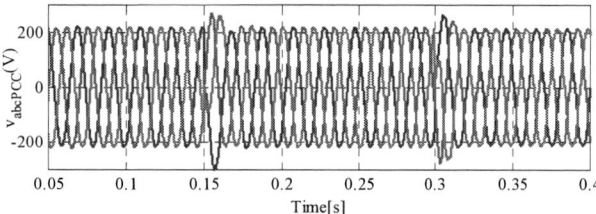

Fig. 10. Three phase voltages at the point of common coupling

V. CONCLUSION

A multi-loops hierarchical controller is proposed for MG system consisting the inner current and voltage loops based on the obtained dynamic model. Local voltage controller with a proper power sharing performance is obtained using the proposed method. The proposed inner current and voltage loops are utilized to accurately form the switching functions of interfaced-inverters and the reference values of DG unit currents, respectively. In order to have a perfect performance for the voltage control loop, a new droop controller is designed.

Also, to suppress the nonlinear loads impacts on PCC voltages, the extraction of harmonic components of PCC voltages is accomplished in the proposed method. In the propose method, a feed-forward compensation technique with a new fundamental current-based *dq* droop method (with harmonics compensation), is employed. The proposed droop curves are only based on currents of DG units. MATLAB/SIMULINK environment has been used to verify the performance of the proposed control technique in the MG system structure. The simulation results show a good transient performance which is maintained specially during non-linear load changes.

REFERENCES

[1] S. Chowdhury, S. P. Chowdhury, and P. Crossley, "Microgrids and active distribution networks." Institution of Engineering and Technology, 2009.

[2] J. M. Guerrero, J. C. Vasquez, J. Matas, V. de, x00F, L. G. a, and M. Castilla, "Hierarchical Control of Droop-Controlled AC and DC Microgrids; A General Approach Toward Standardization," Ind. Electron. IEEE Trans., vol. 58, no. 1, pp. 158–172, 2011.

[3] J. M. Guerrero, L. Garcia De Vicuna, J. Matas, M. Castilla, and J. Miret, "A wireless controller to enhance dynamic performance of parallel

inverters in distributed generation systems," Power Electron. IEEE Trans., vol. 19, no. 5, pp. 1205–1213, 2004.

[4] J. M. Guerrero, L. GarciadeVicuna, J. Matas, M. Castilla, and J. Miret, "Output Impedance Design of Parallel-Connected UPS Inverters With Wireless Load-Sharing Control," IEEE Trans. Ind. Electron., vol. 52, no. 4, pp. 1126–1135, 2005.

[5] K. De Brabandere, B. Bolsens, J. Van den Keybus, A. Woyte, J. Driesen, and R. Belmans, "A Voltage and Frequency Droop Control Method for Parallel Inverters," Power Electron. IEEE Trans., vol. 22, no. 4, pp. 1107–1115, 2007.

[6] H. Bevrani and S. Shokoohi, "An Intelligent Droop Control for Simultaneous Voltage and Frequency Regulation in Islanded Microgrids," Smart Grid, IEEE Trans., vol. 4, no. 3, pp. 1505–1513, 2013.

[7] E. Pouresmaeil, M. Mehrasa, and J. P. S. Catalao, "A Multifunction Control Strategy for the Stable Operation of DG Units in Smart Grids," Smart Grid, IEEE Trans., vol. 6, no. 2, pp. 598–607, 2015.

[8] M. Mehrasa, E. Pouresmaeil, and J. P. S. Catalao, "Direct Lyapunov Control Technique for the Stable Operation of Multilevel Converter-Based Distributed Generation in Power Grid," Emerg. Sel. Top. Power Electron. IEEE J., vol. 2, no. 4, pp. 931–941, 2014.

[9] E. Pouresmaeil, D. Montesinos-Miracle, and O. Gomis-Bellmunt, "Control Scheme of Three-Level NPC Inverter for Integration of Renewable Energy Resources Into AC Grid," Syst. Journal, IEEE, vol. 6, no. 2, pp. 242–253, 2012.

[10] E. Pouresmaeil, C. Miguel-Espinar, M. Massot-Campos, D. Montesinos-Miracle, and O. Gomis-Bellmunt, "A Control Technique for Integration of DG Units to the Electrical Networks," Ind. Electron. IEEE Trans., vol. 60, no. 7, pp. 2881–2893, 2013.

[11] M. Mehrasa, M. E. Adabi, E. Pouresmaeil, and J. Adabi, "Passivity-based control technique for integration of DG resources into the power grid," Int. J. Electr. Power Energy Syst., vol. 58, pp. 281–290, 2014.

7th Power Electronics, Drive Systems & Technologies Conference (PEDSTC 2016)
16-18 Feb. 2016, Iran University of Science and Technology, Tehran, Iran

Developing a New Fault Location Topology for DC Microgrid Systems

Reza Kheirollahi
Department of Electrical and Computer Engineering
Tarbiat Modares University
Tehran, Iran
rkheirollahi@gmail.com

Ehsan Dehghanpour
Department Of Electrical Engineering
Shahid Beheshti University
Tehran, Iran
e.dehghanpour@mail.sbu.ac.ir

Abstract— a new fault location topology implemented in DC microgrid systems is presented in this paper. Due to existing high level fault current in DC microgrids, the detection, isolation, and location of faults are in great degrees of importance. The main structure of proposing fault location unit is an H-bridge controlled by control electronic platform. Based on the selective switching frequency of the semiconductor switches and the amplitude of input DC voltage, the proposed scheme can be used in structurally different microgrids. Low cost design, portability, simple structure, high performance and the acceptable accuracy of the output voltage are the main advantages of the proposed method. The error estimation of the presenting topology in terms of fault position and fault resistance is compared with lastly reported methods. The correctness of the proposed fault location topology is verified through computer simulation.

Keywords— DC microgrids; fault locator unit; power probe unit; switch mode power supply

I. INTRODUCTION

Microgrid is defined as an active distribution network which includes Distributed Generation Resources (DERs) and electrical loads. DERs in Micro grids are renewable generally, and not only do they lead to decrease greenhouse gases, but also reduce using fossil fuels. Nowadays, based on several advantages such as capability of operation in both connected and island mode, improvement in power quality, more flexibility, and using renewable natural resources, microgrids attract a great degree of attention from academia and industrial points of view.

Microgrids can be divided into two categories: AC microgrids and DC microgrids. Using of Low cost power converters, low power losses, benefiting from whole cable's cross-section, and the lack of skin effect all are the advantages of DC microgrids. On the other hand, DC microgrids suffer from the absence of enough standard definition and expensive protective equipment preventing them to be widely implemented in electrical networks.

Fast fault detection and fault isolation play a key role in DC microgrids, and it is of interest to several researches. From the last decade, different approaches and methods have been developed by researches including zero current and zero voltage switching hybrid DC circuit breakers [1], integrating pulse by pulse current limiting technique [2], detecting

abnormal fault current and separating the faulted segment based on segment controllers [3], DC ring bus protection by using intelligent electronic devices [4], high speed current differential fault detection [5]. Added to fault detection and fault isolation operations, fault location is of the great importance owing to the fact that duration times which the electric network is lost should be decreased. To solve the fault location problem and improve the accuracy of output information, several methods have been reported [6-9]; these are: wavelet transform (WT) based Multi-Resolution Analysis (MRA) [6], Artificial Neural Network (ANN) based methods [7]-[8], online fault location and traveling-wave-based methodology for fault location [9]. Another approach is using power probe unit based on extracting the information of fault location from a probe current. This method is capable of locating fault in a pilot test to determine the temporarily existing faults before circuit breaker reclosing [4].

In this paper, a new topology to detect faults in microgrids is presented. To prove the correct functionality of proposal topology, a DC microgrid with annular-ring bus is taken into account. The proposed fault detection topology is verified through analytical formula and simulation results. It is shown that the proposed topology benefits from low cost design, portable usage, simple structure, high reliability, high performance and acceptable accuracy. The paper is organized as follows: in the next section, faults type existing in dc microgrids are described; sections 3 presents the proposed topology to distinguish faults in microgrids; simulation results are presented in section 4.

II. DC BUS MICROGRIDS

A. Possible Fault

Possible fault types in the DC microgrids are divided into two categories as shown in Fig. 1. The first one includes faults which are happen by creating a path between positive or negative line and ground. This type of ground can be regarded as low impedance or high impedance depending on the location of fault. The line-to-ground fault are known as the commonly types of faults in distribution systems. The second categories consists of faults where short-circuit is occurred between the positive and negative lines.

These types of fault usually have low fault resistance [3], [4], and 10]. Generally, both categories of fault may happen in the

978-1-5090-0376-1/16 $31.00 © 2016 IEEE   297

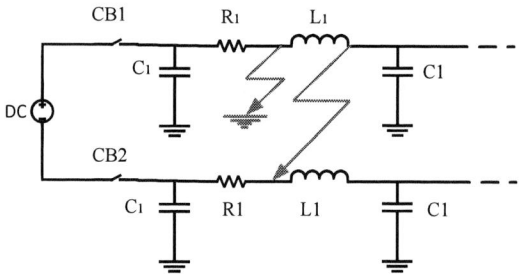

Fig. 1. Two types of faults in DC microgrids

feeders or buses, so Importance of the fault location can be varied for structurally different DC microgrids.

B. Fault Location Techniques

To satisfy DC fault location requirements, existing techniques benefit from rate of current, current's magnitude, current oscillation pattern [11], continuous wavelet transform [12]-[14], distributed parameter line model [15], reference voltage based iterative estimation [16], artificial neural networks [13], [17]. Dependence on two terminal measurement, sensors error, and communication delay are all limitation preventing them to be commonly used in industrial distributed systems [4]. The mainly drawback related to these method is locating fault based on the fault's information extracting at the time of fault. This matter is not acceptable due to the fact that the magnitude of current is risen to a high level fast, and it should be isolated from other part of the DC microgrid as soon as possible. Park et all (2013) have introduced a method based on probe power unit to detect fault location in microgrids; the equivalent circuit of the faulted bus segment with the probe power unit proposed in [4] are shown in Fig. 2.

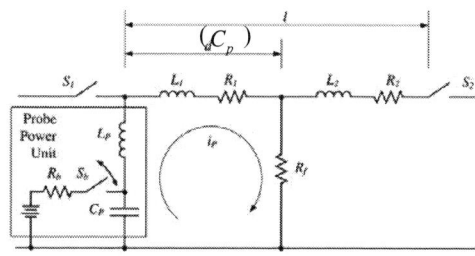

Fig. 2. The equivalent circuit and power probe unit [4]

The operating principles of this method is as follows: firstly, the input capacitor (C_p) is charged, and the energy of capacitor is discharged as an attenuated waveform through internal inductance (L_p), line inductance (L_l) and line resistance (C_l), and the fault resistance (R_f). Secondly, the information of passing current is sampled and held, and a Fast Fourier Transformation (FFT) is executed to provide amplitude and phase of the harmonic spectra of the current. Lastly, the harmonic with the maximum energy in harmonic spectra is distinguished, and based on these information, the fault distance and the value of fault resistance can be

calculated. The accuracy of fault location is depicted in Fig. 3 in terms of fault position and fault resistance.

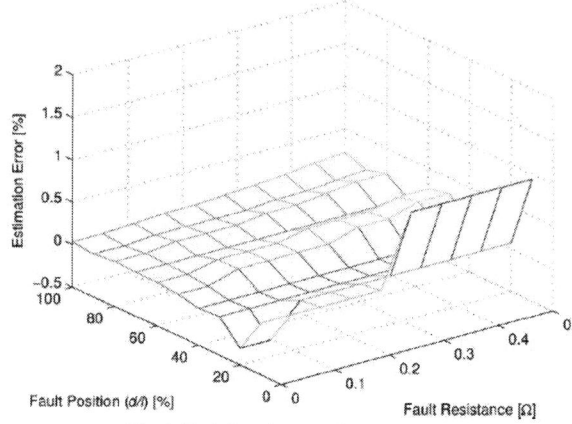

Fig. 3. Fault-location error in terms of fault position and fault resistance [4]

In spite of the fact that this method operates well in the case of short distance and low fault resistance, its accuracy is affected by increasing the length of line and increasing the value of fault resistance. Also, the effects of line's capacitances on probe power unit operation was not considered, and it only can be used in under attenuation state of operation.

III. PROPOSED FAULT LOCATION TOPOLOGY

In this section, the operational principles of proposed topology is presented. Figure .4 shows the general structure of fault locator unit, electrical cable and fault resistance. The power stage of the fault locator unit is mainly constructed based on an H-bridge topology with four semiconductor switches operating in a symmetrical state.

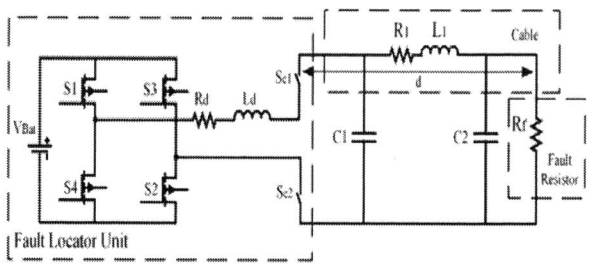

Fig. 4. The main structure of fault location topology

The left hand side of the H-bride is connected to a battery, and the right hand side of the H-bridge is connected to the electrical cable through an R-L circuit. Whenever the switches S_1 and S_2 are turned on and S_3 and S_4 are turned off simultaneously, the electrical cable is connected to the positive polarity of the battery. On the other hand, the electrical cable is connected to the negative polarity of the battery when the switches S_3 and S_4 are turned on and S_1 and S_2 are turned off. The rate of switching frequency can be varied by control

978-1-5090-0376-1/16 $31.00 © 2016 IEEE   298

electronic platform's processor which generates well-formed gating pulses to the switches. R_d and L_d known as internal resistance and inductance of fault location device are the setting parameters which can be chosen based on normal conditions of electrical network. Switches S_{c1} and S_{c2} are used to connect and disconnect fault locator to an ordinary DC microgrid. The electrical cable is implemented based on the π model. This part models the characteristics of electrical cable implemented in microgrids and electrical networks. These parameters are defined as follows:

R_l: it portrays the cable's resistance between the fault location device and a place where the fault has been located. It is defined as $Ru*d$ which Ru introduces the unit resistance of the electrical cable, and parameter d presents the fault distance from fault location unit.

L_l: this parameter describes the cable's inductance between the fault location unit and a place where the fault has been happened. This parameter is defined as $Lu*d$ which Lu refers to the unit inductance of the electrical cable.

C_1 and C_2: these parameters designate the capacitance features of electric cable between two specific points. In Fig.1, these parameters define cable capacitance effect between fault detection device and a place where the fault is located. The value of these two capacitances are determined as $Cu*d$ which Cu shows the unit capacitance of electric cable.

R_f: this parameter indicates value of the fault resistance.

The operational principles of the proposed fault detector are described in a stepwise procedure as follows:

Step 1: the battery provides a DC voltage as the H-bridge's input voltage.

Step 2: the rate of switching frequency is determined based on a specific table proposing a well-tuned frequency related to the cable's length. Selecting proper switching frequency can eliminate the effect of line's parallel capacitances and promote the accuracy of output results.

Step 3: based on the chosen switching frequency and the amplitude of DC output voltage of the battery, a full-wave square pulse is produced and is applied to the electric cable through internal R-L circuit.

Step 4: the amplitude of the current passed through the cable depends on the specifications of the full-wave square voltage generated by H-bridge circuit, the internal resistance and inductance of the fault locator unit, the resistance and inductance of the electric cable, and value of the fault resistance.

Step 5: the current waveform is sampled, and the results are transferred to the control electronic board. This board includes a high performance microcontroller which can analyze the input data and execute the fault location algorithm to calculate location of the fault.

Step 6: FFT algorithm is executed based on the following information: 1) the amplitude and frequency of the full-wave square pulse generated by the H-bridge; 2) the data resulted from sampled current waveform. After the FFT algorithm is executed, following information are provided:

$$v(j\omega_{sw}) = V\angle\alpha \qquad (1)$$

$$i(j\omega_{sw}) = I\angle\beta \qquad (2)$$

The Eq. (1) and (2) show resulted amplitude and phase of the voltage and current waveform, respectively. In this case, the frequency (f_{sw}) points to the switching frequency of the H-bridge.

Step 7: depend on the phase and amplitude of the voltage and current in the switching frequency of the H-bridge, the impedance is obtained. This impedance consists of the internal impedance of the fault location device, the cable's impedance, and the fault resistance. Following equations portrays the relations well:

$$Z(j\omega_{sw}) = \frac{V_{out}(j\omega_{sw})}{i(j\omega_{sw})} = \frac{V\angle\alpha}{I\angle\beta} = Z\angle\theta = R_{cal} + j\omega_{sw}L \qquad (3)$$

$$Z(j\omega_{sw}) = (R_d + j\omega_{sw}L_d) + (R_l + j\omega_{sw}L_l) + R_f \qquad (4)$$

Which ω_{sw} indicates the angular frequency in the switching frequency, R_{cal} shows value of the calculated resistance based on the sampled data, L_{cal} portrays value of the inductance depend on the sampled information. It should be noted that the cable capacitance effect is not considered to calculate cable impedance in Eq. (3) and (4) because the switching frequency is chosen in a way that cable capacitance to be neglected without significant error.

Step 8: by considering the content of equation (3) and (4), and by equaling real and imaginary part of them, it is possible to calculate the fault resistance and fault distance. To find the distance between location device and a place where the fault is located, imaginary part of the both equation are taken into account:

$$j\omega_{sw}L_d + j\omega_{sw}L_l = j\omega_{sw}L_{cal} \qquad (5)$$

The fault distance (d) is also determined based on the calculated inductance value ($L_l = L_u \times d$) and the unit inductance (L_u).

Step 9: by providing the fault distance (d) and by taking the real part of the Eq. (3) and (4), the value of fault resistance is given.

$$R_l = R_u \times d \qquad (6)$$

$$R_f = R_{cal} - R_l - R_d \qquad (7)$$

IV. SIMULATION RESULTS

To verify the correctness of proposing fault location strategy, the computer simulation was performed using MATLAB® software package. Simulation network is illustrated in Fig. 5.

978-1-5090-0376-1/16 $31.00 © 2016 IEEE

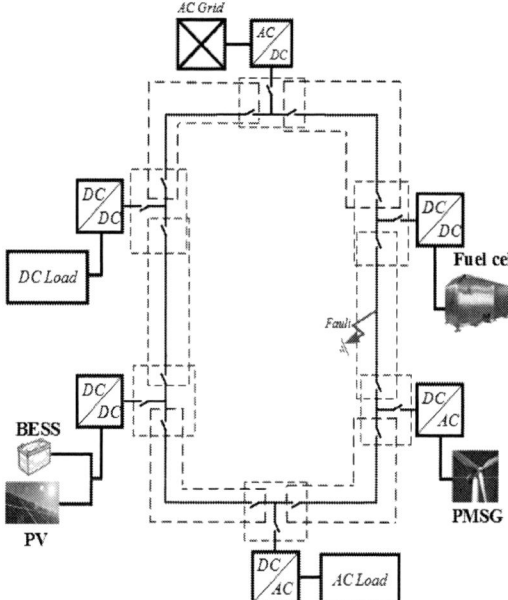

Fig. 5. DC microgrid consist of Permanent Magnet Synchronous Generator (PMSG), Photo Voltaic (PV) panel, Battery Energy Storage System (BESS), Fuel Cell and electrical loads.

As shown in Fig. 5, whenever a fault happens in one section of DC microgrid, relays on both side of faulted section are opened. Then, fault locator can be applied to recognize the fault location and fault resistance.

The performance of the proposed topology should be evaluated from two points of view. First, the estimation error of the fault location is analyzed based on various input frequencies and different amounts of fault distance for a predetermined value of fault resistance. This step explicitly describes the importance of chosen switching frequency of the H-bridge in determining the performance of the proposed topology. As mentioned before, reduction of line's parallel capacitances effects depends on the frequency of the input square-wave pulses. Second, based on the proper switching frequency, the estimation error is obtained in terms of fault resistance and fault position; the output information portrays the correctness of proposing fault location topology.

Table I. Simulation parameter

Parameters	Value
Cable cross-section area	241.9 mm2
Unit resistance R_u	121 $m\Omega/km$
Unit inductance L_u	0.97 mH/km
Fault locator unit R_d	0.1 Ω
Fault locator unit L_d	1 mH
Fault locator unit V_{Bat}	5 V

Simulation parameters can be found in table I. Fig. 6 through 7 show the simulation results for two aforementioned

steps. The estimation error for 0.1 ohm fault resistance is given in Fig. 6; as it can be seen, choosing lower switching frequency leads to lower output estimation error owing to the fact that the effect of line parallel capacitances are decreased. Besides, the accuracy of output results are not affected by changing the amount of fault resistance and increasing the length of line. Fig. 7 shows estimation error of fault locator unit in terms of fault resistance and fault position for 50 Hz input frequency. Deeply studying the output results reveals that the accuracy of proposed topology for a wide range of fault resistance and fault position is better than other method having already been reported.

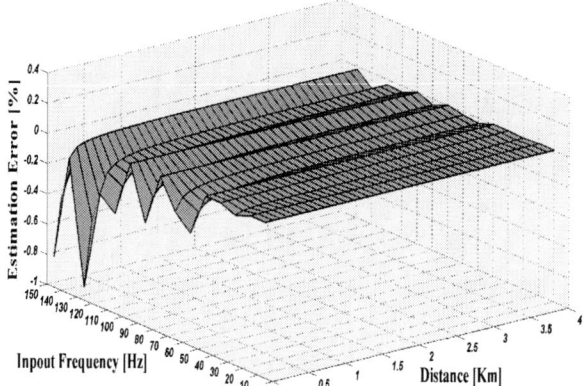

Fig. 6. The estimation error in terms of input switching frequency and fault position

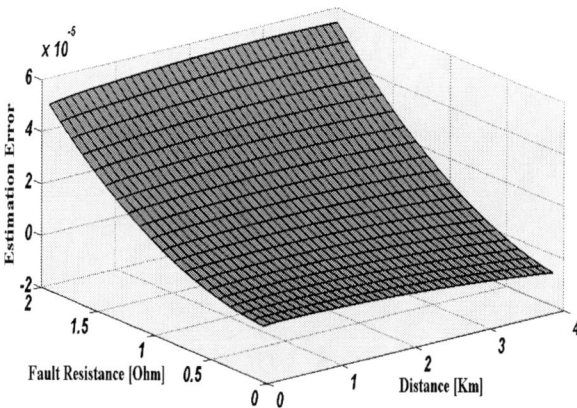

Fig. 7. The estimation error in terms of fault resistance and fault position

V. CONCLUSION

A new fault location strategy for the DC microgrids systems has been presented in this paper. Proposed topology consists of a battery to provide input DC voltage, an H-bridge structure as a switch mode power supply, and an internal R-L circuit. The presenting fault location is capable of locating faults in structurally different DC microgrid systems because of the fact that setting parameters such as the switching frequency of semiconductor switches and the amplitude of input DC voltage can be reset in the control electronic platform. Portability, simple structure, high reliability, high performance and acceptable accuracy of output results are the important advantages of the proposed fault location unit. The effectiveness and correctness of the proposed topology have been shown using computer simulation.

REFERENCES

[1] P. van Gelder and J.A. Ferreira, "Zero volt switching hybrid DC circuit breakers," Industry Applications Conference, Rome, vol. 5, pp. 2923-2927, October 2000.

[2] J. Chunlian and R. Dougal, "Current limiting technique based protection strategy for an industrial DC distribution system," IEEE International Symposium on Industrial Electronics, Montreal, vol. 2, pp. 820-825, July 2006.

[3] J-D. Park and J. Candelaria, "Fault detection and isolation in low-Voltage DC-Bus microgrid system," IEEE Trans. Power Delivery, vol. 28, pp. 779-787, April 2013.

[4] Jae-Do. Park, J. Candelaria, M. Liuyan and K. Dunn, "DC ring-bus microgrid fault protection and identification of fault location" IEEE Trans. Power Delivery, vol. 28, pp. 2574-2584, October 2013.

[5] S.D.A. Fletcher, P.J. Norman, K. Fong, S.J. Galloway and G.M. Burt, "High-speed differential protection for smart DC distribution systems," IEEE Trans. Smart Grid, vol. 5, pp. 2610-2617, September 2014.

[6] L. Weilin, L. Min, A. Monti and F. Ponci, "Wavelet based method for fault detection in Medium Voltage DC shipboard power systems," IEEE International Instrumentation and Measurement Technology Conference, Graz, pp. 2155-2160, May 2012.

[7] N.K. Chanda and F. Yong, "ANN-based fault classification and location in MVDC shipboard power systems," North American Power Symposium, Boston, pp. 1-7, August 2011.

[8] C.S. Chang, S. Kumar, B. Liu and A. Khambadkone, "Real-time detection using wavelet transform and neural network of short-circuit faults within a train in DC transit systems," IEE Proceedings Electric Power Applications, vol. 128, pp. 251-256, May 2001.

[9] S. Azizi, M. Sanaye-Pasand, M. Abedini, and A. Hassani, "A traveling-wave-based methodology for wide-area fault location in multiterminal DC systems," IEEE Trans. Power Delivery, vol. 29, pp. 2552-2560, December 2014.

[10] D. Salomonsson, L. Soder and A. Sannino, "Protection of low-voltage DC microgrids," IEEE Trans. Power Delivery, vol. 24, pp. 1045-1053, July 2009.

[11] L. Tang and B. Ooi, "Protection of VSC-multi-terminal HVDC against DC faults," in Proc. IEEE 33rd Annu. Power Electron. Specialist Conf., vol. 2, pp. 719–724., November 2002.

[12] O. Nanayakkara, A. Rajapakse and R. Wachal, "Location of DC line faults in conventional HVDC systems with segments of cables and overhead lines using terminal measurements," IEEE Trans. Power Del., vol. 27, pp. 279–288, January 2012.

[13] C. Chang, S. Kumar, B. Liu, and A. Khambadkone, "Real-time detection using wavelet transform and neural network of short-circuit faults within a train in DC transit systems," Proc. Inst. Elect. Eng., Elect. Power Appl., vol. 148, pp. 251–256, May 2001.

[14] W. Li, M. Luo, A. Monti, and F. Ponci, "Wavelet based method for fault detection in medium voltage DC shipboard power systems," in Proc. IEEE Int. Instrum. Meas. Technol. Conf., pp. 2155–2160, May 2012.

[15] J. Suonan, S. Gao, G. Song, Z. Jiao, and X. Kang, "A novel fault location method for HVDC transmission lines," IEEE Trans. Power Del., vol. 25, pp. 1203–1209, April 2010.

[16] J. Yang, J. Fletcher, and J. O. Reilly, "Short-circuit and ground fault analyses and location in VSC-based DC network cables," IEEE Trans. Ind. Electron., vol. 59, pp. 3827–3837, October 2012.

[17] N. Chanda and Y. Fu, "ANN-based fault classification and location in MVDC shipboard power systems," in Proc. North Amer. Power Symp., pp. 1–7, August 2011.

7th Power Electronics, Drive Systems & Technologies Conference (PEDSTC 2016)
16-18 Feb. 2016, Iran University of Science and Technology, Tehran, Iran

Three-Phase PFC rectifier with High Efficiency and Low Cost for Small PM Synchronous Wind Generators

Ghasem Rezazadeh, Farzad Tahami, Senior Member, IEEE, Hamed Valipour, Student Member, IEEE

Electrical Engineering Department, Sharif University of Technology, Tehran, Iran

Abstract— Small Permanent Magnet Synchronous Generators (PMSG's) are widely used in low power wind turbines. In order to inject the electrical power, generated by PMSG, to the grid, a back-to-back AC/AC power electronic converter is required. In this paper, a novel low cost efficient AC/DC converter is proposed for rectifier stage to obtain the maximum power per ampere of PMSG by using Power Factor Correction method. The new structure is based on DCM SEPIC converter. Reducing the number of semi-conductor switches has decreased converter cost. Additionally, other advantages of this Converter are employing easy control method to obtain the maximum power per ampere and using Synchronous inductance of PMSG as a converter element. Unlike Conventional DCM Boost rectifier, the proposed converter has a continuous input current that leads to lower conduction power loss in PMSG. Principles of operation are presented as well as the closed-loop simulation results for a 500W/100V prototype.

Keywords— *Permanent Magnet Synchronous Generator; Power Factor Correction Rectifier; SEPIC Converter*

I. INTRODUCTION

The global wind energy capacity has increased rapidly in the past few years and became the fastest developing renewable energy technology [1]. The wind generators' structure can be divided into four categories as follows [1], [2]: fixed-speed induction generators with cage rotor; variable-speed wounded rotor induction generator with rotor resistance control (dynamic slip control); variable-speed doubly fed induction generator; and variable-speed Permanent Magnet Synchronous Generator (PMSG). Most of the direct-drive turbines (gearless generator systems) being sold at the moment have synchronous generators with permanent magnet rotor. This gearless structure leads to failure reduction in gearboxes and less maintenance problems [3]. In particular, active power control will play an important role both during grid faults (low voltage ride-through capability and controlled current injection) and in normal conditions (reserve function and frequency regulation) [4], [5].

Variable-speed fixed-pitch wind turbines with permanent magnet synchronous generators are very popular for small-scale wind power plants because of their direct-drive capability, simple construction, and maintenance-free operation. An AC-DC-AC power electronic converter is utilized to convert the variable-frequency variable-voltage

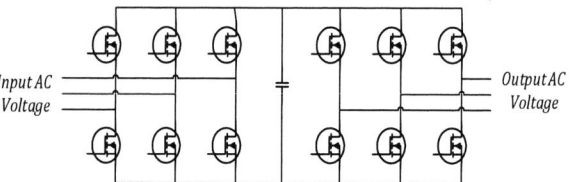

Fig. 1. Conventional back-to-back AC/AC converter

generator output to the fixed-frequency fixed-voltage grid. Conventional back-to-back AC/AC converter used for this task, is demonstrated in Fig.1[1] but there are also some other structures for back-to-back AC/AC converter [6].

In this paper a new structure has been proposed that can convert the variable voltage and frequency of PMSG to the constant voltage and frequency of the grid. It can also convey the maximum possible power from PMSG to the grid for a particular input current level. Some of the main characteristics of the proposed circuit include the elements reduction specially the semiconductor switches, the economical structure, simple structure that leads to the simplicity of analyzing, and high efficiency. Because of the existence of tank elements, soft-switching techniques can be utilized in the proposed structure. Some of the techniques have been discussed in [7] and [8].

The goal is to absorb the maximum possible power from the internal voltage of generator, hence, it is required that PFC rectifier is used in the first stage of the converter. Therefore, the input current shape should follow the input voltage waveform without any phase delay, thus, the input power factor is needed to be close to unity. Utilizing PFC's could lead to reduction of the maximum amplitude of the input current in a particular power level, so it can cause the conduction loss to be decreased. In addition, it can reduce the input current THD which serves to shrink the input filter size [9],[10].

Active PFC rectifiers have two different categories [11]: Phase Modular Systems, which are combined of three single-phase PFC's, in which isolation is needed. The component count is the disadvantage of Phase Modular systems, and their advantage is the capability of controlling each phase in a different mode. Direct Three-Phase Systems, in which a single three-phase structure is used, is better in sense of component count and the lack of the transformer. Direct three-phase systems can be divided into two categories namely, Boost-type, and Buck-type. In order to accomplish

978-1-5090-0376-1/16 $31.00 © 2016 IEEE

the project requirements, Boost-type PFC systems is utilized. DCM approach has been used because of the lower number of the switches and the elements, and simpler control circuitry.

The aforementioned categories leads to three-phase single-switch DCM Boost. Although this circuit seams so simple and efficient, it has some drawbacks that makes it not suitable for PMSG application. Some of the disadvantages include: DCM input current that could lead to increase in current amplitude and RMS that could harm the generator and cause considerable loss. DCM current is composed of high frequency harmonics that can cause higher power loss in the generator due to skin effect.

The proposed circuit does not have the aforementioned problems and has a simple power and control circuitry. The proposed circuit is based on SEPIC DCM and is a Direct Three-Phase PFC System, which can regulate the output voltage, and also correct the power factor. In addition, it is simply controlled like other DCM systems but it does not have discontinuous input current. Nonetheless, there are some papers that have introduced some structures based on DCM SEPIC converter. For instance, in [12] a Phase Modular PFC converter has been proposed that has some drawbacks like non-isolated structure and component count. In [13] and [14] a similar structure has been proposed which needs three power switches and nine diodes, which is higher than what is needed in the presented circuit in this paper.

In section II the proposed circuit will be discussed. Different modes of operation will be analyzed in this section profoundly. In section III the simulation results will be presented and discussed, and eventually, in section IV the conclusion of the presented structure will be reviewed.

II. PRINCIPLES OF THE OPERATION

In this section the proposed circuit will be introduced. Different modes of operation will be discussed, and finally, designed circuit with proper elements values will be illustrated.

The proposed rectifier is illustrated in Fig. 2. As stated above, this circuit is basically a direct three-phase PFC system, and is a single-switch DCM type converter. This circuit is based on a DCM SEPIC converter, nevertheless, in a three-phase structure.

The voltage source and the series inductor represented in Fig. 2 are the simplest circuit model of PMSG from the output terminals. The basic single-phase SEPIC converter is represented in Fig. 3. In order to maintain the similarity between three-phase and single-phase structures, the connections in three-phase structure should be star-type. Hence, the values of the elements of two structures will be the same.

A. Modeling DCM SEPIC converter

Large signal model of a DCM SEPIC converter has been represented in [15], in which has been shown that the converter can be modelled as a resistor from the input side. Therefore, PFC feature can be obtained from this kind of

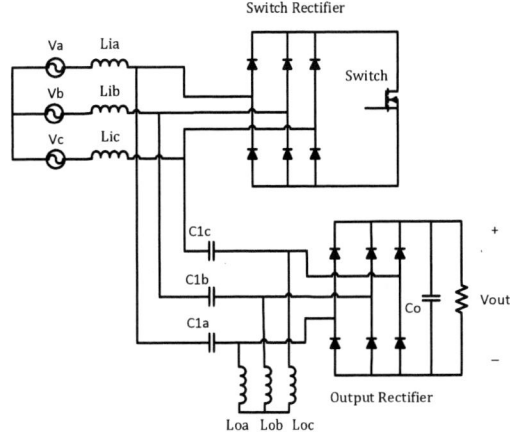

Fig. 2. The proposed three-phase PFC based on SEPIC DCM converter

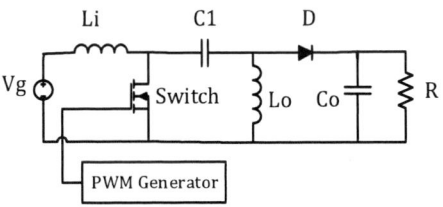

Fig. 3. Equivalent single-phase structure of the proposed three-phase structure

circuit. The equations will be represented briefly in this part, and at last, the little difference between single-phase and three-phase DCM SEPIC will be discussed.

Single-phase DCM SEPIC converter includes three modes of operation, which the equivalent circuit for each have been shown in Fig. 4. However, three-phase structure which is proposed in this paper, has five different modes of operation. The first and the second modes are the same in both single-phase and three-phase converters. The only difference is in the third mode, in which it takes different time for each phase to conclude this mode and enter the next mode, due to different input voltages (Vg) in each phase. As a result, the third mode in single-phase converter can be mapped into three other modes in three-phase structure. Because of the different input voltage in each phases, the time interval of the second mode of operation will be different in each phase, consequently, the starting moment of the third mode will be different. Nonetheless, these differences in the number and the timing of the intervals do not have any influences on the input characteristics of the three-phase circuit, because the input equivalent resistor in DCM SEPIC converter is only related to the duration of the first mode of operation and some of the intrinsic characteristics of the circuit which is common in three-phase and single-phase structures. The equivalent resistor that is seen from the input, is represented in (1). As claimed above, (1) is dependent solely on the duration of the first mode of operation and is valid for both single and three phase of DCM SEPIC converter. The only constraint that should be

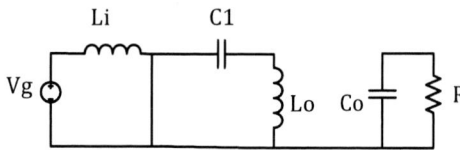

(a) The first mode of operation

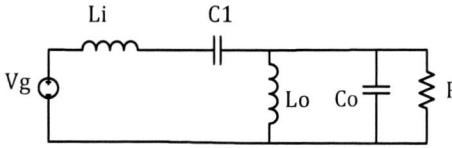

(b) The second mode of operation

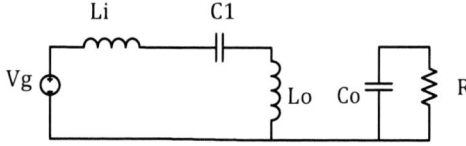

(c) The third mode of operation

Fig. 4. Equivalent circuit of different modes of operation of a DCM SEPIC converter. (a) The switch is ON & the diode is OFF; (b) The switch is OFF & the diode is ON; (c) The switch is OFF & the diode is OFF

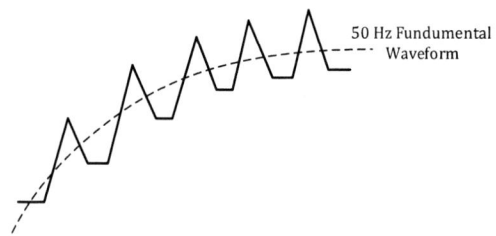

Fig. 5. Input current waveform of DCM SEPIC converter

established for the correction of (1), is DCM operation, which means that the output diode current should reach zero before the end of one period of switching.

$$R_{eq} = \frac{d^2 T_S}{2 L_{eq}} \tag{1}$$

Ts is the period of the switching frequency, d equals to the ratio t_{on}/Ts in which t_{on} is the time of the first interval, and $L_{eq} = \frac{L_i L_o}{L_i + L_o}$. One of the interesting features of DCM SEPIC converter is that it can have PFC in the input by just having a DCM current in the output diode. This means that in PFC mode of DCM SEPIC converter the input current will be continuous and this would help to extend the reliability of PMSG and decrease the resistive loss. Input current of DCM SEPIC converter is illustrated in Fig. 5.

B. Boundary between CCM and DCM in SEPIC converter

In order to obtain the benefits of DCM, the boundary between DCM and CCM should be recognized. This can help to design the converter to operate in desirable condition despite the variation of the input parameters. It has been shown in [15] that the condition which is supposed to be true to maintain the converter in DCM is:

$$K < K_{crit.} \Rightarrow L_{eq} < \frac{R T_S K_{crit.}}{2} \tag{2}$$

In which:

$$K = \frac{2 L_{eq}}{R T_S} \tag{3}$$

$$K_{crit.} = \frac{1}{2(1+M)^2} \tag{4}$$

$R = \frac{V_O}{I_O}$ is representative of the output load, and

$M = \frac{V_O}{V_{in}}$ that $V_{in} = V_m |Sin(\omega t)|$.

C. Designing the elements

Input is connected to the output terminals of a PMSG with variable output RMS voltage between 0 and 48V in each phase. The equivalent series inductor of PMSG is about 5 mH with negligible variation. In the proposed circuit, the series inductor of PMSG has been used as the input inductance of the circuit. Therefore, the component count has been reduced. Since the shunt value of Li and Lo is effective in (1) and the value of Lo is much lower than Li, these variations in the input inductance of the converter do not have significant effect on the functionality of the circuit. The output (inverter output) of the proposed structure should be connected to the grid (220 V AC), thus, the output voltage of the rectifier stage is 311 V. However, low input voltage of the converter could lead to drastic growth in d that could cause numerous problems such as CCM mode of operation, noise, and switches defects. In order to overcome these difficulties, a high-frequency three-phase transformer can be used in the rectifier structure, nonetheless, in the proposed circuit the aforementioned transformer has not been used.

For the purpose of laboratory prototype testing, the above-mentioned transformer has not been used, and the output voltage has been set 100 V. The input and output are as follows:

- Nominal power of the rectifier: 500W

- Desirable output voltage in an equivalent single-phase structure: $V_{O-1ph} = \frac{V_{O-3ph}}{\sqrt{3}} = 57.73 V$

- Maximum amplitude of the input voltage: $V_m = 48\sqrt{2} = 67.88 V$

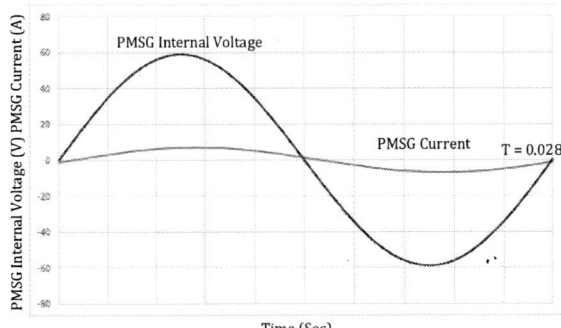

(a)

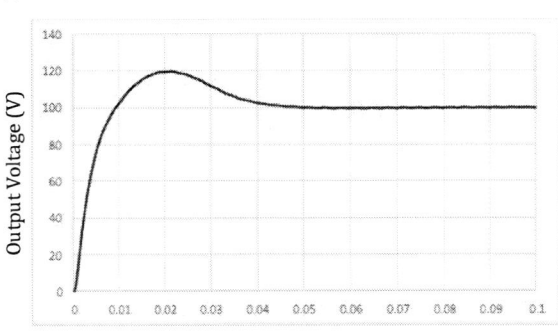

(b)

(c)

Fig. 6. Simulation results with PSIM pre-provided PMSG blocks in case of internal voltage of 41.7 V RMS and 35.5 Hz (a) PMSG voltage and current, in which power factor is 98.7%. (b) Output voltage which is regulated at 100 V. (c) Output diode current waveform, in which the current reaches zero before the end of the period.

• Switching frequency: 50 kHz

The circuit should be designed in worst case condition, in which the ratio of the output to input voltage is minimum. The converter should remain in DCM even in this critical situation.

$$M_{min} = \frac{V_{O-1ph}}{V_m} = 0.85 \tag{5}$$

$$K_{crit.} = \frac{1}{2(1+M_{min})^2} = 0.146 \tag{6}$$

$$L_{eq} < \frac{RT_S K_{crit.}}{2} \Rightarrow L_{eq} = 0.05 mH \tag{7}$$

Now it is possible to determine the elements' values:

$$L_{eq} = 0.05 mH, \ L_i = 5 mH \Rightarrow L_O \approx 0.05 mH \tag{8}$$

For an adequate functionality of the circuit, the resonance frequency of LC tank in the circuit should be somewhat between the input frequency and the switching frequency:

$$f_{in} < f_r = \frac{1}{2\pi\sqrt{L_O C_1}} < f_S \Rightarrow C_1 = 0.5 \mu F \tag{9}$$

D. C1 variations effect on the functionality of the circuit

As stated in (9) there is a degree of freedom in choosing the value of C_1 in Fig. 2. Is there an optimized value for C_1. In order to answer the pre-mentioned question, the effect of the variation in capacitance of C_1 in the functionality of the converter, should be examined. In [15] this examination has been done thoroughly. Therefore, some of the results are presented here:

• As the capacitance of C_1 decreases, the voltage ripple of this capacitor increases. Since the average value of V_{c1} is equal to the input voltage, by decreasing the capacitance of C_1, voltage ripple and the maximum voltage of this capacitor will increase and a capacitor with higher voltage rating is needed.

• As the capacitance of C_1 increases, the level of the harmonics in input current will increase too. This means that the circuit might malfunction.

III. SIMULATION OF THE PROPOSED CIRCUIT

Simulation of the proposed circuit has been done with PSIM software. PMSG can be modelled by a voltage source and a series impedance.

A. Simulation of the circuit with a PSIM pre-defined block diagram as PMSG

The real PMSG characteristics have been used in this part of simulations. A simple control strategy has been utilized for the purpose of closed-loop simulation. Control circuit includes a traditional analog PWM circuit that is a great advantage of the proposed structure. In this method, the output voltage is compared with the reference value and the result will be sent to the controller section (PI controller) that compensates the difference between the feedback value and the desired value. After the compensator, there is a PWM section that consists of a saw-tooth waveform and a comparator that generates the gate signal like the conventional methods in textbooks.

Mechanical input is needed for PMSG block of PSIM, which has been given a wind turbine to resemble the simulation results to the real world, as much as possible. As well as PMSG, the block of PSIM has been used for wind turbine modeling. Wind speed and the blade angle are needed as inputs for wind turbine and the corresponding output is mechanical torque; these factors have a great influence on the output torque due to the corresponding equations stated in [16], [17] and [18]. In addition, these are the only environmental factors that play an important role in the aforementioned equations. Simulation waveforms are illustrated in Fig. 6, which are based on 15 m/s wind speed and a blade angle of 20 degree.

B. Interpretation of the variation in converter behavior due to the variation of the input parameters

In this part, two input parameters (input voltage amplitude and frequency) will be changed, and their effect on important quantities like output voltage and input power factor will be discussed.

1) Input frequency affection on the converter behavior

In this part it is assumed that the RMS value of the input voltage is constant and is equal to 48 V, however, its frequency is not constant and is altering from 5 Hz to 50 Hz with 5 Hz steps. Corresponding results are illustrated in Fig. 7. It should be mentioned that the average value of the output voltage is constant at 100 V and output ripple is changing just due to the input frequency variation.

2) Input voltage influence on the converter behavior

In this part, input frequency is being held constant at 50 Hz and the input voltage RMS has been changed from 5 V to 48 V in 5 V steps. Results have been shown in Fig. 8. As illustrated in Fig. 8, the converter cannot function properly in some low levels of the input voltage RMS. That is, in these values the duty cycle of the switch is stuck at 100% and the converter is doing its best to perform in a suitable fashion, however this situation cannot yield the desirable behavior. This case study is based on assuming the equality of input and output power and simplest equivalent circuit model is used for PMSG modelling.

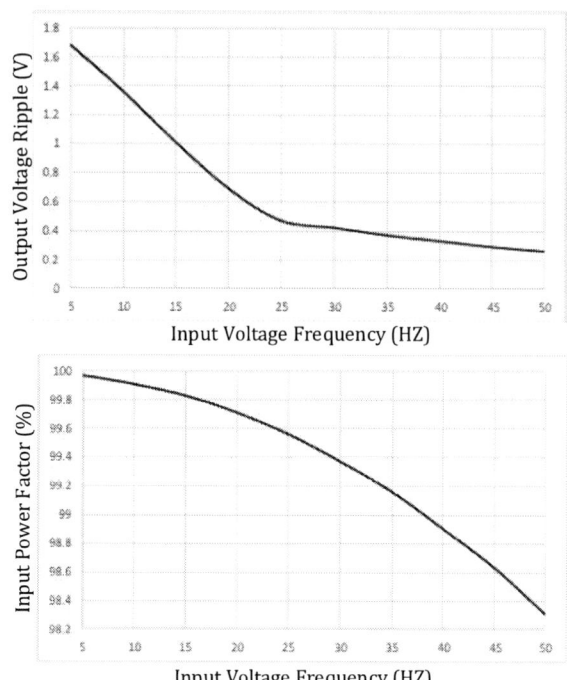

Fig. 7. Output voltage ripple (above) and input power factor (below) variations due to the variation of the input voltage frequency

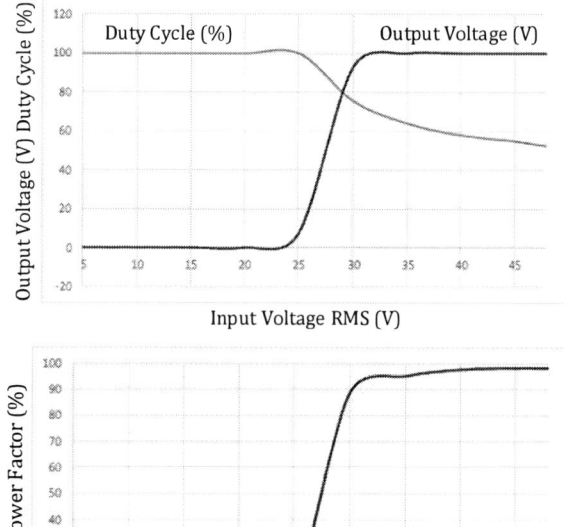

Fig. 8. Output voltage and the switch duty cycle (above) and input power factor (below) variations due to the variation of the input voltage

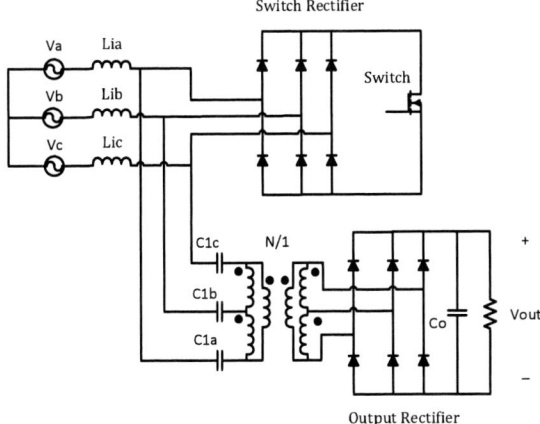

Fig. 9. Proposed DCM three-phase SEPIC converter with added high frequency three-phase transformer which is supposed to increase the level of the input voltage in order to guarantee the great performance of the proposed circuit in low levels of input voltage magnitude

In order to extend the great functionality of the converter into these low values of input voltage magnitude, some changes can be done as stated before. For instance, a three-phase high frequency transformer can be added to the proposed structure as illustrated in Fig. 9. In the proposed structure depicted in Fig. 9, the magnetizing inductance of the three-phase transformer can be utilized to play the role of the required output inductor in SEPIC converter. Nonetheless, it should be emphasized that the leakage inductance of the so-called transformer should be low enough no to have a great impact on the performance of the converter.

IV. CONCOLUSION

A novel PFC three-phase single-switch structure based on DCM SEPIC converter for PMSG and the grid intermediary has been proposed in this paper. The proposed converter has some advantages such as: reduction of the elements, simple structure and obtaining high power factor across the PMSG internal voltage source. The modes of operation discussed, as well as the closed-loop simulation for a 500 W prototype based on the characteristics of a real laboratory system. In addition, some suggestions was made to enhance the great performance of the proposed structure.

REFRENCES

[1] Zhe Chen; Guerrero, J.M.; Blaabjerg, F., "A Review of the State of the Art of Power Electronics for Wind Turbines," in Power Electronics, IEEE Transactions on , vol.24, no.8, pp.1859-1875, Aug. 2009

[2] Carrasco, J.M.; Franquelo, L.G.; Bialasiewicz, J.T.; Galvan, E.; Guisado, R.C.P.; Prats, Ma.A.M.; Leon, J.I.; Moreno-Alfonso, N., "Power-Electronic Systems for the Grid Integration of Renewable Energy Sources: A Survey," in Industrial Electronics, IEEE Transactions on , vol.53, no.4, pp.1002-1016, June 2006

[3] Polinder, H.; van der Pijl, F.F.A.; de Vilder, G.-J.; Tavner, P.J., "Comparison of direct-drive and geared generator concepts for wind turbines," in Energy Conversion, IEEE Transactions on , vol.21, no.3, pp.725-733, Sept. 2006

[4] Rodriguez, P.; Timbus, A.V.; Teodorescu, R.; Liserre, M.; Blaabjerg, F., "Flexible Active Power Control of Distributed Power Generation Systems During Grid Faults," in Industrial Electronics, IEEE Transactions on , vol.54, no.5, pp.2583-2592, Oct. 2007

[5] Conroy, J.F.; Watson, R., "Low-voltage ride-through of a full converter wind turbine with permanent magnet generator," in Renewable Power Generation, IET , vol.1, no.3, pp.182-189, September 2007

[6] Rahnamaee A.; Riazmontazer H.; Mojab A.; Zefran M.; and Mazumder S.K.; "A discontinuous PWM strategy optimized for high-frequency pulsating-dc link inverters" in the 30th Annual IEEE Applied Power Electronics Conf. and Exposition (APEC), pp. 849–853. Mar. 2015

[7] Smith, K.M., Jr.; Smedley, K.M., "A comparison of voltage-mode soft-switching methods for PWM converters," in Power Electronics, IEEE Transactions on , vol.12, no.2, pp.376-386, Mar 1997

[8] Hui, S.Y.R.; Cheng, K.W.E.; Prakash, S.R.N., "A fully soft-switched extended-period quasi-resonant power-factor-correction circuit," in Power Electronics, IEEE Transactions on , vol.12, no.5, pp.922-930, Sep 1997

[9] Redl, R.; Kislovski, A.S., "Telecom power supplies and power quality," in Telecommunications Energy Conference, 1995. INTELEC '95., 17th International , vol., no., pp.13-21, 29 Oct-1 Nov 1995

[10] Tajfar, A.; Riazmontazer, H.; Mazumder, S.K.; "Harmonics analysis for a high-frequency-link (HFL) inverter" Energy Conversion Congress and Exposition (ECCE) , 2014 IEEE, pp-2335-2341, 14-18 Sept. 2014

[11] Kolar, J.W.; Friedli, T., "The essence of three-phase PFC rectifier systems," in Telecommunications Energy Conference (INTELEC), 2011 IEEE 33rd International , vol., no., pp.1-27, 9-13 Oct. 2011

[12] Tibola, G.; Barbi, I., "Isolated Three-Phase High Power Factor Rectifier Based on the SEPIC Converter Operating in Discontinuous Conduction Mode," in Power Electronics, IEEE Transactions on , vol.28, no.11, pp.4962-4969, Nov. 2013

[13] Tibola, G.; Barbi, I., "A single-stage three-phase high power factor rectifier with high-frequency isolation and regulated DC-bus based on the DCM SEPIC converter," in Circuits and Systems (ISCAS), 2011 IEEE International Symposium on , vol., no., pp.2773-2776, 15-18 May 2011

[14] de Freitas, T.R.S.; Antunes, H.M.A.; de Freitas Vieira, J.L.; Ferreira, R.T.; Simonetti, D.S.L., "A DCM three-phase SEPIC converter for low-power PMSG," in Industry Applications (INDUSCON), 2012 10th IEEE/IAS International Conference on , vol., no., pp.1-5, 5-7 Nov. 2012

[15] Lyrio Simonetti, D.S.; Sebastian, J.; Uceda, J., "The discontinuous conduction mode Sepic and Cuk power factor preregulators: analysis and design," in Industrial Electronics, IEEE Transactions on , vol.44, no.5, pp.630-637, Oct 1997

[16] Heier, S., "Grid Integration of Wind Energy Conversion System," Chichester, U.K: Wiley, 1998.

[17] Muyeen, S.M.; Takahashi, R.; Murata, T.; Tamura, J., "A Variable Speed Wind Turbine Control Strategy to Meet Wind Farm Grid Code Requirements," in Power Systems, IEEE Transactions on , vol.25, no.1, pp.331-340, Feb. 2010

[18] Wasynczuk, O.; Man, D. T.; Sullivan, J. P., "Dynamic Behavior of a Class of Wind Turbine Generators during Random Wind Fluctuations," in Power Engineering Review, IEEE , vol.PER-1, no.6, pp.47-48, June 1981

7th Power Electronics, Drive Systems & Technologies Conference (PEDSTC 2016)
16-18 Feb. 2016, Iran University of Science and Technology, Tehran, Iran

A High Step-Up Switched-Capacitor Converter with Zero Current Switching Technique for Using in Solar System Applications

SeyedMohammad Mousavi[1], Reza Rezaii[2], Reza Beiranvand[3], Ali Yazdian Varjani[4]
Department of Electrical and Computer Engineering
Tarbiat Modares University
Tehran, Iran
[1]Mohamad.mousavi@modares.ac.ir
[2]reza.rezaii@modares.ac.ir
[3]beiranvand@modares.ac.ir
[4]yazdian@modares.ac.ir

Abstract— Small size, low cost, high efficiency, and step-up voltage-ratio converters are the most significant features required for solar system applications. In this paper, a high step-up zero current switching switched-capacitor converter is introduced. Its high power density, low cost, and soft switching features meet all the solar system's requirements. To supply a single phase ac load, a half-bridge inverter is used in the second stage. Mathematical analyses of the proposed converter are given, in detail. To verify the given analyses, a 17 V to 68 V switched-capacitor converter has been simulated and implemented at 50 kHz switching frequency. The prototype converter has been tested under different conditions. The experimental and simulated results are in good agreement with the mathematical analyses.

Keywords—Inverter, photovoltaic (PV), resonant power conversion, switched-capacitor converter

I. INTRODUCTION

Requiring new sources of energy is increasing due to technological advancements. Fossil energies will be replaced by renewable energies during time due to their restriction and high emission. Between renewable energies, the sun light is one of the most significant sources of renewable energies [1], [2]. One of the most important applications of the photovoltaic systems is supplying ac grid and load. In this regard, the output voltage of solar module should be increased in order to facilitate connecting PV modules to the network. For achieving this purpose, one can use a step-up converter with a bulk inductance or cascade some PV modules together [3]. In power generation systems such as solar power plant, using one or several central inverters leads to generate power increasing more than 1 MW. Domestic systems which are connected to PV systems, dependence or independence from network, consume electrical power varied from 1 to 10 kW, and a two-level or several one-level inverters are used in this application. These kinds of systems store their energy in batteries when they are separated from network in order to use this stored energy during night time [4].

In domestic consumption, since the output voltage of solar panels changes because of different conditions such as temperature, radiation, etc., there need a boost converter to compensate the dropped voltage of dc link required for inverter supply. Inverter topologies used in solar system can be divided into two separately parts of one-level and two-level inverters, which are based on the number of levels [5]. Central inverter topology can be termed one-level solar system. In this configuration, PV modules are connected to an inverter in order to deliver power to the network or to the load. In single stage topology, as shown in Fig. 1a, PV modules are connected to a dc/ac inverter to deliver power to the grid. In this case, inverter should track the MPP, and it should control the grid current, simultaneously. Consequently, it needs relatively more complicated controller scheme [6].

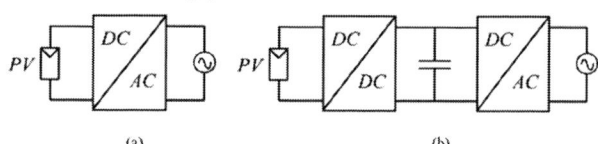

(a)   (b)

Fig. 1 (a) one-level inverter and (b) two-level inverter

On the other hand, two-stage topology uses a dc/dc converter to increase voltage and simultaneously to track MPP. This topology is shown in Fig. 1(b) [7]. Inverter used in this topology controls network or load current. In contrast to aforementioned advantages, two-level system has some disadvantages such as low efficiency, high cost, low reliability, and large size. However, because this system is simpler than previous one, it is more applicable. In this paper, to reduce both cost and size of solar systems, a step-up switched-capacitor converter (SCC) is proposed. The advantages and restrictions of traditional SCCs have been discussed, in details, in [8].

978-1-5090-0376-1/16 $31.00 © 2016 IEEE

Many efforts on topologies have been done to develop high-step voltage ratio SCCs. In [9] SCCs such as Dickson, Ladder, Fibonacci, and doublers topologies have been introduced to achieve a high voltage-ratio. However, they suffer from large number of switching elements and current spike, in addition to low efficiency. A combination of switched-capacitor and PWM converters has been introduced as hybrid SCCs in [10]. Although, these converters benefit from fast transient response and regulated output voltage, they lost high power density which is one of the most significant features of the SCCs. A multilevel modular capacitor clamped DC–DC converter (MMCCC) with the voltage ratio of four has been introduced, too [11], [12]. It improves the efficiency because of reducing conduction losses as compared to the flying-capacitor converter. However, it suffers from high switching losses and large number of switching devices (10 switches for 4 voltage-ratio conversion). Recently, a zero-current switching (ZSC) MMCCC has been proposed which uses stray inductances in each module to realize the soft switching feature [13], [14]. It improves the converter efficiency. But, it employs 3N-2 switching devices rather than 2N in the flying-capacitor topology. In addition, the extra switches have to endure more voltage stress [15].

The proposed converter and its principle operations are introduced in details in Sec. II. DC link control approach is described in Sec. III. In Sec. IV, a 60 W ZCS-exponential (2²) SCC prototype has been implemented to examine the given mathematical analyses and simulation results. Finally, Sec. V concludes the paper.

II. ZCS STEP-UP VOLTAGE- RATIO SWITCHED- CAPACITOR CONVERTER

A. Proposed converter topology and principle operation modes)

The proposed ZCS exponential SCC is indicated in Fig. 2 (a). Its voltage-ratio is equal to $V_{out}/V_{in}= 4$. Compared with other converters, as proposed in [16] and [13], this converter benefits from soft-switching feature. The employed coreless inductance restricts currents spikes during transient, and it provides ZCS condition. C_1 and C_2 are the same and they operate as resonant capacitors. Also, C_3 and C_4 are the same and operate as constant dc voltages. The proposed converter gate drive signals are shown in Fig. 2 (b). Q_2 and Q_3 have the same timing diagram, which is complementary with Q_1 and Q_4 gate drive signals. These power switches have the same gate drive signals, too. Fig. 3 shows the main operation modes of the proposed converter. Operation modes of the ZCS-SCC are described as follows:

Mode I [t_0, t_1]: Fig. 3 (a) shows the converter status during this operation mode. During this time interval Q_1 and Q_4 transistors and D_1 and D_4 diodes are turned on at t_0. When Q_1 is turned on some part of the stored charge in C_1 capacitor is transferred to C_3 and C_4 capacitors through Q_4, D_4, and D_1 devices during a short time interval. Then, combination of C_1- C_2 and L_r, forms a resonant circuit. Thus, the power devices currents can reach to zero after traversing a resonant

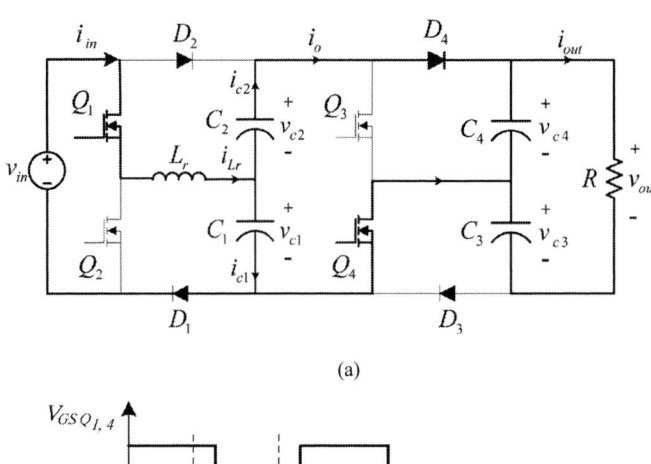

(a)

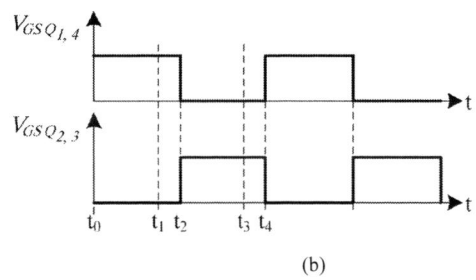

(b)

Fig. 2 (a) Proposed ZSC SCC and (b) MOSFETs gate drive signals.

circulation and before turning the transistors off by the gate-drive signals. So, Q_1, Q_4 transistors and D_1, D_4 diodes can be turned off under the ZCS condition. Although, turn-off switching losses are not dominant as compared to the turn-on losses in the power MOSFETs, but, the diode reverse recovery problem is removed, here. The converter equivalent circuit is described by the following equations during operation mode I:

$$\begin{cases} i_o\left(t\right)=i_{C_2}\left(t\right) \\ i_{in}\left(t\right)=i_{C_1}\left(t\right)+i_{C_2}\left(t\right) \end{cases} \quad (1)$$

$$V_{C_3}\left(t_0\right)+\frac{1}{C_3}\int_{t_0}^{t} i_{C_3}\left(\tau\right)d\tau+V_{C_4}\left(t_0\right)-\frac{1}{C_4}\int_{t_0}^{t} i_{C_4}\left(\tau\right)d\tau=V_{out} \quad (2)$$

Mode II [t_1, t_2]: the resonant operation is ended before this operation mode. During this time interval, as shown in Fig. 3 (b), Q_1 and Q_4 transistors can be turned off under the ZCS condition. D_1 and D_4 diodes are off. During this time interval all of the components currents are equal to zero and voltages across the capacitors C_1, C_2, C_3, and C_4 are kept constant. The output power is delivered by C_3 and C_4. The converter equivalent circuit is described by the following equations:

$$\begin{cases} v_{C_1}\left(t\right)=v_{C_1}\left(t_1\right) \\ v_{C_2}\left(t\right)=v_{C_2}\left(t_1\right) \\ v_{C_3}\left(t\right)=v_{C_3}\left(t_1\right) \\ v_{C_4}\left(t\right)=v_{C_4}\left(t_1\right) \\ i_{in}=0,\ i_o=0 \end{cases} \quad (3)$$

978-1-5090-0376-1/16 $31.00 © 2016 IEEE

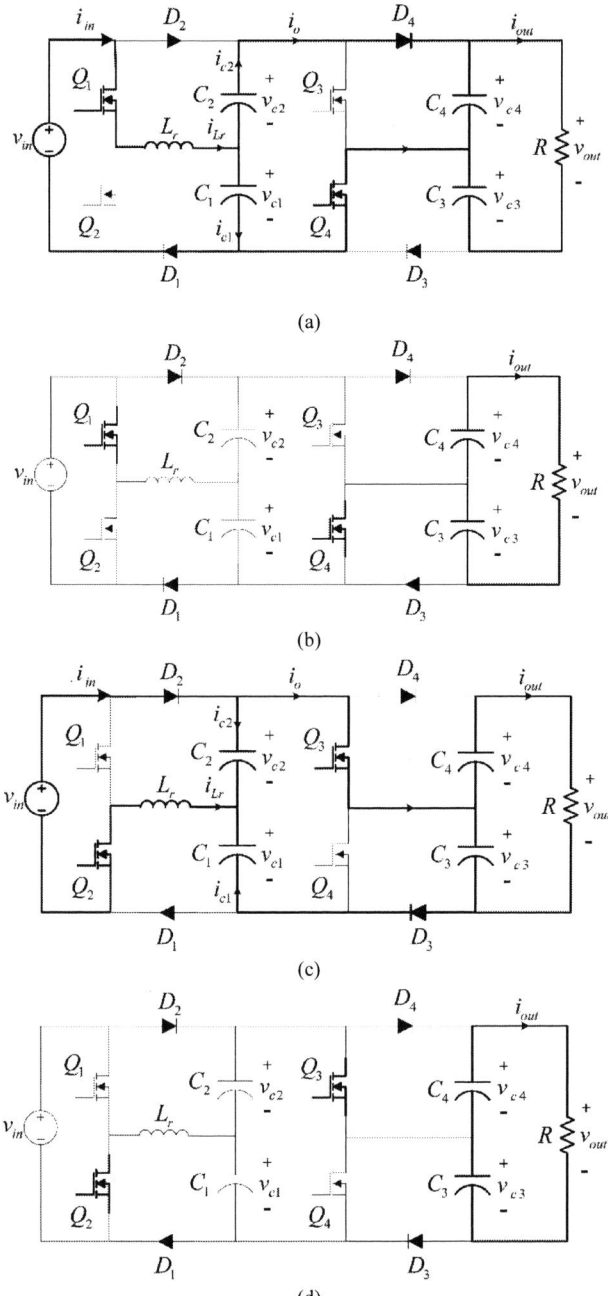

(a)

(b)

(c)

(d)

Fig. 3 Four main operation modes of the converter: (a) mode *I*, (b) mode *II*, (c) mode *III*, and (d) mode *IV*.

Mode III [t_2, t_3]: As shown in Fig. 3 (c), Q_2, Q_3, D_2 and, D_3 devices are turned on at t_2. This operation mode is similar to the first operation mode. The converter equivalent circuit during mode *III* is described by the following equations:

$$\begin{cases} i_o(t) = i_{C_1}(t) \\ i_{in}(t) = i_{C_1}(t) + i_{C_2}(t) \end{cases} \qquad (4)$$

$$V_{C_3}(t_0) - \frac{1}{C_3}\int_{t_2}^{t} i_{C_3}(\tau)d\tau + V_{C_4}(t_0) + \frac{1}{C_4}\int_{t_2}^{t} i_{C_4}(\tau)d\tau = V_{out} \qquad (5)$$

Mode IV [t_3, t_4]: as shown in Fig. 3 (d), this operation mode is similar to mode *II*, i.e., the voltages across the capacitors are kept constant and resonant operation is finished. The converter equivalent circuit is described by the following equations:

$$\begin{cases} v_{C_1}(t) = v_{C_1}(t_3) \\ v_{C_2}(t) = v_{C_2}(t_3) \\ v_{C_3}(t) = v_{C_3}(t_3) \\ v_{C_4}(t) = v_{C_4}(t_3) \\ i_{in} = 0, \, i_o = 0 \end{cases} \qquad (6)$$

Fig. 4 illustrates the simulation waveforms of the converter under the steady state condition.

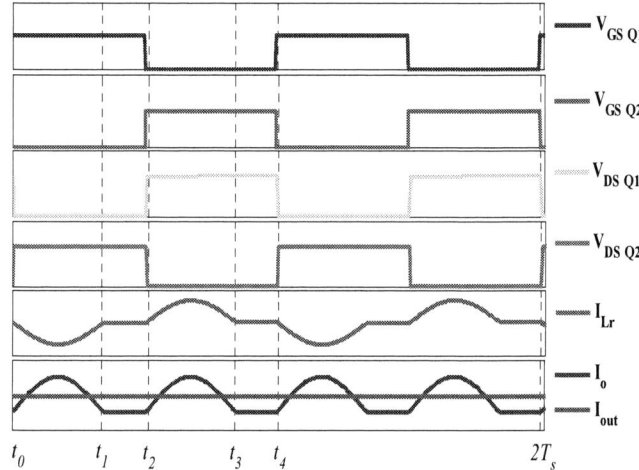

Fig. 4 Typical voltage and current waveforms of the converter, top to bottom: gate-source voltages, drain-source voltages, inductor and output currents, respectively

B. Power devices ZSC operation

Applying KVL in the resonant loop during mode *I*, the following equations can be derived:

$$\begin{cases} v_{C_1}(t) + v_{L_r}(t) - v_{in}(t) = 0 \\ i_{L_r}(t) = 2C_1 \dfrac{dv_{C_1}(t)}{dt} \end{cases} \qquad (7)$$

Considering (7), the converter resonant inductor current in Laplace domain can be calculated:

$$I_{L_r}(s) = \frac{i_{L_r}(0)L_r s + V_{in} - v_{C_1}(0)}{L_r\left(s^2 + \dfrac{1}{2C_1 L_r}\right)} \qquad (8)$$

Therefore, considering $i_{L_r}(0) = 0$ and $v_{C_1}(0) = v_{C_1 max}$, the converter resonant inductor current is obtained:

$$i_{L_r}(t) = \frac{V_{in} - v_{C_1 max}}{L_r} \sin \omega_r (t - t_x) \quad (9)$$

Where t_x is equal to t_0 and t_2 for the first and third operation modes, respectively. Also, angular resonant frequency is given as:

$$\omega_r = \frac{1}{\sqrt{2C_1 L_r}} \quad (10)$$

Considering the given assumptions in [19], to achieve input-output power balancing in the whole switching period, half of the output power should be delivered to the output load during the first operation mode. So, the following equations can be extracted during *I* and *II* operation modes:

$$V_{out} I_{out} \frac{T_s}{2} = V_{in} \int_{t_0}^{t_1} i_{L_r}(t) dt \quad (11)$$

Considering (9), (11), inductor current initial value, i.e., $i_{Lr}(t_0)=0$, and $V_{out}/V_{in}=M$, the input and inductor currents are identified:

$$i_{in}(t) = -i_{L_r}(t) = \frac{\pi M I_{out}}{2 F_n} \sin \omega_r (t - t_0) \quad (12)$$

Like conventional *ZCS* SCCs [17], [18]- [22], currents flowing through the power switches increase from zero in sinusoidal manners and return back to zero when the switching frequency f_s is chosen lower than the resonant frequency f_r. So, to realize the *ZCS* condition, we must have:

$$f_s \le f_r = \frac{1}{2\pi\sqrt{2C_1 L_r}} \quad (13)$$

After a half period, Q_2 and Q_3 are tuned off. Then, Q_1 and Q_4 are turned on. Q_4 and Q_3 are in series with the output and capacitor C_3. So their currents are equal to the output current. Subsequently, Q_1 and Q_2 have the same current value with different sign. Hence, all switches currents can be expressed as follows:

$$\begin{cases} i_{Q_1} = i_{in}(t) = \dfrac{\pi M I_{out}}{2 F_n} \sin \omega_r (t - t_0) \\[2mm] i_{Q_2} = i_{in}(t) = \dfrac{\pi M I_{out}}{2 F_n} \sin \omega_r (t - t_2) \\[2mm] i_{Q_3} = \dfrac{i_{in}(t)}{2} = \dfrac{\pi M I_{out}}{4 F_n} \sin \omega_r (t - t_2) \\[2mm] i_{Q_4}(t) = \dfrac{i_{in}(t)}{2} = \dfrac{\pi M I_{out}}{4 F_n} \sin \omega_r (t - t_0) \end{cases} \quad (14)$$

(14) clearly illustrates that all switch currents increase from zero and return back to zero before switches are turned off by the drive signals. Therefore, this converter operates under ZCS condition.

III. DC Link Control Approach

Considering that the ideal relation between output and input voltages in SCCs is a linear relationship, one can trace MPP by controlling the inverter. According to the proposed converter, the output voltage of dc link is four times greater than the module voltage. Consequently, value of the dc link voltage, as a set point for inverter controller, can be extracted as follows:

$$V_{dc\,Link}^* = (V_{MPP} \times 4) - \Delta V \quad (15)$$

Where, ΔV is equal to the dropped voltage in the circuit. In this way, inverter can control the dc link voltage by controlling its output current. Block diagram of the inverter controller for controlling dc link voltage and achieving MPP is shown in Fig. 5.

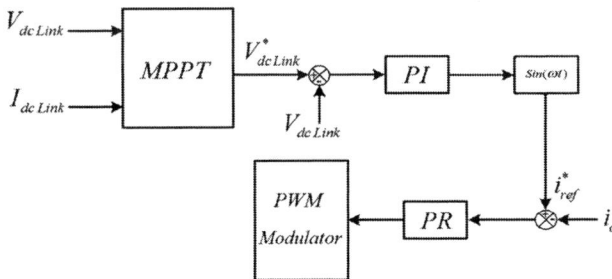

Fig. 5 Block diagram of the control approach used to achieve MPP

It should be mentioned that $V_{dc\,Link}^*$ can be derived by Incremental Conductance (INC) algorithm. In this algorithm, $V_{dc\,Link}^*$ is increased or decreased according to differences between the latest power and previous power values.

IV. Experimental Result

To confirm the practicability and superiority of the proposed converter, a 60 W prototype converter has been implemented to convert 17 V dc to 68 V dc. The solar panel model *M55,* constructed by SIEMENS has been used, here. Voltage and current values of this panel at maximum power point are equal to 17.3 V, and 3.05 A, respectively. The prototype converter photograph has been illustrated in Fig. 6. Also, the main components specifications have been given in table I.

Fig. 6 Prototype of the proposed converter

TABLE I. KEY PARAMETERS VALUES AND MAIN COMPONENTS OF THE PROTOTYPE CONVERTER

Parameter	Value	
	Experimental	Simulated
Input voltage V_{in}	48 V_{dc}	48 V_{dc}
Q_1-Q_4	IRF3205	MOSFET with R_{on}=8 mΩ
D_1-D_4	30CTQ030	V d on=0.35 V and Ron= 9 mΩ
C_1 and C_2	47 μF	47 μF
C_3 and C_4	7×1 μF	7 μF
L_r	400 nH	400 nH
Output power	53 W	53 W
Switching frequency	50 kHz	50 kHz
Maximum efficiency	93.6%	94%

Resonant frequency of the prototype converter is approximately equal to 65 kHz and its switching frequency is set lower than this value, but close to it. The experimental waveforms are shown in Figs. 7-10. Here, the switching frequency is 50 kHz. Gate-source drive signals of both Q_1 and Q_2 are shown in Fig. 7. To avoid transistors short-circuited problem, 325 ns dead-time has been considered, here. Q_1 and Q_2 transistors are turned on and off under the ZCS conditions, as can be seen in Figs. 8. Q_3, and Q_4 transistors current and voltage waveforms are similar to Q_1, and Q_2 waveforms with different amplitudes. The converter output voltage has low voltage ripple, as shown in Fig. 9 due to its dual-phase configuration. It should be mentioned that the difference between the output voltage in ideal and experimental waveforms is mainly due to diodes forward voltage drops, parasitic elements, and output capacitor equivalent resistance and stray inductance values. Different components conduction losses reduce the output voltage to 66.8 V rather than 68 V, in practice. Here, input voltage and output power are respectively equal to 17 V and 53 W.

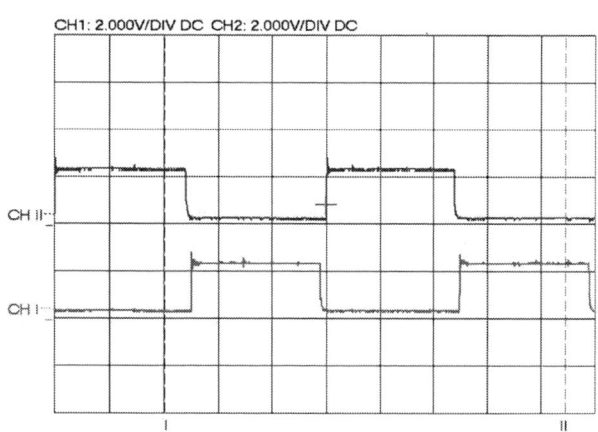

Fig. 7. Gate-source drive voltages of Q_1 and Q_2 transistors

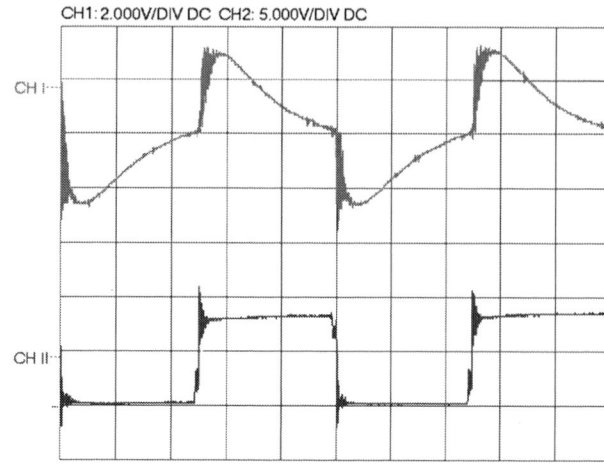

Fig.8 Resonant inductor current (up) and drain-source voltage of Q_1 transistor (down)

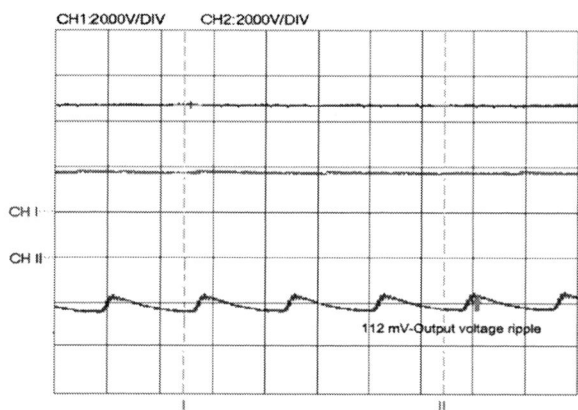

Fig.9 Output (up) and input (middle) voltages and output voltage ripple (down)

The inverter output current is shown in Fig. 10. A current feedback loop has been used to control the output current waveform. Consequently, output current THD less than 2% has been achieved.

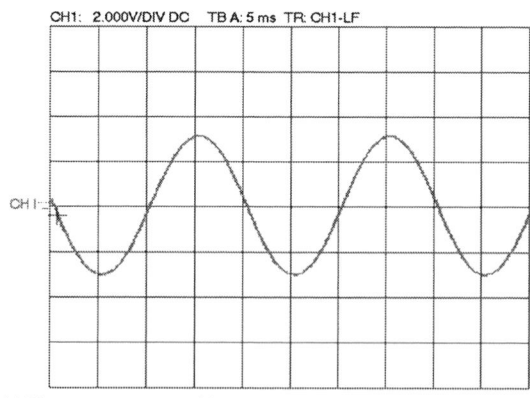

Fig. 10 The output current of inverter

V. CONCLUSION

In this paper, a step-up SCC for using in solar cell applications is proposed. The MPP is traced by INC algorithm, and dc link voltage is controlled by an inverter. The proposed converter benefits from low cost, high efficiency, and low size in comparison with the other topologies. In contrast to the PWM converters, it can be integrated and implemented on solar cell structures. Consequently, it can increase both the voltage level and efficiency of the solar panels.

REFERENCES

[1] Maki, A., & Valkealahti, S. (2012). Power losses in long string and parallel-connected short strings of series-connected silicon-based photovoltaic modules due to partial shading conditions. *Energy Conversion, IEEE Transactions on,27*(1), 173-183.

[2] Park, Y., Sul, S. K., Lim, C. H., Kim, W. C., & Lee, S. H. (2013). Asymmetric control of DC-link voltages for separate MPPTs in three-level inverters. *Power Electronics, IEEE Transactions on, 28*(6), 2760-2769.

[3] Makowski, M. S., & Maksimovic, D. (1995, June). Performance limits of switched-capacitor DC-DC converters. In *Power Electronics Specialists Conference, 1995. PESC'95 Record., 26th Annual IEEE* (Vol. 2, pp. 1215-1221). IEEE.

[4] Cervera, A., & Mordechai Peretz, M. (2015). Resonant switched-capacitor voltage regulator with ideal transient response. *Power Electronics, IEEE Transactions on, 30*(9), 4943-4951.

[5] Parastar, A., & Seok, J. K. (2015). High-gain resonant switched-capacitor cell-based DC/DC converter for offshore wind energy systems. *Power Electronics, IEEE Transactions on, 30*(2), 644-656.

[6] Rosas-Caro, J. C., Mayo-Maldonado, J. C., Mancilla-David, F., Valderrabano-Gonzalez, A., & Carbajal, F. B. (2015). Single-inductor resonant switched capacitor voltage multiplier with safe commutation. *IET Power Electronics,8*(4), 507-516.

[7] Xu, M., Sun, J., & Lee, F. C. (2006, March). Voltage divider and its application in the two-stage power architecture. In *Applied Power Electronics Conference and Exposition, 2006. APEC'06. Twenty-First Annual IEEE* (pp.1-7). IEEE.

[8] Hamo, E., Cervera, A., & Peretz, M. M. (2015). Multiple conversion ratio resonant switched-capacitor converter with active zero current detection. *Power Electronics, IEEE Transactions on, 30*(4), 2073-2083

[9] Lei, Y., & Pilawa-Podgurski, R. C. N. (2013, June). Analysis of switched-capacitor DC-DC converters in soft-charging operation. In *Control and Modeling for Power Electronics (COMPEL), 2013 IEEE 14th Workshop on* (pp. 1-7). IEEE.

[10] Gu, L., Jin, K., Ruan, X., Xu, M., & Lee, F. C. (2014). A family of switching capacitor regulators. *Power Electronics, IEEE Transactions on, 29*(2), 740-749.

[11] Khan, F. H., & Tolbert, L. M. (2007). A multilevel modular capacitor-clamped DC-DC converter. *Industry Applications, IEEE Transactions on, 43*(6), 1628-1638.

[12] Khan, F. H., & Tolbert, L. M. (2009). Multiple-load–source integration in a multilevel modular capacitor-clamped DC-DC converter featuring fault tolerant capability. *Power Electronics, IEEE Transactions on, 24*(1), 14-24.

[13] Cao, D., & Peng, F. Z. (2010). Zero-current-switching multilevel modular switched-capacitor DC-DC converter. *Industry Applications, IEEE Transactions on, 46*(6), 2536-2544.

[14] Cao, D., & Peng, F. Z. (2011). Multiphase multilevel modular DC-DC converter for high-current high-gain TEG application. *Industry Applications, IEEE Transactions on, 47*(3), 1400-1408.

[15] Qian, W., Cao, D., Cintrón-Rivera, J. G., Gebben, M., Wey, D., & Peng, F. Z. (2012). A switched-capacitor DC-DC converter with high voltage gain and reduced component rating and count. *Industry Applications, IEEE Transactions on, 48*(4), 1397-1406.

[16] Xiong, S., Wong, S. C., Tan, S. C., & Tse, C. K. (2014). A Family of Exponential Step-Down Switched-Capacitor Converters and Their Applications in Two-Stage Converters. *Power Electronics, IEEE Transactions on, 29*(4), 1870-1880.

[17] Ye, Y., Cheng, K. W. E., Liu, J., & Xu, C. (2014). A family of dual-phase-combined zero-current switching switched-capacitor converters. *Power Electronics, IEEE Transactions on, 29*(8), 4209-4218.

[18] He, L. (2014). A novel quasi-resonant bridge modular switched-capacitor converter with enhanced efficiency and reduced output voltage ripple. *Power Electronics, IEEE Transactions on, 29*(4), 1881-1893.

[19] Gebben, M. L., Cintron-Rivera, J. G., Qian, W., Cao, D., Pei, X., & Peng, F. Z. (2011, September). A zero-current-switching multilevel switched capacitor DC-DC converter. In *Energy Conversion Congress and Exposition (ECCE), 2011 IEEE* (pp. 1291-1295). IEEE.

[20] Sano, K., & Fujita, H. (2011). Performance of a high-efficiency switched-capacitor-based resonant converter with phase-shift control. *Power Electronics, IEEE Transactions on, 26*(2), 344-354.

[21] Yeung, Y. P. B., Cheng, K. W., Ho, S. L., Law, K. K., & Sutanto, D. (2004). Unified analysis of switched-capacitor resonant converters. *IEEE Transactions on Industrial Electronics, 51*(4), 864-873.

[22] Stauth, J. T., Seeman, M. D., & Kesarwani, K. (2013). Resonant switched-capacitor converters for sub-module distributed photovoltaic power management. *Power Electronics, IEEE Transactions on, 28*(3), 1189-1198.

[23] Jain, S., & Agarwal, V. (2007). A single-stage grid connected inverter topology for solar PV systems with maximum power point tracking. *Power Electronics, IEEE Transactions on, 22*(5), 1928-1940.

7th Power Electronics, Drive Systems & Technologies Conference (PEDSTC 2016)
16-18 Feb. 2016, Iran University of Science and Technology, Tehran, Iran

Space Vector PWM Algorithm for Three-Phase 16-Level Class B-2 Converter

O. Salari*
Omid.Salari@ee.kntu.ac.ir

M.J. Mojibian*
Mojibian@ee.kntu.ac.ir

M. Tavakoli Bina*
Tavakoli@eetd.kntu.ac.ir

*K.N.Toosi University of Technology, Tehran, Iran

Abstract— Multilevel converters offer a wide range of advantages; however, they suffer from their own prevalent issues such as modulation techniques for higher levels. In this paper, a novel Space Vector Modulation (SVM) technique is proposed to be implemented on a three phase Class B-2 converter. Although the SVM has the advantage of digital implementation, as well as offering redundant switching states, at least finding the tip of the reference vector in each triangle is a critical issue. The proposed SVM technique for Class B-2 converter transfers the reference vector to the first sextant, and finds the number of triangles containing the tip of the reference vector and then using angular shifts transfers the triangles of the first sextant to their original sextants. Waveforms of line voltage, load current, and their THDs are discussed.

Keywords—modulation; SVM; multilevel inverters; class B-2 Converters

I. INTRODUCTION

During the 21st century, power electronics has served as a principal player in providing power systems with vital equipment to enjoy exploiting renewable energies. Inverters are among the forerunner modules for interfacing an AC system with a DC one [1-2]. Surveying the background of inverters reveals that their first generations were mainly two level inverters, which suffered from unacceptable voltage waveforms at the AC side [3]. However, the conducted studies during recent decades, have introduced multilevel inverters as a new generation of them [4-5]. In [6], some attempts have been done for selection of the optimal place and number of multilevel inverters as FACTS devices. Although multilevel inverters offer a wide range of merits such as lower voltage THD at AC side, lower switching stresses, lower switching frequency etc., complexity exists in appropriate modulation techniques for better exploitation of them. Among very common modulation techniques such as Level-Shifted Pulse Width Modulation (LSPWM), Phase-Shifted Pulse Width Modulation (PSPWM) [7], and Space Vector Modulation (SVM) [8], the SVM offers some promising features. LSPWM and PSPWM provide simplicity of implementation, but the maximum modulation index for these techniques is 87% [8]. On the other hand, the SVM suggests higher modulation indices up to 100%, exploiting redundant switching states, and digital implementation using DSP or FPGA. Another advantage for the SVM is that it could be combined with other controllers like hysteresis controller for current control or

reduction in switching losses, and high frequency harmonics [9-10]. Some modulation techniques have been proposed in [11-13].

In [14] a general classification for multilevel inverters has been done, and finally class B-2 converter as a new topology has been introduced. Since class B-2 is a promising topology, which is capable of generating higher voltage levels with lower number of switches in comparison with traditional topologies, this paper offers a SVM technique for this topology; however, the proposed technique is capable of being implemented on any type of topologies with any number of voltage levels.

II. CLASS B-2 CONVERTER

The case study system in this paper is a Class B-2 (CB-2) multilevel inverter. For multilevel inverters various topologies with specific characteristics have been introduced up to now. Among the abundant existing topologies, CB-2 provides the advantage of efficient number of switches, and gate drivers in comparison with the number of generated output voltage levels.

The 16-level CB-2 converter is indicated in Fig.1. As can be seen in this figure, each phase of this converter has two modules ($m=2$), and each module consists of three equivalent sources ($n=3$) with two bidirectional and two unidirectional switches.

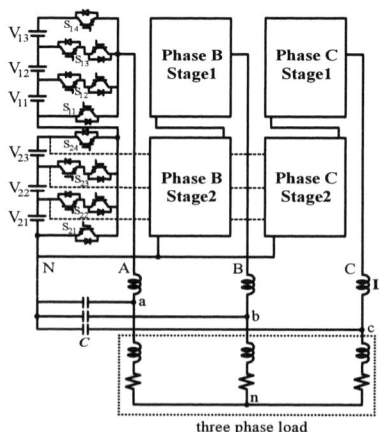

Fig. 1. Three-Phase 16-level CB-2 Converter.

978-1-5090-0376-1/16 $31.00 © 2016 IEEE
314

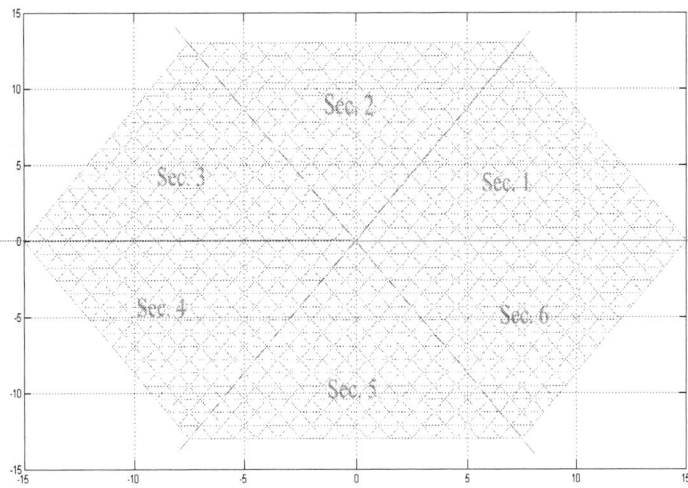

Fig. 2. The resulted nested hexagons from applying the Park's Transformation to the output voltage vectors.

To produce higher numbers of output voltage levels, the upper and lower sources should have a ratio of 1:4 in respective modules. Using this ratio, the maximum output voltage in each phase to the neutral point N, and the number of built voltage levels could be obtained through (1) and (2).

$$V_{peak} = (3 \times V_{11} + 3 \times V_{21}) = \sum_{i=1}^{2} (n_i \times V_{i1}) = 15 V_{dc} \qquad (1)$$

$$N_{level} = \prod_{i=1}^{m=2} (n_i + 1) = 4 \times 4 = 16 \qquad (2)$$

The first stage of CB-2 is capable of producing 4 voltage levels with steps equal to V_{dc}, and the second one generates voltage levels with steps equal to $4 V_{dc}$. Consequently, there would be 16 various voltage levels form 0 to 15 V_{dc}, which are defined in Table I [10].

III. SVM Strategy for CB-2 Converter

As mentioned earlier, the case study CB-2 converter is an asymmetrical converter. Asymmetrical structures produce higher numbers of output levels in comparison with the symmetrical structures of their own family; however, modulation schemes are more complicated for them [10]. Thus, it is essential to address this matter with an efficient modulation technique, which is capable of responding the requested demands of the designer.

The main goal of the SVM technique is to approximate a reference voltage with the reserved voltage vectors that the inverter is capable of producing. Thus, the modulating scheme should be fast and exact enough to tackle this objective.

Each phase of CB-2 converter has 16 different voltage levels. The numbers of switching states for the three-phase structure are equal to $(SS_{ph})^3$ where (SS_{ph}) is the number of switching states of one phase; thus, there are $16^3 = 4096$

switching states. Among the existing switching states, there are 720 switching states that introduce voltage vectors with different amplitudes and angles. The remained 3376 voltage vectors are redundant voltage vectors, which produce the same output voltages like their main voltage vectors. However, they produce different DC or AC side currents. Thus, they can be used for other controlling purposes. Each switching state has been symbolized with a number, which is the same as the ratio of V_{XN}/V_{DC} in Table I, where X could be A, B, or C. For example, the switching state 10 11 12 means that the output voltage of phase A, B, and C are equal to 10 V_{DC}, 11 V_{DC}, and 12 V_{DC} respectively.

Park's equation (3), should be applied to each of these voltage vectors. The result of applying (3) to voltage vectors, is 15 nested hexagons while each of them is the candidate for one level. The zero point in the middle of these hexagons indicates the zero vector. Consequently, there would be 15+1=16 layers. Fig. 2 implies the aforementioned sentences vividly.

$$\begin{bmatrix} v_\alpha \\ v_\beta \end{bmatrix} = \begin{bmatrix} 1 & -1/2 & -1/2 \\ 0 & -\sqrt{3}/2 & \sqrt{3}/2 \end{bmatrix} \begin{bmatrix} V_a \\ V_b \\ V_c \end{bmatrix} \qquad (3)$$

Each sector of this nested hexagonal contains 225 triangles. The method of tagging numbers of these triangles is illustrated in Fig. 3(a). Rotating these tagged triangles by the products of *60-degree* results in the numbers of other triangles in the remaining sectors. Fig. 3 indicates results of rotating the triangles of the first sextant. The switching states that form each triangle have been tagged to them in Fig. 3 above the red spots.

The tip of the reference voltage locates inside one of these triangles in each instant. Thus, finding the exact number of each triangle to select the best switching state in order to synthesize the reference voltage in the best form is vital.

978-1-5090-0376-1/16 $31.00 © 2016 IEEE   315

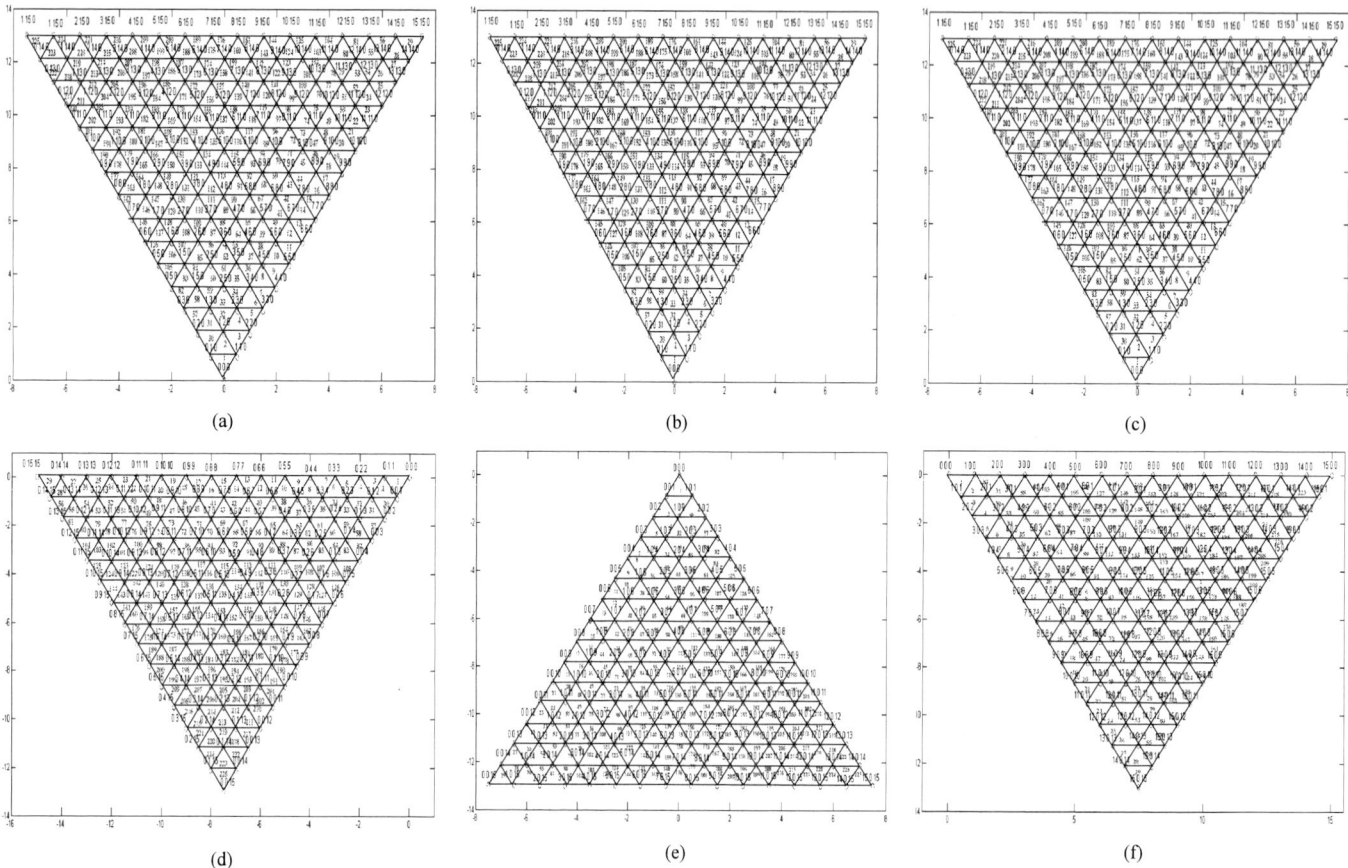

Fig. 3. Numbering process of the triangles of each sextant, (a) Sextant 1, (b) Sextant 2, (c) Sextant 3, (d) Sextant (4), (e) Sextant 5, (f) Sextant 6.

TABLE I. SWITCHING STATES AND THEIR CORESPONDING OUTPUT VOLTAGE LEVELS

Switches states								V_{AN}/V_{DC}
Stage1				Stage2				
S_{11}	S_{12}	S_{13}	S_{14}	S_{21}	S_{22}	S_{23}	S_{24}	
1	0	0	0	1	0	0	0	0
0	1	0	0	1	0	0	0	1
0	0	1	0	1	0	0	0	2
0	0	0	1	1	0	0	0	3
1	0	0	0	0	1	0	0	4
⋮	⋮	⋮	⋮	⋮	⋮	⋮	⋮	⋮
0	0	1	0	0	0	0	1	14
0	0	0	1	0	0	0	1	15

$$V1 = |V_{ref}| \cos(\theta_N) \qquad (4)$$

$$V2 = |V_{ref}| \cos(\theta_N - 60) \qquad (5)$$

First, the number of sextant containing the reference voltage should be specified. Table II indicates the condition for determination of the sextants. Second, no matter which sextant contains the reference vector, it should be transferred to the first sextant. The new angle symbolized by θ_N has been illustrated in Table II either.

Now, based on the transferred reference vector to the first sector, the exact number of the triangle, which contains the tip of the reference vector could be indicated based on (4) and (5) where (4) and (5) are the projection of the reference vector on axis L1 and L2 respectively. L1 and L2 are the axes, which make the borders of the first sextant.

According to the values of V1 and V2, a look -up table could be established to find the number of triangle in which the tip of the reference vector is located. Table III comprises the sufficient conditions for some of the triangles of the first sextant. Now, the numbers of the sextant, as well as the number of triangle that contains the tip of the transformed reference vector to first sextant are available.

As Fig.3 indicates, each triangle in the first sextant has a corresponding one in the other sextants. Thus, the exact position of the reference vector could be determined in the $\alpha\beta$ plane. Each three voltage vectors that make a triangle are corresponding to three different switching states. This means that the switching states that synthesize the reference vector are determined.

The last step is to calculate the duty cycle for each switching state to finish the process of approximating the reference vector.

TABLE II. Determination of Sectors and Transferring Them to a Sector 1

$\theta = Arctg(V_{ref_\beta} / V_{ref_\alpha})$	Sector Number	θ_N
$0 \le \theta \le 60$	1	θ
$60 \le \theta \le 120$	2	$\theta - 60$
$120 \le \theta \le 180$	3	$\theta - 120$
$-180 \le \theta \le -120$	4	$\theta + 180$
$-120 \le \theta \le -60$	5	$\theta + 120$
$-60 \le \theta \le 0$	6	$\theta + 60$

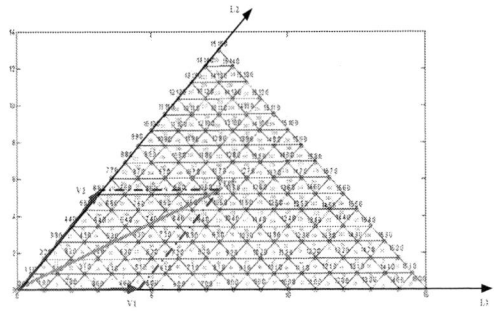

Fig. 4. Transferred vector to sector 1 and the resultant vectors from its projection on L1 and L2.

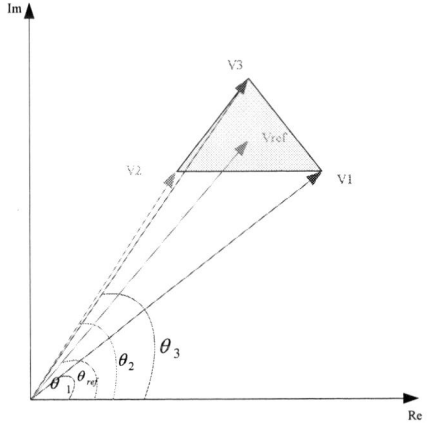

Fig. 5. Reference vector in an assumptive triangle with its adjacent vectors.

Solving (6) for D_1, D_2 and D_3 leads to duty cycles of each switching states.

$$\begin{cases} (V_1 D_1 \cos(\theta 1) + V_2 D_2 \cos(\theta 2) + V_3 D_3 \cos(\theta 3)) = V_{ref} \cos(\theta ref) \\ (V_1 D_1 \sin(\theta 1) + V_2 D_2 \sin(\theta 2) + V_3 D_3 \sin(\theta 3)) = V_{ref} \sin(\theta ref) \\ D_1 + D_2 + D_3 = 1 \end{cases} \quad (6)$$

where V_1, V_2, V_3, V_{ref}, $\theta 1, \theta 2$, $\theta 3$, and θref are illustrated in Fig. 5 for an assumptive triangle in sector 1.

IV. SIMULATION

In order to validate the theory, Matlab/Simulink environment has been used. Line voltage, load current, and voltage and current THD are discussed in this part as the results. Simulation parameters are summarized in Table IV. The load is a three-phase one, connected in Y style, and its neutral point is grounded. Power factor of load is 0.6. As mentioned earlier, the ratio of the upper module voltage sources to lower ones, should be 1:4. In this case, upper ones are chosen 5 V while the lower ones are 20 V.

One of the basic advantages of the SVM over other modulation schemes is its ability to work in full modulation index (M). In other words, LSPWM or PSPWM offer a maximum modulation index of 0.87; however, SVM is capable of performing in M=1. To validate this claim, M=1 was chosen as the first step of simulation part. For M=1 or other modulation indices near it, the voltage waveforms should introduce the maximum number of levels. In other words, if the maximum number of phase voltage level is n, then in line voltage there should be $2n$-1 voltage levels. The studied CB-2 is a 16 level inverter; thus, for line voltage during the full modulation index, there should be 31 voltage levels. Fig. 6 (a), introduces a 31 voltage levels, which is in agreement with the theory.

Line current has been indicated in Fig. 6 (b). Voltage and current THDs, could be followed in Fig. 6(c), Fig. 6 (d) respectively. Voltage THD is 2.11%, and current THD is 1.66 %. Therefore, they are in an acceptable domain. A SVM strategy should be able to respond appropriately in situations where a sudden change is going to happen. In this case, three respective changes in modulation index were applied to the system. At the beginning, M was 0.3, after 0.05 sec, M increased to 0.5, then in t= 0.1 sec M increased to 0.87, and its final value was 0.1 in t= 0.15 sec. Line voltage, and current for this part of the simulations is indicated in Fig. 7 (a), (b) respectively. Voltage levels and amplitudes change in response to change in modulation index. Increase in the value of M results in increase in the number and amplitude of synthesized voltage.

TABLE III. Conditions of Triangle Determination Based on Decomposed Vectors

V1	V2	V1+V2	Tri. No.
$0 \le V1 < 1$	$0 \le V2 < 1$	$V1 + V2 \le 1$	1
$0 \le V1 < 1$	$0 \le V2 < 1$	$1 \le V1 + V2 \le 2$	2
$9 \le V1 < 10$	$0 \le V2 < 1$	$9 \le V1 + V2 \le 10$	20
$5 \le V1 < 6$	$1 \le V2 < 2$	$6 \le V1 + V2 \le 7$	40
$9 \le V1 < 10$	$3 \le V2 < 4$	$12 \le V1 + V2 \le 13$	100
$0 \le V1 < 1$	$14 \le V2 < 15$	$14 \le V1 + V2 \le 15$	225

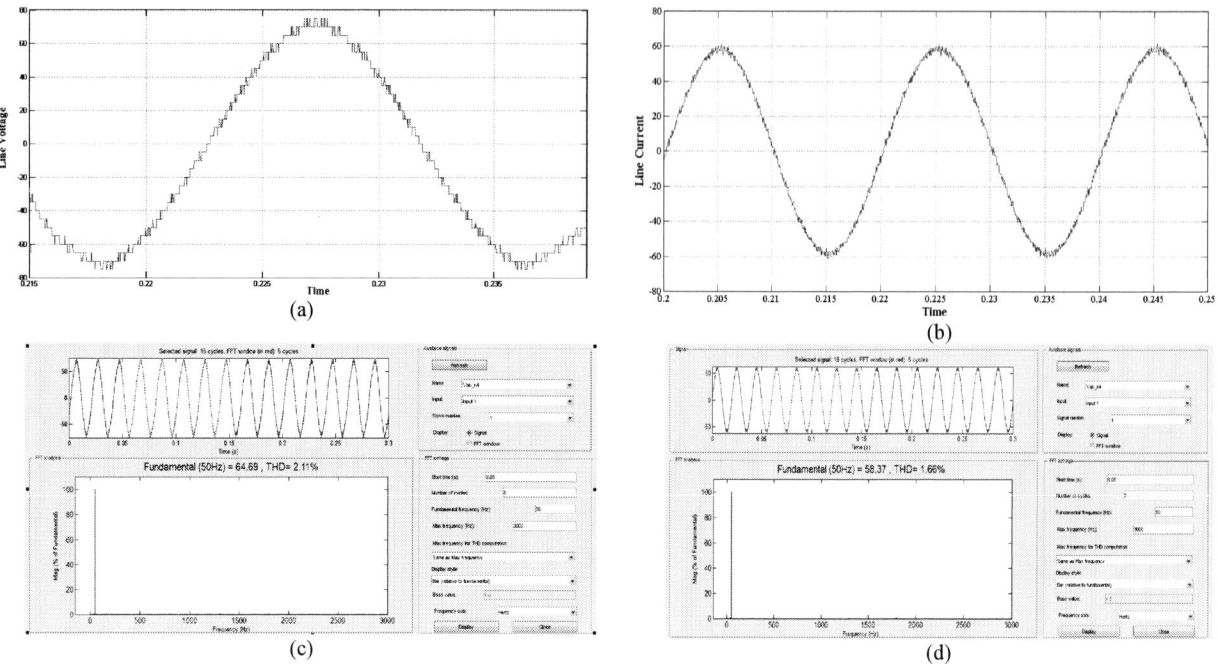

Fig. 6. Simulation results for M=100%, (a) line voltage, (b) load Current, (c) voltage THD, (d) current THD.

TABLE IV. SPECIFICATION OF THREE- PHASE CB-2 CONVERTER

Simulation Parameters	Values
Voltage source ratings of the upper modules	5 v
Voltage source ratings of the lower modules	20 v
Switching Frequency	8K Hz
Fundamental Frequency	50 Hz
Load power factor	0.6
Converter Switch type	IGBT

TABLE V. SVM VS. OTHER MODULATION METHODS

Modulation technique	Implementation Method	THD (%)		Maximum modulation index (%)
		V_{ab}	I_a	
LS-PWM	Analog	4.25	5.6	87
PS-PWM	Analog	4.34	6.1	87
SVM	Digital	2.1	1.66	100

On the other hand, decrease in M results in decreasing the number of output voltage levels and amplitudes.

A summary of implementation of LS-PWM, PSPWM, and SVM on the CB-2 converter has been gathered in Table V. As one may consider, SVM offers a wide range of superiority over LS-PWM and PSPWM; however, it is far harder to implement.

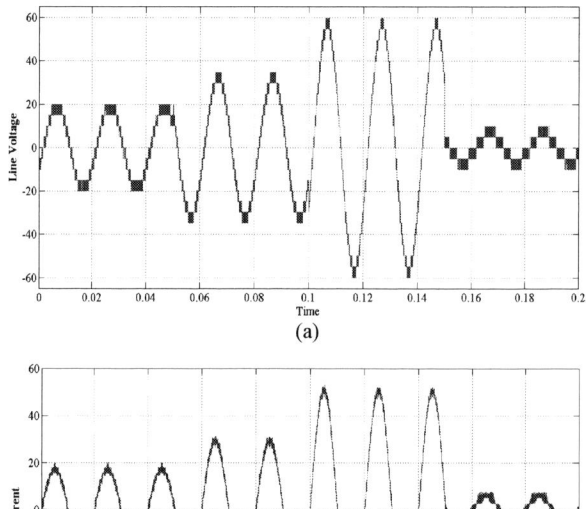

Fig. 7. Simulation results for sudden change in M. (a) line voltage (b) load Current.

978-1-5090-0376-1/16 $31.00 © 2016 IEEE

V. CONCLUSION

Multilevel inverters are experiencing a wide range of development because of their inherent advantages over traditional inverters. As the number of levels increase in these structures, some issues such as modulation strategies emerge. In this paper a SVM strategy was proposed for a novel 16 level multilevel inverter known as CB-2. First, the procedure of implementation was explained, and then it was implemented using Matlab/Simulink on a three-phase CB-2 inverter. Voltage, current, and their THD waveforms were plotted and explained. As current and voltage THD showed, the case study inverter is capable of working without the need for bulk filters.

REFERENCES

[1] J. Rodriguez, S. Bernet, B. Wu, J. O. Pontt, and S. Kouro, "Multilevel voltage-source-converter topologies for industrial medium-voltage drives" *IEEE Trans. Ind. Electron.*, vol. 54, no. 6, pp. 2930–2945, Dec. 2007.

[2] L. G. Franquelo, J. Rodriguez, J. I. Leon, S. Kouro, R. Portillo, and M. A. M. Prats, "The age of multilevel converters arrives" *IEEE Ind. Electron.*, vol. 2, no. 2, pp. 28–39, Jun. 2008.

[3] S. Kouro, M. Malinowski, K. Gopakumar, J. Pou, L. G. Franquelo, "Recent Advances and Industrial Applications of Multilevel Converters" *IEEE Trans. Ind. Electron.*, vol. 57, no. 8, Agu. 2010.

[4] Pirouz, H.M., Bina, M.T., "Modular multilevel converter based STATCOM topology suitable for medium-voltage unbalanced systems", Journal of Power Electronics, Vol. 10, No. 5, September 2010, pp. 572-578

[5] Rahimzadeh, S., Tavakoli Bina, M., Viki, A.H., "Simultaneous application of multi-type FACTS devices to the restructured environment: Achieving both optimal number and location", *IET Generation, Transmission and Distribution*, Volume 4, Issue 3, March 2010, pp. 349 – 362

[6] J. Rodriguez, J.S. Lai, and F.Z.Peng, "Multilevel inverters: A survey of topologies, control, and applications" *IEEE Trans. Ind. Electron.*, vol. 49, no.4, pp. 724- 738, Aug. 2002.

[7] B. P. McGrath, D. G. Holmes, "Multicarrier PWM strategies for multilevel inverters", *IEEE Trans. Ind. Electron.*, vol. 49, no. 4, 2002, pp. 858-867.

[8] M. Saeedifard, P. M. Barbisa, P.K. Steimer, "Operation and Control of a Hybrid Seven- Level Converter" *IEEE Trans, Ind. Electron*, vol.27 , no. 2, 2012.

[9] A. Nazemi, O.Salari, M. Tavakoli Bina, " Design of a three level hystersis controller for a four leg inverter in $\alpha\beta$ 0 Frame", 6th Power Electronics, Drive Systerms and Technologies (PEDSTC), Tehran, Iran, 2015.

[10] Vahedi, H., Sheikholeslami, A., Tavakoli Bina, M., Vahedi, M., "Review and simulation of fixed and adaptive hysteresis current control considering switching losses and high-frequency harmonics", Advances in Power Electronics, May 2011

[11] J. Hu, K. Chen, T. Shen, and C. Tang, " Analytical solutions of multilevel space vector PWM for multiphase voltage source inverters" *IEEE Trans. Power Electron.*, vol.26, no.5, pp. 1489- 1502, May 2011.

[12] B. Jacob, M.R. Baiju, "A New Space Vector Modulation Scheme for Multilevel Inverters Which Directly Vector Quantize the Reference Space Vector" *IEEE Trans, Ind. Electron*, vol. 62, no.1; pp. 88- 95, 2015.

[13] O. Lopez, J. Alvarez, J. Doval-Gandoy, and F. D. Freijedo, "Multilevel multiphase space vector PWM algorithm with switching state redundancy" *IEEE Trans. Ind. Electron.*, vol. 56, no. 3, pp. 792–804, Mar. 2009.

[14] M. J. Mojibian and M.T.Bina, "Classification of multilevel converters with a modular reduced structure: implementing a prominent 31-level 5

kVA class B converter" *IET Power Electron.*, vol. 8, no. 1, pp. 20–32, Jan. 2015.

7th Power Electronics, Drive Systems & Technologies Conference (PEDSTC 2016)
16-18 Feb. 2016, Iran University of Science and Technology, Tehran, Iran

An Interleaved High-Power Two-Switch Flyback Inverter with a Fast and Robust Maximum Power Point Tracker

Saleh Mohammadi
Dept. of Electrical Engineering
Bojnord Branch, Islamic Azad University
Bojnord, Iran
Saleh.67mohammadi@yahoo.com

H.Abootorabi Zarchi
Dept. of Electrical Engineering
Ferdowsi University of Mashhad
Mashhad, Iran
abootorabi@um.ac.ir

Abstract— **The major drawback of the two switch flyback topology is its low-power applications such as microinverter. This study proposes a high-power two-switch flyback inverter and shows its excellent performance as a string-type PV inverter. The suggested inverter system is based on interleaving technique and operated in discontinuous conduction mode. Moreover, a fast and robust maximum power point tracking method is proposed. The fast dynamics and robustness of the proposed method is achieved using variable structure control approach. Simulation results with a 2 kW inverter system confirm the excellent performance of the suggested scheme. The total harmonic distortion and power factor are measured as 2% and 0.99 respectively. Therefore, it is shown that the performance of the suggested system is comparable to the common isolated PV inverters available in the market.**

Keywords—two switch flyback; variable structure control; THD; power factor; MPPT.

I. INTRODUCTION

Nowadays, environmental concerns and freely available nature of photovoltaic (PV) energy have caused extensive applications of this important source of energy [1]. In photovoltaic systems, the flyback converter is considered as an appropriate topology due to advantages of fewer component count, simplicity and isolation between the PV modules and grid line [2]. One of the biggest drawbacks of the flyback inverter is the high voltage and current stresses of the switches which suffers them. A switch with higher voltage rating is required to withstand the turn-off transient voltage. Consequently the size and cost of the switch is increased. Moreover the on-resistance of the switch is increased, resulting in increased conduction losses [3].

The two switch flyback converter is an appropriate solution to reduce the voltage stress of the switches. The switch voltage stress in this converter is decreased to the dc input voltage, reducing the switching and conduction losses. The additional switch and the main one operate homogeneously, therefore the control logic is very simple [4], [5].

The design of a transformer with large energy storage capability is always a challenge. The air gap where the energy

is stored is very large in high-power flyback converter. As a result of this, the leakage flux is increased and efficiency of the converter is decreased. Chiefly this reason most often causes that the flyback converters are designed for low power applications. However, if we apply modern and optimal designing schemes properly, this converter topology can be applied in high-power too. Interleaving of flyback inverters assists designing high-power flyback inverters. Furthermore, when the number of interleaved cells increased, the frequency of harmonics at the waveform are increased as well. This feature makes it easy to filter out the harmonics by using smaller sized filtering elements. Therefore the cost of passive elements are reduced [6], [7].

In contrast to pulse width modulated (PWM) based maximum power point tracking (MPPT), we propose a sliding mode controlled MPPT. Variable structure or sliding mode controlled MPPT presents a fast response to variations in radiation. In addition, the implementation of the control is simple and requires low cost hardware. This work may be viewed as an extension of the work by Levron and Shmilovitz [8]. The authors proposed an MPPT controller based on an inner sliding-mode loop which uses inductor's current as the main state variable. This method is generally more robust and stable, but the switching frequency cannot be determined exactly. Also, the maximum power point (MPP) voltage affects the switching frequency, thus, the switching frequency changes with the temperature and the design of the controller is complicated and may cause instability. A MPPT method is suggested in [9] based on variable structure control approach and P-I characteristic of the PV panel which has a slow response to variation in radiation.

The main goal of this study is to design and suggest a grid-tied, isolated and string-type inverter based on the two-switch flyback converter topology at 2 kW, which is not available in today's PV market. Moreover, a new fast and robust MPPT technique is presented which is based on variable structure control approach. The suggested control system adjusts the switching frequency precisely and independent to MPP voltage. So the switching frequency doesn't change in case of

978-1-5090-0376-1/16 $31.00 © 2016 IEEE

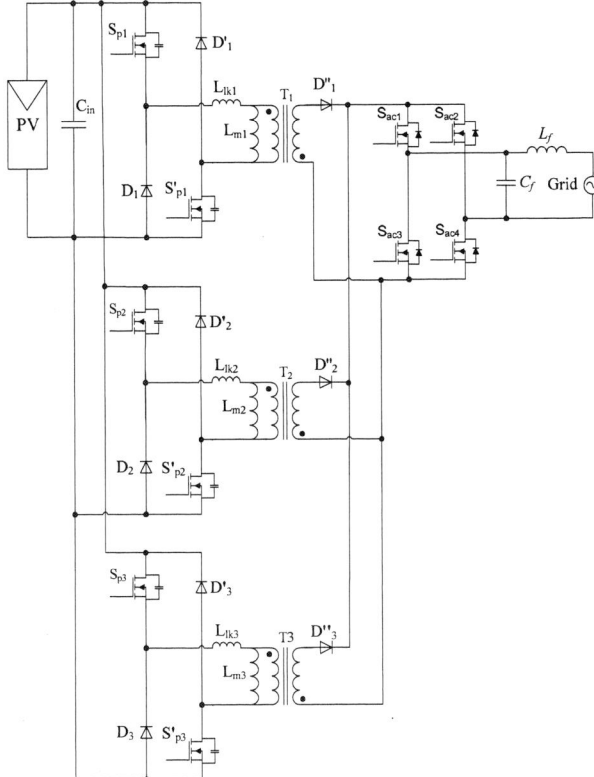

Fig. 1: Schematic diagram of the suggested PV inverter system

variation in temperature. The proposed system has performed effectively according to the main specifications such as the power factor and the THD of the grid current.

This paper has five main parts. Part II introduces the proposed topology. Design steps are stated in part III and design equations are derived. The proper operation of the introduced scheme is confirmed through simulations presented in part IV. The last part provides the conclusions.

II. CONVERTER DESCRIPTION AND ANALYSIS

A. Proposed Structure

Fig. 1 shows the proposed PV inverter system. The proposed system consists of five main parts: first-phase converter, second-phase converter, third-phase converter, unfolding bridge, and C–L filter. S_{Pi}, $S'_{Pi\,(i=1,\,2,\,3)}$ are the main power switches; D''_i is the rectifier diode; L_{lki} is the Leakage inductance, and L_{mi} is the magnetizing inductance. Unfolding the rectified sinusoidal waveform for attaching to the grid is done with a current source inverter formed by S_{ac1}-S_{ac4}. D_i and D'_i are clamping diodes.

B. Operational Analysis of the Proposed Structure

Steady-state performance of the system is composed of four operational modes in a switching period. The simplicity of the control encouraged us to apply discontinuous conduction mode (DCM) in design of the proposed scheme. The principle of the operation of each stage is explained according to the equivalent circuits of the first-phase converter shown in Fig. 2 and the voltage and current waveforms of the converter shown in Fig. 3.

Stage 1 $(t_0 < t < t_1)$: At the beginning of this stage, switches S_{P1} and S'_{P1} are turned on. The PV panel voltage is across the primary winding. So, the current of magnetizing and leakage inductances increases linearly. The current of L_{m1} and L_{lk1} can be stated as:

$$i_{Lm1} = i_{Llk1} = \frac{V_{pv}}{L_{m1} + L_{lk1}}(t - t_0) + i_{Lm1}(t_0) \qquad (1)$$

Where $i_{Lm1}(t_0) = 0$. The equation of the peak current of L_{m1} can be written as follows:

$$i_{Lm1}(t_1) = \frac{V_{pv}\, d\, T_s}{L_{m1} + L_{lk1}} \qquad (2)$$

In (2), d and dT_s are the duty cycle and on time of the main switches S_{P1} and S'_{P1} in a particular switching cycle respectively. The peak current of the magnetizing inductance has a sinusoidal waveform. So we have:

$$d(t) = d_{\max}\sin(\omega t) \qquad (3)$$

d_{max} refers to maximum duty cycle. At the end of this stage the switches S_{P1} and S'_{P1} are turned off.

Stage2 $(t_1 < t \le t_2)$: In this stage, the reflected output voltage $-nV_g$ $(0 \le V_g \le 220\sqrt{2})$ is put across the primary winding and the voltage across the switches S_{P1} and S'_{P1} is limited to V_{pv}. The energy stored in L_{lk1} charges the input capacitance C_{in} through diodes D_1 and D'_1. So, we have:

$$i_{Lm1} = i_{Llk1} = -\frac{V_{pv}}{L_{m1} + L_{lk1}}(t - t_1) - i_{Lm1}(t_1) \qquad (4)$$

$i_{Lm1}(t_1)$ refers to the initial current of the magnetizing inductance at $t = t_1$. At the end of this stage, i_{Llk1} drops to zero. Thus, clamping diodes D_1 and D'_1 are turned off.

Stage3 $(t_2 < t \le t_3)$: In this stage, all the energy stored in L_{m1} is transferred to the secondary side. An equivalent circuit for this stage is depicted in Fig. 2(c). The voltage of $-nV_g$ is put across the magnetizing inductance. So, the current passes through the magnetizing inductance is equal to:

$$i_{Lm1} = -\frac{nV_g}{L_{m1}}(t - t_2) + i_{Lm1}(t_2) \qquad (5)$$

i_{Lm1} (t_2) refers to the initial current of L_{m1} at time t_2. If we consider that the switches S_{P1}, S'_{P1} are identical, the voltages across them is represented as follows:

$$V_{SP1} = V_{S'P1} = \frac{V_{pv} + nV_g}{2} \tag{6}$$

Stage4 $(t_3 < t \le t_4)$: In this stage, the energy stored in C_f and L_f is transferred to the grid. The voltage across S_{P1} and S'_{P1} is represented as follows:

$$V_{SP1} = V_{S'P1} = \frac{V_{pv}}{2} \tag{7}$$

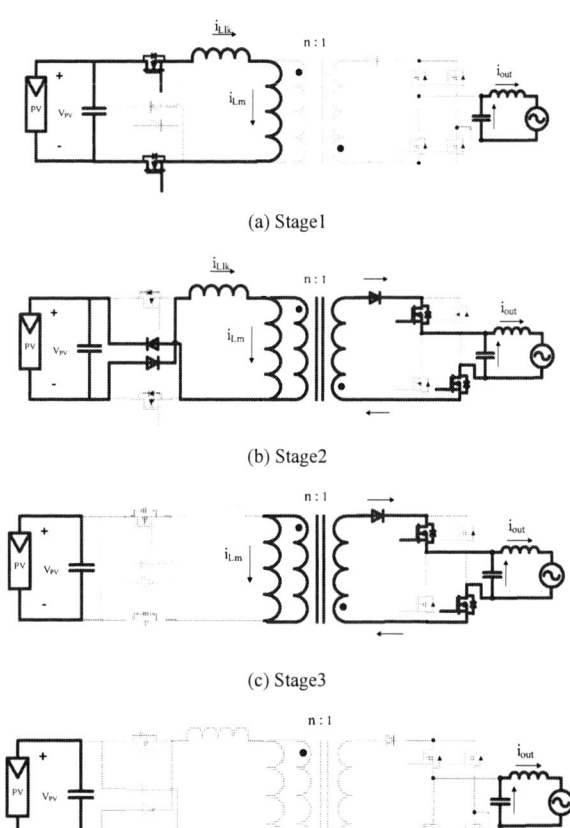

(a) Stage1

(b) Stage2

(c) Stage3

(d) Stage4

Fig.2 Equivalent circuits of the proposed system in steady-state operation

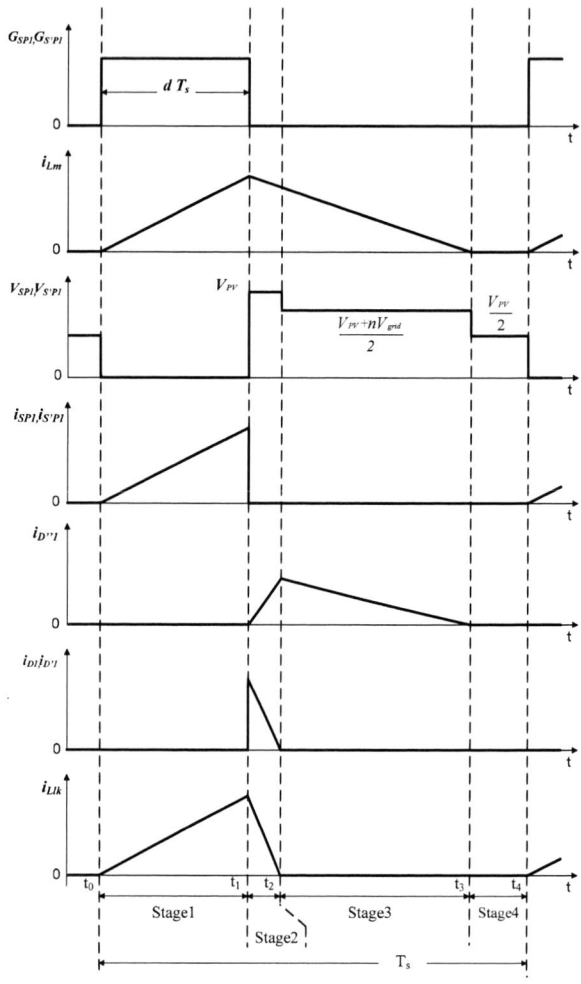

Fig. 3 Voltage and current waveforms of the proposed structure

III. DESIGN PROCEDURE

A. Parameters Design

The main specifications for designing the proposed inverter system is summarized in Table I. When the switches S_{Pi} and S'_{Pi} are turned off, we should prevent conduction of D_i, D'_i and also prevent transferring the energy stored in magnetizing inductance energy to the input capacitor. So, two important limitations in designing the inverter system should be considered as follows:

$$nV_m < V_{pv} \tag{8}$$

$$d_{max} < 0.5 \tag{9}$$

In (8), V_m refers to the peak value of the output voltage. Two-thirds of V_{pv} is an appropriate peak value of reflected output voltage nV_m [10].

TABLE I. DESIGN SPECIFICATIONS

Design parameters	Specifications
PV model and maximum power	JC250M,250 W
Open circuit voltage and short circuit current per panel	37.4 V,8.83 A
Maximum power current and voltage per panel and per the selected panel group arrangement	30.1 V,8.31 A,240.8 V,8.31 A
Total maximum dc power from the PV array	2000 W
Grid characteristics	Single-phase, 220 V, 50 Hz
Grid current percent THD	<5%
Power factor	>0.99
Switching frequency	20 kHz
Number of interleaved cells	3

Following equation should be considered for determining maximum duty cycle d_{max} [7]:

$$d_{max} \le (1 + \frac{V_{pv}}{nV_m})^{-1} \quad (10)$$

For $V_m = 220\sqrt{2}$ V and $V_{pv} = 240.8$ V, turn ratio should be n < 0.77. We select n=0.33 considering the above limitations.

In [11], the average primary current value is described as follows:

$$I_{P,avg} = \frac{n_{cell} V_{pv} T_s d_{max}^2}{4 L_m} \quad (11)$$

The input power of interleaved flyback inverter is:

$$P_{pv} = V_{pv} I_{p,avg} = \frac{n_{cell} V_{pv}^2 T_s d_{max}^2}{4 L_m} \quad (12)$$

For P_{PV}=2000 W, V_{pv}=240.8 V and d_{max}=0.3, the maximum transformer magnetizing inductance is equal to $L_{m,max}$=97.8 µH.

B. Reference Current design

In order to regulate the output current, the reference signal for each phase is obtained as follows[7]:

$$i_{refi} = \sin(\omega t)\sqrt{\frac{4P_o}{n_{cell} L_m f_s}} \quad (13)$$

C. Proposed MPPT control method

The PV generators (PVG) exhibit nonlinear $I–V$ and $P–V$ characteristic curves. The maximum power produced depends on both irradiance and temperature. Reference [8] with the aid of I-V curves of PV panel, defines a switching surface that satisfies MPPT regarding irradiance. The proposed surface in [8] is:

$$S(V, i) = a\,i - b\,V + ref = 0 \quad (14)$$

Where V and i are PV panel output voltage and current, and a, b, and ref define the switching surface. Fig. 4, shows the proposed surface. The switching frequency is dependent on MPP voltage and has different value when the temperature changes according to (15) [8].

$$f_{sw} = \frac{aV_{mpp}}{\Delta.L}(1 - \frac{V_{mpp}}{V_o}) \quad (15)$$

Moreover, the designed switching frequency has a 20% error in comparison with experimental result. Thus, the design of the controller system can be a complicated one. By comparison, in this work we design the switching frequency precisely and independent on MPP voltage.

The schematic diagram of the system in presence of proposed control scheme is illustrated in Fig. 5. The main part of it is "linear and variable structure control" (LVSC) which comprises the sliding-mode and linear controller. This controller takes advantage of the best specifications of the linear controller like, smooth operation and the best features of variable structure control that are, robustness to perturbations and modeling uncertainties. In the proposed system, the MPPT control system consists of sliding-mode controller with an adaptive switching surface which makes the system to operate along a line in close proximity of the MPP loci. Obviously, MPPT would be satisfied if $S(V, i)$ can be preserved in zero value, as shown in Fig. 4.

The controller produces the magnitude of the reference current i_{ref} as given by:

$$i^* = (K_P + K_I/s)(e_s + K\,\text{sgn}(S)) \quad (16)$$

In (16), K_I and K_P represent the PI controller gains and K is the VSC gain. Using PV voltage and current, switching surface is calculated and then an error is defined as:

$$e_s = S^* - S \quad (17)$$

The intercept, ref, is set by a conventional MPPT control algorithm. So, P&O algorithm is selected because of its simplicity. A proper selection of slope introduced by constants a and b, can effectively reduce the MPPT's convergence time. So, we use least square estimate (LSE) method to the set of maximum power points correspond to different irradiation levels as an appropriate choice of slope. During transients, $e_s > K\,\text{sgn}(S)$, and the Linear property is dominant. In the steady-state, error is very small and the switching characteristic plays an important role. Also the K gain determines the ripple value. Adequate balance between

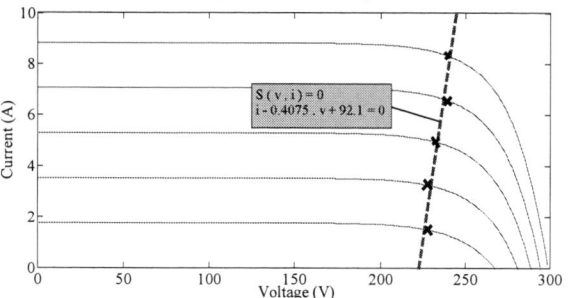

Fig.4 proposed switching surface [8]

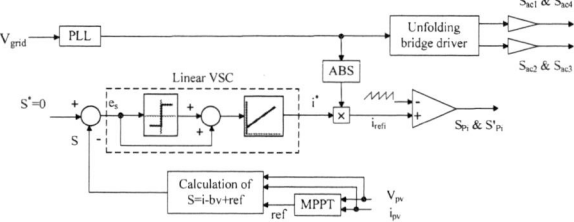

Fig.5 Block diagram of the proposed control system

linear and switching characteristic is easily obtained by proper gain selection. PI gains are chosen so that the linear control provides the desired dynamic response, while the VSC gain determines the robustness in steady-state operation. Phase locked loop (PLL) is used to detect the phase angle, amplitude and frequency of the grid voltage accurately and quickly to synchronize the reference (primary) current with the grid voltage and control the H-bridge inverter for unfolding purpose. The reference current is compared with a sawtooth waveform to produce switching signals for the main switches (S_{Pi} & S'_{Pi}). So, we have a fixed frequency to draw maximum power from PVG.

IV. PERFORMANCE EVALUATION

To evaluate the effectiveness of the suggested inverter and MPPT control system, several tests have been conducted on PSIM 9.0 software integrated with MATLAB software. The main parameters of the simulated system are listed in Table II. The appropriate switching surface is selected by applying LSE to the MPP loci. The resulting slope (normalized by a) is $a = 1$, $b = 0.4075$ and $ref = 92.1$. Waveforms of the simulated system during the grid period are plotted in Fig. 6. Reference current irefl in (13) is followed by current waveform isp1 depicted in Fig. 6(a). Voltage waveform Vsp1 is plotted in Fig. 6(b). As it is depicted, the switch voltage is clamped to the dc input voltage V_{PV}. Waveforms of the simulated system during the switching period are depicted in Fig. 7. Switching signals GSP1, GSP2 and GSP3 are given to the switches SP1, SP2 and SP3 respectively shifted at 120° to have lower ripple [7] which has been plotted in Fig. 7(a). Current waveforms iSP1, iSP2 and iSP3 and output rectifier diode currents iD"1, iD"2 and iD"3 are shown in Fig. 7(b) and 7(c) respectively.

Clamping diode currents iD1, iD2 and iD3 are displayed in Fig. 7(d). Fig. 7(e) illustrates voltage waveform VSP1. According to the simulation results, the predicted waveforms of Fig. 3 are confirmed with the simulation waveforms. Fig.8 shows the simulation results of MPPT control. The irradiation changes from 1000 W/m² to 800 W/m². The sliding-mode MPPT, is compared with an ordinary "PWM based MPPT". The sliding mode MPPT is seen to converge within 7 ms with the MPPT step size of $ref = 0.4$, whereas the PWM based MPPT converges within 0.5 s. Fig. 9 shows switching surface values. Based on these results, it can be concluded that the proposed MPPT control has a fast and robust response in comparison with PWM based MPPT. Fig.10 shows the simulated waveforms of the grid voltage and current. The grid current has 2% THD and the power factor of 99.96. The waveforms demonstrate the success of the proposed control system and the proposed PV inverter system in achieving the high quality energy transfer into the grid.

TABLE II. SIMULATION PARAMETERS OF THE PROPOSED SYSTEM

Parameter	Value	Unit
Switching frequency f_{sw}	20	kHz
Grid frequency f_{grid}	50	Hz
Transformers turn ratio (Np/Ns)	0.33	
Nominal power	2000	W
Magnetizing inductance L_m	78	µH
Leakage inductance L_{lk}	0.78	µH
Input capacitance C_{in}	5.3	mF
Filter inductance Lf	0.3	mH
Filter capacitance C_f	0.35	µF

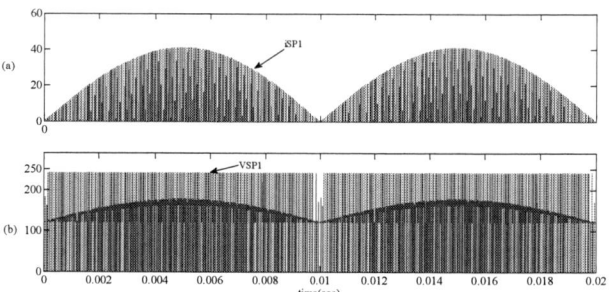

Fig.6. Performance of the suggested system under full load condition during grid period

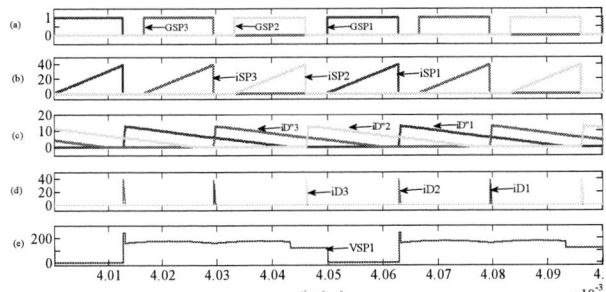

Fig.7. Performance of the suggested system under full load condition during switching period

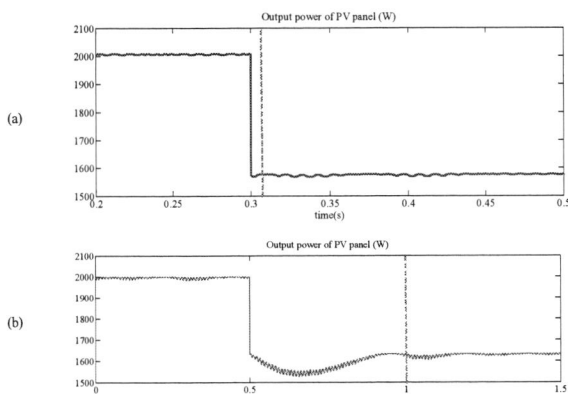

Fig.8: Output power of PV Panel, (a) proposed variable structure control, (b) PWM based MPPT

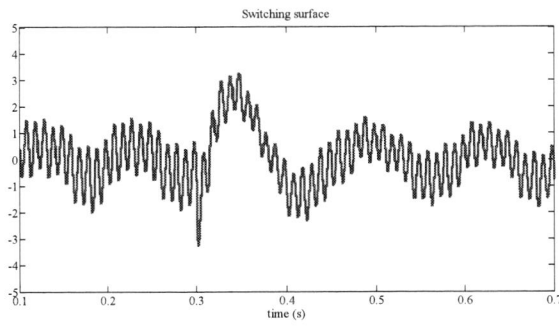

Fig.9: Switching surface values

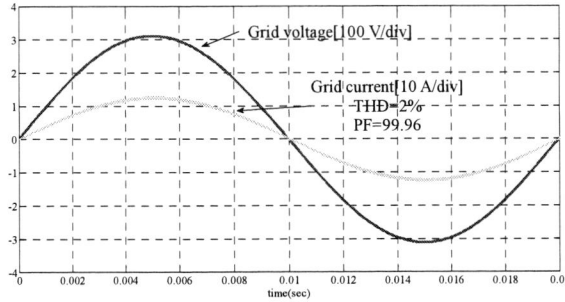

Fig .10 simulated waveform of the grid voltage and current

V. CONCLUSION

A string-type PV inverter for small electric power system applications rated at 2 kW is designed and simulated based on the two switch flyback converter topology. This inverter system consists of three interleaved two switch flyback cells each rated at 700 W limiting the high-voltage transients of the switches to the dc input voltage. Also a linear variable structure MPPT control was presented which can fix the switching frequency, while the fast response and robustness merits of the sliding mode controller. The simulation results demonstrate the successful operation of the inverter and compliance to the specifications. The THD of the grid current is measured as 2% and the power factor is 0.996, which are verifying the high power quality interface to the grid. Therefore, it is shown that interleaved two switch flyback topology is an appropriate topology at high power as a string-type PV inverter.

REFERENCES

[1] S. B. Kjaer, J. K. Pedersen, and F. Blaabjerg, "A review of single-phase grid-connected inverters for photovoltaic modules,"IEEE Trans. Ind.Appl., vol. 41, no. 5, pp. 1292–1306, Sep. 2005

[2] Young-Ho Kim, Young-Hyok Ji, Jun-Gu Kim, Yong-Chae Jung, and Chung-Yuen Won , "A New Control Strategy for Improving Weighted Efficiency in Photovoltaic AC Module-Type Interleaved Flyback Inverters, lIEEE Trans Power Electron, VOL. 28, NO. 6, JUNE 2013

[3] Murthy-Bellur and Kazimierczuk, M.K. "Zero-current-transition TWO-Switch flyback pulse-width modulated DC-DC converter," Power Electronics, IET, P. 288–295 March 2011.

[4] Dakshina Murthy Bellur, "Hard-switching and soft- switching two-switch flyback PWM DC-DC converters and winding loss due to harmonics in high-frequency transformers," Ph.D. dissertation, WRIGHT STATE university, 2010.

[5] Mohammadi, S.; Zarchi, H.A.; Amiri, M., "Interleaved two-switch flyback microinverter for grid-tied photovoltaic applications," Power Electronics, Drives Systems & Technologies Conference (PEDSTC), 2015 6th, vol., no., pp.59, 64, 3-4 Feb. 2015

[6] Tamyurek, B.; Kirimer, B., "An Interleaved High-Power Flyback Inverter for Photovoltaic Applications," Power Electronics, IEEE Transactions on, vol.30, no.6, pp.3228, 324, June 2015

[7] Z. Zhang, X.-F. He, and Y.-F. Liu, "An optimal control method for photovoltaic grid-tied-interleaved flyback microinverters to achieve high efficiency in wide load range,"IEEE Trans. Ind. Appl., vol. 28, no. 11, pp.5074–5087, Nov. 2013.

[8] Levron, Y.; Shmilovitz, D., "Maximum Power Point Tracking Employing Sliding Mode Control," in Circuits and Systems I: Regular Papers, IEEE Transactions on, vol.60, no.3, pp.724-732, March 2013

[9] Ravari, F.K.; Zarchi, H.A., "A fast and robust maximum power point tracker for photovoltaic systems using variable structure control approach," in Electrical Engineering (ICEE), 2015 23rd Iranian Conference on , vol., no., pp.1647-1652, 10-14 May 2015

[10] Abraham I. Pressman, Keith Billings and Taylor Morey, Switching Power Supply Design, 3rd Ed. McGraw-Hill, 2009, pp.157-160.

[11] Kyritsis, A.Ch.; Tatakis, E.C.; Papanikolaou, N.P., "Optimum Design of the Current-Source Flyback Inverter for Decentralized Grid-Connected Photovoltaic Systems," Energy Conversion, IEEE Transactions on , vol.23, no.1, pp.281,293, March 2008

7th Power Electronics, Drive Systems & Technologies Conference (PEDSTC 2016)
16-18 Feb. 2016, Iran University of Science and Technology, Tehran, Iran

An Improved Method for Power Management and Voltage Control of PV Unit in DC Microgrid

S. Soori[1], S. A. Saremi Hasari[2], A. Salemnia[3], S. Khosrogorji[4]

Department of Electrical and Computer Engineering

Shahid Beheshti University

Tehran, Iran

[1]sepehr.soori70@gmail.com, [2]A_Saremi@sbu.ac.ir, [3] Salemnia@pwut.ac.ir, [4] Soheyl_Gorji1991@ace.sbu.ac.ir

Abstract—**According to growing of DC loads such as LEDs and computers, DC Microgrids will play an undeniable role in power systems in future. One of important power sources in DC Microgrids is photo voltaic (PV) unit which usually is composed of PV panel, super-capacitor (SC) and battery. In this paper, using models of PV panel, SC and battery, an improved method is suggested for providing the load power by participation of PV, SC and battery with considering the SOC of battery and SC. The final aim of power management is controlling the DC link voltage at desired amount and improving its quality when some disturbances like load current or PV output power changing are occurred. The performance of the presented method is verified using simulations carried out in MATLAB/SIMULINK software.**

Keywords— photo voltaic; DC Microgrid; modelling; power sharing; voltage control.

Nomenclature

V_{batt}	Output voltage of Battery
V_r	Rated voltage of PV
V_g	Voltage Saturation of PV
E_0	Open circuit voltage of Battery
E_{full}	Battery fully charge voltage
E_{exp}	Battery voltage at the end of exponential area
E_{nom}	Battery voltage at the end of nominal area
I_{cell}	Output Current of solar cell
I_{pv}	Output current of PV panel
I_d	PV panel diode current
I_{sh}	PV panel shunt resistor current
I_r	Rated current of PV panel
I_{sc}	Short circuit of PV panel
I_L	Load current
α	Temperature coefficient of short-circuit current
β	Approximate effect of temperature on power
A	exponential zone amplitude
B	exponential zone time constant inverse
K	Polarization resistance
G	Insolation
G_r	Nominal Insolation
T_r	Environment Temperature
Q	Battery capacity
Q_{max}	Battery capacity at the end of exponential area
Q_{cut}	Battery capacity at cut-off
Q_{nom}	Battery capacity at the end of nominal area
R_L	Load resistance
R_s	PV series Impedance

I. INTRODUCTION

Nowadays because of some disadvantages of traditional power systems like: environment pollution, transmission loss, transmission cost and many other drawbacks, using of traditional power systems is faced with question. For solving these issues the theory of Microgrids has been suggested [1].

Microgrids are local grids that built up from distributed generation (DGs), energy storage system (ESS) and renewable sources like wind or PV cell or etc. Microgrids can work generally in two modes grid-connected mode or islanded mode [2, 3]. Microgrids can be used in backup role [4] or grid's power quality improvement like harmonic determination or voltage regulation [5-7].

In general there is three type of Microgrid: AC Microgrid, DC Microgrid and hybrid Microgrid [8, 9] which is combination of two latter types. Nowadays because of arising growth of DC loads like LEDs and communication systems in power grids, usage of DC Microgrids is more economical compared with two other types. The other benefit of DC Microgrids is simplicity of control system in comparison with AC or hybrid Microgrids.

Some control method for controlling DC Microgrid is presented in [10-12]. Many renewable energy sources like PV cells or fuel cells have DC output. A comparative study of hybrid photovoltaic-fuel cell and hybrid wind-fuel cell system in Coimbatore of India was presented in [13] but control system wasn't introduced. In [14], different topology for connecting battery bank and ultra-capacitor bank is introduced and studied. A control technique for power management between batteries and power management between ultra-capacitors is investigated but power management between battery bank and ultra-capacitor bank isn't presented. In [15], connection between a PV cell and battery and ultra-capacitor and wind turbine is studied and a control technique was investigated, but battery or ultra-capacitor state of charge (SOC) isn't considered in controlling technique. In [16], steady state performance of a grid connected PV cell with battery is studied and a technique for unit sizing is investigated. But, because of weak dynamic response of battery, this system can't response effectively in fast dynamic faults.

In this paper, in an islanded DC Microgrid that is composed of a PV, SC, battery and load, charging and discharging of battery and super-capacitor is controlled by considering SOC of them, amount of generated power by PV and demanded power by load. The purpose of power management is balancing the generated and demanded power so that the DC link voltage could be fixed at set point with a fast dynamic response. So this paper is organized as follows: PV cell modeling of PV cell, SC and battery and required equations are described in section II, power management method and voltage control strategy are described in section III and finally, the performance of the presented strategy is verified using simulations carried out in MATLAB/SIMULINK software in section IV and conclusion is presented in section V.

II. PROPOSED DC MICROGRID

Proposed DC Microgrid is shown in Fig. 1 In this scheme each PV cell is paralleled with battery and SC and finally a PV unit is made which will be controlled by proposed centralized control.

PV array is non dispatch able, so battery is used for balancing the generated and demanded power. To have high power quality and a stable DC link voltage, control system must have ability to response quickly both fast dynamic and slow dynamic errors. Because of slow dynamic response of battery, a SC which has fast dynamic response is required too. So, because of slow dynamic response of battery, it will provide load current in long load turbulence duration or PV turbulence and SC will control DC link voltage in transient states and for short time intervals because of fast dynamic response and lack of stored energy. Both of SC and battery help to regulate load voltage and balance generated and demanded power.

The components of PV unit are described in the following section.

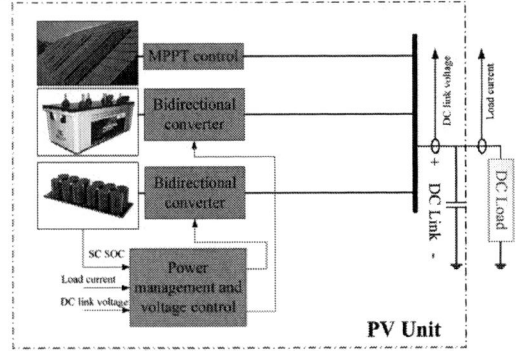

Fig. 1. proposed DC microgrid

A. PV modelling

Some models for PV cell are presented in [17-19] and the model that is presented in [18] is used in this paper. This model is figured out in Fig. 2.

According to Fig. 2 and Kirchhoff's current law, PV cell output current I_{cell} can be defined such as (1):

$$I_{cell} = I_{ph} - I_d - I_{sh} \qquad (1)$$

As we know PV output current is dependent on temperature and irradiation. there is an inverse relation between PV output voltage and Load current. According to [18] overall PV output current and voltage can be defined as (2-4):

$$I_{cell} = I_r + \left[\alpha \left(\frac{G}{G_r} \right)(T_c - T_{cr}) + \left(\frac{G}{G_r} - 1 \right) I_{sc} \right] \qquad (2)$$

$$V_{cell} = -\beta(T_c - T_{cr}) - R_s \Delta I + V_r \qquad (3)$$

$$\Delta I = \left[\alpha \left(\frac{G}{G_r} \right)(T_c - T_{cr}) + \left(\frac{G}{G_r} - 1 \right) I_{sc} \right] \qquad (4)$$

The parameters that is used in PV modelling has been introduced in first section.

In [20-23] many MPPT techniques like constant voltage method, beta method, system oscillation method, ripple correlation method, perturb and observe method (P&O) and incremental conductance method (IC) are introduced. IC method is used in this work because of method effectivity and simplicity.

B. Battery modeling

In stand-alone converters, batteries have an important role, because solar energy exists during the day and the load don't need whole of energy all the time. So, a storage system is needed for converter. In this paper, a lead-acid battery is used for storage system with the model in references [24, 25]. Battery model can be constructed by a variable voltage source and a series resistor. Math equations for lead-acid battery are as follows:

$$\text{Discharge}: V_{batt} = E_0 - R.i - K \frac{Q}{Q - it}(it + i^*) + Exp(t) \qquad (5)$$

$$\text{Charge}: V_{batt} = E_0 - R.i - K \frac{Q}{it - 0.1Q} i^* - K \frac{Q}{Q - it}.it + Exp(t) \qquad (6)$$

$$\overset{\bullet}{Exp}(t) = B.|i(t)|.(-Exp(t) + A\,u(t)) \qquad (7)$$

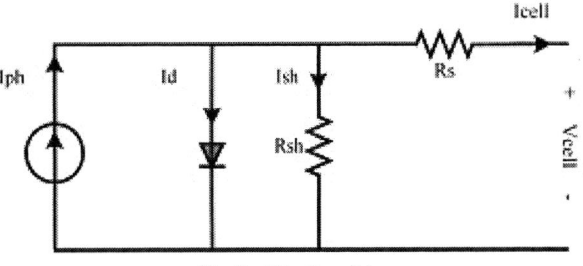

Fig. 2. PV cell model

In (5) and (6), R is output resistance of battery. According to these equations, batteries have two operation modes that are named discharge and charge. Block diagram of battery model in the discharge mode is shown in Fig. 3 that contains battery model. The important feature of this model is using less number of parameters in comparison with other models.

C. SC modelling

SC can be modeled by four resistive-capacitive branches which is shown in Fig. 4 [26]**Error! Reference source not found.**. In this figure, power losses in SC is modeled by R_Leakage, long dynamic response of SC is modeled by R_Long and C3, middle dynamic response is modeled by R_middle and C2 and finally, fast dynamic response of SC is modeled by R_fast and a variable capacitor that composed of C1 and C0.

For simplicity, long dynamic response branch and middle dynamic response branch can be neglected [27]. Simplified model has been shown in Fig. 5.

D. Bidirectional converter

Bidirectional converter is shown in Fig. 6. In this converter capacitor or battery is able to operate in charge mode or discharge mode. According to DC link voltage error switching scenario will be changed which is shown in Fig. 8. Switch S1 for SC charging and S2 is used for SC discharging.

III. PROPOSED CONTROL METHOD

In this section the strategy for power management and DC link voltage regulation will be described.

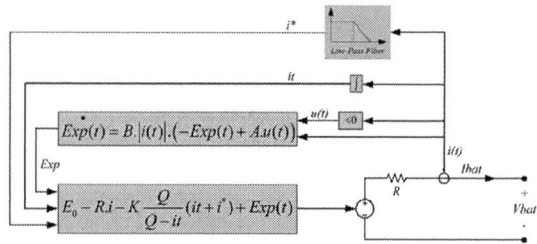

Fig. 3. Battery model

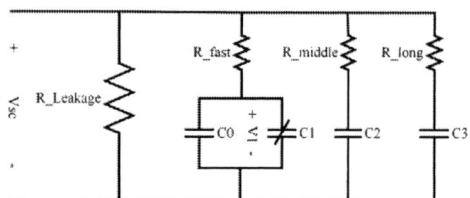

Fig. 4. SC general model according to [26]

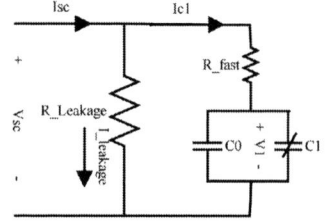

Fig. 5. simple model of SC according to [27]

A. power sharing method

As it said earlier this method is centralized method so battery and SC current reference are produced by local measurement which is shown in Fig. 7. According to Fig. 7, the amount of PV power and load power are calculated using DC link voltage, PV current and Load current. Injected power to the DC link by battery and SC is equal to difference of load power and PV power.

For having effective SC action in any situation, SOC of SC must be in an acceptable range. For this purpose SOC of SC is considered when battery and SC must absorb excessive power from DC link. In other words, for excessive power injected to DC link, SC will be charged at first, and with increasing the SOC of SC, the amount of the battery charging current increases too. It should be noticed that SOC of SC is considered just for charging ESS, not for discharging them.

B. DC link voltage regulating

Control method for battery converter has been depicted in Fig. 8 the aims of this method are accurate control of DC link voltage, less DC link voltage deviation and fast recovery of it for sudden load change. For achieving these goals both of DC link voltage error and battery current error are used for charging or discharging the battery, as shown in Fig. 8.

The control system of SC is like battery's system. So, both of SC and battery are used for regulating the voltage of DC link and making the generated power equal to consumption. But in proposed strategy, main duty of SC is regulating the DC link voltage and main duty of battery is balancing the power. Because of this reason, the amount of coefficients in PIv controller is greater than PIc for SC control system and is smaller for battery control system.

IV. SIMULATION RESULT

For verifying the accuracy of purposed method, two scenarios will be studied. The characteristics of Microgrid components are shown in table 1.

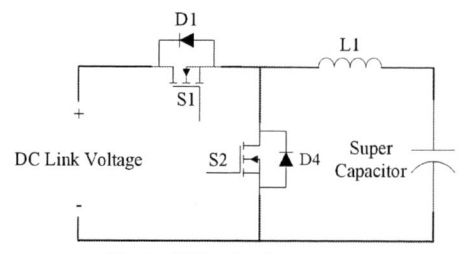

Fig. 6. Bidirectional converter

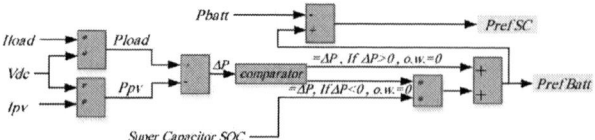

Fig. 7. Reference power production using local measurement

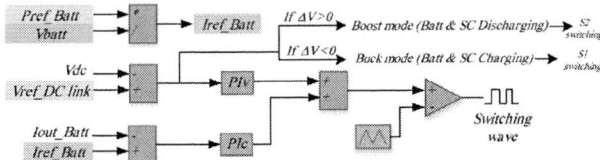

Fig. 8. Block diagram of battery control system

TABLE I.   CHARACTERISTICS OF DC MICROGRID COMPONENTS

Components	Characteristics
DC link	Vdc=100v DC link capacitor=1000uF Rload=40 Ω
PV cells	Ns=4, Np=4, Vr=17.1, Ir=3.5, Isc=3.8
Super capacitor	16v, 5F
Battery	24v, 10Ah
Battery converter	L=1mH, PIv: (kp=0.1, ki=0) , PIc: (kp=1.5, ki=0)
SC converter	L=100mH, PIv: (kp=10, ki=0), PIc: (kp=1.5, ki=0)

A. First scenario

At first scenario, PV power generation will be decreased because of sun insolation decrement between 0.8s to 1.7s. Battery and SC has to provide lack of generation in this interval.

Load current, SC current and battery current are shown in Fig. 9. As it can be seen, load current is kept constant when PV generation is decreased. There are three sections in Fig. 9.

In section 1 and 3 of Fig 9, generated power by PV is greater than load power, so battery and SC charging process takes place. According to control method at first SC is charged and when SOC of SC increased to desired value, battery charging process will be start. This process is shown in section 3, as it can be seen in t=2.6s SC current is decreased which means that SOC of SC has increased to desired value and battery charge process could start.

In section 2 of Fig. 9, at first PV power generation decreases and because of SC fast dynamic response and battery slow dynamic response, SC starts to compensate the lack of power by injecting power to DC link until battery could increase its generation. It could be seen in this section that after PV power generation increment, because of battery low dynamic response, battery can't decrease power injection immediately so SC starts to charge. By SC charging in this state, load power and DC link voltage will be kept constant.

DC link voltage has been shown in Fig. 10. As can be seen DC link voltage is kept constant, although PV generation has changed. When PV power is decreased, voltage is decreased just 0.5% that is acceptable for DC link voltage.

For showing the ability of purposed method in regulating the DC link voltage, large deviation of DC link voltage is presented in fig 11 when voltage error is not used in battery and SC control system.

B. Second scenario

In second scenario PV power generation is constant and load current is increased at two steps and finally decreased to the first value. First load amount is 40Ω and at t = 0.2s it

changes to 20 ohm and at t = 0.6s it changes to 13.33 ohm. Finally at t = 1s, it changes to first value 40 ohm.

Load current and PV current is shown in Fig. 12. As can be seen, there are four sections in this scenario. In section 1 and 4 PV generation is greater than load power, so charge process is started like first scenario. In section 2, PV generation is equal to load power, so it's not required any generation by SC or battery. It can be seen that battery and SC currents are zero in this section. In section 3, load power is greater than PV generation, so both of SC and battery start to inject power to load. In all sections, because of SC fast dynamic response and battery slow dynamic response, at first, SC is injecting power, after a while, battery begins to inject power and after that, SC power gradually decreases to zero.

DC link voltage is shown in Fig. 13. As can be seen in this figure, maximum voltage deviation is 2% that is cleared after 0.03s.

V. CONCLUSION

In this work by accurate models of PV, battery and SC an effective power management and control method was investigated and verified by results of simulations.

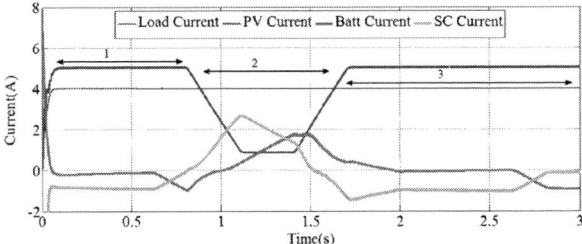

Fig. 9. PV, SC , load and battery current in first scenario

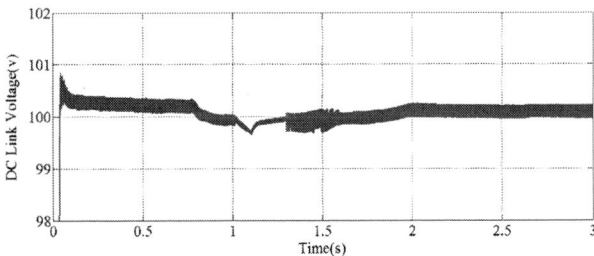

Fig. 10. DC link voltage with proposed control in first scenario

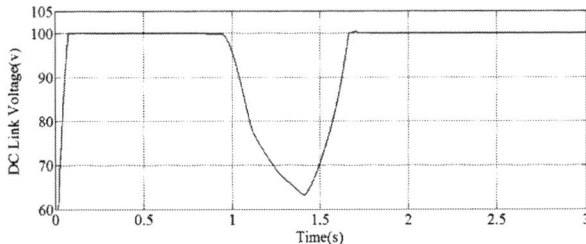

Fig. 11. DC link voltage in current control method

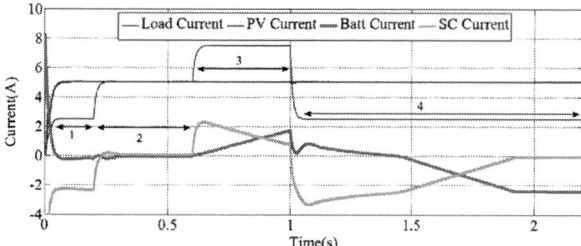

Fig. 12. PV, SC , load and battery current in second scenario

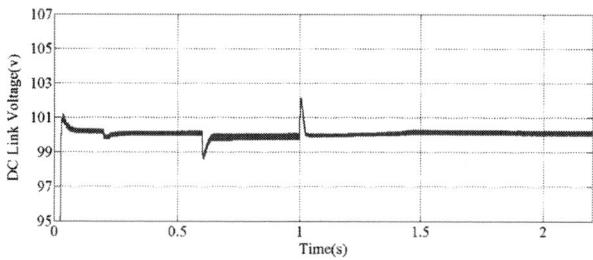

Fig. 13. DC link voltage in second scenario

For verifying the accuracy of purposed method, two scenarios were studied. In first scenario, for a constant load, generated power of PV was decreased because of sun insolation decrement. Results of simulations showed an acceptable DC link voltage regulation and deviation less than 0.5%. In second scenario, generated power of PV was constant and load power was changed. In this scenario DC link voltage was kept nearly reference voltage and deviated less than 3%.

REFERENCES

[1] H. Kakigano, Y. Miura, T. Ise, T. Momose, and H. Hayakawa, "Fundamental characteristics of DC microgrid for residential houses with cogeneration system in each house," in *Power and Energy Society General Meeting - Conversion and Delivery of Electrical Energy in the 21st Century, 2008 IEEE*, 2008, pp. 1-8.

[2] J. M. Guerrero, M. Chandorkar, T. Lee, and P. C. Loh, "Advanced Control Architectures for Intelligent Microgrids Part I: Decentralized and Hierarchical Control," *Industrial Electronics, IEEE Transactions on*, vol. 60, pp. 1254-1262, 2013.

[3] J. M. Guerrero, L. Poh Chiang, L. Tzung-Lin, and M. Chandorkar, "Advanced Control Architectures for Intelligent Microgrids Part II: Power Quality, Energy Storage, and AC/DC Microgrids," *Industrial Electronics, IEEE Transactions on*, vol. 60, pp. 1263-1270, 2013.

[4] K. Kurohane, T. Senjyu, A. Yona, N. Urasaki, E. B. Muhando, and T. Funabashi, "A high quality power supply system with DC smart grid," in *Transmission and Distribution Conference and Exposition, 2010 IEEE PES*, 2010, pp. 1-6.

[5] S. Chakraborty, M. D. Weiss, and M. G. Simoes, "Distributed Intelligent Energy Management System for a Single-Phase High-Frequency AC Microgrid," *Industrial Electronics, IEEE Transactions on*, vol. 54, pp. 97-109, 2007.

[6] M. Prodanovic and T. C. Green, "High-Quality Power Generation Through Distributed Control of a Power Park Microgrid," *Industrial Electronics, IEEE Transactions on*, vol. 53, pp. 1471-1482, 2006.

[7] J. C. Vasquez, R. A. Mastromauro, J. M. Guerrero, and M. Liserre, "Voltage Support Provided by a Droop-Controlled Multifunctional Inverter," *Industrial Electronics, IEEE Transactions on*, vol. 56, pp. 4510-4519, 2009.

[8] P. Sung-Hwan, C. Jin-Young, and W. Dong-Jun, "Cooperative control between the distributed energy resources in AC/DC hybrid microgrid,"

in *Innovative Smart Grid Technologies Conference (ISGT), 2014 IEEE PES*, 2014, pp. 1-5.

[9] L. Xiong, W. Peng, and L. Poh Chiang, "A Hybrid AC/DC Microgrid and Its Coordination Control," *Smart Grid, IEEE Transactions on*, vol. 2, pp. 278-286, 2011.

[10] C. Dong and X. Lie, "Autonomous DC Voltage Control of a DC Microgrid With Multiple Slack Terminals," *IEEE Transactions on Power Systems,*, vol. 27, pp. 1897-1905, 2012.

[11] T. Dragicevic, J. M. Guerrero, J. C. Vasquez, and D. Skrlec, "Supervisory Control of an Adaptive-Droop Regulated DC Microgrid With Battery Management Capability," *IEEE Transactions on Power Electronics,* vol. 29, pp. 695-706, 2014.

[12] J. M. Guerrero, J. C. Vasquez, J. Matas, V. de, x00F, L. G. a, *et al.*, "Hierarchical Control of Droop-Controlled AC and DC Microgrids_A General Approach Toward Standardization," *IEEE Transactions on Industrial Electronics,*, vol. 58, pp. 158-172, 2011.

[13] K. Balachander, S. Kuppusamy, and P. Vijayakumar, "Comparative study of hybrid photovoltaic-fuel cell system/hybrid wind-fuel cell system for smart grid distributed generation system," in *Emerging Trends in Science, Engineering and Technology (INCOSET), 2012 International Conference on*, 2012, pp. 462-466.

[14] Z. Haihua, T. Bhattacharya, T. Duong, T. S. T. Siew, and A. M. Khambadkone, "Composite Energy Storage System Involving Battery and Ultracapacitor With Dynamic Energy Management in Microgrid Applications," *IEEE Transactions on Power Electronics*, vol. 26, pp. 923-930, 2011.

[15] A. Tani, M. B. Camara, and B. Dakyo, "Energy management in the decentralized generation systems based on renewable energy sources," in *International Conference on Renewable Energy Research and Applications (ICRERA)*, 2012, pp. 1-6.

[16] F. Giraud and Z. M. Salameh, "Steady-state performance of a grid-connected rooftop hybrid wind-photovoltaic power system with battery storage," *IEEE Transactions on Power Electronics*, vol. 16, pp. 1-7, 2001.

[17] I. H. Altas and A. M. Sharaf, "A Photovoltaic Array Simulation Model for Matlab-Simulink GUI Environment," in *International Conference on Clean Electrical Power, ICCEP '07, 2007.* , 2007, pp. 341-345.

[18] C. Keles, B. B. Alagoz, M. Akcin, A. Kaygusuz, and A. Karabiber, "A photovoltaic system model for Matlab/Simulink simulations," in *Fourth International Conference on Power Engineering, Energy and Electrical Drives (POWERENG), 2013*, 2013, pp. 1643-1647.

[19] H.-L. Tsai, "Insolation-oriented model of photovoltaic module using Matlab/Simulink," *Solar Energy*, vol. 84, pp. 1318-1326, 2010.

[20] S. K. Dash, D. Verma, S. Nema, and R. K. Nema, "Comparative analysis of maximum power point (MPP) tracking techniques for solar PV application using MATLAB simulink," in *Recent Advances and Innovations in Engineering (ICRAIE), 2014*, 2014, pp. 1-7.

[21] M. A. G. de Brito, L. P. Sampaio, G. Luigi, G. A. e Melo, and C. A. Canesin, "Comparative analysis of MPPT techniques for PV applications," in *International Conference on Clean Electrical Power (ICCEP), 2011*, 2011, pp. 99-104.

[22] S. Jain, A. Vaibhav, and L. Goyal, "Comparative analysis of MPPT techniques for PV in domestic applications," in *Power India International Conference (PIICON), 2014 6th IEEE*, 2014, pp. 1-6.

[23] A. F. Murtaza, H. A. Sher, M. Chiaberge, D. Boero, M. De Giuseppe, and K. E. Addoweesh, "Comparative analysis of maximum power point tracking techniques for PV applications," in *16th International Multi Topic Conference (INMIC), 2013* 2013, pp. 83-88.

[24] O. Tremblay, L. A. Dessaint, and A. I. Dekkiche, "A Generic Battery Model for the Dynamic Simulation of Hybrid Electric Vehicles," in *Vehicle Power and Propulsion Conference*, Arlington, TX, 2007, pp. 284-289.

[25] O. Tremblay and L.-A. Dessaint, "Experimental Validation of a Battery Dynamic Model for EV Applications," *World Electric Vehicle Journal*, vol. 3, 2009.

[26] L. Zubieta and R. Bonert, "Characterization of double-layer capacitors for power electronics applications," *IEEE Transactions on Industry Applications*, vol. 36, pp. 199-205, 2000.

[27] I. San Martín, A. Ursúa, and P. Sanchis, "Integration of fuel cells and supercapacitors in electrical microgrids: Analysis, modelling and experimental validation," *International Journal of Hydrogen Energy,* vol. 38, pp. 11655-11671, 9/10/ 2013.

7th Power Electronics, Drive Systems & Technologies Conference (PEDSTC 2016)
16-18 Feb. 2016, Iran University of Science and Technology, Tehran, Iran

A New DPC Method For Single VSC Based DFIG Under Unbalanced Grid Voltage Condition

Ali Izanlo*, S. Asghar Gholamian**, and Mohammad Verij Kazemi***

Faculty of Electrical and Computer Engineering, Babol University of Technology, Babol, Iran
* izanlo_ali@yahoo.com
** sagholamian@gmail.com
*** mohammad_v_kazemi@yahoo.com

Abstract—**This paper proposes a new Direct Power Control (DPC) strategy for a single Voltage Source Converter (VSC) based Doubly Fed Induction Generator (DFIG) by employing Four Switch Three Phase Inverter (FSTPI) instead of Six Switch Three Phase Inverter (SSTPI). Reduce number of active switch not only improves reliability but also decreases cost, and conduction losses. The new switching table for FSTPI is based on principle of similarity between FSTPI and SSTPI i.e. by using a technique the four unbalanced voltage vectors which generated by FSTPI will be converted to six balanced voltage vectors. In addition, the behavior of the DFIG under unbalanced grid voltage is investigated and then alternative DFIG control targets, such as reducing stator current unbalance, torque, and power pulsation minimization, are identified. Simulation results using Matlab/Simulink are presented for a 1-MW DFIG Wind generation system to validate the proposed control scheme.**

Keywords—*Doubly Fed Induction Generator (DFIG); Direct Power Control (DPC); Four Switch Three Phase Inverter (FSTPI); Six Switch Three Phase Inverter (SSTPI); Unbalanced Voltage*

I. INTRODUCTION

The wind energy system using a DFIG have some advantage due to variable speed operation and four quadrant active and reactive power capabilities compared with fixed speed induction generators. The stator of DFIG is connected direct to the grid and the rotor links the grid by a bi-directional converter. The rotor converter objective aims to the DFIG active and reactive power control between the stator and supply. But, in this paper Grid Side Converter (GSC) has been removed and instead it used from a Battery Energy Storage System (BESS). Fig. 1 shows the configuration of proposed DFIG for a Wind Energy Conversion System (WECS) connected to the Grid.

DPC method for DFIG presented for first time in [1]. This method base on the estimated stator flux was proposed. Since the stator voltage is relatively harmonic-free with fixed frequency, a DFIG's estimated stator flux accuracy can be guaranteed. Switching vectors were selected from the optimal switching table using the estimated stator flux position, and the errors of the active and reactive power. Thus, control of the system is very simple, and the impact of machine parameters on system performance was found to be negligible. However a conventional DPC has switching frequency that varies significantly with active and reactive power variations, machine operating speed, and the power controller's hysteresis bandwidth. This topic complicates the AC filter design.

Several modified DPC strategies with constant switching frequency have also been proposed for the DFIG [2], [3]. Nowadays, since DFIG based WECSs are mainly installed in remote and rural areas with long transmission lines to the load centers, the voltage unbalance is one common and severe disturbances in the power supply. Thus, in order to improve the operation performance of DFIG system on the unbalanced voltage, several solution for DFIGs have been designed and proposed. In [4] and [5], new kinds of reference power generation technique for DPC were proposed to eliminate the torque, power pulsations or stator current unbalance. In [6], an unbalanced DPC-SVM scheme was designed for the DFIG under transient grid faults, but only simulation studies were carried out and no experimental results were proposed. In order to improve the applications quality, it is required to reduce the number of active switches which results in improvement of reliability, reduce of costs, and less conduction losses. Several researchers are done on a reduced switch count structure, due to its many benefits. Among them the FSTPI was introduced with four IGBT switches instead of six in a SSTPI. In [7], the modulation technique such as Hysteresis Band current control (HB), SPWM, and SVM is proposed for a FSTPI. To obtain the simple, effective performance, the fast control of torque and flux in a DTC system for FSTPI induction motor has been proposed [8].

In this paper used from FSTPI instead of SSTPI in the structure of a standalone DFIG. This paper is organized as follows: in section II a mathematical model of the DFIG is explained. Response of DFIG under unbalanced condition is proposed in section III whereas the power compensation strategies are illustrated in section IV. The new switching table for FSTPI is presented in section V. Section VI discusses the simulation results on a 1-MW DFIG system to demonstrate the efficacy and performance of the proposed method, and finally the conclusions are presented in section VII.

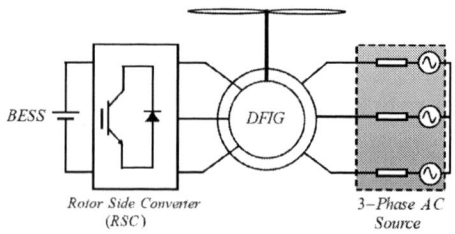

FIG. 1. SCHEMATIC DIAGRAM OF THE PROPOSED DFIG

978-1-5090-0376-1/16 $31.00 © 2016 IEEE

II. Mathematical Model of The DFIG System

The equivalent circuit of a DFIG expressed in the stator stationary frame is shown in Fig. 2. The mathematical equations for a DFIG can be expressed as [1]

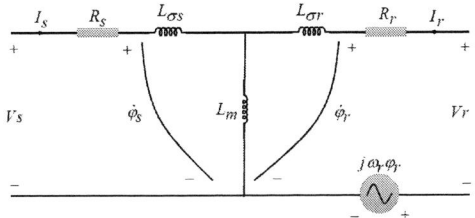

FIG. 2. Equivalent circuit of a DFIG in the stator stationary frame

$$V_s = R_s I_s + \frac{d\varphi_s}{dt} \tag{1}$$

$$V_r = R_r I_r + \frac{d\varphi_r}{dt} - j\omega_r \varphi_r \tag{2}$$

$$\varphi_s = L_s I_s + L_m I_r \tag{3}$$

$$\varphi_r = L_m I_s + L_r I_r \tag{4}$$

$$P_s = \frac{3}{2}\operatorname{Re}\left(V_s I_s^*\right) = \frac{3}{2}\left(V_{s\alpha}I_{s\alpha} + V_{s\beta}I_{s\beta}\right) \tag{5}$$

$$Q_s = \frac{3}{2}\operatorname{Im}\left(V_s I_s^*\right) = \frac{3}{2}\left(V_{s\beta}I_{s\alpha} - V_{s\alpha}I_{s\beta}\right) \tag{6}$$

$$T_e = \frac{3}{2}p\operatorname{Im}\left(\varphi_s^* I_s\right) = \frac{3}{2}p\left(\varphi_{s\alpha}I_{s\beta} - \varphi_{s\beta}I_{s\alpha}\right) \tag{7}$$

Where V_s and V_r stator and rotor voltage vector, φ_s and φ_r stator and rotor flux vector, P_s and Q_s active and reactive power, and T_e is electromagnetic torque respectively.

III. Response of DFIG Under Unbalanced Grid Voltage

According to the symmetrical component theory, if the grid supply is unbalanced, any three phase quantity, e.g., voltage, current or flux, can be decomposed to three balanced symmetric three phase systems; that is, the zero, positive and negative sequence components. Assuming no zero sequence components, the three phase quantities may be decomposed into positive and negative sequence components. In the stationary ($\alpha\beta$) reference frame, the voltage, current, and flux, denoted as a vector F can be decomposed into positive and negative sequence components as [4]

$$F_s = F_s^+ + F_s^- = (F_{s\alpha}^+ + F_{s\alpha}^-) + j(F_{s\beta}^+ + F_{s\beta}^-) \tag{8}$$

Substituting (8) into (5) and (6), we get

$$P_s = (A_P + B_P + C_P + D_P) \tag{9}$$

$$Q_s = (A_Q + B_Q + C_Q + D_Q) \tag{10}$$

Where

$$A_P = \frac{3}{2}\operatorname{Re}\left(V_s^+ \cdot I_s^{+*}\right) = \frac{3}{2}\left(V_{s\alpha}^+ I_{s\alpha}^+ + V_{s\beta}^+ I_{s\beta}^+\right) \tag{11}$$

$$B_P = \frac{3}{2}\operatorname{Re}\left(V_s^- \cdot I_s^{-*}\right) = \frac{3}{2}\left(V_{s\alpha}^- I_{s\alpha}^- + V_{s\beta}^- I_{s\beta}^-\right) \tag{12}$$

$$C_P = \frac{3}{2}\operatorname{Re}\left(V_s^+ \cdot I_s^{-*}\right) = \frac{3}{2}\left(V_{s\alpha}^+ I_{s\alpha}^- + V_{s\beta}^+ I_{s\beta}^-\right) \tag{13}$$

$$D_P = \frac{3}{2}\operatorname{Re}\left(V_s^- \cdot I_s^{+*}\right) = \frac{3}{2}\left(V_{s\alpha}^- I_{s\alpha}^+ + V_{s\beta}^- I_{s\beta}^+\right) \tag{14}$$

$$A_Q = \frac{3}{2}\operatorname{Im}\left(V_s^+ \cdot I_s^{+*}\right) = \frac{3}{2}\left(V_{s\beta}^+ I_{s\alpha}^+ - V_{s\alpha}^+ I_{s\beta}^+\right) \tag{15}$$

$$B_Q = \frac{3}{2}\operatorname{Im}\left(V_s^- \cdot I_s^{-*}\right) = \frac{3}{2}\left(V_{s\beta}^- I_{s\alpha}^- - V_{s\alpha}^- I_{s\beta}^-\right) \tag{16}$$

$$C_Q = \frac{3}{2}\operatorname{Im}\left(V_s^+ \cdot I_s^{-*}\right) = \frac{3}{2}\left(V_{s\beta}^+ I_{s\alpha}^- - V_{s\alpha}^+ I_{s\beta}^-\right) \tag{17}$$

$$D_Q = \frac{3}{2}\operatorname{Im}\left(V_s^- \cdot I_s^{+*}\right) = \frac{3}{2}\left(V_{s\beta}^- I_{s\alpha}^+ - V_{s\alpha}^- I_{s\beta}^+\right) \tag{18}$$

Also with substituting the positive and negative sequence components of the stator current and flux into the expression (7), we obtain

$$T_e = \frac{3}{2}p\operatorname{Im}\left(\psi_s^{+*}I_s^+ + \psi_s^{+*}I_s^- + \psi_s^{-*}I_s^+ + \psi_s^{-*}I_s^-\right) \tag{19}$$

In addition, stator flux can be expressed using the stator voltage from (1) as

$$\psi_s = \frac{j}{\omega_s}\left(R_s I_s - V_s\right) \tag{20}$$

Substituting the flux from expressions (20) in torque equation (19), we get

$$T_e = \frac{3}{2\omega_s}p\operatorname{Re}\left(\begin{array}{c} V_s^+ I_s^{+*} - V_s^- I_s^{-*} - V_s^- I_s^{+*} + V_s^+ I_s^{-*} \\ + R_s\left(\left|I_s^-\right|^2 - \left|I_s^+\right|^2\right) \end{array}\right) \tag{21}$$

According to the active power expression (9), the torque equation (21) become

$$T_e = \frac{3p}{2\omega_s}\left(A_P - B_P + C_P - D_P + T_0\right) \tag{22}$$

Where

$$T_0 = R_s\left(\left|I_s^-\right|^2 - \left|I_s^+\right|^2\right) \tag{23}$$

Notice that A_P, A_Q and B_P, B_Q are constant in steady state since they are composed of the same sequence product. However, the terms C_P, C_Q and D_P, D_Q oscillate with $2\omega_s$ pulsation, because they are composed of positive and negative sequence products.

IV. POWER COMPENSATION STRATEGIES UNDER UNBALANCED CONDITION

In this section, the control strategies that injection sinusoidal and symmetric stator current into the grid, and produces constant active and reactive power and electromagnetic torque will be analyzed.

A. Active and reactive power oscillation cancellation strategy (target 1)

This control target is to allow the existence of negative sequence current components but eliminate the stator output active and reactive power ripples.

In order to obtain constant active and reactive power, the both of them reference must be kept constant, thus the active and reactive power ripples must be zero, i.e. [5]

$$C_P + D_P = 0 \tag{24}$$

$$C_Q + D_Q = 0 \tag{25}$$

As a result, the required powers for compensation become

$$P_{ref} = P_{const} = A_P + B_P \tag{26}$$

$$Q_{ref} = Q_{const} = A_Q + B_Q \tag{27}$$

According to the (22) and (24), even if the stator active and reactive power pulsation can be removed by the unbalanced DPC, the torque pulsation will still exist.

B. Torque oscillation cancellation strategy (target 2)

This objectives is to mitigate the torque oscillation when the network is unbalanced. According to (22), only way to achieve constant electromagnetic torque is by imposing [4]

$$C_P - D_P = 0 \tag{28}$$

As a result, the required active power for compensation become

$$P_{ref} = P_{const} + 2C_P = P_{const} + 2D_P \tag{29}$$

Since the condition for torque oscillation cancellation in (28) is not related to reactive power, no compensation is needed for the reactive power; that is, $Q_{ref}=0$.

C. Sinusoidal and balanced stator current exchanged with grid (target 3)

To obtain sinusoidal and balanced stator current, the negative sequence current must be eliminated. Therefore, the stator output power components containing the negative sequence current must be zero. In other words, $I_{s\alpha}^-$ and $I_{s\beta}^-$ have to be zero if C_P and C_Q are to be maintained as zero. Therefore, If D_P and D_Q are added to the original power references to obtain the new power references, C_P and C_Q can be forced to zero; that is, the negative sequence current is eliminated.

Therefore, the compensating power pulsations become [5]

$$P_{ref} = P_{const} + D_P \tag{30}$$

$$Q_{ref} = Q_{const} + D_Q \tag{31}$$

In the above equation, P_{const} and Q_{const} are equal to original power references under balanced grid voltage condition.

V. THE PRINCIPLE OF THE NEW DPC METHOD FOR DFIG

Before explaining the new DPC method, the characteristics of the FSTPI be checked. The schematic diagram of the FSTPI is shown in Fig. 3. As can be seen one of the three output phases is connected to the middle of the dc link capacitors. A semiconductor reduction of 1/3 is obtained, and two drive circuits can be omitted because only two inverter legs are used. Also, the conduction losses can be reduced with 1/3 because only two inverter legs will conduct compared with a SSTPI, where three legs will conduct. However, the higher voltage rating of the switches will add some extra losses. However, this inverter has some disadvantages as well. For example, because the dc link voltage in this inverter should be considered double dc link voltage in SSTPI, the switching losses rises. The voltage is doubled, which gives approximately 1/3 extra losses in the FSTPI when the same current flows to a load. As well as, the voltage vectors that generated by FSTPI is shown in Fig. 4. These vectors have unbalanced amplitudes and are shifted by an angle of $\pi/2$. Indeed, vectors V_1 and V_3 have an amplitude of $V_{dc}/\sqrt{6}$, while vectors V_2 and V_4 have an amplitude of $V_{dc}/\sqrt{2}$.

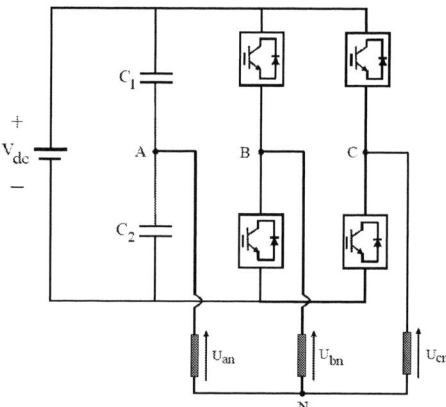

FIG. 3. SCHEMATIC DIAGRAM OF THE FSTPI

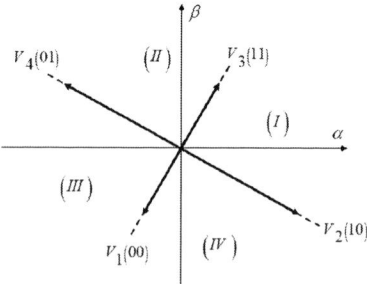

FIG. 4. UNBALANCED ACTIVE VOLTAGE VECTORS GENERATED BY THE FSTPI

In the event that, these vectors using for DPC of a DFIG, high ripples be created in the active and reactive power. Also, another one of problems of the four switch converter is that despite the six switch converter, this converter has no zero vectors. In result it is obvious from the circuit of Fig. 3 that there is no way the three pole voltages can be at the same potential during the operation. To fix these problems, the four unbalanced voltage vectors somehow be combined together to generate six balanced voltage vectors similar to voltage vectors produced by SSTPI. The generated vectors have the same amplitude and angular shift as those of the SSTPI. The amplitude of active voltage vectors generated from SSTPC are equal to $\sqrt{\frac{2}{3}}\, V_{dc}$. If the voltage vector V_1 (or V_3) to be used two times in succession, the voltage vector V_{11} (or V_{33}) is produced with a amplitude

$$V_{11} = V_{33} = \frac{V_{dc}}{\sqrt{6}} + \frac{V_{dc}}{\sqrt{6}} = \sqrt{\frac{2}{3}}\, V_{dc} \qquad (32)$$

Also, if the consecutive voltage vectors V_1 and V_2 (or V_2 and V_3 or V_3 and V_4 or V_4 and V_1) to be used in succession, the voltage vector V_{12} (or V_{23} and V_{34} and V_{41}) is produced with a domain to amount

$$V_{ij} = V_{12} = V_{23} = V_{34} = V_{41} = \sqrt{V_i^2 + V_j^2} = \sqrt{\frac{2}{3}}\, V_{dc} \qquad (33)$$

Also, in order to generate the zero voltage vectors, the sums of two opposite intrinsic voltage vectors V_1 and V_3 or V_2 and V_4 are used. New voltage vectors and the voltage vectors generated by SSTPI shown in Fig. 5a and Fig. 5b, respectively. Seen that, each new voltage vector is equal to a voltage vector in SSTPI. The similarity between space vectors of SSTPI and FSTPI is presented in Table I.

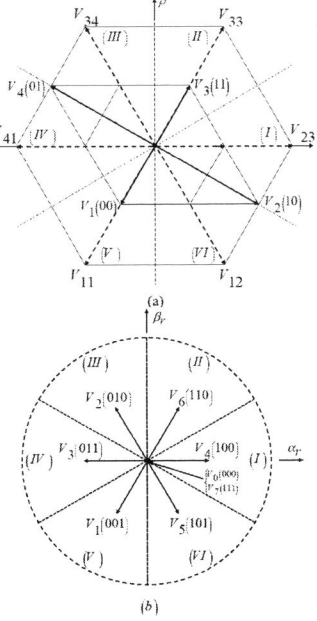

FIG. 5. A) GENERATED BALANCED VOLTAGE VECTORS BY COMBINING OF FOUR UNBALANCED VOLTAGE OF THE FSTPI B) VOLTAGE VECTORS GENERATED BY SSTPI

TABLE I.   SIMILARITY BETWEEN SPACE VECTORS OF FSTPI AND SSTPI

Used voltage space vectors for SSTPC	Used voltage space vectors for FSTPC
V_1	V_{11}
V_2	V_{34}
V_3	V_{41}
V_4	V_{23}
V_5	V_{12}
V_6	V_{33}
$V_{0,7}$	$V_{13,24}$

The switching table for DPC of DFIG is presented in [1]. In result, if you use the SSTPI the switching table will be modified according to the Table I and will be as Table II.

TABLE II.   MODIFIED SWITCHING TABLE FOR FSTPI

	Sector	I	II	III	IV	V	VI
$S_q=1$	$S_p=1$	V_{12}	V_{23}	V_{33}	V_{34}	V_{41}	V_{11}
	$S_p=0$	V_{23}	V_{33}	V_{34}	V_{41}	V_{11}	V_{12}
	$S_p=-1$	V_{33}	V_{34}	V_{41}	V_{11}	V_{12}	V_{23}
$S_q=0$	$S_p=1$	V_{11}	V_{12}	V_{23}	V_{33}	V_{34}	V_{41}
	$S_p=0$	V_{13}	V_{24}	V_{13}	V_{24}	V_{13}	V_{24}
	$S_p=-1$	V_{34}	V_{41}	V_{11}	V_{12}	V_{23}	V_{33}
$S_q=-1$	$S_p=1$	V_{11}	V_{12}	V_{23}	V_{33}	V_{34}	V_{41}
	$S_p=0$	V_{41}	V_{11}	V_{12}	V_{23}	V_{33}	V_{34}
	$S_p=-1$	V_{34}	V_{41}	V_{11}	V_{12}	V_{23}	V_{33}

But, the question that arises here is that, how can generated balanced voltage vectors shown in Fig. 5a. If two voltage vectors V_i and V_j to be used during two successive sampling periods ($2T_s$), the voltage vector V_{ij} will be created ($1 \le i \le 4$ and $1 \le j \le 4$). Notice that the inputs of switching table i.e. error from the hysteresis controllers (Sp & Sq) and stator flux position in the rotor reference frame (sector) should be maintained during $2T_s$. To better understand this topic, refer to the Fig. 6.

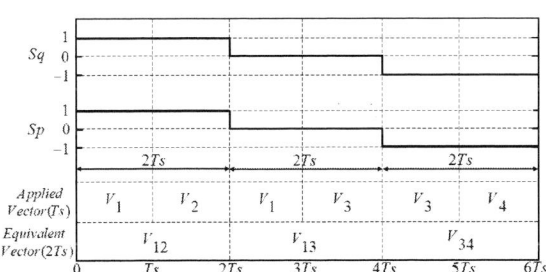

FIG. 6. HOW TO PRODUCE THE EQUIVALENT VOLTAGE VECTOR

VI. SIMULATION REAULTS

In this section, the proposed control strategies are validated. Different simulation results are presented in order to validate the theoretical performance. For simulation results, MATLAB/Simulink software has been used together with the Power System Block-set library tool.

The main characteristics of the simulated system are shown in Table III. The simulated generator is a 1-MW DFIG, and the scheme of the implemented system is shown in Fig. 7. During simulation, a sampling frequency of 20 KHZ was used for the proposed control strategy and the bandwidths of the active and reactive power hysteresis controllers were set at ±4% of the rated generator power of 1 MW.

The performance of the system will be analyzed under steady-state operation condition. The grid voltage unbalance programmed for this experiment is: (Vsa=394.25<0, Vsb=344.45<-113, Vsc=410.85<-231). The DFIG was assumed to be in speed control, i.e., the rotor speed is set externally, as the large inertia of the wind turbine results in slow change of rotor speed. During the period of 0.2-0.7 s, the rotor speed decrease from 1.2 to 0.8 p.u. various power steps are also applied, i.e., active and reactive power reference are changed from -1 to -0.4 MW at 0.5 s and from -0.33 to 0.33 MVar at 0.4 s, respectively.

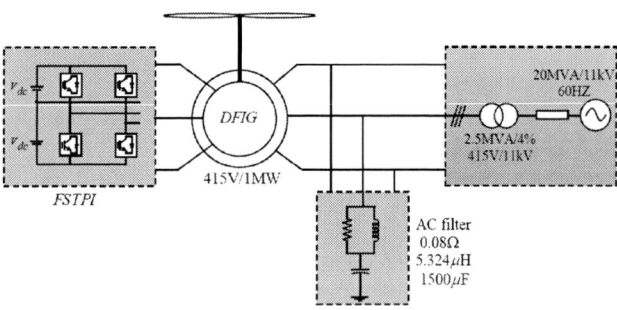

FIG. 7. SCHEME OF THE SIMULATED SYSTEM

TABLE III.   SYSTEM PARAMETERS

Rated Power	1MW
Line to line voltage (RMS)	415V
R_s	0.015pu
R_r (referred to the stator)	0.0167pu
L_m	4.467pu
L_{ls}	0.141pu
L_{lr} (referred to the stator)	0.152pu
Number of pole pairs	2
DC link voltage (V_{dc})	600V
Sampling period	5e-5

In the Fig. 8, The Target 1, i.e., Active and reactive power oscillation cancellation strategy, was selected as the control objective. From Fig. 8(a) Can see that, active and reactive power tracking their reference value, but when is used from FSTPC slightly has risen amount of ripples. Fig. 8(b) shows the three phase stator current that are unbalanced and nonsinusoidal. Fig. 8(c) shows the electromagnetic torque that has a lot of ripples, because not used from target 2.

In the Fig. 9, The Target 2, i.e., electromagnetic torque oscillation cancellation strategy, was selected as the control objective. Fig. 9(a) shows the active power tracking behavior in order to cancel the electromagnetic torque oscillation. The active power is oscillating from 0.4 s, because the terms of $2C_P$ or $2D_P$ have added to it. It can be seen from Fig. 9(b), the stator currents from 0.4s are still unbalanced, but have became sinusoidal. Fig. 9(c) show that the electromagnetic torque oscillations have been removed from 0.4s.

In the Fig. 10, The Target 3 of the power compensation scheme is implemented. As shown in Fig. 10(a), it can be observed that there are oscillating components in both the

active and reactive powers because of the two oscillating terms D_P and D_Q added to the original constant active and reactive powers. However, these eliminate the negative sequence current. As a result, the stator current become quite sinusoidal and symmetric, thus power quality is improved significantly. If look carefully to the Fig. 10(c), you can see that decreased the torque oscillation, but completely has not gone away.

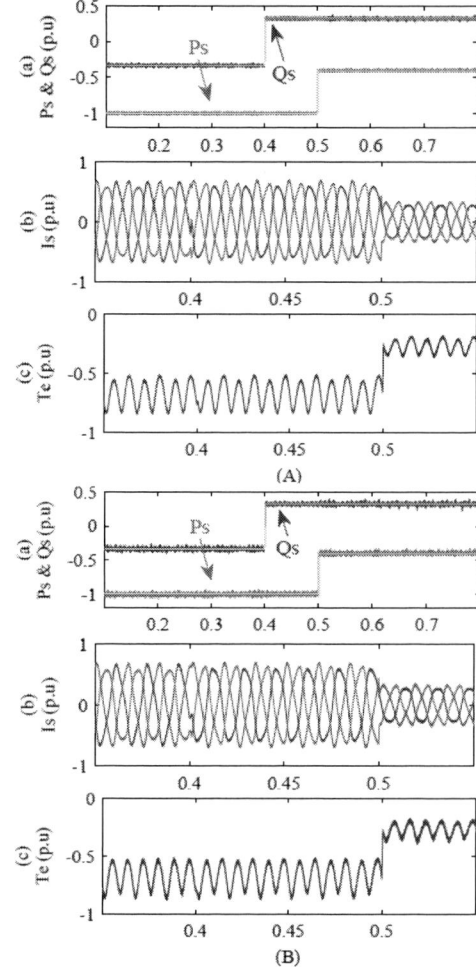

FIG. 8. COMPARISON BETWEEN SIMULATION RESULTS BY APPLYING TARGET 1 IN TWO CASES: A) SSTPI B) FSTPI IN: A) ACTIVE AND REACTIVE POWER B) STATOR CURRENTS C) ELECTROMAGNETIC TORQUE

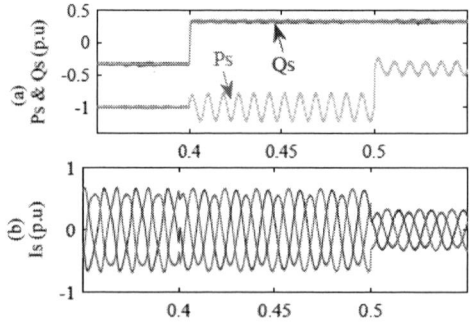

978-1-5090-0376-1/16 $31.00 © 2016 IEEE

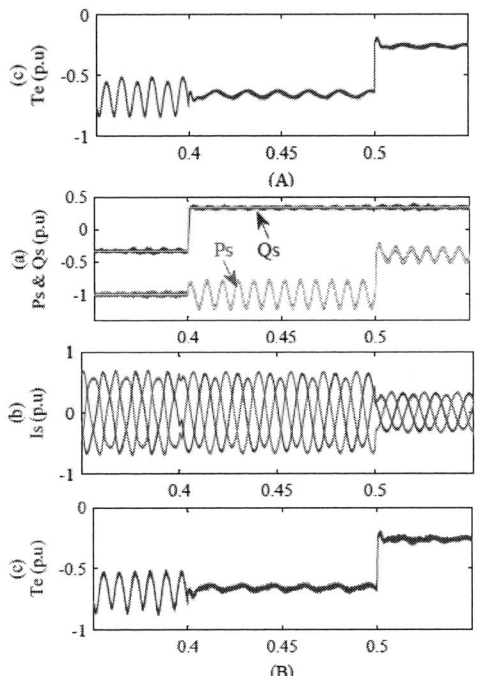

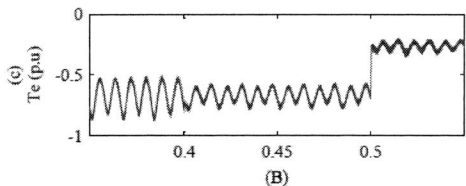

FIG. 10. COMPARISON BETWEEN SIMULATION RESULTS BY APPLYING TARGET 3 IN TWO CASES: A) SSTPI B) FSTPI IN: A) ACTIVE AND REACTIVE POWER B) STATOR CURRENTS C) ELECTROMAGNETIC TORQUE

VII. CONCLUSION

This paper has presented a new DPC for DFIG-based wind energy generation system operating under unbalanced network condition. In this method is used from FSTPI instead of SSTPI in rotor side.   It was shown that this structure has some advantages compared to conventional back to back converters such as lower number of active semiconductors, low cost and simple control system. The proposed DPC strategy is based on the emulation of the operation of the conventional SSTPI. This has been achieved by virtue suitable combinations of the four unbalanced voltage vectors intrinsically generated by the FSTPI, leading to the synthesis of the six balanced voltage vectors yielded by the SSTPI. Then, according to new voltage vectors the new switching tables is presented for FSTPI. Also, three selectable control targets is raised for compensating under unbalanced voltage condition. Choosing this control target is highly dependent on the design of the turbine system and the operation of the network. These targets are used for compensation of power, torque and stator current. The effectiveness of the proposed DPC is validated by a series of simulation results from a 1-MW DFIG system.

References

[1] L. Xu, and P. Cartwright, "Direct active and reactive power control of DFIG for wind energy application," IEEE Trans. Energy Convers., vol. 21, no. 3, pp. 750-758, Sep. 2006.

[2] D. Zhi, L. Xu, and B. W. Williams, "Model-based predictive direct power control of doubly fed induction generators," IEEE Trans. Power Electron., vol. 25, no. 2, pp. 341-351, Feb. 2010.

[3] Mohammad Verij Kazemi, Ahmad Sadeghi Yazdankhah, and Hossein Madadi Kojabadi, "Direct power control of DFIG based on discrete space vector modulation," vol. 35, Issue 5, pp. 1033–1042, May 2010.

[4] G. Abad, M. A. Rodriguez, G. Iwanski, and J. Poza, "Direct power control of doubly-fed-induction-generator based wind turbines under unbalanced grid voltage," IEEE Trans. Power Electron., vol. 25, no. 2, pp. 442-452, Feb. 2010.

[5] Jiefeng Hu, Jianguo Zhu, and David G. Dorrell, "Model-predictive direct power control of foubly fed induction generators under unbalanced grid voltage conditions in wind energy applications ," IET Renew. Power Gener., vol. 8, Iss. 6, pp. 687-695, 2014.

[6] P. Zhou, Y. He, and D. Sun, "Improved direct power control of a DFIG-based wind turbine during network unbalance," IEEE Trans. Power Electron., vol. 24, no. 11, pp. 2465-2474, Nov. 2009.

[7] M. Momfared, H. Rastegar, and H. M. Kojabadi, "Overview of modulation techniques for the four switch converter topology," in PECon 08, Dec. 2008, Johor Baharu, Malaysia.

[8] Mohamed Azab, and A. L. Orille, "Novel flux and torque control of IM drive using FSTPI," in IECON 01, 2001, pp. 1268-1273.

FIG. 9. COMPARISON BETWEEN SIMULATION RESULTS BY APPLYING TARGET 2 IN TWO CASES: A) SSTPI B) FSTPI IN: A) ACTIVE AND REACTIVE POWER B) STATOR CURRENTS C) ELECTROMAGNETIC TORQUE

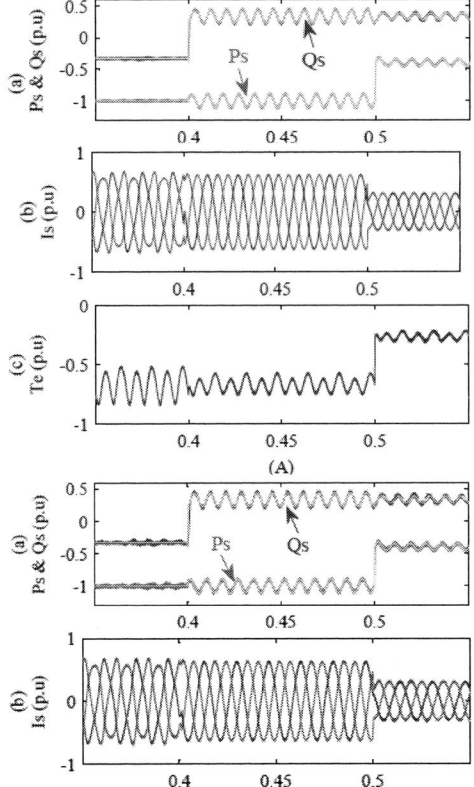

7th Power Electronics, Drive Systems & Technologies Conference (PEDSTC 2016)
16-18 Feb. 2016, Iran University of Science and Technology, Tehran, Iran

Energy Management of Dual-Source Propelled Electric Vehicle using Fuzzy Controller Optimized via Genetic Algorithm

S. Khoobi Arani[1,2]   A. Halvaei Niasar[1]   A. Haji Zadeh[3]

[1]Department of Electrical & Computer Engineering, University of Kashan, Kashan, Iran
[2]Department of Electrical Engineering, University of Shahrood, Shahrood, Iran
[3]Department of Energy Technology, Aalborg University, Denmark
s.khoobi@grad.kashanu.ac.ir   halvaei@kashanu.ac.ir   aminhajizadeh@gmail.com

Abstract— Energy and power distribution between multiple energy sources of electric vehicles (EVs) is the main challenge to achieve optimum performance from EV. Fuzzy inference systems are powerful tools due to nonlinearity and uncertainties of EV system. Design of fuzzy controllers for energy management of EV relies too much on the expert experience and it may lead to sub-optimal performance. This paper develops an optimized fuzzy controller using genetic algorithm (GA) for an electric vehicle equipped with two power bank including battery and super-capacitor. The model of EV and optimized fuzzy controller are simulated in ADVISOR software. Developed method has been implemented on standard driving cycles and simulation results show the decrease on consumed power by developed controller compared with standard fuzzy controller.

Keywords—Electric Vehicle (EV); Energy Management; Fuzzy Controller; Genetic Algorithm (GA), ADVISOR.

I. Introduction

Increasing fuel consumption in transportation system demands the management and control to run this system. Traditionally, fossil fuels were the major energy resources for transportation system. However, price uncertainties, political issues of oil provider countries and environmental problems of fossil fuels resulted in a need to find other energy resources [1-3]. The transportation system as one of the major energy sectors is changing the internal combustion engines. Electric vehicles as one of the best possible option have been developed at the last decade and new development in battery and storage devices, charge and discharge infrastructures has led to relatively high penetration of these vehicles [4,5]. According to above mentioned issues, optimum design of EVs is an important task. Optimal modeling and simulation of EVs leads to energy consumption and cost minimization. There are lots of simulation software to model and simulate the EVs in which ADVISOR seems to be more accurate.

Recently, due to the importance of EVs there has been an augmented interest in energy and power management and control field. A control strategy to reduce the energy consumption in super capacitor and fuel cell based EVs have been developed in [6,7]. Moreover, fuel consumption

optimization has been modeled in [8-9] for super capacitor and fuel cell based EVs. The developed method in [8] has also been examined on real EVs. In [9] the super capacitor's duty is to supply electric power in case of high power consumption of EV, especially in acceleration mode. Adaptive control method for EVs with parallel pattern has been studied in [10]; the developed strategy in this article has been adapted with a driving schematic. Reference [11] has modeled an EV system with just one convertor while [12] has used a fuzzy logic for optimum design of energy consumption in EVs.

In this paper an optimum energy consumption pattern for EV enabled with super capacitor, battery and fuel cell based on fuzzy model has been developed. For efficient and optimum performance of EVs the sustain control system should be designed since storage system selection and charging pattern is a very important task. It has been shown that fuzzy controller has better performance in comparison with other linear or nonlinear controllers. The objective of this study is to design the fuzzy based controller to control and manage EVs' power sources. In a fuzzy system membership function definition and fuzzy rules is the most important task. Hence to improve the controller operation, Genetic Algorithm (GA) is used to determine the membership functions of input and output parameter and variables as well as fuzzy rules.

This paper is organized as follows; sections II introduce the mathematical model of the EV system, and section III develops the fuzzy controller for energy management of EV. The configuration of the simulation and obtained results are provided in section IV and finally, concluding remarks are given in section V.

II. Mathematical Model of Electric Vehicle

The schematic diagram of studied EV has been shown in Fig. 1. The power transmission system contains battery, super capacitor bank, DC/DC convertor, inventor and electric motor. Battery and super capacitor have parallel structure in order to make DC link. DC/DC converter regulates DC link voltage. the controlled DC voltage converts to AC by inverter to start electric motor. Transmission system which contains a gearbox,

978-1-5090-0376-1/16 $31.00 © 2016 IEEE   338

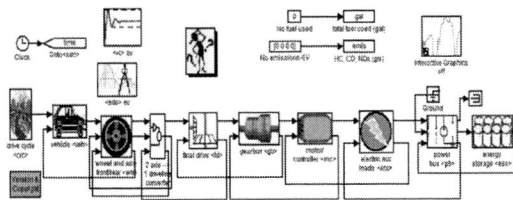

Fig. 1. Block diagram EV with dual energy source in ADVISOR

multiplies motors torque to conversion coefficient. In case of high power need, battery, super capacitor, gearbox and electric motor produce the needed power and supplies the vehicle.

In driving stage, energy storage system (ESS) supplies power needed to run the car. Considering the drag the car in road, mechanical force as much as F1 is needed, with respect to vehicle dynamic theory, F1 includes four terms: air resistance (F_d), wheel friction (F_r), upward drag force (F_c) and accelerating force (F_a) as:

$$F1 = F_d + F_r + F_c + F_a \qquad (1)$$

Aerodynamic force (F_d) is resulted by interaction between vehicle body and air by car movement. This force itself has three terms each of which created by an aerodynamic phenomena: shell traction friction, induced force created behind the vehicle and finally natural pressure force that is rational to front of car and driving speed. The third term is greater than two other terms, and so aerodynamic power could rewritten as:

$$F_d = 0.5\rho \times C_d \times A \times V^2 \qquad (2)$$

which C_d is aerodynamic traction coefficient ρ is air density, A is car front area and V is the current speed. Rotational resistance force that is because of while deformation explained as:

$$F_r = M \times g \times C \qquad (3)$$

that M is mass of car, g is gravity and C is rotational resistance coefficient. The upward drag force is simple and obeys:

$$F_c = M \times g \times \sin(\alpha) \qquad (4)$$

where α is ramp degree. The acceleration force to drive the car can be derived form:

$$F_a = M \times \frac{dV}{dt} \qquad (5)$$

Hence total force can be rewritten as:

$$F_1 = \frac{0.5\rho \times C_d \times A \times V^2 + M \times g \times C + M \times g \times \sin(\alpha) + M \times dV}{dt} \qquad (6)$$

Multiplying this force to velocity, needed power to drive the car is calculated by:

$$P_{Load} = \frac{(0.5\rho \times C_d \times A \times V^2) \times V}{3600} + \frac{(M \times g \times C + M \times g \times \sin(\alpha) + M \times dV) \times V}{3600} \qquad (7)$$

Considering efficiency of studied electric vehicle electric, power demand to drive the car is obtained by:

$$P_{Req} = P_{Load} \times \eta \qquad (8)$$

To have enough power at all time reliably ESS should have enough power and in the other hand considering $K_{Bat}(t)$ and $K_{sc}(t)$ as battery and super capacitor power factors, the amount of power that is extracted from each power sources can be calculated from:

$$\begin{aligned} P_{Bat} &= K_{Bat}(t) \times P_{Req}(t) \\ P_{SC} &= K_{SC}(t) \times P_{Req}(t) \end{aligned} \qquad (9)$$

Also sum of power factors for battery and super capacitor should be unit at all time. Moreover, if SOC of battery and super capacitor rich to more than selected amount, their ability to absorb regenerative breaking power will reduce. On the other hand if SOC become very low, ESS may have lower power than the needed amount in acceleration stage. To increase the life of battery and super capacitor, the stored energy should be in predefined range. Energy resource of EVs charging constrains are as following.

$$\begin{aligned} SOC_{Bat}^{Min} &\leq SOC_{Bat}(t) \leq SOC_{Bat}^{Max} \\ SOC_{SC}^{Min} &\leq SOC_{SC}(t) \leq SOC_{SC}^{Max} \end{aligned} \qquad (10)$$

In this paper, the lower and upper level of capacitor and battery storage limits are 0.2 and 0.8 respectively. Considering all above mentioned notes, mathematical model to manage energy of these to sources can be formulated as following.

$$\begin{aligned} Min \quad & Energy_{Req} \left| K_{Bat}(t), K_{SC}(t) \right. \\ St. \quad & \\ & P_{Req} = P_{Load} \times \eta \\ & SOC_{Bat}^{Min} \leq SOC_{Bat}(t) \leq SOC_{Bat}^{Max} \\ & SOC_{SC}^{Min} \leq SOC_{SC}(t) \leq SOC_{SC}^{Max} \end{aligned} \qquad (11)$$

978-1-5090-0376-1/16 $31.00 © 2016 IEEE   339

III. MIXED INTEGER GENETIC ALGORITHM

MIGA is an optimization technique for solving the integer and mixed integer constrained optimizations. Here, this technique is modified to find a stable solution of the protection coordination problem using FCLs. Different functions of the basic GA, including the Power mutation and Laplace crossover functions have been modified and an especial truncation procedure has been used to cope with the integrality constraints. A parameter free penalty approach is used to handle the problem constraints. In the following formulation n is the number of variables indexed by i.

The modifications with respect to the original GA algorithm are presented here. The Laplace crossover function that was originally proposed by [13] is the first function that is modified here. According to [13], two offspring y^1 and y^2 are generated from two parents x^1 and x^2 based on (12), where β_i is a random number that satisfies the Laplace distribution based on (13). In (13), u_i and r_i, both between 0 and 1, are two uniformly distributed random variables and a and $b > 0$ are location and scaling parameters, respectively. For the continuous variables $b=b_{real}$ and for the integer variables $b=b_{int}$.

$$
\begin{aligned}
y_i^1 &= x_i^1 + \beta_i \cdot \left| x_i^1 - x_i^2 \right| \\
y_i^2 &= x_i^2 + \beta_i \cdot \left| x_i^1 - x_i^2 \right|
\end{aligned}
\tag{12}
$$

$$
\beta_i = \begin{cases} a - b \cdot \log(u_i), & r_i \le 0.5 \\ a + b \cdot \log(u_i), & r_i > 0.5 \end{cases}
\tag{13}
$$

Reference [14] proposed a power distribution for mutation process. Solution x is generated in the vicinity of a parent solution $\bar{x}$ using (14). The variable s follows the power distribution ($s=w^p$). Variable w is a uniformly distributed random variable between 0 and 1. The index of mutation is denoted by p. The values of this index for integer and continuous variables are p_{int} and p_{real}, respectively. In (14), r is a random variable with uniform distribution. Notations x^l and x^u are used for the lower and upper bounds of the decision variable x, respectively.

$$
x = \begin{cases} \bar{x} - s(\bar{x} - x^l), & t < r \\ \bar{x} + s(x^u - \bar{x}), & t \ge r \end{cases}
$$

$$
t = \frac{(\bar{x} - x^l)}{(x^u - \bar{x})}
\tag{14}
$$

In order to select the individuals from the population to be inserted into the mating pool, the same selection technique as [15] is also used here. To ensure that after crossover and mutation operations, the integrality constraints are satisfied, a truncation procedure is adopted. If variable x_i does not have an integer value after these operations, the value of this variable is changed to $[x_i]$ or $[x_i]+1$ each with the probability of 0.5,

where $[x_i]$ is the correct part of x_i.

In order to find a solution that satisfies the problem constraints, a parameter free penalty function approach is adopted [16]. The fitness value for jth solution is calculated using (15), where f_w is the value of objective function for the worst feasible solution available in the current population. Under such setup, the fitness value of a solution depends on the degree of constraint violation and also the population in hand. In the case where there is no feasible solution so far, f_w is set to zero. Based on [16], such constraint handling technique pushes the infeasible solutions towards the problem feasible region. k^{th} inequality constraint and the penalty function regarding to this constraint is shown in (16) for solution j. In order to satisfy the equality constraints in the final solution these constraints are managed to be changed into two inequality constraints using a constraint satisfaction tolerance.

$$
fit(X_j) = \begin{cases} f(X_j), & \text{if } X_j \text{ is a feasible solution} \\ f_w + \sum_{k=1}^{K} \left| \phi_k(X_j) \right|, & \text{otherwise} \end{cases}
\tag{15}
$$

$$
\begin{aligned}
g_k(X) &\le G_k \\
\phi_k(X_j) &= G_k - g_k(X_j)
\end{aligned}
\tag{16}
$$

IV. FUZZY CONTROLLER DESIGN

In order to manage the energy stored in battery and super capacitor sustainably, in each time step a sustainable decision for power factors should be make. According to energy management rules and constraints between two energy storage sources, the optimization problem is nonlinear, and therefore conventional optimization approaches cannot be employed. Fuzzy control is one of best approaches to tackle this optimization problem. Fuzzy control can be easily used for nonlinear problems [17].

A. Fuzzy Systems

A simple block diagram of energy management fuzzy control is illustrated in Fig. 2. Battery and super capacitor charging state are input of this control system and output is battery power factor. Once battery power factor is determined, the super capacitor power factor could be easily determined, since the sum of these factors is unit at all times. Inputs and output membership functions are illustrated in Fig. 3.

B. Proposed Optimized Fuzzy Controller based on Genetic Algorithm

In spite of the capability of fuzzy systems, the design of fuzzy systems need to knowledge of expert human that may not be available. To solve this problem, in this paper, Genetic algorithm is used to optimize the design of fuzzy controller. As shown in Fig. 3, the membership function of the required power is symmetrical around zero. This membership function can be determined individually by three parameters. These parameters are named z_1, z_2 and z_3 as shown in Fig. 4. Similarly, to determine membership function of battery and super capacitor, parameters z_4 and z_5 are needed. Also output membership function should be determined by z_6 and z_7. In fact, these seven parameters are the optimization variables.

There are three states for battery SOC and three states for super capacitor SOC and seven states for required power state. Hence there are 63 if and then rules in the fuzzy system. These 63 rules introduce 63 decision variables; these variables are integer. Moreover, there are 5 state variables for the output. As mentioned earlier GA is employed to optimally determine the fuzzy rules and fuzzy membership of inputs and output. There are continuous and integer variables in this optimization problem. GA is used to properly handle these variables and determine these variables such that the fuzzy system has the best performance. Once the variables are determined, the fuzzy system can be used to optimally determine the battery and super capacitor power factor in different times in the driving cycle.

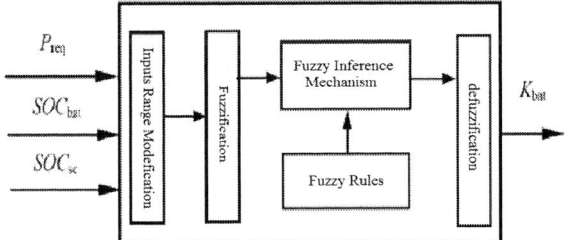

Fig. 2. Fuzzy control system block diagram

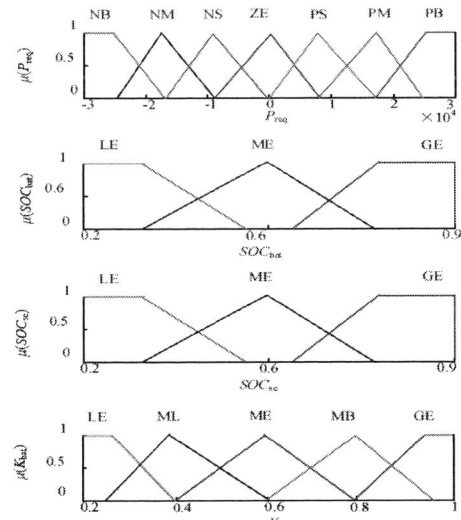

Fig. 3. Input and output membership functions

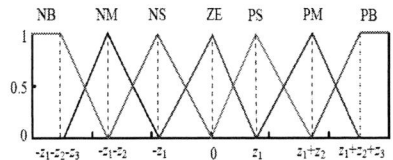

Fig. 4. Parameters of required membership function

V. CASE STUDY

Developed optimal fuzzy controller is employed to control the battery and super capacitor SOC in EV. Some simulation results are provided and compared with ordinary fuzzy controller results. For generation of driving force, EV uses battery and capacitor. To evaluate suggested method, a standard driving cycle has been used. Fig. 5 shows the characteristics of this cycle.

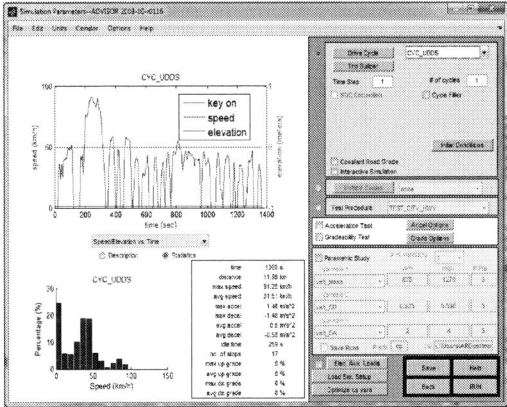

Fig. 5. Driving cycle in ADVISOR software

By simulating this driving cycle in ADVISOR, total and time frame amount of required power can be excluded. The required motor power in each second is shown in Fig. 6. In order to evaluate the developed method in this paper, first energy management results by ordinary fuzzy controller have been calculated and then the results of developed fuzzy control method is presented.

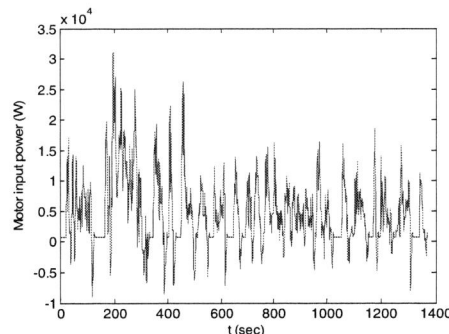

Fig. 6. Needed power in each second

978-1-5090-0376-1/16 $31.00 © 2016 IEEE

A. Ordinary fuzzy control

Using ordinary fuzzy control and simulation of driving cycle second by second, with respect to required power, SOC of super capacitor and battery can be calculated. Fig. 7 shows the battery quota of generated power. Hence super capacitor quota simply can be calculated. Fig. 7 and Fig. 8 show the super capacitor and battery SOCs. Primitive value of SOCs are considered to be 80% of maximum capacity. As shown in Fig. 8 and Fig. 9, the minimum power charge constrains are not met. Total supplied power by battery and capacitor is 6.71 MJ.

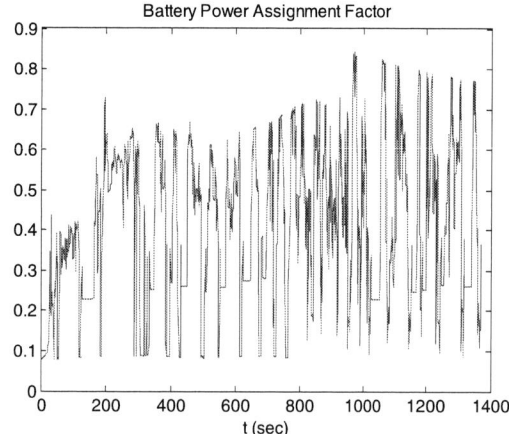

Fig. 7. Battery quota of required power resulted from ordinary fuzzy control

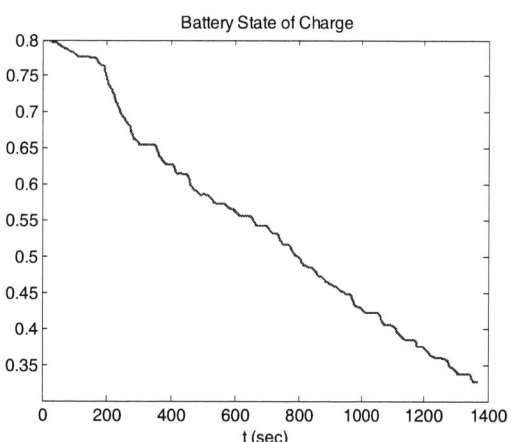

Fig. 8. Battery SOC resulted by ordinary fuzzy control

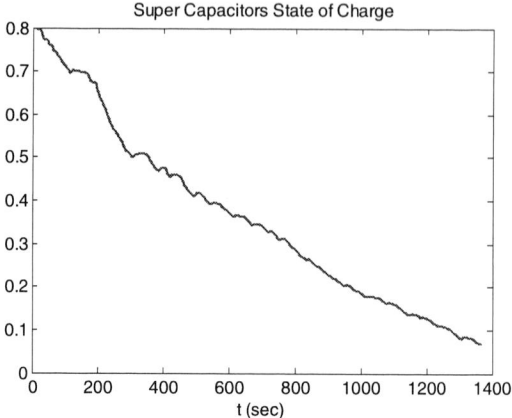

Fig. 9. Super capacitor SOC resulted by fuzzy control

B. Optimal fuzzy control

Using optimal input and output membership function and if and then rules and simulating driving cycle second by second SOC of super capacitor and battery can be extracted. Fig. 10 shows batteries portion of required power. Fig. 11 and Fig. 12 show the super capacitor and battery SOC in driving cycle. It is shown that minimum stored energy constraint cannot be met by ordinary fuzzy control. By optimal control the minimum energy constraint has been met. Also total energy consumption by optimal fuzzy control is 6.55 MJ.

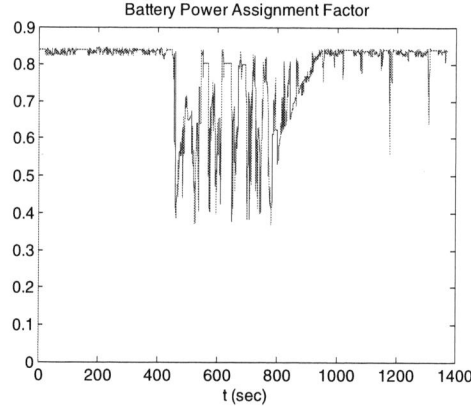

Fig. 10. Battery portion of consumed power.

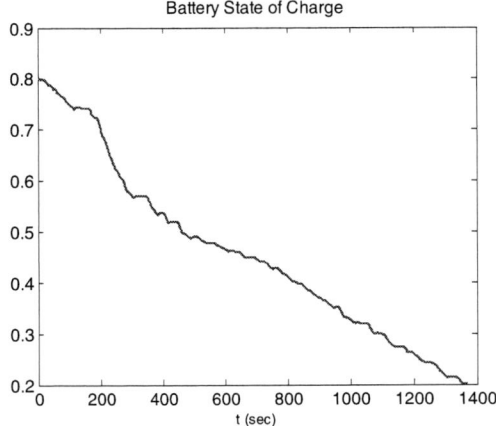

Fig. 11. SOC of battery in driving cycle resulted by optimal fuzzy controller

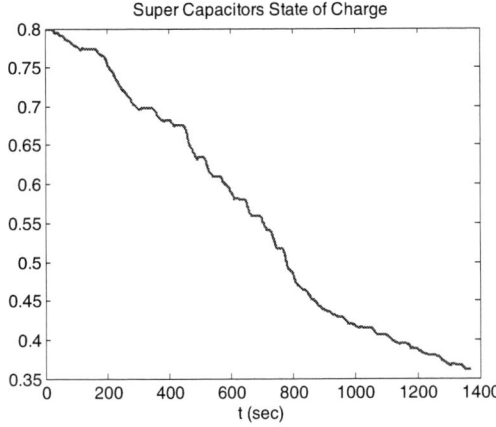

Fig. 12. SOC of super capacitor in driving cycle resulted by optimal fuzzy controller

VI. CONCLUSION

Optimal energy management control in electric vehicle using conventional and GA-based fuzzy controllers have been investigated in this paper. Standard driving cycle has been employed to evaluate the proposed energy management control scheme. Obtained results demonstrated that conventional fuzzy controller cannot manage the energy sources as effectively as GA-based fuzzy controller such that using conventional fuzzy controller the minimum SOC constraints of battery and supercapacitor are violated. Moreover, total energy consumption has shown 2.5 percent reduction in the proposed GA-based fuzzy controller compared with the conventional fuzzy controller. Obtained results demonstrate the importance of optimal input and output fuzzy membership functions as well as fuzzy rules.

REFERENCES

[1] J. Eastin, R. Grundmann, A. Prakash, "The two limits debates: Limits to Growth and climate change", Futures, vol. 43, no.1, pp.16-26, 2011.

[2] Y. Cancino-Solo´rzano, et al, "Electricity sector in Mexico: current status. Contribution of renewable energy sources", Renewable and Sustainable Energy Reviews, vol. 14, no. 1 , pp. 454-461, 2010.

[3] AM. Omer, "Energy, environment and sustainable development", Renewable and Sustainable Energy Reviews, vol. 12, no.9, pp. 2265–2300, 2008.

[4] X. Oua, X. Zhanga, and S. Changa, "Scenario analysis on alternative fuel/vehicle for china's future road transport: Life-cycle energy demand and GHG emissions", Energy Policy, vol. 38, no. 8, pp. 3943-3956, 2010.

[5] R. Sioshansi and P. Denholm, "Emissions impacts and benefits of plugin hybrid electric vehicles and vehicle-to-grid services", Environ. Sci. Technol., vol. 43, no.4, pp.1199–1204, 2009.

[6] T. Azib, G. Remy, O. Bethoux, and C. Marchand, ''Control Strategy with Saturation Management of a Fuel Cell/Ultracapacitors Hybrid Vehicule ",IEEE Vehicle Power and Propulsion Conference (VPPC), pp. 1– 6,2010.

[7] P. Thounthong, S. Rael, and B. Davat, "Control strategy of fuel cell/ultracapacitors hybrid power sources for electric vehicle", Elsevier, Journal of Power Sources, vol. 158, no.1, pp. 806 – 814, 2006.

[8] Z. Jiang, L. Gao, M.J. Blackwelder, and R.A. Dougal, "Design and experimental tests of control strategies for active hybrid fuelcell/batterypower sources'', Journal of Power Sources, vol. 130, no. 1-2, pp. 163- 171, 2004.

[9] T. Azib, O. Bethoux, G. Remy, C. Marchand, "Structure and Control Strategy for a Parallel Hybrid Fuel Cell/Ultracapacitors Power Source", IEEE Vehicle Power and Propulsion Conference (VPPC'09), pp.1858-1863, 2009.

[10] A. Chasse, A. Sciarretta, and J. Chauvin, "Online optimal control of a parallel hybrid with vehicles adaptation", Proceedings of the 6th IFAC Symposium "Advances in Automotive Control", pp.99-104, 2010.

[11] T. Azib, O. Bethoux, G. Remy, C. Marchand, and E. Berthelot "An Innovative Control Strategy of a Single Converter for Hybrid FuelCell/SupercapacitorsPowerSource", IEEE Trans. Industrial Electronics, vol. 57, no.12, pp.4024 – 4031, Dec. 2010.

[12] DaweiGao, ZHenhua Jin, Qingchun Lu. "Energy managementstrategy based on fuzzy logic for a fuel cell hybrid bus", Journal of Power Sources, vol.185, no. 1, pp. 311-317, 2008.

[13] K. Deep, and M. Thakur, "A new crossover operator for real coded genetic algorithms," Applied Mathematics and Computation, vol.188, pp. 895–912, 2007.

[14] K. Deep, and M. Thakur, "A new mutation operator for real coded genetic algorithms," Applied Mathematics and Computation, vol. 193 pp. 211–230, 2007.

[15] D.E. Goldberg, and K. Deb, "A comparison of selection schemes used in genetic algorithms," Foundations of Genetic Algorithms 1, FOGA-1, vol. 1, pp. 69–93, 1991.

[16] K. Deb, "An efficient constraint handling method for genetic algorithms", Computer Methods in Applied Mechanics and Engineering, vol. 186, pp. 311–338, 2000.

[17] M. P. Kevin, and Y. Stephen, Fuzzy control, Addison Wesley Longman, Menlo Park, CA, 1998.

Three Phase Photovoltaic Grid-Tied Inverter Based on Feed-Forward Decoupling Control Using Fuzzy-PI Controller

Faramarz Karbakhsh
Electrical Engineering Department
Amirkabir University of Technology (Tehran Polytechnic)
Tehran, Iran
f.karbakhsh@aut.ac.ir

G. B. Gharehpetian
Electrical Engineering Department
Amirkabir University of Technology (Tehran Polytechnic)
Tehran, Iran
grptian@aut.ac.ir

Jafar Milimonfared
Electrical Engineering Department
Amirkabir University of Technology (Tehran Polytechnic)
Tehran, Iran
monfared@aut.ac.ir

Armin Teymoori
Electrical Engineering Department
Amirkabir University of Technology (Tehran Polytechnic)
Tehran, Iran
armin.teymoori@aut.ac.ir

Abstract—This paper presents a two-stage photovoltaic grid-connected inverter that features fuzzy control. The dq synchronous rotating reference frame is used to model the inverter and feed-forward decoupling control strategy is used. The proposed controller provides decoupled control of active and reactive power while simplicity merit of dq reference frame is preserved. The combination of the flexible and intelligent fuzzy logic with the classical PI controller makes the control of the system more feasible and accurate. In this paper the initial parameters of fuzzy logic controller are obtained from its linear counterpart. The effectiveness and precision of the proposed control strategy are verified by the simulation of a grid-tied three-phase photovoltaic inverter.

Keywords— Photovoltaic systems, feed-forwad decoupling, fuzzy control, dq synchronous rotating frame

I. INTRODUCTION

Although solar panels are relatively expensive, photovoltaic (PV) power systems have been commercialized all over the world due to its long term economic prospects, silent operation, low maintenance and the concerns over the environment [1]. The PV generator (PVG) system has two modes of operation, which are independent operation mode and grid-connected operation mode. Nowadays, the grid connected operation mode is of greater importance and is widely used [2]-[3]. Therefore the grid-connected inverter is of great importance. In the grid-connected inverter systems, the DC voltage outer loop and AC current inner loop are conventionally implemented using fixed-gain proportional-integral (PI) or proportional-integral-derivative (PID) controllers. Simplicity and ease of implementation are the merits of linear PI controllers but its dependence on the system parameters is the drawback of these controllers. The fuzzy logic controller (FLC) is a solution to overcome this problem. The best feature of FLC is that it does not require the knowledge of mathematical model of the plant; however, the procedure of designing and gain tuning is controversial. In this

paper the initial parameters of FLC are obtained from the parameters of its well-tuned linear counterpart [4]. In this paper fuzzy-PI controllers are used instead of conventional PI controllers. Moreover, in order to facilitate the compensator design the control variables are studied under the rotating dq reference frame. In steady state condition, the control variables are DC quantities under this reference frame [5]. Here, in order to control the active and reactive power independently, the feed-forward decoupling control is used.

II. MATHEMATICAL MODEL OF THE INVERTER

A. Operating Principle

Fig. 1 depicts the schematic diagram of the system. It is consisted of a boost stage, DC-Link, three-phase inverter, and the power grid model.

B. Feed-forward Decoupling Control

Assuming the power grid is balanced, it can be defined as follows [5]:

$$\begin{cases} v_{sa} = E\cos(\omega t) \\ v_{sb} = E\cos(\omega t - 2\,{}^{\pi}\!/_{3}) \\ v_{sc} = E\cos(\omega t + 2\,{}^{\pi}\!/_{3}) \end{cases} \quad (1)$$

Where E is the amplitude of the grid voltage, and ω is the angular frequency. Considering Fig.1 The fundamental equation of inverter is defined as follows:

$$\begin{pmatrix} \frac{di_a}{dt} \\ \frac{di_b}{dt} \\ \frac{di_c}{dt} \end{pmatrix} = \begin{pmatrix} -\frac{R}{L} & 0 & 0 \\ 0 & -\frac{R}{L} & 0 \\ 0 & 0 & -\frac{R}{L} \end{pmatrix} \begin{pmatrix} i_a \\ i_b \\ i_c \end{pmatrix} + \frac{1}{L}\begin{pmatrix} v_{ta}-v_{sa} \\ v_{tb}-v_{sb} \\ v_{tc}-v_{sc} \end{pmatrix} \quad (2)$$

978-1-5090-0376-1/16 $31.00 © 2016 IEEE

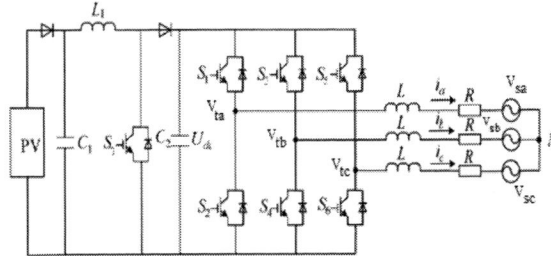

Fig. 1. Schematic diagram of the system

Where v_{ta}, v_{tb}, and v_{tc} represent the output voltage of the inverter, respectively. v_{sa}, v_{sb}, and v_{sc} stand for three-phase grid voltage. i_a, i_b, and i_c represent three-phase injected current to the grid, respectively. L is the sum of the output filter inductance and the grid inductance.

Using coordinate transformation from *abc* frame to dq synchronous rotating reference frame, (2) can be represented as follows [6]:

$$\begin{pmatrix} \frac{di_d}{dt} \\ \frac{di_q}{dt} \end{pmatrix} = \frac{1}{L}\begin{pmatrix} -R & \omega L \\ -\omega L & -R \end{pmatrix}\begin{pmatrix} i_d \\ i_q \end{pmatrix} - \frac{1}{L}\begin{pmatrix} v_{sd} \\ v_{sq} \end{pmatrix} + \frac{1}{L}\begin{pmatrix} v_{td} \\ v_{tq} \end{pmatrix} \quad (3)$$

Where i_d and i_q are defined as d-axis component and q-axis component of the injected current to the grid, respectively. v_{sd} and v_{sq} are the d-axis and q-axis components of the three-phase grid voltage, v_{td} and v_{tq} are defined as the d-axis and q-axis components of the inverter output terminal.

Equation (3) can be rewritten as follows:

$$\begin{cases} v_{td} = L\frac{di_d}{dt} + Ri_d - L\omega_0 i_q + v_{sd} \\ v_{tq} = L\frac{di_q}{dt} + Ri_q + L\omega_0 i_d + v_{sq} \end{cases} \quad (4)$$

Considering the dq mathematical model described with (4), due to the presence of the term $L\omega_0$, dynamics of i_d and i_q are coupled. This leads to difficulty in the control design process. The voltage-sourced converter (VSC) can be modeled as follows:

$$\begin{cases} v_{td} = m_d(t)\frac{V_{DC}}{2} \\ v_{tq} = m_q(t)\frac{V_{DC}}{2} \end{cases} \quad (5)$$

Decoupling the mentioned dynamics can be achieved by defining $m_d(t)$ and $m_q(t)$ as follows [6]:

$$\begin{cases} m_d(t) = \frac{2}{V_{DC}}(u_d - \omega L i_q + v_{sd}) \\ m_q(t) = \frac{2}{V_{DC}}(u_q + \omega L i_d + v_{sq}) \end{cases} \quad (6)$$

Where u_d and u_q are two new control inputs. Substituting for $m_d(t)$ and $m_q(t)$ in (4) and (5), respectively from (6), it is deduced:

$$\begin{cases} L\frac{di_d}{dt} = -Ri_d + u_d \\ L\frac{di_q}{dt} = -Ri_q + u_q \end{cases} \quad (7)$$

Therefore the control equations can be written as follows:

$$\begin{cases} v_{td} = (k_p + \frac{k_i}{s})(i_d^* - i_d) - \omega L i_q + v_{sd} \\ v_{tq} = (k_p + \frac{k_i}{s})(i_q^* - i_q) + \omega L i_d + v_{sq} \end{cases} \quad (8)$$

C. Power decoupling control

Considering the instantaneous power theory [6], under *dq* synchronous rotating reference frame we can get:

$$\begin{cases} P = \frac{3}{2}(v_{sd}i_d + v_{sq}i_q) \\ Q = \frac{3}{2}(v_{sq}i_d - v_{sd}i_q) \end{cases} \quad (9)$$

Where P and Q are the active and reactive power, respectively.

If d-axis is positioned as grid voltage synthesized vector, namely $e_q = 0$, (9) can be declared as bellow:

$$\begin{cases} P = \frac{3}{2}v_{sd}i_d \\ Q = -\frac{3}{2}v_{sd}i_q \end{cases} \quad (10)$$

Considering (10), the active power P and reactive power Q can be controlled via d-axis current and q-axis current, respectively.

III. FUZZY LOGIC CONTROLLER DESIGN

The ability to handle nonlinearities, working with imprecise inputs, and not needing to have accurate mathematical model are the merits of FLCs [7]-[9]. The general FLC structure is shown in Fig. 2. The FLC is consisted of fuzzification, membership function, rule base, fuzzy interference engine and defuzzification [10].

A. Comparetive design for conventioanal FLC

In the case of designing and tuning the scaling gains of fuzzy logic controllers, there is not any methodical approach. Ref. [4] proposes a method that initial parameters of FLC are extracted from its linear PI controller counterpart.

PI, PD, and PID are three different types of controllers in linear control theory. PI type FLC, PD type FLC, and PID type FLC are the analogous ones in fuzzy logic control [11]-[12]. The PI type FLC configuration is shown in Fig.3.

If a fuzzy controller employs a linear rule base, it is linear in the linguistic domain and is called a fuzzy linear controller [4].

A two dimensional linear rule base with seven labels for each input and output is used. The employed rule base is shown in Table I. The membership functions are uniformly distributed triangular. As shown in Fig. 4 the membership functions consist of Negative Big, Negative Medium, Negative Small, Zero, Positive Small, Positive Medium, and Positive Big which are represented by NB, NM, NS, ZR, PS, PM, and PB respectively.

Control surface of a linear rule base with respect to Table I is shown in Fig. 5. Ref. [13] proposes a method in order to analyze the system stability with this three dimensional curve.

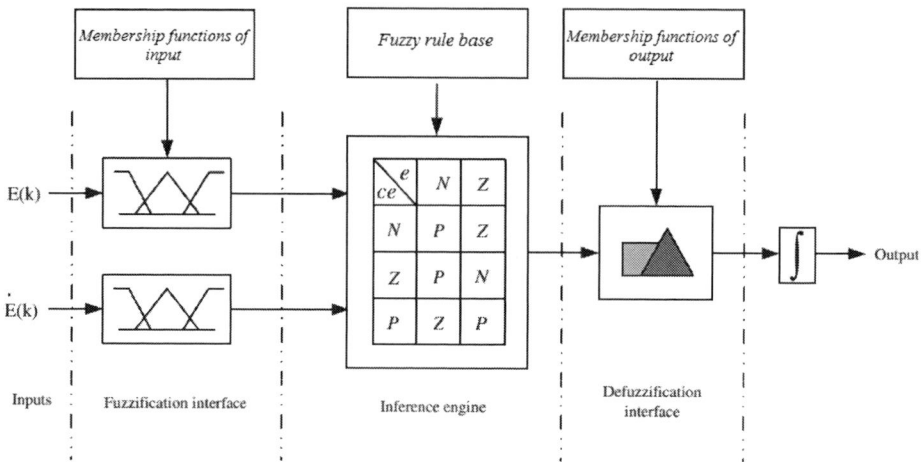

Fig. 2. Basic configuration of FLC

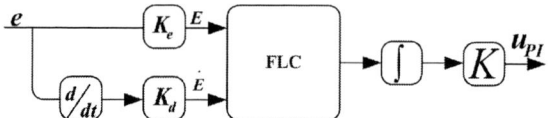

Fig. 3. Basic configuration of PI FLC

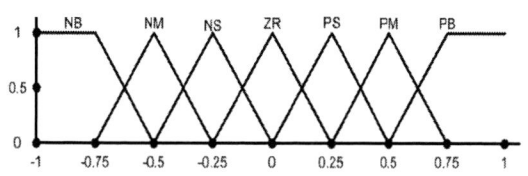

Fig. 4. Input and output membership function

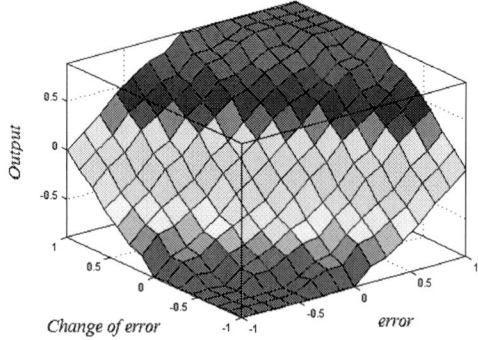

Fig. 5. Control surface of linear rule base

TABLE I.   A LINEAR TWO-DIMENSIONAL RULE BASE

		e(t)						
		NB	NM	NS	ZR	PS	PM	PB
$\dot{e}(t)$	PB	ZR	PS	PM	PB	PB	PB	PB
	PM	NS	ZR	PS	PM	PB	PB	PB
	PS	NM	NS	ZR	PS	PM	PB	PB
	ZR	NB	NM	NS	ZR	PS	PM	PB
	NS	NB	NB	NM	NS	ZR	PS	PM
	NM	NB	NB	NB	NM	NS	ZR	PS
	NB	NB	NB	NB	NB	NM	NS	ZR

IV. THE OVERALL PROPOSED CONTROL STRATEGY

The whole control system consists of MPPT loop, DC voltage outer loop, and AC current inner loop shown in Fig. 6. Because of the introduced boost circuit, MPPT loop is independent with DC voltage outer loop and AC current inner loop.

According to [6], in order to prevent over modulation the following inequality must be ensured:

$$V_t(t) \leq \frac{V_{DC}}{2} \qquad (11)$$

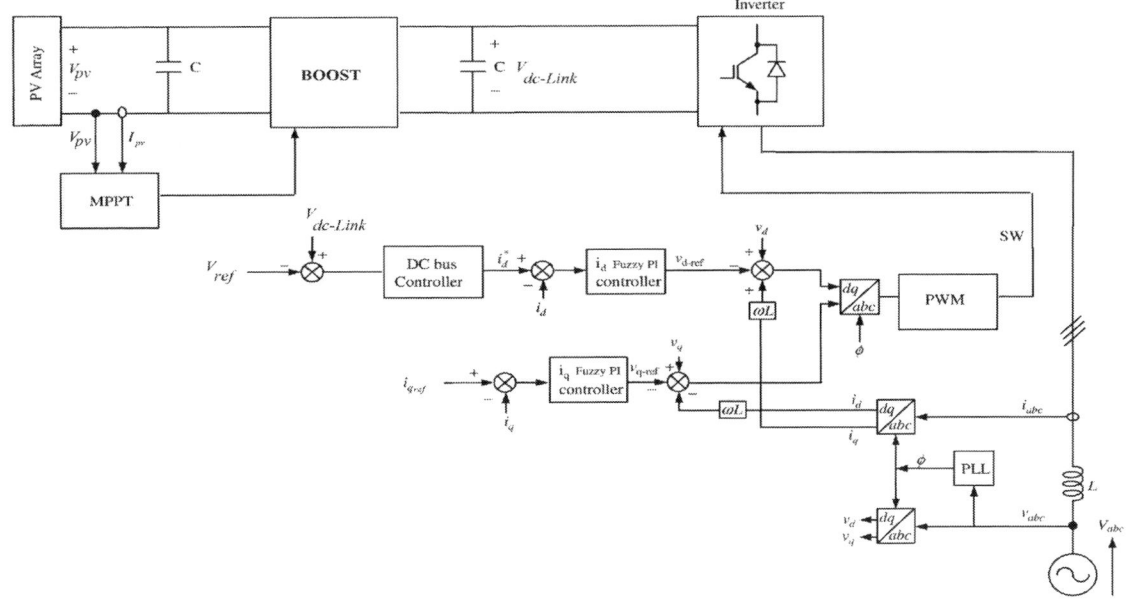

Fig. 6. Block diagram of the overall system

DC voltage outer loop gives out a given value of grid-connected power that ensures the net power transferred by the DC-link capacitance is zero.

V. SIMULATION

In order to verify the correctness of the proposed fuzzy control strategy, a simulation platform based on PSIM 9.0 software integrated with MATLAB is established. In the case of power electronics PSIM is one of the fastest simulators while retaining excellent simulation accuracy.

MATLAB/Simulink presents a graphical simulation environment suitable for controller design and simulation of dynamic systems. The effectiveness and accuracy of PSIM-Simulink co-simulation is verified in [14]-[15]. Table II presents the detailed specifications of the simulation platform.

A. The simulation results

In Fig. 7 the three-phase injected current into grid is shown in full load condition.

In Fig. 8 the output current of phase a, and the sampling of grid voltage are shown in full load condition at different power factors which is determined via the reactive power reference value. In Fig. 8 (a) the current is in phase with the voltage grid that verifies unity power factor as the reactive power reference is zero. The injected current with lagging and leading power factors are shown in Fig.8 (b) and (c).

Fig. 9 shows the DC-Link voltage that effectively tracks the reference voltage.

TABLE II.   PARAMETERS OF SIMULATION

Parameter	Symbol	Value	Unit
Grid frequency	f_{grid}	50	Hz
Switching frequency	f_s	5	kHz
Input capacitance	C_{in}	0.8	mF
DC-Link capacitance	$C_{dc\text{-}link}$	0.2	mF
Output filter inductance	L_f	5	mH
PV array maximum power	P_{max}	4900	W
PV array voltage at P_{max}	V_{mpp}	299	V
PV array current at P_{max}	I_{mpp}	16.4	A

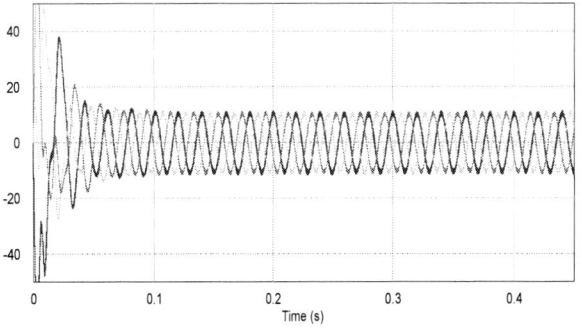

Fig. 7. The three-phase injected current into grid

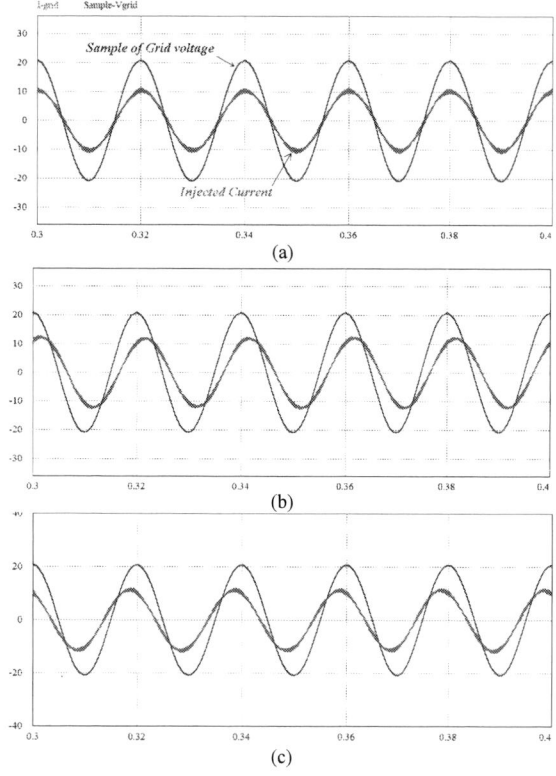

Fig. 8. Injected current and sample of grid voltage. (a) Unity power factor. (b) Lagging power factor. (c) Leading power factor

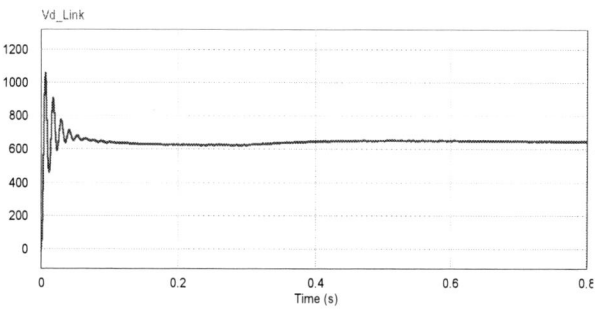

Fig. 9. DC-Link Voltage

VI. CONCLUSION

In this paper a control strategy based on feed-forward decoupling method using fuzzy-PI controller was proposed to control a three-phase photovoltaic two stage inverter. The proposed strategy features decoupled control of active and reactive power, while simplicity and accuracy merits of fuzzy control are preserved. The performance of this control strategy employing fuzzy-PI controller was verified using some simulation results.

REFERENCES

[1] M. A. Eltawi and Z. Zhao, "MPPT techniques for photovoltaic applications," ELSEVIER renewable and sustainable energy reviews 25, pp. 793-813, 2013.

[2] T. Abeyasekera, C. M. Johnson, D. J. Atkinson, and M. Armstrong. "Suppression of line voltage related distortion in current controlled grid connected inverters," IEEE Trans. Power Electron., vol. 20, no. 6, pp. 1393-1401, Nov. 2005.

[3] B. M. T. Ho, and H. S. Chung. "An integrated inverter with maximum power tracking for grid-connected PV systems," IEEE Trans. Power Electron., vol. 20, no. 4, pp. 953-962, Jul. 2005.

[4] H. X. Li, "A Comparative Design and Tuning for Conventional Fuzzy Control," IEEE Trans. Syst., Man, Cybern., vol. 27, Oct. 1997, pp. 884–889.

[5] T. Huang, X. Shi, Y. Sun, and D. Wan, "Three-phase photovoltaic grid-connected inverter based on feedforward decouplnig control," ICMREE, Vol. 2, Aug. 2013, pp. 476-480.

[6] Amirnaser Yazdani, Reza Iravani, "Voltage-Sourced converters in power systems," John wiley & sons, INC, 2012.

[7] D.Vijaya Bandhavi and S.R.Thilaga, "Fuzzy Based Hill-Climbing Method for Maximum Power Point in Microgrid Standalone Photovoltaic System," International Journal of Communications and Engineering Volume 03– No.3, Issue: 04 March2012.

[8] Basil M. Hamed and Mohammed S. El-Moghany, "Fuzzy Controller Design Using FPGA for Photovoltaic Maximum Power Point Tracking," International Journal of Advanced Research in Artificial Intelligence, Vol. 1, No. 3, 2012.

[9] Leonid Reznick, "Fuzzy Controllers," Newnes, 1997.

[10] M. Rosyadi, S.M. Muyeen, R. Takahashi, J. Tamura, "Transient Stability Enhancement of Variable Speed Permanent Magnet Wind Generator using Adaptive PI-Fuzzy Controller," in Proc. Trondheim PowerTech. Conf., 2011, pp.1-6.

[11] H. X. Li, nad H. B. Gatland, "Enhanced methods of fuzzy logic control," in Proc. 4th IEEE Int. Conf. Fuzzy Systems, Japan, Mar. 1995, pp. 331–336.

[12] H. X. Li, nad H. B. Gatland, "Conventional fuzzy control and its enhancement," IEEE Trans. Syst., Man, Cybern., vol. 26, pp. 791–796, Oct. 1996.

[13] K. Guesmi,N. Essounbouli,A. Hamzaoui,"Systematic design approach of fuzzy PID stabilizer for DC–DC converters," Energy Conversion and Management, 2008, pp. 2880–2889.

[14] G. Acciani, F. Vacca, and S. Vergura, "Time domain analysis of switching circuits by using simulink-based co-simulation," EUROCON, May 2009, pp. 256-263.

[15] S. Khader, A. Hadad, and A. Abu-Aisheh, "The application of PSIM & MATLAB/Simulink in power electronics courses," EDUCON, 2011 IEEE, pp. 118-121.

7th Power Electronics, Drive Systems & Technologies Conference (PEDSTC 2016)
16-18 Feb. 2016, Iran University of Science and Technology, Tehran, Iran

Analysis of the Boost Converter Under the DCM Condition to Reduce the MIC Volume to Mitigate Partial Shading Effects in PV Arrays

Reza Rezaii[1], Mohammad Amin Abolhasani[3], Ali Yazdian Varjani[3], Reza Beiranvand[4]

Faculty of Electrical and Computer Engineering, Tarbiat Modares University, Tehran, Iran

[1]reza.rezaii@modares.ac.ir, [2]m.abolhassani@modares.ac.ir, [3]yazdian@modares.ac.ir, [4]beiranvand@modares.ac.ir

Abstract—This paper analyses the conventional boost converter to mitigate partial shading effects in Photovoltaic (PV) panels. Analysis of a string of several boost converters connected in series in the structure of Module Integrated Converters (MICs) has been developed for better combating the problems of solar energy harvesting. Also, small signal ac analysis of the converter in Discontinuous Conduction Mode (DCM) is proposed. . This operation mode has been used to reduce dimensions of the proposed MIC structure by reducing its inductors' volumes. Moreover, exact model of the converter is used to control its input voltage by calculating controller's coefficients to track Maximum Power Point (MPP) of the module. A particular control approach has been used to regulate the DC link voltage which supplies a voltage source inverter. To verify the given control approach and mathematical analyses, a prototype converter has been simulated and implemented, in practice. The experimental and simulated results are in good agreement with the given mathematical calculations.

Keywords— Boost converter, module integrated converter, maximum power point tracking controller, solar energy

I. INTRODUCTION

Renewable energies are getting wider use in the world due to pollution and limitations of the fossil fuel supplies, and among all renewable energy sources, solar energy is one of the most available endless supplies which is considered to be influential in following years [1], [2]. On the other hand, because of the high installation and maintenance costs of PV systems, obtaining maximum power from solar cells is an absolute necessity [3]. Partial shading is one of the most unfavorable problems in PV structures that causes enormous power losses [4]. According to an official statistics of European Union from a wide number of Building Integrated Photovoltaic (BIPV) systems, partial shading dissipates 5-10% of the achievable solar energy annually, and investigations in Germany proves that 10% of the energy is lost due to partial shading [5], [6].

For increasing output voltage in PV arrays, it is necessary to connect several modules in series [7]. If, in a solar array, one or more modules become shaded ascribable to natural phenomena like snow, pollution and dust, bird droppings, trees and neighbor panels' shadows, this shading causes power dissipation in V-I curve of the modules owing to mismatch occurred in power delivery of cells [6], [8]. When mismatch is occurred, current of the shaded module is reduced, and because of almost constant current of the string a negative voltage is applied across the shaded module, as shown in Fig. 1(a) Subsequently, some part of the produced power of other modules is dissipated in the shaded module, and it may be dangerous due to hot spot effect [7], [9].

The easiest way for eliminating the problem of partial shading effect and preventing damage to the module is using bypass diodes connected in parallel to each module [10], [11]. In this method, the shaded module is detached from the string when its parallel bypass diode is turned on [12]. The P-V curve of the array when shading effect is occurred is shown in Fig. 1(b). However, using bypass diodes results in two problems: 1) the whole power of the shaded module is dissipated, and 2) it causes multiple peaks to appear in the P-V characteristic of the array. So, it is too difficult to track the global maximum power point [13]-[15], [27].

Using MICs is another different approach which has been introduced to replace the bypass diodes and to overcome their problems [16]-[18]. MICs are consisted as group of cascaded dc-dc converters connected to each module and make a modular structure. Using modular structure eliminates the need for a wide and plain roof for domestic systems, and MIC systems assign a plug and play capability to PV arrays. The most advantages of using MICs are better protection of panels, higher safety in installation and maintenance [7], and boosting the output voltage.

Among conventional power electronic converters, buck, boost, buck-boost, and Cuk converters can be used in cascaded form in the MIC structure. Also, between these converters, buck and boost topologies are more common due to their higher efficiency and components lower stresses. Buck converters are more suitable for longer strings with high

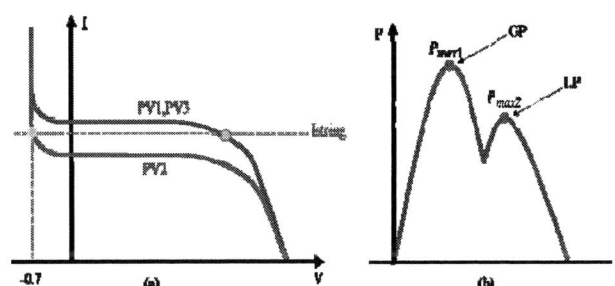

Fig. 1. a) I-V curve b) P-V curve of a PV array.

978-1-5090-0376-1/16 $31.00 © 2016 IEEE

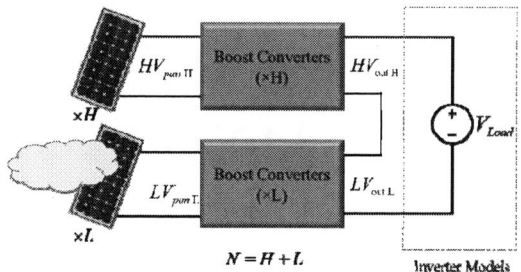

Fig. 2. A string of N MIC converters connected in series.

number of modules, and boost converters are better for short strings with fewer modules [7].

The most advantages of the boost over buck converter are: 1) rms current flows in boost is much less than buck, 2) switches with lower current stress can be used in boost converters, and 3) when sun light is low, grid current will not flow through PV array and causes leakage losses [19].

In this paper, small signal analysis of the boost converter and its control block diagram to control current of the module are investigated to track the MPP in MIC structure. Also, in small signal analysis of the converter under the DCM condition, the inductor of the converter is involved in the analysis, which distinguishes this paper from the literature. For more exact analysis and dynamic investigation of the system and circuit's transition behavior according to the designed PI coefficients, it is necessary to employ a detailed model of the converter. Moreover, by calculating the operating boundary of the voltage of inverter's DC link, this voltage is kept constant in the presented control method by using the Park Transform. Thus, in Sec. II calculating the output voltage of each boost converter in different irradiance conditions is investigated. Then, in Sec. III, analysis of boost converter is conducted. Finally, the presented control method for managing DC link is explained in IV section. Simulated and experimental results are given in Sections V and VI, respectively, and the paper is concluded in Sec. VII.

II. CALCULATING THE OUTPUT VOLTGAE OF EACH CONVERTER

Using first harmonic approximation, the voltage source DC/AC inverter's model can be considered as a voltage source V_{LOAD} along with a R_{LOAD} resistance. Actually, a PV connected inverter can regulate its input voltage by changing its current in a specific range [20]. Considering a string consisted of N modules, each one connected to a boost converter, as shown in Fig. 3, the total power of the string is calculated as follows:

$$P_{tot} = \sum_{i=1}^{N} P_{pan\,i} = \sum_{i=1}^{N} V_{pan\,i} \cdot I_{pan\,i} = \sum_{i=1}^{N} V_{pan\,i} \cdot f\left(V_{pan\,i}\right). \quad (1)$$

If L modules in this string be subjected to low irradiance level and H modules to full irradiance ($N=H+L$), such that H and L represent the level of irradiation on the module, according to Fig. 2, the behavior of the DC system is explained by [21]:

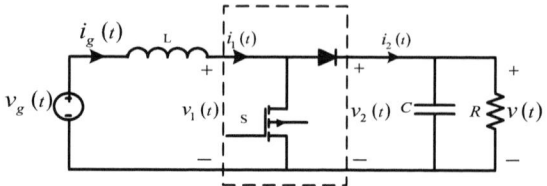

Fig. 3. The network of swithes highlighted in boost converter [22].

$$\begin{cases} H V_{out\,H} + L V_{out\,L} = V_{LOAD} \\ V_{out\,H} = M(D_{H})V_{pan\,H} \\ I_{out\,H} = I_{out\,L} \end{cases}. \quad (2)$$

Because, all converters are connected in series in a string, by keeping the V_{LOAD} constant, the output voltage of each converter is depended on its output power and also on the total output power value. Voltage of the converter number X is calculated as follows:

$$V_{out\,X} = V_{LOAD} \frac{P_{out\,X}}{H P_{out\,H} + L P_{out\,L}} \quad (3)$$

Where, X can be either equal to H or L. Also, $P_{out\,H}$ and $P_{out\,L}$ are the output power of each converter which is under the full or partial irradiation, respectively [21]. According to (3), it is possible in some circumstances that the output voltage of the converter increases; for instance, if $P_{out\,L} \approx 0$ (L modules are completely in shadow) while $P_{out\,H} \neq 0$, the value of the voltages are $V_{out\,L} \approx 0$ and $V_{outH} = V_{LOAD}/H$. Thus, by decreasing the number of H, $V_{out\,H}$ and consequently switches' voltage stress will increase. To prevent the boost converter's excessive voltage increasing from $V_{dc,max}$, there should be a limit on duty cycle in converter's control [21].

III. BOOST CONVERTER ANALYSIS

A. Average Values of the Current and Voltage of the Switch

Considering Fig. 3, average of the switch network voltages are calculated:

$$\begin{cases} \langle v_1(t) \rangle_{T_s} = \langle v_g(t) \rangle_{T_s} - \langle v_L(t) \rangle_{T_s} = \langle v_g(t) \rangle_{T_s} \\ \langle v_2(t) \rangle_{T_s} = \langle v(t) \rangle_{T_s} \end{cases} \quad (4)$$

Also, average values of the input and output currents are:

$$\begin{cases} \langle i_1(t) \rangle_{T_s} = \langle i_L(t) \rangle_{T_s} = R_e \dfrac{\langle v_1(t) \rangle_{T_s} \cdot \langle v(t) \rangle_{T_s}}{\langle v(t) \rangle_{T_s} - \langle v_1(t) \rangle_{T_s}} \\ \langle i_2(t) \rangle_{T_s} = R_e \dfrac{\langle v_1(t) \rangle_{T_s}^2}{\langle v_2(t) \rangle_{T_s} - \langle v_1(t) \rangle_{T_s}} \end{cases} \quad (5)$$

in which R_e is the effective resistance seen from primary side of the switch network [22] and is defined by:

$$R_e = \frac{d_1^2 \cdot Ts}{2L}. \quad (6)$$

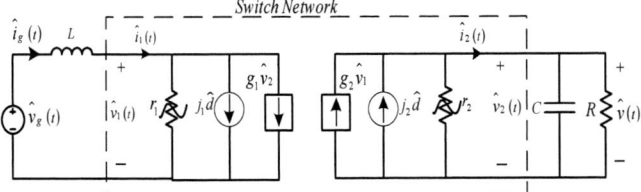

Fig. 4. Small signal model of boost converter in DCM [22].

B. Small Signal Model of the Boost Converter in DCM

To obtain ac small signal model in an operating point it should be assumed that the average values of the variables consists of a constant dc and an ac variant terms. Thus, using Taylor series would lead to linearization of switch network currents as:

$$\langle i_1(t)\rangle_{T_s} = f_1\left(\langle v_1(t)\rangle_{T_s}, \langle v_2(t)\rangle_{T_s}, d(t)\right). \qquad (7)$$

Thus, the current explanation is composed of an ac and a dc term, and by eliminating dc term, the ac term can be like equation (8) for input current i_1 and output current i_2 [22]. The undefined parameters are defined in (9).

$$\begin{cases} \hat{i}_1(t) = \hat{v}_1(t)\left(\dfrac{1}{r_1}\right) + \hat{v}_2(t)g_1 + \hat{d}(t)j_1 \\[2mm] \hat{i}_2(t) = \hat{v}_1(t)g_2 + \hat{v}_2(t)\left(\dfrac{-1}{r_2}\right) + \hat{d}(t)j_2 \end{cases} \qquad (8)$$

$$\begin{cases} r_1 = R_e\dfrac{(M-1)^2}{M} \\[2mm] g_1 = \dfrac{1}{(M-1)^2.R_e}\,, \\[2mm] j_1 = \dfrac{2MV_1}{D(M-1)R_e} \end{cases} \begin{cases} g_2 = \dfrac{2M-1}{(M-1)^2R_e} \\[2mm] r_2 = (M-1)^2R_e \\[2mm] j_2 = \dfrac{2V_1}{D.(M-1).R_e} \end{cases} \qquad (9)$$

Small signal model of boost converter in DCM based on the aforementioned equations is depicted in Fig. 4 [22].

C. Converter's Transfer Functions

Owing to the fact that in MIC structure each module should separately operate in its own MPP, the need for controlling the input current or voltage of the converter is essential. Since in PV modules the current and voltage are related, by controlling just the input current of the converter, the maximum power can be extracted from the module. Taking input control $\hat{d}(s)$ and input voltage $\hat{v}_g(s)$ as ac variables and input current $\hat{i}_g(s)$ as desired output for designing, equation (10) can be extracted, and the input current can be related to $\hat{d}(s)$ and $\hat{v}_g(s)$ by (11) and (12) respectively.

$$\hat{i}_g(s) = G_{id}(s)\hat{d}(s) + G_{iv}(s)\hat{v}_g(s) \qquad (10)$$

$$G_{id}(s) = \left.\frac{\hat{i}_g(s)}{\hat{d}(s)}\right|_{\hat{v}_g(s)=0} \qquad (11)$$

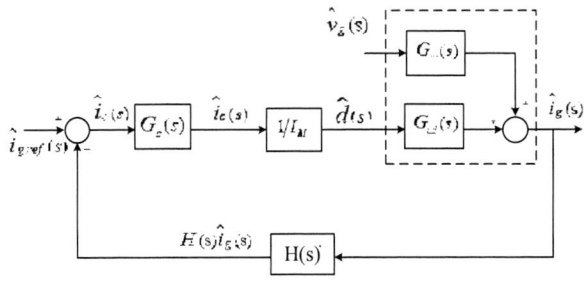

Fig. 5. Control block diagram of boost converter to track MPP.

$$G_{iv}(s) = \left.\frac{\hat{i}_g(s)}{\hat{v}_g(s)}\right|_{\hat{d}(s)=0} \qquad (12)$$

To gain the correct expressions for $G_{id}(s)$ and $G_{iv}(s)$, (10) should be realized by the group of equations (8) and (13).

$$\begin{cases} \hat{v}_1(t) = \hat{v}_g(t) - Ls\times\hat{i}_1(t) \\[2mm] \hat{v}_2(t) = \dfrac{R}{RCs+1}\hat{i}_2(t) \end{cases} \qquad (13)$$

According to (8) and (13) the expression of $\hat{i}_1(t)$ can be modified as:

$$\hat{i}_1(t) = \hat{i}_g(t) = \frac{\left(g_1g_2a + r_1^{-1}\right)\hat{v}_g(t) + \left(g_1j_2a + j_1\right)\hat{d}(t)}{g_1g_2aLs + \dfrac{Ls}{r_1} + 1} \qquad (14)$$

in which:

$$a = \frac{Rr_2}{Rr_2Cs + R + r_2} \qquad (15)$$

To find the expression for $G_{iv}(s)$, $\hat{d}(t)$ in (14) should be equal to zero, so $G_{iv}(s)$ is:

$$G_{id}(s) = \left.\frac{\hat{i}_g(s)}{\hat{d}(s)}\right|_{\hat{v}_g(s)=0} = \frac{g_1j_2a + j_1}{g_1g_2aLs + \dfrac{Ls}{r_1} + 1} \qquad (16)$$

Also, $G_{id}(s)$ can be achieved by equating $\hat{v}_g(t)$ to zero in (14), so $G_{id}(s)$ is:

$$G_{iv}(s) = \left.\frac{\hat{i}_g(s)}{\hat{v}_g(s)}\right|_{\hat{d}(s)=0} = \frac{g_1g_2a + r_1^{-1}}{g_1g_2aLs + \dfrac{Ls}{r_1} + 1} \qquad (17)$$

Therefore, it can be shown that $i_g(t)$ is a function of input voltage, duty cycle, and load current, and block diagram of the converter's control loop can be depicted, as shown in Fig. 5. The input current $i_g(t)$ is sampled with a current sensor with a gain of $H(s)$ and is fed back to the block. The output of the current sensor is $i_g(s)H(s)$, and is compared to the reference current. The difference between these two values is the error signal which is applied to the compensator. At the end, the output signal of the compensator is compare to a sawtooth

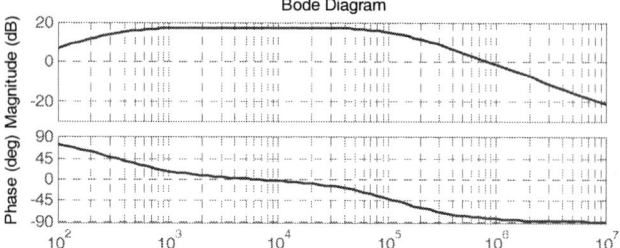

Fig. 6. The converter loop gain bode diagram without PI compensator.

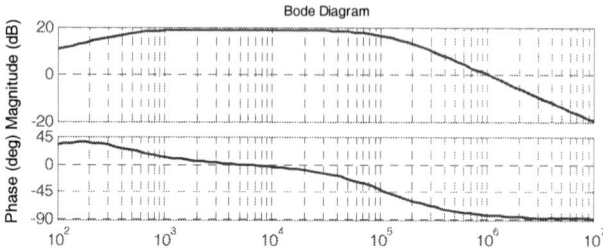

Fig. 7. The converter loop gain bode diagram in presence of PI compensator.

wave to produce the duty cycle signal. Loop gain $T(s)$ is the product of small signal gains in the forward and feedback routes. Each one of disturbance transfer functions to the output has a $1/(1+T(s))$ factor, so by reducing the loop gain $T(s)$, the effect of disturbance on the output will reduce. Moreover, the system feedback is more stable when phase margin of $T(s)$ becomes positive, and increasing the phase margin leads to improve the transients of the circuit. Thus, the circuit's loop gain would be (18). Thus, by employing a PI compensator the cut off frequency of the loop gain can be enhanced [22].

$$T(s) = \frac{HG_cG_{id}}{I_M} \quad (18)$$

D. System Stability Analysis of T(s)

Before adding PI controller, the stability of loop gain should be evaluated in order to analyze the stability of the whole system. Loop gain can be calculated by (18), in which H obtains from the ratio of reference input current to I_g. Firstly, assuming G_c and I_M are equal to one, the loop gain is analyzed. For $I_g=3$ and $I_{gref}=2.5$ the value of gain is calculated and is $H=0.833$, and with input and output voltages as $V_g=17$ and $V=51$, and considering circuit parameters as $R=50\Omega$, $L=100\mu H$, and $C=150\mu F$, transfer function of loop gain is equal to:

$$T(s) = \frac{HG_cG_{id}}{I_M} = 0.833 \frac{25.5s + 0.18}{0.025e^{-3}s^2 + 2.9s + 952}. \quad (19)$$

By plotting bode diagram of loop gain $T(s)$,,as given in Fig. 6, cut off frequency of the circuit is gained $8.4e^5 rad/s$, and in this frequency, operating frequency of the circuit is $1.25e^5$ rad/s. It may be needed to increase the cut off frequency after hardware implementing the circuit because of possible difference between measured and calculated values of gain frequency, so PI compensation is a necessity. Consequently, initial desired cut off frequency is considered $\omega_x=1e^6$ rad/s; then, the amplitude of the transfer function in the noted frequency is calculated. The following equation should be solved in order to obtain the cut off frequency of the circuit after adding PI compensator.

$$\left\|T(s)\right\|_{\omega_x} \times \left\|PI\right\|_{\omega_x} = 1 \quad (20)$$

at ω_x, the amplitude of the transfer function is equal to 1.47 dB, so by replacing it in (20) the following equation is achieved.

$$\left\|\frac{k_i + jk_p\omega}{j\omega}\right\|_{\omega_x} = \frac{1}{1.47\,dB} \quad (21)$$

By selecting $K_i=100$ in (21), $K_p=1.186$ is obtained in cut off frequency of $1e^6(rad/s)$, as can be seen in Fig. 7.

In many references [22], due to small inductor value under the DCM operation mode, the boost inductor is ignored to simplify the analysis. However, to observe the dynamic of the system, transient analysis, and calculate the PI compensator coefficients, more exact model of the circuit's transfer function is needed. Moreover, more exact analysis and correct PI compensator factors are required because for tracking the MPP, reducing tracking time is necessary, and transients should be studied in more details.

IV. STRING VOLTAGE CONTROL

A. Keeping String Voltage Constant

One of the most desirable issues in MIC structures is keeping the string voltage V_{string} constant to connect several strings in parallel. Also, connecting the system to a battery bank would be easier. The voltage and current characteristics of two converters in MIC structure at constant power is shown in Fig. 8. It is evident from the photograph that each converter operates in its own constant power, and all converters operate at a constant current. The summation of converters' output voltages is constant depends on the converter's power, and string current, which is uniform for all converters and it is automatically tuned. It is apparent from the curve that by controlling the string current I_{string} , properly, the voltage of the string can be kept constant. By employing a proper control loop, the inverter's input voltage can be kept constant at a desired value more than the ac peak voltage [24].

B. Proposed Control Method

In the proposed control method, to regulate DC link voltage at a desirable value, a PI controller has been used. The error signal is due to differences between desired and measured

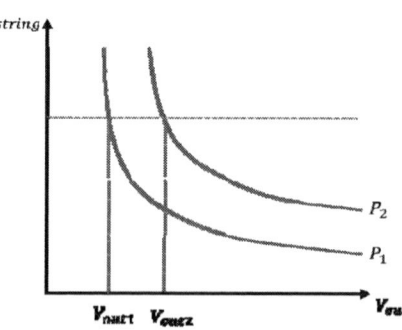

Fig. 8. String current versus output voltage for fixed output powers.

978-1-5090-0376-1/16 $31.00 © 2016 IEEE

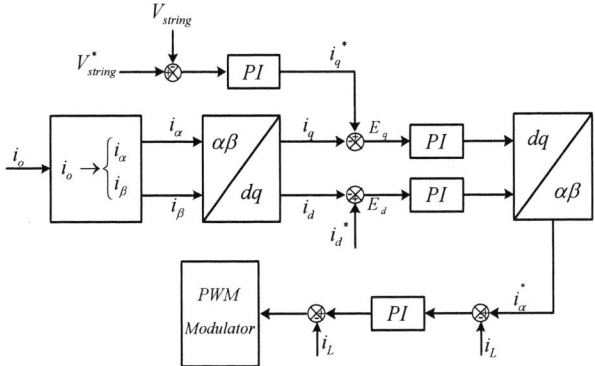

Fig. 9. Proposed control method block diagram for regulating output voltage.

values of the DC link voltage. Also, to achieve an alternative output current, a park transform has been employed to change the alternative current of the output into two DC components. Following equation expresses the park transform in synchronous reference frame.

$$\begin{bmatrix} i_d(t) \\ i_q(t) \end{bmatrix} = \begin{bmatrix} \cos(\omega_f t) & -\sin(\omega_f t) \\ \sin(\omega_f t) & \cos(\omega_f t) \end{bmatrix} \begin{bmatrix} i_\alpha(t) \\ i_\beta(t) \end{bmatrix} \quad (22)$$

From (22), two components for park transform α and β should be created for single phase park transform, and they have 90 degrees phase shift. There are various methods for creating these two perpendicular components, and one of the best methods is utilizing an all pass filter [25], which is a simple method without the problem of weak dynamic response of other methods [26]. The all pass filter transfer function from input to the output is:

$$\frac{v_\beta(s)}{v_\alpha(s)} = \frac{\omega_f - S}{\omega_f + S}. \quad (23)$$

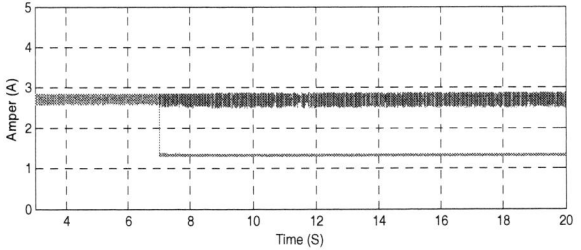

Fig. 10. Modules currents.

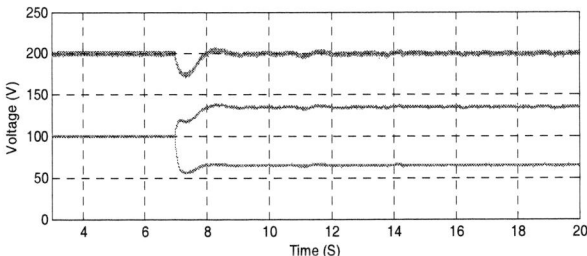

Fig. 11. Output voltage of each converter along with DC link voltage.

Fig. 12. Experimental setup.

Table I. PV Module Characteristics.

Characteristic	Value
V_{mp}	17.4
I_{mp}	3.05
P_{max}	53

Here, ω_f is the main angular frequency of the circuit. The complete block diagram of the proposed control method is drawn in Fig. 9, and obviously it is comprised of an inner current control loop and a feed forward voltage route for eradicating the permanent DC link voltage error. To control the voltage, the current error from measured string current which passes through PI compensator is transferred to an imaginary two phase current by Inverse Park Transform, and only $i^*_{L\alpha}$ which states the real part of this current is employed. Besides, to improve the transient and steady state performance, the inner feedback of inductor current is exercised.

V. SIMULATION RESULTS

In simulation which is conducted in MATLAB/ SIMULINK, two boost converter each one linked to a single module, are connected in series. Every module is separately tracking the maximum power using DMPPT control, and both have the same characteristics which is provided in Table I. In this simulation at $t=7s$ the second module experiences a partial shading, and its irradiation is reduced by half, so its current diminishes. Fig. 10 shows the currents of two modules, and output voltages of two converters linked to these modules. DC link voltage are plotted in Fig. 11. t $t=7s$, by reducing the irradiation on the second converter, its output voltage declines. However, on the grounds that both converters are operating at their MPPT and deliver their maximum power to the output, by decreasing the string current due to constant power of the first converter, its voltage increases, and altogether, DC link voltage remains relatively constant.

Table II. Boost Converter Components.

Component	Value
Filter Inductor	500 μH
Filter Capacitor	22 μF
Boost Inductor	100 μH
Boost Capacitor	150 μF
Power MOSFET	IRFP460
Schottky Diode	DSEC 30-04A
MOSFET Gate Driver	HCPL3120
Current Sensor	LTS 25-NP
Voltage Sensor	HCPL7840

978-1-5090-0376-1/16 $31.00 © 2016 IEEE   353

VI. EXPERIMENTAL RESULTS

Here, the experimental results are given in accordance with the analyses, as done before. Table II represents the elements used in the boost converter and inverter structure which is comprised of power switches and energy saving elements. Control algorithm is implemented by using an ARM microcontroller produced by STMicroelectronics, STM32F407VG. Fig.12 shows the prototype system.

The switching frequencies of the boost converters are set at *20 kHz* and the inverter switches are driven by *5 kHz*. Fig. 13 shows the output signal of the microcontroller's Pulsed Width Modulation (PWM) driving boost switches, which is due to the comparison between a saw-tooth wave and carrier produced by the difference of the MPP and PV voltages in the PI compensator.

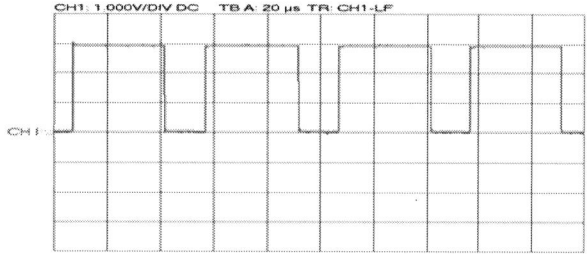

Fig. 13. Output waveform of PWM.

As stated in boost converter analysis section, for reducing the circuit's volume and size, the converter should operate under the DCM conditions. Thus, the inductor becomes smaller. Fig. 14 depicts the DCM current flowing through the inductor. It can be seen that during the first subinterval of length d_1T_s the inductor current increases. During the second subinterval of length d_2T_s it is decreased. During the third subinterval of length d_3T_s the inductor current is zero.

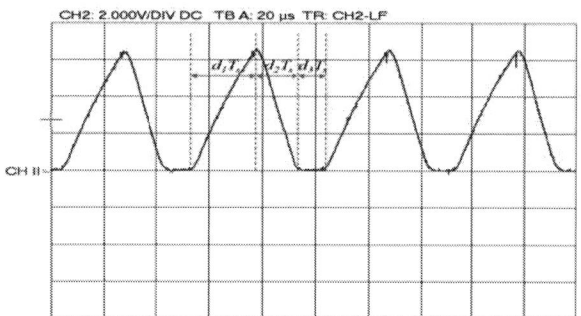

Fig. 14. The inductor current under the DCM condition.

To observe the proposed controller operation's accuracy in keeping DC link voltage in a constant value in order to mitigate partial shading, a scenario is devised. In this experiment, two solar panels are used. At *t=4s* the second one is exposed to shading. As shown in Fig. 15 the DC link is kept constant during this experiment. The current drawn by the inverter is reduced due to the shading effect. Since the output power of each converter is constant owing to tracking MPP, and because their output currents are equal to each other, the output power of the converter is decreased to a constant power, and the other one's voltage is increased as depicted in Fig. 16.

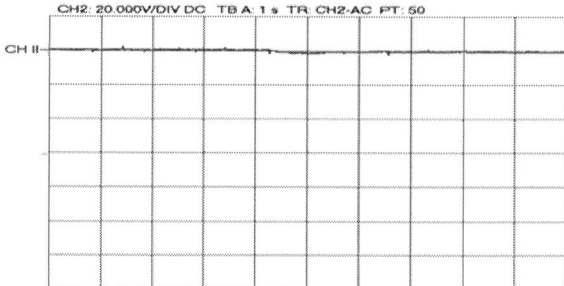

Fig. 15. DC link output voltage when shading occurs.

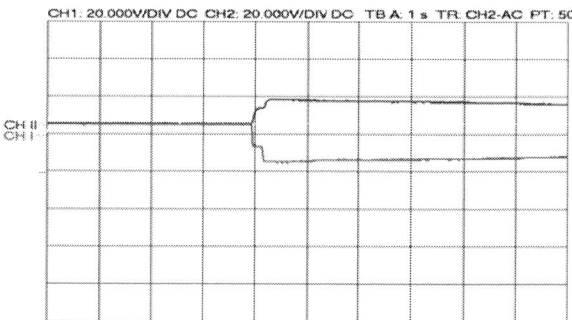

Fig. 16. Output voltage of boost converters when shading occurs.

Efficiency of the prototype at a fixed switching frequency has been plotted versus load resistance, as illustrated in Fig. 17. Maximum efficiency equal to 95.2% is achieved at *20 W*. But, by increasing the load current it is reduced due to voltage drop on parasitic resistances of the circuit. The output waveform of the inverter has been plotted in Fig. 18. Thanks to the current control loop Total Harmonic Distortion (THD) of the converter is reduced.

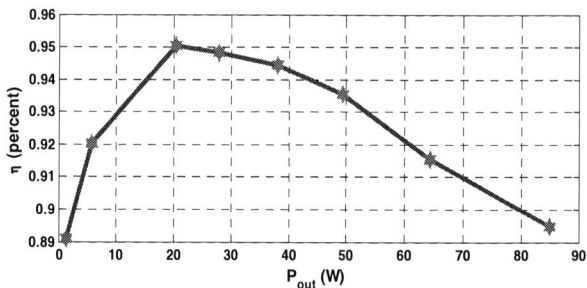

Fig. 17. Efficiency curve versue load changes.

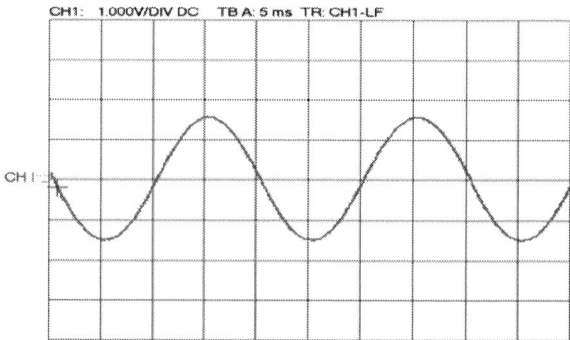

Fig. 18. PWM gate drive signal.

VII. CONCLUSION

Owing to the high cost of photovoltaic structures and the need for mitigating partial shading effects by using MIC structure, reducing the cost of the implemented converter is necessary. DCM operation mode has been used to reduce dimensions of the proposed MIC structure by reducing its inductors' volumes. Using the modified converter's formulas, performance of the boost converter is investigated at the presence of partial shading effects. This analysis can be useful to select the converter components, in practice. Besides, an experimental prototype is designed to confirm the presented structure. The setup uses two 53 W boost converters which accurately track the MPP of their modules with 95.2% efficiency. The given experimental results show that using the DCM operation mode reduces the converter total size, effectively.

REFERENCES

[1] Park, Yongsoon, et al. "Asymmetric control of DC-link voltages for separate MPPTs in three-level inverters." Power Electronics, IEEE Transactions on 28.6 (2013): 2760-2769.

[2] Maki, Anssi, and Seppo Valkealahti. "Power losses in long string and parallel-connected short strings of series-connected silicon-based photovoltaic modules due to partial shading conditions." Energy Conversion, IEEE Transactions on27.1 (2012): 173-183.

[3] Sundareswaran, Kinattingal, Sankar Peddapati, and Sankaran Palani. "MPPT of PV systems under partial shaded conditions through a colony of flashing fireflies." Energy Conversion, IEEE Transactions on 29.2 (2014): 463-472.

[4] Levron, Yoash, et al. "Control of Submodule Integrated Converters in the Isolated-Port Differential Power-Processing Photovoltaic Architecture." Emerging and Selected Topics in Power Electronics, IEEE Journal of 2.4 (2014): 821-832

[5] M. Drif, P. J. Perez, J. Aguilera, and J. D. Aguilar, "A new estimation method of irradiance on a partially shaded PV generator in gridconnected photovoltaic system," Renew. Energy, vol. 33, no. 9, pp. 2048–2056, Feb. 2008

[6] El-Dein, MZ Shams, Mehrdad Kazerani, and M. M. A. Salama. "Optimal photovoltaic array reconfiguration to reduce partial shading losses." Sustainable Energy, IEEE Transactions on 4.1 (2013): 145-153.

[7] Walker, Geoffrey R., and Paul C. Sernia. "Cascaded DC-DC converter connection of photovoltaic modules." Power Electronics, IEEE Transactions on 19.4 (2004): 1130-1139.

[8] Wang, Yu-Jiu, and Po-Chi Hsu. "Analytical modelling of partial shading and different orientation of photovoltaic modules." Renewable Power Generation, IET 4.3 (2010): 272-282.

[9] E. Molenbroek, D. W. Waddington, and K. A. Emery, "Hot spot susceptibility and testing of PV modules," in Proc. Conf. Rec. 22nd IEEE Photovoltaic Spec. Conf., 1991, vol. 1, pp. 547–552.

[10] Lavado Villa, Luiz Fernando, Bertrand Raison, and Jean-Christophe Crebier. "Toward the Design of Control Algorithms for a Photovoltaic Equalizer: Detecting Shadows Through Direct Current Sampling."

Emerging and Selected Topics in Power Electronics, IEEE Journal of 2.4 (2014): 893-906.

[11] Sharma, Parmanand, and Vivek Agarwal. "Maximum power extraction from a partially shaded PV array using shunt-series compensation." Photovoltaics, IEEE Journal of 4.4 (2014): 1128-1137.

[12] S. Silvestre, A. Boronat, and A. Chouder, "Study of bypass diodes configuration on PV modules," Appl. Energy, vol. 86, no. 9, pp. 1632–1640, Sep. 2009.

[13] G. Petrone, G. Spagnuolo, and M. Vitelli, "Analytical model of mismatched photovoltaic fields by means of Lambert W-function," Sol. Energy Mater. Sol. Cells, vol. 91, no. 18, pp. 1652–1657, Nov. 2007.

[14] Gao, Lijun, et al. "Parallel-connected solar PV system to address partial and rapidly fluctuating shadow conditions." Industrial Electronics, IEEE Transactions on 56.5 (2009): 1548-1556.

[15] W. Xiao, N. Ozog, and W. G. Dunford, "Topology study of photovoltaic interface for maximum power point tracking," IEEE Trans. Ind. Electron., vol. 54, no. 3, pp. 1696–1704, Jun. 2007.

[16] E. Roman, R. Alonso, P. Ibanez, S. Elorduizapatarietxe, and D. Goitia, "Intelligent PV module for grid-connected PV systems," IEEE Trans. Ind. Electron., vol. 53, no. 4, pp. 1066–1073, Aug. 2006.

[17] E. Roman, R. Alonso, P. Ibanez, S. Elorduizapatarietxe, and D. Goitia, "Intelligent PV module for grid-connected PV systems," IEEE Trans. Ind. Electron., vol. 53, no. 4, pp. 1066–1073, Aug. 2006.

[18] [18] W. Xiao, N. Ozog, and W. G. Dunford, "Topology study of photovoltaic interface for maximum power point tracking," IEEE Trans. Ind. Electron., vol. 54, no. 3, pp. 1696–1704, Jun. 2007.

[19] Weidong Xiao, Nathan Ozog, William G. Dunford, "Topology study of photovoltaic interface for maximum power point tracking", IEEE Trans. on Industrial Electronics, June 2007, vol. 54, pp. 1696-1704.

[20] Femia, N., et al. "Guidelines for the optimization of the P&O technique in grid-connected double-stage photovoltaic systems." 2007 IEEE International Symposium on Industrial Electronics. 2007.

[21] Femia, Nicola, et al. "Distributed maximum power point tracking of photovoltaic arrays: Novel approach and system analysis." Industrial Electronics, IEEE Transactions on 55.7 (2008): 2610-2621.

[22] Erickson, Robert W., and Dragan Maksimovic. Fundamentals of power electronics. Springer Science & Business Media, 2007.

[23] Alonso, Ricardo, et al. "Analysis of inverter-voltage influence on distributed MPPT architecture performance." Industrial Electronics, IEEE Transactions on 59.10 (2012): 3900-3907.

[24] Linares, Leonor, et al. "Improved energy capture in series string photovoltaics via smart distributed power electronics." Applied Power Electronics Conference and Exposition, 2009. APEC 2009. Twenty-Fourth Annual IEEE. IEEE, 2009.

[25] R. Y. Kim, S. Y. Choi, and I. Y. Suh, "Instantaneous control of average power for grid tie inverter using single phase D-Q rotating frame with all pass filter," in Proc. IEEE Annu. Conf. Ind. Electron. Soc., Nov. 2004, pp. 274–279.

[26] R. Zhang, M. Cardinal, P. Szczesny, and M. Dame, "A grid simulator withcontrol of single-phase power converters in D-Q rotating frame," in Proc. IEEE Power Electron. Spec. Conf., 2002, pp. 1431–1436.

[27] Petrone, Giovanni, Giovanni Spagnuolo, and M. Vitelli. "Analytical model of mismatched photovoltaic fields by means of Lambert W-function." Solar Energy Materials and Solar Cells 91.18 (2007): 1652-1657.

7th Power Electronics, Drive Systems & Technologies Conference (PEDSTC 2016)
16-18 Feb. 2016, Iran University of Science and Technology, Tehran, Iran

A Noninvasive On-line Failure Prediction Technique for Aluminum Electrolytic Capacitors in Photovoltaic Grid-connected Inverters

Amir Sepehr[*], Mehdi Saradarzadeh[**], Shahrokh Farhangi[*]

* School of Electrical and Computer Engineering, University of Tehran, Tehran, Iran
**Electrical and Computer Engineering Department, Jundi Shapur University of Technology, Dezful, Iran

Abstract— Since the electrolytic capacitors are mainly responsible for breakdowns in power electronics converters, they usually determine the overall lifetime. Their failure is due to the deterioration of their dielectric, the production of gases, and, eventually, their explosion. The deterioration process leads to a decrease in its capacitance value and simultaneously, an increase in the capacitor equivalent series resistance (ESR). In this paper, a novel noninvasive technique for failure prediction of the electrolytic capacitors in grid-connected photovoltaic inverters is presented, which can easily be applied online and even in the real time. The technique is based on the DC-link voltage measurement, current calculation, and extracting the current integral and the voltage change during switching intervals. Then, the capacitance can be calculated using the recursive least squares method. Although the estimation results are influenced by the temperature at which the measurement is performed, and the results are dependent on the operating conditions, but the proposed method is modified to obviate the above distorting factors effectively. The proposed technique is simple to apply, accurate and does not need to any extra hardware. Also it can be adapted for any other inverter based applications. The performance of the proposed method is verified by the simulation and experimental results.

Keywords— *aluminum electrolytic capacitors, equivalent circuit estimation, equivalent series resistance, capacitance measurement, failure analysis, life estimation, photovoltaic inverters.*

I. INTRODUCTION

Reliability is a basic feature for the correct operation of all the components used in electrical and electronic devices. The reliability of the electrical components is studied at the design stage and can be improved by means of well-known techniques, such as preventive and corrective maintenance. The aim of this study is to focus on the online predictive maintenance and health diagnosis of electrolytic capacitors when working as the DC-link in a grid-connected photovoltaic inverter.

Due to the large capacity and low cost, the aluminum electrolytic capacitor is one of the most common capacitors used in the power electronics converters in the industrial applications, such as the output voltage regulators in the power factor correctors, the DC-link power balancer for the photovoltaic inverters, the energy buffer for switched-capacitor

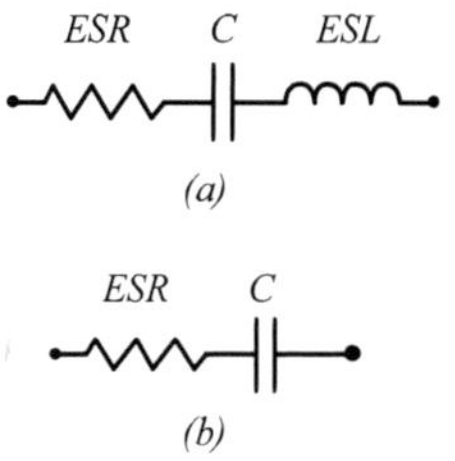

Fig. 1. Capacitor Equivalent Circuit

circuits, and the main DC-bus capacitors in the multilevel converters [1]-[14]. Unfortunately, among the power stage components of the switching-mode power converters, the electrolytic capacitors are the major affected elements by the aging effect and responsible for a large amount of breakdowns in the mentioned converters [2]. Characteristics of the electrolytic capacitor can be found in many papers [3]–[8], by neglecting the capacitor leakage current, the equivalent circuit for the electrolytic capacitor can be drawn as shown in Fig. 1(a), where C is its capacitance, ESL is its equivalent series inductance, and ESR is its equivalent series resistance. Since the ESL is very small, at the operating frequency of the electrolytic capacitor in switching-mode power converters, the equivalent circuit can be further simplified as the one shown in Fig. 1(b) [3], [5].

The performance of the electrolytic capacitor is highly affected by its operating conditions such as voltage, current, frequency, and temperature. Moreover, as the operating temperature and frequency increase, the capacitor's ESR will decrease. On the other hand, the ESR will increase when the operating hours, the operation voltage, and the electrolyte leakage increase. Most manufacturers define the end of life of these capacitors when the ESR doubles or the capacitance reduces by 20%, compared with the initial values [5]. Consequently, the electrolytic capacitor plays a very important role for the switching-mode converter's quality and reliability. Thus, it is of paramount importance to perform an online predictive maintenance, and use of a built-in test equipment (BITE) that is able to provide a real-time diagnosis of a monitored electrolytic capacitor, and allows increasing the reliability and decreasing the costs. This feature is recommended in the power electronic applications which are connected to the grid such as inverter based DGs (photovoltaic

978-1-5090-0376-1/16 $31.00 © 2016 IEEE   356

inverters) and FACTS devices (Static Synchronous Series Compensators) [11], [12]. In this paper, a novel noninvasive technique for failure prediction of the electrolytic capacitors in grid-connected photovoltaic inverters is presented, which can be applied easily online and even in the real time. The technique is based on the DC-link voltage measurement, current calculation, and extracting the current integral and the voltage change during switching intervals. Then, the capacitance can be calculated using the recursive least squares method.

Several works can be found in the literature related to the fault detection of aluminum electrolytic capacitors subject. Those techniques and their drawbacks are discussed in the next section. The new on-line approach, which overtakes the previous disadvantages and is able to estimate the parameters of the DC-link electrolytic capacitor, and consequently, prediction of failure, is proposed in the section III. Then, its performance is validated by simulation and experimental results.

II. BACKGROUND: FAULT DIAGNOSTIC TECHNIQUES

Many papers have proposed different methods or algorithms to determine the ESR of electrolytic capacitors [2]–[9].

A state average method to derive a formula for ESR, which is a ratio of the capacitor voltage to the capacitor or load current and a diagnostic method to determine the value of ESR for a boost converter has been proposed in [6], which is based on the relation between the mean value of the output voltage ripple and the capacitor current ripple.

In the same way, Imam *et al.* [7] presented a method using a digital signal processor to perform the fast Fourier transform of the capacitor ripple voltage and capacitor current. By these items the capacitor impedance is determined, then by performing the time-average technique a smooth capacitor impedance variation curve is obtained, where comparing to a predefined set point, the faulty capacitor is recognized. Gasperi [8] developed a flowchart to determine the ESR by measuring the capacitor parameters, including the electrolyte volume, the currents ripple, and the ambient temperature. The ESR measurement result could be accurate but the calculation burden is very heavy.

In order to justify the aging status of the electrolytic capacitor, comparing the estimated ESR, by measuring the capacitor' voltage and current, with the pre-calculated ESR, is presented in [3]. Another approach also is proposed in [8], to use the capacitor temperature to predict the electrolyte volume, then to justify the increase of ESR. However, the temperature measurement may be affected by the environment.

A decision system that can determine the ESR value of the output capacitor by taking account of the relevant modifications about input voltage and temperature has been proposed by [9]. However, this method needs a complicated preliminary test under a certain operation condition to collect the modification information. For different applications, the modification information may be different, and the experimental test needs to start over again.

The aforementioned failure prediction techniques are complex and are developed to suit a specific application. In

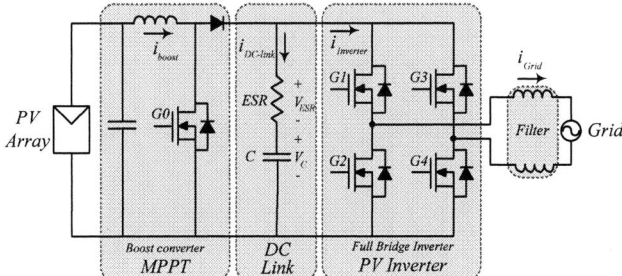

Fig. 2. Power Circuit of a Residential Grid-Connected Photovoltaic Inverter [1]

Fig. 3. Catastrophic Failure of a Electrolytic Capacitor [15]

order to simplify the process and simultaneously create a universal failure prediction technique, this paper proposes a new noninvasive online approach. However, to make this process more precise while simultaneously avoiding the measurement errors that can be resulted from the electrical noises, Recursive Least Square (RLS) method is used. In addition, it can be implemented in a grid-connected photovoltaic inverter without demounting the capacitor from the board. Furthermore, this paper presents a method which has the capability of implementing on all types of switching power electronics converters.

III. NOVEL NONINVASIVE ONLINE APPROACH TO ESTIMATE THE CAPACITANCE AND THE ESR

Commonly used power circuit as a residential grid-connected photovoltaic inverter is demonstrated in Fig. 2, which shows the boost and the full-bridge converters as the MPPT and the inverter respectively [1], [13]. The DC-link capacitor decouples the two converters and also serves as an energy storage element. The DC-link capacitance is selected and designed with respect to the allowable DC-link voltage ripple, the inverter power and the switching frequency. The DC-link capacitance and its equivalent resistance are associated with high impact of the quality and performance of the inverter. Reduce in the capacitance or increase in the equivalent resistance, would increase the DC-link voltage ripple and decrease the output waveform quality which causes an undesired operation. On the other hand, with rise of the resistance, the capacitor loss also increases leading to an increase in the capacitor and the inverter temperature. The exceeded temperature rise results in damage to the semiconductor switches and the other electrolytic capacitors which are used in the power circuit or control circuit. Fig. 3

TABLE I. INPUT CURRENT OF THE FULL-BRIDGE INVERTER ACCORDING TO THE GRID CURRENT'S HALF CYCLE AND STATES OF ITS SWITCHES

State of the grid current	G1	G3	Full-bridge input current
Positive half cycle	0	0	zero
Positive half cycle	0	1	Minus Grid Current
Positive half cycle	1	0	Grid Current
Positive half cycle	1	1	zero
Negative half cycle	0	0	zero
Negative half cycle	0	1	Minus Grid Current
Negative half cycle	1	0	Grid Current
Negative half cycle	1	1	Zero

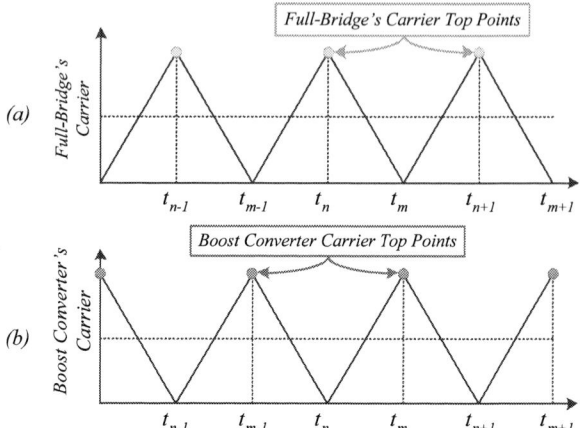

Fig. 4. The Triangular Carriers, (a) Full-Bridge Inverter's Carrier, (b) Boost Converter's Carrier

shows an example of a damaged electrolytic capacitor in a photovoltaic inverter. Since the useful lifetime of photovoltaic modules is estimated to be more than ten years, so it is important that the grid-connected inverters have a reliable long-term performance.

In customary modulation methods for photovoltaic inverters, such as bipolar and unipolar modulation, the two switches on the same leg have opposite switching commands, thus, using the grid current half cycle and two switching commands, the full-bridge inverter input current can be identified. In Fig. 2, the ESR and the capacitor voltage and current are specified and the following relationships are established:

$$i_{DC-link} = \bar{G}_0 \cdot i_{boost} - i_{inverter}$$
$$i_{inverter} = f_{(G_1,G_2,G_3,G_4,i_{Grid})} \times i_{Grid}$$
(1)

Where G_0 to G_4 are the gate signals of the semiconductor switches, $i_{inverter}$ is the full-bridge input current and $i_{DC-link}$ is the DC-link current. Current sensors are usually embedded in the grid side and the boost converter side in grid-connected photovoltaic inverters, however, using a current sensor in the DC-link is not suitable due to the increase in the current path length and its leakage inductance. According to the states of the full-bridge converter switches and using (1), Table I can be achieved. The DC-link current can be calculated at any moment easily using this look-up table. Estimating the equivalent resistance and the DC link capacitance is possible by applying the basic electric circuit relations to the circuit shown in Fig. 2 as follows:

$$V_{DC-link} = V_C + V_{ESR}$$
$$V_{DC-link(t)} = ESR \times i_{DC-link(t)}$$
$$+ \frac{1}{C}\int_{t_0}^{t} i_{DC-link} dt + V_{C_0}$$
(2)

Where V_{C0} is the capacitor C voltage at time t_0. The capacitance of the DC-link capacitor can be estimated using the injected current and its voltage, and it can be concluded from (2) that when the DC-link current is zero, the DC-link

voltage equals to the capacitor voltage. Accordingly, the DC-link capacitance can be calculated using its voltage in two different instants in which the DC-link current is zero, and the integral of the DC-link current between these two instants. t_{n-1} and t_n are the instants in which the DC-link current is zero and are shown in Fig. 4.

The relationship between the integral of the current and the voltage change of the capacitor is given by:

$$if\ i_{DC-link} = 0 \Rightarrow V_{DC-link} = V_C$$
$$\Delta V_{C_{t_n}} = \frac{1}{C}\int_{t_{n-1}}^{t_n} i_{DC-link} dt$$
$$= V_{DC}\big|_{t=t_n, i_{DC-link}=0} - V_{DC}\big|_{t=t_{n-1}, i_{DC-link}=0}$$
(3)
$$C = \frac{\int_{t_{n-1}}^{t_n} i_{DC-link} dt}{\Delta V_{C_{t_n}}}$$

Estimating the capacitance and measuring the voltage and current of the DC-link when its current value is nonzero, can easily determine the ESR value as well as the following:

$$ESR = \frac{V_{DC-link(t_m)} - (\frac{1}{C}\int_{t_n}^{t_m} i_{DC-link} dt + V_{C(t_n)})}{i_{DC-link(t_m)}}$$
(4)

Where at the moment t_m, the DC link current is nonzero. By setting the switching states of the boost converter and the full-bridge inverter, t_m and t_n can be determined in such a way that the above conditions will be met, and moreover, no switching will be done at these moments in the circuit to measure and estimate the capacitance and the ESR free from switching noises. In photovoltaic applications the boost converter duty cycle is normally between 10 to 80 percent, furthermore, the maximum modulation index for full-bridge inverter is considered to be equal to 95 percent [10]. Consequently, according to Fig. 4, where the triangular carriers of the boost converter and the full-bridge inverter are π rad different from

978-1-5090-0376-1/16 $31.00 © 2016 IEEE   358

each other, on the top points of the full-bridge's carrier, the full-bridge input current is zero, and the boost converter's switch is surely on. Therefore, the DC-link current is zero, and the capacitance estimation is possible in two consecutive top points of the full-bridge inverter carrier. At the top points of the boost converter carrier, the switch of the boost converter is off and the full-bridge converter input current is zero. In this case, the DC-link current is equal to boost inductor current. Therefore, using (3) and (4) the capacitance and the ESR can be obtained at t_ns and t_ms respectively. The estimation of the ESR is carried out every sampling period using its voltage and current at t_m instants. Since the ESR and the capacitance value calculated in (3) and (4) usually have ripple components, the RLS method can be applied for a more reliable estimation. Fig. 6 demonstrates the system block diagram estimating the ESR and the capacitance.

The RLS identification algorithm minimizes a least square cost function, allowing the estimated parameter of the system to be updated at each sampling interval whenever new data becomes available. Moreover, it has the advantages of simple calculation and good convergence properties for system parameter identification.

The error cost function for $\hat{C}_{[n]}$ is expressed as:

$$\int_{t_{n-1}}^{t_n} i_{DC-link}\, dt = C_{[n]}(V_{DC-link_{[n]}} - V_{DC-link_{[n-1]}})$$

$$e_{[n]}^2 = \left\{ \int_{t_{n-1}}^{t_n} i_{DC-link}\, dt - \hat{C}_{[n]}(V_{DC-link[n]} - V_{DC-link[n-1]}) \right\}^2 \quad (5)$$

Where $\hat{C}_{[n]}$ is the estimated value of the capacitance. The gradient of the error with respect to $\hat{C}_{[n]}$ value is given by:

$$\frac{\partial e_{[n]}^2}{\partial \hat{C}_{[n]}} = -2(V_{DC-link[n]} - V_{DC-link[n-1]}) \times$$

$$\left\{ \int_{t_{n-1}}^{t_n} i_{DC-link}\, dt - \hat{C}_{[n]}(V_{DC-link[n]} - V_{DC-link[n-1]}) \right\} \quad (6)$$

Which means that the parameter $\hat{C}_{[n]}$ should be updated, so that the right-side terms in (6) can minimize the error cost function $e_{[n]}^2$. Thus, the RLS updates the estimated parameter to $\hat{C}_{[n+1]}$.

$$\hat{C}_{[n+1]} = \hat{C}_{[n]} + \alpha_{[n]}(V_{DC-link[n]} - V_{DC-link[n-1]}) \times$$

$$\left\{ \int_{t_{n-1}}^{t_n} i_{DC-link}\, dt - \hat{C}_{[n]}(V_{DC-link[n]} - V_{DC-link[n-1]}) \right\} \quad (7)$$

Where $\alpha_{[n]}$ is an adjustment gain and can be selected as a constant scalar gain (9e-4) by a trial and error method. The initial value of $\hat{C}_{[0]}$ can be the nominal value of the DC-link capacitance.

The error cost function for $\hat{ESR}_{[n]}$ is expressed as:

$$ESR \cdot i_{DC-link[m]} =$$

$$V_{DC-link[m]} - V_{DC-link[n]} - \frac{1}{C_{[n+1]}}\int_{t_n}^{t_m} i_{DC-link}\, dt \quad (8)$$

$$e_{[m]}^2 = \left\{ \begin{array}{l} V_{DC-link[m]} - V_{DC-link[n]} - \dfrac{1}{C_{[n+1]}}\displaystyle\int_{t_n}^{t_m} i_{DC-link}\, dt \\[2mm] - \hat{ESR}\cdot i_{DC-link[m]} \end{array} \right\}^2 \quad (9)$$

The gradient of the error with respect to $\hat{ESR}_{[n]}$ value is given by:

$$\frac{\partial e_{[m]}^2}{\partial \hat{ESR}_{[m]}} = -2i_{DC-link[m]} \times$$

$$\left\{ \begin{array}{l} V_{DC-link[m]} - V_{DC-link[n]} \\[2mm] - \dfrac{1}{C_{[n+1]}}\displaystyle\int_{t_n}^{t_m} i_{DC-link}\, dt - \hat{ESR}\cdot i_{DC-link[m]} \end{array} \right\} \quad (10)$$

The RLS updating routine for the parameter $\hat{ESR}_{[n]}$, is given by:

$$\hat{ESR}_{[m+1]} = \hat{ESR}_{[m]} + \beta_{[m]}2i_{DC-link[m]} \times$$

$$\left\{ \begin{array}{l} V_{DC-link[m]} - V_{DC-link[n]} \\[2mm] - \dfrac{1}{C_{[n+1]}}\displaystyle\int_{t_n}^{t_m} i_{DC-link}\, dt - \hat{ESR}\cdot i_{DC-link[m]} \end{array} \right\} \quad (11)$$

Where $\beta_{[n]}$ is another adjustment gain, which can be chosen by the trial and error as a constant scalar gain (8e-5).

A. Temperature Sensing and Parameter Estimation Modification

In order to measure the operating temperature, the surface temperature is assumed to be equal to the capacitor core temperature and the temperature sensor can be tightly fixed to the top surface of the capacitor. The influence of the temperature and operating frequency on the capacitor characteristics are depicted in Fig. 5. As the operating temperature increases, the DC-link capacitance increases and its ESR decreases. Fig. 5(a) shows the relationship between the ESR and the temperature, Fig. 5(b) shows the relationship between the capacitance and the temperature, and both relationships can be implemented as look-up tables in the photovoltaic inverter central microcontroller which is able to modify the estimated values of the capacitance and the ESR.

B. Advantages of Proposed Technique

A literature search on the state of the art in Section II showed that the existing methods suffered from the following subjects:

1) Complicated or costly because they require additional hardware.
2) Provide an inconsistent estimation of ESR due to temperature variation.

The proposed predictive maintenance strategy can be used to estimate the ESR and the capacitance automatically and overcomes both limitations of complexity or cost and

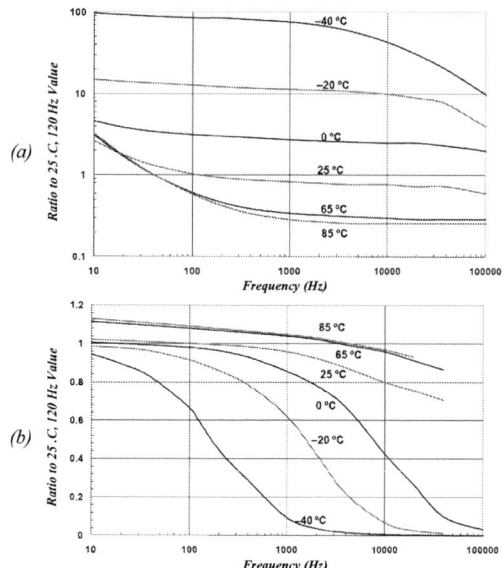

Fig. 5. Electrolytic Capacitor Characteristics vs. Temperature and Frequency, (a) ESR vs. Temperature and Frequency for a 2400µF 350V Capacitor, (b) Capacitance vs. Temperature and Frequency for a 2400µF 350V Capacitor [16].

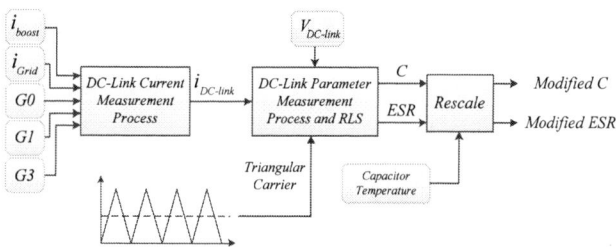

Fig. 6. The Block Diagram for the ESR and the Capacitance Estimation System

inconsistency issues. First of all, the proposed method is a simple and low cost solution, because the inverter is used for obtaining an automated estimation of ESR and capacitance.

Moreover, there are no requirements for additional hardware because the dc-link voltage and current are available by using the grid and the boost converter sensors. The estimation is accurate due to sampling at the instants, which are free of the switching noise, and also using RLS algorithm. In addition, it is able to provide frequent assessment of the DC-link capacitor condition.

IV. SIMULATION AND EXPERIMENTAL VERIFICATION

MATLAB simulations for a 3600W grid-connected photovoltaic inverter, using *unipolar PWM switching method*, were carried out to verify the effectiveness of the proposed technique. The sampling time of the DC-link voltage, and source currents was 32µs, which corresponds to the 15.625 kHz switching frequency of the boost converter and the full-bridge converter. The DC-link capacitor contained two identical sets of capacitance and ESR where the total capacitance and ESR was set to 2,400µF and 18mΩ, respectively. Fig. 7 shows the performance of the Grid-

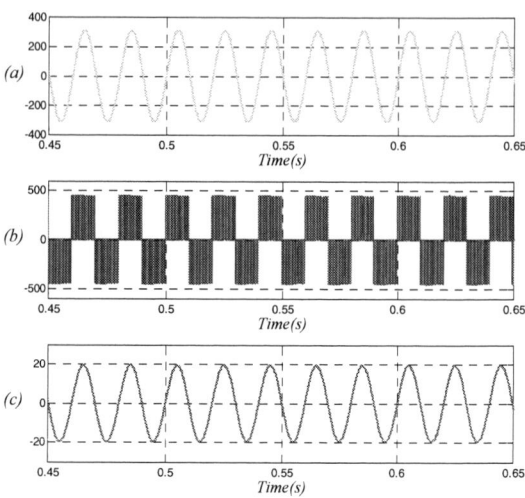

Fig. 7. Simulation Results for a 3600W Grid-connected Photovoltaic Inverter, (a) Grid Voltage, (b) Inverter Voltage, (c) Inverter Terminal Current

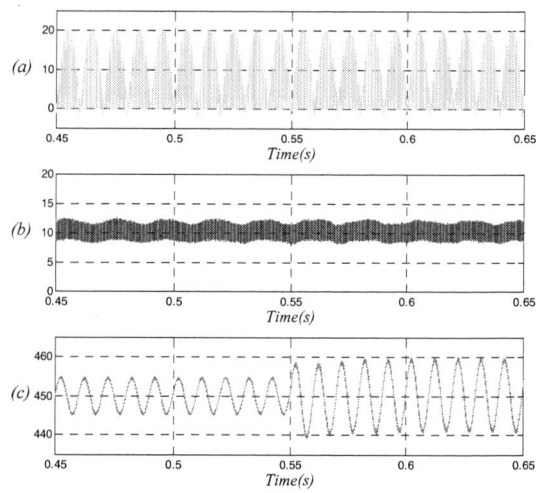

Fig. 8. Simulation Results for a 3600W Grid-connected Photovoltaic Inverter, (a) Full-bridge Input Current, (b) Boost Converter Output Current, (c) DC-link Voltage.

connected photovoltaic inverter. The waveform of the boost converter and the full-bridge current is shown in Fig. 8(a), (b). The DC-link voltage is shown in Fig. 8(c), where the 100Hz double frequency voltage ripple is explicit. The DC-link voltage for the estimation is sampled at the zero current instants every 32µs.

The DC-link current is sampled every 6.4µs. Then, Fig. 9(a) and 9(b) show the capacitance and the ESR estimation performance, respectively when one of the parallel sets of capacitors with ESR of 36mΩ is disconnected at 0.55sec. When the ESR value increases, the ESR voltage would be increased as well, since the current flowing into the DC-link capacitor is constant. The difference in voltage ripple can be clearly observed in Fig. 8 as well.

The above scenario has been implemented on a grid connected photovoltaic inverter with a nominal power of 3600W, in which the DC-link capacitor contains two sets of 1410µF capacitance in parallel (as shown in Fig. 10). The main processor of the inverter is TMS320F2812 which estimates the capacitance value. Capacitance estimation performance in stable operation and when half of the DC link capacitors were disconnected of the DC-link have been successfully implemented and similar results to the simulations have been obtained.

V. CONCLUSION

In this paper, a simple on-line fault detection technique is presented that is able to prevent failures in aluminum electrolytic capacitors used in the DC-link of grid-connected photovoltaic inverters. The proposed fault detection technique is able to estimate the ESR and the capacitance value during converter operation, the temperature effect is also considered, since the ESR and the capacitance change with temperature. Moreover, it is a noninvasive technique that can be implemented online and in real time. Although, in this paper, special emphasis was given to grid-connected photovoltaic inverters, the proposed technique can be extended to any other configuration or application using aluminum electrolytic capacitors. Simulation results and hardware experiments show that the proposed electrolytic capacitor failure prediction technique can be applied successfully.

REFERENCES

[1] M. Moosavi, S. Farhangi, H. Iman-Eini, and A. Haddadi, "An LCL-based interface connecting photovoltaic back-up inverter to load and grid," in *Power Electronics, Drive Systems and Technologies Conference (PEDSTC), 2013 4th*, 2013, pp. 465-470.

[2] G. M. Buiatti, Marti, x, J. A. n-Ramos, C. H. R. Garcia, A. M. R. Amaral, *et al.*, "An Online and Noninvasive Technique for the Condition Monitoring of Capacitors in Boost Converters," *Instrumentation and Measurement, IEEE Transactions on*, vol. 59, pp. 2134-2143, 2010.

[3] M. Makdessi, A. Sari, P. Venet, P. Bevilacqua, and C. Joubert, "Accelerated Ageing of Metallized Film Capacitors Under High Ripple Currents Combined With a DC Voltage," *Power Electronics, IEEE Transactions on*, vol. 30, pp. 2435-2444, 2015.

[4] K. Abdennadher, P. Venet, G. Rojat, J. M. Retif, and C. Rosset, "A Real-Time Predictive-Maintenance System of Aluminum Electrolytic Capacitors Used in Uninterrupted Power Supplies," *Industry Applications, IEEE Transactions on*, vol. 46, pp. 1644-1652, 2010.

[5] A. G. Abo-Khalil and L. Dong-Choon, "DC-Link Capacitance Estimation in AC/DC/AC PWM Converters Using Voltage Injection," *Industry Applications, IEEE Transactions on*, vol. 44, pp. 1631-1637, 2008.

[6] A. M. R. Amaral and A. J. M. Cardoso, "On-line fault detection of aluminium electrolytic capacitors, in step-down DC-DC converters, using input current and output voltage ripple," *Power Electronics, IET*, vol. 5, pp. 315-322, 2012.

[7] A. M. Imam, D. M. Divan, R. G. Harley, and T. G. Habetler, "Electrolytic Capacitor Failure Mechanism Due to Inrush Current," in *Industry Applications Conference, 2007. 42nd IAS Annual Meeting. Conference Record of the 2007 IEEE*, 2007, pp. 730-736.

[8] M. L. Gasperi, "Life prediction modeling of bus capacitors in AC variable-frequency drives," *Industry Applications, IEEE Transactions on*, vol. 41, pp. 1430-1435, 2005.

[9] W. Huiqing, X. Weidong, W. Xuhui, and P. Armstrong, "Analysis and Evaluation of DC-Link Capacitors for High-Power-Density Electric

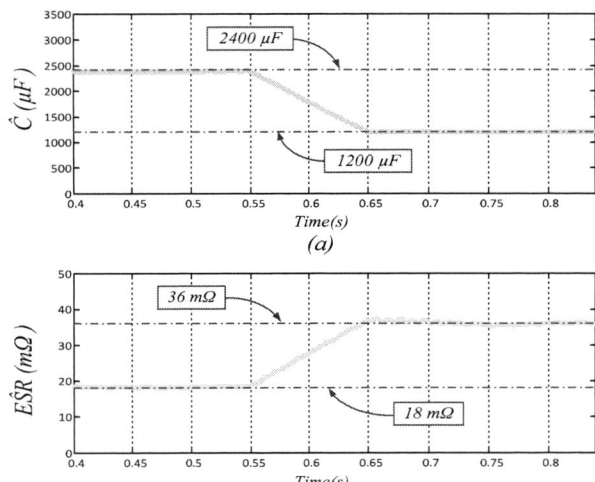

Fig. 9. Simulation Results for a 3600W Grid-connected Photovoltaic Inverter, (a) DC-link Estimated Capacitance, (b) DC-link Estimated ESR.

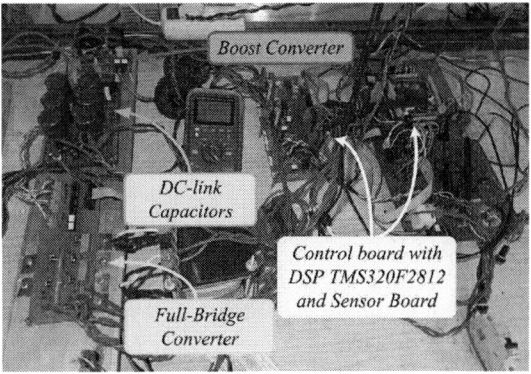

Fig. 10. 3600W Grid-Connected Photovoltaic Inverter in Which the Estimation Technique Has Been Implemented.

Vehicle Drive Systems," *Vehicular Technology, IEEE Transactions on*, vol. 61, pp. 2950-2964, 2012.

[10] Esram, T., Chapman, P.L., "Comparison of Photovoltaic Array Maximum Power Point Tracking Techniques," Energy Conversion, IEEE Transactions on, vol.22, no.2, pp.439-449, June 2007.

[11] Saradarzadeh, M.; Farhangi, S.; Schanen, J.L.; Jeannin, P.-O.; Frey, D., "Application of cascaded H-bridge distribution-static synchronous series compensator in electrical distribution system power flow control," in Power Electronics, IET , vol.5, no.9, pp.1660-1675, November 2012.

[12] Saradarzadeh, M.; Farhangi, S.; Schanen, J.L.; Jeannin, P.-O.; Frey, D., "Combination of power flow controller and short-circuit limiter in distribution electrical network using a cascaded H-bridge distribution-static synchronous series compensator," in Generation, Transmission & Distribution, IET , vol.6, no.11, pp.1121-1131, November 2012.

[13] E. Afshari, R. Rahimi, B. Farhangi, and Sh. Farhangi, "Analysis and Modification of the Single Phase Transformerless FB-DCB Inverter Modulation for Injecting Reactive Power," in Proc. Energy Conversion (CENCON), 2015 IEEE Conference on , pp.1-6, 19-20 Oct. 2015.

[14] R. Rahimi, B. Farhangi, and Sh. Farhangi, "New Topology to Reduce Leakage Current in Three-Phase Transformerless Grid-Connected Photovoltaic Inverters ," in Proc. Power Electronics Drive Systems and Technologies Conference (PEDSTC), 7th 2016, pp. 1-6, 16-18 Feb. 2016.

[15] www.konstant.in

[16] www.cde.com

7th Power Electronics, Drive Systems & Technologies Conference (PEDSTC 2016)
16-18 Feb. 2016, Iran University of Science and Technology, Tehran, Iran

Single Stage DC-AC Boost Converter

Ali Nahavandi
Faculty of Engineering
Malayer University
Malayer, Iran
E-mail:
Ali.nahavandi@malayeru.ac.ir

Mehdi Roostaee
Faculty of Engineering
Malayer University
Malayer, Iran
E-mail:
Mehdi.roostaee@stu.malayeru.ac.ir

Mohammad Reza Azizi
Faculty of Engineering
Malayer University
Malayer, Iran
E-mail:
Azizi.malayeru@gmail.com

Abstract—**This paper presents a new DC-AC converter which not only acts as inverter but also boosts the output voltage with respect to input. This topology is cost-effective due to reduced switch number and it is suitable for compact design. Switching strategy in this topology is similar to conventional inverter and in each half cycle, boost operation is done by one of the two boost converter. Variable duty cycle is applied for generating sinusoidal output voltage. Theoretical analysis and operation principle are presented together with simulated results to validate the proposed concepts.**

Keywords—Photovoltaic system; single-stage DC-AC converter; full bridge Inverter; boost converter.

I. INTRODUCTION

The growing use of renewable energy sources brings new challenges to the energy conversion technology. One of this challenges is related to the fact that some devices that store or produce electric energy (e.g. batteries, solar panels,...) are built using low-voltage cells, Usually connected in series in order to attain a reasonable voltage. The connection of large number of cells in series will increase the complexity of the system and may reduce its performance because of differences between cells and different working conditions. So, the output voltage of source (PV cells) need to be boosted with high conversion ratio and then must be inverted to AC for practical applications.

As a consequence, when a higher output voltage level is needed, a DC-DC boost converter must be used between the DC source and inverter as shown in figure 1. Depending on the power and voltage levels involved, this solution can result in high volume, weight, cost and reduced efficiency.

The proposed structure in this paper as shown in figure 2 uses one circuit for DC-DC boost converter and DC-AC conversion. This structure introduces a new single stage topology based on a full-bridge DC-AC inverter together with two additional diodes and one input inductor to implement two boost converters that share the same input inductor. By such a topology in recluse places that there is no achievement to AC grid, AC voltage in different level and frequency can be generated. Similar circuit is analyzed in [1]. Other DC-AC boost converter achieves DC-AC conversion, by connecting the load differentially across two DC-DC boost converters and modulating the DC-DC converter output voltages sinusoidally. This concept has been discussed in [2-5]. Several authors have worked on expanded circuit with transformer [6-8].In [9] and [10] the DC source is first boosted and then inverted to AC.

In this structure the duty cycle is variable and alternative [11].One advantage of this topology is using the minimum number of switches and consequently less power loss and other advantage is the ability of producing AC voltage in different level with different desired frequency.

In the next part, proposed structure is analyzed and different modes (operation principle) are illustrated and then the simulation results verify converter analyzes and at last the conclusion is added.

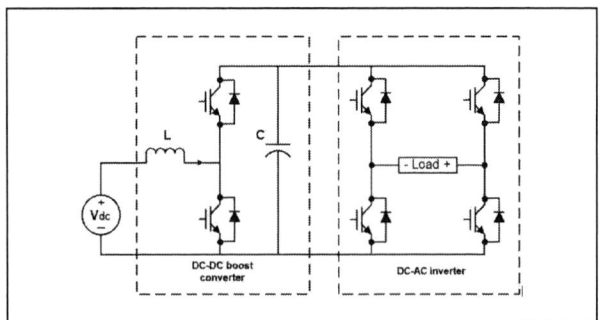

Fig.1. Traditional DC-AC boost converter (consisted of a DC-DC boost converter and an inverter)

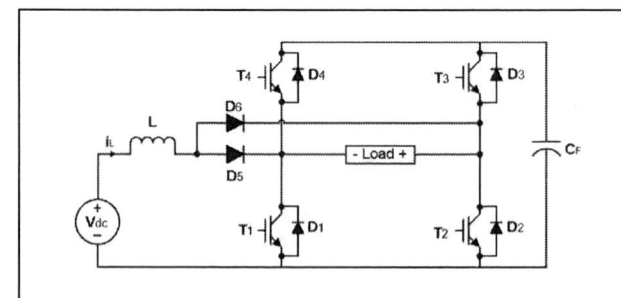

Fig. 2. Proposed structure (DC-AC boost converter)

978-1-5090-0376-1/16 $31.00 © 2016 IEEE

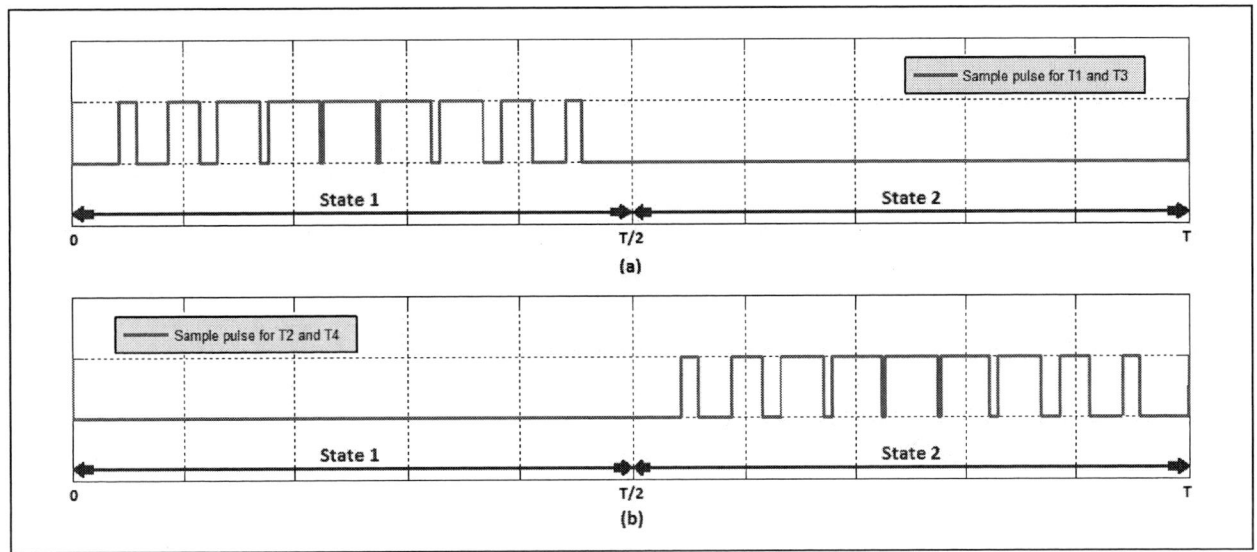

Fig. 3. a. T_1, T_3 are active in first half cycle with variable duty cycle b. T_2, T_4 are active in second half cycle with variable duty cycle. (For a better view number of pulses have been decreased)

II. PROPOSED STRUCTURE

As shown in figure 2 this topology has low number of switches as possible, one inductor and capacitor and a DC source which represents a PV cell and formed a DC-AC boost converter in one circuit. In this case, as an assumption the AC output frequency is 50Hz and the switching frequency is 10kHz with variable duty cycle. The switching strategy is similar to full bridge inverter, except in each half cycle boost operation must be done. So, two state for inverter and in each state two stage for boost operation will be defined. For obtaining a sinusoidal output voltage the duty cycle in this structure must be variable. Figure 3 shows the state of each of switches.

In state 1, T_1 and T_3 are switched with the frequency of 10kHz and variable duty cycle, while T_2 and T_4 are completely off. This state has two stages, first one occurs when T_1 and T_3 are on and second when T_1 and T_3 are off. Figure 4 shows stage 1 of state 1 and figure 5 shows stage 2 of this state.

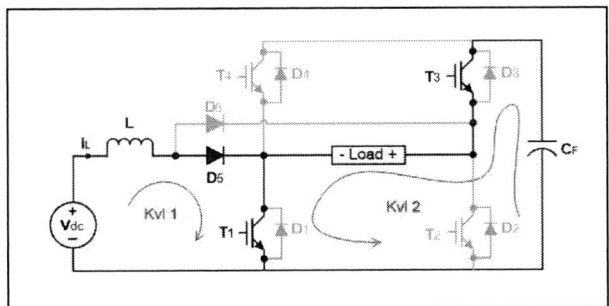

Fig. 4. Stage 1 of state 1($T_1, T_3 \rightarrow$ on $\qquad T_2, T_4 \rightarrow$ off)

In stage 1, two equations can be attained as below:

$$T_1, T_3 \rightarrow \text{ on} \qquad T_2, T_4 \rightarrow \text{ off}$$

kvl(1)

$$
\begin{cases}
V_{dc} = V_L \\[2mm]
V_L = L\dfrac{di_L}{dt} \\[2mm]
i_L = i_0 + \dfrac{V_{dc}}{L} t_n
\end{cases}
\tag{1}
$$

kvl(2)

$$V_C = V_{C0} - \frac{1}{C}\int i_C \, dt = V_{Load} \tag{2}$$

As the duty cycle is variable, consequently t_n in different period will be variable and then i_L and V_C will be so.

By using a kvl in stage 2 another equation will be attained.

978-1-5090-0376-1/16 $31.00 © 2016 IEEE   363

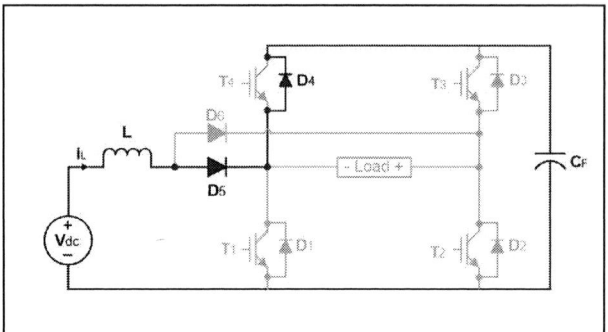

Fig. 5: Stage 2 of state 1 (T_1, T_3 → off T_2, T_4 → off)

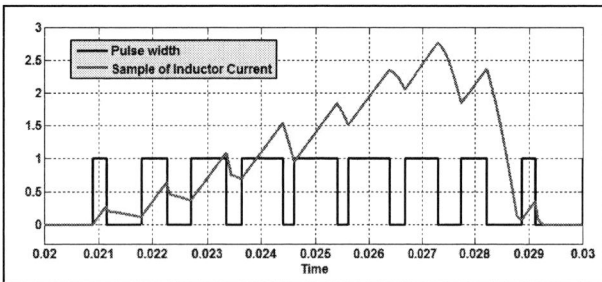

Fig. 6. Variable pulse width and sample of inductor current for different period

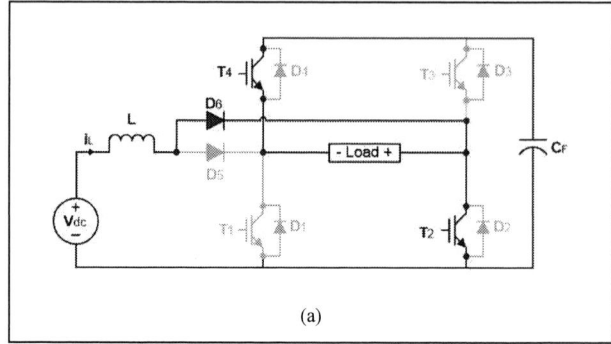

(a)

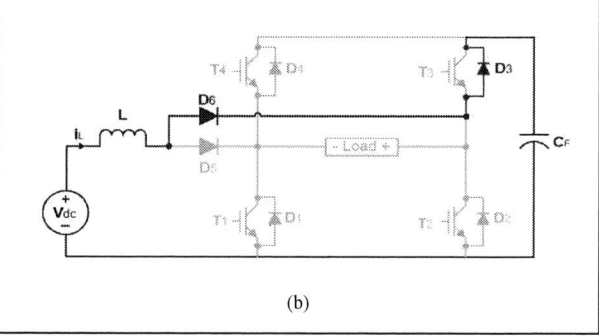

(b)

Fig. 7. a . Stage 1 of state 2 (T_2, T_4 → on T_1, T_3 → off)
       b . Stage 2 of state 2 (T_2, T_4 → off T_1, T_3 → off)

From

T_1, T_3 → off T_2, T_4 → off

$$
\begin{cases}
V_L + V_C = V_{dc} \\
i_L = i_C \\
L\dfrac{di_L}{dt} + \dfrac{1}{C}\int i_L\,dt = V_{dc}
\end{cases}
\tag{3}
$$

$$
\Rightarrow \qquad \frac{di_L^2}{dt} + \frac{i_L}{LC} = 0
\tag{4}
$$

Where

$$
\begin{cases}
i_L(0) = i_0 \\
\dfrac{di_L}{dt}(0) = \dfrac{V_{dc}}{L}
\end{cases}
\tag{5}
$$

$$
\Rightarrow \qquad i_L(t) = i_0 \cos \omega_0 t + V_{dc}\sqrt{\frac{L}{C}} \sin \omega_0 t
\tag{6}
$$

$$
\Rightarrow \qquad V_L(t) = L\frac{di_L(t)}{dt} = -i_0\sqrt{\frac{L}{C}}\sin \omega_0 t + \cos \omega_0 t
\tag{7}
$$

$$
\Rightarrow \qquad V_C(t) = \frac{1}{C}\int i\,dt = i_0\sqrt{\frac{L}{C}}\sin \omega_0 t - V_{dc}\cos \omega_0 t
\tag{8}
$$

Figure 6 shows inductor current in state 1. As shown in this figure, in stage 1 which T_1 and T_3 are on, inductor current is linear with variable duty cycle. In stage 2 which T_2, T_4 are off inductor current can be expressed by equation (6).

State 2 is completely similar to state 1, except the path of capacitor discharging and active switches are changed. This state also has two stages, stage 1 occur when T_2 and T_4 are on, which inductor is charged and capacitor is discharged. Second stage occurs when T_2 and T_4 are off. In this stage capacitor is charged and inductor is discharged. Figure 7 illustrate these stages.

Boost operation in this state in done by this two stage with high frequency and variable duty cycle.

III. SWITCHING STRATEGY

Switching strategy is in the way that T_1 and T_3 are switched in each positive half cycle while T_2 and T_4 are completely off in this state. Also, in negative half cycle T_2 and T_4 are switched while T_1 and T_3 are completely off. Consequently the output voltage of converter not only is boosted but also is inverted to AC. As the desired output voltage in each half cycle is a sinusoidal, the duty cycle in each state for each pair of switches must be variable. Figure 8 shows the variable duty cycle in one period (T=0.02s) for each pair of switches.

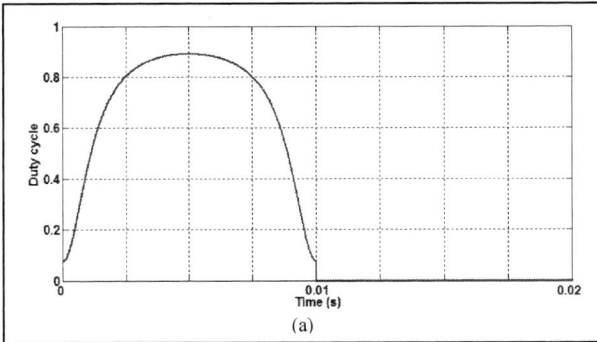

(a)

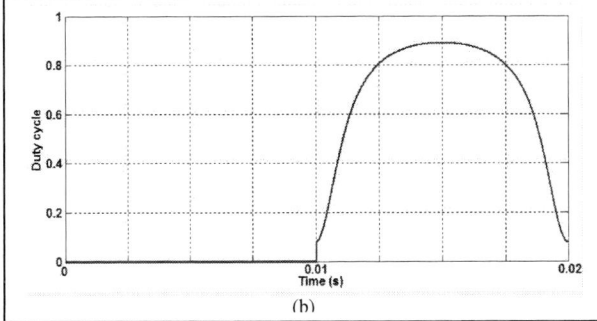

(b)

Fig. 8. a. Variable duty cycle of T_1 and T_3 in one period (T_1 and T_3 are switched in positive half cycle)
b. Variable duty cycle of T_2 and T_4 in one period (T_2 and T_4 are switched in positive half cycle)

IV. SIMULATION RESULT

In this part, by using of a DC-AC boost converter with $L = 10mH$, $C_F = 400\mu F$ from a 12V DC source (PV cell) an AC sinusiodal 48V is obtained. The current which is drawn from the DC source is similar to inductor current of the DC-DC boost converter which is in CCM mode. As the duty cycle of each boost converter is variable in time duration of a period T=0.02s and as inductor current is equal to the sum of two boost converter current, inductor current is alternative with T=0.01s, which has tiny fluctuations of switching with the frequency of 10khz. Figure 9 shows this current. Ripple of inductor current is depended on inductor value and switching frequency.

Figure 10 shows the capacitor voltage. As the stored energy in inductor in each stage is given to the capacitor and as the duty cycle is variable, like inductor current this voltage is variable with T=0.01s and has tiny fluctuation of switching. The ripple of this voltage is depended on capacitor value and switching frequency.

As explained in proposed structure $|V_{Load}| = V_C$. Since T_1, T_2, T_3 and T_4 operate like inverter switches, in a first half cycle (T_1, $T_3 \rightarrow$ on), $V_{Load} = V_C$ and in the second half cycle(T_2, $T_4 \rightarrow$ on), $V_{Load} = -V_C$. With variable duty cycle for boost operation in each half cycle V_{Load} will be AC and sinusoidal. By using a LC filter V_{Load} will be smoother with higher level. Figure 11 shows the output voltage with $L = 10mH$ and $C_F = 500\mu F$ as a load filter.

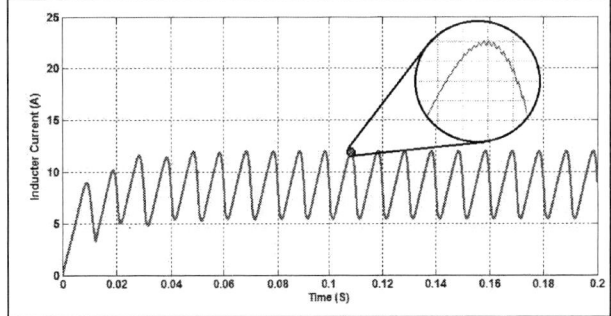

Fig. 9. Inductor current in CCM mode for $L = 10mH$ and $V_{dc} = 12v$

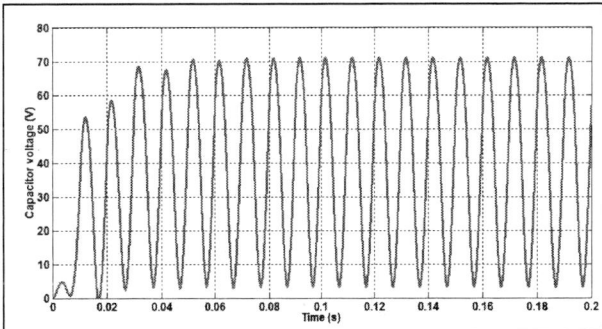

Fig. 10. Capacitor voltage

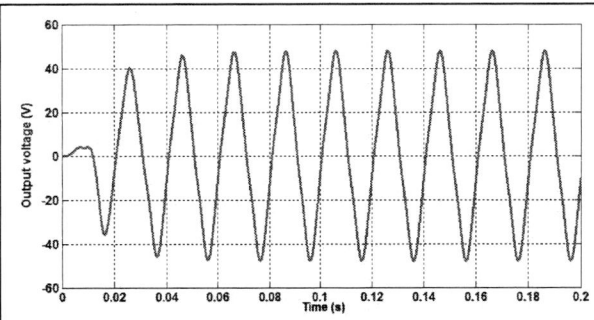

Fig. 11. Sinusoidal output voltage with $R_{Load} = 10\Omega$

Figure 12 shows output voltage, capacitor voltage and inductor current in a period. In state one, first the duty cycle increases. In this case, by increasing the pulse width, inductor will be charged more and more and the capacitor will be discharged. In the following of this state the duty cycle decreases. By decreasing the pulse width, capacitor will be charged more and more and inductor will be discharged. It should be noted that these charge and discharge occur with a little of delay.

As the duty cycle for both states is similar. Inductor current and capacitor voltage in state 2 as shown in figure 12 is similar to state 1. Output voltage in state 2 will be negative because T_2 and T_4 will change the path of circuit and the result is AC output voltage.

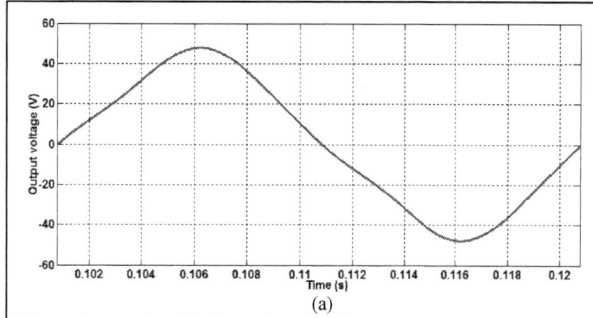

(a)

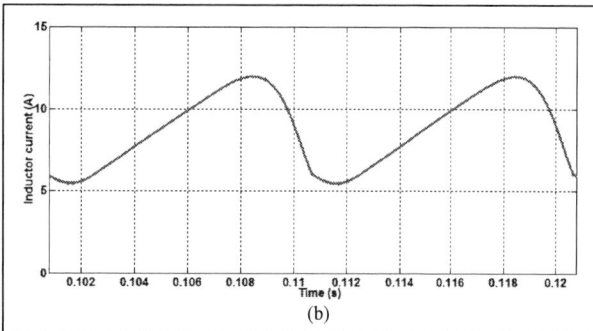

(b)

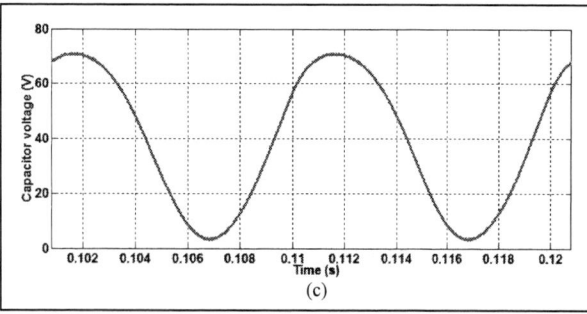

(c)

Fig. 12. a. Output voltage in a complete period b. Inductor current (current of two boost converter, one in positive half cycle and other in negative half cycle) c. Capacitor voltage (Output voltage of two boost converter)

V. CONCLUSION

In this paper, a DC-AC converter in one circuit was presented. By using the low number of switches as possible, from a low-voltage DC source, AC voltage with higher amplitude and desired frequency is generated. It was shown that AC voltage is achievable by applying variable duty cycle and changing the path of capacitor discharge. Also, it was shown that in each half cycle one of two DC-DC boost converter do the boost operation. Consequently, in a complete period the boost operation is done two times. So, capacitor voltage and inductor current are alternative with twice frequency of output voltage.

VI. REFRENCE

[1] H. Ribeiro, A. Pinto, and B. Borges, "Single-Stage DC-AC Converter for Photovoltaic Systems," In Energy Conversion Congress and Exposition (ECCE), 2010 IEEE, pp. 604-610. IEEE, 2010.

[2] R. O. Caceres, and I. Barbi. "A boost DC-AC converter: analysis, design, and experimentation." Power Electronics, IEEE Trans. on. vol. 14, pp.134-141 no. 1, JANUARY 1999.

[3] M. Coppola, S. Daliento, P. Guerriero, D. Lauria, and, E. Napoli, "On the design and the control of a coupled-inductors boost DC-AC converter for an individual PV panel," In Power Electronics, Electrical Drives, Automation and Motion (SPEEDAM), 2012 International Symposium on, pp. 1154-1159. IEEE, 2012.

[4] S. Danyali, S. H.Hosseini, and, G. B. Gharehpetian. "New Extendable Single-Stage Multi-input DC–DC/AC Boost Converter," Power Electronics, IEEE Trans. on. vol. 29, pp.775-788 no. 2, FEBRUARY 2014.

[5] P. Sanchis, E. Gubía, and L. Marroyo. "Design and experimental operation of a control strategy for the buck-boost DC-AC Inverter." In Electric Power Applications, IEE Proc.-Electr. Power Appl, vol. 152, no. 3, pp. 660-668. IET, May 2005.

[6] N. C. Foureaux, L. Adolpho, S.M. Silva, Brito de S, Jose Antonio, and De J. Cardoso Filho. "Application of solid state transformers in utility scale solar power plants," In Photovoltaic Specialist Conf. (PVSC), 2014 IEEE 40th, pp. 3695-3700. IEEE, 2014.

[7] M. Jong, and V. G. Agelidis. "A minimum power-processing-stage fuel-cell energy system based on a boost-inverter with a bidirectional backup battery storage," Power Electronics, IEEE Trans. on. vol. 26, pp. 1568-1577, no. 5, MAY 2011.

[8] D. Patil, and V. Agarwal. "Multi-input DC-AC converter for renewable energy applications," In ECCE Asia Downunder (ECCE Asia), pp. 429-435. 2013.

[9] H. S. Krishnamoorthy, S. Essakiappan, P. N. Enjeti, R. S. Balog, and S. Ahmed. "A new multilevel converter for Megawatt scale solar photovoltaic utility integration," IEEE Transl. In Applied Power Electronics Conference and Exposition (APEC), pp. 1431-1438, February 2012.

[10] Y. J. Song, and P. N. Enjeti. "A high frequency link direct DC-AC converter for residential fuel cell power systems," In Power Electronics Specialists Conference, 2004. PESC 04. vol. 6, pp. 4755-4761, June 2004.

[11] S. Pouresmaeil, B. Eskandari, and M. T. Bina. "Optimized offset modulation technique for three-phase DC-AC boost-converter," In Power Electronics, Drive Systems and Technologies Conference (PEDSTC), pp. 193-198. February 2014.

7th Power Electronics, Drive Systems & Technologies Conference (PEDSTC 2016)
16-18 Feb. 2016, Iran University of Science and Technology, Tehran, Iran

A Comparison Between Buck and Boost Topologies as Module Integrated Converters To Mitigate Partial Shading Effects on PV Arrays

Mohammad Amin Abolhasani[1], Reza Rezaii[2], Reza Beiranvand[3], Ali Yazdian Varjani[4],

Faculty of Electrical and Computer Engineering, Tarbiat Modares University, Tehran, Iran
[1]m.abolhassani@modares.ac.ir, [2]reza.rezaii@modares.ac.ir, [3]beiranvand@modares.ac.ir, [4]yazdian@modares.ac.ir

Abstract—**Module Integrated Converters (MICs) are one of the most reliable solutions for the partial shading effects problem in Photovoltaic (PV) systems. There are number of converters employed as DC-DC converters in MIC structures. Among them, buck and boost converters are more favorable because of their simple implementation, reduced part counts, high efficiency and easy control. In this paper a comparison between buck and boost converters in MIC structures is given. Advantages and disadvantages of both topologies are investigated, and experimental results are provided to verify operation of these two converters.**

Keywords— boost converter; buck converter; DC-DC power converters; module integrated converter; solar energy harvesting

I. INTRODUCTION

World-wide demand for clean electric energy is rapidly increasing during last few years. Among all renewable energy sources, solar energy is the most favorable endless resource, and is most available and easy to use. The global expectation is that the Photovoltaic (PV) energy sources will become the biggest renewable energy source to produce electricity by 2040 [1].

To modify the operation of the photovoltaic interfaces so that the load and the photovoltaic array match their operating characteristics at the Maximum Power Point (MPP), and in order to extract the maximum power delivered by a PV module, there are Maximum Power Point Tracking (MPPT) control techniques. P—V characteristic of a typical PV panel under full solar irradiation have a single power peak (MPP) which is shown in Fig. 1 [2]. However, there are some non-ideal conditions that cause problems in determining and tracking the MPP. One challenging issue is effects of the partial shading occurrence on PV panels.

When this phenomenon is occurred in a PV module, current of the un-shaded cells passes through the shaded cells and causes massive power dissipation in the module resulting in hotspot and cell breakdown. By using bypass diodes in parallel with each module, hotspot breakdown on PV cells due to partial shading can be avoided [3]. However, they cause presence of multiple peaks in the P—V characteristics of PV arrays [4], and they eliminate the available amount of power that shaded modules can deliver even in the presence of partial shading. Fig. 2 shows how the P—V characteristics of the array with bypass diodes changes due to the partial shading.

There are number of different methods that have been employed and developed over the last years to overcome the shading problem [5]. Among all techniques, using MICs has attracted a lot of attention from researchers due to its specific characteristics. MIC systems present "plug and play" concept and greatly optimizes the energy yield. Considering these advantages, the MIC concept has become the trend for future PV system development, but challenges remain in terms of cost, reliability, and stability for the grid connection [6]. Compared with the conventional methods, using MIC structure can increase harvested power from PV modules up to 30% when mismatch occurs [7]-[9].

In order to reduce the cost and power dissipation, and increase the efficiency in MICs, the topology of the converter connected to each module should be ameliorated. There are many types of converters employed in MIC structure including Buck, Boost, Buck-Boost, Flyback, Current-Fed Push-Pull, and Resonant Converters [10]. Most of these topologies are derived

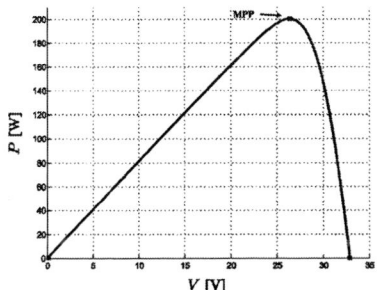

Fig. 1. P-V characteristics of a module under full solar irradiation

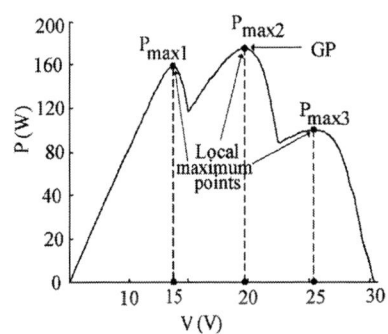

Fig. 2. P-V characteristics of PV modules when partial shading occurs [2]

978-1-5090-0376-1/16 $31.00 © 2016 IEEE

from buck and boost structures. However, buck and boost themselves have high efficiencies compared to other converters. In this paper, a comparison between these two converters in MIC structure is presented, and pros-and-cons of each topologies are investigated.

The characteristics and operation of the used MICs are discussed in Sec. II. In Sec. III characteristics of the buck and boost converters in MIC structure are demonstrated and a comparison between these two topologies is given. Then, experimental results are given in Sec. IV, and the conclusion is made in Sec. V.

II. MODULE INTEGRATED CONVERTERS

In MIC topology, a DC-DC converter is combined with the PV module such that the module is no longer connected in series with neighboring modules directly. The converter and the module both constitute an integrated element which can be connected to other elements in the whole array by output terminals of the converter, and construct a series-connected string. In this topology, each converter is controlled by a local autonomous MPPT controller in other to extract the maximum possible amount of power from each single module [5].

By using MICs, power dissipation resulting from each unbalancing between modules is eliminated, and the module level MPPT is provided so that it contributes to higher efficiency than the central inverters. In different current levels of every single module (either the module is prone to the shading or different level of irradiance) performance of the system will not be changed, and the current control will be facilitated because series connection of MICs has a high degree of controllability. General view of the MIC structure [7] used in this paper is shown in Fig. 3. In corresponding situations compared to other topologies MICs have more capability of harvesting solar energy. Under different irradiation. It is demonstrated that MICs are capable of delivering the output power by 30% more when a single MIC per panel is used and 45% more when there are two MICs per panel [7].

The multiple peaks problem is avoided by using MICs because instead of bypassing the module, the converter can process the reduced amount of power until there is no serious harm for cells. Thus, a simple algorithm can be exploited to track the single MPP. For this purpose the conventional Hill Climbing technique is used in this paper. In this approach by measuring current and voltage of a single module, periodically, and calculating the corresponding power, the duty cycle for converter switches can be determined so that the PV module always deliver its maximum available power [11].

However, because they add more elements to the systems which extract power from PV arrays, DC-DC converters used in this structure should have little power dissipation so that the balance between price and efficiency justifies the increasing amount and volume of the system. Thus, DC-DC converters should be selected by care and accuracy. Therefore, in this paper buck and boost converters are defined as most suitable converters according to [7], and their characteristics of them are investigated in this structure.

A. Series and Parallel connection of MICs

To connect PV arrays detached to MICs to the grid, 220 V output voltage is needed. Given to low output voltage of PV modules, the interface converter should have a high voltage gain by using a transformer or a large inductor. However, the drawbacks of large inductive elements are increasing converter size and cost, and decreasing efficiency. On the contrary, by connecting output of low gain converters in series to each other can eliminate the need for huge inductive elements. Also, voltage stresses on power switches, cost, and volume will be moderated, and efficiency of the system will increase.

Buck and boost topologies are regularly employed in this structure because of their higher efficiencies. As a case in point, in series connection boost converter will operate in its finest condition because it is not necessary to increase its voltage gain higher than its proper value. Nevertheless, if this converter is used in parallel connection, it will need to increase its output voltage farthest than its normal condition; consequently, its efficiency will diminish. Also, it should be considered that in worse cases, for example in higher value of shading, boost converter is rarely able to work at MPP of the shaded module, so consideration based on many factors is mandatory. According to circuit's theory, converters which acts like a voltage source should be used in series structure, and the one which is more like a current source is better to be employed in parallel structure.

B. Cascaded Multilevel Inverters in MIC structure

In the past few years multilevel inverters are used in PV applications [12]. CHB inverter are employed as MICs which is shown in Fig. 4, and by this construction the output filter's size will reduce. Though, these inverters can only reduce their

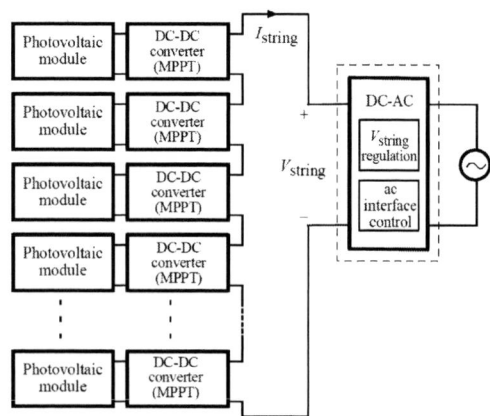

Fig. 3. Module Integrated Converters structure [7]

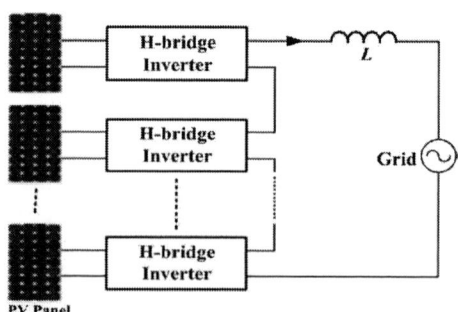

Fig. 4. CHB inverters in MIC structure [12]

output voltage, so many inverters are needed to connect the output of PV system to the grid. In this case, connecting many inverters to each other will endanger the reliability of the system.

To eliminate the problem of increasing the output voltage, Z-source inverters (ZSI) are used in MIC structure as shown in Fig. 5. On the grounds that ZSIs are immune to the inverter short circuit conditions in each legs, so reliability is higher in this configuration [13]. Although cascaded connection of inverters needs no DC link for a string of PV arrays, this structure has some disadvantages: 1) control and filtering of each inverter is sophisticated, 2) an input capacitor should be connected in parallel to each PV module to protect it from probable inrush current, and 3) cost, volume, and power dissipation of the system is increased drastically and reliability is reduced.

III. BUCK AND BOOST CONVERTERS IN MIC STRUCTURE

A. Buck Converter

According to the fact that each module should be connected in series with others to increase the total output voltage, buck converters, despite their reducing voltage instinct, are suitable choices in these structure, in that, their volume, cost, and efficiency is optimized, and can be controlled easily.

In a buck converter output voltage is always less than input voltage, and output current is more than input current. Owing to the fact that in a series connection of MICs the output current is an imposed current of all converter's output currents and is common to all converters, buck converters are more reliable. That is, if a single module considering as the first module is shaded and others are under full irradiation, the input current of the first converter connected to the shaded one is reduced, while other converter's input currents is at their nominal values.

However, the string current is almost constant, so the first converter will increase its duty cycle to compensate its input deficiency. Since the input current is less than the output current, the converter is capable of increasing its output current. But, if the converter's instinct was so that its output current was less than the input current, it would not perform properly in lower levels of shading. However, buck converter limitation is its reduced output voltage since to obtain needed voltage for DC link large number of buck converters should be connected in series.

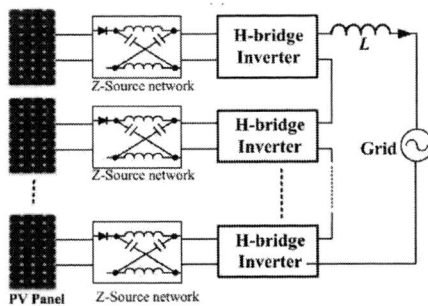

PV Panel   Z-Source network

Fig. 5. Z-Source inverters in MIC structure [13]

B. Boost Converter

The boost convert, also, benefits the advantage of having low part count and simple design. Owing to its capability of enhancing output voltage, boost converters are one of the most applicable candidates in MIC structure. In fact, capability of increasing the output voltage with simple structure and without using high-cost and heavy transformers makes the boost converter the most popular converter for MIC applications among all step-up converters.

This topology presents the ability of obtaining needed DC link voltage by less PV modules. If the desired converter output voltage is designed so high that the open circuit voltage of the PV panel sets below it, a boost characteristic ensures that the maximum power point (MPP) will always be tracked. The boost inductor which is located on the PV module side will reduce electromagnetic interference (EMI) noises and losses resulting from current ripple in the converter. The series connection of boost converters causes the output current for all converters to be the same.

Also, it allows every converter to reduce its voltage boosting factor, so it will reduce the voltage stress on the power switches and contribute to higher efficiency and better performance of the whole system. Since the output currents of neighboring MICs are equal, their output voltage is proportionate to their delivered power. However, boost converter limitation is that their output current is less than their input current. Thus, at high percent of shading their input current would be so that their output current force the string current to be that low.

C. Comparison Between Buck and Boost Converters in MIC Structure

When *shading* occurs on a PV module in MIC structure, it will not bypassed by a diode, so all of its power capability at that moment is extractable by its connected DC-DC converter. According to the aforementioned matters, when buck converters are employed in MIC structure, the string current is always bigger than every converter's input current. Thus, in case of high input current reduction, a wide range of this mismatch can be compensated.

As an illustration, consider a system in which output current of a PV module connected to a buck MIC is 3 A, and the output current of that converter which is equal to the string current is 5 A at normal conditions. When partial shading is occurred on module number 1 and other modules be under full irradiation, the input current of converter 1 becomes 1 A. The current of the string is held approximately at 5 A because 1) there are many converters in series and their output current is 5 A, and 2) the shaded module's buck converter is able to increase its current gain by changing its duty cycle. Consequently output voltage of that converter will reduced.

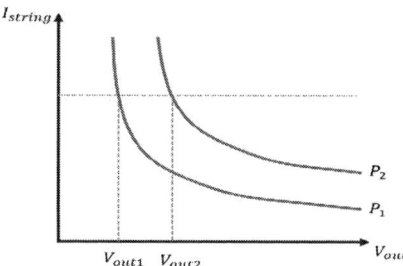

Fig. 6. I-V characteristics of MIC converters

However, input current of a boost converter is always bigger than the string current, so the range in which mismatch is compensable is limited in boost MICs. As a case in point, consider the above string with boost converters as MICs, and the string current is now 1 A. If shading occurs similar to what had happened previously, and input current of boost converter becomes 1 A, the converter is not able to deliver power. Thus, the converter should be removed from the string, and MIC structure is useless. Still, by controlling the DC link which encompass this string this problem can be mitigated.

That is, DC link should be forced to extract less current from the string so that shaded module could be remain in the system. However, control of the system will become slightly difficult because current sensors should be employed in the system. Also, other converters connected to unshaded modules should increase their duty cycle and output voltage to reduce their output current. Although increasing their output voltage compensates the reduction of converter 1's output voltage, the boost converter's operation will weaken by increasing its duty cycle very much, and its efficiency will reduce.

Fig. 6 shows the I-V characteristics of two converters in MIC structure. In this figure $P1$ curve is the output power of the converter connected to an unshaded module, and $P2$ curve is for shaded one. It shows that by reducing the string current DC link voltage is regulated since incrementing the output voltage of unshaded boost MIC compensates the reduction of the output voltage of the shaded one. Output voltage of each boost converter in MIC string is:

$$V_{out\,X} = V_{LOAD}\frac{P_{out\,X}}{H\,P_{out\,H} + L\,P_{out\,L}}. \tag{1}$$

Here, $H+L$ is the total number of PV modules H of which are under full irradiation, and L modules are shaded partially. X can be either among H converters or L converters.

Also, P_{outH} and P_{outL} are these converter's output power. It can be noticed from (1) that the output voltage of each converter is subjected to drastic increase in some

Table I Buck converter elements

Component	Value
Buck Inductor	20 μH
Buck Capacitor	33 μF
Power MOSFET	STP40NF10
Schottky Diode	MBR20100CT
MOSFET Gate Driver	HCPL3120
Voltage Sensor	HCPL7840

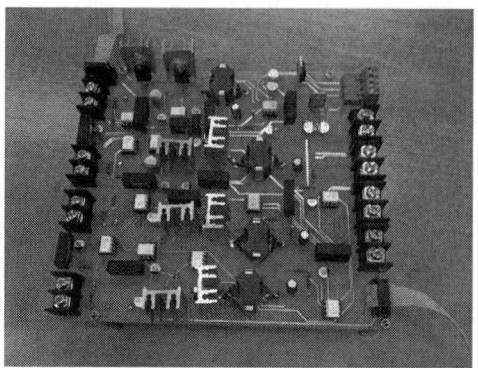

Fig. 7. Implementation of buck MICs

circumstances. For instance, if L modules become highly shaded, (i.e. $P_{outL}\approx0$), while $P_{outH}\neq0$, output voltages of L modules is approximately zero ($V_{outL}\approx0$), and output voltage of other converters are:

$$V_{out\,H} = \frac{V_{Load}}{H}. \tag{2}$$

Therefore, if H is small, V_{outH} will be so high, and voltage stresses on power switches will be very great. Consequently, to prevent excessive output voltage increase in boost MICs, some limitations should be considered [14].

On the other hand, capability of increasing output voltage is a remarkable advantage of boost converters. In places like building roofs where there is a limitation on number of PV modules, connecting system to the grid is a primary issue. If buck converters are used in these situations, a transformer with tap changer is needed to increase the voltage and provide the suitable input source for inverters. This will reduce the system's efficiency dramatically. Thus, using boost converters as MICs despite their weak performance in high levels of shading is unavoidable, and will ensure a high efficiency operation of the system at least in uncritical conditions.

In sum, what is concluded from the whole analysis is that in places where lots of PV modules are available like photovoltaic power stations, it is better to use buck converters in MIC structure. Since solar modules are numerous in PV farms, so much buck converters can be series connected to supply an inverter's DC link. What is achieved is low part counts in the system, high efficiency and reliability, and wide range of operation under mismatch conditions. In contrast, in domestic

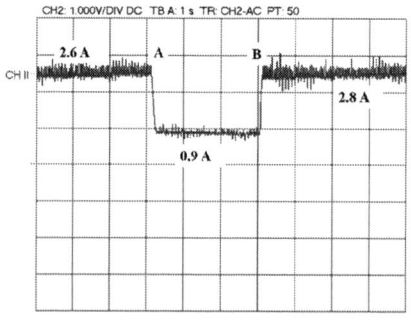

Fig. 8. Input current of a buck MIC when shading occurs on its module

usages, where PV module numbers are limited, using boost MICs are more beneficial since buck converters cannot increase the output voltage of the converters and reach the needed input voltage of inverters.

IV. EXPERIMENTAL RESULTS

Here, the experimental results of implementing buck and boost converters in MIC structure are presented. The results confirms the claims asserted in section III truly.

A. Buck Converter Implementing

In the first test, to investigate the capability of buck converter in tracking the maximum power point of PV arrays due to mitigating partial shading effects, a prototype with four module integrated buck converters is implemented. Table I shows the elements of a single converter and the whole system is shown in Fig. 7. Output of these converters are connected in series to feed the DC link of an H-Bridge inverter. The inverter is connected to the single phase grid to represent the domestic applications.

In order to extract maximum power from the PV modules, each converter senses its own output voltage and changes the duty cycle so that it leads to maximizing its output voltage. Since the string current of the converters is constant instantaneously, maximizing the output voltage leads to maximizing the output power, and consequently extracting maximum power from PV module all the time [15]. Converters are controlled by an ARM7 microcontroller, named NXP/LPC2138, which is a fast, economic, and accurate microcontroller. Gate drivers of each switches are fired by a *100 kHz* switching frequency which is created by microcontroller's PWM generator.

Fig. 8 illustrates the input current of the first converter when it experiences partial shading at point *A*. Then, partial shading is removed at point *B*. Fig. 9 shows the string current of converters at this situation. Other converters are under full irradiation. It can be noticed from the figure that the string current is reduced by a little value when shading occurs. In theory, the string current should not change; however, due to the buck's inductor, the output current decrease by a little amount. Also, it is obvious that by reducing input current of the first converter from *2.8 A* to *0.9 A*, change in its output current is very little, from *4.97 A* to *4.65 A*, because the string is imposing its current.

Fig. 10 depicts the output voltage of each converter over the aforementioned transition. It is obvious in the figure that because of constant output current of the converters the converter connected to the shaded module (Buck 1) changes its duty cycle so that it increases its output current to reach the string current. Thus, Buck 1's output voltage is decreased, and because the string current is reduced by a little value, the output voltages of other converters are increased.

In the next test, two boost converters connecting to similar modules of the previous test to constitute a series of boost MICs as shown in Fig 11. Table II represents the elements used in boost converters. Control algorithm is, also, implemented on an ARM microcontroller. The switching frequency of boost converters are *20 kHz*, and they track the MPP by a

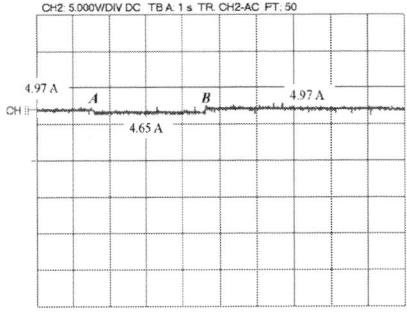

Fig. 9. Output current of a buck MIC when shading occurs on its module

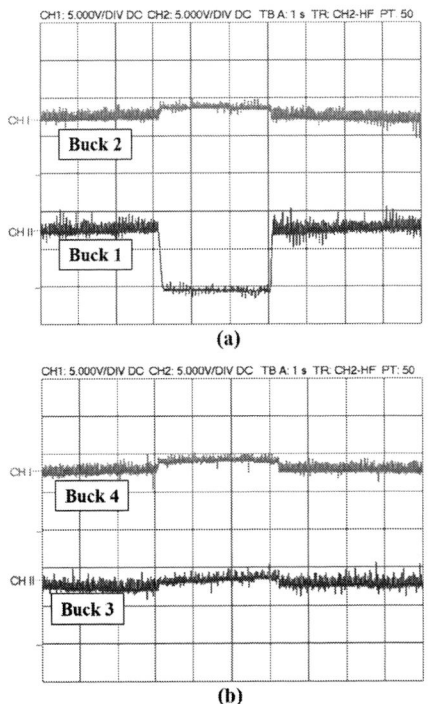

Fig. 10. Output voltage of four series connected buck MICs when shading occurs on module of first buck

Fig. 11. Implementation of boost MICs

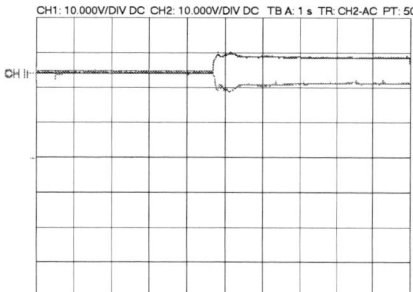

CH1: 10.000V/DIV DC CH2: 10.000V/DIV DC TB A: 1 s TR: CH2-AC PT: 50

Fig. 12. Output voltage of two boost MICs befor and after a light partial shading

Table II Boost converter elements

Component	Value
Boost Inductor	100 µH
Boost Capacitor	150 µF
Power MOSFET	IRFP460
Schottky Diode	DSEC 30-04A
MOSFET Gate Driver	HCPL3120
Current Sensor	LTS 25-NP
Voltage Sensor	HCPL7840

conventional Hill Climbing method. Output of these two converters are connected to an inverter which is controlled to keep DC link voltage at a constant value. Fig. 12 shows the output voltage of two converters when the module connected to one of them is shaded at the middle of sampling time. It is observable in the figure that shaded module's converter decreases its output voltage, and unshaded one increases its output voltage to compensate the deficiency.

Fig. 13 depicts the output voltages of two boost MICs when shading on the first module is very high. It can be seen that the converter which its module is not shaded increases its output voltage to an extremely high value, and the other one's output voltage is approximately zero. Thus, the efficiency of the system is reduced intensely since both converters are operating under unsuitable conditions.

V. CONCLUSION

Buck and boost topologies are the most regular converters in MIC structure, and many other converters are derived from these two topologies. They are easily implemented and controlled, and have suitable efficiency and operation on a wide range. In this paper a comparison between merits and drawbacks of buck and boost MICs was presented, and experimental results of testing these structures under different conditions was investigated. Finally, we have made this conclusion that in places where large number of PV panels are available, buck converters have better efficiency and operation because they facilitate the operation of the second stage converter by cheaper solutions. However, less PV modules at hand leads us to choose boost converter because the output voltage cannot be suitable for connecting system to the grid if buck converters are used.

REFERENCES

[1] European Renewable Energy Council (2004, May). Renewable Energy Scenario to 2040[Online]. Available: http://www.erec-renewables.org/ documents/targets_2040/EREC_Scenario%202040.pdf .

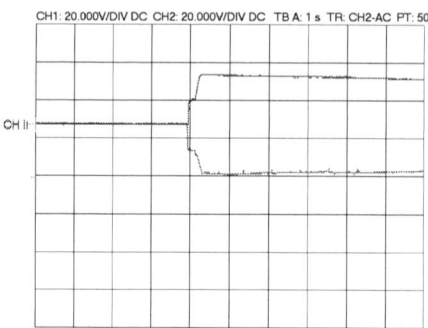

CH1: 20.000V/DIV DC CH2: 20.000V/DIV DC TB A: 1 s TR: CH2-AC PT: 50

Fig. 13. Output voltage of two boost MICs befor and after a heavy partial shading

[2] Villalva, M. Gradella, and J. R Gazoli. "Comprehensive approach to modeling and simulation of photovoltaic arrays." Power Electronics, IEEE Transactions . vol. 24, pp. 1198-1208, Jan-Feb,2009.

[3] S. Silvestre, A. Boronat, and A. Chouder. "Study of bypass diodes configuration on PV modules." Applied Energy . vol. 86, no. 9, pp. 1632-1640, September, 2009.

[4] H. Patel, and V. Agarwal. "MATLAB-based modeling to study the effects of partial shading on PV array characteristics." Energy Conversion, IEEE Transactions . vol. 23, no. 1, pp. 302-310, March 2008.

[5] A. Bidram, A. Davoudi, and R. S. Balog. "Control and circuit techniques to mitigate partial shading effects in photovoltaic arrays." Photovoltaics, IEEE Journal , Vol. 2, No. 4, October 2012.

[6] G. R. Walker and J. C. Pierce, "Photovoltaic dc–dc module integrated converter for novel cascaded and bypass grid connection topologies— Design and optimisation," in Proc. IEEE PESC, pp. 3094–3100, 2006.

[7] L. Linares, R. Erickson, S. MacAlpine, and M. Brandemuehl, "Improved energy capture in series string photovoltaic via smart distributed power electronics," in IEEE 2009 Applied Power Electronics Conference, pp. 904-905, 2009.

[8] S. Kjaer, J. Pedersen and F. Blaabjerg, "A review of single-phase gridconnected inverters for photovoltaic modules," IEEE Transactions on Industry Applications, vol. 41, no. 5, pp. 1292-1306, Sept. 2005.

[9] Chris Deline, Bill Marion, Jennifer Granata, Sigifredo Gonzalez, "A performance and economic analysis of distributed power electronics in photovoltaic Systems", Nation renewable energy lab report, NREL Report, No. TP-5200-50003, Jan. 2011.R. Nicole, "Title of paper with only first word capitalized," J. Name Stand. Abbrev., in press.

[10] Liang, Zhigang, et al. "A high-efficiency PV module-integrated DC/DC converter for PV energy harvest in FREEDM systems." Power Electronics, IEEE Transactions on 26.3 (2011): 897-909.

[11] B. Subudhi, and P. Raseswari, "A comparative study on maximum power point tracking techniques for photovoltaic power systems." Sustainable Energy, IEEE Transactions . Vol. 4, No. 1, Jan. 2013.

[12] E. Villanueva, P. Correa, J. Rodriguez and M. Pacas, "Control of a single-phase cascaded Hbridge multilevel inverter for grid-connected photovoltaic systems," IEEE Trans. Ind.Electron., vol. 56, no. 11, pp. 4399-4406, Nov. 2009.

[13] Yi Huang, Miaosen Shen, F.Z. Peng and Jin Wang, "Z-source inverter for residential photovoltaic systems," IEEE Trans. Power Electron., vol. 21, no. 6, pp. 1776-1782, Nov. 2006.

[14] Femia, Nicola, et al. "Distributed maximum power point tracking of photovoltaic arrays: Novel approach and system analysis." Industrial Electronics, IEEE Transactions on 55.7 (2008): 2610-2621.

[15] Pilawa-Podgurski, Robert CN, and David J. Perreault. "Submodule integrated distributed maximum power point tracking for solar photovoltaic applications." Power Electronics, IEEE Transactions on 28.6 (2013): 2957-2967.

7th Power Electronics, Drive Systems & Technologies Conference (PEDSTC 2016)
16-18 Feb. 2016, Iran University of Science and Technology, Tehran, Iran

Output Power Smoothing of PMSG-Based Wind Energy Conversion System Equipped with Matrix Converter

Koosha Mehdizadegan
Department of Electrical Engineering
Islamic Azad University – south Tehran Branch
Tehran, Iran
Koosha_mehdizadegan@yahoo.com

Karim Abbaszadeh
Department of Electrical Engineering
Islamic Azad University – south Tehran Branch
Tehran, Iran
abbaszadeh@eetd.kntu.ac.ir

Abstract— **Due to the growing of the electrical energy demand and depletion of the fossil-fuel resources, especial attention has been made to utilize the renewable-energy sources such as wind. Nowadays, variable-speed wind turbine (VSWT) is employed to increase the amount of the extracted wind power and also to mitigate the stress imposed on the turbine. The PMSG-based VSWT equipped with the conventional back to back converter is one of the most popular variable-speed structures. In this paper, the conventional back to back converter is replaced by a matrix type. Furthermore, it is discussed the extracting the maximum power of wind may lead to appearance of fluctuation in the output power. In this paper, a novel and simple scheme is introduced, which not only mitigate the fluctuation of output power, but also preserve the peak power tracking capability. Comparison between proposed and conventional scheme confirm the effectiveness of the proposed scheme.**

Keywords—Matrix Converter; Wind Turbine; Permanent Magnet Synchronous Generator; Maximum Power Point Tracking; Output Power Smothing; Low Pass Filter

I. INTRODUCTION

Due to the depletion of fossil fuels and its environmental pollution, special attention has been made for using renewable energy such as wind. Wind turbine can harness the wind power in different ranges of wind velocity [1]. As shown in Fig. 1, below cut-in wind speed, i.e., V_{W1}, the turbine does not operate. However, below rated wind speed, i.e., V_{Wr}, obtaining the optimal power of wind is useful. In this condition, the pitch angle is fixed and by adjusting the rotational speed of turbine, the optimum available power of turbine is extracted. Activating the pitch control system during high wind speed condition, i.e. above rated wind speed, is essential for preventing damages imposed on the turbine and generator. In the present study, it is assumed that the wind speed is below rated speed and therefore, pitch control is fixed and tracking maximum power point (MPP) of turbine is useful. To achieve the maximum power point tracing (MPPT) capability, it is essential to adjust the rotational speed of turbine on the optimal value.

Using the variable-speed wind turbine (VSWT) with power electronic interface is indispensable for achieving MPPT [2]. VSWT structures generally use double fed induction generator or permanent magnet synchronous generator (PMSG). the PMSG based VSWT has the advantageous of higher efficiency and better power factor condition. Furthermore, the gearbox can be eliminated from this structure and therefore, the reliability and losses are improved.

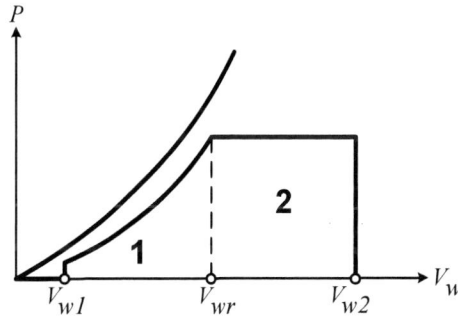

Fig. 1.   Variation of the turbine power versus wind speed

In a PMSG based VSWT, the series full rated inverter is employed to adjust the voltage and frequency and therefore, the control of the rotational speed is obtained [3]. In [4], vector control of PMSG equipped with a conventional back to back converter has been studied. The back to back converter consists of two converters and DC-link voltage. The converter connected to the PMSG is named machine side converter (MSC) and the other is known as the grid side converter (GSC). The MSC adjusts the flux and rotational speed of PMSG, while, the GSC is used for controlling the DC-voltage and the injected reactive power to the grid [4]. These converters can be controlled by pulse with modulation (PWM) techniques.

In the present paper, the matrix converter (MC) is employed as an interface between the PMSG and grid. The MC is an AC/AC converter which any storage energy

978-1-5090-0376-1/16 $31.00 © 2016 IEEE   373

equipment is not used in its structure. Therefore, the size of the converter is reduced and its reliability is improved [5]. Furthermore, the life-time of the converter is extended and the control is applied only on one converter, which results in decreasing the complexity of the implementation. Sinusoidal waveform for the input current and output voltage is achieved, and the power factor can be set on one. More comparisons between MC and conventional back to back converter can be found in [6].

Many works have been carried out for tracking the MPP of turbine. Using the optimal power curve is one of the popular methods. As it will be shown, this method imposes further fluctuation in the output power. The fluctuation of output power causes many problems in the power system. Furthermore, the additional fluctuation may impose stress on the generator. In the present paper, the conventional power curve method is improved. Consequently, the MPP is tracked and the output power is smoothed. Comparisons with the conventional power curve method confirm the effectiveness of the proposed scheme.

This paper is structured as follows; the turbine and generator are modeled in the section II. In the section III, the matrix converter and its space vector modulation are studied. The conventional power curve method for obtaining MPP of turbine is reviewed in the section IV. The proposed method is introduced in the next section V. In the section VI, simulation results are given. Finally, conclusion is given in the section VII.

II. VSWT EQUIPPED WITH THE PMSG AND MC

As shown in Fig. 2, mechanical power of turbine is converted to the electric form by the PMSG. Then, the electric power can be achieved through stator winding of the generator. The PMSG is connected to the main grid by MC. For eliminating harmonics, employing the input filter is essential [7]. By switching the MC, the rotational speed of turbine is controlled and consequently the capability of MPPT can be obtained. In this section, the controls of turbine, PMSG and MC are reviewed.

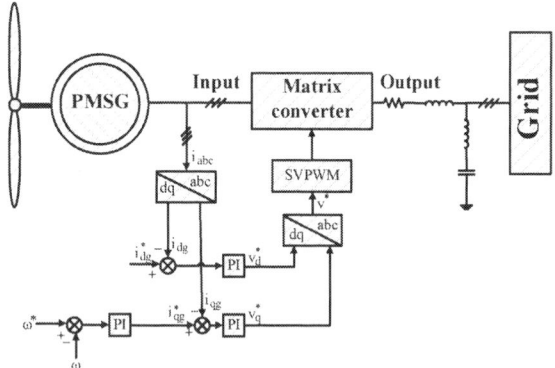

Fig. 2. Control of wind turbine equipped with the PMSG and MC

A. Wind turbine modeling

The mechanical steady-state model of turbine is as follows [8, 9]:

$$P_m = \frac{1}{2}\rho A C_p\left(\lambda,\beta\right)v_w^3 \tag{1}$$

where, P_m is the mechanical power of turbine in Watt, C_p is the aerodynamic efficiency of turbine, ρ is the air density in Kg/m³, λ is the tip speed ratio, β_p is the pitch angle in Deg and v is the wind speed in m/s. The equation corresponds to the aerodynamic efficiency of turbine is as follows [10]:

$$c_p\left(\lambda,\beta_p\right) = c_1\left(\frac{c_2}{\lambda_i}-c_3\beta_p-c_4\right)e^{\left(-\frac{c_5}{\lambda_i}\right)}+c_6\lambda;$$

$$\frac{1}{\lambda_i} = \frac{1}{\lambda+0.08\beta_p}-\frac{0.035}{\beta_p^3+1} \tag{2}$$

The mechanic torque (T_m), electrical torque (T_e) and rotational speed (ω_m) are given as follows:

$$T_m = \frac{P_m}{\omega_m} \quad ; \quad T_e = \frac{P_e}{\omega_e} \tag{3}$$

$$\omega_m = \int \frac{T_m - T_e}{J}\,dt \tag{4}$$

where J is the inertia constant in Kg/m², P_e is the electrical power in Watt, P_m is the mechanical torque in Watt and ω_e is the electrical speed of rotor in rad/s.

B. Modeling of the generator

In the reference frame is aligned with the rotor flux, the model of PMSG can be written as follows:

$$i_d R_s - v_d - L_d \frac{di_d}{dt}+\omega_r L_q i_q \tag{5}$$

$$i_q R_s - v_q - L_q \frac{di_q}{dt}+\omega_r\left(L_q i_q + \psi_{pm1}\right) \tag{6}$$

where, V_d and V_q are the voltage of d and q axes, respectively. i_d and i_q are the current of dq axes, R_s is the stator resistance. L_d and L_q are the linkage inductance for the d and q axis, respectively, ω_r is the generator speed and ψ_{pm1} is the flux of the permanent magnetic.

The electromagnetic torque can be expressed as follows:

$$T_e = P i_q \left[\left(L_d - L_q \right) i_d \right] \qquad (7)$$

where, P is the pole pair number. The above equations can be re-written as follows:

$$v_d = L_d \frac{di_d}{dt} - \omega_r L_q i_q + i_d R_s \qquad (8)$$

$$v_q = L_q \frac{di_q}{dt} - \omega_r \left(L_q i_q + \psi_{pm1} \right) + i_q R_s \qquad (9)$$

According to the above equations, by controlling the d and q components of the stator current, the electrical torque and the reactive power can be controlled, respectively. It should be noted that the control of the PMSG connected to the MC is the same as the control of the MSC of the conventional back-to-back converter [11]. The vector control of PMSG equipped with the MC is given in Fig. 2.

III. MATRIX CONVERTER

The MC is a direct Ac/Ac converter which is a good candidate for replacing the conventional back-to-back ones [12]. This converter includes 18 switches used for connecting the load to grid. The size and weight of this converter are reduced and it has better lifetime and reliability [13]. The main characteristics of a MC can be expressed as: 1) the MC has a simple and compact electrical circuit; 2) controlling the frequency and amplitude of the output voltage can be achieved; 3) The sinusoidal waveform for the input and output currents is obtained; 4) The unity power factor can be obtained. Due to these characteristics, back-to-back converter can be replaced with the Mc. In this paper, switching of the MC is applied by the space vector modulation technique.

IV. MAXIMUM POWER POINT TRACKING OF TURBINE USING OPTIMAL POWER CURVE METHOD

The optimal rotational speed of turbine can be determined by using the optimal power curve which can be implemented by a lookup table [14]. General Electric has been used this method for the MPPT [15], [16]. The maximum available power of turbine can be expressed as follows:

$$P_{max} = \frac{1}{2} \rho A_r \frac{R^3 C_{p\,max}}{\lambda_{opt}^3} \omega_t^3 = K_{opt} \omega_t^3 \qquad (10)$$

where, $C_{p\,max}$ is the maximum value of turbine aerodynamic efficiency. In this condition, the TSR has an optimal value. According to the above equation, the optimal rotational speed of turbine can be calculated. The reference rotational speed is tracked by the vector control scheme shown in Fig. 2. As it will be shown, tracking the optimal rotational speed imposes additional fluctuations in the output power.

V. THE PROPOSED SCHEME

Assume that the reference of the rotational speed is suddenly increased. For tracking the reference speed, according to equation (4), the electrical power is reduced. However, this leads to appearance of fluctuations in the output power.

The proposed scheme is depicted in Fig. 3. As it can be seen, the calculated reference speed by the power curve method flows through a low pass filter (LPF). By employing LPF, the variation of the reference speed is significantly mitigated which results in smoothing of the output power. It should be noted that the implementation of the proposed scheme is very simple.

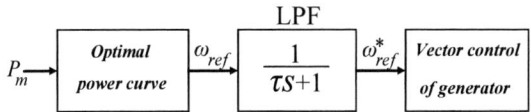

Fig. 3. The proposed MPPT scheme

VI. CASE STUDIES

A. case1: Step variation of the wind speed

Figure 4 illustrates the variation of the wind speed. The results are depicted in Fig. 5. The parameters of the grid, wind turbine and PMSG are given in the Appendix. As shown in Fig. 4, at t=5 s, the wind speed decreases from 11 to 10 m/s. Figure 5-a, compares the variation of the rotational speed in the cases of using the conventional and proposed scheme. It is clear that the variation of the rotational speed in the case of using the proposed scheme significantly reduces. The optimal value of the rotational speed for wind velocity of 10 m/s is 1.05 pu, while it is 1.15 pu for the case of 11 m/s. It is obvious that both schemes track the optimal reference of speed rapidly.

Fig.5-b depicts the variation of the generator output power. In the conventional scheme, significant variations appear in the output power. Due to the power fluctuation, the output power exceeds from the nominal value and consequently, the generator may be damaged. Clearly, in the proposed scheme, the variation of output power is smooth.

Figures 5-c and 5-d illustrate the variation of the aerodynamic efficiency and the turbine mechanical power, respectively. These values for the case of using conventional speed are greater than those of proposed method.

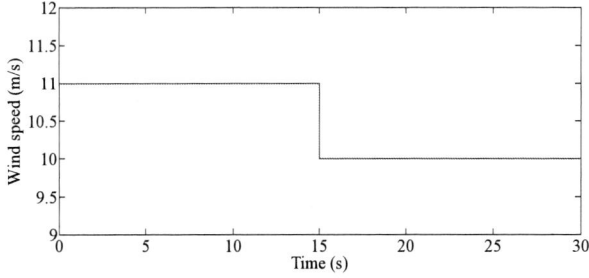

Fig. 4. Variation of the wind speed

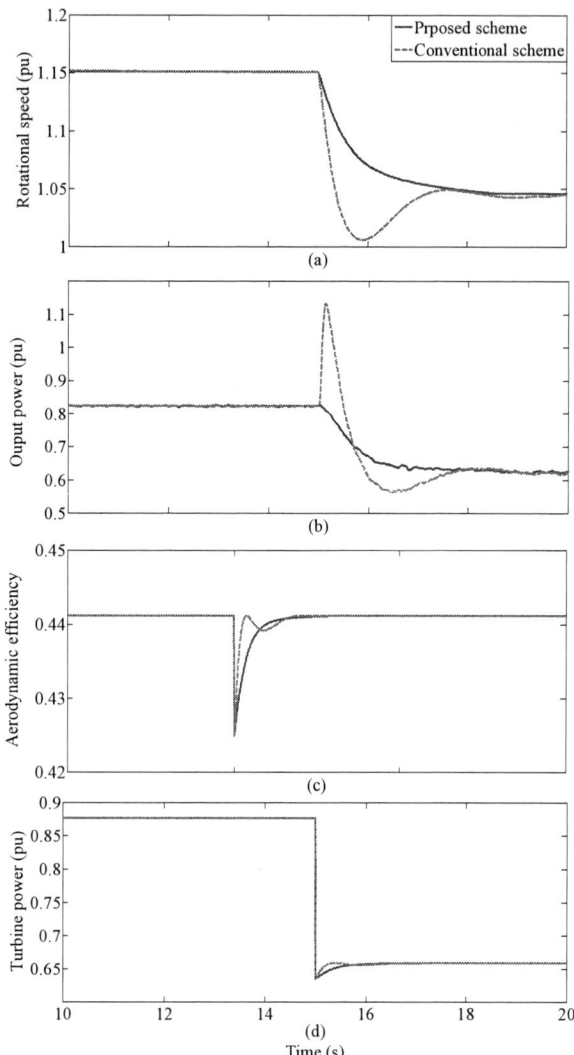

Fig. 5. Result of the first: case a) rotational speed, b) Ouput power, c) aerodynamic efficiency and d) turbine mechanical power

B. case2: real variation of the wind speed

Fig. 6 illustrates the variation of the wind speed. Weibull distribution is used for modeling the variation of the wind speed.

Figures 7-a and 7-b depict the variation of the rotational speed and the output power of the generator, respectively. It is clear that, in the conventional scheme, the fluctuation of the output power and the rotational speed is higher.

The variation of the aerodynamic efficiency and mechanical power of turbine is depicted in Figs. 7-c and 7-d, respectively. These values are almost equal in both schemes.

For better comparison, it is useful to employ the mean and standard deviation of variables. The mean and standard deviation values of the output electrical power, mechanical power, rotational speed and aerodynamic efficiency of turbine

are given in Table I. The mean value of variables, for both schemes, are equal. Therefore, the proposed scheme preserves the MPPT and efficiency. However, the standard deviation of variables for the case of using the proposed scheme is lower than those of the conventional scheme. Consequently, the proposed scheme significantly mitigates the fluctuation of variables.

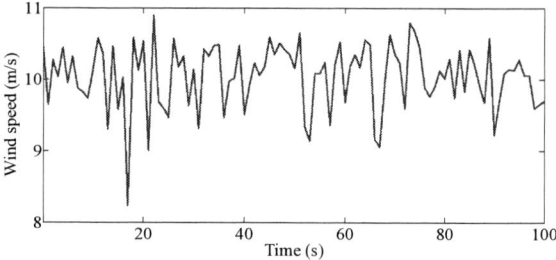

Fig. 6. Variation of the wind speed

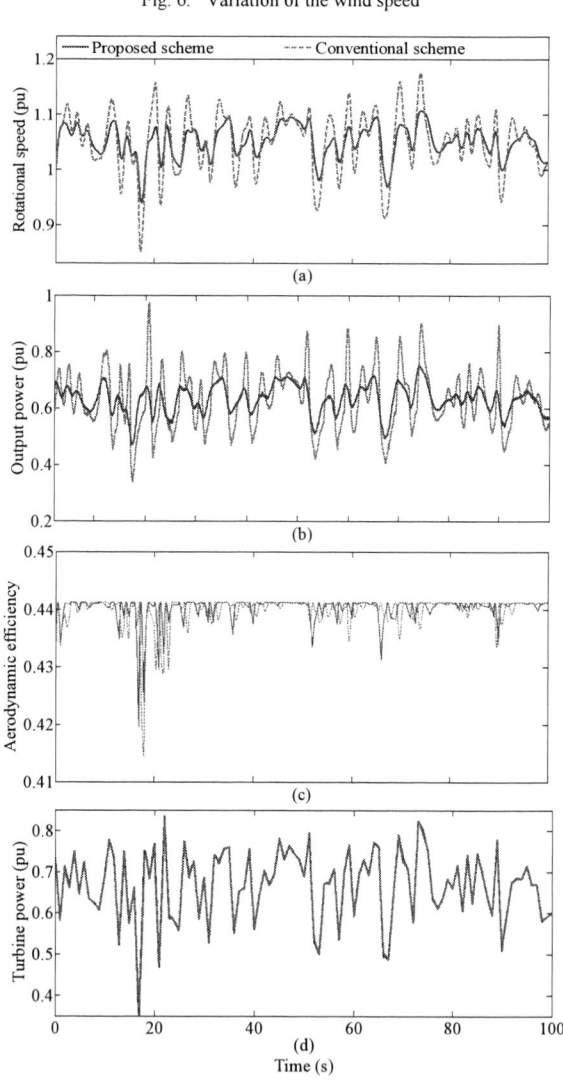

Fig. 7. Result of the first: case a) rotational speed, b) Ouput power, c) aerodynamic efficiency and d) turbine mechanical power

Comparing the standard deviation of the output electrical power with the input mechanical power reveals that the proposed scheme mitigates the fluctuation of power. While, the conventional scheme deteriorates the quality of the output power and imposes additional fluctuation in the power.

TABLE I.   COMPARISON BETWEEN PROPOSED AND CONVENTIONAL METHOD

	Proposed method		Conventional method	
	Average	Standard deviation	Average	Standard deviation
p_{out}	0.105	0.63	0.05	0.63
ω_r	0.053	1.05	0.028	1.05
C_p	0.002	0.44	0.002	0.44
P_t	0.072	0.67	0.072	0.67

VII. CONCLUSION

In this paper, due to the advantages of the MC, the PMSG based-VSWT has been connected to the gird through the MC. Controlling the rotational speed and therefore, tracking the MPP is achieved by appropriate control of the MC. As it has been shown, using the conventional scheme to track the optimal rotational speed causes additional fluctuation in the output power. In the proposed scheme, a low pass filter has been used for mitigating the variation of the reference rotational speed. Consequently, the fluctuations of output power have been significantly reduced while the efficiency has been preserved. Furthermore, the proposed scheme has very simple structure.

VIII. REFRENCES

[1] Jan Machowski, Janusz W.Bialek and James R.Bumby, "Power System Dynamics Stability and control", 2rd ed.,WILEY 2008.

[2] Chinchilla, M., Arnaltes, S.: 'Control of permanent magnet generators applied to variable speed wind energy systems connected to grid', IEEE Trans. Energy Convers., 2006, 21, (1), pp. 130–135

[3] Nasr El-Khoury, C., Kanaan, H. Y., Mougharbel, I. (2014). A review of modulation and control strategies for matrix converters applied to PMSG based wind energy conversion systems. Paper presented at the Industrial Electronics (ISIE), 2014 IEEE 23rd International Symposium on

[4] Enamul Haque, Md., Negnevitsky, M.: 'A novel control strategy for a variable speed wind turbine with a permanent magnet synchronous generator', IEEE Trans. Ind. App., 2010, 46, (1), pp. 331–339

[5] Kumar V., Joshi R.R., and Bansal R.C, 'Optimal Control of Matrix Converter Based WECS for Performance Enhancement and Effuciency Optimization', IEEE Trans. Energy. Convers, vol. 24,no. 1, pp. 264-713, march 2009

[6] Friedli, T., Kolar, J.W., Rodriguez, J., Wheeler, P.W.: 'Comparative evaluation of three phase AC-AC matrix converter and voltage DC link Back-to- Back converter systems', IEEE Trans. Ind. Electron., 2012, 59, (12), pp. 4487–4510

[7] Wheeler, P.W., Clare, J., Weinstein, A.: 'Matrix converters: a technology review', IEEE Trans. Ind. Electron., 2002, 49, (2), pp. 276–287

[8] Bhadra, S.N., Kastha, D., Banerjee, S.: 'Wind electrical systems' (Oxford University Press, 2005)

[9] Kesraoui, M., Korichi, N., Belkadi, A.: 'Maximum power point tracker of wind energy Conversion system', Elsevier – Renew. Energy, 2011, 36, (10), pp. 2655–2662

[10] Qiao W., 'Dynamic Modeling and Control of Doubly Fed Induction Generators Driven by Wind Turbines', in Proc. IEEE PESC conf, pp. 1-8, 2009

[11] H. Hojabri, H. Mokhtari and L. Chang, "Reactive power control of permanent-magnet synchronous wind generator with matrix converter", IEEE Trans. Power Delivery., vol. 28, no. 2, pp. 575- 584, April 2013

[12] Tazil M., Kumar V., Bansal R.C., Kong S., Dong Z.Y., and Freiyas W., 'Three phase doubly fed induction generators: an overview' ,IET Electr. Power Appl., vol.4, no.2, pp.75-89, 2010.

[13] Kumar V., Joshi R.R., and Bansal R.C., 'Optimal Control of Matrix-Converter-Based WECS for Performance Enhancement and Efficiency Optimization' ,IEEE Trans. Energy convers., vol.24, no.1, pp.264-713, march 2009.

[14] H. Camblong, I. M. de Alegria, M. Rodriguez, and G. Abad, "Experimental evaluation of wind turbines maximum power point tracking controllers," Energy Convers. Manag., vol. 47, nos. 18–19, pp. 2846–2858, Nov. 2006

[15] R. W. Delmerico, P. L. Jansen, R. Qu, Z. Tan, C. Wang, and X. Yuan, "System and method for controlling torque ripples in synchronous machines," U.S. Patent 7 847 526, Dec. 7, 2010.

[16] R. W. Delmerico and E. V. Lersen, "HVDC connection of wind turbine," U.S. Patent 2 416 466, Feb. 8, 2012.

IX. APPENDIX

Parameters of the grid, wind turbine and PMSG are given in Table II.

TABLE II.   PARAMETER OF STUDIED VSWT

GENERATOR	$S = 1.5$ MW, $V = 575$ v, $P = 24$, $R_s = 0.06\ \Omega$, $L_d = L_q = 0.3$ mH, $J = 35000$ kg.m^2, $\psi_{pm1} = 1.48$ V.s
TURBINE	$P_m = 1.5$ Mw, $C_p^{max} = 0.44$, $\lambda_{opt} = 6.91$, $K_{opt} = 0.5753$, $v_{nom} = 11.5$ m/s
INPUT FILTER	$R = 0.003$ pu, $L = 0.3$ pu

978-1-5090-0376-1/16 $31.00 © 2016 IEEE

7th Power Electronics, Drive Systems & Technologies Conference (PEDSTC 2016)
16-18 Feb. 2016, Iran University of Science and Technology, Tehran, Iran

Control of Storage System in Series Collection Grid in Offshore Wind Farm for Limiting DC/DC Converter Overvoltaging

Diana Flórez Rodríguez *, Ehsan Enferad[†] and Christophe Saudemont*

* HEI, L2EP Laboratory, Lille, France

[†] Université Lille 1, L2EP Laboratory, Lille, France

diana.florez@hei.fr, ehsan.enferad@etudiant.univ-lille1.fr, christophe.saudemont@hei.fr

Abstract—DC series collection system is a low cost collection grid for integrating offshore wind power to the HVDC grid. The main advantage is to step-up the MVDC voltage level to HVDC level without using extra equipment. However, an effective control system to enable extract and deliver the fluctuation wind power to series collection grid. Previously, a control system has been proposed by authors to this aim, while the propose control system causes overvoltage on DC/DC converter of wind units. In this study a solution is investigated to this limitation. In this paper, utilization of storage system has been considered as a solution for avoiding overvoltage on DC/DC converter. To this aim, the control system is developed for both managing storage system and delivering fluctuating power of wind unit to series collection grid.

Index Terms—DC Series Collection System, Energetic Macro-scopic Representation (EMR), Inversion Based Control, HVDC, Offshore Wind Farm, Storage System.

I. INTRODUCTION

Offshore wind power is a fast growing renewable energy form, which will have significant contribution in grid mix in coming future. The power generation capacity of offshore wind power is expected to reach 150 GW by 2030 which will be 14% of the demand by the time [1][2]. However the higher fed of offshore wind power to the grid is important in term of fossil fuel consumption reduction and environmental aspects, this issue will be challenging in point of view of grid management and its stabilization. The problem is due to the fluctuating nature of power generated by the offshore wind (the well-known phenomena for all renewable sources) .

Nowadays, the wind power plants can only deliver a small share of their installed capacity. The mains reason is the grid stability, since power supply and demand have to be equilibrium at any moment in the grid and the technical limitations, like as grid overload or grid instabilities, which restricts the fed of offshore wind power to the network. In order to higher injection of the offshore wind power to the grid, a technical equipment or technical method is rudimentary to fed offshore wind power to grid with capability of smoothing power fluctuations. One technical solution for this challenge is utilization of storage systems [3].

To date, different storage system with different physical principal has been proposed for integrating with offshore wind power or other renewable plants [4] [5]. The working

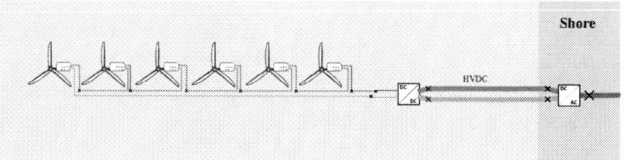

Figure 1. Overall Schematics of Series Collection Grid

principals are mostly on mechanical, electrical, chemical, potential storages and storage technologies differ in conversion efficiency, storage duration, power density, installed location and condition [6] [7]. One of well-known storages systems are Pumped-Storage Hydro Power with conversion efficiency of the 80% (based on potential principal). Well-known mechanical storage is flywheel with almost no power losses with storage volume of 5-100 kW/h and limited storage time. Batteries are another example which have diverse technologies and the efficiency varies between 70-95% depending the technology and discharging capacity is 10-20 MW. Hydrogen storage is another technology which store energy in compressed hydrogen form by the electrolysis phenomena [8]. The stored hydrogen can be re-electrified or directly market, while the efficiency of re-electrifying is very low near 30%.

The aim of this paper is not to choose the proper storage system for offshore wind farm but to investigate the technical advantages of the integrating storage system for the offshore wind power collection grid. The so called Series collection grid has been considered for this study (Fig. 1) [9]. The Series collection grid is an offshore grid by which several wind turbines are connected to each other in series branch, in order to increase voltage level. The Series collection grid functions based on DC voltage form of electrical power and the offshore power is delivered to the onshore grid via HVDC (High Voltage Direct Current) transmission lines. The overall control system for controlling series collection grid has been studied by authors previously. This study focuses on using storage system in each wind unit where the storage is connected to collection grid via a DC/DC converter.

The objective of this paper is to use storage system for limiting overvoltage of DC/DC converter of wind turbine during power oscillation in the wind power generation. Con-

978-1-5090-0376-1/16 $31.00 © 2016 IEEE   378

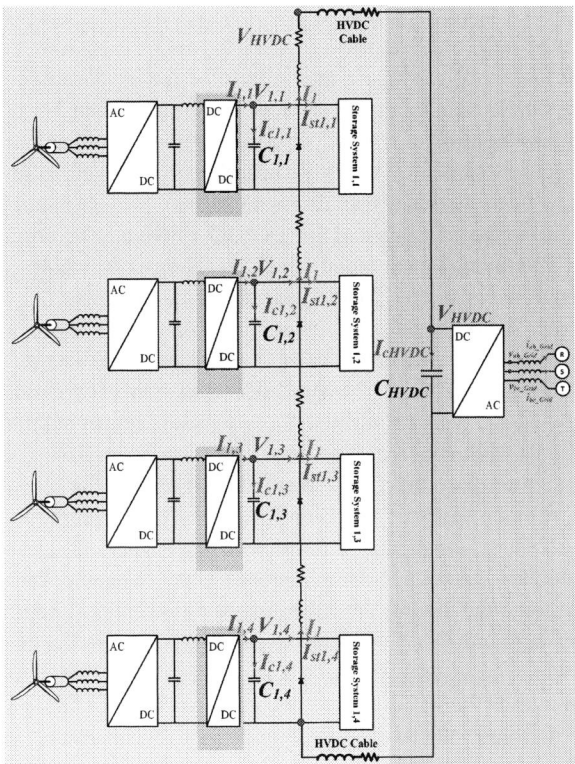

Figure 2. Dynamic Model of Monopolar Series Collection System

sequently, the control system of the series collection grid has been develop to manage overvoltage in collection grid by storing energy. Energetic Macroscopic Representation (EMR) has been used to model the series collection grid and its element. Also, Inversion Based Control method has been used to develop control system of the power system. According to developed model and control system, some simulation tests have been done in Matlab/Simulink. The results show to well performance of control system and storage system in limiting voltage of DC/DC converter.

II. SERIES COLLECTION GRID AND STORAGE SYSTEM

A. Series collection grid and its elements

In order to develop investigate the series collection system, the system shown in Fig. 2 is modelled dynamically. As seen in the figure, the output capacitor of each wind unit corresponds to the current source DC-DC converter. The HVDC cable is modelled as a series inductor and resistor inside the series collection system (The capacitive behavior of the HVDC cable is negligible in when compared to the output capacitor of DC-DC converter). In Fig. 2, the L_{ij} represents both the smoothing reactor and the inductance of the cable for the ij^{th} wind turbine, the R_{ij} represents the damping resistor and the resistance of the cable connected to the ij^{th} wind turbine. The C_{ij} is the equivalent capacitance of current source DC-DC converter seen from the output of the the ij^{th} wind turbine's

converter. For simplicity the wind turbine is considered as an ideal current source whose current value I_{ij} is defined as;

$$I_{i,j} = \frac{P_{i,j}}{V_{i,j}} \quad (1)$$

where, $P_{i,j}$ represents the power of the ij^{th} wind unit, and $V_{i,j}$ represents the output voltage of the DC-DC converter for the same wind unit.

The main challenge in connecting wind units into series collection grid is the fluctuating output power of the wind turbines. This is explained by the fact that, in a series collection grid the current of the series branch should be the same for all the wind units. Moreover the output voltage of the series branch should respect the HVDC link voltage in order to avoid any disturbance of the HVDC voltage balance. As a consequence of having several wind turbines with possibly different power levels, integration and injection of wind power is a challenge merged with series collection grid.

Concerning the power control in series branches, in a previous study the auther presented a control method for extracting fluctuating power of the wind units to be injected into the series branch by taking into account the two aforementioned constrains the unique current in the whole series branch and to comply the HVDC voltage. The control aim was to control the series branch current in a way that the series branch voltage keeps the desired HVDC voltage level, while the voltage of each wind unit fluctuates according to its power level. In other words, in a series grid, the wind unit that had more power production (in comparison to other wind units in the same series branch) had more contribution on the series branch voltage and each wind unit with lower power production had lower voltage at the output.

The drawback of previous study was overvoltage of the DC/DC converter. As power production of any wind unit increases suddenly in series collection grid, its voltage contribution was increasing. Adding this fact that the DC/DC converter of wind unit has its own limit in tolerating overvoltage, consequently, a proper solution is presented in this paper by introducing utilizing storage system. The objective is to develop previously introduced control system for injecting variable power of wind units to collection grid as well as limiting the voltage variation of each wind unit by proper controlling of storage system in collection grid.

B. Storage system of wind units

In this study, the storage system has been considered in the output of the each wind unit. As it is seen in Fig. 2, the storage system are connected to Medium Voltage DC grid. The connection of storage system connected to the collection grid can have other utilizations like as managing unbalanced between power collection grid and onshore grid or frequency support of onshore grid. While in this study just effect of the storage on limiting over voltage of wind units is studied. The storge system can be super capacitor, battery or combination of them and use of a DC/DC converter maybe inevitable to connect each storage to collection grid. The objective in this

978-1-5090-0376-1/16 $31.00 © 2016 IEEE

study is to control the collection grid in a way that the required storage capacity be as small as possible.

C. EMR of system and its control system

In order to carry out the control system of series collection system, the series collection grid and storage systems are modelled according to the Energetic Macroscopic Representation (EMR) in this section. The EMR is an effective modeling tool to model and carry out the controller of the complex energetic systems [10] [11]. Since in series collection grid there are multiple wind units which are multiple input of the energetic system of series branch, the EMR facilitates to graphical inspection of the system and aid in finding out the control path. Fig. 3 shows the EMR of series branch when connected to the onshore grid Fig. 2. For the sake of simplicity, only four wind units have been considered in the series branch and modeled in EMR.

Fig. 4 shows the control system developed by inversion based control and EMR of the system in Fig. 3. The objective of the proposed control method is the output voltage control of each wind unit with appropriate tuning of the output current of the onshore DC/AC converter meanwhile when the voltage limit is achieved by wind unit the power is directed to storage system of that wind unit (for avoiding overvoltage on wind unit power converter). The tuning path for satisfying this control condition is highlighted on Fig. 3 by green color. According to the tuning path and aforementioned control strategy, the maximal control of path of one series branch is achieved according to the EMR and Inversion-Based Control concept (see Fig. 3 blue pictograms).

The realization of this control system uses PI controllers as shown in Fig. 4. There are three controller levels for the system, a controller is used to control the voltage of each wind unit output capacitor in series branch ($V_{C_{i,n}}$) (four controllers for four series wind units), a controller for current of smoothing reactor and inductance of HVDC cable (I_{L_i}) , and one to control the voltage in input capacitance of DC/AC onshore inverter ($V_{C_{HVDC}}$). Since the case of power variation in each wind units is a target of this study, thus the maximal control is mandatory and maximal measurement is necessary. A comparator always compares the output voltage of each wind unit with the DC/DC converter voltage limit, if the voltage has not exceed the limit, the control is done by just acting on onshore converter (as aforementioned). In case the voltage of wind unit reaches to voltage limit, the control loop for storage system is activated and the power flow to storage instead of series collection grid in order to not to exceed voltage limit of power converter.

As seen in Fig. 4 to control the series branch one strategy block is necessary. The strategy block receives the voltage and current of each wind unit, the series branch current and HVDC voltage. The outputs of the strategy block are the reference values of the voltages for each wind units, as shown in Fig. 3 and described in Eq. ??.

III. SIMULATION RESULTS AND DISCUSSIONS

In order to investigate the validity of the proposed control method. A simulation bench has been carried out by using Matlab Simulink in which four wind units with nominal power generation capability of 5MW have been connected in series. It has been assumed that the wind turbine is well controlled by an active rectifier and its DC/DC converter for ensuring the Maximum Power Point Tracking. Hence the whole wind turbine is considered as an ideal current source controlled by power. Also, the DC/AC inverter is considered as an ideal current source with input capacitor and output inductor. The storage system for each wind unit has been carried out by using Nickel Metal Hydride battery which is connected the via ideal DC/DC converter to series collection grid. The system simulations have been implemented according to Fig. 2 and its control loop designed based on Inversion-Based control in Fig. 4. All the accomplished controllers are PI based. The cables, smoothing reactor and damping resistor have been modeled according to the models described in previous sections. The tests have been designed concerning power generation fluctuation of wind units series connected and injection of this variable power into the series collection grid, where the extra power of each unit is injected to the collection by increasing units voltage level. It has been assumed that the DC/DC converter of wind units has overvoltage capability of 5% of nominal voltage value. The overvoltage effect of DC/DC converter for wind units is compared for both with and without storage system.

For investigation of advantage of storage system in series collection grid, a simulation has been done for the case of different power level of wind turbines. In fact this phenomenon occurs during wind speed variation when extracted power by wind turbines becomes different. For this case a simulation bench is arranged so the wind units $WU_{1,1}$, $WU_{1,2}$, $WU_{1,3}$ and $WU_{1,4}$ were supposed to operate initially at 80 percent of their nominal value (0.2 per unit), then due to the increase of the wind speed the power generation of all wind units increases to their nominal value (0.25 per unit). Since the offshore wind units are physically far from each other they experience power variation with a time delay. As seen in Fig. 5.a. the generated power varies initially on $WU_{1,1}$ (Fig. 5.a solid blue line), after a time span on $WU_{1,2}$ (Fig. 5.a doted blue line), then on $WU_{1,3}$ (Fig. 5.a dashed blue line), and finally on $WU_{1,4}$ (Fig. 5.a dot dashed blue line).

The initial reaction for power variation of each wind turbine comes from onshore converter which is commanded by the proposed strategy level of proposed control method. Regarding Fig. 6.a, during time interval of 4-7 increase of the first wind unit power the series branch current (solid blue line) increases due to the controlling onshore converter. The delivered current of the $WU_{1,1}$ is higher than the series branch (solid black line) current during this interval. This extra power is injected to the capacitor of $WU_{1,1}$ (Fig. 6.b blue solid line) and causes the voltage of $WU_{1,1}$ increase in the series branch (Fig. 7.a (blue solid line). When the voltage of $WU_{1,1}$ ($V_{1,1}$) reaches to the DC/DC converter voltage limit ($V_{DC/DC_{Limit}}$), the extra

978-1-5090-0376-1/16 $31.00 © 2016 IEEE

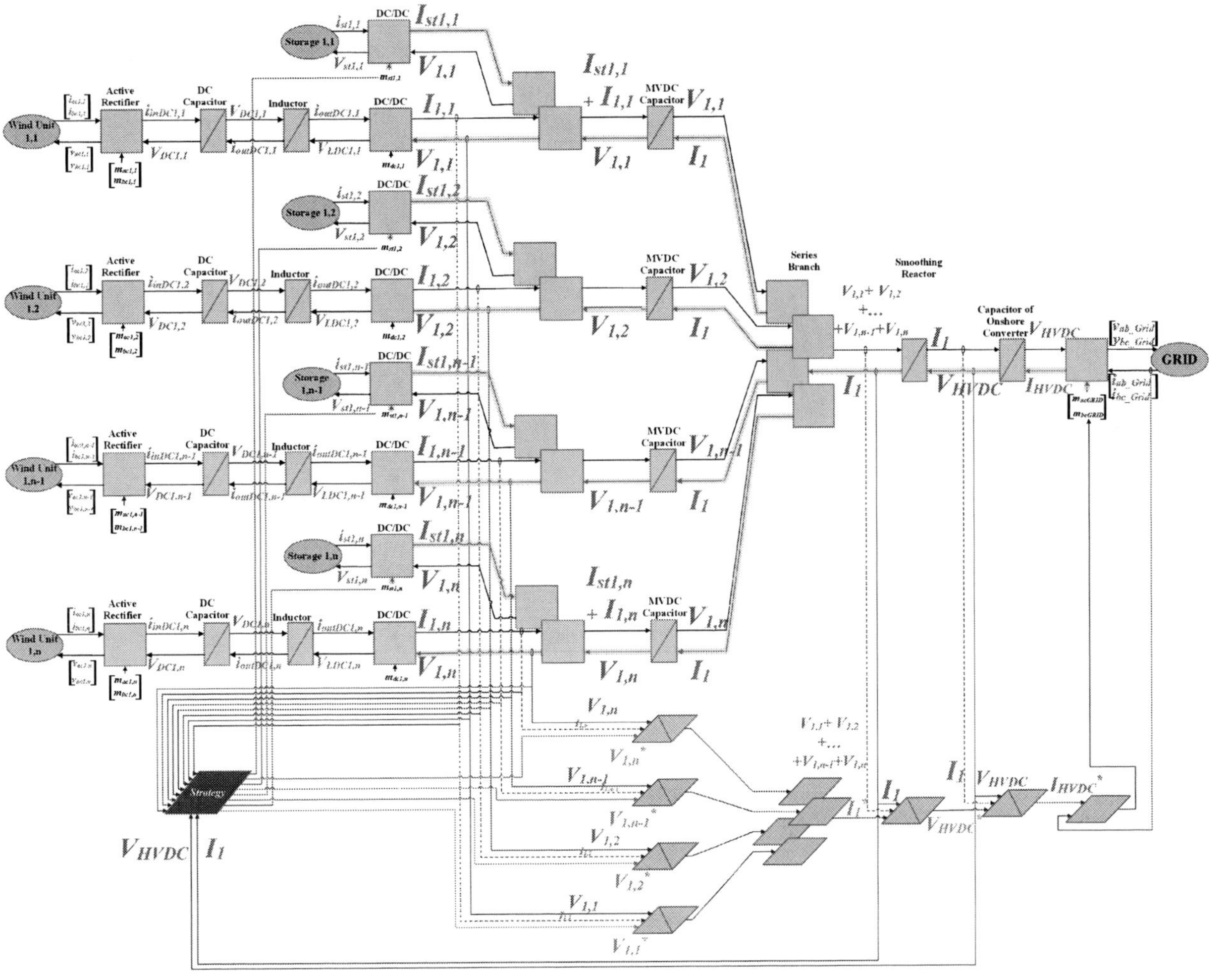

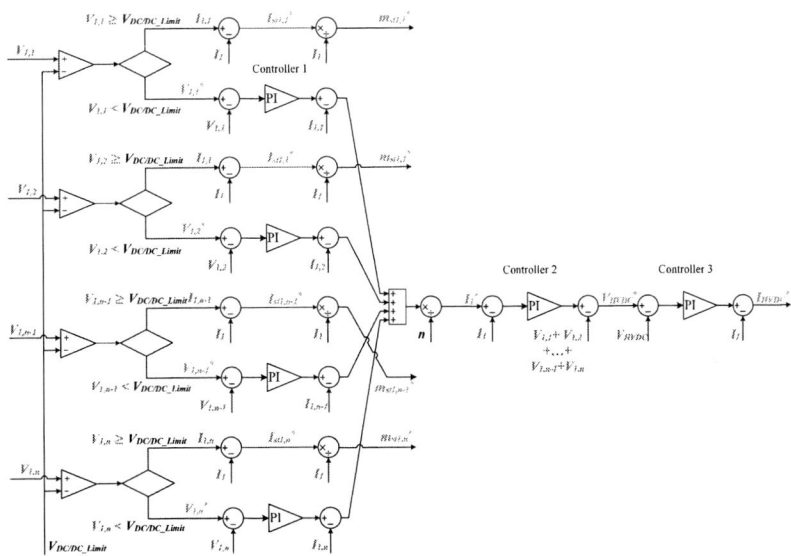

Figure 3. EMR of wind units and their storage system connected to series collection grid and their Inversion Based Control

Figure 4. Controller Block Diagram of Series Collection Grid with Storage

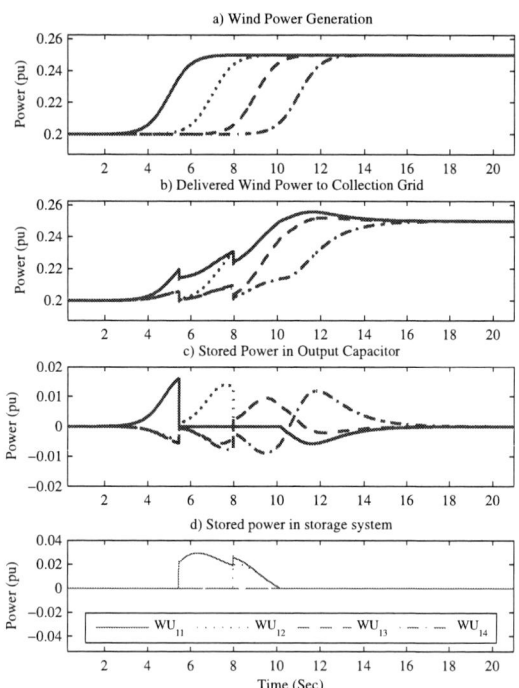

Figure 5. Power plot of series connected wind turbines equipped with storage system during wind speed increase **a)** Generated wind power by each wind turbine **b)** Delivered wind power to series branch **c)** Delivered Power to the output capacitor of each wind unit **d)** Stored power in storage of each wind unit

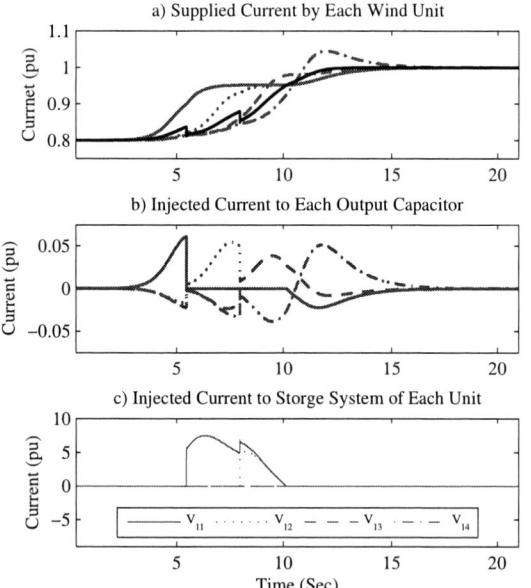

Figure 6. Current Plot **a)** Current of each wind unit and series branch current (solid black line) during power wind speed increase **b)** Current flow inside each output capacitor

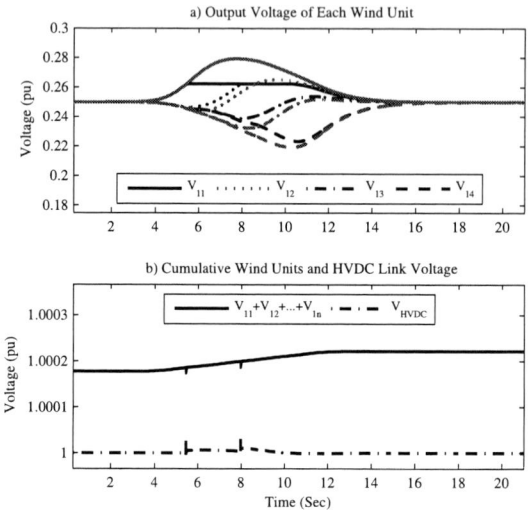

Figure 7. Voltage plot of series connected wind turbines during wind speed increase **a)**Output voltage of each wind unit in collection grid (black figure are for the case of using storage system and the blue ones are for the case of without using storage) **b)** HVDC link voltage and cumulative voltage of wind units

current of wind unit $WU_{1,1}$ is directed to the storage of that unit ($I_{st1,1}$). In this simulation case this event happens in the time 5.5 sec for $WU_{1,1}$. As it is illustrated in Fig. 5.c solid blue line, by increase of power generation in this unit, a part of this extra power is delivered power to the capacitor $C_{1,1}$. At the time 5.5 second, despite the power generation is still increasing, the power delivery for $C_{1,1}$ is stopped because the voltage limit has been achieved. After this moment the extra power is delivered to the storage system of $WU_{1,1}$, the solid blue line in Fig. 5.d clarifies this point. It should be highlighted that at the same time interval, when the voltage of the wind unit $WU_{1,1}$ is increasing, the delivered current by wind units $WU_{1,2}$, $WU_{1,3}$ and $WU_{1,4}$ are lower than the series branch current (controlled current by onshore converter) as seen in Fig. 5.a. In these wind units, lacking current is supplied by the capacitors of the mentioned units Fig.5.b and it results in a voltage reduction of these wind units outputs as illustrated in Fig. 7.b.

The same phenomenon occurs sequently in later time intervals on second wind unit $WU_{1,2}$. While the event is a bit different for wind units $WU_{1,3}$ and $WU_{1,4}$. In this simulation case this two wind unit do not store any energy in their storages. The reason is due to increase of power from $WU_{1,1}$ to $WU_{1,4}$, when the power generation of $WU_{1,3}$ increases the its voltage is very lower than wind units $WU_{1,1}$ and $WU_{1,2}$. Consequently the power generation increase cause this unit return back to its equilibrium voltage level in the collection grid and never reaches to DC/DC voltage limit. So that this unit never store energy in its storage in this simulation case. The same explanation is valid for the $WU_{1,4}$.

The interesting feature is the well behavior of the control system in keeping the HVDC voltage in balanced condition

978-1-5090-0376-1/16 $31.00 © 2016 IEEE

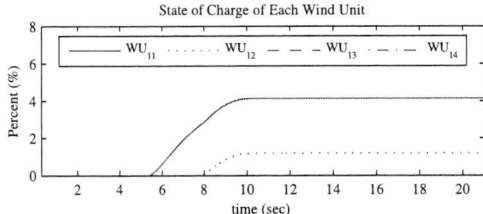

Figure 8. State of Charge (SOC) in the storages of the wind units

as in Fig. 7.b (dot dashed line). The solid line in the same figure shows the increased voltage of all wind units due to increased power production increasing. According to the results presented in here, it can strongly be concluded that the proposed control method allows the maximum power point tracking of each wind turbine inside the series connected grid with respecting the voltage limitation on DC/DC converter.

According to Fig. 7.a, comparison can be made between the control system of series collection control without storage (blue curves) and with storage system (black curve). As it is apparent in this figure, the use of storage system and its control system well limits the over violating of DC/DC converter meanwhile it manages to deliver generated power to collection grid. Moreover, use of storage system reduces the amplitude of voltage variation between wind units during power increase. It can be concluded from smaller difference between maximum and minimum voltage of wind units with storage (black curve) in comparison to wind units without storage (blue curves). Last not the least, Fig. 8 shows the state of charge of storage system for each wind units (the storage capacity of batteries are 12 kwh). As it is shown in this figure and as it was described earlier, there is stored energy in storage of $WU_{1,1}$ and $WU_{1,2}$. While the storages of $WU_{1,3}$ and $WU_{1,3}$ is not used in this simulation case.

IV. CONCLUSION

In this paper, the use of storage system has been proposed to limit the voltage variation on the DC/DC converter. Current control of series branch has been proposed by authors to feasible Maximum Power Point Tracking of wind units and delivering power to series collection grid. The technic was very useful while it cause overvoltage on DC/DC converter. The use of storage system and development was one of the possibility to solve the issue for series collection grid. The simulation result proves that this approach is effective to limit the overvoltage of DC/DC converter. It should be highlighted since this storage system is used for a very short time, hence its size is very small comparing to full power generation capacity of wind units. The existence of storage system and its management in collection grid of offshore wind farm can be used to other application as well.

REFERENCES

[1] Roadmap for maritime spatial planning: Achieving common principles in the eu. Technical report, Commission of the European, Brussel, 2008.

[2] Jie Wu, Zhi-Xin Wang, Lie Xu, and Guo-Qiang Wang. Key technologies of VSC-HVDC and its application on offshore wind farm in china. *Renewable and Sustainable Energy Reviews*, 36:247–255, August 2014.

[3] Lingfeng Wang, J.-Y. Yu, and Y.-T. Chen. Dynamic stability improvement of an integrated offshore wind and marine-current farm using a flywheel energy-storage system. *Renewable Power Generation, IET*, 5(5):387–396, September 2011.

[4] Alexander H. Slocum. Symbiotic offshore energy harvesting and storage systems. *Sustainable Energy Technologies and Assessments*, 11:135–141, September 2015.

[5] Yajun Fan, Anle Mu, and Tao Ma. Study on the application of energy storage system in offshore wind turbine with hydraulic transmission. *Energy Conversion and Management*, 110:338–346, February 2016.

[6] Haoran Zhao, Qiuwei Wu, Shuju Hu, Honghua Xu, and Claus Nygaard Rasmussen. Review of energy storage system for wind power integration support. *Applied Energy*, 137:545–553, January 2015.

[7] A. H. Slocum, G. E. Fennell, Gunhan Dundar, B. G. Hodder, J. D. C. Meredith, and M. A. Sager. Ocean Renewable Energy Storage (ORES) System: Analysis of an Undersea Energy Storage Concept. *Proceedings of the IEEE*, 101(4):906–924, April 2013.

[8] Daniel Kroniger and Reinhard Madlener. Hydrogen storage for wind parks: A real options evaluation for an optimal investment in more flexibility. *Applied Energy*, 136:931–946, December 2014.

[9] E. Veilleux and P.W. Lehn. Interconnection of Direct-Drive Wind Turbines Using a Series-Connected DC Grid. *IEEE Transactions on Sustainable Energy*, 5(1):139–147, January 2014.

[10] A. Bouscayrol, P. Delarue, F. Giraud, X. Guillaud, X. Kestelyn, B. Lemaire-Semail, and W. Lhomme. Teaching drive control using energetic macroscopic representation-expert level. In *Power Electronics and Applications, 2009. EPE '09. 13th European Conference on*, pages 1–9, Sept 2009.

[11] A. Bouscayrol, P. Delarue, X. Guillaud, W. Lhomme, and B. Lemaire-Semail. Simulation of a wind energy conversion system using energetic macroscopic representation. In *Power Electronics and Motion Control Conference (EPE/PEMC), 2012 15th International*, pages DS3e.8-1–DS3e.8-6, Sept 2012.

978-1-5090-0376-1/16 $31.00 © 2016 IEEE   383

7th Power Electronics, Drive Systems & Technologies Conference (PEDSTC 2016)
16-18 Feb. 2016, Iran University of Science and Technology, Tehran, Iran

Analysis and Control of Single-Phase Converters for Integration of Small-Scaled Renewable Energy Sources into the Power Grid

Majid Mehrasa
Babol (Noshirvani) Univ.
of Technology, Babol,
Iran

Mohamad Rezanejhad
Mazandaran Univ. of
Science and Tech., Babol,
Iran

Edris Pouresmaeil and João P. S. Catalão
FEUP, Porto, UBI, Covilha, and
INESC-ID, IST, Univ. Lisbon, Lisbon,
Portugal

Sasan Zabihi
ABB,
Darwin,
Australia

Abstract—**A comprehensive dynamic model based on Direct-Quadrature (DQ) rotating frame is proposed in this paper that is used along with a capability curve (CC) based on the active and reactive power to control a grid-connected single-phase voltage-source inverter (SPVSI). With the proposed dynamic model, a droop-passivity based controller can be designed for the grid-connected inverter in the presence of nonlinear loads. Stability analysis of the proposed control technique is also discussed in the paper as well as design principles. Moreover, an accurate performance area of SPVSI active and reactive power in dynamic transitions is achieved using the CC. Furthermore, an effective harmonic compensation scheme along with a proper active and reactive power sharing algorithm are performed by a well-designed reference waveform generation process. Performance of the grid-connected SPVSI, under the proposed controller, is thoroughly evaluated in the Matlab/Simulink environment.**

Keywords— grid-connected; single-phase voltage-source inverter; passivity based controller; Droop Control; active and reactive power sharing.

I. INTRODUCTION

Integration of renewable energy resources into power grid via power electronic interfaced-converters has been widely investigated by numerous literatures from a diverse viewpoints including accurate performance of active and reactive power sharing, harmonic compensation, unity power factor achievement and so on [1-2]. Single-phase voltage-source inverter (SPVSI) has been employed in distributed generation and grid-connected applications as an applicable solution. In these applications, a properly obtained dynamic model and a well-designed control technique are substantial parts of SPVSI use process. In [3], a simple control method is proposed for the purpose of stabilization of a grid-connected inverter in both grid-connected and stand-alone modes, whereas in another study, harmonic impacts of nonlinear loads on grid currents are eliminated by adding the load current into the filter inductor current loop [4]. In [5], a synchronous reference frame PI controller (SRFPI) and a multi-resonant harmonic compensator are utilized to regulate the output current with zero steady state error and to provide an efficient attenuation of the grid voltage harmonic distortion respectively. Also, an average power controller is used to control the active and reactive power sharing [5].

For compensating current harmonics and electromagnetic interferences, a current-mode asynchronous sigma-delta modulation (CASDM) is employed for single phase grid-tied photovoltaic (PV) inverters [5]. Also, for an effective control of reactive power, a cost-effective micro-controller is proposed [5]. An adaptive control technique is proposed in [6] for single-phase grid-connected PV inverters to compensate lower order harmonics introduced by the core saturation induced distorted magnetizing current of the transformer and the dead-time of the inverter and so on. Also, a Proportional-Resonant-Integral (PRI) controller is used to eliminate the dc component in the control system.

In this paper, a comprehensive Direct-Quadrature (DQ) rotating frame based dynamic model is proposed for a SPVSI. Then, a droop-passivity [7-8] based controller is used considering the harmonic effects of nonlinear loads. The complete design process and the system stability analysis are discussed in detailed. A CC is proposed to assess SPVSI active and reactive power areas in dynamic operation conditions. Also a well-designed reference generation process is utilized for the power sharing purposes. Finally Matlab/Simulink environment is employed to verify the validity of the proposed controller.

II. DYNAMIC MODEL OF GRID-CONNECTED SPVSI

The proposed grid-connected SPVSI is illustrated in Fig.1. The inverter consists of a dc link, and the output resistance and inductance of R_c and L_c respectively. Utility is connected to the point of common coupling (PCC) as depicted in Fig.1. In addition, a nonlinear load draws harmonic current from grid that should be compensated by SPVSI [9-11]. Renewable energy sources of PV and wind turbine system (WTS) utilize dc-dc and ac-dc converters respectively to generate a dc voltage. As shown in Fig.1, i_{dc} and C_{dc} are the dc-link current and the output capacitor, respectively. According to Fig.1, the mathematical model of a grid-connected SPVSI is obtained as,

$$L_c \frac{di_c}{dt} + R_c i_c + v_c + v_{dc} u_c = 0$$

$$C_{dc} \frac{dv_{dc}}{dt} = i_c u_c + i_{dc}$$

(1)

978-1-5090-0376-1/16 $31.00 © 2016 IEEE

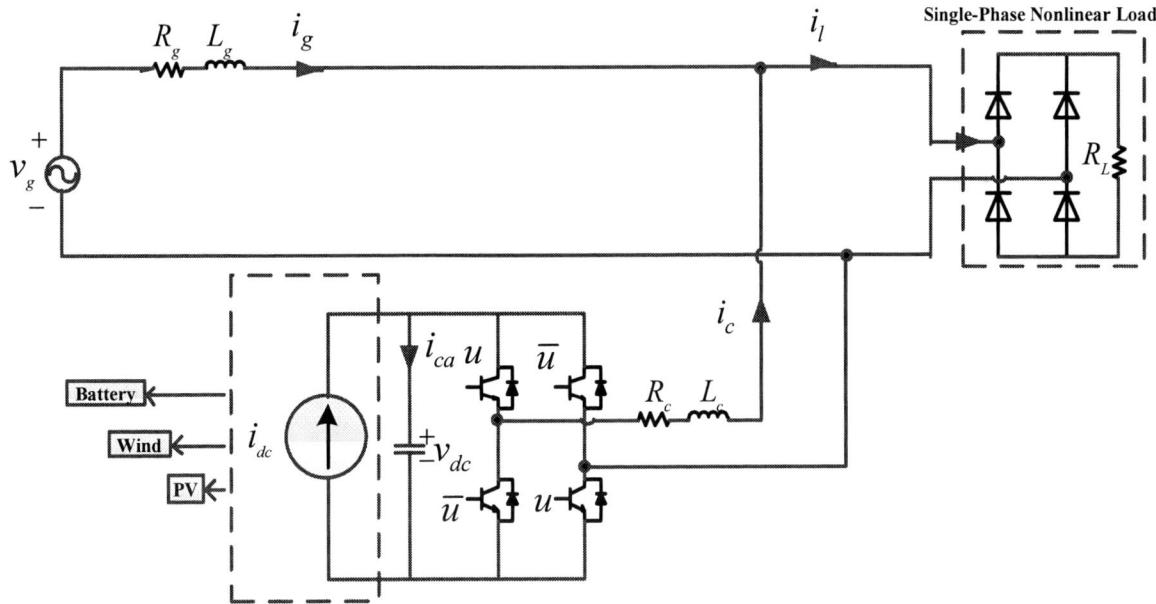

Fig. 1. The proposed grid-connected SPVSI

where i_c and v_c are the output inverter current and PCC voltage, respectively, and u_c is SPVSI switching function. If the real state variable of SPVSI is equal to:

$$x_r = x_m \sin\left(\omega t + \varphi\right) \tag{2}$$

where x_m, is the state variable, φ is the initial phase and ω is the angular frequency. Also, the imaginary state variable of the proposed inverter is:

$$x_i = x_m \cos\left(\omega t + \varphi\right) \tag{3}$$

where $x_{(ir)} \in \left\{i_c, v_{dc}\right\}$. Considering equations (2) and (3), the d-q frame transformation matrix can be written as:

$$\begin{bmatrix} x_d \\ x_q \end{bmatrix} = \begin{bmatrix} \sin\left(\omega t\right) & \cos\left(\omega t\right) \\ \cos\left(\omega t\right) & -\sin\left(\omega t\right) \end{bmatrix} \begin{bmatrix} x_r \\ x_i \end{bmatrix} \tag{4}$$

Using (1), (2) and (3), the real-imaginary coordinate-based dynamic model of SPVSI is derived as,

$$\begin{bmatrix} L_c & 0 & 0 & 0 \\ 0 & L_c & 0 & 0 \\ 0 & 0 & C_{dc} & 0 \\ 0 & 0 & 0 & C_{dc} \end{bmatrix} \begin{bmatrix} \dot{i}_{cr} \\ \dot{i}_{ci} \\ \dot{v}_{dcr} \\ \dot{v}_{dci} \end{bmatrix} + \begin{bmatrix} R_c & 0 & 0 & 0 \\ 0 & R_c & 0 & 0 \\ 0 & 0 & 0 & 0 \\ 0 & 0 & 0 & 0 \end{bmatrix} \begin{bmatrix} i_{cr} \\ i_{ci} \\ v_{dcr} \\ v_{dci} \end{bmatrix}$$
$$+ \begin{bmatrix} 0 & 0 & u_{cr} & 0 \\ 0 & 0 & 0 & u_{ci} \\ -u_{cr} & 0 & 0 & 0 \\ 0 & -u_{ci} & 0 & 0 \end{bmatrix} \begin{bmatrix} i_{cr} \\ i_{ci} \\ v_{dcr} \\ v_{dci} \end{bmatrix} + \begin{bmatrix} 0 \\ 0 \\ i_{dcr} \\ i_{dci} \end{bmatrix} = 0 \tag{5}$$

$v_{dc(ri)}$ and $i_{dc(ir)}$ are the real and imaginary components of SPVSI dc-link voltage and current, respectively.

To maintain the dc link voltage and current at a constant value in both states, therefore $v_{dc(ri)} = v_{dc}$ and $i_{dc(ri)} = i_{dc}$. Therefore, by applying the matrix of (4) to (5), the d-q dynamic model of SPVSI is obtained as,

$$\begin{bmatrix} L_c & 0 & 0 \\ 0 & L_c & 0 \\ 0 & 0 & C_{dc} \end{bmatrix} \begin{bmatrix} \dot{i}_{cd} \\ \dot{i}_{cq} \\ \dot{v}_{dc} \end{bmatrix} + \begin{bmatrix} R_c & 0 & 0 \\ 0 & R_c & 0 \\ 0 & 0 & 0 \end{bmatrix} \begin{bmatrix} i_{cd} \\ i_{cq} \\ v_{dc} \end{bmatrix} + \tag{6}$$
$$\begin{bmatrix} 0 & \omega L_c & 0 \\ -\omega L_c & 0 & 0 \\ 0 & 0 & 0 \end{bmatrix} \begin{bmatrix} i_{cd} \\ i_{cq} \\ v_{dc} \end{bmatrix} +$$
$$\begin{bmatrix} 0 & 0 & u_{cd} \\ 0 & 0 & u_{cq} \\ -u_{cd}/2 & -u_{cq}/2 & 0 \end{bmatrix} \begin{bmatrix} i_{cd} \\ i_{cq} \\ v_{dc} \end{bmatrix} + \begin{bmatrix} v_d \\ v_q \\ -i_{dc} \end{bmatrix} = 0$$

III. PROPOSED CONTROL METHOD

An effective harmonic compensation with unity power factor as well as a good dynamic execution of active and reactive power sharing in the grid-connected SPVSI requires design of an appropriate closed-loop controller with fast dynamic responses. The proposed controller is discussed in details in this section.

A. Current-based closed-loop controller

In this section, a passivity-based controller including injection of damping resistances and shaping energy is used to design proposed control technique. To use the proposed controller for the system model shown in Fig.1, a closed-loop dynamic model based on the error state variables to be achieved, so initially (6) is rewritten as following matrix demonstration:

$$\dot{IX} + RX + WX + UX + S = 0$$

$$I = \begin{bmatrix} L_c & 0 & 0 \\ 0 & L_c & 0 \\ 0 & 0 & C_{dc} \end{bmatrix}, X = \begin{bmatrix} i_{cd} \\ i_{cq} \\ v_{dc} \end{bmatrix}, W = \begin{bmatrix} \omega L_c & 0 & 0 \\ 0 & -\omega L_c & 0 \\ 0 & 0 & 0 \end{bmatrix},$$

$$R = \begin{bmatrix} R_c & 0 & 0 \\ 0 & R_c & 0 \\ 0 & 0 & 0 \end{bmatrix}, U = \begin{bmatrix} 0 & 0 & u_{cd} \\ 0 & 0 & u_{cq} \\ -u_{cd}/2 & -u_{cq}/2 & 0 \end{bmatrix},$$

$$S = \begin{bmatrix} v_d \\ v_q \\ -i_{dc} \end{bmatrix} \tag{7}$$

By defining the error state variables vector as (8),

$$E = X - X^* = \begin{bmatrix} e_1 & e_2 & e_3 \end{bmatrix}^T =$$

$$\begin{bmatrix} i_{cd} - i_{cd}^* & i_{cq} - i_{cq}^* & v_{dc} - v_{dc}^* \end{bmatrix}^T \tag{8}$$

X^* is the reference values vector of the proposed system state variables. Thus, the error closed-loop dynamic model of the proposed SPVSI can be obtained by substituting (8) into (7) as,

$$\dot{IE} + RE + WE + UE = \tag{9}$$

$$-S - \left(\dot{IX}^* + RX^* + WX^* + UX^* \right)$$

In order to have a faster transient response and zero steady state error in different operational conditions, a damping resistance matrix of R_d is added to (9) as,

$$\dot{IE} + \left(R + R_d \right) E + WE + UE = \tag{10}$$

$$-S - \left(\dot{IX}^* + RX^* + WX^* + UX^* - R_d E \right)$$

where R_d is considered as,

$$R_d = \begin{bmatrix} R_{d1} & 0 & 0 \\ 0 & R_{d2} & 0 \\ 0 & 0 & R_{d3}^{-1} \end{bmatrix} \tag{11}$$

As the first step for the design of the proposed controller, the SPVSI output currents and dc-link voltage should track their reference values that lead to $E = 0$. Thus, based on (10), (12) is obtained as,

$$\dot{IE} + \left(R + R_d \right) E + WE + UE = 0 \tag{12}$$

The asymptotical stable operation of the grid-connected SPVSI along with the proposed controller can be proved by the use of system total saved energy function developed based on the error state variables. The function is defined as,

$$H(\overline{e}) = \frac{1}{2} L_c e_1^2 + \frac{1}{2} L_c e_3^2 + \frac{1}{2} C_{dc} e_3^2 \tag{13}$$

According to Lypunov stability theory, the total saved energy of a stable system has to be minimized.

Therefore, by making a derivative of (13) and using (12), the following equation is yield.

$$\dot{H}(\overline{e}) = L_c \dot{e}_1 e_1 + L_c \dot{e}_2 e_2 + C_{dc} \dot{e}_3 e_3 = E^T \dot{IE} = \tag{14}$$

$$-E^T \left(\left(R + R_d \right) E + WE + UE \right)$$

It is noticed that the terms related to the damping resistances in (14) are much larger than other terms. Thus, (14) can be rewritten as,

$$\dot{H}(\overline{e}) = -E^T \left(R + R_d \right) E = -\left(R_c + R_{d1} \right) e_1^2 \tag{15}$$

$$-\left(R_c + R_{d2} \right) e_2^2 - R_{d3}^{-1} e_3^2$$

Equation (15) verifies that the designed controller based on the closed-loop error dynamic model presented in (10) is able to make the proposed system stable and to reach zero errors. Substitution of (12) into (10) results in:

$$U = \left(-S - \dot{IX}^* - RX^* - WX^* + R_d E \right) X^{*-1} \tag{16}$$

Switching functions obtained from (16) are utilized to control the grid-connected SPVSI during various operating conditions.

B. Calculation of SPVSI reference currents

SPVSI is responsible for injecting the required harmonic components of both active and reactive power of nonlinear loads. Also, in order to achieve high quality grid currents and power, the whole reactive power of the load has to be supplied by the SPVSI. Thus, an accurate calculation of SPVSI currents in d-q frame is necessary to meet the addressed targets.

According to Fig.1, the relation between the grid, SPVSI and load currents can be written as,

$$\begin{bmatrix} i_{cd} \\ i_{cq} \end{bmatrix} = \begin{bmatrix} \sin(\omega t) & \cos(\omega t) \\ \cos(\omega t) & -\sin(\omega t) \end{bmatrix} \begin{bmatrix} i_{gr} + i_{lr} \\ i_{gi} + i_{li} \end{bmatrix} = \begin{bmatrix} i_{gd} + i_{ld} \\ i_{gq} + i_{lq} \end{bmatrix} \tag{17}$$

i_g and i_l are the grid and load currents, respectively. Since the load generates harmonic components in the proposed system, the d-q component of load currents can be stated as,

$$i_{ldq} = I_{ldq1} + \sum_{n=2}^{\infty} i_{ldqhn} \tag{18}$$

I_{ldq1} and $\sum_{n=2}^{\infty} i_{ldqhn}$ are the fundamental-frequency and harmonic components of the load current. The grid-connected SPVSI has to supply the whole harmonic parts of the nonlinear load current. Also, the q-component of grid current to be zero in order to achieve the unity power factor, so $i_{gq} = 0$.

Consequently, the SPVSI currents should be equal to,

$$i_{cd} = \alpha I_{ld1} + \sum_{n=2}^{\infty} i_{ldhn} \tag{19}$$

$$i_{cq} = i_{lq}$$

α shows a respective portion of d-component of load current at main frequency to be injected by SPVSI. Thus, grid should generate $(1-\alpha)I_{ld1}$ during controller operation. Considering (19), reference currents of SPVSI can be achieved as,

$$i_{cd}^{*} = \left(i_{cd} - \alpha I_{ld1} - \sum_{n=2}^{\infty} i_{ldhn} \right) \left(k_{pd} + k_{id}/s \right) \quad (20)$$

$$i_{cq}^{*} = \left(i_{cq} - i_{lq} \right) \left(k_{pq} + k_{iq}/s \right)$$

Where k_{pdq} and k_{idq} are the proportional and integral coefficients of the PI controllers used in the reference current calculation process of SPVSI that are effective in the convergence rate of zero state variables errors.

C. Droop controller

A new droop control method is presented in this section for the grid-connected SPVSI which is based on d and q components of its current. This feature is added to the controller in order to complete the passivity-based controller of (16) with an accurate estimation of S vector. Considering conventional droop controller equations as follows,

$$\omega = \omega^{*} - m_{p}P \quad (21)$$

$$E = E^{*} - m_{q}Q$$

where ω^{*} and E^{*} are the desired value of angular frequency and voltage magnitude of the inverter, respectively. m_{p} and m_{q} are also the conventional droop coefficients of frequency and voltage magnitude, respectively. The instantaneous SPVSI active and reactive power with respect to the d-q components of the inverter current and voltage can be written as,

$$p = v_{d}i_{cd} + v_{q}i_{cq} \quad (22)$$

$$q = v_{q}i_{cd} - v_{d}i_{cq}$$

By an acceptable assumption, the active and reactive power of SPVSI at fundamental frequency can be obtained as,

$$P = v_{m}i_{cd1}, \mathrm{Q} = -v_{m}i_{cq1} \quad (23)$$

where v_{m} and i_{cdq1} are the reference values of the inverter voltage magnitude and the fundamental frequency components of SPVSI current, respectively. By substituting (23) into (21), the new droop equation can be obtained as,

$$\omega = \omega^{*} - m_{p}'i_{cd1} \quad (24)$$

$$E = E^{*} + m_{q}'i_{cq1}$$

where $m_{p}' = m_{p}v_{m}$ and $m_{q}' = m_{q}v_{m}$.

IV. CAPABILITY CURVE OF SPVSI

A significant criterion for evaluating a good performance of SPVSI in the grid-connected mode is a proper execution of active and reactive power sharing in presence of dynamic load changes.

In this case, knowing the maximum and minimum active and reactive power of SPVSI can assist the proposed controller to perform in a more effective manner.

The grid-connected SPVSI switching functions can be achieved through the two first terms of (6) as,

$$u_{cd} = -\left(LI_{avcd} + Ri_{cd} + \omega Li_{cq} + v_{d} \right)/v_{dc} \quad (25)$$

$$u_{cq} = -\left(LI_{avcq} + Ri_{cq} - \omega Li_{cd} + v_{q} \right)/v_{dc}$$

By substitution of (25) into the last term of (6) and also after simplifying the equation, (26) can be achieved as,

$$\left(i_{cd} + \frac{LI_{avcd} + v_{d}}{2R} \right)^{2} + \left(i_{cq} + \frac{LI_{avcq} + v_{q}}{2R} \right)^{2} = \quad (26)$$

$$\frac{\left(LI_{avcd} + v_{d} \right)^{2} + \left(LI_{avcq} + v_{q} \right)^{2} + 8Ri_{dc}v_{dc} - 8RC_{dc}\overline{v}_{dc}v_{dc}}{4R^{2}}$$

I_{avdq} and $\overline{v}_{dc}$ are the average values of the inverter currents and dc-link voltages, respectively. By substituting (22) in (25), and with an acceptable approximation and also considering $v_{q} = 0$, the active and reactive power based curve is obtained as,

$$\left(p + \frac{LI_{avcd}v_{d} + v_{d}^{2}}{2R} \right)^{2} + \left(q - \frac{LI_{avcq}v_{d}}{2R} \right)^{2} = \quad (27)$$

$$\frac{\left(LI_{avcd}v_{d} + v_{d}^{2} \right)^{2} + \left(LI_{avcq}v_{d} \right)^{2} + 8Ri_{dc}v_{dc}v_{d}^{2} - 8RC_{dc}\overline{v}_{dc}v_{dc}v_{d}^{2}}{4R^{2}}$$

Equation (27) is plotted in Fig. 2. As evident, the maximum and minimum active and reactive power of SPVSI can be achieved by the means of center coordinates and radius value. It can be understood that the active and reactive power of the inverter are dependent of the dc and ac link parameters of SPVSI.

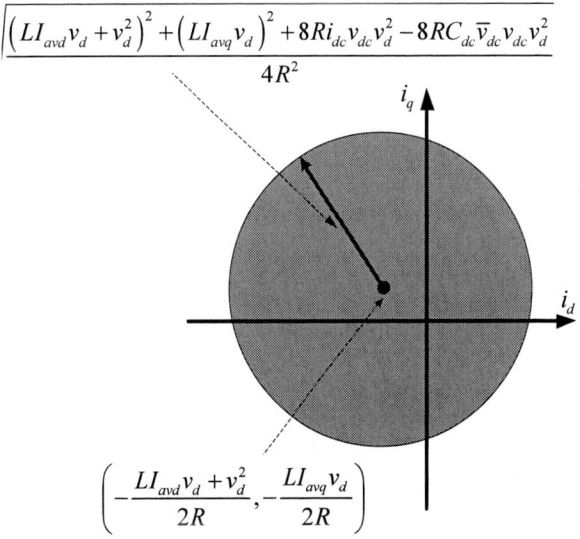

Fig. 2. Capability curve of SPVSI active and reactive power

V. SIMULATION RESULTS

The assessment of the proposed control technique is carried out based on the overall structure presented in Fig.3. The Matlab/Simulink environment is used to verify the excellent performance of the grid-connected SPVSI during various operating conditions. The system parameters are given in Table I.

TABLE I
SIMULATION MODEL PARAMETERS

Parameter	Value
Load of diode rectifier	20+j0.3
dc-link voltage set-point (v_{dc})	200 V
ac voltage	120V
Fundamental frequency	50 Hz
Switching/Sampling frequency	10 kHz
SPVSI resistance	0.1 mΩ
SPVSI inductance	1 mH

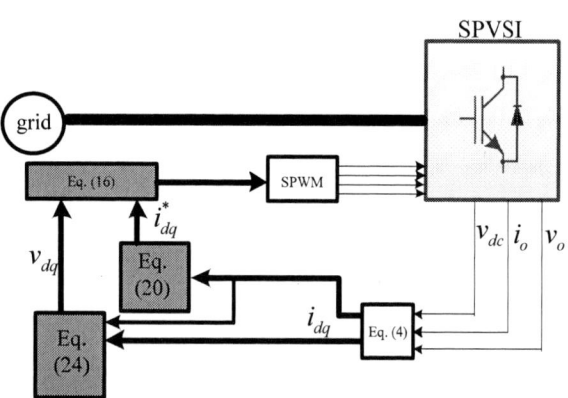

Fig. 3. Overall structure of the proposed controller

A. The SPVSI active and reactive power evaluation

Firstly, the grid supplies a single-phase full-bride diode rectifier as a nonlinear load in the time interval of [0, 0.05].

Then, SPVSI is connected to the point of common coupling at t=0.05s. Active and reactive power of SPVSI, grid, and load are illustrated in Fig.4.

As can be seen from the figure, when the SPVSI is not connected, both active and reactive power of the load are withdrawn from the grid and its respective powers for the inverter are zero. Then, after SPVSI connection, the whole reactive power of the load is supplied by the inverter and also the total harmonic components of active power and a fraction of active power at fundamental frequency are produced by SPVSI as depicted in Fig. 4.

This test verifies that the proposed controller can accomplish an effective active and reactive power contribution during any sudden changes.

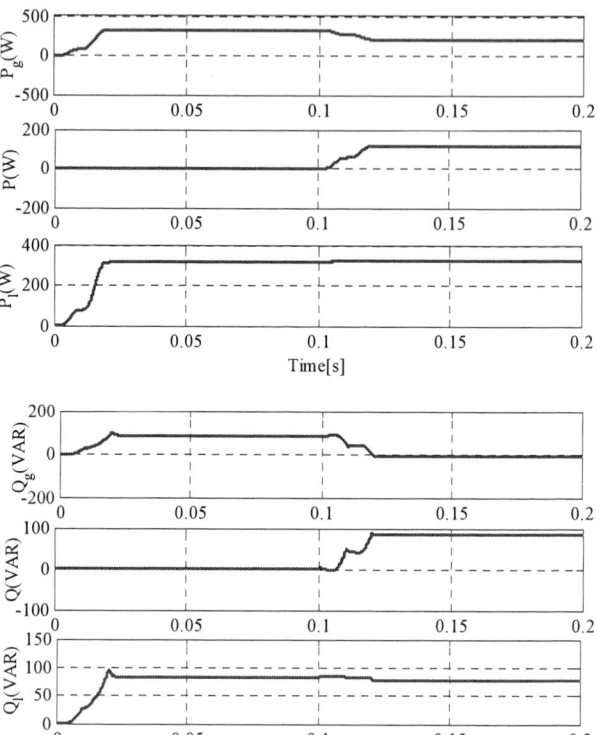

Fig. 4. Active and reactive power of grid, load and SPVSI

B. Analysis of the grid current

In order to assess the merit of the proposed controller in enhancement of power quality and compensation of harmonic components, the upstream grid current is analyzed in this section.

Fig. 5 shows the grid current and voltage, load current, and SPVSI current. As can be seen in Fig.5, once SPVSI is connected to the grid, the grid current will become sinusoidal and also in phase with the grid voltage that demonstrates a complete generation of load reactive power through SPVSI.

Fig. 6 illustrates the harmonic spectrum of grid current during SPVSI connection. This figure shows that the SPVSI is able to properly inject the required harmonic components of the load and make the grid current pure sinusoidal with a low THD.

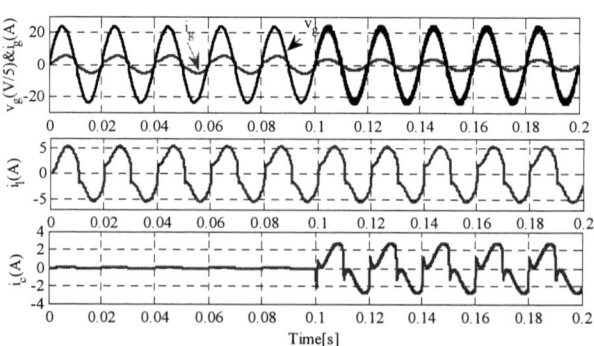

Fig. 5. Grid current and voltage, load current, and SPVSI current

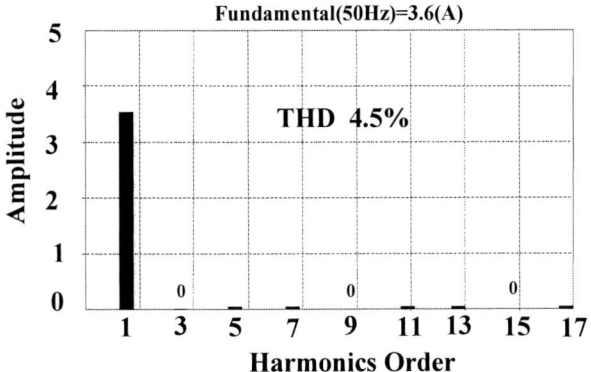

Fig. 6. Harmonic spectrum of grid current

VI. CONCLUSION

A comprehensive Direct-Quadrature (DQ) rotating frame based dynamic model for a grid-connected SPVSI was developed in this paper to design a droop-passivity based controller for a better performance in presence of nonlinear loads. Also, details of the controller design principles and its stability issues were thoroughly discussed. Moreover, an accurate operation area for SPVSI active and reactive power was obtained that was used as a capability curve for achieving a more precise power sharing performance. By using a well-designed reference generation scheme, an effective harmonic compensation was attained in the proposed controller. Finally, the proposed controller impacts on the grid-connected SPVSI operation was evaluated in Matlab/Simulink environment.

ACKNOWLEDGEMENT

This work was supported by FEDER funds (European Union) through COMPETE, and by Portuguese funds through FCT, under Projects FCOMP-01-0124-FEDER-020282 (Ref. PTDC/EEA-EEL/118519/2010), UID/CEC/50021/2013 and SFRH/BPD/102744/2014. Also, the research leading to these results received funding from the EU Seventh Framework Programme FP7/2007–2013 under GA no. 309048.

REFERENCES

[1] M. Mehrasa, E. Pouresmaeil, B. N. Jørgensen, and J. P.S. Catalão, "A control plan for the stable operation of microgrids duringgrid-connected and islanded modes." Electric Power Systems Research., vol. 129, pp. 10–22, 2015.

[2] E. Pouresmaeil, M. Mehrasa, and J. P. S. Catalão, "A Multifunction Control Strategy for the Stable Operation of DG Units in Smart Grids." Smart. Grid. IEEE Trans., vol. 6, no. 2, pp. 598-607, 2015.

[3] Z. Yao, and L. Xiao, "Control of Single-Phase Grid-Connected Inverters With Nonlinear Loads." Ind. Electron. IEEE Trans., vol. 60, no. 4, pp. 1384–1389, 2013.

[4] S. Golestan, M. Monfared, Member, and J. M. Guerrero, "A Novel Control Technique for Single-Phase Grid-Connected Inverters," 2011 International Conference on Electrical Engineering and Informatics, Bandung, Indonesia.

[5] C-H Chang, Y-H Lin, Y-M Chen, and Y-R Chang, "Simplified Reactive Power Control for Single-Phase Grid-Connected Photovoltaic Inverters." Ind. Electron. IEEE Trans., vol. 61, no. 5, pp. 2286 - 2296, 2013.

[6] A. Kulkarni, and V. John, "Mitigation of Lower Order Harmonics in a Grid Connected Single Phase PV Inverter." Power. Electron. IEEE Trans., vol. 28, no. 11, pp. 5024 - 5037, 2013.

[7] M. Mehrasa, M. E. Adabi, E. Pouresmaeil,and J. Adabi, "Passivity-based control technique for integration of DG resources into the power grid." International Journal of Electrical Power & Energy Systems, vol. 58, pp. 281–290, 2014.

[8] H Komurcugil, "Improved passivity-based control method and its robustness analysis for single-phase uninterruptible power supply inverters." Power. Electron. IET., vol. 8, no. 8, pp. 1558 – 1570, 2015.

[9] Zabihi, Sasan and Zare, Firuz.; "A New Adaptive Hysteresis Current Control with Unipolar PWM Used in Active Power Filters" [online]. Australian Journal of Electrical & Electronics Engineering, Vol. 4, No. 1, 2008: 9-16.

[10] Zabihi, S.; Zare, F., "Active Power Filters with Unipolar Pulse Width Modulation to Reduce Switching Losses," in Power System Technology, 2006. PowerCon 2006. International Conference on , vol., no., pp.1-5, 22-26 Oct. 2006

[11] Zabihi, Sasan; Zare, Firuz; "An Adaptive Hysteresis Current Control Based on Unipolar PWM for Active Power Filters," Proceedings of the 2006 Australasian Universities Power Engineering Conference, 10-13 December 2006, Australia, Victoria, Melbourne.

978-1-5090-0376-1/16 $31.00 © 2016 IEEE

Comparison of Single Loop Based Control Strategies for a Grid Connected Inverter in a Photovoltaic System

Mohammad Hossein Mahlooji   Hamid Reza Mohammadi   Mohsen Rahimi

Department of electrical and computer engineering
University of Kashan
Kashan, Iran

Abstract— **Nowadays many of the distributed generation (DG) sources such as photovoltaic systems are connected to the grid via inverters. One of the main problems of the inverters is the injection of current harmonics to the grid. For attenuating the injected current harmonics, low pass filters such as LCL filter are used in the output of the inverters. The quality of the injected current to the grid has a significant importance. The control method for grid connected inverters with LCL filter is mainly based on the output current feedback. But there is another method that is based on the inverter current feedback. In this paper, moreover designing the LCL filter parameters the controller design for both methods are discussed. Also by using the output impedance criterion, the performance of these methods are compared with respect to voltage disturbance rejection and satisfying the power quality criteria and finally the better one is introduced. The analytical results are verified by simulation results using MATLAB/SIMULINK software.**

  Keywords— photovoltaic systems; LCL filter; controller design; power quality.

I. INTRODUCTION

Nowadays regarding to energy crisis issue and fuel decaying over time, the application of renewable energy resources are widely increased. Photovoltaic systems are also one of them that have much popularity. Photovoltaic systems are often grid connected and the inverters are used for connecting these sources to the grid. The main problem of inverters is the harmonic components of the output current which are appeared due to switching operation.

The passive filters are used in the output of inverters to overcome these drawbacks. Reference [1] describes the designing procedure of an LCL filter in a three phase rectifier, step by step. The design procedure of an LCL filter and a cascade controller are discussed in [2]. Reference [3] analyses the effect of filter parameters on the utility grid output current affecting by the resonance. Reference [4] deals with L and LCL filter design and also expresses L filter problems such as much volume and size in satisfying grid standards requirements such as IEEE 1547 standard. Commonly third order LCL filter is used in place of conventional L filter that can satisfy the grid standards requirements. An LCL filter has low size and weight, cost saving and higher harmonic components attenuation. Non-recursive LCL filter design methodology using an approximated harmonic analysis is used

in [5]. Reference [6] presents the modeling of grid connected PWM inverters that have been controlled using direct current control method and also analyses the damping resistor effect in the system stability. In [7] an LCL filter is designed for a grid connected photovoltaic system and also the PI controller design using Ruth-Herwitz stability criterion is presented.

In this paper moreover designing LCL filter, the controller design for a grid connected inverter is presented by using two different strategies. These two strategies are: 1- grid current feedback based control and 2- inverter current feedback based control. By using the output impedance criterion, the performances of these two strategies are compared with respect to THD and voltage disturbance rejection and finally the better one is introduced.

II. SYSTEM MODELING AND LCL FILTER DESIGN

Fig. 1 illustrates the topology of a grid connected inverter in a photovoltaic system. The state equations of the LCL filter are derived with the assumption that the capacitor is large enough and consequently the dc link voltage is constant. The equations in dq reference frame can be written as follows:

$$\dot{i1}_{dq} = \frac{1}{L_1}V_{inv_{dq}} - \frac{1}{L_1}(R_1 \cdot i1_{dq} + V_{c_{dq}}) - j\,\omega_g \cdot L_1 \cdot i1_{dq}$$

$$\dot{i2}_{dq} = \frac{1}{L_2}V_{c_{dq}} - \frac{1}{L_2}U_{s_{dq}} - \frac{1}{L_2}R_2 i2_{dq} - j\,\omega_g \cdot L_2 \cdot i2_{dq} \qquad (1)$$

$$\dot{V}_{c_{dq}} = \frac{1}{C}(i1_{dq} - i2_{dq}) - j\,\omega_g \cdot V_{c_{dq}}$$

Where each typical variable such as F can be written as $F_{dq}=F_d+jF_q$ and $\omega_g = 2\pi.f_g$. The LCL filter block diagram with respect to (1) is plotted in Fig 2.

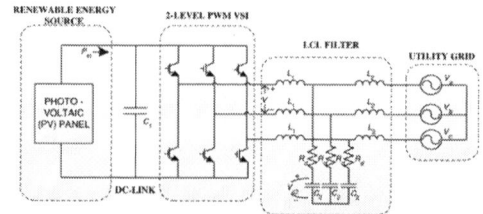

Fig. 1. Topology of a grid connected inverter in a photovoltaic system

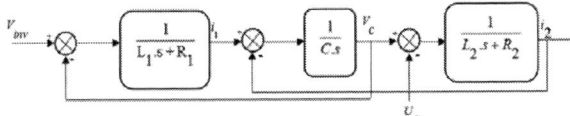

Fig 2. LCL filter block diagram

A. Filter Design Procedure

In filter design procedure, the parameters such as current ripple, filter size and harmonic attenuation are considered. The calculation of filter components can be performed by using the system parameters such as inverter output voltage, grid frequency and rated power. The system parameters are provided in table 1.

The base impedance and base capacitor values are calculated as follows:

$$Z_{base} = \frac{v_{LL}^2}{P_n} \tag{2}$$

$$C_{base} = \frac{1}{\omega_g \cdot Z_{base}} \tag{3}$$

Where v_{LL} is inverter line to line output voltage and P_n is rated power. The output current ripple is used as a criterion for determining the inverter inductor L_1. To finding the proper value of L_1 the maximum output current ripple should be calculated. The maximum output current ripple is calculated by (4).

$$\Delta I_{L,max} = \frac{2V_{DC}}{3L_1 f_{sw}} m(1-m) \tag{4}$$

Where m is modulation index and f_{sw} is switching frequency. The output current ripple is maximized for m=0.5 and thus L1 value can be obtained as follows:

$$L_1 = \frac{V_{DC}}{6\Delta I_{Lmax} f_{sw}} \tag{5}$$

Usually current ripple is considered 5 to 25 percent of the maximum current and the maximum current is represented by (6).

$$I_{Lmax} = \frac{P_n \times \sqrt{2}}{3v_{ph}} \tag{6}$$

Where v_{ph} is phase to neutral RMS voltage. The maximum permissible variation of reactive power in the point of common coupling is used to determine the value of the filter's capacitor.

TABLE 1. SYSTEM PARAMETERS

Parameter	Value
Rated power	100 Kw
Line-Line Voltage	380 v
System frequency	60 Hz
Switching frequency	1.7 KHZ

The Reactive power exchanged by this capacitor is considered 5 to 7.5 percent of the rated power. Thus the capacitor value is determined as follows:

$$C = 0.05C_{base} \tag{7}$$

The inverter current ripple reduction is the designing criteria for design of L_2. The grid current to the inverter current ratio is defined as:

$$\frac{i_g(h)}{i_i(h)} \square K_a = \frac{1}{\left|1 + r[1 - L_1 c_f \omega_{sw}^2]\right|} \tag{8}$$

Where K_a is the desired attenuation and $r = \dfrac{L_2}{L_1}$. According to Fig. 3 (the plot curve of Ka versus r) the value of r is determined for a desired attenuation K_a. By determining r value, the value of L2 can be determined as follows:

$$L_2 = r.L_1 \tag{9}$$

By using (1-9) the parameters of the LCL filter are calculated and the results are shown in table 2. It should be noted that in this design process the value of Ka is selected equal to 0.26.

B. Resonance and Frequency Response of filter

Resonance frequency of the LCL filter is given in (10) that the equivalent inductor is equal to the parallel connection of L_1 and L_2.

$$f_{res} = \frac{1}{2\pi\sqrt{(L_1 \square L_T).C}} = \frac{1}{2\pi}\sqrt{\frac{L_1 + L_T}{L_1.L_T.C}} \tag{10}$$

The resonance frequency should not be interfere with bandwidth of the control system and also it is better that be far from switching frequency, hence the resonance frequency range should be considered to satisfy (11).

$$10f_g \langle f_{res} \langle 0.5 f_{sw} \tag{11}$$

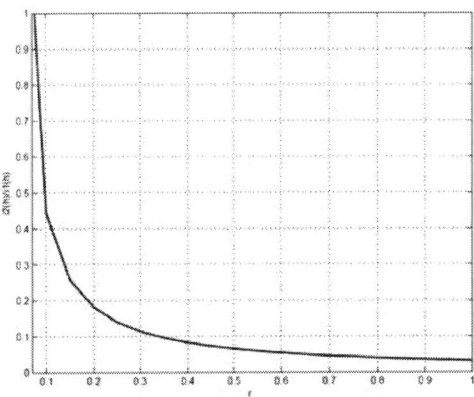

Fig 3. Plot curve of Ka versus r

TABLE 2. FILTER PARAMETERS

Parameter	Value
Inverter side inductor	3.2 mH
Grid side inductor	0.48 mH
Capacitor	91.8 mF

The bode diagram of the LCL filter using i_2 to V_{inv} transfer function which is given in (12) is illustrated in Fig. 4.

$$\frac{i_2}{v_{inv}} = \frac{1}{L_1 L_2 C s^3 + (L_1 + L_2) s} \quad (12)$$

It can be seen that the magnitude diagram has a peak value at resonance frequency that lead to oscillation and instability in the system. There are some ways for damping the resonance oscillation includes passive and active damping. Passive damping uses a damping passive resistor. There are different possible places for the resistor (in series or parallel with L1 or C) that usually it is placed in series with capacitor. The value of damping resistor can be calculated as follow [8]:

$$R_d = \frac{1}{3\omega_{res}.C} \quad (13)$$

III. THE CONTROLLER DESIGN FOR GRID CONNECTED INVERTER

The schematic diagram of the grid connected VSI and its control is illustrated in Fig. 5. Commonly the grid current feedback strategy is used for control system although there is another strategy that is based on inverter current feedback which is not considered in literatures well.

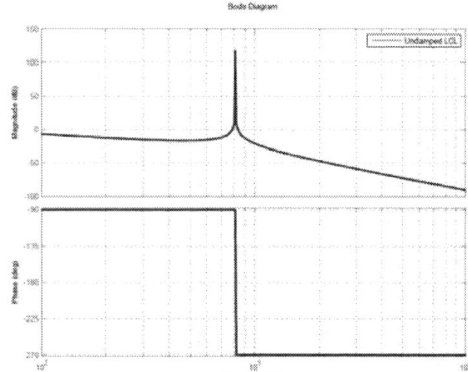

Fig 4. The bode diagram of the LCL filter transfer function

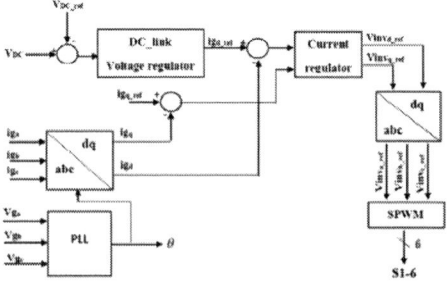

Fig 5. Schematic diagram of the grid connected VSI and its control

In this paper moreover designing the controller based on grid current feedback strategy, the controller design based on inverter current feedback strategy is also depicted and finally the results of these two strategies are compared using output impedance criterion.

A. Controller Design Based On Grid Current Feedback Strategy

The block diagram of the control system based on grid current feedback strategy is shown in Fig. 6. It should be noted that the variables are transferred to dq synchronoeus reference frame. The grid current is compared with the reference current and the error signal is processed by a PI controller to make the steady-state error equal to zero. The output of the PI controller is the reference value for the inverter output voltage.

Due to the superposition rule, the grid current is derived from two inputs (inverter output voltage and grid voltage). In this block diagram the grid current to inverter output voltage transfer function is represented by $G_{i_2}^{ol}$ and the grid current to grid voltage transfer function is also represented by $Y_{o-inv}^{i_2}$ which is equal to inverter output admittance. The $G_{i_2}^{ol}$ transfer function is given by (14).

$$G_{i_2}^{ol} = \frac{i_2}{V_{inv}}\bigg|_{Vg=0} = \frac{A.s + 1}{B_0.s^3 + B_1.s^2 + B_2.s + B_3} \quad (14)$$

Where the numerator and denumerator coefficients are represented as follows:

$A = R_d.C$

$B_0 = L_1.L_2.C$

$B_1 = C(R_d.(L_1 + L_2) + L_1.R_2 + L_2.R_1)$

$B_2 = L_1 + L_2 + C(R_1.R_d + R_1.R_2 + R_2.R_d)$

$B_3 = R_1 + R_2$

$Y_{o-inv}^{i_2}$ is the inverter output admittance and its transfer function is given by (15):

$$Y_{o-inv}^{i_2} = \frac{i_2}{V_g}\bigg|_{Vinv=0} = \frac{C'.s^2 + D.s + 1}{B_0.s^3 + B_1.s^2 + B_2.s + B_3} \quad (15)$$

Where B0-B3 coeficients are same as (14) and C', D coefficients are represented as follows:

C'=L$_1$.C

D=R$_d$.C

In the controller design process, the phase angle of the $G_{i_2}^{ol}$ transfer function in a frequency corresponding to bandwidth is derrieved using it's bode diagram in Fig. 7. Since the resonance frequency (ω_{res}) is 898 Hz, hence the bandwidth of close loop control system is given by (16).

$$f_{cl} = \frac{1}{5} f_{res} = 179.6\,Hz \quad (16)$$

978-1-5090-0376-1/16 $31.00 © 2016 IEEE

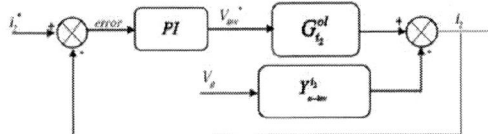

Fig 6. Block diagram of the grid current feedback based control strategy

The system phase angle in bandwidth corresponding frequency (179.6Hz) is -90 degree as can be seen in fig. 7. The phase margin is selected to be 80 degree and hence for achieving this phase margin, the PI controller should compensate for -10 degree.

The controller design process is done using following steps:
1. Cutoff frequency of the PI controller is considered too smaller than the close loop bandwidth.

$$\omega_{cut-off} \ll \omega_{cl} \Rightarrow \frac{K_I}{K_P} \ll \omega_{cl} \tag{17}$$

2. Open loop transfer function is calculated in crossover frequency that is equal to close loop bandwidth.

$$G_{sys}^{ol}(s)\Big|_{s=j\omega_{cl}} = (K_P + \frac{K_I}{j\omega_{cl}}).G_{i_2}^{ol}(s)\Big|_{s=j\omega_{cl}} \tag{18}$$

Due to $\frac{K_I}{\omega_{cl}} \ll K_P$, the $\frac{K_I}{\omega_{cl}}$ term can be neglected against K_P.

$$G_{sys}^{ol}(s)\Big|_{s=j\omega_{cl}} = K_P .G_{i_2}^{ol}(s)\Big|_{s=j\omega_{cl}} \tag{19}$$

The magnitude of transfer function in crossover frequency is equal to one and by using this relation the K_p can be calculated as in (20).

$$G_{sys}^{ol}(s)\Big|_{s=j\omega_{cl}} = 1$$

$$\rightarrow K_P.\left|G_{i_2}^{ol}(s)\right|_{s=j\omega_{cl}} = 1 \tag{20}$$

$$\rightarrow K_P = \frac{1}{\left|G_{i_2}^{ol}(s)\right|_{s=j\omega_{cl}}}$$

3. The phase margin of the close loop system is equal to 80 degree as mentioned earlier. Using this matter it can be written as follows:

$$-180 + 80 = \angle PI + \angle G_{i_2}^{ol}(s) \tag{21}$$

Where the angle of PI controller is given by (22)

$$\angle PI = -90 + tg^{-1}\frac{K_P\omega_{cl}}{K_I} \tag{22}$$

Using (21) and (22) the K_I can be calculated as shown in (23).

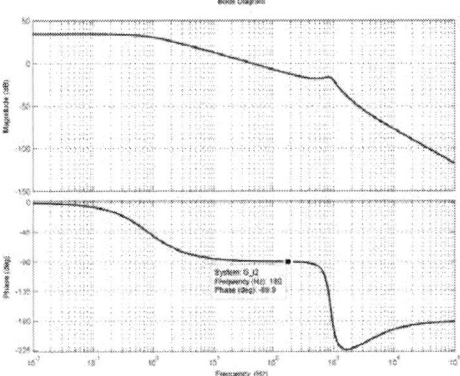

Fig. 7. Bode diagram of the $G_{i_2}^{ol}$

$$-180 + 80 = -90 + tg^{-1}\frac{K_P\omega_{cl}}{K_I} + \angle G_{i_2}^{ol}(s)$$

$$\Rightarrow \frac{K_P\omega_{cl}}{K_I} = tg(-\frac{\pi}{18} - \angle G_{i_2}^{ol}(s)) \tag{23}$$

$$\Rightarrow K_I = \frac{K_P\omega_{cl}}{tg(-\frac{\pi}{18} - \angle G_{i_2}^{ol}(s))}$$

By using the designed controller, the bode diagram of open loop transfer function is plotted in Fig. 8. It can be seen that in the crossover frequency (180 Hz), the desired phase margin is achieved which verifies the designing method.

B. Controller Design Based On Inverter Current Feedback Strategy

The block diagram of the control system based on inverter current feedback strategy is shown in Fig.9. The output of DC link voltage controller is equal to the inverter reference current in d axis (i_{1d}*). It is compared with the inverter actual current and then processed by a PI controller to achieving zero steady-state error same as grid current feedback strategy. Due to the superposition rule, the inverter current is derived from two inputs (inverter output voltage and grid voltage). In this block diagram the inverter current to inverter voltage transfer function is represented by $G_{i_1}^{ol}$ and the inverter current to grid voltage transfer function is also represented by Y_{12-inv}. The $G_{i_1}^{ol}$ transfer function is given by (24).

$$G_{i_1}^{ol} = \frac{i_1}{V_{inv}}\Big|_{Vg=0} = \frac{C''.s^2 + D.s + 1}{B_0.s^3 + B_1.s^2 + B_2.s + B_3} \tag{24}$$

Where $C'' = L_2.C$ and other coefficients are same as (15).

Also Y_{12-inv} Admittance is given by (25).

978-1-5090-0376-1/16 $31.00 © 2016 IEEE   393

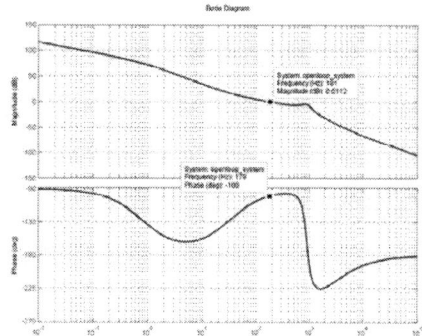

Fig 8. Bode diagram of the open loop transfer function in output current feedback strategy considering the designed PI controller

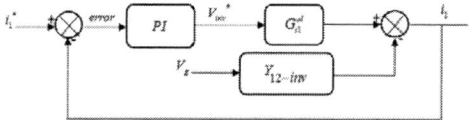

Fig 9. Block diagram of the inverter current feedback based control strategy

$$Y_{12-inv} = \left.\frac{i_1}{V_g}\right|_{V_{inv}=0} = \frac{A.s+1}{B_0.s^3 + B_1.s^2 + B_2.s + B_3} \qquad (25)$$

Where all coefficients are same as coaficients in (14).

Using bode diagram of $G_{i_1}^{ol}$, it can be seen that the phase angle of $G_{i_1}^{ol}$ in close loop bandwidth frequency (180Hz) is almost 90 degree, and if PI controller compensate 10 degree, the desired phase margin that is 80 degree is achieved. In this control strategy, the controller design process is exactly same as grid current feedback based control strategy and previous three steps are done here. Finally, the PI controller coefficients which are obtained by this design process are given in table 3. By using the designed controller, the bode diagram of open loop transfer function is plotted in Fig. 10. It can be seen that in the crossover frequency (180 Hz), the desired phase margin is achieved that verify designing method.

As well as, because of $\dfrac{K_I}{\omega_{cl}} = 0.18 K_P$ the early assumption of

$\dfrac{K_I}{\omega_{cl}} \square K_P$ is realized.

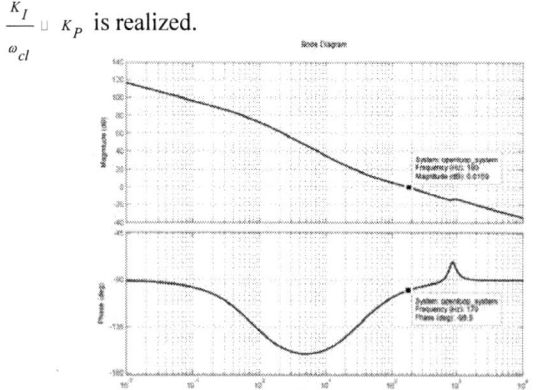

Fig 10. Bode diagram of the open loop transfer function in output current feedback strategy considering the designed PI controller

TABLE 3. PI Controller parameters in the two strategies

	Grid current feedback strategy	Inverter current feedback strategy
Controller coefficients	K_P=3.085	K_P=4.03
	K_I=766.03	K_I=782.57

IV. OUTPUT IMPEDANCE

The output impedance of the inverter is a proper criterion for evaluation of the inverter performance in case of nonsinusoidal grid voltage condition. The output impedance is represented in (26).

$$Z_o = \left.\frac{V_g}{i_g}\right|_{ig^*=0} \qquad (26)$$

Using (26) and block diagrams in Figures 6 and 9 the output impedance in these two control strategies are calculated and their bode diagrams are illustrated in Fig. 11. It can be seen that the magnitude of the output impedance in the inverter current feedback strategy is larger than the grid current feedback strategy. The greater output impedance results in lower current harmonic in case of nonsinusoidal grid voltage condition. Therefor the inverter current feedback strategy has better performance regarding voltage disturbance rejection issue.

V. SIMULATION RESULTS

A photovoltaic system with a grid connected inverter is considered and two aforementioned single loop based control strategies have been tested through extensive case study simulations using MATLAB/SIMULINK software. The performances of these two strategies are analyzed and compared in sinusoidal and nonsinusoidal grid voltage conditions.

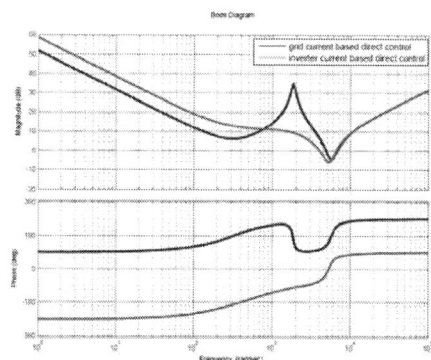

Fig 11. Bode diagram of the output impedance in two strategies

A. Sinusoidal Grid Voltage Condition

The simulation results for grid current and inverter current feedback strategies include grid voltage and injected grid current are shown in figures 12 and 13 respectively. The THD of the injected grid current for these two strategies are given in table 4. It can be seen that the THD of the injected grid current in the inverter current feedback strategy is lower than other strategy.

B. Nonsinusoidal Grid Voltage Condition

In this case, the grid voltage is composed of fundamental and 5, 7 and 11 harmonic components. The grid voltage and injected grid current for these two strategies are shown in figures 14 and 15. The THD of the injected grid current for these two strategies are also given in table 5. It can be seen that the inverter current feedback strategy has better performance and smaller THD than other strategy. This result is pereviously expected by output impedance comparison between these two strategies. The inverter current feedback strategy has bigger output impedance and consequently has better voltage disturbance rejection property.

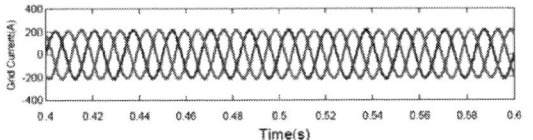

Fig 12. Simulation results for grid current feedback strategy in sinusoidal grid condition

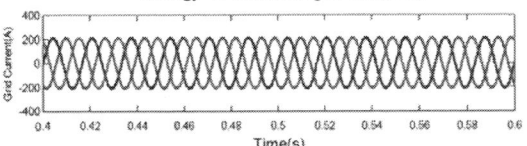

Fig 13. Simulation result for inverter current feedback strategy in sinusoidal grid condition

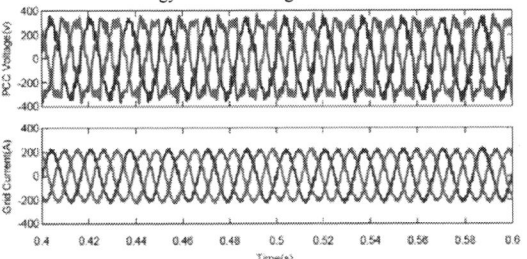

Fig 14. Simulation result for grid current feedback strategy in nonsinusoidal grid voltage condition

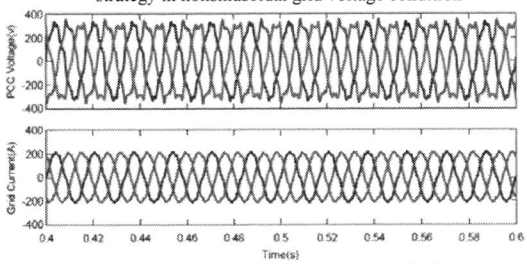

Fig 15. Simulation result for inverter current feedback strategy in nonsinusoidal grid voltage condition

TABLE 4. Injected grid current THD in two control strategies in sinusoidal grid voltage condition

Control strategy	Current THD
Inverter current based feedback	4.8%
Grid current based feedback	3.3%

TABLE 5. Injected grid current THD in two control strategies in nonsinusoidal grid voltage condition

Control strategy	Current THD
Inverter current based feedback	8.1%
Grid current based feedback	5.1%

VI. CONCLUSION

The grid connected inverters with LCL lowpass filter play an important role as interface circuit between distributed generation (DG) sources such as photovoltaic systems and grid. Hence the proper control of these inverters has significant importance for satisfying power quality criteria. In this paper moreover designing the LCL filter, the comparison of single loop control strategies for a grid connected inverter in a photovoltaic system is done. The controller design for two different control strategies based on inverter current and grid current feedback are carried out. By comparison of simulation results, it was shown that the THD of the grid injected current in the inverter current feedback strategy is lower than other strategy. Also the inverter current feedback strategy has better performance with respect to grid voltage disturbance rejection. These results are anlytically verified by using the output impedance criterion.

REFRENCES

[1] M. Liserre, F.Blaabjerg, and S. Hansen, "Design and Control of an LCL-filter Based three phase active rectifier" IEEE Trans. Industry Applications, vol.41, pp. 1281-1291, 2005.

[2] B.G.Cho, S.K.Sul, H.Yoo and S.M.Lee, "LCL filter design and control for gridconnected PWM converter", 2011 IEEE 8th International Conference on PowerElectronics and ECCE Asia (ICPE & ECCE), pp.756-763 , 2011.

[3] [24] F. Liu, X. Zha, Y. Zhou and S. Duan, "Design and Research on Parameter of LCL Filter in Three-Phase Grid-Connected Inverter", IEEE 6th International Power Electronics and Motion Control Conference, pp.2174-2177 , 2009

[4] M.A. Elsaharty, H.A.Ashour , "Passive L and LCL Filter Design Method for Grid-Connected Inverters" 2014 IEEE Innovative Smart Grid Technologies Conference - Asia (ISGT Asia), pp.13-18 , 20-23 May 2014.

[5] K.Y. Choi, T.R. Kim, R.Y. Kim " Non-recursive LCL Filter Design Methodology for a Grid-connected PWM Inverter using an Approximated Harmonic Analysis," IEEE Vehicle Power and Propulsion Conference, pp.364-369 , 9-12 October 2012.

[6] X.G. Guo, W.Y. Wu, and H.R. Ju, "Modeling and simulation of direct output current control for LCL-interfaced grid-connected inverters with parallel passive damping" simulation modeling practice and theory, vol.18, pp. 946-956, 2010.

[7] Y. Chen, F. Liu, " Design and Control for Three-phase Grid-connected Photovoltaic Inverter with LCL Filter," IEEE Circuits and Systems International Conference on Testing and Diagnosis, pp.1-4 , 28-29 April 2009.

[8] S. V. Araújo, A. Engler, B. Sahan, V. U. Kassel, F. Luiz, and M.Antunes, "LCL Filter design for grid-connected NPC inverter offshore wind turbines," in the 7th International Conference on Power Electronics, 2007, pp. 1133–1138.

7th Power Electronics, Drive Systems & Technologies Conference (PEDSTC 2016)
16-18 Feb. 2016, Iran University of Science and Technology, Tehran, Iran

THD Minimization in Variable Input Cascaded H-Bridge Multi-level Inverters via State Table

Reza Emamalipour, Behzad Asaei, Babak Farhangi
Department of Electrical and Computer Engineering
College of Engineering, University of Tehran
Tehran, Iran
r.emamalipour@ut.ac.ir, basaei@ut.ac.ir, b.farhangi@ieee.org

Abstract— This paper proposes a method for achieving the optimum Total Harmonic Distortion (THD) of the output voltage in the cascaded multilevel inverter in case that dc cells' voltages vary during operation. It considers switching angles and voltage steps as optimization variables in order to obtain the minimum THD for the 5, 7, 9, 11, and 13-level cases. The proposed method studies +20% and -20% variations in the optimum dc voltage levels in 7-level cascaded inverters. It obtains new optimum switching angles according to cells' voltages by building a state table of the possible output vectors and applying the new switching angles from a lookup table by means of a microcontroller. Furthermore, it examines fault conditions in the optimized voltages leading a drop of 50% and 80% for each cell. By building the fault state table, the proposed method results in voltage THD reduction for each state after re-optimizing the switching angles. Also, it compares the output voltage THD before and after applying the new switching angle values from a lookup table. Up to 12% THD reduction during fault conditions is achieved by applying the proposed optimization method. The analytical results are verified experimentally.

Keywords - Total Harmonic Distortion, THD minimization, Multilevel inverter, H-bridge, Cacaded H-Bridge Inverters

I. INTRODUCTION

Multi-level inverters have been introduced since 1975 and are popular for several applications [1] including electric drives[2, 3], Flexible alternating current transmission systems (FACTS)[4-6], HVDC systems[7, 8], Electric Vehicles [9, 10] and high power applications[11]. They also condition the renewable energy sources such as photovoltaic plants, wind turbines and fuel cells [12-14]. These converters enable high voltage applications, due to the use of multiple similar cells with available low voltage switches. The switching frequency can be reduced compared to traditional multi-level converters. This reduces the switching losses. Multi-level inverters came in various topologies like Diode clamped and flying capacitor; cascaded H-Bridge inverter is one of the most popular multi-level converter topologies. [15, 16].

Total Harmonic Distortion (THD) of the output waveform is an important design aspect for the multilevel converters. Increasing the dc voltage levels in multilevel inverters improves the total harmonic distortion. In the original cascaded H-bridge inverter patent [17], the input voltages were assumed identical. In contrary, the input voltages may vary when batteries or photovoltaic panels are used as the input sources. [18, 19] has proposed asymmetric voltage ratio between the dc sources and showed that integer ratio (2:1 or 3:1) achieves better output waveforms. [18] also proposed non-integer dc voltage ratios in order to reduce the THD for three phase applications. In [20], symmetric switching angles were proposed and unequal voltage sources were investigated in order to reduce the output THD. In [21-23], both dc voltage sources and switching angles have been proposed as optimization variables. [21, 24] used the same technique and the harmonic elimination method in order to minimize the output THD.

In case the dc sources are powered by photovoltaic panels (PV) or batteries, the input voltages may not stay constant and they may vary during operation. As an example, if PV is used as the dc source, the output voltage may vary due to partial shading, the change in irradiance and hence the change in operating point of the MPPT (Maximum Power Point Tracker) which leads to voltage variation in cells' terminal. [25]

Also in case that batteries are used as dc sources, this voltage variation can happen. In battery powered vehicles [10], the batteries' voltages may vary due to the unbalanced operation of multi-level converters during charging and discharging periods or due to tolerances. This affects voltage harmonics and increases the voltage THD. In battery powered multi-level converters, a large number of batteries are usually connected in series to build a battery pack. The operational differences in electro chemical characteristics can cause state of charge imbalance. In some cases, a fault can occur between several batteries and terminal voltage can drop significantly [26].

This voltage variation can have a severe effect on output voltage THD of the multi-level converter. This paper utilizes the effects of both the switching angles and the dc voltages in order to reduce the output voltage THD when the ratios between dc sources are non-integers. Also, it investigates the case that the voltage of the dc sources varies during different operation modes and examines voltage variation of +20% and -20% for each cell during charging and discharging states and obtains the optimum switching angles in order to achieve the minimum THD according to cells' voltages.

978-1-5090-0376-1/16 $31.00 © 2016 IEEE

When the switching angles need to be reconfigured, the new switching angles may be calculated either online or offline. In online method, the microcontroller gets the cells' voltages via sensors and runs the PSO algorithm online to obtain switching angles. However, the complexity of the calculations and optimization algorithm and its dependence on the initial values as well as random parameters can lead to the divergence of the algorithm, especially when the number of levels and optimization variables increases. In offline method, a state table is formed including cell's voltage variations and optimized switching angles and unlike the online method, the calculations and optimization technique are developed and implemented in a computer rather than microcontroller with higher accuracy and the lower chance of divergence. Then, the table including the converged results of optimization algorithm for different states is fed to the microcontroller. In this case, the task of microcontroller is just comparing the sensed voltages to the predicted voltage vectors in the fed table and applies the corresponding switching angles which takes much less time and the calculations are much simpler without the possibility of divergence.

This paper utilizes the offline method to apply the new switching angles. In addition, it studies a fault condition in dc cells which may be caused by partial shading in PV panels or occurring fault between battery cells leading to the voltage drop of 50% and 80% form the nominal voltage. Considering fault states results in building a state table of different voltage waveforms in fault and normal working states. The state table includes the new optimized THD and new switching angles for each state that are applied via Microcontroller.

II. THD OPTIMIZATION

For a K-cell cascaded inverter, the number of optimization variables is 2K-1. K-1 parameters are Cells' dc voltages and the other K variables are switching angles. Optimum dc voltages and switching angles for different levels are given in Table I. The number of levels are considered from 5 to 13. For each case, both the cell's voltage values and switching angles are taken into account as optimization variables and the maximum number of 100 harmonic order is considered in PSO algorithm. Table II compares minimum output voltage THD considering different parameters as optimization variables for 5- to 13- level cascaded H-bridge inverters.

In the 1st case study, the dc values are equal and the switching angles are considered to be symmetric. (e.g., for a 5-level cascaded inverter the symmetric switching angles are π/6 and π/3 and for a 7-level they are π/8, π/4 and 3π/8 respectively).

TABLE II
OUTPUT THD OBTAINED THD CONSIDERING DIFFERENT OPTIMIZATION VARIABLES

Number of Levels	THD with equal P symmetric α	THD with optimized P symmetric α	THD with equal P optimized α	THD with optimized P optimized α
5	31.50%	26.16%	15.90%	15.84%
7	25.12%	17.57%	11.00%	10.93%
9	21.72%	13.00%	8.51%	8.29%
11	19.68%	10.24%	6.73%	6.63%
13	18.27%	8.33%	5.58%	5.46%

In the 2nd case study, switching angles are again symmetric and the dc voltages are considered as optimization variables. In the 3rd case study, the dc sources are equal and the switching angles can acquire optimized amounts. In the 4th case study both the switching angles and cell's voltage values are taken into account as optimization variables in order to minimize the output voltage THD. The output waveform and THD for a 5- and 7- level cascaded inverter with optimized dc voltage values as well as switching angles are shown in Fig. 1 and Fig. 2.

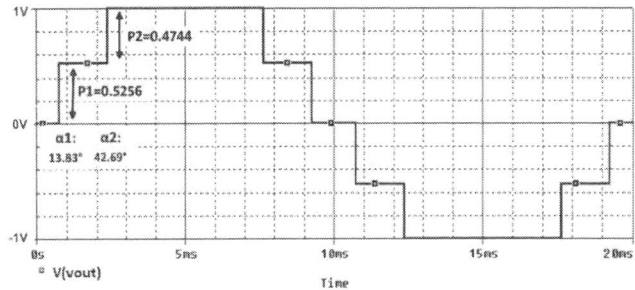

Total Harmonic Distortion = 1.584579E+01 Percent

Fig. 1. THD and output waveform specification for a 5-level cascaded H-Bridge inverter with optimum steps and switching angles

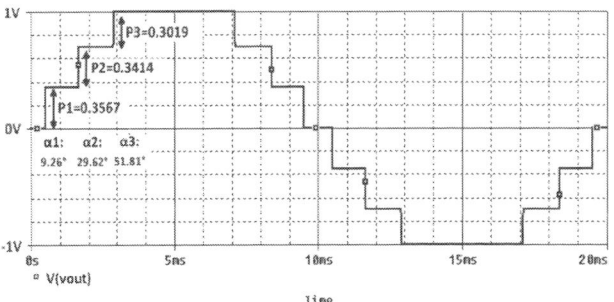

Total Harmonic Distortion = 1.093134E+01 Percent

Fig. 2. THD and output waveform specification for a 7-level cascaded H-bridge inverter with optimum steps and switching angles

TABLE I
OPTIMIZED DC VOLTAGES AND SWITCHING ANGLES FOR A 5-, 7-, 9-, 11-, 13-LEVEL CASCADED H-BRIDGE INVERTER

Number of Levels	P1	P2	P3	P4	P5	P6	α₁ (deg)	α₂ (deg)	α₃ (deg)	α₄ (deg)	α₅ (deg)	α₆ (deg)
5	0.5256	0.4744					13.83	42.69				
7	0.3567	0.3414	0.3019				9.26	29.62	51.81			
9	0.2652	0.2533	0.2632	0.2183			7.13	21.43	38.32	57.22		
11	0.2248	0.2131	0.2076	0.1980	0.1565		6.47	18.37	31.56	45.67	62.77	
13	0.1830	0.1846	0.1809	0.1607	0.1610	0.1298	5.16	15.70	26.48	37.09	50.22	65.52

III. THE PROPOSED METHOD

If PVs or batteries are used as the cells' dc sources, they may acquire values different from the nominal voltage during their operation. For instance terminal voltage of photovoltaic panels can vary due to the partial shading, the change in irradiance or the operating point of the MPPT. Another example is in battery powered vehicles. The battery in a vehicle with nominal voltage of 12 V can have maximum amount of 14.5 V during charging mode and a minimum voltage of 9 V during discharge operation mode. Hence, the voltage level can vary from -20% to +20% from nominal voltage. As a result of these changes in batteries' voltages, the THD will be different from the expected value; consequently, in order to reduce the THD, the switching angles should be reconfigured according to these variations in voltages.

The flowchart of the proposed algorithm is given in the below figure.

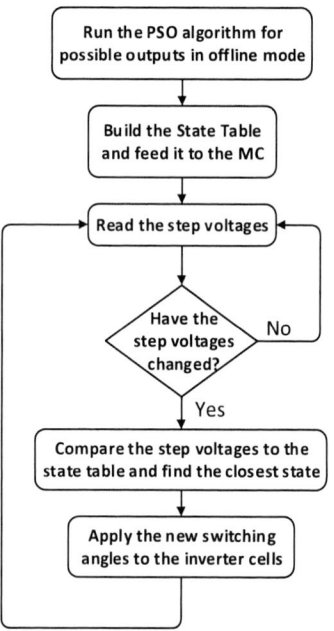

Fig. 3. Flowchart of the proposed method

The total state count of voltage variations could be expanded to a large number depending on the desired accuracy. As an example, in a 3-cell 7-level inverter, considering 21 states for each cell, including nominal voltage, +2%, +4%... +20% over voltages and -2%, -4%, …, -20% under voltages, accumulates the total number of voltage vectors to 21 states for each cell and 9261 possible vectors for all the 3 cells. In order to study the proposed method, the state table is simplified, hence, for each cell, 3 states are considered as 0.8Vb, Vb and 1.2Vb where Vb is the cell's nominal voltage. This results in the total 27 voltage vectors for a 7-level cascaded H-bridge as listed in Table III.

As it is shown in Table III, in voltage vector 25, the 1st step voltage is equal to 1.2Vb while the 2nd and 3rd steps are equal to 0.8Vb. This yields the THD of 12.508% with the switching

angles before any change in cells' voltages; however, if we apply new switching angles according to the new step voltages, the THD will be reduced to 11.537% with almost 1% reduction. This improvement can be crucial in some quality sensitive applications.

TABLE III
STATE TABLE FOR DC STEP VOLTAGE VARIATIONS INCLUDING OBTAINED THD AND NEW OPTIMIZED SWITCHING ANGLES FOR A 7-LEVEL INVERTER

Voltage Vector	3rd cell Voltage (P_3)	2nd cell Voltage (P_2)	1st cell Voltage (P_1)	Opt α_1 (deg)	Opt α_2 (deg)	Opt α_3 (deg)	THD without α optimizing	Opt THD value
1	120%	120%	120%	9.26	29.62	51.82	10.932%	10.932%
2	120%	120%	100%	8.54	27.19	51.30	11.243%	11.021%
3	120%	120%	80%	7.11	25.07	49.72	12.325%	11.343%
4	120%	100%	120%	9.95	29.88	51.38	11.077%	11.039%
5	120%	100%	100%	9.26	29.62	51.81	11.221%	11.059%
6	120%	100%	80%	7.66	25.09	49.37	12.207%	11.279%
7	120%	80%	120%	10.65	28.94	49.78	11.577%	11.369%
8	120%	80%	100%	9.26	29.62	51.81	11.564%	11.313%
9	120%	80%	80%	9.26	29.62	51.81	12.446%	11.790%
10	100%	120%	120%	10.31	31.27	53.68	11.170%	11.027%
11	100%	120%	100%	8.80	29.35	53.31	11.090%	11.041%
12	100%	120%	80%	8.13	27.44	51.63	11.759%	11.348%
13	100%	100%	120%	10.65	31.57	53.05	11.265%	11.027%
14	100%	100%	100%	9.26	29.62	51.82	10.932%	10.932%
15	100%	100%	80%	9.26	29.62	51.81	11.391%	11.078%
16	100%	80%	120%	11.28	31.65	52.34	11.725%	11.337%
17	100%	80%	100%	10.07	29.90	51.22	11.146%	11.089%
18	100%	80%	80%	8.81	26.71	50.40	11.372%	11.093%
19	80%	120%	120%	10.55	32.60	56.12	11.926%	11.383%
20	80%	120%	100%	9.26	29.62	51.81	11.500%	11.362%
21	80%	120%	80%	8.24	28.94	53.52	11.732%	11.557%
22	80%	100%	120%	11.96	33.46	54.59	12.032%	11.328%
23	80%	100%	100%	10.38	31.53	53.73	11.278%	11.069%
24	80%	100%	80%	8.71	29.32	53.41	11.171%	11.101%
25	80%	80%	120%	12.26	33.66	54.38	12.508%	11.537%
26	80%	80%	100%	10.80	31.80	53.11	11.428%	11.088%
27	80%	80%	80%	9.27	29.62	51.82	10.932%	10.932%

The simulation and experimental results prove the theoretical results. The output waveform without and with α reconfiguration for the voltage vector 25, including harmonic orders and voltage THD, obtained using P-Spice software is shown in Fig. 4 and Fig. 5 respectively.

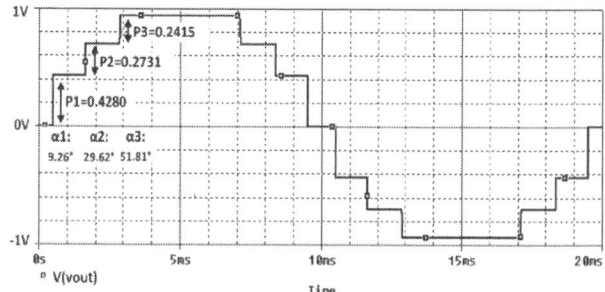

Total Harmonic Distortion = 1.250551E+01 Percent

Fig. 4. THD and output waveform specification for voltage vector 25 without α reconfiguration

978-1-5090-0376-1/16 $31.00 © 2016 IEEE

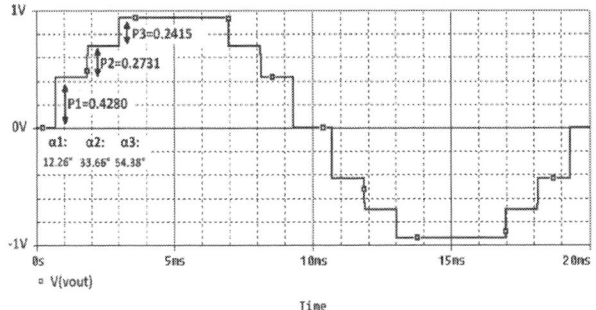

Total Harmonic Distortion = 1.153644E+01 Percent

Fig. 5. THD and output waveform specification for voltage vector 25 after α reconfiguration

The same table can be obtained for a 9-, 11- and 13-level inverter. For a 9-level inverter the total number of states will be 3^4 which is equal to 81. 9 rows of the state table for a 9-level inverter is given in Table IV.

TABLE IV
STATE TABLE FOR DC STEP VOLTAGE VARIATIONS INCLUDING OBTAINED THD AND NEW OPTIMIZED SWITCHING ANGLES FOR A 9-LEVEL INVERTER

Voltage Vector	P_4	P_3	P_2	P_1	Opt α_1 (deg)	Opt α_2 (deg)	Opt α_3 (deg)	Opt α_4 (deg)	Not Opt THD	Opt THD value
1	120%	120%	120%	120%	7.13	21.43	38.32	57.22	8.292%	8.292%
2	120%	120%	120%	100%	6.42	20.09	37.03	57.48	8.545%	8.370%
3	120%	120%	120%	80%	7.13	21.48	38.32	57.20	9.410%	8.587%
4	120%	120%	100%	120%	7.41	21.24	36.59	56.89	8.465%	8.410%
5	120%	120%	100%	100%	6.70	19.85	35.88	56.64	8.796%	8.447%
6	120%	120%	100%	80%	5.27	17.89	34.39	55.43	9.798%	8.660%
7	120%	120%	80%	120%	7.72	21.31	35.59	56.52	8.908%	8.613%
8	120%	120%	80%	100%	6.68	19.35	34.48	55.75	9.350%	8.678%
9	120%	120%	80%	80%	7.13	21.48	38.32	57.20	10.511%	8.882%

As is clear, in the 9th voltage vector the THD has been reduced by 1.63% after reconfiguration of the switching angles. For an 11-level inverter the algorithm is the same and the total number of states will be 3^5 that is 243. 9 rows of the state table for an 11-level inverter is given in Table V.

TABLE V
STATE TABLE FOR DC STEP VOLTAGE VARIATIONS INCLUDING OBTAINED THD AND NEW OPTIMIZED SWITCHING ANGLES FOR AN 11-LEVEL INVERTER

Voltage Vector	P_5	P_4	P_3	P_2	P_1	Opt α_1 (deg)	Opt α_2 (deg)	Opt α_3 (deg)	Opt α_4 (deg)	Opt α_5 (deg)	Not Opt THD	Opt THD value
1	120%	120%	120%	120%	120%	6.44	18.38	31.63	45.73	62.76	6.635%	6.635%
2	120%	120%	120%	120%	100%	5.28	16.76	30.20	44.56	62.03	6.872%	6.647%
3	120%	120%	120%	120%	80%	4.94	15.65	29.41	44.15	61.67	7.672%	6.784%
4	120%	120%	120%	100%	120%	6.56	17.84	30.48	44.99	62.17	6.751%	6.696%
5	120%	120%	120%	100%	100%	5.44	16.22	29.54	44.15	61.72	7.135%	6.680%
6	120%	120%	120%	100%	80%	4.96	15.23	27.50	44.07	62.02	8.121%	6.842%
7	120%	120%	120%	80%	120%	6.75	17.43	28.81	44.06	60.71	7.127%	6.859%
8	120%	120%	120%	80%	100%	5.55	15.68	27.80	43.58	61.24	7.686%	6.858%
9	120%	120%	120%	80%	80%	4.85	14.47	26.12	42.54	60.45	8.866%	6.938%

As can be seen, again in the 9th voltage vector, the THD has been reduced by 1.93% after reconfiguring the switching angles. Due to the size of the state table for a 13-level inverter, it hasn't been given in this paper. However, the algorithm for obtaining the table is the same except the number of the total states will be 729 (3^6).

It is unlikely that the step voltages obtain the exact amounts that are supposed in the state table. In this case, the microcontroller acquires the new switching angles from the existing information. It calculates the relation between the measured step voltages, then, compares it to the relation between the steps in each of the 27 states. Afterwards, it applies the switching angles of the state with the closest relation. The interpolation technique can also be utilized in order to make the calculation more precise. As an example, for a 7-level inverter, if the steps were all 0.9Vb, the relation between the steps would be exactly same as the state where all the steps are Vb. Also, if the steps were 1.1, 0.9 and 1.1Vb respectively, the closest state would be voltage vector 4 where the voltages are 1.2, 1 and 1.2Vb respectively.

This method can also be used to improve the output voltage waveform when a fault occurs in one or more dc cells of the multilevel inverter. Since each dc voltage is produced by several batteries in series, as a result of short circuit between some batteries, voltage at the dc source terminal can drop dramatically. In this case, with the previous switching angles, not only the output voltage amplitude but also the quality of output power will decrease and THD will rise. However with α reconfiguration according to the cell's voltage values, THD can be improved and the fault effects can be reduced.

Table VI is a state table for a 7-level cascaded inverter considering 3 state for each step including 0%, -50% and -80% voltage variations for each cell that makes total voltage vectors equal to 27. Again for more accuracy, the total number of voltage vectors can be expanded and different variation levels can be considered e.g. 100% voltage decrease instead of 80%.

As can be seen in voltage vector 9, with a severe fault in the 1st and the 2nd cells, the THD rises from 10.93% to about 30%. Yet after reconfiguration of switching angles, it will decrease about 12% to 18.35%. This means that changing the switching angles in order to reduce the THD level and improving the output quality could be crucial in some conditions. The output waveform without and with α reconfiguration for the voltage vector 9, including harmonic orders and THD, obtained using P-Spice software is shown in Fig. 6 and Fig. 7 respectively. As can be seen, the THD is improved form 30.01% to 18.35% when the 2nd and 3rd cells have experienced under voltage faults. The advantage of this method in multilevel inverters compared to other inverters is the simplicity of the control strategy and its flexibility in THD reduction by adjusting the switching angles especially for the MV and HV applications. Also, by means of this method, the THD of the output voltage can stay within the limits which can be vital in EV applications e.g. its application in smart grids especially when the EV is injecting power into the grid.

978-1-5090-0376-1/16 $31.00 © 2016 IEEE

TABLE VI
STATE TABLE FOR DC VOLTAGE SOURCES FAULT CONDITION INCLUDING OBTAINED THD AND NEW OPTIMIZED SWITCHING ANGLES

Voltage Vector	3rd cell Voltage (P_3)	2nd cell Voltage (P_2)	1st cell Voltage (P_1)	Opt α_1 (deg)	Opt α_2 (deg)	Opt α_3 (deg)	THD without α optimization	Opt THD value
1	100%	100%	100%	9.26	29.62	51.82	10.932%	10.932%
2	100%	100%	50%	5.52	22.96	48.16	14.355%	11.966%
3	100%	100%	20%	2.93	17.84	45.61	20.830%	13.910%
4	100%	50%	100%	9.26	29.62	51.81	12.528%	12.042%
5	100%	50%	50%	7.10	21.35	44.97	15.378%	12.631%
6	100%	50%	20%	0.00	15.41	40.40	23.930%	15.188%
7	100%	20%	100%	12.66	29.37	46.30	15.680%	13.976%
8	100%	20%	50%	8.59	19.95	40.61	19.029%	14.928%
9	100%	20%	20%	3.94	11.53	35.14	30.016%	18.354%
10	50%	100%	100%	10.96	35.21	57.01	13.229%	11.973%
11	50%	100%	50%	7.01	28.34	53.69	13.205%	12.803%
12	50%	100%	20%	2.95	22.93	51.30	18.341%	15.293%
13	50%	50%	100%	13.82	37.12	54.58	15.053%	12.791%
14	50%	50%	50%	9.26	29.62	51.82	10.932%	10.932%
15	50%	50%	20%	4.81	21.40	48.15	16.082%	12.510%
16	50%	20%	100%	16.13	37.92	51.74	18.735%	15.394%
17	50%	20%	50%	12.14	29.64	47.93	13.356%	12.524%
18	50%	20%	20%	6.52	19.07	42.31	18.453%	13.842%
19	20%	100%	100%	12.95	37.58	52.59	16.902%	14.301%
20	20%	100%	50%	9.26	29.62	51.81	15.922%	15.495%
21	20%	100%	20%	9.26	29.62	51.81	19.872%	19.057%
22	20%	50%	100%	15.85	43.17	59.15	19.827%	15.107%
23	20%	50%	50%	11.95	36.58	58.91	14.269%	12.510%
24	20%	50%	20%	6.58	28.08	55.34	14.698%	14.125%
25	20%	20%	100%	18.92	44.87	54.27	24.477%	18.797%
26	20%	20%	50%	12.84	30.57	50.99	17.384%	15.617%
27	20%	20%	20%	9.26	29.62	51.82	10.932%	10.932%

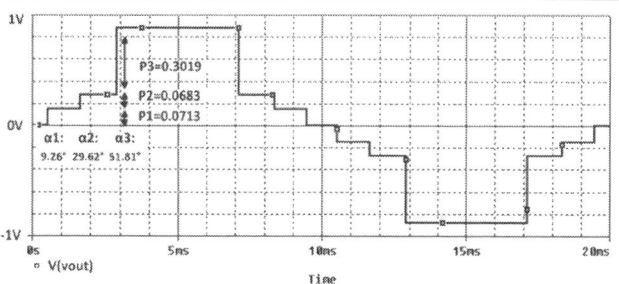

Total Harmonic Distortion = 3.002033E+01 Percent

Fig. 6. THD and output waveform specification for fault voltage vector 9 without α reconfiguration

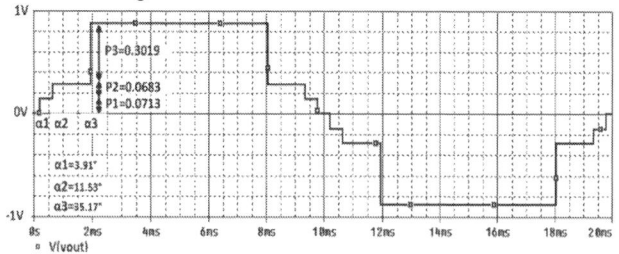

Total Harmonic Distortion = 1.835385E+01 Percent

Fig. 7. THD and output waveform specification for fault voltage vector 9 after α reconfiguration

IV. EXPERIMENTAL VERIFICATION

In order to verify the analysis, the proposed modulation method is tested on a 7-level cascaded H-bridge inverter. The dc Cells are supplied by laboratory power supplies (HYELEC HY2005E). In practice, the dc sources can be implemented by an ac source via transformer and rectifiers or by a dc source such as battery or photovoltaic panels. The STM32F4 ARM is used to generate the switching angles.

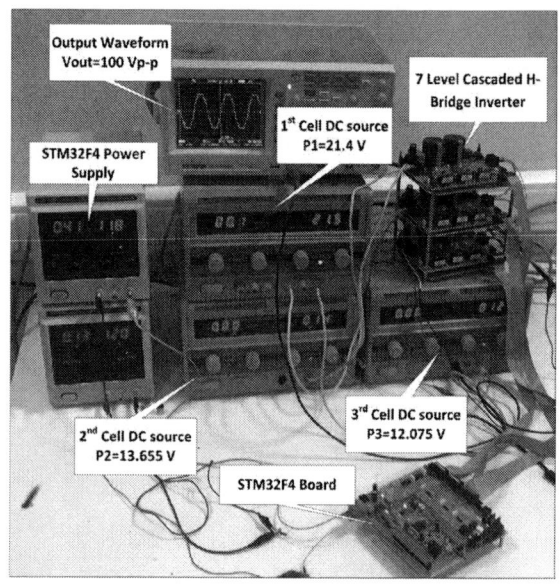

Fig.8. Laboratory prototype of the 7-level H-bridge inverter.

TABLE VII
PARAMETERS FOR EXPERIMENTAL VERIFICATION

Case Study	P_1	P_2	P_3	Opt α_1 (deg)	Opt α_2 (deg)	Opt α_3 (deg)	THD Simulation	THD Experimental
Opt	P_{1opt}= 17.83 v	P_{2opt}= 17.07 v	P_{3opt}= 15.09 v	9.26	29.62	51.81	10.932%	10.968%
1	1.2 P_{1opt}	0.8 P_{2opt}	0.8 P_{3opt}	9.26	29.62	51.81	12.508%	12.640%
2	1.2 P_{1opt}	0.8 P_{2opt}	0.8 P_{3opt}	12.26	33.66	54.38	11.537%	11.346%
3	P_{1opt}	0.2 P_{2opt}	0.2 P_{3opt}	9.26	29.62	51.81	30.016%	30.500%
4	P_{1opt}	0.2 P_{2opt}	0.2 P_{3opt}	3.91	11.53	35.17	18.350%	18.690%

First, the 7-level inverter were supplied with optimized voltage values of P1=17.835 V, P2=17.07 V and P3=15.095 V and switching angles of α1=9.26°, α2=29.62° and α3=51.81° in order to achieve the minimum THD. This yielded a THD of 10.968% calculated by PSPICE from the data collected by GWINSTEK GDS-2000 oscilloscope, which is very close to theoretical value (10.932%). Fig. 9 shows the experimental output waveform of a 7-level cascaded inverter with optimum steps and switching angles.

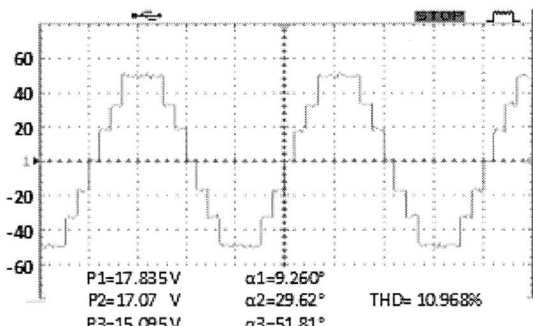

Fig. 9. THD and output voltage waveform with optimized step voltages and switching angles

It should be noted that the amplitude of the output voltage is considered 50 V for 20% voltage variation experiments before any voltage variation and for the fault experiments the amplitude of the output voltage is set to 100 V before any voltage drop.

In the first case study, we examined +20% voltage variation in the 1st voltage step and -20% in the 2nd and the 3rd voltage steps with the previous switching angles. In this case, the output THD went up to 12.64%. Afterwards, the new optimized switching angles equal to α1=12.26°, α2=33.66° and α3=54.38° that were obtained according to the voltage step variations, were applied to Inverter. The new switching angles enhanced the THD to 11.346% which is very close to the theoretical value (11.537%). The THD value is very slightly lower than the theoretical amount which can be related to the accuracy of the HYELEC HY2005E power supplies, the GWINSTEK GDS-2000 and the probes. Experimental output waveforms for the voltage vector 25 in Table III before and after α reconfiguration are shown in Fig.10 and Fig.11 respectively.

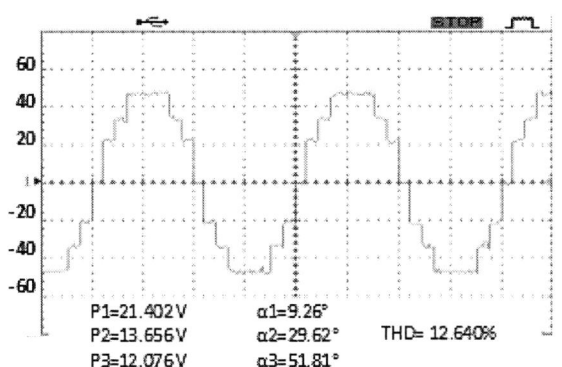

Fig. 10. THD and output voltage waveform for voltage vector 25 in Table III before α reconfiguration.

In the next case study, a 80% voltage drop is examined in the 2nd and the 3rd voltage steps with optimal voltage in the 1st step. This fault condition distorts the output waveform and the output voltage THD increases to 30.5%.

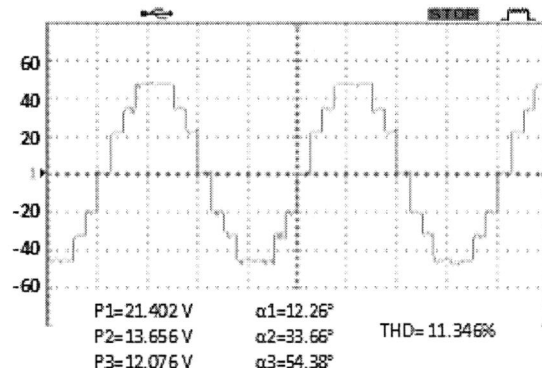

Fig. 11. THD and output voltage waveform for voltage vector 25 in Table III after α reconfiguration

Modifying the switching angles to the optimized value of α1=3.91°, α2=11.53° and α3=35.17° results in the optimum THD. The new switching angles decreased the THD by almost 12% to 18.69% which is very close to the theoretical value of 18.35% that was shown in the simulation case study of Fig. 6 and Fig. 7. The experimental output waveforms for voltage vector 9 in Table VI before and after α reconfiguration are shown in Fig. 12 and Fig. 13 respectively.

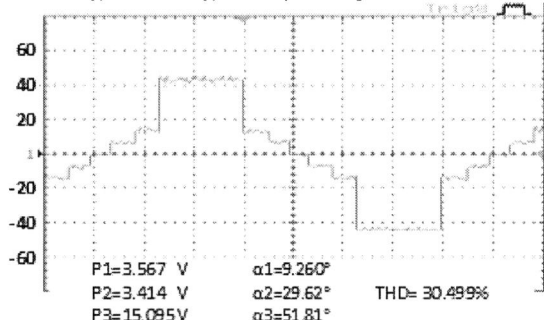

Fig. 12. THD and output voltage waveform for fault voltage vector 9 in Table VI before α reconfiguration.

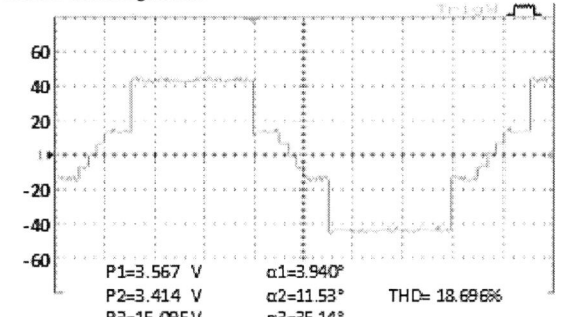

Fig. 13. THD and output voltage waveform for fault voltage vector 9 in Table VI after α reconfiguration.

As can be seen, the THD is improved form 30.49% to 18.69% when the 2nd and 3rd cells have experienced under voltage faults. The experimental results verify the analyses and simulation results with very good accordance.

978-1-5090-0376-1/16 $31.00 © 2016 IEEE

V. CONCLUSION

This paper proposed a method for reducing the output voltage THD in a cascaded H-Bridge inverter when the dc voltage sources are variable and acquire different values during various operating modes. The optimum THD values, considering dc steps and switching angles as optimization variables have been obtained for a 5- 7- 9- 11- 13-level H-bridge inverter. Also, +20% and -20% variations in optimum nominal dc voltages for a 7-level H-bridge inverter are studied in case that dc cells' voltages are variable such as photovoltaic panels or batteries . The THD obtained from voltage variations with previous switching angles were compared to the THD after modifying of the switching angles. A state table of output voltage waveforms was built including the new switching angles and calculated THD values before and after the switching angles re-optimization. The results showed that the THD level can be reduced up to 1% according to the case studies. Furthermore, the proposed method were applied to the condition that the dc cells may experience under voltage faults. A voltage drop of 50% and 80% from nominal steps were examined. The THD with previous switching angles in fault conditions were compared to the THD after reconfiguration of the switching angles. Afterwards, a state table of output voltage vectors was built including the new switching angles and calculated THD values before and after the switching angles re-optimization. The output voltage THD were improved up to 12% in the demonstrated experiment. This result can improve the fault tolerant operation of the load that is connected to the multi-level inverter.

REFERENCES

[1] J. Rodriguez, L. Jih-Sheng, and P. Fang Zheng, "Multilevel inverters: a survey of topologies, controls, and applications," Industrial Electronics, IEEE Transactions on, vol. 49, pp. 724-738, 2002.

[2] S. Le, Z. Wu, W. Ma, X. Fei, C. Xinjian, and Z. Liang, "Analysis of the DC-Link Capacitor Current of Power Cells in Cascaded H-Bridge Inverters for High-Voltage Drives," Power Electronics, IEEE Transactions on, vol. 29, pp. 6281-6292, 2014.

[3] Z. Xiaoming, X. Lan, G. Jinwu, and L. Fei, "Cascaded multilevel converter for medium-voltage motor drive capable of regenerating with part of cells," Power Electronics, IET, vol. 7, pp. 1313-1320, 2014.

[4] L. K. Haw, M. S. A. Dahidah, and H. A. F. Almurib, "A New Reactive Current Reference Algorithm for the STATCOM System Based on Cascaded Multilevel Inverters," Power Electronics, IEEE Transactions on, vol. 30, pp. 3577-3588, 2015.

[5] H. Law Kah, M. S. A. Dahidah, and H. A. F. Almurib, "SHE-PWM Cascaded Multilevel Inverter With Adjustable DC Voltage Levels Control for STATCOM Applications," Power Electronics, IEEE Transactions on, vol. 29, pp. 6433-6444, 2014.

[6] D. Sixing, L. Jinjun, L. Jiliang, and H. Yingjie, "A Novel DC Voltage Control Method for STATCOM Based on Hybrid Multilevel H-Bridge Converter," Power Electronics, IEEE Transactions on, vol. 28, pp. 101-111, 2013.

[7] A. Nami, L. Jiaqi, F. Dijkhuizen, and G. D. Demetriades, "Modular Multilevel Converters for HVDC Applications: Review on Converter Cells and Functionalities," Power Electronics, IEEE Transactions on, vol. 30, pp. 18-36, 2015.

[8] L. Wuhua, J. Qun, M. Ye, L. Chushan, D. Yan, and H. Xiangning, "Modular Multilevel DC/DC Converters With Phase-Shift Control Scheme for High-Voltage DC-Based Systems," Power Electronics, IEEE Transactions on, vol. 30, pp. 99-107, 2015.

[9] L. M. Tolbert, F. Z. Peng, and T. G. Habetler, "Multilevel inverters for electric vehicle applications," in Power Electronics in Transportation, 1998, 1998, pp. 79-84.

[10] Z. Zedong, W. Kui, X. Lie, and L. Yongdong, "A Hybrid Cascaded Multilevel Converter for Battery Energy Management Applied in Electric Vehicles," Power Electronics, IEEE Transactions on, vol. 29, pp. 3537-3546, 2014.

[11] D. E. Soto-Sanchez, R. Pena, R. Cardenas, J. Clare, and P. Wheeler, "A Cascade Multilevel Frequency Changing Converter for High-Power Applications," Industrial Electronics, IEEE Transactions on, vol. 60, pp. 2118-2130, 2013.

[12] S. Daher, J. Schmid, and F. L. M. Antunes, "Multilevel Inverter Topologies for Stand-Alone PV Systems," Industrial Electronics, IEEE Transactions on, vol. 55, pp. 2703-2712, 2008.

[13] H. Ziar, M. Nouri, B. Asaei, and S. Farhangi, "Analysis of Overcurrent Occurrence in Photovoltaic Modules With Overlapped By-Pass Diodes at Partial Shading," Photovoltaics, IEEE Journal of, vol. 4, pp. 713-721, 2014.

[14] H. Ziar, S. Farhangi, and B. Asaei, "Modification to Wiring and Protection Standards of Photovoltaic Systems," Photovoltaics, IEEE Journal of, vol. 4, pp. 1603-1609, 2014.

[15] M. A. Perez, S. Bernet, J. Rodriguez, S. Kouro, and R. Lizana, "Circuit Topologies, Modeling, Control Schemes, and Applications of Modular Multilevel Converters," Power Electronics, IEEE Transactions on, vol. 30, pp. 4-17, 2015.

[16] A. Nabae, I. Takahashi, and H. Akagi, "A New Neutral-Point-Clamped PWM Inverter," Industry Applications, IEEE Transactions on, vol. IA-17, pp. 518-523, 1981.

[17] R. H. Baker, "Electric power converter," U.S. Patent 3867643, Feb. 18,1975.

[18] Q. Jiang and T. A. Lipo, "Switching angles and DC link voltages optimization for multilevel cascade inverters," in Power Electronic Drives and Energy Systems for Industrial Growth, 1998. Proceedings. 1998 International Conference on, 1998, pp. 56-61 Vol.1.

[19] P. Wheeler, L. Empringham, and D. Gerry, "Improved output waveform quality for multi-level H-bridge chain converters using unequal cell voltages," in Power Electronics and Variable Speed Drives, 2000. Eighth International Conference on (IEE Conf. Publ. No. 475), 2000, pp. 536-540.

[20] H. Ziar, E. Afjei, A. Siadatan, and S. Mansourpour, "Optimization of Voltage Levels in Multilevel Inverters," in Universities' Power Engineering Conference (UPEC), Proceedings of 2011 46th International, 2011, pp. 1-5.

[21] B. Diong, H. Sepahvand, and K. A. Corzine, "Harmonic Distortion Optimization of Cascaded H-Bridge Inverters Considering Device Voltage Drops and Noninteger DC Voltage Ratios," Industrial Electronics, IEEE Transactions on, vol. 60, pp. 3106-3114, 2013.

[22] N. Farokhnia, H. Vadizadeh, S. H. Fathi, and F. Anvariasl, "Calculating the Formula of Line-Voltage THD in Multilevel Inverter With Unequal DC Sources," Industrial Electronics, IEEE Transactions on, vol. 58, pp. 3359-3372, 2011.

[23] C. Younghoon, T. LaBella, L. Jih-Sheng, and M. K. Senesky, "A Carrier-Based Neutral Voltage Modulation Strategy for Multilevel Cascaded Inverters Under Unbalanced DC Sources," Industrial Electronics, IEEE Transactions on, vol. 61, pp. 625-636, 2014.

[24] A. Marzoughi and H. Imaneini, "Optimal selective harmonic elimination for cascaded H-bridge-based multilevel rectifiers," Power Electronics, IET, vol. 7, pp. 350-356, 2014.

[25] H. Ziar, S. Mansourpour, E. Afjei, and M. Kazemi, "Bypass diode characteristic effect on the behavior of solar PV array at shadow condition," in Power Electronics and Drive Systems Technology (PEDSTC), 2012 3rd, 2012, pp. 229-233.

[26] L. Maharjan, T. Yamagishi, H. Akagi, and J. Asakura, "Fault-Tolerant Operation of a Battery-Energy-Storage System Based on a Multilevel Cascade PWM Converter With Star Configuration," Power Electronics, IEEE Transactions on, vol. 25, pp. 2386-2396, 2010.

978-1-5090-0376-1/16 $31.00 © 2016 IEEE

7th Power Electronics, Drive Systems & Technologies Conference (PEDSTC 2016)
16-18 Feb. 2016, Iran University of Science and Technology, Tehran, Iran

New Switching Strategy for Single-Phase Multilevel Quasi-Z-Source Inverter

Saman Radman/Mostafa Shahnazari/Hamidreza Toodeji
Dept. of Electrical engineering
University of Yazd
Yazd, Iran
saman.radman@gmail.com

Abstract— A modified pulse width modulation strategy for single-phase multilevel quasi-Z-Source inverters has been proposed in this paper, which reduces switching action applying to traditional simple boost control or maximum constant boost control. Therefore, this strategy yields lower switching loss and higher system efficiency. A Seven-level quasi-Z-Source inverter, based stand-alone photovoltaic system with proposed switching strategy, is designed and simulated in MATLAB/SIMULINK. Simulation results verify the proposed modulation strategy.

Keywords— *Z-Source inverters (ZSIs), quasi-Z-Source inverters (qZSIs), simple boost control (SBC), photo voltaic (PV), pulse width modulation (PWM).*

I. INTRODUCTION

Z-Source inverters (ZSIs) and quasi-Z-Source inverters (qZSI) attract increasing interests in application to PV power systems [1]-[3]. To date, numerous studies have paid close attention to achieving higher efficiency in PV inverters so that the competition is on increasing efficiency by every 0.1% [4]. Due to the ability of ZSI [5] and qZSI [6] to deal with the wide range of voltage variations and improve the inverter reliability in a single-stage topology, they were proposed to overcome the barriers of traditional voltage source inverters in 2003 and 2008, respectively. In ZSI, two switches in one bridge leg can gate on meanwhile, hence, the dead time between switching is removed and the inverter reliability as well as output waveform quality is highly improved. qZSIs provide several improvement in contrast with ZSIs, like drawing a continues current at input inductor, reducing capacitor voltage range and existence of a common dc rail between the input and output of the qZSI, which is easy to assemble and diminish electromagnetic interference problems [7].

More recently, various studies of ZSIs/qZSIs have been devoted in; 1) their topologies to improve the voltage gain of the basic ZSI/qZSI [8]–[12]; 2) Z-source PWMs to achieve buck/boost [13]–[15]; 3) feedback control and parameter design by state space averaging or dynamic modeling [16]–[17]; 4) application on cascade multilevel inverter (CMI) [18]–[20]; 5) analyzing and eliminating low frequency ripples [21]-[23] and ac equivalent model [24]; 6) the ability to achieve distributed maximum power point tracking (MPPT) [25]-[26], and etc. Different kinds of PWM and space vector

modulation (SVM) [27]-[28] techniques have been proposed to achieve wider modulation range, lower voltage stress on power devices, and easier real-time implementation. In [29] a PWM strategy of ZSIs has been proposed with minimum inductor current ripple. A method based on saw-tooth carrier has been presented in [20] which compared with the triangular carrier-based method, results in lower switching loss and improved system efficiency during boost operation.

Classic technique for controlling ZSIs/qZSIs voltage boost called simple boost control (SBC) [5] compares carrier signals with two straight line shape references signal to determine duty cycle of shoot-through state. For achieving wider modulation range, maximum boost control (MBC) [30] strategy was proposed. In this controlling method all traditional zero states are replaced by the shoot-through states, causing a low-frequency ripple in the qZS components. So the maximum constant boost control (MCBC) method is proposed to overcome SBC's low voltage gain and MBC's low-frequency ripples [31]. In MCBC, carrier signal is compared with two sinusoidal shape references, which are upper and lower than the envelope of fundamental wave in order to determine duty cycle of shoot-through state.

The novelty of this paper is to propose a modified PWM technique for the qZS inverters based PV power system to minimize the number of switching actions, which reduces switching loss and improves system efficiency. The proposed strategy can be applied to both SBC and MCBC and also it is compatible with multilevel single-phase qZS inverters. The paper is organized as follows: an overview of ZS/qZS inverter presented in Section II; Section III focuses on the proposed switching strategy and compare it with traditional modulation; a seven-level qZSI based stand-alone PV system is simulated and controlled completely in Section IV; at last, a conclusion is made in Section V.

II. PRINCIPLE OF ZS/QZS INVERTERS

Topologically, ZSIs/qZSIs have an impedance network in their DC-link that is responsible for store and transmit the energy for buck/boost capability. Impedance network consist of two inductors, two capacitors and one diode. Fig. 1 shows the Z-Source and quasi-Z-Source inverters. When DC source voltage is insufficient to supply the output voltage, the traditional voltage source inverters (VSI) require an extra DC–

978-1-5090-0376-1/16 $31.00 © 2016 IEEE

DC converter in their Dc-link that cause increasing cost and system complexity. While ZS/qZS inverters fulfill buck/boost ability in single-stage without extra switch. To achieve voltage boosting, ZSI/qZSI should goes to shoot-through state at certain duty cycle. Shoot-through state occurs when at least one bridge leg be short circuit, which will be discussed further in section III. Both ZSIs and qZSIs have advantages like ability to cope with the wide range of input voltage, removing dead time between upper and lower switches, reducing loss and improving output voltage quality as well as system reliability. Moreover qZSI provides several improvement in ZSI like drawing a continues current at input inductor, reducing capacitor size and existence of a common dc rail between the input and output of the qZSI, which is easy to assemble and diminish electromagnetic interference problems. So in this paper we will continue to use qZSI.

The boost factor of qZSIs is defined as:

$$\frac{V_{dc}}{V_{in}} = \frac{1}{1-2D} \tag{1}$$

Where D represents the shoot-through duty ratio and defines as: $D = T_{st} / T_s$. Where T_{st} is the shoot-through time interval and T_s is switching cycle. So at modulation index of m, voltage gain G is expressed as:

$$G = \frac{m}{1-2D} \tag{2}$$

For achieving higher voltage gain, some ZSI topologies have been used additional elements like capacitors, diode or transformers [8]–[12].

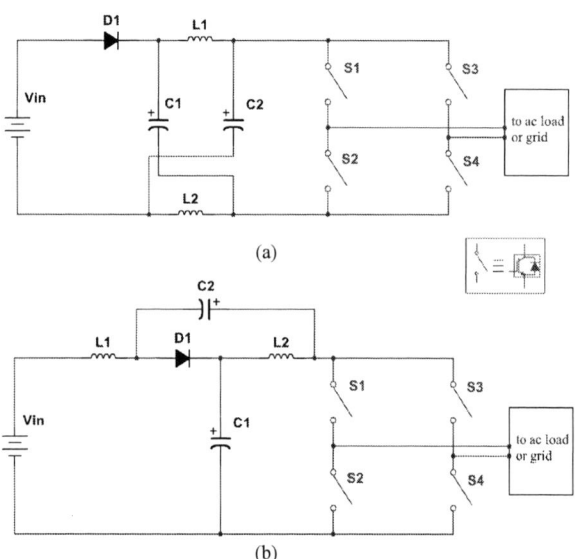

(a)

(b)

Fig. 1. (a) Z-Source inverter; (b) quasi-Z-Source inverter.

The voltage stress on C1 and C2 are defined as follows:

$$V_{C1} = \frac{1-D}{1-2D} \cdot V_{in} \tag{3}$$

$$V_{C2} = \frac{D}{1-2D} \cdot V_{in} \tag{4}$$

can be seen that in qZSI V_{C2} is very smaller than V_{C1} because D is usually less than 0.5 while in ZSI $V_{C1} = V_{C2}$. Thus, ZSI needs higher voltage range capacitor incomparison with qZSI.

III. SWITCHING STRATEGY

A. Various PWM modulations

Traditional unipolar PWM modulation for conventional VSI has two states; active state and zero state. But in qZSIs there is another state called shoot-through. Shoot-through state occurs in part of zero state's time interval for boosting capability. As mentioned, D is the shoot-through duty ratio that determines duty cycle of inverter legs short-circuit. To avoid overlapping, the sum of D and M must be smaller than one (M+D≤1). When D=0 the qZSI operates as conventional inverters. Control of boosting in PWM modulation is done by comparing two reference signals with carrier signal. When the carrier C is higher than the top shoot-through reference (SP) or lower than the bottom shoot-through reference (SN), inverter legs short-circuit must happen. It means S1-S2 or S3-S4 should conduct at this time. Otherwise depend on comparison of C with sinusoidal reference (R) or reference inverse (RI), the active or zero state of inverter can be formed. In SBC method, two reference signals have a straight line shape, as shown in fig. 2(a). In order to achieve wider modulation range, MBC strategy is proposed. In MBC controlling method all traditional zero states are replaced by the shoot-through states, causing a low-frequency ripple in the qZS components. So, the MCBC method is proposed to overcome SBC's low voltage gain and MBC's low-frequency ripples, fig. 2(b).

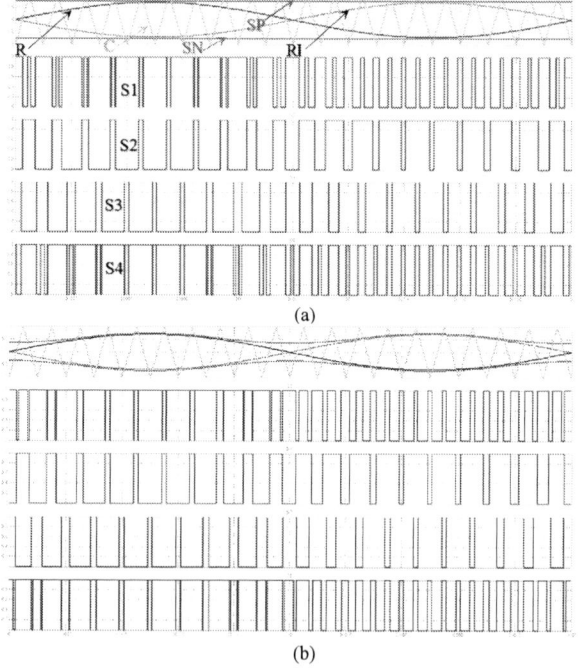

(a)

(b)

Fig. 2. (a) SBC method; (b) MCBC method.

B. Proposed Switching Strategy

As seen in fig.2, total switching frequency is higher than conventional PWM method that causes more switching loss. To reduce total switching actions we propose modified PWM strategy using one shoot-through reference. Fig. 3 shows the proposed switching strategy. Compared to the previous model, total switching action is reduced to half. i.e. at 10 KHz carrier signal and 50 Hz sinusoidal reference wave, total switching actions is 1200, while in proposed method it is just 600 per fundamental cycle without any distortion in output voltage quality or boosting capability. To achieve wider modulation range shoot-through reference can be also sinusoidal shape, like MCBC. In cascade multilevel qZSI structure switching strategy for each cell is the same and just $2\pi/n$ phase shift of carrier signal is needed to apply each cell, where n is number of cascaded qZSI modules. As seen in fig. 3, during positive half cycle, S3 is kept OFF, S4 is kept ON and when the reference (R) is greater than the carrier or carrier is greater than SP switch S1 is turned on, and when carrier is greater than (R) switch S2 is turned on. In negative half cycle S1 is kept OFF, S2 is kept ON and when the reference inverse (RI) is greater than the carrier or carrier is greater than SP switch S3 is turned on, and when carrier is greater than (RI) switch S4 is turned on.

Therefore qZSI has five operation modes. Mode1: the qZSI operates in positive active state (S1 and S4 are ON). Mode2: zero-state (S2 and S4 are ON). Mode3: shoot-through state (S1, S2 and S4 are ON). Mode4: negative active state (S2 and S3 are ON). Mode5: shoot-through state (S2, S3 and S4 are ON). Fig. 4 shows switching patterns under different operation modes.

C. Power Loss Calculation

Power loss in qZSIs consists of H-bridge device loss, qZS diode loss, inductor and capacitor losses of qZS network. For simplicity of power loss calculation, positive half fundamental cycle was considered because there are the same states in the negative half cycle. By using analytical model in [24], device currents and time intervals of qzsi module in one switching cycle T_s for positive half fundamental cycle can be determined as shown in table I.

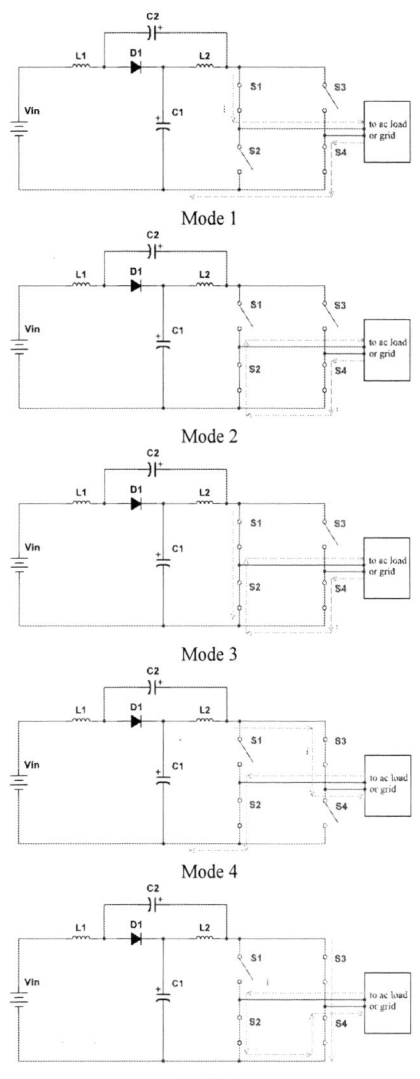

Fig. 4. Switching patterns under different operation modes.

TABLE I. DEVICE CURRENTS AND TIME INTERVALS OF QZSI MODULE FOR POSITIVE HALF FUNDAMENTAL CYCLE.

devices	Conducting time	Conducting current
S1	$m.T_s$	i_{load}
	$D.T_s$	$2i_{l1}$
S2	$D.T_s$	$2i_{l1}$
S4	$m.T_s$	i_{load}
d2(antiparallel diode)	$(1-m).T_s$	i_{load}
D1(qZS diode)	$(1-D-m).T_s$	$2i_{l1}$
	$m.T_s$	$2i_{l1} - i_{load}$
L1,L2	T_s	i_{l1}
C1,C2	$D.T_s$	$-i_{l1}$
	$m.T_s$	$i_{l1} - i_{load}$
	$(1-D-m).T_s$	i_{l1}

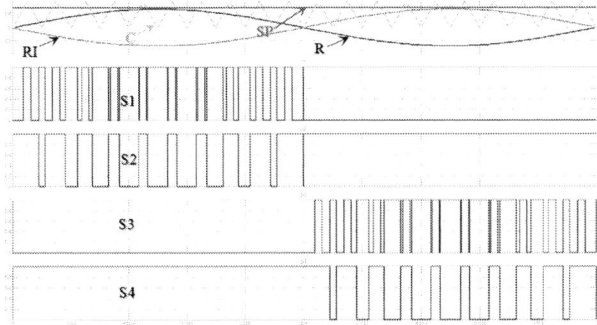

Fig. 3. Proposed Switching Strategy for each module.

1) H-bridge device power loss

There are four IGBTs in qZS module, each with an anti-parallel diode. IGBT and diode power losses, as well as power losses in any semiconductor component, can be divided in two groups: conduction losses and switching losses. [32]

a) Conduction loss

In positive half cycle the conduction loss includes the switch loss (S1, S2 and S4) and free-wheeling diode (d2). And there is:

$$P_{con-S1} = P_{con-tr-S1} + P_{con-st-S1}$$
$$= \frac{1}{\pi}\int_0^\pi [(u_{CE}.i_{load} + r_c.i^2_{load}).m]d\omega t \qquad (5)$$
$$+2(u_{CE}.2i_{l1} + r_c.(2i_{l1})^2).D$$

$$P_{con-S2} = P_{con-st-S2}$$
$$= 2(u_{CE}.2i_{l1} + r_c.(2i_{l1})^2).D \qquad (6)$$

$$P_{con-S3} = 0 \quad \text{(in positive half cycle doesn't conduct)}$$

$$P_{con-S4} = P_{con-tr-S4}$$
$$= \frac{1}{\pi}\int_0^\pi [(u_{CE}.i_{load} + r_c.i^2_{load}).m]d\omega t \qquad (7)$$

$$P_{con-d2} = \frac{1}{\pi}\int_0^\pi [(u_{d0}.i_{load} + r_d.i^2_{load}) \atop .(1-m)]d\omega t \qquad (8)$$

Where P_{con-tr} is conduction loss in traditional state, P_{con-st} is conduction loss in shoot-through state, u_{CE} is on-state zero-current collector-emitter voltage, r_c is collector-emitter on-state resistance, u_{d0} and r_d are the on-state zero-current voltage and the on-state resistance of the anti-parallel diode, respectively. Total conduction power loss in full fundamental cycle can be calculated using:

$$P_{con_{Total}} = 2(P_{con_{S1}} + P_{con_{S2}} + P_{con_{S4}} \atop + P_{con\,d2}) \qquad (9)$$

Comparing with conventional switching method that the current passes through two diode and four switch in each half cycle, conduction power loss is reduced roughly 0.05% by proposed strategy.

b) Switching loss

The switching losses in the IGBT and the diode are the product of switching energies and the switching frequency (f_{sw}).

$$P_{sw} = (E_{onT} + E_{offT} + E_{onD}).f_{sw} \qquad (10)$$

Where E_{onT} and E_{offT} are the turn-on and turn-off energy losses per pulse of the IGBT; E_{onD} is the reverse recovery energy of the antiparallel diode. With proposed switching strategy, two switch is kept ON or OFF in half fundamental cycle so total switching action is reduced to half. Which causes significant switching loss reduction.

2) qZS diode power loss

The shoot-through action of the H-bridge switches will lead to the qZS diode being turned off, otherwise the diode turns on, which causes a loss. The qZS diode blocks the dc-link peak voltage V_{dc}, but the diode current depends on the operating status of the H-bridge switches. According to table I, conduction loss of qZS diode in half fundamental cycle is

$$P_{con-D} = \frac{1}{\pi}\int_0^\pi [(u_{D0}.2i_{l1} + r_D.(2i_{l1})^2) \atop .(1-D-m)]d\omega t$$
$$+ \frac{1}{\pi}\int_0^\pi [(u_{D0}.(2i_{l1} - i_{load}) \atop + r_D.(2i_{l1} - i_{load})^2)].m\, d\omega t \qquad (11)$$

And the reverse recovery loss is

$$P_{rr-D} = (Q_{rr-D} \times V_{dc}).f_{sw} \qquad (12)$$

Where u_{D0} and r_D are the on-state zero-current voltage and the on-state resistance of the qZS network diode, respectively. Q_{rr-D} is the reverse recovery charge of the qZS diode. So total qZS diode power loss in whole fundamental cycle can be obtained by:

$$P_{D,total} = 2(P_{con-D} + P_{rr-D}) \qquad (13)$$

3) qZS inductor power loss

The inductor power loss is composed of the core loss and winding loss. In the quasi-Z-source network, the current through the inductors contains dc, double-frequency, and high-frequency ac components. The dc and low-frequency ac components mainly cause the winding loss. The high frequency ac current is related to the core loss. The inductors total loss can be calculated as

$$P_{ind,total} = 2P_{copper,ind} + P_{core,ind} \qquad (14)$$

$$P_{copper,ind} = i_{l1}^2.R_l \qquad (15)$$

$$P_{core,ind} = kf_{sw}^m B^n W_t \qquad (16)$$

Where R_l resistance of the inductor's copper winding. f_{sw} is the inductor current ripple frequency; B is the high-frequency ac flux density; W_t is the weight of the core; the coefficients k, m and n for different core materials can be found in [33].

4) qZS capacitor power loss

Total power loss of qZS capacitors can be calculate as:

$$P_{C,total} = 2.R_{esr}.i_{C1}^2 \qquad (17)$$

Where R_{esr} is the capacitor's equivalent series resistance And I_{C1} is the rms value of the capacitor current [24], according table I, I_{C1} can be calculated as

$$I_{C1} = \sqrt{\begin{array}{c} (-i_{l1})^2.D \\ + \frac{1}{\pi}\int_0^\pi (i_{l1} - i_{load})^2).m\, d\omega t \\ + \frac{1}{\pi}\int_0^\pi (i_{l1})^2.(1-D-m)]d\omega t \end{array}} \qquad (18)$$

IV. SIMULATION RESULT

In order to verify the proposed switching strategy, the simulation is performed by using MATLAB/SIMULINK. Three cascaded qZSIs are connected to resistance load through LCL filter. Each qZS module is supplied by single PV panel. The P-V characteristic of panel is shown in Fig. 5. The system is simulated in non-uniform irradiations. First PV array is irradiated by 1000 W/m², second and third PV array is irradiated by 850 W/m² and 750 W/m² respectively. Each qZS module must be controlled independently. Index SP and M are controller parameters. The MPPT adjusts the PV voltage by regulating the index D (where SP=1-D) and defines shoot-through duty ratio. Output voltage is kept to 220 volt rms by comparing DC-link capacitor voltage (VC1) of each module and reference voltage (Vref), where Vref is adjusted by output power regulation. Finally M index can be achieved through PI controller. Fig. 6 shows SP and M index of each module during 0.35s simulation time. It is obvious that SP3>SP2>SP1 to keep PV voltage at maximum power point. In second 0.1s DC-link controller is involved and M indexes are regulated to keep output voltage at 220 Vrms. For maximum power point tracking (MPPT) of PV panel, Hill-climbing algorithm has been used.

Fig. 7 shows configuration of whole system. The inductances, L_1 and L_2, and the capacitances, C_1 and C_2, of each qZS network are 470 µH and 100 µF, respectively. Four IGBT are chosen as power switch. Carrier frequency 10 kHz with 120 degree phase shifting for each module. The system is in stand-alone mode with 110 Ω resistance load. The filter parameters are $L_{f1} = L_{f2} = 1mH$, $C_f = 2.2\mu F$. In control scheme, index n defines module number and can be 1, 2 or 3.

Fig. 8 shows Z-Source network Capacitors voltage. As seen in Fig. 8, DC-link voltage of each module is kept the same after transient state. Fig. 9 shows load voltage. Fast Fourier transform (FFT) spectrums of output voltage is obtained at this condition and open-loop control condition. Total harmonic distortion (THD) of voltage in close-loop condition is 4.7%, while in conditions where SPn=Mn=0.85, THD is 3.2%. Fig. 10 shows output voltage of 7-level qZSI before filters. Results show that proposed switching method in addition to reducing switching power loss, keeps power quality and boosting capability.

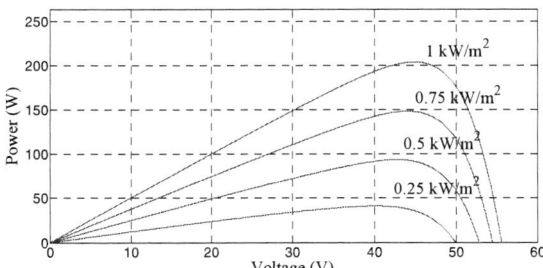

Fig. 5. P-V characteristic of PV panel in different irradiations.

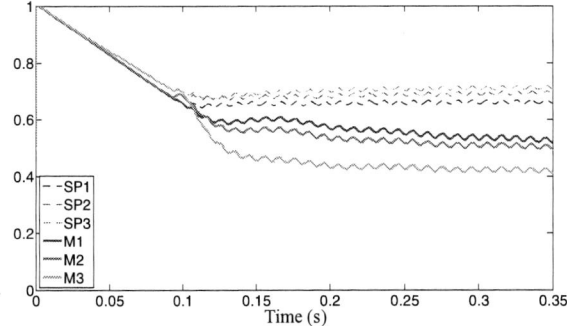

Fig. 6. Control indexes

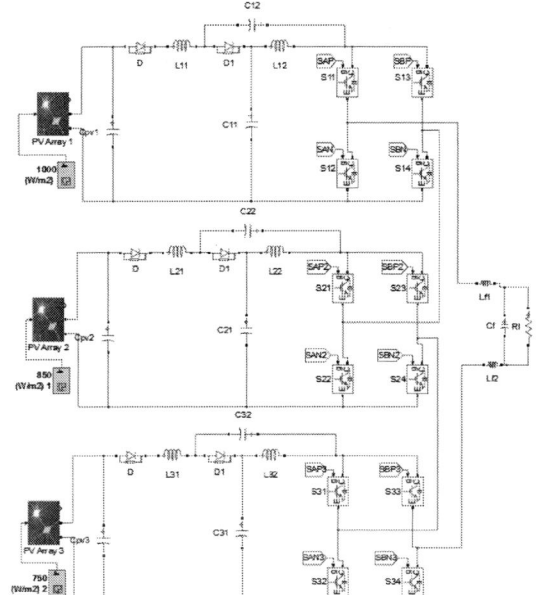

Fig. 7. 7-level qZSI configuration.

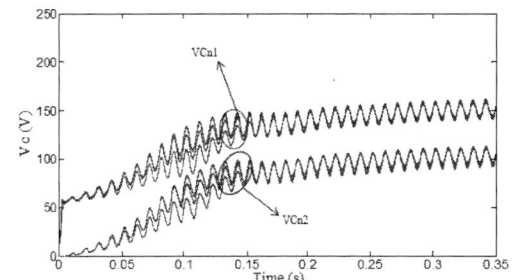

Fig. 8. Capacitors voltage.

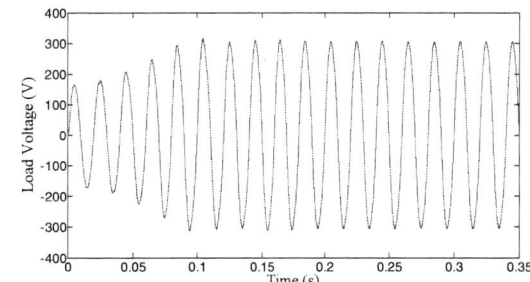

Fig. 9. Load voltage.

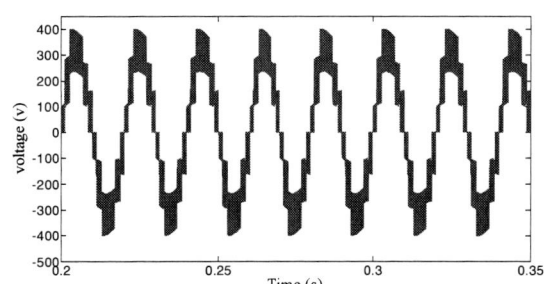

Fig. 10. Output voltage of 7-level qZSI before filter.

V. CONCLUSION

In this paper, a modified PWM switching strategy based SBC method was proposed for 7-level qZS single-phase inverter. With the proposed switching technique, the number of switching actions and total conduction losses of H-bridge devices are greatly reduced. H-bridge device power loss reduction was about 0.55 % causing total system efficiency improvement. The proposed switching technique is also compatible with MCBC method. Power loss of qZSI is calculated in detail and finally a 7-level qZSI with proposed switching strategy based PV system was simulated in MATLAB/SIMULINK. Independent MPPT control and DC-link voltage regulation are applied to system. Simulink results verify proposed switching strategy.

REFERENCES

[1] B. Ge, H. Abu-Rub, F. Peng, Q. Lei, de Almeida A., Ferreira F., D. Sun,Y. Liu, "An energy stored quasi-Z-source inverter for application to photovoltaic power system," *IEEE Trans. Ind. Electron.*, vol.60, no.10, pp.4468-4481, Oct. 2013.

[2] B. Farhangi, S. Farhangi, "Comparison of Z-source and boost-buck inverter topologies as a single phase transformer-less photovoltaic grid-connected power conditioner," *37th IEEE Power Electronics Specialists Conference*, (PESC'06), pp.1-6, 18-22 June. 2006.

[3] D. Sun, B. Ge, D. Bi, and F. Z. Peng, "Analysis and control of quasi-Z source inverter with battery for grid-connected PV system," *International Journal of Electrical Power & Energy Systems*, vol. 46, pp. 234-240, March. 2013.

[4] B. Sahan, S. V. Araujo, T. Kirstein, L. Menezes, and P. Zacharias, "Photovoltaic converter topologies suitable for SiC-JFETs," in Proc. PCIM Europe, 2009, pp. 431-437.

[5] F. Z. Peng, "Z-source inverter," *IEEE Trans. Ind. Applicat.*, vol. 39, no. 2, pp. 504–510, April. 2003.

[6] J. Anderson, F. Z. Peng, "Four quasi-ZSource inverters," in 39th *IEEE Annu. Power Electronics Specialists Conf., (PESC'08)*, June 2008, pp. 2743–2749.

[7] Y. Liu,H. A.Rub, B. Ge, "Z-Source/Quasi-Z-Source Inverters: Derived Networks, Modulations, Controls, and Emerging Applications to Photovoltaic Conversion," *IEEE Trans. Ind. Electron*, vol. 8, no.4, pp. 32-44, December. 2014.

[8] C. J. Gajanayake, F.-L. Luo, H. B. Gooi, P. L. So, and L. K. Siow, "Extended-boost Z-source inverters," *IEEE Trans. Power Electron.*, vol. 25, no. 10, pp. 2642–2652, Oct. 2010.

[9] W. Qian, F. Z. Peng, and H. Cha, "Trans-Zsource inverters," *IEEE Trans. Power Electron.*, vol. 26, no. 12, pp. 3453–3463, Dec. 2011.

[10] M. Adamowicz, J. Guzinski, R. Strzelecki, F. Z. Peng, and H. Abu-Rub, "High step-up continuous input current LCCT-Z-source inverters for fuel cells," in *Proc. 2011 IEEE Energy Conversion Congress Exposition (ECCE)*, Sept. 17–22, pp. 2276–2282.

[11] M. Adamowicz, R. Strzelecki, F. Z. Peng, J. Guzinski, and H. Abu-Rub, "New type LCCT-Zsource inverters," in *Proc. 14th European Conf. Power Electronics Applications (EPE 2011)*, Aug. 30–Sept. 1, pp. 1–10.

[12] R. Strzelecki, W. Bury, M. Adamowicz, and N. Strzelecka, "New alternative passive networks to improve the range output voltage regulation of the PWM inverters," in *Proc. 2009 IEEE Applied Power Electronics Conf. Exposition (APEC)*, vols. 1–4, pp. 857–863.

[13] H. Rostami and D. A. Khaburi, "Voltage gain comparison of different control methods of the Z-source inverter," in *Proc. Int. Conf. Electrical Electronics Engineering (ELECO 2009)*, Nov. 5–8, pp. I-268–I-272.

[14] O. Ellabban, J. Van Mierlo, and P. Lataire, "Experimental study of the shoot-through boost control methods for the Z-source inverter," *Eur. Power Electron. Drives Assoc. J.*, vol. 21, no. 2, pp. 18–29, June. 2011.

[15] Y. Liu, B. Ge, H. Abu-Rub, and F. Z. Peng, "Overview of space vector modulations for three-phase Z-source/quasi-Z-source inverters," *IEEE Trans. Power Electron.*, vol. 29, no. 4,pp. 2098–2108, Apr. 2014.

[16] Y. Li, S. Jiang, J. Cintron-Rivera, and F. Peng, "Modeling and control of quasi-Z-source inverter for distributed generation applications," *IEEE Trans. Ind. Electron.*, vol. 60, no. 4, pp. 1532–1541, Apr. 2013.

[17] C. J. Gajanayake, D. M. Vilathgamuwa, and P. C. Loh, "Development of a comprehensive model and a multiloop controller for Z-source inverter DO systems," *IEEE Trans. Ind. Electron.* vol. 54, pp. 2352–2359, Aug. 2007.

[18] E. Villanueva, P. Correa, J. Rodriguez, and M. Pacas, "Control of a single-phase cascaded H-bridge multilevel inverter for grid-connected photovoltaic systems," *IEEE Trans. Ind. Electron.*, vol. 56, no. 11, pp. 4399–4406, Nov. 2009.

[19] M. Malinowski, K. Gopakumar, J. Rodriguez, and M. A. Pérez, "A survey on cascaded multilevel inverters," *IEEE Trans. Ind. Electron.*, vol. 57, no. 7, pp. 2197–2206, Jul. 2010.

[20] Y. Zhou, L. Liu, and H. Li, "A high performance photovoltaic module-integrated converter (MIC) based on cascaded quasi-z-source inverters (qZSI) using eGaN FETs," *IEEE Trans. Power Electron.*, vol.28, no.6, pp. 2727-2738, Jan. 2013.

[21] Y. Yu, Q. Zhang, B. Liang, and S. Cui, "Single-phase Z-source inverter: analysis and low-frequency harmonics elimination pulse width modulation," *IEEE ECCE'2011*, Sept. 2011, pp. 2260-2267.

[22] Z. Gao, Y. Ji, Y. Sun, and J. Wang, "Suppression of voltage fluctuation on DC link voltage of Z-source," *Journal of Harbin University of Science and Technology*, vol.16, no.4, pp.86-89, Aug. 2011.

[23] Liu, Y.; Ge, B.; Abu-Rub, H.; Sun, D, "Comprehensive Modeling of Single-Phase Quasi-Z-Source Photovoltaic Inverter to Investigate Low-Frequency Voltage and Current Ripple," *IEEE Trans. Ind. Electron.*, vol. 62, no. 7, pp. 4194–4202, Jul. 2015.

[24] D. Sun, B. Ge, X. Yan, D. Bi, H.Zhang, Y. Liu, H. A. Rub, L. B. Brahim, F.Z. Peng, "Modeling, Impedance Design, and Efficiency Analysis of Quasi-Z Source Module in Cascaded Multilevel Photovoltaic Power System," *IEEE Trans. Ind. Electron.*, vol. 61, no. 11, pp. 6108–6117, June. 2014.

[25] Y. Liu, B. Ge, H. Abu-Rub, F. Z. Peng, "An effective control method for quasi-Z-source cascade multilevel inverter-based grid-tie single-phase photovoltaic power system," *IEEE Trans. Ind. Informat.*, vol.10, no.1, pp.399-407, Feb. 2014.

[26] H. Abu-Rub, A. Iqbal, S. M. Ahmed, F. Peng, Y. Li, B. Ge,"Quasi-Z-source inverter-based photovoltaic generation system with maximum power tracking control using ANFIS," *IEEE Trans. Sustain. Energy*, vol.4, no.1, pp.11-20, Jan. 2013.

[27] Y. Liu, B. Ge, F. J. T. E. Ferreira, A. T. de Almeida, and H. Abu-Rub, "Modeling and SVM control of quasi-Z-source inverter," in *Proc. 11th Int. Conf. Electrical Power Quality Utilization (EPQU)*, Oct. 17–19, 2011, pp. 1–7.

[28] U. S. Ali and V. Kamaraj, "A novel space vector PWM for Z-source inverter," in *Proc. 1st Int. Conf. Electrical Energy Systems (ICEES)*, Jan. 3–5, 2011, pp. 82–85.

[29] Y. Tang, Sh. Xie, J. Ding, "Pulse width Modulation of Z-Source Inverters with Minimum Inductor Current Ripple," *IEEE Trans. Ind. Electron.*, vol. 61, no. 1, pp. 98–106, January. 2014.

[30] F. Z. Peng, M. Shen, and Z. Qian, "Maximum boost control of the Z-source inverter," *IEEE Trans. Power Electron.*, vol. 20, no. 4, pp. 833–838, July. 2005.

[31] M. Shen, J. Wang, A. Joseph, F. Z. Peng, L. M. Tolbert, and D. J. Adams, "Constant boost control of the Z-source inverter to minimize current ripple and voltage stress," *IEEE Trans. Ind. Applicat.*, vol. 42, no. 3, pp. 770–778, May/June. 2006.

[32] D. Graovac, M.Purschel, "IGBT Power Losses Calculation Using the Data-Sheet Parameters," *Automotive Power*, Application Note, vol.1.1, Jul. 2009.

[33] W. McLyman, *Transformer and Inductor Design Handbook*, 3rd ed. Idyllwild, CA: Kg Magnetics, Inc., 2004.

7th Power Electronics, Drive Systems & Technologies Conference (PEDSTC 2016)
16-18 Feb. 2016, Iran University of Science and Technology, Tehran, Iran

Reliability Evaluation of Two-Stage Interleaved Boost Converter Interfacing PV Panels Based on Mode of Use

Farid Hamzeh Aghdam
Faculty of Electrical and Computer
Engineering
University of Tabriz, Iran
f.hamzeh93@ms.tabrizu.ac.ir

Mehrdad Tarafdar Hagh
Faculty of Electrical and Computer
Engineering
University of Tabriz, Iran
tarafdar@tabrizu.ac.ir

Mehdi Abapour
Faculty of Electrical and Computer
Engineering
University of Tabriz, Iran
abapour@tabrizu.ac.ir

Abstract— By rapid penetration of renewable energy sources, such as photovoltaic (PV) and wind generation systems, many issues about utilization of these units have emerged. To achieve a sustainable source of electrical energy, a reliable generation unit is required. In PV systems, the DC–DC boost converters are responsible for maximum power point tracking (MPPT). Consequently reliability of these converters should be increased to have a reliable source of electrical energy. The aim of this paper is to evaluate reliability of interleaved boost converters interfacing a PV panel based on mode of use. A comparative study on reliability of a two-stage interleaved boost converter has been done based on different operation modes; redundant operation or parallel operation. For parallel mode of use, two scenarios are considered. In low power generation of PV panels, one healthy stage would be adequate to process the power, however in nominal power generation, operation of both stages would be obligatory. Effects of converter components residential and switching power losses, are contemplated in reliability calculations. Without considering off time of PV panels during night hours, previous studies performed on reliabilities of PV panels power electronic interfaces are pessimistic. Due to this fact, actual time of operation in PV panels, are considered in reliability calculation of the converter, by a clustering method. Finally output data of a PV panel installed in the campus is used to get the actual reliability relations.

Keywords—Renewable Energy, PV Generation Systems, Interleaved DC-DC Boost Converters, Reliability.

I. INTRODUCTION

Environmental concerns, and limited amount of fossil fuels and nuclear energy sources, have caused that renewable energy sources to gather so much attention for replacing conventional power plants to generating electricity [1],[2].

The energy produced by photovoltaic solar cells can be considered as the most important and essential energy resource due to abundance and sustainability of solar energy. Furthermore, solar power causes no emission and destruction. These advantages have led to rapid growth of solar power installed capacity, which has reached at least 177 Giga watts (GW) by the end of 2014 [3].

Low voltage direct current distribution system, is a promising solution for the restoration of MV branch lines with

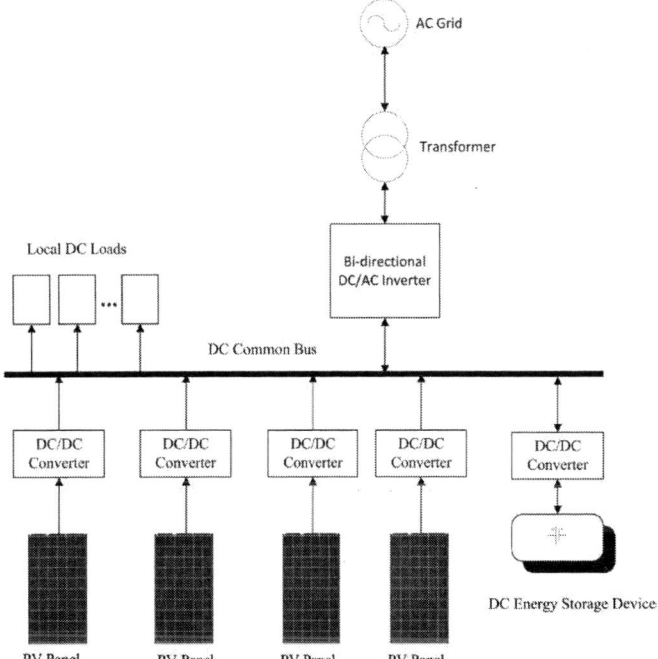

Fig. 1. The grid-connected PV generation system with energy storage elements, local dc loads, dc–dc converters, common DC bus, dc–ac inverters and utility grid.

benefits of large power transfer capacity with low voltage, high cost saving potential and improvements in reliability and voltage quality [4]. Most of distributed energy resources, like PV generation units, generate DC power and also energy storage devices supply DC power. Furthermore most of the consumer loads such as LEDs in lighting systems and digital devices like PCs, use DC power [5]. Hence DC interface seems more efficient than AC, in units with DC voltage generation, like PV panels, due to reduction in conversion loss in DC to AC and AC to DC stages [6].

DC-DC converters are essential parts in DC distribution systems for connecting PV panels and energy storage devices to the DC bus [7]. These converters are responsible for maximum power point tracking (MPPT) and regulating voltage on the DC bus, despite variation of solar radiation and

978-1-5090-0376-1/16 $31.00 © 2016 IEEE

changes in size of the load. Fig. 1 shows a complete structure of a PV generation system consisting of energy resources, energy storage elements, loads, DC–DC converters, DC bus, DC/AC inverters and utility grid.

PV power systems could fail as a result of accidental events or occasional failures in their components. So, it seems necessary to evaluate reliabilities of DC-DC converters used in PV systems, to get a perspective of the reliability of the whole system.

There has been a little research on reliabilities of power electronic devices. The following references are introduced shortly. Ref [8] presents a comprehensive review of reliability assessment and improvement of power electronic systems. A survey of determination of the industrial requirements and expectations of reliability in power electronic converters, systems, and components is presented in [9]. Reference [10] presents a reliability evaluation and comparison of conventional and interleaved DC-DC boost converters. Reference [11] analyzes reliability of the power electronic converter for a grid-connected PMSG wind turbine, based on the semiconductor power loss. Ref [12] has done a research on reliability analysis of a push-pull converter designed for connecting to a 125-W photovoltaic (PV) panel. Reference [13] studies failure rates of PV modules, inverters, and capacitors used in a grid connected PV power system. Ref. [14] introduces an algorithm for accurate calculation of reliability of PV systems and number of PV modules and batteries to satisfy a desired loss of load probability (LOLP). Also ref. [15] analyses thermal histories of inverter components which are collected from operating inverters from several manufacturers to determine thermal profiles and the dependence on local conditions, as well as to assess the effect on inverter reliability

Interleaved boost converters have multiple stages that each stage consists of a diode, inductor and switch. Having more stages has led to an increment in power processing capability and improving reliability of power electronic system in interleaved boost converters. Consequently they have gathered so much interest recently.

Aim of this paper is to evaluate reliability of a two-stage boost converter used in a PV panel, based on mode of use. First, parameters affecting the failure rate of a power electronic device will be discussed and then power loss and temperature effects on failure rate will be studied. Next section is dedicated to reliability calculations according to Markov model of converter. Reliability relation will be computed, based on different modes of operation; in one mode, one stage remains redundant and the other will have duty of power processing; in the other mode, stages will operate simultaneously. For parallel mode of use, two scenarios are considered; in low power generation of PV panels, one healthy stage would be adequate to process the power, however in nominal power generation, operation of both stages is obliged. Finally, reliability and mean time to failure (MTTF) of both modes, based on clustered actual data, will be calculated, and compared to each other.

TABLE I
FAILURE RATE MODELS

MOSFET	$\lambda_{MOSFET} = \lambda_b \pi_T \pi_A \pi_Q \pi_E$
Diode	$\lambda_{Diode} = \lambda_b \pi_T \pi_S \pi_C \pi_Q \pi_E$
Inductor	$\lambda_{Inductor} = \lambda_b \pi_T \pi_Q \pi_E$

TABLE II
TEMPERATURE FACTORS

MOSFET	$\pi_T = \exp\left[-1925\left(\dfrac{1}{T_j + 273} - \dfrac{1}{298}\right)\right]$
Diode	$\pi_T = \exp\left[-3091\left(\dfrac{1}{T_j + 273} - \dfrac{1}{298}\right)\right]$
Inductor	$\pi_T = \exp\left[\dfrac{-0.11}{8.617 \times 10^{-5}}\left(\dfrac{1}{T_{HS} + 273} - \dfrac{1}{298}\right)\right]$

II. PARAMETERS AFFECTING FAILURE RATE

In this section, parameters influencing failure rate of converter components are discussed.

According to reference [16] the failure rate for a microelectronic part is

$$\lambda_{part} = \lambda_b \prod_{i=1}^{n} \pi_i \qquad (1)$$

where n indicates the number of factors that affect the failure rate of components.

Table I, shows the failure rates of boost converter components. From the mentioned relations in table I, temperature effect on the failure rate is obvious. The temperature factor π_T is function of the device junction temperature, T_j, for MOSFET and diode, or the hot spot temperature, T_{HS}, for inductor as listed in table II [10]. With an increment in temperature of the elements, the failure rate increases.

Beside temperature, failure rate also depends on the application of the device in the circuit, the quality and the type of materials used in the fabrication of the device and environment of operation which are considered in failure rate relations by π_Q and π_E respectively [16]. The electrical stress factor π_S for a diode is affected by its voltage stress ratio which is the ratio of reverse voltage applied to rated reverse voltage. Also for a diode, the contact construction factor will be $\pi_C = 1$.

Except temperature factor, all the other factors are assumed to be constant in considered converter circuit. In the next section a detailed analysis of temperature factor of converter components is presented.

III. TEMPERATURE AND POWER LOSS EFFECTS

In this section, power loss and temperature effects on the junction temperature, T_j, and thereby on failure rate of the part is discussed in details.

According to [10], junction temperature for semi-conductor devices are calculated based on maximum power dissipation, P_D, and the device junction to case thermal resistance, θ_{JC}.

TABLE III
POWER LOSS RELATIONS OF BOOST CONVERTER COMPONENTS

MOSFET	$P_D = R_{DS(on)}(DI_{in})^2 + \frac{1}{2}V_{out}I_{in}(t_{rise} + t_{fall})f_s$ $+ \frac{1}{2}C_{DS}f_sV_{out}^2$
Diode	$P_D = V_fI_{out} + R_DI_{out}^2 + \frac{1}{2}C_Df_sV_{out}^2$
Inductor	$P_D = R_LI_{in}^2$

With a detailed thermal analysis of the components and using equations (2)-(5), components junction temperatures are calculated [10].

$$T_j = T_C + \theta_{JC}P_D \qquad (2)$$

where, T_j is junction temperature, θ_{JC} is junction-to-case thermal resistance, P_D is power dissipation and T_C is the case temperature and is expressed as

$$T_C = T_a + \theta_{CA}P_D \qquad (3)$$

where, θ_{CA} is the thermal resistance between the case and ambient and T_a, is ambient temperature.

Also the hot spot temperature, T_{HS} of an inductor is a function of its power dissipation, P_D, and radiating surface area, A, of the case. From ref. [10] the relations are:

$$T_{HS} = T_a + 1.1\Delta T \qquad (4)$$

$$\Delta T = 125\,P_D/A \qquad (5)$$

The power loss is the only parameter that affects the temperature factor in test case systems. Hence the relations of boost converter components power losses, are presented in table III. Both switching and resistance power losses of components have been considered in equations which are shown in the table. As the power flowing from the device increases, its corresponding current and power dissipation will rise. Next section is dedicated to Markov reliability development.

IV. MARKOV RELIABILITY MODEL DEVELOPMENT OF 2-STAGE INTERLEAVED BOOST CONVERTER IN DIFFERENT MODES OF OPERATION

One of the important techniques for evaluating reliability of systems is Markov approach.

For any given system, a Markov model consists of a list of that system's possible states, the possible transition paths between those states, and the rate parameters of those transitions. In reliability analysis, the transitions usually consist of failures and repairs [17].

Fig. 2 shows the functional diagram of a two-stage interleaved converter, which composed of two boost converters. Each stage consists of a diode, inductor and switch. Interleaved converters are more efficient and reliable comparing with conventional converters [10].

In this section, Markov reliability model of a two-stage interleaved DC-DC boost converter is presented according to modes of use. Fig. 3 depicts the Markov chain diagram of a two-stage interleaved configuration. Both stages are assumed to be identical. In Markov chain, in state 11 both stages are

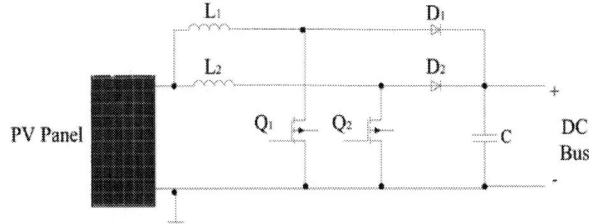

Fig. 2. A 2-stage interleaved DC-DC Boost Converter connected to a PV panel

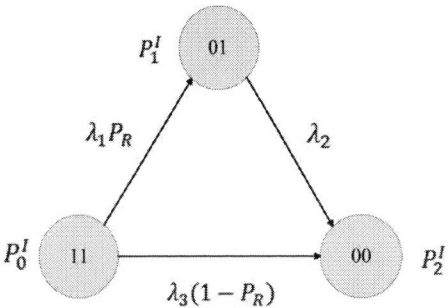

Fig. 3. Markov chain of 2-stage interleaved DC-DC boost converter

healthy; in state 01 one of the stages has failed; in state 00 both stages have failed.

In this study, two modes of operation are supposed: In first mode, one stage will be used in all power levels, until it fails operating. After its failure, the second stage will start power processing. In the second mode, both stages will work simultaneously. In this mode, two scenarios would be considered: In the first scenario, it is assumed that in low power generation of PV panel, after the failure of one of the stages, the other stage continues operating. The second scenario occurs when PV panel is generating in its nominal rating. In this scenario, when one stage fails the other stage cannot handle the input power and whole system fails.

In Markov chain diagram, the rate of going from stage 11 to stage 01, is $\lambda_1 P_R$ and the rate of transitioning from stage 01 to stage 00 is λ_2, and transitioning from stage 11 to stage 00 is $\lambda_3(1 - P_R)$. P_R is the probability that the fault management mechanism can manage the fault in case of fault occurrence.

A) Redundant Mode: For first mode of operation, with a redundant stage, equation (6) shows occupational probabilities of states in which just state 00 is failure state. By assuming the first state to be initial state, as in equation (7), and solving the differential equation (6), we will have the reliability of the converter, which is sum of the occupational probabilities of states 11 and 01, as shown in equation (8).

We have $\lambda_1 = \lambda_2 = \lambda_3 = \lambda_L + \lambda_D + \lambda_Q$ in this mode of operation. λ_L, λ_D and λ_Q are failure rates of inductor, diode and MOSFET respectively.

$$\frac{d}{dt}\begin{bmatrix} P_0^{I-2} \\ P_1^{I-2} \\ P_2^{I-2} \end{bmatrix} = \begin{bmatrix} -\lambda_1 P_R - \lambda_3(1 - P_R) & 0 & 0 \\ \lambda_1 P_R & -\lambda_2 & 0 \\ \lambda_3(1 - P_R) & \lambda_2 & 0 \end{bmatrix}\begin{bmatrix} P_0^{I-2} \\ P_1^{I-2} \\ P_2^{I-2} \end{bmatrix} \qquad (6)$$

$$P(0) = \begin{bmatrix} 1 & 0 & 0 \end{bmatrix} \qquad (7)$$

$$R^{I-2}(t) = e^{-\lambda t} + \lambda P_R t e^{-\lambda t} \qquad (8)$$

In above equations, P_i^{I-2} and P_R are occupational probability of state i and probability of correct operation of fault detection system respectively.

B) Simultaneous Mode: For second mode of operation, two scenarios are assumed.

First scenario: In low power generation, when a fault occurs and one of the stages fails, the other stage continues to operate. The occupational probabilities of the states, are just like previous mode of operation as presented in equation (6) with different failure rates, due to different power passage in healthy stage, before and after failure. So we will have $\lambda_1 = 2(\lambda_L + \lambda_D + \lambda_Q)$, $\lambda_2 = \lambda_L + \lambda_D + \lambda_F$ and $\lambda_3 = 2(\lambda_L + \lambda_D + \lambda_Q)$, in this scenario. By solving the differential equation (6), we will have the reliability of the converter, which is sum of the occupational probabilities of states 11 and 01, as shown in equation (9).

$$R^{I-2}(t) = 1 - P_2^{I-2}$$
$$= -\frac{(\lambda_2 - \lambda_3(1 - P_R))e^{-(\lambda_1 P_R + \lambda_3(1-P_R))t} - \lambda_1 P_R e^{-\lambda_2 t}}{\lambda_1 P_R + \lambda_3(1 - P_R) - \lambda_2} \qquad (9)$$

Second scenario: In this scenario, when PV panel produces in its nominal rating, both stages are working in their full-power modes. As a result, when a fault occurs and one stage fails operating, the other stage fails to deliver its input power to the output. By considering states 01 and 00 as failure states and assuming the first state to be initial state, the reliability equation leads to below equations. Also $\lambda_1 = \lambda_3 = 2(\lambda_L + \lambda_D + \lambda_Q)$, in this scenario.

$$\frac{d}{dt}\begin{bmatrix} P_0^{I-2} \\ P_1^{I-2} \\ P_2^{I-2} \end{bmatrix} = \begin{bmatrix} -\lambda_1 P_R - \lambda_3(1 - P_R) & 0 & 0 \\ \lambda_1 P_R & 0 & 0 \\ \lambda_3(1 - P_R) & 0 & 0 \end{bmatrix}\begin{bmatrix} P_0^{I-2} \\ P_1^{I-2} \\ P_2^{I-2} \end{bmatrix} \qquad (10)$$

$$P(0) = [1 \quad 0 \quad 0] \qquad (11)$$
$$R^I(t) = P_0^{I-2} = e^{-(\lambda_1 P_R + \lambda_3(1-P_R))t} \qquad (12)$$

Each scenario will take part in reliability relation, by its probability of occurrence.

In next section, reliability of a 2-stage interleaved converter in different modes of use, will be discussed based on output data of a practical PV panel installed in the campus.

V. RELIABILITY EVALUATION OF A TWO-STAGE INTERLEAVED CONVERTER IN A PRACTICAL SYSTEM

As discussed in the previous sections, failure rate of power electronic devices, depend on the mission time and their power dissipation. In this section reliability relations of converter in different modes would be calculated and quantified by mean time to failure (MTTF) index to get a better view of systems reliability as presented in equation (13).

$$MTTF = \int_0^\infty R(t)dt \qquad (13)$$

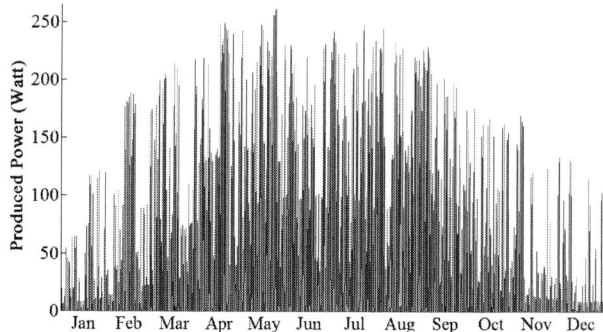

Fig. 4. Hourly produced power of the PV Panel

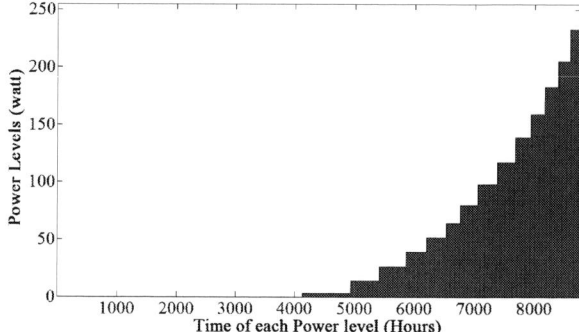

Fig. 5. Clustered output data of PV panel constructed with clustered

One year output power data for a PV panel installed in the campus is depicted in fig. 4. Also fig. 5 shows the clustered output data of the PV panel. Reliability of the converter is output data of PV panel by means of a K-means clustering method [18] which penetrates in calculating of mean values of failure rates by equation (14).

$$\lambda_T = \sum_{i=1}^K p_i \lambda_i \qquad (14)$$

Where p_i is probability of cluster i and λ_i is failure rate in power level i.

A) Redundant operation mode: For first mode of operation, with redundant stage, table IV, shows failure rates of MOSFET, diode, and inductor for each power level.

Expected values of failure rates for MOSFET, diode and inductor are $\lambda_{eQ} = 3.6755 \times 10^{-6}$, $\lambda_{eD} = 3.1803 \times 10^{-5}$ and $\lambda_{eL} = 1.1384 \times 10^{-8}$, respectively.

Considering mentioned values for failure rates, and equation (8) the reliability relation for the 2-stage interleaved boost converter is given in equation (15).

$$R_1^{I-2}(t) = e^{-3.5489 \times 10^{-5}t} + 3.5489 \times 10^{-5}t e^{-3.5489 \times 10^{-5}t} \qquad (15)$$

According to equations (13) and (15), MTTF of the 2-stage interleaved converter in redundant mode of operation is as follows.

$$MTTF = \int_0^\infty e^{-3.5489 \times 10^{-5}t} + 3.5489 \times 10^{-5}t e^{-3.5489 \times 10^{-5}t} dt = 5.6 \times 10^4 \text{(hours)} \qquad (16)$$

B) Simultaneous operation mode: In this mode of operation, as discussed in previous section, there will be two scenarios for reliability calculation. Each scenario will participate in reliability calculation by probability of its occurrence. In this study, we have considered power levels below 150 watts, in first scenario calculations and power levels above that, in second scenario calculations.

Table V shows failure rates of MOSFET, diode, and inductor in the converter, while there is no fault in stages. Also table VI presents failure rates of components, when a fault has occurred in one of the stages; so all of the input power goes through just one stage. Table VII, shows failure rates of components while system is in second scenario and proper working of both stages is obligatory. From the tables V, VI and VII we can calculate the λs for reliability calculation in both scenarios. Expected values of λs are considered in calculation.

In first scenario, expected values for failure rates of components are $\lambda_{LH} = 1.796 \times 10^{-8}$, $\lambda_{LF} = 4.6123 \times 10^{-8}$, $\lambda_{DH} = 7.006 \times 10^{-6}$, $\lambda_{DF} = 5.0704 \times 10^{-5}$, $\lambda_{QH} = 2.3606 \times 10^{-5}$, $\lambda_{QF} = 2.6476 \times 10^{-5}$, $\lambda_1 = \lambda_3 = 2(\lambda_{LH} + \lambda_{DH} + \lambda_{QH}) = 6.126 \times 10^{-6}$, and $\lambda_2 = \lambda_{LF} + \lambda_{DF} + \lambda_{QF} = 7.7226 \times 10^{-6}$ respectively.

In second scenario, expected values for failure rates are $\lambda_L = 1.83 \times 10^{-9}$, $\lambda_D = 1.94 \times 10^{-6}$, $\lambda_Q = 6.57 \times 10^{-7}$, and $\lambda_1 = \lambda_3 = 2(\lambda_L + \lambda_D + \lambda_Q) = 2.6 \times 10^{-6}$, respectively.

Equation (17)-(18) shows reliability relation and MTTF of converter, in second operating mode, respectively considering both scenarios with their probability of occurrence.

$$R_2^{l-2}(t) = 0.9 \times \left[4.6 \times e^{-6.126 \times 10^{-6}t} - 3.6 \times e^{-7.723 \times 10^{-6}t}\right] + 0.1 \times e^{-2.6 \times 10^{-6}t} \quad (17)$$

$$MTTF = \int_0^\infty \left(0.9 \times \left[4.6 \times e^{-6.126 \times 10^{-6}t} - 3.6 \times e^{-7.723 \times 10^{-6}t}\right] + 0.1 \times e^{-2.6 \times 10^{-6}t}\right)dt = 2.9474 \times 10^5 \text{(hours)} \quad (18)$$

It is obvious from MTTF, that simultaneous mode of operation leads to higher lifetime and reliability. Fig. 8 shows the reliabilities of two-stage interleaved boost converter, in different operation modes.

Fig. 6 depicts reliability curve of two-stage converter based on different modes of operation.

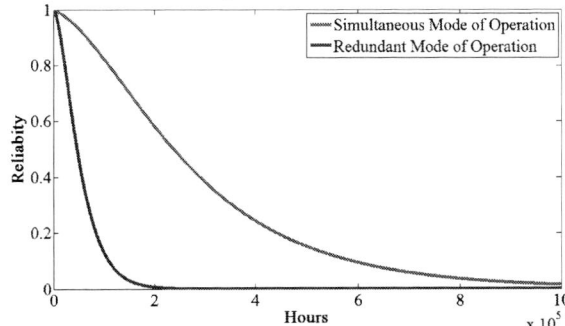

Fig. 6. Reliability of different modes of operation in interleaved boost converter

TABLE IV
POWER LOSS, FAILURE RATE AND PROBABILITY VALUES OF EACH LEVEL OCCURRENCE FOR MOSFET WITH REDUNDANT OPERATING STAGE

Cluster i	μ_i (w)	p_i	λ_Q (failure/hour)	λ_D (failure/hour)	λ_L (failure/hour)
1	0	0.471	0	0	0
2	2.9	0.0925	5.23e-6	3.7e-7	2.64e-9
3	14.05	0.0555	5.27e-6	6.0e-7	2.84e-9
4	26.3	0.0498	5.36e-6	1.02e-6	3.4e-9
5	39.3	0.0396	5.51e-6	1.8e-6	4.52e-9
6	52.2	0.0374	5.72e-6	3.23e-6	6.28e-9
7	64.13	0.0259	5.97e-6	5.4e-6	8.65e-9
8	79.9	0.0334	6.4e-6	1.03e-5	1.3e-8
9	97.9	0.0369	7.03e-6	2.02e-5	2.0e-8
10	116.85	0.0347	7.9e-6	3.85e-5	2.91e-8
11	138.8	0.0295	9.2e-6	7.44e-5	4.14e-8
12	159.5	0.0264	1.08e-5	1.3e-4	5.38e-8
13	183.1	0.0258	1.31e-5	2.2e-4	6.78e-8
14	205.36	0.0232	1.59e-5	3.35e-4	8.03e-8
15	232.94	0.0185	2.03e-5	5.27e-4	9.44e-8

TABLE V
POWER LOSS, FAILURE RATE AND PROBABILITY VALUES OF EACH LEVEL OCCURRENCE FOR MOSFET, DIODE AND IN FIRST SCENARIO FOR SIMULTANEOUS OPERATION WHILE BOTH STAGES WORK PROPERLY

Cluster i	μ_i (w)	p_i	λ_Q (failure/hour)	λ_D (failure/hour)	λ_L (failure/hour)
1	0	0.471	0	0	0
2	2.9	0.0925	5.22e-6	3.5e-7	2.63e-9
3	14.05	0.0555	5.23e-6	4.42e-7	2.68e-9
4	26.3	0.0498	5.26e-6	5.75e-7	2.81e-9
5	39.3	0.0396	5.3e-6	7.65e-7	3.05e-9
6	52.2	0.0374	5.35e-6	1.02e-6	3.4e-9
7	64.13	0.0259	5.41e-6	1.33e-6	3.83e-9
8	79.9	0.0334	5.51e-6	1.89e-6	4.59e-9
9	97.9	0.0369	5.66e-6	2.81e-6	5.76e-9
10	116.85	0.0347	5.84e-6	4.24e-6	7.42e-9
11	138.8	0.0295	6.10e-6	6.72e-6	9.95e-9

TABLE VI
POWER LOSS, FAILURE RATE AND PROBABILITY VALUES OF EACH LEVEL OCCURRENCE FOR MOSFET, DIODE AND INDUCTOR IN FIRST SCENARIO FOR SIMULTANEOUS OPERATION WHILE ONE STAGE HAS FAILED

Cluster i	μ_i (w)	p_i	λ_Q (failure/hour)	λ_D (failure/hour)	λ_L (failure/hour)
1	0	0.471	0	0	0
2	2.9	0.0925	5.23e-6	3.7e-7	2.64e-9
3	14.05	0.0555	5.27e-6	6.0e-7	2.84e-9
4	26.3	0.0498	5.36e-6	1.02e-6	3.4e-9
5	39.3	0.0396	5.51e-6	1.8e-6	4.52e-9
6	52.2	0.0374	5.72e-6	3.23e-6	6.28e-9
7	64.13	0.0259	5.97e-6	5.4e-6	8.65e-9
8	79.9	0.0334	6.4e-6	1.03e-5	1.3e-8
9	97.9	0.0369	7.03e-6	2.02e-5	2.0e-8
10	116.85	0.0347	7.9e-6	3.85e-5	2.91e-8
11	138.8	0.0295	9.2e-6	7.44e-5	4.14e-8

TABLE VII

POWER LOSS, FAILURE RATE AND PROBABILITY VALUES OF EACH LEVEL OCCURRENCE FOR MOSFET, DIODE AND INDUCTOR IN SECOND SCENARIO FOR SIMULTANEOUS OPERATION

Cluster i	μ_i (w)	p_i	λ_Q (failure/hour)	λ_D (failure/hour)	λ_L (failure/hour)
12	159.5	0.0264	6.39e-6	1.02e-5	1.30e-8
13	183.1	0.0258	6.78e-6	1.6e-5	1.73e-8
14	205.36	0.0232	7.23e-6	2.4e-5	2.20e-8
15	232.94	0.0185	7.87e-6	3.8e-5	2.89e-8

VI. CONCLUSION

This paper has investigated reliability of two-stage boost converter interfacing a PV panel and DC bus, based on different modes of operation, using a clustering method to reduce consideration of all hours for calculating reliability. Results has shown that using stages simultaneously leads to higher reliability and more MTTF than redundant mode of operating. Redundant mode of operation results a 6.4 years of MTTF, while simultaneous mode has 33.6 years of MTTF. Obviously a better strategy in controlling the converter, leads to almost 5 times more MTTF.

REFERENCES

[1] O. M. Toledo, D. O. Filho, and A. S. A. C. Diniz, "Distributed photovoltaic generation and energy storage systems: A review," Renewable Sustainable Energy Rev., Vol. 14, No. 1, pp. 506–511, 2010.

[2] M. Bragard, N. Soltau, S. Thomas, and R. W. De Doncker, "The balance of renewable sources and user demands in grids: Power electronics for modular battery energy storage systems," IEEE Trans. Power Electron., Vol. 25, No. 12, pp. 3049–3056, Dec. 2010.

[3] "Global Market Outlook for Photovoltaics until 2016" (PDF). EPIA. 2012. pp. 9, 11, 12, 64.

[4] T. Hakala, T. Lähdeaho, and P. Järventausta,x "Low Voltage DC Distribution – Utilization Potential in a Large Distribution Network Company", IEEE Trans on power delivery, Vol. 30, No. 4, pp. 1694 – 1701, 2015.

[5] Manoranjan Sahoo, and Siva Kumar K, "High Gain Step Up DC-DC Converter For DC Micro-Grid Application", Information and Automation for Sustainability (ICIAfS), 2014 7th International Conference.

[6] Gab-Su Seo, Student Member, IEEE, Kyu-Chan Lee, Student Member, IEEE, and Bo-Hyung Cho," A New DC Anti-Islanding Technique of Electrolytic Capacitor-Less Photovoltaic Interface in DC Distribution Systems", IEEE Trans. on power electronics, Vol. 28, No. 4, pp. 1632-1641, 2013.

[7] Zhan Wang, and Hui Li, "An Integrated Three-Port Bidirectional DC–DC Converter for PV Application on a DC Distribution System", IEEE Trans on power electronics, Vol. 28, no. 10, pp. 4612-4624, 2013.

[8] Y. Song, B. Wang, "Survey on Reliability of Power Electronic Systems, IEEE Trans. Power electronics, Vol. 28, No. 1, pp. 591-604, 2013.

[9] Shaoyong Yang, Angus Bryant, Philip Mawby, Dawei Xiang, Li Ran, and Peter Tavner, "An Industry-Based Survey of Reliability in Power Electronic Converters " IEEE Trans. Industry applications, Vol. 47, No. 3, pp. 1441-1451, 2011.

[10] Alireza Khosroshahi, Mehdi Abapour, and Mehran Sabahi, "Reliability Evaluation of Conventional and Interleaved DC-DC Boost Converters" IEEE Trans. Power electronics, Vol 30, No. 10, pp. 5821-5828, 2015.

[11] Md. Arifujjaman, "Reliability comparison of power electronic converters for grid-connected 1.5kW wind energy conversion system", Elsevier Renewable Energy, Vol. 57, pp. 348-357, 2013.

[12] Susana Estefany De Leon-Aldaco, Hugo Calleja, Freddy Chan, and Humberto R. Jimenez-Grajales, "Effect of the Mission Profile on the Reliability of a Power Converter Aimed at Photovoltaic Applications— A Case Study", IEEE Trans on power electronics, Vol. 28, No. 6, pp. 2998-3008, 2013.

[13] Peng Zhang, Yang Wang, Weidong Xiao, and Wenyuan Li, "Reliability Evaluation of Grid-Connected Photovoltaic Power Systems", IEEE Trans on sustainable energy, Vol. 3, No. 3, pp. 379-389, 2012.

[14] Marios Theristis, and Ioannis A. Papazoglou, "Markovian Reliability Analysis of Standalone Photovoltaic Systems Incorporating Repairs", IEEE journal of photovoltaics, Vol. 4, No. 1, pp. 414-422, 2014.

[15] N. Robert Sorensen, Edward V. Thomas, Michael A. Quintana, Stephen Barkaszi, Andrew Rosenthal, Zhen Zhang, and Sarah Kurtz, "Thermal Study of Inverter Components", IEEE journal of photovoltaics Vol. 3, No. 2, pp. 807- 813, 2013.

[16] U. S. o. A. D. o. Defense, Military Handbook: Reliability Prediction of Electronic Equipment: MIL-HDBK-217F: 2 December 1991: Department of defense, 1991. [17]   Roy Billinton, Ronald N. Allan, "Reliability Evaluation of Engineering Systems: Concepts and Techniques", Springer US, 2012.

[18] W. Li, "Risk Assessment of Power Systems: Models, Methods, and Applications", New York, IEEE Wiley, 2005.

7th Power Electronics, Drive Systems & Technologies Conference (PEDSTC 2016)
16-18 Feb. 2016, Iran University of Science and Technology, Tehran, Iran

A Comparative Study of Different Multilevel Converter Topologies for High Power Photovoltaic Applications

A. Delavari
Electrical Engineering Department
Laval University

I. Kamwa, *Fellow, IEEE*
IREQ, Chief of power systems and mathematics

A. Zabihinejad
Electrical Engineering Department
Laval University

Abstract— **This paper investigates the modern topology of multilevel converters, which are suitable for use in high power photovoltaic applications with the focus on achieving lower total harmonic distortion and better efficiency. Multilevel converters offer several advantages compared to conventional types. Multilevel converters provide high quality output while using the low switching frequency. It affects the switching losses, size of semiconductor switches and harmonic filters. This research investigates various topologies of multilevel converter for high power photovoltaic applications and compares their THD, efficiency, number of required semiconductors and other important characteristics. All topologies are simulated using MATLAB/Simulink in the same operating conditions. Finally, the more suitable multilevel topology is selected with respect to the simulation results.**

Keywords—photovoltaic; Multilevel converter; qualitative study; high power application

I. INTRODUCTION

Recent years have seen a growing trend for generating electric power from renewable energies sources [1,2]. At the same time, the increase in the power rating of wind turbines, photovoltaic power plants and other renewable equipment has been accelerated sharply [3,4]. In this context, high demand for medium and high power converters has made multilevel converters a timely and interesting subject in the field of power electronics [3,4,5]. Researchers strive to propose new multilevel topologies able to provide lower THD and higher efficiency [5], especially at high power level.

Conventional PV plant consisted of a large number of PV modules connected in series and parallel to form strings and sub-arrays, which are combined to feed the inverters. The inverters are then connected to the medium-voltage (MV) electric grid through a low-frequency (LF) transformer [5,6,7,8]. The trend in the industry is to design and utilize the higher inverter ratings since pricing analysis proves that the inverter cost per watt decreases by increasing the inverter power rating. Therefore, inverters with power ratings up to a few megawatts are now being offered on the market [5,6,7].

Also, designers prefer to use higher nominal voltages for both the DC and AC side of the inverter, which leads to reduce wire costs and power losses. These design choices also result in smaller cross-section cables, fewer generator connection boxes

and less cabling at the DC end, which is important in balance-of-system costs [5,9]. Therefore, topologies for medium-voltage grid integration of megawatt-scale PV inverters are moving toward multilevel structures.

Researchers have recently proposed many different multilevel topologies for PV applications [2,5,9,11-14]. Neutral point Clamped converter (NPC) [1], cascaded H-bridge [10], Y-Connected Hybrid Cascaded [15], Capacitor Clamped [2], Z-source [16] and quasi Z-source [17] are important topologies, which are proposed to use with the PV modules. It is possible to investigate these topologies from different point of views. As this work is concerned, to find the most appropriate structure for the PV modules, our investigation is organized in two stages; dealing with quantitative and qualitative study respectively. Quantitative study investigates the output specification of the converter, which is analyzed using Matlab/Simulink. The important parameters, which should be evaluated, are line voltage and current, THD, losses and efficiency. Qualitative study verifies the characteristics, which are important to implement the converter. However; converter reliability, modularity, scalability and functionality are the important issues in qualitative study.

II. MULTILEVEL TOPOLOGY REVIEW

In this section, a brief review of the most common topologies is presented. The topologies considered in this paper are shown in Fig.1.

A. Diode-Clamped Topology (NPC)

According to records the first multi-level inverter was a cascaded one which was designed in 1975 with diodes blocking the source [12]; this inverter was later driven into the diode clamed multilevel inverter proposed in [1].This topology is shown in Figure 1(a). Each of the three-phase outputs of the inverter is connected to a common DC bus voltage divided into three levels over two DC bus capacitors. Existing A high number of clamping diodes results in high cost and different limitations for high-voltage level applications [18]. In addition, a special control is required to balance the capacitor voltages. Consequently, most of practical applications for a diode clamped multilevel inverter are limited to lower than five levels [12,19].

978-1-5090-0376-1/16 $31.00 © 2016 IEEE

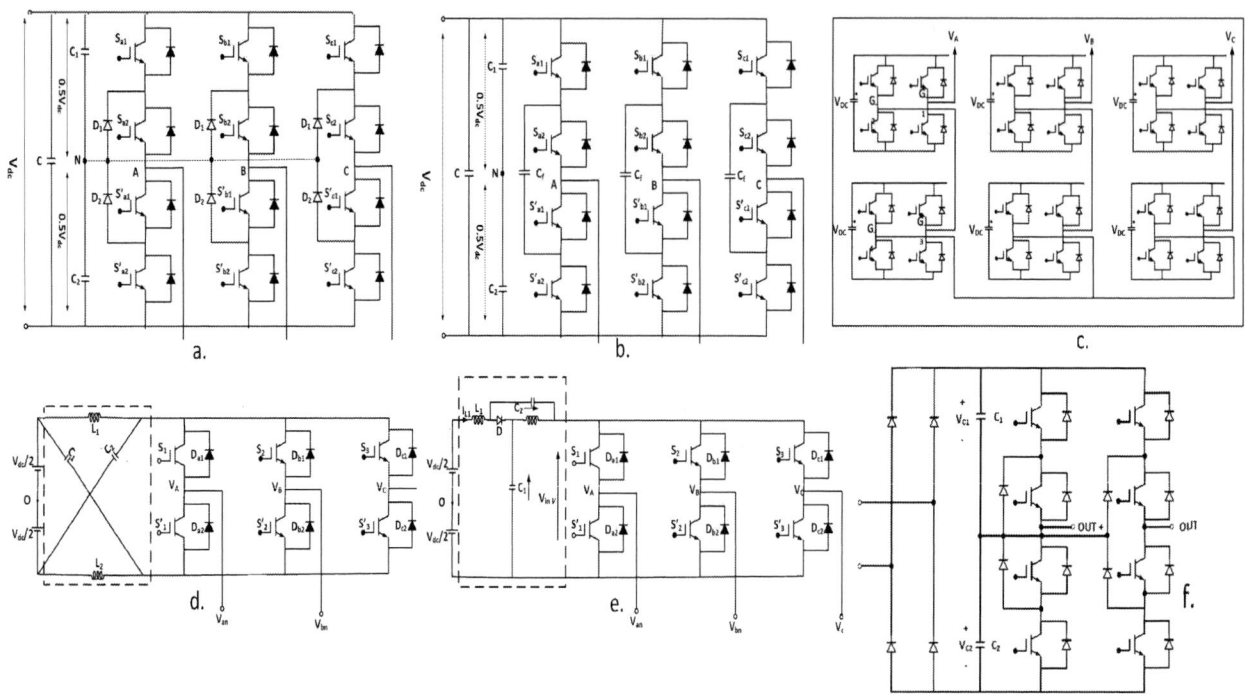

Fig 1: a) NPC   b) Capacitor clamped   c) Cascade   d) Z-source   e) Quasi Z-source   f) Hybrid

B. Capacitor Clamped Topology

Another type of multi-level inverter which has similar topology to the NPCMLI topology is named the flying capacitor inverter or capacitor clamped multilevel inverter [2,14,18] as shown in Fig.1(b). However; instead of using clamping diodes it uses capacitors to keep the voltages to the favored values.[18] It is considered as a good substitute of NPC topology to dominate some of its shortcomings according to [20] and [21].

C. Cascade H-Bridge Topology

Cascaded Multi-Cell Inverter (CMCI), which is proposed in [10] is different in several ways from NPCMLI and CCMLI, especially in how to build the multilevel voltage waveform. It creates the step waveform by using cascaded full-bridge inverters with separate DC-sources, as shown in Fig.1(c) [22]. The cascade topology allows utilizing dc sources with dissimilar voltage values, and high-resolution multilevel waveforms can be reached with a fairly low number of components [23,24].

D. Z-source Topology

The impedance source or Z-source inverter was proposed for the first time by [16] and is shown in Fig.1(d). Z-source inverters distinguished themselves from other conventional types of inverters by providing voltage boost capability in common inverters. The conventional inverters are always buck converter; because of generating the output voltage lower than the DC input voltage [18]. In addition, the upper and lower power switch cannot conduct all together; if not, the DC

source will short-circuit. Therefore, a dead band is provided purposefully between the switching on and switching off of the complimentary power switches of the identical leg, consequently some deformations in the output current are caused by this dead band. These deficiencies are overcome in the Z-source inverter [18]. Comprehensive discussions on the Z-source inverters are given in [25,26,28].

E. Quasi Impedance Source or QZSI Topology

Fig.1(e) presents the QZSI topology which was proposed in [17] as a derivative of the original Z-source inverter; so it contains all the benefits of the ZSI. The impedance source or Z-source inverter has the weakness of discontinuous input (DC) current during boost mode, high voltages across the capacitors, and higher stress on power switches [18, 26]. These limitations are overcome by QZSI [26,27]. Drawing continuous current from DC source, decreasing the voltage across the capacitor C2, lower elements count and therefore high reliability as well as putting lower voltage stress on the power switches are considered as the major advantages of a QZSI [18].

F. Y-Connected Hybrid Cascaded Topology

This kind of topology is obtained by substituting the conventional two-level leg in the H-bridge module of the CMI with diode clamped or capacitor clamped multilevel leg in order to diminish the number of separate DC sources. Each module of this topology can output three-level voltage and each phase contains a cascaded NPC-based H-bridge module [15, 19]. The number of switching devices in the conversion system will be reduced by taking hybrid topologies.

978-1-5090-0376-1/16 $31.00 © 2016 IEEE

III. RESULTS DISCUSSION

A. Quantitative study

In this section, the most common topologies of multilevel converters, which are connected to PV array, are scrutinized in six case studies. By comparing their output wave forms and their characteristics, the most suitable inverter configuration is found. All scenarios have been done in identical situations using the same PV array source and loads while all switches are modeled as IGBT ones. The PV array module is called Canadian solar load CS5C90M with 40 parallel strings and 10 series connected modules per strings, with irradiation of rate 1000, temperature of 25^{oc} and a three phase resistive load of R=10(Ω).

a) Three level NPC PV source inverter

Fig.2 illustrates a three level NPC PV source inverter model in Matlab. The inverter is connected to the pre-defined PV array and load. The voltage and current wave forms of this simulation are shown in Fig.3.

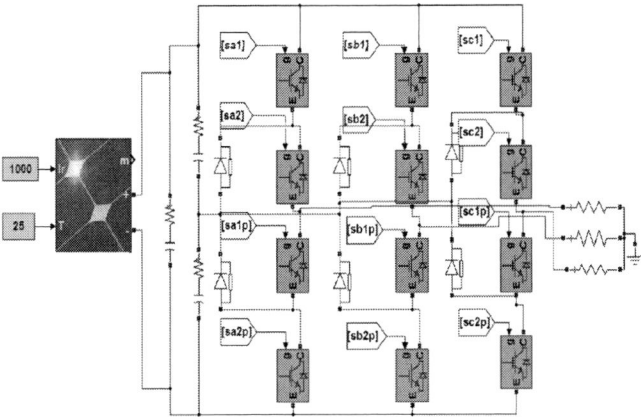

Fig.2:. Three level NPC, PV source model in Matlab/Simulink.

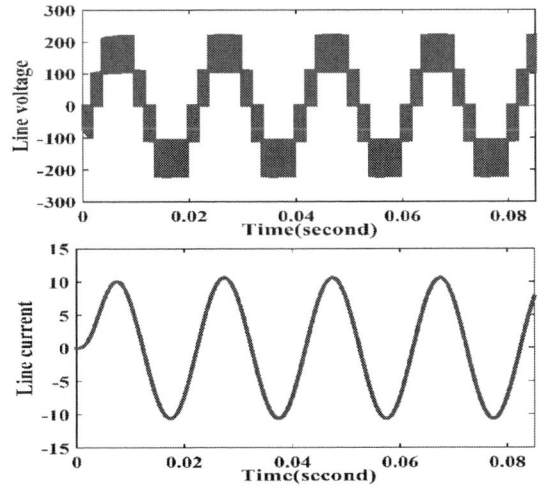

Fig.3. Three level NPC inverter voltage and current waveforms.

The total harmonic distortion (THD) value of each waveform is calculated by Matlab/Simulink. In this way the capacitor

values are considered 2200µF while THD of line voltage is 36.22% for this case study; in addition, the efficiency of inverter is calculated as η = 98.93%

b) Capacitor clamped three level PV source inverter

Three level capacitor clamped PV source inverter model is shown in Fig.4. Capacitor values are 1000µF. The voltage and current waveforms of this simulation are shown in Fig.5. THD line voltage is 49.89% for this inverter topology; and the efficiency is calculated as η = 98.65%.

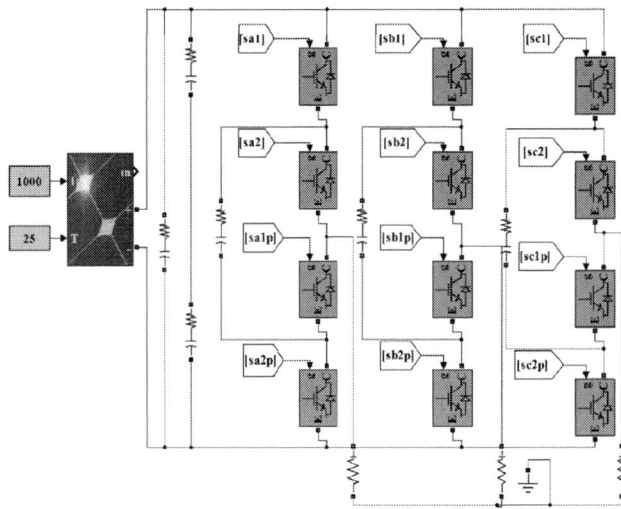

Fig. 4. Three level Capacitor clamped, PV source model in Matlab/Simulink.

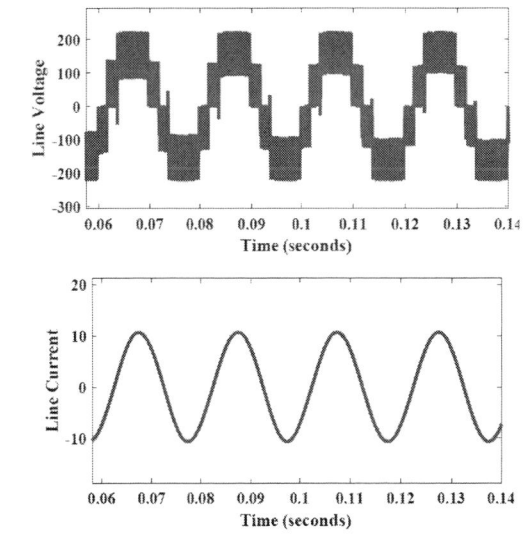

Fig. 5. Three level Capacitor clamped voltage and current waveforms.

c) Three level Cascaded PV source inverter

Fig.6 represents a three level cascaded PV source inverter model in Matlab, and its voltage and current waveforms are depicted in Fig.7. THD line voltage is obtained 47.18% for this model; and the efficiency is calculated as η = 83.33%.

d) Three level Z-source PV connected inverter

Three level Z-source PV connected inverter as well as its output wave forms are shown in Fig.8 and 9. The inductance values are assumed to be the same equal to 0.5mH as are the capacitor values 0.4mF. THD of this modeled is measured 42.19% and its efficiency is calculated as η=99.48%.

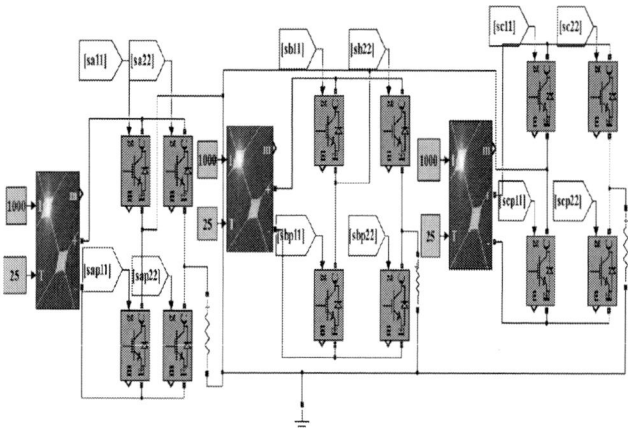

Fig. 6. Three level Cascaded PV source model in matlab/simulink.

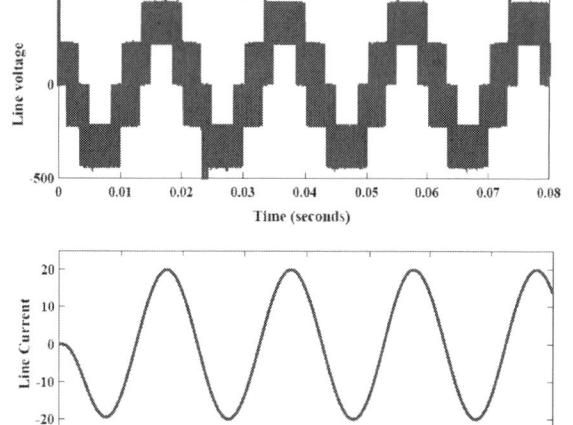

Fig. 7. Voltage and current waveforms of three level cascaded inverter.

e) Three level Quasi-Z source PV source inverter

The Quasi-Z source model is done according to Fig.10, and its output waveforms are shown in Fig.11. The inductance values are assumed to be the same equal to 0.5mH as are the capacitor values 0.4mF. Line voltage THD as well as efficiency for this model are 41.49% and η=98.95% respectively.

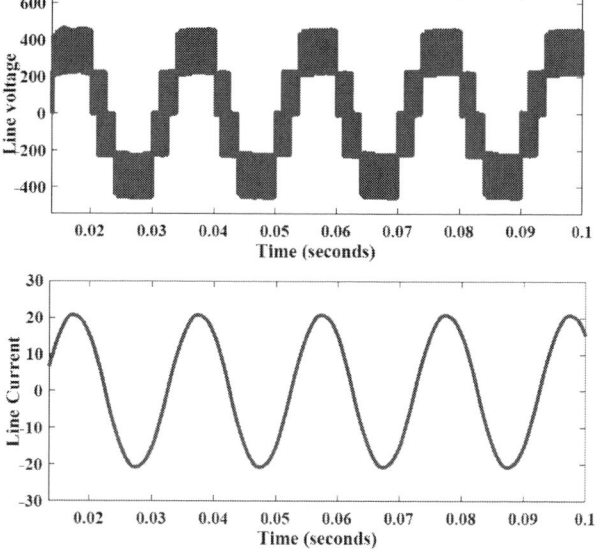

Fig. 9. Voltage and current waveforms of three level Z source inverter.

f) Y-Connected three level Hybrid Cascaded PV source inverter (CMI)

Three level hybrid cascaded NPC PV source inverter model as well as its voltage and current waveforms are shown in Fig.12 and 13, respectively. Capacitors values are 2200µF while THD is 37.57% and efficiency is η=81.8%. Normally hybrid topologies are used for creating high level output voltage. This concept was introduced in [19] by presenting 17-level CMI as the most suited for application to PV power generation. The simulation result of this topology confirms the low THD rate of this topology.

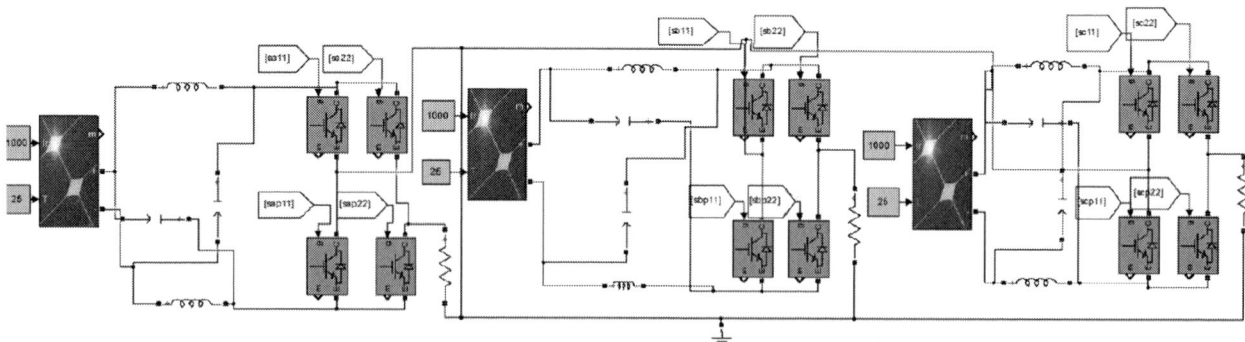

Fig. 8. Three level Z-source PV connected inverter in matlab/simulink.

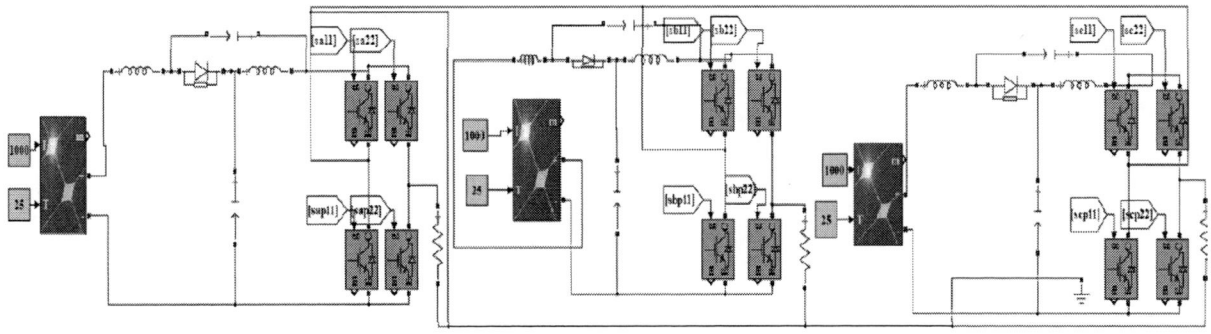

Fig. 10. Three level Quasi Z source PV connected model in matlab/simulink.

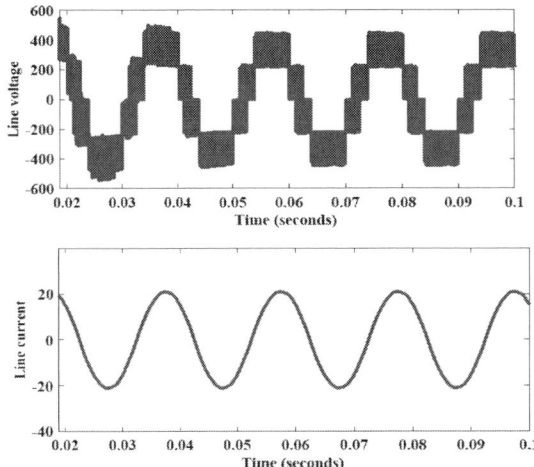

Fig. 11. Voltage and current waveforms of three level Quasi Z source.

B. Qualitative Study

Table I, represents quantitative results of case studies; in addition, the characteristics of different topologies are summarized in Table III in accordance with Table II .As it can be seen in Table II, a lot of clamping diodes in NPC topology make it very expensive and raise different issues in high-voltage level applications, Therefore, according to table III practical uses of diode clamped multilevel inverters are limited to lower than five levels [18].

The second inverter, which has a quite similar topology to the first one is the capacitor clamped topology. The major dissimilarity is the use of clamping capacitors in place of clamping diodes, and the number of switching combinations rises as capacitors do not block reverse voltages [17,18].

According to table III, both NPC and capacitor clamped topologies are single input inverters, however other types of topologies are modular so they reaches the higher reliability in comparison with NPC and capacitor clamped because of its modular topology [9].

Quasi-Z source inverter is introduced as a derivative of Z-source inverter by having the ability of solving some Z-Source topology problems such as high voltages across the capacitors,

and higher stress on power switches [18] and therefore reaching a reduced value of THD. In addition the efficiency of Z source as well as Quasi-Z source inverters are superior among other types of multi-level inverters. Hybrid multi level inverter is also considered as a suitable case in THD rate according to Table I; however, its efficiency is lower than Z types.

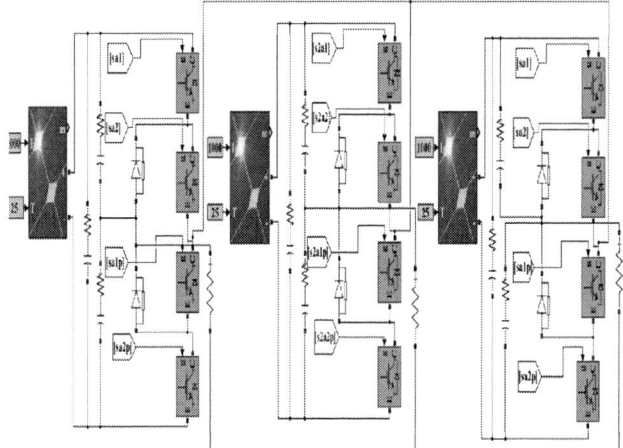

Fig. 12. Three level Hybrid PV source inverter in matlab/simulink.

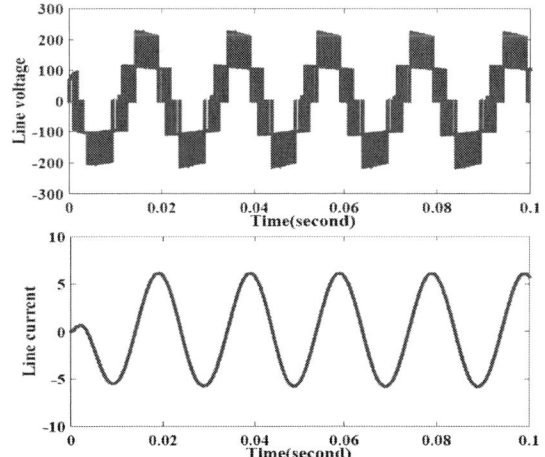

Fig. 13. Voltage and current waveforms of three level hybrid model.

TABLE I.   THD AND EFFICIENCY OF DIFFERENT TOPOLOGIES

R	THD and Total Efficiency		
	Converter Topology (3level)	Efficiency %	THD %
1	NPC	98.93	36.22
2	Capacitor clamped	98.65	49.89
3	Cascaded	83.33	47.18
4	Hybrid	81.08	37.57
5	Z-source	99.48	42.19
6	Quasi Z source	98.95	41.49

TABLE II.   TABLE STYLES

R	Parameter Identification	
	Id	Inverter Characteristics
1	M1	Number of power Switch
2	M2	Number of Capacitor
3	M3	Number of Inductance
4	M4	Number of Diode
5	M5	Single source input
6	M6	Suitable to implement high level voltage
7	M7	Reliability (Medium/High)
8	M8	Bidirectional (Yes/No)

TABLE III.   DESIGN DATA AND PERFORMANCE

R	Parameters							
	Converter Topology (3level)	M1	M2	M3	M4	M5	M6	M7	M8
1	NPC	12	3	0	6	Y	N	M	Y
2	Capacitor clamped	12	6	0	0	Y	N	M	Y
3	Cascaded	12	0	0	0	N	Y	H	Y
4	Hybrid	12	9	0	6	N	Y	H	Y
5	Z-source	12	6	6	0	N	Y	H	Y
6	Quasi Z source	12	6	6	3	N	Y	H	Y

IV. CONCLUSION

The price analysis of the converter shows that multilevel converters are more economic than conventional types in the case of medium and high power applications. In This research, different multilevel converter topologies have been investigated and compared in order to find the most suitable topology, which is appropriate to use in the PV applications. Six multilevel topologies, which were proposed in the literature, have been investigated. The investigation was done via quantitative and qualitative study. In quantitative study, important output parameters of proposed multilevel topologies were evaluated using Matlab/Simulink at the same operating point. Also, a qualitative analysis has been performed to investigate some advantages and disadvantages of each topology, which cannot be considered in the simulation. The results prove that quasi Z-source converter has better performance in comparison with other types.

REFERENCES

[1] A. Nabae, I. Takahashi and H. Akagi, "A new neutral point clamped PWM inverter", IEEE Trans. Ind. Appl., IA-17 (5) 518–523, 1981.

[2] T. A. Meynard, H. Foch, P. Thomas, J. Couralt, R. Jakob, and m. Naherstaedt, "Multicel converters: Basic consepts and industry application", IEEE Trans. Ind. Electron., 49 (5), 955-964, 2002.

[3] M. F. Escalante, J. C. Vannier, and A. Arzande, "Flying capacitor multilevel inverters and DTC motor drive applications", IEEE Trans. Ind. Elect., 49 (4), 809–815, 2002.

[4] S. S. Fazel, S. Bernet, D. Krug and K. Jalili, "Design and comparison of 4 kV Neutral-pointclamped, flying capacitor and series-connectd H-bridge multilevel converters", IEEE Trans. Ind. Appl., 43(4), 1032-1040, 2007.

[5] J. V. Núñez, "Multilevel Topologies: Can New Inverters Improve Solar Farm Output? " Solar industry journal, 5, 12, 2013.

[6] A. Zabihinejad, P. Viarouge, "Design of Direct Power Controller for a High Power Neutral Point Clamped Converter using Real time Simulator", World Academy of Science, Engineering and Technology, Energy and Power Engineering, 1(1), 187, 2014.

[7] R. Badin, Y. Huang, F. Z. Peng, and H. G. Kim, "Grid interconnected Z-source PV system", Proc. IEEE PESC'07, Orlando, FL, June, pp. 2328–2333, 2007.

[8] F. Z. Peng. "Z-source networks for power conversion". 23rd Ann. IEEE App. Power Elect. Conf. Exp, APEC2008, 24–28 February, Austin, TX, pp. 1258–1265, 2008.

[9] M. Malinowski, K.Gopakumar, Jose Rodriguez, and A. Perez, "A Survey on Cascaded Multilevel Inverters". IEEE Transaction on Industrial Electronics, 57(7), 2010.

[10] P. W. Hammond,." A new approach to enhance power quality for medium voltage AC drives", IEEE Trans.Ind.Appl., 33 (1), 202-208, 1997.

[11] T. A. Meynard, M. Fadel, N .Aouda, "Modelling of multilevel converters". IEEE Trans. Industrial Electronics 44(3), 356–364,1997

[12] A. Nordvall, "Multilevel Inverter Topology Survey", Master of Science Thesis in Electric Power Engineering, Department of Energy and Environment Division of Electric Power Engineering Chalmers University of technology Göteborg, Sweden, 2011.

[13] S. Chakraborty, M. G. Simões, W. E. Kramer "Power electronics for renewable and distributed energy systems", Springer, 2013.

[14] N. S. Cho, "A general circuit topology of multilevel inverter," in Proc. IEEE Power Electron. Specialists Conf., Cambridge, 96–103, 1991.

[15] F. Khoucha, S. Mouna Lagoun, K. Marouani, A. Kheloui and M. Benbouzid. "Hybrid Cascaded H-Bridge Multilevel-Inverter Induction-Motor-Drive Direct Torque Control for Automotive Applications". IEEE Transactions on Industrial Electronics, Institute of Electrical and Electronics Engineers, 57 (3), 892-899, 2010.

[16] F. Z .Peng. "Z-source inverter. IEEE Trans". Ind. Appl., 39(2), 504–510, 2003.

[17] Y. Li, J. Anderson, F. Z.Peng, and D.Liu, "Quasi-Z-source inverter for photovoltaic power generation systems". 24th Ann. IEEE Appl. Power Elect. Conf. Exp., APEC 2009, 15–19 February, Washington, DC pp. 918–924, 2009

[18] H. Abu-Rub, A. Iqbal, J.Guzinski, "High performance control of AC driveswith Matlab#simulink models" J. Wiley & S. Ltd IEEE express, 2012.

[19] H. Abu-Rub, M. Malinowski, K. Alhaddad, power electronics for renewable energy systems, transportation, and industrial applications, John Wiley & Sons Ltd,A co-publication of IEEE Press, 2014.

[20] J. Huang and K. A. Corzine, "Extended operation of flying capacitor multilevel inverters," IEEE Trans. Power Electron., 21 (1)140– 147, 2006.

[21] S. Sirisukprasert, "Optimized harmonic stepped-waveform for multilevel inverter," M.S. thesis, Dept. Elect. Eng., Virginia Polytechnic Inst. State Univ., Blacksburg, VA, 1999.

[22] B. S.Jin, W. K. Lee, T. J.Kim, D. W.Kang, and D. S. Hyun, " A study on the multi-carrier PWM methods for voltage balancing of flying capacitor in the flying capacitor multilevel converter". Proc. 31st IEEE Ind. Elect. Conf. IECON, 6–10 November. North Carolina, 721–726, 2005.

[23] S. Mariethoz and , A. Rufer, "Design and control of asymmetrical multilevel inverters," in Proc. Int. Conf. Ind. Electron. Control Instrum., Seville, Spain, pp. 840–845, 2002

[24] J. Dixon and L. Moran, "High-level multistep inverter optimization using a minimum number of power transistors," IEEE Trans. Power Electron., vol. 21, no. 2, pp. 330–337, 2006.

[25] M. Shen, J. Wang, A. Joseph , F. Z. Peng, L. M.Tolbert and D. J. Adams, "Constant boost control of the Z-source inverter to minimize current ripple and voltage stress". IEEE Trans. Ind. Appl, 42(3), 770–778, 2006.

[26] J. Park, H. Kim, E. Nho, T. Chun, , and J. Choi, " Grid-connected PV system using a quasi-Zsource inverter". 24th Ann. IEEE AplL. Power Elect. Conf. Exp., APEC 2009, 15–19 February, Washington, DC pp. 925–929, 2009

[27] F. Z. Peng, , A. Joseph, J. Wangetal. " Z-Source inverter for motor drives". IEEE Trans. Power Elect., 20(4), 857–863, 2005.

[28] F. Z.Peng, M. Shen and Z. Qian, " Maximum boost control of the Z-source inverter". IEEE Trans. Power Elect, 20(4), 833–838, 2005.

7th Power Electronics, Drive Systems & Technologies Conference (PEDSTC 2016)
16-18 Feb. 2016, Iran University of Science and Technology, Tehran, Iran

New Topology to Reduce Leakage Current in Three-Phase Transformerless Grid-Connected Photovoltaic Inverters

Ramin Rahimi[1], Babak Farhangi[2], Shahrokh Farhangi[3]

Department of Electrical and Computer Engineering, University of Tehran, Tehran, Iran.

Emails: [1] raminrahimi@ut.ac.ir, [2] b.farhangi@ece.ut.ac.ir, [3] farhangi@ut.ac.ir

Abstract— **Three phase converters are favorable for high power grid connected systems. In photovoltaic (PV) application, it is possible to remove the transformer from the PV system in order to reduce size, losses, and cost; and improve the efficiency. In other hand, presence of the parasitic capacitor between the panel's metal frame and cells causes the leakage current issue. The leakage current increases current harmonics injected into the utility grid, the radiated and conducted electromagnetic interference, and losses. In this paper, a novel grid-connected inverter topology with an optimum modulation technique is proposed to address leakage current issue. The proposed topology also improves output current THD in three-phase transformerless PV inverters. Moreover, the proposed topology is compared by traditional three phase inverter in terms of common mode leakage current and output current THD. In reference to MATLAB/SIMULINK simulation results, the proposed topology is able to reduce the common mode ground leakage current and improve the output current THD by 2.17%.**

Keywords— Photovoltaic, PV, transformerless inverter, full-bridge inverter, common mode voltage, leakage current, parasitic capacitor, sinusoidal pulse width modulation, SPWM.

I. INTRODUCTION

Electricity production from Photovoltaic (PV) systems has become more cost effective and efficient in recent years. Renewable energy recourses supplied 19.1% of the global energy in 2013. Renewables continue to grow in capacity and generation in 2014. Solar PV plays a substantial role in electricity generation in some countries as rapidly falling costs have made unsubsidized solar PV-generated electricity cost-competitive with fossil fuels. In 2014, solar PV marked another record year for growth, with an estimated 40 GW installed for a total global capacity of about 177 GW such as seen in Fig. 1 [1]. PV generation has a large field of study in various application such as PV based water desalination systems [2, 3].

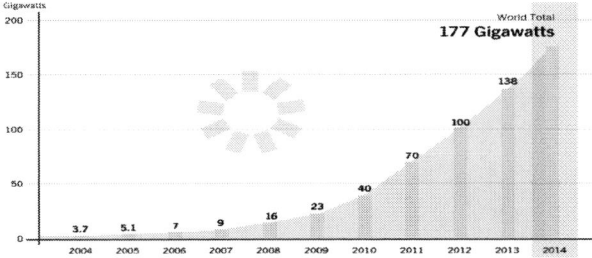

Fig. 1. Solar PV Global Capacity, 2004–2014 [1]

In PV single-phase PV systems, there is a pulsating ac power on the output, whilst the input is a smooth dc. Large dc capacitors are required that decrease the lifetime and reliability of the whole system [4-12]. On the other hand, in a three-phase system, there is constant ac power on the output, which means that there is no need for large capacitors, leading to smaller cost, and a higher reliability and lifetime of the whole system [13-16].

The majority of the PV inverters use isolation transformers such as high-frequency or line-frequency transformers. On the other hand, the high-frequency isolated topologies usually include several power stages, which reduces the system efficiency and increases the system complicacy [17-22]. Also line-frequency isolated topologies have a weighty and big transformer that makes the whole system large and difficult to install. Consequently, it is possible to remove the transformer from the PV system in order to improve the efficiency and reduce size, losses, and cost in PV application. The direct conductive connection of the dc source and the grid through transformerless PV inverter, introduce a leakage current, which parasitic capacitance between the PV panels and the earth caused to this leakage current. This current increases current harmonics injected into the utility grid, the radiated and conducted electromagnetic emissions, and losses. Amplitude of the leakage current depends on the converter topology and switching strategy [4, 5, 23, 24]. In transformerless PV inverters, common mode leakage current reduction and efficiency increment, have been widely studied in many papers. One approach is to connect the middle point of the dc-link to the neutral of the grid that has been applied to neutral point clamped (NPC) and FB topologies. Another solution is to separate the PV array from the grid. In FB inverters, this method is applied [25, 26]. In the HERIC topology, the ac side is disconnected during zero voltage vectors. In H5 and H7 topologies [27], PV side is disconnected from the grid on DC side of the inverter. A new conversion topology has been proposed in [28] with extra two switches and twelve diodes. Several additional components are required in [28]; this significantly increase the cost and losses. Reduced CMV PWM (RCMV-PWM) methods such as active zero-state PWM (AZPWM) [29], near-state PWM (NSPWM) [30] and remote-state PWM (RSPWM) [31], are proposed recently. RCMV-PWM methods are able to reduce the CMV to reduce the leakage current without zero vectors. However, In RCMV-PWM methods, output bipolar voltage generates overvoltage transients [30, 32, 33], large current ripples across the filter

978-1-5090-0376-1/16 $31.00 © 2016 IEEE   421

inductors, and high switching losses that reduces overall efficiency of the system [27]. As a summary, there are three methods to eliminate leakage current issue including circuit parameters matching, hardware topology modification, and modulation mode improvement [34].

In this paper, an improved grid-connected inverter topology with an optimum modulation technique is proposed to reduce leakage current and improve output current THD in three-phase transformerless PV systems. This new topology is called "H8 topology". Also this paper evaluates the FB topology. MATLAB/SIMULINK simulation results of these two topologies are compared. According to the simulation results, that the proposed H8 topology and novel modulation technique achieves the best performance; both leakage current and the output current THD are reduced.

The rest of this paper is organized as follows: In Section II, a common mode leakage current model will be presented in order to analyze the leakage current issue. The proposed three-phase topology will be introduced in Section III. In section IV, simulation results of the FB inverter and H8 inverter will be reported. Finally, the conclusion will be presented in Section V.

II. COMMON MODE LEAKAGE CURRENT MODEL

Fig. 2 shows schematic of the transformerless grid-connected three-phase PV system with FB topology for inverter; where, C_{PE} is the stray capacitor between the PV panel and ground. In transformerless PV systems, the PV system is directly connected to the grid and the common mode voltage can generate large leakage ground current flowing through the parasitic capacitance C_{PE}. The leakage ground current behavior is strongly influenced by the converter topology and the PWM strategy. Thus, the common mode leakage current model is a helpful tool to describe how the common mode voltage (CMV) contributes to generation of the currents through stray capacitor of the PV panel. Common-mode and differential-mode voltages in the three phase PV systems, can be analyzed by using two phases [35].

The common-mode voltage is defined as the average of the sum of inverter output voltages. In this case, the common reference is taken to be the negative terminal of the dc link (marked with N). Fig. 3 shows the simplified model of common-mode voltage for phase "a" and phase "b". In this model, the common-mode voltage for phase "a" and phase "b" (V_{cm-ab}), L_{c-ab}, and L_{g-ab} are defined by (1). In (1) L_{c-a} and L_{g-a} are inverter side inductor and grid side inductor of the LCL filter, respectively.

$$
\begin{aligned}
V_{cm-ab} &= \frac{V_{aN} + V_{bN}}{2} \\
L_{c-ab} &= L_{c-a} \parallel L_{c-b} \\
L_{g-ab} &= L_{g-a} \parallel L_{g-b}
\end{aligned}
\tag{1}
$$

The total common-mode voltage is defined by (2).

$$
V_{cm-total} = \frac{V_{cm-ab} + V_{cm-bc} + V_{cm-ca}}{3} = \frac{V_{aN} + V_{bN} + V_{cN}}{3}
\tag{2}
$$

The common mode voltage of the three-phase inverter is calculated in (2). This common mode voltage charges and discharges the stray capacitance C_{PE}, which means that the variations in this voltage will produce a voltage transient over the stray capacitance. This voltage transient results in leakage current through the C_{PE} between PV array and ground. The magnitude of this leakage ground current depends on the amplitude and frequency of the voltage across the stray capacitances. Therefore, the common mode voltage is a major concern and its variations should be reduced. The common mode model for grid-connected three-phase PV systems is shown in Fig. 4.

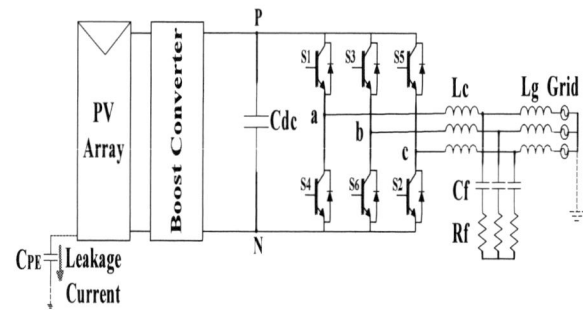

Fig. 2. Grid-connected three-phase PV system with FB topology

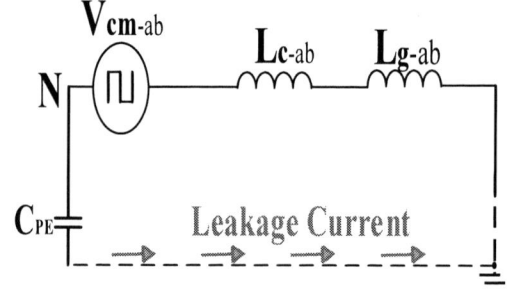

Fig. 3. Simplified common mode leakage mode for phase "a" and phase "b"

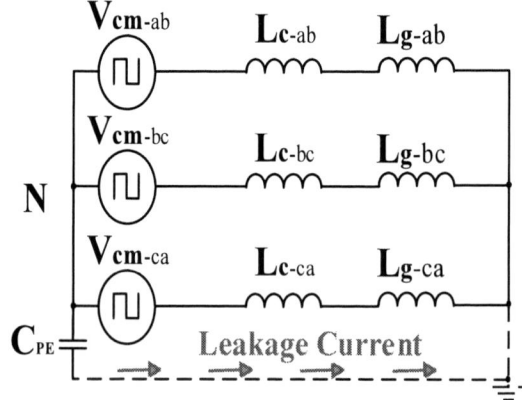

Fig. 4. Simplified common mode leakage mode for three-phase system

III. PROPOSE TOPOLOGY

As mentioned in introduction, one of the techniques to reduce the common mode leakage current is separating the PV array from the grid. In case of the full-bridge inverter, the grid is disconnected from the PV array when the zero voltage vectors are applied. In this paper, this solution is practiced and a new transformerless three-phase topology is proposed that is called "H8 topology". An improved sinusoidal pulse width modulation (SPWM) strategy is used for this proposed topology that is able to improve the common mode behavior without compromising the harmonic distortion of the whole system. The H8 topology is shown in Fig. 5. This topology has eight switches, the switches S7 and S8 are the dc isolating stage.

In general, eight state vectors are applied to the output along the grid period, three odd active vectors (i.e. V_1, V_3, and V_5), three even active vectors (i.e. V_2, V_4, and V_6), and two null vectors (i.e. V_0 and V_7). Fig. 6 shows the proposed modulation technique for H8 topology. In this figure Va*, Vb*, and Vc* are voltage references. These reference voltages are generated by close loop control system. The carrier is a 16 kHz triangular waveform; g1 to g8 are switch gate pulses. The switch gate pulses g7 and g8 can be obtained from (3) as follows:

$$g_7 = \overline{g_1 g_3 g_5}$$
$$g_8 = g_1 + g_3 + g_5 \tag{3}$$

Where, g7 is NAND of g1, g2, and g3, and g5; and g8 is OR of g1, g3, and g5. The proposed three-phase PV topology with proposed modulation strategy, works as explained below:

1. During odd active vectors (V_1, V_3, and V_5), S7 and S8 are on to generate the desired output voltage and the corresponding CMV becomes VDC/3.

2. During even active vectors (V_2, V_4, and V_6), S7 and S8 are on to generate the desired output voltage and the corresponding CMV becomes 2VDC/3.

3. During zero vector V_7, all the upper switches S1, S3 and S5 are turned on and connected to positive (P) of the DC-link. At this moment, S7 is turned off to disconnect the PV from the grid. Therefore, there is no path to allow leakage current to flow.

4. During zero vector V_0, all the lower switches S4, S6 and S2 are turned on and connected to negative (N) of the DC-link. At this moment, S8 is turned off to disconnect PV from the grid. Therefore, leakage current finds no path to flow.

The switches S7 and S8 allow to isolate PV panels from the grid during zero voltage vectors. As a consequence, the leakage ground current does not flow through the stray capacitances during zero vectors. Therefore, stray capacitor voltage variations are reduced. In other words, common mode leakage current is reduced. Table I illustrates operation of the switches according to the proposed modulation method. The proposed modulation strategy doesn't remove zero vector voltages in order to reduce leakage current. In this situation, due to the use of zero voltage vectors, the output voltage pattern does not generate overvoltage transients, current ripples across the filter inductors, and high frequency losses are reduced.

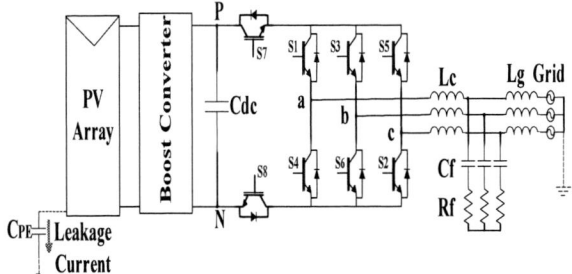

Fig. 5. Proposed H8 three-phase transformerless PV topology

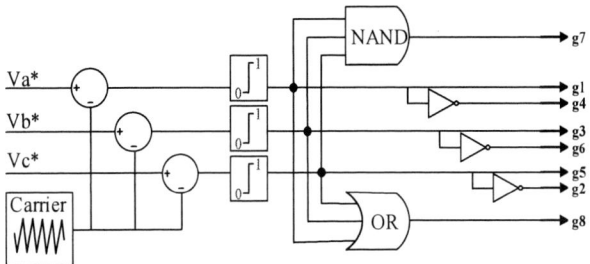

Fig. 6. Proposed modulation technique for H8 topology

IV. SIMULATION RESULTS

Tab. II shows Circuit parameters that are used in the simulations. Two case studies are demonstrated to verify the proposed topology and proposed modulation strategy. Case studies are performed with closed loop controls. The control method includes dc link voltage loop control and grid current loop control. The dc-link voltage is controlled by the proportional integral (PI) controller. The current injected to the grid is controlled by the proportional resonant (PR) controller. The benefit characteristic of PR controller is the feasibility of implementing harmonic compensator without intervening with control dynamics, getting a high quality current delivered to the grid [4, 36]. Simulation results of the FB inverter and H8 topologies will be presented in the next sub-sections. Boost DC/DC converter has been chosen as the maximum power point tracking (MPPT) controller. Perturb and observe (P&O) method is applied for the MPPT controller of a two-stage PV grid-connected converter system. The maximum power point was reached in 0.05 seconds. The waveforms have been reported in the time period 0.1-0.14 second after the MPPT is reached.

A. Three-Phase FB Topology

This topology is shown in Fig. 2. This topology is the simplest and most widely used for general applications with three-phase system. Fig. 7, Fig. 8, Fig. 9 and Fig. 10 show simulation results for this topology when SPWM strategy is implemented. In these figures, the sinusoidal grid currents, output current's FFT, stray capacitor voltage, and the leakage ground current are depicted. The output current THD is 9.52% and leakage current is 344.8 mA (RMS). This leakage current avoids the VDE 0126-1-1 standard [4]. The maximum level for the leakage ground current should not exceed 300 mA. Hence, this topology is not suitable for three-phase transformerless PV application.

978-1-5090-0376-1/16 $31.00 © 2016 IEEE

TABLE I. SWITCHING SEQUENCES OF THE ROPOSED MODULATION STRATEGY FOR H8 TOPOLOGY

Vector	S1	S2	S3	S4	S5	S6	S7	S8
V_1	ON	OFF	OFF	OFF	ON	ON	ON	ON
V_2	ON	ON	OFF	OFF	OFF	ON	ON	ON
V_3	OFF	ON	OFF	ON	OFF	ON	ON	ON
V_4	OFF	ON	ON	ON	OFF	OFF	ON	ON
V_5	OFF	OFF	ON	ON	ON	OFF	ON	ON
V_6	ON	OFF	ON	OFF	ON	OFF	ON	ON
V_0	OFF	OFF	OFF	ON	ON	ON	OFF	ON
V_7	ON	ON	ON	OFF	OFF	OFF	ON	OFF

TABLE II. TECHNICAL SPECIFICATION OF THE THREE-PHASE TRANSFORMERLESS PV SYSTEM

Grid frequency	50 Hz
PWM carrier frequency	16 kHz
Nominal power	1980 W
Phase grid voltage (RMS)	220 V
DC link voltage	700 V
Inverter side inductor (Lc)	6.5 mH
Grid side inverter (Lg)	650 uH
Capacitor filter (Cf)	2.2 uF
Damping Resistor (Rf)	5.6 Ω
Parasitic Capacitance (C_{PE})	20 uF
DC link Capacitor (C_{DC})	340 uF
Boost Inductor	6.5 mH

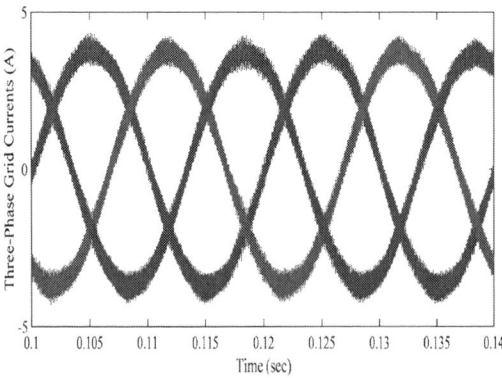

Fig. 7. Three-phase grid currents for FB topology

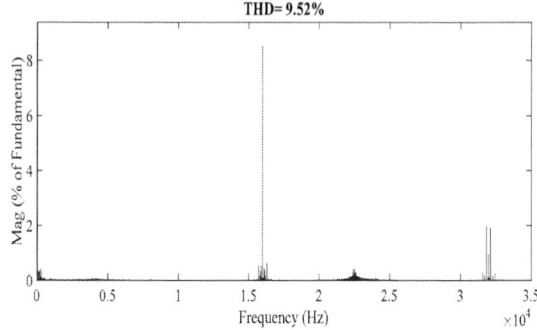

Fig. 8. Output current in FFT in FB topology

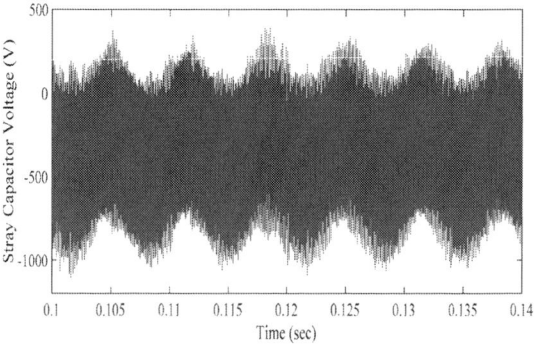

Fig. 9. Stray capacitor voltage for FB topology

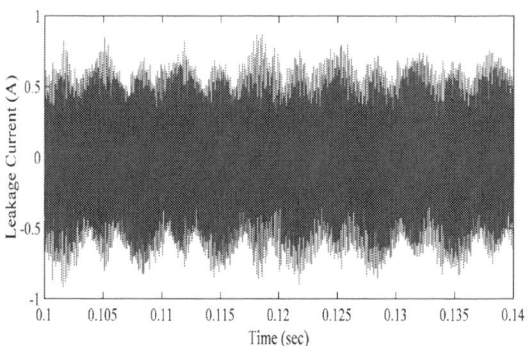

Fig. 10. Leakage ground current for FB topology

B. Three-Phase Proposed Topology (H8 Topology)

This topology is shown in Fig. 5. The performance of H8 converter is evaluated in this section. Fig. 11 and Fig. 12 show the grid currents and its FFT, respectively. As can be seen, grid current THD is 7.35%. Fig. 13 and Fig. 14 show stray capacitor voltage and leakage current, respectively. In this case, leakage current is 255 mA (RMS). The measured leakage ground current is lower than permissible level of 300 mA limit on the VDE 0126-1-1 standard. According to Fig. 9 and Fig. 13, stray capacitor voltage variations is reduced in H8 topology that results in lower leakage current. Thus, this topology is suitable for three-phase transformerless PV application.

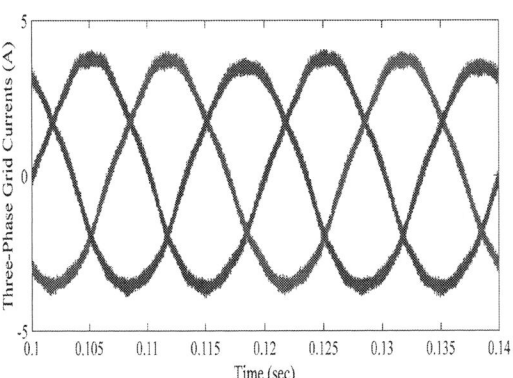

Fig. 11. Three-phase grid currents for H8 topolog

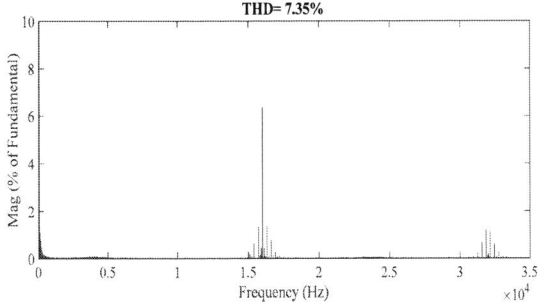

Fig. 12. Output current in FFT in H8 topology

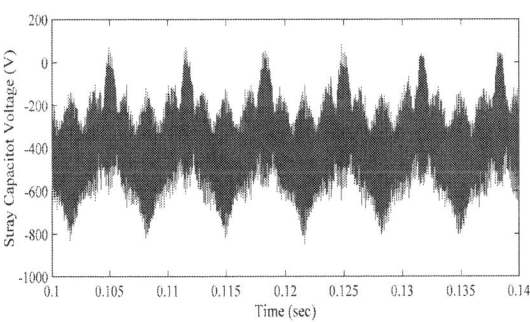

Fig. 13. tray capacitor voltage for H8 topology

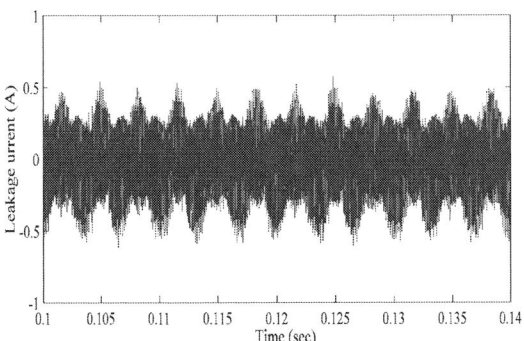

Fig. 14. Leakage ground current for H8 topology

V. CONCLUSION

In this paper, a novel grid-connected inverter topology with an optimum modulation technique was proposed to reduce leakage current and improve output current THD in three-phase transformerless PV systems. Performance of the three phase transformer-less FB PV inverter was evaluated. Although the output current THD is acceptable, FB topology does not satisfy the standards in term of the common mode leakage current. In contrats, the proposed H8 topology with proposed optimum modulation technique satisfies the required common mode leakage current limit. This is achieved by reducing the stray capacitor voltage fluctuations.The proposed modulation method separates the PV side from the grid side during zero voltage vectors. As the bottom line, the H8 topology was superior compared to FB topology as a three phase transformerless PV inverter.

ACKNOWLEDGMENT

The authors wish to thank the Renewable Energy Organization of Iran for financial support of this project through contract number 94/17.

REFERENCES

[1] REN21, "Renewables 2014: Global Status Report (GSR)," [Online]. Avaiable: http://www.ren21.net/,Jun.2015. J. Clerk Maxwell, A Treatise on Electricity and Magnetism, 3rd ed., vol. 2. Oxford: Clarendon, 1892, pp.68-73.

[2] H. Mehrabian-Nejad, B. Farhangi, and Sh. Farhangi, S.Vaez-Zadeh, " DC-DC Converter for Energy Loss Compensation and Maximum Frequency Limitation in Capacitive Deionization Systems," in Proc. Power Electronics Drive Systems and Technologies Conference (PEDSTC), 7th 2016, pp. 1-6, 16-18 Feb. 2016.

[3] H. Mehrabian-Nejad, B. Farhangi, and Sh. Farhangi, "Application of PV and Solar Energy in Water Desalination System," in Proc. 2nd Int. Conf. and Exhibition on Solar Energy, 2015, ICESE-1078.

[4] R. Rahimi, E. Afshari, B. Farhangi, and Sh. Farhangi, "Optimal Placement of Additional Switch in the Photovoltaic Single-phase Grid-Connected Transformerless Full Bridge Inverter for Reducing Common Mode Leakage Current", in Proc. Energy Conversion (CENCON), 2015 IEEE Conference on, pp. 1-5, 19-20 Oct. 2015.

[5] E. Afshari, R. Rahimi, B. Farhangi, and Sh. Farhangi, "Analysis and Modification of the Single Phase Transformerless FB-DCB Inverter Modulation for Injecting Reactive Power," in Proc. Energy Conversion

[6] M. H. Jahanbakhshi, B. Asaei, and B. Farhangi, "A novel deadbeat controller for single phase PV grid connected inverters," in Electrical Engineering (ICEE), 2015 23rd Iranian Conference on, 2015, pp. 1613-1617.

[7] M. Hamzeh, S. Farhangi, and B. Farhangi, "A new control method in PV grid connected inverters for anti-islanding protection by impedance monitoring," in Control and Modeling for Power Electronics, 2008. COMPEL 2008. 11th Workshop on, 2008, pp. 1-5.

[8] B. Farhangi and S. Farhangi, "Comparison of z-source and boost-buck inverter topologies as a single phase transformer-less photovoltaic grid-connected power conditioner," in Power Electronics Specialists Conference, 2006. PESC'06. 37th IEEE, 2006, pp. 74-79.

[9] B. Farhangi and S. Farhangi, "Application of Z-Source Converter in Photovoltaic Grid-Connected Transformer-Less Inverter," Electrical Power Quality and Utilisation, Journal, vol. 12, pp. 41-45, 2006.

[10] B. Farhangi, "A novel modified deadbeat controller for vehicle to grid application," in Power Electronics, Drives Systems & Technologies Conference (PEDSTC), 2015 6th, 2015, pp. 47-52.

[11] B. Farhangi, H. A. Toliyat, and A. Balaster, "High impedance grounding for onboard plug-in hybrid electric vehicle chargers," in Power Engineering, Energy and Electrical Drives (POWERENG), 2013 Fourth International Conference on, 2013, pp. 609-613.

[12] A. Sepehr, M. Saradarzadeh, and Sh. Farhangi, " A Noninvasive On-line Failure Prediction Technique for Aluminum Electrolytic Capacitors in Photovoltaic Grid-connected Inverters," in Proc. Power Electronics Drive Systems and Technologies Conference (PEDSTC), 7th 2016, pp. 1-6, 16-18 Feb. 2016.

[13] T. Kerekes, M. Liserre, R. Teodorescu, C. Klumpner, and M. Sumner, "Evaluation of Three-Phase Transformerless Photovoltaic Inverter Topologies," Power Electronics, IEEE Transactions on, vol. 24, pp. 2202-2211, 2009.

[14] S. Farhangi, B. Vafakhah, B. Farhangi, P. Kanaan, and S. Maneshipoor, "A 5kW grid-connected system with totally home-made components in Iran," in 19th European Photovoltaic Solar Energy Conference and Exhibition, 2004, pp. 2988-2991.

[15] J. Farhang, M. Eydi, B. Asaei, and B. Farhangi, "Flexible strategy for active and reactive power control in grid connected inverter under unbalanced grid fault," in Electrical Engineering (ICEE), 2015 23rd Iranian Conference on, 2015, pp. 1618-1623.

[16] B. Farhangi and K. Butler-Purry, "Transient study of DC Zonal Electrical Distribution System in Next Generation Shipboard Integrated Power Systems using PSCAD™," in North American Power Symposium (NAPS), 2009, 2009, pp. 1-8.

[17] S. Farhangi, B. Vafakhah, B. Farhangi, P. Kanaan, and S. Maneshipoor, "A 5kW grid-connected system with totally home-made components in Iran," in 19th European Photovoltaic Solar Energy Conference and Exhibition, 2004, pp. 2988-2991.

[18] B. Farhangi and H. A. Toliyat, "Modeling and Analyzing Multiport Isolation Transformer Capacitive Components for Onboard Vehicular Power Conditioners," Industrial Electronics, IEEE Transactions on, vol. 62, pp. 3134-3142, 2015.

[19] B. Farhangi and H. A. Toliyat, "Modeling Isolation Transformer Capacitive Components in a Dual Active Bridge Power Conditioner," in Energy Conversion Congress and Exposition (ECCE), 2013 IEEE, 2013, pp. 5476-5480.

[20] B. Farhangi and H. Toliyat, "A Novel Vehicular Integrated Power System Realized with Multi-port Series Ac Link Converter," in Applied Power Electronics Conference and Exposition, 2015. APEC 2015. 30th Annual IEEE, 2015, pp. 1353-1359.

[21] B. Farhangi and H. A. Toliyat, "Piecewise Linear Model for Snubberless Dual Active Bridge Commutation," Industry Applications, IEEE Transactions on, vol. 51, pp. 4072-4078, 2015.

[22] B. Farhangi and H. A. Toliyat, "Piecewise linear modeling of snubberless dual active bridge commutation," in Energy Conversion Congress and Exposition (ECCE), 2014 IEEE, 2014, pp. 2065-2071.

[23] Mehrabian-Nejad, H.; Mohammadi, S.; Farhangi, B., "Novel control method for reducing EMI in shunt active filters with level shifted random modulation," in Power Electronics, Drives Systems & Technologies Conference (PEDSTC), 2015 6th , vol., no., pp.585-590, 3-4 Feb. 2015.

[24] A. Ramezani, Sh. Farhangi, H. Iman-Eini, and B. Farhangi, "High Efficiency Wireless Power Transfer System Design for Circular Magnetic Structures ", in Proc. Power Electronics Drive Systems and Technologies Conference (PEDSTC), 7th 2016, pp. 1-6, 16-18 Feb. 2016.

[25] E. Gubía, P. Sanchis, A. Ursúa, J. Lopez, and L. Marroyo, "Groundcurrents in single-phase transformerless photovoltaic systems," Prog. Photovolt., Res. Appl., vol. 15, no. 7, pp. 629–650, 2007.

[26] T. Kerekes, R. Teodorescu, and U. Borup, "Transformerless photovoltaic inverters connected to the grid," in Proc. APEC, Feb. 25– Mar. 1, 2007, pp. 1733–1737.

[27] K. S. F. Tan, N. A. Rahim, H. Wooi-Ping, and C. Hang Seng, "Modulation Techniques to Reduce Leakage Current in Three-Phase Transformerless H7 Photovoltaic Inverter," Industrial Electronics, IEEE Transactions on, vol. 62, pp. 322-331, 2015.

[28] G. Vazquez, T. Kerekes, J. Rocabert, P. Rodriquez, R. Teodorescu, and D. Aguilar, "A photovoltaic three-phase topology to reduce commonmode voltage," in Proc. IEEE ISIE, Jul. 2010, pp. 2885-2890.

[29] G. Oriti, A. L. Julian, and T. A. Lipo, "A new space vector modulation strategy for common mode reduction," in Proc. IEEE PESC 1997, Jun. 1997, pp. 1541-1546.

[30] E. Ün, and A. M. Hava, "A near-state PWM method with reduced switching losses and reduced common-mode voltage for three-phase voltage source inverters," IEEE Trans Ind. Appl., vol. 45, no. 2, pp. 782-793 , Mar./Apr. 2009.

[31] M. Cacciato, A. Consoli, G. Scarccella, and A. Testa, "Reduction of common-mode currents in PWM inverter motor drives," IEEE Trans. Ind. Appl., vol. 35, no. 2, pp. 469-476, Mar./Apr. 1999.

[32] A. M. Hava, and E. Ün, "Performance analysis of reduced commonmode voltage PWM methods and comparison with standard PWM methods for three-phase voltage-source inverters," IEEE Trans. Power Electron., vol. 24, no. 1, pp. 241-252, Jan. 2009.

[33] A. M. Hava, and E. Ün, "A high-performance PWM algorithm for common-mode voltage reduction in three-phase voltage source inverters," IEEE Trans. Power Electron., vol. 26, no. 7, pp. 1998-2008, July 2011.

[34] G. Xiaoqiang, H. Ran, J. Jiamin, L. Zhigang, S. Xiaofeng, and J. M. Guerrero, "Leakage Current Elimination of Four-Leg Inverter for Transformerless Three-Phase PV Systems," Power Electronics, IEEE Transactions on, vol. 31, pp. 1841-1846, 2016.

[35] T. Kerekes, R. Teodorescu, and M. Liserre, "Common mode voltage in case of transformerless PV inverters connected to the grid", in Industrial Electronics, 2008. ISIE 2008. IEEE International Symposium on, 2008, págs. 2390-2395.

[36] Li, Bin; Zhang, Ming; Huang, Long; Hang, Lijun; Tolbert, Leon M., "A robust multi-resonant PR regulator for three-phase grid-connected VSI using direct pole placement design strategy," In Proc. Twenty- Eighth Annual IEEE , Applied Power Electronics Conference and Exposition (APEC), 2013, pp.960-966, March 2013.

7th Power Electronics, Drive Systems & Technologies Conference (PEDSTC 2016)
16-18 Feb. 2016, Iran University of Science and Technology, Tehran, Iran

Efficiency Optimization and Power Management in a Stand-Alone Photovoltaic (PV) Water Pumping System

Behzad Mirshekarpour
Department of Electrical Engineering
Sahid Rajaee Teacher Training University
Tehran, Iran
b.mirshekarpour@gmail.com

S. Alireza Davari
Department of Electrical Engineering
Sahid Rajaee Teacher Training University
Tehran, Iran
davari@srttu.edu

Abstract— **In this paper a power management strategy for the stand alone PV based system for water pumping is proposed. The mentioned system is practical for off-grid rural areas. The main part of the system includes PV panels, three phase induction motor and the battery. PV panel voltage is delivered to the inverter by a boost converter. For the battery charging and discharging a simple bidirectional converter is used. Efficiency optimization of these systems is very important. In this paper the efficiency of motor in light loads is increased by Rosenbrock method. Also, the PV maximum power point Tracking (MPPT) is done by perturb and observe (P&O) algorithm. The mentioned system is evaluated under different conditions. This paper proposed a control strategy for power management. Simulation results, using MATLAB/SIMULINK, confirm the validity of the system, and verify MPPT, DC link voltage regulation and optimal charging/discharging of the battery.**

Keywords: Stand-Alone PV system, Efficiency optimization, MPPT.

I. INTRODUCTION

Nowadays, with the increasing of the world population, limited resources and environmental adverse effects resulting from the consumption of conventional fossil fuel resource, increases the attention to the renewable forms of energy. Solar energy is one of the renewable energy sources that can be used as a source of stand-alone generation [1]. In rural areas where there is no access to the grid, using solar cells is particularly important. One of the main applications of solar cells in these regions is water pumping. A system contains solar cells, converter, motors and energy storage source can be used, especially in those areas. Reducing losses in mentioned system is very important. The loss reduction can be contained of increasing motor, converter efficiency and maximum power point tracking of solar cells. There are several methods to reach these goals. In [2], design of a single-stage single switch dc/dc converter for a PV-battery-powered water pump system is presented. By using the variable-frequency control, the main functions such as MPPT and driving the motor with specific

speed can be realized. In [3], a converter for PV water pumping with having low cost, high efficiency and robustness was presented. In [4], a stand-alone generation system, with a reduced number of energy-processing stages, is proposed, allowing for the implementation of a high-efficiency PV system.

In this paper a novel control strategy has been proposed for the stand-alone solar water pumping systems. The control of every part is performed by considering the condition of the whole system. The control strategy should reach to the following goals:

- All day maximum feasible power of the pump
- Controlling dc link voltage
- MPPT of the PV
- Controlling the battery charging/discharging
- Reducing the motor loss

For reaching these goals, a combination of the perturb and observe (P&O) algorithm for Maximum Power Point Tracking (MPPT) and Rosenbrock algorithm for increasing motor efficiency is used. For the battery charging/discharging control, a simple bidirectional dc–dc power converter is used. The whole process is driven by a central power management system.

This paper is organized as follows. In section II, the structure of the mentioned system is described. In section III, the control strategy is explained. Section IV defines the operating modes and in section V the simulation results are presented. Finally, conclusions of this research are given in section VI.

II. THE STRUCTURE OF A STAND-ALONE PHOTOVOLTAIC WATER PUMPING SYSTEM

In this section the structure of a stand-alone photovoltaic water pumping system studied and has been investigated. Fig.1 illustrates the simplified block diagram of the mentioned system.
The main parts of this system are:
- PV panel
- DC/DC converter

978-1-5090-0376-1/16 $31.00 © 2016 IEEE   427

- Bidirectional converter and battery
- Inverter
- Pump motor

The output power of a PV panel depends on radiation, ambient temperature and voltage terminal. Therefore, the control operating point of a PV panel is important in order to attract the maximum power. Thus, an appropriate algorithm should be used for the maximum power tracking. In order to obtain a good dynamic performance from the MPPT, different control algorithms were developed [5]–[9]. The output voltage of the PV panels is usually low. Therefore, a boost converter is required for connecting the PV panels to system. Also a boost converter is used for maximum power point tracking. The output voltage of boost converter will be applied to the DC link capacitors. The Inverter provides the AC voltage for three-phase induction motor. To avoiding a boost converter between the PV panel and the motor, the majority of commercial systems use low voltage dc motors. However, dc motors have high losses and high maintenance cost compared to induction motors. Therefore, dc motors are not suitable for rural areas, because of operating and maintenance problems. Besides, the induction motor has the following advantages: higher efficiency, accessibility in local markets, and need to less maintenance [4].

In this paper, direct torque control (DTC) method is used for motor control. DTC is a direct method which is used in the variable frequency drives to control the torque and the speed of three-phase AC electric motors. This involves calculating an estimation of the motor's magnetic flux and torque based on the measured voltage and current of the motor. Dynamic response of the torque is very fast in DTC and complexity processing requirements is low. Inverter switching modes are selected using hysteresis based lookup table [10].

The motor efficiency could be increased at low load condition by changing the stator flux reference. When motor operate in low load condition, stator flux reduction reduces the motor current and it will lead to increase the efficiency of the motor [11].

In some conditions, the power produced by the PV panel is greater than the required motor power. In this case, in order to use the maximum power of PV panel, a source of energy storage such as a battery is needed. Battery stores the extra energy and delivered it in low PV power conditions. Also, the DC link voltage will be more stable when a battery is connected to the DC link.

The battery cannot be connected directly to DC link. Therefore, a simple bidirectional converter is necessary. Battery charging/discharging control is the main task of the bidirectional converter.

III. PROPOED POWER MANAGEMENT STRTEGY

The aim of this research is reaching to a thorough power management strategy for the standalone solar power pumping systems. This strategy controls every part of the system by considering the situation of the others. The maximum power tracking of the PV panel, efficiency optimization of the

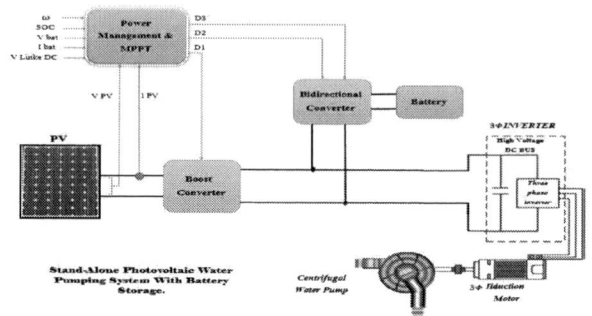

Fig.1. Structure of a stand-alone photovoltaic (PV) water pumping systems.

induction motor and control of charging/discharging of the battery is performed in an interconnected way. Fig. 2 shows the mentioned system circuits and blocks. The power flow of the PV and the battery is controlled in order to fulfill the motor control and the optimized battery charging/discharging. In order to reduce the power demand, an efficiency optimization method is used for motor control. Also, the PV panel may be used in a power less than the maximum power for optimized charging of the battery. In different conditions the power management strategy manage the power between the components of the system. To achieve the mentions goals, the duty cycle of boost converter (D1) and duty cycle of bidirectional converter (D2, D3) changes according to the system condition. In order to determine the optimal value of the duty cycle of the boost converter (D1) and the duty cycle of the bi-directional converter (D2, D3), the following constrains should be considered:

- The accurate control of the motor
- Ideal voltage for DC link with respect to load variation
- Optimal charge and discharge of the battery
- Delivering the maximum power by the PV panels

In the studied system, the pump motor is controlled by inverter with direct torque control method. Main features of DTC is absence of mechanical transducers, very simple control, low computational time and reduced parameter sensitivity [10].

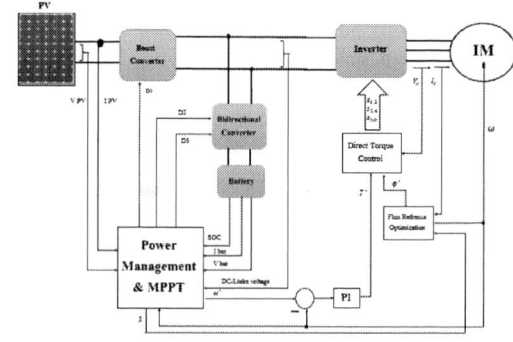

Fig.2. Circuits and blocks of the stand-alone PV water pumping systems.

Also, efficiency of the induction motor can be improved by making the stator flux amplitude as a function of the load. Many search algorithms of efficiency optimization are presented in the literature. Search algorithms are based on input power measurement. Their main principle is composed of changing the flux level in small steps to keep the output power of the motor constant and find the minimum of input power [11]. In this paper, Rosenbrock algorithms is used. This algorithm is explained as follow:

$$\Delta\phi = f(\Delta p_{abs}) \tag{1}$$

$$\Delta\phi = f(\Delta I_S) \tag{2}$$

$$
\begin{cases}
\Delta\phi_k = \phi_k - \phi_{k-1} & \phi_{k+1} = \phi_k + k\,\Delta\phi_k \\
\Delta I_S = I_{S(k)} - I_{S(k-1)} & k = 1 \xrightarrow{for} \Delta I_S \langle 0
\end{cases} \tag{3}
$$
$$k = -\frac{1}{2} \xrightarrow{for} \Delta I_S \rangle 0$$

For PV panel control, among several algorithms proposed for accessing the maximum power point in the PV panel, P&O algorithm is very common and used extensively. The advantage of P&O algorithm is simple control algorithm, Low cost and easy implementation [5]. In this simulation, the PV voltage is perturbed in an arbitrary path and the power levels of two successive samples are compared. According to sign power, the path for further perturbation is decided.

Battery is connected to the DC-link by a simple bidirectional converter [4],[12],[13]. This bidirectional converter acts as a buck converter in the battery charging mode. The reason behind this is that DC link voltage must be reduced for optimal battery charge. Also, the bidirectional converter acts as a boost converter in the battery discharging mode. The reason behind this is that when the battery is connected to DC-link the voltage of the battery must be increased.

Under different environmental conditions for the mentioned system, several operation modes are possible. This operation modes can be summarized as follows:
1. *The PV panel only charges the battery.*
2. *The battery and the motor are fed by the PV panel simultaneously.*
3. *The PV panel alone feeds the motor.*
4. *The motor is fed by the PV panel and the battery simultaneously.*
5. *The motor is fed by the battery only, and the power of the PV panel is zero.*

The mentioned modes are elaborated as follows.

A. *Operation Mode 1*

In this mode motor is not connected to the system, the battery is charged with the PV panel. In this mode, the bidirectional converter operates as a buck converter to charge the battery. The optimal charging current battery is considered 0.9 of battery capacity. The PV panel works at maximum power point by P&O algorithm. If the PV panel power is higher than the needed power of the battery, the PV panel should not work at maximum power to avoid damaging the battery. To resolve this problem, at first the PV is placed in the maximum power point. Afterward, the DC-link voltage is changed with the aim of the optimal charge of the battery via changing the duty cycle of the bidirectional converter (D2). Then, the battery current is measured. If the charging current of the battery is more than the optimum one, the PV power is reduced by changing the boost converter duty cycle (D₁). By reducing the PV panel power, battery charge current is also reduced. The diagram of this mode is illustrated in Fig.3.

B. *Operation Mode 2*

If the load power value is lower than the PV panel power, a part of the energy produced by the PV panel is used to charge the battery, and another part is used to supply the motor. The control algorithm maintains the sum of the battery power and the load power equal to the PV panel power. The bidirectional converter operates as a buck converter to charge the battery. The challenge in this mode is that the DC link voltage is changed when the battery and the motor are fed by PV panel simultaneously. In this case, the duty cycle of the boost converter is fixed and the DC link voltage variation is transferred to the PV panel voltage. Thus, the required power is drawn from the solar panel. The diagram of this mode is illustrated in Fig.4.

C. *Operation Mode 3*

If the load power value is higher than the PV panel power, all power produced by the PV panel is supplied to the load. The remained power that is necessary for the load must be supplied by the battery. In this mode, PV panel must be in MPP. The control algorithm maintains the PV panel power at the MPP by P&O algorithm. The bidirectional converter operates as a boost converter and the battery is discharged during this operation mode. If the battery charge level is not suitable for supply the pump motor, the battery is switched off and the reference of the motor speed is reduced until the PV panel feed induction motor singly. The diagram of this mode is illustrated in Fig.5.

D. *Operation Mode 4*

When the load power is equal to the PV power, the power processed by the bidirectional converter is zero. In this case, the battery is switched off. The diagram of this mode is illustrated in Fig.6.

E. *Operation Mode 5*

In this operation mode, the battery supplies the load without solar irradiation or under shading. The battery is discharged during this operation mode, and the bidirectional converter operates as a boost converter. The control system must turn off the bidirectional converter when the energy of the battery is lower than a certain limit. The diagram of this mode is illustrated in Fig.7.

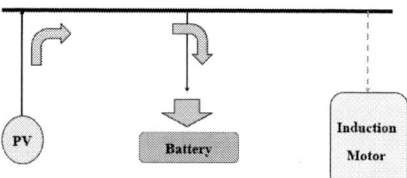

Fig.3. Operation mode 1.

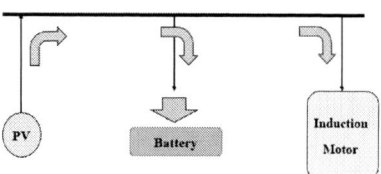

Fig.4. Operation mode 2.

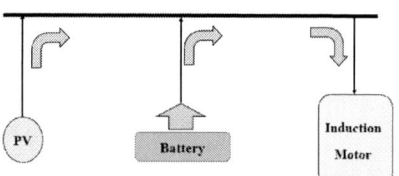

Fig.5. Operation mode 3.

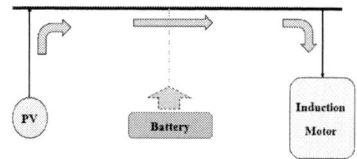

Fig.6. Operation mode 4.

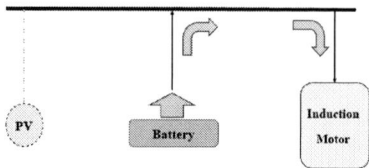

Fig.7. Operation mode 5.

TABLE I.   FEATURE OF SOLAR PANEL

TYPE	JAM5(L) 72-195/SI
Rated Maximum Power at STC (W)	195
Open Circuit Voltage (Voc/V)	45.56
Maximum Power Voltage (Vmp/V)	36.66
Short Circuit Current (Imp/A)	5.6
Maximum Power Current (Imp/A)	5.32
Module Efficiency [%]	15.27
STC Irradiance 1000W/m², Module Temperature 25⁰C , Air Mass 1.5	

TABLE II.   FEATURE OF INDUCTION MOTOR

TYPE	Three Phase IM
Nominal Power (HP)	5
Voltage (V)	460
Frequency (Hz)	36.66
Nominal Speed (RPM)	1750
Pole Pairs	2

TABLE III.   FEATURE OF BATTERY

TYPE	Lead-Acid FT-1201500
Nominal Voltage (V)	12
Capacity (Ah)	200
Weight Approx. (Kg)	2.44

IV. SIMULATION RESULTS

The simulation results are used to evaluate the proposed strategy. The model number of the simulated PV is JAM5(L)72-195/SI. The PV model specifications are tabulated in Table I. Also, the specifications of the induction motor and the battery are shown in Tables II and III.

Six similar solar panels are connected in series and each of these branches is paralleled with the other three branches. In this system, 8 batteries are connected in series and a battery pack with 96 V is made. Due to change in the temperature and irradiance the PV maximum power changes.

The PV power, DC-link voltage, motor speed, electromagnetic torque and battery current waveforms for changes in the irradiance is shown in Figs.8-12. At first, irradiance is 1000 W/m2 and PV panel alone feeds the induction motor (mode 4). At t=2.5 sec, irradiance is reduced from 1000 to 500 W/m2. In this case, the motor speed do not follows the reference speed and the battery is connected to the circuit (mode 3). At t=4.5 sec., irradiance is increased from 500 to 800 W/m2. At this moment the control scheme draws the maximum power from the PV panel and the battery is discharged in a longer time.

In Figs.13-17 the other operating modes are discussed. In this case, irradiance is 1000 W/m2 and IM motor works at light load. So, the battery can be charged by means of bidirectional converter. At t=0-4 sec, The PV panel provides motor power (mode 4). At t=4-7 sec, the battery is connected to the system and PV panel power delivery increases by varying the duty cycle of bi-directional converter (mode 2). At t=7-10 sec, PV power delivery increases again by varying the duty cycle of bi-directional converter (D2) and battery charge current can be controlled with these changes (mode 2).

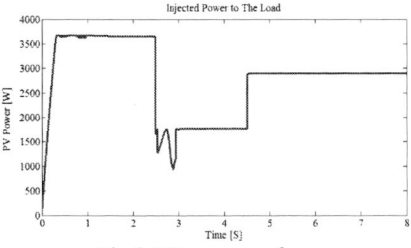
Fig. 8. PV power waveform.

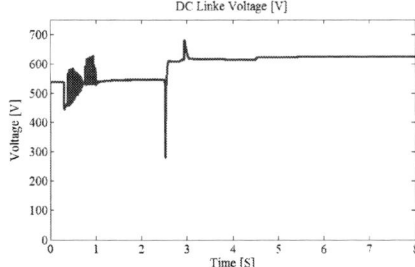
Fig. 9. DC-linke voltage waveform.

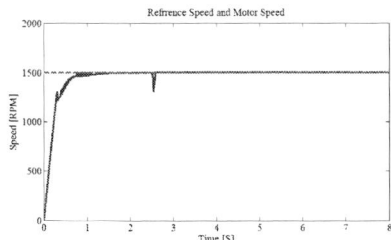
Fig. 10. Reffrence speed and motor speed waveform.

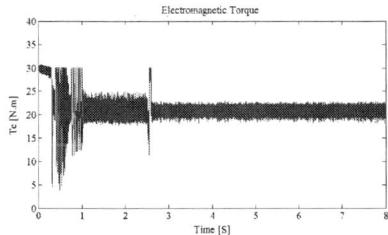
Fig. 11. Electromagnetic torque waveform.

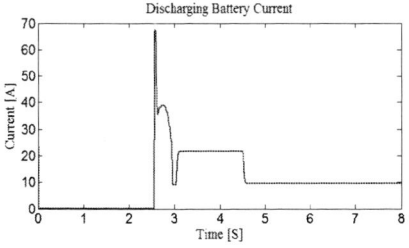
Fig. 12. Battery current waveform.

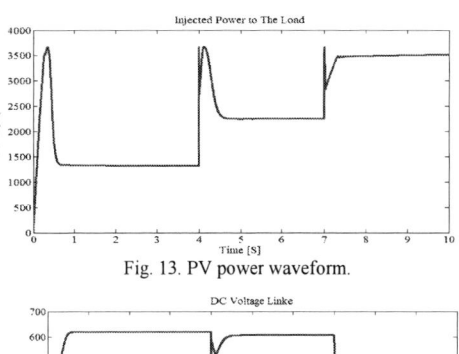
Fig. 13. PV power waveform.

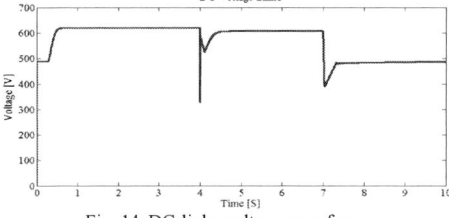
Fig. 14. DC-linke voltage waveform.

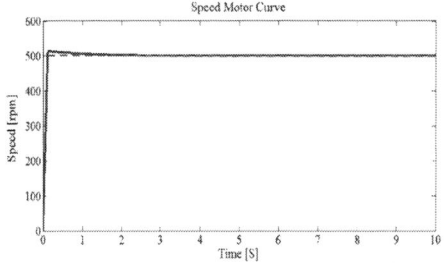
Fig. 15. Reffrence speed and motor speed waveform.

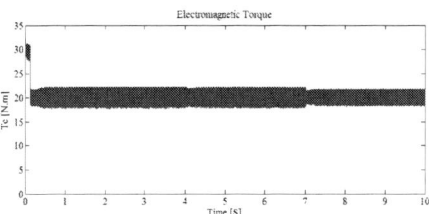
Fig. 16. Electromagnetic torque waveform.

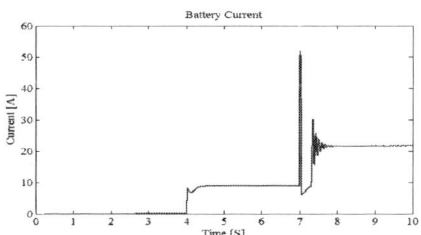
Fig. 17. Battery current waveform.

To evaluate the performance of systems due to the improving the efficiency of an induction motor, mode 5 is studied elaborately. In this mode, the battery alone feeds the motor. In this mode, 15%, 30%, 45%, and 60% of motor nominal load is simulated and battery output power has been studied. The battery power in mentioned loads is shown in Fig.18.

TABLE IV. EFFICIENCY OPTIMIZATION AT DIFFERENT LOADS

% Load	Battery power without efficiency optimization (W)	Battery power with efficiency optimization (W)	% efficiency optimization
٪ 15	830	740	٪ 10
٪ 30	1400	1320	٪ 6
٪ 45	1970	1930	٪ 2
٪ 60	2570	2540	٪ 1.16

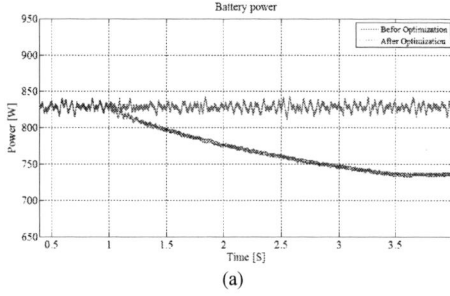

(a)

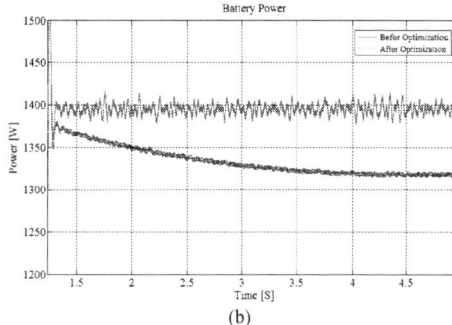

(b)

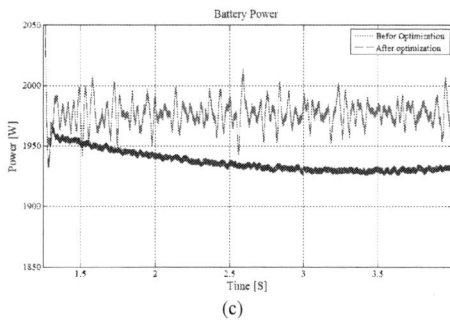

(c)

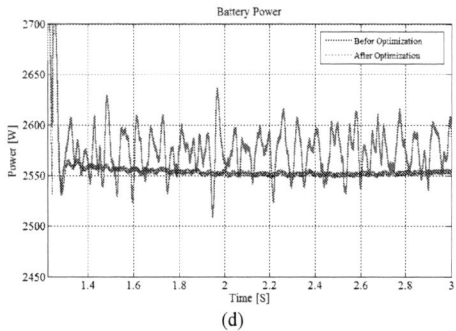

(d)

Fig.18. Waveform battery power with efficiency optimization and without efficiency optimization; (a): 15% motor nominal load; (b): 30% nominal load engine; (c): 45% nominal load engine; (d): 60% nominal load engine.

Table IV shows the battery power and efficiency optimization in the different percentage of loads in summary.

V. CONCLUSION

A power management strategy for the stand alone PV based system for water pumping is proposed in this paper. The aim of the power management strategy is feeding the induction motor and optimal charging/discharging of the battery with respective of power availability from PV. The maximum power of the PV panel was obtained by using the P&O algorithm. Control of the speed of induction motor based on DTC method and efficiency optimization of the induction motor was carried out by using Rosenbrock method. Charging/discharging of the battery was performed by using a simple bidirectional converter. Simulation with MATLAB/SIMULINK is done in order to show the validity of the proposed strategy. The simulation results of system validate that the mentioned system has the fast and effective response under variation irradiance levels and demonstrate that the performance of the power management strategy is satisfying under steady state and dynamic conditions.

REFERENCES

[1] R.Marouani,F.Bacha,"A maximum-power-point tracking algorithm applied to a photovoltaic water-pumping system",*EPE Chapter 'Electric Drives' Joint Symposium*, 1-3 July 2009.

[2] Le An and Dylan Dah-Chuan Lu, "Design of a Single-Switch DC/DC Converter for a PV-Battery-Powered Pump System With PFM+PWM Control," *IEEE Transactions on Industrial Electronics,* vol. 62, no. 2, February 2015.

[3] J. M. Caracas, G. D. C. Farias, L. M. Teixeira, and L. D. S. Ribeiro, "Implementationof a high-efficiency, high-lifetime,and low-cost converter foran autonomous photovoltaic water pumping system," *IEEE Trans. Ind. Appl.*, vol. 50, no. 1, pp. 631–641, Jan./Feb. 2014.

[4] Roger Gules, Juliano De Pellegrin Pacheco, Hélio Leães Hey and Johninson Imhoff, "A Maximum Power Point Tracking System With Parallel Connection for PV Stand-Alone Applications, " *IEEE Tans. Ind. Electron.*, vol. 55, NO. 7, July 2008.

[5] W. Xiao, W. G. Dunford, P. R. Palmer, and A. Capel, "Application of centered differentiation and steepest descent to maximum power point tracking," *IEEE Trans. Ind. Electron.*, vol. 54, no. 5, pp. 2539–2549, Oct. 2007.

[6] N. Femia, G. Petron, G. Spagnuolo, and M. Vitelli, "Optimization of perturb and observe maximum power point tracking method," *IEEE Trans. Power Electron.*, vol. 20, no. 4, pp. 963–973, Jul. 2005.

[7] M. G. Simões and N. N. Franceschetti, "Fuzzy optimisation based control of a solar array system," *Proc. Inst. Electr. Eng.—Electric Power Applications*, vol. 146, no. 5, pp. 552–558, Sep. 1999.

[8] J. A. Abu-Qahouq, H. Mao, H. J. Al-Atrash, and I. Batarseh, "Maximum efficiency point tracking (MEPT) method and digital dead time control

implementation," *IEEE Trans. Power Electron.*, vol. 21, no. 5, pp. 1273– 1281, Sep. 2006.

[9] N. Mutoh, M. Ohno, and T. Inoue, "A method for MPPT control while searching for parameters corresponding to weather conditions for PV generation systems," *IEEE Trans. Ind. Electron.*, vol. 53, no. 4, pp. 1055– 1065, Aug. 2006.

[10] I. Takashi and T. Noguchi, "A new quick-response and high-effiency control strategy of an induction motor." *IEEE Trans. Ind. Applicat.*, IA-22, pp. 820-827,Sept./Oct. 1986.

[11] S.Ghozzi, K. Jrlassi,X.Roboam," Energy optimization of induction motor drives",*IEEE International Conference on Industrial Technotogy*,2004.

[12] R. J. Wai, R. Y. Duan, and K. H. Jheng, "High-efficiency bidirectionaldc–dc converter with high-voltage gain," *IET Power Electron.*, vol. 5,no. 2, Feb. 2012.

[13] R. J. Wai, R. Y. Duan, and K. H. Jheng, "High-efficiency bidirectionaldc–dc converter with high-voltage gain," *IET Power Electron.*, vol. 5,no. 2, Feb. 2012.

7th Power Electronics, Drive Systems & Technologies Conference (PEDSTC 2016)
16-18 Feb. 2016, Iran University of Science and Technology, Tehran, Iran

Increasing the Battery Life of the PMSG Wind Turbine by Improving Power Division of the Hybrid Energy Storage System

Mohammad Eydi, Javad Farhang, Behzad Asaei, Reza Emamalipour
School of Electrical and Computer Engineering
Factulty of Engineering, University of Tehran, Tehran, Iran
m.eydi@ut.ac.ir, j.farhang@ut.ac.ir, basaei@ut.ac.ir, r.emamalipour@ut.ac.ir

Abstract— **Nowadays, with an increase in the number of wind power plants, the rules for connecting these plants to the grid have been changed. Under the new standards, to maintain a frequency stability, the power variations of wind farms must be limited. For this reason, the use of the energy storage sources along with wind turbines have been proposed and many publications have been published on this subject. One of the energy storage systems is the hybrid energy storage system with battery and capacitor. In this paper, the control of the hybrid energy storage system with battery and capacitor for the PMSG based wind turbines has been investigated and a method is proposed to control and divide the power between the battery and the capacitor. By eliminating the unnecessary charges and discharges of the battery, which existed in the conventional approaches, this method improves the performance of the energy storage system and increases the battery life compared to the conventional methods. Finally, the functionality of the proposed method is investigated by simulation in MATLAB/Simulink Software.**

Keywords— ***Battery lifetime; Hybrid energy storage system; Output Power smoothing; Permanent magnet synchronous generator; Wind turbine.***

I. INTRODUCTION

The increasing growth of the world population and industrialization in developing countries have increased the use of the fossil fuels. Excessive consumption of the fossil fuels results in a reduction in the reservoirs and an increase in air pollution and the price of fuels. Thus, considering the problems of fossil fuels and the increasing global need for energy, the use of renewable energy sources has been taken into consideration more and more. Among renewable energy sources, wind energy has been considered very largely due to its high production capacity up to several megawatts as well as its fast return on the investment. Today, most distributed generation power plants which use wind energy produce a small portion of the total power that is injected into the power system. In this case, network variations affect the turbine, however, the turbine has no significant effect on the grid stability because the variations in a small portion of the grid's power doesn't affect the grid frequency. Due to the growing use of the wind turbines in the world, this will change soon and as a result, the grid frequency will be affected by the wind turbines. Hence, if the injected power of these turbines changes a lot, the grid frequency will change. These variations will result in the power system instability. In order to avoid grid instability, a limit is imposed on the output power of these sources. The amount of this limit is different in various countries and depends on the wind turbine production capacity in each system. For example, Japan and China have set the power variation limit to be 2% and 7% of the nominal power in each minute and each 30 minutes respectively [1]. Due to the limitations, it is clear that these sources must produce and inject smooth power into the grid. Output power smoothing is done in two methods. The 1st method is to absorb a smooth power from the wind. This is done by changing the speed of the turbine and the angle of the blades. The 2nd method is the use of the energy storage sources. In this case, the turbine tracks the maximum power point. A smooth power is delivered to the grid and the power fluctuations are absorbed by the energy storage system. Different energy storage sources can be used to absorb the power fluctuations. [2] has utilized batteries, [3] has used capacitor and [1,4,5] has taken advantage of both batteries and capacitors simultaneously. Among the energy storage systems mentioned above, the simultaneous use of batteries and capacitors has a better performance compared to others [5]. In this method, the fluctuating power that must be compensated by the energy storage system is passed through a filter. The high frequency fluctuations are compensated by the capacitor and the low frequency variations are compensated by the batteries. If the power isn't divided properly between the battery and the capacitor, the lifetime of these sources will be reduced. Factors such as temperature, peak current and variations in charge amount affect the battery lifetime. In [4-6] fluctuations are divided between batteries and capacitors by means of a filter. The high frequency fluctuations are passed through the capacitor and the low frequencies are passed through the battery. In [7], the difference between the input and output power is compensated by the capacitor and the battery carries out the power compensation at specific times. In [8], the power division between the battery and the capacitor is determined by the PSO algorithm. In [9], by performing optimization techniques, the power variations are divided among different kinds of capacitors and batteries.

In this paper, a method is proposed in which the power reference of the battery and the capacitor is determined without the use of any optimization algorithm. Therefore, some of the problems of the optimization methods such as response divergence and long response time don't exist in this method .

978-1-5090-0376-1/16 $31.00 © 2016 IEEE

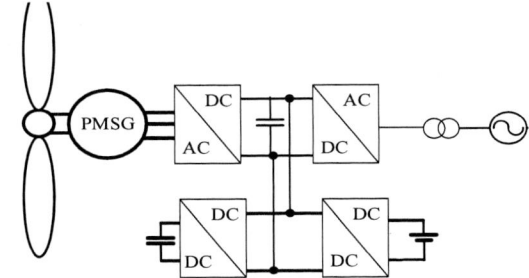

Fig.1. Overall structure of the system

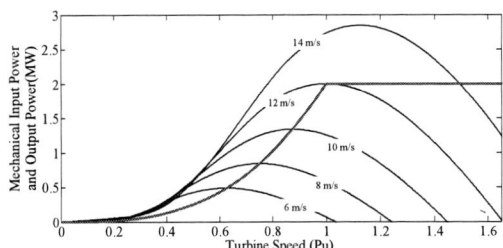

Fig.2.
Black lines: The turbine's input power against turbine speed for different wind velocities
Brown line : The turbine's output power against turbine speed

Moreover, in this method, power division is done according to the state of the system while in power division via filters, system state isn't taken into account.

The rest of this paper is as follows: section 2 presents a general overview of the system. The proposed control algorithm is presented in section 3. To evaluate the performance of the proposed method some simulation has been done using Matlab/Simulink. Section 4 is assigned to the simulation result. Finally section 5 concludes the paper.

II. SYSTEM'S MODEL

Fig. 1 shows a simplified block diagram of a wind turbine equipped with a permanent magnet generator and a hybrid energy storage system. Wind turbine absorbs the power from the wind according to the wind speed and the rotational speed of the turbine and delivers this power to the DC bus. The system can be divided into 2 parts. The mechanical and the electrical part. In the following, the operation of each part has been explained.

A. The Mechanical Part

The task of the mechanical part is to absorb power from the wind and deliver it to the generator shaft. The absorbed power and the torque can be calculated by the following formula.

$$P_w = \frac{1}{2}C_p(\lambda, \beta)\rho\pi R^2 V_w{}^3 \tag{1}$$

$$T_w = \frac{1}{2}C_p(\lambda, \beta)\rho\pi R^3 \frac{V_w{}^2}{\lambda} \tag{2}$$

Where C_p is the turbine's power factor, ρ is the air density, R is the blade radius, V_w is the wind speed. As can be seen in (1) and (2), C_p is a function of β and λ, where β is the blade angle and λ is the ratio between the velocity of the blades' tip and the wind speed. Hence, λ can be calculated as follows:

$$\lambda = \frac{\omega_t R}{V_w} \tag{3}$$

Where ω_t is the rotational speed of the turbine shaft. Fig. 2 shows the power produced by the turbine in terms of the rotational speed of the turbine for different wind speeds. As can

be seen in this figure, at any wind speed, the maximum power that can be

absorbed from the wind is achieved at a particular rotational speed called the maximum power point.

At nominal wind speed and maximum power point, the turbine absorbs nominal power. When the wind speed increases and becomes greater than the rated wind speed, at maximum power point, the power that is absorbed by the turbine will become greater than its rated power. Hence, if the turbine is planned to operate at wind speeds greater than the rated speed, it shouldn't be at the maximum power point. Thus at wind speeds greater than the nominal value, turbine rotates at a speed to absorb the rated power [3].

In Fig. 2, the output power variation curve is shown in terms of different wind speeds (Brown line). As can be seen, when the wind speed is less than the nominal value, the power that is produced by the turbine is at its maximum possible amount. When the wind speed becomes higher than the rated value, the nominal power is absorbed from the wind. The turbine output power curve in terms of different wind speeds is obtained in Fig. 2.

B. The Electrical Part

The 1st part of the electrical system is the generator. The generator used in the system presented in Fig. 1 is a permanent magnet. The voltage and the current equations of a permanent magnet generator in the rotating reference frame are as below:

$$V_d = -R_s I_d - L_d \frac{dI_d}{dt} + \omega_r L_q I_q \tag{4}$$

$$V_q = -R_s I_q - L_q \frac{dI_q}{dt} - \omega_r L_d I_d + \omega_r \Psi_f \tag{5}$$

Where V_d, V_q, I_d and I_q are the d and q components of the voltage and the current of the machine. L_d and L_q are the d and q components of the machine inductance respectively in rotating reference frame. ω_r is the rotational speed of the rotor and Ψ_f is the magnetizing flux. The power and the torque equations can be obtained as follows:

$$P = \frac{3}{2}(V_d I_d + V_q I_q) \tag{6}$$

$$T = -\frac{3}{2}P_n \Psi_f I_q \tag{7}$$

As can be seen in (7), in order to control the motor's torque, I_q must be controlled. In the following, it is explained how the control system of the generator-side and the grid-side converter operate [10].

B-1. *Control of the Generator-side Converter*

The task of the generator-side converter is to deliver the power of the generator to the DC bus and adjust the turbine speed. In this regard, first, the reference is determined. If the wind speed is less than the rated value, the generator's speed reference will be determined by the maximum power point tracker (MPPT). Otherwise, the speed reference will be calculated in a way that the turbine absorbs the nominal power. Then, the generator speed is compared to the speed reference. In case the generator speed were less than the speed reference, the q axis current of the generator would decrease. The result is a reduction in electrical torque and eventually an increase in generator speed. In case the turbine speed were higher than the speed in maximum power point, the operation would be the same. The task of the generator-side converter is to absorb this current reference from the generator and deliver the produced power to the DC bus. Fig .3 shows the control algorithm of the mentioned method.

B-2. *Control of the Grid-side Converter*

The task of this converter is to deliver the power from the DC bus to the grid. This transmission must be in such a way that meet the grid standards. Also, the voltage of the DC bus must be controlled if possible. Furthermore, this converter has to control the output voltage by controlling the reactive power that is injected into the grid. The relationship between the output active and reactive power of the grid-side converter is as follows.

$$V_{dinv} = RI_d + L\frac{dI_d}{dt} - L\omega_s I_q + V_d \qquad (8)$$

$$V_{qinv} = RI_q + L\frac{dI_q}{dt} + L\omega_s I_d + V_q \qquad (9)$$

Where V_d, V_q, V_{dinv} and V_{qinv} are the d and q components of the grid and the inverter voltages. I_d and I_q are the d and q components of the current respectively in the rotating reference frame. R and L are the resistance and the inductance of the filter and the line and ω_s is the angular speed. The active and reactive power that the converter exchanges with the grid are calculated by the following formula [10].

$$P = \frac{3}{2}V_d I_d \qquad (10)$$

$$Q = \frac{3}{2}V_d I_q \qquad (11)$$

B-3. *Control of the Energy Storage System*

The generator-side and the grid-side converters are separated by the DC link capacitor. If there were a difference between the input and the output power of the DC link, this difference would lead to charges and discharges of the DC link capacitor and thus variations in its voltage. Hence, a system must adjust the voltage of the DC link at the nominal value.

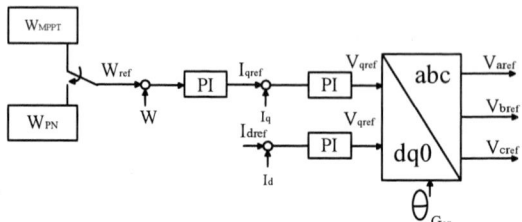

Fig.3. Control system of PMSG

For this purpose, two DC to DC converters are used. One side of these converters is connected to the DC bus and the other side is connected to the battery and the capacitor. In this case, the DC bus voltage is compared to a reference voltage and the difference passes through a PI controller. The output of the PI controller is the power reference of the energy storage system. Finally, this power must be divided between the battery and the capacitor. In order to determine the battery and the capacitor power references, a low pass filter is used to separate the low and high frequency fluctuations [8].

In order to prevent the capacitor from over charges and over discharges, a proportional controller is used. The output of this controller produces a current reference which is added to the current reference of the battery by a feed-forward.

The proper charge of the capacitor is considered to be in the middle of the allowed operating zone in order to allow the capacitor to absorb the power fluctuations with the lowest frequency and also to inject these fluctuations into the grid. In this case, if the capacitor state of charge (SOC) were higher than this proper charge, a current would be injected to the battery and if the capacitor's voltage were lower than this amount, a current would be drawn from the battery to bring the capacitor's voltage closer to the ideal value. The amount of this current is proportional to the difference between the capacitor SOC and the determined charge.

This problem also exists for the battery. The battery may also experience over charge or over discharge conditions. To solve this problem, in case of a low battery state of charge, the output power is reduced so that a share of the produced power charges the battery. In case the battery state of charge were high, the output power would increase (considering the allowed limits) to deliver a share of the battery's stored energy to the grid.

In this control method when the power is divided by a low pass filter, it can be seen often that although the capacitor has a high state of charge, the battery injecting power into the grid or while the capacitor has a low state of charge, both the battery and the capacitor absorb power according to the difference between the output power and the power produced by the turbine. However, after providing the required power by the energy storage system, a current is injected or drawn from the battery to bring the capacitor state of charge closer to its ideal value. The unnecessary charging and discharging of the battery, reduces the system's lifetime. If the adjustment of the voltage of the capacitor bank to the ideal value is done by the input power, then the battery life will increase. In the following, a method is proposed for dividing the power between the battery and the capacitor to minimize the problems that are mentioned above and to increase the battery life compared to the conventional methods [11].

III. THE PROPOSED METHOD

In this method, first, the charging and discharging conditions of the energy storage system and the voltage of the capacitor are investigated. If the energy storage system were being charged and the capacitor's SOC were less than 50%, the capacitor would receive the power alone until its SOC reaches 50%. If the capacitor's SOC were higher than 50% and the capacitor were absorbing the power alone, it would become fully charged quickly. After being fully charged, all of the power would be provided by the battery. Therefore, when the capacitor is in this area, the power is divided between the battery and the capacitor by a low pass filter. When the capacitor's SOC is more than 50% and the system is injecting power into the grid, the power will be supplied through the capacitor until the SOC reaches 50%. With this, not only the time that was required by the battery to supply power is reduced but also it takes less time that the capacitor's energy reaches its reference amount. When the capacitor's SOC reaches 50% and the power division changes from the capacitor alone to the conventional method or vice versa, there will be rapid changes in the battery current (since the battery's power reference changes as stairs). Therefore, to avoid these rapid changes, a low pass filter is used with a frequency of 10 times as much as the cur-off frequency of the power division system. This value is chosen so that the 2nd filter has no impact on the power division. In general, according to the state of charge of the energy storage system and the voltage of the capacitor, the following operation areas can be considered for the system.

- **The capacitor SOC is greater than its ideal value and the current is positive:** In this case, the voltage of the capacitor is high and the system is absorbing power. In this area, the capacitor operates like the conventional method and considering the amount of the energy that the capacitor can store, the power division between the battery and the capacitor is performed.

- **The capacitor SOC is less than its ideal value and the current is negative:** In this case, the voltage of the capacitor is low and the system must inject power into the grid. In this area, the capacitor operates like the conventional method and considering the amount of power that the capacitor can supply to the grid, the power division between the battery and the capacitor is performed.

- **The capacitor SOC is less than its ideal value and the current of the energy storage system is positive:** In this case, the capacitor is partially charged and the energy storage system is being charged. First, the current reference of the battery is set to zero in order to avoid the battery and the capacitor to charge simultaneously and then the battery discharges into the capacitor and hence prevents a reduction in the battery life. This is done as long as the capacitor state of charge is located in its lower

half. Once the capacitor state of charge reaches 50%, the system state changes and the power is divided between the battery and the capacitor.

- **The capacitor's SOC is greater than its ideal value and the current of the energy storage system is negative:** In this case, the capacitor state of charge is higher than 50% and the system is being discharged. In conventional methods, both of the systems are discharged first. After a change in the current's polarity, the battery charges so that the capacitor reaches its ideal point. In this method, when this happens, all of the current is drawn from the capacitor. This continues until either the current's polarity changes or the capacitor's SOC becomes less than 50%. In this case, the battery's power changes from zero to a new value which is determined by the filter.

The figure below shows the control algorithm of the proposed method. As can be seen, first, the power of the Hybrid energy storage system passes through a low pass filter. In the conventional method, the output of the filter determines the battery power. On the other hand, it is examined whether the power of the energy storage system is positive or negative. Also, the capacitor state of charge is investigated. If the capacitor SOC were low and the power of the energy storage system were positive, X would become zero. Another case where X becomes zero is when the voltage of the capacitor is high and the power of the energy storage system is negative. In other cases, X equals to 1. This amount is multiplied by the power reference which is calculated by the filter. As a result, the power of the battery will become zero, if the capacitor's voltage is high and the power of the energy storage system is negative. Also, this will be valid in case the capacitor's voltage is low and the power of the energy storage system is positive. The result of the X times the output of the LPF1 passes through another low pass filter (LPF2) with a cut-off frequency 10 times as much as the cut-off frequency of the first filter. So, the rapid changes of the battery power during this filter is considered to be high, it doesn't have any impact on the power division. Another point that must be noted is that because of the new proposed control method, the proportional the variations of X is overcome. Since the cut-off frequency of controller cannot be omitted since the proportional controller is applicable to all cases. However, the new controller is applied when a specific condition prevails over the system.

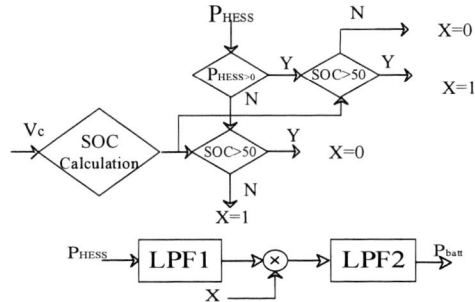

Fig. 4: Proposed method approach schematic

IV. SIMULATION RESULTS

In order to evaluate the functionality of the proposed method, the system is simulated in MATLAB software. The parameters of the simulated system are given in table I [3, 4].

Figure 5.a shows the wind speed variation curve. As illustrated, the average wind speed is 8 meters per second. The maximum power that can be absorbed in this condition is 600 MW. The wind speed varies between 4 to 12 (m/s). In this range, the turbine produces between zero and rated power. Applying this wind speed to the turbine and tracking the maximum power point, the absorbed power will be as Fig 5.b. As can be seen, when the wind speed is close to 12 m/s, the turbine produces the nominal power. The input cut-off speed is considered to be 5 m/s. Hence, when the wind speed becomes less than 5 m/s, the turbine doesn't produce any power. By applying the output power variation limits according to the grid standards, the output power waveform of Fig. 5.b is achieved. As shown, a relatively constant power is delivered to the grid. The difference between these two powers is supplied by the energy storage system. Figure 5.c shows the power variation curve of the energy storage system. Figure 5.d represents the power variations of the battery in both the proposed and conventional methods. As it is clear, in the proposed method, the power passes through the battery has a lower variation range. The reason is that the time that the capacitor has a high SOC and the battery is being discharged or the capacitor has a low SOC and the battery is being charged is reduced. Variations in the power that passes through the capacitor are shown in Fig. 5.e for both the proposed and conventional methods. As can be seen in Fig 5.d, in moments like the 22nd to 25th seconds and 55th to 57th seconds, the passing power through the battery cancels, becouse capacitor has a low voltage and the power of the battery is positive. In these moments, in order to bring the voltage of the capacitor closer to the ideal value, the whole power is absorbed by the capacitor. Between the 55th and 57th seconds, the battery power doesn't become zero. this difference

Table I   Simulation parameters

Generator parameters	
Rated output power	2MW
Resistance	50uΩ
d axis Inductance	5.5mH
q axis Inductance	5.5mH
Number of Pair Pole	11
Field Flux	136.25 V.s/rad
Equivalent Inertia	100000 Kg.m^2
HESS parameters	
Battery Capacity	50Kwh
Battery Bank Voltage	1kV
Initial SOC	50
Rated Capacity	50Ah
Internal resistance	0.2Ω
HESS Capacitance	0.2F
Rated Capacitor Bank Voltage	7kV
Other parameters	
DC Link Voltage	8kV
DC Link Capacitance	0.2F
Line Filter Inductance	3mH
Filter Cut off Frequency	0.05 Hz

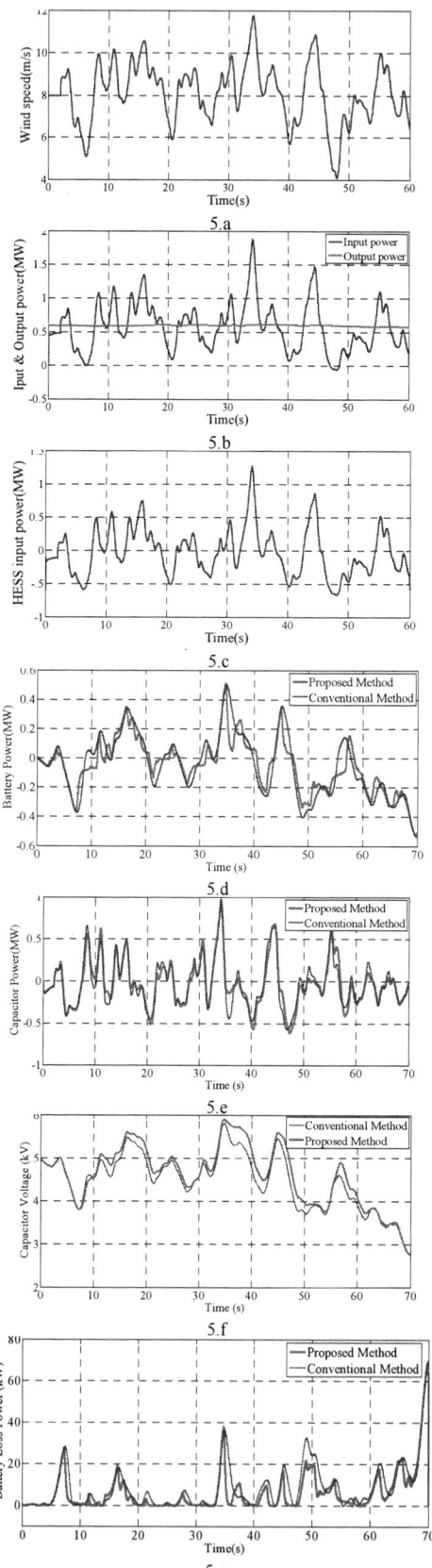

5.a

5.b

5.c

5.d

5.e

5.f

5.g

978-1-5090-0376-1/16 $31.00 © 2016 IEEE

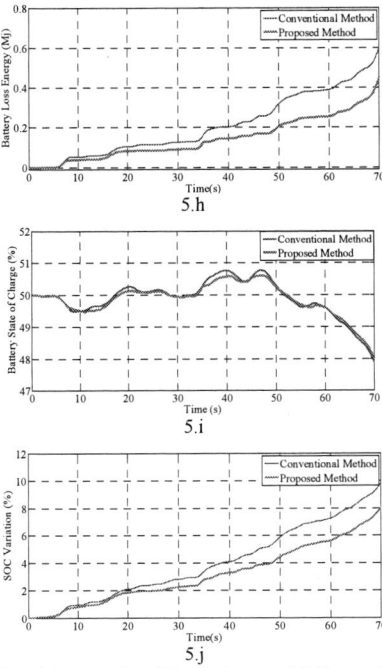

Fig. 5. Simulation results. (a) wind speed,(b)Input & output power, (c)HESS input power,(d) battery power,(e) capacitor power,(f) capacitor voltage,(g)battery SOC,(h) battery SOC variation (i) Battery loss power (j) Battery loss energy

is due to the feed-forward current which is drawn from the battery by the proportional controller. The capacitor voltage variation waveform is shown in Fig. 5.f. As can be seen, the voltage across the capacitor reaches its ideal amount earlier than that of in conventional method. For example, in 30th and 55th seconds, the voltage of the capacitor is identical in both methods. After a few seconds, it can be seen that the voltage of the capacitor gets closer to the ideal value in the proposed method. The ideal amount in this case is considered to be 5 kV. Fig 5.g shows the dissipation waveform of the battery for both methods. As it is clear, the dissipation in the proposed method is less than that of in the conventional methods. The dissipated energy from the beginning is shown in Fig 5.h for both methods. In this case, the average loss is reduced from 8.58 kW to less than 6.3 kW. This means that the battery works in lower temperature. To fully assess the status of the battery, the SOC variations of the battery are shown in Fig 5.i. As it is clear, the SOC variation of the battery in the proposed method is less than the conventional method which means an increase in battery lifetime. Figure 5.j shows the SOC variation of the battery. As illustrated, the SOC variation range has decreased by 20% which means the battery life has increased by 25% for the presented wind spectrum just by the SOC variations.

V. CONCLUSION

In this paper, the output power smoothing of the PMSG based wind turbine was studied by means of a hybrid energy storage system. For this purpose, different parts of the PMSG wind turbine along with its control system was described. After that, considering the goal of this paper, the problems of the conventional energy storage systems with battery and capacitor were described. It was observed that in the conventional methods, at intervals, the battery experienced unnecessary charges and discharges which in the long run reduces the battery life. In order to overcome this problem, a new control method was proposed. The main goal of the proposed method was to distribute the power between the battery and capacitor such that the unnecessary charge and discharge of the battery was reduced. Which resulted in a longer battery lifetime.

REFRENCE

[1]. Quanyuan Jiang and Haisheng Hong "Wavelet-Based Capacity Configuration and Coordinated Control of Hybrid Energy Storage System for Smoothing Out Wind Power Fluctuations" IEEE TRANSACTIONS ON POWER SYSTEMS, VOL. 28, NO. 2, MAY 2013

[2]. Quanyuan Jiang and Haijiao Wang "Two-Time-Scale Coordination Control for a Battery Energy Storage System to Mitigate Wind Power Fluctuations" IEEE TRANSACTIONS ON ENERGY CONVERSION, VOL. 28, NO. 1, MARCH 2013

[3]. Nicholas P. W. Strachan and Dragan Jovcic "Improving Wind Power Quality using an Integrated Wind Energy Conversion and Storage System (WECSS)" Power and energy Society General Meeting july 2008 Pittsburgh

[4]. A.Esmaili, B.Novakovic, A.Nasiri, and O. Abdel-Baqi," A Hybrid System of Li-Ion Capacitors and Flow Battery for Dynamic Wind Energy Support ",IEEE TRANSACTIONS ON INDUSTRY APPLICATIONS, VOL. 49, NO. 4, JULY/AUGUST 2013.pp.1649-1657

[5]. Wei Li ,Géza Joós and Jean Bélanger "Real-Time Simulation of a Wind Turbine Generator Coupled With a Battery Supercapacitor Energy Storage System" IEEE TRANSACTIONS ON INDUSTRIAL ELECTRONICS, VOL. 57, NO. 4, APRIL 2010

[6]. N.Mendis, K.M.Muttaqi and S.Perera "Management of Battery-Supercapacitor Hybrid Energy Storage and Synchronous Condenser for Isolated Operation of PMSG Based Variable-Speed Wind Turbine Generating Systems" IEEE TRANSACTIONS ON SMART GRID

[7]. Gao Chen, Qiang Yang, Ting Zhang, Zhejing Bao, and Wenjun Yan "Real-time wind power stabilization approach based on hybrid energy storage systems" 2013 Sixth International Conference on Advanced Computational Intelligence October 2013 pp124-129

[8]. T. Zhang, Z. J. Bao, G. Chen, Q. Yang, W. J. Yan "Control Strategy for a Hybrid Energy Storage System to Mitigate Wind Power Fluctuations" 2013 Sixth International Conference on Advanced Computational Intelligence , October 2013 pp 27-32

[9]. Mid-Eum Choi, Seong-Woo Kim, and Seung-Woo Seo "Energy Management Optimization in a Battery/Supercapacitor Hybrid Energy Storage System" IEEE TRANSACTIONS ON SMART GRID, VOL. 3, NO. 1, MARCH 2012

[10]. S.Li, T.A. Haskew, RP.Swatloski, and W.Gathings, "Optimal and Direct-Current Vector Control of Direct-Driven PMSG Wind Turbines ",IEEE TRANSACTIONS ON POWER ELECTRONICS, VOL. 27, NO. 5, MAY 2012.pp.2325-2337

[11]. Anthony M. Gee, Francis V. P. Robinson and R.W. Dunn "Analysis of Battery Lifetime Extension in a Small-Scale Wind-Energy System Using Supercapacitors" IEEE TRANSACTIONS ON ENERGY CONVERSION, VOL. 28, NO. 1, MARCH 2013

978-1-5090-0376-1/16 $31.00 © 2016 IEEE

7th Power Electronics, Drive Systems & Technologies Conference (PEDSTC 2016)
16-18 Feb. 2016, Iran University of Science and Technology, Tehran, Iran

Fast and Simple Open-Circuit Fault Detection Method for Interleaved DC-DC Converters

Mahmoud Shahbazi, Mohammad Reza Zolghadri, Saeed Ouni

Center of Excellence in Power System Management & Control (CEPSMC), Sharif University of Technology, Tehran, Iran

Mahmoudshahbazi@outlook.com

Abstract—**Interleaved DC-DC boost converters are interesting choices in applications like fuel cells and photovoltaic systems. Although this converter offers low current ripple, but an open-circuit switch fault can lead to unacceptable current ripples. In this paper, a very fast and simple method is proposed to detect an open-circuit switch fault and its location. This method doesn't need any additional sensors, is efficient in CCM and DCM modes of operation, and can detect the fault in less than one switching period. Moreover, this method is suitable for implementation on an FPGA, due to the use of simple math and state machine blocks. Simulations are carried out to validate the effectiveness of this method.**

Keywords— *Power switch fault detection, open-circuit fault, interleaved converter, boost converter, Field-Programmable Gate Array (FPGA).*

I. INTRODUCTION

Interleaved DC-DC converters are being used in a variety of applications like photovoltaic (PV) and fuel cell systems. This is mostly because they offer higher reliability, efficiency and modularity, as well as lower current ripple [1-4] .however, like other power electronic converters, these converters are also sensitive to a switch failure. While this converter has an inherent fault-tolerant capability, however an Open-Circuit switch Fault (OCF) can lead to higher current ripples in the input and output of the converter, which may be beyond the allowed limits. Therefore, OCF detection can be useful in order to achieve the required performance and reliability of the system.

Many papers have studied Fault Detection (FD) and fault tolerant operation in power electronic converters. For non-isolated DC-DC and conventional two-level AC-DC converters, fast detection methods are proposed in [5, 6]. An OCF detection method for a parallel connected single active bridge DC-DC converter is presented in [7]. In [8] fault detection for DC-DC converters is carried out by monitoring the magnetic component (inductor or transformer) voltage and switch gate signals. A model-based estimator is proposed in [9] and is studied for real-time fault detection in a DC-DC converter. OCF detection in an isolated full-bridge converter using the primary voltage of the transformer is presented in [10]. The frequency information of the measured waveform of magnetic near field is used for fault detection in buck and full bridge DC-DC converters in [11].

However, few researches have been carried out on the fault detection and fault-tolerant operation of interleaved converters. FPGA-based fault-tolerant control of an interleaved boost converter for fuel cell electric vehicle application is studied in [12]. It is shown that an open-circuit fault in a semiconductor switch can lead to drastic increase of fuel cell current ripples, which in turn can lead to degradation of the fuel cell in long term. PWM gate signals of the healthy switches are modified to improve the post-fault performance [12]. However, no detail on the fault-detection method is provided. Interactions between the fuel cell and the converter in case of an OCF are also studied [13]. A fault detection method for an interleaved boost DC-DC converter for a PV application is presented in [1]. Here, the DC-link current is monitored and therefore no extra sensor is needed for fault detection. This method, however, is very complicated and does not seem suitable for practical implementation. Moreover, its performance is studied just for Continuous Conduction Mode (CCM), and not for Discontinuous Conduction Mode (DCM). It should be noted that while the approaches of [6] is applicable for interleaved converters, it need one current sensor per phase.

In this paper, a fast yet simple fault detection algorithm for interleaved DC-DC converters is presented and tested for a three-phase boost converter. This method needs no extra sensors, therefore it is cost effective, while maintaining high detection speed and accuracy. The proposed method uses the information already available in the closed loop control of the converter, to get an estimation of the current of each leg of the converter, and also the input source's current. This estimation is then compared with the measured input current, and their difference is used as an error signal in fault detection unit. Simulations are carried out to evaluate the effectiveness of this method. It is shown that this method not only is simple, but is several times faster than the existing methods. Moreover, due to the use of simple mathematical, logic and state machine units, this method is suitable for FPGA implementation.

In the following, first an interleaved boost converter is briefly reviewed in section II. Then the proposed method is explained in details in section III. Finally, simulation results are provided in section IV.

978-1-5090-0376-1/16 $31.00 © 2016 IEEE

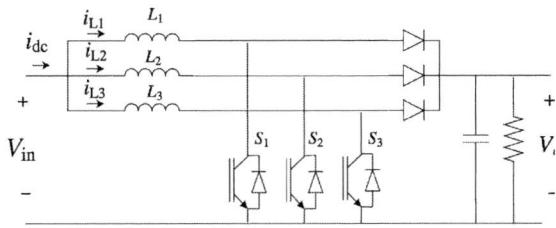

Fig. 1. Interleaved three-phase boost converter.

II. INTERLEAVED BOOST CONVERTER

Without lack of generalization, a three-phase (three-leg) interleaved boost converter is studied here. Fig. 1 shows circuit diagram of this converter. A voltage control loop is normally used for voltage regulation. The inner current control loops on each phase control the phase inductor currents [3]. In this configuration, one current sensor is needed per phase, which limits the number of phases and increases the cost. It is however also possible to use a single current loop for the DC-link current, and use input current estimation or DCM operation mode control to balance the phase currents [1, 2].

In order to reduce the current ripples, PWM signals of the switches have phase shifts of $2\pi/n$ from each other, where n is the number of phases. Each phase of the converter is a boost converter that can operate in CCM or DCM, or the boundary mode between them. Fig. 2 presents the inductor waveforms in CCM and DCM, and defines the periods T_1, T_2, T_3 which are later used in the FD algorithm. The gate commands of the three switches are shown by g_i ($i = \{1,2,3\}$. During T_1 period, the corresponding switch of the phase is on and the inductor current increases. When the switch is turned off, the T_2 period starts, and the inductor current starts to decrease. In CCM the switch is turned on before the inductor current reaches zero, but in DCM this current reaches zero and remains zero until the next switching. T_3 period in DCM is the period when the inductor current is equal to zero.

III. PROPOSED FAULT DETECTION METHOD

Figure 3 shows the proposed fault detection method. It is consisted of simple blocks that make its implementation on

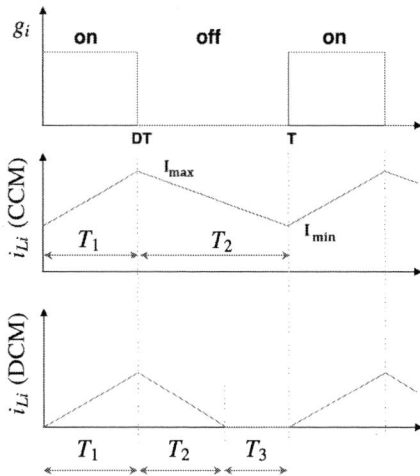

Fig. 2. Inductor current waveforms in CCM and DCM.

digital targets like FPGA easy. Inspired by the author's previous fault detection method [6], this method uses the slope of the inductor currents and compares them with their estimated values to detect a fault. As shown in Fig. 2, in T_1 period the inductor current increase with the slope of V_{in}/L. In T_2 the current decreases with a slope of $(V_{in} - V_o)/L$, and in T_3 in DCM mode it remains at zero. If the current measurement is carried out for each phase, then the same method as of [6] can be applied for fast fault detection. Here, however, the case where only one current sensor is used in control is considered. This current sensor measures the DC-link current i_{dc}. On the other hand, a current estimation block uses the information that is already available in the control to estimate the inductor currents, and to construct the estimated DC-link current ($i_{dc,es}$) out of them:

$$i_{dc} = i_{iL1} + i_{L2} + i_{L3} \qquad (1)$$

The measured and estimated values of DC-link current are then passed through a derivative function and then compared to see if there is any difference between them. Since these derivatives have normally very high values, before comparing them they are multiplied by a constant value of $k =$

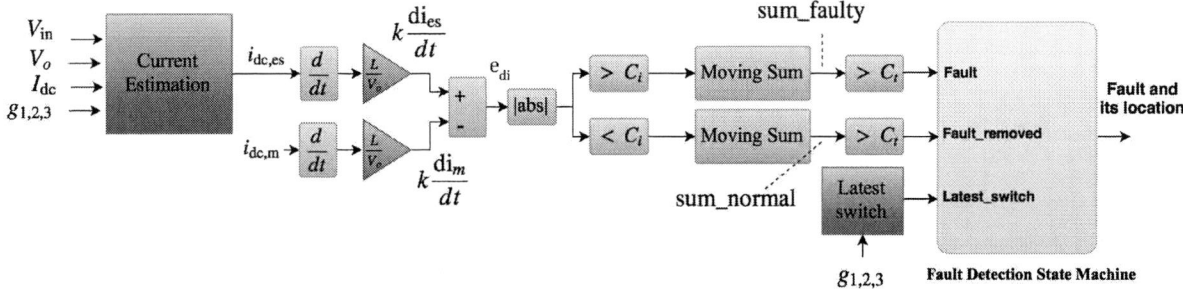

Fig. 3. Proposed fault detection method.

$L/V_{o,nominal}$ to have more practical values.

The absolute value of the error between measured and estimated current slopes (e_{di}) is then fed to two comparators with a threshold value of C_i. These comparators determine if the error between measured and estimated values is large enough to be considered as a sign of a fault or not. It is necessary to have these comparators, as in practice there are always errors in measurement, and also in our current estimation.

Since in practice, there are also unavoidable delays in the control and measurement systems, a time-based condition must be used as well to account for the probable time mismatches to avoid false detection. Therefore, the outputs of the C_i comparators are then passed through Moving Sum (MS) blocks. Moving sum, also known as the running sum, is the simplest form of a Finite Impulse Response (FIR) filter, and is defined as the sum of element over a moving window of values with length N, as shown in equation (2).

$$y(n) = x(n) + x(n-1) + \cdots + x(n-N+1) \qquad (2)$$

Here, moving sums are carried out for 15 sampling periods (equal to a window length of 15 μs when sampling frequency is 1MHz). The length of this observation window is chosen based on the typical delays in the control, drivers and switching devices of the studied power electronic converter and is chosen to be sufficiently larger than all these delays, but still small enough to guaranty fast fault detection. The upper moving sum block in Fig. 3 calculates in how many of the past 15 samples (the observation window) there has been a considerable error between measured and estimated values. If this value (*sum_faulty*) is greater than a constant C_t, then the detection algorithm can conclude that a fault has occurred somewhere. Here, C_t is chosen equal to 10. The lower MS does the reverse

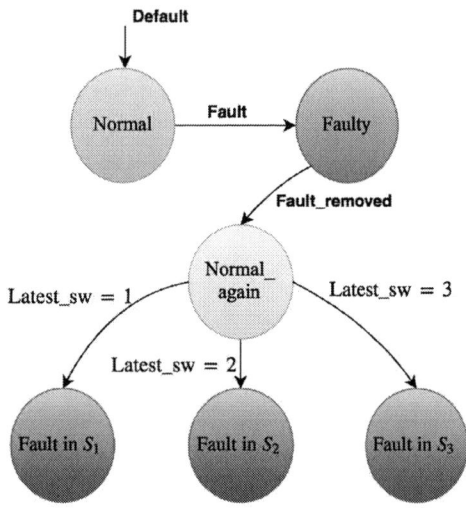

Fig. 4. Fault detection state machine.

and calculates for how many samples in the observation window there has not been a considerable error. If the output of this block is larger than C_t, the algorithm can conclude that there is no fault in the system, or the fault has been cleared out. The two moving sums and their following comparators are actually used to account for the delays in the control loop.

It is worth mentioning that clearly it is possible to use lower sampling rates, and change the N and C_t values accordingly. For example, if the sampling frequency is chosen equal to 333 kHz (sampling period of 3µs), choosing N=5 will result to the same observation window of 15µs. In this case C_t can be chosen equal to 3, so when at least four of the last five samples are indicating the existence of a considerable error between measured and estimated values, the fault will be declared. Also the mentioned sampling frequencies can be easily achieved by a wide range of microcontrollers or by many low-cost analog to digital conversion devices.

It can also be noted that here it is possible to use only the upper MS block, and set *Fault_removed* if $sum\_faulty < (N - C_t)$, which will result in the same outcomes. However the two MSs here are used to produce *sum_faulty* and *sum_normal* signals which can be useful in explaining and understanding the simulation results. Moreover, a MS is very easy to implement and doesn't need much hardware resources.

A. Fault location identification

After fault detection, it is necessary to detect the fault location as well. Here, a simple yet effective method is used based on the fact that when the command of the faulty switch goes back to zero, the error will also disappear, because the converter will act normally again. The Fault Detection State Machine (FDSM) of Fig. 3 is responsible for the detection of fault's location. Figure 4 shows the details of the FDSM.

When $sum\_faulty > C_t$, the state of FDSM changes from *Normal* to *Faulty* and the SM waits for the fault removal signal (*Fault_removed*). When the switching command of the faulty switch goes back to zero, the converter starts to act normally again and at some point *sum_normal* will be larger than C_t. Then, the state machine will go to the *Normal_again* state. Now it is just necessary to investigate which switching command has been put to zero, which can be easily carried out by monitoring the switching commands. For example if the latest switch command that has put to zero is g_2, then a fault in S_2 is declared.

B. Current estimation

A simple current estimation method is used here to build the current waveforms of each inductor, using only the information that is already available in the control. Figure 5 shows the SM that is used for current estimation. In each state, the current is calculated as:

978-1-5090-0376-1/16 $31.00 © 2016 IEEE   442

$$\begin{cases} in\ T_1: & i_{es} = \dfrac{I_{dc}}{3} - \dfrac{I_{max} + I_{min}}{2} + \displaystyle\int_0^t \dfrac{V_{in}}{L}\,dt \\[2mm] in\ T_2: & i_{es} = I_{max} + \displaystyle\int_0^t \dfrac{V_{in} - V_o}{L}\,dt \\[2mm] in\ T_3: & i_{es} = 0 \end{cases} \qquad (3)$$

where the periods T_1, T_2, T_3 are as explained before in Fig.2. I_{dc} is the average value of the DC-link current. I_{min} and I_{max} are sampled and the end of T_1 and T_2 periods respectively. If the current state is T_1 and the gate signal becomes '0', the T_2 state will be activated. This state will remain active until the gate signal goes back to '1' which points to a CCM operation, or the estimated current reaches zero, which means the converter is working in DCM. In these cases, the states of T_1 or T_3 will be activated respectively. This current estimation method is simple, and does not take into account the non-ideal parameters like inductor resistances and diode voltage drops. However, this method is later shown to be adequate for the fault detection algorithm. Nevertheless, any more accurate current estimation may be used to further improve the estimation accuracy, which may be also useful in the control of the converter.

It is worth mentioning that in this paper, open-circuit faults are considered. For short-circuit switch faults, normally using fast acting fuses the converter topology will become similar to that after an open-switch fault [6, 12], or special supplementary hardware is needed to detect the fault, as the software methods are not fast enough to detect the short-circuit switch faults. Nonetheless, this is not in the scope of this paper

IV. SIMULATION RESULTS

Simulations are carried out in Matlab/Simulink to evaluate the proposed method. A three-phase interleaved boost converter is simulated, and the parameters are provided in Table I. Since fault detection is so much faster compared with the control system, changes in the duty cycle during the fault detection are negligible. Therefore the converter can be simulated in open loop control. Non-ideal parameters like series resistances of inductors, IGBT switching delays and diode voltage drops are also included to verify the robustness of the FD method.

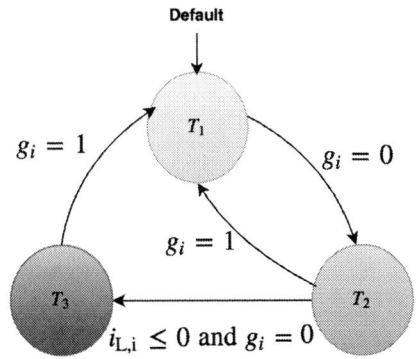

Fig. 5. The state machine for the estimation of phase currents.

First an OCF in S_2 at $t = 0.4s$ is investigated. The converter is in CCM, with duty cycle of 50%. The results are shown in Fig. 6. Figure 6(a) shows the measured and estimated values of the DC-link current. Note that while the estimated current successfully repeats the waveform of the measured current, there is a steady state error between them. This error is partly due to non-ideal parameters that are not taken into account in the current estimation. However, since the detection method uses the derivative of these currents. A fixed steady state error is not important, as its derivative is zero. Therefore, while the current estimation method can be adjusted to produce more accurate results, it is not necessary for the fault detection and is deliberately left unchanged here. The scaled-down derivatives of the currents are shown in Fig. 6(b). It can be verified that prior to fault occurrence, the estimated value follows the real one successfully, while upon fault occurrence there is considerable error between them. It is worth mentioning that sharp differences in the switching moments are as well a result of the non-ideal estimation of the DC-link current. But using the proper values for C_i and C_t these spikes can be filtered out. The outputs of the MSs are shown in Fig. 6(c). After the fault occurrence, the *sum_faulty* starts to increase, until it reaches its maximum of 15. This clearly tells the FDSM that a fault is present in the system, and the state *Faulty* in the FDSM will be activated. When the gate signal of S_2 is put to zero, *sum_normal* starts increasing, which is a sign that the fault has been removed. Gating signals for the three

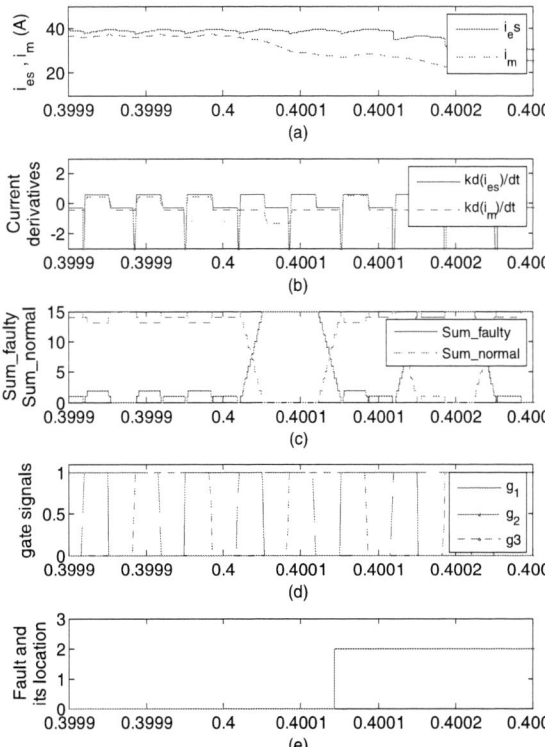

Fig. 6. Simulation results for a fault in S_2 at t=0.4s in CCM.

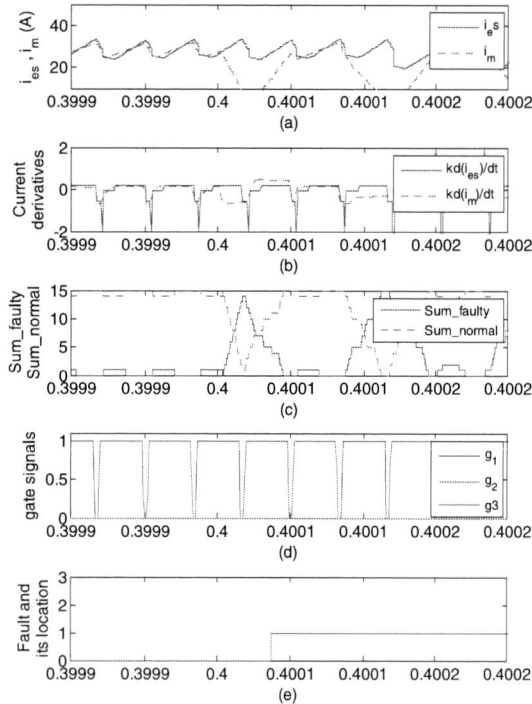

Fig. 7. Simulation results for a fault in S_1 at t=0.4s in DCM.

The results confirm that the proposed method is capable of fast and accurate detection of an OCF and its location. Non-ideal parameters can affect the current estimation accuracy, but they have limited effect on the fault detection algorithm. It is also worth mentioning that since the fault detection is very fast and detects the fault in less than one switching period, load dynamics have no effect on the performance of the fault detection algorithm and can be neglected.

V. CONCLUSION

In this paper, a fast and simple fault detection method for interleaved DC-DC converters is proposed. This method is based on the estimation of the DC-link current and comparing its derivative with that of its measured value. Two sets of comparators along with two moving sum calculators account for the probable errors in measurement and estimation due to parameter mismatch, model inaccuracy or natural delays of the system. A three-phase boost converter is studied and simulated. The simulation results show that the proposed method is capable of detecting the fault and finding its location in less than one switching cycle, which is several times faster than the currently available methods. This method is robust, and is not affected by the unaccounted non-ideal parameters such as series resistances or voltage drops. Moreover, due to use of simple blocks, this method is suitable for implementation on FPGA, to guarantee the fast detection speed.

switches are depicted in Fig. 6(d). when *Fault_removed* signal is applied to the FDSM, it looks up to see what switch has been put to zero more recently, and then the *'Fault in S_2'* state will be activated, and the output of the FDSM will be set to *'2'*. The fault detection and determining its location is carried out in less than one switching period, which is several times faster than those reported in [1].

Then the performance in DCM operation is studied. Duty cycle in this case is set to 30%, and a smaller value is used for inductances. It should be noted that the duty cycle is less than 1/3, therefore switching patterns in this case are different than the previous one [1]. Fault is introduced at S_1 in $t = 0.4s$. Figure 7 shows the simulation results in this case. Same analysis can be carried out as the previous case, and visibly the fault is detected in less than one switching period (less than $40\mu s$).

REFERENCES

[1] Ribeiro, Eraldo, Antonio J. Marques Cardoso, and Chiara Boccaletti. "Open-circuit fault diagnosis in interleaved DC–DC converters." Power Electronics, IEEE Transactions on, vol. 29, no.6, pp. 3091-3102, 2014.

[2] Ni, Liqin, Dean J. Patterson, and Jerry L. Hudgins. "High power current sensorless bidirectional 16-phase interleaved DC-DC converter for hybrid vehicle application." Power Electronics, IEEE Transactions on, vol. 27, no.3, pp. 1141-1151, 2012.

[3] Somkun, Sakda, Chatchai Sirisamphanwong, and Sukruedee Sukchai. "A DSP-based interleaved boost DC–DC converter for fuel cell applications." International Journal of Hydrogen Energy, vol. 40, pp. 6391-6404, 2015.

[4] Benyahia, N., et al. "MPPT controller for an interleaved boost DC–DC converter used in fuel cell electric vehicles." International Journal of Hydrogen Energy vol. 39, no. 27, pp. 15196-15205, 2014.

[5] M. Shahbazi, M. R. Zolghadri, P. Poure, S. Saadate, "Fast detection of open-switch faults with reduced sensor count for a fault-tolerant three-phase converter", Power Electronics, Drive Systems and Technologies Conference (PEDSTC), pp. 546-550, 2011.

[6] M. Shahbazi, E. Jamshidpour, P. Poure, S. Saadate, M. R. Zolghadri, "Open and Short-Circuit Switch Fault Diagnosis for Non-Isolated DC-DC Converters Using Field Programmable Gate Array", IEEE Trans. on Industrial Electronics, vol. 60, no. 9, pp. 4136-4146,Sep. 2013.

[7] Park, Kiwoo, and Zhe Chen. "Open-circuit fault detection and tolerant operation for a parallel-connected SAB DC-DC converter." Applied Power Electronics Conference and Exposition (APEC), 2014 Twenty-Ninth Annual IEEE. IEEE, 2014.

[8] Nie, Songsong, et al. "Fault Diagnosis of PWM DC–DC Converters Based on Magnetic Component Voltages Equation." Power Electronics, IEEE Transactions on, vol. 29, no.9, pp. 4978-4988, 2014.

[9] Poon, Jason, et al. "Real-time model-based fault diagnosis for switching power converters." Applied Power Electronics Conference and Exposition (APEC), 2015 IEEE. IEEE, 2015.

TABLE I.   SYSTEM PARAMETERS

Parameter	description	value
V_{in}	Input DC voltage	100 V
f_{sw}	Switching frequency	10 kHz
L_{CCM}	Converter inductor in CCM	1 mH
L_{DCM}	Converter inductor in DCM	0.1 mH
D_{CCM}	Duty cycle in CCM	0.5
D_{DCM}	Duty cycle in DCM	0.3

[10] X. Pei, S. Nie, Y. Chen, and Y. Kang, "Open-circuit fault diagnosis and fault-tolerant strategies for full-bridge DC–DC converters," IEEE Trans. Power Electron., vol. 27, no. 5, pp. 2550–2565, May 2012.

[11] Y. Chen, X. Pei, S. Nie, and Y. Kang, "Monitoring and diagnosis for the DC–DC converter using the magnetic near field waveform," IEEE Trans. Ind. Electron., vol. 58, no. 5, pp. 1634–1647, May 2011.

[12] Guilbert, Damien, et al. "FPGA based fault-tolerant control on an interleaved DC/DC boost converter for fuel cell electric vehicle applications." International Journal of Hydrogen Energy, vol. 40, pp.15815-15822, 2015.

[13] Guilbert, Damien, et al. "Investigation of the interactions between proton exchange membrane fuel cell and interleaved DC/DC boost converter in case of power switch faults." International Journal of Hydrogen Energy, vol. 40, no.1, pp. 519-537, 2015.

7th Power Electronics, Drive Systems & Technologies Conference (PEDSTC 2016)
16-18 Feb. 2016, Iran University of Science and Technology, Tehran, Iran

A New Hybrid Method of MPPT for Photovoltaic Systems Based on FLC and Three Point-Weight Methods

M. Bahrami, M. Zandi, R. Gavagsaz
Renewable energy department
Shahid Beheshty University
Tehran, Iran
Milad.bahrami@mail.sbu.ac.ir

B. Nahid-Mobarakeh, S. Pierfederici
GREEN
Université de Lorraine
Lorraine, France

Abstract—**In this paper, a hybrid method is proposed for maximum power point tracking (MPPT) of solar photovoltaic system. This method combines three point-weight and Fuzzy Logic Control (FLC) methods. First, the direction of change in the voltage or no change compared to the state before is determined based on three point-weight method. Then a change in the voltage magnitude is done based on FLC. Three point-weight, FLC and hybrid algorithms are simulated in Matlab software for comparison of their performance. The mean square error is used for a better comparison.**

Keywords—Maximum power point tracking, Photovoltaic, Fuzzy logic control, Three point-weight, Matlab.

I. INTRODUCTION

Recently, both interest and demand about the substitute energy systems have been increased. Among the variety of clean energy sources, solar energy is used because it is more environmentally friendly and clean [1-2]. The photovoltaic (PV) system is preferred because the output is extracted to the useful electric energy. However, the (I-V) characteristic of PV depends on the weather conditions. The output characteristics of this system are under influence of humidity, dust, wind speed, irradiation level, module temperature and etc. [3-7]. Among these, both irradiation level and cell temperature make the electrical output be varied largely. So, it is generally hard to develop the control algorithm and its related controllers for the PV because of the weather conditions changed randomly [8-9].

The paper is organized as follows: Photovoltaic (PV) modeling in section II, Maximum Power Point Tracking (MPPT) methods in section III, simulation and results in section IV and the conclusion in section V.

II. PV MODELING

The traditional equivalent circuits of a solar PV cell represented by a current source in parallel with one or two diodes are shown in FIGURE 1a and FIGURE 1b, respectively. The single-diode model (FIGURE 1a) includes five components: a photo current source, a diode parallel to the source and two

resistors. In double-diode model [FIGURE 1b], an additional diode is added for better curve fitting [10].

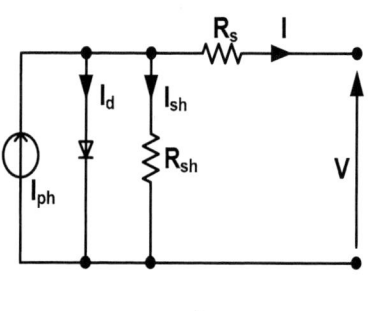

a)

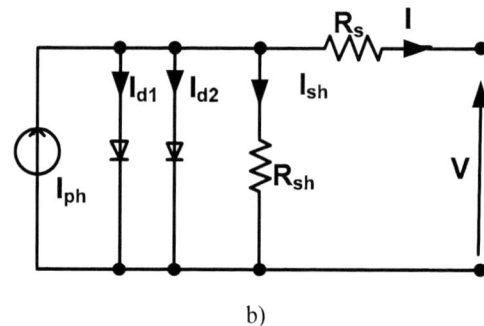

b)

FIGURE 1- Euivalent circuit of the PV with a: one diode b: two diodes

In the equivalent circuit of the PV, the photo source is for modeling the irradiation. The parallel diode is for modeling the cell polarization phenomena. The two resistors (series and shunt) are for modeling the losses [11].

The equation which describes the I-V characteristic of the equivalent circuit of the PV with one diode is [12]:

$$I = I_{ph} - I_d - I_{sh} = I_{ph} - I_o \left[e^{\frac{q(V+IR_s)}{nKT}} - 1 \right] - \frac{V+IR_s}{R_{sh}}$$

(1)

978-1-5090-0376-1/16 $31.00 © 2016 IEEE   446

Where V and I represent the output voltage and current of the PV, respectively; R_s and R_{sh} are the series and shunt resistance of the cell; q (= $1.6.10^{-19}$ C) is the electronic charge; I_{ph} is the light-generated current; I_o is the reverse saturation current; n is a dimensionless factor; K(= 1.38×10^{-23} J/K) is the Boltzman constant, and T_k (°K) is the ambient temperature [12].

In this paper, the model described in [12], is used to simulate the PV characteristics. In the variation of temperatures and irradiation values, I– V characteristic of the PV can be varied, as shown in FIGURE 2.

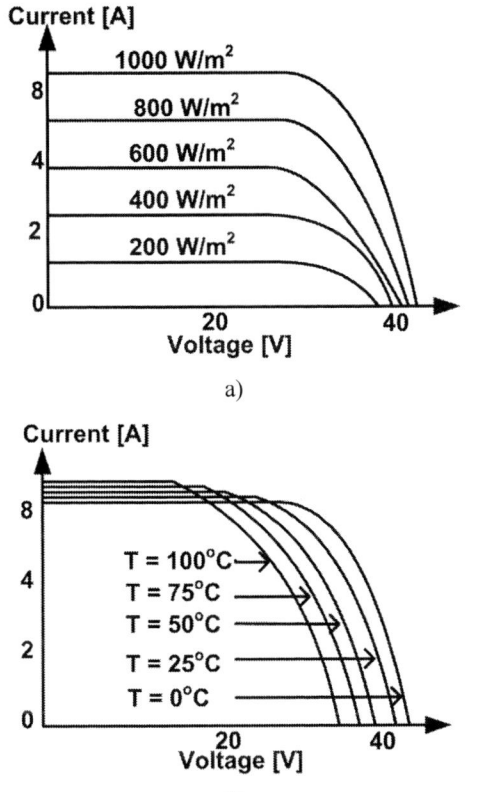

a)

b)

FIGURE 2- Impact of environmental condition on PV characteristics. a: impact of insulation, b: impact of ambient temperature

FIGURE 2a shows with the higher irradiation level, both voltage and current increases. FIGURE 2b shows with higher temperature level, the voltage decreases, but current is increased very slightly [13, 9]. I-V characteristic of PV shows that in each whether condition, the I-V has just one maximum. This point is named Maximum power point (MPP). The position of this point on the PV characteristic depends strongly on the solar irradiation and the cell temperature.

Operation of the PV generator at its MPP involves matching the impedance of the load to that of the generator. For this purpose, an electronic device, normally a power conditioning unit, capable of performing the function of an MPPT has to be connected between PV generator and the load. This is a control unit. Many different techniques have been developed to provide Maximum Power Point Tracking (MPPT)

of PV generators. These techniques can be classified as either direct or indirect methods. The direct methods are based on a searching algorithm to determine the maximum of the power curve without interruption of the normal operation of the PV generator. The indirect methods use an outside signal to estimate the MPP. Such outside signals may be given by measuring the solar radiation, the module temperature, the short circuit current, or the open circuit voltage of a reference PV cell. A set of physical parameters has to be given, and the MPP set point is derived from the monitored signal [14].

Many MPPT methods and algorithms have been developed and implemented. The methods and algorithms vary in complexity, sensors required, convergence speed, cost, range of effectiveness, implementation hardware, popularity, and in other respects [15-16].

III. MPPT METHODS

A. Perturb and Observe (P&O) method

The MPP tracker with P&O algorithm operates by periodically incrementing or decrementing the solar array voltage. If a given perturbation leads to an increase (decrease) the output power of the PV, then the subsequent perturbation is generated in the same (opposite) direction.

There are many methods for the P&O algorithm in the references. The hill-climbing method [17], the three-point weight comparison method [18], the extremum-seeking method [19] and modified methods [20-22] are some of them, the three-point weight comparison method is proposed to avoid having to move rapidly the operating point, when the solar radiation is varying quickly or when a disturbance or a data reading error occur. Restated, the MPPT can be traced accurately when the solar radiation is stable and power loss is low. The three-point weight comparison method is run periodically by perturbing the solar array terminal voltage and comparing the PV output power on three points of the V-P curve. The three points are the current operating point (A), a point (B), which is perturbed from point (A) and a point(C) with doubly perturbed in the opposite direction from point (B).

B. Fuzzy Logic Control (FLC) Method

The Fuzzy logic controller (FIGURE 3) is divided into four sections: Fuzzification, Rule-Base, Inference and Defuzzification. The FLC inputs turn to fuzzy language in fuzzyfication section. Then the outputs in inference section are made based on laws of rule-based. In the Defuzzification section, quantifiable results produce from the outputs of inference section.

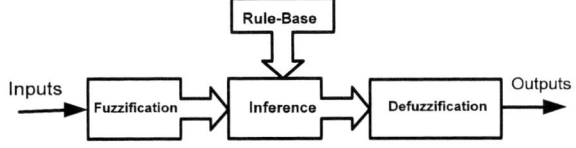

FIGURE 3- Block diagram of the Fuzzy controller

In [23] change in the power and change current are considered as inputs of FLC. The change of power dp and change of power with respect to change of voltage dp/dv have been selected as the inputs in [24]. In [25] error (dP/dI) and change of error have been considered as input variables. In this study, the inputs to the Fuzzy logic controller (FLC) are considered change in PV array Power (ΔP) and change in PV array voltage (ΔV). The two inputs are processed by the Fuzzy controller and the output of the Fuzzy controller is considered the increment in duty cycle (Δd). The universe of discourse for inputs and outputs are shown in TABLE 1.

TABLE 1- the rule base for FLC

Rule No.	If ΔP is	And ΔV is	Then Δd is	Rule No.	If ΔP is	And ΔI is	Then Δd is
1	PB	P	PB	11	NM	N	PM
2	PM	P	PM	12	NS	N	PS
3	PS	P	PS	13	PB	Z	PM
4	PB	N	NB	14	PM	Z	PS
5	PM	N	NM	15	NB	Z	NM
6	PS	N	NS	16	NM	Z	NS
7	NB	P	NB	17	ZZ	P	PS
8	NM	P	NM	18	ZZ	N	NS
9	NS	P	NS	19	ZZ	Z	ZZ
10	NB	N	PB				

C. Hybrid Method

By combining these two methods (Three point-weight and FLC), decision-making has been improved in transient conditions and thus accuracy and convergence speed have been increased. This algorithm is shown in the FIGURE 4. As shown in this figure, the algorithm is based on the three point weight algorithm, but in two states that are associated with changes in the voltage, the FLC method has been applied.

IV. SIMULATION AND RESULTS

Specifications of Conergy PH 255m solar photovoltaic panel have been used for simulation. These specifications are shown in TABLE 2.

TABLE 2- PV characteristics

Rated Power (P_{max}) at STC	255W
Power tolerance	+3%
Module efficiency	15.5%
Maximum Power Voltage (V_{MPP})	30.68V
Maximum Power Current (I_{MPP})	8.33A
Open Circuit Voltage (V_{OC})	38.4V
Short Circuit Current (I_{SC})	8.69A

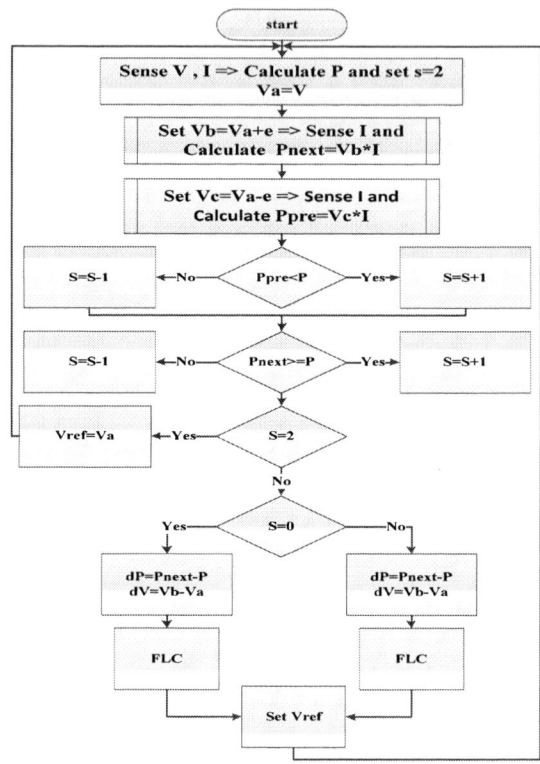

FIGURE 4- Algorithm of proposed method

Three methods are simulated and implemented simultaneously using Matlab software. Results of simulation are demonstrated in FIGURE 5. As seen in this figure, the proposed method, better than the other two methods has been tracked the maximum power point. It should be noted that the number of repetitions and the size of the change in the voltage magnitude, in addition to the nature of the algorithm, have been affected on the MPPT. If the changing the voltage (or the change in the duty ratio) is small, more number of repetitions is needed for adequate MPPT. In the case of large size of the change in the voltage magnitude, accuracy is low. A constant value has been considered for the size of the change in the voltage. This value can be multiplied according to the output of FLC. Consequently, all methods are completely comparable.

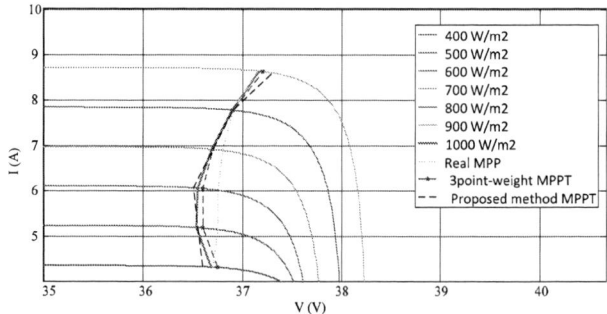

FIGURE 5- simulation of three MPPT methods. Yellow: FLC, Green: Three point-weight and blue: proposed method.

Considering optimized values for the size of the change in the voltage and the number of repetitions, the simulation is performed and the result is demonstrated in FIGURE 5. The mean square error is used for a better comparison in this paper. These values are shown in TABLE 3. As seen in this table, the least mean square error is related to proposed method.

TABLE 3- mean square error of three methods related to FIGURE 5

Three Point-weight	FLC	Proposed method
0.1148	0.0494	0.0096

If the number of repetitions is reduced to half, MPP tracked by three-point weight method is intensively moved away from real MPP and the error of FLC method is increased. But the proposed method has good accuracy in this case (FIGURE 6). This is indicated that the proposed method has higher speed and thus power losses, especially in transient conditions will be lower. Mean square error in this case is shown in TABLE 5 which indicates that the proposed method is the best method in comparison to the other two methods.

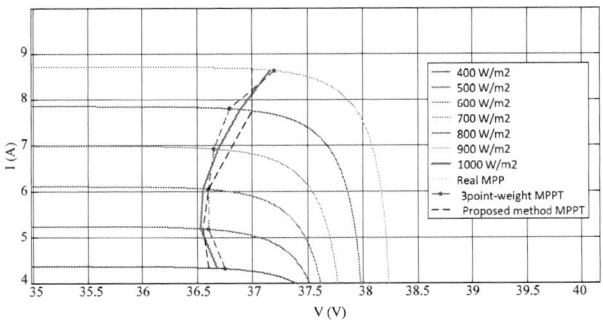

FIGURE 6- Simulation of three MPPT methods with half number of repetitions. Yellow: FLC, Green: Three point-weight and blue: proposed method.

TABLE 5- mean square error of three methods related to FIGURE 6.

Three Point-weight	FLC	Proposed method
Very high	0.0639	0.0191

It is considering that MPPT methods should be examined in another point of the view.

A. The Comparison of the MPPT Algorithms

The ease of implementation is an important factor in deciding which The MPPT algorithm to use. Some of the algorithms might be more familiar with analog circuitry and some of them with digital circuitry.

The number and type of the sensors required to implement also affects the decision. The monetary cost of the MPPT algorithm hardware is an important factor for comparison between the algorithms [13]. The MPPT algorithms are compared in the TABLE 5.

V. CONCLUSION

The main purpose of this paper is to propose a new technique for PV MPPT. Simulation results show that the proposed MPPT can track the MPP faster when compared to FLC or three point-weight methods. This method prevents of power losses in transient condition and when the condition is changed. In the future, this method will be implemented in the real PV system.

REFERENCES

[1] Xu, Dan, Qi Yu Li, Hui Chen, and Bing Gang Cao. "Review on Method of Maximum Power Point Tracking in Photovoltaic Generation System." InApplied Mechanics and Materials, vol. 494, pp. 1710-1714. 2014.

[2] Chiu, Yi-Hsun, Yu-Shan Cheng, Yi-Hua Liu, Shun-Chung Wang, and Zong-Zhen Yang. "A novel asymmetrical FLC-based MPPT technique for photovoltaic generation system." In Power Electronics Conference (IPEC-Hiroshima 2014-ECCE-ASIA), 2014 International, pp. 3778-3783. IEEE, 2014.

[3] Skoplaki, E., and J. A. Palyvos. "On the temperature dependence of photovoltaic module electrical performance: A review of efficiency/power correlations." Solar energy 83, no. 5 (2009): 614-624.

[4] Kaldellis, John K., Marina Kapsali, and Kosmas A. Kavadias. "Temperature and wind speed impact on the efficiency of PV installations. Experience obtained from outdoor measurements in Greece." Renewable Energy 66 (2014): 612-624.

[5] Skoplaki, E., A. G. Boudouvis, and J. A. Palyvos. "A simple correlation for the operating temperature of photovoltaic modules of arbitrary mounting." Solar Energy Materials and Solar Cells 92, no. 11 (2008): 1393-1402.

[6] Rouholamini, Amin, Hamed Pourgharibshahi, Roohollah Fadaeinedjad, and Morteza Abdolzadeh. "Temperature of a photovoltaic module under the influence of different environmental conditions–experimental investigation."International Journal of Ambient Energy ahead-of-print (2014): 1-7.

[7] Dash, Soubhagya Kumar, R. Akhil Raj, S. Nema, and R. K. Nema. "Development of photovoltaic (PV) cell/module/array and non-uniform irradiance effect based on two-diode model by using PSPICE simulator." In Nascent Technologies in the Engineering Field (ICNTE), 2015

TABLE 4- the comparison table of the MPPT algorithms

MPPT algorithm	PV array dependent	True MPPT?	Analog or digital	Periodic Tuning	Convergence Speed	Implementation complexity	Sensed Parameters
3 point-weight	NO	YES	Both	NO	Varies	Low	V , I
FLC	YES	YES	Digital	YES	Fast	High	Varies
Proposed	No	Yes	Digital	Yes	Fast	High	V,I

978-1-5090-0376-1/16 $31.00 © 2016 IEEE   449

International Conference on, pp. 1-6. IEEE, 2015.

[8] Babu, B. Chitra, and Shyoram Gurjar. "A novel simplified two-diode model of photovoltaic (PV) module." Photovoltaics, IEEE Journal of 4, no. 4 (2014): 1156-1161.

[9] Aissou, S., D. Rekioua, N. Mezzai, T. Rekioua, and S. Bacha. "Modeling and control of hybrid photovoltaic wind power system with battery storage." Energy Conversion and Management 89 (2015): 615-625.

[10] Ma, Tao, Hongxing Yang, and Lin Lu. "Solar photovoltaic system modeling and performance prediction." Renewable and Sustainable Energy Reviews 36 (2014): 304-315.

[11] Rodríguez, Juan David Bastidas, Carlos Andrés Ramos-Paja, and Edison Franco Mejia. "Modeling and parameter calculation of photovoltaic fields in irregular weather conditions." Ingeniería 17, no. 1 (2012): 37-48.

[12] M, Zandi, A, Payman, J-P, Martin, S, Pierfederici, and B. Davat. "The Novel Coefficients for Modeling the Photovoltaic Systems and MPPT Control" ICREDG10-RE2-150, First Iranian Conference on Renewable Energies and Distributed Generation, Birjand, (2010).

[13] Rahim, Nasrudin Abd, Hew Wooi Ping, and Jeyraj Selvaraj. "Photovoltaic Module Modeling using Simulink/Matlab." Procedia Environmental Sciences 17 (2013): 537-546.

[14] Reisi, Ali Reza, Mohammad Hassan Moradi, and Shahriar Jamasb. "Classification and comparison of maximum power point tracking techniques for photovoltaic system: A review." Renewable and Sustainable Energy Reviews19 (2013): 433-443.

[15] Trishan Esram and Patrick L. Chapman, " Comparison of Photovoltaic Array Maximum Power Point Tracking Techniques ", the project report of Center for Electric Machinery and Electromechanics at the University of Illinois at Urbana-Champaign., the National Science Foundation ECS-01-34208.

[16] Kamarzaman, Nur Atharah, and Chee Wei Tan. "A comprehensive review of maximum power point tracking algorithms for photovoltaic systems."Renewable and Sustainable Energy Reviews 37 (2014): 585-598.

[17] Tan, C. Y., Nasrudin Abd Rahim, and Jeyraj Selvaraj. "Improvement of hill climbing method by introducing simple irradiance detection method." In Clean Energy and Technology (CEAT) 2014, 3rd IET International Conference on, pp. 1-5. IET, 2014.

[18] Jiang, J., T. Huang, Y. Hsiao, and C. H. Chen. "Maximum power tracking for photovoltaic power systems." Tamkang Journal of Science and Engineering 8, no. 2 (2005): 147.

[19] R. LEYVA, C. ALONSO, I. QUEINNEC, A. CID-PASTOR, D. LAGRANGE, L. MARTÍNEZ-SALAMERO, " MPPT of Photovoltaic Systems using Extremum–Seeking Control ", Aerospace and Electronic Systems, IEEE Transactions, Vol. 42, Issue: 1, pp. 249- 258 , Jan. 2006 04.

[20] Sokolov, Michael, and Doron Shmilovitz. "Power Engineering Letters." IEEE Transactions on Energy Conversion 23, no. 4 (2008): 1105.

[21] Sodhi, Paloma, Dhruv Kapoor, and Mahesh S. Illindala. "An enhanced MPPT strategy for a grid-connected PV station under rapidly varying environmental conditions." In Power Electronics, Drives and Energy Systems (PEDES), 2012 IEEE International Conference on, pp. 1-6. IEEE, 2012.

[22] Elgendy, Mohammed, Bashar Zahawi, and David J. Atkinson. "Assessment of perturb and observe MPPT algorithm implementation techniques for PV pumping applications." Sustainable Energy, IEEE Transactions on 3, no. 1 (2012): 21-33.

[23] Takun, Pongsakor, Somyot Kaitwanidvilai, and Chaiyan Jettanasen. "Maximum power point tracking using fuzzy logic control for photovoltaic systems." InProceedings of International MultiConference of Engineers and Computer Scientists. Hongkong. 2011.

[24] Chim, Chia Seet, Prabhakaran Neelakantan, P. Yoong, and Kenneth Tze Kin Teo. "Fuzzy logic based MPPT for photovoltaic modules influenced by solar irradiation and cell temperature." In Computer Modelling and Simulation (UKSim), 2011 UkSim 13th International Conference on, pp. 376-381. IEEE, 2011.

[25] Algazar, Mohamed M., Hamdy Abd EL-halim, and Mohamed Ezzat El Kotb Salem. "Maximum power point tracking using fuzzy logic control." International Journal of Electrical Power & Energy Systems 39, no. 1 (2012): 21-28.

Stabilizing a Photovoltaic Plant Power Output by Employing an Auxiliary Power Source

Naier Mahdinejad
Graduate Program in Electrical Engineering,
Federal University of Minas Gerais (UFMG)
Belo Horizonte, Brazil
mahdinejad@ufmg.br

Luiz Machado
Program of Post-Graduation in Mechanical Engineering,
Federal University of Minas Gerais (UFMG)
Belo Horizonte, Brazil
Luizm@demec.ufmg.br

Ricardo Nicolau Nassar Koury
Program of Post-Graduation in Mechanical Engineering,
Federal University of Minas Gerais (UFMG)
Belo Horizonte, Brazil
Koury@demec.ufmg.br

Ramon Molina Valle
Program of Post-Graduation in Mechanical Engineering,
Federal University of Minas Gerais (UFMG)
Belo Horizonte, Brazil
Ramon@demec.ufmg.br

Abstract—**The use of compressed air energy storage (CAES) systems instead of conventional energy storage systems in large scale grid connected photovoltaic (PV) plans has already been proposed and investigated thermo-economically, resulting to very satisfactory outcomes. On the other hand, city gate stations (CGS), in which high pressure natural gas is expanded to much lower presser levels, has been proved to be a very suitable place for producing free electricity by employing turbo-expanders instead of conventional throttling valves. In this work, the feasibility of employing a CGS power output for improving the performance of a grid connected PV plant accompanied with a CAES system and enhancing its power output stability is studied. Comprehensive energy analysis and economic assessment on the proposed configuration is carried out and the results are discussed thoroughly. Finally, the performance of this hybrid configuration is compared with the PV plant and the CGS station while working individually. Internal rate of return (IRR) method as an authentic economic evaluation approach is used for comparing the considered systems economically.**

Keywords— *PV Plant, CAES, CGS, Power Stabilizer, Energy Analysis*

I. INTRODUCTION

Among all renewable energy sources, the sun is the most plentiful and available. The radiated energy from the sun is 3.8×10^{23} kW out of which almost 1.8×10^{14} kW is received by the earth. This amount of energy is almost well over 7500 times the world's total energy demand [1]. PV panels can generate electricity employing the sun clean energy; therefore, they are environmentally friendly. Furthermore, low restriction in installation location and low maintenance costs are other advantages of such panels [2]. PV panels can be employed to generate power in either small or large scales. Extensive research has been carried out, during the last decades, on PV technology experimentally and theoretically. Mendez et al. [3] studied the use of standalone PV systems in places far from electricity distribution grids. Economic and environmental impact aspects of this work were later assessed by Wies et al. [4]. In another work, King et al. [5] developed an electrical simulation model for PV cell modules for analyzing the performance of PV cell arrays. Chenni et al. [6] also developed a computer simulation model that revealed the PV system features by changing solar irradiation intensity and ambient temperature. Today, in addition to the huge capacity of power production by standalone PV systems, there are numerous grid connected PV farms with capacities from below 1 MWp up to hundreds of MWp all around the world.

The negative point of grid connected PV plants is that the source of energy is intermittent and as a result, instantaneous variations of electricity demand in the grid could not be accurately responded. For overcoming this problem, employing energy storage systems seems to be the best measure by now. Among various energy storage systems, for large scale applications, the CAES has attracted more attention over the recent years due to its lower capital cost, flexibility in the required size and also being environmentally friendly. After introducing the CAES technology for the first time in the 1970s, many studies have been carried out to improve its efficiency and configuration. In terms of novel applications of CAES technology, numerous works have been addressed in the literature focusing on renewable energy source power generation systems. For example, Denholm [7] proposed combining a wind farm with a CAES unit and biofuels. In another work, in order to compensate the power output fluctuations of a wind farm, Cavallo [8] suggested a hybrid wind-CAES configuration. In one of the last works in this area, Arabkoohsar et al. [9] proposed a new configuration of a large scale grid connected PV plant accompanying with a CAES unit. In this work, which was the first study on dynamic modelling of a CAES system in a grid connected renewable energy source power plant, they indicated the size of all components in the CAES system and specified the best power sales strategy for the whole power plant.

978-1-5090-0376-1/16 $31.00 © 2016 IEEE

On the other hand, CGS is a place in natural gas transmission system in which the pressure of natural gas from a very high level is reduced to a much lower value [10]. In fact, considering the long distance that the natural gas stream must pass from the refinery to the consumption points like cities and factories, its pressure should be so high that can overcome the losses along the path. Near the consumption points, CGSs are located to regulate the natural gas pressure to much lower levels. Therefore, before the expansion process at CGSs, there is considerable amount of exergy along the natural gas stream due to its high mass flow rate and high pressure. In the conventional configuration of CGS, a throttling value accomplished the pressure drop mission and as a result, all the exergy along with the natural gas flow was destructed [11]. However, replacing the throttling valve by a turbo-expander and an electricity generator, in order to hire the natural gas exergy and produce free power, has recently been proposed and studied [12].

In this work, the feasibility of employing the power output of a CGS equipped with a turbo-generator set for stabilizing the power output and enhancing the power generation efficiency of a grid connected PV plant accompanied with a CAES system is investigated. In this way, the power sales strategy of the PV plant can be optimized and consequently, the revenue of the power plant increases significantly. Also, the intermittent power produced by the PV farm can be stabilized considerably.

II. THE PROPOSED SYSTEM

Fig. 1 illustrates the schematic diagram of the proposed system.

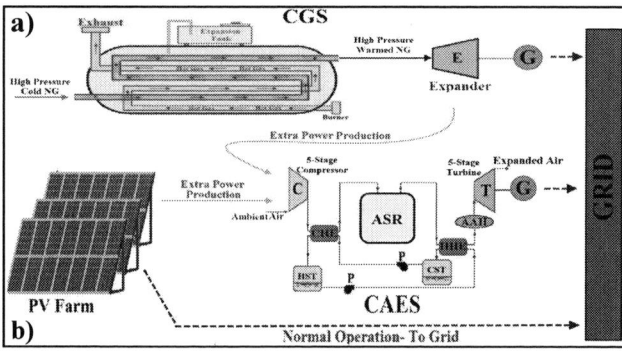

Fig. 1. The schematic of the proposed system; C: compressor set, CHE: cooling heat exchanger set, HST: hot storage tank, ASR: air storage reservoir, T: turbine set, HHE: cooling heat exchanger set, CST: cold storage tank, G: generator; E: expander

According to the figure, the PV farm produces power employing solar irradiation. Based on the amount of power that is supposed to be sold to the grid, there may be either extra produced electricity or lack of electricity in the system. Note that the amount of power that is going to be sold to grid by the PV plant should be specified in advance. Therefore, one should find the best power sales strategy of the power plant based on energy-economic considerations. In case of producing extra power by the PV farm, this extra power could

be employed by the CAES unit to produce compressed air. This compressed air, then, could be employed to produce extra electricity when the PV farm is not able to cover the grid demand solely. On the other hand, in this configuration, there is a CGS station that can generate free electricity employing a turbo-generator set. The power produced by the CGS could also be either sold directly to the grid or used by the CAES system to produce compressed air. In this way, not only the power sales strategy of the plant is improved significantly, but also the energy storage potential of the plant increases considerably. This, in fact, means that the capability of the energy storage unit of the plant to offset solar irradiation ramps increases dramatically.

A. CGS and case study

As it was explained, in a CGS, the pressure of natural gas should be reduced down considerably. Due to the positive Joule-Thompson coefficient, the pressure reduction process causes the natural gas temperature to fall significantly. On the other hand, there is a restriction for the minimum allowable temperature of natural gas which is called hydrate forming in which the water droplets suspended in the natural gas stream begin to freeze. The hydrate forming temperature is a functional of the natural gas compositions and consequently, it is in different zones for various natural gases. In order to not approach hydrate forming zone, as can be seen in fig. 1 (a), before the expansion process, the natural gas stream is heated up employing a heater.

In this work, in order to have an accurate and purposive simulation, one of CGSs that supply Natal city natural gas, where the PV farm along with a CAES unit was previously proposed to be built, is selected as the case study of this work. Also, table 1 details the technical information about this CGS.

TABLE I. CGS TECHNICAL INFORMATION

Characteristics	Information
Total Capacity of Station (Sm³/h)	400,000
Number of Line Heaters	4
Individual Heater Maximum Capacity (Sm³/h)	100,000
Hydrate Temperature (°C)	5 °C
Nominal Inlet Pressure (kPa)	7100
Nominal Outlet Pressure (kPa)	1700
Relative Density (kg/Sm³)	0.63

B. PV plant

Fig. 1 (b) shows the schematic of a PV farm taking advantage of a CAES unit. The PV farm, which includes thousands of connected PV cells, generates electricity employing sunshine. Depending on the amount of power that is to be sold to the grid in every moment, there may be extra power produced or electricity shortage in the system. If there is extra power, it is utilized by the CAES system to produce compressed air. In this case, the five stage compressor (C) intakes ambient air to produce compressed air. These compressors are equipped with inter-cooling heat exchangers (CHE) to enhance the efficiency of the compression process as well as collecting the generated heat in the air flow through the compressors. This heat is then stored in the hot storage tank (HST) in form of a hot oil. Subsequently, the cooled compressed air is kept in an air storage reservoir (ASR).

On the other hand, when there is electricity shortage in the system, i.e. when the PV farm cannot cover the grid electricity demand, the compressed air is used by the five stage turbine (T) to be expanded and produce rotational work that could be hired by an electricity generator (G). Naturally, the compressed air flow must be heated up to the required temperatures before being expanded. For this objective, the air flow, first, passes through the heating heat exchangers (HHE), located between the expander stages, which is supplied by the hot oil existed in the hot storage tank (HST). Then, the rest of the required heat is provided by an auxiliary air heater (AAH). The cooled working fluid (oil) outgoing from the heating heat exchangers is then kept in the cold storage tank (CST). As an important design target in this system, if there is still extra compressed air in the air reservoir at the end of the day, it is utilized by the expanders to produce power at peak power consumption hours (from 8 pm to 12 pm).

The effective parameter on the operational conditions and the size of CAES system is the pattern based on which the power plant sells electricity to the grid. Clearly, the power plant benefit depends on the value of vendible power to the grid and, on the other hand, the power plant must, beforehand, specify how much power can provide for the grid in every moment of each day. In case of any delinquency from the PV farm side for any reason; there will be huge financial penalties for the power plant. Therefore, there should be a thorough assessment on selecting the best pattern of power sales of the power plant. Generally, two different strategies are possible namely constant daily power value and daily-time dependent power value. As the names imply, in the first strategy, the power plant pledges to provide a constant amount of power to the grid during the day while in the second pattern the amount of vendible power to the grid varies during the day based on a time dependent function. Fig. 2 explains these approaches schematically.

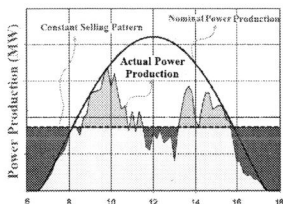 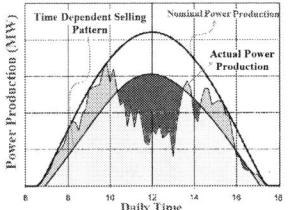

Fig. 2. Power sales strategies; a) constant; b) time dependent

In this figure, the yellow, the red and the green areas show respectively the amount of power sold to the grid directly, the amount of power shortage relative to the grid demand and the amount of extra electricity produced by the PV farm. Note that, by the time of energy shortage in the system, the CAES system must offset this power shortage; otherwise, the power plant is fined financially. It is also noteworthy that in power sales strategy selection assessment, the daily and nightly consumption electricity prices and the penalty rates need to be taken into account.

III. MATHEMATICAL MODELING OF THE SYSTEM

Considering the information given for the CGS, the maximum obtainable work from the natural gas stream by the expander could be calculated by [11]:

$$\dot{W}_{E\text{-rev}} = \dot{m}_{NG} \left(\psi_{NG\text{-i}} - \psi_{NG\text{-e}} \right) \tag{1}$$

Where, ψ is the summation of physical, chemical, potential and kinetic exergies the natural gas stream. It's worth mentioning that the kinetic and potential exergies are considered to be negligible in this work. Also, as the chemical components of the natural gas don't vary along the expander, the chemical exergy variation is also zero. The physical exergy is defined as follow:

$$\psi_{ph} = (h - h_o) - T_o(s - s_o) \tag{2}$$

Where, h and s are the enthalpy and entropy of the natural gas in the given temperature and h_o and s_o refer to the enthalpy and entropy of the stream in dead state (T_o=298.15 K and P_o=1.01325 bar), respectively.

Having the expander exergetic efficiency (ε_E), one could calculate the actual producible work by the turbo-expander as:

$$\dot{W}_E = \varepsilon_E \cdot \dot{W}_{E\text{-rev}} \tag{3}$$

According to the previous study of the authors, the exergetic efficiency of turbo-expanders suitable for this purpose can be considered equal to 85% [12]. Knowing the electricity generator conversion efficiency (η_g, 95%), one can compute the total producible power in the CGS station as:

$$\overline{P}_{CGS} = \eta_g \cdot \dot{W}_E \tag{4}$$

For the PV farm, the amount of receivable solar irradiation on a sloped cell with 1 m^2 area can be calculated by:

$$I_T = I_b R_b + I_d \left(\frac{1+\cos\beta}{2} \right) + I \cdot \rho_g \left(\frac{1-\cos\beta}{2} \right) \tag{5}$$

In which, I_b, I_d, ρ_g and β are the beam and diffuse components of solar radiation, the sunlight reflection coefficient and the slop angle, respectively. R_b is also a functional of incident angle (the angle between the sunlight and the cell surface, θ). This equation could be used for calculating the amount of receivable solar irradiation by the cells with various tracking modes where the slop angle and the incident angle (θ) for each case are different. The incident angle and the instantaneous slop angle for the selected tracking mode after evaluating simulation results (north-south axis parallel to the earth's axis rotator) are computed as [13]:

$$R_b = \frac{\cos\theta}{\cos\theta_z}; \quad \theta = \delta; \quad \beta = \tan^{-1}\left(\frac{\tan\varphi}{\cos\gamma} \right) \tag{6}$$

Where, γ is the surface azimuth angle.

Also, the employed PV cells present an average energy conversion efficiency of 19.3%. The other important point to mention is that the average energy conversion efficiency of the ground mounted PV arrays, due to the numerous sources of losses in the ground mounted PV arrays, which is considered equal to 80% of the average efficiency of an individual PV cell operating in normal conditions. Therefore, the average energy conversion efficiencies of the PV farm should be considered as 15.44%.

According to the operation explanation presented for the CAES system, two different operational periods are possible: when the compression facilities are working and when the air expansion process is to be accomplished. In the air compression stage, the five stage compressors and the inter-cooling heat exchangers work continuously and the hot storage tank and the air reservoir are charged. One important note here is the arrangement of the compressors that changes from parallel to series based on the air reservoir pressure. Naturally, based on the operational conception of the CAES system in the PV plant, the compressor set total work is equal to:

$$\dot{W}_C = \left(\overline{P}_{PV} + \overline{P}_{CGS}\right) \times \eta_C \tag{7}$$

In which, P_{PV}, P_{CGS} and η_C are respectively the extra power produced by the PV farm and the turbo-generator set in the CGS and the compressor overall efficiency. On the other hand, for calculating the compressors inlet temperatures which are equal to the cooling heat exchanger outlet temperatures, the cooling heat exchangers should be analyzed contemporary with the compressors. Employing technical correlations related to co-axial counter flow heat exchangers and having the heat exchangers effectiveness (E_{CHE}), one could calculate the working fluid (f) and air (a) outlet temperatures by:

$$T_{e-a} = T_{i-a}\left(1 - E_{CHE}\right) + E_{CHE} \cdot T_{i-f} \tag{8}$$

$$T_{e-f} = \frac{E_{CHE} \cdot \dot{m}_a \cdot c_{p-a} \cdot \left(T_{i-a} - T_{i-f}\right)}{\dot{m}_f \cdot c_{p-f}} + T_{i-f} \tag{9}$$

As it was mentioned before, while the compressors and inter-cooling heat exchangers work, the hot storage tank and the air reservoir are charged. From the mass conservation law and the first law of thermodynamics for the hot storage tank, one has:

$$m_{HST}^{\lambda+1} = m_{HST}^{\lambda} + \left(\sum \dot{m}_i - \sum \dot{m}_e\right)_{HST}^{\lambda} \cdot \Delta t \tag{10}$$

$$T_{HST}^{\lambda+1} = \frac{\left(\sum \dot{m}_i c_f T_i - \sum \dot{m}_e c_f T_e\right)^{\lambda} \Delta t + m_{HST}^{\lambda} \cdot c_f \cdot T_{HST}^{\lambda}}{m_{HST}^{\lambda+1} \cdot c_f} \tag{11}$$

Where, c_{HST}, m_{HST} and T_{HST} are the working fluid specific heat, mass and temperature within the hot storage tank, respectively. Similarly, for the air reservoir, the mass conservation law could be written as:

$$m_{ASR}^{\lambda+1} = m_{ASR}^{\lambda} + \left(\sum \dot{m}_i - \sum \dot{m}_e\right)_{ASR}^{\lambda} \cdot \Delta t \tag{12}$$

Where, m_{ASR} is the mass of air in the reservoir. In the expansion unit, the formulation about the heating heat exchangers and the cold storage tank are similar to the formulation presented for the compression unit. Also, the turbine set total work is given by:

$$\dot{W}_T = \frac{\overline{P}_{sh}}{\eta_{T-G}} \tag{13}$$

Where, P_{sh} is the energy shortage in the system that should be offset by the CAES system and η_{T-G} is the turbo-generator set overall efficiency.

IV. RESULTS AND DISCUSSIONS

In CGS modeling Fig. 3 shows the daily average pressure of natural gas incoming into/outgoing from the Natal CGS station during the year. It is reminded that the natural gas pressure is also opted according to inlet and outlet pressures of typical CGSs all over Brazil and taking ambient temperature changes into account. As can be seen, the inlet pressure is always around 5 MPa and the outlet pressure is normally in range of 1.45-1.65 MPa.

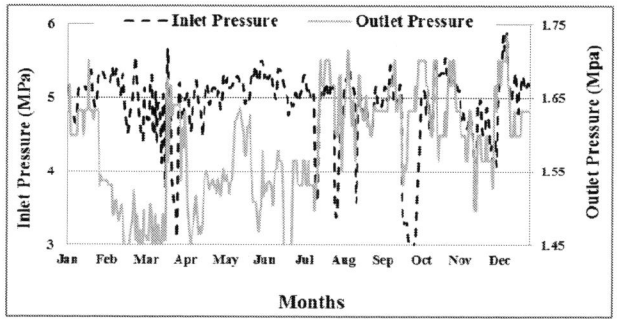

Fig. 3. The inlet and outlet natural gas pressures into/from the CGS

Fig. 4 shows the monthly averaged producible power from the natural gas stream in this station in case of employing a turbo-generator pack. According to the figure, the maximum power at the CGS could be produced in February which is almost 11 MW and the least value of power which is approximately 8.8 MW is produced in September.

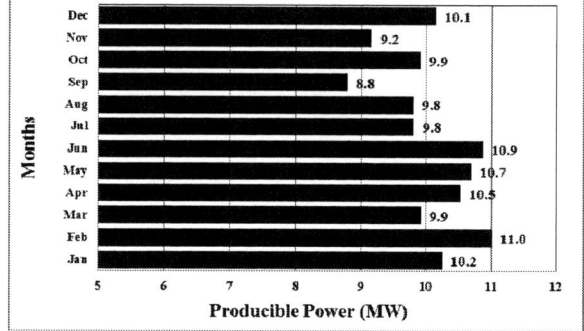

Fig. 4. The producible power at the CGS station by the turbo-generator pack

Note that the receivable solar irradiation by a horizontal surface with 1 m² area in Natal has been measured in minutely gaps and are available for all days of the year in 2012. Selecting the tracking mode and calculating the theoretical obtainable solar irradiation by such a tracking system, one could calculate the actual receivable solar energy by the tracking system. Fig. 5 shows the actual minutely-monthly averaged solar irradiation absorbable by a tracker cell with 1 m² area in Natal in 2012.

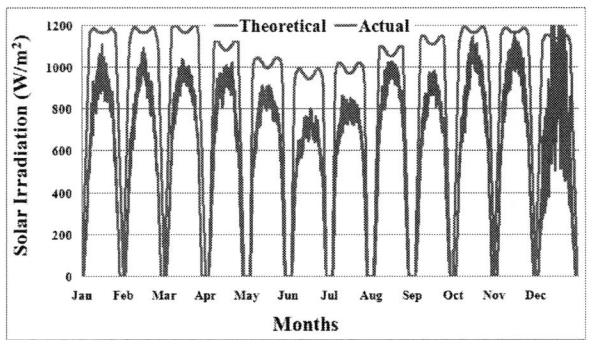

Fig. 5. Minutely-monthly averaged solar irradiation expected to be received by a tracking surface in Natal

As it was explained, the capacity of the PV plant is to be 100 MWp. According to fig. 5, the maximum theoretically solar irradiation receivable by each tracking PV cell is 1220 W/m² in March. Considering the utilized PV cells efficiency (19.3%) and consequently the expected overall ground mounted PV cell modules efficiency (15.44%), a maximum of 188.4 W/m² electricity is producible during the year. Therefore, almost 530000 m² PV cells are required for the PV farm. Fig. 6 shows the minutely-monthly averaged electricity producible by the PV farm for the first and second six months of the year, respectively.

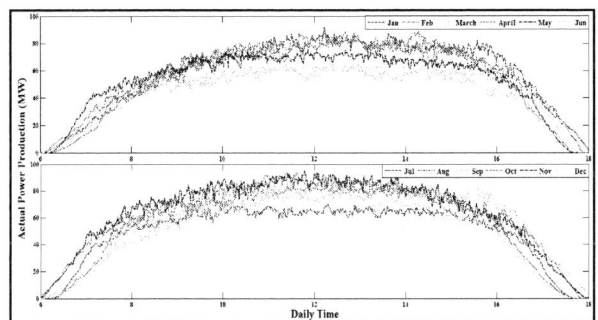

Fig. 6. Minutely-monthly averaged producible power by the PV farm

The next step is presenting the total producible power in the system including the PV farm and the CGS. Naturally, presenting the monthly averaged data, in this step, cannot be illustrative enough as averaging process prohibits well showing the solar energy ramps effects on the PV farm output and the effect of utilizing the CGS power output as the stabilizer source of system. Therefore, the producible power of the whole system is presented for three sample days during the year in this step. Note that the sample days are so selected that one could observe the effects of the proposal in all possible cases i.e. a very sunny day, a very cloudy day and one temperate day. Fig. 7 illustrates the total producible power of the whole system during these three sample days. According to the figure, March 4th is among top days of the year in terms of available solar irradiation as well as the least solar energy ramps while June 22nd is among the worst days of the year in this aspect and January 18th is a moderate level day.

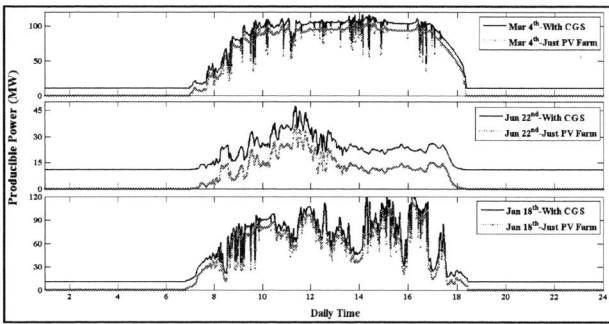

Fig. 7. Producible power in the PV farm in presence/absence of the CGS power production potential in three sample days

Fig. 8 shows the results of power sales strategy selection for various maximum capacities of the CAES system as a portion of the maximum extra produced power during the year. In this figure, CV stands for daily constant value strategy and TD refers to time dependent pattern. The percentages in front of these letters indicate the level of vendible power as a portion of the averaged producible power by the PV farm (monthly averaged for CV and instantaneously-monthly averaged for TD). For example, TD-80% refers to the case that the power plant sells power to the grid based on 80% of the instantaneously-monthly averaged producible power by the PV farm. Also, the phrase CAES-40%, for instance, shows the capacity of the CAES system equal to 40% of the maximum annual extra power produced by the whole power plant. According to the figure, overall, a time dependent power sales pattern results to more satisfactory economic outcomes. In time dependent category also, the power plant shows the best economic performance if: 1) it sells power to the grid based on 90% of the minutely-monthly averaged producible power of the PV farm, 2) the capacity of its CAES system is 40% of the maximum extra producible power by the PV farm during the year (30 MW).

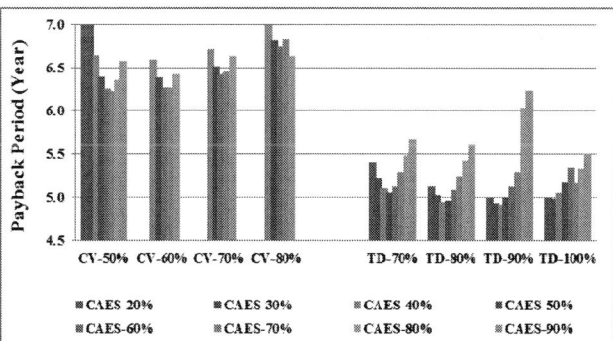

Fig. 8. Payback period of the power plant based on various power sales strategies and CAES maximum capacity; CV: constant value, TD: time dependent

In order to investigate the effect of the CAES system operation on the performance of the PV plant plus the CGS power production unit during the day (from 6 am to 6 pm), Fig. 9 is given for the three selected sample days. Note that this figure is presented to observe the level of power ramp offset by the CAES system, the level of power slumps that could not be recovered (fine levels) and the order of extra power production in the PV farm in each day. In this figure set, the brown area shows the amount of power that is sold to the grid directly by the PV farm. The red area also refers to the amount of power that is supplied to the system by the CAES system to offset the power ramps of the system. Also, the green area is the level of extra power that is produced by the PV farm and could be employed by the CAES system. Clearly, the value of this parameter is less than the total extra power that is produced by the PV farm as the CAES capacity is restricted to 30 MW. Finally, the blue zone is where the steep solar radiation slumps not only cause the PV farm to not be able to cover the grid demand, it cause the CAES system to fail in offsetting the power shortage in the system. Evidently, for such cases, financial fines are considered for the power plant. According to the figure, in March 4th, the situation is very satisfactory and no sharp power ramp occurs during the day. In a sharp contrast, in June 22nd, the system is not able to cover the grid electricity demand and as a result, the power plant is subjected to huge financial penalties, though a portion of this shortage is offset by the CAES system. On the other hand, January 18th is a moderate day in which solar radiation slumps are recovered by the CAES system and there is no penalty for the power plant.

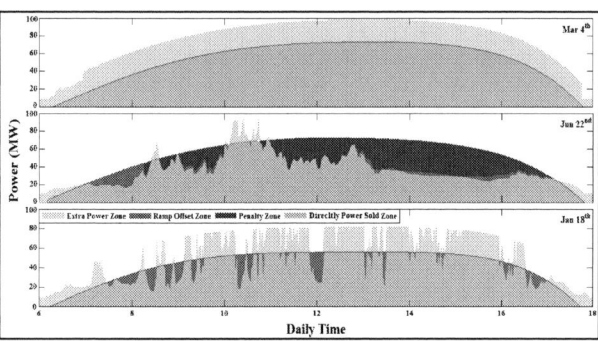

Fig. 9. Overall effect of the CAES system on the performance of the PV farm and the CGS power production station during the three sample days

To assess the overall annual effect of the CAES system on the PV plant and the CGS power production station performance, Fig. 10 presents a detailed report of the whole system performance including the total annual stored energy by the CAES system, the total annual electricity shortage compensated by the CAES system, the total annual peak consumption power produced by the CAES system and CGS power generation station, the total amount of electricity shortage remained in the system for which the power plant is fined financially, annual power that is wasted as the CAES capacity is limited and the total power that is produced by the CGS station in other hours of the day (from 24 to 6 am).

Finally, the investigations show that by this configuration, a total daily and nightly power of 243.5 GWh and 53.2 GWh could be sold to the grid over a whole year and taking the Brazilian electricity prices for daily and nightly consumptions, which are respectively 0.306 USD/kWh and 0.459 USD/kWh, the annual income of the power plant will be almost 99 million USD which is a great income for such a power plant and it could result to great sustainability in electricity production

industry and electricity distribution tends surely to use the power supplied by such power plant types.

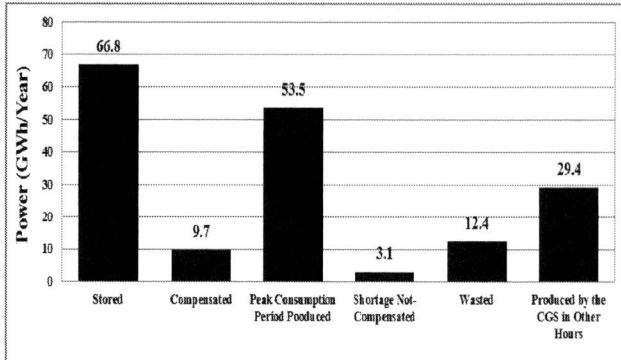

Fig. 10. Total annual performance of the whole system

REFERENCES

[1] A. Arabkoohsar, M. Farzaneh-Gord, M. Deymi-Dashtebayaz, L. Machado, R. N. N. Koury, Energy and Exergy Analysis of Natural Gas Pressure Reduction Points Equipped with Solar Heat and Controllable Heaters, Renewable Energy, 2014, 72, pp: 258-270.

[2] A. Mellit, A. Massi Pavan, V. Lughi, Short-term forecasting of power production in a large-scale photovoltaic plant, Solar Energy, Volume 105, July 2014, Pages 401-413.

[3] Mendez L, Naravarte L, Maninach A G, Izquierdo P, Carrasco L M, Eyras R, Centralized stand-alone PV system in micro grid in Morocco. In: Proc. 3rd world conference on photovoltaic energy conversion, Osaka, Japan, 2003, 3, pp: 2326-8.

[4] Wies R W, Johnson R A, Agrawal A N, Chubb T J, Simulink model for economic analysis and environmental impacts on a PV with diesel-battery system for remote villages. IEEE Transactions on Power Systems, 2005, 20, pp: 692-700.

[5] King D L, Dudley J K, Boyson W E, PVSIMC: a simulation program for photovoltaic cells, modules, and arrays. In: Proc, Twenty Fifth IEEE Photovoltaic Specialists Conference, 1996, 1, pp: 1295-7.

[6] Chenni R, Makhlouf M, Karabacho T, Bouzid A, A detailed modeling method for photovoltaic cells, Energy, 2007, 32, pp: 1724-30.

[7] Denholm P, Improving the technical, environmental and social performance of wind energy systems using biomass-based energy storage, Renew Energy, 2006, 31(9), pp: 1355–70.

[8] Cavallo A, Controllable and affordable utility-scale electricity from intermittent wind resources and compressed air energy storage (CAES), Energy, 2007, 32(2), pp:120–127.

[9] A. Arabkoohsar, L. Machado, M. Farzaneh-Gord, R. N. N. Koury, Thermo-economic analysis and sizing of a PV plant equipped with a compressed air energy storage system, Renewable Energy, 2015, 83, pp: 491–509.

[10] A. Arabkoohsar, M. Farzaneh-Gord, M. Deymi Dashtebayaz, L. Machado, R.N.N. Koury, A New Design for Natural Gas Pressure Reduction Points by Employing a Turbo-Expander and a Solar Heating Set, Renewable Energy 81, 239-250.

[11] M Kargaran, A Arabkoohsar, SJ Hagighat-Hosini, V Farzaneh-Kord, Mahmood Farzaneh-Gord, The second law analysis of natural gas behavior within a vortex tube, Thermal Science 17 (4), 1079-1092.

[12] Wojciech J. Kostowski, Jacek Kalina, Pawel Bargiel, Pawel Szuflenski, Energy and exergy recovery in a natural gas compressor station - A technical and economic analysis, Energy Conversion and Management, Volume 104, 1 November 2015, Pages 17-31.

[13] A Soteris. Kalogirou (2004), Solar thermal collectors and applications, Progress in Energy and Combustion Science, 30, 231–295.

7th Power Electronics, Drive Systems & Technologies Conference (PEDSTC 2016)
16-18 Feb. 2016, Iran University of Science and Technology, Tehran, Iran

The Effect of Energy Efficiency Increase on a PV Plant with a Non-conventional Energy Storage System Income

Ali Moallemi
Program of Post-Graduation in Mechanical Engineering,
Federal University of Minas Gerais (UFMG)
Belo Horizonte, Brazil
Alimoallemi@gmail.com

Luiz Machado
Program of Post-Graduation in Mechanical Engineering,
Federal University of Minas Gerais (UFMG)
Belo Horizonte, Brazil
Luizm@demec.ufmg.br

Ricardo Nicolau Nassar Koury
Program of Post-Graduation in Mechanical Engineering,
Federal University of Minas Gerais (UFMG)
Belo Horizonte, Brazil
Koury@demec.ufmg.br

Fabricio Jose Pacheco Pujatti
Program of Post-Graduation in Mechanical Engineering,
Federal University of Minas Gerais (UFMG)
Belo Horizonte, Brazil
pujatti@demec.ufmg.br

Abstract—**This Photovoltaic (PV) plants are widely used to produce power in either large or small scales all around the world. In addition, compressed air energy storage (CAES) system has attracted considerable attention as one of the most efficient candidates for large scales energy storage applications in the recent years. In this work, detailed energy and exergy analysis of a 100 MWp grid connected PV plant equipped with a CAES system is carried out. The PV plant is assumed to be located in Brazil. The formulations related to the first and the second laws of thermodynamic for all components as well as detailed solar engineering formulations for both the PV farm and the solar heating unit are presented. The performance of the power plant is comprehensively investigated for one entire year in real circumstances. The results revealed that the annual average exergy and energy efficiencies of the power plant are 17.9% and 16.2%, respectively and for 1% energy efficiency enhancement the power plant, the annual income increases almost 4 million USD.**

Keywords— PV Farm, CAES, Solar Heating System, Energy Analysis, Exergy Investigation

I. INTRODUCTION (*HEADING 1*)

PV farms are also built in a wide range from 1 MWp up to hundreds of MWp. The largest PV plant in the world is located in Arizona and has a capacity of 290 MWp. The main problem with PV farms is the sharp ramps in their power production which is clearly because of solar irradiation intensity fluctuations. As power distribution grids always need to make a balance between the power required and the power produced, the amount of power that the power plant sells to the grid should be defined in advance based on a mutual contract between the grid and the power plant. In case of any ramp in the power delivered to the grid relative to the value defined in the contract, the PV plant will be penalized financially. Therefore, the solar energy fluctuations must be predicted and offset to minimize these fines. As accurate long term forecast of solar energy fluctuations is not applicable yet, a second measure should be taken. In this regards, the only effective measure seems to be the use of energy storage systems [1]. Among all energy storage system candidates, the CAES is the most promising technology for large scale applications due to being environmentally friendly and its lower capital cost [2]. The CAES technology was proposed and extensively investigated in the 1970s and it is still in development stage. The first CAES plant was built for a gas turbine plant in Germany in 1978. The next CAES system was built in 1991 for McIntosh power plant with the power production capacity of 110 MW for 26 h [3]. In one of the most recent works in this area, a comprehensive thermo-economic study on a large scale grid connected PV plant equipped with a CAES system and an ancillary solar heating system in Brazil was presented [4]. In this work, the best location for building the power plant was selected and selecting the most efficient tracking mode for the PV cells, sizing of all components in the CAES system and finding the best power sales strategy for the power plant were carried out [4]. In the current work, a comprehensive thermodynamic analysis on the PV plant is presented. For this objective, the performance of the power plant over an entire year is assessed and detailed energy-exergy performance is presented.

II. SYSTEM CONFIGURATION

Fig. 1 illustrates the understudy system configuration.

978-1-5090-0376-1/16 $31.00 © 2016 IEEE

According to the figure, the power plant comprises two main subsystems, i.e. the PV farm and the CAES system.

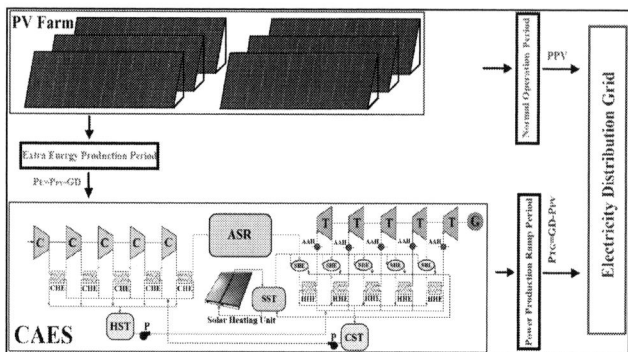

Fig. 1. The proposed system schematic; C: compressor, CHE: cooling heat exchanger, HHE: heating heat exchanger, HST: hot storage tank, CST: cold storage tank, ASR: air storage reservoir, T: turbine, SCS: solar collector series, SST: solar storage tank, AAA: auxiliary air heater, P: pump, G: generator

The PV farm including thousands of connected PV cells is supposed to produce power. Depending on the amount of power that is to be sold to the grid in every moment, there may be extra produced power or electricity shortage in the system. In case of the existence of any extra power (PE) in the system, the CAES unit can hire it to produce compressed air. In this case, as it can be seen, the compression part of the CAES system, including a multiple-stage compressor (C) with inter-cooling heat exchangers (CHE) and the hot storage tank (HST), is in operation state. In this state, the multiple-stage compressor is used to produce compressed air employing the extra power produced by the power plant during off peak periods. This air is stored in the air storage reservoir (ASR). The cooling heat exchangers are also used as intercoolers between the compressor stages. In this way, not only the harvested heat from the hot air is stored and could be used when required, but also the compressor efficiencies can increase significantly. The compressed air remains in the cavern until producing extra electricity is required. This is mainly by the time that intensive solar ramps occur and as a result, the PV farm is not able to cover the grid demand (GD). At this time, the compressed air is reclaimed to be expanded through the multiple-stage turbine (T). Before each step of expansion, the air stream should be heated up to the required temperature. The heating process is done in three stages. The first stage is employing solar heat exchangers (SHE) supported by hot water provided by the flat plate collectors. The heat harvested from the compressed air during compression process is used in the next stage by employing the heating heat exchangers (HHE). Finally, the warmed high pressure air is heated up to the desired temperature by auxiliary air heaters (AAH). Note that the cooled working fluid outgoing from the heating heat exchangers is then kept in the cold storage tank (CST).

It should be mentioned that, as it was explained, the configuration shown in Fig. 1 has been proposed and analyzed thermo-economically in reference [4]. Table 1 shows the results of thermo-economic analysis and sizing assessment on the understudy system accomplished in this reference.

TABLE I. POWER PLANT TECHNICAL AND SIZING INFORMATION

Characteristics	Information
PV Cell Type	Monocrystalline Silicon
Tracking Mode	North-South Axis Parallel to the Earth's Axis Rotation
PV Farm Average Efficiency	15.44 %
Total PV Cell Area	530000 m^2
Power Plant Power Sales Strategy	70% of Averaged Power Production
CAES Power Production Capacity	50 MW
Air Reservoir Volume	20000 m^3
Air Reservoir Maximum Pressure	120 barg
Thermal Storage Tanks Volume	5500 m^3
The Heat Exchangers Working Fluid	Industrial Oil
Number of Compression/Expansion Stages	5
Compression/Expansion Pressure Ratios	2.88
Number of Collectors in Solar Heating Unit	2500

III. PERFORMANCE INVESTIGATION OF THE SYSTEM

In this section, detailed formulation about energy and exergy analysis of the system is presented.

A. Energy Performance Investigation

In the configuration shown in Fig. 1, five separate control volumes could be specified, namely, the PV farm, the compression unit, the air storage reservoir, the solar heating system and the expansion unit.

In the first control volume (the PV farm), the first law efficiency for the PV farm could be defined as the ratio of the produced power by the PV farm (P'_{PV}) to the incident solar irradiation on the farm ($I_{T,PV}$). Therefore:

$$\xi_{PV} = \frac{P'_{PV}}{I_{T,PV}} \tag{1}$$

Based on isotropic model, the solar irradiation on a horizontal surface with 1 m^2 area outside the earth's atmosphere (I_o) and on the earth's surface (I) could be given by the following equations [5]:

$$I_o = \frac{12 \times 3600}{\pi} \times G_{sc}\left(1 + 0.033 \cdot \cos\frac{360\,n}{365}\right)$$
$$\left[\cos\varphi \cdot \cos\delta\,(\sin\omega_2 - \sin\omega_1) + \frac{\pi(\omega_2 - \omega_1)}{180} \times \sin\varphi \cdot \sin\delta\right] \tag{2}$$
$$I = I_o \times K_T$$

Where, φ, δ, n, ω and G_{sc} are local latitude angle, the inclination angle of the sun, the day number in the year, the time angle and the solar constant (1367 W/m^2), respectively. As it was explained before, the PV farm is supposed to hire north-south axis parallel to the earth's axis sun tracker for which the incident solar irradiation on the cell surface with 1 m^2 area ($I_{T,cell}$) could be calculated as [5]:

$$I_{T,cell} = I_b R_b + I_d\left(\frac{1+\cos\beta}{2}\right) + I \cdot \rho_g\left(\frac{1-\cos\beta}{2}\right)$$
$$R_b = \frac{\cos\theta}{\cos\theta_z}; \quad \theta = \delta; \quad \beta = \tan^{-1}\left(\frac{\tan\varphi}{\cos\gamma}\right) \tag{3}$$

Where, I_b, I_d, ρ_g, β, θ, θ_z and γ are the beam and diffuse components of solar radiation, sunlight reflection coefficient, the instantaneous slop angle of the cell, the solar irradiation incident angle, the zenith angle and the surface azimuth angle, respectively.

The second control volume (the compression unit) includes three main components, namely, the compressor set, the CHEs and the HST. In the compressor set, the first law of thermodynamic for each compressor could be written as:

$$\dot{Q}_C + \dot{W}_C + \dot{m}_a (h_i - h_e)_C = 0 \tag{4}$$

In which, Q_C, W_C, $\dot{m}_a$, h_i and h_e are the compressor heat transfer and work, the air mass flow rate, inlet and outlet enthalpies, respectively. The total work of the compressor set is equal to the extra produced electricity (P_E) multiplied by the compressor set overall efficiency (η_C).

$$\dot{W}_{C,t} = P_E \times \eta_C \tag{5}$$

The second component in this control volume is the cooling heat exchanger set. The first law of thermodynamic for each heat exchanger could be written as:

$$\dot{m}_a (h_i - h_e)_a + \dot{m}_f (h_i - h_e)_f = 0 \tag{6}$$

In which, the subscript f represents the working fluid in the heat exchanger. Defining the heat exchanger effectiveness of each cooling heat exchanger (E_{CHE}) as:

$$E_{CHE} = \frac{U_{CHE} A_{CHE} / C_{min}}{1 + U_{CHE} A_{CHE} / C_{min}}; \quad C_{min} = \dot{m}_a \cdot c_{p,a} \tag{7}$$

In the last equations, A_{HE}, U_{CHE}, $c_{p,a}$ and c_f are the heat transfer area, the overall heat transfer coefficient, the constant pressure specific heat capacity of air and the working fluid specific heat, respectively.

The last component in the second control volume is the hot storage tank for which the first law of thermodynamic could be written as follow:

$$u_{HST}^{\lambda+1} = \frac{(m \cdot u)_{HST}^{\lambda} + \Delta t \cdot \left(\sum \dot{m}_{f,i} h_{f,i} - \sum \dot{m}_{f,e} h_{f,e} \right)^{\lambda}}{m_{HST}^{\lambda+1}}; \tag{8}$$

$$u_{HST} = c_f \cdot T_{HST}$$

Where, u_{HST}, m_{HST}, T_{HST} and c_f are the internal energy, the mass, the temperature and the specific heat of working fluid within the hot storage tank, respectively.

The third control volume is the air storage reservoir. Considering the assumptions made for the air storage reservoir, the first law of thermodynamic for it could be written as below:

$$u_{ASR}^{\lambda+1} = \frac{(m \cdot u)_{ASR}^{\lambda} + T_o \cdot c_{p,a} \left(\sum \dot{m}_{a,i} - \sum \dot{m}_{a,e} \right)^{\lambda} \cdot \Delta t}{m_{ASR}^{\lambda+1}}; \tag{9}$$

$$u_{ASR} = c_{v,a} \cdot T_o$$

In which, m_{ASR} is the mass of air encapsulated in the reservoir. Finally, the reservoir pressure (P_{ASR}) could be obtained by:

$$P_{ASR} = m_{ASR} \times R \times T_o / V_{ASR} \tag{10}$$

The fourth control volume in the system is the solar heating unit which includes the flat plate solar collector modules as well as the solar storage tank. The energy balance on each flat plate solar collector could be written as:

$$I_{T,fc} - \dot{Q}_{\tau\alpha} - \overbrace{\dot{m}_w (h_e - h_i)_w}^{\dot{Q}_w} = 0 \tag{11}$$

Where, $I_{T,fc}$ is the receivable solar irradiation by fixed sloped surface that could be calculated by Eq. 6 where the slop angle is constant and equal to 12°. $\dot{Q}_{\tau\alpha}$ is the solar irradiation losses due to not perfect emittance coefficient of the cover and absorption coefficient of the absorber surface. Also, Q_w is the amount of obtainable heat by the working fluid (water) while passing through the collector which could be computed as follow [5]:

$$\dot{Q}_w = A_p \cdot FR \cdot \left[\bar{S} - U_l (T_{w,i} - T_o) \right] \tag{12}$$

Where, A_p, FR, S, U_l, $T_{w,i}$ and T_o are the collector absorber surface area, the collector removal factor, the absorbed solar flux, the overall loss coefficient of collector, the inlet water temperature and ambient temperature, respectively.

The fifth control volume in the system is the expansion unit. This control volume comprises the solar heat exchangers, the heating heat exchangers, the expanders and the cold storage tank. The first law of thermodynamic for each expander could be written as below:

$$\dot{W}_T + \dot{m}_a (h_i - h_e)_T = 0 \tag{13}$$

In which, W_T is the work done with each expander. Naturally, the total work of the expander set should provide enough power to offset the PV farm power slumps during the days and produce extra power during peak consumption periods (if applicable). Therefore:

$$\dot{W}_{T,t} = \frac{P_R}{\eta_{TG}} \tag{14}$$

Where, P_R and η_{TG} are the amount of power required to be produced and the turbo-generator set overall efficiency, respectively.

Finally, the last control volume is the ancillary air heater. Considering the coils transmitting the air through the heater as the control volume, the first law of thermodynamic is written as:

$$\dot{Q}_h + \dot{m}_a (h_i - h_e)_h = 0 \tag{15}$$

Where, Q_h is the amount of auxiliary heat required to be provided by the air heater. The amount of fuel required to be burnt by the heater to provide this heat is also computed as:

$$\dot{V}_{fu} = \frac{\dot{Q}_h}{LHV \cdot \eta_h} \tag{16}$$

In which, LHV and η_h are lower heating value of the fuel and the heater thermal efficiency (50%), respectively.

Regarding the formulation presented above, the total daily fist law efficiency for the power plant (ξ_{PP}, including both the CAES system and the PV farm) and the CAES system individually (ξ_{CAES}) could respectively be defined as:

$$\xi_{CAES} = \frac{SPS}{\sum\limits_{t=6}^{24} \left(I_{T,fc} + P_E + P_{fu}\right)} \qquad (17)$$

$$\xi_{PP} = \frac{PS}{\sum\limits_{t=6}^{24} \left(I_{T,fc} + I_{T,PV} + P'_{fu}\right)} \qquad (18)$$

In which, PS and SPS are the total daily power sold to the grid and the total daily power produced by the CAES system, respectively. P'_{fu} is also the result of multiplying the fuel mass flow rate by the fuel LHV. The variable t also refers to hourly time steps from 6 am to 12 pm.

B. Exergy Performance Investigation

Similar to the energy analysis section, the first control volume in the system is the PV farm for which the exergy analysis is presented. The second law efficiency for the PV farm is defined as the ratio of the amount of useful exergy gained, i.e. the net power produced (P_{PV}), to the solar exergy received (Ψ_{PV}).

$$\varepsilon_{PV} = \frac{P_{PV}}{\Psi_{PV}} \qquad (19)$$

The amount of solar exergy received by 1 m^2 area of a PV cell, on the other hand, could be calculated as [6]:

$$\Psi_{cell} = I_{T,cell} \cdot \left[1 - \frac{4}{3}\left(\frac{T_o}{T_{sun}}\right) + \frac{1}{3}\left(\frac{T_o}{T_{sun}}\right)^4 \right] \qquad (20)$$

In which, T_{sun} is the effective temperature of the sun (4350 K).

In the second control volume, for the compressor set, considering the assumptions taken for the compressors, the entropy balance on each compressor could be written as:

$$\dot{\sigma}_C = \dot{m}_a \left(s_e - s_i\right)_C - \frac{\dot{Q}_C}{T_{C,s}} \qquad (21)$$

Where, σ_C, Q_C and $T_{C,s}$ are the rate of entropy generation through the compressor, the rate of heat transfer from the compressor to the environment and the compressor surface average temperature, respectively. Finally, defining the second law efficiency for a compressor as the ratio of exergy increase in the fluid through the compressor to the actual work of compressor, the compressor second law efficiency could be calculated as:

$$\varepsilon_C = 1 - \frac{T_o\left(s_e - s_i - \frac{\dot{Q}_C}{T_C}\right)_C}{\left(h_e - h_i\right)_C - \dot{Q}_C} \qquad (22)$$

For the heat exchangers, on the other hand, regarding the assumptions taken, from the second law of thermodynamic, one has:

$$\dot{\sigma}_{CHE} = \dot{m}_a \left(s_e - s_i\right)_{CHE,a} + \dot{m}_f \left(s_e - s_i\right)_{CHE,f} \qquad (23)$$

Defining the second law efficiency for a heat exchanger as the ratio of exergy increase in the cold fluid to exergy increase in the hot fluid, one has:

$$\varepsilon_{CHE} = \frac{\dot{m}_f \left[\left(h_e - h_i\right) - T_o\left(s_e - s_i\right)\right]_{CHE,f}}{\dot{m}_a \left[\left(h_i - h_e\right) - T_o\left(s_i - s_e\right)\right]_{CHE,a}} \qquad (24)$$

The third component in the second control volume is the hot storage tank. The entropy balance on the storage tank may be written as:

$$S_2 - S_1 = \left(\sum \dot{m}_i s_i - \sum \dot{m}_e s_e + \dot{\sigma}\right)_{HST} \cdot \Delta t \qquad (25)$$

The second law efficiency for the hot storage tank can only be defined for a period of performance including a full charging step as well as a complete discharging step and it is written as below:

$$\varepsilon_{HST} = \left.\frac{\Delta\Psi_{Discharging}}{\Delta\Psi_{Charging}}\right)_{HST} = \left.\frac{\left[\Delta H - T_o \Delta S\right]_{Discharging}}{\left[\Delta H - T_o \Delta S\right]_{Charging}}\right)_{HST} \qquad (26)$$

For the third control volume (the air storage reservoir), the entropy balance results to:

$$S_2 - S_1 = \left(\sum \dot{m}_i s_i - \sum \dot{m}_e s_e + \dot{\sigma}\right)_{ASR} \cdot \Delta t \qquad (27)$$

The second law efficiency of the reservoir should also be defined for a period of performance including a full charging step as well as a complete discharging step.

$$\varepsilon_{ASR} = \left.\frac{\Delta\Psi_{Discharging}}{\Delta\Psi_{Charging}}\right)_{ASR} = \left.\frac{\left[\Delta H - T_o \Delta S\right]_{Discharging}}{\left[\Delta H - T_o \Delta S\right]_{Charging}}\right)_{ASR} \qquad (28)$$

The next control volume is the solar heating unit. In this control volume, for the solar collector module, the second law efficiency is defined as the ratio of the net exergy obtained from the collector by the working fluid (Ψ_w) to the received solar exergy by the collector (Ψ_{fc}) [6].

$$\varepsilon_{fc} = \frac{\Psi_w}{\Psi_{fc}} \qquad (29)$$

The amount of solar exergy received by the flat plate solar collector could be calculated by:

$$\Psi_{fc} = \eta_o \cdot I_{T,fc} \cdot A_p \cdot \left[1 - \left(\frac{T_o}{T_{sun}}\right)\right]; \quad \eta_o = \frac{S}{I_{T,fc}} \qquad (30)$$

Where, η_o is the optical efficiency of the collector. The rate of net exergy gained by water while passing through the collector could also be given by:

$$\Psi_w = \dot{m}_w \left[\left(h_e - h_i\right) - T_o\left(s_e - s_i\right)\right]_w \qquad (31)$$

For calculating the storage tank entropy balance, exergy efficiency, the same correlations as the formulation presented for the cold storage tank in the second control volume could be employed.

The fifth control volume is the expansion unit in the CAES system in which the first component to be analyzed is the expanders set. The entropy balance on each expander could be presented as:

$$\dot{\sigma}_T = \dot{m}_a \left(s_e - s_i\right)_T \qquad (32)$$

The second law efficiency of each expander is defined as the ratio of the actual work of the turbine to the amount of decrease in the availability of air while passing through that:

$$\varepsilon_T = \frac{1}{1 + \frac{T_o(s_e - s_i)_T}{c_{p,ave}(T_i - T_e)_T}} \tag{33}$$

For exergy analysis on the other components in this control volume such as the solar heat exchangers, the heating heat exchangers and the cold storage tank the same formulation as previous sections could be presented.

The last control volume is the auxiliary air heater for which the second law of thermodynamic could be written as:

$$\dot{\sigma}_h = \dot{m}_a (s_e - s_i)_h \tag{34}$$

Naturally, the heater inlet exergy ($\Psi_{h,i}$) is the summation of the inlet air exergy and the chemical exergy along with the fuel.

$$\Psi_{h,i} = \Psi_{fu} + \dot{m}_a [(h_i - h_o) - T_o (s_i - s_o)]_a \; ; \tag{35}$$
$$\Psi_{fu} = \dot{n}_{fu} [(\hat{h} - \hat{h}_o) - T_o (\hat{s} - \hat{s}_o) + \hat{\psi}^{ch}]_{fu}$$

Note that the fuel used by the heater is assumed to be pure methane. The second law efficiency of the heater is also calculated as:

$$\varepsilon_h = \frac{\dot{m}_a [(h_e - h_i) - T_o (s_e - s_i)]_a}{\Psi_{h,i}} \tag{36}$$

Finally, the total daily second law efficiency of the PV farm (ε_{PV}), the total daily second law efficiency of the power plant (ε_{PP}) and the total daily second law efficiency of CAES unit (ε_{CAES}) could be calculated by the following equations, respectively:

$$\varepsilon_{CAES} = \frac{SPS}{\sum_{t=6}^{24} (\Psi_{fc} + P_E + \Psi_{fu})} \tag{37}$$

$$\varepsilon_{PP} = \frac{PS}{\sum_{t=6}^{24} (\Psi_{fc} + \Psi_{PV} + \Psi_{fu})} \tag{38}$$

Where, PPV is the total daily power that is sold directly to the grid by the PV farm.

IV. RESULTS AND DISCUSSIONS

In this section, the results of the simulation accomplished on the power plant are presented. Fig. 2 shows the monthly averaged theoretically receivable solar irradiation by a tracking PV cell and a sloped flat plate collector both with 1 m2 area.

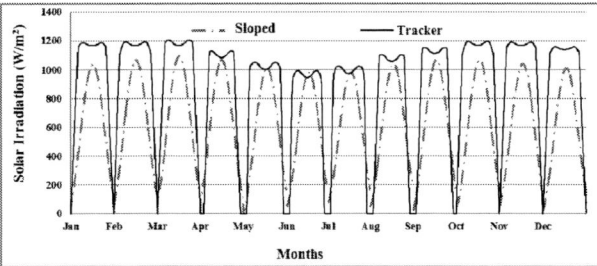

Fig. 2. Instantaneously-monthly averaged theoretical solar irradiation on a PV cell and a flat plate collector with 1 m^2 area

Fig. 3 gives information about the monthly-instantaneously averaged actual producible power by the PV farm. It is reminded that the PV farm takes advantage of 530000 m^2 north-south axis parallel to the earth's axis tracker PV cells with overall efficiency of 15.44%.

Fig. 4 shows the extra power produced by the PV farm that could be used by the CAES system in March 23rd, April 30th and May 24th. Clearly, January 30th is one of the best; April 21st is a moderate and May 23rd is one of the worst days of the year in terms of available solar energy. According to the figure, the maximum extra power to be employed by the compressors is 50 MW.

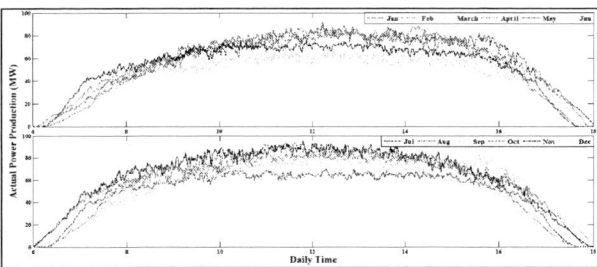

Fig. 3. Instantaneously-monthly averaged actual producible power by the PV farm

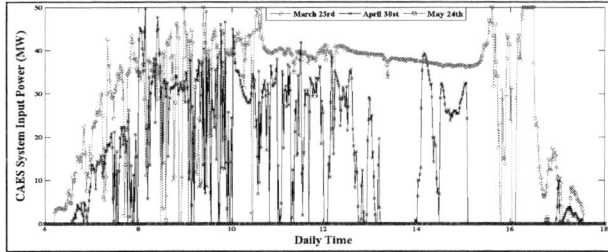

Fig. 4. The CAES system compressor set work for three sample days (MW)

Fig. 5 also shows the amount of power shortage in the power plant relative to the grid demand based on the chosen characteristics (the CAES unit capacity) and power sales strategy for the system during the three sample days. This energy shortage, in fact, should be offset by the CAES system, if possible. Expectedly, the energy shortage in May 24th is so much so that even the CAES system may not be able to offset it. In a sharp contrast, the energy shortage during March 23rd is very trifle. The power shortage in April 30th is always in moderate levels.

978-1-5090-0376-1/16 $31.00 © 2016 IEEE

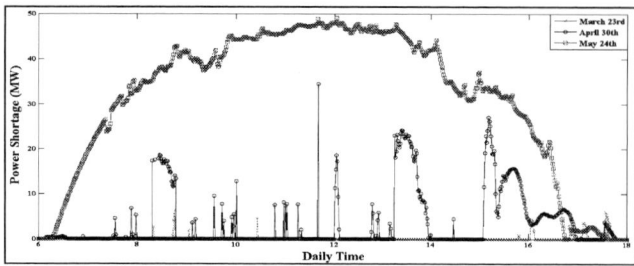

Fig. 5. Power shortage in the system that the CAES unit should offset during the three sample days (MW)

Fig. 6 shows the total daily solar energy and exergy received by the PV farm.

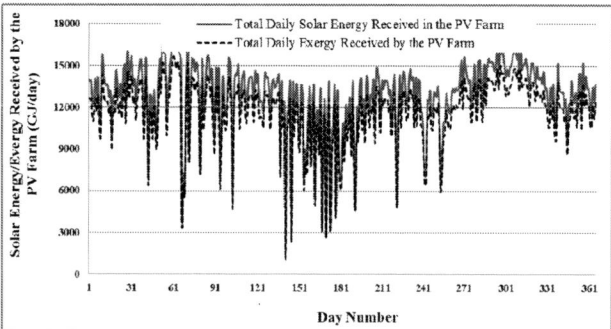

Fig. 6. Total daily solar exergy and energy received by the PV farm

Fig. 7 gives information about total daily energy performance of the CAES system such as the total daily energy produced by the CAES system and the total daily power received by/available for the compressor set in the CAES system.

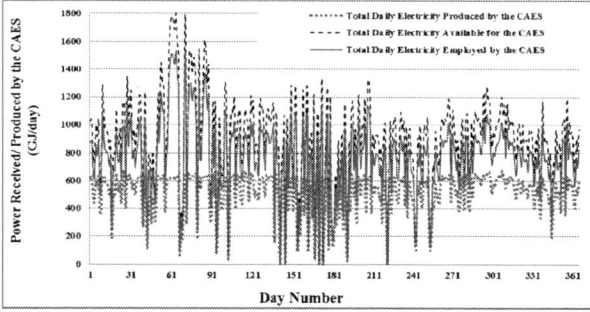

Fig. 7. Total daily energy performance of the CAES unit

Fig. 8 shows the total daily exergy and energy efficiency of the CAES system including the compression set, the expansion set, the air heater and the solar heating unit.

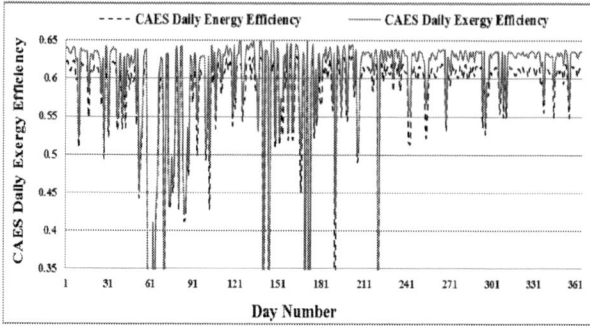

Fig. 8. Total daily exergy and energy efficiency of the CAES system

Fig. 9 reveals how much power is totally sold by the power plant in each day, how much power ramps could (not) finally be offset and how much power is produced at peak consumption periods of each day.

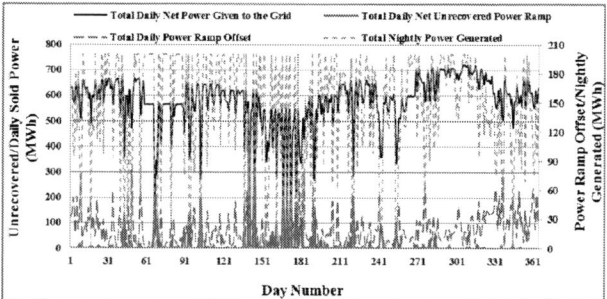

Fig. 9. Total daily power sold/offset/un-recovered and nightly produced in the power plant

Fig. 10 shows the total daily energy and exergy efficiencies of the whole power plant.

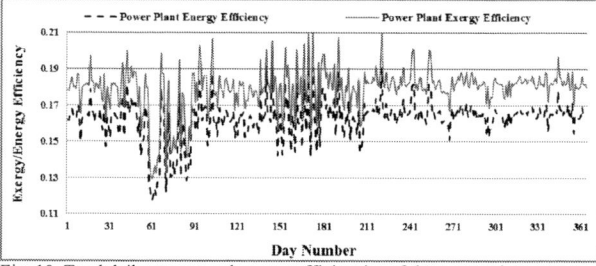

Fig. 10. Total daily energy and exergy efficiencies of the power plant

According to the figure, overall, the power plant exergy efficiency is more than its energy efficiency in all days of the year. The exergy efficiencies up to 21% are achievable in the plant while it doesn't fall below 13% in any day over the entire year. On the other hand, the energy efficiency of the plant varies from 12% in the first days of March to almost 19% in the middle days of May and June. The annual average exergy and energy efficiency of the power plant with all its peripheral equipment are 17.9% and 16.2%, respectively.

In the end, as a result, taking the total daily and nightly vendible power of the power plant to the grid which are correspondingly 151.22 GWh and 41.74 GWh and Brazilian electricity prices for daily and nightly consumptions, which are respectively 0.306 USD/kWh and 0.459 USD/kWh, increasing the power plant energy conversion efficiency for 1 percent, leads to an overall income enhancement of well over 4 million USD per year. Therefore, if this achievement is obtained, a great sustainability in electricity production industry will occur and electricity distribution tends to be more supported by such power plant types.

References

[1] Paul Breeze, Chapter 10-Power System Energy Storage Technologies, Power Generation Technologies (Second Edition), 2014, Pages 195-221.

[2] V Gadhamshetty, V G Gude, N Nirmalakhandan, Thermal energy storage system for energy conservation and water desalination in power plants, Energy, Volume 66, 2014, Pages 938-949.

[3] M Raju, Siddhartha K Khaitan, Modelling and simulation of compressed air storage in caverns: A case study of the Huntorf plant, Applied Energy, Volume 89, Issue 1, January 2012, Pages 474-481.

[4] A. Arabkoohsar, L. Machado, M. Farzaneh-Gord, R. N.N Koury, Thermo-Economic Analysis and Sizing of a CAES Unit Equipped with an Ancillary Solar Heating System for a PV Farm, Renewable Energy, 2015.

[5] A Soteris. Kalogirou (2004), solar thermal collectors and applications, Progress in Energy and Combustion Science, 30, 231–295.

[6] A Arabkoohsar, M Farzaneh-Gord, M Deymi-Dashtebayaz, L Machado, R N.N Koury, Energy and Exergy Analysis of Natural Gas Pressure Reduction Points Equipped with Solar Heat and Controllable Heaters, Renewable Energy, Volume 72, December 2014, Pages 258-270.

7th Power Electronics, Drive Systems & Technologies Conference (PEDSTC 2016)
16-18 Feb. 2016, Iran University of Science and Technology, Tehran, Iran

Signal Flow Graph Modeling and Disturbance Observer based Output Voltage Regulation of an Interleaved Boost Converter

Majid Abbasi
Faculty of Electrical Engineering,
Malek-Ashtar University of
Technology (MUT) Tehran, Iran.
Mabbasi_iust86@yahoo.com

Ahmad Afifi
Faculty of Electrical Engineering,
Malek-Ashtar University of
Technology (MUT) Tehran, Iran.
ah_afifi@iust.ac.ir

Mohammad Reza Alizadeh Pahlavani
Faculty of Electrical Engineering,
Malek-Ashtar University of
Technology (MUT) Tehran, Iran.
mr_alizadehp@iust.ac.ir

Abstract- **In this paper, a disturbance observer (DOB) algorithm is proposed to solve the output voltage regulation of an Interleaved Boost Converter (IBC) in presence of unknown disturbance and parameter uncertainty. IBCs are improved model of boost converters which provide better operation than conventional converters. Also, these converters are rather complicated, therefore, the modeling of them are more cumbersome. To solve this problem, the Signal Flow Graph (SFG) method is implemented for a 2-cells IBC. The main advantage of SFG is obtaining unified dynamic mode of converter. Beside the modeling problem, unknown disturbances such as the input voltage ripple, mismatching of duty cycles and load variations are inevitable in real circuit and degrade the system performance.**
To obtain a high performance regulation in presence of disturbances, a DOB is integrated with an optimal controller. According to our models, the duty cycle variation is limited so an optimal control technique is implemented to achieve a good tracking. Furthermore, the DOB can be estimate and compensate the unknown disturbances. Simulation results for a real model of a 2-cells IBC indicates that application of DOB and SFG modeling is quite effective in correct performance of IBC in presence of unknown disturbances.

Keywords: Disturbance observer, Interleaved boost converter, Signal flow graph.

I. INTRODUCTION

Boost converters are widely used in power electronics applications such as increasing DC voltage level, photovoltaic (PV) systems [1] and power factor correction [2]. For high current applications, interleaved boost converters are preferable, since the currents through the switches are just fractions of the input current [3]. In addition, interleaved boost converters can also reduce input current ripple and switching

losses, provide higher efficiency, lower complexity rather than cascaded converters, lower input current and output voltage ripple [4]. Interleaving also can double the effective switching frequency, reliability by distributed power losses evenly [5], [6].

Interleaved boost converter (IBC) consists of N number of basic boost cells connected in parallel that cells have $2\pi/N$ phase difference of driving pulses in a cycle. Each cell in this converter can operate in discontinuous and continuous inductor current mode, denoted by DCM and CCM respectively. However, more phases in the IBC increase the number of components, such as inductors, and active and passive switches. The dimension of state and control input also becomes higher and it is more difficult to analyze and investigate the operating characteristics at both steady and transient states.

In the purpose of controlling of converters the general models, which describe all behaviors, are needed. State-space averaging method is the most popular approach used for modeling of the DC/DC switching converters [7]. This method is sometimes prolix, especially when the numbers of elements in converters are rather high. Furthermore, this method does not predict the large signal behavior of the system. A large signal modeling tool is necessary to study the global dynamic behavior of switching converters to design robust system.

Switching converters are variable structure systems with linear subsystems. In [8] a method was introduced that is based on signal flow graph (SFG) to model nonlinear behavior of pulse-width-modulated (PWM) switching converters that each subsystem is represented by a flow graph. The advantages of this method are surveyed in [9]. In [9] SFG

978-1-5090-0376-1/16 $31.00 © 2016 IEEE

method is used for modeling a 2-cell interleaved boost converter. It can be shown in the some switching intervals, the power switches cells are on or off simultaneously which is not considered in [9]. The model is correct for only duty cycle (D) equals 0.5. The proposed model in this paper has good performance for operation mode $D>0.5$.

To achieve the good tracking of desired value, there exist some algorithms such as state feedback, optimal control, model predictive control and etc. In [10], a model predictive controller is proposed for an IBC in two operation mode. In this work, according to limitation of duty cycle variation, an optimal control technique is implemented. Another problem that influences the voltage regulation in IBC is the need of robustness against disturbances such as the ripple of input voltage, mismatching of duty cycles and load variations. In order to compensate for external disturbances and improve the performance of system, Ohnishi [11] proposed the concept of DOB. Since the DOB is able to improve the system robustness and dynamic performance, it is widely applied in motion control systems, such as robotic systems, hard disk, electric load simulator and etc [12]. Simple disturbance observers are proposed in [13] which unlike adaptive and robust methods, don't require heavy computation or knowledge of the norm bound on unknown disturbances. It seems that the application of DOB in voltage regulation of IBCs is almost a new design.

The rest of the paper is arranged as follows; in section II, the principle of IBC is reviewed. In section III, the SFG modeling of IBC is presented. Section IV explains the design process of an optimal controller and a disturbance observer to the ability of disturbance attenuation.

II. THE PRINCIPLE OF IBC AND MODELING

A two-phase conventional interleaved boost converter is shown in Fig.1 which has allocated many applications in electrical systems. Each cell can operate in CCM or in DCM mode. In CCM the input current and output voltage ripple is lower than DCM but hard switching occurs that lead to an increase in switching losses. Farther more, the reverse-recovery problems of boost diodes exist too [14]. The operation mode is chosen by duty cycle (D) and the values of parameters such as inductor, output capacitor and load resistance. The inductors current and the sub-circuits in CCM, $D>0.5$ are shown in fig. 2 and Fig. 3 respectively.

For each state, a SFG is shown separately which denoted by G_1 to G_4. Branches that exist in G_1 but not in $G_2... G_4$ are

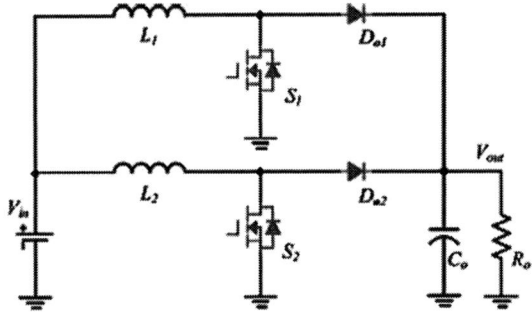

Fig. 1 Two-phase conventional interleaved boost converter

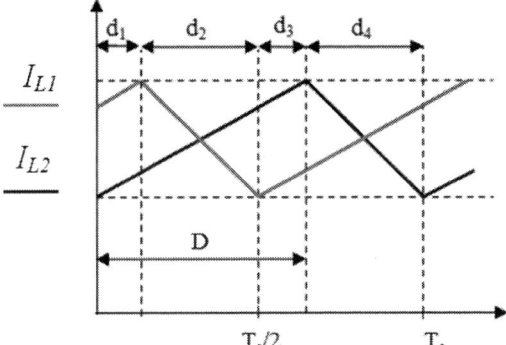

Fig. 2 The current of two inductors, CCM (D>0.5)

replaced by K_1 branches and we do the same for three remain sub-circuits respectively. The four SFGs are combined to form a simplified SFG and the resulting graph topology can be mathematically written as:

$$G = \sum_{i=1}^{4} K_i G_i \qquad (1)$$

where K_1, K_2, K_3 and K_4 are the switching functions defined as:

$$K_i = \begin{cases} 1 & t_{i-1} < t < t_i \\ 0 & otherwise \end{cases} \qquad (2)$$

Finally, the Simplified SFG of a 2-cell interleaved boost converter is obtained shown in Fig 4. As shown in Fig 4, only two branches remain that are denoted by K_2 and K_4.

In this method and in large signal model the output signal of the switches $y_i(t)$ from i'Th branch is defined as product of its input signal $x(t)$ and duty cycle function $d(t)$ as follows:

$$y_i(t) = x_i(t)d_i(t) \qquad (3)$$

The Small signal flow graph of 2-cell interleaved boost converter is shown in Fig. 5 where $D' = 1 - D_1$, $D'' = 1 - D_2$ and D_1, D_2 are the duty cycles corresponding to each cells. The parameters a and B are obtained as:

$$a = -I_{in} + V_o \left(\frac{D''^2}{L_1 s + r_1} + \frac{D'^2}{L_2 s + r_2} \right) \qquad (4)$$

$$B = \frac{R(1 + sCR_c)}{1 + sC(R + R_c)} \qquad (5)$$

Where V_o and I_{in} are the value of output voltage and input current in the steady state respectively. Fig. 4 illustrate the SFG modeling of a 2-cell IBC based on relationships between output voltage variation $\hat{v}_o$, duty cycle variation $\hat{d}$ and input voltage variation $\hat{v}_g$ as disturbance. State-space representation of IBC in CCM mode can be writing as follow:

$$\dot{x}(t) = Ax(t) + Bu(t) + N\hat{v}_g$$
$$y(t) = Cx(t) \qquad (6)$$

where $x = [\hat{\imath}_1, \hat{\imath}_2, \hat{\imath}_o, \hat{v}_o]^T$ is state vector, $u = \hat{d}$ and $y = \hat{v}_o$. The matrices A, B, N, C are as follow:

$$A = \begin{bmatrix} -\frac{r}{L} & 0 & 0 & -\frac{D''}{L} \\ 0 & -\frac{r}{L} & 0 & -\frac{D'}{L} \\ 0 & 0 & -\frac{r}{L} & -\frac{2D'D''}{L} \\ 0 & 0 & A_{43} & A_{44} \end{bmatrix}, B = \begin{bmatrix} 0 \\ 0 \\ B_{31} \\ B_{41} \end{bmatrix}, N = \frac{r}{L}\begin{bmatrix} 1 \\ 1 \\ D'+D'' \\ N_{41} \end{bmatrix},$$

$$C = [0 \quad 0 \quad 0 \quad 1] \qquad (7)$$

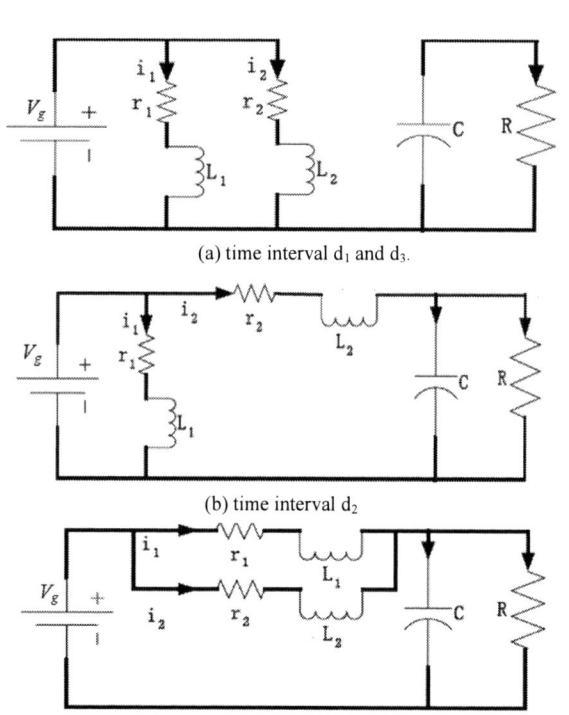

(a) time interval d_1 and d_3.

(b) time interval d_2

(C) time interval d_4

Fig.3 The subcircuites of a 2-cells IBC in CCM and D>0.5 corresponding to time intervales shown in Fig. 2

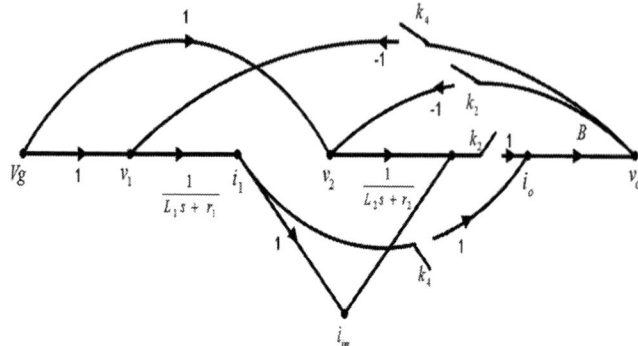

Fig.4 Simplified signal flow graph of 2-cell interleaved boost converter in CCM (D>0.5)

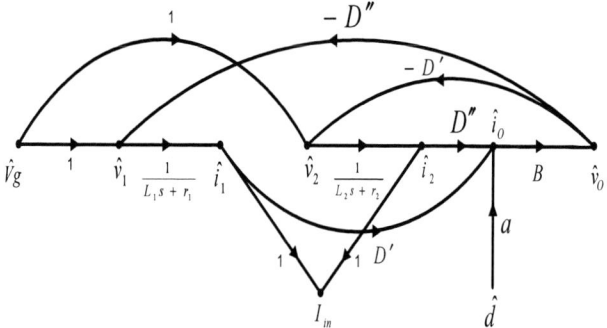

Fig. 5 Small signal SFG model, CCM (D>0.5).

where

$$A_{43} = \frac{1}{C(R+R_c)}\left(R - \frac{rR_cRC}{L}\right), \quad A_{44} = \frac{-1}{C(R+R_c)}\left(1 + 2D'D''\frac{R_cRC}{L}\right),$$

$$B_{31} = \frac{V_o\left(D'^2+D''^2\right)-rI_{in}}{L} \quad , \quad B_{41} = \frac{R_cR}{L(R+R_c)}\left[V_o\left(D'^2 + D''^2\right) - rI_{in}\right], N_{41} = \frac{R_cR(D'+D'')}{R+R_c}.$$

III. DISTURBANCE OBSERVER AND OPTIMAL CONTROLLER DESIGN

Fig. 5 illustrates the proposed control scheme in this paper. Therefore, first design a optimal controller using the state-space equation (6); then, a disturbance observer is presented to estimate the unknown disturbances and then is integrated with control input of optimal controller u^*.

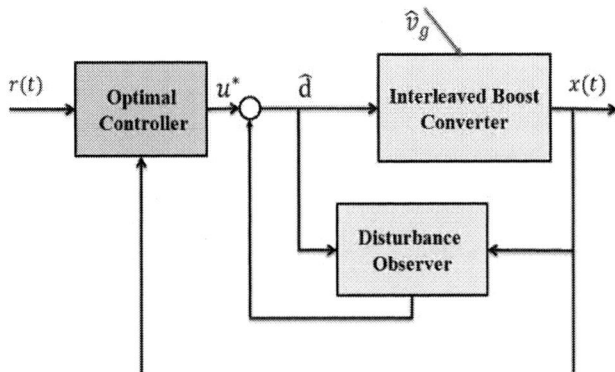

Fig. 5 Block diagram of proposed scheme

A. Disturbance observer design

DOB has been widely utilized for high-precision and high-speed control applications. The structure of the DOB for a 2-cell IBC is shown in Fig. 6 where $H_1(s) = C(sI - A)^{-1}B$ and $H_2(s) = C(sI - A)^{-1}N$. In Fig. 6, the transfer function from two external inputs $u^*, \hat{v}_g$ to output y is obtained as follows:

$$y = H_1(s)u^* + [1 - Q(s)]H_2(s)\hat{v}_g \qquad (8)$$

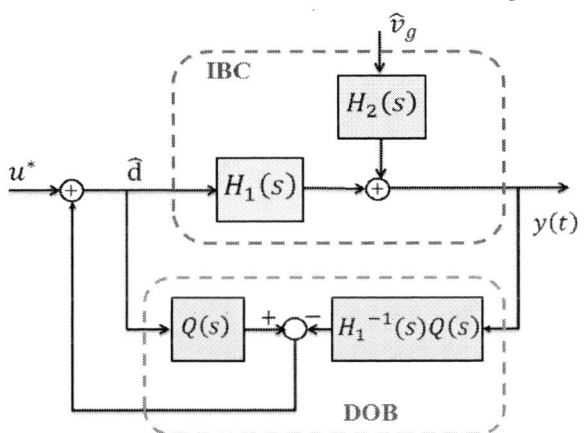

Fig. 6 structure of propose DOB.

The characteristics of DOB is determined by the filter time constant, numerator order and denominator order (or relative degree) of $Q(s)$ filter. If $|Q(s)| \approx 1$ in low frequencies, then the DOB system (8) can be approximated in the following form $y = H_1(s)u^*$ and the disturbances can be rejected. First, we assume the Q filter of the following form:

$$Q_{mn}(s) = \frac{\sum_{i=0}^{n} a_{mi}(\tau s)^i}{(\tau s + 1)^m} \qquad (9)$$

where τ is the filter time constant, $a_{mi} = m!/(m - i)! \, i!$ the binomial coefficient, m the denominator order and n the

numerator order. As shown in [14], the larger the numerator order lead to the better the disturbance rejection performance. Moreover, small τ means wider frequency range for disturbance rejection.

B. Optimal controller design

Let the state equation are

$$\dot{x}(t) = Ax(t) + Bu(t) \qquad (10)$$

and the performance measure to be minimized is

$$J = \frac{1}{2} \int_{t_0}^{t_f} ([x(t) - r(t)]^T H[x(t) - r(t)] + Ru^2) \, dt \qquad (11)$$

where $r(t)$ is the desired or reference value of the state vector. H, R are real symmetric and positive definite matrices. The Hamiltonian is given by

$$\mathcal{H}(x(t), u(t), p(t), t) = \frac{1}{2}[x(t) - r(t)]^T H[x(t) - r(t)] + \frac{1}{2}Ru^2 + p(t)^T[AX(t) + Bu(t)] \qquad (12)$$

where $p(t)$ is the Lagrange multipliers vector. The costate equations and the algebraic relations are as follow:

$$\dot{p}^*(t) = -\frac{\partial \mathcal{H}}{\partial x} \quad , \quad \frac{\partial \mathcal{H}}{\partial x} = 0. \qquad (13)$$

Solving for $p^*(t)$ and $u^*(t)$ yields

$$p^*(t) = Kx^*(t) + s(t) \qquad (14)$$

$$u^*(t) = -R^{-1}B^T p^*(t) \qquad (15)$$

where state feedback K and $s(t)$ are solution of the equations [15]:

$$KA + A^T K - KBR^{-1}B^T K + H = 0 \qquad (16)$$

$$\dot{s}(t) = -(A^T - KBR^{-1}B^T)s(t) + Hr(t). \qquad (17)$$

The block diagram of the optimal controller is shown in Fig. 7

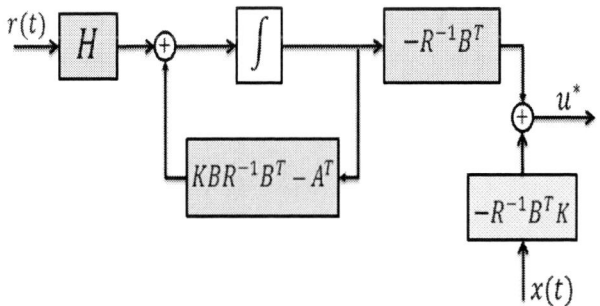

Fig. 7 Block diagram of optimal controller.

978-1-5090-0376-1/16 $31.00 © 2016 IEEE

IV. SIMULATION AND RESULTS

In this section a 2-cell interleaved boost converter is simulated and the SFG results in CCM are compared with equivalent switching circuit in MATLAB/SIMULINK. The parameters chosen for this converter are: $V_g = 10$, $L_1 = 30\mu H$, $L_2 = 30\,\mu H$, $C = 60\,\mu H$, $r_1 = 0.1\Omega$, $r_2 = 0.1\Omega$, $R_C = 0.3\,\Omega$, $R = 10\,\Omega$ and $D' = D'' = 0.7$. The corresponding results are plotted in Figs. 8(a) and 8(b). The results are closely matching and the slight discrepancies in the switching model and SFG modeling results are mainly due to output capacitor. The input voltage ripple is such as disturbance.

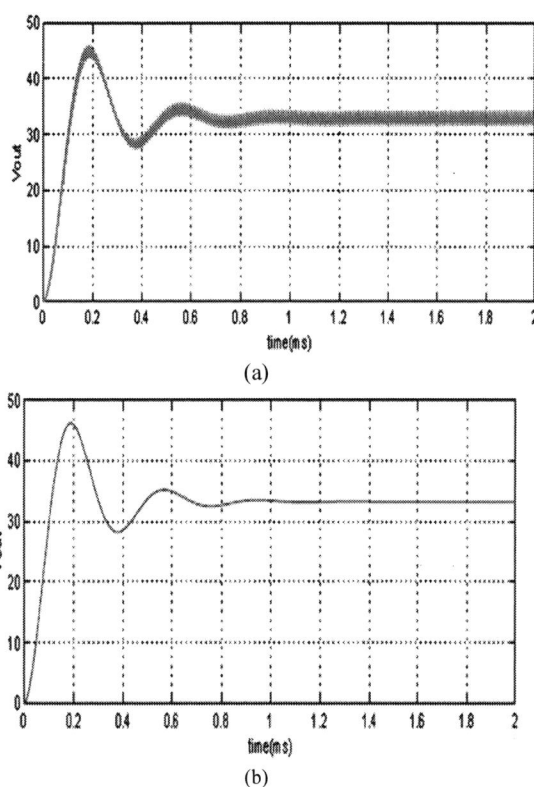

Fig. 8 Output voltage of a 2-cells IBC (a) output voltage of SFG in CCM, D>0.5.(b) Output voltage of switching model CCM, D>0.5.

The input disturbances are sinusoidal wave with amplitude of 2 volt and frequency of 50 hertz. These simulation results indicate that the performance of the optimal controller under the disturbance is significantly improved and the DOB achieves good disturbance attenuation ability. Fig. 9(a) shows the performance of proposed controller with and without DOB. Moreover, control input (duty-cycle variation) is shown in Fig. 9(b). The bode diagram of filter $Q(s)$ is shown in Fig. 10.

V. CONCLUSION

A new controlling method of interleaved boost converter is proposed in this paper. The Signal Flow Graph method is implemented for a 2-cells interleaved boost converter and a disturbance observer algorithm is proposed to solve its output voltage regulation.

The simulation results of modeling are in close agreement with switching model, thus validating the signal flow graph modeling of the interleaved boost converter system. These simulation results indicate that the performance of the optimal controller under the disturbance is significantly improved and the DOB achieves good disturbance attenuation ability.

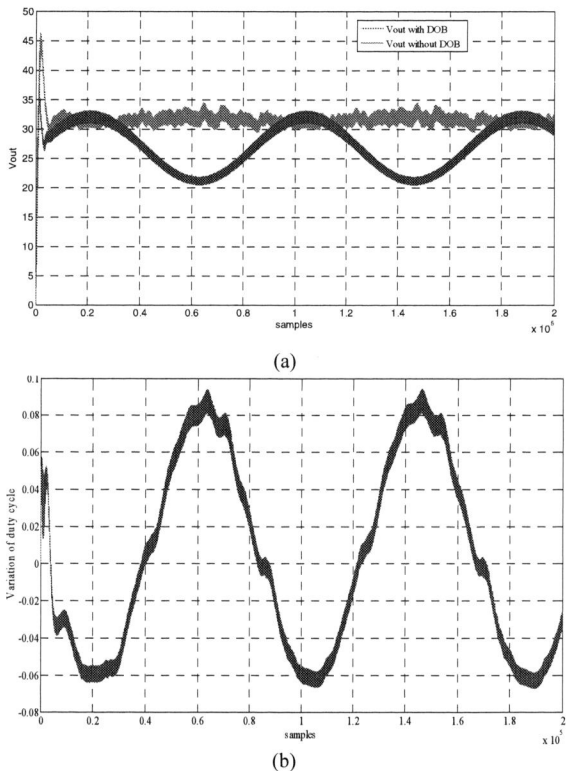

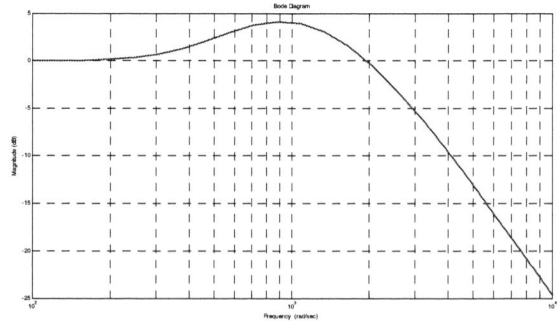

Fig .9 (a) output voltage regulation with and without DOB. (b) Control input (duty-cycle variation) of integrated optimal controller with DOB.

Fig. 10 Bode diagram of $Q(s)$.

REFERENCES

[1] H. van der Broeck and I. Tezcan, "1 KW dual interleaved boost converter for low voltage applications," in *Proc. CES/IEEE 5th IPEMC*, vol. 3, pp. 1–5, Aug. 2006.

[2] Garinto. D, "Interleaved Boost Converter System for Unity Power Factor Operation," *European Conference on Power Electronics and Applications*, pp 1 - 7, Sept. 2007.

[3] W. Li, Y. Deng, R. Xie, X. He, "A Non-Isolated Interleaved ZVT Boost Converter with High Step-Up Conversion Derived from its Isolated Counterpart," *in Proc. IEE EPE '07*, pp. 1-8, 2007.

[4] Michael O'Loughlin, "An Interleaving PFC Pre Regulator for High-Power Converters," Texas Instruments.

[5] Yanshen Hu, Yunxiang Xie, Huamin Xu, and Tian Hao, "Design and Implementation of Two-Channel Interleaved Boost Converters with Integrated Coupling Inductors," *IEEE EPEPEMC*, pp. 625-630, Aug. 2006.

[6] Abbasi, M.; Mortazavi, N.; Rahmati, A., "A novel ZVS interleaved boost converter," in Power Electronics, Drive Systems and Technologies Conference (PEDSTC), 2014 5th , vol., no., pp.535-538, 5-6 Feb. 2014.

[7] Rashid M.H., Power Electronics handbook, Academic Press, Aug. 2001, chapter 19.

[8] Smedley, K., and Cuk, S, "Switching flow-graph nonlinear modeling technique," *IEEE Trans. Circuits Syst*, Vol. 9, pp. 245 251, July 1994.

[9] Veerachary M, "General rules for signal flow graph modeling and analysis of DC-DC converters," *IEEE Transactions on Aerospace and Electronic Systems*, Vol. 40, pp. 251-271. Jan 2004.

[10] C. S. Babu, M. J. Veerachary, "Predictive Controller for Interleaved Boost Converter" *Proc. of IEEE ISIE*, pp.577-581, June 2005.

[11] K. Ohnishi "A New Servo Method in Mechatronics," *Trans Japanese Society of Electrical Engineering*, vol. 107, pp.83-86, 1987.

[12] S. K. Park, and S. H. Lee, "Disturbance Observer Based Robust Control for Industrial Robots with Flexible Joints," *International Conference on Control, Automation and Systems*, COEX, Seoul, Korea, 2007.

[13] Uran, S., Milanovic M. , "Advanced Control of Buck Converter," *IEEE International Conference on Industrial Technology*, Vol. 1, pp. 602-607, 2003.

[14] Y. Choi, K. Yang, W. K. Chung, H. R. Kim, and I. H. Suh, "On the robustness and performance of disturbance observers for second-order systems," *IEEE Trans. on Automatic Control*, Vol.48, No.2315-320, 2003.

[15] Kirk, D. E, Optimal Control Theory: An Introduction. *Lebedev, L. P, and Cloud*, M. J. 2003.

Design Procedure and Experimental Validation of a 30kVA DSP-Based PWM Rectifier

Mohammad Pichan
Department of Electrical Converters & Power Systems
IRIEE, ACECR
Tehran, Iran
m_pichan@yahoo.com

Adib Abrishamifar
Department of Electrical Engineering
Iran University of Science and Technology
Tehran, Iran
Abrishamifar@iust.ac.ir

Amir Mirzabayati
Department of Electrical Converters & Power Systems
IRIEE, ACECR
Tehran, Iran
amirmirzabayati@gmail.com

Mehdi Fazeli
Department of Electrical Converters & Power Systems
IRIEE, ACECR
Tehran, Iran
mfazeli@jdevs.com

Abstract— **Nowadays DSPs have been used widely in power electronic applications because of powerful calculation ability, several peripherals and serial interfaces. Three phase PWM rectifiers (TPPR) have been used widely since they can provide unity power factor (PF), a sinusoidal grid current and also higher DC voltage than grid voltage amplitude. Besides, current vector control is a powerful strategy to be used for three phase PWM rectifier to regulate input active and reactive powers separately. Therefore, in this paper, precise design procedure is presented firstly. Secondly, simulation and experimental evaluation of the three phase PWM rectifier is considered based on a digital signal processor (DSP)-based digital controller. An expanded simulation and experimental results are done to verify the effectiveness of complete control system. The results show the low-ripple adjustable DC voltage between 650 to 750v. Moreover, the THD value of the input current is about 2.4% which satisfy international standards such as IEEE Standard 1547.**

Keywords— *three phase PWM rectifier; current vector control; Park transformation; DSP processors*

I. INTRODUCTION

Nowadays, several applications need DC voltage such as adjustable AC drives and uninterruptible power supplies (UPSs). A simple solution to provide DC voltage is using a conventional six pulse rectifiers based on a diode bridge. Although, it has a very simple structure but, it suffers from several disadvantages like constant and unregulated output DC voltage, low input power factor (PF) and high total harmonic distortion (THD) value of the input current. Recently, TPPR have found more attention especially in UPS applications. They can provide unity PF, low-THD sinusoidal input current and adjustable output DC-voltage with low ripple. Also, the output DC voltage is higher than the peak amplitude of the grid voltage which may be beneficial for some applications. Therefore, their several advantages and an increasing demand stimulate the researcher to develop high performance control strategies for them [1].

Generally, the control strategies of the TPPR can be divided into indirect and direct control strategies [2-5]. The direct control strategy which is mainly direct power control (DPC), the active and reactive powers are used as control variables. Since one of the main goals in rectifiers is to control the active and reactive powers, the DPC proposes to remove all internal current controllers. The DPC strategies can be implemented either based on switching look-up table [3] or PWM scheme [4-6]. Since the DPC directly controls the active and reactive powers, it can guarantee good dynamic performance [6]. But, it suffers from some disadvantages like variable switching frequency, high sampling frequency for digital implementation and also needing large input inductance [7-12].

Indirect control strategies are mainly current control or current vector orientation according to input line voltage called voltage oriented control (VOC). Due to an internal current control loop, this method shows good dynamic and steady state performance for current regulating. This method decouples the input current into two components which can regulate the active and reactive powers, respectively [13].

In this paper, current vector control method is used to control TPPR. It is mainly related to this matter that for UPS applications, unity PF and low THD value at input (grid side) are so important. Besides, this paper deals with digital implementation of the current vector control for TPPR based on DSP processors.

II. SYSTEM MODELLING AND PRINCIPLE OF THE CURRENT VECTOR CONTROLLER

The schematic of the three phase PWM rectifier is shown in Fig.1. The voltage equations for each phase are as follow:

$$L\frac{d}{dt}i_a = -ri_a - V_{aN} + E_a$$

$$L\frac{d}{dt}i_b = -ri_b - V_{bN} + E_b \qquad (1)$$

$$L\frac{d}{dt}i_c = -ri_c - V_{cN} + E_c$$

If the (1) is rewritten in matrix form, the (2) will be resulted in which I_3 is the 3*3 identity matrix.

$$\frac{d}{dt}\begin{bmatrix} i_a \\ i_b \\ i_c \end{bmatrix} = -\frac{r}{L}I_3\begin{bmatrix} i_a \\ i_b \\ i_c \end{bmatrix} - \frac{1}{3L}\begin{bmatrix} 2 & -1 & -1 \\ -1 & 2 & -1 \\ -1 & -1 & 2 \end{bmatrix}\begin{bmatrix} V_a \\ V_b \\ V_c \end{bmatrix} + \frac{1}{L}I_3\begin{bmatrix} E_a \\ E_b \\ E_c \end{bmatrix} \qquad (2)$$

A. Park transformation

The park transformation is a well known linear transformation with time varying coefficients. This is used mainly to control the three phase machines and power converters. This transformation maps the three phase variables into two-axis (dq) synchronous reference frame.

For TPPR, the speed of the synchronous rotating reference frame is equal to the grid angular frequency (ω). Using the transformation matrix (3), the variables in the abc reference frame will transfer to the synchronous reference frame.

$$f_{dq} = T_{dq}f_{abc}$$

$$T_{dq} = \frac{2}{3}\begin{bmatrix} \cos\omega t & \cos(\omega t - 120°) & \cos(\omega t + 120°) \\ \sin\omega t & \sin(\omega t - 120°) & \sin(\omega t + 120°) \end{bmatrix} \qquad (3)$$

Based on (2) and (3), the three phase PWM rectifier equations in the dq reference frame are given by (4) where I_2 is 2*2 identity matrix.

$$\frac{d}{dt}\vec{i}_{dq} = \begin{bmatrix} -\dfrac{r}{L} & \omega \\ -\omega & -\dfrac{r}{L} \end{bmatrix}\vec{i}_{dq} - \frac{1}{L}I_2\vec{V}_{dq} + \frac{1}{L}I_2\vec{E}_{dq} \qquad (4)$$

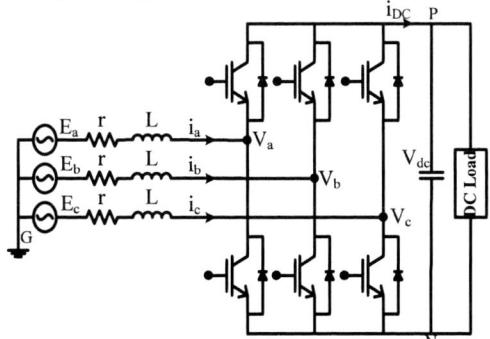

Fig. 1. The schematic of the three phase PWM rectifier.

One of the most advantages of the park transformation is that sinusoidal signals with angular frequency ω will be constant signals in the dq reference frame. Therefore, a common PI controller can be used to guarantee zero steady state error thanks to the built in integral action.

B. The current vector control strategy

In order to simplify the overall equations, the phase locked loop (PLL) is employed to fix the d-axis of the rotating reference frame to follow the d-axis of the grid voltage. Consequently, the q component of the grid voltage vector will be equal to zero. The grid voltage vector can be expressed as follow:

$$\begin{cases} E_d = E_m \\ E_q = 0 \end{cases} \qquad (5)$$

where E_m is the amplitude of the grid phase voltage. Substituting the (5) in (4) will result:

$$L\frac{d}{dt}i_d + ri_d = V_d - E_d + \omega Li_q$$

$$L\frac{d}{dt}i_q + ri_q = V_q - \omega Li_d \qquad (6)$$

Also, the power equations in the dq reference frame can be expressed as follow:

$$P = E_d i_d + E_q i_q$$

$$Q = E_d i_q - E_q i_d \qquad (7)$$

Substituting the (5) in (7) we have:

$$P = E_d i_d$$

$$Q = E_d i_q \qquad (8)$$

Since the amplitude and frequency of the grid voltage are constant for normal operation, the E_d can be assumed as a constant value. Hence, the active and reactive powers are proportional to the i_d and i_q, respectively. It means that in order to regulate the active power, the i_d should be controlled. On the other hand, controlling the i_q, the reactive power can be regulated. Therefore, if the i_d and i_q can be controlled separately, the active and reactive powers can be regulated separately.

According to (4), the d and q current components have coupling with each other so, they should be decoupled. In order to decouple these components, new inputs are defined as follow:

$$\begin{cases} V_{dl}^* = V_d - E_d + \omega Li_q \\ V_{ql}^* = V_q - \omega Li_q \end{cases} \qquad (9)$$

Substituting (9) in (4) will result in (10).

$$\begin{cases} L\dfrac{di_d}{dt} + Ri_d = V_{dl}^* \\ L\dfrac{di_q}{dt} + Ri_q = V_{ql}^* \end{cases} \qquad (10)$$

According to (10), it is evident that decoupled active (i_d) and reactive (i_q) current components controlling is achieved.

The relative current to voltage transfer function of the dq component is given in (11).

$$G_{dq} = \frac{I_{dq}(s)}{V_{dql}(s)} = \frac{1/R}{(L/R)s+1} \tag{11}$$

C. The Model of the DC-Link Voltage Loop

Assuming ignorable power loss, the output power will be equal to the input power as follow:

$$3E_{rms}I_{rms} = i_{dc}v_{dc} \tag{12}$$

where $E_a = E_b = E_c = E$ and $i_a = i_b = i_c = I$. Applying KCL law at the output node and based on the park transformation, the (13) and (14) can be concluded.

$$i_{dc} = c\frac{d}{dt}v_{dc} + \frac{v_{dc}}{R} \tag{13}$$

$$i_d = \sqrt{2}I_{1rms} \tag{14}$$

Replacing (13) and (14) in (12) will result (15).

$$\frac{3E_{1,rms}i_d}{\sqrt{2}} = \frac{1}{2}C\frac{d}{dt}v_{dc}^2 + \frac{v_{dc}^2}{R} \tag{15}$$

Finally, linearization of the (15) around operating point and after some simplification, the DC-Link voltage loop transfer function is concluded as follow:

$$\frac{V_{dc}(s)}{I_d(s)} = \frac{3E_{1,rms}R}{2\sqrt{2}V_{dc}^*}\frac{1}{\frac{RC}{2}s+1} \tag{16}$$

III. THE CURRENT AND VOLTAGE CONTROLLER DESIGN

The parameters of the three phase PWM rectifier which are used for hardware implementation are given in Table.I.

A. The current controller design

The simplified current loop is depicted in Fig. 2. According to the (11) and Table. I, the G_{dq} can be calculated as follow:

$$G_{d,q} = \frac{20}{0.024s+1} \tag{17}$$

The common PI controller can be used to achieve reference current tracking with zero steady state error. In order to avoid aliasing effect, the closed loop bandwidth should be lower

TABLE I.   THE PARAMETERS OF THE THREE PHASE PWM RECTIFIER.

Rectifier power	15 kVA
Grid voltage (E_s)	380 v (LL, rms)
DC-Link voltage (V_{DC})	690 v
Input inductor (L_s)	3.35 mH
DC-Link capacitance (C_{dc})	23.8 mF
Sampling frequency	6 kHz
Switching frequency	3 kHz

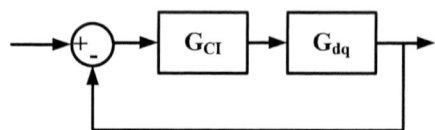

Fig. 2. The simplified current loop

enough than switching frequency. On the other hand, to remove disturbances and track the reference precisely, the closed loop bandwidth should be selected as high as enough. As a result, the closed loop bandwidth is selected as 20% of the switching frequency to satisfy both limitations. Also, to consider PWM and calculation delays, the phase margin is selected 60°. The bode plot of the $G_{CI}(s)G_{d,q}(s)$ is shown in Fig.3. To achieve mentioned specifications, the K_P and K_I of the current controller are selected as 3.8 and 8374.

B. The DC-Link voltage controller design

Since the three phase PWM rectifier should provide adjustable DC voltage, a well performance voltage controller should be used. In this paper, the PI controller is used to control the DC-Link voltage with zero steady state error and fast transient response. Also, this controller must reduce the voltage ripple. The overall control loop of the PWM rectifier is depicted in Fig. 4.

According to (16) and Table. I, the relative DC-Link transfer function is given in (18).

$$G_V = \frac{2.178}{0.0833s+1} \tag{18}$$

For this controller, the controller bandwidth is selected as 10% of the current loop bandwidth. Consequently, the time constant of the current loop is higher enough the voltage loop and as a result, the current loop can track the reference current (output of the voltage loop controller) precisely. Also, again the phase margin is selected 60°. For these conditions, the PI voltage controller coefficients, the K_P and K_I are selected as 13 and 3218, relatively. The bode plot of the open loop voltage controller, $G_{CV}(s)G_V(s)$, is shown in Fig. 5.

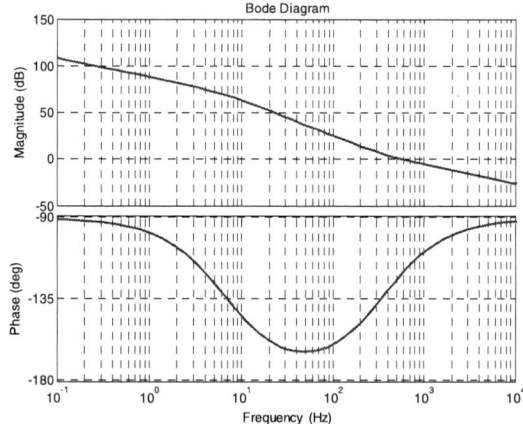

Fig. 3. The bode plot of the current open loop.

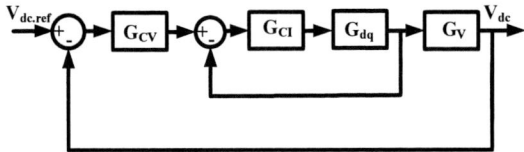

Fig. 4. The overall control loop of the PWM rectifier.

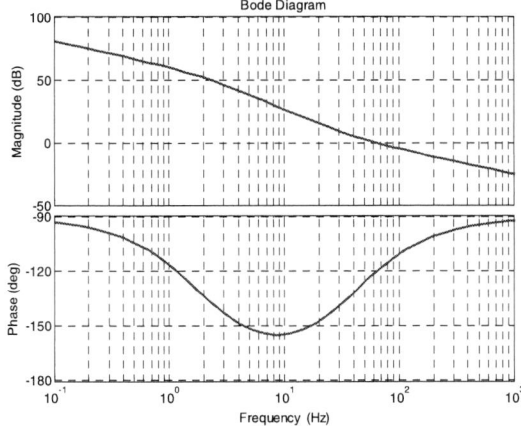

Fig. 5. The bode plot of the open loop voltage cotroller.

Based on (8), input active power is proportional to the i_d. On the other side, if the DC load changes, the DC-Link voltage will be changed relatively. In order to compensate the DC-Link voltage, the input power should be changed. Hence, it can be concluded that the output of the DC-Link voltage controller generates the d component of the input reference current. Again, based on (8), the input reactive power is proportional to the i_q. Commonly, it is desired that rectifiers don't consume any reactive power or in other word, rectifiers should provide unity PF at input or grid side. Therefore, the q component of the input reference current should be set to zero. The schematic of the control system is shown in Fig. 6.

According to Fig. 6, the outputs of the PI current controllers are V_{dq}^* which should be added to decoupling term to generate the final reference voltages in dq reference frame.

IV. SIMULATION RESULTS

To validate the effectiveness of the current vector control strategy and design procedure, various simulations are done in SIMULINK. The parameters of the three phase PWM rectifier are given in Table. I. Also these values are used for hardware implementation. To ensure unity PF, PLL plays an important role in control system and it should work properly even at non-ideal grid voltage, such as harmonic polluted grid voltage. In this paper, a simple structure three phase PLL with good performance is used [14].

A. Steady state performance

The steady state performance of the three phase PWM rectifier with current vector control is shown in Fig.7. According to this figure, the grid current is synchronized with grid voltage to provide unity PF. Also, the grid current THD

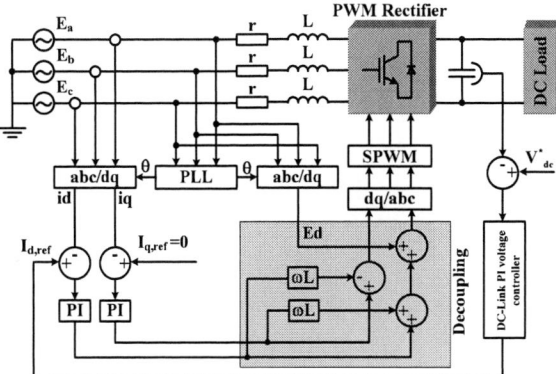

Fig. 6. The schematic of the control system.

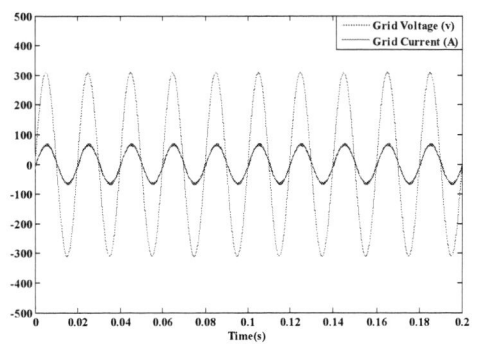

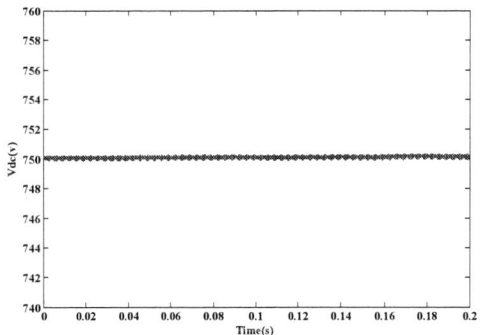

Fig. 7. The steady state performance of the three phase PWM rectifier.

value is 2.2% at this condition which satisfies international standards like IEEE Standard 1547. One of the main goals of the rectifier is to make a low ripple constant DC voltage. For example, to charge the battery bank in the UPSs, a relative high voltage ripple reduces the life time of the battery and also causes additional loss and heat. According to the Fig.7, it is evident that rectifier provides constant DC voltage with voltage ripple below 0.3%.

B. Dynamic response

To evaluate the dynamic performance of the three phase PWM rectifier, both rising and falling 50% step changes in load power are simulated and the result is shown in Fig.8. The designed current vector control strategy tracks the reference

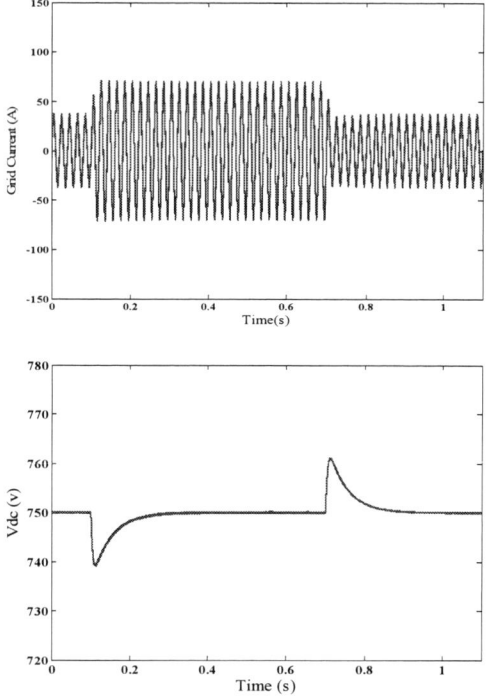

Fig. 8. The dynamic performance with load step change.

values with fast response in few milliseconds. Although there is a voltage change at the DC-Link voltage at stepping time but, the DC-Link voltage controller restores the voltage at nominal value very fast.

V. EXPERIMENTAL RESULTS

The power set-up which is similar to Fig.1 is provided to validate simulation results with experimental results. The control system is implemented with DSP development board based on a TMS320f28335 processor. The DSP clock is 150MHz and the control loop is implemented with 6kHz sampling frequency. A development board is used to generate gate signals for IGBTs. The performance of system in steady state condition is shown in Fig.9. As shown in this figure, the DC-Link voltage controller set the voltage at the nominal value. The vector current controller tracks the reference current and the grid current THD value is almost 2.3% which is in standard range. Also, the PF is greater than 0.99.

For some application such as battery chargers, it is essential that PWM rectifier provides variable DC voltage. To show controllable DC voltage, various set points are selected and the results are shown in Fig. 10. At all of these set points, the low-ripple DC voltage is achieved.

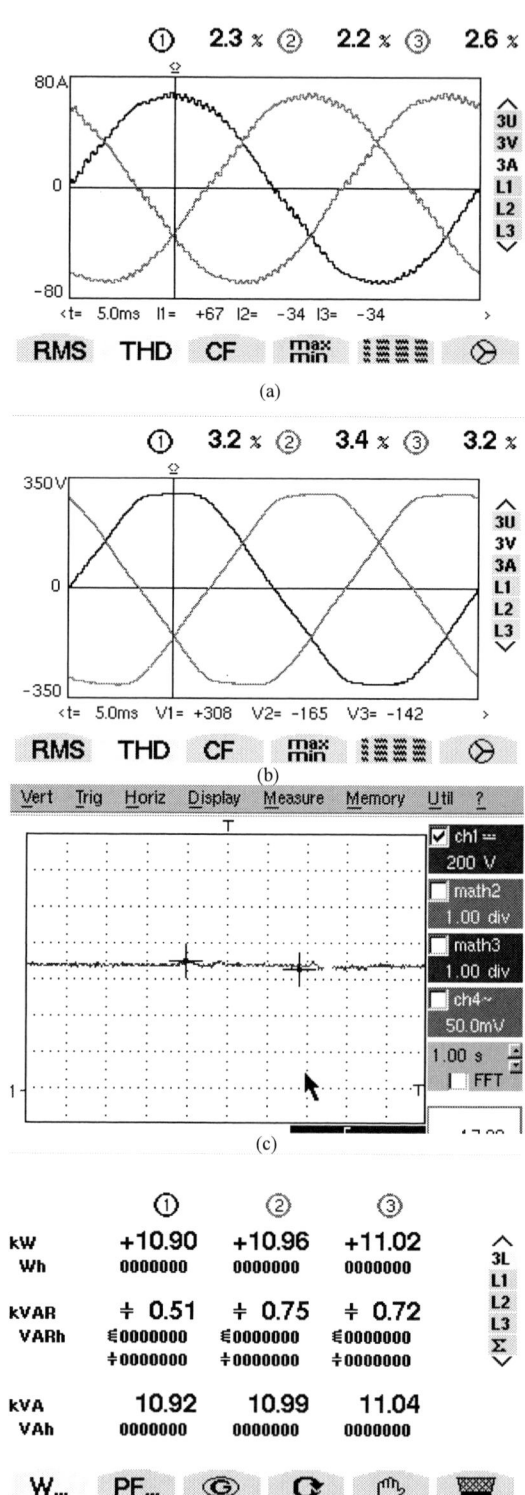

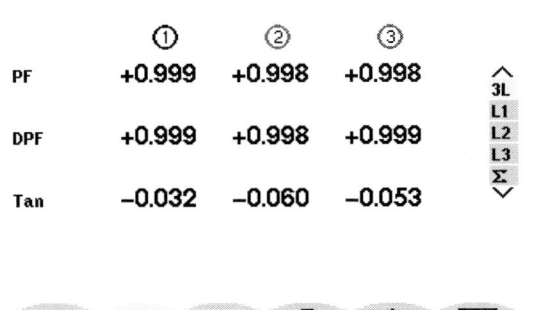

	①	②	③
PF	+0.999	+0.998	+0.998
DPF	+0.999	+0.998	+0.999
Tan	−0.032	−0.060	−0.053

(e)

Fig. 9. The experimental result under steady state condition: a) Grid current, b) Grid voltage, c) DC-Link voltage, d) Total power, e) PF.

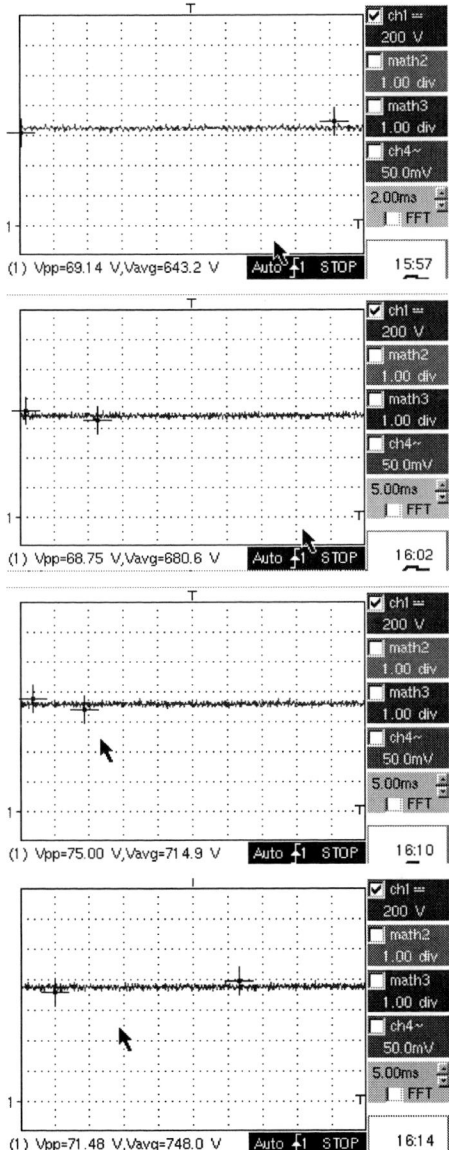

Fig. 10. Experimental results of output DC voltage at various set points.

VI. CONCLUSION

In this paper, simulation and experimental evaluations of a three phase PWM rectifier for UPS application were presented. The whole control system is based on indirect vector current controller in dq reference frame. Also, the PI controller is used to fix the DC-Link voltage. Due to the several advantages of DSPs, the DSP development board is used as digital controller. Expanded simulations were done in MATLAB/SIMULINK software to verify the effectiveness of control system in steady state and dynamic condition. Also, the power set-up was provided to verify simulation results and all experimental results validate simulation results. All of these results show the effectiveness of the control system including low input THD value, unity PF and constant and low ripple DC voltage. All of these are almost essential for UPS applications.

References

[1] J. R. Rodriguez, J. W. Dixon, J. R. Espinoza, J. Pontt, and P. Lezana, "PWM regenerative rectifiers: State of the art," *IEEE Trans. Ind. Electron.*, vol. 52, no. 1, pp. 5–22, Feb. 2005.

[2] Y. Lo, T. Song, and H. Chiu, "Analysis and elimination of voltage imbalance between the split capacitors in half-bridge boost rectifier," *IEEE Trans. Ind. Electron.*, vol. 49, no. 5, pp. 1175–1177, Oct. 2002.

[3] M. Malinowski, M. P. Kazmierkowski, S. Hansen, F. Blaabjerg, and G. D. Marques, "Virtual-flux-based direct power control of three-phase PWM rectifiers," *IEEE Trans. Ind. Appl.*, vol. 37, no. 4, pp. 1019–1027, Jul./Aug. 2001.

[4] T. Ohnishi, "Three phase PWM converter/inverter by means of instantaneous active and reactive power control," in *Proc. IECON*, 1991, vol. 1, pp. 819–824.

[5] M. Malinowski, M. Jasinski, and M. P. Kazmierkowski, "Simple direct power control of three-phase PWM rectifier using space-vector modulation (DPC-SVM)," *IEEE Trans. Ind. Electron.*, vol. 51, no. 2, pp. 447–454, Apr. 2004.

[6] P. R. M. Rodriguez, G. Escobar, A. A. V. Fernandez, M. H. Gomez, J. M. Sosa, " Direct power control of a three-phase rectifier based on positive sequence detection" *IEEE Trans. Ind. Electron.*, vol. 61, no. 8, Aug 2014.

[7] M. Cichowlas, M. Malinowsk, M. P. Kazmierkowski, D. L. Sobczuk, P. Rodriguez, J. Pou," Active filtering function of three-phase PWM boost rectifier under different line voltage conditions" IEEE Trans Ind Electron, vol. 52, no.2, 2005.

[8] M. Malinowski, M. Jasinski, M. P. Kazmierkowski," Simple direct power control of three-phase PWM rectifier using space–vector modulation (DPC-SVM)" IEEE Trans Ind Electron, vol. 51, no.2, 2004.

[9] P. Cortes, J. Rodriguez, P. Antoniewicz, M. Kazmierkowski," Direct power control of an AFE using predictive control" IEEE Trans Power Electron, vol. 23, no.5, 2008.

[10] A. Bouafia, F. Krim, J. P. Gaubert," Design and implementation of high performance direct power control of three-phase PWM rectifier, via fuzzy and PI controller for output voltage regulation" Energy Convers Manage, vol.50, no.1, 2009.

[11] A. Chaoui, F. Krim, J. P. Gaubert, L. Rambault," DPC controlled three-phase active filter for power quality improvement" Int J Electric Power Energy Syst, vol.30, no.8, 2008.

[12] M. Monfared, H. Rastegar, H. M. Kojabadi," High performance direct instantaneous power control of PWM rectifiers" Energy Convers Manage, vol.51, 2010.

[13] S. Buso, P. Mattavelli," Digital control in power electronics" USA: Morgan and Claypool, 2006.

[14] C.H. Silva, R. R. Pereira, L. E.B. Silva, G.L. Torres, B.K. Bose, S. U. Ahn, "A Digital PLL Scheme for Three-Phase System Using Modified Synchronous Reference Frame", *IEEE Trans on. Ind Elect*, vol. 57, no. 11, NOVEMBER 2010.

7th Power Electronics, Drive Systems & Technologies Conference (PEDSTC 2016)
16-18 Feb. 2016, Iran University of Science and Technology, Tehran, Iran

Virtual Flux Based Direct Power Control of a Three-Phase Rectifier Connected to an LCL Filter with Sensorless Active Damping

Mehran Maghamizadeh
M.Sc. Student
Department of Electrical Engineering
Amirkabir University of Technology
Tehran, Iran
mmaghamizadeh@aut.ac.ir

S. Hamid Fathi
Professor
Department of Electrical Engineering
Amirkabir University of Technology
Tehran, Iran
fathi@aut.ac.ir

Abstract— **This paper proposes a simple method for connecting an LCL filter to a three-phase pulse-width modulation (PWM) rectifier with direct power control space vector modulation (DPC-SVM) strategy. The goal is to improve the transient response and reduce the harmonic distortion, as well as size and cost reduction of the input filter. Since the proposed method is sensorless, the AC line voltage sensors are eliminated and the virtual flux block would be used instead. The cost and size of the filter could be reduced since the LCL-filter has lower inductance values and also the harmonic distortions could be reduced as well. The only drawback of the LCL-filter is producing resonance oscillations due to the resonance frequency of its third-order transfer function. In order to eliminate the oscillations, a simple method of active damping is utilized. Finally, simulation results have shown an excellent dynamic response compared with a three-phase converter that uses a conventional L-filter.**

Keywords—PWM Rectifier; Virtual Flux; Direct Power Control; DPC-SVM; LCL-filter; Active Damping

I. INTRODUCTION

There are two common control strategies for a three-phase PWM rectifier, voltage oriented control (VOC) and direct power control (DPC) [1]. In VOC, the current is controlled directly and guarantees a good dynamic response, but the response is largely dependent on the design of control parameters. In addition, the current components are not decoupled from each other and it makes the control loop more complicated. On the other hand, DPC has a very simple algorithm compared with VOC. In DPC, the instantaneous active and reactive powers are controlled directly and very simple, but the main drawback of DPC is the variable and high switching frequency, which is caused by the hysteresis controllers and the switching table.

Both mentioned strategies need to a large input inductance to have a low current ripple and THD. Therefore, an LCL filter can be used, which can reduce the cost and size of the filter, as well as the current ripple and harmonic distortion. The major problem with LCL filters is the produced oscillations caused by the resonance frequency. This problem can be solved by damping the resonance, but a fixed switching frequency is needed to do so, otherwise the control strategy

would be very complicated. Therefore, another control strategy called direct power control with space vector modulation (DPC-SVM), proposed by Malinowski et al. is used, which has the advantages of VOC and DPC together [2].

In order to overcome the drawback of the resonance oscillations of LCL-filter, the damping techniques can be applied. There are two main methods for this goal: passive damping (PD) and active damping (AD). In passive damping, one or more passive elements is added to the LCL filter. This element is resistive (a resistor in the simplest way) and causes power loss. However, PD has its own popularity because of its simplicity and reliability [3-5].

Therefore, the active damping would be a better choice since it can stabilize the system without any power loss and only by a change in the control system [6]. There are several methods presented for the active damping in the literature. Among them is virtual resistor method, proposed by Dahono [7], which assumes there is a resistance in the filter (similar to the resistance in passive damping method), however it needs extra current sensors to implement. Another method adds a lead-lag controller to the control system [8], but the performance of damping would depend on the design of the controller. Serpa et al. proposed a method in [9] to damp the inverter current oscillations with the conventional DPC strategy, but the method is complicated and needs to extra sensors as well. Liserre et al. proposed a sensorless method based on genetic algorithm [10], which makes the control system difficult to implement.

In this paper, a method proposed by Malinowski et al. [11] will be applied to DPC-SVM strategy and virtual flux concept is utlized to remove the voltage sensors. The main advantage of the proposed method in comparison with other active damping methods is that it is sensorless and very simple to implement.

II. CONTROL STRATEGY

A. Direct Power Control

The direct power control strategy was proposed by Noguchi [12]. It uses a switching table to determine the switching states and there is no PWM block and current

978-1-5090-0376-1/16 $31.00 © 2016 IEEE

control loop either. The instantaneous active and reactive powers are compared with a reference value. The reference active power is calculated from the DC-link voltage controller output and the reference reactive power is set to zero for unity power factor. The error passes a hysteresis controller, which limits the error in a specific range. The outputs of the hysteresis controllers determine the switching states using a switching table. It is very important to have a fast estimation for the active and reactive powers.

The active and reactive powers are calculated by scalar and vector product of the line current and line voltage, respectively. However, we can use virtual flux instead of the line voltage for the sensorless purpose. Therefore, the estimated active and reactive powers can be easily calculated from the following equations [13]:

$$p = \omega \cdot (\Psi_{L\alpha} i_{L\beta} - \Psi_{L\beta} i_{L\alpha}) \tag{1}$$

$$q = \omega \cdot (\Psi_{L\alpha} i_{L\alpha} + \Psi_{L\beta} i_{L\beta}) \tag{2}$$

Fig. 1 shows a DPC-SVM rectifier connected to an LCL-filter. Since the control signals are DC values, the errors can be controlled easily and the result will be transformed to stationary coordinate. A phase locked loop (PLL) could be used to find the angle of transformation block, since the disturbances and noises of the line voltage can affect the voltage angle. However, the virtual flux angle can be free of any distortion because of its inherent low-ass filter. One of the other advantages of DPC-SVM strategy is the utilization of SVM block instead of switching table, which leads to a fixed switching frequency and eases the active damping of the LCL-filter.

B. Sensorless Operation

Virtual flux concept in DPC was introduced by Malinowski with the goal of removing the AC line voltage sensors [14]. There are many advantages in using virtual flux instead of voltage sensors such as cost reduction, reliability and simplification. The virtual flux is defined as integration of the line voltage:

$$\underline{\Psi}_L = \begin{bmatrix} \Psi_{L\alpha} \\ \Psi_{L\beta} \end{bmatrix} = \begin{bmatrix} \int u_{L\alpha} dt \\ \int u_{L\beta} dt \end{bmatrix} \tag{3}$$

Where

$$\underline{u}_L = \begin{bmatrix} u_{L\alpha} \\ u_{L\beta} \end{bmatrix} = \sqrt{\frac{2}{3}} \begin{bmatrix} 1 & \frac{1}{2} \\ 0 & \frac{\sqrt{3}}{2} \end{bmatrix} \begin{bmatrix} u_{ab} \\ u_{bc} \end{bmatrix} \tag{4}$$

And we have

$$\underline{u}_L = \underline{u}_S + \underline{u}_I \tag{5}$$

And similarly

$$\underline{\psi}_L = \underline{\psi}_S + \underline{\psi}_I \tag{6}$$

The converter input voltage can be estimated using the measured DC-link voltage and the converter switching states:

$$u_{S\alpha} = \sqrt{\frac{2}{3}} U_{dc} (S_a - \frac{1}{2}(S_b + S_c)) \tag{7}$$

$$u_{S\beta} = \frac{1}{\sqrt{2}} U_{dc} (S_b - S_c) \tag{8}$$

Where U_{dc} is the measured DC-link voltage and S_a, S_b and S_c are the switching states. By substituting the (7) and (8) in (6), finally the virtual flux can be calculated as:

$$\Psi_{L\alpha} = \int (u_{S\alpha} + L \frac{di_{L\alpha}}{dt}) dt \tag{9}$$

$$\Psi_{L\beta} = \int (u_{S\beta} + L \frac{di_{L\beta}}{dt}) dt \tag{10}$$

In practice, the ideal integration in the virtual flux cannot be used since it produces a DC offset and causes undesired impact on the control system. One solution, proposed by Malinowski [14], is using a low-pass filter (Fig. 2-(a)) where the frequency of filter must be lower than the frequency of grid and it is usually set as one tenth of the frequency.

Although a simple low-pass filter can relieve the problem of DC offset, there is another problem of phase lag with it too. Since the filter frequency is one tenth of the grid frequency, the phase angle would be 84.2 degrees lagged and occurred in -20 dB of magnitude (Fig. 3-(a)), while it should be 90 degrees to guarantee the stability of the system and a phase lag causes a delay in the output.

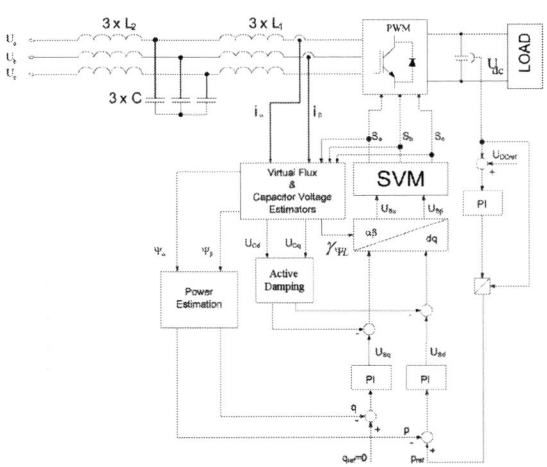

Figure 1. Block diagram of the proposed method

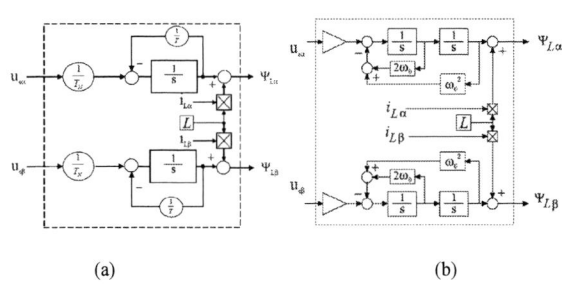

(a)   (b)

Figure 2. Low-pass filter that is used in the virtual flux; (a) conventional first-order LPF (b) proposed second-order LPF

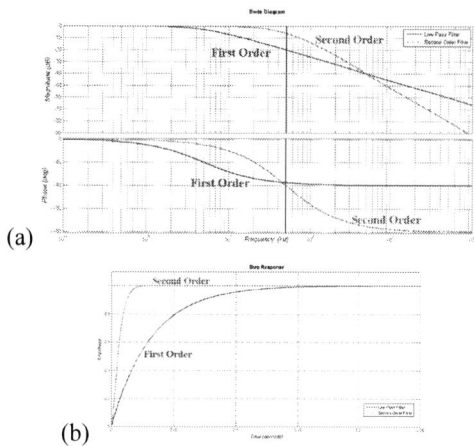

(a)

(b)

Figure 3. Comparison of the first-order and second-order LPF by (a) bode diagram (b) step response

Therefore, a second-order low-pass filter, proposed by Kulka [15], can be used instead. A second-order low-pass filter can bring a faster response with a phase shift of 90 degrees and -6 dB attenuation in magnitude and the cut-off frequency is the same grid frequency, which leads to a faster time response (Fig. 3-(b)). The transfer function of the proposed filter is:

$$H_1(s) = \frac{\omega_0^2}{(s+\omega_0)^2}$$

(11)

It should be noticed that the voltage of the (dq/αβ) block should be specified, which is done by a PLL in the conventional DPC strategy, since the voltage angle can be influenced by disturbances. But the virtual flux angle $\gamma_{\psi L} = tan^{-1}(\psi_{L\alpha}/\psi_{L\beta})$ is purely sinusoidal and can bring a good response without using any PLL. Additionally it is very easier to use virtual flux angle than designing a PLL precisely.

III. THE THREE-PHASE RECTIFIER WITH AN LCL-FILTER

A. LCL-Filter Design

Fig. 4 shows the model of a three-phase filter. In practice, the values of the resistances are very small and can be neglected in the calculations. A step-by-step design procedure of the LCL-filter has been explained at [6] by Liserre et al. In order to find the parameter values, firstly the converter-side inductance (L_1) must be found from the following equation:

$$L_1 = \frac{V_{dc}}{12 f_{sw} \Delta i_{max}}$$

(12)

Where V_{dc} is the DC-link voltage, f_{sw} is the switching frequency and Δi_{max} is the permitted ripple of the line current, which is usually chosen between 10 to 25 percent of the nominal current. The coefficient of 12 in the denominator has been chosen by the designer. Actually this coefficient is usually chosen between 3 and 12, but we selected the biggest possible value in order to have a smaller inductance. Also the

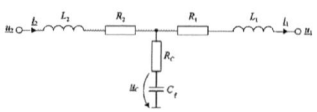

Figure 4. One-phase model of a three-phase LCL filter

permitted current ripple has been chosen 25 percent of the nominal current.

For the grid-side inductance, we assume $L_2=L_1$ and for the purpose of a large power factor, between 0.95 and 1, the capacitor of the filter must be chosen a value to have a reactive power not more than 5% of the rated power. Therefore, the maximum value for the capacitance is 5% of the base capacitance and can be written as:

$$C_f = \frac{0.05 \, P_n}{\omega V_{ph}^2}$$

(13)

Where P_n is the rated power of the system, ω is the angular frequency of the grid and V_{ph} is the phase voltage of the grid.

B. Active Damping

The transfer function of the LCL-filter can be written as:

$$G_{LCL}(s) = \frac{1}{C_f L_1 L_2 s^3 + (L_1+L_2)s}$$

(14)

Therefore, the resonance frequency is obtained as:

$$f_{res} = \frac{1}{2\pi} \sqrt{\frac{L_1+L_2}{L_1 L_2 C_f}}$$

(15)

The transfer function of a LCL-filter is third-order and has three poles, while the transfer function of a simple L-filter has only one pole. The bode diagram of LCL-filter together with L-filter can be seen in Fig. 5. The LCL-filter is equivalent to an L-filter with the inductance value of L_1+L_2 in low frequencies. Impact of the capacitor is observed at the higher frequencies, especially above the resonance frequency, where the magnitude of LCL-filter is lower than L-filter, which leads to attenuate the harmonics better than L-filter. The major drawback occurs in the resonance frequency, where the magnitude raises suddenly and brings oscillations to the system.

The active damping can compensate the sudden magnitude rise at resonance frequency by a negative peak at the same point (Fig. 5) and the result would be a filter that treats like an L-filter in the low frequencies and has a good harmonic attenuation capability in the high frequencies. In this paper, a high-pass filter (HPF) has been utilized for the active damping (Fig. 6). The method was proposed by Malinowski et al. for VOC strategy [11] and we apply that in DPC-SVM. In this method, the capacitor voltage passes an HPF to filter out the high frequency components of the signal and the output signal will be deducted from the modulator input.

978-1-5090-0376-1/16 $31.00 © 2016 IEEE

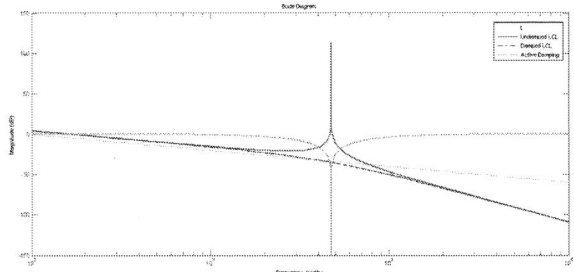

Figure 5. The bode diagram of the L-fitler, undamped LCL-filter, active damping, and damped LCL-filter

This results in eliminating the high frequency harmonics from the modulator signal and the oscillated signal becomes a pure signal without any harmonics. The frequency of the high-pass filter equals to resonance frequency of LCL-filter and its transfer function is:

$$\text{HPF(s)} = \frac{s^2}{s^2 + \omega_0^2} \tag{16}$$

C. Controller Design

Since the impedance of capacitor is very high in low frequencies, the capacitor current can be assumed as zero. Consequently, we can assume that the currents of the grid-side and converter-side inductances equal to each other and use the sum of inductances as the equivalent inductance ($L_{12} = L_1 + L_2$). The additional capacitor in the filter makes another a minor change in the sensorless operation as well, it can be calculated using the converter estimated voltage and the converter-side inductance voltage:

$$\underline{u}_C = \underline{u}_S + \underline{u}_{L1} \tag{17}$$

According to (17), as well as (9) and (10), the virtual flux can be easily calculated from the following equations:

$$\Psi_{Line\_\alpha} = \int \left(\sqrt{\frac{2}{3}} U_{dc} \left(S_a - \frac{1}{2}(S_b + S_c) \right) \right) dt + Li_{L\alpha} \tag{18}$$

$$\Psi_{Line\_\beta} = \int \left(\frac{1}{\sqrt{2}} U_{dc} (S_b - S_c) \right) dt + Li_{L\beta} \tag{19}$$

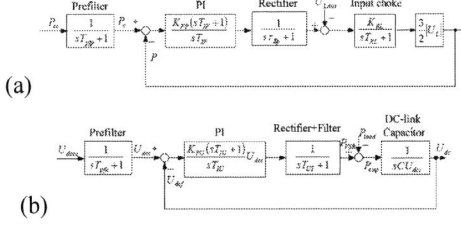

(a)

(b)

Figure 7. Control loop of the DPC-SVM with LCL-filter for (a) active power; (b) DC-link voltage

In order to design the controller parameters, symmetric optimum (SO) has been used [16]. Fig. 7-(a) shows the control loop of the active power, since there is no difference between the active and reactive power controlling, the control loop remains with no change for the reactive power as well.

After calculating the closed-loop transfer function of the control loop and with the help of symmetric optimum, the PI controller parameters can be obtained as:

$$K_{PP} = \frac{T_{RL}}{2\tau_{\Sigma p} K_{RL}} \frac{2}{3|U_L|} \tag{26}$$

$$T_{IP} = 4\tau_{\Sigma p} \tag{20}$$

Where $T_{\Sigma p} = T_s + T_{PWM}$, $T_{RL} = (L_1 + L_2)/R$ and $K_{RL} = 1/R$ are the converter time constant, the input filter time constant and the proportional coefficient of the equivalent L-filter of the LCL-filter respectively. T_s is the sampling time and T_{PWM} is the modulator delay, which usually equals to $0.5T_s$. The PI controller, although is able to damp the oscillations, has a minor overshoot in its step response because of the zero it has. Therefore, the prefilter block is added to the control loop to reduce this overshoot. The prefilter time constant is $T_{pfp} = 6T_s$.

Similarly, the parameters of the DC-link PI controller can be obtained from the DC-link voltage control loop (Fig. 7-(b)):

$$K_{PU} = \frac{C}{2T_{UT}} U_{dcc} \tag{21}$$

$$T_{IU} = 4T_{UT} \tag{22}$$

Where $T_{UT} = T_U + T_{IT}$ and U_{dcc} is the reference DC-link voltage. T_U is the time constant of the low-pass filter in the DC-link and equals to 0.003 (s). T_{IT} is the time constant of the inner active power control loop, $T_{IT} = 1.5T_s$, and the prefilter time constant is $T_{pfu} = T_{UT}$.

IV. SIMULATION RESULTS

The simulation of the proposed method was performed in Matlab/Simulink and Table I shows the parameters used in the simulation.

Fig. 8 shows the simulation results of DPC-SVM strategy with the conventional L-filter and in steady-state. The

TABLE I.   DATA USED IN THE SIMULATION MODEL

	Parameter	Symbol	Value
AC-side	Line voltage	U_L	380 v
	Grid frequency	f	50 Hz
L-filter	Line resistance	R	0.1 Ω
	Line inductance	L	10 mH
LCL-filter	Converter-side inductance	L_1	3 mH
	grid-side inductance	L_2	3 mH
	Line resistance	R_2	0.1 Ω
	Filter capacitor	C_f	30 μF
	Switching frequency	f_{sw}	5 KHz
DC-side	DC-link voltage	U_{DC}	750 v
	DC-link capacitor	C	1 mF
	Load resistance	R_L	100 Ω

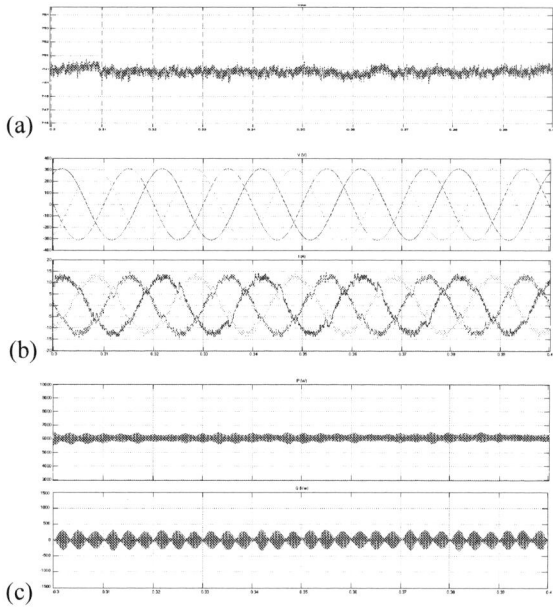

Figure 8. Steady-state waveforms of DPC-SVM strategy with conventional L-filter (a) DC-link voltage; (b) from top: line voltage, line current with THD=6.11; (c) from top: instantaneous active power, instantaneous reactive power

reference voltage has been set on 750v and the DC-link voltage ripple is lower than 0.15%, which shows the good performance of the DC-link voltage controller. The line voltage and line current have remained balanced with THD=6.11%. The active and reactive powers have also remained almost constant and tracked the reference power values. The reactive power has remained in zero, which guarantees the unity power factor.

Note that the second-order LPF in virtual flux leads to a precise tracking of the active and reactive powers, it means if the first-order LPF was used, there would be a small offset in the power values.

Fig. 9 shows transient operation of the DPC-SVM with conventional L-filter. For the transient state, a double-step of 200Ω resistance load has been added to the output load at t=0.1s and terminated at t=0.2s. Again, the active and reactive powers have tracked the reference values. There is a small overshoot in the active power at the entrance and termination point of the load, which is the result of active power PI controller. This overshoot would have a larger value if the prefilter block was not added to the active power control loop.

We can see a minor coupling between the active and reactive powers, which is not existed in the conventional DPC strategies with switching table. The reason is the high sampling frequency in switching table methods. The sampling frequency is ten times bigger than the switching frequency in a switching tables-based method, while the sampling and switching frequencies in a SVM-based method are equal to each other as well as a lower fixed value. In this case, the sampling and switching frequencies are 5 KHz. During the transient operation, the line voltage and line current have

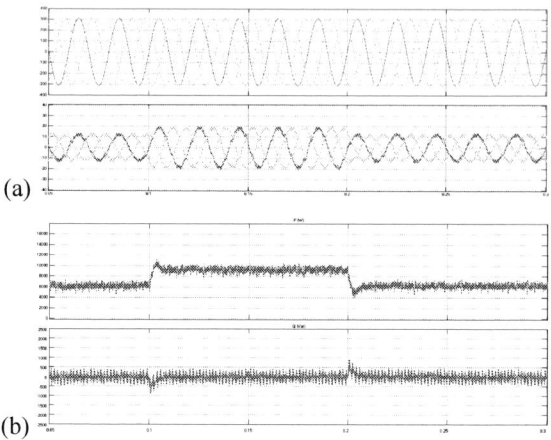

Figure 9. Transient operation of DPC-SVM strategy with conventional L-filter (a) from top: line voltage, line current; (c) from top: instantaneous active power, instantaneous reactive power

remained in-phase and without distortions, which shows the unity power factor. The unity power factor can be observed from the zero reactive power as well.

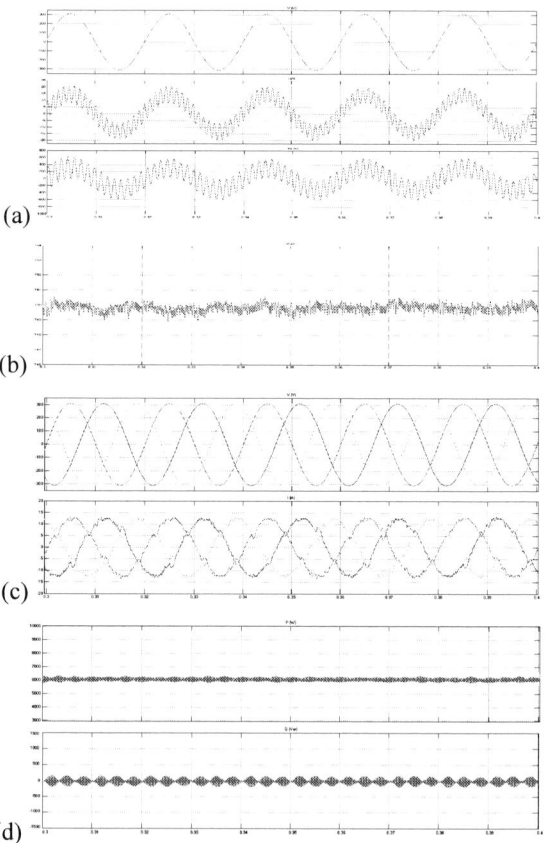

Figure 10. Steady-state waveforms of DPC-SVM strategy with LCL-filter (a) without damping, from top: line voltage, line current, capacitor voltage; (b) and with proposed active damping method, the DC-link voltage; (c) from top: line voltage, line current with THD=3.32; (d) from top: instantaneous active power, instantaneous reactive power

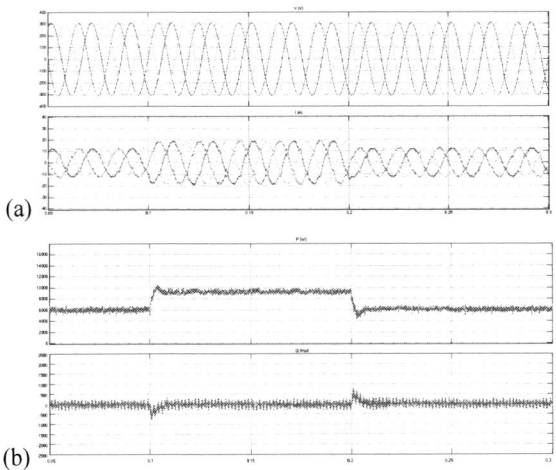

(a)

(b)

Figure 11. Transient operation of DPC-SVM strategy with LCL-filter and proposed active damping method (a) from top: line voltage, line current; (c) from top: instantaneous active power, instantaneous reactive power

Fig 10-(a) Shows the operation of the three-phase rectifier when the LCL-filter has been connected. Since there is no active damping used in the control system, the resonance oscillations can be seen clearly in the line current and capacitor voltage. The harmonic order of the oscillations equals to resonance frequency of the LCL-filter. As it can be seen, the oscillations made the line current increase to 20A, which could be harmful for the system, in comparison with 12A current of DPC-SVM with L-filter.

With the utilization of active damping, according to Fig. 11, the result would be elimination of the resonance oscillations, as well as the harmonics attenuation. In Fig. 11-(b), the DC-link voltage is similar to Fig. 8-(a), that is because of the identical DC-link voltage PI controllers of them. In Fig. 8-(c), the high-order harmonics has been attenuated by the LCL-filter in line current and made it smooth with the lower THD of 3.32%. Moreover, the active and reactive powers oscillate in a lower range in comparison with Fig. 8-(c). Finally the transient operation of DPC-SVM with LCL-filter can be seen at Fig. 11.

V. CONCLUSION

In this paper, a simple sensorless active damping method for the DPC-SVM of a three-phase rectifier connected to an LCL-filter was proposed. The method was based on subtracting the high-frequency harmonics from the modulator input signal. This was done by filtering the oscillated capacitor voltage by means of a high-pass filter with the cut-off frequency of the LCL-filter resonance frequency. It was shows that the proposed method has the capability of harmonic attenuation without any power loss. The line current THD with LCL-filter was reduced by a half with a sum of inductances noticeably smaller than in L-filter, which leads to size and cost reduction of the filter.

Moreover, an improved low-pass filter was replaced with the conventional filter of the virtual flux method, which made

the active and reactive powers track the reference values precisely. The control loops were designed with the help of symmetric optimum, which makes the controllers easy to tune. There are several advantages for the proposed method in comparison with the other methods, including being sensorless, easy to implement, low-cost, reliable, and with fixed switching frequency.

References

[1] M. Malinowski, P. Kazmierkowski and A. Trzynadlowski, "Review and comparative study of control techniques for three-phase PWM rectifiers," Mathematics and computers in simulation, 63(3), 2003, pp. 349-361.

[2] M. Malinowski, Jasiński and M. P. Kazmierkowski, "Simple direct power control of three-phase PWM rectifier using space-vector modulation (DPC-SVM)," Industrial Electronics, IEEE Transactions on, 51(2), 2004, 447-454..

[3] R. Pena-Alzola, M. Liserre, F. Blaabjerg, R. Sebastián, J. Dannehl and F. W. Fuchs, "Analysis of the passive damping losses in LCL-filter-based grid converters," Power Electronics, IEEE Transactions on, 28(6), 2013, pp. 2642-2646.

[4] W. Wu, Y. He, T. Tang and F. Blaabjerg, "A new design method for the passive damped LCL and LLCL filter-based single-phase grid-tied inverter," Industrial Electronics, IEEE Transactions on, 60(10), 2013, pp. 4339-4350.

[5] W. Zhao and G. Chen, "Comparison of active and passive damping methods for application in high power active power filter with LCL-filter," In Sustainable Power Generation and Supply, 2009. SUPERGEN'09. International Conference on, April 2009, pp. 1-6.

[6] M. Liserre, F. Blaabjerg and S. Hansen, "Design and control of an LCL-filter-based three-phase active rectifier," Industry Applications, IEEE Transactions on, 41(5), 2005, pp. 1281-1291.

[7] P. A. Dahono, "A control method for DC-DC converter that has an LCL output filter based on new virtual capacitor and resistor concepts," In Power Electronics Specialists Conference, 2004. PESC 04. 2004 IEEE 35th Annual Vol. 1, June 2004, pp. 36-42.

[8] V. Blasko and V. Kaura, "A novel control to actively damp resonance in input LC filter of a three-phase voltage source converter," Industry Applications, IEEE Transactions on, 33(2), 1997, pp. 542-550.

[9] L. A. Serpa, S. Ponnaluri, P. M. Barbosa and J. W. Kolar, "A modified direct power control strategy allowing the connection of three-phase inverters to the grid through LCL filters," Industry Applications, IEEE Transactions on, 43(5), 2007, pp. 1388-1400.

[10] M. Liserre, A. D. Aquila and F. Blaabjerg, "Genetic algorithm-based design of the active damping for an LCL-filter three-phase active rectifier," Power Electronics, IEEE Transactions on, 19(1), 2004, pp. 76-86.

[11] M. Malinowski and S. Bernet, "A simple voltage sensorless active damping scheme for three-phase PWM converters with an filter," Industrial Electronics, IEEE Transactions on, 55(4), 2008, pp. 1876-1880.

[12] T. Noguchi, H. Tomiki, S. Kondo and I. Takahashi, "Direct power control of PWM converter without power-source voltage sensors," Industry Applications, IEEE Transactions on, 34(3), 1998, pp. 473-479.

[13] M. Malinowski, M. P. Kazmierkowski, S. Hansen, F. Blaabjerg and G. D. Marques, "Virtual-flux-based direct power control of three-phase PWM rectifiers," Industry Applications, IEEE Transactions on, 37(4), 2001, pp. 1019-1027.

[14] M. Malinowski, "Sensorless control strategies for three-phase PWM rectifiers," Rozprawa doktorska, Politechnika Warszawska, Warszawa, 2001.

[15] A. Kulka, "Sensorless digital control of grid connected three phase converters for renewable sources," Ph.D. Thesis, Norwegian University of Science and Technology, Trondhein, 2009.

[16] J. W. Umland and M. Safiuddin, "Magnitude and symmetric optimum criterion for the design of linear control systems: what is it and how does it compare with the others?," Industry Applications, IEEE Transactions on, 26(3), 1990, pp. 489-497.

7th Power Electronics, Drive Systems & Technologies Conference (PEDSTC 2016)
16-18 Feb. 2016, Iran University of Science and Technology, Tehran, Iran

Dynamic Formation of Time-Dependent Duty Cycles for a Three-Phase Boost-Type DC-AC Converter Based on Averaging Model

B. Eskandari, J. Javidi Hagh, J. shojaee
Faculty of Engineering, Malayer University
Malayer, Iran
b.eskandary@gmail.com

M. Tavakoli Bina
Faculty of Electrical Engineering, K. N. Toosi University
Tehran, Iran
tavakoli@kntu.ac.ir

Abstract—**Power electronic converters are considered as an expensive part of energy harvesting systems while their reliability is low. Most of these converters consist of more than one stage such as DC/AC or DC/DC/AC converters that each one has its own power semiconductor switches. Increasing the number of switches is costly and leads to lower the reliability. Among different types of converters there is an especial topology with minimal structure and nonlinear behavior. Because of nonlinearity, the generation of three-phase SPWM for this converter is so complicated. In this paper a dynamic model for the boost-converter has been extracted from the basic switching functions; hence a precise modulation technique is obtained.**

Keywords—DC/DC; DC/AC; dynamic modeling; modulation;

I. INTRODUCTION

There are lots of different topologies to convert a DC voltage source to three-phase AC type by boosting level at the same time. The simple way is combination of conventional three level Voltage source converter along with an iron core transformer [1] (DC/AC/AC) while the large volume and weight of transformer is a negative point. Placing a DC/DC boost type regulator between source and three level converter [2-3] may lead to remove the transformer but it led to an increase in the number of switches. The current source converter [4-5] can also increase the amplitude of output AC voltage but the size of DC inductor and stability of DC current are the negative points of this converter. The Z-source converter [6] is consist of two conventional voltage source and current source converter (VSC and CSC) while the maximum gain of this converter is limited (less than 2) by restricted shoot-thru. There is a unified converter which produce boosted three-phase AC voltage simultaneously while its structure is minimal [7]. This converter consists of only six, two-quadrant, switches (Fig.1) and the rest of passive elements have little size. Despite the higher reliability and low cost, its behavior is nonlinear and so it is difficult to modulation and control [8-9].

The ultimate goal of this paper is to achieve an analytical modulation technique for boosted three-phase converter while efficiency is improved [10]. To reach the final aim, taking a series of steps is required. First of all dynamic equations of DC/DC boost converter is extracted and the authenticity of equations is evaluated through simulation. Solving the

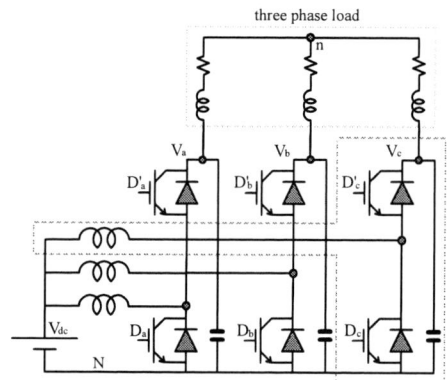

Fig. 1. Three-phase DC/AC boost type converter.

nonlinear equations to achieve the variable sinusoidal gain, as second step, leads to solving the modulation problem.

II. DYNAMIC MODELING OF BOOST CONVERTER

A. Deriving the Average Equations

By applying the averaging model, the dynamic response of DC/DC switching converters can be derived. As illustrated in Fig. 2(a), for a simple boost DC/DC converter, there are two possible states in continues conduction mode (CCM). During the first stage current of inductor is increase by the positive slope while the capacitor discharged. This process is reversed at second stage. Figure .2(b) shows the slope of variations in each stage. The average equation of inductor current and capacitor voltage, in each switching period, could be written as:

$$
\left.
\begin{aligned}
t_1 = DT_s : \quad & L\frac{dI_1}{dt_1} = E - rI_1 \\
t_2 = D'T_s : \quad & L\frac{dI_2}{dt_2} = E - rI_2 - V_2
\end{aligned}
\right\} \Rightarrow L\frac{dI}{dt} = E - rI - D'V
$$

$$
\left.
\begin{aligned}
t_1 = DT_s : \quad & C\frac{dV_1}{dt_1} = -\frac{V_1}{R} \\
t_2 = D'T_s : \quad & C\frac{dV_2}{dt_2} = I_2 - \frac{V_2}{R}
\end{aligned}
\right\} \Rightarrow C\frac{dV}{dt} = D'I - \frac{V}{R} \quad (1)
$$

$$ t_1 + t_2 = T_s $$

978-1-5090-0376-1/16 $31.00 © 2016 IEEE   482

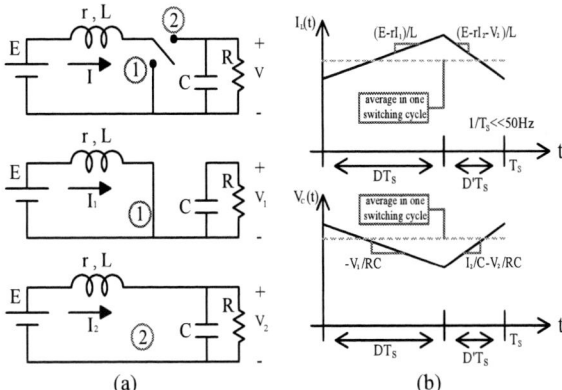

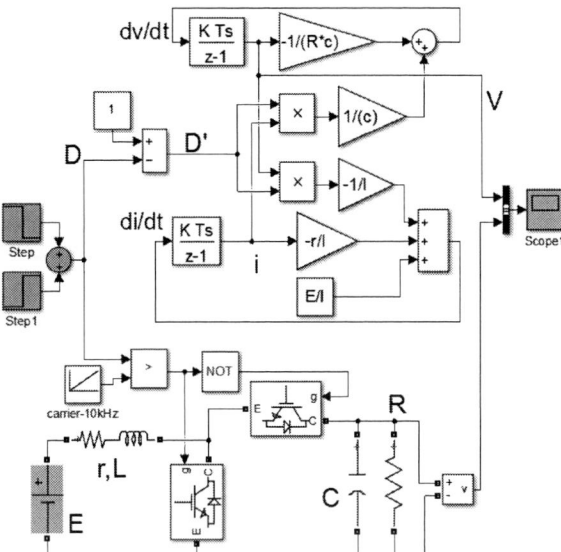

Fig. 2. (a) Two possible stage for boost converter in CCM, (b) current and voltage variations in each stage.

Fig. 3. Confirmation of equations by means of two independent simulations.

B. Confirmation of Derived Equations

The derived equation (1) is verified truth analysis of step response. As illustrated in Fig. 3, two simulations have been run independently, one as nonlinear differential equation and another as the actual circuit. All parameters are identical in both simulations and are as follows:

$$T_s = 100\mu s \quad R = 6\ \Omega \quad r = 0.1\ \Omega \quad E = 40\ V$$
$$L = 50\mu H \quad C = 600\mu F \quad \omega = 100 \cdot \pi \quad A = 50\ V$$

The parameters A and ω will be used in the following sections. The results of both simulations have been shown in Fig. 4. Both step responses are exactly the same. This confirms the validity of derived average equation.

III. SOLVING THE NONLINEAR STATE EQUATION

Equation (1) can be used to calculate the pulse width, $D\ (t)$, for a desired variable voltage, especially the sinusoidal wave form. Because of characteristic curve of the boost converter there are two possible answers for (1). This concept is expressed in Fig. 5. The first answer is located in the stable region and acceptable. The second answer comes from unstable region and is not acceptable.

A. Solving for Single Boost Converter

Because of characteristic of boost converter the minimum of output voltage of it must be greater than input DC source. There for the output desire voltage could be as follow:

$$V = A \cdot \sin(\omega t + \varphi) + A + E \tag{2}$$

and there after:

$$\frac{dV}{dt} = A \cdot \omega \cdot \cos(\omega t + \varphi) \tag{3}$$

By considering (2), (3) and other certain parameters a set of non-linear differential equation (4) is obtained.

$$\begin{cases} D' = (E - rI - L\dfrac{dI}{dt})/V \\ D'I = C\dfrac{dV}{dt} + \dfrac{V}{R} \end{cases} \tag{4}$$

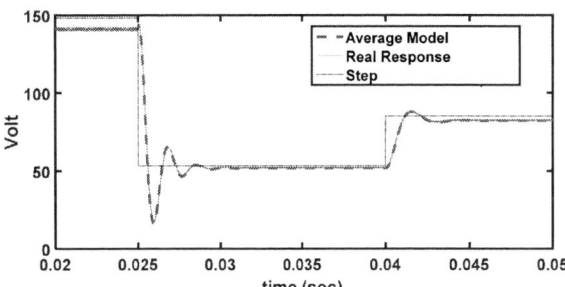

Fig. 4. Step response of two independent simulations.

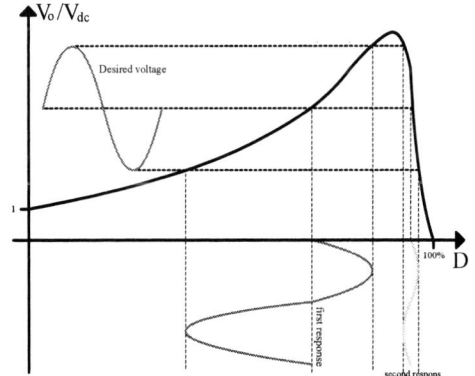

Fig. 5. Characteristic curve of the boost converter and two possible answers.

Solving the (4) leads to determining the patterns of $D\ (t)$ and $I\ (t)$, Fig. 6. Given the pattern of $D\ (t)$, sinusoidal voltage could be generated. It should be noted that the appropriate answer is selected. The difference between stable and unstable responses is their amplitude. The amplitude of unstable response is very narrow (i.e. $0.9 < D\ (t) < 1$). It's shown in Fig. 5.

978-1-5090-0376-1/16 $31.00 © 2016 IEEE   483

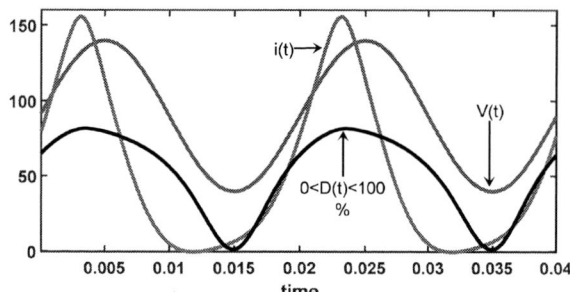

Fig. 6. The result of solving (4) for certain sinusoidal $V(t)$.

B. Solving for three-phase

It is clear that the dc voltage impregnated with sine distortion is not beneficial, at all. However presence of two other converter eliminate the DC offset in line voltages. One of three boost converters along with delta load is shown in Fig. 7. Assuming the desired phase voltages as follows:

$$\begin{cases} V_{aN} = A.\sin(\omega t) + A + E \\ V_{bN} = A.\sin(\omega t + 120^\circ) + A + E \\ V_{cN} = A.\sin(\omega t + 240^\circ) + A + E \end{cases} \quad (5)$$

There after the line voltages will be:

$$\begin{cases} V_{ab} = V_{aN} - V_{bN} = \sqrt{3}A.\sin(\omega t - 30^\circ) \\ V_{bc} = V_{bN} - V_{cN} = \sqrt{3}A.\sin(\omega t + 90^\circ) \\ V_{ca} = V_{cN} - V_{aN} = \sqrt{3}A.\sin(\omega t - 150^\circ) \end{cases} \quad (6)$$

Here, the equation (1) is true as well, except that in the load current the DC component has been removed. This content can be proved as follows:

$$\begin{cases} L\dot{I}_a = E - rI_a - D'_a V_a \\ C\dot{V}_a = D'_a I_a - \dfrac{V_a - V_b}{R} - \dfrac{V_a - V_c}{R} \end{cases}$$

$$V_{abc} = A \cdot \sin(100\pi t + \varphi_{abc}) + V_{offset} \quad (7)$$

$$\varphi_{abc} = 0^\circ, +120^\circ, -120^\circ \quad V_{offset} > A + E$$

Which the sum of V_{abc} is equal to three times of offset value.

$$V_a + V_b + V_c = 3V_{offset} \quad (8)$$

By substituting in (7):

$$C\dot{V}_a = D'_a I_a - \frac{V_a - V_{offset}}{R/3} \quad (9)$$

Elimination the DC component of load current is obvious. Now by solving (7) for certain V_{abc} the pattern of $D_{abc}(t)$ will be obtained, Fig. 9. The results of (7) for one of three converters is shown in Fig. 8. The inductor current has found a negative part while V_{abc} is like as before.

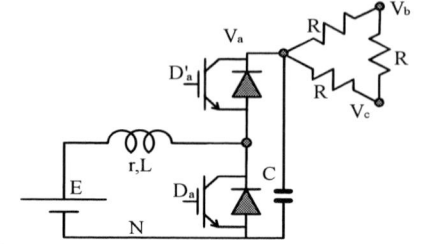

Fig. 7. One of three boost converters along with delta load.

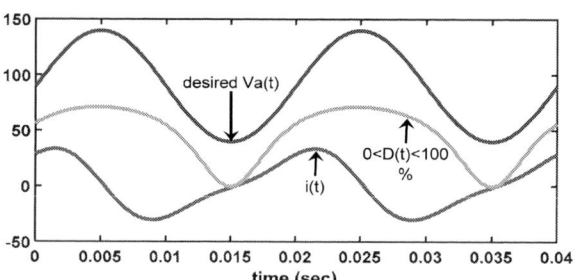

Fig. 8. The results of solving (7) for phase "a".

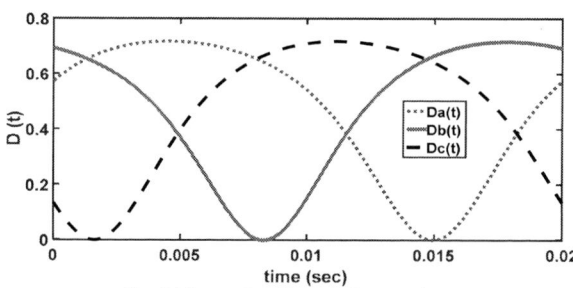

Fig. 9. The results of $D_{abc}(t)$ for tree-phase.

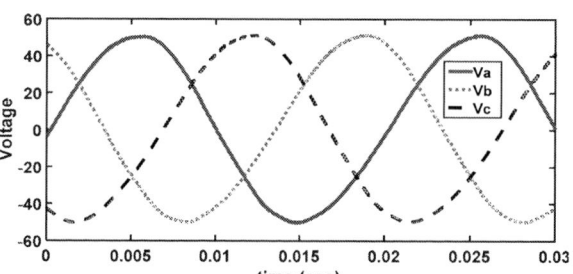

Fig. 10. The results of power simulation, three-phase voltages.

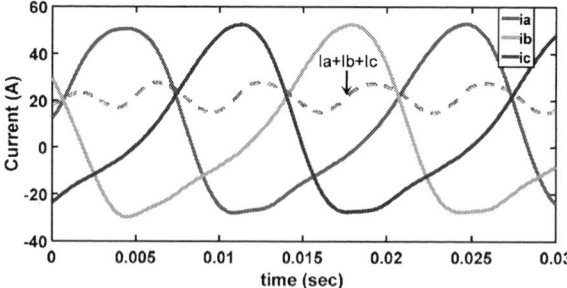

Fig. 11. The results of power simulation, inductor currents and their sum.

The results of power simulation have been illustrated in Figs. 10 and 11. Output voltages are very similar to the desired wave form and the measured currents are also similar to the calculated values. The interesting point is the sum of the three inductor currents, which is the current of DC source. The current of DC source has a DC component which is impregnated with third harmonic while the amplitude of inductor current is twice the amplitude of source current.

IV. CONCLUSION

Here it is introduced modeling of a non-linear three-phase boosted converter as a cost-effective and more reliable power electronic device. Contrary to Buck-type converter, the boosted three phase converter has a non-linear behavior and an analytical procedure of its modulation is presented. It is shown that the proposed modulation technique is consistent with dynamic behavior of the converter. The validity of all statements and equations have been evaluated by detailed simulations.

REFERENCES

[1] O. Ojo, "The generalized discontinuous PWM scheme for three-phase voltage source inverters," IEEE Trans on Ind. Electron, vol. 51, pp. 1280-1289, 2004.

[2] B. D. Reddy, N. K. Anish, M. P. Selvan, S. Moorthi, "Embedded Control of n-Level DC–DC–AC Inverter," IEEE Trans on Power. Electron, vol. 30, pp. 3703-3711, 2015.

[3] Pirouz, H.M., Bina, M.T., "Modular multilevel converter based STATCOM topology suitable for medium-voltage unbalanced systems", Journal of Power Electronics, Vol. 10, No. 5, September 2010, pp. 572-578.

[4] Ye. Yang, M. Kazerani,V. H. Quintana, "Current-source converter based STATCOM: modeling and control," IEEE Trans on power Del, vol. 20, pp. 795-800, 2005.

[5] Vahedi, H., Sheikholeslami, A., Tavakoli Bina, M., Vahedi, M., "Review and simulation of fixed and adaptive hysteresis current control considering switching losses and high-frequency harmonics", Advances in Power Electronics, May 2011

[6] C.J. Gajanayake, Lin Luo Fang, Beng Gooi Hoay, Lam So Ping, Kian Siow Lip, "Extended-Boost Z -Source Inverters," IEEE Trans on Power electron, vol. 25, pp. 2642-2652, 2010.

[7] Chan-Hee Choi, Kyung-Rae Cho, and Jul-Ki Seok, "Inverter Nonlinearity Compensation in the Presence of Current Measurement Errors and Switching Device Parameter Uncertainties," IEEE Trans on Power Electron, vol. 22, pp. 576-583 , 2007.

[8] B. Eskandari, M. Tavakoli Bina, "Support vector regression-based distortion compensator for three-phase DC–AC boost-inverters: analysis and experiments," IET Power Electronic, vol. 7, pp. 251-258, 2014.

[9] B. Eskandari, M. Tavakoli Bina, "New concept on sinusoidal modulation for three-phase DC/AC converters: analysis and experiments," IET Power Electronic, vol. 7, pp. 357-365, 2014.

[10] Firouz, Y., Bina, M.T., Eskandari, B., "Efficiency of three-level neutral-point clamped converters: Analysis and experimental validation of power losses, thermal modeling and lifetime prediction", IET Power Electronics, Volume 7, Issue 1, January 2014, pp. 209 – 219

7th Power Electronics, Drive Systems & Technologies Conference (PEDSTC 2016)
16-18 Feb. 2016, Iran University of Science and Technology, Tehran, Iran

A Modified SVM Switching Pattern for Z-Source Inverter

Mostafa Abarzadeh, Hossein Fathi kivi, Hossein Madadi Kojabadi
Renewable Energy and Advanced Power Electronic Research Center
Sahand University of Technology
Tabriz, Iran
m_abarzadeh@sut.ac.ir, h.fathi.1990@gmail.com, hmadadi64@yahoo.ca

Abstract— Because of significant capabilities of Space vector pulse width modulation (SVPWM) implementation in discrete systems, this method has been widely applied in Z-source inverters. ZSVM is one of the switching methods for Z-source inverters. Couple of disadvantages of this method can be mentioned as 1) Fluctuation of inductor current causes earsplitting noises phenomenon, 2) During shoot-through period, the switches of one leg turn on and this leads to extra losses and higher voltage drop in the switches. 3) Stochastic spot of shoot-through implementation during one switching period causes irregularity in DC bus voltage. In this paper, a new modified ZSVM (M-ZSVM) switching method for z-source inverter has been proposed. In this switching method by utilizing fixed shoot-through spot, irregularity of DC bus voltage is diminished and this leads to elimination of low frequency oscillation in inductor current. Moreover, because all switches are turned on during shoot-through period, switch losses and voltage drop decrease and causes equal distribution of heat between switches. The performance of the proposed switching strategy is analyzed in detail. Simulation and experimental results of proposed method verify the effectiveness and feasibilities of suggested strategy.

Keywords— *Z-source Inverter; Space vector pulse width modulation (SVM); inductor current ripple; shoot-through duty cycle*

I. INTRODUCTION

Z-source inverter overcomes disabilities and disadvantages of voltage source inverters (VSI) and current source inverters (CSI). This breed of inverters can operate in buck and boost modes. Also, there is not any menace about shoot-through and open circuit conditions in mentioned inverters [1].

Various switching methods have been provided for Z-source inverters. For instance, simple boosting control [1], maximum boosting control [2], constant boosting control [3] and space vector pulse width modulation (SVM) based control have been proposed [4-9]. Shoot-through period calculation is indispensable matter in Z-source inverter switching. This matter depends on choosing switching method. Also, essential issue in applying each modulation method is invariability of time length of active vectors. Attention to this matter is essential for preserving output wave harmonic effects and minimizing the size of the output filter [1]. The simple boosting strategy has been proposed on sinusoidal pulse width modulation (SPWM) method. In order to utilize this method on

Z-source inverter, in addition to three reference signals, two auxiliary signals are applied to create shoot-through states. Significant advantage of this method is easy implementation and having fixed boost coefficient for fixed amplitude of auxiliary reference signals. In order to increase the voltage, the value of modulation index (M) must be decrease as much as possible. This operation will increase the voltage stress on the switches. In order to decrease the voltage stress on switches, boosting control method has been proposed [1]. The main disadvantage of maximum boosting control is appearance of low frequency ripple in capacitor voltage and inductor current of impedance network that derives from output fundamental waveform. This fact causes earsplitting noises. In order to decrease low frequency ripples of capacitor voltage and inductor current of impedance network, larger capacitor and inductor must be used [2]. For prevailing disadvantages of two mentioned approaches, constant maximum boosting control has been proposed. The merits of this approach are elimination of low frequency oscillations of impedance network and significant decreasing of voltage stress. There is no ability of sequence switching control of switches in the above-mentioned approaches that causes increasing switching losses [3]. In space vector pulse width modulation (SVM) method, in order to minimizing switching losses, only one leg switches have been changed during transferring from one vector to next vector. This function leads to high current stress and asymmetrical heating of switches. The provided method for Z-source inverter switching by SVM has been involved in couple of problems which affect the performance of this approach. Z-source inverter switching has been occurred in zero section. Depends on different angles, this section has various intervals that will causes oscillation in inductor current ripple and irregularity in shoot-through applying. Also, in this method, during shoot-through period, only one leg is short circuited and all inverter current will pass through it that will lead in increased amount of losses and increased inverter voltage drop [4, 5, 7, 8].

In order to overcome SVM switching method disadvantages in Z-source inverter, in this paper, the new method has been proposed. In this method, fixed shoot-through location causes decreasing irregularity of DC bus waveform and elimination of inductor current low frequency ripple. These functions lead to decreasing earsplitting noises. Furthermore, since all switches are turned on simultaneously, so, switching losses are decreased, emanated heating derive from losses are

978-1-5090-0376-1/16 $31.00 © 2016 IEEE

symmetrical divided between all switches and also, voltage drops on the switches are reduced.

II. SVM SWITCHING FOR Z-SOURCE INVERTER

The space vector pulse width modulation (SVM) has significant capabilities of implementation in discrete systems. Therefore, this method has been widely utilized in Z-source inverters. As previously mentioned, for voltage source inverters (VSI), eight vectors have been defined in SVM switching method. SVM method concept has been depicted in Fig. 1. $d - q$ Frame has been divided to six 60-degree sectors. In each sector, switching function and time have been defined by two vectors adjacent to reference vector. For instance, in sector 1, by using (1), reference vector V_{ref} has been defined by V_1 and V_2 vectors.

$$V_{ref} = V_1 \frac{T_1}{T_S} + V_2 \frac{T_2}{T_S} \tag{1}$$

where T_1, T_2 and T_0 values are defined by (2), (3) and (4), i is sector number, θ is angle between V_{ref} and V_1 vectors, and M is modulation index.

$$T_1 = T_S M sin[\frac{\pi}{3} - \theta + \frac{\pi}{3}(i - 1)] \tag{2}$$

$$T_2 = T_S M sin[\theta - \frac{\pi}{3}(i - 1)] \tag{3}$$

$$T_0 = T_S - T_1 - T_2 \tag{4}$$

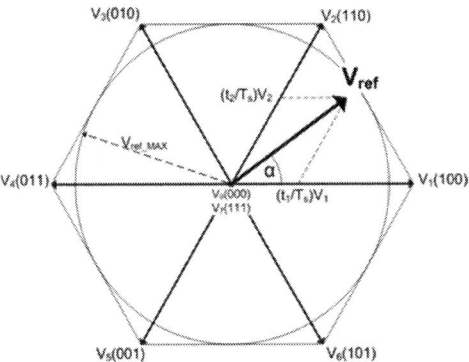

Fig. 1. Space vectors in SVM method

In conventional SVM method, on and off states of upper and lower switches of each leg is complement. Hence, for six switches, three switching states have been defined. For sector 1, switching is depicted in Fig. 2. Since, for voltage boosting in Z-source inverter, switching occurs in zero sector. So, maximum and mean values of this sector are important. Mean value which is defined by (5) is time interval of zero sector for each sector per switching time duration [8].

$$D_{MAX} = \frac{T_0}{T_S} = \frac{1}{\pi/3} \int_0^{\pi/3} [1 - Msin(\pi/3 + \theta)]\, d\theta \\ = 1 - M \tag{5}$$

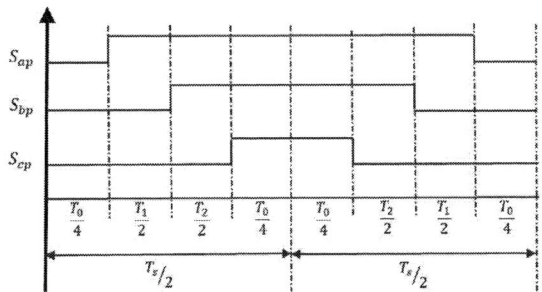

Fig. 2. Switching algorithm in SVM method

A. ZSVM switching method for Z-source inverter

As mentioned before, in SVM switching method, there are eight vectors that six vectors are related to active section and two vectors are related to zero section. Voltage boosting switching for Z-source inverter must not influence on active state, though this switching has been applied on zero state. In [3], four SVM switching approaches for Z-source inverter are mentioned as ZSVM1, ZSVM2, ZSVM4 and ZSVM6. The switching methods of ZSVM1, ZSVM2, ZSVM4 and ZSVM6 are depicted in Fig. 3, Fig. 4, Fig. 5 and Fig. 6, respectively [8].

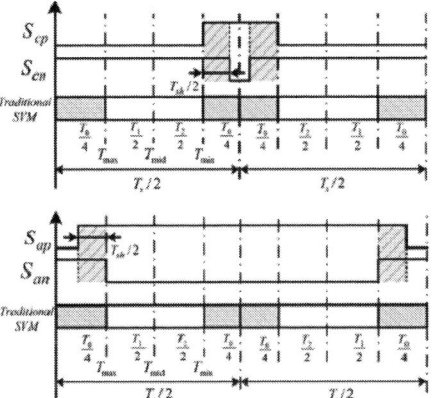

Fig. 3. ZSVM1 switching method [8]

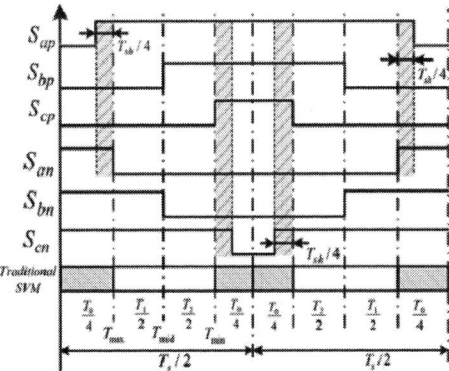

Fig. 4. ZSVM2 switching method [8]

978-1-5090-0376-1/16 $31.00 © 2016 IEEE   487

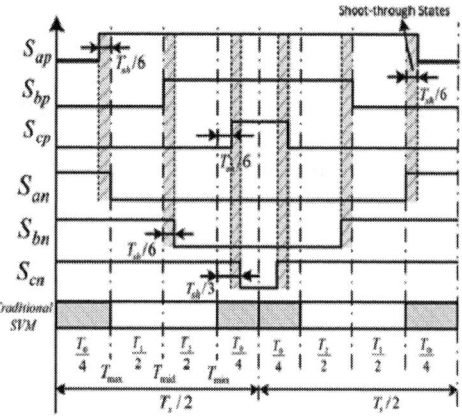

Fig. 5. ZSVM4 switching method [8]

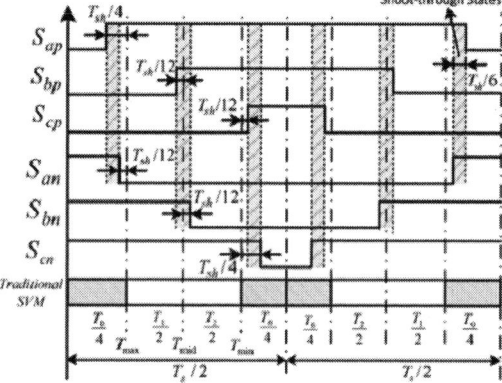

Fig. 6. ZSVM6 switching method [8]

Considering depicted figures, it is vivid that ZSVM2 and ZSVM6 methods have had maximum exploitation of zero section and it is probable to claim that boosting gain in these approaches are more than the other methods. In the aspect of current ripple, because shoot-through occurs during one switching period, ZSVM4 and ZSVM6 methods are better than the other three methods. Hence, in these methods, inductor size is smaller than the other three approaches. In order to minimize the switching losses in SVM switching method, during transition of one switching vector to another vector, only one leg switches status have been changed. This matter causes high current stress and asymmetrical switches heating distribution. In Z-source inverter, forbearing switching losses versus resistance losses though high current at shoot-through moment, has preference. Z-source inverter switching has been occurred in zero section and this section has various values depends on different angles. The provided approach for Z-source inverter switching with SVM method has couple of problems that can effect on its performances. Couple of disadvantages can be mentioned such as asynchronous firing of switches during shoot-through period, low frequency ripple in inductor current and disorderliness in shoot-through applying. In order to eliminate frequency oscillations on inductor current that causes earsplitting noises, the size of inductor must be selected bigger [8].

III. PROPOSED M-ZSVM2 METHOD

Disadvantages of ZSVM2 switching method have been emerged since shoot-through section is stood between zero section and active section. In order to prevail over this problem, the new M-ZSVM2 switching algorithm has been provided. Despite previous mentioned method, in this method zero and active section always have been stood sequentially and after this section, active section will be active. This issue causes more arrangement of shoot-through applying and elimination of low frequency oscillations in inductor current ripple. This method has been implemented by logical circuits. Since, all inverter switches are short circuited simultaneously, thus current, losses and voltage drop on each switch will be reduced. The proposed method has been shown in Fig. 7. The algorithm of proposed method has been depicted in Fig. 8 and Fig. 9. Based on proposed algorithm, the reference signals of S_{ap}, S_{bp} and S_{cp} are the same as references signals of conventional SVM method for VSI inverter. In fact, these signals are as same as traditional SVM method's output signals. In this proposed method, only two shoot-through signals are added. These signals are defined by (6) and (7). By using logical circuits which are depicted in Fig. 8, these signals have been converted to considered signals.

$$Ref_{sh1} = \frac{T_{sh}}{4} \qquad (6)$$

$$Ref_{sh2} = \frac{T_s}{2} - \frac{T_{sh}}{4} \qquad (7)$$

Where T_s is switching time period and T_{sh} is shoot-through time period.

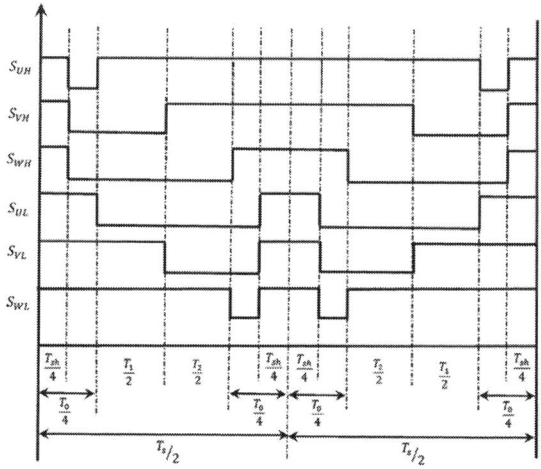

Fig. 7. Proposed switching method for Z-source inverter

IV. PROPOSED METHOD SWITCHING ANALYSIS

Z-source inverter has been shown in Fig. 10. Inductor voltage during shoot-through condition can be expressed as [10]:

$$V_{L1} = V_{C1} = V_{L2} = V_{C2} \qquad (8)$$

Inductor voltage during non shoot-through condition can be written as

978-1-5090-0376-1/16 $31.00 © 2016 IEEE   488

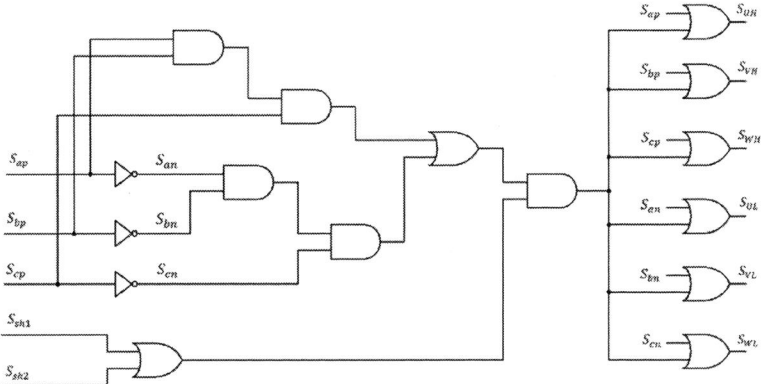

Fig. 8. Switching method implementation by logical circuits

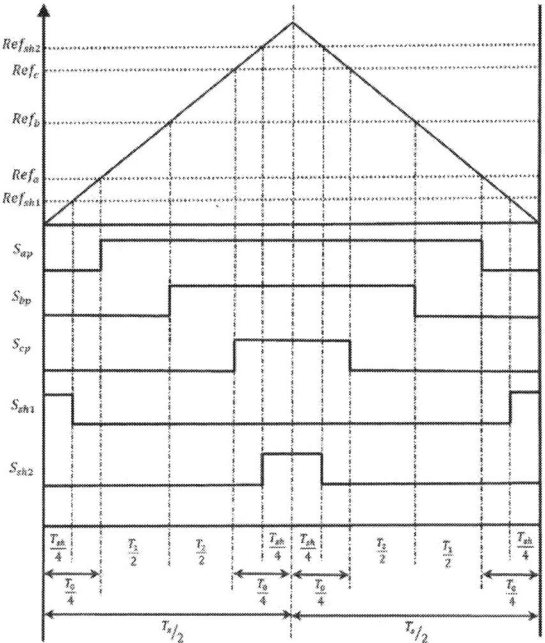

Fig. 9. Proposed switching method switching signals generation algorithm.

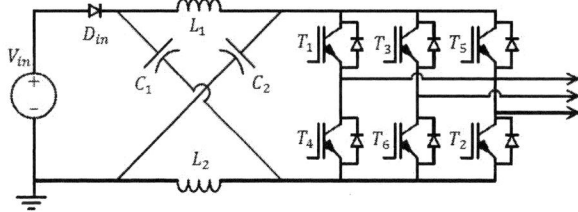

Fig. 10. Z-source inverter

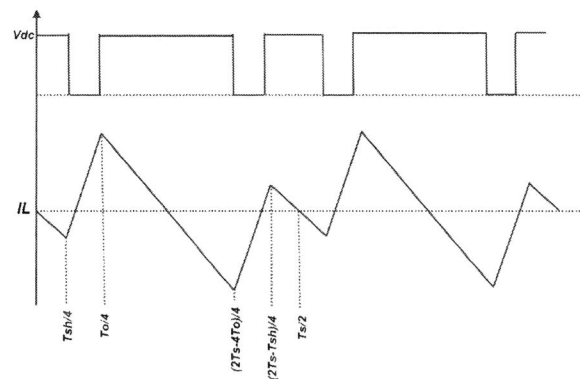

Fig. 11. Inductor current waveform during one switching period in ZSVM2 method

$$V_{L1-OFF} = V_{L2-OFF} = -\frac{D}{1-D}V_{C1} = -\frac{D}{1-D}V_{C2} \qquad (9)$$

With regards to correlation between inductor voltage and current, the Δi_L can be expressed as

$$\Delta i_L = \frac{\Delta t}{L}V_L \qquad (10)$$

Mean value of inductor current can be written as

$$i_{L-av} = \frac{P_{out}}{V_{IN}} \qquad (11)$$

In Fig. 11 and Fig. 12 inductor current waveforms in ZSVM2 and proposed M-ZSVM2 method are shown, respectively. Based on Fig. 11 and with regards to (8), (9) and (10), maximum and minimum values of inductor current for ZSVM2 method can be expressed as (12) and (13).

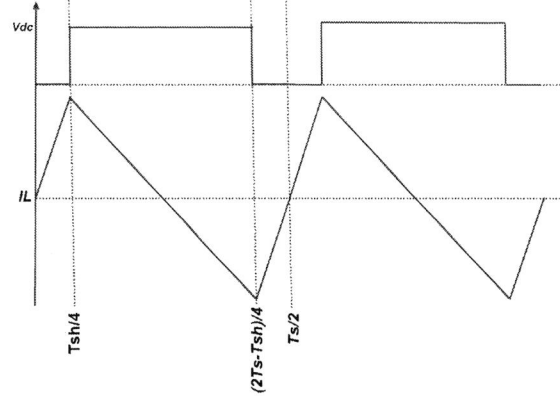

Fig. 12. Inductor current waveform during one switching period in proposed M-ZSVM2 method

$$i_{L-max} = i_{L-av} + \frac{V_c T_{sh}}{4L} + \left(\frac{V_{in} - V_c}{L} \left(\frac{T_s}{4} - \frac{T_{sh}}{4} - \left(\frac{T_s}{4} \left(1 - \frac{T_{sh}}{T_s} \right) \sin \left(\frac{\pi}{3} + \theta \right) \right) \right) \right) \quad (12)$$

$$i_{L-min} = i_{L-av} + \frac{(V_{in} - V_c)T_s}{2L} - \frac{(2V_{in} - 3V_c)T_{sh}}{4L} - \left(\frac{V_{in} - V_c}{L} \left(\frac{T_s}{4} - \frac{T_{sh}}{4} - \left(\frac{T_s}{4} \left(1 - \frac{T_{sh}}{T_s} \right) \sin \left(\frac{\pi}{3} + \theta \right) \right) \right) \right) \quad (13)$$

With regards to Fig. 12 and refer to (8), (9) and (10), maximum and minimum values of inductor current for proposed M-ZSVM2 can be expressed as

$$i_{L-max} = i_{L-av} + \frac{V_c T_{sh}}{4L} \quad (14)$$

$$i_{L-min} - zsvm2 = i_{L-av} + \frac{(V_{in} - V_c)T_s}{2L} - \frac{(2V_{in} - 3V_c)T_{sh}}{4L} \quad (15)$$

With regards to (12) and (13) for ZSVM2 switching method, it has been concluded that maximum and minimum values of inductor current depend on θ angle. Hence, in this method, inductor current is pendulous and its oscillation frequency is six times more than inverter's output frequency. Since inverter's output frequency has low value, hence, oscillation frequency of inductor current has been occurred in low frequencies. This issue leads to emanate earsplitting noises. Thus, for elimination of these noises, inductor size must be selected larger.

Considering (14) and (15) for proposed M-ZSVM2 switching method, it has been concluded that on the contrary to previous method, maximum and minimum value of inductor current independent of θ angle. Hence, it can be concluded that proposed method can eliminate inductor current oscillations. In Fig. 13 maximum and minimum values of inductor current for ZSVM2 and proposed M-ZSVM2 method during one period of output frequency are shown. Refer to Fig. 13, it is vivid that maximum and minimum values of inductor current are fixed during one output period.

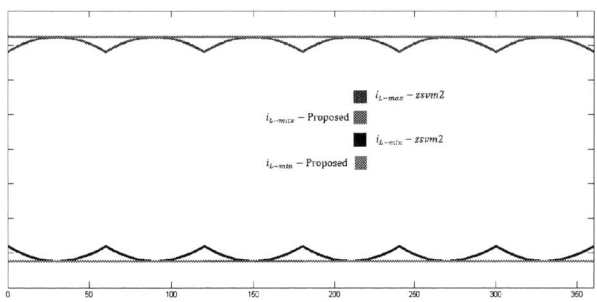

Fig. 13. Maximum and minimum values of inductor current for ZSVM2 and proposed M-ZSVM2 switching method versus variation of space vector angle.

With regards to ZSVM2 switching algorithm, in this method, during shoot-through section only one leg of inverter is turned on, thus, all amount of shoot-through current is passed through one leg that causes asymmetrical switches heating distribution. On the other hand, in the proposed M-ZSVM2 method, during shoot-through moment all switches of inverter are turned on, hence, shoot-through current divided between three legs that leads to reducing conducting losses of switches and causes symmetrical switches heating distribution. Also, because of all switches are turned on during shoot-through moment, voltage drop in system are decreased.

V. SIMULATION RESULTS

In order to compare ZSVM2 and proposed M-ZSVM2 methods, both of these methods have been implemented in Z-source inverter. Z-source inverter for $D = 0.2$ and $M = 0.8$ has been simulated under following conditions
1. All inductors are same with the value of 700 µH.
2. All capacitors are same with the value of 500 µF.
3. Switching frequency is 10 kHz.
4. Load is balanced, star and with the value of 60 Ω.
5. Power supply voltage is 170 V.
6. All elements are assumed to be ideal.

Simulation results of ZSVM2 switching method for Z-source inverter have been shown in Fig. 14. Refer to depicted results, inductor current ripple is very disordered and number of DC bus shoot-through occurrences is high and disordered. Simulation results of proposed M-ZSVM2 method have been depicted in Fig. 15. With regards to depicted results, in proposed method, it is vivid that inductor current oscillation has been eliminated and shoot-through occurrences are more ordered in comparison with ZSVM2 method.

VI. EXPERIMENTAL RESULTS

Traditional ZSVM2 and proposed M-ZSVM2 switching methods have been implemented based on parameters that have been mentioned in section V. Experimental results of ZSVM2 and proposed M-ZSVM2 switching methods for Z-source inverter have been shown in Fig. 16 and Fig. 17, respectively.

VII. CONCLUSION

In this paper, a new modified ZSVM (M-ZSVM) switching method for z-source inverter has been proposed. In this switching method by utilizing fixed shoot-through spot, irregularity of DC bus voltage is diminished and this will lead in elimination of low frequency oscillation in inductor current. Moreover, because of all switches are turned on during shoot-through period, switch losses and voltage drop decrease and causes equal distribution of heat between switches. The performance of the proposed switching strategy is analyzed in detail. Simulation and experimental results of proposed method verify the effectiveness and feasibilities of suggested strategy.

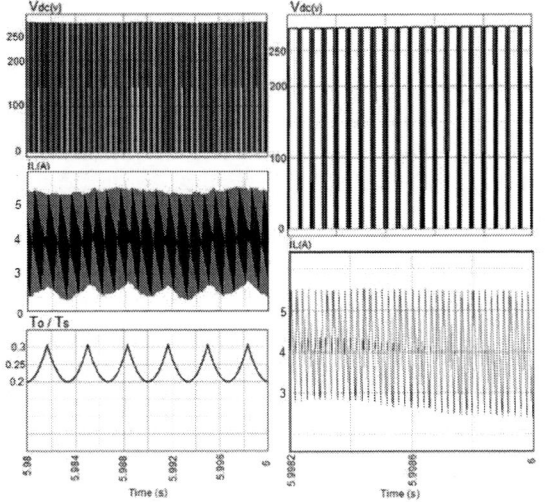

Fig. 14. Simulation results of ZSVM2 switching method for Z-source inverter

Fig. 15. Simulation results of proposed M-ZSVM2 switching method for Z-source inverter

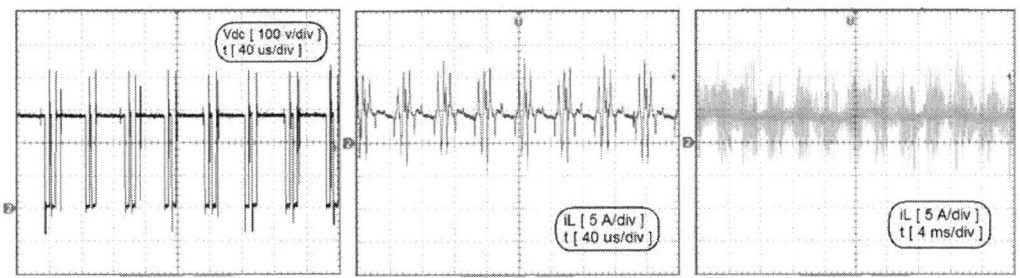

Fig. 16. Experimental results of ZSVM2 switching method for Z-source inverter for $D = 0.2$ and $M = 0.8$

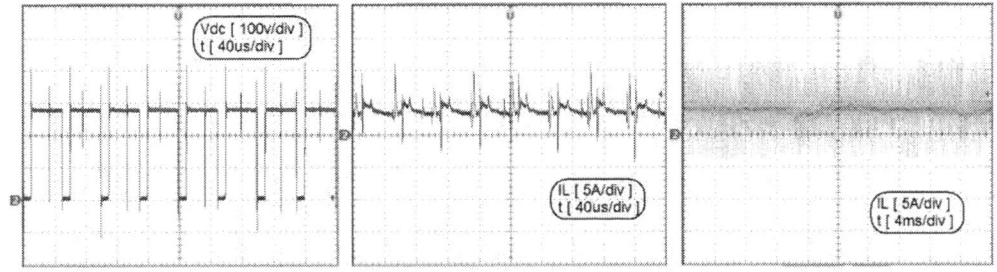

Fig. 17. Experimental results of proposed M-ZSVM2 switching method for Z-source inverter for $D = 0.2$ and $M = 0.8$

REFERENCES

[1] F. Z. Peng, "Z-source inverter," *Industry Applications, IEEE Transactions on,* vol. 39, pp. 504-510, 2003.

[2] F. Z. Peng, M. Shen, and Z. Qian, "Maximum boost control of the Z-source inverter," *IEEE Transactions on power electronics,* vol. 20, pp. 833-838, 2005.

[3] M. Shen, J. Wang, A. Joseph, F. Z. Peng, L. M. Tolbert, and D. J. Adams, "Constant boost control of the Z-source inverter to minimize current ripple and voltage stress," *Industry Applications, IEEE Transactions on,* vol. 42, pp. 770-778, 2006.

[4] U. S. Ali and V. Kamaraj, "A modified space vector PWM for bi-directional z-source inverter," in *Emerging Trends in Electrical and Computer Technology (ICETECT), 2011 International Conference on,* 2011, pp. 342-345.

[5] U. S. Ali and V. Kamaraj, "A novel space vector PWM for Z-source inverter," in *Electrical Energy Systems (ICEES), 2011 1st International Conference on,* 2011, pp. 82-85.

[6] Q. Lei, D. Cao, and F. Z. Peng, "Novel SVPWM switching pattern for high efficiency 15KW current-fed quasi-Z-source inverter in HEV motor drive application," in *Applied Power Electronics Conference and Exposition (APEC), 2012 Twenty-Seventh Annual IEEE,* 2012, pp. 2407-2420.

[7] J. Liu, J. Hu, and L. Xu, "A modified space vector PWM for Z-source inverter-modeling and design," in *Electrical Machines and Systems, 2005. ICEMS 2005. Proceedings of the Eighth International Conference on,* 2005, pp. 1242-1247.

[8] Y. Liu, B. Ge, and H. Abu-Rub, "Theoretical and experimental evaluation of four space-vector modulations applied to quasi-Z-source inverters," *IET Power Electronics,* vol. 6, pp. 1257-1269, 2013.

[9] P. C. Loh, D. M. Vilathgamuwa, C. J. Gajanayake, Y. R. Lim, and C. W. Teo, "Transient modeling and analysis of pulse-width modulated Z-source inverter," *Power Electronics, IEEE Transactions on,* vol. 22, pp. 498-507, 2007.

[10] H. Fathi Kivi, and H. Madadi Kojabadi, "Enhanced-boost Z-source inverters with switched Z-impedance, " *IEEE Transactions on Indus. Electronics,* Issue 99, 2015.

Reduced Size Single-Phase PHEV Charger with Output Second-Order Voltage Harmonic Elimination Capability

Hamid Rezaie, Hassan Rastegar and Mohammad Pichan
Department of Electrical Engineering
Amirkabir University of Technology
Tehran, Iran
h.rezaie@aut.ac.ir, rastegar@aut.ac.ir and m.pichan@aut.ac.ir

Abstract—**The single-phase PWM rectifier is one of the converter types applicable for the Plug-in Hybrid Electric Vehicles (PHEVs) chargers. A primary challenge with the single-phase rectifiers is the input alternative power with twice the grid frequency, which should be filtered out properly. The lack of proper filtering leads to significant second-order voltage harmonic at the charger output. This results overheating and decreasing of the battery lifetime and performance. In this paper, an active method to filter out the second-order voltage harmonic is proposed. In the proposed method, the input alternative power of the charger is absorbed by an inductive auxiliary energy storage element. Using the proposed method, the size, weight and implementation cost of the charger are decreased, as well as its power density and reliability are increased. The effectiveness of the proposed method has been verified by detailed simulation studies in MATLAB/Simulink.**

Keywords— Single-phase PHEV charger, pulsating power, second-order voltage harmonic, auxiliary energy storage.

I. Introduction

Using the internal combustion vehicles, leads to increase air pollution and global warming and also rapid evacuation of the earth's oil resources. So, interest in the use of electric vehicles is increasing, day by day. The battery can be considered as the most important part of the electric vehicles, since the battery restrictions is the most important factor that limits the use of the electric vehicles. The battery charging time and its lifetime greatly depends on the characteristics of the charger. Therefore, the battery charging system plays an important role in development of the electric vehicles [1]. Generally, from different aspects, the chargers can be categorized as follows [2]; 1) Single-phase or three-phase 2) On-board or off-board 3) Unidirectional or bidirectional.

Each of the mentioned categories has its specific advantages and disadvantages. For example, the on-board chargers enable the electric vehicle to recharge in several places. But, considering the volume and weight of the vehicle, the output power of the on-board chargers is limited. While the off-board chargers have no restriction on size, weight and output power and its output power is limited only by the limitations of the battery current. The unidirectional chargers only allow charging vehicles, whereas using bidirectional

chargers, the possibility of benefit from vehicle-to-grid capability for Plug-in Hybrid Electric Vehicles (PHEVs), is also provided [3,4].

Different types of battery chargers can be used in the electric vehicles [5]. In this paper, the single-phase full-bridge PHEV charger is studied and an active method is proposed to eliminate its output second-order voltage harmonic caused by the input alternative power. In the proposed method, the second-order harmonic is absorbed by an inductive auxiliary energy storage element. Using the proposed method, the function of the DC-link capacitor is only elimination of high-order harmonics such as harmonics caused by switching. Hence, the required capacitance for the charger output filter is reduced significantly. The decrement in the DC-link capacitance leads to reduce the size, weight and implementation cost of the charger and increase its power density. In addition, since the conventional high capacity capacitors for this purpose, such as Aluminum-Electrolytic capacitor, usually have a short lifetime, using the proposed allows using a small capacitor with long lifetime like Film-capacitor at the DC-link, and leads to increase the charger reliability [6, 7].

The rest of this paper is organized as follows:

The single-phase full-bridge converter is studied and the elimination method of the input power fluctuating part is described in section 2. In section 3, the charger control system is explained. The strength of the proposed method is confirmed by system simulations in MATLAB/Simulink, in section 4.

II. The Single-Phase Full-Bridge Charger and the Elimination Method of the Second-Order Voltage Harmonic

Fig. 1 shows the typical block diagram of a single-phase full-bridge AC/DC converter. Suppose that the AC source voltage is as follow:

$$v_S(t) = V_S \sin(\omega t) \qquad (1)$$

978-1-5090-0376-1/16 $31.00 © 2016 IEEE

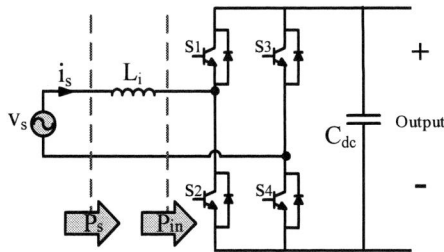

Fig. 1. The typical block diagram of a single-phase full-bridge AC/DC converter

Where V_S is the source voltage amplitude and ω is the angular frequency. Assuming unity power factor operation, the input current must be as (2).

$$i_S(t) = I_S \sin(\omega t) \qquad (2)$$

Where I_S is the source current amplitude. According to the source voltage and current, the source power is given in (3).

$$P_S = v_S(t)i_S(t) = \frac{V_S I_S}{2} - \frac{V_S I_S}{2}\cos(2\omega t) \qquad (3)$$

Multiplying the voltage and current of the L_i, the instantaneous power of the L_i can be expressed as bellow:

$$P_{L_i} = L_i \frac{di_S}{dt} \times i_S = \frac{1}{2}L_i I_S^2 \omega \sin(2\omega t) \qquad (4)$$

Which L_i is the input inductance. So, the input power of the converter can be obtained from (5).

$$P_{in} = P_S - P_{L_i} \qquad (5)$$

Substituting (3) and (4) in (5), the input instantaneous power of the converter can be calculated as follow:

$$P_{in} = \frac{V_S I_S}{2} - \frac{V_S I_S}{2}\cos(2\omega t) - \frac{1}{2}L_i I_S^2 \omega \sin(2\omega t) = P_{const.} + P_r \quad (6)$$

According to (6), the converter input power is composed of two parts; a constant part and a time-varying part which pulsates with twice the grid frequency. In the following, an active method to eliminate the time-varying part is proposed.

The proposed topology for the charger which is able to remove second-order voltage harmonic is shown in Fig. 2. Noted that in practice, considering the battery characteristics this topology may need a DC-DC converter at the output to regulate the output voltage and satisfy the battery requirements.

Pulsating power absorption by auxiliary inductor which is highlighted in Fig. 2, prevent injection of the second-order

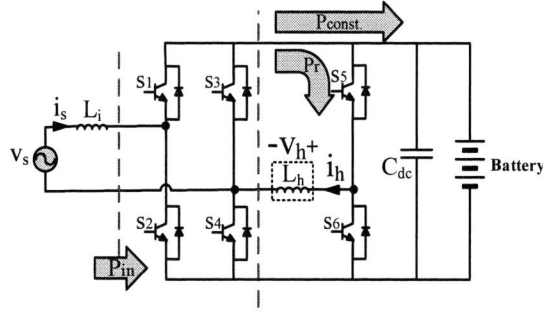

Fig. 2. The proposed topology for the charger with second-order harmonic elimination capability

voltage harmonic to the DC-link. Consequently, the required output filter capacitance is significantly reduced and prevents from overheating and decrement in the battery lifetime and its performance. In order that, the auxiliary inductance L_h be able to absorb the pulsating power effectively, its instantaneous power should be equal to the alternative part of the input power.

$$P_r = P_{L_h} \qquad (7)$$

Considering that the P_r pulsates at twice of the grid frequency and it should be equal to the instantaneous power of the auxiliary inductor. Thus, the reference current of the L_h should be sinusoidal at the grid frequency. Suppose that the current of the L_h is:

$$i_h(t) = I_h \sin(\omega t + \varphi) \qquad (8)$$

Where I_h and φ are the L_h current amplitude and the angle between the source voltage and current of L_h, respectively. So, the instantaneous power of the auxiliary inductor can be given by:

$$P_{L_h} = L_h \frac{di_h}{dt} \times i_h = \frac{1}{2}L_h I_h^2 \omega \sin(2\omega t + 2\varphi) \qquad (9)$$

Using the trigonometric equations, the P_r can be rewritten as follows:

$$P_r = -\frac{1}{2}\sqrt{\left(L_i I_S^2 \omega\right)^2 + \left(V_S I_S\right)^2}\,\sin(2\omega t + \gamma) \qquad (10)$$

Where

$$\gamma = \tan^{-1}\left(\frac{V_S}{L_i I_S \omega}\right) \qquad (11)$$

Substituting (9) and (10) in (7), will result in (12).

978-1-5090-0376-1/16 $31.00 © 2016 IEEE   493

$$-\frac{1}{2}\sqrt{\left(L_i I_S^2 \omega\right)^2 + \left(V_S I_S\right)^2}\sin(2\omega t + \gamma) \qquad (12)$$
$$= \frac{1}{2}L_h I_h^2 \omega \sin(2\omega t + 2\varphi)$$

Which can be rewritten as:

$$\sqrt{\left(L_i I_S^2 \omega\right)^2 + \left(V_S I_S\right)^2}\sin(2\omega t + \gamma) \qquad (13)$$
$$= L_h I_h^2 \omega \sin(2\omega t + 2\varphi + \pi)$$

According to (13), the amplitude and phase of the reference current in order to eliminate the output voltage second-order harmonic can be achieved from following equations:

$$I_h = \sqrt[4]{\frac{\left(V_S I_S\right)^2 + \left(L_i \omega I_S^2\right)^2}{\left(L_h \omega\right)^2}} \qquad (14)$$

$$\varphi = \frac{\gamma - \pi}{2} \qquad (15)$$

According to the described equations and by neglecting the switching losses, if the auxiliary inductor current controlled according to the obtained amplitude and phase in (14) and (15), L_h will be able to fully absorb the second-order harmonic and prevent from its transmission to the DC-link. It should be mentioned that, according to (14), to decrease the current stress of the third leg, a larger inductor must be chosen. However, an inductor with high inductance leads to high volume and implementation cost of the charger. Thus, a tradeoff in selection between the inductance value and the current stress must be made.

III. CHARGER CONTROL SYSTEM

The desired characteristics of a single-phase charger are: 1) Adjustable output voltage with minimum ripple, despite of source voltage probable fluctuations, 2) Input current with low distortion and harmonic contents, and 3) unity power factor operation [8]. In addition, in the proposed method, the input pulsating power should be filtered by proper switching method of the additional leg. One of the advantages of the proposed method is that the conventional charger control system can remain unchanged and the system is able to work properly, with or without the additional leg. The schematic control system of the proposed charger is presented in Fig. 3.

Part i, is the conventional control system of the single-phase PWM battery charger and part ii shows the control method of the third leg. The control system of the single-phase PWM charger which is shown in part i of Fig. 3, consists of two parts; an outer-loop voltage controller and an inner-loop current controller. The DC-link voltage is regulated by a PI compensator. Using a Phase-Locked Loop (PLL), an unit amplitude synchronous signal with the source voltage is generated.

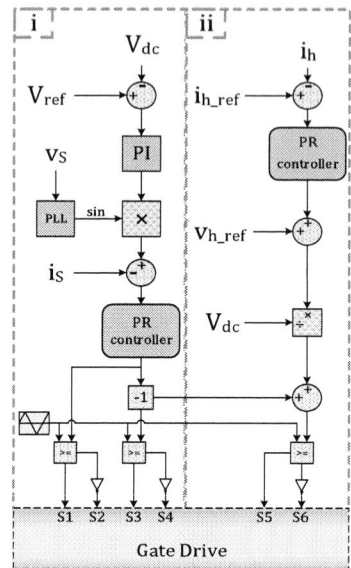

Fig. 3. The schematic control system of the proposed charger

The PLL is a closed-loop system that synchronizes its output signal in frequency and phase with its input signal [9]. Multiplying the outputs of the PI controller and PLL, the reference source current is achieved. Then, the source current is controlled according to its reference and the control signal for PWM generator is produced. As the reference current is an AC waveform, a Proportional-Resonant (PR) controller is able to track it correctly [10]. In this work, the unipolar switching method is used, to increase the effective switching frequency and reduce the output voltage ripple [11].

The control system for the third leg which conducts the input pulsating power into the auxiliary energy storage element is presented in part ii of Fig. 3. Since the reference current of the auxiliary inductor is a sinusoidal waveform, again a PR controller is used for its regulation. Sinusoidal reference waveform provides simpler control method compared with other methods which the reference current or voltage of the auxiliary energy storage element to eliminate the input pulsating power is a non-sinusoidal signal [12]. Considering the relatively high switching frequency, the converter averaged model is applicable and the control signal for the third leg switching can be obtained from (16) [13].

$$V_h = (d_c - d_b)V_{dc} \qquad (16)$$

That can be rewritten as:

$$d_c = \frac{V_h}{V_{dc}} + d_b \qquad (17)$$

Where d_b and d_c are the duty ratio of the S3 and S5, respectively. The control structure shown in part ii of Fig. 3 is based on (17) and v_{h_ref} is achieved by multiplying auxiliary inductance and time derivative of the its current.

IV. SIMULATION RESULTS

The system simulations has been performed using one of the standard values (120 V, 15 A) of the PHEV single-phase chargers [14]. The used parameters in simulation of the proposed charger are listed in Table I.

To evaluate the charger performance, the battery is modeled as a resistive load [15]. Fig. 4 shows the charger output voltage. The charger input voltage and current are shown in Figure 5. As it is clear in Fig. 4 and Fig. 5, the charger control system has suitable performance in the output voltage stabilization, providing input current with minimum disturbances and unity power factor operation. It should be noted that, using FFT analysis tool, the total harmonic distortion (THD) of the input current has been achieved 3.78%, which is an acceptable value.

To better show the effectiveness of the proposed method to remove the second-order voltage harmonic, the charger output voltage with and without the proposed method and with 200 μF DC-link capacitance is presented in Fig. 6.

TABLE I.   MAIN PARAMETERS

parameter	variable	value
Source voltage	v_S	120 Vrms
Source current	i_S	15 Arms
Frequency	f_S	50 Hz
Input inductance	L_i	1 mH
Auxiliary inductance	L_h	7 mH
DC-link capacitance	C_{dc}	200 μF
Charger nominal power	P_n	1.8 kW
Switching frequency	f_{sw}	10 kHz

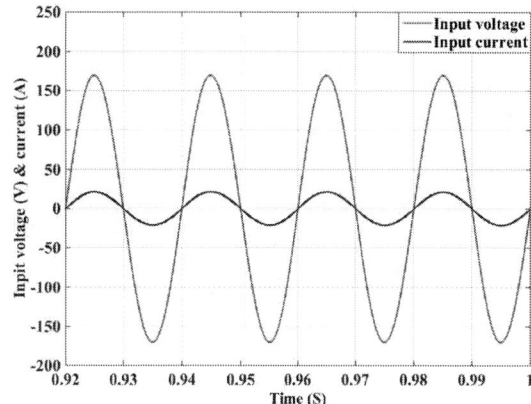

Fig. 5. The input voltage and current of the proposed charger

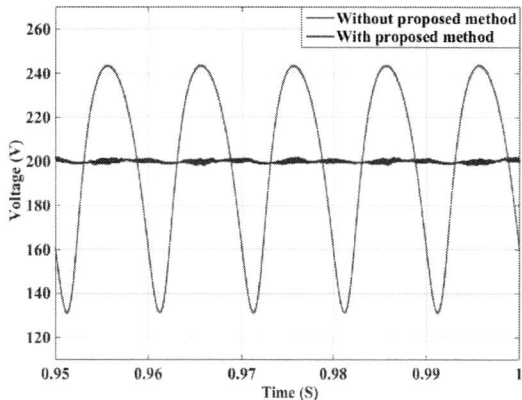

Fig. 6. The charger output voltage, with and without the proposed method, and with 200 μF DC-link capacitance

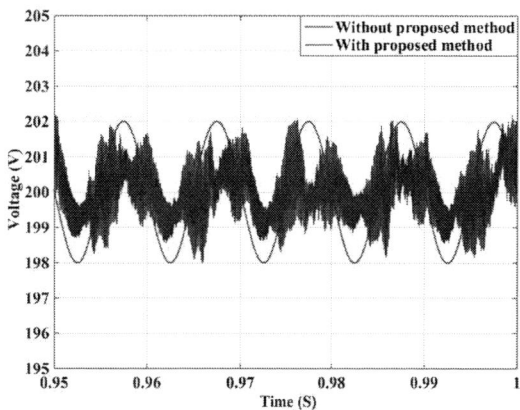

Fig. 7. The charger output voltage with proposed method and 200 μF DC-link capacitance, and without proposed method and 7200 μF DC-link capacitance.

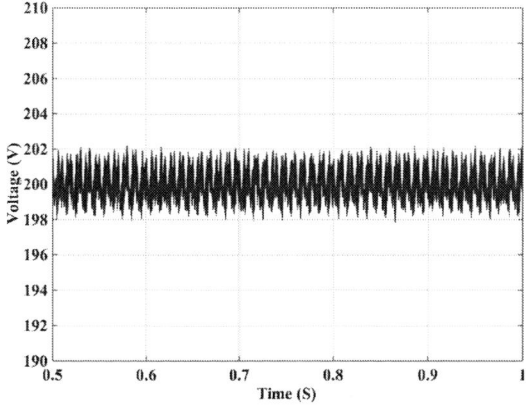

Fig. 4. The output voltage of the proposed charger

Also, the charger output voltage with proposed method and 200 μF DC-link capacitance, and without proposed method and 7200 μF DC-link capacitance are shown in Fig.7, simultaneously. As it is evident in Fig. 6, the peak-to-peak output voltage ripple, without the proposed method is 113 V, i.e. 56.5% of the reference voltage. While, its value using proposed method, has decreased to 4 V, i.e. 2% of the reference voltage. Fig. 7 indicates that, without the proposed method applying, in order to achieve to 4 V voltage ripple, the output capacitor must be 36 times larger than used capacitor with the proposed method applying. Noted that, due to the use of a large DC-link capacitor, without the proposed method applying, the charger power density is reduced, and its size and weight are increased, significantly. Moreover, since the conventional high capacity capacitors have relatively a short lifetime, the charger reliability is also reduced. On the other hand, using the small capacitor, without the proposed method applying, leads to high output voltage ripple and input current THD. The high output voltage ripple results in battery overheating, and decreasing of the battery lifetime and performance. The charger Output voltage ripple (peak-to-peak) and its 100-Hz component, with and without proposed method, are given in Table II. The input current harmonics expressed as a percentage of the fundamental component, in three different cases are listed in Table III; Case 1: with proposed method and 200 μF DC-link capacitance, Case 2: without proposed method and 7200 μF DC-link capacitance, and Case 3: without proposed method and 200 μF DC-link capacitance. According to Table III without applying proposed method to achieve acceptable value of THD a large DC-link capacitance is required. The effect of the proposed method on the input current harmonics reduction is clear in Table III. Noted that using unipolar switching method leads to increase the effective switching frequency and the dominant switching harmonics appear around $2m_f$ ($m_f = f_{sw}/f_s$).

Study of the charger dynamic and its capability to adjust the output voltage and current, has been done in Fig. 8 and Fig. 9. In Fig. 8, the charger output voltage and current, with changes in the reference output voltage is shown. At 1 S, the reference voltage increased suddenly from 200 V to 220 V, and at 2 S, decreased suddenly from 220 V to 180 V. Fig. 9 shows the charger output voltage and current, with changes in the output power (current), with fixed output voltage. At 1 S, the charger output power (current), increased suddenly from 1.8 kW (9 A) to 2.4 kW (12 A), and at 2 S, decreased suddenly from 2.4 kW (12 A) to 1.44 kW (7.2 A).

TABLE II.   THE CHARGER OUTPUT VOLTAGE RIPPLE (PEAK-TO-PEAK) AND ITS 100-HZ COMPONENT, WITH AND WITHOUT PROPOSED METHOD

	Output voltage ripple (peak-to-peak)	100-Hz component at the output voltage
With proposed method, with small DC-link capacitor	4	0.5297
Without proposed method, with large DC-link capacitor	4	2.003
With proposed method, with large DC-link capacitor	113	52.75

According to Fig. 8 and Fig. 9, the charger has suitable dynamic, and is able to adjust the output voltage and current, considering to the requirements, in a short time.

TABLE III.   THE PERCENTAGE OF THE INPUT CURRENT HARMONICS; CASE 1: WITH PROPOSED METHOD AND C_{DC}=200 μF CASE 2: WITHOUT PROPOSED METHOD AND C_{DC}=7200 μF CASE 3: WITHOUT PROPOSED METHOD AND C_{DC}=200 μF

	Case 1	Case 2	Case 3
3rd harmonic	0.57	1.27	26.85
5th harmonic	0.31	0.21	6.88
7th harmonic	0.18	0.13	1.99
9th harmonic	0.12	0.15	0.79
Max harmonics around m_f	0.22	0.02	0.00
Max harmonics around $2m_f$	2.15	2.16	2.65
THD	3.78	3.92	28.10

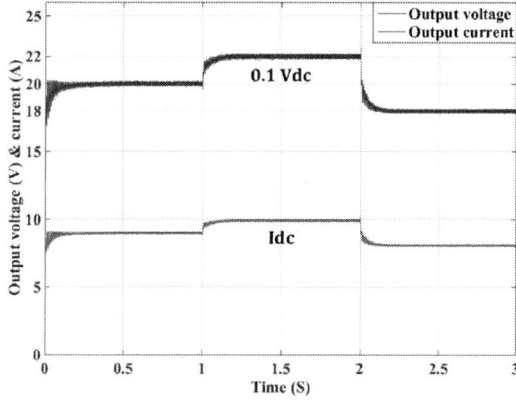

Fig. 8. The charger output voltage and current, with changes in the reference output voltage

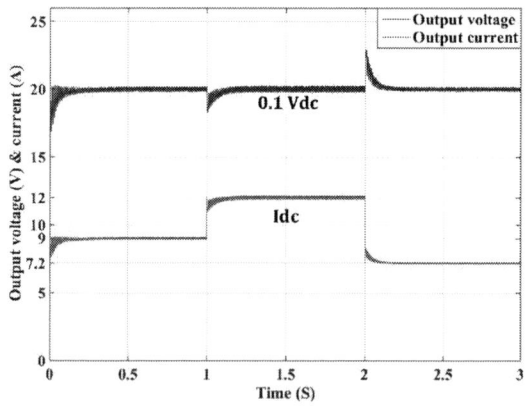

Fig. 9. The charger output voltage and current, with changes in the output power (current), with fixed output voltage.

V. CONCLUSION

In this paper, an active method to eliminate the output second-order voltage harmonic of the single-phase full-bridge PHEV chargers was proposed. In the proposed method, the second-order harmonic caused by the input alternative power, was absorbed by an inductive energy storage element. The second-order harmonic elimination, led to a significant reduction in the required capacitance of the charger output filter. Due to the use of a small capacitor at the charger output, the charger size, weight and cost was reduced, and its power density was increased. In addition, since the conventional high capacity capacitors, for this purpose, have relatively a short lifetime, the proposed method led to increase the charger reliability. The system simulations in MATLAB/Simulink verified the effectiveness of the proposed method.

References

[1] Gomez, J.C. and Morcos, M.M., "Impact of EV Battery Chargers on the Power Quality of Distribution Systems," IEEE Transactions on., vol.18, Issue: 3, pp. 975–981, July 2003.

[2] Grenier, M., Hosseini Aghdam, M.G. and Thiringer, T., "Design of On-Board Charger for Plug-in Hybrid Electric Vehicle," Power Electronics, Machines and Drives (PEMD 2010), 5th IET International Conference on., Brighton, UK., pp. 1-6, 19-21 April 2010.

[3] Kramer, Bill, Sudipta Chakraborty, and Benjamin Kroposki. "A review of plug-in vehicles and vehicle-to-grid capability." Industrial Electronics, 2008. IECON 2008. 34th Annual Conference of IEEE. IEEE, 2008.

[4] Salari, O., Nazemi, A., and Shamlou, S. "A new multiple input bidirectional HEV battery charger." In Smart Grid Conference (SGC), 2014, pp. 1-5. IEEE, 2014.

[5] Yilmaz, Murat, and Philip T. Krein. "Review of battery charger topologies, charging power levels, and infrastructure for plug-in electric and hybrid vehicles." Power Electronics, IEEE Transactions on 28.5 (2013): 2151-2169.

[6] R. Wang, F. Wang, D. Boroyevich, R. Burgos, R. Lai, P. Ning, and K. Rajashekara, "A High Power Density Single-Phase PWM Rectifier With Active Ripple Energy Storage", IEEE Trans. On Power Electron., vol. 26, no. 5, pp. 1430-1443, May 2011.

[7] H. Li, K. Zhang, H. Zhao, S. Fan, and J. Xiong, "Active Power Decoupling for High-Power Single-Phase PWM Rectifiers", IEEE Trans. On Power Electron., vol. 28, no. 3, pp. 1308-1319, Mar. 2013.

[8] Naouar, Mohamed Wissem, B. Ben Hania, Ilhem Slama-Belkhodja, Eric Monmasson, and Ahmad Ammar Naassani. "FPGA-based sliding mode direct control of single phase PWM boost rectifier." Mathematics and Computers in Simulation 91 (2013): 249-261.

[9] Shinnaka, Shinji. "A robust single-phase PLL system with stable and fast tracking." Industry Applications, IEEE Transactions on 44.2 (2008): 624-633.

[10] Herman, Leopold, Igor Papic, and Bostjan Blazic. "A proportional-resonant current controller for selective harmonic compensation in a hybrid active power filter." Power Delivery, IEEE Transactions on 29.5 (2014): 2055-2065.

[11] Mohan, Ned, and Tore M. Undeland. Power electronics: converters, applications, and design. John Wiley & Sons, 2007.

[12] M. Su, X. Long, Y. Sun, and J. Yang, "An active power decoupling method for single-phase AC/DC converters," IEEE Trans. Industrial Informatics, Volume: 10, Issue: 1, May. 2013.

[13] Yazdani, Amirnaser, and Reza Iravani. Voltage-sourced converters in power systems: modeling, control, and applications. John Wiley & Sons, 2010.

[14] Yilmaz, Murat, and Philip T. Krein. "Review of battery charger topologies, charging power levels, and infrastructure for plug-in electric and hybrid vehicles." Power Electronics, IEEE Transactions on 28.5 (2013): 2151-2169.

[15] Ketsingsoi, Supasit, and Yuttana Kumsuwan. "An Off-line Battery Charger based on Buck-boost Power Factor Correction Converter for Plug-in Electric Vehicles." Energy Procedia 56 (2014): 659-666.

7th Power Electronics, Drive Systems & Technologies Conference (PEDSTC 2016)
16-18 Feb. 2016, Iran University of Science and Technology, Tehran, Iran

A Novel Application of H_∞ Robust Controller for a Single Phase Inverter in Uninterruptible Power Supply

Ahmad Irani[1], Mahdi Sojoodi[2], Mustafa Mohamadian[3]
Department of Electrical and Computer Engineering
Tarbiat Modares University
Tehran, Iran
[1]a.irani@modares.ac.ir, [2]sojoodi@modares.ac.ir, [3]mohammadian@modares.ac.ir

Abstract— **This paper focuses on a robust state feedback controller design strategy for a single phase inverter with an output lowpass filter. State space representation of the single phase DC/AC converter was presented by assuming the load as an exsanguinous disturbance input modeled as a current source in parallel with an uncertain linear load. According to norm-bounded linear differential inclusion (NLDI) notation, robust stability, and linear matrix inequalities (LMI) techniques, sufficient conditions are gathered to design the robust state feedback. Furthermore, based on internal model principle, the steady-state error is removed by adding a resonant controller. The Simulation results for linear and nonlinear load are presented.**

Keywords— DC/AC power convertors; power filters; robust state feedback; convex optimization.

I. INTRODUCTION

Uninterruptible power supplies (UPS) used to deliver reliable, and high quality power for sensitive loads [1], [2] in many applications despite Power failures and distortions. UPS systems provide backup power, trusty and high-quality power for these sensitive loads. based on industrial standards, a UPS system objective is to deliver a little tracking error, an excellent-regulated sinusoidal output voltage with low total harmonic distortion (THD), and rapid sinusoidal waveform recovering time[3]–[5].

several attempts, like repetitive control[6], [7], sliding mode control[8]–[10], multi-loop feedback[11], robust and dissipative based state feedback control [12]–[16] have been made to single phase inverter in UPS to attain the aforementioned specification. In [17] an extended Lyapunov- function- based control strategy is designed and the global asymptotically stability of system is investigated. The performance of system is good. In [18] an observer-based optimal voltage control is used and load current is estimated. Although using an observer the proposed algorithm does not eliminate hardware complexity for current sensor, the specification of inverter is good.

According to LMI's theory, H_∞ robust approach was used in [13] to design PID controller which has a sensitive response in nonlinear load condition. By using Park

transformation, according to LMI, H_∞ robust state feedback control is designed for three-phase system in[14]. Uncertainties are assumed polytopic in [14] and nonlinear load has high THD. Furthermore, [15] recommend a systematic approach for LMI formulation of multiple resonant controllers design. In that approch each sub harmonic has a separate resonant controller, and represents good regulation and tracking. The uncertainties are considered polytopic and the controller which designed, is complicated and has zero which stables it and suffers from its complexity in implementation.

In this paper, uncertain load is composed of a bounded linear resistor with current source in parallel. In contrast to [13]–[15], the inverter is modeled as norm-bounded linear differential inclusion (NLDI). In order to elimination of tracking error and rejection of disturbances, we combined the state space representation of plant with a resonant controller, and then we presented a theorem to design H_∞ state feedback scheme based on D-stability theorem to improve performance. An algorithm is introduced to obtain and implement the controller. In order to solve this eigenvalue optimization problem, several software packages were used to solve obtained LMI constraints such as LMI Lab, YALMIP, and CVX toolboxes of MATLAB. The computer simulation results, simulated by Simpower toolbox of Simulink, demonstrate the capability of designed control strategy to powerfully assure stringent specification of standards.

This paper begins by the preliminary background and the problem statement presented in Section 2. It then goes on to, in section 3, developing the norm-bounded linear

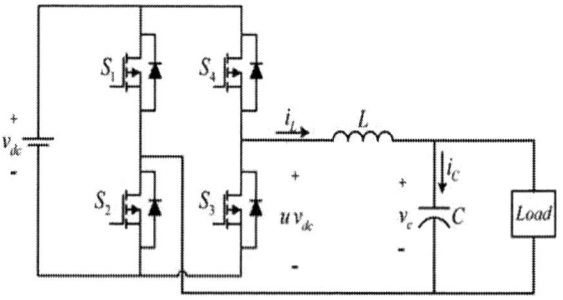

Fig. 1. Single phase inverter with low pass filter.

978-1-5090-0376-1/16 $31.00 © 2016 IEEE

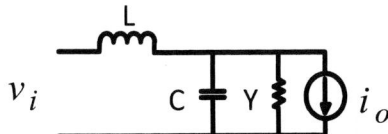

Fig. 2. Equivalent circuit for single phase inverter

differential inclusion model of DC to AC converter assuming that load is uncertain and then using a resonant controllers which eliminate the steady state error, and finally getting the LMI formulation of the system to design the robust state feedback. Section 4 is devoted to a numerical example to investigate the theoretical results. Finally, the paper is concluded in section 5.

Notation: throughout this paper the term R^n will refer to the n-dimensional Euclidean space, $R^{n \times m}$ denotes the real matrices set, the Euclidean vector norm is represented as $\|.\|$, I_n denotes the $n \times n$ identity matrix, and $0_{n \times m}$ is the $n \times m$ zero matrix. The term $.^T$ used to the matrices transpose. In the real matrices $Q > 0$ ($Q < 0$) used to refer that Q is positive definite (negative definite) and symmetric and $\otimes$ used to represent the Kronecker product of matrices.

II. Problem Statement and Preliminaries

Three major parts of online single phase UPS system are: power factor correction, battery bank as DC link, and static inverter including output lowpass LC filter. The performance of system is mostly affected with the last two issues and fluctuation of DC link which can be modeled as exogenous disturbance. These parts are illustrated in Fig.1.

In this system, the output voltage selected as a variable should be controlled. Uncertainties are imposed to system by uncertain load, might be nonlinear in most cases, and fluctuations of DC link. Current source in parallel with linear resistance is used as model of nonlinear load, which increasing the accuracy.

By using the average model of single phase DC to AC converter, full bridge inverter can be modeled as constant gain which is multiplied by control effort and can be considered one. So the dynamic of the converter is indicated by lowpass LC filter and uncertain load which represented in Fig.2.

According to Kirchhoff rules, we can derive system dynamic as:

$$\dot{x}(t) = A(\Delta)x + B_u u + E i_d$$
$$y = Cx \tag{1}$$

where $x = [x_1, x_2]^T = [i_L \quad v_C]^T$ is state vector, u is control input, i_d is exogenous input, and i_L is the current of inductor and v_C is the voltage of capacitor. $\Delta = Y_0$ is the admittance of uncertain load, satisfying the following condition:

$$0 \leq Y_0 \leq 1 \tag{2}$$

and:

$$A(\Delta) = \begin{bmatrix} 0 & \dfrac{-1}{L} \\ \dfrac{1}{C} & \dfrac{-\Delta}{C} \end{bmatrix} \quad B = \begin{bmatrix} \dfrac{1}{L} \\ 0 \end{bmatrix}$$

$$E = \begin{bmatrix} 0 \\ \dfrac{-1}{C} \end{bmatrix} \quad C = [0 \quad 1] \tag{3}$$

where C denotes capacity of capacitor and L represents the inductance of inductor.

Lemme1: [19] (Schure Complement): Consider the following symmetric matrix:

$$M = \begin{bmatrix} A & B & C \\ B^T & D & 0 \\ C^T & 0 & E \end{bmatrix} \tag{4}$$

By applying Schure complement to diagonal blocks including D and E we have:

$$M > 0 \leftrightarrow A - BD^{-1}B^T - CE^{-1}C^T > 0 \text{ and } D > 0 \text{ and } E > 0 \tag{5}$$

Lemma 2: S procedure consider symmetric matrices $T_0, \dots, T_P \in \Re^{n \times n}$. We assume $T_0, \dots, T_P$ satisfy the following condition:

$$\forall \xi \neq 0 \; \xi^T T_0 \xi > 0, \xi^T T_i \xi \geq 0, i = 1, \dots, p \tag{6}$$

Now, If there exist $\tau_1 \geq 0, \dots, \tau_p \geq 0$ such that:

$$T_0 - \sum_{i=1}^{p} \tau_i T_i > 0 \tag{7}$$

Then (6) holds[20]–[22].

III. Robust Controller Design

In the current part, the design of H_∞ robust state feedback controller is discussed for a single-phase full bridge inverter including output lowpass LC filter augmented with a resonant controller according to NLDI notation, D-stability, and LMI techniques.

According to the internal mode principle, steady state error is eliminated by adding a multiple resonant controller in feed forward path. Model of multiple resonant controller is represented by:

$$\dot{x}_c(t) = \begin{bmatrix} 0 & 1 \\ -\omega^2 & 0 \end{bmatrix} x_c + \begin{bmatrix} 0 \\ \omega \end{bmatrix} e \tag{8}$$

where **e** is the difference between reference and output voltage, and $x_c = [x_{1c} \quad x_{2c}]^T$ and ω denote states of the controller and network frequency, respectively. We can write:

$$e = r - y = r - Cx \tag{9}$$

Replacing (8) into (9), we conclude that:

$$\dot{x}_c = A_c x_c - B_c C x + B_c r \qquad (10)$$

where periodic reference input is **r** that should be tracked, x are plants states and:

$$A_c = \begin{bmatrix} 0 & 1 \\ -\omega^2 & 0 \end{bmatrix} \quad B_c = \begin{bmatrix} 0 \\ \omega \end{bmatrix} \qquad (11)$$

By augmenting (10) and (1), the following state space representation for the system is derived:

$$\begin{bmatrix} \dot{x} \\ \dot{x}_c \end{bmatrix} = \begin{bmatrix} A & 0 \\ -B_c C & A_c \end{bmatrix} \begin{bmatrix} x \\ x_c \end{bmatrix} + \begin{bmatrix} B \\ 0 \end{bmatrix} u + \begin{bmatrix} E \\ 0 \end{bmatrix} i_d + \begin{bmatrix} 0 \\ B_c \end{bmatrix} r \quad (12)$$

Now, (12) can be written as:

$$\begin{bmatrix} \dot{x}_1 \\ \dot{x}_2 \\ \dot{x}_{1c} \\ \dot{x}_{2c} \end{bmatrix} = \begin{bmatrix} 0 & -\frac{1}{L} & 0 & 0 \\ \frac{1}{C} & -\frac{\Delta}{C} & 0 & 0 \\ 0 & 0 & 0 & 1 \\ 0 & -\omega & -\omega^2 & 0 \end{bmatrix} \begin{bmatrix} x_1 \\ x_2 \\ x_{1c} \\ x_{2c} \end{bmatrix} + \begin{bmatrix} \frac{1}{L} \\ 0 \\ 0 \\ 0 \end{bmatrix} u$$
$$+ \begin{bmatrix} 0 \\ -1 \\ C \\ 0 \\ 0 \end{bmatrix} i_d + \begin{bmatrix} 0 \\ 0 \\ 0 \\ \omega \end{bmatrix} r \qquad (13)$$

In order to separate the uncertainty, (13) can be written as:

$$\begin{bmatrix} \dot{x}_1 \\ \dot{x}_2 \\ \dot{x}_{1c} \\ \dot{x}_{2c} \end{bmatrix} = \begin{bmatrix} 0 & -\frac{1}{L} & 0 & 0 \\ \frac{1}{C} & 0 & 0 & 0 \\ 0 & 0 & 0 & 1 \\ 0 & -\omega & -\omega^2 & 0 \end{bmatrix} \begin{bmatrix} x_1 \\ x_2 \\ x_{1c} \\ x_{2c} \end{bmatrix} + \begin{bmatrix} 0 \\ -\frac{1}{C} \\ 0 \\ 0 \end{bmatrix} \Delta x_2$$
$$+ \begin{bmatrix} \frac{1}{L} \\ 0 \\ 0 \\ 0 \end{bmatrix} u + \begin{bmatrix} 0 \\ -1 \\ C \\ 0 \\ 0 \end{bmatrix} i_d + \begin{bmatrix} 0 \\ 0 \\ 0 \\ \omega \end{bmatrix} r \qquad (14)$$

Now, By considering:

$$\Delta x_2 = p \qquad (15)$$

And

$$q = \begin{bmatrix} 0 & 1 & 0 & 0 \end{bmatrix} \begin{bmatrix} x_1 \\ x_2 \\ x_{1c} \\ x_{2c} \end{bmatrix}, \qquad (16)$$

Therefore, $p = \Delta q$ is obtained.

Considering $\|\Delta\| \le 1$, we can derive the NLDI representation of (12) as following:

$$\dot{x} = A_{cl} x + B_p p + B_u u + E i_d + B_r r \qquad (17)$$

$$y = C_z x \qquad (18)$$

$$q = C_q x \qquad (19)$$

$$pp^T \le qq^T \qquad (20)$$

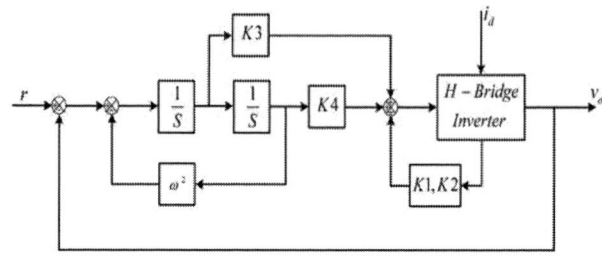

Fig. 3. Controlled system with state feedback and resonant controller

where:

$$A_{cl} = \begin{bmatrix} 0 & -\frac{1}{L} & 0 & 0 \\ \frac{1}{C} & 0 & 0 & 0 \\ 0 & 0 & 0 & 1 \\ 0 & -\omega & -\omega^2 & 0 \end{bmatrix}, B_p = \begin{bmatrix} 0 \\ -\frac{1}{C} \\ 0 \\ 0 \end{bmatrix}$$

$$B_u = \begin{bmatrix} \frac{1}{L} \\ 0 \\ 0 \\ 0 \end{bmatrix}, \ B_r = \begin{bmatrix} 0 \\ 0 \\ 0 \\ \omega \end{bmatrix}, \quad E = \begin{bmatrix} 0 \\ -1 \\ C \\ 0 \\ 0 \end{bmatrix} \qquad (21)$$

$$C_z = \begin{bmatrix} 0 & 1 & 0 & 0 \end{bmatrix}, C_q = \begin{bmatrix} 0 & 1 & 0 & 0 \end{bmatrix}$$

Equations (17) denote the NLDI representation of system.

Theorem 1: System (17) is robustly D-stable with the robust state gain K for a given α, if there exist matrix Y, scalars γ and μ and positive definite matrix Q with appropriate dimensions such that the following eigenvalue problem holds:

$$\min \gamma$$
$$subject\ to \begin{bmatrix} \Gamma_{11} & * & * \\ \Gamma_{21} & -\gamma^2 I & * \\ \Gamma_{31} & 0 & -\tau I \end{bmatrix} \le 0, \qquad (22)$$
$$L_1 \otimes Q + M_1 \otimes (A_{cl} Q + B_u Y) + M_1^T \otimes (A_{cl} Q + B_u Y)^T < 0$$

where:

$$\Gamma_{11} = A_{cl} Q + Q A^T_{cl} + B_u Y + Y^T B_u^T + EE^T + \tau B_p B_p^T$$
$$\Gamma_{21} = C_z Q + D_{zu} Y$$
$$\Gamma_{31} = C_q Q + D_{zu} Y$$
$$L_1 = 2\alpha$$
$$M_1 = 1$$

The gain of robust state feedback control:

$$K = YQ^{-1} \qquad (23)$$

Assurances us that:

1) All poles of the closed-loop are stable, with real parts smaller than $-\alpha$.

2) The closed-loop system attenuate the effect of disturbance w in the output y, by minimizing γ.

Proof: for simplicity let's assume that $D_{qp} = 0$. We define the L_2 gain of the NLDI (17) as the quantity

$$\sup_{\|w\|_2 \neq 0} \frac{\|z\|_2}{\|w\|_2} \tag{24}$$

Where L_2 norm of u is $\|u\|_2^2 = \int_0^\infty u^T u\, dt$ and suprimume is taken over all nonzero trajectories of the NLDI with initial condition $x(0) = 0$.

Let us suppose that there exist a quadratic Lyapunov function

$$V(x) = x^T P x, P > 0 \tag{25}$$

and $\gamma > 0$ such that for all t

$$\frac{d}{dt} V(x) + z^T z - \gamma^2 w^T w \leq 0 \tag{26}$$

Then L_2 gain of the NLDI is less than γ. To show this, (26) is integrated from 0 to T, with the initial condition $x(0) = 0$, to get

$$V\big(x(T)\big) + \int_0^T (z^T z - \gamma^2 w^T w)dt \leq 0 \tag{27}$$

Since $V\big(x(T)\big) \geq 0$, this implies

$$\frac{\|z\|_2}{\|w\|_2} \leq \gamma \tag{28}$$

Now reconsidering (25) and (26) and some algebraic operations, due to controllability of system, we yield:

$$\begin{bmatrix} \Gamma'_{11} & * & * \\ E^T P & -\gamma^2 I & * \\ B_p^T P & 0 & 0 \end{bmatrix} \leq 0 \tag{29}$$

Where:

$$\Gamma'_{11} = A^T P + PA + K^T B_u^T P + P B_u K \\ + (C_Z + D_{zu}K)^T (C_Z + D_{zu}K) \tag{30}$$

Integral quadratic constraint (20) can be written as:

$$\begin{bmatrix} -(C_q + D_{zu}K)(C_q + D_{zu}K)^T & 0 & 0 \\ 0 & 0 & 0 \\ 0 & 0 & I \end{bmatrix} \leq 0 \tag{31}$$

Now we apply Lemma 2 to (29) and (31). So we have:

$$\begin{bmatrix} \Gamma'_{11} + \mu(C_q + D_{zu}K)(C_q + D_{zu}K)^T & * & * \\ E^T P & -\gamma^2 I & * \\ B_p^T P & 0 & -\mu I \end{bmatrix} \\ \leq 0 \tag{32}$$

Now we can multiply (32) pre and post in:

$$\begin{bmatrix} -\gamma P^{-1} & 0 & 0 \\ 0 & 0 & 0 \\ 0 & 0 & 0 \end{bmatrix} \tag{33}$$

Considering $Y = KQ^{-1}$ and $Q = \gamma P^{-1}$, we derive:

$$\begin{bmatrix} \Gamma''_{11} + \mu(QC_q + D_{zu}Y)\gamma^2(QC_q + D_{zu}Y)^T & * & * \\ E^T \gamma & -\gamma^2 I & * \\ B_p^T \gamma & 0 & -\mu I \end{bmatrix} \\ \leq 0 \tag{34}$$

Where:

$$\Gamma''_{11} = QA^T + AQ + Y^T B_u^T + B_u Y \\ + (C_Z + D_{zu}Y)^T \gamma^2 (C_Z + D_{zu}Y) \tag{35}$$

Considering Lemma 1 and $\tau = \mu\gamma^2$, we reach to (22) and the proof completes. ∎

Algorithm 1: To design the proposed controller, following steps can be used:

Step1: derive NLDI model of system

Step2: obtain the LMI according Theorem 1.

Step3: use software package to solve this LMI and derive robust state feedback gains such as CVX and YALMIP.

Step4: discretize the robust state feedback and implement it. To implement the controller, shift the first branch of resonant controller to forward.

IV. SIMULATION RESULTS

In the current section, the results of simulating the proposed controller in Theorem 1 by SIMPOWER toolbox of MATLAB for a single phase DC/AC converter are reported using the parameters mentioned in Table I and are compared with those of the multi-loop PID controller. The parameters of nonlinear load R_l and C_l are defined as bellow:

$$R_l = \frac{(1.22V_{in})^2}{0.66P} \quad , \quad C_l = \frac{7.5}{f.R_l} \tag{36}$$

Where V_{in} and P are the nominal *RMS* of reference voltage in inverter and apparent power of the single phase full bridge inverter respectively, and f represents the fundamental frequency of reference voltage.

Considering the parameters of Table I, R_l and C_l are derived as 21.8 Ω and 6880μF, respectively.

To derive robust state feedback controller, the results of Theorem 1 is solved, using YALMIP in MatLAB 2015 for

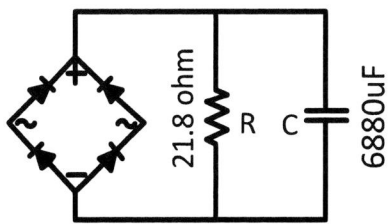

Fig. 4 Nonlinear load.

Table I. Nominal parameters

DC link	400
Reference voltage	311V (220 V in RMS)
Frequency of output	314 rad/s
frequency of switching	20 kHz
Inductor of filter	1.0mH
Capacitor of filter	300μF
Rated power of output	5kVA

Table II. Robust state feedback gains

k_1	-21
k_2	-118.9
k_3	83924
k_4	339

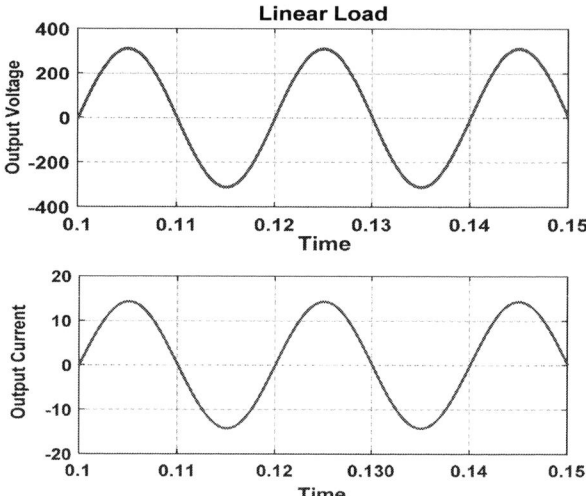

Fig.5. Output voltage and current for linear load

$\alpha = 500$ and $M_1 = 1$. The obtained gains of the robust state feedback controller are represented in Table II.

In Table III simulation results are reported for nonlinear and linear load. In the case of linear load, the THD is obtained 0.09. Fig. 5 depictures the output current and voltage for linear load case. THD for nonlinear load and its crest factor reported 1.95 and 3 respectively in the worst case according to (33). Figure 6 represents the output current and output voltage. Fig. 7 illustrate the output current and output voltage behavior at step linear load condition. We used a fluctuation of amplitude $\pm 40\ v$ which is rejected by our control algorithm, in DC bus.

V. CONCLUSIONS

The paper aimed to form a new modeling of single phase full bridge inverter with an output lowpass LC filter for UPS application. According to LMIs, convex optimization problem was defined to derive the robust state feedback controller. An algorithm was given to design and implement the robust state feedback controller. For the proposed controller, the performances were illustrated with a 5-kVA single phase inverter. Simulation results illustrated that the proposed controller satisfied the IEC62040-3 and IEEE 1547 specifications. The designed controller's THD was 1.95 for nonlinear load.

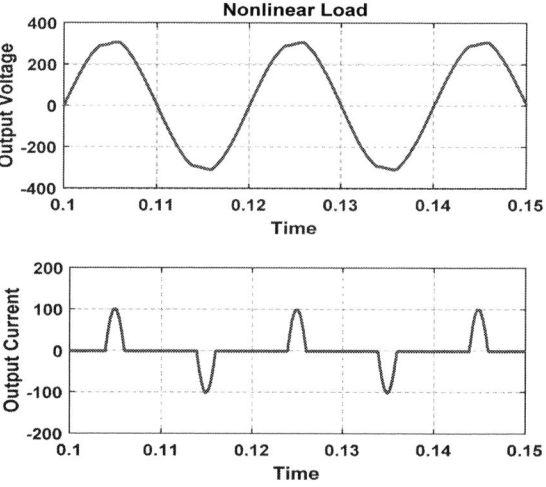

Fig. 6. Output current and voltage for nonlinear load

Table III. Analyses of output voltage for linear and nonlinear load

	Output Voltage	THD	CF
Linear load	311 (219.9 RMS)	0.09	1.4
Nonlinear load	311.1(220 RMS)	1.95	3

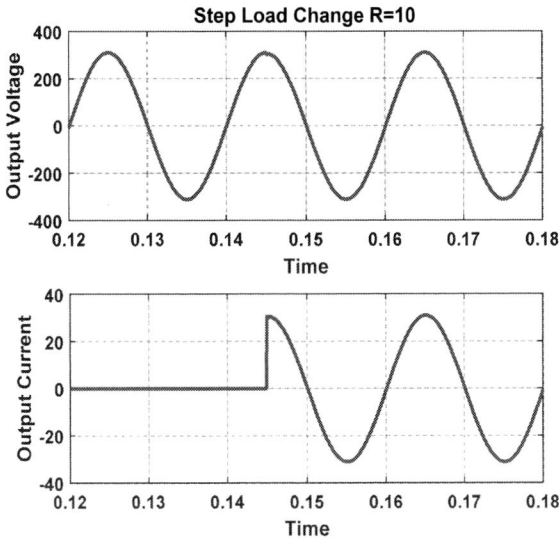

Figure 7: linear step load

REFERENCES

[1] A. Nasiri, *Power Electronics Handbook*. Elsevier, 2011.

[2] N. Mohan, T. M. Undeland, and W. P. Robbins, "Power Electronics. Converters, Application and Design. John Willy & Sons," *INC, NY*, 1995.

[3] "IEEE Recommended Practice and Requirements for Harmonic Control in Electric Power Systems," *IEEE Std 519-2014 (Revision of IEEE Std 519-1992)*. pp. 1–29, 2014.

[4] "Uninterruptible power systems (UPS)-part3: method of specifying the performance and test requirements. IEC 62040-3." IEC, Switzerland, 2003.

[5] "IEEE Recommended Practice for the Application and Testing of Uninterruptible Power Supplies for Power Generating Stations," *ANSI/IEEE Std 944-1986*. p. 0_1, 1986.

[6] K. Zhang, Y. Kang, J. Xiong, and J. Chen, "Direct repetitive control of SPWM inverter for UPS purpose," *Power Electronics, IEEE Transactions on*, vol. 18, no. 3. pp. 784–792, 2003.

[7] C. Rech, H. Pinheiro, H. A. Grundling, H. L. Hey, and J. R. Pinheiro, "A modified discrete control law for UPS applications," *Power Electronics, IEEE Transactions on*, vol. 18, no. 5. pp. 1138–1145, 2003.

[8] A. Abrishamifar, A. A. Ahmad, and M. Mohamadian, "Fixed Switching Frequency Sliding Mode Control for Single-Phase Unipolar Inverters," *Power Electronics, IEEE Transactions on*, vol. 27, no. 5. pp. 2507–2514, 2012.

[9] H. Komurcugil, "Rotating-Sliding-Line-Based Sliding-Mode Control for Single-Phase UPS Inverters," *Industrial Electronics, IEEE Transactions on*, vol. 59, no. 10. pp. 3719–3726, 2012.

[10] T.-L. Tai and J.-S. Chen, "UPS inverter design using discrete-time sliding-mode control scheme," *Industrial Electronics, IEEE Transactions on*, vol. 49, no. 1. pp. 67–75, 2002.

[11] N. M. Abdel-Rahim and J. E. Quaicoe, "Analysis and design of a multiple feedback loop control strategy for single-phase voltage-source UPS inverters," *Power Electronics, IEEE Transactions on*, vol. 11, no. 4. pp. 532–541, 1996.

[12] A. Nasiri, "24 - Uninterruptible Power Supplies," M. H. B. T.-P. E. H. (Third E. Rashid, Ed. Boston: Butterworth-Heinemann, 2011, pp. 627–641.

[13] G. Willmann, D. F. Coutinho, L. F. A. Pereira, and F. B. Libano, "Multiple-Loop H-Infinity Control Design for Uninterruptible Power Supplies," *Industrial Electronics, IEEE Transactions on*, vol. 54, no. 3. pp. 1591–1602, 2007.

[14] J. S. Lim, C. Park, J. Han, and Y. Il Lee, "Robust Tracking Control of a Three-Phase DC–AC Inverter for UPS Applications," *Industrial Electronics, IEEE Transactions on*, vol. 61, no. 8. pp. 4142–4151, 2014.

[15] L. F. Alves Pereira, J. Vieira Flores, G. Bonan, D. Ferreira Coutinho, and J. M. Gomes da Silva Junior, "Multiple Resonant Controllers for Uninterruptible Power Supplies—A Systematic Robust Control Design Approach," *Industrial Electronics, IEEE Transactions on*, vol. 61, no. 3. pp. 1528–1538, 2014.

[16] V. F. Montagner and S. P. Ribas, "State feedback control for tracking sinusoidal references with rejection of disturbances applied to UPS systems," *Industrial Electronics, 2009. IECON '09. 35th Annual Conference of IEEE*. pp. 1764–1769, 2009.

[17] H. Komurcugil, N. Altin, S. Ozdemir, and I. Sefa, "An Extended Lyapunov-Function-Based Control Strategy for Single-Phase UPS Inverters," *Power Electronics, IEEE Transactions on*, vol. 30, no. 7. pp. 3976–3983, 2015.

[18] E. Kim, F. Mwasilu, H. H. Choi, and J. Jung, "An Observer-Based Optimal Voltage Control Scheme for Three-Phase UPS Systems," *Industrial Electronics, IEEE Transactions on*, vol. 62, no. 4. pp. 2073–2081, 2015.

[19] C. Scherer and S. Weiland, "Linear matrix inequalities in control," *Lect. Notes, Dutch Inst. Syst. Control. Delft, Netherlands*, 2000.

[20] L. E. Ghaoui and S. I. Niculescu, *Advances in Linear Matrix Inequality Methods in Control*. Society for Industrial and Applied Mathematics, 2000.

[21] S. Boyd, L. El Ghaoui, E. Feron, and V. Balakrishnan, *Linear matrix inequalities in system and control theory*, vol. 15. SIAM, 1994.

[22] G. Dullerud and F. Paganini, *Course in Robust Control Theory*. Springer-Verlag New York, 2000.

7th Power Electronics, Drive Systems & Technologies Conference (PEDSTC 2016)
16-18 Feb. 2016, Iran University of Science and Technology, Tehran, Iran

Design and Implementation of an FPGA-based Real-Time Simulator for H-Bridge Converter

Morteza Rezaei Larijani, Mohammad-Reza Zolghadri, Member, IEEE, and Mahmoud Shahbazi

Department of Electrical Engineering, Sharif University of Technology, Tehran, Iran

mrl.larijani@gmail.com, zolghadr@sharif.edu, mahmoudshahbazi@outlook.com

Abstract—This paper presents a methodology for implementing of the mathematical model of H-Bridge converter in an FPGA-based Real-time simulator. Furthermore, it introduces a new method for choosing parameters of the Associate Discrete Circuit (ADC) model of semiconductor switches. The ADC-based model allows obtaining a fixed topology irrespective of switches states for the power electronic converters in the digital simulation. Backward-Euler based discretized state space matrix (SSM) of the circuit used for ADC parameter. Choosing appropriate switch parameter is based on 1) reducing the distance of SSM eigenvalues from origin in z-Plane to reduce settling-time of system response; and 2) reducing the angle of SSM eigenvalues to increase damping factor of the system. Modified nodal analysis (MNA) approach is used for solving the circuit model. Verilog code to simulate the behavior of the circuit is developed and implemented on a SPARTAN-6 FPGA. Simulation step times as low as 200ns is easily achieved. Experimental result presented confirms the performance of applied method for Real-time simulation of the H-bridge converter.

Keywords—*Real-time Simulation, field-programmable gate array (FPGA), associate discrete circuit (ADC), modified nodal analysis (MNA), H-bridge converter, state space matrix (SSM).*

I. INTRODUCTION

With development of power electronic science and with the existence of different power electronic converters in this field, the need for improving their operation and the fault detection methods of these converters, the need for simulation and test of them is strongly felt. However, experimental constrain such as damage risks, reliability, costs, and so on obviate the need for making the converter and the need for testing it and its controller. Real-time simulation is a usual practice for testing the protection platform and converter controller in a hardware-in-the-loop configuration [1]. Control/protection platform needs to be tested and its functionality should be verified prior to installation and commissioning. Testing in the Real-time simulator environment is practically the only option to safely and thoroughly verify the design integrity and evaluate its functionality and performance. Moreover, a Real-time simulator is also required to conduct statistical switching studies, as it substantially reduces the total run-time of the study [2]. This paper proposes and develops a generalized methodology for implementing a model of the power electronic in an FPGA environment.

The studied power electronic converter is an H-Bridge converter. ADC method is used to develop the model of this converter. Using ADC method, each switch is represented by an inductor and a capacitor when switch is on and off respectively. Then, the discrete model of switch is obtained as a conductance and shunt current source. Advantage of this method is the fact that irrespective of switch state, with selection of appropriate time-step the conductance can be constant, so the topology of the circuit can be fixed [3], [4], [5]. With backward-Euler integration method, which has inherent damping [6], elements of the converter circuit will be discrete and the matrix of the circuit is obtained. Solving of the circuit model is based on MNA approach. Solution of the discrete circuit model leads to nodal voltages and some branch currents. Using ADC method not only the network matrix would be fixed and the need for calculation of the matrix inversion can be obviating, but also the occupied area in FPGA would be reduced. However, the ADC-based model produces artificial losses and transients, which are not real. Errors due to ADC-based model are treated as numerical errors and those are given no physical concept [4], [5], [7].

The developed methodology enables Real-time operation for statistical switching studies. Also, Based on the developed methodology, an FPGA-based Real-time simulator is developed and tested. It enables the use of a nanosecond range simulation time-step to simulate large systems in Real-time. Thus, it is also able to provide a wide frequency bandwidth for simulation results. Furthermore, due to small time-step the corrective algorithm can be obviating to solve the problem of inter-simulation time-step switching [2]. Also, the created system can be used for the fault detection of power electronic converters in a nanosecond range.

Easily programmable hardware based rapid calculation capability of FPGAs make them a suitable choice for application has the need for simulation of the converters with high speed. Also, FPGA has a parallel adventure for sum or multiply operation whose its feature cause to it be used in works which calculation speed is important and hardware can be used up to implementation and configuration of a controller or a power electronic converter. [2],[8].

The rest of this paper is organized as follows. Section II describes modeling of H-Bridge converter for Real-Time simulation and explanation of the problem definition. Section III describes the proposed method for optimum selection of model parameter to solving that problem. Simulation results using the proposed method are presented in section IV. Section

978-1-5090-0376-1/16 $31.00 © 2016 IEEE

V describes implementation of the model in an FPGA board and finally conclusion of the work is described in section VI.

II. ADC MODEL OF H-BRIDGE CONVERTER

Real-time simulation of H-Bridge converter needs to an accurate modeling. In the former articles variety of models, three of which are functional, has been used. Two-valued switch model, averaged model, and switching function model [2], [9], [10]. In the first method each switch is represented from others separately, so each switch can be represented by: 1) the ideal switch model, 2) the small/large resistor model, and 3) the ADC-based model [11].

In this paper the ADC-based method has been used for modeling the converter switches. The ADC-based switch model shows a switch by a small inductance (L) when on and by a small capacitance (C) when OFF [11]. Fig. 1 shows the continuous and discontinuous model of an on switch with ADC-based model.

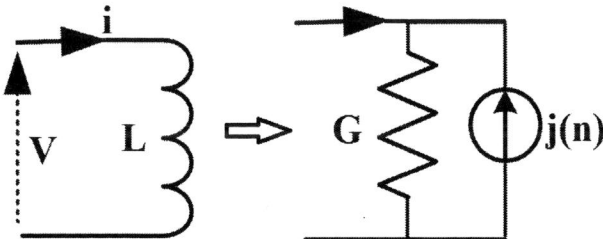

Fig. 1. An on switch model (inductance) and its ADC model

Fig. 2 shows the continuous and discontinuous model of an off switch with ADC-based model.

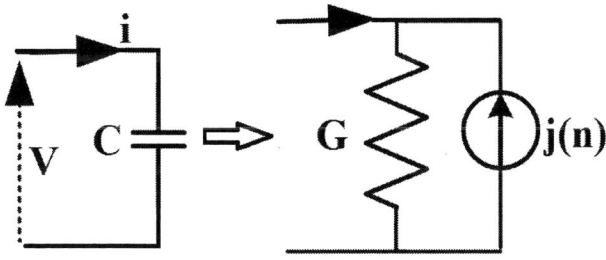

Fig. 2. An off switch model (capacitance) and its ADC model

The value of resistance and current source depend on switch state. However, with appropriate choice of time-step the value of the resistance would be the same for both switch states. By doing so, a constant conductance matrix which is not depend on switch states be obtained. To set this limitation for circuit time-step to achieve a constant matrix, reducing complication of Real-time simulator algorithm and the occupied area of the FPGA. These are the adventure of ADC method.

For digital simulation, discrete form of the circuit should be used. In this case Backward Euler Method is used because of its inherent damping. Table 1 shows the value of shunt dependent current source and the conductance of discrete switch model.

Table 1. Parameters of discrete switch model by ADC method

Switch state	G(conductance)	Current source
Switch is ON	$G = \dfrac{T}{L}$	$j(n) = -i_s(n-1)$
Switch is OFF	$G = \dfrac{C}{T}$	$j(n) = G_c V_s(n-1)$

In table 1 $j(n)$ is dependent current source in nth time-step. Also, $v_s(n)$ and $i_s(n)$ are the voltage and current of switches in nth time-step, respectively. The dependent current source whose value is depending on switch state compute at each time-step from the former time-step voltage and current of switch. G, the conductance of switches, which is related to state of switch, is not constant so far. In order to set a constant value for conductance of switch the time-step should be chosen as $T = \sqrt{LC}$. By doing so, we have a constant $G=G_s$ and a constant matrix.

Therefore, in discrete form, the ADC-based method represents each switch by a resistance, or conductance in parallel with a dependent current source, as shown in Fig. 3.

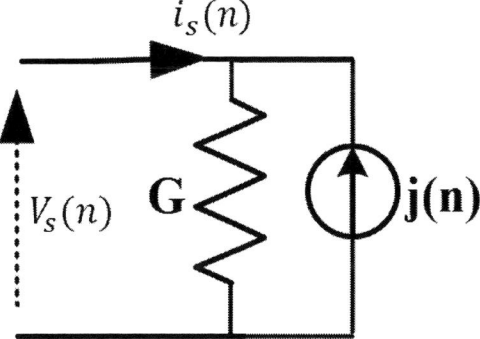

Fig. 3. ADC-based model of the switch.

Effect of switch anti parallel diode (if exist) is modeled as a controlling effect. The state of diode updated at each time-step as follows equation (1).

$$D(n) = D(n-1).(I_D(n-1) \geq 0) + \overline{D(n-1)}.(V_D(n-1) > 0) \quad (1)$$

$D(n)$ is diode state in nth time-step. $V_D(n-1)$, $I_D(n-1)$, and $D(n-1)$ are respectively voltage, current, and diode state in n-1th time-step. To achieve circuit currents as well as voltages, modified nodal analysis (MNA) is used. [4], [5].

Circuit equations can be written in the form shown in (2).

$$Y.X(n) = B(n) \quad (2)$$

Where Y and its inverse are computed offline and the vector of voltages and currents ($X(n)$) are calculated at each time-step by solving the equation (3).

$$X(n) = Y^{-1}.B(n) \quad (3)$$

Altering switches state affects the value of current source which affects $B(n)$ [7].

The Calculation of voltages and currents are done in parallel. The multiplication operations to construct sum of product are independent and can be carried out in parallel [2].

The computation of circuit variable is done in two stages: first calculating $X(n)$ and second stage is updating the dependent current sources, which is related to former values of voltages and currents.

H-bridge, with a resistive load shown in Fig. 4 is used for this study.

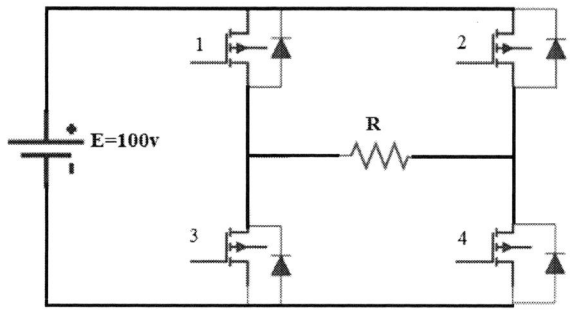

Fig. 4. Schematic of H-bridge converter.

The discrete equivalent circuit of H-Bridge converter is shown in Fig. 5.

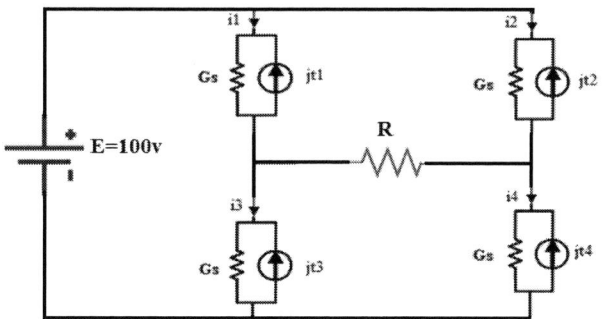

Fig. 5. ADC model of H-bridge converter (discrete of Fig. 2).

Simulink-Matlab is used to simulate the circuit of fig. 5 with G_s=1. The switching frequency is 2 KHz, R=10Ω, E=100v, and T (time-step) is 200 nanosecond. Simulation result is shown in Fig. 6.

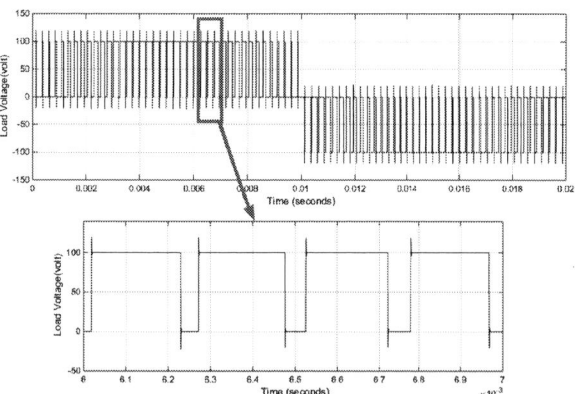

Fig. 6. Load voltage waveform with G_s=1.

It can be seen that the voltage across the load has a small transient spike. This is mainly due to the resonance of the LCs added in ADC model instead of switches. In the next section a Z domain eigenvalue analysis is proposed to minimize this problem. This method based on analysis of state space matrix eigenvalues in z-Plane.

III. SELECTION OF SWITCH PARAMETER BY EIGENVALUE ANALYSIS

Proposer selection of G can affect the performance of the transient response of the ADC model. To investigate it, eigenvalue analysis in z-Plane is used. First, the state space equation be obtained, second by using backward Euler method continues equations convert to discrete equations. Finally, the eigenvalues of that SSM calculated and by offered method the optimum G_s be obtained. This method based on analysis of eigenvalues of state space matrix of the circuit whose matrix is discrete by backward Euler method in z-Plane.

Using continuous form state space equations form these equations is shown in (4).

$$X'(t) = A.X(t) + B.U(t) \qquad (4)$$

The discrete state space equation of (4) is achieved with backward Euler method is shown in (5) [12].

$$X(n+1) - X(n) = ATX(n+1) + BTU(n+1) \qquad (5)$$

$$\xrightarrow{yields} X(n+1) = \acute{A}.X(n) + \acute{B}.U(n)$$

$$\acute{A} = [(I - AT)^{-1}], \ \acute{B} = [(I - AT)^{-1}BT]$$

The relationship between z and s by Backward Euler is shown.

$$z = \frac{1}{1 - TS} \qquad (6)$$

T is time-step and $s = \sigma + j\omega$.

Mapping the Left-half of the s-Plan into the z-Plane by the backward Euler integration method leads to a circle centered at 0.5+0j and its radius is 0.5, which is shown in Fig. 7.

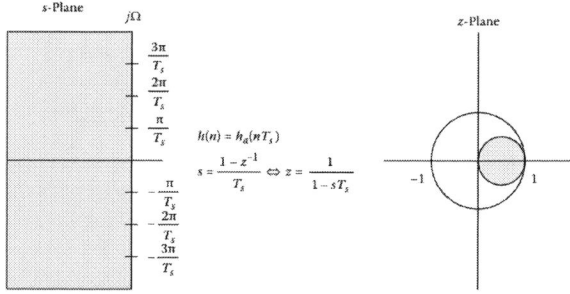

Fig. 7. Mapping the left half of the s-Plane into the z-Plane by the backward Euler method [13].

To achieve acceptable transients output from Real-time simulator settling-time and overshoot of the output is considered.

• Settling-time

Moving the poles away from origin in s-Plan to the left half-plane (LHP) results in shorter settling-time [14]. The

978-1-5090-0376-1/16 $31.00 © 2016 IEEE   506

relation between settling-time (T_s) and real section of pole (σ) is shown in (7) [12].

$$T_s \cong -\frac{4}{\sigma} \tag{7}$$

The (7) for a discrete system in z-Plan can be converted using (6). By doing so, this relation is shown in (8).

$$T_s \cong \frac{4TZ}{1-Z} \tag{8}$$

According to (8) the $\lim_{z\to0} Ts=0$ and $\lim_{z\to1} Ts=\infty$, so moving the poles in z-Plan close to the origin result in decreasing settling-time. By doing so, the minimum settling-time occurred when the poles of the system are close to the origin. The specified path for poles in the Fig. 8 reducing Ts of the system.

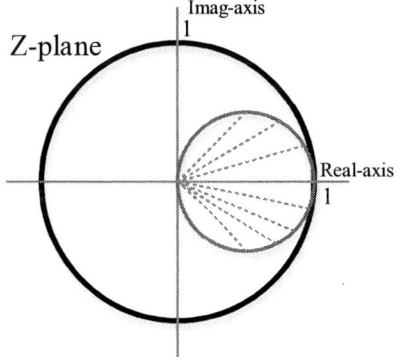

Fig. 8.The dashed lines are direction of poles near the origin for minimum settling-time.

The first condition done is minimizing the distance of poles from origin for reducing settling-time.

- Overshoot

The second condition is considering the relationship between damping factor (ζ) and overshoot. As shown in Fig. 4 the load voltage has a transient overshoot. To obviate them, the damping ratio should be increased. By doing so, the decreasing overshoot would be obtained.

$s = -\zeta\omega_0 + j\sqrt{1-\zeta^2}\omega_0$, and using (6) the relationship between z and ζ is obtained as in (9).

$$z = \frac{1+\zeta\omega_0T+j(\omega_0T\sqrt{1-\zeta^2})}{1+(\omega_0T)^2+(2\zeta\omega_0T)} \tag{9}$$

To getting minimum overshoot, the damping of the system should be high. To do so, the value of ζ should be 1 or the angel of poles should be 0. The angel of poles is drawn versus ζ. First, ω_0T in (9) assumed 1 and according to it the angle of Z ($\tan^{-1}\frac{Imag(z)}{Real(z)}$) is drawn, which is shown in Fig. 9 (dimension of angle is radian).

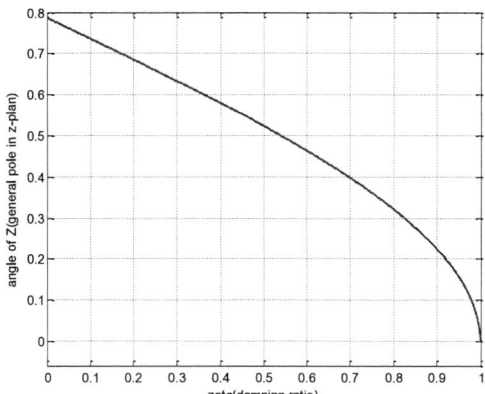

Fig. 9.Waveform of pole angel versus ζ.

According to Fig. 9 when the angel of poles is zero, ζ is 1. Thereby, for removing overshoot the location of poles should be near the real-axis of z-Plane.

IV. CASE STUDY

H-bridge converter, which is shown in Fig. 2, is considered. The SSM of the circuit is obtained and then it be discretized by backward Euler integration method. H-Bridge has 4 switches, so it has $2^4 = 16$ topologies. However, the states of switches in each leg are opposite of each other. Also, two of states are equal and results same topologies. Therefore, the converter has two different topologies totally. These are shown in Table 2.

Table 2. States of the converter according to the switches.

Switch Number	1	2	3	4	Output voltage
State 1	ON	OFF	OFF	ON	E=100
State 2	ON	ON	OFF	OFF	0

Due to L and C, which are series to each other, both of topologies of the circuit have identical property, so one of them can be considered. State 1, which the switches 1 and 4 are on and 2 and 3 are off is assigned to be considered. ON switches are represented with inductor and OFF switches are represented with capacitor. Therefore, the circuit contains two inductor and capacitor. The circuit of the state 1 is shown in Fig. 10.

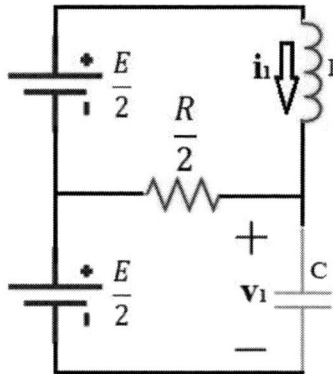

Fig. 10. Modified schematic of converter for state 1.

State space matrix of the circuit of Fig. 10 is shown in (10).

$$A = \begin{bmatrix} 0 & \dfrac{-1}{L} \\ \dfrac{1}{C} & \dfrac{-2}{RC} \end{bmatrix} \qquad (10)$$

According to (5) the discrete state space matrix can be achieved according to G, which is shown in (11).

$$\acute{A} = \begin{bmatrix} 1 & G \\ \dfrac{-1}{G} & 1 + \dfrac{2}{RG} \end{bmatrix}^{-1} \qquad (11)$$

- Reducing the settling-time

The matrix $\acute{A}$ has two eigenvalues. First, the absolute values of the eigenvalues to consider distance of them from origin are shown in Fig. 11 and Fig. 12.

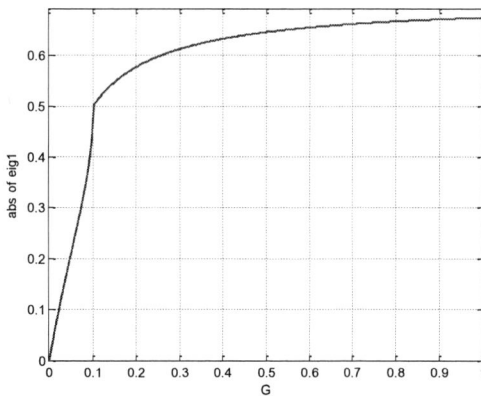

Fig. 11. Absolute value of the first eigenvalue versus G_s is plotted.

As shown in Fig. 11, with increasing G_s, the absolute value of eigenvalue increases. Also, when the poles moving far from the origin, the settling-time increases. In fact, the location of the first eigenvalue for small amount of G_s is good.

Absolute value of the second eigenvalue of discrete state space matrix versus G_s is shown in Fig. 12.

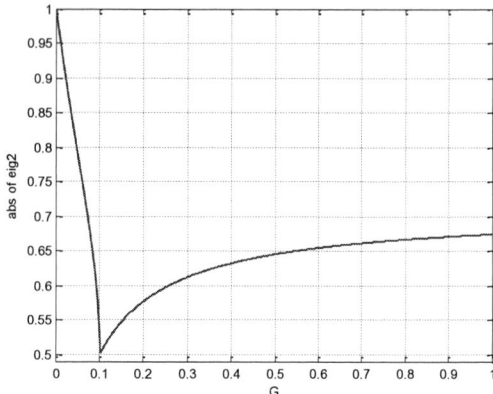

Fig. 12. Absolute value of the second eigenvalue, which is versus G_s, is plotted.

As shown in Fig. 12, by increasing G_s the absolute value of eigenvalue decrease as far as $G_s = 0.1$, and Afterwards the absolute value of eigenvalues increases. In fact, the location of second eigenvalue for small and big amount of G_s is not acceptable.

The best location of eigenvalues should be found for obtaining the minimum settling-time of system response. To do so, a function which is sum of square of eigenvalues absolute value is defined. This function is shown in (12).

$$F(G_s) = \sum_{i=1}^{n} |\lambda_i|^2 \qquad (12)$$

The G_s which causes the minimum value of (12), gives the best value of G_s, which is suitable and the output has minimum settling-time.

Due to the use of Square in the (12) is that the location of first eigenvalue is acceptable and there is no need for moving it away. Therefore, the square absolute used rather than absolute to decreasing sensitivity of first eigenvalue, which is near origin. Accordingly, the $F(G_s)$ is shown in Fig. 13.

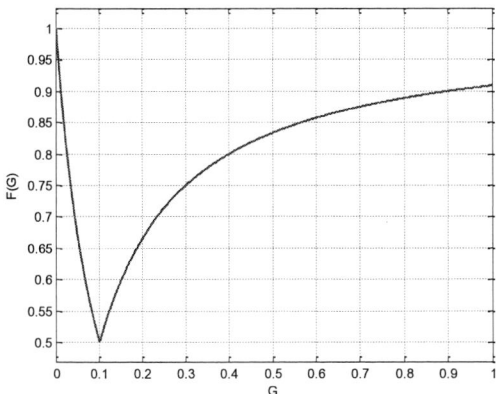

Fig. 13. F(G_s) function, or sum of square of eigenvalues absolute value.

According to Fig. 13 the minimum value of F(G_s) happened when G_s is 0.1 and the output of the system has minimum settling-time. Thereby, the optimum value of G_s is 0.1 till this moment.

- Reducing the overshoot

To obviate the spike of output voltage the angle of poles should be zero. Both poles, the angle of one of which is shown in Fig. 14, are conjugate.

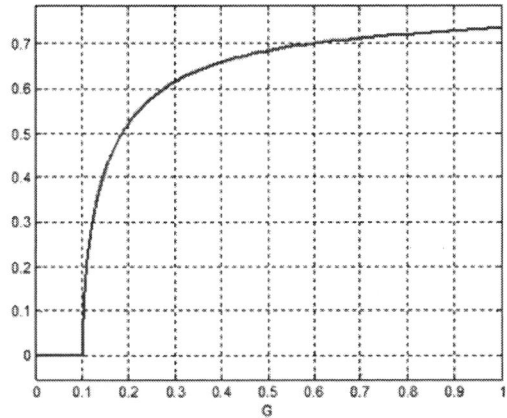

Fig. 14.Angle of one of the poles versus G_s.

As shown in Fig. 14, the angle for G_s less than 0.1 is zero. To have minimum overshoot the amount of G_s should be less than or equal 0.1. So there are two conditions for G_s. Finally, with combination of them the optimum value of G_s, which is 0.1 will be obtained. This is shown in (15).

$$\begin{cases} 1) G_s = 0.1 \\ 2) G_s \le 0.1 \end{cases} \Rightarrow G_s = 0.1 \qquad (13)$$

Now the ADC model of H-Bridge with $G_s = 0.1$ is simulated and the load voltage is plotted. The expectation is that the load voltage should not contain any overshoot.

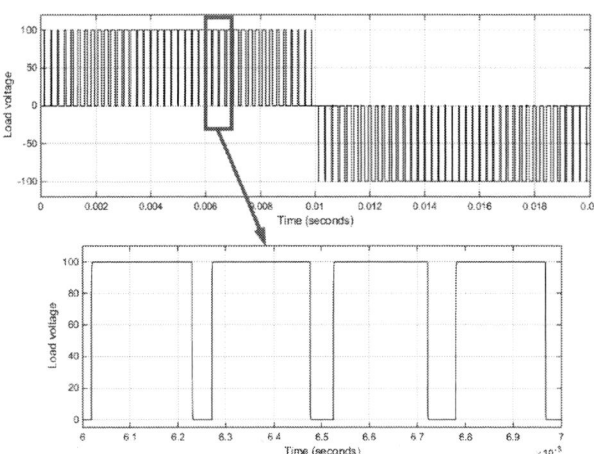

Fig. 15. Load voltage waveform with G_s=0.1 in Simulink.

As shown in Fig. 15, the load voltage does not have any transient spike because of appropriate choice of conductance of switch (G_s). The selection of conductance was done correctly, so the implantation of converter on FPGA begins.

V. IMPLEMANTION OF CONVERTER MODEL ON THE FPGA AND VALIDATION OF IT

Solution of the H-Bridge model equations based on the modified nodal analysis (MNA). After creation of admittance matrix of discrete circuit the code of converter model will be written. The model of the power electronic converter under study is described using the Verilog (Hardware Description Language). Verilog is a high level programming language that has been specially designed for describing the behavior of digital electronic systems [2]. The developed Verilog code is based on a fixed-point, 32-bit, signed Q (32.20) representation, i.e., 32 bits for the total length, 11 bits for integer part, and 20 bits for the fractional part.

This FPGA-based Real-time simulator solves the mathematical model of the converter at each time-step according to Eq. 3. The code of H-bridge model is implanted on the FPGA with ISE 14.5 (Xilinx software) on SPARTAN-6 FPGA. Also, for generate the PWM signal for Converter we used the system generator of Xilinx (blocks in Simulink). By using the system generator we join the converter code which is written manually with PWM blocks. Finally, the final Verilog code will be product. Then, by ISE software the bitstream of Verilog code would be obtained.

When using a Xilinx FPGA target (SPARTAN-6, XC6SLX9-2tqg144) for the implementation of the H-Bridge converter model, the resources consumption is evaluated at 6% the Number of slice Registers over 11440 and 41% the Number of slice Look-up tables(LUTs) over 5720. The frequency of FPGA clock is 50 MHz.

The AD7302BNZ and SN74HC573N ICs are used as DAC and buffering respectively. Figure of the FPGA platform is shown in Fig. 16.

978-1-5090-0376-1/16 $31.00 © 2016 IEEE   509

Fig. 16. FPGA platform for Real-time simulator.

By using JTAG cable the bitstream is implemented into FPGA. The output of board which is load voltage given to the DAC and is displayed on the oscilloscope.

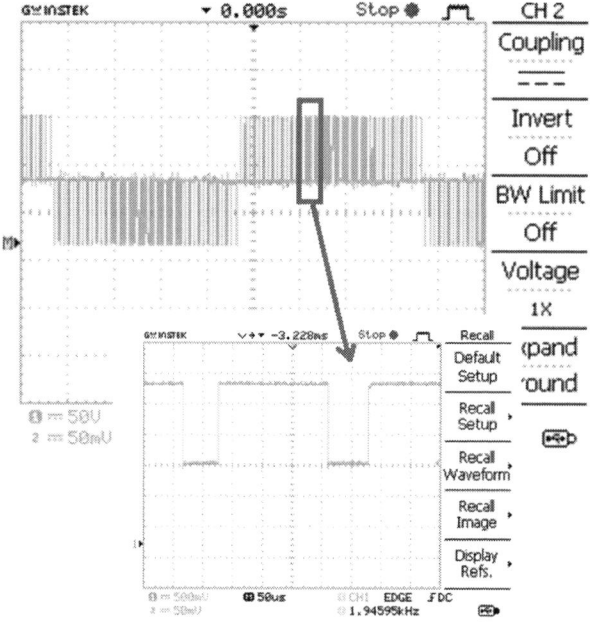

Fig. 17. Voltage of load is displayed by Real-time simulator.

As shown in Fig. 17, load voltage is simulated with Real-time simulator very well and implantation of H-Bridge converter is done sufficiently.

VI. CONCLUSION

This paper introduces a manner to choose parameter of discrete switch model for Real-time simulation. The switches of converter modeled correctly should be suitable for Real-time simulation. The method for modeling switches is ADC method. After modeling the circuit by analysis of eigenvalues of state space matrix which is discrete with BEM the optimum value of conductance was obtained. Finally with ISE the

Verilog code of converter was written and the code is implemented into FPGA board. The oscilloscope shows the load voltage of H-Bridge converter.

Reference

[1] Y. Chen and V. Dinavahi, "FPGA-Based Real-Time EMTP," *IEEE Transactions on Power Delivery,* vol. 24, no. 2, pp. 892-902, 2009.

[2] M. Matar and R. Iravani, "FPGA Implementation of the Power Electronic Converter Model for Real-Time Simulation of Electromagnetic Transients," *IEEE TRANSACTION ON POWER DELIVERY,* vol. 25, no. 2, April 2010.

[3] H. W. Dommel, EMTP Theory Book. Microtran Power System Analysis Corporation, 1996.

[4] P. Pejovic and D. Maksimovic, "A Method for Fast Time-Domain Simulation of Networks with Switches," *IEEE TRANSACTIONS ON POWER ELECTRONICS,* vol. 9, no. 4, July 1994.

[5] S. Y. R. Hui and S. Morrall, "Generalised Associated Discrete Circuit Model for Switching Devices," *IEE Proceedings -Science, Measurement and Technology,* vol. 141, no. 1, pp. 57-64, 1994.

[6] C. W. Gear, Numerical Initial Value Problems in Ordinary Diferential, Englewood Cliffs, NJ: Prentice Hall, 1971.

[7] R. Razzaghi, C. Foti and M. Paolone, "A Novel Method for the Optimal Parameter Selection of Discrete-Time Switch Model," in *conferance Power Systems Transients (IPST2013),* Vancouver, Canada, July 18-20, 2013.

[8] R. C. Restle, "choosing between DSPs, FPGAs, UPs, and ASICs to implement digital signal processing," in *Conference proceedings of ICSPAT: DSP world,* 2000.

[9] T. O. Bachir, J.-P. David, C. Dufour and J. Belanger, "Effective FPGA-based Electric Motor Modeling with Floating-Point Cores," *IEEE,* 2010.

[10] K. L. Lian and P. W. Lehn, "Real Time Simulation of Voltage Source Converters based on Time Average Method," *IEEE Trans on power electronic,* vol. 20, no. 1, 2005.

[11] M. Dagbagi, L. Idkhajine, E. Monmasson and I. Slama-Belkhodja, "FPGA Implementation of Power Electronic Converter Real-Time Model," in *International Symposium on Power Electronics, Electrical Drives, Automation and Motion,* 2012.

[12] G. F.Franklin, J. D. Powel and M. L.Workman, Digital Control of Dynamic Systems, THIRD ed., Adison wesley Longman, 1998.

[13] C. Wai-Kai, The Circuits and Filters Handbook Analog and VLSI Circuits, 3rd ed., Chicago: CRC Press, 2009.

[14] M. E. Schlarmann and R. L. Geiger, "Relationship Between Amplifier Settling Time and Pole-Zero Placements for Second-Order Systems," in *Circuits and Systems,* Lansing, MI, 2000.

7th Power Electronics, Drive Systems & Technologies Conference (PEDSTC 2016)
16-18 Feb. 2016, Iran University of Science and Technology, Tehran, Iran

A New SVM-based Voltage Balancing method for Five-Level NPC Inverter

[1]Pouria Qashqai, [1]Abdolreza Sheikholeslami
[2]Hani Vahedi, *Student, IEEE*, [2]Kamal Al-Haddad, *Fellow, IEEE*
[1]Babol Noshirvani University of Technology, Babol, Iran
[2]Ecole de Technologie Superieure, University of Quebec, Montreal, Canada
pouria.qashqai@gmail.com, asheikh@nit.ac.ir, hani.vahedi@etsmtl.ca, kamal.al-haddad@etsmtl.ca

Abstract—**This paper presents a new voltage-balancing method of five-level neutral-point clamped inverter, based on space-vector modulation redundant switching states. The proposed method uses measurement of DC-link capacitors and output phase currents in order to determine the best fitted redundancies. Due to its severe approach, the proposed method provides relatively low voltage ripples using low-capacitance DC-link capacitors. This advantage leads to reduction in size of the inverter and cost of manufacturing. Finally, some simulations in MATLAB/Simulink environment have been done in order to validate the dynamic performance of the proposed method.**

Index Terms—**Multilevel Inverters; Space Vector Modulation; Voltage-balancing; NPC; Power Quality.**

I. INTRODUCTION

Multilevel inverters (MLI's) have been studied and developed since past three decades due to their significant advantages over conventional two-level inverters. Advantages such as: 1- lower harmonic distortion, 2-generation of higher voltages with medium voltage semiconductors, 3-lower dv/dt, 4-smoother output waveform, 5-lower stress on semiconductors which results in increment of their lifetime [1-7].

Among different modulation techniques [8-17] for multilevel inverters, pulse width modulations (PWM) are the most popular ones due to their high frequency switching which provides smoother output waveforms and consequently reduction of harmonic distortions. Two major types of PWM techniques are: 1-carrier based techniques such as Sinusoidal PWM (SPWM) and 2-Space Vector PWM or simply Space Vector Modulation (SVM). SPWM is simple and efficient but it can't provide maximum voltage utilization, it is incapable of self-balancing for n>3 where n is the number of levels. Moreover, it needs higher number of carrier signal generators and controllers, especially when the number of levels increase. SVM switching technique, on the other hand, has resolved the above mentioned problems [3, 18, 19].

Neutral-Point Clamped (NPC) which is also known as Diode-Clamped topology [20], is one of the most popular multilevel topologies mostly because of its flexibility, cost-effectiveness and easy self-balancing feature [8, 16, 21, 22]. NPC topology utilizes only a single DC source in its input and generates different voltage levels using clamping diodes and DC-link capacitors. This advantage has a major side-effect which is voltage deviation of the DC-link capacitors [8, 15, 16, 23, 24].

Many studied have been done in the literature on voltage balancing of NPC inverter. While some methods have proposed using auxiliary circuits for voltage balancing of three-level [25], four-level [26] and five-level [27] NPC, other methods have used redundant switching states for that purpose on three-level [28, 29], five-level [30, 31] and n-level [32] NPC inverter. Some methods used g-h coordinate [15] for voltage balancing, other methods [33] have used optimization of DC-link capacitors energy, based on redundant vectors of SVM technique. In [13], a method to set the maximum ripple of the DC-link capacitors has been introduced but it needed external integrators, external Schmitt triggers and there were no optimization on choosing switching states. In [23] an SVM voltage balancing based on 3-level NPC DC-link capacitor voltages and output phase currents has been introduced which also reduces common-mode voltage as a side benefit. However, capacitors sizes have been never mentioned in that paper. In [29] a new method of voltage balancing based on sensation of the neutral-point current has been introduced. In that method, current direction of the neutral point in three-level NPC were used to select the best fitted redundant states for switching.

In his paper, firstly in section II, basic knowledge about SVM technique and five-level NPC have been presented. Afterwards in section III a voltage balancing using redundant vectors has been introduced which is based on DC-link capacitor voltages and output phase currents of the inverter. This method provides very low voltage ripples which leads to using low-capacity, smaller and cheaper capacitors. Finally, in section IV, some simulations have been done in MATLAB/Simulink environment in order to validate the dynamic performance of the proposed schemes.

II. FIVE-LEVEL NEUTRAL POINT CLAMPED INVERTER

Structure of a five-level three-phase NPC inverter has been illustrated in Fig. 1. This topology uses only a single DC source as its input and generates different voltage levels using DC-link capacitors and clamping diodes. All valid switching combinations in a leg of three-level NPC inverter have been listed in Table I. where 0,1...4 are symbols of their corresponding switching states.

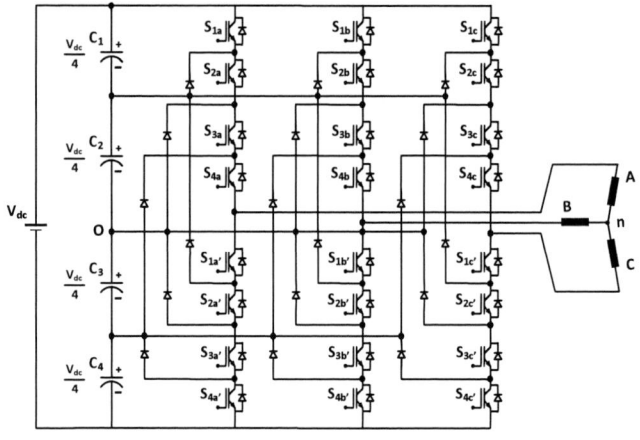

Fig. 1. Structure of 5-level NPC inverter

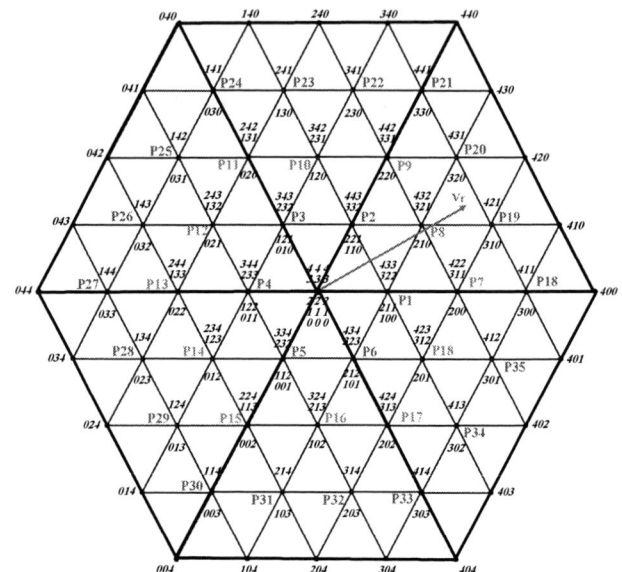

Fig. 2. SVM diagram of 5-level NPC

Since there are 5 valid switching states per phase, overall number of switching states of a three-phase NPC inverter would be $5^3 = 125$, which they can be displayed as a three-digit number $\overline{xyz}$ where x, y, z $\in$ {0,1,2,3,4}as. Each of these phase switching states (x,y and z) generate a predetermined voltage at their corresponding output phase according to Table I. Thus the overall three-digit state generates a three phase voltage at the output that can be represented as a voltage vector in $\alpha\beta$ plane using Eq. (1).

$$\begin{bmatrix} v_\alpha(t) \\ v_\beta(t) \end{bmatrix} = \frac{2}{3} \begin{bmatrix} 1 & -\frac{1}{2} & -\frac{1}{2} \\ 0 & \frac{\sqrt{3}}{2} & -\frac{\sqrt{3}}{2} \end{bmatrix} \begin{bmatrix} v_{ao}(t) \\ v_{bo}(t) \\ v_{co}(t) \end{bmatrix} \tag{1}$$

TABLE I
SWITCHING STATES OF NPC INVERTER (PHASE A)

Switching State	S_{1a}	S_{2a}	S_{3a}	S_{4a}	$S_{1a'}$	$S_{2a'}$	$S_{3a'}$	$S_{4a'}$	Inverter Terminal Voltage (V_{AO})
4	1	1	1	1	0	0	0	0	$\frac{V_{dc}}{2}$
3	0	1	1	1	1	0	0	0	$\frac{V_{dc}}{4}$
2	0	0	1	1	1	1	0	0	0
1	0	0	0	1	1	1	1	0	$-\frac{V_{dc}}{4}$
0	0	0	0	0	1	1	1	1	$-\frac{V_{dc}}{2}$

Where V_{AO}, V_{BO} and V_{CO} are the output phase voltages of the inverter. Variables in the left side of that equation can be considered as real and imaginary parts of a complex number P using Eq. (2).

$$\vec{P}_i(t) = v_\alpha(t) + jv_\beta(t) \qquad i = 0.1.2.\dots.60 \tag{2}$$

Converting all 125 different switching states using the aforementioned equations, 61 unique vectors would be calculated. By drawing the vectors on a two dimensional plane, the 5-level SVM hexagram as depicted in Fig. 2 would be achieved.

As it can be seen, some switching states lead to identical vectors in the diagram. Such redundant switching states play a major role in DC capacitor voltage balancing techniques [15, 23, 27, 34]. Except zero vectors at the origin (000,111…,444), other redundant vectors generate identical voltages at the output but their current paths are different. The proposed method in this paper selects the best redundant state in each vector to ensure that the voltage ripple of the capacitors remains in a predetermined band.

The idea of space vector modulation is based on approximation of the reference vector with three nearest vectors. To do so, conversion of the three-phase reference vector into a two-dimensional vector using Eq. (1) is inevitable. As shown in Fig. 2, the reference vector (V_r) is surrounded by three adjacent vectors (Let's say V_1, V_2 and V_3). Applying the volt-second principle in Eq. (3) and considering the fact that sampling time (T_s) is equal to summation of switching time intervals (T_1, T_2 and T_3) as observable in Eq. (4), all the time intervals in every sampling period would be calculated.

$$V_r T_s = V_1 T_1 + V_2 T_2 + V_3 T_3 \tag{3}$$

$$T_s = T_1 + T_2 + T_3 \tag{4}$$

$$T_s = \frac{1}{f_s} \tag{5}$$

Solving equations (3) and (4) leads to:

$$M_a = \frac{\sqrt{3}v_r}{V_{dc}} \tag{7}$$

$$T_1 = T_s[1 - 2M_a sin\theta] \tag{8}$$

$$T_2 = T_s\left[2M_a \sin\left(\frac{\pi}{3} + \theta\right) - 1\right] \tag{9}$$

978-1-5090-0376-1/16 $31.00 © 2016 IEEE

$$T_3 = T_s - T_1 - T_2 = T_s[1 - 2M_a \sin\left(\frac{\pi}{3} - \theta\right)] \qquad (10)$$

Where M_a is the modulation index, V_{dc} is the DC-link voltage and θ is the angle between the reference vector and horizontal axis.

Finding the adjacent vectors and calculating the time-intervals of them have been widely studied in the literature [9-12, 15, 35-37]. In this paper the method in [11] has been used. In the following section, an effective method is introduced which utilizes redundant states for arbitrary voltage balancing of five-level NPC in a way that relatively low voltage ripples occur to DC-link capacitors.

III. PROPOSED VOLTAGE BALANCING METHOD

As shown in Fig. 1, there are four DC-link capacitors in a five-level NPC. Thus the voltage of each capacitor would be $V_{dc}/4$ in steady state operation where V_{dc} is the DC source voltage (V_{dc}). But voltage of those capacitors are open to change because of non-zero currents which are flowing through them. Voltage deviation of the DC-link capacitors is a major concern in NPC inverters. In [38] a study has shown that average value of capacitors currents in five-level NPC when the inverter is delivering non-zero real power, is not equal to zero thus voltage deviation phenomenon would undoubtedly occur.

Owing to redundant states in SVM technique, voltages can be balanced without any auxiliary circuit. In this section a new method has been introduced which uses capacitor voltages and phase currents for arbitrary voltage balancing of five-level NPC.

A. Creating the list of priorities

First step in the proposed method is finding the most deviated capacitor voltage among others. To do so, voltages of all the capacitors (V_{C1}-V_{C4}) have to be measured and compared with the reference value ($V_{dc}/4$). The error matrix "Err" indicates the absolute error between each measured voltage and the reference value as shown in Eq. (12).

$$Err = [E_1 \quad E_2 \quad E_3 \quad E_4] \qquad (11)$$

$$E_i = V_{Ci} - \frac{V_{dc}}{4} \qquad (12)$$

By sorting the error matrix Err, a new matrix called "Matrix of priorities" which is demonstrated by Eq. (13) would be achieved.

$$Cerr = [CE_1 \quad CE_2 \quad CE_3 \quad CE_4] \qquad (13)$$

In every sampling period, the capacitor whose absolute error or in other words, its corresponding element in Err matrix is the greatest, would be considered as the most deviated capacitor which is equal to the first element of Cerr matrix (CE_1). Other elements may be used if there would be no way to minimize the CE_1 deviation.

In continue, the proposed method would try to reduce the voltage deviation of that capacitor, using redundant switching states.

B. Representation of the redundancies

Considering the SVM diagram of five-level NPC shown in Fig. 2, it is obvious that except the vectors on the outer hexagram, all other vectors contain at least one redundant states. Each vector P_i with redundant states can be represented as:

$$P_i = \sum_{j=1}^{n}[x_j \quad y_j \quad z_j]U_j \qquad (14)$$

Where n is the number of all redundant switches. For example, n for the zero vector at the origin is equal to 5. Xj, Yj, and Zj are the corresponding switching states of the jth redundancy. And finally redundancy selection coefficient, Uj is either 1 or 0 and implies which redundant state is active.

Take P_1 as an example. This vector can be written as:

$$P_1 = [4 \quad 3 \quad 3]U_1 + [3 \quad 2 \quad 2]U_2 + [2 \quad 1 \quad 1]U_3 + [1 \quad 0 \quad 0]U_4 \qquad (15)$$

Since there are four redundant states in this vector (n=4), thus there would be four redundancy selection coefficients U_1-U_4. It is worth to mention that only one of those coefficient can be 1 at a time and others must remain 0 to ensure that only one redundancy is selected.

C. Selecting the best redundant state

As mentioned earlier, redundant states generate identical voltage vectors but with different current paths. There are three types of non-zero vectors containing redundant states:

Type I: P1 is an example of this type. There are four redundancies for that vector. Each of those redundancies and their corresponding current paths have been demonstrated in Fig. 3.

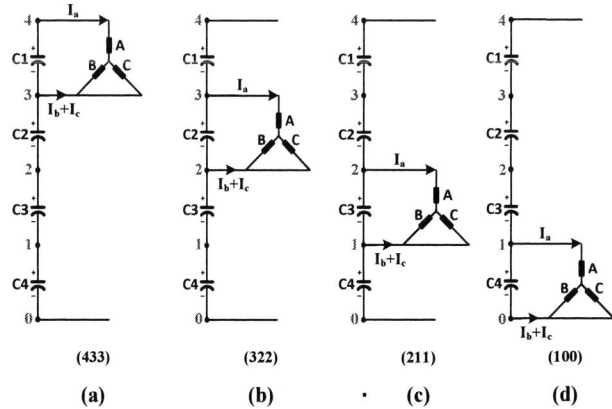

Fig. 3. Redundant switching states of P1: 433(a); 322(b); 211(c); and 100(d)

As it is clear, the number used to display switching states, also implies the connection pattern of that state. First digit means which point has been connected to phase A, second and third digits also display that which points have been connected to phase B and C respectively. As shown in Fig. 3, I_a flows through a capacitor in each state and $I_b + I_c$ or $-I_a$ flows through other capacitors. Thus it's always possible to reduce voltage deviation of CE_1 by charging/discharging it when it is undercharged/overcharged. But it is important to remember that charge/discharge of CE_1, will result in discharge/charge of the other capacitors. So the best case scenario would be to select a redundancy in which, the most deviated capacitor (CE_1) and the second most deviated one (CE_2), both get balanced. The flowchart of such procedure has been demonstrated in Fig. 4. Where I_x is equal to I_a for P1 but it takes different values for other 5 vectors of the inner hexagram (P2-P6). And i,j,k and l are the corresponding indexes showing which capacitors are the most deviated ones. For example, if second capacitor is the most deviated one and the 4th capacitor is in the next rank, then i=2 and j=4 respectively.

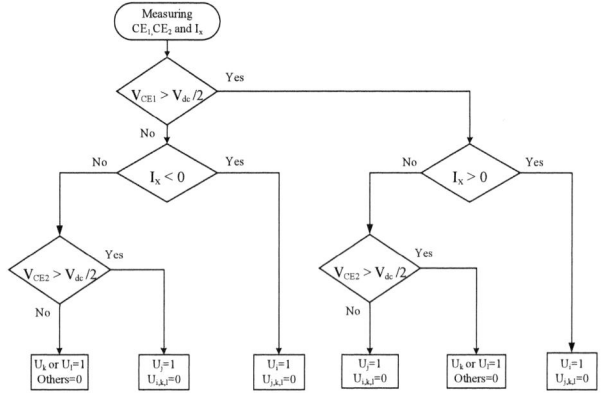

Fig. 4.   Flowchart of redundancy selection of vector P1.

In case that CE_1 could be balanced, then always its corresponding U has to be selected. But if it couldn't be balanced directly, then flowing I_x through the second most deviated capacitor would balance both capacitors. I_x for all 6 vectors of type I has been listed in Table II. Due to significant increase in complexity, checking further capacitors is no longer needed and one of two remaining redundancies can be selected. For modulation indexes under $1/2\sqrt{3}$ that the reference vector lies inside the inner hexagram, voltage balancing would be always possible.

TABLE II
I_X CURRENT FOR ALL TYPE I VECTORS

Switching Vector	P1	P2	P3	P4	P5	P6
I_x	I_a	$-I_c$	I_b	$-I_a$	I_c	$-I_b$

Type II vectors have three redundant states. Take vector P7 as an example of this type. Fig. 5 illustrates various states of that vector and their corresponding current paths. As it can be seen, in all cases there are two pairs of capacitors that change together. This dependence of capacitor voltages makes the voltage balancing process harder in some conditions.

When that type of vectors gets involved, the proposed method tries to reduce deviation of CE_1 and it is always to do so. But it may result in over/under charging another capacitor. So among all possible redundancies, the one who leads to the least voltage deviation have to be selected. To do so, CE_2 always have to be checked and if it is possible, a state in which both CE_1 and CE_2 get balanced have to be selected.

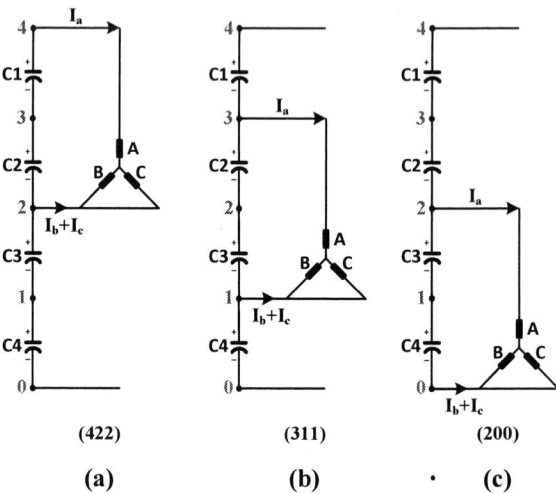

Fig. 5.   Redundant switching states of P1: 422(a); 311(b); and 200(c);

The I_x current for P7 and all other vectors of type II that are $n \times 60°$ rotated with respect to P7 where n=1, 2,...,5 has been listed in Table III.

TABLE III
I_X CURRENT FOR TYPE II VECTORS, P7 GROUP

Switching Vector	P7	P9	P11	P13	P15	P17
I_x	I_a	$-I_c$	I_b	$-I_a$	I_c	$-I_b$

But for P8 and all other 5 vectors that are $n \times 60°$ rotated with respect to it, it is different because each phase is connected to a different point. Thus two currents I_{x1} and I_{x2} have to be checked where are used to balance CE_1 and CE_2 respectively. These currents for P8 group of type II vectors have been listed in Table IV.

TABLE IV
I_X CURRENT FOR TYPE II VECTORS, P8 GROUP

Switching Vector	P8	P10	P12	P14	P16	P18
I_{x1}	I_a-I_b	I_b-I_a	I_b-I_c	I_c-I_b	I_c-I_a	I_a-I_c
I_{x2}	I_b-I_c	I_a-I_c	I_c-I_a	I_b-I_a	I_a-I_b	I_c-I_b

For outermost hexagram with redundancies, type III would be defined which contains three groups: P18, P19 and P20. Members of each group are $n \times 60°$ rotated with respect to each other. Tables V and VI has listed I_x currents that must be checked in those vectors for voltage balancing.

TABLE V
I_X CURRENT FOR TYPE III VECTORS, P18 GROUP

Switching Vector	P18	P21	P24	P27	P30	P33
I_x	$-I_a$	I_c	$-I_b$	I_a	$-I_c$	I_b

TABLE VI
I_X CURRENT FOR TYPE III VECTORS, P19 AND P20 GROUP

Switching Vector	P19	P22	P25	P28	P31	P34
I_{x1}	I_a-I_b	I_b-I_a	I_b-I_c	I_c-I_b	I_c-I_a	I_a-I_c
I_{x2}	I_b-I_c	I_a-I_c	I_c-I_a	I_b-I_a	I_a-I_b	I_c-I_b

As seen, I_x in Table V is negative of I_x in Table II. The reason is because in this group, each 3 capacitors are in series and for example in P18, Ia flows through them. Since controlling one capacitor is simpler, it can be considered equal as if –Ia where flowing through the remaining capacitor. Same analogy is valid for Table IV and VI.

978-1-5090-0376-1/16 $31.00 © 2016 IEEE   514

D. Operation limits

According to [32] arbitrary voltage balancing for n-level NPC has some operation limits dictated by modulation index and AC-side current angle. Fig. 6 has demonstrated these limits for four and five-level NPC. For significantly high number of levels, the voltage-balancing limits reaches the shaded area.

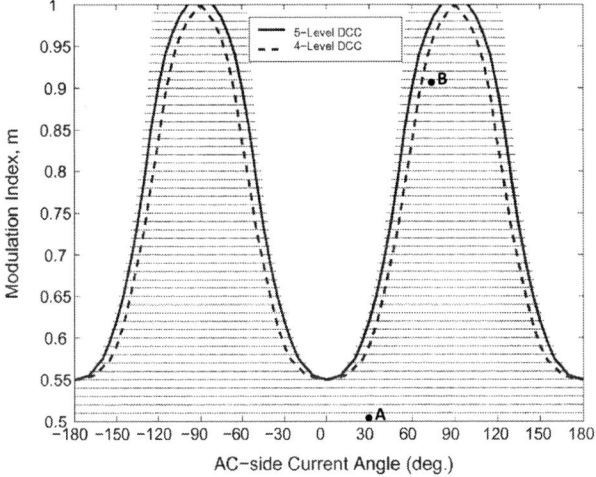

Fig. 6. Operation limits of the proposed SVM-based balancing method for four-level and five-level NPC.

IV. SIMULATION RESULTS

In this section, some simulations have been done in order to investigate the dynamic performance of the proposed voltage-balancing method. Configuration of the test system has been listed in Table VII.

TABLE VII
CONFIGURATION OF THE TEST SYSTEM

Output Frequency	50Hz
Sampling Frequency	2000Hz
DC Link Voltage	400V
DC Link Capacitors	500uF
Load Impedance for PF=0.86	40Ω , 20mH
Load Impedance for PF=0.34	40Ω , 250mH

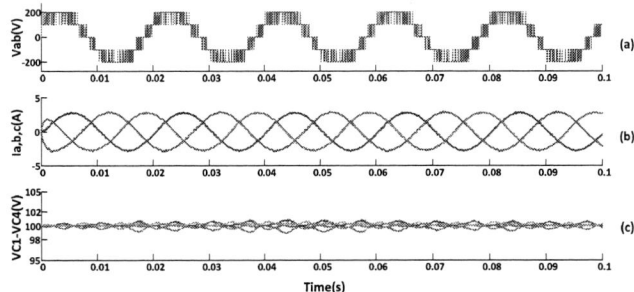

Fig. 7. Output line voltage (a); output phase currents (b); and DC-capacitor voltages (c); under PF=0.86 and Ma=0.5 condition.

Considering diagram of operation limits in Fig. 6, two points have been selected to investigate. Point A which is in the normal operation area, is feeding a typical load with PF=0.86 using a mediocre modulation index (Ma=0.5). output waveforms and DC-link capacitor voltages of this point have been demonstrated in Fig. 7. As seen, the voltage ripples are around 2% which considering the low-capacitance DC-capacitors, it is quite acceptable.

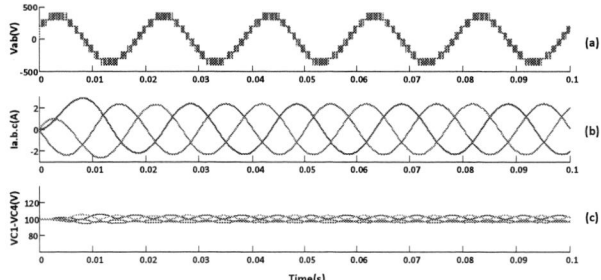

Fig. 8. Output line voltage (a); output phase currents (b); and DC-capacitor voltages (c); under PF=0.34 and Ma=0.9 condition.

As an example of critical operation conditions, waveforms of point B have been illustrated in Fig. 8. Due to connection of a highly inductive load (PF=0.34), arbitrary voltage balancing is valid even for relatively high modulation indexes. But as it is clear in Fig. 8-c, there are some offsets on DC-capacitor voltages. Nevertheless, voltage ripples remain under 5% which is also fairly acceptable considering the low capacity of the DC-link Capacitors.

V. CONCLUSION

Multilevel inverters play a crucial role in modern power electronics especially when it comes to using renewable energy sources. NPC inverter is one of the most popular multilevel inverter topologies due to its cost-effectiveness and robustness. That configuration utilizes some DC-link capacitors to clamp the input DC voltage. That capacitors suffer voltage-deviation phenomenon. SVM modulation method not only overcomes drawbacks of conventional SPWM, but also has an inherent capability of arbitrary voltage balancing.

Many studies have been done on SVM-based voltage balancing of three-level NPC. In this paper a new voltage balancing method of five-level NPC has been introduced which is based on measurement of DC-link capacitor voltages and output currents. Due to severe decision process of the proposed method, voltage ripples of the DC-link capacitors would be relatively low even by connection low-capacitance capacitors. This is the main advantage of the proposed method which leads to implementation of smaller and cheaper capacitors that results in reduction of manufacturing size and price.

Some simulations have been done and shown the good dynamic performance of the proposed method. Based on predecessor sturdies, five-level NPC has some practical limitations when it comes to SVM-based voltage balancing.

978-1-5090-0376-1/16 $31.00 © 2016 IEEE

These limitations have been investigated by simulations in this paper and validity of those studies have been proved.

REFERENCES

[1] M. Malinowski, K. Gopakumar, J. Rodriguez, and M. A. Perez, "A survey on cascaded multilevel inverters," *Industrial Electronics, IEEE Transactions on,* vol. 57, pp. 2197-2206, 2010.

[2] B. P. McGrath and D. G. Holmes, "Multicarrier PWM strategies for multilevel inverters," *Industrial Electronics, IEEE Transactions on,* vol. 49, pp. 858-867, 2002.

[3] J. Rodriguez, J.-S. Lai, and F. Z. Peng, "Multilevel inverters: a survey of topologies, controls, and applications," *Industrial Electronics, IEEE Transactions on,* vol. 49, pp. 724-738, 2002.

[4] P. Qashqai, A. Sheikholeslami, H. Vahedi, and K. Al-Haddad, "A Review on Multilevel Converter Topologies for Electric Transportation Applications," in *Vehicle Power and Propulsion Conference (VPPC), 2015 IEEE,* 2015, pp. 1-6.

[5] H. Vahedi and K. Al-Haddad, "Real-Time Implementation of a Packed U-Cell Seven-Level Inverter with Low Switching Frequency Voltage Regulator," *IEEE Trans. Power Electron.,* 2015.

[6] H. Vahedi, K. Al-Haddad, Y. Ounejjar, and K. Addoweesh, "Crossover Switches Cell (CSC): A new multilevel inverter topology with maximum voltage levels and minimum DC sources," in *Industrial Electronics Society, IECON 2013-39th Annual Conference of the IEEE,* 2013, pp. 54-59.

[7] H. Vahedi, S. Rahmani, and K. Al-Haddad, "Pinned mid-points multilevel inverter (PMP): three-phase topology with high voltage levels and one bidirectional switch," in *Industrial Electronics Society, IECON 2013-39th Annual Conference of the IEEE,* 2013, pp. 102-107.

[8] M. Sharifzadeh, H. Vahedi, A. Sheikholeslami, P.-A. Labbé, and K. Al-Haddad, "Hybrid SHM-SHE Modulation Technique for Four-Leg NPC Inverter with DC Capacitors Self-Voltage-Balancing," *IEEE Trans. Ind. Electron.,* vol. 62, pp. 4890-4899, 2015.

[9] N. Celanovic and D. Boroyevich, "A fast space-vector modulation algorithm for multilevel three-phase converters," *Industry Applications, IEEE Transactions on,* vol. 37, pp. 637-641, 2001.

[10] Y. Deng, Y. Wang, K. Teo, and R. G. Harley, "A Simplified Space Vector Modulation Scheme for Multilevel Converters," *Power Electronics, IEEE Transactions on,* vol. PP, pp. 1-1, 2015.

[11] A. K. Gupta and A. M. Khambadkone, "A general space vector PWM algorithm for multilevel inverters, including operation in overmodulation range," *Power Electronics, IEEE Transactions on,* vol. 22, pp. 517-526, 2007.

[12] S. Jae Hyeong, C. Chang Ho, and H. Dong-Seok, "A new simplified space-vector PWM method for three-level inverters," *Power Electronics, IEEE Transactions on,* vol. 16, pp. 545-550, 2001.

[13] R. Rojas, T. Ohnishi, and T. Suzuki, "An improved voltage vector control method for neutral-point-clamped inverters," *Power Electronics, IEEE Transactions on,* vol. 10, pp. 666-672, 1995.

[14] M. Saeedifard, R. Iravani, and J. Pou, "A space vector modulation strategy for a back-to-back five-level HVDC converter system," *Industrial Electronics, IEEE Transactions on,* vol. 56, pp. 452-466, 2009.

[15] J. Yang, F. C. Lee, and L. Sizhao, "Space Vector Modulation for Three-Level NPC Converter With Neutral Point Voltage Balance and Switching Loss Reduction," *Power Electronics, IEEE Transactions on,* vol. 29, pp. 5579-5591, 2014.

[16] W. Yue, L. Ning, L. Su, C. Wulong, L. Wanjun, and W. Zhao'an, "Research on DC capacitor voltage self-balancing space vector modulation strategy of five-level NPC converter," in *Applied Power Electronics Conference and Exposition (APEC), 2014 Twenty-Ninth Annual IEEE,* 2014, pp. 2694-2699.

[17] R. Zgheib and K. Al-Haddad, "Neural Network Controller to Manage the Power Flow of a Hybrid Source for Electric Vehicles," in *Vehicle Power and Propulsion Conference (VPPC), 2015 IEEE,* 2015, pp. 1-6.

[18] B. Wu, *High-power converters and AC drives*: John Wiley & Sons, 2006.

[19] H. Vahedi, P.-A. Labbé, and K. Al-Haddad, "Sensor-Less Five-Level Packed U-Cell Inverter (PUC5) Operating in Stand-Alone and Grid-Connected Modes," *IEEE Trans. Ind. Informat.,* 2015.

[20] R. H. Baker, "Switching circuit," ed: Google Patents, 1980.

[21] M. Sharifzade, H. Vahedi, A. Sheikholeslami, H. Ghoreishy, and K. Al-Haddad, "Selective Harmonic Elimination Modulation Technique Applied on Four-Leg NPC," in *ISIE 2014-23rd IEEE International Symposium on Industrial Electronics,* Turkey, 2014, pp. 2163-2168.

[22] M. Sharifzade, H. Vahedi, A. Sheikholeslami, H. Ghoreyshi, and K. Al-Haddad, "Modified selective harmonic elimination employed in four-leg NPC inverters," in *IECON 2014-40th Annual Conference of the IEEE Industrial Electronics Society,* 2014, pp. 5196-5201.

[23] A. Choudhury, P. Pillay, and S. S. Williamson, "Modified DC-bus Voltage Balancing Algorithm for a 3-Level Neutral Point Clamped (NPC) PMSM Inverter Drive with Reduced Common-Mode Voltage," *Industry Applications, IEEE Transactions on,* vol. PP, pp. 1-1, 2015.

[24] H. Vahedi, P.-A. Labbe, H. Y. Kanaan, H. F. Blanchette, and K. Al-Haddad, "A new five-level buck-boost active rectifier," in *Industrial Technology (ICIT), 2015 IEEE International Conference on,* 2015, pp. 2559-2564.

[25] A. Von Jouanne, S. Dai, and H. Zhang, "A simple method for balancing the DC-link voltage of three-level inverters," in *Power Electronics Specialists Conference, 2001. PESC. 2001 IEEE 32nd Annual,* 2001, pp. 1341-1345.

[26] K. A. Corzine, J. Yuen, and J. R. Baker, "Analysis of a four-level DC/DC buck converter," in *Industry Applications Conference, 2001. Thirty-Sixth IAS Annual Meeting. Conference Record of the 2001 IEEE,* 2001, pp. 1882-1888.

[27] A. A.-M. Ibrahim, "A practical method for capacitor voltage balancing of diode clamped multilevel inverters," Heriot-Watt University, 2004.

[28] R. M. Tallam, R. Naik, and T. Nondahl, "A carrier-based PWM scheme for neutral-point voltage balancing in three-level inverters," *Industry Applications, IEEE Transactions on,* vol. 41, pp. 1734-1743, 2005.

[29] J. Pou, R. Pindado, D. Boroyevich, and P. Rodriguez, "Evaluation of the low-frequency neutral-point voltage oscillations in the three-level inverter," *Industrial Electronics, IEEE Transactions on,* vol. 52, pp. 1582-1588, 2005.

[30] C. Newton and M. Sumner, "Novel technique for maintaining balanced internal DC link voltages in diode clamped five-level inverters," in *Electric Power Applications, IEE Proceedings-,* 1999, pp. 341-349.

[31] H. Zhang, S. Finney, T. Lim, A. Massoud, J. Yang, and B. Williams, "DC-link capacitor voltage balancing for a five-level diode-clamped active power filter using redundant vectors," *Australian Journal of Electrical & Electronics Engineering,* vol. 10, p. 137, 2013.

[32] J. Pou, R. Pindado, and D. Boroyevich, "Voltage-balance limits in four-level diode-clamped converters with passive front ends," *Industrial Electronics, IEEE Transactions on,* vol. 52, pp. 190-196, 2005.

[33] M. Saeedifard, H. Nikkhajoei, and R. Iravani, "A space vector modulated STATCOM based on a three-level neutral point clamped converter," *Power Delivery, IEEE Transactions on,* vol. 22, pp. 1029-1039, 2007.

[34] H. Vahedi, K. Al-Haddad, and H. Y. Kanaan, "A new voltage balancing controller applied on 7-level PUC inverter," in *Industrial Electronics Society, IECON 2014-40th Annual Conference of the IEEE,* 2014, pp. 5082-5087.

[35] A. K. Gupta and A. M. Khambadkone, "A space vector PWM scheme for multilevel inverters based on two-level space vector PWM," *Industrial Electronics, IEEE Transactions on,* vol. 53, pp. 1631-1639, 2006.

[36] B. P. McGrath, D. G. Holmes, and T. Lipo, "Optimized space vector switching sequences for multilevel inverters," *Power Electronics, IEEE Transactions on,* vol. 18, pp. 1293-1301, 2003.

[37] S. Wei, E. Wu, F. Li, and C. Liu, "A general space vector PWM control algorithm for multilevel inverters," in *Applied Power Electronics Conference and Exposition, 2003. APEC'03. Eighteenth Annual IEEE,* 2003, pp. 562-568.

[38] M. Saeedifard, R. Iravani, and J. Pou, "Analysis and control of DC-capacitor-voltage-drift phenomenon of a passive front-end five-level converter," *Industrial Electronics, IEEE Transactions on,* vol. 54, pp. 3255-3266, 2007.

7th Power Electronics, Drive Systems & Technologies Conference (PEDSTC 2016)
16-18 Feb. 2016, Iran University of Science and Technology, Tehran, Iran

Model Predictive Control of Classic Bidirectional DC-DC Converter for Battery Applications

A. Pirooz, R. Noroozian

Department of Electrical Engineering
University of Zanjan
Zanjan, Iran
a.pirooz@znu.ac.ir, noroozian@znu.ac.ir

Abstract— **In this paper, controlling of a DC-DC bidirectional converter is investigated using model predictive control (MPC) technique. The converter is supposed to connect a battery bank to the DC bus. The main purpose of MPC method is to maintain the DC bus voltage in a predefined rated range, which means injecting active power to the DC bus to increase bus voltage and discharging the battery, and on the contrary absorbing active power from the DC bus to decrease bus voltage and charging the battery. The control strategy also disconnects the battery bank while DC bus voltage is in the predefined rated range. Simulations have been done in PSCAD/MATLAB software interfacing and results show the feasibility of the proposed controlling strategy to conduct active power on both directions.**

Keywords—bettery bank; model predictive control; bidirectional DC-DC converter; DC voltage control

I. INTRODUCTION

Nowadays distributed generation (DG) units are highly integrated into DC and AC distribution networks. Renewable energy sources as photovoltaic and wind power units cannot supply the needs of power continuously in a distribution network due to their dependencies to environmental conditions [1]. A battery bank is used to compensate the power difference between loads and generation units [2]. Obviously batteries are not infinite sources of power and discharge in a rather short period. In these cases also surplus power in the system is used to charge the battery banks. As a result, the converter connecting the battery bank to the system must allow power flow between two buses on both directions. Bidirectional converters are also used in DC uninterruptable power supplies (UPS), electrical vehicles and ships [3-5].

Basically, unidirectional DC converters as buck and boost, doesn't allow reverse current conductance since they have diodes in their topology. By replacing diodes with controllable switches they could be turned into bidirectional converters [6]. The classic bidirectional converter which is shown in Fig. 1 is a combination of buck and boost converter which has the same structure in both cases. The inductor is on the low voltage (LV) side which the battery is connected to. Existence of the inductor on LV side decreases current ripple which is advantageous for charging and discharging the battery and makes this converter suitable for battery applications.

Controlling bidirectional DC-DC converters is another challenge that has been the subject of many researches in literature. Controlling method must provide fast dynamic response and follow the reference signal in both power flow directions. Pulse width modulation (PWM) and phase-shift control [7], hysteresis current control (HCC) [8] and fuzzy control [9] are different controlling techniques that could be applied to bidirectional DC-DC converters.

Model predictive control (MPC) is a derivative of predictive control family which uses a predefined cost function in order to make system variables follow their reference values. An accurate mathematical discrete-time model of the system is needed to predict future behavior of the variables. On each time step, subject to the predicted values and the cost function, an optimization problem is solved online over a finite prediction horizon which leads to optimal controlling actions. Basic idea of MPC is easy to understand and it could be applied to every system with known mathematical models. A large number of calculations needs to be done online, but with powerful microprocessors available today, this cannot be a challenge for implementing MPC into power electronic converters which operate with fast dynamic behavior [10]. This research applies MPC to a classic bidirectional DC-DC converter. It takes advantage of independent buck and boost operational modes of the converter. Low voltage (LV) and high voltage (HV) sides of the converter are connected to a battery bank and a DC distribution network bus respectively. A predefined rated range is provided for the DC bus voltage which has a lower and upper limit. A controlling strategy is proposed based on a voltage feedback signal from DC bus to discharge the battery while DC bus voltage value is below the predefined rated range, to charge the battery while DC bus voltage value exceeds the predefined rated range and to disconnect the battery (idle mode) while DC bus voltage value is in the predefined rated range. The battery itself is modeled by its internal voltage V_{BAT} and resistance R_{BAT}. Accurate modeling of the battery bank is out of the subject of this research.

The rest of this paper is organized as follows. Section II includes modeling of the converter. Predictive control of the converter is discussed in section III. Section IV includes proposed controlling strategy for charging/discharging the battery. Finally, simulation results using PSCAD/MATLAB software interfacings are demonstrated in section V.

978-1-5090-0376-1/16 $31.00 © 2016 IEEE

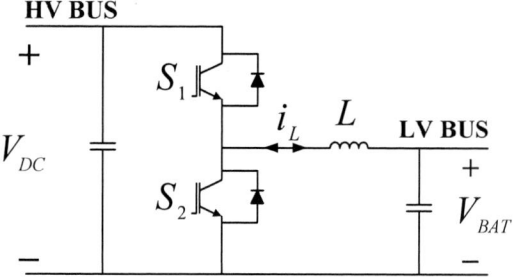

Fig. 1. Classic bidirectional DC-DC converter topology.

II. CONVERTER MODELING

The bidirectional DC-DC converter shown in Fig. 1 is a combination of buck and boost converters. While S_1 is switching and S_2 is constantly off, the converter operates in independent buck mode, DC bus voltage level is decreased to battery voltage level, and power flows from the DC bus to charge the battery. While S_2 is switching and S_1 is constantly off, the converter operates in independent boost mode, battery voltage level is increased to DC bus voltage level, and power injected to the DC bus discharges the battery. Due to the specific application and battery behavior, the converter is supposed to operate only in continuous current mode (CCM).

A. Buck Mode

Buck mode topology is shown in Fig. 2 (a). There are two semiconductors, controllable switch S_1 and a diode. The inductor L is used to store and release electrical energy depending on switch S_1 state. When S_1 is on (S_1=1), energy is stored in the inductor L through DC bus as shown in Fig. 2 (b). When S_1 is off (S_1=0), energy is released from the inductor L to the battery as shown in Fig. 2 (c). Consequently, continuous-time state space equations of the converter becomes as below [11]:

$$\frac{dx(t)}{dt} = (A_1 + A_2 u(t))x(t) + Bu(t)V_{DC} \tag{1}$$
$$y(t) = Cu(t)x(t)$$

where state variables are inductor current $i_L(t)$ and battery voltage $v_{BAT}(t)$:

$$x(t) = [i_L(t) \quad v_{BAT}(t)]^T \tag{2}$$

system matrices A_1, A_2, B and C are defined as below. $u(t)$ is status of the switch S_1. If S_1 is on, $u(t)$=1 and if S_1 is off, $u(t)$=0. $y(t)$ is also the output of the system.

$$A_1 = \begin{bmatrix} 0 & 0 \\ 0 & -\dfrac{1}{R_{BAT}C} \end{bmatrix}, A_2 = \begin{bmatrix} 0 & -\dfrac{1}{L} \\ \dfrac{1}{C} & 0 \end{bmatrix}, B = \begin{bmatrix} \dfrac{1}{L} \\ 0 \end{bmatrix}, C = \begin{bmatrix} 0 & 1 \end{bmatrix}$$

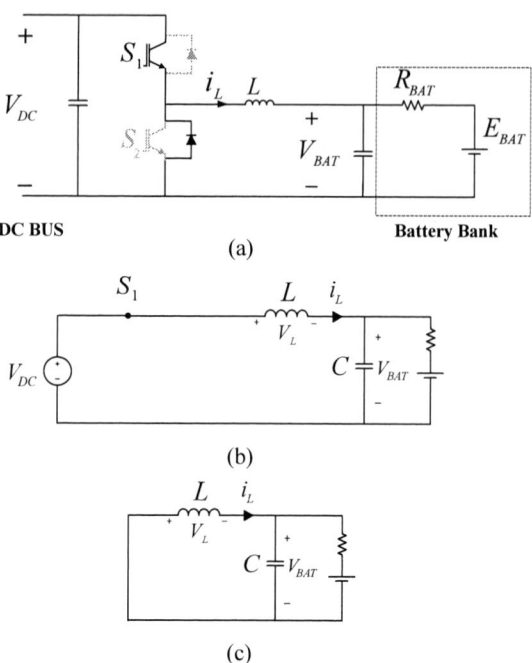

Fig. 2. Buck mode operation: a) buck mode topology b) switch on S_1=1 c) switch off S_1=0.

To obtain an accurate equation for prediction block, discrete-time model of the system is needed due to inherent procedure of MPC. By using Euler forward method, discrete-time model of first order systems could be obtained by approximation of derivatives as below:

$$\frac{dx}{dt} = \frac{x(k+1) - x(k)}{T_s} \tag{3}$$

where T_s is the sampling time. However, for higher order systems this method introduces significant error, thus exact discretization must be done [10]. By using (3), discretization of (1) is resulted as:

$$x(k+1) = \begin{cases} E_1 x(k) + FV_{DC}, y(k) = Gx(k) & \text{for } S_1 = 1 \\ E_2 x(k), y(k) = 0 & \text{for } S_1 = 0 \end{cases} \tag{4}$$

the matrices E_1, E_2, F and G are also defined as below:

$$E_1 = \mathbf{1} + (A_1 + A_2)T_s, \ E_2 = \mathbf{1} + A_1 T_s$$
$$F = BT_s, \ G = C$$

note that $\mathbf{1}$ is the identity matrix.

B. Boost Mode

Boost mode topology is shown in Fig. 3 (a). There are two semiconductors, controllable switch S_2 and a diode. The inductor L is used to store and release electrical energy depending on switch S_2 status. When S_2 is on (S_2=1), energy is stored in the inductor L through DC bus as shown in Fig. 3 (b). When S_2 is off (S_2=0), energy is released from the inductor L to the battery as shown in Fig. 3 (c). Consequently, continuous-time state space equations of the converter becomes as below [11]:

$$\frac{dx(t)}{dt} = (A_1 + A_2 u(t))x(t) + BV_{BAT} \tag{5}$$
$$y(t) = Cx(t)$$

where state variables are inductor current $i_L(t)$ and DC bus voltage $v_{DC}(t)$:

$$x(t) = [i_L(t) \quad v_{DC}(t)]^T \tag{6}$$

system matrices A_1, A_2, B and C are defined as below. $u(t)$ is status of the switch S_2. If S_2 is on, $u(t)$=1 and if S_2 is off, $u(t)$=0. $y(t)$ is also the output of the system.

$$A_1 = \begin{bmatrix} 0 & -\dfrac{1}{L} \\ \dfrac{1}{C} & -\dfrac{1}{R_{DC}C} \end{bmatrix}, A_2 = \begin{bmatrix} 0 & \dfrac{1}{L} \\ -\dfrac{1}{C} & 0 \end{bmatrix}, B = \begin{bmatrix} \dfrac{1}{L} \\ 0 \end{bmatrix}, C = \begin{bmatrix} 0 & 1 \end{bmatrix}$$

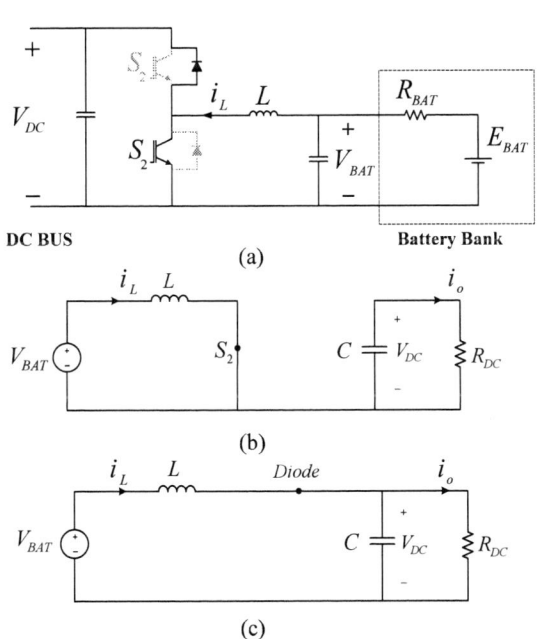

DC BUS

(a)

(b)

(c)

Fig. 3. Boost mode operation: a) boost mode topology b) switch on S_2=1 c) switch off S_2=0.

where R_{DC} is the equivalent resistance seen by the DC bus. By using (3), discretization of (5) is resulted as:

$$x(k+1) = \begin{cases} E_1 x(k) + FV_{BAT} & \text{for } S=1 \\ E_2 x(k) + FV_{BAT} & \text{for } S=0 \end{cases} \tag{7}$$
$$y(k) = Gx(k)$$

the matrices E_1, E_2, F and G are also defined as below:

$$E_1 = \mathbf{1} + (A_1 + A_2)T_s, E_2 = \mathbf{1} + A_1 T_s$$
$$F = BT_s, G = C$$

note that $\mathbf{1}$ is the identity matrix.

III. MODEL PREDICTIVE CONTROL OF THE CONVERTER

MPC scheme of buck and boost converters are presented in this section. Discrete-time modeling of the converter has been achieved on previous section. MPC is generally classified into discrete-time modeling, prediction and optimization blocks. Measurements are imported to the algorithm on prediction stage, based on previous modelings, all possible future behavior of controlled variables are predicted as below:

$$x_{pi}(t_{k+1}) = f_p\{x(t_k), S_i\} \tag{8}$$
$$i = 1,...,n$$

future predicted values of control variables $x_{pi}(t_{k+1})$ are a function of current values of control variables $x(t_k)$ subject to switching state of the converter S_i. i is number of possible states for switchings. For example, there are two possible switchings (on or off) for boost mode as there is only one controllable switch. Equations (4) and (7) are used as prediction function for buck and boost modes respectively. The output of prediction stages is a two dimensional array for each operation mode. All predictions are forwarded to the optimization stage. MPC defines the output control actions subject to a predefined cost function:

$$g_i = f_g\{x^*(t_{k+1}), x_{pi}(t_{k+1})\} \tag{9}$$

g_i is a function of predicted values imported from prediction block $x_{pi}(t_{k+1})$ and future values of reference signal $x^*(t_{k+1})$. Since the reference signal is a DC value, it can be supposed that it remains constant during sampling time T_s:

$$x^*(t_{k+1}) \approx x^*(t_k) \tag{10}$$

g_i is usually defined as error between variable values and their reference value. Consequently the cost function needs to be minimized by MPC scheme. On each sampling interval T_s, the switching state that minimizes g_i is chosen by MPC scheme to be applied on next sampling interval. The cost function for buck mode is defined as below:

$$J_{buck} = \begin{cases} |V^*_{BAT}(k+1) - V_{BAT,S_1=1}(k+1)| & \text{for } S_1 = 1 \\ |V^*_{BAT}(k+1) - V_{BAT,S_1=0}(k+1)| & \text{for } S_1 = 0 \end{cases} \tag{11}$$

and also for boost mode:

$$J_{boost} = \begin{cases} |V^*_{DC}(k+1) - V_{DC,S_2=1}(k+1)| & \text{for } S_2 = 1 \\ |V^*_{DC}(k+1) - V_{DC,S_2=0}(k+1)| & \text{for } S_2 = 0 \end{cases} \tag{12}$$

978-1-5090-0376-1/16 $31.00 © 2016 IEEE

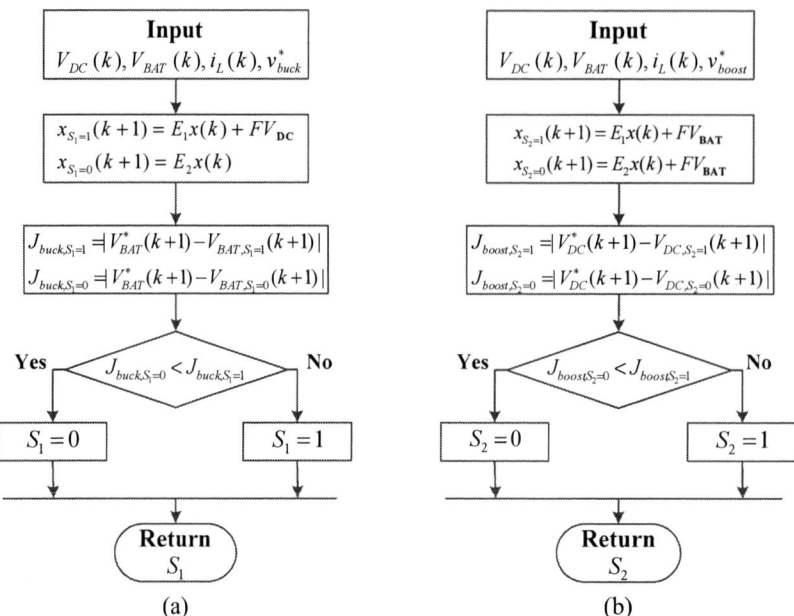

Fig. 4. Simplified flowcharts of MPC scheme a) buck operation b) boost operation.

Simplified flowcharts of MPC scheme for buck and boost operational modes of bidirectional DC-DC converter are shown in Fig. 4.

IV. PROPOSED CONTROL STRATEGY

The converter aims to regulate DC bus voltage V_{DC} and keep it in the predefined rated range. Upper and lower limits are defined for the control scheme as shown in Fig. 5. As DC bus voltage V_{DC} drops below the lower limit, bidirectional converter switches to boost mode, injects power to the DC bus by discharging the battery in order to keep V_{DC} on lower limit. On the contrary, as the DC bus voltage V_{DC} exceeds the upper limit, bidirectional converter switches to buck mode, absorbs power from the DC bus by charging the battery in order to keep V_{DC} on upper limit and also switches to idle mode and disconnects the battery from the DC bus as DC bus voltage goes into the allowed range. Fig. 6 demonstrates the whole control scheme. Bidirectional converter is placed between battery bank bus and DC bus. The main control idea is based on a voltage feedback signal V_{DC} coming from the DC bus. It is passed through a low pass filter to attenuate voltage ripple. Buck and boost predictive controllers operate independently. For the boost operation, DC bus voltage $V_{DC}(k)$, inductor current $i_L(k)$ and battery voltage $V_{BAT}(k)$ are imported to the prediction stage. A two dimensional array $V_{DC}(k+1)$ is sent to the optimization block from prediction stage. This array contains two predicted DC bus voltage values of the next sampling interval $t_{k+1}=t_k+T_s$ subject to switching states. Reference value $V_{DC}^*(k) \approx V_{DC}^*(k+1)$ is also imported to the optimization block. V_{DC}^* is lower limit of the predefined rated voltage range of DC bus. Cost function (12) is evaluated for each predicted value and the switching state S_{Boost} that

minimizes cost function is selected and sent to the converter to be applied on the next sampling interval. For the buck operation, DC bus voltage $V_{DC}(k)$, inductor current $i_L(k)$ and battery voltage $V_{BAT}(k)$ are imported to the prediction stage. A two dimensional array $V_{BAT}(k+1)$ is sent to the optimization block from prediction stage. This array contains two predicted battery voltage values of the next sampling interval $t_{k+1}=t_k+T_s$ subject to switching states. Reference value $V_{BAT}^*(k) \approx V_{BAT}^*(k+1)$ is generated from an external loop as below:

$$V_{BAT}^*(k) = \frac{i_{DC}(k).V_{Buck}^*}{i_L(k)}$$

(13)

where $i_{DC}(k)$ is the current passing through DC bus and V_{Buck}^* is upper limit of the predefined rated voltage range of DC bus. By using (13) for calculating reference value of buck operation, the converter controls its output voltage V_{BAT} to keep its input voltage V_{DC} on the upper limit of allowed range.

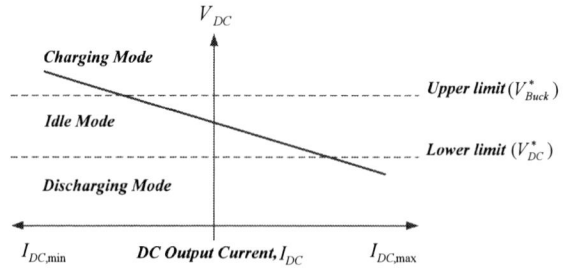

Fig. 5. Three operational modes based on DC bus voltage droop.

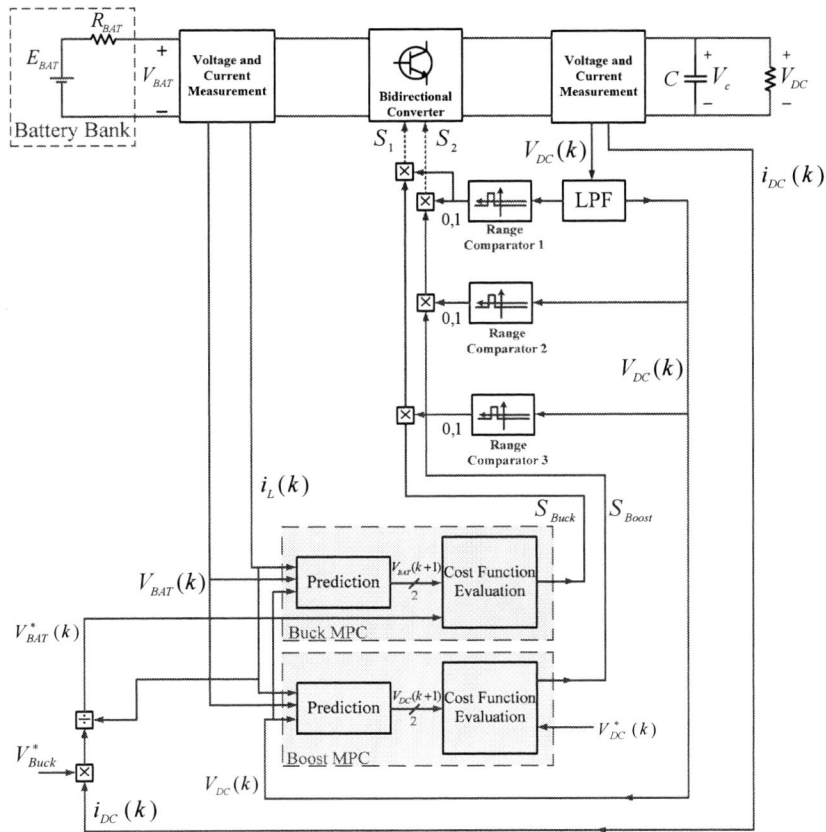

Fig. 6. Block diagram of the proposed control strategy.

Cost function (11) is evaluated for each predicted value and the switching state S_{Buck} that minimizes cost function is selected and sent to the converter to be applied on the next sampling interval. The outputs of predictive controllers S_{Boost} and S_{Buck} are headed toward the bidirectional converters switches. But three range comparators (RC) decide which operational mode to be activated. DC bus voltage V_{DC} will enter into three range comparators after passing low pass filter. The output of each comparator is 0 or 1. Table I shows how range comparators control operational modes. RC1 controls the idle mode, RC2 controls the boost mode and RC3 controls the buck mode.

TABLE I.   OPERATIONAL MODE CONTROL

Mode/RC	Range Comparator 1	Range Comparator 2	Range Comparator 3
Idle	0	0	0
Buck/Charging	1	0	1
Boost/Discharging	1	1	0

V. SIMULATION RESULTS

Steady state buck and boost operations and transition from boost to buck operation through idle mode by increasing DC bus voltage have been implemented in PSCAD/MATLAB interfacing. Fig. 7 and Fig. 8 show the steady state operations of the converter in buck and boost modes respectively. System parameters are also demonstrated in Table II.

TABLE II.   SYSTEM PARAMETERS

Parameter	Value
Voltage Range Lower Limit (V_{DC}^{*})	100 V
Voltage Range Upper Limit (V_{Buck}^{*})	150 V
Battery Rated Voltage	50 V
Internal Battery Resistance (R_{BAT})	0.01 Ω
Converter Inductance (L)	8 mH
Converter Capacitance (C)	1000 μF
Sampling Time (T_s)	33 μs

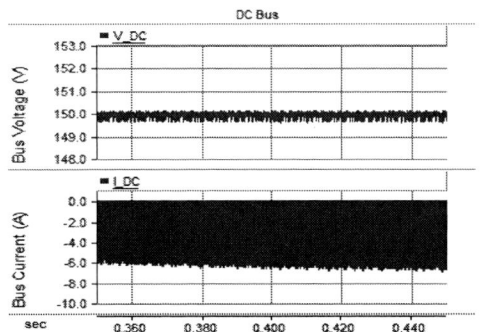

Fig. 7. DC bus voltage and current in buck mode.

978-1-5090-0376-1/16 $31.00 © 2016 IEEE

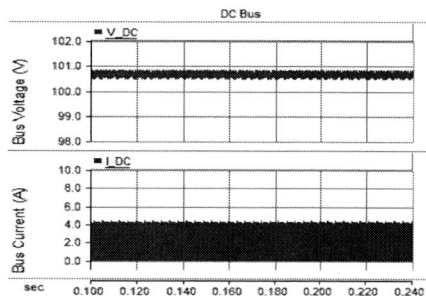

Fig. 8. DC bus voltage and current in boost mode.

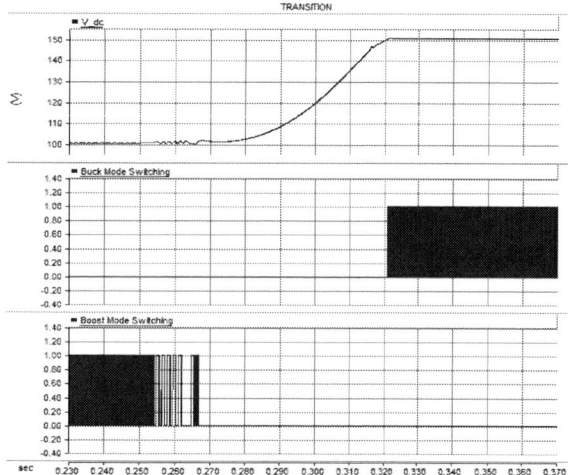

Fig. 9. Transition between three operational modes.

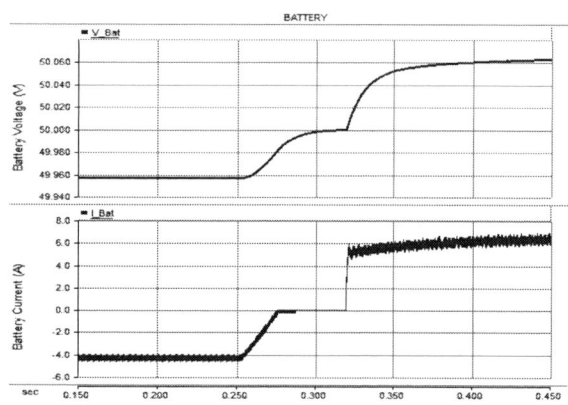

Fig. 10. Battery voltage and current during boost, idle and buck mode.

Transition from boost to buck mode is shown in Fig. 9. During $0.23 < t < 0.25$ the battery is discharging and converter operates in boost mode and bus voltage is kept on 100 V which is the lower limit of the predefined allowed range. Then during $0.25 < t < 0.32$ DC bus voltage is increased to the allowed range and all switchings are off since the converter is in idle mode. As the voltage reaches its predefined upper limit

150 V, converter starts to charge the battery in buck mode operation to prevent the voltage from exceeding voltage upper limit.

As shown in Fig. 10, during battery discharge, about 4 amperes of current flows to the DC bus from the battery. On idle mode, no current is exchanged between battery and DC bus, consequently during charging mode, current direction is reversed and battery absorbs 6 amperes of current.

VI. CONCLUSION

This research applies model predictive control to a bidirectional DC-DC converter. Mode activation control strategy is proposed to control the battery charge, discharge and idle modes. Each of the three modes operate independently due to the converters structure. The main control strategy is based on a voltage feedback signal from DC bus. A predefined rated range is considered for the bus voltage value thus the controller defines which mode to be activated. Simulation results show the feasibility of MPC and proposed controlling strategy to be used in battery applications due to proper reference tracking and fast dynamic response during transitions between modes.

References

[1] Lin, C. C., Yang, L. S., & Wu, G. W. (2013). Study of a non-isolated bidirectional DC–DC converter. *IET Power Electronics, 6*(1), 30-37.

[2] Noroozian, R., Abedi, M., Gharehpetian, G. B., & Hosseini, S. H. (2009). Combined operation of DC isolated distribution and PV systems for supplying unbalanced AC loads. *Renewable Energy, 34*(3), 899-908.

[3] Zhang, Z., Thomsen, O. C., Andersen, M. A., Schmidt, J. D., & Nielsen, H. R. (2009, February). Analysis and design of bi-directional DC-DC converter in extended run time DC UPS system based on fuel cell and supercapacitor. In *Applied Power Electronics Conference and Exposition, 2009. APEC 2009. Twenty-Fourth Annual IEEE* (pp. 714-719). IEEE.

[4] Abedi, M., Song, B. M., & Kim, R. Y. (2011, September). Nonlinear-model predictive control based bidirectional converter for V2G battery charger applications. In *Vehicle Power and Propulsion Conference (VPPC), 2011 IEEE* (pp. 1-6). IEEE.

[5] Zahedi, B., Nebb, O. C., & Norum, L. E. (2012, June). An isolated bidirectional converter modeling for hybrid electric ship simulations. In *Transportation Electrification Conference and Expo (ITEC), 2012 IEEE* (pp. 1-6). IEEE.

[6] Karshenas, H. R., Bakhshai, A., Safaee, A., Daneshpajooh, H., & Jain, P. (2011). *Bidirectional dc-dc converters for energy storage systems.* INTECH Open Access Publisher.

[7] Xu, D., Zhao, C., & Fan, H. (2004). A PWM plus phase-shift control bidirectional DC-DC converter. *Power Electronics, IEEE Transactions on, 19*(3), 666-675.

[8] Cernat, M., Scortaru, P., Tanase, A., & Iov, F. (2007, September). Hysteretic current controlled ZVS dc/dc converter for automobiles. In *Power Electronics and Applications, 2007 European Conference on* (pp. 1-10). IEEE.

[9] Hazil, O., Bououden, S., Chadli, M., & Filali, S. (2014). Fuzzy model predictive control of dc-dc converters. In *AETA 2013: Recent Advances in Electrical Engineering and Related Sciences* (pp. 423-432). Springer Berlin Heidelberg.

[10] Rodriguez, J., & Cortes, P. (2012). Predictive control of power converters and electrical drives (Vol. 40). John Wiley & Sons.

[11] Erickson, R. W., & Maksimovic, D. (2007). *Fundamentals of power electronics.* Springer Science & Business Media.

7th Power Electronics, Drive Systems & Technologies Conference (PEDSTC 2016)
16-18 Feb. 2016, Iran University of Science and Technology, Tehran, Iran

A Novel Method for Real-time Selective Harmonic Elimination in Five-Level Converters

Adib Abrishamifar
School of Electrical Engineering, Iran
University of Science & Technology
Tehran, Iran
abrishamifar@iust.ac.ir

Mohammad Arasteh
University of Science and Culture
Tehran, Iran
marasteh@iust.ac.ir

Farzad Golshan
School of Electrical Engineering, Iran
University of Science & Technology
Tehran, Iran
fgolshan@iust.ac.ir

Abstract— **Multilevel converters are attractive solution for high-power high-voltage applications such as motor drive systems, active filters, and flexible AC transmission systems (FACTS). Several modulation techniques are proposed for multilevel converters. Among them, selective harmonic elimination (SHE) offers high quality voltage waveform with low switching frequency which is especially suitable for high-power applications. This paper proposes a novel method for real-time implementation of SHE modulation technique in five-level converters. Using this method, all possible sets of solutions and the exact value of switching angles are determined by a very simple formulation. Furthermore, this technique systematically suggests three-level output waveform to extend the minimum value of modulation index to zero for each eliminated harmonic. To verify the validity of the proposed algorithm, simulation and experimental results for a single phase five-level diode-clamped inverter are reported.**

Keywords— *Selective Harmonic Elimination (SHE), Multilevel Converters, Diode-Clamped Inverter, Real-time implementation.*

I. INTRODUCTION

Multilevel converters provide an attractive solution to overcome the restriction of conventional two-level converters in high-voltage and high-power applications. Three conventional topologies of multilevel converters are: diode clamped, cascaded H-bridge, and capacitor clamped. Among them, diode clamped converter is the most commonly topology in industry [1-4].

Several different modulation methods such as sinusoidal PWM, space vector modulation, and selective harmonic elimination (SHE) have been proposed for multilevel converters in the literature. Among them, SHE method is the most suitable switching strategy for high-voltage high-power applications due to its low frequency switching [1,5,6].

A fundamental problem with such method is to obtain the arithmetic solution of nonlinear transcendental equations which contain trigonometric terms. Several algorithms such as the Newton–Raphson method, resultant theory [7], genetic algorithm (GA) [8], particle swarm optimization (PSO) [9], and Colonial Competitive Algorithm [10] have been reported in the technical literature to solve these transcendental equations [6,7].

The real-time implementation of aforementioned methods is too hard and sometimes impossible. So the switching angles of SHE equations are calculated offline and saved in a memory

as a lookup table [11]. recently, a new analytical method have been introduced in [12] for five-level converters which is able to calculate the exact amount of switching angles for elimination of each selected harmonic. This technique can be implemented in real time but the authors assume the scheme of output voltage waveform has always five levels like Fig. 1. However, there exists another possible switching scheme with three-level output waveform which may work over different ranges of the modulation index.

This paper has proposes a novel simple analytical technique to solve SHE problem for five-level converters suitable for real-time implementation. This technique, not only determines the exact value of switching angles by a very simple formulation, but also systematically suggests three-level output waveform to extend the minimum value of modulation index to zero for each harmonic as well. Using this new approach, all possible sets of solutions and the exact boundaries of all valid modulation index intervals can be determined for each harmonic. The simulation and experimental results agree well with the analytical results and prove the validity of the proposed technique.

The rest of paper is organized as follows: First, an overview of standard SHE problem in five-level converters is presented in section II. A new expression for SHE problem in five-level converters is described in section III which is suitable for real-time implementation. Section IV presents some simulation results on a 6 kV five-level Diode-clamped converter and finally, experimental results are presented in section V for a 600 V prototype to confirm the simulation results.

II. STANDARD SHE PROBLEM IN FIVE-LEVEL CONVERTERS

Assuming both quarter and half wave symmetries, Fig. 2 shows a staircase voltage waveform for a five-level converter. The output voltage can be illustrated by adding two quasi-square waveforms with the amplitude of V_{DC}. Due to odd quarter wave symmetry, the Fourier series of these quasi-square waveforms can be expressed by (1):

$$v(\omega t) = \sum_{n=1}^{\infty} V_n \sin(n\omega t) \qquad (1)$$

where n is the harmonic number and V_n represents the amplitude of harmonic order n which can be calculated by (2).

$$V_n = \frac{4V_{DC}}{n\pi} \cos(n\theta) \quad , \quad n = 1,3,5,7,\dots \qquad (2)$$

978-1-5090-0376-1/16 $31.00 © 2016 IEEE

As the output waveform decomposes to two quasi-square waveforms, the amplitude of odd order harmonics is calculated by (3).

$$V_n^{Out} = \frac{4V_{DC}}{n\pi}(\cos(n\theta_1) + \cos(n\theta_2)) \ , \ n = 1,3,5,7,\dots \quad (3)$$

The following restrictions must be satisfied by switching angles [7,8].

$$0 \leq \theta_1 \leq \theta_2 \leq \pi/2$$

There should be two equations because there exist only two degree of freedom. In other words, The SHE problem is formulated by equations (4)

$$\begin{cases} \cos(\theta_1) + \cos(\theta_2) = V_1^{Out}/(4V_{DC}/\pi) \\ \cos(n\theta_1) + \cos(n\theta_2) = 0 \end{cases} \quad (4)$$

where V_1^{Out} is the desired value of first order harmonic and n is the harmonic order which is going to be eliminated.

By defining the modulation index M in equation (5), the SHE problem can be formulated by equations (6).

$$M = \frac{V_1^{Out}}{8V_{DC}/\pi} \quad (0 \leq M \leq 1) \quad (5)$$

$$\begin{cases} \cos(\theta_1) + \cos(\theta_2) = 2M \\ \cos(n\theta_1) + \cos(n\theta_2) = 0 \end{cases} \quad (6)$$

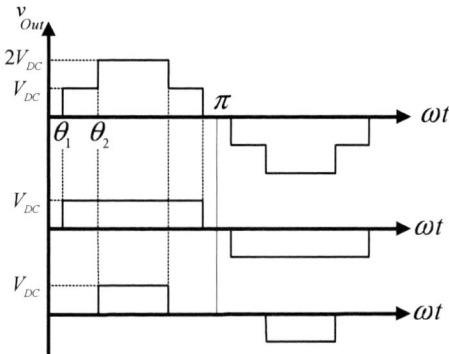

Fig. 1. The output voltage waveform of a five-level converter (standard expression).

III. NEW EXPRESSION FOR SHE PROBLEM IN FIVE-LEVEL CONVERTERS

In this section, we propose a new expression for overcoming SHE problem in five-level converters. In standard SHE problem, as discussed in previous section, the output waveform decomposes to two quasi-square waveforms with different switching angles (θ_1 and θ_2) as shown in Fig. 1, but another expression is to provide the output waveform by subtracting two quasi-square waveforms with the same switching angles (θ) and different phases (φ) like Fig. 2.

Using (2), the fundamental harmonic amplitude of both quasi-square waveforms is determined by (7)

$$V_1 = \frac{4V_{DC}}{\pi}\cos(\theta) \quad (7)$$

but they have a phase difference of φ. Therefore, the fundamental harmonic amplitude of output waveform (V_1^{Out}) is calculated as follows:

$$v_1^{Out}(t) = V_1\sin(\omega t) - V_1\sin(\omega t - \varphi)$$

$$\Rightarrow v_1^{Out}(t) = V_1[2\sin(\varphi/2)\cos(\omega t - \varphi/2)]$$

$$\Rightarrow v_1^{Out}(t) = [2V_1\sin(\varphi/2)]\cos(\omega t - \varphi/2)$$

$$\Rightarrow V_1^{Out} = \frac{8V_{DC}}{\pi}\cos(\theta)\sin(\varphi/2)$$

Using equation (5):

$$\cos(\theta)\sin(\varphi/2) = M \quad (8)$$

In a similar way, the nth order harmonic amplitude of the output waveform can be written as:

$$v_n^{Out}(t) = V_n\sin(n\omega t) - V_n\sin[n(\omega t - \varphi)]$$

$$\Rightarrow v_n^{Out}(t) = V_n[2\sin(n\varphi/2)\cos(n\omega t - n\varphi/2)]$$

$$\Rightarrow v_n^{Out}(t) = [2V_n\sin(n\varphi/2)]\cos(n\omega t - n\varphi/2)$$

$$\Rightarrow V_n^{Out} = \frac{8V_{DC}}{n\pi}\cos(\theta)\sin(n\varphi/2) \quad (9)$$

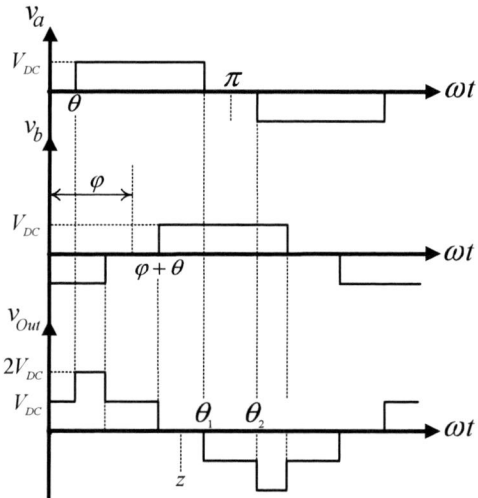

Fig. 2. The output voltage waveform of a five-level converter (new expression, five-level waveform).

According to equations (8) and (9), the SHE problem for five-level converters is redefined by two following equations:

$$\begin{cases} \cos(\theta)\sin(\varphi/2) = M \\ \cos(\theta)\sin(n\varphi/2) = 0 \end{cases} \quad (10)$$

Solving these new equations compared to solving standard SHE problem is simpler. By choosing phase difference φ according to (11), the desired n^{th} order harmonic is eliminated completely.

$$\sin(n\varphi/2) = 0 \ \Rightarrow \ n\varphi/2 = k\pi \ \Rightarrow \ \varphi = 2k\pi/n \quad (11)$$

978-1-5090-0376-1/16 $31.00 © 2016 IEEE

In (11), k is a positive integer number and all results less than π are acceptable. For attaining desired amplitude of fundamental harmonic, it is enough to calculate switching angle θ by (12).

$$\theta = \cos^{-1}\left(\frac{M}{\sin(\varphi/2)}\right) \qquad (12)$$

It is noticeable that the output waveform does not have five levels in all situations. Fig. 3 illustrates that the output waveform has three levels if $\theta \geq \varphi/2$. The equations (11) and (12) are validated in this condition too. According to (12), the solution exists if the following inequation is satisfied.

$$M \leq \sin(\varphi/2)$$

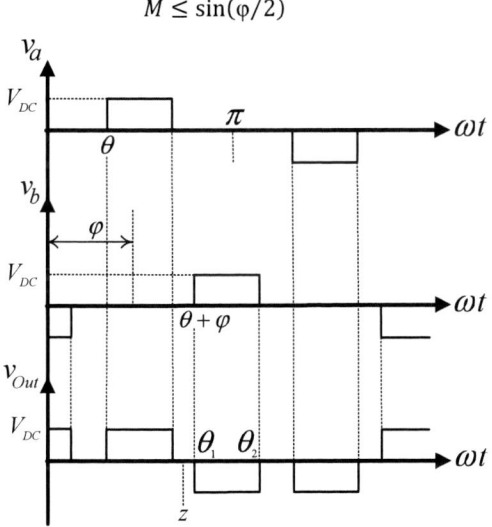

Fig. 3. The output voltage waveform of a five-level converter (new expression, three-level waveform).

By choosing $\theta = \varphi/2$ the border of modulation index for five-level and three-level waveform is determined. Using (10):

$$M = \cos(\varphi/2)\sin(\varphi/2)$$

Therefore, equation (13) shows the feasible modulation index intervals for five-level and three-level output waveform for an acceptable phase difference φ.

$$\begin{cases} 0 \leq M \leq \cos(\varphi/2)\sin(\varphi/2) \\ \qquad \Rightarrow \text{three} - \text{level output waveform} \\ \cos(\varphi/2)\sin(\varphi/2) \leq M \leq \sin(\varphi/2) \\ \qquad \Rightarrow \text{five} - \text{level output waveform} \end{cases} \qquad (13)$$

In fact, this technique is able to systemically propose another acceptable output waveform scheme with desired fundamental harmonic amplitude and suppressed selected harmonic for small value of the modulation index. This waveform has only three levels contrary to usual staircase waveform.

In order to clarify this method, the Elimination of third and fifth harmonics are discussed in detail.

A. Elimination of Third Harmonic ($n = 3$)

Using (11)
$$n = 3 \Rightarrow \varphi = 2k\pi/3$$

The only acceptable answer is $\varphi = 2\pi/3$ (by considering $k = 1$). The switching angle θ is determined by (12).

$$\theta = \cos^{-1}\left(\frac{M}{\sin(\pi/3)}\right) = \cos^{-1}\left(\frac{2M}{\sqrt{3}}\right)$$

In this condition, the feasible modulation index interval for three-level and five-level output waveform is obtained as follows:

$$\begin{cases} 0 \leq M \leq \cos(\pi/3)\sin(\pi/3) \\ \qquad \Rightarrow \text{three} - \text{level output waveform} \\ \cos(\pi/3)\sin(\pi/3) \leq M \leq \sin(\pi/3) \\ \qquad \Rightarrow \text{five} - \text{level output waveform} \end{cases}$$

$$\Rightarrow \begin{cases} 0 \leq M \leq 0.4330 \Rightarrow three - level\ output\ waveform \\ 0.433 \leq M \leq 0.8660 \Rightarrow five - level\ output\ waveform \end{cases}$$

B. Elimination of Fifth Harmonic ($n = 5$)

$$n = 5 \Rightarrow \varphi = 2k\pi/5$$

There are two acceptable answers and therefore, two different switching angles are achieved. The feasible modulation index interval for each answer is demonstrated as follows:

$$k = 1 \Rightarrow \varphi' = 2\pi/5$$

$$\Rightarrow \begin{cases} 0 \leq M \leq \cos(\pi/5)\sin(\pi/5) \\ \qquad \Rightarrow \text{three} - \text{level output waveform} \\ \cos(\pi/5)\sin(\pi/5) \leq M \leq \sin(\pi/5) \\ \qquad \Rightarrow \text{five} - \text{level output waveform} \end{cases}$$

$$\Rightarrow \begin{cases} 0 \leq M \leq 0.4755 \Rightarrow \text{three} - \text{level output waveform} \\ 0.4755 \leq M \leq 0.5878 \Rightarrow \text{five} - \text{level output waveform} \end{cases}$$

$$k = 2 \Rightarrow \varphi'' = 4\pi/5$$

$$\Rightarrow \begin{cases} 0 \leq M \leq \cos(2\pi/5)\sin(2\pi/5) \\ \qquad \Rightarrow \text{three} - \text{level output waveform} \\ \cos(2\pi/5)\sin(2\pi/5) \leq M \leq \sin(2\pi/5) \\ \qquad \Rightarrow \text{five} - \text{level output waveform} \end{cases}$$

$$\Rightarrow \begin{cases} 0 \leq M \leq 0.2939 \Rightarrow \text{three} - \text{level output waveform} \\ 0.2939 \leq M \leq 0.9511 \Rightarrow \text{five} - \text{level output waveform} \end{cases}$$

These results illustrate that there may be more than one acceptable solution for desirable modulation index. For example, there exist two different output waveforms for modulation index $M = 0.4$ without fifth harmonic which are determined by:

$$\varphi' = 2\pi/5 \Rightarrow \theta' = \cos^{-1}\left(\frac{0.4}{\sin(\pi/5)}\right) \approx 0.8223$$

$$\varphi'' = 4\pi/5 \Rightarrow \theta'' = \cos^{-1}\left(\frac{0.4}{\sin(2\pi/5)}\right) \approx 1.1367$$

It is noticeable that the output waveform of the first answer has only three levels but the second answer produces a five-level output waveform. The simulation and experimental results of this condition will be considered in the following sections.

Elimination of other odd harmonics is also possible by using this technique. Fig. 4 shows all acceptable phase differences and their modulation index intervals for each harmonic. It is noticeable that in all conditions, the minimum value of modulation index is zero due to generating three-level waveform in low modulation index range.

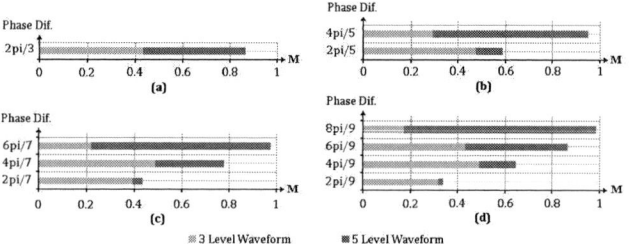

Fig. 4. Acceptable phase difference and their modulation index intervals for (a) third harmonic. (b) fifth harmonic. (c) seventh harmonic. (d) ninth harmonic.

Final step is to attain the relationship between (θ, φ) and (θ_1, θ_2). It is noticeable that the output switching angles (θ_1, θ_2) on Fig. 1 are defined relative to the new origin (point z) on Fig. 2 and Fig. 3. The coordinates of z is determined by:

$$z = \pi/2 + \varphi/2$$

According to Fig. 2:

$$\begin{aligned}\theta_1 &= (\pi - \theta) - z = \pi/2 - \theta - \varphi/2 \\ \theta_2 &= (\pi + \theta) - z = \pi/2 + \theta - \varphi/2\end{aligned} \quad (14)$$

According to Fig. 3:

$$\begin{aligned}\theta_1 &= (\varphi + \theta) - z = \theta + \varphi/2 - \pi/2 \\ \theta_2 &= (\varphi + \pi - \theta) - z = \pi/2 - \theta + \varphi/2\end{aligned} \quad (15)$$

Therefore, if $\theta < \varphi/2$, equations (14) is used to determine output switching angles (θ_1, θ_2) and otherwise, equations (15) is used to calculate them.

Using the equations (11), (12), (14), and (15), the real-time implementation of proposed method is practicable without the need of any lookup table. Following steps have been used for real-time implementation of proposed method.

1) Calculation of the phase difference φ by (11)
2) Calculation of the switching angle θ by (12)
3) if $\theta < \varphi/2$, equations (14) is used to determine output switching angles (θ_1, θ_2) and θ_2 is rising edge.
4) if $\theta \geq \varphi/2$, equations (15) is used to determine output switching angles (θ_1, θ_2) and θ_2 is falling edge.

IV. SIMULATION RESULTS

In this part, the validity of the proposed method is verified by simulations on a five-level diode-clamped inverter in

MATLAB/SIMULINK. The characteristics of the converter are as follows:
- total DC input voltage equals to 6 kV ($V_{DC} = 1.5\ kV$)
- output frequency is 50Hz

Several cases are examined in which the selected harmonic is eliminated and the desired fundamental amplitude is achieved.

Case 1: Elimination of third harmonic for $M = 0.65$

As shown in Fig. 4(a), only one solution exists whose output waveform has five levels. The output voltage waveform and its FFT spectrum are shown in Fig. 5.

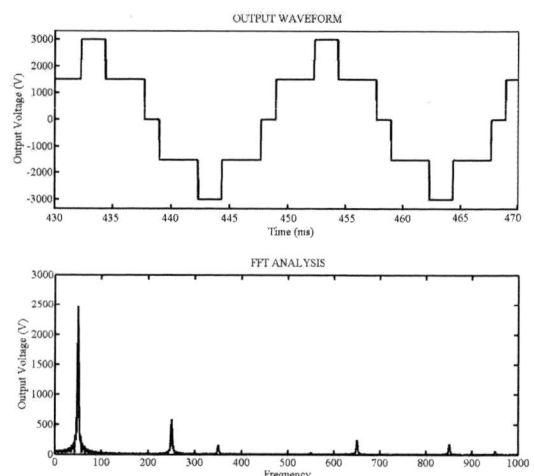

Fig. 5. Elimination of third harmonic for M=0.65, using phase difference $\varphi = 2\pi/3$ (simulation results).

Case 2: Elimination of third harmonic for $M = 0.3$

According to Fig. 4(a), there exists only one solution with three-level output waveform. Fig. 6 shows the output waveform and its FFT spectrum.

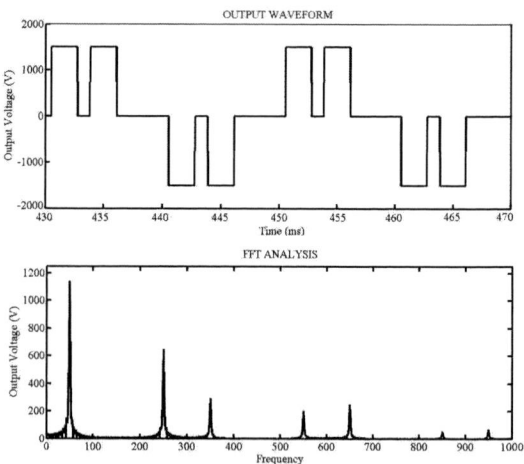

Fig. 6. Elimination of third harmonic for M=0.3, using phase difference $\varphi = 2\pi/3$ (simulation results).

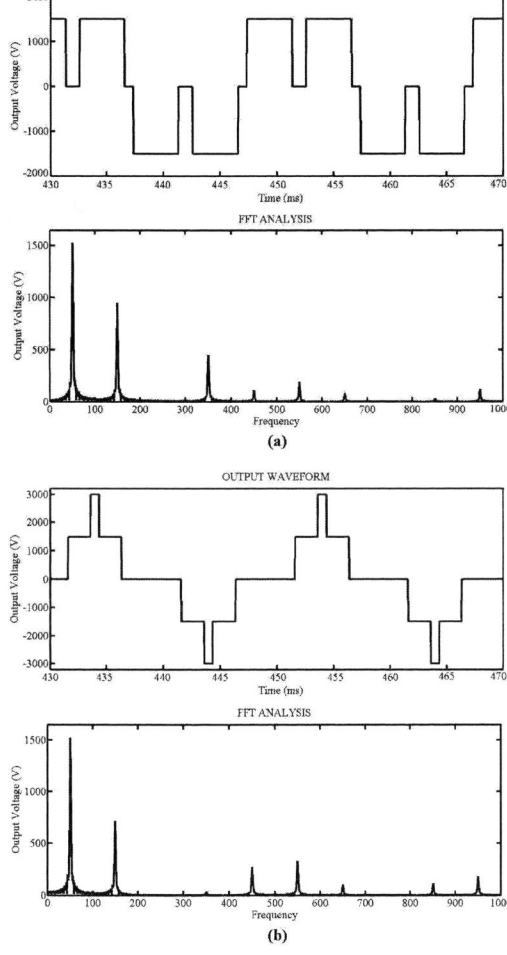

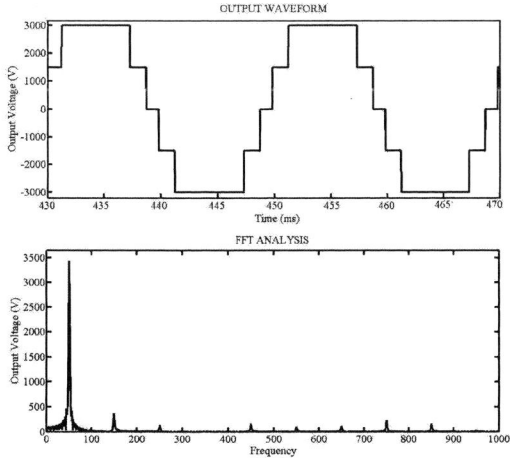

Fig. 7. Elimination of fifth harmonic for *M=0.4* (a) $\varphi = 2\pi/5$. (b) $\varphi = 4\pi/5$. (simulation results).

Fig. 8. Elimination of seventh harmonic for *M=0.9*, using phase difference $\varphi = 6\pi/7$ (simulation results).

Case 3: *Elimination of fifth harmonic for M = 0.4*

According to Fig. 4(b), two solutions exist, one of which for $\varphi = 2\pi/5$ with three-level output waveform and the other for $\varphi = 4\pi/5$ with five-level output waveform. Output voltage waveforms and their FFT spectrum are shown in Fig. 7.

Case 4: Elimination of seventh harmonic for M = 0.9

Fig. 8 shows the simulation results of this condition. There exists only one acceptable answer for $\varphi = 6\pi/7$ which has a five-level output waveform.

V. EXPERIMENTAL RESULTS

To further verify the validity of the proposed method, the simulation results are checked by experiments on a five-level single-phase diode-clamped inverter prototype with the input DC voltage of 600V ($V_{DC} = 150\,V$) and 50Hz output frequency. The results demonstrate the effectiveness of the proposed SHE method.

Case 1: Elimination of third harmonic for M = 0.65

As shown in Fig. 9, the output voltage waveform and its FFT spectrum are similar to Fig. 5. The base scales for FFT analysis are $100\,V_{rms}/div$ and $f = 62.5\,Hz/div$.

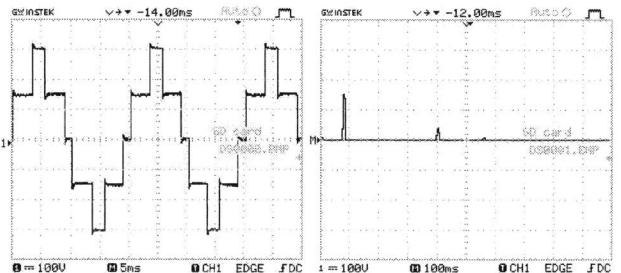

Fig. 9. Elimination of third harmonic for *M=0.65*, using phase difference $\varphi = 2\pi/3$ (experimental results).

Case 2: Elimination of third harmonic for M = 0.3

The output voltage waveform and its FFT spectrum are shown in Fig. 10. The voltage and frequency scales for FFT analysis are $50\,V_{rms}/div$ and $f = 62.5\,Hz/div$ respectively.

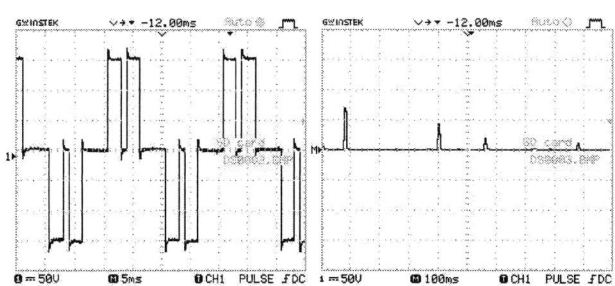

Fig. 10. Elimination of third harmonic for *M=0.3*, using phase difference $\varphi = 2\pi/3$ (experimental results).

978-1-5090-0376-1/16 $31.00 © 2016 IEEE

Case 3: *Elimination of fifth harmonic for M = 0.4*

According to Fig. 11(a) the output waveform for $\varphi = 2\pi/5$ has only three levels and the base scale for its FFT analysis are $50\ V_{rms}/div$ and $f = 62.5\ Hz/div$ but for $\varphi = 4\pi/5$, the output waveform has five levels and the base scale for its FFT analysis are $100\ V_{rms}/div$ and $f = 62.5\ Hz/div$.

Case 4: Elimination of seventh harmonic for M = 0.9

Fig. 12 shows the experimental results of this condition. The base scales for both FFT analysis are $100\ V_{rms}/div$ and $f = 62.5\ Hz/div$.

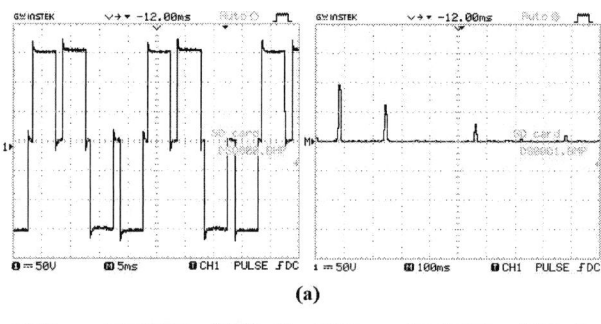

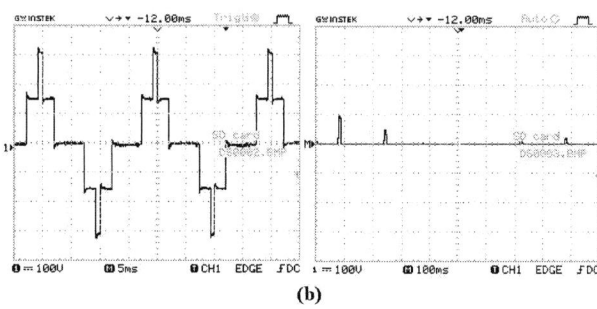

Fig. 11. Elimination of fifth harmonic for *M=0.4*
(a) $\varphi = 2\pi/5$. (b) $\varphi = 4\pi/5$. (experimental results).

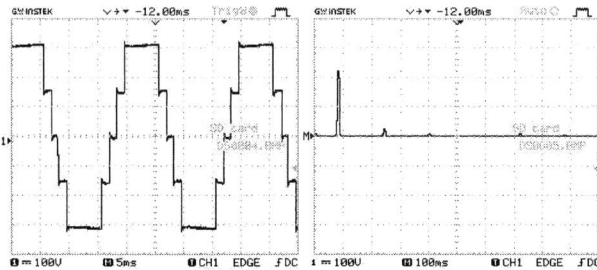

Fig. 12. Elimination of seventh harmonic for *M=0.9*, using phase difference $\varphi = 6\pi/7$ (experimental results).

The experimental results not accurate but totally, there exists a good adaptation between the simulation and experimental FFT spectrum for all cases.

VI. CONCLUSION

This paper proposes a novel and simple analytical method for SHE problem in five-level converters which is suitable for real-time implementation by a simple microprocessor or DSP.

This technique, not only determines the exact value of switching angles by a very simple formulation, but also systematically suggests three-level output waveform to extend the minimum value of modulation index to zero for each harmonic as well. All possible sets of solutions and the feasible modulation index range for each solution are achieved with less computational burden. The validity of the proposed method is investigated through simulation and implementation of a five-level diode-clamped inverter.

ACKNOWLEDGMENT

The authors wish to thank the invaluable support provided by Dr. S. M. Sadegh Mirghafourian and Borna Electronics Company.

REFERENCES

[1] H. Abu-Rub, J. Holts, J. Rodriguez, and G. Baoming, "Medium voltage multilevel converters, state of the art, challenges and requirements in industrial applications," *IEEE Trans. Ind. Electron.*, vol. 57, no. 8, pp. 2581–2596, August 2010.

[2] J. Rodriguez, S. Bernet, P. K. Steimer, I. E. Lizama, "A Survey on Neutral-Point-Clamped Inverters," *IEEE Trans. Ind. Electron.*, vol. 57, no. 7, pp. 2219–2230, July 2010.

[3] S. R. Pulikanti, G. Konstantinou, and V. G. Agelidis, "Hybrid seven-level cascaded active neutral-point-clamped-based multilevel converter under SHE-PWM," *IEEE Trans. Ind. Electron.*, vol. 60, no. 11, pp. 4794-4804, 2013.

[4] R. Abdullah, N. A. Rahim, S. R. Sheikh Raihan, and A. Z. Ahmad, "Five-Level Diode-Clamped Inverter with Three Level Boost Converter," *IEEE Trans. Ind. Electron.*, vol. 61, no. 10, pp. 5155-5163, 2014.

[5] Y. Deng, K. H. Teo, C. Duan, T. G. Habetler, and R. G. Harley, "A fast and generalized space vector modulation scheme for multilevel inverters," *IEEE Trans. Power Electron.*, vol. 29, no. 10, pp. 5204–5217, 2014.

[6] M. R. Banaei and P. A. Shayan, "Solution for selective harmonic optimisation in diode-clamped inverters using radial basis function neural networks," *IET Power Electronics*, vol. 7, no. 7, pp. 1797-1804, 2014.

[7] M. S. A. Dahidah, G. Konstantinou, and V. G. Agelidis, "A Review of Multilevel Selective Harmonic Elimination PWM: Formulations, Solving Algorithms, Implementation and Applications," *IEEE Trans. Power Electron.*, vol. 30, no. 8, pp. 4091-4106, 2015.

[8] N. Farokhnia, S. H. Fathi, R. Salehi, G. B. Gharehpetian, and M. Ehsani, "Improved selective harmonic elimination pulse-width modulation strategy in multilevel inverters," *IET Power Electronics*, vol. 5, no. 9, pp. 1904-1911, 2012.

[9] K. Shen, D. Zhao, J. Mei, L. M. Tolbert, J. Wang, M. Ban, Y. Ji, and X. Cai, "Elimination of harmonics in a modular multilevel converter using particle swarm optimization-based staircase modulation strategy," *IEEE Trans. Ind. Electron.*, vol. 61, no. 10, pp. 5311–5322, October 2014.

[10] M. H. Etesami, N. Farokhnia, and S. H. Fathi, "Colonial Competitive Algorithm Development Toward Harmonic Minimization in Multilevel Inverters, " *IEEE Trans. on Ind. Infor.*, vol. 11, no. 2, pp. 459-466, 2015.

[11] Y. Zhang, Y. W. Li, N. R. Zargari, and Z. Cheng, "Improved selective harmonics elimination (SHE) scheme with online harmonic compensation for high-power PWM converters," *IEEE Trans. Power Electron.*, vol. 30, no. 7, pp. 3508-3517, 2015.

[12] C. Buccella, C. Cecati, M. G. Cimoroni, and K. Razi, "Analytical Method for Pattern Generation in Five Levels Cascaded H-bridge Inverter using Selective Harmonics Elimination," *IEEE Trans. Ind. Electron.*, vol. 61, no. 11, pp. 5811–5819, 2014.

Elimination of Low Order Harmonics in Nine-level Cascaded H-bridge Converter

Adib Abrishamifar
School of Electrical Engineering, Iran
University of Science and Technology
Tehran, Iran
abrishamifar@iust.ac.ir

Mohammad Arasteh
University of Science and Culture
Tehran, Iran
marasteh@iust.ac.ir

Farzad Golshan
School of Electrical Engineering, Iran
University of Science and Technology
Tehran, Iran
fgolshan@iust.ac.ir

Abstract— **MULTILEVEL converters compare with the conventional two-level converters provide noticeable advantages such as lower harmonic distortion, lower switching frequency, lower switching losses, and lower stress across the semiconductor switches. To improve the output quality of multilevel converters, different modulation techniques have been suggested in the literature. The most suitable modulation technique for high-power high-voltage applications is selective harmonic elimination (SHE) due to having fundamental switching frequency. In this paper, a simple method for elimination of low order harmonics in nine-level CHB converter is proposed. Using this technique, there is no need to solve complex trigonometric equations and therefore, this technique can be implemented in real time using a simple processor. Furthermore, all CHB modules operate at the fundamental switching frequency and power consumption is shared identically among them. To demonstrate the validity of the proposed method, simulation for a three phase nine-level CHB inverter is presented.**

Keywords— Selective Harmonic Elimination (SHE), Multilevel Converters, cascaded H-bridge Inverter, Modulation Techniques.

I. INTRODUCTION

In recent years, multilevel converters attract much attention in high-voltage high-power applications such as adjustable-speed drives, flexible ac transmission systems (FACTS), and high voltage direct current lines. The output voltage of multilevel converters is generated by the combination of multiple low dc voltage sources used at the input side. As the number of dc sources increases, the output voltage will be closer to sinusoidal waveform. The main advantage of multilevel converters compare with traditional two-level converters is their lower output THD. Also, they can operate with lower switching frequency, thus reducing the switching power loss [1,2]. The most known topologies of multilevel converters are diode-clamped, capacitor-clamped, and the cascaded H-bridge (CHB) multilevel converters. Among them, the cascaded multilevel converter has received special attention due to its modularity and simplicity of control. In addition, CHB converter needs lowest number of components compare with other topologies [1-3].

Several modulation and control techniques have been developed for multilevel converters including high frequency PWM techniques such as sinusoidal PWM (SPWM), selective harmonic elimination PWM (SHE-PWM) and, space vector PWM (SVPWM) and low frequency switching techniques such as space vector control (SVC) and selective harmonic

elimination (SHE). The SVC and SHE methods are fundamental frequency switching methods and perform one or two commutations of the power semiconductors during one cycle of the output voltages to generate a staircase waveform [4-7].

The efficiency parameters of a multilevel converter such as switching losses and THD principally depend on the modulation strategies used to control the converter. Therefore, high frequency switching techniques are not suitable for high power converters. The switching frequency must be kept low in high power and efficient applications because high power semiconductor switches are not able to work in high frequency switching and also the switching frequency is directly proportional to the switching losses [7,8].

Selective harmonic elimination is the most suitable switching strategy for high-voltage high-power applications. The aim of this method is the elimination of low order harmonics by choosing appropriate switching angles. To determine the switching angles, it is necessary to solve nonlinear transcendental equations which contain trigonometric terms. The main challenge in the SHE method is how to solve these nonlinear equations. Several iterative numerical techniques such as Newton-Raphson method have been used to solve SHE problem but these techniques need a good initial guess which should be very close to the exact solution. Resultant theory is another approach to solve SHE problem. Using this technique, the SHE problem is converted into an equivalent set of polynomial equations whose degree is proportionate to the number of switching angles. Therefore, the complexity increases dramatically with an increase in the number of switching angles [7].

In addition, stochastic optimization techniques such as genetic algorithm (GA) [9], the particle swarm optimization (PSO) [10], and simulated annealing have been suggested to solve SHE equations. Optimization techniques can also introduce optimum angles for infeasible modulation indices. However, these algorithms may trap a local optima and not reach all sets of solutions [7,11].

Regretfully, there is no analytical method to solve SHE transcendental equations for n-level converters. Therefore, to implement the SHE modulation technique, the switching angles of SHE equations are calculated offline and saved in a memory as a lookup table. In other words, the real time implementation of SHE technique is too hard and sometimes impossible [7].

978-1-5090-0376-1/16 $31.00 © 2016 IEEE

This paper proposes a novel technique to eliminate low order harmonics in nine-level three-phase CHB converter. Contrary to standard SHE method, this technique does not deal with trigonometric equations and can be implemented in real time. Also, all CHB modules have the same working time and operate at the fundamental switching frequency in all conditions. Simulation results prove the validity of the proposed technique.

The rest of paper is organized as follows: section II describes Cascaded H-bridge converter and its operation principle. In section III, the standard SHE problem is explained. This is followed by introducing a novel technique for elimination of low order harmonics in section IV and finally, simulation results are presented in section V for a nine-level three-phase CHB inverter.

II. CASCADED H-BRIDGE CONVERTER

Cascaded H-bridge multilevel converter is made from a series of H bridge (single-phase full-bridge) converter units as shown in Fig. 1. Each H-bridge unit generates a quasi-square waveform with three different voltage level output. The ac output of each H-bridge is connected in series such that the output voltage waveform is the sum of all of the individual H-bridge outputs.

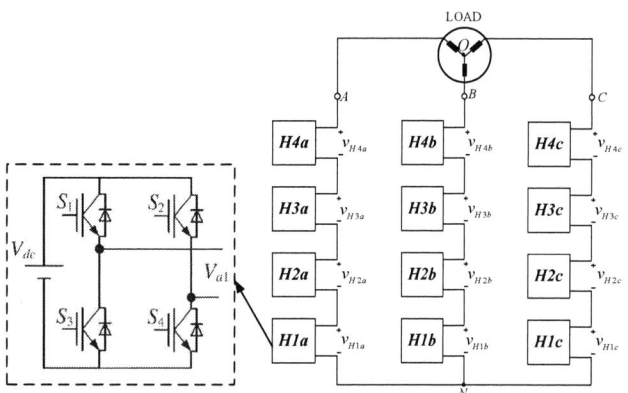

Fig. 1. A three-phase nine-level CHB converter

Fig. 2 shows the output phase voltage waveform of a 9-level inverter. θ_1, θ_2, θ_3, and θ_4 are the switching angles of the CHB converters. It is noticeable that the working time of CHB modules are not equal. Therefore, the power dissipation is not the same in CHB modules [7,9].

III. STANDARD SHE PROBLEM IN MULTILEVEL CONVERTERS

As shown in Fig. 2, each CHB module produces a quasi-square waveform with different switching angles. The aim of SHE method is to choose switching angles in a way that low order harmonics are eliminated completely.

The Fourier series of a quasi-square waveform can be expressed by (1):

$$v_H(\omega t) = \sum_{n=1}^{\infty} V_n \sin(n\omega t) \qquad (1)$$

where n is the harmonic number and V_n represents the amplitude of harmonic order n which can be calculated by (2). Note that due to odd quarter wave symmetry, the amplitude of all even order harmonics is zero.

$$V_n = \frac{4V_{DC}}{n\pi} \cos(n\theta_i) \quad , \quad n = 1,3,5,7,\dots \qquad (2)$$

As the output voltage waveform is the sum of all of the individual H-bridge outputs. The harmonic amplitude of output voltage is determined by (3)

$$V_n^{Out} = \frac{4V_{DC}}{n\pi} \sum_{i=1}^{N} \cos(n\theta_i) \quad , \quad n = 1,3,5,7,\dots \qquad (3)$$

where V_n^{Out} represents the amplitude of harmonic order n and N is the number of CHB modules per phase. The modulation index M is defined by equation (4).

$$M = \frac{V_1^{Out}}{4NV_{DC}/\pi} \quad (0 \leq M \leq 1) \qquad (4)$$

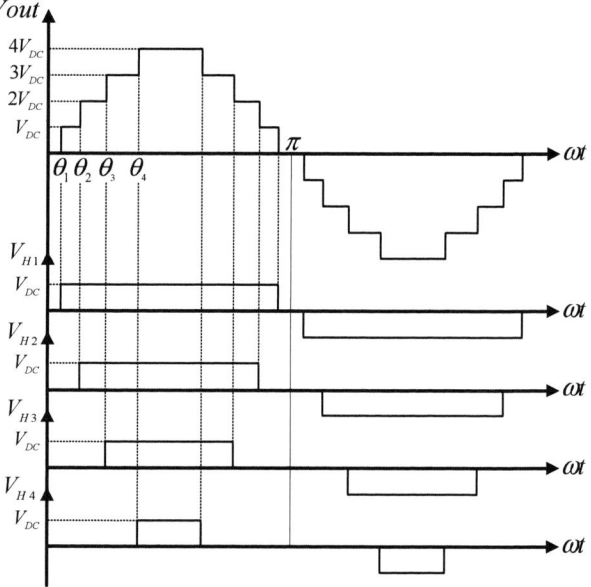

Fig. 2. Nine-level stepped voltage waveform (standard expression).

The switching angles should be determined in such that the low order harmonics of output waveform are eliminated and at the same time, the amplitude of fundamental harmonic becomes equal to the desired value. For a nine-level converter as shown in Fig. 2, the SHE problem can be formulated by following equations. It is noticeable that the triplen harmonics are not considered because they are naturally canceled out in the line-to-line voltages of a three-phase system.

$$\begin{cases} \sum_{i=1}^{4} \cos(\theta_i) = 4M & \left(M = \frac{V_1^{Out}}{16V_{DC}/\pi} \right) \\ \sum_{i=1}^{4} \cos(5\theta_i) = 0 \\ \sum_{i=1}^{4} \cos(7\theta_i) = 0 \\ \sum_{i=1}^{4} \cos(11\theta_i) = 0 \end{cases} \qquad (5)$$

The following restrictions must be satisfied by switching angles [11-14].

$$0 \leq \theta_1 \leq \theta_2 \leq \cdots \leq \theta_4 \leq \pi/2 \tag{6}$$

To determine the switching angles, the transcendental equations in (5) should be solved. Several methods such as Newton-Raphson method, Resultant theory, genetic algorithm, and particle swarm optimization are suggested to solve the SHE equations. These methods are not able to find the exact value of switching angles and cannot suppress the low order harmonics completely [8].

Depending on the modulation index M, there are two regions: (i) the feasible region in which there are a set or a multiple set of solutions, and (ii) the infeasible region in which there is no solution for the equations in (5). Therefore, the standard SHE technique has a narrow modulation region and do not have solution for the entire modulation range. To overcome this drawback, we need more degree of freedom. In other words, the number of switching angles should be more than the number of harmonics which are selected to eliminate. Some articles consider more switching angles and some others ignore the least significant equation in the SHE problem[9].

IV. A NOVEL TECHNIQUE FOR ELIMINATION OF LOW ORDER HARMONICS IN CHB MULTILEVEL CONVERTERS

In this section, we propose a new technique for overcoming SHE problem in CHB multilevel converters. In standard SHE problem, as discussed in previous section, the output waveform decomposes to N quasi-square waveforms with different switching angles $(\theta_1 - \theta_N)$ as shown in Fig. 2. Another expression is to provide the output waveform by adding N quasi-square waveforms with the same switching angles (θ) but in different phases.

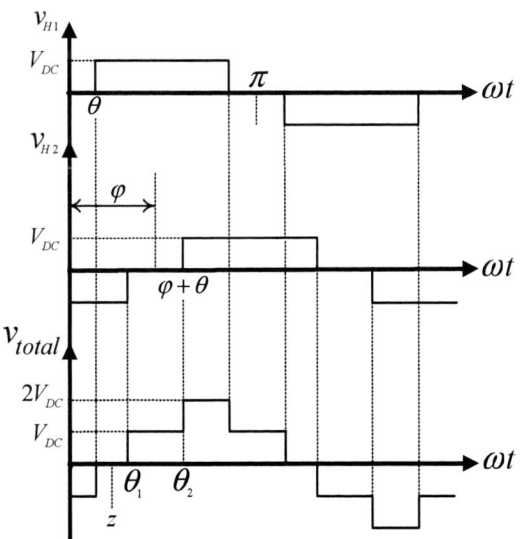

Fig. 3. Five-level stepped voltage waveform (new expression).

Due to having four CHB modules in each phase, at first, we focus on two CHB modules (for example $H1$ and $H2$) which both generate a quasi-square waveform with the same switching angles (θ) and different phase (φ) like Fig. 3.

Using (2), the fundamental harmonic amplitude of both quasi-square waveforms is determined by (7)

$$V_1 = \frac{4V_{DC}}{\pi}\cos(\theta) \tag{7}$$

but they have a phase difference of φ. Therefore, the fundamental harmonic amplitude of the total waveform (V_1^{total}) is calculated as follows:

$$v_1^{total}(t) = V_1 \sin(\omega t) + V_1 \sin(\omega t - \varphi)$$
$$\Rightarrow v_1^{total}(t) = V_1[2\cos(\varphi/2)\sin(\omega t - \varphi/2)]$$
$$\Rightarrow v_1^{total}(t) = [2V_1 \cos(\varphi/2)]\sin(\omega t - \varphi/2)$$
$$\Rightarrow V_1^{total} = \frac{8V_{DC}}{\pi}\cos(\theta)\cos(\varphi/2) \tag{8}$$

In a similar way, the nth order harmonic amplitude of the total waveform can be written as:

$$v_n^{total}(t) = V_n \sin(n\omega t) + V_n \sin[n(\omega t - \varphi)]$$
$$\Rightarrow v_n^{total}(t) = V_n[2\cos(n\,\varphi/2)\sin(n\omega t - n\varphi/2)]$$
$$\Rightarrow v_n^{total}(t) = [2V_n \cos(n\,\varphi/2)]\sin(n\omega t - n\varphi/2)$$
$$\Rightarrow V_n^{total} = \frac{8V_{DC}}{n\pi}\cos(\theta)\cos(n\,\varphi/2) \tag{9}$$

According to equation (9), to eliminate the nth order harmonic, it is enough to choose phase difference φ according to (10).

$$\cos(n\,\varphi/2) = 0 \Rightarrow n\varphi/2 = (2k-1)\,\pi/2$$
$$\Rightarrow \varphi = (2k-1)\pi/n \tag{10}$$

In (10), k is a positive integer number and all results less than π are acceptable.

Using equation (8), for attaining desired amplitude of fundamental harmonic, it is enough to calculate switching angle θ by (11).

$$\theta = \cos^{-1}\left(\frac{V_1^{total}}{(8V_{DC}/\pi)\cos(\varphi/2)}\right) \tag{11}$$

For instance, to eliminate fifth harmonic:

$$n = 5 \Rightarrow \varphi = (2k-1)\pi/5 \tag{12}$$

There are two acceptable answers and therefore, two different switching angles are achieved. These results illustrate that there may be more than one acceptable solution for desirable modulation index.

$$k = 1 \Rightarrow \varphi = \pi/5 \Rightarrow \theta = \cos^{-1}\left(\frac{V_1^{total}}{(8V_{DC}/\pi)\cos(\pi/10)}\right)$$

$$k = 2 \Rightarrow \varphi = 3\pi/5 \Rightarrow \theta = \cos^{-1}\left(\frac{V_1^{total}}{(8V_{DC}/\pi)\cos(3\pi/10)}\right)$$

According to above explanation, the fifth harmonic of the total output waveform can be eliminated by choosing appropriate phase difference in accordance with (12). Also, the desired fundamental harmonic can be achieved by choosing the proper switching angles for both CHB modules according to (11).

Two other CHB modules ($H3$ and $H4$) can generate this waveform, too. In other words, we have two total waveforms in each phase without fifth harmonic ($H1$ and $H2$ generate *total1* waveform and $H3$ and $H4$ generate *total2* waveform). The output voltage of each phase is the sum of these total waveforms and it does not contain fifth harmonic consequently.

Consider that these total waveforms have phase difference φ' as shown in Fig. 4. The fundamental harmonic amplitude of both total waveforms is determined by (8). Therefore, the fundamental harmonic amplitude of the output waveform is calculated as follows:

$$v_1^{out}(t) = V_1^{total} \sin(\omega t) + V_1^{total} \sin(\omega t - \varphi')$$

$$\Rightarrow v_1^{out}(t) = V_1^{total}[2\cos(\varphi'/2)\sin(\omega t - \varphi'/2)]$$

$$\Rightarrow v_1^{out}(t) = [2V_1^{total}\cos(\varphi'/2)]\sin(\omega t - \varphi'/2)$$

$$\Rightarrow V_1^{out} = 2V_1^{total}\cos(\varphi'/2) \qquad (13)$$

Also, the nth order harmonic amplitude of the output waveform can be written as:

$$v_n^{out}(t) = V_n^{total}\sin(n\omega t) + V_n^{total}\sin[n(\omega t - \varphi)]$$

$$\Rightarrow v_n^{out}(t) = V_n^{total}[2\cos(n\,\varphi'/2)\sin(n\omega t - n\varphi'/2)]$$

$$\Rightarrow v_n^{out}(t) = [2V_n^{total}\cos(n\,\varphi'/2)]\sin(n\omega t - n\varphi'/2)$$

$$\Rightarrow V_n^{out} = 2V_n^{total}\cos(n\,\varphi'/2) \qquad (14)$$

In accordance with (14), to suppress the nth order harmonic, it is enough to choose phase difference φ' according to (15), with in which k' is a positive integer number and all results less than π are acceptable.

$$\cos(n\,\varphi'/2) = 0 \Rightarrow n\,\varphi'/2 = (2k'-1)\,\pi/2$$

$$\Rightarrow \varphi' = (2k'-1)\pi/n \qquad (15)$$

To eliminate seventh harmonic from the output waveform ($n = 7$), there exist three different answers.

$$n = 7 \Rightarrow \varphi' = (2k'-1)\pi/7 \qquad (16)$$

Using equation (13):

$$k' = 1 \Rightarrow \varphi' = \pi/7 \Rightarrow V_1^{total} = \frac{V_1^{out}}{2\cos(\pi/14)}$$

$$k' = 2 \Rightarrow \varphi' = 3\pi/7 \Rightarrow V_1^{total} = \frac{V_1^{out}}{2\cos(3\pi/14)}$$

$$k' = 3 \Rightarrow \varphi' = 5\pi/7 \Rightarrow V_1^{total} = \frac{V_1^{out}}{2\cos(5\pi/14)}$$

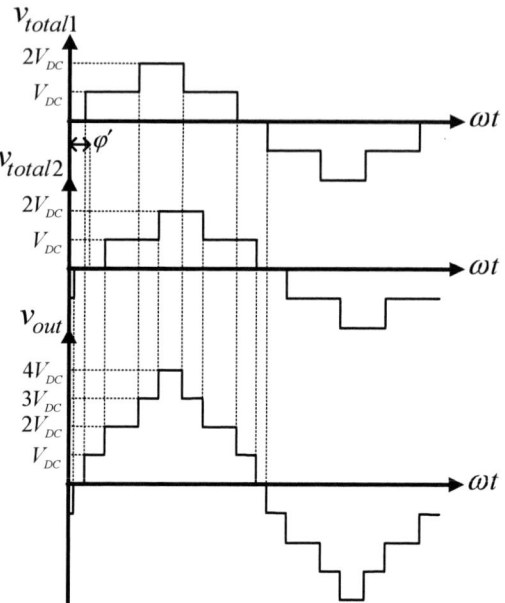

Fig. 4. Nine-level stepped voltage waveform generated by two total waveform.

These results demonstrate that there exist different output waveforms with desired fundamental harmonic amplitude and without fifth and seventh harmonic. Fig. 5 shows the flowchart of this new technique. This flowchart can be implemented in real time using a simple processor or DSP without the need of any lookup table. It is noticeable that all CHB modules operate at the fundamental switching frequency. Using this technique, there is no need to solve complex trigonometric equations. All CHB modules have the same working time as shown in Fig.6 and consequently, the power dissipation is shared identically among them.

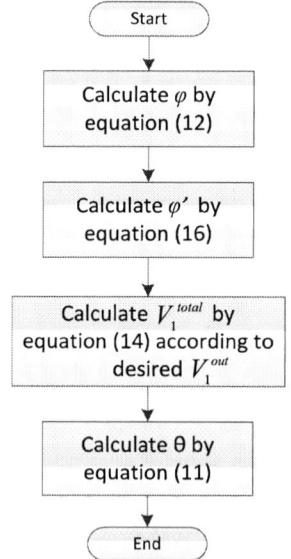

Fig. 5. The flowchart of the proposed method.

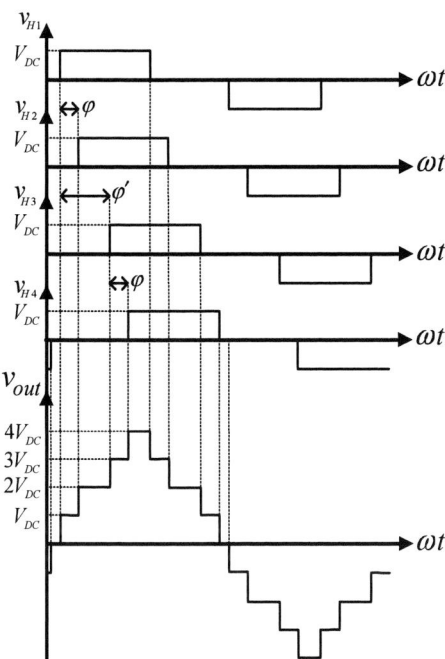

Fig. 6. Nine-level stepped voltage waveform (new expression).

V. SIMULATION RESULTS

To investigate the validity of the proposed method, a nine-level cascaded H-bridge inverter is simulated in MATLAB/SIMULINK. Several cases are examined in which the fifth and seventh harmonics are eliminated completely and the desired fundamental amplitude is achieved. In all simulated cases, the DC input voltage of each CHB module equals to 1 kV.

Case 1: The desired fundamental harmonic is ($V_1^{out} = 4330$) obtained by ($\varphi = \pi/5$, $\varphi' = \pi/7$, and $\theta = 0.4110$)

The output phase voltage, output line voltage and their FFT spectrum are shown in Fig. 7. It is noticeable that the fifth and seventh order harmonics are eliminated from phase and line output voltage completely. Also, the triplen harmonics are suppressed in the FFT spectrum of line voltage.

Case 2: The desired fundamental harmonic is ($V_1^{out} = 2500$) obtained by ($\varphi = \pi/5$, $\varphi' = \pi/7$, and $\theta = 1.014$)

The simulation results of this condition are shown in Fig. 8 which indicates that the fifth and seventh order harmonics are eliminated from the phase and line output voltage.

Case 3: The desired fundamental harmonic is ($V_1^{out} = 2500$) obtained by ($\varphi = \pi/5$, $\varphi' = 3\pi/7$, and $\theta = 0.851$)

As mentioned before, there exists more than one solution for some conditions. Case 3 and case 2 have the same fundamental amplitude without fifth and seventh order harmonics. It is noticeable that the output phase voltage of this condition is a seven-level waveform; in spite of all CHB modules cooperate to generate this waveform with the same

switching angle. The simulation results of this condition are shown in Fig. 9.

Case 4: The desired fundamental harmonic is ($V_1^{out} = 1000$) obtained by ($\varphi = 3\pi/5$, $\varphi' = 3\pi/7$, and $\theta = 1.12$)

Fig. 10 shows the simulation results of this condition. The output phase voltage has only five levels due to small value of desired fundamental harmonic.

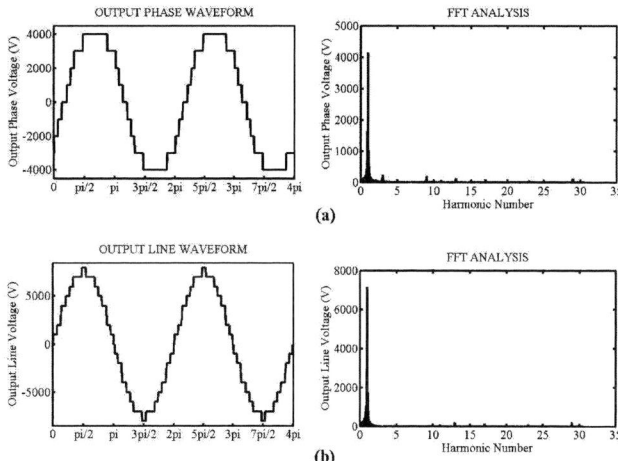

Fig. 7. The output phase and line voltage and their FFT spectrum for case 1.

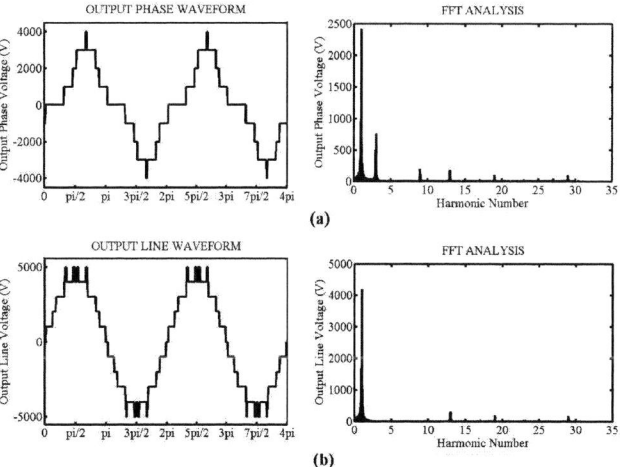

Fig. 8. The output phase and line voltage and their FFT spectrum for case 2.

VI. CONCLUSION

This paper proposes a new method to eliminate low order harmonics in nine-level three-phase CHB converter. However, it is important to point out that this technique can be extended for higher number of levels and multiple harmonic eliminations. Using this technique, there is no need to solve complex trigonometric equations and therefore, this technique is suitable for real-time implementation. Furthermore, All CHB modules have the same working time and therefore, the power dissipation is shared identically among them. The validity of

978-1-5090-0376-1/16 $31.00 © 2016 IEEE

the proposed method is investigated through the simulation of a nine-level CHB inverter.

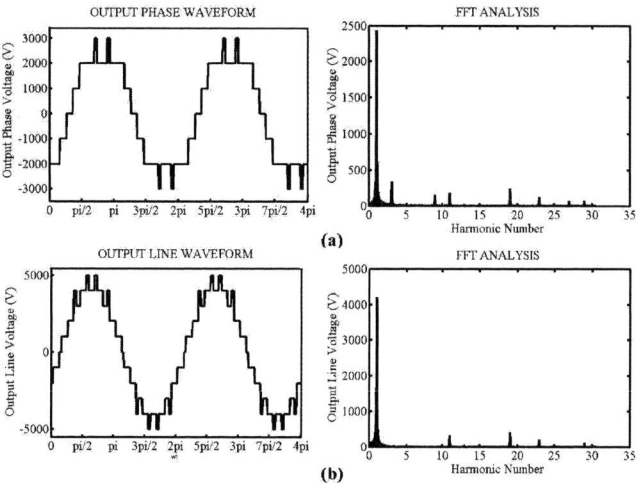

Fig. 9. The output phase and line voltage and their FFT spectrum for case 3.

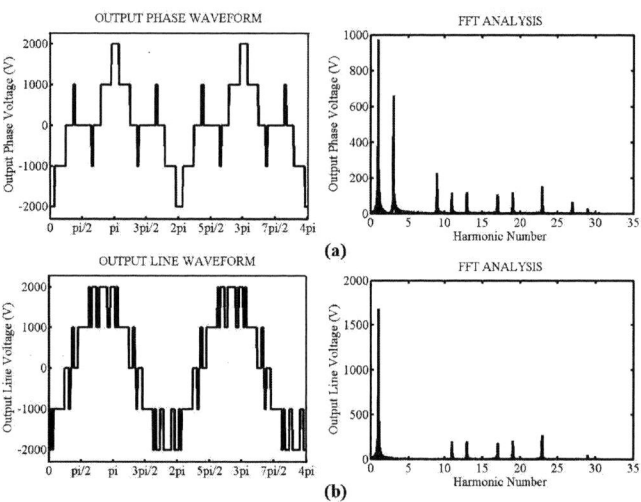

Fig. 10. The output phase and line voltage and their FFT spectrum for case 4.

ACKNOWLEDGMENT

The authors gratefully acknowledge the invaluable support provided by Dr. S. M. Sadegh Mirghafourian and Borna Electronics Company.

REFERENCES

[1] H. Abu-Rub, J. Holts, J. Rodriguez, and G. Baoming, "Medium voltage multilevel converters, state of the art, challenges and requirements in industrial applications," *IEEE Trans. Ind. Electron.*, vol. 57, no. 8, pp. 2581–2596, August 2010.

[2] S. Kouro, M. Malinowski, K. Gopakumar, J. Pou, L. G. Franquelo, B.Wu, J. Rodriguez, M. A. Perez, and J. I. Leon, "Recent advances and industrial applications of multilevel converters," *IEEE Trans. Ind. Electron.*, vol. 57, no. 8, pp. 2553–2580, August 2010.

[3] S. S. Fazel, S. Bernet, D. Krug, and K. Jalili, "Design and comparison of 4-kv neutral-point-clamped, flying-capacitor, and series-connected H-

bridge multilevel converters," *IEEE Trans. Ind. Appl.*, vol. 43, no. 4, pp. 1032–1040, July/August 2007.

[4] C. M. Young and S. F. Wu, "Selective harmonic elimination in multi-level inverter with zig-zag connection transformers," *IET Power Electronics*, vol. 7, no. 4, pp. 876-885, 2014.

[5] Z. Du, L.M. Tolbert, J.N. Chiasson, and B. Ozpineci, "Reduced Switching-Frequency Active Harmonic Elimination for Multilevel Converters," *IEEE Trans. Ind. Electron.*, vol. 55,no. 4, pp. 1761-1770, April 2008.

[6] Y. Deng, K. H. Teo, C. Duan, T. G. Habetler, and R. G. Harley, "A fast and generalized space vector modulation scheme for multilevel inverters," *IEEE Trans. Power Electron.*, vol. 29, no. 10, pp. 5204–5217, 2014.

[7] M. S. A. Dahidah, G. Konstantinou, and V. G. Agelidis, "A Review of Multilevel Selective Harmonic Elimination PWM: Formulations, Solving Algorithms, Implementation and Applications," *IEEE Trans. Power Electron.*, vol. 30, no. 8, pp. 4091-4106, 2015.

[8] M. R. Banaei and P. A. Shayan, "Solution for selective harmonic optimisation in diode-clamped inverters using radial basis function neural networks," *IET Power Electronics*, vol. 7, no. 7, pp. 1797-1804, 2014.

[9] N. Farokhnia, S. H. Fathi, R. Salehi, G. B. Gharehpetian, and M. Ehsani, "Improved selective harmonic elimination pulse-width modulation strategy in multilevel inverters," *IET Power Electronics*, vol. 5, no. 9, pp. 1904-1911, 2012.

[10] K. Shen, D. Zhao, J. Mei, L. M. Tolbert, J. Wang, M. Ban, Y. Ji, and X. Cai, "Elimination of harmonics in a modular multilevel converter using particle swarm optimization-based staircase modulation strategy," *IEEE Trans. Ind. Electron.*, vol. 61, no. 10, pp. 5311–5322, October 2014.

[11] M. H. Etesami, N. Farokhnia, and S. H. Fathi, "Colonial Competitive Algorithm Development Toward Harmonic Minimization in Multilevel Inverters, " *IEEE Trans. on Ind. Infor.*, vol. 11, no. 2, pp. 459-466, 2015.

[12] Y. Zhang, Y. W. Li, N. R. Zargari, and Z. Cheng, "Improved selective harmonics elimination (SHE) scheme with online harmonic compensation for high-power PWM converters," *IEEE Trans. Power Electron.*, vol. 30, no. 7, pp. 3508-3517, 2015.

[13] H. Zhao, T. Jin, S. Wang, D. Wu, and L. Sun, "A real-time selective harmonic elimination based on a transient-free, inner closed-loop control for cascaded multilevel inverters," *IEEE Trans. Power Electron.*, vol. 31, no. 2, pp. 1000-1014, 2016.

[14] C. Buccella, C. Cecati, M. G. Cimoroni, and K. Razi, "Analytical Method for Pattern Generation in Five Levels Cascaded H-bridge Inverter using Selective Harmonics Elimination," *IEEE Trans. Ind. Electron.*, vol. 61, no. 11, pp. 5811–5819, 2014.

Robust Control of the DC-DC Ćuk Converter in Discontinuous Conduction Mode

Vadood Hajbani
Department of Engineering
Ardabil Branch, Islamic Azad
University
Ardabil, Iran
v.hajbani@iauardabil.ac.ir

Mahdi Salimi
Department of Engineering
Ardabil Branch, Islamic Azad
University
Ardabil, Iran
m.salimi@iauardabil.ac.ir

Jafar Soltani
Faculty of Electrical Engineering
Isfahan University of Technology
Isfahan, Iran
jsoltani@iaukhsh.ac.ir

Abstract—In this paper a robust sliding mode controller is designed for DC-DC Ćuk converter. Sliding surface of the controller includes allof the system state variables. Hence, proposed controller has four-loop structure.Reference values of the state variables are calculated according to output voltage references using feedforward and feedback control loops. Designed controller canregulateoutput voltage of the converter in both continues and discontinues conduction modes in a wide range of the load and line variations. The accuracy and effectiveness of the proposed control approach is verified in some testsusing MATLAB/Simulink software.

Keywords—sliding mode controller; steady-state error; four-loop controller; Continuous Conduction Mode (CCM);Discontinuous Conduction Mode (DCM).

I. INTRODUCTION

Application of the renewable energy sources such as photovoltaic (PV), wind and fuel-cell (FC) has been increased exponentially in recent year. For example in Germany, average capacity of the installed PV systems has been doubled every 18 months between 1990and 2012. Costs of the roof-top grid connected PV generators have deceased from 5000euro per kW in 2006 to around 1200 euro in 2015 [1].

It is completely well known that power electronics converters play a vital role in power conditioning of the PV systems; especially DC-DC converters are employed for maximum power point tracking (MPPT) of the solar arrays. Among different topologies, in DC-DC boost and Ćuk choppers, converter inductor is directlconnected into input power source. Hence, the current which is drawn from PV source can be non-pulsating and it is possible to regulate operating point of the input power source in the maxim power point.

Considering simplicity of modeling and circuit design, standard DC-DC boost converter are widely used for MPPT of the PV systems. On the other hand, Ćuk topology is step-up/step-down DC-DC converter and obviously can be employed in a wide range of operation compared with boost chopper.

In spite of different advantages of the Ćuk converter, from controller design view point, the analysis of the closed loop control system is completely complicated. The converter includes more storage passive elements; hence state space variable of the system is more than standard DC-DC converters.

Considering nonlinear nature of the DC-DC converters, application of the linear controllers is not a good choice in a wide range of operation. Among different nonlinear controllers, sliding mode controller (SMC) is more proffered in closed loop control of the power electronics systems due to simplicity of design and robustness. Application of the SMC is studied in closed loop control of the standard DC-DC converters in more detail[2-4].

In [5], fourth-order Ćuk converter has been controlled using self-oscillating SMC. However, the designed controller in [5] has variable switching frequency and hence, it is not easy to employ it in practical DC-DC regulators.

A constant-frequency SMC controller for Ćuk converters is proposed in [6] using equivalent control theory. The various issues of the SMC are studied including theoretical analysis and experimental verification. It is stated that the proposed controller has stable and robust response in a wide range of operation. However, considerable output voltage error is seen in[6] which may be unacceptable in different applications.

A simple reduced-order SMC for power factor correction is designed for Ćuk DC-DC converters in CCM operation [7, 8]. In the paper, systematic technique is introduced for selection of sliding surface coefficients. According to Pade's approximation technique, model of the converter is reduced to simplify the SMC design. Hence, it clear that the designed controller will be reliable if the approximation conditions satisfied. Briefly, it is not possible to employ the controller of [7, 8]in a wide range of operation.

Linear and robust SMC controllers are combined for output voltage regulation of the fourth-order Ćuk converter in the CCM operation using equivalent control approach[9]. The minimum and non-minimum phase nature of the Ćuk converter are studied under step variations of model parameters. It is stated that the proposed controller has an acceptable output voltage error. However, the proposed controller can only be employed and obviously, it is not possible to regulate output voltage of the Ćuk chopper in a wide range of operation.

978-1-5090-0376-1/16 $31.00 © 2016 IEEE

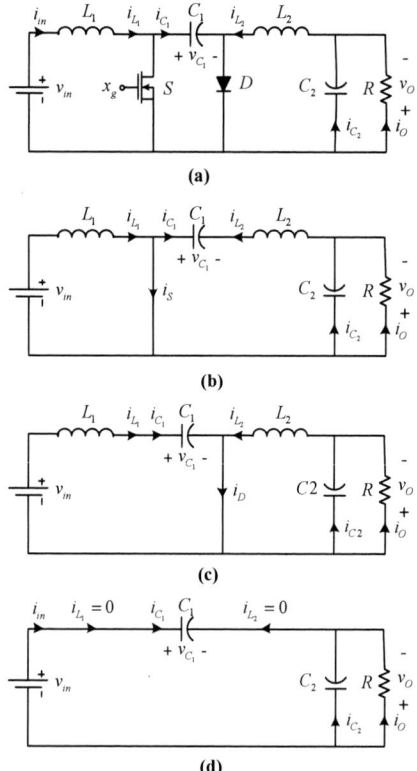

Fig. 1. Topology of the DC-DC Ćuk converter in different switching intervals. (a) Power circuit of the DC-DC Ćuk converter. (b) Equivalent circuit of the converter for $0 \leq t < DT$. (c) Equvalent circuit of the converter for $DT \leq t < (D + D')T$. (d) Equvalent circuit of the converter for $(D + D')T \leq t < T$.

Desired closed loop controller for DC-DC Ćuk converter should satisfy the following conditions:

- Output voltage regulation in both CCM and DCM operations with zero steady-state error.

- Fast transient response with respect to line and load changes.

- Full order SMC design using all of the state variables in sliding surface.

To cover all of these items satisfactorily, a novel SMC is designed for DC-DC Ćuk in a wide range of operation. The controller is developed according to equivalent control theory; hence switching frequency of the converter is completely constant. All of state variables of the converter are included in the controller design in a four-loop structure. It is shown the developed controller can be applied in a wide range of operation in both CCM and DCM operations.

II. STEADY STATE OPRATION OF THE CONVERTER

Power circuit of the DC-DC Ćuk converter in different operating mode is illustrated in Fig. 1. Typical waveform of an inductor current in DC-DC Ćuk converter is shown in Fig. 2 for DCM operation. It is clear that if the equation $(1 - D -$

$D' = 0)$ is satisfied, the operating mode of the controller will be CCM.

During steady-state operation, inductors voltages are shown in Fig. 3. It is well-known that the average value of the inductors voltages is zero over a switching period. Hence, the following equations can be written easily:

$$V_{in}DT_S + (V_{in} - V_{C_1})D'T_S = 0 \implies V_{C_1} = \frac{D+D'}{D'}V_{in} \quad (1)$$

and

$$(V_{C_1} - V_O)DT_S + (-V_O)D'T_S = 0 \implies V_{C_1} = \frac{D+D'}{D}V_O \quad (2)$$

By equaling the above equations:

$$V_O = \frac{D}{D'}V_{in} \quad (3)$$

Equation (5) describes voltage characteristic of the DC-DC Ćuk Converter in DCM operation. In CCM operation, it is clear that $D' = 1 - D$ and converter voltage gain will be as $\frac{V_o}{V_{in}} = \frac{D}{1-D}$.

III. AVERAGED STATE SPACE MODEL OF DC TO DC ĆUK CONVERTER

Considering presence of different energy storage elements in the converter, it is possible to define converter state variables as:

$$\mathbf{X}^T = (x_1, x_2, x_3, x_4) = \left(i_{L_1}, i_{L_2}, v_{C_1}, v_o\right) \quad (4)$$

When the power switch is on, the following equation can be written by consideration of the Fig. 1-b for $0 \leq t < DT_S$:

$$\dot{\mathbf{X}} = \mathbf{A}_{on}\mathbf{X} + \mathbf{B}_{on} = \begin{bmatrix} 0 & 0 & 0 & 0 \\ 0 & 0 & \frac{1}{L_2} & -\frac{1}{L_2} \\ 0 & -\frac{1}{C_1} & 0 & 0 \\ 0 & \frac{1}{C_2} & 0 & -\frac{1}{RC_2} \end{bmatrix} \begin{bmatrix} i_{L_1} \\ i_{L_2} \\ v_{C_1} \\ v_o \end{bmatrix} + \begin{bmatrix} \frac{v_{in}}{L_1} \\ 0 \\ 0 \\ 0 \end{bmatrix} \quad (5)$$

When the power switch is turned off and inductors are currents not zero, the following state equation can be obtained using Fig. 1-c for $DT_S \leq t < (D + D')T_S$:

$$\dot{\mathbf{X}} = \mathbf{A}_{off}\mathbf{X} + \mathbf{B}_{off}$$

$$= \begin{bmatrix} 0 & 0 & -\frac{1}{L_1} & 0 \\ 0 & 0 & 0 & -\frac{1}{L_2} \\ \frac{1}{C_1} & 0 & 0 & 0 \\ 0 & \frac{1}{C_2} & 0 & -\frac{1}{RC_2} \end{bmatrix} \begin{bmatrix} i_{L_1} \\ i_{L_2} \\ v_{C_1} \\ v_o \end{bmatrix} + \begin{bmatrix} \frac{v_{in}}{L_1} \\ 0 \\ 0 \\ 0 \end{bmatrix} \quad (6)$$

And finally for DCM operation, when the power switch is off and inductors currents are zero, the following state equation can be written for $(D + D')T_S \leq t < T_S$:

$$\dot{\mathbf{X}} = \mathbf{A}_\Delta \mathbf{X} + \mathbf{B}_\Delta = \begin{bmatrix} 0 & 0 & 0 & 0 \\ 0 & 0 & 0 & 0 \\ 0 & 0 & 0 & 0 \\ 0 & 0 & 0 & -\frac{1}{RC_2} \end{bmatrix} \begin{bmatrix} i_{L_1} \\ i_{L_2} \\ v_{C_1} \\ v_o \end{bmatrix} + \begin{bmatrix} \frac{v_{in}}{L_1} \\ 0 \\ 0 \\ 0 \end{bmatrix} \quad (7)$$

Considering state space averaging method, general model of the DC-DC Ćuk converter can be obtained as follows:

$$\dot{\mathbf{X}} = \mathbf{A}_{avg}\mathbf{X} + \mathbf{B}_{avg}$$

$$= \begin{bmatrix} 0 & 0 & -\frac{u_B}{L_1} & 0 \\ 0 & 0 & \frac{u}{L_2} & -\frac{u+u_B}{L_2} \\ \frac{u_B}{C_1} & -\frac{u}{C_1} & 0 & 0 \\ 0 & \frac{u+u_B}{C_2} & 0 & -\frac{1}{RC_2} \end{bmatrix} \begin{bmatrix} i_{L_1} \\ i_{L_2} \\ v_{C_1} \\ v_o \end{bmatrix} + \begin{bmatrix} \frac{u+u_B}{L_1}v_{in} \\ 0 \\ 0 \\ 0 \end{bmatrix} \quad (8)$$

where$u$ and u_B are average values of the$D$ and D' respectively.

IV. CONTROLLER DESIGN

In this section, sliding mode control of the converter in DCM operation is presented in detail. First sliding mode control theory is reviewed.

A. Theory of the applied sliding mode controller

Suppose that, the n-dimensional nonlinear system model is considered as below. It is assumed that the origin of the coordinate is the steady-state operating point of the system.

$$\dot{z} = \mathbf{A}z + u\mathbf{B}z \quad (9)$$

where$\mathbf{A}$ and $\mathbf{B}$ (square $n \times n$) matrices are defined with real components. The scalar control function$u$ takes values between 0 and 1. In the sliding mode control method, $u$can be defined as [10]:

$$u = \frac{1}{2}(1 + sgnS(z)) \quad (10)$$

where in this equation, sgn is the sign function and $S(z)$ is named sliding surface. According to (10):

$$if S(z) > 0 \Rightarrow u = 1 \ and if S(z) < 0$$
$$\Rightarrow u = 0 \quad (11)$$

Necessary and sufficient conditions for the existence of sliding motion on the sliding surface can be written as[10]:

$$lim \, S . \dot{S} < 0 \quad (12)$$

The smooth control function for nonlinear system which its model is written in (9) and adopts sliding surface as a local integral manifold, is known as equivalent control and is shown byu_{eq}. Equivalent control can be extracted by setting derivative of the sliding surface into zero[10]:

$$\frac{dS}{dt} = 0 \Rightarrow \left[\frac{\partial S}{\partial z}\right]^T \dot{z} = 0 \quad (13)$$

Considering that the value of $\dot{z}$ is given in (9), the controller can be calculated according to (13):

$$u_{eq} = -\frac{\left[\frac{\partial S}{\partial z}\right]^T Az}{\left[\frac{\partial S}{\partial z}\right]^T Bz} \quad (14)$$

The sliding motion exists locally on the sliding surface, if and only if, u_{eq} satisfies the following condition[10]:

$$0 < u_{eq} < 1 \quad (15)$$

B. Calculation of the u_B in Ćuk converter

It is clear that according to system model in(8), it would be difficult to calculate the controller directly because of the presence of u_B in final control equation. That's why we try to cancel this parameterfrom controller.

According to Fig. 2, the average value of the inductor current can be found from the following equation:

$$I_{L_1} = \frac{1}{T_S}\int_0^{T_S} i_{L_1}(t)dt = \frac{D+D'}{2}\hat{I}_{L_1} \quad (16)$$

whereI_{L_1} is average value of a inductor current. Also the value of $\hat{I}_{L_1}$ can be simply obtained as follows:

$$i_{L_1} = \frac{1}{L_1}\int v_{L_1}dt \rightarrow \hat{I}_{L_1} - 0 = \frac{1}{L_1}\int_0^{DT_S} v_{in}dt$$
$$\rightarrow \hat{I}_{L_1} = \frac{1}{L_1}v_{in}DT_S \quad (17)$$

By substituting (17) in equation (16); the average value of the inductor current can be written as:

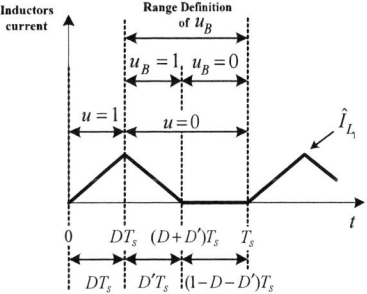

Fig. 2. Inductor current of the DC-DC converter in DCM operations.

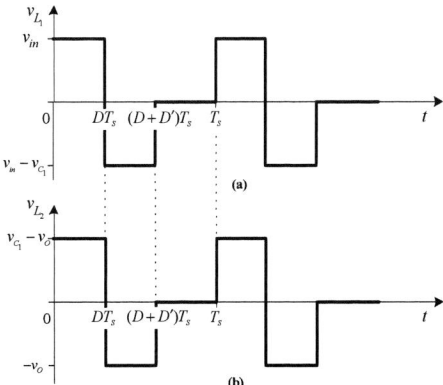

Fig. 3. Inductors voltage in DCM operation of the DC-DC Ćuk onverter.

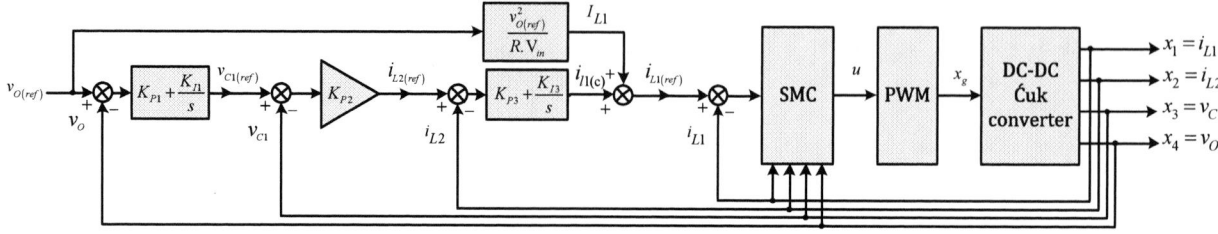

Fig. 4. Four-loop voltage control of the Ćuk converter.

$$I_{L_1} = \frac{V_{in}}{2L_1} DT_S (D + D') \tag{18}$$

If it is assumed that the converter is lossless; the following equation can be obtained:

$$P_{out} = P_{in} \quad \rightarrow \quad V_o I_o = V_{in} I_{in} \tag{19}$$

where in the above equation, $I_{L_1} = I_{in}$. Hence the following equation can be obtained:

$$V_o I_o = V_{in} I_{L_1} \quad \rightarrow \quad V_o I_o = V_{in} \left(\frac{V_{in}}{2L_1} DT_S (D + D') \right) \tag{20}$$

By replacing the value of D' from equation (3) in (20):

$$\frac{V_o^2}{R} = \frac{V_{in}^2}{2L_1} DT_S \left(D + \frac{V_{in}}{V_o} D \right) \tag{21}$$

or

$$D = D_{DCM} = \sqrt{\frac{2L_1}{RT_S} \frac{V_o}{V_{in}} \sqrt{\frac{V_{in}}{V_{in} + V_o}}} \tag{22}$$

and

$$D' = \sqrt{\frac{2L_1}{RT_S} \sqrt{\frac{V_{in}}{V_{in} + V_o}}} \tag{23}$$

Also, the duty cycle of converters in CCM is calculated as follow:

$$D_{CCM} = \frac{V_o}{V_{in} + V_o} \tag{24}$$

It is obvious that the maximum interval in which u_B is equal to one is related to the case which the converter operates in CCM. So:

$$D'_{max} T_S = (1 - D_{CCM}) T_S \tag{25}$$

and we know that in DCM operation, $u_B = 1$ is true for $D' T_S$ duration. That's why u_{Beq} can be defined as follows. It should be noted that $0 < u_{Beq} < 1$ and is called the equivalent virtual switching function.

$$u_{Beq} = \frac{D' T_S}{D'_{max} T_S} \tag{26}$$

Also, u_{Beq} can be calculated as follows using equations (24) and (26):

$$u_{Beq} = \sqrt{\frac{2L_1}{RT_S}} \sqrt{1 + \frac{V_o}{V_{in}}} \tag{27}$$

C. Sliding surface selection

Proposed four-loop SMC for DC-DC Ćuk converter is shown in Fig. 4. The switching manifold [11] for the sliding mode current controllercan be written as:

$$S(z) = \alpha_1 z_1 + \alpha_2 z_2 + \alpha_3 z_3 + \alpha_4 z_4 \tag{28}$$

where α_1, α_2, ,α_3, and α_4 are constant values. During development of the sliding mode controller, the selected state variables should have zero steady-state values. For this reason, the capacitorvoltages(v_{C1} and v_o) and inductor currents (i_{L1} and i_{L2}) errors are considered as state variablesof the system:

$$\begin{cases} z_1 = x_{1(ref)} - x_1 = i_{L_1(ref)} - i_{L_1} \\ z_2 = x_{2(ref)} - x_2 = i_{L_2(ref)} - i_{L_2} \\ z_3 = x_{3(ref)} - x_3 = v_{C_1(ref)} - v_{C_1} \\ z_4 = x_{4(ref)} - x_4 = v_{O(ref)} - v_O \end{cases} \tag{29}$$

In this equations,$i_{L_1(ref)}$, $i_{L_2(ref)}$, and $v_{C_1(ref)}$are reference values of the outer loops and will be defined as follows:

$$\begin{cases} i_{L_1(ref)} = I_{L_1} + i_{l1(c)} \\ i_{L_2(ref)} = K_{P2} [v_{C_1(ref)} - v_{C_1}] \\ v_{C_1(ref)} = K_{P3} [v_{O(ref)} - v_o] + K_{I3} \int [v_{O(ref)} - v_o] dt \end{cases} \tag{30}$$

Also in (30); $I_{L_1}$and $i_{l1(c)}$ can be written:

$$I_{L_1} = \frac{v_{O(ref)}^2}{R.v_{in}} \tag{31}$$

$$i_{l1(c)} = K_{P1} [i_{L_2(ref)} - i_{L_2}] + K_{I1} \int [i_{L_2(ref)} - i_{L_2}] dt \tag{32}$$

D. Equivalent controller desgin

Considering the performance of Ćuk converter in DCM operation, the dynamic model of the systemcan be derived by regarding the basic rules of the electrical circuit and the equations (28) and (29).Finallyu_{eq} will be calculated as follows by setting$\dot{S} = 0$ in (14):

978-1-5090-0376-1/16 $31.00 © 2016 IEEE   538

$$u_{eq} = \frac{1}{\{K_{i_{L_2}} i_{L_2} + K_{v_{C_1}} v_{C_1} + K_{v_o} v_o + K_{v_{in}} v_{in}\}} \times \Big\{K'_{i_{L_2}} i_{L_2} + K'_{v_{C_1}} v_{C_1} +$$
$$K'_{v_o} v_o + \Big[K''_{i_{L_1}} i_{L_1} + K''_{i_{L_2}} i_{L_2} + K''_{v_{C_1}} v_{C_1} + K''_{v_o} v_o +$$
$$K''_{v_{in}} v_{in}\Big] \Big(\sqrt{1 + \frac{v_o}{v_{in}}}\Big) + K_{e_{v_o}} \big[v_{O(ref)} - v_o\big] + K_{ie_{v_o}} \int \big[v_{O(ref)} -$$
$$v_O\big] dt\Big\} \qquad (33)$$

where

$$K_{i_{L_2}} = (-\alpha_1 K_{P1} K_{P2} C_2 + \alpha_1 K_{P1} K_{P2} K_{P3} C_1 + \alpha_2 K_{P2} K_{P3} C_1 - \alpha_2 K_{P2} C_2 +$$
$$\alpha_3 K_{P3} C_1 - \alpha_3 C_2 + \alpha_4 C_1)/(C_1 C_2) \qquad (34)$$

$$K_{v_{C_1}} = -K_{v_o} = \frac{\alpha_1 K_{P1} + \alpha_2}{L_2} \qquad (35)$$

$$K_{v_{in}} = \frac{\alpha_1}{L_1} \qquad (36)$$

$$K'_{i_{L_2}} = -\alpha_1 K_{I1} \qquad (37)$$

$$K'_{v_{C_1}} = -\alpha_1 K_{I1} K_{P2} \qquad (38)$$

$$K'_{v_o} = (\alpha_1 K_{P1} K_{P2} K_{P3} + \alpha_2 K_{P2} K_{P3} + \alpha_3 K_{P3} + \alpha_4)/(RC_2) \qquad (39)$$

$$K''_{i_{L_1}} = -\sqrt{\frac{2L_1}{RT_S}} (\alpha_1 K_{P1} K_{P2} + \alpha_2 K_{P2} + \alpha_3)/(C_1) \qquad (40)$$

$$K''_{i_{L_2}} = -\sqrt{\frac{2L_1}{RT_S}} (\alpha_1 K_{P1} K_{P2} K_{P3} + \alpha_2 K_{P2} K_{P3} + \alpha_3 K_{P3} + \alpha_4)/(C_2) \qquad (41)$$

$$K''_{v_{C_1}} = -K''_{v_{in}} = \frac{\alpha_1}{L_1} \sqrt{\frac{2L_1}{RT_S}} \qquad (42)$$

$$K''_{v_o} = \frac{\alpha_1 K_{P1} + \alpha_2}{L_2} \sqrt{\frac{2L_1}{RT_S}} \qquad (43)$$

$$K_{e_{v_o}} = \alpha_1 K_{P1} K_{P2} K_{I3} + \alpha_1 K_{I1} K_{P2} K_{P3} + \alpha_2 K_{P2} K_{I3} + \alpha_3 K_{I3} \qquad (44)$$

$$K''_{v_o} = \frac{\alpha_1 K_{P1} + \alpha_2}{L_2} \sqrt{\frac{2L_1}{RT_S}} \qquad (45)$$

V. SIMULATION

DC-DC Ćuk converter is simulated in CCM and DCM operations according to the designed SMC robust controller (equation (33)) using MATLAB/Simulink. In this section, in order to build the controller, all of the state variables are measured. Due to switching, sampled values may have a considerable ripple. Large values of the ripple may ruin stability of the closed-loop controller. Hence. low-pass filters are employed to calculate average of the state variables. Power circuit elements values and controller gains are listed in TABLE I. Also, maximum step size for all of the simulations is assumed to be 100ns.

Test 1- Considering nominal values of the closed loop system in TABLE I, steady state response of the proposed robust controller is illustrated in Fig.5. It is clear that the converter is in DCM operation and has zero steady state error. In this test, waveforms of the different state variables are shown with respect to control signal. It is seen that both of the inductors currents switching intervals are completely similar.

Test 2- In this test, response of the proposed controller is investigated during load variations. Considering nominal parameters of the system, load value is stepped from 100 ohm to 50 ohm in t=0.3s and from 50 ohm to 33.3 ohm in t=0.5.s. From Fig.6, it is seen that, in spite of variation of the converter operating mode from DCM to CCM, proposed converter is stable. Also, it is obvious that steady state error of the proposed robust controller is zero for different loads. Also,the developed SMC has acceptable dynamic response.

TABLE I. NOMINAL SPECIFICATIONS OF THE DC-DC ĈUK CONVERTER AND CONTROLLER.

Description	Parameter	Nominal value
Input voltage	v_{in}	25 V
Converter inductor	L_1	1 mH
Converter inductor	L_2	1 mH
Converter capacitor	C_1	850 μF
Output capacitor	C_2	2000μF
Load resistance	R	100 Ω
Reference output voltage	v_o	24V
Switching frequency	f_s	20 kHz
control parameters	α_1	70
control parameters	α_2	20
control parameters	α_3	60
control parameters	α_4	32
control parameters	K_{P1}	0.1
control parameters	K_{P2}	32
control parameters	K_{P3}	0.25
control parameters	K_{I1}	100
control parameters	K_{I3}	90

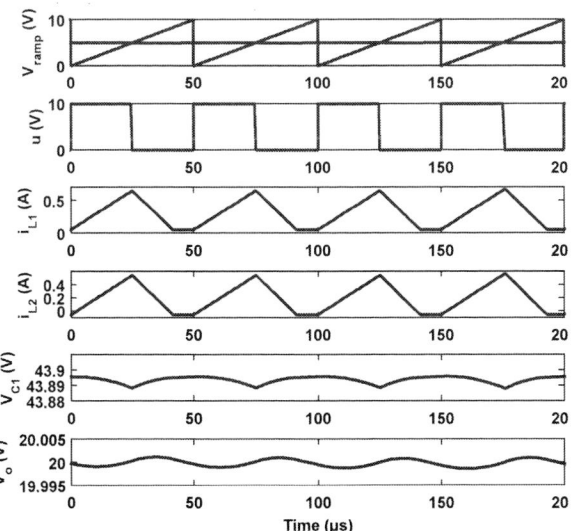

Fig. 5. Steady state response of the proposed robust controller for nominal parameters.

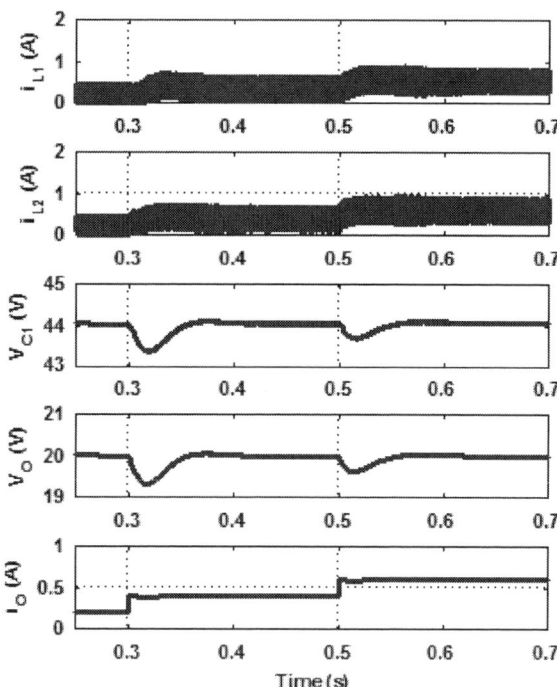

Fig. 6.   response of the proposed controller during load variations.

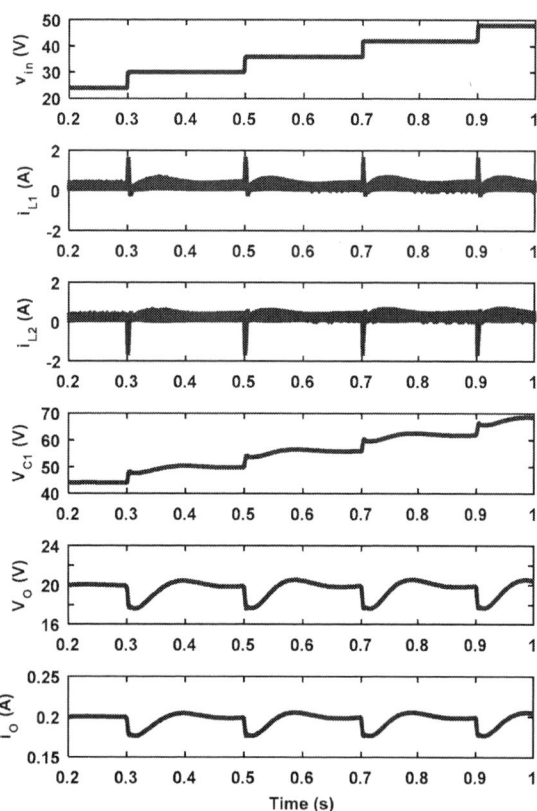

Fig. 7.   Response of the closed loop controller for input voltage.

Test 3- Step change of the input voltage from 24V to 30V, 36V, 42V and 48V in different instants is studied in this test. Response of the closed loop controller is shown in Fig.7. Transient response of the system is illustrated considering systems state variables. Again, the controller is completely stable, has good dynamic response with zero steady state error.

VI. CONCLUSION

In this paper, robust SMC is proposed to regulate output voltage of the DC-DC Ćuk converter in both CCM and DCM operations. The controller is developed according to different control loop. All of the converter state variables are controlled in a cascade structure to achieve fast transient response and stable behavior. Reference values of the state variables are calculated using feedforward and feedback signals in the designed four loop controller. Proposed controller includes integral of the output voltage error in the final control law; hence steady state error of the system is equal to zero in different conditions. Application of the equivalent SMC guarantees fixed switching frequency of the system. Designed controller is simulated using MATLAB/ Simulink toolbox for different tests. Simulation results show that the designed SMC has fast transient response. In spite of large variations of the input voltage and load resistance, developed controller is completely fast and stable.

REFERENCES

[1] *Solar power in Germany.* Available: https://en.wikipedia.org/wiki/Solar_power_in_Germany. 2015.

[2] M. Salimi, J. Soltani, A. Zakipour, and N. R. Abjadi., Hyper-plane sliding mode control of the DC–DC buck/boost converter in continuous and discontinuous conduction modes of operation. *IET Power Electronics8(8),* 1473-1482, 2015.

[3] M. Salimi, J. Soltani, A. Zakipour, and V. Hajbani, "Sliding mode control of the DC-DC flyback converter with zero steady-state error," in *Power Electronics, Drive Systems and Technologies Conference (PEDSTC), 2013 4th,* pp. 158-163, 2013.

[4] M. Salimi and V. Hajbani, "Sliding-mode control of the DC-DC flyback converter in discontinuous conduction mode," in *Power Electronics, Drives Systems & Technologies Conference (PEDSTC), 2015 6th,* pp. 13-18, 2015.

[5] L. Martinez-Salamero, H. Valderrama-Blavi, R. Giral, C. Alonso, B. Estibals, and A. Cid-Pastor, "Self-oscillating DC-to-DC switching converters with transformer characteristics," *Aerospace and Electronic Systems, IEEE Transactions on,* vol. 41, pp. 710-716, 2005.

[6] S. C. Tan and Y. M. Lai, "Constant-frequency reduced-state sliding mode current controller for Cuk converters," *Power Electronics, IET,* vol. 1, pp. 466-477, 2008.

[7] M. G. Umamaheswari, G. Uma, and K. M. Vijayalakshmi, "Design and implementation of reduced-order sliding mode controller for higher-order power factor correction converters," *Power Electronics, IET,* vol. 4, pp. 984-992, 2011.

[8] M. G. Umamaheswari, G. Uma, and K. M. Vijayalakshmi, "Analysis and design of reduced-order sliding-mode controller for three-phase power factor correction using cuk rectifiers," *Power Electronics, IET,* vol. 6, pp. 935-945, 2013.

[9] C. Zengshi, "PI and Sliding Mode Control of a Cuk Converter," *Power Electronics, IEEE Transactions on,* vol. 27, pp. 3695-3703, 2012.

[10] H. J. Sira-Ramirez and M. Ilic, "A geometric approach to the feedback control of switch mode DC-to-DC power supplies," *Circuits and Systems, IEEE Transactions on,* vol. 35, pp. 1291-1298, 1988.

[11] V. Utkin, J. Guldner, and M. Shijun, *Sliding Mode Control in Electromechanical Systems*: Taylor & Francis, 1999.

978-1-5090-0376-1/16 $31.00 © 2016 IEEE

Indirect Voltage Regulation of Double Input Y-Source DC-DC Converter Based on Sliding Mode Control

Soheil Ahmadzadeh

Dep. of Computer Science and Electronic Engineering
University of Shahrekord,
Shahrekord, Iran
soheil.a@stu.sku.ac.ir

Gholamreza Arab Markadeh

Dep. of Computer Science and Electronic Engineering
University of Shahrekord,
Shahrekord, Iran
arab-gh@eng.sku.ac.ir

Abstract—**In This paper, a sliding mode controller is designed for regulating the output voltage of a double input and high step-up DC-DC converter with three coupled inductors called Y-source impedance network, which can provide a very high boost at lower shoot-through duty cycle of switch and has one more degree of freedom to achieve voltage boost. In the proposed converter, the set of input sources hand over power to the load at the same time or separately, so joining of different new dc energy sources such as wind turbine, solar array, or fuel cell with a battery can be applied as input sources. The proposed controller is robust and stable against variation of load demands and uncertainty of parameters. Simulation results confirm the capability and effectiveness of the proposed control approach.**

Keywords—*Y-source impedance network, sliding mode control; high step-up dc-dc converter; double input converter.*

I. INTRODUCTION

In recent years, the renewable energy sources such as photovoltaic and wind due to facility of access, cheapness and lack of environmental pollution have applied widely by name distributed power generation. As a result to merge more than one energy source such as solar array, fuel cell and small scale wind turbines, the large number of multi input converter configurations have been proposed [1]-[4]. In the other hand, these distributed generators need a dc-dc converter with high step up gain to boost the output voltage than the dc link voltage required for interfacing to the utility grid [5]-[7]. Hence comprising these properties lead to discussion about a dc-dc converter which has high boost function and double sources at the input side.

Double input Y-source converter such as double input dc-dc converter explained in [8]-[10] must be able to deliver power to the load at the same time or individually. In addition proposed converter, in comparison with Double input Z-source converter, has a preference which provides higher boost gain with same duty cycle. Also despite of high-efficiency and high sep-up dc-dc converter design is usually a critical point due to high current at input side and high voltage at output side, according to power conservation, we propose double input and high step up converter using Y shaped impedance network called Y-source converter.

Operating the converter at the lower duty cycle compared to other step-up converter such as conventional boost, Z and Quasi Z-source, T and Trans-Z-source and Γ-source converters, reduces the shoot through interval time in each switching period and increases the efficiency. The

aforementioned converter provides the flexibility to choose the the turns ratio of the coupled inductor and range of shoot through time. As the consequence, it gives two degree of freedom in the boost function and eliminates the requirement for a high turns ratio of the coupled inductor or isolation transformer. Because using large turns ratio transformer increase the cost and size, it is not a good method for voltage increment. Output voltage enhancement by implementing a multi level converter and a voltage multiplier or cascading topologies entails many active and passive elements. Using large number of devices increases the total cost of the system as well as reduces the power density and efficiency [7],[11]-[12]. Hence proposed converter overcomes these limitations by means of high voltage boost gain with minimum number of passive elements and only one active switch. In addition due to existence of two independent input sources, it is appropriate for power generation applications due to connect a battery and an energy source (small scale wind turbines, solar array, wind turbine or fuel cell). Due to dc input voltage variation of the converter in different conditions, an appropriate controller is required in order to prepare robustness towards input changes and parameter uncertainties. In this paper sliding mode control approach is employed for dc output voltage regulation versus any change in input, load and other system characteristics.

In order to control nonlinear systems a kind of closed loop control approach called sliding mode control (SMC) was proposed initially [13]. The distinguished characteristic of this type of controller is the robustness and stability towards parameters uncertainties, system input, and output variation. Due to bounded switching frequency of power converters, ideal SMC cannot be used. Indeed all of SMC approaches are quasi-sliding mode control. Hence various sliding surfaces are proposed to control output voltage of conventional dc-dc converters mainly buck, boost and cuk converters in many papers [14]-[19]. Equivalent control is utilized to SMC to provide fixed switching frequency [17]-[19]. In order to design static controllers, conventional SMC employs the Ackermann's Formula and does not consider dynamic of the system [18]-[20]. It essentially relates to how the sliding coefficients on the desired dynamic properties were selected [21].

The operating principle of double input Y-source converter is presented in Section II. Sliding mode control is presented in Section III. Simulation results and conclusion are presented in Section IV and section V respectively.

978-1-5090-0376-1/16 $31.00 © 2016 IEEE

II. OPERATING PRINCIPLE AND TOPOLOGY OF PROPOSED CONVERTER

The proposed converter contains double voltage source in the input side, a Y–source impedance network whose function is to boost the voltage at the dc-link and inductive load in the output side. In Fig. 1 the circuit diagram of the double input Y-source converter is illustrated.

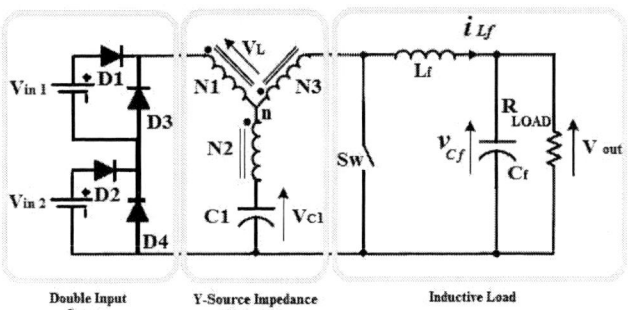

Fig.1. Proposed Double Input Y-Source DC-DC Converter

According to active or inactive operation states of dc input sources, four distinct states is Available. Each source independently can be applied either individually or simultaneously and hand over power to the load through the converter. If one of them is in active state, dc input source feeds the converter individually and the converter will operate as does a PWM converter. Operation states summary of the proposed dc-dc converter is presented in Table I [8]-[10]. So input stage is modeled by a diode and a single input source with various values according to table I. Input voltage source realization (V_{in}) in this converter is presented in Fig. 2.(a).

The proposed converter operates as follows: when the switch turns on, the equivalent diode is inversely polarized (shoot-through state). This situation generates a circuit topology demonstrated in Fig. 2.(b). During this interval, stored energy in the inductor L_f was transferred to the system load resistor. When the switch is in OFF state, the equivalent diode is directly polarized (non shoot-through state). In a similar way, this situation makes a circuit topology which is demonstrated in Fig 2.(c). In the second interval, the equivalent voltage source V_{in} and Y-source network generates the current of inductor L_f [22].

TABLE I. INPUT SOURCE STATES OF PROPOSED CONVERTER

State	Input Sources State		Diodes State				Input Voltage Source Value (V_{in})
	V_{in} 1	V_{in} 2	D1	D2	D3	D4	
1	Active	Active	On	On	Off	Off	V_{in} 1 + V_{in} 2
2	Active	Inactive	On	Off	Off	On	V_{in} 1
3	Inactive	Active	Off	On	On	Off	V_{in} 2
4	Inactive	Inactive	Off	Off	On	On	0

According to volt-second balance equation for V_L, one has

$$\frac{n_{12}}{n_{12}+1}(V_{in} - V_{C1})(1-d) + \frac{n_{12}n_{13}}{n_{12}-n_{13}}V_{C1}d = 0 \tag{1}$$

Where n_{12} = N1/N2, n_{13}=N1/N3. The peak of output voltage during the non shoot-through interval describes as,

$$\widehat{v_{Cf}} = \widehat{v_{out}} = \frac{V_{in}}{\left(1 - \frac{1+n_{13}}{1-n_{23}}d\right)} \tag{2}$$

So, the average output voltage define as:

$$V_{out} == \frac{(1-d)V_{in}}{\left(1 - \frac{1+n_{13}}{1-n_{23}}d\right)} \tag{3}$$

Forasmuch as $v_{c1}(t) \cong V_{c1}$ (as shown in [23]), v_{c1} suppose constant as follows:

$$V_{C1} = V_{in}(1-d)/(1 - \frac{n_{12}(1+n_{13})}{n_{12}-n_{13}}d) \tag{4}$$

Since v_{c1} is constant, we present state equations in terms i_{Lf} and v_{Cf} so if the KVL and KCL apply to the two circuit diagram of Fig. 2.(b) and (c), the differential equations describing the system in shoot-through state is the following:

$$L_f \frac{di_{Lf}}{dt} + v_{Cf} = 0 \tag{5}$$

$$i_{Lf} = C_f \frac{dv_{Cf}}{dt} + \frac{v_{Cf}}{R} \tag{6}$$

And differential equations describing the system in non shoot-through state describe as follows:

$$L_f \frac{di_{Lf}}{dt} + v_{Cf} = E \tag{7}$$

$$i_{Lf} = C_f \frac{dv_{Cf}}{dt} + \frac{v_{Cf}}{R} \tag{8}$$

As,

$$E = V_{C1} + \frac{(1-n_{32})(V_{in}-V_{C1})}{(1+n_{12})} \tag{9}$$

And n_{32} = N3/N2.

So, according to (5-9), the following average model obtains from differential equations:

$$L_f \frac{di_{Lf}}{dt} = -v_{Cf} + (1-d)E \tag{10}$$

$$C_f \frac{dv_{Cf}}{dt} = -i_{Lf} - \frac{v_{Cf}}{R} \tag{11}$$

When the average model of the proposed converter is obtained as mentioned in (10, 11) we make straightforward changes in the time variable (t) and scales which measure the magnitudes of the state variables (i_{Lf}, v_{Cf}). The normalized system equations with the new set of variables are defined as follows:

$$\begin{pmatrix} x_1 \\ x_2 \end{pmatrix} = \begin{pmatrix} \frac{1}{E}\sqrt{L_f/C_f} & 0 \\ 0 & 1/E \end{pmatrix} \begin{pmatrix} i_{Lf} \\ v_{Cf} \end{pmatrix} \tag{12}$$

$$\tau = \frac{t}{\sqrt{L_f \times C_f}} \quad , \quad u = 1-d$$

Now average normalized model obtains as:

$$\frac{dx_1}{d\tau} = -x_2 + u \tag{13}$$

$$\frac{dx_2}{d\tau} = x_1 - Qx_2 \tag{14}$$

Where, the parameter Q is the quality factor of the load circuit. It is defined with the load resistance R, inductance L_f and capacitance C_f by means of the relation:

$$Q = \frac{1}{R}\sqrt{L_f/C_f} \tag{15}$$

In the steady state regime, for the description of the average normalized model, all time derivatives of the state variables equals zero. These constant values are corresponding to equilibrium points applied to sliding mode control approach as the current reference or voltage reference. In (16, 17) relating equations were expressed.

$$\bar{x}_1 = Q\bar{x}_2 = Qu \qquad (16)$$
$$\bar{x}_2 = u \qquad (17)$$

On the other hand, in the steady state condition, the equilibrium values are expressed by desired average output voltage.

$$\bar{x}_1 = QV_d \qquad (18)$$
$$\bar{x}_2 = V_d \qquad (19)$$

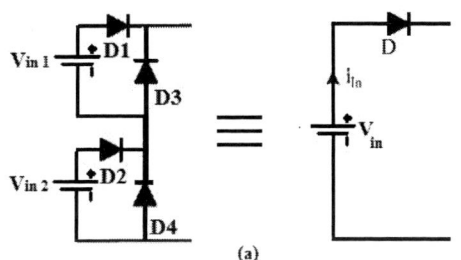

(a)

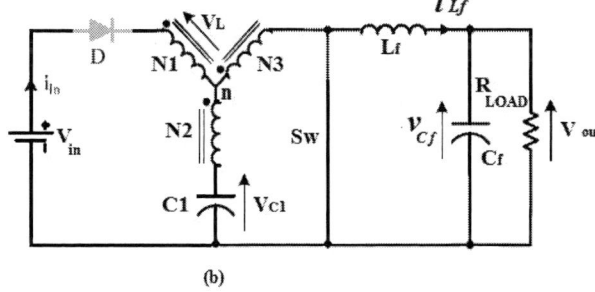

(b)

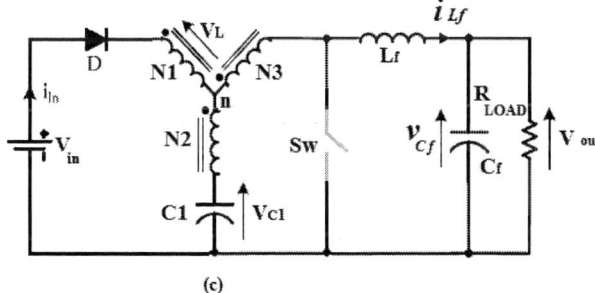

(c)

Fig.2. Exhibition of proposed converter ideal switching (a) Equivalent circuit of double input voltage source corresponding to V_{in} in Table I. (b)Switch state at d=1. (c)Switch state at d=0.

III. SLIDING MODE CONTROL

With normalized averaged model of the converter, as explained in (13, 14), we present the sliding mode control approach for the double input Y-source converter. In the context of the general notation,

$$\dot{x} = f(x) + g(x)u , \quad y = h(x) \qquad (20)$$

We have,

$$f(x) = \begin{pmatrix} -x_2 \\ x_1 - Qx_2 \end{pmatrix}, g(x) = \begin{pmatrix} 1 \\ 0 \end{pmatrix} \qquad (21)$$

Control objective in the indirect voltage regulation, is to attain the normalized average current x_1 to the current reference value $\bar{x}_1$ corresponding to desired output voltage. In the following, we express (22) as sliding surface coordinate function.

$$h(x) = x_1 - \bar{x}_1 = x_1 - Q\bar{x}_2 \qquad (22)$$

The following invariance condition is satisfied by the coordinate function $h(x)$ then,

$$\dot{h}(x) = \frac{\partial h}{\partial x}\dot{x} = \frac{\partial h}{\partial x}\big(f(x) + g(x)u_{eq}\big) = 0 \qquad (23)$$

In other words,

$$\mathcal{L}_f h(x) + \mathcal{L}_g h(x)u_{eq} = 0 \qquad (24)$$

So, the equivalent control is represented as (25).

$$u_{eq} = -\frac{\mathcal{L}_f h(x)}{\mathcal{L}_g h(x)} \qquad (25)$$

From (21) and (25), one has

$$u_{eq} = x_2 \qquad (26)$$

With (2) and denormalizing of x_2, we will prove that the equivalent switching control function (u_{eq}) is shorter than one. $0 \le u_{eq} \le 1$.

$$u_{eq} = x_2 = \frac{v_{Cf}}{E} \le \frac{\widehat{v_{Cf}}}{E} = \frac{\dfrac{V_{in}}{\left(1 - \dfrac{1 + n_{13}}{1 - n_{23}}d\right)}}{V_{C1} + \dfrac{(1 - n_{32})(V_{in} - V_{C1})}{(1 + n_{12})}}$$

$$= \frac{\dfrac{V_{C1}}{1 - d}}{V_{C1} + \dfrac{(1 - n_{32})(V_{in} - V_{C1})}{(1 + n_{12})}}$$

$$= \frac{\dfrac{V_{C1}}{1 - d}}{V_{C1} - \dfrac{(1 - n_{32})n_{13}dV_{C1}}{(n_{12} - n_{13})(1 - d)}}$$

$$= 1 \qquad (27)$$

As it was explained in [23], $\widehat{v_{Cf}}$ or $\widehat{v_O}$ is peak of output voltage obtained in non shoot-through state. Thus $0 \le x_2 \le 1$ or in the other words $0 \le u_{eq} \le 1$.

In the x_2 space, we recommend the following Lyapunov function candidate, which clearly specify the ideal sliding dynamics as (28).

$$V(x_2) = \frac{1}{2}(x_2 - \bar{x}_2)^2 \qquad (28)$$

Considering the ideal sliding dynamics from (13), (14), (16) and (26), which correspond to $h(x)=0$, the time derivative of (28) can be defined by (29).

$$\dot{V}(x_2) = (x_2 - \bar{x}_2)\dot{x}_2 = (\bar{x}_1 - Qx_2)(x_2 - \bar{x}_2)$$
$$= -Q(x_2 - \bar{x}_2)^2 < 0 \qquad (29)$$

Evidently, the last expression is negative definite around the equilibrium point $\bar{x}_2$. The desired voltage gives an asymptotically stable equilibrium point exhibited by the ideal sliding dynamics. Also we demonstrate that $h(x)=0$

assumption corresponding to the ideal sliding dynamics is true.

$$V(x_1) = \frac{1}{2}(x_1 - \bar{x}_1)^2 \tag{30}$$

$$\dot{V}(x_1) = \dot{x}_1(x_1 - \bar{x}_1) = (-x_2 + u)(x_1 - \bar{x}_1) \tag{31}$$

Using the switching policy according to (27) and (32), $\dot{V}(x_1)$ is negative definite and the sliding surface is accessible:

$$u = \begin{cases} 0 & x_1 - \bar{x}_1 > 0 \\ 1 & x_1 - \bar{x}_1 < 0 \end{cases} \tag{32}$$

And therefore, the control effort is represented in (33).

$$u = \frac{1}{2}[1 + sign(x_1 - \bar{x}_1)] \tag{33}$$

The desired regulation is provided by (33). It yields a system with internal stability. This expression is represented by (34) in non normalized form:

$$u = \frac{1}{2}[1 + sign(i_{Lf} - I_{ref})] \tag{34}$$

As i_{Lf} and I_{ref} are the real and reference current of inductor L_f. From (10) the control effort for proposed converter is given by (35).

$$d = 1 - u \tag{35}$$

IV. SIMULATION RESULTS

A double input Y-source dc-dc converter with an inductive load using SMC approach is simulated by PSIM to confirm above analysis. Referring to Fig. 1, the system characteristics were demonstrated in table II.

TABLE II. CHARACTERISTIC OF PROPOSED CONVERTER

Parameter	Description	Value
$V_{in\,1}$	First Input Voltage	10 V
$V_{in\,2}$	Second Input Voltage	50 V
C_1 & C_f	Capacitance	470 µF
$N_1:N_2:N_3$	Turns Ratio	80:16:48 = 5:1:3
L_f	Inductance	330 µH
R	Resistance	25 Ω

Fig.3 demonstrates the power circuit which provides proposed converter simulation results. This figure illustrates uncertainty of the parameters L_f, C_f and R. Switching Off/On the power switches $S_2 - S_4$ can perform the parameter step changes. Also, double input stage is modeled by additional circuit. The converter can be supplied by each dc source, individually or simultaneously, as previously mentioned. The converter state variables are measured and plotted by selecting a sampling time step of 1.639×10^{-6}s in each switching period. In order to show independence of input sources from each other, four different states are considered for simulations.

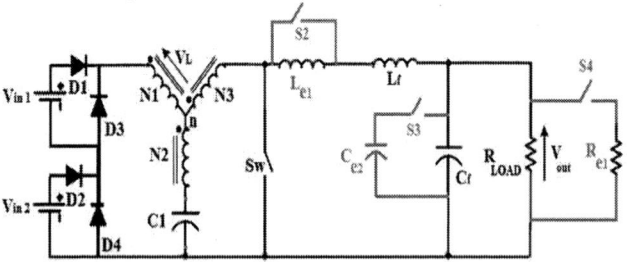

Fig.3. Power circuit of double input Y-Source converter

A. State 1: Both input source 1 and 2 be active

Test 1)

During simulation time in this test, both input sources are active. The converter produces 127.5 V in the output side with 15% duty cycle. Fig. 4 shows the output voltage in this situation.

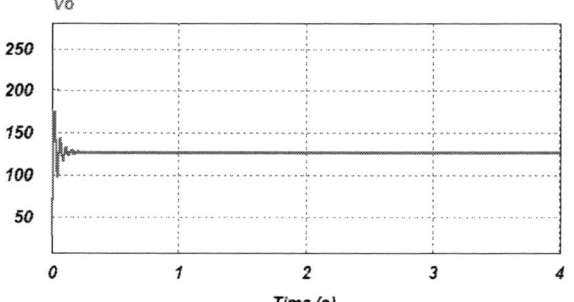

Fig.4. Output Voltage in State 1 with D=0.15

Test 2)

The converter operation, in this test, was considered in the steady-state regime with defined parameters as table II. At $t=1.2$ s. output desired voltage of the converter is stepped up from $V_{ref}=+80$ V to $V_{ref} = +240$ V. The simulation result is demonstrated in Figure 5.

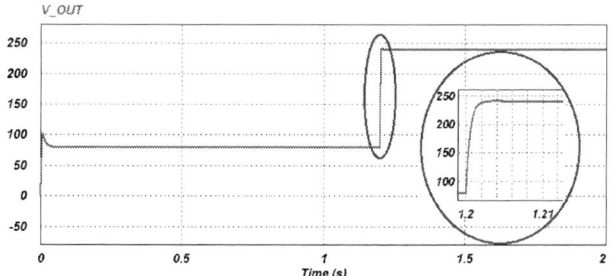

Fig.5. Step response of sliding mode controller in state 1

Test 3)

Fig. 6 shows simulation results related to load resistor uncertainty with reference voltage value 80 V. $R_{e1} = 25\ \Omega$ at $t = 0.8$ s is connected and then load resistor back to its initial value at $t=1.6$ s with disconnection of R_{e1}. Output voltage and current of R_{e1} are shown in Fig. 6.

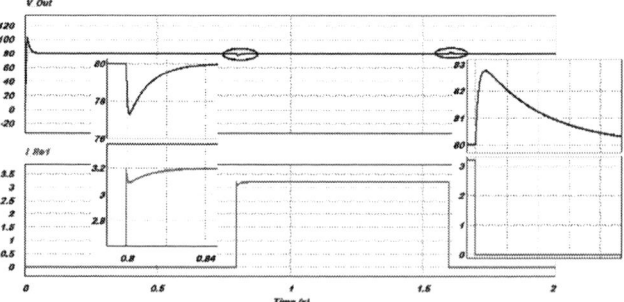

Fig.6. Load resistor uncertainty

Test 4)

Considering $L_{e1}=100~\mu H$ and $C_{e2}=47~\mu F$ with reference voltage value 80 V, the effect of simultaneous changes in parameters is studied in this test. At $t=1~s$ and $t=1.5~s$, S_2 and S_3 become close respectively. So L_{e1} will be shorted and C_f is stepped up from its initial value to C_f+C_{e2}. Simulation results are shown in Fig. 7.

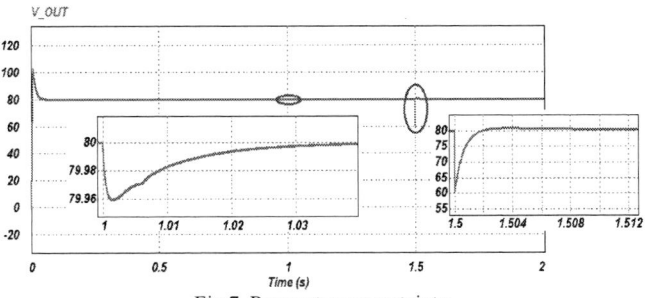

Fig.7. Parameters uncertainty

B. *State 2: input source 1 becomes inactive at t=1 s*

Test 5)

In this test, we have investigated the simultaneous change in input side. Related Simulation results were shown in Fig. 8 by 15% duty cycle. So output average voltage decreases from initial value 127.5 V to 106.25 V.

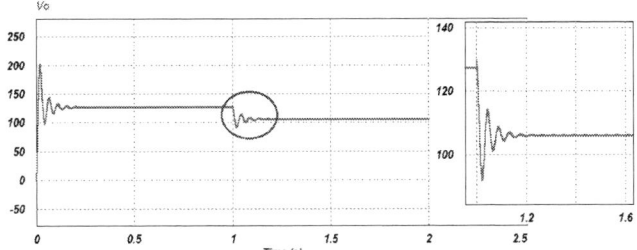

Fig.8. Simultaneous changes in input side by turning off the source1

Test 6)

In this test with consideration $V_{Ref}=80$ V, load resistance, inductance L_f, capacitance C_f and input voltage change simultaneously. Input source 1 becomes inactive at $t=1~s$. At $t=0.5, 0.75$ and $1.25s$, S_4, S_2 and S_3 become close and at $t=1.5s$, S_4 will be open Again. So the equivalent resistance changes between 25Ω and 12.5Ω. Fig. 9 demonstrates the results.

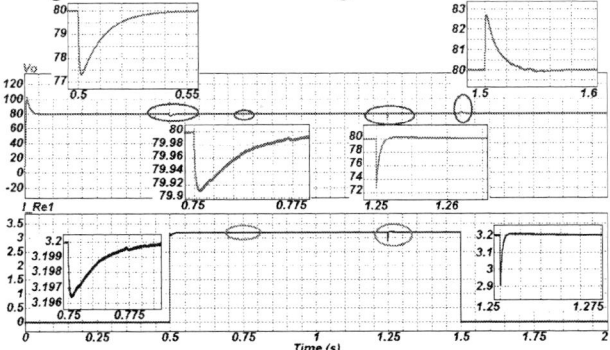

Fig.9. Effect of input voltage and parameters changes

C. *State 3: input source 2 becomes inactive at t=1 s*

Test 7)

For this state, input source 2 becomes inactive at $t=1~s$, converter produced 21.25 V by 15% duty cycle. Fig. 10 demonstrates the output voltage in this situation.

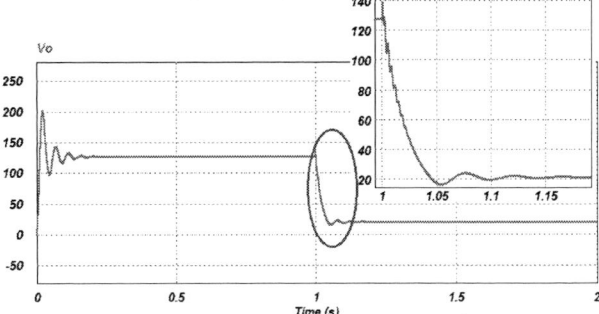

Fig.10. Simultaneous changes in input voltage

Test 8)

Considering $V_{Ref}=80$ V output capacitor, load resistance and input voltage have simultaneously changed and these effects were investigated. At $t=0.4~s$ and $t=1.4~s$, S_4 and S_3 become close respectively. At $t=1~s$ input source 2 becomes inactive. The simulation results are illustrated in Fig. 11.

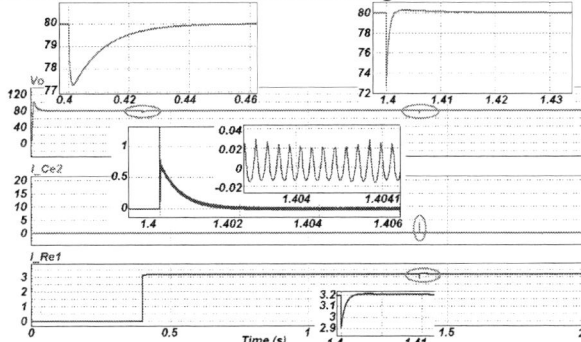

Fig.11. Input voltage, output capacitor and load resistance simultaneous changes.

Test 9)

In this test, output desired voltage and input voltage have simultaneously changed. The effects of these changes are studied. The output desired voltage is stepped up at $t=0.5~s$ from +80 V to +200 V, and input source 2 becomes inactive at $t=1~s$. In Fig. 12, the simulation results show output voltage versus changing in desired output voltage.

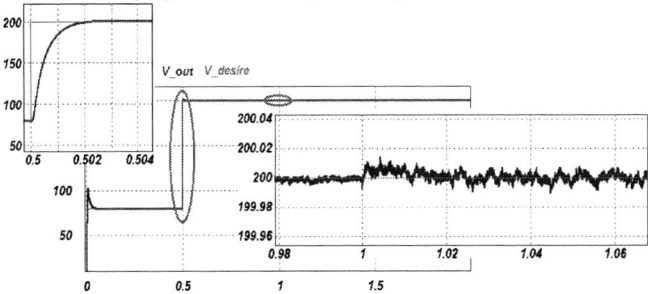

Fig.12. Step response using SMC

D. State 4: source 1 become inactive at t=0.5s and source 2 become inactive at t=1.5 s

Test 10)

In this state both input sources become inactive finally .So we do not consider it as a profitable operation state. At *t=0.5 s* and *t=1.5 s*, source 1 and 2 become inactive respectively. Indeed, converter and load cannot be supplied because no active source is available. But the simulation was done for this state by 15% duty cycle and the results are shown in Fig. 13.

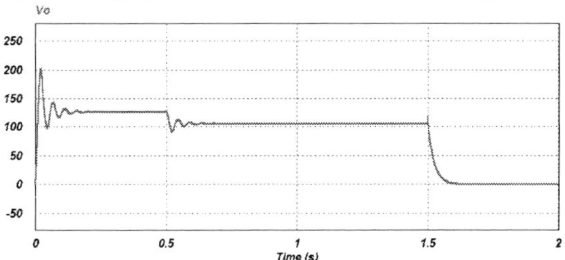

Fig.13. Deactivation of input sources.

V. CONCLUSION

In this paper, a nonlinear control approach was utilized for voltage regulation of DC-DC double input Y-source converter. It has been developed by means of a SMC. Due to the independence of each input source, four distinguished states have been considered and specific simulation has been done for each state. Simulation results verify capability and effectiveness of the SMC. According to these results, proposed control approach clearly confirms the robustness and stability of the converter against variations of input voltage, uncertainty of parameters and load disturbances.

REFERENCES

[1] E.Muljadi and H.F.Mckenna, "Power Quality issues in a hybrid power sysytem" IEEE Trans Ind Appl vol 38 pp 803-809 May/Jun 2002

[2] F Giroud and Z.M. Salameh "Steady State performance of a grid connected rooftop hybrid wind-photovoltaic power system with battery storage" IEEE Trans Energy Convers vol 16 no.1 pp 1-7 march 2001

[3] Yaow Ming Chen, Yuan Cuan Liu, Cheng Sien Lin "Double Input PWM DC-DC Converter for High/Low Voltage Source" IEEE Trans Ind. Electron, vol. 53 no.5 October 2005

[4] Yuan Chuan Liu, Yaow Ming Cheng "A systematic approach to synthesizing Mul input DC_DC converter" IEEE Trans Power Electron.vol.24 no.1 January 2009

[5] F. Blaabjerg, Z. Chen and S. B. Kjaer, "Power Electronics as Efficient Interface in Dispersed Power Generation Systems," IEEE Trans. Power Electron., vol. 19, no. 5, pp. ll84-ll94, Sept. 2004.

[6] E. H. Ismail, M. A. AI-Saffar, A. J. Sabzali and A. A. Fardoun, "A Family of Single-Switch PWM Converters With High Step-Up Conversion Ratio," IEEE Trans. on Circuits and Systems, vol. 55, no. 4, pp. 1159-1171,2008

[7] Yam P. Siwakoti, Graham E. Town, Poh Chiang Loh, Frede blaabjerg "Y Source Impedance-Network -Based Isolated Boost DC/DC Converter" the International Power Electronics Conference pp1801-1805, 2014

[8] Samthosh Yamsani, Sympath Kumar Bioni, "DC motor Fed with double-input Z-source DC_DC converter" International Journal of modern Engineering Research vol.2 Issue 6, pp 4144-4148, Nov/Dec 2012

[9] Fazad Sedaghati and Ebrahim Baberi, " Double input Z-Source DC-DC Converter", *Proceeding of IEEE 2011 2nd Power Electronics, Drive System and Technologies Conferences*, pp. 581-586.

[10] K.Subramanian V. Kavitha "Simulation Study of Double Input Z-Source DC-to-DC Converter" International Conference on Renewable Energy and Sustainable Energy [ICRESE], 2013

[11] W. Li and X. He, "Review of Nonisolated High-Step-Up DC/DC Converters in Photovoltaic Grid-Connected Applications," IEEE Trans. Ind. Electron., vol. 58, no. 4, pp. 1239-1250, April 2011.

[12] Y. Du, S. Lukic, B. Jacobson and A. Huang, "Review of high power isolated bi-directional DC-DC converters for PHEV/EV DC charging infrastructure," in Proc. Energy Conversion Congress and Exposition (ECCE), pp. 553,560, 17-22 Sept. 2011.

[13] V. Utkin, J. Guldner, and J. X. Shi, "Sliding Mode Control in Electromechanical Systems " . London, U.K.: Taylor and Francis, 1999.

[14] L. Malesani, L. Rossetto, G. Spiazzi, and P. Tenti, "Performance optimization of cuk converters by sliding-mode control," *IEEE Trans. Power Electron.*, vol. 10, no. 3, pp. 302--309, May 1995.

[15] S. C. Tan, Y. M. Lai, M. K. H. Cheung, and C. K. Tse, "On the practical design of a sliding-mode voltage controlled buck converter," *IEEE Trans. Power Electron.*, vol. 20, no. 2, pp. 425-- 437, Mar. 2005.

[16] S. C. Tan, Y. M Lai, and C. K. Tse, "Adaptive feedforward and feedback control schemes for sliding mode controlled power converters," *IEEE Trans. Power Electron.*, vol. 21, no. 1, pp. 182-- 192, Jan. 2006.

[17] S. C. Tan, Y. M. Lai, C. K. Tse, and M. K. H. Cheung, "A fixedfrequency pulse width-modulation-based quasi-sliding-mode controller for buck converters," *IEEE Trans. Power Electron.*, vol. 20, no. 6, pp. 1379--1392, Nov. 2005.

[18] Y. He, and F. L. Luo, "Sliding-mode control for dc-dc converters with constant switching frequency," IEE Proc.-Control Theory Appl., Vol. 153, No.1, Jan 2006

[19] S. C. Tan, Y. M. Lai, and C. K. Tse, "A unified approach to the design of PWM-based sliding-mode voltage controllers for basic dc-dc converters in continuous conduction mode," *IEEE Trans. Circuits Syst..*, vol. 53, no. 8, pp. 1816--1827, Aug. 2006.

[20] J. Ackermann, and V. Utkin, "Sliding-mode control design based on Ackermann's formula," *IEEE Trans. Autom. Contr.*, vol. 43, no. 2,pp. 234--237, Feb. 1998.

[21] A. H. Rajaei, S. Kaboli, and A. Emadi, "Sliding-mode control of z-source inverter," in *Proc. 34th Annu. IEEE Ind. Electron. Conf*, pp.947-952, 10-13 Nov. 2008.

[22] Y. P. Siwakoti, P. C. Loh, F. Blaabjerg and G. E. Town, "Effects of Leakage Inductances on Magnetically-Coupled Impedance-Source Networks" *IEEE Trans. on Power Electron. (Letter)*, vol. 29, no. 11, pp. 5662-5666, Nov. 2014.

[23] Y. P. Siwakoti, P.C. Loh, F. Blaabjerg and G. Town, "Y-Source Impedance Network," *IEEE Trans. Power Electron. (Letter)*, vol. 29, no. 7, pp. 3250-3254, Jul. 2014.

7th Power Electronics, Drive Systems & Technologies Conference (PEDSTC 2016)
16-18 Feb. 2016, Iran University of Science and Technology, Tehran, Iran

Flatness-Based Control Method:

A Review of its Applications to Power Systems

M. Soheil-Hamedani [1], M. Zandi[1]
[1]Department of Energy Engineering, Shahid Beheshti
University, Tehran–Iran
mary.soheil@gmail.com

R. Gavagsaz-Ghoachani [1,2], B. Nahid-Mobarakeh[2],
SenoirMember, IEEE, S. Pierfederici[2]
[2]GREEN, Université de Lorraine–France

Abstract— **Flatness systems comprise controllable linear systems, as well as nonlinear systems by static and dynamic feedback call endogenous dynamic feedback. Systems which are differently flat have several advantages which can be utilized to generate effective control strategies for nonlinear systems. In order to highlight the improvement of this control method, this paper provides an overview of this approach and its usage in some power systems including hybrid systems, power electronics and electrical drives.**

Keywords—Flatness, DC-DC converter, electrical hybrid system, fuel cell, ultracapacitor, inverter, photovoltaic, wind turbine, power systems, control.

I. INTRODUCTION

For nonlinear systems, the most common control is to invert the system dynamics to calculate the inputs required for a specific task.

In most control techniques, a mathematical model of systems is exploited to obtain a solution to the inverse dynamics and feedback regulation problems. The most common structure is linear, in which the properties of linear control problems may be used. By using different linearizations about different operation points, best control results may be obtained. However, as the systems become more complex, using only the linear structure may not be adequate [1]. In order to overcome these difficulties, a particular class of systems, called differentially flat system is introduced [2-3].

Differential flatness is a concept introduced in 1992 by [2-3]. It is a new approach for the analysis and design of nonlinear continuous-time control systems. A system is differentially flat if there is a basic variable (a flat output) in which to describe the complete motion of the system [4]. The essential condition for flatness is that the system must be controllable and linearizable by endogenous feedback, either static or dynamic state feedback [5-6].

The flatness property is used to control of nonlinear systems with good performances in term of tracking trajectory. By the use of reference trajectories of the flat outputs, a safe operation during the start-up and steady-state can be obtained. Furthermore, Flat controllers use the reference reactions instead of disturbance reactions in order to reduce the noise impact. Moreover, the transient states can be analytically anticipated by generating feasible trajectories, which cannot be achieved with classical approaches, by using flatness property [7]. To highlight the improvement of the flatness-based control, this paper provides an overview of this control method in some power systems, including hybrid systems [8-21], power electronics [22-34], and electrical drives [35-45].

This paper is organized as follows: Section II describes the flatness theory. Section III provides a literature review of systems which are differentially flat. An overview of differential *control systems based on differential flatness theory* for various applications is given in Section VI. The conclusions of the study are given in Section V.

II. FLATNESS THEORY

The main feature of the differential flatness theory is the presence of a fictitious output, called flat output denoted by $y = (y_1, y_2, ..., y_m)$, such that [46] the system variables may be expressed as a function of the flat output components and their time-derivatives; the flat output components may be expressed as a function of the system variables and their time-derivatives; and the number m of components y is equal to the number of input (control) variables.

In other word, assumed a non-linear system such as:

$$x = f(x, u); \ x \in R^n; u \in R^m \qquad (1)$$

Where $x = [x_1, x_2, ..., x_n]^T$ and $u = [u_1, u_2, ..., u_m]^T$ denote state and input vectors, respectively.

This system is differentially flat, if, there exist a vector $y \in R^m$ called flat output, in the form $y = \psi(x, u, ..., u^{(\beta)})$ such that [2], [5-6]:

$$\begin{cases} x = \varphi_1(y, \dot{y}, ..., y^{(\alpha)}) \\ u = \varphi_2(y, \dot{y}, ..., y^{(\alpha+1)}) \end{cases} \qquad (2)$$

α and β are finite numbers of derivatives.

Those properties, which are enhanced most often by the existence of a flat output with a clear engineering and physical meaning, make their control rather easy. They permit a straightforward, open loop path tracking. The equivalence of the system with a linear controllable one yields feedback stabilization around the desired trajectory [7].

III. SYSTEM DESCRIPTION AND EQUATIONS

The theory of flatness is well developed and has made accessible by several books [5] and [47], in parallel, a high number of applications have studied. In section III, we present an overview for the use of differential flatness theory-based control of the three main categories of applications including hybrid systems, power electronics, and electrical drives.

A. Electrical hybrid systems

The hybridization concept is to combine two or more electrochemical or electromechanical devices with at least one storage element in order to gain the respective aims of each one while minimizing their disadvantages. The flatness property of hybrid systems is introduced in [8-21] in order to satisfy the stabilization problem, dc bus voltage regulation as

978-1-5090-0376-1/16 $31.00 © 2016 IEEE   547

well as energy management. Some of the researchers working on this technique are shown in Table I.

In [8] a flatness-based control method is proposed to control a multisource/multiload Electrical Hybrid System (EHS). As illustrated in Fig. 1, the EHS is composed of a fuel cell (FC) and a supercapacitor (SC) as the main and auxiliary sources. They supply two independent loads that are fed by buck converters. As can be seen in Fig. 1, the electrostatic energy stored in the capacitors marked in green, the DC-bus and the output capacitor voltages marked in blue and the power of loads and supercapacitors marked in red are the output, state and input flat components, respectively. The advantages of this control approach are to manage the energy in the system without algorithm commutations. In [11], the control of an FC/Ultracapacitor hybrid system is presented based on flatness control approach. The ultracapacitor is connected in parallel to the FC through a bidirectional converter. This association supplies a load through another dc/dc converter. As presented in Table I, the electrostatic energy stored in the dc-bus capacitors considered as the system output to control the power flow in the system. Furthermore, the minimum value for C_0 and C_2 are measured during the worst case to reduce cost and increase the stability of the system. This control method is applied in [12] to stabilize the dc bus and energy management between sources for a hydrogen/solar power plant. Moreover, this method is compared with Linear proportional-integral during a load step and it was illustrated that the convergence via the new operating point with flatness control method is faster than PI controller. [13] is applied the combination of flatness control technique (FCT) and fuzzy logic control (FLC) for the FC/BATs/SCs hybrid power source. By using the specific trajectories for the system output, the power flow management and the output voltage regulation are obtained in different operating mode. To reach these objectives, it was considered the energies stored in output capacitor C_2 and the total energies stored on C_1 and C_2 as flat output variables (illustrated in Table I). This control approach is implemented in [15] to manage the power flow in a BATs/SCs hybrid systems. These two sources are connected to the DC bus via a DC-DC current bidirectional converter. The advantages of this control are the regulation of DC bus voltage as well as convincing the power demand. [16] is used flatness control method in a renewable energy power plant. As illustrated in Fig. 2, a reduced order of a hybrid power plant composed of photovoltaic (PV), wind turbine (WT), and FC as the main source and a SC as an auxiliary device is presented. To simplify the system, only static losses in the converters and sources are considered. The FC, PV, and WT converters have 4-phase parallel boost converters and the SC converter has 4-phase parallel bidirectional converters. By choosing the dc-bus capacitive energy and the total energies stored on C_{Bus} and C_{SC} as flat output components marked in green, the power flow between sources and network and voltage regulation of dc bus have been obtained. Also, the state and input variables are shown in blue and red, respectively.

B. Power electronic systems

The flatness property of converters is presented in [22-31] in order to investigate the power and voltage regulation besides stabilization of the whole system during transient and steady-state. Some of the researchers working on this technique are shown in Table II.

In [22], the control of dc–dc converter with a high voltage ratio based on flatness control is implemented. As shown in Fig. 3, in order to optimize the efficiency of the converter with high voltage and small input-current undulation, cascaded converter structure is proposed. Therefore, the two-interleaved boost converter and three-level boost converter are used as the first and second subconverter, respectively. As illustrated in Fig. 3, the energy stored on capacitor C and the total energy stored in capacitors C_1 and C_2 marked with green, are considered as flat output to control of the whole system. Moreover, the state and input variables can be seen with blue and red, respectively. A flatness control method is applied in [23] to control a non-ideal DC/DC boost converter. Also, an online estimator is used to track the evolution of the converter losses and the load variations. As can be seen in TABLE I, to control the total energy stored in the system, the energy stored in C and L are candidates flat output. This control approach is also applied for a DC/DC buck-power-converter and DC-motor system [24]. It was used in the buck converter to require a voltage profile v for the dynamic model of the DC motor, in order to follow the angular speed. As illustrated in TABLE II, the output voltage between the capacitor terminals is chosen as flat output to reach the desired voltage profile. In [25], a new control law based on the flatness approach for nonlinear switching-power converters applied in a fuel cell (FC) generator is presented. Only the inner FC power regulation loop is studied and multiphase boost converters with an interleaving switching technique are utilized as power converter. To operate the power regulation loop at very high dynamics, input power of each cell is considered as the flat output (as illustrated in Table II).

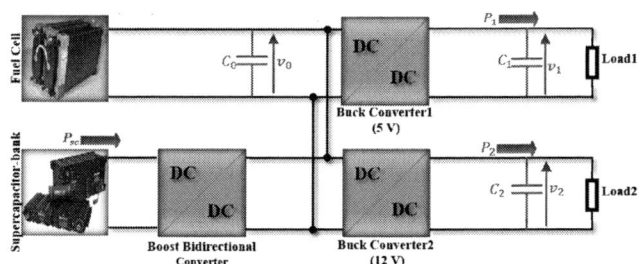

Fig. 1. Structure of an electrical hybrid system composed of FC/SC.

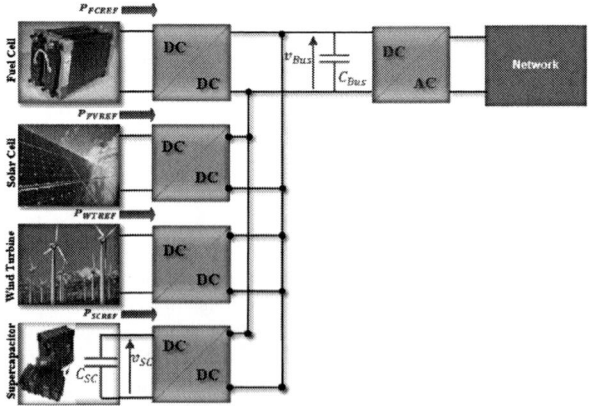

Fig. 2. Architecture of a renewable energy hybrid power plant

Table I. STUDIED HYBRID SYSTEMS BASED ON FLATNESS CONTROL.

Ref	Studied system	State Variables		Input Variables		Output Variables	
8	Electrical hybrid system (FC and SCB)	v_0	-DC-bus capacitor voltage C_0	P_1	- Power of load 1	y_0	- Energy stored on C_0
		v_1	- Output capacitor voltage C_1	P_2	- Power of load 2	y_1	- Energy stored on C_1
		v_2	- Output capacitor voltage C_2	P_{sc}	- Power of supercapacitor	y_2	- Energy stored on C_2
11	Electrical hybrid system (FC and ultracapacitor)	v_0	-Capacitor voltage C_0	P_1	- Power of ultracapacitor	y_0	Energy stored on C_0
		v_2	- Capacitor voltage C_2	P_2	- Power of dc-link	y_2	Energy stored on C_2
9	Distributed dc generation (FC and supercapacitor)	v_{Bus}	-DC-bus capacitor voltage C_{Bus}	P_{SCREF}	-Supercapacitor power reference	y_{Bus}	- DC-bus capacitive energy
		v_{SC}	- Supercapacitor voltage C_{SC}	P_{FCREF}	- FC power reference	y_T	- Total energies stored on C_{Bus} and C_{SC}
10	FC/supercapacitor hybrid power source	v_{Bus}	DC bus capacitor voltage C_{Bus}	P_{SCREF}	Supercapacitor power reference	y_{Bus}	Energy stored on C_{Bus}
13	Electrical hybrid power source (FC, SCs and BATs)	v_1	- Output capacitor voltage C_1	P_2	-Power of converter 2	y_2	- Energies stored in output capacitor C_2
		V_2	- Output capacitor voltage C_2	P_{st}	-Sum of the storage power in SCs and BATs	y_s	- Total energies stored on C_1 and C_2
12	Hybrid energy system (PV, PEMFC and a Li-ion)	v_{Bus}	DC-bus capacitor voltage C_{Bus}	P_{Bat}	Battery power	y_{Bus}	DC bus capacitive energy
14	Hybrid source system (FC and SCB)	i_L	- Inductor current L	d	Duty cycle	y	Sum of energies stored in the L and C
		v_o	- Output capacitor voltage C				
15	Multi-source systems (BAT, SC)	v_{bus}	DC bus capacitor voltage C_{bus}	P_{sc}	Power of supercapacitor	y_{Bus}	Energy stored in the DC bus
17	DC hybrid power sources (PEMFC and SCs)	v_{bus}	DC bus capacitor voltage C_{bus}	P_{sc}	Power of supercapacitor	y_{Bus}	Energy stored in the DC bus
16	Renewable energy hybrid power plant (PV, WT, FC and SC)	v_{Bus}	- DC-bus capacitor voltage C_{Bus}	P_{SCREF}	- Supercapacitor power reference	y_{Bus}	- DC-bus capacitive energy
		v_{SC}	- Supercapacitor voltage C_{SC}	P_{TREF}	- Total power from the PV array, WT, and FC	y_T	- Total energies stored on C_{Bus} and C_{SC}
18	Solar cell/hydrogen energy power plant	v_{Bus}	DC-bus capacitor voltage C_{Bus}	P_{BatREF}	Battery power reference	y	DC-bus capacitive energy

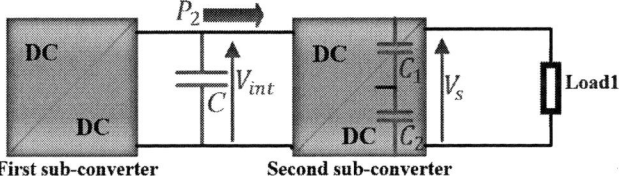

Fig. 3. Schema of a cascaded converter.

The control of this kind of converters for a PV power source based on the flatness based control is also investigated in [26]. As illustrated in Fig. 4, the input power of each cell of PV p_L marked with green is considered the flat output component. The inductor Current i_L with blue and the duty cycle of the pulse width modulation (PWM) converter d with red are also assumed as the state and the control variable, respectively. Using flatness property, performance and stabilization of the system besides voltage regulation of dc bus have been achieved. In [27] a combination of sliding mode and flatness control are applied to control a current-fed isolated boost converter. This converter is composed of an inductor and a controllable full-bridge converter, which is connected to a full-bridge rectifier via a high-frequency transformer. The paper is used flatness-based control for the outer voltage loop and the sliding mode for the inner loop. As can be seen in Table II, by choosing the energy stored in the output capacitor C as the flat output, flatness control could not only satisfy the responses during start-up but also could guarantee the system in steady-state operation. In [29], a flatness control method is used to control the three-port bidirectional full-bridge dc-dc converter for a FC/SC hybrid system. The advantages of using this control method are also

to estimate the behavior of the system in both transient and steady-state operation. To achieve these properties, the energy stored in the input and output capacitors C_0 and C_1 are chosen as the flat output variables. Furthermore, A combination of flatness control method and sliding mode are used for a new current sharing technique on a general case of N paralleled DC–DC boost converters in [30]. The first-loop controller is based on flatness method to control the voltage regulation and the second one is based on the sliding mode to ensure the current regulation for each individual converter. The benefits of the proposed control strategy are high dynamics, regulation and system robustness against parameter variations.

Three phase voltage source inverters (VSI) are being controlled based on the flatness technique by many researchers [32-34] as depicted in Table III. These inverters fed from DC or AC voltage source with diode rectifier. To make the input voltage constant, a large capacitor is connected at the input terminals. An output LC filter is needed in VSI which causes more losses and control complexity. VSI can be operated as a stepped wave inverter or a pulsewidth modulated (PWM) inverter and is frequently met in the electric power system, such as in the connection of photovoltaics with the rest of the grid [48-49]. For this kind of inverters to be controlled, flatness-based control method can be utilized [32-34]. In [32], this control approach is applied to investigate the behavior of the system state variables in transient and steady-state in the presence of any cases of load. To simplify the calculation, loss of dynamical information is supposed to be negligible. A one loop controller based on flatness technique is also applied for a non-isolated power supply composed of N

Table II. STUDIED POWER CONVERTERS BASED ON FLATNESS CONTROL.

Ref	Studied system	State Variables		Input Variables		Output Variables	
22	Non isolated dc–dc converter with high voltage ratio	V_{int} V_s	-Intermediate voltage -Sum of the voltages across the capacitors C_1 and C_2	P_1 P_2	- Source power - First subconverter power	y_1 y_2	- Energy Stored on capacitor C - Total energy stored on capacitors C_1 and C_2
25	Multiphase Interleaved DC–DC Converter	i_L	Inductor current	d	Duty cycle of the pulsewidth modulation (PWM) converter	p_L	- Input power of each cell of FC
27	Current-fed isolated boost converter	V_2	Output capacitor voltage (C)	p_0	Power from the primary side	y	Energy stored in the output capacitor C
28	Current-fed full-bridge DC-DC converter	v_2	Output capacitor voltage (C_{eq})	p_0	Power from the primary side	y	Energy stored in the capacitor C_{eq}
26	Multiphase Interleaved DC–DC Converter	i_L	Inductor current	d	Duty cycle of the pulsewidth modulation (PWM) converter	p_L	Input power of each cell of PV
29	Three-port Bidirectional DC-DC converter	V_0 V_{ch}	- Input capacitor voltage C_0 -Output capacitor voltage C_1	p_{10} p_{20}	-Power transferred from each 2 port -Power transferred from each 2 port	y_0 y_1	- Energy stored in C_0 - Energy stored in C_1
31	Isolated three-port Bidirectional dc-dc converter	V_0 V_{ch}	- Input capacitor voltage C_0 -Output capacitor voltage C_1	p_{10} p_{20}	-Power transferred from each 2 port -Power transferred from each 2 port	y_0 y_1	- Energy stored in C_0 - Energy stored in C_1
23	Non-ideal DC/DC Boost Converter	i_L v_s	- Inductor current L - Output capacitor voltage C	d	Duty cycle of the pulsewidth modulation (PWM) converter	y	Total energy stored in C and L
24	DC/DC Buck-Power-Converter	v i	- Output voltage between the capacitor terminals - Inductor current	u_{av}	Switch position function	v	Output voltage between the capacitor terminals
30	N paralleled DC–DC boost converters	V_0	Output voltage	p_{in}	Input power	y	Energy stored in C_0

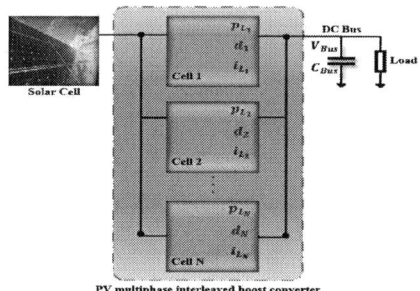

Fig. 5. PV multiphase interleaved boost converter.

parallel inverters with the LC output filter and common DC bus [33] and [34]. This technique prepared the highest dynamic response and enhanced a reliable operation of the parallel system when one or more inverters of the parallel system are in faulty mode. The benefits of using flatness are the reductions of the circulating currents and high power quality at the Point of Common Coupling (PCC).

C. AC and DC drives

The flatness property of AC and DC drives is also introduced in [35-45] in order to guarantee power and frequency regulation. It can also minimize the copper losses, avoid the system from the uncontrollable behavior and prepare system robustness during perturbations Some papers using this control approach are presented in Table IV.

In [35] this control method is used to minimize the copper losses excising in permanent magnet synchronous motor (PMSM) at all operating points. The flat output components of this synchronous motor are the load torque T_e, position θ and the current in d-axes i_d. It was shown that all the variables of the system, including the mechanical speed Ω, current in d-axes i_d, voltage in d-axes v_d and voltage in q-axes v_q can be expressed in terms of output components. In [36], the combination of flatness control and PI are used to satisfy torque and speed tracking with respect to parameter variations in PMSM. A cascaded flatness control is also used for PMSM

in [37] to eliminate the static errors of the system state variables. The d-q axes stator currents and the angle of flux orientated coordinate system are candidate as the flat output components. The flatness control of this kind of machine fed by bidirectional Quasi Z-source inverter (Q-ZSI) is also presented in [38]. The three flat output components are the stored energy in the ZSI, the mechanical speed and the d-axis flux. The advantages of this control approach are avoiding the system from the uncontrollable behavior and preparing robustness during perturbations. The concept of flatness-based control method is used in [39] in order to control a synchronous machine. To simplify the calculation, it was assumed that the generator terminals were connected to the infinite bus and the dynamics of the static excitation system are neglected. It was considered the load angle δ and the mechanical power p_m as the flat output components The system was subjected to the three-phase short circuit fault for a period of three cycles. Using the flatness property, the stabilization of the sample power system under different fault conditions and parameter variations was achieved. This approach is extended in [40] in order to control frequency and power flow for an automatic generation control (AGC) of a multi-machine system. Flatness-based control method is implemented on a 3-area, 10- machine and the 39-bus system. All the state, control and output components of flat can be seen in Table IV. The advantage of this control approach is the n-machine system is split into n-linear controllable systems. In [41], a fuzzy differential flatness-based controller is implemented to improve the control performance of an induction motor. The model of motor is considered a three-phase voltage source inverter-fed squirrel cage induction motor. The flatness control is utilized to generate a suitable output, whereas the fuzzy logic is used to eliminate the effects of the time-varying nonlinear system. This control approach is also applied to the induction machine fed by a voltage source

978-1-5090-0376-1/16 $31.00 © 2016 IEEE

Table III. STUDIED INVERTERS BASED ON FLATNESS CONTROL.

Ref	Studied system	State Variables		Input Variables		Output Variables	
32	Three-phase inverter with output LC filter	V_{cd}	- d-axes voltage value across the filter capacitor C_f	V_d	- Inverter d-voltage	y_d	- Electrostatic energy in d-axes
		V_{cq}	- d-axes voltage value across the filter capacitor C_f	V_q	- Inverter q-voltage	y_q	- Electrostatic energy in d-axes
33	parallel voltage-source inverters	V_{cd}	- d-axes output voltage of the output LC filter	V_{di}	- Output d-voltages for each inverter	y_c	- Energy stored in the AC capacitor of the output LC
		V_{cq}	- q-axes output voltage of the output LC filter	V_{qk}	- Output q-voltages for each inverter	y_{zk}	- Current errors referred to the first inverter
34	parallel voltage-source inverters	V_{cd} V_{cq}	- d-voltages at the PCC - q-voltages at the PCC	V_{dn}	- Output d-voltages for each inverter	y_c	- Energy stored in the AC capacitor filters
				V_{qn}	- Output q-voltages for each inverter	y_z	- Balance the power between the first inverter and a kth inverter

Table IV. STUDIED ELECTRICAL DRIVES BASED ON FLATNESS CONTROL.

Ref	Studied system	State Variables		Input Variables		Output Variables	
38	Permanent Magnet Synchronous Machine (PMSM)	i_d i_q Ω v_C i_L	- d-axis current machine - q-axis current machine - Mechanical Speed - Capacitive voltage $v_{C_1}+v_{C_2}$ - Inductive current $i_{in}+i_z$	V_d V_q	- Three-phase reference voltages in d-axes - Three-phase reference voltages in q-axes	y_Ω y_d	- Mechanical speed - d-axes flux
39	Synchronous Machine	i_d i_q Ω	- d-axis current machine - q-axis current machine - Mechanical Speed	V_d V_q d	-Three-phase reference voltages in d-axes - Three-phase reference voltages in q-axes - Duty cycle of Q-ZSI command	y_e y_Ω y_d	- Stored energy in the Q-ZSI - Mechanical speed - d-axes flux
37	Permanent Magnet-Excited Synchronous Motor	i_{sd} i_{sq} v_s	- d-axes stator current - q-axes stator current - Angle of flux orientated coordinate system	u_{sd} u_{sq} ω_s	- d-axes stator voltage - q-axes stator voltage - Stator circuit velocity,	i_{sd} i_{sq} v_s	- d-axes stator current - q-axes stator current - Angle of flux orientated coordinate system
43	Power System Generator	x_1 x_2 x_3	- Load angle - Generator speed deviation - Stator voltage	E_{fd}	Excitation control input	F	Load angle
24	DC-motor system	i_a ω	- Armature current - Angular speed	v	Voltage in the motor armature terminal	ω	Angular speed
36	Permanent magnet synchronous motor (PMSM)	i_d i_q Ω	- Stator d-current - Stator q-current - Rotor Speed	V_{dn} V_{qn}	- Stator d-voltage reference - Stator q-voltage reference	y_Ω y_d	- Electrostatic Energy in d-axes - Flux in d-axes
40	Multi-machine system	δ_i ω_i P_{gvi} P_{mi}	- Rotor electrical angle - Rotational speed of rotor - Governor power - Mechanical power	P_i^{ref}	Power of Supercapacitor	δ_i	- Rotor electrical angle, i denotes the ith generator

converter in which the flatness-based control is used for the inner current and outer flux and speed loops.

IV. CONTROL STRATEGY

Although flatness-based design could be applied to any control scheme because flatness is such a general property, the main two objectives of flatness-based control are trajectory generation and trajectory tracking. These aspects can be combined with other control systems such as predictive, sliding mode, feedforward and optimal control systems [4]. More detailed information about how flatness can improve and control these systems will discuss in future work.

V. CONCLUSION

In this paper, differential flatness theory-based methods of control using various applications are reviewed. Based on the study, the main conclusions are as follows: The stabilization problem, dc bus voltage regulation, robustness, stability as well as energy management can be achieved in hybrid systems using flatness control method. The flatness property of converters can satisfy the power and voltage regulation besides stabilization of the whole system during transient and

steady-state. This technique prepares the highest dynamic response and enhanced a reliable operation of the parallel inverters in faulty mode. Moreover, the reductions of the circulating currents besides high power quality at the Point of Common Coupling (PCC) are the achievement of using this control method. Also, the flatness property of AC and DC drives is guarantee the power and frequency regulation. Besides, It can minimize the copper losses, avoid the system from the uncontrollable behavior and prepare system robustness during perturbations.

REFERENCES

[1] Ph. Martin, R. M. Murray, P. Rouchon. Flatness based design. Notes for EOLSS, January 2002

[2] Fliess M, Levine J, Martin P, Rouchon P. Flatness and defect of nonlinear systems: introductory theory and examples. Int J Control 1995;61:1327-61.

[3] M. Fliess, J. Lévine, P. Martin, and P. Rouchon. On differentially flat nonlinear systems. In Proc. IFAC-Symp. Nonlinear Control Systems, pages 408–412, 1992.

[4] J.F. Stumper. Flatness-based predictive and optimal control for electrical drives, Theses, 04.12.2012, Munich Technical University.

[5] Levine J. Analysis and control of nonlinear systems: a flatness-based approach. Springer; 2009.

[6] Fliess M, Levine J, Martin P, Rouchon P. A lie-bäcklund approach to equivalence and flatness of nonlinear systems. IEEE Trans Automatic Control 1999;44:922-34

[7] M. van Nieuwstadt and R.M. Murray. Real-time trajectory generation for differentially flat systems. In Int. J. of Robust and Nonlinear Control, Vol. 8, pages 995–1020, 1998

[8] A. Payman, S. Pierfederici, and F. Meibody-Tabar. Energy Management in a Fuel Cell/Supercapacitor Multisource/Multiload Electrical Hybrid System. IEEE Trans. Ind. Electron, vol. 24, no. 12, december 2009.

[9] P. Thounthong, S. Pierfederici, B. Davat. Analysis of Differential Flatness-Based Control for a Fuel Cell Hybrid Power Source. IEEE Trans. Energy Conversion, vol. 25, no. 3, september 2010

[10] P. Thounthong, S. Pierfederici, J.P. Martin, M. Hinaje, B. Davat, ''Modeling and Control of Fuel Cell/Supercapacitor Hybrid Source Based on Differential Flatness Control'', IEEE 2010.

[11] A. Payman, S. Pierfederici, F. Meibody-Tabar, B. Davat," An Adapted Control Strategy to Minimize DC-Bus Capacitors of a Parallel Fuel Cell/Ultracapacitor Hybrid System",. IEEE 2011.

[12] P. Thounthong, S. Sikkabut, P. Mungporn, L. Piegari, B. Nahid-Mobarakeh, S. Pierfederici, B. Davat. DC Bus Stabilization of Li-Ion Battery Based Energy Storage for a Hydrogen/Solar Power Plant for Autonomous Network Applications. IEEE Trans. Industry Applications, vol. 51, no. 4, July/August 2015

[13] M. Zandi, A. Payman, J.-P. Martin, S. Pierfederici, B. Davat, F. Meibody-Tabar, "Energy Management of a Fuel Cell/Supercapacitor/Battery Power Source for Electric Vehicular Applications,", IEEE 2011.

[14] A. Shahin, M. Zandi, M. Phattanasak, H. Renaudineau, J.-P. Martin, B. Nahid-Mobarakeh, S. Pierfederici, Bernard Davat. Flatness Based Control Of Hybrid Systems For Fuel Cell Applications.

[15] M. Benaouadj, A. Aboubou, R. Saadi, M.Y. Ayad, M. Becherif, M. Bahri, O. Akhrif. Flatness Control of Batteries/Supercapacitors Hybrid Sources for Electric Traction. 13-17 May 2013.

[16] P. Thounthong, S. Sikkabut, P. Mungporn, P. Sethakul, S. Pierfederici, and B. Davat. Differential Flatness Based-Control of Fuel Cell/Photovoltaic/Wind Turbine/Supercapacitor Hybrid Power Plant. IEEE. 2013.

[17] R. Saadi, M. Benaouadj, O. Kraa, M. Becherif, M.Y. Ayad, A. Aboubou, M. Bahri, A. Haddi. Energy Management of Fuel cell/ Supercapacitor Hybrid Power Sources Based on The Flatness Control. 4th International Conference on Power Engineering, Energy and Electrical Drives, 13-17 May 2013.

[18] P. Thounthong, S. Sikkabut, P. Mungporn, B. Nahid-Mobarakeh, S. Pierfederici, B. Davat. Differential Flatness Control Approach for Fuel Cell/Solar Cell Power Plant with Li-Ion Battery Storage Device for Grid-Independent Applications. 2014 International Symposium on Power Electronics, Electrical Drives, Automation and Motion. IEEE.

[19] M. Benaouadj, A. Aboubou, M.Y. Ayad, M. Becherif. Nonlinear Flatness Control Applied to Supercapacitors Contribution in Hybrid Power Systems using Photovoltaics Source and Batteries. The International Conference on Technologies and Materials for Renewable Energy, Environment and Sustainability, TMREES 14. Energy Procedia 50 (2014) 333-341.

[20] C. Jia, T. Bai, X. Shan, F. Cui, S. Xu. Cloud Neural Fuzzy PID Hybrid Integrated Algorithm of Flatness Control. Journal of Iron and Steel Research, International. 2014, 21 (6): 559-564.

[21] H. Sira-Ram'ırez, R. Silva-Ortigoza. On the Control of the Resonant Converter: A Hybrid-Flatness Approach. Departamento de Ingenier'ia El'ectrica

[22] A. Shahin, M. Hinaje, J-P. Martin, S. Pierfederici, S. Raël, B. Davat. High Voltage Ratio DC–DC Converter for Fuel-Cell Applications. IEEE Trans. Industry Applications, VOL. 57, NO. 12, DECEMBER 2010

[23] M. Zandi, R. Gavagsaz-Ghoachani, M. Phattanasak, J-P Martin, B. Nahid-Mobarakeh, S. Pierfederici, B. Davat, A. Payman. Flatness Based Control of a non-ideal DC/DC Boost Converter. IEEE 2011.

[24] A. Shahin, M. Hinaje, J-P. Martin, S. Pierfederici, S. Raël, B. Davat. High Voltage Ratio DC–DC Converter for Fuel-Cell Applications. IEEE Trans. Industry Applications, VOL. 57, NO. 12, DECEMBER 2010

[25] P. Thounthong, S. Pierfederici. A New Control Law Based on the Differential Flatness Principle for Multiphase Interleaved DC–DC Converter. IEEE Trans. Circuits and Eystems—II: express briefs, vol. 57, no. 11, november 2010

[26] P. Mungporn, S. Sikkabut, B. Yodwong, C. Ekkaravarodome, S. Toraninpanich, B. Nahid-Mobarakeh, S. Pierfederici, B. Davat6, P. Thounthong. Photovoltaic Power Control Based on Differential Flatness Approach of Multiphase Interleaved Boost Converter for Grid Connected Applications. IEEE. 2015

[27] M. Phattanasak, W. Kaewmanee, P. Thounthong, Panarit Sethakul, R. Gavagsaz-Ghoachani, J.-P. Martin, S. Pierfederici, B. Davat, and M. Zandi, Current-fed DC-DC converter with Flatness based control for Renewable Energy. IEEE, 2014.

[28] M. Phattanasak, W. Kaewmanee, P. Mungporn, S. Sikkabut, B. Yodwong, A. Boonseng, P. Thounthong, P. Sethakul, R. Gavagsaz-Ghoachani, J.- P. Martin, S. Pierfederici, B. Davat. Current-fed full-bridge DC-DC converter with nonlinear control scheme. IEEE 2014.

[29] M. Phattanasak, R. Gavagsaz-Ghoachani, J-P. Martin, S. Pierfederici, B. Davat. Flatness Based control of an Isolated Three-port Bidirectional DC-DC converter for a Fuel cell hybrid source. IEEE 2011.

[30] H. Renaudineau, A. Houari, A. Shahin, J-P. Martin, S. Pierfederici. Efficiency Optimization Through Current-Sharing for Paralleled DC–DC Boost Converters With Parameter Estimation IEEE Trans. Power Electronics, Vol. 29, No. 2, February 2014.

[31] M. Phattanasak, R. Gavagsaz-Ghoachani, J-P. Martin, B. Nahid-Mobarakeh, S. Pierfederici, B. Davat. Comparison of two nonlinear control strategies for a hybrid source system using an isolated three-port bidirectional DC-DC converter. IEEE 2011.

[32] A. Houari, H. Renuadineau, J-P. Martin, S. Pierfederici, and F. Meibody-Tabar, "Flatness-based control of threephase inverter with output LC filter," IEEE Trans. Ind. Electron., vol.59, no.7, pp. 2890–2897, Jul. 2012.

[33] A. Shahin, S. Eskander, H. Moussa, J.-P. Martin, B. Nahid-Mobarakeh, S. Pierfederici. A New Approach Based on Flatness Control to Improve Reliability of Parallel Connected Inverters. IEEE 2015.

[34] A. Houari, A. Battiston, J-P Martin, S. Pierfederici, F. Meibody-Tabar. Flatness-Based-Control for Parallel Operation of N Voltage-Source Inverters.

[35] E. Delaleau and A.M. Stankovic, Flatness-based hierarchical control of the PM synchronous motor. Proceeding of the 2004 American Control Conference, Boston, Massachusetts, pp. 65-70, 2004.

[36] A. Fezzani, S. Drid, A. Makouf, L. Chrifi Alaoui. Speed Sensorless Flatness-Based Control of PMSM Using a Second Order Sliding Mode Observer. 2013 Eighth International Conference and Exhibition on Ecological Vehicles and Renewable Energies (EVER).

[37] P. Tam Thanh, N.D. That. Nonlinear Flatness-Based Controller for Permanent Magnet-Excited Synchronous Motor. (ISARC 2014)

[38] A. Battiston, J-P. Martin, E-H. Miliani, B. Nahid-Mobarakeh, S. Pierfederici1, F. Meibody-Tabar. Control of a PMSM Fed by a Quasi Z-Source Inverter Based on Flatness Properties and Saturation Schemes.

[39] E.C. Anene, U.O. Aliyu, J. Levine and G. K. Venayagamoorthy. Flatness-based feedback linearization of a synchronous machine model with static excitation and fast turbine valving. IEEE Power Engineering Society General Meeting, Tampa, FL, pp. 1-6, 2007.

[40] M. Hassani-Variani, K. Tomsovic. Distributed Automatic Generation Control Using Flatness-Based Approach for High Penetration of Wind Generation. IEEE 2013.

[41] L. Fan and L. Zhang. Fuzzy based flatness control of an induction motor. Procedia Engineering, vol .23, pp. 72-76, 2011.

[42] A. Battiston, E. Hadj Miliani, J.-P. Martin, B. N. Mobarakeh, S. Pierfederici, F. Meibody-Tabar. A Control Strategy for Electric Traction Systems Using a PM-Motor Fed by a Bidirectional Z-Source Inverter. IEEE Trans. Veh. Techn. , vol. 63, no. 9, november 2014.

[43] H.A. Yousef, M. Hamdy, M. Shafiq. Flatness-based adaptive fuzzy output tracking excitation control for power system generators. Journal of the Franklin Institute 350 (2013) 2334–2353.

[44] J. Dannehl and F.W. Fuchs. Flatness-based control of an Induction Machine Fed via Voltage Source Inverter-concept, Control Design and Performance Analysis. IEEE Industrial Electronics IECON 2006-32nd Annual Conference, Paris, pp. 5125- 5130, 2006.

[45] M. Hassani-Variani, K. Tomsovic. Two-Level Control of Doubly Fed Induction Generator Using Flatness-Based Approach. IEEE Trans. Power Systems. 2015.

[46] Fliess M, Martin P, Cluistopher I. Byrnes, DavidS. Gilliam, Clyde F. Martin, Biswa N. Datta. Controlling Nonlinear Systems by Flatness. Springer: 1997

[47] H. Sira-Ramírez and S.K. Agrawal. Differentially flat systems. Control engineering series. New York: Marcel Dekker, 2004.

[48] Suresh L., G.R.S. Naga Kumar, And M.V. Sudarsan. Modeling And Simulation Of Z-Source Inverter. From the SelectedWorks of suresh L. January 2012.

[49] G. G. Rigatos, Nonlinear Control and Filtering Using Differential Flatness Approaches. Springer; 2015.

7th Power Electronics, Drive Systems & Technologies Conference (PEDSTC 2016)
16-18 Feb. 2016, Iran University of Science and Technology, Tehran, Iran

Space Vector PWM Method for Two-Phase Three-Leg Inverters

Maedeh Mirazimi
Faculty of Electrical and Computer
Engineering,
University of Tabriz,
Tabriz, Iran
E-mail: maedehazimi66@gmail.com

Ebrahim Babaei, *Member, IEEE*
Faculty of Electrical and Computer
Engineering,
University of Tabriz,
Tabriz, Iran
E-mail: e-babaei@tabrizu.ac.ir

Mohammad Bagher Banna Sharifian
Faculty of Electrical and Computer
Engineering,
University of Tabriz,
Tabriz, Iran
E-mail: sharifian@tabrizu.ac.ir

Abstract— **In this paper, the space vector pulse width modulation (SVPWM) method is developed to control two-phase inverter with three legs. To implement in digital signal processors (DSP), the adjustability of switching sequences to reduce the harmonic contents of the output voltages and switching losses is very important. For this reason, the proposed control method is very suitable in comparison with the other modulation techniques that produce output voltages with variable frequency and magnitude. The proposed control scheme allows independent control of the magnitude and quadrature phase angle for two output phases. Thus, the presented two-phase inverter and its modulation scheme make the two-phase inverter commensurate with both balanced and unbalanced two-phase loads. Simulation results by using MATLAB/SIMULINK reconfirm the feasibility of the structure based on SVPWM method for the inverter.**

Keywords—Space vector; pulse width modulation; two-phase three-leg inverters

I. INTRODUCTION

The voltage source inverter is a power electronic device that converts a dc voltage to ac voltage of variable frequency and magnitude. Variable output voltage in an inverter is achieved by varying the dc input voltage or inverter's gain (the ratio of ac output voltage to dc input voltage) which is controlled by using pulse width modulation (PWM) techniques [1-3]. Furthermore SVPWM methods have several advantages over the other PWM techniques such as lower total harmonic distortion (THD), higher ac gain, and simpler implementation in DSP based control systems [4]. In recent years, there is an increasing trend of using SVPWM in variable ac drives because of their easier digital realization and better dc bus utilization.

Based on the number of phases of output voltage of inverters, they are categorized into three types including single-phase, two-phase and three-phase. The single-phase induction motor with two unbalanced windings is widely used at the lower power level, especially in household applications where a three-phase supply is not available. In such application the motor operates at low efficiency and power factor. Single phase induction motor can be considered as two-phase induction motor with main and auxiliary winding and it is possible to apply vector control of induction motor or any

other control technique to increase its efficiency [5-8]. Traditional rectifier-inverter based structures for two-phase drive systems are typically categorized with two-leg and three-leg circuit configurations, depending on a neutral point connection of loads. In the two-leg inverters, the neutral terminal of two-phase loads is linked to a center-tap of the series connected dc-link capacitors. On the other hand, the three-leg inverters assign one redundant leg to connect the load neutral point [9]. The three-leg configurations have shown advantages over the two-leg counterparts, including no requirement of accessing the capacitor mid-point, no harmonic current flow into the dc-link capacitor, and less harmonic components in load currents [10]. In this paper, the two phase inverter based on three-leg structure and SVPWM method has been simulated by MATLAB/SIMULINK. The simulation results reconfirm the capability of the mentioned inverter to feed sinusoidal output currents for both balanced and unbalanced two-phase loads.

II. THREE-LEG TWO-PHASE VSI

The typical model for a three-leg two-Phase voltage source inverter (VSI) is shown in Fig. 1. This inverter consists of six unidirectional switches (s_{ap}, s_{bp}, s_{cp}, s_{an}, s_{bn} and s_{cn}) that connect the dc input voltage source to the output loads and being able to produce three voltage levels on each output. As it is shown in Fig. 1, each unidirectional switch consists of an IGBT and an anti-parallel fast recovery diode. As illustrated in Table. I, there are eight possible combinations of on and off patterns for the three upper power switches [12]. The on and off states of the lower power devices are opposite to the upper ones.

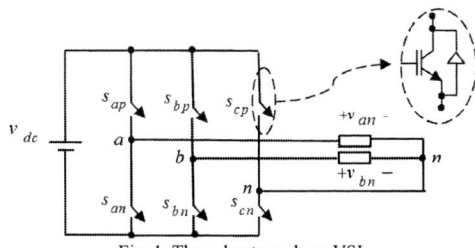

Fig. 1. Three-leg two-phase VSI.

978-1-5090-0376-1/16 $31.00 © 2016 IEEE   553

TABLE I. OUTPUT VOLTAGES OF A THREE-LEG TWO-PHASE INVERTER FOR DIFFERENT SWITCHING STATES

States	s_{ap}	s_{bp}	s_{cp}	s_{an}	s_{bn}	s_{cn}	v_{an}	v_{bn}
1	0	0	0	1	1	1	0	0
2	1	0	0	0	1	1	v_{dc}	0
3	1	1	0	0	0	1	v_{dc}	v_{dc}
4	0	1	0	1	0	1	0	v_{dc}
5	0	1	1	1	0	0	$-v_{dc}$	0
6	0	0	1	1	1	0	$-v_{dc}$	$-v_{dc}$
7	1	0	1	0	1	0	0	$-v_{dc}$
8	1	1	1	0	0	0	0	0

Considering Fig. 1 and Table I, the output voltages can be expressed by the following equations:

$$v_{an} = (S_{ap} - S_{cp})v_{dc}$$
$$v_{bn} = (S_{bp} - S_{cp})v_{dc} \tag{1}$$

In the aforementioned equation, S_{ap}, S_{bp} and S_{cp} are switching functions of s_{ap}, s_{bp} and s_{cp}, switches respectively. They are defined as follows:

$$S_{ip} = \begin{cases} 1 & s_{ip}\ ON \\ 0 & s_{ip}\ OFF \end{cases} \quad for \ \ i = a,b,c \tag{2}$$

III. PRINCIPLE OF SVPWM

The desired output voltages of a two-phase inverter are sinusoidal and orthogonal just as follows:

$$v_{an,ref}(t) = V_m \cos(\omega t)$$
$$v_{bn,ref}(t) = V_m \cos\left(\omega t + \frac{\pi}{2}\right) \tag{3}$$

where V_m and ω denote the amplitude and angular frequency of the desired output voltages, respectively.

In SVPWM method for each switching states of Table I, a reference voltage vector (v_o^*) is defined. As the output voltages of a two-phase inverter (v_{an}, v_{bn}) is quadrature in phase, v_o^* is defined as:

$$v_o^* = v_{an} + jv_{bn} \tag{4}$$

where v_{an} and v_{bn} denote output voltages of the inverter. By substitution of (1) into (4), v_o^* can be rewritten as the following form:

$$v_o^* = (S_{ap} - S_{cp})v_{dc} + j(S_{bp} - S_{cp})v_{dc} \tag{5}$$

Considering (5), eight space vectors can be achieved for the switching states presented in Table I. The voltage space

vectors are shown in Fig. 3 and denoted by V_0, V_1, V_2, V_3, V_4, V_5, V_6 and V_7. The binary numbers on the figure indicate the on or off states of upper switches of inverter legs. The most significant bit is for leg a, the least significant bit is related to leg n and the middle one is for leg b. V_0 and V_7 related to switching states that all the upper switches of inverter legs are on or off, respectively, and produce zero vectors and apply zero voltage to the load. Meanwhile the remaining voltage space vectors are non-zero vectors. The amplitudes of V_2 and V_5 are $\sqrt{2}\ v_{dc}$ while the amplitude of V_1, V_3, V_4 and V_6 is v_{dc}. As a result, related to switching states presented in Table I, six non-zero vectors and two zero vectors are possible. Six nonzero vectors shape the axes of an irregular hexagon and six sectors numbered 1 to 6 as shown in Fig. 2. The angle between two adjacent non-zero vectors in sectors number 1, 2, 4 and 5 is 45 degrees. Meanwhile, the angle between two adjacent non-zero vectors in sectors number 3 and 6 is 90 degrees.

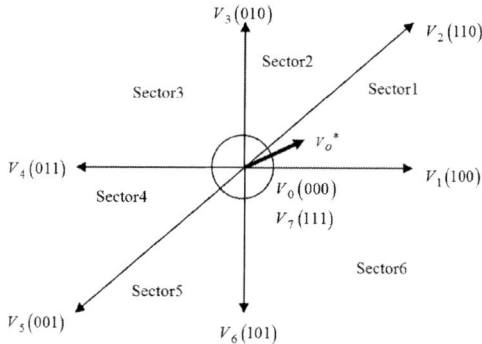

Fig. 2. Basic switching vectors and sectors.

Based on the switching states presented in Table I and equation (5), the angle and magnitude of voltage space vectors are obtained in details as follows:

$$state = 1 : V_0 = (0-0)v_{dc} + j(0-0)v_{dc} = 0$$
$$state = 2 : V_1 = (1-0)v_{dc} + j(0-0)v_{dc} = v_{dc}\angle 0°$$
$$state = 3 : V_2 = (1-0)v_{dc} + j(1-0)v_{dc} = \sqrt{2}\ v_{dc}\angle 45°$$
$$state = 4 : V_3 = (0-0)v_{dc} + j(1-0)v_{dc} = v_{dc}\angle 90°$$
$$state = 5 : V_4 = (0-1)v_{dc} + j(1-1)v_{dc} = v_{dc}\angle 180° \tag{6}$$
$$state = 6 : V_5 = (0-1)v_{dc} + j(0-1)v_{dc} = \sqrt{2}\ v_{dc}\angle 225°$$
$$state = 7 : V_6 = (1-1)v_{dc} + j(0-1)v_{dc} = v_{dc}\angle 270°$$
$$state = 8 : V_7 = (1-1)v_{dc} + j(1-1)v_{dc} = 0$$

This section discusses the SVPWM principle. It is seen that a two-phase VSI generates eight switching states and consequently eight voltage space vectors. Using (3) and (4), the desired reference voltage vector at the output of the inverter (v_o^*) is achieved in each switching period. This vector represents two-phase sinusoidal voltages while it has an arbitrary amplitude and angle. The objective of SVPWM technique is to approximate the reference voltage vector (v_o^*) by using the eight voltage space vectors. The conventional method of approximation is to generate the average output of the inverter in each switching period by using two nearest non-zero voltage space vectors (based on the sector that v_o^* lies in according to its angle (α) as it is shown in Fig. 3 and is obtained by using (7) and a zero voltage space vector. Assuming v_o^* to be lying in sector 1, the adjacent voltage vectors to v_o^* in this sector (V_1 and V_2) are used to produce v_o^*.

$$\alpha = 2\pi ft \tag{7}$$

where f represents the frequency of output voltages of the inverter.

The reference voltage vector (v_o^*) within T_s can be constructed by vector sum of its two adjacent vectors as shown in Fig. 3. So, the following equation can be obtained:

$$v_o^* T_s = V_1 T_1 + V_2 T_2 \tag{8}$$

where T_s is the switching cycle that is determined according to the maximum switching frequency of the inverter.

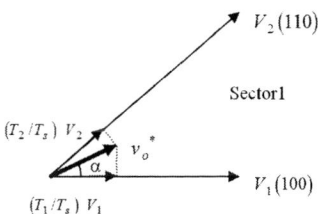

Fig. 3. Output voltage space vector of the inverter and its components when it lies in sector 1.

As an illustration for an inverter with the switching frequency of 10 kHz, T_s is 0.1ms. So in order to produce output voltages with the fundamental frequency of 50 Hz, there are 200 switching cycles in each period of the output voltages. In addition T_1 and T_2 are the duty cycle periods of V_1 and V_2 (the adjacent space voltage vectors to v_o^*), respectively, while the sum of these times is less than or equal to T_s as the following equation:

$$T_s \geq T_1 + T_2 \tag{9}$$

The on-time for the zero space vectors (V_0 or V_7) is obtained as follows:

$$T_0 = T_s - (T_1 + T_2) \tag{10}$$

Considering equations (5) and (9) and representing each vector into matrix notation, T_1, T_2 and T_0 can be calculated as follows:

$$T_s \left| v_{ref} \right| \begin{bmatrix} \cos\alpha \\ \sin\alpha \end{bmatrix} = T_1 v_{dc} \begin{bmatrix} 1 \\ 0 \end{bmatrix} + \sqrt{2} T_2 v_{dc} \begin{bmatrix} \cos\dfrac{\pi}{4} \\ \sin\dfrac{\pi}{4} \end{bmatrix} \tag{11}$$

The aforementioned equation can be rewritten in the following form:

$$T_s \left| v_{ref} \right| \cos\alpha = T_1 v_{dc} + \sqrt{2} T_2 v_{dc} \cos\frac{\pi}{4} \tag{12}$$

$$T_s \left| v_{ref} \right| \sin\alpha = \sqrt{2} T_2 v_{dc} \sin\frac{\pi}{4} \tag{13}$$

Considering (10), (12) and (13), the time intervals of application of V_1, V_2, (V_0 or V_7) are achieved as follows:

$$T_1 = \sqrt{2} T_s m_v \sin\left(\frac{\pi}{4} - \alpha\right) \tag{14}$$

$$T_2 = m_v \sin\alpha \tag{15}$$

$$T_0 = T_s - (T_1 + T_2) \tag{16}$$

where m_v is the modulation index or the gain of the inverter and is defined as follows:

$$m_v = \frac{v_{ref}}{v_{dc}} \tag{17}$$

Similarly if the reference voltage vector (v_o^*) lies in sector 2, it can be constructed by vector sum of its two adjacent vectors (V_2 and V_3) as shown in Fig. 4. So, the following equation can be obtained:

$$v_o^* T_s = V_2 T_2 + V_3 T_3 \tag{18}$$

where T_2 and T_3 are the duty cycles of V_2 and V_3.

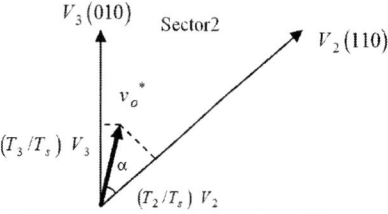

Fig. 4. Output voltage space vector of the inverter and its components when it lies in sector 2.

Considering (18) and representing each vector into matrix notation, T_2, T_3 and T_0 can be calculated as follows:

$$T_s \left| v_{ref} \right| \begin{bmatrix} \cos\alpha \\ \sin\alpha \end{bmatrix} = \sqrt{2} T_2 v_{dc} \begin{bmatrix} \cos\dfrac{\pi}{4} \\ \sin\dfrac{\pi}{4} \end{bmatrix} + T_3 v_{dc} \begin{bmatrix} 0 \\ 1 \end{bmatrix} \tag{19}$$

$$T_s \left| v_{ref} \right| \cos \alpha = \sqrt{2} T_2 v_{dc} \cos \frac{\pi}{4} \qquad (20)$$

$$T_s \left| v_{ref} \right| \sin \alpha = \sqrt{2} T_2 v_{dc} \sin \frac{\pi}{4} + T_3 v_{dc} \qquad (21)$$

$$
\begin{aligned}
T_2 &= T_s m_v \cos(\alpha) \\
T_3 &= \sqrt{2} m_v \sin\left(\alpha - \frac{\pi}{4}\right) \\
T_0 &= T_s - (T_1 + T_2)
\end{aligned}
\qquad (22)
$$

Similarly if the reference voltage vector ($v_o{}^*$) lies in the i-th sector, it can be constructed by vector sum of its two adjacent vectors (V_i and V_{i+1}). Hence, the application time of non-zero vectors at each sector (V_i and V_{i+1}) and also the application time of zero vectors (V_0 or V_7) are summarized in Table. II.

TABLE II. THE DUTY CYCLES OF THE ADJACENT VECTORS TO $v_o{}^*$ AT EACH SECTOR

Sector	Angular position of $v_o{}^*$ (α)	T_i	T_{i+1}
1	$0 \le \alpha < \frac{\pi}{4}$	$\sqrt{2} T_s m_v \sin\left(\frac{\pi}{4} - \alpha\right)$	$T_s m_v \sin(\alpha)$
2	$\frac{\pi}{4} \le \alpha < \frac{\pi}{2}$	$T_s m_v \cos(\alpha)$	$\sqrt{2} T_s m_v \sin\left(\alpha - \frac{\pi}{4}\right)$
3	$\frac{\pi}{2} \le \alpha < \pi$	$T_s m_v \sin(\alpha)$	$-T_s m_v \cos(\alpha)$
4	$\pi \le \alpha < \frac{5\pi}{4}$	$\sqrt{2} T_s m_v \sin\left(\frac{3\pi}{4} + \alpha\right)$	$-T_s m_v \sin(\alpha)$
5	$\frac{5\pi}{4} \le \alpha < \frac{3\pi}{2}$	$-T_s m_v \cos(\alpha)$	$\sqrt{2} T_s m_v \sin\left(\frac{3\pi}{4} + \alpha\right)$
6	$\frac{3\pi}{2} \le \alpha < 2\pi$	$T_s m_v \cos(\alpha)$	$T_s m_v \sin(\alpha)$

IV. SWITCHING PATTERNS

The output reference voltage vector can be calculated in terms of its adjacent space vectors and also one of zero vectors (V_0 or V_7) within a switching period in each sector. Assuming that the switching period is divided into odd parts (for instance 5) to have a symmetrical switching pattern and on the other hand, in order to obtain the switching pattern with low switching

frequency and as a result low switching losses, each leg should change its state only once in one switching period or minimum number of switching transitions is desired. As an illustration if $v_o{}^*$ lies in sector 1 and the switching period is divided into five parts, the first and the second parts will be dedicated to $V_1 (100)$ and $V_2 (110)$ for half of their application times $\left(\frac{T_1}{2} \right.$ and $\left. \frac{T_2}{2} \right)$, respectively. In addition, the next part is specified for application of the zero vector ($V_7 (111)$) for T_0 while the last two parts of the switching period are the same as two first parts to have a symmetrical switching pattern. The switching sequences in other sectors are summarized in Table. III.

TABLE III. SWITCHING SEQUENCES FOR DIFFERENT SECTORS

		$\frac{T_1}{2}$	$\frac{T_2}{2}$	T_0	$\frac{T_2}{2}$	$\frac{T_1}{2}$
Sector 1	S_{ap}	ON	ON	ON	ON	ON
	S_{bp}	OFF	ON	ON	ON	OFF
	S_{cp}	OFF	OFF	ON	OFF	OFF
Sector 2	S_{ap}	ON	OFF	OFF	OFF	ON
	S_{bp}	ON	ON	OFF	ON	ON
	S_{cp}	OFF	OFF	OFF	OFF	OFF
Sector 3	S_{ap}	OFF	OFF	ON	OFF	OFF
	S_{bp}	ON	ON	ON	ON	ON
	S_{cp}	OFF	ON	ON	ON	OFF

		$\frac{T_1}{2}$	$\frac{T_2}{2}$	T_0	$\frac{T_2}{2}$	$\frac{T_1}{2}$
Sector 4	S_{ap}	OFF	OFF	OFF	OFF	OFF
	S_{bp}	ON	OFF	OFF	OFF	ON
	S_{cp}	ON	ON	OFF	ON	ON
Sector 5	S_{ap}	OFF	ON	ON	ON	OFF
	S_{bp}	OFF	OFF	ON	OFF	OFF
	S_{cp}	ON	ON	ON	ON	ON
Sector 6	S_{ap}	ON	ON	OFF	ON	ON
	S_{bp}	OFF	OFF	OFF	OFF	OFF
	S_{cp}	ON	OFF	OFF	OFF	ON

V. SIMULATION RESULTS

To demonstrate the validity of the SVPWM method for the two-phase inverter based on the presented model in Fig. 1, the inverter has been simulated with a two-phase R-L load (2.5Ω and 7.5mH) by using MATLAB/SIMULINK. Figs. 5, 6 and 7 show the simulation results for the SVPWM based three-leg two-phase inverter with balanced output voltage references. Identical magnitude commands, 60V have been applied to the two output phases (v_{anref} and v_{bnref}). Moreover the output frequency reference was set to 50 Hz. Fig. 5 shows the output phase voltages generated by the inverter while Fig. 6 represents their harmonic spectrum. As it is obvious, the maximum amplitude of the first output voltage occurs in fundamental frequency is 59.68V and the maximum amplitude of second output voltage occurs in fundamental frequency is 58.58V. So, their magnitudes are equal and other harmonic distortions are negligible. Fig. 7 shows the output currents' waveforms. It is seen that the output currents are sinusoidal with equal amplitude while the phase displacement between two output currents is 90 degrees with respect to each other. The sinusoidal waveforms of output currents are due to the inductive characteristics of load. As a result the load acts as a low pass filter that passes low order harmonics of output voltages and filters high order harmonics.

Additional simulation results for unbalanced output voltage references are presented in Figs. 8, 9 and 10. The amplitudes of the output voltage references are unequal while they are 60V and 80V, respectively. Based on Figs. 8 and 9, the maximum amplitude of first output voltage occurs in fundamental frequency is 58.53V and maximum amplitude of second output voltage occurs in fundamental frequency is 78.64V. Fig. 10 depicts the output currents' waveforms. It is shown that the proposed two-phase inverter can produce output currents with independent control of the amplitudes for two output phases. In addition the inverter keeps 90 degrees shift between the output currents independent of unequal magnitudes of output voltage references.

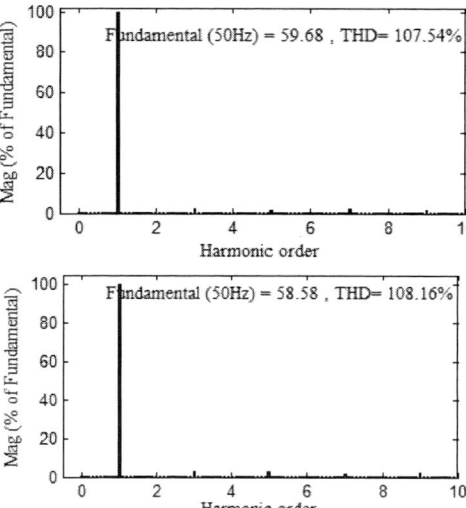

Fig. 6. Harmonic spectrum of output voltages of three-leg two-phase inverter with balanced output voltage references (top figure for phase a and the bottom figure for phase b).

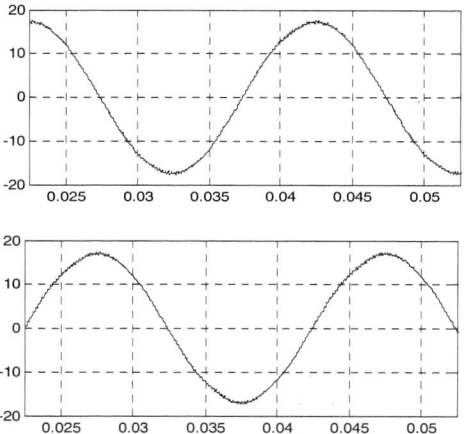

Fig. 7. Output currents of three-leg two-phase inverter with balanced output voltage references (top figure for phase a and the bottom figure for phase b).

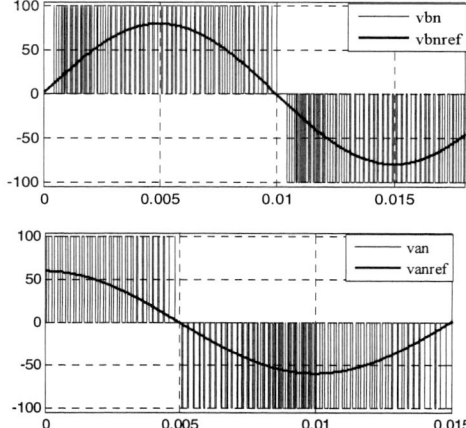

Fig. 8. Output voltages of three-leg two-phase inverter with unbalanced output voltage references (top figure for phase a and the bottom figure for phase b).

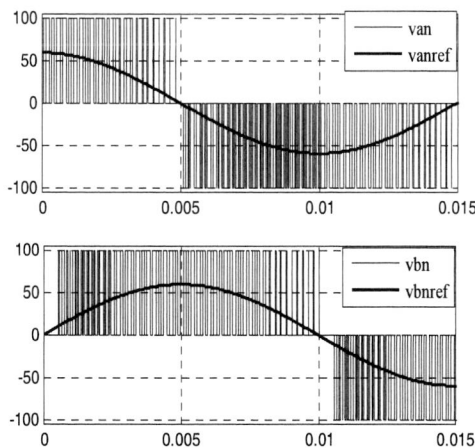

Fig. 5. Output voltages of three-leg two-phase inverter with balanced output voltage references (top figure for phase a and the bottom figure for phase b).

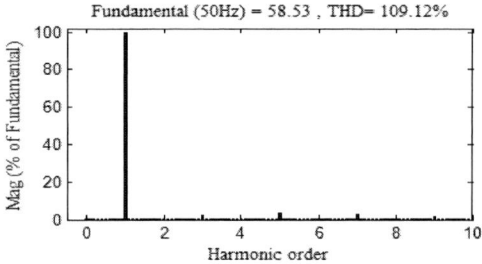

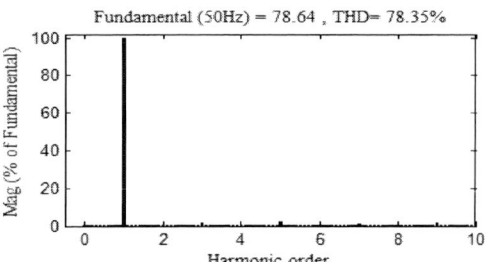

Fig. 9. Harmonic spectrum of output voltages of three-leg two-phase inverter with unbalanced output voltage references (top figure for phase a and the bottom figure for phase b).

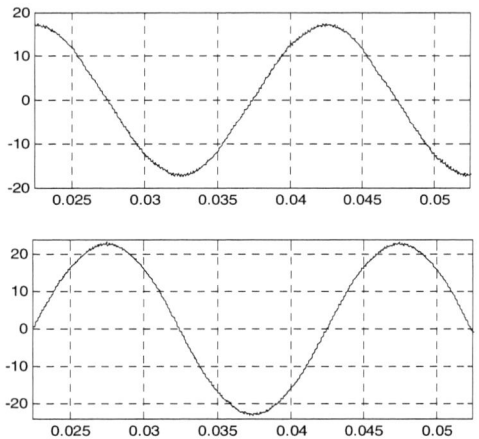

Fig. 10. Output currents of three-leg two-phase inverter with unbalanced output voltage references (top figure for phase a and the bottom figure for phase b).

VI. CONCLUSIONS

In this paper, the SVPWM method for the two-phase three-leg inverter has been presented and simulated. In the proposed method, the mentioned inverter can generate sinusoidal output currents with independently control of magnitudes while the phase angle between them is set to 90 degrees, regardless of unequal output magnitudes. In addition the generated output voltages by the inverter follow their reference signals. The simulation results reconfirm that the SVPWM method allows the two-phase three-leg inverter to feed sinusoidal output currents for both balanced and unbalanced two-phase loads.

REFERENCES

[1] E. Babaei and M. Sabahi, "Development of pulse width modulation technique for controlling inverters under balanced and unbalanced operations," *Arabian Journal for Science and Engineering (AJSE)*, vol. 39, no. 4, pp. 2941-2951, April 2014.

[2] S. Golabi, E. Babaei, M.B. Bannae Sharifian, Z. Golabi, "Application of speed, rotor flux, electromagnetic, load torque observers and diagnostic system in a vector controlled high-power traction motor drive," *Arabian Journal for Science and Engineering (AJSE)*, vol. 39, no. 4, pp. 2979-2996, April 2014.

[3] E. Babaei, "A new pulse width modulation technique for inverters," *Arabian Journal for Science and Engineering (AJSE)*, vol. 39, no. 8, pp. 6235-6247, Aug. 2014.

[4] J. Espinoza and G. Joos, "Modelling and implementation of space vector PWM techniques in active filter applications," *IEEE Power Elec.Workshop*, 1996.

[5] L.C. Tomaselli, I.B. Lazzarin, D.C.Martins, and I. Barbi, "Application of the Vector Modulation in the Symmetrical Two-Phase Induction Machine Drive," in *Proc. PESC*, 2005, Recife, Brazil, pp. 1253 -1258.

[6] M. Frivaldsky, B. Dobrucky, G. Scelba. P.Spanik, and P. Drgona, "Bidirectional Step-up/step-down DC-DC Converter with Magnetically Coupled Coils, " in *CSL Communications - Scientific Letters of University of Zilina*, vol. 15, no. 3, pp. N/A, ISSN 1335-4205, 2013.

[7] B. Dobrucky, J. Michalik, P. Spanik, and V. Bobek, "Virtual HF Injection Method (VHFIM) of Rotor Position Estimation of PMSM under Field Oriented Control " in *Proc. SPEEDAM*, 2006, Taormina, Italy, pp. SI 28-30.

[8] P. Zaskalicky and L. Schreier, "Using Fourier analysis for torque estimation of a two-Phase induction motor supplied by a half-bridge inverter with PWM Control, " in *CSL Communications -Scientific Letters of University of Zilina*, vol. 15, no. 3, pp. N/A, ISSN 1335-4205, 2013.

[9] D. H. Jang and D. Y. Yoon, "Space-vector PWM technique for two-phase inverter-fed two-phase induction motors, " *IEEE Trans. Ind. Appl.*, vol. 39, no. 2, pp. 542-549, Mar./Apr 2003.

[10] S. Kwak, T. Kim, and O. Vodyakho, "Space vector control methods for two-leg and three-leg based direct ac to ac converters for two-phase drive systems," in *Proc. IECON*, 2008, Orlando, Florida, pp. 959-964.

[11] F.Z. Peng, J.W. McKeever, D.J. Adams, "A power line conditioner using cascade multilevel inverters for distribution systems, " *IEEE Trans. Ind. Appl.*, vol. 34, no. 6, pp.1293-1298, Nov./Dec 1998.

[12] J. Do-Hyan, "PWM methods for two-phase inverters," *IEEE Industry Applications Magazine*, vol. 13, pp. 50-61, 2007.

Improved Equations of Switching Loss and Conduction Loss in SPWM Multilevel Inverters

Abolfazl Babaie
School of Electrical Engineering
Iran University of Science &
Technology (IUST)
Tehran, Iran
a.babaiee@yahoo.com

Bagher Karami
School of Electrical Engineering
Iran University of Science &
Technology (IUST)
Tehran, Iran
bagher.karami@yahoo.com

Adib Abrishamifar
School of Electrical Engineering
Iran University of Science &
Technology (IUST)
Tehran, Iran
abrishamifar@iust.ac.ir

Abstract— Loss is a very important parameter in the analysis of the power electronic systems, and accurate calculation of this parameter directly effects on the economic and technical evaluation. Junction temperature, heat-sink sizing and cooling system, failure rate and MTTF[1] are some examples of the basic parameters in designing a multilevel inverter, which all depends on the loss. Through previous methods, the loss and power equations were general and the results were limited to the simulations, where this paper proposed estimated equations for the switching loss and conduction loss. Previously, calculating the loss of an IGBT, estimating the conduction and the switching times in an on/off cycle, were a major obstacle. This paper utilizes a simpler method which provides more accurate results to calculate the conduction and switching times. Simulations are performed for a 5-level NPC[2] inverter. All of the results confirm high accuracy of proposed equations. Also, these equations can be extended to the n-level SPWM[3] inverters.

Keywords—Conduction Loss; Switching Loss; Junction Temperature; SPWM; Mean Time to Failure.

I. INTRODUCTION

Multilevel inverters have many industrial applications, including: controlling and driving motor [1, 2], UPS systems, transmission lines, and etc. Fewer switching harmonics at low frequencies and consequently less switching loss and also less EMI[4] are quite exciting among the advantages of these systems. Basically, there are three main commercial topologies of the multilevel voltage-source inverters well established in industry: Neutral-Point Clamped (NPC-MLI) Cascaded H-bridge (CHB-MLI), and Flying Capacitor (FC-MLI). Among these topologies, NPC-MLI is so widely applied in industry for high power applications. It is used in high voltage and high power levels. It has a reliable modular

structure compared to the CHB-MLI and FC-MLI. Studying the behavior of power semiconductor switches is necessary, according to the important role of the multi-level inverters in the industrial applications. Power loss is the most important parameter in operation and protection of switches. Practically, loss evaluation in the multilevel inverter is not an easy task and much more challenging compared to the basic two-level inverters. This is due to the fact that the current differs in each power switches in the inverter. Power losses of the power devices consist of the conduction losses and the switching losses. Different methods have been suggested in the literature works to calculate the power loss in the multilevel inverters. Some of these methods are based on an online model calculation from the simulated circuit and some are done based on deep mathematical analysis and calculations. In [3], a general procedure proposed for calculating the switching and conduction losses of the power semiconductors in the modular multilevel converter (MMC). As In [4] switching loss is just related to EON, EOFF, and switching frequency, it seems to be wrong. It is assumed that the power switch is conducting in the whole cycle leads to a very higher value of loss than the actual value, while the power switches in the multilevel inverters are only active in a few levels, so the conduction and switching losses have to be calculated in that time scale. Switching functions have been implemented in [6] to model the inverter losses for three phase nine level cascaded H-bridge inverter in which the load was assumed to be mixed RL load at a modulation index of 0.85. All the previous papers applied online modeling for calculating the losses by applying curve fitting to characterize the IGBT based on the actual device datasheet. In [7], the losses of multilevel inverter were calculated using a mathematical model which the voltage across the switch is modeled using threshold voltage and a series resistance. In [5,8], the conduction and switching losses are calculated using general formulas and the results are limited to simulation outcomes. In [10], to calculate the conduction loss a coefficient is introduced to the equation of conduction loss which considers the percentage of being active through a period. Assuming a sinusoidal waveform for the inverter output is not true since this assumption for the multilevel inverter is not permissible. Although the problem arises from the point that the output of the power electronic

---

[1] Mean Time To Failure
[2] Neutral Point Converter
[3] Sinusoidal Pulse Width Modulation
[4] electromagnetic interference

inverters is in PWM waveform. Through all of the methods proposed for loss calculation, results are limited to the simulation outputs while just general equations are introduced in [10]. In this paper, an accurate method is presented for calculating the conduction and switching losses. The switching time and the conduction time are calculated based on simpler equations. These equations can be extended to the n-level SPWM multilevel inverters. Also, to evaluate the proposed equations, a large number of simulations are performed on a 5 level NPC multilevel inverter.

II. MODULATION TECHNIQUE

There are various modulation methods such as Selective Harmonic Elimination (SHE), SPWM, Space Vector Control (SVC), Space Vector Modulation (SVM) used in the multilevel inverters. Among these, SVM and SPWM are used at high frequency, and SVC and SHE are used in low-frequency systems [9,11,12]. The SPWM is a common modulation method in the multilevel inverters which has two different types: phase shift and level shift as illustrated in figure 1. Between these two methods, the level-shift method is used in this paper as it is the most common way.

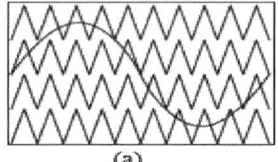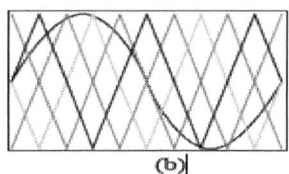

Figure 1: SPWM with (a) amplitude-shift (b) phase-shift [1]

III. CONDUCTION LOSS

The conduction loss in a power switch is generated in ON state. the general equation for P_{Con} is as (1) [5-9]:

$$P_{CON} = V_{CE(ON)} \times I_{C(ON)} \qquad (1)$$

$I_{C(ON)}$ and $V_{CE(ON)}$ are current and Collector-Emitter voltage drop in ON state, respectively. Equation (1) has been commonly used in papers to calculate the conduction loss, but this is a general equation. Usually, in some of the literatures the conduction loss is calculated based on simulation results and they don't care about theoretical calculations. While this paper has focused on the conduction loss calculations and proposed equations for estimating the conduction loss with high accuracy. The conduction time is a major parameter in calculating the conduction loss which its calculation is too difficult. In this paper, the conduction loss is calculated through (2).

$$P_{CON} = V_{CE(ON)} \times I_{C(ON)} \times D_{CON} \qquad (2)$$

In (2), D_{con} is the correction coefficient of the conduction loss, and is calculated through (3):

$$D_{CON} = \frac{t_{Level-i}}{T} \qquad (3)$$

In (3), T is the period and $t_{Level-i}$ is the total ON state time of the switch in I^{th} level. This is illustrated in figure (2) for a 5-level inverter. By integrating intervals that are in red color, the total conduction time for each switch is calculated.

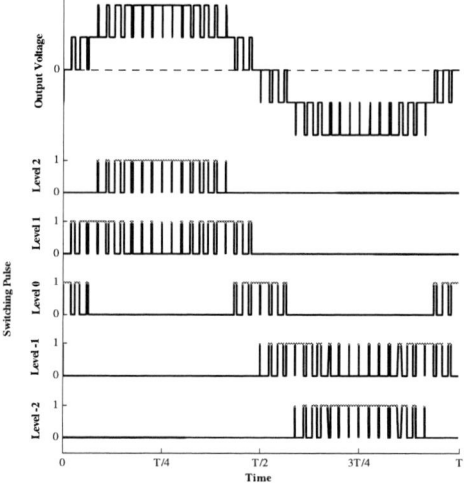

Figure 2: Conduction time for each level in a sample 5-level inverter

Due to the fact that the switching frequency is high, according to figure 2, the calculation of conduction time will be too difficult. So we have suggested a method for calculating the intervals of the conduction time for each switch in a period. According to figure 3, the conduction time for $(i)^{th}$ level is between the $(i-1)^{th}$ and $(i+1)^{th}$ levels. As a result, we should calculate the conduction time in these intervals. Due to the fact that the level shift SPWM is a modulation technique in this paper, each level can be estimated by a sinusoidal wave according to figure 4 and TABLE I. Therefore, we should calculate the percentage of being active in $(i)^{th}$ level. Hence, We need the calculation of the average area of each level that is calculated by (4). According to (4) S_i is the average of being active in each level.

978-1-5090-0376-1/16 $31.00 © 2016 IEEE   560

$$S_i = \frac{1}{t_i - t_{i-1}} \int_{t_{i-1}}^{t_i} (A_r \sin wt - i + 1)\, dt \qquad (4)$$

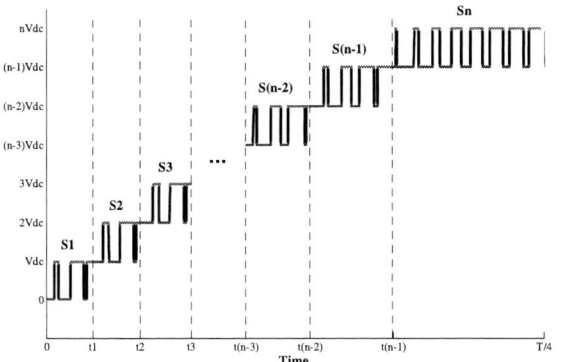

Figure 3: How to find S for each level

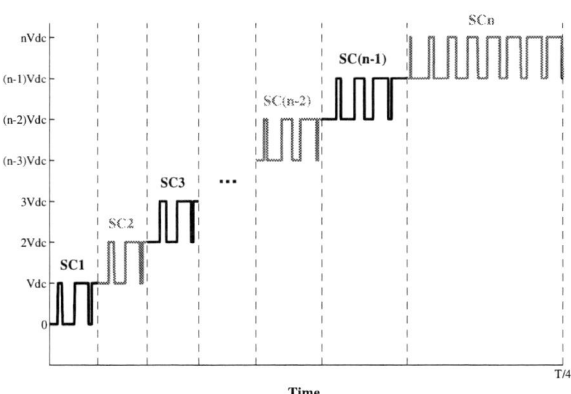

Figure 4: Calculating S with sinusoidal estimation

TABLE I. ESTIMATION OF (6) FOR CALCULATING S IN VARIOUS LEVELS

Staircase	Approximate Equation	Output Level
SC_1	$\dfrac{A_r \sin wt}{t_1}$	0, 1
SC_2	$\dfrac{A_r \sin wt - 1}{t_2 - t_1}$	1, 2
$\vdots$	$\vdots$	$\vdots$
SC_{n-1}	$\dfrac{A_r \sin wt - n + 2}{t_{n-1} - t_{n-2}}$	n-2, n-1
SC_n	$\dfrac{A_r \sin wt - n + 1}{\dfrac{T}{4} - t_{n-1}}$	n-1, n

Where A_r is the amplitude of the sinusoidal waveform in SPWM. According to figure 5 t_i is the start of switching between $(i)^{th}$ and $(i+1)^{th}$ levels and is calculated using (5).

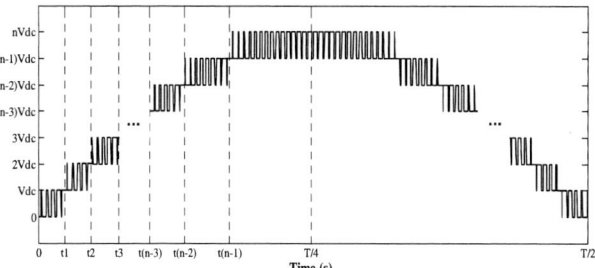

Figure 5: t_i for n-level inveter

$$t_i = \frac{\theta_i \times T}{2\pi} \qquad (5)$$

Where θ_i is the angle of t_i and is calculated by (6):

$$\theta_i = \arcsin \frac{i}{A_r} \qquad (6)$$

Where A_r is calculated using (7):

$$M = \frac{A_r}{a} \qquad , \ a-1 < A_r < a \qquad (7)$$

Where (a) is the number of voltage source and M is the modulation index. So the conduction time for (level 1) and (level i) are calculated through the (8) and (9) after calculation S_i.

$$t-level1 = 2\left[S_1(t_1 - t_0) + (1 - S_2)(t_2 - t_1) \right] \qquad (8)$$

Where in (8) $S_1(t_1 - t_0)$ is the percentage of being active in (level-1) between (level- 0) and (level- 1) and $(1 - S_2)(t_2 - t_1)$ is the percentage of being active in (level- 1) between (level- 1) and (level- 2).

$$t-leveli = 2\left[S_i(t_i - t_{i-1}) + (1 - S_{i+1})(t_{i+1} - t_i) \right] \qquad (9)$$

IV. SWITCHING LOSS

The switching loss in the IGBT is the product of the switching energies and the switching frequency[5-9]:

$$P_{SW} = F_{sw} \times (E_{sw(on)} + E_{sw(off)}) \quad (10)$$

Equation (10) is suitable for 2 or 3 level inverters. We should calculate the switching loss in the interval of the switching time since the switching time in these inverters is through the whole cycle interval. But the switching time is limited in multilevel inverters for each switch. The calculation of the switching time is challenging when the switching frequency is high. Consequently, the best way to calculate the switching loss is proposed through (11).

$$P_{SW} = F_{sw} \times (E_{sw(on)} + E_{sw(off)}) \times D_{SW} \quad (11)$$

Where D_{sw} is calculated using (12):

$$D_{SW} = \frac{t_{sw}}{T} \quad (12)$$

Where t_{sw} it the total time of switching for a switch. According to figure 6, switching time can be estimated so accurately. For instance, according to figure 6, one switch for a 5-level inverter is active in two levels, in level 1 and level 2, the switching time for this switch is calculated by (13):

$$t_{sw} = 2(\frac{T}{4} - t_1) \quad (13)$$

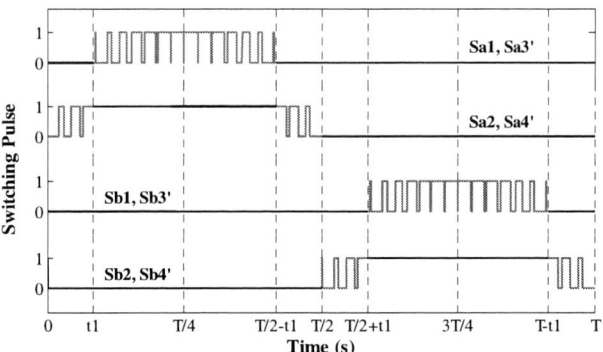

Figure 6: Switching signals for switches of a 5-level inverter

Also, according to figure 7 the current ($i_c(t)$) and voltage ($v_{ce}(t)$) waveforms should be considered for calculating E_{ON} and E_{OFF} during the switching time. So the E_{OFF} and E_{ON} are calculated by (14) and (15)[8]:

$$E_{sw(off)} = \int_{t_0}^{t_2} v_{ce}(t) \times i_c(t) \quad (14)$$

$$E_{sw(on)} = \int_{t_0}^{t_4} v_{ce}(t) \times i_c(t) \quad (15)$$

According to the figure 7, t2 is the total time of the delay when the switch becomes turned off and t4 is the total time of the delay when the switch becomes turned on.

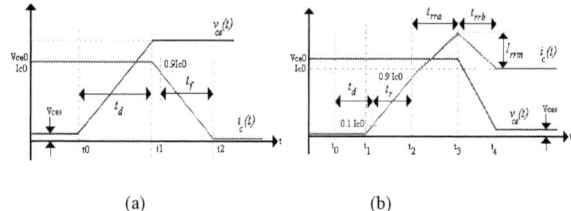

(a)   (b)

Figure 7: (a): Approximations of IGBT turn off waveforms, (b) approximations of IGBT turn on waveforms[8]

Due to the dependency of the E_{ON} and E_{OFF} to the case temperature and considering that the initial value of case temperature is not definite at the beginning, the results of (11) will be wrong if there is an error in the considered value for the initial case temperature. To avoid this, we can use repeating method for calculating the case temperature. In this method, an initial value is supposed for the case temperature in the beginning. The derived E_{ON} and E_{OFF} are based on the case temperature. Thus, using (16) a new case temperature is derived. Repeating this method, the case temperature arrives at a constant value that is the base of the E_{ON} and E_{OFF} calculations, leading to the accurate calculation of switching loss through the (11).

$$T_c = T_A + \left(\left((E_{on} + E_{off}) \times F_{sw} \times D_{sw}\right) + P_{con}\right) \times Z_{th(c-A)}\right) \quad (16)$$

Where T_A is the ambient temperature and $Z_{th(c-A)}$ is the thermal impedance between the case to ambient.

V. DISCUSSING SIMULATION RESULTS

The first theoretical calculations are performed to evaluate the estimated equations for two switches Sa_1 and Sa_2 in the NPC shown in figure 10. The NPC is simulated in PLECS software where the NPC of figure 10 includes: two equal 300 V voltage sources, one 12 Ohms resistive load, and eight IGBT with an antiparallel diode from Fairchild FGH60N60SMD are chosen. The IGBT datasheet is available in [14] and the switching frequency is 15 kHz. Also, the thermal impedance is considered based on [15] is shown in TABLE II.

978-1-5090-0376-1/16 $31.00 © 2016 IEEE

TABLE II. THERMAL IMPEDANCE BETWEEN JUNCTION TO AMBIENT

Thermal Impedance				
$Z_{th(j-c)}$	$Z_{th(c-s)}$		$Z_{th(s-a)}$	
0.15	$R_{th(c-s)}$	$C_{th(c-s)}$	$R_{th(s-a)}$	$C_{th(s-a)}$
	0.2	0.95	0.63	2.4

Anyway, the values of D_{SW} and D_{con} are the first step in theoretical calculations. Therefore, D_{sw} calculated For the 5-level NPC based on (12) which are listed as table III. Also, the value of modulation index is 1 in all of the results of D_{sw} and D_{con}.

TABLE III. CORRECTION COEFFICIENT BETWEEN DIFFERENT LEVELS FOR SWITCHING LOSS

LEVELS	2 , 1	1 , 0	0 , -1	-1 , -2
$D^{'}$	0.334	0.165	0.165	0.334

Figure 8 illustrates the simulation results for the correction coefficient of the switching loss in 5-level NPC, where (a) is the value of D_{sw} for (S_{a1}, S_{a3}) and (b) is the value of D_{sw} for (S_{a2}, S_{a4}). Also, (c) and (d) are the simulation results for the value of D_{sw} for (S_{b1}, S_{b3}) and (S_{b2}, S_{b4}).

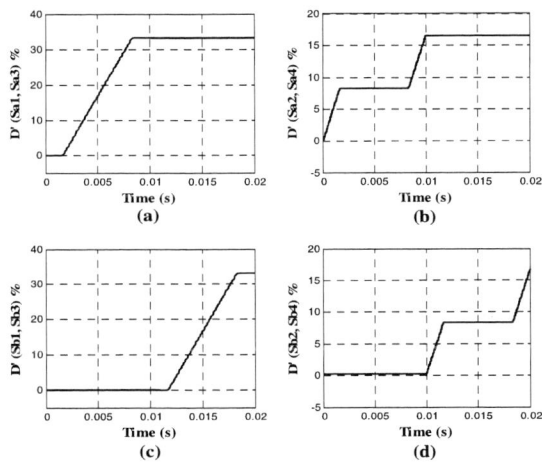

Figure 8: The simulation of correction coefficient for switching loss in a 5-level inverter

Also, D_{con} resulted from (3) is available in table IV.

TABLE IV. CORRECTION COEFFICIENT BETWEEN DIFFERENT LEVELS FOR CONDUCTION LOSS

D_2	D_1	D_0	D_{-1}	D_{-2}
0.218	0.201	0.162	0.201	0.218

Moreover, after the theoretical calculation of the correction coefficient of the conduction loss, the simulation results for this parameter is provided in figure 9. In figure 9 (a) to (e) are related to +2, +1, 0, -1, and -2 levels in a 5-level NPC inverter. Comparing the theoretical results and the simulation results validate the accuracy of the estimated equations. After calculating the correction coefficients, the conduction loss is calculated through (2). For calculating the switching loss, the E_{ON} and E_{OFF} are require, therefore, they are calculated based on repeat method according to TABLE V.

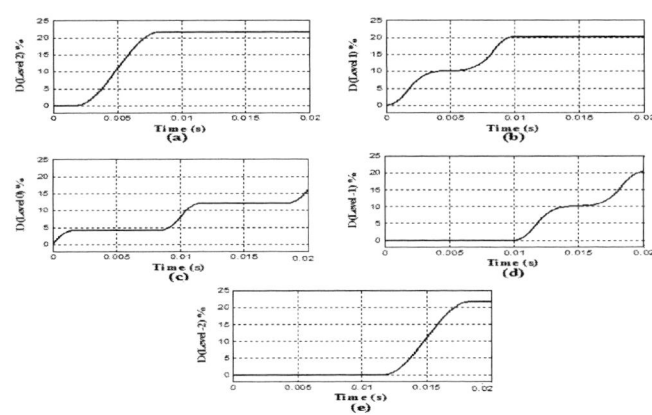

Figure 9: The simulation of correction coefficient for conduction loss in a 5-level inverter

TABLE V. RESULTS OF CALCULATING E_{ON} AND E_{OFF}

a_1			
T_C	E_{on}	E_{off}	T_C
125	1.295	0.459	76.84
76.84	1.075	0.3628	75.13
75.13	1.0672	0.35825	75.06
a_2			
TC	Eon	Eoff	TC
125	0.564	0.186	76.92
76.92	0.4339	0.1306	76.42
76.42	0.4325	0.13	76.41

After the theoretical calculations, the NPC is simulated in PLECS according to figure 10. The results are shown in table VI where the theoretical results are similar to the simulation results that show high accuracy of the estimated equations in calculating the conduction loss and the switching loss. The advantages of proposed equations in this paper are simplicity and being extendable to the n-level SPWM inverters.

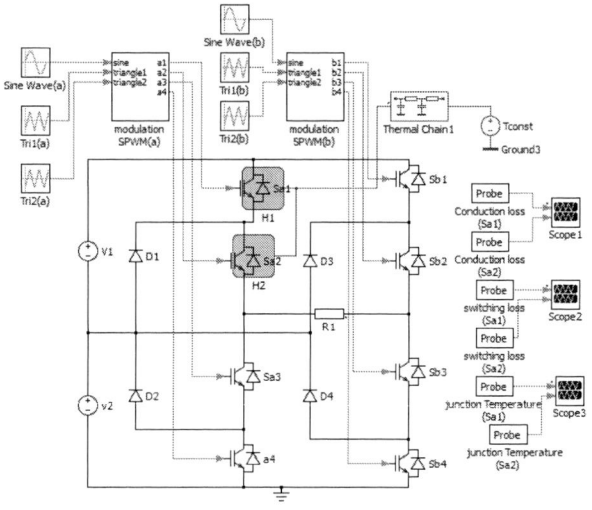

Figure 10: simulated system in PLECS

TABLE VI. THE THEORETICAL AND SIMULATION RESULTS OF A 5-LEVEL NPC

Calculation Method	Parameter	Switches		Parameter	Error (%)	
		S_{a1}	S_{a2}		S_{a1}	S_{a2}
Theoretical	$P_{SW}(w)$	7.25	1.39	P_{SW}	0.6	0.7
	$P_{Con}(w)$	20.7	27.7			
	$T_j(C^0)$	86	86.5			
Simulation	$P_{SW}(w)$	7.27	1.4	P_{Con}	0	0
	$P_{Con}(w)$	20.7	27.7			
	$T_j(C^0)$	87	87.4			

VI. CONCLUSIONS

The accurate calculations of the power and junction temperature of the switches are vital for reliable, technical and financial analysis of a power electronic system. The accurate calculation of the power depends on the switching loss and the conduction loss calculations. Previously, the switching loss and the conduction loss were calculated using some complicated equations, while this paper proposes some simpler estimated equations. In this method, the switching time and the conducting time are calculated based on simpler mathematical equations. Also, the equations presented here can be extended to the n-level SPWM inverters. The simulation has been done using PLECS software to evaluate the proposed methods of this paper. All of the results show the accuracy of proposed equations.

VII. REFERENCES

[1] J. Rodriguez, J. S. Lai, and F. Z. Peng, "Multilevel inverters: A survey of topologies, controls, and applications," IEEE Trans. Ind. Electron., vol. 49, no. 4, pp. 724–738, Aug. 2002.

[2] R.Ribeiro, C.jacobina, A. Lima, and E. da Silva, " A Strategy for improving reliability of motor drive systems using a four-leg three phase converter," IEEE proc. Of APEC'01, vol.1,pp.385-391.

[3] Haitian Wang, Guangfu Tang, Zhiyuan He, Junzheng Cao, Xiaoping Zhang, "Analytical approximate calculation of losses for modular multilevel converters", IET Gener. Transm. Distrib., 2015, Vol. 9, Iss. 16, pp. 2455–2465.

[4] U. Drofenik, J.W. Kolar, "A general scheme for calculating switching- and conduction losses of power semiconductors in numerical circuit simulations of power electronic system", in: Proc. of the 5th Int. Power Electron.conference, Niigata, Japan, 2005.

[5] Alamri,B, Darwish,M. "precise modeling of switching and conduction losses in cacade H-BRIDGE multilevel inverters" Power Engineering Conference (UPEC), 2014 49th International Universities, ISBN: 978-1-4799-6556-4, 2-5 Sept. 2014.

[6] M.G.H. Aghdam, S.S. Fathi, A. Ghasemi, "The analysis of conduction and switching losses in three-phase OHSW multilevel inverter using switching functions", International Conference on Power Electronics and Drives Systems, 2005, vol. 1, pp. 209-218.

[7] J. Ramu, S. Parkash, K. Srinivasu, R. Ram, M. Prasad, Md. Hussain, "Comparison between symmetrical and asymmetrical single phase seven level cascaded h-bridge multilevel inverter with PWM topology", International Journal of Multidisciplinary Sciences and Engineering 3 (4) (2012) 16-20.

[8] A. Farzaneh, J. Nazarzadeh, "Precise loss calculation in cascaded multilevel inverters", in: Second International Conference on Computer and Electrical Engineering, Dec. 28-30, 2009, vol. 2, pp.563-568.

[9] Chinnathambi Govindaraju, Kaliaperumal Baskaran "Power Loss Minimizing Control of Cascaded Multilevel Inverter with Efficient Hybrid Carrier Based Space Vector Modulation", International Journal of Electrical and Computer Engineering Systems, Volume 1, Number 1, June 2010.

[10] P. Satish Kumar, K. Ramakrishna, Ch. Lokeshwar Reddy, G. Sridhar, "Minimization of Switching Loss in Cascaded Multilevel Inverters Using Efficient Sequential Switching Hybrid Modulation Techniques", International Journal of Electrical, Computer, Energetic, Electronic and Communication Engineering Vol:8, No:3, 2014.

[11] Surin Khomfoi, Leon M. Tolbert, "Multilevel Carrier-Based Pulse Width Modulation Techniques Applied to a Diode-Clamped Converter for Use as a Universal Power Conditioner", The University of Tennessee, 1999.

[12] ON SEMICONDUCTOR, Appl.Note AND9140/D, "Thermal Calculations for_IGBTs" April, 2014.

[13] Mwinyiwiwa, B, Wolanski, Z., Ooi, B.-T. "Microprocessorimplemented SPWM for multiconverters with phase-shifted triangle carriers", IEEE Trans. Industry Applications., 1998, 34, pp. 487–495.

[14] Fairchild Corporation: Datasheet of IGBT "FGH60N60 SMD" published at http://www.fairchild.com/.

[15] T. Hopkins, R. Tiziani "Transient Thermal Impedance Consideration in Power Semiconductor Applications" IEEE Automotive Power Electronics ,pp. 89 – 97, 28-29 Aug 1989.

High Efficiency Wireless Power Transfer System Design for Circular Magnetic Structures

A. Ramezani[1], Sh. Farhangi, H. Iman-Eini ,B. Farhangi

School of Electrical and Computer Engineering, College of Engineering University of Tehran
Tehran, Iran
[1] aliramezani@ut.ac.ir

Abstract—In this paper, a new approach is proposed to achieve maximum coupling coefficient for the windings in circular Wireless Power Transfer (WPT) systems. The designed magnetic structure has the maximum magnetic coupling coefficient in a specific air gap in order to maximize the system efficiency. Series resonant compensator network is analyzed in frequency domain and proper operating point is selected to achieve constant voltage gain in different loads. This system controls output voltage through a closed loop control via controlling inverter's duty cycle. The design procedure is explained for a 5 kW system. This system can transfer power from 100mm to 200mm air gap by resonant network and magnetic circular coupling that has a 480mm diameter. Efficiency results for output power in various air gaps are presented, where efficiency of the nominal load for 150mm air gap is 93%.

Keywords— wireless power transfer; circular magnetic structures; high efficiency; loss analysis; maximum coupling coefficient ; finite element analysis

I. INTRODUCTION

With the development of power electronic devices, wireless power transfer has found various applications in industry. Wireless power transfer has many advantages such as:

- Safety

- Robustness against environmental conditions like rain

- High reliability

- It does not require copper cable and plugs

- It does not require isolating transformer

Since a magnetic coupling is used in the wireless power transfer system's structure, it has an inherent isolation, and thus it does not need any isolating transformers so cost, weight and dimensions of the system can reduce considerably [2-5].

A wireless power transfer network is shown in Fig. 1. First, an AC/DC converter receives electrical energy from power grid, then a high-frequency inverter generates a high frequency square wave. With the design of a resonant compensator network, where its various structures are introduced in [6], a resonance circuit with magnetic structure is formed. In fact, these magnetic structures are a high frequency transformer with specific magnetic coupling in the specified air gap. Resonant circuit transfers electrical energy with high frequency through the air gap by inducing voltage on the secondary winding. Secondary circuit comprises a rectifier and a resistive load, in its simplest form.

In recent years with increasing problems of fossil fuels like contamination and environment destruction and increasing cost, pure and hybrid electrical vehicles have attracted attentions. Conductive charging of electrical vehicles requires a connection to the charging station through cable, isolation transformer, and control equipment [7-13]. The vehicle has to be parked and stay stationary. The conductive connection can be hazardous for the user [14]. In contrast, wireless charging can potentially improve safety and reduce the equipment weight. Moreover, wireless charging introduce more flexibility; vehicles can even be charged while traveling. New method for charging electrical vehicles is using wireless power transfer system. In this method, user does not interfere in charging stages and overcomes disadvantages of conventional method. In addition, charging electric vehicles along highways allows the electrical vehicle to go further distances, which is studied in [17-19]. Another type of chargers based on wireless power transfer is used in charge stations and home garages, which is shown in Fig. 1 [1, 20 and 21].

Various structures have been introduced to be used in magnetic coupling in wireless power transfer [22]. Among these structures, circular magnetic structure has attracted more attentions and has many advantages like simplicity of design, implementation and higher magnetic coupling coefficient in a specific level compared to other structures [24]. According to (1) where r_1 and r_2 are the primary and secondary winding's resistance, angular frequency of ω, R_{ac} is equivalent resistance of the rectifier and M is the mutual inductance of the windings, the efficiency of the WPT system is calculated as [21]:

$$\eta = \cfrac{R_{ac}}{R_{ac} + r_1 + r_2 \cfrac{\left(R_{ac} + r_2\right)^2 + \left(\omega L_2 - \cfrac{1}{\omega C_2}\right)^2}{\omega^2 M^2}} \quad (1)$$

According to (1) the efficiency of the WPT is a function of load, frequency and mutual inductance. Mutual inductance of the windings is affected by the air gap. So, in a specific air gap the mutual inductance should be maximized to achieve

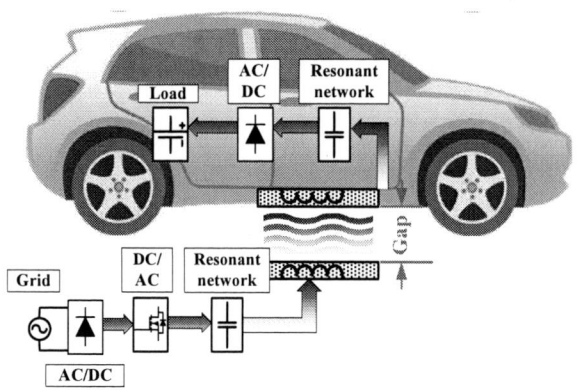

Fig. 1. Wireless power transfer for EV's charger application

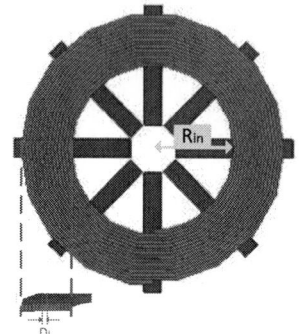

Fig. 2. Circular magnetic structure

maximum efficiency. In section II such objective is followed by properly designing circular structures. Circular magnetic pads have different structures for various applications. Each magnetic structure has specific characteristics, which have been explained and compared in [22, 26]. One effective factor in designing magnetic structure is the magnetic coupling coefficient between primary and secondary windings. Magnetic structures should also have minimum dimensions so that they can be used for applications such as charging an electric vehicle. Design parameters and optimization of circular magnetic couplers has been well studied in [24, 26].

Circular magnetic structure has been used in this paper because its implementation and design is very simple. This structure comprise a number of ferrite strips which are stacked vertically, flat circular winding and an air gap between primary and secondary. Magnetic coupling coefficient of these two magnetic pads varies with air gap changes, which directly affects maximum transferred power.

II. DESIGNING MAGNETIC STRUCTURE

In this paper, a circular magnetic structure with 480mm diameter is designed. In order to avoid increase in windings' losses and its induced voltage, number of turns should not be large. If operating frequency of the winding is 200 kHz, then skin depth (δ) is calculated according to (2) and number of the strands for Litz wire of the winding is calculated from (3). Parameters D_s, A_s and N_s are respectively the diameter, area and number of strands of Litz wire. Parameter A_{cu} is the required area of wire based on current density (J_{cu}). D_L is the equivalent diameter of wire.

TABLE I.   Parameter Value Of Magnetic Couplers

Parameter	Value
Number of ferrite bars	8
Ferrite bar's thickness (mm)	15
Ferrite bar's width (mm)	30
Ferrite bar's length (mm)	200
Wire diameter (mm)	6
Number of both winding turns	14

$$\delta = \sqrt{\frac{2\rho}{\omega\mu}} = 0.167600716 \text{ mm} \tag{2}$$

$$D_s \approx \delta \rightarrow A_s = \pi\frac{D_s^2}{4} = 0.02206183441 mm^2 \tag{3}$$

$$A_{cu} = \frac{I}{J_{cu}} = \frac{20}{4} = 5mm^2 \tag{4}$$

$$N_s = \frac{A_{cu}}{A_s} = \frac{5}{0.02206183441} \cong 226 \tag{5}$$

$$D_L = 1.3\sqrt{N_s} \times D_s = 4.64mm \tag{6}$$

So, considering the insulator between each layer and equivalent diameter of Litz wire, 6mm diameter is chosen for each turn of the windings. Designed magnetic structure is presented in Table I.

In order to find optimum inner radius of the winding, finite element analysis (FEA) is used with ANSYS Maxwell by keeping number of turns constant and changing the inner radius from 40 to 160mm. Variations of magnetic coupling coefficient for primary and secondary windings are calculated in a 150mm air gap. The results are shown in Fig. 3, it shows that for a specific radius, magnetic coupling coefficient is maximum, which is the optimum point of design, thus by choosing inner radius equal to 140mm, maximum magnetic coupling coefficient is achievable.

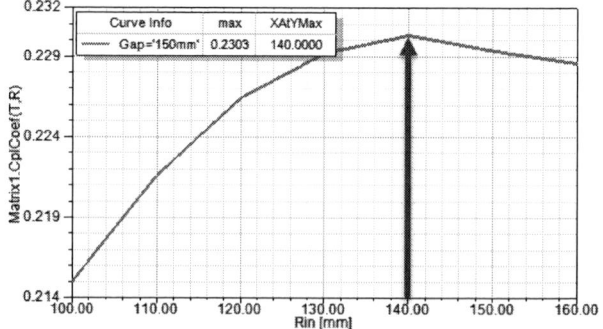

Fig. 3. Magnetic coupling coefficient as function of inner radius

Fig. 4. Magnetic flux vector

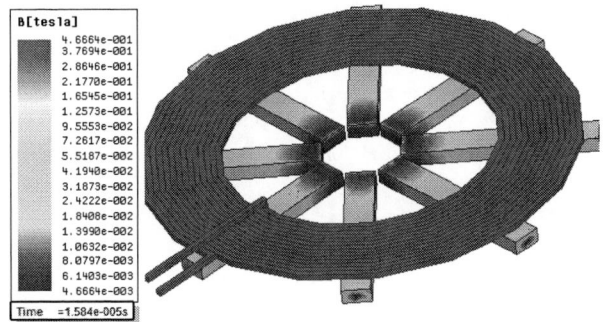

Fig. 7. Flux density in ferrite bars.

A. Effect Of Air Gap On Magnetic Coupling Coefficient

Increasing air gap between magnetic couplers decreases mutual inductance. By getting magnetic pads closer to each other there is little impact on self-inductance. Because in short distances, secondary pad's ferrite gets close to primary winding and self-inductance increases. For 140mm inner radius, variations of magnetic coupling coefficient for primary and secondary windings are shown in Fig. 5 for air gap variations form 100 to 200 mm.

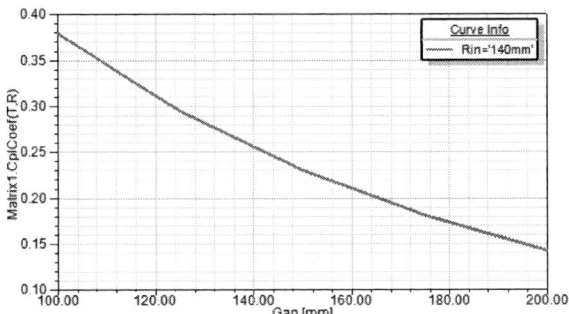

Fig. 5. Magnetic coupling coefficient as function of air gap

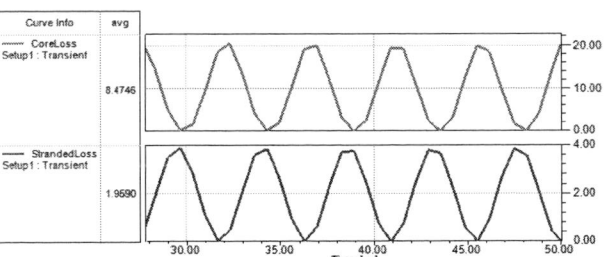

Fig. 6. Core and strand losses of the designed circular magnetic structure

B. Loss Analysis

Magnetic coupler's losses include core loss, winding losses due to proximity effect and skin effect. For proposed circular magnetic structure of Table I, these losses are shown in Fig. 6 with their mean value for the nominal current excitation in primary and secondary in 130 kHz.

III. CIRCUIT TOPOLOGY AND ANALYSIS

A resonant circuit with series resonant network in primary and secondary are proposed for this charger as shown in Fig. 8. This resonant circuit is designed such that series capacitor C_1 resonates with inductance L_1 in a specific frequency. Since difference between self-inductance of primary and secondary is negligible, resonance frequency of primary and secondary can be considered equal. Fundamental harmonic analysis of the circuit allows to consider rectifier and load as an equivalent resistive load and inverter as an AC voltage source. Such approximation simplifies analysis and provides the possibility of designing circuit elements. Since output voltage of the inverter is a square wave, Fourier series of the inverter's output voltage is equal to:

$$V_{inv} = \frac{4 \times V_{dc}}{\pi} \sum_{n=1}^{\infty} \frac{1-(-1)^n}{2n} Sin(n\omega t) \qquad (7)$$

$$V_{ac} = V_1 = \frac{4 \times V_{dc}}{\pi} Sin(\omega t) \qquad (8)$$

$$V_1^{rms} = 4 \times V_{dc} / \pi\sqrt{2} \qquad (9)$$

Similarly equivalent resistance on the rectifier's output voltage in the secondary can be written as:

$$R_{ac} = \frac{8}{\pi^2} R_L \qquad (10)$$

Thus circuit of Fig. 8 can be shown simply for fundamental harmonic as Fig.9. In this circuit, according to variations of the mutual inductance, magnetic coupling coefficient varies for different distances. Thus, charger circuit is analyzed in frequency domain for various magnetic coupling coefficients to choose operating frequency and values of circuit elements.

A. Circuit Design

Considering the magnetic structure with circular pad

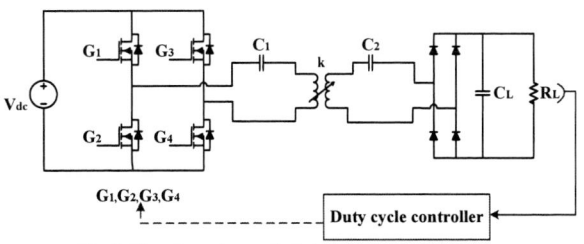

Fig. 8. Circuit topology of wireless power transfer system

978-1-5090-0376-1/16 $31.00 © 2016 IEEE   567

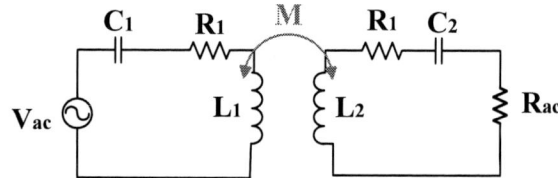

Fig. 9. Equivalent system model

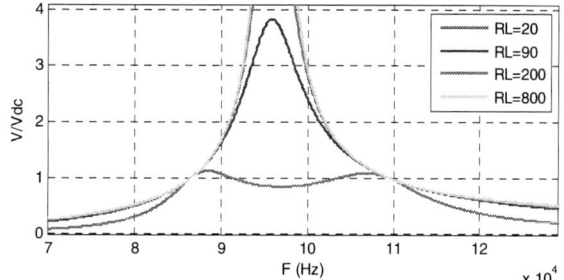

Fig. 10. Voltage gain function of frequency for different loads

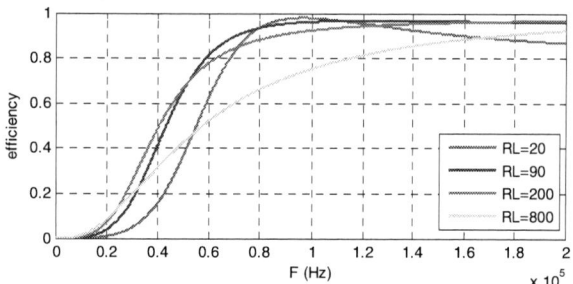

Fig. 11. System efficiency function of frequency for different loads

introduced in section II, self-inductance's value of primary and secondary are known. If series capacitor of primary and secondary chose large values, resonance frequency and total efficiency decrease but high resonance frequency increases switching frequency. Also in order to meet ZVS condition, operating frequency of the circuit should be higher than resonance frequency of the resonant network. This operating frequency should be chosen such that the series compensator network's voltage gain is the same for various loads such that circuit works like an ideal voltage source with maximum efficiency [21]. Resonant frequency of this series resonant network is calculated according to (11).

$$f_0 = \frac{1}{2\pi\sqrt{L_1C_1}} \cong \frac{1}{2\pi\sqrt{L_2C_2}} = 95.97 \quad kHz \qquad (11)$$

In this design, range of resistive load is considered to be between 20Ω to 800Ω for 5kW to 0.11kW output power. Variations of the output voltage gain are shown in Fig. 10. It is clear that voltage gain is the same for various resistive loads in 110 kHz frequency. This point is chosen as the nominal operating point of the circuit.

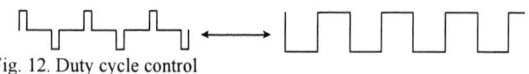

Fig. 12. Duty cycle control

B. Control

Considering variations of the input DC link voltage or output load's voltage when the load is a battery, a PI type closed loop controller is proposed to control output voltage by duty cycle of the input inverter's switches, which is shown in Fig. 12. Controller gains are chosen such that the desired time response is obtained. Choosing duty cycle control method causes circuit elements to be chosen for a specific frequency and controlling with this method is much easier than frequency control, because in frequency control, system's response is non-linear.

IV. SIMULATION RESULTS

Results of DC to DC efficiency in operating frequency for various values of output power are shown in Fig. 13 to Fig. 16. By increasing air gap, system's efficiency and its maximum output power for similar resistive load has decreased. The reason, why output power decreases is because of voltage gain curve of the resonant circuit does not reach 1pu in operating frequency, that is why for air gaps larger than 150mm, duty cycle controller of the switches are set about 50% such that maximum output voltage is obtained. Fig. 14 shows the results of power transfer for nominal air gap equal to 150mm, which indicates that in small loads, efficiency will drop below 85%, but in nominal load and values close to that, efficiency is higher than 90%. System efficiency for different air gaps between 125mm to 200mm respectively presented in Fig. 13 to Fig 16.

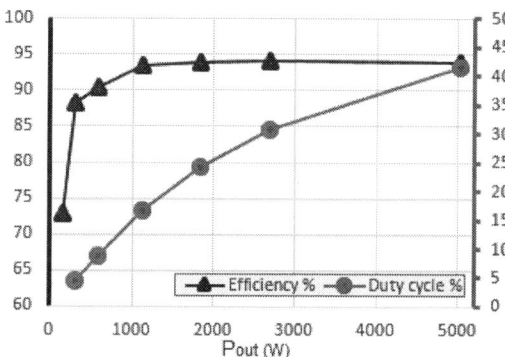

Fig. 13. System efficiency and duty cycle versus output power with 125mm air gap

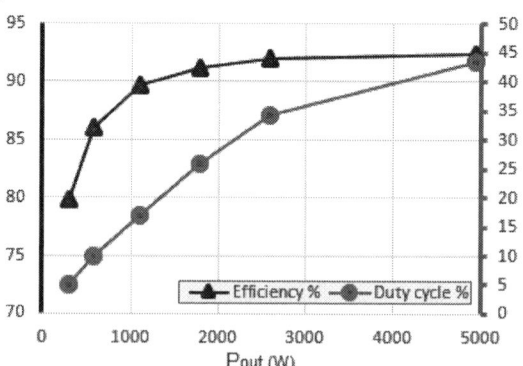

Fig. 14. System efficiency and duty cycle versus output power with 150mm air gap

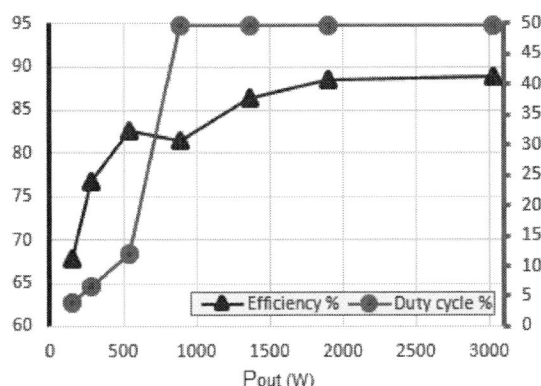

Fig. 15. System efficiency and duty cycle versus output power with 175mm air gap

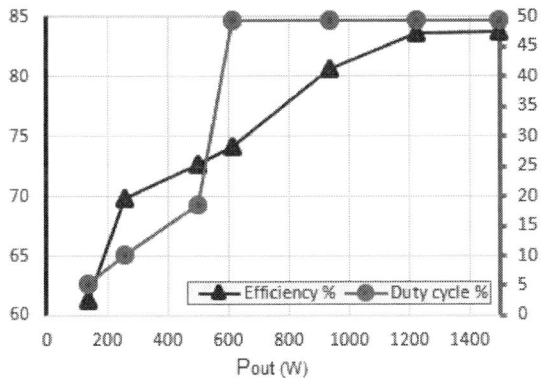

Fig. 16. System efficiency and duty cycle versus output power with 200mm gap

V. CONCLUSION

In this paper, a circular magnetic structure with total diameter of 480mm was introduced. Based on the proposed design method, coupling coefficient of the windings are calculated as a function of the winding radius in order to find the maximum k. This structure is designed such that its magnetic coupling coefficient for a specific air gap reaches the maximum value. Designed diameter of the applied conductor is determined based on skin depth and magnetic circular pad's losses in nominal load. Core losses and conduction losses are negligible compared to the transferred power, which guarantees high efficiency. Power transfer circuit includes H bridge inverter with 110 kHz frequency that generates a square wave with 400V amplitude. Series compensator network is designed with capacitors of primary and secondary being the same and equal to 20nf. According to the values of self-inductance and series capacitor, resonance frequency of the circuit would be 95.97 kHz. Voltage gain's curve for different resistive loads from 20 to 800Ω is plotted to determine the operating frequency of the circuit. This operating frequency is determined such that the resonant network obtains the same voltage gain for different loads and meets ZVS conditions. The efficiency results and set values for duty cycles shows higher than 90% overall efficiency for this wireless power transfer system.

APPENDIX

TABLE II. Nominal system specification

Parameter	Value
L_1 (uH)	137.503
L_2 (uH)	137.71
M (hH)	31.687
C_1 (nF)	20
C_2 (nF)	20
R_1 (ohm)	0.1
R_2 (ohm)	0.1

REFERENCES

[1] H. H. Wu, A. Gilchrist, K. D. Sealy, and D. Bronson, "A high efficiency 5 kW inductive charger for EVs using dual side control," Industrial Informatics, IEEE Transactions on, vol. 8, pp. 585-595, 2012.

[2] F. Musavi, M. Edington, and W. Eberle, "Wireless power transfer: A survey of EV battery charging technologies," in Energy Conversion Congress and Exposition (ECCE), 2012 IEEE, 2012, pp. 1804-1810.

[3] R. Rahimi, E. Afshari, B. Farhangi, Sh. Farhangi, "Optimal placement of additional switch in the photovoltaic single-phase grid-connected transformerless full bridge inverter for reducing common mode leakage current", in Proc. Energy Conversion (CENCON), 2015 IEEE Conference on, pp. 1-5, 19-20 Oct, 2015.

[4] E. Afshari, R. Rahimi, B. Farhangi, Sh. Farhangi, "Analysis and Modification of the Single Phase Transformerless FB-DCB Inverter Modulation for Injecting Reactive Power," in Proc. Energy Conversion (CENCON), 2015 IEEE Conference on , pp.1-6, 19-20 Oct. 2015.

[5] R. Rahimi, B. Farhangi, Sh. Farhangi, " A Novel Topology to Reduce Leakage Current in Three-Phase Transformerless Grid-Connected Photovoltaic Inverter", in Proc. Power Electronics Drive Systems and Technologies Conference (PEDSTC), 7th 2016, pp. 1-6, 16-18 Feb. 2016.

[6] R. L. Steigerwald, "A comparison of half-bridge resonant converter topologies," Power Electronics, IEEE Transactions on, vol. 3, pp. 174-182, 1988.

[7] B. Farhangi and H. A. Toliyat, "Modeling and Analyzing Multiport Isolation Transformer Capacitive Components for Onboard Vehicular Power Conditioners," Industrial Electronics, IEEE Transactions on, vol. 62, pp. 3134-3142, 2015.

[8] B. Farhangi and H. Toliyat, "A Novel Vehicular Integrated Power System Realized with Multi-port Series Ac Link Converter," in Applied Power Electronics Conference and Exposition, 2015. APEC 2015. 30th Annual IEEE, 2015, pp. 1353-1359.

[9] B. Farhangi, "A novel modified deadbeat controller for vehicle to grid application," in Power Electronics, Drives Systems & Technologies Conference (PEDSTC), 2015 6th, 2015, pp. 47-52.

[10] B. Farhangi and H. A. Toliyat, "Modeling Isolation Transformer Capacitive Components in a Dual Active Bridge Power Conditioner," in Energy Conversion Congress and Exposition (ECCE), 2013 IEEE, 2013, pp. 5476-5480.

[11] B. Farhangi and H. A. Toliyat, "Piecewise Linear Model for Snubberless Dual Active Bridge Commutation," Industry Applications, IEEE Transactions on, vol. 51, pp. 4072-4078, 2015.

[12] B. Farhangi and H. A. Toliyat, "Piecewise linear modeling of snubberless dual active bridge commutation," in Energy Conversion Congress and Exposition (ECCE), 2014 IEEE, 2014, pp. 2065-2071.

[13] B. Farhangi and K. Butler-Purry, "Transient study of DC Zonal Electrical Distribution System in Next Generation Shipboard Integrated Power Systems using PSCAD™," in North American Power Symposium (NAPS), 2009, 2009, pp. 1-8.

[14] B. Farhangi, H. A. Toliyat, and A. Balaster, "High impedance grounding for onboard plug-in hybrid electric vehicle chargers," in Power Engineering, Energy and Electrical Drives (POWERENG), 2013 Fourth International Conference on, 2013, pp. 609-613.

[15] D. Kobayashi, T. Imura, and Y. Hori, "Real-time coupling coefficient estimation and maximum efficiency control on dynamic wireless power transfer for electric vehicles," in Emerging Technologies: Wireless Power (WoW), 2015 IEEE PELS Workshop on, 2015, pp. 1-6.

[16] G. A. Covic and J. T. Boys, "Modern Trends in Inductive Power Transfer for Transportation Applications," Emerging and Selected Topics in Power Electronics, IEEE Journal of, vol. 1, pp. 28-41, 2013.

[17] J. Huh, S. W. Lee, W. Y. Lee, G. H. Cho, and C. T. Rim, "Narrow-Width Inductive Power Transfer System for Online Electrical Vehicles," Power Electronics, IEEE Transactions on, vol. 26, pp. 3666-3679, 2011.

[18] G. A. Covic, J. T. Boys, M. L. G. Kissin, and H. G. Lu, "A Three-Phase Inductive Power Transfer System for Roadway-Powered Vehicles," Industrial Electronics, IEEE Transactions on, vol. 54, pp. 3370-3378, 2007.

[19] G. Covic, G. Elliott, O. Stielau, R. Green, and J. Boys, "The design of a contact-less energy transfer system for a people mover system," in Power System Technology, 2000. Proceedings. PowerCon 2000. International Conference on, 2000, pp. 79-84.

[20] J. M. Miller, O. C. Onar, and M. Chinthavali, "Primary-Side Power Flow Control of Wireless Power Transfer for Electric Vehicle Charging," Emerging and Selected Topics in Power Electronics, IEEE Journal of, vol. 3, pp. 147-162, 2015.

[21] Z. Cong, L. Jih-Sheng, C. Rui, W. E. Faraci, Z. Ullah Zahid, G. Bin, et al., "High-Efficiency Contactless Power Transfer System for Electric Vehicle Battery Charging Application," Emerging and Selected Topics in Power Electronics, IEEE Journal of, vol. 3, pp. 65-74, 2015.

[22] G. Covic and J. T. Boys, "Inductive power transfer," Proceedings of the IEEE, vol. 101, pp. 1276-1289, 2013.

[23] R. Bosshard, J. Muhlethaler, J. W. Kolar, and I. Stevanovic, "Optimized magnetic design for inductive power transfer coils," in Applied Power Electronics Conference and Exposition (APEC), 2013 Twenty-Eighth Annual IEEE, 2013, pp. 1812-1819.

[24] J. T. Boys, G. A. Covic, and A. W. Green, "Stability and control of inductively coupled power transfer systems," Electric Power Applications, IEE Proceedings -, vol. 147, pp. 37-43, 2000.

[25] W. Shuo and D. Dorrell, "Review of wireless charging coupler for electric vehicles," in Industrial Electronics Society, IECON 2013 - 39th Annual Conference of the IEEE, 2013, pp. 7274-7279.

[26] M. Budhia, G. A. Covic, and J. T. Boys, "Design and Optimization of Circular Magnetic Structures for Lumped Inductive Power Transfer Systems," Power Electronics, IEEE Transactions on, vol. 26, pp. 3096-3108, 2011.

7th Power Electronics, Drive Systems & Technologies Conference (PEDSTC 2016)
16-18 Feb. 2016, Iran University of Science and Technology, Tehran, Iran

Minimum Weight Wireless Power Transfer Coil Design

Adel Moradi, Farzad Tahami, Amirreza Poorfakhraei
Department of Electrical Engineering
Sharif university of Technology
Tehran, Iran

Abstract-**In optimum designing of wireless power transfer (WPT) coils, maximum efficiency and maximum power transfer capability of coils should be considered.**

In this paper, a novel technique for optimum design of WPT coils based on maximum efficiency with considering required power transfer capability has been given. Optimum selection of system parameters including coil and wire dimensions, number of turns of coils and compensation capacitor selection based on minimum weight of copper mass is also given.

For evaluation purpose a 100W experimental sample is designed and tested in 20cm range.

Keywords: Wireless power transfer, minimum weight algorithm, Maximum efficiency, Copper mass optimization.

I. INTRODUCTION

After that a promising report was published in MIT in 2007 [1], investigations on wireless power transfer technology have been widely promoted to make old dreams of engineers from Nicola Tesla era [2] come true.

The reasons behind inventing wireless power transfer (WPT) technology are same as those that motivate engineers to enhance wireless communications. In fact applying this technology is very helpful and effective in applications that using wires is impractical and occupant, such as electric vehicle charging [3] and biomedical applications [4].

The main concept of inductively coupled resonant power transfer is based on compensating large leakage inductances of air-cored coils by capacitors in a special frequency which is called resonant operation of converter and enlarging mutual impedance between coils by increasing operating resonant frequency as large as possible unless it is harmful for human body or electrical components [5].

In addition to the system efficiency, the power transfer capability of the coupled coils is of concern to many applications. Without being compensated appropriately for minimizing the power dissipation, the operating distance between a given pair of coupled coils can hardly be increased [6].

In [7] a new ICPT (Inductively Coupled Power Transfer) optimization method based on minimum copper mass and proper stability conditions has been introduced and a new

design factor K_D is proposed to select the optimum configuration among four most common compensation topologies (Series-Series, Series-Parallel, Parallel-Series, Parrallel-Parallel).

In [8], [9] a design methodology based on maximum efficiency for series-loaded, series-resonant contactless converter is described and also it is shown that for practical values of the separation distance, the leakage inductance, being part of the resonant inductor, remains almost unchanged. Nevertheless, the current distribution between primary and secondary windings changes significantly due to the large variation of the magnetizing inductance.

In this paper a novel step by step algorithm including a mathematical transformation technique to formulate optimum design of a series compensated wireless power system will be introduced. This approach is based on maximum efficiency criterion for transferring a specified power in a determined distance by simultaneously considering required power transfer capability of the coils.

In addition, an optimization algorithm for determining the properties of coils based on minimum copper weight has been introduced.

Moreover in section IV, minimum design weight variation versus required efficiency and distance between coils, using the given algorithm have been simulated and studied.

Finally an experimental case study with power rating of 100W and efficiency of 80% is presented for approving the design approach.

II. THEORY OF WPT SYSTEM

In Fig.1, T-model of two windings of a WPT coils is shown. In this model M is the mutual inductance of two windings and L_1, R_1 and L_2, R_2 are respectively self-inductance and internal resistance of each winding and R_s is the voltage source resistance.

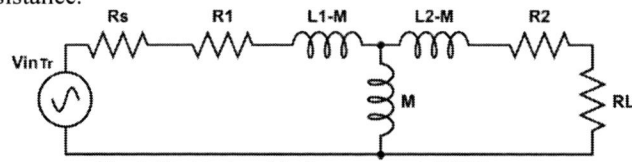

Fig. 1. Electrical model of uncompensated WPT

978-1-5090-0376-1/16 $31.00 © 2016 IEEE

The efficiency of the uncompensated system shown in Fig.1 is calculated as below:

$$\eta_{uncomp} = \frac{R_L}{R_L + R_2 + \frac{(R_1+R_s)(R_2+R_L)^2 + (\omega L_2)^2(R_1+R_s)}{(\omega M)^2}} \tag{1}$$

By definition of new parameters:

$$Q_1 = \frac{\omega.L_1}{R_s+R_1}, Q_2 = \frac{\omega.L_2}{R_L+R_2}, Q_{1,int} = \frac{\omega.L_1}{R_1}, Q_{2,int} = \frac{\omega.L_2}{R_2} \text{ and}$$

$$k = \frac{M}{\sqrt{L_1.L_2}} \tag{2}$$

The uncompensated system efficiency can be rewritten as below:

$$\eta_{uncomp} = \frac{k^2 Q_1 Q_2 - k^2 Q_1 Q_2^2 / Q_{2,int}}{1 + k^2 Q_1 Q_2 + Q_2^2} \tag{3}$$

If the series compensation is applied to the secondary side of the circuit then efficiency of the system is given by:

$$\eta_{SScomp} = \frac{k^2 Q_1 Q_2 - k^2 Q_1 Q_2^2 / Q_{2,int}}{1 + k^2 Q_1 Q_2} \tag{4}$$

It can be easily seen that the only difference in the series compensated system is that a term in the denominator is removed (Q_2^2). This term is usually a large value compared to the other terms and compensating it will result in higher efficiency.

A. Series Compensation for Maximum efficiency

Fig.2 shows a T-Model of series compensated transformer in which the series reactance is denoted by X. In this model it is assumed that $X_1 = X_2 = X$. The reason behind this assumption is that even if L_1 and L_2 weren't equal then by selecting the compensation elements i.e. C_1 and C_2 in a special frequency it will result in an equal reactance for the two windings. The efficiency of the circuit in terms of X and other circuit parameters (including winding losses) has been calculated as below:

$$\eta = \frac{R_L}{R_L + R_2 + R_1 \frac{(X+M\omega)^2 + (R_2+R_L)^2}{(M\omega)^2}} \tag{5}$$

The maximum efficiency is found by setting the derivative of (5) equal to zero. It will happen when $X_{MaxEff} = -M\omega$ and the maximum efficiency will be:

$$\eta_{Max} = \frac{R_L}{R_L + R_2 + R_1 \frac{(R_2+R_L)^2}{(M\omega)^2}} \tag{6}$$

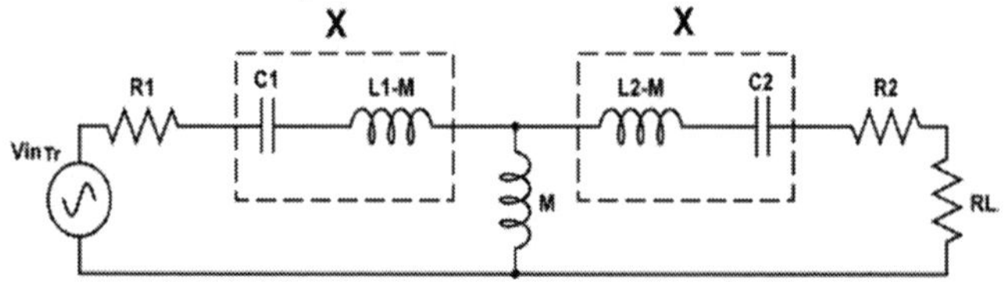

Fig. 2. T-Model of series compensated transformer

The frequency in which $X_{MaxEff} = -M\omega$ is called the resonance frequency:

$$\omega_{res} = \omega = \frac{1}{\sqrt{L_1 C_1}} = \frac{1}{\sqrt{L_2 C_2}} \tag{7}$$

Considering the maximum efficiency formula in (6) it is obvious that for achieving higher efficiency, the resonant frequency should be practically increased as high as possible.

B. Series Compensation for Maximum Power Transfer Capability

to determine the transferable power to the load, the output power is calculated as below:

$$P_{out} = \frac{R_L(M\omega)^2 V_{inTr}^2}{(-X^2 - 2M\omega X + R_1 R_2 + R_1 R_L)^2 + (M\omega+X)^2(R_1+R_2+R_L)^2} \tag{8}$$

The maximum transferable power to the load is determined by derivation of (8) with respect to X:

$$\frac{dP_{out}}{dX} = 0 \rightarrow if \ M\omega < R_c \rightarrow X_{MaxPower} = -M\omega \rightarrow$$

$$P_{out\_Max} = \frac{R_L V_{inTr}^2}{\left(M\omega + \frac{R_1 R_2 + R_1 R_L}{M\omega}\right)^2} \tag{9}$$

Where a critical resistance is defined as:

$$R_c = \sqrt{\frac{R_1^2 + R_2^2 + R_L^2}{2} + R_2 R_L} \tag{10}$$

$$if \ M\omega > R_c \rightarrow X_{MinPower} = -M\omega \rightarrow P_{out\_Min} =$$

$$\frac{R_L V_{inTr}^2}{(M\omega + \frac{R_1 R_2 + R_1 R_L}{M\omega})^2} \tag{11}$$

And

$$\begin{cases} X_{MaxPower1} = -M\omega + \sqrt{(M\omega)^2 - (R_c)^2} \\ X_{MaxPower2} = -M\omega - \sqrt{(M\omega)^2 - (R_c)^2} \end{cases} \rightarrow$$

$$\begin{cases} P_{outMax1} = \frac{R_L V_{inTr}^2}{(R_1+R_2+R_L)^2 \left[1 - \frac{\frac{1}{2}((R_c)^2 - R_1 R_2 - R_1 R_L)}{(M\omega)^2}\right]} \\ P_{out\_Max2} = \frac{R_L V_{inTr}^2}{(R_1+R_2+R_L)^2 [1 - \frac{\frac{1}{2}((R_c)^2 - R_1 R_2 - R_1 R_L)}{(M\omega)^2}]} \end{cases} \tag{12}$$

From the above equations, if $M\omega < R_c$ then the maximum transferable power as well as the maximum efficiency is obtained when $X = -M\omega$. So in this condition by using ($M\omega < R_c$), the maximum possible resonant frequency can be calculated so that the efficiency to be maximized and the desired power to be transferred. From (12) it is obvious that if the resonant frequency is increased such that $M\omega \gg R_c$ then the maximum transferable power take places at two different points. The first point is when $X \approx -2M\omega$ and the second one is for $X \approx 0$ which both are different from the condition obtained for maximum efficiency in the previous section ($X_{MaxEff} = -M\omega$). So to obtain the best power transfer capability of the series compensated WPT system with the highest efficiency it is required to design coils with the below condition:

$$M\omega < R_c \tag{13}$$

978-1-5090-0376-1/16 $31.00 © 2016 IEEE

Actually in this section, by finding the optimum values of X, it has been proven that in series compensated WPT system the maximum efficiency of system is in resonant frequency found in (7) and the maximum power transfer capability of the coils is maximized when (13) is held, otherwise the maximum power transfer capability of the coils move to two other frequencies. It should be emphasized that the above mentioned theory is for optimum designing of the coils of WPT system for constant load and constant mutual inductance between coils and for maximizing efficiency of the coils with load and distance variations, other techniques should be applied which is out of this paper scope.

C. Normalization of Transferred Power

Power is normalized based on $\frac{X}{M\omega}$, by introducing new variables:

$$P_\circ = \frac{V_{inTr}^2}{R_L}, \; Y = \frac{X}{M\omega}, \; Z = \frac{R_L}{M\omega}, \; U_1 = \frac{R_1}{M\omega}, \; U_2 = \frac{R_2}{M\omega} \quad (14)$$

The expression is reduced as:

$$\frac{P_{out}}{P_\circ} = \frac{Z^2}{(Y^2+2Y-U_1Z-U_1U_2)^2+(1+Y)^2(Z+U_1+U_2)^2} \quad (15)$$

The normalized output power variations with respect to Y and Z has been depicted in Fig.3, assuming that $U_1, U_2 = 0$.

At the resonant condition with $Y = -1$ the above expression is summarized as below:

$$\frac{P_{out}}{P_\circ} = \frac{Z^2}{(1+U_1Z+U_1U_2)^2} \quad (16)$$

The symmetrical curves demonstrated in Fig.3 for an extreme case ($U_1, U_2 = 0$), are useful in optimum designing of WPT system parameters to determine the required mutual inductance of the coils with considering the required output power.

In the first iteration with considering $U_1, U_2 = 0$ from (14),(16):

$$Z_d = \frac{R_L}{M_d\omega_0} = \sqrt{\frac{P_{out}}{P_\circ}} \quad (17)$$

This results in:

$$M_d = \frac{R_L}{a_0\omega_0} \quad (18)$$

Where $a_0 = \sqrt{\frac{P_{out}}{P_\circ}} = \frac{V_{out}}{V_{in}}$ and M_d is the required designed mutual inductance for transferring P_{out} to the load resistance R_L with constant output voltage.

The maximum efficiency in (6) can also be rewritten in the following form:

$$\eta_{Max} = \frac{1}{1+\frac{U_2}{Z_d}+\frac{U_1}{Z_d}(U_2+Z_d)^2} \quad (19)$$

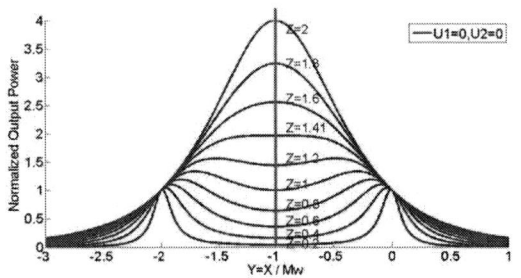

Fig. 3. Normalized output power of the ideal coils versus $Y = \frac{X}{M\omega}$ assuming

Knowing Z_d from (17) and the required maximum efficiency, multiple candidates of the coils with internal resistances of R_{d1}, R_{d2} can be calculated from (14), (18) and (19), which finally the lightest one among them will be selected as the optimum design.

It should be noted that by calculating U_1 and U_2, in this step, the exact value of the Z_d can be calculated again from (16) :

$$Z_d = \frac{a_0+a_0U_1U_2}{1-a_0U_1} \quad (20)$$

Then the corresponding values of U_1 and U_2 can be obtained and this approach can be iterated to find the exact values of them.

From (18), it is obvious that when $a_0 = \sqrt{\frac{P_{out}}{P_\circ}} = \frac{V_{out}}{V_{in}}$ is increased then the value of M_d will be smaller which corresponds to a lighter weight for fabrication of the coils. So the weight of the designed coils also depends on the ratio of the selected input and output voltages of the converter and it can be further optimized by arbitrary selecting of this ratio, but in the real applications, the input and output voltage are dictated by the system requirements. Furthermore if $Z_d < \sqrt{2}$, then the maximum power transfer capability of the coils will encounter in two other frequencies as can be seen in Fig.3 which was mentioned in the previous section.

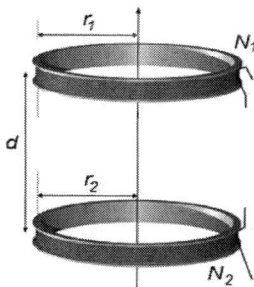

Fig. 4. Parallel circular coaxial coils

III. DESIGN ALGORITHM

A. Calculation of the mutual and self-inductances

For two parallel circular coaxial coils as shown in Fig.4, the self and the mutual inductances can be calculated by equations (21) and (22) where the coil parameters are; r_1, r_2: radii of the coils, N_1, N_2: number of turns of the coils, d: distance between center of the coils, a_1, a_2: radii of the coil wires.

$$M = \mu_0 N_1 N_2 \frac{2r_1 r_2}{\sqrt{(r_1+r_2)^2+d^2}} \int_0^{\frac{\pi}{2}} \frac{2sin^2\varphi-1}{\sqrt{1-4r_1 r_2 sin^2\varphi/(d^2+(r_1+r_2)^2)}} d\varphi \tag{21}$$

$$L_1 = \mu_0 r_1 N_1^2 \left(Ln\left(\frac{8r_1}{a_1}\right) - 2\right)$$

$$L_2 = \mu_0 r_2 N_2^2 \left(Ln\left(\frac{8r_2}{a_2}\right) - 2\right) \tag{22}$$

B. Optimum design algorithm

In this section, an optimum design procedure for transferring a given amount of power with the required efficiency with given input and output voltages will be introduced. Then, another procedure for optimum design of coils will be provided.

A full-bridge converter in which the output voltage is controlled by phase shifting between two legs is shown in Fig.5. The switching pattern of the converter is given in [15].

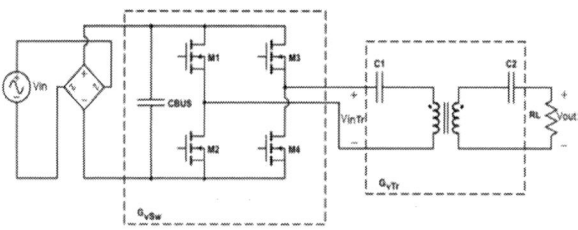

Fig. 5. Phase shifted control converter circuit

A step by step design algorithm for the coils of converter is as follows:

1-Input design data: $V_{in}, V_{out}, P_{out\_Max}, \eta_{Max\_d}, d$ (23)

Where:

d: distance between coils, V_{in}: input voltage, V_{out}: output voltage, P_{out_Max}: maximum transferred output power, η_{Max_d}: desired WPT maximum efficiency.

2-For maximum efficiency, according to (6) the converter should be compensated so that:

$$X = -M\omega \rightarrow Y = \frac{X}{M\omega} = -1 \tag{24}$$

It means that converter should be designed so that its operating point to be always on the $Y = -1$ axis as shown in Fig.3. This condition is satisfied when:

$$\omega_0 = \omega_{res} = \frac{1}{\sqrt{LC}} \tag{25}$$

3-Based on the input data, R_L is calculated as below:

$$R_L = \frac{V_{out}^2}{P_{out\_Max}} \tag{26}$$

4-According to (14), and (17) for the required power transfer capability criterion:

$$Z_d = \sqrt{\frac{P_{out}}{P_\circ}} \tag{27}$$

and $M_d = \frac{R_L}{a_0 \omega_0}$ (28)

Then from (19) and the desired maximum efficiency (η_{Max}) and the given Z_d from (27), a set of combinations of U_1, U_2 will be obtained as below:

$$U_1 = \frac{(1-\eta_{Max})Z_d - \eta_{Max}U_2}{\eta_{Max}(Z_d+U_2)^2} \tag{29}$$

Which the related values of R_{d1}, R_{d2} for two coils can be calculated from (14).
In this step by some iteration the real values of M_d, R_{d1} and R_{d2} can be calculated.

5-The next step is to design two optimum coils so that in distance (d) between coils, the desired mutual inductance (M_d) and desired internal resistances (R_{d1}), (R_{d2}) in previous section are met, it consists of selecting physical parameters of the coils including optimum dimension of wires, geometry and number of turns, considering minimum weight of the used copper as the optimization parameter, the problem is to find the lightest coils based on the input parameters ($M_d, R_{d1}, R_{d2}, \omega_0$) given in the previous step.

By considering the predetermined radii of the coils, a set of pairs of coils which in distance d, satisfies M_d will be calculated regardless of diameters of wires (a_1, a_2), because according to (21), mutual inductance isn't related to the diameters of wires. After that by considering the skin effect and the proximity effect coefficients [10] for satisfying R_{d1} and R_{d2} criteria and considering multiple combinations of these pairs of internal resistances from (29), the wire diameters will be calculated and finally the lightest one among them will be selected.

6-After determining the dimensions and the other properties of coils, the values of L_1 and L_2 also can be determined according to (22). The values of C_1 and C_2 will be obtained from (25).

IV. SAMPLE DESIGN AND SIMULATION

TABLE I. OPTIMUM DESIGN OF WPT SYSTEM FOR $P_{out\_max} = 100w$, $d_{max} = 0.2$ AND VARIOUS DESIRED EFFICIENCIES

η_{Max_d}	0.8	0.9	0.98	0.99
$M_d(uH)$	186	222	223	224
$R_{d1}(\Omega)$	97.48	53.6	8.7	4.59
$R_{d2}(\Omega)$	1.43	1.13	0.221	0.102
$min.weight(kg)$	0.092	0.35	3.7	14.6
$r(m)$	0.25	0.25	0.25	0.25
N_1	74	75	74	79
N_2	16	17	18	17
$a_1(mm)$	0.106	0.193	0.640	1.187
$a_2(mm)$	0.268	0.559	1.781	3.685

A sample design for $\eta_{Max\_d} = 0.80$ based on the given algorithm has been performed as an example. The input voltage of dc bus is considered to be 310V ac and the output dc voltage after rectification is 48V. The optimum design parameters have been tabulated in TABLE I. In this optimization the radius of both coils has been restricted to $r = 0.25m$.

In Fig.6, the required mutual inductance and its associated internal resistance of optimum design of coils versus the variations in efficiency and distance between the coils have been depicted. By increasing the desired efficiency, the value of the required mutual inductance doesn't change, but the value of the required internal resistance is decreased as shown in Fig.6.a. The optimum value of the mutual inductance and its internal resistance does not depend on different distances between coils as shown in Fig.6.b.

After calculating required values of the mutual inductance and its internal resistance for a given efficiency, the variations

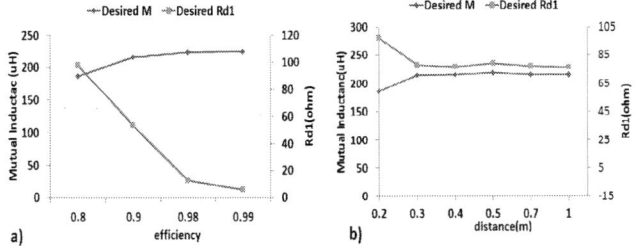

Fig. 6. Optimum required mutual inductance and optimum required internal resistance of primary coil **a)** versus desired maximum efficiency for Pout=100W and d=0.2m **b)** versus distance variation for Pout=100W and $\eta_{Max\_d} = 0.80$

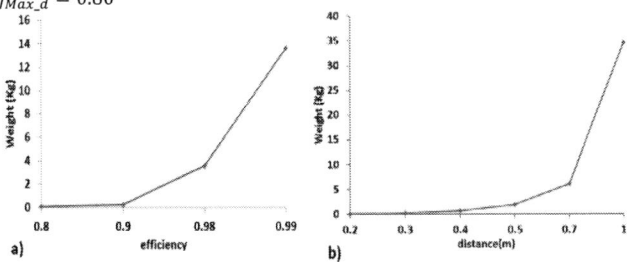

Fig. 7. Minimum weight of winding copper **a)** versus desired efficiency for Pout=100W and d=0.2m **b)** versus distance of two coils for Pout=100W and $\eta_{Max\_d} = 0.80$

of the minimum copper weight versus the desired efficiency, and distance between coils have been also calculated and depicted in Fig.7. Briefly it can be concluded that in designing WPT coils, the amount of transferred power determines the required mutual inductance between the coils and the efficiency has impact on specifying of the internal resistances of the coils. The optimum copper weight is increased by increasing in desired efficiency and distance between coils as shown in Fig.7.a. As can be seen in Fig.6.b, by increasing the distance between the coils, the required mutual inductance and its resistance don't change, but since the radius of the wires is increased, the weight of copper is also increased, as depicted in Fig.7.b.

V. EXPERIMENTAL VERIFICATION

For verification of the given algorithm an experimental sample with Pout=100W for $\eta = 0.8$, $d = 0.2$ and with output dc resistance of $R_{L\_DC} = 25$ has been designed and practically implemented, which by considering rectifier model [11]:

$$R_{L\_DC} = \frac{\pi^2}{8} R_L \qquad (30)$$

The coils have been fabricated with Litz wires and the width of window for winding the wires has been considered to be 20cm.

The design parameters of the converter coils are given in Table II. The experimental and simulated efficiency of the designed WPT coils versus distance has been shown in Fig. 9. The measured efficiency of the coils in d=0.2 (the target design) is $\eta = 0.78$. By decreasing distance between the coils, the efficiency is increased but as shown in Fig. 10, the maximum output transferable power for constant load ($R_{L\_DC} = 25\Omega$) is decreased. On the other hand by increasing the distance, even though the efficiency is decreased but the maximum transferable output power is increased more than designed value (100W).

In Fig.9, the output power for distances more than 0.2m was regulated to the design value by phase shift controlling of the converter, so the efficiency values in Fig.9, are correspondent to 100W output power and for distances less than 0.2m maximum output powers shown in Fig. 10, are considered for calculating the efficiency.

Fig. 8. Prototype of WPT system for maximum power of 100W in 20cm distance with 80% efficiency

978-1-5090-0376-1/16 $31.00 © 2016 IEEE

TABLE II
WPT PARAMETERS FOR $\eta_{Max\_d} = 0.8$, $P_{out\_max} = 100W$ FOR d=0.2

	Calculated	Experimental
$M_d(uH)$	186	180
$R_{1d}(\Omega)$	97.48	93.5(@100kHz)
$R_{2d}(\Omega)$	1.43	1.2(@100kHz)
$min.\,weight(kg)$	0.092	0.107
$r(m)$	0.25	0.25
N_1	86	86
N_2	13	13
$a_1(mm)$	0.097	0.1(1x0.1)
$a_2(mm)$	0.312	0.346 (3x0.2)
$L_1(uH)$	4282	4072
$L_2(uH)$	97.8	94
$C_1(nF)$	0.591	0.68
$C_2(nF)$	25.9	27
$f_{res}(Hz)$	100000	97000
$f_{sw}(Hz)$	100000	97000

For approval of the proposed algorithm and the results given in Table II, the Q factors and the mutual coupling coefficient is calculated for the third case study:

$$Q_1 = \frac{L_1\omega_{res}}{R_1} = 26, Q_2 = \frac{L_2\omega_{res}}{R_2} = 48, k_{mutual} = \frac{M}{\sqrt{L_1 L_2}} = 0.29$$

these results meet the Q factor and coupling coefficient requirements for efficiency of 0.8 given in Fig.3 and equation (13) in [12].

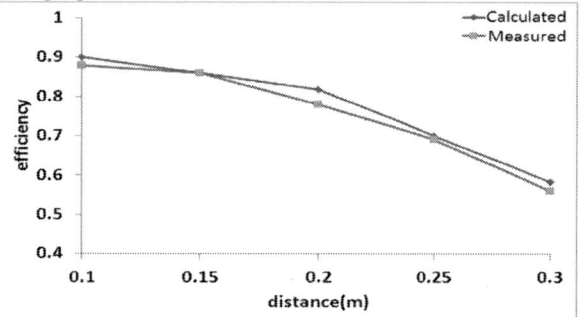

Fig. 9.efficiency of experimental setup versus distance variation with constant load

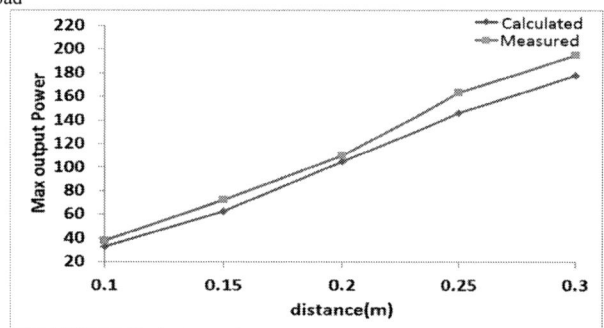

Fig. 10. Maximum output power of experimental setup versus distance variation with constant load

VI. CONCLUSION

In this paper, a novel and simple efficiency based approach for designing wireless power transfer coils was proposed. In this approach, the amount of transferred power determines the required mutual inductance between the coils and the required efficiency has impact on specifying of internal resistances of the coils. Moreover, this design approach is independent from self-inductance values of the coils and only in the final step their values are compensated by appropriate amount of series capacitors.

Furthermore, an algorithm based on minimum copper mass for designing optimal coils for transferring determined amount of power in a specified distance with a desired efficiency was given. Finally an experimental verification of the given algorithms was successfully tested for transferring 100W in maximum 20cm distance range with 80% efficiency.

REFERENCES

[1] Kurs A, Karalis A, Moffatt R, et al. "Wireless power transfer via strongly coupled magnetic resonances," Science, 2007, 317(5834):83-86.

[2] "Tesla Has Fired the Spark Flashed Round the World", The New York Journal, Sunday, August 8, 1897.

[3] U. K. Madawala and D. J. Thrimawithana "A bidirectional inductive power interface for electric vehicles in V2G systems", IEEE Trans. Ind. Electron., vol. 58, no. 10, pp.4789 -4796 2011.

[4] D. Ahn and S. Hong, "Wireless Power Transmission With Self-Regulated Output Voltage for Biomedical Implant," IEEE Trans. Ind. Electron., vol. 61, no. 5, pp. 2225-2235, May 2014.

[5] S. Li and C. C. Mi , "Wireless power transfer for electric vehicle applications" , IEEE Journal of Emerging and Selected Topics in Power Electronics , vol. 3 , no. 1 , pp.4 -17 , 2015.

[6] C. Chih-Jung, Tah-Hsiung, C. Ling,Z. Jou,"A study of loosely coupled coils for wireless power transfer", IEEE Trans. Circuits Syst. II, Exp. Briefs, vol. 57, no. 7, pp.536 -540 2010

[7] J. Sallen , J. L. Villa , A. Llombart and J. F. Sanz "Optimal design of ICPT systems applied to electric vehicle battery charge", IEEE. Trans. Ind. Electron., vol. 56, no. 6, pp.2140 -2149 2009

[8] Valtchev, S. Borges, B. Brandisky, K. Klaassens, J. B. "Resonant contactless energy transfer with improved efficiency," Power Electronics, IEEE Transactions on. vol.24, NO. 3, pp.685-699, Mar. 2009.

[9] W. Zhang , S. C. Wong , C. K. Tse and Q. Chen "Design for efficiency optimization and voltage controllability of series-series compensated inductive power transfer systems", IEEE Trans. Power Electron., vol. 29, no. 1, pp.191 -200 Jan. 2014 .

[10] P. L. Dowell, "Effects of Eddy Currents in Transformer Windings", Proc. IEE, Vol 113 No. 8, August 1966.

[11] R. W. Erickson, "Fundamentals of power electronics", Springer Science & Business Media, 2001.

[12] S. Li and C. Mi "Wireless power transfer for electric vehicle applications", IEEE J. Emerg. Sel. Topics Power Electron. Vol.3, no.1, March 2015.

7th Power Electronics, Drive Systems & Technologies Conference (PEDSTC 2016)
16-18 Feb. 2016, Iran University of Science and Technology, Tehran, Iran

A New Pulsed Power Generator Topology for Corona Discharge

Mohammad Kebriaei, Abolfazl HalvaeiNiasar, Abbas Ketabi

Faculty of Electrical &Computer Engineering
University of Kashan
Kashan, Iran
m.kebriaei@grad.kashanu.ac.ir

Abstract—**Using effective and harmless methods to remove contaminants from the environment and water resources have been considered by many researchers. Among the modern methods, corona discharge for creating free radicals and ozone, which is a strong and unstable pollution treatment, is the best method. The corona discharge is created by applying narrow high-voltage pulses by pulsed power generators on two electrodes. In this paper, a new pulsed power generator topology is proposed that is based on combining voltage source and current source to produce high-voltage pulse. The proposed topology, which is supplied by a DC voltage source with low voltage, can produce the flexible pulses. In order to show the functionality of the system, some simulations have also been performed in Matlab/Simulink.**

Keywords—corona discharge, pulsed power generator, Marx topology, current source

I. INTRODUCTION

Nowadays, the use of high-voltage electric pulses for treatment and microbial removal of the water and air is the interest of many researches and industrialists. Generally, these pulses which are produced between two electrodes can cause purification and microbial removal in two ways: using pulsed electric field for electroporation and pulsed corona discharge for treatment. The electrical breakdown of water/gas or streamer discharge is generated, due to the high voltage pulses between the electrodes. This phenomenon causes the advanced oxidation processes in which active components in the environment are decomposed to carbon dioxide and water .A streamer discharge is seen in the water using electric discharge. Its wavelength changes with the change of water conductivity. The wavelength increases by increasing the pulse-width. In the meantime, a spectrum of radicals H, OH, and O is produced [1]. Fig. 1 shows an emission spectrum produced by pulsed discharge in water [2]. Electric voltage discharge is directly used in the water or in the gas phase above water in order to break peroxide, Molecular oxygen and hydrogen, hydroxyl, hydro-peroxide, hydrogen, oxygen and other radicals; by adding air or oxygen in high-voltage electrode, ozone will be created which will act as a strong microbial removal and antiseptic. Other applications of high-energy discharges are pulsed arcs in water that use for simulation of underwater explosions, metal forming, rock fragmentation, shock wave lithotripsy, such biomedical engineering applications as surgery and skin treatment[3].

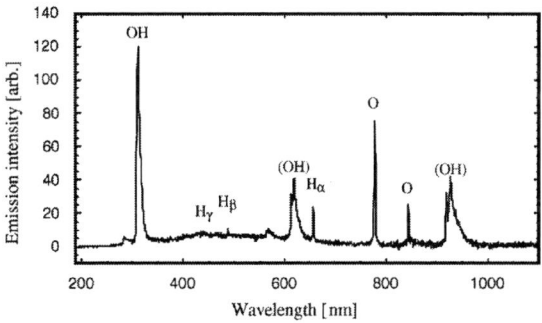

Fig. 1. Emission spectrum from the pulsed discharge in water with point-to-plane electrode configuration [2]

Pulsed electric field, which occurs due to applying high-voltage pulses lower than the required voltage for discharging, will destroy bacteria because of electroporation mechanism.This is a topic of research for scientists working on food treatment [4-5].

This paper reviews the characteristics and requirements of corona discharge and some of electrical model for this phenomenon. Then, the proposed topology is introduced and functional modes are described. Finally, the proposed topology with two stages Marx generator is simulated to evaluate of its performance.

II. CHARACTERISTICS AND REQUIREMENTS OF CORONA DISCHARGE

Dielectric ionization is created due to increase in voltage gradient to get a value higher than dielectric threshold voltage, and then forms corona. Under electric discharge in gas such as air and in gas bubbles in liquids and solids, corona can be easily observed. Corona phenomenon has observable signs like visible light, hearable noise, and ozone gas smell [6]. To discharge and to form corona in pure water containing no bubbles, a voltage of almost 2.5 Megavolt/cm is needed; whereas, this voltage for air is about 30 kV/cm [7]. To lower the required voltage for making electric discharge in water, air or gas bubbles can be injected in the water [8], plasma in gas phase on the water surface can be created [9], and/or water can be sprayed in the gas phase plasma [10]. To electrically investigate corona, the electrical equivalent circuit is needed. The researchers have presented various models for this phenomenon. Corona discharge can be simply modeled as a

978-1-5090-0376-1/16 $31.00 © 2016 IEEE

resistor that variables with time [11-12]. Some authors have modeled it as a resistor in parallel with a capacitor [13-14].Ref. [13] showed this phenomenon by a resistor in parallel with a capacitor but authors in [14] believe that neither a single variable resistor nor a single variable capacitor can show the discharge behavior. The simplest circuit is a variable resistor in parallel with a capacitor. In this model, after that the voltage has reached its breakdown point, the resistor drops to a value which is a little greater than zero. Researchers of some paper have modeled this phenomenon with two back-to-back Zener diodes [15]. In [16]it is tried to provide a model for discharge in room temperature, in dry air and proposed equivalent circuit includes a variable resistor in series with variable capacitor. Various papers have measured and calculated discharge time current. According to the measured values, pulse current can be divided into three parts. In the beginning, the current has a small value. Then, it gets a maximum value rapidly and eventually decreases by decreasing pulse voltage. These three parts are common in all models [11-16]. However, according to the electrodes' geometry and the type of applied pulse, there are different forms of the rise and fall shape. Variable resistance, capacitance, and even inductance are effective on the current pulse rise and fall time and its shape. It is obvious that there is a high resistance between electrodes before producing corona; after formation of corona, the resistance will then decrease dramatically. Therefore, in this paper a variable resistor R in parallel with a capacitor C_g is used as shown in Fig. 2. This model is the simplest electrical one to show dielectric breakdown phenomenon, which has also been used in this paper.

III. PULSE PRODUCTION FOR CORONA

To produce the high-voltage pulses, various pulsed power supplies can be used. Among high-voltage pulse generators, the Marx is an appropriate candidate due to its simple structure, high flexibility, and easy use of power semiconductor switches, that has been employed by many researches. The base of producing the high voltage in Marx generator, which is indeed a voltage source, is to charge capacitors in parallel through relatively low voltage, and then arranging these capacitors in series to create high voltage in order to apply to the load [17-20].

A. Proposed generator

Since, corona has resistor-capacitor characteristic, using a current source and an open switch for making pulse in the load is more appropriate. However, to avoid using the open switches and for using typical power semiconductor switches in this paper, employing Marx generator is suggested as voltage source and current source for load supply. Fig. 3 shows the proposed structure of the generator.

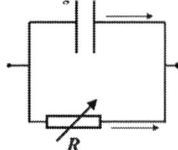

Fig. 2. Simple equivalent circuit for corona discharge

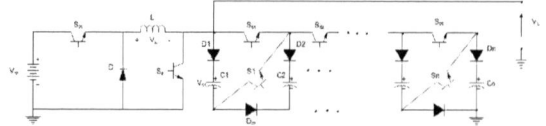

Fig. 3. Proposed generator

The inductor L is charged by the input DC voltage through switches S_{in} and S_s. The amount of current of the inductor can be controlled by the switch S_{in}. Diode D provides current flow of inductor when S_{in} goes to off. The capacitors C_1, C_2,...,C_n represent the Marx generator, and at the times the switches S_{in} and S_s are off, they will be charged through diodes D_1, D_2,...,D_n, D_{c1},D_{c2},...,D_{cn} and switches S_{c1}, S_{c2},...,S_{cn}. The load voltage is equal to the voltage of capacitor C_1 just before the load starts to discharge. The energy saved in the inductor, charges the capacitors. Finally, by changing the states of S_{c1} ~S_{cn} to off and then turning switches S_1, S_2,...,$S_n$to on, the sum of capacitor voltages will drop on the load and it starts to discharge.

The proposed generator is an appropriate candidate to produce corona in various environment with different corona chamber. It is capable of:

- Producing high voltage from low DC voltage

- Using current source along with voltage source to supply RC load

- Maintaining modular structure for voltage generator

- Keeping input and output separately when pulse is applied (because switch Sin is off)

- Being flexible in output voltage while changing switching.

Fig. 4 shows the combination of current source and a two-stage Marx generator as a simplified model which will be explained in details. The results are extensible for each unit of Marx generator.

B. Functional Modes

According to the switches status, the performance of circuit can be explained in 4 modes as following:

1) First mode: charging the inductor

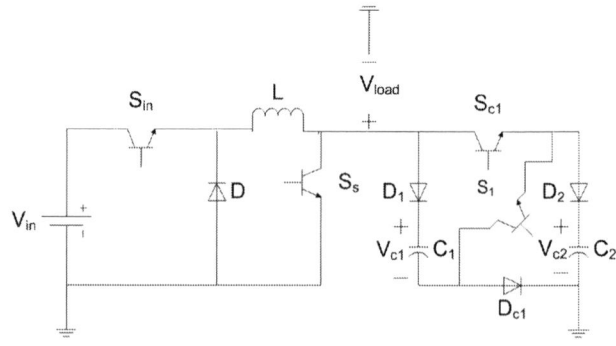

Fig. 4. Proposed generator with two stage

978-1-5090-0376-1/16 $31.00 © 2016 IEEE

If among all switches, only S_{in} and S_s are on, the input voltage applies on the inductor, and then the inductor current is increasing through the circuit shown in Fig. 5. Regardless of the voltage drop on the switches, the current of inductor is calculated as:

$$V_L = V_{in} \tag{1}$$

$$V_L = L\frac{di}{dt} = L\frac{\Delta i}{\Delta t_L} \tag{2}$$

With assumption the inductor's initial current is zero, the maximum inductor current can be expressed as follows:

$$I_{L,max} = \frac{V_L \Delta t_L}{L} \tag{3}$$

that, $I_{L,max}$ is the maximum current of inductor and Δt_L is the duration of inductor charging.

2) Second mode: Inductor current flow

When the switch S_{in} is off, diode D makes the current flows into inductor. In this step, the switch and diode's internal resistors and the voltage drop on them cause losses, and hence inductor current decreases. By ignoring these losses, it is assumed that the inductor current remains constant in this mode through the circuit shown in Fig. 6.

3) Third mode: Charging the capacitors of Marx generator

When the switch S_{c1} goes on and Switch S_s goes off, the inductor current passes through capacitors C_1 and C_2 and makes them to be charged as shown in Fig. 7. If the size of these capacitors is much larger than the equivalent capacitor of the load, the impact of the load capacitor can be ignored, and the voltage of capacitors will then be calculated as follows (it is assumed that the current is divided equally between two capacitors):

$$C_{eq} = C_1 + C_2 \tag{4}$$

$$i = C_{eq}\frac{\Delta V_{ceq}}{\Delta t_c} \tag{5}$$

That C_{eq} is the equivalent capacitor, ΔV_{ceq} is its voltage, I is its current, and Δt_c is the duration of charging. If the capacitors' initial voltage to be zero:

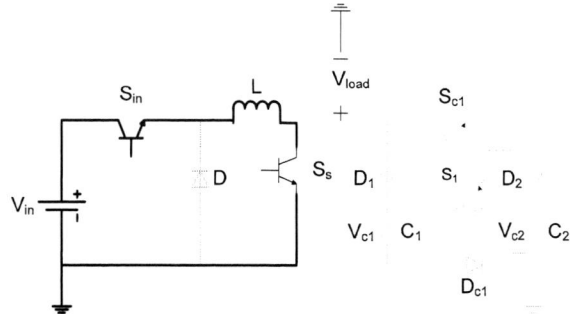

Fig. 5. The switching state of first mode

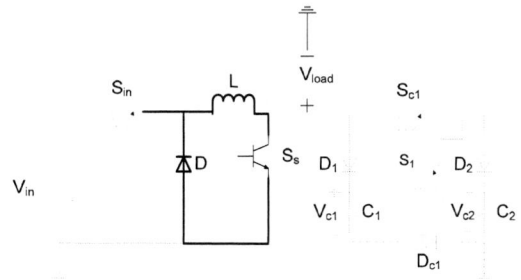

Fig. 6. The switching state of second mode

$$V_{c1} = V_{c2} = \Delta V_{ceq} \tag{6}$$

If during capacitors charging the inductor current is constant, the voltage across the capacitors can be expressed as follows:

$$V_{c1} = V_{c2} = \frac{I_{L,max}\Delta t_c}{C_{eq}} \tag{7}$$

The load resistor is large enough to be ignored in above equations.

4) Fourth mode: Corona discharge

When the switch S_{c1} goes off and switch S_1 goes on, the voltage of the capacitors are summed together:

$$V_{load} = V_{c1} + V_{c2} \tag{8}$$

If this voltage is larger than the threshold value, it starts discharging, load resistor rapidly drops, and capacitors and inductor both give energy to load simultaneously, and then is discharged. After finishing the process of load-feeding, it can be started from the beginning and the circuit can be re-charged. When a discharge finishes till the next discharge begins, a minimum amount of time is required to charging inductor and capacitors. This time is calculated as follows:

$$\Delta t_{min} = \Delta t_L + \Delta t_c \tag{9}$$

$$\Delta t_L = \frac{L I_{L,max}}{V_L} \tag{10}$$

$$\Delta t_c = \frac{C_{eq}\cdot V_{c1}}{I_{L,max}} \tag{11}$$

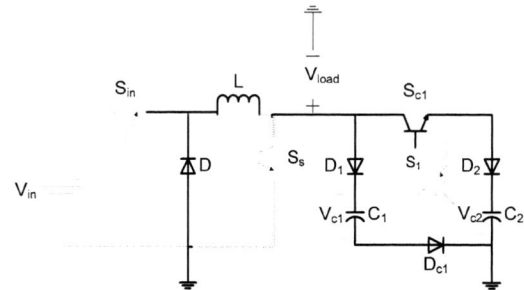

Fig. 7. The switching state of third mode

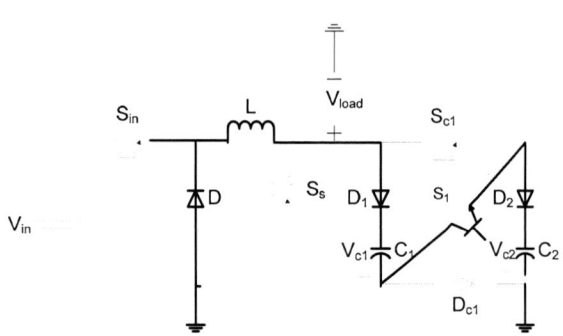

Fig. 8. The switching state of fourth mode

To evaluate the functionality of mentioned circuit, some simulations are performed in Matlab/Simulink. For simulations, the values in Table I have been considered for different elements. R_{1l} and R_{2l} are the load resistance before and after discharge and C_l denotes the capacitance of the load.

TABLE I.   VALUE USED IN SIMULATION

V_{in}	L	C_1	C_2	R_{1l}	R_{2l}	C_l
100	.1mH	10nF	10nF	1M	20	100pF

The maximum current of the inductor is 25A during charge time, and the maximum voltage of each capacitors 1500V during charge time. Fig. 9 shows the simulated inductor current.

From Fig. 9, the inductor current increases almost linearly. During the time of charging capacitors and during pulse, current decreases exponentially with two different time constants. The load voltage is equal to voltage of C_1 before the pulse. But during the pulse, load voltage is determined by the sum of capacitor voltage and inductor current. Because of high resistance of the load before pulse, the discharge in load resistor is very small until the beginning of pulse. The inductor current and capacitor voltages are almost constant at this time.

When the pulse ends, a voltage remains on the load and the current remains in the inductor. These voltage and current can be used in next cycle. Therefore, no energy is wasted in this topology. The load voltage before the pulse improves the circuit performance. It reduces the rise time and causes the corona begins faster.

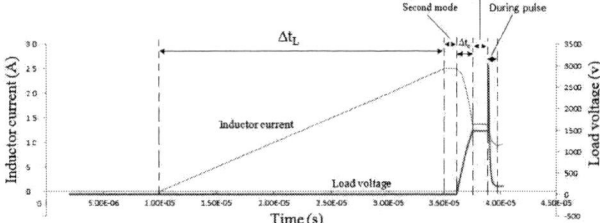

Fig. 9.   Current of inductor and output voltage during simulation

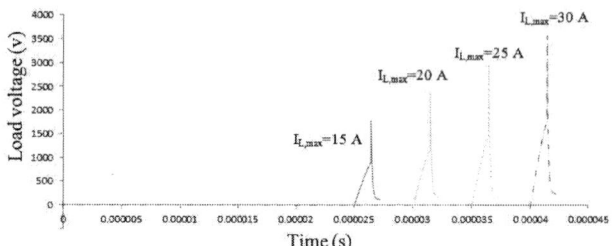

Fig. 10. Load voltage for various Inductor current

The load voltage can vary by different switching. It is possible to change the output voltage by changing the inductor charging time that change the maximum current of inductor or by changing the capacitor charging time. The simulation results of the load voltage by four different inductor currents are shown in Fig. 10. The capacitor charging time is kept constant for these simulations.

IV.   CONCLUSIONS

This paper proposes a high-voltage pulse generator for creating corona. The presented generator includes combination of Marx generator, using power semiconductor switches, with current source. The current source has been utilized to charge Marx capacitors and to feed the load. The performance of the circuit has been evaluated using Matlab/Simulink. Due to its modular structure and also change of the switching, the topology used in this paper is capable of working in a wide range of voltages which makes it useful for plasma applications.

REFERENCES

[1] M. Sato, "Environmental and biotechnological applications of high-voltage pulsed discharges in water",Plasma Sources Science and Technology,vol. 17, no. 2, 2008, p.024021

[2] B. Sun, S. Masayuki, and J. S. Clements, "Optical study of active species produced by a pulsed streamer corona discharge in water",Journal of Electrostaticsvol, vol. 39, no. 3, 1997,pp.189-202

[3] B. R. Locke, M. Sato, P. Sunka, M. R. Hoffmann, and J-S. Chang, "Electrohydraulic discharge and nonthermal plasma for water treatment", Industrial & engineering chemistry research, vol. 45, no. 3, 2006, pp. 882-905.

[4] P. Sharma, P. Bremer, I. Oey, D. Everett, "Bacterial inactivation in whole milk using pulsed electric field processing", International Dairy Journal, vol. 35, no.1, 2014, pp. 49-56

[5] A. Guionet, V. Joubert, D. Packan, C. Cheype, J□P. Garnier, F. David, C. Zaepffel, R. Leroux, J. Teissié, and V. Blanckaert, "Effect of nanosecond pulsed electric field on Escherichia coli in water: inactivation and impact on protein changes",Journal of applied microbiology,vol. 117, no. 3, 2014, pp.721-728.

[6] M. Abdel-Salam, High-Voltage Engineering: Theory and Practice, Revised and Expanded. CRC Press, 2000, pp:175-177

[7] A. Abou-Ghazala, S. Katsuki, K. H. Schoenbach, F. C. Dobbs, and K. R. Moreira, "Bacterial decontamination of water by means of pulsed-corona discharges", IEEE Trans. on Plasma Science vol. 30, no. 4, 2002, pp.1449-1453.

[8] P. Baroch, V. Anita, N. Saito, and O. Takai, "Bipolar pulsed electrical discharge for decomposition of organic compounds in water", Journal of Electrostatics, vol. 66, no. 5, 2008, pp. 294-299.

[9] P. Lukes, E. Dolezalova, I. Sisrova, and M. Clupek, "Aqueous-phase chemistry and bactericidal effects from an air discharge plasma in contact with water: evidence for the formation of peroxynitrite through a pseudo-second-order post-discharge reaction of H2O2 and HNO2.", Plasma Sources Science and Technology, vol. 23, no. 1, 2014: 015019.

[10] J. Song, Y. Wen, and K. Liu, "Investigation of pulsed dielectric barrier discharge system on water treatment by liquid droplets in air.", IEEE Trans. on Dielectrics and Electrical Insulation, vol. 22, no. 4, 2015, pp. 1866-1871.

[11] Y. D. Korolev, O. B. Frants, V. G. Geyman, V. S. Kasyanov, and N. V. Landl, "Transient processes during formation of a steady-state glow discharge in air.", IEEE Trans. on Plasma Science, vol. 40, no. 11, 2012, pp. 2951-2960.

[12] M. Simek, and M. Clupek, "Efficiency of ozone production by pulsed positive corona discharge in synthetic air.", Journal of Physics D: Applied Physics, vol. 35, no. 11, 2002, pp. 1171.

[13] Y. Lee, W. Jung, Y. Choi, J. Oh, S. Jang, Y. Son, M. Cho et al, "Application of pulsed corona induced plasma chemical process to an industrial incinerator.", Environmental science & technology, vol. 37, no. 11, 2003, pp. 2563-2567.

[14] A. V. Pipa, J. Koskulics, R. Brandenburg, and T. Hoder, "The simplest equivalent circuit of a pulsed dielectric barrier discharge and the determination of the gas gap charge transfer.", Review of Scientific Instruments, vol. 83, no. 11, 2012, pp. 115112.

[15] Q. Hu, T. Li, L. Shu, X. Jiang, and B. Luo, "Dynamic characteristics of the corona discharge during the energised icing process of conductors.", Generation, Transmission & Distribution, IET, vol. 7, no. 4, 2013, pp. 366-373.

[16] D. Raouti, S. Flazi, and D. Benyoucef, "Electrical Modelling of a Positive Point to Plane Corona Discharge at Atmospheric Pressure.", Contributions to Plasma Physics, vol. 54, no. 10, 2014, pp. 851-858.

[17] S. Zabihi, F. Zare, G. Ledwich, A. Ghosh, and H. Akiyama. "A new family of Marx generator based on resonant converter.", In Energy Conversion Congress and Exposition (ECCE), 2010 IEEE, pp. 3841-3846. IEEE, 2010.

[18] A. Caiafa, V. B. Neculaes, A. L. Garner, Y. Jiang, S. Klopman, A. Torres, and N. LaPlante, "Compact solid state pulsed power architecture for biomedical workflows: Modular topology, programmable pulse output and experimental validation on Ex vivo platelet activation.", IEEE International Conference In Power Modulator and High Voltage (IPMHVC), pp. 35-40, 2014.

[19] H. Shi, Y. Lu, T. Gu, J. Qiu, and K. Liu, "High-voltage pulse waveform modulator based on solid-state Marx generator.", IEEE Trans. on Dielectrics and Electrical Insulation, vol. 22, no. 4, 2015, pp. 1983-1990.

[20] Y. Wang, Y. Lu, J. Qiu, and K. Liu, "Repetitive high voltage all-solid-state Marx generator for dielectric barrier discharge pulsed plasma.", IEEE International Conference In Power Modulator and High Voltage (IPMHVC), pp. 648-651. IEEE, 2014.

978-1-5090-0376-1/16 $31.00 © 2016 IEEE

7th Power Electronics, Drive Systems & Technologies Conference (PEDSTC 2016)
16-18 Feb. 2016, Iran University of Science and Technology, Tehran, Iran

A Review of Predictive Control Techniques for Matrix Converters - Part I

M. Rivera, P. Wheeler, A. Olloqui, and D.A. Khaburi

Abstract—The predictive strategy is a promising alternative to control matrix converters and due to its simplicity and flexibility to include additional aspects in the control, its implementation can be extended to different topologies and industrial applications. This paper presents an overview of the predictive control principles applied to matrix converters and other derived topologies including cascade systems in different industrial applications, such as renewable energy and multi-drive systems.

Index Terms—Matrix Converter, Direct Matrix Converter, Indirect Matrix Converter, Power Control, Predictive Control, Modulation Schemes

NOMENCLATURE

$\mathbf{i}_s$	Source current	$[i_{sA}\ i_{sB}\ i_{sC}]^T$
$\mathbf{v}_s$	Source voltage	$[v_{sA}\ v_{sB}\ v_{sC}]^T$
$\mathbf{i}_i$	Input current	$[i_A\ i_B\ i_C]^T$
$\mathbf{v}_i$	Input voltage	$[v_A\ v_B\ v_C]^T$
i_{dc}	dc-link current	
v_{dc}	dc-link voltage	
$\mathbf{i}_o$	Output current	$[i_a\ i_b\ i_c]^T$
$\mathbf{v}_o$	Output voltage	$[v_a\ v_b\ v_c]^T$
$\mathbf{i}_s^*$	Source current reference	$[i_{sA}^*\ i_{sB}^*\ i_{sC}^*]^T$
$\mathbf{i}_o^*$	Output current reference	$[i_a^*\ i_b^*\ i_c^*]^T$
$\mathbf{v}_o^*$	Output voltage reference	$[v_a^*\ v_b^*\ v_c^*]^T$
C_f	Input filter capacitor	
L_f	Input filter inductor	
R_f	Input filter resistor	
R_L	Load resistance	
L_L	Load inductance	

I. INTRODUCTION

The matrix converter (MC) is a simple and compact power circuit that directly connects the ac-source with any arbitrary ac-load without the need for large storage elements, making this topology suitable for many applications where weight and size are important issues. With this converter, generation of output voltage with different amplitude and frequency, sinusoidal input and output current waveforms, as well as operation with unity displacement power factor and regenerative capability are made possible. One challenge of MCs used to be the commutation of bidirectional switches but this issue has been solved with multi-step commutation techniques and the use of new technologies in power elements. Due to these characteristics, in recent years MCs have shown continuous and fast development related to the development of new topologies and control methods, including industrial applications with standard units for high and medium voltage using cascade connections [1]–[3].

Different modulation and control methods for MCs are found in the literature and also in the industry. As studied in [4], the most used techniques nowadays are Venturini, carrier-based pulse width modulation (CB-PWM), space vector modulation (SVM) and direct torque control (DTC). Other methods that have been applied to MCs in specific applications are fuzzy control, neural networks and genetic algorithms. Predictive control has shown to be a very interesting alternative for control MCs, specially because the use of the discrete nature of power converters and its simplicity for implementation and intuitive approach. The main objective of this work is to provide a review of the main contributions and trends of predictive control in MCs: the topologies, the different control strategies, and the applications where it has been implemented. The second part of this paper presents an overview of different control strategies and applications for MCs where predictive control techniques are applied.

II. MATHEMATICAL MODEL OF THE MC

The power topology of the MC is presented in Fig. 1. It consists of bidirectional switches to directly connect the input side with the load side without any intermediate dc-link storage element. An input filter is connected at the input side of the converter with two purposes [1]:

- To avoid over-voltage due to short-circuiting the impedance of the power supply, by cause of the fast commutation of currents $\mathbf{i}_i$.
- To eliminate high-frequency harmonics in the input currents $\mathbf{i}_s$.

Such as in each converter, the operation of the direct MC (DMC) is restricted to some operation constraints: the load current cannot be interrupted abruptly, due to the inductive nature of the load, and the operation of the

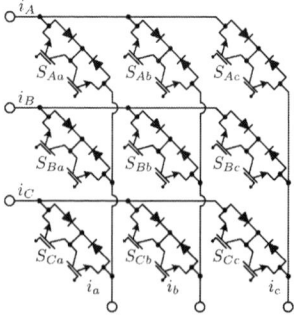

Fig. 1. Power topology of the conventional direct matrix converter.

978-1-5090-0376-1/16 $31.00 © 2016 IEEE

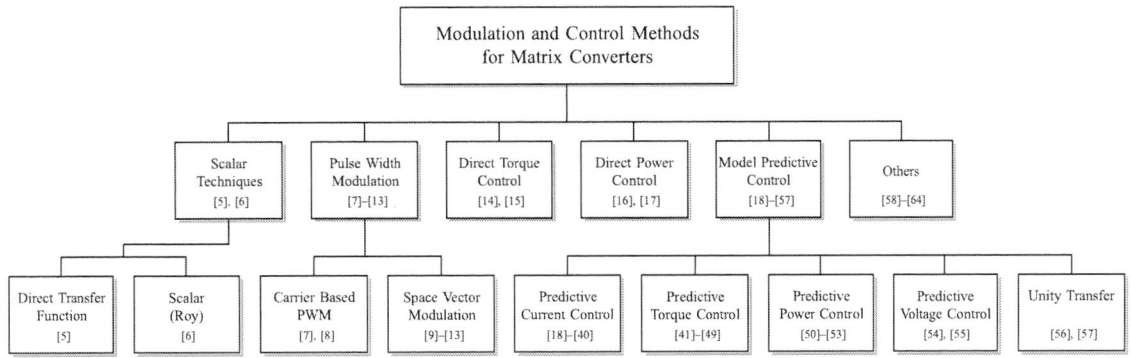

Fig. 2. Summary of modulation and control methods for matrix converters.

switches cannot short-circuit two input lines, owing to the presence of capacitors in the input filter. These restrictions can be expressed by:

$$S_{Ay} + S_{By} + S_{Cy} = 1, \quad \forall \, y = a, \, b, \, c \quad (1)$$

The relations between the input and output variables of the MC are given by:

$$\mathbf{v}_o = \mathbf{T}(S_{ij}) \, \mathbf{v}_i \quad (2)$$

$$\mathbf{i}_i = \mathbf{T}(S_{ij})^T \mathbf{i}_o \quad (3)$$

where T is the instantaneous transfer matrix defined as:

$$\mathbf{T}(S_{ij}) = \begin{bmatrix} S_{Aa} & S_{Ba} & S_{Ca} \\ S_{Ab} & S_{Bb} & S_{Cb} \\ S_{Ac} & S_{Bc} & S_{Cc} \end{bmatrix} \quad (4)$$

Equations (3) and (4) are the basis of all modulation and control methods, which consist of selecting appropriate combinations of on and off switches to achieve the desired output voltages.

III. CLASSICAL MODULATION AND CONTROL TECHNIQUES FOR MCs

Fig. 2 presents a summary of the more relevant modulation and control methods applied to MCs. As described in [4], these methods have different explanations and different levels of complexity, with a dynamic behavior acceptable for various applications. The first methods applied to MCs where Venturini and Roy's techniques which present a complex mathematical development [5], [6]. The pulse width modulation (PWM) technique is the simplest approach to modulate MCs [7], [8], [12], [13]. The space vector modulation (SVM) [9]–[13] and direct torque control (DTC) [14], [15] are the most robust and used techniques for drives control in industrial applications but they are complex and not intuitive.

More advanced techniques, such as model predictive control, have recently been introduced to simplify the complexity of MC control with reliable and fast performance in both steady and transient states [18]–[57]. Other techniques that have been applied to MCs are direct power control [16], [17], fuzzy control, neural networks, genetic algorithms, among others [58]–[64].

IV. PRINCIPLE OF PREDICTIVE CONTROL IN MCs

Model Predictive Control (MPC) is a relatively new control technique applied for the control of power electronic converters. This method utilizes the mathematical model of the controlled system in order to predict, at each sampling instant k, its behavior at $k+1$. For selecting an optimal state of the power converter, a cost function is defined. This function is composed of several constrains and control conditions. It usually contain the differences between the reference values and the predicted values of the variables being controlled. Many other components of this function represent specific constrains, such as limitation of the switching frequency, or other nonlinearities. As an example the predictive current control (PCC) for the DMC is presented in this section.

The PCC scheme is shown in Fig. 3. It shows the switching state selection of the converter, which provides the controlled variables to the nearest respective references at the end of the sampling period. This control approach utilizes the converter and load models in order to predict the future value of the currents. A simple and representative model of the load can be expressed as:

$$\frac{d\mathbf{i}_o}{dt} = \frac{1}{L_o}\mathbf{v}_o - \frac{R_o}{L_o}\mathbf{i}_o \quad (5)$$

knowing the nature of the load (first order in our case), a first order discrete approximation allows predicting the future load current:

$$\mathbf{i}_o(k+1) = \frac{T_s \mathbf{v}_o(k+1) + L_o \mathbf{i}_o(k)}{L_o + R_o T_s} \quad (6)$$

where T_s corresponds to the sampling time.

A cost function is defined in order to determine the error between the current references $\mathbf{i}_o^*$, and their respective current predictions $\mathbf{i}_o^p$, given by:

$$g(k+1) = |i_a^* - i_a^p| + |i_b^* - i_b^p| + |i_c^* - i_c^p| \quad (7)$$

As reported in [26]–[29], [32], [34]–[37], [65] this strategy performs well with a very good behavior in both steady and transient state showing to be a very good alternative to classical control strategies.

978-1-5090-0376-1/16 $31.00 © 2016 IEEE

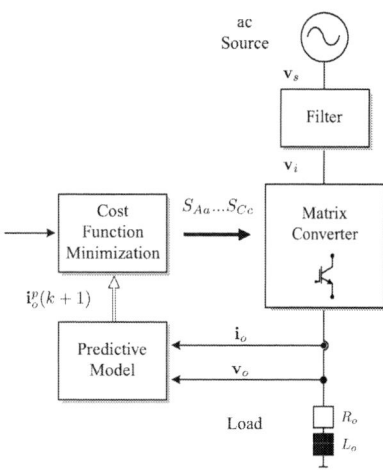

Fig. 3. Block diagram of the predictive current control strategy.

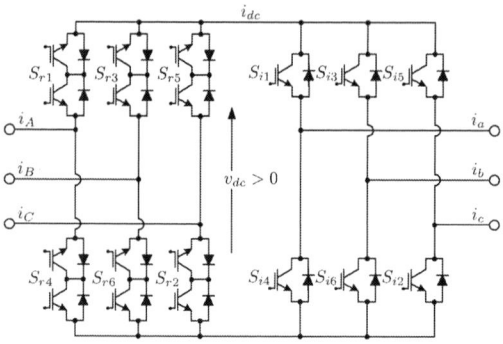

Fig. 4. Conventional indirect matrix converter (common collector).

V. PREDICTIVE CONTROL IN DIFFERENT TOPOLOGIES OF MATRIX CONVERTERS

As presented in [2], there are several topologies of MCs. The main differences between them are given by the number of switches, operation constraints and applications. The most important advantages of these extensions are: the increment of output voltage control range and the reduction of the switching frequency harmonics, losses and common mode voltage. The most common and used topology where predictive control has been implemented is the DMC, shown in Fig. 1. One challenge of this converter is the safe commutation of the nine switches (eighteen IGBTs in total) and its modulation, which is very complex. Additionally, in the operation of a DMC with predictive control, a critical issue has been the high sampling frequency needed, but nowadays this problem has been solved due to the technological progress in fast and powerful microcontrollers. Many authors have used this converter in [19], [22], [24], [32]–[35], [37]–[39], [42], [43], [47], [48], [50], [54], [56], [65]–[67] to apply different techniques for a large number of applications. For this converter, twenty seven different switching states must be evaluated every sampling instant in order to select the one that minimizes the cost function. The current control on the output side of the converter is a very well studied issue, specially for motor drive and grid interconnection applications. Some aspects considered in the control of the DMC are the amplitude and phase control of the input currents in order to operate with unity, capacitive or inductive power factor. Another relevant issue that has been objective for study with predictive control, in consideration for the safe operation of the DMC, is the reduction of the distortion in the input currents produced by input filter resonances due to the commutation and several perturbations in the ac-supply. Due to the large number of power semiconductors of the DMC, some authors have studied also the increment of the efficiency of the converter by reducing the switching losses and frequency. The most important contribution in all the

works done with predictive control, is the simplicity for the safe operation of the DMC, eliminating complex transformations and modulations which are required in PWM and SVM techniques and also the capability to use all the available switching vectors of the converter which is not possible with classical techniques, making this topology an effective alternative when size and weight are important requirements. In [54], [55], are proposed two predictive controllers for a single-phase MC (SPMC) where the topology, along with a resonant circuit, a HF-transformer and a diode bridge, is used as a dc power supply for high-power radio frequency (RF) applications, mainly used in some industrial applications such as microwaves for mineral extraction, medical imaging, television transmission and also research applications (mainly particle physics research). Additionally, this converter is meant to be used in cascaded configurations for high power applications which are connected to the ac source by a common multi-pulse transformer [3]. In [68], a recent work has been published to control a SPMC where is also discussed a possible use of this type of converters in grid interconnection systems, where a medium frequency transformer is required to isolate the grids. In [18], an interesting application of predictive control is found for a three-to-five leg DMC, where complex modulations and three-dimensional transformations are avoided with only a predictive model of the load and source currents. The main challenge in the implementation of predictive control in this converter is the large number of available switching states (243 valid switching states) that must be taking into consideration which requires a high computational cost. The indirect matrix converter (IMC) shown in Fig. 4 is other topology where important contributions of predictive control have been done, such as reported in [20], [23], [25]–[29], [31], [36], [40], [44]–[46], [62]. The IMC, in contrast to the DMC, presents a more simple modulation and commutation known as zero dc-link current strategy [3], [40], which allows to reduce the commutation losses and thus increase the efficiency of the converter. The main challenge in this topology is to ensure a positive dc-link voltage while working with a unity displacement power factor at the input side [23], [26]–[29], [31], [36], [40], [44], [46]. For this converter there are seventy-two valid switching states to be evaluated in the cost function each sampling time, nine given by the rectifier side and eight by the inverter side. But as only a positive dc-link voltage is allowed at any time, the number of valid switching states in the rectifier side that can be applied at any specific time are reduced to only three, thus, the total number of valid switching states that are evaluated in the cost function are reduced to twenty-four. Similarly to the DMC, predictive control in an IMC has been implemented for motor drives in military, aerospace and renewable energy applications where size and weight are relevant issues [25], [62]. Moreover the mitigation of resonances on the input filter due to perturbations of the AC source and due to the commutations of the switches has been considered [23], [26], [28], [31]. The utilization of an IMC as a shunt active power filter operating with a predictive current control strategy was proposed in [20], where the output reference currents are obtained using P-Q theory. An important aspect observed in [20] was the fast dynamic response of the predictive controller, allowing to obtain almost sinusoidal source currents, eliminating the effect of non-linear load currents. As reviewed in [2], [69], there are different topologies derived from the IMC with reduced number of switches and switching states for specific applications. Predictive control has also been applied to these topologies such as reported in [21], [30], [41], [51], [57], [70], where are used a sparse matrix converter (SMC) (Fig. 5), ultra sparse matrix converter (USMC), or with extended number of switches as the IMC with four and six legs, and a hybrid indirect matrix converter (HIMC) (Fig. 6), respectively. As shown in Fig. 5, the SMC utilizes twelve IGBTs and thirty diodes. This reduction in terms of switches allow a more simple topology functionally equivalent to the conventional IMC. Again, the main challenges for the safe operation of this converter are to generate maximum voltage in the dc-link while maintaining sinusoidal currents and unity displacement input power factor [21]. Special attention must be taken into consideration while working with this converter because, as similar to the IMC, it is necessary synchronize the commutation of the rectifier and inverter switches. The inverter must be switched into a free-wheeling state in order to commutate the rectifier at zero current in the dc-link. In [21], the predictive technique is mixed with a space vector pulse with modulation (SVPWM) allowing the operation with fixed switching frequency. In this particular application only the load current is operated by the predictive controller avoiding the use of linear current controllers. As only the load currents are controlled by the predictive algorithm only eight valid switching states are evaluated in the cost function and thus the optimal selected switching vector generates the reference voltage for the modulator. The PWM technique is in charge to ensure unity power factor operation on the input side while following the voltage reference imposed by the predictive controller. A modified topology of an ultra sparse matrix converter (USMC) is proposed in [51]. The main difference between this new

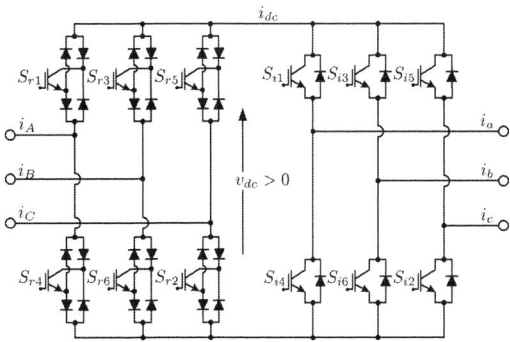

Fig. 5. Sparse matrix converter.

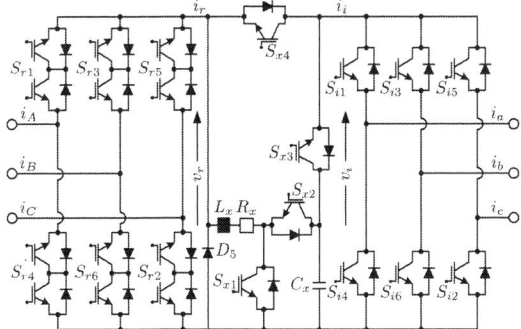

Fig. 6. Hybrid indirect matrix converter.

topology and the classical USMC is that in the proposed architecture, the output stage has bidirectional switches, allowing the transfer of energy from the dc-link to the load and viceversa. Here, the authors used predictive control to handle the electrical power of the micro-turbine grid by the manipulation of the fuel flow, being very effective for both small step changes and large random changes. In [70] predictive control has been applied to a four-leg indirect matrix converter (4L-IMC). The control objectives are load current control and minimization of the instantaneous reactive input power or also the load voltage control by including an LC filter in the output side [30]. As in the previously analyzed indirect matrix converter topologies, the main challenge is given by the safe commutation of the switches in both rectifier and inverter stages. Additionally, the large number of IGBTs and thus the large number of available switching states, make this converter difficult to control requiring 3D modulation techniques which are very complex and non intuitive to understand. One relevant aspect of predictive control is its simplicity and intuitive concept allowing a very simple control for this kind of power converter topologies. As it has been demonstrated in these papers, the predictive strategy performs well with a good performance in both steady and transient states, using only the mathematical model of the converter and load, with all the control objectives merged in only one cost function. In [41] a predictive control for a multi-drive system has been proposed to control two induction machines feeded

978-1-5090-0376-1/16 $31.00 © 2016 IEEE

by a six-leg IMC. Multi-drive systems are very useful nowadays in aerospace applications, exploration and military vehicles, tractors, mining trucks, conveyor belts, among others. Traditionally these systems have the same number of converters than loads (generally ac motors), but recently it has been proposed the use of multi-phase matrix converters to feed several induction machines (IM) with a single converter and thus, to reduce the weight and size of the multi-drive system. The main challenges of multi-phase matrix converters are the increment of the available switching states that can be applied to the converter and must be taking into consideration every sampling time for the optimization algorithm. But also is very relevant to ensure a correct synchronization in the commutation of the switches in order to operate the converter correctly. It is well known that a high sampling frequency is required for a good performance of the predictive control in its implementation. But, is necessary to decrease the sampling frequency in order to be able to evaluate all the valid switching states at every sampling instant which make worse the control behavior. In order to solve this aspect, some recent studies have considered some redundances in the model and switching states, in order to reduce the number of states under evaluation. Despite this, in [41] it has been demonstrated the feasibility for implementation of this control strategy, obtaining an effective control of both IMs operating at the same speed but at different load torque. Also recent work on IMC-based topologies redefine the valid states of the rectifier side whether the maximum dc-link voltage at any specific time is needed or not, reducing to two thirds the total number of valid switching states that are evaluated in the cost function. As highlighted in this paper, the MC presents different advantages in term of size and weight, allowing the operation with sinusoidal source and load currents, regeneration capability among others, but despite all the rewards, there exists two main disadvantages in the MC topology (as long as the maximum voltage transfer ratio is not reached [71]): the output voltage is limited to 86% of the input, and secondly, any perturbation in the supply deteriorates the quality of the load side, due to the absence of storage elements.

In order to solve this issue, in [57] has been proposed the implementation of a hybrid power converter which connects an auxiliary voltage source in the dc-link of the IMC, obtaining unity voltage transfer capability even under severe distortions in the source voltage as shown in Fig. 6. The main challenges of this topology are the control of the input side and the commutation of the switches in the rectifier side as well as the control of the auxiliary circuit connected to the dc-link and finally the control on the output side of the converter.

A predictive current control strategy is proposed in this work for the auxiliary voltage source, where the current reference is given by a PI linear controller and the predictive control generates the duty cycle for the pulse-width modulator. With this predictive controller and the proposed architecture is ensured balanced power in the

converter and unity voltage ratio.

In summary, in all the previous cases, predictive control demonstrated a very good performance, being a very simple method for implementation. Predictive control has been implemented in several topologies of MCs to overcome disadvantages of the MC technology over the two-level voltage dc-link back-to-back converter (V-BBC) in specific applications, demonstrating to be a very flexible and useful technique, introducing a new and promising alternative for electric power conversion for low-voltage and low-power (≤ 100kW) [71].

VI. Conclusions

The main contribution of this paper has been to present an overview of different topologies of matrix converters where predictive control techniques are applied. A detailed description about the constraints, limitations and challenges of each topology for the implementation of predictive control have been presented.

Based on the review given by the authors, predictive control has a very high impact in the control of matrix converters, due its simplicity and intuitive approach, making this control strategy a real alternative in power electronics. Each power converter topology has its own requirements and difficulties, specially those cases where there are multiple control objectives for the same load and multiple motors driven by the same power converter. The main challenge of multi-phase matrix converters is the increment of the available switching states. On one hand, high sampling frequency is required to improve the performance. On the other hand, it requires enough time to evaluate all possible states. Despite the difficulties, recent studies demonstrated the feasibility for implementation of this control strategy in multi-drive systems, obtaining an effective control of two induction motors operating at the same speed but at different load torque.

Acknowledgments

This publication was made possible by the Newton Picarte Project EPSRC: EP/N004043/1: New Configurations of Power Converters for Grid Interconnection Systems / CONICYT DPI20140007 and British Council through the Institutional Skills Development Newton Picarte Project ISCL 2015006.

References

[1] P. W. Wheeler, J. Rodriguez, J. C. Clare, L. Empringham, and A. Weinstein, "Matrix converters: a technology review," *Industrial Electronics, IEEE Transactions on*, vol. 49, no. 2, pp. 276–288, 2002.

[2] T. Friedli and J. W. Kolar, "Comprehensive comparison of three-phase ac-ac matrix converter and voltage dc-link back-to-back converter systems," in *Power Electronics Conference (IPEC), 2010 International*, 21-24 June 2010 2010, pp. 2789–2798.

[3] J. W. Kolar, T. Friedli, J. Rodriguez, and P. W. Wheeler, "Review of three-phase pwm ac-ac converter topologies," *Industrial Electronics, IEEE Transactions on*, vol. 58, no. 11, pp. 4988–5006, 2011.

[4] J. Rodriguez, M. Rivera, J. W. Kolar, and P. W. Wheeler, "A review of control and modulation methods for matrix converters," *Industrial Electronics, IEEE Transactions on*, vol. 59, no. 1, pp. 58–70, 2012.

[5] S. Lopez Arevalo, P. Zanchetta, P. W. Wheeler, A. Trentin, and L. Empringham, "Control and implementation of a matrix-converter-based ac ground power-supply unit for aircraft servicing," *Industrial Electronics, IEEE Transactions on*, vol. 57, no. 6, pp. 2076–2084, 2010.

[6] G. Roy and G. E. April, "Cycloconverter operation under a new scalar control algorithm," in *Power Electronics Specialists Conference, 1989. PESC '89 Record., 20th Annual IEEE*, 26-29 Jun 1989 1989, pp. 368–375 vol.1.

[7] T. D. Nguyen and L. Hong-Hee, "Generalized carrier-based pwm method for indirect matrix converters," in *Sustainable Energy Technologies (ICSET), 2012 IEEE Third International Conference on*, 24-27 Sept. 2012 2012, pp. 223–228.

[8] P. G. Potamianos, E. D. Mitronikas, and A. N. Safacas, "Open-circuit fault diagnosis for matrix converter drives and remedial operation using carrier-based modulation methods," *Industrial Electronics, IEEE Transactions on*, vol. 61, no. 1, pp. 531–545, 2014.

[9] M. Jussila and H. Tuusa, "Comparison of simple control strategies of space-vector modulated indirect matrix converter under distorted supply voltage," *Power Electronics, IEEE Transactions on*, vol. 22, no. 1, pp. 139–148, 2007.

[10] H. M. Nguyen, L. Hong-Hee, and C. Tae-Won, "Input power factor compensation algorithms using a new direct-svm method for matrix converter," *Industrial Electronics, IEEE Transactions on*, vol. 58, no. 1, pp. 232–243, 2011.

[11] A. M. Bozorgi, M. Monfared, and H. R. Mashhadi, "Optimum switching pattern of matrix converter space vector modulation," in *Computer and Knowledge Engineering (ICCKE), 2012 2nd International eConference on*, 18-19 Oct. 2012 2012, pp. 89–93.

[12] T. D. Nguyen and H. H. Lee, "A new svm method for an indirect matrix converter with common-mode voltage reduction," *Industrial Informatics, IEEE Transactions on*, vol. PP, no. 99, pp. 1–1, 2013.

[13] W. Xingwei, L. Hua, S. Hongwu, and F. Bo, "A research on space vector modulation strategy for matrix converter under abnormal input-voltage conditions," *Industrial Electronics, IEEE Transactions on*, vol. 59, no. 1, pp. 93–104, 2012.

[14] D. Xiao and M. F. Rahman, "A novel sensorless direct torque control for matrix converter-fed ipm synchronous machine," in *Electrical Machines and Systems (ICEMS), 2011 International Conference on*, 20-23 Aug. 2011 2011, pp. 1–6.

[15] A. Yousefi-Talouki, S. A. Gholamian, M. Yousefi-Talouki, R. Ilka, and A. Radan, "Harmonic elimination in switching table-based direct torque control of five-phase pmsm using matrix converter," in *Humanities, Science and Engineering Research (SHUSER), 2012 IEEE Symposium on*, 24-27 June 2012 2012, pp. 777–782.

[16] T. Noguchi and A. Sato, "Direct power control based indirect ac to ac power conversion system," in *Power Electronics and Applications, 2009. EPE '09. 13th European Conference on*, 8-10 Sept. 2009 2009, pp. 1–8.

[17] J. Monteiro, J. F. Silva, S. F. Pinto, and J. Palma, "Matrix converter-based unified power-flow controllers: Advanced direct power control method," *Power Delivery, IEEE Transactions on*, vol. 26, no. 1, pp. 420–430, 2011.

[18] S. K. M. Ahmed, A. Iqbal, H. Abu-Rub, and P. Cortes, "Model predictive control of a three-to-five phase matrix converter," in *Predictive Control of Electrical Drives and Power Electronics (PRECEDE), 2011 Workshop on*, 14-15 Oct. 2011 2011, pp. 36–39.

[19] J. D. Dasika and M. Saeedifard, "An on-line fault detection and a post-fault strategy to improve the reliability of matrix converters," in *Applied Power Electronics Conference and Exposition (APEC), 2013 Twenty-Eighth Annual IEEE*, 17-21 March 2013 2013, pp. 1185–1191.

[20] A. A. Heris, E. Babaei, and S. H. Hosseini, "A new shunt active power filter based on indirect matrix converter," in *Electrical Engineering (ICEE), 2012 20th Iranian Conference on*, 15-17 May 2012 2012, pp. 581–586.

[21] E. Lee, L. Kyo-Beum, L. Jae-Sik, L. Youngil, and S. Joong-Ho, "Predictive current control for a sparse matrix converter," in *Power Electronics and Motion Control Conference (IPEMC), 2012 7th International*, vol. 1, 2-5 June 2012 2012, pp. 36–40.

[22] F. Morel, J. M. Retif, L.-S. Xuefang, B. Allard, and P. Bevilacqua, "A predictive control for a matrix converter-fed permanent magnet synchronous machine," in *Power Electronics Specialists Conference, 2008. PESC 2008. IEEE*, 15-19 June 2008 2008, pp. 15–21.

[23] M. Rivera, P. Correa, J. Rodriguez, I. Lizama, and J. Espinoza, "Predictive control of the indirect matrix converter with active damping," in *Power Electronics and Motion Control Conference, 2009. IPEMC '09. IEEE 6th International*, 17-20 May 2009 2009, pp. 1738–1744.

[24] M. Rivera, P. Correa, J. Rodriguez, I. Lizama, J. Espinoza, and C. Rojas, "Predictive control with active damping in a direct matrix converter," in *Energy Conversion Congress and Exposition, 2009. ECCE 2009. IEEE*, 20-24 Sept. 2009 2009, pp. 3057–3062.

[25] M. Rivera, J. L. Elizondo, M. E. Macias, O. M. Probst, O. M. Micheloud, J. Rodriguez, C. Rojas, and A. Wilson, "Model predictive control of a doubly fed induction generator with an indirect matrix converter," in *IECON 2010 - 36th Annual Conference on IEEE Industrial Electronics Society*, 7-10 Nov. 2010 2010, pp. 2959–2965.

[26] M. Rivera, J. Rodriguez, W. Bin, J. R. Espinoza, and C. A. Rojas, "Current control for an indirect matrix converter with filter resonance mitigation," *Industrial Electronics, IEEE Transactions on*, vol. 59, no. 1, pp. 71–79, 2012.

[27] M. Rivera, J. Rodriguez, J. Espinoza, and W. Bin, "Reduction of common-mode voltage in an indirect matrix converter with imposed sinusoidal input/output waveforms," in *IECON 2012 - 38th Annual Conference on IEEE Industrial Electronics Society*, 25-28 Oct. 2012 2012, pp. 6105–6110.

[28] M. Rivera, J. Rodriguez, J. R. Espinoza, and H. Abu-Rub, "Instantaneous reactive power minimization and current control for an indirect matrix converter under a distorted ac supply," *Industrial Informatics, IEEE Transactions on*, vol. 8, no. 3, pp. 482–490, 2012.

[29] M. Rivera, J. Rodriguez, J. R. Espinoza, T. Friedli, J. W. Kolar, A. Wilson, and C. A. Rojas, "Imposed sinusoidal source and load currents for an indirect matrix converter," *Industrial Electronics, IEEE Transactions on*, vol. 59, no. 9, pp. 3427–3435, 2012.

[30] M. Rivera, J. Rodriguez, C. Garcia, R. Pena, and J. Espinoza, "A simple predictive voltage control method with unity displacement power factor for four-leg indirect matrix converters," in *Power Electronics and Motion Control Conference (EPE/PEMC), 2012 15th International*, 4-6 Sept. 2012 2012, pp. DS2c.5–1–DS2c.5–6.

[31] M. Rivera, J. Rodriguez, M. Lopez, and J. Espinoza, "Control of an induction machine fed by an indirect matrix converter with unity displacement power factor operating with an unbalanced ac-supply," in *Power Electronics and Motion Control Conference (EPE/PEMC), 2012 15th International*, 4-6 Sept. 2012 2012, pp. DS2c.4–1–DS2c.4–8.

[32] M. Rivera, J. Rodriguez, P. W. Wheeler, C. A. Rojas, A. Wilson, and J. R. Espinoza, "Control of a matrix converter with imposed sinusoidal source currents," *Industrial Electronics, IEEE Transactions on*, vol. 59, no. 4, pp. 1939–1949, 2012.

[33] M. Rivera, C. Rojas, Rodri, x, J. guez, P. Wheeler, W. Bin, and J. R. Espinoza, "Predictive current control with input filter resonance mitigation for a direct matrix converter," *Power Electronics, IEEE Transactions on*, vol. 26, no. 10, pp. 2794–2803, 2011.

[34] M. Rivera, C. Rojas, J. Rodriguez, and J. Espinoza, "Methods of source current reference generation for predictive control in a direct matrix converter," *Power Electronics, IET*, vol. 6, no. 5, 2013.

[35] J. Rodriguez, J. Espinoza, M. Rivera, F. Villarroel, and C. Rojas, "Predictive control of source and load currents in a direct matrix converter," in *Industrial Technology (ICIT), 2010 IEEE International Conference on*, 14-17 March 2010 2010, pp. 1826–1831.

[36] J. Rodriguez, J. Kolar, J. Espinoza, M. Rivera, and C. Rojas, "Predictive current control with reactive power minimization in an indirect matrix converter," in *Industrial Technology (ICIT), 2010 IEEE International Conference on*, 14-17 March 2010 2010, pp. 1839–1844.

[37] C. Rojas, M. Rivera, J. Rodriguez, A. Wilson, J. Espinoza, F. Villarroel, and P. Wheeler, "Predictive control of a direct matrix converter operating under an unbalanced ac source," in *Industrial Electronics (ISIE), 2010 IEEE International Symposium on*, 4-7 July 2010 2010, pp. 3159–3164.

[38] R. Vargas, U. Ammann, and J. Rodriguez, "Predictive approach to increase efficiency and reduce switching losses on matrix converters," *Power Electronics, IEEE Transactions on*, vol. 24, no. 4, pp. 894–902, 2009.

[39] R. Vargas, J. Rodriguez, U. Ammann, and P. W. Wheeler, "Predictive current control of an induction machine fed by a matrix

converter with reactive power control," *Industrial Electronics, IEEE Transactions on*, vol. 55, no. 12, pp. 4362–4371, 2008.

[40] P. Correa, J. Rodriguez, M. Rivera, J. R. Espinoza, and J. W. Kolar, "Predictive control of an indirect matrix converter," *Ieee Transactions on Industrial Electronics*, vol. 56, no. 6, pp. 1847–1853, June 2009, iEEE Trans. Ind. Electron.

[41] M. Lopez, M. Rivera, C. Garcia, J. Rodriguez, R. Pena, J. Espinoza, and P. Wheeler, "Predictive torque control of a multi-drive system fed by a six-leg indirect matrix converter," in *Industrial Technology (ICIT), 2013 IEEE International Conference on*, 25-28 Feb. 2013 2013, pp. 1642–1647.

[42] C. Ortega, A. Arias, and J. Espina, "Predictive vector selector for direct torque control of matrix converter fed induction motors," in *Industrial Electronics, 2009. IECON '09. 35th Annual Conference of IEEE*, 3-5 Nov. 2009 2009, pp. 1240–1245.

[43] C. Ortega, A. Arias, and et al., "Predictive direct torque control of matrix converter fed permanent magnet synchronous machines," in *Industrial Electronics (ISIE), 2010 IEEE International Symposium on*, 4-7 July 2010 2010, pp. 1451–1455.

[44] J. Rodriguez, J. Kolar, and et al., "Predictive torque and flux control of an induction machine fed by an indirect matrix converter with reactive power minimization," in *Industrial Electronics (ISIE), 2010 IEEE International Symposium on*, 4-7 July 2010 2010, pp. 3177–3183.

[45] J. Rodriguez, J. Kolar, J. Espinoza, M. Rivera, and C. Rojas, "Predictive torque and flux control of an induction machine fed by an indirect matrix converter," in *Industrial Technology (ICIT), 2010 IEEE International Conference on*, 14-17 March 2010 2010, pp. 1857–1863.

[46] S. M. M. Uddin, S. Mekhilef, M. Rivera, and J. Rodriguez, "A fcs-mpc of an induction motor fed by indirect matrix converter with unity power factor control," in *Industrial Electronics and Applications (ICIEA), 2013 8th IEEE Conference on*, 19-21 June 2013 2013, pp. 1769–1774.

[47] R. Vargas, U. Ammann, B. Hudoffsky, J. Rodriguez, and P. Wheeler, "Predictive torque control of an induction machine fed by a matrix converter with reactive input power control," *Power Electronics, IEEE Transactions on*, vol. 25, no. 6, pp. 1426–1438, 2010.

[48] R. Vargas, M. Rivera, J. Rodriguez, J. Espinoza, and P. Wheeler, "Predictive torque control with input pf correction applied to an induction machine fed by a matrix converter," in *Power Electronics Specialists Conference, 2008. PESC 2008. IEEE*, 15-19 June 2008 2008, pp. 9–14.

[49] L. Zakaria and K. Barra, "Predictive direct torque and flux control of an induction motor drive fed by a direct matrix converter with reactive power minimization," in *Networking, Sensing and Control (ICNSC), 2013 10th IEEE International Conference on*, 10-12 April 2013 2013, pp. 34–39.

[50] P. Gamboa, J. F. Silva, S. F. Pinto, and E. Margato, "Predictive optimal matrix converter control for a dynamic voltage restorer with flywheel energy storage," in *Industrial Electronics, 2009. IECON '09. 35th Annual Conference of IEEE*, 3-5 Nov. 2009 2009, pp. 759–764.

[51] M. Ortega, F. Jurado, and J. Carpio, "Control of indirect matrix converter with bidirectional output stage for micro-turbine," *Power Electronics, IET*, vol. 5, no. 6, pp. 659–668, 2012.

[52] S. Yusoff, L. De Lillo, P. Zanchetta, and P. Wheeler, "Predictive control of a direct ac/ac matrix converter power supply under non-linear load conditions," in *Power Electronics and Motion Control Conference (EPE/PEMC), 2012 15th International*, 4-6 Sept. 2012 2012, pp. DS3c.4–1–DS3c.4–6.

[53] S. Yusoff, L. De Lillo, P. Zanchetta, P. Wheeler, P. Cortes, and J. Rodriguez, "Predictive control of a direct ac/ac matrix converter for power supply applications," in *Power Electronics, Machines and Drives (PEMD 2012), 6th IET International Conference on*, 27-29 March 2012 2012, pp. 1–6.

[54] D. J. Cook, M. Catucci, P. W. Wheeler, J. C. Clare, J. S. Przybyla, and B. R. Richardson, "Development of a predictive controller for use on a direct converter for high-energy physics applications," *Industrial Electronics, IEEE Transactions on*, vol. 55, no. 12, pp. 4325–4334, 2008.

[55] E. F. Reyes, A. J. Watson, J. C. Clare, and P. W. Wheeler, "Comparison of predictive control strategies for direct resonant high voltage dc power supply," in *Power Electronics, Machines and Drives (PEMD 2012), 6th IET International Conference on*, 27-29 March 2012 2012, pp. 1–6.

[56] R. Vargas, J. Rodriguez, C. Rojas, and P. Wheeler, "Predictive current control applied to a matrix converter: An assessment with the direct transfer function approach," in *Industrial Technology (ICIT), 2010 IEEE International Conference on*, 14-17 March 2010 2010, pp. 1832–1838.

[57] T. Wijekoon, C. Klumpner, P. Zanchetta, and P. W. Wheeler, "Implementation of a hybrid ac-ac direct power converter with unity voltage transfer," *Power Electronics, IEEE Transactions on*, vol. 23, no. 4, pp. 1918–1926, 2008.

[58] L. Hong Hee, D. Phan Quoc, P. Le Minh, and K. Le Dinh, "A new artificial neural network controller for direct control method for matrix converters," in *Power Electronics and Drive Systems, 2009. PEDS 2009. International Conference on*, 2-5 Nov. 2009 2009, pp. 434–439.

[59] H. Karaca, R. Akkaya, and H. Dogan, "A novel compensation method based on fuzzy logic control for matrix converter under distorted input voltage conditions," in *Electrical Machines, 2008. ICEM 2008. 18th International Conference on*, 6-9 Sept. 2008 2008, pp. 1–5.

[60] P. Zanchetta, J. C. Clare, P. Wheeler, D. Katsis, M. Bland, and L. Empringham, "Control design of a three-phase matrix converter mobile ac power supply using genetic algorithms," in *Power Electronics Specialists Conference, 2005. PESC '05. IEEE 36th*, 16-16 June 2005 2005, pp. 2370–2375.

[61] P. Zanchetta, M. Sumner, J. C. Clare, and P. W. Wheeler, "Control of matrix converters for ac power supplies using genetic algorithms," in *Industrial Electronics, 2004 IEEE International Symposium on*, vol. 2, 4-7 May 2004 2004, pp. 1429–1433 vol. 2.

[62] C. F. Calvillo, F. Martell, J. L. Elizondo, A. Avila, M. E. Macias, M. Rivera, and J. Rodriguez, "Rotor current fuzzy control of a dfig with an indirect matrix converter," in *IECON 2011 - 37th Annual Conference on IEEE Industrial Electronics Society*, 7-10 Nov. 2011 2011, pp. 4296–4301.

[63] T. S. Sivarani, S. J. Jawhar, and C. A. Kumar, "Novel intelligent hybrid techniques for speed control of electric drives fed by matrix converter," in *Computing, Electronics and Electrical Technologies (ICCEET), 2012 International Conference on*, 21-22 March 2012 2012, pp. 466–471.

[64] W. Xiaohong, G. Quanxue, and T. Lianfang, "A novel adaptive fuzzy control for output voltage of matrix converter," in *Power Electronics and Motion Control Conference (IPEMC), 2012 7th International*, vol. 1, 2-5 June 2012 2012, pp. 53–58.

[65] M. Rivera, A. Wilson, C. A. Rojas, J. Rodriguez, J. R. Espinoza, P. W. Wheeler, and L. Empringham, "A comparative assessment of model predictive current control and space vector modulation in a direct matrix converter," *Industrial Electronics, IEEE Transactions on*, vol. 60, no. 2, pp. 578–588, 2013.

[66] F. Villarroel, J. Espinoza, C. Rojas, C. Molina, and E. Espinosa, "A multiobjective ranking based finite states model predictive control scheme applied to a direct matrix converter," in *IECON 2010 - 36th Annual Conference on IEEE Industrial Electronics Society*, 7-10 Nov. 2010 2010, pp. 2941–2946.

[67] F. Villarroel, J. Espinoza, C. Rojas, C. Molina, and J. Rodriguez, "Application of fuzzy decision making to the switching state selection in the predictive control of a direct matrix converter," in *IECON 2011 - 37th Annual Conference on IEEE Industrial Electronics Society*, 7-10 Nov. 2011 2011, pp. 4272–4277.

[68] M. Rivera, J. Munoz, C. Baier, J. Rodriguez, J. Espinoza, V. Yaramasu, B. Wu, and P. Wheeler, "A simple predictive current control of a single-phase matrix converter," in *4th IEEE International Conference on Power Engineering, Energy and Electrical Drives, POWERENG (2013, Turkey)*, 13-17 May 2013 2013.

[69] J. W. Kolar, T. Friedli, F. Krismer, and S. D. Round, "The essence of three-phase ac/ac converter systems," in *Power Electronics and Motion Control Conference, 2008. EPE-PEMC 2008. 13th*, 1-3 Sept. 2008 2008, pp. 27–42.

[70] M. Rivera, I. Contreras, J. Rodriguez, R. Pena, and P. Wheeler, "A simple current control method with instantaneous reactive power minimization for four-leg indirect matrix converters," in *Power Electronics and Applications (EPE 2011), Proceedings of the 2011-14th European Conference on*, 2011, pp. 1–9.

[71] T. Friedli, J. W. Kolar, J. Rodriguez, and P. W. Wheeler, "Comparative evaluation of three-phase ac-ac matrix converter and voltage dc-link back-to-back converter systems," *Industrial Electronics, IEEE Transactions on*, vol. 59, no. 12, pp. 4487–4510, 2012.

978-1-5090-0376-1/16 $31.00 © 2016 IEEE

7th Power Electronics, Drive Systems & Technologies Conference (PEDSTC 2016)
16-18 Feb. 2016, Iran University of Science and Technology, Tehran, Iran

A Review of Predictive Control Techniques for Matrix Converters - Part II

M. Rivera, P. Wheeler, A. Olloqui, and D.A. Khaburi

Abstract—The second part of this paper presents an overview of different control strategies and applications for matrix converters (MCs) where predictive control techniques are applied. It will be shown that predictive control is a promising alternative to control MCs due to its simplicity and flexibility to include different constrains in the control for different industrial applications such as renewable energies, grid interconnection, multi-drives systems control, among others. In addition, some limitations and weaknesses of predictive control in MCs such as variable switching frequency, high dependence on the predictive model quality and high computational cost, will be discussed as well as some future trends in control strategies and its application to other topologies or loads.

Index Terms—Matrix Converter, Direct Matrix Converter, Indirect Matrix Converter, Power Control, Predictive Control, Control Strategies, AC-AC conversion, Modulation Schemes

I. INTRODUCTION

Model predictive control (MPC) offers a flexible and better alternative for the control of electrical energy with power electronics converters. This new approach takes into consideration the discrete and nonlinear nature of power converters and drives and promises to have a strong impact on control in power electronics in the near future.

In the second part of this paper, a review of several predictive control schemes proposed for different topologies of MCs and their applications will be presented. It will be demonstrated that these techniques can be easily implemented by taking advantage of the available technologies of digital signal processors. Also limitations and/or weaknesses in a comparison to conventional control concepts as well as open questions and future trends are discussed in the final part of this paper.

II. PREDICTIVE CONTROL STRATEGIES AND APPLICATIONS FOR MATRIX CONVERTERS

Based on the review done by the authors, there are several implementations of predictive control applied to MCs. As indicated in Fig. 1, the most relevant techniques correspond to predictive current control (PCC) and predictive torque control (PTC). It is possible also to find some implementations of predictive reactive/active power control (PPC), and predictive voltage control (PVC) where an *LC* filter is considered in the output side of the converter [1].

A. Predictive Current Control

In the group of PCC, different have been the implementations which are focused to specific applications and objectives. The basic PCC strategy has been presented in the first part of this paper, where only the load current is controlled [2]. In this case, the cost function includes only the error between the load current references and their respective predictions but the source currents are highly distorted, which is an undesired performance for the MC. For this reason, in order to improve the input current behavior, a new term is included in the cost function, where an instantaneous reactive input power minimization is considered, as reported in [3]–[17]. The control scheme with this implementation is detailed in Fig. 2, where the cost function is now defined as:

$$g(k+1) = \triangle i_o + \gamma_q \triangle Q \qquad (1)$$

with $\triangle i_o = |i_a^* - i_a^p| + |i_b^* - i_b^p| + |i_c^* - i_c^p|$, and $\triangle Q = |Q^* - (v_{s\alpha}i_{s\beta} - v_{s\beta}i_{s\alpha})|$. By including this new term, a significant improvement of the input current performance is obtained while maintaining the good behavior of the load current. An important issue here is the weighting factor selection. There are some guidelines to determine this value which can be reviewed in [18]. By including the instantaneous reactive power minimization on the input side, is possible to obtain unity power factor. However, one interesting issue that have been observed is that the PCC with reactive power minimization method is very sensitive to the distortion of the source voltage and the resonance of the input filter and thus, this affects the output waveform distortion level, particularly when harmonic distortion is existing in the source voltage. One drawback in the operation of a MC with predictive control is the variable switching frequency. Because at every sampling time is selected a switching state to be applied to the converter, is possible to have the same optimal state for a while, which will produce a variable switching frequency and thus, a spread spectrum. This variable switching frequency, along with disturbances in the source voltage, could produce a resonance in the input filter generating high distortion on the input current which are also reflected in the output current due to the direct connection between input and output sides of the converter. In order to solve this problem, in [19]–[22], an input filter resonance mitigation have been proposed, based on the control scheme presented in Fig. 3. Active damping is a control approach used for achieving an attenuation of the resonance, which avoid the disadvantages of using

978-1-5090-0376-1/16 $31.00 © 2016 IEEE

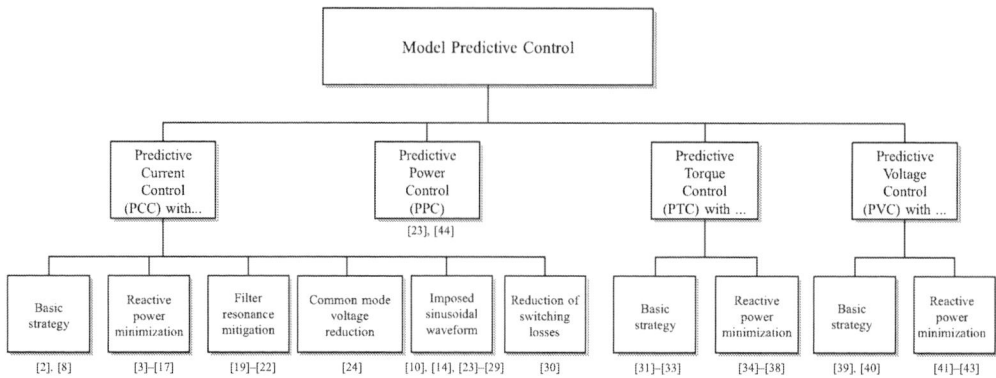

Fig. 1. Predictive control strategies applied to MCs.

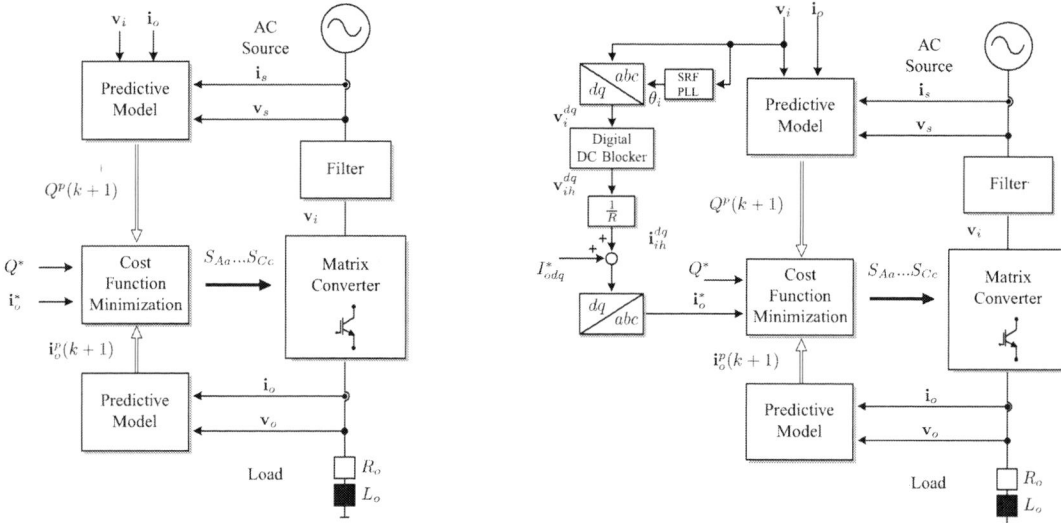

Fig. 2. Block diagram of the predictive current control strategy with instantaneous reactive input power minimization.

Fig. 3. Block diagram of the predictive current control strategy with active damping implementation.

passive damping which has a negative impact on system efficiency. The main idea of this implementation is to emulate a virtual resistor, which is immune to system parameter variations, in parallel to the capacitor of the input filter for reducing the harmonics level without affecting the first harmonic component. The power converter draws a damping current that is proportional to the capacitor voltage. Then, the active damping approach in the matrix converter is achieved by passing the harmonic component effects existing in the input side to the output side of the converter, adding this effect on the reference value of the output current. For this case, is used the same cost function defined in eq. (1), and the only difference is given by the way that the output current reference is obtained. As reported in [19]–[22], the performance of the system is improved, mitigating the resonance of the input filter and obtaining a more sinusoidal source current. This strategy reduces also the power losses comparing to the method with resistive damping. The cost of such improvement is an inclusion of additional virtual resistor

in the predictive control algorithm. Another alternative, that minimizes the instantaneous reactive power without the use of active damping, is to force the source current to follow a sinusoidal reference value, regardless the distortion level at the input side. The block diagram of this new control method is presented in Fig. 4 and the cost function is defined as:

$$g(k+1) = \triangle i_o + \gamma_i \triangle i_s \qquad (2)$$

where $\triangle i_s = |i_{sA}^* - i_{sA}^p| + |i_{sB}^* - i_{sB}^p| + |i_{sC}^* - i_{sC}^p|$. Results shown in [10], [14], [23]–[29] demonstrated that by imposing a given waveform for the source current, it is possible to obtain a better performance than an instantaneous reactive power minimization, reducing the total harmonic distortion (THD) of both input and load currents, reducing the resonance of the input filter and consequently, extending the capacitors life-span. The source current reference and its amplitude are obtained such as reported in [10], [14], [23]–[29], and they are

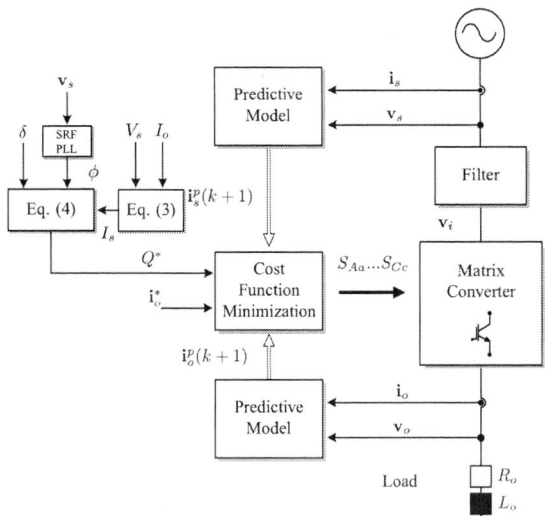

Fig. 4. Block diagram of the predictive current control strategy with imposed sinusoidal input waveform.

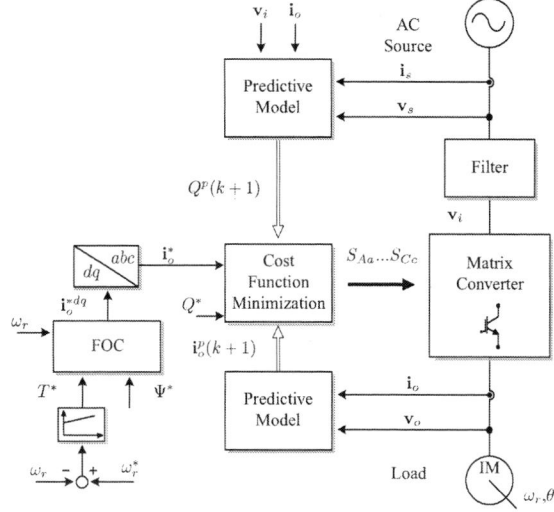

Fig. 5. Block diagram of the predictive current control strategy for torque and flux control with minimization of the input reactive power.

defined by:

$$I_s = \frac{-\lambda V_s \pm \sqrt{(\lambda V_s)^2 - 4\lambda R_f R_o I_o^{*2}/\eta}}{-2\lambda R_f} \quad (3)$$

where $\lambda = 1 - 8\pi^2 f_s^2 C_f L_f$.

$$\begin{aligned} i_{sA}^* &= I_s \sin(w_s t + \theta) \\ i_{sB}^* &= I_s \sin(w_s t - 2\pi/3 + \theta) \\ i_{sC}^* &= I_s \sin(w_s t + 2\pi/3 + \theta) \end{aligned} \quad (4)$$

In [15] is presented an interesting contribution where a different approach is done to control an induction machine with PCC with instantaneous reactive power minimization. The block diagram of this implementation is shown in Fig. 5. There is possible to differentiate a predictive stage that performs PCC and a classic stage that controls the speed, flux, and torque control using field oriented control (FOC), which provides the reference currents for the predictive control stage. The cost function is defined as such as eq. (1). With this strategy, predictive control has been demonstrated as a very powerful tool opening new possibilities in the control of power converters in a very simple way. Similarly, a PCC for an induction machine with an increment of the efficiency and a reduction of switching losses of the converter is proposed in [30]. The control scheme is the same as the diagram shown in Fig. 5, because only the cost function is modified. In order to reduce the switching frequency and thus increment the efficiency of the converter, the idea of the method consists in to include in the cost function the number of commutations needed to transit from the actual switching state being evaluated. This is represented as:

$$g(k+1) = \triangle i_o + \gamma_q \triangle q + \gamma_{sw} \, n \quad (5)$$

where n is the number of switches commutations considered by choosing the evaluated state. This new term in the cost function means that changing the state of the

converter will have a cost. Therefore, the controller will select as optimal, the switching state that involves and produces less commutations. With the goal to reduce the switching losses of the converter, a variation for the PCC method was proposed also in [30]. The idea of this technique is to predict not only the number of commutations, but also the switching losses that the switching of the power converter would produce, and then, include that prediction in the cost function, which is defined as:

$$g(k+1) = \triangle i_o + \gamma_q \triangle q + \gamma_{sl} \sum_{i=1}^{18} \triangle i_c^{(i)} \triangle v_{ce}^{(i)} \quad (6)$$

where $\triangle i_c^{(i)}$ and $\triangle v_{ce}^{(i)}$ are variations of the collector current and collector-emitter voltage of the power transistor i, respectively. Note that 18 correspond to the eighteen switches of a DMC. As reported in [30], by considering these terms in the cost function, an important increment in the efficiency of the converter is obtained. As previously demonstrated, one advantage of model predictive control is the possibility to include several control objectives in the cost function, which could be the current, voltage, reactive power, switching frequency, switching losses, torque, flux, etc. By considering this approach, in [24], a PCC with imposed source current is proposed and this strategy is enhanced with a common-mode voltage (CMV) reduction. The control strategy for this implementation is shown in Fig. 6 and the considered cost function is given by:

$$g(k+1) = \triangle i_o + \gamma_i \triangle i_s + \gamma_v \, |v_{cm}(k+1)| \quad (7)$$

where the CMV is defined as $v_{cm} = (v_a + v_b + v_c)/3$. The results showed in [24], demonstrate that just by including a new term in the cost function is possible to have simultaneous control of the input and output (source and load) currents with waveforms according their references and with reduction of the CMV.

978-1-5090-0376-1/16 $31.00 © 2016 IEEE

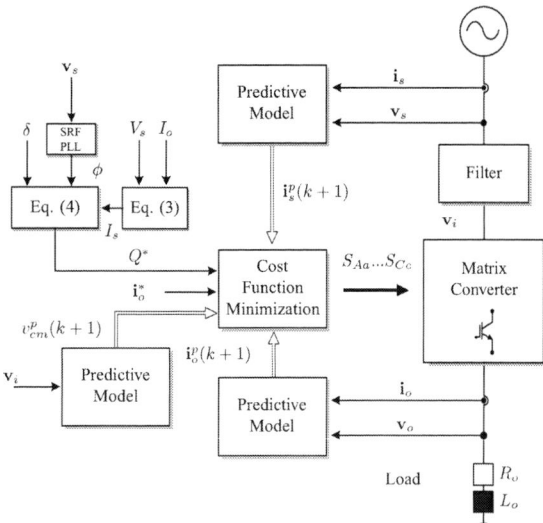

Fig. 6. Block diagram of the predictive current control strategy with common-mode voltage minimization and imposed sinusoidal input waveform.

B. Predictive Torque Control

In Fig. 7 is shown the control scheme for the basic predictive torque strategy. This control method has been introduced in [31]–[33], [45]. Similar to the PCC method, predictive torque control (PTC) consists of selecting, at fixed sampling periods, one of the possible switching states of the matrix converter. Again, the selection of the switching state for the next time instant is performed by using a predefined cost function with proper minimization technique. This cost function g represents the evaluation criteria in order to select the best switching state for the next sampling interval. For the computation of the cost function g, the input current $\mathbf{i}_s$, the electromagnetic torque T_e, and the stator flux $\psi_\mathbf{s}$ are predicted in the future sampling period, using the mathematical model of the input filter and the induction machine (IM). A PI controller is adopted to generate the reference torque T_e^* for the predictive control algorithm. A mathematical discrete-time model is elaborated to predict the behavior of the system under a particular switching state, based on dynamic model for the IM [31]–[33], [45]. This model is used in our case to predict the stator flux and the electromagnetic torque produced by the machine during the next sampling period. In this case, the cost function is composed of the absolute errors of the predicted torque, flux magnitude and reactive input power, resulting in:

$$ g = \triangle T_e(k+1) + \gamma_\psi \triangle \psi(k+1) \qquad (8) $$

where γ_q and γ_ψ are weight factors selected to define the relation between reactive input power, torque and stator flux conditions. In Fig. 8 is presented the same strategy but enhanced with instantaneous reactive power minimization in order to improve the performance of the converter in the input side. For this case, the cost function

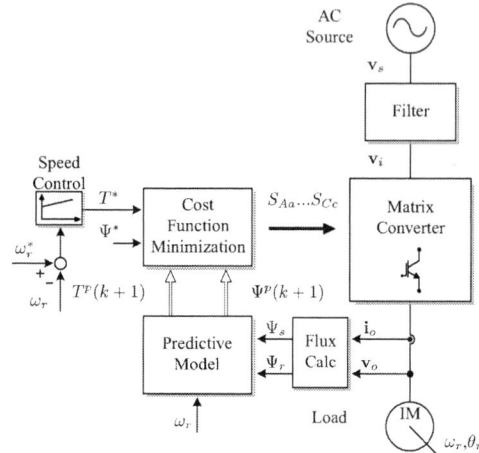

Fig. 7. Block diagram of the predictive torque and flux control strategy.

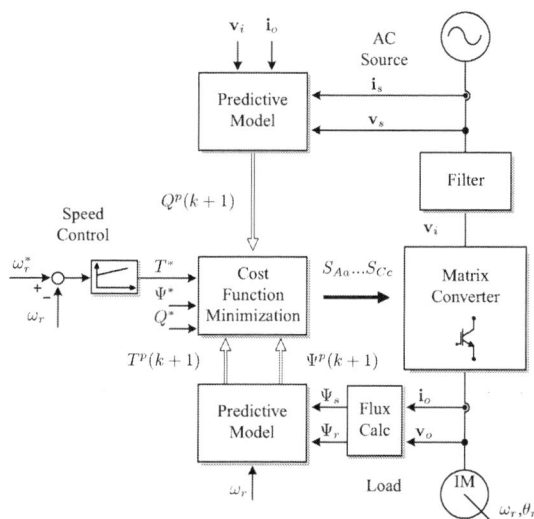

Fig. 8. Block diagram of the predictive torque and flux control strategy with minimization of the instantaneous input reactive power.

is given as

$$ g = \triangle T_e(k+1) + \gamma_\psi \triangle \psi(k+1) + \gamma_q \triangle q_s(k+1) \quad (9) $$

As reported in [34]–[38], the method is very simple and effective with a very fast dynamic response of the electrical torque and a decoupled control respect to the stator flux control maintaining, at the same time, unity displacement power factor for both motoring and regeneration operation modes. All these remarkable characteristics make PTC a suitable alternative to the well known direct torque control (DTC) having the advantages of be simpler and consider all the available switching states of the converter. As indicated in Fig. 1, predictive power control [23], [44] and predictive voltage control [39]–[43] have been also implemented in MCs but due to space limitations, these techniques are not analyzed in this work.

978-1-5090-0376-1/16 $31.00 © 2016 IEEE

III. Limitations, Weaknesses, Open Questions and Future Trends

As well it has been demonstrated that predictive control is a very powerful alternative to control MCs, this technique also presents some limitations and weaknesses. Some of them are:

- Variable switching frequency.
- High dependence on the predictive model quality.
- High sampling frequency.
- Sensitive to variation on the parameters.
- High computational cost.

As highlighted in Part I of this paper, the MC presents different advantages in term of size and weight, allowing the operation with sinusoidal source and load currents, regeneration capability among others, but despite all the rewards, there exists two main disadvantages in the MC topology (as long as the maximum voltage transfer ratio is not needed by the load [46]): the output voltage is limited to 86% of the input, and secondly, any perturbation in the supply deteriorates the quality of the load side, due to the absence of storage elements. But, as mentioned in the first part of this paper, in order to solve this issue, in [47] has been proposed the implementation of a hybrid power converter which connects an auxiliary voltage source in the dc-link of the indirect matrix converter (IMC), obtaining unity voltage transfer capability even under severe distortions in the source voltage. Additionally, a PCC strategy is proposed in this work for the auxiliary voltage source, where the current reference is given by a PI linear controller and predictive control selects the duty cycle for the pulse-width modulator, ensuring balanced power in the converter and unity voltage ratio. Predictive control can select be used to select the optimum state of the converter to control directly the primary control variable or it can be used to calculate the duty cycle to be applied for a set of space vectors to control indirectly the primary control variable by a space vector modulation strategy. The MC topology has also limited capability to perform power factor correction or active damping, since the reactive power is directly proportional to the modulation index, amplitude of the input line voltage and output line current, and inversely proportional to the current-to-voltage displacement angle at the converter. Special hybrid modulation schemes applied to the sparse matrix converter topology have been studied to manage reactive power for a purely reactive loads [48]. However, if the instantaneous output currents of the MC are equal to zero, then no reactive input power can be provided with any modulation scheme [46]. In MC topologies, the energy storage is located at the input filter (despite this is very small), therefore the MC topology suffer from an aging process, limiting the lifetime of the converter. There is a relation between the capacitor lifetime and the capacitance value, calculated in terms of the volume of the dielectric material and the current ripple at a given operating voltage. For an specific type of dielectric material, an operating temperature and rated voltage, the capacitance value, and therefore, the lifetime of the input

filter capacitors is given by the maximum input current ripple. The losses and capacitor value are significantly lower in a MC than for an equally sized dc-link capacitor in a two-level voltage dc-link back-to-back converter (V-BBC) [46]. However, a minimum internal energy storage needs to be provided in order to meet the dynamic requirements imposed by the load. For high-load dynamics, extended ridethrough capability, or unconstrained reactive power compensation, the MC has not yet proved to be the most appropriate solution. The matrix converter should hence be selected for high frequency applications, where the conduction losses are dominant. However, the cause of electromagnetic interference (EMI) problems in power electronics is fast switching of high currents and voltages within the power converter systems. Other causes of EMI are supply voltage interruptions, dips, unbalance, surges and fluctuations that are also reflected in the load side because of the direct topology [19], [20]. In order to reduce the input current ripple, the input displacement power factor and other causes of EMI, many solutions have been reported. For example, it is possible to select a proper filter resonant frequency, which may limit the converter performance because the LC resonant frequency is a function of the system impedance. This impedance varies with the operating conditions of the system. Also, it is possible to adopt a high switching frequency. The first solution results in lower THD of the output currents, however the converter losses and the EMI increase significantly and finally, the converter efficiency with both solutions decreases [21], [22]. Another suitable method consists of emulating a damping resistor placed in parallel with the filter capacitor in such way that the harmonic currents caused by the resonances flow through this resistor. Active damping improves high quality of the input currents and the instantaneous reactive power minimization on the input side. However, this method only mitigates higher current harmonics resulted from the switching operation and cannot ensure a sinusoidal input current, particularly when the source voltage is distorted. To overcome these issues, in the implementations done with predictive control, the term that minimizes the reactive power at the input side is replaced by a direct control of the source currents. This approach forces the input currents to follow the sinusoidal value, regardless of the distortion level at the input side [25], [49]. In all the studied cases where predictive control has been applied, it has been demonstrated a very good performance, being a very simple method for implementation. But in each case, only a second order LC filter has been used. An industrial type system with a MC should be considered with an EMI input filter (not only single-stage LC filter), in order to provide results which are immediately useful to the industry area. Additionally, there are two major reasons for using an EMI input filters. The first reason is to prevent electromagnetic interference of the power electronic converter with the integrated electronic systems, and the second reason is to avoid the malfunction of the power converter operation caused by sources of electromagnetic noise in the surrounds. Also there are some requirements that the EMI filter has to fulfill:

- Minimum filter attenuation requirement at given frequencies.
- Minimization of input current fundamental displacement factor.
- Limitation of the volume/energy stored in the filter components.
- Sufficient passive damping for minimum losses, in order to avoid oscillations also for no-load operation.
- Avoiding of filter resonances at the multiples of switching frequency.
- Minimization of the filter output impedance, in order ensures system stability and minimizes control design constrains

As it has been showed, predictive control has been successfully implemented in MCs. But, there are still some open topics for research as well as some trends in the work with this type of power converters and control strategy. Finally, an exhaustive comparison between predictive control and classical control techniques must be taking into consideration in order to demonstrate the simplicity and feasibility of the strategy.

IV. CONCLUSIONS

The main contribution of this paper has been to presents an overview of different topologies, control strategies and applications for MCs where predictive control techniques are applied. The main limitations of simple predictive control applied to traditional MCs fall on its power factor correction or active damping capability and variable switching frequency producing resonance in the input filter. This resonance generates high distortion on the input current which are also reflected in the output current due to the direct connection between input and output sides of the converter. The performance of predictive control strategies depends on the digital platform and the predictive strategy since it is based on an iterative and finite-horizon optimization of the load model.

Named limitations and weaknesses of predictive control in matrix converters, constitute future topics for research. It has been shown that predictive control can be recognized as an attractive control approach, with significant benefits such as flexibility, versatility, and performance, with real applications of the power converters and electric drives. The advantages of predictive control can be enhanced with commutation strategies to avoid variable switching frequency and resonance. Moreover, other numeric integration algorithms can be implemented to overcome limitations of the sampling time in the digital platform.

ACKNOWLEDGMENTS

This publication was made possible by the Newton Picarte Project EPSRC: EP/N004043/1: New Configurations of Power Converters for Grid Interconnection Systems / CONICYT DPI20140007 and British Council through the Institutional Skills Development Newton Picarte Project ISCL 2015006.

REFERENCES

[1] J. Rodriguez, M. Rivera, J. W. Kolar, and P. W. Wheeler, "A review of control and modulation methods for matrix converters," *Industrial Electronics, IEEE Transactions on*, vol. 59, no. 1, pp. 58–70, 2012.

[2] C. F. Calvillo, A. Olloqui, F. Martell, J. L. Elizondo, A. Avila, M. E. Macias, M. Rivera, and J. Rodriguez, "Comparison of model based predictive control and fuzzy logic control of a dfig with an indirect matrix converter," in *IECON 2012 - 38th Annual Conference on IEEE Industrial Electronics Society*, 25-28 Oct. 2012 2012, pp. 6063–6068.

[3] P. Correa, J. Rodriguez, M. Rivera, J. R. Espinoza, and J. W. Kolar, "Predictive control of an indirect matrix converter," *Industrial Electronics, IEEE Transactions on*, vol. 56, no. 6, pp. 1847–1853, 2009.

[4] J. D. Dasika and M. Saeedifard, "An on-line fault detection and a post-fault strategy to improve the reliability of matrix converters," in *Applied Power Electronics Conference and Exposition (APEC), 2013 Twenty-Eighth Annual IEEE*, 17-21 March 2013 2013, pp. 1185–1191.

[5] A. A. Heris, E. Babaei, and S. H. Hosseini, "A new shunt active power filter based on indirect matrix converter," in *Electrical Engineering (ICEE), 2012 20th Iranian Conference on*, 15-17 May 2012 2012, pp. 581–586.

[6] E. Lee, L. Kyo-Beum, L. Jae-Sik, L. Youngil, and S. Joong-Ho, "Predictive current control for a sparse matrix converter," in *Power Electronics and Motion Control Conference (IPEMC), 2012 7th International*, vol. 1, 2-5 June 2012 2012, pp. 36–40.

[7] F. Morel, J. M. Retif, L.-S. Xuefang, B. Allard, and P. Bevilacqua, "A predictive control for a matrix converter-fed permanent magnet synchronous machine," in *Power Electronics Specialists Conference, 2008. PESC 2008. IEEE*, 15-19 June 2008 2008, pp. 15–21.

[8] M. Rivera, J. L. Elizondo, M. E. Macias, O. M. Probst, O. M. Micheloud, J. Rodriguez, C. Rojas, and A. Wilson, "Model predictive control of a doubly fed induction generator with an indirect matrix converter," in *IECON 2010 - 36th Annual Conference on IEEE Industrial Electronics Society*, 7-10 Nov. 2010 2010, pp. 2959–2965.

[9] M. Rivera, J. Rodriguez, J. R. Espinoza, and H. Abu-Rub, "Instantaneous reactive power minimization and current control for an indirect matrix converter under a distorted ac supply," *Industrial Informatics, IEEE Transactions on*, vol. 8, no. 3, pp. 482–490, 2012.

[10] M. Rivera, C. Rojas, J. Rodriguez, and J. Espinoza, "Methods of source current reference generation for predictive control in a direct matrix converter," *Power Electronics, IET*, vol. 6, no. 5, 2013.

[11] M. Rivera, R. Vargas, J. Espinoza, J. Rodriguez, P. Wheeler, and C. Silva, "Current control in matrix converters connected to polluted ac voltage supplies," in *Power Electronics Specialists Conference, 2008. PESC 2008. IEEE*, 15-19 June 2008 2008, pp. 412–417.

[12] M. Rivera, A. Wilson, C. A. Rojas, J. Rodriguez, J. R. Espinoza, P. W. Wheeler, and L. Empringham, "A comparative assessment of model predictive current control and space vector modulation in a direct matrix converter," *Industrial Electronics, IEEE Transactions on*, vol. 60, no. 2, pp. 578–588, 2013.

[13] J. Rodriguez, J. Kolar, J. Espinoza, M. Rivera, and C. Rojas, "Predictive current control with reactive power minimization in an indirect matrix converter," in *Industrial Technology (ICIT), 2010 IEEE International Conference on*, 14-17 March 2010 2010, pp. 1839–1844.

[14] C. Rojas, M. Rivera, J. Rodriguez, A. Wilson, J. Espinoza, F. Villarroel, and P. Wheeler, "Predictive control of a direct matrix converter operating under an unbalanced ac source," in *Industrial Electronics (ISIE), 2010 IEEE International Symposium on*, 4-7 July 2010 2010, pp. 3159–3164.

[15] R. Vargas, J. Rodriguez, U. Ammann, and P. W. Wheeler, "Predictive current control of an induction machine fed by a matrix converter with reactive power control," *Industrial Electronics, IEEE Transactions on*, vol. 55, no. 12, pp. 4362–4371, 2008.

[16] R. Vargas, J. Rodriguez, C. Rojas, and P. Wheeler, "Predictive current control applied to a matrix converter: An assessment with the direct transfer function approach," in *Industrial Technology (ICIT), 2010 IEEE International Conference on*, 14-17 March 2010 2010, pp. 1832–1838.

[17] L. Yulong, C. Nam-Sup, C. Honnyong, and F. Z. Peng, "Carrier-based predictive current controlled pulse width modulation for matrix converters," in *Power Electronics and Motion Control Conference, 2009. IPEMC '09. IEEE 6th International*, 17-20 May 2009 2009, pp. 1009–1014.

[18] P. Cortes, S. Kouro, B. La Rocca, R. Vargas, J. Rodriguez, J. Leon, S. Vazquez, and L. Franquelo, "Guidelines for weighting factors design in model predictive control of power converters and drives," 2009, pp. 1–7, Gippsland, Australia.

[19] M. Rivera, P. Correa, J. Rodriguez, I. Lizama, and J. Espinoza, "Predictive control of the indirect matrix converter with active damping," in *Power Electronics and Motion Control Conference, 2009. IPEMC '09. IEEE 6th International*, 17-20 May 2009 2009, pp. 1738–1744.

[20] M. Rivera, P. Correa, J. Rodriguez, I. Lizama, J. Espinoza, and C. Rojas, "Predictive control with active damping in a direct matrix converter," in *Energy Conversion Congress and Exposition, 2009. ECCE 2009. IEEE*, 20-24 Sept. 2009 2009, pp. 3057–3062.

[21] M. Rivera, J. Rodriguez, W. Bin, J. R. Espinoza, and C. A. Rojas, "Current control for an indirect matrix converter with filter resonance mitigation," *Industrial Electronics, IEEE Transactions on*, vol. 59, no. 1, pp. 71–79, 2012.

[22] M. Rivera, C. Rojas, Rodri, x, J. guez, P. Wheeler, W. Bin, and J. R. Espinoza, "Predictive current control with input filter resonance mitigation for a direct matrix converter," *Power Electronics, IEEE Transactions on*, vol. 26, no. 10, pp. 2794–2803, 2011.

[23] S. K. M. Ahmed, A. Iqbal, H. Abu-Rub, and P. Cortes, "Model predictive control of a three-to-five phase matrix converter," in *Predictive Control of Electrical Drives and Power Electronics (PRECEDE), 2011 Workshop on*, 14-15 Oct. 2011 2011, pp. 36–39.

[24] M. Rivera, J. Rodriguez, J. Espinoza, and W. Bin, "Reduction of common-mode voltage in an indirect matrix converter with imposed sinusoidal input/output waveforms," in *IECON 2012 - 38th Annual Conference on IEEE Industrial Electronics Society*, 25-28 Oct. 2012 2012, pp. 6105–6110.

[25] M. Rivera, J. Rodriguez, P. W. Wheeler, C. A. Rojas, A. Wilson, and J. R. Espinoza, "Control of a matrix converter with imposed sinusoidal source currents," *Industrial Electronics, IEEE Transactions on*, vol. 59, no. 4, pp. 1939–1949, 2012.

[26] J. Rodriguez, J. Espinoza, M. Rivera, F. Villarroel, and C. Rojas, "Predictive control of source and load currents in a direct matrix converter," in *Industrial Technology (ICIT), 2010 IEEE International Conference on*, 14-17 March 2010 2010, pp. 1826–1831.

[27] F. Villarroel, J. Espinoza, C. Rojas, C. Molina, and E. Espinosa, "A multiobjective ranking based finite states model predictive control scheme applied to a direct matrix converter," in *IECON 2010 - 36th Annual Conference on IEEE Industrial Electronics Society*, 7-10 Nov. 2010 2010, pp. 2941–2946.

[28] F. Villarroel, J. Espinoza, C. Rojas, C. Molina, and J. Rodriguez, "Application of fuzzy decision making to the switching state selection in the predictive control of a direct matrix converter," in *IECON 2011 - 37th Annual Conference on IEEE Industrial Electronics Society*, 7-10 Nov. 2011 2011, pp. 4272–4277.

[29] F. Villarroel, J. R. Espinoza, C. A. Rojas, J. Rodriguez, M. Rivera, and D. Sbarbaro, "Multiobjective switching state selector for finite-states model predictive control based on fuzzy decision making in a matrix converter," *Industrial Electronics, IEEE Transactions on*, vol. 60, no. 2, pp. 589–599, 2013.

[30] R. Vargas, U. Ammann, and J. Rodriguez, "Predictive approach to increase efficiency and reduce switching losses on matrix converters," *Power Electronics, IEEE Transactions on*, vol. 24, no. 4, pp. 894–902, 2009.

[31] M. Lopez, M. Rivera, C. Garcia, J. Rodriguez, R. Pena, J. Espinoza, and P. Wheeler, "Predictive torque control of a multi-drive system fed by a six-leg indirect matrix converter," in *Industrial Technology (ICIT), 2013 IEEE International Conference on*, 25-28 Feb. 2013 2013, pp. 1642–1647.

[32] C. Ortega, A. Arias, and J. Espina, "Predictive vector selector for direct torque control of matrix converter fed induction motors," in *Industrial Electronics, 2009. IECON '09. 35th Annual Conference of IEEE*, 3-5 Nov. 2009 2009, pp. 1240–1245.

[33] C. Ortega, A. Arias, and et al., "Predictive direct torque control of matrix converter fed permanent magnet synchronous machines," in *Industrial Electronics (ISIE), 2010 IEEE International Symposium on*, 4-7 July 2010 2010, pp. 1451–1455.

[34] J. Rodriguez, J. Kolar, and et al., "Predictive torque and flux control of an induction machine fed by an indirect matrix converter with reactive power minimization," in *Industrial Electronics (ISIE), 2010 IEEE International Symposium on*, 4-7 July 2010 2010, pp. 3177–3183.

[35] S. M. M. Uddin, S. Mekhilef, M. Rivera, and J. Rodriguez, "A fcs-mpc of an induction motor fed by indirect matrix converter with unity power factor control," in *Industrial Electronics and Applications (ICIEA), 2013 8th IEEE Conference on*, 19-21 June 2013 2013, pp. 1769–1774.

[36] R. Vargas, U. Ammann, B. Hudoffsky, J. Rodriguez, and P. Wheeler, "Predictive torque control of an induction machine fed by a matrix converter with reactive input power control," *Power Electronics, IEEE Transactions on*, vol. 25, no. 6, pp. 1426–1438, 2010.

[37] R. Vargas, M. Rivera, J. Rodriguez, J. Espinoza, and P. Wheeler, "Predictive torque control with input pf correction applied to an induction machine fed by a matrix converter," in *Power Electronics Specialists Conference, 2008. PESC 2008. IEEE*, 15-19 June 2008 2008, pp. 9–14.

[38] L. Zakaria and K. Barra, "Predictive direct torque and flux control of an induction motor drive fed by a direct matrix converter with reactive power minimization," in *Networking, Sensing and Control (ICNSC), 2013 10th IEEE International Conference on*, 10-12 April 2013 2013, pp. 34–39.

[39] D. J. Cook, M. Catucci, P. W. Wheeler, J. C. Clare, J. S. Przybyla, and B. R. Richardson, "Development of a predictive controller for use on a direct converter for high-energy physics applications," *Industrial Electronics, IEEE Transactions on*, vol. 55, no. 12, pp. 4325–4334, 2008.

[40] E. F. Reyes, A. J. Watson, J. C. Clare, and P. W. Wheeler, "Comparison of predictive control strategies for direct resonant high voltage dc power supply," in *Power Electronics, Machines and Drives (PEMD 2012), 6th IET International Conference on*, 27-29 March 2012 2012, pp. 1–6.

[41] P. Gamboa, J. F. Silva, S. F. Pinto, and E. Margato, "Predictive optimal matrix converter control for a dynamic voltage restorer with flywheel energy storage," in *Industrial Electronics, 2009. IECON '09. 35th Annual Conference of IEEE*, 3-5 Nov. 2009 2009, pp. 759–764.

[42] S. Yusoff, L. De Lillo, P. Zanchetta, and P. Wheeler, "Predictive control of a direct ac/ac matrix converter power supply under non-linear load conditions," in *Power Electronics and Motion Control Conference (EPE/PEMC), 2012 15th International*, 4-6 Sept. 2012 2012, pp. DS3c.4–1–DS3c.4–6.

[43] S. Yusoff, L. De Lillo, P. Zanchetta, P. Wheeler, P. Cortes, and J. Rodriguez, "Predictive control of a direct ac/ac matrix converter for power supply applications," in *Power Electronics, Machines and Drives (PEMD 2012), 6th IET International Conference on*, 27-29 March 2012 2012, pp. 1–6.

[44] M. Ortega, F. Jurado, and J. Carpio, "Control of indirect matrix converter with bidirectional output stage for micro-turbine," *Power Electronics, IET*, vol. 5, no. 6, pp. 659–668, 2012.

[45] J. Rodriguez, J. Kolar, J. Espinoza, M. Rivera, and C. Rojas, "Predictive torque and flux control of an induction machine fed by an indirect matrix converter," in *Industrial Technology (ICIT), 2010 IEEE International Conference on*, 14-17 March 2010 2010, pp. 1857–1863.

[46] T. Friedli, J. W. Kolar, J. Rodriguez, and P. W. Wheeler, "Comparative evaluation of three-phase ac-ac matrix converter and voltage dc-link back-to-back converter systems," *Industrial Electronics, IEEE Transactions on*, vol. 59, no. 12, pp. 4487–4510, 2012.

[47] T. Wijekoon, C. Klumpner, P. Zanchetta, and P. W. Wheeler, "Implementation of a hybrid ac-ac direct power converter with unity voltage transfer," *Power Electronics, IEEE Transactions on*, vol. 23, no. 4, pp. 1918–1926, 2008.

[48] F. Schafmeister and J. W. Kolar, "Novel modulation schemes for conventional and sparse matrix converters facilitating reactive power transfer independent of active power flow," in *Power Electronics Specialists Conference, 2004. PESC 04. 2004 IEEE 35th Annual*, vol. 4, 2004 2004, pp. 2917–2923 Vol.4.

[49] M. Rivera, J. Rodriguez, J. R. Espinoza, T. Friedli, J. W. Kolar, A. Wilson, and C. A. Rojas, "Imposed sinusoidal source and load currents for an indirect matrix converter," *Industrial Electronics, IEEE Transactions on*, vol. 59, no. 9, pp. 3427–3435, 2012.

7th Power Electronics, Drive Systems & Technologies Conference (PEDSTC 2016)
16-18 Feb. 2016, Iran University of Science and Technology, Tehran, Iran

Optimized Current Control of Vienna Rectifier Using Finite Control Set Model Predictive Control

Ali R. Izadinia
Dept. of Elec. and Computer Eng.
Isfahan University of Tech.
a.izadinia@ec.iut.ac.ir

Hamid R. Karshenas
Dept. of Elec. and Computer Eng.
Isfahan University of Tech.
karshen@cc.iut.ac.ir

Abstract. **This paper is concerned with finite control set model predictive control (FCS-MPC) method for the current control of a Vienna rectifier. Vienna rectifier has received a lot of attention in recent years as a practical high-power rectifier. This topology presents good performance in terms of input current quality with lower number of controlled switches as compared to other active rectifier structures. Such high-power rectifiers always use current control schemes in order to achieve fast and reliable performance. Among various current control methods, model predictive control (MPC) method is well suited for implementation with microprocessors. Furthermore, by using finite control set model predictive control (FCS-MPC) method, many performance criteria can be defined as cost functions and optimized simultaneously. This will result in very good performance without increasing controller complexity. The basic concepts of Vienna rectifier switching strategy with space-vector modulation (SVM) technique is presented. The implementation of FCS-MPC for Vienna rectifier is described. The selection of optimum control law with the aim of optimizing a cost function is explained. The theoretical analysis is verified using a laboratory-type 500 W experimental setup.**

Keywords- Vienna Rectifier; Model Predictive Control; Current Control

I. INTRODUCTION

The Applications of high power rectifiers in industry have been expanded in recent decades. These rectifiers are employed in applications such as railway traction systems, electrochemical systems and dc arc furnaces [1]. High power rectifiers should be reliable and cost-effective to be viable in industry. In this regards, phase-controlled rectifiers with multi-pulse structure have become industry standard due to their reliability, cost-effectiveness and high power density[2].

In the recent years, more stringent standards have been enacted for enhancing power quality in power systems. Considering that traditional phase-controlled rectifiers suffer from low power factor and harmonic-rich input current, substantial research has been carried out to present solutions without the mentioned drawbacks. Among various solutions, the so-called active rectifiers which are based on forced commutated switches have received noticeable attention both from structure and control point of view [3]-[4]. One of these active structure that has been widely considered in recent years

is Vienna rectifier [5].

Vienna rectifier has some advantages in comparison with other PWM-controlled rectifier topologies. This topology has only 3 active switches which decreases the complexity of structure and control [6]. In addition, the voltage stress on switching devices in this topology is reduced resulting in reduction in switching losses [7]. Moreover, the rating of inductive element is reduced due to the harmonic level of main currents as opposed to 2-level rectifier topologies [8].

Current control of three phase rectifiers has always been an important and challenging topic in power electronic converters. In past decades, the availability and development of powerful and fast microprocessors has evolved the way for the implementation of digital current control in power converters. One of the current control techniques with some interesting features is model predictive control (MPC) scheme.

Predictive control comprises a wide family of controllers with different approaches [9]. In all predictive control methods, control system uses the model of converter and system to predict the future behavior of the controlled variable, e.g. line current. Power converters are nonlinear systems in general, consisting of linear and nonlinear components and finite number of switching devices. Furthermore, the control system of power converter usually has to consider some inevitable constraints like maximum voltage, current or modulation index. PID controllers for nonlinear, multi-input, multi output (MIMO) systems with constraints are usually slow and cumbersome to design [10]. MPC has developed in 1970s to overcome PID controller's problems.

Finite control set model predictive control (FCS-MPC) is a control approach that can be categorized in predictive methods and has been introduced for active front-end rectifiers [11], [12], matrix converters[13], and multilevel inverters [9]. FCS-MPC can be easily implemented to solve complex and diverse challenges in control systems [9].In FCS-MPC, control system minimizes a cost function that is usually the error between reference signal and control variables and then selects the optimum switching state of system. This control scheme is able

978-1-5090-0376-1/16 $31.00 © 2016 IEEE

to consider different constraints like minimizing switching frequency [14], switching losses or balancing the dc link voltage in bipolar DC bus configuration [15]. In this control scheme, online optimization can be achieved thanks to fast and powerful microprocessors. Furthermore, by eliminating the modulator in FCS-MPC, the control system complexity is reduced [9]. This paper proposes a FCS-MPC algorithm for the current control of Vienna rectifiers. The procedure for calculating and minimizing a predefined cost function to obtain the sinusoidal line current with adjustable phase shift is described. The rest of the paper is organized in the following manner. In section II, Vienna rectifier model is elaborately described to be apply in FCS-MPC and operating constraints for Vienna rectifier will be defined. The procedure to implementing FCS-MPC that has improved speed with reduced number of calculation cycle described in section III. The control system which is used to be implemented for experimental setup on DSP is investigated in section IV. The performance of the FCS-MPC controlled Vienna rectifier is verified with a 500 W experimental set-up in section V. Finally, the conclusions are drawn in section VI.

II. VIENNA RECTIFIER TOPOLOGY

Fig. 1 shows the structure of a Vienna rectifier [5]. In this structure, the voltages on the rectifier poles (v_{rabc}) depend on the line currents polarity and the states of active switches ($T_{1,2,3}$). If an active switch is turned on, the center point of DC link bus will be connected to the corresponding input phase voltage. Leaving the active switches off allows the rectifier input phase voltage to be connected to either positive or negative dc bus rail via diodes depending on line currents polarity. The upper and lower diodes are used to prevent dc bus short circuit during active switches turn-on.

As a results, the ac-side voltage in this structure has a three-level voltage characteristic.

A. space vector modulation of 3-level converters

The three-level voltage characteristic of Vienna rectifier leads to 27 switching states. By eliminating the 8 redundant states, i.e. those with similar voltage space vectors, 19 non-repetitive states remain with 19 independent voltage space vectors in 6 sectors (Fig.2). In each sector, Vienna rectifier can pick 7 independent space vectors and states [16]. For example, in sector one ($I_a>0$, $I_b<0$, $I_c<0$), the vectors V_0, V_1, V_2, V_6, V_7, V_8, V_{18} can be acquired to shape the line currents. This vectors are located in the shaded area (Sector 1) shown in Fig.2.

B. basic equations

Writing KVL in rectifier pole in Fig.1 yields [11],

$$v_{sabc} = L_s \frac{di_{abc}}{dt} + R_s i_{abc} + v_{rabcN} - v_{nN} \qquad (1)$$

writing (1) in space-vector form yields

$$\bar{V}_s = \frac{2}{3}(v_{sa} + \alpha v_{sb} + \alpha^2 v_{sc}) \qquad (2)$$

$$\bar{V}_s = L_s \frac{d}{dt}(\overline{I_s}) + R_s(\overline{I_s}) + \overline{V_{rN}} - \overline{V_{nN}} \qquad (3)$$

where $\overline{I_s}$, $\overline{V_{rN}}$ and $\overline{V_{nN}}$ are the space vectors associated with input line current, input phase voltage to dc center point voltage and dc center point to neutral voltage respectively. It should be noted that the last term in (3) is equal to zero.

$$\frac{2}{3}(\overline{V_{nN}}) = \frac{2}{3}v_{nN}(1 + \alpha + \alpha^2) = 0 \qquad (4)$$

The rectifier input voltage in (3) can be defined by switching states and DC link voltage,

$$\overline{V_{rN}} = \overline{S_r} * V_{DC} \qquad (5)$$

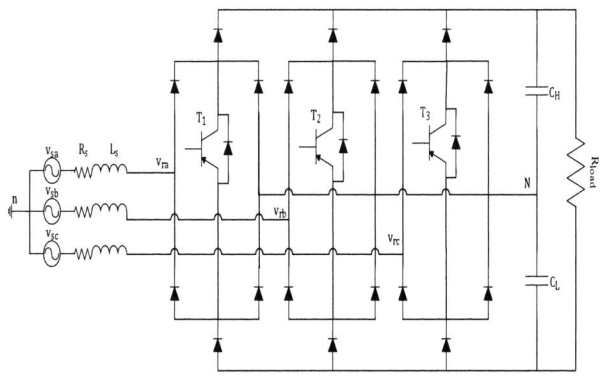

Fig. 1.Three Phase/Level/Switch Vienna Rectifier

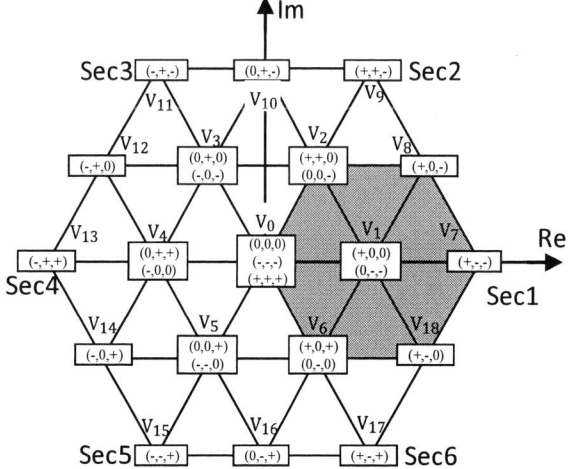

Fig. 2. Vienna Rectifier 19 space vectors, presented by their electrical states [17]

978-1-5090-0376-1/16 $31.00 © 2016 IEEE   597

where $\bar{S}_r$ is the switching state vector of the rectifier and V_{DC} is DC link voltage. The switching state vector of the rectifier is defined by

$$\bar{S}_r = \frac{2}{3}(S_1 + \alpha S_2 + \alpha^2 S_3) \qquad (6)$$

where S_1, S_2 and S_3 are the switching states of each rectifier leg and will have the value of 1, 0 and -1 depending on current polarity and T_1, T_2 and T_3 state.

III. FINITE CONTROL SET MODEL PREDICTIVE CONTROL (FCS-MPC) STRATEGY

As explained in the introduction, FCS-MPC is used where the traditional PID controller is slow and the system is multi-objective and non-linear. To implementing this strategy, the following steps must be taken in design stage [18]:

1. Modeling the converter topology including all possible switching state,

2. Defining the cost function based on the required performance,

3. Discretizing the system model to predict the future state of input variables such that the cost function defined in 2 is optimized.

In this strategy, online optimization can be achieved and system constraints can be easily included in control law. Furthermore, there is no need to modulator in this control scheme so system complexity is reduced. This is done at the expense of variable switching frequency. A basic block diagram of this control strategy used to implement current control for a Vienna rectifier is shown in Fig.3.

In this control scheme, by discretizing the converter model and applying the measured line currents ($i_s(k)$) and main voltages ($v_s(k)$) as well as seven possible vectors that can be selected in next sampling interval, seven possible predicted current ($i_s(k+1)$) are generated. The predicted current is then compared with reference current ($i_s^*(k+1)$) which is created by outer control loop, e.g. DC link regulation loop.

The minimization step in the control system takes place in every single sampling interval and is a challenging part of this control scheme due to the large number of calculations that is needed for comparing, minimizing and selecting optimum state. That is why it is very important to optimize the microprocessor programming required for these stages.

The cost function for current prediction technique can be expressed in orthogonal coordinates as

$$g = |i_{s\alpha}^* - i_{s\alpha}| + |i_{s\beta}^* - i_{s\beta}| \qquad (7)$$

where $i_{s\alpha}^*$ and $i_{s\beta}^*$ are the real and imaginary parts of reference current and $i_{s\alpha}$ and $i_{s\beta}$ are the real and imaginary parts of predicted current respectively.

IV. CONTROL SYSTEM IMPLEMENTATION

A. Rectifier Model Discretization

The first step for implementing the FCS-MPC is to discretize (3) and substituting (4) and (5) into it. Using Euler forward method yields

$$\bar{I}_s(k+1) = \frac{T_s}{L_s}[\bar{V}_s(k) - \overline{V_{rN}}] + (1 - \frac{R_s L_s}{T_s})\bar{I}_s(k) \qquad (8)$$

where T_s is the sampling time.

By using Clarke's transformation to switching states vector (6), the switching states of rectifier can be obtained in $\alpha\beta$ reference frame, as shown in Table I for sector 1.

By applying seven possible vectors for the next sampling interval to (9) and (10), all seven predicted currents in orthogonal coordinates can be obtained.

$$i_{s\alpha}(k+1) = \frac{T_s}{L_s}[v_{s\alpha}(k) - S_{r\alpha}(k) * \frac{V_{DC}}{2}] + (1 - \frac{R_s L_s}{T_s})i_{s\alpha}(k)$$

$$(9)$$

$$i_{s\beta}(k+1) = \frac{T_s}{L_s}[v_{s\beta}(k) - S_{r\beta}(k) * \frac{V_{DC}}{2}] + (1 - \frac{R_s L_s}{T_s})i_{s\beta}(k)$$

$$(10)$$

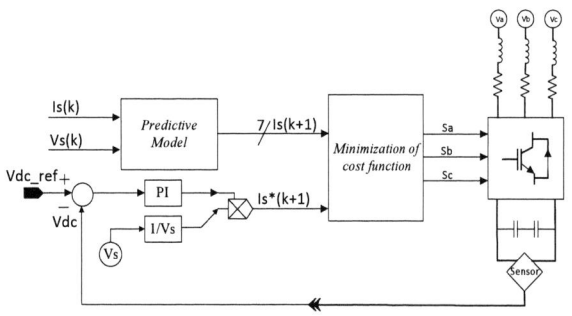

Fig. 3. Basic principle of predictive current control [19]

TABLE I. SWITCHING STATES AND CORRESPONDING VECTORS IN SECTOR 1

Vector	T_1	T_2	T_3	S_1	S_2	S_3	$S_{r\alpha}$	$S_{r\beta}$
V_7	0	0	0	+1	-1	-1	$4/3$	0
V_{18}	0	0	1	+1	-1	0	1	$-\sqrt{3}/3$
V_8	0	1	0	+1	0	-1	1	$\sqrt{3}/3$
V_1	0	1	1	+1	0	0	$2/3$	0
V_1(redundant)	1	0	0	0	-1	-1	$2/3$	0
V_6	1	0	1	0	-1	0	$1/3$	$-\sqrt{3}/3$
V_2	1	1	0	0	0	-1	$1/3$	$\sqrt{3}/3$
V_0	1	1	1	0	0	0	0	0

As discussed in Sec. III, the calculation of all predicted currents for next sampling intervals is a challenging part of this control scheme due to the large number of calculations for selecting the optimum state. It should be noted that if the calculation time is comparable to the sampling time, a delay time will be generated in applying the next optimum state. This delay time can lead to a significant ripple in line current. Therefore, the programming code must be as optimized as possible. Another method to avoid this problem is to properly model the time delay and compensate it in the control system [20].

By looking at (9) and (10), one can realize that by applying the switching states presented in Table I to the cost function, all 19 predicted currents corresponding to 19 independent space vectors can be calculated by a finite number of algebraic expressions in αβ coordinates. It can be shown that these expressions can be categorized into 5 expressions in α axis and 3 expressions in β axis. These algebraic expressions have shown in Table II. Furthermore, Table III shows how each current vectors in sector 1 is calculated by the expressions given in Table II.

B.   Generating Reference Current

The target that was followed in this study was generating sinusoidal current with adjustable phase shift regard to grid voltage by Vienna rectifier. Therefore, synchronizing the reference current with grid voltage had considered in the control system and the phase locked loop (PLL), shown in Fig. 4, had implemented in control loop.

TABLE II.   ALGEBRAIC EXPRESSIONS FOR VECTORS IN SECTOR 1 WITH $X \stackrel{\text{def}}{=} \frac{T_S}{L}$ AND $Y \stackrel{\text{def}}{=} 1 - X * R$

Predicted Expression	Definition
$i_{\alpha 0}[k+1]$	$X * v_{s\alpha}[k] + Y * i_\alpha[k]$
$i_{\alpha 1}[k+1]$	$i_{\alpha 0}[k+1] - \frac{1}{3} X * \frac{v_{DC}}{2}$
$i_{\alpha 2}[k+1]$	$i_{\alpha 0}[k+1] - \frac{2}{3} X * \frac{v_{DC}}{2}$
$i_{\alpha 3}[k+1]$	$i_{\alpha 0}[k+1] - \frac{3}{3} X * \frac{v_{DC}}{2}$
$i_{\alpha 4}[k+1]$	$i_{\alpha 0}[k+1] - \frac{4}{3} X * \frac{v_{DC}}{2}$
$i_{\beta 0}[k+1]$	$X * v_{s\beta}[k] + Y * i_\beta[k]$
$i_{\beta 1}[k+1]$	$i_{\beta 0}[k+1] - \frac{\sqrt{3}}{3} X * \frac{v_{DC}}{2}$
$i_{\beta 2}[k+1]$	$i_{\beta 0}[k+1] + \frac{\sqrt{3}}{3} X * \frac{v_{DC}}{2}$

TABLE III.   PREDICTED CURRENT VECTORS IN SECTOR 1

Vector	Corresponding Current Vector
V_7	$\bar{I}_s(k+1) = i_{\alpha 0}[k+1] + i_{\beta 0}[k+1]$
V_{18}	$\bar{I}_s(k+1) = i_{\alpha 2}[k+1] + i_{\beta 1}[k+1]$
V_8	$\bar{I}_s(k+1) = i_{\alpha 2}[k+1] + i_{\beta 1}[k+1]$
V_1	$\bar{I}_s(k+1) = i_{\alpha 1}[k+1] + i_{\beta 0}[k+1]$
V_6	$\bar{I}_s(k+1) = i_{\alpha 3}[k+1] + i_{\beta 2}[k+1]$
V_2	$\bar{I}_s(k+1) = i_{\alpha 3}[k+1] + i_{\beta 1}[k+1]$
V_0	$\bar{I}_s(k+1) = i_{\alpha 0}[k+1] + i_{\beta 0}[k+1]$

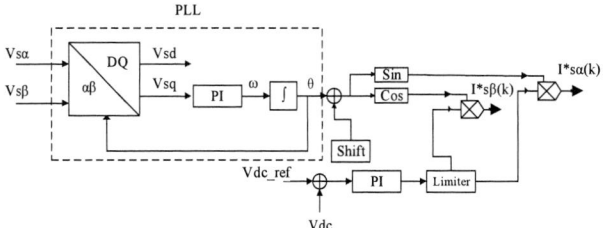

Fig. 4. Block Diagram of PLL

In this PLL, which is based on Park transformation, the Q axis quantity become zero after passing through the PI controller and therefore θ controlled in such a way that grid voltage becomes in phase with D axis. The detail analysis of this PLL scheme had discussed in [21].

To obtain the optimum state that should be selected for next intervals, (11) has to be minimized. Therefore, control system has to predict the reference current for next interval by using only the present and past values of reference current.

$$g = |i_{s\alpha}^*(k+1) - i_{s\alpha}(k+1)| + |i_{s\beta}^*(k+1) - i_{s\beta}(k+1)| \tag{11}$$

A promising solution is to applying Lagrange extrapolation to obtain the predicted reference current [9]. The 2th-order of this extrapolation for reference current can be expressed as,

$$i^*(k+1) = \sum_{m=0}^{2}(-1)^{2-m}\begin{bmatrix}2+1\\m\end{bmatrix}i^*(k+m-2) = \\ 3i^*(k) - 3i^*(k-1) + i^*(k-2) \tag{12}$$

It should be noted that for sufficient small sampling time (T_s) the error between $i_s^*(k+1)$ and $i_s^*(k)$ can be neglected.

V.   EXPERIMENTAL RESULTS

A.   Experimental Setup

In order to validate the performance of FCS-MPC controller on Vienna rectifier, a laboratory type 500 W experimental setup was implemented. The experimental setup parameters are listed in Table V.

A DSP control board is used to implement the control law which is connected to the host computer through a serial bus to transmit various control variables in real-time. Fig. 5 shows different parts of experimental setup.

TABLE IV.   EXPERIMENTAL SETUP PARAMETERS

Parameter	Description	Value
V_s	Supply Line to Line Voltage	400 V_{rms}
f_s	Supply Frequency	50 Hz
V_{DC}	Reference DC-Link Voltage	160 V
R_{load}	DC Load Resistor	50 Ω
$C_{H,L}$	DC Link Capacitors	2 mF
T_S	Sampling Time	50 μs
$\frac{N_1}{N_2}$	Input Transformer Turns Ratio	6
L_s	Input Filter Inductance	3 mH
R_s	Input Filter Resistance	0.2 Ω

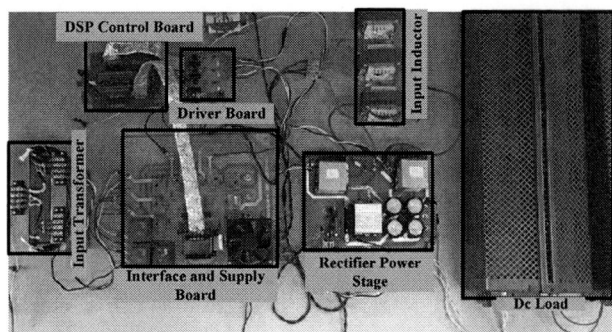

Fig. 5. Laboratory Prototype of Vienna rectifier

B. Experimental Result

Fig. 6 shows the 7 errors obtained by comparing 7 predicted currents with reference signals. The selected error, which is the smallest error between all 7 errors, is shown in bold black. This error is the result of applying the optimum switching state to the cost function defined in (7). Fig. 7 shows the reference and actual line current αβ coordinates. As can be seen, a good tracking performance has been achieved. Fig. 6 and 7 are captured with the real-time monitoring capability of controller variables in experimental setup. Fig. 8 illustrates three phase line currents and scaled phase a voltage. In this figure, the phase angle between line current and phase voltage has been set to zero to achieve unity power factor operation. Fig. 9 shows line current and DC bus voltage with step change in DC bus voltage reference (Fig. 9(a)) and step load change (Fig. 9(b)).

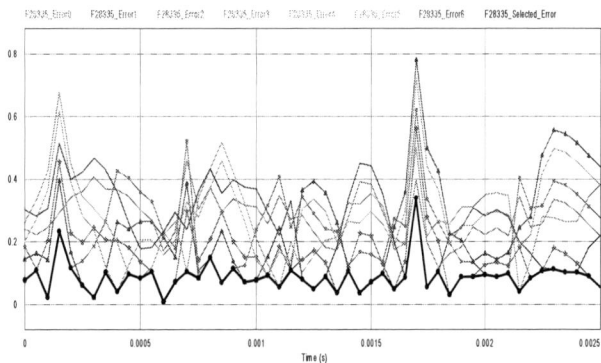

Fig. 6. The 7 error corresponding to the 7 predicted current and the selected error in every time step (T_s=50 μs).

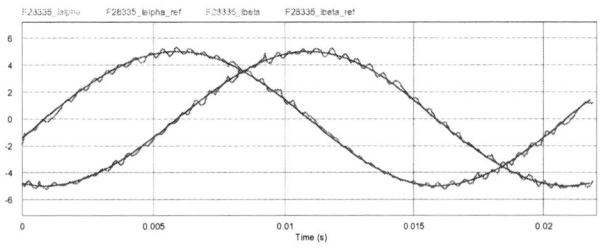

Fig. 7. Line currents and reference current in αβ coordinates

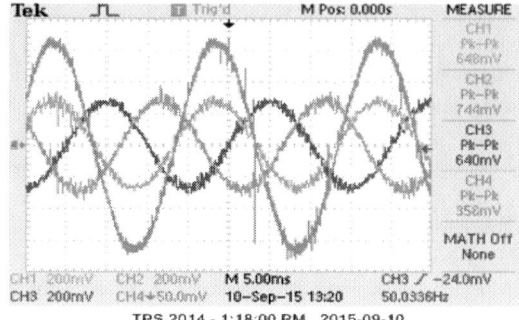

Fig. 8. Three phase line current and phase A voltage

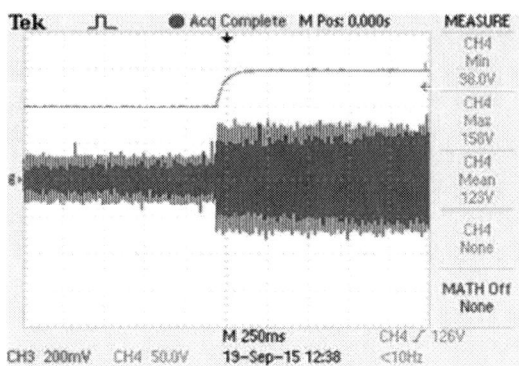

(a)

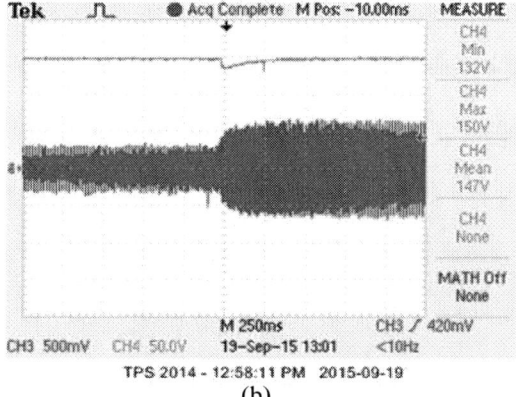

(b)

Fig. 9. DC bus voltage and line current with (a) DC bus reference step change, (b) Step load change

VI. CONCLUSION

In this paper the current control of Vienna rectifiers with finite control set model predictive control (FCS-MPC) strategy was discussed. In this strategy there is no need to add any type of modulator in the control system and gate signals are directly generated by minimizing a given cost function. The efficient approach to implementing this strategy was studied. It was shown that to achieve the online optimization and generating 19 space vectors associated with space vector modulation (SVM), there is only need to calculate 5 algebraic expressions in α coordinates and 3 algebraic expressions in β coordinates. The proposed method has a good dynamic response and can closely track the reference signal. The current control strategy presented in this paper employs a discrete time model of converter and the control law is obtained by solving the equations in discrete form. Therefore, this strategy is well suited for implementation in digital control platforms exploiting high-performance DSP. The results obtained from a laboratory type experimental setup show that the FCS-MPS is a promising solution for controlling the linear and non-linear power converter compared with classical methods.

REFRENCES

[1]   S. Yongsug and P. K. Steimer, "Application of IGCT in High-Power Rectifiers," *Industry Applications, IEEE Transactions on,* vol. 45, pp. 1628-1636, 2009.

[2]   R. L. Alves and I. Barbi, "Analysis and Implementation of a Hybrid High-Power-Factor Three-Phase Unidirectional Rectifier," *Power Electronics, IEEE Transactions on,* vol. 24, pp. 632-640, 2009.

[3]   R. Ghosh and G. Narayanan, "Control of Three-Phase, Four-Wire PWM Rectifier," *Power Electronics, IEEE Transactions on,* vol. 23, pp. 96-106, 2008.

[4]   H. Yoo, K. Jang-Hwan, and S. Seung-Ki, "Sensorless Operation of a PWM Rectifier for a Distributed Generation," *Power Electronics, IEEE Transactions on,* vol. 22, pp. 1014-1018, 2007.

[5]   J. W. Kolar and F. C. Zach, "A novel three-phase utility interface minimizing line current harmonics of high-power telecommunications rectifier modules," in *Telecommunications Energy Conference, 1994. INTELEC '94., 16th International,* 1994, pp. 367-374.

[6]   N. B. H. Youssef, F. Fnaiech, and K. Al-Haddad, "Small signal modeling and control design of a three-phase AC/DC Vienna converter," in *Industrial Electronics Society, 2003. IECON '03. The 29th Annual Conference of the IEEE,* 2003, pp. 656-661 vol.1.

[7]   N. Bel Haj Youssef, K. Al-Haddad, and H. Y. Kanaan, "Implementation of a New Linear Control Technique Based on Experimentally Validated Small-Signal Model of Three-Phase Three-Level Boost-Type Vienna Rectifier," *Industrial Electronics, IEEE Transactions on,* vol. 55, pp. 1666-1676, 2008.

[8]   J. W. Kolar and H. Ertl, "Status of the techniques of three-phase rectifier systems with low effects on the mains," in *Telecommunication Energy Conference, 1999. INTELEC '99. The 21st International,* 1999, p. 16 pp.

[9]   V. Yaramasu, B. Wu, M. Rivera, and J. Rodriguez, "Predictive current control and DC-link capacitor voltages balancing for four-leg NPC inverters," in *Industrial Electronics (ISIE), 2013 IEEE International Symposium on,* 2013, pp. 1-6.

[10]   P. Karamanakos, T. Geyer, N. Oikonomou, F. D. Kieferndorf, and S. Manias, "Direct Model Predictive Control: A Review of Strategies That Achieve Long Prediction Intervals for Power Electronics," *Industrial Electronics Magazine, IEEE,* vol. 8, pp. 32-43, 2014.

[11]   M. Parvez, S. Mekhilef, N. M. L. Tan, and H. Akagi, "An improved active-front-end rectifier using model predictive control," in *Applied Power Electronics Conference and Exposition (APEC), 2015 IEEE,* 2015, pp. 122-127.

[12]   D. E. Quevedo, R. P. Aguilera, M. A. Perez, P. Cortes, and R. Lizana, "Model Predictive Control of an AFE Rectifier With Dynamic References," *Power Electronics, IEEE Transactions on,* vol. 27, pp. 3128-3136, 2012.

[13]   M. Rivera, J. Rodriguez, W. Bin, J. R. Espinoza, and C. A. Rojas, "Current Control for an Indirect Matrix Converter With Filter Resonance Mitigation," *Industrial Electronics, IEEE Transactions on,* vol. 59, pp. 71-79, 2012.

[14]   J. Holtz and S. Stadtfeld, "A predictive controller for the stator current vector of ac machines fed from a switched voltage source," in *JIEE IPEC-Tokyo Conf,* 1983, pp. 1665-1675.

[15]   F. Rojas-Lobos, R. Kennel, and R. Cardenas-Dobson, "Current control and capacitor balancing for 4-leg NPC converters using finite set model predictive control," in *Industrial Electronics Society, IECON 2013 - 39th Annual Conference of the IEEE,* 2013, pp. 590-595.

[16]   R. Burgos, L. Rixin, P. Yunqing, W. Fei, D. Boroyevich, and J. Pou, "Space Vector Modulator for Vienna-Type RectifiersBased on the Equivalence BetweenTwo- and Three-Level Converters:A Carrier-Based Implementation," *Power Electronics, IEEE Transactions on,* vol. 23, pp. 1888-1898, 2008.

[17]   H. Lijun, L. Bin, Z. Ming, W. Yong, and L. M. Tolbert, "Equivalence of SVM and Carrier-Based PWM in Three-Phase/Wire/Level Vienna Rectifier and Capability of Unbalanced-Load Control," *Industrial Electronics, IEEE Transactions on,* vol. 61, pp. 20-28, 2014.

[18]   J. Rodriguez and P. Cortes, "Model Predictive Control," in *Predictive Control of Power Converters and Electrical Drives,* ed: Wiley-IEEE Press, 2012, pp. 31-39.

[19]   J. Rodriguez, J. Pontt, C. Silva, P. Cortes, U. Amman, and S. Rees, "Predictive current control of a voltage source inverter," in *Power Electronics Specialists Conference, 2004. PESC 04. 2004 IEEE 35th Annual,* 2004, pp. 2192-2196 Vol.3.

[20]   P. Cortes, J. Rodriguez, P. Antoniewicz, and M. Kazmierkowski, "Direct Power Control of an AFE Using Predictive Control," *Power Electronics, IEEE Transactions on,* vol. 23, pp. 2516-2523, 2008.

[21]   M. Ebrahimi, H. R. Karshenas, and M. Hassanzahraee, "Comparison of orthogonal quantity generation methods used in single-phase grid-connected inverters," in *IECON 2012 - 38th Annual Conference on IEEE Industrial Electronics Society,* 2012, pp. 5932-5937.

978-1-5090-0376-1/16 $31.00 © 2016 IEEE

7th Power Electronics, Drive Systems & Technologies Conference (PEDSTC 2016)
16-18 Feb. 2016, Iran University of Science and Technology, Tehran, Iran

Predictive Control of a Five-Level NPC Inverter Using a Three-Phase Coupled Inductor

Fazel Seyed Saeed
Iran University of Science and Technology (IUST)
School of Railway Engineering (SRE)
Tehran, Iran
fazel@iust.ac.ir

Piryaei Hamid Reza
Iran University of Science and Technology (IUST)
School of Railway Engineering (SRE)
Tehran, Iran
hamidpiryaei@gmail.com

Abstract— **The predictive control is presented for a five-level neutral point clamped inverter using a three-phase coupled inductor. This topology produces the five-level output voltage with only 12 switches that in comparison with other five-level counterparts, it uses half. The predictive method uses the discrete model of the inverter and load to predict the future load current and capacitor voltages for all possible switching states. Then chooses the best switching state in each sampling period by minimizing a cost function. In fact, the switching state that generates the best reference current tracking and also the best DC link voltage balance, and produces the minimum value of the common mode voltage, will be selected. The controller obtains good reference tracking with less current harmonic distortion. Also, ensures fast dynamic response. The control method has been simulated in The MATLAB/SIMULINK. The feasibility of the predictive control is verified by showing a good performance.**

Keywords—Coupled inductor inverter (CII), Predictive control, Five-level inverter, reference tracking, DC link voltage balance, Common mode voltage elimination

I. INTRODUCTION

Today, multilevel power converters are more desirable due to their greater number of voltage levels. The main features of multilevel converters that caused to be more demanding are:

1) "They can generate output voltages with extremely low distortion and lower dv/dt.
2) They draw input current with very low distortion.
3) They generate smaller common-mode (CM) voltage, thus reducing the stress in the motor bearings.
4) They can operate with a lower switching frequency" [8].

Recently, multilevel inverters with coupled inductors have drawn some researchers' interest and a halfbridge 3-level inverter has been intended using two power switches, two diodes, and two coupled inductors [10]–[13].

One of the multilevel voltage source inverter which is almost new, is a five-level neutral point clamped inverter using a three-phase coupled inductor that has been developed in [3]. In comparison to other five-level counterparts, this configuration uses half the number of power switches (only 12 switches) and eliminates the requirement for the dead time to avoid shoot-through current that are its benefits [2].

Current control of a three-phase inverter is one of the most important and classical subjects in power electronics and has been extensively studied in the last decades. Nonlinear methods, like hysteresis control and linear methods, like proportional-integral controllers using pulse width modulation (PWM) are presented in literatures [5].

Predictive control presents several advantages that make it suitable for the control of power converters: the concepts are intuitive and easy to understand; it can be applied to a variety of systems; constraints and nonlinearities can be easily included; multivariable cases can be considered; and the resulting controller is easy to implement. It requires a high number of calculations, compared to a classical control scheme, but the fast microprocessors available today make possible the implementation of predictive control [1].

Among the different kinds of predictive control, the finite control set model predictive control (FCS-MPC) is almost the best one with these advantages to other predictive control types [1]:

- No modulator
- Online optimization
- Low complexity (N=1)
- Constraints can be included.

The finite control set model predictive control with a prediction horizon of one sample time is a simple and very flexible control scheme that does not require internal current control loops and modulators[9].

This paper presents the finite control set model predictive control (FCS-MPC) for control a three phase five-level neutral point clamped coupled inductor inverter (NPC-CII). A brief description of the inverter topology and the predictive method including the system model used for current prediction and the cost function used for switching state selection is presented. Also the process of selecting the optimum switching state to obtain the neutral point voltage balance and common mode voltage reduction is explained.

978-1-5090-0376-1/16 $31.00 © 2016 IEEE

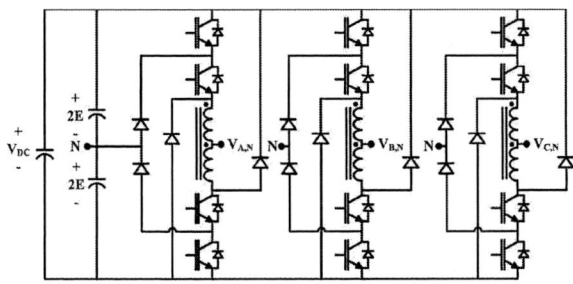

Fig. 1. Five-level Three-phase Neutral Point Clamped Coupled Inductor Inverter (NPC-CII) Topology [7].

Simulation results are presented and verify good performance of the FCS-MPC method in controlling this topology by showing features such as reference current tracking.

II. COUPLED INDUCTOR INVERTER TOPOLOGY

The five-level NPC-CII topology is shown in Fig. 1. Split-wound coupled inductors are placed at the NPC inverter outputs between the upper and lower middle switches in each inverter leg [2]. Each leg is composed of four active switches, two clamping diodes connected to the neutral point, and two free wheel diodes connected to the dc bus [2]. In comparison to standard NPC (three-level NPC), NPC-CII has two additional state ($\pm(1/4)\,V_{dc}$) with two states charge and discharge in coupled inductor winding for each as shown in Table I. So it has 7 possible states for each phase and 343 switching states when considering the three-phase structure. The various possible switching states for one inverter leg are shown in Fig. 2. The switching states in Fig. 2(iv) and (v) produce the extra voltage level of $+(1/4)\,V_{dc}$, and the switching states in Fig. 2(vi) and (vii) produce the extra voltage level of $-(1/4)\,V_{dc}$. The additional $\pm(1/4)\,V_{dc}$ voltage levels are produced by the center-tap connnection of the coupled inductors producing a voltage midway between the split capacitor dc rail voltages [2].

III. PREDICTIVE CURRENT CONTROL METHOD

First, the load model and then the discrete-time model of system is needed. So the system should be discretized with forward Euler method and T_s as sampling period. finally, a cost function will be defined to fulfill the control requirements.

A. Load model

The load is three phase which consists of resistor (R) and inductor (L) in star shape. For simplicity, the three phase equations are written in a vector equation form. So the equation for the load will be:

$$v = Ri + L\frac{di}{dt} \qquad (1)$$

v, i are the inverter terminal voltage vector and phase currents vector respectively.

$$i = \frac{2}{3}(i_a + ai_b + a^2 i_c) \qquad (2)$$

$$v = \frac{2}{3}(v_{aN} + av_{bN} + a^2 v_{cN}) \qquad (3)$$

where $a = e^{\frac{j2\pi}{3}} = -\frac{1}{2} + j\sqrt{3}/2$.

In result, for any possible switching states, the inverter output voltage can be obtained.

B. Discrete-time model of system

The system model will be discretized with replacing the load current derivative di/dt by forward Euler approximation as follows:

$$\frac{di}{dt} \approx \frac{i(k+1)-i(k)}{T_s} \qquad (4)$$

Therefore, by substituting it in (1), the expression for the predicted load current will be:

$$i(k+1) = \left(1 - \frac{RT_s}{L}\right)i(k) + \frac{T_s}{L}v(k) \qquad (5)$$

where k is represents now and k+1 represents the next sampling instant.

C. Cost function

The cost function for reference tracking is generally in this form:

$$g_1 = |i_\alpha^*(k+1) - i_\alpha(k+1)|$$
$$+|i_\beta^*(k+1) - i_\beta(k+1)| \qquad (6)$$

where $i_\alpha(k+1)$ and $i_\beta(k+1)$ are the real and imaginary parts of the load current that is predicted by the system model. If the sampling frequency is high, it is possible to assume that the refrence current in one sampling period is constant; in other words: $i_\alpha^*(k+1) = i_\alpha^*(k)$ and $i_\beta^*(k+1) = i_\beta^*(k)$. Therefore, the cost function will be:

$$g_1 = |i_\alpha^*(k) - i_\alpha(k+1)| + |i_\beta^*(k) - i_\beta(k+1)| \qquad (7)$$

One of the main challenges in multilevel inverters, is DC link voltage unbalanced; which increases the voltage stress on the switches and create low-order harmonics. It also makes a difference in the life of the capacitors.

The other challenge, is the common mode voltage that circulating a leakage current and can cause motor bearings failures and electromagnetic interference (EMI) [6].

Both issues can be easily fulfilled with predictive control by adding additional terms to the cost function as follows:

$$g_2 = \lambda_{dc} * |V_{c1}(k+1) - V_{c2}(k+1)| \qquad (8)$$

$V_c(k+1)$ is the predicted capacitor voltage. To obtain it, the same approximation of the derivative considered in (4), can be used for the capacitor voltages for a sampling time T_s,

$$\frac{dV_c}{dt} = \frac{V_c(k+1)-V_c(k)}{T_s} \qquad (9)$$

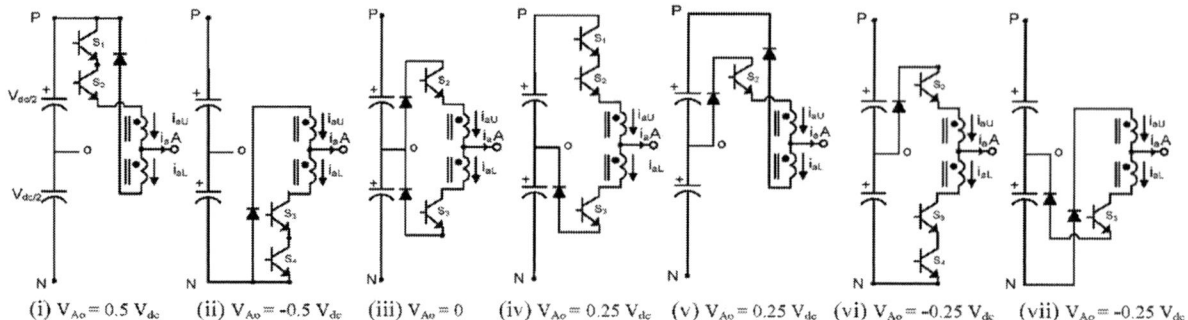

(i) $V_{Ao} = 0.5\ V_{dc}$ (ii) $V_{Ao} = -0.5\ V_{dc}$ (iii) $V_{Ao} = 0$ (iv) $V_{Ao} = 0.25\ V_{dc}$ (v) $V_{Ao} = 0.25\ V_{dc}$ (vi) $V_{Ao} = -0.25\ V_{dc}$ (vii) $V_{Ao} = -0.25\ V_{dc}$

Fig. 2. Effective switching states and the corresponding terminal voltages for a single inverter leg [2].

So the predicted value will be :

$$V_c(k + 1) = V_c(k) + \frac{1}{c} i_c(k) T_s \qquad (10)$$

where $V_c(k)$ is the capacitor voltage and $i_c(k)$ is the capacitor current which can be calculated by considering the load currents and present switching state and there is no need to measurement [4]. Also, the related term to the reduction the common mode voltage will be:

$$g_3 = \lambda_{cm} * |V_{cm}(k + 1)| \qquad (11)$$

The common mode voltage will be:

$$V_{cm}(k + 1) = \frac{1}{3}(v_{aN} + v_{bN} + v_{cN}) \qquad (12)$$

where v_{aN}, v_{bN} and v_{cN} are the voltages of terminals a, b and c with respect to the neutral.

In (8) and (10), λ_{dc} and λ_{cm} are weighting factors, which can be adjusted according to the desired performance, and the greater value for each of them, means that issue is more important. Finally the total cost function will be:

$$g = g_1 + g_2 + g_3 \qquad (13)$$

According to Table I., there are 343 switching state that some of them produce identical voltage vectors. The controller repeats testing all switching states to find the minimum value of cost function in any sampling period. After distinguish the best switching state, it is selected and applied to switches in next sampling period.

IV. SIMULATION RESULTS

Simulation results of applying the FCS-MPC method on the five-level CII topology are presented. The results have been taken by considering the sampling period $T_s = 100\ \mu s$ and DC link voltage $V_{dc} = 300$ V and $C = 4700\ \mu F$. The coupled inductor upper and lower winding inductances are 1.5 mH each, and the coupling factor is 0.99 in each phase winding. The inverter is connected to a three phase load with $R = 15\ \Omega$, $L = 10$ mH. The reference currents are sinusoidal with 10 A amplitude and 50 Hz frequency.

The output three phase load currents that are tracking their refrences are presented in Fig. 3(a). It can be seen that the FCS-MPC control method is very successful in reference tracking. In comparison with classical methods such as PI-PWM, this method does the reference tracking in lower switching frequency. In fact, if both methods are compared at the same switching frequency, the predictive strategy reveals a lower tracking error [1]. Fig. 3(b) illustrates the currents ripple of phase-a more clearly. Also, the current frequency spectrum has been shown in Fig. 3(c). The THD of phase-a output current is calculated until 5000 Hz (100th harmonic order). This THD value is easily a satisfied value under the IEEE Std 519-1992. As seen, the most significant harmonic order is 19th and it's magnitude is very small. The inverter phase-a voltage and Line-to-line terminal voltage has been shown in Figs. 4(a) and (b).

TABLE I.   SWITCHING STATES FOR ONE PHASE [2].

Switching type	S_1	S_2	S_3	S_4	Switching states in Fig. 2	Inverter terminal voltage (V_{Ao})	Coupled inductor winding
P	1	1	0	0	(i)	$0.5\ V_{dc}$	Short-Circuit
M_{P-C}	1	1	1	0	(iv)	$0.25\ V_{dc}$	Charge Mode
M_{P-D}	0	1	0	0	(v)	$0.25\ V_{dc}$	Discharge Mode
O	0	1	1	0	(iii)	0	Short-Circuit
M_{N-C}	0	1	1	1	(vi)	$-0.25\ V_{dc}$	Charge Mode
M_{N-D}	0	0	1	0	(vii)	$-0.25\ V_{dc}$	Discharge Mode
N	0	0	1	1	(ii)	$-0.5\ V_{dc}$	Short-Circuit

978-1-5090-0376-1/16 $31.00 © 2016 IEEE

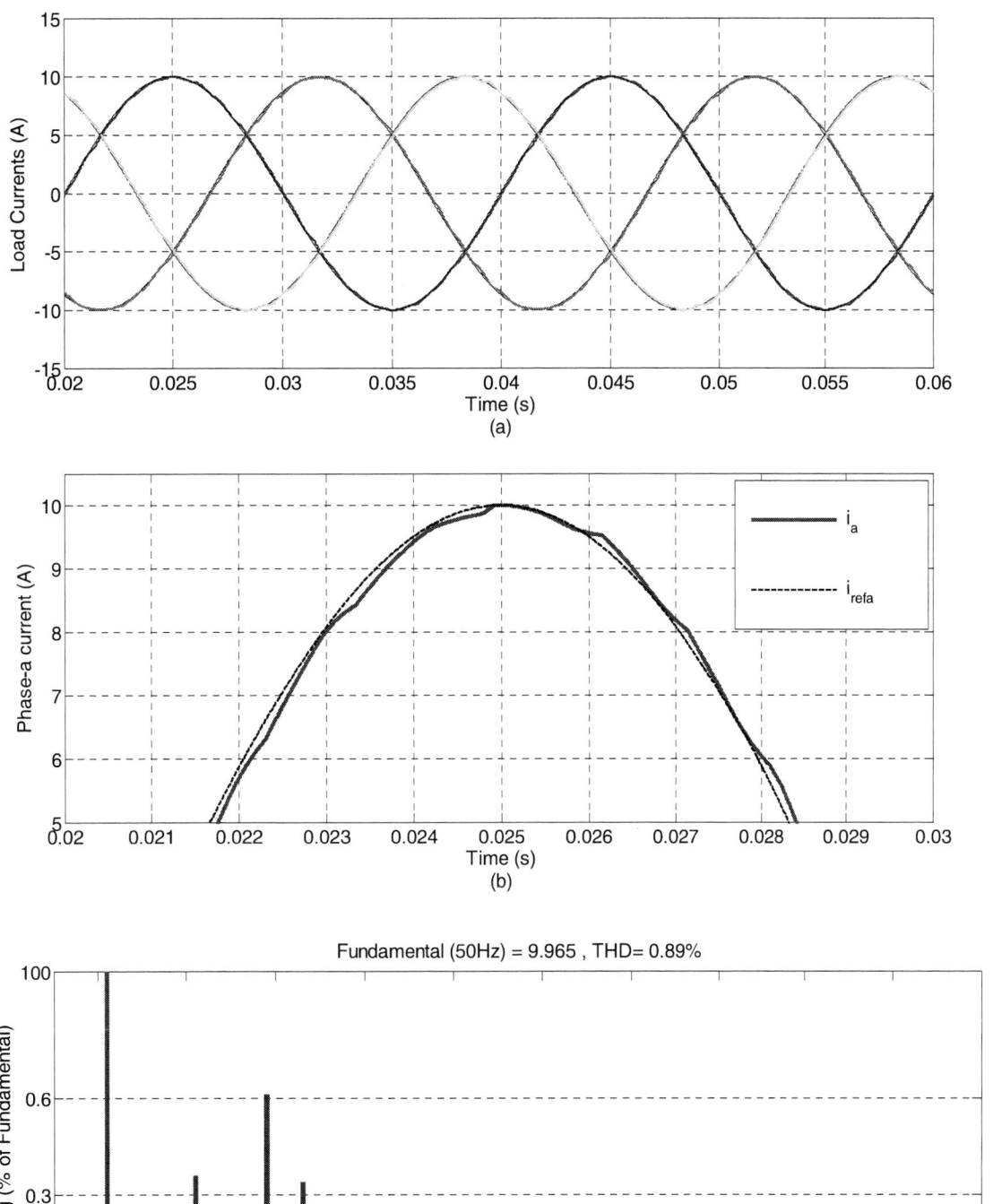

Fig. 3. Simulation results for reference tracking: (a) three phase load currents, (b) focused view of phase-a current, (c) Phase-a current frequency spectrum.

978-1-5090-0376-1/16 $31.00 © 2016 IEEE

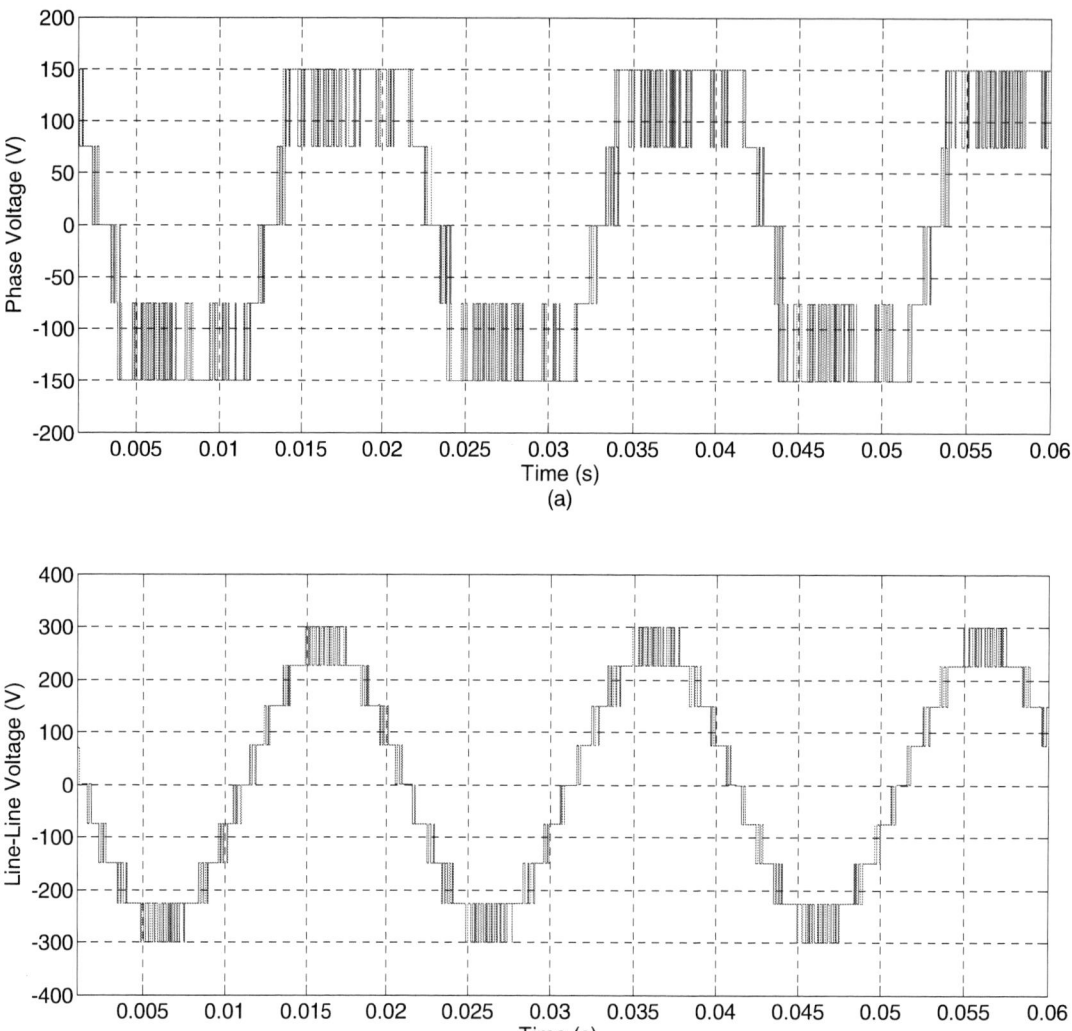

Fig. 4. Simulation results for output voltages: (a) phase-a voltage (V_{aN}), (b) line-to-line voltage.

V. CONCLUSION

The control of a five-level NPC inverter using three phase coupled inductor by one of the predictive control methods that known as FCS-MPC, has been presented. Although this inverter topology uses the half number of switches in comparison with other five level topologies, it has all of the advantages of a multilevel inverter.

The FCS-MPC method is simple and one of it's greatest advantages compared with the traditional controllers, is the capability to control different variables, constraints and additional system requirements (such as common-mode voltages, switching losses, voltage unbalance, etc.). The controller tests all 343 possible switching states and then choose the best one to comply the control requirements. The main control requirement is the load currents reference tracking. The subsequent requirements are DC link capacitor voltage balancing and reducing the common mode voltage.

From the simulation results, it was concluded that applying the FCS-MPC method on this topology ensures: a) good reference current tracking, b) low harmonic content as a result of the large number of voltage vectors participating in control. Also the Neutral point voltage balancing and eliminating the common mode voltage is done and their results will be brought in future papers.

978-1-5090-0376-1/16 $31.00 © 2016 IEEE

REFERENCES

[1] J. Rodriguez and P. Cortes, Predictive control of power converters and electrical drives, John Wiley & Sons, 2012.

[2] B. Vafakhah, J. Ewanchuk, and J. Salmon, "Multicarrier interleaved PWM strategies for a five-level NPC inverter using a three-phase coupled inductor," IEEE Trans. Ind. Appl., vol. 47, no. 6, pp. 2549–2558, 2011.

[3] J. Ewanchuk, B. Vafakhah, and J. Salmon, "A five-/nine-level twelveswitch neutral point clamped inverter for high speed electric drives," in Conj Rec. IEEE ECCE., Sep. 12-16, 2010, pp. 2333-2340.

[4] R. Vargas, P. Cortés, U. Ammann, J. Rodríguez, and J. Pontt, "Predictive control of a three-phase neutral-point-clamped inverter," IEEE Trans. Ind. Electron., vol. 54, no. 5, pp. 2697–2705, Oct. 2007.

[5] J. Rodríguez, S. Member, J. Pontt, C. A. Silva, P. Correa, P. Lezana, P. Cortés, S. Member, and U. Ammann, "Predictive Current Control of a Voltage Source Inverter," vol. 54, no. 1, pp. 495–503, 2007.

[6] J. Rodríguez, J. Pontt, P. Correa, P. Cortés, and C. Silva, "A new modulation method to reduce common-mode voltages in multilevel inverters," IEEE Trans. Ind. Electron., vol. 51, no. 4, pp. 834–839, Aug. 2004.

[7] J. Ewanchuk, R. Ul Haque, a Knight, and J. Salmon, "Three phase common-mode winding voltage elimination in a three-limb five-level coupled inductor inverter," Energy Convers. Congr. Expo. (ECCE), 2012 IEEE, pp. 4794–4801, 2012.

[8] J. Rodriguez, L. Jih-Sheng, and P. Fang Zheng, "Multilevel inverters: A survey of topologies, controls, and applications," IEEE Trans. Ind. Electron., vol. 49, no. 4, pp. 724-738, Aug. 2002.

[9] M. Rivera, J. Rodriguez, V. Yaramasu, and B. Wu, "Predictive load voltage and capacitor balancing control for a four-leg NPC inverter," in 2012 15th International Power Electronics and Motion Control Conference (EPE/PEMC), 2012, pp. DS3c.8–1–DS3c.8–5.

[10] A. M. Knight, J. Ewanchuk, and J. C. Salmon, "Coupled three-phase inductors for interleaved inverter switching," IEEE Trans.Magn., vol. 44, no. 11, pp. 4199–4122, Nov.08.

[11] J. Salmon, A. Knight, and J. Ewanchuk, "Single phase multi-level PWM inverter topologies using coupled inductors," in Proc. IEEE Power Electron. Spec. Conf. (PESC), 08, pp. 802–808.

[12] J. Salmon, J. Ewanchuk, and A. M. Knight, "PWM inverters using split wound coupled inductors," IEEE Trans. Ind. Appl., vol. 45, no. 6, pp. 2001–2009, Nov. 09.

[13] C. Chapelsky, J. Salmon, and A. M. Knight, "Design of the magnetic components for high-performance multilevel half-bridge inverter legs," IEEE Trans. Magn., vol. 45, no. 10, pp. 4785–4788, Oct. 09.

7th Power Electronics, Drive Systems & Technologies Conference (PEDSTC 2016)
16-18 Feb. 2016, Iran University of Science and Technology, Tehran, Iran

A Predictive Control Strategy for a Single-Phase AC-AC Converter

M. Rivera, S. Rojas
Universidad de Talca
Curico, CHILE
Email: marcoesteban@gmail.com
http://www.utalca.cl

P. Wheeler
The University of Nottingham
Nottingham, U.K.
Email: pat.wheeler@nottingham.ac.uk
http://www.nottingham.ac.uk

J. Rodriguez
Universidad Nacional Andrés Bello
Santiago, CHILE
Email: jose.rodriguez@unab.cl
http://www.unab.cl

Abstract—This paper presents a predictive control strategy to control the load current in a single-phase AC-AC converter. The proposed technique consists on a prediction estimation to choose the available switching state of the converter to be applied one step ahead. By considering a cost function, the best switching state to be applied into the converter the next sampling period is selected. All this is done with the objective to obtain a good performance of the load currents. The proposal is verified by both simulation and experimental results, which show a very good dynamic and stationary performance.

Nomenclature

Variable	Description	
$\mathbf{v_i}$	Input voltage	$[v_A \ v_B \ v_C]^T$
$\mathbf{i_i}$	Input current	$[i_A \ i_B \ i_C]^T$
$\mathbf{v}$	Load voltage	$v^+ - v^-$
i_o	Load current	
C_f	Input filter capacitor	

I. Introduction

The AC-AC matrix converter (MC) is formed by an array of bidirectional switches. They are used to connect the AC source directly to the load without the necessity of any large energy storage element [1]. The MCs is characterized because it is a very compact and simple power converter, it has the possibility of load voltage generation with arbitrary amplitude and frequency, it can generate sinusoidal input and load currents, it can operates with unity power displacement factor and the converter has regeneration capability [2], [3]. All these characteristics have been the reason for the research interest in this converter. Venturini and Alesina started the work on matrix converters in 1980 [2], where it was provided a detailed mathematical background about how the low-frequency behavior of the currents and voltages are obtained at the output and input. One of the main problems in the safe operation of this power topolog̣u was the commutation of the bidirectional switches [4]. This issue was solved with the use of intelligent and soft commutation techniques, providing new momentum to research in the area. After more than three decades of research, this power topology is considering in industrial application [5]. In fact, there is one company (Yaskawa) which offers standard units for up to several megawatts and medium voltage using a cascade connection with rated power (and voltages) of 9-114 kVA (200 V and 400 V) for low voltage MC, and 200-6.000

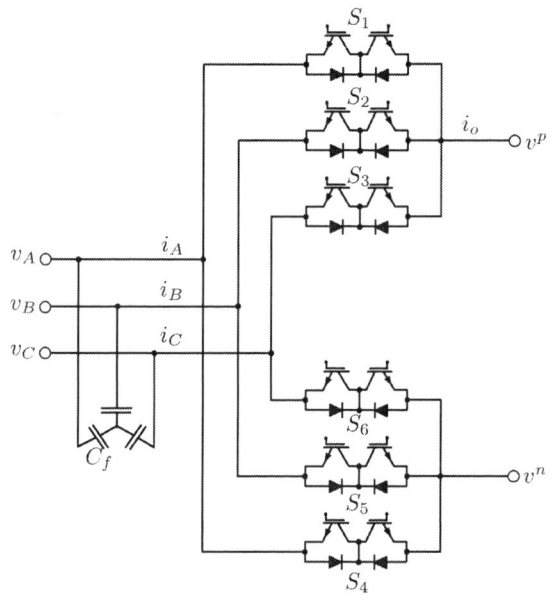

Fig. 1. Topology of the single-phase matrix converter.

kVA (3.3 kV, 6.6 kV) for medium voltage [1]. At the same time, since several years of continuous effort, researchers have been focused to the implementation of different modulation and control techniques for matrix converters [4], [6]–[12].

Regarding the main applications of matrix converters there are two main approaches: (1) speed variation on fans and pumps that require strict control of harmonic distortion and on the other hand (2) applications with high mechanical inertia as centrifuges, cranes and mechanical stairs, in order to employ the regenerative capacity [1].

Predictive control is considered one of the latest and most successful strategies for the effective control of matrix converters [13]. The main objective of this control strategy is the selection of the best switching state (from all the feasible ones) that minimizes an objective function, which is proposed based on the control goals. This control strategy uses the mathematical model of the full system to determine its performance in the future, and, based on this prediction, the optimal switching state is selected to be applied to the converter [14], [15].

TABLE I
FEASIBLE SWITCHING STATES OF THE SINGLE-PHASE MATRIX CONVERTER

Switching State #	S_1	S_2	S_3	S_4	S_5	S_6	v^p	v^n	i_A	i_B	i_C
1	0	0	1	0	0	1	v_C	v_C	0	0	0
2	0	1	0	0	1	0	v_B	v_B	0	0	0
3	1	0	0	1	0	0	v_A	v_A	0	0	0
4	0	0	1	0	1	0	v_C	v_B	0	$-i_o$	i_o
5	0	0	1	1	0	0	v_C	v_A	$-i_o$	0	i_o
6	0	1	0	0	0	1	v_B	v_C	0	i_o	$-i_o$
7	0	1	0	1	0	0	v_B	v_A	$-i_o$	i_o	0
8	1	0	0	0	0	1	v_A	v_C	i_o	0	$-i_o$
9	1	0	0	0	1	0	v_A	v_B	i_o	$-i_o$	0

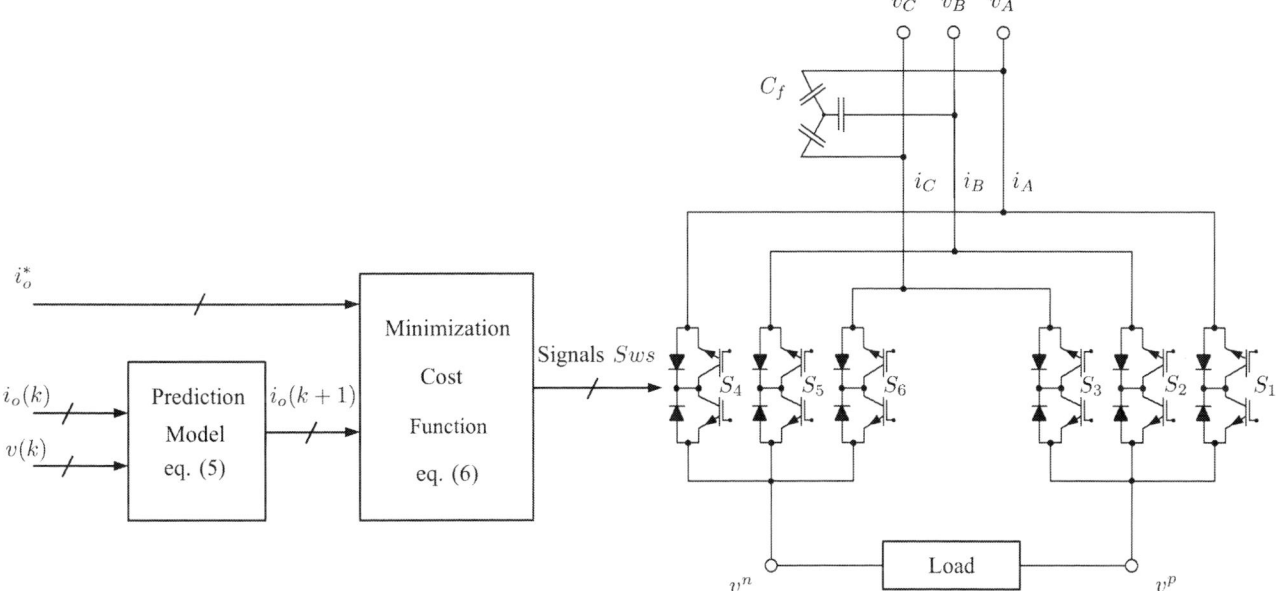

Fig. 2. Proposed control scheme for the single-phase matrix converter.

II. TOPOLOGY AND MATHEMATICAL MODEL OF THE SINGLE-PHASE MATRIX CONVERTER

The single-phase matrix converter topology is depicted in Fig. 1. From this topology it is possible to obtain the mathematical model of the converter which is given by the following equations:

$$v^p = \begin{bmatrix} S_1 & S_2 & S_3 \end{bmatrix} \mathbf{v_i}, \qquad (1)$$

$$v^n = \begin{bmatrix} S_4 & S_5 & S_6 \end{bmatrix} \mathbf{v_i}. \qquad (2)$$

$$\mathbf{i_i} = \begin{bmatrix} S_1 - S_4 \\ S_2 - S_5 \\ S_3 - S_6 \end{bmatrix} i_o. \qquad (3)$$

All these previous equations define the nine feasible commutation states of the power topology [16]. By considering the constrains of no short circuits in the input and no open lines in the output. Table I shows the suitable switching states of

the converter. Finally, the dynamic model of the load is given as:

$$\frac{di_o}{dt} = \frac{1}{L}v - \frac{R}{L}i_o. \qquad (4)$$

III. PREDICTIVE CURRENT CONTROL FOR THE SINGLE-PHASE MATRIX CONVERTER

In Fig. 2 is represented the predictive control scheme for the topology. The proposal pursues the optimal selection of the commutation state of the converter that generates the output current closest to its respective reference at the end of the sampling instant.

The mathematical model of the converter is used to predict the performance of the variable that will be controlled in the next sampling period for each suitable commutation state. The prediction is thus considered to optimize a cost function which deals with the control objective. Among the nine available switching states, the one that generates the lowest value of the cost function is selected to be applied in the following sampling instant.

978-1-5090-0376-1/16 $31.00 © 2016 IEEE

TABLE II
SIMULATION PARAMETERS

Variables	Description	Value
T_s	Sampling time	10, 20, 40 kHz
V_s	Source voltage	56, 112 V
f_s	Source frequency	50 Hz
R	Load resistor	10 Ω
L	Load inductor	10 mH

A. Prediction model

A forward Euler approximation in eq. (4) is used to obtain the load current prediction:

$$i_o(k+1) = d_1 v(k) + d_2 i_o(k), \qquad (5)$$

where, $d_1 = T_s/L$ and $d_2 = 1 - RT_s/L$ are constants dependent on load parameters and the sampling time T_s [17].

B. Cost function

The cost function is defined as:

$$g(k+1) = (i_o^*(k+1) - i_o(k+1))^2, \qquad (6)$$

where the error between the reference and the predicted value of the load current is considered. The objective of this cost function is to obtain g value very close to zero. The commutation state that minimizes this cost function is selected and then applied to the converter at the next sampling period.

IV. SIMULATION AND EXPERIMENTAL RESULTS

The parameters shown in Table II have been used to simulate the performance of the predictive control strategy. The same parameters have been considered for the experimental validation. The figures are divided in a) load current (red), b) load voltaje (blue) and source voltage (purple).

In Fig. 3 simulation are presented in steady state where an output current amplitude of 6 Apk @ f_o=50 Hz has been imposed. The same conditions are evaluated experimentally as depicted in Fig. 4. Both simulation and experimental results are obtained for a sampling frequency of $f_s = 20$ kHz. In Fig. 5 and Fig. 6 are presented simulation and experimental results for a sampling frequency of $f_s = 40$ kHz, respectively. In all these cases is observed a very good tracking of the load current to its reference under the different conditions.

One important issue that it is observed in the experimental results is the lower switching frequency in respect to the simulations which is evident in the figures. Transient analysis in both simulation and experimental have also been done. In Fig. 7 and Fig. 8 are shown simulation results for a step change in amplitude and frequency, respectively. The amplitude change is from 3 Apk to 6 Apk @ $f_o = 50$ Hz and the frequency change is from $f_o = 50$ Hz to $f_o = 25$ Hz at a sampling frequency of $f_s = 10$ kHz. The same analysis is done in experimental implementation as shown in Fig. 9 and Fig. 10, respectively. Again, in all these cases a very good tracking of the load current to its reference is obtained with a very fast dynamic response.

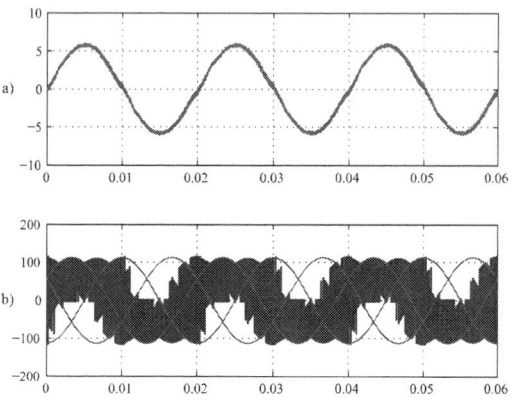

Fig. 3. Simulation results in steady state: i_o = 6Apk; f_o = 50 Hz; f_s = 20 kHz; v_i = 112 Vpk;

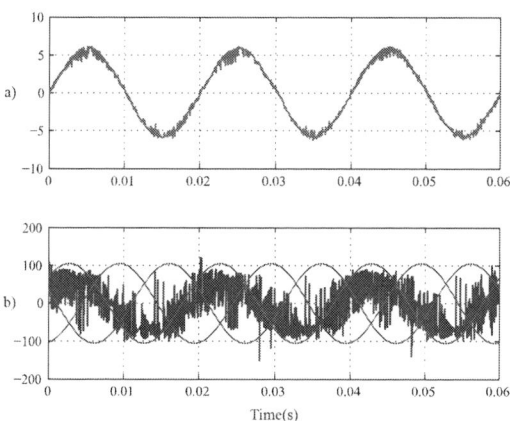

Fig. 4. Experimental results in steady state: i_o = 6 Apk; f_o = 50 Hz; f_s = 20 kHz; v_i = 112 Vpk;

In order to assess the performance of the predictive current control scheme, two parameters are defined: the mean current tracking error and the output current THD. The percentage mean absolute current reference tracking error $\%e_{io}$ is defined as the absolute difference between the reference and load currents (for m number of samples) with respect to the *rms* value of load current [18], [19]:

$$\%e_{i_o} = \frac{\frac{1}{m}\sum_{k=0}^{m}|i_o^*(k) - i_o(k)|}{rms(i_o(k))}. \qquad (7)$$

In the case of the output current THD, it is defined as follows:

$$\% \text{THD} = \frac{\sqrt{s_2^2 + s_3^2 + .. + s_n^2}}{s_1}, \qquad (8)$$

where s_n and s_1 are n^{th} order harmonic and fundamental components of the signal, respectively.

Table III shows the mean average error of the load current for different sampling frequency and references evaluated in simulation and experiments. As expected, in the experiments a higher error is obtained due to some unknown parameters

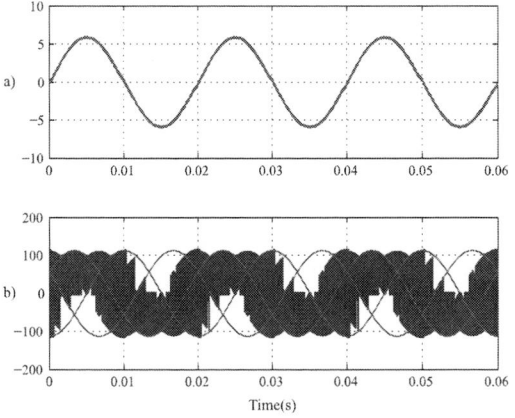

Fig. 5. Simulation results in steady state: i_o = 6 Apk; f_o = 50 Hz; f_s = 40 kHz; v_i = 112 Vpk;

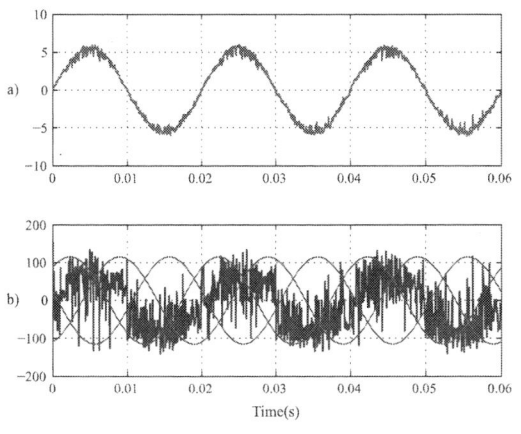

Fig. 6. Experimental results in steady state: i_o = 6 Apk; f_o = 50 Hz; f_s = 40 kHz; v_i = 112 Vpk;

Fig. 7. Simulation results in transient state: i_o=3-6 Apk; f_o = 50 Hz; f_s = 10 kHz; v_i = 112 Vpk;

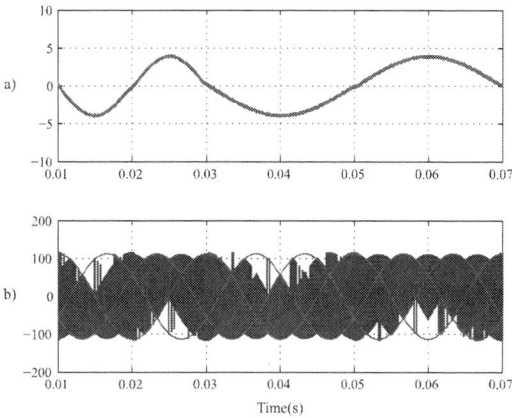

Fig. 8. Simulation results in transient state: i_o = 4 Apk; f_o = 50-25 Hz; f_s = 20 kHz; v_i = 112 Vpk;

TABLE III
MEAN AVERAGE ERROR OF i_{ref} VERSUS i_o A UNA f_o = 50 Hz

Sampling (f_s)	Amplitude (i_o)	Error (e_{sim})	Error (e_{exp})
10 kHz	2 Apk	6,994%	8,294%
10 kHz	6 Apk	4,732%	6,640%
20 kHz	2 Apk	4,192%	5,541%
20 kHz	6 Apk	2,869%	4,796%
40 kHz	2 Apk	2,097%	5,181%
40 kHz	6 Apk	1,425%	5,112%

TABLE IV
TOTAL HARMONIC DISTORTION OF THE LOAD CURRENT @ f_o = 50 Hz

Sampling (f_s)	Amplitude (i_{ref})	THD Sim. (i_o)	THD Exp. (i_o)
10 kHz	2 Apk	12,534%	11,572%
10 kHz	6 Apk	7,235%	10,121%
20 kHz	2 Apk	6,608%	7,830%
20 kHz	6 Apk	4,387%	6,923%
40 kHz	2 Apk	3,465%	10,307%
40 kHz	6 Apk	2,376%	8,837%

which are not considered in the model. The smaller error is observed when a sampling frequency of $f_s = 20kHz$ is considered. Table IV shows the THD values for simulation and experiments for the load current. Similarly, here is also observed that the lower THD value is observed for a sampling frequency of $f_s = 20$ kHz.

From Fig. 11 to Fig. 14 are presented simulation and experimental results under frequency and amplitude variations. There is observed a very good dynamic responde without any significant overshoot or delay.

V. CONCLUSION

A predictive current control strategy for a single-phase matrix converter was presented. The predictive control strategy uses the predicted values of the load currents to select the best-suited converter switching state taking into consideration the load current error in a given cost function. Predictive control does not need the use of complex modulation schemes or internal control loops. The gate drive signals for the devices are generated directly by the predictive controller.

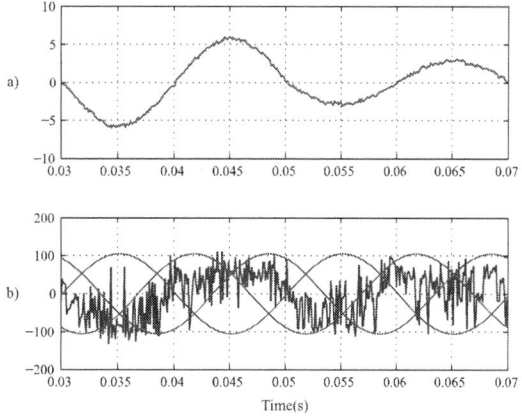

Fig. 9. Experimental results in transient state: i_o = 3-6 Apk; f_o = 50 Hz; f_s = 10 kHz; v_i = 112 Vpk.

Fig. 11. Simulation results in transient state: i_o = 3-6 Apk; f_o = 50 Hz; f_s = 40 kHz; v_i = 112 Vpk;

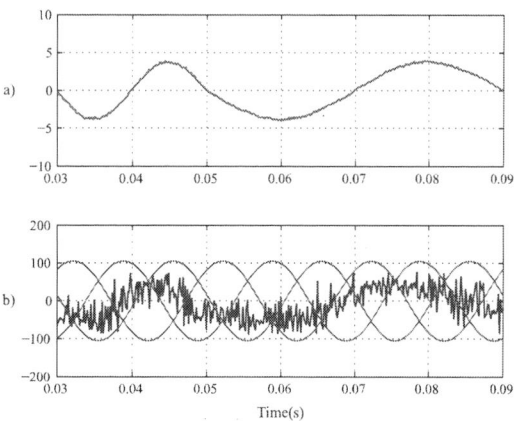

Fig. 10. Experimental results in transient state: i_o = 4 Apk; f_o = 50-25 Hz; f_s = 20 kHz; v_i = 112 Vpk;

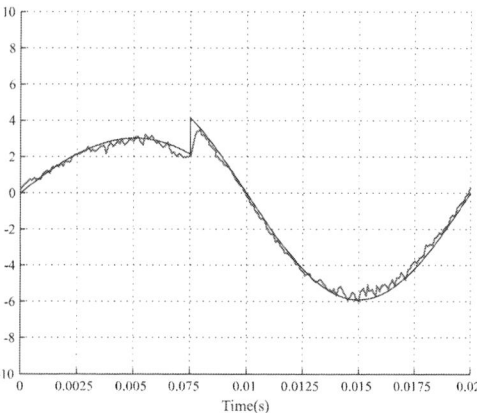

Fig. 12. Experimental results in transient state: i_o = 3-6 Apk; f_o = 50 Hz; f_s = 40 kHz; v_i = 112 Vpk;

ACKNOWLEDGMENTS

This publication was made possible by the Newton Picarte Project EPSRC: EP/N004043/1: New Configurations of Power Converters for Grid Interconnection Systems / CONICYT DPI20140007 and British Council through the Institutional Skills Development Newton Picarte Project ISCL 2015006.

REFERENCES

[1] E. Yamamoto, T. Kume, H. Hara, T. Uchino, J. Kang, and H. Krug, "Development of matrix converter ans its applications in industry," *35th Annual Conference of the IEEE Industrial Electronics Society IECON 2009, Porto, Portugal,* 2009.

[2] M. Venturini, "A new sine wave in sine wave out, conversion technique which eliminates reactive elements," *Powercon 7, 1980,* pp. E3/1–E3/15, Mar. 2001.

[3] J. Rodriguez, E. Silva, F. Blaabjerg, P. Wheeler, J. Clare, and J. Pontt, "Matrix converter controlled with the direct transfer function approach: analysis, modelling and simulation," *Taylor and Francis-International Journal of Electronics,* vol. 92, no. 2, pp. 63 –85, Feb. 2005.

[4] L. Zhang, C. Watthanasarn, and W. Shepherd, "Control of ac-ac matrix converters for unbalanced and/or distorted supply voltage," *Power Electronics Specialists Conference, 2001. PESC. 2001 IEEE 32nd Annual,* vol. 2, pp. 1108 –1113 vol.2, 2001.

[5] E. Yamamoto, H. Hara, T. Uchino, M. Kawaji, T. Kume, J. K. Kang, and H.-P. Krug, "Development of mcs and its applications in industry [industry forum]," *Industrial Electronics Magazine, IEEE,* vol. 5, no. 1, pp. 4 –12, march 2011.

[6] G. Roy and G.-E. April, "Cycloconverter operation under a new scalar control algorithm," *Power Electronics Specialists Conference, 1989. PESC '89 Record., 20th Annual IEEE,* pp. 368 –375 vol.1, Jun. 1989.

[7] J. Rodriguez, "High performance dc motor drive using a pwm rectifier with power transistors," *Electric Power Applications, IEE Proceedings B,* vol. 134, no. 1, p. 9, Jan. 1987.

[8] L. Huber and D. Borojevic, "Space vector modulated three-phase to three-phase matrix converter with input power factor correction," *Industry Applications, IEEE Transactions on,* vol. 31, no. 6, pp. 1234 –1246, Nov. 1995.

[9] F. Blaabjerg, D. Casadei, C. Klumpner, and M. Matteini, "Comparison of two current modulation strategies for matrix converters under unbalanced input voltage conditions," *Industrial Electronics, IEEE Transactions on,* vol. 49, no. 2, pp. 289 –296, Apr. 2002.

[10] I. Takahashi and T. Noguchi, "A new quick response and high efficency control strategy for an induction motor," *Industrial Electronics, IEEE Transactions on,* vol. 22, no. 5, pp. 820 –827, Sep. 1986.

[11] S. Muller, U. Ammann, and S. Rees, "New time-discrete modulation scheme for matrix converters," *Industrial Electronics, IEEE Transactions on,* vol. 52, no. 6, pp. 1607 – 1615, Dec. 2005.

[12] M. Rivera, R. Vargas, J. Espinoza, and J. Rodriguez, "Behavior of the predictive dtc based matrix converter under unbalanced ac-supply,"

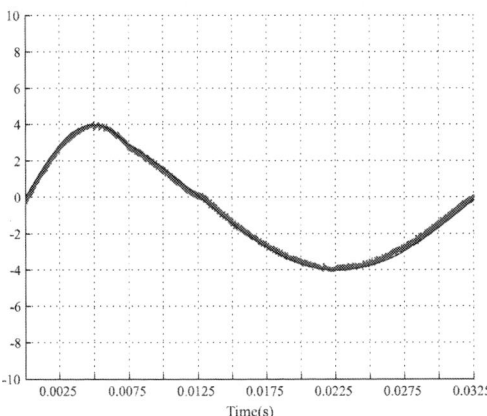

Fig. 13. Simulation results in transient state: i_o = 4 Apk; f_o = 50-25 Hz; f_s = 40 kHz; v_i = 112 Vpk;

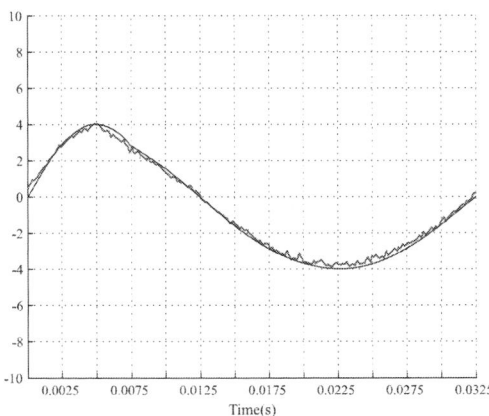

Fig. 14. Experimental results in transient state: i_o = 4 Apk; f_o = 50-25 Hz; f_s = 40 kHz; v_i = 112 Vpk;

current control of two-level four-leg inverters - part II: Experimental implementation and validation," *IEEE Trans. Power Electron.*, vol. 28, no. 7, pp. 3469–3478, Jul. 2013.

Power Electronics Specialists Conference, 2008. PESC 2008. IEEE, pp. 202 –207, Sep. 2007.

[13] M. Rivera, C. Rojas, J. Rodriguez, P. Wheeler, B. Wu, and J. Espinoza, "Predictive current control with input filter resonance mitigation for a direct matrix converter," *IEEE Trans. Power Electron.*, vol. 26, no. 10, pp. 2794–2803, Oct. 2011.

[14] S. Kouro, P. Cortes, R. Vargas, U. Ammann, and J. Rodriguez, "Model predictive control-A simple and powerful method to control power converters," *IEEE Trans. Ind. Electron.*, vol. 56, no. 6, pp. 1826–1838, Jun. 2009.

[15] J. Rodriguez, M. P. Kazmierkowski, J. R. Espinoza, P. Zanchetta, H. Abu-Rub, H. A. Young, and C. A. Rojas, "State of the art of finite control set model predictive control in power electronics," *IEEE Trans. Ind. Informat.*, vol. 9, no. 2, pp. 1003–1016, May. 2013.

[16] C. Rojas, J. Rodriguez, A. Iqbal, H. Abu-Rub, A. Wilson, and S. Moin Ahmed, "A simple modulation scheme for a regenerative cascaded matrix converter," in *IECON 2011 - 37th Annual Conference on IEEE Industrial Electronics Society*, nov. 2011, pp. 4361 –4366.

[17] J. Rodriguez and P. Cortes, *Predictive Control of Power Converters and Electrical Drives*, 1st ed. Chichester, UK: IEEE Wiley press, Mar. 2012.

[18] V. Yaramasu, M. Rivera, B. Wu, and J. Rodriguez, "Model predictive current control of two-level four-leg inverters - part I: Concept, algorithm and simulation analysis," *IEEE Trans. Power Electron.*, vol. 28, no. 7, pp. 3459–3468, Jul. 2013.

[19] M. Rivera, V. Yaramasu, J. Rodriguez, and B. Wu, "Model predictive

978-1-5090-0376-1/16 $31.00 © 2016 IEEE

7th Power Electronics, Drive Systems & Technologies Conference (PEDSTC 2016)
16-18 Feb. 2016, Iran University of Science and Technology, Tehran, Iran

Predictive Torque Control of a Permanent Magnet Synchronous Motor fed by a Matrix Converter without weighting factor

Mohsen Siami[1], Hamed Kiani Savadkoohi[1], Alireza Abbaszadeh[1], D. A. Khaburi[1], Jose Rodriguez[2], Marco Rivera[3]

[1]Center of Excellence for Power Systems Automation and Operation, Department of Electrical Engineering, Iran University of Science and Technology, Tehran, Iran
[2]Faculty of Engineering, Universidad Andres Bello, Santiago, Chile
[3]Department of Industrial Technologies, Universidad de Talca, Curico, Chile

Abstract— **Finite control set model predictive control is a flexible control scheme with fast dynamic. The standard form of this controller is based on the minimization of a cost function. The cost function needs weighting factors that is dependent to the system parameters and operating point. The tuning of the weighting factors requires nontrivial process. This paper presents a predictive direct torque and stator flux control of a permanent magnet synchronous motor fed by a matrix converter without weighting factor. The proposed method uses a multiobjective optimization based on a ranking approach instead of a cost function. Therefore, the tuning of weighting factors is unnecessary. Simulation results are presented to confirm the good performance of the proposed method.**

Keywords— *Predictive control; matrix converter; weigting factor; ranking approach*

I. INTRODUCTION

Two widely used control schemes in commercial are field-oriented control (FOC) and direct torque control (DTC). DTC technique was developed for induction motor drivers in the middle of 1980s [1]. It was applied to permanent magnet drives in [2],[3]. The main advantage of DTC in comparison with FOC is a faster dynamic torque response. Furthermore, DTC is independent of motor parameters except for stator resistance. There are some disadvantages such as torque ripple, current distortion and mainly needing a high sampling frequency for digital implementation. Some studies have been done to solve these problems [4]-[5].

Development of powerful and fast microprocessors allows implementation of more complex control schemes method i.e., model predictive control (MPC). This kind of controller was developed at the end of the 1970s [6]. Due to the technique's qualities such as fast dynamic torque response, low torque ripple, and reduced switching frequency, the application of this control techniques for torque and flux control of induction machines (IMs) and PMSMs, has received attention from researchers [7]–[10]. In [11] different approaches of predictive method were used for current control of PMSM. .A Comparative study between DTC and predictive method was done in [12]. Finite control set MPC (FCS-MPC) is a flexible kind of MPC to control different variables. FCS-MPC uses the discrete nature of the power converters is considered. The behavior of motor is estimated for every feasible voltage vector of the converter based on predictions obtained from a discrete time model of the machine. Then the voltage vector that minimizes the cost function is actuated in the next time interval.

The cost function needs weighting factors that are used to determine the importance of different variables and to normalize the different control objectives. However, the tuning of the factors requires nontrivial process. Several methods using offline and online search procedures have been presented in the literature, but they are dependent on the system parameters and are only formulated for two control objectives [13], [14]. When more objectives are considered, the weighting factors are usually obtained using trial and error procedures and running time-consuming simulations [15]–[17]. However, the cost function with weighting factor is not the only possible solution to solve the optimization problem [18]. In [19] a simple multiobjective optimization method has been presented in order to get rid of the requirement of weighting factors. The optimization problem is solved using a multiobjective ranking-based approach.

The limited number of voltage vectors from traditional inverters makes the torque ripple problem more challenging. So, a number of researchers turns to utilization of multilevel inverters that develop a higher number of voltage vectors [20]. Recently, Matrix Converters (MCs) due to the higher number of voltage vectors have received considerable attention as an attractive alternative to the conventional voltage-source inverter (VSI) [21]. The absence of large capacitors or inductances allows the MC to give a compact design. Modulation strategies for MCs are reviewed in [22]. These can be classified into two main groups: scalar and space vector methods. In [23] a FCS-MPC method was introduced an IM fed by a MC. In this model the selection of the switching state of the MC is performed by means of a cost function with weighting factors to control different objectives.

The objective of this paper is to predictive control of torque and stator flux of a PMSM fed by a matrix converter using a multiobjective ranking-based approach. The main goal is the evaluation of each control objective for every voltage vectors of matrix converter independently and, then, the computation of a ranking of each possible solution using a sorting algorithm.

978-1-5090-0376-1/16 $31.00 © 2016 IEEE

Simulation results which confirm the good performance of the proposed methods are presented.

II. MATRIX CONVERTER

A Matrix Converter (MC) is an ac–ac single-stage power converter with m × n bidirectional switches, which connects a m-phase voltage source to a n-phase load. The most widely used configuration is the three-phase MC, 3 × 3-switches, shown in Fig. 1. It connects a three-phase voltage source to a three-phase load directly without using any intermediate dc link circuit. The input filter attenuates the high-frequency switching components in the input current.

The corresponding switching function of each nine bidirectional switches is S_{xy} with $x \in \{A, B, C\}$ and $y \in \{a, b, c\}$, as shown in Fig. 2. In order to achieve the safe operation of the converter, two basic rules must be observed. Normally the matrix converter is fed by a voltage source and therefore, the input terminals should not be short circuited. The load has typically an inductive nature and for this reason output phase must never be opened. Considering these rules, switching functions should fulfill, at all times, the following equation:

$$S_{Ay} + S_{By} + S_{Cy} = 1 \qquad \forall y \in \{a, b, c\} \qquad (1)$$

Where

$$S_{xy} = \begin{cases} 0 & if \quad S_{xy} \quad is \quad open \\ 1 & if \quad S_{xy} \quad is \quad close \end{cases} \qquad (2)$$

The state of the converter switches can be represented by means of the following matrix:

$$T = \begin{bmatrix} S_{Aa}(t) & S_{Ba}(t) & S_{Ca}(t) \\ S_{Ab}(t) & S_{Bb}(t) & S_{Cb}(t) \\ S_{Ac}(t) & S_{Bc}(t) & S_{Cc}(t) \end{bmatrix} \qquad (3)$$

Considering the equations (1), (2) there are 27 valid switching states for a 3×3 MC. According to the kind of output voltage vector, these 27 switching configurations can be grouped into three groups as follows:

1) *Zero vectors*: All three output phases are linked to the one input phase.

2) *Space vectors with varying amplitude and fixed direction*: Two output phases are linked to one input phase, and the other output phase is linked to a different input phase.

3) *Rotational space vectors*: Each output phase is connected to a different input phase. These vectors have constant amplitudes, but their angles change at the source frequency.

In Fig. 1, the output voltage and the input current space vectors of MC can be expressed as (4) and (5), respectively.

$$v_o = \frac{2}{3} \left(v_a + a \cdot v_b + a^2 \cdot v_c \right) \qquad (4)$$

$$i_e = \frac{2}{3} \left(i_A + a \cdot i_B + a^2 \cdot i_C \right) \qquad (5)$$

Where $a = e^{j(2\pi/3)}$ and v_a, v_b and v_c are output phase voltages

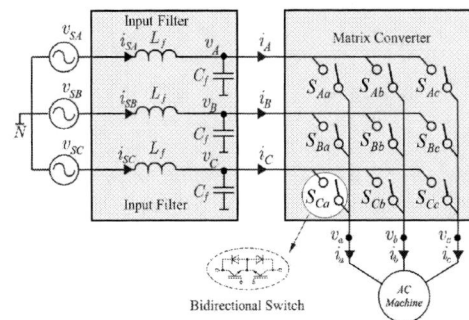

Fig. 1. 3 × 3 MC

and i_a, i_b and i_c are input phase currents of the MC. A similar expression can be defined for the source current vector i_s, the source voltage vector v_s and the output current i_o.

III. PREDICTIVE TORQUE CONTROL

The standard predictive torque control (PTC) selects one of the 27 feasible switching states of the MC, at fixed time intervals, based on the minimization of a cost function (CF). Actually, the cost function defines the evaluation criteria to choose the best switching configuration for the next time interval. For the computation of the CF, the electromagnetic torque T_e, and the stator flux φ_s on the next time interval are predicted, supposing the application of each valid switching state in the next time interval, by a mathematical discrete-time model of the PMSM. These predicted values are compared with their reference values in CF through the use of weighting factor. The factor is dependent on the operating point and system parameters. Therefore, its choice is not an easy task. Furthermore, the weighting factor decides the relative importance of the torque and stator flux and thus, affects the performance of the controller. The PTC algorithm can be executed in three steps: measuring variables, prediction of the system behavior, and optimization of the cost function.

A. Models Used to Obtain Predictions

1) *Model of Matrix Converter*: Due to the instantaneous power transfer of MCs, voltages and currents at any moment in one side may depend on the voltages and currents in the other side. Because the MC is connected to the source, the input line-to-neutral voltages are known; therefore, the output line-to-neutral output voltages are obtained as follows:

$$\begin{bmatrix} v_a \\ v_b \\ v_c \end{bmatrix} = \underbrace{\begin{bmatrix} S_{Aa} & S_{Ba} & S_{Ca} \\ S_{Ab} & S_{Bb} & S_{Cb} \\ S_{Ac} & S_{Bc} & S_{Cc} \end{bmatrix}}_{T} \cdot \begin{bmatrix} v_A \\ v_B \\ v_C \end{bmatrix} \qquad (7)$$

So the output voltages applied to the load are dependent on the switching functions, reflected in matrix T, and the input voltages.

The output currents are resulted from applying these output voltages to a given load. By measuring the output currents, the input currents can be easily found as

$$\begin{bmatrix} i_A \\ i_B \\ i_C \end{bmatrix} = \underbrace{\begin{bmatrix} S_{Aa} & S_{Ab} & S_{Ac} \\ S_{Ba} & S_{Bb} & S_{Ac} \\ S_{Ca} & S_{Cb} & S_{Cc} \end{bmatrix}}_{T^T} \cdot \begin{bmatrix} i_a \\ i_b \\ i_c \end{bmatrix} \qquad (8)$$

2) Stator Flux and Electromagnetic torque Prediction: To predict the response of the system for every switching state, a mathematical discrete-time model is obtained on the basis of the dynamic equations of a PMSM.

The stator voltage equation of a PMSM in the stationary reference frame can be expressed as

$$v_s = R_s i_s + \frac{d\varphi_s}{dt} \qquad (9)$$

Where v_s, i_s, φ_s and R_s are the stator voltage, current, flux and resistance, respectively. The stator flux prediction at the sampling step $k+1$ (φ_s^{k+1}) with sampling time T_s is obtained by means of the forward-Euler discretization of (9)

$$\varphi_s^{k+1} = \varphi_s^k + T_s v_s^k - T_s R_s i_s^k \qquad (10)$$

The quadrature component of stator currents is obtained as

$$\frac{di_q}{dt} = \frac{v_q}{L_s} - \frac{R_s i_q}{L_s} - \omega i_d - \frac{\omega \varphi_m}{L_s} \qquad (11)$$

Where i_q and i_d are quadrature and direct components of stator currents, respectively. v_q is quadrature component of stator currents and φ_m is rotor magnet flux. The prediction of the quadrature component of stator current is obtained using the forward-Euler discretization of (11)

$$i_q^{k+1} = i_q^{k+1} + \frac{T_s}{L_s}(v_q^k - R_s i_q^k - \omega i_d^k - \omega \varphi_m) \qquad (12)$$

The electromagnetic torque prediction is obtained as

$$T_e^{k+1} = \frac{2}{p} \varphi_m i_q^{k+1} \qquad (13)$$

Equations (10) and (13) are used to obtain predictions of torque and stator flux for each of the 27 feasible voltage vectors.

B. Cost Function

The evaluation criteria used to determine which voltage vector is the best to be applied in next time interval, are defined by the cost function. The cost function is made-up of the absolute error of the predicted torque and the absolute error of the predicted stator flux magnitude as follows

$$CF = \left| T_e^* - T_e^{k+1} \right| + \lambda_\varphi \left\| \varphi_s^* \right| - \left| \varphi_s^{k+1} \right\| \qquad (14)$$

Where the references values are shown by the superscript "*". The cost function must be computed for each of the 27 possible voltage vectors. The vectors that produces the minimum value will be selected and exerted during the next time interval. λ_φ is a weighting factor that manages the relationship between torque and stator flux situations. The determination of the λ_φ is a hard process, and it has a high effect on the performance of the controller. At the present state of the art, this weighting factor is determined analytically [11] with genetic algorithms [17] and empirically [15], [16], [24].

C. Multiobjective Ranking-Based Optimization

For using the multiobjective optimization approach in PTC algorithm the two different cost functions should be minimized:

$$CF_1 = \left| T_e^* - T_e^{k+1} \right| \qquad (15)$$

$$CF_2 = \left\| \varphi_s^* \right| - \left| \varphi_s^{k+1} \right\| \qquad (16)$$

The proposed multiobjective ranking-based approach evaluates these tow cost functions for each feasible voltage vector of the matrix converter. The obtained values of each cost function are sorted separately. Then, a ranking value is assigned to each value.

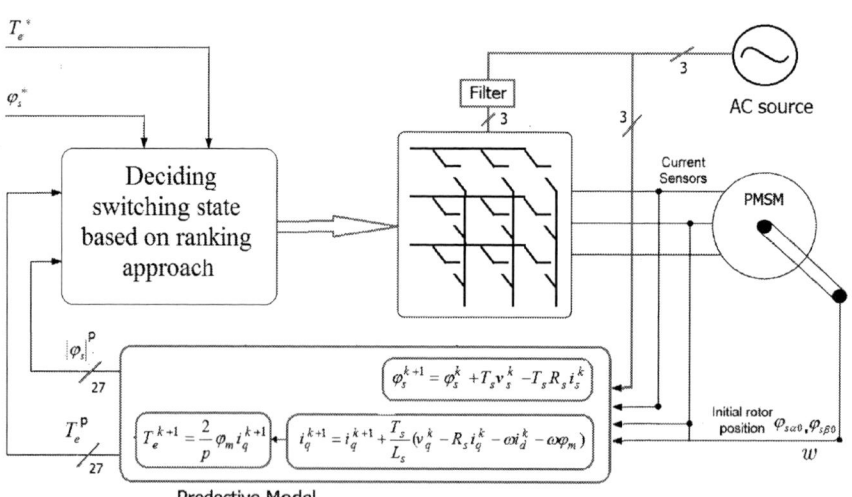

Fig. 2. Block diagram for proposed predictive torque control with matrix converter.

The voltage vectors with lower value are assigned a lower ranking, while the voltage vectors with higher value are assigned a higher ranking. Now, sum of the ranks of these tow cost functions for each voltage vector is used to decide which voltage vector is the best to be applied. The voltage vector that minimizes this value is selected [19].

Table I illustrates the proposed algorithm for a matrix converter with 27 voltage vectors. The utilization of the zero voltage vector is alternating between three zero voltage vectors and for this reason, just one zero vector is considered in the calculation (v_0). CF_1 and CF_2 are computed for different voltage vectors and a ranking is assigned to each voltage vector according to these values. For example, considering that the CF_1 for v_{24} is 0.004, the ranking value assigned to v_{24} is 1 because it produces the lowest value of CF_1, while the CF_1 associated with v_{13} is 0.546, and then, the ranking assigned is 25 because v_{13} produces the highest value of CF_1. The same procedure is performed for the CF_2 with its respective ranking assignation. Now, the sum of these two ranking for every voltage vector results that the voltage vector v3 gives the lowest value (Rank CF_1 + Rank CF_2 = 4) and applied in the next time interval.

IV. SIMULATION RESULTS

A MATLAB SIMULINK model has been used in order to verify the performance of the proposed predictive control. Simulation parameters have been listed in Table II. The block diagram of the proposed predictive torque control has been shown in Fig. 2. As shown in this figure, unlike DTC this control method doesn't need switching table, and the proper switching state in each time interval is selected according to the flowchart that has been shown in Fig. 3. To predict the torque and stator flux for each possible switching state, the discrete time model of the machine presented in previous part, are used.

TABLE I.   EXAMLPE OF VOLTAGE VECTOR SELECTION

V_0	CF_1	CF_2	Rank CF_1	Rank CF_2	Rank CF_1 + Rank CF_2
V_0	0.271	0.0008	13	10	23
V_1	0.147	0	7	1	8
V_2	0.421	0.0018	21	20	41
V_3	0.377	0.0006	18	7	25
V_4	0.253	0.0002	12	4	16
V_5	0.527	0.0016	24	18	32
V_6	0.143	0.0011	6	13	19
V_7	0.018	0.0002	2	2	4
V_8	0.293	0.002	15	21	36
V_9	0.289	0.0018	14	19	33
V_{10}	0.165	0.001	8	12	20
V_{11}	0.44	0.0028	22	24	46
V_{12}	0.395	0.0016	19	17	36
V_{13}	0.546	0.0026	25	23	48
V_{14}	0.167	0.0021	9	22	31
V_{15}	0.038	0.0012	3	14	17
V_{16}	0.312	0.0031	16	25	41
V_{17}	0.249	0.0004	11	5	16
V_{18}	0.124	0.0013	5	15	20
V_{19}	0.399	0.0006	20	8	28
V_{20}	0.355	0.0006	17	6	23
V_{21}	0.23	0.0015	10	16	26
V_{22}	0.505	0.0006	23	9	32
V_{23}	0.121	0.0002	4	3	7
V_{24}	0.004	0.001	1	11	12

TABLE II.   PMSM PARAMETERS IN SIMULATION MODEL

Parameter	Description	Value
p	number of pole pairs	4
R_s / Ω	stator resistance	0.2
L_s / mH	stator inductance	8.5
φ_m / Wb	rotor magnet flux	0.175
$j /(kg.m^2)$	rotor inertia	0.089
B	damping coefficient	0.005
V_{rms} / V	Source voltage	100
ω_r / rpm	rated speed	300
$T_N /(N.m)$	rated torque	11

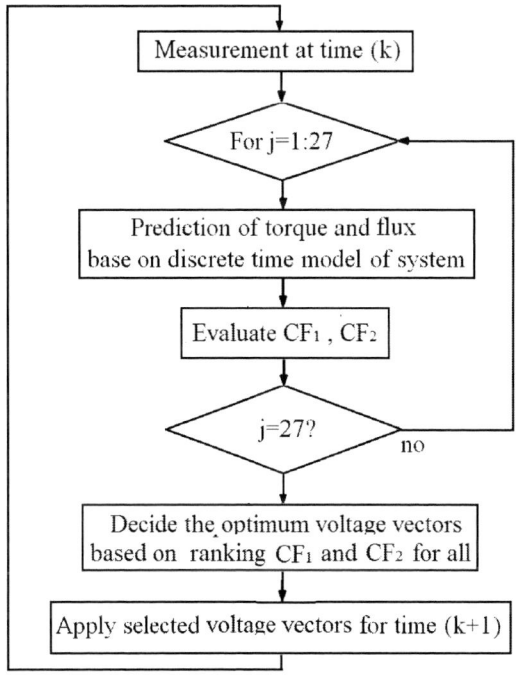

Fig. 3. Flowchart of the proposed predictive control.

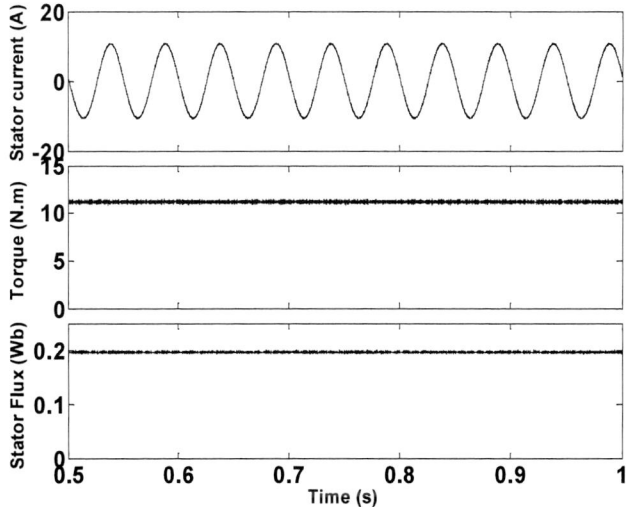

Fig. 4. Steady state performance of the proposed method

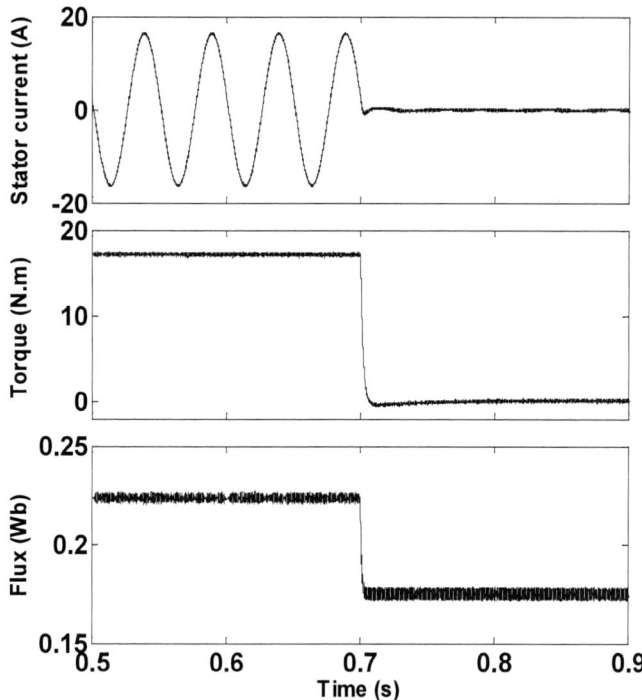

Fig. 6. Dynamic performance of the standard PTC.

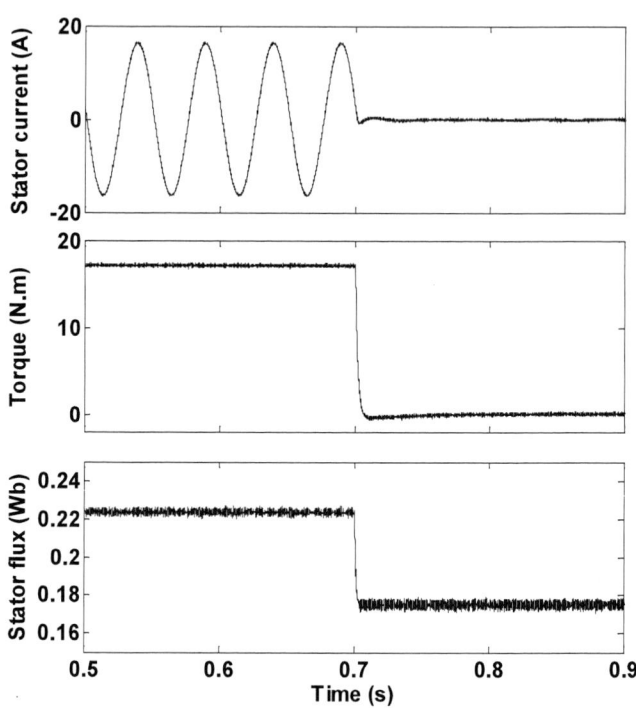

Fig. 5. Dynamic performance of the proposed method.

It is noticeable that the process of selecting switch for every time interval is done at the previous sampling interval. Thus, for determining sample time, the sufficient time for this process should be taken into account.

In Fig. 4, the steady state performance of the proposed method has been shown. In this case the machine is operating at a nominal speed at 300 r/min with 11 N.m load torque. As this figure depicts the stator flux produces good current waveforms and, as a consequence, good torque performance in steady state. Such behavior can be achieved by standard PTC with the offline tuned weighting factor. However, in the proposed method in this paper, no additional offline optimization or weighting factor computation is needed.

The dynamic behavior of the motor has been shown in Fig. 5, and 6, for the proposed ranking approach and the standard PTC, respectively. The load torque changes from 17 N.m to 0 N.m at t=0.7s. It can be seen that the proposed method in this paper has a better performance in torque and stator flux ripple reduction rather than the standard PTC. It is due to the fact that the weighting factor in this case has not been tuned and it is fixed at the obtained optimized value in steady state with 11 N.m load toque. However, with additional computation effort to offline optimization of weighting factor for different load torque, the dynamic performance of the standard PTC can be same as the proposed ranking approach. But, the proposed approach in this paper doesn't need any additional computation for different load torque.

V. CONCLUSION

A new predictive torque control (PTC) with matrix converter to control torque and stator flux of a PMSM has been presented in this article. In contrast with the standard PTC that uses a cost function with weighting factor, the proposed method in this paper doesn't need any weighting factor. The tuning of

weighting factor is a very complex task but, it is replaced by a multiobjective ranking-based approach and so, weighting factor calculation is avoided. The approach is based on the fact that the selected voltage vector should minimize all the objective functions, in this case torque and stator flux tracking. Simulation results confirm the good performance of the proposed method in both steady state and dynamic operation of the motor. Furthermore, comparing the simulation results of the proposed method with standard PTC, shows an improvement in torque and stator flux in dynamic operation without any offline optimizations such as is done in the standard PTC method.

REFERENCES

[1] I. Takahashi, T. Noguchi, "A new quick-response and high-efficiency control strategy of an induction motor," *IEEE Trans. Ind. Appl.*, vol. IA-22, no. 5, pp. 820–827, Sep. 1986.

[2] L. Zhong, M. F. Rahman, W. Y. Hu, and K. W. Lim, "Analysis of Direct Torque Control in Permanent Magnet Synchronous Motor Drives," *IEEE Trans. Power Electeron.*, vol. 12, no. 3, pp. 528 – 536, May, 1997.

[3] M. Siami, S. A. Gholamian, "Application of Direct Torque Control Technique for Three Phase Surface Mounted AFPM Synchronous Motors," *International Journal of Science and Advanced Technology*, vol. 1, no. 10, pp. 15-20, Dec. 2011.

[4] J. Faiz, M.B.B Sharifian, "Comparison of different switching patterns in direct torque control technique of induction motors," *Electric Power Systems Research*, vol. 60, no. 2, pp. 63-75, Dec. 2001.

[5] M.Pacas, J.Weber, "Predictive Direct Torque Control for the PM Synchronous Machine," *IEEE Trans. Ind. Electron.*, vol. 52, no. 5, pp. 1350-1356, Oct. 2005.

[6] E. Camacho and C. Bordons, Model Predictive Control. Berlin, Germany: Springer-Verlag, 1999.

[7] M. Siami, D. A. Khaburi, M. Yousefi and J. Rodriguez "Improved predictive torque control of a permanent magnet synchronous motor fed by a matrix converter," in *Proc. 6th IEEE Power Electron., Drive Syst. Technol. Conf.*, 2015, pp. 369 – 374..

[8] J. Holtz, "Advanced Pulse width Modulation and Predictive Control– An Overview," *IEEE Trans. Ind. Electron.*, Accepted for publication, DOI: 10.1109/TIE.2015.2504347.

[9] M. Siami, S. A. Gholamian, "Predictive Torque Control of three phase Axial Flux Permanent Magnet Synchronous Machines," *Majlesi Journal of Electrical Engineering*, vol. 6, no. 2, pp. 7-13, Jun. 2012.

[10] M. Siami, A. Abbaszadeh, D. A. Khaburi and J. Rodriguez, "Robustness Improvement of Predictive Current Control Using Prediction Error Correction for Permanent Magnet Synchronous Machines," *IEEE Trans. Ind. Electron.*, Accepted for publication, DOI: 10.1109/TIE.2016.2521734.

[11] F. Morel, X. Lin-Shi, J-M. Rétif, B. Allard and C. Buttay, "A Comparative Study of Predictive Current Control Schemes for a Permanent-Magnet Synchronous Machine Drive," *IEEE Trans. Ind. Electron.*, vol. 56, no. 7, pp. 2715–2728, Jul. 2009.

[12] M. Siami, S. A. Gholamian, M. Yousefi, "A Comparative Study Between Direct Torque Control and Predictive Torque Control for Axial Flux Permanent Magnet Synchronous Machines," *Journal of Electrical Engineering*, vol. 64, no. 6, pp. 346-353, Dec, 2013.

[13] S. A. Davari, D. A. Khaburi, and R. Kennel, "An improved FCS-MPC algorithm for induction motor with imposed optimized weighting factor," *IEEE Trans. Power. Electron.*, vol. 27, no. 3, pp. 1540–1551, Mar. 2012..

[14] T. J. Vyncke, S. Thielemans, T. Dierickx, R. Dewitte, M. Jacxsens, and J. A. Melkebeek, "Design choices for the prediction and optimization stage of finite-set model based predictive control," in Proc. Workshop PRECEDE , 2011, pp. 47–54.

[15] P. Cortes, S. Kouro, B. La Rocca, R. Vargas, J. Rodriguez, J. Leon, S. Vazquez, and L. Franquelo, "Guidelines for weighting factors design in model predictive control of power converters and drives," in Proc. IEEE ICIT , 2009, pp. 1–7.

[16] R. Vargas, J. Rodriguez, M. Rivera, C. Rojas, and M. Rivera, "Predictive Control of an Induction Machine Fed by a Matrix Converter With Increased Efficiency and Reduced Common-Mode Voltage," *IEEE Trans. Energy Conv.*, vol.29, no.2, pp.473-485, June 2014.

[17] P. Zanchetta, "Heuristic multi-objective optimization for cost function weights selection in finite states model predictive control," in Proc. Workshop PRECEDE , 2011, pp. 70–75.

[18] Y. Zhang, H. Yang, S. Member "Model-Predictive Flux Control of Induction Motor Drives with Switching Instant Optimization," *IEEE Enegy Conv.*, Accepted for publication, 2015, DOI: 10.1109/TEC.2015.2423692.

[19] C. A. Rojas, , J. Rodríguez, F. Villarroel, J. R. Espinoza, C. A. Silva and M, Trincado, "Predictive Torque and Flux Control Without Weighting Factors," *IEEE Trans. Ind. Electron.*, vol. 60, no. 2, pp. 681–690, Feb. 2013.

[20] K. B. Lee, J. H. Song, J. H. I. Choy, J. Y. Yoo and "Torque ripple reduction in DTC of induction motor driven by three-level inverter with low switching frequency," *IEEE Trans. Power. Elec.*, vol. 17, no. 2, pp. 255–264, Mar. 2000.

[21] P. Wheeler, J. R odriguez, J. Clare, L. Empringham, and A. Weinstein, "Matrix converters: A technology review," *IEEE Trans. Ind. Electron.*, vol l. 49, no. 2, pp. 276–288, Apr. 2002.

[22] L. Helle, K. B. Larsen, A. H. Jorgensen, S. M. Nielsen, and F. Blaabjerb, "Evaluation of modulation schemes for three-phase to three-phase matrix converters," *IEEE Trans. Ind. Electron.*, vol. 51, no. 1, pp. 158–170, Feb. 2004.

[23] J. Rodriguez. M, Rivera. J.W. Kolar and P.W. Wheeler, "A Review of Control and Modulation Methods for Matrix Converters", *IEEE Trans. Ind. Electron.*, vol. 59, no. 1, pp. 58 – 70, Jan. 2012.

[24] R. Vargas, J. Rodríguez, U. Ammann, and P. Wheeler, "Predictive current control of an induction machine fed by a matrix converter with reactive power control," *IEEE Trans. Ind. Electron.*, vol. 55, no. 12, pp. 4362-4371, Dec. 2008.

[25] E. Bećirović1, J. Osmić, M. Kušljugić and N. Perić3 "Analysis And Synthesis Of Model Reference Controller For Variable Speed Wind Generators Inertial Support," *Journal of Electrical Engineering*, vol. 66, no. 1, pp. 3-10, Mar, 2015.

[26] V. Veselý and A. Ilka, "Robust Gain–Scheduled PID Controller Design For Uncertain LPV Systems," *Journal of Electrical Engineering*, vol. 66, no. 1, pp. 19-25, Mar, 2015.

[27] Y. Errami, M. Ouassaid, M. Cherkaoui, and M. Maaroufi, "Variable Structure Sliding Mode Control and Direct Torque Control of Wind Power Generation System Based on the Pm Synchronous Generator," *Journal of Electrical Engineering*, vol. 66, no. 3, pp. 121-131, Jul, 2015.

[28] N. Ghaffarzadeh, "Water Cycle Algorithm Based Power System Stabilizer Robust Design for Power Systems," *Journal of Electrical Engineering*, vol. 66, no. 2, pp. 91-96, May, 2015.

[29] Ž. Despotović and V. Šinik, "The Simulation and Experimental Results of Dynamic Behaviour of Torque Motor Having Permanent Magnets," *Journal of Electrical Engineering*, vol. 66, no. 2, pp. 97-102, May, 2015.

[30] A. H. Jafari, H. S. Shahhoseini, "A Reinforcement Routing Algorithm with Access Selection in the Multi–Hop Multi–Interface Networks," *Journal of Electrical Engineering*, vol. 66, no. 2, pp. 70-78, May, 2015.

7th Power Electronics, Drive Systems & Technologies Conference (PEDSTC 2016)
16-18 Feb. 2016, Iran University of Science and Technology, Tehran, Iran

One Step Model Predictive Control of Five Level ANPC Permanent Magnet Motor Drive

Mohammad Niliyan[1], Mustafa Mohamadian[2], Ali Yazdian Varjani[3]

Department of Electrical and Computer Engineering, Tarbiat Modares University, Tehran, Iran

[1] m.niliyan@modares.ac.ir
[2] mohamadian@modares.ac.ir
[3] yazdian@modares.ac.ir

Abstract— **Five-level active neutral-point clamped (5L-ANPC) converter is an interesting topology for high-power medium-voltage motor drives. In addition to a neutral point potential in the DC-link, it has a flying capacitor in each phase. Controlling these internal inverter voltages while producing desired torque and flux for the motor is a challenging control problem. In published articles that use model predictive control (MPC) to control the ANPC inverter as an induction motor drive, prediction horizon was equal to two steps. In this paper, using MPC with one step prediction horizon and a cost function with variable weighting factors, not only the four internal voltages of the inverter is controlled and the desired torque and magnetic flux for permanent magnet motor is produced, but also the magnitude of common mode voltage (CMV) is limited to one sixth of DC-link voltage. In this paper according to one step prediction horizon, the number of calculations and the complexity is less than other articles while the quality of motor and inverter variables are the same as other articles.**

Keywords—MPC; ANPC; common mode voltage; pm motor;

I. INTRODUCTION

Five-level active neutral-point clamped (ANPC) converter is a desirable multilevel topology, which is more appropriate for high-performance medium-voltage motor drives [1]–[8]. Its DC-link is split up into two parts and just four switches and one flying capacitor are needed for clamping per phase. The main problem of this topology is the disparate voltage stresses of switches and voltage balancing of the DC-link and flying capacitors [1]–[6]. A new medium-voltage drive using this inverter is explained in [2]. The direct torque control method based on space-vector PWM (SVPWM) is used to control this converter. A five-level virtual-flux direct power control for the five-level ANPC converter was described in [3].

A phase-disposition PWM (PD-PWM) with zero-sequence voltage injection method was proposed in [4] to control the NP potential of the 5L-ANPC converter. In [5], a control strategy based on selective harmonic elimination PWM was proposed and the voltage across the flying capacitor was balanced by exchanging the switching patterns. The NP voltage was regulated by adding or subtracting a relatively small pulse to the switching pulse signals [5], [6]. In [9],

controlling the neutral point and flying capacitors was presented using phase-shifted PWM.

With improvement of more powerful microprocessors, new control methods have been proposed. Along with these new control methods, predictive control seems to be an attractive substitute for the control of power converters and drives. Predictive control covers a wide type of controllers with different methods. A very strong predictive control method that has been used in recent times for power electronics is model predictive control [10].

In [11] internal voltages of inverter were controlled by using MPC and this method utilized to control the 5L-ANPC inverter as an IM motor in [12]. In [11] and [12] to controll the variables, predictive control with prediction horizon equal to two and a fairly complex algorithm was used. This paper proposes a simple MPC with one step prediction horizon to control the 5L-ANPC drive system and limit the common mode voltage to one sixth of DC-link voltage. Because of using one step prediction horizon the number of calculation is less than [11] and [12] and the algorithm is more simple.

II. 5L- ANPC DRIVE SYSTEM

A. 5L-ANPC Inverter Topology

The circuit diagram of a 5L-ANPC is displayed in Fig. 1 [12]. The inverter can generate 61 line-to-line voltage vectors. These vectors are created from the 125 three-phase voltage vectors. The DC-link is divided into an upper and a lower DC-link capacitors (C_{dc}). The potential of the neutral point describe with $V_n = 0.5\left(V_{dc\ low} - V_{dc\ up}\right)$, where $V_{dc\ low}$ and $V_{dc\ up}$ representing the voltages over the lower and the upper DC-link halves. With different switching state, it can produce five voltage levels. Table I summarizes the relationship between switching states, phase voltage, and the influence on the flying capacitor (C_f) voltage [1]. The nominal voltage of DC-link capacitors is $V_{dc}/2$ and the flying capacitor voltage is $V_{dc}/4$.

978-1-5090-0376-1/16 $31.00 © 2016 IEEE   620

Table 1. Switching state and their effect on phase voltage and flying capacitor [1].

S_1	S_5	S_6	Phase Voltage	Effect on C_f		Switching state
				I>0	I<0	
0	0	0	$-V_{dc}/2$	-	-	V0
0	0	1	$-V_{dc}/4$	Discharge	Charge	V1
0	1	0	$-V_{dc}/4$	Charge	Discharge	V2
0	1	1	0	-	-	V3
1	0	0	0	-	-	V4
1	0	1	$-V_{dc}/4$	Discharge	Charge	V5
1	1	0	$+V_{dc}/4$	Charge	Discharge	V6
1	1	1	$+V_{dc}/2$	-	-	V7

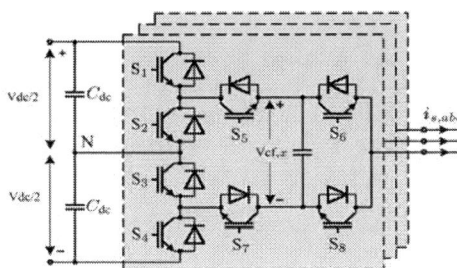

Fig. 1: Circuit diagram of 5L-ANPC [12].

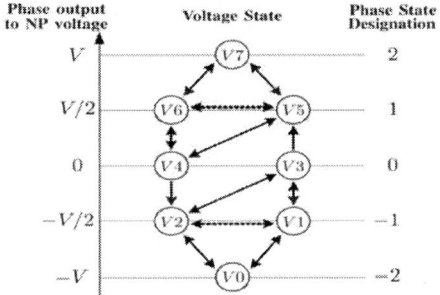

Fig. 2: Switching state Restrictions of the ANPC-5L and output voltage [11].

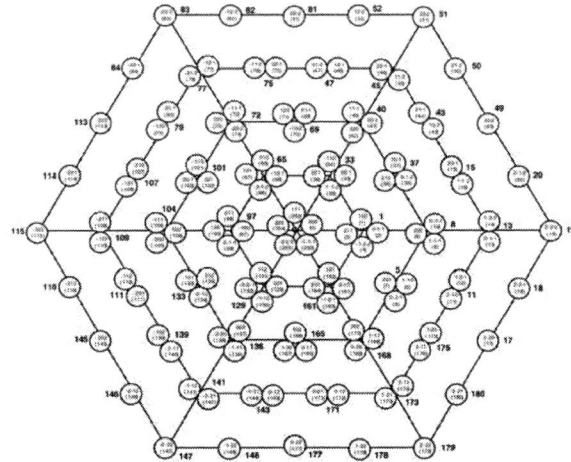

Fig. 3: Five-level inverter voltage vectors [12].

B. Switching constraints

Several switching constraints are inflicted in the 5L-ANPC topology. The allowed single-phase switching transitions are shown in Fig. 2. switching state should change in a way that just one phase voltage level up or down [11]. With this restriction, the maximum usable switching states is 64 instead of 512 in each step.

C. Dynamics of the Internal Voltages of Inverter

The voltage of flying capacitor in phase x, with $x \in$ {a, b, c}, is explained as equation (1) [16].

$$\frac{dV_{cfx}}{dt} = \frac{i_x}{C_f}(s_5 - s_6) \tag{1}$$

And the dynamic of the neutral point potential is given by:

$$\frac{dV_n}{dt} = \frac{1}{2C_{dc}}\sum H_x . i_x \tag{2}$$

$$H_x = \begin{cases} 1 & if \ V_x \in \{0,1,6,7\} \\ 0 & else \end{cases} \tag{3}$$

It should be Noted that the voltage of flying capacitor only depends on the switch state and current of own phase but the neutral point potential depends on all three switch positions and all three-phase currents[12].

And The CMV of three-phase 5L-ANPC converters can be describe by the phase output voltages as equation (4).

$$V_{cm} = \frac{v_{aN} + v_{bN} + v_{cN}}{3} \tag{4}$$

D. Permanent magnet Synchronous motor

For modeling PM motor, it is better to use rotor synchronous coordinate because all the parameters become constant.
Motor equations in rotor synchronous coordinate are as follows:

$$u_d = R_s i_d + L_d \frac{di_d}{dt} - \omega L_q i_q a \tag{5}$$

$$u_q = R_s i_d + L_q \frac{di_q}{dt} + \omega L_d i_d + \omega \psi_f \tag{6}$$

Where R_s is stator resistance, L_d, L_q and u_q, u_d and i_d, i_q are d and q axes inductance, stator voltage and stator current, ψ_f is PM flux, ω is electrical rotor speed. The torque is expressed as:

$$T_e = 1.5p\left(\left(L_d i_d + \psi_f\right)i_q - \left(L_q i_q\right)i_d\right) \tag{7}$$

If $L_q = L_d$, T_e becomes simple as equation (8).

$$T_e = 1.5p i_q \psi_f \tag{8}$$

Where p is the number of pole pairs. And the magnitude of motor flux is expressed as

$$|\psi| = \sqrt{\left(L_q i_q\right)^2 + \left(L_d i_d + \psi_f\right)^2} \tag{9}$$

III. CONTROL SCHEME

A. Model Predictive Control

MPC is an optimization problem that consist of minimizing the cost function, for a predefined horizon in time, subject to the model of the system and the restrictions of the system. The result is a sequence of M optimal actuations. The controller will apply only the first element of the sequence and the process is repeated at the consecutive sampling time[10],[15].

B. Internal Controller Model

To predict future values of the variables, their dynamic equations should be considered and by using forward Euler Approximation, calculate the next step value.
According to Equation (1)-(3) and (5)-(9) the future amounts of internal voltages of inverter and motor torque and flux can be obtained as follows.

$$V_{cfx}(t+T_s) = V_{cfx}(t) + \frac{T_s}{C_f} i_x (s_5 - s_6) \tag{10}$$

$$V_n(t+T_s) = V_n(t) + \frac{T_s}{2C_{dc}} \sum H_x \cdot i_x \tag{11}$$

$$i_d(t+T_s) = i_d(t) + \frac{T_s}{L_d}\left(u_d + \omega L_q i_q - R_s i_d\right) \tag{12}$$

$$i_q(t+T_s) = i_q(t) + \frac{T_s}{L_q}\left(u_q - \omega L_d i_d - \omega \psi_f - R_s i_q\right) \tag{13}$$

$$T_e(t+T_s) = 1.5 p \psi_f i_q(t+T_s) \tag{14}$$

$$\left|\psi(t+T_s)\right| = \sqrt{\left(L_q i_q(t+T_s)\right)^2 + \left(L_d i_d(t+T_s) + \psi_f\right)^2} \tag{15}$$

Where T_s is sampling time.

C. Cost Function whith variable wighting factor

A cost function that represents the desired behavior of the system and containing the variables needs to be defined [10]. In this article the following cost function is considered.

$$
\begin{aligned}
g = & \lambda_{Te}\left|T_e(t+T_s) - T_e^{ref}\right| + \lambda_\psi \left\|\psi(t+T_s)\right| - \psi^{ref}\right| + \\
& \lambda_{cfa}\left|V_{cfa}(t+T_s) - V_{dc}/4\right| + \lambda_{cfb}\left|V_{cfb}(t+T_s) - V_{dc}/4\right| + \\
& \lambda_{cfc}\left|V_{cfc}(t+T_s) - V_{dc}/4\right| + \lambda_n\left|V_n(t+T)\right|
\end{aligned} \tag{16}
$$

Coefficients λ are Weighting factors that have a significant impact on the control variables.
Since the main purpose of control are torque and flux, their coefficient are greater than the rest. To adjust the weighting factors of torque and flux, it is better to divide them to their nominal values[10].
If the weighting factors of internal voltages of inverter be fixed numbers, It is difficult to select the right ratio and also it causes to increase the switching frequency because at any moment switch state would change to control the variables exactly at their reference values. But the variable weighting factor can be chosen to control the variables in their reference bands and the switching frequency can be prevented from

rising. In this paper, the following weighting factor have been used.

$$\lambda_{Te} = 10 \ , \ \lambda_\psi = \frac{10 T_e^{base}}{\left|\psi^{base}\right|} \tag{17}$$

Where ψ^{base} and T_e^{base} are nominal value of flux and torque.

$$\lambda_n = \begin{cases} 1 & if \ \left|V_n\right| > 0.07 \times V_{dc}/2 \\ 0 & else \end{cases} \tag{18}$$

$$\lambda_{fcx} = \begin{cases} 1 & if \ \left|V_{cfx} - V_{dc}/4\right| > 0.05 \times \left(V_{dc}/4\right) \\ 0 & else \end{cases} \tag{19}$$

Equation (4) can be added to cost function to control the common mode voltage, But by doing this, control quality of other variables will reduce. But rather than reducing common mode voltage, it could be limited to specify amount by adding λ_{cm} to cost function as Equation (20).

$$\lambda_{cm} = \begin{cases} \infty & if \ \dfrac{v_{aN} + v_{bN} + v_{cN}}{3} > \dfrac{V_{dc}}{6} \\ 0 & else \end{cases} \tag{20}$$

The simulation results show that by adding λ_{cm}, other variables will be controlled as well as they were.

D. Control performance

Control Performance is as follow :
At the beginning of each sampling time by using measured values (flying capacitors voltages, neutral point potential and motor currents) and equations (10)-(15), the future values of variables are calculated and by minimizing the cost function (equations (16) plus equations (20)), optimum switching state will be selected.

IV. SIMULATION RESULT

In this section, simulation results of mentioned method is given. A PI speed controller is used to produce the reference torque and flux reference is calculated by utilizing the method of Maximum Torque Per Ampere (MTPA) as equation (21) [14] :

$$\psi^{ref} = \sqrt{\psi_f^2 + \left(\frac{T_e^{ref}}{1.5 p \psi_f} L_q\right)^2} \tag{21}$$

The speed reference is 75 rad/s, load torque is 35 KN.m and sample time is 200 μs.
Motor and inverter parameters are shown in Table 2.
Figures are based on P.U and horizontal axes are time except figure 8 that is frequency. Base values are as follows :
$V_{base} = 3000$ v , $I_{base} = 600$ A , $T_{e\,base} = 35000$ N.m
It can be seen from Fig. 4 that peak to peak torque ripple is less than %5 of its nominal value. The amplitude of neutral point potential never go upper than %8 of half of the DC-link voltage, that means voltage ripple across the DC-link capacitors is always less than %8 of their nominal value

Table 2. Motor and inverter parameters.

Motor parameter		Inverter parameter	
L_q=L_d=9.8 mH	V = 4200 v	C_{dc}	2000 µf
ψ_f=9.75web	I = 430 A	C_{fc}	2000 µf
R_s = 0.024	Ferquency = 50 Hz	V_{dc}	6000 v
P = 4			

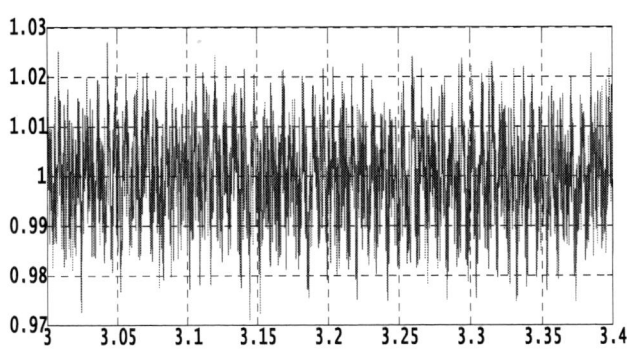

Fig. 4: Electromagnetic torque.

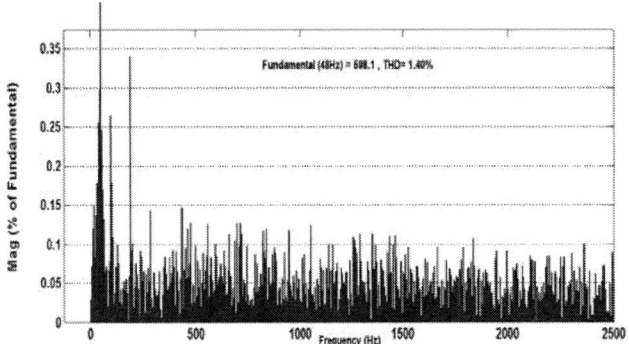

Fig. 7: Stator current spectrum.

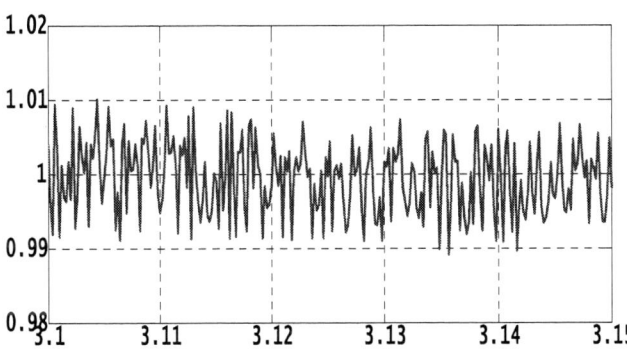

Fig. 5: Magnitude of flux.

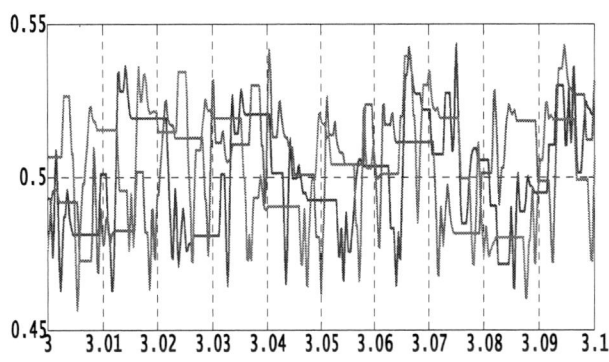

Fig. 8: Flying capacitor voltages.

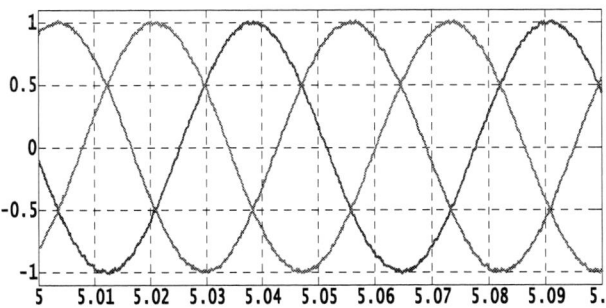

Fig. 6: Three phase stator current.

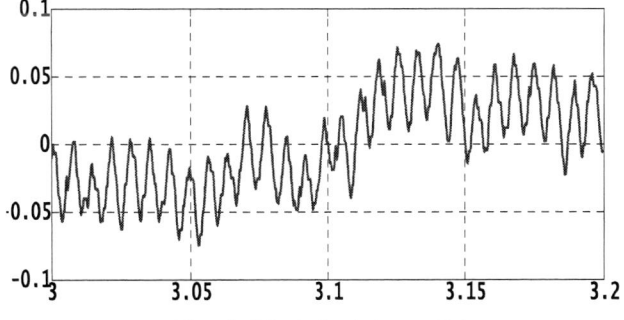

Fig. 9: Neutral point potential.

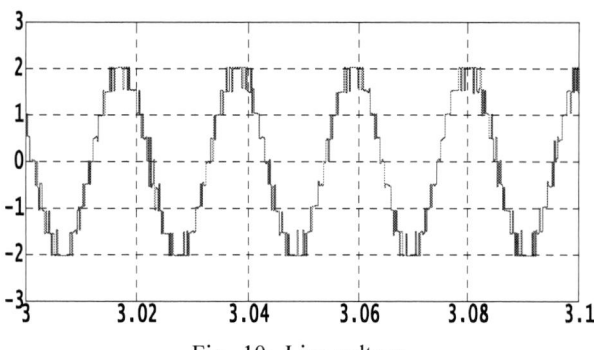

Fig. 10: Line voltage.

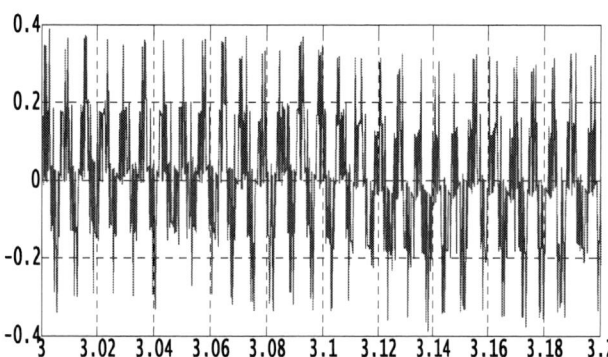

Fig. 11: Common mode voltage.

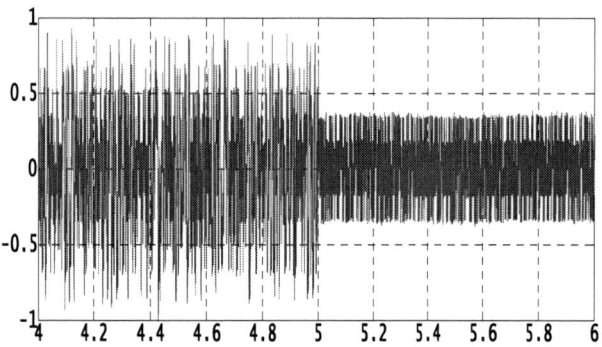

Fig. 12: Common mode voltage without and with λ_{cm}.

(given that DC-link is connected to an active front end rectifier and the suming up of two DC-link capacitor voltages is a fix value). According to Fig. 7 the THD of output current is about %1.4. Fig. 11 shows that the maximum amplitude of common mode voltage is about 0.33 in P.U ($V_{dc}=2$ p.u); that means its amplitude is limited to one-sixth of DC-link voltage. Also common mode voltage waveform is shown in Fig. 12 at low speed (speed reference is 30 rad/s) that λ_{cm} is added to cost function after t=5s.

V. CONCLUSIONS

In this article, predictive control method with one step prediction horizon is used to control the 5L-ANPC inverter as PM motor drive.

In addition to observe the switching limitation (according to figure 2) and control four internal voltages of inverter and produce the desired flux and torque of pm motor, common-mode voltage is limited to one-sixth of DC-link voltage. Moreover ,because of low number of calculation in this method, there is no need to very powerful microprocessor to implementation that.

References

[1] P. Barbosa, P. Steimer, J. Steinke, L. Meysenc, M. Winkelnkemper, and N. Celanovic, "Active neutral-point-clamped (ANPC)multilevel converter Technology," in *Proc. Conf. Rec. Eur. Conf. Power Electron. Appl.*, 2005, CDROM, pp. 1–10.

[2] F. Kieferndorf, M. Basler, L. A. Serpa, J. H. Fabian, A. Coccia, and G. A. Scheuer, "A new medium voltage drive system based on ANPC-5L technology," in *Proc. Conf. Rec. IEEE Int. Conf. Ind. Technol.*, 2010, Pp. 643–649.

[3] L. A. Serpa, P. M. Barbosa, P. K. Steimer, and J. W. Kolar, "Five-level Virtual-flux direct power control for the active neutral-point clamped Multilevel inverter," in *Proc. IEEE Power Electron. Spec. Conf.*, 2008, Pp. 1668–1674.

[4] K. Wang, Z. Zedong, Y. Li, K. Liu, and J. Shang, "Neutral-point potential balancing of a five-level active neutral-point-clamped inverter," *IEEE Trans. Ind. Electron.*, vol. 60, no. 5, pp. 1907–1918, May 2013.

[5] S. R. Pulikanti and V. G. Agelidis, "Hybrid flying-capacitor-based activeneutral- point-clamped five-level converter operated with SHE-PWM," *IEEE Trans. Ind. Electron.*, vol. 58, no. 10, pp. 4643–4653, Oct.2011.

[6] S. R. Pulikanti and V. G. Agelidis, "Control of neutral point and flying Capacitor voltages in five-level SHE-PWM controlled ANPC converter," In *Proc. IEEE 4th Conf. Ind. Electron. Appl.*, 2009, pp. 172–177.

[7] J. Li, A. Q. Huang, Z. Liang, and S. Bhattacharya, "Analysis and design Of active NPC (ANPC) inverters for fault-tolerant operation of high-power Electrical drives," *IEEE Trans. Power Electron.*, vol. 27, no. 2, pp. 519–533, Feb. 2012.

[8] S. R. Pulikanti, G. S. Konstantinou, and V. G. Agelidis, "Generalisation Of flying capacitor-based active-neutral-point-clamped multilevel converter Using voltage-level modulation," *IET Power Electron.*, vol. 5, no. 4, Pp. 456–466, 2012.

[9] Wang, Kui, Lie Xu, Zedong Zheng, and Yongdong Li. "Capacitor Voltage Balancing of a Five-Level ANPC Converter Using Phase-Shifted PWM." *Power Electronics, IEEE Transactions on* 30, no. 3 (2015): 1147-1156.

[10] Rodriguez, Jose, and Patricio Cortes. *Predictive control of power converters and electrical drives*. Vol. 40. John Wiley & Sons, 2012.

[11] Kieferndorf, Frederick, Petros Karamanakos, Philipp Bader, Nikolaos Oikonomou, and Tobias Geyer. "Model predictive control of the internal voltages of a five-level active neutral point clamped converter." In *Energy Conversion Congress and Exposition (ECCE), 2012 IEEE*, pp. 1676-1683. IEEE, 2012.

[12] Geyer, Tobias, and Silvia Mastellone. "Model predictive direct torque control of a five-level ANPC converter drive system." *Industry Applications, IEEE Transactions on* 48, no. 5 (2012): 1565-1575.

[13] M. M. Renge and H. M. Suryawanshi, "Five-level diode clamped inverter to eliminate common mode voltage and reduce dv/dt in medium voltage rating induction motor drives," *IEEE Trans. Power Electron.*, vol. 23, no. 4, pp. 1598–1607, Jul. 2008

[14] Zhang, Yongchang, and Xianglong Wei. "Torque ripple RMS minimization in model predictive torque control of PMSM drives." In

Electrical Machines and Systems (ICEMS), 2013 International Conference on, pp. 2183-2188. IEEE, 2013.

[15] J. M. Maciejowski. *Predictive Control*. Prentice Hall, 2002.

[16] Pulikanti, Sridhar R., and Vassilios G. Agelidis. "Control of neutral point and flying capacitor voltages in five-level SHE-PWM controlled ANPC converter." In *Industrial Electronics and Applications, 2009. ICIEA 2009. 4th IEEE Conference on*, pp. 172-177. IEEE, 2009.

7th Power Electronics, Drive Systems & Technologies Conference (PEDSTC 2016)
16-18 Feb. 2016, Iran University of Science and Technology, Tehran, Iran

Improved Direct Torque Control of Induction Motor with the Model Predictive Solution

Mohammad Reza Nikzad, Seyed omid Ahmadi
Dept. of computer and electrical Engineering
University of Tehran
Tehran, Iran
nikzad@ut.ac.ir

Behzad Asaei
Dept. of computer and electrical Engineering
University of Tehran
Tehran, Iran
basaei@ut.ac.ir

Abstract— Today, **The Direct Torque Control (DTC) is a common solution in modern induction motor drives. Because of the low number of available active voltage vectors, the DTC method has high flux and torque ripple.**
In this paper, an improved model predictive DTC method will be proposed to decrease both torque and flux ripples without the use of weighting factor. This method is based on the combination of the both conventional DTC and Finite Control Set Model Predictive Control (FCS-MPTC) schemes.
The performance of the proposed method will be validated by simulation results.

   Keywords— Direct torque control (DTC); Finite Control Set Model Predictive Control (FCS-MPTC); induction machine; variable speed drives.

Introduction

Nowadays, the direct torque control (DTC) is known as a popular control method for high performance induction motor drives [1]. The simple structure and fast response are the main advantages of this control scheme.

In DTC scheme, in order to directly control stator flux and torque separately and keep them within their hysteresis band, one voltage vectors (VV) among all available VVs that could be generated by inverter is selected in each sampling period. The proper VV is selected based on the torque and stator flux errors and according to the predefined switching table.

In this algorithm, there is no need for the PWM pulse generation, current regulators , and coordinate transformation.

However, the traditional DTC method has some well-known drawbacks such as high sampling requirement for digital implementation, considerable torque and current ripples due to the limited number of available VVs and variable switching frequency [2].
In order to directly control both of the torque and stator flux, finite control set model predictive torque control (FCS-MPTC) have been so attractive during the last decade [3, 4].
FCS-MPTC can predict the influence of each possible VVs on torque and stator flux by using the system model. Then, the VV that minimizes a single cost function including the torque and flux errors is selected as the output of the controller.

The intuitive concept and straightforward implementation are the main advantages of the FCS-MPTC method.

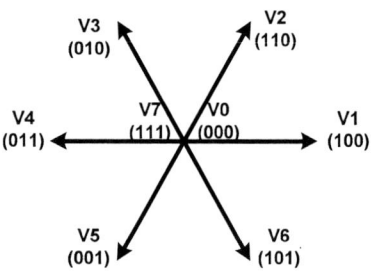

Fig. 1. Eight possible VVs in three phase voltage source inverter (VSI).

Furthermore, in MPTC, the nonlinearity of the variables and impose constraints could be included in the cost function [3]. Tuning of the weighting factor is the main drawbacks of this method. Furthermore, due to the limited number of the available VVs, this method has a relatively high torque and flux ripples.

In this paper, an improved model predictive DTC (MP_DTC) scheme to decrease torque and flux ripples without the use of weighting factor will be presented.

The basic idea of the proposed method is too simple. Firstly, in order to compensate the torque and flux errors, by using a switching table, an appropriate active voltage vector $\vec{V_n}$ is selected. Then, several VVs with the same direction and different duty cycle will be generated from original vector $\vec{V_n}$. Finally, between the new VVs the VV that minimizes a single variable cost function is selected by the controller and generated by the inverter for the next sampling period.

The paper is organized as follows. Principle of DTC and FCS-MPTC methods are explained in section II and section III respectively. Section IV explains the proposed DTC Method with Model Predictive Solution. Section V gives the simulation results, and performance validities.

Direct Torque Control Principle

The control Block Diagram of DTC is shown in Fig. 2. A close loop estimator estimates the values of the stator flux and torque at each sampling time. A three level hysteresis comparator is used for torque controller and a two level hysteresis controller control the stator flux error. By selecting

978-1-5090-0376-1/16 $31.00 © 2016 IEEE   626

the proper inverter VV, among six active VVs and two zero VVs, the Stator flux and torque can be controlled directly and independently [1]. As shown in Table I., a predefined switching table is used to select the proper VV based on the flux and torque errors as well as stator flux position.

FCS-MPTC FOR IMs

Fig. 3 shows the block diagram of the FCS-MPTC. This algorithm includes a measurement and estimation stage, a prediction of the outputs and an optimization stage.

Current sensors measure the motor current. Then, by using the motor internal dynamic model, the flux and torque are estimated for the next sampling time. Among the available VVs , the VV that minimize the cost function will be select for applying in the next switching period.

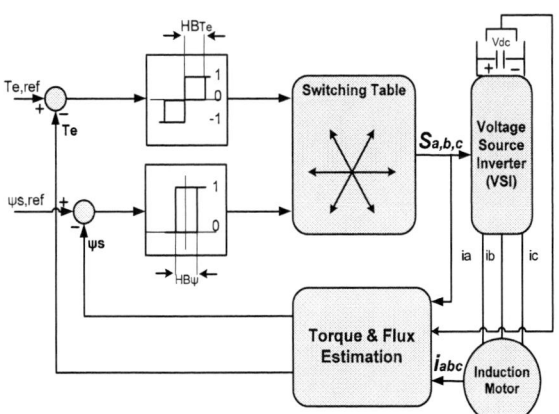

Fig. 2. Basic DTC scheme

TABLE I. DTC SWITCHING TABLE

sector		I	II	III	IV	V	VI
$\Psi s \uparrow$	$\tau \uparrow$	V2	V3	V4	V5	V6	V1
	τ --	V7	V0	V7	V0	V7	V0
	$\tau \downarrow$	V6	V1	V2	V3	V4	V5
$\Psi s \downarrow$	$\tau \uparrow$	V3	V4	V5	V6	V1	V2
	τ --	V0	V7	V0	V7	V0	V7
	$\tau \downarrow$	V5	V6	V1	V2	V3	V4

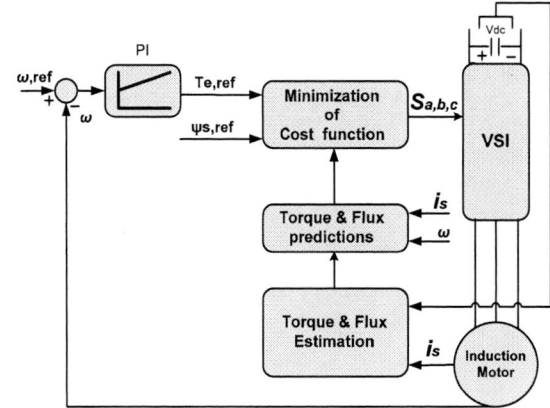

Fig. 3. Block diagram of FCS-MPTC.

Discrete Model of IMs

In FCS-MPTC, an internal model of the electric drive is required to predict the future values of the controlled output variables over a next sampling period T_s.

By means of normal forward Euler approximation, the discrete equations of the rotor and stator flux estimation (at the present sampling step k) are

$$\vec{\psi}_r(k) = \frac{L_r}{L_r + T_s.R_r} \vec{\psi}_r(k-1) + \frac{L_m}{\frac{\tau_r}{T_s}+1}.\vec{\iota}_s(k) \qquad (1)$$

$$\vec{\psi}_s(k) = \vec{\psi}_s(k-1) + T_s.\vec{v}_s(k) - R_s.T_s.\vec{\iota}_s(k) \qquad (2)$$

Where $\tau_r = L_r/R_r$ and T_s corresponds to the sampling time [6].

The electromagnetic torque can be predicted by using the stator flux and stator current at the next sampling time according to

$$\hat{T}(k+1) = \frac{3}{2}p.\Im m \left\{ \vec{\psi}_s(k+1)^*.\vec{\iota}_s(k+1) \right\} \qquad (3)$$

The prediction equations of the stator flux and stator current   at the instant $k+1$ are

$$\vec{\psi}_s(k+1) = \vec{\psi}_s(k) + T_s.\vec{v}_s(k) - R_s.T_s.\vec{\iota}_s(k) \qquad (4)$$

$$\vec{\iota}_s(k+1) = \left(1 - \frac{T_s}{\tau_\sigma}\right).\vec{\iota}_s(k) + \frac{T_s}{\tau_\sigma}$$
$$\cdot \left\{ \frac{1}{R_\sigma}.\left(\left(\frac{k_r}{\tau_r} - j.k_r.\omega\right).\vec{\psi}_r(k) + \vec{v}_s(k) \right) \right\} \qquad (5)$$

Where ω corresponds to the rotor speed, $k_r = L_m/L_r$ $R_\sigma = R_s + k_r{}^2 R_r$, $\tau_\sigma = \frac{\sigma L_s}{R_\sigma}$ and $\sigma = 1 - (L_m{}^2/L_sL_r)$.

978-1-5090-0376-1/16 $31.00 © 2016 IEEE   627

Optimization of the Cost Function

The next step is to minimize cost function. The cost is a control low and defined in a manner to select the appropriate VVs, among the available, in order to track torque and flux at sampling time T_s with the following structure:

$$g_i = |T_{ref} - \hat{T}(k+1)_i| + \lambda_0 . \left| \left| \psi_{sref} \right| - \left| \left| \vec{\psi}_s (k+1)_i \right| \right| \right| \quad (6)$$

Where $i/1 \triangleleft \square \triangleleft 7$ denotes the feasible switching states of a two level inverter. T_{ref} and ψ_{sref} are the torque and flux references and g_i is the cost value corresponding to the i_{th} VV at the next sampling time.

The factor $\square_0$ denotes a weight factor. This factor could define based on the importance of the control objectives. With the same importance, this factor would define to the ratio between the nominal magnitudes of the torque and stator flux

$$\lambda_0 = \frac{T_{nom}}{|\psi_{sn}|} \quad (7)$$

Finally, after the optimization stage, the controller selects the proper VV as the optimal switching state and applies it to the electric motor via the inverter for the next sampling period $k+1$.

PROPOSED DTC METHOD WITH MODEL PREDICTIVE SOLUTION

As discussed above, due to the limited number of available VVs (seven VVs) the DTC and FCS-MPTC schemes have considerable torque and current ripples [3]. In order to decrease the ripples, sampling rate must be increased respectively [3]. Increasing the sampling frequency leads to increasing inverter switching frequency and also the calculation effort for the control system.

In this paper, an improved direct torque control method is proposed to decrease torque and flux ripples with constant switching frequency. This method is based on the combination of the both conventional DTC and FCS-MPTC schemes. Fig. 4 shows The block diagram of the proposed model predictive direct torque control (MP-DTC). The proposed algorithm is explained as follows.

Initially, like the DTC algorithm, the values of stator flux and torque are estimated and compared with reference values at each sampling period. The Stator flux can be calculated by using (2) and the torque can be estimated according to

$$\hat{T}(k) = \frac{3}{2} p . \Im m \left\{ \vec{\hat{\psi}}_s (k)^* . \vec{i}_s (k) \right\} \quad (8)$$

In the next stage, the switching table is used to select the appropriate $\vec{V_n}$ ($n/1 \triangleleft \square \triangleleft 6$) to compensate the torque and flux errors. In this method, the switching table is similar to the DTC one. The only difference is, when the estimated torque is equal to reference value, the active VV that compensate the flux error is selected instead of zero vectors in DTC.

In the next stage, the selected active voltage vector $\vec{V_n}$ will be divided into m subdivisions $k/m . \vec{V_n}$ ($k = 1, ..., m$). Therefore, several vector with same direction and different duty cycle extracted from the VV $\vec{V_n}$ will be extracted. In this paper in order to simplicity the voltage vector $\vec{V_n}$ is divided into four subdivisions, $1/4 \vec{V_n}$, $2/4 \vec{V_n}$, $3/4 \vec{V_n}$ and $\vec{V_n}$. Then, by using (1) to (5), the stator flux and torque will be predicted for the new VVs as well as the zero voltage vector for the next sampling period $k+1$.

For instance, Assuming the stator flux vector lying in sector I, if the estimated flux and torque are smaller than (or equal to) reference values, the vector $\vec{V_2}$ (110) will be selected from switching table. Effect of the each new extracted VVs ($1/4 \vec{V_2}$, $2/4 \vec{V_2}$, $3/4 \vec{V_2}$ and $\vec{V_2}$) on the stator flux vector $\vec{\psi}_{s1}$ is shown in Fig. 5.

The aim of the proposed algorithm is to minimize the torque ripple and maintain the stator flux within the predefined limits. Because the selected voltage vector $\vec{V_n}$ (and its subdivisions) has a proper direction to compensate flux error, a simple cost function with single variable (torque) could satisfy the control strategy requirements and the flux variable

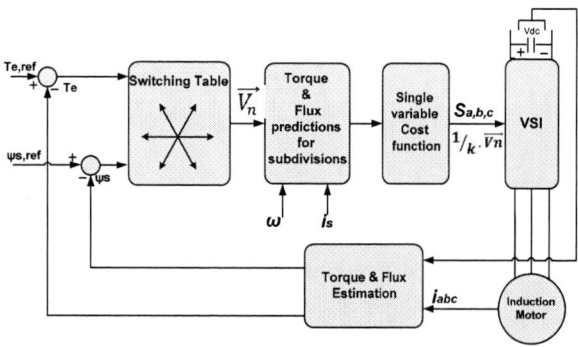

Fig. 4. Block diagram of the proposed MP-DTC.

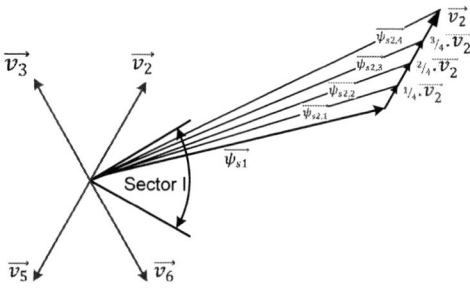

Fig. 5. Effect of the each new extracted VVs on the stator flux vector.

is included as a constraint in the cost function. Therefore, the following is used as the cost function:

$$g_i = \left| T_{ref} - \hat{T}(k+1)_i \right| + \psi_{max} \qquad (9)$$

Where

$$\psi_{max} = \begin{cases} \infty & if & \left| \vec{\psi}_s(k+1) \right| < \psi_{s\,min} \\ 0 & if & \psi_{s\,min} \leq \left| \vec{\psi}_s(k+1) \right| \geq \psi_{s\,max} \\ \infty & if & \left| \vec{\psi}_s(k+1) \right| > \psi_{s\,max} \end{cases} \qquad (10)$$

Finally, the optimization step is done, and between the new VVs and zero VV the VV that minimizes (9) is selected and applied as the optimal switching state for the next sampling period $k+1$.

SIMULATION STUDY

In this section, firstly, the dynamic behavior of the proposed MP-DTC will be examined. Then, the performance of conventional DTC and FCS-MPTC will be compared with the proposed method. The parameters of machine are listed in Table II.

For the proposed the proposed MP-DTC the sampling time is set to 100 $\square$s (3.3 kHz switching frequency) while the sampling time of the DTC and FCS-MPTC is 40 $\square$s.

In the first simulation, different speed reference steps are applied to evaluate the dynamic performance of the proposed MP-DTC method. Initially, the 200 rad/s speed reference is applied; the second speed reference step is from 200 to 100 rad/s at 0.4 s. The simulated results are shown in Fig. 6. As shown in Fig. 6, the reference torque is tracked very well by the proposed MP-DTC during both startup and reversal torque reference. Thus, the actual mechanical speed tracks the speed command very well. Furthermore, Fig. 6 shows the low current harmonics and flux ripples of the proposed method.

The comparative study is shown in Fig. 7. At the speed of 100 rad/s and a torque load of 20 N.m with simulation time of $t = 0.5$ s, it can be seen that the proposed MP-DTC has better torque performance than DTC and MPTC. Furthermore, the stator current and flux ripples are much lower in the proposed method.

Considering that, the sampling frequency of the proposed MP-DTC is only less than half of the other two methods. Furthermore, it is clear that by increasing the subdivision the better performance will be achieved. Thus, it is clear that the performance of the proposed MP-DTC is better than two prior methods.

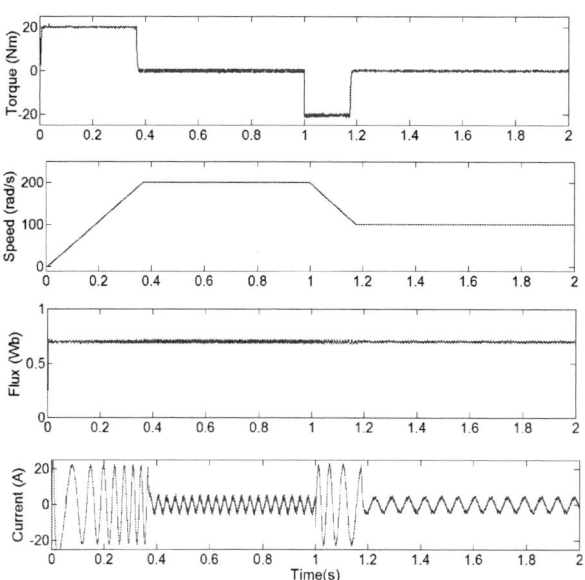

Fig. 6. Dynamic control behavior of the proposed MP-DTC.

TABLE II

INDUCTION MACHINE SPECIFICATIONS

DC-bus voltage	520 V
Rated torque	20 Nm
Number of pole pairs	2
Stator resistance	2.1 Ω
Mutual inductance	0.198 H
Flux amplitude reference	0.7 Wb
Stator inductance	0.212 H
Rotor inductance	0.212 H

CONCLUSION

In this paper, an improved direct torque control method has been proposed to decrease torque and flux ripples. This method is based on the combination of the both conventional DTC and FCS-MPTC schemes.

The dynamic behavior of the proposed MP-DTC was validated with simulation results. Furthermore, the performance of the proposed MP-DTC was compared with two prior DTC and FCS-MPTC methods. The simulation results show the better performance and lower flux and torque ripples and less current harmonics of the proposed method compared with the two prior schemes.

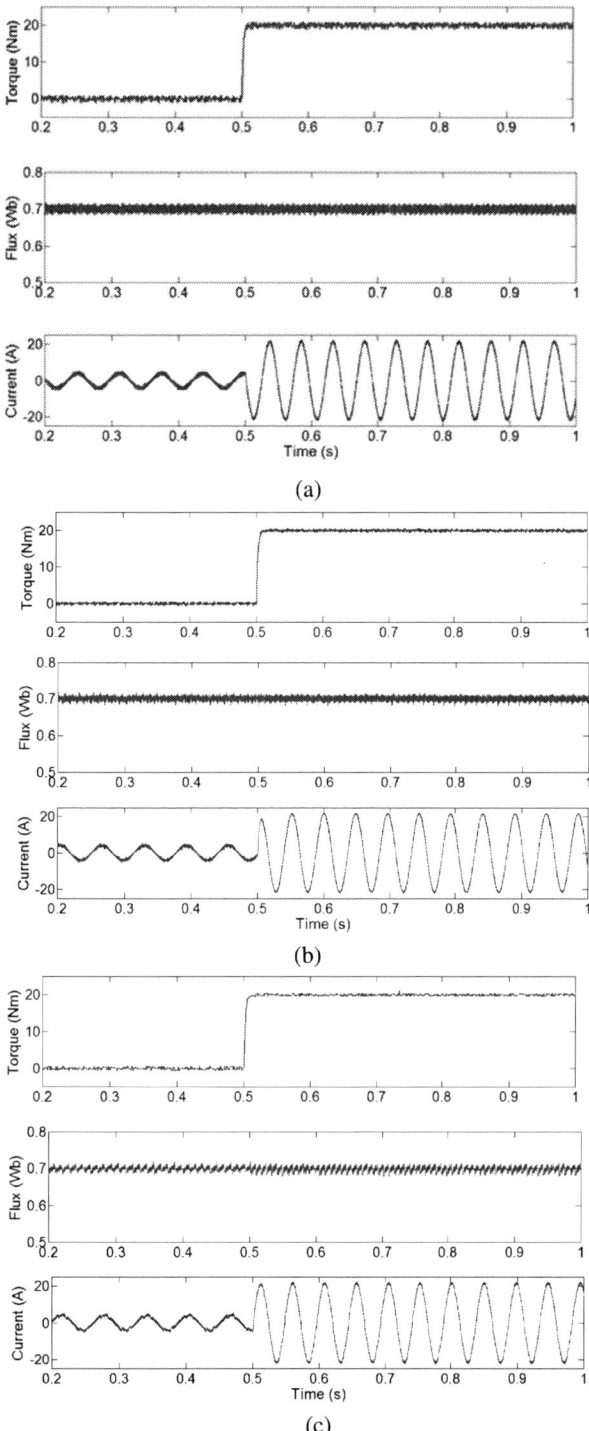

Fig. 7. Simulated responses of torque, stator flux and stator current with Sudden load change for (a) conventional DTC, (b) FCS-MPTC and (c) proposed MP-DTC.

REFERENCES

I. Takahashi and T. Noguchi, "A new quick-response and high-efficiency control strategy of an induction motor," IEEE Trans. Ind. Appl., pp. 820-827, 1986.

G. S. Buja and M. P. Kazmierkowski, "Direct torque control of PWM inverter-fed AC motors-a survey," IEEE Trans. Ind. Electron., vol. 51, pp. 744-757, 2004.

D. Casadei, F. Profumo, G. Serra, and A. Tani, "FOC and DTC: two viable schemes for induction motors torque control," IEEE Trans. Power Electron., vol. 17, pp. 779-787, 2002.

J. Rodriguez, M. P. Kazmierkowski, J. R. Espinoza, P. Zanchetta, H. Abu-Rub, H. Young, et al., "State of the art of finite control set model predictive control in power electronics," IEEE Trans. Ind. Informat.,vol. 9, pp. 1003-1016, 2013.

J. Rodríguez, R. M. Kennel, J. R. Espinoza, M. Trincado, C. Silva, and C. Rojas, "High-performance control strategies for electrical drives: An experimental assessment," IEEE Trans. Ind. Electron.,vol. 59, pp. 812-820, 2012.

P. Cortés, S. Kouro, B. La Rocca, R. Vargas, J. Rodríguez, J. León, et al., "Guidelines for weighting factors design in model predictive control of power converters and drives," in Industrial Technology, 2009. ICIT 2009. IEEE International Conference on, 2009, pp. 1-7.

J. Rodriguez, J. Pontt, C. Silva, P. Cortés, S. Rees, and U. Ammann, "Predictive direct torque control of an induction machine," in *Proc. Power Electron. Motion Control Conf*, 2004, pp. 2-4.

978-1-5090-0376-1/16 $31.00 © 2016 IEEE

7th Power Electronics, Drive Systems & Technologies Conference (PEDSTC 2016)
16-18 Feb. 2016, Iran University of Science and Technology, Tehran, Iran

Minimum Slope Model Predictive Control with Double Margins

Reza Fotouhi, Hui Fang, Ralph Kennel
Institute for Electrical Drive Systems and Power Electronics
Technical University of Munich
Munich, Germany

Abstract—**In Finite Control Set Model Predictive Control, FCS-MPC, with single step prediction horizon; absolute value of control variable errors corresponding to each of input sets are precalculated and the input sets which satisfies an objective function are selected to be applied in the next sampling time. This paper proposes a scheme to utilize predicted values in order to precalculate not only the absolute error, but also the slope and the direction of control variables. Finally, these predictions are used to reduce switching frequency and torque ripple in an induction motor drive.**

Keywords—Model Predictive Control; FCS-MPC; Induction Motor Drive

I. INTRODUCTION

Power electronics society has shown a high tendency to apply Model Predictive Control to Electrical Drives over the last decade [1], [2]. The foundation stone of MPC for electrical drives was introduced in 1983 [3], in which a current error vector is monitored with a high sampling rate. A boundary is considered for reference current and whenever a boundary condition occurs, the switching state of the converter is changed with the aim of maximizing the time between switching instances, or in other words minimizing the switching frequency. Availability of higher calculation power together with some further research works [4] escalated the research on predictive control in power converters and electrical drives in recent years.

The emerging field of MPC for electrical drives is introduced by replacing inner current control loops of Field Oriented Control (FOC) with Model Predictive Control [5] in which online optimization of cost function was a challenge. In an attempt to simplify the control problem, Finite Control Set Model Predictive Control (FCS-MPC) was proposed in [6], [7] for induction machine. In the later the modulator is omitted and the control problem is solved simply by enumeration of cost function for each feasible voltage vector. The switching states are defined based on the voltage vector which corresponds to the minimum value of cost function. This approach is very simple and flexible for modification and extension. Besides, nonlinearities of power converters and different constraints can be easily considered. Moreover, its

performance can be highly improved and losses can be minimized by choosing long prediction horizons. However, its application for long prediction horizons and multi-level converters encounters calculation time limits.

Reducing calculation effort of MPC has become an attractive topic for research in recent years. To increase the prediction horizon a candidate rejection technique is proposed in [8] and [9] for Direct Torque Control and in [10] and [11] for Direct Current Control of induction machine. In another attempt, a heuristic is used to reduce the number of switching sequences for longer horizons [12]. More recently, sphere decoding is used to increase the prediction horizon of model predictive control for multilevel converters [13], [14].

In literature, research on FCS-MPC with a single step prediction horizon is mainly limited to cost function formulation and weighting factor calculation[15], [16], stability issues [17] and its application on different topologies such as Matrix Converter [18], UPS [19], four-Leg converters [20] and etc. There is a common principle in all of these applications where the absolute value of control variable error, corresponding to each applicable input state for the next sampling time, are precalculated. Then the input state which satisfies the objective or cost function is selected to be applied in the next sampling time.

In principle, by predicting the control variable values based on each applicable input set, not only the absolute value of the errors are predicted but also the direction and slope of control variable trajectories are precalculated as well. This last piece of information can be also included into the objective function to achieve secondary goals such as torque ripple reduction.

The main motivation of this paper is to use the slope and direction of trajectories of control variables in order to reduce switching frequency in steady state for an induction motor which is fed by a 2-Level inverter.

In the next section, the model of induction motor and converter which are used for precalculations are proposed. Section III includes the Minimum Slope MPC with Double Margins that utilizes absolute error, slopes and direction information in different operating points in order to reduce the switching frequency. Finally, experimental results are presented in V.

978-1-5090-0376-1/16 $31.00 © 2016 IEEE   631

II. MODEL FOR PRECALCULATIONS

A. Two Level Inverter

First, a two-level voltage source inverter and its voltage vectors are shown in Fig. 1. The switching states of the inverter, S, is expressed as following:

$$S = \frac{2}{3}(S_a + aS_b + a^2 S_c) \tag{1}$$

where $a = e^{j2\pi/3}$, $S_i = 1$ corresponds to on state of the switch S_i, $\overline{S}_i = 0$ corresponds to its off state and $i = a,b,c$. The voltage vector v of the switching state S is:

$$v = u_{dc} S \tag{2}$$

where u_{dc} is the dc-link voltage.

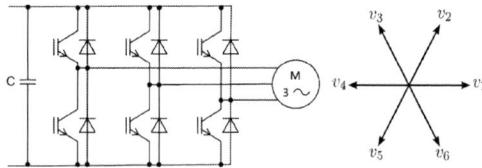

Fig. 1. Left: two-level voltage source inverter; right: voltage vectors

B. Induction Motor

A squirrel-cage IM can be described by a well-known set of complex equations using stator reference frame [21]:

$$\frac{d\psi_s}{dt} = \frac{-R_s}{\sigma L_s}\psi_s + \frac{R_s L_m}{\sigma L_r L_s}\psi_r + v_s \tag{3a}$$

$$\frac{d\psi_r}{dt} = \frac{R_r L_m}{\sigma L_r L_s}\psi_s - \frac{R_r}{\sigma L_r}\psi_r + j\omega_m\psi_r \tag{3b}$$

$$T = \frac{L_m}{\sigma L_s L_r}|\psi_r \times \psi_s|_z \tag{3c}$$

where v_s denotes stator voltage vector, ψ_s and ψ_r represent the stator flux and rotor flux, respectively. R_s and R_r are the stator and rotor resistance. L_s, L_r and L_m are stator, rotor and mutual inductance. $\sigma = 1 - L_m^2 / L_s L_r$ is the total leakage factor. ω_m is the mechanical speed and T denotes the electromagnetic torque.

Using (3) a discrete time model for precalculations is given by:

$$\psi_s(k+1) = (1 + \frac{T_s R_s}{\sigma L_s})\psi_s + \frac{T_s R_s L_m}{\sigma L_r L_s}\psi_r + T_s v_s \tag{4a}$$

$$\psi_r(k+1) = \frac{T_s R_r L_m}{\sigma L_r L_s}\psi_s + (1 + \frac{T_s R_r}{\sigma L_r})\psi_r + j\omega_m\psi_r \tag{4b}$$

$$T(k+1) = \frac{L_m}{\sigma L_s L_r}|\psi_r(k+1) \times \psi_s(k+1)|_z \tag{4c}$$

where T_s is the sampling time.

III. MINIMUM SLOPE MPC WITH DOUBLE MARGINS

The proposed scheme aims to directly control electromagnetic torque and magnitude of stator flux of an induction motor with a high dynamic and at the same time reduce the switching frequency. In order to illustrate controller's principle, reference and estimated values of torque are shown in the Fig. 2, where three regions are defined around the reference torque. The first region is an area where reference torque and estimated torque are both within a certain boundary which is determined by maximum permissible absolute value of torque error. When torque is in region I, it means that the preliminary control objective is satisfied. Thus no extra switching shall take place; Nevertheless, if a switching is necessary due to another control variable (ie.: when amplitude of stator flux trajectory exceeds a boundary), it is preferred to minimize the slope of control variable while switching. This is depicted at point k_2 in Fig. 2.

The second region is a narrow area on the boarders of the inner region. When the control variable enters this area it means that the absolute value of torque error has exceeded the inner boundaries. Therefore, a switching shall take place in order to bring it back to region I. The voltage vector with minimum slope which is directing to the reference point shall be selected in this case. This is depicted at point k_1 in Fig. 2. Using minimum slope trajectories at this point increases the interval time between two successive switchings and hence, reduces the switching frequency.

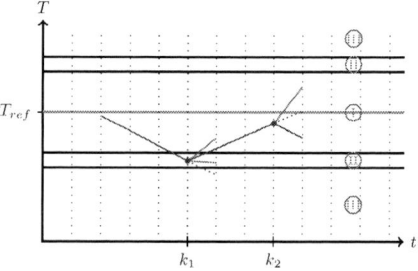

Figure 2: Torque trajectory in minimum slope model predictive control with double margins

When the control variable is located in regions I or II, the machine is operating in steady-states. Thus the slope and direction of control variable trajectories are used to keep the machine at this operating point with minimum number of switching. If the control variable enters region III, the machine is in transient mode so the absolute value of control variable error is used to take the control variables to the new steady state point with the highest dynamic.

Accordingly, the following sequence of steps are developed for this scheme:

978-1-5090-0376-1/16 $31.00 © 2016 IEEE

$$J_i = \lambda_1 (\lambda_2 \underbrace{((T(k+1)-T(k))^2 + \lambda_3 (|\psi_s(k+1)| - |\psi_s(k)|)^2)}_{\text{Minimum Slope Term}} + \overline{\lambda_2} \underbrace{((T(k+1)-T_{ref})^2 + \lambda_4 (|\psi_s(k+1)| - |\psi_{s,ref}|)^2))}_{\text{Fast Dynamic Term}} \quad (5)$$

$$\lambda_1 = \begin{cases} 1 & \text{if stator flux amplitude or torque trajectory for all voltage vectors directing outside inner bands} \\ & \text{or located in region III} \\ \infty & \text{elseif stator flux amplitude or torque trajectory directing outside inner regions} \\ 1 & \text{else} \end{cases}$$

$$\lambda_2 = \begin{cases} 0 & \text{if stator flux amplitude or torque trajectory for all voltage vectors directing outside inner bands} \\ & \text{or located in region III} \\ 1 & \text{else} \end{cases}$$

1. The amplitude of stator flux, $|\psi_s|$ and electromagnetic torque, T are estimated based on the measured signals, flux estimator and equations (3).

2. Based on feasible voltage vectors of the converter, predicted values for torque, $T(k+1)$ and amplitude of stator flux, $|\psi_s(k+1)|$ are calculated according to (4).

3. If previously applied voltage vector keeps torque and stator flux amplitude in region I at $k + 1$, no action is performed in this sample time. Otherwise next step shall be followed.

4. Cost function, J, is calculated based on (5) for all feasible voltage vectors and the voltage vector corresponding to its minimum value is applied.

In (5), J_i denotes the value of cost function for voltage vector i, λ_1 and λ_2 are weighting factors to decide between FCS and minimum slope approaches and λ_3 and λ_4 are weighting factors which are considered as tuning parameters.

It is worthy to note that when all predicted values for stator flux amplitude or torque trajectory are located in region III the machine is in transients. Setting $\lambda_1 = 1$ and $\lambda_2 = 0$ omits the minimum slope term from cost function in this case. Hence, the cost function is only consists of absolute error which leads to a fast dynamic response.

Setting $\lambda_1 = 1$ and $\lambda_2 = 1$ omits the fast dynamic term from the objective function. This happens during steady state operation in which both control variables are located either in region I or II.

On the other hand, when at least a single precalculation for stator flux amplitude and torque trajectory is located in inner regions, all other trajectories which are directing to or are located in region III are canceled by setting $\lambda_1 = \infty$. In fact, λ_1 is a factor to reject voltage vectors which has a wrong direction or are located in a wrong region.

As a result, in steady state only the voltage vectors with right direction and minimum slope are chosen in inner region.

Simulation results are shown for an induction motor at rated torque and 50% of nominal speed in Fig. 3. It can be seen that several switchings are taking place while a variable

is in region I due to the fact that the other variable exceeds the boundaries. It is worthy to mention that, although considering wide boundaries for control variables always reduce the switching frequency, but it does not guarantee to choose the best voltage vector in case of switching. The reason is that the definition of best is different in steady-state and transient. In transient best means fast dynamic and in steady state it refers to low switching frequency in our case. Therefore, the secondary bounds are considered to distinguish between switching in steady state and transients. Thus, the switching frequency can be set to a low value without compromising the dynamics.

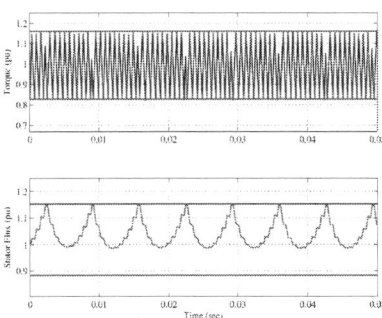

Figure 3: Simulation results of torque and amplitude of stator flux at 50% of nominal speed

IV. EXPERIMENTAL RESULTS

The experimental tests are carried out on an IM fed by a two-level inverter to evaluate the effectiveness of the proposed method. Also a dSpace system with ds1006 processor is employed to realize the control algorithm. In all tests the speed of IM is either kept constant or changed via a load machine and the controller objective is to regulate the torque and stator flux magnitude.

The performance of the proposed scheme is illustrated and compared with conventional FCS-MPC via two tests: Torque step and speed reversal test. Fig. 4 shows a torque step from 10 Nm to -10 Nm with the conventional FCS-MPC and Fig. 6 illustrates the proposed method's performance in the same conditions. It's worthy to note that although the flux bandwidth is considered to be relatively large to have the minimum number of switchings. As a result, the flux trajectory is very similar to a DSC control and stator currents are highly distorted. But even in such an extreme case, torque has relatively low ripple. It is due to the fact that torque slope is considered in the objective function. Besides, such a low switching has led to no sacrifice on system dynamics.

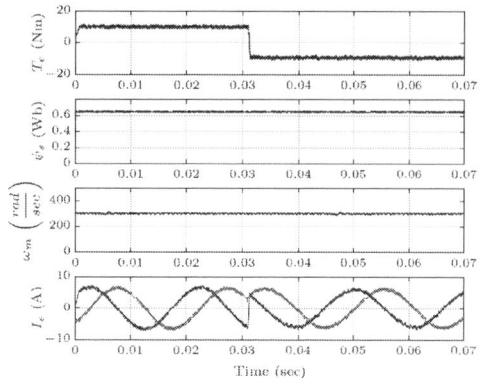

Figure 4: Torque step with FCS-MPC

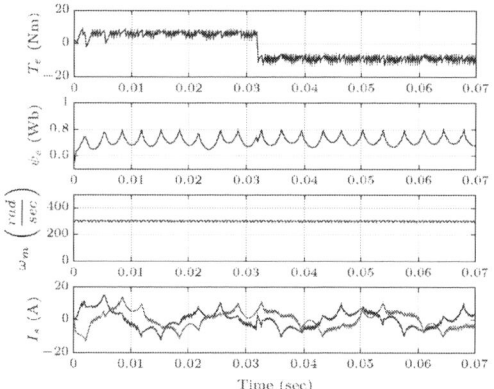

Figure 6: Torque step with the proposed method

The results during speed reversal are shown in Fig. 5 for FCS-MPC and in Fig. 7 for Minimum Slope MPC with Double Marines. It demonstrates the stability of the controller

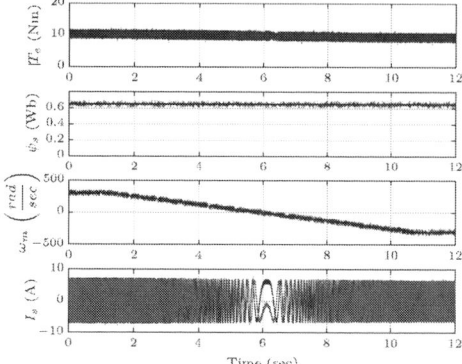

Figure 5: Speed reversal test with FCS-MPC

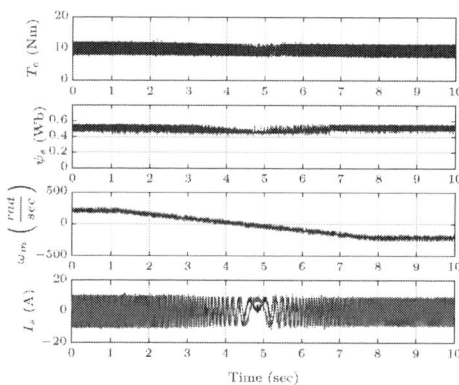

Figure 7: Speed reversal test with the proposed method

during speed reversal test. Also in this case smaller bandwidth is selected for stator flux magnitude and bigger bandwidth is selected for torque in comparison to the dynamic tests. This has led to a better current waveform which proves that current waveform is highly affected by stator flux bandwidth.

The results of Fig. 5 and Fig. 7 are magnified in Fig. 8 in order to compare the switching frequency of both methods. By applying the proposed algorithm the switching frequency reduces from 8.3 kHz to 5 kHz. It shall be noted that the proposed method is a hysteresis top controller. Hence it has a bigger torque ripple and lower switching frequency. On the other hand, FCS-MPC has a deadbeat behavior with a higher switching frequency and lower torque ripple. The figures are given for reference; however, comparing switching frequencies of these two methods is not fair.

V. CONCLUSION

In this work a new cost function is introduced for predictive torque control of induction motor. The aim was to emphasis on the role of objective function in the design of a model predictive controller. Cost function is the core heart of MPC and the controller performance highly relies on that;

however, in most cases only a reference tracking cost function is used. This can be changed in the future and this work is a contribution in that direction.

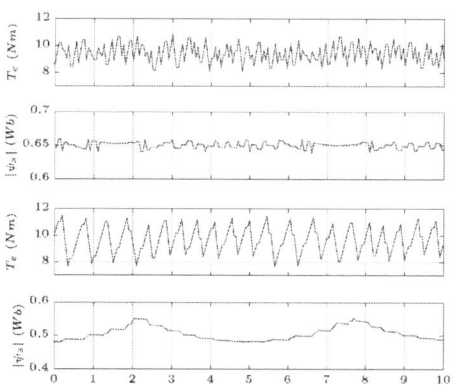

Figure 8: Torque and stator flux amplitude. top: FCS-MPC, bottom: the proposed method

References

[1] P. Cortes, M. Kazmierkowski, R. Kennel, D. Quevedo, and J. Rodriguez, "Predictive control in power electronics and drives," Industrial Electronics, IEEE Transactions on, vol. 55, no. 12, pp. 4312–4324, Dec 2008.

[2] J. Rodriguez, M. Kazmierkowski, J. Espinoza, P. Zanchetta, H. Abu-Rub, H. Young, and C. Rojas, "State of the art of finite control set model predictive control in power electronics," Industrial Informatics, IEEE Transactions on, vol. 9, no. 2, pp.1003–1016, May 2013.

[3] J. Holtz and S. Stadtfeldt, "A predictive controller for the stator current vector of AC machines fed from a switched voltage source," in Proc. IEEE Int. Power Electron. Conf. IPEC, vol. 2, Mar. 27-31, 1983, pp. 1665–1675.

[4] R. Kennel, A. Linder, and M. Linke, "Generalized predictive control (gpc)-ready for use in drive applications?" in Power Electronics Specialists Conference, 2001. PESC. 2001 IEEE 32nd Annual, vol. 4, 2001, pp. 1839–1844 vol. 4.

[5] A. Linder and R. Kennel, "Model predictive control for electrical drives," in Power Electronics Specialists Conference, 2005. PESC '05. IEEE 36th, June 2005, pp. 1793–1799.

[6] J. Rodriguez, J. Pontt, C. Silva, M. Salgado, S. Rees, U. Ammann, P.Lezana, R. Huerta, and P. Cortes, "Predictive control of three-phase inverter," Electronics Letters, vol. 40, no. 9, pp.561–563, April 2004.

[7] J. Rodriguez, J. Pontt, P. Cortes, and R. Vargas, "Predictive control of a three-phase neutral point clamped inverter," in Power Electronics Specialists Conference, 2005. PESC '05. IEEE 36th, June 2005, pp. 1364–1369.

[8] T. Geyer, G. Papafotiou, and M. Morari, "Model predictive direct torque control - concept, algorithm, and analysis," Industrial Electronics, IEEE Transactions on, vol. 56, no. 6, pp.1894–1905, June 2009.

[9] G. Papafotiou, J. Kley, K. Papadopoulos, P. Bohren, and M. Morari, "Model predictive direct torque control - implementation and experimental evaluation," Industrial Electronics, IEEE Transactions on, vol. 56, no. 6, pp. 1906–1915, June 2009.

[10] T. Geyer, "Model predictive direct current control: Formulation of the stator current bounds and the concept of the switching horizon," Industry Applications Magazine, IEEE, vol. 18, no. 2, pp. 47–59, March 2012.

[11] J. C. Ramirez Martinez, R. Kennel, and T. Geyer, "Model predictive direct current control," in Industrial Technology (ICIT), 2010 IEEE International Conference on, March 2010, pp. 1808–1813.

[12] P. Stolze, P. Landsmann, R. Kennel, and T. Mouton, "Finite-set model predictive control with heuristic voltage vector preselection for higher rediction horizons," in Power Electronics and Applications (EPE 2011), Proceedings of the 2011-14th European Conference on, Aug 2011, pp. 1–9.

[13] T. Geyer and D. Quevedo, "Multistep direct model predictive control for power electronics - part 1: Algorithm," in Energy Conversion Congress and Exposition (ECCE), 2013 IEEE, Sept 2013, pp. 1154–1161.

[14] ——, "Multistep direct model predictive control for power electronics - part 2: Analysis," in Energy Conversion Congress and Exposition (ECCE), 2013 IEEE, Sept 2013, pp. 1162–1169.

[15] P. Cortes, S. Kouro, B. La Rocca, R. Vargas, J. Rodriguez, J. Leon, S. Vazquez, and L. Franquelo, "Guidelines for weighting factors design in model predictive control of power converters and drives," in Industrial Technology, 2009. ICIT 2009. IEEE International Conference on, Feb 2009, pp. 1–7.

[16] S. Alireza Davari, D. Khaburi, and R. Kennel, "An improved fcs-mpc algorithm for an induction motor with an imposed optimized weighting factor," Power Electronics, IEEE Transactions on, vol. 27, no. 3, pp. 1540–1551, March 2012.

[17] R. Aguilera and D. Quevedo, "On stability of finite control set mpc strategy for multicell converters," in Industrial Technology (ICIT), 2010 IEEE International Conference on, March 2010, pp. 1277–1282.

[18] R. Vargas, U. Ammann, and J. Rodriguez, "Predictive approach to increase efficiency and reduce switching losses on matrix converters," Power Electronics, IEEE Transactions on, vol. 24, no. 4, pp. 894–902, April 2009.

[19] P. Cortes, J. Rodriguez, S. Vazquez, and L. Franquelo, "Predictive control of a three-phase ups inverter using two steps prediction horizon," in Industrial Technology (ICIT), 2010 IEEEInternational Conference on, March 2010, pp. 1283–1288.

[20] F. Rojas-Lobos, R. Kennel, and R. Cardenas-Dobson, "Current control and capacitor balancing for 4-leg npc converters using finite set model predictive control," in Industrial Electronics Society, IECON 2013 - 39th Annual Conference of the IEEE,Nov 2013, pp. 590–595.

[21] J. Holtz, "The dynamic representation of ac drive systems by complex signal flow graphs," in Industrial Electronics, 1994. Symposium Proceedings, ISIE '94., 1994 IEEE International Symposium on, May 1994, pp. 1–6.

7th Power Electronics, Drive Systems & Technologies Conference (PEDSTC 2016)
16-18 Feb. 2016, Iran University of Science and Technology, Tehran, Iran

The Challenges of Predictive Control to reach acceptance in the Power Electronics Industry

Margarita Norambuena
Departamento de Electronica UTFSM
Valparaiso, Chile
Fakultat IV
Technische Universität Berlin
Berlin, Germany
e-mail: margarita.norambuena@gmail.com.

Cristian Garcia
Departamento de Electronica
Universidad Técnica Federico Santa María
Valparaíso, Chile
e-mail: cristian.garciap@alumnos.utfsm.cl

Jose Rodriguez
Universidad Nacional Andrés Bello
Santiago, Chile
e-mail: jose.rodriguez@unab.cl

Abstract—**Model Predictive Control (MPC) has emerged as a very attractive alternative for the control of electrical energy. This paper discusses the challenges faced by (MPC) to reach industrial acceptance. Some of these issues are operation with fixed switching frequency and simple calculation of the weighting factors.The main conclusions are that the algorithm must remain simple and must demonstrate that it brings operating advantages in comparison to the classical linear control with Pulse Width Modulation. These problems have not been solved yet.**

I. INTRODUCTION

The control and transformation of electrical energy using power semiconductors has been one of the most active research topics in the last decades. The development of new application areas like electric transportation and renewable energies have increased the use of power converters and made the control of electrical energy even more timely and attractive for research. In addition, the availability of fast, cheap and very powerful microprocessors originated the development of new and more intelligent control techniques [1]. Model Predictive Control (MPC) is one of these intelligent techniques that is being studied and considered for power electronics systems, [2], [3]. MPC has several attractive features like: it is simple to understand and to implement and can include easily any type of restrictions and nonlinearities. Preliminary results have shown that this strategy works well specially in dynamic behavior [4]–[6]. However, to reach a wider acceptance, it must improve certain operating aspects, some of these are:

i It must avoid the operation with variable switching frequency. In most applications, fixed switching frequency is highly desirable [7].

ii It must improve the method to calculate the weighting factors [8], [9].

iii It must keep its simplicity.

iv It must keep a reduced computational cost, specially for multilevel inverters [10]–[15].

In addition, to be accepted by industry, MPC must demonstrate that it offers advantages over linear controllers an classical modulation strategies in terms of losses, operating performance and costs. These issues are discussed in the following pages of this paper.

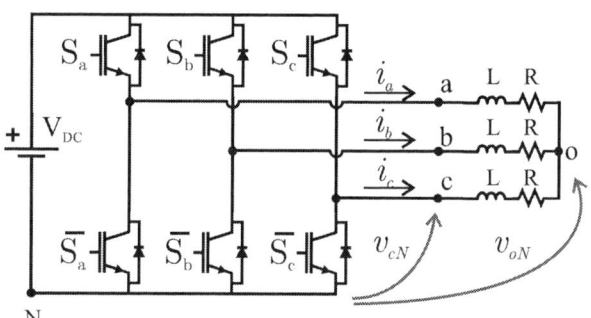

Fig. 1: Power circuit: 2-level VSI

II. OPERATION PRINCIPLE

One of the most widespread converter topologies found in the industry is the 2-level Voltage Source Inverter (VSI) shown in Fig. 1. The 2-level VSI consists of three phase legs, each one characterized by two power switches that connect the corresponding load terminal to $+V_{DC}$ or to the neutral point of the converter N. S_a, S_b and S_c are the switching signals, which can take only two values, $S_x = \{1, 0\}$ where $x = \{a, b, c\}$. When $S_x = 1$ the power switch S_x is in ON state and when $S_x = 0$ the power switch is in OFF state, $\bar{S}_x$ denotes the negated state of S_x.

Since there are three power switches that can take only two possible states, there are a total of $2^3 = 8$ possible switching states in the whole system.

Voltages v_{aN}, v_{bN} and v_{cN} can be modeled in discrete-time by the Eq. (1)

$$v_{xN}^k = S_x^k V_{DC}, \tag{1}$$

where $x \, \epsilon \{a, b, c\}$ and k is the sampling instant.

The equations for the load in a discrete-time model are described by:

$$i_x^{k+1} = (v_{xN}^k - v_{oN}^k)\frac{T}{L} + i_x^k \left(1 - \frac{RT}{L}\right), \tag{2}$$

$$v_{oN}^k = \frac{v_{aN}^k + v_{bN}^k + v_{cN}^k}{3}, \tag{3}$$

978-1-5090-0376-1/16 $31.00 © 2016 IEEE

where T is the sampling period and i_x^k denotes the measured values of the output currents at instant k.

Finite Control Set Model Predictive Control (FCS-MPC) explicitly takes into account the power switches as constraint in the optimal problem and chooses the switching state to be applied at the next sampling instant $(k + 1)$ through minimization of a cost function [2]. One possible cost function is given by tracking the error in the output currents and the number of commutations in the switches of the inverter:

$$g^k = \left(i_a^* - i_a^{k+2}\right)^2 + \left(i_b^* - i_b^{k+2}\right)^2 + \left(i_c^* - i_c^{k+2}\right)^2 \quad (4)$$
$$+ \lambda_1 \left(C_{sw}^* - C_{sw}\right)^2$$

Where i_x^* is the reference for current i_x, i_x^{k+2} is the current prediction at the sampling instant $k+2$ and λ_1 is a weighting factor to govern the switching frequency. C_{sw}^* is the reference for the number of commutations and C_{sw} is the actual value.

The block diagram of the control is shown in Fig. 2. The load current is measured at instant k and used as input for a predictive model that calculates the values of the current at the next sampling time for all possible switching states of the inverter. The cost function is evaluated for all possible switching states and the switching state which minimizes (4), it is applied at the begin of the next sampling time.

Fig. 3 shows the flow diagram of the predictive control applied in a 2-level VSI.

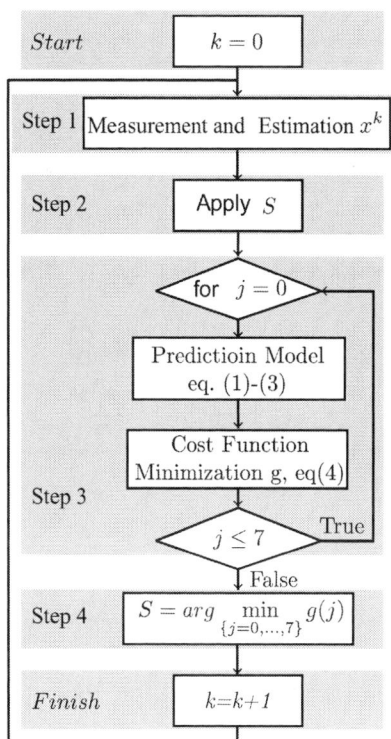

Fig. 3: Flow diagram of MPC

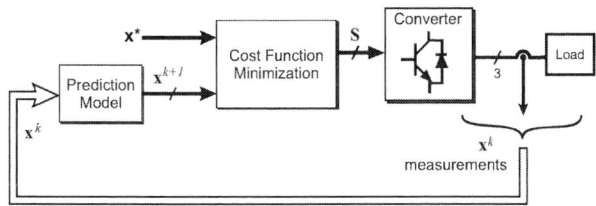

Fig. 2: Block diagram of MPC

III. SIMPLICITY BEHIND THE CONTROL PRINCIPLE

The main characteristic of MPC is its simplicity. MPC uses the equations that describe the system to establish a work framework and the controller defines a cost function to evaluate the performance. The simplicity of MPC is not only behind the theory of the control principle, but also in how the cost function is defined.

The following sections show two examples of application of MPC. The first is the speed control of an induction machine and the second is the control of a multilevel inverter.

A. Induction Machine

The induction machine feed by 2-Level voltage source inverter is a typical industrial application.

During the past decades, two main control approaches have gained widespread use in control of high performance electrical drives and they have become a standard in industrial applications. These are Field Oriented Control (FOC) [16] and Direct Torque Control (DTC) [17], [18].

FCS-MPC is an attractive alternative for the control of the machine thanks to high dynamics response and easy conceptual application [4]. In this section a Predictive Torque Control (PTC) is presented. PTC has two objectives: to control the electric torque (T) and the stator flux (Ψ_s). The scheme for PTC is shown in Fig. 4. The cost function to achieve the objectives is the following:

$$g = (T^* - T^p)^2 - \lambda_\Psi \left(\Psi_s^* - \Psi_s^p\right)^2, \quad (5)$$

where T^* and Ψ_s^* are the references for torque and flux respectively. T^p and Ψ_s^p are the prediction of the electric torque and stator flux using the mathematical model of the electrical machine. λ_Ψ is the weighting factor to adjust both control objectives [4].

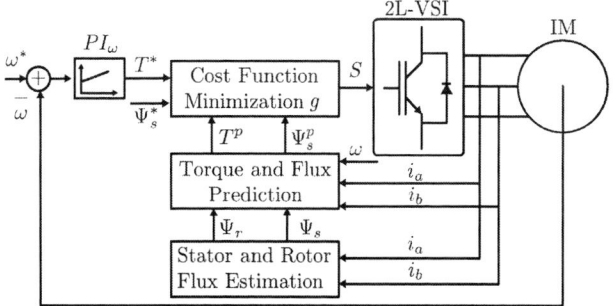

Fig. 4: Predictive Torque Control Scheme.

The dynamic behavior of predictive torque control is demon-

978-1-5090-0376-1/16 $31.00 © 2016 IEEE   637

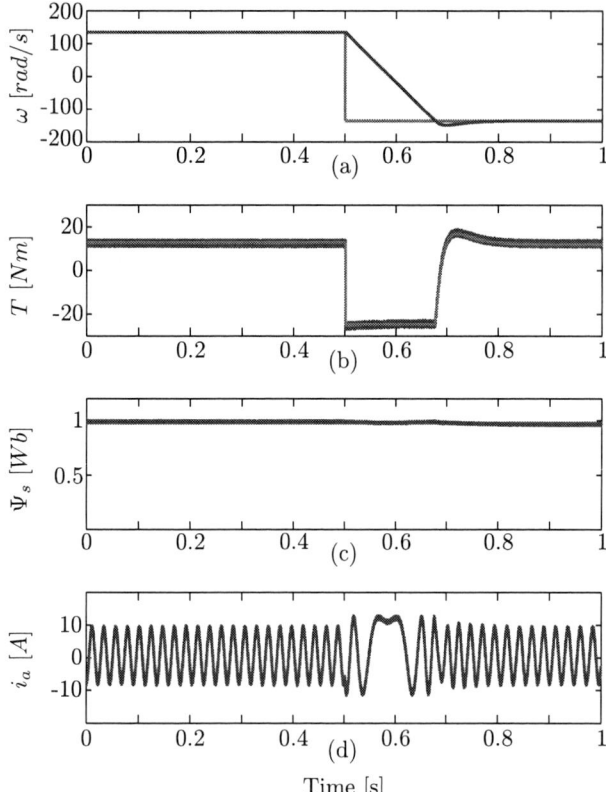

Fig. 5: Simulation results of Predictive Torque Control. (a) Speed; (b) Electric torque; (c) Stator flux; (d) Phase stator current.

strated through a reversing speed maneuver shown in Fig. 5. The machine has a constant load torque of 12.5[Nm]. At $t = 0.5[s]$, a speed reversal is applied from +90% to -90% of the rated speed of the machine. The electric torque correctly tracks the reference.

B. Multilevel Converter

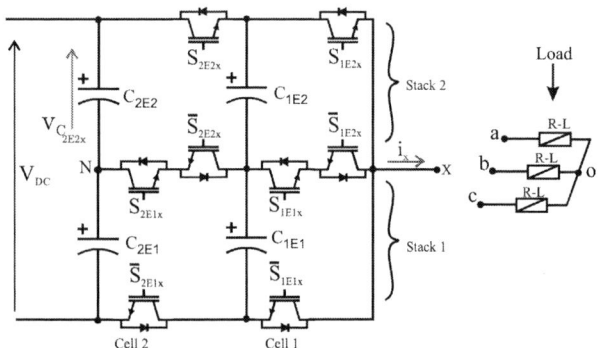

Fig. 6: Power circuit of Multilevel Converter

Fig.6 shows the power circuit of a Stacked Multilevel Convert (SMC), in this case, the variables of interest are the inner voltages (the voltage in every capacitor) and the output

current. It is necessary to have balanced the inner voltages to achieve a correct performance of the converter with the desired number of levels.

A cost function that can achieve current control of the SMC while controlling simultaneously the capacitor voltages is given by

$$g^k = g_a^k + g_b^k + g_c^k \qquad (6)$$

$$
\begin{aligned}
g_x^k = {} & (i_x^* - i_x^{k+2})^2 + \lambda_{v_{C2i}}(v_{C_{2E2x}}^* - v_{C_{2E2x}}^{k+2})^2 \\
& + \lambda_{v_{C2i}}(v_{C_{2E1x}}^* - v_{C_{2E1x}}^{k+2})^2 \\
& + \lambda_{v_{C1i}}(v_{C_{1E2x}}^* - v_{C_{1E2x}}^{k+2})^2 \\
& + \lambda_{v_{C1i}}(v_{C1E1x}^* - v_{C1E1x}^{k+2})^2 \qquad (7) \\
& x \epsilon \{a, b, c\},
\end{aligned}
$$

where $\lambda_{v_{C2i}}$ and $\lambda_{v_{C1i}}$ are weighting factors, which give the control designer two degrees of freedom to improve the tracking of capacitor voltages and/or output currents. Variables with superscript * are reference values corresponding to the next sampling time ($k + 2$).

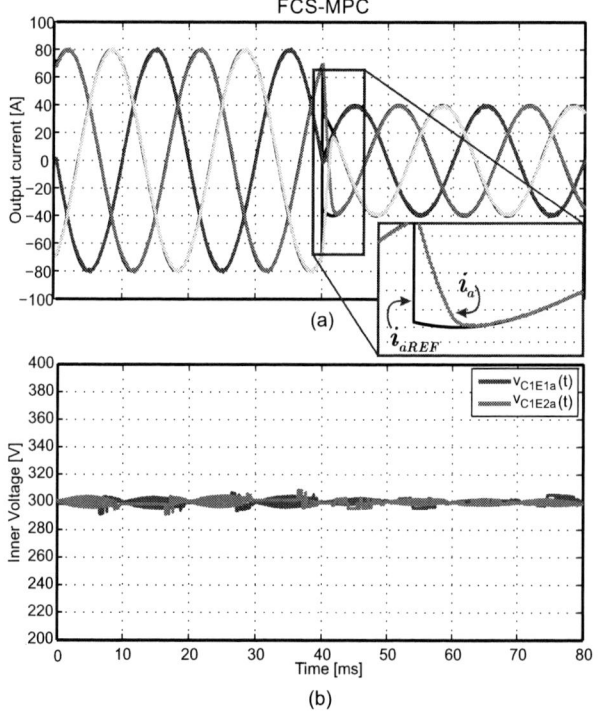

Fig. 7: SMC performance using Model Predictive Control: (a) Output currents; (b) Inner voltages.

Fig.7 and Fig.8 show the performance of MPC and linear controller, respectively, to control the main variables in the stacked multilevel inverter. These figures show that both strategies are able to control the load currents and the capacitor voltages. However, MPC shown in Fig. 7 has better dynamic behavior and better balance of the capacitor voltages.

IV. FIXED SWITCHING FREQUENCY

One of the drawbacks of MPC is its variable switching frequency. A variable switching frequency generates a wide

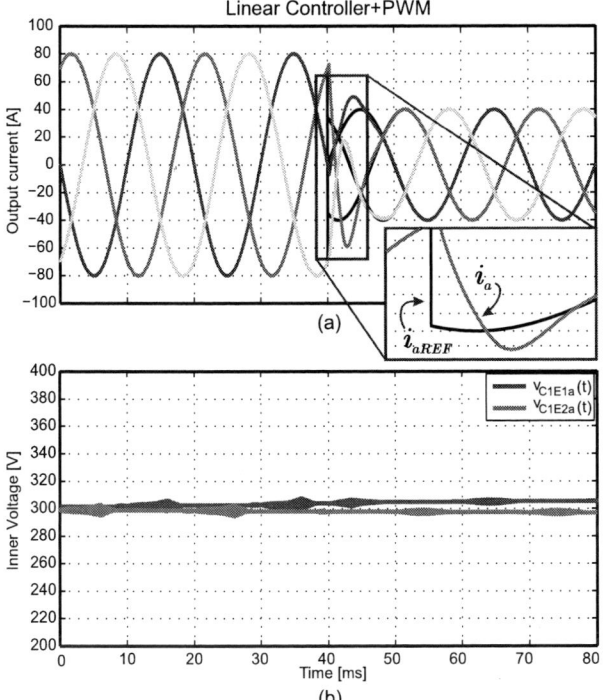

Fig. 8: SMC performance using Linear Controller with Pulse Width Modulation: (a) Output currents; (b) Inner voltages.

frequency spectrum. This can excite non expected resonances in the system and complicate the filter design to reduce the harmonic content.

Because MPC has a variable switching frequency, the possible application of this control is more limited. For example, it is not easy to apply MPC in converters connected to the grid, because the performance of the control can generate some problems. To use MPC in converters connected to the grid it is necessary to do some considerations in the cost function to achieve a more concentrated frequency spectrum.

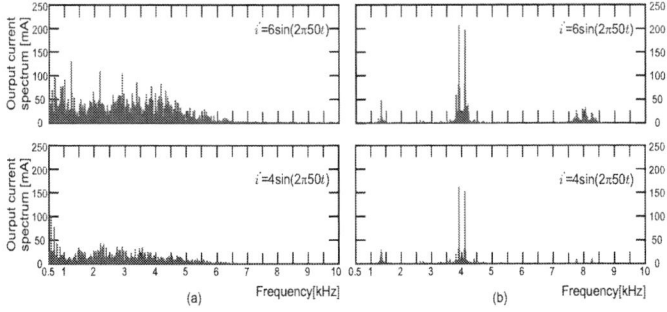

Fig. 9: Output current spectrum: (a) Using MPC; (b) Using linear controller with PWM.

Fig.9 shows the output current spectrum for sinusoidal reference of 6A and 4A respectively. It can be seen that the output current spectrum obtained with MPC is distributed over the range of 500-5000Hz and the spectrum changes significatively

with the current reference.

V. WEIGHTING FACTOR DESIGN

Another problem with MPC it is how to define the weighting factor. The weighting factor must be properly designed in order to achieve the desired performance. Unfortunately, there are no analytical, numerical methods or control design theories to adjust these parameters.

Currently, the weighting factor is determined based on empirical procedures and heuristic methods. The weighting factor tuning procedure depends of the type of terms present in the cost function [19].

Fig. 10 shows the effect of different weighting factors in a cost function. It can be observed that the value of the weighting factor plays an important role in the behavior of the converter.

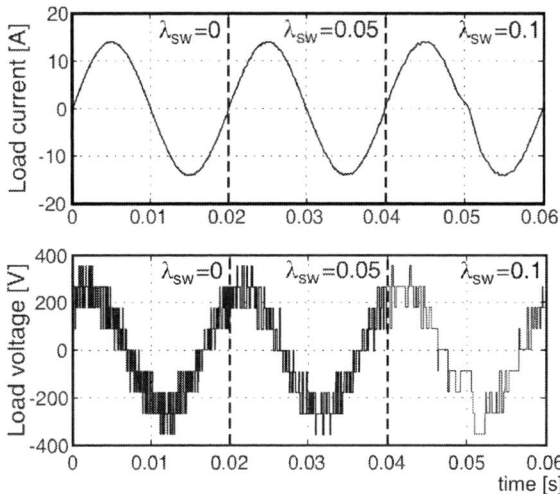

Fig. 10: Weighting factor influence over the current error and the switching frequency.

VI. COMPUTATIONAL COST

MPC requires a large amount of calculations, which can be an obstacle for its application, especially in multilevel converters.

Table I: Computational cost for 3x2 stacked multilevel converter.

Control Scheme	Number of iterations	Number of calculations
Finite Control Set MPC	19683	393676

Table I shows for example, the number of necessary calculations in a Stacked Multilevel Converter of 3-cells and 2-stacks (see Fig. 6). Today, none microcontroller can handle this amount of calculations, for this reason in recent years some efforts have been made to reduce the calculation cost [10]–[15].

978-1-5090-0376-1/16 $31.00 © 2016 IEEE

VII. PRODUCT DEVELOPMENT

Very important in the discussion is the point of view from the industry. This was presented in [20], by people of an important manufacturer of equipment for application in power electronics. To reach the market, an idea has the following steps: i) basic research, ii) technology development, iii) product development and iv) maintenance and support. The activity in MPC is today in the first stage. In effect, an interesting research effort can be observed from the side of academia to demonstrate that this technology offers some advantages over established alternatives. The economic impact of MPC is today unknown, but promising.

VIII. CONCLUSIONS

In the last decade, MPC has shown an increasing activity. It has proved that it works well and that it can be used in a wide variety of applications, using a very simple and intuitive approach. MPC avoids the use of sophisticated modulation strategies, that now appear as non necessary, to control the operation of the converters. Works have been presented to improve some operational aspects, like the variable switching frequency and the total harmonic distortion in steady state. These works have demonstrate that an improvement is possible. On the other hand , the calculation of the weighting factors is an open problem that needs to be addressed. Probably, the solution will have to consider an operation without weighting factors. In general, MPC continues to improve its performance, even though it has not reached yet the maturity for industrial use.

ACKNOWLEDGEMENT

The authors acknowledge the support of university Andres Bello, the support of Universidad Tecnica Federico Santa Maria and the support of the Chilean Research Council (CONICYT) under grant "Doctorado Nacional 2014" (21140574) and "Doctorado Nacional 2013" (1120683), the Advanced Center for Electrical and Electronic Engineering (AC3E) and FONDECYT under grant 1150829.

REFERENCES

[1] C. Buccella, C. Cecati, and H. Latafat, "Digital control of power converters - a survey," *Industrial Informatics, IEEE Transactions on*, vol. 8, no. 3, pp. 437–447, Aug 2012.

[2] J. Rodriguez, J. Pontt, C. A. Silva, P. Correa, P. Lezana, P. Cortes, and U. Ammann, "Predictive current control of a voltage source inverter," vol. 54, no. 1, pp. 495–503, Feb. 2007.

[3] J. Rodriguez, M. P. Kazmierkowski, J. R. Espinoza, P. Zanchetta, H. Abu-Rub, H. A. Young, and C. A. Rojas, "State of the Art of Finite Control Set Model Predictive Control in Power Electronics," *IEEE Transactions on Industrial Informatics*, vol. 9, no. 2, pp. 1003–1016, 2013.

[4] J. Rodriguez, R. Kennel, J. Espinoza, M. Trincado, C. Silva, and C. Rojas, "High-performance control strategies for electrical drives: An experimental assessment," *Industrial Electronics, IEEE Transactions on*, vol. 59, no. 2, pp. 812–820, Feb 2012.

[5] A. Darba, F. De Belie, P. D haese, and J. Melkebeek, "Improved dynamic behaviour in bldc drives using model predictive speed and current control," *Industrial Electronics, IEEE Transactions on*, vol. PP, no. 99, pp. 1–1, 2015.

[6] B. Riar, T. Geyer, and U. Madawala, "Model predictive direct current control of modular multilevel converters: Modeling, analysis, and experimental evaluation," *Power Electronics, IEEE Transactions on*, vol. 30, no. 1, pp. 431–439, Jan 2015.

[7] S. Vazquez, J. Leon, L. Franquelo, J. Carrasco, O. Martinez, J. Rodriguez, P. Cortes, and S. Kouro, "Model predictive control with constant switching frequency using a discrete space vector modulation with virtual state vectors," in *Industrial Technology, 2009. ICIT 2009. IEEE International Conference on*, Feb 2009, pp. 1–6.

[8] P. Cortes, S. Kouro, B. La Rocca, R. Vargas, J. Rodriguez, J. Leon, S. Vazquez, and L. Franquelo, "Guidelines for weighting factors design in model predictive control of power converters and drives," in *Industrial Technology, 2009. ICIT 2009. IEEE International Conference on*, Feb 2009, pp. 1–7.

[9] S. Alireza Davari, D. Khaburi, and R. Kennel, "An improved fcs-mpc algorithm for an induction motor with an imposed optimized weighting factor," *Power Electronics, IEEE Transactions on*, vol. 27, no. 3, pp. 1540–1551, March 2012.

[10] C. Xia, T. Liu, T. Shi, and Z. Song, "A simplified finite-control-set model-predictive control for power converters," *IEEE Transactions on Industrial Informatics*, vol. 10, no. 2, pp. 991–1002, 2014.

[11] M. Vatani, B. Bahrani, M. Saeedifard, and M. Hovd, "Indirect finite control set model predictive control of modular multilevel converters," *IEEE Transactions on Smart Grid*, vol. 6, no. 3, pp. 1520–1529, 2015.

[12] J.-W. Moon, J.-S. Gwon, J.-W. Park, D.-W. Kang, and J.-M. Kim, "Model predictive control with a reduced number of considered states in a modular multilevel converter for HVDC system," *IEEE Transactions on Power Delivery*, vol. 30, no. 2, pp. 608–617, 2015.

[13] S. Kwak and J.-C. Park, "Switching strategy based on model predictive control of vsi to obtain high efficiency and balanced loss distribution," *IEEE Transactions on Power Electronics*, vol. 29, no. 9, pp. 4551–4567, 2014.

[14] T. Geyer and D. E. Quevedo, "Performance of Multistep Finite Control Set Model Predictive Control for Power Electronics," *IEEE Transactions on Power Electronics*, vol. 30, no. 3, pp. 1633–1644, 2015.

[15] M. Norambuena, S. Kouro, S. Dieckerhoff, and J. Rodriguez, "Finite Control Set-Model Predictive Control of a Stacked Multicell Converter with Reduced Computational Cost," in *IEEE Industrial Electronics, IECON 2015 - 41nd Annual Conference on*, 2015.

[16] F. Blaschke, "The principle of field-orientation applied to the transvector closed-loop control system for rotating field machines," *Siemens*, p. Vol 3, 1972.

[17] I. Takahashi and T. Noguchi, "A New Quick-Response and High-Efficiency Control Strategy of an Induction Motor," *IEEE Transactions on Industry Applications*, vol. IA-22, no. 5, pp. 820–827, Sep. 1986.

[18] M. Depenbrock, "Direct self-control (DSC) of inverter-fed induction machine," *IEEE Transactions on Power Electronics*, vol. 3, no. 4, pp. 420–429, 1988.

[19] P. Cortes, S. Kouro, B. La Rocca, R. Vargas, J. Rodriguez, J. I. Leon, S. Vazquez, and L. G. Franquelo, "Guidelines for weighting factors design in model predictive control of power converters and drives," in *Proc. IEEE International Conference on Industrial Technology ICIT 2009*, 10–13 Feb. 2009, pp. 1–7.

[20] G. Papafotiou, G. Demetriades, and V. Agelidis, "Integration of model-predictive control in medium and high-voltage power electronics products: An industrial perspective on gaps and progress required," in *Industrial Electronics Society, IECON 2015 - 41th Annual Conference of the IEEE*, Nov 2015.

978-1-5090-0376-1/16 $31.00 © 2016 IEEE

Distributed Secondary Control in DC Microgrids with Low-Bandwidth Communication Link

Saeed Peyghami-Akhuleh, and Hossein Mokhtari
Dept. of Electrical Engineering
Sharif University of Technology
Tehran, Iran
saeed_peyghami@ee.sharif.edu, mokhtari@sharif.edu

Poh Chiang Loh, and Frede Blaabjerg
Dept. of Energy Technology
Aalborg University
Aalborg, Denmark
pcl@et.aau.dk, fbl@et.aau.dk

Abstract— In this paper, a distributed secondary power sharing approach with low bandwidth communication network is proposed for low voltage direct current (LVDC) microgrids. Conventional droop control causes voltage drop in the grid and also a mismatch on the current of converters in the case of consideration of the line resistances. Proposed control system carry out the current value of the other converters to reach the accurate current sharing and suitable voltage regulation as well. Voltage and current controllers locally regulate the voltage and current of converters as a secondary controller. Secondary controller is realized locally and the communication network is only used to transfer the data of dc currents. Therefore, the secondary controller can regulate the average voltage by only using the data of currents. The proposed approach is verified with simulations based on PLECS.

Keywords—secondary control, droop control, dc microgrid, distributed power sharing control

I. INTRODUCTION

The concept of ac/dc microgrids has been proposed in recent years [1], [2] to increase reliability, power quality and decrease losses and pollution. Both ac and dc power systems have been studied and implemented for years. However, due to the advances in the power electronics technology, dc-based power networks have been used in industrial applications such as data centers [3], space applications [4], offshore wind farms [5], ships [6], electric vehicles [7] and HVDC transmission systems [8]. Many power sources and loads, such as photovoltaic (PV) modules, fuel cell units, batteries, motor driven loads, and full converter based generators (i.e. micro-turbines and wind turbines) have a natural dc coupling [9]. Therefore, it is a more efficient and reliable method to integrate these sources and loads into a dc-based system by using dc-dc power electronic converters [10].

Power sharing control is a key issue in a stable and efficient operation of a microgrid. Many studies have been carried out to attain proper power management in ac and dc power systems at all three levels of the hierarchical control of a microgrid. In all of these studies, both tertiary and secondary control are done by communication links [11]–[14]. However, primary control is usually done with distributed droop methods.

The main objectives of the control system (primary and secondary) in a dc microgrid are voltage regulation at all buses and proportional load sharing among dispatchable sources

[15]–[19]. Despite the simplicity of implementation, conventional droop methods suffer from poor voltage regulation and load sharing, particularly when the line impedances are not negligible [20]–[22]. The primary reason for this poor voltage regulation is the voltage drop caused by the virtual impedance. Another factor is the output voltage mismatch among different converters, which is crucial for the natural power flow in the dc systems. Possible solutions to the aforementioned issues have been reviewed in [14]. These solutions are either centralized or require a communication network throughout the microgrid [23]. A centralized secondary control in [24] measures the microgrid voltage, calculates a voltage restoration term, and sends the restoration term to all sources. Decentralized secondary controllers also, require communication link to regulate the average voltage and also increase the current sharing among the converters[14]. Therefore, a complicated communication network is needed to transmit data of voltages and currents between converters. Moreover, the delay of the communication link for both current and voltage data transmission affects the stability.

In this paper, conventional droop method is used to control the current sharing of converters. A distributed secondary controller is proposed to compensate the voltage drop due to the droop gains and line resistances and also to eliminate the mismatch of currents dispatched between converters. A low-band width communication link is used to transfer the data of currents of converters and the secondary controller locally controls the voltage and current of each converter. The main advantage of the proposed controller is locally regulating the voltage and current of converters by only using the data of currents.

This paper is organized as follows. Section II explains the details of the power management system and the main idea of the proposed control system as a secondary controller. Small signal stability analysis is discussed in section III. Section IV presents the simulation results of the control scheme using PLECS. Finally, section VI summarizes the conclusions of this paper.

II. PROPOSED CONTROL APPROCH

Proposed distributed secondary control system includes in two controllers for current sharing and voltage restoring in microgrid. The control system and single line diagram of a typical microgrid is depicted in Fig. 1. Current Regulator

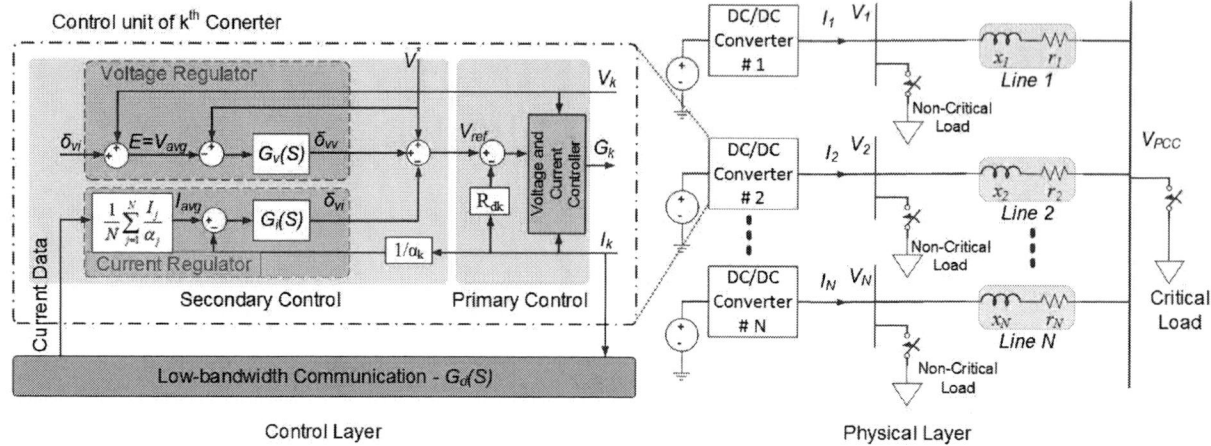

Fig. 1. Proposed control approach; Voltage Regulator G_v, Current Regulator G_i, and low- bandwidth communication link with delay function of $G_d(s) = \dfrac{1}{1+\tau S}$.

is used to increase the accuracy of the dispatched current with conventional droop controller and Voltage Regulator restores the average voltage of the microgrid to the nominal value. Both regulators are described in the following.

A. Current Regulator

Current sharing among converters is conventionally performed by droop gain R_{dk} which can be defined as (1) for k^{th} converter.

$$R_{dk} = \frac{V_{max} - V_{min}}{I_{nk}} \tag{1}$$

where V_{max} and V_{min} is the maximum and minimum allowable voltage range and I_{nk} is the nominal current of kth converter. However, because of the differences in line resistances, the currents are not proportionally dispatched among converters. The current regulator calculates the weighted average currents of all converters and tries to regulate the output current proportional to the nominal current of each converter. The average current (I_{avg}) can be calculated as (2).

$$I_{avg} = \frac{1}{N}\sum_{j=1}^{N} I_j / \alpha_j \quad j = 1:N \tag{2}$$

where N is the number of converters, I_j is the measured current and α_j is the sharing coefficient of jth converter respectively.

A simplified single line model of a dc microgrid with two converter is shown in Fig. 2. At steady state, droop gain acts as a series resistor (R_{d1} & R_{d2}). The secondary current regulator behaves as a small positive/negative resistor (r_{d1} & r_{d2}) such that the total resistance of each line becomes proportional to the rated current of each converter. Hence, the relation between rated current (I_{nj}) and sharing coefficient (α_j) and total line resistance between j^{th} converter and Point of Common Coupling (PCC) can be described as:

$$\frac{I_{ni}}{I_{nj}} = \frac{\alpha_j}{\alpha_i} = \frac{R_{dj} + r_{dj} + r_j}{R_{di} + r_{di} + r_i}; \quad i,j = 1:N, i \neq j \ . \tag{3}$$

where r_j is the resistance of the line connected to j^{th} converter.

Effect of current regulator in power sharing system is schematically described in Fig. 3. The blue graph shows the

effect of conventional droop gain. The secondary current regulator changes the slope of this droop characteristics to reach the same current between two converters. Sharing coefficients are assumed 1 for both converters. Therefore, the accurate current sharing is obtained with droop controller and current regulator. However, the average voltage through the microgrid is decreased and a distributed voltage regulator is required to restore the average voltage of the microgrid.

B. Voltage Regulator

This regulator compensates the voltage drop though the droop gain. From Fig. 2, the output voltage of each converter (V_1 & V_2) can be calculated as:

$$\begin{aligned} V_1 &= V_r - R_{d1}I_1 - \delta_{vi1} \\ V_2 &= V_r - R_{d2}I_2 - \delta_{vi2} \end{aligned}. \tag{4}$$

where V_r is the reference value of the voltage loop. If sharing coefficients are assumed to be 1, then the droop gains have to be equal ($R_{d1}=R_{d2}=R_d$) and also, at steady states $I_1=I_2$. The output of the current regulators are:

$$\begin{aligned} \delta_{vi1} &= (\frac{I_1+I_2}{2} - I_1)G_dG_i(s) = (\frac{I_2-I_1}{2})G_dG_i(s) \\ \delta_{vi2} &= (\frac{I_1+I_2}{2} - I_2)G_dG_i(s) = (\frac{I_1-I_2}{2})G_dG_i(s) \end{aligned} \tag{5}$$

where $G_i(s)$ is the PI controller and $G_d(s)$ is the delay of communication link. At steady state $\delta_{vi1}+\delta_{vi2}=0$. Therefore the average voltage of the microgrid is:

$$V_{avg} = \frac{1}{2}(V_1 + V_2) = V_r - R_d I \ . \tag{6}$$

Applying primary droop controller and secondary current regulator causes average voltage drop equal to R_dI. From the single line model of the microgrid depicted in Fig. 2, internal voltages (i.e. E_1 & E_2) are equal to the average voltage calculated by (6). Therefore, the distributed voltage regulator can estimate the internal voltage and regulate it at the reference value. In fact, the correction term (δ_{vv}) shifts up the droop characteristics in Fig. 3 to restore the average voltage of the microgrid which can be calculatad as (7), where V* is the rated voltage of the microgrid.

$$\delta_{vv1} = (V^* - (V_1 + \delta_{vi1}))\,G_v(s)$$
$$\delta_{vv2} = (V^* - (V_2 + \delta_{vi2}))\,G_v(s) \tag{7}$$

III. SMALL SIGNAL STABILITY

In this section, analysis of small signal stability of the simplified dc microgrid depicted in Fig. 2 is described. The relation between converter voltage and PCC voltage can be find as:

$$V_1 - V_{PCC} = r_1 I_1$$
$$V_2 - V_{PCC} = r_2 I_2 \tag{8}$$

where

$$V_{PCC} = R_L(I_1 + I_2) \tag{9}$$

The set point value for primary controller is:

$$V_{ref1} = V^* + \delta_{vv1} - \delta_{vi1}$$
$$V_{ref2} = V^* + \delta_{vv2} - \delta_{vi2} \tag{10}$$

And the set point value for the inner voltage loop can be determined by primary controller as:

$$V^*_1 = V_{ref1} - R_d I_1$$
$$V^*_2 = V_{ref2} - R_d I_2 \tag{11}$$

Substituting (5), (7) and (10) in (8) gives:

$$V^*_1 = V^* - \delta_{vi1} - \frac{R_d I_1}{1 + G_v(s)}$$
$$V^*_2 = V^* - \delta_{vi2} - \frac{R_d I_2}{1 + G_v(s)} \tag{12}$$

This equations show that the term of R_dI which is related to the primary controller can be eliminated in low frequencies i.e. in the secondary controller frequency bandwidth. Therefore, primary controller tries to dispatch the current between converters based on droop gain and secondary controller tries to reduce the mismatch in current sharing as well as decrease the voltage drop because of the droop gain.

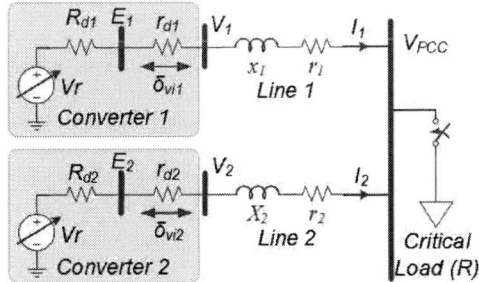

Fig. 2. Simplified model of 2-converter based dc microgrid.

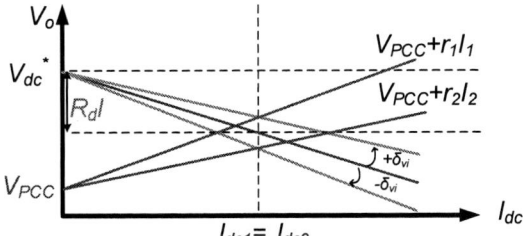

Fig. 3. Proposed droop characteristic adjustment.

Combining equations (5), (7), (8), (9) and (10) gives the system representation in state space as (13).

$$\begin{bmatrix} 0.5 G_i G_d - \dfrac{R_d}{1+G_v} - R_L - r_1 & -0.5 G_i G_d - R_L \\[2mm] -0.5 G_i G_d - R_L & 0.5 G_i G_d - \dfrac{R_d}{1+G_v} - R_L - r_2 \end{bmatrix} \begin{bmatrix} I_1 \\ I_2 \end{bmatrix}$$
$$= \begin{bmatrix} -V^* \\ -V^* \end{bmatrix} \tag{13}$$

The zeros of this matrix determines the poles of the system. For a simple system described in TABLE I, the root loci are shown in Fig. 4. The inner voltage loop poles are faster than primary controller and secondary controller as well. Therefore, the primary controller regulates the output currents based on droop gains. After current sharing, secondary Current Regulator eliminates the small mismatches between currents. At the end, the secondary Voltage Regulator compensates the average voltage drop.

The effect of communication delay on dominant poles (i.e., primary and secondary poles) is shown in Fig. 5. Communication delay varies from 2 to 30 ms. The closed loop poles are located on the left half of the plane at different communication delay.

TABLE I.   SPECIFICATIONS OF MICROGRID AND CONTROL SYSTEM

Definition	Symbol	Value
Impedance of line 1	$r_1(\Omega)/L_1(\mu H)$	0.5/600
Impedance of line 2	$r_2(\Omega)/L_2(\mu H)$	1-2/900
Rated current of Converters	I_{n1}/I_{n2} (A)	7/7
DC link voltage	V_{dc} (V)	700
Maximum Voltage Variation	ΔV (%-V)	5% -35
Voltage-Current droop gain	R_d (V/A)	5
Loads	R_{L1}/R_{L2} (Ω)	100/100
Communication link delay	τ (ms)	5
Current regulator	$G_i = k_{pi} + k_{ii}/s$	3+25/s
Voltage regulator	$G_v = k_{pv} + k_{iv}/s$	1.5+15/s

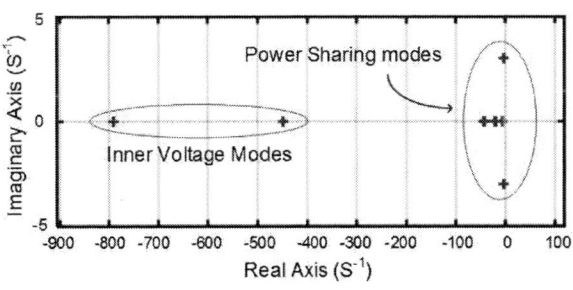

Fig. 4. Root loci of closed loop system, inner voltage loop poles, and Power Sharing controller poles. (Inner current controller poles are not shown.)

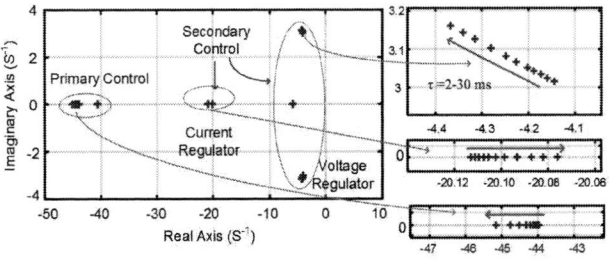

Fig. 5. Power sharing controller poles. Effect of communication delay (τ) on root loci is very small.

978-1-5090-0376-1/16 $31.00 © 2016 IEEE   643

IV. SIMULATION RESULTS

A simplified dc microgrid with two converters like one in Fig. 2 is considered to validate the proposed control approach. The system parameters are given in TABLE I. Boost topology is selected for both dc converters with input voltage of 540 V and output voltage of 700 V. Each boost converter has $L=2mH$, $C=500\mu F$, switching frequency $f_s=20\ kHz$.

In Case I, the performance of proposed control method is validated and compared to the conventional droop controller without secondary controller. In this case a different line resistance are considered as well. In Case II, the performance of proposed controller is demonstrated at different loading conditions. Voltage and current waveforms are exhibited the performance of the controller.

A. Case I: Comparison with Conventional Droop Method

In this case, the performance of the proposed controller is demonstrated in the considered microgrid in Fig. 2. The load of 100 Ω is connected to the PCC. At first, the conventional droop method controls the current and voltage of the converters. As depicted in Fig. 6, the currents are not equally dispatched between converters. However, after applying the secondary controller the output current of both converters are the same. Proposed controller regulates the average voltage of the microgrid at the nominal value (i.e. 700 V). As shown in Fig. 7, after applying the proposed controller, the average voltage – yellow one – is settled at 700 V. Load voltage is around 680 V with conventional droop method, however, after applying the proposed controller, it regulates at 697 V.

In order to demonstrate the further performance of the control system, different line resistances are considered. Here, r_1 is fixed at 0.5 Ω and r_2 is set to 1 Ω and 2 Ω respectively. The output current for both resistances are depicted in Fig. 6 and Fig. 8 respectively and also, the voltage of converters are shown in the Fig. 7 and Fig. 9 respectively. The equal current sharing and suitable voltage regulation is achieved with both resistances.

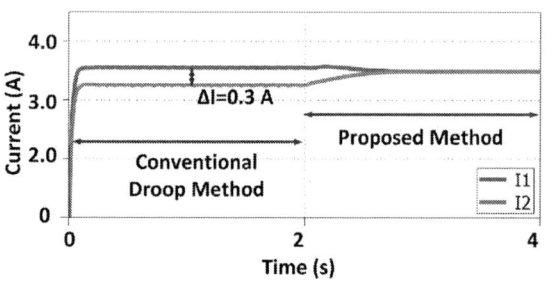

Fig. 6. DC current of converters with $r_1=0.5$ and $r_2=1\ \Omega$, applying proposed control causes accurate current sharing.

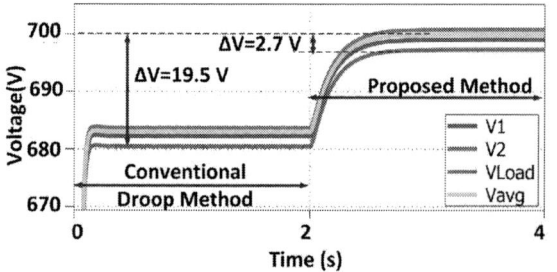

Fig. 7. DC voltage of converters with $r_1=0.5$ and $r_2=1\ \Omega$, applying proposed control regulates the average voltage of microgrid.

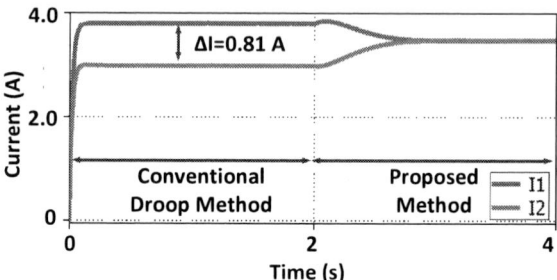

Fig. 8. DC current of converters with $r_1=0.5$ and $r_2=2\ \Omega$, applying proposed control causes accurate current sharing.

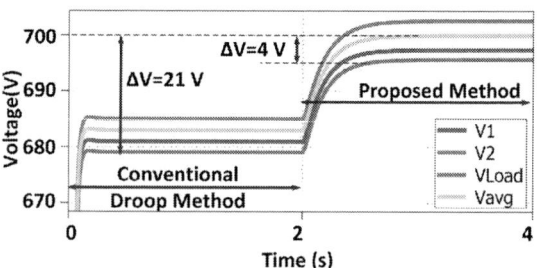

Fig. 9. DC voltage of converters with $r_1=0.5$ and $r_2=2\ \Omega$, applying proposed control regulates the average voltage of microgrid.

B. Case II: Different Loading Condition

In this case, the applicability of the proposed control system is demonstrated in different loading conditions. The resistive load of 100 Ω is connected into the PCC. At t = 2 s the other load with 100 Ω resistance is turned on and at t = 4 s is turned off. The current and voltage waveforms of both converters are shown in Fig. 10 and Fig. 11 respectively. The equal and accurate current sharing at both 100 Ω and 50 Ω loading conditions are achieved. The acceptable transient response of the secondary controller is obtained as well and the currents and voltages are settled at 0.5 second. The average voltage of the grid is regulated at the nominal value in both loading conditions. And load voltage deviation from the rated value is acceptable.

From the dc power flow theory, in case of no control on dc voltage, the currents have to be dispatched based on line resistances. Therefore, for the case with $r_1=0.5\ \Omega$ and $r_2=1$ Ω the output current of first converter has to be 2 times more than the second one. However, all simulation results show that the primary controller dispatch the currents regarding to the droop gain and the secondary controller reduces the mismatch between currents and also restores the voltage drop as well.

The viability of the proposed control system at different line resistances and different loading conditions are represented by simulation results.

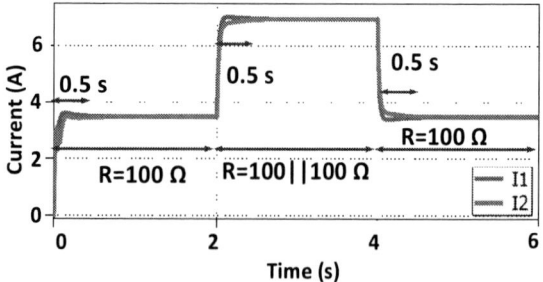

Fig. 10. DC current of converters with $r_1=0.5$ and $r_2=1\ \Omega$, at both loads of 100 Ω and 50 Ω, accurate current sharing between converters is achieved.

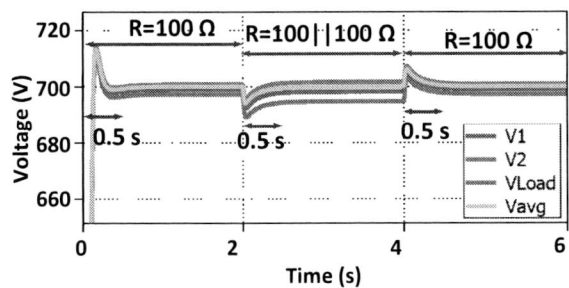

Fig. 11. DC voltage of converters with $r_1 = 0.5$ and $r_2 = 1$ Ω, at both loads of 100 Ω and 50 Ω, proposed control can regulate the average voltage of microgrid at 0.5 second.

V. CONCLUSION

This paper has presented a reliable secondary control with low-bandwidth communication link for power management in LVDC microgrids to improve the voltage regulation and current sharing accuracy. Secondary control regulates the average voltage of microgrid and also, proportionally controls the output current of converters by carrying out the information of the other converter currents, without communicating the voltage information, and then, leads to increase of reliability. The viability of proposed control system is ensured for different line resistances and different loading condition. The proposed approach is verified by simulations based on PLECS. Future research will be focused on generalizing the proposed secondary control approach using the information of neighbor converters for multi converter microgrids.

REFERENCES

[1] F. Katiraei, R. Iravani, N. Hatziargyriou, and A. Dimeas, "Microgrids Management," *IEEE Power Energy Mag.*, vol. 6, no. 3, pp. 54–65, 2008.

[2] N. Hatziargyriou, H. Asano, R. Iravani, and C. Marnay, "Microgrids," *IEEE Power Energy Mag.*, vol. 5, no. 4, pp. 78–94, 2007.

[3] E. Alliance, "380 Vdc Architectures for the Modern Data Center," pp. 1–29, 2013.

[4] S. P. Barave and B. H. Chowdhury, "Hybrid AC/DC Power Distribution Solution for Future Space Applications," *IEEE Power Eng. Soc. Gen. Meet.*, vol. 65401, 2007.

[5] J. Robinson, D. Jovcic, and G. Joos, "Analysis and Design of an Offshore Wind Farm Using a MV DC Grid," *IEEE Trans. Power Del.*, vol. 25, no. 4, pp. 2164–2173, Oct. 2010.

[6] J. G. Ciezki and R. W. Ashton, "Selection and Stability Issues Associated with a Navy Shipboard DC Zonal Electric Distribution System," *IEEE Trans. Power Del.*, vol. 15, no. 2, pp. 665–669, 2000.

[7] C. C. Chan, "An Overview of Electric Vehicle Technology," *in Proc. IEEE*, vol. 81, no. 9, pp. 1202–1213, 1993.

[8] N. G. Hingorani, "High-Voltage DC Transmission: A Power Electronics Workhorse," *in IEEE Spectr.*, vol. 33, no. 4, pp. 63–72, 1996.

[9] D. Manz, R. Walling, N. Miller, B. LaRose, R. D'Aquila, and B. Daryanian, "The Grid of the Future: Ten Trends That Will Shape the Grid Over the Next Decade," *IEEE Power Energy Mag.*, vol. 12, no. 3. pp. 26–36, 2014.

[10] M. E. Baran and N. R. Mahajan, "DC Distribution for Industrial Systems: Opportunities and Challenges," *IEEE Trans. Ind. Appl.*, vol. 39, no. 6, pp. 1596–1601, 2003.

[11] T. Zhou and B. François, "Energy Management and Power Control of a Hybrid Active Wind Generator for Distributed Power Generation and Grid Integration," *IEEE Trans. Ind. Electron.*, vol. 58, no. 1, pp. 95–104, 2011.

[12] J. Rocabert, A. Luna, F. Blaabjerg, and P. Rodriguez, "Control of Power Converters in AC Microgrids," *IEEE Trans. Power Electron.*, vol. 27, no. 11, pp. 4734–4749, 2012.

[13] P. C. Loh and F. Blaabjerg, "Autonomous Control of Distributed Storages in Microgrids," *in Proc. IEEE ICPE-ECCE Asia*, pp. 536–542, 2011.

[14] S. Anand, B. G. Fernandes, and J. M. Guerrero, "Distributed Control to Ensure Proportional Load Sharing and Improve Voltage Regulation in Low-Voltage DC Microgrids," *IEEE Trans. Power Electron.*, vol. 28, no. 4, pp. 1900–1913, 2013.

[15] D. Chen, L. Xu, and L. Yao, "DC Voltage Variation Based Autonomous Control of DC Microgrids," *IEEE Trans. Power Del.*, vol. 28, no. 2, pp. 637–648, 2013.

[16] X. Lu, K. Sun, J. M. Guerrero, J. C. Vasquez, and L. Huang, "State-of-Charge Balance Using Adaptive Droop Control for Distributed Energy Storage System in DC Microgrid Applications," *IEEE Trans. Ind. Electron.*, vol. 61, no. 6, pp. 2804–2815, 2014.

[17] P. C. Loh, D. Li, Y. K. Chai, and F. Blaabjerg, "Autonomous Operation of Hybrid Microgrid with AC and DC Subgrids," *IEEE Trans. Power Electron.*, vol. 28, no. 5, pp. 2214–2223, 2013.

[18] Y. Ito, Y. Zhongqing, and H. Akagi, "DC Microgrid Based Distribution Power Generation System," *in Proc. IEEE IPEMC*, vol. 3, pp. 1740–1745, 2004.

[19] P. Frack, P. Mercado, M. Molina, E. Watanabe, R. De Doncker, and H. Stagge, "Control-Strategy for Frequency Control in Autonomous Microgrids," *IEEE J. Emerg. Sel. Top. Power Electron.*, vol. 6777, no. c, pp. 1–1, 2015.

[20] H. Nikkhajoei and R. Iravani, "Steady-State Model and Power Flow Analysis of Electronically-Coupled Distributed Resource Units," *IEEE Power Eng. Soc. Gen. Meet. PES*, vol. 22, no. 1, pp. 721–728, 2007.

[21] Y. W. Li and C.-N. Kao, "An Accurate Power Control Strategy for Power-Electronics-Interfaced Distributed Generation Units Operating in a Low-Voltage Multibus Microgrid," *IEEE Trans. Power Electron.*, vol. 24, no. 12, pp. 2977–2988, Dec. 2009.

[22] J. He and Y. W. Li, "Analysis, Design, and Implementation of Virtual Impedance for Power Electronics Interfaced Distributed Generation," *IEEE Trans. Ind. Appl.*, vol. 47, no. 6, pp. 2525–2538, 2011.

[23] T. L. Vandoorn, S. Member, and B. Meersman, "A Control Strategy for Islanded Microgrids With DC-Link Voltage Control," *IEEE Trans. Power Del.*, vol. 26, no. 2, pp. 703–713, 2011.

[24] J. M. Guerrero, J. C. Vasquez, J. Matas, L. G. De Vicuña, and M. Castilla, "Hierarchical Control of Droop-Controlled AC and DC Microgrids - A General Approach toward Standardization," *IEEE Trans. Ind. Electron.*, vol. 58, no. 1, pp. 158–172, 2011.

7th Power Electronics, Drive Systems & Technologies Conference (PEDSTC 2016)
16-18 Feb. 2016, Iran University of Science and Technology, Tehran, Iran

Voltage Unbalance and Harmonic Compensation in Microgrids by Cooperation of Distributed Generators and Active Power Filters

Mohammad M. Hashempour, Mehdi Savaghebi, *Senior Member, IEEE,* Juan C. Vasquez, *Senior Member, IEEE*, and Josep M. Guerrero, *Fellow, IEEE*

Abstract— In this paper, the power quality of microgrids is addressed. To achieve the desired level of power quality, a strategy based on the coordinated control between DGs and APFs is proposed. In this regard, hierarchical control is applied where primary control consists of power droop controller of DGs, selective virtual impedance and voltage/current regulators. Based on the secondary control, at first voltage harmonic compensation and voltage unbalance compensation of point of common coupling (PCC), that might includes sensitive loads, is carried out by DGs. Voltage compensation of PCC by DGs may cause severe voltage distortion at DGs terminals. Thus, the coordinated control is used to mitigate the voltage distortion to the defined maximum allowable value at DGs terminals. Evaluation of the proposed hierarchical control is carried out by a simulation study.

Index Terms—Active power filter, Distributed Generator, Hierarchical control, Microgrid, Voltage unbalance/harmonic compensation.

I. INTRODUCTION

DISTRIBUTED generators (DGs) are usually connected to microgrids (MGs) by power electronic interface converters. Regulating voltage/frequency of DG terminal is accomplishable by proper control of the interface inverters [1],[2]. Furthermore, many strategies have been suggested for improving power quality of MGs based on DGs inverters control [3]-[14].

Unbalanced voltage might be produced due to asymmetrical transmission lines or loads. It might cause serious problems such as increase of power losses in equipment, disturbing sensitive loads performance and even instability of system. As a common problem in three-phase MGs, voltage unbalance has been addressed in some previous works. A well-known strategy addresses voltage unbalance compensation (VUC) of point of common coupling (PCC) or DG terminal by proper control of DGs interface inverters [3]-[6]. To compensate unbalanced voltage of Sensitive Load Bus (SLB), an extra control loop is devised in [5] as secondary control. Although

SLB voltage is improved in [5], DGs terminal might be distorted. This strategy is also applied in [6], moreover, an extra control loop is contrived to distribute the distortion rate among DGs terminal in an optimized way. However, in severe unbalance conditions, one or more DGs terminal might become distorted severely by this method.

With high penetration of nonlinear loads and power electronic equipment, harmonic pollution is considered as an important power quality problem. Many efforts have been done for voltage harmonic compensation (VHC). Like VUC, DGs inverters are usually used for VHC. The common strategy is making resistance emulation at harmonic frequencies [7]-[14]. The methods proposed in [11] and [12] address VUC and VHC of PCC while compensation sharing is considered too. Furthermore, [11] is built on selective harmonic compensation approach and in [12], transient state is regarded too. Again, DG terminal might become distorted severely by these methods. To simultaneously compensate voltage harmonics of the both points (PCC and DGs buses), coordinated control of DGs inverters and Active Power Filter (APF) is suggested in [13] and [14]. In [13], satisfactory voltage quality of multi-area MG (with different voltage quality requirement) is obtained while in [14], DGs inverters rated power is considered in the coordinated control too.

In line with previous efforts regarding power quality improvement of MG by DGs inverters and APFs, in the present paper, voltage unbalance mitigation is considered while VHC is carried out too.

II. PROPOSED HIERARCHICAL CONTROL SCHEME

A typical MG is represented in Fig. 1. Note that there might be more than one DG connected to DG(s) bus. Two points are represented in the system as nodes and PCC where nodes are DGs terminals and PCC is the point that there might be high amount of loads (including sensitive loads). Note that there should be very low voltage harmonic distortion (VHD) and voltage unbalanced factor (VUF) at PCC while satisfactory power quality of nodes (according to the nodes voltage quality requirement) is considered. Meanwhile, since there is no constraint defined for voltage quality of a typical node, well-known power quality indexes should be considered.

Fig. 2 shows general scheme of the proposed hierarchical control. The proposed control contains two levels. Primary control is DGs local control that power sharing is considered in this level. Secondary control is for PCC voltage quality improvement by DGs interface inverters. However, due to PCC voltage compensation by DGs, nodes voltages might

---

This work was supported by the Technology Development and Demonstration Program (EUDP) through the Sino-Danish Project "Microgrid Technology Research and Demonstration" (meter.et.aau.dk).

M. M. Hashempour is with the Department of Electrical Engineering, Karaj Branch, Islamic Azad University, Iran (e-mail: hashempourmehdi@gmail.com).

M. Savaghebi, J. C. Vasquez and J. M. Guerrero are with the Department of Energy Technology, Aalborg University, Aalborg East DK-9220, Denmark (e-mail: mes@et.aau.dk, juq@et.aau.dk, joz@et.aau.dk).

978-1-5090-0376-1/16 $31.00 © 2016 IEEE

become distorted [14]. In this situation, APF is used for voltage distortion mitigation of nodes by making APF cooperated with DGs for PCC compensation.

According to Fig. 2, at first PCC voltage is analyzed by "PCC voltage analysis" block. In this block PCC voltage components in dq frame ($v_{dq}^{h,1-}$) is extracted by synchronous reference frame-phase lock loop (SRF-PLL) [15] extraction method. Note that both positive and negative sequences of individual harmonics is considered in this block. Then $v_{dq}^{h,1-}$ is transferred to secondary control by low bandwidth communication (LBC) link. As it is shown in Fig. 2, all the signals communicated to secondary control are LBC signals so secondary control as a central control can be far from DGs and APF. In secondary control, voltage distortion rate of PCC ($VDR_{PCC}^{h\pm,1-}$) is calculated. Due to nonlinear-unbalance loads in the system, both negative and positive sequences of individual harmonics ($VHD_{PCC}^{h\pm}$) and VUF of PCC (VUF_{PCC}) are considered to be very low; so these parameters are calculated in "PCC compensation rate cal.". $VDR_{PCC}^{h\pm,1-}$ is compared with its reference value, since there is any violation from the reference value, proper signals ($C_{dq}^{h,1-}$) are generated and sent to primary control of DGs for removing the violation and compensating PCC. Note that $C_{dq}^{h,1-}$ is the same for all DGs and sharing compensation among DGs is explained in follow.

As it is proved in Section V, PCC compensation by DGs might result severe VDR at nodes ($VDR_N^{h\pm,1-}$). This distortion is directly dependent on $VDR_{PCC}^{h\pm,1-}$; by increase of $VDR_{PCC}^{h\pm,1-}$, DG should tolerate more efforts for PCC compensation so the corresponding $VDR_N^{h\pm,1-}$ increases. In this situation, APF is active to help DGs for improving PCC and reducing $VDR_N^{h\pm,1-}$ to its reference value ($VDR_{N_{ref}}^{h\pm,1-}$). To make the cooperation, individual $VDR_N^{h\pm,1-}$ are calculated and transferred to this level by LBC (see Fig. 1). In the cooperation block, $VDR_N^{h\pm,1-}$ is compared with its reference values since there is any violation, according to the proposed cooperation policy between DGs and APF, proper signals are sent to primary control of the relative DGs ($T_i^{h\pm,1-}$) for reducing those DGs efforts. Moreover, proper signals ($T^{h\pm,1-}$) are sent to control stage of APF to include APF in compensation, partially. In follow, detailed descriptions of the hierarchical control is offered.

A. Primary Control

As mentioned before, in primary control power sharing is considered. Since there might be power circulation between parallel DGs, droop control is used to share fundamental component of power. As the main drawback of droop control, nonlinear and unbalance load sharing is not considered in the droop scheme. To share unbalance and harmonic current, using virtual impedance is suggested to provide resistance behavior toward harmonic components and the negative sequence of fundamental component of output current. Remember that different sequences of harmonic components should be regarded in virtual impedance since we have unbalance load condition in the system. Noteworthy that multiple second-order generalized integrator-frequency lock loop (MSOGI-FLL) is used in this paper to extract fundamental and harmonic components in stationary

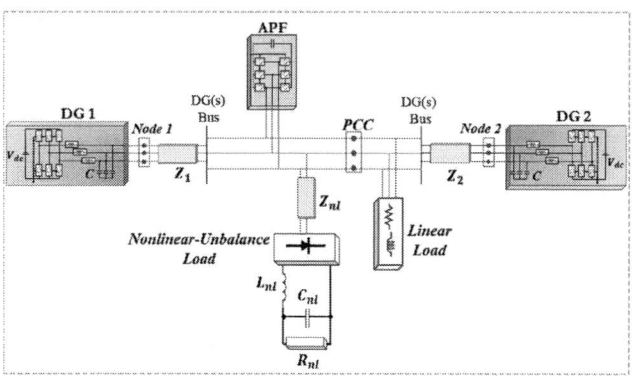

Fig. 1. General scheme of microgrid.

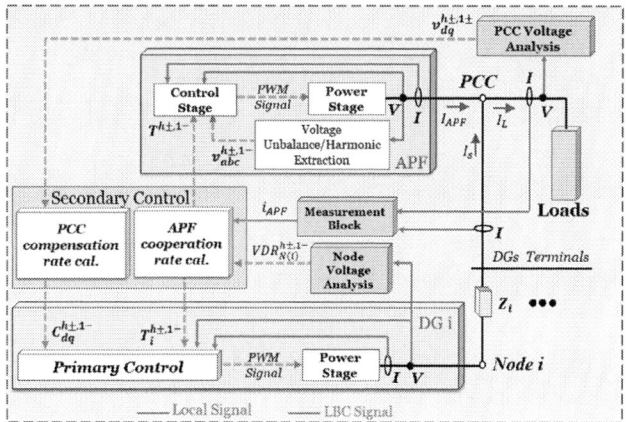

Fig. 2. General scheme of the proposed hierarchical control.

frameworks [15]. Note that PCC compensation by DGs is shared between them in compensation effort controller of primary control. Comprehensive explanations of different parts of primary control is available in [1], [5] and [14].

B. Secondary(Central) Control

As mentioned before, in secondary control PCC compensation is carried out by DGs, firstly. Fig. 3 shows PCC compensation rate calculation block of secondary control. As it is shown in Fig. 3, VDR is calculated like below:

$$VHD_V^{h\pm} = \frac{v_{rms_{o\alpha}}^{h\pm}}{v_{rms_{o\alpha}}^{1+}} \quad and \quad VUF = \frac{v_{rms_{o\alpha}}^{1-}}{v_{rms_{o\alpha}}^{1+}} \quad (1)$$

The VDR is compared with the reference value (VDR_{ref}), if there is any violation, depending on the violated rate, proper signal is produced for individual sequence of harmonics ($c_{dq}^{h,1-}$) by using a Proportional-Integrator (PI) controller. Then $C_{dq}^{h,1-}$ is sent to compensation effort controller of all DGs to mitigate PCC distortion to the reference value. The PI controller should be tuned so that the reference quality is achieved in short time while stability margin is considered too. It is worth noting that the "dead band" block is used to prohibit secondary control operation since PCC voltage distortion is lower that the reference value. Remember that the coordination step is also included in secondary control that is described in Section V.

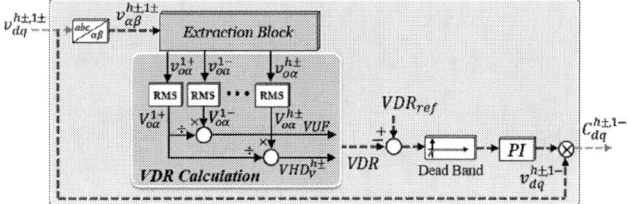

Fig. 3. Block diagram of compensation rate calculation.

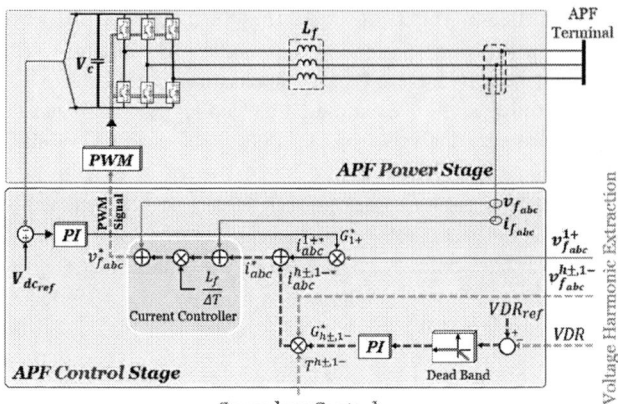

Fig. 4. APF power stage and control structure.

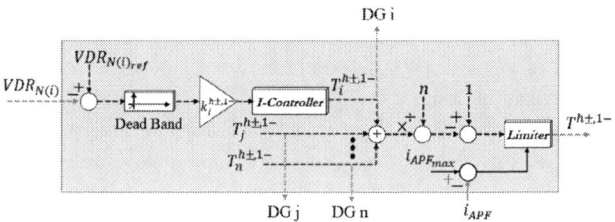

Fig. 5. Block diagram of APF cooperation rate calculation.

III. ACTIVE POWER FILTER, CONTROL SCHEME

Shunt APF is used in this paper to compensate both unbalance and harmonic distortion. The general control approach of the applied APF is extracted from [16]. The APF suppresses VDR by making resistive behavior toward fundamental negative sequence and harmonic components of current. The current command is determined as the follow equation:

$$i_{abc}^* = \sum_{h\pm,1-} G_{h\pm,1-}^* \cdot v_{fabc}^{h\pm,1-}. \tag{2}$$

where h is the harmonic order. In Eq. (2), $v_{fabc}^{h\pm}$ and v_{fabc}^{1-} are h^{th} harmonic component (including positive and negative sequences) and the fundamental negative sequence of voltage at the APF installation point (that is PCC in this paper, according to Fig. 1), respectively. Furthermore, $G_{h,1-}^*$ is the conductance command that might be different for fundamental and individual harmonic components. In fact, $G_{h,1-}^*$ is tuned based on the violation rate of VDR from the reference value by using a PI controller. As a result, APF can regulate compensation according to nonlinear/unbalance load condition. Note that the PI controller is designed exactly like that used in secondary control to make the better cooperation. Finally, the current regulator produces the following voltage command at the APF terminal [16]:

$$v_{fabc}^* = v_{fabc} - \frac{L_f}{\Delta T}\left(i_{abc}^* - i_{fabc}\right). \tag{3}$$

where L_f and ΔT are the APF inductance and sampling period, respectively [16]. According to above explanations, Fig. 3 shows the APF structure. It is worth noting that a PI controller is used for fixing the dc link of the filter and the dead band is applied to inhibit compensation if PCC compensation is not required.

IV. PROPOSED COORDINATION SCHEME

As mentioned before, the coordination is contrived for the situations that DG(s) is not sufficient for PCC compensation. As a result, the coordinated control manages compensation by using APF as auxiliary compensator. It is happened in "APF cooperation rate cal." of secondary control (see Fig. 2). The main policy of the proposed hierarchical control is that PCC compensation is carried out with the least cost investment and applying auxiliary compensator is avoided as far as possible so DGs play the main role in compensation. However, nodes voltage might become distorted severely. This can damage possible loads around nodes. Furthermore, when there is high amount of VDR at PCC, DGs have to devote high their capacities for compensation. It might diminish their efficiency as generators and even it might result instability. As a result, it is essential limiting DGs effort as compensators and provide satisfactory voltage quality at nodes.

The coordination scheme is represented in Fig. 5. As shown, at first $VDR_{N(i)}$ of all nodes is compared with the reference value; if there is any violation, proper signals ($T_i^{h\pm,1-}$) are sent to the primary control of the corresponding DG to obtain the reference voltage quality at the respective node. $T_i^{h\pm,1-}$ is tuned by using an integrator controller with the initial condition set to 1. In fact, $T_i^{h\pm,1-}$ changes between one and zero, as a result, DG compensation effort can be changed from 100% to 0%, depending on the violated rate. The integrator controller should be tuned so that the DG is able to tolerate the possible overshoot produced due to fast response while the time-response is not very long. Note that $T_i^{h\pm,1-}$ is individual for each voltage distortion including voltage unbalance of fundamental component and positive and negative sequences of voltage harmonic components. As it is shown in Fig. 5, VDR of all nodes are considered in the coordination while compensation effort reduction is carried out just for those DGs that their terminals are distorted severely. It is achieved via the dead band block in the coordination block (see Fig. 5).

However, by reducing DGs efforts for PCC compensation, PCC voltage will be distorted depending on the total compensation effort reduced by DGs. For this, APF is used to cooperate with DGs. In this line, it is needed calculating the total compensation effort reduction. Therefore, the following equation is used to determine the required cooperation rate:

$$T^{h\pm,1-} = 1 - \frac{\sum_{i=1}^n T_i^{h\pm,1-}}{n}. \tag{4}$$

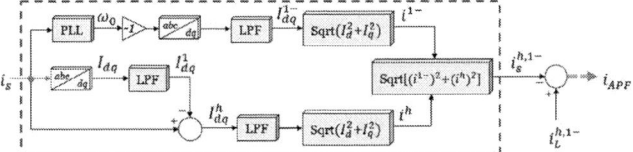

Fig. 6. APF current measurement.

where n is the number of DGs. In Eq. (4), it is assumed that PCC required compensation for achieving the reference voltage quality is 1 (or 100%). APF cooperation is obtained through multiplying $T^{h\pm,1-}$ to $G^*_{h\pm,1-}$ (see Fig. 4).

Another factor important in the cooperation is APF operation situation that can be determined by measuring its output current. Simply, since APF is overloaded, its cooperation should be limited (see Fig. 5). To measure APF current (i_{APF}), adaptive noise canceling technology (ANCT) is used [17]. According to ANCT, i_{APF} can be measured based on the following equation:

$$i_{APF} = i_L^{h\pm,1-} - i_s^{h\pm,1-}. \tag{5}$$

where $i_L^{h\pm,1-}$ and $i_s^{h\pm,1-}$ are harmonic component (including both positive and negative sequences) and fundamental negative sequence of load and DGs current, respectively (see Figs. 2&6). $i_L^{h\pm,1-}$ and $i_s^{h\pm,1-}$ extraction is taken place in measurement block of Fig. 2. This block is represented in Fig. 6. As shown, SRF-PLL extraction method is used in this block. Based on measurement block, harmonic component is measured by using Low Pass Filters (LPFs) and subtracting total current from fundamental component and the fundamental negative sequence is extracted by using PLL [18]. Note that $i_L^{h\pm,1-}$ is measured like $i_s^{h\pm,1-}$.

V. SIMULATION STUDY

The test MG is shown in Fig. 1. To provide high power quality for the loads at PCC, voltage quality requirement of PCC is set to higher levels than that of nodes (reference values $VHD_{PCC_{ref}}$ and $VUF_{PCC_{ref}}$ =0.5%). It is worth noting that compensation of 3rd, 5th and 7th harmonics (the main orders) of nodes and PCC voltage is concerned in this paper.

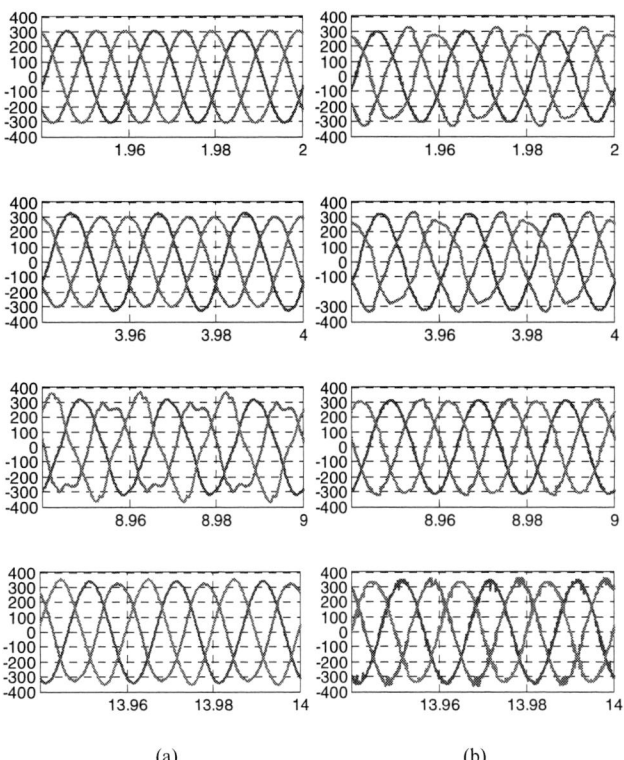

(a)   (b)
Vertical axis: voltage(V), Horizontal axis: time(s)
Fig. 7. Voltage waveforms: (a) Node1, (b) PCC.

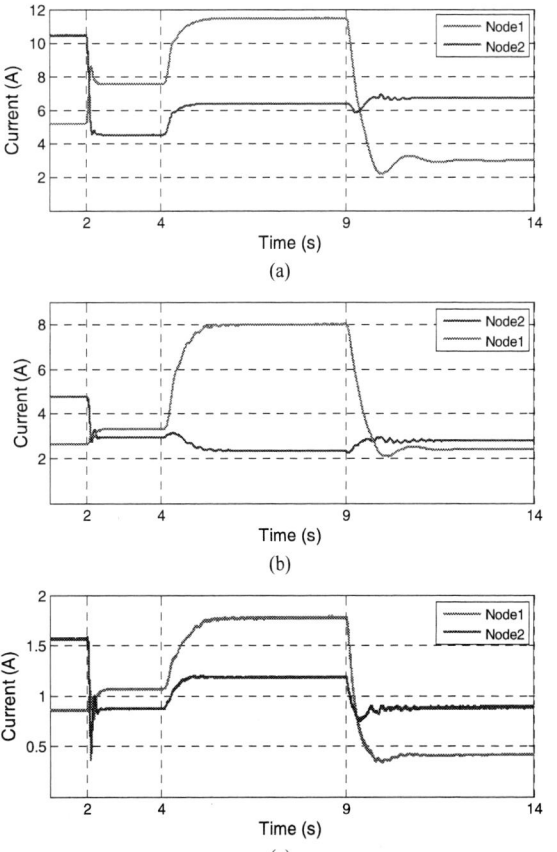

Fig. 8. Current distortion rate of nodes (RMS value in $\alpha\beta$-frame): (a) Fundamental negative sequence, (b) 3rd harmonic, (c) 5th harmonic.

TABLE I
TEST SYSTEM PARAMETERS

Power Stage Parameters								
APF Power Stage		Distribution Lines			nonlinear load			linear load
L_F (mH)	C_F (μF)	$Z_1(\Omega)$	$Z_2(\Omega)$	$Z_{nl}(\Omega)$	$C_{nl}(\mu F)$	$R_{nl}(\Omega)$	$L_{nl}(mH)$	$Z_{ll}(\Omega)$
15	2200	0.2+j1.131	0.1+j0.565	0.2+j1.005	235	20	0.084	50+j6.238
Control Stage Parameters								
APF Capacitor-PI Controller			Cooperation stage					
K_p		K_i	k_1^{1-}	$k_1^{3\pm}$	$k_1^{5\pm}$	$k_1^{7\pm}$		
0.16		0.02	0.5	0.45	0.4	0.35		

TABLE II
SIMULATION TIME PERIODS

Time (s)	0 < t < 2	2 < t < 4	4 < t < 9	9 < t < 14
Control loop	Droop Control And Power Sharing	Current sharing	Secondary Control (PCC compensation by DGs)	Secondary Control (PCC compensation by the cooperation)

978-1-5090-0376-1/16 $31.00 © 2016 IEEE

Required data of power and control stages of the system is available in Table I and the data concerning primary and secondary controls can be found in [11] and [14]. Based on DGs power droop characteristics, DG1 is twice of DG2.

To test different parts of the proposed hierarchical control clearly, Table II shows simulation process and the following explanations are based on this table. It is worth noting that MATLAB/Simulink is used for evaluating the proposed hierarchical control.

Two DGs are in the system so power, harmonic and unbalance current sharing should be considered. It is assumed that voltage quality reference of the node 1 is Total Harmonic Distortion (THD) equal to 3% ($THD_{N(1)_{ref}}$ =3%) and $VUF_{N(1)_{ref}}$ =1% while the maximum permissible VDR is considered for node 2 according to IEEE Standards 519 and 141 ($THD_{N(2)_{max}}$ =5% and $VUF_{N(2)_{max}}$ =2%) [19],[20].

Since APF is only cooperated with DG1, Fig. 7 shows voltage waveforms of node 1 and PCC according to simulation time periods (Table II). As it is represented, the node voltage is completely sinusoidal in the first period, demonstrating perfect performance and well design of droop control and selective virtual impedance. However, PCC voltage waveform is distorted. It is for the voltage drop produced through $Z_{1,2}$ (see Fig. 1). As it is shown in Fig. 7, voltage waveform of node 1 and PCC are severely distorted while current unbalance and harmonic sharing between DGs is obtained by virtual impedance (see Fig. 8). It can be seen in Fig. 8 that harmonic and unbalance components in DG2-current is higher than that of DG1 before $t = 2s$ while after this time, current sharing is happened based on DGs rated power. As shown in Fig. 7, PCC voltage is significantly improved since secondary control is initiated but node 1 voltage is severely distorted. It shows that DG1 has main role in PCC compensation. However, the cooperation is required because $VDR_{N(1)}$ is very higher than the reference value, according to Fig. 10. Since the cooperation is initiated, it can be seen that $VDR_{N(1)}$ is reduced while PCC voltage is remained unchanged. It proves that the cooperation is designed accurately.

Fig. 8 shows current distortion at nodes. As shown, since secondary control is initiated, node 1 current is severely distorted. However, as the cooperation is initiated, it can be seen that current distortion in node 1 is reduced while this parameter is nearly unchanged in node 2. It is because APF is only cooperated with DG1.

To be more illustrative, Fig. 9 shows current waveforms of nodes according to the simulation time periods. It can be seen that node2-current is more close to sinusoidal waveform since virtual impedance is initiated and current sharing is taken place. On the other hand, node1-current is more distorted. As PCC compensation is occurred by DGs, it is shown that node1-current is even more distorted than before because DG1 plays the main role in compensation. However, by initiation of the cooperation at $t = 9s$, node1-current waveform is nearly sinusoidal due to the fact that a part of PCC compensation is carried out by APF.

To test the proposed hierarchical control more accurately, Fig. 10 shows VHD curves of both positive and negative sequences of individual harmonic components and VUF of the nodes and PCC. Based on Fig. 10, nodes and PCC-VDR are

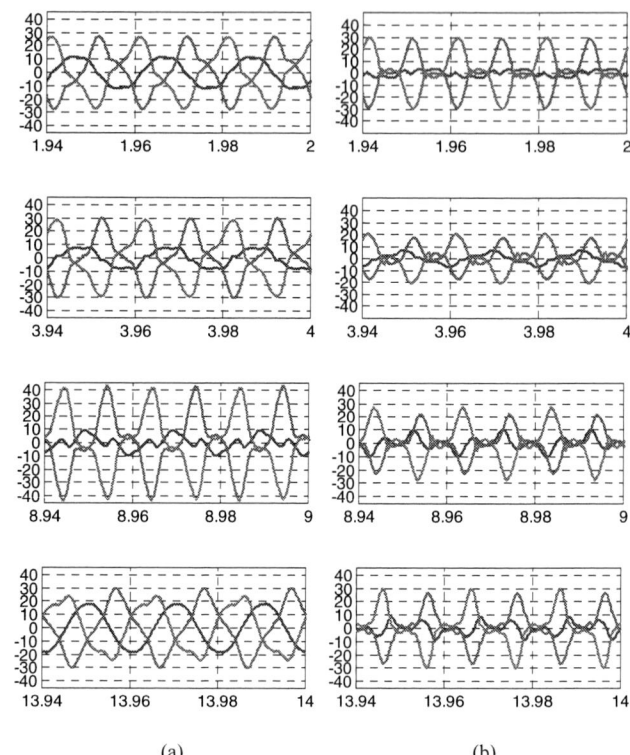

(a)   (b)

Vertical axis: current(A), Horizontal axis: time(s)

Fig. 9. Current waveforms: (a) Node1, (b) Node2.

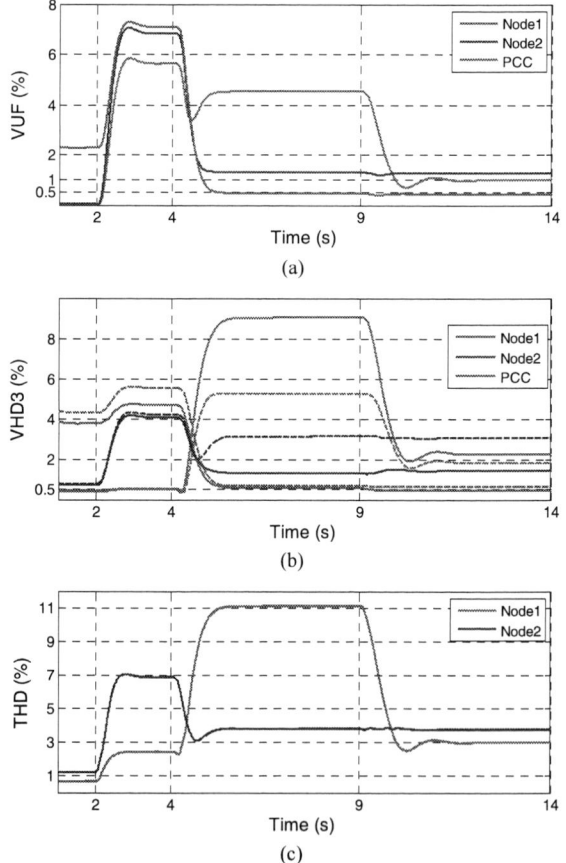

Fig. 10. Voltage Distortion Rate (dashed: positive sequence, solid: negative sequence): (a) VUF, (b) 3ᵗʰ harmonic, (c) THD.

increased since current sharing is happened. According to Fig. 10, when secondary control is active, PCC-VDR is significantly reduced and $VDR_{PCC_{ref}}$ is achieved but node1-VDR is severely increased and the violation from $VDR_{N(1)_{ref}}$ is occurred. In the final period, the cooperation between APF and DG1 is initiated and $VDR_{N(1)_{ref}}$ is achieved. It is shown in Fig. 10 that $VDR_{N(2)}$ and VDR_{PCC} (both positive and negative sequences of voltage harmonics and VUF) are remained unchanged when APF is participated in compensation. It shows perfect performance of the coordination and precise design of its parameters.

VI. CONCLUSION

A hierarchical control scheme to improve voltage harmonic distortion and voltage unbalance factor of microgrid is proposed. The hierarchical structure includes two levels. In the primary control, power and current sharing is carried out. Secondary level compensates PCC by controlling DG(s) inverters. Compensation of PCC by DG(s) might increase voltage distortion at DG(s) terminal. Thus, a coordinated control between DGs and APF is designed to share compensation. The coordinated control is based on the required power quality of each DG terminal and the APF capacity.

REFERENCES

[1] J. M. Guerrero, P. C. Loh, M. Chandorkar, and T.-L. Lee, "Advanced Control Architectures for Intelligent MicroGrids, Part I: Decentralized and Hierarchical Control Advanced Control Architectures for Intelligent Microgrids," *IEEE Trans. Ind. Electron.*, vol. 60, pp. 1254-1262, 2013.

[2] J. M. Guerrero, P. C. Loh, T.-L. Lee, and M. Chandorkar, "Advanced control architectures for intelligent microgrids—part II: power quality, energy storage, and AC/DC microgrids," *IEEE Trans. Ind. Electron.*, vol. 60, no. 4, pp. 1263-1270, Apr. 2013.

[3] M. Hojo, Y. Iwase, T. Funabashi, and Y. Ueda, "A method of three-phase balancing in microgrid by photovoltaic generation systems," in *Power Electronics and Motion Control Conference, 2008. EPE-PEMC 2008. 13th*, 2008, pp. 2487-2491.

[4] P.-T. Cheng, C.-A. Chen, T.-L. Lee, and S.-Y. Kuo, "A cooperative imbalance compensation method for distributed-generation interface converters," *IEEE Trans. Ind. Appl.*, vol. 45, pp. 805-815, 2009.

[5] M. Savaghebi, A. Jalilian, J. C. Vasquez, and J. M. Guerrero, "Secondary control scheme for voltage unbalance compensation in an islanded droop-controlled microgrid," *IEEE Trans. Smart Grid*, vol. 3, pp. 797-807, 2012.

[6] L. Meng, F. Tang, M. Savaghebi, J. C. Vasquez, and J. M. Guerrero, "Tertiary Control of Voltage Unbalance Compensation for Optimal Power Quality in Islanded Microgrids," *IEEE Trans. Energy Conv.*, vol. 29, pp. 802-815, 2014.

[7] T. L. Lee, and P. T. Cheng, "Design of a new cooperative harmonic filtering strategy for distributed generation interface converters in an islanding network," *IEEE Trans. Power Electron.*, vol. 22, no. 5, pp. 1919-1927, Sept. 2007.

[8] M. Cirrincione, M. Pucci, and G. Vitale, "A single-phase DG generation unit with shunt active power filter capability by adaptive neural filtering," *IEEE Trans. Ind. Electron.*, vol. 55, no. 5, pp. 2093-2110, May 2008.

[9] M. Savaghebi, M. M. Hashempour, and J. M. Guerrero, "Hierarchical coordinated control of distributed generators and active power filters to enhance power quality of microgrids," in *Power and Electrical Engineering of Riga Technical University (RTUCON), 2014 55th International Scientific Conference on*, 2014, pp. 259-264.

[10] N. Pogaku and T. C. Green, "Harmonic mitigation throughout a distribution system: a distributed-generator-based solution," *Generation, Transmission and Distribution, IEE Proceedings-*, vol. 153, pp. 350-358, 2006.

[11] M. Savaghebi, A. Jalilian, J. C. Vasquez, and J. M. Guerrero, "Secondary control for voltage quality enhancement in microgrids," *IEEE Trans. Smart Grid*, vol. 3, no. 4, pp. 1893-1902, Dec 2012.

[12] J. He, Y. W. Li, and F. Blaabjerg, "An Enhanced Islanding Microgrid Reactive Power, Imbalance Power, and Harmonic Power Sharing Scheme," *IEEE Trans. Power Electron.*, vol. 30, pp. 3389-3401, 2015.

[13] M. M. Hashempour, M. Savaghebi, J. C. Vasquez, and J. M. Guerrero, "Hierarchical control for voltage harmonics compensation in multi-area microgrids," in *Diagnostics for Electrical Machines, Power Electronics and Drives (SDEMPED), 2015 IEEE 10th International Symposium on*, 2015, pp. 415-420.

[14] M. M. Hashempour, M. Savaghebi, J. C. Vasquez, and J. M. Guerrero, "A Control Architecture to Coordinate Distributed Generators and Active Power Filters Coexisting in a Microgrid," *IEEE Trans. Smart Grid*, Early Access, pp. 1-12, 2015.

[15] P. Rodriguez, A. Luna, I. Candela, R. Mujal, R. Teodorescu, and F. Blaabjerg, "Multiresonant frequency-locked loop for grid synchronization of power converters under distorted grid conditions," *IEEE Trans. Ind. Electron.*, vol. 58, no. 1, pp. 127-138, Jan. 2011.

[16] T. L. Lee, J. C. Li, and P. T. Cheng, "Discrete frequency tuning active filter for power system harmonics," *IEEE Trans. Power Electron.*, vol. 24, no. 5, pp. 1209-1217, May 2009.

[17] B. Widrow and S. D. Stearns, "Adaptive signal processing," *Englewood Cliffs, NJ, Prentice-Hall, Inc.*, 1985, 491 p., vol. 1, 1985.

[18] Q. Wang, N. Wu, and Z. Wang, "A neuron adaptive detecting approach of harmonic current for APF and its realization of analog circuit," *IEEE Trans. Instrumentation and Measurement*, vol. 50, pp. 77-84, 2001.

[19] IEEE Recommended Practice and Requirements for Harmonic Control in Electric Power Systems, IEEE Standard 519-2014, 2014.

[20] IEEE Recommended Practice for Electric Power Distribution for Industrial Plants, ANSI/IEEE Std. 141-1993, (Red Book).

Method for Load Sharing and Power Management in a Hybrid PV/Battery Source Islanded Microgrid

Yaser Karimi, Hashem Oraee

Dept. of Electrical Engineering, Sharif University of
Technology, Tehran, Iran
y_karimi@ee.sharif.edu, oraee@sharif.edu

Josep M.Guerrero, Juan C.Vasquez, Mehdi Savaghebi

Dept. of Energy Technology, Aalborg University.
Aalborg, Denmark
{joz, juq, mes}@et.aau.dk

Abstract—**This paper presents a decentralized load sharing and power management method for an islanded microgrid composed of PV units, battery units and hybrid PV/battery units. The proposed method performs all the necessary tasks such as load sharing among the units, battery charging and discharging and PV power curtailment with no need to any communication among the units. The proposed method is validated experimentally.**

Keywords—islanded microgrid; decentralized power management; hybrid PV/battery; battery SOC.

I. INTRODUCTION

Due to environmental concerns and continuous decrease in the price, utilization of photovoltaic (PV) generation systems has been increasing in the recent years. Because of the intermittent nature of the PV power output, storage batteries are integrated to PV systems and a set of PV sources, batteries and loads can form a microgrid. Battery storage can be connected through an inverter as a separate unit to the microgrid or can be combined with the PV unit forming a hybrid source unit [1], [2].

In islanded mode of operation, the control system objective is to share the load among different units and balance the power in the microgrid while considering power rating of the units and State of Charge (SOC) of the batteries. For eliminating communication among units, decentralized methods are used for controlling microgrids. In previous works, several decentralized control strategies are proposed. In [1-3] three power management methods are proposed based on frequency signaling but the application of these methods are limited to the microgrids composed of only one energy storage unit. In [4] a frequency based energy management strategy is proposed for a microgrid with distributed battery storage but it is only valid for systems with separate battery units not hybrid source units. Similarly the method proposed in [5] is only applicable to separate battery units. The strategy proposed in [6] considers a separate battery storage unit and a PV unit connected to a droop controlled microgrid while the decentralized power management strategy proposed by the same authors in [7] considers a single PV/battery hybrid unit connected to a droop controlled islanded microgrid. This strategy is very useful for a single hybrid unit connected to a microgrid that contains only droop controlled sources, however if there are multiple hybrid units or there are other battery storage units in the microgrid, this strategy will not be applicable.

This paper proposes a decentralized method for power management and load sharing in an islanded microgrid consisting of different PV units, battery storage units and hybrid PV/battery units. Unlike previous works, the proposed method is not limited to the systems with separate PV and battery units or systems with only one hybrid unit. A modified droop method is used in the proposed method that uses the frequency as trigger for different state changes in each unit. The main contributions of the proposed method are:

• When the microgrid load is more than total PV generation, all PV sources (in both separate and hybrid units) operate in Maximum Power Point (MPP) and all the batteries (in both separate and hybrid units) supply the surplus load power. The droop equations are chosen such that the surplus power is shared among the batteries such that batteries with higher SOC have higher discharging power.

• If some batteries have not reached to charge limiting mode and the total PV generation is more than microgrid load, the batteries are being charged with the excess PV power. The excess power is shared among the batteries such that batteries with lower SOC absorb more charging power.

• When the microgrid load is less than total PV generation and all batteries are in charge limiting mode, PV power curtailment is performed and the load is shared among the units that have PV source based on the inverter capacity and considering their available PV power.

It is worth noting that the proposed method is applicable to both single phase and three phase microgrids but for simplicity single phase microgrid is considered in this paper. Moreover, other sources other than PV can be used in the units.

The rest of the paper is organized as follows: in section II the general structure of the hybrid source single phase microgrid is presented. A microgrid with three hybrid units is considered in this study. In section III the proposed method is presented in detail and different operating modes of the whole microgrid and operating states of each unit in the microgrid are presented. The proposed method is validated experimentally in section IV. Section V concludes the paper.

II. SINGLE PHASE MICROGRID STRUCTURE

A general single phase microgrid system consisting of PV units, battery units and hybrid source units is depicted in Fig. 1. For being comprehensive, a microgrid with three hybrid units is considered in the following analysis. The proposed

978-1-5090-0376-1/16 $31.00 © 2016 IEEE

method can be easily applied to separate PV and battery units with minor changes. All the microgrid loads are centralized in a single load.

By using an inductance in the output filter of each unit and by implementing virtual inductance [8], it is assumed that the output impedances of units are mainly inductive and a modified P-f, Q-E droop method is utilized in the proposed control method. The inner voltage and current loops are studied in other papers and they are not addressed in this paper.

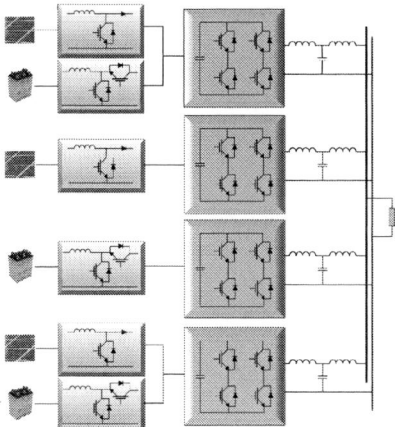

Fig. 1. Single phase microgrid system

III. PROPOSED METHOD

In this section, first, the general operating modes of the whole microgrid is described; then, the operating states of each hybrid unit and criteria for changing of states are described. It is worth mentioning that battery power, P_{Bat}, is positive in discharging mode and is negative in charging mode.

A. General Operating Modes of the Whole Microgrid

Depending on the total maximum available PV power, total charging capacity of the batteries and total load in the microgrid, it can operate in three main modes:

Mode I) Battery Discharging Mode
In this mode, the total microgrid load is more than total PV maximum power, i.e.

$$P_{LD} > \sum_{i=1}^{m} P_{PV-MP_i} \tag{1}$$

where P_{LD} is the load power, P_{PV-MP_i} is the maximum PV power of Unit$_i$ and m is the number of units in the microgrid. P_{PV-MP_i} depends on the solar irradiance and temperature of the PV array.

In this mode, all PVs work at MP and the batteries supply the surplus load power, $P_{LD} - \sum_{i=1}^{m} P_{PV-MP_i}$. The dc-link voltage, V_{dc} is controlled by battery boost converter. In order to share the surplus load power among the units, load is shared such that the required total discharging power of the batteries is

shared among units based on the State of Charge (SOC) of the unit's battery. The droop coefficient is selected to be inversely proportional to SOC of the battery similar to method proposed in [9, 10]for DC microgrids. To achieve this battery power sharing strategy, following droop equation is used:

$$f = f_0 + m_p (P_{PV-MP} - P_{out}), m_p \propto \frac{1}{SOC^n} \tag{2}$$

where n is selected as described in [9, 10]. In this strategy, considering that discharging power of the battery is

$$P_{Bat} = P_{out} - P_{PV}, \tag{3}$$

if m_p is same for all units, the battery discharging power of all units are equal regardless of their PV power (unless output power reaches inverter rating limit which is discussed in the next section.)

Mode II) Battery Charging Mode
In this mode, the total microgrid load is less than total PV maximum power and at least the battery of one unit has the capability to absorb the surplus power, i.e. one of the units has not reached to battery charge limiting mode. In this mode, all PVs work at MP. In units that are not in charge limiting mode, V_{dc} is controlled by battery boost converter, in addition, the inverter is controlled in Voltage Control Mode (VCM) and the P-f droop is selected similar to Mode I. The droop coefficient is selected to be proportional to SOC of the battery similar to method proposed in [10] for charging mode so that batteries with higher SOC absorb less power:

$$f = f_0 + m_p (P_{PV-MP} - P_{out}), m_p \propto SOC^n \tag{4}$$

In units that enter to charge limiting mode, V_{dc} is controlled by controlling output power, in addition, the inverter is controlled in Power Control Mode (PCM) and output power is controlled with following equations,

$$P_{ref} = (K_{V-P} + \frac{K_{V-I}}{s})(V_{dc} - V_{dc}^*) \tag{5}$$

$$f = f_0 + (K_{P-P} + \frac{K_{P-I}}{s})(P_{ref} - P_{out}) $$

where P_{ref} is the reference value of output power, V_{dc}^* is reference value of dc-link voltage, K_{V-P} and K_{V-I} are proportional and integral gains of V_{dc} controller and K_{P-P} and K_{P-I} are proportional and integral gains of output power controller. Note that in steady state,

$$P_{ref} = P_{PV-MP} - |P_{Bat-ChLimit}| \tag{6}$$

where $P_{Bat-ChLimit} < 0$ is battery charging power in charge limiting mode. $P_{Bat-ChLimit}$ is zero when battery reaches SOC_{max} and has varying value in constant current and constant voltage charging modes. This control strategy ensures charging power distribution between all batteries. It is worth mentioning that each battery can be charged with PV power of other units.

Fig. 2-a shows the P-f characteristics of a two hybrid unit microgrid system in Modes I and II. It is assumed that $SOC_1 > SOC_2$; therefore, the discharging and charging powers of Unit1 battery are higher and less than that of Unit2 battery, respectively. In this figure, f_I and f_{II} are sample equilibrium

978-1-5090-0376-1/16 $31.00 © 2016 IEEE

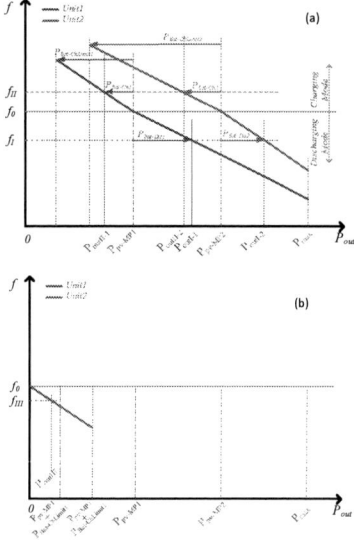

Fig. 2. P-f characteristics of a two hybrid unit microgrid system in a) Modes I and II assuming $SOC_1 > SOC_2$ b) Mode III

frequencies in discharging and charging modes, P_{outI} and P_{outII} are corresponding output powers, and $P_{Bat-Dis}$ and P_{Bat-Ch} are corresponding discharging and charging powers, respectively.

Mode III) PV Power Curtailment Mode

In this mode, the sum of total microgrid load and total charging capacity of the batteries is less than total PV maximum power, i.e.,

$$P_{LD} + \sum_{i=1}^{m} P_{Bat-ChLimit_i} < \sum_{i=1}^{m} P_{PV-MP_i} \qquad (7)$$

therefore, PV power curtailment should be performed and some PVs must deviate out of their maximum power. In this mode all batteries are charged with maximum power, units with low PV maximum power work at MP and are controlled in PCM, units with high PV maximum power are controlled in VCM and PV boost converter controls the dc-link voltage. The simple P-f droop is used for VCM units in this mode:

$$f = f_0 - m_p P_{out}, m_p = \frac{\Delta f_{max}}{P_{max}} \qquad (8)$$

Fig. 2-b shows the P-f characteristics of a two hybrid unit microgrid system in Mode III. It is assumed that m_p is equal for two units. In this figure, f_{III} is a sample equilibrium frequency and P_{outIII} is its corresponding output power for two units.

B. Operating States of Each Unit in the Microgrid

The general operating modes of the microgrid was described in previous section. In this section, the detailed operating states of a single unit in the microgrid and criteria for different state changes are presented. In the following, "this unit" refers to the single unit under control.

Each unit in the microgrid can be in five states:

State 1) VCM, battery charge/discharge, V_{dc} is controlled by battery boost converter

This state is associated with Mode I or VCM condition of Mode II of the described operating modes. In this state, PV works at MPP and battery boost converter controls V_{dc} by either charging or discharging of the battery. If total PV maximum power is less than load power and battery is in discharging mode, (2) is used for controlling output power; otherwise, (4) is used. Note that in (4), if P_{PV-MP} is small or zero, P_{out} can be negative which means that battery is charged with power delivered by the microgrid.

Criteria for exiting from State 1:

- If SOC of the battery reaches to its minimum value, SOC_{min}, state is changed to State 4.
- If output power reaches inverter rating limit, $P_{out-max}$, state is changed to State 5.
- If battery is completely charged or battery reaches its maximum charging power due to decrease in load or increase in PV generation; in other words, when $SOC=SOC_{max}$ or $P_{Bat}=-P_{Bat-ChLimit}$ or $I_{Bat}=I_{Bat-max}$ or $V_{Bat} = V_{Bat-max}$, state is changed to State 2.

State 2) PCM, Battery charge limit, V_{dc} is controlled by output power

This state is associated with PCM condition of Mode II of the described operating modes. In this state, PV works at MPP, battery is in charge limiting mode and V_{dc} is controlled by (5).

Criteria for exiting from State 2:

- If because of decrease in load, increase in total PV generation or decrease in total battery charging power (as a result of reaching SOC_{max} or entering constant voltage charging mode) all units enter this state from State 1, due to using PI controller in (5) f increases in all units until it saturates to f_{max}. At this point all units change to State 3 to reduce PV power generation and keep the power balance.

- If at least one of other units is in State 1 in battery charging mode and the others—including this unit—are in State 2, (i.e. all batteries are in charging mode but some of them are in charge limiting mode) if load is increased or total PV generation is decreased, charging power of units in State 1 decreases and according to (4) f decreases. When charging power of units in State 1 multiplied by m_p of the units is less than a determined ratio of this unit's corresponding value, unit must return to State 1. In other words, when

$$\left| m_{p-i} P_{Bat-S1-i} \right| < K_{ch} \left| m_p P_{Bat} \right|, \qquad (9)$$

if state is returned to State 1 and the total charging power is shared again based on (4), this unit will no longer enter battery charge limit mode and return to State 2. $P_{Bat-S1-i}$ is charging power of ith unit which is in State 1 and $K_{ch}<1$ is a margin used for preventing unwanted changing of state because of error in measuring power. Note that according to (4) $m_{p-i}P_{Bat-S1-i}$ are equal for all units in State 1. As P_{Bat} is negative in charging mode, (9) can be written as

$$m_{p-i} P_{Bat-S1-i} > K_{ch} m_p P_{Bat} \qquad (10)$$

978-1-5090-0376-1/16 $31.00 © 2016 IEEE   654

Since $m_{p-i}P_{Bat-S1-i}$ is not measurable directly, it can be determined according to f. As PI controller is used in (5), the unit's frequency follows the frequency of other units that are in State 1; therefore, after some manipulation using (3), (4) and (9) the criterion for returning from State 2 to State 1 is:

$$f < f_0 - K_{ch}m_pP_{Bat} \text{ and preState} = 1. \qquad (11)$$

In the worst case that the unit is in SOC_{max} and P_{Bat} =0, if $f < f_0$ it means that other units are in discharging mode and this unit starts to discharge.

- If because of increase in load or decrease in total PV generation all units enter this state from State 3, due to using PI controller in (5) f decreases in all units until it saturates to f_{min}. At this point all units change to State 1 in order to reduce the battery charging power or enter battery discharging mode.

If at least one of other units is in State 3 and the others—including this unit—are in State 2, when load is decreased or output power of this unit is increased because of increase in the unit's PV generation, output power of units that are in State 3 decrease and according to (8) f increases. When output power of the units that are in State 3 multiplied by m_p of the units is less than a determined ratio of this unit's corresponding value, unit must return to State 3. In other words, when

$$m_{p-i}P_{out-S3-i} < K_{pc}m_pP_{out}, \qquad (12)$$

if state is returned to State 3 and load is shared again based on (8), this unit will no longer return to State 2 because of insufficient PV power. $P_{out-S3-i}$ is output power of ith unit which is in State 3 and $K_{pc}<1$ is a margin used for preventing unwanted changing of state because of error in measuring power. Note that in State 2, P_{out} is equal to P_{ref} determined by (6) in steady-state; furthermore, according to (8) $m_{p-i}P_{out-S3-i}$ are equal for all units in State 3. Since $m_{p-i}P_{out-S3-i}$ is not measurable directly, it can be determined according to f. As PI controller is used in (5), the unit's frequency follows the frequency of other units that are in State 3. Since $P_{out}>0$ for units that are in State 3, criterion (12) can be written as,

$$f_0 - m_{p-i}P_{out-S3-i} > f_0 - K_{pc}m_pP_{out} \qquad (13)$$

and based on (8) the criterion for returning from State 2 to State 3 is:

$$f > f_0 - K_{pc}m_pP_{out} \text{ and preState} = 3. \qquad (14)$$

State 3) VCM, PV power curtailment, V_{dc} is controlled by PV boost converter

This state is associated with Mode III of the described operating modes in which total PV maximum power is more than total power required by load and charging of the batteries. In this state, the unit's battery is charged with maximum power, output power is controlled by (8) and PV boost converter controls V_{dc}.

Criterion for exiting from State 3:

- If the PV maximum power is less than sum of output power determined by (8) and battery charging power, state is changed to State 2. This criterion happens when the unit is in State 3 and either load is increased or PV generation is decreased, or when all units enter to State 3 from State 2 but PV power is not sufficient for this unit.

State 4) PCM, Battery disconnection, V_{dc} is controlled by output power

When SOC of the battery reaches to SOC_{min} the unit enters this state. In this state, battery is disconnected to prevent damage due to its deep discharging, PV works at MPP and V_{dc} is controlled by (5). Since $P_{Bat}=0$, in the steady-state

$$P_{out} = P_{PV-MP}. \qquad (15)$$

Criterion for exiting from State 4:

- If all units enter this state and load power is more than total PV maximum power, f decreases until it reaches the critical minimum frequency. At this point load shedding is inevitable and some non-critical loads must be disconnected. Load shedding is out of scope of this paper.
- According to (2) and (4), if $f > f_0$ it means that $P_{Bat}<0$ in units that are in State 1 and they are in battery charging mode. Therefore, this unit can return to State 1 to charge the battery.

State 5) PCM, Output power limiting, V_{dc} is controlled by battery boost converter

When output power of the unit reaches $P_{out-max}$, the unit enters this state to limit its output power. In this state, PV works at MPP and V_{dc} is controlled by battery boost converter by discharging. Output power is controlled by (5) with $P_{ref}= P_{out-max}$.

Criteria for exiting from State 5:

- If all units enter this state and load power is more than total ratings of the units, f decreases until it reaches the critical minimum frequency. At this point load shedding is inevitable and some non-critical loads must be disconnected.
- If the unit is in output power limiting and load is decreased such that battery discharging power of other units in State 1 multiplied by m_p of the units is less than a determined ratio of this unit's corresponding value, unit must return to State 1. In other words, when

$$m_{p-i}P_{Bat-S1-i} < K_{pl}m_pP_{Bat} \qquad (16)$$

if state is returned to State 1 and the load is shared again based on (2), this unit will no longer enter output power limiting. $K_{pl}<1$ is a margin used for preventing unwanted changing of state because of error in measuring power. Similar to previous discussions, this criterion can be determined indirectly by f and after some manipulation the criterion can be written as,

$$f > f_0 - K_{pl}m_pP_{Bat}. \qquad (17)$$

Different states of each unit have been summarized in Table I. In addition, criteria for changing the states have been depicted in Fig. 3.

Table I Summary of states of each unit

State	1 Battery Charge/Discharge	2 Battery Charge Limit	3 PV Power Curtailment	4 Battery Disconnect	5 Output Power Limit
Control mode	VCM	PCM	VCM	PCM	PCM
V_{dc} Control	Battery	Pref	PV	Pref	Battery
PV Power	MPP	MPP	<MPP	MPP	MPP

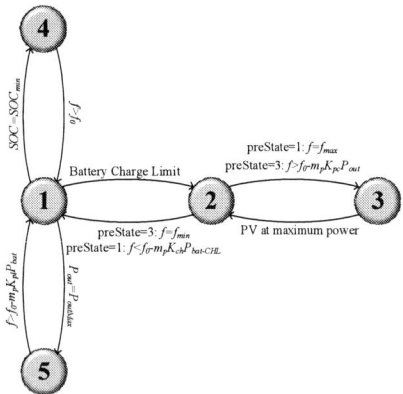

Fig. 3. Criteria for changing the states of each single unit in the microgrid

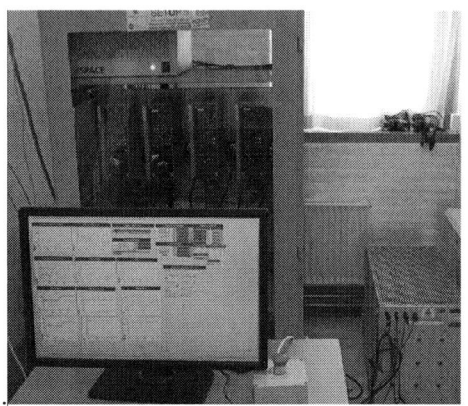

Fig. 4. Experimental setup

IV. EXPERIMENTAL RESULTS

The proposed method has been evaluated experimentally using the experimental setup shown in Fig. 4. It consists of three Danfoss three phase inverters used as single phase inverters, a real-time dSPACE1006 platform, LCL filters and load. Batteries and PVs are emulated as Hardware In the Loop (HIL).

Fig. 5. shows the state, output power, PV power, battery power and output frequency of each unit in different load and PV generation conditions. In this experiment m_p is considered equal for all units for clarity of the results. The maximum PV powers of Units 1, 2 and 3 are considered 300W, 500W and 600W, respectively and the maximum charging powers of Units 1, 2 and 3 are considered 400W, 300W and 150W, respectively.

At start all units are in State 1 and load power which is 1700W is shared between units such that the battery discharging power of all units are equal. All PVs work at MP in this state.

At t=20s, load is decreased to 1400W, therefore discharging power of all batteries decrease to 5W.

At t=40s, load is decreased to 1100W and since total PV generation is more than load, all batteries enter to charging mode with equal charging power.

At t=60s, load is decreased to 800W, as a result, charging power of all units increase. Since Unit3 reaches its maximum charging power, it changes to State 2 with constant output power equal to $P_{pv-MP}+P_{Bat-ChLimit}$ in steady state. The remaining charging power is shared between Units 1 and 2 equally based on (4).

At t=80s, load is decreased to 500W, as a result, charging power of Units 1 and 2 increase. First Unit 2 reaches its maximum charging power and changes to State 2 and in a short time, Unit 1 also reaches maximum charging power and changes to State 2. Since all units are in State 2, frequency increases until it saturates to f_{max}. At this point all units change to State 3; however, since PV power is not sufficient for supplying both battery charging power and output power determined by (8) in Units 1 and 2, they return to State 2. Note that output power of Unit1 is negative at State 2 which means it is absorbing power from the microgrid to charge its battery.

At t=100s, load is decreased to 200W, since Units 1 and 2 are controlled in constant power, output power of Unit3 decreases and consequently f increases based on (8). After 3s criterion (14) is validated in Unit2 and it changes to State3 and decreases its PV generation.

At t=120s, irradiance of unit1 PV is increased and therefore unit1 PV maximum power increases from 300W to 600W; consequently output power of Unit1 increases to 600W. Since load power is constant it results in decreasing output power of Unit3 and increasing in f based on (8). After 3s criterion (14) is validated in Unit1 and it also changes to State3. At this point all units are in State 3 having same output power, battery charging with maximum power and deviating out of their maximum PV power.

At t=140s, load is increased to 500W and output power of all units increase equally. Since none of them reach maximum PV power, they remain at State3.

At t=160s, load is increased to 800W. Output power of all units increase but Units 1 and 2 reach their maximum PV power and change to State 2.

At t=180s, load is increased to 1100 W. Unit3 reaches its maximum power and it also changes to State 2. Since all units are in State 2, frequency decreases until it saturates to f_{min}. At this point all units change to State 1; however, since Unit3 battery charging power determined by (4) is more than its maximum value, it returns to State 2.

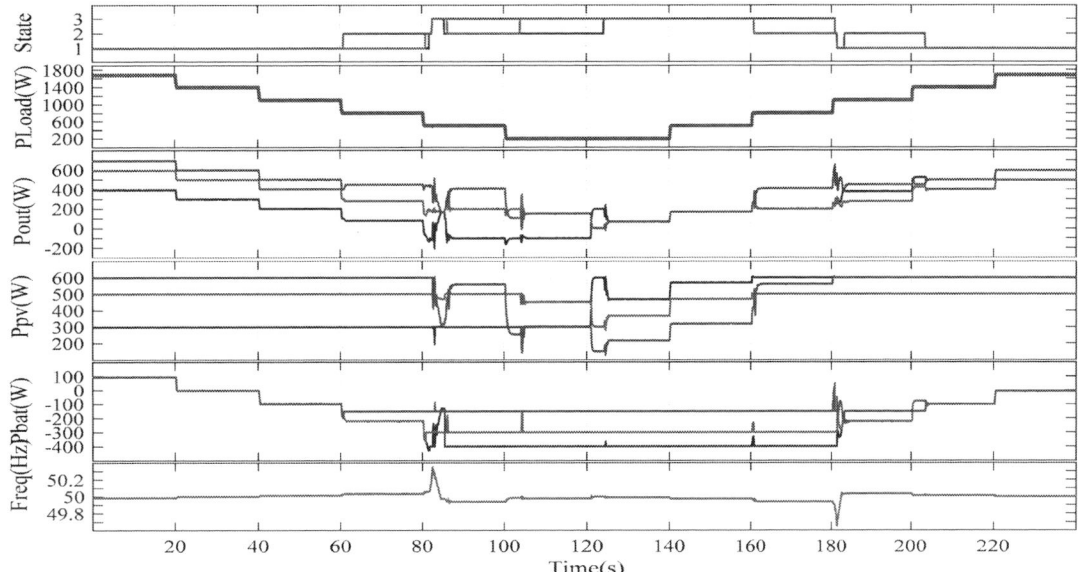

Fig. 5. Experimental results of a three unit single phase microgrid in different load and PV generation conditions. Blue: Unit1 Red: Unit2, Green: Unit3

At t=200s, load is increased to 1400W. Output power of Units 1 and 2 increase resulting in decrease in f based on (4). After 3s, criterion (11) is validated in Unit3 and it also changes to State 1. At this point all units are in State 1 having same battery charging power and PVs at their maximum power.

At t=220s, load is increased to 1700W and the charging power of all units decrease equally to -4W.

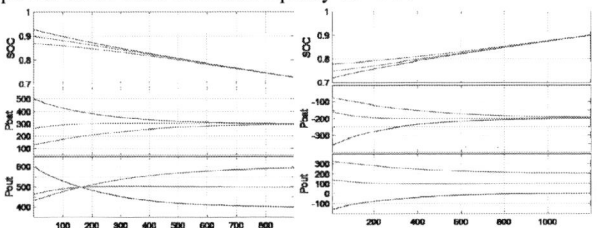

Fig. 6. SOC balancing in a) discharging mode b) charging mode

Fig. 6-a. shows SOC balancing of the batteries in discharging mode. All the units are in State 1 and the discharging power of the batteries are determined by (2) with n=15. Load power is 1500W, PV_1=100W, PV2=200W, PV3=300W. Fig. 6-b. shows SOC balancing of the batteries in charging mode. The charging power of the batteries are determined by (4) with n=15.

V. CONCLUSION

A decentralized method was proposed in this paper for load sharing and power management in an islanded microgrid with PV units, battery units and hybrid PV/battery units. The proposed method uses frequency as trigger for different state changes in each unit. The proposed method was validated experimentally with a three unit setup.

REFERENCES

[1] E. Serban and H. Serban, "A Control Strategy for a Distributed Power Generation Microgrid Application With Voltage- and Current-Controlled Source Converter," *Power Electronics, IEEE Transactions on,* vol. 25, pp. 2981-2992, 2010.

[2] W. Dan, T. Fen, T. Dragicevic, J. C. Vasquez, and J. M. Guerrero, "Autonomous Active Power Control for Islanded AC Microgrids With Photovoltaic Generation and Energy Storage System," *Energy Conversion, IEEE Transactions on,* vol. 29, pp. 882-892, 2014.

[3] J. G. de Matos, F. S.F.e Silva, and L. A. de S Ribeiro, "Power Control in AC Isolated Microgrids With Renewable Energy Sources and Energy Storage Systems," *Ind. Elec, IEEE Trans. on,* vol.62, pp.3490-3498, 2015.

[4] A. Urtasun, E. L. Barrios, P. Sanchis, and L. Marroyo, "Frequency-Based Energy-Management Strategy for Stand-Alone Systems With Distributed Battery Storage," *Power Electronics, IEEE Transactions on,* vol. 30, pp. 4794-4808, 2015.

[5] W. Dan, T. Fen, T. Dragicevic, J. C. Vasquez, and J. M. Guerrero, "A Control Architecture to Coordinate Renewable Energy Sources and Energy Storage Systems in Islanded Microgrids," *Smart Grid, IEEE Transactions on,* vol. 6, pp. 1156-1166, 2015.

[6] H. Mahmood, D. Michaelson, and J. Jin, "Strategies for Independent Deployment and Autonomous Control of PV and Battery Units in Islanded Microgrids," *Emerging and Selected Topics in Power Electronics, IEEE Journal of,* vol. 3, pp. 742-755, 2015.

[7] H. Mahmood, D. Michaelson, and J. Jin, "Decentralized Power Management of a PV/Battery Hybrid Unit in a Droop-Controlled Islanded Microgrid," *Power Electronics, IEEE Transactions on,* vol. 30, pp. 7215-7229, 2015.

[8] J. C. Vasquez, J. M. Guerrero, M. Savaghebi, J. Eloy-Garcia, and R. Teodorescu, "Modeling, Analysis, and Design of Stationary-Reference-Frame Droop-Controlled Parallel Three-Phase Voltage Source Inverters," *Industrial Electronics, IEEE Transactions on,* vol. 60, pp. 1271-1280, 2013.

[9] L. Xiaonan, S. Kai, J. M. Guerrero, J. C. Vasquez, and H. Lipei, "State-of-Charge Balance Using Adaptive Droop Control for Distributed Energy Storage Systems in DC Microgrid Applications," *Industrial Electronics, IEEE Transactions on,* vol. 61, pp. 2804-2815, 2014.

[10] L. Xiaonan, S. Kai, J. M. Guerrero, J. C. Vasquez, and H. Lipei, "Double-Quadrant State-of-Charge-Based Droop Control Method for Distributed Energy Storage Systems in Autonomous DC Microgrids," *Smart Grid, IEEE Transactions on,* vol. 6, pp. 147-157, 2015.

978-1-5090-0376-1/16 $31.00 © 2016 IEEE

Author Index

A

Mehdi Abapour			69(409-414)
Mostafa Abarzadeh			82(486-491)
Majid Abbasi			78(464-469)
Alireza Abbaszadeh			103(614-619)
Karim Abbaszadeh			63(373-377)
Mohammad Amin Abolhasani	35(204-210)	59(349-355)	62(367-372)
H. Abootorabi Zarchi			54(320-325)
Seyed Adib Abrishamifar	37(217-221)	79(470-475)	88(523-528)
		89(529-534)	94(559-564)
Masoumeh Adham Haghighpour			23(129-133)
Ehsan Adib	21(116-122)	24(134-139)	26(147-152)
Ahmad Afifi			78(464-469)
Seyed Ebrahim Afjei		2(7-12)	8(41-47)
M.R. Agha Kashkooli		10(53-57)	11(58-63)
Mostafa Ahmadi Darmani			3(13-18)
Saeed Ahmadi			40(235-240)
Seyed Omid Ahmadi			105(626-630)
Soheil Ahmadzadeh			91(541-546)
Kamal Al-Haddad			86(511-516)
M. Alinaghizadeh Ardestani			38(222-228)
Mohammad Reza Alizadeh Pahlavani			78(464-469)
Masoume Amirbande			28(159-164)
Davood Arab Khaburi	9(48-52)	12(64-69)	14(76-81)
	16(87-92)	98(582-588)	99(589-595)
	103(614-619)		
Gholamreza Arab Markadeh			91(541-546)
Mohammad Arasteh	18(99-104)	88(523-528)	89(529-534)
Masoud Arefian			37(217-221)
Mahdi Asadi		17(93-98)	46(273-278)
Behzad Asaei	48(285-290)	67(396-402)	73(434-439)
	105(626-630)		
Mohammad Reza Azizi			61(362-366)

B

Ebrahim Babaei	29(165-170)	32(185-190)	33(191-196)
	93(553-558)	94(559-564)	
Alfred Baghramian		23(129-133)	28(159-164)
M. Bahrami			75(446-450)
Amin Banaiemoqadam			34(197-203)
Mahdi Banejad			49(291-296)
Mohammad Bagher Banna Sharifian			93(553-558)
Seyed Masoud Barakati		22(123-128)	31(177-184)
A. Baraston			47(279-284)
Farhad Barati			13(70-75)
Reza Beiranvand	35(204-210)	52(308-313)	59(349-355)
	62(367-372)		
Frede Blaabjerg			108(641-645)
Concettina Buccella			29(165-170)

C

João P.S. Catalão			65(384-389)
Carlo Cecati			32(185-190)

D

Ali Dastfan			40(235-240)
Seyed Alireza Davari			72(427-433)
S.M. Dehghan	15(82-86)	30(171-176)	38(222-228)
Ehsan Dehghanpour			50(297-301)
A. Delavari			70(415-420)
M. Delhommais			47(279-284)
Azadeh Doulatshah			6(29-34)

E

H. Ebrahimirad			46(273-278)
Nasrin Einabadi			24(134-139)
Abdolhossein Ejlali			9(48-52)
Reza Emamalipour		67(396-402)	73(434-439)
Ehsan Enferad			64(378-383)
B. Eskandari			81(482-485)
Moretza Esteki		21(116-122)	24(134-139)
Mohammad Eydi		48(285-290)	73(434-439)

F

Esmaeil Fallah-Chulabi			4(19-23)
Hui Fang			106(631-635)
Javad Farhang		48(285-290)	73(434-439)
Babak Farhangi	36(211-216) 71(421-426)	48(285-290) 95(565-570)	67(396-402)
Shahrokh Farhangi	36(211-216) 95(565-570)	60(356-361)	71(421-426)
Ebrahim Farjah			39(229-234)
Hosein Farzanehfard	21(116-122)	24(134-139)	26(147-152)
Hossein Fathi Kivi			82(486-491)
Seyed Hamid Fathi		80(476-481)	34(197-203)
Seyed Saeed Fazel			101(602-607)
Mahdi Fazeli		37(217-221)	79(470-475)
Mahmoud Fekri			26(147-152)
Mojtaba Forouzesh			28(159-164)
Reza Fotouhi			106(631-635)

G

Cristian Garcia			107(636-640)
R. Gavagsaz-Ghoachani		75(446-450)	92(547-552)
Ali Ghaffarpour			13(70-75)
Arsham Ghanbari			1(1-6)
Teymoor Ghanbari			39(229-234)
Pooya Ghani			18(99-104)
G.B. Gharehpetian			58(344-348)
Masoud Ghodsi		22(123-128)	31(177-184)
S. Asghar Gholamian			56(332-337)
Farzad Golshan		88(523-528)	89(529-534)

Shahin Goodarzi			35(204-210)

H

Reza Haghighian			7(35-40)
Vadood Hajbani			90(535-540)
A. Haji Zadeh			57(338-343)
M.R. Hajimoradi			25(140-146)
Hossein Hajisadeghian			17(93-98)
Abolfazl Halvaei Niasar	1(1-6)	57(338-343)	97(577-581)
Farid Hamzeh Aghdam			69(409-414)
S. Hasanzadeh			30(171-176)
Mohammad M. Hashempour			109(646-651)
Hadi Hosseini Kordkheili			49(291-296)
Mohammad Ali Hosseinzadeh		29(165-170)	32(185-190)

I

H. Iman-Eini			95(565-570)
Ahmad Irani			84(498-503)
Ali R. Izadinia			100(596-601)
Ali Izanlo			56(332-337)

J

Alireza Jalilian			44(261-267)
Sadegh Jamali			44(261-267)
J. Javidi Hagh			81(482-485)
Josep M. Guerrero			110(652-657)

K

I. Kamwa			70(415-420)
Bagher Karami			94(559-564)
Faramarz Karbakhsh			58(344-348)
Mohammad-Sadegh Karbasforooshan			43(253-260)
Yaser Karimi			110(652-657)
Hamid Reza Karshenas			100(596-601)
Mohammad Kebriaei			97(577-581)
Ralph Kennel			106(631-635)
Abbas Ketabi			97(577-581)
Seyed Reza Khayam Hosseini			39(229-234)
Reza Kheirollahi			50(297-301)
S. Khoobi Arani			57(338-343)
S. Khosrogorji		42(248-252)	55(326-331)
Hamed Kiani Savadkoohi			103(614-619)
Ricardo Nicolau Nassar Koury		76(451-456)	77(457-463)

L

Poh Chiang Loh			108(641-645)

M

Luiz Machado		76(451-456)	77(457-463)
Hossein Madadi Kojabadi			82(486-491)
Seyed M. Madani		10(53-57)	11(58-63)

Mehran Maghamizadeh			80(476-481)
Naier Mahdinejad			76(451-456)
Hossein Mahdinia Roudsari			44(261-267)
Mohammad Hossein Mahlooji			66(390-395)
M. Mansourian			15(82-86)
Mohamad Reza Mazinanian			7(35-40)
Koosha Mehdizadegan			63(373-377)
Hamed Mehrabian-Nejad			36(211-216)
Majid Mehrasa			65(384-389)
Jafar Milimonfared			58(344-348)
Maedeh Mirazimi			93(553-558)
Seyyed Mehdi Mirimani			3(13-18)
Behzad Mirshekarpour			72(427-433)
Amir Mirzabayati			79(470-475)
Ali Moallemi			77(457-463)
Hasan Moghbeli			17(93-98)
Mustafa Mohamadian	35(204-210)	37(217-221)	45(268-272)
	84(498-503)	104(620-625)	
Mohammad Mohammadi Firozjaee			5(24-28)
Hamid Reza Mohammadi			66(390-395)
Mehdi Mohammadi			21(116-122)
Saleh Mohammadi			54(320-325)
M.J. Mojibian			53(314-319)
Hossein Mokhtari		25(140-146)	108(641-645)
Mohammad Monfared			43(253-260)
Adel Moradi			96(571-576)
A. Mosallanejad			2(7-12)
Ali Mostaan			23(129-133)
Seyed Mohammad Mousavi			52(308-313)

N

Peyman Naderi			6(29-34)
Ali Nahavandi			61(362-366)
B. Nahid-Mobarakeh		75(446-450)	92(547-552)
M. Nezamabadi		2(7-12)	8(41-47)
Mohammad Reza Nikzad			105(626-630)
Mohammad Niliyan			104(620-625)
N. Noori			14(76-81)
Margarita Norambuena			107(636-640)
R. Noroozian			87(517-522)

O

A. Olloqui		98(582-588)	99(589-595)
Hashem Oraee		13(70-75)	110(652-657)
Saeed Ouni			74(440-445)

P

Saeed Peyghami-Akhuleh			108(641-645)
Mohammad Pichan		79(470-475)	83(492-497)
S. Pierfederici			75(446-450)
A. Pirooz			87(517-522)
Hamid Reza Piryaei			101(602-607)

Amirreza Poorfakhraei			96(571-576)
Javad Poshtan			20(111-115)
Edris Pouresmaeil			65(384-389)
Fabricio Jose Pacheco Pujatti			77(457-463)

Q

Pouria Qashqai			86(511-516)

R

Saman Radman			68(403-408)
Elham Rahimi			40(235-240)
Mohsen Rahimi			66(390-395)
Ramin Rahimi			71(421-426)
A. Ramezani			95(565-570)
Hassan Rastegar			83(492-497)
Morteza Rezaei Larijani			85(504-510)
Hamid Rezaie			83(492-497)
Reza Rezaii	35(204-210) 62(367-372)	52(308-313)	59(349-355)
Mohammad Rezanejad		27(153-158)	65(384-389)
Ghasem Rezazadeh			51(302-307)
Marco Rivera	12(64-69) 99(589-595)	16(87-92) 102(608-613)	98(582-588) 103(614-619)
Diana Florez Rodriguez			64(378-383)
Jose Rodriguez	102(608-613)	103(614-619)	107(636-640)
S. Rojas		102(608-613)	61(362-366)
Mehdi Roostaee			

S

Ramtin Sadeghi		10(53-57)	11(58-63)
Sajad Sadr			16(87-92)
O. Safdarzadeh		2(7-12)	8(41-47)
Simin Sakhavati			33(191-196)
O. Salari			53(314-319)
Sina Salehi Dobakhshari			34(197-203)
S.M. Salehi			30(171-176)
Nima Salehiyan Zandi			23(129-133)
A. Salemnia		42(248-252)	55(326-331)
Mahdi Salimi			90(535-540)
Emad Samadaei			27(153-158)
Pooya Samanipour			20(111-115)
Mostafa Sanikhani			5(24-28)
Mehdi Saradarzadeh			60(356-361)
Ali Sarajian			12(64-69)
Maryam Sarbanzadeh		29(165-170)	32(185-190)
S.A. Saremi Hasari		42(248-252)	55(326-331)
Christophe Saudemont			64(378-383)
Mehdi Savaghebi		109(646-651)	110(652-657)
J.L. Schanen			47(279-284)
Amir Sepehr			60(356-361)
Seyed Hamid Shahalami			4(19-23)

Mahmoud Shahbazi	45(268-272)	74(440-445)	85(504-510)
Mostafa Shahnazari			68(403-408)
Mehdi Shahrdad			37(217-221)
Abdolreza Sheikholeslami			86(511-516)
Abbas Shiri			19(105-110)
J. Shojaee			81(482-485)
Javad Shokrollahi Moghani			34(197-203)
Mohsen Siami			103(614-619)
M. Soheil-Hamedani			92(547-552)
Mahdi Sojoodi			84(498-503)
Javad Soleimani			9(48-52)
Jafar Soltani			90(535-540)
S. Soori		42(248-252)	55(326-331)

T

Farzad Tahami	41(241-247)	51(302-307)	96(571-576)
Mehrdad Tarafdar Hagh			69(409-414)
Nima Tashakor			39(229-234)
M. Tavakoli Bina		53(314-319)	81(482-485)
Hamid Reza Tayebi			18(99-104)
Armin Teymoori			58(344-348)
Hamidreza Toodeji			68(403-408)

V

Sadegh Vaez-Zadeh			36(211-216)
Abolfazl Vahedi		5(24-28)	7(35-40)
Hani Vahedi			86(511-516)
Hamed Valipour			51(302-307)
Ramon Molina Valle			76(451-456)
Juan C. Vasquez		109(646-651)	110(652-657)
Mohammad Verij Kazemi			56(332-337)
Mahdi Vizheh			27(153-158)

W

P. Wheeler	98(582-588)	99(589-595)	102(608-613)

Y

Keyvan Yari			28(159-164)
Farzad Yazdani			41(241-247)
Ali Yazdian Varjani	45(268-272)	52(308-313)	59(349-355)
	62(367-372)	104(620-625)	

Z

Sasan Zabihi			65(384-389)
A. Zabihinejad			70(415-420)
P. Zanchetta			47(279-284)
M. Zandi		75(446-450)	92(547-552)
Amir Zare-Bazghaleh			4(19-23)
Mohammad-Reza Zolghadri		74(440-445)	85(504-510)

IEEE
445 Hoes Lane
Piscataway, NJ 08854-4141

ISBN 978-1-5090-0376-1